er Art: George Herman, Untitled.

United States of America

4 3 2 1

0-6

& SCHUSTER CUSTOM PUBLISHING
treet/Needham Heights, MA 02194
uster Education Group

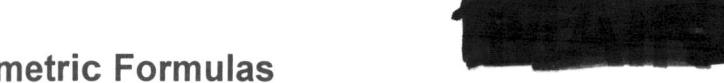

Geometric Formulas

Rectangle

Area: $A = \ell w$
Perimeter: $P = 2\ell + 2w$

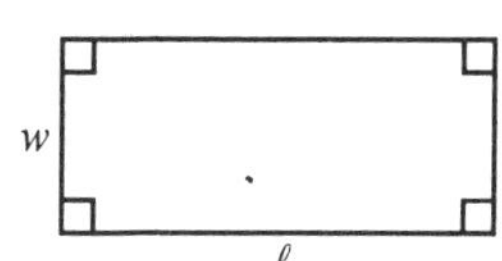

Square

Area: $A = s^2$
Perimeter: $P = 4s$

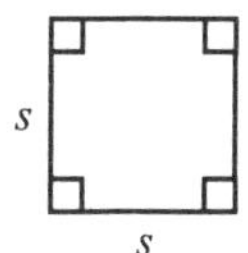

Parallelogram

Area: $A = bh$

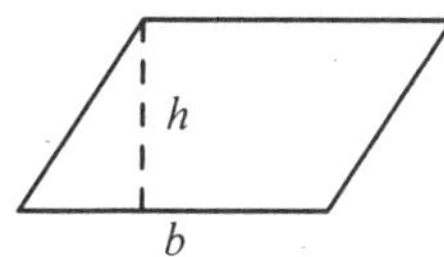

Trapezoid

Area: $A = \frac{1}{2}h(a+b)$

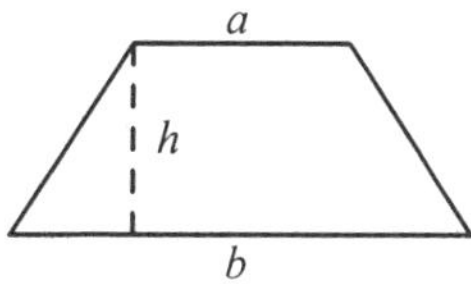

Triangle

Area: $A = \frac{1}{2}bh$

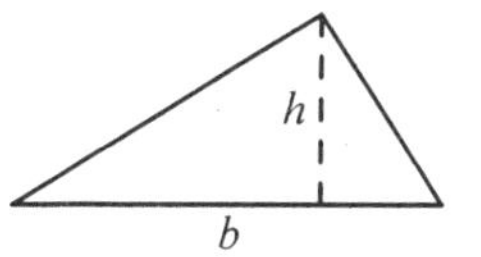

or

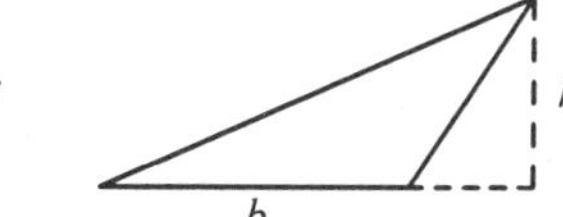

Area: $A = \sqrt{s(s-a)(s-b)(s-c)}$,
where $s = \frac{1}{2}(a+b+c)$
Angle sum: $A + B + C = 180^\circ$

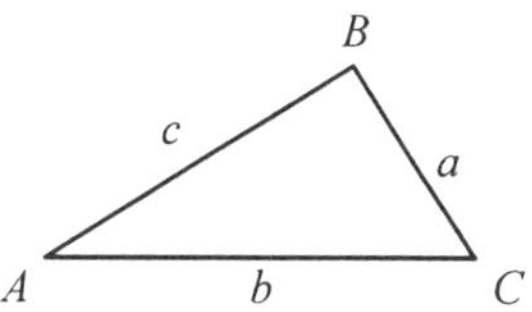

Right Triangle

Pythagorean theorem: $a^2 + b^2 = c^2$,

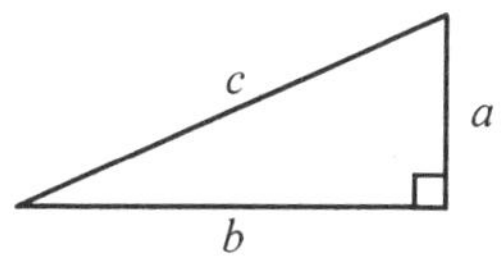

Circle

Area: $A = \pi r^2$
Circumference: $C = \pi d = 2\pi r$

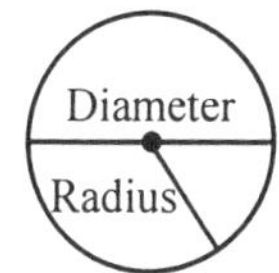

ESSENTIA
of Precalculus, Alg
and Trigonometry

SIXTH EDITION

Dennis

Christy

NASSAU COMMUNITY C

Robert

Rosenfeld

SIMON & SCHUSTER CUSTOM P

Printed in the

10 9 8 7 6 5

ISBN 0-536-0026
BA 97409

SIMON
160 Gould
Simon & Sc

To Margaret and Leda

Contents

Preface

Audience

This book is intended for students who need a concrete approach to mathematics. Throughout the text, two key principles are applied to make the material less abstract to the student.

1. Topics are developed in the context of meaningful problems by first illustrating an application for a topic before exploring the specifics of the mathematics that is involved.

2. Topics are reinforced with graphical and numerical examples to support the algebraic analysis so that the student will understand a topic from algebraic, geometric, and numeric viewpoints.

The presentation assumes that the student has a background that includes two years of high-school mathematics. However, a detailed review of all necessary ideas is given early in the text so that students whose basic skills need improvement have a wealth of helpful material.

● ● ●

Approach

This book is written with the belief that current textbooks must provide a *balanced approach* to learning precalculus, which incorporates many of the changes recommended by various reform programs, while maintaining the sound pedagogical features that have proven to be effective in traditional texts. As a minimum, a modern text must provide a dynamic approach to problem solving that allows students to monitor their progress and that effectively integrates graphics and calculator use. Our approach in these four areas is explained next.

Problem-Solving Approach Our experience is that students who take precalculus, algebra, and trigonometry learn best by "doing." Examples and exercises are crucial, since it is usually in these areas that the students' main interactions with the material take place. The problem-solving approach contains brief, precisely formulated paragraphs, followed by many detailed examples. A relevant word problem introduces *every* section of the text, and word problems or other motivational problems are included in *every* section's exercise set.

Graphics A major component of a problem-solving approach with intuitive concept development is a strong emphasis on graphics. Students are given, and are encouraged to draw for themselves, visual representations of the concepts they are analyzing and the problems they are solving. Effective use of color enhances the many images in concept development and in exercise sets.

Interactive Approach Because students learn best by doing, Progress Check exercises are associated with each example problem in the text. By doing these exercises, students obtain immediate feedback on their understanding of the concept being discussed. In actual practice we have also found that the Progress Check exercises are an excellent source for homework problems.

Calculator Use The text encourages the use of graphing calculators and discusses how they can be used effectively throughout the exercises and explorations. For these discussions, it is assumed that students have at least a basic graphing calculator such as a Texas Instruments TI-81, and possibly a more advanced model such as a TI-85. Because the Texas Instruments family of graphing calculators is widely used, we have chosen to use feature names and to show calculator screens associated with this family of calculators. However, the exercises and explorations are not machine-specific, so a variety of graphers may be used.

The main limitation for students will be dictated by the features of their calculator. For example, a calculator that has equation-solving features provides the student with much more support than a basic graphing calculator.

● ● ●

Organization

All the essential precalculus, algebra, and trigonometry topics are covered in an organization that differs significantly from the somewhat standard arrangement of topics. As indicated in the Table of Contents, the text is divided into two parts. Part 1 covers the fundamentals of algebra along with an introduction to functions and trigonometry. The early introduction of functions, graphing, and trigonometry injects new and central topics at the beginning of the course (before the more customary rehash of algebraic manipulations) and helps provide a more integrated treatment of algebra and trigonometry, one that can take advantage of modern technology. Students enrolled in engineering or technology programs benefit especially from the early discussion of trigonometry. The trigonometric expressions are carried along simultaneously with the work in algebra, so that students can see that algebraic and trigonometric expressions are handled in basically the same way. They will find this insight particularly useful when working with trigonometric equations and identities, which is one of the most crucial areas in trigonometry.

Part 2 uses an "elementary functions" approach, with the function concept playing the unifying role in the study of linear, quadratic, polynomial, rational, exponential, logarithmic, and circular functions. The text concludes with the traditional chapters on analytic geometry, systems of equations, and discrete algebra.

● ● ●

Pedagogy

Section Introductions In the spirit of problem solving, each section opens with a problem that should quickly involve students and teachers in a discussion of an important section concept. This problem is later worked out as an example in each section.

Systematic Review Students benefit greatly from a systematic review of previously learned concepts. At the end of each chapter, there is a detailed chapter summary that includes: a checklist of objectives illustrated by example problems, a chapter overview organized in a section-by-section format, abundant chapter review exercises, and a chapter test.

"Explore" Exercises Each exercise set is preceded by a set of "Explore" exercises. Through these problems, students can use modern technology and work individually or in groups to discover properties and learn about additional features of graphing calculators that are useful for precalculus. These exercises will help students to develop critical-thinking skills and will increase their ability to use technology.

"Think About It" Exercises Each exercise set is followed immediately by a set of "Think About It" exercises. Although some of these problems are challenging, this section is not intended as a set of "mind bogglers." Instead, the goal is to help students develop critical-thinking skills by asking them to create their own examples, express concepts in their own words, extend ideas covered in the section, and analyze topics slightly out of the mainstream. These exercises are an excellent source of nontemplate problems and problems that can be assigned for group work.

● ● ●

Changes to the New Edition

Authors can usually anticipate that by the time a text reaches a sixth edition, only minor revisions will be needed. However, current trends in mathematics education and advances in technology have necessitated extensive rewriting for this edition. There are three main areas where changes have been made.

1. Graphing calculator coverage is now integrated throughout the text, and more graphical and numerical analysis is provided, to supplement algebraic developments.

2. Adjustments in organization have been made to reflect the increased emphasis on using technology, and to provide a smoother transition between ideas. For example, the ordering of topics in Chapter 1 on "Functions and Graphs," has been changed significantly.

3. Section-opening problems, stated section objectives, titled example problems, Progress Check exercises, Technology Link segments, Explore exercises, and an Objectives checklist that is illustrated by example problems are new features of the text.

● ● ●

Features

Section Openers Each section opens with a motivational problem that is solved later as an example problem.

Section Objectives The specific objectives of each section are listed to help students focus on the basic goals of the section.

Titled Examples Each example problem is titled so that students can quickly see the point of the example.

Progress Check Exercises Each example problem has an associated Progress Check exercise, so that students can obtain immediate feedback on their understanding of the concept being discussed.

Remarks "Note" and "Caution" remarks provide helpful insights and point out potential student errors.

Concept Highlight Boxes with labels highlight important definitions and rules.

Technology Links Discussions about using graphing calculators are included often, so that students can benefit from using technology. This ability can help the student to understand topics from geometric, numeric, and algebraic viewpoints.

Applications Application problems are included in the examples and exercise sets of *every* section. This edition contains many new and realistic applications, including many problems that are based on real data.

Explore Exercises These exercises can promote discovery and cooperative learning and increase the student's proficiency with graphing calculators.

Think About It Exercises Students work individually or in groups to develop critical-thinking skills by creating their own examples, expressing concepts in their own words, and/or extending ideas covered in the section.

Objectives Checklist Each Chapter Summary begins with a checklist of specific objectives that are illustrated by example problems.

● ● ●

Instructional Package

For the Instructor

Instructor's Resource Manual The *Instructor's Resource Manual* contains three tests for each chapter, three final exams, and the answers to the "Think About It" exercises, the "Explore" exercises, and the even-numbered text exercises.

For the Student

Student Solutions Manual The *Student Solutions Manual* contains detailed solutions to all odd-numbered text exercises. This manual is written by Dr. John R. Garlow who teaches at Tarrant County Junior College.

● ● ●

Acknowledgment

We wish to thank the many users and reviewers of our books who have suggested improvements. At this point, it is hard to separate our original ideas from the many valuable observations they made, and we are indebted to all of them. For help with this and related projects we are especially grateful to Dr. John R. Garlow, Tarrant County Junior College, who wrote the *Student Solutions Manual*; Theresa Grutz and Earl McPeek, who provided editorial guidance with Wm. C. Brown Publishers; Chris Shenk and Dennis Ricci, who brought this project to fruition at Simon & Schuster Publishing; and Deborah Levine and Gene Zirkel, Nassau Community College, who provided significant support at various stages of the project. Our wives, Margaret and Leda, once again helped in the many special ways that were needed to keep our writing efforts going. So, to our families, our colleagues at Nassau Community College, the staff at Simon & Schuster Publishing, and the many users and reviewers, thank you.

Dennis Christy
Robert Rosenfeld

Part 1

Algebra with an Introduction to Trigonometry

• • •

Chapter

Review of Basic Algebra

TIAA-CREF is one of the nation's largest pension programs for college educators. Each participant receives a yearly benefits report that shows the total accumulation in his or her account, with various illustrations of first-year retirement income at age 65. One set of illustrations assumes that no additional premiums are paid and shows hypothetical rates of return until retirement of 3 percent, 6 percent, 9 percent, and 12 percent. Use these assumptions and find the hypothetical retirement accumulations at age 65 for a teacher whose accumulation at age 55 is $312,573, by evaluating

$$312{,}573(1+r)^{10}$$

where r equals 0.03, 0.06, 0.09, and 0.12. If your calculator has a Last Entry button, then use this feature to simplify the computation. (See Example 12 of Section R.1).

Source: TIAA-CREF Annuity Benefits Report, January 1997

Photo Courtesy of ***David Powers*** *of Stock Boston, Boston, MA*

A common student lament goes something like, "I understand the new concepts, but the algebra is killing me!" In this chapter, we hope to remedy this problem by reviewing *in detail* some basic rules in algebra about real numbers, exponents, and first-degree equations and inequalities. We then expand the algebra review in Chapters 2 and 3 to include factoring, fractions, radicals, and a wide variety of equations and inequalities. Success in these algebra chapters will go a long way toward ensuring success in this course and in higher mathematics.

In this text we take a problem-solving approach which emphasizes that one learns mathematics by *doing* mathematics, while *thinking* mathematically. That is, you need to actively work through the problems (with pencil and paper), while *focusing on the definitions, relationships, and procedures* that link together all steps in the solution. In this spirit of problem solving we open each section with a problem. Most are applications, some are puzzles, and a few are proofs. Taken together, they illustrate the varied nature of problem solving. Since none of them requires a lot of sophisticated mathematics, we hope you will take a stab at an answer either initially or after covering the relevant section in the text.

• • •

R.1 Real Numbers and Calculator Computation

Objectives

1. Express quotients of integers as repeating decimals, and vice versa.
2. Identify integers, rational numbers, irrational numbers, and real numbers.
3. Identify the properties of real numbers.
4. Graph and order real numbers.
5. Find the absolute value of a real number and operate on real numbers.
6. Evaluate numeric expressions on a graphing calculator.

Mathematics is a basic tool in analyzing concepts in every field of human endeavor. In fact, the primary reason you have studied this subject for at least a decade is that mathematics is the most powerful instrument available in the search to understand the world and to control it. Mathematics is essential for full comprehension of technological and scientific advances, economic policies and business decisions, and the complexities of social and psychological issues. At the heart of this mathematics is algebra. Calculus, statistics, and computer science are but a few of the areas in which a knowledge of algebraic concepts and manipulations is necessary.

Algebra is a generalization of arithmetic. In arithmetic we work with specific numbers, such as 5. In algebra, we study numerical relations in a more general way by using symbols, such as x, that may be replaced by a number from some collection of numbers. Since the symbols represent numbers, they behave according to the same rules that numbers must follow. Consequently, instead of studying specific numbers, we study symbolic representations of numbers and try to define the laws that govern them.

We begin our study of algebra by giving specific names to various sets* of numbers. The collection of the counting numbers, zero, and the negatives of the counting numbers is called the **integers**. Thus, the set of integers may be written as

$$\{\ldots, -3, -2, -1, 0, 1, 2, 3, \ldots\}.$$

The set of fractions with an integer in the top of the fraction (numerator) and a nonzero integer in the bottom of the fraction (denominator) is called the **rational numbers**. Symbolically, a rational number is a number that may be written in the form a/b, where a and b are integers, with b not equal to ($\neq$) zero. The numbers $\sqrt{2}/3$ and $2/\pi$ are fractions, but they are not rational numbers because they cannot be written as the quotient of two integers. All integers are rational numbers because we can think of each integer as having a 1 in its denominator. For example, $4 = 4/1$.

Our definition for rational numbers specified that the denominator cannot be zero. To see why, you need to know that

$$\frac{8}{2} = 4 \text{ is equivalent to saying that } 8 = 4 \cdot 2 \text{ and}$$
$$\frac{55}{11} = 5 \text{ is equivalent to saying that } 55 = 5 \cdot 11.$$

If $8/0 = a$, where a is some rational number, this would mean that $8 = a \cdot 0$. But $a \cdot 0 = 0$ for any rational number. There is no rational number a such that $a \cdot 0 = 8$. Thus we say that $8/0$ is *undefined.*

Now consider $0/0 = a$. This is equivalent to $0 = a \cdot 0$. But $a \cdot 0 = 0$ for *any* rational number. Thus, there is not just one number a that will solve the equation—any a will. Since $0/0$ does not

---

* A **set** is simply a collection of objects, and we may describe a set by listing the objects or members of the collection within braces.

name a particular number, it is also undefined. Consequently, division by zero is undefined in every case, so the denominator in a rational number cannot be zero.

To define our next set of numbers, we now consider the decimal representation of numbers. We may convert rational numbers to decimals by long division. Consider the following examples of repeating decimals. A bar is placed above the portion of the decimal that repeats.

$$\frac{3}{4} = 0.7500\ldots \qquad \frac{2}{3} = 0.6666\ldots \qquad \frac{8}{7} = 1.1428571\ldots$$

$$= 0.75\overline{0} \qquad = 0.\overline{6} \qquad = 1.\overline{142857}$$

$1.\overline{142857}$
$7\overline{)8.000000}$
7
10
7
30
28
20
14
60
56
40
35
50
49
1

Figure R.1

The decimals repeat because at some point we must perform the same division and start a cycle. For example, when converting $8/7$, the only possible remainders are 0, 1, 2, 3, 4, 5, and 6. In performing the division, as shown in Figure R.1, we had remainders of 1, 3, 2, 6, 4, and 5. In the next step we must obtain one of these remainders a second time and start a repeating cycle, or obtain 0 as the remainder, which results in repeating zeros. Thus, if a/b is a rational number, it can be written as a repeating decimal.

It is also true that any repeating decimal may be converted to a ratio between two integers, as shown in Example 1.

Example 1: Converting a Repeating Decimal to a Ratio of Integers

Express the repeating decimal $0.\overline{17}$ as the ratio of two integers.

Solution: First, let $x = 0.1717\ldots$. Multiplying both sides of this equation by 100 moves the decimal two places to the right, so we obtain

$$\begin{aligned} 100x &= 17.1717\ldots \\ x &= 0.1717\ldots \\ \hline \text{now subtracting yields} \quad 99x &= 17 \qquad \text{or } x = \frac{17}{99}. \end{aligned}$$

Thus, the repeating decimal $0.\overline{17}$ is equivalent to the fraction $17/99$.

Note: In this example we multiplied by 100 because the decimal repeated after every two digits. If the decimal repeats after one digit, we multiply by 10; if it repeats every three digits, we multiply by 1,000, and so on.

PROGRESS CHECK 1

Express the repeating decimal $0.\overline{6}$ as the ratio of two integers. ■

We have illustrated that we may define a rational number either as the quotient of two integers or as a repeating decimal. There are decimals that do not repeat, and this set of numbers is called the **irrational numbers**. The number π, which represents the ratio between the circumference and the diameter of a circle, is an example of a number with a nonrepeating decimal form. Other examples of irrational numbers are $\sqrt{2}$, $\sqrt{3}$, $\sqrt{5}$, and $\sqrt[3]{2}$. It is not obvious whether or not certain numbers are irrational, and the result must be proved mathematically. Early Greek mathematicians were able to prove that $\sqrt{2}$ is irrational (and a proof that $\sqrt{2}$ cannot be written as the quotient of two integers is considered in Question 5 of the "Think About It" exercises). However, the proof that π is irrational was obtained only in 1767 by the Swiss mathematician Johann Lambert.

Since an irrational number is a nonrepeating decimal, and a rational number is a repeating decimal, there is no number that is both rational and irrational. The set of numbers that are either repeating decimals (rational numbers) or nonrepeating decimals (irrational numbers) constitutes the **real numbers**. Real numbers are used extensively in this text, and unless it is stated otherwise, you may assume that the symbols in algebra (such as x) may be replaced by any real number. Consequently, the rules that govern real numbers determine our methods of computation in algebra. A graphical illustration of the relationships among the various sets of numbers is given in Figure R.2. Note that all of the sets we have discussed are subsets* of the set of real numbers.

Real numbers

Rational numbers	Irrational numbers
$\frac{2}{3}$ $0.\overline{17}$	$\sqrt{2}$
Integers: -3, 0, $1{,}000$	$\sqrt{7}$, π

Figure R.2

Example 2: Classifying Numbers

For each number, select all correct classifications from the following categories: real number, irrational number, rational number, integer, or none of these.

a. $\sqrt{225}$ **b.** $\sqrt{-4}$ **c.** $0.505005\ldots$ **d.** $\dfrac{22}{7}$

Solution:

a. Since $\sqrt{225} = 15$, the number $\sqrt{225}$ is an integer, a rational number, and a real number. Note from the diagram in Figure R.2 that any integer is automatically a rational number and a real number.

b. No real number is the square root of a negative number, such as -4, because the product of two equal real numbers is never negative. Thus, $\sqrt{-4}$ belongs to none of the listed categories of numbers.

c. The number $0.505005\ldots$ is a nonrepeating decimal. Therefore, it is an irrational number and a real number.

d. The number $22/7$ is a ratio of two integers (not involving division by zero), so it is a rational number and a real number. Because $22/7$ is a common approximation for π, some students mistakenly think that $22/7$ is irrational. It is often useful to write rational number approximations for irrational numbers, but it is important to keep in mind the difference between them.

Technology Link

In part **a** of Example 2, a calculator can be used to determine that $\sqrt{225}$ equals 15, to help classify $\sqrt{225}$, but there are limitations to this calculator method. For instance, $\sqrt{3.6101520016}$ is rational, but $\sqrt{3.6101520017}$ is not, and many calculators cannot distinguish them. In part **b** trying computing $\sqrt{-4}$ on your calculator to see if an error message results. In Section 2.5 we will extend the number system beyond real numbers to complex numbers, which will enable us to work with square roots of negative numbers. If your calculator returns (0, 2) or $2i$ for $\sqrt{-4}$, then it is programmed to output complex number results.

PROGRESS CHECK 2

For each number select all correct classifications from the following categories: real number, irrational number, rational number, integer, or none of these.

a. $\sqrt{-7}$ **b.** $-\sqrt{7}$ **c.** 3.14 **d.** 0.05

---

* Set A is a **subset** of set B if every element of A is an element of B.

We now list the most important properties of the real numbers with respect to addition and multiplication. These properties are the basis for the justification of many algebraic manipulations.

Properties of the Real Numbers

Let a, b, and c be real numbers.

	Addition	**Multiplication**
Closure Properties	$a+b$ is a unique real number	ab is a unique real number
Commutative Properties	$a+b=b+a$	$ab=ba$
Associative Properties	$(a+b)+c=a+(b+c)$	$(ab)c=a(bc)$
Identity Properties	There exists a unique real number 0 such that $a+0=0+a=a$	There exists a unique real number 1 such that $a\cdot 1=1\cdot a=a$
Inverse Properties	For every real number a, there is a unique real number, denoted by $-a$, such that $a+(-a)=(-a)+a=0$	For every real number a except zero, there is a unique real number, denoted by $1/a$, such that $a\cdot 1/a=1/a\cdot a=1$
Distributive Properties	$a(b+c)=ab+ac$	$(a+b)c=ac+bc$

Example 3: Naming Properties of Real Numbers

Name the property of real numbers illustrated in each statement.

a. $(2+3)+4=2+(3+4)$ **b.** $5+(-5)=0$

c. $2(x+4)=2\cdot x+2\cdot 4$ **d.** $0\cdot x=x\cdot 0$

Solution:

a. This statement illustrates the associative property of addition. This property indicates that we obtain the same result if we change the grouping of the numbers in an addition problem.

b. This statement illustrates the addition inverse property. The number $-a$ is called the **negative** or **additive inverse** of a. Note that there is a difference between a negative number and the negative of a number. Although -5 is a negative number, the negative of -5 is 5.

c. This statement illustrates the distributive property (or more technically, the distributive property of multiplication over addition).

d. This statement illustrates the commutative property of multiplication. This property indicates that the order in which we write numbers in a multiplication problem does not affect their product.

PROGRESS CHECK 3

Name the property of real numbers illustrated in each statement.

a. $5\left(\frac{1}{5}\right)=1$ **b.** $(a+b)(a-b)=(a-b)(a+b)$

c. $-1(2x)=(-1\cdot 2)x$ **d.** $-1(x+2)=(-1)(x)+(-1)(2)$ ■

Real Number Line

The real numbers may be interpreted geometrically by considering a straight line. Every point on the line can be made to correspond to a real number, and every real number can be made to correspond to a point. The first point that we designate is zero. It is the dividing point between positive and negative real numbers. Any number to the right of zero is called positive, and any number to the left of zero is called negative. Figure R.3 is the result of assigning a few positive and negative real numbers to points on the line.

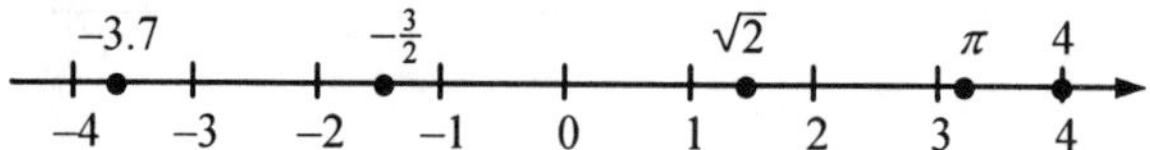

Figure R.3

Order

The **trichotomy property** states that if a and b are real numbers, then either a is less than b, a is greater than b, or a equals b. By definition, a is **less than** b, written $a < b$, if and only if $b - a$ is positive; a is **greater than** b, written $a > b$, if and only if $a - b$ is positive. We also define $a \le b$ (read "a is **less than or equal to** b") to mean $a < b$ or $a = b$. Alternately, we can write $b \ge a$ and say that b is **greater than or equal to** a. Relations of "less than" and "greater than" can be seen very easily on the number line, as shown in Figure R.4. The point representing the larger number will be to the right of the point representing the smaller number. (*Note*: The definitions use the phrase "if and only if." This phrase is commonly used in mathematical definitions and theorems, and the statement "p **if and only if** q" means "if p then q, and conversely, if q then p.")

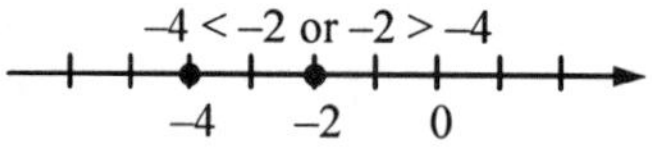

$-4 < -2$ or $-2 > -4$

−4 −2 0

Figure R.4

Example 4: Graphing and Ordering Real Numbers

Insert the property symbol (<, >, or =) to indicate the correct order.

a. -1 ________ -3 **b.** -3 ________ $\sqrt{2}$ **c.** $-\sqrt{9}$ ________ -3

Solution: The given numbers are graphed in Figure R.5.

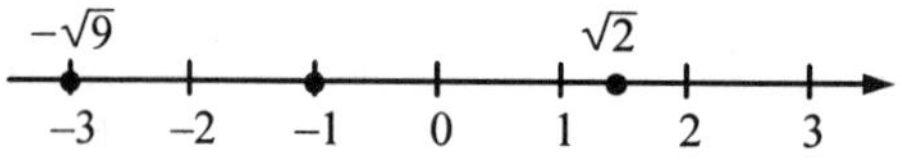

$-\sqrt{9}$ $\sqrt{2}$

−3 −2 −1 0 1 2 3

Figure R.5

a. -1 is greater than -3, written $-1 > -3$, because -1 is to the right of -3 on the number line.

b. $-3 < \sqrt{2}$ because -3 is graphed to the left of $\sqrt{2}$. In general, any negative number is less than any positive number.

c. $-\sqrt{9} = -3$ since both numerals represent the same number.

PROGRESS CHECK 4

Insert the proper symbol (<, >, =) to indicate the correct order.

a. -2 ________ 0 **b.** $\frac{2}{3}$ ________ $\frac{6}{9}$ **c.** $-\sqrt{2}$ ________ $-\sqrt{5}$ ■

Absolute Value

Positive and negative numbers in mathematics designate direction or indicate whether a result is above or below some reference point. However, sometimes the numerical size of a number is more important than its sign, and in such cases, the concept of absolute value is useful. The **absolute value** of a real number a, denoted $|a|$, is the distance between a and 0 on the number line. For example, $|3| = 3$ and $|-3| = 3$, as shown in Figure R.6. Note that if $a \ge 0$, then $|a| = a$, but if $a < 0$, then $|a| = -a$ since $-a$ is positive in this case. Thus, algebraically we can define the absolute value of a real number a in tabular form as follows:

$$|a| = \begin{cases} a & \text{if } a \ge 0 \\ -a & \text{if } a < 0 \end{cases}$$

The concept of absolute value is discussed in detail in Section 3.5.

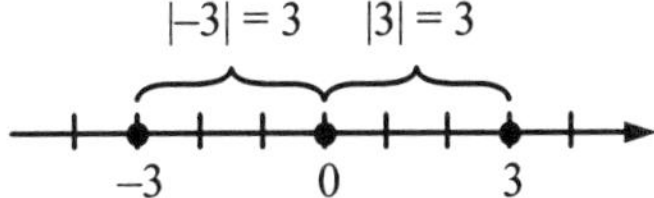

Figure R.6

Addition

Addition of two real numbers may be interpreted as movement on the number line. The initial starting point is zero. To add a positive number, we draw an arrow to the right the desired distance. To add a negative number, we draw an arrow to the left. Consider Figure R.7, which illustrates the procedure. The sum $2+(-4)$ is shown on the real number line by drawing an arrow with a length of 2 units, starting at zero and pointing to the right. From the tip of the arrow at 2 we draw an arrow pointing to the left with a length of 4 units. Since the tip of the second arrow is at –2, we conclude that $2+(-4)=-2$. As another example, the diagram in Figure R.8 may be used to show geometrically that $-2+(-4)=-6$.

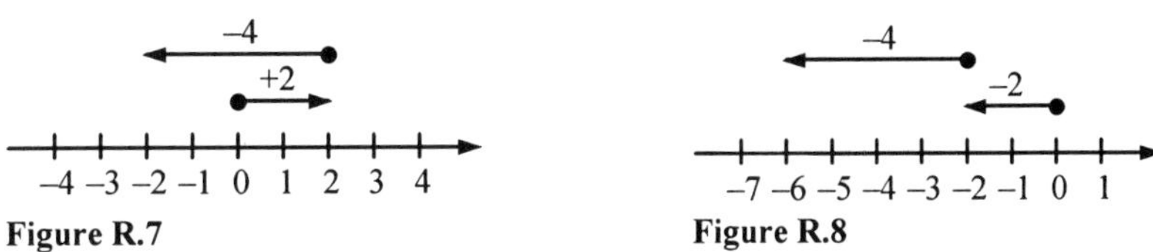

Figure R.7

Figure R.8

The number line method will help you to envision the addition of real numbers, but it is not an efficient computational method. Instead, the concept of absolute value may be used to add real numbers, as described next. Observe that the two methods are in agreement with the number line procedure, since the lengths of the arrows are given by the absolute values of the numbers.

Adding Real Numbers

> **Like Signs** To add two real numbers with the same sign, add their absolute values and attach the common sign.
> **Unlike Signs** To add two real numbers with different signs, subtract the smaller absolute value from the larger and attach the sign of the number with the larger absolute value.

Example 5: Adding Real Numbers

Find each sum.

a. $-9+(-18)$ **b.** $-7+52$ **c.** $-\frac{5}{7}+\frac{1}{7}$

Solution:

a. $-9+(-18)=-27$ *Think*: $|-9|=9$, $|-18|=18$, $9+18=27$, and the common sign is negative.

b. $-7+52=45$ *Think*: $|-7|=7$, $|52|=52$, $52-7=45$, and the positive number 52 has the larger absolute value.

c. $-\frac{5}{7}+\frac{1}{7}=-\frac{4}{7}$ *Think*: $\left|-\frac{5}{7}\right|=\frac{5}{7}$, $\left|\frac{1}{7}\right|=\frac{1}{7}$, $\frac{5}{7}-\frac{1}{7}=\frac{4}{7}$, and the negative number $-\frac{5}{7}$ has the larger absolute value.

PROGRESS CHECK 5
Find each sum.

a. $-25+(-19)$ **b.** $\frac{7}{9}+\left(-\frac{2}{9}\right)$ **c.** $-18+10$ ■

Subtraction

We define subtraction in terms of addition. If a and b are any real numbers, then

$$a-b=a+(-b)$$

To subtract b from a, we add to a the negative of b.

Example 6: Subtracting Real Numbers

Find each difference.

a. $-5-2$ **b.** $5-\left(-\frac{1}{2}\right)$ **c.** $-4.4-(-1.5)$

Solution: Subtract by converting to addition.

a. $-5-2=-5+(-2)=-7$ **b.** $5-\left(-\frac{1}{2}\right)=5+\frac{1}{2}=\frac{10}{2}+\frac{1}{2}=\frac{11}{2}$

c. $-4.4-(-1.5)=-4.4+1.5=-2.9$

PROGRESS CHECK 6
Find each difference.

a. $3-7$ **b.** $-9-\left(-\frac{5}{9}\right)$ **c.** $1.8-(-1.8)$ ■

Multiplication

It seems natural to define the product of two positive numbers as positive. Thus, $4\cdot 3=12$. To determine what sign to use for the product of a positive number and a negative number, we use the distributive property.

$$\begin{aligned} 5[2+(-2)] &= 5\cdot 2+5(-2) \\ 5\cdot 0 &= 10+5(-2) \\ 0 &= 10+? \end{aligned}$$

We must define $5(-2)$ so that $10+5(-2)=0$. Therefore, $5(-2)$ must equal -10. In every case the product of a positive number and a negative number is negative.

To determine the sign of the product of two negative numbers, consider this problem.

$$\begin{aligned} -5[2+(-2)] &= (-5)2+(-5)(-2) \\ (-5)0 &= -10+(-5)(-2) \\ 0 &= -10+? \end{aligned}$$

We must define $(-5)(-2)$ so that $-10+(-5)(-2)=0$. Therefore, $(-5)(-2)$ must equal 10. In every case the product of two negative numbers is positive.

Since multiplication and division of real numbers are related in that

$$(-5)(-2)=10 \text{ implies } \frac{10}{-5}=-2 \text{ or } \frac{10}{-2}=-5$$

the sign rules just determined also apply to quotients. The above examples suggest the following procedures:

Products and Quotients of Nonzero Real Numbers

Same Sign: To multiply (or divide) two real numbers with the same sign, multiply (or divide) their absolute values, and make the sign of the product (or quotient) positive.
Different Signs: To multiply (or divide) two real numbers with different signs, multiply (or divide) their absolute values, and make the sign of the product (or quotient) negative.

Example 7: Multiplying and Dividing Real Numbers

Find each product or quotient.

a. $(-6)(-9)$ **b.** $\dfrac{-12.6}{3}$ **c.** $(-4)(-4)(-4)$

Solution:

a. $(-6)(-9) = 54$ *Think*: $6 \cdot 9 = 54$ and the sign of the product is positive.

b. $\dfrac{-12.6}{3} = -4.2$ *Think*: $\dfrac{12.6}{3} = 4.2$ and the sign of the quotient is negative.

c. $(-4)(-4)(-4) = -64$ *Think*: $4 \cdot 4 \cdot 4 = 64$ and the sign of the product is negative.

PROGRESS CHECK 7

Find each product or quotient.

a. $(-8)(7)$ **b.** $\dfrac{-70.5}{-5}$ **c.** $(-3)(-3)(-3)(-3)$ ■

If two or more numbers are multiplied together, each number is a **factor** of the product. In Example 7c the factor –4 is used three times. An alternate way of writing $(-4)(-4)(-4)$ is $(-4)^3$. The number $(-4)^3$, or –64, is called the third **power** of –4. In general, by a^n, where n is a positive integer, we mean to use a as a factor n times.

$$a^n = \underbrace{a \cdot a \cdot a \cdots a}_{n \text{ factors}}$$

In the expression a^n, n is called the **exponent**, and a is called the **base**. When a product contains many factors the following sign rule is helpful.

Sign Rule

A product of nonzero factors is positive if the number of negative factors is even. The product is negative if the number of negative factors is odd.

Example 8: Evaluating a Power

Multiply

a. $(-3)^5$ **b.** $(-2)^4$ **c.** -2^4

Solution:

a. $(-3)^5 = -243$ *Think*: $3^5 = 243$ and the sign of the product is negative because there are an odd number of negative factors.

b. $(-2)^4 = 16$ *Think*: $2^4 = 16$ and the sign of the product is positive because there are an even number of negative factors.

c. $-2^4 = -16$ *Think*: $2^4 = 16$ and the negative of 2^4 is –16.

Caution: In this example, note the difference in meaning of $(-2)^4$ and -2^4. The expression $(-2)^4$ denotes the fourth power of -2, while -2^4 denotes the negative of 2^4.

PROGRESS CHECK 8
Multiply

a. $(-5)^3$ **b.** -2^6 **c.** $(-2)^6$ ■

When the product of two numbers is 1, then the numbers are called **reciprocals** of each other. The reciprocal of the nonzero number a is $1/a$, and the reciprocal of the nonzero fraction a/b is b/a. Using the concept of a reciprocal, division may be defined in terms of multiplication as follows.

Definition of Division

If a and b are real numbers with $b \neq 0$, then

$$a \div b = a \cdot \frac{1}{b}$$

To divide a by b, we multiply a by the reciprocal of b.

Example 9: Dividing by Using Reciprocals

Divide using the reciprocal definition of division.

a. $35 \div (-5)$ **b.** $\left(-\frac{7}{18}\right) \div \left(-\frac{4}{3}\right)$

Solution:

a. $35 \div (-5) = 35\left(-\frac{1}{5}\right) = -7$ *Think*: Product of 35 and the reciprocal of -5.

b. $\left(-\frac{7}{18}\right) \div \left(-\frac{4}{3}\right) = \left(-\frac{7}{18}\right) \cdot \left(-\frac{3}{4}\right) = \frac{7}{24}$ *Think*: Product of $-\frac{7}{18}$ and the reciprocal of $-\frac{4}{3}$.

PROGRESS CHECK 9
Divide using the reciprocal definition of division.

a. $(-54) \div (-9)$ **b.** $\left(-\frac{7}{3}\right) \div \left(\frac{11}{9}\right)$ ■

Without a priority for performing operations, different values may be possible for an expression involving more than one operation. To avoid this uncertainty, use this agreed-upon order of operations.

Order of Operations

1. Perform all operations within grouping symbols, such as parentheses, first. If there is more than one symbol of grouping, simplify the innermost symbol of grouping first, and simplify the numerator and denominator of a fraction separately.
2. Evaluate powers of a number.
3. Multiply or divide working from left to right.
4. Add or subtract working from left to right.

Example 10: Using the Order of Operations

Evaluate $1 + 7(2-5)^2$.

Solution: Follow the order of operations given above.

$$\begin{aligned} 1+7(2-5)^2 &= 1+7(-3)^2 && \textit{Think}\text{: Operate within parentheses.} \\ &= 1+7(9) && \textit{Think}\text{: Evaluate powers.} \\ &= 1+63 && \textit{Think}\text{: Multiply.} \\ &= 64 && \textit{Think}\text{: Add.} \end{aligned}$$

PROGRESS CHECK 10

Evaluate $8-5(6-10)^2$. ■

Calculator Computation

A graphing calculator has three distinct capabilities useful in this course.

1. It can do numeric calculations.
2. It can display graphs.
3. It is programmable.

At this time we focus on some of the more frequently used features for numeric calculations. You should learn to use these features on your own calculator.

a. The subtraction key is different from the key for making a negative sign.

b. There is a special key, usually marked [^], for raising numbers to powers. For example, 3^5 is entered as 3 [^] 5. Special keys may be used for squaring $[x^2]$, finding square roots $[\sqrt{\ }]$, and computing reciprocals $[x^{-1}]$.

c. There is usually a menu choice called **abs** for finding the absolute value of a number.

d. There is usually a key marked [ANS], which will reproduce the last answer.

e. There is usually a feature for converting decimal results (within certain limits) to fractions.

f. In some expressions multiplication is implied, which means that it is not always necessary to enter a multiplication symbol.

g. There are special keys for parentheses. Proper use of parentheses is crucial for correct evaluation of expressions.

h. There is a way to reproduce and edit the last line that was entered. This allows you to make slight changes easily without having to enter the entire expression again. In this editing procedure, you will often use keys marked [DEL] (for delete) and [INS] (for insert).

Example 11: Using Parentheses to Evaluate Expressions

Evaluate on a graphing calculator.

a. $\dfrac{3-(-5)}{-1-2}$ **b.** $\sqrt{8^2+15^2}$

Solution: A calculator display for these evaluations is shown in Figure R.9.

```
(3 - -5)/(-1 - 2)
            -2.66666666667
Ans▸Frac
                      -8/3
√(8² + 15²)
                        17
```

Figure R.9

a. Group the numerator and denominator in parentheses because the given problem may be expressed as $[3-(-5)]/(-1-2)$. Note that parentheses are not needed around –5, and note that the subtraction sign and the negative sign look different from one another on your calculator. The resulting decimal display is an approximate answer. The next line on the screen shows the result of converting the answer to fractional form. Thus, the exact value of the expression is $-8/3$.

b. Use the square root key and group the sum in parentheses. The resulting display shows $\sqrt{8^2+15^2}=17$.

PROGRESS CHECK 11

Evaluate on a graphing calculator.

a. $\dfrac{3-(-1)}{-5-2}$ **b.** $|4-3^2|$ ■

Example 12: Using Varying Return Rates in an Annuity Benefits Report

Solve the problem in the Chapter introduction on page 3.

Solution: In the given expression, replace r with 0.03 and determine

$$312{,}573(1+0.03)^{10}\approx 420{,}071.97$$

as shown in the first two lines of Figure R.10. Note that a tenth power was determined by using the power key [^] and that the multiplication symbol is not needed in front of the left parenthesis.

```
312573(1+.03)^10
           420071.974441
312573(1+.06)^10
           559770.637052
312573(1+.09)^10
           739973.965858
312573(1+.12)^10
```

Figure R.10

For the next calculation, since only the value of r changes, you can reproduce the original entry and edit it to change the .03 to .06. Then you press ENTER to get the new answer. You can repeat this procedure for $r=0.09$ and $r=0.12$ also. These steps give

$$312{,}573(1+0.06)^{10}\approx 559{,}770.64$$
$$312{,}573(1+0.09)^{10}\approx 739{,}973.97$$
$$312{,}573(1+0.12)^{10}\approx 970{,}804.29.$$

Observe that the total accumulation varies significantly as the rate of return increases in increments of 3 percent, and that the difference associated with each 3 percent increase is getting larger.

PROGRESS CHECK 12

Redo the problem in Example 12 assuming that the teacher has accumulated \$97,018 at age 35. You must evaluate $97{,}018(1+r)^{30}$ where r equals 0.03, 0.06, 0.09, and 0.12. ■

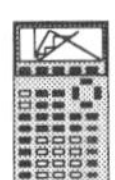

EXPLORE R.1

1. **a.** Explain why $12^{15}+1-12^{15}$ must equal 1.
   b. Now enter this expression into your calculator. Does it get the right answer? If not, explain why.
2. **a.** Explain why 111, 222, 333, 444, 555, 666 divided by 111, 222, 333, 444, 555, 000 does not equal 1.
   b. Now enter this expression into your calculator. Does it get the right answer? If not, explain why.
3. The following example shows how you can use a calculator to convert an improper fraction to a mixed number.
   Steps 1 through 3 illustrate the procedure for the improper fraction $215/7$.
   1. Divided 215 by 7 to get an approximate decimal answer.
   2. Subtract the integer part, which is 30.
   3. Use the decimal-to-fraction feature to get $5/7$.

   Thus, $215/7$ equals $30\ 5/7$.
   Use this procedure to express $312/11$ as a mixed number.
4. **a.** On most calculators if you divide 1 and 17 to find the repeating decimal form for $1/17$, the repeating part of the decimal is too long to fit on the display. However, by cleverly combining the information from several displays, you can determine the repeating block. By looking at these 4 results, for example, you can quickly determine that the repeating decimal form for $1/17$ is $.\overline{0588235294117647}$.

   $$\frac{1}{17}=.058823529412 \qquad \frac{2}{17}=.117647058824$$
   $$\frac{3}{17}=.176470588235 \qquad \frac{4}{17}=.235294117647$$

   Use these same displays to find the repeating decimal form for $2/17$.
   b. Using a calculator, express $2/23$ as a repeating decimal.

EXERCISES R.1

In Exercises 1–8 express each rational number as a repeating decimal

1. $\frac{4}{5}$ **2.** $\frac{5}{4}$ **3.** $\frac{5}{11}$

4. $\frac{2}{9}$ **5.** $\frac{17}{6}$ **6.** $\frac{26}{11}$

7. $\frac{10}{7}$ **8.** $\frac{100}{99}$

In Exercises 9–18 express each repeating decimal as the ratio of two integers

9. $0.\overline{2}$ **10.** $0.\overline{07}$ **11.** $0.\overline{321}$

12. $0.\overline{6332}$ **13.** $2.\overline{73}$ **14.** $1.\overline{6}$

15. $5.\overline{9}$ **16.** $4.\overline{81}$ **17.** $2.14\overline{3}$

18. $2.\overline{143}$

In Exercises 19–30 select all correct classifications from the following categories: real number, irrational number, rational number, integer, or none of these.

19. 0 **20.** $\frac{0}{3}$ **21.** $\frac{3}{0}$

22. $\frac{3}{4}$ **23.** -9 **24.** $\sqrt{9}$

25. $-\sqrt{9}$ **26.** $\sqrt{-9}$ **27.** $\sqrt{7}$

28. π **29.** $0.\overline{01}$ **30.** $0.101001\ldots$

In Exercises 31–50 name the property illustrated in the statement.

31. $2+7=7+2$ **32.** $11+0=11$

33. $4(5\cdot 11)=(4\cdot 5)11$ **34.** $6(4+3)=6\cdot 4+6\cdot 3$

35. $\sqrt{2}\cdot 1=\sqrt{2}$ **36.** $-7.3+7.3=0$

37. $(2.5)3=3(2.5)$ **38.** $17\left(\frac{1}{17}\right)=1$

39. $\pi\cdot 3+\pi\cdot 8=\pi(3+8)$

40. $(5+3)+(2+1)=[(5+3)+2]+1$

41. $z(xy)=(zx)y$ **42.** $(xy)z=z(xy)$

43. $x+0=0+x$ **44.** $(-z)+z=0$

45. $y(z+x)=(z+x)y$ **46.** $y\cdot 1=1\cdot y$

47. $x\left(\frac{1}{x}\right)=1$ if $x\neq 0$ **48.** $ax+ay=a(x+y)$

49. $y+(x+z)=(x+z)+y$

50. $(x+z)+y=x+(z+y)$

In Exercises 51–62 use the proper symbol (<, >, or =) between the pairs of numbers to indicate their correct order.

51. -7________1 **52.** 42.8________-91

53. $|14|$________$|-16|$ **54.** $|1.46|$________$|-1.46|$

55. -0.0001________-0.00001

56. 0.0001________0.00001

57. $\sqrt{5}$________$\sqrt{7}$ **58.** $-\sqrt{5}$________$-\sqrt{7}$

59. $\frac{5}{8}$________$\frac{2}{3}$ **60.** $\frac{3}{7}$________$\frac{5}{9}$

61. π________3.14 **62.** $\frac{1}{9}$________$0.\overline{1}$

In Exercises 63–106 evaluate each expression by performing the indicated operations.

63. $(-4)+(-3)$ **64.** $43+(-21)$

65. $|3-8|$ **66.** $|0-6|$

67. $(-6)-(-4)$ **68.** $(-12)-5$

69. $\left(-\frac{1}{5}\right)+3$ **70.** $7-\left(-\frac{3}{8}\right)$

71. $(-0.87)+0.33$ **72.** $0.17-(-0.48)$

73. $(-0.1)-4$ **74.** $(-3.3)+(-0.67)$

75. $(-11)6$ **76.** $(-8)(-3)$ **77.** $(-14)\div(-7)$

78. $89\div(-89)$ **79.** $-\left(\frac{1}{8}\right)\div 4$

80. $\left(-\frac{2}{3}\right)\div\left(-\frac{4}{15}\right)$ **81.** $(-5)\left(-\frac{4}{9}\right)$

82. $\left(\frac{3}{16}\right)(-2)$ **83.** $0\div(-4)$

84. $(-4)\div 0$ **85.** $0.2(-0.2)$

86. $(-0.64)\div(-0.16)$ **87.** $(-32.4)\div(-3)$

88. $10.1(-11)$ **89.** $(-2)\div(6-5)$

90. $(5-3)(5+2)$ **91.** $-4+2(3-8)$

92. $(-4+2)(3-8)$ **93.** $(2-4)(3-6)$

94. $2-4(3-6)$ **95.** $(-3)^3$

96. $(-2)^4$ **97.** -2^4 **98.** $-4\cdot 5^2$

99. $2\left(-\frac{1}{2}\right)^5$ **100.** $1+2\cdot 3^4$ **101.** $2-8(3-7)^2$

102. $(-5+2)^2+(-2-6)^2$

103. $-5(11-6)-[7-(11-19)]$

104. $7-3[2(13-5)-(5-13)]$

105. $3(2)(-1)+(-1)(-3)(2)+(6)(1)(-3)-6(2)(2)$
$-3(-3)(-3)-(-1)(1)(-1)$

106. $3[2(-1)-(-3)(-3)]-(-1)[1(-1)-(-3)(2)]$
$+6[1(-3)-2(2)]$

In Exercises 107 and 108 use a calculator to evaluate the given expression. If your calculator can convert decimals to fractions, express any rational answers in fractional form.

107. **a.** $\frac{2+3}{5-1}$ **b.** $\sqrt{3^2+5}$

c. $\frac{-2-(-8)}{1+3^2}$ **d.** $\left|5-4^2\right|$

108. **a.** $\frac{-3-1}{4+5}$ **b.** $\sqrt{5^2-4^2}$

c. $\frac{22-8}{(1+3)^2}$ **d.** $|5|-|-4|^2$

109. Evaluate $\left(1+\frac{1}{n}\right)^n$ for $n=1$, 10, 100, 1,000, and 10,000.

110. Evaluate $\left(1+\frac{2}{t}\right)^t$ for $t=2$, 20, 200, 2,000, and 20,000.

111. If \$2000 is deposited in a bank account when a baby is born, then its value 21 years later is given by $2000(1+r)^{21}$ where r is the annual interest rate. Evaluate the expression for $r = 0.03$, 0.06, and 0.09.

112. If \$100 is deposited at the end of every month in a savings account that pays monthly interest, then the value of the account after 240 deposits (20 years) is given by $100\left[\frac{(1+(r/12))^{240}-1}{r/12}\right]$. Evaluate this expression for $r = 0.03$, 0.06, and 0.09.

113. A glass vase has a temperature of 2,000 degrees Fahrenheit when it is removed from the glassblower's furnace. As it cools to room temperature (80 degrees), its temperature at any time x is given by the formula $T = 80 + 2000(0.942)^x$ where x is the number of minutes elapsed since removal from the furnace. Find the temperature of this vase after 15, 30, 60 and 90 minutes have elapsed.

114. A researcher worked out a formula that gives the area A in square centimeters of a growing colony of bacteria to be

$$A = \frac{0.252}{0.0051 + 0.119^x}$$

where x is the number of days elapsed from the initial measurement. Find the area after 1 and 2 days have elapsed.

115. Find, by two methods, the area of the rectangle and explain how the rectangle can be used to illustrate geometrically the distributive property.

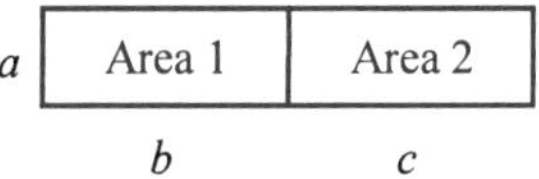

116. Explain in terms of multiplication why $1/0$ is undefined.

117. If $xy = x$, what are the possible values for x and y?

118. If $x^2 + y^2 = 0$, what are the possible values for x and y?

In Exercises 119–128 answer true or false. If false, give a specific counterexample.

119. All rational numbers are integers.

120. All rational numbers are real numbers.

121. All real numbers are irrational numbers.

122. All integers are irrational numbers.

123. The quotient of two integers is always an integer.

124. The quotient of two integers is always a rational number.

125. Every real number is either a rational number or an irrational number.

126. A number that can be written as a repeating decimal is called a rational number.

127. The nonnegative integers are the positive integers.

128. The nonpositive real numbers are the negative real numbers and zero.

THINK ABOUT IT R.1

1. A common student error is to assume that $-a$ must represent a negative number. Explain why this assumption is incorrect.
2. Give an example in which the sum of two irrational numbers is a rational number.
3. Use a dictionary to find meanings for the words *commute*, *associate*, and *distribute* that are consistent with the concepts expressed by the commutative, associative, and distributive properties discussed in this section.
4. What is the 1998th digit to the right of the decimal point in the decimal expansion of $22/7$?
5. A prime number is a positive integer other than 1 that is exactly divisible only by itself and 1.
   a. The *fundamental theorem of arithmetic* states that every integer greater than 1 can be written as the product of prime factors in exactly one way, if we disregard the order of the factors. For example, the prime factorization of 12 is $2 \cdot 2 \cdot 3$. This means that if 2 numbers are equal, they must have the same prime factorization. Suppose that n is any integer greater than 1. Explain why n^2 must have an even number of prime factors, and why $2n^2$ must have an odd number of prime factors.
   b. Explain (using the results of part **a**) how the assumption that $\sqrt{2}$ may be written in the form a/b, where a and b are positive integers, contradicts the fundamental theorem of arithmetic.

● ● ●

R.2 Algebraic Expressions and Geometric Formulas

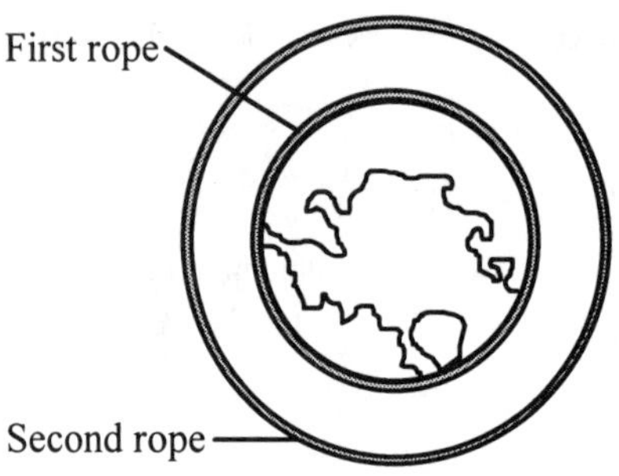

A rope is wrapped tightly around the equator of the earth. A second rope is suspended in the air 1 ft directly above the first all the way around the earth. Is the difference between the rope lengths about 6 ft, 600 ft, 60,000 ft, or 6 million ft? Justify your answer.
(See Example 6.)

Objectives

1. Evaluate algebraic expressions.
2. Simplify algebraic expressions by combining like terms.
3. Apply geometric formulas.
4. Translate between verbal expressions and algebraic expressions.

In algebra, two types of symbols are used to represent numbers: variables and constants. A **variable** is a symbol that may be replaced by different numbers in a particular problem, while a **constant** is a symbol that represents the same number throughout a particular problem. An expression that combines variables and constants using the operations of arithmetic is called an **algebraic expression**. For example,

$$3x^2 - 5x + 7 \qquad \frac{-b + \sqrt{b^2 - 4ac}}{2a}, \text{ and } \frac{1}{2}gt^2$$

are algebraic expressions.

If we are given numerical values for the symbols, we can evaluate the expression by substituting the given values and performing the indicated operations. Such substitutions are based on the **substitution property** that states that if a and b are real numbers and $a = b$, then either may replace the other without affecting the truth value of the statement.

Example 1: Evaluating an Algebraic Expression

Evaluate each expression, given that $x = -2$, $y = 3$, and $z = -4$.

a. $5y - x$ **b.** $-x^2 + 5yz^2$

Solution:

a. $5y - x = 5(3) - (-2)$ Replace y by 3 and x by -2.

$= 15 + 2$

$= 17$

b. $-x^2 + 5yz^2 = -(-2)^2 + 5(3)(-4)^2$ Replace x by -2, y by 3 and z by -4.

$= -(4) + 5(3)(16)$

$= -4 + 240$

$= 236$

Caution: Pay close attention to how the expression $-x^2$ was evaluated in part **b**. Because powers precede multiplication in the order of operations, $-x^2$ means that you must first square x and then take the negative of your result. In general, $-a^n$ denotes the negative of a^n.

Technology Link

On some calculators, variable names may be used to name memory cells in which you can save numbers or expressions. Figure R.11 shows a calculator display for the evaluation considered in Example 1**a**.

Figure R.11

To obtain this type of display, you should understand the Store and Alphabetic features of your calculator.

PROGRESS CHECK 1

Evaluate each expression, given that $x=-2$, $y=3$, and $z=-4$.

a. $4x-3y$ **b.** $-y^2+2xz^3$ ■

Example 2: Evaluating an Expression Involving Subscripts

If

$$m=\frac{y_2-y_1}{x_2-x_1}$$

evaluate m when $x_1=2$, $y_1=-5$, $x_2=-1$, and $y_2=3$. (*Note*: In the symbols, x_1, y_1, x_2 and y_2 the numbers 1 and 2 are called **subscripts**. In this case x_1 and x_2 are used to denote various x values.)

Solution:

$$m=\frac{y_2-y_1}{x_2-x_1}$$

so when $x_1=2$, $y_1=-5$, $x_2=-1$, and $y_2=3$, we have

$$m=\frac{(3)-(-5)}{(-1)-(2)}=\frac{8}{-3}=-\frac{8}{3}.$$

PROGRESS CHECK 2

Let m be defined as in Example 2, and evaluate m when $x_1=3$, $y_1=-4$, $x_2=7$, and $y_2=-10$. ■

We often need to combine algebraic expressions that are added or subtracted. Those parts of an algebraic expression separated by plus (+) signs are called **terms** of the expression. The following example will help you recognize terms.

Example 3: Recognizing Terms of an Algebraic Expression

Identify the term(s) in the following expressions.

a. $3x^2-5x+7$ **b.** πr^2h

Solution:

a. $3x^2 - 5x + 7$ may be written as $3x^2 + (-5x) + 7$ and is an algebraic expression with three terms, $3x^2$, $-5x$, and 7.

b. Because $\pi r^2 h$ is a product (instead of a sum or difference), it is an algebraic expression with one term, $\pi r^2 h$.

PROGRESS CHECK 3

Identify the terms in the following expressions.

a. $-4x^3 + x^2 - 3x + 7$ **b.** $\dfrac{-7x}{2}$ ■

If a term is the product of some constants and variables, the constant factor is called the (**numerical**) **coefficient** of the term. For example, the coefficient of the term $2x$ is 2, and the coefficient of $-7x^2$ is -7. Every term has a coefficient. If the term is x, the coefficient is 1, since $x = 1 \cdot x$. Similarly, if the term is $-x$, the coefficient is -1, since $-x = -1 \cdot x$. If two terms have identical variable factors (such as $3x^2y$ and $-2x^2y$), they are called **like terms**. The distributive property indicates that we combine like terms by combining their coefficients.

Example 4: Combining Like Terms

Simplify by combining like terms if possible.

a. $3x + 7x$ **b.** $yz^2 - 10yz^2$ **c.** $8x - y + 2y - 3x$ **d.** $-3x^2y + xy^2 - 7xy$

Solution:

a. $3x + 7x = (3 + 7)x = 10x$

b. $yz^2 - 10yz^2 = (1 - 10)yz^2 = -9yz^2$

c. $8x - y + 2y - 3x = (8 - 3)x + (-1 + 2)y = 5x + y$

d. There are no like terms in $-3x^2y + xy^2 - 7xy$, so we cannot simplify the expression.

PROGRESS CHECK 4

Simplify by combining like terms if possible.

a. $x - 8x$ **b.** $3 - 8x$ **c.** $2p - 3p + 4p$ **d.** $-x^2 + 6x + 6x^2 - x$ ■

To combine algebraic expressions, it is sometimes necessary to remove the parentheses or brackets that group certain terms together. Parentheses are removed by applying the distributive property; that is, we multiply each term inside the parentheses by the factor in front of the parentheses. If the grouping is preceded by a minus sign, the factor is -1, so the sign of each term inside the parentheses must be changed. If the grouping is preceded by a plus sign, the factor is 1, so the sign of each term inside the parentheses remains the same. If there is more than one symbol of grouping, it is usually better to remove the innermost symbol of grouping first.

Example 5: Removing Grouping Symbols and Simplifying

Remove the symbols of grouping and combine like terms.

a. $-(x - y) - 2(3x - y)$ **b.** $7 - 2[4x - (1 - 3x)]$

Solution:

a. $-(x - y) - 2(3x - y)$

$= -1(x - y) - 2(3x - y)$ Use $-(x - y) = -1(x - y)$

$= -x + y - 6x + 2y$

$= -7x + 3y$

b. $7-2[4x-(1-3x)]$
$=7-2[4x-1+3x]$ Distributive property (remove parentheses).
$=7-2[7x-1]$
$=7-14x+2$ Distributive property (remove brackets).
$=-14x+9$

PROGRESS CHECK 5
Remove the symbols of grouping and combine like terms.
a. $-(x+y)+3(x-y)$ **b.** $5x-4[x-(1+x)]$ ■

An application of combining or evaluating algebraic expressions often occurs when we apply formulas. A **formula** is an equality statement that expresses the relationship between two or more variables. For this course, a useful category of formulas is found in geometry, and Figures R.12 and R.13 provide some of these formulas for reference purposes.

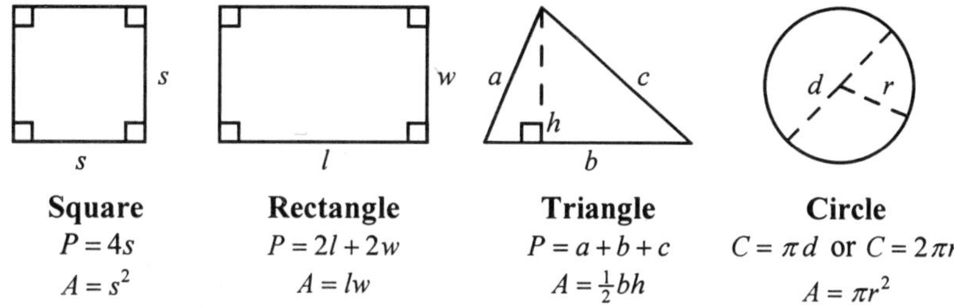

Square	**Rectangle**	**Triangle**	**Circle**
$P=4s$	$P=2l+2w$	$P=a+b+c$	$C=\pi d$ or $C=2\pi r$
$A=s^2$	$A=lw$	$A=\frac{1}{2}bh$	$A=\pi r^2$

Figure R.12
Two-dimensional figures: Formulas for perimeter P, circumference C, and area A.

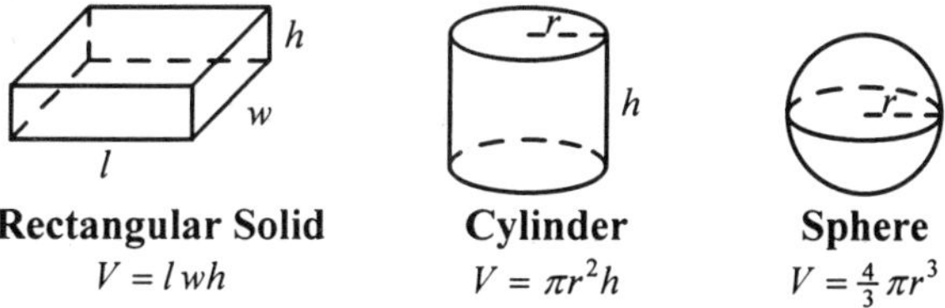

Rectangular Solid	**Cylinder**	**Sphere**
$V=lwh$	$V=\pi r^2h$	$V=\frac{4}{3}\pi r^3$

Figure R.13
Three-dimensional figures: Formulas for volume V.

Example 6: A Popular Geometric Puzzle

Solve the problem in the section introduction on page 18.

Solution: Although you might expect the difference between the rope lengths to be very large, they only differ by about 6 ft. Since the equator of the earth may be approximated by a circle with diameter d, the circumference formula, $C=\pi d$, tells us that the length of the first rope is πd, while the length of the second rope is $\pi(d+2)$. Then

$$\begin{aligned}\text{difference} &= \pi(d+2)-\pi d\\ &= \pi d+2\pi-\pi d\\ &= 2\pi\text{ ft.}\end{aligned}$$

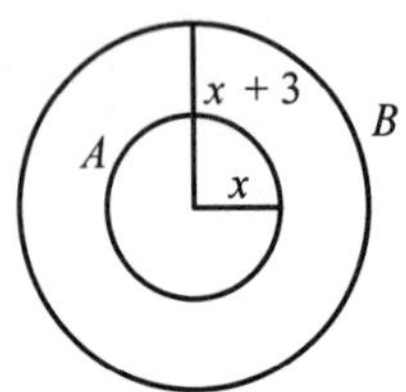

Figure R.14

Since $\pi \approx 3.14$, the difference is about 6 ft. Note that *any* two circles that differ by 2 ft in diameter differ by 2π ft in circumference.

PROGRESS CHECK 6

Circle A has a radius of x ft and circle B has a radius of $x+3$ ft, as shown in Figure R.14. How much larger is circumference B than circumference A? ■

Example 7: Finding the Volume and Weight of a Bowling Ball

A bowling ball has an 8-in. diameter.

a. Find its volume to the nearest cubic inch. Disregard any holes in the ball.

b. If the material weights 0.9 oz/in.^3, find the weight of the ball to the nearest pound. (There are 16 oz. in 1 lb.)

Solution:

a. Since the shape of the ball is a sphere, the formula to use is $V = (4/3)\pi r^3$. Because $d = 8$ in. and $r = (1/2)d$, we replace r by 4. If available, use the calculator key labeled π in the computation. Otherwise, replace π by 3.14159.

$$\begin{aligned} V &= \frac{4}{3}\pi r^3 \\ &= \frac{4}{3}\pi(4^3) \\ &= 268.08257\ldots \end{aligned}$$

To the nearest cubic inch, the volume is 268 in.^3.

b. Each cubic inch weighs 0.9 oz, so the ball weighs 0.9(268), or 241.2 oz. Since there are 16 oz in 1 lb, this weight to the nearest pound is 15 lb.

Technology Link

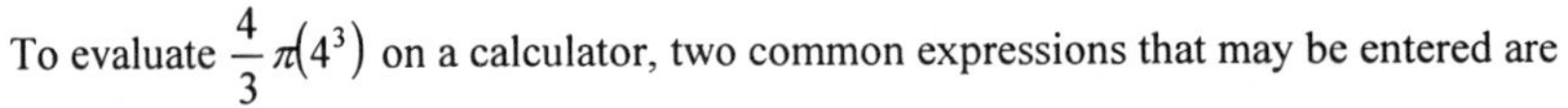

To evaluate $\frac{4}{3}\pi(4^3)$ on a calculator, two common expressions that may be entered are

$$\left(\frac{4}{3}\right)\pi(4^3) \text{ and } 4\pi\frac{4^3}{3}.$$

It is incorrect to enter $4/3\pi(4^3)$ here, because a calculator interprets this expression to mean

$$\frac{4}{3\pi(4^3)}.$$

PROGRESS CHECK 7

A circular backyard aboveground pool has a diameter of 20 ft.

a. What volume of water (to the nearest cubic foot) will it take to fill the pool to a depth of 4.3 ft?

b. Use the fact that one cubic foot of water holds about 7.5 gallons to express the volume in gallons of water. ■

To develop proficiency in the language of algebra, you must be able to translate between verbal expressions and algebraic expressions. The following chart shows some algebraic expressions that may be used to translate typical verbal expressions that involve arithmetic operations.

Operations	Verbal Expression	Algebraic Expression
Addition	The sum of a number x and 1	$x+1$, or $1+x$
	A number y plus 2	$y+2$, or $2+y$
	A number w increased by 3	$w+3$, or $3+w$
	4 more than a number b	$b+4$, or $4+b$
	Add 5 and a number d	$d+5$ or $5+d$
Subtraction	The difference of a number x and 6	$x-6$
	A number y minus 5	$y-5$
	A number w decreased by 4	$w-4$
	3 less than a number b	$b-3$
	Subtract 11 from a number d	$d-11$
	Subtract a number d from 11	$11-d$
Multiplication	The product of a number x and 5	$5x$
	4 times a number t	$4t$
	A number y multiplied by 7	$7y$
	Twice a number w	$2w$
	1/3 of a number n	$(1/3)n$
	25 percent of a number P	$0.25P$
Division	The quotient of a number x and 8	$x \div 8$, or $x/8$
	A number y divided by 2	$y \div 2$, or $y/2$
	The ratio of a number a to a number b	$a \div b$, or a/b, or $a{:}b$

Example 8: Translating to an Algebraic Expression

Translate each statement to an algebraic expression.

a. \$3 less than the list price p. **b.** 6 percent of the sum of x and y.

Solution:

a. \$3 less than p is expressed as $p-3$ dollars. Note that $3-p$ dollars is not correct here.

b. 6 percent is written as 0.06, the sum of x and y is written as $x+y$, and in this context the word *of* implies multiplication. Thus, the given expression translates to $0.06(x+y)$.

PROGRESS CHECK 8

Translate each statement to an algebraic expression.

a. \$5 more than the retail price p. **b.** 8 percent of the product of a and b. ■

Example 9: Translating to a Verbal Expression

Translate each algebraic expression to a verbal expression.

a. $ab+ac$ **b.** $a(b+c)$

Solution:

a. The sum $ab+ac$ may be read as "the sum of a times b and a times c."

b. The product $a(b+c)$ may be stated as "a times the sum of b and c."

Note: In this type of example, other translations are possible. For instance, $a(b+c)$ may be read as "the product of a and the sum of b and c." However, care must be taken to avoid ambiguous statements like "a times b plus c" where it is not clear whether $ab+c$ or $a(b+c)$ is intended.

PROGRESS CHECK 9

Translate each algebraic expression to a verbal expression

a. $ab+c$ **b.** $\dfrac{a}{a+5}$ ■

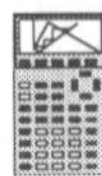

EXPLORE R.2

1. Determine on your calculator which of these sequences correctly evaluates $60/[3(2)]$.
   a. $60 \div (3 \cdot 2)$ **b.** $60 \div 3 \cdot 2$ **c.** $60 \div 3 \div 2$
2. Determine on your calculator which of these sequences correctly evaluates $60/3x$ when $x = 2$. Begin by storing 2 for x on your calculator.
   a. $60 \div 3x$ **b.** $60 \div 3 \cdot x$ **c.** $60 \div 3 \div x$
3. Most graphing calculators have a List feature, which can provide a list of values for substitution into an expression. Usually, this feature is engaged by using braces to group numbers. The figure below illustrates a typical screen for this procedure. Note that the list is first stored as a variable, and then that variable appears in the desired expression. The screen shows one way to evaluate $3x + 8$ for $x = 5$, 1, and 7.

```
{5,1,7}→X
                    {5 1 7}
3X+8
                 {23 11 29}
```

   a. Use this method to evaluate $-3x^2 + 1$ for $x = -1, -.4$ and 2.5.
   b. Use this method to evaluate $-4x^3 + x^2 - 3x - 2$ for $x = -2$, 0.01, and $1/5$.

EXERCISES R.2

In Exercise 1–20 evaluate each expression, given that $x = -2$, $y = 3$, and $z = -4$.

1. $x + y + z$ **2.** $4x - z$

3. x^3x^2 **4.** $\dfrac{y^3}{y^2}$

5. $5z - xy^2$ **6.** $20 - xyz^3$

7. $(-x)^2 + 2y$ **8.** $-x^2 + 2y$

9. $3y^3 - 2y^2 + y - 2$ **10.** $2x^3 + 3x^2 - x + 1$

11. $x - 2(3y - 4z)$ **12.** $(x - 2)(3y - 4z)$

13. $(z - x)(y - 3z)$ **14.** $z - x(y - 3z)$

15. $(2y - x)x - z$ **16.** $(2y - x)(x - z)$

17. $(x + y + z)(x + y - z)$

18. $x + y + z(x + y - z)$

19. $(x + 2)(y - 3) + (3 - z)(4 + x)$

20. $(5 - y)(z + 1) - (y - 2)(3 + x)$

In Exercises 21–26 $m = \dfrac{y_2 - y_1}{x_2 - x_1}$. Evaluate m for the following values of x_1, y_1, x_2, and y_2.

	x_1	y_1	x_2	y_2
21.	5	4	1	2
22.	1	6	4	8
23.	–3	–1	–6	–4
24.	–5	–3	–2	–6
25.	2	–1	–6	3
26.	0	3	–2	–4

In Exercises 27–32 identify the term(s) in the given expressions.

27. $2n + 4w$ **28.** $-m + 4x$

29. $x^2 - 3x - 2$ **30.** $2y^2 + 3y - 4$

31. $\dfrac{1}{2}bh$ **32.** $(2a)(3b)$

In Exercises 33–56 simplify each expression by combining like terms. Remove symbols of grouping when necessary.

33. $6a + 4a + 9a$ **34.** $-3b - b + 7b$

35. $8a - 3b + 6a - 5b$

36. $5k - 6m - 7 + 3k - 4m + 2$

37. $5xy - 4cd - 2xy + 9cd$

38. $4p - 5q + 4p + q + 4p - 2q$

39. $2x^3 + 6y^2 - 5x^3 + 2y^2$

40. $2a^2b + 3ab - 4ab^2$ **41.** $x - 2(x + y) + 3(x - y)$

42. $k + (m - k) - (m - 2k)$

43. $-2(a - 2b) - 7(2a - b)$

44. $2a - (7a + 3) + (4a - 6)$

45. $-(x + y) + 4(x - y) + 7x - 2y$

46. $a(b - c) - b(a + c)$

47. $(2x^3 + 7x^2y^2 + 9xy^3) + (6x^2y^2 - 3x^3y)$

48. $(3c^2d + 2cd - 5d^3) - (9d^3 - 6c^2d - 2cd)$

49. $(3y^3 - 2y^2 + 7y - 1) + (4y^3 - y^2 - 3y + 7)$

50. $(2n - 4n^3 + 3n^2 + 4) - (6 - 2n^2 + 7n - n^3)$

51. $(b^3 - 2b^2 + 3b + 4) - (b^2 + 2b - 1)$

52. $(4x^3 + 7x^2y^2 - 2y^3) + (-7x^3 + 2x^2y^2 + 2y^3)$

53. $2[a + 5(a + 2)] - 6$ **54.** $10 - 4[3x - (1 - x)]$

55. $3a - [3(a - b) - 2(a + b)]$

56. $-[x - (x + y) - (x - y) - (y - x)]$

57. The side of the inside square is x; the side of the outside square is $x + 2$. How much larger is the perimeter of the outside square than the perimeter of the inside square?

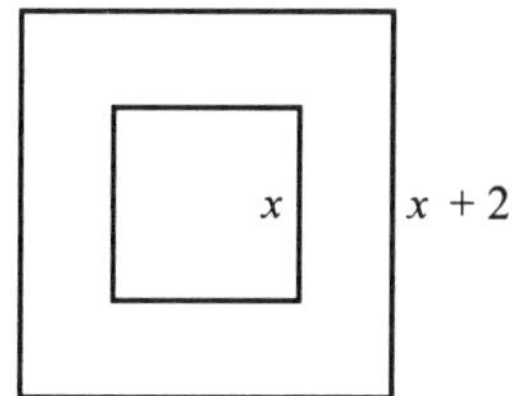

58. One solid ball with an 8-inch diameter is made of material weighing 0.8 oz./in.^{3}. A second ball is the same size, but made of material weighing 0.7 oz./in.^{3}. How much heavier is the first ball?

59. Find and simplify an expression for the perimeter of the figure shown. Use 3.14 for π. The curve is a semicircle.

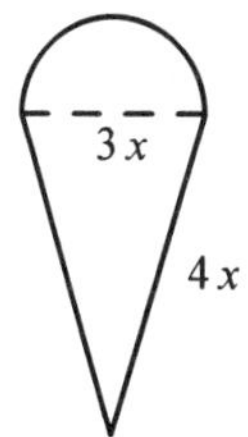

60. Find and simplify an expression for the perimeter of the figure shown. Use 3.14 for π. The curves are semicircles that have the same diameter.

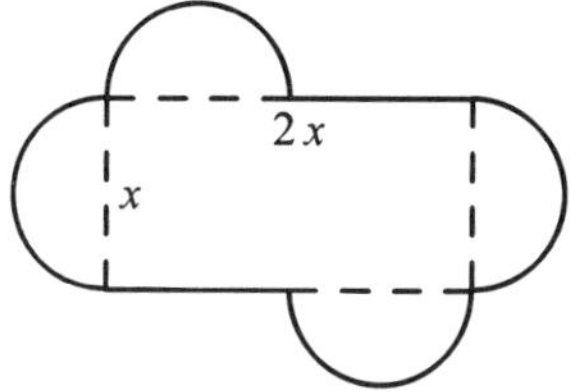

61. Find and simplify an expression for the area of the figure shown. The curves are semicircles. Use 3.14 for π.

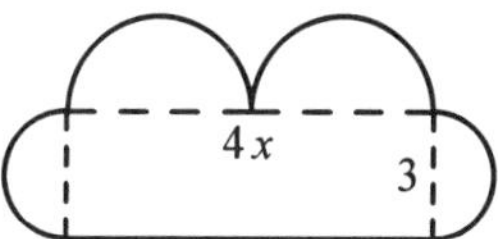

62. Find and simplify an expression for the area of the figure shown.

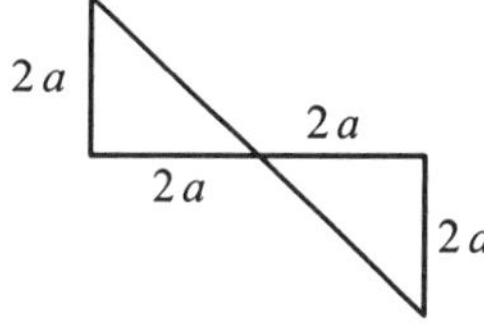

63. A cylindrical storage vat has a diameter of 8 feet and a depth of 7 feet.

a. If it is filled with liquid to within 1 ft of the top, how much liquid does it hold? Give the answer to the nearest cubic foot.

b. Express the volume in gallons if 1 ft^3 of liquid holds about 7.5 gallons. Give the answer to the nearest 10 gallons.

64. Redo Exercise 63, but now assume that the vat is filled to within 2 ft of the top.

In Exercise 65–72 translate the given statement to an algebraic expression.

65. 4.5 percent of the amount of sales t.

66. 10 percent of the rental fee f.

67. $1 more than the cost of the ticket c.

68. $9.95 less than the list price l.

69. The sum of the price p and the sales tax, which is 4 percent of the price.

70. The list price l minus the 30 percent discount on the list price.

71. The sum of a and the product of m and n.

72. The quotient of x and the sum of 2 and x.

In Exercise 73–76 translate each algebraic expression into a verbal expression

73. $4y + x$

74. $4(y + x)$

75. $\frac{a}{bc}$

76. $\frac{a}{b} \cdot c$

THINK ABOUT IT R.2

1. Explain the difference in meaning between $-a^2$ and $(-a)^2$.
2. **a.** Use the distributive property to show that $2x - 2x$ equals 0 for all values of x.
   b. Use the distributive property to show that $2 - 2x$ does not equal 0 unless $x = 1$.
3. Simplify $(2/3)\{(1/2) - (1/8)[4x - 2(8x + 3)]\}$. Check by letting $x = 2$.
4. **a.** Make up two terms that are like terms with respect to $3xy^2$. Why are they like terms?
   b. Give two terms that are unlike terms with respect to $3xy^2$. Explain why they are unlike terms.
5. In parts **a** and **b** try the number trick with at least one number. Then show algebraically why the trick works.
   a. Choose a number.
   Add three.
   Multiply by two.
   Add six.
   Divide by two.
   Subtract your original number.
   Your result is six.
   b. Choose a number.
   Triple it.
   Add the number that is one larger than your original number.
   Add seven.
   Divide by four.
   Subtract two.
   The result is your original number.

● ● ●

R.3 Integer Exponents

Photo Courtesy of ***Chuck Mason*** *of International Stock*

The equal-monthly-payment formula is used by banks when they finance many common loans. The problem is to find the equal monthly installment E that will pay off a loan of P dollars with interest in n months. The monthly interest rate is r and interest is charged at each payment period only on the unpaid balance. The formula used by the bank is

$$E = \frac{Pr}{1-(1+r)^{-n}}.$$

Suppose you obtain an $11,000 home improvement loan for 36 months with 9 percent annual interest charged on the unpaid balance. Determine the amount of the equal monthly installment that will pay off this debt. (See Example 5.)

Objectives

1. Evaluate expressions containing integer exponents.
2. Simplify expressions involving integer exponents.
3. Simplify expressions involving literal exponents.
4. Convert numbers from normal notation to scientific notation, and vice versa.
5. Solve problems by calculator that involve scientific notation.

Recall that a positive integer exponent is a shortcut way of expressing a repeating factor. That is, a^n, where n is a positive integer, is shorthand for

$$\underbrace{a \cdot a \cdot a \cdots a}_{n \text{ factors}}.$$

From this definition it is easy to obtain the following laws of exponents, which we will illustrate in specific terms in Example 1.

Laws of Positive Integer Exponents

Let a and b be any real numbers and m and n be any positive integers.

1. $a^m \cdot a^n = a^{m+n}$
2. $(a^m)^n = a^{mn}$
3. $(ab)^n = a^n b^n$
4. $\left(\frac{a}{b}\right)^n = \frac{a^n}{b^n} \quad (b \neq 0)$
5. $\frac{a^m}{a^n} = \begin{cases} a^{m-n} & \text{if } m > n \\ \frac{1}{a^{n-m}} & \text{if } n > m \end{cases} \quad (a \neq 0)$

Example 1: Using the Laws of Positive Integer Exponents

Simplify by the laws of exponents and check your result.

a. $2^2 \cdot 2^3$ **b.** $(2^3)^2$ **c.** $(2 \cdot 3)^2$ **d.** $\frac{2^5}{2^2}$ **e.** $\frac{2^2}{2^5}$

Solution:	*Check*
a. $2^2 \cdot 2^3 = 2^{2+3} = 2^5 = 32$	$2^2 \cdot 2^3 = 4 \cdot 8 = 32$
b. $(2^3)^2 = 2^{3 \cdot 2} = 2^6 = 64$	$(2^3)^2 = 8^2 = 64$

c. $(2\cdot 3)^2 = 2^2\cdot 3^2 = 4\cdot 9 = 36$ $\quad (2\cdot 3)^2 = 6^2 = 36$

d. $\frac{2^5}{2^2} = 2^{5-2} = 2^3 = 8$ $\quad \frac{2^5}{2^2} = \frac{32}{4} = 8$

e. $\frac{2^2}{2^5} = \frac{1}{2^{5-2}} = \frac{1}{2^3} = \frac{1}{8}$ $\quad \frac{2^2}{2^5} = \frac{4}{32} = \frac{1}{8}$

PROGRESS CHECK 1

Simplify by the laws of exponents and check your result.

a. $3^4\cdot 3$ b. $(3^4)^2$ c. $(4\cdot 2)^3$ d. $\frac{3^4}{3^2}$ e. $\frac{3^2}{3^4}$ ■

Example 2: Using the Laws of Positive Integer Exponents

Simplify by the laws of exponents

a. $(6x^5y^3)(-2xy^7)$ b. $\frac{18a^3b^2c^2}{12ab^4c}$

Solution:

a. $(6x^5y^3)(-2xy^7) = 6(-2)x^{5+1}y^{3+7} = -12x^6y^{10}$

b. $\frac{18a^3b^2c^2}{12ab^4c} = \frac{18a^{3-1}c^{2-1}}{12b^{4-2}} = \frac{3a^2c}{2b^2}$

PROGRESS CHECK 2

Simplify by the laws of exponents.

a. $(-4x)^2(9x^5)$ b. $\frac{14xy^5}{21x^4y^4}$ ■

We now wish to extend our definition of exponents to zero and negative integers. Note that it is meaningless to use x as a factor either zero times or a negative number of times, so we must define these exponents in a different manner. However, our guideline in these new definitions is to retain the laws of exponents developed for positive integers.

We start by considering the first law of exponents.

$$a^m\cdot a^n = a^{m+n}$$

If $n=0$, we have

$$a^m\cdot a^0 = a^{m+0} = a^m.$$

When we multiply a^m by a^0, our result is a^m. Thus, a^0 must equal 1, and we make the following definition.

Zero Exponent

> If a is a nonzero real number, then
>
> $$a^0 = 1.$$

Example 3: Evaluating an Expression with a Zero Exponent

Evaluate each expression. Assume $x\neq 0$.

a. 9.64×10^0 b. $3x^0$ c. -4^0

Solution: Apply the zero exponent definition.

a. Because $10^0 = 1$, $9.64 \times 10^0 = 9.64(1) = 9.64$

b. $3x^0$ is equivalent to $3^1 \cdot x^0$, so $3x^0 = 3(1) = 3$.

c. Recall that the form $-a^n$ denote the negative of a^n. Thus, $-4^0 = -(4^0) = -1$.

PROGRESS CHECK 3

Evaluate each expression. Assume $x \neq 0$.

a. $-5x^0$ **b.** -5^0 **c.** 4.72×10^0 ■

To obtain a definition for exponents that are negative integers, we will again consider the first law of exponents.

$$a^m \cdot a^n = a^{m+n}$$

If $m = 5$ and $n = -5$, we have

$$a^5 \cdot a^{-5} = a^{5+(-5)} = a^0 = 1.$$

When we multiply a^5 by a^{-5}, the result is 1. Thus, a^{-5} is the reciprocal of a^5, or $a^{-5} = 1/a^5$. In general, our previous laws of exponents may be extended by making the following definition.

Negative Exponent

If a is a nonzero real number and n is an integer, then

$$a^{-n} = \frac{1}{a^n}.$$

Example 4: Evaluating an Expression with a Negative Exponent

Evaluate each expression.

a. 5^{-2} **b.** $(-4)^{-3}$ **c.** $\left(\frac{3}{2}\right)^{-4}$

Solution: Use the negative exponent definition.

a. $5^{-2} = \frac{1}{5^2} = \frac{1}{25}$ **b.** $(-4)^{-3} = \frac{1}{(-4)^3} = \frac{1}{-64} = -\frac{1}{64}$

c. $\left(\frac{3}{2}\right)^{-4} = \frac{1}{(3/2)^4} = \frac{1}{81/16} = 1 \cdot \frac{16}{81} = \frac{16}{81}$

Note: Part **c** shows the evaluation of a fraction raised to a negative power. This type of evaluation may also be done by using

$$\left(\frac{a}{b}\right)^{-n} = \left(\frac{b}{a}\right)^n \quad a, b \neq 0.$$

For instance,

$$\left(\frac{3}{2}\right)^{-4} = \left(\frac{2}{3}\right)^4 = \frac{2^4}{3^4} = \frac{16}{81}.$$

Technology Link

Integer exponents are generally evaluated on a calculator with the Power key in the usual way. For example, Figure R.15 displays the computation for $(-4)^{-3}$ in which we also used the feature for converting a rational number to fraction form. When entering such expressions, be careful to use the key [(–)] for making a negative sign instead of the subtraction key [–].

```
(-4)^-3
                -.015625
Ans▸Frac
                   -1/64
```

Figure R.15

An exponent of –1 is best computed with the Reciprocal key, which is labeled x^{-1} on some calculators, and $1/x$ on others.

PROGRESS CHECK 4

Evaluate each expression.

a. 7^{-1} **b.** $(-3)^{-4}$ **c.** $\left(\frac{2}{5}\right)^{-3}$ ■

Example 5: Calculating Monthly Installment Payments

Solve the problem in the section introduction on page 27.

Solution: The annual interest rate is 9 percent, so the monthly interest rate is 0.09/12, or 0.0075. We substitute $r = 0.0075$, $P = 11{,}000$ and $n = 36$ in the given formula, and then simplify with the aid of a calculator.

$$\begin{aligned} E &= \frac{Pr}{1-(1+r)^{-n}} && \text{Equal monthly payment formula.} \\ &= \frac{11{,}000(0.0075)}{1-(1+0.0075)^{-36}} && \text{Substitution.} \\ &= 349.80 && \text{By calculator to the nearest cent.} \end{aligned}$$

The monthly payment will be \$349.80. Note that this schedule gives a total of \$12,592.80, of which \$1,592.80 is interest.

PROGRESS CHECK 5

Redo Example 5, but assume that the annual interest rate is 12 percent. ■

Integer exponents have been defined so that all previous properties of exponents continue to apply. We will therefore restate the laws of exponents as they apply to all integers. In particular, note that property 5 (the quotient property) can now be stated simply as $a^m/a^n = a^{m-n}$ because negative integer and zero exponents are now sensible.

Laws of Integer Exponents

Let a and b be any nonzero real numbers, and m and n be any integers.

1. $a^m \cdot a^n = a^{m+n}$
2. $(a^m)^n = a^{mn}$
3. $(ab)^n = a^n b^n$
4. $\left(\dfrac{a}{b}\right)^n = \dfrac{a^n}{b^n}$
5. $\dfrac{a^m}{a^n} = a^{m-n}$
6. $\left(\dfrac{a}{b}\right)^{-n} = \left(\dfrac{b}{a}\right)^n$

Example 6: Using the Laws of Integer Exponents

Simplify and write the result using only positive exponents. Assume $x \neq 0$ and $y \neq 0$.

a. $2^{-5} \cdot 2^3$ **b.** $5x \cdot x^{-4}$ c. $\left(\dfrac{9}{y}\right)^{-2}$ **d.** $(3x^{-2})^{-3}$ **e.** $\dfrac{3^{-2}x^2y^{-3}}{3x^{-5}y}$

Solution:

a. $2^{-5} \cdot 2^3 = 2^{-5+3} = 2^{-2} = \dfrac{1}{2^2} = \dfrac{1}{4}$ **b.** $5x \cdot x^{-4} = 5x^{1+(-4)} = 5x^{-3} = 5 \cdot \dfrac{1}{x^3} = \dfrac{5}{x^3}$

c. $\left(\dfrac{9}{y}\right)^{-2} = \left(\dfrac{y}{9}\right)^2 = \dfrac{y^2}{9^2} = \dfrac{y^2}{81}$ **d.** $(3x^{-2})^{-3} = 3^{-3}(x^{-2})^{-3} = \dfrac{1}{3^3} \cdot x^6 = \dfrac{x^6}{27}$

e. $\dfrac{3^{-2}x^2y^{-3}}{3x^{-5}y} = 3^{-2-1}x^{2-(-5)}y^{-3-1} = 3^{-3}x^7y^{-4} = \dfrac{1}{3^3} \cdot x^7 \cdot \dfrac{1}{y^4} = \dfrac{x^7}{27y^4}$

Technology Link

A calculator may be used to support the results in this example by evaluating both the original expression and the simplified version for at least one value of x or y. For instance, Figure R.16 shows that when $x = 2$, both $5x \cdot x^{-4}$ and $5/x^3$ are equal to .625.

```
2→X
                2
5X*X^-4
             .625
5/X^3
             .625
```

Figure R.16

PROGRESS CHECK 6

Simplify and write the result using only positive exponents. Assume $x \neq 0$ and $y \neq 0$.

a. $2^{-5} \div 2^3$ **b.** $3y^{-5} \cdot y^4$ c. $\left(\dfrac{2x}{5}\right)^{-3}$ **d.** $(6y^{-4})^{-1}$ **e.** $\dfrac{3^2x^{-4}y^{-5}}{3^{-1}x^4y^{-3}}$ ■

When working with negative exponents, keep in mind the important principle that any *factor* of the numerator may be made a factor of the denominator (and vice versa) by changing the sign of the exponent. For example,

$$\frac{3^{-2}}{3^{-4}} = \frac{3^4}{3^2} \text{ and } \frac{3^{-2}x^2y^{-3}}{3x^{-5}y} = \frac{x^2x^5}{3 \cdot 3^2y^3y}.$$

Such transformations may then be easier to simplify, as shown in the next example. When this method is used, it is important to note that this principle applies only to factors.

Example 7: Simplifying by Rewriting Without Negative Exponents

Simplify and write the result using only positive exponents. Assume $x \neq 0$ and $y \neq 0$.

a. $\dfrac{3^{-2}}{3^{-4}}$ **b.** $\dfrac{3^{-2}x^2y^{-3}}{3x^{-5}y}$

Solution: Begin by rewriting the expressions without negative exponents, as discussed above.

a. $\dfrac{3^{-2}}{3^{-4}} = \dfrac{3^4}{3^2} = 3^{4-2} = 3^2 = 9$ **b.** $\dfrac{3^{-2}x^2y^{-3}}{3x^{-5}y} = \dfrac{x^2x^5}{3 \cdot 3^2y^3y} = \dfrac{x^{2+5}}{3^{1+2}y^{3+1}} = \dfrac{x^7}{3^3y^4} = \dfrac{x^7}{27y^4}$

Note: Observe that Examples 6e and 7b show alternate methods for simplifying the same expression. Check that the simplified results for Progress Check 6e and 7b are the same also.

PROGRESS CHECK 7

Simplify and write the result using only positive exponents. Assume $x \neq 0$ and $y \neq 0$.

a. $\dfrac{2}{2^{-3}}$ **b.** $\dfrac{3^2x^{-4}y^{-5}}{3^{-1}x^4y^{-3}}$ ■

In algebra it is also common for variables to appear as exponents. We can apply the exponent properties of this section if we assume that the variables in the exponents represent integers. Letters used as exponents are called **literal exponents**.

Example 8: Simplifying an Expression Involving Literal Exponents

Simplify the given expression. Assume $x \neq 0$, $b \neq 0$, and variables as exponents denote integers.

a. $\dfrac{2x^{n+3}}{-4x^n}$ **b.** $b^x \cdot b^{-x}$ **c.** $\dfrac{b^x}{b^{2x}}$

Solution:

a. $\dfrac{2x^{n+3}}{-4x^n} = \dfrac{2x^{(n+3)-n}}{-4} = -\dfrac{x^3}{2}$ **b.** $b^x \cdot b^{-x} = b^{x+(-x)} = b^0 = 1$

c. $\dfrac{b^x}{b^{2x}} = \dfrac{1}{b^{2x-x}} = \dfrac{1}{b^x}$

PROGRESS CHECK 8

Simplify the given expression. Assume $x \neq 0$, $a \neq 0$, and variables as exponents denote integers.

a. $a^x \div a^{-x}$ **b.** $3^n \cdot 3^n$ **c.** $\dfrac{x^{3n}}{x^{2n}}$ ■

Scientific Notation

Many numbers that appear in scientific work are either very large or very small. For example, the average distance from the earth to the sun is approximately 93,000,000 mi, and the mass of an atom by hydrogen is approximately

$$0.0000000000000000000000017 \text{ g}.$$

To work conveniently with these numbers, we often write them in a form called **scientific notation**.

A number is expressed in scientific notation when it is written in the form

$$N = m \times 10^k, \text{ where } 1 \leq |m| < 10 \text{ and } k \text{ is an integer}.$$

For example,

$$93{,}000{,}000 = 9.3(10{,}000{,}000) = 9.3 \times 10^7$$
$$0.00103 = 1.03\frac{1}{1{,}000} = 1.03 \times 10^{-3}.$$

To convert a positive number from normal notation to scientific notation, use the following procedure.

To Convert Positive Numbers to Scientific Notation

1. Immediately after the first nonzero digit of the number, place an apostrophe (').
2. Starting at the apostrophe, count the number of places to the decimal point. If you move to the right, your count is expressed as a positive number, if you move to the left, the count is negative.
3. The apostrophe indicates the position of the decimal in the factor between 1 and 10; the count represents the exponent to be used in the factor, which is a power of 10.

The following examples illustrate how this procedure is used for positive numbers. (*Note*: The arrow indicates the direction of the counting.)

Number	=	Number from 1 to 10	×	Power of 10
9'3000000. (→)	=	9.3	×	10^7
0.0001'36 (←)	=	1.36	×	10^{-4}
6'.2	=	6.2	×	10^0

To express a negative number in scientific notation, you simply use the procedure above and write a negative sign before the number.

Example 9: Converting to Scientific Notation

Write each number in scientific notation.

a. 27 billion **b.** –0.00000031

Solution: Use the procedure outlined above.

a. $27{,}000{,}000{,}000 = 2.7 \times 10^{10}$ *Think*: 2'7,000,000,000. (→)

b. $-0.00000031 = -3.1 \times 10^{-7}$ *Think*: –0.0000003'1 . (←)

PROGRESS CHECK 9

Write each number in scientific notation.

a. 615 million **b.** –0.09 ■

To convert a number from scientific notation to normal notation, just carry out the indicated multiplication. When the power of 10 has a positive exponent, move the decimal point to the right and insert zeros, as needed, to indicate the final position of the decimal point. When the power of 10 has a negative exponent, the decimal point is moved to the left instead.

Example 10: Converting to Normal Notation

Express each number in normal notation.

a. 8.07×10^6 **b.** 4.3×10^{-3}

Solution:

a. To multiply by 10^6, move the decimal point 6 places to the right.
$8.07 \times 10^6 = 8,070,000$ *Think*: $8.\underbrace{070000}_{\text{6 places}}$

b. To multiply by 10^{-3}, move the decimal point 3 places to the left.
$4.3 \times 10^{-3} = 0.0043$ *Think*: $0.\underbrace{004}_{\text{3 places}}3$

PROGRESS CHECK 10

Express each number in normal notation.

a. 9.2×10^9 **b.** -2.7×10^{-6} ■

Graphing calculators are programmed to work with scientific notation, and this feature is useful for many applied problems. To enter a positive number in scientific notation, first enter the significant digits of the number from 1 to 10, press [EE] or [EXP], and finally, enter the exponent of the power of 10. For example, to enter 7.3×10^{-6} on a Texas Instruments model, press

7.3 [EE] [(−)] 6.

The display looks like

[7.3E − 6].

You should consult the owner's manual to your calculator to learn its scientific notation capabilities. In particular, here are two common features you need to check for.

1. If a calculation results in an answer that either exceeds the display range or is less than a certain number (0.001 on Texas Instruments models) in absolute value, then the calculator automatically displays the answer in scientific notation. The computations of 40^8 and $3/10,000$ on a Texas Instruments model are shown in Figure R.17 to illustrate this feature.

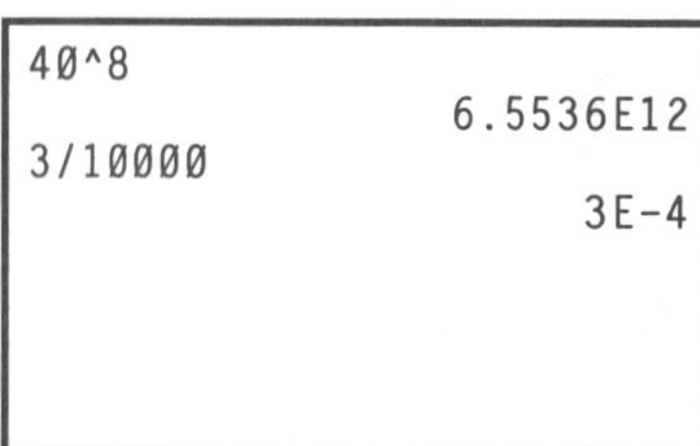

Figure R.17

2. By use of the Mode button, the calculator can be put in scientific notation display format, which forces all numeric results to appear in scientific notation.

Example 11: Number of Hemoglobin Molecules

If one red blood cell contains about 270 million hemoglobin molecules, about how many molecules of hemoglobin are there in 30 trillion red blood cells (which is the approximate number of red blood cells in a typical adult)?

Solution: 270 million $= 2.7 \times 10^8$, and 30 trillion $= 3 \times 10^{13}$. Find the product of these numbers (with these keystrokes on a Texas Instruments model).

2.7 EE 8 × 3 EE 13 ENTER

The answer in the display reads 8.1E21, which means there are about 8.1×10^{21} hemoglobin molecules in the total supply of red blood cells in a typical adult.
Note: In order to interpret large numbers, it is useful to know the following names.

$$1 \text{ million} = 10^6 = 1{,}000{,}000$$
$$1 \text{ billion} = 10^9 = 1{,}000{,}000{,}000$$
$$1 \text{ trillion} = 10^{12} = 1{,}000{,}000{,}000{,}000$$
$$1 \text{ quadrillion} = 10^{15} = 1{,}000{,}000{,}000{,}000{,}000$$

Caution: In this type of problem, students sometimes answer that there are 8.1E21 molecules. You should convert 8.1E21 to 8.1×10^{21} when expressing the final result.

PROGRESS CHECK 11
In a certain year the U.S. government spent about \$298 billion for national defense, out of a total budget of \$1.14 trillion. To the nearest tenth of a percent, what percent of the total budget was spent on national defense? ■

EXPLORE R.3

1. Use the Power key and parentheses to compare the results for $(2^3)^4$ and $2^{(3^4)}$. Do you get either of those results if you omit all the parentheses?
2. Try each of these on your calculator and comment on the result.
   a. 1^0 **b.** $(-1)^0$ **c.** -1^0 **d.** $(1 \times 10^{-30})^0$ **e.** 0^0 **f.** 0^{-1}
3. If your calculator has an Engineering mode, figure out how it works, and what its relationship is to scientific notation.

EXERCISES R.3

In Exercise 1–24 evaluate each expression and check your result with a calculator.

1. $3^2 \cdot 3^3$ **2.** $(3^2)^3$ **3.** 7^0

4. -7^0 **5.** 3^{-2} **6.** $(-2)^{-3}$

7. 2.04×10^{-3} **8.** 4.12×10^0

9. $(2^3 \cdot 3)^2$ **10.** $(3^2 \cdot 4)^3$ **11.** $(4/3)^{-1}$

12. $(2/5)^{-3}$ **13.** $4^5 \cdot 4^{-2}$ **14.** $5^{-4} \cdot 5$

15. $(-3)^2 \cdot (-3)^3$ **16.** $(-2)^{-1} \cdot (-2)^{-3}$

17. $\dfrac{2}{2^4}$ **18.** $\dfrac{3^6}{3^2}$ **19.** $\dfrac{1}{2^{-3}}$

20. $\dfrac{2}{2^{-2}}$ **21.** $\dfrac{3^{-1}}{3}$ **22.** $\dfrac{5^{-2}}{5^{-4}}$

23. $\dfrac{6^{-1} \cdot 2^{-4}}{3^{-2} \cdot 2^{-3}}$ **24.** $\dfrac{2^{-3} \cdot 7^0}{4^{-1}}$

In Exercises 25–64 use the laws of exponents to perform the operations and write the result in the simplest form that contains only positive exponents.

25. $x^4 \cdot x^3$ **26.** $x \cdot x^5$ **27.** $(x^3)^4$

28. $(t^5)^2$ **29.** $(-2p)^2$ **30.** $(-y)^3$

31. $(-5y^5z)^3$ **32.** $(3x^3y^2)^4$ **33.** $c^9 \div c^3$

34. $8k^8 \div 2k^2$ **35.** $9x^2 \div x^3$ **36.** $a \div a^7$

37. $(x+y) \div (x+y)^4$ **38.** $(a+b)^2(a+b)^6$

39. $(-3x^2)(-3x)^2$ **40.** $(-4s^2c)^3(3sc^2)^5$

41. $x^5 \cdot x^{-3}$ **42.** $3y \cdot y^{-4}$ **43.** $t^{-1} \div t$

44. $a^{-2} \div a^{-3}$ **45.** $\left(\frac{y}{5}\right)^{-2}$ **46.** $\left(\frac{3}{x}\right)^{-1}$

47. $\left(\frac{1}{x^{-1}}\right)^{-1}$ **48.** $(2x^{-3})(-3x^{-2})$

49. $(4x^0)(-2x^{-3})$ **50.** $(2a^{-3})^{-2}$

51. $\left[3(x+h)^{-1}\right]^2$ **52.** $(1-n)^{-1}(1-n)$

53. $\left(\frac{2a^{-1}}{x}\right)^{-2}$ **54.** $\left(\frac{3x^{-2}}{y}\right)^{-1}$ **55.** $\frac{2a^{-1}}{a}$

56. $\frac{-x^{-2}}{x^{-3}}$ **57.** $\frac{(-x^3)^2(4yz)}{(-2x^2)(2y^2z^3)^3}$

58. $\frac{(4x^2yz^3)^4}{(-20x^6yz^2)^2}$ **59.** $\frac{3^{-1}x^2y^{-3}}{9x^{-2}y^{-3}}$

60. $\frac{2^2x^{-3}y^{-4}}{2^{-1}x^3y^{-2}}$ **61.** $\frac{(xyz)^{-1}}{x^{-2}yz^{-3}}$ **62.** $\frac{x^0y^{-2}z^3}{(xy^{-1}z^{-3})^{-1}}$

63. $\frac{(2ax^2)^{-2}(a^3x^{-1})^2}{2(ax)^{-1}(ax^5)}$ **64.** $\frac{\left[a^2(a+y)^3\right]^{-1}}{a(a+y)^2}$

In Exercise 65–88 use the laws of exponents to simplify each expression. Variables as exponents denote integers.

65. $2^x \cdot 2^y$ **66.** $5^a \cdot 5$ **67.** $\frac{2}{2^n}$

68. $\frac{5^{2x}}{5^x}$ **69.** $(5^x)^{2x}$ **70.** $3^a \cdot 3^{-a}$

71. $y^{2a} \cdot y^{5a}$ **72.** $x^{b+2} \cdot x^{2b}$

73. $(a-b)^x(a-b)^y$ **74.** $\left[(b-a)^x\right]^y$

75. $a^x \cdot a^{-x}$ **76.** $2x^{1-a} \cdot x^{a-1}$

77. $x \div x^{-x}$ **78.** $a^{-x} \div a^0$ **79.** $\left[(y^x)^{2x}\right]^3$

80. $(2x^a)^c(3x^c)^b$ **81.** $\frac{y^x}{y^{x+1}}$

82. $\frac{x^{2n+2}}{x^2}$ **83.** $(x^a)^{p-1} \cdot x^{a-1}$

84. $(b^x)^{1-a} \div b^x$ **85.** $\frac{x^{2a}y^{a+1}}{x^ay^{a+2}}$

86. $\frac{x^{a+b}y^b}{x^{a-b}y^{b-1}}$ **87.** $\frac{(1-x)^{2a}(1-x)^2}{(1-x)^a}$

88. $\frac{(y+2)^{1+m}}{(y+2)^m(y+2)}$

89. Use the equal-monthly-payment formula in Example 5 to determine the monthly installment that will pay off a \$10,000 car loan in 48 months with a 10.8 percent annual interest rate charged on the unpaid balance.

90. Redo Exercise 89, but assume that the annual interest rate is 8.4 percent.

In Exercises 91–100 express each number in scientific notation.

91. 42 **92.** 0.6 **93.** 34,251 **94.** 7.21

95. A light-year (that is, the distance light travels in 1 year) is about 5,900,000,000,000 mi.

96. A human body contains about 10 quadrillion cells.

97. A certain radio station broadcasts at a frequency of about 1,260,000 hertz (cycles per second).

98. A certain computer can perform an addition in about 0.000014 second.

99. The weight of an oxygen molecule is approximately 0.000000000000000000000053 g.

100. The wavelength of yellow light is about 0.000023 in.

In Exercises 101–110 express each number in normal notation.

101. 9.2×10^4 **102.** 3×10^{-1}

103. 4.21×10^1 **104.** 6.3×10^0

105. The earth travels about 5.8×10^8 mi in its trip around the sun each year.

106. The number of atoms in 1 oz of gold is approximately 8.65×10^{21}.

107. One coulomb is equal to about 6.28×10^{18} electrons.

108. An atom is about 5×10^{-9} in. in diameter.

109. The mass of a molecule of water is about 3×10^{-23} g.

110. The wavelength of red light is approximately 6.6×10^{-5} cm.

111. If light travels about $186{,}000 \text{ mi/second}$, about how far will it travel in 50 minutes?

112. If one red blood cell contains about 270,000,000 hemoglobin molecules, about how many molecules of hemoglobin are there in 2 million red cells?

113. If the mass of one electron is about 0.0000000000000000000000000009 g, what is approximately the mass of 400 electrons?

114. If a certain computer can process an instruction in a nanosecond (a billionth of a second), then how long will it take this computer to process 25,000 instructions?

115. The annual budget of the U.S. government is currently between \$1 trillion and \$2 trillion. Represent 1 trillion and 1 million as powers of 10, and then determine how many times the government has to spend a million dollars in order to spend a trillion dollars.

116. The U.S. federal deficit in 1992 was about \$4 trillion. If you spent \$1,000 per second, how long would it take to spend \$4 trillion? Express the answer in seconds in scientific notation. Then express the answer in years.

117. One liter of water is made up of 55.6 moles of water molecules, where 1 mole means 6.02×10^{23} things (just as 1 dozen means 12 things).

a. How many water molecules are in one liter of water?

b. One liter of distilled water contains 10^{-7} moles of hydrogen ions. What percent of the water is made up of hydrogen ions?

THINK ABOUT IT R.3

1. Why is a^{-n} the multiplicative inverse of a^n if $a \neq 0$?

2. If $2^a = b$, then 2^{a+3} equals

a. $b+3$ **b.** b^3 **c.** $6b$ **d.** $8b$

3. A student reasoned that $2^3 \cdot 5^2$ must equal 10^5, because "when you multiply, you add the exponents." What crucial idea did the student forget?

4. The Greek mathematician Archimedes (c. 250 B.C.) is usually considered one of the greatest mathematicians of all time. Archimedes considered his most important achievement to be the discovery that whenever a sphere is circumscribed by a cylinder, the ratio of their volume is 3:2. He even asked that this figure and ratio be engraved on his tombstone. If the volume of a cylinder is $\pi r^2 h$, and the volume of a sphere is $(4/3)\pi r^3$, verify Archimedes' discovery.

5. a. The correct measurement in an experiment is 3.2×10^3. Which is a more serious error, to record it as 4.2×10^3 or as 3.2×10^4?

b. When using scientific notation, why might a scientist be more concerned about the power of 10 than about the other part of the number?

● ● ●

R.4 Products of Algebraic Expressions

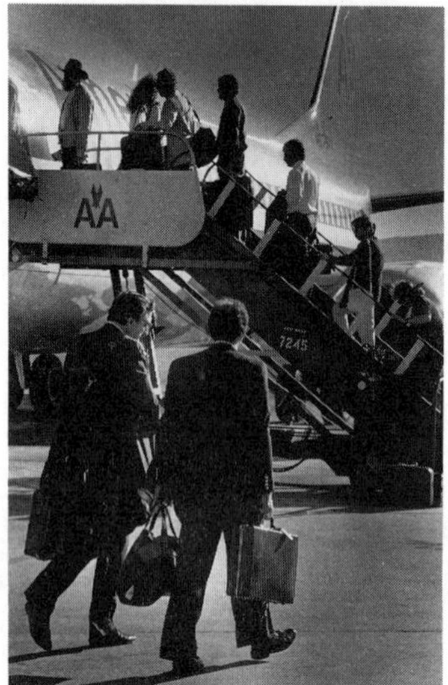

Photo Courtesy of ***Spencer Grant*** *of The Picture Cube*

An airline offers a charter flight at a fare of \$250 per person, if 120 passengers sign up. For each passenger above 120, the fare for each passenger is reduced by \$2. Explain why the total revenue for the flight is given by $(120+x)(250-2x)$, where x represents the number of passengers above 120. Then rewrite the expression for the total revenue by performing the indicated multiplication.
(See Example 6.)

Objectives

1. Use the distributive property to multiply algebraic expressions.
2. Use the FOIL method to multiply two expressions that each contain two terms.
3. Use special product formulas to find certain products.

To multiply algebraic expressions, we use the laws of exponents and the distributive property in the forms

$$a(b+c)=ab+ac \text{ and } (a+b)c=ac+bc$$

or in an extended form such as

$$a(b+c+\ldots+n)=ab+ac+\ldots+an.$$

Example 1 shows the basic procedure.

Example 1: Using the Distributive Property

Find each product.

a. $3x(x^2-2x+1)$ **b.** $(2c^2+t)ct$

Solution:

a. $3x(x^2-2x+1)=(3x)(x^2)+(3x)(-2x)+(3x)(1)=3x^3-6x^2+3x$

b. $(2c^2+t)ct=(2c^2)(ct)+(t)(ct)=2c^3t+ct^2$

PROGRESS CHECK 1

Find each product.

a. $-2x^2(3x^2-x+9)$ **b.** $(7y-x)3y$ ■

Since division is defined in terms of multiplication, a similar procedure is used to simplify division problems in which the divisor is a single term. For example, we may divide $a+b$ by c as follows:

$$(a+b)\div c=(a+b)\cdot\frac{1}{c}=\frac{a}{c}+\frac{b}{c}.$$

This result shows that

$$\frac{a+b}{c}=\frac{a}{c}+\frac{b}{c}$$

Example 2: Dividing by a Single Term Divisor

Divide $\dfrac{2xh+h^2+2h}{h}$.

Solution: $\dfrac{2xh+h^2+2h}{h}=\dfrac{2xh}{h}+\dfrac{h^2}{h}+\dfrac{2h}{h}=2x+h+2$

PROGRESS CHECK 2
Divide $-12n^2x^2+2nx^2$ by $4nx$. ■

If both factors in the multiplication contain more than one term, the distributive property must be used more than once. For example, no matter what expression is inside the parentheses

$$(\blacksquare)(x+2) \text{ means } (\blacksquare)x+(\blacksquare)2.$$

Thus,

$$(x+3)(x+2) \text{ means } (x+3)x+(x+3)2.$$

Using the distributive property the second time, we get

$$(x+3)x+(x+3)2=x^2+3x+2x+6=x^2+5x+6$$

Therefore,

$$(x+3)(x+2)=x^2+5x+6$$

Example 3: Using the Distributive Property Twice

Find each product.
a. $(2t+3)(5t-1)$ **b.** $(x^n-2)(x^n-1)$

Solution:
a. $(2t+3)(5t-1)=(2t+3)(5t)+(2t+3)(-1)=10t^2+15t-2t-3=10t^2+13t-3$
b. $(x^n-2)(x^n-1)=(x^n-2)(x^n)+(x^n-2)(-1)=x^{2n}-2x^n-x^n+2=x^{2n}-3x^n+2$

PROGRESS CHECK 3
Find each product.
a. $(3y-4)(7y-2)$ **b.** $(a^x+3)(a^x-5)$ ■

Notice that this method of multiplication is equivalent to multiplying each term of the first factor by each term of the second factor, and then combining like terms. This observation leads to the arrangement in Example 4, which is good for multiplying longer expressions since like terms are placed under each other.

Example 4: Multiplying in a Vertical Format

Multiply $(c^2 - 5c + 25)(c + 3)$.

Solution: Multiply each term of $c^2 - 5c + 25$ by each term of $c + 3$. Arrange like terms in the products under each other; then add.

$$\begin{array}{lrrrl} & c^2 - 5c + 25 & & & \\ & c + 3 & & & \\ \hline & c^3 - 5c^2 + 25c & & & \text{This line equals } c(c^2 - 5c + 25). \\ \text{add} & 3c^2 - 15c + 75 & & & \text{This line equals } 3(c^2 - 5c + 25). \\ \hline & c^3 - 2c^2 + 10c + 75 & & & \end{array}$$

The product is $c^3 - 2c^2 + 10c + 75$.

PROGRESS CHECK 4
Multiply $(4x^2 + 3x - 2)(x + 3)$. ■

When you multiply expressions that contain two terms, a mental shortcut may be used, as shown in Examples 5–7. This method is called the FOIL method, and the letters, F, O, I, and L denote the products of the **first**, **outer**, **inner**, and **last** terms, respectively.

Example 5: Using the FOIL Method

Multiply using the FOIL method: $(2x - 1)(3x + 4)$.

Solution: Note that the outer and inner pairs are like terms in this example.

$$\begin{aligned} & \quad\ \text{F} \quad + \quad \text{O} \quad + \quad \text{I} \quad + \quad \text{L} \\ (2x - 1)(3x + 4) &= 2x(3x) + 2x(4) + (-1)(3x) + (-1)(4) \\ &= 6x^2 + 8x - 3x - 4 \\ &= 6x^2 + 5x - 4 \end{aligned}$$

Thus, $(2x - 1)(3x + 4) = 6x^2 + 5x - 4$.

Technology Link

As in the previous section, a calculator may be used to support the results here by evaluating both the original expression and the expanded product for at least one set of values for the variables. For instance, Figure R.18 shows that when $x = 5$, both $(2x - 1)(3x + 4)$ and $6x^2 + 5x - 4$ are equal to 171.

```
5→X
                    5
(2X-1)(3X+4)
                  171
6X²+5X-4
                  171
```

Figure R.18

PROGRESS CHECK 5
Multiply using the FOIL method: $(5y - 4)(4y - 3)$. ■

Example 6: Revenue for a Charter Flight

Solve the problem in the section introduction on page 38.

Solution: Because revenue = (number of passengers) · (fare per passenger), we first write algebraic expressions for these two factors. The variable x represents the number of passengers above 120 so

$$\text{number of passengers} = 120 + x.$$

For each passenger above 120, the fare for each passenger is reduced by \$2, so

$$\text{fare per passenger} = 250 - 2x.$$

Thus,

$$\text{revenue} = (120 + x)(250 - 2x)$$

Now multiply using the FOIL method.

$$\begin{aligned} & \quad\ \text{F} \quad + \quad \text{O} \quad + \quad \text{I} \quad + \quad \text{L} \\ (120 + x)(250 - 2x) &= 120(250) + 120(-2x) + x(250) + x(-2x) \\ &= 30{,}000 - 240x + 250x - 2x^2 \\ &= 30{,}000 + 10x - 2x^2 \end{aligned}$$

The revenue for the flight is given by $30{,}000 + 10x - 2x^2$ dollars.

PROGRESS CHECK 6

A manufacturer plans to introduce a new model of computer at a list price of \$3,000 and expects to sell 20,000 units. The sales department forecasts that for each \$100 the price is cut, sales will go up by 700 units. Explain why the formula for estimated revenue is then $(20{,}000 + 700x)(3{,}000 - 100x)$, where x represents the number of \$100 price cuts. Then rewrite the expression for the estimated revenue by performing the indicated multiplication. ■

Example 7: Using the FOIL Method

Multiply using the FOIL method.

a. $(a+b)(a-b)$ **b.** $(a+b)^2$ **c.** $(a-b)^2$

Solution:

a.

$$\begin{aligned} & \quad \text{F} \quad + \quad \text{O} \quad + \quad \text{I} \quad + \quad \text{L} \\ (a+b)(a-b) &= a(a) + a(-b) + b(a) + b(-b) \\ &= a^2 - ab + ab - b^2 \\ &= a^2 - b^2 \end{aligned}$$

Thus $(a+b)(a-b) = a^2 - b^2$.

b. $(a+b)^2 = (a+b)(a+b)$. Then,

$$\begin{aligned} & \quad \text{F} \quad + \quad \text{O} \quad + \quad \text{I} \quad + \quad \text{L} \\ (a+b)(a+b) &= a(a) + a(b) + b(a) + b(b) \\ &= a^2 + 2ab + b^2 \end{aligned}$$

Thus $(a+b)^2 = a^2 + 2ab + b^2$

c. $(a-b)^2 = (a-b)(a-b)$. Then,

$$\begin{aligned} &\quad\ \ \text{F} \quad + \quad \text{O} \quad + \quad \text{I} \quad + \quad \text{L} \\ (a-b)(a-b) &= a(a) + a(-b) + (-b)(a) + (-b)(-b) \\ &= a^2 - 2ab + b^2 \end{aligned}$$

Thus $(a-b)^2 = a^2 - 2ab + b^2$

PROGRESS CHECK 7

Multiply using the FOIL method.

a. $(k-8)^2$ **b.** $(x+h)^2$ **c.** $(3y+5)(3y-5)$ ■

Example 7 illustrates products that occur so frequently that you should memorize and use these results as special product formulas.

Special Product Formulas

	Comment
1. $(a+b)(a-b) = a^2 - b^2$	The product of the sum and the difference of two terms is the square of the first term minus the square of the second term.
2. $(a+b)^2 = a^2 + 2ab + b^2$	The square of the sum of two terms is the square of the first term, plus twice the product of the two terms, plus the square of the second term.
3. $(a-b)^2 = a^2 - 2ab + b^2$	The square of the difference of two terms is the square of the first term, minus twice the product of the two terms, plus the square of the second term.

Example 8: Using Special Product Formulas

Use a special product formula to find each product.

a. $(4x+5)(4x-5)$ **b.** $(3x+4)^2$ **c.** $(x^3 - y^3)^2$

Solution:

a. Replace a by $4x$ and b by 5 in the formula for the product of the sum and difference of two terms.

$$(a+b)(a-b) = a^2 - b^2$$
$$\downarrow \quad \downarrow$$
$$(4x+5)(4x-5) = (4x)^2 - 5^2 = 16x^2 - 25$$

b. Use the formula for $(a+b)^2$ with $a = 3x$ and $b = 4$.

$$(a+b)^2 = a^2 + 2a\ b + b^2$$
$$\downarrow \quad \downarrow \ \downarrow \quad \downarrow$$
$$(3x+4)^2 = (3x)^2 + 2(3x)(4) + (4)^2 = 9x^2 + 24x + 16$$

c. Use the formula for $(a-b)^2$ with $a = x^3$ and $b = y^3$.

$$(a-b)^2 = a^2 - 2\ a\ \ b + b^2$$
$$\downarrow \quad \downarrow \ \downarrow \quad \downarrow$$
$$(x^3 - y^3)^2 = (x^3)^2 - 2(x^3)(y^3) + (y^3)^2 = x^6 - 2x^3y^3 + y^6$$

Note: It should be your goal to become so proficient with these special products that the work displayed above is done *mentally*, and you can just write down the answer.

PROGRESS CHECK 8

Use a special product formula to find each product.

a. $(2x+7)^2$ **b.** $(2x+7)(2x-7)$ **c.** $(3m-5n)^2$ ■

Example 9: Combining Multiplication and Division Methods

Show that $\dfrac{2(x+h)^3-1-(2x^3-1)}{h}$ simplifies to $6x^2+6xh+2h^2$.

Solution: First, we determine that

$$\begin{aligned}(x+h)^3 &= (x+h)(x+h)(x+h)\\ &= (x^2+2xh+h^2)(x+h)\\ &= x^3+3x^2h+3xh^2+h^3\end{aligned}$$

Then,

$$\begin{aligned}\frac{2(x+h)^3-1-(2x^3-1)}{h} &= \frac{2x^3+6x^2h+6xh^2+2h^3-1-2x^3+1}{h}\\ &= \frac{6x^2h+6xh^2+2h^3}{h}\\ &= 6x^2+6xh+2h^2\end{aligned}$$

PROGRESS CHECK 9

Simplify $\dfrac{3(x+h)^2+7-(3x^2+7)}{h}$ ■

EXPLORE R.4

1. Check that $(x-3)(x+3)$ equals x^2-9 by evaluating both expressions for several values of x. Use the List or Table feature if available.
2. Show that $3x(x+4)$ does *not* equal $3x^2+4$ by evaluating both expressions for several values of x.

EXERCISES R.4

In Exercise 1–40 perform the multiplication or division and combine like terms.

1. $2(x-y)$ **2.** $-6(z+y)$

3. $-5x(x^3-x^2-x)$ **4.** $y(3y^2+5y-6)$

5. $-2xyz(4x-y+7z)$

6. $4x^2yz^3(3x^3-2yz+5xz^4)$

7. $(p^2+q^2)p^2q^2$ **8.** $(a^2-b^2)ab$

9. $(24n^2x+18nx^2)\div 6nx$

10. $(6dt^2-12d^2t)\div 2dt$ **11.** $\dfrac{2hx+h^2}{h}$

12. $\dfrac{6xh+3h^2}{h}$ **13.** $2x^4\left(\dfrac{3}{x}+\dfrac{4}{x^2}-\dfrac{5}{x^3}\right)$

14. $-12x^2\left(3x^2-\dfrac{1}{3}+\dfrac{3}{4x}-\dfrac{1}{x^2}\right)$

15. $(a+3)(a+4)$ **16.** $(z-2)(z-6)$

17. $(x+7)(x-5)$ **18.** $(y-8)(y-2)$

19. $(x+4)(x-4)$ **20.** $(3-t)(3+t)$

21. $(3x-4)(2x-1)$ **22.** $(7h+6)(5h-3)$

23. $(2t+5c)(6t-c)$ **24.** $(5y-4x)(3y+x)$

25. $(k-2)^2$ **26.** $(x+7)^2$ **27.** $(2x+3y)^2$

28. $(4c-a)^2$ **29.** $(x-1)^3$ **30.** $(x+h)^3$

31. $(y-4)(y^2+5y-1)$ **32.** $(2a-1)(a^3+a+1)$

33. $(x-y)(x^2+xy+y^2)$

34. $(2y+3z)(y^2-2yz+z^2)$

35. $(2y-1)(3y+2)(1-y)$

36. $(x+1)^2(2-x)$ **37.** $[(x-y)-1]^2$

38. $[3+(4a-b)]^2$

39. $(x^2-2xy+y^2)(x^2+2xy+y^2)$

40. $(x^2-x-1)(x^2+x+1)$

In Exercises 41–60 use a special product formula to find each product.

41. $(y+3)(y-3)$ **42.** $(x-8)(x+8)$

43. $(5n-7)(5n+7)$ **44.** $(3t+4)(3t-4)$

45. $(6x+y)(6x-y)$ **46.** $(4a-10b)(4a+10b)$

47. $(a+1)^2$ **48.** $(y+9)^2$ **49.** $(x-7)^2$

50. $(1-k)^2$ **51.** $(3c+5)^2$ **52.** $(4n+8)^2$

53. $(2-7x)^2$ **54.** $(7x-2)^2$ **55.** $(5x+4y)^2$

56. $(6r+2h)^2$ **57.** $(10a-5b)^2$ **58.** $(3s-11t)^2$

59. $[(x-y)-1]^2$ **60.** $[(x+h)+1]^2$

In Exercises 61–70 find each product. Assume that variables as exponents denote integers.

61. $(x^n+5)(x^n+2)$ **62.** $(3x^n-2)(x^n-1)$

63. $(z^a+3)^2$ **64.** $(t^b-7)^2$

65. $(x^a-y^b)(x^a+y^b)$ **66.** $(x^a+y^b)^2$

67. $(a^{bx}+a^{-bx})^2$ **68.** $(a^{bx}+a^{-bx})(a^{bx}-a^{-bx})$

69. $(x^n+y^n)(x^{2n}-x^ny^n+y^{2n})$

70. $(x^n-y^k)(x^{2n}+x^ny^k+y^{2k})$

71. Show that $\dfrac{(x+h)^2+1-(x^2+1)}{h}$ simplifies to $2x+h$.

72. Show that $\dfrac{\left[1-(-3+h)^2\right]-\left[1-(-3)^2\right]}{h}$ simplifies to $6-h$.

73. Simplify $\dfrac{(x+h)^3-x^3}{h}$.

74. Simplify $\dfrac{(2+h)^3-2^3}{h}$.

75. Simplify $\dfrac{[(x+h)+1]^2-(x+1)^2}{h}$.

76. Simplify $\dfrac{\left[128(t+h)-16(t+h)^2\right]-\left[128t-16t^2\right]}{h}$.

77. A manufacturer has been selling about 900 Model X notebook computers every month at the price of $900 each, for a gross monthly income of $810,000. The sales department estimates that for each $50 the price is cut, sales will rise by 50 units per month, so the formula

$$r=(900-50x)(900+50x)$$

gives the estimated revenue, where x represents the number of $50 price cuts. Multiply the two factors on the right side, and rewrite the result as a polynomial expression. Explain, according to this expression, why cutting the price is not a good idea.

78. For the company described in Exercise 77, if each $50 cut results in 100 new sales, the formula for expected revenue becomes

$$r=(900-50x)(900+100x).$$

Find a polynomial expression for the right-hand side, and explain why raising the price is a good idea in this case. [Hint: show that there are some values of x that will yield higher revenue than $810,000.]

79. **a.** Write a polynomial expression for the area of the rectangle.

15 − x

15 + x

b. For what value of x will the area be maximum?

80. **a.** Write a polynomial expression for the area of the rectangle.

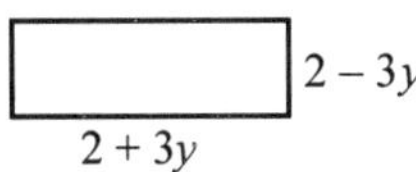

b. For what value of y will the area be a maximum?

81. Find a formula for the shaded area.

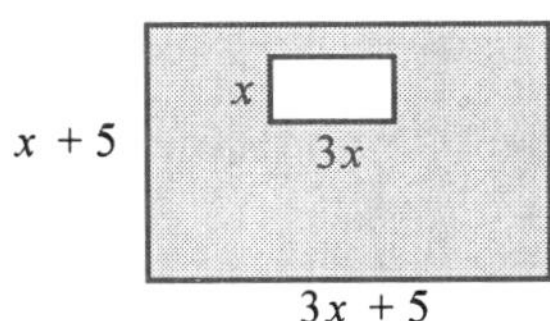

82. Find a formula for the shaded area.

$2x + 2$
$2x + 5$
$3x - 2$
$3x + 5$

THINK ABOUT IT R.4

1. A common student error is to expand $(a+b)^2$ as a^2+b^2. Explain how the square in the diagram with side length $a+b$ may be used to illustrate geometrically the correct expansion of $(a+b)^2$.

a b
a 1 2
b 3 4

2. Show that for any three consecutive integers the square of the middle one is always one more than the product of the smallest and the largest.
3. **a.** What do the letters FOIL stand for?
   b. Would the FILO method give the same answer?
   c. How many different ways can you arrange the letters {F, O, I, L}?
4. Find these products.
   a. $(x-1)(x+1)$
   b. $(x-1)(x^2+x+1)$
   c. $(x-1)(x^3+x^2+x+1)$
   d. Assume that the pattern above holds, and find $(x-1)(x^{20}+x^{19}+x^{18}+\cdots+x+1)$.
   e. This product can be found the hard way or the easy way. Do the calculation both ways. Find $(d^4+e^4)(d^2+e^2)(d+e)(d-e)$.
5. The Pythagorean relation is one of the most famous ideas in mathematics. Several hundred different proofs of this theorem have been recorded, and the National Council of Teachers of Mathematics published a book, *The Pythagorean Proposition* (1968), that presents 370 demonstrations of this statement. Early proofs, which were geometric in nature, gradually gave way to analytical proofs that used algebra to verify geometric ideas. Two well-known proofs of this variety follow.
   a. Consider the square with side length $a+b$. Calculate the area of the square in two ways and establish the Pythagorean relation $c^2=a^2+b^2$. (**Note:** It must be shown that $\theta = 90°$.)

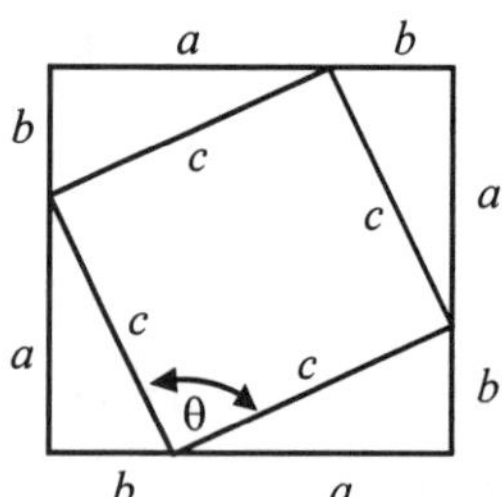

b. The trapezoid in the diagram was used by James Garfield (the twentieth president of the United States) to prove the Pythagorean theorem. Once again the idea is to find the area in two ways and establish $c^2 = a^2 + b^2$. Try it.

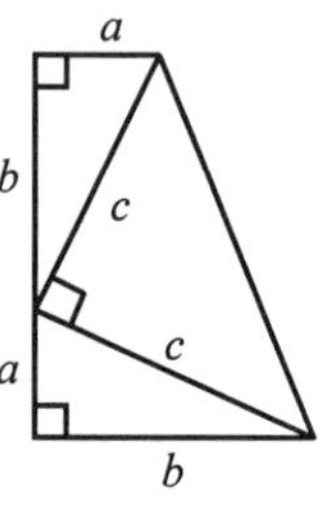

● ● ●

R.5 Linear Equations and Literal Equations

Photo Courtesy of Archive Photos, New York

A hydrogen bomb is exploded that releases 6.7×10^{16} joules of energy, the equivalent of 16 million tons of TNT. Use Einstein's formula $E = mc^2$ to find m and determine how much matter was used to create this energy. If we take c, the speed of light, to be 3.0×10^8 m/second, then the units for mass will be kilograms. Use 1 kg $\approx$ 2.2 lb to also find the answer in pounds.
(See Example 6.)

Objectives

1. Solve linear equations.
2. Find the value of a variable in a formula when given values for the other variables.
3. Solve a given formula or literal equation for a specified variable.

If you can solve equations, you can answer the section-opening problem in a straightforward manner. To solve an equation means to find the set of *all* values for the variable that make the equation a true statement. This set is called the **solution set** of the equation. Equations may be always true or always false, or their truth may depend on the value substituted for the variable. If the equation is a true statement for all admissible values of the variable, as is $2x + x = 3x$, then it is called an **identity**. A **conditional equation** is an equation in which some replacements for the variable make the statement true, while others make it false. For example, the conditional equation

$x+5=21$ is true if $x=16$ and false otherwise. We call 16 the **solution** or **root** of the equation. The concept of solving equations is usually associated with conditional equations.

One of the simplest equations to solve is a linear equation. By definition, a **linear equation in one variable** is an equation that can be written in the form

$$ax+b=0,$$

where a and b are real numbers with $a \neq 0$. This type of equation is called **linear** because (as shown in Section 1.1) it is associated with a graph that is a straight line. We solve a linear equation by isolating the variable on one side of the equation, as described next.

Solution of Linear Equations

Two equations are **equivalent** when they have the same solution set. We may change a linear equation to an equivalent equation of the form x = number, by performing any combination of the following steps:

1. Adding or subtracting the same expression to (from) both sides of the equation.
2. Multiplying or dividing both sides of the equation by the same (nonzero) number.

Example 1: Solving a Linear Equation

Solve the equation $5x+3=33$.

Solution: We first isolate the term involving the variable by subtracting 3 from both sides of the equation.

$$\begin{aligned} 5x+3&=33 \\ 5x+3-3&=33-3 \\ 5x&=30 \end{aligned}$$

Next we want the coefficient of x to be 1. Therefore, divide both sides of the equation by 5:

$$\frac{5x}{5}=\frac{30}{5}$$
$$x=6$$

Thus, 6 is the solution of the equation, and the solution set is $\{6\}$. We can check the solution by replacing x by 6 in the original equation.

$$\begin{aligned} 5x+3&=33 \\ 5(6)+3&\stackrel{?}{=}33 \\ 33&\stackrel{\checkmark}{=}33 \end{aligned}$$

PROGRESS CHECK 1

Solve the equation $-5x+9=64$ ■

Example 2: Solving a Linear Equation

Solve the equation $6(x+1)=14x+2$.

Solution: The form x = number may be obtained as follows.

$$\begin{aligned} 6(x+1) &= 14x+2 \\ 6x+6 &= 14x+2 \\ 6x+6-6 &= 14x+2-6 && \text{Subtract 6 from both sides.} \\ 6x &= 14x-4 \\ 6x-14x &= 14x-14x-4 && \text{Subtract } 14x \text{ from both sides.} \\ -8x &= -4 \\ \frac{-8x}{-8} &= \frac{-4}{-8} && \text{Divide both sides by } -8. \\ x &= \frac{1}{2} \end{aligned}$$

We verify that $1/2$ is a solution next.

$$6\left(\frac{1}{2}+1\right) \stackrel{?}{=} 14\left(\frac{1}{2}\right)+2$$
$$9 \stackrel{\checkmark}{=} 9$$

Thus, the solution set is $\{1/2\}$.

Technology Link

A calculator check that both sides in this equation equal 9 when $x = 1/2$ is shown in Figure R.19.

```
1/2 X
                    .5
6(X+1)
                     9
14X+2
                     9
```

Figure R.19

PROGRESS CHECK 2
Solve the equation $2(5-3y) = 1-8y$. ■

Example 3: Solving a Linear Equation Involving Fractions

Solve $\frac{x-1}{2} = \frac{2x}{5}$.

Solution: Although equations that contain fractions are discussed in detail in Section 3.2, it is useful to solve some simple examples of these equations now. Here we can simplify by multiplying both sides of the equation by the least common denominator of the denominators, which is 10.

$$\begin{aligned} \frac{x-1}{2} &= \frac{2x}{5} \\ 10\left(\frac{x-1}{2}\right) &= 10\left(\frac{2x}{5}\right) && \text{Multiply both sides by 10.} \\ 5x-5 &= 4x \\ x-5 &= 0 && \text{Subtract } 4x \text{ from each side.} \\ x &= 5 && \text{Add 5 to both sides.} \end{aligned}$$

Thus, the solution set is $\{5\}$.

Now check by replacing x by 5 in the original equation.

$$\frac{5-1}{2} \stackrel{?}{=} \frac{2(5)}{5}$$
$$2 \stackrel{\checkmark}{=} 2$$

Thus, the solution set is $\{5\}$.

PROGRESS CHECK 3

Solve $\frac{2x-5}{6} = \frac{x-6}{9}$. ■

Note that in Examples 1–3 each linear equation had exactly one solution. We can anticipate this result since the general form of a linear equation, $ax+b=0$ with $a \neq 0$, may be solved as follows:

$$ax+b=0$$
$$ax=-b \quad \text{Add } -b \text{ to both sides.}$$
$$x=\frac{-b}{a} \quad \text{Divide both sides by } a \neq 0.$$

Therefore, the linear equation $ax+b=0$, $a \neq 0$, has exactly one solution, $-b/a$. The requirement that $a \neq 0$ is important, as shown in Example 4.

Example 4: Identifying a False Equation

Solve $3(x+3)=3x$.

Solution: If we proceed in the usual way, we have

$$3(x+3)=3x$$
$$3x+9=3x$$
$$9=0.$$

When we subtract $3x$ from both sides of $3x+9=3x$, the result is $9=0$. This statement is never true and indicates that the original equation has no solution. When the solution set contains no elements, we say the solution set is $\varnothing$, called the **empty set**. Note that when an equation may be written in the form $ax+b=0$, with $a=0$, then either the equation has no solution (if $b \neq 0$) or the equation is an identity (if $b=0$).

PROGRESS CHECK 4

Solve $x=-(1-x)$. ■

To analyze relationships, we often use **literal equations**, which are equations that contain two or more letters. The letters may represent any mix of variables and constants. Common examples of literal equations are formulas, such as the perimeter formula $P=4s$, and general forms of equations such as $ax+b=0$. Because literal equations and formulas are types of equations, we may use the equation-solving techniques developed in this section to analyze them.

Example 5: Finding the Time in a Simple Interest Investment

When an original principal P is invested at simple interest for t years at annual interest rate r, then the amount A of the investment is given by $A=P(1+rt)$. Find the value of t when $A=\$5{,}920$, $P=\$4{,}000$, and $r=0.06$.

Solution: Substitute the given values into the formula and then solve for t.

$$
\begin{aligned}
A &= P(1+rt) && \text{Given formula.}\\
5{,}920 &= 4{,}000(1+0.06t) && \text{Substitute the given values.}\\
5{,}920 &= 4{,}000+240t && \text{Distributive property.}\\
1{,}920 &= 240t && \text{Subtract 4,000 from both sides.}\\
8 &= t && \text{Divide both sides by 240.}
\end{aligned}
$$

Check the solution by replacing each letter by its value. You will get the identity $5{,}920 = 5{,}920$, proving that $t = 8$ years is correct.

Caution: In this example a common student error is to replace $1+0.06t$ with $1.06t$. Recall that 1 and $0.06t$ are not like terms and cannot be combined. The expression $1.06t$ would result from $t+0.06t$.

Technology Link

Some calculators have an equation-solving feature that you will find useful when working with formulas, in a wide variety of math-dependent disciplines. For instance, Figure R.20 displays the solution to the problem in Example 5 on a particular calculator with such a feature.

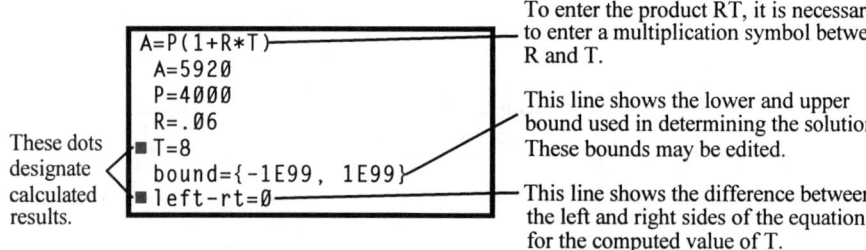

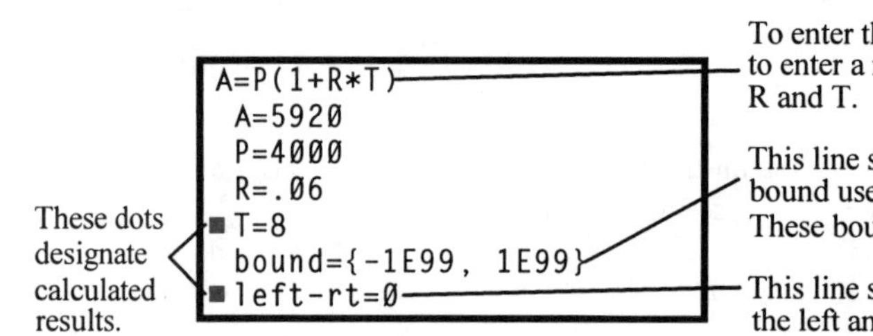
A=P(1+R*T)
A=5920
P=4000
R=.06
T=8
bound={-1E99, 1E99}
left-rt=0
These dots designate calculated results.
To enter the product RT, it is necessary to enter a multiplication symbol between R and T.
This line shows the lower and upper bound used in determining the solution. These bounds may be edited.
This line shows the difference between the left and right sides of the equation for the computed value of T.

Figure R.20

A nice benefit of an equation-solving feature is that it allows you to investigate quickly how a change in one variable affects another variable.

PROGRESS CHECK 5

Find the value of r in the formula $A = P(1+rt)$ if $A = \$6{,}240$, $P = \$4{,}000$, and $t = 7$ years. ■

Example 6: Finding the Mass in a Hydrogen Bomb Explosion

Solve the problem in the section introduction on page 46.

Solution: First, solve for m using the given information.

$$
\begin{aligned}
E &= mc^2 && \text{Given formula.}\\
6.7\times10^{16} &= m(3\times10^8)^2 && \text{Substitute the given values.}\\
\frac{6.7\times10^{16}}{(3\times10^8)^2} &= m && \text{Divide both sides by } (3\times10^8)^2.
\end{aligned}
$$

Now compute the quotient on a calculator. One possible sequence of keystrokes is

6.7 [EE] 16 [÷] 3 [EE] 8 [x^2] [ENTER].

To the nearest hundredth, the computed result in the display rounds off to 0.74, so about 0.74 kg of matter were used. Since 1 kg is about 2.2 lb, this explosion was generated from about 1.6 lb of matter!

PROGRESS CHECK 6

The largest hydrogen bomb every exploded (the former Soviet Union, 1961) released 2.4×10^{17} joules of energy, the equivalent of 57 million tons of TNT. Use the information in Example 6 and determine how much matter was used to create this energy. Express the answer in both kilograms and pounds. ■

In Example 6 you may find it convenient to first rewrite

$$E = mc^2 \text{ as } m = \frac{E}{c^2}$$

before substituting the values given for E and c. This rearrangement is permitted because $E = mc^2$ may be "solved for m" by dividing both sides by c^2. It is often useful to solve a literal equation for a specified variable to obtain a more helpful version of the equation, and Examples 7–9 illustrate a few such conversions.

Example 7: Rearranging a Statistics Formula

The Z-score formula in statistics is $Z = (x - m)/s$, where x is a raw score, m is the mean, and s is the standard deviation. Solve the formula for the raw score (x).

Solution: To solve for x, we need to isolate x on one side of the equation.

$$\begin{aligned} Z &= \frac{x-m}{s} \\ sZ &= s\left(\frac{x-m}{s}\right) && \text{Multiply both sides by } s. \\ sZ &= x - m \\ sZ + m &= x - m + m && \text{Add } m \text{ to both sides.} \\ sZ + m &= x \end{aligned}$$

The formula $x = sZ + m$ gives the raw score in terms of s, Z, and m.

PROGRESS CHECK 7

Solve $m = \dfrac{y-b}{x}$ for y. ■

Example 8: Rearranging a Temperature Conversion Formula

The formula $C = (5/9)(F - 32)$ is usually used to convert from degrees Fahrenheit to degrees Celsius. Solve this formula for F.

Solution: We will solve for F so that the result is the common formula for converting from degrees Celsius to degrees Fahrenheit.

$$\begin{aligned} C &= \frac{5}{9}(F-32) \\ \frac{9}{5}C &= \frac{9}{5}\cdot\frac{5}{9}(F-32) && \text{Multiply both sides by } \frac{9}{5}. \\ \frac{9}{5}C &= F - 32 \\ \frac{9}{5}C + 32 &= F - 32 + 32 && \text{Add 32 to both sides.} \\ \frac{9}{5}C + 32 &= F \end{aligned}$$

The result, $F = (9/5)C + 32$, gives a formula for F in terms of C.

Note: In this example it is also logical to clear fractions as a first step by multiplying both sides by 9. This approach leads to the result

$$F = \frac{9C + 160}{5}.$$

When you rearrange formulas, there may be several acceptable answer forms, and you may need to consult with your instructor to see if your answer is correct.

PROGRESS CHECK 8

Solve $S = t - \frac{1}{2}gt^2$ for g. ■

In Chapter 1 we will sometimes find it useful to solve equations of the form $ax + by = C$ for either x or y. Example 9 illustrates this type of problem.

Example 9: Solving for a Specified Variable

Solve the equation $2x - 3y = 9$ for the indicated letter.

a. For x **b.** For y

Solution:

a. The equation may be solved for x as follows.

$$\begin{aligned} 2x - 3y &= 9 \\ 2x &= 3y + 9 && \text{Add } 3y \text{ to both sides.} \\ x &= \frac{3y + 9}{2} && \text{Divide both sides by 2.} \end{aligned}$$

b. The equation may be solved for y as follows.

$$\begin{aligned} 2x - 3y &= 9 \\ -3y &= 9 - 2x && \text{Subtract } 2x \text{ from both sides.} \\ y &= \frac{9 - 2x}{-3} && \text{Divide both sides by } -3. \\ y &= \frac{2x - 9}{3} && \text{Multiply numerator and denominator by } -1. \end{aligned}$$

Note: In Example 9b we choose as a last step to multiply numerator and denominator by -1 to avoid an answer with a negative number in the denominator. However, this conversion is optional, and all of the following equations are acceptable answers here.

$$y = \frac{2x - 9}{3};\ y = \frac{2}{3}x - 3;\ y = \frac{9 - 2x}{-3};\ y = \frac{-9 + 2x}{3};\ y = -3 + \frac{2}{3}x$$

PROGRESS CHECK 9

Solve $5x - 2y = 30$ for the indicated letter.

a. For x **b.** For y ■

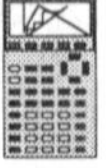

EXPLORE R.5

If your calculator has an equation solver, do the first three explorations to familiarize yourself with its behavior.

1. Enter the equation $A = P(1 + RT)$ [**Note:** For some solvers you enter $0 = A - P(1 + R \times T)$]. Set P equal to 1,000, and R equal to 0.05.

Then find T when $A = 2{,}500$, $3{,}000$, $4{,}000$, and $5{,}000$.
This approach is another way to use the calculator to evaluate an expression for several values of one or more variables.

2. A calculator Equation Solver searches for solutions between 2 given bounds. Enter the equation $y = 2x - 8$, then set y equal to 0.
   - **a.** Set the bounds at -10 and 10, then solve for x. What happens?
   - **b.** Next, set the bounds at -10 and 0. Now try to solve for x, and explain what happens.
3. What happens when you use the Equation Solver to solve a false equation? Enter the equation $x = x + 1$, and solve for x. Explain the result.
4. In Example 6 we converted 0.74 kg to pounds. If your calculator has a conversion feature, use this feature to do the conversion automatically.
5. At one particular temperature the degree measure is the same in Fahrenheit and Centigrade.
   - **a.** Find this temperature by using a guess and check approach with any calculator method for evaluating expressions rapidly. The formula is $F = (9/5)C + 32$.
   - **b.** Now find this temperature by setting up and solving a linear equation.

EXERCISES R.5

In Exercises 1–30 solve each question and check your answer.

1. $6x = 18$ **2.** $-3x + 3 = 0$

3. $3x + 2 = 2x + 8$ **4.** $7y - 9 = 6y - 10$

5. $2z + 12 = 5z + 15$ **6.** $-6x + 7 = 5x - 4$

7. $3y - 14 = -4y + 7$ **8.** $3z - 8 = 13z - 9$

9. $8x - 10 = 3x$ **10.** $15 - 6y = 0$

11. $9 - 2y = 27 + y$ **12.** $15x + 5 = 2x + 14$

13. $4x - x = 3x$ **14.** $x - 1 = x + 3$

15. $2(x+4) = 2(x-1) + 3$

16. $2(x+3) - 1 = 2x + 5$ **17.** $7x = 2x$

18. $3y - 7 = 2y - 7$ **19.** $4 - 3y = 5(4 - y)$

20. $7x - 2(x - 7) = 14 - 5x$

21. $5(x - 6) = -2(15 - 2x)$

22. $18(z + 1) = 9z + 10$

23. $3[2x - (x - 2)] = -3(3 - 2x)$

24. $2y - (3y - 4) = 4\left(y + \frac{3}{4}\right)$

25. $\frac{x-2}{7} = -1$ **26.** $\frac{2y-7}{9} = 5$

27. $\frac{x-9}{3} = 3x - 11$ **28.** $\frac{-14y-17}{-5} = 2 + 3y$

29. $\frac{9x}{4} = \frac{18+x}{2}$ **30.** $\frac{7x-5}{6} = \frac{x-44}{9}$

In Exercises 31 – 40 find the value of the indicated variable in each formula.

31. $d = \frac{1}{2}at^2$; $d = 144$, $t = 3$. Find a.

32. $C = \frac{5}{9}(F - 32)$; $C = -25$. Find F.

33. $S = \frac{1}{2}gt^2 + vt$; $g = 32$, $t = 2$, $S = 144$. Find v.

34. $F = \frac{9}{5}C + 32$; $F = 212$. Find C.

35. $R = \frac{kL}{d^2}$; $R = 32$, $L = 100$, $d = 5$. Find k.

36. $a = p(1 + rt)$; $a = 3{,}000$, $p = 2{,}000$, $r = 0.05$. Find t.

37. $P = 2(l + w)$; $P = 76$, $l = 27$. Find w.

38. $V = \frac{1}{3}\pi r^2 h$; $V = 51\pi$, $r = 3$. Find h.

39. $T = mg - mf$; $T = 80$, $m = 10$, $g = 14$. Find f.

40. $S = \frac{1}{2}n(a + L)$; $S = 85$, $n = 17$, $a = -7$. Find L.

In Exercises 41–60 solve each formula for the variable indicated.

41. $F = ma$ for m **42.** $D = r \cdot t$ for r

43. $d = \frac{1}{2}at^2$ for a

44. $V = \frac{1}{3}\pi r^2 h$ for h

45. $P = mgh$ for m

46. $i = p \cdot r \cdot t$ for r

47. $A = \frac{a+b+c}{3}$ for b

48. $T = mg - mf$ for f

49. $S = \frac{1}{2}n(a + L)$ for L

50. $A = \frac{1}{2}h(b + c)$ for b

51. $a = p(1 + rt)$ for t

52. $a = p + prt$ for t

53. $A = \pi(R^2 - r^2)$ for R^2

54. $R = \frac{kL}{d^2}$ for d^2

55. $S = \frac{a}{1-r}$ for r

56. $C = \frac{E}{R+S}$ for R

57. $n = \frac{A+D}{A}$ for D

58. $\frac{I_1}{I_2} = \frac{r_1^2}{r_2^2}$ for I_2

59. $I = \frac{2E}{R+2r}$ for r

60. $C = \frac{nE}{1+nr}$ for E

In Exercises 61–68 solve each equation for x.

61. $2x = a$

62. $x - 7 = b$

63. $3az - x = 1 + t$

64. $2t - 2x = 3c$

65. $3x + b = c - x$

66. $5x - a = 2x + a$

67. $\frac{2x-a}{b} = 5$

68. $\frac{4-3x}{c} = 1$

In Exercises 69–76 solve for the indicated letter.

a. For x **b.** For y

69. $x + 2y = 4$

70. $7x - y = 2$

71. $-x - y = 2$

72. $-4x + 2y = 5$

73. $3x + 4y = 6$

74. $3x - 2y = -6$

75. $-\frac{3}{5}x + \frac{2}{5}y = 10$

76. $\frac{4}{3}x - \frac{1}{3}y = 1$

77. The formula $h = \frac{w+200}{5}$ shows the approximate relationships of "normal" weight to height for adult females, where h is height in inches and w is weight in pounds.
   a. Solve this formula for w.
   b. How many pounds should a 5-ft. 6-in. woman "normally" weigh?

78. The formula for converting centimeters (c) to inches (i) is $i = 0.3937c$.
   a. Solve the formula for c.
   b. A yardstick is 36 in. long. Find this length to the nearest centimeter.

79. Radio waves travel at the speed of light, which is about 299,800 km/second. How long does it take for a radio message from an astronaut to travel from the moon to the earth, which is about 3.844×10^8 m away? Use the distance formula, $d = rt$.

80. Use the formula $E = mc^2$ to find m (in kilograms) if c, the speed of light, is 3×10^8 m/second and E is 4.2×10^{16} joules. This will give you the amount of matter used to create the explosive force of the first hydrogen bomb ever exploded. (This force is equivalent to 10 million tons of TNT and was released by the U.S. in 1952 at Eniwetok.)

81. The shaded area equals 55 in.^{2}. Find the area of the larger rectangle.

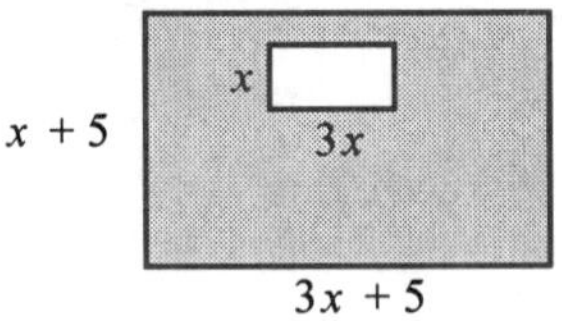

82. The shaded area equals 98 cm^2. Find the area of the smaller rectangle.

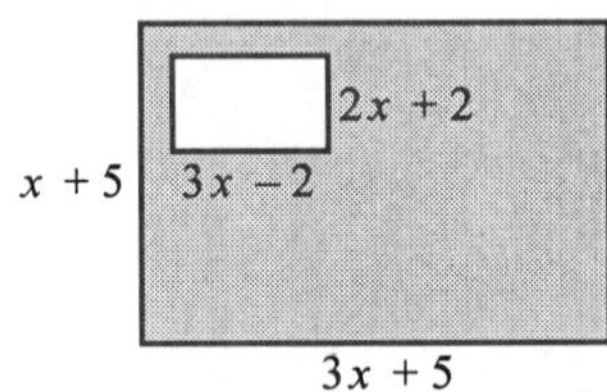

THINK ABOUT IT R.5

1. **a.** Write in words what this equation says, and then explain why it must be an identity: $x + x = 2x$.
   b. Write in words what this equation says. Is it an identity or a false equation? $x - 1 = x + 1$.

2. The formula $S = (1/2)n(n+1)$ gives the sum of the integers from 1 to n.
   a. To see that the formula is reasonable, let $n = 3$ and use the formula to show that $1+2+3$ is equal to 6.
   b. Find the sum of the integers from 1 to 999.
   c. Explain why these three versions of the formula are equivalent. Then use each version to find the sum of the numbers from 1 to 100.
      i. $S = [(n+1)/2]\cdot n$ ii. $S = (n+1)\cdot(n/2)$ iii. $S = [(n+1)(n)] \div 2$
3. Assume $a \neq 0$ and solve each equation for x.
   a. $ax + a^2 = 4a^2$ b. $4ax - 3a^{-2} = 9a^{-2}$
4. In the equation $ax = b$, a and b are constants. What combination of values for a and b results in an equation with exactly one solution? No solution? Infinitely many solutions?
5. Consider this illustration of a piece of a broken chariot wheel found by a group of archaeologists. What is the radius of the wheel?

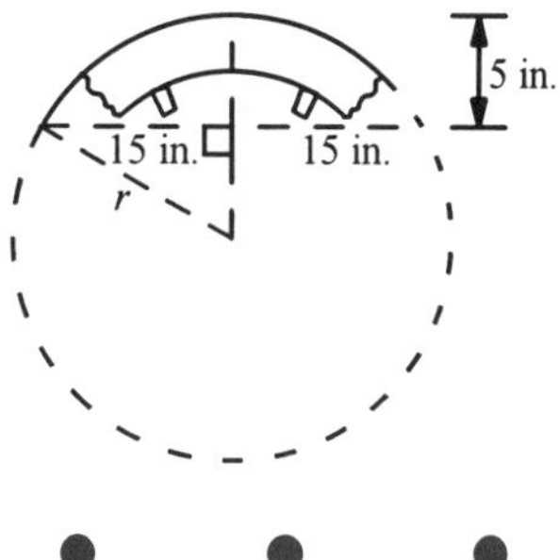

● ● ●

R.6 Applications of Linear Equations

Photo Courtesy of Archive Photos, New York

In a vacation community the real estate salespeople get a 12 percent commission on rentals. An owner wants to rent out a vacation house for the summer and end up with \$10,000 income. Find the required rental price for the house.
(See Example 1.)

Objectives

1. Solve word problems by translating phrases and setting up and solving equations.
2. Solve problems involving geometric figures, annual interest, uniform motion, liquid mixtures, and proportions.

Many problems that are analyzed by mathematics begin with writing an equation to express some key relationship. Then the solution to the equation is used to solve the problem. In this section we discuss how to solve several basic types of word problems that require us to write and solve linear equations. Consider carefully the following steps, which are recommended by both mathematicians and reading specialists as a general approach to solving word problems.

To Solve a Word Problem

1. **Read the problem several times.** The first reading is a preview and is done quickly to obtain a general idea of the problem. The objective of the second reading is to determine exactly what you are asked to find. Write this down. Finally, read the problem carefully and note what information is given. If possible, display the given information in a sketch or chart.
2. **Let a variable represent an unknown quantity** (which is usually the quantity you are asked to find). Write down precisely what the variable represents. If there is more than one unknown, represent these unknowns in terms of the original variable, when possible.
3. **Set up an equation** that expresses the relationship between the quantities in the problem.
4. **Solve the equation.**
5. **Answer the question.**
6. **Check the answer** by interpreting the solution in the context of the word problem.

Example 1: Finding the Rental Price for a Vacation House

Solve the problem in the section introduction on page 55.

Solution: We need to find the required rental price for the house. If x = rental price, then the sales commission is 12 percent of x, or $0.12x$, so

$$\underset{x}{\underset{\downarrow}{\text{rental price}}}\quad \underset{-}{\underset{\downarrow}{\text{minus}}}\quad \underset{0.12x}{\underbrace{\text{sales commission}}}\quad \underset{=}{\underset{\downarrow}{\text{equals}}}\quad \underset{10{,}000}{\underbrace{\text{amount for owner.}}}$$

Solve the Equation:

$$\begin{aligned} 0.88x &= 10{,}000 && \text{Combine like terms.} \\ x &= \frac{10{,}000}{0.88} && \text{Divide both sides by 0.88.} \\ x &= 11{,}363.64 && \text{Simplify.} \end{aligned}$$

Answer the Question: The rental price should be \$11,363.64. Of course, in practice the owner usually rounds the number up a little, to say \$11,400.

Check the Answer: If the rental price is \$11,363.64, then

$$\text{sales commission} = 0.12(\$11{,}363.64) = \$1{,}363.64$$
$$\text{income from rental} = \$11{,}363.64 - \$1{,}363.64 = \$10{,}000$$

The solution checks.

PROGRESS CHECK 1

The total cost (including tax) of a new car is \$14,256. If the sales tax rate is 8 percent, how much is paid in taxes? ■

Historically, the study of algebra has always been connected with problems about number relations and properties. The next example illustrates a type of number problem that traditionally is used to build proficiency in writing and solving equations.

Example 2: Translating Phrases to Solve a Problem

The sum of two consecutive even integers is 238. Find the integers.

Solution: If x represents the first even integer, then $x+2$ represents the next even integer. Now **set up an equation**.

$$\underbrace{\text{The sum of two consecutive even integers}}_{x+(x+2)} \quad \underset{=}{\overset{\text{is}}{\downarrow}} \quad \underset{238}{\overset{238.}{\downarrow}}$$

Solve the Equation:

$$\begin{aligned} 2x+2 &= 238 && \text{Combine like terms.} \\ 2x &= 236 && \text{Subtract 2 from both sides.} \\ x &= 118 && \text{Divide both sides by 2.} \end{aligned}$$

Answer the Question: The smaller integer is 118, and the next consecutive even integer is 120.

Check the Answer: The sum of 118 and 120 is 238, and they are consecutive even integers, so the solution checks.

PROGRESS CHECK 2
The sum of two consecutive integers is 147. Find the integers. ■

To develop skill in problem solving, it is useful to build a supply of model problems that you understand. Then, when a new situation arises, you may find that it is similar to a problem you already know how to solve, and you will quickly know some approaches that might work. To build this base, we now consider word problems that involve geometric figures, annual interest, uniform motion, liquid mixtures, and proportions.

Geometric Problems

To solve problems about geometric figures, we often need perimeter, area, and volume formulas, and Section R.2 contains the formulas you need for now. When triangles and angle measures are involved, then you should know that any two angles whose measures add up to 90° are called **complementary**, and any two angles whose measures add up to 180° are called **supplementary**. Also, the sum of the angle measures in a triangle is 180°, and a **right triangle** is a triangle that contains a 90° angle.

Example 3: Solving a Geometric Problem

The length of a rectangular solar panel is three times its width. If the perimeter is 224 in., determine the area of the solar panel.

Solution: To find the dimensions of the solar panel, let

$$x = \text{width}$$

so

$$3x = \text{length.}$$

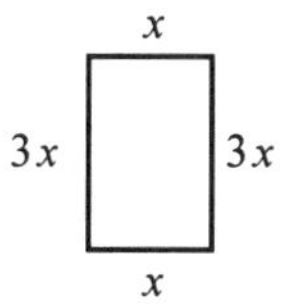

Figure R.21

Now illustrate the given information as in Figure R.21.

The perimeter is 224 in., and we can **set up an equation** by using $P = 2l + 2w$, the formula for the perimeter of a rectangle.

$$\begin{aligned} P &= 2l + 2w \\ 224 &= 2(3x) + 2x \quad \text{Replace } P \text{ by 224, } l \text{ by } 3x\text{, and } w \text{ by } x. \end{aligned}$$

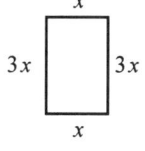

Solve the Equation:

$$224 = 8x \quad \text{Combine like terms.}$$
$$28 = x \quad \text{Divide both sides by 8.}$$

Answer the Question: The area is given by $A = lw$ where $w = 28$ in., and $l = 3(28) = 84$ in. So the area is 2,352 in.^{2}.

Check the Answer: The perimeter is $2(28) + 2(84) = 224$ in., while 84 in. is three times as long as 28 in. The area is $28(84) = 2{,}352$ in.^{2}, so the solution checks.

PROGRESS CHECK 3
In a right triangle the measure of one of the acute angles is 36° greater than the other. What is the measure of the larger acute angle? ■

Annual Interest Problems

When a problem involves annual interest, use $I = Pr$, where P represents principal, r represents the annual interest rate, and I represents the amount of interest earned in one year. This formula is a special case of $I = Prt$ where $t = 1$.

Example 4: Solving an Annual Interest Problem

How should a $160,000 investment be split so that the total annual earnings are $14,000, if one portion is invested at 6 percent annual interest and the rest at 11 percent?

Solution: To find the amount invested at each rate, let

$$x = \text{amount invested at 6 percent}$$

so

$$160{,}000 - x = \text{amount at 11 percent.}$$

In general, note that if the sum of two numbers is n, then one number can be called x and the other $n - x$. Now, use $I = Pr$ and analyze the investment in a chart format.

Investment	Principal ·	Rate =	Interest
1st account	x	0.06	$0.06x$
2nd account	$160{,}000 - x$	0.11	$0.11(160{,}000 - x)$

Set up an equation using 14,000 as the desired amount of total interest.

$$\underbrace{0.06x}_{\text{Interest from 1st account}} \; \underset{\text{plus}}{+} \; \underbrace{0.11(160{,}000 - x)}_{\text{interest from 2nd account}} \; \underset{\text{equals}}{=} \; \underbrace{14{,}000}_{\text{total interest.}}$$

Solve the Equation:

$$0.06x + 17{,}600 - 0.11x = 14{,}000 \quad \text{Distributive property.}$$
$$-0.05x + 17{,}600 = 14{,}000 \quad \text{Combine like terms.}$$
$$-0.05x = -3{,}600 \quad \text{Subtract 17,600 from each side.}$$
$$x = 72{,}000 \quad \text{Divide both sides by } -0.05.$$

Answer the Question: Invest \$72,000 at 6 percent and \$160,000 − \$72,000, or \$88,000, at 11 percent.

Check the Answer: \$72,000 + \$88,000 = \$160,000 and the first account earns 6 percent of \$72,000, or \$4,320, while the second account earns 11 percent of \$88,000, or \$9,680, for a total of \$14,000 interest. The solution checks.

PROGRESS CHECK 4
How should a \$200,000 investment be split so that the total annual earnings are \$18,000, if one portion is invested at 10 percent annual interest and the rest at 6 percent? ■

Uniform Motion Problems

To solve uniform motion problems you need the formula $d = rt$, where d represents the distance traveled in time t by an object moving at a constant rate r. This formula applies to objects moving at a **constant** (or uniform) speed and to objects whose **average** speed is involved. A chart format is recommended for this type of problem, and a sketch of the situation may also be helpful.

Example 5: Solving a Uniform Motion Problem

One jet takes 1 hour 15 minutes for a certain flight between two airports. Under the same conditions another jet makes the trip in 1 hour by going 100 mi/hour faster. What is the distance between airports?

Solution: The key here is to find the rate for one of the jets. If

$$x = \text{rate for slower jet,}$$

then

$$x + 100 = \text{rate for faster jet.}$$

Now analyze the problem in a chart format using $d = rt$. Note that 1 hour 15 minutes is equivalent to $1 + 15/60$, or 1.25 hours.

Plane	Rate (mi/hour) ·	Time (hours) =	Distance (mi)
Slower jet	x	1.25	$1.25x$
Faster jet	$x + 100$	1	$1(x + 100)$

To **set up an equation**, sketch the situation as in Figure R.22. From the sketch observe that both jets travel the same distance.

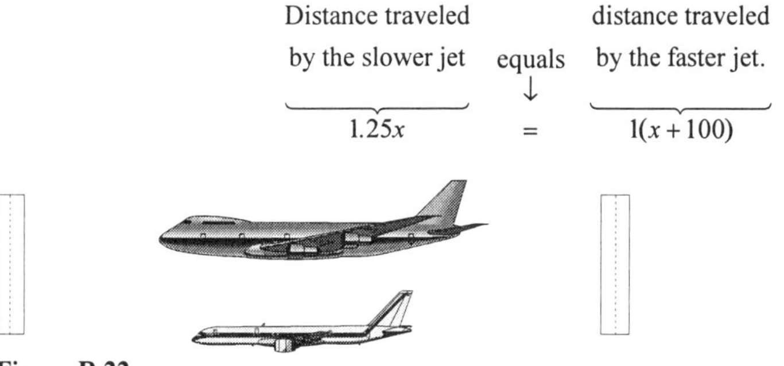

Distance traveled by the slower jet
equals
distance traveled by the faster jet.
1.25x
=
1(x + 100)

Figure R.22

Solve the Equation:

$1.25x = x + 100$	Distributive property.
$0.25x = 100$	Subtract x from both sides.
$x = 400$	Divide both sides by 0.25.

Answer the Question: The slower jet travels at 400 mi/hour for 1.25 hours, so the distance between airports is 400(1.25) mi, or 500 mi.

Check the Answer: If the distance between airports is 500 mi, then a rate of 400 mi/hour is required to complete the trip in 1.25 hours, and a rate of 500 mi/hour is required to complete the trip in 1 hour. Since 500 is 100 more than 400, the solution checks.

PROGRESS CHECK 5
On a video display, an air traffic controller notices two planes 120 mi apart and flying toward each other on a collision course. One plan is flying at 500 mi/hour; the other is flying at 300 mi/hour. How much time is there for the controller to prevent a crash? ■

Liquid Mixture Problems

To solve problems about liquid mixtures, you need to apply the concept of percentage in the following context.

$$\begin{pmatrix}\text{percent of}\\ \text{an ingredient}\end{pmatrix}\cdot\begin{pmatrix}\text{amount of}\\ \text{solution}\end{pmatrix}=\begin{pmatrix}\text{amount of}\\ \text{the ingredient}\end{pmatrix}.$$

For example, the amount of acid in 10 liters of a solution that is 30 percent acid is 0.30(10), or 3 liters. Once again a chart is recommended for this type of problem.

Example 6: Solving a Liquid Mixture Problem

A chemist has two acid solutions, one 30 percent acid and the other 70 percent acid. How much of each solution must be used to obtain 100 ml of a solution that is 41 percent acid?

Solution: To find the correct mixture, let

$$x = \text{amount used of 30 percent solution,}$$

so

$$100 - x = \text{amount used of 70 percent solution.}$$

As recommended, we analyze the problem with a chart, and illustrate this information as in Figure R.23.

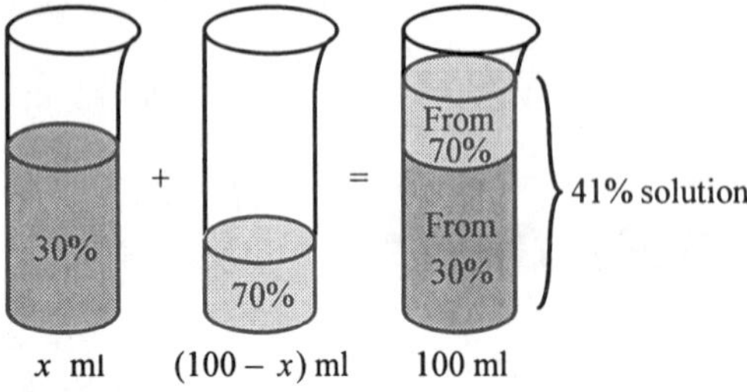

Figure R.23

Solution	Percent Acid	·	Amount of Solution (ml)	=	Amount of Acid (ml)
1st solution	30		x		$0.3x$
2nd solution	70		$100-x$		$0.7(100-x)$
New solution	41		100		$0.41(100)$

To **set up an equation**, we reason that the amount of acid in the new solution is the sum of the amounts contributed by the 1st and 2nd solutions.

$$\underbrace{0.3x}_{\text{Amount of acid in 1st solution}} \underset{\text{plus}}{+} \underbrace{0.7(100-x)}_{\text{amount of acid in 2nd solution}} \underset{\text{equals}}{=} \underbrace{0.41(100)}_{\text{amount of acid in new solution.}}$$

Solve the Equation:

$$\begin{aligned} 0.3x + 70 - 0.7x &= 41 && \text{Distributive property.} \\ -0.4x + 70 &= 41 && \text{Combine like terms.} \\ -0.4x &= -29 && \text{Subtract 70 from both sides.} \\ x &= 72.5 && \text{Divide both sides by } -0.4. \end{aligned}$$

Answer the Question: Mix 72.5 ml of the 30 percent solution with $100-72.5$, or 27.5, ml of the 70 percent solution.

Check the Answer: The new solution contains $72.5+27.5$, or 100 ml. Also, 72.5 ml from the lst solution contains $0.3(72.5)=21.75$ ml of acid, while 27.5 ml from the 2nd solution contains $0.7(27.5)=19.25$ ml of acid. Thus, the new mixture contains 41 ml of acid in 100 ml of solution. Because $41/100=0.41=41$ percent, the new mixture does contain 41 percent acid, and the solution checks.

PROGRESS CHECK 6

One metal contains 30 percent gold by weight and the rest silver. Another contains 50 percent gold by weight and the rest silver. They will be melted down and mixed together to form a new alloy that is 35 percent gold. How much of each should be used to form 5 lb of the new alloy? ■

Proportion Problems

A **ratio** is a comparison of two quantities by division, and a **proportion** is a statement that two ratios are equal. For example, the ratios $2/3$ and $6/9$ are equal and form a proportion that may be written as $2/3=6/9$ (read: 2 is to 3 as 6 is to 9) or as $2:3=6:9$. When we work with proportions, it is easier to write the ratios as fractions and then use the techniques we have developed for solving an equation.

Example 7: Solving a Proportion Problem

A cylindrical tank holds 480 gal when it is filled to its full height of 8 ft. When the gauge shows that it contains water at a height of 3 ft 1 in., how many gallons are in the tank?

Solution: If we let x gal represent the unknown amount of water in the tank and *set up two equal ratios* that compare like measurements, we have the proportion

$$\frac{x \text{ gal}}{480 \text{ gal}} = \frac{37 \text{ in.}}{96 \text{ in.}} \qquad \textit{Note: } 3 \text{ ft } 1 \text{ in.} = 37 \text{ in. and } 8 \text{ ft} = 96 \text{ in.}$$

Solve the Equation:

$$x = \left(\frac{37}{96}\right) \cdot 480 \quad \text{Multiply both sides by 480.}$$
$$x = 185 \quad \text{Simplify.}$$

Answer the Question: There are 185 gal of water in the tank when the water is at a height of 3 ft 1 in.

Check the Answer: In the context of the problem, 185 gal at 37 inches is a rate of 5 gal/inch, and 480 gal at 96 inches is also a rate of 5 gal/inch. The solution checks.
Note: It is possible to set up the proportion in other ways, as long as both sides of the equation express similar ratios. For instance, as suggested in the Check, a logical proportion in this example would be

$$\frac{x \text{ gal}}{37 \text{ in.}} = \frac{480 \text{ gal}}{96 \text{ in.}}.$$

PROGRESS CHECK 7
If an idling car uses 35 oz of gasoline in 50 minutes, how long to the nearest minute must it idle to use 1 gal of gas? One gallon is 128 oz. ■

EXPLORE R.6

For the following exercises:

a. Use a guess and check approach with a calculator to solve the problem. Use a calculator technique of your choice for rapidly evaluating expressions.
b. Solve the problem algebraically by setting up and solving an equation.

1. A student has 3 test grades: 80, 84, and 92. What grade on the fourth test will result in an average grade of 85?
2. The total cost of a car, including a 5 percent sales tax, is $18,732. What is the price of the car before the tax is added?
3. A restaurant patron would like to spend no more than $15 for a meal, including a 15 percent tip and a 4.5 percent tax. To the nearest dollar, what is the most expensive menu item the patron can order?

EXERCISES R.6

Solve each problem by first setting up an appropriate equation.

1. During a test run, machine A produced x cans; machine B turned out three times as many; machine C produced 10 more cans than A. The total production during the run was 5,510. How many cans did each machine produce?

2. A magazine and a newsletter cost $1.10. If the magazine cost $1.00 more than the newsletter, how much did each cost?

3. A stockbroker, after deducting a commission, gives a client $7,650. If the commission is 10 percent of the selling price of the stock, how much was the customer's stock worth?

4. Suppose that 5 percent of your salary is deducted for a retirement fund. If your total deduction for one year is $1,725, what is your annual salary?

5. A car dealer offers a used car at a price of $6,104. This selling price represents a profit of 12 percent on the amount paid for the car. How much did the car cost the dealer?

6. If the total cost (including tax) of a new car is $13,054, how much do you pay in taxes if the sales tax rate is 7 percent?

7. The final bill for the sofa you purchased totaled \$1,349. If this included 8.25 percent sales tax plus a \$50 delivery fee, what was the original price of the sofa? The delivery fee was not taxed.

8. The population of a city increased 2 percent in one year to reach about 3 million people. To the nearest thousand people, about what was the population at the beginning of the year?

9. Multiplying a number by 4 gives the same result as adding 6 to the number. What is the number?

10. Taking one-half of a number gives the same result as adding 3 to the number. What is the number?

11. If a number is decreased by 5, the result is twice the original number. What is the number?

12. Multiplying a number by 4 and adding 2 to the product gives the same result as dividing the number by –2. What is the number?

13. The sum of two consecutive integers is 89. Find the larger integer.

14. The sum of two integers is 35. One of the integers is four times the other. Find the integers.

15. The sum of the three angles in a triangle is 180°. The second angle is 20° more than the first, and the third angle is twice the first. What is the measure of each of the angles in the triangle?

16. The length of a rectangle is three times its width. If the perimeter is 160 in., determine the area of the rectangle.

17. In a right triangle the measure of one of the acute angles is 3 times the measure of the other. What is the measure of the larger acute angle?

18. The sum of the angles of a quadrilateral is 360°. If the measures of the angles of a quadrilateral are four consecutive odd integers, find them.

19. Two angles are supplementary. If the measure of one of them is 5 more than 4 times the other, what are the measures of the angles?

20. If the measure of two complementary angles is represented by $2x+3$ and $3x-8$, what are the measures of these angles?

21. A certain amount of money is invested at 12 percent interest per year, and twice that amount is invested at 9 percent annual interest. Together, the two accounts earned annual interest of \$5,400. How much is invested in each account?

22. Money in one account is earning 7 percent annually, and another account with \$2,000 more earns 11 percent annually. Altogether, the two accounts earn \$2,700 a year. How much money is invested in each account?

23. The trustees of a \$375,000 fund that provides scholarship monies need to generate \$26,000 worth of income each year. Their advisors recommend investing in two different financial instruments, one very safe but yielding only 5 percent annually, and another yielding 12 percent annually but more risky.

a. To the nearest thousand dollars, how much should be allocated to each investment?

b. What should be the split if they also want to generate an additional \$10,000 for investment.

24. An accountant recommends two investments to a client: one earning 10 percent annually and one earning 15 percent annually. These investments should cover the cost of the client's rent (\$900 per month) for a year. If there is \$86,000 to invest, how should the money be split between the two investments?

25. Two fishing boats leave the harbor at the same time and using the same float plan. One averages 17 knots (nautical miles) per hour, and the other averages 14.5 knots per hour. How long will it take for the boats to be 10 nautical miles apart?

26. A moving van makes the trip along the total length of Highway 4 in about 6 hours. Another, newer moving van makes the same trip in 5.5 hours by traveling 4 mi/hour faster. How fast does each van travel? How long is Highway 4?

27. Two bicyclists make the same trip. The first takes about 3.5 hours, and the second, traveling about 1 mi/hour faster, only takes 3 hours. How many miles is the trip?

28. One plane takes 2 hours and 45 minutes for a particular flight. With a headwind, the same plane goes 20 mi/hour slower and requires 3 hours to reach its destination. How far apart are the airports?

29. A spacecraft moves through intergalactic space at a rate of 15,000 mi/hour. Unknown to the crew, a huge asteroid traveling at 7,000 mi/hour is hurtling directly toward the spacecraft. The craft and the asteroid are presently 165,000 mi apart. How long will it take for them to be within scanning range (22,000 mi) of each other, at which time the crew can make a course correction to avoid impact?

30. On a video display an air traffic controller notices two planes 120 mi apart and flying toward each other on a collision course. One plan is flying at 500 mi/hour; the other is flying 300 mi/hour. How much time is there for the controller to prevent a crash?

31. A machine shop has two large containers that are each filled with a mixture of oil and gasoline. Container A contains 2 percent oil (and 98 percent gasoline). Container B contains 6 percent oil (and 94 percent gasoline). How much of each should be used to obtain 18 quarts (qt) of a new mixture that contains 5 percent oil?

32. A chemist has 5 liters of a 25 percent sulfuric acid solution. She wishes to obtain a solution that is 35 percent acid by adding a solution of 75 percent acid to her original solution. How much of the more concentrated acid must be added to achieve the desired concentration?

33. One bar of tin alloy is 35 percent pure tin, and another bar is 10 percent pure tin. How many pounds of each must be used to make 95 lb of a new alloy that is 20 percent pure tin?

34. An alloy of copper and silver weighs 40 lb and is 20 percent silver. How much silver must be added to produce a metal that is 50 percent silver?

35. A 20 percent antifreeze solution is a solution that consists of 20 percent antifreeze and 80 percent water. If we have 20 qt of such a solution, how much water should be added to obtain a solution that contains 10 percent antifreeze?

36. One cup of vinegar is mixed with 3 cups of a solution that is 15 percent vinegar and 85 percent oil. What is the percentage of vinegar in the final mixture?

37. In a certain concrete mix the ratio of sand to cement is 4:1. At this ratio, how many pounds of cement are needed to be mixed with 50 lb of sand?

38. A spring is stretched 6 in. by a weight of 2 lb. How much weight is needed to stretch the spring 15 in.?

39. A car travels 225 mi in 5 hours. At the same rate, how long will it take the car to travel 405 mi?

40. An object that weighs 48 lb on earth weighs 8 lb on the moon. How much will a person who weighs 174 lb on earth weigh on the moon?

41. A stake 12 ft high casts a shadow 9 ft long at the same time that a tree casts a shadow of 33 ft. What is the height of the tree?

42. The gable end of a building has the dimensions illustrated in the drawing on the left below. If a carpenter wishes to keep the same proportions, what would be the vertical height of the structure illustrated in the drawing on the right?

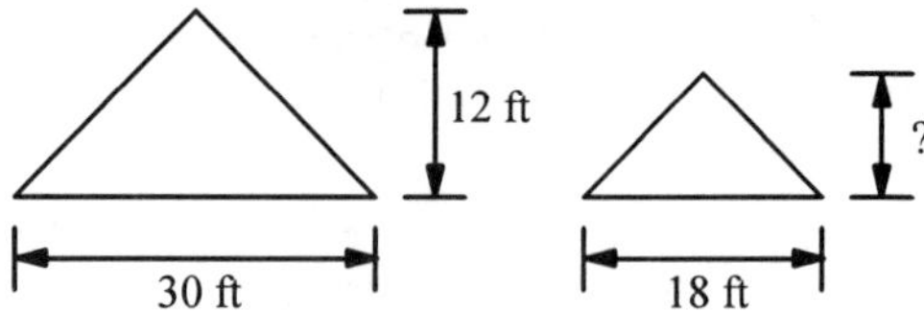

43. On a test, you received a grade of 114 on a scale of 0–120. Convert your score to a scale of 0–100.

44. A cylindrical tank holds 370 gal when it is filled to its full height of 9 ft. When the gauge shows that it contains oil at a height of 6 ft 7 in., how many gallons (to the nearest gallon) are in the tank?

45. If 72 gal of water flow through a feeder pipe in 40 minutes, how many gallons of water will flow through the pipe in 3 hours.

46. A map has a scale of 1.5 in. = 7 mi. To the nearest tenth of a mile, what distance is represented by 20 in. on the map?

47. A bus will travel 67 mi on 8 gal of gas. To the nearest mile, how far will the bus travel on 11 gal of gas?

48. If 2 g of hydrogen unite with 16 g of oxygen to form water, how many grams of oxygen are needed to produce 162 g of water? (Assume that no loss takes place during the reaction.)

49. A certain antifreeze mixture calls for 5 qt of antifreeze to be mixed with 2 gal of water. How many quarts of antifreeze are needed for a total mixture of 65 qt?

50. A team of sociologists wishes to do a survey in a city where the census report indicates that the ratio of Protestants to Catholics is 5:2. They also determine that a suitable sample size is 420 people. How many people of each religion should they interview if the sample is to have the same religious distribution as the city's population?

51. The power and beauty of mathematics is that one problem analyzed abstractly and mathematically can be applied to many different physical situations. For example, consider the following applications, all of which refer to the drawing shown.

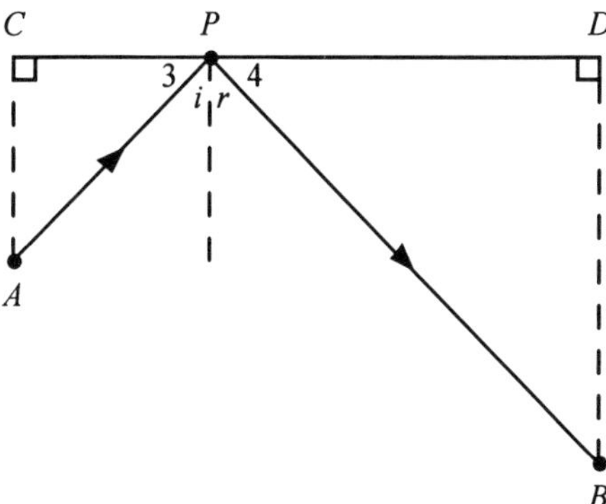

a. One of the most important phenomena in our physical world is light. The simplest principle in the mathematical approach to the theory of this subject is the **law of reflection**. This law states that a ray of light that strikes a reflecting surface is reflected so the angle of incidence (i) equals the angle of reflection (r). Many optical instruments, such as telescopes, work on this principle. For a better lighting effect a photographer positioned at A aims her flash at position P on a reflecting surface to take a picture of a client at B. If $\overline{AC} = 4$ ft, $\overline{CP} = 3$ ft, and $\overline{PD} = 6$ ft, determine $\overline{AP} + \overline{PB}$, which is the distance the light travels from the flash to the client.

b. The same principle applies to a hard body that bounces off an elastic surface. That is, it rebounds at the same angle at which it strikes. A billiards player wishes to make a ball at A strike the cushion at P to hit another ball at B. If $\overline{AC} = 2$ ft, $\overline{CD} = 6$ ft, and $\overline{DB} = 3$ ft, determine $\overline{CP}$ so that point P may be found.

c. Suppose that a company owns two plants located at A and B, and that line segment CD represents a railroad track. The company wishes to build a station at P to ship its merchandise. Naturally it wants the total trucking distance $\overline{AP} + \overline{BP}$ to be minimal. Mathematically, it can be shown that $\overline{AP} + \overline{BP}$ is minimal when angle 3 equals angle 4. If $\overline{AC} = 3$ mi, $\overline{CD} = 10$ mi, and $\overline{DB} = 5$ mi, determine $\overline{DP}$ so that the most desirable spot for the station may be found.

52. One of the classic problems from the history of mathematics is the ingenious method used by the Greek mathematician Eratosthenes to estimate the circumference of the earth in about 200 B.C. Eratosthenes obtained his estimate by noting the following information, which is illustrated here. Alexandria (where Eratosthenes was librarian) was 500 mi due north of the city of Syene. At noon on June 21 he knew from records that the sun cast no shadow at Syene, which meant that the sun was directly overhead. At the same time in Alexandria he measured from a shadow that the sun was 7.5° south from the vertical. By assuming that the sun was sufficiently far away for the light rays to be parallel to the earth, he then determined that angle AOB measured 7.5°. Why? Complete the line of reasoning and obtain Eratosthenes' estimate for the circumference. You will find the result remarkably close to the modern estimate of 24,900 mi, which is obtained with the same basic procedure but with more accurate measurements.

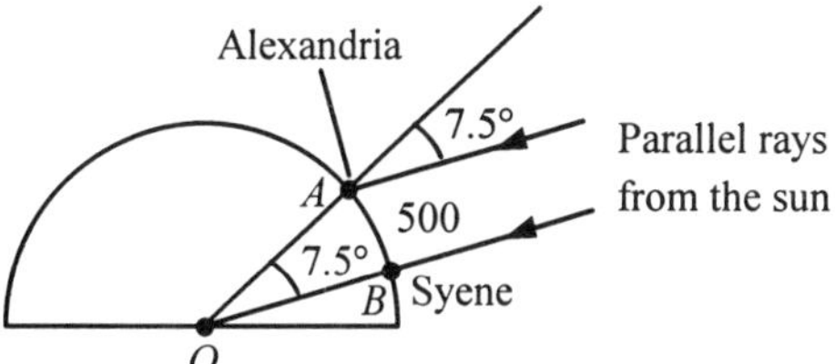

THINK ABOUT IT R.6

1. Create a word problem that may be solved by the equation

$$1000 - x = 4x$$

where x represents the unknown requested in the problem.

2. Give two reasons why we check the answer in a word problem by interpreting the solution in the context of the word problem, instead of checking in the equation that has been set up to solve the problem.

3. The following question is a favorite of problem solvers because it makes an important point. A car leaves Washington and heads for New York at 55 mi/hour. Two hours later a bus leaves New York and heads for Washington at 60 mi/hour. When the car and the bus meet, which is closer to New York? What is the point made by this problem?

4. A chemist has two mixtures, 30 percent acid and 20 percent acid. Can they be mixed to form 6 liters of a 40 percent solution? Set up the equation for this and see what the solution tells you. What is the difficulty?
5. You have 60 in. of string to shape into a rectangle with a perimeter equal to 60 in. Make a chart that shows various possibilities for the length and width. Do they all have the same area? What dimensions appear to give the maximum area? The minimum area?

● ● ●

R.7 Linear and Compound Inequalities

Photo Courtesy of **Eric Roth**, The Picture Cube

Water turns to steam when the temperature is at least 100 degrees Celsius. At what temperature in degrees Fahrenheit will water turn to steam if the temperature scales are related by the formula $C = (5/9)(F - 32)$? (See Example 3.)

Objectives

1. Specify solution sets of linear inequalities by using graphs and interval notation.
2. Solve linear inequalities by applying properties of inequalities.
3. Solve compound inequalities that specify numbers between two numbers.
4. Solve compound inequalities involving *or* statements.

The solution to a problem often requires us to use the inequality signs ($<$, $\leq$, $>$, $\geq$) when expressing relationships. For instance, in the section-opening problem we will determine the temperature in degrees Fahrenheit at which water turns to steam by solving the inequality

$$\frac{5}{9}(F-32) \geq 100.$$

As with equations, a **solution** of an inequality is a value for the variable that makes the inequality a true statement, and the set of all such solutions is called the **solution set**. Thus, the solution set of the inequality $x < 3$ is the set of all real numbers less than 3, and this set is written in set-builder notation* as $\{x\colon\ x < 3\}$. Another way to describe this infinite set of numbers is to graph it on the number line, as shown in Figure R.24(a). The parenthesis at 3 in this figure means that 3 is not included in the solution set, and the arrow specifies all real numbers less than 3. To graph a set like $\{x\colon\ x \leq 3\}$, we show that 3 is included in the solution set by putting a bracket at this point, as shown in Figure R.24(b).

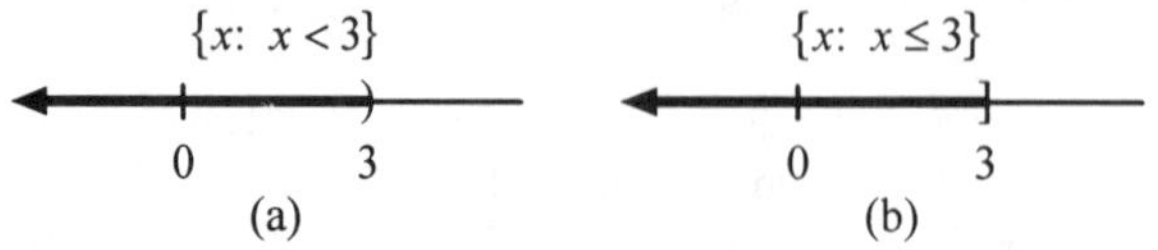

Figure R.24

* **Set-builder notation** writes sets in the form $\{x\colon x \text{ has property } P\}$, which is read, "the set of all elements x such that x has property P." The colon (:) is read "such that."

Sets of real numbers that may be represented graphically as half lines or the entire number line are examples of **intervals** that may be expressed conveniently by using **interval notation**, as outlined in the following chart.

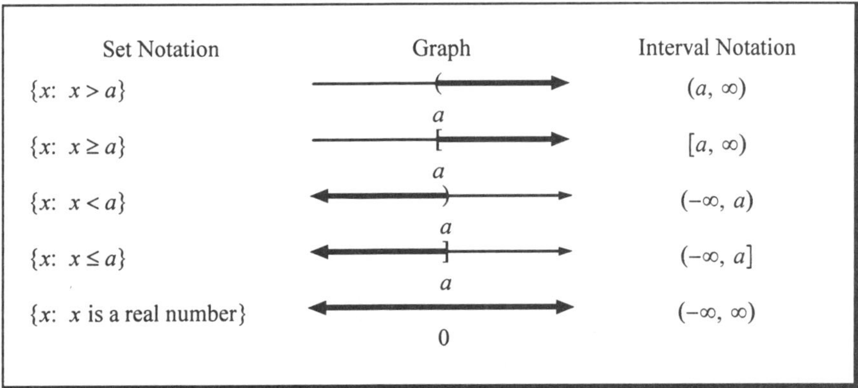

Set Notation	Graph	Interval Notation
$\{x\colon\ x > a\}$	a	(a, ∞)
$\{x\colon\ x \geq a\}$	a	$[a, \infty)$
$\{x\colon\ x < a\}$	a	$(-\infty, a)$
$\{x\colon\ x \leq a\}$	a	$(-\infty, a]$
$\{x\colon\ x$ is a real number$\}$	0	$(-\infty, \infty)$

Note that the symbols ∞, read "infinity," and $-\infty$ are not real numbers but are convenient symbols that help us designate intervals that are unbounded in a positive or negative direction.

Example 1: Using Interval Notation and Graphs to Specify Solution Sets

Write the solution set to each inequality in interval notation, and graph the interval.

a. $x \geq -1$ **b.** $x < \dfrac{3}{2}$

Solution:

a. Draw a number line. Place a bracket at -1, and then draw an arrow from the bracket to the right, as in Figure R.25(a). The bracket means that the number -1 is a member of the solution set, and the arrow specifies all real numbers greater than -1. In interval notation $[-1, \infty)$ represents this set of numbers.

b. Figure R.25(b) shows the graph of $\{x\colon\ x < 3/2\}$, which is written as $(-\infty, 3/2)$ in interval notation. Note that in both the graph and the interval notation, the parenthesis indicates that $3/2$ is not a member of the solution set.

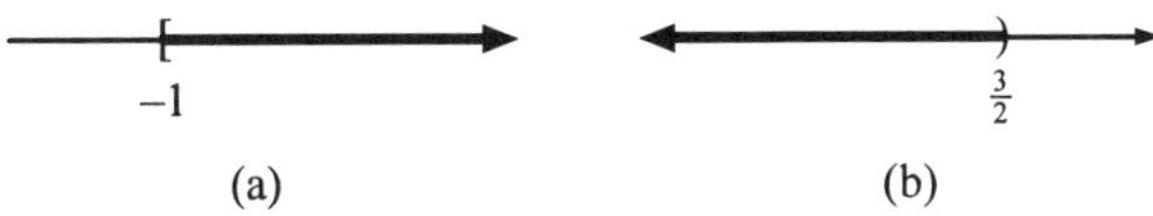

Figure R.25

Note: As shown in Figure R.26, it is also common notation to use an open circle in a graph instead of a parenthesis to indicate an endpoint that is not included in the solution set, and to use a solid dot instead of a bracket when the endpoint is included. In this text we choose to use parentheses and brackets because this method reinforces writing intervals in interval notation and is also more common in higher-level mathematics.

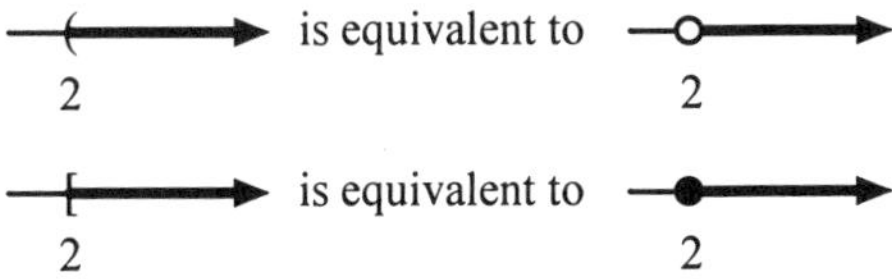

Figure R.26

PROGRESS CHECK 1

Write the solution set of each inequality in interval notation, and graph the interval.

a. $x \le 4$ **b.** $x > -\dfrac{3}{2}$ ■

By analogy to equations, a **linear inequality** results if the equal sign in a linear equation is replaced by one of the inequality symbols. For example,

$$-5x + 1 < 16,\ x \ge 2, \text{ and } \frac{5}{9}(F - 32) \ge 100$$

are linear inequalities. The procedures for solving linear equalities are similar to the procedures for solving linear equations. The key idea is to create equivalent but simpler inequalities at each step until the solution set is clear. However, to create equivalent inequalities, we apply the following properties of inequalities. Although these properties are given only for <, similar properties may be stated for the other inequality symbols.

Properties of Inequalities

Let a, b, and c be real numbers.

		Comment
1.	If $a < b$, then $a + c < b + c$ and $a - c < b - c$.	The sense of the inequality is preserved when the same number is added to (or subtracted from) both sides of an inequality.
2.	If $a < b$ and $c > 0$, then $ac < bc$ and $a/c < b/c$.	The sense of the inequality is preserved when both sides of an inequality are multiplied (or divided) by the same positive number.
3.	If $a < b$ and $c < 0$, then $ac > bc$ and $a/c > b/c$.	The sense of the inequality is reversed when both sides of an inequality are multiplied (or divided) by the same negative number.
4.	If $a < b$ and $b < c$, then $a < c$.	This property is called the **transitive property**.

Note that multiplying or dividing on both sides of an inequality demands special care because there are two cases; one for positive multipliers and one for negative multipliers. To see why, notice that the true inequality $-5 < 2$ leads to a true inequality if both sides are multiplied by *positive* 3.

$-5(3) < 2(3)$	Multiply both sides by 3.
$-15 < 6$	True inequality.

But $-5 < 2$ leads to a false inequality when both sides are multiplied by *negative* 3.

$-5(-3) < 2(-3)$	Multiply both sides by -3.
$15 < -6$	False inequality.

To obtain a true statement when multiplying both sides by –3, the direction of the inequality must be reversed.

$-5 < 2$	Original inequality.
$-5(-3) > 2(-3)$	Multiply both sides by -3 and *reverse* the inequality symbol.
$15 > -6$	True inequality.

Always remember to reverse the direction of the inequality when multiplying or dividing by a negative number on both sides of an inequality. This step is also called reversing the ***sense*** of the inequality. A formal proof for this property may be found by consulting "Think About It" Exercise 5.

Example 2: Solving a Linear Inequality

Solve $2x+1<5x-8$. Express the solution set graphically and in interval notation.

Solution: We isolate x on one side of the inequality in the following sequence of equivalent (same solution set) inequalities.

$$2x+1<5x-8$$

$$2x<5x-9 \quad \text{Subtract 1 from both sides.}$$

$$-3x<-9 \quad \text{Subtract } 5x \text{ from both sides.}$$

$$\frac{-3x}{-3}>\frac{-9}{-3} \quad \text{Divide both sides by } -3 \text{ and change the sense of the inequality.}$$

$$x>3$$

Thus, all real numbers greater than 3 make the original inequality a true statement. The solution set is written as $(3, \infty)$ and is graphed as shown in Figure R.27.

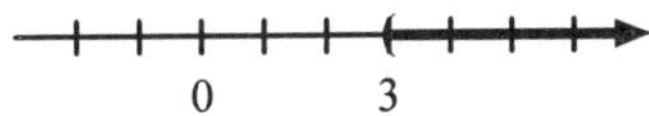

Figure R.27

PROGRESS CHECK 2

Solve $-7(x-1)\geq 11x-29$. Express the solution set graphically and in interval notation. ■

Example 3: Finding Temperatures at Which Water Turns to Steam

Solve the problem in the section introduction on page 66.

Solution: If water turns to steam when $C\geq 100$, then by the given formula this event occurs when the Fahrenheit temperature satisfies

$$\frac{5}{9}(F-32)\geq 100.$$

The easiest way to solve this inequality is to multiply both sides by 9/5, to clear the fractions. Doing this, we have

$$\frac{9}{5}\cdot\frac{5}{9}(F-32)\geq\frac{9}{5}\cdot 100$$

$$F-32\geq 180$$

$$F\geq 212.$$

Therefore, water turns to steam when the Fahrenheit temperature is at least 212 degrees. We express this result in a graph as shown in Figure R.28.

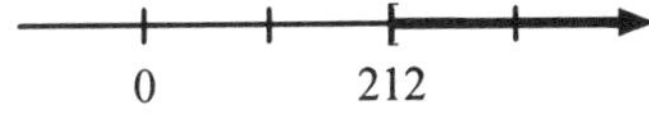

Figure R.28
Fahrenheit temperatures for steam

PROGRESS CHECK 3

A certain antifreeze product does not guarantee freeze-up protection for temperatures below –85 degrees Fahrenheit. What is this unprotected temperature range in degrees Celsius? Use $F=(9/5)C+32$. ■

The two inequalities $3<8$ an $8>3$ have the same meaning, and in general, the inequality

$a<b$ is equivalent to $b>a$.

This equivalence means that the left and right sides of an inequality are interchangeable, provided that the sense of the inequality is reversed. As illustrated in Example 4, sometimes one version is preferable because it is easier to visualize.

Example 4: Solving a Linear Inequality

Solve $8-2x<5x$, and graph the solution set.

Solution: For this inequality an efficient method is to isolate x on the right side of the inequality.

$$\begin{aligned} 8-2x &< 5x \\ 8 &< 7x && \text{Add } 2x \text{ to both sides.} \\ \frac{8}{7} &< x && \text{Divide both sides by 7.} \end{aligned}$$

Then, the answer is easier to visualize if we interchange sides and reverse the inequality symbol to get

$$x>\frac{8}{7}.$$

Substituting any number greater than 8/7 will show that our answer is reasonable. The solution set $(8/7,\infty)$ is graphed in Figure R.29.

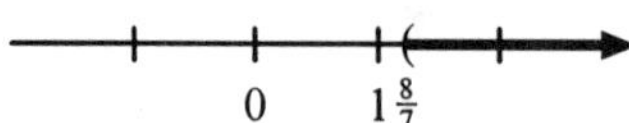

Figure R.29

PROGRESS CHECK 4
Solve $2x<5x$ and graph the solution set. ■

Some inequalities are true for all real numbers, while others can never be true. The next example illustrates one of the latter cases.

Example 5: An Inequality That is Always False

Solve $x>x+1$.

Solution: Because no number is greater that the number than is 1 bigger than it, this inequality is never true, and the solution set is $\varnothing$. If this result is not apparent when $x>x+1$ is examined, then a useful next step is to subtract x from both sides.

$$\begin{aligned} x &> x+1 \\ 0 &> 1 && \text{Subtract } x \text{ from both sides.} \end{aligned}$$

Because $0>1$ is never true, it follows that $x>x+1$ is never true.

PROGRESS CHECK 5
Solve $3+2x\le 2(x+1)$. ■

When two inequalities are joined by the word *and* or *or*, the result is called a **compound inequality**. For an and statement to be true, both inequalities must be true *simultaneously*. Compound inequalities involving *and* are often used to describe an interval between two numbers where the endpoints of the interval may also be included. For example, suppose an average (a) from 80 up to but not including 90 results in a grade of B for a course. This interval may be described by the pair of inequalities

$$80 \le a \text{ and } a < 90.$$

The inequality $80 \le a$ specifies that the average must be at least 80, while $a < 90$ specifies that the average must be below 90. As shown in Figure R.30, the two inequalities are both true only for averages between 80 and 90, including 80 and excluding 90.

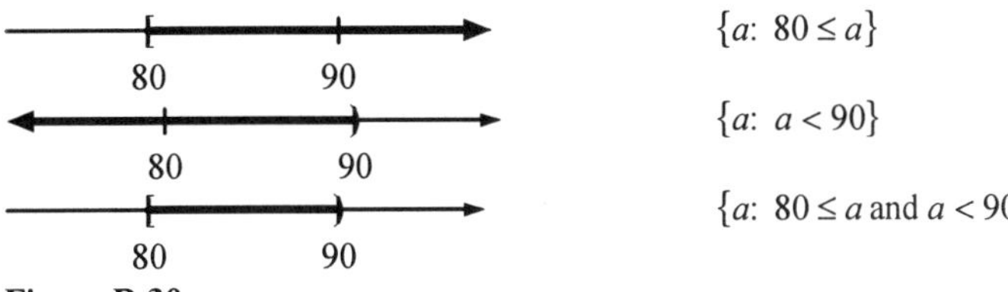

Figure R.30

The statement "$80 \le a$ and $a < 90$" is usually expressed in compact form and written $80 \le a < 90$. As suggested by Figure R.31 this set is written as [80,90) in interval notation. The following chart shows the different methods for expressing intervals that may be represented graphically as line segments

Type of Interval	Set Notation	Graph	Interval Notation
Open interval	$\{x\colon a < x < b\}$	a b	(a, b)
Closed interval	$\{x\colon a \le x \le b\}$	a b	$[a, b]$
Half-open interval	$\{x\colon a \le x < b\}$	a b	$[a, b)$
	$\{x\colon a < x \le b\}$	a b	$(a, b]$

Example 6: Solving a Compound Inequality in Compact Form

Solve $-4 < 5 - 3x < 4$ and express the solution set graphically and in interval notation.

Solution: Although the given inequality is actually an abbreviated form of a pair of inequalities, $-4 < 5 - 3x$ and $5 - 3x < 4$, the statement can remain in compact form provided that we use the properties of inequalities carefully. The goal is to isolate x in the middle member of the compound inequality.

$$-4 < 5 - 3x < 4$$

$$-9 < -3x < -1 \quad \text{Subtract 5 from each member.}$$

$$3 > x > \frac{1}{3} \quad \text{Divide each member by } -3 \text{ and change the sense of the inequalities.}$$

$$\frac{1}{3} < x < 3 \quad a < b \text{ is equivalent to } b > a.$$

Thus, the solution set is the interval $(1/3, 3)$, as graphed in Figure R.31. Note that we do not write $(3, 1/3)$, since a must be less than b when writing (a, b).

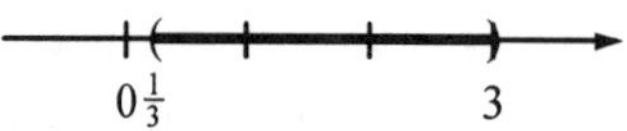

Figure R.31

PROGRESS CHECK 6

Solve $-24 \le 32 - 4t < 24$ and express the solution set graphically and in interval notation. ■

Example 7: Finding Test Results that Lead to Certain Grades

Suppose a student's average grade on four tests must be in the interval [69.5, 79.5) to yield a grade of C. If the student has grades of 83, 66, and 71 so far, find all possible grades for the last test that will result in a grade of C.

Solution: A grade of C results if the average (a) satisfies $69.5 \le a < 79.5$. Since the average is the sum of the four test grades divided by 4, letting x represent the grade on the last test yields

$$69.5 \le \frac{83 + 66 + 77 + x}{4} < 79.5$$

$$69.5 \le \frac{220 + x}{4} < 79.5$$

$$278 \le 220 + x < 318 \qquad \text{Multiply each member by 4.}$$

$$58 \le x < 98 \qquad \text{Subtract 220 from each member.}$$

Any grade in the last test from 58 up to but not including 98 results in a grade of C. In interval notation, the answer is [58,98).

PROGRESS CHECK 7

An average from 79.5 up to but not including 89.5 earns a grade of B. If your first three grades are 76, 93 and 88, find all possible grades on the fourth exam that will result in a grade of B. ■

To solve compound inequalities involving *or*, first note that the *union* of two sets A and B is denoted by $A \cup B$ and is the set of all elements belong to A or B or both. Because an *or* statement is true when at least one of the statements is true, we find the solution set to a compound inequality involving *or* by finding the union of the solution sets of its component inequalities.

Solution of *Or* Inequalities

To solve a compound inequality involving *or*

1. Solve separately each inequality in the compound inequality.
2. Find the union of the solution sets of the separate inequalities.

Example 8: Solving a Compound Inequality Involving *Or*

Solve $4x - 5 < -3$ or $4x - 5 > 3$. Express the solution set graphically and in interval notation.

Solution: First solve each inequality separately.

$$4x - 5 < -3 \quad \text{or} \quad 4x - 5 > 3$$

$$4x < 2 \quad \text{or} \quad 4x > 8$$

$$x < \frac{1}{2} \quad \text{or} \quad x > 2$$

The union of the solution sets of these two inequalities is graphed in Figure R.32. In interval notation this solution set is written as $(-\infty, 1/2) \cup (2, \infty)$.

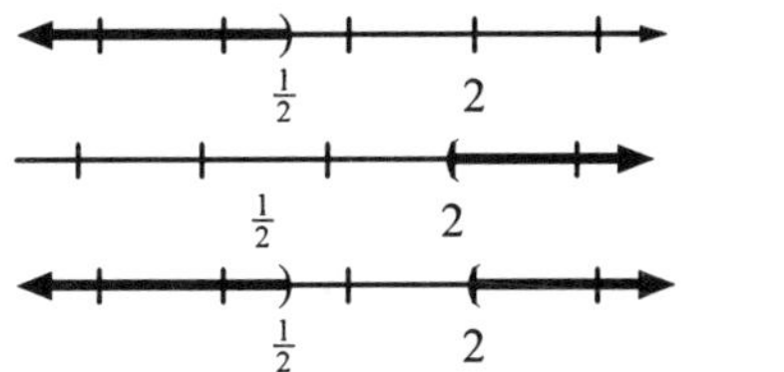

$\left\{x\colon\ x < \frac{1}{2}\right\}$

$\{x\colon\ x > 2\}$

$\left\{x\colon\ x < \frac{1}{2} \text{ or } x > 2\right\}$

Figure R.32

Caution: Only an *and* compound inequality can be written in compact form. Common student errors occur when statements like "$x < 1/2$ or $x > 2$" are shortened to

WRONG **WRONG**

$$\frac{1}{2} > x > 2 \quad \text{or} \quad 2 < x < \frac{1}{2}.$$

PROGRESS CHECK 8

Solve $3x - 7 \le -13$ or $3x - 7 \ge 13$. Express the solution set graphically and in interval notation. ■

EXPLORE R.7

Many calculators include the ability to enter inequality symbols, often through a key or menu marked Test. If your calculator can express inequalities, try the following explorations.

1. Enter a true inequality, such as $3 < 5$. What happens? In general, how does the Test feature work with true inequalities?
2. Enter a false inequality, such as $5 < 1$. What happens? In general, how does the Test feature work with false inequalities?
3. Predict the value of $(3 < 5) + (5 > 2)$. Then use the calculator to check your prediction.
4. Use the calculator to evaluate these expressions. In each case first predict the calculator output.

   a. $3 < 3$ **b.** $4 \ge 4$ **c.** $(3 < 5) - (3 > 0)^{(1 > -1) - (-3 > 0)}$
5. Determine which of these values {–2, 0, 3, 5} are in the solution set of the inequality $3x - 4 < 9$. Store –2, then enter $3x - 4 < 9$. If your answer is 1, then –2 is in the solution set. Repeat for the other values.

EXERCISES R.7

In Exercises 1–30 solve the inequality. Express the solution set graphically and in interval notation.

1. $x - 8 > 0$ **2.** $x + 5 < 0$ **3.** $-4x < -8$

4. $-9x \le 81$ **5.** $4x \le 3x + 1$

6. $10x < 11x - 3$ **7.** $2x + 9 < 0$

8. $3 - 4x > 0$ **9.** $-4x - 5 \ge 6x + 5$

10. $4 - 6x \ge -7 - 9x$ **11.** $13 - 2x < 7 - 3x$

12. $16x - 7 \le 17x - 4$ **13.** $-2x + 5 > 5x - 7$

14. $1 - 4x \ge 2x + 3$ **15.** $7 - x > 5 - 2x$

16. $214 + 6x \le -4x + 6$ **17.** $2(x - 3) \ge 6(x + 1)$

18. $4(3 + x) > 8(x - 4)$ **19.** $\frac{x}{2} + 4 < 8 + x$

20. $24 - 13x \ge -12.5x + 26$

21. $x > x - 1$ **22.** $x < x - 1$ **23.** $3x < 4x$

24. $-x \ge 2x$ **25.** $2(x + 3) \ge 7 + 2x$

26. $7 - 4y \le 4(7 - y)$ **27.** $1 - 3y \le 1 - 4y$

28. $5x+4>x+4$ **29.** $2(2-x)+1>5-2x$

30. $2(2-x)+1\geq 5-2x$

In Exercises 31–40 complete the chart and provide the alternative designations for the given set of real numbers

	Interval Notation	Set Notation	Graph
31.	$[1, 4)$		
32.	$(-\infty, 1)$		
33.		$\{x\colon x\leq 3\}$	
34.		$\{x\colon -1<x\leq 1\}$	
35.			−3 0
36.			
37.	$[0, \infty)$		
38.		$\{x\colon 0<x\}$	
39.			−2 2
40.			 −2 2

In Exercises 41–58 solve each inequality. Express the solution set graphically and in interval notation.

41. $-2<x-4<2$ **42.** $0<x+7<10$

43. $-5<-15+10t<5$ **44.** $-8<4x+10<8$

45. $-5<2-x\leq 7$ **46.** $-0.1\leq 1-x\leq 0.1$

47. $-16\leq 4-8x\leq 16$ **48.** $-1\leq 1-2x<1$

49. $-1\leq\dfrac{5-2x}{3}<\dfrac{1}{2}$ **50.** $-\dfrac{9}{4}<\dfrac{3x+2}{-2}\leq-\dfrac{1}{4}$

51. $x>6$ or $x<1$ **52.** $x\geq 3$ or $x>8$

53. $x+2<-1$ or $x+2>1$

54. $x-1\leq -2$ or $x-1>-3$

55. $3x-2>2$ or $2x-3>3$

56. $5x+1\geq x+1$ or $3x-2<2x+1$

57. $3-2x<2x$ or $4+3x<5-x$

58. $5>-2x+6$ or $-5<2x-6$

Intervals appear in a natural way whenever we analyze data based on measurements. Measurements of one kind or another are essential to both scientific and nontechnical work. Weight, distance, time, volume, and temperature are but a few of the quantities for which measurements are required. However, any reading is an approximation that can only be as accurate as the measuring instruments allow. Thus, if we record a man's weight as 164 lb, we guarantee only that his exact weight is somewhere in the interval [163.5 lb, 164.5 lb). Use this concept to answer Exercises 59 and 60.

59. The circumference of the earth through the North and South Poles is about 25,000 mi. Use interval notation to write an interval that contains the earth's exact polar circumference. If we use a more accurate measuring system and record the polar circumference as 24,860 mi, which interval now contains the exact distance?

60. To the nearest inch, the height of a woman is 69 in. Use interval notation to write an interval that contains the woman's exact height. If we use a more accurate measuring device and record her height as 69.3 in., which interval now contains the woman's exact height?

61. In Lobachevskian geometry the sum of the three angles in a triangle is always less than 180°. If the second angle is 20° more than the first, and the third is twice the first, how many degrees might there be in the first angle?

62. If we categorize any temperature in adults above 98.6 degrees Fahrenheit as a "fever," then what is the temperature range of a fever in degrees Celsius? Use $F=\dfrac{9}{5}C+32$.

63. A publishing company plans to produce and sell x textbooks. The cost of producing x books is given by $161,100+9x$, which represents a setup cost of \$161,000 plus a variable cost of \$9 per book. The publisher receives \$23 on the sale of each book, so the revenue for selling x books is given by $23x$. For what values of x is revenue greater than cost?

64. A recording company plans to produce and sell x compact discs of a particular album. The production cost includes a one-time "up-front" recording cost of \$150,000 followed by a duplicating cost of \$2 per CD. The company receives \$8 on the sale of each CD. For what number of CDs sold is the revenue greater than the cost?

65. The element mercury is a liquid between –38.87° and 356.58°C. This range of temperature values is therefore given by $-38.87 < C < 356.58$. Express this in degrees Fahrenheit by using the formula $C = \frac{5}{9}(F-32)$ and solving the resulting compound inequality. Round values to the nearest hundredth.

66. A ski resort advertises that the temperature this weekend will range from –20° to –5° Celsius. Let $-20 \le C \le -5$ represent this interval and find the corresponding range in degrees Fahrenheit. Use $C = \frac{5}{9}(F-32)$.

67. Suppose a student's average grade on four tests must be in the interval [74.5, 79.5) to yield a grade of C^+. If the student has grades of 81, 70, and 76 so far, find all possible grades for the last test that result in a grade of C^+.

68. Scores (X) on a national exam are "standardized" by using the formula $Z = \frac{X-400}{120}$. Standardized scores between 2 and 3 are classified as "good." What interval of exam scores does this correspond to?

69. If you can purchase a discount care for \$20 from a video store that entitles you to 15 percent off all purchases, then how much must you purchase for the card to save you over \$25?

70. World grain production between 1950 and 1992 is displayed in the bar graph. (Data from *Vital Signs, 1993*, Worldwatch Institute). This pattern may be modeled approximately by the formula $y = 580.916 + 28.468x$, where x is the number of years elapsed since 1950, and y is total grain production in millions of tons.

a. According to this model, in what year did total world grain production first reach at least 1 billion tons?

b. According to the bar graph, in what year did total world grain production first reach at least 1 billion tons?

c. If the model continues to hold beyond 1992, when will total world grain production first reach at least 2 billion tons?

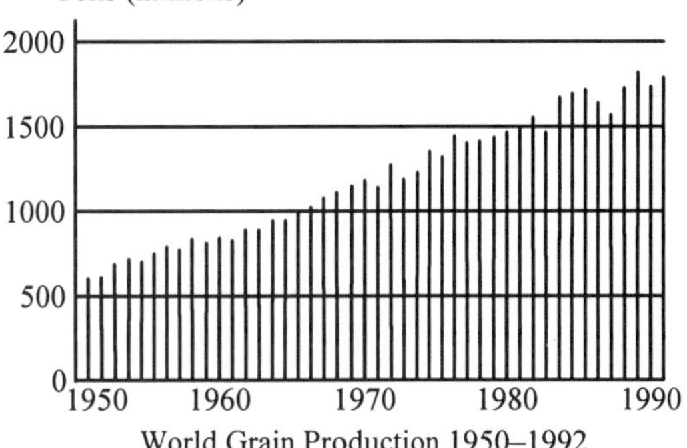

World Grain Production 1950–1992

THINK ABOUT IT R.7

1. Give examples of values for the unknowns so that the following statements are *false*.

a. If $a < b$ then $ac < bc$.
b. If $a < b$ then $a^2 < b^2$.
c. If $a > 0$ then $a^2 \ge a$.
d. If $a < 0$ and $b > 1$, then $ab < -1$.

2. Solve $x < 2x - 7 \le 10$.

3. Take two identical sheets of notebook paper. Form two cylinders from these rectangular $(y > x)$ sheets: one with the length as the base [diagram (a)], the other with the width as the base [diagram (b)]. Show that the cylinder formed in (a) will always have the greater volume. (*Note*: $V = \pi r^2 h$.)

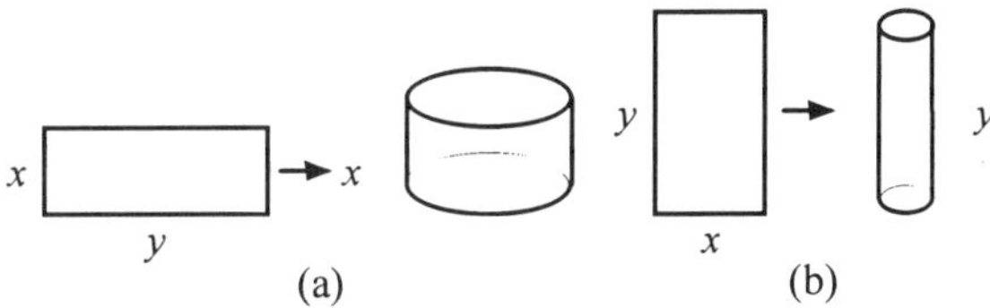

4. In solving $a < bx < c$ for x, explain why you should not divide by b unless some other information is given. What is the needed information?

5. The properties of inequalities given in this section may be formally proved by using the definition of "less than" as given in Section R.1 and the fact that the sum and product of two positive numbers is positive. For example, to prove that if $a < b$ and $c < 0$, then $ac > bc$, we recall that $a < b$ if and only if $b - a$ is positive. Similarly, $c < 0$ implies that $0 - c$, or $-c$, is positive. Then

$$\underbrace{(b-a)}_{\text{pos}}\underbrace{(-c)}_{\text{pos}} = \underbrace{-bc + ac.}_{\text{pos}}$$

Thus, $ac - bc$ is positive, so $bc < ac$, or alternatively $ac > bc$, which proves the property.

Use similar proofs to prove each of the following statements. Assume that *a*, *b*, *c*, and *d* are real numbers.

a. If $a < b$ then $a + c < b + c$. **b.** If $a < b$ and $c > 0$, then $ac < bc$.
c. If $a < b$ and $b < c$, then $a < c$. **d.** If $a < b$ and $c < d$, then $a + c < b + d$.

● ● ●

CHAPTER R SUMMARY

OBJECTIVES CHECKLIST

Specific chapter objectives are summarized below along with numbered example problems from the text that should clarify the objectives. If you do not understand any objectives, or do not know how to do the selected problems, then restudy the material.

R.1: Can you:

1. **Express quotients of integers as repeating decimals, and vice versa?** Express the repeating decimal $0.\overline{17}$ as the ratio of two integers.

[Example 1]

2. **Identify integers, rational numbers, irrational numbers, and real numbers?** For each number select all correct classifications from the following categories: real number, irrational number, rational number, integer, or none of these.

a. $\sqrt{225}$ **b.** $\sqrt{-4}$ **c.** 0.505005... **d.** $\frac{22}{7}$

[Example 2]

3. **Identify the properties of real numbers?** Name the property of real numbers illustrated in each statement.

a. $(2+3)+4 = 2+(3+4)$ **b.** $5+(-5)=0$
c. $2(x+4) = 2\cdot x + 2\cdot 4$ **d.** $0\cdot x = x\cdot 0$

[Example 3]

4. **Graph and order real numbers?** Insert the proper symbol (<, >, or =) to indicate the correct order.

a. -1 ________ -3 **b.** -3 ________ $\sqrt{2}$ **c.** $-\sqrt{9}$ ________ -3

[Example 4]

5. **Find the absolute value of a real number and operate on real numbers?** Evaluate $1+7(2-5)^2$.

[Example 10]

6. **Evaluate numeric expressions on a graphing calculator?** Evaluate on a graphing calculator.

a. $\dfrac{3-(-5)}{-1-2}$ **b.** $\sqrt{8^2+15^2}$

[Example 11]

R.2: Can you:

1. **Evaluate algebraic expressions?** Evaluate each expression given that $x = -2$, $y = 3$, and $z = -4$.

a. $5y - x$ **b.** $-x^2 + 5yz^2$

[Example 1]

2. **Simplify algebraic expressions by combining like terms?** Simplify by combining like terms if possible.
   a. $3x + 7x$ **b.** $yz^2 - 10yz^2$ **c.** $8x - y + 2y - 3x$ **d.** $-3x^2y + xy^2 - 7xy$
   [Example 4]

3. **Apply geometric formulas?** A bowling ball has an 8-in. diameter.
   a. Find its volume to the nearest cubic inch. Disregard any holes in the ball.
   b. If the material weighs 0.9 oz/in.^3, find the weight of the ball to the nearest pound. (There are 16 oz in 1 lb.)
   [Example 7]

4. **Translate between verbal expressions and algebraic expressions?** Translate each statement to an algebraic expression.
   a. $3 less than the list price *p*. **b.** 6 percent of the sum of *x* and *y*.
   [Example 8]

R.3: Can you:

1. **Evaluate expressions containing integer exponents?** Simplify by the laws of exponents and check your result.
   a. $2^2 \cdot 2^3$ **b.** $(2^3)^2$ **c.** $(2 \cdot 3)^2$ **d.** $\frac{2^5}{2^2}$ **e.** $\frac{2^2}{2^5}$
   [Example 1]

2. **Simplify expressions involving integer exponents?** Simplify and write the result using only positive exponents. Assume $x \neq 0$ and $y \neq 0$.
   a. $2^{-5} \cdot 2^3$ **b.** $5x \cdot x^{-4}$ **c.** $\left(\frac{9}{y}\right)^{-2}$ **d.** $(3x^{-2})^{-3}$ **e.** $\frac{3^{-2}x^2y^{-3}}{3x^{-5}y}$
   [Example 6]

3. **Simplify expressions involving literal exponents?** Simplify the given expression. Assume $x \neq 0$, $b \neq 0$ and variables as exponents denote integers.
   a. $\frac{2x^{n+3}}{-4x^n}$ **b.** $b^x \cdot b^{-x}$ **c.** $\frac{b^x}{b^{2x}}$
   [Example 8]

4. **Convert numbers from normal notation to scientific notation, and vice versa?** Write each number in scientific notation.
   a. 27 billion **b.** –0.00000031.
   [Example 9]

5. **Solve problems by calculator that involve scientific notation?** If one red blood cell contains about 270 million hemoglobin molecules, about how many molecules of hemoglobin are there in 30 trillion red blood cells (which is the approximate number of red blood cells in a typical adult)?
   [Example 11]

R.4: Can you:

1. **Use the distributive property to multiply algebraic expressions?** Find each product.
   a. $3x(x^2 - 2x + 1)$ **b.** $(2c^2 + t)ct$
   [Example 1]

2. **Use the FOIL method to multiply two expressions that each contain two terms?** Multiply using the FOIL method: $(2x - 1)(3x + 4)$.
   [Example 5]

3. **Use special product formulas to find certain products?** Use a special product formula to find each product.

   a. $(4x+5)(4x-5)$ **b.** $(3x+4)^2$ **c.** $\left(x^3-y^3\right)^2$

[Example 8]

R.5: Can you:

1. **Solve linear equations?** Solve the equation $6(x+1)=14x+2$.

[Example 2]

2. **Find the value of a variable in a formula when given values for the other variables?** When an original principal P is invested at simple interest for t years at annual interest rate r, then the amount A of the investment is given by $A=P(1+rt)$. Find the value of t when $A=\$5{,}920$, $P=\$4{,}000$, and $r=0.06$.

[Example 5]

3. **Solve a given formula or literal equation for a specified variable?** The Z-score formula in statistics is $Z=(x-m)/s$, where x is a raw score, m is the mean, and s is the standard deviation. Solve the formula for the raw score (x).

[Example 7]

R.6: Can you:

1. **Solve word problems by translating phrases and setting up and solving equations?** The sum of two consecutive even integers is 238. Find the integers.

[Example 2]

2. **Solve problems involving geometric figures, annual interest, uniform motion, liquid mixtures, and proportions?** How should a \$160,000 investment be split so that the total annual earnings are \$14,000, if one portion is invested at 6 percent annual interest and the rest at 11 percent?

[Example 4]

R.7: Can you:

1. **Specify solution sets of linear inequalities by using graphs and interval notation?** Write the solution set to each inequality in interval notation, and graph the interval.

   a. $x \geq -1$ **b.** $x < \dfrac{3}{2}$

[Example 1]

2. **Solve linear inequalities by applying properties of inequalities?** Solve $2x+1<5x-8$. Express the solution set graphically and in interval notation.

[Example 2]

3. **Solve compound inequalities that specify numbers between two numbers?** Solve $-4<5-3x<4$ and express the solution set graphically and in interval notation.

[Example 6]

4. **Solve compound inequalities involving *or* statements?** Solve $4x-5<-3$ or $4x-5>3$. Express the solution set graphically and in interval notation.

[Example 8]

KEY CONCEPTS AND PROCEDURES

Section	Key Concepts and Procedures to Review
R.1	• Definitions of integers, rational numbers, irrational numbers, real numbers, $a<b$, $a>b$, $a \leq b$, absolute value, subtraction, factor, power, exponent and division • Relationships among the various sets of numbers • Statements of basic properties of real numbers • Methods to add, subtract, multiply, and divide real numbers

- Order of operations
- Methods for performing numeric calculations on a graphing calculator

R.2

- Definitions of variable, constant, algebraic expression, terms, (numerical) coefficient, and like terms
- Methods to evaluate an algebraic expression
- Methods to add or subtract algebraic expressions

R.3

- Laws of exponents (m and n denote integers)

  1. $a^m \cdot a^n = a^{m+n}$ **2.** $\left(a^m\right)^n = a^{mn}$ **3.** $(ab)^n = a^n b^n$

  4. $\left(\frac{a}{b}\right)^n = \frac{a^n}{b^n} \quad (b \neq 0)$ **5.** $\frac{a^m}{a^n} = a^{m-n} = \frac{1}{a^{n-m}} \quad (a \neq 0)$

  6. $a^0 = 1 \quad (a \neq 0)$ **7.** $a^{-n} = \frac{1}{a^n} \quad (a \neq 0)$

- Methods to convert a number from normal notation to scientific notation, and vice versa
- Guidelines for working with scientific notation on a graphing calculator

R.4

- Methods to multiply various types of algebraic expressions
- FOIL multiplication method
- Special product formulas:

  1. $(a+b)(a-b) = a^2 - b^2$

  2. $(a+b)^2 = a^2 + 2ab + b^2$

  3. $(a-b)^2 = a^2 - 2ab + b^2$

R.5

- Definitions of equation, conditional equation, identity, false equation, equivalent equation, solution set, linear equation, and literal equation
- Methods to obtain an equivalent equation
- Methods to solve formulas and literal equations for a given variable

R.6

- Guidelines to setting up and solving word problems
- Methods for solving problems involving geometric figures, annual interest, uniform motion, liquid mixtures, and proportions

R.7

- Definition of solution of an inequality, and representation by graphs and interval notation
- Properties of inequalities
- Methods to solve linear inequalities
- Methods to solve compound inequalities involving linear inequalities

CHAPTER R REVIEW EXERCISES

1. Evaluate these expressions both with and without the use of a calculator.

a. $1-3(2-4)$ **b.** $1+\frac{3-11}{(4-8)^2}$

2. Is –3 a solution of the equation $\frac{2}{2-c}+\frac{c}{c-2}=1$?

3. If $m=\frac{y_2-y_1}{x_2-x_1}$, evaluate m for $x_1=2$, $x_2=-5$, $y_1=-1$, and $y_2=4$.

4. Find the value of the expression $-x^2+2xy^3$ when $x=-1$ and $y=3$.

5. If $x=-2$ and $y=5$, then evaluate $|2x|+|y|$.

6. Find the value of t in the formula $a = p(1 + rt)$ if $a = 3{,}000$, $p = 1{,}500$, and $r = 0.05$.

7. Combine like terms: $2(x - y) + (x + y) - (3 - y)$.

8. Solve for y' (read "y prime"): $2x + 2y' = 0$.

9. Express the rational number $\frac{13}{11}$ as a repeating decimal.

10. Use inequalities and interval notation to describe the sets of numbers illustrated.

a. 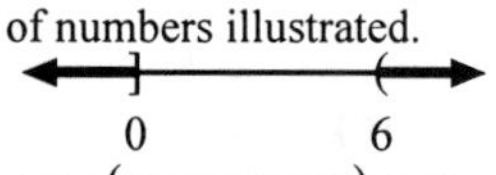

b. 

11. Simplify $x - [-x - (x - 1)]$.

12. Simplify $(5x^2 - 3x + 1) - (2x^2 - 2x + 7)$.

13. Name the property of real numbers illustrated in the given statements.
a. $(x + y) + z = z + (x + y)$
b. $\frac{1}{9} \cdot 8 + \frac{1}{9} \cdot 2 = \frac{1}{9}(8 + 2)$

14. Use the correct symbol (<, >, or =) between the following pairs of expressions.
a. -0.1________ -0.01
b. 2%________ 0.2
c. $|-3 + 7|$________ $|-3| + |7|$
d. $|-2a|$________ $|-2| \cdot |a|$

15. The number 3.14 is a member of which of the following sets of numbers; reals, rationals, irrationals, integers?

16. If $x = -3$, find
a. the absolute value of x,
b. the reciprocal of x, and
c. the negative of x.

17. Evaluate $2x^0 + x^{-1}$ when $x = -\frac{1}{4}$.

18. A mutual fund company manages \$28 billion in assets. Express this number in scientific notation.

19. The wavelength of violet light is about 4.0×10^{-5} cm. Express this number in normal notation.

20. Rewrite $\frac{2^{-3}x^2}{4y^{-2}}$ with only positive exponents.

In Exercises 21–32 use the laws of exponents to perform the operations, and write the answer in the simplest form that contains only positive exponents

21. $(-3x^3) \cdot (-3x)^3$ **22.** $(2st)^2 \cdot (3tc)^3$

23. $2 \cdot 2^x$ **24.** $2^x \cdot 2^{-x}$

25. $(3^x)^{2x}$ **26.** $x^{2n+2} \cdot x^2$

27. $(-3)^{-1} \cdot (-3)^{-3}$ **28.** $5 \div 5^{-1}$

29. $\frac{3}{3^n}$ **30.** $\frac{(3x)^2}{y} \cdot \left(\frac{2zy}{x^2}\right)^3$

31. $\frac{3^2 x^{-4} y^{-3}}{3^{-1} x^4 y^{-2}}$ **32.** $\frac{(2ax^2)^2}{2(ax)^{-1}}$

In Exercises 33–42 perform the multiplication or division and combine like terms.

33. $x^2y(2x - 3y + 4)$ **34.** $(a - 4b)a^3b^2$

35. $\frac{cs^2 - 2c^3s}{cs}$ **36.** $\frac{2xh + h^2 + 2h}{h}$

37. $(1 - k)^2$ **38.** $(4x - y)^2$

39 $(x^a - y^b)^2$ **40.** $(x - 3)(x + 2)(x - 4)$

41. $(a + b)(a^2 - ab + b^2)$ **42.** $[(x + h) - 2]^2$

In Exercises 43–54 solve each equation or inequality.

43. $3(x - 6) = -2(12 - 3x)$

44. $1 - x = 1 + x$ **45.** $1 - x \le 1 + x$

46. $-x + 1 > 3$ **47.** $6{,}000 = 1{,}000(1 + 0.05t)$

48. $5 - 2(a - 2) = 2 - 3a$

49. $6x < 7x - 1$ **50.** $-2(x + 4) \le -2x + 4$

51. $2(x + 4) = 2x + 4$ **52.** $\frac{20 - x}{3} = \frac{x}{5}$

53. $-3 < 4 - 2x < 3$ **54.** $x + 5 \le -1$ or $x + 5 \ge 1$

In Exercises 55–60 solve each equation for the letter indicated.

55. $C = 2\pi r$ (for r) **56.** $5x + 3y = 6$ (for y)

57. $ax + b = n$ (for x) **58.** $2x + 2y' = 5$ (for y')

59. $P = \frac{x - y - z}{10}$ (for y) 60. $\frac{a_1}{c_1} = \frac{a_2}{c_2}$ (for c_2)

In Exercises 61–70 answer true or false. If false, give a specific counterexample.

61. All real numbers are rational numbers.

62. All integers are rational numbers.

63. There is no number equal to $\frac{0}{4}$.

64. If $ab = ac$, then $b = c$.

65. If $a < b$, then $a + c < b + c$.

66. If $a < b$, then $ac < bc$.

67. $|-a| = a$ 68. $|a + b| = |a| + |b|$

69. $|ab| = |a| \cdot |b|$ 70. $|a - b| = |b - a|$

71. Find the volume of a cylinder whose radius and height are each 6 inches.

72. A part of a microchip is a cube with side 0.001 mm. Calculate its surface area.

73. On a test you received a grade of 123 on a scale of 0–150. Convert your score to a scale of 0–100.

74. A brick weights 6 lb plus half a brick. What does one brick weigh?

75. The total cost (including tax) of a new car is $9,222. If the sales tax is 6 percent, how much do you pay in taxes?

76. To the nearest inch, the height of a woman is 64 in. Use interval notation to write an interval that contains the woman's exact height. If we use a more accurate measuring device and record her height as 63.9 in., which interval now contains the woman's exact height?

77. Ohm's law states that the current in a circuit (I) is equal to the voltage (E) divided by the resistance (R). Write a formula for the resistance in terms of the current and the voltage.

78. You have 20 qt of a solution that is 10 percent antifreeze. How much pure antifreeze must be added to obtain a solution that is 30 percent antifreeze?

79. A map has a scale of $1.5 \text{ in.} = 11 \text{ mi}$. To the nearest tenth of a mile, what distance is represented by 7 in. on the map?

80. A student needs an average of 90.0 or better to get an A in a course. Her first four test grades were 94, 91, 88, and 97. What possible grades on her fifth test will result in an A grade?

81. A real estate broker's commission for selling a house is 6 percent of the selling price. If you want to sell your house and receive $80,000, then what must be the selling price of the house?

82. Triangle, ABC is similar to triangle EDC. If $\overline{BD} = 25$ ft, $\overline{AB} = 9$ ft, and $\overline{DE} = 11$ ft, then find $\overline{BC}$ to the nearest foot.

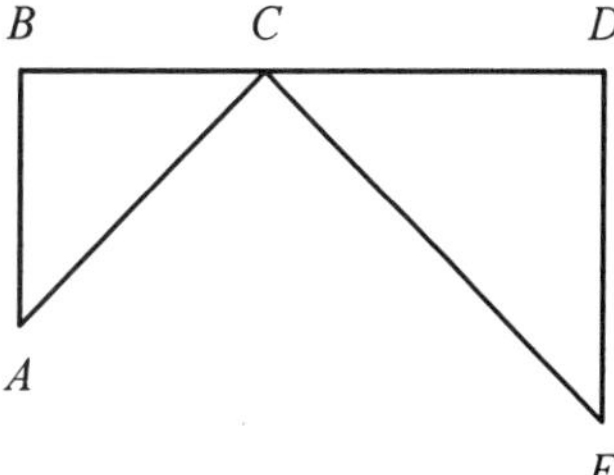

In Exercises 83–92 select the choice that answers the question or completes the statement.

83. Which statement illustrates the commutative property of multiplication?
a. $a(b + c) = (b + c)a$ **b.** $a(b + c) = ab + ac$
c. $(ab)c = a(bc)$ **d.** $a \cdot 1 = a$

84. Which number is a rational number?
a. $\sqrt{2}$ **b.** $\sqrt{-9}$ **c.** $-\sqrt{9}$ **d.** π

85. If $2^a = x$, then 2^{a+3} equals
a. $x + 3$ **b.** x^3 **c.** $6x$ **d.** $8x$

86. The side of a square is given by $4x^3$. The area of the square is given by
a. $12x^3$ **b.** $16x^5$ **c.** $16x^6$ **d.** $4x^9$

87. The side of a square is given by $4 + x^3$. The area of the square is given by
a. $x^6 + 8x^3 + 16$ **b.** $x^9 + 8x^6 + 16$
c. $x^9 + 16$ **d.** $x^6 + 16$

88. Pick the statement that describes the set of numbers illustrated.

a. $\{x: \ -1 < x < 3\}$ **b.** $\{x: \ 3 < x < -1\}$
c. $\{x: \ x < -1 \text{ and } x > 3\}$ **d.** $\{x: \ x < -1 \text{ or } x > 3\}$

89. If $a < 0$, then $|a|$ equals

a. a b. a^2 c. $-a$ d. $\frac{1}{a}$

90. If the width of a rectangle is one-fourth the length, and the perimeter is 70 cm, what is the area?

a. 140 cm^2 b. 196 cm^2 c. 392 cm^2 d. 784 cm^2

91. Which expression is undefined when $x = -1$?

a. $\frac{x}{x+1}$ b. $\frac{x+1}{x}$ c. x^0 d. x^{-1}

92. The inequality $-x < 1$ is equivalent to

a. $x < 1$ b. $x < -1$ c. $x > 1$ d. $x > -1$

CHAPTER R TEST

1. List all correct classifications of the number –5 from the following categories: real number, irrational number, rational number, integer.

2. Number the property of real numbers illustrated by $8x + 2x = (8+2)x$.

In Questions 3 and 4 evaluate each expression.

3. $5 - 9(2-7)^2$ 4. $(-3)^{-1} \cdot (-3)^{-3}$

In Questions 5 and 6 simplify each expression and write the result using only positive exponents.

5. $\left(\frac{10x^{-2}}{a}\right)^{-3}$ 6. $\frac{x^n}{x^{n+2}}$

In Questions 7–10 solve each equation or inequality.

7. $2(4-y) = 26 + y$ 8. $4 - 5x \geq x$

9. $-3(x+2) < -3x + 1$ 10. $\frac{x}{4} = \frac{10-x}{8}$

11. Multiply $(5x+7)(3x-1)$.

12. Simplify $x - 2[5x - (1-x)]$.

13. Evaluate $\frac{y_2 - y_1}{x_2 - x_1}$ if $x_1 = 4$, $y_1 = -2$, $x_2 = -6$, and $y_2 = 3$.

14. Find the value of F in the formula $C = \frac{5}{9}(F - 32)$ if $C = -15$.

15. Solve $0 \leq 2x - 6 < 10$ and express the answer graphically and in interval notation.

16. Solve for x: $3x + 2y = 5$.

17. Solve for c: $A = \frac{1}{2}h(b+c)$.

18. Simplify $\frac{(x+h)^2 + 5 - (x^2 + 5)}{h}$.

19. During a certain day about 172 million shares of stock were sold on the New York Stock Exchange. Express this number in scientific notation.

20. The total cost (including tax) of a new car is $13,482. If the sale tax rate is 7 percent, how much is paid in taxes?

• • •

Chapter

1

Functions and Graphs

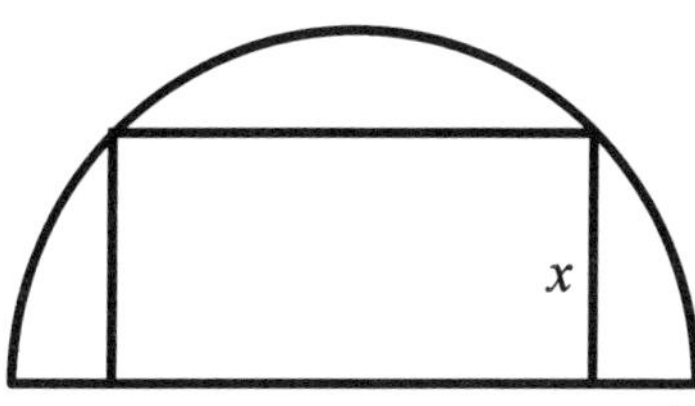

A rectangular insert of clear glass is placed in a colored semicircular glass window of radius 1 m to improve the lighting near a stairway. To find the dimensions of the rectangular insert of greatest area that most improves the lighting, it is necessary to first find a formula for the area of the insert in terms of side dimension x, as shown. What is this formula? Find the domain and range of this function.
(See Example 12 of Section 1.1.)

We now introduce the idea of a function and its graph. This concept will then become an important theme in the text, as we investigate the behavior of the trigonometric functions in Chapters 2 and 8 and the behavior of linear, quadratic, polynomial, rational, exponential, and logarithmic functions in Chapters 5, 6 and 7. In addition, graphical analysis will be used to reinforce further algebraic techniques in Chapter 3 and to provide a picture of the solutions to equations and inequalities in Chapter 4. This chapter starts us on the road to analyzing relationships with formulas, tables, and graphs, and this three-point approach will help you to understand topics from algebraic, numeric, and geometric viewpoints.

● ● ●

1.1 Functions and Graphs

Objectives

1. Find a formula that defines a function and determine its domain and range.
2. Determine ordered pairs that are solutions of an equation.
3. Determine if a set of ordered pairs is a function.
4. Graph an ordered pair and determine the coordinates of a point.
5. Graph a function by the point-plotting method or by using a grapher.

One of the most important considerations in mathematics is determining the relationship between two variables. For example:

- The postage required to mail a package is a function of the weight of the package.
- The bill from an electric company is a function of the number of kilowatt-hours of electricity that are purchased.
- The current in a circuit with a fixed voltage is a function of the resistance in the circuit.
- The demand for a product is a function of the price charged for the product.
- The perimeter of a square is a function of the length of the side of the square.

In each of these examples we determine the relationship between the two variables by finding a rule that establishes a correspondence between values of each variable. For example, if we know the length of the side of a square, we can determine the perimeter by the formula $P = 4s$. In this case the rule is a formula or equation. Sometimes the rule is given in tabular form. For example, consider the formula table below. This rule assigns to each final average (a) a final grade for the course.

$$\text{Final grade} = \begin{cases} \text{A} & \text{if} \quad 90 \le a \le 100 \\ \text{B} & \text{if} \quad 80 \le a < 90 \\ \text{C} & \text{if} \quad 70 \le a < 80 \\ \text{D} & \text{if} \quad 60 \le a < 70 \\ \text{F} & \text{if} \quad 0 \le a < 60 \end{cases}$$

Since formulas and tables are not always applicable, it is sometimes best to give the rule verbally or to make a list that shows the correspondence. The relationship between students and their Social Security numbers is such a case. Finally, you are undoubtedly familiar with the type of graph shown in Figure 1.1, which is commonly used to specify relationships in a quick and vivid way.

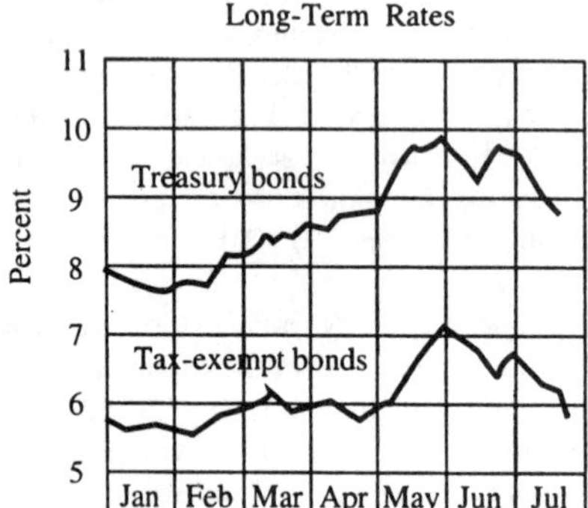

Figure 1.1

Whether the rule is given by formula or table, graphically, or by a list, the rule is most useful if we obtain *exactly one* answer whenever we use it. For example, the rule in the formula table above assigns to each final average exactly one final grade. Once we compute the final average, the rule tells us exactly what grade to assign for the course. Some typical assignments are shown in Figure 1.2. Assignments are represented as arrows from the points that represent final averages to the points that represent final grades. However, some correspondences do not always give us exactly one answer. For example, if we reverse the assignments in Figure 1.2, then we cannot determine a unique final average when the final grade is A, as shown in Figure 1.3.

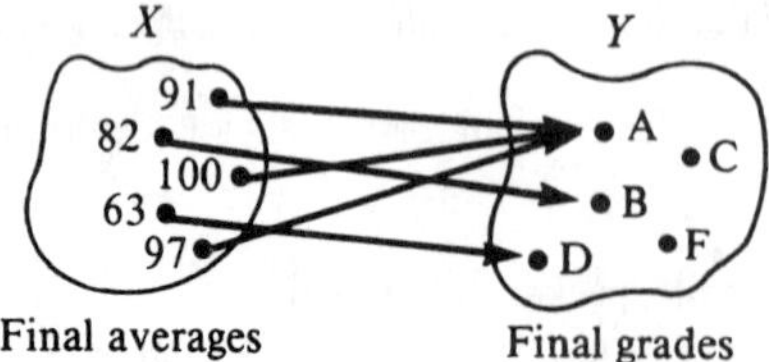

Figure 1.2
Function

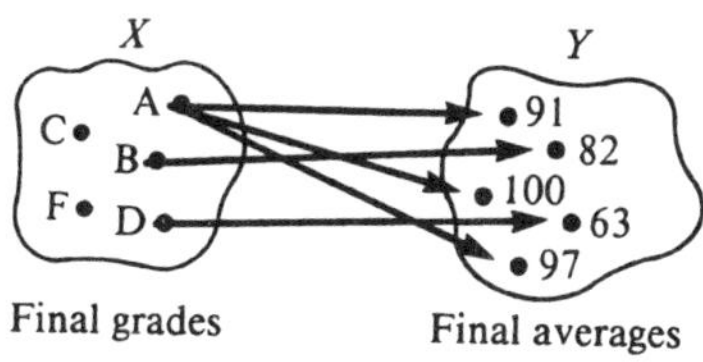

Figure 1.3
Not a Function

We wish to define a function so that the correspondence in Figure 1.2 is a function, while the correspondence in Figure 1.3 is not a function. We do this as follows:

Definition of a Function

> A **function** is a rule that assigns to each element x in a set X exactly one element y in a set Y. In this definition, set X is called the **domain** of the function, and the set of all elements of Y that correspond to elements in the domain is called the **range** of the function.

The analogy between a function and a computing machine may help to clarify the concept. Consider Figure 1.4, which shows a machine processing domain elements (x values) into range elements (y values). In goes an x value, out comes exactly one y value. With this in mind, you should have a clear image of the three features of a function: (1) the domain (the input), (2) the range (the output), and (3) the rule (the machine). When y is a function of x, as just described, the value of y depends on the choice for x, so we call x the **independent variable** and y the **dependent variable**.

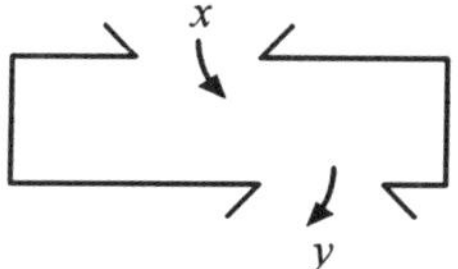

Figure 1.4

It is common practice to refer to a function by stating only the rule, no domain or range is specified. Thus we say, "Consider the circumference function $C = \pi d$." The domain is then assumed to be the collection of values for d that are interpretable in the problem. Since the length of the diameter of a circle must be positive in this case, the domain is the set of positive real numbers. For functions defined by algebraic equations such as $y = 1/\sqrt{x-1}$, the domain is the set of all real numbers for which a real number exists in the range. Thus we exclude from the domain values for the independent variable (x) that result in an even root of a negative number or in division by zero. For $y = 1/\sqrt{x-1}$ we want $x - 1 > 0$ so that the domain is the interval $\{x\colon\ x > 1\}$, or $(1, \infty)$ in interval notation. We illustrate these ideas in the following example.

Example 1: Describing a Function for Bank Charges

Express the monthly cost (c) for a checking account as a function of the number (n) of checks serviced that month if the bank charges 10 cents per check plus a \$3.50 maintenance charge. Describe the domain and range of this function.

Solution: To help us see a pattern, we will first analyze the problem numerically and determine the cost of writing a specific number of checks, say 20. If the bank charges 10 cents per check plus a \$3.50 charge, then the cost of 20 checks is given by

$$c = 0.10(20) + 3.50 = \$5.50.$$

To generalize and obtain a formula for the function, let the variable n replace the specific number 20 to obtain

$$c = 0.10n + 3.50.$$

Since the cost depends on the number of checks written, c is the dependent variable, and n is the independent variable. Thus, the domain is the set of values for n that are meaningful in this context, and the range is the corresponding set of costs. Using set notation we specify these sets as follows.

$$\text{Domain} = \{0, 1, 2, 3, \ldots\}$$
$$\text{Range} = \{\$3.50, \$3.60, \$3.70, \$3.80, \ldots\}$$

Note that 0 is included in the domain because there is a monthly charge even when no checks are serviced.

PROGRESS CHECK 1

Express the cost (c) of a shipment of CDs as a function of the number (n) of CDs ordered if the charge is \$9 per CD plus \$2.95 for shipping and handling. Describe the domain and range of this function. ■

Functions as Ordered Pairs

In mathematical notation we use **ordered pairs** to show the correspondence in a function. For example, consider the equation $y = 2x + 1$.

If x Equals	Then $y = 2x + 1$	Thus, the Ordered Pairs Are
2	$2(2) + 1 = 5$	(2, 5)
1	$2(1) + 1 = 3$	(1, 3)
0	$2(0) + 1 = 1$	(0, 1)
–1	$2(-1) + 1 = -1$	(–1, –1)
–2	$2(-2) + 1 = -3$	(–2, –3)

In the pairs that represent the correspondence, we list the values of the independent variable first and the values of the dependent variable second. Thus the order of the numbers in the pair is significant. The pairing (2, 5) indicates that when $x = 2$, $y = 5$; (5, 2) means when $x = 5$, $y = 2$. In the equation $y = 2x + 1$, (2, 5) is an ordered pair that makes the equation a true statement; so (2, 5) is said to be a solution of the equation; (5, 2) is not a solution of this equation.

Example 2: Ordered Pair Solutions

Which of the following ordered pairs are solutions of the equation $y = 3x - 2$?

a. $(-1, -5)$ **b.** $(-5, -1)$

Solution:

a. $(-1, -5)$ means that when $x = -1$, $y = -5$. Then

$$y = 3x - 2$$
$$(-5) \stackrel{?}{=} 3(-1) - 2$$
$$-5 \stackrel{\checkmark}{=} -5$$

Thus, $(-1, -5)$ is a solution

b. $(-5, -1)$ means that when $x=-5$, $y=-1$. Then

$$y = 3x - 2$$
$$(-1) \stackrel{?}{=} 3(-5) - 2$$
$$-1 \neq -17.$$

Thus, $(-5, -1)$ is not a solution.

PROGRESS CHECK 2

Which of the following ordered pairs are solutions of the equation $y = 2x - 3$?

a. $(-2, -7)$ **b.** $(-7, -2)$ ■

The representation of a correspondence as a set of ordered pairs gives us a different perspective on the function concept. We call any set of ordered pairs a relation, and the following definitions show that a function is a special type of relation.

Relation and Function Definitions

A **relation** is a set of ordered pairs. The set of all first components of the ordered pairs is called the **domain** of the relation. The set of all second components is called the **range** of the relation. A **function** is a relation in which no two different ordered pairs have the same first component.

In this section we have defined a function in terms of (1) a rule and (2) ordered pairs. Since the function concept is so important, you should consider both definitions and satisfy yourself that these definitions are equivalent.

Example 3: Determining Functions and Finding Domain and Range

Determine if the relation is a function. Also specify the domain and the range.

a. $\{(-1, 0), (0, 0), (1, 0)\}$ **b.** $\{(0, -1), (0, 0), (0, 1)\}$

Solution:

a. This relation is a function because the first component in each of the ordered pairs is different. The domain is $\{-1, 0, 1\}$ and the range is $\{0\}$.

b. This relation is not a function because the number 0 is the first component in more than one ordered pair. The domain is $\{0\}$ and the range is $\{-1, 0, 1\}$.

PROGRESS CHECK 3

Determine if the relation is a function. Also specify the domain and the range.

a. $\{(-2, 1), (-2, 0), (2, 0)\}$ **b.** $\{(-2, 1), (2, 1), (1, -2)\}$ ■

Functions As Graphs

One of the sources of information and insight about a relationship is a picture that describes the particular situation. Pictures or graphs are often used in business reports, laboratory reports, and newspapers to present data quickly and vividly. Similarly, it is useful to have a graph that describes the behavior of a particular function, for this picture helps us see the essential characteristics of the relationship.

We can pictorially represent a function by using the **Cartesian coordinate system**. This system was devised by the French mathematician and philosopher René Descartes and is formed from the intersection of two real number line at right angles. The values for the independent variable (usually x) are represented on a horizontal number line, and values for the dependent variable (usually y) on a vertical number line. These two lines are called **axes**, and they intersect at their common zero point, which is called the **origin** [see Figure 1.5 (a)].

This coordinate system divides the plane into four regions called **quadrants**. The quadrant is which both x and y are positive is designated the first quadrant. The remaining quadrants are labeled in a counterclockwise direction. Figure 1.5(b) shows the name of each quadrant as well as the sign of x and y in that quadrant.

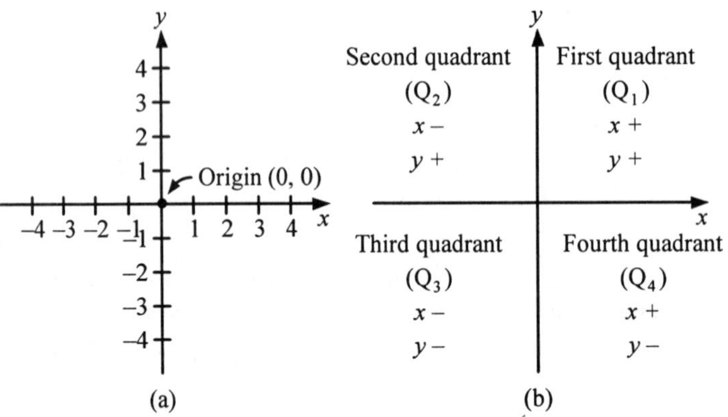

Figure 1.5

Any ordered pair can be represented as a point in this coordinate system. The first component indicates the distance of the point to the right or left of the vertical axis. The second component indicates the distance of the point above or below the horizontal axis. These components are called the **coordinates** of the point.

Example 4: Plotting Points

Represent the ordered pairs $(0, -3)$, $(-4, -2)$, $(\pi, -5/2)$ and $\left(-4/3, \sqrt{2}\right)$ as points in the Cartesian coordinate system.

Solution: See Figure 1.6.

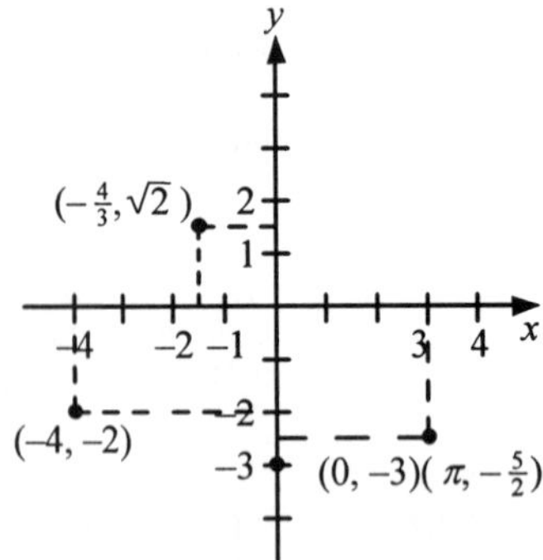

Figure 1.6

PROGRESS CHECK 4

Represent the ordered pairs $(-2, 0)$, $(-2, -4)$, $(3/2, \pi)$ and $\left(-5/3, \sqrt{5}\right)$ as points in the Cartesian coordinate system.

We now consider one of the most useful ideas in mathematics. With the Cartesian coordinate system we can represent any ordered pair of real numbers by a particular point in the system. This enables us to draw a geometric picture (graph) of a relation.

Graph

> The graph of a relation is the set of all points in a coordinate system that correspond to ordered pairs in the relation.

There are many techniques associated with determining the graph of a relation. One technique is simply to assign values to the independent variable and obtain a list of ordered-pair solutions. By plotting enough of these solutions, we can establish a trend and then complete the graph by following the established pattern. However, we cannot possibly list all the solutions of most equations, because they are an infinite set of ordered pairs. Determining how many and which points to plot is a difficult decision. Therefore, as we proceed in this section and succeeding chapters, we develop the more efficient method of determining the essential characteristics of the graph from the form of the equation. We may also use a graphing calculator or computer software to quickly obtain the graph of a relation. In Examples 5–7, we begin our development of graphing techniques by considering the point-plotting method for obtaining a graph.

Example 5: Graphing a Line by Plotting Points

Graph the function $y = -2x + 1$.

Solution: Make a list of ordered-pair solutions by replacing x with integer values from, say, 2 to –2.

If x Equals	Then $y = -2x + 1$	Thus, the Ordered Pairs Are
2	$-2(2) + 1 = -3$	$(2, -3)$
1	$-2(1) + 1 = -1$	$(1, -1)$
0	$-2(0) + 1 = 1$	$(0, 1)$
–1	$-2(-1) + 1 = 3$	$(-1, 3)$
–2	$-2(-2) + 1 = 5$	$(-2, 5)$

These ordered pairs are graphed in Figure 1.7, where they appear to all lie in a straight line. In fact, the graph of $y = -2x + 1$ is the straight line show in Figure 1.8. Note that the given equation is of the form $y = mx + b$ with $m = -2$ and $b = 1$. In Section 5.4 we will prove that the graph of every function of the form $y = mx + b$, where m and b are real number constants, is a straight line.

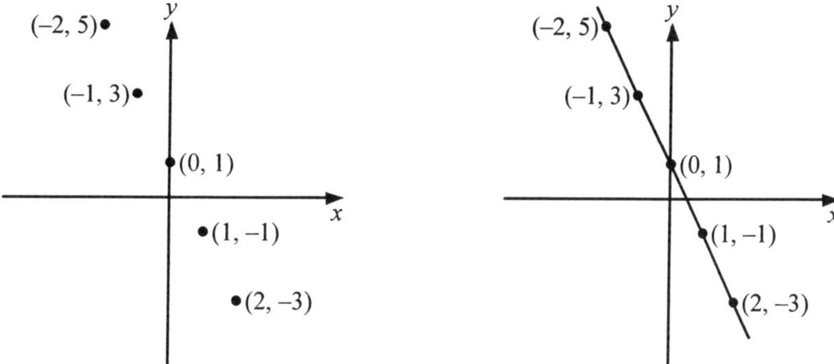

Figure 1.7 **Figure 1.8**

Technology Link

Most graphing calculators have a Table feature or a List feature for creating a table of ordered-pair solutions. If such features are available on your calculator, then use them to confirm the ordered-pair solutions shown for the function in Example 5.

PROGRESS CHECK 5

Graph the function $y = 2x - 1$. ■

We have said that a graph helps us see the essential characteristics of a function. However, the benefits derived from such a picture are directly related to your ability to read the graph. Essential features such as the domain and range of a function stand out, but you have to know what you are looking for. For example, consider Figure 1.9, which specifies the domain and range for $y = -2x + 1$.

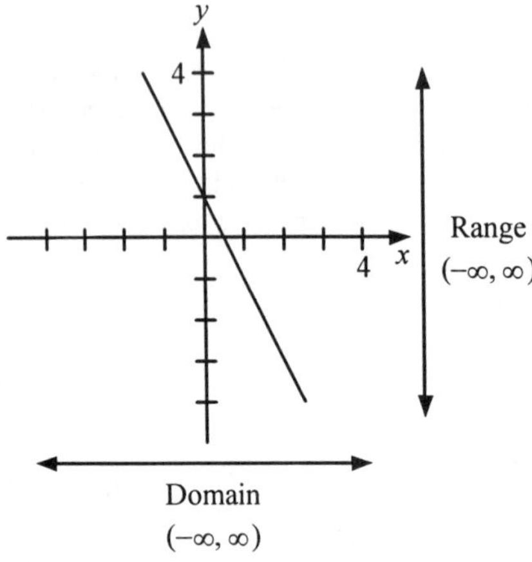

Figure 1.9

Because x may be replaced by any real number, in this equation, the domain is the set of all real numbers. This result can be realized geometrically by noticing that there is no limit to the horizontal extent of the graph; the line continues both to the right and the left without end. Similarly, because there is no limit to the vertical extent of the graph, the range is also the set of all real numbers. In general, the domain of a graph is given by the variation in the horizontal direction, while the range is given by the variation in the vertical direction.

Example 6: Graphing a Parabola by Plotting Points

Graph $y = 4 - x^2$ and indicate the domain and range of the function on the graph.

Solution: Begin by substituting integer values for x from, say, 3 to –3 and make a list of ordered-pair solutions.

If $x =$	Then $y = 4 - x^2$	Thus, the Ordered Pairs Are
3	$4-(3)^2=-5$	(3, −5)
2	$4-(2)^3=0$	(2, 0)
1	$4-(1)^2=3$	(1, 3)
0	$4-(0)^2=4$	(0, 4)
−1	$4-(-1)^2=3$	(−1, 3)
−2	$4-(-2)^2=0$	(−2, 0)
−3	$4-(-3)^2=-5$	(−3, −5)

Now graph these ordered pairs and draw a smooth curve through them. The resulting graph of $y = 4 - x^2$ is shown in Figure 1.10, and we read from the graph that the domain is the set of all real numbers, and the range is the interval $(-\infty, 4]$. Not that the cuplike curve in Figure 1.10 is called a **parabola**. In every case, the graph of a function that may be written in the form $y = ax^2 + bx + c$, where a, b, and c are real numbers with $a \neq 0$, is a parabola. We discuss these functions in detail in Section 5.5.

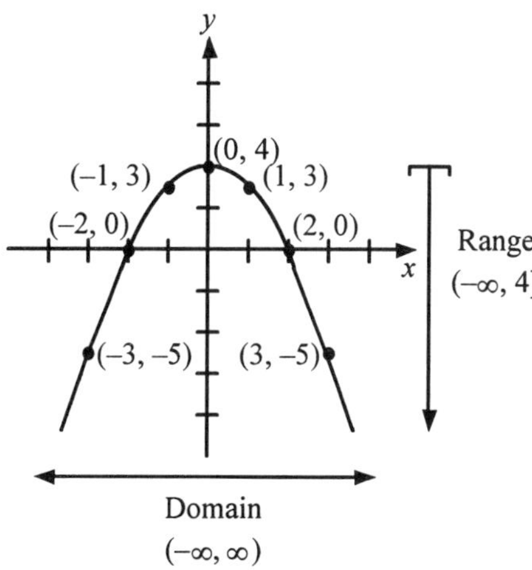

Figure 1.10

PROGRESS CHECK 6

Graph $y = (x-2)^2$ and indicate the domain and range of the function on the graph. ■

Example 7: Graphing a Horizontal Line by Plotting Points

Graph $y = 4$ (or $y = 4 + 0x$). Indicate the domain and range of the function on the graph.

Solution: Create a table of ordered pair solutions as in the chart below.

If $x =$	Then $y = 4 + 0 \cdot x$	Thus, the Ordered Pairs Are
3	$4 + 0(3) = 4$	(3, 4)
1	$4 + 0(1) = 4$	(1, 4)
−2	$4 + 0(-2) = 4$	(−2, 4)
−3	$4 + 0(-3) = 4$	(−3, 4)

Observe that no matter what number we substitute for x, the resulting value for y is 4. Thus, this function graphs as shown in Figure 1.11, and the domain is the set of all real numbers, while the range is $\{4\}$. Since the value of the function does not change, $y = 4$ is called a **constant function**. In every case, the graph of a constant function is a horizontal line.

PROGRESS CHECK 7

Graph $y = -3$ (or $y = -3 + 0x$). Indicate the domain and range of the function on the graph. ■

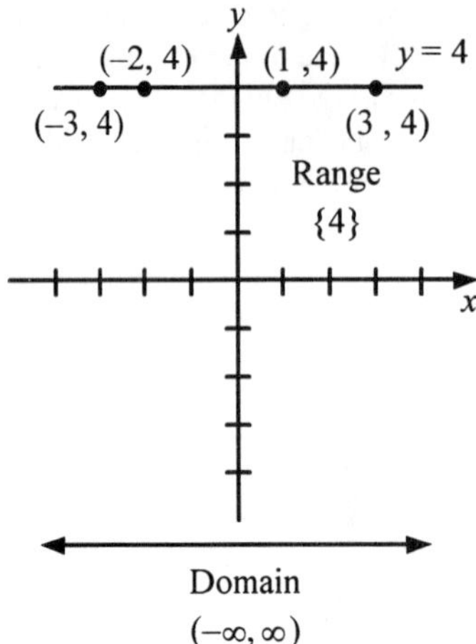

Figure 1.11

Obtaining Graphs by Graphing Calculator

A great strength of a graphing calculator is its ability to plot many points accurately in a few seconds. These points are then usually connected by line segments to obtain a better visual image of the relation. To obtain the graph of a function by calculator two important steps are necessary.

1. Enter an expression to define a function in the equation list.
2. Establish lower and upper bounds for x and y for the calculator display, and then activate the graphing routine.

The limits set in step 2 define a **viewing window**, and this procedure is often called setting the window or range for x and for y. Note that this use of the word "range" is different from the range of a function. For some graphing calculators, the **Standard Viewing Window** is obtained by letting both x and y vary from –10 to 10, where scale markers are 1 unit apart, and we will adopt this convention for this text. In the next example we use the ideas just discussed to redraw the graph of $y = 4 - x^2$ with the aid of a graphing calculator.

Example 8: Obtaining a Graph by Calculator

Graph $y = 4 - x^2$ using a graphing calculator.

Solution: Figures 1.12(a), (b), and (c) show typical calculator screens that result from entering this function in the equation list, setting the standard window, and then obtaining the graph, respectively. Observe that the resulting graph of $y = 4 - x^2$ is in agreement with the parabola we obtained by plotting points in Example 6.

Caution: If the display is entirely blank when you expect to see a graph, the most likely cause is that the window settings are not suitable. Another possible cause is that the function to be graphed has been inadvertently turned off. For instance, on Texas Instruments models the equal sign is highlighted for a selected function, and not highlighted for an unselected function. You should learn how to turn a function "on" or "off" on your calculator.

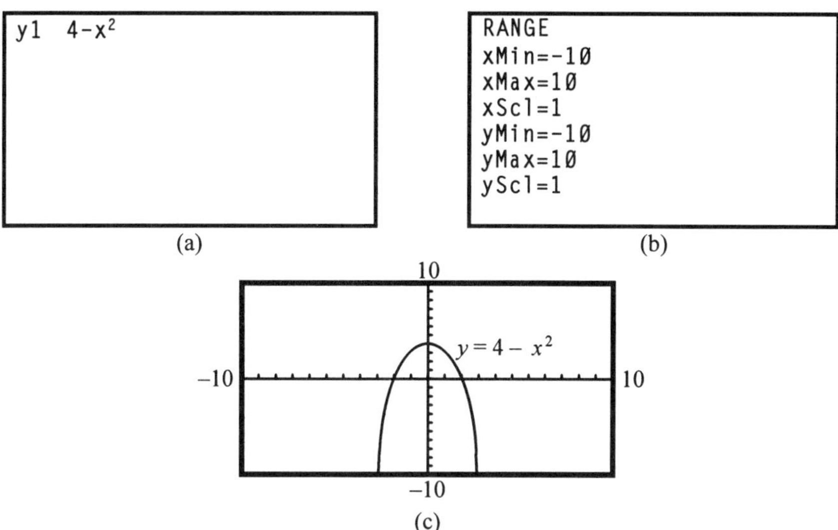

Figure 1.12

PROGRESS CHECK 8
Graph $y = (x-2)^2$ using a graphing calculator. ■

When graphing functions we usually do not want to think of a function in terms of a single picture. Instead, many pictures are possible depending on how the viewing window is defined. It is assumed that when we are directed to graph a function, a complete graph is requested. A **complete graph** is a graph that shows all the significant features of a function. The next two examples involve adjusting viewing windows to obtain a complete graph.

Example 9: Adjusting Viewing Windows to Find a Complete Graph

Graph $y = (x-10)^2 - 15$ using the following viewing windows. Which picture shows a complete graph?

a.	**b.**
$x\text{Min} = -10$	$x\text{Min} = -5$
$x\text{Max} = 10$	$x\text{Max} = 20$
$x\text{Scl} = 1$	$x\text{Scl} = 5$
$y\text{Min} = -10$	$y\text{Min} = -50$
$y\text{Max} = 10$	$y\text{Max} = 100$
$y\text{Scl} = 1$	$y\text{Scl} = 10$

Solution: The function is graphed using the two viewing windows in Figure 1.13(a) and Figure 1.13(b), respectively. Because $y = (x-10)^2 - 15$ may be expressed in the form $y = ax^2 + bx + c$ with $a \neq 0$, the graph is a parabola. Observe that Figure 1.13(b) shows a complete graph of the function, while Figure 1.13(a) does not illustrate all the significant features of the graph.

PROGRESS CHECK 9
Graph $y = 60 - x$ using the following viewing windows. Which picture shows a complete graph?

a.	**b.**
$x\text{Min} = -10$	$x\text{Min} = -70$
$x\text{Max} = 70$	$x\text{Max} = 10$
$x\text{Scl} = 10$	$x\text{Scl} = 10$
$y\text{Min} = -10$	$y\text{Min} = -70$
$y\text{Max} = 70$	$y\text{Max} = 10$
$y\text{Scl} = 10$	$y\text{Scl} = 10$

■

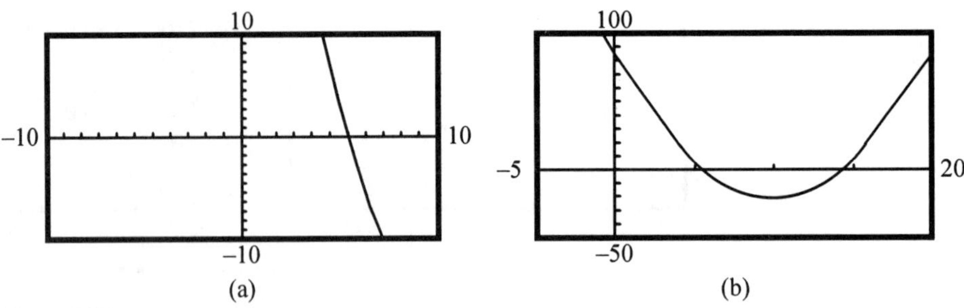

Figure 1.13

Example 10: Finding Domain and Range Algebraically and Graphically

Determine the domain and range of $y = \sqrt{x+12}$. Use both algebraic and geometric methods.

Solution: Algebraic Analysis: For the output of $\sqrt{x+12}$ to be a real number, x must satisfy

$$x + 12 \geq 0$$
$$x \geq -12.$$

Thus, the domain is $\{x\colon\ x \geq -12\}$, or $[-12, \infty)$, in interval notation. The radical sign $\sqrt{\ }$ denotes the nonnegative square root, so if $x \geq -12$, then $\sqrt{x+12}$ is greater than or equal to 0. Therefore, the range is $\{y\colon\ y \geq 0\}$, or alternatively $[0, \infty)$.

Geometric Analysis

First, set $y1 = \sqrt{x+12}$. Then we use $[-15, 15]$ by $[-10, 10]$ to define a viewing window and graph the function, as shown in Figure 1.14. This graph suggests that the domain is $[-12, \infty)$ and the range is $[0, \infty)$. With this graphical approach, analysis of the equation $y = \sqrt{x+12}$ is required to determine that $(-12, 0)$ is a left end point of this graph, and that this graph is unbounded to the right.

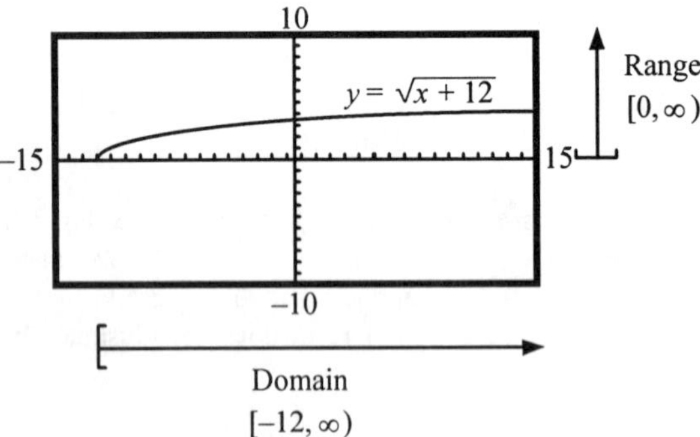

Figure 1.14

Caution: On a calculator, remember to enter the function input (argument) inside of parentheses when the input is more than one term. For instance, the keystroke

$$\sqrt{\ }\ x + 12 \text{ means } \sqrt{x} + 12$$
$$\text{while } \sqrt{\ }\ (x + 12) \text{ means } \sqrt{x+12}.$$

PROGRESS CHECK 10

Determine the domain and range of $y = -\sqrt{18 - x}$. Use both algebraic and geometric methods. ∎

The Zoom-in feature and Trace feature on a graphing calculator are particularly useful for graphing operations, as shown next.

Example 11: Using Zoom and Trace Features

Graph $y = -3x^2 + 10x - 5$ and estimate the coordinates of the highest point on the graph to the nearest tenth.

Solution: First, graph $y = -3x^2 + 10x - 5$ in the standard viewing window, as shown in Figure 1.15(a). Next, zoom in on the part of the picture that displays the highest point. Figure 1.15(b) shows a typical calculator screen that results when the cursor is located near the highest point, and the Zoom-In feature (set to zoom factors of 4) is applied twice. Finally, we activate the Trace feature and try to stop the cursor on the highest point, to obtain a display like that in Figure 1.15(c). From this result we estimate (to the nearest tenth) that the highest point is located at (1.7, 3.3). The Zoom-In and Trace Procedures may be repeated as many times as you wish, to improve the estimate.

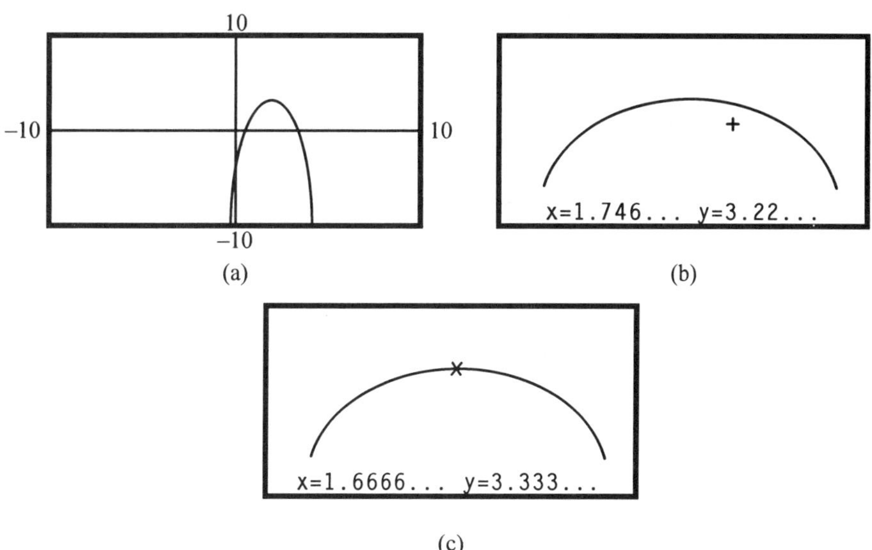

Figure 1.15

PROGRESS CHECK 11

Graph $y = 3x^2 + 8x$ and estimate the coordinates of the lowest point on the graph to the nearest tenth. ∎

Example 12: Finding a Formula for a Function

Solve the problem in the chapter introduction on page 83.

Solution: First, consider the sketch of the situation in Figure 1.16 and note that we strategically placed the radius of the semicircle so as to form a right triangle inside the rectangle. We may then use the Pythagorean theorem to write z in terms of x as follows:

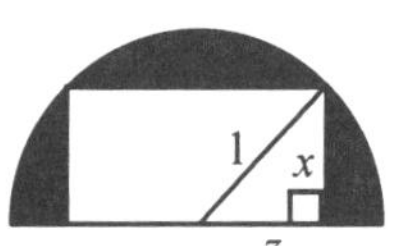

Figure 1.16

$$x^2 + z^2 = 1^2$$
$$z^2 = 1 - x^2$$
$$z = \sqrt{1 - x^2} \quad \text{(since } z \text{ is positive).}$$

Then the area formula in terms of x is

$$A = 2xz$$
$$= 2x\sqrt{1 - x^2}.$$

In the context of this problem, the meaningful replacements for x are the real numbers between 0 and 1, so the domain is the interval (0, 1).

To determine the range, use [0, 1] by [0, 2] to define a viewing window, and then graph the function as shown in Figure 1.17. In this graph, note that the y-axis represents values of the area A. Using the Zoom and Trace features you should determine that $y \approx 1$ at the highest point in the graph. (In fact, check by calculator that when $x = \sqrt{1/2}$, then $y = 1$.) The lowest y value is 0, but $y > 0$ because of context. Thus, the range is the interval (0, 1].

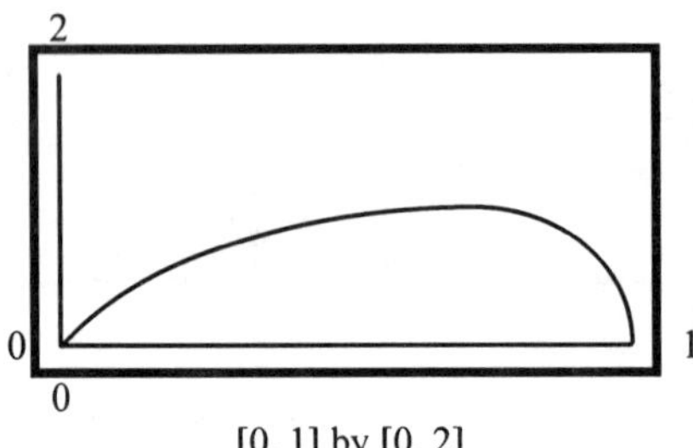

[0, 1] by [0, 2]

Figure 1.17

PROGRESS CHECK 12

Suppose the radius of the semicircle in Example 12 is 2 ft. Find a formula for the area of the rectangular insert as a function of the side dimension x, and determine the domain and the range. ■

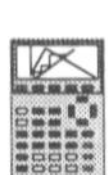

EXPLORE 1.1

1. Earlier in this section of the text, it was stated that the graphs of equations that have the form $y = mx + b$ are straight lines.
   a. Investigate the effect on the graph when the value of b is changed. For instance, compare the graphs for $y = 2x + b$ for these values of b: $\{-3, -2, -1, 0, 1, 2, 3\}$. What does b control in the graph of the line?
   b. Investigate the effect on the graph when the value of m is changed. For instance, compare the graphs for $y = mx + 2$ for these values of m: $\{-3, -2, -1, 0, 1, 2, 3\}$. What does m control in the graph of the line?
2. You may find it useful to be able to control how far the cursor steps when you move it by pressing an arrow key one time. The display screen has room for a certain number of steps, according to the manufacture of the calculator. On the T185, for instance, there are 126 horizontal steps and 62 vertical steps in the display. The sizes of each horizontal and vertical step are therefore determined by $(x\text{Max} - x\text{Min})/126$ and $(y\text{Max} - y\text{Min})/62$, respectively. As another example, the TI-82 has 94 horizontal steps and 62 vertical steps. Determine the number of steps in each direction for *your* calculator, and then use this result in the following explorations.

a. Find several window settings that will give a horizontal step equal to 1.
b. Find several window settings that will give a horizontal step equal to 0.1
c. Find several window settings that will give a vertical step equal to 1.
d. Find several window settings that will give a vertical step equal to 0.1.

EXERCISES 1.1

In Exercises 1–10, find a formula that defines the functional relationship between the two variables; in each case indicate the domain of the function.

1. Express the area (A) of a square in terms of the length of its side (s).

2. Express the area (A) of a circle as a function of its radius (r).

3. Express the length of the side (s) of a square in terms of its perimeter (P)

4. Express the area (A) of a square as a function of the perimeter (P) of the square.

5. Express the earnings (e) of an electrician in terms of the number (n) of hours worked if the electrician makes $28 per hour.

6. Express the earnings (e) of a real estate agent who receives a 6 percent commission as a function of the sale price (p) of a house.

7. As a rule, in oil spills, the number of tons spilled can be multiplied by 7 to estimate the number of barrels spilled. Each barrel contains 42 gallons. Express the number of gallons (g) spilled as a function of the number of tons (t) spilled.

8. Express the monthly cost (c) for a checking account as a function of the number (n) of checks serviced that month if the bank charges 10 cents per check plus a 75-cent maintenance charge.

9. The total cost of producing a certain product consists of paying $400 per month rent plus $5 per unit for material. Express the company's monthly total costs (c) as a function of the number of units (x) it produces that month.

10. Express the monthly cost of renting a computer as a function of the number (n) of hours the computer is used if the company charges $200 plus $100 for every hour the computer is used during the month.

In Exercises 11–12 determine which of the ordered pairs are solutions of the equation.

	Equation	Ordered Pairs
11.	$y = 3x + 1$	$(4, 1), (-2, -5), (1, 4), (-5, -2)$
12.	$y = \lvert x \rvert$	$(-1, -1), (1, 1), (-1, 1), (1, -1)$

In Exercises 13–16 find five ordered pairs that are solutions of each formula or equation.

13. $y = 4 - x$ **14.** $y = 2x - 7$

15. $y = x^2 + 2x - 1$ **16.** $y = x^3 - 1$

17. Fill in the missing component in each of the following ordered pairs so they are solutions of the equation $y = -3x + 7$: $(0, \ ), (\ , 0), (-5, \ ), (\ , 5)$.

18. Fill in the missing component in each of the following ordered pairs so they are solutions of the equation $y = \dfrac{5x - 3}{6}$: $(0, \ ), (\ , 0), (\ , -2), (-3, \ )$.

19. If $(a, -1)$ is a solution of the equation $y = -2x + 9$, find the value of a.

20. If $(-2, b)$ is a solution of the equation $y = 7x + 5$, then find the value of b.

In Exercises 21–26 state which relations are functions. Specify the domain and range.

21. $\{(1, -3), (2, 0), (3, 1)\}$

22. $\{(-3, 1), (0, 2), (1, 3)\}$

23. $\{(2, -1), (3, 0), (4, -1)\}$

24. $\{(-1, 2), (0, 3), (-1, 4)\}$

25. $\{(2, 1), (2, 2), (2, 3)\}$

26. $\{(1, 2), (2, 2), (3, 2)\}$

In Exercises 27–30 determine whether or not the given correspondence is a function

27.

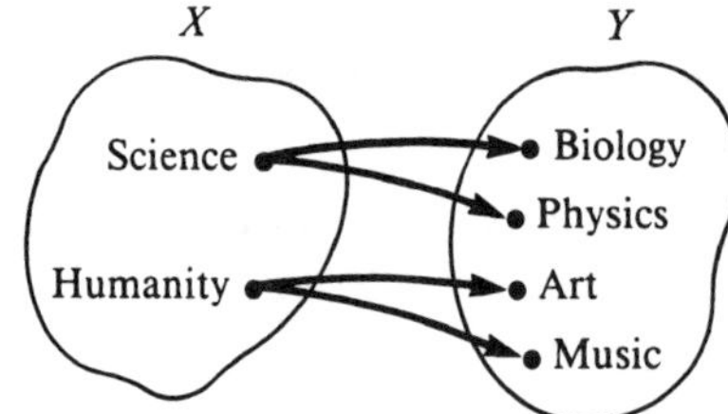

28.

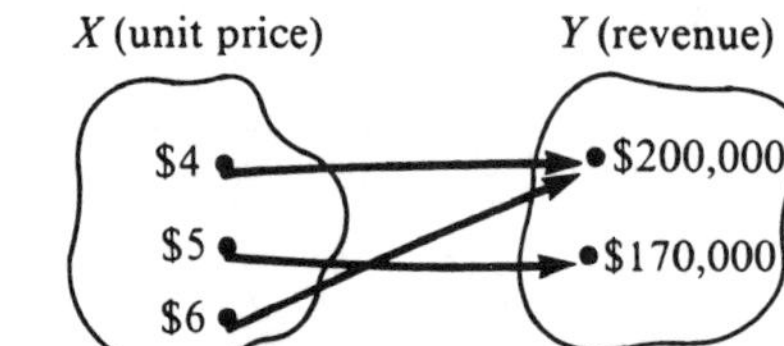

29.

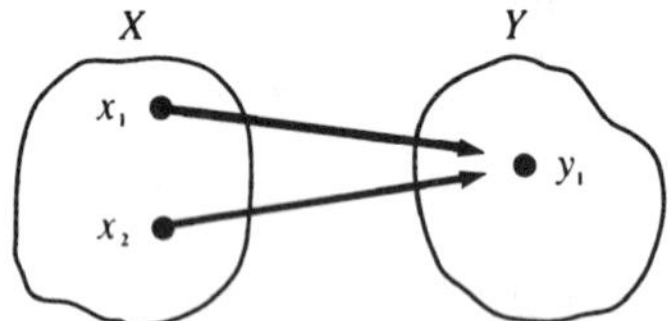

30.

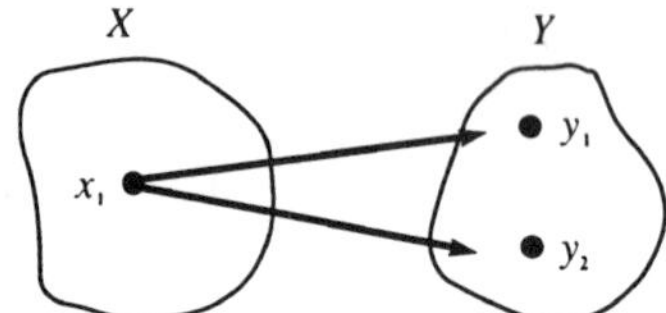

31. Graph the following ordered pairs.

a. (3, 1) **b.** (−3, 4) **c.** (0, 2)
d. (−3, 0) **e.** (−1, −2) **f.** (2, −3)
g. (−2, −3) **h.** (−1, 2)
i. $(\sqrt{2}, 3)$ **j.** $(1, -\pi)$

32. Approximate (use integers) the ordered pairs corresponding to the points shown in the graph.

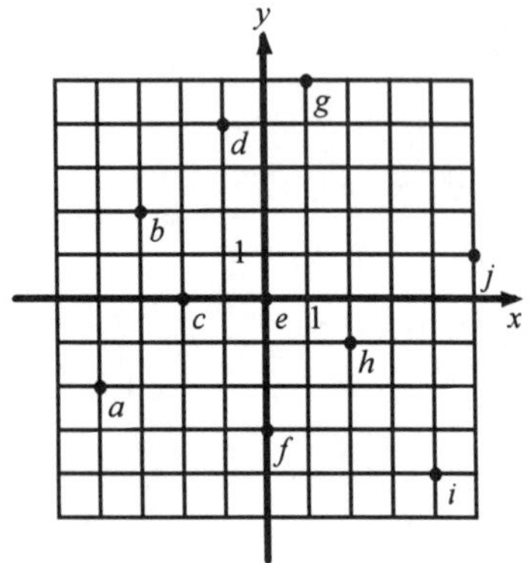

In Exercises 33–46 draw the graph by the point-plotting method. Confirm your answer with a grapher. Indicate the domain and range on the graph.

33. $y = x + 3$ **34.** $y = 3x + 1$ **35.** $y = -3x - 2$

36. $y = 2 - x$ **37.** $y = 4$ **38.** $y = -3$

39. $y = -\sqrt{3}$ **40.** $y = \pi$ **41.** $y = x^2 + 1$

42. $y = x^2 + x$ **43.** $y = (x - 1)^2$ **44.** $y = 5 - x^2$

45. $y = 3x - x^2$ **46.** $y = x^2 - 2x + 1$

In Exercises 47–52 graph these essential functions, and make note of their names. They will be referred to by name later in the text.

47. The squaring function: $y = x^2$

48. The square root function: $y = \sqrt{x}$

49. The cubing function: $y = x^3$

50. The identity function: $y = x$

51. The constant function: $y = c$, where c is some constant.

52. The absolute value function: $y = |x|$. Note: On many graphing calculators there is a key or menu choice labeled "abs" for absolute value. So $|x|$ is entered as abs x.

In Exercises 53–64 use a grapher to graph the function with the given viewing window, shown as [xMin, xMax]; [yMin, yMax].

53. $y = 3x + 1$ [−10, 10]; [−10, 10]

54. $y = -2x + 2$ [−10, 10]; [−10, 10]

55. $y = 3 - 2x^2$ [−5, 5]; [−5, 5]

56. $y = 3x^2 - 2$ [−5, 5]; [−5, 5]

57. $y = x^2 + x + 10$ [−10, 10]; [−5, 25]

58. $y = x^2 - 10x - 10$ [−5, 15]; [−40, 15]

59. $y = \sqrt{x + 5}$ [−10, 10]; [−5, 5]

60. $y = \sqrt{x - 5}$ [−1, 15]; [−2, 5]

61. $y = x^3 - 3x^2 - 3$ [−5, 10]; [−15, 10]

62. $y = -x^3 - 3x^2 - 3$ [−5, 5]; [−20, 20]

63. $y = |x + 5|$ [−15, 5]; [−5, 15]

64. $y = |x - 5|$ [−5, 15]; [−5, 15]

In Exercises 65–68 decide which viewing window shows a complete graph.

65. $y = (x + 12)^2 - 9$
a. [−10, 10]; [−10, 10]
b. [−20, 10]; [−20, 10]

66. $y = 20 + (x - 12)^2$
a. [−10, 10]; [−10, 10]
b. [−5, 20]; [−5, 40]

67. $y = x - 20$
a. [−10, 40]; [−40, 10]
b. [−40, 10]; [−10, 40]

68. $y = 3x + 20$
a. [−5, 50]; [−20, 10]
b. [−20, 10]; [−50, 50]

In Exercises 69–72 find a viewing window that produces the given display.

69. $y = (x - 8)^2 + 3$

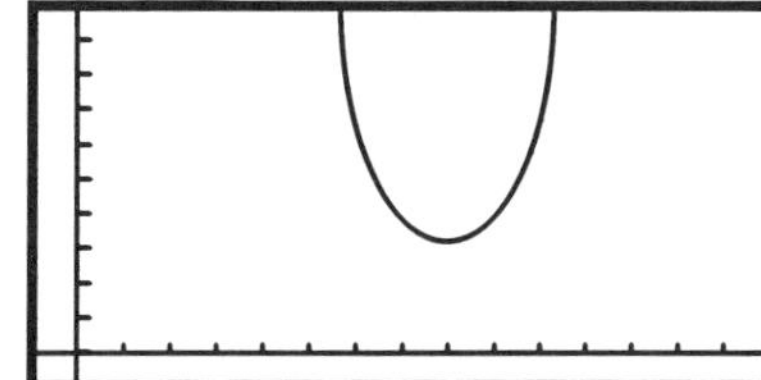

70. $y = (x + 8)^2 + 3$

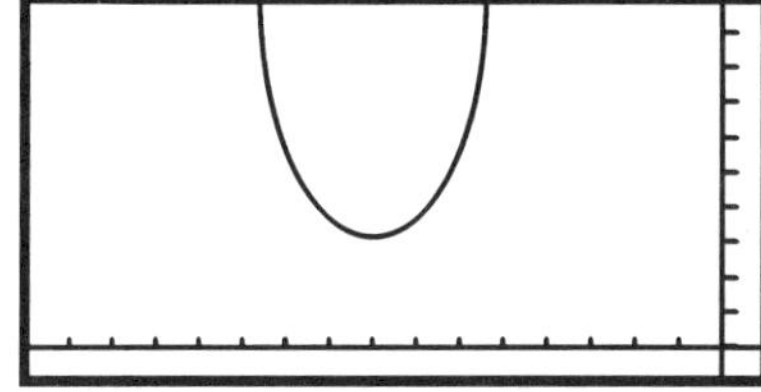

71. $y = 15 - x$

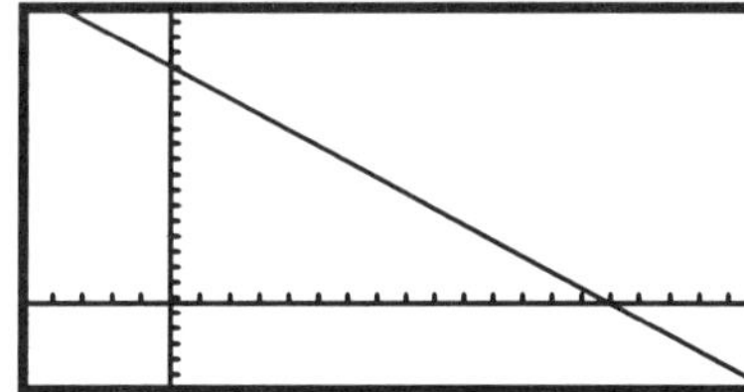

72. $y = -15 - x$

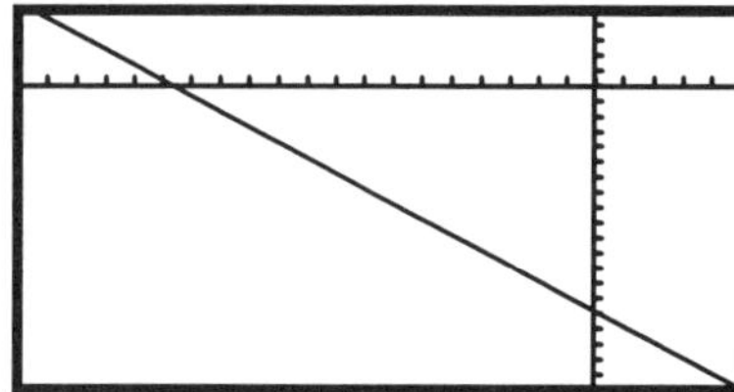

In Exercises 73–82 determine the domain and range of the given function. Use both algebraic and geometric methods.

73. $y = \sqrt{x + 6}$

74. $y = -\sqrt{10 - x}$

75. $y = 5 + \sqrt{3x - 9}$

76. $y = 4 - \sqrt{2x + 10}$

77. $y = |x - 4| + 2$

78. $y = |x + 4| - 1$

79. $y = 7$

80. $y = -8$

81. $y = \dfrac{1}{x + 1}$

82. $y = \dfrac{1}{\sqrt{x + 5}}$

In Exercises 83–88 use the Zoom and Trace features of a grapher to estimate the coordinates of the desired point to the nearest tenth.

83. $y = -x^2 + 5x - 3$; highest point

84. $y = -x^2 - 3x + 5$; highest point

85. $y = 5x^2 + 2x - 2$; lowest point

86. $y = 5x^2 - 3x + 2$; lowest point

87. $y = (3x - 10)^2 + 20$; lowest point

88. $y = -(10x + 3)^2 - 10$; highest point

89. A projectile fired vertically upward from the ground with an initial velocity of 128 ft/second will hit the ground 8 seconds later, and the speed of the projectile in terms of the elapsed time t equals $|128 - 32t|$. Graph the function $s = |128 - 32t|$. Describe the domain and range.

90. The height (y) above water of a diver t seconds after the diver steps off a platform 100 ft high is given by the formula $y = 100 - 16t^2$. Graph this function. Describe the domain and range.

91. A reservoir contains 10,000 gal. of water. If water is being pumped from the reservoir at a rate of 50 gal/minute, write a formula expressing the amount (a) of water remaining in the reservoir as a function of the number (n) of minutes the water is being pumped. Describe the domain and range.

92. Oil is leaking from a tanker at the rate of 40,000 gallons per hour. Originally, the tanker was carrying 2 million gallons. Write a function that expresses the amount (a) of oil that remains in the tanker as a function of the number of hours (h) elapsed since the spill began. Describe the domain and range.

93. Express the federal income tax (t) for a single person as a function of taxable income (i) if the person's taxable income is between \$115,000 and \$250,000 inclusive, and the tax rate is \$31,172 plus 36 percent of the excess over \$115,000. Describe the domain and range.

94. A rule in a certain state income tax form says that if a person's federal tax is between \$3,400 and \$13,100, then the state tax is \$952 plus 31 percent of the excess over \$3,400. Express this state tax (s) as a function of the federal tax (t). Describe the domain and range.

95. For the given figure find a formula that expresses the area (A) of the shaded triangle as a function of the distance labeled as x. The outer figure is a rectangle. Describe the domain and range of the function.

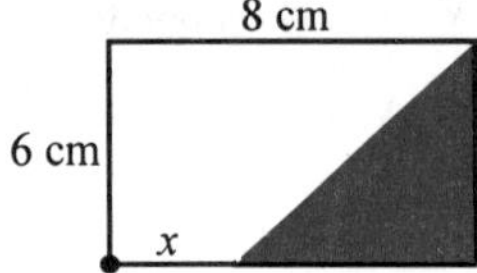

96. A long strip of galvanized sheet metal 12 in. wide is to be formed into an open gutter by bending up the edges to form a gutter with a rectangular cross-section. Write the cross-sectional area of the gutter as a function of the depth (x). Describe the domain and range of the function.

97. For the given figure find a formula that expresses the area (A) of the rectangle as a function of the height (x). The outer figure is a semicircle with radius 3 inches. Describe the domain and range of the function.

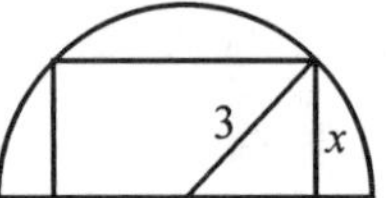

98. Express the shaded area as a function of the x coordinate of the vertical line. Describe the domain and range.

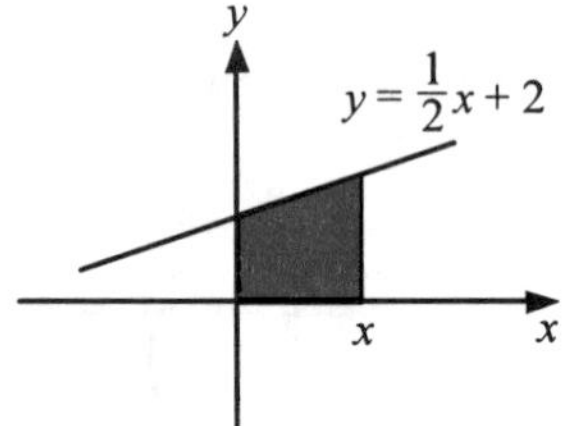

THINK ABOUT IT 1.1

1. A graphing calculator displays a graph by darkening dots on the display screen. Each tiny area of the screen (each "point") that can be darkened is called a *pixel*. This method imposes certain limitations on the graphs. Think about it by writing an answer to these questions. When you use the standard viewing window, [−10, 10]; [−10, 10], why do the lines $y = 4$ and $y = 4.01$ look the same? How can you find a viewing window that will show that they are not the same?
2. Describe a realistic situation that can be analyzed by the given function.
   a. $y = 3x$, where $x > 0$ **b.** $y = 2x + 10$, where x is a nonnegative integer
3. **a.** The relation "west of" on the set {Denver, San Diego, Boston} is given by {(San Diego, Denver), (Denver, Boston), (San Diego, Boston)}. Is this relation a function?
   b. The relation "is 2 more than" on the set {0, 1, 2, 3, 4} is given by what set of ordered pairs? Is this set a function?
4. For homework you need to solve every other odd-numbered section exercise, starting with Exercise 1. Will you need to solve Exercise 67? How about Exercise 93? Write a formula that shows the numbers of the exercises that must be solved. Give a verbal description for this rule.
5. It is often useful to formulate (at least mentally) a verbal description of a function. For example, the function defined by the equation $y = 2x + 5$ may be expressed in words as follows: "For each real number, double it, and then add 5 to the result." Write verbal descriptions for the following functions.

a. $y = x^2 + 3$ **b.** $y = (x+3)^2$ **c.** $y = 1 - x^3$

d. $y = -|x|$ **e.** $y = \sqrt{x+2}$ **f.** $y = \dfrac{1}{x-4}$

● ● ●

1.2 Functional Notation and Piecewise Functions

Photo Courtesy of ***Reuters/Roy Letkey*** *of Archive Photos, New York*

If the value V of a particular work of art is given by the function

$$V = f(x) = 50{,}000(1.07)^x,$$

where x is the number of years since its purchase at \$50,000, then find and interpret $f(9)$. (See Example 2.)

Objectives

1. Evaluate functions using functional notation.
2. Write the difference quotient for a function in simplest form.
3. Find function values and graphs for piecewise functions.
4. Read from a graph of function f the domain, range, function values, values of x for which $f(x) = 0$, $f(x) < 0$, or $f(x) > 0$, intercepts, and intervals where f is increasing, decreasing, or constant.

A useful notation commonly used with functions allows us to represent more conveniently the value of the dependent variable for a particular value of the independent variable. In this notation a letter such as f is used to name a function, and then an equation such as

$$y = 2x + 5 \text{ is written as } f(x) = 2x + 5.$$

The dependent variable y is replaced by $f(x)$, with the independent variable x appearing in parentheses. The expression $f(x)$ is read "f of x" or "f at x" and means the value of the function (the y value) corresponding to the value of x. Similarly, $f(7)$ is read "f of 7" or "f at 7" and means the function value when $x = 7$. To find $f(7)$ in this example, we substitute 7 for x in the equation $f(x) = 2x + 5$.

$f(x) = 2x + 5$	Given equation.
$f(7) = 2(7) + 5$	Replace x by 7.
$= 19$	Simplify.

The result $f(7) = 19$ says that when $x = 7$, $y = 19$. The notation $f(x)$ originated with the Swiss mathematician Leonhard Euler (1734), and in this context note that $f(x)$ does not mean *f* times *x*.

Example 1: Using Functional Notation

If $y = f(x) = 2x^2 - x + 4$, find $f(2)$, $f(15)$, and $f(-3)$.

Solution: In each case replace all occurrences of *x* by the number inside the parentheses and then simplify.

$$y_{\text{when } x=2} = f(2) = 2(2)^2 - 2 + 4 = 10$$
$$y_{\text{when } x=15} = f(15) = 2(15)^2 - 15 + 4 = 439$$
$$y_{\text{when } x=-3} = f(-3) = 2(-3)^2 - (-3) + 4 = 25$$

Thus, $f(2) = 10$, $f(15) = 439$, and $f(-3) = 25$.

Technology Link

Most graphing calculators have special features to find many function values quickly. These features may include a functional notation capability, an Evaluate function, a Table feature, or a List feature. If your calculator has such features, then you should use them to redo the problem in Example 1 and compare your result to the text's answers.

PROGRESS CHECK 1

If $y = f(x) = 5x - x^2$, find $f(4)$, $f(20)$, and $f(-5)$. ■

Example 2: Interpreting Functional Notation

Solve the problem in the section introduction on page 101.

Solution: *f*(9) gives *V* when $x = 9$. Using $V = f(x) = 50{,}000(1.07)^x$ gives

$$V_{\text{when } x=9} = f(9) = 50{,}000(1.07)^9 = 91{,}922.96$$

Thus, the value of this particular work of art in 9 years is about \$92,000 (to the nearest thousand dollars).

PROGRESS CHECK 2

Use $V = 50{,}000(1.07)^x$ and find $f(6)$. Interpret the meaning of $f(6)$ in the context of Example 2.

■

Example 3: Evaluating Two Functions

If $f(x) = x - 1$ and $g(x) = x^2 + 1$, find $3f(-1) - 4g(2)$.

Solution: The expression given above means that you are to find the difference of 3 times "*f* of −1" and 4 times "*g* of 2." First, determine $f(-1)$ and $g(2)$.

$$\begin{array}{ll} f(x) = x - 1 & g(x) = x^2 + 1 \\ f(-1) = (-1) - 1 & g(2) = (2)^2 + 1 \\ f(-1) = -2 & g(2) = 5 \end{array}$$

Then

$$\begin{aligned} 3f(-1) - 4g(2) &= 3(-2) - 4(5) \\ &= -26 \end{aligned}$$

PROGRESS CHECK 3
If $f(x)=1-x$ and $g(x)=x^2+2$, find $4f(-1)-2g(3)$. ■

Example 4: Testing for Function Properties

If $f(x)=x^2$, show that $f(a+b)$ does not equal $f(a)+f(b)$ for all a and b.

Solution: To determine $f(a+b)$, $f(a)$, and $f(b)$, replace x in the function $f(x)=x^2$ by $a+b$, a, and b, respectively.

$$\begin{aligned} f(a+b) &= (a+b)^2 = a^2+2ab+b^2 \\ f(a) &= a^2 \\ f(b) &= b^2 \end{aligned}$$

Since $a^2+2ab+b^2 \neq a^2+b^2$, $f(a+b)\neq f(a)+f(b)$.

PROGRESS CHECK 4
If $f(x)=3x$, show that $f(a+b)$ *does* equal $f(a)+f(b)$ for all a and b. ■

Example 5: Finding a Difference Quotient

The difference quotient of a function $y=f(x)$ is defined as

$$\frac{f(x+h)-f(x)}{h},\ h\neq 0.$$

Computing this ratio is an important consideration when you are analyzing the rate of change of a function. Find the difference quotient for $f(x)=x^2+2x$ in simplest form.

Solution: If $f(x)=x^2+2x$, we have

$$f(x+h)=(x+h)^2+2(x+h)=x^2+2xh+h^2+2x+2h.$$

Then

$$\begin{aligned} \frac{f(x+h)-f(x)}{h} &= \frac{\left(x^2+2xh+h^2+2x+2h\right)-\left(x^2+2x\right)}{h} \\ &= \frac{x^2+2xh+h^2+2x+2h-x^2-2x}{h} \\ &= \frac{2xh+h^2+2h}{h} \\ &= \frac{2xh}{h}+\frac{h^2}{h}+\frac{2h}{h} \\ &= 2x+h+2. \end{aligned}$$

PROGRESS CHECK 5
Find the difference quotient for $f(x)=x^2+3x$ in simplest form. ■

Piecewise Functions

A **piecewise function** is a function in which different rules apply for different intervals of domain values. An example of an everyday situation that leads to a piecewise function is considered next.

Example 6: A Piecewise Function

Each month a salesperson earns \$500 plus 7 percent commission on sales above \$2,000. Find a rule that expresses the monthly earnings (e) of the salesperson in terms of the amount (a) of merchandise sold during the month.

Solution: If the salesperson sells less than or equal to \$2,000 worth of merchandise for the month, he earns \$500. Thus,

$$e = \$500 \text{ if } \$0 \le a \le \$2{,}000.$$

If the salesperson sells above \$2,000, she earns \$500 plus 7 percent of the amount above \$2,000. Thus

$$e = \$500 + 0.07(a - \$2{,}000) \text{ if } a > \$2{,}000.$$

The following rule may then be used to determine the monthly earnings of the salesperson when we know the amount of merchandise sold:

$$e = \begin{cases} \$500 & \text{if } \$0 \le a \le \$2{,}000 \\ \$500 + 0.07(a - \$2{,}000) & \text{if } a > \$2{,}000. \end{cases}$$

The domain of the function is $\{a\text{: } a \ge \$0\}$, and the range is $\{e\text{: } e \ge \$500\}$.

PROGRESS CHECK 6

Express the monthly earnings (e) of a salesperson in terms of the cash amount (a) of merchandise sold if the salesperson earns \$600 per month plus 8 percent commission on sales above \$10,000. ■

Examples 7 and 8 consider how to evaluate and how to graph a piecewise function.

Example 7: Evaluating a Piecewise Function

If $f(x) = \begin{cases} 2 & \text{if } x < 0 \\ x+1 & \text{if } 0 \le x < 3 \end{cases}$, find

a. $f(2)$ **b.** $f(-2)$ **c.** $f(3)$

Solution:

a. Since 2 is in the interval $0 \le x < 3$, use $f(x) = x + 1$, so $f(2) = 2 + 1 = 3$.

b. Since –2 is less than 0, use $f(x) = 2$, so $f(-2) = 2$.

c. Since 3 is not in the domain of the function, $f(3)$ is undefined.

PROGRESS CHECK 7

If $f(x) = \begin{cases} x-1 & \text{if } 0 \le x < 5 \\ 0 & \text{if } x \ge 5 \end{cases}$ find

a. $f(3)$ **b.** $f(-3)$ **c.** $f(5)$ ■

Example 8: Graphing a Piecewise Function

Graph the function defined as follows:

$$f(x) = \begin{cases} -1 & \text{if } x < 1 \\ x+2 & \text{if } x \ge 1 \end{cases}$$

Indicate the domain and range on the graph.

Solution: If $x < 1$, $f(x) = -1$, which is a constant function whose graph is a horizontal line with such ordered pairs as $(0, -1)$, $(-1, -1)$, $(-2, -1)$, and so on. If $x \ge 1$, $f(x) = x + 2$, which graphs as a line with ordered pairs (1,3), (2,4), (3,5), and so on. The graph is given in Figure 1.18. The vertical axis is labeled $f(x)$, and the domain and range are indicated. Note that we draw a solid circle at $(1,3)$ and an open circle at $(1,-1)$ to show that $(1,3)$ is part of the graph, while $(1,-1)$ is not.

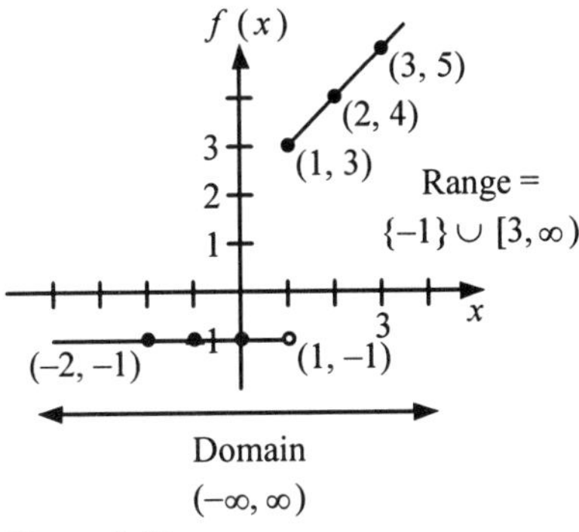

Figure 1.18

Technology Link

It is possible to use some graphing calculators to graph a piecewise function. For example, to graph the function in Example 8 on Texas Instruments models, it is first recommended that the calculator be switched from connected mode to dot mode. The dot mode prevents the calculator from connecting points that are in separate pieces of the graph. Then enter the following expression for this piecewise function in the equation list and obtain the graph in the usual way. In this expression the symbols < and ≥ may be entered through a key or menu marked TEST.

$$y1 = -1(x < 1) + (x + 2)(x \geq 1)$$

The rationale for entering this expression is that the calculator returns a 1 when an expression containing a relational operation (such as <) is true, and a 0 when it is false. Thus, if you enter the expression $(x+2)(x \geq 1)$, then the value of the expression in the right-hand parentheses is 1 when x is greater than or equal to one and 0 otherwise. So the graph of $y = (x+2)(x \geq 1)$ is the line $y = x + 2$ when x is greater than or equal to one and the line $y = 0$ otherwise. Note that the line $y = 0$ graphs as the x-axis, so it cannot be seen in the display.

PROGRESS CHECK 8
Graph the function defined as follows:

$$f(x) = \begin{cases} 2 & \text{if } x < -2 \\ x + 1 & \text{if } x \geq -2 \end{cases}$$

Indicate the domain and range on the graph. ■

The next example illustrates that it is sometimes helpful to convert to a piecewise function to analyze a graph.

Example 9: Converting to a Piecewise Function

Graph $y = \dfrac{|x-10|}{x-10}$. Indicate the domain and range on the graph.

Solution: From Section R.1 we know

$$|x-10| = \begin{cases} x-10 & \text{if } x-10 \geq 0 \\ -(x-10) & \text{if } x-10 < 0 \end{cases}$$

Since division by 0 is undefined, $x \neq 10$. If $x > 10$, we have

$$\frac{|x-10|}{x-10} = \frac{x-10}{x-10} = 1.$$

Similarly, if $x < 10$, we have

$$\frac{|x-10|}{x-10} = \frac{-(x-10)}{x-10} = -1$$

Thus, $y = \dfrac{|x-10|}{x-10}$ is equivalent to $y = \begin{cases} 1 & \text{if } x > 10 \\ -1 & \text{if } x < 10. \end{cases}$

The function is graphed in Figure 1.19 and the domain and range are indicated. Note the open circles above and below $x = 10$, which indicate that the function is not defined at this point.

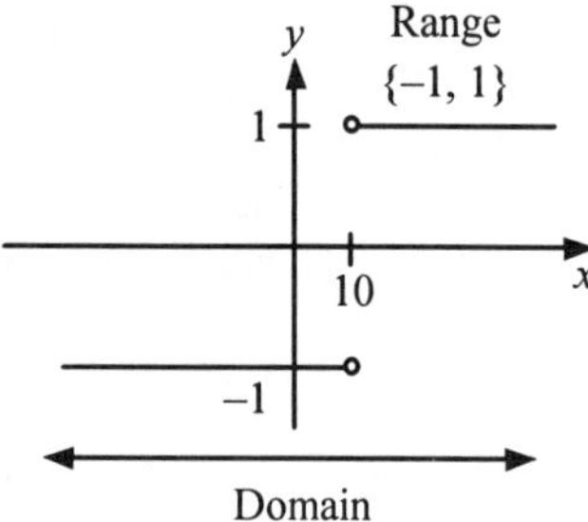

Figure 1.19

PROGRESS CHECK 9

Graph $y = \dfrac{|x+2|}{x+2}$. Indicate the domain and range on the graph. ■

Reading Graphs

It is important to be able to read information about a relation from a graph of a function *f*. Some common objectives associated with reading graphs are outlined next.

1. Read the domain and range.
2. Read function values and values of *x* for which $f(x) = 0$, $f(x) < 0$, or $f(x) > 0$.
3. Read all points where the graph intersects an axis. Such a point is called an **intercept** of the graph.
4. Read intervals where a function is increasing, decreasing, or constant, as outlined in the following chart. For this analysis, let *f* be defined on an interval, and let x_1 and x_2 be any two *x*-values in this interval, with $x_1 < x_2$.

Function on the interval	Possible Graph	Geometric Viewpoint	Algebraic Viewpoint
Increasing	$f(x)$, $f(x_2)$, $f(x_1)$, x_1, x_2, x	From left to right, graph rises.	$x_1 < x_2$ implies $f(x_1) < f(x_2)$. That is, as x increases, y increases.
Decreasing	$f(x)$, $f(x_1)$, $f(x_2)$, x_1, x_2, x	From left to right, graph falls.	$x_1 < x_2$ implies $f(x_1) > f(x_2)$. That is, as x increases, y decreases.
Constant	$f(x)$, $f(x_1) = f(x_2)$, x_1, x_2, x	From left to right, graph is horizontal.	$f(x_1) = f(x_2)$. That is, as x increases, y remains constant.

Example 10: Reading a Graph

Consider the graph of $y = f(x)$ in Figure 1.20

a. What is the domain of *f*?

b. What is the range of *f*?

c. Determine $f(0)$.

d. For what values of *x* does $f(x) = 0$?

e. For what values of *x* is $f(x) < 0$?

f. Solve $f(x) > 0$.

g. Find all intercepts.

h. Find the open intervals on which *f* is increasing, decreasing, or constant.

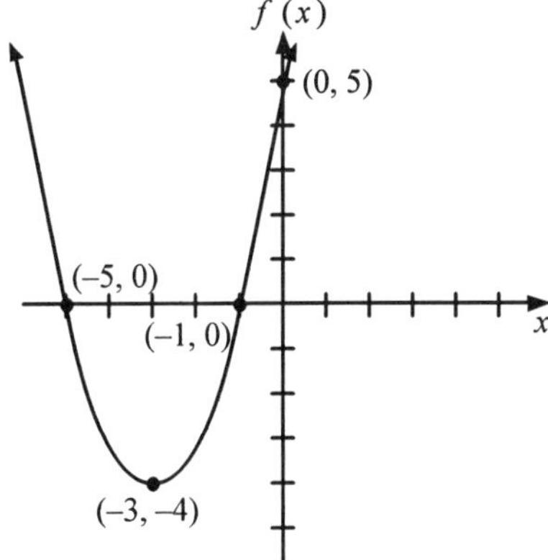

Figure 1.20

Solution:

a. The graph is unbounded to the left and to the right, so the domain is the set of all real numbers, or $(-\infty, \infty)$ in interval notation.

b. The minimum y value is –4, and the graph extends indefinitely in the positive y direction, so there is no maximum value. Thus, the range is $\{y\colon\ y \geq -4\}$, or $[-4, \infty)$ in interval notation.

c. To determine $f(0)$ requires finding the y value where $x = 0$. Using the ordered pair (0,5) gives $f(0) = 5$.

d. From the ordered pairs $(-5, 0)$ and $(-1, 0)$, we know $f(x)$ or y equals 0 when $x = -5$ or $x = -1$. The solution set is therefore $\{-5, -1\}$.

e. The y values are less than zero when the graph is below the x-axis. As indicated in color, $f(x) < 0$ for $-5 < x < -1$, so the solution set is the interval $(-5, -1)$.

f. The y values are greater than zero when the graph is above the x-axis. Thus, $f(x) > 0$ when $x < -5$ or $x > -1$, so the solution set in interval notation is $(-\infty, -5) \cup (-1, \infty)$.

g. An intercept in a graph is a point where the graph intersects an axis, so the intercepts are $(-5, 0)$, $(-1, 0)$, and (0,5).

h. From left to right, the graph falls to the point $(-3, -4)$ and rises thereafter. Thus, the function decreases for the interval $(-\infty, -3)$ and increases for the interval $(-3, \infty)$.

PROGRESS CHECK 10

Answer the questions in Example 10 for the graph of $y = f(x)$ in Figure 1.21. ■

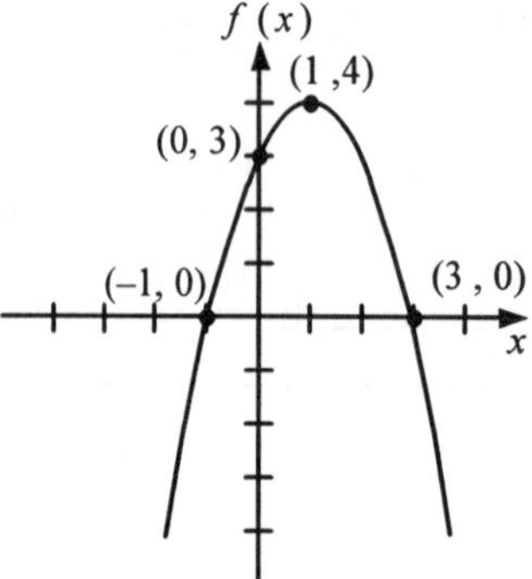

Figure 1.21

In conclusion, here are some important ideas to keep in mind about functional notation.

Functional Notation Concepts

1. If $y = f(x)$ and a is in the domain of f, then $f(a)$ means the value of y when $x = a$. Thus, evaluating $f(a)$ often requires nothing more than a *substitution* of the value a for x.
2. $f(a)$ is a y value, a is an x value. Hence, ordered pairs for the function defined by $y = f(x)$ all have the form $(a, f(a))$.
3. In functional notation we use the symbols f and x more out of custom than necessity, and other symbols work just as well. The notations $f(x) = 2x$, $f(t) = 2t$, $g(y) = 2y$, and $h(z) = 2z$ all define exactly the same function if x, t, y, and z may be replaced by the same numbers.

EXPLORE 1.2

Many calculators include the ability to enter inequality symbols, often through a key or menu marked Test. If your calculator can express inequalities, try the following explorations. The results may be best seen in dot mode.

1. Graph each of the following. In each case, first predict the calculator output. Then explain the basis for the graph that results.

   a. $y = (x > 2)$ **b.** $y = (x > 3)(x < 5)$ **c.** $y = (x > 2) + (x > 3)$

2. Find an expression that may be set equal to y on a graphing calculator to produce this graph.

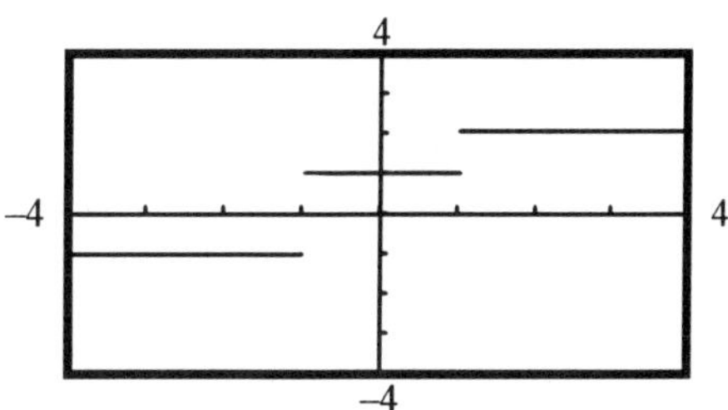

EXERCISES 1.2

1. If $f(x) = x - 2$, find $f(0)$, $f(-1)$, and $f(4)$.

2. If $g(t) = t - 2$, find $g(0)$, $g(-1)$, and $g(4)$.

3. If $h(x) = -2x + 7$, find $h(-2)$, $h(1)$, and $h(5)$.

4. If $f(t) = t^2 + 1$, find $f(-1)$, $f(0)$, and $f(1)$.

5. If $g(x) = 2x^2 - x + 4$, find $g(3)$, $g(0)$, and $g(-1)$.

6. If $h(t) = -t^2$, find $h(5)$, $h(1)$, and $h(-5)$.

7. If $f(x) = \dfrac{x+1}{x-2}$, find $f(-3)$, $f(1)$, and $f(2)$.

8. If $h(r) = \dfrac{r}{r+1}$, find $h(1)$, $h(0)$, and $h(-1)$.

9. If $f(x) = 5$, find $f(1)$, $f(0)$, and $f(a)$.

10. If $g(x) = 2x - 5$, find $g(a)$ and $g(m)$.

11. If $f(x) = 3x$, find $f(a+1)$.

12. If $f(x) = x^2 - x$, find $f(a+3)$.

13. If $f(x) = x + 1$, find
    a. $f(2)$ b. $f(-2)$
    c. Does $f(-2) = -f(2)$?

14. If $g(x) = x^2$, find
    a. $g(2)$ b. $g(-2)$
    c. Does $g(x) = g(-x)$ for all values of x?

15. You are about to throw out an old banjo when a friend tells you it is now "highly collectible." An appraiser says the value of the banjo has increased about 15 percent per year and gives you the function $V = f(x) = (\text{original cost})(1.15)^x$ for its value, where x is the number of years since it was purchased. If your grandmother originally paid \$45 for it, what is it worth today (50 years later)? That is, find and interpret $f(50)$.

16. A function that approximates the value V of a machine purchased for \$30,000 and which decreases in value 10 percent per year, is $V = f(x) = 30{,}000(0.90)^x$, where x is the number of years since the machine was purchased. Find $f(4)$ and $f(7)$ and interpret their meaning.

17. The height y above the water, in meters, of a diver t seconds after stepping off a diving tower 10 m high is given by the function $y = f(t) = -4.9t^2 + 10$. Find and interpret $f(1)$ and $f(1.42)$.

18. During exercise a person's maximum target heat rate is a function of age. The recommended number of beats per minute is given by $y = f(x) = -0.85x + 187$, where x represents age in years and y represents the recommended number of beats per minute. Find and interpret $f(19)$ and $f(38)$.

19. The function $M(p) = 2^p - 1$, where p is a prime number, gives results that are called *Mersenne numbers*. Some Mersenne numbers themselves are prime. Find $M(2)$, $M(3)$, and $M(5)$ and see if they are prime.

20. The function $T(n) = \dfrac{1}{2}n(n+1)$, where n is a positive integer, gives results that are called triangular numbers. Find the first three triangular numbers, $T(1)$, $T(2)$, and $T(3)$. What is the interpretation of $T(100)$?

21. If $f(x)=x+1$ and $g(x)=x^2$, find

a. $f(1)+g(0)$ **b.** $f(-2)-g(-3)$

c. $4g(0)+5f(1)$ **d.** $2f(3)-3g(2)$

e. $f(2)\cdot g(1)$ **f.** $\dfrac{f(1)}{g(2)}$ **g.** $[f(1)]^2$

22. If $f(x)=x+3$ and $g(x)=x^3$, find

a. $f(-1)+g(0)$ **b.** $f(0)+g(-1)$

c. $3f(-1)-g(3)$ **d.** $-2g(-2)-3f(-3)$

e. $f(3)\cdot g(3)$ **f.** $\dfrac{f(3)}{g(-3)}$ **g.** $[f(-4)]^2$

23. If $f(x)=3x^2-7$ and $g(x)=\sqrt{\dfrac{x+7}{3}}$, find

a. $f(0)+g(-7)$ **b.** $f(2)-g(5)$

c. $f(1)\cdot g(-4)$ **d.** $\dfrac{f(3)}{g(20)}$

24. If $f(x)=(x+1)^2$ and $g(x)=x^2+1$, find

a. $f(-1)+g(-1)$ **b.** $f(2)-g(2)$

c. $f(3)\cdot g(3)$ **d.** $\dfrac{g(-2)}{f(-2)}$

In Exercise 25 and 26, find

a. $f(2)$ **b.** $f(3)$ **c.** $f(2+3)$

d. Does $f(2+3)=f(2)+f(3)$?

e. $f(a)$ **f.** $f(b)$ **g.** $f(a+b)$

h. Does $f(a+b)=f(a)+f(b)$?

25. $f(x)=x+1$ **26.** $f(x)=5x$

27. If $f(x)=x+4$ show that $f(a+b)\neq f(a)+f(b)$ for all values of a and b.

28. If $f(x)=-12x$ show that $f(a+b)=f(a)+f(b)$ for all values of a and b.

29. If $f(x)=x^3$ show that $f(a+b)\neq f(a)+f(b)$ for all values of a and b.

30. If $f(x)=x^2$ show that $f\left(\dfrac{a}{b}\right)=\dfrac{f(a)}{f(b)}$ for all values of a and b except $b=0$.

31. If $f(x)=2x^2-1$, find

a. $f(x+h)$ **b.** $f(x+h)-f(x)$

c. $\dfrac{f(x+h)-f(x)}{h}$, if $h\neq 0$

32. If $f(x)=1-x$, find

a. $f(1+h)$ **b.** $f(1+h)-f(1)$

c. $\dfrac{f(1+h)-f(1)}{h}$, if $h\neq 0$

In Exercises 33–40 write the difference quotient for each function in simplest form (see Example 5)

33. $f(x)=x^2+x$ **34.** $f(x)=x^2+4x$

35. $f(x)=7x-5$ **36.** $f(x)=2x+3$

37. $g(x)=2$ **38.** $g(x)=7$

39. $f(x)=1-x^2$ **40.** $f(x)=3-2x^2$

41. A tax bill on certain luxury items is computed as \$400 plus 28 percent of the excess in price over \$1000. Find a rule that expresses the tax (t) as a function of the price (p) of the item.

42. For a week's work on a construction job, a carpenter charges \$680 plus \$27 per hour for any time over 40 hours. Find a rule that expresses the charge (c) as a function of the hours worked (h).

43. Express the monthly cost (c) of an electric bill in terms of the number (n) of electrical units purchased (the unit of measure is the kilowatt-hour) if the customer used no more than 48 units and the electric company has the following rate schedule:

Amount	Charge
First 12 units or less	\$5.25
Next 36 units at	12.82 cents/unit

44. Express the cost (c) of a phone call in terms of the length (m) of the call in minutes according to this table.

Length	Charge
First minute	\$0.58
Each additional minute	\$0.29

45. An instruction for income tax in Vermont reads as follows:

> Vermont's base personal income tax rate … is 28% of a taxpayer's federal income tax liability. In addition to the base rate, there is a surtax of 3% on federal tax liability between \$3,400 and \$13,100, and a 6% surtax on federal income tax over \$13,100.

Using x to represent the federal tax and s to represent the state tax, express s as a function of x by a piecewise function that has 3 pieces.

46. An electric utility company charges \$16 per month plus 10 cents for every kilowatt-hour from 150 to 300 and 15 cents for every kilowatt-hour over 300. Express this relationship as a piecewise function with 3 pieces, using c for the monthly cost and x for the kilowatt-hours of electricity used.

In Exercises 47–56 find the given function values.

47. $f(x)=\begin{cases} x-1 & \text{if } x<0 \\ -1 & \text{if } x>0 \end{cases}$

a. $f(-1)$ **b.** $f(1)$ **c.** $f(0)$

48. $f(x)=\begin{cases} 1 & \text{if } x\ge 0 \\ -1 & \text{if } x<0 \end{cases}$

a. $f(3)$ **b.** $f(0)$ **c.** $f(-3)$

49. $f(x)=\begin{cases} -1 & \text{if } x>0 \\ 0 & \text{if } x=0 \\ 1 & \text{if } x<0, \end{cases}$

a. $f(1)$ **b.** $f(0)$
c. $f(-1)$ **d.** $f(0.0001)$

50. $h(x)=\begin{cases} x & \text{if } 0\le x\le 1 \\ -1 & \text{if } x>1, \end{cases}$

a. $h(3)$ **b.** $h(1)$ **c.** $h\left(\frac{1}{2}\right)$
d. $h(0)$ **e.** $h(-3)$

51. $h(x)=\begin{cases} 25 & \text{if } 0\le x\le 4 \\ 25+5(x-4) & \text{if } x>4 \end{cases}$

a. $h(3)$ **b.** $h(1)$ **c.** $h(-1)$
d. $h(5)$ **e.** $h(4)$

52. $g(x)=\begin{cases} 30 & \text{if } 0\le x\le 10 \\ 30+6(x-10) & \text{if } x>10 \end{cases}$

a. $g(2)$ **b.** $g(10)$ **c.** $g(11)$
d. $g(6.4)$ **e.** $g(12.5)$

53. $f(x)=\begin{cases} 0 & \text{if } 0\le x<1 \\ 1 & \text{if } 1\le x<2 \\ 2 & \text{if } 2\le x<3 \\ \vdots & \\ n & \text{if } n\le x<n+1 \quad \begin{pmatrix}\text{where } n \text{ is a} \\ \text{positive integer}\end{pmatrix} \\ \vdots & \end{cases}$

a. $f(0)$ **b.** $f\left(\frac{1}{2}\right)$
c. $f(1)$ **d.** $f(1.8)$
e. $f(-1)$
f. Explain what this function does.
g. Graph the function.

54. $f(x)=\begin{cases} 1 & \text{if } 0.5\le x<1.5 \\ 2 & \text{if } 1.5\le x<2.5 \\ 3 & \text{if } 2.5\le x<3.5 \\ \vdots & \\ n & \text{if } n-0.5\le x<n+0.5 \quad \begin{pmatrix}\text{where } n \text{ is a} \\ \text{positive integer}\end{pmatrix} \\ \vdots & \end{cases}$

a. $f(.5)$ **b.** $f(.6)$
c. $f(1.12)$ **d.** $f(3.64)$
e. $f(0)$
f. Explain what this function does.
g. Graph the function.

55. The charge for the first 3 minutes of a Monday station-to-station call from New York to Los Angeles depends on the time of day when the call is made. The following formula table indicates the charge in terms of 24-hour time (that is, 1700 hours is equivalent to 5 P.M).

$$\text{Charge}=\begin{cases} \$0.85 & 0 \text{ hours}\le t<0800 \text{ hours} \\ \$1.45 & 0800 \text{ hours}\le t<1700 \text{ hours} \\ \$0.85 & 1700 \text{ hours}<t\le 2400 \text{ hours} \end{cases}$$

Graph this function.

56. Because of the need for energy conservation, the local power company advertises ways to "save a watt." To this end, it calculates the efficiency to air conditioners by finding the ratio between the cooling ability of the machine (Btu) and the amount of watts of electricity that it requires (that is, efficiency = Btu/watt). The rating in terms of the efficiency (e) of the air conditioner is as follows:

$$\text{Rating}=\begin{cases} 0(\text{flunk}) & \text{if } 0\le e<6 \\ 1(\text{pass}) & \text{if } 6\le e<8 \\ 2(\text{good}) & \text{if } 8\le e<10 \\ 3(\text{very good}) & \text{if } 10\le e. \end{cases}$$

Graph this function.

In Exercises 57–68 graph the function and indicate the domain and range of the function on the graph.

57. $g(x)=\begin{cases} 1 & \text{if } x\ge 0 \\ -1 & \text{if } x<0 \end{cases}$

58. $h(x)=\begin{cases} x & \text{if } x\ge 1 \\ 1 & \text{if } x<1 \end{cases}$

59. $f(x)=\begin{cases} x & \text{if } -2\le x<0 \\ 0 & \text{if } x=0 \\ -x & \text{if } x>0 \end{cases}$

60. $y = \begin{cases} -x & \text{if } 0 < x < 3 \\ 2 & \text{if } x = 0 \\ x+1 & \text{if } x < 0 \end{cases}$

61. $y = \begin{cases} x^2 & \text{if } 0 \le x \le 1 \\ x & \text{if } x > 1 \end{cases}$

62. $f(x) = \begin{cases} -x & \text{if } x \le 1 \\ -x^2 & \text{if } x > 1 \end{cases}$

63. $g(x) = \begin{cases} 1 & \text{if } x < 0 \\ x^2+1 & \text{if } x \ge 0 \end{cases}$

64. $h(x) = \begin{cases} |x| & \text{if } x < 0 \\ \sqrt{x} & \text{if } x \ge 0 \end{cases}$

65. $y = \dfrac{|3x+6|}{3x+6}$

66. $y = \dfrac{1-x}{|1-x|}$

67. $y = \dfrac{|3x|}{x}$

68. $y = \dfrac{2x}{|2x|}$

In Exercises 69 and 70 answer these questions by interpreting the graph of $y = f(x)$.

a. What is the domain of the function?
b. What is the range of the function?
c. True or False? The function is never decreasing?
d. For how many values of x does $f(x) = 0$?
e. True or false? $f(0) > 0$.

69.

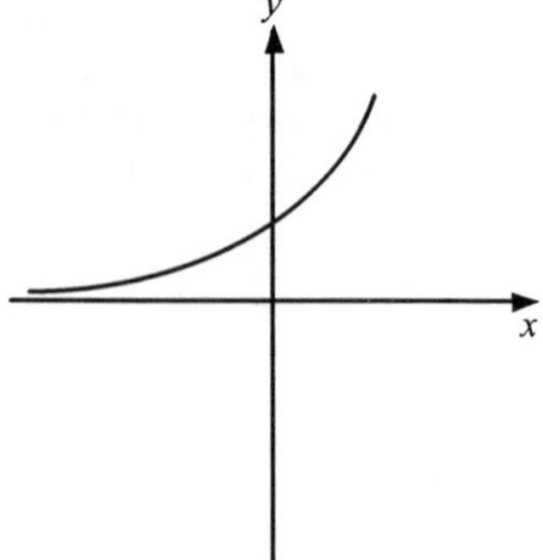

70.

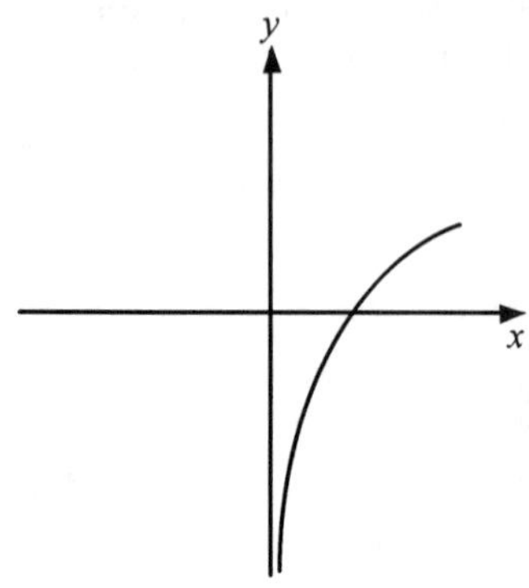

71. Consider the graph of $y = f(x)$ given here.

a. What is the domain of the function?
b. What is the range of the function?
c. Determine $f(0)$ and $f(5)$.
d. For what value(s) of x does $f(x) = 4$?
e. Solve $f(x) = 0$.
f. For what value(s) of x is $f(x) > 0$?
g. For what value(s) of x is $f(x) < 0$?
h. For what value(s) of x is $|f(x)| < 5$?
i. Find all intercepts.
j. Find the open intervals on which f is increasing, decreasing, or constant.

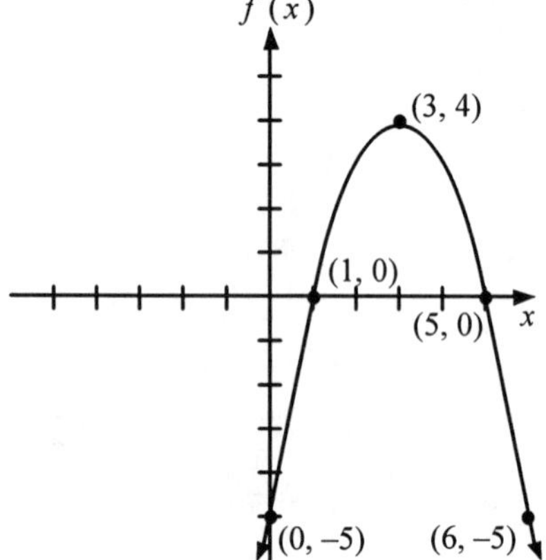

72. Consider the graph of $y = f(x)$ given here.

a. What is the domain of the function?
b. What is the range of the function?
c. Determine $f(a)$ and $f(c)$.
d. Solve $f(x) = a$.
e. For what value(s) of x does $f(x) = 0$?
f. For what value(s) of x is $f(x) < 0$?
g. For what value(s) of x is $f(x) > 0$?
h. For what value(s) of x is $|f(x)| < a$?
i. Find all intercepts.
j. Find the open intervals on which f is increasing, decreasing, or constant.

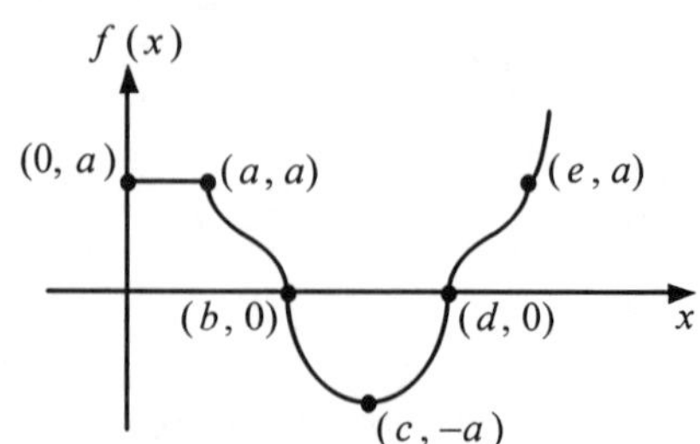

73. The bracket notation $[x]$ means the greatest integer less than or equal to x. For example $[4.3]=4$, $[\pi]=3$, $\left[-\frac{1}{2}\right]=-1$ and $[2]=2$. If $f(x)=[x]$, which is called the **greatest integer function**, find

a. $f\left(\frac{1}{2}\right)$ **b.** $f(\sqrt{2})$

c. $f(-1.4)$ **d.** $f(-\pi)$

e. the graph of this function for $-2 \le x < 3$

Note Some calculators have this function built in where $[x]$ is entered as $\text{int}(x)$.

74. The sign function, often written as SIGN(x) is defined as follows.

$$\text{SIGN}(x)=\begin{cases}1 & \text{if } x>0\\ 0 & \text{if } x=0\\ -1 & \text{if } x<0\end{cases}$$

Let $f(x)=\dfrac{\text{SIGN}(x)}{2}+\dfrac{1}{2}$. Find $f(2)$, $f(-2)$, $f(0)$ and show the graph of *f* for $-5 \le x < 5$.

75. Use the greatest integer function (See Exercise 73) to find a formula for postage (p) in terms of weight (x) given that postage is 32 cents for the first ounce plus 23 cents for each additional ounce.

76. An interesting function based on those defined in Exercises 73 and 74 is $f(x)=[x]\text{SIGN}(x)$. Find $f(2)$, $f\left(-\frac{1}{2}\right)$, $f\left(\frac{1}{2}\right)$ and show the graph of *f* for $-10 \le x \le 10$.

77. This graph describes the distance along a straight road that a traveler has come from home, as time goes by.

a. What does $f(1.5)$ represent?

b. Explain what is happening to the traveler on each of these intervals: (0,1); (1,2); (2,3).

c. How can you tell that the person never reversed direction on the road?

d. How can you tell that the traveler was slowing down during the last half-hour?

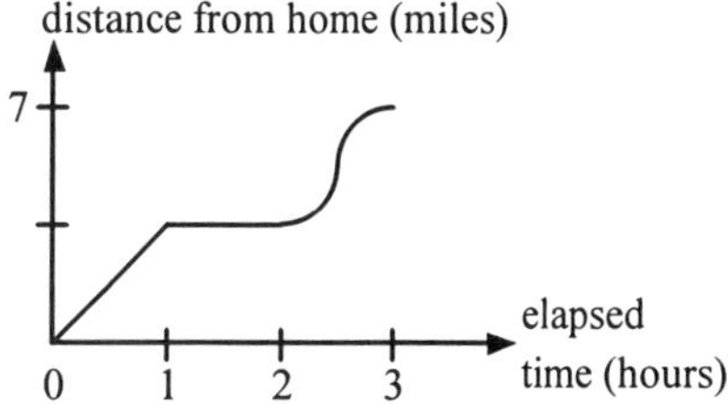

78. Make up an applied problem that corresponds to each graph shown. (See Exercise 77 for an illustration.)

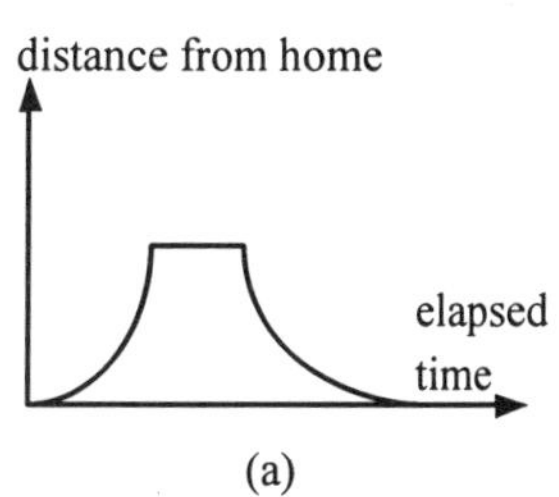

(a)

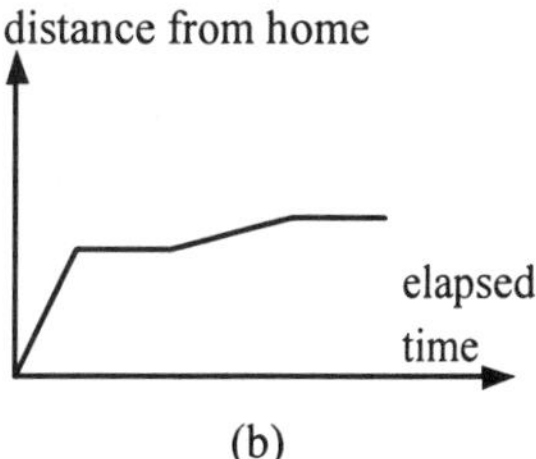

(b)

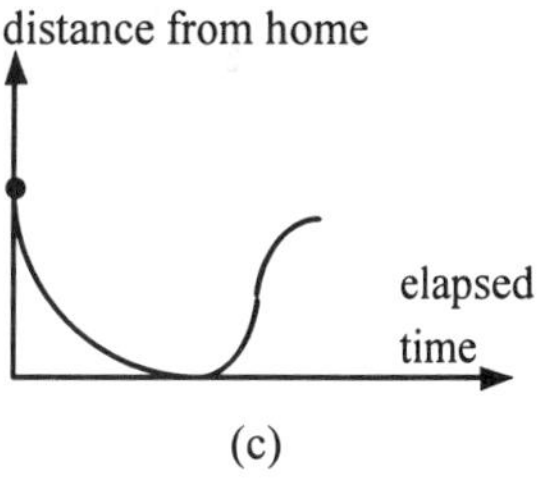

(c)

THINK ABOUT IT 1.2

1. Denote $f[f(a)]$ by $f^{(2)}(a)$, and $f(f[f(a)])$ by $f^{(3)}(a)$.

a. If $f(x)=x^2$, what is $f^{(3)}(a)$? **b.** If $f(x)=-x$, what is $f^{(3)}(a)$?

c. If $f(n)=(-1)^n$, what is $f^{(2)}(100)$?

2. a. Let $f(x)=2x+3$, and let $g(x)=(x-3)/2$. Check that $f[g(4)]$ and $g[f(4)]$ are each equal to 4. Check that $f[g(n)]=n$, and that $g[f(n)]=n$ for all values of n.

b. Let $f(x)=3x-2$. Find a formula for $g(x)$ so that $f[g(n)]$ and $g[f(n)]$ are each equal to n for all values of n.

c. Explain the relationship between *f* and *g* that was illustrated in parts a and b.

3. Give two examples of a function for which $f(-x)=f(x)$ for all real numbers x. In each case, graph *f* and describe how the graph of *f* is related to the y-axis.

4. Graph $y=|x|$ and $y=-|x|$ on the same coordinate system. How are the graphs related? In general, what is the relationship between the graphs of $y=f(x)$ and $y=-f(x)$.
5. A function may be simple to represent by a formula but impossible to graph in a meaningful way. Compare these three functions and describe why the first two are easy to represent graphically but the third is not. Describe the domain and range of each, and give the value of $f(1)$, $f(2)$, $f(1.4)$, and $f(\sqrt{2})$ for each function.

   a. $f(x)=\begin{cases}1 & \text{if } x \geq 0\\ -1 & \text{if } x<0\end{cases}$

   b. $f(x)=\begin{cases}1 & \text{if } x \text{ is even}\\ -1 & \text{if } x \text{ is odd}\end{cases}$

   c. $f(x)=\begin{cases}1 & \text{if } x \text{ is rational}\\ -1 & \text{if } x \text{ is irrational}\end{cases}$

● ● ●

1.3 Graphing Techniques

Photo Courtesy of Camerique, Inc., The Picture Cube

A projectile fired vertically up from the ground with an initial velocity of 128 ft/second will hit the ground 8 seconds later, and the speed (s) of the projectile in terms of the elapsed time t is given by

$$s=32|t-4|.$$

Graph this function.
(See Example 5.)

Objectives

1. Graph a function using translation, reflection, stretching, or shrinking.
2. Prove or classify a function to be even, odd, or neither of these.
3. Complete the graph of an even function or odd function given the graph for $x \geq 0$.

There are many techniques associated with graphing functions. In the previous sections we concentrated primarily on obtaining graphs by using the point-plotting method or by using a grapher. We now mention some useful general approaches that revolve around two central themes: (1) how we can graph variations of familiar functions by somehow adjusting a known curve and (2) how we can use the idea of symmetry to "cut in half" the job of graphing unfamiliar functions. In Exercises 47–52 of Section 1.1 we considered the graphs shown in Figure 1.22. From this point on, it is important that you memorize the graphs of these basic functions. Knowledge of these graphs will also facilitate use of a grapher, particularly when you need to adjust the viewing window to obtain a complete graph.

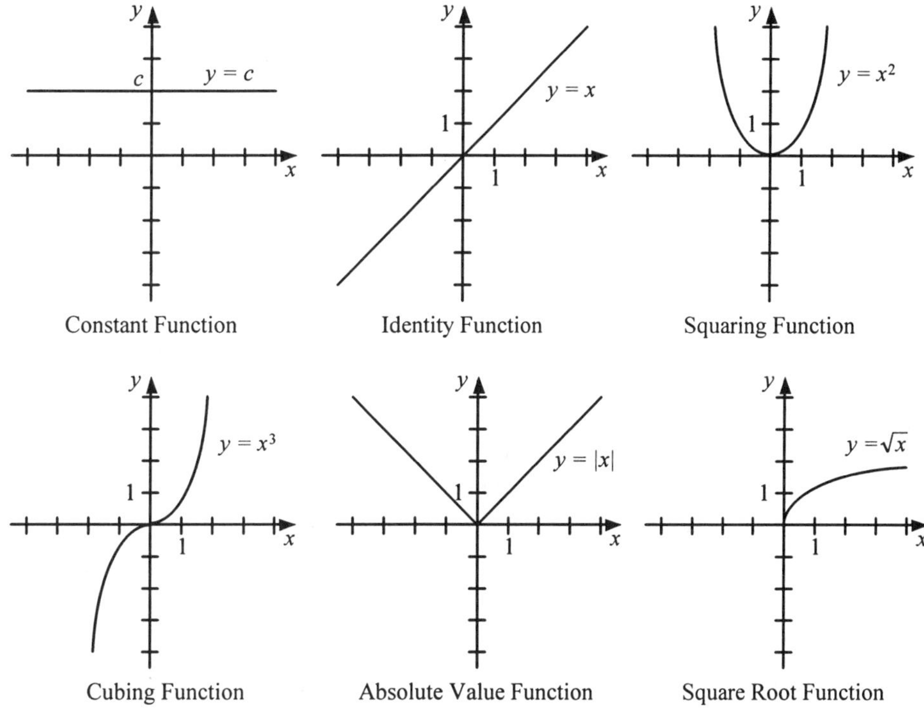

Figure 1.22

Translations

Many graphs may be sketched quickly if you learn to graph variations of basic functions by properly adjusting a familiar graph. For instance, use a grapher or the point-plotting method to graph the four variations of the absolute value function $f(x)=|x|$ that are shown in Figure 1.23. Do you see how to predict the resulting graph from the form of the equation?

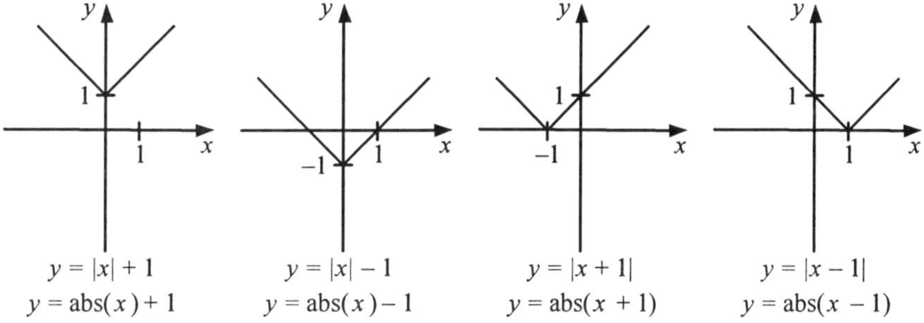

$y=|x|+1$ $y=\text{abs}(x)+1$ | $y=|x|-1$ $y=\text{abs}(x)-1$ | $y=|x+1|$ $y=\text{abs}(x+1)$ | $y=|x-1|$ $y=\text{abs}(x-1)$

Figure 1.23

All four graphs have the basic $\vee$ shape that characterizes the absolute value function. Our job is merely to translate the $\vee$ to the right spot. Note that when we add or subtract the 1 *after* applying the absolute value rule, the effect is to move the $\vee$ up (if adding) or down (if subtracting) a distance of 1 unit. If we add or subtract the 1 *before* applying the absolute value rule, the $\vee$ moves to the left (if adding) or to the right (if subtracting) a distance of 1 unit. Be careful with the left-right shifts—they can be deceiving. The graph of $y=|x-1|$ is 1 unit to the right (not the left) of $y=|x|$, because x must be one larger in $y=|x-1|$ than in $y=|x|$ to produce the same y

value. For example, $y = 0$ when $x = 1$, in $y = |x-1|$, while $y = 0$ when $x = 0$ in $y = |x|$. These ideas generalize to other functions and constants and provide us with the following guidelines.

Vertical and Horizontal Shifts

Let c be a positive constant.

1. The graph of $y = f(x) + c$ is the graph of f raised c units.
2. The graph of $y = f(x) - c$ is the graph of f lowered c units.
3. The graph of $y = f(x + c)$ is the graph of f shifted c units to the left.
4. The graph of $y = f(x - c)$ is the graph of f shifted c units to the right.

Example 1: Graphing Functions Using Translation

Use the graph of $y = x^2$ to graph each of the following functions.
a. $y = x^2 + 3$ **b.** $y = (x-2)^2$ **c.** $y = (x+1)^2 - 2$

Solution: The graph of the squaring function $y = x^2$ is a parabola. The rules above tell us to move this basic shape as follows (see Figure 1.24):

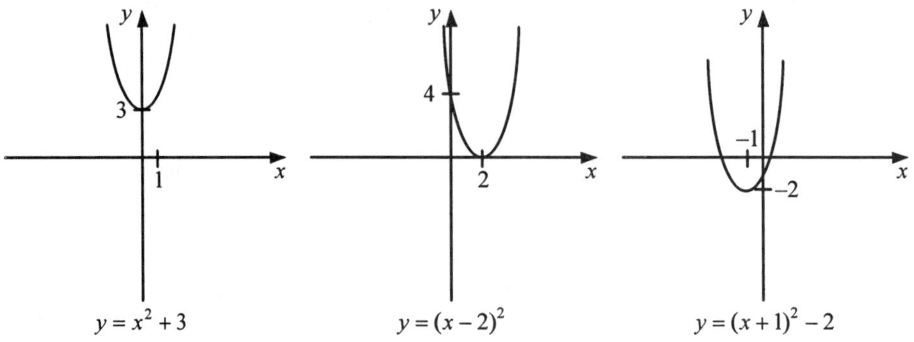

Figure 1.24

a. The constant 3 is added *after* the squaring rule, which means we move the parabola up 3 units.
b. The constant 2 is subtracted *before* the squaring rule, which moves the parabola 2 units to the right.
c. In this case we move the parabola 1 unit to the left and 2 units down.

PROGRESS CHECK 1
Use the graph of $y = x^3$ to graph each of the following functions.
a. $y = x^3 + 3$ **b.** $y = (x-2)^3$ **c.** $y = (x+1)^3 - 2$ ■

Example 2: Using Trace to Explore Translations

Let $f(x) = x^2$, $g(x) = x^2 + 3$, and $h(x) = x^2 - 5$.
a. Graph f, g, and h in the standard viewing window without erasing.
b. Describe how to obtain the graph of g and the graph of h from the graph of f.
c. Use the Trace feature to check that your descriptions are accurate.

Solution:
a. The requested graphs are shown in Figure 1.25.
b. The graph of g is the graph of f shifted 3 units up, while the graph of h is the graph of f shifted 5 units down.
c. To check these descriptions, use the Trace feature to display the coordinates of a point in the graph of f, possibly (0,0). If you move the cursor up to the graph of g, the display reads $x = 0$,

$y = 3$, so this point in g is 3 units higher than the corresponding point in f. Similarly, if you move the cursor down to the graph of h, the display reads $x = 0$, $y = -5$, so this point in h is 5 units lower then the corresponding point in f. These types of results will occur if the cursor initially starts at any point in the graph of f, so the descriptions appear to be accurate.

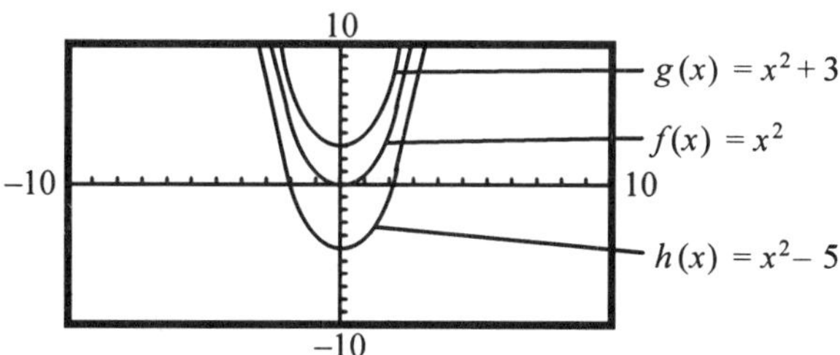

Figure 1.25

Note: On many calculators the up-down arrow keys are used to move from one graph to another when more than one graph is displayed, and the cursor movement is based on the order of the equations in the equation list (not their appearance in the viewing window). On different calculators, the Trace feature will stop at different x-coordinates, but in all cases the vertical distance between the y-coordinates will conform to our general observations. It may appear to you that the graphs in Figure 1.25 tend to get closer as $|x|$ increases, but this is an optical illusion.

PROGRESS CHECK 2

Let $f(x) = |x|$, $g(x) = |x| - 7$, and $h(x) = |x| + 2$ and answer the questions in Example 2. ■

Reflecting, Stretching and Shrinking

Consider Figures 1.26 and 1.27, which illustrate the graphs of $y = |x|$, $y = -|x|$, $y = 2|x|$, and $y = (1/2)|x|$. These graphs show what happens when $|x|$ is multiplied by some constant. The most dramatic effect comes when the sign of the function is changed. As shown in Figure 1.26, switching from $|x|$ to $-|x|$ reflects the graph about the x-axis. Multiplying by a positive constant, such as 2, causes the curve to climb faster, while multiplying by a constant like $1/2$ flattens out the graph (see Figure 1.27). We summarize this pattern in the following rule.

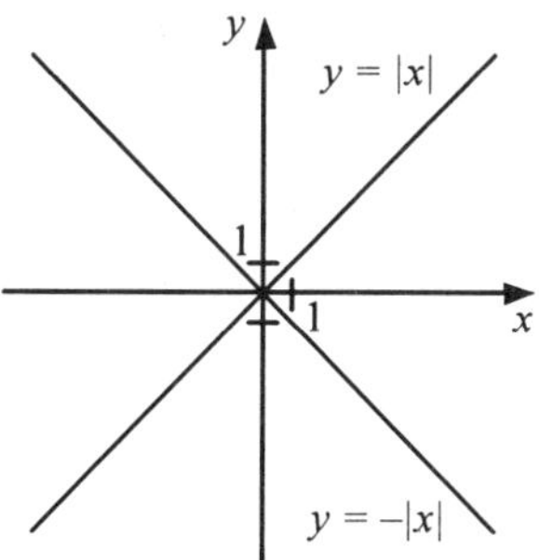

Figure 1.26

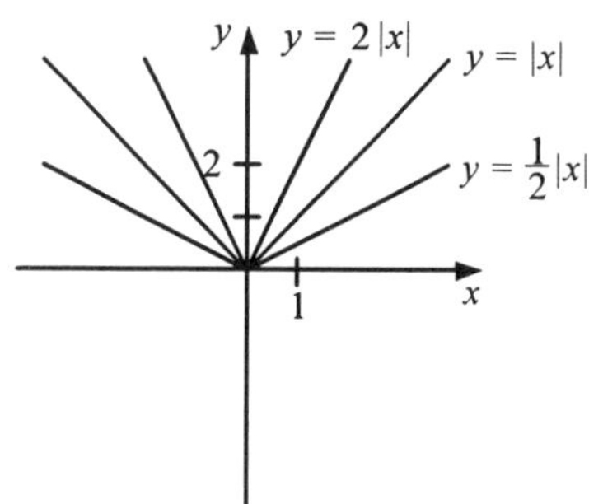

Figure 1.27

To Graph $y = cf(x)$

Reflecting: The graph of $y = -f(x)$ is the graph of f reflected about the x-axis.
Stretching: If $c > 1$, the graph of $y = cf(x)$ is the graph of f stretched by a factor of c.
Shrinking: If $0 < c < 1$, the graph of $y = cf(x)$ is the graph of f flattened out by a factor of c.

Example 3: Combining Reflection and Translation

Use the graph of $y = x^2$ to graph $y = 2 - x^2$.

Solution: To graph $y = -x^2$, we reflect the basic ∪ shape about the x-axis as in Figure 1.28(a). Now $2 - x^2$ is the same as $-x^2 + 2$, so we raise the graph in Figure 1.28(a) up 2 units to get our answer [see Figure 1.28(b)].

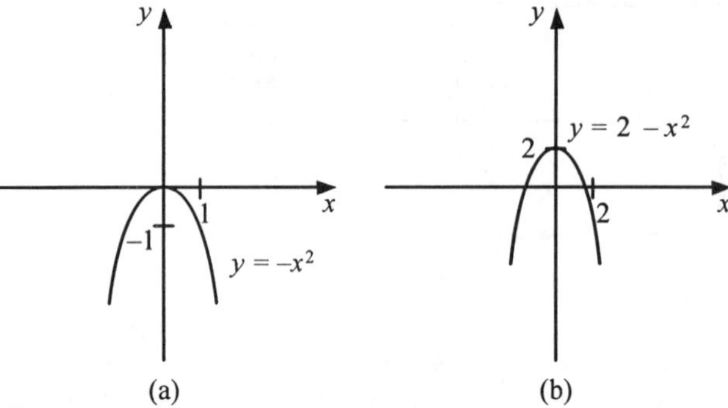

Figure 1.28

PROGRESS CHECK 3

Use the graph of $y = x^2$ to graph $y = 4 - x^2$. ■

Example 4: Graphing Functions Using Stretching or Shrinking

Consider the graph of $y = f(x)$ in Figure 1.29. Use this graph to sketch $y = 2f(x)$ and $y = (1/2)f(x)$.

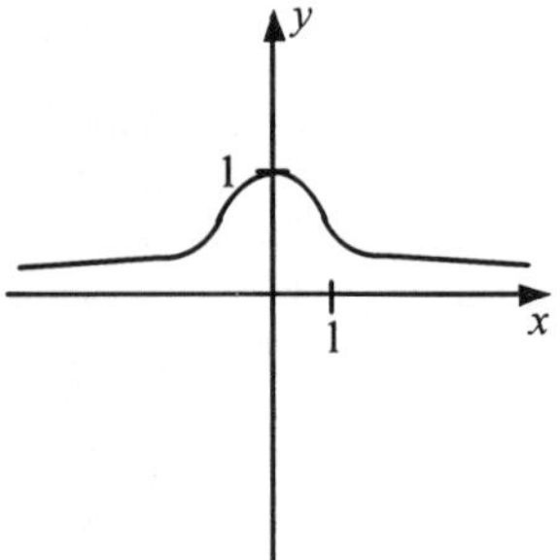

Figure 1.29

Solution: To graph $y = 2f(x)$, we just double each y value in f. Note in particular that the maximum y in f is 1, while the biggest y in $2f$ is 2. Similarly, we graph $y = (1/2)f(x)$ by halving the y values in f. Both graphs are shown in Figure 1.30.

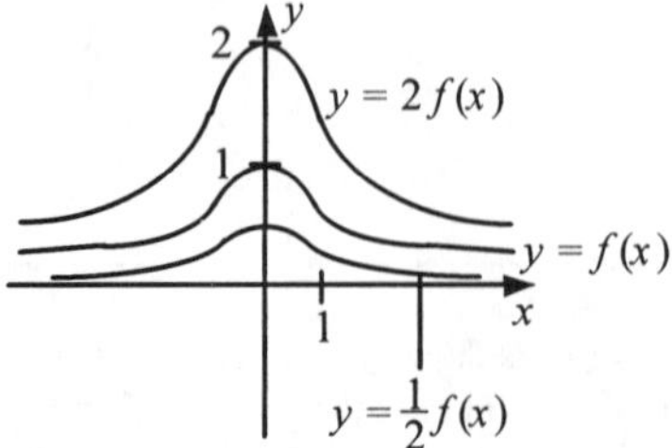

Figure 1.30

PROGRESS CHECK 4

Use the graph of $y = f(x)$ in Figure 1.31 to sketch $y = 2f(x)$ and $y = (1/2)f(x)$. ■

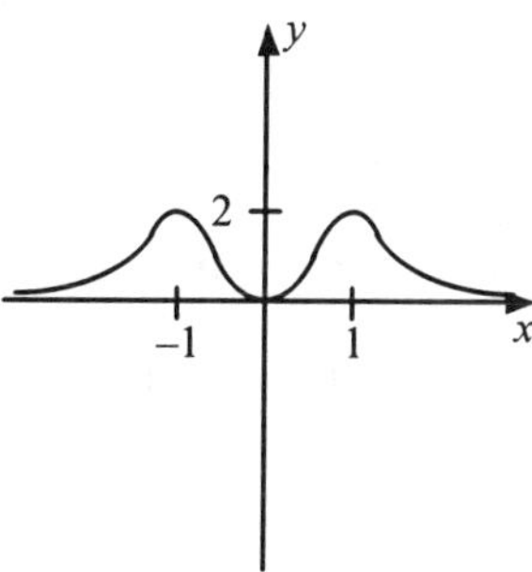

Figure 1.31

Example 5: Combining Translation and Stretching

Solve the problem in the section introduction on page 114.

Solution: To graph $s = 32|t-4|$, shift the graph of $s = |t|$ to the right 4 units, and then stretch the resulting graph by a factor of 32. This graph is shown in Figure 1.32. The given formula is meaningful starting at $t = 0$ seconds and ending at $t = 8$ seconds (when the projectile hits the ground), so the domain is the interval [0 seconds, 8 second]. We then use this domain restriction and identify appropriate units on both axes to graph the function as shown in Figure 1.33.

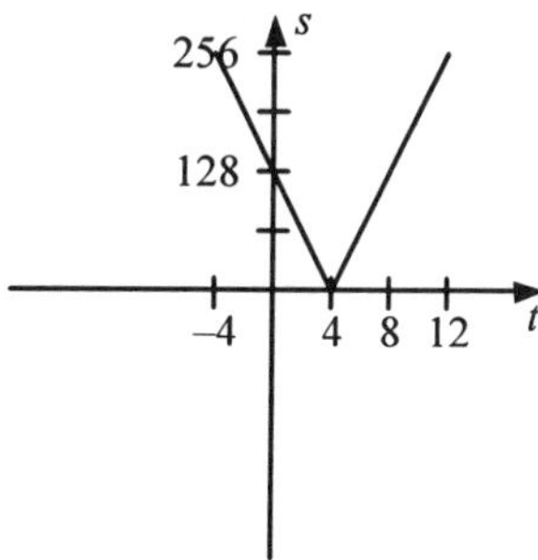

Figure 1.32

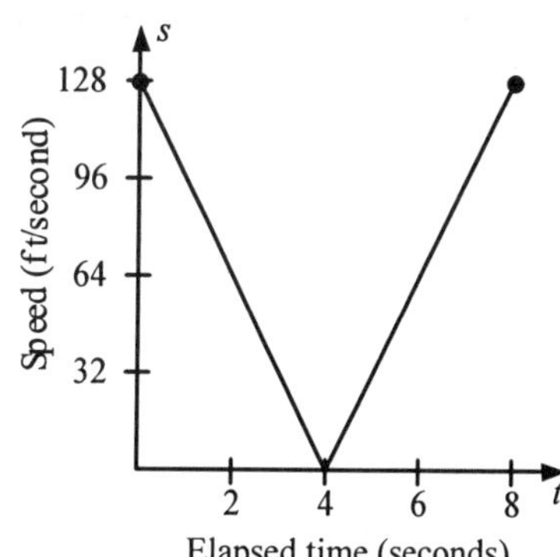

Figure 1.33

Many applications involve graphs that will not be sensible in the standard viewing window, and you should use your knowledge of graphing techniques to set appropriate ranges. In this example you can tell from the discussion above that a meaningful graph is obtained in the viewing window defined by [0,8] by [0, 128]. Figure 1.34 shows a graph that results when these range settings are used.

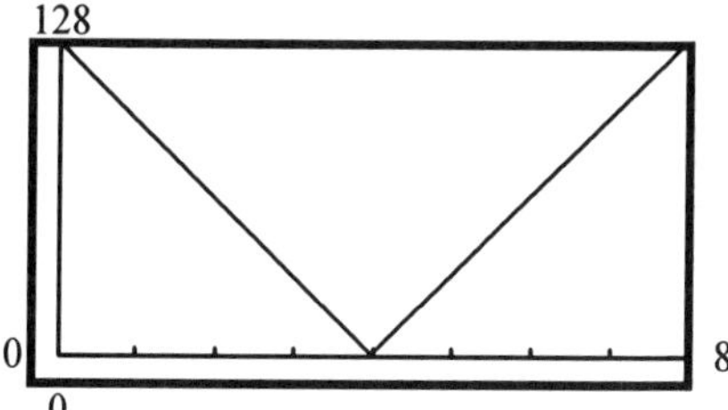

Figure 1.34

PROGRESS CHECK 5

If a stone is dropped from a bridge at a point that is 400 ft above water, then $y = -16t^2 + 400$ gives the height y above water in feet after t seconds have elapsed. Graph this function. ■

Symmetry

Consider the graph of $y = x^2$ in Figure 1.35. Notice that the y-axis divides the parabola into two segments such that if the curve were folded over on this line, its left half would coincide with its right half. In such a case we say $y = x^2$ has **symmetry about the *y*-axis**. When a curve has such symmetry, graphing it is simplified. We have to plot only the curve for $x \geq 0$ and then reflect our result over to the other side of the y-axis to complete the graph.

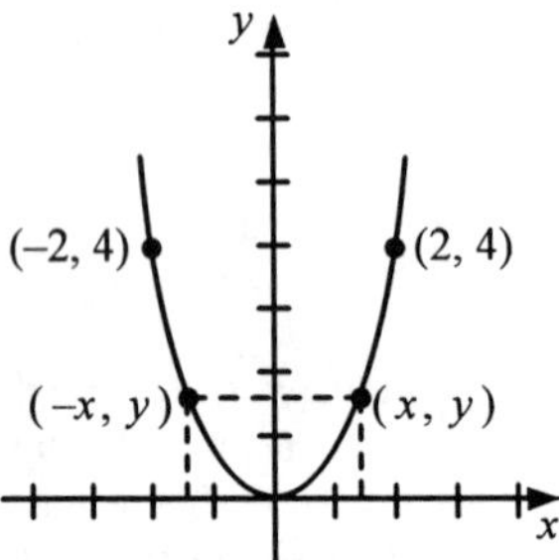

Figure 1.35

Example 6: Using *y*-axis Symmetry

Complete the graph in Figure 1.36 for $x < 0$ if f is symmetric about the y-axis.

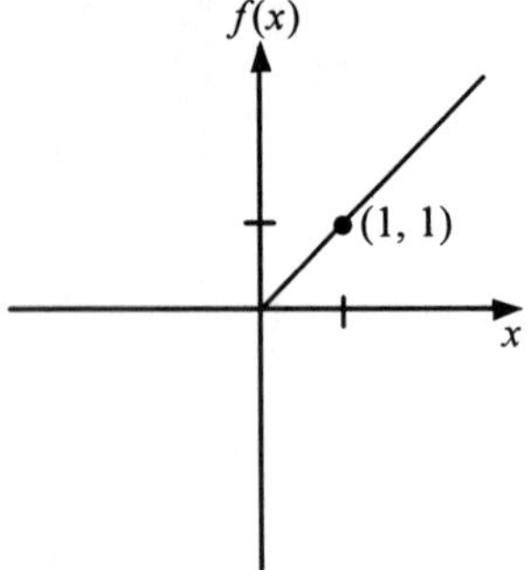

Figure 1.36

Solution: Just reflect the given segment over to quadrant 2 to obtain the graph for $x < 0$. The completed graph is shown in Figure 1.37.

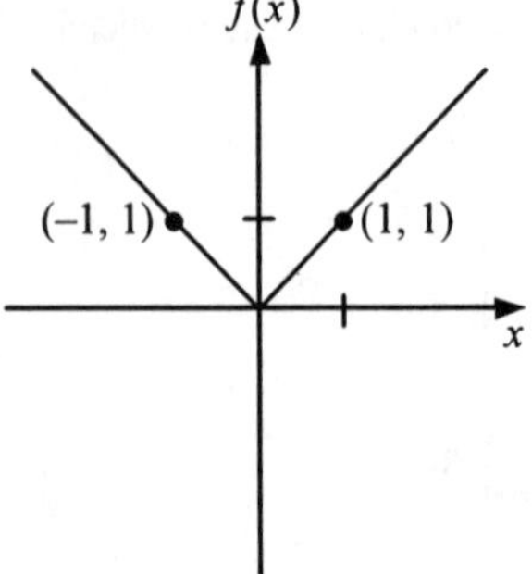

Figure 1.37

PROGRESS CHECK 6
Complete the graph in Figure 1.38 for $x < 0$ if *f* is symmetric about the *y*-axis. ■

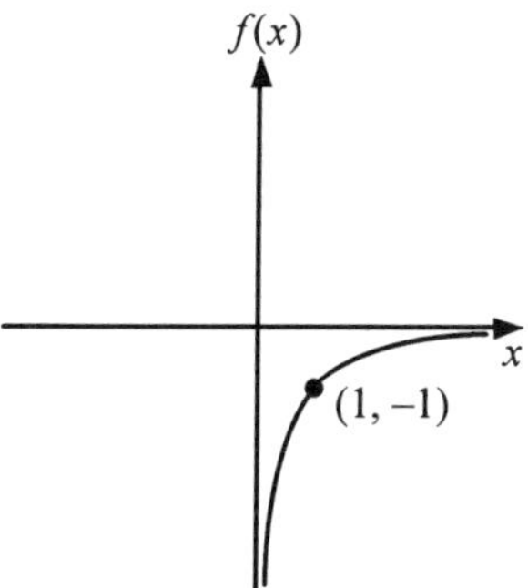

Figure 1.38

Now the question is how can we anticipate *y*-axis symmetry by looking at an equation? Consider Figure 1.35 again and note that *y*-axis symmetry results when *x*-values that are opposite in sign share the same *y* value. For example, in the case of $y = x^2$, values of both 2 and –2 for *x* lead to a *y* value of 4, and in general, both (x, y) and $(-x, y)$ are on the graph. This observation leads us to the following rule for recognizing *y*-axis symmetry.

Even Function

A function *f* is called an *even* function when $f(-x) = f(x)$ for all *x* in the domain of *f*. The graph of an even function is symmetric about the *y*-axis.

Example 7: Proving a Function is Even

Show that the absolute value function $f(x) = |x|$ is an even function.

Solution:

$$f(x) = |x| \text{ and } f(-x) = |-x| = |x| = f(x)$$

Since $f(-x) = f(x)$ for any *x*, the absolute value function is an even function. In case you did not recognize it, Example 6 is the graph of $y = |x|$, which has *y*-axis symmetry.

PROGRESS CHECK 7
Show that the squaring function $f(x) = x^2$ is an even function. ■

Another important type of symmetry is shown in the graph of $y = x^3$ in Figure 1.39(a). In this graph every point (x, y) in the first quadrant has a matching point $(-x, -y)$ in the third quadrant equidistant from the origin. In such a case we say $y = x^3$ has **symmetry about the origin**. Origin symmetry is a little more subtle than y-axis symmetry in that we must reflect the given segment twice in our shortcut graphing approach. As shown in Figure 1.39 (b), we reflect the curve for $x \geq 0$ about both the *x*- and *y*-axes (in either order) to complete the graph.

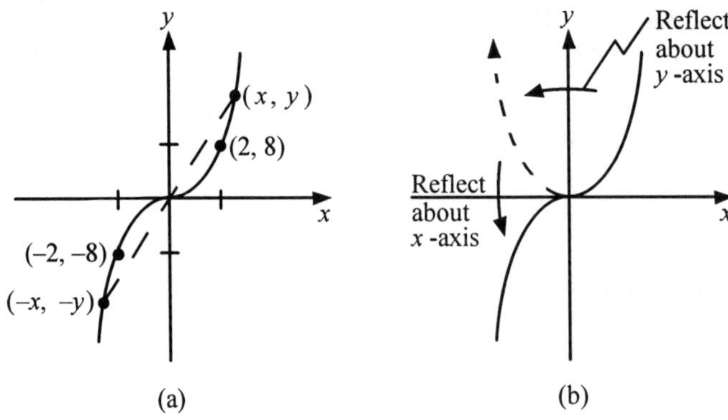

Figure 1.39

Example 8: Using Origin Symmetry

Complete the graph in Figure 1.40 for $x < 0$ if f is symmetric about the origin.

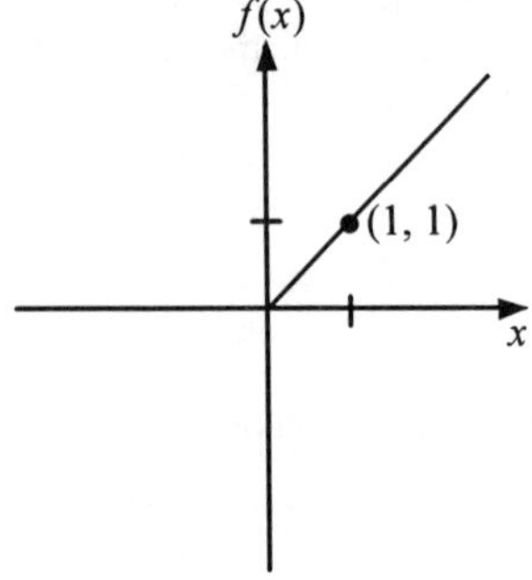

Figure 1.40

Solution: Reflect the given segment over to quadrant 2 and then reflect the result about the x-axis and down to quadrant 3. The completed graph is shown in Figure 1.41.

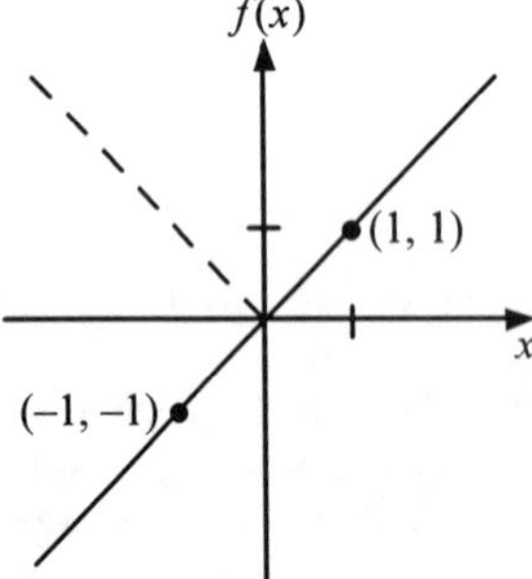

Figure 1.41

PROGRESS CHECK 8

Complete the graph in Figure 1.42 for $x < 0$ if f is symmetric about the origin. ■

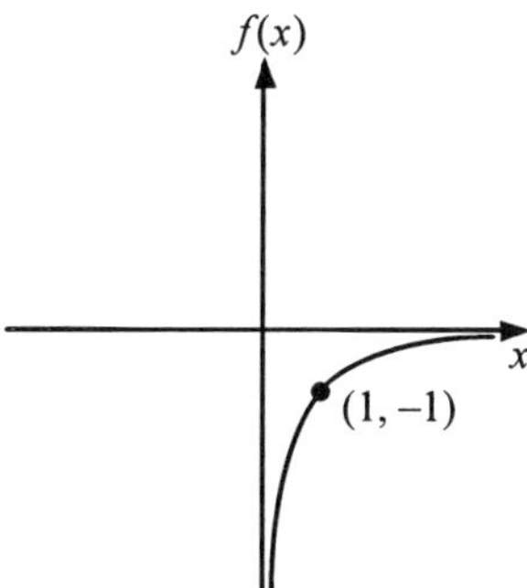

Figure 1.42

We can anticipate origin symmetry because it results when x values that are opposite in sign produce y values that are opposite in sign. For example, in the case of $y = x^3$, values of 2 and –2 for x produce y values of 8 and –8 respectively, and in general, both (x, y) and $(-x, -y)$ are on the graph. Thus, we have the following rule for recognizing origin symmetry.

Odd Function

> A function f is called an **odd** function when $f(-x) = -f(x)$ for all x in the domain of f. The graph of an odd function is symmetric about the origin.

Example 9: Proving a Function is Odd

Show that the identity function $f(x) = x$ is an odd function.

Solution:

$$f(x) = x \text{ and } f(-x) = -x = -f(x)$$

Thus, $f(-x) = -f(x)$ for any x, and the identity function is an odd function. Example 8 is the graph of the identity function that has origin symmetry.

PROGRESS CHECK 9

Show that the cubing function $f(x) = x^3$ is an odd function. ■

Example 10: Using Symmetry to Graph a Function

Is the function $y = (x^2 - 1)/x^3$ an even function, an odd function, or neither? Figure 1.43 is the graph of this function for $x \geq 0$. Complete the graph and check your result with a grapher.

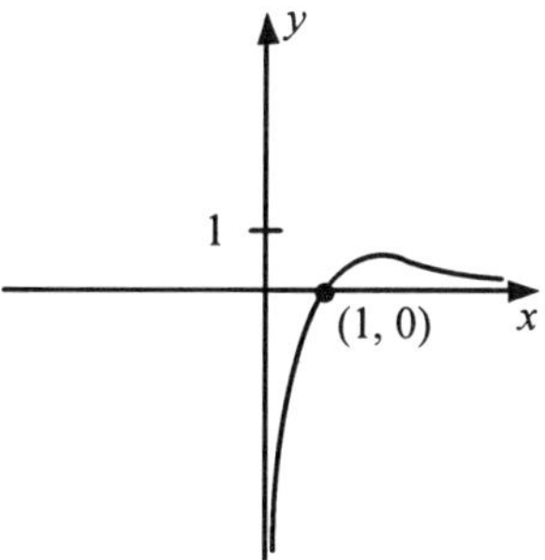

Figure 1.43

Solution:

$$f(x)=\frac{x^2-1}{x^3}$$

and

$$f(-x)=\frac{(-x)^2-1}{(-x)^3}=-\frac{x^2-1}{x^3}=-f(x)$$

Since $f(-x)=-f(x)$, f is an odd function that is symmetric about the origin. Reflecting the given curve about both the x- and y-axes gives the completed graph in Figure 1.44. This result checks with the display in a grapher, as shown in Figure 1.45.

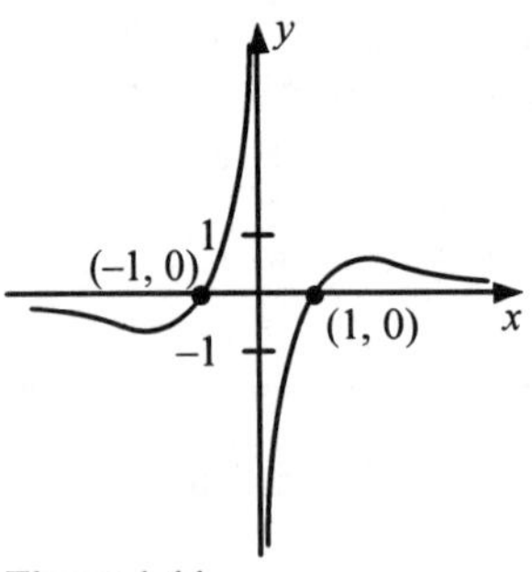

Figure 1.44

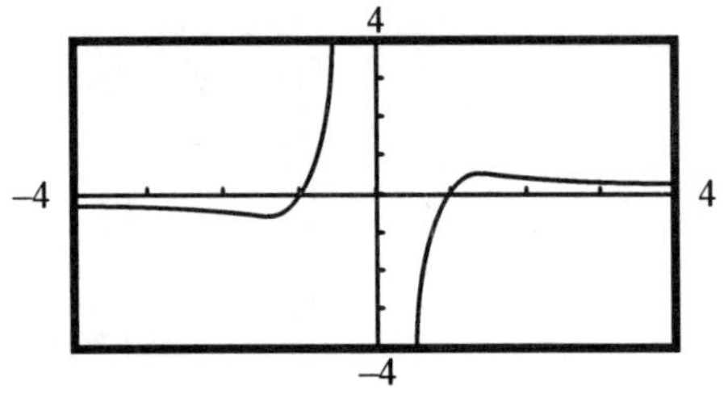

Figure 1.45

PROGRESS CHECK 10

Is the function $y=5x/(x^2+1)$, whose graph is given for $x\geq 0$ in Figure 1.46, an even function, an odd function, or neither? Complete the graph and check your result with a grapher. ■

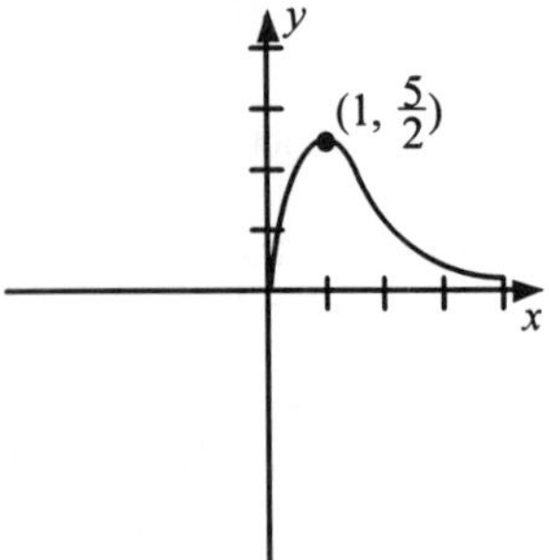

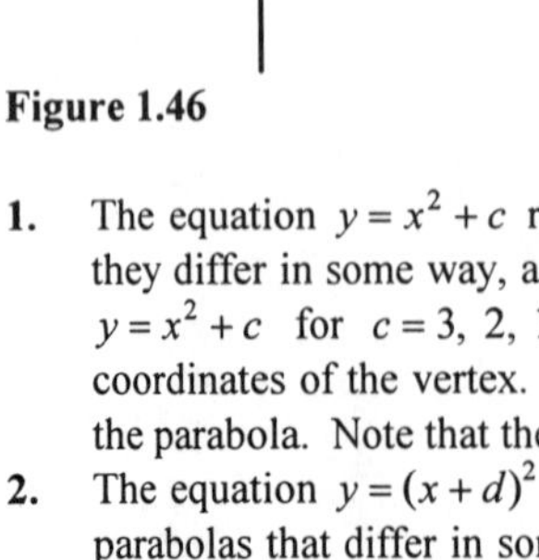

Figure 1.46

EXPLORE 1.3

1. The equation $y=x^2+c$ represents a *family* of functions. All the graphs are parabolas, but they differ in some way, according to the value of c. Use a grapher to examine the graphs of $y=x^2+c$ for $c=3, 2, 1, 0, -1, -2$ and -3. For each of these parabolas give the coordinates of the vertex. Write a few sentences that describe the effect of c on the graph of the parabola. Note that the constant c is added after the *squaring* is carried out.
2. The equation $y=(x+d)^2$ also represents a *family* of functions, and again all the graphs are parabolas that differ in some way, according to the value of d. Use a grapher to examine the graphs of $y=(x+d)^2$ for $d=3, 2, 1, 0, -1, -2$, and -3. For each of these parabolas give

the coordinates of the vertex. Write a few sentences that describe the effect of d on the graph of the parabola. Note that now the constant d is added *before* the squaring is carried out.

3. Given the results of the previous explorations, describe the family of functions defined by $y = (x+d)^2 + c$. Check your answer by looking at the graph of $y = (x+3)^2 + 4$.

EXERCISES 1.3

In Exercises 1–4 use the graph of $y = |x|$ to graph the given function.

1. $y = |x| + 2$ **2.** $y = |x + 3|$

3. $y = |x - 2| - 1$ **4.** $y = |x + 1| + 3$

In Exercises 5–8 use the graph of $y = x^2$ to graph the given function.

5. $y = (x+2)^2$ **6.** $y = x^2 - 3$

7. $y = (x-1)^2 + 2$ **8.** $y = (x+3)^2 - 1$

In Exercises 9–12 use the graph of $y = \sqrt{x}$ to graph the given function.

9. $y = 3 + \sqrt{x}$ **10.** $y = \sqrt{x+5}$

11. $y = \sqrt{x-1} + 1$ **12.** $y = -3 + \sqrt{x+2}$

In Exercises 13 and 14 use the graph of $y = f(x)$ that follows to graph the given function.

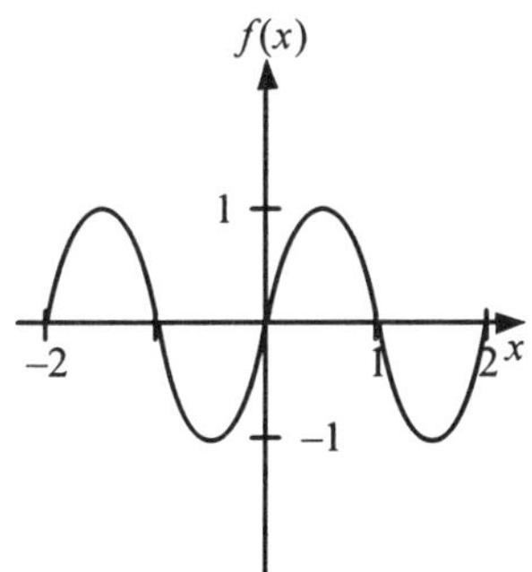

13. $y = f(x) + 1$ **14.** $y = f(x-1)$

In Exercises 15 and 16 find a viewing window that includes all three graphs on the same display, then describe how the graphs in parts **b** and **c** are related to the graph in part **a**. Use the Trace feature to check that your descriptions are accurate.

15. **a.** $y = x^3$ **b.** $y = (x+20)^3$ **c.** $y = x^3 + 15$

16. **a.** $y = \sqrt{x}$ **b.** $y = \sqrt{x+10}$ **c.** $y = \sqrt{x} - 10$

In Exercises 17–20 find equations for each of the graphs shown, which were all determined by translating the graph of $y = x^2$.

17. 

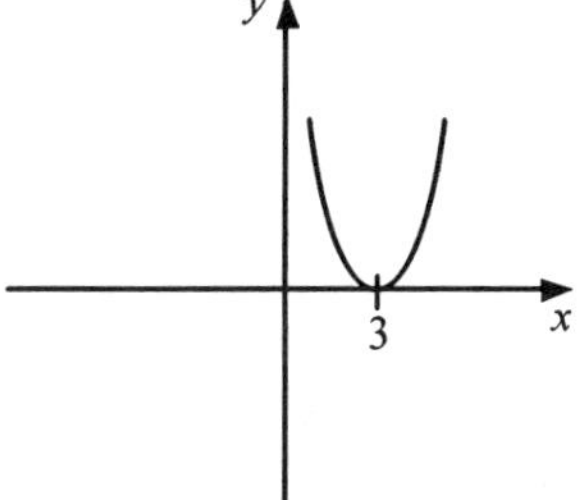

18. 

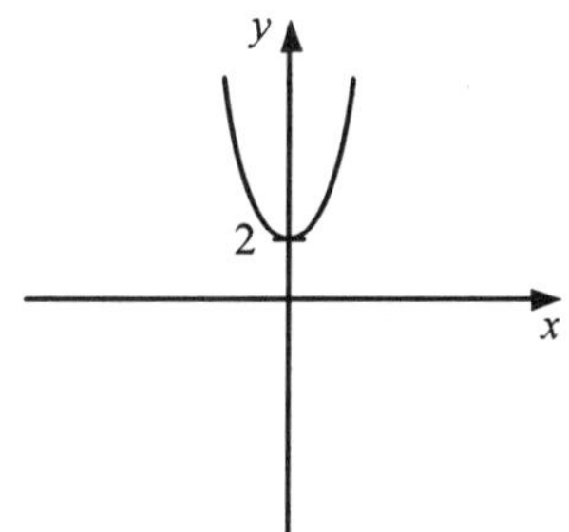

19. 

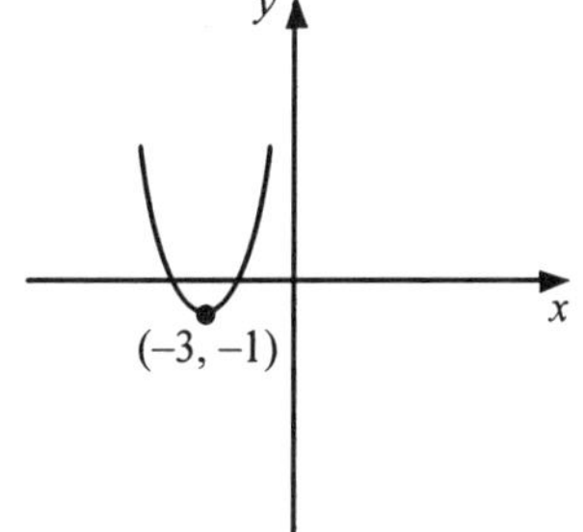

20.

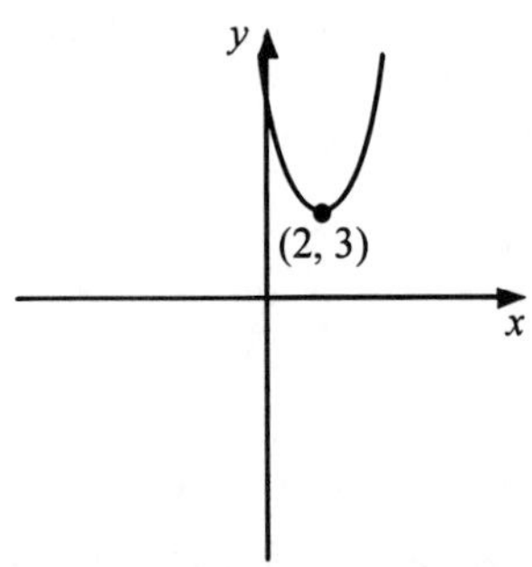

In Exercise 21–24 use the given graph of $y = f(x)$ to name the graphs shown.

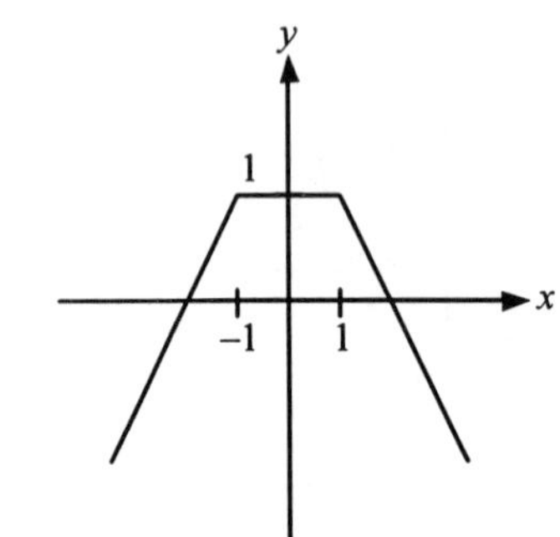

21.

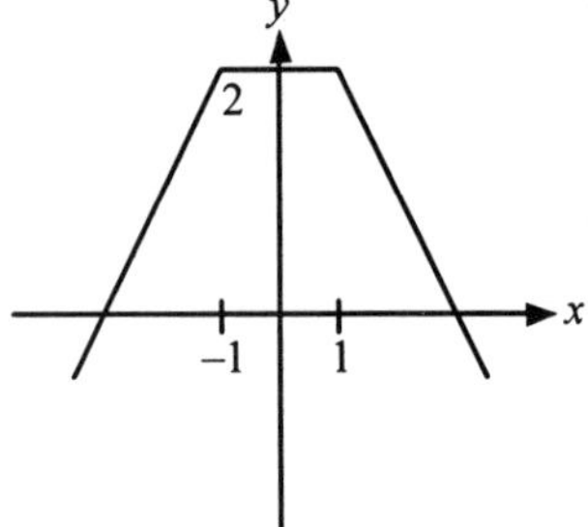

22.

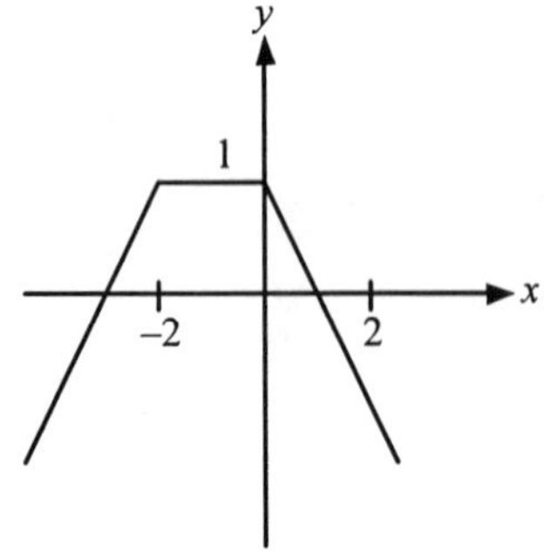

23.

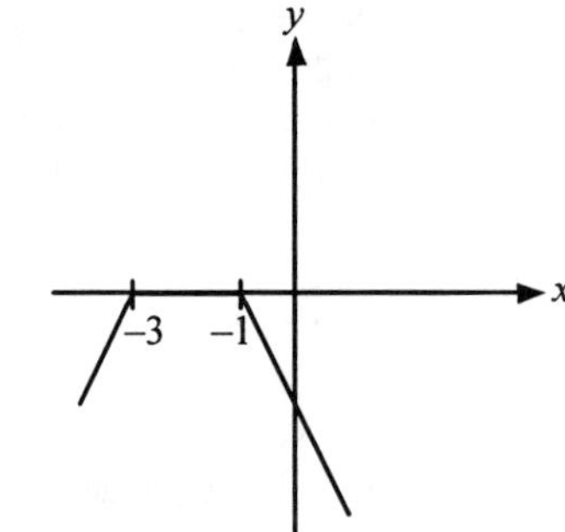

24.

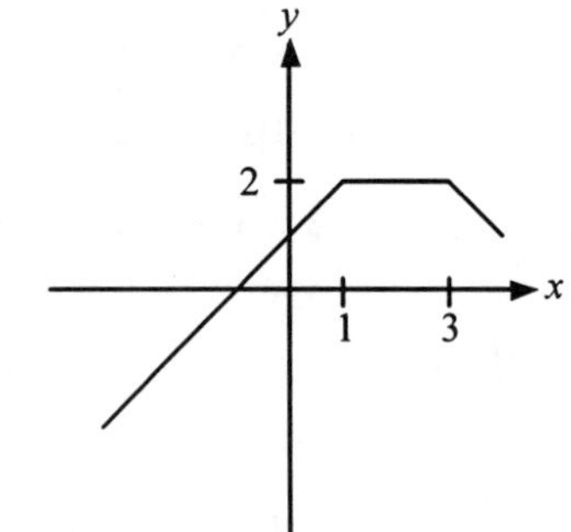

In Exercise 25–28, us the graph of $y = |x|$ to graph the given functions

25. $y = -|x|$ **26.** $y = 2 - |x|$

27. $y = -1 - |x+5|$ **28.** $y = 1 - |x+1|$

In Exercises 29–32 use the graph of $y = x^2$ to graph the given functions.

29. $y = -x^2$ **30.** $y = 3 - x^2$

31. $y = 2 - (x+3)^2$ **32.** $y = -(x-1)^2 - 4$

In Exercises 33–36 use the graph of $y = \sqrt{x}$ to graph the given functions.

33. $y = -\sqrt{x}$ **34.** $y = -1 - \sqrt{x}$

35. $y = -2 - \sqrt{x-2}$ **36.** $y = 2 - \sqrt{x+2}$

In Exercises 37–40 find equations for each of the graphs shown, which were all determined by translating and/or reflecting the graph of $y = x^3$.

37.

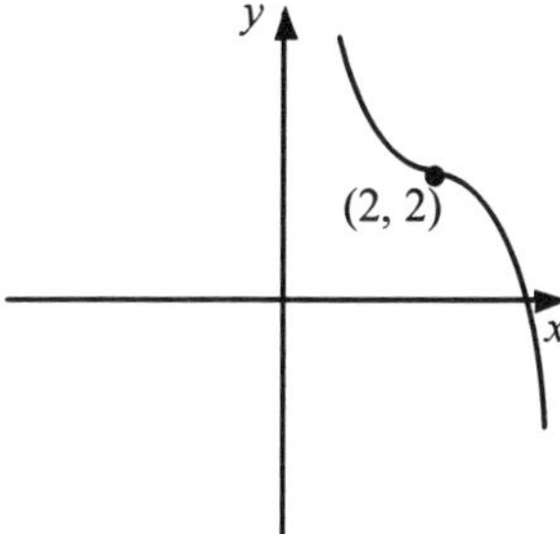

38.

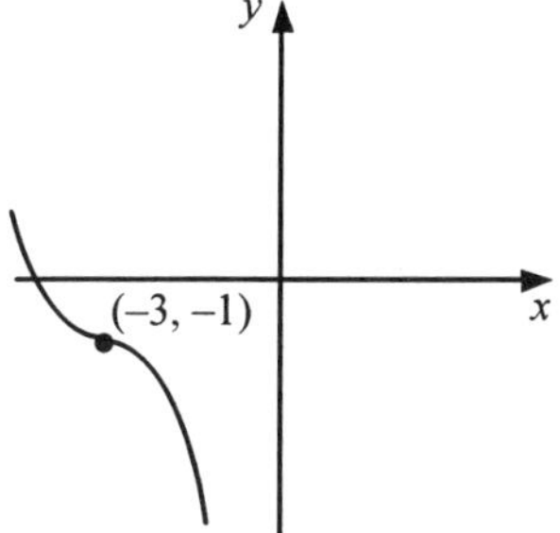

39.

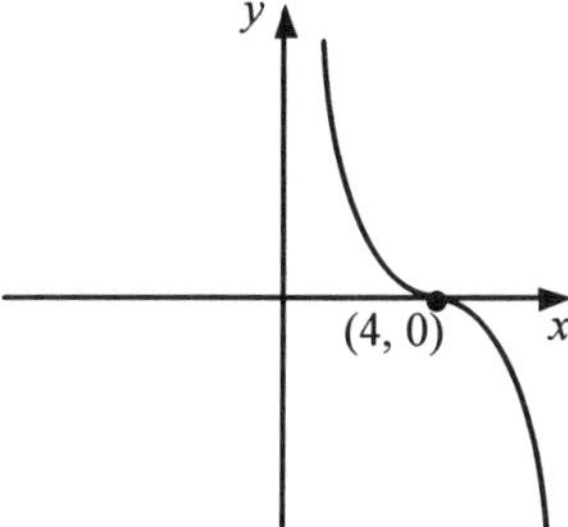

40.

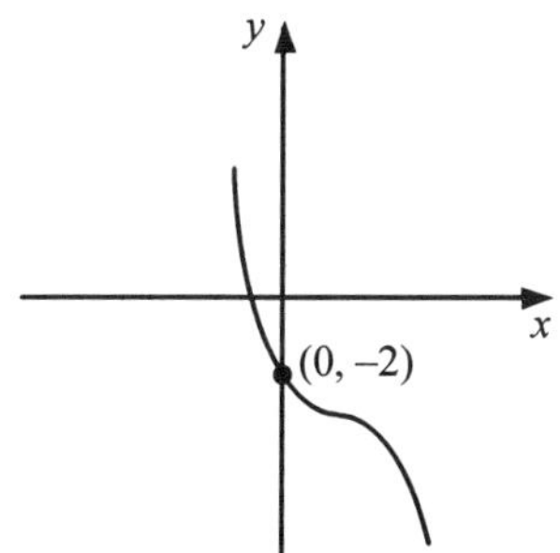

In Exercises 41–44 use the graph of $y = f(x)$ that follows to graph each given function. Describe the relationship of each graph to the given one.

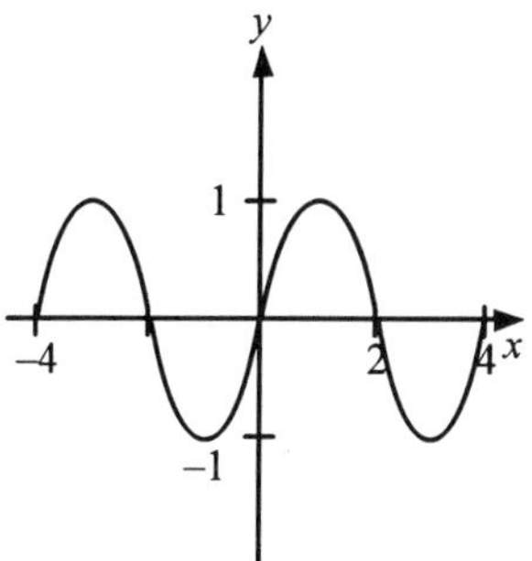

41. $y = f(x) + 2$ **42.** $y = -f(x)$

43. $y = f(x+2)$ **44.** $y = 2 - f(x)$

In Exercises 45–48 use the graph of $y = g(x)$ to name the given graphs.

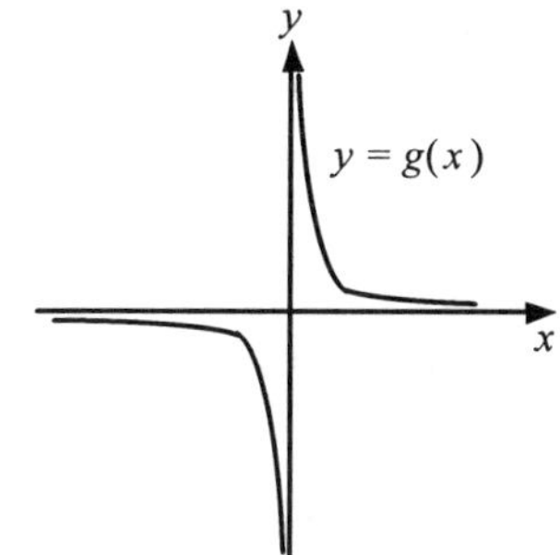

45.

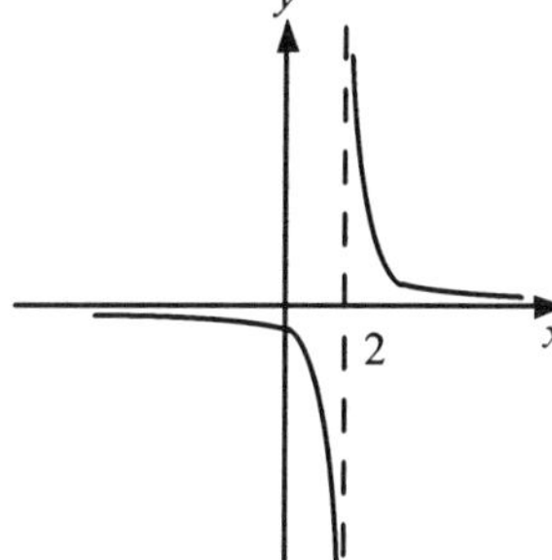

46.

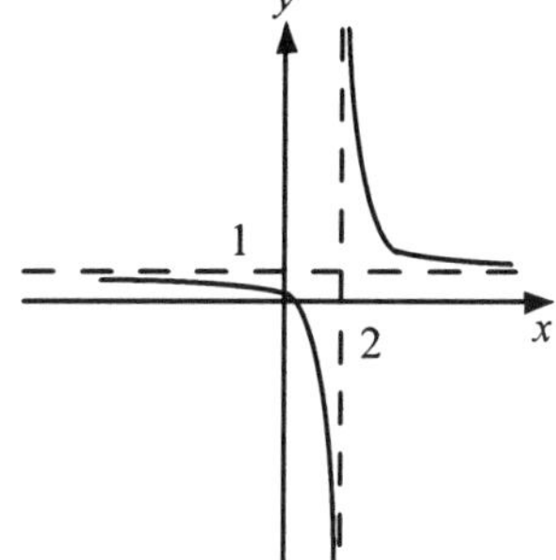

47.

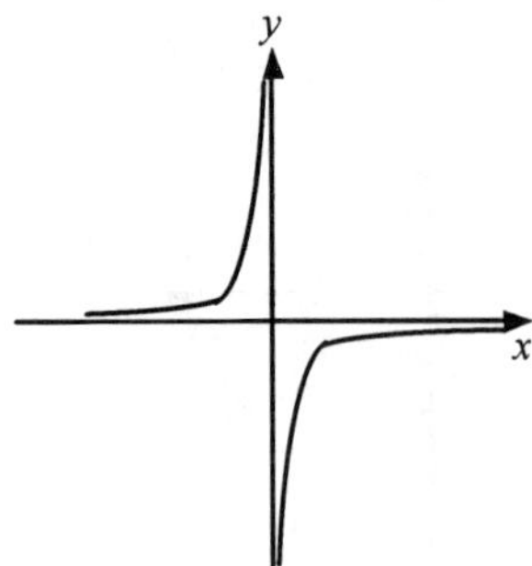

48.

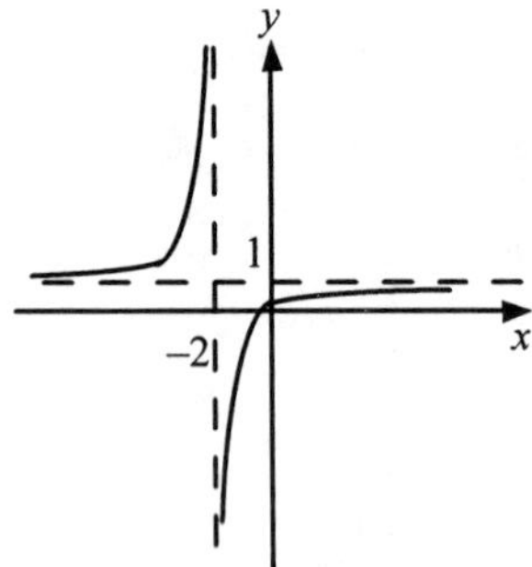

In Exercise 49–52 use the graph of $y=|x|$ to graph the given functions.

49. $y=2|x|$

50. $y=-\frac{1}{2}|x|$

51. $y=1-3|x-5|$

52. $y=\frac{1}{4}|x+2|-3$

In Exercise 53 and 54 find a viewing window that displays complete graphs for both functions in the same display.

53. $y=\sqrt{x}$; $y=8\sqrt{x}$

54. $y=\dfrac{(x-12)^2}{4}$; $y=-4x^2+12$

55. If the projectile in Example 5 has an initial velocity of 256 ft/sec, then its flight will last 16 seconds, and during flight its speed is given in terms of elapsed time t by $s=32|t-8|$. Graph this function, giving its domain and range.

56. A hammer falls to the ground from a high girder of a skyscraper. Its height above the ground after t seconds is given by $y=-16t^2+300$. Graph this function, giving its domain and range.

In Exercises 57–60 complete the graphs for $x<0$ if (a) f is an even function and (b) f is an odd function.

57.

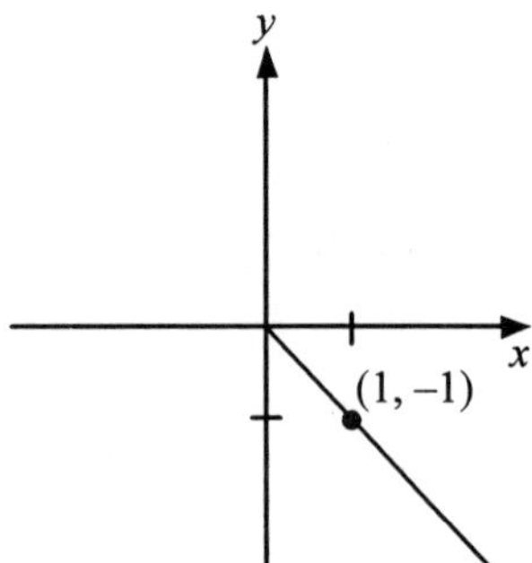

58.

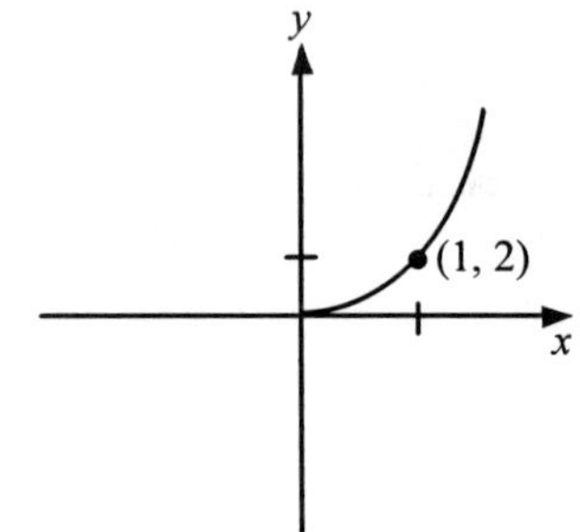

59.

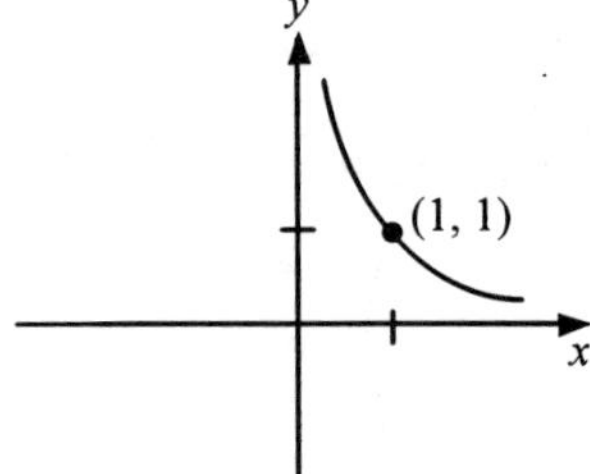

60.

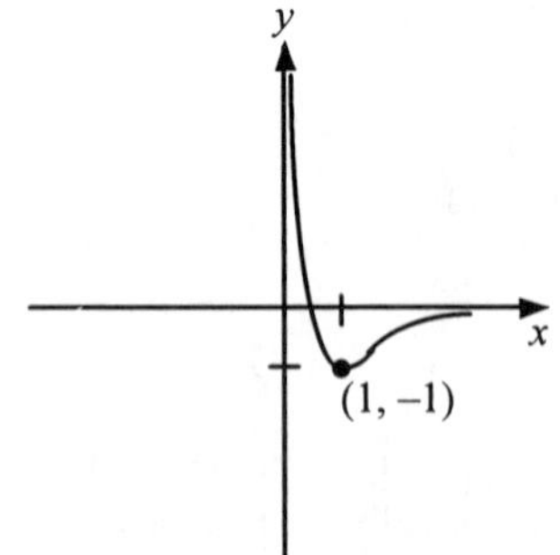

In Exercises 61–66 classify each of the functions in the figures shown as even, odd, or neither of these.

61.

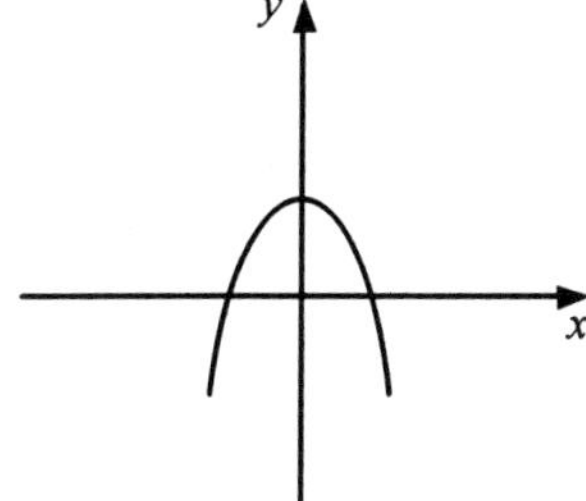

62.

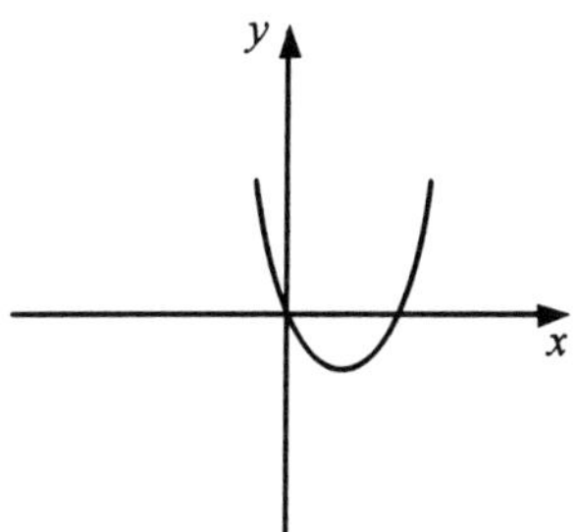

63.

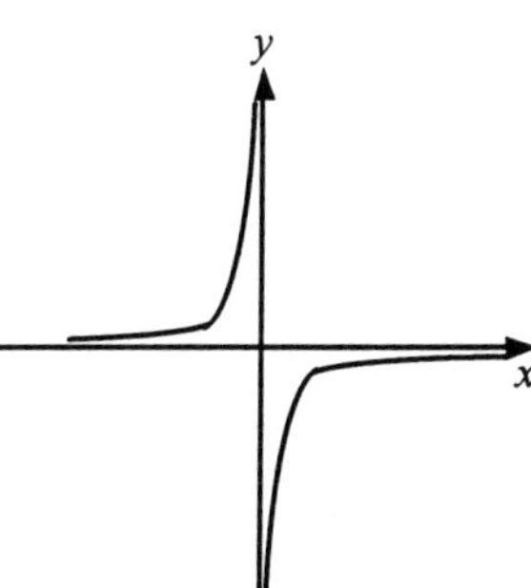

64.

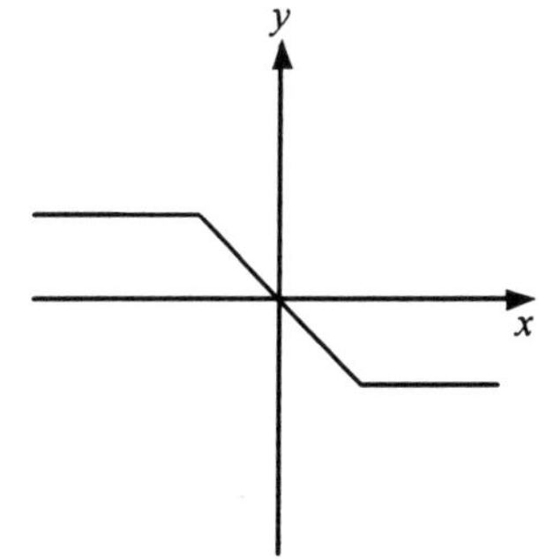

65.

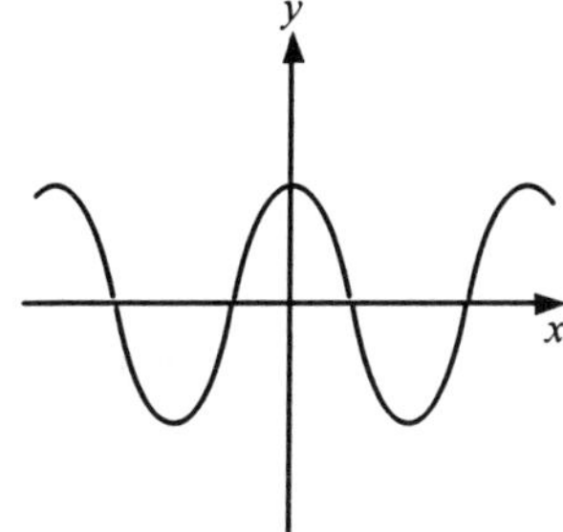

66.

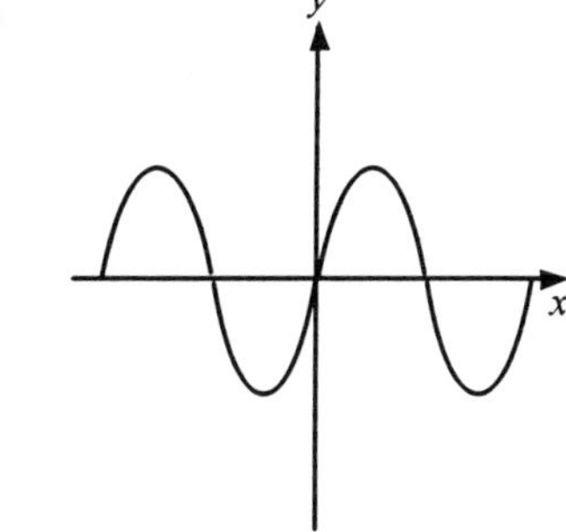

In Exercises 67–70 prove algebraically that the given function is an even function, by showing that $f(-x)$ is equal to $f(x)$. Confirm your proof by drawing the graph.

67. $f(x)=|3x|$ **68.** $f(x)=3x^2$

69. $f(x)=7$ **70.** $f(x)=\dfrac{1}{x^2+1}$

In Exercises 71–74 prove algebraically that the given functions are odd, by showing that $f(-x)$ equals $-f(x)$. Confirm your proof by drawing the graph.

71. $f(x)=3x$ **72.** $f(x)=2x^3$

73. $f(x)=\dfrac{-2x^3-x}{x^2}$ **74.** $f(x)=\dfrac{1}{x}$

In Exercises 75–78 classify the given function as even, odd, or neither of these. Justify your answer algebraically and geometrically.

75. $y=2-|x|$ **76.** $y=\dfrac{1}{x^3+1}$

77. $y=x^2-x$ **78.** $y=x+1$

79. Besides looking for symmetry, here is another way a grapher can be used to test whether a function is even or odd. Enter expressions for $y1$, $y2$, and $y3$ based on the following scheme.

$$y1 = f(x);\ y2 = f(-x);\ y3 = y1/y2$$

Then, graph so that the display shows only the graph of $y3$.

a. What graph results when f is an even function? Why?

b. What graph results when f is an odd function? Why?

c. Use this method to see if $f(x) = \dfrac{2x}{x^2 - 2}$ is even, odd, or neither.

80. Exercise 79 illustrates that if f is an even function then $\dfrac{f(x)}{f(-x)} = 1$ for all values for x where this ratio is defined. Thus the graph of $y = \dfrac{f(x)}{f(-x)}$ is the straight line $y = 1$. Use this idea and a graphing calculator to see which of the given functions is even.

a. $g(x) = x^2 + 1$ **b.** $h(x) = (x+1)^2$

81. Exercise 79 illustrates that if f is an odd function then $\dfrac{f(x)}{f(-x)} = -1$ for all values of x where this ratio is defined. Thus the graph of $y = \dfrac{f(x)}{f(-x)}$ is the straight line $y = -1$. Use this idea and a graphing calculator to see which of the given functions is odd.

a. $g(x) = x^3 + 1$ **b.** $h(x) = x^3 - x$

THINK ABOUT IT 1.3

1. Classify the following function as even, odd, or neither of these.

$$f(x) = \begin{cases} 1 & \text{if} \quad x > 1 \\ x & \text{if} \quad -1 \le x \le 1. \\ -1 & \text{if} \quad x < -1 \end{cases}$$

2. Show that the function $f(x) = 0$ is **both** odd and even. Explain why this is the only function that is both odd and even.
3. The point (–4, 3) lies on the graph of $f(x) = (x+2)^2 - 1$. What are the translated coordinates of this point if the graph of f is shifted so that the minimum point of the parabola is at the origin?
4. If the point (a, b) lies on the graph of $y = f(x)$, then what are the translated coordinates of this point if the graph of f is shifted to graph $y = f(x-2) - 3$?
5. The graph of $y = x^2$ is a parabola that has vertex at the origin and includes the point (1, 1). Find a formula for a parabola with vertex at the origin that includes the point (2, 1). Find a formula for a parabola with vertex at the origin that includes the point (a, b).

1.4 The Distance Formula and Circles

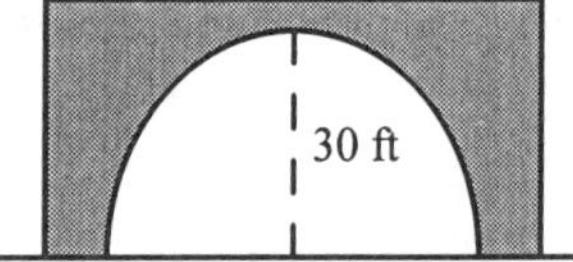

The cross-section of a tunnel is the semicircle shown in the figure. How high above the road is the tunnel ceiling 15 ft on each side of the center? Approximate the answer to the nearest tenth of a foot.
(See Example 5.)

Objectives

1. Find the distance between two points.
2. Graph and write equations for circles and semicircles with centers at the origin.
3. Determine if a graph or equation defines y as a function of x.

When working with the Cartesian coordinate system, it is often necessary to determine the distance between two points in the system. It is easiest to calculate the distance between two points on a horizontal or vertical line. If the points lie on the same horizontal line, we find the distance between them by taking the absolute value of the difference between their x-coordinates. If two points lie on the same vertical line, we calculate the distance between them by taking the absolute value of the difference between their y-coordinates.

Example 1: Horizontal and Vertical Distance

Find the distance between the given points.

a. $(-4, 3)$ and $(2, 3)$ **b.** $(1, 1)$ and $(1, -3)$

Solution: See Figure 1.47

a. The points $(-4, 3)$ and $(2,3)$ have the same y-coordinate, so

$$d = |x_2 - x_1| = |2 - (-4)| = |6| = 6.$$

b. The points $(1, 1)$ and $(1, -3)$ have the same x-coordinate, so

$$d = |y_2 - y_1| = |-3 - 1| = |-4| = 4.$$

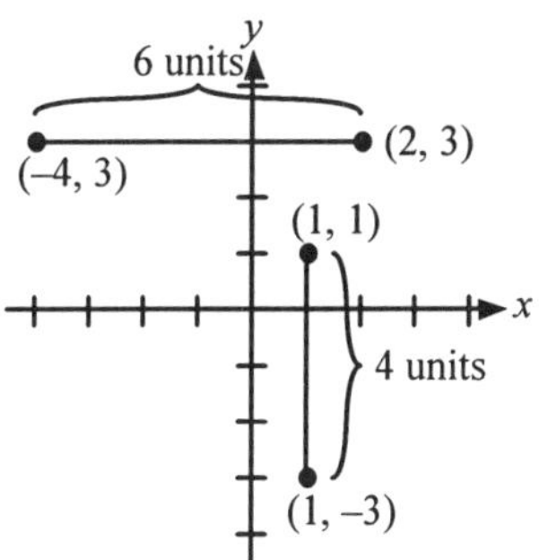

Figure 1.47

PROGRESS CHECK 1

Find the distance between the given points.

a. $(-3, 4)$ and $(1, 4)$ **b.** $(2, 2)$ and $(2, -1)$ ■

If two points do not lie on the same horizontal or vertical line, we can find the distance between them by drawing a horizontal line through one point and a vertical line through the other, as illustrated in Figure 1.48. The x-coordinate at the point where the two lines intersect will be x_2. The y-coordinate will be y_1. The distance between (x_1, y_1) and (x_2, y_1) will be $|x_2 - x_1|$ because they lie on the same horizontal line. The distance between (x_2, y_2) and (x_2, y_1) will be $|y_2 - y_1|$

because they lie on the same vertical line. The vertical and horizontal lines meet at a right angle (90°), and thus the resulting triangle is a right triangle.

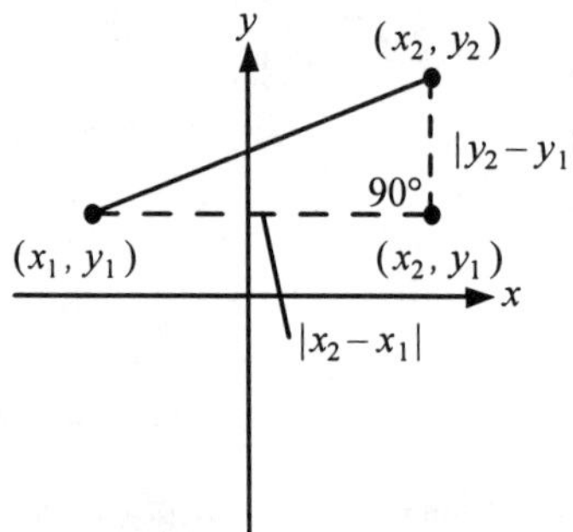

Figure 1.48

The distance d is the length of the hypotenuse in the right triangle, and by the Pythagorean theorem we get

$$d^2 = |x_2 - x_1|^2 + |y_2 - y_1|^2$$
$$d = \sqrt{(x_2 - x_1)^2 + (y_2 - y_1)^2}.$$

Note that we no longer need the absolute value symbol since the square of any real number is never negative. Thus, we have the following formula.

Distance Formula

> The distance d between the points $P_1(x_1, y_1)$ and $P_2(x_2, y_2)$ is given by
>
> $$d = \overline{P_1P_2} = \sqrt{(x_2 - x_1)^2 + (y_2 - y_1)^2}.$$

Example 2: Using the Distance Formula

Find the distance between $(-3, -2)$ and $(-1, 4)$.

Solution: Let $x_1 = -3$, $y_1 = -2$, $x_2 = -1$, and $y_2 = 4$, and use the distance formula,

$$\begin{aligned} d &= \sqrt{(x_2 - x_1)^2 + (y_2 - y_1)^2} \\ &= \sqrt{[-1-(-3)]^2 + [4-(-2)]^2} \\ &= \sqrt{(2)^2 + (6)^2} \\ &= \sqrt{4 + 36} \\ &= \sqrt{40} \quad \text{(See Figure 1.49.)} \end{aligned}$$

PROGRESS CHECK 2

Find the distance between $(-3, -4)$ and $(0, 2)$. ■

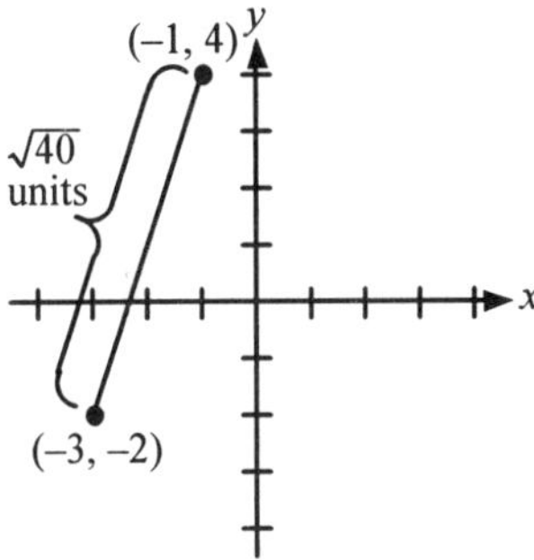

Figure 1.49

Example 3: An Application of the Distance Formula

Show that the points (–4, – 2), (–3, 2), and (1, 1) are vertices of an isosceles triangle. Is this triangle a right triangle?

Solution: Label (–4, – 2) as point A, (–3, 2) as B, and (1, 1) as C. Then

$$\overline{AB} = \sqrt{[-4-(-3)]^2 + (-2-2)^2} = \sqrt{17}$$
$$\overline{BC} = \sqrt{(-3-1)^2 + (2-1)^2} = \sqrt{17}$$
$$\overline{AC} = \sqrt{(-4-1)^2 + (-2-1)^2} = \sqrt{34}.$$

Since the lengths of two sides are equal, the triangle is isosceles. Since $(\sqrt{34})^2 = (\sqrt{17})^2 + (\sqrt{17})^2$, the Pythagorean theorem holds and ABC is a right triangle with angle B as the right angle (see Figure 1.50).

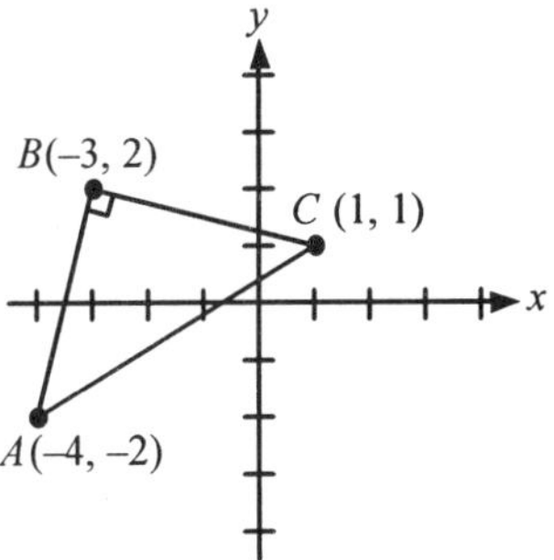

Figure 1.50

PROGRESS CHECK 3

Show that the points (3, 5), (3, – 3) and (–1, 1) are vertices of an isosceles triangle. Is this triangle a right triangle? ■

The distance formula may be used to derive equations for circles. At this time we will consider only a circle with center at the origin and radius r as shown in Figure 1.51. Note that (x, y) is a point on the circle if and only if the distance from the (x, y) to the origin is r. Thus, by the distance formula, the equation for the specified circle is

$$\sqrt{(x-0)^2 + (y-0)^2} = r$$

or, after squaring both sides of the equation,

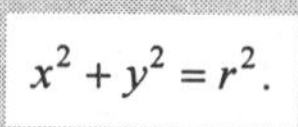

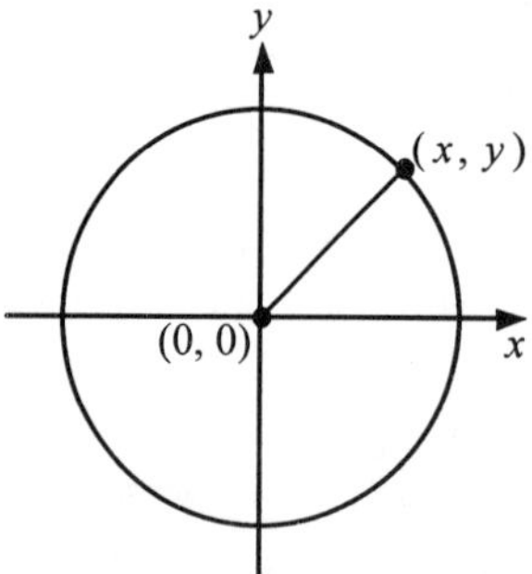

Figure 1.51

This result allows us to graph without difficulty the types of equations given in the next example.

Example 4: Graphing Circles and Semicircles Centered at the Origin

Graph each equation and indicate the domain and range of each relation.

a. $x^2 + y^2 = 25$ **b.** $y = \sqrt{25 - x^2}$ **c.** $y = -\sqrt{25 - x^2}$

Solution:

a. The equation $x^2 + y^2 = 25$ fits the form $x^2 + y^2 = r^2$, so the graph is a circle centered at the origin. Also, $r^2 = 25$ so the radius is $\sqrt{25}$, or 5. By drawing a circle that satisfies these conditions, we obtain the graph in Figure 1.52 (a). Both the domain and the range are given by $[-5, 5]$.

b. We start with $y = \sqrt{25 - x^2}$ and square both sides of the equation to obtain $y^2 = 25 - x^2$, or $x^2 + y^2 = 25$, which is the equation from **a**. However, in the original equation, y is restricted to positive numbers or zero, so the graph is a semicircle, as shown in Figure 1.52 (b). From the graph we read that the domain is $[-5, 5]$ and the range is $[0, 5]$.

c. As in **b**, the graph of $y = -\sqrt{25 - x^2}$ is a semicircle. But y is restricted to negative numbers or zero in this case, so we draw the semicircle in Figure 1.52(c), which shows that the domain is $[-5, 5]$ and the range is $[-5, 0]$.

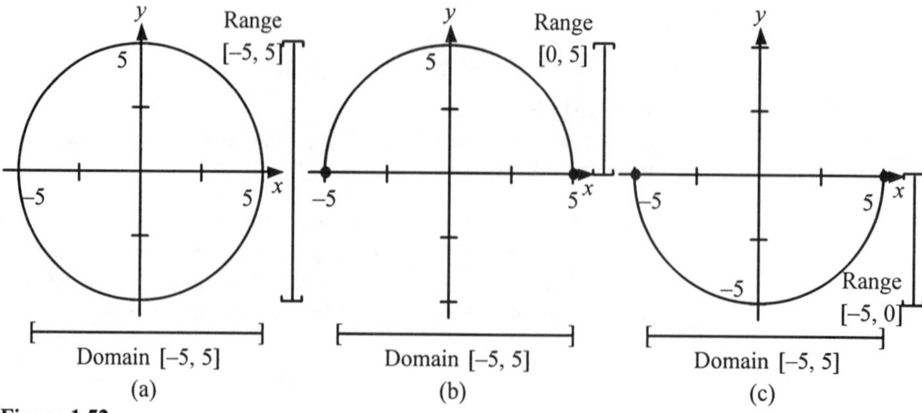

Figure 1.52

Technology Link

To graph $x^2 + y^2 = 25$ on a graphing calculator, we graph the top semicircle given by $y = \sqrt{25 - x^2}$ and the bottom semicircle given by $y = -\sqrt{25 - x^2}$ in the same display. This method is required because it is necessary to enter an expression that defines a function on a graphing calculator. Also, the ranges must be adjusted to make the graph appear round instead of oval. Although the exact adjustment depends on the particular calculator display, many calculators have a square setting feature that defines the viewing window as needed. For example, in Figure 1.53 we show the calculator screens that may result when the square setting feature is invoked to graph $y1 = \sqrt{25 - x^2}$ and $y2 = -y1$ (or $y2 = -\sqrt{25 - x^2}$) on a particular calculator.

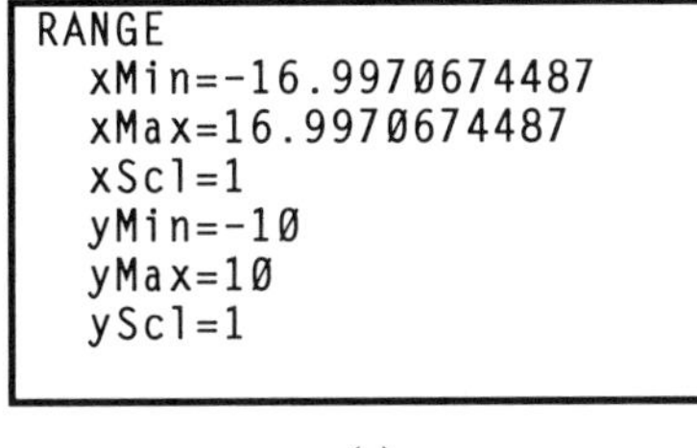

(a)

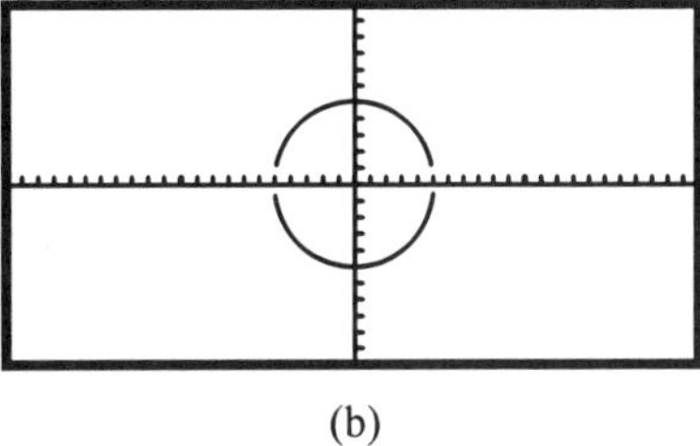

(b)

Figure 1.53

PROGRESS CHECK 4

Graph each equation and indicate the domain and the range of each relation.

a. $x^2 + y^2 = 9$ **b.** $y = \sqrt{9 - x^2}$ **c.** $y = -\sqrt{9 - x^2}$ ■

Example 5: Ceiling Heights in a Tunnel

Solve the problem in the section introduction on page 131.

Solution: We are free to place the specific figure in any convenient position in a coordinate system, so we start by diagramming the problem as shown in Figure 1.54.

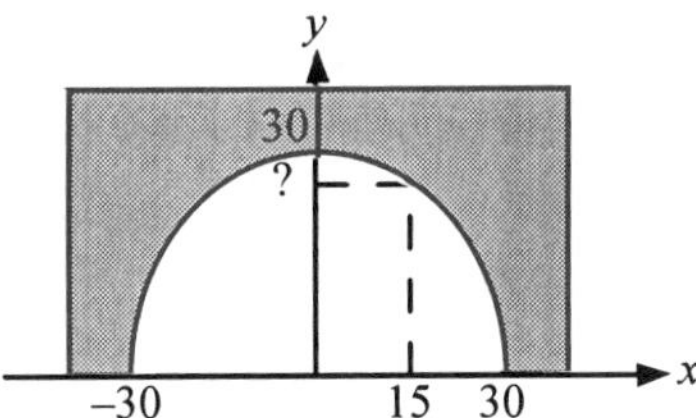

Figure 1.54

The radius is 30, so an equation for the entire circle is

$$x^2 + y^2 = (30)^2.$$

Then this equation implies

$$y^2 = 900 - x^2$$
$$y = \pm\sqrt{900 - x^2},$$

and the equation for the semicircle in the diagram is

$$y = \sqrt{900 - x^2}.$$

Now we find y when $x = 15$ to determine the requested height.

$$y = \sqrt{900-(15)^2} = \sqrt{675} \approx 26.0$$

The tunnel is about 26.0 ft above the road at the specified point.

PROGRESS CHECK 5
If the tunnel described in Example 5 reaches a maximum height of 20 ft above the road, then to the nearest tenth of a foot how high is the tunnel ceiling 14 ft on each side of center? ■

The graph of a circle is the first graph we have drawn in this chapter that is not the graph of a function. It is easy to determine if a given graph is the graph of a function because no two ordered pairs in a function may have the same first component. Graphically, this means that a function cannot contain two points that lie on the same vertical line. This observation leads to the following simple test for determining if a graph represents a function.

Vertical Line Test

> Imagine a vertical line sweeping across the graph. If the vertical line at any position intersects the graph in more than one point, the graph is not the graph of a function.

Example 6: Using the Vertical Line Test

Use the vertical line test to determine which graphs in Figure 1.55 represent the graph of a function.

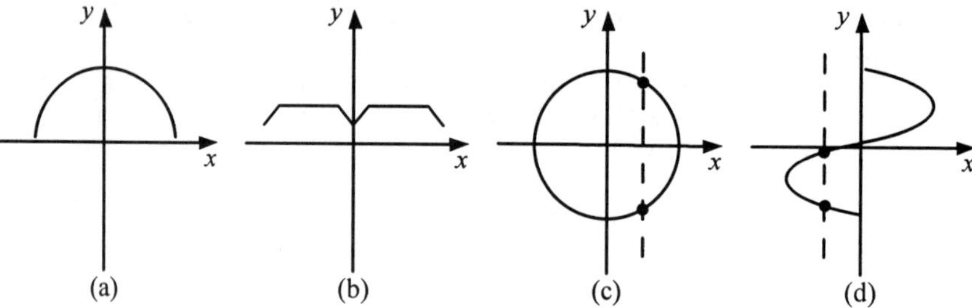

Figure 1.55

Solution: By the vertical line test, graphs (a) and (b) represent functions, whereas (c) and (d) do not.

PROGRESS CHECK 6
Use the vertical line test to determine which graphs in Figure 1.56 are the graph of a function. ■

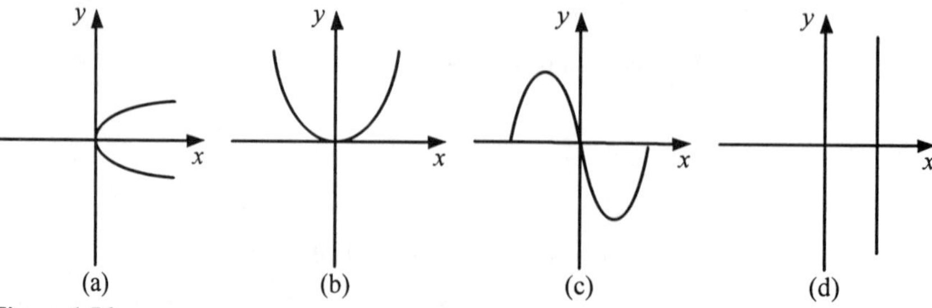

Figure 1.56

The next example considers whether certain relations are functions from both the algebraic and the geometric viewpoint.

Example 7: Determining Functions Algebraically and Graphically

Does the given rule determine y as a function of x? Use both algebraic and geometric methods.

a. $x + y = 4$ **b.** $x^2 + y^2 = 4$

Solution:

a. $x + y = 4$ implies $y = 4 - x$. The assignment of any real number to x results in exactly one output for y, so the equation determines y as a function of x. In Figure 1.57(a) this result is confirmed graphically by the vertical line test.

b. Since $x^2 + y^2 = 4$ implies $y = \pm\sqrt{4 - x^2}$, it is possible for an x value to correspond to two different y values. For example if $x = 0$, then $y = 2$ or $y = -2$. Thus, $x^2 + y^2 = 4$ does not determine y as a function of x. The graph of this equation is the circle in Figure 1.57(b), and the vertical line test indicates that a circle is never the graph of a function.

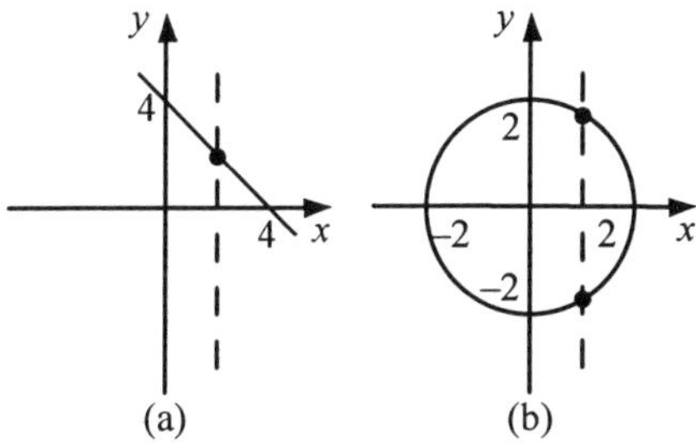

Figure 1.57

PROGRESS CHECK 7

Does the given rule determine y as a function of x? Use both algebraic and geometric methods.

a. $y^2 = x$ **b.** $y = 4$ ■

In the first four sections of this chapter we have considered many ideas associated with graphing relations. Some of the important principles that have been mentioned are summarized next.

1. Coordinate (or analytic) geometry bridges the gap between algebra and geometry. The fundamental principle in this scheme is that every ordered pair that satisfies an equation corresponds to a point in its graph, and every point in the graph corresponds to an ordered pair that satisfies the equation. Thus, a picture may be drawn for each equation relating x and y, which helps us see the essential characteristics of the relationship. On the other hand, a graph can often be represented as an equation, so that the powerful methods of algebra can be used to analyze these curves. This simple, yet brilliant, idea is one of the most useful in mathematics.
2. All necessary information must be shown in the graph, so that you can see the relationship without consulting any supplementary material. In particular, be sure to identify the appropriate units on both axes. A graph must be able to "stand" by itself!
3. It is not necessary to scale the two axes in the same manner. In fact, different scales are frequently desirable. It is important that you be aware that different scales can have a dramatic effect on the apparent behavior of a relationship.
4. In this section we discussed the Cartesian (or rectangular) coordinate system. There are other important systems that can be used to obtain a picture of a relationship, such as a polar

coordinate system. No one system is ever more correct than another, just more advantageous for the particular situation being analyzed.

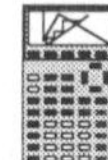

EXPLORE 1.4

1. See if your calculator has a built-in distance calculator for finding the distance between 2 points. If it does, use it to find the distance between (1, 1) and (3, 3). Then confirm the calculator result by using the distance formula.
2. Examine the graph of the circle $x^2 + y^2 = 9$ under all these windows and explain the results.
   a. [−10, 10]; [−10, 10] **b.** [−200, 200]; [−200, 200] **c.** [−0.01, 0.01]; [2.99, 3.01]
3. Usually when you change one value in an equation by just a little, the graph also changes by just a little. Confirm this for the graphs of $x^2 + y^2 = 1$ and $x^2 + y^2 = 1.1$. But this is not always true. Describe and compare the graphs of $x^2 + y^2 = c$ for $c = 0.1$, 0, and −0.1.

EXERCISES 1.4

In Exercises 1–12 represent the ordered pairs as points in the Cartesian coordinate system and find the distance between them.

1. (−1, 3), (−1, 1) **2.** (2, −4), (2, −1)

3. (4, 1), (0, 1) **4.** (−3, −2), (2, −2)

5. (−12, 5), (0, 0) **6.** (0, 0), (−4, 1)

7. (4, −1), (−2, −3) **8.** (0, 2), (4, −3)

9. (−1, −1), (−2, −2) **10.** (5, 2), (1, 3)

11. (−2, 5), (4, −3) **12.** (6, −2), (−6, 3)

13. Three vertices of a square are (3, 3), (3, −1), and (−1, 3). Find the ordered pair corresponding to the fourth vertex. What is the perimeter of the square?

14. Three vertices of a rectangle are (4, 2), (−5, 2), and (−5, −4). Find the ordered pair corresponding to the fourth vertex. What is the perimeter of the rectangle?

15. Show that the points (3, 5), (−3, 7), and (−6, −2) are vertices of a right triangle.

16. Show that the points (−8, −2), (−11, 3) and (3, 8) are vertices of an isosceles triangle. Is this triangle a right triangle?

17. Find the radius of a circle whose diameter extends from (−2, 1) to (3, 4).

18. Find the radius of a circle with center at (0, 0) and which passes through the point (2, 3).

19. Find the area of a circle whose radius extends from (−1, 1) to (2, 1).

20. Find the area of a circle whose diameter extends from (−1, 3) to (3, 1).

21. Find a point (x, y) not on the x- or y-axis whose distance from the origin is 3 units. How many such points are there?

22. Find a point (x, y) with $x \neq 1$, $y \neq 1$ whose distance from (1, 1) is 3 units. How many such points are there?

Graph each equation and indicate the domain and range of each relation on the graph.

23. **a.** $x^2 + y^2 = 1$ **b.** $y = \sqrt{1 - x^2}$
c. $y = -\sqrt{1 - x^2}$

24. **a.** $x^2 + y^2 = 5$ **b.** $y = \sqrt{5 - x^2}$
c. $y = -\sqrt{5 - x^2}$

25. **a.** $x^2 + y^2 = \frac{1}{4}$ **b.** $y = \sqrt{\frac{1}{4} - x^2}$
c. $y = -\sqrt{\frac{1}{4} - x^2}$

26. **a.** $x^2 + y^2 = 36$ **b.** $y = \sqrt{36 - x^2}$
c. $y = -\sqrt{36 - x^2}$

27. **a.** $x^2 + y^2 = 1.44$ **b.** $y = \sqrt{1.44 - x^2}$
c. $y = -\sqrt{1.44 - x^2}$

28. **a.** $2x^2 + 2y^2 = 50$ **b.** $y = \sqrt{\frac{50 - 2x^2}{2}}$
c. $y = -\sqrt{\frac{50 - 2x^2}{2}}$

29.

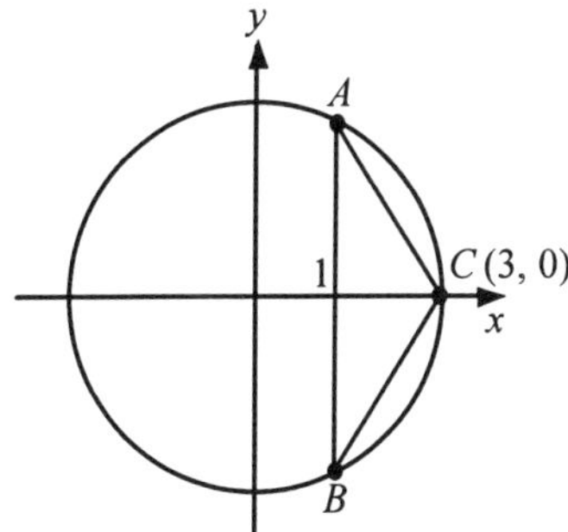

a. Find the equation of the circle shown.
b. Find the lengths of AB and AC.
c. Find the area of triangle ABC.

30. Find an equation for the semicircle shown. The length of AB is 10, and the length of BC is 12.

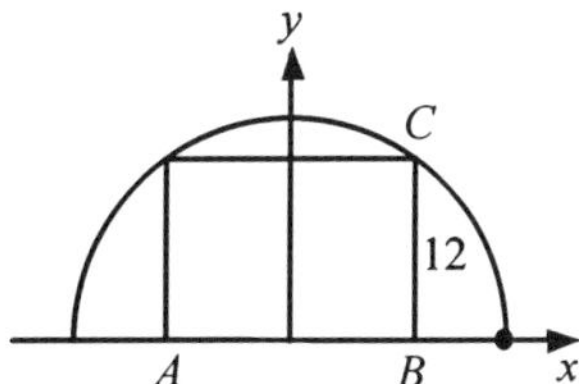

31. A pendulum is suspended from a bar and swings back and forth as indicated in the figure.

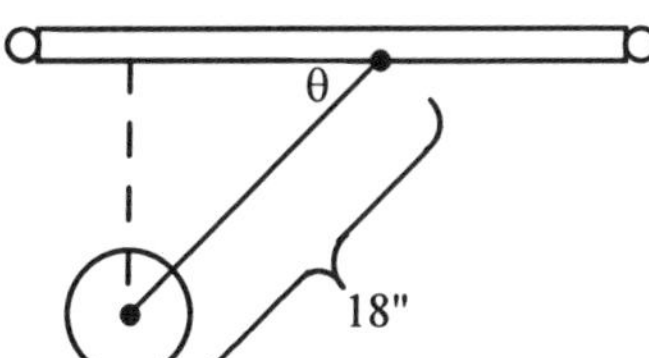

a. Why is the path of the center of the pendulum bob part of a semicircle? Find an equation for this semicircle using the point of attachment to the bar as the origin.
b. The left-most position occurs when angle θ is 45°. To the nearest tenth of an inch find the vertical distance from the center of the bob to the bar at that position.

32. A small satellite is put into a circular orbit around an asteroid at a distance of 1000 feet above the surface. If the diameter of the asteroid is 10,000 feet, find an equation for the orbit using the center of the asteroid as the origin.

33. Use a grapher to graph the relation $2x^2 + y^2 = 16$. Solve for y and graph the top and bottom halves separately. Because the coefficients of x^2 and y^2 are not equal, the graph is not a circle but an ellipse. Use the viewing window $[-5, 5]$ by $[-5, 5]$ and describe the domain and range of this relation. Estimate to the nearest tenth where the ellipse crosses the x- and y-axes.

34. Use a grapher to graph the relation $x^2 - y^2 = 10$. Solve for y and graph the top and bottom halves separately. Because the left side expresses the *difference* between x^2 and y^2 instead of the sum, the graph is not a circle but a hyperbola. Use the viewing window $[-10, 10]$ by $[-10, 10]$ and describe the domain and range of this relation. Estimate to the nearest tenth where the hyperbola crosses the x- and y-axes.

In Exercises 35-44 use the vertical line test to determine which graphs represent the graph of a function.

35.

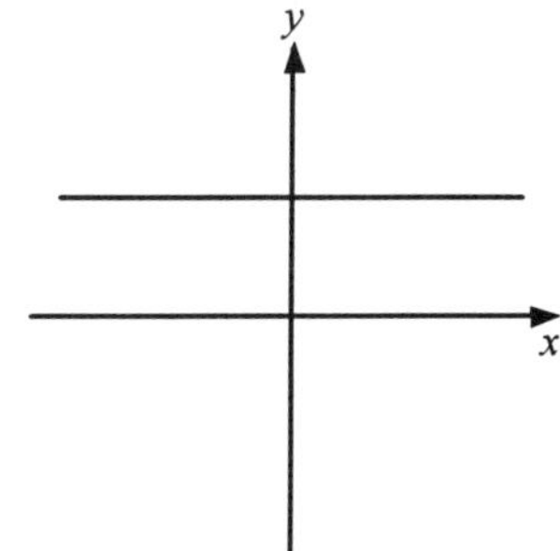

36.

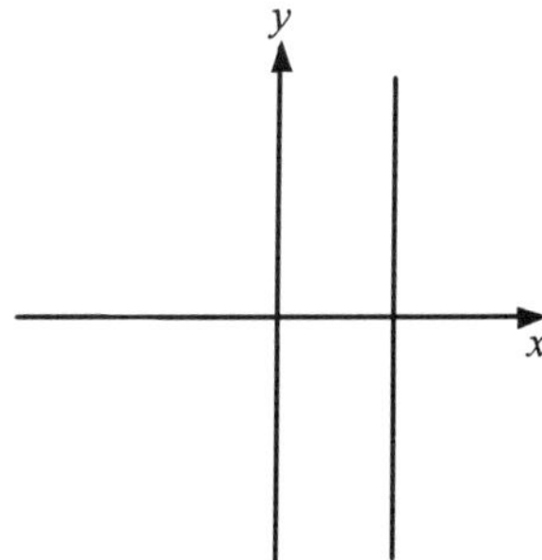

37.

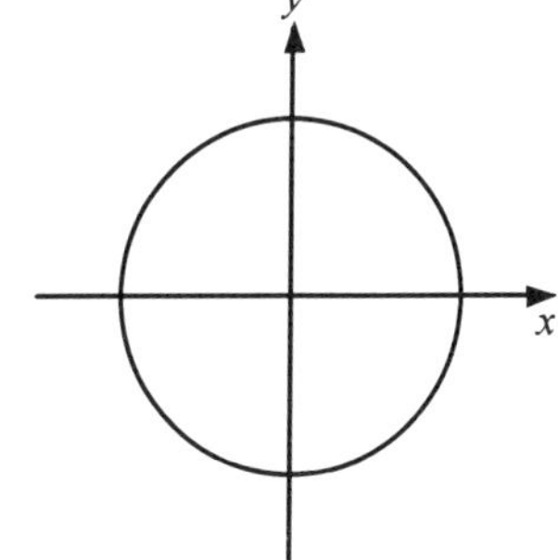

38.

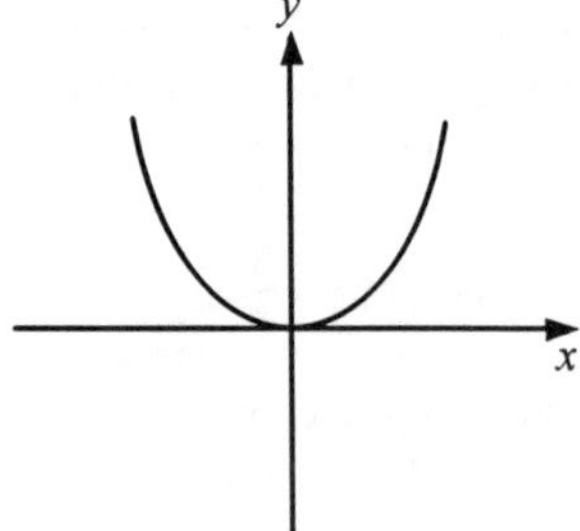

39.

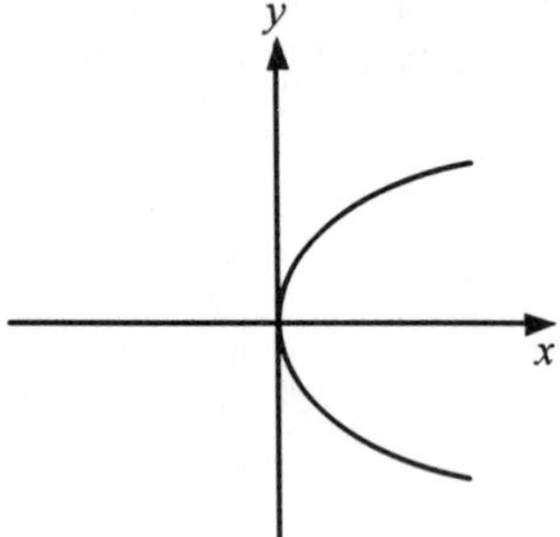

40.

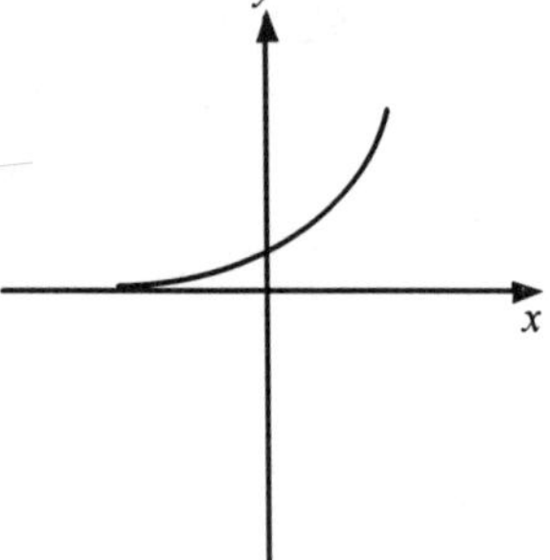

41.

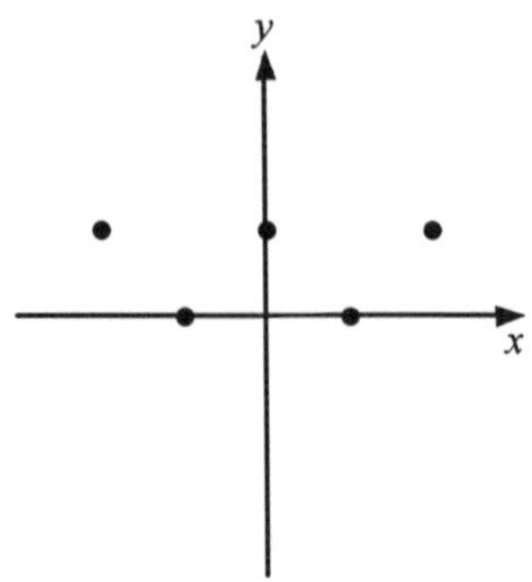

42.

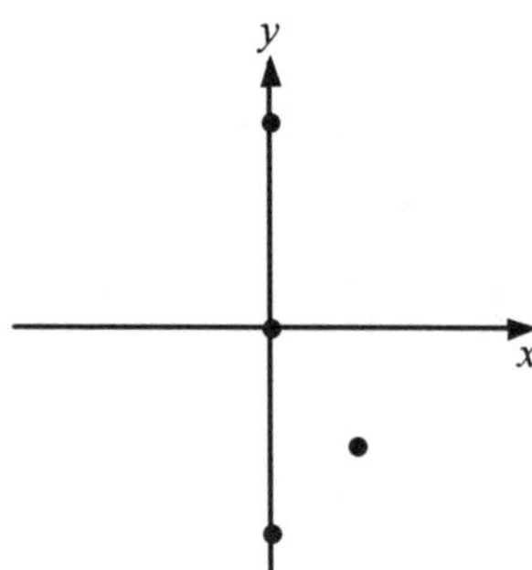

43.

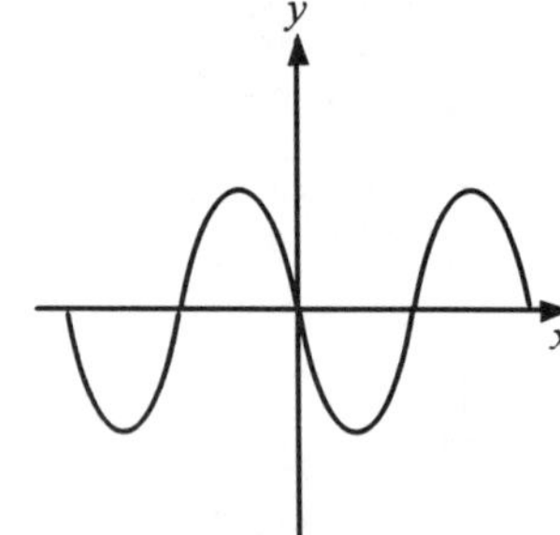

44.

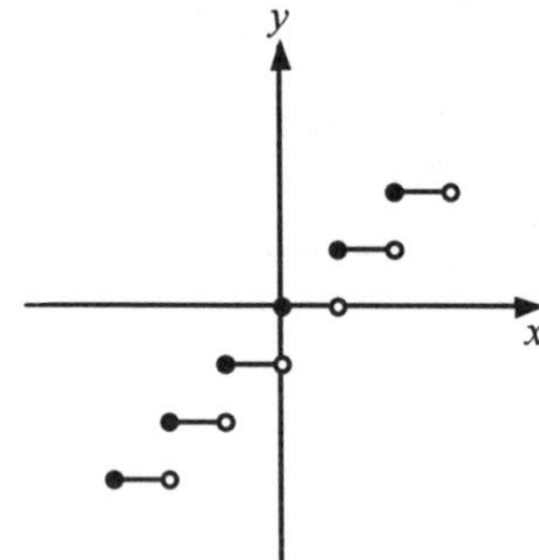

In Exercises 45–54 use both algebraic and geometric methods to see if the given rule determines y as a function of x.

45. $y = 9$ **46.** $x = 9$ **47.** $x + y = 9$

48. $x^2 + y = 9$ **49.** $x^2 + y^2 = 9$

50. $x + y^2 = 9$ **51.** $x^3 + y = 9$

52. $x + y^3 = 9$ **53.** $x^2 + y^3 = 9$ **54.** $x^3 + y^2 = 9$

THINK ABOUT IT 1.4

1. Find x, given that the point $(x, 0)$ is equidistant from $(-4, -4)$ and $(2, 3)$.

2. The graphs of ordered *pairs* of variables are conveniently drawn on a *two*-dimensional coordinate system. By analogy, the graphs of ordered *triples* of variables may be drawn on a *three*-dimensional coordinate system like the one shown. The graph of the point (1, 2, 3) is drawn. Graph these points.

a. (2, 3, 4) **b.** (3, 5, 0)

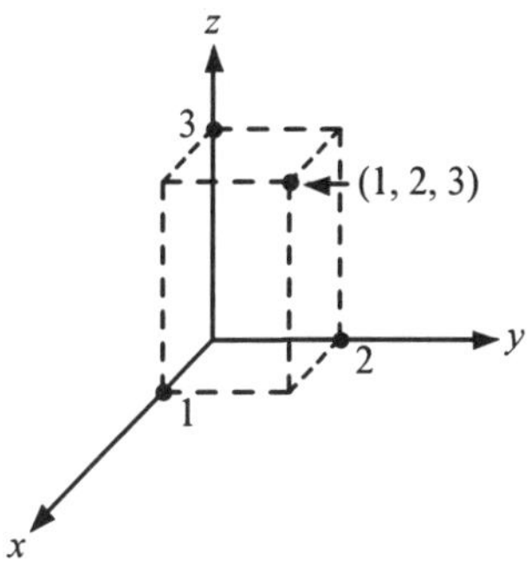

c. In 3-dimensional coordinate systems the distance formula becomes

$$d = \sqrt{(x_2 - x_1)^2 + (y_2 - y_1)^2 + (z_2 - z_1)^2}\,.$$

Find the distance between (2, 3, 4) and (3, 5, 0).

3. **a.** Graph the line $y = 3$.

b. How many points are on the line $y = 3$ between (0, 3) and (1, 3)?

c. Why is the following question difficult? Are there twice as many points on the line $y = 3$ between (0, 3) and (2, 3) as there are between (0, 3) and (1, 3)? A mathematician who made a major contribution to answering this question is Georg Cantor (1845–1918). A lively account of his life and work is given in *Men of Mathematics* by E. T. Bell, Simon and Schuster, 1937.

4. Explain why when you repeatedly zoom in on a point on a circle eventually the graphs looks like a straight line.

5. The graph of $y = -\sqrt{9 - x^2}$ is a semicircle.

a. What is the radius?

b. By adding k the graph is translated up k units, as represented by $y = -\sqrt{9 - x^2} + k$. What value of k raises the semicircle so that it intersects the x-axis in exactly one point?

c. Find an equation for the whole circle that results from completing the semicircle in part **b**.

● ● ●

1.5 Simultaneous Equations

Photo Courtesy of ***Barry M. Winiker****, The Picture Cube*

Pat is offered two part-time jobs selling exercise equipment. The first company offers a straight 20 percent commission, whereas the second company offers a salary of $70 per week plus a 10 percent commission.

a. Write a formula for each offer that expresses Pat's income (y) as a function of sales (x).

b. Sketch graphs of both functions on the same axes.

c. How should Pat decide which offer is better financially?

(See Example 1.)

Objectives

1. Solve a system of linear equations by the graphing method, the substitution method, or the addition-elimination method.

2. Solve applied problems by setting up and solving a system of linear equations.
3. Solve linear equations and linear inequalities in one variable graphically.

To understand what is meant by a set of simultaneous equations and its solution, we begin by analyzing the section-opening problem.

Example 1: Comparing Two Salary Offers

Solve the problem in the section introduction on page 141.

Solution:

a. x stands for the amount of sales, so Pat's income (y) for each offer is given by

First offer (20% commission): $y = 0.2x$
Second offer: ($70 + 10% commission): $y = 70 + 0.1x$.

b. A benefit comparison between these two offers can be made by constructing the following tables and then graphing the two lines on the same axes, as shown in Figure 1.58.

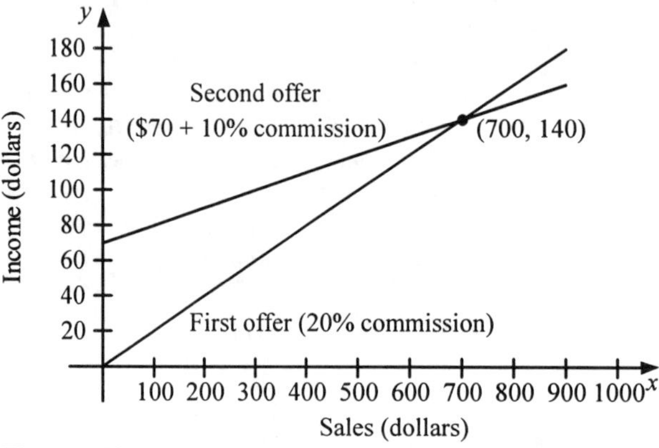

Figure 1.58

First offer (20% commission): $y = 0.2x$

Sales (in dollars) x	100	200	300	400	500	600	700	800	900	1,000
Income (in dollars) y	20	40	60	80	100	120	140	160	180	200

Second offer ($70 + 10% commission): $y = 70 + 0.1x$

Sales (in dollars) x	100	200	300	400	500	600	700	800	900	1,000
Income (in dollars) y	80	90	100	110	120	130	140	150	160	170

c. By examining the graph and the table, we can see that Pat earns $140 from either job when $700 of equipment are sold, so the two offers are financially equivalent at this point. We also see that when sales exceed $700, the larger commission (first offer) is the better offer. Thus, Pat should decide as follows.

Anticipated Weekly Sales	**Decision**
Above $700	Choose 1st offer
Below $700	Choose 2nd offer
Exactly $700	Choose either offer

PROGRESS CHECK 1

Redo the problem in Example 1, but assume that the second company offers \$150 per week plus a 5 percent commission.

The pair of equations in the opening problem is an example of a set of *simultaneous equations*. Note that there are many ordered pairs that satisfy $y = 0.2x$ and many ordered pairs that satisfy $y = 70 + 0.1x$, but there is only one ordered pair that satisfies both equations. In general, the **solution set** of a system of equations consists of all ordered pairs that satisfy all equations in the system at the same time (simultaneously). In this section we limit our coverage to a pair of equations with straight lines for graphs. Such a set of simultaneous equations is called a **system of linear equations**, and its solution is given by all the points where the lines intersect. There are three possible cases, and they are represented in Figure 1.59.

Case 1 The equations represent two lines that intersect at one point and so have 1 point in common. This system is called **consistent**.

Case 2 The equations represent two distinct lines that are parallel and do not intersect at all, and so have no points in common. This system is called **inconsistent**.

Case 3 Both equations represent the same line and so have all the points of that line in common. This system is called **dependent** (and consistent).

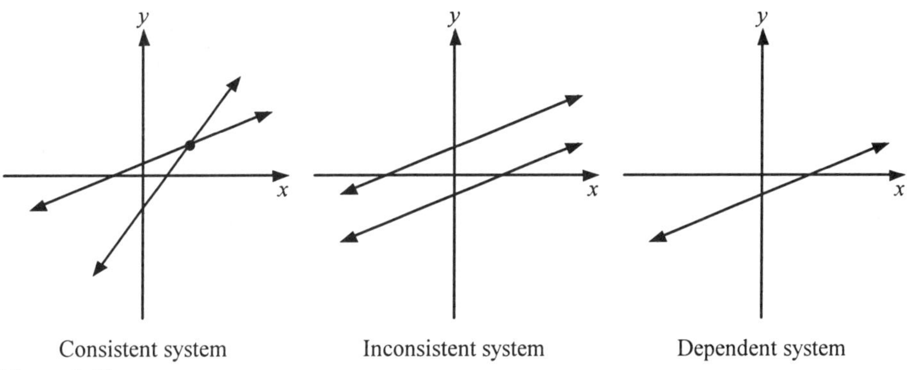

Figure 1.59

The most useful situation is usually when the graphs of the two equations intersect at exactly one point, as in our next example.

Example 2: Solving a Linear System Graphically

Solve by graphing:
$$y = x + 2$$
$$y = 2x - 3.$$

Solution: In Figure 1.60 we graph both of these equations on the same coordinate system, and it appears that the lines meet at (5,7). We check this apparent solution by replacing x by 5 and y by 7 in **both** equations.

$$y = x + 2 \qquad\qquad y = 2x - 3$$
$$7 \stackrel{?}{=} 5 + 2 \qquad\qquad 7 \stackrel{?}{=} 5 + 2$$
$$7 \stackrel{\checkmark}{=} 7 \text{ True} \qquad\qquad 7 \stackrel{\checkmark}{=} 7 \text{ True}$$

Thus, the solution is (5,7)

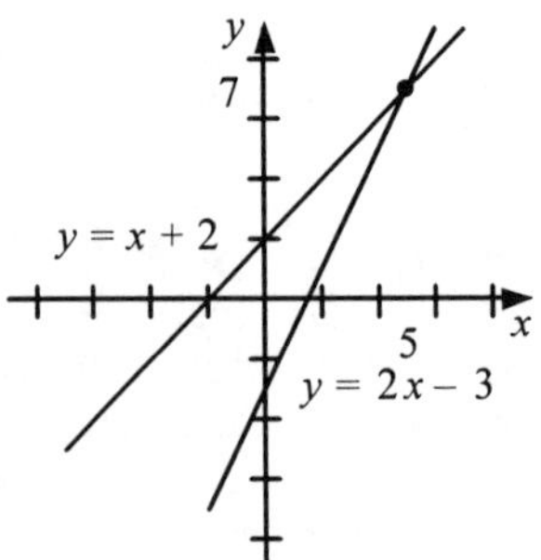

Figure 1.60

Technology Link

To obtain an approximate solution in this example using the graphing method and a graphing calculator, set $y1 = x + 2$ and $y2 = 2x - 3$. Then graph both equations in the standard viewing window and use the Trace feature to estimate the coordinates of the intersection point, as shown in Figure 1.61(a). Estimates obtained by this method may be improved as needed by repeated use of the Zoom and Trace features.

In addition, some calculators have a built-in operation that finds an intersection of two functions in an interval. For example, on one such calculator the Intersection operation may be utilized to display where two graphs intersect, as shown in Figure 1.61(b). We will use this feature throughout the text when simultaneous equations are displayed graphically, because this operation gives very accurate answers for x and y. But keep in mind that repeated use of Zoom and Trace is a workable alternative, if this special feature is not available on your calculator.

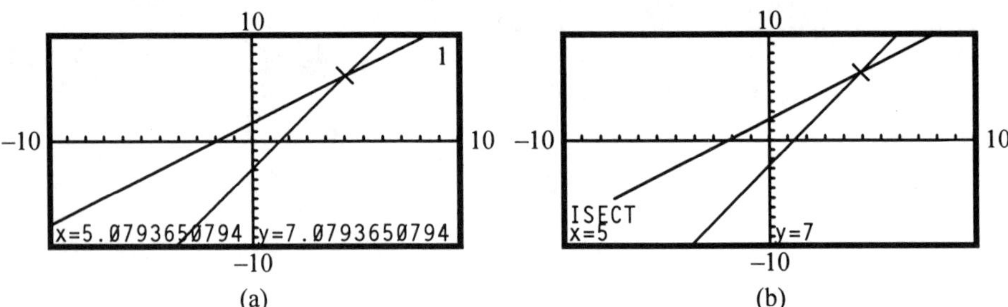

Figure 1.61

PROGRESS CHECK 2

Solve by graphing: $\begin{aligned} y &= -2x + 5 \\ y &= 3x - 10. \end{aligned}$ ■

The graphing method for solving simultaneous equations is good for illustrating the principle involved and for estimating the solution. However, for finding exact solutions, algebraic methods are usually preferable. When at least one equation in a linear system is solved for one of the variables, then the system may be solved efficiently by the **method of substitution**, as shown in the next two examples.

Example 3: Solving a Linear System by Substitution

Solve by substitution: $\begin{aligned} y &= x + 2 \\ y &= 2x - 3. \end{aligned}$

Solution: First, note that the previous example showed how to solve this system by the graphing method. For an algebraic solution, observe that both equations are solved by y. We choose to substitute $x+2$ for y in the bottom equation and solve for x.

$$\begin{aligned} x+2 &= 2x-3 \\ 2 &= x-3 \\ 5 &= x \end{aligned}$$

Thus, the x-coordinate of the solution is 5. To find the y-coordinate, substitute 5 for x in either of the given equations.

$$\begin{aligned} y &= x+2 \\ &= (5)+2 \\ &= 7 \end{aligned} \quad \text{or} \quad \begin{aligned} y &= 2x-3 \\ &= 2(5)-3 \\ &= 7 \end{aligned}$$

Thus, the solution is (5, 7). (**Note:** A good check of your result is to find the y-coordinate by substituting the x-coordinate in both equations.)

PROGRESS CHECK 3

Solve by substitution: $\begin{aligned} 5x-4y &= 6 \\ y &= 1.1x. \end{aligned}$ ■

Example 4: Finding a Break-Even Point

The total cost of producing necklaces consists of \$300 per month for rent plus \$4 per unit for material. If the selling price for the necklaces is \$9, how many units must be made and sold per month for the company to break even?

Solution: If x represents the number of necklaces made and sold, then

$$\begin{aligned} \text{total cost} &= 300+4x \\ \text{total revenue} &= 9x. \end{aligned}$$

We wish to find the value for x that makes the total cost equal to the revenue, or equivalently, we want to know the value for x that makes $300+4x$ equal to $9x$.

$$\begin{aligned} 300+4x &= 9x \\ 300 &= 5x \\ 60 &= x \end{aligned}$$

Thus, the company will break even when it makes and sells 60 necklaces. To check the result, substitute 60 in the original equations (see Figure 1.62).

$$\begin{aligned} \text{Total cost} &= 300+4x \\ &= 300+4(60) \\ &= 540 \end{aligned} \quad \text{or} \quad \begin{aligned} \text{Total revenue} &= 9x \\ &= 9(60) \\ &= 540 \end{aligned}$$

PROGRESS CHECK 4

One video store rents tapes for an annual membership of \$12 plus \$2.75 per rental. For nonmembers the cost is \$3.15 per tape. How many tapes must be rented annually for the cost to be the same for members and nonmembers? ■

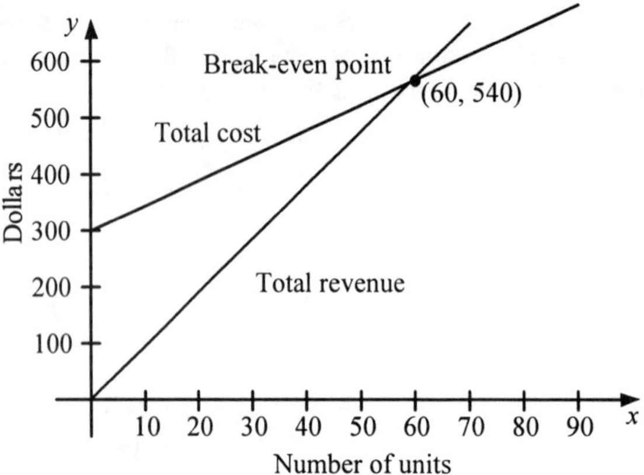

Figure 1.62

Another algebraic method for solving simultaneous equations is called the **addition-elimination method**. This method is based on the property that

$$\begin{aligned} \text{if } A &= B \\ \text{and } C &= D \\ \text{then } A + C &= B + D \end{aligned}$$

In words, adding equal quantities to equal quantities results in equal sums. For this method to result in the elimination of a variable, the coefficients of either x or y must be opposites, as in the system of Example 5.

Example 5: Solving a Linear System by Addition-Elimination

Solve by addition-elimination: $\begin{aligned} x + y &= 9 \\ x - y &= 5. \end{aligned}$

Solution: This method attempts to eliminate one of the variables by adding the two equations together. In this example the coefficients for y are $+1$ and -1. If we add the equations, the result will be an equation that contains only x.

$$\begin{array}{rcrcrl} x & + & y & = & 9 & \\ x & - & y & = & 5 & \\ \hline 2x & & & = & 14 & \text{Add the equations.} \\ x & & & = & 7 & \end{array}$$

Thus, the x-coordinate of the solution is 7.

To find the y-coordinate, substitute 7 for x in either of the given equations.

$$\begin{array}{rcr} x + y = 9 & \text{or} & x - y = 5 \\ (7) + y = 9 & & 7 - y = 5 \\ y = 2 & & y = 2 \end{array}$$

Thus, the solution is (7,2).

PROGRESS CHECK 5

Solve by addition-elimination: $\begin{aligned} 3x - 4y &= 7 \\ -2x + 4y &= 2. \end{aligned}$ ■

Example 6: Solving a Linear System by Addition-Elimination

Solve by addition-elimination: $\begin{aligned} 3x - 2y &= 27 \\ 2x + 5y &= -1. \end{aligned}$

Solution: If we form equivalent equations by multiplying the top equation by –2 and the bottom equation by 3, we can eliminate the x variable.

$$\begin{array}{rcrcr} -6x & + & 4y & = & -54 \\ 6x & + & 15y & = & -3 \\ \hline & & 19y & = & -57 \\ & & y & = & -3 \end{array} \quad \text{Add the equations.}$$

Thus, the y-coordinate of the solution is –3.

To find the x-coordinate, substitute –3 for y in either of the given equations.

$$\begin{aligned} 3x - 2y &= 27 \\ 3x - 2(-3) &= 27 \\ x &= 7 \end{aligned} \quad \text{or} \quad \begin{aligned} 2x + 5y &= -1 \\ 2x + 5(-3) &= -1 \\ x &= 7 \end{aligned}$$

Thus, the solution is $(7, -3)$.

PROGRESS CHECK 6

Solve by addition-elimination: $\begin{aligned} 4x + 3y &= -2 \\ 3x + 2y &= 1. \end{aligned}$ ■

Example 7 points out what happens when an algebraic method is applied to inconsistent systems and dependent systems.

Example 7: An Inconsistent System and a Dependent System

Solve each system.

a. $\begin{aligned} y &= 2x + 1 \\ y &= 2x + 3 \end{aligned}$ **b.** $\begin{aligned} -2x + y &= 3 \\ -4x + 2y &= 6 \end{aligned}$

Solution:

a. The substitution method yields

$$\begin{aligned} 2x + 1 &= 2x + 3 \\ 1 &= 3. \end{aligned}$$

The false equation $1 = 3$ indicates that there is no solution and that the system is inconsistent. The solution set for every inconsistent system is Ø. Graphing these two equations as in Figure 1.63 reveals that they represent distinct parallel lines, which never intersect.

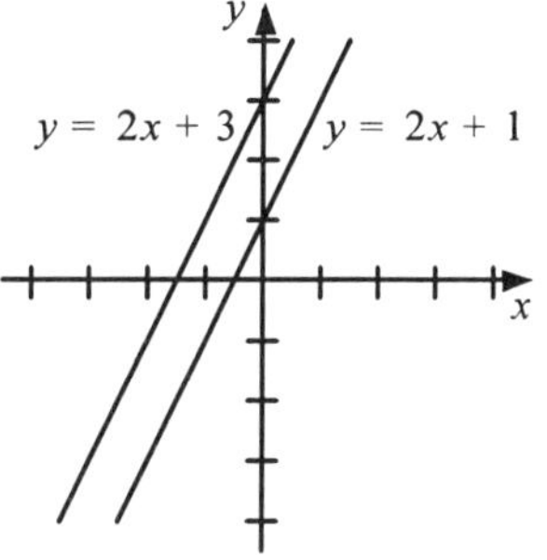

Figure 1.63

b. If we multiply both sides of the top equation by –2 and add the result to the bottom equation, both x and y are eliminated.

$$\begin{aligned} 4x - 2y &= -6 \\ -4x + 2y &= 6 \\ \hline 0 &= 0 \end{aligned}$$

The equation $0 = 0$, which is always true, indicates that the system is dependent. Thus the same line is the graph of both equations, as shown in Figure 1.64. Graphically, the solution set is the set of all points on that line, which may be specified by

$$\{(x, y)\colon -2x + y = 3\}$$

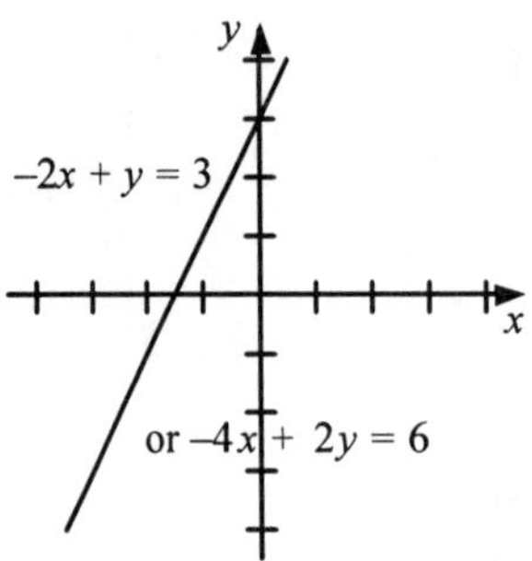

Figure 1.64

PROGRESS CHECK 7
Solve each system.

a. $y = -\frac{1}{2}x + 5$, $x + 2y = 10$ **b.** $2x - y = 5$, $-2x + y = 5$ ■

Many word problems that were solved in Chapter R using one variable are also readily solved using two variables and a system of linear equations. To illustrate, the next example reconsiders the problem in Example 4 of Section R.6 and uses the addition-elimination method to solve the problem. Take the time to compare the solution methods in these two examples.

Example 8: Managing Investments

How should a \$160,000 investment be split so that the total annual earnings are \$14,000 if one portion is invested at 6 percent annual interest and the rest at 11 percent?

Solution: To find the amount invested at each rate, let

$$x = \text{amount invested at 6 percent}$$

and

$$y = \text{amount invested at 11 percent.}$$

Use $I = Pr$ and analyze the investment in a chart format.

Investment	Principle ·	Rate =	Interest
1st account	x	0.06	$0.06x$
2nd account	y	0.11	$0.11y$

The two principals add up to $160,000 and the sum of the interests is $14,000 so the required system is

$$\begin{aligned} x + y &= 160{,}000 \quad (1) \\ 0.06x + 0.11y &= 14{,}000 \quad (2) \end{aligned}$$

To solve this system, first multiply both sides of equation (2) by 100 to clear decimals.

$$\begin{aligned} x + y &= 160{,}000 \quad (1) \\ 6x + 11y &= 1{,}400{,}000 \quad (3) \end{aligned}$$

To eliminate x, we can multiply both sides of equation (1) by -6 and then add the result to equation (3).

$$\begin{array}{rcrcl} -6x & - & 6y & = & -960{,}000 \\ 6x & + & 11y & = & 1{,}400{,}000 \\ \hline & & 5y & = & 440{,}000 \quad \text{Add the equations.} \\ & & y & = & 88{,}000 \end{array}$$

To find x, substitute 88,000 for y in equation (1).

$$\begin{aligned} x + y &= 160{,}000 \\ x + 88{,}000 &= 160{,}000 \\ x &= 72{,}000 \end{aligned}$$

Thus, the investment should be split so that $72,000 is invested at 6 percent and $88,000 is invested at 11 percent. Check this solution (as in Example 4 of Section R.6).

PROGRESS CHECK 8

How should a $50,000 investment be split so that the total annual earnings are $5,000, if one portion is invested at 8.5 percent annual interest and the rest at 10.5 percent? ■

The methods of this section may be used together with a graphing calculator to estimate solutions to linear equations and inequalities. This approach has the advantage of transforming questions about equations or inequalities to questions about graphs, which can then be answered (if only approximately) by looking at the calculator screen.

Example 9: Solving Linear Equations and Inequalities Graphically

Solve each equation or inequality graphically. Approximate the solution to the nearest tenth.

a. $2(4 - x) = 3x - 6$ **b.** $2(4 - x) < 3x - 6$

Solution:

a. Solving $2(4 - x) = 3x - 6$ is equivalent to finding the x-coordinate of the point where the graphs of $y = 2(4 - x)$ and $y = 3x - 6$ intersect. We therefore begin by displaying the graphical solution of the system defined by these equations, as shown in Figure 1.65. We see "$x = 2.8$," and an algebraic check confirms that 2.8 is an exact solution. Thus, the solution set is $\{2.8\}$.

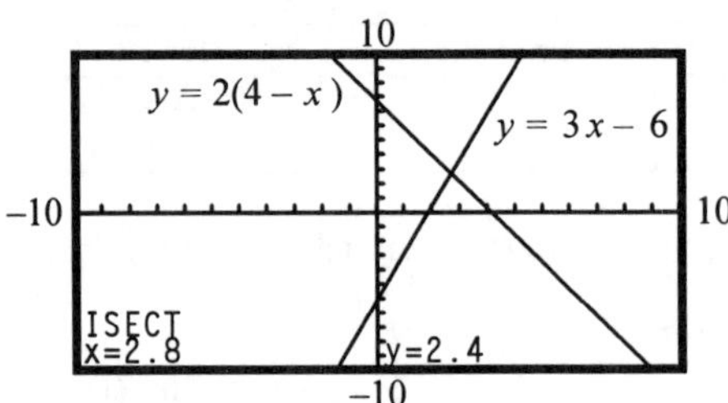

Figure 1.65

b. The graph of $y = 2(4 - x)$ lies **below** the graph of $y = 3x - 6$, to the right of the point of intersection. Thus, $2(4 - x) < 3x - 6$ is true provided $x > 2.8$. In interval notation the solution set is $(2.8, \infty)$.

PROGRESS CHECK 9

Solve each equation or inequality graphically. Approximate the solution to the nearest tenth.

a. $2x - 1 = -3(x + 1)$ **b.** $2x - 1 < -3(x + 1)$ ■

In this section we have shown several different methods for solving a linear system of equations. Keep in mind the following guidelines concerning the efficient use of each method.

Graphing: The graphing method is useful for *estimating* the solution and for *comparing* visually the linear equations in a system. When solving systems, we usually use it in conjunction with one of the algebraic methods.

Substitution: The substitution method is most efficient when at least one equation is solved for one of the variables. It is also a good choice when the system contains a variable with a coefficient of 1 or −1. We avoid this method when it leads to significant work with fractions.

Addition-Elimination: The addition-elimination method is usually the easiest when neither equation is solved for one of the variables. If you have trouble choosing, select this method.

EXPLORE 1.5

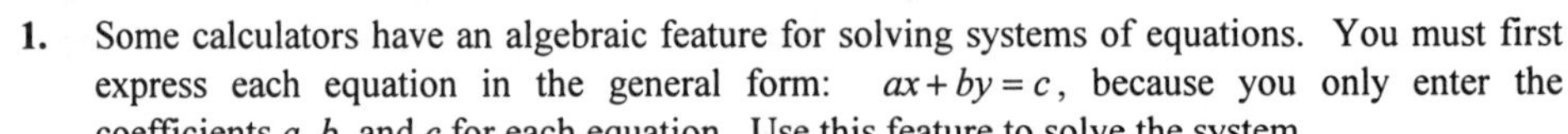

1. Some calculators have an algebraic feature for solving systems of equations. You must first express each equation in the general form: $ax + by = c$, because you only enter the coefficients *a*, *b*, and *c* for each equation. Use this feature to solve the system

$$y = 2x - 2$$
$$y = 4 + 3x.$$

2. One way to solve the inequality $2x + 1 < x - 2$ is to look at the graph of

$$y1 = (2x + 1) < (x - 2).$$

The graph will be the horizontal line $y = 1$ whenever the inequality is true, and the line $y = 0$ whenever it is false. Thus, the solution set consists of all the values of *x* for which $y = 1$. Use this method to see that the solution set is $\{x\colon\ x < -3\}$. For more details on using relational operations (such as <) on a calculator, see Example 8 in Section 1.2.

3. If your calculator has an Intersect feature, see what happens when you use it to find a point of intersection for parallel lines. Try it on this system:

$$y = 2x + 4$$
$$y = 2x + 1$$

EXERCISES 1.5

In Exercises 1–20 solve the given system.

1. $y = x - 2$
$y = 5x + 6$

2. $y = -x - 3$
$y = 2x + 3$

3. $y = 3x + 4$
$y = -x - 2$

4. $y = 4x + 7$
$y = 3x + 5$

5. $y = x + 4$
$y = 2x + 4$

6. $y = -5x + 2$
$y = 4x - 7$

7. $y = x$
$x - 2y = 6$

8. $y = -3x$
$2x + 3y = -21$

9. $y = \frac{1}{3}x - 5$
$x - 3y - 15 = 0$

10. $y = -\frac{5}{2}x + 3$
$5x + 2y + 6 = 0$

11. $x + y = 25$
$6x - y = 3$

12. $2x + 3y = 8$
$2x - 7y = -32$

13. $5x + 3y = -2$
$x - 2y = -3$

14. $-2x - y = -5$
$5x + 2y = -17$

15. $3x - 2y = 1$
$6x - 4y = 5$

16. $2x - 3y = 1$
$6x - 9y = 3$

17. $2x - 5y = 5$
$4x + 3y = 23$

18. $4x + 2y = 2$
$6x - 5y = 27$

19. $7x - 2y - 19 = 0$
$3x + 5y + 14 = 0$

20. $6x + 10y - 7 = 0$
$15x - 4y - 3 = 0$

In Exercises 21–28 set up a system of linear equations and solve by an appropriate method illustrated in this section.

21. The sum of two numbers is 70, and their difference is 22. What are the numbers?

22. Find two complementary angles whose difference is 20°.

23. Find two supplementary angles whose difference is 100°.

24. A piece of lumber is 120 in. long. Where must it be cut for one piece to be four times longer than the other piece?

25. A container holding a liquid weighs 500 g. If one-half the liquid is poured out, the weight is 350 g. What is the weight of the empty container?

26. If four black metal balls and one red metal ball are placed on a scale, they balance a weight of 100 g. A weight of 90 g will balance two black and three red balls. Find the weight of each kind of metal ball.

27. The velocity of a particle that accelerates at a uniform rate is linearly related to the elapsed time by the equation $v = v_0 + at$, where v_0 is the initial velocity and a the acceleration. If $v = 36$ ft/second when $t = 2$ seconds and $v = 4$ ft/second when $t = 3$ seconds, find values for v_0 and a.

28. In Exercise 27 find the values of v_0 and a if $v = 45$ ft/second when $t = 2$ seconds and if $v = 72$ ft/second when $t = 5$ seconds.

In Exercises 29–30 two physics principles about systems in equilibrium are used. First, the sum of the forces pointing up must equal the sum of the forces pointing down. Second, the products, $F_1 d_1$ and $F_2 d_2$ must be equal.

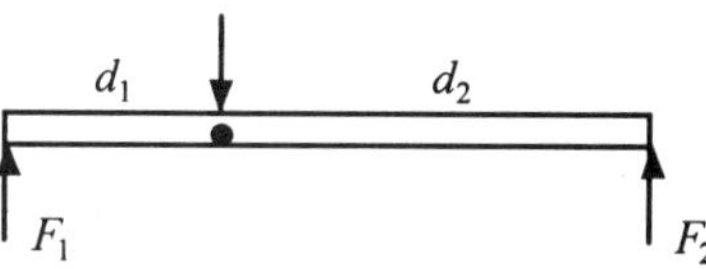

29. Find the forces F_1 and F_2 that achieve equilibrium for the beam in the following diagram.

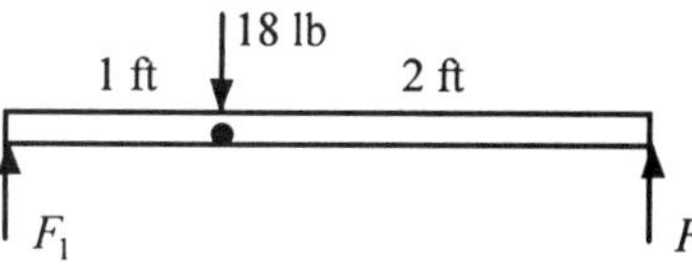

30. Find the forces F_1 and F_2 that achieve equilibrium for the beam in the following diagram.

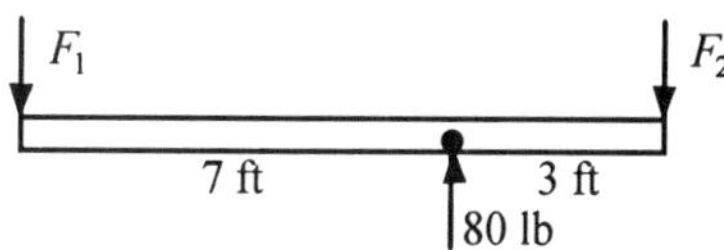

31. Two acid mixtures, one 6 percent acid and the other 8 percent acid, are to be combined to make 8 gal of a 6.5 percent acid solution. How much of each should be used?

32. A chemist has a 10 percent solution and a 25 percent solution of alcohol. How much of each should be mixed together to obtain 100 gal of a 20 percent solution?

33. A college wants to invest \$10,000 to have annual income of \$700 for a scholarship. The college plans to invest part of the money in a bank that yields 6 percent interest and the remainder in a speculative fund that promises to yield 11 percent interest. How much should be invested in each to obtain the desired income?

34. An accountant recommends two investments to a client: one earning 10 percent annually and one earning 15 percent annually. These investments should cover the cost of the client's rent (\$900 per month) for a year. If there is \$86,000 to invest, how should the money be split between the two investments.

35. You are trying to decide between two positions as a salesperson. The first offer is a straight 15 percent commission, while the second offer pays a salary of \$60 per week plus a 10 percent commission.

   a. Write a formula for each offer that expresses pay (y) as a function of sales (x).

   b. Sketch graphs of both functions on the same axes.

   c. How should you decide which offer is better financially?

36. A manufacturer wants to know whether it will pay to buy and install a special machine to turn out cartons. that until now have been purchased from an outside supplier for \$1 each. The machine will cost \$500 per year and will be able to produce cartons for 50 cents each.

   a. Write a formula for each system of paying for cartons, where cost (y) is given as a function of the number of cartons used (x).

   b. Sketch graphs of both functions on the same axes.

   c. How should the manufacturer decide which machine is better financially?

37. In economics an analysis of the law of supply and demand involves the intersection of two curves. Basically, as the price for an item increases, the quantity of the product that is supplied increases while the quantity that is demanded decreases. The point at which the supply and demand curves intersect is called the **point of market equilibrium**. This principle is shown in the figure below where, for illustrative purposes, we assume the supply and demand equations to be linear.

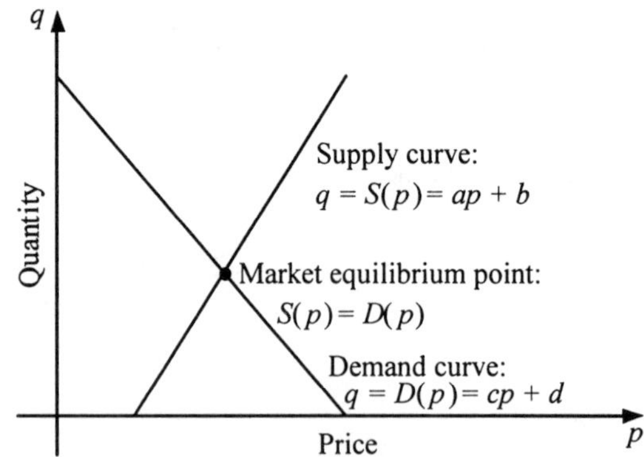

The equilibrium price is the value for p at which supply equals demand. Using the figure above, determine the equilibrium price (p) in terms of the constants a, b, c, and d.

38. In electronics the analysis of a circuit often utilizes several laws that apply at the same time. This situation leads naturally to simultaneous equations. By applying Kirchhoff's laws to the circuit shown in the diagram, we obtain the following equations:

$$R_1I_1 + R_3(I_1 + I_2) = E_1$$
$$R_2I_2 + R_3(I_1 + I_2) = E_2.$$

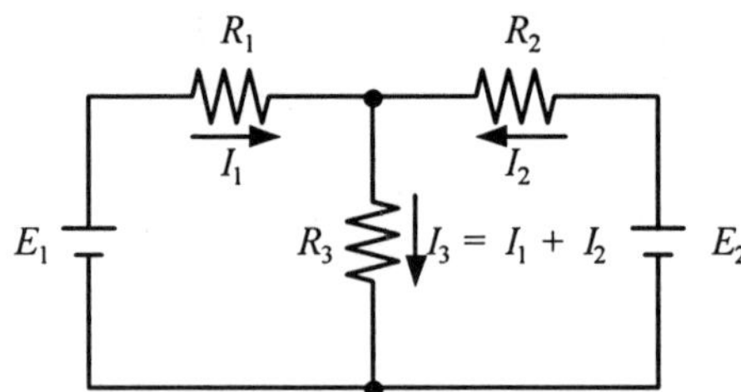

Determine the value of the currents I_1 and I_2 if $E_1 = 8$ volts, $E_2 = 5$ volts, $R_1 = 6$ ohms, $R_2 = 10$ ohms, and $R_3 = 3$ ohms.

In Exercises 39–44 solve each equation or inequality graphically. If necessary, approximate the solution to the nearest tenth.

39. **a.** $2x - 4 = 5x + 1$ **b.** $2x - 4 < 5x + 1$

40. **a.** $x = 1 - x$ **b.** $x > 1 - x$

41. **a.** $2(3 + x) = 6 + x$ **b.** $2(3 + x) \geq 6 + x$

42. **a.** $1 - 2x = 3 - 2x$ **b.** $1 - 2x \leq 3 - 2x$

43. **a.** $x = x + 2$ **b.** $x > x + 2$

44. **a.** $x + 1 = \frac{6x}{4} - \frac{x - 2}{2}$ **b.** $x + 1 < \frac{6x}{4} - \frac{x - 2}{2}$

THINK ABOUT IT 1.5

1. Describe the graph and the solution set of a system of linear equations in two variables for each type of system.
   a. inconsistent system **b.** dependent system

2. 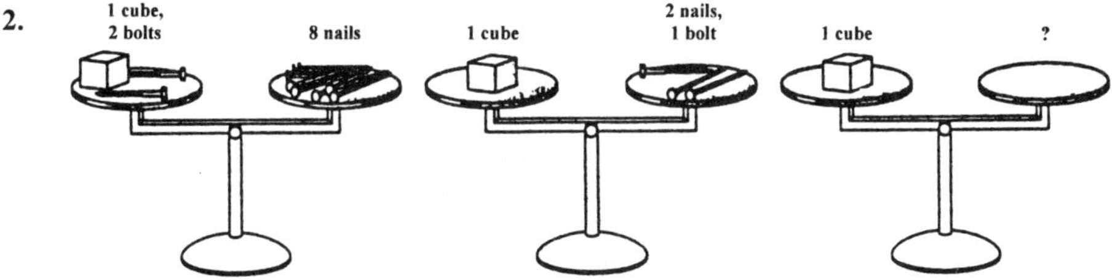

   Consider the figure. How many nails are needed to balance the cube?

3. The coordinate system we use in algebra is called rectangular because the x- and y-axes are perpendicular. Other systems are possible. For instance, let the y-axis be a vertical line, and let the x-axis make a 45° angle with it, as shown in the figure below.

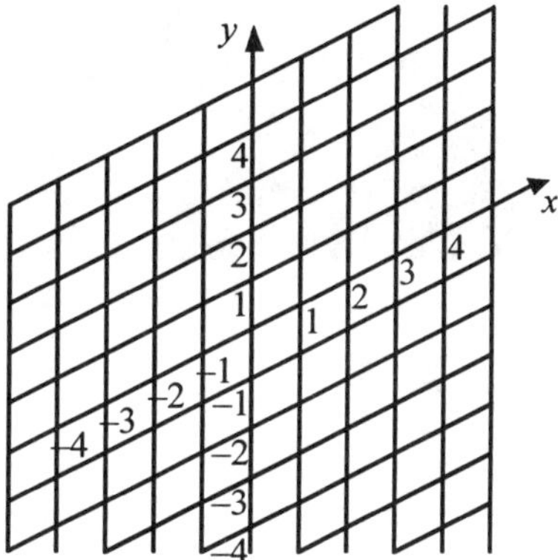

   a. Use the new system to graph this system: $\begin{matrix} x+y=2 \\ x+y=3 \end{matrix}$. On a rectangular coordinate system the equations represent parallel lines. Are the lines still parallel in the new system?

   b. Use the new system to graph this system: $\begin{matrix} x+y=7 \\ y-x=1 \end{matrix}$. On a rectangular coordinate system the lines intersect at $(3, 4)$. Do they still intersect at $(3, 4)$ in the new system?

   c. Explain why the choice of coordinate system cannot change the solution set of the system of linear equations.

The following problems are usually considered in the third semester of accounting. In these problems we find several important business considerations defined in terms of each other. Thus, they have interlocking solutions that lead naturally to simultaneous equations.

4. A company operates in a state that levies a 10 percent tax on the income that remains after paying the federal tax. Meanwhile, the federal tax is 50 percent of the income that remains after paying the state tax. If during the current year a corporation has \$400,000 in taxable income, determine the state and federal income taxes.

5. A company is to give a performance bonus to one of its top managers. The income in the manager's division is \$100,000 and the bonus rate is 10 percent. The bonus is based on profit, so it is based on income after both taxes and the bonus have been subtracted from the \$100,000. The taxes amount to 30 percent of the taxable income, with the bonus counting as a tax deduction. What is the manager's bonus?

● ● ●

1.6 Variation

Photo Courtesy of Archive Photos, New York

Johannes Kepler (1571–1630) was the first person to give a convincing argument that the orbits of the planets around the sun are elliptical rather than circular. He formulated several laws that describe the motion of the planets. Kepler's third law states that for any planet, the square of the time (T) needed to complete one orbit varies directly with the cube of the average distance (d) of the planet from the sun.

a. Express this relationship with an equation.

b. Taking the average distance of earth from the sun as 1 unit, it is known that the average distance of Mars from the sun is 1.524 units. To the nearest hundredth of a year, find the time it takes Mars to orbit the sun one time

(See Example 7.)

Objectives

1. Solve problems involving direct variation.
2. Solve problems involving inverse variation.
3. Solve problems involving combined variation or variation of powers of variables.

In many scientific laws the functional relationship between variables is stated in the language of variation. The statement "y varies *directly* as x" means that there is some nonzero number k such that $y = kx$. The constant k is called the **variation constant**. In some applications the relationship $y = kx$ is also described by saying that y is **proportional** to x, and that k is the **constant of proportionality**.

Example 1: Expressing Direct Variation by an Equation

Write a variation equation for the given relation and determine the value of the variation constant if it is known.

a. The perimeter P of a square varies directly as the side s.

b. The sales tax T on a purchase varies directly as the price p of the item.

Solution:

a. P varies directly as s means that $P = ks$. In this case we know that $k = 4$.

b. T varies directly as p means that $T = kp$. The value of k depends on the sales tax rate in a given location.

PROGRESS CHECK 1
Write a variation equation for the given relation and determine the value of the variation constant if it is known.

a. The circumference C of a circle varies directly as the diameter d.
b. The property tax T on a house varies directly as the assessed value v of the house. ■

Figure 1.66 shows the graph of the variation equation $y = kx$, $k > 0$. Note that the graph is a straight line through the origin, and that as x increases, y increases. We may determine the value of k if one ordered pair in the relation, other than (0, 0), is known. This value of k may then be used to find other corresponding values of the variables.

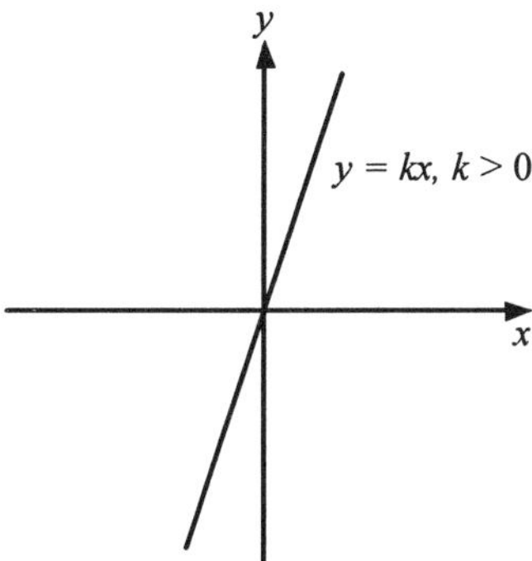

Figure 1.66

Example 2: Finding and Using k in a Direct Variation Relation

If y varies directly as x and $y = 21$ when $x = 3$, write y as a function of x. Determine the value of y when $x = 10$.

Solution: Since y varies directly as x, we have

$$y = kx.$$

To find k, replace y by 21 and x by 3.

$$21 = k \cdot 3$$
$$7 = k$$

Thus, $y = 7x$. When $x = 10$, $y = 7(10) = 70$.

PROGRESS CHECK 2
If y varies directly as x, and $y = 3$ when $x = 4$, write y as a function of x. Find y when $x = 12$. ■

Example 3: Hooke's Law

Hooke's law states that the distance (d) a spring is stretched varies directly as the force (F) applied to the spring (see Figure 1.67). The value of the constant of variation depends on the particular spring. Suppose that for one such spring a force of 20 lb stretches the spring 6 in.

a. Express Hooke's law using an equation of variation that gives d as a function of F, and find the value of k.
b. Graph the function from part **a** and give its domain.
c. Using the equation from part **a** or the graph from part **b**, determine how far the spring will be stretched by a force of 13 lb.

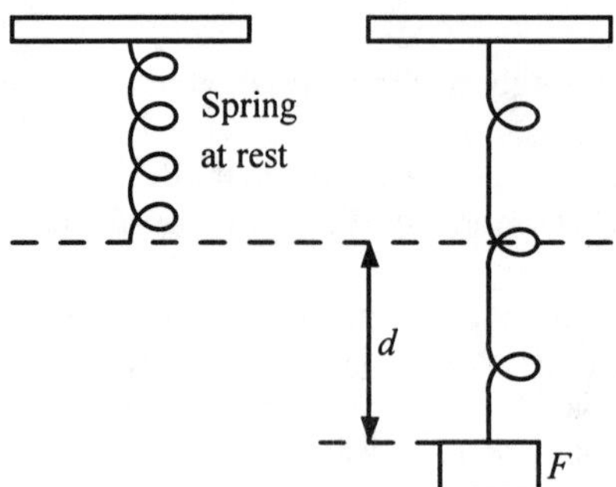

Figure 1.67

Solution:

a. Since d varies directly as F, we can write

$$d = kF.$$

To find k, replace d by 6 and F by 20.

$$6 = k \cdot 20$$
$$0.3 = k$$

Thus, the desired equation is $d = 0.3F$.

b. Note that in this application there is a natural restriction of the domain of the function, which represents the force on the spring, to nonnegative values. So we take the domain to be all nonnegative real numbers. This interval is clearly an oversimplification, because in reality tiny weights may not make any observable stretch, and extremely heavy weights will stretch the spring to its full length. The graph is shown in Figure 1.68.

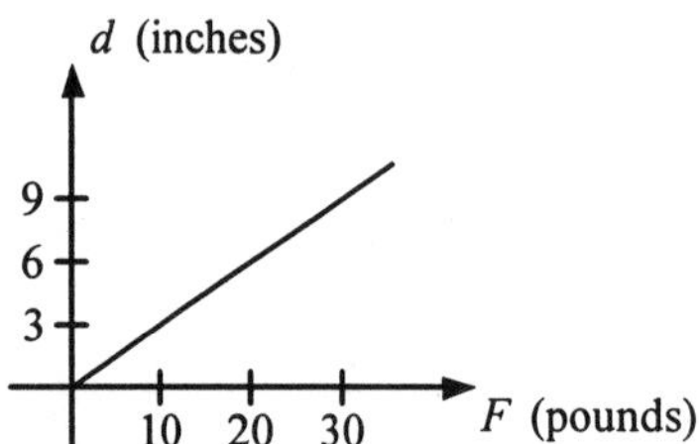

Figure 1.68

c. When $F = 13$, we get $d = 0.3(13) = 3.9$. A 13-lb force will stretch this spring 3.9 inches.

PROGRESS CHECK 3

The weight of an object on the moon w_m is proportional to its weight on Earth w_e. An object that weights 204 pounds on the Earth weighs 34 pounds on the moon.

a. Express the moon weight as a function of the Earth weight using an equation of variation, and find the value of the constant of proportionality.

b. Graph the function from part **a** and give its domain.

c. Using the equation from part **a** or the graph from part **b**, determine what a person who weighs 138 pounds on the Earth will weigh on the moon. ■

In some relationships one variable decreases as another increases. If this happens in such a way that the product of the two variables is constant, then we say that the variables **vary inversely**, or that one is **inversely proportional** to the other. The statement "y varies inversely as x" means that there is some nonzero number k (the variation constant) such that

$$xy = k \text{ or } y = \frac{k}{x}.$$

Figure 1.69 shows the graph of the variation equation $y = k/x$ for $k > 0$. The graph in this figure, which is called a **hyperbola**, consists of two disconnected curves, known as **branches**. The graph shows that the variation equation $y = k/x$ is meaningless if $x = 0$ or $y = 0$, and that on each branch as x increases, y decreases.

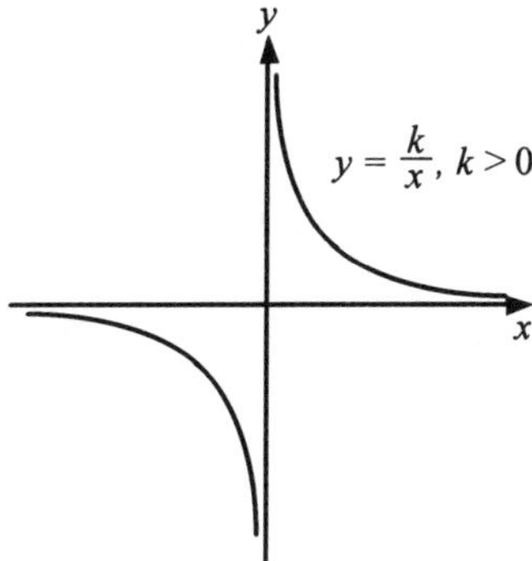

Figure 1.69

Example 4: Finding and Using k in an Inverse Variation Relation

If y varies inversely with x and $y = 27$ when $x = 2$, write y as a function of x. Find y when $x = 9$.

Solution: Since y varies inversely as x, we have

$$y = \frac{k}{x}.$$

To find k, replace y by 27 and x by 2.

$$27 = \frac{k}{2}$$
$$54 = k$$

Thus, $y = 54/x$. When $x = 9$, $y = 54/9 = 6$.

PROGRESS CHECK 4

If y varies inversely with x and $y = 13$ when $x = 5$, write y as a function of x. Find y when $x = 1$. ■

Example 5: Speed of a Gear

The speed (S) of a gear varies inversely as the number (n) of teeth. Gear A, which has 12 teeth, makes 400 rpm. How many revolutions per minute are made by a gear with 32 teeth that is connected to gear A?

Solution: Since S varies inversely with n, we have

$$S = \frac{k}{n}.$$

To find k, replace S by 400 and n by 12.

$$400 = \frac{k}{12}$$
$$4{,}800 = k$$

Thus, $S = 4{,}800/n$. When $n = 32$, $S = 4{,}800/32 = 150$ rpm.

PROGRESS CHECK 5
The speed of a pulley is inversely proportional to the radius of the pulley. The speed of the larger of two pulleys, which has a 2-inch radius, is 45 rpm. What is the speed of a $1/2$-inch pulley connected to it. ■

We may extend the concept of variation to include direct and inverse variation of variables raised to specified powers and relationships that involve more than two variables.

Example 6: Finding and Using k in a Combined Variation Relation

If y varies directly as x^2 and inversely as z^3, and $y = 18$ when $x = 3$ and $z = 2$, determine the value of y when $x = 1$ and $z = 5$.

Solution: Since y varies directly as x^2 and inversely as z^3, we have

$$y = \frac{kx^2}{z^3}.$$

To find k, replace y by 18, x by 3, and z by 2.

$$18 = \frac{k(3)^2}{(2)^3}$$
$$18 = \frac{9k}{8}$$
$$16 = k$$

Thus, $y = 16x^2/z^3$. When $x = 1$ and $z = 5$,

$$y = \frac{16(1)^2}{(5)^3} = \frac{16}{125}.$$

PROGRESS CHECK 6
If y varies directly as x^3 and inversely as z^2, and $y = 10$ when $x = 2$ and $z = 5$, determine the value of y when $x = 5$ and $z = 2$. ■

Example 7: Orbits of Planets

Solve the problem in the section introduction on page 154.

Solution:
a. Using k as the variation constant we can write

$$T^2 = kd^3$$

b. We use the data for Earth to find the value of k. For Earth we have $T = 1$ year (since it takes 1 year for Earth to orbit the sun), and $d = 1$ unit as given, so

$$\begin{aligned} 1^2 &= k(1)^3 \\ 1 &= k(1) \\ k &= 1. \end{aligned}$$

To find the time for an orbit of Mars, use $k = 1$ and $d = 1.524$ units, and solve for T.

$$\begin{aligned} T^2 &= 1(1.524)^3 \\ T &= \sqrt{(1.524)^3} \approx 1.88 \end{aligned}$$

So it takes Mars 1.88 years (about 687 days) to orbit the sun.

PROGRESS CHECK 7
The average distance from Mercury to the sun is 0.387 units. Find the time it takes Mercury to orbit the sun one time. ■

Example 8: Gravitational Attraction Between Two Objects

Newton's law of gravitation states that the gravitational attraction between two objects varies directly as the product of their masses, and inversely as the square of the distance between their centers of mass. What will be the change in attraction between the two objects if both masses are doubled and the distance between their centers is cut in half?

Solution: If F represents the gravitational attraction, m_1 and m_2 represent the masses of the objects, and d represents the distance between the centers of mass, then algebraically, we write Newton's law as

$$F = \frac{km_1m_2}{d^2}.$$

If the masses are doubled and the distance is cut in half, we have

$$\begin{aligned} F &= \frac{k(2m_1)(2m_2)}{[(1/2)d]^2} \\ &= \frac{4km_1m_2}{(1/4)d^2} \\ &= \frac{16km_1m_2}{d^2}. \end{aligned}$$

Thus, the gravitational attraction becomes 16 times as great.
Note: If y varies directly as the product of other variables, say x and w, we write $y = kxw$, and we say that y **varies jointly as** x and w. In this example an alternative way to say that F varies directly as the product m_1m_2 is to say that F varies jointly as m_1 and m_2.

PROGRESS CHECK 8
Using Newton's law of gravitation from Example 8, what will be the change in attraction between two objects if both masses are cut in half and the distance between them is doubled? ■

Finally, we point out that any variation problem may also be solved by using a proportion. For example, if y varies directly as x, we know that $y = kx$. Thus, $y/x = k$ and there is a constant ratio between any corresponding values of x and y. This means that

$$\frac{y_1}{x_1} = \frac{y_2}{x_2} \text{ and } \frac{y_1}{y_2} = \frac{x_1}{x_2}.$$

The equation $y_1/y_2 = x_1/x_2$ is called a **proportion** and may be used to solve the problem. It is for this reason that the variation constant k is often called the constant of proportionality, and the expression "varies directly as" is often replaced by "is proportional to." However, the language of variation usually provides a more convenient and informative statement of a relationship.

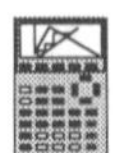

EXPLORE 1.6

1. The set of variation functions given by $y = kx^n$ $(x > 0)$ for various integer values of n form a family of functions. Use a grapher with window $[0, 5]$ by $[0, 5]$ to examine the graphs of $y = 2x^n$ for $n = -3, -2, -1, 0, 1, 2,$ and 3. Describe in a few sentences the effect that the value of n has on the graph. What values of n yield graphs that are increasing, decreasing, or constant? Which graph increases the fastest? Which decreases the fastest? Is there any point common to all 7 graphs?
2. The set of variation functions given by $y = k/x$ $(x > 0)$ for positive values of k forms a family of functions. Use a grapher with window $[0, 5]$ by $[0, 5]$ to examine the graphs of $y = k/x$ for $k = 4, 3, 2, 1, 0.5, 0.1$. Describe in a few sentences the effect that the value of k has on the graph. Is there any point common to all 6 graphs? Find where each graph intersects the line $y = x$.

EXERCISES 1.6

In Exercises 1–6 write a variation equation for the given relation and determine the value of the variation constant if it is known.

1. The perimeter P of an equilateral triangle varies directly as the side s.

2. The circumference C of a circle varies directly as the radius r.

3. The sales tax T on a car varies directly as the price of the car.

4. The air resistance R on a falling object varies directly as its velocity v.

5. According to Boyle's law, the volume V of a given mass of gas at a constant temperature varies inversely as the pressure P exerted on it.

6. In a manufacturing process the cost per unit c varies inversely as the number n of units produced.

7. If y varies directly as x, and $y = 14$ when $x = 6$, write y as a function of x. Determine the value of y when $x = 10$.

8. If y varies directly as x, and $y = 2$, when $x = 5$, write y as a function of x. Determine the value of y when $x = 11$.

9. If y varies directly as the square of x, and $y = 3$ when $x = 2$, write y as a function of x. Find y when $x = 4$.

10. If y varies directly as x^3, and $y = 5$ when $x = 3$, write y as a function of x. Find y when $x = 2$.

11. If y varies inversely as x, and $y = 9$ when $x = 8$, write y as a function of x. Find y when $x = 24$.

12. If y varies inversely as x, and $y = 3$ when $x = 7$, write y as a function of x. Find y when $x = 21$.

13. If y varies inversely as x^3, and $y = 3$ when $x = 2$, write y as a function of x. Find y when $x = 3$.

14. If y varies inversely as the square of x, and $y = 2$ when $x = 4$, write y as a function of x. Find y when $x = 8$.

15. If y varies directly as x and z, and $y = 105$ when $x = 7$ and $z = 5$, find y when $x = 10$ and $z = 2$.

16. If y varies directly as x and inversely as z, and $y = 10$ when $x = 4$ and $z = 3$, find y when $x = 7$ and $z = 15$.

17. If y varies directly as x and inversely as z^2, and $y = 36$ when $x = 4$ and $z = 7$, find y when $x = 9$ and $z = 9$.

18. If y varies inversely as x^2 and z^3 and $y = 0.5$ when $x = 3$ and $z = 2$, find y when $x = 2$ and $z = 3$.

19. In a spring to which Hooke's law applies (see Example 3), a force of 15 lb stretches the spring 10 in. How far will the spring be stretched by a force of 6 lb?

20. The weight of an object on the moon varies directly with the weight of the object on Earth. An object that weighs 114 lb on Earth weighs 19 lb on the moon. How much will a person who weighs 174 lb on Earth weigh on the moon?

21. The amount of garbage produced in a given location varies directly with the number of people living in the area. It is known that 25 tons of garbage are produced by 100 people in 1 year. If there are 8 million people in New York City, how much garbage is produced by New York City in 1 year?

22. Property tax varies directly as assessed valuation. The tax on property assessed at $12,000 is $400. What is the tax on property assessed at $40,000?

23. If the area of a rectangle remains constant, the length varies inversely as the width. The length of a rectangle is 9 in. when the width is 8 in. If the area of the rectangle remains constant, find the width when the length is 24 in.

24. The speed of a gear varies inversely as the number of teeth. Gear A with 48 teeth makes 40 rpm. How many revolutions per minute are made by a gear with 120 teeth that is connected to gear A.

25. The speed of a pulley varies inversely as the diameter of the pulley. The speed of pulley A, which has an 8-in. diameter, is 450 rpm. What is the speed of a 6-in. diameter pulley connected to pulley A.

26. The time required to complete a certain job varies inversely as the number of machines that work on the job (assuming each machine does the same amount of work). It takes five machines 55 hours to complete an order. How long will it take 11 machines to complete the same job?

27. The volume of a sphere varies directly as the cube of its radius. The volume is 36π cubic units when the radius is 3 units. What is the volume when the radius is 5 units?

28. The distance an object falls due to gravity varies directly as the square of the length of time of the fall. If an object falls 144 ft in 3 seconds, how far did it fall the first second?

29. The weight of an object varies inversely as the square of the distance from the object to the center of the Earth. At sea level (4,000 mi from the center of the Earth) a man weighs 200 lb. Find his weight when he is 200 mi above the surface of the Earth.

30. The intensity of light on a plane surface varies inversely as the square of the distance from the source of light. If we double the distance from the source to the plane, what happens to the intensity?

31. The exposure time for photographing an object varies inversely as the square of the lens diameter. What will happen to the exposure time if the lens diameter is cut in half?

32. The resistance of a wire to an electrical current varies directly as its length and inversely as the square of its diameter. If a wire 100 ft long with a diameter of 0.01 in. has a resistance of 10 ohms (Ω), what is the resistance of a wire of the same length and material but 0.03 in. in diameter.

33. The general gas law states that the pressure of an ideal gas varies directly as the absolute temperature and inversely as the volume. If $P = 4$ atmospheres (atms) when $V = 10\text{ cm}^3$ and $T = 200°$ Kelvin, find P when $V = 30\text{ cm}^3$ and $T = 250°$ Kelvin.

34. The safe load of a beam (the amount it supports without breaking) that is supported at both ends varies directly as the width and the square of the height, and inversely as the distance between supports. If the width and height are doubled and the distance between supports remains the same, what is the effect on the safe load?

35. Coulomb's law states that the magnitude of the force that acts on two charges q_1 and q_2 varies directly as the product of the magnitude of q_1 and q_2 and inversely as the square of the distance between them. If the magnitude of q_1 is doubled, the magnitude of q_2 is tripled, and the distance between the charges is cut in half, what happens to the force?

36. Newton's law of gravitation states that the gravitational attraction between two objects varies directly as the product of their masses, and inversely as the square of the distance between their centers of mass. What will be the change in attraction between the two objects if both masses are cut in half and the distance between their centers is cut in half?

37. Three people invest in a business venture. A puts in \$1000, B invests \$1800, and C invests \$2400. They agree to split the profits each year in proportion to their original investment.

a. Complete this chart:

Year	Profit	A share	B share	C share
1.	\$2,000			
2.	\$3,600			

b. Find formulas for each person's share if the profit is x dollars.

38. Four people invest in a business venture. A puts in \$1000, B invests \$1800, C invests \$2000, and D invests \$2400. They agree to split the profits each year in proportion to their original investment.

a. Complete this chart:

Year	Profit	A share	B share	C share	D share
1.	\$24,000				
2.	\$60,000				

b. Find formulas for each person's share if the profit is x dollars.

39. In some office buildings designers must consider the potential damage to the floor tile due to the pressure of spike-heeled shoes on the floor. The pressure on the floor is given by

$$P = \frac{\text{weight on spike}}{\text{area of spike that touches the floor}}$$

a. Suppose that roughly speaking the spike is a square with side s inches and that $1/4$ of the woman's weight rests on the spike while she is walking. Find a formula for the pressure on the floor. Use w for the person's weight and s for the side of the square.

b. For a woman who weighs 120 pounds, write the formula, graph it for $0 < s \le 1$, showing the domain and range, and describe the relation using the language of variation.

40. Galileo found by experiment that the distance traveled by a falling body varies directly as the square of the elapsed time. We write this relation today as $d = kt^2$. He did not have our modern notation; he expressed the relationship in the language of proportion, which had been in use at least since the time of Euclid. His statement looked more like the following.

If a body falls distance d_1 in time t_1, and if it falls distance d_2 in time t_2, then $d_1{:}d_2 = t_1^2{:}t_2^2$.

a. Show that the variation and proportion versions are equivalent by writing out the variation equation twice, once for d_1 and once for d_2, and equating the resulting values for k.

b. Suppose we use a stopwatch and observe that the body falls 4 feet in the first half second. What is the value of k for this relationship?

41. The graph of an inverse variation relation goes through the point (2, 4). Find an equation for this relation, and give two other points on the graph.

42. The graph of a direct variation relation goes through the point (2, 4). Find an equation for this relation, and give two other points on the graph.

THINK ABOUT IT 1.6

1. **a.** If y varies directly as x, does x vary directly as y?

b. In an example, y varies directly as x, with variation constant equal to 10. Write this as an equation; then rewrite the equation to show that x varies directly as y. What is the new variation constant?

2. **a.** If y varies inversely as x, does x vary inversely as y?

b. In an example, y varies inversely as x, with variation constant equal to 5. Write this as an equation; then rewrite the equation to show that x varies inversely as y. What is the new variation constant?

3. **a.** What happens to the *circumference* of a circle when you triple the radius? Explain this in terms of direct variation.

b. What happens to the *area* of a circle when you triple the radius? Explain this in terms of direct variation.

c. The area of a circle varies directly as the square of the diameter. What is the variation constant in this relation.

4. **a.** Solve the problem in Exercise 20 by setting up a *proportion.*

b. Solve the problem in Exercise 23 by setting up a *proportion.*

5. **a.** Describe two instances not discussed in this section in which one variable varies *directly* as another variable.

b. Describe two instances not discussed in this section in which one variable varies *inversely* as another variable.

● ● ●

CHAPTER 1 SUMMARY

OBJECTIVES CHECKLIST

Specific chapter objectives are summarized below along with numbered example problems from the text that should clarify the objectives. If you do not understand any objectives, or do not know how to do the selected problems, then restudy the material.

1.1: Can you:

1. **Find a formula that defines a function, and determine its domain and range?** Express the monthly cost (c) for a checking account as a function of the number (n) of checks serviced that month, if the bank charges 10 cents per check plus a $3.50 maintenance charge. Describe the domain and range of the function.

   [Example 1]

2. **Determine ordered pairs that are solutions of an equation?** Which of the following ordered pairs are solutions of the equation $y = 3x - 2$?

   a. $(-1, -5)$ **b.** $(-5, -1)$

   [Example 2]

3. **Determine if a set of ordered pairs is a function?** Determine if the relation is a function. Also, specify the domain and the range.

   a. $\{(-1, 0), (0, 0), (1, 0)\}$ **b.** $\{(0, -1), (0, 0), (0, 1)\}$

   [Example 3]

4. **Graph an ordered pair and determine the coordinates of a point?** Represent the ordered pairs $(0, -3)$, $(-4, -2)$, $(\pi, -5/2)$, and $\left(-4/3, \sqrt{2}\right)$ as points in the Cartesian coordinate system.

   [Example 4]

5. **Graph a function by the point-plotting method or by using a grapher?** Graph $y = 4 - x^2$ and indicate the domain and range of the function on the graph.

   [Example 6]

1.2: Can you:

1. **Evaluate functions using functional notation?** If $y = f(x) = 2x^2 - x + 4$ find $f(2)$, $f(15)$, and $f(-3)$.

   [Example 1]

2. **Write the difference quotient for a function in simplest form?** Find the difference quotient for $f(x) = x^2 + 2x$ in simplest form.

   [Example 5]

3. **Find function values and graphs for piecewise functions?** Graph the function defined as follows:

$$f(x) = \begin{cases} -1 & \text{if } x < 1 \\ x + 2 & \text{if } x \geq 1 \end{cases}$$

   Indicate the domain and range on the graph.

   [Example 8]

4. **Read from a graph of function *f* the domain, range, function values, values of *x* for which $f(x)=0$, $f(x)<0$, or $f(x)>0$, intercepts and intervals where *f* is increasing, decreasing, or constant?** Consider the graph of $y=f(x)$ in the figure below.

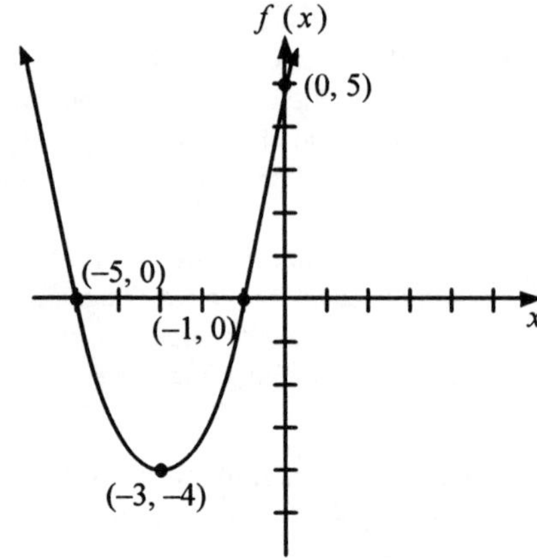

a. What is the domain of *f*?
b. What is the range of *f*?
c. Determine $f(0)$.
d. For what values of *x* does $f(x)=0$?
e. For what values of *x* is $f(x)<0$?
f. Solve $f(x)>0$.
g. Find all intercepts.
h. Find the open intervals on which *f* is increasing, decreasing, or constant.

[Example 10]

1.3: Can you:

1. **Graph a function using translation, reflection, stretching, or shrinking?** Use the graph of $y=x^2$ to graph each of the following functions.

   a. $y=x^2+3$ b. $y=(x-2)^2$ c. $y=(x+1)^2-2$

[Example 1]

2. **Prove or classify a function to be even, odd, or neither of these?** Show that the identity function $f(x)=x$ is an odd function.

[Example 9]

3. **Complete the graph of an even function or odd function given the graph for $x\geq 0$? Is the** function $y=(x^2-1)/x^3$ an even function, an odd function, or neither? The figure below is the graph of this function for $x\geq 0$. Complete the graph and check your result with a grapher.

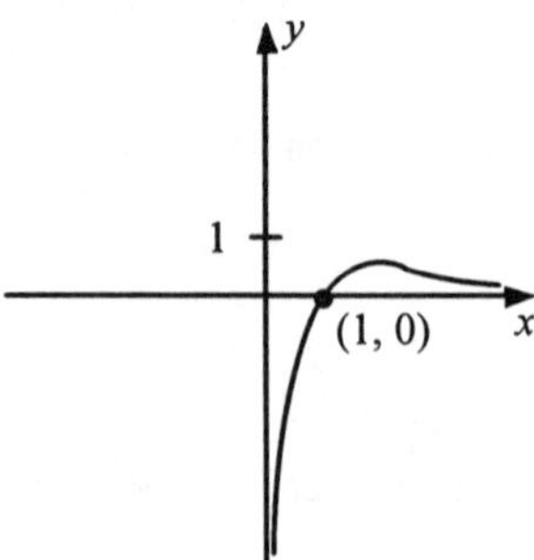

[Example 10]

1.4: Can you:

1. **Find the distance between two points?** Find the distance between $(-3,\ -2)$ and $(-1,\ 4)$.
   [Example 2]

2. **Graph and write equations for circles and semicircles with centers at the origin?** Graph each equation and indicate the domain and range of each relation.
   a. $x^2+y^2=25$ **b.** $y=\sqrt{25-x^2}$ **c.** $y=-\sqrt{25-x^2}$
   [Example 4]

3. **Determine if a graph or equation defines *y* as a function of *x*?** Does the given rule determine *y* as a function of *x*? Use both algebraic and geometric methods.
   a. $x+y=4$ **b.** $x^2+y^2=4$
   [Example 7]

1.5: Can you:

1. **Solve a system of linear equations by the graphing method, the substitution method, or the addition-elimination method?** Solve by addition-elimination: $\begin{aligned}3x-2y&=27\\2x+5y&=-1\end{aligned}$
   [Example 6]

2. **Solve applied problems by setting up and solving a system of linear equations?** How should a $160,000 investment be split so that the total annual earnings are $14,000 if one portion is invested at 6 percent annual interest and the rest at 11 percent?
   [Example 8]

3. **Solve linear equations and linear inequalities in one variable graphically?** Solve each equation or inequality graphically. Approximate the solution to the nearest tenth.
   a. $2(4-x)=3x-6$ **b.** $2(4-x)<3x-6$
   [Example 9]

1.6: Can you:

1. **Solve problems involving direction variation?** If *y* varies directly as *x* and $y=21$ when $x=3$, write *y* as a function of *x*. Determine the value of *y* when $x=10$.
   [Example 2]

2. **Solve problems involving inverse variation?** If *y* varies inversely as *x* and $y=27$ when $x=2$, write *y* as a function of *x*. Find *y* when $x=9$.
   [Example 4]

3. **Solve problems involving combined variation or variation of powers of variables?** If *y* varies directly as x^2 and inversely as z^3, and $y=18$ when $x=3$ and $z=2$, determine the value of *y* when $x=1$ and $z=5$.
   [Example 6]

KEY CONCEPTS AND PROCEDURES

Section	Key Concepts and Procedures to Review
1.1	• Rule definitions of a function and its domain and range. • Ordered-pair definitions of a relation and function and their domain and range. • Definition of the graph of a relation. • Graph of $y=mx+b$ is a straight line. • Graph of $y=ax^2+bx+c\quad(a\neq 0)$ is a parabola. • Graph of constant function is a horizontal line. • Guidelines for graphing functions with a graphing calculator.
1.2	• The term $f(x)$ is read "*f* of *x*" or "*f* at *x*" and means the value of the function (the *y* value) corresponding to the value of *x*. • If *a* is in the domain of *f*, then ordered pairs for the function defined by $y=f(x)$ all have the form $(a, f(a))$. **Note:** *a* is an *x* value and $f(a)$ is a *y* value.

- Definition of a piecewise function
- Methods to construct, evaluate, and graph a piecewise function

1.3
- Graphs of basic functions
  $y = c$: constant function $y = x$: identity function
  $y = x^2$: squaring function $y = x^3$: cubing function
  $y = |x|$: absolute value function $y = \sqrt{x}$: square root function
- Methods to graph variations of a familiar function by using vertical and horizontal shifts, reflecting, stretching, and shrinking.
- Definitions of an even function and an odd function
- The graph of an even function is symmetric about the y-axis.
- The graph of an odd function is symmetric about the origin.

1.4
- Distance formula: $d = \sqrt{(x_2 - x_1)^2 + (y_2 - y_1)^2}$ (for any two points)
- Equations and graphs of circles and semicircles centered at the origin
- Vertical line test
- Statement of the fundamental principle in coordinate (or analytic) geometry

1.5
- Definition of a system of linear equations
- Definitions of consistent, inconsistent and dependent systems
- The solution of a system of linear equations in two variables is the set of all the ordered pairs that satisfy both equations. Graphically this corresponds to the collection of points where the lines intersect.
- Methods to solve a system of equations by the substitution method and by the addition-elimination method.

1.6
- The statement "y varies directly as x" means there is some nonzero number k (variation constant) such that $y = kx$.
- The statement "y varies inversely as x" means there is some nonzero number k such that $xy = k$ or $y = k/x$.

CHAPTER 1 REVIEW EXERCISES

1. Fill in the missing component in each of the following ordered pairs so they are solutions of the equation $y = -2x + 4$: (0,), (, −1), (−1,), (, 0), (3,).

2. Give three ordered pairs in f if $f(x) = 2$.

3. Is the set of ordered pairs {(−1, 3), (0, 3), (1, 3)} a function?

4. Write a formula expressing the circumference of a circle as a function of its radius.

5. Is the following graph the graph of a function?

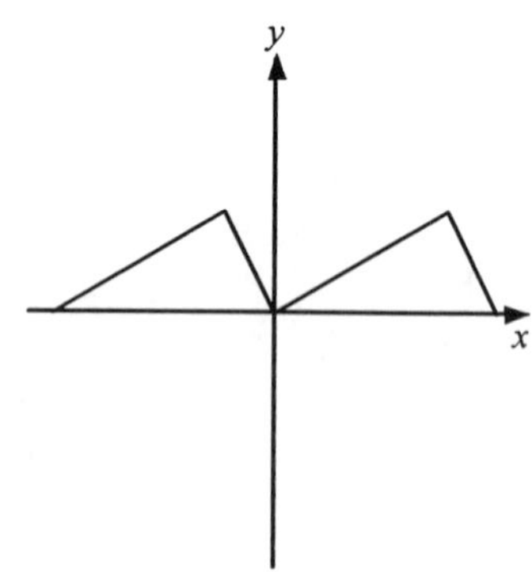

In Exercises 6–8 solve the given system of linear equations using the method indicated.

6. $\begin{array}{l} y = x + 1 \\ y - 2x + 4 = 0 \end{array}$ (graphing)

7. $\begin{array}{l} y = x - 1 \\ y = 4 - 2x \end{array}$ (substitution)

8. $\begin{array}{l} x + y = 5 \\ 2x - 3y = 1 \end{array}$ (addition-elimination)

9. If y varies directly as x and $y = 5$ when $x = 12$, write y as a function of x. Determine the value of y when $x = 20$.

10. Find the distance between $(-2, 1)$ and $(3, -1)$.

In Exercises 11 and 12 state the domain and range of the function.

11. $g(x) = \dfrac{1}{x+4}$ **12.** $y = \sqrt{x+7}$

13. Write a formula that expresses the area of a square as a function of the diagonal. (*Hint*: Use the following drawing and the Pythagorean theorem.)

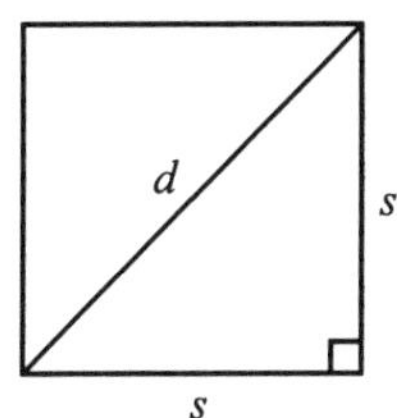

14. The speed of a pulley varies inversely as the diameter of the pulley. The speed of pulley A, which has a 7-in. diameter, is 320 rpm. What is the speed of a 5-in. diameter pulley that is connected to pulley A?

15. Are the following points the vertices of a right triangle: $(1, -3)$, $(8, -7)$ and $(5, -1)$?

In Exercises 16–21 let $f(x) = x - 1$ and $g(x) = 2x$.

16. Evaluate $3f(3) - 4g(-2)$.

17. Does $f[g(1)] = g[f(1)]$?

18. Does $f(a+b) = f(a) + f(b)$?

19. Does $g(a+b) = g(a) + g(b)$?

20. Is f an even function, an odd function, or neither?

21. Is g an even function, an odd function, or neither?

In Exercises 22–25 use the graph of $y = x^2$ to graph each function.

22. $y = (x+1)^2$ **23.** $y = x^2 + 1$

24. $y = -x^2$ **25.** $y = (x-2)^2 - 1$

26. Is the function $y = 3/x^2$ even, odd, or neither of these? The following figure gives the graph of this function for $x \geq 0$. Complete the graph.

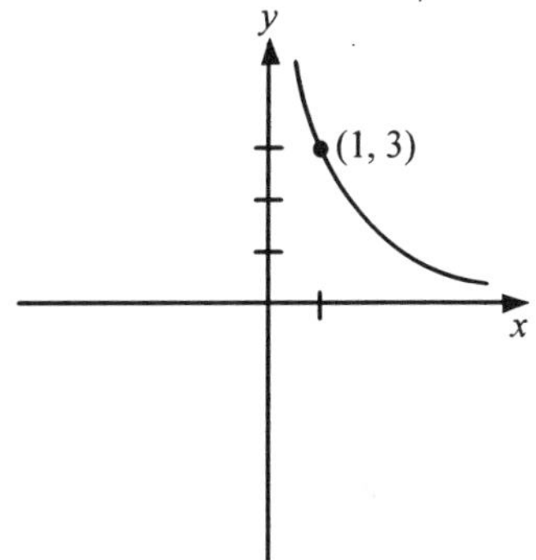

27. The intensity of light on a surface is inversely proportional to the square of the distance from the source of light. You are reading a book 6 ft from an electric light. If you move 2 ft closer, how many times greater is the intensity of light on the book?

28. A circle of radius 5 has its center at $(-1, -1)$. Is the point $(3, 2)$ on this circle?

29. Express the area (A) of a circle as a function of its diameter (d).

In Exercises 30–33 consider the function

$$f(x) = \begin{cases} -x & \text{if} \quad 0 \leq x \leq 3 \\ 3 & \text{if} \quad x > 3. \end{cases}$$

30. Find $f(3)$. **31.** Graph the function.

32. What is the domain? **33.** What is the range?

In Exercises 34–37 use the following graph of $y = f(x)$ to graph each function.

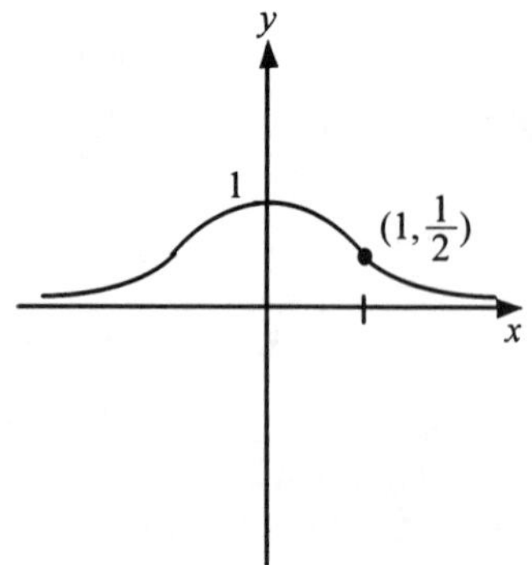

34. $y = 2f(x)$ **35.** $y = -\frac{1}{2}f(x)$

36. $y = f(x-2)$ **37.** $y = f(x) - 2$

In Exercises 38–41 graph the given function.

38. $y = -3$ **39.** $f(x) = 2x - 4$

40. $g(x) = 9 - x^2$ **41.** $y = |x+5|$

In Exercises 42 and 43 classify each function as even, odd, or neither of these.

42. $f(x) = \frac{|x|}{x}$ **43.** $y = x^4 + 1$

44. Is a vertical line the graph of a function, a relation, or both? Is a horizontal line the graph of a function, a relation, or both?

45. Match the graph of $y = x^2$ to the given viewing window:
a. [−1, 1] by [−1, 1]
b. [−10, 10] by [−10, 10]
c. [−100, 100] by [−100, 100]

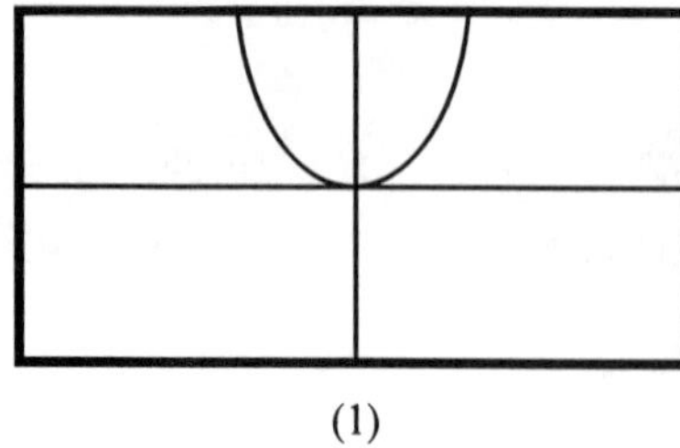

(1)

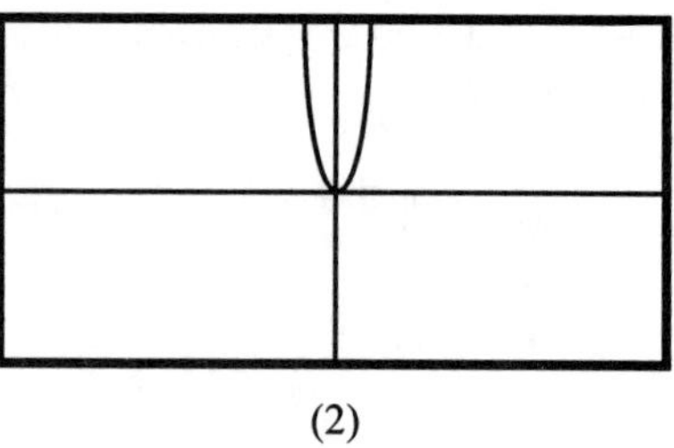

(2)

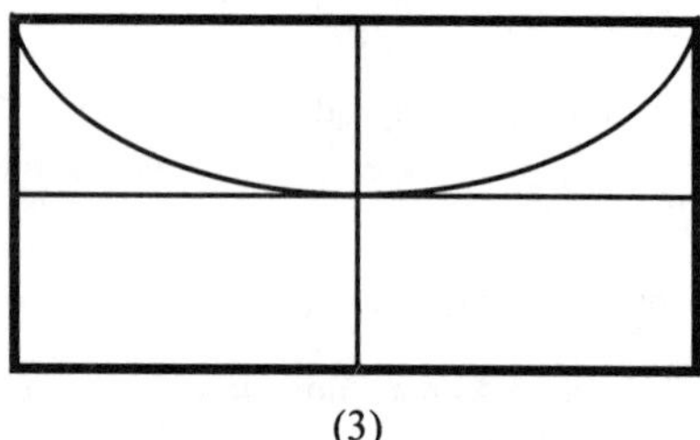

(3)

In Exercises 46–54 use the following figure.

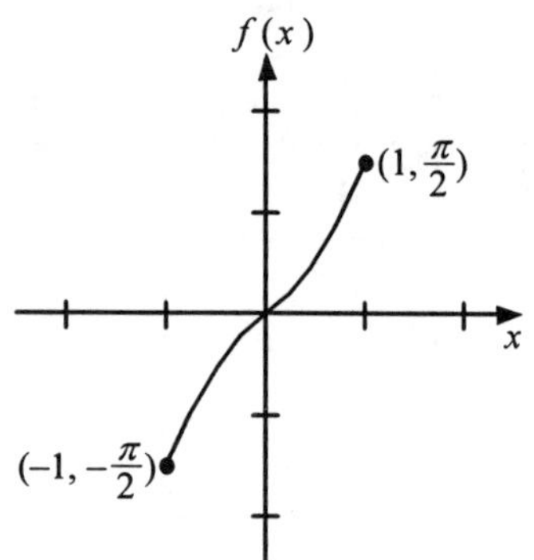

46. What is the domain? **47.** What is the range?

48. If f even, odd, or neither?

49. Find $f(0)$. **50.** Solve $f(x) = 0$.

51. Solve $f(x) > 0$. **52.** Solve $f(x) \leq 0$.

53. Find all intercepts.

54. Find the open intervals on which f is increasing, decreasing, or constant.

In Exercises 55 and 56 compute in simplest form

$$\frac{f(x+h) - f(x)}{h} \text{ with } h \neq 0.$$

55. $f(x) = 3x^2 - 1$ **56.** $f(x) = 3 - x$

57. Solve graphically: $3(2-x) \geq 2x+1$.

58. Find a formula for this semicircle.

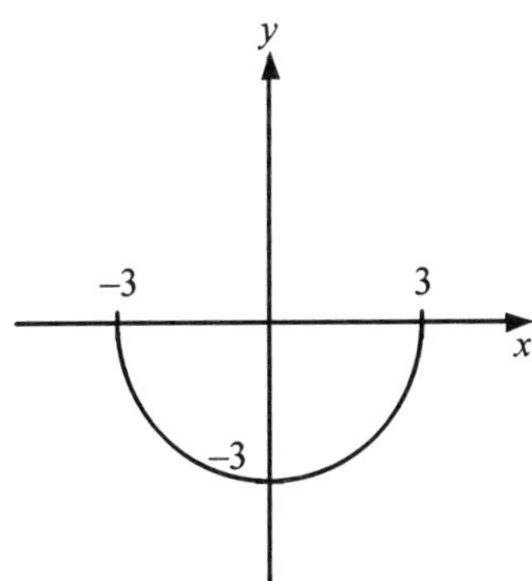

59. A company is trying to decide between two machines for packaging its new product. Machine A will cost \$5,000 per year plus \$2 to package each unit; machine B will cost \$8,000 per year plus \$1 to package each unit.
   a. Write a formula for each machine that expresses cost (y) as a function of the number of units packaged (x).
   b. How many units must be packaged for the cost of the two machines to be the same? If the company plans to package more units than that which machine should it purchase?
   c. If packaging can be subcontracted to another firm at a cost of \$5 per unit, how many units would have to be packaged before the purchase of machine A would be worthwhile?

60. A rectangle of sides x and y units is inscribed in a circle of radius 5 units. Express the area of the rectangle as a function of x. What are the domain and range of the function?

61. Determine if $(5, -2)$ is a solution of the system

$$\begin{aligned} x - y &= 7 \\ 2x + 3y &= 4 \end{aligned}$$

62. A rectangular field is completely enclosed by x feet of fencing. If the length of the rectangular region is twice the width, then express the area of the field as a function of x.

63. Express the monthly earnings (e) of a salesperson in terms of the cash amount (a) of merchandise sold, if the salesperson earns \$500 per month plus 9 percent commission on sales above \$2,000.

64. How should a \$40,000 investment be split so that the total annual earnings are \$2,700, if one portion is invested at 7 percent annual interest and the rest at 5 percent?

In Exercises 65–74 use the following figure.

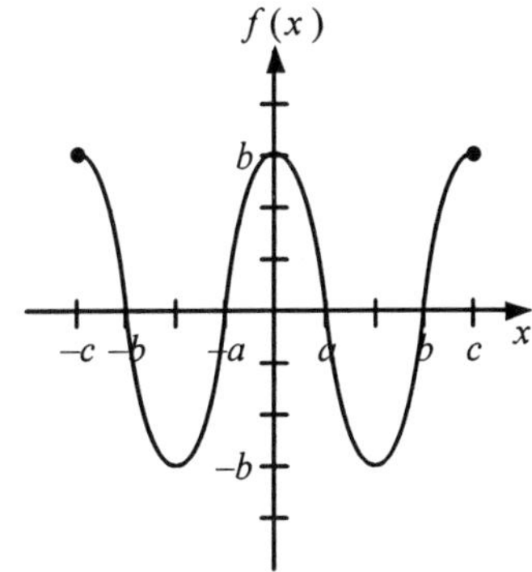

65. What is the domain? 66. What is the range?

67. Is f even, odd, or neither?

68. Find $f(b)$. 69. Solve $f(x)=b$.

70. Solve $f(x)<0$. 71. Graph $y=-\frac{1}{2}f(x)$

72. Graph $y=f(x-c)$. 73. Find all intercepts.

74. Find the open intervals on which f is increasing, decreasing, or constant.

75. Find the domain and range of the functions shown.
   a.

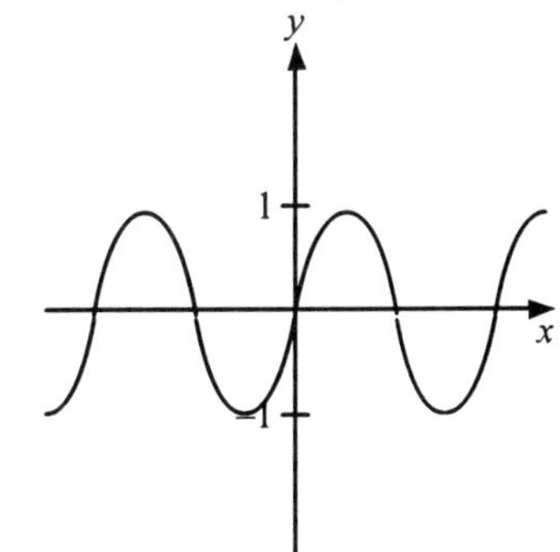

b.

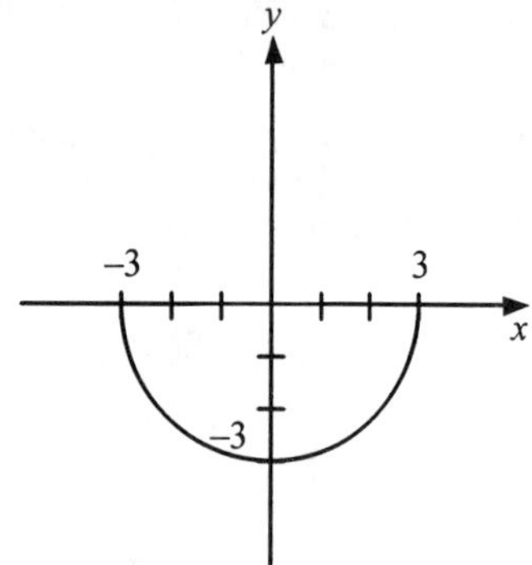

In Exercises 76–85 select the choice that answers the question or completes the statement.

76. Which number is not in the domain of $y = \frac{x+2}{x+3}$?

a. 2 **b.** −2 **c.** −3 **d.** 3

77. The graph of $y = f(x) - 3$ is the graph of $y = f(x)$ shifted 3 units

a. up **b.** down **c.** left **d.** right

78. If $f(x) = 5x^2$, then the difference quotient $\frac{f(x+h) - f(x)}{h}$ with $h \neq 0$ in simplest form is

a. $10x + 5h$ **b.** $5h$
c. 1 **d.** $10xh + 5h$

79. The area of a circle whose radius extends from (−2, 5) to (5, 1) is

a. 11π **b.** 25π **c.** 45π **d.** 65π

80. If y varies inversely with x, then when x is doubled y is

a. decreased by 2 **b.** increased by 2
c. multiplied by 2 **d.** divided by 2

81. If $(a, -1)$ is a solution of $y = -4x + 5$, then a equals

a. 9 **b.** $\frac{3}{2}$ **c.** $-\frac{2}{3}$ **d.** −1

82. The graph of an odd function is symmetric with respect to the

a. x-axis **b.** y-axis **c.** origin

83. If $f(x) = 2x + b$ and $f(-2) = 3$, then b equals

a. 7 **b.** −1 **c.** 1 **d.** −8

84. If a square field is completely enclosed by x feet of fencing, then the area of the field as a function of x equals

a. x^2 **b.** $\frac{x^2}{4}$ **c.** $4x^2$ **d.** $\frac{x^2}{16}$

85. If $f(x) = 2x$, then which one of the following is true?

a. $f(-a) = -f(a)$ **b.** $f(-a) = f(a)$
c. $f(ab) = f(a) \cdot f(b)$ **d.** $f\left(\frac{a}{b}\right) = \frac{f(a)}{f(b)}$

CHAPTER 1 TEST

1. If $f(x) = \frac{2}{x-9}$, find the domain of f.

2. What is the range of the function defined by $y = 6 - |x|$?

3. If $f(x) = x^2$ and $g(x) = 3x - 2$, find $3f(4) + \frac{1}{2}g(-6)$.

4. If $f(x) = 3x^2 - x + 5$, find $f(-2)$.

5. A rectangular area of sides x and y units is enclosed using 800 ft of fencing. Find a formula for the enclosed area as a function of x.

6. Find the area of a circle whose radius extends from $(-1, 2)$ to $(0, -1)$.

7. Is the following graph the graph of a function?

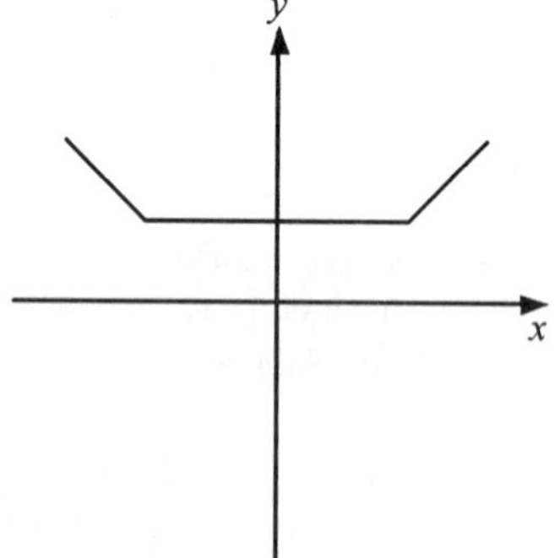

8. Solve the system
$$2x - y = 3$$
$$x + 4y = 5.$$

9. Find the domain and range of the function graphed below. Also, write an equation for this function.

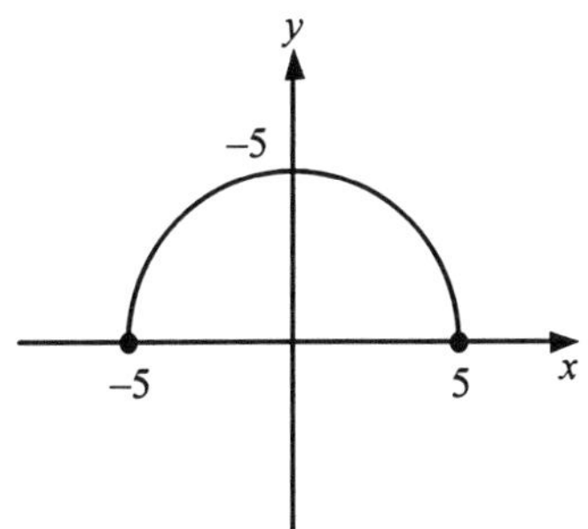

10. Graph $y = \sqrt{x-3}$.

11. Graph $y = (x+2)^2 - 1$.

12. Graph $y = \begin{cases} x & \text{if} \quad x \geq 0 \\ 2 & \text{if} \quad x < 0. \end{cases}$

13. Is the function defined by $f(x) = 3 - x^2$ even, odd, or neither of these?

14. If $(a, \ -2)$ is a solution of $y = -8x + 10$, find the value of a.

15. Is the set of ordered pairs $\{(1, \ 1), \ (0, \ 0), \ (1, \ -1)\}$ a function?

16. How should a \$60,000 investment be split so that the total annual earnings are \$6,000, if one portion is invested at 8.5 percent annual interest and the rest at 10.5 percent?

17. Solve graphically: $3(2-x) = 2x + 1$.

18. If y varies inversely as x, and $y = 8$ when $x = 12$, find y when $x = 4$.

19. In a spring to which Hooke's law applies, a force of 25 lb stretches the spring 4 in. How far will the spring be stretched by a force of 40 lb?

20. The exposure time for photographing an object varies inversely as the square of the lens diameter. What will happen to the exposure time if the lens diameter is tripled?

● ● ●

Chapter 2

Introduction to Trigonometry

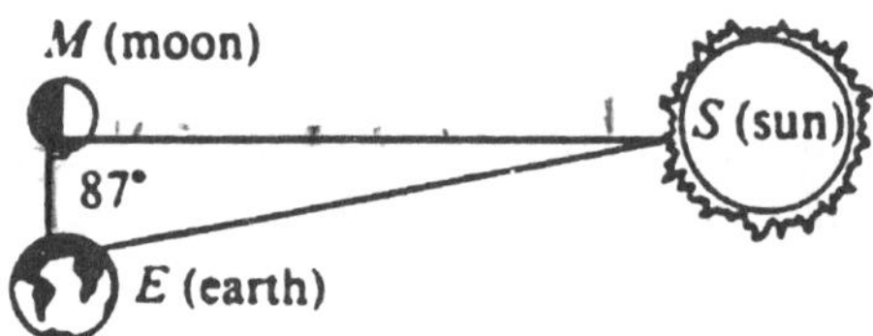

The first major attempt to measure the relative distances of the sun and the moon from Earth was made in about 280 B.C. by the Alexandrian astronomer Aristarchus. Basically, he reasoned that the moon has no light of its own since we do not always see a "full moon." Therefore, the moon, like Earth, must receive its light from the sun and then reflect this light toward Earth. As illustrated in the figure, he also correctly reasoned that at a quarter phase of the moon, when it is half light and half dark, angle *EMS* is a right angle. Using primitive instruments, Aristarchus measured angle *SEM* to be 29/30 of a right angle (or 87°). About how many times more distant is the sun than the moon in Aristarchus' estimate?
(See Example 7 of Section 2.1.)

In this chapter and in Chapters 8 and 9, we turn to trigonometry. This topic can be approached from different viewpoints, and in these chapters we provide three definitions of the trigonometric functions. These definitions are presented in order of abstraction and sophistication. In this section we begin with the rather concrete study of right-triangle trigonometry, and we follow this in Section 2.3 with a general angle approach. Finally, in Section 8.2 we discuss the modern concept of the trigonometric functions of real numbers which is the approach most useful in calculus. In the end, you must merge these definitions for a thorough understanding of trigonometry.

Historically, trigonometry involved the study of triangles for the purpose of measuring angles and distances in astronomy (as illustrated above). In early times this science was the primary concern of scholars who sought to understand God's design of the universe, aesthetically and, practically, sought to obtain a more accurate system of navigation. As is often the case in mathematics, the ideas developed in this study later proved useful in analyzing a wide variety of situations.

● ● ●

2.1 Trigonometric Functions of Acute Angles

Objectives

1. Find the six trigonometric functions of an acute angle in a right triangle, when given at least two side lengths.
2. Find the values of the five remaining trigonometric functions of an acute angle, given one function value.

3. Find the length of any side in a right triangle, given a trigonometric ratio and the length of one side.
4. State exact trigonometric values for 30°, 45°, and 60°.
5. Express any trigonometric function of an acute angle as the cofunction of the complementary angle.
6. Use a calculator to approximate the trigonometric value of a given acute angle.
7. Use a calculator to approximate the measure of an acute angle, given the value of a trigonometric function of that angle.

In the chapter-opening problem, Aristarchus calculated on Earth that the angle between the sun and the moon measures about 87° at a quarter phase of the moon. Before developing the trigonometry necessary to answer the question posed by Aristarchus, we need to review some basic concepts about angles and their measures.

A ray is a half-line that begins at a point and extends indefinitely in some direction. Two rays that share a common endpoint (or vertex) form an angle. If we designate one ray as the **initial ray** and the other ray as the **terminal ray** (see Figure 2.1), the **measure of the angle** is the amount of rotation needed to make the initial ray coincide with the terminal ray. Notice that there are many rotations that will make the rays coincide, since there is no limitation on the number of revolutions made by the initial ray. In fact, it is useful to allow the initial ray to rotate through many revolutions, since the rotating initial ray will demonstrate cyclic behavior that can serve as a model to simulate physical phenomena that occur in cycles. Also, the initial ray can rotate in two possible directions, as shown in Figure 2.1. To show the direction of the rotation, we define the measure of an angle to be positive if the rotation is counterclockwise, and negative if the rotation is clockwise.

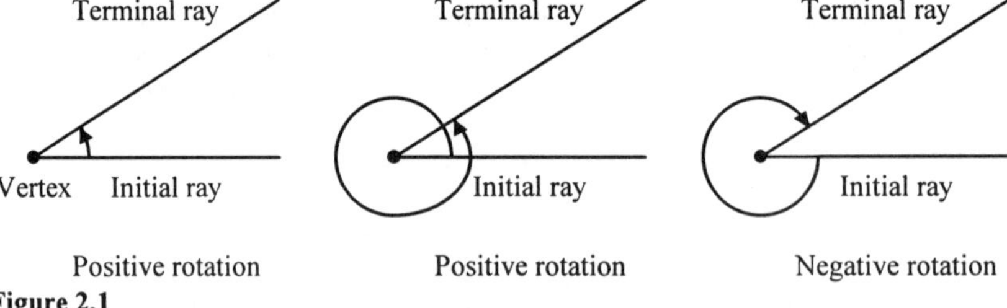

Figure 2.1

A common unit of the measure of an angle is a degree. We define **one degree (1°)** to be $1/360$ of a complete counterclockwise rotation. Equivalently, this means that there are 360° in a complete counterclockwise rotation. Figure 2.2 illustrates this case, a straight angle, and a right angle using the Greek letter θ (theta) to represent the angle measure. An angle is **acute** if it measures between 0° and 90°, and **obtuse** if it measures between 90° and 180°.

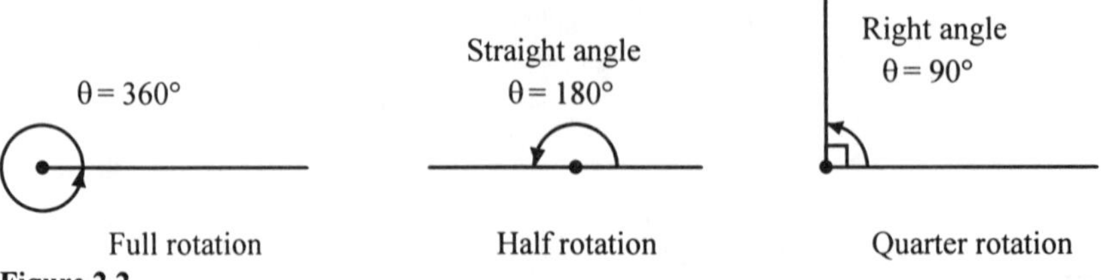

Figure 2.2

Because precise measurements are often needed, the degree is subdivided into 60 smaller units called minutes, with each minute being subdivided into 60 seconds. Thus, **one minute**, written 1', equals $1/60$ of one degree, and **one second**, written 1", equals $1/60$ of one minute—or $1/3600$ of one degree. One degree may also be subdivided by using decimal degrees such as 18.5°, which is a unit more convenient for calculator evaluation.

To define the trigonometric functions with respect to a triangle, we start with a right triangle. A **right triangle** is a triangle that contains a 90° (or right) angle. The two other angles in the triangle are acute angles and the side opposite the right angle is called the **hypotenuse**. In a right triangle, six different ratios can be formed using the measures of the three sides of the triangle, and these ratios are related to the measure of an acute angle in the following definitions.

Definition of the Trigonometric Functions

If θ (theta) is an acute angle in a right triangle, as shown in Figure 2.3, then

Name of Function	Abbreviation	Ratio	
sine of angle θ	$\sin\theta$	$=\dfrac{\text{opposite}}{\text{hypotenuse}}$	reciprocal functions
cosecant of angle θ	$\csc\theta$	$=\dfrac{\text{hypotenuse}}{\text{opposite}}$	
cosine of angle θ	$\cos\theta$	$=\dfrac{\text{adjacent}}{\text{hypotenuse}}$	reciprocal functions
secant of angle θ	$\sec\theta$	$=\dfrac{\text{hypotenuse}}{\text{adjacent}}$	
tangent of angle θ	$\tan\theta$	$=\dfrac{\text{opposite}}{\text{adjacent}}$	reciprocal functions
cotangent of angle θ	$\cot\theta$	$=\dfrac{\text{adjacent}}{\text{opposite}}$	

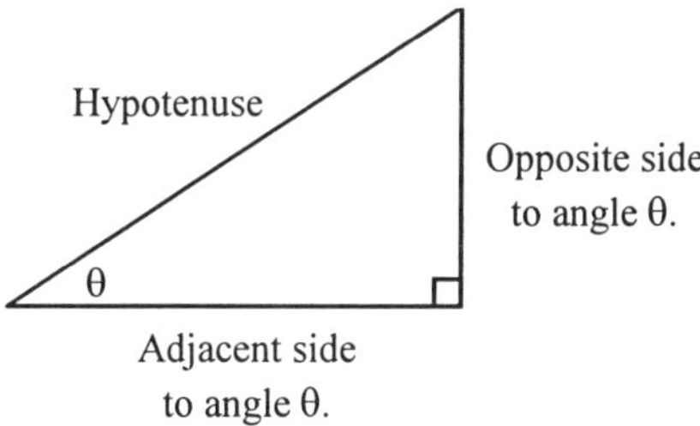

Figure 2.3

Example 1: Finding Trigonometric Functions When Given Side Lengths

Find the values of the six trigonometric functions of angle θ in Figure 2.4.

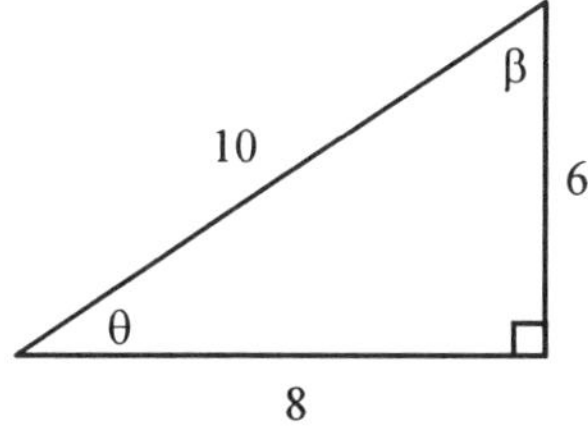

Figure 2.4

Solution: The length of the side opposite angle θ is 6, the adjacent side length is 8, and the length of the hypotenuse is 10. Substituting these numbers in the above definitions gives

$$\sin\theta = \frac{\text{opp}}{\text{hyp}} = \frac{6}{10} = \frac{3}{5} \leftarrow \text{reciprocals} \rightarrow \csc\theta = \frac{\text{hyp}}{\text{opp}} = \frac{10}{6} = \frac{5}{3}$$
$$\cos\theta = \frac{\text{adj}}{\text{hyp}} = \frac{8}{10} = \frac{4}{5} \leftarrow \text{reciprocals} \rightarrow \sec\theta = \frac{\text{hyp}}{\text{adj}} = \frac{10}{8} = \frac{5}{4}$$
$$\tan\theta = \frac{\text{opp}}{\text{adj}} = \frac{6}{8} = \frac{3}{4} \leftarrow \text{reciprocals} \rightarrow \cot\theta = \frac{\text{adj}}{\text{opp}} = \frac{8}{6} = \frac{4}{3}.$$

Note that if we say $\tan\theta = 3/4$, we do not necessarily mean that the opposite side length is 3 and the adjacent side length is 4, but that their lengths are in the ratio of 3 to 4.

PROGRESS CHECK 1
Find the values of the six trigonometric functions of angle β (beta) in Figure 2.4. ■

Example 2: Finding Trigonometric Functions Given One Function Value

Find the values of the remaining trigonometric functions of acute angle θ if $\tan\theta = \frac{2}{5}$.

Solution: First, draw a right triangle, as in Figure 2.5, and label one acute angle θ. Since $\tan\theta = 2/5$, the ratio of the opposite side length to the adjacent side length is 2.5. Although many choices are possible, it is easiest to label the opposite side 2 and the adjacent side 5. The hypotenuse in the triangle is found by Pythagorean relationship,

$$(\text{hypotenuse})^2 = 2^2 + 5^2 = 29$$
$$\text{hypotenuse} = \sqrt{29}.$$

Then

$$\sin\theta = \frac{2}{\sqrt{29}} \leftarrow \text{reciprocals} \rightarrow \csc\theta = \frac{\sqrt{29}}{2}$$
$$\cos\theta = \frac{5}{\sqrt{29}} \leftarrow \text{reciprocals} \rightarrow \sec\theta = \frac{\sqrt{29}}{5}$$
$$\cot\theta = \frac{5}{2}.$$

Note that $\cot\theta = 5/2$ can be determined from its reciprocal relation to $\tan\theta = 2/5$ without constructing the triangle.

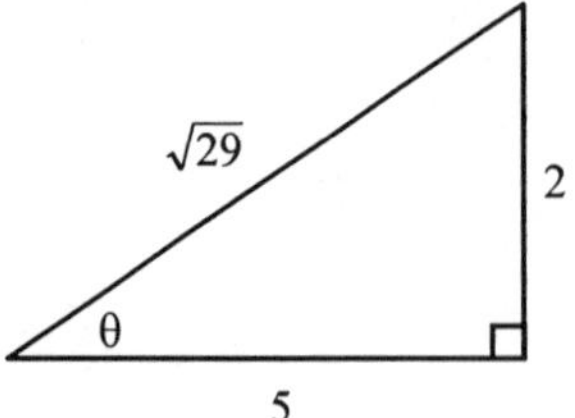

Figure 2.5

PROGRESS CHECK 2

Find the values of the remaining trigonometric functions of acute angle θ if $\cos\theta = \frac{2}{5}$. ■

It is common notation to label a right triangle as in Figure 2.6. Capital letters such as *A*, *B*, and *C* denote angles, while the lengths of the sides opposite these angles are labeled with the corresponding lowercase letters, *a*, *b*, and *c*. The next example illustrates this notation.

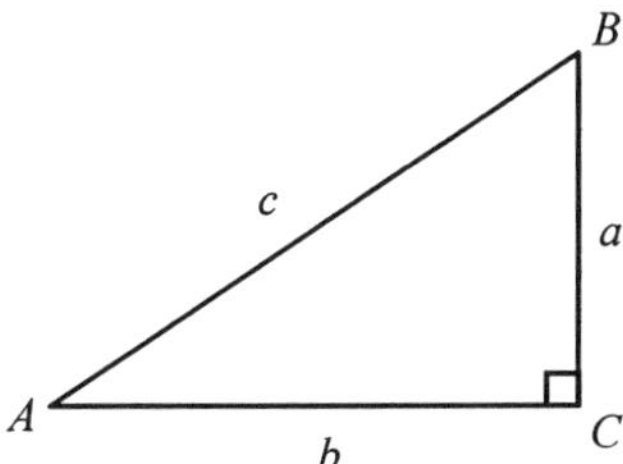

Figure 2.6

Example 3: Using a Trigonometric Ratio to Find a Side Length

In the right triangle *ABC* with $C = 90°$, if $\cos B = \frac{5}{13}$ and $c = 39$, find *a*.

Solution: From Figure 2.6 we have

$$\cos B = \frac{\text{adj}}{\text{hyp}} = \frac{a}{c} = \frac{5}{13}.$$

Substituting $c = 39$ gives

$$\frac{a}{39} = \frac{5}{13}.$$

Then

$$\begin{aligned} a &= 39\left(\frac{5}{13}\right) \\ &= 15. \end{aligned}$$

PROGRESS CHECK 3

In triangle *ABC* with $C = 90°$, if $\sin A = \frac{8}{17}$ and $a = 40$, find *c*. ■

To demonstrate why the trigonometric relationships define functions, consider Figure 2.7. Right triangles *ABC* and *AB'C'* contain the same angle measures. Two triangles with the same angle measures are called **similar triangles**, and one of the properties of similar triangles is that the lengths of corresponding sides are proportional. Therefore,

$$\frac{a}{c} = \frac{a'}{c'}, \quad \frac{b}{c} = \frac{b'}{c'}, \quad \frac{a}{b} = \frac{a'}{b'}, \quad \text{and so on.}$$

This means that the six trigonometric ratios depend only on angle A, and not on the size of the right triangle that contains A. Since there is exactly one number associated with a particular trigonometric ratio of angle A, the term *function* applies.

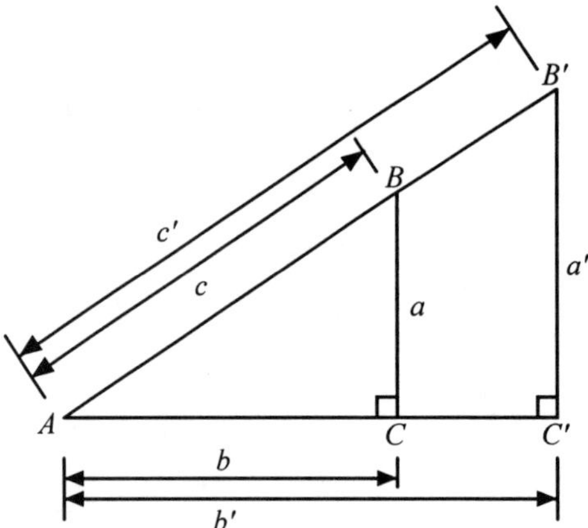

Figure 2.7

So far we have determined the values of the trigonometric functions by applying their definitions to triangles whose side measures were known. A common problem is determining these values if we know the measure of angle θ. First, let us consider the special angles whose trigonometric values are known exactly. Figure 2.8 shows the acute angles 30°, 45°, and 60° contained in two right triangles. The diagonal in a square with a side length of 1 unit forms the right triangle with a 45° angle; the altitude in an equilateral triangle with a side length of 2 units forms the right triangle with angles of 30° and 60°. Using these two triangles, we may determine the value of any of their trigonometric functions.

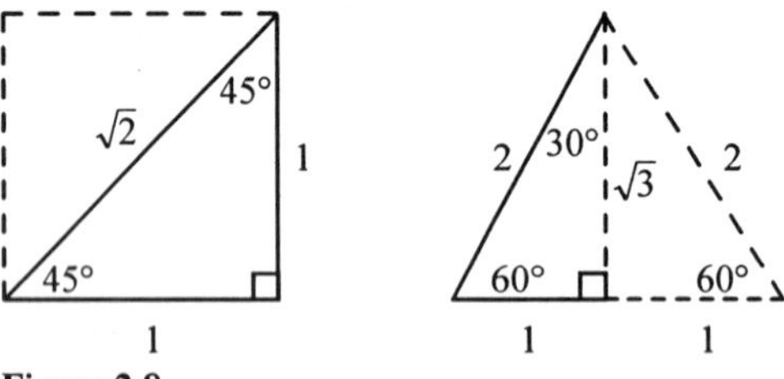

Figure 2.8

Example 4: Exact Trigonometric Values for 30°, 45°, and 60°

Use Figure 2.8 to find the exact values of sin 30°, cos 45°, sec 45°, and tan 60°.

Solution: Applying the definitions of the trigonometric functions to the appropriate triangle in Figure 2.8 gives

$$\sin 30° = \frac{\text{opp}}{\text{hyp}} = \frac{1}{2}, \qquad \cos 45° = \frac{\text{adj}}{\text{hyp}} = \frac{1}{\sqrt{2}},$$

$$\sec 45° = \frac{\text{hyp}}{\text{adj}} = \sqrt{2}, \quad \text{and} \quad \tan 60° = \frac{\text{opp}}{\text{adj}} = \sqrt{3}.$$

PROGRESS CHECK 4

Use Figure 2.8 to find the exact values of sin 60°, cot 30°, csc 30°, and tan 45°. ■

The angles 30°, 45°, and 60° appear often in trigonometry, so it is useful to use the methods shown in Example 4 to create the following table, which lists exact values of all trigonometric functions of these special angles.

θ	$\sin\theta$	$\csc\theta$	$\cos\theta$	$\sec\theta$	$\tan\theta$	$\cot\theta$
30°	$\frac{1}{2}$	2	$\frac{\sqrt{3}}{2}$	$\frac{2}{\sqrt{3}}$	$\frac{1}{\sqrt{3}}$	$\sqrt{3}$
60°	$\frac{\sqrt{3}}{2}$	$\frac{2}{\sqrt{3}}$	$\frac{1}{2}$	2	$\sqrt{3}$	$\frac{1}{\sqrt{3}}$
45°	$\frac{1}{\sqrt{2}}$	$\sqrt{2}$	$\frac{1}{\sqrt{2}}$	$\sqrt{2}$	1	1

This table may be used to provide numerical evidence for an important relation in trigonometry by noting the similarity between the values of the trigonometric functions for 30° and 60°. For example:

$$\sin 30° = \frac{1}{2} \quad \text{and} \quad \cos 60° = \frac{1}{2}$$
$$\tan 30° = \frac{1}{\sqrt{3}} \quad \text{and} \quad \cot 60° = \frac{1}{\sqrt{3}}$$
$$\sec 30° = \frac{2}{\sqrt{3}} \quad \text{and} \quad \csc 60° = \frac{2}{\sqrt{3}}.$$

This similarity results from the fact that in the right triangle, the side opposite the 30° angle is adjacent to the 60° angle. Thus,

$$\sin 30° = \frac{\text{side opposite } 30° \text{ angle}}{\text{hypotenuse}} = \frac{1}{2}$$
$$= \frac{\text{side adjacent to } 60° \text{ angle}}{\text{hypotenuse}} = \cos 60°$$
$$\tan 30° = \frac{\text{side opposite } 30° \text{ angle}}{\text{side adjacent to } 30° \text{ angle}} = \frac{1}{\sqrt{3}}$$
$$= \frac{\text{side adjacent to } 60° \text{ angle}}{\text{side opposite } 60° \text{ angle}} = \cot 60°$$
$$\sec 30° = \frac{\text{hypotenuse}}{\text{side adjacent to } 30° \text{ angle}} = \frac{2}{\sqrt{3}}$$
$$= \frac{\text{hypotenuse}}{\text{side opposite } 60° \text{ angle}} = \csc 60°.$$

Observe that in each case a trigonometric function of 30° is equal to the corresponding cofunction of 60°. The corresponding cofunction is easy to remember since the *co*function of the sine is the *co*sine, the cofunction of the tangent is the *co*tangent, and the cofunction of the secant is the *co*secant. We can generalize from these examples concerning 30° and 60° to any two angles A and B that are **complementary** (that is, $A + B = 90°$), since in any right triangle the side opposite angle A is adjacent to angle B. Thus, a trigonometric function of any acute angle is equal to the corresponding cofunction of the complementary angle. This result may be stated as follows.

Cofunction Properties

For any acute angle θ,

$$\begin{aligned} \sin(90^\circ-\theta) &= \cos\theta & \cos(90^\circ-\theta) &= \sin\theta \\ \tan(90^\circ-\theta) &= \cot\theta & \cot(90^\circ-\theta) &= \tan\theta \\ \sec(90^\circ-\theta) &= \csc\theta & \csc(90^\circ-\theta) &= \sec\theta. \end{aligned}$$

Example 5: Equating Cofunctions of Complementary Angles

Express tan 75° as a function of the angle complementary to 75°.

Solution: Using $\cot(90^\circ-\theta) = \tan\theta$ with $\theta = 75^\circ$ yields

$$\begin{aligned} \tan 75^\circ &= \cot(90^\circ - 75^\circ) \\ &= \cot 15^\circ. \end{aligned}$$

In other words, $\tan 75^\circ = \cot 15^\circ$, since 75° and 15° are complementary angles and the cofunction of the tangent is the cotangent.

PROGRESS CHECK 5
Express cos 34° as a function of the angle complementary to 34°. ■

A fast and accurate method for evaluating trigonometric functions of acute angles is to use a calculator. Because calculators may be set in different modes for measuring angles, you must first consult the owner's manual to your calculator and learn the procedure for operating in Degree mode (instead of Radian mode, which is considered in Section 8.1). Then, the keys Sin, Cos, or Tan may be used to evaluate sine, cosine, or tangent expressions, while the reciprocals of these functions may be evaluated by using

$$\csc\theta = \frac{1}{\sin\theta}, \ \sec\theta = \frac{1}{\cos\theta}, \text{ or } \cot\theta = \frac{1}{\tan\theta}.$$

Example 6: Evaluating Trigonometric Functions by Calculator

Evaluate each expression by calculator to four decimal places.

a. sin 18° **b.** cos 52°20' **c.** cot 35.6°

Solution: See Figure 2.9

a. $\sin 18^\circ \approx 0.3090$

b. $\cos 52^\circ 20' \approx 0.6111$. Note that 20 minutes may be entered as $20/60$ because one minute is $1/60$ of one degree.

c. Using $\cot 35.6^\circ = 1/\tan 35.6^\circ$ yields $\cot 35.6^\circ \approx 1.3968$.

```
sin 18
                .309016994375
cos (52 + 20/60)
                 .61106662153
1/tan 35.6
                1.39678522019
```

Figure 2.9

Note: Many calculators have special features for entering angles in degree-minute-second format and for displaying the degree symbol so that the calculator computes in Degree mode regardless of the current angular mode setting. You should learn how to use these features on your calculator if they are available.

PROGRESS CHECK 6
Evaluate each expression by calculator to four decimal places.
a. tan 62° **b.** sec 16.7° **c.** sin 81°50' ■

Example 7: Relative Distances of the Sun and the Moon

Solve the problem in the chapter introduction on page 173.

Solution: Consider Figure 2.10. The length of the hypotenuse $\left(\overline{ES}\right)$ represents the distance from the Earth to the sun, while the length of the adjacent side $\left(\overline{EM}\right)$ represents the distance from Earth to the moon. The ratio $\overline{ES}/\overline{EM}$ represents how many times more distant is the sun than the moon from Earth. Then

$$\frac{\overline{ES}}{\overline{EM}} = \frac{\text{hypotenuse}}{\text{adjacent}} = \sec 87° \approx 19.11.$$

Thus, Aristarchus estimated that the sun is about 19 times farther away than the moon.

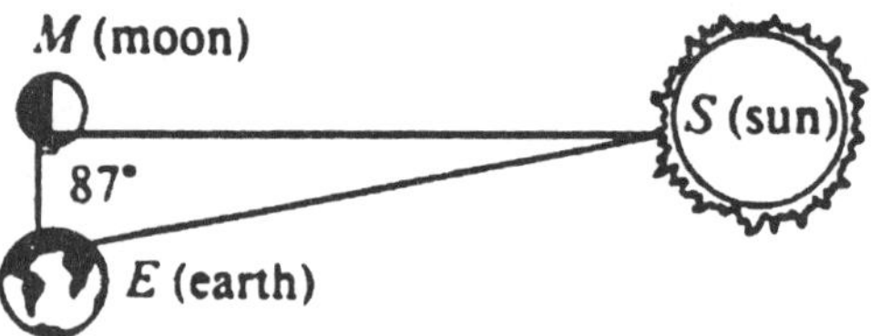

Figure 2.10

PROGRESS CHECK 7
In reality, Aristarchus' estimate is not close to the true ratio. The error is caused by his measurement for angle *SEM*, which should be about 89°50'. On the basis of this measurement for angle *SEM*, about how many times farther away is the sun than the moon? ■

An important theme in problem solving is the notion that certain problems may be viewed as inverse problems. In the context of trigonometry, the following problems state the basic inverse problems with respect to the sine function.

Inverse problems
→ <u>Problem</u> Given θ, find $\sin\theta$.
→ <u>Problem</u> Given $\sin\theta$, find θ.

On the basis of this perspective, the function that reverses the assignments of the sine function is called the **inverse sine function.** On most calculators this function is labeled $\sin^{-1}$. Similar notation is used for the other inverse trigonometric functions, and the following table illustrates comparable information using a trigonometric function and its corresponding inverse trigonometric function.

Trigonometric relation	Relation in inverse notation
$\sin 30° = \frac{1}{2}$	$\sin^{-1}\frac{1}{2} = 30°$
$\tan 45° = 1$	$\tan^{-1} 1 = 45°$
$\cot 60° = \frac{1}{\sqrt{3}}$	$\cot^{-1}\left(\frac{1}{\sqrt{3}}\right) = 60°$

Most calculators have keys for $\sin^{-1}$, $\cos^{-1}$, and $\tan^{-1}$ that may be used to find an acute angle when one of the trigonometric functions is known, as illustrated in the next example. For such problems, it is important to recognize that a superscript -1 is used to denote an inverse function. Do *not* interpret the -1 in this notation as an exponent.

Example 8: Using an Inverse Key on a Calculator

Use a calculator to approximate the acute angle θ that satisfies the given equation. Write solutions to the nearest tenth of a degree and to the nearest 10 minutes.

a. $\sin\theta = 0.7957$ **b.** $\cot\theta = 2.583$

Solution:

a. If $\sin\theta = 0.7957$ and θ is an acute angle, then

$$\theta = \sin^{-1} 0.7957 \approx 52.7°,$$

as shown in Figure 2.11. To convert to degree-minute format, multiply the decimal portion of this answer by 60 minutes, as shown in the last four lines of Figure 2.11. To the nearest 10 minutes, the decimal portion rounds to 40 minutes, so $\sin^{-1} 0.7957 \approx 52°40'$.

```
sin-10.7957
                 52.7214224844
Ans-52
                 .721422484404
Ans*60
                 43.2853490642
```

Figure 2.11

b. Calculators do not have keys for inverse cotangent, secant, or cosecant functions, so first observe that $\cot\theta = 2.583$ is equivalent to $\tan\theta = 1/2.583$. The display in Figure 2.12 then shows that acute angle θ is given by

$$\theta = \tan^{-1}(1/2.583) \approx 21.2°.$$

By the methods discussed in part **a**, the decimal portion of this result rounds to 10 minutes, so we also have $\theta \approx 21°10'$. Many calculators have a special feature for converting from decimal degrees to degree-minute-second format, and the last two lines in Figure 2.12 display such a conversion that supports our result in this example.

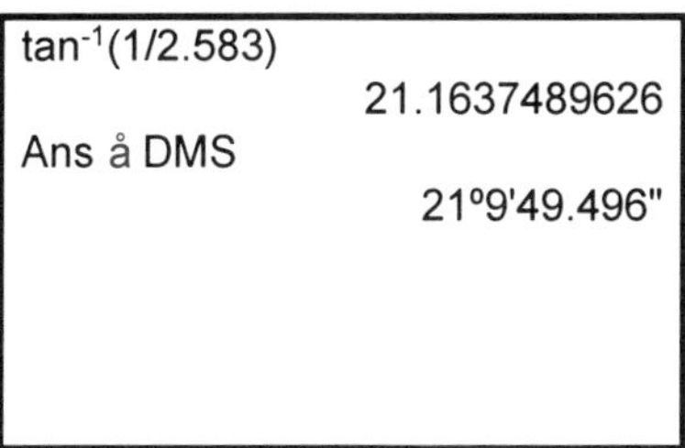

Figure 2.12

PROGRESS CHECK 8

Use a calculator to approximate the acute angle θ that satisfies the given equation. Write solutions to the nearest tenth of a degree and to the nearest 10 minutes. ■

a. $\cos\theta = 0.2728$ **b.** $\csc\theta = 1.448$

Note: For correct calculator evaluation of trigonometric expressions, the calculator Mode must be set properly. At this point in the course, the calculator should be set in Degree mode. Later you may want to switch to Radian mode. Another approach, which allows you to ignore the mode setting, is to always insert the degree symbol, but this is less convenient.

EXPLORE 2.1

1. Use a calculator to evaluate these pairs of expressions, and to decide if the two expressions in each pair are equal. Give each answer to four decimal places. Each angle is given in degrees.
   - **a.** $\cos(40^2)$; $(\cos 40)^2$; True or False? $\cos(x^2) = (\cos x)^2$
   - **b.** $\sin(30^2)$; $(\sin 30)^2$; True or False? $\sin(x^2) = (\sin x)^2$
   - **c.** $\sin(40) + \sin(70)$; $\sin(110)$; True or False? $\sin A + \sin B = \sin(A + B)$
   - **d.** $\cos(20) + \cos(40)$; $\cos(60)$; True or False? $\cos A + \cos B = \cos(A + B)$
2. For each of these values of θ evaluate $(\sin\theta)^2 + (\cos\theta)^2$. Does there appear to be a pattern?
   - **a.** 40° **b.** 55° **c.** 84°
3. Sometimes angles are given in decimal notation as in 10.5°, and sometimes in degree-minute-second notation, as in 10°30'15". Many calculators have commands to convert from decimal to DMS notation. If your calculator has these conversion commands, use them to convert each of the given expressions to DMS notation.
   - **a.** 10.4° **b.** 10.41° **c.** 10.413°

EXERCISES 2.1

In Exercises 1–4 find the six trigonometric functions of the acute angles in the right triangle.

1.

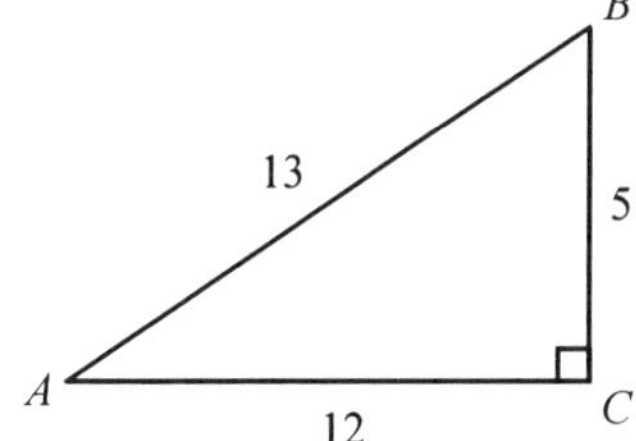

2.

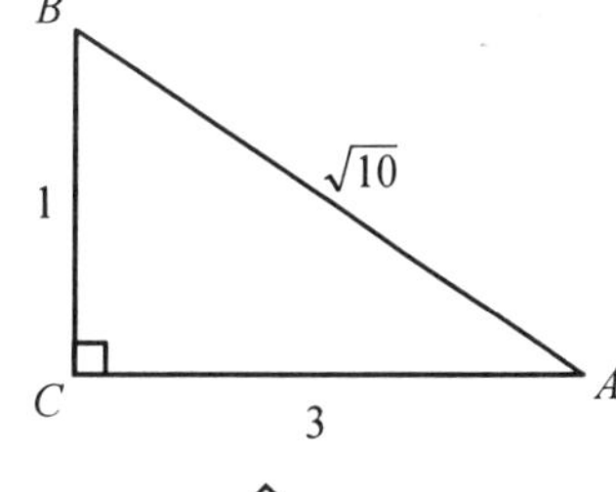

3.

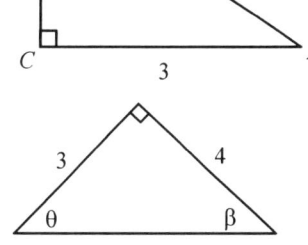

4.

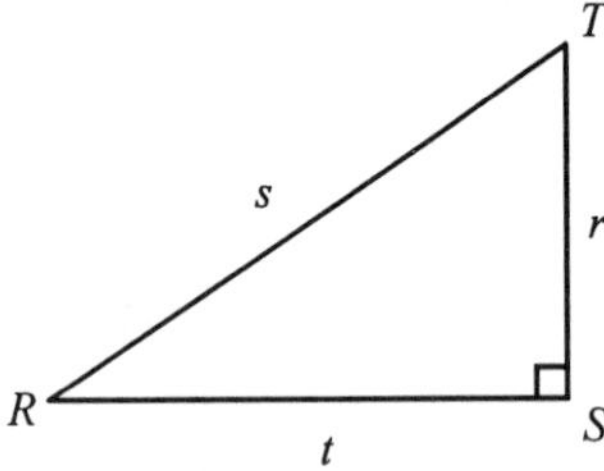

In Exercises 5–12 use these two triangles

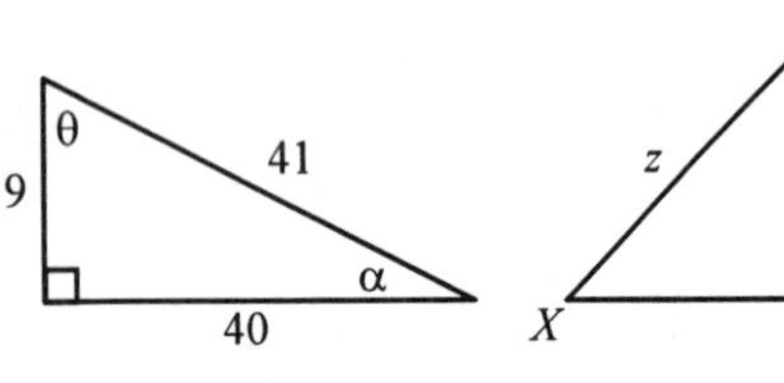

5. Find cos Z.

6. Find sec θ.

7. Find cot α.

8. Find sin α.

9. Find csc X.

10. Find tan θ.

11. Represent the ratio $\frac{y}{z}$ as a trigonometric function of an acute angle.

12. Represent the ratio $\frac{9}{40}$ as a trigonometric function of an acute angle.

In Exercises 13–18 use the reciprocal relations between the functions.

13. If $\sin\theta = \frac{1}{3}$, find csc θ.

14. If $\cos\theta = \frac{3}{4}$, find sec θ.

15. If $\sec\theta = 1.4$, find cos θ.

16. If $\tan\theta = 0.7$, find cot θ.

17. What is the value of $\sin\theta \cdot \csc\theta$ for all acute angles θ?

18. What is the value of $3\tan\theta \cdot \cot\theta$ for all acute angles θ?

In Exercises 19–24 use the given trigonometric ratio to find the other trigonometric functions of acute angle θ. Use the definitions of the functions, not a calculator.

19. $\sin\theta = \frac{1}{2}$

20. $\csc\theta = 2$

21. $\cot\theta = \frac{3}{5}$

22. $\cos\theta = \frac{8}{17}$

23. $\tan\theta = 2$

24. $\sec\theta = 5$

In Exercises 25–30 refer to this triangle

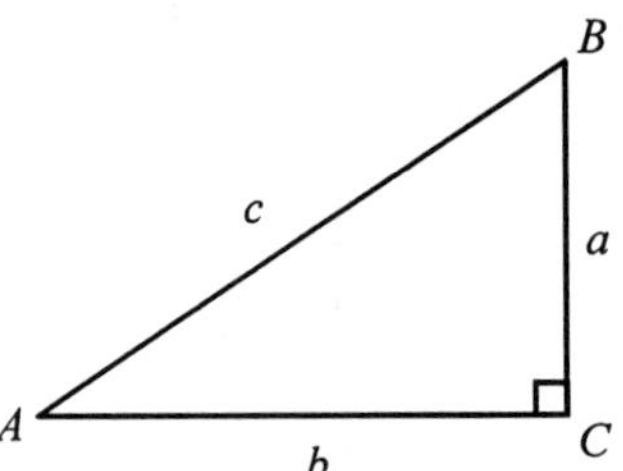

25. If $\cos B = \frac{4}{5}$ and $a = 16$, find c.

26. If $\sin A = \frac{12}{13}$ and $c = 39$, find a.

27. If $\tan A = \frac{7}{12}$ and $a = 21$, find b.

28. If $\csc B = 2$ and $c = 8$, find a

29. If $a = 5$, $b = 12$, and $c = 13$, find the cosine of the smaller acute angle.

30. If $c = 17$ and $a = 8$, find the tangent of the larger acute angle.

In Exercises 31–36 express each term as a function of the angle complementary to the given angle.

31. $\sin 17°$

32. $\cos 64°$

33. $\tan 81°30'$

34. $\sec 33°58'$

35. $\csc 68.1°$

36. $\cot 0.5°$

In Exercises 37–44 use the exact trigonometric values of 30°, 45°, or 60°.

37. Does $2\sin 30° = \sin 60°$?

38. Does $\sin 60° \csc 30° = \tan 60°$?

39. Does $\sin 45° \cos 45° = 1$?

40. Does $\sin 30° = \sqrt{\frac{1 - \cos 60°}{2}}$?

41. Does $(\tan 45°)^2 + 1 = (\sec 45°)^2$?

42. Does $\cos 60° = 1 - 2(\sin 30°)^2$?

43. If the length of the short leg of a 30–60–90 triangle is 2, determine the lengths of the other two sides of the triangle.

44. If the length of a leg of a 45–45–90 triangle is 5, determine the length of the hypotenuse of the triangle.

In Exercises 45–52 use the given figures to find the exact values of the given expression

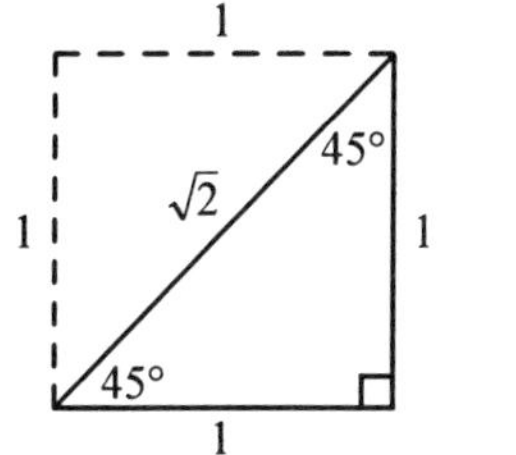

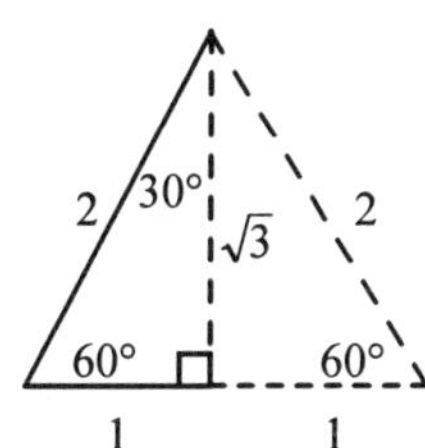

45. cos 30° **46.** sin 45° **47.** sec 30°

48. csc 45° **49.** cot 60° **50.** tan 30°

51. sin 45° **52.** cos 60°

In Exercises 53–66 approximate each using a calculator.

53. cos 7° **54.** sin 42° **55.** cot 83°

56. csc 54° **57.** tan 79°30' **58.** sec 65°10'

59. csc 51°00' **60.** cot 16°40' **61.** sin 12.3°

62. tan 47.6° **63.** cot 89.9° **64.** csc 0.1°

65. cot 2°07' **66.** cos 19°19'

For Exercises 67–72 use the figure given with Exercises 45–52 to find the acute angle given.

67. $\cos^{-1}\frac{1}{2}$ **68.** $\sec^{-1}\sqrt{2}$ **69.** $\sin^{-1}\frac{\sqrt{3}}{2}$

70. $\csc^{-1}\frac{2}{\sqrt{3}}$ **71.** $\tan^{-1}1$ **72.** $\cot^{-1}\sqrt{3}$

In Exercises 73–84 approximate the measure of angle θ. Write solutions to the nearest 10 minutes and to the nearest tenth of a degree by calculator.

73. $\sin\theta = 0.7071$ **74.** $\cos\theta = 0.8660$

75. $\tan\theta = 0.7907$ **76.** $\cot\theta = 2.699$

77. $\sec\theta = 1.781$ **78.** $\csc\theta = 49.11$

79. $\cot\theta = 0.7651$ **80.** $\tan\theta = 0.0402$

81. $\csc\theta = 16.00$ **82.** $\sec\theta = 1.549$

83. $\sin\theta = 0.9973$ **84.** $\cos\theta = 0.9513$

85. The formula for the horizontal distance traveled by a projectile, neglecting air resistance, is

$$d = V^2 \frac{\sin A \sin B}{16}$$

where V is the initial velocity, A the angle of elevation, and $B = 90° - A$. In the 16-lb shot-putting event, an athlete releases the ball at an angle of elevation of 42°, with an initial velocity of 47 ft/second. Determine the distance of the throw. (The maximum distance is attained when the release angle is 45°.)

86. Answer the question posed in Exercise 85 if the ball is released at an angle of elevation of 48° with an initial velocity of 47 ft/second.

87. A major principle in the theory of light is the **law of refraction.** Refraction is the bending of light as it passes from one medium to another. For example, consider the following diagram, which shows a ray of light bending toward the perpendicular as it passes from air to water. The bending is caused by the change in speed of the light ray as it slows down in the water, which is the denser medium. The mathematical relation between the angle of incidence (i) and the angle of refraction (r) is a trigonometric ratio called **Snell's law**. The law is

$$\frac{\sin i}{\sin r} = \frac{v_i}{v_r},$$

where $\frac{v_i}{v_r}$ is the ratio between the velocities of light in the two media.

Diver's conception of the direction of the sun.
Everything above water appears to lie within this cone.
Sun's ray
i
Air
Water
Refracted ray
r
Diver

a. As light passes from air to water, $\frac{v_i}{v_r}$ is about $\frac{4}{3}$.

Find the angle of refraction if $i = 68.5°$.

b. When the sun is near the horizon, the angle of incidence approaches 90° and the angle at which the sun's rays penetrate the water approaches a limiting value. Determine this value of r. It is interesting to note that this restriction on the angle of refraction causes an optical illusion for a diver under water, as shown in the diagram. As she looks up, the world above the surface appears to be in the shape of a cone. This distorted perspective is called the "fish-eye view of the world."

88. The ratio $\frac{v_i}{v_r}$ given in Exercise 87 is called the Index of Refraction when one medium is air. Most ordinary glass has a refraction index of about 1.5. Refer to the figure in Exercise 87, but assume that instead of water, the bottom medium is glass.

a. Find the angle of refraction if $i = 50°$.

b. Find r, the angle of refraction, when $i = 90°$.

THINK ABOUT IT 2.1

1. Explain why the values of $\sin\theta$ range between 0 and 1 when θ is an acute angle.
2. Explain why the values of $\csc\theta$ are always greater than 1 when θ is an acute angle.
3. The formula for the volume of an oblique cylinder (see diagram below) is $V = \pi R^2 H$. Find the volume of an oblique cylinder in which $R = 0.5$ m, $\theta = 75°$, and the slant height is 2.1 m.

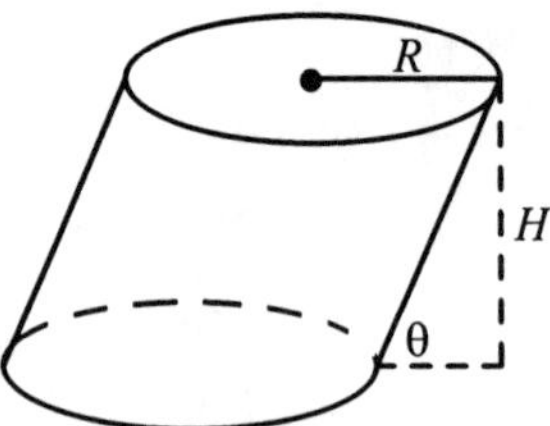

4. In the figure below find the exact values for m, n, p, and q.

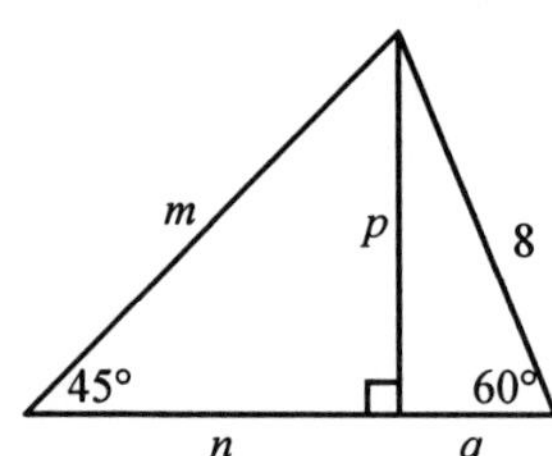

5. In the figure below find θ_1, θ_2, θ_3, θ_4 and θ_5. Which measures are exact and which are approximations?

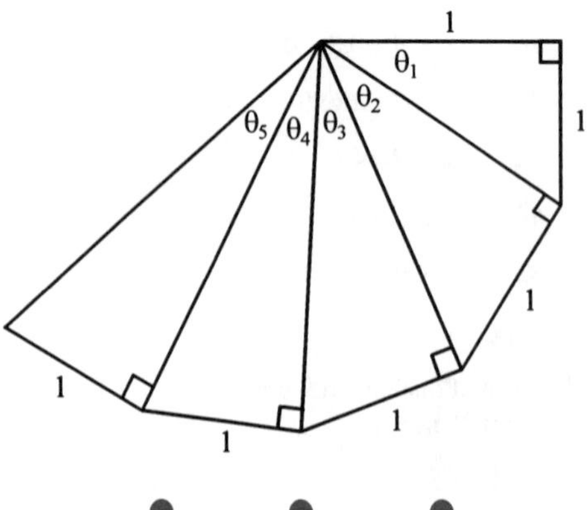

• • •

2.2 Right Triangle Applications

Photo Courtesy of ***Frank Siteman,*** *The Photo Cube, Inc.*

For maximum safety the distance between the base of a ladder and a building should be one-fourth of the length of the ladder. Find the angle that the ladder makes with the ground when it is set up in the safest position.
(See Example 6.)

Objectives

1. Solve a right triangle, given one side and one acute angle.
2. Solve a right triangle, given two sides.
3. Solve applied problems involving right triangles.

Solving triangles is a basic goal in trigonometry. To solve a right triangle means to find the measures of the two acute angles and the lengths of the three sides of the triangle. To accomplish this, at least two of these five values must be known, and one or more must be a side length. In this section we follow the standard practice of simplifying the notation by always labeling the angles of the triangle as A, B, and C, with C designating the right angle. Before attempting these problems, it is suggested that you study the topic of significant digits in the Appendix. **Note:** As discussed in the Appendix, a bar above a zero, $\bar{0}$, is used to avoid ambiguity and indicates a zero that is a significant digit.)

Example 1: Solving a Right Triangle: Angle-Side Case

Solve the right triangle ABC in which $A = 30°$ and $c = 1\bar{0}0$ ft.

Solution: First, sketch Figure 2.13. We find angle B, since angles A and B are complementary.

$$\begin{aligned} A + B &= 90° \\ 30° + B &= 90° \\ B &= 60° \end{aligned}$$

Second, we can find side length a by using the sine function.

$$\begin{aligned} \sin A &= \frac{a}{c} \\ \sin 30° &= \frac{a}{100} \\ 100(\sin 30°) &= a \\ 5\bar{0} \text{ ft} &= a \quad \text{(two significant digits)} \end{aligned}$$

Third, we can find b by using the cosine function.

$$\cos A = \frac{b}{c}$$
$$\cos 30° = \frac{b}{100}$$
$$100(\cos 30°) = b$$
$$87 \text{ ft} = b \quad \text{(two significant digits)}$$

To summarize, we found that $B = 60°$, $a = 5\overline{0}$ ft, and $b = 87$ ft.

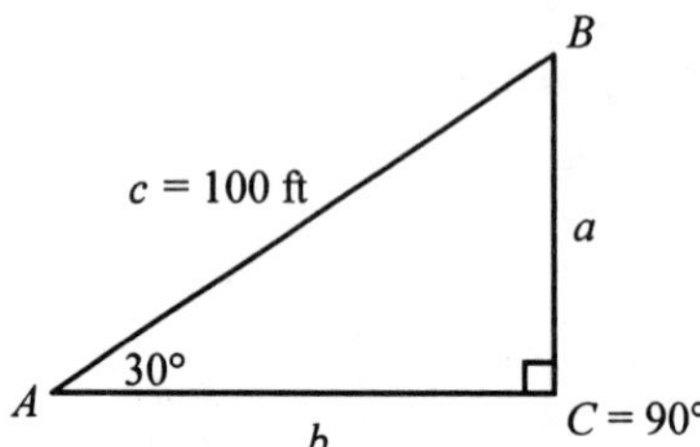

Figure 2.13

PROGRESS CHECK 1
Solve the right triangle *ABC* in which $B = 58°$ and $a = 35$ ft. ■

In applications of trigonometry our values are only as accurate as the devices we use to measure the data. However, although our answers are approximations, the symbol for equality (=) is generally used instead of the more precise symbol for approximation (≈). Computed results should not be used to determine other parts of a triangle, since the given data produce more accurate answers. The results are usually rounded off as follows:

Accuracy of Sides	Accuracy of Angle
Two significant digits	Nearest degree
Three significant digits	Nearest 10 minutes or tenth of a degree
Four significant digits	Nearest minute or hundredth of a degree

Example 2: Solving a Right Triangle: Angle-Side Case

Solve the triangle *ABC* in which $B = 47°10'$ and $a = 45.6$ ft.

Solution: First, sketch Figure 2.14. Now find *A*.

$$A + B = 90°$$
$$A + 47°10' = 90°$$
$$A = 42°50'$$

Second, we can find the length of the hypotenuse *c* by using the secant function.

$$\sec B = \frac{c}{a}$$
$$\sec 47°10' = \frac{c}{45.6}$$
$$45.6(\sec 47°10') = c$$
$$67.1 \text{ ft} = c \quad \text{(three significant digits)}$$

Third, we can find b by using the tangent function.

$$\tan B = \frac{b}{a}$$
$$\tan 47°10' = \frac{b}{45.6}$$
$$45.6(\tan 47°10') = b$$
$$45.6(1.079) = b$$
$$49.2 \text{ ft} = b \quad \text{(three significant digits)}$$

To summarize, we determined that $A = 42°50'$, $c = 67.1$ ft, and $b = 49.2$ ft.

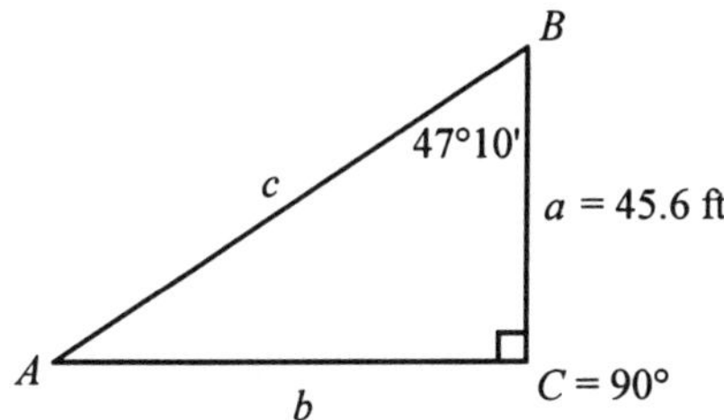

Figure 2.14

PROGRESS CHECK 2
Solve the triangle ABC in which $A = 9°40'$ and $c = 7.52$ ft. ∎

Example 3: Solving a Right Triangle: Two Sides Case

Solve the right triangle ABC in which $a = 11.0$ ft and $b = 5.00$ ft.

Solution: First, sketch Figure 2.15. Now find the length of the hypotenuse by the Pythagorean theorem.

$$c^2 = a^2 + b^2$$
$$= 11^2 + 5^2$$
$$= 146$$
$$c = \sqrt{146}$$
$$= 12.1 \text{ ft} \quad \text{(three significant digits)}$$

Second, we can find angle A by using the tangent function.

$$\tan A = \frac{a}{b}$$
$$= \frac{11}{5}$$
$$A = \tan^{-1}\frac{11}{5} = 65°30' \quad \text{(nearest 10 minutes)}$$

Third, we can find angle B, since angle A and angle B are complementary.

$$A + B = 90°$$
$$65°30' + B = 90°$$
$$B = 24°30'$$

To summarize, we found that $c = 12.1$ ft, $A = 65°30'$, and $B = 24°30'$.

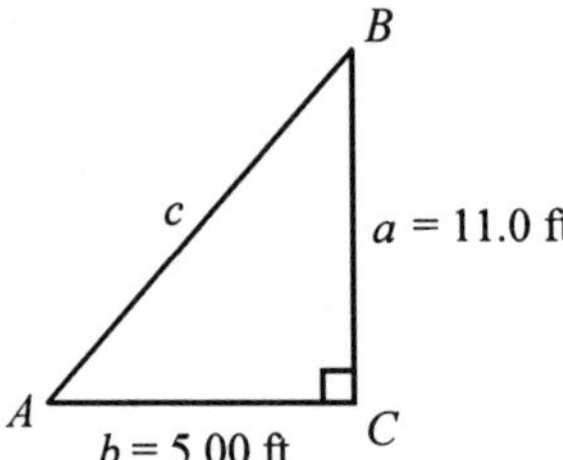

Figure 2.15

PROGRESS CHECK 3
Solve the triangle ABC in which $b = 12.0$ ft and $c = 19.0$ ft. ■

In many practical applications of right triangles, an angle is measured with respect to a horizontal line. This measurement is often accomplished by means of a transit. (By centering a bubble of air in a water chamber, the table of this instrument may be horizontally leveled.) The sighting tube of the transit is then tilted upward or downward until the desired object is sighted. This measuring technique will result in an angle that is described as either an angle of elevation or an angle of depression (see Figure 2.16). The measurement results in an **angle of elevation** if the object being sighted is above the observer, and the measurement results in an **angle of depression** if the object being sighted is below the observer.

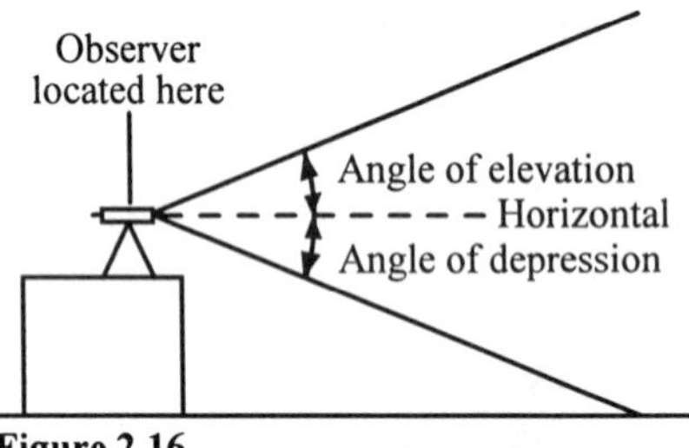

Figure 2.16

Example 4: Height of a Smokestack

It is necessary to determine the height of a smokestack to estimate the cost of painting it. At a point 225 ft from the base of the stack, the angle of elevation is 33.0°. How high is the smokestack?

Solution: First, sketch Figure 2.17. In right triangle ABC we can find x, the height of the smokestack, by using the tangent function.

$$\tan A = \frac{\text{opposite}}{\text{adjacent}}$$
$$\tan 33.0° = \frac{x}{225}$$
$$225(\tan 33.0°) = x$$
$$146 \text{ ft} = x \quad \text{(three significant digits)}$$

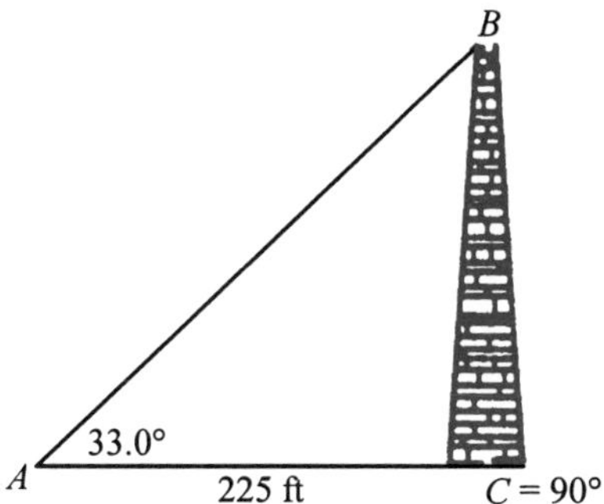

Figure 2.17

PROGRESS CHECK 4

A ladder leans against the side of a building and makes an angle of 72.0° with the ground. If the ladder is 25.0 ft long, then find the height the ladder reaches on the building. ■

Example 5: Distance to a Buoy

The measure of the angle of depression of a buoy from the platform of a radar tower that is 85 ft above the ocean is 15°. Find the distance of the buoy from the base of the tower. (See Figure 2.18.)

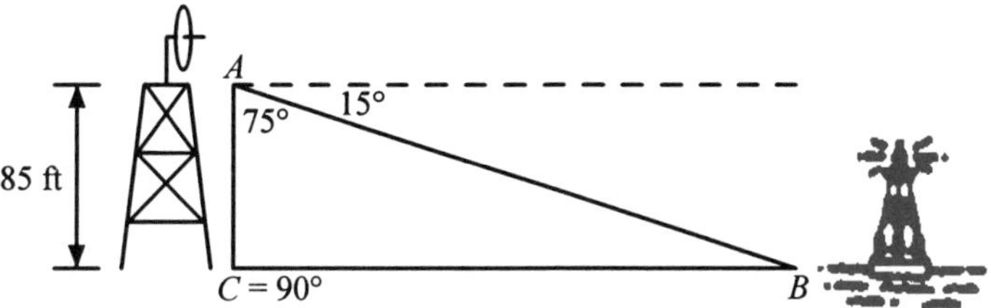

Figure 2.18

Solution: In right triangle ABC we can find the measure of angle A, since angle A and the angle of depression are complementary.

$$A + 15° = 90°$$
$$A = 75°$$

We know the side length adjacent to angle A and we need to find x, which is the opposite side length. Thus

$$\tan 75° = \frac{x}{85}$$
$$85(\tan 75°) = x$$
$$320 \text{ ft} = x \quad \text{(two significant digits).}$$

PROGRESS CHECK 5

A surveyor stands on a cliff 175 ft above a river. If the angle of depression to the water's edge on the opposite bank is 8.4°, how wide is the river at this point? ■

Example 6: Setting Up a Ladder in the Safest Position

Solve the problem in the section introduction on page 187.

Solution: First, sketch Figure 2.19. The length of the hypotenuse is x, while the length of the side adjacent to θ is $(1/4)x$. Thus,

$$\cos\theta = \frac{(1/4)x}{x} = \frac{1}{4}$$

so,

$$\theta = \cos^{-1}\frac{1}{4} \approx 75.5°.$$

For maximum safety, a ladder should make an angle of 75.5° with the ground.

Figure 2.19

PROGRESS CHECK 6

A ladder is set up so that the distance between the base of the ladder and the building against which it leans is one-fifth the length of the ladder . Find the angle the ladder makes with the ground. ■

EXPLORE 2.2

1. In a right triangle with fixed base (as shown in the figure) if you double base angle A, does side a also double? Given that $c = 10$ and $A = 20°$, find a. Then, double A so that $A = 40°$. Does a also double?

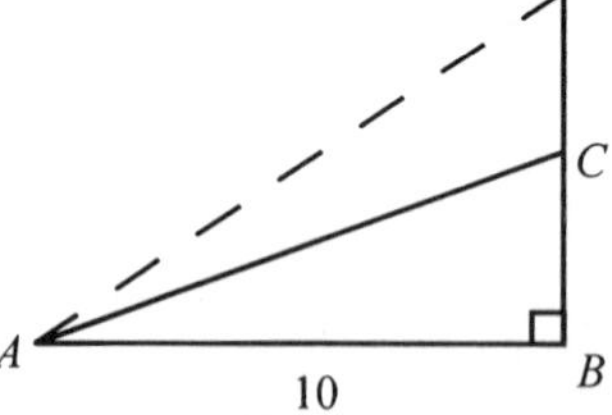

2. If you double an angle, do you double its sine? Its cosine? Its tangent? To investigate, compare the trigonometric functions for 25° and 50°.
3. **a.** The tangent of an acute angle A is 3. Find A. Find another acute angle B such that the tangent of B is twice as large as the tangent of A. Is B twice as large as A?

   b. The cotangent of an acute angle A is 4. Find A. Find another acute angle B such that the cotangent of B is twice as large as the cotangent of A. Is B twice as large as A?

EXERCISES 2.2

In Exercises 1–20 solve each right triangle ABC ($C = 90°$) for the given data. Round off answers using the guidelines given in this section. (**Note:** As discussed in the Appendix, a bar above a zero, $\bar{0}$ is used to avoid ambiguity and indicates a zero that is a significant digit.)

1. $A = 30°,\ a = 5\bar{0}$ ft **2.** $B = 45°,\ a = 85$ ft

3. $A = 60°,\ c = 15$ ft **4.** $A = 22°,\ b = 62$ ft

5. $B = 71°,\ c = 25$ ft **6.** $A = 19°,\ a = 17$ ft

7. $A = 55°,\ c = 25$ ft **8.** $B = 10.3°,\ a = 24.5$ ft

9. $A = 45.5°,\ a = 86.6$ ft

10. $A = 84°50',\ c = 12.4$ ft

11. $B = 52°40',\ c = 625$ ft

12. $A = 31.5°,\ b = 29.7$ ft

13. $B = 88°10',\ a = 31.2$ ft

14. $A = 10.8°,\ a = 49.2$ ft

15. $a = 6.0$ ft, $c = 15$ ft

16. $b = 1.0$ ft, $c = 2.0$ ft

17. $a = 1.0$ ft, $b = 1.0$ ft

18. $a = 7.00$ ft, $c = 11.0$ ft

19. $b = 12.0$ ft, $c = 26.0$ ft

20. $a = 5.00$ ft, $b = 4.00$ ft

In Exercises 21–44 solve each problem by making a careful diagram and using right triangles.

21. An escalator from the first floor to the second floor of a building is $5\bar{0}$ ft long and makes an angle of 30° with the floor. Find the vertical distance between the floors.

22. A ladder leans against the side of a building and makes an angle of 60° with the ground. If the foot of the ladder is $1\bar{0}$ ft from the building, find the height the ladder reaches on the building.

23. A road has a uniform elevation of 6°. Find the increase in elevation in driving $5\bar{0}0$ yd along the road.

24. If the angle of elevation of the sun at a certain time is 40°, find the height of a tree that casts a shadow of 45 ft.

25. To find the width of a river, a surveyor sets up her transit at C and sights across the river to point B (both B and C are at the water's edge, as shown below). She then measures off $2\bar{0}0$ ft from C to A such that C is a right angle. If she determines that angle A is 24°, how wide is the river?

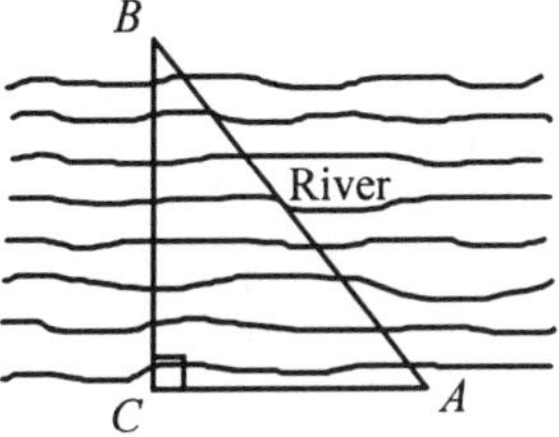

26. The distance from ground level to the underside of a cloud is called the "ceiling." At an airport, a ceiling light projector throws a spotlight vertically on the underside of a cloud. At a distance of $6\bar{0}0$ ft from the projector, the angle of elevation of the spot of light on the cloud is 58°. What is the ceiling?

27. A lighthouse built at sea level is 180 ft high. From its top, the angle of depression of a buoy is 24°. Find the distance from the buoy to the foot of the lighthouse.

28. A pilot in an airplane at an altitude of $4,\bar{0}00$ ft observes the angle of depression of an airport to be 12°. How far is the airport from the point on the ground directly below the plane?

29. A surveyor stands on a cliff $5\bar{0}$ ft above the water of the river below. If the angle of depression to the water's edge on the opposite bank is 10°, how wide is the river at this point?

30. A person looks down from the edge of the roof of a skyscraper to the nearest edge of the roof of another building. If the angle of depression is 15.6°, and the distance between the buildings is 205 yd, how much taller is the skyscraper than the other building?

31. At an airport, cars drive down a ramp 85 ft long to reach the lower baggage claim area 15 ft below the main level. What angle does the ramp make with the ground at the lower level?

32. Suppose the distance between the base of a ladder and a building is one-third of the length of the ladder. Find the angle that the ladder makes with the ground when it is set up in this position.

33. In building a warehouse, a carpenter checks the drawings and finds the roof span to be $4\overline{0}$ ft, as shown in the sketch below. If the slope of the roof is 17°, what length of 2-by 6-in. stock will he need to make rafters if a 12-in. overhang is desired?

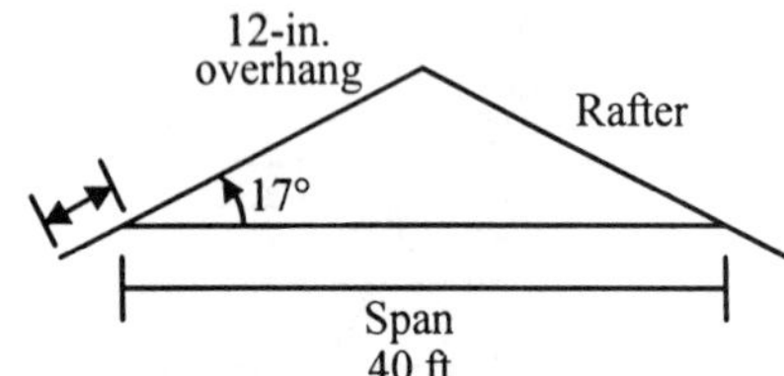

34. A welder is to weld vertical supports for a 25-ft conveyor so that it will operate at a 14° angle with respect to the horizontal. What is the length of the supports?

35. A carpenter is to build a concrete ramp with an 18° slope leading up to a platform. If the platform is 38 in. above the ground, how far from the base of the platform should she start to lay the concrete forms?

36. A tinsmith forms an angle of 120° for a $1\overline{0}$-in flashing, as shown. Give the height of the bend in inches.

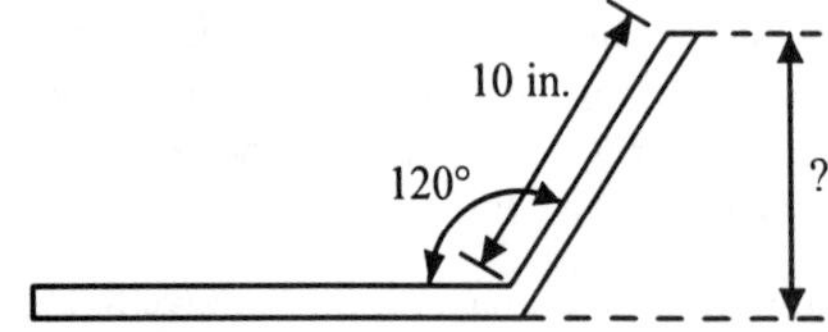

37. A right triangle, called an **impedance triangle**, is used to analyze alternating-current (a.c.) circuits. Consider diagram (a), which shows a resistor and an inductor in series. As current flows through these components, it encounters some resistance, and the total effective resistance is called the **impedance**. We determine the impedance by making the resistances of the two circuit components the measures of the legs of a right triangle. As shown in diagram (b), the hypotenuse of the triangle then represents the impedance. The degree to which the voltage and current are in phase is given by angle θ, which is called the **phase angle**. From the data in this figure, determine the impedance and the phase angle.

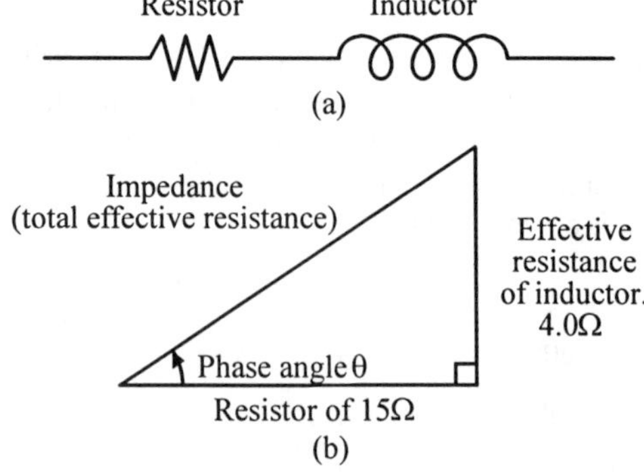

38. The sides of a rectangle are 18 and 31 ft. Find the angle that the diagonal makes with the shorter side.

39. In an isosceles triangle each of the equal sides is $7\overline{0}$ in., and the base is $8\overline{0}$ in. Find the angles of the triangle.

40. An important principle in the mathematical analysis of light is the **law of reflection**. This law states that a ray of light that strikes a reflecting surface is reflected so that the angle of incidence (i) equals the angle of reflection (r). In the diagram below, a photographer positions his flash at A. For a better lighting effect he aims this flash at position P on a reflecting surface to take a picture of a subject at B. If $\overline{AC} = 4.0$ ft, $\overline{CD} = 12$ ft and $i = 41°$, find the length of BD.

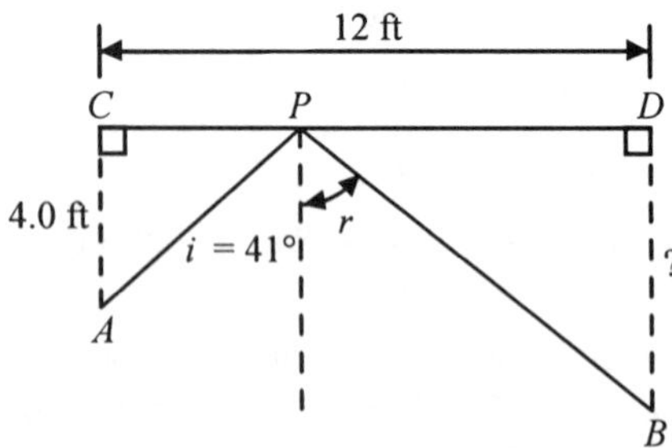

41. An observer on the third floor of a building determines that the angle of depression of the foot of a building across the street is 28°, and the angle of elevation of the top of the same building is 51°. If the distance between the two buildings is $5\overline{0}$ ft, find the height of the observed building.

42. An observer on the top of a hill 350 ft above the level of a road spots two cars due east of her. Find the distance between the cars if the angles of depression noted by the observer were 16° and 27°.

43. An artillery spotter in a plane at an altitude of 950 m observes two tanks in a line due east of the plane. If the angles of depressions of the two tanks measures 62° and 44°, then find the distance between the tanks.

44. A circular disc 24.0 in diameter is to have five equally spaced holes as shown. Determine the correct setting (x) for the dividers to space these holes.

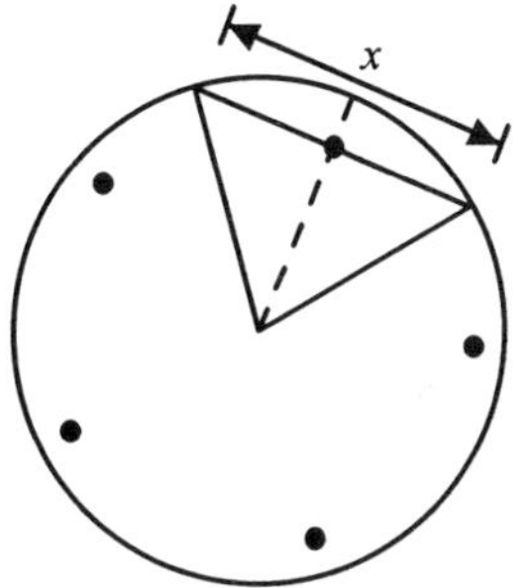

45. In pipe fitting there is a measure called the Take Out of an elbow. See the figure below, which shows the case of a 50° elbow. If the inner radius of the elbow shown is 2.0 in. and the outer radius is 4.0 in., find the take out. (Idea for this exercise was taken from Pipe Fitter's Math Guide by J.E. Hamilton, Construction Trades Press).

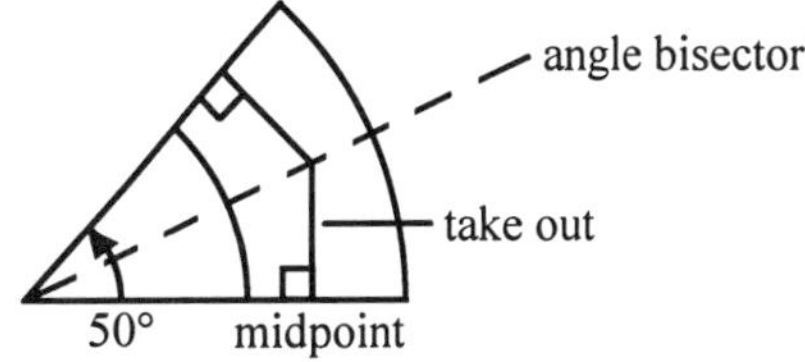

46. Refer to Exercise 45. If the inner radius is changed to 3.0 in., find the new take out.

THINK ABOUT IT 2.2

1. In the figure below show that $\tan(\theta/2) = \sin\theta/(1+\cos\theta)$.

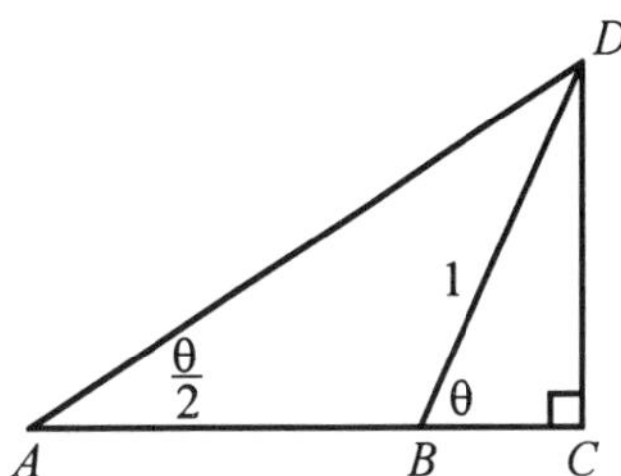

2. A machinist is given a 5.00-in.-diameter steel rod with instructions to make a tapered pin 12.0 in. long. The pin must have diameters of 4.00 and 2.00 in. What angle of taper should be used to obtain the right dimensions? (**Hint:** Make use of the dashed line parallel to the center axis in the diagram.)

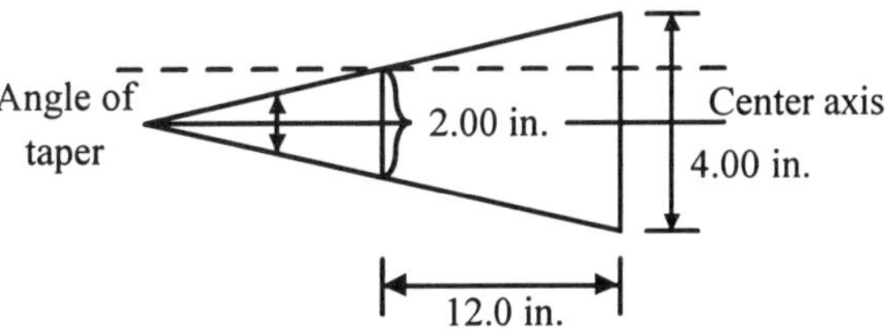

3. Consider the diagram below, which illustrates that twilight lasts until the sun is 18° below the horizon. From this, estimate the height (h) of the atmosphere.

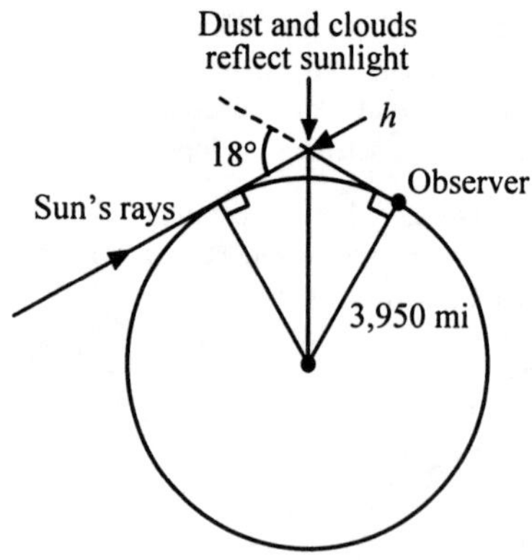

4. In the figure below, derive a formula for h in terms of d, cot θ, and cot α.

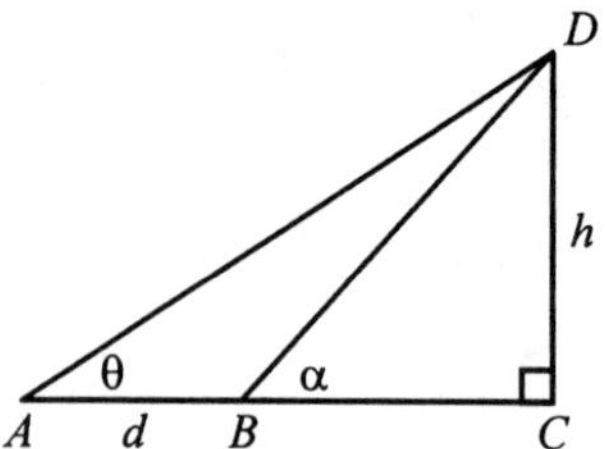

5. Create a word problem that may be solved by using the diagram and formula from Question 4. Then, find the solution to the problem you created.

● ● ●

2.3 Trigonometric Functions of General Angles

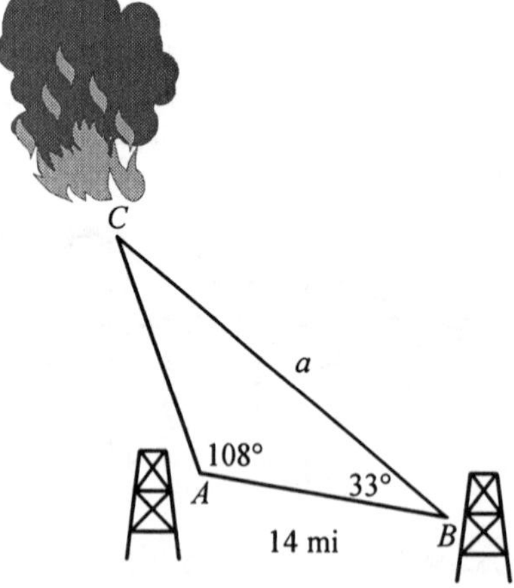

Two fire towers, A and B, are located 14 mi apart. A fire is sighted at point C, and observers in towers A and B measure angles A and B to be 108° and 33°, as illustrated. In accordance with the law of sines (see Section 9.1), the distance a between tower B and the fire is given by

$$a = \frac{c \sin A}{\sin C}$$

To the nearest mile, find a. (See Example 9.)

Objectives

1. Find the values of the six trigonometric functions of an angle, given the terminal ray of the angle.
2. Find the values of the five remaining trigonometric functions of θ, given one function value and the quadrant containing the terminal ray of θ.

3. Determine the exact value of any trigonometric function of a quadrantal angle.
4. Determine the exact value of any trigonometric function of angles with reference angles of 30°, 45°, or 60°.
5. Use a calculator to approximate the trigonometric value of a given angle.

Right-triangle trigonometry is useful but limited. To solve triangles that might not contain a right angle, we need trigonometric functions that are defined for $0° < \theta < 180°$ as illustrated in the section-opening problem. Also, although triangle trigonometry is important, an approach emphasizing the repetitive characteristics of the trigonometric functions is very useful today. The rhythmic motion of the heart, the sound waves produced by musical instruments, and alternating electric current are only a few of the physical phenomena that occur in cycles and can be studied through the aid of the trigonometric functions.

To begin this general study, we use the Cartesian coordinate system as the frame of reference for defining the trigonometric functions. First, we put angle θ in **standard position**, as shown in Figure 2.20.

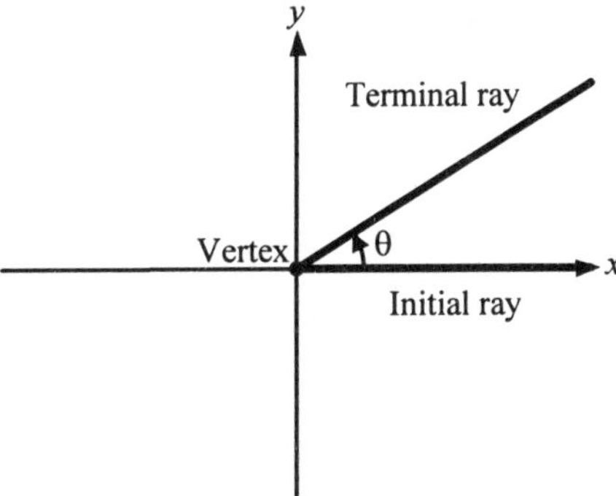

Figure 2.20

1. We place the vertex of the angle at the origin (0, 0).
2. We place the initial ray of the angle along the positive x-axis.

When we have the angle in standard position, we can define the trigonometric functions of an angle by considering any point on the terminal ray of θ [except (0, 0)]. Three numbers can be associated with the location of this point (see Figure 2.21):

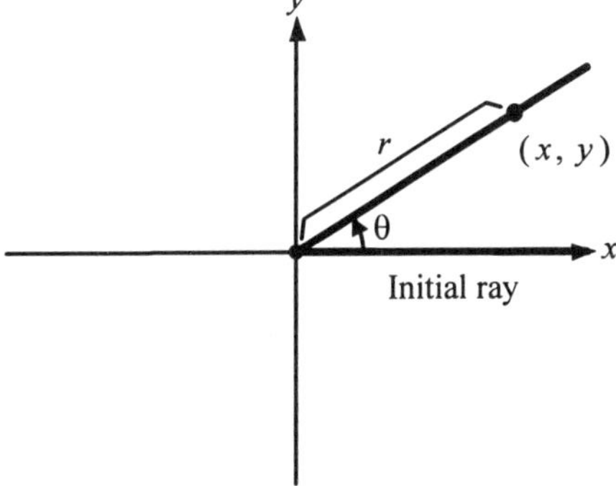

Figure 2.21

1. The x-coordinate of the point
2. The y-coordinate of the point
3. The distance r between the point and the origin

Since r represents the distance from the origin (0, 0) to the point (x, y), we can find the relationship among x, y, and r by using the distance formula.

$$r = \sqrt{(x-0)^2 + (y-0)^2}$$
$$r = \sqrt{x^2 + y^2}$$

or

$$r^2 = x^2 + y^2$$

If we consider the number of ratios that can be obtained from the three variables x, y, and r, we find that there are six. It is these six ratios that define the six *trigonometric functions.*

Definition of the Trigonometric Functions

If θ is an angle in standard position, and if (x, y) is any point on the terminal ray of θ [except (0, 0)], then

Name of Function	Abbreviation	Ratio	
sine of angle θ	$\sin\theta$	$=\frac{y}{r}$	reciprocal functions
cosecant of angle θ	$\csc\theta$	$=\frac{r}{y}\ (y \neq 0)$	
cosine of angle θ	$\cos\theta$	$=\frac{x}{r}$	reciprocal functions
secant of angle θ	$\sec\theta$	$=\frac{r}{x}\ (x \neq 0)$	
tangent of angle θ	$\tan\theta$	$=\frac{y}{x}\ (x \neq 0)$	reciprocal functions
cotangent of angle θ	$\cot\theta$	$=\frac{x}{y}\ (y \neq 0)$	

Example 1: Finding Trigonometric Functions Using Their Definition

Find the value of the six trigonometric functions of an angle θ if $(2, -5)$ is a point on the terminal ray of θ.

Solution: Figure 2.22 illustrates that the terminal ray of θ lies in quadrant 4, and that $x = 2$ and $y = -5$. Then

$$r = \sqrt{x^2 + y^2} = \sqrt{2^2 + (-5)^2} = \sqrt{29},$$

so by definition

$$\sin\theta = \frac{y}{r} = \frac{-5}{\sqrt{29}} \leftarrow \text{reciprocals} \rightarrow \csc\theta = \frac{r}{y} = \frac{\sqrt{29}}{-5}$$

$$\cos\theta = \frac{x}{r} = \frac{2}{\sqrt{29}} \leftarrow \text{reciprocals} \rightarrow \sec\theta = \frac{r}{x} = \frac{\sqrt{29}}{2}$$

$$\tan\theta = \frac{y}{x} = \frac{-5}{2} \leftarrow \text{reciprocals} \rightarrow \cot\theta = \frac{x}{y} = \frac{2}{-5}.$$

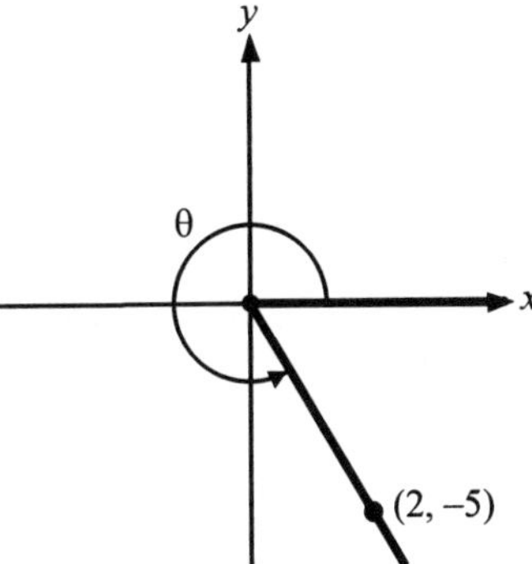

Figure 2.22

PROGRESS CHECK 1
Find the value of the six trigonometric functions of angle θ if $(-3, -1)$ is a point on the terminal ray of θ. ■

Note two important features of the definition above. First, (x, y) may be any point (except the origin) on the terminal ray of θ. Figure 2.23 shows that if we pick different points on the terminal ray of θ, say (x_1, y_1) and (x_2, y_2), they determine similar triangles. It follows, as in Section 2.1 that corresponding side lengths are proportional, so the trigonometric ratios are uniquely determined by the terminal ray of θ. This means that in addition to picking any point on the terminal ray of θ, we may also choose any of the angle measures associated with the angle.

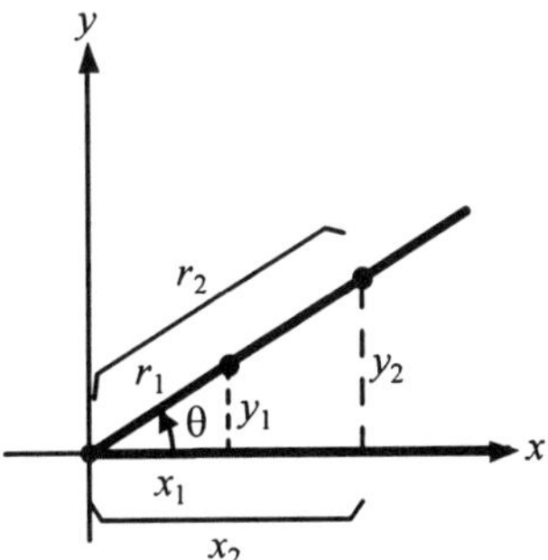

Figure 2.23

Second, note in Figure 2.24 that if θ is a positive acute angle in standard position with the point (x, y) on its terminal ray, then a right triangle may be formed with opposite side of length y, adjacent side of length x, and hypotenuse of length r. Thus, right triangle trigonometry is a special case of the more general definition of the trigonometric functions, for it arises when we are dealing with an acute angle whose terminal ray lies in the first quadrant.

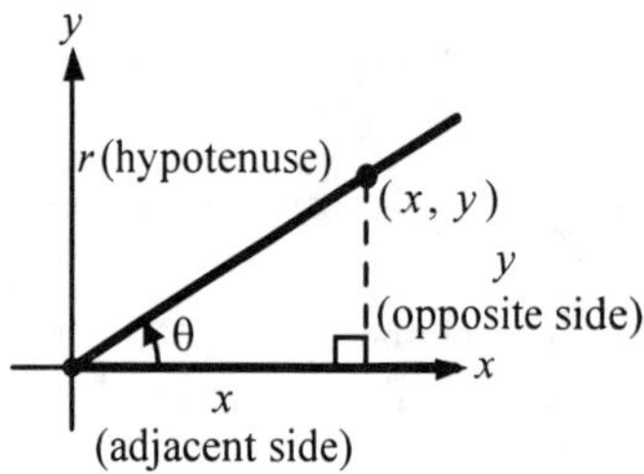

Figure 2.24

By considering the definition of the trigonometric functions and the signs of x and y in the various quadrants, we can construct a chart of the signs of the trigonometric ratios, as shown in Figure 2.25. To illustrate, because $\cos\theta = x/r$ with $r > 0$, the sign in the cosine ratio depends on the sign of x. Thus, cos θ is positive in Q_1 and Q_4 and negative in Q_2 and Q_3. As a memory aid, the underlined first letters in the chart in Figure 2.25 correspond to the first letters in the mnemonic "*A*ll *S*tudents *T*ake *C*alculus." One use for this chart is shown in the next example.

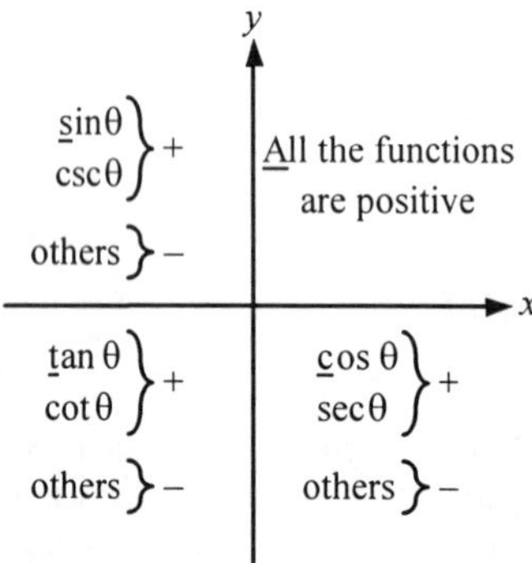

Figure 2.25

Example 2: Finding Trigonometric Functions Given One Function Value

Find the value of the remaining trigonometric functions if $\tan\theta = \dfrac{2}{3}$ and $\sec\theta < 0$.

Solution: First, determine the quadrant that contains the terminal ray of θ. $\tan\theta = 2/3$, which is positive. Therefore, θ is in Q_1 or Q_3; $\sec\theta < 0$ means that sec θ is negative and θ is in Q_2 or Q_3. Only Q_3 satisfies both conditions. Therefore, the terminal ray of θ is in Q_3, where x is negative and y is negative.

Second, determine appropriate values for x, y, and r.

$$\tan\theta = \frac{y}{x} = \frac{2}{3}$$

Since both x and y are negative in Q_3, let $x = -3$ and $y = -2$; then find r.

$$\begin{aligned} r &= \sqrt{x^2 + y^2} \\ &= \sqrt{(-3)^2 + (-2)^2} \\ &= \sqrt{9+4} \\ &= \sqrt{13} \end{aligned}$$

Third, calculate the values of the different trigonometric functions. If $x=-3$, $y=-2$, and $r=\sqrt{13}$, then

$$\sin\theta=\frac{y}{r}=\frac{-2}{\sqrt{13}} \leftarrow \text{reciprocals} \rightarrow \csc\theta=\frac{r}{y}=\frac{\sqrt{13}}{-2}$$
$$\cos\theta=\frac{x}{r}=\frac{-3}{\sqrt{13}} \leftarrow \text{reciprocals} \rightarrow \sec\theta=\frac{r}{x}=\frac{\sqrt{13}}{-3}$$
$$\tan\theta=\frac{y}{x}=\frac{-2}{-3}=\frac{2}{3} \leftarrow \text{reciprocals} \rightarrow \cot\theta=\frac{x}{y}=\frac{-3}{-2}=\frac{3}{2}.$$

Note once again that the value of a trigonometric function is a ratio. $\tan\theta=2/3$ did not mean that $y=2$ and $x=3$. Also, in the second step we could have proceeded by drawing a right triangle with opposite side 2 and adjacent side 3. With this method we then determine the hypotenuse, set up the trigonometric ratios, and affix the correct sign depending on the function value in Q_3.

PROGRESS CHECK 2
Find the value of the remaining trigonometric functions if $\sin\theta=-3/4$ and $\tan\theta>0$. ■

We now consider the problem of evaluating a trigonometric function for any angle. It is easy to use a calculator for such evaluations, since we merely need to extend the methods shown in Section 2.1 for evaluating trigonometric functions of acute angles. Although most calculators limit the size of the angle you can enter, you will rarely exceed this input range. In case you do, we will discuss how to reduce large angles later in this section. Thus, a calculator should easily give you a number associated with a trigonometric evaluation.

What is missing in this simple calculator method, however, is an understanding of some important concepts in trigonometry. Reference and coterminal angles, the signs of the functions in the various quadrants, and special angles with exact trigonometric values that are easily found and often used are just some of the ideas associated with a noncalculator approach. So keep in mind that a final numerical answer is only part of the objective of this section. We begin by using the definition of the trigonometric functions in the following example to evaluate functions for a special angle.

Example 3: Evaluating Trigonometric Functions by Definition

Find the six trigonometric functions of 90° using the definition of the trigonometric function.

Solution: We have shown that we may pick any point on the terminal ray of the angle when applying the definition. The terminal ray of a 90° angle is the positive *y*-axis, and a choice of $r=1$ determines the point (0, 1), as shown in Figure 2.26. Then, since $x=0$, $y=1$, and $r=1$, we have

$$\sin 90°=\frac{y}{r}=\frac{1}{1}=1 \leftarrow \text{reciprocals} \rightarrow \csc 90°=\frac{r}{y}=\frac{1}{1}=1$$
$$\cos 90°=\frac{x}{r}=\frac{0}{1}=0 \leftarrow \text{reciprocals} \rightarrow \sec 90°=\frac{r}{x}=\frac{1}{0} \text{ undefined}$$
$$\tan 90°=\frac{y}{x}=\frac{1}{0} \text{ undefined} \leftarrow \text{reciprocals} \rightarrow \cot 90°=\frac{x}{y}=\frac{0}{1}=0.$$

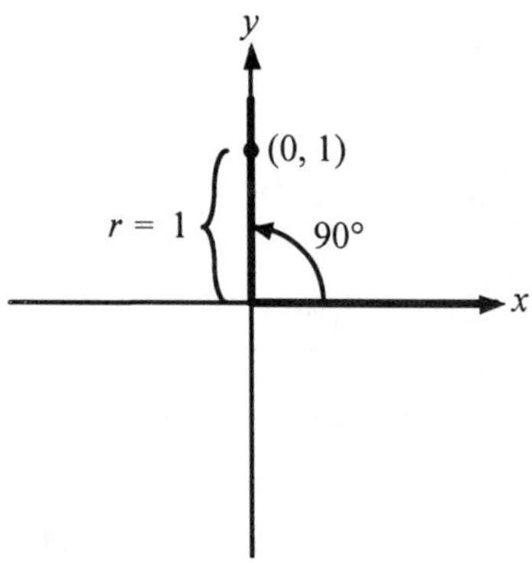

Figure 2.26

You should check these results by calculator and note the cases in which the calculator is unable to produce an answer.

PROGRESS CHECK 3

Find the six trigonometric functions of 180°, using the definition of the trigonometric functions. ■

An angle of 90° is "special" because its terminal ray coincides with one of the axes. Such angles are called **quadrantal angles**, and any quadrantal angle can be expressed as the product of 90° and some integer. For example, $270° = 90° \cdot 3$ and $-180° = 90° \cdot (-2)$. We can repeat the procedure from Example 3 when evaluating functions for any quadrantal angle, and Figure 2.27 shows the four possible positions for the terminal ray of such angles. There are other quadrantal angles besides 0°, 90°, 180°, and 270° (which are shown in the figure), but their trigonometric values must be the same as one of the four listed. For example, the trigonometric values for 360° will be the same as the trigonometric values for 0°, since the terminal ray for both angles is the same (positive x-axis). In general, if two angles have the same terminal ray, they are called **coterminal**, and the trigonometric functions of coterminal angles are equal. The following table summarizes the values of the trigonometric functions for quadrantal angles.

θ	$\sin\theta$	$\csc\theta$	$\cos\theta$	$\sec\theta$	$\tan\theta$	$\cot\theta$
0°	0	undefined	1	1	0	undefined
90°	1	1	0	undefined	undefined	0
180°	0	undefined	−1	−1	0	undefined
270°	−1	−1	0	undefined	undefined	0

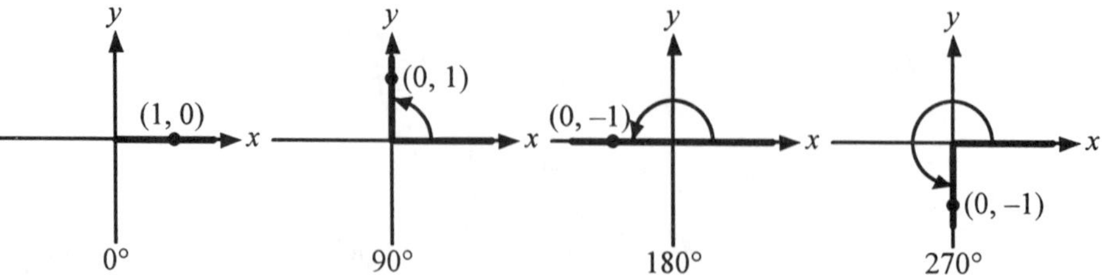

Figure 2.27

Example 4: Finding Trigonometric Functions of Quadrantal Angles

Find the exact value of each expression.

a. $\sin 540°$ **b.** $\cos(-270°)$

Solution:

a. $540°$ is a quadrantal angle that is coterminal with $180°$, as shown in Figure 2.28. Therefore, $\sin 540° = \sin 180° = 0$.

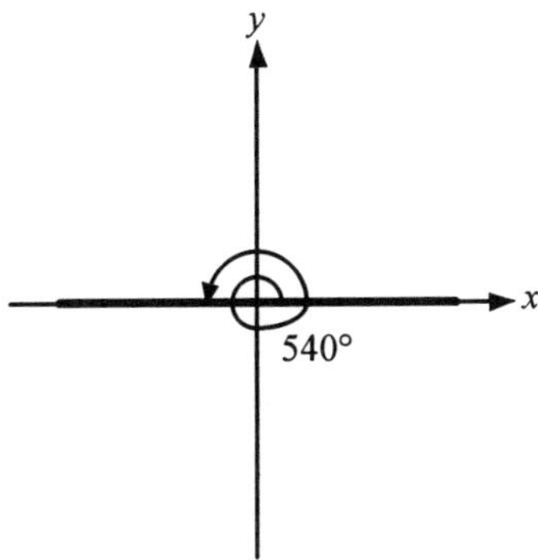

Figure 2.28

b. Recall that a negative angle measure indicates a rotation in a clockwise direction. So $-270°$ is a quadrantal angle that is coterminal with $90°$, as shown in Figure 2.29. Thus, $\cos(-270°) = \cos 90° = 0$.

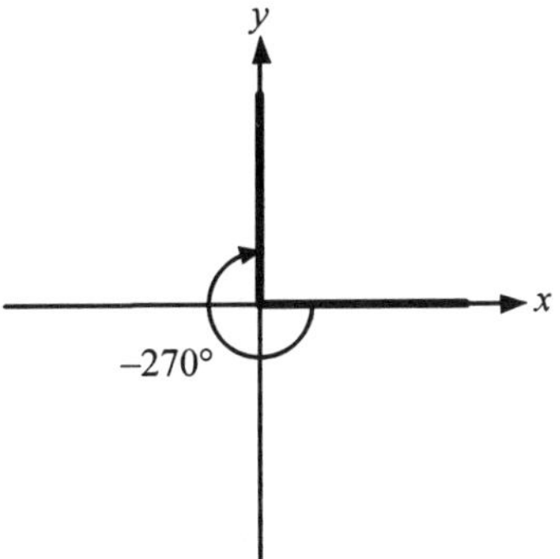

Figure 2.29

PROGRESS CHECK 4

Find the exact value of each expression.

a. $\cos 720°$ **b.** $\cot(-90°)$ ■

To evaluate trigonometric functions for angles that are not quadrantal angles, we introduce the concept of a **reference angle**. The reference angle for an angle θ is defined to be the positive acute angle formed by the terminal ray of θ and the horizontal axis. For example, the reference angle for $150°$ is $30°$, since the closest segment of the horizontal axis (negative x-axis) may correspond to a rotation of $180°$ and $|180° - 150°| = 30°$, as shown in Figure 2.30.

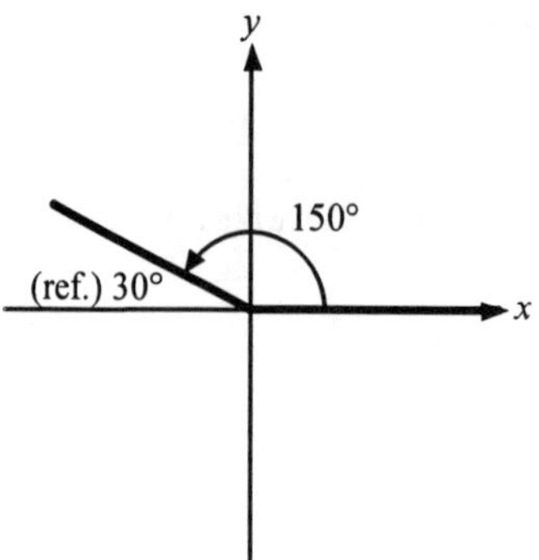

Figure 2.30

Now consider the angles 30°, 150°, 210°, and 330°, where the reference angle for each of these angles is 30°. Note that the points on the terminal ray for each of these angles differ only in the sign of the x-coordinate or the y-coordinate. Therefore, the values of the trigonometric functions (which are ratios among, x, y, and r) of these angles can differ only in their sign. For instance, by considering the appropriate points in Figure 2.31, we can determine $\sin 30° = 1/2$, $\sin 150° = 1/2$, $\sin 210° = -1/2$, and $\sin 330° = -1/2$. Thus, the reference angle provides us with a method for relating an angle in any quadrant to some acute angle.

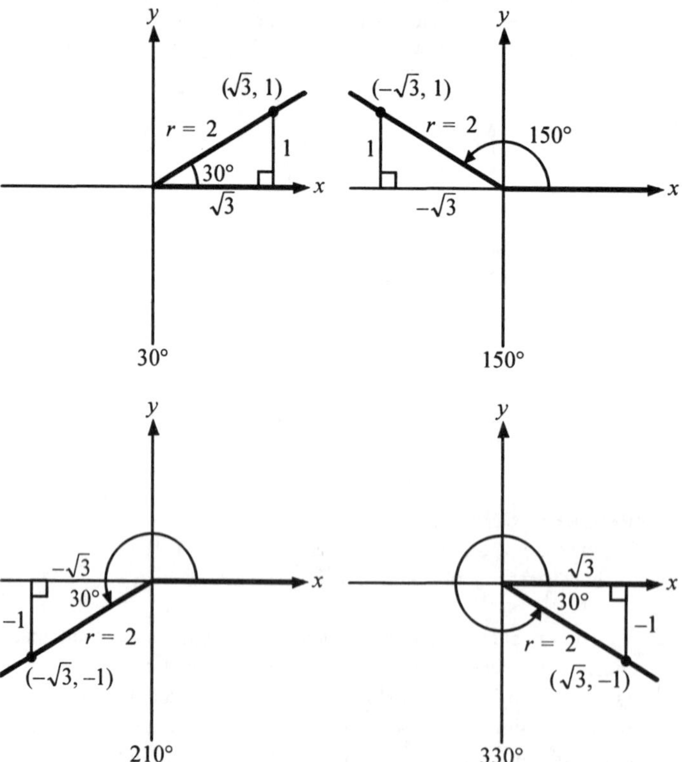

Figure 2.31

Although the trigonometric values of an angle and its reference angle may differ in sign, we can determine the correct sign using the chart for the signs of the trigonometric ratios that was given in Figure 2.25. A systematic method for using this chart and the concept of a reference angle to evaluate trigonometric functions for nonquadrantal angles is shown in the next example.

Example 5: Evaluations Involving 30°, 45°, or 60° Reference Angles

Find the exact value of tan 135°.

Solution: First, determine the reference angle.

$$|180° - 135°| = 45° \quad \text{(See Figure 2.32.)}$$

Second, determine tan 45°.

$$\tan 45° = 1 \quad \text{(See Figure 2.33)}$$

Third, determine the correct sign.

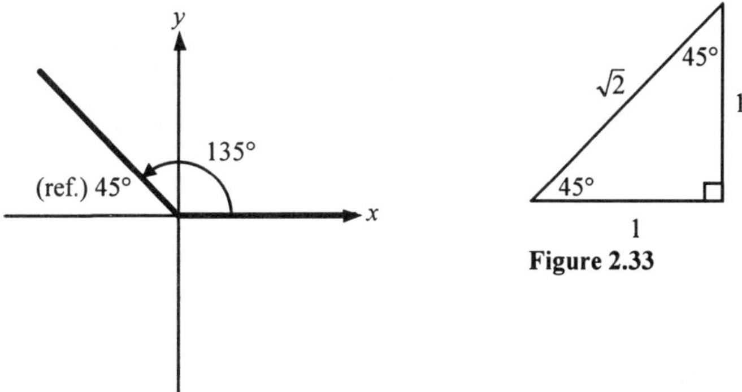

Figure 2.33

Figure 2.32

135° is in Q_2, where the value of the tangent function is negative. Therefore, $\tan 135° = -1$. Check this result by calculator.

PROGRESS CHECK 5
Find the exact value of sin 300°. ■

In general, we can find the trigonometric value of nonquadrantal angles by doing the following.

To Evaluate Trigonometric Functions

For nonquadrantal angles

1. Find the reference angle for the given angle.
2. Find the trigonometric value of the reference angle using the appropriate function. If the reference angle is 30°, 45°, or 60°, the exact answer is preferable.
3. Determine the correct sign according to the terminal ray of the angle and the chart in Figure 2.25.

Example 6: Evaluations Involving 30°, 45°, or 60° Reference Angles

Find the exact value of cos 300°.

Solution: First, determine the reference angle.

$$|360° - 300°| = 60° \quad \text{(See Figure 2.34)}$$

Second, determine cos 60°.

$$\cos 60° = \frac{1}{2} \quad \text{(See Figure 2.35)}$$

Third, determine the correct sign.

300° is in Q_4, where the cosine function is positive.

Therefore, $\cos 300° = \cos 60° = 1/2$. Check this result by calculator.

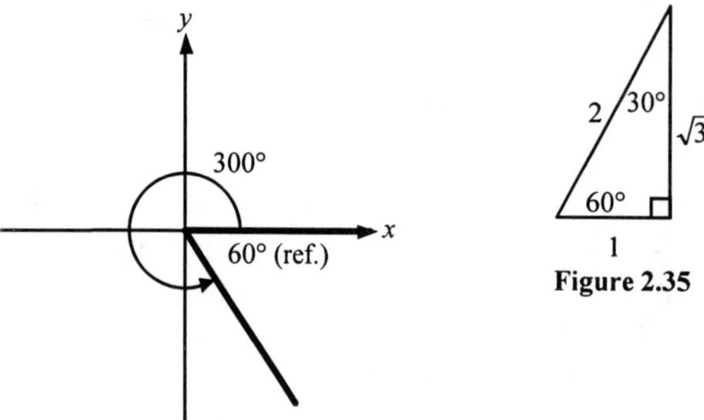

Figure 2.35

Figure 2.34

PROGRESS CHECK 6
Find the exact value of sin 315°. ■

Example 7: Evaluations Involving 30°, 45°, or 60° Reference Angles

Find the exact value of sec 510°.

Solution: First, determine the reference angle.

$$|540° - 510°| = 30° \quad \text{(See Figure 2.36)}$$

Second, determine sec 30°.

$$\sec 30° = \frac{2}{\sqrt{3}} \quad \text{(See Figure 2.35)}$$

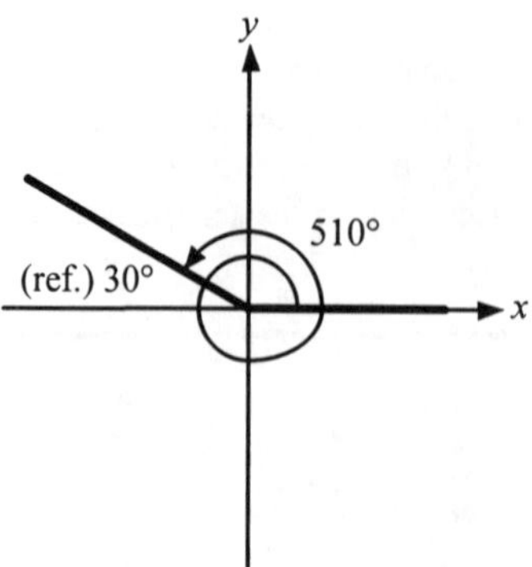

Figure 2.36

Third, determine the correct sign.

$510°$ is in Q_2, where the secant function is negative.

Therefore, $\sec 510° = -\sec 30° = -2/\sqrt{3}$.

A calculator check that supports this result, using $\sec\theta = 1/\cos\theta$, is shown in Figure 2.37.

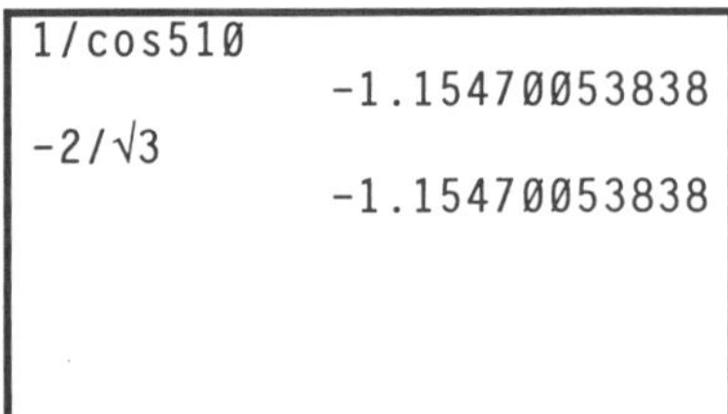

Figure 2.37

PROGRESS CHECK 7
Find the exact value of cot 660°. ■

Quadrantal angles and angles with reference angles of 30°, 45°, or 60° may be considered special cases because exact values may be found for trigonometric functions of these angles. For other angles, we can easily approximate their function values by using a calculator, as shown in the next two examples.

Example 8: Approximating Trigonometric Values by Calculator

Find the approximate value of each expression. Round off answers to four decimal places.
a. $\cos 227°20'$ **b.** $\csc(-412°)$

Solution: See Figure 2.38.
a. $\cos 227°20' \approx -0.6777$. Note that 20 minutes may be entered as $20/60$ because one minute is $1/60$ of one degree.
b. Using $\csc(-412°) = 1/\sin(-412°)$ yields $\csc(-412°) \approx -1.2690$.

```
cos(227 + 20/60)
              -.677731999146
1/sin(-412)
              -1.26901821507
```

Figure 2.38

PROGRESS CHECK 8
Find the approximate value of each expression. Round off answers to four decimal places.
a. $\sin 96°40'$ **b.** $\sec(-518°)$ ■

Example 9: Distance to a Forest Fire

Solve the problem in the section introduction on page 196.

Solution: Consider the sketch of the problem in Figure 2.39. The angle measures sum to 180° in a triangle, so

$$\begin{aligned} A+B+C &= 180^\circ \\ 108^\circ+33^\circ+C &= 180^\circ \\ C &= 39^\circ. \end{aligned}$$

Then replacing c by 14, A by 108°, and C by 39° in $a = c\sin A/\sin C$ yields

$$a = \frac{14\sin 108^\circ}{\sin 39^\circ} \approx 21.1574 \quad \text{(by calculator).}$$

To the nearest mile, the fire is 21 miles from tower B.

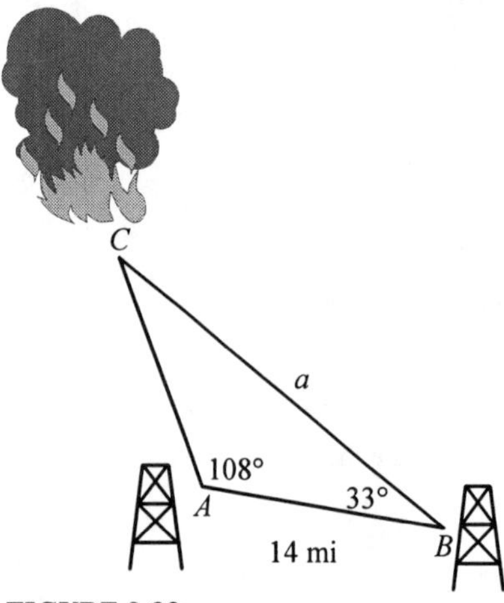

FIGURE 2.39

PROGRESS CHECK 9

Redo the problem in Example 9 assuming that $A = 97^\circ$, $B = 29^\circ$, and $\overline{AB} = 12$ mi. ■

EXPLORE 2.3

1. Compare the values of these trigonometric expressions, and then explain any pattern that you see.
   - **a.** sin(40°); sin(400°); sin(4000°); sin(40000°)
   - **b.** cos(50°); cos(500°); cos(5000°); cos(50000°)
   - **c.** tan(70°); tan(700°); tan(7000°); tan(70000°)
2. According to the law of cosines (discussed in Section 9.2) the following equation is true for any triangle ABC.

$$\cos A = \frac{b^2+c^2-a^2}{2bc}$$

   - **a.** In a triangle with $a=2$, $b=3$, and $c=4$, find angle A. Make a rough sketch of this triangle.
   - **b.** In a triangle with $a=1$, $b=2$, and $c=3$, find angle A. Why do you get a strange answer?
3. As indicated by Figure 2.25 in this section, if you know the signs (+ or –) for $\sin\theta$ and $\cos\theta$, you can identify the quadrant for θ. Use this approach with a calculator to determine the quadrant for each of these angles.

   a. 987° **b.** –1,234° **c.** 1 million°

EXERCISES 2.3

In Exercises 1–10 find the values of the six trigonometric functions of the angle θ that is in standard position and satisfies the given condition.

1. $(3, -4)$ is on the terminal ray of θ.
2. $(-1, -1)$ is on the terminal ray of θ.
3. $(-12, 5)$ is on the terminal ray of θ.
4. $(2, 4)$ is on the terminal ray of θ.
5. $(9, -12)$ is on the terminal ray of θ.
6. $(-7, -13)$ is on the terminal ray of θ.
7. The terminal ray of θ lies in Q_1 on the line $y = x$.
8. The terminal ray of θ lies in Q_3 on the line $y = x$.
9. The terminal ray of θ lies on the line $y = -2x$, and θ is in Q_2.
10. The terminal ray of θ lies on the line $y = -2x$, and θ is in Q_4.

In Exercises 11–16 determine in which quadrant the terminal ray of θ lies.

11. $\sin\theta$ is negative, $\cos\theta$ is positive.
12. $\cot\theta$ is positive, $\csc\theta$ is positive.
13. $\tan\theta$ is positive, $\sec\theta$ is negative.
14. $\sec\theta > 0$, $\csc\theta < 0$.
15. $\tan\theta < 0$, $\sin\theta > 0$.
16. $\cos\theta < 0$, $\cot\theta < 0$.

In Exercises 17–24 find the values of the remaining trigonometric functions of θ.

17. $\sin\theta = -\dfrac{3}{5}$, terminal ray of θ is in Q_3.
18. $\sec\theta = \dfrac{13}{12}$, terminal ray of θ is in Q_4.
19. $\cos\theta = \dfrac{1}{2}$, $\tan\theta$ is negative.
20. $\tan\theta = \dfrac{2}{3}$, $\sin\theta$ is negative.
21. $\csc\theta = -\dfrac{4}{3}$, $\cot\theta$ is positive.
22. $\cot\theta = 1$, $\csc\theta < 0$.
23. $\sin\theta = \dfrac{1}{3}$, $\sec\theta > 0$.
24. $\csc\theta = 2$, $\cos\theta < 0$.

In Exercises 25–30 find the six trigonometric functions of the given angle by using the definitions of the trigonometric functions

25.	−90°	**26.**	−180°	**27.**	360°
28.	270°	**29.**	−540°	**30.**	450°

In Exercises 31–36 find the exact value of the given expression.

31.	sin(450°)	**32.**	cos(630°)	**33.**	sec(720°)
34.	csc(−450°)	**35.**	cot(990°)	**36.**	tan(−900°)

In Exercises 37–64 find the *exact* value of each expression.

37.	sin 210°	**38.**	tan 225°	**39.**	sec 330°
40.	cos 135°	**41.**	cot 315°	**42.**	csc120°
43.	sin 150°	**44.**	tan 300°	**45.**	cos 225°
46.	cot 240°	**47.**	sec 390°	**48.**	tan 420°
49.	cot 690°	**50.**	cos 840°	**51.**	sin 1,035°
52.	csc 675°	**53.**	cos 495°	**54.**	sin 570°
55.	sin(−60°)	**56.**	cos(−45°)	**57.**	tan(−120°)
58.	sec(−225°)			**59.**	cot(−315°)
60.	csc(−210°)			**61.**	sec(−330°)
62.	cos(−480°)			**63.**	cot(−495°)
64.	sin(−1,050°)				

In Exercises 65–94 find the approximate value of each expression. Use a calculator, and round off answers to four decimal places.

65.	sin 212°	**66.**	cos 307°	**67.**	tan 254°
68.	cot 115°			**69.**	sec 301°20′
70.	csc 163.4°			**71.**	cos 148°50′
72.	sin 354.5°			**73.**	cot 298°10′

74. $\tan 190°40'$ **75.** $\sin 177°10'$

76. $\cos 252°20'$ **77.** $\csc 672°40'$

78. cot 392.1° **79.** $\sin 626°40'$

80. sin 531.5° **81.** $\cos 952°20'$

82. $\sec 452°20'$ **83.** $\tan 738°30'$

84. $\sin 521°50'$ **85.** $\cos(-81°)$

86. $\sin(-25°)$ **87.** $\tan(-131°)$

88. $\csc(-322°40')$ **89.** $\cos(-61°)$

90. $\sin(-251.5°)$ **91.** $\tan(-214°10')$

92. $\sec(-400°)$ **93.** $\cot(-512°)$

94. $\csc(-938°20')$

95. The cross-section of a teepee is shown in the figure. If the base angles are each 70.0° and the diameter of the floor is 16.0 ft, determine the minimum pole length needed for the side if the pole is to extend at least 1 ft beyond the peak. Solve this problem in two ways.

a. Use the law of sines (discussed in Section 9.1), which states that $a = \dfrac{b\sin A}{\sin B}$.

b. Split the given triangle into two right triangles, and apply right triangle trigonometry.

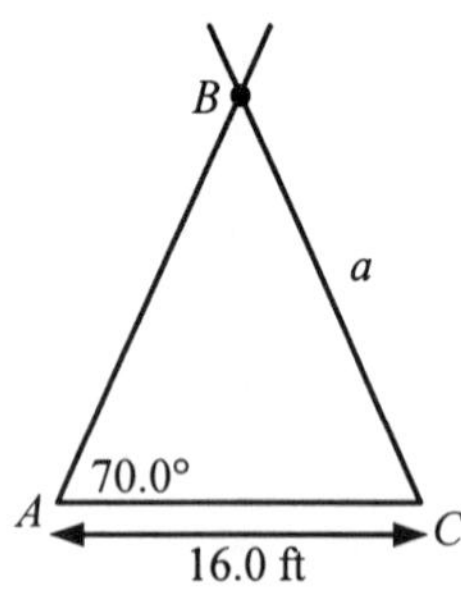

96. Refer to the figure in Exercise 95. Solve the problem again, but assume that the base angles are reduced to 65.0°.

97. A straight tunnel is to be cut through a mountain as indicated in the figure. Sighting from point A shows that angle A is 115.2°, the distance from A to B is 1.4 miles, and the distance from point A to point C is 2.1 miles. According to the law of cosines (discussed in Section 9.2),

$$a^2 = b^2 + c^2 - 2bc\cos A.$$

Use this relationship to determine the distance from B to C to the nearest tenth of a mile.

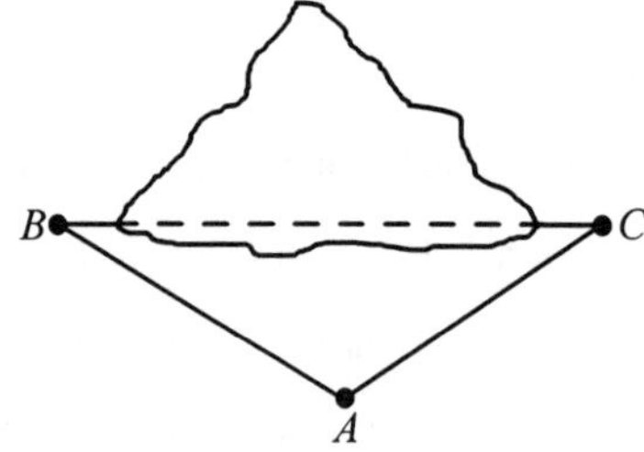

98. Refer to the figure in Exercise 97. Solve the problem again, but assume that angle A is 98.6°.

99. A formula for the horizontal distance traveled by a projectile, neglecting air resistance, is $d = \dfrac{1}{32}V^2 \sin 2\theta$. If a ball is kicked with an initial velocity V of 76 ft/second, when θ is 35°, how far will it go? When V is given in ft/second, then d is in feet. Determine the answer to the nearest foot.

100. Refer to Exercise 99. If the angle θ is changed to 40°, how far will the ball go?

THINK ABOUT IT 2.3

1. **a.** Give two examples of specific values for θ_1 and θ_2 so that $\sin(\theta_1 + \theta_2)$ does not equal $\sin\theta_1 + \sin\theta_2$. Verify your answers without using a calculator.
   b. Give two examples of specific values for θ_1 and θ_2 so that $\sin(\theta_1 + \theta_2)$ does equal $\sin\theta_1 + \sin\theta_2$. Verify your answers without using a calculator.
2. Evaluate $\sin 1° + \sin 2° + \cdots + \sin 360°$.
3. If $\cos\theta = a$ and $270° < \theta < 360°$, express the values of the other trigonometric functions in terms of a.
4. Consider the unit circle in the accompanying figure, in which each of the six trigonometric functions can be represented as a line segment. For example, since $\overline{OC} = 1$ in right triangle OAC, we have $\sin\theta = \overline{AC}/\overline{OC} = \overline{AC}/1 = \overline{AC}$.

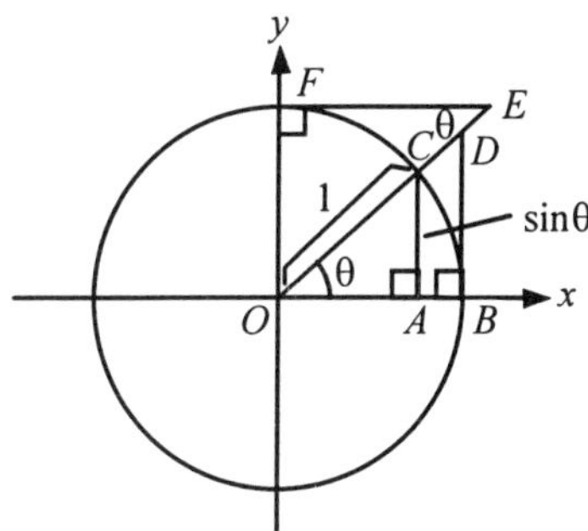

Notice that we obtained the desired line segment by selecting a right triangle where the denominator in the defining ratio is 1. Determine the line segments representing the five remaining trigonometric functions in the figure.

5. A simple method to obtain a rough estimate of the values of the trigonometric functions is to use construction. Start by using a compass to draw a circle with a radius of 20 spaces on a piece of graph paper. (**Note:** Do not be concerned if the circle goes slightly off the paper.) Label the graph so that a distance of 2 spaces corresponds to one-tenth of a unit, and the radius of 20 spaces corresponds to 1 unit (so $r = 1$). Now use a protractor to find the point on the circle that corresponds to the following rotations, and on the basis of your estimate for their x or y components, approximate to two significant digits the value of the given trigonometric functions.

   a. sin 30°, sin 150°, sin 210°, sin 330° **b.** cos 30°, cos 150°, cos 210°, cos 330°
   c. cos 45°, cos 135°, cos 225°, cos 315° **d.** sin 45°, sin 135°, sin 225°, sin 315°
   e. sin 62°, sin 118°, sin 242°, sin 298° **f.** cos 62°, cos 118°, cos 242°, cos 298°
   g. sec 15°, sec 165°, sec 195°, sec 345° **h.** csc 15°, csc 165°, csc 195°, csc 3345°

● ● ●

2.4 Introduction to Trigonometric Equations

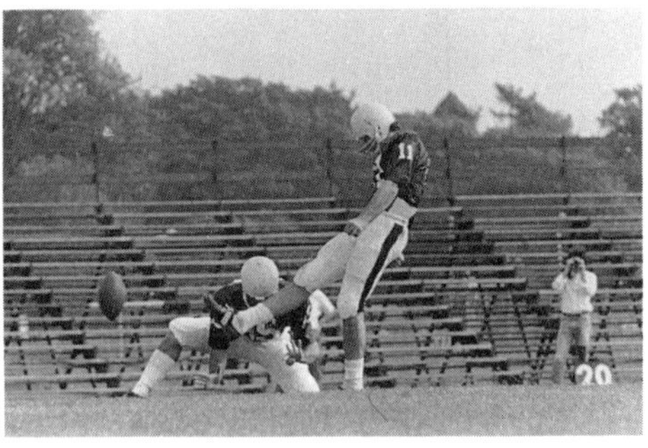

Photo Courtesy of ***Frank Siteman*** *of The Picture Cube*

A formula for the horizontal distance traveled by a projectile, neglecting air resistance, is

$$d = \frac{1}{32}V^2 \sin 2\theta$$

where θ and V measure the angle of elevation and the initial velocity in ft/second of the projectile, respectively. If a professional field-goal kicker boots a football with an initial velocity of 76 ft/second, and the ball travels 180 ft, find θ to the nearest degree. (See Example 6.)

Objectives

1. Find the exact solution to certain trigonometric equations for $0° \le \theta < 360°$.
2. Find the approximate solution to a trigonometric equation for $0° \le \theta < 360°$.
3. Find all solutions to a trigonometric equation, specifying exact solutions where possible.
4. Solve applied problems involving trigonometric equations.

Equations that involve trigonometric expressions commonly appear in two types: identities and conditional equations. An identity is an equation such as

$$\csc\theta = \frac{1}{\sin\theta}$$

that is true for all values of θ for which the expressions are defined (this topic is discussed in detail in Sections 8.7–8.9). Conditional trigonometric equations differ from identities in that they are true only for certain values of θ. For instance, in the section-opening problem we need to find all meaningful values for θ that satisfy

$$180 = \frac{1}{32}(76)^2 \sin 2\theta,$$

and we will find that there are just two possible solutions. Before we attempt to find a procedure for writing all the solutions to a particular equation, it is helpful to first establish a method for finding solutions where $0° \le \theta < 360°$. A method for finding such solutions is shown in Examples 1–4.

Example 1: Finding Exact Solutions Where $0° \le \theta < 360°$

Find the exact values of $\theta (0° \le \theta < 360°)$ for which the equation $\sin\theta = 1/2$ is a true statement.

Solution: First, determine the quadrant that contains the terminal ray of θ.

$$\sin\theta = \frac{1}{2}, \text{ which is a positive number.}$$

The terminal ray of θ could be in either quadrant 1 or 2, since the sine function is positive in both quadrants.
Second, determine the reference angle. Using Figure 2.40 we see that

$$\sin^{-1}\frac{1}{2} = 30°.$$

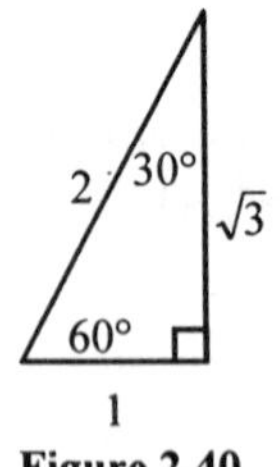

Figure 2.40

Third, determine the appropriate values of θ (see Figure 2.41).

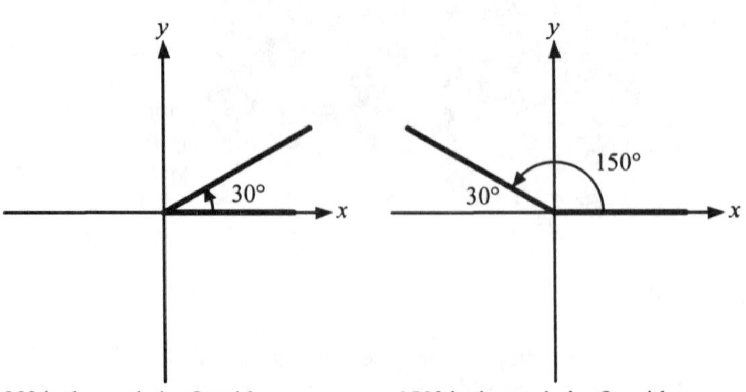

30° is the angle in Q_1 with a reference angle of 30°.

150° is the angle in Q_2 with a reference angle of 30°.

Figure 2.41

Therefore, 30° and 150° make the equation a true statement, and the solution set in the interval $[0°, 360°)$ is $\{30°, 150°\}$.

Technology Link

Although formal coverage of the graphs of the trigonometric functions is presented in Chapter 8, you should find that it is easy to use a grapher to create a picture of the solution in this example, as shown in Figure 2.42. To obtain this display a grapher was first set to Degree mode. Then, the equations $y1 = \sin x$ and $y2 = 1/2$ were graphed in the viewing window shown, and the Intersection operation on a grapher was used to estimate the points of intersection. (If your calculator does not have this feature, then the Zoom and Trace features may be used to estimate these points.) Observe that the x-coordinates of the intersection points are in agreement with the solution set $\{30°, 150°\}$.

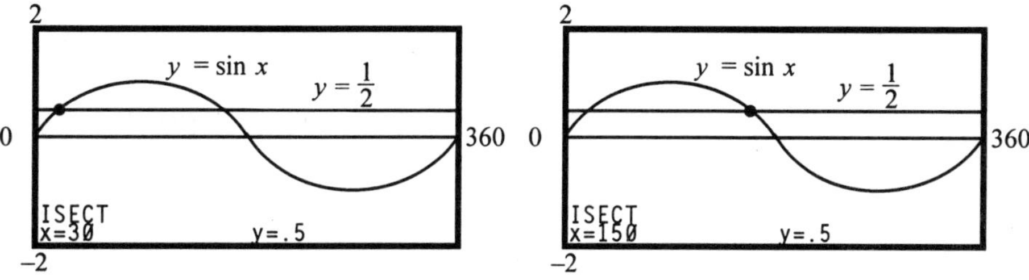

Figure 2.42

PROGRESS CHECK 1

Find the exact value of $\theta(0° \le \theta < 360°)$ for which the equation $\cos\theta = \sqrt{3}/2$ is a true statement. ■

Example 2: Finding Exact Solutions Where $0° \le \theta < 360°$

Find the exact values of $\theta(0° \le \theta < 360°)$ for which the equation $2\sin\theta + \sqrt{3} = 0$ is a true statement.

Solution: First, solve the equation for sin θ.

$$2\sin\theta + \sqrt{3} = 0$$
$$2\sin\theta = -\sqrt{3}$$
$$\sin\theta = \frac{-\sqrt{3}}{2}$$

Second, determine the quadrant that contains the terminal ray of θ.

$$\sin\theta = \frac{-\sqrt{3}}{2}, \text{ which is a negative number.}$$

The terminal ray of θ could be in either Q_3 or Q_4 since the sine function is negative in both quadrants.

Third, determine the reference angle. Discard the sign on the trigonometric ratio and use Figure 2.40 once again to determine that

$$\sin^{-1}\frac{\sqrt{3}}{2} = 60°.$$

Therefore, 60° is the reference angle.

Fourth, determine the appropriate values of θ (see Figure 2.43)

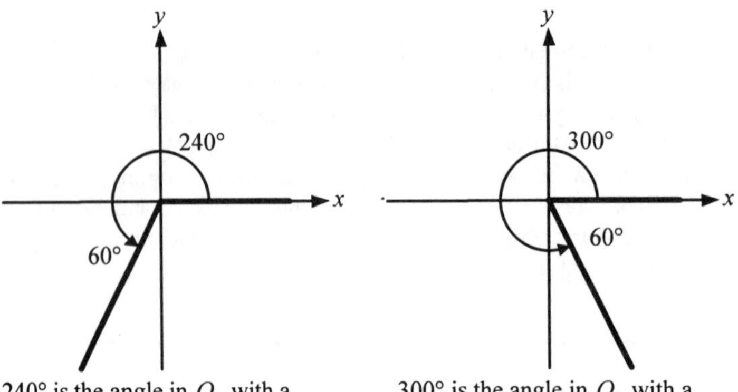

240° is the angle in Q_3 with a reference angle of 60°.

300° is the angle in Q_4 with a reference angle of 60°.

Figure 2.43

Therefore, 240° and 300° make the equation a true statement, and the solution set in the interval [0°, 360°) is {240°, 300°}.

Figure 2.44 provides graphical evidence that supports this result.

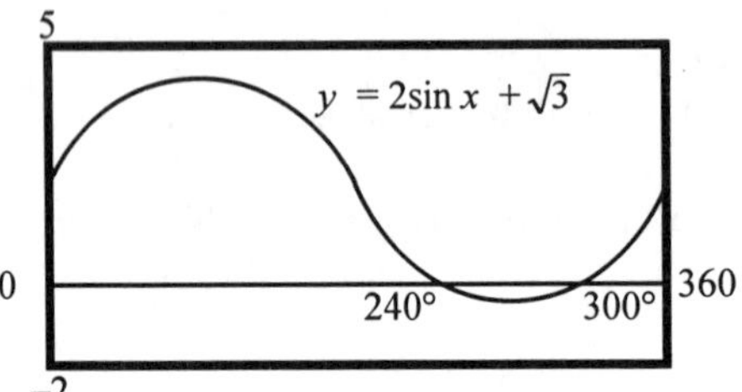

Figure 2.44

PROGRESS CHECK 2

Find the exact values of $\theta (0° \le \theta < 360°)$ for which the equation $2\cos\theta + 1 = 0$ is a true statement.

■

A solution in each of the four quadrants was illustrated in Examples 1 and 2. Observe that solutions in each of the quadrants are determined as follows:

Quadrant	Solution
1	reference angle
2	180° – reference angle
3	180° + reference angle
4	360° – reference angle

By using the inverse trigonometric function keys on a calculator, reference angles may be found exactly in the cases of 30°, 45°, and 60°, and approximately in other cases. The use of a calculator to determine the approximate value of a reference angle is illustrated in the next three examples.

Example 3: Finding Approximate Solutions Where $0° \le \theta < 360°$

To the nearest 10 minutes, approximate the values of $\theta (0° \le \theta < 360°)$ for which the equation $\cos\theta = -0.7969$ is a true statement.

Solution: First, determine the quadrant that contains the terminal ray of θ.

$$\cos\theta = -0.7969, \text{ which is a negative number.}$$

The terminal ray of θ could be in either quadrant 2 or 3, since the cosine function is negative in both quadrants.

Second, determine the reference angle. Discard the sign on the trigonometric ratio and use a calculator to determine that

$$\cos^{-1} 0.7969 \approx 37.165^\circ \approx 37^\circ 10'.$$

To the nearest 10 minutes, the reference angle is $37^\circ 10'$.

Third, determine the appropriate values of θ (see Figure 2.45).

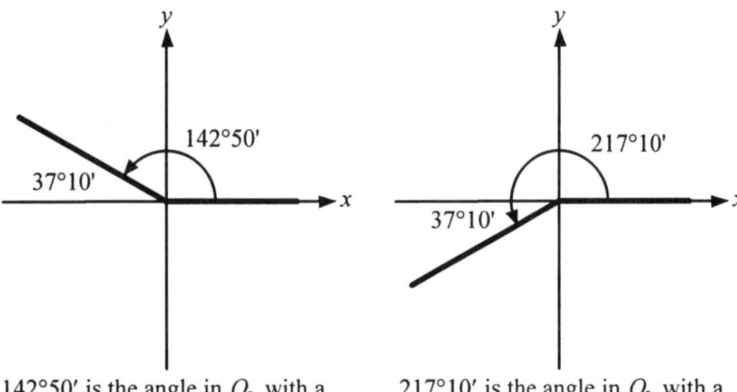

$142^\circ 50'$ is the angle in Q_2 with a reference angle of $37^\circ 10'$.

$217^\circ 10'$ is the angle in Q_3 with a reference angle of $37^\circ 10'$.

Figure 2.45

Therefore, $\{142^\circ 50', 217^\circ 10'\}$ is the solution set for $\theta (0^\circ \leq \theta < 360^\circ)$. A graphical check that supports this result is shown in Figure 2.46.

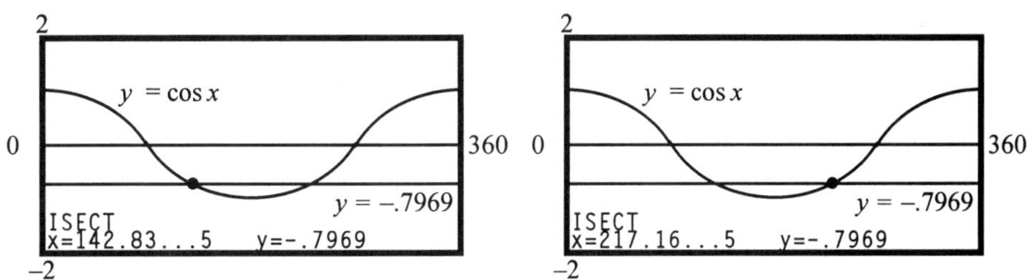

Figure 2.46

PROGRESS CHECK 3

To the nearest 10 minutes, approximate the values of $\theta (0^\circ \leq \theta < 360^\circ)$ for which the equation $\sin\theta = -0.4928$ is a true statement. ■

Example 4: Finding Approximate Solutions Where $0^\circ \leq \theta < 360^\circ$

Solve $4\tan\theta = 1$ for $0^\circ \leq \theta < 360^\circ$. Round off answers to the nearest tenth of a degree.

Solution: The graphical solution to this equation necessitates a small adjustment in our methods, so we will deal with this issue first. The difficulty arises because the tangent function is not defined in the interval $[0^\circ, 360^\circ)$ at 90° and 270°, so it is recommended that the calculator be set in

Dot mode, not Connected mode, when drawing graphs involving tan θ. The Dot mode prevents the calculator from connecting points that are in separate pieces of the graph (as discussed in Section 1.2). Figure 2.47 shows graphs drawn in this mode that indicate that the solution set is $\{14.0°, 194.0°\}$ for $0° \leq \theta < 360°$.

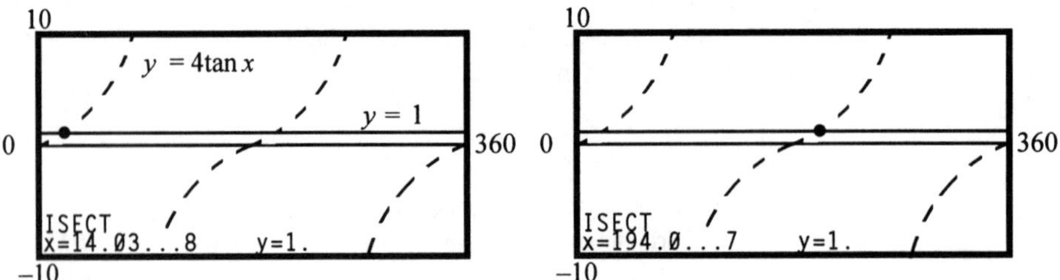

Figure 2.47

To confirm this result analytically, first solve the equation for tan θ.

$$4\tan\theta = 1$$

$$\tan\theta = \frac{1}{4}$$

Second, determine the quadrant that contains the terminal ray of θ.

$$\tan\theta = \frac{1}{4}, \text{ which is a positive number.}$$

The terminal ray of θ could be in either Q_1 or Q_3, since the tangent function is positive in both quadrants.

Third, determine the reference angle.

$$\tan^{-1}\frac{1}{4} = 14.0°$$

Therefore, 14.0° is the reference angle.

Fourth, determine the appropriate values of θ (see Figure 2.48).

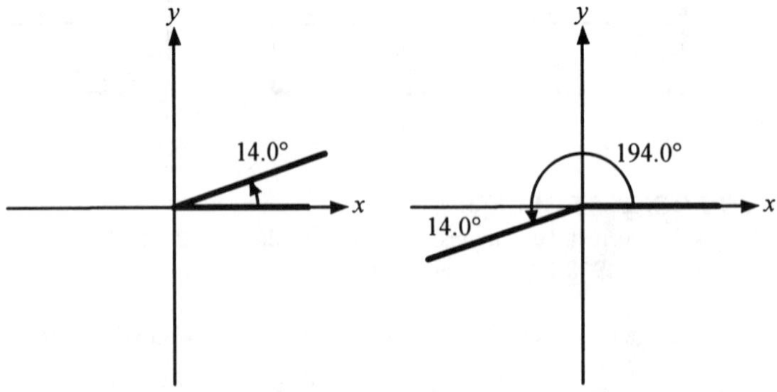

14.0° is the angle in Q_1 with a reference angle of 14.0°.

194.0° is the angle in Q_3 with a reference angle of 14.0°.

Figure 2.48

Therefore, $\{14.0°, 194.0°\}$ is the solution set for $0° \leq \theta < 360°$, which confirms the results we obtained initially through graphical analysis.

PROGRESS CHECK 4
Solve $3\tan\theta = -5$ for $0° \leq \theta < 360°$. Round off answers to the nearest tenth of a degree. ■

Once we are able to find the solutions of a trigonometric equation that are between 0° and 360°, we can determine *all* the solutions by finding the angles that are coterminal with our results. That is, we wish to find all the angles that have the same terminal ray as the solutions that are between 0° and 360°. The problem arises that we cannot possibly list all the angles that are coterminal with a given angle. Therefore, we indicate some rule by which coterminal angles can be found. We can generate all the angles coterminal to a given angle of, say, 30° by adding to 30° the multiples of 360°. Thus,

$$30° + (0)360° = 30°$$
$$30° + (1)360° = 390°$$
$$30° + (2)360° = 750°$$
$$30° + (-1)360° = -330°$$
$$30° + (-2)360° = -690° \quad \text{etc.}$$

In general, $\theta + k360°$, where k is an integer, will generate all the angles that have the same terminal ray as θ.

Example 5: Approximating All Solutions to a Trigonometric Equation

Approximate all the solutions to $10\cot\theta = 3$, to the nearest 10 minutes.

Solution: First solve the equation for cot θ. Then, for calculator purposes, use the reciprocal relation between the tangent and cotangent functions to rewrite the equation in terms of tan θ.

$$10\cot\theta = 3$$
$$\cot\theta = \frac{3}{10}$$
$$\tan\theta = \frac{10}{3}$$

Second, determine which quadrant contains the terminal ray of θ.

tan θ is a positive number.

The terminal ray of θ could be in either Q_1 or Q_3, since the tangent function is positive in both quadrants.

Third, determine the reference angle.

$$\tan^{-1}\frac{10}{3} \approx 73°20'$$

Therefore, $73°20'$ is the reference angle.

Fourth, determine the solutions between 0° and 360° (see Figure 2.49).

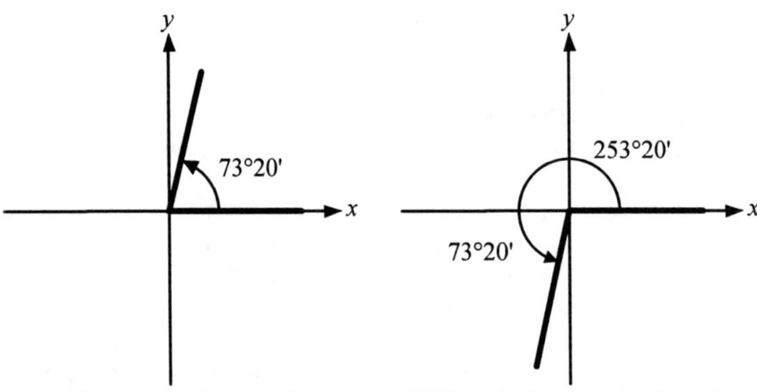

$73°20'$ is the angle in Q_1 with a reference angle of $73°20'$.

$253°20'$ is the angle in Q_3 with a reference angle of $73°20'$.

Figure 2.49

Therefore, $73°20'$ and $253°20'$ make the equation a true statement.

Fifth, indicate how angles coterminal to the above angles may be generated.

$$\left.\begin{array}{r} 73°20' + k360° \\ 253°20' + k360° \end{array}\right\} \text{ or equivalently } 73°20' + k180°,$$

where k is an integer, generates all the solutions to the equation. Thus, the solution set is $\{\theta\colon \theta = 73°20' + k180°,\ k \text{ any integer}\}$.

A graphical check that supports this result for $0° \le \theta < 360°$ is shown in Figure 2.50. Note that we use $\cot\theta = 1/\tan\theta$ to obtain this display, since $y = 10\cot x$ is graphed by entering the equation $y = 10(1/\tan x)$.

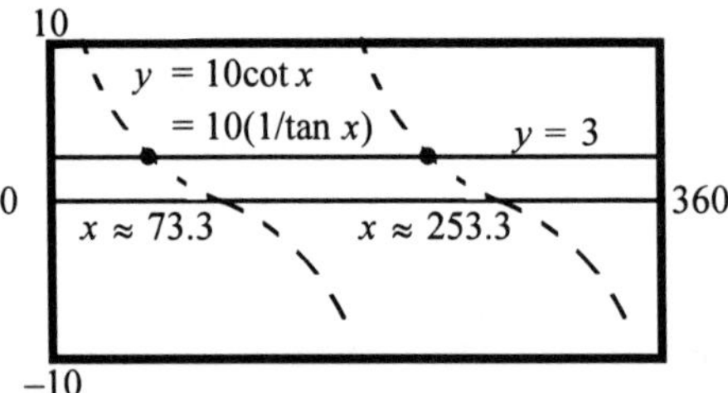

Figure 2.50

PROGRESS CHECK 5

Approximate all the solutions to $2\sec\theta + 7 = 1$, to the nearest 10 minutes. ■

Applications in trigonometry sometimes involve functions of multiple angles such as 2θ or $\pi\theta/6$. We illustrate how to solve equations involving such angles in the next example.

Example 6: Projectile Motion

Solve the problem in the section introduction on page 211.

Solution: Consider carefully how this problem may be solved using either a graphical approach or an analytic approach.

Graphic Approach: We will first try to obtain a picture of the solution through graphical analysis. Using

$$d = \frac{1}{32} V^2 \sin 2\theta$$

we replace d by 180 (since the ball traveled 180 ft) and V by 76 (since the initial velocity is 76 ft/second) to obtain

$$180 = \frac{1}{32}(76)^2 \sin 2\theta$$

Given the context of the problem, θ is meaningful for $0° < \theta < 90°$, so we choose a viewing window of [0, 90] by [0, 200] to obtain a picture of the situation in the problem, as shown in Figure 2.51. Then, zooming in on the maximum point in the graph and finding the intersection point shown in Figure 2.52 indicates that one solution (to the nearest degree) is 43°. Repeating the intersection process to the right of the maximum point yields a second solution at 47°. Thus, the angle of elevation of the kick was either 43° or 47°.

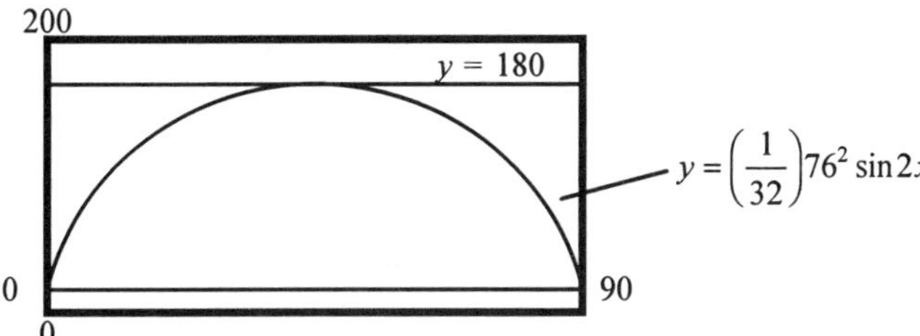

Figure 2.51

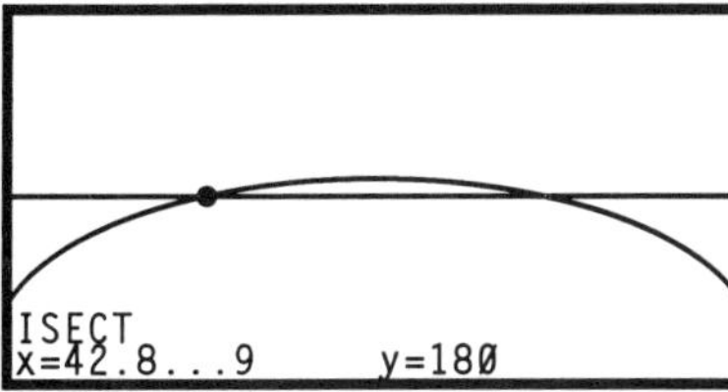

Figure 2.52

Analytic Approach: To obtain the solution analytically, follow these steps, as outlined in the previous examples.

1. $180 = \frac{1}{32}(76)^2 \sin 2\theta$ implies $\sin 2\theta = \frac{5760}{5776}$.
2. The sine function is positive in quadrants 1 and 2.
3. The reference angle is 86° since $\sin^{-1}\frac{5760}{5776} \approx 86°$.
4. Q_1 solution: $2\theta =$ reference angle $= 86°$
   Q_2 solution: $2\theta = 180° -$ reference angle $= 180° - 86° = 94°$

Finally, θ may now be found since

$$2\theta = 86° \text{ implies } \theta = 43°$$

and

$$2\theta = 94° \text{ implies } \theta = 47°$$

Thus, we have confirmed analytically that the angle of elevation of the kick was either 43° or 47°.

PROGRESS CHECK 6

A projectile is to be shot at an initial velocity of 450 ft/second at a target 6000 ft away, as shown in Figure 2.53. Use the formula in Example 6 to find all settings for θ, to the nearest tenth of a degree, such that the projectile hits the target. ■

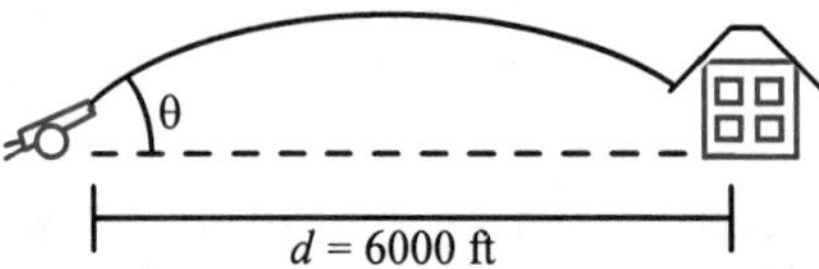

Figure 2.53

EXPLORE 2.4

1. With a calculator set in Degree mode, you can generate a graph or table for the trigonometric functions. The properties of the graphs are discussed more fully in Chapter 8.
   a. Draw each of the following graphs:
   $y = \sin x (0° \le x \le 360°)$; $y = \cos x (0° \le x \le 360°)$; $y = \tan x (0° \le x \le 360°)$
   Experiment with the range settings (say by using Zoom Trig or Zoom Decimal) so that Trace causes the cursor to jump by convenient increments.
   b. If your calculator has a Table feature, generate tables for the sine, cosine, and tangent functions using increment of 10 degrees.
2. Trigonometric equations can be solved with the aid of a grapher by finding the intersection of appropriate graphs. Use this technique to solve the following equations. To the nearest tenth of a degree, estimate all solutions between 0° and 360°.
   a. $2\sin(3x) = 1.4$ (6 solutions) **b.** $\sin(2x) = \cos x$ (4 solutions)
3. Use a grapher to show that these equations have no solution.
   a. $\sin x = 3$ **b.** $4\cos(2x) = 5 - 2\sin(2x)$

EXERCISES 2.4

In Exercises 1–20 find the exact values of θ between 0° and 360° that make the equation a true statement.

1. $\cos\theta = \frac{1}{2}$ **2.** $\cos\theta = -\frac{1}{2}$

3. $\tan\theta = -1$ **4.** $\tan\theta = 1$

5. $\sin\theta = \frac{1}{\sqrt{2}}$ **6.** $\sin\theta = \frac{-1}{\sqrt{2}}$

7. $\sec\theta = \frac{-2}{\sqrt{3}}$ **8.** $\sec\theta = \frac{2}{\sqrt{3}}$

9. $\cot\theta = \sqrt{3}$ **10.** $\cot\theta = -\sqrt{3}$

11. $\csc\theta = -\sqrt{2}$ **12.** $\csc\theta = \sqrt{2}$

13. $2\tan\theta = 2\sqrt{3}$ **14.** $-2\cos\theta = \sqrt{3}$

15. $2\sin\theta + \sqrt{3} = 0$ **16.** $\csc\theta - 2 = 0$

17. $4\sin\theta + 3 = 1$ **18.** $3\sec\theta - 7 = -1$

19. $2\tan\theta + 5 = 7$ **20.** $3\cot\theta + 4 = 1$

In Exercises 21–40 find the approximate values of θ between 0° and 360° that make the equation a true statement.

21. $\sin\theta = 0.1219$ **22.** $\sin\theta = -0.1219$

23. $\cos\theta = -0.5125$ **24.** $\cos\theta = 0.5125$

25. $\cot\theta = 2.457$ **26.** $\cot\theta = -2.457$

27. $\sec\theta = -1.058$ **28.** $\sec\theta = 1.058$

29. $\tan\theta = 3.145$ **30.** $\tan\theta = -3.145$

31. $5\cot\theta = -1$ **32.** $7\sin\theta = 2$

33. $3\cos\theta + 2 = 0$ **34.** $3\tan\theta - 1 = 0$

35. $2\tan\theta - 7 = 0$ 36. $5\sin\theta + 2 = 0$

37. $4\csc\theta + 9 = 0$ 38. $\cos\theta - 3 = 0$

39. $\sin\theta + 2 = 0$ 40. $3\sec\theta - 1 = 1$

In Exercises 41–50 find five angles that are coterminal with the given angle.

41. 45° 42. 90° 43. 22°10'

44. 84.4° 45. 120° 46. 215°

47. –30° 48. –60° 49. –100°50'

50. –312°20'

In Exercises 51–60 find all the values of θ for which the given trigonometric equation is a true statement. Where possible, find exact solutions.

51. $\sqrt{2}\sin\theta = 1$ 52. $\sqrt{3}\csc\theta = 2$

53. $3\sin\theta + 2 = 1$ 54. $5\tan\theta + 2 = -3$

55. $10\cos\theta + 7 = 3$ 56. $2\cot\theta - 3 = 0$

57. $5\sec\theta + 1 = 3$ 58. $\sin\theta = 1$

59. $\cos\theta = -1$ 60. $\cos\theta = 2$

61. A formula for the horizontal distance traveled by a projectile, neglecting air resistance, is

$$d = \frac{1}{32}V^2\sin 2\theta$$

where θ and V measure the angle of elevation and the initial velocity in ft/second of the projectile, respectively. If a projectile can be launched with $V = 100$ ft/second, at which two angles can it be launched so that it travels horizontally 200 ft? Round your answer to the nearest degree.

62. Refer to Exercise 61. If the projectile can be launched with $V = 100$ ft/second, at what angle can it be launched so that it travels horizontally 312.5 ft?

63. For a projectile launched from the ground at angle θ, its height above the ground (in feet) at time t (seconds) is given by

$$h = V(\sin\theta)t - 16t^2.$$

Suppose a projectile is launched at 1000 ft/second. To the nearest degree, what angle of elevation should be used so that it achieves a height of 1500 feet when 3 seconds have elapsed?

64. Refer to Exercise 63. Suppose the projectile must be 1500 feet above the ground when 2 second have elapsed. What value of θ will achieve this result?

THINK ABOUT IT 2.4

1. Explain in terms of the general definition of the trigonometric functions why the solution set of the equation $\sin\theta = 2$ is $\varnothing$.
2. If the point $\left(-\sqrt{3},\ -1\right)$ is on the terminal ray of θ, give one possible measure of θ.
3. Find all angles θ between 0° and 360° for which the absolute value of $\sin\theta$ equals the absolute value of $\sin(2\theta)$.
4. Find a value for k such that $\sin\theta = k + \cos\theta$ has exactly one solution between 0° and 360°.
5. In astronomy there are several ways of locating a star in the sky. One system is called **horizon coordinates**, and it uses altitude and azimuth. If you stand facing north, then the altitude tells you how far up in degrees to look, and the azimuth tells you how far around from north to turn to locate the star. Another system is called **equatorial coordinates**. In this system "how far up" is given by "declination," and "how far around" is given by the "hour-angle" In this second system, measurements are given relative to a plane through the equator of the Earth, so it is also necessary to give the observer's geographical latitude. (It is assumed that the observer is in the northern hemisphere.) To make conversions between these two systems of star location, the following formulas are used.

$$\sin a = \sin\delta\sin\phi + \cos\delta\cos\phi\cos H$$

$$\cos A = \frac{\sin\delta - \sin\phi\sin a}{\cos\phi\cos a}$$

where altitude is a, azimuth is A, declination is δ, and hour-angle is H. The observer's latitude is given by ϕ.

Use these equations to find the altitude and azimuth for a star whose hour-angle is 87.933°, and whose declination is 23°13', for an observer at latitude 52°N. [**Note:** if $\sin H > 0$ then the azimuth is taken to be $360° - A$.] (The idea for this exercise came from *Practical Astronomy with Your Calculator*, P. Duffett-Smith, Cambridge University Press.)

● ● ●

2.5 Vectors

Photo Courtesy of Archive Photos, New York

A ship is headed due each at $2\bar{0}$ knots (nautical miles per hour) while the current carries the ship due south at 5.0 knots. Find

a. the speed of the ship and

b. the direction (course) of the ship.

(See Example 2.)

Objectives

1. Find the resultant, or vector sum, of two vectors, using right-triangle trigonometry.
2. Resolve a vector into components.
3. Find the resultant, or vector sum, of two or more vectors, using components.
4. Solve applied problems using vectors.

An application of trigonometry occurs with the study of physical quantities that act in a definite direction. For example, when meteorologists describe wind, they mention both the speed of the wind and the direction from which the wind is blowing. Similarly, quantities such as forces, weights, and velocities must be described in such a way that both the strength (magnitude) and the direction of the quantity can be determined. Mathematically, we represent such a quantity by a line segment with an arrowhead at one end. This directed line segment is called a **vector**. The direction in which the arrowhead is pointing represents the direction in which the quantity is acting, while the length of the line segment is proportional to the magnitude of the quantity.

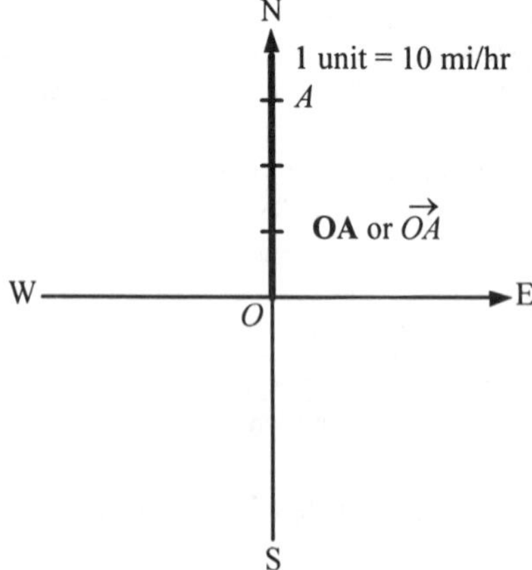

Figure 2.54

For instance, Figure 2.54 illustrates how to use a vector to represent graphically wind that is blowing due north at a speed of 40 mi/hour. Note that a vector that starts at **O** and ends at **A** is labeled **OA**, while a vector that starts at **A** and ends at **O** is a different vector and is labeled **AO**. In handwritten work, a vector such as **OA** is written $\overrightarrow{\mathbf{OA}}$.

Frequently, there are two (or more) forces acting on a body from different directions, and their net effect is a third force with a new direction. This new force is called the **resultant**, or **vector sum**, of the given forces. For example, in Figure 2.55 two forces of 20 lb and 30 lb are both acting on a body with an angle of 30° between the two forces. It can be shown that the resultant of vectors **OA** and **OB** is vector **OC**, which is the diagonal of a parallelogram formed from the given vectors. The magnitude of the resultant **OC** can be determined by the length of the line segment from *O* to *C*. The direction of the resultant is the same as the direction of the arrowhead on vector **OC** and is described in terms of the original vectors by finding either angle *AOC* or angle *BOC*. The next two examples illustrate this procedure for finding the resultant of two forces in cases that involve only right-triangle trigonometry.

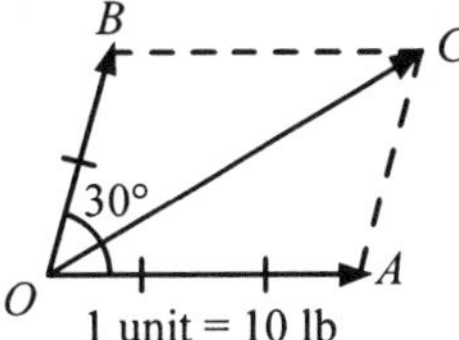

Figure 2.55

Example 1: Vector Analysis Involving a Right Triangle

A force of 3.0 lb and a force of 4.0 lb are acting on a body with an angle of 90° between the two forces. Find

a. the magnitude of the resultant and

b. the angle between the resultant and the larger force.

Solution: Let **OA** represent the 4.0-lb force and **OB** the 3.0-lb force. Then vector **OC** represents the resultant (see Figure 2.56).

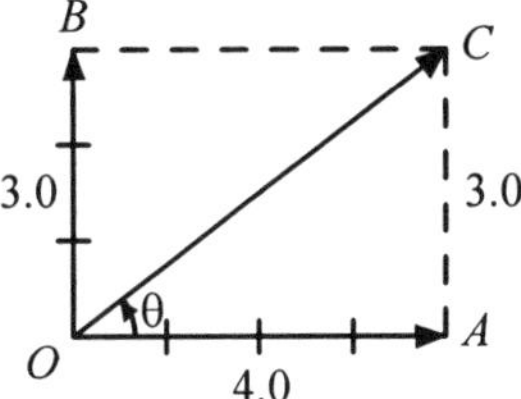

Figure 2.56

a. We can find the length of **OC** by using the Pythagorean theorem in right triangle *OAC*. Notice that line segments *OB* and *AC* have the same length.

$$\begin{aligned}\left(\overline{OC}\right)^2 &= (4)^2 + (3)^2 \\ &= 25 \\ \overline{OC} &= 5.0\end{aligned}$$

Thus, the magnitude of the resultant is 5.0 lb.

b. We can find the angle between the resultant and the 4.0-lb force by using the tangent function in right triangle *OAC*.

$$\tan\theta = \frac{3}{4}$$
$$\theta = 37° \quad \text{(to the nearest degree)}$$

Thus, the angle between the resultant and the larger force is 37°.

PROGRESS CHECK 1

Two forces one 45 lb and the other 24 lb, act on the same object at right angles to each other. Find

a. the magnitude of the resultant and

b. the angle between the resultant and the smaller force. ■

Example 2: Speed and Course of a Ship

Solve the problem in the section introduction on page 222.

Solution: Let vector **OA** represent the velocity of the ship in the easterly direction Let vector **OB** represent the velocity of the ship in the southerly direction due to the current. Let vector **OC** represent the actual velocity of the ship. (See Figure 2.57.)

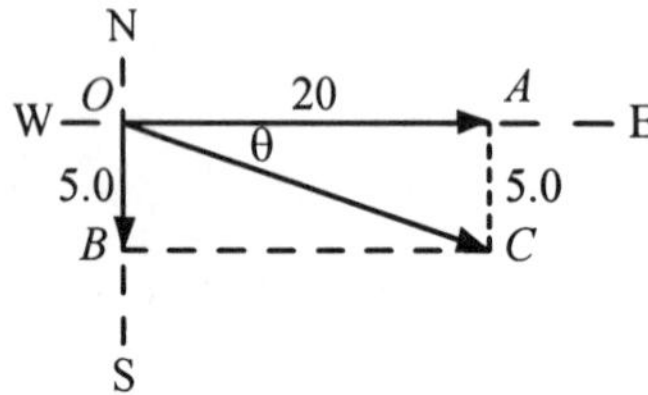

Figure 2.57

a. We can find the length of **OC** by using the Pythagorean theorem in right triangle *OAC*.

$$\left(\overline{OC}\right)^2 = (20)^2 + (5)^2$$
$$= 425$$
$$\overline{OC} = \sqrt{425} \approx 21$$

Thus, the speed of the ship is 21 knots.

b. We can find the angle between the resultant and vector **OA**, which is due east, by using the tangent function in right triangle *OAC*.

$$\tan\theta = \frac{5}{20}$$
$$\theta = 14° \quad \text{(to the nearest degree)}$$

Thus, the ship is heading in a direction that is 14° south of east.

PROGRESS CHECK 2

An airplane can fly $56\overline{0}$ mi/hour in still air. If it is heading due south in a wind that is blowing due east at a rate of 75.0 mi/hour, find

a. the distance the plane can fly in 1 hour and

b. the angle that the resultant force will make with respect to due east. ■

In the previous examples we considered how the net effect of having two forces acting on a body produced a third force, the resultant. However, the reverse situation often arises, where we are given a single force that we think of as the resultant, and we need to calculate two forces, which are called **components**, that produce the resultant. This process of expressing a single force in terms of two components, which are usually at right angles to each other, is called **resolving a vector**. The following examples illustrate the usefulness of resolving a vector into components.

Example 3: Effective Force Lowering a Window

A man pulls with a force of $4\bar{0}$ lb on a window pole, in an effort to lower a window. What part of the man's force lowers the window if the pole makes an angle of 20° with the window?

Solution: Let vector **OA** represent the pull on the window pole of $4\bar{0}$ lb. The resolution of vector **OA** results in the vertical component vector **OC** (the force lowering the window) and the horizontal component vector **OB** (wasted force) (see Figure 2.58).

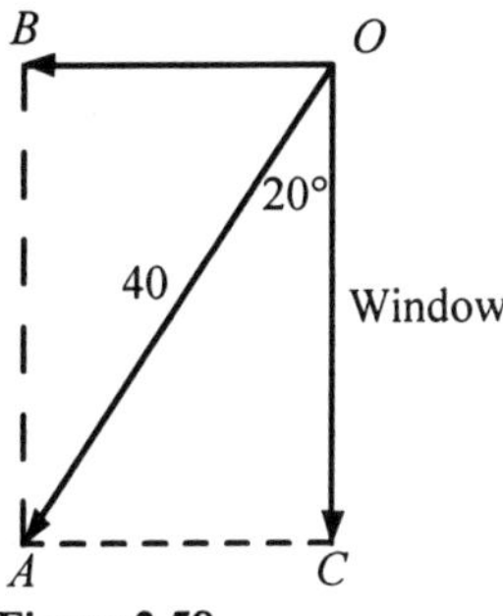

Figure 2.58

We can find the magnitude of vector **OC** by using the cosine function in the right triangle *ACO*.

$$\cos 20° = \frac{\overline{OC}}{40}$$
$$40(\cos 20°) = \overline{OC}$$
$$38 = \overline{OC} \quad \text{(two significant digits)}$$

Thus, a force of 38 lb is lowering the window.

PROGRESS CHECK 3
A woman pulls with a force of 55 lb on a window pole, in an effort to lower a window. How much force is wasted if the pole makes an angle of 16° with the window? ■

Example 4: Component Forces Affecting a Plane's Velocity

An airplane, pointed due west, is traveling 10.0° north of west at a rate of $40\bar{0}$ mi/hour. This resultant course is due to a wind blowing north. Find

a. the velocity of the plane if there were no wind (that is, the vector pointing due west) and
b. the velocity of the wind (that is, the vector pointing due north).

Solution: Let vector **OC** represent the resultant velocity of the plane. The resolution of vector **OC** results in a westerly component vector **OA** (the velocity of the plane if there were no wind) and northerly component vector **OB** (the velocity of the wind) (see Figure 2.59).

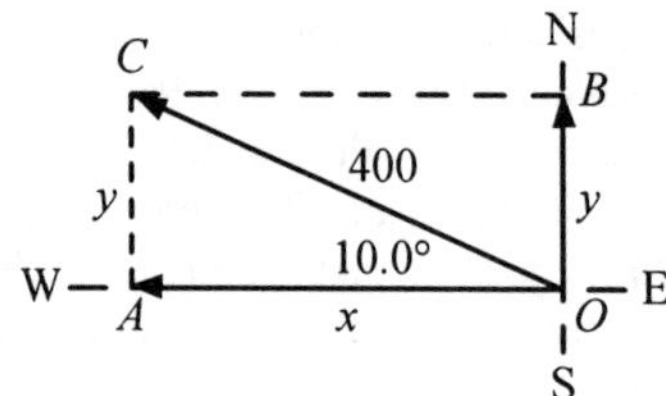

Figure 2.59

a. We can find the velocity of the plane if there were no wind by using the cosine function in right triangle OAC.

$$\cos 10.0° = \frac{x}{400}$$
$$400(\cos 10.0°) = x$$
$$394 = x \quad \text{(three significant digits)}$$

Thus, the velocity of the plane if there were no wind would be 394 mi/hour due west.

b. We can find the velocity of the wind by using the sine function in right triangle OAC.

$$\sin 10.0° = \frac{y}{400}$$
$$400(\sin 10.0°) = y$$
$$69.5 = y \quad \text{(three significant digits)}$$

Thus, the velocity of the wind would be 69.5 mi/hour due north.

PROGRESS CHECK 4

An airplane, pointed due east, is traveling 6.0° south of east at a rate of $46\bar{0}$ mi/hour. The resultant course is due to a wind blowing south. Find

a. the velocity of the plane if there were no wind and

b. the velocity of the wind. ■

Example 5: Resolving a Vector Into Components

Find the horizontal and vertical components of a vector that has a magnitude of 25 lb and makes an angle of 40° with the positive x-axis.

Solution: Let vector OA be the given vector. The resolution of vector **OA** results in the horizontal component, vector **OB**, and the vertical component, vector **OC** (see Figure 2.60).

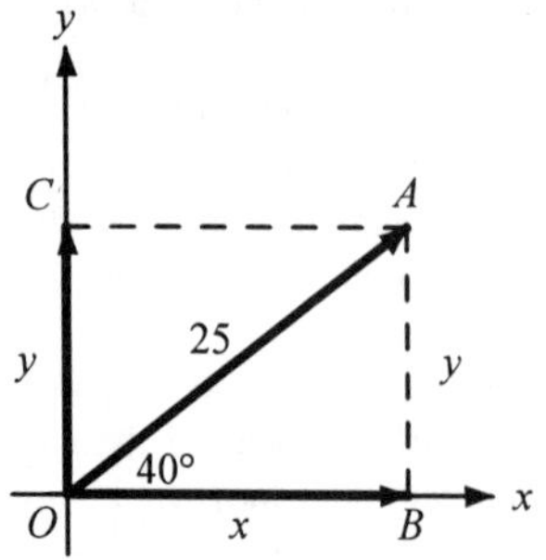

Figure 2.60

We can find the magnitude of the horizontal component, vector **OB**, by using the cosine function in right triangle OBA.

$$\begin{aligned} \cos 40° &= \frac{x}{25} \\ 25(\cos 40°) &= x \\ 19 &= x \quad \text{(two significant digits)} \end{aligned}$$

We can find the magnitude of the vertical component, vector **OC**, by using the sine function in right triangle OBA.

$$\begin{aligned} \sin 40° &= \frac{y}{25} \\ 25(\sin 40°) &= y \\ 16 &= y \quad \text{(two significant digits)} \end{aligned}$$

Thus, the magnitude of the horizontal component is 19 lb, and the magnitude of the vertical component is 16 lb.

PROGRESS CHECK 5
Find the horizontal and vertical components of a vector that has a magnitude of 42 lb and makes an angle of 28° with the positive x-axis. ■

Example 5 shows how to resolve a vector into horizontal (or x) and vertical (or y) components. This breakdown can be very useful if we want to add two (or more) vectors. That is because the resultant of two (or more) vectors may be found by breaking each vector into its x and y components. The sum of the x components of each vector is the x component of the resultant (labeled R_x). Similarly, the sum of the y components of each vector is the y component of the resultant (labeled R_y). By the Pythagorean theorem, the magnitude of the resultant is then

$$R = \sqrt{(R_x)^2 + (R_y)^2},$$

and the angle the resultant makes with the positive x-axis (labeled θ_R) is determined from the equation

$$\tan \theta_R = \frac{R_y}{R_x}.$$

Example 6 illustrates this approach to vector addition.

Example 6: Finding the Resultant Using Components

Vector **A** has a magnitude of 5.0 lb and makes an angle of 35° with the positive x-axis. Vector **B** has a magnitude of 9.0 lb and makes an angle of 160° with the positive x-axis. Find the resultant (or vector sum) of **A** and **B**.

Solution: We organize the procedure described above with the following table.

Vector	Magnitude	Director	x Component	y Component
A	5.0	35°	5.0 cos 35° = 4.096	5.0 sin 35° = 2.868
B	9.0	160°	9.0 cos 160° = –8.457	9.0 sin 160° = 3.078
R			R_x = –4.361	R_y = 5.946

The magnitude of the resultant is

$$\begin{aligned} R &= \sqrt{(R_x)^2 + (R_y)^2} \\ &= \sqrt{(-4.361)^2 + (5.946)^2} \\ &= \sqrt{54.37} \\ &= 7.4. \end{aligned}$$

The angle the resultant makes with the positive x-axis is found by solving

$$\tan \theta_R = \frac{R_y}{R_x} = \frac{5.946}{-4.361} = -1.363.$$

Since the y component is positive and the x component is negative, θ_R is in Q_2. By calculator we determine that the reference angle is 54°. Then

$$\begin{aligned} \theta_R &= 180° - 54° \\ &= 126°. \end{aligned}$$

Thus, the resultant has magnitude 7.4 lb and makes an angle of 126° with the positive x-axis. (See Figure 2.61.)

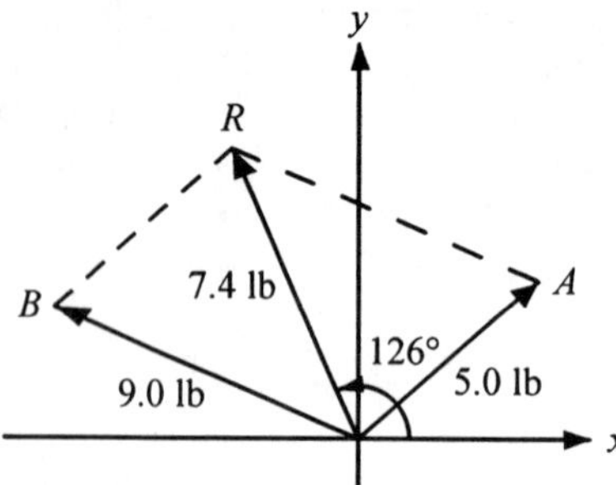

Figure 2.61

Technology Link

Some calculators have the capability to find a vector sum in both rectangular form and polar form, which are often displayed as $[x \ \ y]$ and $[r\angle\theta]$, respectively. To illustrate, Figure 2.62 shows a display on one such calculator that indicates the solution to the problem in Example 6. You should learn to use this feature if it is available on your calculator.

```
[5 ∠ 35] + [9 ∠ 160]
[-4.36147336563   5.94...
Ans ▸ Pol
[7.37415220404 ∠ 126.2...
```

Figure 2.62

PROGRESS CHECK 6

Vector **A** has a magnitude of 7.0 lb and makes an angle of 62° with the positive x-axis. Vector **B** has a magnitude of 4.0 lb and makes an angle of 175° with the positive *x*-axis. Find the resultant (or vector sum) of **A** and **B**. ■

EXPLORE 2.5

Some calculators have special commands for working with vectors. On such calculators you may have several format options for entering vectors. One way is to enter the length and the angle (with respect to the positive *x*-axis) for the vector inside special bracket symbols; this method will involve a special angle symbol. Another way is to enter the horizontal and vertical components of the vector, separated by a comma, inside the bracket symbols. Be careful to distinguish between available methods and to know what form the calculator uses for answers. You can probably control the format through the Mode menu.

1. The operation to find the resultant, or vector sum, of two vectors is given by the ordinary plus sign. If your calculator has this feature, use it to find the resultant for each given pair of vectors. (See Figure 2.62 for one illustration.) Then draw a sketch of the two given vectors and the resultant to illustrate the computation.
   a. $[5,\ 30°]$ and $[5,\ 30°]$ **b.** $[5,\ 30°]$ and $[5,\ -30°]$ **c.** $[5,\ 30°]$ and $[0,\ 10°]$
2. Use a calculator to find the resultant for three vectors, then draw a sketch to illustrate the computation: $[1,\ 10°]+[2,\ 20°]+[3,\ 30°]$.
3. If we say for three vectors that $A+B=C$ then it is reasonable to expect that $A=C-B$. Use a calculator to find $[5,\ 45°]-[2,\ 10°]$. Then check the answer by addition.

EXERCISES 2.5

In Exercises 1–10 find the magnitude of all forces to two significant digits and all angles to the nearest degree.

1. A force of 5.0 lb and a force of 12 lb are acting on a body with an angle of 90° between the two forces. Find
   a. the magnitude of the resultant
   b. the angle between the resultant and the larger force
2. A force of 12 lb and a force of 16 lb are acting on a body with an angle of 90° between the two forces. Find.
   a. the magnitude of the resultant
   b. the angle between the resultant and the smaller force
3. Two forces, one $1\bar{0}$ lb and the other 15 lb, act on the same object at right angles to each other. Find
   a. the magnitude of the resultant
   b. the angle between the resultant and the larger force
4. Two velocities, one $2\bar{0}$ mi/hour north and the other $3\bar{0}$ mi/hour east, are acting on the same body. Find
   a. the speed of the resultant velocity
   b. the angle of the resultant velocity with respect to a direction of due east
5. Two velocities, one 5.0 mi/hour south and the other 15 mi/hour west, are acting on the same body. Find
   a. the speed of the resultant velocity
   b. the angle of the resultant velocity with respect to a direction of due west
6. A ship is headed due south at 18 knots (nautical miles per hour) while the current carries the ship due west at 5.0 knots. Find
   a. the speed of the ship
   b. the direction (course) of the ship
7. An airplane can fly $50\bar{0}$ mi/hour in still air. If it is heading due north in a wind that is blowing due east at a rate of 80.0 mi/hour, find

a. the distance the plane can fly in 1 hour
b. the angle that the resultant velocity will make with respect to due east

8. An object is dropped from a plane that is moving horizontally at a speed of $3\overline{0}0$ ft/second. If the vertical velocity of the object in terms of time is given by the formula $v = 32t$, 5 seconds later:
a. What is the speed of the object?
b. What angle does the direction of the object make with the horizontal?

9. A pilot wishes to fly due east at $30\overline{0}$ mi/hour when a 70.0 mi/hour wind is blowing due north.
a. How many degrees south of east should the pilot point the plane to attain the desired course?
b. What airspeed should the pilot maintain?

10. A ship wishes to travel due south at 25 knots in an easterly current of 7.0 knots.
a. How many degrees west of south should the navigator direct the ship?
b. What speed must the ship maintain?

11. An airplane pointed due east is traveling 7.0° north of east at a rate of $35\overline{0}$ mi/hour. The resultant course is due to a wind blowing north. Find
a. the velocity of the plane if there were no wind (that is, the vector pointing due east)
b. the velocity of the wind (that is, the vector pointing due north)

12. Answer the questions posed in Exercise 11 if the plane is traveling 14.0° north of east at 358 mi/hour.

13. A woman pushes with a force of $4\overline{0}$ lb on the handle of a lawn mower that makes an angle of 33° with the ground. How much force pushes the lawn mower forward?

14. A man pulls with a force of 25 lb on a window pole, in an effort to lower a window. How much force is wasted if the pole makes an angle of 15° with the window?

15. A car weighing 3,500 lb is parked in a driveway that makes an angle of 10° with the horizontal. Find the minimum brake force that is needed to keep the car from rolling down the driveway. (Use the illustration below and assume no friction.)

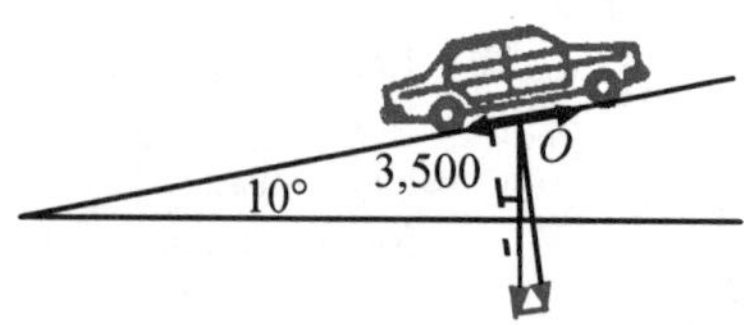

16. The answer to Exercise 15 is 608 pounds. If the driveway angle is cut in half to 5°, then what will the minimum brake force be?

17. Find the force needed to keep a barrel weighing $1\overline{0}0$ lb from rolling down a ramp that makes an angle of 15° with the horizontal (assume no friction).

18. A box is resting on a ramp that makes an angle of 18° with the horizontal. What is the force of friction between the box and the ramp if the box weights $8\overline{0}$ lb?

19. A body is acted on by two forces with magnitudes of 16 lb and 25 lb, which act at an angle of 30° with each other. Find
a. the magnitude of the resultant
b. the angle between the resultant and the larger force

20. Two force with magnitudes of 45 lb and $9\overline{0}$ lb are applied to the same point. If the angle between them measures 72°, find
a. the magnitude of the resultant
b. the angle between the resultant and the smaller force

21. Forces with magnitudes of 126 lb and 198 lb act simultaneously on a body in such a way that the angle between the forces is 14°50′. Find
a. the magnitude of the resultant
b. the angle between the resultant and the larger force

22. Two forces with magnitudes of 15 lb and $2\overline{0}$ lb act on a body in such a way that the magnitude of the resultant is 28 lb. Find the angles that the three forces make with each other.

23. Two forces act on a body to produce a resultant of 75 lb. If the angle between the two forces is 56° and one of the forces is $6\overline{0}$ lb, find the magnitude of the other force.

24. City B is located 50° north of east of city A. There is a $3\bar{0}$ mi/hour wind from the west and a pilot wishes to maintain an airspeed of 450 mi/hour. How many degrees north of east should the pilot head the plane to arrive directly at city B from city A?

25. Two forces with magnitudes of 42 lb and 71 lb are applied to the same object. If the magnitude of the resultant is 85 lb, find the angles that the three forces make with each other.

26. A force **A** of $4\bar{0}0$ lb and a force **B** of $6\bar{0}0$ lb act at a point. Their resultant, **R**, makes an angle of 42° with **A**. Find the magnitude of **R**.

In Exercises 27–30 find the horizontal and vertical component of each vector.

27. A magnitude of $1\bar{0}0$ lb; makes an angle of 30° with the positive x-axis.

28. A magnitude of 75 lb; makes an angle of 45° with the positive x-axis.

29. A magnitude of 18 lb; makes an angle of 27° with the positive x-axis.

30. A magnitude of 125 lb; makes an angle of 72° with the positive x-axis.

In Exercises 31–36 find the resultant (or vector sum) of the given vectors by resolving each vector into its x and y components.

	Vector	*Magnitude*	*Direction (with respect to positive x-axis)*
31.	**A**	5.0 lb	26°
	B	3.0 lb	84°
32.	**C**	6.0 lb	95°
	D	9.0 lb	15°
33.	**C**	12 lb	110°
	D	15 lb	180°
34.	**A**	4.0 lb	90°
	B	3.0 lb	190°
35.	**A**	$1\bar{0}$ lb	42°
	B	$2\bar{0}$ lb	140°
	C	$3\bar{0}$ lb	240°
36.	**A**	5.0 lb	60°
	B	2.0 lb	210°
	C	1.0 lb	270°

THINK ABOUT IT 2.5

1. Two vectors **AB** and **CD** are said to be equal, and we write **AB** = **CD** if and only if they have the same length and direction. Use this definition to explain why the two methods shown below give the same result for the vector sum **A** + **B**.

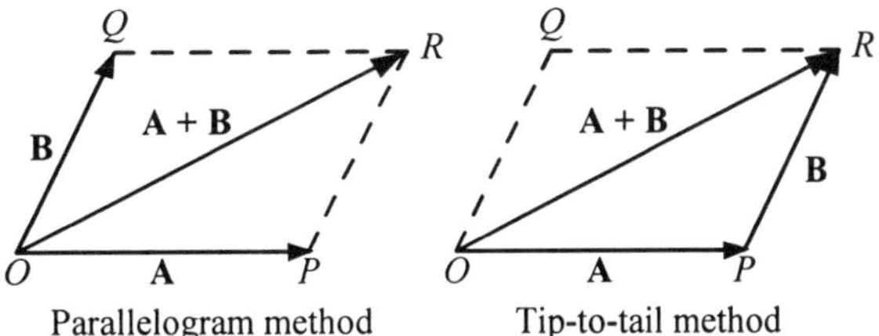

2. Use the tip-to-tail method of vector addition and show in a diagram that

$$\mathbf{A} + \mathbf{B} = \mathbf{B} + \mathbf{A} \quad \text{(commutative property)}$$

3. Use the tip-to-tail method of vector addition and show in a diagram that

$$(\mathbf{A} + \mathbf{B}) + \mathbf{C} = \mathbf{A} + (\mathbf{B} + \mathbf{C}) \quad \text{(associate property)}$$

4. If the magnitude of vector **A** is 6.0 lb and the magnitude of vector **B** is 8.0 lb, then show in a diagram how the vectors may be combined so that the vector sum has the given magnitude.
a. 14 lb **b.** 2.0 lb. **c.** 10 lb **d.** 7.2 lb

5. During a tournament, a golfer requires three putts to sink the ball in the hole. In respective order, the three putts displaced the ball 18.00 ft due east, 6.00 ft in a direction 30.0° north of west, and 1.00 ft in a direction 45.0° south of east. Find the magnitude and the direction of the putt that would have been needed to sink the ball on the first putt.

● ● ●

CHAPTER 2 SUMMARY

OBJECTIVES CHECKLIST

Specific chapter objectives are summarized below, along with numbered example problems from the text that should clarify the objectives. If you do not understand any objectives, or do not know how to do the selected problems, then restudy the material.

2.1: Can you:

1. **Find the six trigonometric functions of an acute angle in a right triangle, when given at least two side lengths?** Find the six trigonometric functions of angle θ in the figure.

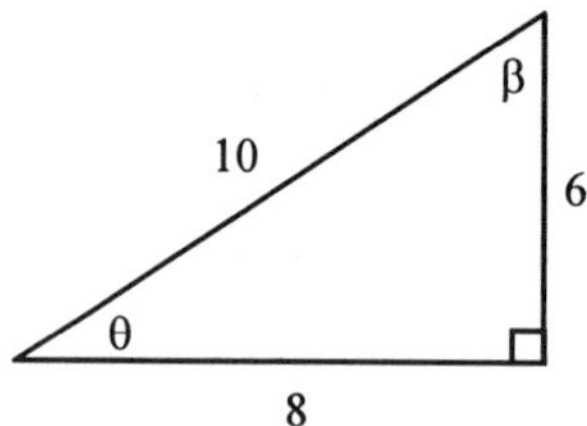

[Example 1]

2. **Find the values of the five remaining trigonometric functions of an acute angle, given one function value?** Find the values of the remaining trigonometric functions of acute angle θ if $\tan\theta = 2/5$.

[Example 2]

3. **Find the length of any side in a right triangle, given a trigonometric ratio and the length of one side?** In right triangle ABC with $C = 90°$, if $\cos B = 5/13$ and $c = 39$, find a.

[Example 3]

4. **State exact trigonometric values for 30°, 45°, and 60°?** Use the figure to find the exact values of sin 30°, cos 45°, sec 45°, and tan 60°.

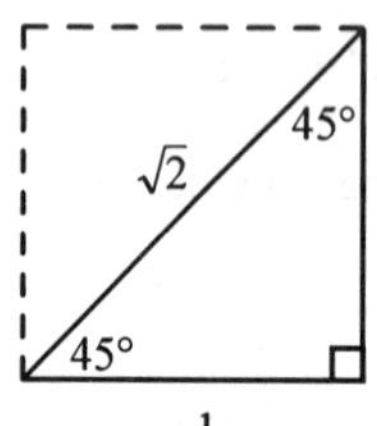

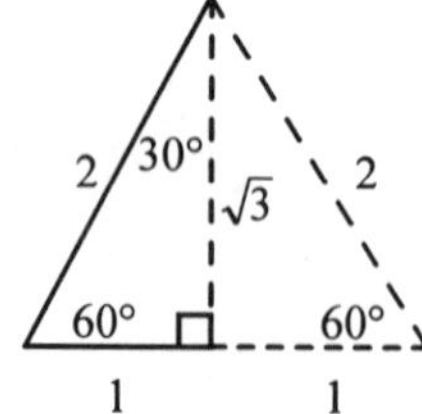

[Example 4]

5. **Express any trigonometric function of an acute angle as the cofunction of the complementary angle?** Express tan 75° as a function of the angle complementary to 75°.

[Example 5]

6. **Use a calculator to approximate the trigonometric value of a given acute angle?** Evaluate sin 18° by calculator to four decimal places.
[Example 6a]

7. **Use a calculator to approximate the measure of an acute angle, given the value of a trigonometric function of that angle?** Use a calculator to approximate the acute angle θ that satisfies the equation $\sin\theta = 0.7957$. Write the solution to the nearest tenth of a degree and the nearest 10 minutes.
[Example 8a]

2.2: Can you:

1. **Solve a right triangle, given one side and one acute angle?** Solve the right triangle *ABC* in which $A = 30°$ and $C = 1\bar{0}0$ ft.
[Example 1]

2. **Solve a right triangle given two sides?** Solve the right triangle *ABC* in which $a = 11.0$ ft and $b = 5.00$ ft.
[Example 3]

3. **Solve applied problems involving right triangles?** It is necessary to determine the height of a smokestack to estimate the cost of painting it. At a point 225 ft from the base of the stack, the angle of elevation is 33.0°. How high is the smokestack?
[Example 4]

2.3: Can you:

1. **Find the value of the six trigonometric functions of an angle, given the terminal ray of the angle?** Find the value of the six trigonometric functions of an angle θ if $(2, -5)$ is a point on the terminal ray of θ.
[Example 1]

2. **Find the values of the five remaining trigonometric functions of θ, given one function value and the quadrant containing the terminal ray of θ?** Find the value of the remaining trigonometric functions if $\tan\theta = 2/3$, and $\sec\theta < 0$.
[Example 2]

3. **Determine the exact value of any trigonometric function of a quadrantal angle?** Find the exact value of sin 540°.
[Example 4]

4. **Determine the exact value of any trigonometric function of angles with reference angles of 30°, 45°, or 60°?** Find the exact value of tan 135°.
[Example 5]

5. **Use a calculator to approximate the trigonometric value of a given angle?** Find $\cos 227°20'$. Round the answer to four decimal places.
[Example 8]

2.4: Can you:

1. **Find the exact solution to certain trigonometric equations for $0° \le \theta < 360°$?** Find the exact value of $\theta (0° \le \theta < 360°)$ for which the equation $2\sin\theta + \sqrt{3} = 0$ is a true statement.
[Example 2]

2. **Find the approximate solution to trigonometric equations for $0° \le \theta < 360°$?** Solve $4\tan\theta = 1$ for $0° \le \theta < 360°$. Round off answers to the nearest tenth of a degree?
[Example 4]

3. **Find all solutions to a trigonometric equation, specifying exact solutions where possible?** Approximate all the solutions to $10\cot\theta = 3$, to the nearest 10 minutes.
[Example 5]

4. **Solve applied problems involving trigonometric equations?** A formula for the horizontal distance traveled by a projectile, neglecting air resistance, is

$$d = \frac{1}{32}V^2 \sin 2\theta,$$

where θ and V measure the angle of elevation and the initial velocity in ft/second of the projectile, respectively. If a professional field-goal kicker boots a football with an initial velocity of 76 ft/second, and the ball travels 180 ft, find θ to the nearest degree.

[Example 6]

2.5: Can you:

1. **Find the resultant, or vector sum, of two vectors, using right-triangle trigonometry?** A force of 3.0 lb and a force of 4.0 lb are acting on a body with an angle of 90° between the two forces. Find the magnitude of the resultant and the angle between the resultant and the larger force.

[Example 1]

2. **Resolve a vector into components?** Find the horizontal and vertical components of a vector that has a magnitude of 25 lb and makes an angle of 40° with the positive x-axis.

[Example 5]

3. **Find the resultant, or vector sum, of two or more vectors, using components?** Vector **A** has a magnitude of 5.0 lb and makes an angle of 35° with the positive x-axis. Vector **B** has a magnitude of 9.0 lb and makes an angle of 160° with the positive x-axis. Find the resultant (or vector sum) of **A** and **B**.

[Example 6]

4. **Solve applied problems using vectors?** A ship is headed due east at $2\overline{0}$ knots (nautical miles per hour) while the current carries the ship due south at 5.0 knots. Find the speed and direction of the ship.

[Example 2]

KEY CONCEPTS AND PROCEDURES

Section — **Key Concepts and Procedures to Review**

2.1

- Definitions of right triangle, acute angle, hypotenuse, similar triangles, cofunctions, complementary angles, one degree, one minute (written 1'), and reciprocal functions
- Definition of the trigonometric functions of an acute angle θ of a right triangle
- The side lengths in a 30–60–90 triangle are in the ratio of $1{:}\sqrt{3}{:}2$
- The side lengths in a 45–45–90 triangle are in the ratio of $1{:}1{:}\sqrt{2}$
- A trigonometric function of any acute angle is equal to the corresponding cofunction of the complementary angle.
- Guidelines for using a calculator to evaluate a trigonometric function of an acute angle and to evaluate an inverse trigonometric function

2.2

- Definitions of angle of elevation and of angle of depression
- Methods to solve a right triangle
- Guidelines for accuracy in computed results

2.3

- Definitions of standard position of angle θ, quadrantal angle, coterminal angle, and reference angle
- Definition of the trigonometric functions of an angle in standard position
- Chart summarizing the signs of the trigonometric ratios
- Methods to determine (if defined) approximate trigonometric values for any angle and exact values in special cases

2.4

- Methods to solve trigonometric equations. Reference numbers and graphical calculator methods may be used.
- Determine solutions between 0° and 360° as follows:

Quadrant	Soultion
1	reference angle
2	$180°-$reference angle
3	$180°+$reference angle
4	$360°-$reference angle

- If θ is a solution of a trigonometric equation, then $\theta + k360°$ (where k is any integer) is also a solution of the equation.

2.5

- Definitions of vector, resultant of two (or more) vectors, and components of a vector
- Methods to determine the resultant of two (or more) vectors
- Methods to determine the horizontal and vertical components of a vector
- Resultant formulas: $R = \sqrt{(R_x)^2 + (R_y)^2}$ and $\tan\theta = R_y/R_x$

CHAPTER 2 REVIEW EXERCISES

1. Give five angles that are coterminal with 100°. Include at least two negative angles.

2. Determine the exact value of sec 135°.

3. If $\sin\theta = a$, find csc θ.

4. If $\tan\theta = \frac{3}{4}$ and $\cos\theta < 0$, find sin θ.

5. Express 22° as a function of the angle complementary to 22°.

6. Determine an acute angle θ for which $\sin\theta = \cos 56°40'$.

7. Find the sine of the acute angle whose cosine is $\frac{2}{5}$.

8. What is the reference angle for 261°?

9. If the point $(1, -1)$ is on the terminal ray of θ, find cos θ.

10. If the point $(-1, 1)$ is on the terminal ray of θ, give one possible measure of angle θ.

In Exercises 11–16 answer true for false.

11. $\sin 30° + \sin 60° = \sin 90°$

12. $\sin(30° + 60°) = \sin 30° + \sin 60°$

13. $\sin 45° = \cos 45°$

14. $\sin 30° = \frac{1}{2}\sin 60°$

15. $(\sin 30°)^2 = 2$

16. $\sin 30° \cdot \csc 30° = 1$

In Exercises 17–20 evaluate the given expression.

17. $\sin(-210°)$

18. $\cos 626°40'$

19. $3\cos 720° + \tan(-540°)$

20. $\sin 2\theta$ if $\theta = 45°$

In Exercises 21–24 solve each equation for the given interval.

21. $3\cot\theta - 1 = 0$; all solutions

22. $3\csc\theta - 7 = -13$; $0° \le \theta < 360°$

23. $2\sin\theta = -3$; all solutions

24. $2\sin\theta = -2$; $0° \le \theta < 720°$

25. In a 30°, 60°, 90° triangle, the shortest side is 1 cm long. Find the lengths of the other two sides.

26. In right triangle *ABC*, $C = 90°$. If $\sin A = .8021$, find cos *B*.

27. Solve the triangle ABC, if $C = 90°$, $B = 47°$, and $a = 45$ ft.

28. In right triangle ABC with $C = 90°$, if $b = 5.00$ and $c = 8.00$, determine B.

29. If $\cos\theta > 0$ and $\sin\theta < 0$, then in which quadrant is the terminal ray of θ?

30. What is the minimum value of $3 - \cos x$?

31. What is the maximum value of $2 + \sin x$?

32. True or False? In any triangle ABC, $\sin\frac{1}{2}A = \cos\frac{1}{2}(B + C)$.

33. Find the area of triangle ABC if $C = 90°$, $A = 40°$, and $b = 3.0$ ft.

34. If $\sin\theta = 0.7$ and θ is an acute angle, then find θ to the nearest tenth of a degree.

35. Consider Snell's law (see Exercise 87 in Section 2.1). As light passes from air to water, v_i/v_r is about $4/3$. Find the angle of incidence of a ray of light if the angle of refraction is $25°40'$.

36. A carpenter has to build a stairway. The total rise is 8 ft 6 in. and the angle of rise is 30°, as shown in the illustration. What is the shortest piece of $2\text{-}\times 12\text{-in.}$ stock that can be used to make the stringer?

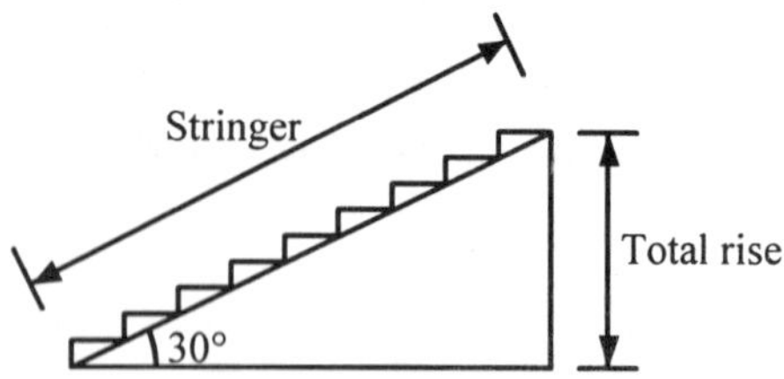

37. Two points, A and B, are $1\bar{0}0$ yd apart. Point C across a canyon s located so that angle CAB is 90° and angle CBA is 50°. Compute the distance $\overline{BC}$ across the canyon.

38. What is the vertical component of a vector with a magnitude of $5\bar{0}$ lb which makes an angle of 25° with the positive x-axis?

39 A road rises 25 ft in a horizontal distance of $4\bar{0}0$ ft. Find the angle that the road makes with the horizontal.

40. A force of 12 lb and a force of 15 lb are acting on a body with an angle of 90° between the two forces. Find the magnitude of the resultant and the angle between the resultant and the larger force.

41. From the top of a building $4\bar{0}$ ft tall, the angle of depression to the foot of a building across the street is 60° and the angle of elevation to the top of the same building is 70°. How tall is the building?

42. In a 3–4–5 right triangle find the smaller acute angle. Round to the nearest tenth of a degree.

43. A force of $4\bar{0}$ lb and a force of $3\bar{0}$ lb act on a body so that their resultant is a force of 38 lb. Find the angle between the two original forces.

44. To determine the radius of the sun, an observer on Earth at point O measures angle θ to be 16', as illustrated by the diagram. If the distance from Earth to the sun is about 93,000,000 mi, what is the radius of the sun?

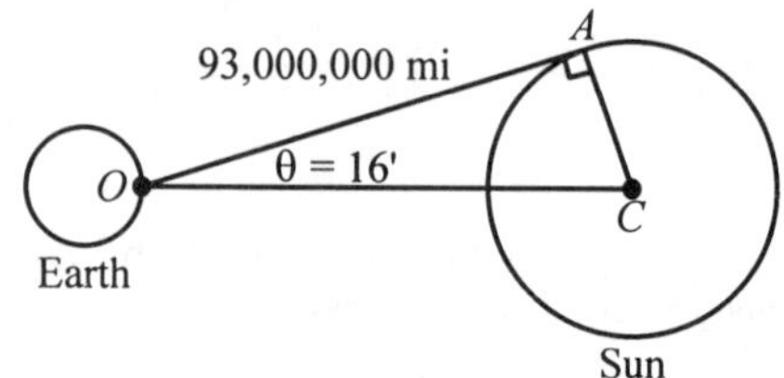

45. Vector **A** has a magnitude of 6.0 lb and makes an angle of 55° with the positive x-axis. Vector **B** has a magnitude of 2.0 lb and makes an angle of 110° with the positive x-axis. Find the resultant (or vector sum) of **A** and **B**.

46. The area of an equilateral triangle is $\sqrt{3}$. What are the lengths of the sides?

In Exercises 47–56 select the choice that completes the statement or answers the question.

47. The value of $\sin 100°$ is equal to
a. $\sin 10°$ **b.** $\sin 80°$
c. $-\sin 80°$ **d.** $-\sin 10°$

48. Which angle is a solution of $\tan\theta = -\sqrt{3}$?
a. 120° **b.** 150° **c.** 210° **d.** 240°

49. The x component of a vector with a magnitude of 25 lb which makes an angle of 67° with the x-axis is
a. 9.8 lb **b.** 12 lb **c.** 17 lb **d.** 23 lb

50. If the terminal ray of θ lies on the line $y = 2x$ and θ is in Q_3 then one angle measure for θ is
a. 225° **b.** 210° **c.** 207° **d.** 243°

51. Find sec R in the following right triangle.

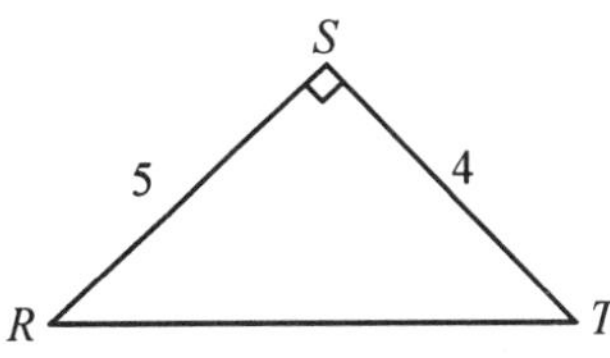

a. $\frac{3}{5}$ **b.** $\frac{4}{3}$ **c.** $\frac{\sqrt{41}}{5}$ **d.** $\frac{\sqrt{41}}{4}$

52. Solve $\frac{\sin A}{a} = \frac{\sin B}{b}$ for a.

a. $a = \frac{\sin A}{\sin B}$ **b.** $a = b\sin A - \sin B$

c. $a = b$ **d.** $a = \frac{b\sin A}{\sin B}$

53. In a right triangle, the length of the hypotenuse is 17 and the longer side length is 15. The sine of the smaller acute angle is

a. $\frac{15}{17}$ **b.** $\frac{8}{17}$ **c.** $\frac{15}{8}$ **d.** $\frac{17}{15}$

54. As angle θ increases from 180° to 360°, the sine of that angle

a. increases **b.** decreases

c. increases, then decreases

d. decreases, then increases

55. Which of the following is an identity?

a. $\sin\theta \cdot \cos\theta = 1$ **b.** $\sec\theta \cdot \csc\theta = 1$

c. $\cos\theta \cdot \csc\theta = 1$ **d.** $\tan\theta \cdot \cot\theta = 1$

56. For $0° < \theta < 45°$ which statement must be true?

a. $\sin\theta = \cos\theta$ **b.** $\sin\theta < \cos\theta$

c. $\sin\theta > \cos\theta$ **d.** $\tan\theta < \sin\theta$

CHAPTER 2 TEST

1. Find the exact value of sin 60°.

2. In right triangle ABC with $C = 90°$, if $c = 17$ and $a = 15$, find the tangent of the smaller acute angle.

3. If $\cot\theta = \frac{4}{5}$ and θ is an acute angle, find cos θ.

4. Find the exact value of cos 225°.

5. Evaluate $132°50'$ to four significant digits.

6. Find three angles coterminal to 150°. Include at least one negative angle measure.

7. If $\csc\theta = 3.85$ and θ is an acute angle, find θ to the nearest degree.

8. Solve $4\cos\theta + 3 = 0$ for $0° \leq \theta < 360°$ to the nearest degree.

9. Find all solutions to $\tan\theta + 1 = 0$.

10. If $(3, -2)$ is a point on the terminal ray of θ, find sin θ.

11. True or False: In triangle ABC if $\cot A = 0$, then ABC is a right triangle.

12. In right triangle ABC with $C = 90°$, $\sin A = \frac{1}{2}$. Find $(\sin A)^2 + (\cos A)^2$.

13. Find the hypotenuse of right triangle ABC with $C = 90°$ if $A = 45°$ and $a = 55$ ft.

14. In a right triangle, the hypotenuse is 3 times as long as the shortest side. Find the measure of the smaller acute angle, to the nearest tenth of a degree.

15. If $\tan\theta > 0$ and $\cos\theta < 0$, what quadrant contains the terminal side of θ?

16. What is the vertical component of a vector with a magnitude of 56 lb which makes an angle of 37° with the positive x-axis?

17. Find the exact values of θ between 0° and 360° that make $\cos\theta = -\frac{1}{2}$ a true statement.

18. In right triangle ABC with $C = 90°$, $b = 35$ ft and $c = 45$ ft. Solve the triangle.

19. A road has a uniform elevation of 4.0°. Find the increase in elevation in driving 95 m along this road.

20. A force of 28 lb and a force of 12 lb are acting on a body with an angle of 28° between the two forces. Find the magnitude of the resultant force.

Chapter 3

Further Algebraic Techniques

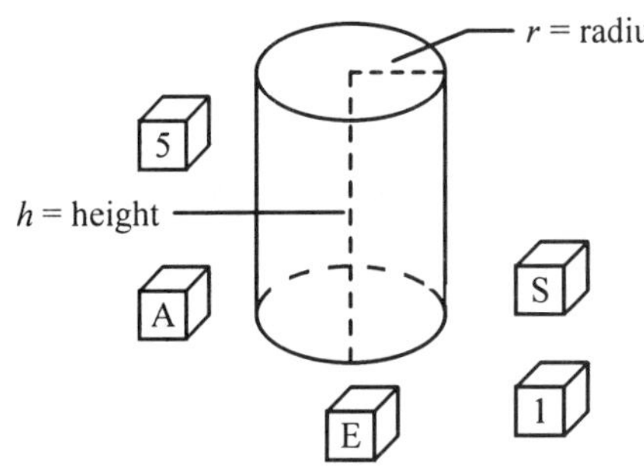

A manufacturer chooses to use a cylindrical container to package its new product, as shown in the illustration. The total surface area for such a container is given by

$$A = 2\pi r^2 + 2\pi rh.$$

Write a common alternative formula for the surface area of a cylinder by factoring the right side of this equation.
(See Example 2 of Section 3.1)

In Chapter R we discussed some essential topics in algebra that have served our purposes to this point. However, to continue our study of functions, we need more algebraic techniques. In this chapter we will learn to factor expressions, to perform the basic operations with fractional expressions, radicals, and complex numbers, and to simplify expressions containing rational number exponents. Since factoring is a key technique in many algebraic manipulations, we begin with this topic.

● ● ●

3.1 Factoring

Objectives

1. Factor out the greatest common factor (GCF).
2. Factor by grouping.
3. Factor trinomials of the form $ax^2 + bx + c$.
4. Factor out the GCF and then factor trinomials.
5. Use $b^2 - 4ac$ to determine if $ax^2 + bx + c$ can be factored.

The process of factoring undoes the process of multiplication, as shown below.

If we change a product into a sum, we are multiplying.

$$(x+5)(x+1)\xrightarrow{\text{multiplying}} x^2+6x+5$$

If we change a sum into a product, we are factoring.

$$(x+5)(x+1)\xleftarrow{\text{factoring}} x^2+6x+5$$

Both processes are important because in different situations one form may be more useful than the other. In general, the procedures for factoring are not as straightforward as those for multiplication, and we limit our discussion of factoring primarily to algebraic expressions called polynomials. Examples of polynomials are

$$64x^6 - y^6,\ 9x^2+24x+16, \text{ and } 12ct^2+9c^2t,$$

and a **polynomial** is defined as an algebraic expression that may be written as a finite sum of terms that contain only nonnegative integer exponents on the variables. Thus, algebraic expression such as

$$4x^{-2},\ \sqrt{x}-1, \text{ and } 5x^{1/3}+5$$

are not polynomials. A polynomial with just one term is called a **monomial**; one that contains exactly two terms is a **binomial**; and one with exactly three terms is a **trinomial**. The **degree of a monomial** is the sum of the exponents on the variables in the term. The **degree of a polynomial** is the same as the degree of its highest monomial term. Thus, the degree of $5x^4y^2$ is 6, the degree of $2xy^2$ is 3, and the degree of $5x^4y^2+2xy^2$ is 6, since $5x^4y^2$ is the highest-degree term.

There are several techniques that enable us to factor certain polynomials, and you need to consider carefully each of the factoring methods that follow in this section and in Section 3.2.

Common Factors

The first method that should be employed in factoring a polynomial is to attempt to find a factor that is common to each of the terms. For example:

Polynomial	Common Factor
$9s+6t$	3
$15x^2+(-5x)$ (or $15x^2-5x$)	$5x$
$2a^2b+8ab^2$	$2ab$

In each case we attempt to pick the greatest common factor (**GCF**) that divides into each term of the polynomial. Therefore, although 2 is a common factor of $2a^2b+8ab^2$, a preferable common factor is $2ab$, since $2ab$ is the largest factor that divides both terms. With respect to variables, the GCF is the product of all common variable factors, with each variable appearing the fewest number of times it appears in any one of the terms. After we determine the greatest common factor, we use the distributive property to factor it out. For instance, the GCF of the terms in $3x^2+6x$ is $3x$, and we factor it out as follows.

$$\begin{aligned} 3x^2+6x &= 3x\cdot x+3x\cdot 2 && \textit{Think: } \frac{3x^2}{3x}=x,\ \frac{6x}{3x}=2 \\ &= 3x(x+2) && \text{Distributive property} \end{aligned}$$

Thus, the factored form of $3x^2 + 6x$ is $3x(x+2)$. Because factoring undoes multiplying, you should check factoring answers by multiplication. For this example,

$$3x(x+2) = 3x \cdot x + 3x \cdot 2 = 3x^2 + 6x.$$

which checks. Use this example to clarify the following factoring procedure.

To Factor Out the GCF

1. Find the GCF of the terms in the polynomial.
2. Express each term in the polynomial as a product with the GCF as one factor.
3. Factor out the GCF using the distributive property.
4. Check the answer through multiplication.

You will notice that the directions in factoring problems use the phrase "factor completely." This expression directs us to continue factoring until the polynomial contains no factors of two or more terms that can be factored again. The restrictions we place on the form of the factors will determine whether a polynomial is factorable, so it is important to note that, unless otherwise specified, we are interested only in polynomial factors with integer coefficients.

Example 1: Factoring Out the GCF

Factor completely.

a. $21x^2 - 7x$ **b.** $s(c+2) - t(c+2)$

Solution:

a. The GCF of $21x^2$ and $7x$ is $7x$.

$$\begin{aligned} 21x^2 - 7x &= 7x \cdot 3x - 7x \cdot 1 \\ &= 7x(3x-1) \quad \text{Distributive property} \end{aligned}$$

b. The greatest common factor is the binomial $c+2$. Factoring out the GCF using the distributive property gives

$$s(c+2) - t(c+2) = (c+2)(s-t).$$

PROGRESS CHECK 1

Factor completely.

a. $8y^3 + 4y^2$ **b.** $2x(x+3) + 5(x+3)$ ■

Example 2: Using Factoring to Rewrite a Formula

Solve the problem in the chapter introduction on page 239.

Solution: A common alternative formula for $A = 2\pi r^2 + 2\pi rh$ may be obtained by observing that $2\pi r$ is the GCF of the expression on the right side of this equation. Then

$$\begin{aligned} 2\pi r^2 + 2\pi rh &= 2\pi r \cdot r + 2\pi r \cdot h \\ &= 2\pi r(r+h) \end{aligned}$$

So the surface area of the cylindrical container is also given by

$$A = 2\pi r(r+h).$$

PROGRESS CHECK 2

The total surface area of a right circular cone, as shown in Figure 3.1, is given by $A = \pi r^2 + \pi rs$. Write an alternative formula by factoring the right side of this equation.

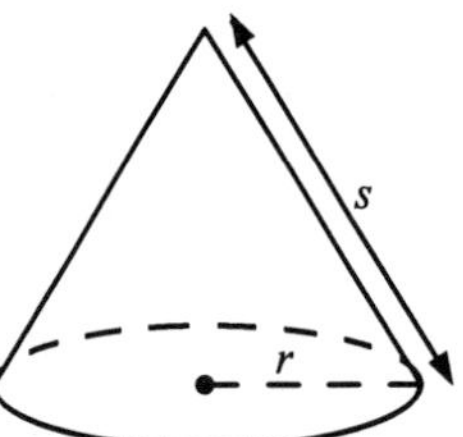

Figure 3.1 ■

Trigonometric expressions are factored in the same manner as algebraic expressions. Throughout the chapter, we will often present problems in two parallel forms: the algebraic form in part **a** and a similar trigonometric expression in part **b**. The techniques for both problems are the same since a symbol in algebra, such as s, represents a real number; while an expression in trigonometry, such as $\sin\theta$, represents a ratio that is also a real number. Thus, the rules that govern both algebraic and trigonometric expressions are the rules that govern the real numbers. When using exponents with functions, it is common notation to avoid parentheses and to put the exponent between the name of the function and the symbol for the independent variable. For example, $(\sin\theta)^3$, which means $(\sin\theta)(\sin\theta)(\sin\theta)$, is written as $\sin^3\theta$.

Example 3: Contrasting Factorizations of Algebraic and Trigonometric Expressions

Factor completely.

a. $12ct^2 + 9c^2t$ **b.** $12\cos\theta\tan^2\theta + 9\cos^2\theta\tan\theta$

Solution:

a. The GCF of $12ct^2$ and $9c^2t$ is $3ct$.

$$\begin{aligned} 12ct^2 + 9c^2t &= 3ct \cdot 4t + 3ct \cdot 3c \\ &= 3ct(4t + 3c) \end{aligned}$$

b. The GCF of $12\cos\theta\tan^2\theta$ and $9\cos^2\theta\tan\theta$ is $3\cos\theta\tan\theta$.

$$\begin{aligned} 12\cos\theta\tan^2\theta + 9\cos^2\theta\tan\theta &= 3\cos\theta\tan\theta \cdot 4\tan\theta + 3\cos\theta\tan\theta \cdot 3\cos\theta \\ &= 3\cos\theta\tan\theta(4\tan\theta + 3\cos\theta) \end{aligned}$$

PROGRESS CHECK 3

Factor completely.

a. $6s^2 - 9s$ **b.** $6\sin^2\theta - 9\sin\theta$ ■

Factoring by Grouping

In Example 1b we factored out a common factor that was a binomial. This situation often occurs when factoring a polynomial with four terms by a method called **factoring by grouping**. For example,

$$3xy + 2y + 3xz + 2z$$

can be factored by rewriting the expression as

$$(3xy + 2y) + (3xz + 2z).$$

If we factor out the common factor in each group, we have

$$y(3x + 2) + z(3x + 2).$$

Now $3x + 2$ is a common factor, and the final result is

$$(3x + 2)(y + z).$$

You should check here that an alternate grouping of the original expression, such as

$$(3xy + 3xz) + (2y + 2z),$$

also achieves the same result.

Example 4: Factoring by Grouping

Factor by grouping.

a. $3x^2 + 6x + 4x + 8$ **b.** $x^3 - 5x^2 - 3x + 15$

Solution:

a. $3x^2 + 6x + 4x + 8$

$= \overbrace{(3x^2 + 6x)}^{\substack{\text{common factor} \\ 3x}} + \overbrace{(4x + 8)}^{\substack{\text{common factor} \\ 4}}$ Group terms with common factors.

$= 3x(x + 2) + 4(x + 2)$ Factor in each group.

$= (x + 2)(3x + 4)$ Factor out the common binomial factor.

b. Note in the following solution that we factor out –3 instead of 3 from the grouping on the right to reach the goal of obtaining a common binomial factor.

$x^3 - 5x^2 - 3x + 15$

$= \overbrace{(x^3 - 5x^2)}^{\substack{\text{common factor} \\ x^2}} + \overbrace{(-3x + 15)}^{\substack{\text{common factor} \\ -3}}$ Group terms with common factors.

$= x^2(x - 5) - 3(x - 5)$ Factor in each group.

$= (x - 5)(x^2 - 3)$ Factor out the common binomial factor.

PROGRESS CHECK 4

Factor by grouping.

a. $2x^2 + 4x + 3x + 6$ **b** $x^3 - 6x^2 - 6x + 36$ ■

Factoring Trinomials (Leading Coefficient 1)

In some cases trinomials of the form $ax^2 + bx + c$ can be factored (with integer coefficients) by reversing the FOIL multiplication process shown earlier. Example 4 discusses this factoring procedure in the simplest case when the leading coefficient a equals 1.

Example 5: Factoring $ax^2 + bx + c$ with $a = 1$

Factor completely $x^2 - 5x - 6$.

Solution:

The first term, x^2, is the result of multiplying the first terms in the FOIL method. Thus,

$$x^2 - 5x - 6 = (x + ?)(x + ?).$$

The last term, –6, is the result of multiplying the last terms in the FOIL method. Since –6 equals $(-6)(1)$, $(6)(-1)$, $(-3)(2)$ and $(3)(-2)$, we have four possibilities. We want the pair whose sum is –5, which is the coefficient of the middle term. The combination of –6 and 1 satisfies this condition and produces the middle term of $-5x$.

$$(x-6)(x+1)$$

$$x+(-6x)=-5x$$

Thus, $x^2-5x-6=(x-6)(x+1)$. We summarize our method in this example with the factoring formula

$$x^2+(m+n)x+mn=(x+m)(x+n),$$

which you can use as a model for factoring such expressions.

Technology Link

A graphing calculator can often be useful when factoring a polynomial in one variable. To see how, consider Figure 3.2, which shows a graph of the function associated with the polynomial in this example. Do you see a relation between the factors of the polynomial and the points where the graph crosses the x-axis?

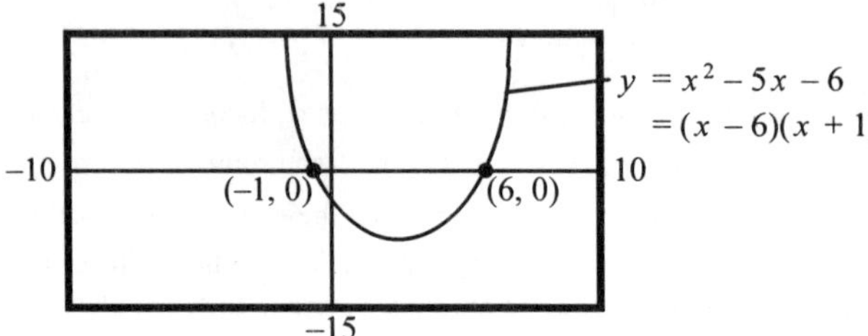

Figure 3.2

Observe that the graph crosses the x-axis at $x=6$ and at $x=-1$, while $x-6$ and $x+1$ are factors of the polynomial. This result illustrates the general principle that if the graph of $y=f(x)$ intersects the x-axis at $x=a$, then $x-a$ is a factor of $f(x)$. Try to use this principle to factor the polynomial in the Progress Check exercise that follows.

PROGRESS CHECK 5

Factor completely x^2-x-20. ■

Factoring Trinomials (General Case)

When the coefficient of the squared term is not 1, the trinomial is harder to factor. The reason for this difficulty can be seen in the following product.

$$(px+m)(qx+n)=(pq)x^2+(pn+qm)x+mn$$

In factoring these more complicated trinomials, if you keep in mind that the middle term is the sum of the inside product and the outside product, then a little trial and error will eventually produce the answer. Although there are often many combinations to consider, experience and practice will help you *mentally* eliminate many of the possibilities.

Example 6: Factoring $ax^2 + bx + c$ with $a \neq 1$

Factor completely.

a. $4c^2 - 12c + 5$ **b.** $4\cos^2\theta - 12\cos\theta + 5$

Solution:

a. The first term, $4c^2$, is the result of multiplying the first terms in the FOIL method. Thus,

$$4c^2 - 12c + 5 \stackrel{?}{=} \begin{matrix} (4c + ?)(c + ?) \\ \text{or} \\ (2c + ?)(2c + ?) \end{matrix}$$

The last term, 5, is the result of multiplying the last terms in the FOIL method. There is only one possibility for 5, $(-5)(-1)$, since we can eliminate $(5)(1)$ because the middle term in negative.

$$4c^2 - 12c + 5 \stackrel{?}{=} \begin{matrix} (4c - 5)(c - 1) \\ (4c - 1)(c - 5) \\ (2c - 5)(2c - 1) \end{matrix}$$

The middle term, $-12c$, is the sum of the inner and the outer terms.

$$(4c - 5)(c - 1) \qquad (2c - 5)(2c - 1) \qquad (4c - 1)(c - 5)$$

$$-4c - 5c = -9c \qquad -2c - 10c = -12c \qquad -20c - c = -21c$$

Thus, $4c^2 - 12c + 5 = (2c - 5)(2c - 1)$ is the correct choice.

b. In a similar way, $4\cos^2\theta - 12\cos\theta + 5 = (2\cos\theta - 5)(2\cos\theta - 1)$.

Technology Link

When a factorization is to be limited to polynomial factors with integer coefficients, then the general principle given in the technology link associated with Example 4 needs to be customized as follows.

> If the graph of $y = f(x)$ intersects the x-axis at $x = a/b$, where a/b is a rational number, then $bx - a$ is a factor of $f(x)$.

For instance, we see in Figure 3.3 that the graph of $y = 4x^2 - 12x + 5$ intersects the x-axis at $x = 5/2$ and $x = 1/2$. Therefore, $2x - 5$ and $2x - 1$ are factors of $4x^2 - 12x + 5$.

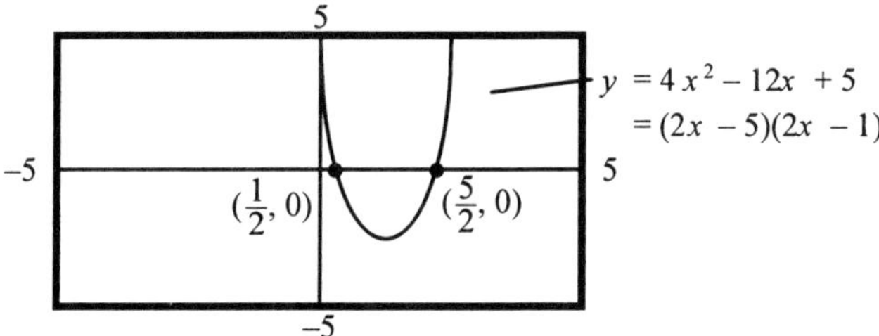

Figure 3.3

PROGRESS CHECK 6

Factor completely.

a. $4t^2 - 4t - 15$ **b.** $4\tan^2\theta - 4\tan\theta - 15$ ■

Example 7: Factoring $ax^2 + bx + c$ with $a \neq 1$

Factor completely.

a. $9\sin^2\theta + 12\sin\theta + 4$ **b.** $9s^2 + 12s + 4$

Solution:

a. The first term, $9\sin^2\theta$, is the result of multiplying the first terms in the FOIL method.

$$9\sin^2\theta + 12\sin\theta + 4 \stackrel{?}{=} \begin{matrix} (9\sin\theta + ?)(\sin\theta + ?) \\ \text{or} \\ (3\sin\theta + ?)(3\sin\theta + ?) \end{matrix}$$

The last term, 4, is the result of multiplying the last terms in the FOIL method. Since 4 equals (4)(1) and (2)(2), we have two possibilities. We eliminate negative factors of 4 since the middle term is positive.

$$9\sin^2\theta + 12\sin\theta + 4 \stackrel{?}{=} \begin{matrix} (9\sin\theta + 1)(\sin\theta + 4) \\ (9\sin\theta + 4)(\sin\theta + 1) \\ (9\sin\theta + 2)(\sin\theta + 2) \\ (3\sin\theta + 4)(3\sin\theta + 1) \\ (3\sin\theta + 2)(3\sin\theta + 2) \end{matrix}$$

The middle term, $12\sin\theta$, is the sum of the inner and the outer terms.

$$(3\sin\theta + 2)(3\sin\theta + 2)$$

$$6\sin\theta + 6\sin\theta = 12\sin\theta$$

Thus, $9\sin^2\theta + 12\sin\theta + 4 = (3\sin\theta + 2)^2$.

b. In a similar way, $9s^2 + 12s + 4 = (3s + 2)(3s + 2)$ or $(3s + 2)^2$.

Note: Very often when the first term and the last term of the trinomial to be factored are perfect squares, the factored form is also a perfect square. Try this possibility first. The factoring models in such cases (called perfect square trinomials) are

$$a^2 + 2ab + b^2 = (a + b)^2$$
$$a^2 - 2ab + b^2 = (a - b)^2.$$

You can spot a perfect square trinomial in one variable on a graphing calculator by looking for a graph that intersects the x-axis in exactly one point. For instance, Figure 3.4 shows that the graph of $y = 9x^2 + 12x + 4$ intersects the x-axis only at $x = -2/3$, so only $3x + 2$ is a factor of $9x^2 + 12x + 4$ (if integer coefficients are required).

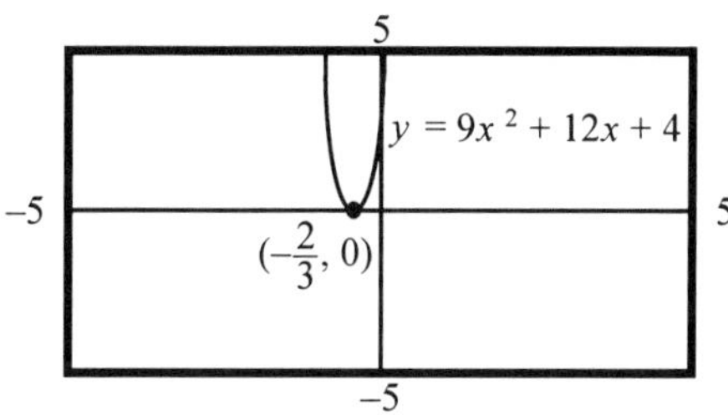

Figure 3.4

PROGRESS CHECK 7
Factor completely.
a. $4\cos^2\theta - 4\cos\theta + 1$ **b.** $4c^2 - 4c + 1$. ■

Factoring by grouping is a key step in the ***ac* method** for factoring the trinomial $ax^2 + bx + c$. Example 8 explains in detail the steps for this method, which is especially useful if our previous method of reversing FOIL results in many possible factorizations to consider.

Example 8: Using the *ac* method

Factor completely $12x^2 - x - 20$.

Solution: Follow the steps below.
Step 1: Find two integers whose product is *ac* and whose sum is *b*. In the trinomial $a = 12$, $b = -1$, and $c = -20$. Thus $ac = 12(-20) = -240$, and we look for two integers whose product is –240 and whose sum is –1. With a little trial and error we find that

$$-16(15) = -240 \text{ and } -16 + 15 = -1,$$

so the required integers are –16 and 15.

Step 2: Replace *b* by the sum of the two integers from step 1 and then distribute *x*.

$$\begin{aligned} 12x^2 - x - 20 &= 12x^2 + (-16 + 15)x - 20 \\ &= 12x^2 - 16x + 15x - 20 \end{aligned}$$

Step 3: Factor by grouping.

$$\begin{aligned} \left(12x^2 - 16x\right) + (15x - 20) &= 4x(3x - 4) + 5(3x - 4) \\ &= (3x - 4)(4x + 5) \end{aligned}$$

Thus, $12x^2 - x - 20 = (3x - 4)(4x + 5)$.

PROGRESS CHECK 8
Factor completely $12x^2 + 19x - 18$. ■

In the next example more than one factoring method is required. As a general strategy, it is recommended that you factor out any common factors as your first factor procedure.

Example 9: Combining Factoring Methods

Factor completely $18x^3 + 15x^2 - 12x$.

Solution: First look for common factors. The GCF is $3x$, so factor it out.

$$18x^3 + 15x^2 - 12x = 3x(6x^2 + 5x - 4)$$

Now suppose we try to factor $6x^2 + 5x - 4$ as $(2x + 1)(3x - 4)$.

$$(2x + 1)(3x - 4)$$

$$-8x + 3x = -5x$$

This result is the opposite of the middle term we want, so we switch signs on the factors of -4, from $(1)(-4)$ to $(-1)(4)$.

$$(2x - 1)(3x + 4)$$

$$8x - 3x = 5x$$

Thus, $6x^2 + 5x - 4 = (2x - 1)(3x + 4)$, and so the complete factorization is

$$18x^3 + 15x^2 - 12x = 3x(2x - 1)(3x + 4)$$

PROGRESS CHECK 9
Factor completely $6y^4 - 15y^3 - 9y^2$. ■

It should be noted that many trinomials cannot be factored with integer coefficients, and if you suspect that you are trying to find a factorization that may not exist, then apply the following test.

Factoring Test for Trinomials

A trinomial of the form $ax^2 + bx + c$, where a, b, and c are integers, is factorable into binomial factors with integer coefficients if and only if $b^2 - 4ac$ is a perfect square.

The expression $b^2 - 4ac$ is called the **discriminant**, and when we derive the discriminant in Section 4.1, the rationale for this test will be easy to understand.

Example 10: Using a Factoring Test for Trinomials

Calculate $b^2 - 4ac$ and determine if $x^2 - 10x - 30$ can be factored into binomial factors with integer coefficients.

Solution: In this trinomial $a = 1$, $b = -10$, and $c = -30$. Therefore,

$$\begin{aligned} b^2 - 4ac &= (-10)^2 - 4(1)(-30) \\ &= 220. \end{aligned}$$

Since 220 is not a perfect square, $x^2 - 10x - 30$ cannot be factored into binomial factors with integer coefficients.

PROGRESS CHECK 10
Calculate $b^2 - 4ac$ and determine if $9x^2 - 9x - 4$ can be factored into binomial factors with integer coefficients. ■

A polynomial is said to be **irreducible** over the set of integers if it cannot be written as the product of two polynomials of positive degree with integer coefficients. Thus, we say $x^2 - 10x - 30$ is irreducible over the set of *integers*. Note, however, that $x^2 - 10x - 30$ is not irreducible over the set of *real numbers*, since we can show that

$$x^2 - 10x - 30 = \left[x - \left(5 + \sqrt{55}\right)\right]\left[x - \left(5 - \sqrt{55}\right)\right]$$

by using techniques that we will develop. Thus, whether or not a polynomial is irreducible depends on the number system from which the coefficients may be selected.

EXPLORE 3.1

1. Factoring one expression, $f(x)$, produces an equivalent expression, $g(x)$. By comparing the graphs of $y = f(x)$ and $y = g(x)$ you can check whether a factorization is correct. Use this technique to decide which of these factorizations are correct
   a. $x^2 + x \stackrel{?}{=} x(x+1)$ **b.** $x^2 + 9 \stackrel{?}{=} (x+3)^2$
   c. $x^3 - x^2 + 1 \stackrel{?}{=} x\left(x^2 - x + 1\right)$ **d.** $x^2 - 8x - 20 \stackrel{?}{=} (x-10)(x+2)$
   e. $\sin 5x^2 - \sin 3x \stackrel{?}{=} x(\sin 5x - \sin 3)$
2. If $g(x)$ is the factored version of $f(x)$, then a table of values for the two expressions should match exactly. Since many calculators easily produce lists or tables, this provides another way to use a calculator to check factorizations. Use this approach to check the factorizations in Explore Exercise 1, by completing tables for both expressions using $x = -2, -1, 0, 1$, and 2.
3. Refer to the two Technology Links in this section for suggestions on using a calculator to find the factors of a polynomial. Then use those ideas to determine the factors of these polynomials.
   a. $x^2 - 6x - 16$ **b.** $x^2 - 2x - 143$
   c. $x^3 + 5x^2 - 12x - 36$ **d.** $6x^2 - x - 12$
4. CAS utilities do factoring automatically. If you have such a utility use it to confirm the results in Explore Exercises 1 and 3.

EXERCISES 3.1

In Exercises 1–20, factor out the greatest common factor.

1. $yx + yz$ **2.** $ab - ac$ **3.** $8x - 12$

4. $9y + 15$ **5.** $b - b^2$ **6.** $x^2 + x$

7. $5x^3y - 10x^2y^2$ **8.** $12ct^2 - 21c^2t$

9. $9a^2x^2 + 3ax$ **10.** $22x^3y^3 - 2x^2y^2$

11. $2x^2z^2 + 4xy - 5x^2y^2$

12. $3a^{10}x^6 - 9a^7x^7 + 6a^9x^4$

13. $x(x+1) + 2(x+1)$ **14.** $x(x-2) - 3(x-2)$

15. $7t(t-4) - (t-4)$ **16.** $(1-y) - y(1-y)$

17. $b(a-c) + d(a-c)$ **18.** $2x(y+z) - 5y(y+z)$

19. $(x-5)^2 + 7(x-5)$ **20.** $(x+3)^2 - (x+3)$

In Exercises 21–32 factor by grouping.

21. $ax + ay + 5x + 5y$ **22.** $2a + 2b + ka + kb$

23. $3xy + 6x + y + 2$ **24.** $1 + x + c + cx$

25. $4x^2 - 8x + 3x - 6$ **26.** $12y^2 - 18y + 8y - 12$

27. $a^2 - 3a - 3a + 9$ **28.** $4t^2 - 6t - 6t + 9$

29. $x^5 + x^3 + 5x^2 + 5$ **30.** $y^4 - 5y^3 + 2y - 10$

31. $x^3 + 7x^2 - 7x - 49$ **32.** $x^3 - 3x^2 - x + 3$

In Exercises 33–60 factor each polynomial.

33. $a^2 + 5a + 4$ **34.** $y^2 + 7y + 10$

35. $x^2 - 9x + 14$
36. $r^2 - 13r + 12$
37. $y^2 + 13y - 30$
38. $p^2 - 3p - 18$
39. $c^2 - 3c - 10$
40. $x^2 - x - 6$
41. $x^2 - 6x + 9$
42. $y^2 + 2y + 1$
43. $t^2 + 10t + 25$
44. $c^2 - 8c + 16$
45. $x^2 + 8xy + 15y^2$
46. $y^2 - 10ay + 16a^2$
47. $a^2 + 3ab - 4b^2$
48. $x^2 - bx - 20b^2$
49. $7y^2 - 13y + 6$
50. $2x^2 + 9x + 10$
51. $3x^2 - 5x - 2$
52. $5r^2 - 4r - 12$
53. $6k^2 - 7k + 2$
54. $4a^2 + 13a - 12$
55. $9a^2 - 12a + 4$
56. $4x^2 - 12x + 9$
57. $9b^2 - 25b - 6$
58. $12x^2 + 19x - 18$
59. $12x^2 - 29xy + 10y^2$
60. $18t^2 - 3kt - 10k^2$

In Exercises 61–70 factor each expression completely.

61. $3k^2 - 6k - 24$
62. $5y^2 - 15y - 50$
63. $6t^2 + 12t - 48$
64. $4x^2 - 12x + 8$
65. $3x^3 - 12x^2 + 9x$
66. $2k^3 - 18k^2 + 40k$
67. $9y^3 - 21y^2 + 10y$
68. $16a^3 - 4a^2 - 6a$
69. $24x^3y + 40x^2y^2 + 16xy^3$
70. $72a^3b - 12a^2b^2 - 60ab^3$

In Exercises 71–92 factor each expression completely.

71. $7\sin\theta + 7\cos\theta$
72. $3\tan\theta - 3\sin\theta$
73. $21\sin^2\theta - 14\sin\theta$
74. $7\tan\theta - 35\tan^3\theta$
75. $15\cos\theta\tan^2\theta - 21\cos^2\theta\tan\theta$
76. $20\tan^2\theta - 15\tan\theta\sin\theta$
77. $\cos\theta(\sin\theta - 5) + \tan\theta(\sin\theta - 5)$
78. $\sin\theta(2\tan\theta + 1) - \cos\theta(2\tan\theta + 1)$
79. $\tan^2\theta - 3\tan\theta + 2$
80. $\tan^2\theta - 8\tan\theta + 12$
81. $\tan^2\theta - 9\tan\theta + 20$
82. $\sin^2\theta - 8\sin\theta - 20$
83. $3\sin^2\theta - 8\sin\theta - 3$
84. $2\tan^2\theta - 7\tan\theta + 6$
85. $6\sin^2\theta + 7\sin\theta + 2$
86. $4\cos^2\theta - 13\cos\theta + 3$
87. $20\sin^2\theta - 43\sin\theta - 12$
88. $18\cos^2\theta - 57\cos\theta + 24$
89. $4\cos^2\theta - 9\cos\theta\tan\theta + 2\tan^2\theta$
90. $2\sin^2\theta + 5\sin\theta\cos\theta - 7\cos^2\theta$
91. $4\sin^2\theta - 4\sin\theta + 1$
92. $4\tan^2\theta + 20\tan\theta + 25$

In Exercises 93–100 determine whether or not each expression can be factored into binomial factors with integer coefficients by calculating $b^2 - 4ac$.

93. $6s^2 + 7s + 2$
94. $3\sin^2\theta + \sin\theta - 1$
95. $t^2 - 5t + 3$
96. $2\tan^2\theta - \tan\theta - 3$
97. $2\cos^2\theta - 5\cos\theta - 7$
98. $18x^2 + 19x - 12$
99. $12x^2 - 11x - 18$
100. $12s^2 - s + 1$

In Exercises 101–104 write an alternative formula by factoring the right side of the equation.

101. The surface area of the rectangular solid shown. $S = 2b^2 + 4ab$

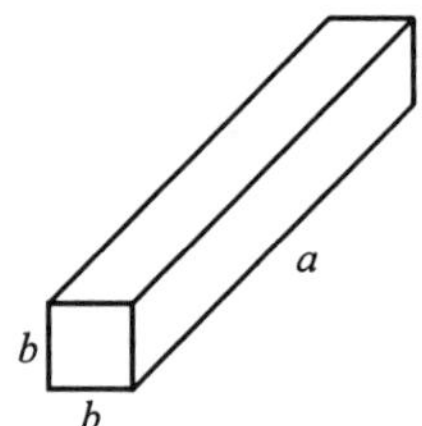

102. The total surface area of the metal gutter shown (not including edges) $S = 4ac + 2bc$.

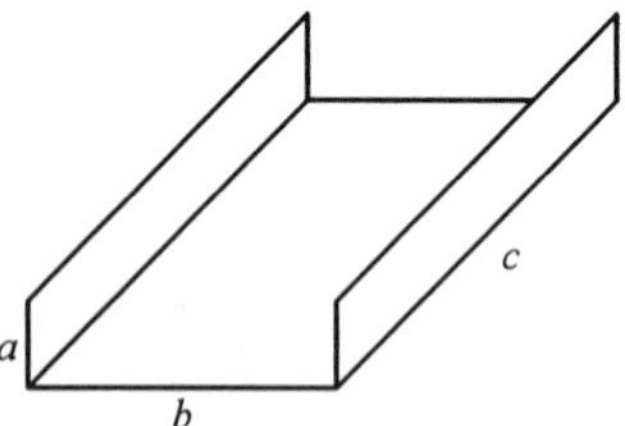

103. The area of a ring where the small radius is half the large radius, R. $A = \pi R^2 - \pi\left(\frac{R}{2}\right)^2$

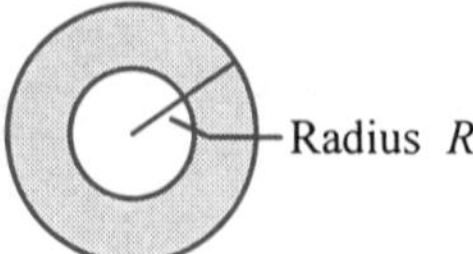

104. The shaded area given that the small radii are half the large radius, R. $A = \pi R^2 - \pi\left(\frac{R}{2}\right)^2 - \pi\left(\frac{R}{2}\right)^2$

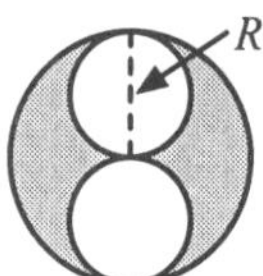

105. If a storekeeper raises a price 30 percent and one week later cuts this new price by 30 percent, the final price can be represented as $(x+0.30x)-0.30(x+0.30x)$, where x is the original price. Factor out $x+0.30x$; then simplify and interpret the result. Is the final price more or less than the original price?

106. If a storekeeper raises a price 70 percent and one week later cuts this new price by 70 percent, the final price can be represented as $(x+0.70x)-0.70(x+0.70x)$, where x is the original price. Factor out $x+0.70x$; then simplify and interpret the result. Is the final price more or less than the original price?

107. What is the overall result of raising a price by 10 percent and then raising the resulting price by 20 percent? Factor out $x+0.10x$ in $(x+0.10x)+0.20(x+0.10x)$ and simplify to find out.

108. What is the overall result of raising a price by 20 percent and then raising the resulting price by 10 percent? Factor out $x+0.20x$ in $(x+0.20x)+0.10(x+0.20x)$ and simplify to find out.

109. There are two numbers for which the number plus twice its square equals 55. They can be found by solving $x+2x^2-55=0$. Factor the left side of this equation.

110. A ball is thrown down from a roof 160 feet high with an initial velocity of 48 ft/sec. To find the number of seconds it takes to hit the ground you can solve $16t^2+48t-160=0$. Factor the left side of this equation.

THINK ABOUT IT 3.1

1. **a.** What does the word *common* mean in *greatest common factor*?
   b. Look up *factor* in the dictionary. What does it have to do with *factory*? English use of this word in algebra dates back to the seventeenth century.
2. If c is a prime number and x^2+bx+c is factorable, express b in terms of c.
3. Factor each expression completely. Assume that variables as exponents denote integers.
   a. $x^{n+1}+x^n$ **b.** $y^{2n}-y^n$ **c.** $x^{2m}+6x^m+9$
   d. $10x^{2n}-9x^n+2$ **e.** $y^{2n+1}+y^{n+1}-20y$ **f.** $x^{2m+2}+x^{m+2}-30x^2$
4. Find two expressions whose greatest common factor is the following.
   a. $5x^2$ **b.** $3x-1$
5. An odd number is a positive integer that can be written in the form $2k+1$ where k is an integer. Show that the product of two odd numbers is an odd number.

• • •

3.2 Special Factoring Models and a Factoring Strategy

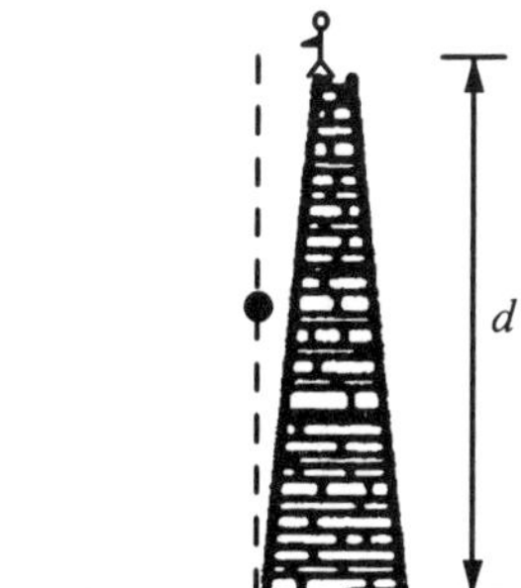

A formula for the height (y) above ground of an object dropped d ft above the earth's surface is $y=d-16t^2$, where t is the elapsed time in seconds.

a. What is the formula for an object dropped from 625 ft?

b. Express the result from part **a** in factored form.

c. Show that the object hits the ground in 6.25 seconds.

(See Example 3.)

Objectives

1. Factor an expression that is the difference of squares.
2. Factor an expression that is the sum or difference of cubes.
3. Apply a general strategy to factor expressions systematically.

In this section we rely on certain models to factor three special types of binomials. All of these models may be verified through multiplication and they should all be memorized.

Difference of Squares

An important method of factoring comes from reversing the following special product that was considered in Section R.4.

$$(a+b)(a-b)=a^2-b^2$$

Note that the factors on the left differ only in the operation between a and b, while the result of the multiplication is two perfect squares with a minus sign between them. Thus, in factoring two perfect squares with a minus sign between them, we obtain two factors that consist of the sum and the difference of the square roots of each of the squared terms.

Factoring Model for a Difference of Squares

$$a^2-b^2=(a+b)(a-b)$$

Keep in mind that even powers of variables such as x^2, t^4, y^6, and so on are perfect squares.

Example 1: Factoring a Difference of Squares

Factor completely.

a. x^2-81 **b.** $25y^6-49n^4$

Solution:

a. Substitute x for a and 9 for b in the given model.

$$\begin{array}{c} a^2-b^2=(a+b)(a-b) \\ \downarrow \quad \downarrow \qquad \downarrow \ \downarrow \ \downarrow \ \downarrow \\ x^2-81=x^2-9^2=(x+9)(x-9) \end{array}$$

b. Because $25y^6=(5y^3)^2$ and $49n^4=(7n^2)^2$, replace a by $5y^3$ and b by $7n^2$ in the factoring formula for a difference of squares.

$$\begin{array}{c} a^2 \quad - \quad b^2 \quad = (a \ + \ b)(a \ - \ b) \\ \downarrow \qquad \downarrow \qquad \downarrow \ \ \downarrow \ \ \downarrow \ \ \downarrow \\ 25y^6-49n^4=(5y^3)^2-(7n^2)^2=(5y^3+7n^2)(5y^3-7n^2) \end{array}$$

Technology Link

Two expressions that are equivalent will produce the same graph. Therefore, one way to test for equivalent expressions on a graphing calculator is to set $y1$ equal to one of the expressions, $y2$ equal to the other, and then draw complete graphs of both functions in the same display. If you see two distinct graphs, then the expressions are not equivalent. Two graphs that appear to be the same will support the conclusion that the expressions are equivalent. To illustrate, visual support for the factorization in Example 1a may be obtained by letting

$$y1=x^2-81$$
$$y2=(x+9)(x-9)$$

and then graphing both functions as shown in Figure 3.5.

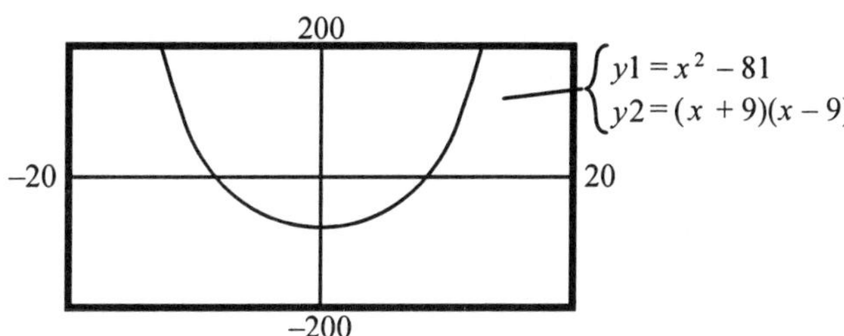

Figure 3.5

PROGRESS CHECK 1
Factor completely.
a. $n^2 - 16$ **b.** $100x^2y^2 - 49z^2$ ■

Example 2: Contrasting Factorizations of Algebraic and Trigonometric Expressions

Factor completely.
a. $16 - 9\tan^2\theta$ **b.** $16 - 9t^2$

Solution:

a. Note that $16 - 9\tan^2\theta = 4^2 - (3\tan\theta)^2$ and apply the factoring formula with $a = 4$ and $b = 3\tan\theta$.

$$16 - 9\tan^2\theta = 4^2(3\tan\theta)^2$$
$$= (4 + 3\tan\theta)(4 - 3\tan\theta)$$

b. In a similar way, $16 - 9t^2 = (4 + 3t)(4 - 3t)$.

PROGRESS CHECK 2
Factor completely.
a. $1 - 4\sin^2\theta$ **b.** $1 - 4s^2$ ■

Example 3: Using Factoring to Rewrite a Formula

Solve the problem in the section introduction page 251.

Solution:

a. Replacing d by 625 in the formula $y = d - 16t^2$ yields

$$y = 625 - 16t^2.$$

b. The right side of the equation above factors as a difference of squares.

$$625 - 16t^2 = (25 + 4t)(25 - 4t)$$

Thus, $y = (25 + 4t)(25 - 4t)$ express the result from part **a** in factored form.

c. To show the object hits the ground in 6.25 seconds, we show that when $t = 6.25$, $y = 0$. Choosing the factored version of the formula we get

$$y = (25 + 4t)(25 - 4t)$$
$$= [25 + 4(6.25)][25 - 4(6.25)]$$
$$= (50)(0)$$
$$= 0.$$

Note that an expression in factored form has a value of 0 if and only if at least one of the factors has a value of 0.

PROGRESS CHECK 3

Because the moon is less massive than the earth, the pull of gravity at its surface is weaker, and so objects fall more slowly on the moon. The formula for the height (y) above ground of an object dropped d ft above the moon's surface is roughly $y = d - 4t^2$, where t is the elapsed time in seconds.

a. What is the formula for an object dropped for 625 ft?

b. Express the result from part **a** in factored form.

c. Show that the object hits the ground in 12.5 seconds. ■

Sum or Difference of Two Cubes

The product of $a + b$ and $a^2 - ab + b^2$ is $a^3 + b^3$, as shown below.

$$(a+b)(a^2 - ab + b^2) = a^3 - a^2b + ab^2 + a^2b - ab^2 + b^3$$
$$= a^3 + b^3$$

Similarly,

$$(a-b)(a^2 + ab + b^2) = a^3 + a^2b + ab^2 - a^2b - ab^2 - b^3$$
$$= a^3 - b^3.$$

By reversing these special products, we obtain factoring models for the sum and the difference of two cubes.

Factoring Models for Sums and Differences of Cubes

$$a^3 + b^3 = (a+b)(a^2 - ab + b^2)$$
$$a^3 - b^3 = (a-b)(a^2 + ab + b^2)$$

To use these models, we identify appropriate replacements for a and b in the expression to be factored and then substitute in these formulas.

Example 4: Factoring a Sum or Difference of Cubes

Factor completely.

a. $8x^3 + 27$ **b.** $y^6 - 125$

Solution:

a. Here we use the formula for the sum of two cubes. Since $8x^3 = (2x)^3$ and $27 = (3)^3$, we replace a with $2x$ and b with 3. The result is

$$a^3 + b^3 = (a + b)(a^2 - a\,b + b^2)$$
$$\downarrow \quad \downarrow \quad \downarrow \quad \downarrow \quad \downarrow \quad \downarrow \quad \downarrow \quad \downarrow$$
$$8x^3 + 27 = (2x)^3 + 3^3 = (2x+3)\left[(2x)^2 - (2x)(3) + (3)^2\right]$$
$$= (2x+3)(4x^2 - 6x + 9)$$

b. To factor $y^6 - 125$, use the formula for the difference of two cubes. Because $y^6 = (y^2)^3$ and $125 = 5^3$, replace a with y^2 and b with 5.

$$\begin{array}{c} a^3 - b^3 = (a - b)(a^2 + ab + b^2) \\ \downarrow \quad \downarrow \quad \downarrow \; \downarrow \; \downarrow \quad \downarrow \; \downarrow \; \downarrow \end{array}$$

$$y^6 - 125 = (y^2)^3 - 5^3 = (y^2 - 5)\left[(y^2)^2 + (y^2)(5) + (5)^2\right]$$
$$= (y^2 - 5)(y^4 + 5y^2 + 25)$$

PROGRESS CHECK 4

Factor completely.

a. $8y^3 - 1$ **b.** $x^9 + 27y^3$ ■

The next example points out that we often need to apply more than one factoring procedure to factor completely.

Example 5: Combining Factoring Methods

Factor completely.

a. $10x^4 + 10x$ **b.** $y^4 - x^4$

Solution:

a. First, factor out the common factor, $10x$; then apply the factoring model for the sum of two cubes.

$$10x^4 + 10x = 10x(x^3 + 1) \qquad \text{Factor out } 10x.$$
$$= 10x(x + 1)(x^2 - x + 1) \qquad \text{Sum of cubes.}$$

b. First, factor $y^4 - x^4$ as the difference of squares.

$$y^4 - x^4 = (y^2 + x^2)(y^2 - x^2)$$

Then, $y^2 + x^2$ does not factor, but $y^2 - x^2$ is a difference of squares, so

$$y^4 - x^4 = (y^2 + x^2)(y + x)(y - x)$$

represents the complete factorization.

PROGRESS CHECK 5

Factor completely.

a. $5x^3 - 40$ **b.** $16m^4 - 81n^4$ ■

At this point it is useful to summarize all of our factoring models and to state a general strategy to help you to systematically factor a variety of expressions.

Summary of Factoring Models

1. Common factor: $ab + ac = a(b + c)$
2. Trinomial $(a = 1)$: $x^2 + (m + n)x + mn = (x + m)(x + n)$
3. General trinomial: $(pq)x^2 + (pn + qm)x + mn = (px + m)(qx + n)$
4. Perfect square trinomial: $a^2 + 2ab + b^2 = (a + b)^2$
5. Perfect square trinomial: $a^2 - 2ab + b^2 = (a - b)^2$

6. Difference of squares: $a^2 - b^2 = (a+b)(a-b)$
7. Sum of cubes: $a^3 + b^3 = (a+b)(a^2 - ab + b^2)$
8. Difference of cubes: $a^3 - b^3 = (a-b)(a^2 + ab + b^2)$

With the aid of these models you may use the following steps as a guideline for factoring polynomials.

Guidelines to Factoring a Polynomial

1. Factor out any common factors (if present) as the first factoring procedure.
2. Check for factorizations according to the number of terms in the polynomial.
   Two terms: Look for a difference of squares or cubes, or a sum of cubes. Then apply models 6, 7, or 8, if applicable. Remember that $a^2 + b^2$ is irreducible.
   Three terms: If the coefficient of the squared term is 1, try to use model 2. If the coefficient of the squared term is not 1, use FOIL reversal or the *ac* method. Check for the special case of a perfect square trinomial in which models 4 and 5 apply.
   Four terms: Try factoring by grouping.
3. Make sure that no factors of two or more terms can be factored again.

Also remember that trigonometric expressions are factored in the same manner as polynomials and that you can always check your factoring by multiplying out your answer.

Example 6: Factoring a Polynomial Systematically

Factor completely $144 - 16t^2$.

Solution: Although $144 - 16t^2$ is a difference of two squares, we will first factor out the GCF, 16, as recommended in the guidelines above.

$$144 - 16t^2 = 16(9 - t^2)$$

Now by the difference of squares model, $9 - t^2 = (3+t)(3-t)$, so

$$144 - 16t^2 = 16(3+t)(3-t).$$

PROGRESS CHECK 6
Factor completely $64 - 16t^2$. ■

Example 7: Factoring a Trigonometric Expression Systematically

Factor completely $6\sin^2\theta - 19\sin\theta + 10$

Solution: There are no common factors to factor out and this expression matches the general trinomial model. Use FOIL reversal or the *ac* method to obtain

$$6\sin^2\theta - 19\sin\theta + 10 = (3\sin\theta - 2)(2\sin\theta - 5).$$

PROGRESS CHECK 7
Factor completely $6\cos^2\theta - 48\cos\theta + 72$. ■

Example 8: Factoring a Polynomial Systematically

Factor completely $x^6 - a^6$.

Solution: The polynomial contains two terms with no common factors, and we find that $x^6 - a^6$ is both a difference of squares and a difference of cubes. In such cases, apply the difference of squares model first.

$$\begin{aligned} x^6 - a^6 &= \left(x^3\right)^2 - \left(a^3\right)^2 \\ &= \left(x^3 + a^3\right)\left(x^3 - a^3\right) \end{aligned}$$

Now factor $x^3 + a^3$ by the sum of cubes model, and $x^3 - a^3$ by the difference of cubes model. The final factorization is

$$x^6 - a^6 = (x + a)\left(x^2 - ax + a^2\right)(x - a)\left(x^2 + ax + a^2\right).$$

PROGRESS CHECK 8
Factor completely $y^6 - 1$. ■

Example 9: Factoring a Polynomial Systematically

Factor completely $x^2 + y^2 - z^2 - 2xy$.

Solution: The polynomial contains four terms with no common factors, so try factoring by grouping. We need to recognize that if we group together $x^2 - 2xy + y^2$, it factors into the perfect square $(x - y)^2$. We then have $(x - y)^2 - z^2$, which factors as the difference of two squares. Thus,

$$\begin{aligned} x^2 + y^2 - z^2 - 2xy &= \left(x^2 - 2xy + y^2\right) - z^2 \\ &= (x - y)^2 - z^2 \\ &= (x - y + z)(x - y - z). \end{aligned}$$

PROGRESS CHECK 9
Factor completely $c^2 - d^2 + 4c + 4$. ■

EXPLORE 3.2

1. Factoring one expression, $f(x)$, produces an equivalent expression, $g(x)$, and so the graphs of $y = f(x)$ and $y = g(x)$ are the same. Therefore, to help check whether a factorization is correct you can compare the graphs of the two expressions. Use this technique to decide which of these factorizations are correct, and which are not.
   a. $x^2 + 9 = (x - 3)^2$ **b.** $x^4 + x^2 + 1 = \left(x^2 + 1\right)^2$
   c. $x^3 + 3x^2 + 6x + 8 = (x + 2)^3$ **d.** $\sin^2 x - 6 = (\sin x + 3)(\sin x - 3)$
   e. $1 - \tan^2 2x = (1 + \tan 2x)(1 - \tan 2x)$ **f.** $25^x + 2\left(5^x\right) + 1 = \left(5^x + 1\right)^2$
2. In Explore Exercise 1 we stated that graphing can help you check if a factorization is correct. It is not a foolproof approach because two graphs may appear very similar in appearance and yet not be identical. In cases where graphs appear to be the same, the Trace feature is an additional aid because tracing causes the *x*- and *y*-coordinates to appear on the screen. In the two following problems, notice that on a standard viewing window the graphs may appear the same. Use Trace to decide if the functions are identical, then state whether the factorization is correct or not.
   a. $x^2 + 0.02x + 0.04 \stackrel{?}{=} (x + 0.02)^2$ **b.** $8x^3 - x^2 + 0.1 \stackrel{?}{=} x\left(8x^2 - x + 0.1\right)$
3. Recall that if $g(x)$ is the factored version of $f(x)$, then a table of values for the two expressions should match exactly. Use the table or list capabilities of your calculator to check the factorizations in Explore Exercises 1 and 2 by completing tables for both expressions using $x = -2, -1, 0, 1,$ and 2.
4. CAS utilities do factoring automatically. If you have such a utility use it to confirm the results in Explore Exercises 1 and 2.

EXERCISES 3.2

In Exercises 1–10 factor each difference of squares.

1. $n^2 - 9$
2. $x^2 - 64$
3. $36x^2 - 1$
4. $100 - 81y^2$
5. $25p^2 - 49q^2$
6. $s^2t^2 - 9c^2$
7. $(x + y)^2 - 4$
8. $(x - y)^2 - 121$
9. $36r^6 - k^4$
10. $25x^{12} - 36a^{10}$

In Exercises 11–20 factor each sum or difference of cubes.

11. $y^3 + 27$
12. $x^3 + 64$
13. $x^3 - 1$
14. $t^3 - 125$
15. $a^3b^3 + c^3$
16. $8x^3 - 27y^3$
17. $x^6 + 64$
18. $216 - y^9$
19. $(x + 3)^3 - 8$
20. $(x - 2)^3 + 1$

In Exercises 21–50 factor each expression completely.

21. $x - xy^2$
22. $n^3 - n$
23. $16 - 16t^2$
24. $4 - 16t^2$
25. $1 - t^4$
26. $x^4 - 81$
27. $27(x - 1)^2 - 48x^2$
28. $5(k + 7)^2 - 45k^2$
29. $t^4 + t$
30. $n^4 - 27n$
31. $2y^5 - 16y^2$
32. $ax^5 + 8ax^2$
33. $\pi R^2 - \pi r^2$
34. $\frac{4}{3}\pi R^3 - \frac{4}{3}\pi r^3$
35. $9n^2 + 3n + 3$
36. $12a^2 + 10a - 12$
37. $6x^3 - 24x^2 + 18x$
38. $2x^3y - 18x^2y + 40xy$
39. $x^5 + x^3 + x^2 + 1$
40. $y^4 - y^3 - 8y + 8$
41. $4x^4 - 144x^2$
42. $2x^5 - 162x$
43. $x^6 - 1$
44. $y^6 - x^6$
45. $x^8 - y^8$
46. $1 - x^{12}$
47. $x^2 + 4x + 4 - y^2$
48. $a^2 + x^2 - 2ax - 1$
49. $x^2 - y^2 - z^2 - 2yz$
50. $4x^2 + y^2 + 4xy - 1$

In Exercises 51–62 factor each expression completely.

51. $\sin^2\theta - 1$
52. $9 - \tan^2\theta$
53. $4\tan^2\theta - 25\sin^2\theta$
54. $49\sin^2\theta - 16\cos^2\theta$
55. $7\sin^2\theta - 63$
56. $3\tan^2\theta - 75\sin^2\theta$
57. $\cos^3\theta - \cos\theta$
58. $\sin^2\theta - \sin^4\theta$
59. $\tan^4\theta - 1$
60. $\cos^4\theta - \sin^4\theta$
61. $\sin^3\theta - 1$
62. $\tan^3\theta + 1$
63. A hole is cut from a rectangular solid as shown in the figure. The volume of the remaining solid is given by $4a^3 - 4ab^2$. Factor this expression for the volume.

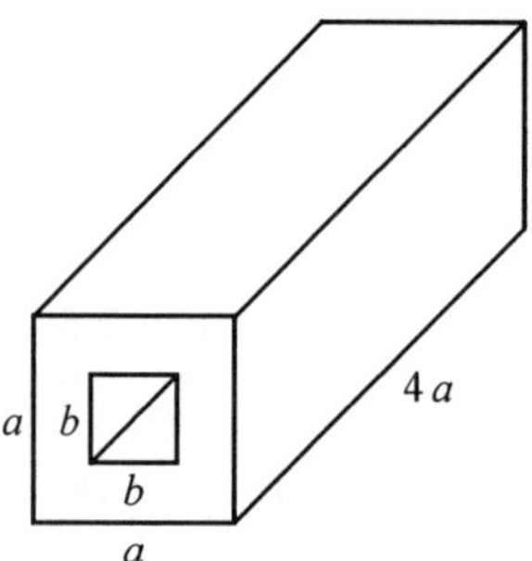

64. A rectangular solid with height h has a square cross-section of sides s. If a corner piece with two sides equal to x is removed, the volume of the solid that remains is $s^2h - x^2h$. Factor this to find another expression for the volume.

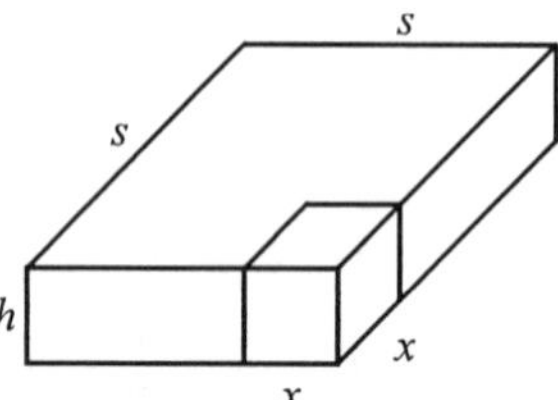

65. The sketch shown consists of three areas whose sum is $c^2 + 2cd + d^2$.
    a. Factor this polynomial to find another expression for this area.

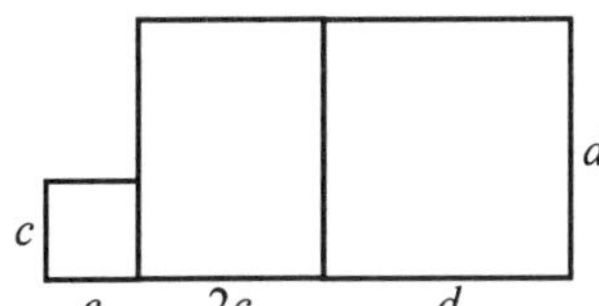

b. Show that the three pieces can be cut up and rearranged to make a square whose dimensions correspond to the two factors from part **a**.

66. The sketch shown consists of three areas whose sum is $r^2 + 2r(2r+1) + (2r+1)^2$.

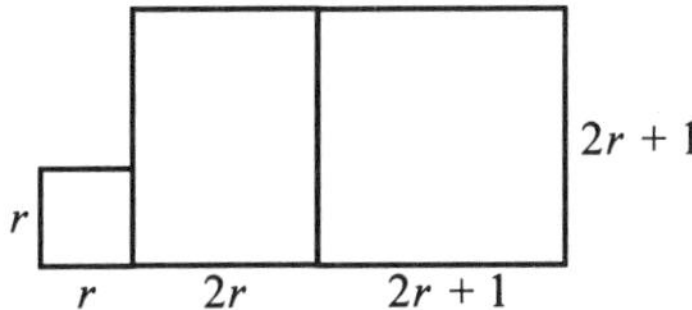

a. Simplify and factor this polynomial to find another expression for this area.
b. Show that the three pieces can be cut up and rearranged to make a square whose dimensions correspond to the two factors from part **a**.

67. The area of a washer can be expressed as $\pi R^2 - \pi r^2$, where R is the radius of the larger circle and r is the radius of the smaller circle. (See the figure below.) A ring-shaped region of this type is called an **annulus**. Factor this polynomial to find another expression for the area.

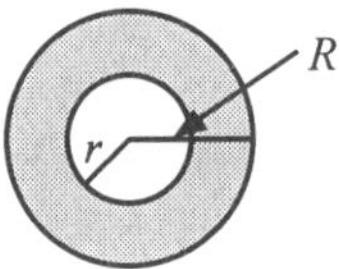

68. The total shaded area shown in the figure is given by $\underbrace{\pi(a+2)^2 - \pi(a+1)^2}_{\text{outside ring}} + \underbrace{\pi \cdot 1^2}_{\text{inside circle}}$.

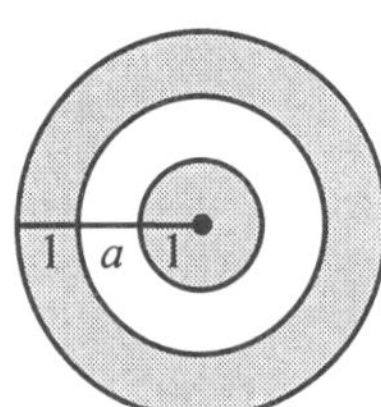

a. Simplify this expression and factor it.
b. Find an expression for the circumference of the largest circle.

69. An approximate formula for the height (y) above the sun's surface of an object dropped from a height of d feet is $y = d - 441t^2$, where t is the elapsed time in seconds, and y is given in feet.
a. What is the formula for an object dropped from a height of 5329 feet? This is a drop of a little more than a mile.
b. Express the result from part **a** in factored form. (Hint: $5329 = 73^2$, and $441 = 21^2$.)
c. Show that the object hits the surface in 73/21 seconds. This is about 3.5 seconds.
d. What is the average speed (in feet per second, and in miles per hour) of the object while it falls this distance?

70. An approximate formula for the height (y) above Pluto's surface of an object dropped from a height of d feet is $y = d - t^2$, where t is the elapsed time in seconds, and y is given in feet.
a. What is the formula for an object dropped from a height of 5329 feet? This is a drop of a little more than a mile.
b. Express the result from part **a** in factored form. (Hint: $5329 = 73^2$.)
c. Show that the object hits the surface in 73 seconds.
d. What is the average speed (in feet per second, and in miles per hour) of the object while it falls this distance?

THINK ABOUT IT 3.2

1. Factor each expression completely. Assume that variables as exponents denote integers.
   a. $x^{2n} - 16$ b. $y^{4n} - y^{2n}$ c. $y^{4n} - 49$
   d. $x^{m+2} - x^m$ e. $1 - t^{3m}$ f. $x^{3n} + y^{3n}$
2. Consider the illustration below.
   a. Explain why the shaded area is given by $a^2 - b^2$. Then, factor this polynomial to find an expression for this area in factored form.
   b. Show that the shaded area can be cut up and rearranged into a rectangle whose length and width are the two factors found in part **a**.

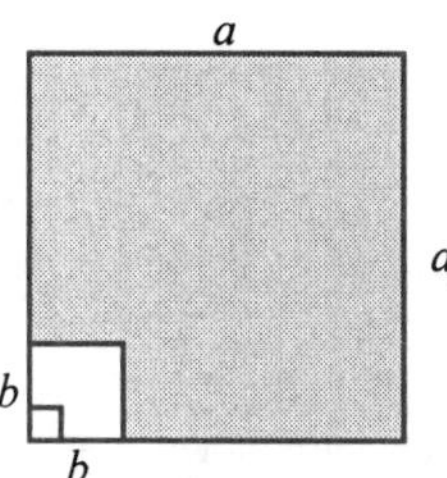

3. If the shaded region in Question 2 is divided into three regions as illustrated, find an expression for the shaded area by adding the areas of the three regions. Then, factor this sum to obtain the same result as in Question **2a**.

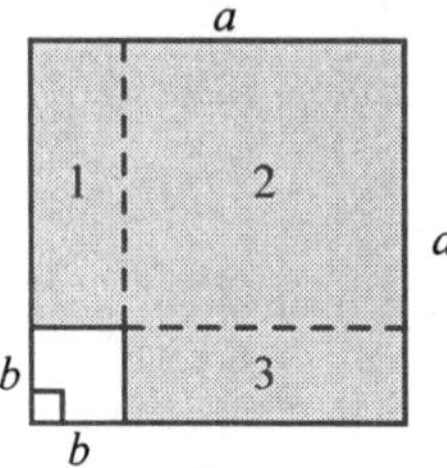

4. **a.** Multiply $(x-1)(x+1)$. **b.** Multiply $(x-1)(x^2+x+1)$.
   c. Multiply $(x-1)(x^3+x^2+x+1)$.
   d. Look at your results from parts **a-c**. Without multiplying, simplify the product $(x-1)(x^7+x^6+x^5+x^4+x^3+x^2+x+1)$.
   e. What is a factoring model for x^n-1?
5. Factor x^6-64 completely by the following methods.
   Method 1 First, factor as a difference of squares, and then use the formulas for the sum and the difference of cubes.
   Method 2 First, factor as a difference of cubes. Then, use the difference of squares formula on *both* factors and match the result from the first method. Applying the difference of squares formula to the trinomial factor requires some ingenuity.

• • •

3.3 Fractional Expressions

Photo Courtesy of Popperfoto/Archive Photos, New York

A race car driver must average at least 150 mi/hour for two laps around a track to qualify for the finals. If a driver averages 180 mi/hour on the first lap, but mechanical trouble reduces the average speed on the second lap to only 120 mi/hour, does the driver qualify for the finals?
(See Example 10. *Hint*: average speed = total distance/total time.)

Objectives

1. Express a fraction in simplest form.
2. Multiply fractions and simplify the product.
3. Divide fractions and simplify the quotient.
4. Add and subtract fractions and write the result in simplest form.
5. Change a complex fraction to a fraction in simplest form.

Algebraic fractions are the quotients of algebraic expressions, and the same principles that govern a fraction in arithmetic also apply when the numerator and the denominator contain algebraic expressions. Two fraction principles of particular importance follow.

Fraction Principles

Let a, b, c, d, and k be real numbers with b, d, and $k \neq 0$.

Equality of fractions $\frac{a}{b} = \frac{c}{d}$ if and only if $ad = bc$

Fundamental principle $\frac{ak}{bk} = \frac{a}{b}$

The fundamental principle can be established from the criterion for the equality of two fractions, since $(ak)b$ is equal to $(bk)a$. As its name suggests, the fundamental principle is applied often, and two common uses of this principle are shown below.

Simplifying fractions $\frac{6}{8} = \frac{3 \cdot 2}{4 \cdot 2} = \frac{3}{4}$

↑ ↑

fundamental principle

↓ ↓

Adding fractions $\frac{1}{8} + \frac{3}{4} = \frac{1}{8} + \frac{3 \cdot 2}{4 \cdot 2} = \frac{1}{8} + \frac{6}{8} = \frac{7}{8}$

When simplifying fractions by the fundamental principle, it is important to recognize that we may divide out only nonzero factors of the numerator and the denominator. Thus, we have a general procedure for expressing a fraction in simplest form.

To Simplify Fractions

1. Factor completely the numerator and the denominator of the fraction.
2. Divide out nonzero factors that are common to the numerator and the denominator according to the fundamental principle.

Example 1 illustrates this procedure.

Example 1: Reducing Algebraic Fractions

Express in simplest form.

a. $\frac{x^2 - 3x - 4}{x^2 - 2x - 8}$ **b.** $\frac{(x^2 + 2) - (a^2 + 2)}{x - a}$

Solution: Factor and use the fundamental principle.

a. $\dfrac{x^2-3x-4}{x^2-2x-8}=\dfrac{(x+1)(x-4)}{(x+2)(x-4)}=\dfrac{x+1}{x+2},\quad x\neq 4$

b. $\dfrac{(x^2+2)-(a^2+2)}{x-a}=\dfrac{x^2-a^2}{x-a}=\dfrac{(x+a)(x-a)}{x-a}=x+a,\quad x\neq a$

Technology Link

In the previous section we discussed a graphing calculator method to test for equivalent expressions, based on the fact that equivalent expressions produce the same graph. A disadvantage of this method is that it may be hard to interpret the resulting display if you are unfamiliar with the types of graphs that are associated with certain equations. One way to overcome this difficulty is to graph the quotient of the expressions in question. This quotient will be 1 for equivalent expressions throughout the domain of this quotient function. To illustrate, the simplification in Example 1a may be checked by letting

$$y1=\frac{x^2-3x-4}{x^2-2x-8}$$
$$y2=\frac{x+1}{x+2}$$
$$y3=\frac{y1}{y2}.$$

Then graph only $y3$ as shown in Figure 3.6.

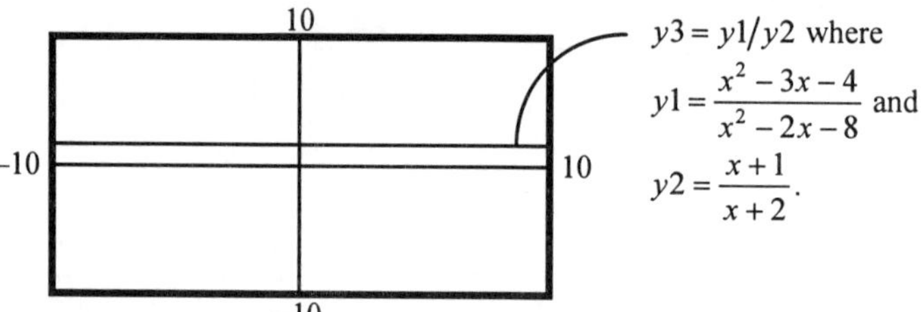

Figure 3.6

This graph, which is a horizontal line where y is fixed at 1, supports our results.

PROGRESS CHECK 1

Express in simplest form.

a. $\dfrac{4x}{8x^2-4x}$ **b.** $\dfrac{(x^2-5)-(c^2-5)}{x-c}$ ■

In Example 1 the restrictions $x\neq 4$ and $x\neq a$ are necessary because we may divide out only nonzero factors, since division by zero is undefined. With the understanding that such restrictions always apply, we will not continue to list them from this point on.

When working with fractions, it is important to keep in mind what happens in addition and subtraction if we change the order of the terms. In addition $a+b=b+a$. However, in subtraction $a-b=-(b-a)$. Thus,

$$\frac{a+b}{b+a}=1,\text{ while }\frac{a-b}{b-a}=\frac{-(b-a)}{b-a}=-1.$$

Although $a-b$ does not equal $b-a$, the expressions differ only by a factor of -1 and can be simplified accordingly.

Example 2: Simplification Involving Factors That Are Opposites

Express in simplest form $\frac{x^2-y^2}{y-x}$.

Solution: Since $y-x=-(x-y)$ we proceed as follows:

$$\frac{x^2-y^2}{y-x}=\frac{(x+y)(x-y)}{(-1)(x-y)}$$
$$=\frac{x+y}{-1} \text{ or } -x-y.$$

PROGRESS CHECK 2

Express in simplest form $\frac{2-3x}{9x^2-4}$. ∎

Multiplication

In arithmetic we know that the product of two or more fractions is the product of their numerators divided by the product of their denominators. In symbols, this principle is

$$\frac{a}{b}\cdot\frac{c}{d}=\frac{ac}{bd} \quad b,\, d\neq 0.$$

Similarly, we will use this procedure for multiplying algebraic fractions. To express products in simplest form, factoring and the fundamental principle are usually required.

Example 3: Multiplying Algebraic Fractions

Simplify $\frac{5x^2+15x}{4-x^2}\cdot\frac{12-6x}{(3+x)^2}$.

Solution: We factor, multiply, and then simplify as shown.

$$\frac{5x^2+15x}{4-x^2}\cdot\frac{12-6x}{(3+x)^2}=\frac{5x(x+3)}{(2+x)(2-x)}\cdot\frac{6(2-x)}{(3+x)^2}$$
$$=\frac{5\cdot 6x(x+3)(2-x)}{(2+x)(2-x)(3+x)(3+x)}$$
$$=\frac{30x}{(2+x)(3+x)}$$

It is not necessary to multiply out expressions like $(2+x)(3+x)$ for the final answer to these problems.

PROGRESS CHECK 3

Simplify $\frac{x-3}{x^2-x}\cdot\frac{x^2-2x+1}{x^2-9}$. ∎

Division

To divide two fractions, we must recall that division is defined in terms of multiplication. That is, to divide a by b, we multiply a by the reciprocal of b. Note that the reciprocal of a fraction can be found by inverting the fraction. For example, the reciprocal of 2/7 is 7/2. Thus, to divide two fractions, we invert the fraction by which we are dividing, to find its reciprocal, and then we multiply.

$$\frac{a}{b} \div \frac{c}{d} = \frac{a}{b} \cdot \frac{d}{c} \qquad b,\ d,\ \frac{c}{d} \neq 0$$

Example 4: Dividing Algebraic Fractions

Simplify $\frac{(y-2)^2}{5y} \div \frac{y^2-4}{15y}$.

Solution: We convert to multiplication and simplify.

$$\begin{aligned} \frac{(y-2)^2}{5y} \div \frac{y^2-4}{15y} &= \frac{(y-2)^2}{5y} \cdot \frac{15y}{y^2-4} \\ &= \frac{(y-2)(y-2)}{5y} \cdot \frac{3 \cdot 5 \cdot y}{(y+2)(y-2)} \\ &= \frac{(y-2)(y-2) \cdot 3 \cdot 5 \cdot y}{5 \cdot y(y+2)(y-2)} \\ &= \frac{3(y-2)}{y+2} \end{aligned}$$

PROGRESS CHECK 4

Simplify $\frac{(n+5)^2}{4n} \div \frac{n^2-25}{12n^2}$. ■

Addition and Subtraction

In arithmetic we now that the sum (or difference) of two or more fractions that have the same denominator is given by the sum (or difference) of the numerators divided by the common denominator. In symbols, the principle is

$$\frac{a}{b} \pm \frac{c}{b} = \frac{a \pm c}{b} \qquad b \neq 0.$$

We add or subtract algebraic fractions in a similar way, as shown in Example 5.

Example 5: Subtracting Fractions with the Same Denominators

Simplify $\frac{2x+1}{x-3} - \frac{x-2}{x-3}$.

Solution: The fractions have the same denominator, so

$$\begin{aligned} \frac{2x+1}{x-3} - \frac{x-2}{x-3} &= \frac{(2x+1)-(x-2)}{x-3} \\ &= \frac{2x+1-x+2}{x-3} \\ &= \frac{x+3}{x-3}. \end{aligned}$$

Caution: Be careful in this and other subtraction problems to remove the parentheses correctly in an expression like $-(x-2)$.

PROGRESS CHECK 5

Simplify $\frac{3x}{x+5} - \frac{2x-5}{x+5}$. ■

When fractions have different denominators, we can change them into equivalent fractions with the same denominator and then add them. In such cases, computation is simpler if we use the smallest possible common denominator, called the **least common denominator** or **LCD**. One way of finding the LCD follows.

To Find the LCD

1. Factor *completely* each denominator.
2. The LCD is the product of all the different factors, with each factor raised to the highest power to which it appears in any one factorization.

The procedure is illustrated in Example 6.

Example 6: Finding the LCD

Find the LCD of $\frac{2}{x^2-1}$ and $\frac{1}{x^2+2x+1}$.

Solution: First, factor each denominator completely.

$$x^2 - 1 = (x+1)(x-1)$$
$$x^2 + 2x + 1 = (x+1)(x+1) = (x+1)^2$$

The LCD will contain the factors $x+1$ and $x-1$. The highest power of $x+1$ is 2, and the highest power of $x-1$ is 1. Thus,

$$\text{LCD} = (x+1)^2(x-1).$$

PROGRESS CHECK 6

Find the LCD of $\frac{1}{x^2+5x}$ and $\frac{x}{3x+15}$. ■

We can now combine several principles of fractions and formulate a general procedure for adding or subtracting two or more fractions.

To Add or Subtract Fractions

1. Completely factor each denominator and find the LCD.
2. For each fraction, obtain an equivalent fraction by applying the fundamental principle and multiplying the numerator and the denominator of the fraction by the factors of the LCD that are not contained in the denominator of that fraction.
3. Add or subtract the numerators and divide this result by the common denominator.

Example 7: Subtracting Fractions with Different Denominators

Simplify $\frac{2a+3}{a^2+a} - \frac{a}{a^2-1}$.

Solution: The denominators a^2+a and a^2-1 factor as $a(a+1)$ and $(a+1)(a-1)$, respectively. Since the highest power of each different factor is 1, the LCD is $a(a+1)(a-1)$. Then,

$$\begin{aligned}\frac{2a+3}{a^2+a}-\frac{a}{a^2-1}&=\frac{2a+3}{a(a+1)}-\frac{a}{(a+1)(a-1)}\\&=\frac{(2a+3)(a-1)}{a(a+1)(a-1)}-\frac{a(a)}{(a+1)(a-1)(a)}\\&=\frac{2a^2+a-3-a^2}{a(a+1)(a-1)}\\&=\frac{a^2+a-3}{a(a+1)(a-1)}.\end{aligned}$$

PROGRESS CHECK 7

Simplify $\frac{b}{b^2+3b+2}+\frac{1}{b^2+4b+3}$. ■

We often need to add or subtract two fractions for which the LCD is simply the product of the denominators. This special situation is handled directly by the following formula.

$$\frac{a}{b}\pm\frac{c}{d}=\frac{ad\pm bc}{bd}\qquad b,\ d\neq 0$$

We can verify this property from our previous methods, since

$$\frac{a}{b}\pm\frac{c}{d}=\frac{a\cdot d}{b\cdot d}\pm\frac{c\cdot b}{d\cdot b}=\frac{ad\pm bc}{bd}.$$

Example 8: Using a Formula to Subtract Two Fractions

Simplify $\frac{x+h+1}{x+h}-\frac{x+1}{x}$.

Solution: The LCD is the product of the denominators, so we choose to apply the above formula, and then we simplify.

$$\begin{aligned}\frac{x+h+1}{x+h}-\frac{x+1}{x}&=\frac{(x+h+1)x-(x+h)(x+1)}{(x+h)x}\\&=\frac{x^2+hx+x-x^2-x-hx-h}{(x+h)x}\\&=\frac{-h}{(x+h)x}\end{aligned}$$

PROGRESS CHECK 8

Simplify $\frac{2y}{y-5}+\frac{3}{y}$. ■

Complex Fractions

A **complex fraction** is a fraction in which the numerator or the denominator, or both, involve fractions. A procedure for simplifying complex fractions follows.

To Simplify Complex Fractions

1. Find the LCD of all the fractions that appear in the numerator and the denominator of the complex fraction.
2. Multiply the numerator and the denominator of the complex fraction by this LCD and simplify your results.

Example 9: Simplifying Complex Fractions

Simplify.

a. $\dfrac{x^{-1}+y^{-1}}{(x/y)-(y/x)}$ **b.** $\left(\dfrac{1}{x+h}-\dfrac{1}{x}\right)\div h$

Solution:

a. $x^{-1}=1/x$, $y^{-1}=1/y$, and the LCD is xy.

$$\frac{\frac{1}{x}+\frac{1}{y}}{\frac{x}{y}-\frac{y}{x}}=\frac{xy\left(\frac{1}{x}+\frac{1}{y}\right)}{xy\left(\frac{x}{y}-\frac{y}{x}\right)}$$

$$=\frac{y+x}{x^2-y^2}$$

$$=\frac{y+x}{(x+y)(x-y)}$$

$$=\frac{1}{x-y}$$

b. The LCD is $x(x+h)$.

$$\frac{\frac{1}{x+h}-\frac{1}{x}}{h}=\frac{x(x+h)\left(\frac{1}{x+h}-\frac{1}{x}\right)}{x(x+h)\cdot h}$$

$$=\frac{x-(x+h)}{x(x+h)h}$$

$$=\frac{-h}{x(x+h)h}$$

$$=\frac{-1}{x(x+h)}$$

PROGRESS CHECK 9

Simplify.

a. $\dfrac{x^{-1}+1}{1-x^{-2}}$ **b.** $\left(\dfrac{1}{(x+h)^3}-\dfrac{1}{x^3}\right)\div h$ ■

Example 10: Finding an Average Speed

Solve the problem in the section introduction on page 260.

Solution: Let d represent the distance around the track. Since $d=rt$, the times required for the first and second laps are denoted by $d/180$ and $d/120$, respectively. Then,

$$\text{average speed}=\frac{\text{total distance}}{\text{total time}}=\frac{2d}{\frac{d}{180}+\frac{d}{120}}.$$

Multiplying each term in the numerator and the denominator by 360 gives

$$\frac{360(2d)}{360(\frac{d}{180}+\frac{d}{120})}=\frac{720d}{2d+3d}=\frac{720d}{5d}=144.$$

Thus, the average speed is 144 mi/hour, and the driver does not qualify for the finals. Not that the average speed is closer to 120 mi/hour because more *time* was needed for the second lap. An interesting follow-up to this question is given in the introduction to Section 4.2.

PROGRESS CHECK 10

A lawyer drives from her home to work at an average speed of 50 mi/hour. On the way home she takes the same route and averages 30 mi/hour. What is her average speed for the round trip? ■

Fractional expressions that involve the trigonometric functions are simplified in the same manner as algebraic fractions as illustrated in the next example.

Example 11: Simplifying Trigonometric Fractional Expressions

Simplify to a single fraction in simplest form.

a. $\frac{5}{3\sin^2\theta}+\frac{7}{6\sin\theta}$ **b.** $\frac{1+(1/\sin\theta)}{1/\sin^2\theta}$

Solution:

a. The LCD of $3\sin^2\theta$ and $6\sin\theta$ is $6\sin^2\theta$. Then,

$$\begin{aligned}\frac{5}{3\sin^2\theta}+\frac{7}{6\sin\theta}&=\frac{5(2)}{3\sin^2\theta(2)}+\frac{7(\sin\theta)}{6\sin\theta(\sin\theta)}\\&=\frac{10}{6\sin^2\theta}+\frac{7\sin\theta}{6\sin^2\theta}\\&=\frac{10+7\sin\theta}{6\sin^2\theta}.\end{aligned}$$

b. The LCD is $\sin^2\theta$. Then,

$$\begin{aligned}\frac{1+\frac{1}{\sin\theta}}{\frac{1}{\sin^2\theta}}&=\frac{\sin^2\theta(1+\frac{1}{\sin\theta})}{\sin^2\theta(\frac{1}{\sin^2\theta})}\\&=\frac{\sin^2\theta+\sin\theta}{1}\text{ or }\sin^2\theta+\sin\theta.\end{aligned}$$

PROGRESS CHECK 11

Simplify to a single fraction in simplest form.

a. $\frac{1}{\tan\theta+2}+\frac{1}{\tan\theta+4}$ **b.** $\frac{\sin\theta-1}{\sin\theta}\div\frac{\sin^2\theta-1}{\sin^2\theta}$ ■

EXPLORE 3.3

1. If $f(x)$ simplifies to $g(x)$, then $y=f(x)$ and $y=g(x)$ have the same graph whenever both functions are defined. Also, because the quotient of two equivalent expressions is identically 1 throughout the domain of the quotient function, the graph of the quotient $y=f(x)/g(x)$ will be a horizontal line at $y=1$.
   a. Check that $5x/8x$ simplifies to $5/8$ by showing that $y_1=f(x)=5x/8x$ and $y_2=g(x)=5/8$ have the same graph, except when $x=0$. Compare the values of $f(0)$ and $g(0)$ as given by the calculator. Find the domain of each function. Let $y_3=y_1/y_2$

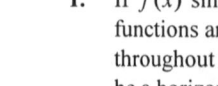

and check that the graph of y_3 is a horizontal line at $y=1$. What is the domain of the quotient function? What is the value of y_3 when $x=0$?

b. Check that $(x^2-3x-4)/(x^2-2x-8)$ simplifies to $(x+1)/(x+2)$ by showing that $y_1=f(x)=(x^2-3x-4)/(x^2-2x-8)$ and $y_2=g(x)=(x+1)/(x+2)$ have the same graph, except when $x=4$. Compare the value of $f(4)$ and $g(4)$. Compare the values of $f(-2)$ and $g(-2)$. Find the domain of each function. Let $y_3=y_1/y_2$ and check that the graph of y_3 is a horizontal line at $y=1$. What is the domain of the quotient function? What is the value of y_3 when $x=4$?, when $x=-2$?

2. Certain mistakes are common in simplification of fractional expressions. By showing that the two expressions produce different graphs, confirm that all the following "simplifications" are wrong. Even though the two expressions are not equivalent, there may some values of x for which they are equal. Find all such values.

a. $\dfrac{x^2+x+5}{5}$; x^2+x+1 **b.** $\dfrac{3x+1}{3}$; $x+1$ **c.** $\dfrac{(2x-4)(4x+6)}{2}$; $(x-2)(2x+3)$

3. Check by graphs whether each of these functions simplifies to 1, −1 , or neither.

a. $\dfrac{x+2}{2+x}$ **b.** $\dfrac{x-2}{2-x}$ **c.** $\dfrac{x^2+2x-3}{-x^2-2x+3}$ **d.** $\dfrac{x^2-4}{x^2+4}$

4. **a.** Find a fractional expression that has the same graph as $y=2x-5$, except when $x=7$. Support your answer by constructing a table of x and y values that includes 7 as one of the x values.

b. Find a fractional expression that has the same graph as $y=2x-5$, except when $x=7$ and when $x=8$. Support your answer by constructing a table of x and y values that includes 7 and 8 as two of the x values.

5. CAS utilities do simplifications of fractional expressions automatically. If you have such a utility, use it to confirm the simplifications in Explore Exercises 1 and 3.

EXERCISES 3.3

In Exercises 1–14, express each fraction in simplest form.

1. $\dfrac{4x-4}{11x-11}$ **2.** $\dfrac{y^2+2y}{y^2+3y}$ **3.** $\dfrac{4b}{12b^2+4b}$

4. $\dfrac{20z^2-5z}{10z}$ **5.** $\dfrac{y-x}{x-y}$ **6.** $\dfrac{-x+a}{x-a}$

7. $\dfrac{x-a}{ax(a-x)}$ **8.** $\dfrac{(x-4)^2}{4-x}$

9. $\dfrac{a^2-3a-4}{a^2-8a+16}$ **10.** $\dfrac{49-14y+y^2}{49-y^2}$

11. $\dfrac{6x^3-3x^2-30x}{2x^2-4x-16}$ **12.** $\dfrac{z^3-z^2-6z}{z^3+z^2-12z}$

13. $\dfrac{x^3-a^3}{x-a}$ **14.** $\dfrac{x^4-a^4}{x-a}$

In Exercises 15–30 express each product or quotient in simplest form.

15. $\dfrac{2x}{5y}\cdot\dfrac{5y^2}{8x^3}$ **16.** $\dfrac{14z}{10y}\cdot\dfrac{15y}{35z}$

17. $\dfrac{5a}{a^2-9}\cdot\dfrac{6a-18}{15a^2}$ **18.** $\dfrac{n^2-n-12}{n^2}\cdot\dfrac{n}{n-4}$

19. $\dfrac{x^2+3x-4}{x^2-3x-4}\cdot\dfrac{x^2-5x+4}{x^2+5x+4}$

20. $\dfrac{2a^2-7a+6}{a^2-a-6}\cdot\dfrac{5a-15}{a^2-4}$

21. $\dfrac{xy}{z}\div\dfrac{x^2y}{z}$ **22.** $\dfrac{zy^2}{2x}\div\dfrac{yz^2}{x}$

23. $\dfrac{n^2+n-2}{n-3}\div(n+2)$ **24.** $\dfrac{4x^2-9}{4x-9}\div(2x-3)$

25. $\dfrac{3b^2+6b-24}{b^2-7b+10} \div \dfrac{3b^2+4b}{b^3-5b^2}$

26. $\dfrac{x^2-7x+10}{5x-25} \div \dfrac{4-2x}{25-x^2}$

27. $\dfrac{x^4-1}{7x+7} \div \dfrac{x^3+x}{x^2-1}$ 28. $\dfrac{1-a}{x^2+x} \div \dfrac{a^2-1}{x^2-1}$

29. $\dfrac{x^3-y^3}{(x-y)^3} \div \dfrac{x^2+xy+y^2}{x^2-2xy+y^2}$

30. $\dfrac{2x^2-2ax+2a^2}{(ax)^3} \div \dfrac{a^3+x^3}{ax^3}$

In Exercises 31–38 find the LCD of the given fractions.

31. $\dfrac{3}{x^2-9}$ and $\dfrac{4}{x^2+6x+9}$

32. $\dfrac{-3}{4x^2-1}$ and $\dfrac{6}{4x^2-8x+3}$

33. $\dfrac{x}{x^2+2x}$ and $\dfrac{1}{x^2-4}$

34. $\dfrac{1}{3x^2-6x}$ and $\dfrac{x}{x^2-4x+4}$

35. $\dfrac{3x+2}{x^2-x-2}$ and $\dfrac{6x}{x^2+4x+3}$

36. $\dfrac{1-2x}{x^2+3x-4}$ and $\dfrac{3}{x^2+2x-8}$

37. $\dfrac{x^2+4}{3x^3-3x^2}$ and $\dfrac{2-x^2}{2x^3-4x^2+2x}$

38. $\dfrac{x^2-2}{2x^3-6x^2}$ and $\dfrac{x+1}{2x^2+8x^2}$

In Exercises 39–54 combine each expression into a single fraction in simplest form.

39. $\dfrac{4a}{xy}+\dfrac{2a}{xy}-\dfrac{3a}{xy}$ 40. $\dfrac{1}{a^2x^2}-\dfrac{5}{a^2x^2}+\dfrac{11}{a^2x^2}$

41. $\dfrac{3x+4}{x+2}-\dfrac{2x+5}{x+2}$ 42. $\dfrac{4y+1}{2y-6}-\dfrac{2y+7}{2y-6}$

43. $\dfrac{2n-3}{n-6}-\dfrac{n+1}{6-n}$ 44. $\dfrac{2x-5}{4-3x}+\dfrac{2-x}{3x-4}$

45. $\dfrac{4a}{5}+a$ 46. $z+\dfrac{1}{z}$ 47. $\dfrac{2}{k^3}-\dfrac{3}{k^2}+\dfrac{5}{k}$

48. $\dfrac{16}{a^2b}+\dfrac{1}{ab}-\dfrac{6}{ab^2}$ 49. $\dfrac{2x+3}{5x}-\dfrac{2x-1}{10x}+\dfrac{4}{x}$

50. $\dfrac{7}{n^2}-\dfrac{5n-2}{n}+6$ 51. $\dfrac{3}{2x-3}-\dfrac{2}{9-4x^2}$

52. $\dfrac{d}{b+d}-\dfrac{b}{b-d}$ 53. $\left(\dfrac{1}{x}+\dfrac{1}{y}\right)\div\left(\dfrac{x}{y}-\dfrac{y}{x}\right)$

54. $\left(4+\dfrac{1}{a-1}\right)\div\left(\dfrac{2}{a-1}+3\right)$

In Exercises 55–64 change the complex fraction to a fraction in simplest form.

55. $\dfrac{3+\frac{1}{4}}{4-\frac{2}{5}}$ 56. $\dfrac{\frac{5}{18}+\frac{11}{12}}{\frac{7}{6}-\frac{2}{9}}$ 57. $\dfrac{1+\frac{1}{x}}{1-\frac{1}{x}}$

58. $\dfrac{x+\frac{1}{a}}{\frac{2}{a}-x}$ 59. $\dfrac{n+1+\frac{2}{n}}{n-1-\frac{2}{n}}$ 60. $\dfrac{z+4+\frac{5}{z}}{z+1-\frac{2}{z}}$

61. $\left(\dfrac{y^2}{x^2}-1\right)\div\left(\dfrac{y}{x}-1\right)$ 62. $\left(k-\dfrac{1}{x^2}\right)\div\left(2-\dfrac{k}{x}\right)$

63. $\dfrac{1+\frac{3}{x+1}}{\frac{4}{x^2-1}}$ 64. $\dfrac{\frac{b}{b-a}-\frac{a}{b+a}}{\frac{b^2+a^2}{b^2-a^2}}$

In Exercises 65–82 perform the indicated operations and/or simply.

65. $\dfrac{x+y-z}{y^2-x^2+xz-yz}$ 66. $\dfrac{a^3-8}{a^3-2a^2+4a-8}$

67. $\dfrac{2(2+h)-3(2+h)}{2h(2+h)}$ 68. $\dfrac{(2+h)^2-4}{(2+h)-2}$

69. $\dfrac{(x^2-x)-(a^2-a)}{x-a}$ 70. $\dfrac{(x^2+2x)-(a^2+2a)}{x-a}$

71. $\dfrac{5}{y+1}-\dfrac{1}{y^2-1}+\dfrac{2}{y-1}$

72. $\dfrac{2a}{a^2-1}-\dfrac{a+1}{a-1}+7$ 73. $\dfrac{y}{y-x}-\dfrac{x}{x+y}+\dfrac{x^2+y^2}{x^2-y^2}$

74. $\dfrac{a}{a^2+ab}-\dfrac{b}{a^2-b^2}+\dfrac{a+b}{ab-b^2}$

75. $\dfrac{1}{h}\left(\dfrac{1}{x+h}-\dfrac{1}{x}\right)$ 76. $\dfrac{1}{h}\left(\dfrac{2}{(x+h)^2}-\dfrac{2}{x^2}\right)$

77. $\left[\dfrac{x+h}{x+h+1}-\dfrac{x}{x+1}\right]\div h$

78. $\left[\frac{3}{(x+h)^2} - \frac{3}{x^2}\right] \div h$

79. $\dfrac{\frac{(x+h)^2+1}{x+h} - \frac{x^2+1}{x}}{h}$

80. $\dfrac{\frac{x+h-4}{x+h+1} - \frac{x-4}{x+1}}{h}$

81. $\dfrac{x^{-1}+y^{-1}}{(xy)^{-1}}$

82. $\dfrac{xy^{-1}+x^{-1}y}{x^{-1}y-xy^{-1}}$

In Exercises 83-102 simplify the given expression.

83. $\dfrac{3\sin\theta+\sin\theta\cos\theta}{\sin\theta}$

84. $\dfrac{5\tan\theta\cos\theta}{\tan\theta+\tan\theta\sin\theta}$

85. $\dfrac{4\sin^2\theta-4\cos^2\theta}{4\sin\theta-4\cos\theta}$

86. $\dfrac{\cos^2\theta+2\sin\theta\cos\theta+\sin^2\theta}{\cos\theta(\sin\theta+\cos\theta)}$

87. $\dfrac{3\tan\theta}{5}\cdot\dfrac{3}{5\tan\theta}$

88. $\dfrac{\tan\theta(\sin\theta-4)}{5(4+\cos\theta)}\cdot\dfrac{10\cos\theta+40}{3\tan\theta}$

89. $\dfrac{\sin\theta-1}{1+\sin\theta}\cdot\dfrac{\sin\theta}{1-\sin\theta}$

90. $\dfrac{\tan^2\theta-9}{3-\tan\theta}\cdot\dfrac{3}{3+\tan\theta}$

91. $\dfrac{\sin\theta}{\tan\theta}\div\sin\theta$

92. $\dfrac{\tan\theta}{\cos\theta}\div\dfrac{2}{\cos\theta}$

93. $\dfrac{\sin\theta-1}{\cos\theta}\div\dfrac{1-\sin\theta}{3\cos\theta}$

94. $\dfrac{\cos^2\theta-\sin^2\theta}{\tan^2\theta-4}\div\dfrac{\sin\theta-\cos\theta}{\tan\theta+2}$

95. $\dfrac{3\sin\theta}{1+3\sin\theta}+\dfrac{1}{1+3\sin\theta}$

96. $\dfrac{1+\sin\theta}{2-\sin\theta}+\dfrac{1-2\sin\theta}{2-\sin\theta}$

97. $\dfrac{3}{2\sin\theta-1}-\dfrac{1}{\sin\theta+2}$

98. $\dfrac{8}{3\sin\theta-6}-\dfrac{-1}{3\sin\theta+12}$

99. $\dfrac{\frac{1}{\sin\theta}+2}{3-\frac{1}{\sin\theta}}$

100. $\dfrac{\frac{1}{\sin\theta}+\frac{1}{\cos\theta}}{\frac{1}{\sin\theta}-\frac{1}{\cos\theta}}$

101. $\dfrac{1}{\frac{1}{\tan\theta}+2}$

102. $\dfrac{1+\tan^2\theta}{1+\frac{2}{\tan\theta}}$

103. In photography the focal length of a lens is given by $f=\dfrac{1}{\frac{1}{d}+\frac{1}{a}}$, where d is the distance from some object to the lens, and a is the distance of its image from the lens. Simplify the complex fraction in this equation.

104. When resistors R_1 and R_2 are connected in parallel, their combined resistance is given by

$$\frac{1}{\frac{1}{R_1}+\frac{1}{R_2}}.$$

Change this complex fraction to a fraction in simplest form.

105. The odds *against* an event occurring equal the probability that it will *not* occur divided by the probability that it *will* occur. Find the odds against event A if the probability that A will occur is y/m, and the probability that A will not occur is $1-(y/m)$.

106. The odds *favor* of an event equal the probability that it will occur divided by the probability that it will *not* occur. Find the odds in favor of event A if the probability that A will occur is a/n, and the probability that A will *not* occur is $1-(a/n)$.

107. A person drives from home to work at an average speed of 30 mi/hour, and returns on the same route at an average speed of 20 mi/hour. What is the average speed for the whole trip?

108. A person drives from home to the beach at an average speed of 35 mi/hour, and returns on the same route at an average speed of 55 mi/hour. What is the average speed for the whole trip?

109. We can generalize the idea in Exercises 87 and 88. If we let s_g represent the speed going, and s_b represent the speed coming back, then the following formula can be used to find the average speed for the round trip.

$$s_{avg}=\frac{2}{\frac{1}{s_g}+\frac{1}{s_b}}$$

Change this complex fraction to a fraction in simplest form.

110. a. The harmonic mean of two numbers a and b is given by

$$\frac{2}{\frac{1}{a}+\frac{1}{b}}.$$

Simplify this expression.

b. Verify that the average speed in Exercise 107 is the harmonic mean of 30 and 20.

c. A teacher invests \$6,000 in a growth-oriented mutual fund at \$10 a share. Six months later another \$6,000 is invested in the same fund at \$15 a share. Find the average per-share purchase price and verify that it is the harmonic mean of 10 and 15.

91. The "masterpiece of ancient geometry" is the formula for the volume of a truncated square pyramid (see figure). We know the Egyptians had the idea about 1850 BC, because it was used in Problem 14 of a famous scroll, called the Moscow Papyrus, which is a major source of current knowledge about ancient Egyptian mathematics.

$$V = \frac{hb^2 + (b^2 - a^2)\frac{ah}{b-a}}{3}$$

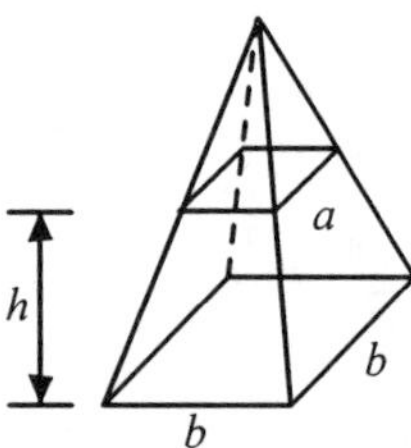

Write an alternative formula by simplifying and factoring the right side of this equation. [*Source: The History of Mathematics*, Burton, McGraw Hill Publishers.]

92. Simplify the formula in the previous exercise for a case in which the pyramid is cut half-way up, so that $a = \frac{b}{2}$.

THINK ABOUT IT 3.3

1. Some people say informally that you reduce $5x/6x$ to $5/6$ by "canceling" the x's. They also say that $5+2x-2x$ equals 5 because $2x$ and $-2x$ "cancel out." Explain *in terms of arithmetic operations* why these two uses of the word *cancel* are not the same. (Consequently, many people prefer to avoid the word together.)

2. A common student error is contained in the following "simplification." Find the mistake and discuss how to avoid such an error.

$$\frac{x^2}{x-1} - \frac{2x+1}{x-1} = \frac{x^2-2x+1}{x-1} = \frac{(x-1)^2}{x-1} = x-1, \text{ if } x \neq 1.$$

3. The Egyptians in ancient times expressed most fractional quantities as sums of unit fractions, which are fractions with numerators equal to 1. For example, the answer given in the Rhind papyrus (1650 B.C.) to the question of how to divide 6 loaves among 10 men is $1/2 + 1/10$ (but we would write $6/10$ or $3/5$).

a. Show that $(x+y)/xy = 1/x + 1/y$.

b. For the equation in part **a**, let $x = 3$ and $y = 5$ to express $8/15$ as a sum of two unit fractions.

c. Use the equality given in part **a** to find two unit fractions whose sum is $2/7$.

4. Consider these expressions.

$$\frac{1}{2+\frac{1}{2}}, \quad \frac{1}{2+\frac{1}{2+\frac{1}{2}}}, \quad \frac{1}{2+\frac{1}{2+\frac{1}{2+\frac{1}{2}}}}$$

a. Show that they simplify to $2/5$, $5/12$, and $12/29$, respectively.

b. If you keep going like this, the expression is called a *continued fraction.* There is a pattern to the fractions in part **a**. The next fraction is $29/70$. What comes after that?

c. If you express each fraction as a decimal and add 1 to each, you may see that the answers are getting closer and closer to $\sqrt{2}$. Find an approximation to $\sqrt{2}$ on a calculator, and see how these answers compare.

5. Simplify $\left(1-\frac{1}{n}\right)\left(1-\frac{1}{n+1}\right)\left(1-\frac{1}{n+2}\right)\cdots\left(1-\frac{1}{n+8}\right)\left(1-\frac{1}{n+9}\right)$.

● ● ●

3.4 Radicals and Rational Exponents

Photo Courtesy of ***Reuters/Tass*** *of Archive Photos, New York*

A dangerous amount of radioactive strontium 90 was released in the Chernobyl accident in Ukraine in 1986. The formula

$$A = A_0 \cdot 2.718^{-0.02424t}$$

approximates how much of an original quantity A_0 of this strontium remains after t years. (The number 2.718 is an approximation to a constant called e that will be considered in Section 7.6.) What percent of the initial amount of the strontium 90 that was released will remain after 50 years? Express the answer to the nearest tenth of a percent.
(See Example 4.)

Objectives

1. Find principal nth roots.
2. Evaluate expressions containing rational exponents.
3. Simplify certain expressions involving rational exponents by using exponent properties.
4. Simplify certain expressions involving rational exponents by factoring out the smallest power of the variable from each term.

In many applications we need to reverse the process of raising a number to a power, and we ask

$$\text{if } x^n = a, \text{ what is } x?$$

To answer this question, it is useful to define an nth root.

Definition of *n*th Root

> For any positive integer n, the number b is an nth root of a if $b^n = a$.

In other words, if raising b to the nth power results in a, then b is an nth root of a. It is important to note the following ideas concerning the root of a number.

1. In the expression $\sqrt[n]{a}$, which is called the **principal *n*th root of *a***, we say that $\sqrt{}$ is the **radical sign**, a is the **radicand**, and n is the **index** of the radical. The index is usually omitted from the square root radical, and $\sqrt[3]{a}$ is called the **cube root of *a***.
2. There are two square roots for a positive number such as 4 since $2^2 = 4$ and $(-2)^2 = 4$. To avoid ambiguity, we define the **principal square root** of a positive number to be its *positive* square root. Thus, the principal square root of 4 is 2, and not –2. The radical sign, $\sqrt{}$, is used to symbolize the principal square root, so $\sqrt{4} = 2$. To symbolize the negative square

root of 4 (which is –2), we use $-\sqrt{4}$. In general, when n is even, $\sqrt[n]{a}$ means the nonnegative nth root of a.

3. No real number is the square root of a negative number such as –4, since the product of two equal real numbers is never negative. In general, when n is even, the nth root of a negative number does not exist in the set of real numbers.
4. The square root of any positive number that is not a perfect square is an irrational number. Consequently, if we wish to express the square root as a decimal, we can only *approximate* the number to some desired number of significant digits. In most cases we will leave these numbers in radical form.

Example 1: Finding Roots of a Real Number

Find each root that is a real number.

a. $\sqrt{64}$ **b.** $\sqrt[3]{27}$ **c.** $\sqrt[3]{-27}$ **d.** $\sqrt[4]{16}$ **e.** $\sqrt[4]{-16}$

Solution:

a. $\sqrt{64}$ denotes the positive square root of 64. Since $8^2 = 64$, $\sqrt{64} = 8$.
b. $\sqrt[3]{27} = 3$, because $3^3 = 27$. Read $\sqrt[3]{27} = 3$ as "the cube root of 27 is 3."
c. $\sqrt[3]{-27} = -3$, because $(-3)^3 = -27$. Note that the cube roots of negative numbers are negative numbers, and in general odd roots of negative numbers are negative.
d. $\sqrt[4]{16}$ denotes the positive fourth root of 16. Since $2^4 = 16$, $\sqrt[4]{16} = 2$.
e. $\sqrt[4]{-16}$ is not a real number. In general, when n is even and a is negative, then $\sqrt[n]{a}$ is not a real number.

Technology Link

Graphing calculators routinely have a Square Root feature and an nth Root feature, and sometimes have a Cube Root feature. You should use these features to redo the problems in Example 1 and compare your results to the text's answers. In Example 1e a complex number result will appear, instead of an error message, if your calculator has a complex number capability. Such numbers will be discussed in detail in Section 3.6.

PROGRESS CHECK 1

Find each root that is a real number.

a. $\sqrt{100}$ **b.** $\sqrt{-100}$ **c.** $\sqrt[3]{-125}$ **d.** $\sqrt[3]{125}$ **e.** $\sqrt[4]{81}$ ■

It is a goal of algebra to be able to use any real number as an exponent, and we can use our work with radicals to give meaning to expressions with rational number exponents such as

$$16^{1/2},\ 16^{-3/4},\ \text{and}\ 8^{2/3}.$$

Recall that the power-to-a-power property for integral exponents is

$$\left(a^m\right)^n = a^{mn}.$$

If this law is to hold for rational exponents, then consider

$$\left(16^{1/2}\right)^2 = 16^{(1/2)2} = 16^1 = 16.$$

We see that squaring $16^{1/2}$ results in 16, so $16^{1/2}$ is a square root of 16. We choose to define $16^{1/2}$ as the positive square root of 16, so

$$16^{1/2} = \sqrt{16} = 4.$$

In general, our previous laws of exponents may be extended by defining $a^{1/n}$ as the principal nth root of a.

Definition of $a^{1/n}$

> If n is a positive integer and $\sqrt[n]{a}$ is a real number, then
>
> $$a^{1/n} = \sqrt[n]{a}.$$

Visual support for this definition in the case when $n=5$ is shown in Figure 3.7, where it appears that $y = x^{1/5}$ and $y = \sqrt[5]{x}$ have identical graphs.

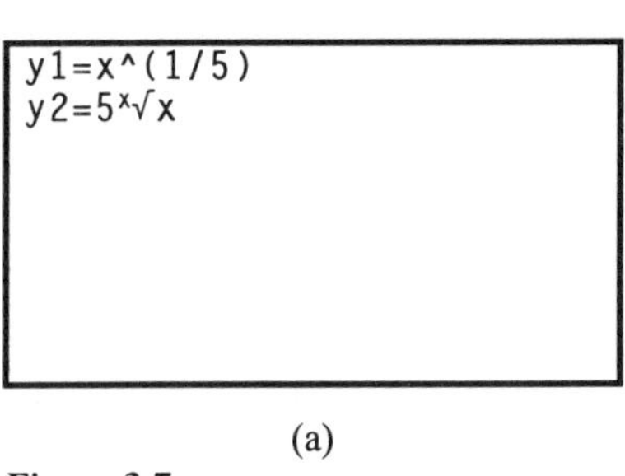

(a)

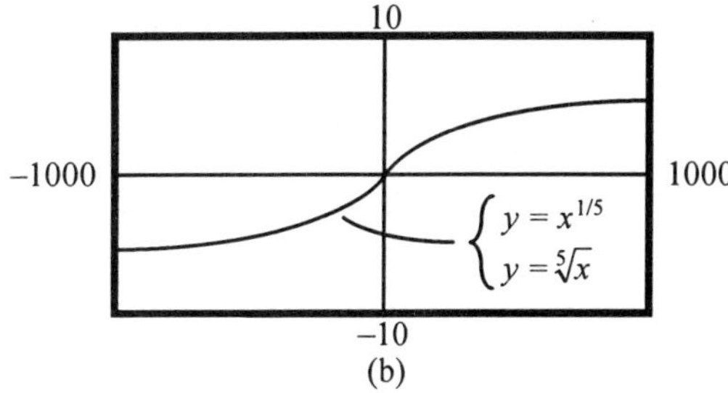

(b)

Figure 3.7

Example 2: Using $a^{1/n} = \sqrt[n]{a}$

Evaluate each expression.

a. $9^{1/2}$ **b.** $8^{1/3}$ **c.** $(-8)^{1/3}$ **d.** $-8^{1/3}$ **e.** $16^{-1/4}$

Solution: Convert to radical form and then simplify

a. $9^{1/2} = \sqrt{9} = 3$ **b.** $8^{1/3} = \sqrt[3]{8} = 2$ **c.** $(-8)^{1/3} = \sqrt[3]{-8} = -2$

d. $-8^{1/3} = -\sqrt[3]{8} = -2$ **e.** $16^{-1/4} = \dfrac{1}{16^{1/4}} = \dfrac{1}{\sqrt[4]{16}} = \dfrac{1}{2}$

PROGRESS CHECK 2

Evaluate each expression.

a. $16^{1/2}$ **b.** $(-16)^{1/2}$ **c.** $-16^{1/2}$ **d.** $64^{-1/3}$ **e.** $32^{1/5}$ ■

The power-to-power property, $a^{mn} = (a^m)^n$, is the basis for extending the definition of rational exponent to the case when the numerator is not 1. For instance, to evaluate $8^{2/3}$, we reason that

$$8^{2/3} = (8^{1/3})^2 = (\sqrt[3]{8})^2 = 2^2 = 4,$$

or

$$8^{2/3} = (8^2)^{1/3} = 64^{1/3} = \sqrt[3]{64} = 4.$$

By both methods $8^{2/3} = 4$, and this example suggests how a rational number exponent should be defined.

Rational Exponent

If m and n are integers with $n > 0$, and if m/n represents a reduced fraction such that $a^{1/n}$ represents a real number, then

$$a^{m/n} = \left(a^{1/n}\right)^m = \left(a^m\right)^{1/n}$$

or equivalently

$$a^{m/n} = \left(\sqrt[n]{a}\right)^m = \sqrt[n]{a^m}.$$

Notice that in this definition we avoid the situation where the base is negative and the exponent is a reduced fraction in which n is an even number. This is done because such a condition deals with numbers that are not real numbers and, in such a case, the laws of exponents do not necessarily hold.

Example 3: Using $a^{m/n} = \left(a^{1/n}\right)^m$

Evaluate each expression.

a. $8^{2/3}$ **b.** $49^{3/2}$ **c.** $(-32)^{4/5}$ **d.** $125^{-2/3}$

Solution: Use $a^{m/n} = \left(a^{1/n}\right)^m$, since each root is a rational number.

a. $8^{2/3} = (8^{1/3})^2 = (\sqrt[3]{8})^2 = (2)^2 = 4$

b. $49^{3/2} = (49^{1/2})^3 = (\sqrt{49})^3 = (7)^3 = 343$

c. $(-32)^{4/5} = [(-32)^{1/5}]^4 = (\sqrt[5]{-32})^4 = (-2)^4 = 16$

d. $125^{-2/3} = \dfrac{1}{125^{2/3}} = \dfrac{1}{(125^{1/3})^2} = \dfrac{1}{(\sqrt[3]{125})^2} = \dfrac{1}{(5)^2} = \dfrac{1}{25}$

Technology Link

Various methods may be used to evaluate expressions of the form $a^{m/n}$ with a graphing calculator. If $a > 0$, then the most straightforward general method is to use the Power key and enclose the exponent within parentheses. For example, an evaluation of $8^{5/3}$ is shown in the first two lines of the display in Figure 3.8.

```
8^(5/3)
                    32
((-8)^(1/3))^5
                   -32
```

Figure 3.8

If $a < 0$, then you may need to use one of the forms for $a^{m/n}$ that was given in the rational exponent definition. For instance, an evaluation of $(-8)^{5/3}$ that uses $a^{m/n} = \left(a^{1/n}\right)^m$ is shown in lines 3 and 4 of the display in Figure 3.8. You may also need a form of the definition of $x^{m/n}$ to graph $y = x^{m/n}$ for a viewing window that involves negative values for x.

PROGRESS CHECK 3

Evaluate each expression

a. $9^{3/2}$ **b.** $(-27)^{2/3}$ **c.** $(-32)^{-3/5}$ **d.** $16^{5/4}$ ■

Example 4: Radioactive Contamination from the Chernobyl Accident

Solve the problem in the section introduction on page 273.

Solution: The percent of the initial amount of released strontium 90 that will still remain after t years is given by A/A_0, so we begin by dividing both sides of the given formula by A_0.

$$A = A_0 \cdot 2.718^{-0.02424t}$$

$$\frac{A}{A_0} = 2.718^{-0.02424t}$$

Then replacing t by 50 and evaluating the resulting expression by calculator yields

$$\frac{A}{A_0} = 2.718^{(-0.02424 \cdot 50)}$$

$$\approx 0.29763888.$$

To the nearest tenth of a percent, about 29.8 percent of the strontium 90 that was released will remain after 50 years (or in 2036).

PROGRESS CHECK 4

To the nearest tenth of a percent, what percent of the strontium 90 that was released at the Chernobyl accident will remain after 75 years. ■

Because all previous laws of exponents hold for rational exponents, we may use these extended properties to simplify expressions with rational exponents.

Example 5: Using Exponent Properties with Rational Exponents

Simplify by performing the indicated operations and writing your results with only positive exponents. Also, express the result in radical form.

a. $6^{7/5} \cdot 6^{8/5}$ **b.** $\left(a^8\right)^{-3/16}$ **c.** $\dfrac{\sqrt{2}}{\sqrt[3]{2}}$ **d.** $\dfrac{2^{-1/2} \cdot 3x^{1/3}}{2 \cdot 3^{1/4} x}$ **e.** $\left(\dfrac{4x^{-1}y^{1/5}}{32x^5y^0}\right)^{1/3}$

Solution:

a. $6^{7/5} \cdot 6^{8/5} = 6^{7/5+8/5} = 6^{15/5} = 6^3 = 216$

b. $\left(a^8\right)^{-3/16} = a^{8(-3/16)} = a^{-3/2} = \dfrac{1}{a^{3/2}}$ or $\dfrac{1}{\sqrt{a^3}}$

c. $\dfrac{\sqrt{2}}{\sqrt[3]{2}} = \dfrac{2^{1/2}}{2^{1/3}} = 2^{1/2-1/3} = 2^{1/6}$ or $\sqrt[6]{2}$

d. $\dfrac{2^{-1/2} \cdot 3x^{1/3}}{2 \cdot 3^{1/4} x} = \dfrac{3^{1-1/4}}{2^{1-(-1/2)} x^{1-(1/3)}} = \dfrac{3^{3/4}}{2^{3/2} x^{2/3}}$ or $\dfrac{\sqrt[4]{3^3}}{\sqrt{2^3}\,\sqrt[3]{x^2}}$

e. $\left(\dfrac{4x^{-1}y^{1/5}}{32x^5y^0}\right)^{1/3} = \left(\dfrac{y^{1/5}}{8x^{5-(-1)} \cdot 1}\right)^{1/3} = \left(\dfrac{y^{1/5}}{8x^6}\right)^{1/3} = \dfrac{y^{(1/5)(1/3)}}{8^{1(1/3)} x^{6(1/3)}} = \dfrac{y^{1/15}}{2x^2}$ or $\dfrac{\sqrt[15]{y}}{2x^2}$

PROGRESS CHECK 5

Simplify by performing the indicated operations and writing your results with only positive exponents. Also, express the result in radical form.

a. $7^{1/4} \cdot 7^{1/4}$ **b.** $\left(8x^6\right)^{-1/3}$ **c.** $\dfrac{\sqrt{5}}{\sqrt[4]{5}}$ **d.** $\dfrac{x^{-1}y}{x^{4/5}y^{-3/2}}$ **e.** $\left(\dfrac{x^{-5/2}y^{3/4}}{xy^{-1/4}}\right)^{1/3}$ ■

Expressions that involve rational exponents sometime need to be simplified in calculus. The next two examples illustrate simplifications of this type.

Example 6: Using the Fundamental Fraction Principle with Rational Exponents

Simplify by performing the indicated operations and writing your results with only positive exponents. Also, express the result in radical form.

a. $\dfrac{x^{1/2}\cdot 2-(2x-1)\cdot \frac{1}{2}x^{-1/2}}{x}$ **b.** $\dfrac{1}{2}x^{-1/2}-x^{-3/2}$

Solution:

a. The main problem is how to take care of the expression $(1/2)x^{-1/2}$ or $1/(2x^{1/2})$. One way to simplify this expression is to multiply it by $2x^{1/2}$, since the resulting product is 1. Thus, in this case we will multiply both the numerator and the denominator of the given fraction by $2x^{1/2}$.

$$\left(\frac{2x^{1/2}}{2x^{1/2}}\right)\left(\frac{x^{1/2}\cdot 2-(2x-1)\cdot \frac{1}{2}x^{-1/2}}{x}\right)=\frac{4x-(2x-1)(1)}{2x^{3/2}}$$
$$=\frac{2x+1}{2x^{3/2}} \text{ or } \frac{2x+1}{2\sqrt{x^3}}$$

b.
$$\frac{1}{2}x^{-1/2}-x^{-3/2}=\frac{1}{2x^{1/2}}-\frac{1}{x^{3/2}}$$
$$=\frac{1}{2x^{1/2}}\left(\frac{x}{x}\right)-\frac{1}{x^{3/2}}\left(\frac{2}{2}\right)$$
$$=\frac{x}{2x^{3/2}}-\frac{2}{2x^{3/2}}$$
$$=\frac{x-2}{2x^{3/2}} \text{ or } \frac{x-2}{2\sqrt{x^3}}$$

PROGRESS CHECK 6

Simplify by performing the indicated operations and writing your results with only positive exponents. Also, express the result in radical form.

a. $\dfrac{x^{1/2}\cdot 3-(3x+1)\cdot \frac{1}{2}x^{-1/2}}{x}$ **b.** $\dfrac{3}{2}x^{1/2}+\dfrac{1}{2}x^{-1/2}$ ■

Expressions with negative exponents are often simplified by factoring out from each term the smallest power of the variable along with other common numerical factors, as discussed in Section 3.1. The next example gives a simple illustration of this technique in part **a** and then provides an alternative solution to Example 6b by using this method in part **b**.

Example 7: Using Factoring with Rational Exponents

Simplify each expression by factoring and then writing the result with only positive exponents

a. $10x^{-4}+15x^{-2}$ **b.** $\dfrac{1}{2}x^{-1/2}-x^{-3/2}$

Solution:

a. The smallest power of x is x^{-4}, and the GCF of 10 and 15 is 5. Thus, we may simplify by factoring out $5x^{-4}$, as follows.

$$\begin{aligned}10x^{-4} + 15x^{-2} &= 5x^{-4} \cdot 2 + 5x^{-4} \cdot 3x^2 \\ &= 5x^{-4}\left(2 + 3x^2\right) \\ &= \frac{5\left(2 + 3x^2\right)}{x^4}\end{aligned}$$

b. In this case $x^{-3/2}$ is the smallest power of x, so

$$\begin{aligned}\frac{1}{2}x^{-1/2} - x^{-3/2} &= x^{-3/2} \cdot \frac{1}{2}x - x^{-3/2} \cdot 1 \\ &= x^{-3/2}\left(\frac{1}{2}x - 1\right) \\ &= \frac{\frac{1}{2}x - 1}{x^{3/2}} \\ &= \frac{x-2}{2x^{3/2}} \text{ or } \frac{x-2}{2\sqrt{x^3}}.\end{aligned}$$

PROGRESS CHECK 7

Simplify each expression by factoring and then writing the result with only positive exponents.

a. $18x^{-1} + 24x^{-3}$ **b.** $\frac{3}{2}x^{1/2} + \frac{1}{2}x^{-1/2}$ ■

EXPLORE 3.4

Typically, in evaluating $a^{b/c}$, a calculator will first divide b by c and then use a preprogrammed general procedure (involving logarithms) to raise the base to a noninteger power. It will not treat b and c as separate integers unless you use parentheses appropriately. The evaluation of rational exponents on a calculator must be done thoughtfully, because the built-in general procedure may not behave the way you expect it to, particularly for negative bases. On some calculators imaginary numbers result (because logarithms of negative numbers are imaginary), or an error message may appear.

1. The following expressions all equal 4. Try them on your calculator and see if any of them produce an error message or cause the machine to enter complex mode. If you get an imaginary number, raise it to the $3/2$ power to see if you get back to -8.

   a. $(-8)^{2/3}$ **b.** $\left[(-8)^2\right]^{1/3}$ **c.** $\left[(-8)^{1/3}\right]^2$

2. Use a grapher to graph $y = x^{(2/3)}$ and $y = \left(x^2\right)^{1,3}$. Explain why these graphs may not be the same. Which graph is the correct graph of $y = x^{2/3}$?
3. In this section of the text we define rational exponents so that if b is very close to c, then a^b is very close to a^c.
   a. Explore this characteristic of rational exponents by evaluating and comparing these 3 expressions: $4^{0.49}$, $4^{1/2}$, $4^{0.51}$.
   b. Graph $y = 4^x$, and trace it as x goes from 0.49 to 0.51. Find the value of y when $x = 0.49$, 0.50, and 0.51.
4. Because the graph of $y = x^2$ is a parabola, we expect the graph of $y = x^{2.1}$ to be somewhat similar. Compare the two graphs and describe how they differ. For what positive integer value of x is the graph of $y = x^{2.1}$ more than 100 units above the other graph for the first time?

EXERCISES 3.4

In Exercises 1–8 find each root that is a real number.

1. $\sqrt{49}$ 2. $\sqrt{121}$ 3. $\sqrt[3]{8}$

4. $\sqrt[3]{216}$ 5. $\sqrt{-16}$ 6. $\sqrt{-144}$

7. $\sqrt[5]{-32}$ 8. $\sqrt[3]{-8}$

In Exercises 9–34 evaluate each expression

9. $25^{1/2}$ 10. $81^{1/2}$ 11. $(-27)^{1/3}$

12. $(-64)^{1/3}$ 13. $16^{1/4}$ 14. $81^{1/4}$

15. $\left(\dfrac{9}{25}\right)^{-1/2}$ 16. $\left(\dfrac{49}{9}\right)^{-1/2}$ 17. $(0.04)^{-1/2}$

18. $(0.027)^{-1/3}$ 19. $-4^{1/2}$

20. $-16^{1/4}$ 21. $9^{3/2}$ 22. $4^{5/2}$

23. $8^{5/3}$ 24. $16^{3/4}$ 25. $(-27)^{2/3}$

26. $(-27)^{-2/3}$ 27. $\left(\dfrac{4}{9}\right)^{-3/2}$ 28. $\left(\dfrac{27}{8}\right)^{-2/3}$

29. $2^{1/2}\cdot 2^{3/2}$ 30. $5^{1/2}\cdot 5^{1/2}$

31. $(-3)^{5/3}\cdot(-3)^{4/3}$ 32. $(-4)^{7/5}\cdot(-4)^{8/5}$

33. $\dfrac{5^{1/2}}{5^{3/2}}$ 34. $\dfrac{6^{1/5}\cdot 6^{3/5}}{6^{-1/5}}$

In Exercises 35–74 simplify each expression by performing the indicated operations and writing your result with only positive exponents. Also show your result in radical form. (**Note:** When the denominator in a fractional exponent is even, assume that the base is a positive number.)

35. $x^{1/5}\cdot x^{2/5}$ 36. $y^{-1/3}\cdot y^{2/3}$ 37. $\dfrac{b^{2/3}}{b^{1/3}}$

38. $\dfrac{a^{1/5}}{a^{2/5}}$ 39. $\dfrac{y^{1/4}}{y}$ 40. $\dfrac{x}{x^{1/2}}$

41. $x^{4/3}\cdot x^{-1}$ 42. $x^{1/2}\cdot x^{2/3}$ 43. $\dfrac{x^{-1/3}}{x^{5/6}}$

44. $\dfrac{x^{-1/4}}{x^{1/2}}$ 45. $\left(8x^3y^6\right)^{1/3}$ 46. $\left(3^{1/2}x^{3/4}\right)^{-2}$

47. $\left(4a^{-2}b^6\right)^{-1/2}$ 48. $\left(2x^{1/3}y^{5/6}\right)^6$

49. $\dfrac{x^{1/2}yz^{1/3}}{x^0y^{-1}z}$ 50. $\dfrac{2x^{4/5}y^{-2/3}}{6^0x^{-1}y}$

51. $(x+1)^{-1/2}(x+1)^{3/2}$ 52. $(x+y)^{5/3}(x+y)^{-2/3}$

53. $\dfrac{x-2}{(x-2)^{1/2}}$ 54. $\dfrac{y+3}{(y+3)^{-1/2}}$

55. $(a-b)^{1/2}\div(a-b)^{3/4}$ 56. $(x-a)^{-2/3}\div(x-a)^{1/4}$

57. $3u^{1/2}\left(2u^{1/2}+u^4\right)$ 58. $-a^{2/3}\left(a^3-a^{1/2}\right)$

59. $\dfrac{\left(4x^2\right)^{1/2}-x^{1/2}}{(16x)^{1/2}}$ 60. $\left(\dfrac{b^{1/3}}{b^{4/3}+b^{7/3}}\right)^{-1}$

61. $\left(\dfrac{x^{2/3}y^{-3/2}}{x^{-1/3}y}\right)^{-1}$ 62. $\left(\dfrac{27x^{-3}}{64y^{-3}}\right)^{-1/3}$

63. $\left(\dfrac{49x^{-4}y^6}{25x^{-2}y^{-10}}\right)^{-3/2}$ 64. $\left(\dfrac{x^{1/4}y^{-2/3}z^0}{x^{3/4}y^{-2/3}z^{1/2}}\right)^{-3/2}$

65. $\dfrac{3x^{1/2}y^{-2/3}}{5y^{-1}}\cdot\dfrac{20y^{1/4}}{27x^{-2}}$ 66. $\dfrac{7^{-1}x^{1/4}}{3^{-2}y^{-1}}\div\dfrac{2\left(3x^{-1/2}\right)^2}{(49y)^0}$

67. $x^{-1/2}-\dfrac{1}{2}x^{-3/2}$ 68. $\dfrac{2}{3}x^{-1/3}+\dfrac{1}{3}x^{-4/3}$

69. $\dfrac{x^{1/3}+x^{-2/3}}{x^{1/3}-x^{-2/3}}$ 70. $\dfrac{x^{1/2}}{2x^{-3/2}+3x^{-1/2}}$

71. $(x+2)\left(-\dfrac{1}{2}x^{-3/2}\right)+x^{-1/2}\cdot 1$

72. $x^{1/2}\cdot\left[-(x+2)^{-2}\right]+(x+2)^{-1}\cdot\dfrac{1}{2}x^{-1/2}$

73. $\dfrac{x^{1/2}\cdot 1-(x+2)\cdot\frac{1}{2}x^{-1/2}}{x}$

74. $\dfrac{1+\frac{1}{2}\left(1+x^2\right)^{-1/2}(2x)}{x+\left(1+x^2\right)^{1/2}}$

In Exercises 75–84 simplify each expression by factoring out the smallest power of the variable from each term and writing your result with only positive exponents.

75. $x^{-5}+x^{-2}$ 76. $x^{-8}+3x^{-4}$

77. $12x^{-1}+8x^{-5}$ 78. $5x^{-2}-10x^{-1}$

79. $3x^{1/3} - 15x^{-2/3}$ **80.** $4x^{-1/3} + 12x^{-4/3}$

81. $2x + \frac{1}{2}x^{-1/2}$ **82.** $x^2 - \frac{1}{3}x^{-2/3}$

83. $x^2(x-5)^{1/2} - x(x-5)^{-1/2}$

84. $-x^4(1-x^2)^{-1/2} + 3x^2(1-x^2)^{1/2}$

Simplify the following expressions. Assume $\sin\theta$, $\cos\theta$, and $\tan\theta$ are positive.

85. $(\sin^2\theta)^{1/2}$ **86.** $(\tan^{1/3}\theta)^3$

87. $(\cos^{3/2}\theta)(\cos^{5/2}\theta)$ **88.** $\sec^{1/2}\theta\sec\theta$

89. $(\sin\theta + \cos\theta)^{3/2} \div (\sin\theta + \cos\theta)^{1/2}$

90. $(\tan^{1/2}\theta - 1)(\tan^{1/2}\theta + 1) \div (\tan\theta + 1)$

91. The accident at the Chernobyl nuclear reactor released dangerous amounts of the radioactive element cesium–137. The formula $A = A_0 \cdot 2.718^{-0.02295t}$ approximates how much of the original quantity A_0 remains after t years. What percent of the cesium that was released will still remain after 60 years? Express the answer to the nearest tenth of a percent.

92. Refer to the previous exercise. What percent of the cesium will remain 100 years after the accident?

93. If the world population is growing at about 1.8 percent per year, then the population after x years may be approximated by $P = P_0 \cdot 2.718^{0.018x}$. How many times larger will the population be in 39 years? Express the answer to the nearest tenth by evaluating P/P_0 when $x = 39$.

94. Repeat the previous exercise if the growth rate drops to 1 percent per year. The formula becomes $P = P_0 \cdot 2.718^{0.01x}$.

95. The third law of planetary motion discovered by the German astronomer Johann Kepler (1571–1630) states that "The square of the time of one complete revolution of a planet about its orbit is proportional to the cube of the orbit's semimajor axis." In symbols, this relation is $T^2 = kx^3$, where T and x represent time and semimajor axis respectively. This formula can be expressed as $T = \sqrt{k}\,x^{3/2}$. For the given planets, use the table to find the approximate time in years for one complete orbit of the sun.

Planet	Semimajor Axis (A.U.)
Earth	1.000
Mercury	0.387
Jupiter	5.202

96. The formula in the previous exercise can be solved for x to get $x = \sqrt[3]{\frac{T^2}{k}}$. For the given planets, use the table to find the approximate length of the semimajor axis.

Planet	Time in Years for One Orbit
Earth	1.000
Mars	1.881
Venus	0.615

97. The volume and the surface area of a cube of side s are given by the following formulas. $V = s^3$ and $A = 6s^2$. Find a formula that expresses the area A as a function of the volume, V. Use this function to find (to the nearest square inch) the surface area of a cube with volume 2000 in^3.

98. The volume and the surface area of a sphere with radius r are given by the following formulas. $V = \frac{4}{3}\pi r^3$ and $A = 4\pi r^2$. Use these formula to express the volume V, as a function of the surface area, A. Find the volume of a sphere with surface area 144 cm^2.

99. **The geometric mean** of n positive numbers is defined as the nth root of their product, so for positive numbers $a_1, a_2, \ldots a_n$,

$$\text{geometric mean} = \sqrt[n]{(a_1)(a_2)\cdots(a_n)}$$

a. Find the geometric mean of 3 and 27.

b. It can be shown that the geometric mean of two positive numbers can never be larger than their arithmetic mean (the ordinary average). Show that this is correct for 3 and 27.

100. If sales at a company over a two-year period had annual growth rates of 40 percent and 10 percent, then annual growth increased by factors of 1.40 and 1.10, respectively. Find and interpret the geometric mean of these two numbers. Give the answer to the nearest tenth of a percent.

THINK ABOUT IT 3.4

1. Explain why the nth root of a negative number does not exist in the set of real numbers when n is even.
2. In this section we defined rational exponents, but not irrational exponents. For instance, 2^{π} and $5^{\sqrt{2}}$ were not defined. We shall assume that it is possible to write irrational exponents and that the regular rules apply to them.
   a. Explain why $\left(5^{\sqrt{2}}\right)^{\sqrt{2}}$ must be equal to 25.
   b. Explain why 2^{π} should be between $2^{3.13}$ and $2^{3.15}$. Approximate π by 3.14159 and compute 2^{π} to three decimal places.
3. The Pythagoreans of ancient Greece studied many types of means. In general, a mean of a and b is some number between them. The three earliest recorded types are the arithmetic mean, $A=(a+b)/2$; the geometric mean, $G=\sqrt{ab}$; and the harmonic mean, $H=2ab/(a+b)$. Compute each of these means when $a=4$ and $b=9$. Numerically, it is always true that A is the largest and H is the smallest. Verify that this is true in this particular example.
4. Determine the fallacy in the following argument: $\left(a^m\right)^{1/n}=a^{m/n}$ implies

$$\left[(-5)^2\right]^{1/2}=(-5)^1=-5, \text{ but } \left[(-5)^2\right]^{1/2}=(25)^{1/2}=\sqrt{25}=5.$$

   Therefore $-5=5$.
5. Graph the function $y=\sqrt[x]{x}$, and estimate (to 3 decimal places) the coordinates of the maximum point. The problem of finding the coordinates exactly is called Steiner's problem.

● ● ●

3.5 Operations with Radicals

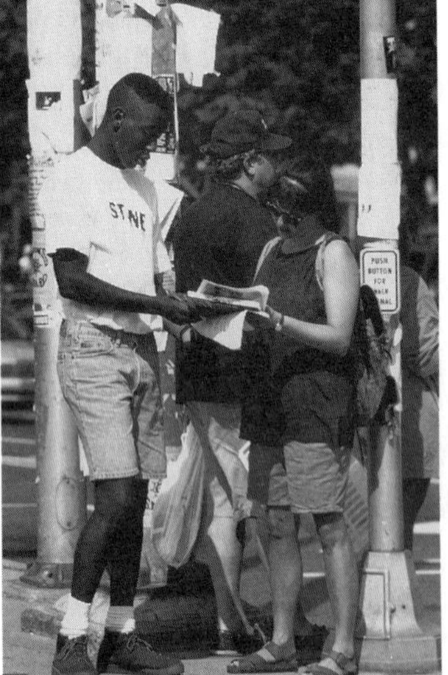

Photo Courtesy of ***John Colletti**, The Picture Cube*

An important expression in statistics is

$$\sqrt{\frac{p(1-p)}{n}},$$

which is used in computing the margin of error for a survey of n people, where p is the proportion who agree on some opinion. Assume $p=0.5$ and simplify the formula. (See Example 5.)

Objectives

1. Express a radical in simplified form.
2. Add and subtract radicals, and simplify where possible.

3. Multiply radicals, and simplify where possible.
4. Divide radicals, and simplify where possible.
5. Rationalize the denominator or the numerator of a fractional expression containing radicals.
6. Simplify expressions containing radicals by using combinations of the basic operations with radicals.

Properties of radicals are used often to simplify radicals and to operate on radicals. To illustrate two such properties, consider

$$\sqrt{4\cdot 9}=\sqrt{36}=6 \text{ and } \sqrt{4}\cdot\sqrt{9}=2\cdot 3=6$$

and

$$\sqrt[3]{\frac{64}{8}}=\sqrt[3]{8}=2 \text{ and } \frac{\sqrt[3]{64}}{\sqrt[3]{8}}=\frac{4}{2}=2.$$

We see that $\sqrt{4\cdot 9}=\sqrt{4}\cdot\sqrt{9}$, and that $\sqrt[3]{64/8}=\sqrt[3]{64}/\sqrt[3]{8}$, which illustrates product and quotient properties of radicals, respectively.

Product and Quotient Properties of Radicals

For real numbers a, b, $\sqrt[n]{a}$, and $\sqrt[n]{b}$:

1. $\sqrt[n]{a\cdot b}=\sqrt[n]{a}\cdot\sqrt[n]{b}$
2. $\sqrt[n]{\frac{a}{b}}=\frac{\sqrt[n]{a}}{\sqrt[n]{b}} \quad (b\neq 0)$

The product property may be proved by converting between radical form and exponential form and using the product-to-a-power-property of exponents.

$$\sqrt[n]{ab}=(ab)^{1/n}=a^{1/n}\cdot b^{1/n}=\sqrt[n]{a}\cdot\sqrt[n]{b}$$

The quotient property is proved in a similar way, and this proof is requested in the exercises. The first application of these properties that we consider is their role in simplifying radicals.

Simplifying Radicals

One condition that must be met for a radical to be expressed in simplified form is that we remove all factors of the radicand whose indicated root can be taken exactly.

Example 1: Removing Constant Factors of the Radicand

Simplify each radical.

a. $\sqrt{48}$ **b.** $\sqrt[3]{81}$

Solution:

a. Rewrite 48 as the product of a perfect square and another factor, and simplify. Both 4 and 16 are perfect square factors of 48. Choosing the *larger* perfect square is more efficient, so

$$\sqrt{48}=\sqrt{16\cdot 3}=\sqrt{16}\cdot\sqrt{3}=4\sqrt{3}.$$

Check that using $\sqrt{48}=\sqrt{4\cdot 12}$ leads to the same result, but with more work.

b. To simplify cube roots, look for factors of the radicand from the perfect cubes 8, 27, 64, 125, and so on. Seeing $81 = 27 \cdot 3$ yields

$$\sqrt[3]{81} = \sqrt[3]{27 \cdot 3} = \sqrt[3]{27} \cdot \sqrt[3]{3} = 3\sqrt[3]{3}.$$

PROGRESS CHECK 1

Simplify each radical.

a. $\sqrt{32}$ **b.** $\sqrt[3]{32}$ ■

The next example discusses how to simplify $\sqrt[n]{a^m}$ when m and n have common factors (other than 1) and a is nonnegative. In parts *c* and *d*, note that the index for the radical has been reduced. A second consideration for expressing a radical in simplified form is that the index of the radical be as small as possible.

Example 2: Using $\sqrt[n]{a^m} = a^{m/n}$ to Simplify Radicals

Simplify each radical. Assume $x \geq 0$, $y \geq 0$.

a. $\sqrt{x^6}$ **b.** $\sqrt[3]{2^6}$ **c.** $\sqrt[6]{y^4}$ **d.** $\sqrt[4]{9}$

Solution: Use $\sqrt[n]{a^m} = a^{m/n}$ and reduce the rational exponent. Then convert back to radical form where necessary.

a. $\sqrt{x^6} = x^{6/2} = x^3$ **b.** $\sqrt[3]{2^6} = 2^{6/3} = 2^2 = 4$

c. $\sqrt[6]{y^4} = y^{4/6} = y^{2/3} = \sqrt[3]{y^2}$ **d.** $\sqrt[4]{9} = \sqrt[4]{3^2} = 3^{2/4} = 3^{1/2} = \sqrt{3}$

Note: In this example we assume $x \geq 0$ and $y \geq 0$ so that expressions such as $\sqrt{x^2}$ simplify to x. Without this assumption it is necessary to use absolute value and write

$$\sqrt{x^2} = |x|.$$

For instance, $\sqrt{(-7)^2} \neq -7$. Instead $\sqrt{(-7)^2} = \sqrt{49} = 7$, so $\sqrt{(-7)^2} = |-7| = 7$. The general rule is

$$\sqrt[n]{a^n} = \begin{cases} a, & \text{if } n \text{ is odd,} \\ |a|, & \text{if } n \text{ is even.} \end{cases}$$

Throughout this section we will restrict radicands involving variables to nonnegative real numbers so that simplifications involving absolute value will not be necessary.

PROGRESS CHECK 2

Simplify each radical. Assume $x \geq 0$ and $y \geq 0$.

a. $\sqrt{11^2}$ **b.** $\sqrt[4]{y^{12}}$ **c.** $\sqrt[12]{25}$ **d.** $\sqrt[9]{x^6}$ ■

Parts **a** and **b** in Example 2 showed how to simplify a radical when the radicand contains a power of a variable that is a multiple of the index. This method is also used in the simplification in the next example.

Example 3: Removing Constant and Variable Factors of the Radicand

Simplify each radical. Assume $x \geq 0$, $y \geq 0$.

a. $\sqrt{8x^9y^6}$ **b.** $\sqrt[3]{x^7y^5}$

Solution:

a. Rewrite $8x^9y^6$ as a product of its largest perfect square factor and another factor, and simplify.

$$\sqrt{8x^9y^6} = \sqrt{(4x^8y^6)(2x)} = \sqrt{4x^8y^6} \cdot \sqrt{2x} = 2x^4y^3\sqrt{2x}$$

b. The largest powers in which exponents are multiples of the index of 3 are x^6 and y^3, so

$$\sqrt[3]{x^7y^5} = \sqrt[3]{(x^6y^3)(xy^2)} = \sqrt[3]{x^6y^3} \cdot \sqrt[3]{xy^2} = x^2y\sqrt[3]{xy^2}.$$

PROGRESS CHECK 3

Simplify each radical. Assume $x \geq 0,\ y \geq 0$.

a. $\sqrt{98x^3y^5}$ **b.** $\sqrt[5]{x^7y^{12}}$ ■

A third consideration for expressing a radical in simplified form is to eliminate any fractions in the radicand. The quotient property of radicals is used in such simplifications, as is shown in the next two examples.

Example 4: Eliminating Fractions in the Radicand

Simplify each radical. Assume $x > 0,\ y > 0$.

a. $\sqrt{\dfrac{3}{2}}$ **b.** $\sqrt[3]{\dfrac{7}{9}}$ **c.** $\sqrt{\dfrac{2x}{y}}$

Solution:

a. Rewrite $3/2$ as an equivalent fraction whose denominator is a perfect square, and simplify.

$$\sqrt{\frac{3}{2}} = \sqrt{\frac{3\cdot 2}{2\cdot 2}} = \sqrt{\frac{6}{4}} = \frac{\sqrt{6}}{\sqrt{4}} = \frac{\sqrt{6}}{2}$$

b. Rewrite $7/9$ as an equivalent fraction whose denominator is a perfect cube, and simplify.

$$\sqrt[3]{\frac{7}{9}} = \sqrt[3]{\frac{7\cdot 3}{9\cdot 3}} = \sqrt[3]{\frac{21}{27}} = \frac{\sqrt[3]{21}}{\sqrt[3]{27}} = \frac{\sqrt[3]{21}}{3}$$

c. Since $\sqrt{y^2} = y$ for $y > 0$,

$$\sqrt{\frac{2x}{y}} = \sqrt{\frac{2x\cdot y}{y\cdot y}} = \sqrt{\frac{2xy}{y^2}} = \frac{\sqrt{2xy}}{\sqrt{y^2}} = \frac{\sqrt{2xy}}{y}$$

PROGRESS CHECK 4

Simplify each radical. Assume $x > 0,\ y > 0$.

a. $\sqrt{\dfrac{2}{5}}$ **b.** $\sqrt[3]{\dfrac{5}{4}}$ **c.** $\sqrt{\dfrac{y}{3x}}$ ■

Example 5: Simplifying an Expression Involving Margin of Error

Solve the problem in the section introduction on page 282.

Solution: Replacing p by 0.5 in

$$\sqrt{\frac{p(1-p)}{n}} \text{ gives } \sqrt{\frac{0.5(1-0.5)}{n}}.$$

This expression with $n > 0$ may be changed to simplified radical form as follows.

$$\sqrt{\frac{0.5(1-0.5)}{n}} = \sqrt{\frac{0.25}{n}} = \sqrt{\frac{0.25 \cdot n}{n \cdot n}} = \frac{\sqrt{0.25} \cdot \sqrt{n}}{\sqrt{n^2}} = \frac{0.5\sqrt{n}}{n}.$$

Thus, when $p = 0.5$, the expression simplifies to $0.5\sqrt{n}/n$.

PROGRESS CHECK 5
Repeat Example 5, but take $p = 0.1$. ■

Addition and Subtraction of Radicals

We can simplify an algebraic expression such as $7x + 2x - x$ by combining like terms. That is,

$$7x + 2x - x = (7 + 2 - 1)x = 8x.$$

Similarly, if x is replaced by $\sqrt{2}$, we have

$$7\sqrt{2} + 2\sqrt{2} - \sqrt{2} = (7 + 2 - 1)\sqrt{2} = 8\sqrt{2},$$

or if x is replaced by $\sqrt[3]{5x}$, we have

$$7\sqrt[3]{5x} + 2\sqrt[3]{5x} - \sqrt[3]{5x} = (7 + 2 - 1)\sqrt[3]{5x} = 8\sqrt[3]{5x}.$$

Thus, to add or subtract radicals, we combine like radicals using the distributive property. By definition, **like radicals** are radicals that have the same radicand and the same index. Note that only like radicals may be combined.

Example 6: Adding and Subtracting Radicals

Simplify where possible. Assume $x > 0,\ y > 0$.

a. $2\sqrt{32} - 4\sqrt{1/2}$ **b.** $\sqrt[3]{-16} + \sqrt[3]{54}$ **c.** $\sqrt{50x^3y} - \sqrt{8xy^3}$

Solution:

a. First, simplify each square root.

$$2\sqrt{32} = 2\sqrt{16 \cdot 2} = 2\sqrt{16}\sqrt{2} = 2 \cdot 4\sqrt{2} = 8\sqrt{2}$$

$$4\sqrt{\frac{1}{2}} = 4\sqrt{\frac{1 \cdot 2}{2 \cdot 2}} = 4\sqrt{\frac{2}{4}} = \frac{4\sqrt{2}}{\sqrt{4}} = \frac{4\sqrt{2}}{2} = 2\sqrt{2}$$

$$\text{Then, } 2\sqrt{32} - 4\sqrt{\frac{1}{2}} = 8\sqrt{2} - 2\sqrt{2} = (8-2)\sqrt{2} = 6\sqrt{2}.$$

b. First, simplify each cube root.

$$\sqrt[3]{-16} = \sqrt[3]{-8 \cdot 2} = \sqrt[3]{-8}\sqrt[3]{2} = -2\sqrt[3]{2}$$

$$\sqrt[3]{54} = \sqrt[3]{27 \cdot 2} = \sqrt[3]{27}\sqrt[3]{2} = 3\sqrt[3]{2}$$

$$\text{Then, } \sqrt[3]{-16} + \sqrt[3]{54} = -2\sqrt[3]{2} + 3\sqrt[3]{2} = (-2+3)\sqrt[3]{2} = \sqrt[3]{2}.$$

c. Simplify each radical and then combine like radicals.

$$\sqrt{50x^3y} = \sqrt{(25x^2)(2xy)} = \sqrt{25x^2}\sqrt{2xy} = 5x\sqrt{2xy}$$
$$\sqrt{8xy^3} = \sqrt{(4y^2)(2xy)} = \sqrt{4y^2}\sqrt{2xy} = 2y\sqrt{2xy}$$
$$\text{Then, } \sqrt{50x^3y} - \sqrt{8xy^3} = 5x\sqrt{2xy} - 2y\sqrt{2xy} = (5x-2y)\sqrt{2xy}\,.$$

Caution: Although $\sqrt[n]{a}\cdot\sqrt[n]{b} = \sqrt[n]{ab}$ and $\sqrt[n]{a}/\sqrt[n]{b} = \sqrt[n]{a/b}$ are properties of radicals, note that

$$\sqrt[n]{a} + \sqrt[n]{b} \text{ does not equal } \sqrt[n]{a+b},$$

except for certain instances. For example,

$$\sqrt{16} + \sqrt{9} = 4 + 3 = 7\text{, while } \sqrt{16+9} = \sqrt{25} = 5,$$

so $\sqrt{16} + \sqrt{9} \neq \sqrt{16+9}$.

PROGRESS CHECK 6
Simplify where possible. Assume $x > 0,\ y > 0$.

a. $\sqrt{75} - \frac{1}{5}\sqrt{\frac{1}{3}}$ **b.** $\sqrt[3]{-54} - \sqrt[3]{128}$ **c.** $2\sqrt{27x^2y} + 5\sqrt{12x^2y}$ ■

Example 7: Testing for Identical Expressions with a Grapher

A common student error is to rewrite $\sqrt{x} + \sqrt{x}$ as $\sqrt{2x}$. Show that these two expressions are not identical by showing that the graphs of $y = \sqrt{x} + \sqrt{x}$ and $y = \sqrt{2x}$ are different. Use the graphs to find all values of x for which the expressions are equal.

Solution: The two functions are graphed in Figure 3.9 using a viewing window that takes into account that the domain for both functions is $x \geq 0$ and that the square root function grows slowly. It is apparent that the two graphs are different, so the expressions are not identical. We can also see that the two graphs intersect only at the origin, so that $\sqrt{x} + \sqrt{x} = \sqrt{2x}$ is true only when $x = 0$.

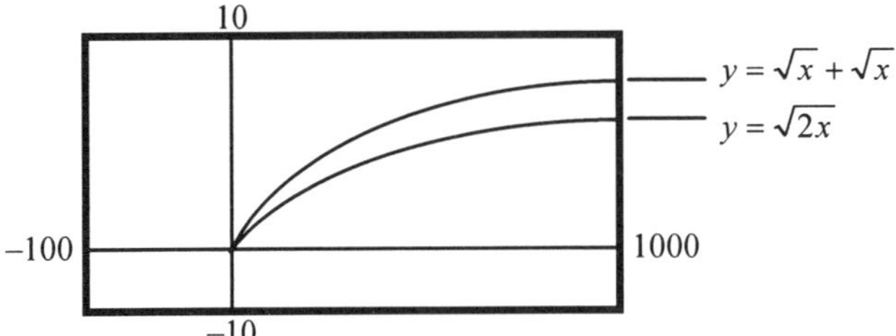

Figure 3.9

PROGRESS CHECK 7
A common student error is to rewrite $\sqrt{x^2+9}$ as $x + 3$. Show that these two expressions are not identical by showing that the graphs of $y = \sqrt{x^2+9}$ and $y = x + 3$ are different. Use the graphs to find all values of x for which the expressions are equal.

Multiplication of Radicals
To multiply expressions containing radicals, we use the product property of radicals in the form

$$\sqrt[n]{a}\cdot\sqrt[n]{b}=\sqrt[n]{ab}.$$

Applying this property sometimes results in expressions that can be simplified, and final answers must be stated in simplified radical form.

Example 8: Multiplying Radical Expressions

Multiply and simplify where possible. Assume $x>0$ and $y>0$.

a. $\sqrt{15}\cdot\sqrt{5}$ **b.** $8\sqrt[3]{6x}\cdot 5\sqrt[3]{3x}$ **c.** $\left(\sqrt{6}+\sqrt{y}\right)\left(\sqrt{6}-\sqrt{y}\right)$

Solution:

a. First, multiply to get

$$\sqrt{15}\cdot\sqrt{5}=\sqrt{15\cdot 5}=\sqrt{75}$$

Now simplify.

$$\sqrt{75}=\sqrt{25\cdot 3}=\sqrt{25}\cdot\sqrt{3}=5\sqrt{3}$$

Thus, $\sqrt{15}\cdot\sqrt{5}=5\sqrt{3}$.

b. Reorder and regroup as shown. Then multiply.

$$8\sqrt[3]{6x}\cdot 5\sqrt[3]{3x}=(8\cdot 5)\left(\sqrt[3]{6x}\cdot\sqrt[3]{3x}\right)=40\sqrt[3]{18x^2}$$

No factor of 18 or x^2 is a perfect cube, so the result does not simplify.

c. Based on the distributive property, multiply each term of $\sqrt{6}+\sqrt{y}$ by each term of $\sqrt{6}-\sqrt{y}$, and then simplify.

$$\begin{aligned}\left(\sqrt{6}+\sqrt{y}\right)\left(\sqrt{6}-\sqrt{y}\right)&=\sqrt{6}\cdot\sqrt{6}+\sqrt{6}\left(-\sqrt{y}\right)+\sqrt{y}\cdot\sqrt{6}+\sqrt{y}\left(-\sqrt{y}\right)\\&=\sqrt{36}-\sqrt{6y}+\sqrt{6y}-\sqrt{y^2}\\&=6-y\end{aligned}$$

Note: A property that is used often when simplifying radicals stems from the definition of $\sqrt[n]{a}$ and states

$$\left(\sqrt[n]{a}\right)^n=a$$

if $\sqrt[n]{a}$ is a real number. For instance, $\sqrt{6}\cdot\sqrt{6}=\left(\sqrt{6}\right)^2=6$, and this result checks, since $\sqrt{6}\cdot\sqrt{6}=\sqrt{36}=6$. It is recommended that you use $\left(\sqrt[n]{a}\right)^n=a$ whenever possible.

PROGRESS CHECK 8

Multiply and simplify where possible. Assume $x>0$ and $y>0$.

a. $\sqrt{7}\cdot\sqrt{14}$ **b.** $9\sqrt[5]{5y}\cdot 6\sqrt[5]{7y}$ **c.** $\left(5+\sqrt{x}\right)\left(5-\sqrt{x}\right)$ ■

The special product formulas from Section R.4 are often used to simplify certain products involving square roots. For instance, the product in Example 8c may be found more easily by using

$$(a+b)(a-b)=a^2-b^2$$

with a replaced by $\sqrt{6}$ and b replaced by $\sqrt{y}$, to obtain

$$\begin{aligned}(\sqrt{6}+\sqrt{y})(\sqrt{6}-\sqrt{y}) &= (\sqrt{6})^2-(\sqrt{y})^2\\ &= 6-y.\end{aligned}$$

Other useful special product formulas from Section R.4 are

$$\begin{aligned}(a+b)^2 &= a^2+2ab+b^2\\ (a-b)^2 &= a^2-2ab+b^2.\end{aligned}$$

Example 9: Using Special Product Formulas

Simplify each expression. Assume $x>0$.

a. $(4\sqrt{3}+\sqrt{2})(4\sqrt{3}-\sqrt{2})$ **b.** $(5+\sqrt{x})^2$

Solution:

a. Replace a with $4\sqrt{3}$ and b with $\sqrt{2}$ in the formula for $(a+b)(a-b)$.

$$\begin{array}{ccc} (\; a \;+\; b)(\; a \;-\; b) & = & a^2 \;-\; b^2\\ \downarrow\quad\downarrow\quad\downarrow\quad\downarrow & & \downarrow\qquad\downarrow\\ (4\sqrt{3}+\sqrt{2})(4\sqrt{3}-\sqrt{2}) & = & (4\sqrt{3})^2-(\sqrt{2})^2\\ & = & 48-2\\ & = & 46\end{array}$$

b. Use $(a+b)^2=a^2+2ab+b^2$ with $a=5$ and $b=\sqrt{x}$.

$$\begin{aligned}(5+\sqrt{x})^2 &= 5^2+2(5)(\sqrt{x})+(\sqrt{x})^2\\ &= 25+10\sqrt{x}+x\end{aligned}$$

PROGRESS CHECK 9

Simplify each expression. Assume $x>0$.

a. $(2-6\sqrt{x})^2$ **b.** $(3+2\sqrt{5})(3-2\sqrt{5})$ ■

Division of Radicals

The quotient property of radicals in the form

$$\frac{\sqrt[n]{a}}{\sqrt[n]{b}}=\sqrt[n]{\frac{a}{b}}$$

may be used to divide expressions containing radicals with the same index. Once again, remember that after applying this property final answers must be stated in simplified radical form.

Example 10: Dividing Radical Expressions

Divide and simplify where possible. Assume $x>0$.

a. $\dfrac{\sqrt{7}}{\sqrt{2}}$ **b.** $\dfrac{50\sqrt{32}}{5\sqrt{2}}$ **c.** $\dfrac{\sqrt[3]{48x}}{\sqrt[3]{2x}}$

Solution:

a. $\frac{\sqrt{7}}{\sqrt{2}}=\sqrt{\frac{7}{2}}=\sqrt{\frac{7\cdot 2}{2\cdot 2}}=\frac{\sqrt{14}}{\sqrt{4}}=\frac{\sqrt{14}}{2}$

b. $\frac{50\sqrt{32}}{5\sqrt{2}}=\frac{50}{5}\sqrt{\frac{32}{2}}=10\sqrt{16}=10\cdot 4=40$

c. $\frac{\sqrt[3]{48x}}{\sqrt[3]{2x}}=\sqrt[3]{\frac{48x}{2x}}=\sqrt[3]{24}=\sqrt[3]{8}\cdot\sqrt[3]{3}=2\sqrt[3]{3}$

PROGRESS CHECK 10

Divide and simplify where possible. Assume $y>0$.

a. $\frac{\sqrt{3}}{\sqrt{10}}$ b. $\frac{10\sqrt[3]{54}}{3\sqrt[3]{2}}$ c. $\frac{\sqrt{27y}}{\sqrt{6y}}$ ■

We have discussed how to simplify a radical so the radicand contains no fractions. A similar procedure may be used to ensure that no radicals appear in a denominator. For example, $1/\sqrt{2}$ can be converted to $\sqrt{2}/2$ as follows.

$$\frac{1}{\sqrt{2}}=\frac{1}{\sqrt{2}}\cdot\frac{\sqrt{2}}{\sqrt{2}}=\frac{\sqrt{2}}{2}$$

One instance when the latter form is more useful occurs with the addition of radicals, as shown below.

$$3\sqrt{2}+\frac{1}{\sqrt{2}}=3\sqrt{2}+\frac{\sqrt{2}}{2}=\frac{7}{2}\sqrt{2}$$

Thus, we adopt the condition that a simplified radical cannot have radicals in the denominator. The process of obtaining a radical-free denominator is called **rationalizing the denominator**, and all of the conditions that have been given in this section for writing a simplified radical are summarized next.

Simplified Radical

To write a radical in simplified form:

1. Remove all factors of the radicand whose indicated root can be taken exactly.
2. Write the radical so that the index is as small as possible.
3. Eliminate all fractions in the radicand and all radicals in the denominator (which is called rationalizing the denominator).

Example 11: Rationalizing Denominators with One Term

Rationalize the denominator. Assume $x>0$.

a. $\frac{2}{\sqrt{5}}$ b. $\frac{7}{4\sqrt{3}}$ c. $\frac{9}{2\sqrt{18}}$ d. $\frac{9}{\sqrt{3x}}$

Solution:

a. $\frac{2}{\sqrt{5}}=\frac{2\sqrt{5}}{\sqrt{5}\sqrt{5}}=\frac{2\sqrt{5}}{5}$

b. $\frac{7}{4\sqrt{3}}=\frac{7\sqrt{3}}{4\sqrt{3}\sqrt{3}}=\frac{7\sqrt{3}}{4\cdot 3}=\frac{7\sqrt{3}}{12}$

c. $\frac{9}{2\sqrt{18}}=\frac{9\sqrt{2}}{2\sqrt{18}\sqrt{2}}=\frac{9\sqrt{2}}{2\sqrt{36}}=\frac{9\sqrt{2}}{2\cdot 6}=\frac{3\sqrt{2}}{4}$

d. $\frac{9}{\sqrt{3x}}=\frac{9\sqrt{3x}}{\sqrt{3x}\sqrt{3x}}=\frac{9\sqrt{3x}}{3x}=\frac{3\sqrt{3x}}{x}$

PROGRESS CHECK 11

Rationalize the denominator. Assume $x > 0$.

a. $\dfrac{1}{\sqrt{2}}$ **b.** $\dfrac{3}{\sqrt{x}}$ **c.** $\dfrac{6}{\sqrt{20}}$ **d.** $\dfrac{\sqrt{5}}{\sqrt{7x}}$ ■

To eliminate radicals in denominators that contain square roots and two terms, consider that for nonnegative a and b

$$\left(\sqrt{a}+\sqrt{b}\right)\left(\sqrt{a}-\sqrt{b}\right)=\left(\sqrt{a}\right)^2-\left(\sqrt{b}\right)^2=a-b.$$

In general, the sum and the difference of the same two terms are called **conjugates** of each other. Binomial denominators involving square roots are rationalized by multiplying the numerator and the denominator by the conjugate of the denominator.

Example 12: Rationalizing a Binomial Denominator

Rationalize the denominator of $\dfrac{10}{3+\sqrt{7}}$.

Solution: The conjugate of $3+\sqrt{7}$ is $3-\sqrt{7}$. Then,

$$\frac{10}{3+\sqrt{7}}=\frac{10}{3+\sqrt{7}}\cdot\frac{3-\sqrt{7}}{3-\sqrt{7}} \quad \text{Multiply using the conjugate of the denominator.}$$

$$=\frac{10\left(3-\sqrt{7}\right)}{(3)^2-\left(\sqrt{7}\right)^2}$$

$$=\frac{10\left(3-\sqrt{7}\right)}{2}$$

$$=5\left(3-\sqrt{7}\right) \quad \text{Express in lowest terms.}$$

$$=15-5\sqrt{7}$$

Note: To express the result in lowest terms, it is recommended that fractions be simplified where possible before removing parentheses in the numerator, as was done in this example.

PROGRESS CHECK 12

Rationalize the denominator of $\dfrac{8}{3-\sqrt{5}}$. ■

In some cases, instead of rationalizing the denominator, it is useful to rationalize the numerator. The idea is now to obtain a radical-free numerator, and the following example illustrates the procedure for a numerator that contains square roots and two terms.

Example 13: Rationalizing the Numerator

Rationalize the numerator: $\dfrac{\sqrt{x}-\sqrt{a}}{x-a}$.

Solution: The conjugate of $\sqrt{x}-\sqrt{a}$ is $\sqrt{x}+\sqrt{a}$. Then,

$$\frac{\sqrt{x}-\sqrt{a}}{x-a} = \frac{(\sqrt{x}-\sqrt{a})(\sqrt{x}+\sqrt{a})}{(x-a)(\sqrt{x}+\sqrt{a})}$$

$$= \frac{x-a}{(x-a)(\sqrt{x}+\sqrt{a})} = \frac{1}{\sqrt{x}+\sqrt{a}}.$$

PROGRESS CHECK 13

Rationalize the numerator of $\dfrac{\sqrt{x}-1}{x-1}$. ■

Expressions that involve radicals sometimes need to be simplified in calculus by using combinations of the basic operations with radicals. The next example illustrates a simplification of this type.

Example 14: Simplifying a Radical Expression Seen in Calculus

Write as a single fraction: $x \cdot \dfrac{2x}{2\sqrt{x^2+5}} + \sqrt{x^2+5}$.

Solution:

$$x \cdot \frac{2x}{2\sqrt{x^2+5}} + \sqrt{x^2+5} = \frac{x^2}{\sqrt{x^2+5}} + \sqrt{x^2+5}\left(\frac{\sqrt{x^2+5}}{\sqrt{x^2+5}}\right)$$

$$= \frac{x^2+(x^2+5)}{\sqrt{x^2+5}}$$

$$= \frac{2x^2+5}{\sqrt{x^2+5}} \text{ or } \frac{(2x^2+5)\sqrt{x^2+5}}{x^2+5}$$

PROGRESS CHECK 14

Write as a single fraction: $h \cdot \dfrac{-2h}{2\sqrt{4R^2-h^2}} + \sqrt{4R^2-h^2}$ ■

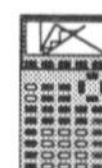

EXPLORE 3.5

1. Use graphs to check the following "simplifications." Some of them are correct, and some are not. If the two given expressions are not equivalent, then find any values of x for which they are equal.
   - **a.** $\sqrt{x^2}$; x **b.** $\sqrt{x^2+9}$; $x+3$ **c.** $\sqrt{x}+\sqrt{x}$; $\sqrt{2x}$
   - **d.** $(5+\sqrt{x})(5-\sqrt{x})$; $25-x$ **e.** $\sqrt{9x}$; $3\sqrt{x}$ **f.** $\sqrt{x^6}$; x^3
2. Use a grapher to find the maximum values of $y=\sqrt{x(1-x)}$. This expression is used in finding the margin of error in a statistical survey. For what value(s) of x does y equal half the maximum value?
3. Graph each of these functions, and determine the domain of each. Note, when the square root is involved, care is needed to determine the domain.
   - **a.** $y=\sqrt{\sqrt{x^2}}$ **b.** $y=\sqrt{x^2}$ **c.** $y=\sqrt{x}$
4. **a.** For what values of x is $\sqrt{x}$ greater than x?
   - **b.** For what values of x is $\sqrt{x}$ greater than $2x$?
   - **c.** For what values of x is $\sqrt{x}$ greater than $3x$?
   - **d.** Use the pattern from parts **a**, **b**, and **c** to determine the values of x for which $\sqrt{x}$ is greater than ax $(a>0)$.

EXERCISES 3.5

In Exercises 1–30 express each radical in simplest form (assume that x and y represent positive real numbers).

1. $\sqrt{108}$
2. $-\sqrt{80}$
3. $5\sqrt{8}$
4. $\frac{1}{3}\sqrt{50}$
5. $\sqrt[3]{-32}$
6. $\sqrt[5]{64}$
7. $\sqrt{12x^7y^4}$
8. $\sqrt{54x^{11}y^5}$
9. $\sqrt[4]{9x^6y^7}$
10. $\sqrt[5]{16x^3y^5}$
11. $\sqrt[6]{7^2}$
12. $\sqrt[8]{3^4}$
13. $\sqrt[4]{25}$
14. $\sqrt[9]{8}$
15. $\sqrt[9]{y^6}$
16. $\sqrt[4]{x^2}$
17. $\sqrt[8]{2^6(x-1)^6}$
18. $\sqrt[6]{y^4(x+y)^2}$
19. $\sqrt[6]{81y^8}$
20. $\sqrt[4]{4x^{10}}$
21. $9\sqrt{\frac{4}{3}}$
22. $-12\sqrt{\frac{20}{27}}$
23. $\sqrt[4]{\frac{5}{8}}$
24. $\sqrt[5]{\frac{2}{9}}$
25. $\sqrt{\frac{9}{x}}$
26. $\sqrt{\frac{5}{3x^2y}}$
27. $\sqrt[3]{\frac{x}{y}}$
28. $\sqrt[4]{\frac{2x}{y^5}}$
29. $\sqrt{x^2-\frac{2x^2}{4}}$
30. $\sqrt{\frac{1}{x^2}+\frac{1}{y^2}}$

In Exercises 31–60 express each sum or difference in simplest form (assume that x and y represent positive real numbers).

31. $4\sqrt{2}+7\sqrt{2}$
32. $-2\sqrt{7}-\sqrt{7}$
33. $\sqrt{27}+\sqrt{12}$
34. $\sqrt{8}-\sqrt{32}$
35. $3\sqrt{72}-5\sqrt{50}$
36. $3\sqrt{28}-10\sqrt{63}$
37. $-2\sqrt{176}-3\sqrt{44}$
38. $3\sqrt{48}-\sqrt{20}$
39. $4\sqrt{12}-\sqrt{27}+2\sqrt{48}$
40. $-4\sqrt{45}+2\sqrt{125}-6\sqrt{20}$
41. $2\sqrt{50}-6\sqrt{\frac{1}{2}}$
42. $4\sqrt{\frac{1}{5}}+2\sqrt{125}$
43. $6\sqrt{\frac{1}{6}}-\sqrt{\frac{2}{3}}$
44. $5\sqrt{\frac{1}{3}}-3\sqrt{\frac{1}{12}}$
45. $3\sqrt{\frac{3}{2}}-\sqrt{\frac{2}{3}}+5\sqrt{6}$
46. $\frac{1}{3}\sqrt{\frac{5}{3}}-\frac{1}{2}\sqrt{60}+\frac{1}{6}\sqrt{15}$
47. $\sqrt[3]{24}+7\sqrt[3]{3}$
48. $2\sqrt[3]{-54}-\sqrt[3]{128}$
49. $\sqrt[3]{\frac{1}{9}}+2\sqrt[3]{-375}$
50. $3\sqrt[4]{32}-5\sqrt[4]{2}$
51. $\sqrt{8x}+\sqrt{72x}$
52. $9\sqrt{3y}-4\sqrt{12y}$
53. $\sqrt{20xy^2}-2\sqrt{45x^3}$
54. $\sqrt{\frac{x^3y}{2}}+\frac{1}{2}\sqrt{32xy^7}$
55. $\sqrt{50x^3y}+x\sqrt{\frac{xy}{2}}-2\sqrt{\frac{9x^3y}{2}}$
56. $2\sqrt{\frac{x}{y}}-\sqrt{\frac{y}{x}}+3\sqrt{\frac{1}{xy}}$
57. $2\sqrt{\frac{2x}{y}}-4\sqrt{\frac{y}{2x^3}}+5\sqrt{\frac{1}{8}x^3y}$
58. $\sqrt[3]{54x^4y}-\sqrt[3]{\frac{xy^4}{4}}$
59. $\sqrt[4]{\frac{x}{y}}-\sqrt[4]{xy^3}$
60. $\sqrt{4+4x}+\sqrt{16+16x}$

In Exercises 61–80 express each product or quotient in simplest form (assume that x and y represent positive real numbers).

61. $\sqrt{6}\cdot\sqrt{2}$
62. $\left(-3\sqrt{5}\right)^2$
63. $\sqrt{24}\div\sqrt{3}$
64. $\sqrt{14}\div\sqrt{5}$
65. $\sqrt[3]{-4}\cdot\sqrt[3]{16}$
66. $\sqrt[4]{2}\div 2\sqrt[4]{2}$
67. $\left(\sqrt{15}+\sqrt{60}\right)\div\sqrt{3}$
68. $-3\sqrt{3}\left(4-2\sqrt{27}\right)$
69. $\left(1+\sqrt{2}\right)\left(1-\sqrt{2}\right)$
70. $\left(1+\sqrt{2}\right)^2$
71. $\left(\sqrt{5}-2\sqrt{6}\right)^2$
72. $\left(5\sqrt{7}-3\sqrt{8}\right)\left(2\sqrt{7}+4\sqrt{2}\right)$
73. $\sqrt{15x}\cdot\sqrt{6x}$
74. $\sqrt{5x^3y}\cdot\sqrt{10y}$
75. $\sqrt[3]{4x^2y}\cdot\sqrt[3]{6x^2y^3}$
76. $4\sqrt[4]{27x^3}\cdot 5\sqrt[4]{3x^5}$
77. $\sqrt{18}\div\sqrt{3x}$
78. $-2\sqrt{54x^3y}\div 4\sqrt{3xy}$

79. $\sqrt[4]{7x^2} \div \sqrt[4]{2x}$ 80. $\sqrt[3]{-3y} \div \sqrt[3]{48y}$

In Exercises 81–90 rationalize the denominator.

81. $\dfrac{2}{\sqrt{3}}$ 82. $\dfrac{7}{2\sqrt{18}}$ 83. $\dfrac{1-\sqrt{3}}{\sqrt{3}}$

84. $\dfrac{\sqrt{3}}{1-\sqrt{3}}$ 85. $\dfrac{2-\sqrt{2}}{2+\sqrt{2}}$ 86. $\dfrac{\sqrt{3}+\sqrt{5}}{2\sqrt{3}-7\sqrt{5}}$

87. $\dfrac{\sqrt{x}}{\sqrt{x}+\sqrt{y}}$ 88. $\dfrac{2\sqrt{y}}{x-2\sqrt{y}}$

89. $\dfrac{x-1}{\sqrt{x^2-1}}$ 90. $\dfrac{x}{\sqrt{2x^3+x}}$

In Exercises 91–96 rationalize the numerator.

91. $\dfrac{\sqrt{x}-2}{x-4}$ 92. $\dfrac{\sqrt{x}-\sqrt{9}}{x-9}$ 93. $\dfrac{\sqrt{x+h}-\sqrt{x}}{h}$

94. $\dfrac{\sqrt{1+h}-1}{h}$ 95. $\dfrac{\sqrt{2x+2h+1}-\sqrt{2x+1}}{h}$

96. $\dfrac{\sqrt{x+h-1}-\sqrt{x-1}}{h}$

In Exercises 97–108 write each expression as a single fraction in simplest form (assume that all variables represent positive real numbers.)

97. $600\sqrt{5}+\dfrac{360{,}000}{120\sqrt{5}}$ 98. $900\sqrt{2}+\dfrac{540{,}000}{300\sqrt{2}}$

99. $\dfrac{\sqrt{\frac{3}{x}}\left(1+\frac{3}{x}\right)}{\frac{\sqrt{3}}{2x^{3/2}}}$ 100. $-\dfrac{1}{2}\sqrt{\dfrac{x}{y}}\cdot\dfrac{x\sqrt{\frac{y}{x}}-y}{x^2}$

101. $1-\dfrac{1}{2\sqrt{x+1}}$ 102. $\dfrac{u}{2\sqrt{u+1}}+\sqrt{u+1}$

103. $a\cdot\dfrac{a}{\sqrt{a^2+b^2}}+b\cdot\dfrac{b}{\sqrt{a^2+b^2}}$

104. $\dfrac{2x}{2\sqrt{x^2-1}}-\dfrac{1}{x\sqrt{x^2-1}}$

105. $\pi\left(h\cdot\dfrac{-2h}{2\sqrt{4R^2-h^2}}+\sqrt{4R^2-h^2}\right)$

106. $\sqrt{R^2-x^2}+(R+x)\dfrac{-x}{\sqrt{R^2-x^2}}$

107. $\dfrac{x\cdot\frac{2x}{2\sqrt{x^2+1}}-\sqrt{x^2+1}}{x^2}$

108. $\dfrac{\sqrt{1-x^2}-x\cdot\frac{-x}{\sqrt{1-x^2}}}{1-x^2}$

In Exercises 109–114 show that the two given expressions are not identical, by showing that they yield different graphs. Use the graphs to find all values of x for which the expressions are equal.

109. $\sqrt{x}+\sqrt{2x}$ and $\sqrt{3x}$

110. $5\sqrt{x}$ and $\sqrt{5x}$

111. $\sqrt[3]{x}+\sqrt[3]{10}$ and $\sqrt[3]{x+10}$

112. $\sqrt[3]{5x^2}-\sqrt[3]{3x^2}$ and $\sqrt[3]{2x^2}$

113. $\sqrt{x^2+16}$ and $x+4$ 114. $\sqrt[3]{x^3-8}$ and $x-2$

In Exercises 115–122 simplify and rationalize any denominators. Assume all radicands are positive.

115. $4\sqrt{\sin\theta}+6\sqrt{\sin\theta}$ 116. $\sqrt{\cos\theta}-5\sqrt{\cos\theta}$

117. $\dfrac{1}{\sqrt{\tan\theta}}$ 118. $\dfrac{3}{\sqrt{2\cot\theta}}$

119. $\sqrt{\tan\theta}\sqrt{\tan\theta}$ 120. $\sqrt{18\sin\theta}\sqrt{2\sin\theta}$

121. $\left(1-\sqrt{\tan\theta}\right)\left(1+\sqrt{\tan\theta}\right)$

122. $\left(3\sqrt{\cos\theta}+\dfrac{\sqrt{3}}{2}\right)\left(3\sqrt{\cos\theta}-\dfrac{\sqrt{3}}{2}\right)$

In Exercises 123–124 give exact answers using radicals in simplest form.

123. If a beam emerging from a laser travels as shown in the diagram, find the distance traveled by the beam, in simplified radical form.

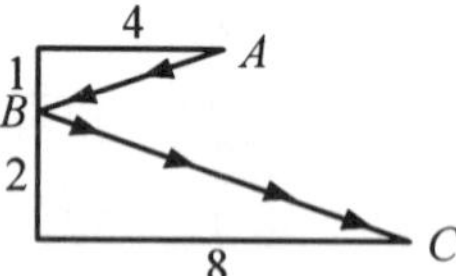

124. If a beam emerging from a laser travels as shown in the diagram, find the distance traveled by the beam, in simplified radical form.

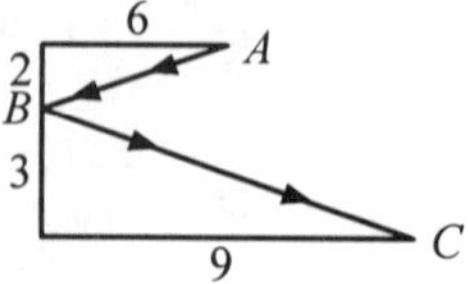

For Exercises 125–126; The period T of a pendulum is the time required for the pendulum to complete one round-trip of motion, that is, one complete cycle. When the period is measured in seconds, then a formula for the period is

$$T = 2\pi\sqrt{\frac{l}{32}},$$

where l is the length in feet of the pendulum.

125. In simplified radical form, what is the period of a pendulum that is 16 ft long? Approximate the answer to the nearest hundredth of a second.

126. In simplified radical form, what is the period of a pendulum that is 9 ft long? Approximate the answer to the nearest hundredth of a second.

127. An important concept in statistics is that for a random survey the margin of error (e) is inversely proportional to the square root of the sample size (n). This relation can be represented as $e = \frac{k}{\sqrt{n}}$.

a. Rewrite this formula after rationalizing the denominator.

b. If the error is 3% when $n = 900$, find k, then find the error when $n = 2500$.

128. The velocity (v) of a satellite in circular orbit around a planet is inversely proportional to the square root of the distance (r) from the center of the planet. This relation can be expressed as $v = \frac{k}{\sqrt{r}}$.

a. Rewrite this formula after rationalizing the denominator.

b. For the planet Earth, when distance is measured in feet and time in seconds, the value of k is $14 \cdot 10^{15}$. Find the velocity needed to keep a satellite in orbit at a distance of 100 miles (528,000 feet).

129. The length (L) of the edge of a regular hexagon can be found from the area (A) by $L = \sqrt{\frac{2A}{3\sqrt{3}}}$. Rewrite this formula by relationalizing the denominator, then find L when $A = 6\sqrt{3}$.

130. The length (L) of the edge of a regular pentagon can be found from the area (A) by $L = \frac{2\sqrt{A\tan 36°}}{\sqrt{5}}$.
Rewrite this formula by relationalizing the denominator, then find L to the nearest hundredth when $A = 16$.

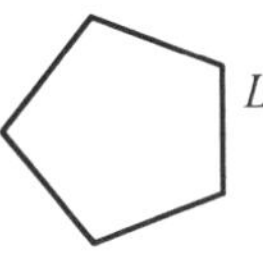

THINK ABOUT IT 3.5

1. This problem demonstrates that visual patterns are not always helpful when adding and subtracting radicals.

a. Is it true that $1 + 49 = 25 + 25$? Will the equality still be correct if you just write a square root symbol in front of each term? In other words, is it true that $\sqrt{1} + \sqrt{49} = \sqrt{25} + \sqrt{25}$?

b. Is it true that $512 + 216 + 1 = 729$? Will the equality still be correct if you just write a cube root symbol in front of each term? In other words, is it true that $\sqrt[3]{512} + \sqrt[3]{216} + \sqrt[3]{1} = \sqrt[3]{729}$?

c. Is it true that $75 + 12 = 48 + 27$? Is it true that $\sqrt{75} + \sqrt{12} = \sqrt{48} + \sqrt{27}$?

d. Given that $a + b = c + d$, which of these statements must also be correct?

i. $\sqrt[n]{a+b} = \sqrt[n]{c+d}$ **ii.** $\sqrt[n]{a} + \sqrt[n]{b} = \sqrt[n]{c} + \sqrt[n]{d}$

2. Which statement is true for all values of x (for which it is defined)?

a. $\left(\sqrt{x} + \sqrt{2}\right)^2 = x + 2$ **b.** $\sqrt{x} + \sqrt{x} = \sqrt{2x}$

c. $\left(\sqrt{x+2}\right)^2 = x + 2$ **d.** $\sqrt{x^2 + 1} = x + 1$

3. The formula for the period of a pendulum, as illustrated in Exercise 125, is $T = 2\pi\sqrt{l/32}$. The formula shows how the period depends on the length.

a. Does the period increase or decrease if you make the pendulum longer?

b. Explain why the period does not double when you double the length.

c. The 32 in the formula represents the force of gravity near the surface of the earth. On the moon the force of gravity would be less than 32. If you moved a pendulum from the earth to the moon, would its period be longer or shorter there?

4. Use the product model

$$(a-b)(a^2+ab+b^2)=a^3-b^3$$

to rationalize the numerator: $\left(\sqrt[3]{x+h}-\sqrt[3]{x}\right)/h$.

5. When drawing a rectangle, what ratio of length to width should one use, to achieve the most satisfying visual effect? The answer is obviously subjective, but there is a leading candidate for the title of most pleasing. This rectangle is called the **golden rectangle**, and the ratio of the length to the width in this figure is called the **golden ratio**. Consider the buildings sketched within rectangles whose side ratios are indicated. Which figure appeals the most to you?

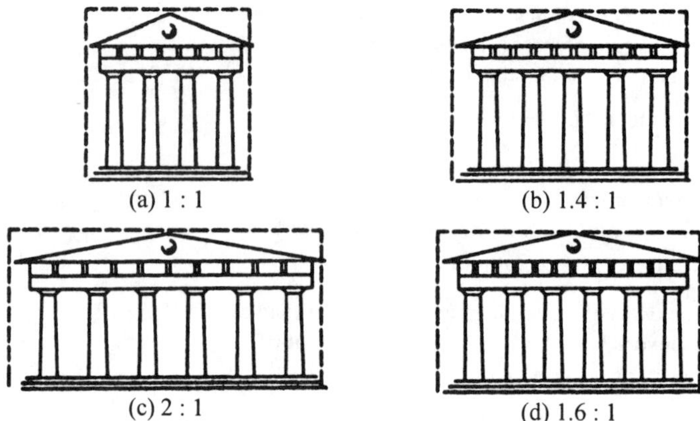

If you selected figure (d) you are in good company, for this figure has the same side ratios as the Parthenon in Athens. This fact and other examples of the golden rectangle in art and architecture are nicely illustrated in the Time-Life book by David Bergamini entitled *Mathematics* (1963, 1970). The ratio 1.6:1 (or simply 1.6) is actually an approximation of the golden ratio. The exact number can be determined by considering the simple geometric method used to construct a golden rectangle, which is shown below. We start with square *ABCD* and extend this square to form golden rectangle *ABEF* by drawing arc *CF*, which is centered at the midpoint *M* of *AD*.

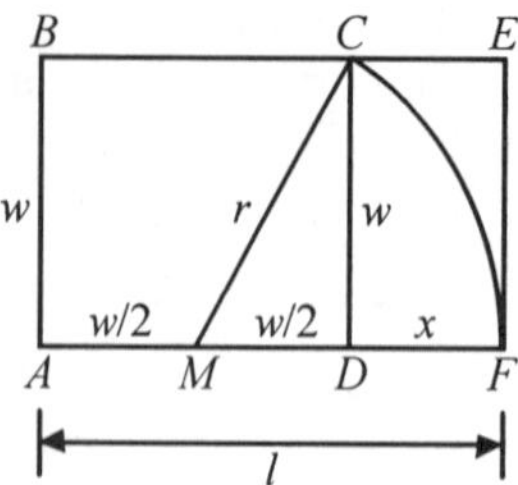

a. Show in rectangle *ABEF* that $l/w=\left(1+\sqrt{5}\right)/2$, which is the exact value of the golden ratio. (*Hint*: Note that $l=(w/2)+r$.)

b. Show that $w/x = (1+\sqrt{5})/2$, which means that rectangle *DCEF* is also a golden rectangle.

c. Subtracting 1 from the golden ratio yields its reciprocal. Verify this.

d. In some literature, the golden ratio is defined as w/l instead of l/w. Using this definition, what is the exact value of the golden ratio in simplest form?

● ● ●

3.6 Complex Numbers

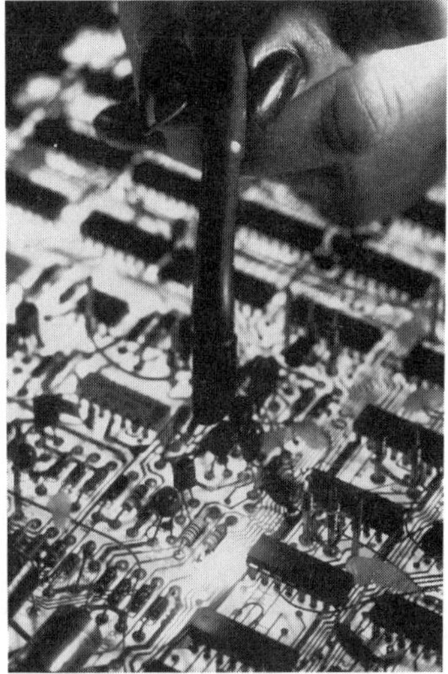

*Photo Courtesy of **Jorge Ramirez**, International Stock*

Ohm's law for alternating current circuits states

$$V = IZ,$$

where V is the voltage in volts, I is the current in amperes, and Z is the impedance in ohms. If the voltage in a particular circuit is $18-21i$ volts, and the current is $3-6i$ amperes, find the complex number that measures the impedance. (See Example 7.)

Objectives

1. Express square roots of negative numbers in terms of i.
2. Add, subtract, and multiply complex numbers.
3. Divide complex numbers.
4. Find powers of i.

In this section we extend the number system beyond real numbers to complex numbers, which have significant applications in mathematics, physics, and engineering. For instance, in the section-opening problem complex numbers are used to describe voltage and other electrical quantities, because such a numerical representation indicates both the strength and time (or phase) relationships of the quantities.

To define a complex number, we first introduce a new set of numbers called **imaginary numbers**, in which square roots of negative numbers are defined. The basic unit in pure imaginary numbers is $\sqrt{-1}$, and it is designed by i. Thus, by definition,

$$i = \sqrt{-1} \text{ and } i^2 = -1.$$

Square roots of negative numbers may now be written in terms of i by defining the principal square root of a negative number as follows.

Principal Square Root of a Negative Number

> If a is a positive number, then
>
> $$\sqrt{-a} = i\sqrt{a}.$$

In this definition we choose to write i in front of any radicals, so that expressions such as $\sqrt{a}\,i$ are not confused with $\sqrt{ai}$.

Example 1: Simplifying Square Roots of Negative Numbers

Express each number in terms of i.

a. $\sqrt{-3}$ **b.** $\sqrt{-4}$ **c.** $\sqrt{-50}$ **d.** $\sqrt{(-1)^2 - 4(2)(2)}$

Solution:

a. $\sqrt{-3} = i\sqrt{3}$

b. $\sqrt{-4} = i\sqrt{4} = 2i$

c. $\sqrt{-50} = i\sqrt{50} = i\sqrt{25}\sqrt{2} = 5i\sqrt{2}$

d. $\sqrt{(-1)^2 - 4(2)(2)} = \sqrt{-15} = i\sqrt{15}$

PROGRESS CHECK 1

Express each number in terms of i.

a. $\sqrt{-9}$ **b.** $\sqrt{-7}$ **c.** $\sqrt{-8}$ **d.** $\sqrt{(-2)^2 - 4(1)(4)}$ ■

Using real numbers and pure imaginary numbers, we may extend the number system to include a number like $3 - 6i$, which was used to describe the current in the section-opening problem. This type of number is called a **complex number**.

Definition of Complex Number

> A number of the form $a + bi$, where a and b are real numbers and $i = \sqrt{-1}$, is called a complex number.

The number a is called the **real part** of $a + bi$, and b is called the **imaginary part** of $a + bi$. Note that both the real part and the imaginary part of a complex number are real numbers. Two complex numbers are **equal** if and only if their real parts are equal and their imaginary parts are equal. That is,

$$a + bi = c + di \text{ if and only if } a = c \text{ and } b = d.$$

If we let a and/or b equal zero, both real numbers and pure imaginary numbers may be expressed in $a + bi$ form. For instance:

Real number → $2 = 2 + 0i$ ↘
Pure imaginary number → $5i = 0 + 5i$ ↗ Complex number in $a + bi$ form

Thus, the complex numbers include the real numbers and the imaginary numbers. Figure 3.10 illustrates the relationship among the various sets of numbers.

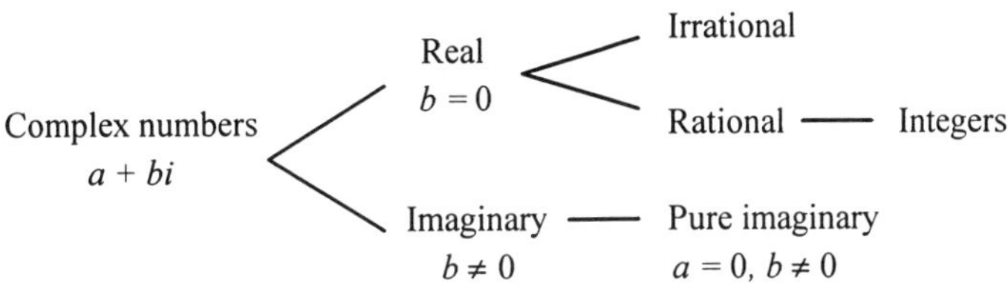

Figure 3.10

Operations with complex numbers are defined so that the properties of real numbers (like the commutative properties) continue to apply, and in many ways computations with complex numbers are similar to computations with polynomials. Consider carefully Examples 2–5, which illustrate how to add, subtract, and multiply with complex numbers.

Example 2: Adding and Subtracting Complex Numbers

Combine the complex number.

a. $\sqrt{-36}+\sqrt{-100}$ **b.** $(7-3i)+(1+6i)$ **c.** $(5+4i)-(9-i)$

Solution:

a. $\sqrt{-36}+\sqrt{-100} = i\sqrt{36}+i\sqrt{100}$
$= 6i + 10i$
$= 16i$

b. $(7-3i)+(1+6i) = (7+1)+(-3+6)i$
$= 8+3i$

c. $(5+4i)-(9-i) = 5+4i-9+i$
$= (5-9)+(4+1)i$
$= -4+5i$

PROGRESS CHECK 2

Combine the complex numbers.

a. $\sqrt{-9}-\sqrt{-25}$ **b.** $(-1-2i)+(8+5i)$ **c.** $(9+3i)-(16-i)$ ■

Example 3: Multiplying Complex Numbers

Multiply the complex numbers.

a. $\sqrt{-25}\sqrt{-4}$ **b.** $(3+4i)(2-6i)$ **c.** $(5+2i)^2$

Solution:

a. $\sqrt{-25}\sqrt{-4} = i\sqrt{25}\,i\sqrt{4}$
$= 5i \cdot 2i$
$= 10i^2$
$= 10(-1)$ Replace i^2 by -1.
$= -10$

b. $(3+4i)(2-6i)$
$= 3(2)+3(-6i)+4i(2)+4i(-6)$ Use FOIL.
$= 6-18i+8i-24i^2$
$= 6-18i+8i-24(-1)$ Replace i^2 by (-1).
$= 30-10i$

c. $(5+2i)^2 = 5^2+2(5)(2i)+(2i)^2$ Use $(a+b)^2 = a^2+2ab+b^2$.
$= 25+20i+4i^2$
$= 25+20i+4(-1)$ Replace i^2 by -1.
$= 21+20i$

Caution: The property of radicals $\sqrt{a}\sqrt{b} = \sqrt{ab}$ does not hold when a and b are both negative. For instance, the solution in part **a** shows

$$\sqrt{-25} \cdot \sqrt{-4} = -10.$$

Therefore,

$$\sqrt{-25} \cdot \sqrt{-4} = \sqrt{(-25)(-4)} = \sqrt{100} = 10 \text{ is } \textbf{wrong}.$$

Always remember to express complex numbers in terms of i before performing computations.

PROGRESS CHECK 3
Multiply the complex numbers.
a. $\sqrt{-16}\sqrt{-9}$ **b.** $(2+3i)(10-i)$ **c.** $(4+7i)^2$ ■

In Chapter 4 we will consider equations with complex number solutions that are not real numbers. The next two examples illustrate how operating on complex numbers can be useful in this context.

Example 4: Verifying a Complex Number Solution to an Equation

Verify that the complex number $3+2i$ is a solution of the equation $(x-3)^2+4=0$.

Solution: We replace x by $3+2i$ and verify.

$$[(3+2i)-3]^2+4 \stackrel{?}{=} 0$$
$$(2i)^2+4 \stackrel{?}{=} 0$$
$$4i^2+4 \stackrel{?}{=} 0$$
$$4(-1)+4 \stackrel{?}{=} 0$$
$$0 \stackrel{\checkmark}{=} 0$$

Note in the check above that $4i^2 = -4$, since $i^2 = -1$. Thus, $3+2i$ is a solution of $(x-3)^2+4=0$.

PROGRESS CHECK 4
Verify that the complex number $-5-3i$ is a solution for the equation $(x+5)^2+9=0$. ■

Example 5: Testing Whether or Not a Complex Number is a Solution

Is $4-3i$ a solution of the equation $x^2-8x+25=0$?

Solution: To check, we replace x by $4-3i$ in the given equation.

$$(4-3i)^2-8(4-3i)+25 \stackrel{?}{=} 0$$
$$16-24i+9i^2-32+24i+25 \stackrel{?}{=} 0$$
$$9i^2+9 \stackrel{?}{=} 0$$
$$-9+9 \stackrel{?}{=} 0$$
$$0 \stackrel{\checkmark}{=} 0$$

Therefore, $4-3i$ is a solution of the equation $x^2-8x+25=0$.

PROGRESS CHECK 5
Is $1+2i$ a solution of the equation $x^2-2x+4=0$? ■

Conjugates and Division

To understand the procedure for dividing complex numbers, we first must define the conjugate of a complex number and then consider some basic properties of conjugates. The complex conjugate or simply **conjugate**, of $a+bi$ is $a-bi$. It is standard notation to denote the conjugate of a number by placing a bar above the number. For example,

$$\overline{3+2i}=3-2i \text{ and } \overline{-4-0i}=-4+0i.$$

We now state some of the basic properties of conjugates that will be needed either here or in Section 6.2.

Properties of Conjugates

If z and w are complex numbers, then

1. $z\cdot\overline{z}$ is a real number.
2. $\overline{z}=z$ if and only if z is a real number.
3. $\overline{z+w}=\overline{z}+\overline{w}$.
4. $\overline{z\cdot w}=\overline{z}\cdot\overline{w}$.
5. $\overline{z^n}=(\overline{z})^n$ for any positive integer n.

Illustrations of these properties with specific complex numbers are given in the exercises, which also outline proofs for properties 2, 3, and 4. (The proof of property 5 requires the use of mathematical induction, which is discussed in the last chapter of the text.) Property 1 is our main interest at this time because it provides the rationale for the procedure for dividing two complex numbers. In words, property 1 states that the product of a complex number and its conjugate is always a real number. To prove this property, we first let $z=a+bi$. Then

$$\begin{aligned} z\cdot\overline{z} &= (a+bi)(a-bi) \\ &= a^2-abi+abi-b^2i^2 \\ &= a^2+b^2. \end{aligned}$$

Since a and b are real numbers, a^2+b^2 is a real number, which proves the property. Because $z\cdot\overline{z}$ is a real number, the quotient w/z can be written in the form $a+bi$ by multiplying both the numerator and the denominator by $\overline{z}$. Examples 6-8 use this procedure.

Example 6: Dividing Complex Numbers

Write the quotient $\dfrac{4+3i}{1-2i}$ in the form $a+bi$.

Solution: As discussed, we divide two complex numbers by multiplying the numerator and the denominator by the conjugate of the denominator. Thus,

$$\frac{4+3i}{1-2i}=\frac{4+3i}{1-2i}\cdot\frac{1+2i}{1+2i}$$
$$=\frac{4+8i+3i+6i^2}{1+2i-2i-4i^2}$$
$$=\frac{-2+11i}{5}=-\frac{2}{5}+\frac{11}{5}i$$

PROGRESS CHECK 6

Write the quotient $\frac{1+5i}{4+3i}$ in the form $a+bi$. ■

Example 7: Using Ohm's Law to Find Impedance

Solve the problem in the section introduction on page 297.

Solution: To find the impedance, which is symbolized by *Z*, first solve $V = IZ$ for *Z*, to obtain

$$Z=\frac{V}{I}.$$

Now replace *V* by $18-21i$ and *I* by $3-6i$ and divide, using the conjugate of the denominator.

$$Z=\frac{18-21i}{3-6i}=\frac{18-21i}{3-6i}\cdot\frac{3+6i}{3+6i}=\frac{54+108i-63i-126i^2}{9+18i-18i-36i^2}=\frac{54+45i-126(-1)}{9-36(-1)}=\frac{180+45i}{45}=4+i$$

The impedance is therefore $4+i$ ohms.

PROGRESS CHECK 7

Find the impedance in an alternating-current circuit in which the voltage is $23-14i$ volts, and the current is $3-4i$ amperes. ■

Example 8: Finding the Reciprocal of a Complex Number

The reciprocal, or multiplicative inverse, of a complex numer z is $1/z$. Write the reciprocal of $4i$ in the form $a+bi$.

Solution: The reciprocal of $4i$ is $1/4i$. Since the conjugate of $0+4i$ is $0-4i$, we have

$$\frac{1}{4i}=\frac{1}{4i}\cdot\frac{-4i}{-4i}=\frac{-4i}{-16i^2}=\frac{-4i}{16}=-\frac{1}{4}i.$$

Thus, the reciprocal of $4i$ is $-(1/4)i$, or $0-(1/4)i$ in standard form. We can verify our result as follows:

$$(4i)\left(-\frac{1}{4}i\right)=-1\cdot i^2=-1(-1)=1.$$

PROGRESS CHECK 8

Find the reciprocal of $1-i$ and write the answer in the form $a+bi$. ■

Powers of *i*

To simplify i^n, where n is a positive integer, consider the cyclic pattern contained in the following simplifications.

$$\begin{aligned} &i^1 = i && i^5 = i^4 \cdot i = 1 \cdot i = i \\ &i^2 = -1 && i^6 = i^4 \cdot i^2 = 1(-1) = -1 \\ &i^3 = i^2 \cdot i = (-1)i = -i && i^7 = i^4 \cdot i^3 = 1(-i) = -i \\ &i^4 = i^2 \cdot i^2 = (-1)(-1) = 1 && i^8 = \left(i^4\right)^2 = 1^2 = 1 \end{aligned}$$

Continuing this pattern to higher powers of i gives the cyclic property shown in Figure 3.11. Note in particular that $i^n = 1$ if n is a multiple of 4, because we may use this observation to simplify large powers of i, as shown in Example 9.

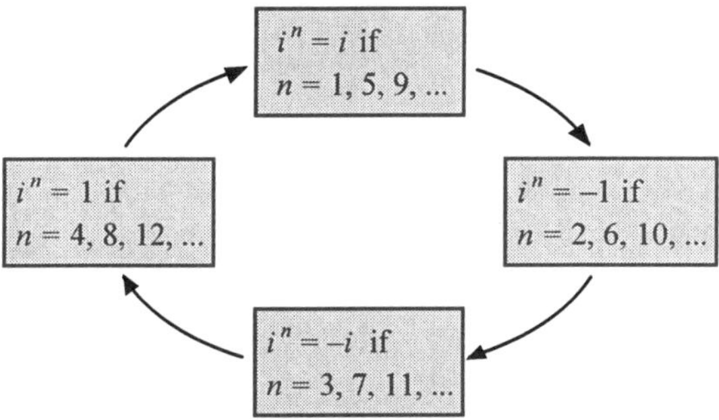

Figure 3.11
Powers of i^n

Example 9: Simplifying Powers of *i*

Simplify each power of i.

a. i^{21} **b.** i^{95}

Solution:

a. The largest multiple of 4 that is less than or equal to the exponent is 20. Therefore, rewrite i^{21} with i^{20} as one factor and then simplify.

$$i^{21} = i^{20} \cdot i = \left(i^4\right)^5 \cdot i = (1)^5 \cdot i = i$$

b. Because $i^{92} = \left(i^4\right)^{23} = 1$, we have

$$i^{95} = i^{92} \cdot i^3 = 1(-i) = -i.$$

PROGRESS CHECK 9

Simplify each power of i.

a. i^{10} **b.** i^{101} ■

In conclusion, note that we have been operating with complex numbers according to the following definitions. You should use the example problems of this section as a basis for understanding these definitions.

Operations on Complex Numbers

1. Two complex numbers are added by adding separately their real parts and their imaginary parts.

$$(a+bi)+(c+di)=(a+c)+(b+d)i$$

2. Two complex numbers are subtracted by subtracting separately their real parts and their imaginary parts.

$$(a+bi)-(c+di)=(a-c)+(b-d)i$$

3. Two complex numbers are multiplied as two binomials are multiplied, with i^2 being replaced by -1.

$$\begin{aligned}(a+bi)(c+di) &= ac+adi+bci+bdi^2 \\ &= (ac-bd)+(ad+bc)i\end{aligned}$$

4. Two complex numbers are divided by multiplying the numerator and the denominator by the conjugate of the denominator.

$$\frac{a+bi}{c+di}=\frac{(a+bi)(c-di)}{(c+di)(c-di)}=\frac{(ac+bd)+(bc-ad)i}{c^2+d^2}$$

Technology Link

A complex number capability is a feature of some graphing calculators. If your calculator can perform operations with complex numbers, then you should redo the problems in Examples 1–9 and compare your results to the text's answers. Be advised that for some calculator models the complex number

$a+bi$ is denoted by (a, b).

To illustrate, Figure 3.12 shows the computations involved in Example 2a, Example 7, and Example 9b, respectively, using a calculator that works with this notation. In the last line of this display, note that the discrepancy from 0 in the real part of the computed result is due to round-off error and should be ignored.

```
√-36 + √-100
                    (0, 16)
(18, -21)/(3, -6)
                     (4, 1)
(0, 1)^95
              (4.5E-12, -1)
```

Figure 3.12

EXPLORE 3.6

These explorations may be worked using a calculator with complex number capability.

1. Evaluate $\sqrt{-1}$. Determine if your calculator produces an error message, the letter i, or the ordered pair $(0, 1)$. In the ordered pair notation, the pair (a, b) represents $a + bi$. For instance, if you enter $(0, 1)^2$ to calculator i^2, you will get $(-1, 0)$ as the answer.
2. a. Confirm by calculator that $\sqrt{-3}\sqrt{-3} = -3$.
   b. Use a calculator to evaluate i^n, for various positive integers n, such as 2, 10, 41, and 100. Are there some values of n for which your calculator does not get the answer exactly?
   c. Use a calculator to evaluate $2^{(i^2)}$, $(2^i)^2$, $(2^2)^i$, and $2^{(2^i)}$. Are any of these expressions equal? Are any of them real numbers?
3. a. There is usually a menu choice to calculate just the real part, or just the imaginary part of a complex number. Use this feature to confirm that the real part of $(3+4i)^3$ is -117, and that the imaginary part is 44.
   b. The *absolute value* of $a + bi$ is defined as $\sqrt{a^2 + b^2}$. This can be calculated automatically by using the Absolute Value feature. Use this feature to find the absolute value of $3 - 4i$. Explain why this definition of the absolute value of a complex number works for real numbers, too.
4. Explain what happens when you graph $y = (x + 2i)(x - 2i)$. Why do you do a better job than the calculator does?

EXERCISES 3.6

In Exercises 1–20 simplify each expression in terms of i.

1. $\sqrt{-25}$ **2.** $\sqrt{-49}$ **3.** $2\sqrt{-16}$

4. $5\sqrt{-121}$ **5.** $\sqrt{-22}$ **6.** $-3\sqrt{-7}$

7. $\frac{1}{2}\sqrt{-12}$ **8.** $-3\sqrt{-98}$ **9.** $2 - \sqrt{-128}$

10. $-3 + \frac{1}{2}\sqrt{-162}$ **11.** $\sqrt{(-4)^2 - 4(1)(5)}$

12. $\sqrt{(12)^2 - 4(9)(8)}$ **13.** $\sqrt{(2)^2 - 4(-1)(-3)}$

14. $\sqrt{(-5)^2 - 4(2)(7)}$

15. $\dfrac{-(-1) - \sqrt{(-1)^2 - 4(2)(3)}}{2(2)}$

16. $\dfrac{-(0) + \sqrt{(0)^2 - 4(1)(2)}}{2(1)}$

17. $\dfrac{-(4) + \sqrt{(4)^2 - 4(1)(5)}}{2(1)}$

18. $\dfrac{-(-2) - \sqrt{(-2)^2 - 4(1)(5)}}{2(1)}$

19. $\dfrac{-(2) - \sqrt{(2)^2 - 4(5)(3)}}{2(5)}$

20. $\dfrac{-(-2) + \sqrt{(-2)^2 - 4(-5)(-3)}}{2(-5)}$

In Exercises 21–40 perform the indicated operations and write each result in the form $a + bi$.

21. $2\sqrt{-9} - 3\sqrt{-9}$ **22.** $4\sqrt{-1} - 2\sqrt{-4}$

23. $\sqrt{-8} + \sqrt{-18} - \sqrt{-50}$

24. $4\sqrt{-27} - 2\sqrt{-48} + 3\sqrt{-75}$

25. $(-1-2i) + (-2+i)$ **26.** $(11+2i) - (-5-7i)$

27. $(3-i) - (-2+5i)$ **28.** $(6+2i) + (2-4i)$

29. $\sqrt{-3} \cdot \sqrt{-12}$ **30.** $\sqrt{-5} \cdot \sqrt{-18}$

31. $(2i)^2$ **32.** $(-i)^2$ **33.** $(4+2i)(1-3i)$

34. $(2-5i)(3-i)$ **35.** $(2+3i)(2-3i)$

36. $(6-5i)(6+5i)$ **37.** $(1-2i)^2$

38. $(5+i)^2$ **39.** $\left(-3+\frac{1}{3}i\right)^2$ **40.** $\left(-1-\frac{1}{2}i\right)^2$

In Exercises 41–50 verify that the given number is a solution of the equation.

41. $x^2+25=0;\ 5i$ **42.** $x^2+9=0;\ -3i$

43. $x^2+8=0;\ -2i\sqrt{2}$ **44.** $x^2+48=0;\ 4i\sqrt{3}$

45. $x^2-2x+2=0;\ 1-i$

46. $x^2-2x+2=0;\ 1+i$

47. $2x^2-x+3=0;\ \dfrac{1}{4}+i\dfrac{\sqrt{23}}{4}$

48. $2x^2-x+3=0;\ \dfrac{1}{4}-i\dfrac{\sqrt{23}}{4}$

49. $x^4-16=0;\ 2i$ **50.** $x^4+13x^2-48=0;\ -4i$

In Exercises 51–60 determine by checking if the given number is a solution of the equation.

51. $x^2+4=0;\ -2i$ **52.** $x^2-4=0;\ 2i$

53. $x^2-4x+5=0;\ 2+i$

54. $x^2-4x+5=0;\ 2-i$

55. $2x^2+x+1=0;\ -1+2i$

56. $2x^2+x+1=0;\ -1-2i$

57. $3x^2-2x+1=0;\ \dfrac{1}{3}-i\dfrac{\sqrt{2}}{3}$

58. $3x^2-2x+1=0;\ \dfrac{1}{3}+i\dfrac{\sqrt{2}}{3}$

59. $x^3-27=0;\ 3i$ **60.** $x^4-81=0;\ -3i$

In Exercises 61–66 find the conjugate of the given complex number.

61. $3+4i$ **62.** $5-7i$ **63.** i

64. $-4i$ **65.** -7 **66.** 7

In Exercises 67–76 write each quotient in the form $a+bi$.

67. $\dfrac{1}{1+2i}$ **68.** $\dfrac{-3}{2-3i}$ **69.** $\dfrac{-2}{5i}$ **70.** $\dfrac{1}{i}$

71. $\dfrac{i}{-1-4i}$ **72.** $\dfrac{-2i}{6-2i}$ **73.** $\dfrac{2+i}{3+i}$

74. $\dfrac{2+5i}{4+2i}$ **75.** $\dfrac{\sqrt{2}-i}{\sqrt{2}+i}$ **76.** $\dfrac{(1/2)+i}{(1/2)-i}$

In Exercises 77–80 find the reciprocal and write the answer in the form $a+bi$.

77. $-i$ **78.** $3i$ **79.** $1+i$ **80.** $-2-2i$

In Exercises 81–88 simplify each power of i.

81. i^3 **82.** i^4 **83.** i^7 **84.** i^9

85. i^{36} **86.** i^{43} **87.** i^{93} **88.** i^{100}

In Exercises 89 and 90 verify that

a. $\overline{z+w}=\overline{z}+\overline{w}$, **b.** $\overline{z\cdot w}=\overline{z}\cdot\overline{w}$, and **c.** $\overline{z^2}=(\overline{z})^2$

for the given values of z and w.

89. $z=1+i,\ w=2-3i$ **90.** $z=4-i,\ w=5+2i$

In Exercises 91–94 let $z=a+bi$ and $w=c+di$, and use the definition of conjugate to prove each of the following properties.

91. $\overline{z+w}=\overline{z}+\overline{w}$ **92.** $\overline{z\cdot w}=\overline{z}\cdot\overline{w}$

93. If z is a real number, then $\overline{z}=z$.

94. If $\overline{z}=z$, then z is a real number.

Ohm's law for alternating current circuits states that $V=IZ$, where V is the voltage in volts, I is the current in amperes, and Z is the impedance in ohms. Use this law to solve Exercises 95–100.

95. Find the complex number that represents the impedance in a particular circuit if the voltage is $14+8i$ volts and the current is $1+2i$ amperes.

96. Find the complex number that represents the impedance in a given circuit if the voltage is $12-5i$ volts and the current is $2-3i$ amperes.

97. Find the complex number that represents the voltage in a given circuit if the impedance is $240-50i$ ohms and the current is $0.4+0.3i$ amperes.

98. Find the complex number that represents the voltage in a given circuit if the impedance is $10+i$ ohms and the current is $5-0.5i$ amperes.

99. Find the complex number that represents the current in a given circuit if the voltage is 12 volts and the impedance is $1-i$ ohms.

100. Find the complex number that represents the current in a given circuit if the voltage is 120 volts and the impedance is $-3i$ amperes.

THINK ABOUT IT 3.6

1. **a.** Complex numbers are not ordered like real numbers. The relations "less than" and "greater than" do not apply to complex numbers. But they can be assigned a magnitude, similar to the concept of the absolute value of a real number. The **absolute value** of the complex number $a+bi$ is defined as $\sqrt{a^2+b^2}$. Find the absolute value of these complex numbers.

$$1+i;\ 1-i;\ i;\ \frac{\sqrt{2}}{2}+\frac{\sqrt{2}}{2}i$$

   b. Just as a real number can be represented by a point on a line, a complex number can be represented by a point in a plane. The number $a+bi$ is represented by the point (a,b). Find the point represented by each of the complex numbers in part **a**. Verify that the absolute value of each number equals its distance from the origin.
2. Not only real numbers but *all* complex numbers (except zero) have two square roots. Show that $\sqrt{2}/2+(\sqrt{2}/2)i$ and $-\sqrt{2}/2-(\sqrt{2}/2)i$ are both square roots of i. To do this, show that the square of each number is i.
3. Determine b so that $1+i$ is a solution of the equation $x^2+2=bx$.
4. If m and n are positive integers and $i^m=i^n$, describe the relation between m and n.
5. Because complex numbers can be represented by ordered pairs, it is possible to define the operations for them entirely in terms of ordered pairs without any reference to i. For instance, the sum of (a,b) and (c,d) is defined as $(a+c,b+d)$. This takes the place of saying that the sum of $a+bi$ and $c+di$ equals $(a+c)+(b+d)i$.
   a. Show that the product of (a,b) and (c,d) should be defined as $(ac-bd, ad+bc)$.
   b. What should the definition be of $(a,b)\div(c,d)$?

● ● ●

CHAPTER 3 SUMMARY

OBJECTIVES CHECKLIST

Specific chapter objectives are summarized below, along with numbered example problems from the text that should clarify the objectives. If you do not understand any objectives, or do not know how to do the selected problems, then restudy the material.

3.1: Can you:

1. **Factor out the greatest common factor (GCF)?** Factor completely: $21x^2-7x$. [Example 1a]
2. **Factor by grouping?** Factor by grouping: $3x^2+6x+4x+8$. [Example 4a]
3. **Factor trinomials of the form ax^2+bx+c?** Factor completely: $4c^2-12c+5$ [Example 6a]
4. **Factor out the GCF and then factor trinomials?** Factor completely: $18x^3+15x^2-12x$. [Example 9]
5. **Use b^2-4ac to determine if ax^2+bx+c can be factored?** Calculate b^2-4ac and determine if $9x^2-9x-4$ can be factored into binomial factors with integer coefficients. [Example 10]

3.2: Can you:

1. **Factor an expression that is the difference of squares?** Factor completely: x^2-81. [Example 1a]
2. **Factor an expression that is the sum or difference of cubes?** Factor completely: $8x^3+27$. [Example 4a]

3. **Apply a general strategy to factor expressions systematically?** Factor completely: $x^6 - a^6$.
[Example 8]

3.3: Can you:

1. **Express a fraction in simplest form?** Simplify $\dfrac{x^2 - y^2}{y - x}$.
[Example 2]

2. **Multiply fractions and simplify the product?** Simplify $\dfrac{5x^2 + 15x}{4 - x^2} \cdot \dfrac{12 - 6x}{(3 + x)^2}$.
[Example 3]

3. **Divide fractions and simplify the quotient?** Simplify $\dfrac{(y-2)^2}{5y} \div \dfrac{y^2 - 4}{15y}$.
[Example 4]

4. **Add and subtract fractions and write the result in simplest form?** Simplify $\dfrac{2a+3}{a^2 + a} - \dfrac{a}{a^2 - 1}$.
[Example 7]

5. **Change a complex fraction to a fraction in simplest form?** Simplify $\dfrac{x^{-1} + y^{-1}}{(x/y) - (y/x)}$.
[Example 9a]

3.4: Can you:

1. **Find principal *n*th roots?** Find $\sqrt[3]{-27}$.
[Example 1c]

2. **Evaluate expressions containing rational exponents?** Evaluate $49^{3/2}$.
[Example 3b]

3. **Simplify certain expressions involving rational exponents by using exponent properties?** Simplify by performing the indicated operations and writing your results with only positive exponents. Also, express the result in radical form.

$$\frac{2^{-1/2} \cdot 3x^{1/3}}{2 \cdot 3^{1/4} x}$$

[Example 5d]

4. **Simplify certain expressions involving rational exponents by factoring out the smallest power of the variable from each term?** Simplify by factoring and then writing the result with only positive exponents.

$$10x^{-4} + 15x^{-2}$$

[Example 7a]

3.5: Can you:

1. **Express a radical in simplified form?** Simplify $\sqrt{8x^9y^6}$. Assume $x \geq 0$, $y \geq 0$.
[Example 1a]

2. **Add and subtract radicals, and simplify where possible?** Simplify where possible: $\sqrt{50x^3y} - \sqrt{8xy^3}$
[Example 6c]

3. **Multiply radicals, and simplify where possible?** Simplify. Assume $x > 0$. $(5+\sqrt{x})^2$
[Example 9b]

4. **Divide radicals, and simplify where possible?** Divide and simplify where possible. Assume $x > 0$.

$$\frac{\sqrt[3]{48x}}{\sqrt[3]{2x}}$$

[Example 10c]

5. **Rationalize the denominator or the numerator of a fractional expression containing radicals?** Rationalize the denominator of $\frac{10}{3+\sqrt{7}}$.
[Example 12]

6. **Simplify expressions containing radicals by using combinations of the basic operations with radicals?** Write as a single fraction: $x \cdot \frac{2x}{2\sqrt{x^2+5}} + \sqrt{x^2+5}$.
[Example 14]

3.6: Can you:

1. **Express square roots of negative numbers in terms of *i*?** Express in terms of i: $\sqrt{-50}$
[Example 1c]

2. **Add, subtract, and multiply complex numbers?** Multiply: $(3+4i)(2-6i)$.
[Example 3b]

3. **Divide complex numbers?** Write the quotient $\frac{4+3i}{1-2i}$ in the form $a+bi$.
[Example 6]

4. **Find powers of *i*?** Simplify i^{21}.
[Example 9a]

KEY CONCEPTS AND PROCEDURES

Section	Key Concepts and Procedures to Review
3.1	• Definitions of polynomial, monomial, binomial, trinomial, degree of a monomial, and degree of a polynomial. • Methods to factor an expression by factoring out the greatest common factor, by factoring using grouping, or by factoring a trinomial using FOIL reversal or the *ac* method. • Method to determine whether or not ax^2+bx+c can be factored into binomial factors with integer coefficients.
3.2	• Methods to factor an expression that is a difference of squares, or that is a sum or difference of cubes. • Summary of factoring models. • Guidelines to factoring a polynomial.
3.3	• Methods to simplify, multiply, and divide fractions • Equality principle: $\frac{a}{b}=\frac{c}{d}$ if and only if $ad=bc$ $(b, d \neq 0)$ • Fundamental principle: $\frac{ak}{bk}=\frac{a}{b}$ $(b, k \neq 0)$

- Multiplication principle: $\frac{a}{b}\cdot\frac{c}{d}=\frac{ac}{bd}$ $(b, d \neq 0)$
- Division principle: $\frac{a}{b}\div\frac{c}{d}=\frac{a}{b}\cdot\frac{d}{c}$ $\left(b, d, \frac{c}{d}\neq 0\right)$
- Definition of least common denominator
- Methods to add and subtract fractions and to find the least common denominator
- Addition or subtraction principles $(b, d \neq 0)$:

$$\frac{a}{b}\pm\frac{c}{b}=\frac{a\pm c}{b} \qquad \frac{a}{b}\pm\frac{c}{d}=\frac{ad\pm bc}{bd}$$

- Definition of a complex fraction
- Methods to simplify a complex fraction

3.4

- Definitions of $a^{1/n}$, index and radicand of a radical, principal square root, and nth root of a number
- If m/n represents a reduced fraction such that $a^{1/n}$ represents a real number, then $a^{m/n}=\left(a^{1/n}\right)^m=\left(a^m\right)^{1/n}$, or equivalently, $a^{m/n}=\left(\sqrt[n]{a}\right)^m=\sqrt[n]{a^m}$.

3.5

- Methods to simplify a radical by removing any factor of the radicand whose indicated root can be taken exactly, by writing the index as small as possible, or by eliminating any fractions in the radicand
- Definition of like radicals
- Methods to add and subtract radicals
- Properties of radicals (a, b, $\sqrt[n]{a}$, $\sqrt[n]{b}$ denote real numbers)

  1. $\sqrt[n]{a\cdot b}=\sqrt[n]{a}\cdot\sqrt[n]{b}$
  2. $\sqrt[n]{\frac{a}{b}}=\frac{\sqrt[n]{a}}{\sqrt[n]{b}}$ $(b\neq 0)$
  3. $\left(\sqrt[n]{a}\right)^n=a$
  4. $\sqrt[n]{a^n}=\begin{cases} a, & \text{if } n \text{ is odd}\\ |a|, & \text{if } n \text{ is even}\end{cases}$

- Definition of conjugates
- Methods to multiply and divide radicals
- Methods to rationalize the denominator or the numerator

3.6

- Definitions of imaginary number, principal square root of a negative number, complex number, equality for complex numbers, and the conjugate of a complex number
- $i=\sqrt{-1}$ and $i^2=-1$
- Relationships among the various sets of numbers
- Properties of conjugates
- Methods to find powers of i
- Methods to add, subtract, multiply, and divide complex numbers

CHAPTER 3 REVIEW EXERCISES

In Exercises 1–60 perform the indicated operations and express your result in the simplest form that contains only positive exponents.

1. $\frac{2}{2^{-1/2}}$
2. $\left(3\sqrt{7}\right)^2$
3. $5\sqrt{12}-2\sqrt{27}$
4. $2^0-\left(\frac{4}{9}\right)^{-3/2}$
5. $\left(1-\sqrt{3}\right)^2$
6. $\sqrt{\frac{4}{3}}+\sqrt{\frac{3}{4}}$
7. $\sqrt{\left(-\pi\sqrt{3}\right)^2+(3\pi)^2}$

8. $(-2)^{-1}\cdot(-2)^{-3/5}$

9. $-6\left(-\frac{\sqrt{3}}{2}\right)-4\left(\frac{\sqrt{3}}{2}\right)(-1)$

10. $1\div\left(3+\sqrt{2}\right)$

11. $\frac{6R^2}{\sqrt{3R^2}},\ R>0$

12. $\frac{a^{-2/3}}{\sqrt[3]{a}}$

13. $\sqrt{\frac{1+(4/5)}{2}}$

14. $\frac{1}{\sqrt{3}}+\frac{2\pi}{3}$

15. $\frac{\left(x^{-2/3}x^{1/6}\right)^{-1}}{x^{1/3}}$

16. $2\pi h\sqrt{\frac{100}{\pi h}},\ h>0$

17. $\frac{\sqrt{3}+\sqrt{2}}{\sqrt{3}}$

18. $\sqrt{18x^7y^4},\ x,y\ge 0$

19. $\frac{2\sqrt{2}h\pi}{\sqrt{2}\pi h^2}$

20. $\left[\frac{\sqrt{3}/2}{\sqrt{1-(3/4)}}-\frac{\pi}{3}\right]\div\frac{3}{4}$

21. $\frac{\sin\theta\cos^3\theta-4\sin^2\theta\cos\theta}{\sin\theta\cos\theta}$

22. $\left(2+\sin^3\theta\right)^2$

23. $\frac{1}{2}-\frac{x-1}{2}$

24. $(2\cos\theta+1)(3\cos\theta-2)$

25. $\left(a^x-3\right)\left(a^x-1\right)$

26. $\frac{(3x)^2}{y}\left(\frac{2zy}{x^2}\right)^3$

27. $\frac{(x+h)^4-x^4}{h}$

28. $(2x+4)\left(4+x-x^2\right)$

29. $\frac{x+1}{x-1}-\frac{4}{1-x}$

30. $\frac{(c-3)(c+3)(c-3)}{(3-c)(-3+c)(3+c)}$

31. $\frac{(\cos\theta-1)^2}{\cos^2\theta-1}$

32. $\frac{2}{x}+\frac{3}{y}$

33. $\frac{10ax^2}{21y^4}\cdot\frac{7y}{5ax}$

34. $\frac{\cos\theta}{\left(1/\cos^2\theta\right)+(2/\cos\theta)}$

35. $\frac{2k+1}{k-1}-\frac{3k-4}{k-1}$

36. $\frac{(a-1)x-(1-a)y}{(a-1)xy}$

37. $\frac{2\sin\theta}{5}-\frac{7}{3\sin\theta}+\frac{2}{\sin\theta}$

38. $\frac{x^2+1}{4x+4}\cdot\frac{(x+1)^2}{x^2+x}$

39. $\frac{1-(1/s)}{1+(1/s)}$

40. $\left(\frac{x}{1+x}-\frac{1-x}{x}\right)\div\left(\frac{x-1}{x}-\frac{x}{x+1}\right)$

41. $\left[\frac{2}{x+h+1}-\frac{2}{x+1}\right]\div h$

42. $\frac{2t}{t^2+5t+6}+\frac{5t}{t^2+2t-3}$

43. $\frac{\sin^2\theta-5\sin\theta+6}{\sin\theta+6}\div(\sin\theta-2)$

44. $\frac{(1+h)^3-1}{(1+h)-1}$

45. $\frac{1}{x^n}+\frac{1}{x^{n+1}}$

46. $x^2+\frac{kx^2}{2x-k}$

47. $\frac{(x+y)^2-4xy}{x^2-y^2}$

48. $\frac{\left(2x^2-1\right)-\left(2a^2-1\right)}{x-a}$

49. $\frac{1}{\tan\theta}-\frac{1}{\tan\theta+1}$

50. $6\sqrt[3]{5x}-2\sqrt[3]{5x}+\sqrt[3]{5x}$

51. $\frac{\left(x^2+x\right)-\left(a^2+a\right)}{x-a}$

52. $\frac{\left(1+x^{-1}\right)\cdot 1-x\left(-x^{-2}\right)}{\left(1+x^{-1}\right)^2}$

53. $\left[\frac{1}{x}-\frac{1}{a}\right]\div(x-a)$

54. $x^{1/3}+\frac{1}{2}x^{-1/3}$

55. $(x-3)\left(\frac{1}{2}x^{-1/2}\right)+x^{1/2}\cdot 1$

56. $\left[(x+h)^{1/3}-x^{1/3}\right]\cdot\left[(x+h)^{2/3}+x^{1/3}(x+h)^{1/3}+x^{2/3}\right]$

57. $\frac{1}{\sqrt{1-x^2}}-x\cdot\frac{-x}{\sqrt{1-x^2}}-\sqrt{1-x^2}$

58. $\sqrt{R^2-y^2}(-1)+(R-y)\frac{-2y}{2\sqrt{R^2-y^2}}$

59. $\frac{u\cdot\left(2u/2\sqrt{u^2+4}\right)-\sqrt{u^2+4}}{u^2}$

60. $\frac{\left(1/\sqrt{x+h}\right)-\left(1/\sqrt{x}\right)}{h}$

In Exercises 61–66 perform the indicated operations and simplify.

61. $3\sqrt{-1}-2\sqrt{-4}$

62. $\frac{-(2)+\sqrt{(2)^2-4(3)(1)}}{2(3)}$

63. $(1+i)^2$

64. $(1-\sqrt{-7})(1+\sqrt{-7})$

65. $\dfrac{3}{3-2i}$

66. i^{92}

In Exercises 67–74 factor each of the expressions completely.

67. $\cos^2\theta + 8\cos\theta + 16$

68. $t^2 + t^2x^3$

69. $t^3 - t^3x^2$

70. $x^2 - 8x + 12$

71. $x^3 + x^2 + x + 1$

72. $4c^2 - 2ct - 12t^2$

73. $2\sin^3\theta\cos\theta - 18\sin^2\theta\cos\theta + 40\sin\theta\cos\theta$

74. $2x(h-k) + (h^2 - k^2) - (h-k)$

75. Verify through multiplication the following factoring model.

$$a^3 + b^3 = (a+b)(a^2 - ab + b^2)$$

76. Calculate $b^2 - 4ac$ for the expression $2c^2 - 3c + 7$. Can the expression be factored into binomial factors with integer coefficients?

In Exercises 77 and 78 determine the least common denominator.

77. $\dfrac{5}{18x^2yz^3}, \dfrac{11}{24xy^2z}$

78. $\dfrac{1}{t^2-4}, \dfrac{t+4}{t^2-4t+4}$

79. Rationalize the numerator in the answer to Exercise 60.

80. Rationalize the denominator: $\dfrac{x-2}{\sqrt{x^2-4}}$.

81. Rationalize the numerator: $\dfrac{\sqrt{x-2}}{x^2-4}$.

82. If $f(x) = x^{3/2} + 2x^0$, find $f(4)$.

83. Verify that $1+i$ is a solution of $x^2 - 2x + 2 = 0$.

84. Find the product of $3+2i$ and its conjugate.

85. If $f(x) = 1$ and $g(x) = \dfrac{x+1}{x-1}$, write $f(x) + g(x)$ as a single fraction in simplest form.

86. Show that $\left[\dfrac{1}{1+(1/n)}\right]^p$ is equivalent to $\dfrac{n^p}{(n+1)^p}$.

87. Show that $\dfrac{(x-1)^3 - (x+1)\cdot 3(x-1)^2}{(x-1)^6}$ simplifies to $\dfrac{-2(x+2)}{(x-1)^4}$.

88. Simplify

$$n\left(\frac{2}{n}\right) + \frac{1}{2}n^2\left(\frac{2}{n}\right)^2 + \frac{1}{4}\left(4n^3 - n\right)\left(\frac{2}{n}\right)^3 + \frac{1}{8}\left(2n^4 - n^2\right)\left(\frac{2}{n}\right)^4.$$

89. Write an alternative formula for the area of the shaded trapezoid by simplifying the right side of the equation. The triangles are equilateral.

$$A = \frac{1}{2}\left(s + \frac{s}{k}\right)\left(\frac{s\sqrt{3}}{2} - \frac{s\sqrt{3}}{2k}\right).$$

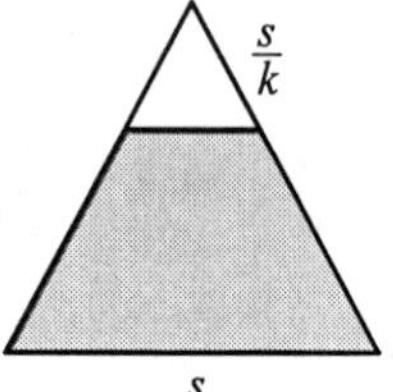

90. A person drives to work at an average speed of 24 miles per hour and returns on the same route at an average speed of 36 miles per hour. What is the average speed for the round trip?

91. An oil spill is dispersing such that the amount remaining after t hours is given approximately by $A = A_0 3^{-0.01t}$. What percent of the original amount is left after 24 hours?

92. Ohm's law for alternating current states that $V = IZ$, where V is voltage in volts, I is current in amperes, and Z is impedance in ohms. Find the impedance in a particular circuit if the voltage is $10+6i$ volts, and the current is $2+3i$ amperes.

In Exercises 93–100 select the choice that completes the statement or answers the question.

93. The fraction $\dfrac{x-y}{4}$ is *not* equal to

a. $\dfrac{y-x}{-4}$ **b.** $-\dfrac{x-y}{-4}$

c. $\dfrac{-x+y}{-4}$ **d.** $\dfrac{y-x}{4}$

94. $\dfrac{(R+r)^2E^2 - RE^2\cdot 2(R+r)}{(R+r)^4}$ simplifies to

a. $\dfrac{rE^2(r-2R)}{(R+r)^4}$ **b.** $\dfrac{E^2(r-2R)}{(R+r)^3}$

c. $\dfrac{E^2(r-R)}{(R+r)^3}$ **d.** $\dfrac{E^2(r-R)^2}{(R+r)^4}$

95. $\left(1-\dfrac{\sin x}{\cos x}\right)\div\left(\cos x-\dfrac{\sin^2 x}{\cos x}\right)$ simplifies to

a. $\cos x-\sin x$ **b.** $\cos x+\sin x$

c. $\dfrac{1}{\cos x+\sin x}$ **d.** $\dfrac{1}{\cos x-\sin x}$

96. The reciprocal of $3+\dfrac{1}{n}$ is

a. $\dfrac{3n+1}{3}$ **b.** $\dfrac{n}{3n+1}$

c. $\dfrac{3n+1}{n}$ **d.** $3+n$

97. The expression $\dfrac{a^{-1}+b^{-1}}{a^{-1}}$ is equivalent to

a. $\dfrac{1}{b}$ **b.** $\dfrac{a+b}{b}$

c. $\dfrac{b}{a+b}$ **d.** $\dfrac{a}{a+b}$

98. $\left(1+\dfrac{1}{n}\right)\left(\dfrac{1}{n+1}-1\right)$ simplifies to

a. -1 **b.** 1

c. n **d.** $n+1$

99. The expression $\dfrac{\sqrt{2}+1}{\sqrt{2}-1}$ is equivalent to

a. 3 **b.** -1

c. $5+\sqrt{2}$ **d.** $3+2\sqrt{2}$

100. The expression $x^{-1/3}$ is equivalent to

a. $\sqrt[3]{x}$ **b.** $-\sqrt[3]{x}$

c. $\dfrac{1}{\sqrt[3]{x}}$ **d.** $\dfrac{-1}{\sqrt[3]{x}}$

CHAPTER 3 TEST

In Questions 1–10 perform the indicated operations and/or simplify.

1. $\dfrac{y^2-7y+12}{y^2-3y}$ **2.** $\dfrac{1/a}{(1/a)+(1/x)}$

3. $(5+\sqrt{3})(5-\sqrt{3})$ **4.** $(\sqrt{x}-\sqrt{a})^2$ for x, $a>0$

5. $\sqrt{56}-\sqrt{\dfrac{2}{7}}$ **6.** $\dfrac{\sin\theta}{\cos\theta}-\dfrac{\cos\theta}{\sin\theta}$

7. $\dfrac{4}{1-x}-\dfrac{x+3}{x-1}$ **8.** $\dfrac{1-x}{9x+6}+\dfrac{x-2}{6x+4}$

9. $\dfrac{4\cos\theta-6}{7}\div\dfrac{6\cos\theta-9}{35}$

10. $\dfrac{(x-1)^2}{(x+2)^2}\cdot\dfrac{x^2-4}{x^2-1}$

In Questions 11–14 factor each expression completely.

11. x^3-x **12.** $27x^3-8y^3$

13. $20\sin^2\theta+2\sin\theta-6$

14. $4t^3-12t^2+9t$

15. Evaluate $(3+2i)(3-2i)$

16. Multiply $a^{1/2}\cdot a^{2/5}$ and write the result in radical form.

17. Rationalize the denominator: $\dfrac{12}{3-\sqrt{3}}$.

18. Find the least common denominator: $\dfrac{1}{6x^2y^2}$, $\dfrac{4}{9xy^3}$.

19. Simplify and write the result using only positive exponents: $\left(\dfrac{xy^{-1/3}}{x^{-3/2}y^{2/3}}\right)^{-1}$.

20. A race car driver averages 180 mi/hour on the first lap around a track, but mechanical trouble reduces the average speed on the second lap to only 90 mi/hour. What is the average speed for the first two laps?

Chapter

4

Equations and Inequalities

Photo Courtesy of S. Tanaka, The Picture Cube

An airline offers a charter flight at a fare of \$99 per person if 100 passengers sign up. The plane has a capacity of 200 seats, so the airline provides a quantity discount and offers to reduce the fare of each passenger by 50 cents for each passenger above 100 who signs up.

a. Explain why the total revenue for the flight is given by

$$R = (100 + x)(99 - 0.5x)$$

where x represents the number of passengers above 100.

b. Use a grapher to graph the equation in part **a** and describe what happens to revenue as the number of passengers increases.

c. Use the Trace feature to estimate all passenger counts that produce a revenue of \$11,040 for the flight.

d. Find algebraically all passenger counts sought in part **c**.

(See Example 11 of Section 4.1).

Setting up and solving equations and inequalities is often an essential part of the mathematical analysis of a problem, and in this chapter we discuss some basic techniques and applications that involve quadratic, fractional, radical, polynomial, and absolute value equations and inequalities. But the topic of equations and inequalities does not end here. In subsequent chapters, when we introduce a new idea (such as a logarithm) you will need to extend your skills in solving equations and inequalities to include this new concept.

● ● ●

4.1 Quadratic Equations

Objectives

1. Solve quadratic equations using a specified method.
2. Find intercepts of graphs of equations of the form $y = ax^2 + bx + c$.
3. Use the discriminant to determine the number of real roots of a quadratic equation or the number of x-intercepts in the graph of $y = ax^2 + bx + c$.
4. Solve quadratic equations after choosing an efficient method.
5. Solve applied problems involving quadratic equations.

In the chapter-opening problem the airline can find all passenger counts that produce a revenue of \$11,040 by solving the equation

$$11{,}040 = (100 + x)(99 - 0.5x).$$

This equation is an example of a **second-degree** or **quadratic equation** because it can be written in the standard form $ax^2 + bx + c = 0$, where a, b, and c are real numbers with $a \neq 0$. Other examples of quadratic equations are

$$3x^2 - 2x - 6 = 0,\ y^2 = 5y, \text{ and } (t - 1)^2 = 3.$$

Depending on the particular equation, different procedures may be used for solving quadratic equations, and you need to consider carefully each of the methods that follow.

The Factoring Method

A technique that can often be used to solve a quadratic equation relies on factoring. For example, the equation $x^2 + 5x + 4 = 0$ may be solved by first factoring on the left-hand side of this equation to obtain

$$(x + 4)(x + 1) = 0.$$

We now have the situation where the product of two factors is zero. In multiplication zero is a special number, as outlined in the following principle.

Zero Product Principle

For any numbers a and b,

$$ab = 0 \text{ if and only if } a = 0 \text{ or } b = 0.$$

By applying this principle, we know

$$(x + 4)(x + 1) = 0 \text{ if and only if } x + 4 = 0 \text{ or } x + 1 = 0.$$

Since $x + 4 = 0$ when $x = -4$ and $x + 1 = 0$ when $x = -1$, the solutions are -4 and -1, and the solution set is $\{-4,\ -1\}$. To catch any mistakes, it is recommended that solutions be checked in the original equation, so

$$\begin{array}{rcl} x^2 + 5x + 4 = 0 & \qquad & x^2 + 5x + 4 = 0 \\ (-4)^2 + 5(-4) + 4 \stackrel{?}{=} 0 & & (-1)^2 + 5(-1) + 4 \stackrel{?}{=} 0 \\ 0 \stackrel{\checkmark}{=} 0 & & 0 \stackrel{\checkmark}{=} 0 \end{array}$$

To summarize, quadratic equations that are factorable (with integer coefficients) are usually solved as follows.

Factoring Method for Solving Quadratic Equations

1. If necessary, change the form of the equation so one side is 0.
2. Factor the nonzero side of the equation.
3. Set each factor equal to zero and obtain the solution(s) by solving the resulting equations.
4. Check each solution by substituting it in the original equation.

Example 1: Solving a Quadratic Equation by Factoring

Solve $4x^2 = 12x$ using the factoring method.

Solution: Rewrite the given equation so that one side is zero. Then, factor and apply the zero product principle.

$$4x^2 = 12x$$

$$4x^2 - 12x = 0 \qquad \text{Rewrite the equation so one side is 0.}$$

$$4x(x-3) = 0 \qquad \text{Factor the nonzero side.}$$

$$4x = 0 \quad \text{or} \quad x - 3 = 0 \qquad \text{Set each factor equal to 0.}$$

$$x = \frac{0}{4} \qquad x = 3 \qquad \text{Solve each linear equation.}$$

$$x = 0$$

Check each possible solution by substituting it in the original equation

$$4x^2 = 12x \qquad\qquad 4x^2 = 12x$$

$$4(0)^2 \stackrel{?}{=} 12(0) \qquad\qquad 4(3)^2 \stackrel{?}{=} 12(3)$$

$$0 \stackrel{\checkmark}{=} 0 \qquad\qquad 36 \stackrel{\checkmark}{=} 36$$

Thus, the solution set is $\{0,\ 3\}$.

Technology Link

Figure 4.1 shows how a grapher may be used to display a picture of the solution in this example. To obtain this display the equations $y1 = 4x^2$ and $y2 = 12x$ were graphed in the viewing window shown, and the Intersection operation on a grapher was used to estimate the points of intersection. (If your calculator does not have this feature, then the Zoom and Trace features may be used to estimate these points.) Observe that the displayed x-coordinates of the intersection points are in agreement with the solution set $\{0,\ 3\}$.

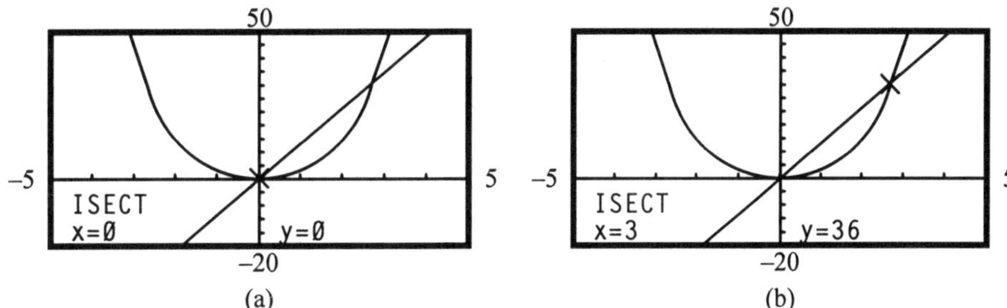

Figure 4.1

PROGRESS CHECK 1

Solve $3x^2 = -21x$ using the factoring method. ■

The Technology Link in Example 1 showed how to picture the real number solutions of a quadratic equation in terms of simultaneous equations. To understand an alternative method that is often used when solving a wide variety of equations, we first need to introduce the concept of an **intercept**. A point where a graph crosses the x-axis is called an ***x*-intercept**. Because an x-intercept lies on the x-axis, *its second component must be zero.* Similarly, a point where a graph crosses the y-axis is called a ***y*-intercept**, and *its first component must be zero.* This leads to a direct way to find intercepts.

To Find Intercepts

1. To find x-intercepts, which have the form $(a,\ 0)$, let $y=0$ and solve for x.
2. To find y-intercepts, which have the form $(0,\ b)$, let $x=0$ and solve for y.

Example 2: Find x- and y-intercepts

Find the intercepts of the graph of $y=f(x)=x^2-2x-3$.

Solution: To find y-intercepts, we let $x=0$ and solve for y.

$$\begin{aligned} y &= (0)^2-2(0)-3 \\ &= -3 \end{aligned}$$

The y-intercept is the point $(0,\ -3)$.

To find x-intercepts, let $y=0$ and solve for x.

$$\begin{aligned} 0 &= x^2-2x-3 \\ 0 &= (x-3)(x+1) \\ x-3=0 \quad &\text{or} \quad x+1=0 \\ x=3 \quad & \qquad\quad x=-1 \end{aligned}$$

The x-intercepts are at $(3,\ 0)$ and $(-1,\ 0)$.

To check these answers, use a grapher to graph f as shown in Figure 4.2. It appears that the graph crosses the y-axis when y is –3, and the graph crosses the x-axis when x is 3 and x is –1. So, our results check graphically.

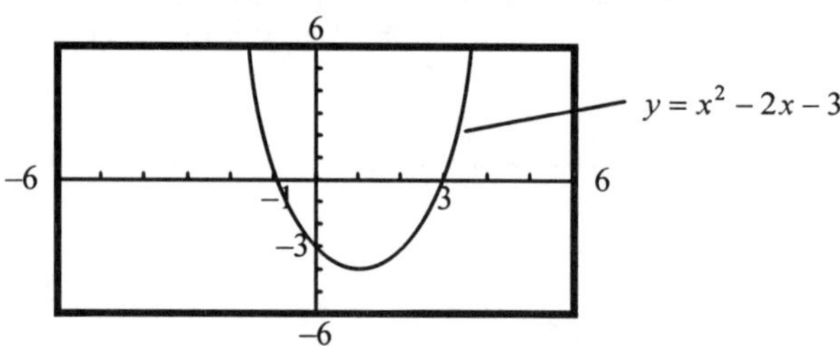

Figure 4.2

PROGRESS CHECK 2

Find the intercepts of the graph of $y=f(x)=x^2+2x-8$. ■

In Example 2 observe that 3 and –1 are both the roots of the equation $0=x^2-2x-3$ and the x-coordinates of the x-intercepts of the graph of $y=x^2-2x-3$. This match illustrates an important connection between roots and x-intercepts.

Connection Between Roots and x-intercepts

The real-number roots of the equation $f(x)=0$ are identical to the x-coordinates of the x-intercepts of the graph of $y=f(x)$.

Thus, for the remainder of the text, we can choose to display on a graphing calculator the real-number solutions of an equation by expressing the equation in the form $f(x)=0$, graphing $y=f(x)$, and then estimating the x-coordinates of the x-intercepts. Note that this method will fail to reveal solutions that are not real numbers, because imaginary numbers are not represented on a number line.

The Quadratic Equation $ax^2+c=0$

The easiest equation to solve is one in which there is no x term and the equation has the form $ax^2+c=0$ with $a \neq 0$. For example, to solve $2x^2-10=0$, we merely transform the equation so that x^2 is on one side of the equation by itself, giving $x^2=5$. We then conclude that the solutions are $\sqrt{5}$ and $-\sqrt{5}$ by applying the following property.

Square Root Property

> If n is any real number, then
>
> $$x^2=n \text{ implies } x=\sqrt{n} \text{ or } x=-\sqrt{n}.$$

This property is a key step in the general solution to quadratic equations, and we can prove this property as follows:

$$x^2=n$$
$$x^2-n=0$$
$$\left(x-\sqrt{n}\right)\left(x+\sqrt{n}\right)=0 \qquad \text{Factoring over the set of complex numbers.}$$
$$x-\sqrt{n}=0 \quad \text{or} \quad x+\sqrt{n}=0 \qquad \text{Zero product principle.}$$
$$x=\sqrt{n} \quad \text{or} \quad x=-\sqrt{n}.$$

We usually abbreviate these solutions as $\pm\sqrt{n}$. Note that there are two real number solutions if $n>0$, two (conjugate) complex number solutions if $n<0$, and one solution (namely, 0) if $n=0$.

Example 3: Solving a Quadratic Equation by the Square Root Property

Solve each equation by using the square root property.

a. $x^2=27$ **b.** $5(x-3)^2+20=0$

Solution:

a. $x^2=27$ implies $x=\pm\sqrt{27}$. Since $\sqrt{27}=\sqrt{9}\sqrt{3}=3\sqrt{3}$, the solution set is $\{\pm 3\sqrt{3}\}$.

To check this answer graphically, we may look at the x-intercepts in the graph of $y=x^2-27$ that is shown in Figure 4.3. We see that the graph crosses the x-axis when x is a little to the right of 5 and when x is a little to the left of –5. Since $\pm 3\sqrt{3}\approx \pm 5.2$, this geometric result is in agreement with the solutions that were obtained algebraically.

b. First, transform the equation so $(x-3)^2$ is on one side of the equation by itself.

$$5(x-3)^2+20=0$$
$$5(x-3)^2=-20$$
$$(x-3)^2=-4$$

Now apply the square root property and simplify.

$$x - 3 = \pm\sqrt{-4}$$
$$x = 3 \pm 2i \quad \left(\text{since } \sqrt{-4} = 2i\right)$$

Thus, the solution set is $\{3 \pm 2i\}$.

Because the solutions to this equation are not real numbers, we cannot check the solution graphically. However, this type of solution does imply that the graph of $y = 5(x-3)^2 + 20$ has no x-intercepts, and the graph in Figure 4.4 shows that this prediction is accurate.

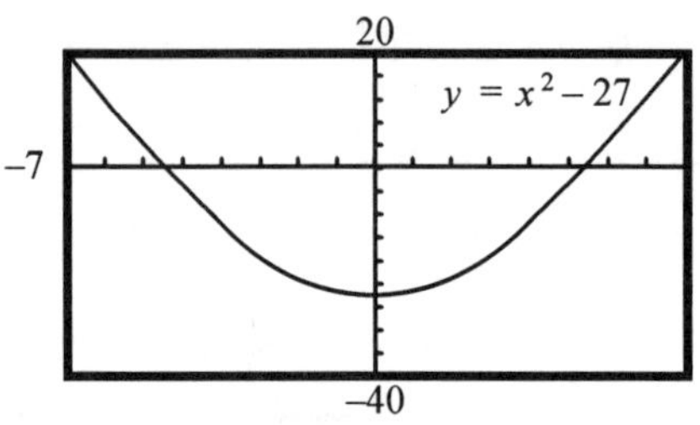

Figure 4.3

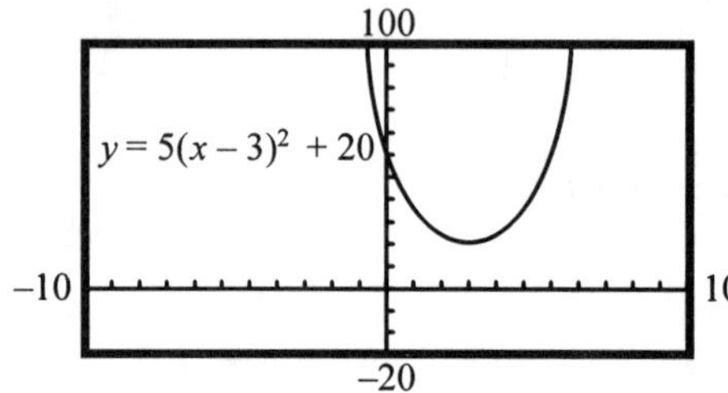

Figure 4.4

Technology Link

Some calculators have a built-in operation that finds an x-intercept of a graph in an interval. For example, on one such calculator the Root operation may be utilized to display the x-intercepts for the graph from part **a** of $y = x^2 - 27$, as shown in Figure 4.5. We will use this feature occasionally in the text to find x-intercepts because this operation gives very accurate answers. But keep in mind that repeated use of Zoom and Trace is a workable alternative, if the Root feature is not available on your calculator.

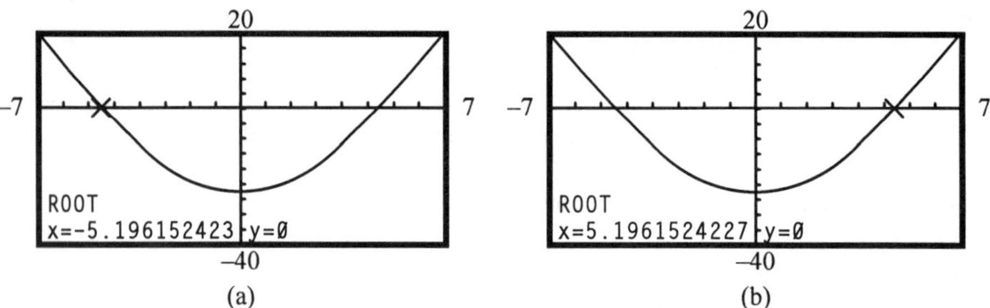

Figure 4.5

PROGRESS CHECK 3

Solve each equation, using the square root property.

a. $x^2 = 18$ **b.** $(x+4)^2 + 7 = 0$ ■

Completing the Square

In Example 3b we used the square root property to solve a quadratic equation of the form

$$(x + \text{constant})^2 = \text{constant}.$$

Actually, any quadratic equation may be placed in this form by a technique called completing the square. Consider an expression such as

$$x^2 + 8x.$$

What constant needs to be added to make this expression a perfect square? Since

$$(x+k)^2 = x^2 + 2kx + k^2$$

we set $2k = 8$, so $k = 4$ and $k^2 = 16$. Thus, adding 16 gives

$$x^2 + 8x + 16 = (x+4)^2.$$

In general, we complete the square for $x^2 + bx$ by adding $(b/2)^2$, which is the square of one-half of the coefficient of x. Example 4 shows how completing the square can solve a quadratic equation.

Example 4: Solving a Quadratic Equation by Completing the Square

Solve the equation $x^2 - 4x - 3 = 0$ by completing the square.

Solution: First, rearrange the equation with the x terms to the left of the equals sign and the constant to the right.

$$x^2 - 4x = 3$$

Now complete the square on the left. Half of –4 is –2, and $(-2)^2 = 4$. Add 4 to both sides of the equation and proceed as follows:

$$\begin{aligned} x^2 - 4x + 4 &= 3 + 4 \\ (x-2)^2 &= 7 \\ x - 2 &= \pm\sqrt{7} \\ x &= 2 \pm \sqrt{7}. \end{aligned}$$

The solutions are $2+\sqrt{7}$ and $2-\sqrt{7}$, and the solution set is abbreviated $\{2 \pm \sqrt{7}\}$.

Figure 4.6 shows a graphical check and a numerical check that support this answer. Observe in the numerical check involving $2-\sqrt{7}$ that our computed result is $-5E-13$, which is scientific notation for –0.0000000000005 This discrepancy from 0 is due to round-off errors and should be ignored.

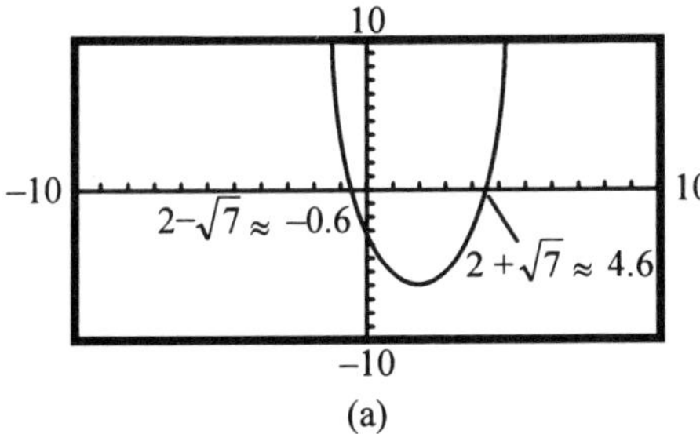

(a)

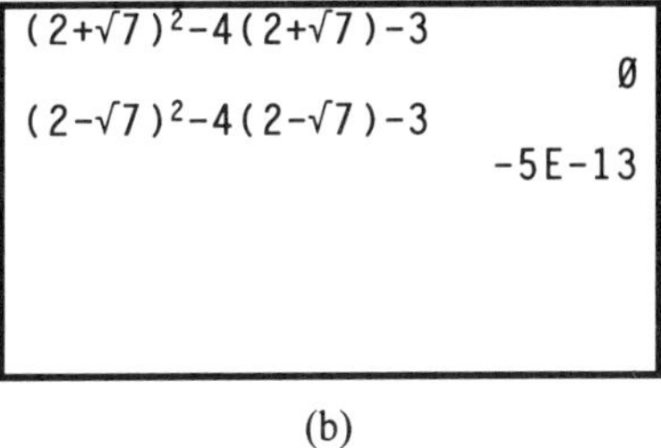

(b)

Figure 4.6

PROGRESS CHECK 4

Solve $x^2 - 12x + 6 = 0$ by completing the square. ■

Quadratic Formula

Completing the square is a useful technique in many situations because it often converts an expression to a standard form that is easy to analyze. In this case, the completing the square method works so well that if we apply it to the general quadratic equation $ax^2 + bx + c = 0$, we obtain a powerful formula (the **quadratic formula**) that solves all quadratic equations. In the following derivation of this formula, we also display the corresponding steps required to solve the particular equation $2x^2 + 5x + 1 = 0$, to illustrate in specific terms what is happening.

General Equation	Particular Equation	Comment
$ax^2 + bx + c = 0,\ a \neq 0$	$2x^2 + 5x + 1 = 0$	Given equation
$x^2 + \frac{b}{a}x + \frac{c}{a} = 0$	$x^2 + \frac{5}{2}x + \frac{1}{2} = 0$	Divide on both sides by the coefficient of x^2.
$x^2 + \frac{b}{a}x = -\frac{c}{a}$	$x^2 + \frac{5}{2}x = -\frac{1}{2}$	Subtract the constant term from both sides.
$x^2 + \frac{b}{a}x + \frac{b^2}{4a^2} = -\frac{c}{a} + \frac{b^2}{4a^2}$	$x^2 + \frac{5}{2}x + \frac{25}{16} = -\frac{1}{2} + \frac{25}{16}$	Add the square of one-half of the coefficient of x to both sides.
$\left(x + \frac{b}{2a}\right)^2 = \frac{b^2 - 4ac}{4a^2}$	$\left(x + \frac{5}{4}\right)^2 = \frac{17}{16}$	Factor on the left and add fractions in the right.
$x + \frac{b}{2a} = \pm\sqrt{\frac{b^2 - 4ac}{4a^2}}$	$x + \frac{5}{4} = \pm\sqrt{\frac{17}{16}}$	Apply the square root property.
$x + \frac{b}{2a} = \pm\frac{\sqrt{b^2 - 4ac}}{2a}$	$x + \frac{5}{4} = \frac{\pm\sqrt{17}}{4}$	Simplify the radical.
$x = \frac{-b}{2a} + \frac{\pm\sqrt{b^2 - 4ac}}{2a}$	$x = \frac{-5}{4} + \frac{\pm\sqrt{17}}{4}$	Isolate x on the left.
$a = \frac{-b \pm \sqrt{b^2 - 4ac}}{2a}$	$x = \frac{-5 \pm \sqrt{17}}{4}$	Combine fractions on the right.

In the particular equation, the two solutions are

$$x = \frac{-5 + \sqrt{17}}{4} \text{ and } x = \frac{-5 - \sqrt{17}}{4},$$

and in the general equation, the two solutions are

$$x = \frac{-b + \sqrt{b^2 - 4ac}}{2a} \text{ and } x = \frac{-b - \sqrt{b^2 - 4ac}}{2a}.$$

The work with the general equation results in the quadratic formula.

Quadratic Formula

If $ax^2 + bx + c = 0$, and $a \neq 0$, then

$$x = \frac{-b + \sqrt{b^2 - 4ac}}{2a}$$

Any quadratic equation may be solved with this formula. The idea is to substitute appropriate values for a, b, and c in the formula and then simplify.

To illustrate the use of this formula, we next resolve the equation in Example 4, to show that the same result is obtained by both methods.

Example 5: Quadratic Formula: Two Real Roots

Solve the equation $x^2 - 4x - 3 = 0$, using the quadratic formula.

Solution: In this equation $a = 1$, $b = -4$, and $c = -3$. Therefore,

$$\begin{aligned} x &= \frac{-b \pm \sqrt{b^2 - 4ac}}{2a} = \frac{-(-4) \pm \sqrt{(-4)^2 - 4(1)(-3)}}{2(1)} \\ &= \frac{4 \pm \sqrt{16 + 12}}{2} \\ &= \frac{4 \pm \sqrt{28}}{2} \\ &= \frac{4 \pm 2\sqrt{7}}{2} && \text{Simplify the radical.} \\ &= 2 \pm \sqrt{7}. && \text{Divide out 2.} \end{aligned}$$

Thus, $x_1 = 2 + \sqrt{7}$, $x_2 = 2 - \sqrt{7}$, and (as in Example 4) the solution set is $\{2 \pm \sqrt{7}\}$.

Once again, Figure 4.6 on page 321 shows a graphical check and a numerical check that support this answer.

Caution: A common student error is to interpret the quadratic formula as

Wrong $\quad$ **Wrong**

$$x = -b \pm \frac{\sqrt{b^2 - 4ac}}{2a} \quad \text{or as} \quad x = \frac{-b}{2a} \pm \sqrt{b^2 - 4ac}.$$

Remember to divide the *entire* expression $-b \pm \sqrt{b^2 - 4ac}$ by $2a$.

PROGRESS CHECK 5

Solve $x^2 - 2x - 1 = 0$ using the quadratic formula. ■

Example 6: Quadratic Formula: One Repeated Root

Solve the equation $4x^2 = 20x - 25$ using the quadratic formula.

Solution: First, express the equation in the form $ax^2 + bx + c = 0$.

$$4x^2 - 20x + 25 = 0$$

In this equation $a = 4$, $b = -20$, and $c = 25$. Therefore,

$$\begin{aligned} x &= \frac{-b \pm \sqrt{b^2 - 4ac}}{2a} = \frac{-(-20) \pm \sqrt{(-20)^2 - 4(4)(25)}}{2(4)} \\ &= \frac{20 \pm \sqrt{400 - 400}}{8} \\ &= \frac{20 \pm \sqrt{0}}{8}. \end{aligned}$$

$$x_1 = \frac{20 + 0}{8} = \frac{20}{8} = \frac{5}{2} \qquad x_2 = \frac{20 - 0}{8} = \frac{20}{8} = \frac{5}{2}$$

The roots x_1 and x_2 both equal $5/2$, so the solution set is $\{5/2\}$.

Since this equation has exactly one real solution, $5/2$, the graph of $y = 4x^2 - 20x + 25$ should have exactly one x-intercept at $(5/2,\ 0)$. Figure 4.7 shows a graph that helps to confirm this observation.

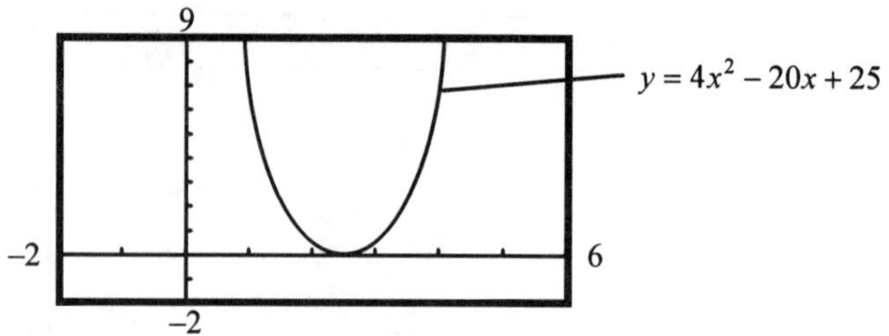

Figure 4.7

PROGRESS CHECK 6
Solve $25x^2 + 16 = 40x$ by means of the quadratic formula. ■

Example 7: Quadratic Formula: No Real Roots

Solve the equation $2x^2 - x + 3 = 0$ using the quadratic formula.

Solution: In this equation $a = 2$, $b = -1$, and $c = 3$. Therefore,

$$\begin{aligned} x &= \frac{-b \pm \sqrt{b^2 - 4ac}}{2a} = \frac{-(-1) \pm \sqrt{(-1)^2 - 4(2)(3)}}{2(2)} \\ &= \frac{1 \pm \sqrt{1 - 24}}{4} \\ &= \frac{1 \pm \sqrt{-23}}{4} \\ &= \frac{1 \pm i\sqrt{23}}{4}. \end{aligned}$$

$$x_1 = \frac{1 + i\sqrt{23}}{4} \qquad x_2 = \frac{1 - i\sqrt{23}}{4}$$

The solutions of this equation are (conjugate) complex numbers, and the solution set is $\{(1 \pm i\sqrt{23})/4\}$.

Figure 4.8 shows that the graph of $y = 2x^3 - x + 3$ has no x-intercepts, and this result is in agreement with the conclusion that $2x^2 - x + 3 = 0$ has no real roots.

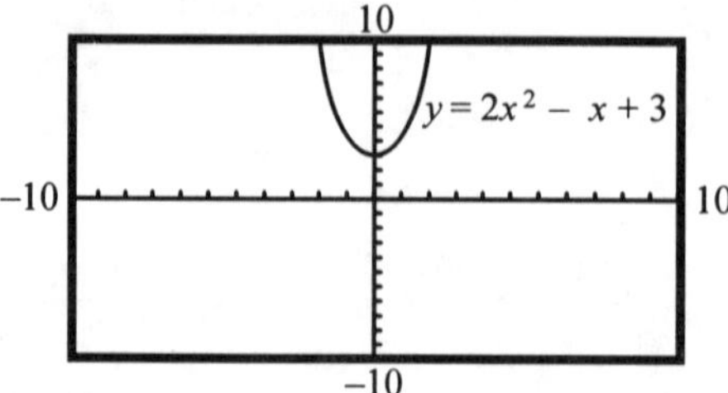

Figure 4.8

PROGRESS CHECK 7
Solve $x^2 + 4 = 3x$ by means of the quadratic formula. ■

Examples 5–7 have illustrated quadratic equations with two real roots, one real root, and no real roots, respectively. Given the quadratic formula, you can see that the number of real roots depends on the value of $b^2 - 4ac$, which is called the **discriminate**. This expression appears under the radical in the quadratic formula and reveals the following about the real roots of $ax^2 + bx + c = 0$.

Discriminate $b^2 - 4ac$	Number of Real Roots of $f(x) = 0$	Possible Graph of $y = f(x)$	Geometric Interpretation
$b^2 - 4ac > 0$	Two	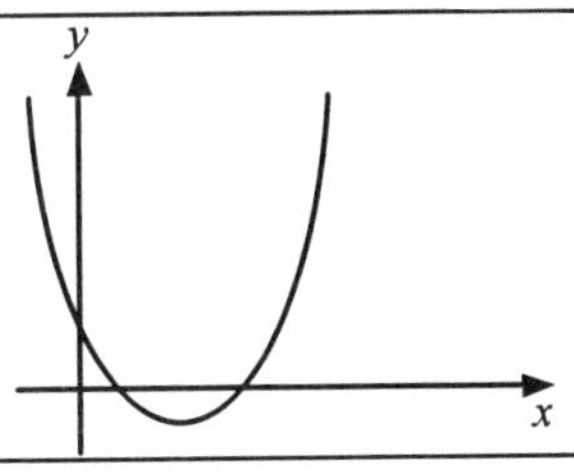	The graph has two x-intercepts.
$b^2 - 4ac = 0$	One	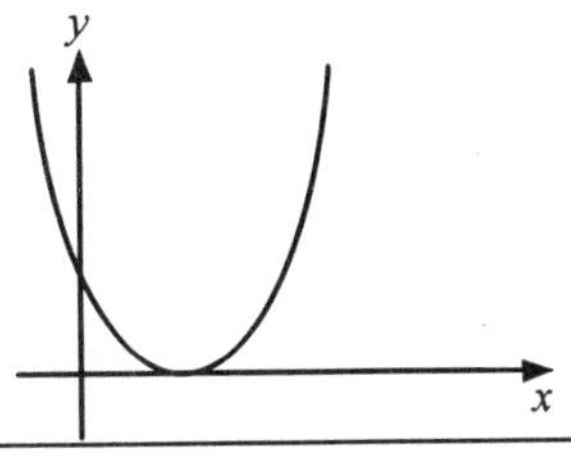	The graph has one x-intercept.
$b^2 - 4ac < 0$	None	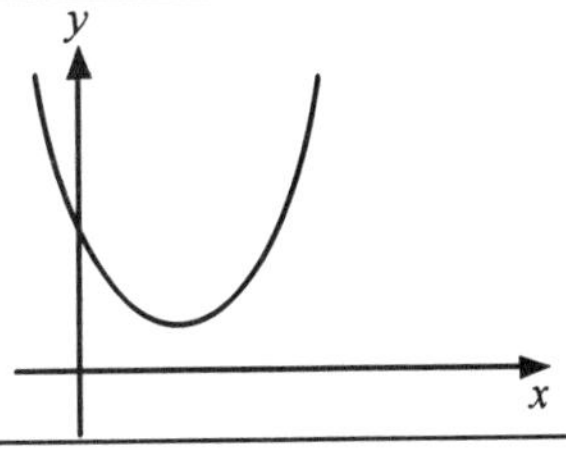	The graph has no x-intercepts.

Example 8: Using the Discriminate

Use the discriminate to determine the number of real roots of the equation $2x^2 - 5x = 3$.

Solution: This equation is equivalent to $2x^2 - 5x - 3 = 0$, in which $a = 2$, $b = -5$, and $c = -3$. The discriminate is

$$b^2 - 4ac = (-5)^2 - 4(2)(-3) = 49.$$

Since 49 is greater than 0, the equation has two real roots. These two roots can be seen in the graph of $y = 2x^2 - 5x - 3$ that is shown in Figure 4.9.

PROGRESS CHECK 8
Use the discriminate to determine the number of real roots of the equation $16x^2 + 25 = 40x$. ■

In this section we have shown several methods for solving a quadratic equation. Selecting an efficient method depends on the particular equation to be solved, and the following guidelines will help in your choice of methods.

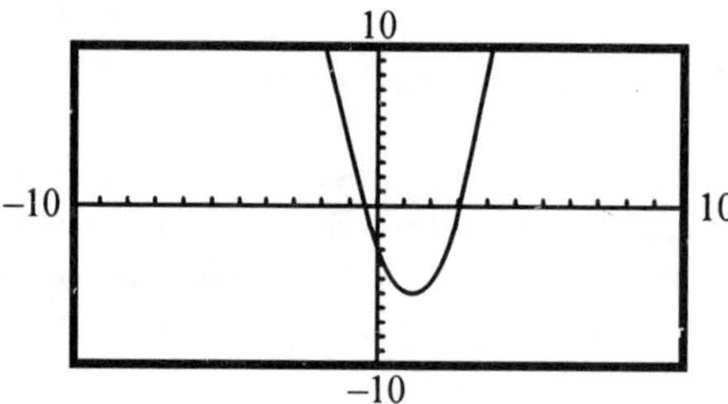

Figure 4.9

Guidelines for Solving a Quadratic Equation

Equation Type	Recommended Method
$ax^2+c=0$	Use the square root property. If ax^2+c is a difference of squares, consider the factoring method.
$ax^2+bx=0$	Use the factoring method.
$(px+q)^2=k$	Use the square root property.
$ax^2+bx+c=0$	First, try the factoring method, which applies when the discriminant is a perfect square. If ax^2+bx+c does not factor or is hard to factor, use the quadratic formula.

Example 9: Solving a Quadratic Equation Efficiently

Solve the equation $80=96t-16t^2$. Choose an efficient method.

Solution: This equation is equivalent to $16t^2-96t+80=0$, in which $a=16$, $b=-96$, and $c=80$. First, we try the factoring method, and we find it can be used.

$$\begin{aligned} 16t^2-96t+80&=0\\ 16\left(t^2-6t+5\right)&=0\\ 16(t-5)(t-1)&=0\\ t-5=0 \quad &\text{or} \quad t-1=0\\ t=5 \quad &\qquad\quad t=1 \end{aligned}$$

Thus, there are two real roots, and the solution set is $\{5,\ 1\}$.

If the factoring step in this solution seemed hard, then an alternate solution method is to use the quadratic formula.

$$t=\frac{-b\pm\sqrt{b^2-4ac}}{2a}=\frac{-(-96)\pm\sqrt{(-96)^2-4(16)(80)}}{2(16)}=\frac{96\pm\sqrt{4096}}{32}=\frac{96\pm 64}{32}$$

Then simplifying each solution separately gives

$$t_1=\frac{96+64}{32}=5 \text{ or } t_2=\frac{96-64}{32}=1.$$

By either method, the solution set is $\{5,\ 1\}$. These two roots can be seen in the graph of $y=16x^2-96x+80$ that is shown in Figure 4.10.

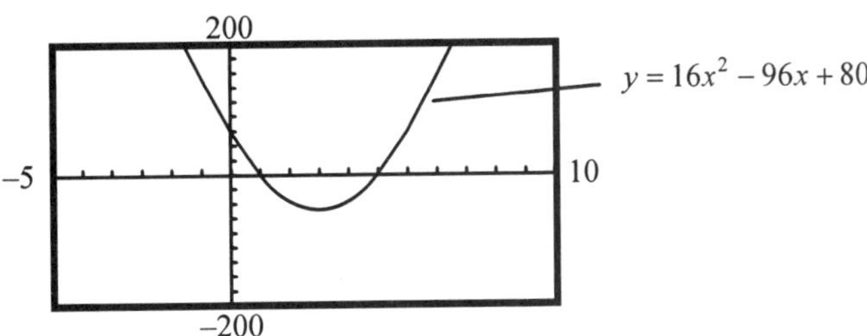

Figure 4.10

Technology Link

A quadratic equation may be viewed in a broader sense as a polynomial equation of degree or order 2. Some calculators have a built-in feature that may be used to solve polynomial equations. For instance, Figure 4.11 shows the coefficient entry screen and the solution screen associated with solving $16x^2 - 96x + 80 = 0$ on a particular calculator with a polynomial root finding feature.

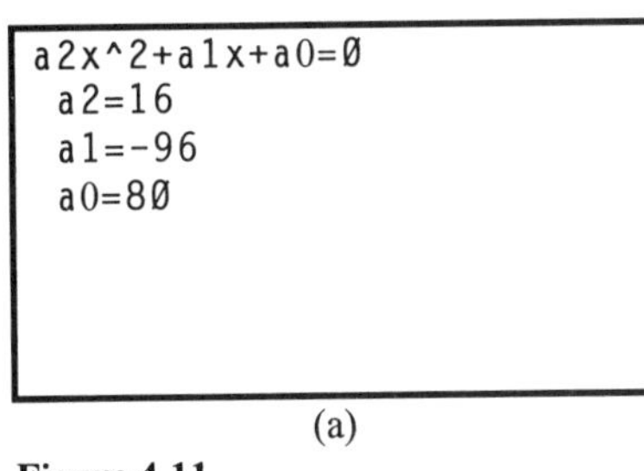

(a)

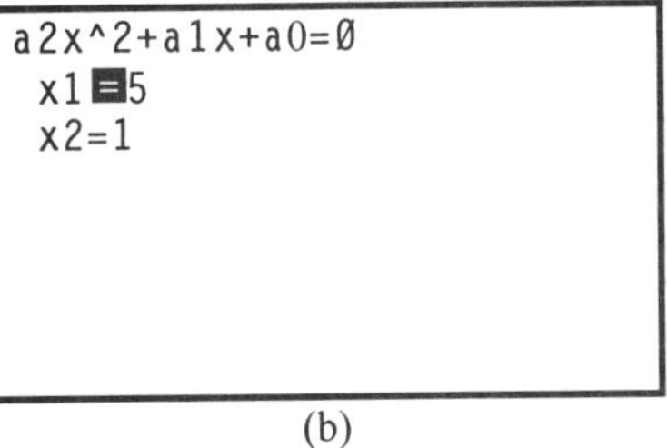

(b)

Figure 4.11

If your calculator does not have this special feature, then you should consider using the programming capabilities of your calculator to solve quadratic equations. For example, the program shown in Figure 4.12 can be copied into a TI-82 calculator for automatic evaluation of the quadratic formula.

PROGRESS CHECK 9

Solve $96 = 80t - 16t^2$. Choose an efficient method. ■

Example 10: Finding Elapsed Time in a Projective Problem

The height (y) of a projectile that is shot directly up from the ground with an initial velocity of 100 ft/second is given by the formula $y = 100t - 16t^2$ where t is the elapsed time in seconds. To the nearest hundredth of a second, when will the projectile initially attain a height of 115 ft?

Solution: Replacing y by 115 in the given formula leads to

$$115 = 100t - 16t^2$$
$$16t^2 - 100t + 115 = 0.$$

Then applying the quadratic formula to the resulting equation in which $a = 16$, $b = -100$, and $c = 115$ gives

$$t = \frac{-(-100) \pm \sqrt{(-100)^2 - 4(16)(115)}}{2(16)} = \frac{100 \pm \sqrt{2640}}{32}.$$

To the neatest hundredth, the two roots are

```
:Lbl 1
:Disp "A IS"
:Input A
:Disp "B IS"
:Input B
:Disp "C IS"
:Input C
:B² – 4AC → D
:If D<0
:Goto 2
:(–B–√D)/2A → R
:(–B+√D)/2A → S
:Disp "ROOTS ARE"
:Disp R
:Disp S
:Goto 1
:Lbl 2
:Disp "NO REAL
 SOLUTIONS"
:Goto 1
```

Figure 4.12

$$t_1 = \frac{100+\sqrt{2640}}{32} \approx 4.73 \text{ and } t_2 = \frac{100-\sqrt{2640}}{32} \approx 1.52.$$

The projectile attains a height of 115 ft after 1.52 seconds (on the way up) or after 4.73 seconds (on the way down). So the projectile *initially* attains this height at 1.52 seconds. Figure 4.13 shows a graphical check of this solution.

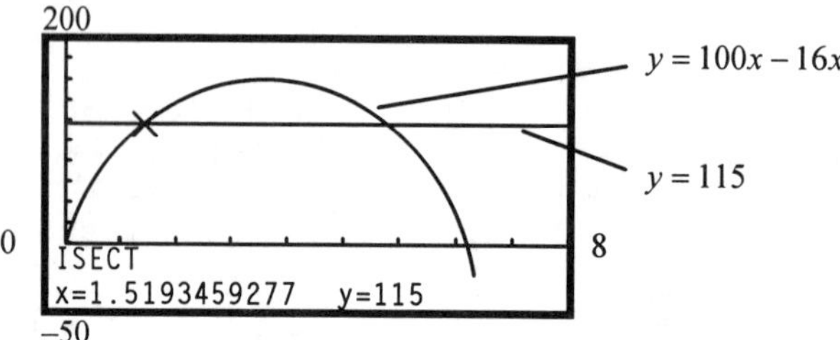

Figure 4.13

PROGRESS CHECK 10

To the nearest hundredth of a second, when will the projectile described in Example 10 initially attain a height of 95 ft? ■

Example 11: Analyzing a Relation Involving Total Revenue

Solve the problem in the chapter introduction on page 315.

Solution:

a. x represents the number of passengers above 100, so the number of passengers is $100+x$. The fare for each of these passengers is $99-0.5x$ since the \$99 fare is reduced by 50 cents for each passenger above 100. Then,

$$\text{total revenue} = (\text{number of passengers})\cdot(\text{fare for each passenger})$$
$$R = (100+x)(99-0.5x).$$

b. Figure 4.14 shows a graph of $y=(100+x)(99-0.5x)$ where y measures revenue. The graph shows that the revenue increases as the passenger count increases, to a point; then as more passengers sign up, the revenue decreases (because of the declining fare).

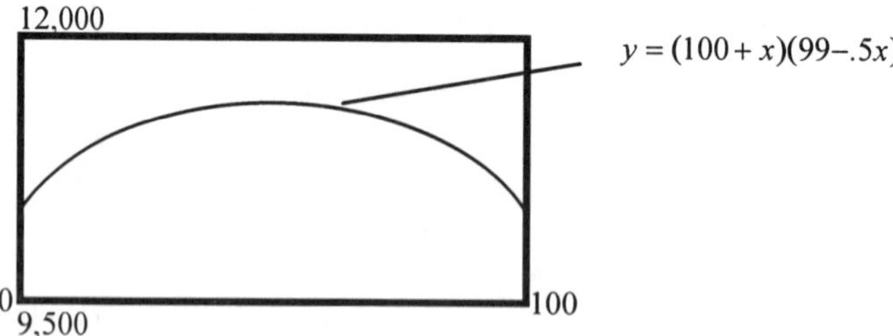

Figure 4.14

c. Tracing along the graph in Figure 4.14 shows that y is 11,040 when x is about 38 or when x is about 60. Therefore, passenger counts of 138 or 160 produce about \$11,040 in revenue for the flight.

d. To find algebraically all values of x at which revenue is 11,040, we solve the equation $11{,}040 = (100 + x)(99 - 0.5x)$. This equation may be converted to the form $ax^2 + bx + c = 0$, as shown next.

$$11{,}040 = (100 + x)(99 - 0.5x)$$
$$11{,}040 = 9900 + 49x - 0.5x^2$$
$$0.5x^2 - 49x + 1140 = 0$$

In the resulting equation, $a = 0.5$, $b = -49$ and $c = 1140$. Then the quadratic formula gives

$$x = \frac{-(-49) \pm \sqrt{(-49)^2 - 4(0.5)(1140)}}{2(0.5)} = \frac{49 \pm \sqrt{121}}{1} = 49 \pm 11.$$

So the two roots are

$$x_1 = 49 + 11 = 60 \text{ and } x_2 = 49 - 11 = 38.$$

This result confirms algebraically that passenger counts of 138 or 160 produce $11,040 in revenue for the flight.

PROGRESS CHECK 11

The total sales revenue for a product is estimated by the formula $R = 17x - x^2$, where x is the unit selling price (in dollars) and R is the total sales revenue measured in units where 1 unit equals $10,000.

a. Use a grapher to graph this equation and describe what happens to total sales revenue as the unit selling price increases.

b. Use the Trace feature to estimate to the nearest integer all unit selling prices at which the estimated revenue is $720,000 (that is, $R = 72$).

c. Find algebraically all the unit selling prices sought in part **b**. ■

EXPLORE 4.1

1. Graphical solutions of the equation $4x^2 = 12x$ can be found by looking for the points of intersection of $y = 4x^2$ and $y = 12x$. Rewriting the equation as $4x^2 - 12x = 0$ shows that the solutions can also be found by looking for points of intersection of $y = 4x^2 - 12x$ and $y = 0$, which is the x-axis. Use both methods to check that the solutions are $x = 0$ and $x = 3$.
2. Two parabolas may have 0, 1, or 2 points of intersection.
   a. Use a grapher to find (to the nearest hundredth) the points of intersection of $y = 2x^2 - 3x - 4$ and $y = -4x^2 + x + 3$. Find the two points exactly by solving this system of equations algebraically.
   b. Create a specific system of equations that illustrates the case of two parabolas with no points of intersection. Check your solution graphically and algebraically.
   c. Repeat part **b**, but for one point of intersection.
3. The number of x-intercepts of the 2nd-degree function $y = ax^2 + bx + c$ is determined by the sign of its discriminant, $b^2 - 4ac$.
   a. To illustrate this relation let $a = 1$, and $b = 4$, and then try several values of c, namely $c = 3$, $c = 4$, and $c = 5$. In each case determine the sign of the discriminant, then determine the number of x-intercepts in the graph. In general, what seems to be the relation between the number of x-intercepts and the sign of the discriminant?
   b. Repeat the procedure with the same values for a and b, but with $c = 3.99$, 4, and 4.01. Discuss what happens when you try to confirm your prediction graphically.

EXERCISES 4.1

In Exercises 1–18, solve each equation using the factoring method.

1. $x(x+5)=0$
2. $x(x-1)=0$
3. $x^2-5x=0$
4. $y^2=2y$
5. $a^2-4=0$
6. $5x^2-5=0$
7. $y^2-2y-8=0$
8. $z^2-3z-18=0$
9. $k^2-3k=4$
10. $b^2+3b=4$
11. $3x^2-16x+5=0$
12. $5y^2-11y+2=0$
13. $3r^2=5r-2$
14. $2x^2=7x+4$
15. $2(x^2-1)=3x$
16. $3(x^2+1)=10x$
17. $2x(x+3)=15-x$
18. $3x(x-2)=6+x$

In Exercises 19–24 find the intercepts of the graph of the function.

19. $y=x^2-4x-5$
20. $y=6x^2-x-2$
21. $g(x)=x^2-4x+4$
22. $g(x)=25x^2-10x+1$
23. $f(x)=6x^2+5x-6$
24. $y=2x^2+3x-2$

In Exercises 25–36 solve each equation using the square root property.

25. $x^2=9$
26. $x^2=36$
27. $x^2=-25$
28. $x^2=-9$
29. $3x^2+24=0$
30. $2y^2+48=0$
31. $(b-5)^2=4$
32. $(x+3)^2=100$
33. $(x+1)^2=-1$
34. $(a-1)^2=-16$
35. $5(k-7)^2+35=0$
36. $3(z+4)^2+42=0$

In Exercises 37–44 solve each equation by completing the square.

37. $x^2-2x+2=0$
38. $x^2+4x+5=0$
39. $x^2+x-3=0$
40. $x^2+4x+4=0$
41. $3x^2-4x-2=0$
42. $5x^2-3x-4=0$
43. $-5x^2=2x+3$
44. $-2x^2=6x+5$

In Exercises 45–56 solve each equation, using the quadratic formula

45. $x^2+5x+4=0$
46. $x^2-3x+2=0$
47. $x^2-6x+9=0$
48. $x^2+2x-1=0$
49. $x^2-4=x$
50. $x^2+2=2x$
51. $3x^2-2x=6$
52. $2x^3+x=14$
53. $3x^2=2x-1$
54. $4x^2=12x-9$
55. $4x^2+x=3$
56. $5x^2-28x=12$

In Exercises 57–66 use the discriminant to determine the number of real roots of the equation.

57. $2x^2-x+2=0$
58. $-3x^2+x+1=0$
59. $x^2+2x+1=0$
60. $5x^2-7x+2=0$
61. $x^2-10x-9=0$
62. $-2x^2+4x+9=0$
63. $4x^2=2x+5$
64. $3x^2=-4x-1$
65. $x^2-5x=-7$
66. $-2x^2-x=-2$

In Exercises 67–78 solve each equation. In all cases, choose an efficient method for solving the particular quadratic equation involved in the problem.

67. $x^2-x-6=0$
68. $x^2-7x+8=0$
69. $5r^2+20=16$
70. $4y^2+25=20y$
71. $3n^2=2n$
72. $4(a+3)^2=9$
73. $t(t-2)=4$
74. $(x+2)^2=9x$
75. $x^2+540=69x$
76. $76=100t-16t^2$
77. $\dfrac{x^2}{2}-5x=0$
78. $\dfrac{x^2}{3}+\dfrac{x}{2}=1$
79. On a sailboat the wind causes a wind pressure gauge to register 4.2 lb/ft^2. If the pressure p in pounds per square foot of a wind blowing at v mi/hour is given by

$$p=0.003v^2,$$

find the wind speed at that moment. (Round to the nearest tenth.)

80. The scientist Galileo (1564–1642) discovered, when he rolled balls down an inclined plane (see figure), that the equation $v^2 = 64h$ relates the velocity of the ball and the *vertical* distance it has covered. Starting from rest, a ball that has dropped h ft has a velocity of v ft/second. It is remarkable that this velocity has nothing to do with the steepness of the plane. Use the given equation to find the velocity of a rolling ball when its vertical height is 2 ft below its starting height.

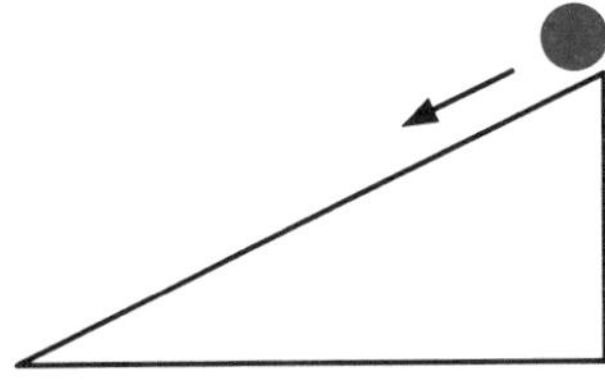

81. The sum of a positive number and its square is 56. What is the number?

82. An ancient problem leading to a quadratic equation is given on an Egyptian papyrus from about 2000 B.C. The problem is as follows: Divide 100 square units into two squares such that the side of one of the squares is three-fourths the side of the other.

83. A rectangular plate is to have an area of 30 in.^{2} and the sides of the plate are to differ in length by 4 in. To the nearest tenth of an inch, find the dimensions of the plate.

84. The perimeter of a rectangle is 100 ft and the area is 400 ft^2. Find the dimensions of the rectangle.

85. The total cost of manufacturing x units of a certain product is given by $C = 0.1x^2 + 0.7x + 3$. How many units can be made for a total cost of \$20?

86. A manufacturer sells a product for \$10 per unit. The cost of manufacturing x units is estimated by the formula $C = 200 + x + 0.01x^2$. How many units must be manufactured and sold to earn a profit of \$1,200? (**Hint:** profit = income − cost.)

87. The total sales revenue for a company is estimated by the formula $S = 8x - x^2$, where x is the unit selling price (in dollars) and S is the total sales revenue (1 unit equals \$10,000). What should be the unit selling price if the company wishes the total revenue from sales to be \$150,000 (that is, $S = 15$)? Why are there two solutions to the problem?

88. The total cost (in dollars) to a company of producing x units of a product in 1 month is estimated by the formula $C = 10x^2 + 150x + 200$.

a. What are the fixed costs of the company (that is, what costs does the company experience even if it produces no units)?

b. If the total cost during a month is \$2,700, how many units did the company produce?

89. The height (y) above water of a diver t seconds after she steps off a platform 10 ft high is given by the formula $y = 10 - 16t^2$. When will the diver hit the water?

90. The height (y) of a projectile that is shot directly up from the ground with an initial velocity of 100 ft/second is given by the formula $y = 100t - 16t^2$. To the nearest hundredth of a second, when will the projectile initially attain a height of 50 ft?

91. A 14-in. piece of wire must be bent as shown in the figure to make a three-sided frame. The area is to be 10 in.^{2}. Show that there are two different solutions to the problem.

x x

$14 - 2x$

92. From a square sheet of metal an open box is made by cutting 2-in. squares from the four corners and folding up the ends. To the nearest tenth of an inch, how large a piece of metal should be used if the box is to have a volume of 100 in.^{3}?

93. When two resistors, R_1 and R_2 are connected in series, their combined resistance is given by $R = R_1 + R_2$. If the resistors are connected in parallel, their combined resistance is given by $1/R = 1/R_1 + 1/R_2$. Find the value of resistors R_1 and R_2 if their combined resistance is 32 ohms when connected in series, and 6 ohms when connected in parallel.

94. Show that the sum of the solutions of the equation $ax^2 + bx + c = 0$ (with $a \neq 0$) is $-\frac{b}{a}$ and that the product of the solution is $\frac{c}{a}$.

95. In the field of sociology, mathematical models are sometimes used to describe human behavior. In some cases the model used is quadratic in form. For instance, one study analyzed the relationship between the length of time a U.S. president had been in office (x, from 0.00 to 4.00) and his approval rating (y) as given by Gallup poll data. In this study the data suggested that two separate functions made sense, one for peacetime and one for wartime. The equations are as follows.

Peacetime: $y = 70.48 + 3.22(2.18 - x)^2 \quad 0 \le x \le 4$

Wartime: $y = 48.08 + 3.45(3.11 - x)^2 \quad 0 \le x \le 4$

a. Use a grapher to compare the two graphs. Describe what happens in each case.

b. Use graphical methods to estimate to the nearest tenth all times the model predicts a 75 percent approval rating. Find algebraically the exact values of these times.

c. Over the given domain, which situation leads to the highest approval rating? When does this highest rating occur?

d. Over the given domain, which situation leads to the lowest approval rating? When does this lowest rating occur?
[Source: *Linear Models in Social Research*, ed. Marsden, Sage, 1981]

THINK ABOUT IT 4.1

1. Create a quadratic equation with the given solution set.
   a. $\{2/3, -5\}$ **b.** $\{4\}$
2. A quadratic equation cannot have three distinct solutions. Explain why not, in terms of the zero product principle.
3. The great equation solver of the sixteenth century, Gerolamo Cardano, perhaps the most bizarre character in the whole history of mathematics, used a technique called "depressing" to put equations in simpler form. When you "depress" an equation, you get rid of the term that has the next to the highest power of the variable. Through this technique Cardano was able to solve cubic equations. You can use it to solve quadratic equations.
   a. Given the equation $ax^2 + bx + c = 0$, rewrite it by replacing each occurrence of x by $y - b/2a$. This will give you a simpler (depressed) equation in y, which you can solve by the square root property. You will see that this is another way to arrive at the quadratic formula. [For more on Cardano see *Journey Through Genius* by William Dunham (Wiley, 1990) or any history of mathematics text.]
   b. Solve the equation $x^2 + 4x + 1 = 0$ by the method of depression.
4. A fascinating number that appears in many areas of mathematics is called the golden ratio. For example, in an isosceles triangle where the base angles are double the vertex angle, it is the ratio of the side to the base. An early appearance is in Euclid's *Elements*, where an exercise says: "Divide a line segment such that the ratio of the large part to the whole is equal to the ratio of the small part to the large." Refer to the sketch, from which we derive golden ratio $l/(l + s) = s/l$. If you set $s = 1$ and solve the resulting quadratic equation, you will determine the exact value of the golden ratio.

5. In Example 8 of this section, the discriminant was used to determine the nature of the solutions of the equation $2x^2 - 5x - 3 = 0$.
   a. Show that the discriminant is not changed if the equation is rewritten as $-2x^2 + 5x + 3 = 0$, which results when both sides are multiplied by -1.
   b. Show that, in general, the discriminant $b^2 - 4ac$ is not changed if a, b, and c are replaced by their opposites.
   c. Show that in the quadratic formula if a, b, and c are replaced by their opposites, then the solutions do not change.

• • •

4.2 Equations That Contain Fractions

Photo Courtesy of **David Jones**/PA News Ltd./ Archive Photos, New York

A race car driver must average at least 100 mi/hour for two laps around a track to qualify for the finals. Because of mechanical trouble, the driver is only able to average 50 mi/hour for the first lap. What minimum speed must the driver average for the second lap to qualify for the finals? (See Example 7.)

Objectives

1. Solve fractional equations that lead to linear equations.
2. Solve fractional equations that lead to quadratic equations.
3. Solve formulas that contain fractions for a specified variable.
4. Solve applied problems involving fractional equations.

Many applications of algebra involve equations or formulas that contain fractions. The usual first step for solving such an equation is to multiply both sides of the equation by the least common denominator of all fractions that appear in it. This results in an equation that does not contain fractions and that may often be solved by methods we have already discussed. For instance, in Examples 1 and 2 this procedure will lead to a linear equation.

Example 1: Solving a Fractional Equation That Leads to Linear Equation

Solve $\frac{2}{3a} = \frac{3}{2a-1}$.

Solution: We first remove the fractions by multiplying both sides of the equation by the LCD, which is $3a(2a-1)$ in this equation. Note that this step requires the restrictions that $a \neq 0$ and $a \neq 1/2$ to ensure that we are multiplying both sides of the equation by a nonzero number.

$$3a(2a-1)\frac{2}{3a} = 3a(2a-1)\frac{3}{2a-1}$$
$$2(2a-1) = 9a$$
$$4a-2 = 9a$$
$$-2 = 5a$$
$$\frac{-2}{5} = a$$

The LCD is nonzero when $a = -2/5$; and the check of this solution is shown next.

$$\frac{2}{3\left(-\frac{2}{5}\right)} \stackrel{?}{=} \frac{3}{2\left(-\frac{2}{5}\right)-1}$$
$$\frac{2}{-\frac{6}{5}} \stackrel{?}{=} \frac{3}{-\frac{9}{5}}$$
$$\frac{5}{-3} \stackrel{\checkmark}{=} \frac{5}{-3}$$

Thus the solution set is $\{-2/5\}$.

Technology Link

Graphical support for the solution of a fractional equation can be complicated because fractional expressions are undefined when the denominator is zero. Although a detailed discussion for graphing rational functions will be given in Section 6.4, we can still use a calculator to draw such graphs at this time. First, it is recommended that the calculator be set in dot mode, not connected mode, when initially drawing such graphs. The dot mode prevents the calculator from connecting points that are in separate pieces of the graph (as discussed in Section 1.2). Figure 4.15 shows three graphs that are associated with the graphical solution of the equation in Example 1. Observe that the x-coordinate of the intersection point of the two graphs agrees with our finding that the root of the equation is $-2/5$.

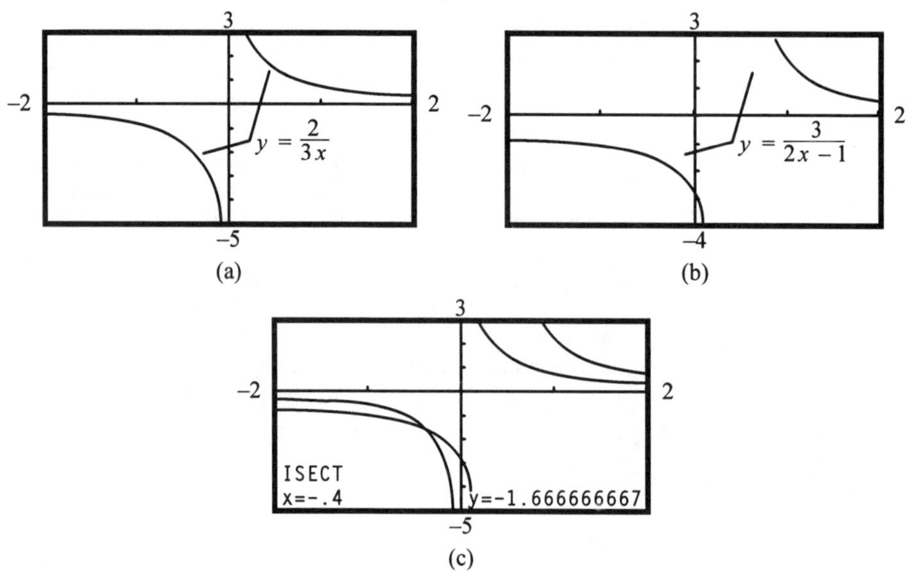

Figure 4.15

PROGRESS CHECK 1

Solve $\dfrac{4-3x}{x}=\dfrac{5}{2x}$. ■

The next example shows that the methods of this section may lead to **extraneous solutions**, which appear to be solutions but do not check in the original equation and are therefore not part of the solution set. When solving equations containing fractions, check for extraneous solutions at values for the variable that make the LCD zero.

Example 2: Solving a Fractional Equation That Leads to a Linear Equation

Solve $\dfrac{2}{x+1}+\dfrac{3}{x-1}=\dfrac{-4}{x^2-1}$.

Solution: Figure 4.16 shows a graph of the equation

$$y=\frac{2}{x+1}+\frac{3}{x-1}+\frac{4}{x^2-1}$$

which may be used to illustrate the solution graphically. Since there are no x-intercepts in the graph, it appears that there are no real roots of the original equation. To confirm this result algebraically, observe that the LCD is $(x+1)(x-1)$. So if $x \neq -1$ and $x \neq 1$, then

$$\begin{aligned}(x+1)(x-1)\left(\frac{2}{x+1}+\frac{3}{x-1}\right) &= (x+1)(x-1)\frac{-4}{x^2-1}\\ 2(x-1)+3(x+1) &= -4\\ 2x-2+3x+3 &= -4\\ 5x+1 &= -4\\ 5x &= -5\\ x &= -1.\end{aligned}$$

The restriction $x \neq -1$ eliminates -1 as a solution, so no number satisfies the original equation, and the solution set is $\varnothing$. We verify that -1 is not a solution in the following check.

$$\begin{aligned}\frac{2}{(-1)+1}+\frac{3}{(-1)-1} &\stackrel{?}{=} \frac{-4}{(-1)^2-1}\\ \frac{2}{0}+\frac{3}{-2} &\neq \frac{-4}{0}\end{aligned}$$

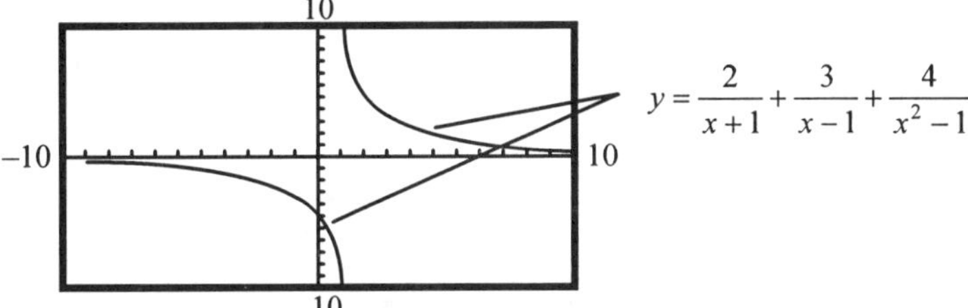

Figure 4.16

PROGRESS CHECK 2

Solve $\frac{2}{x+1}=\frac{1}{x}-\frac{2}{x^2+x}$. ■

Clearing fractions in the next two examples leads to quadratic equations that can be solved by the methods of the previous section.

Example 3: Solving a Fractional Equation That Leads to a Quadratic Equation

Solve $\frac{7}{t-4}-\frac{56}{t^2-16}=1$.

Solution: The LCD is $(t+4)(t-4)$ or t^2-16. So, we begin by multiplying both sides of the equation by t^2-16, assuming $t \neq -4$ and $t \neq 4$.

$$\left(t^2-16\right)\left(\frac{7}{t-4}-\frac{56}{t^2-16}\right)=\left(t^2-16\right)\cdot 1$$
$$7(t+4)-56=t^2-16$$
$$7t+28-56=t^2-16$$
$$0=t^2-7t+12$$
$$0=(t-3)(t-4)$$
$$t-3=0 \quad \text{or} \quad t-4=0$$
$$t=3 \qquad\qquad t=4$$

The solution $t=3$ does check, as shown next, but the restriction $t \neq 4$ means that 4 is an extraneous solution and should not be included in the solution set. Substitution of 4 for t shows that this extraneous solution leads to division by zero.

Check

$$\frac{7}{3-4}-\frac{56}{3^2-16}\stackrel{?}{=}1 \qquad\qquad \frac{7}{4-4}-\frac{56}{4^2-16}\stackrel{?}{=}1$$
$$-7-(-8)\stackrel{?}{=}1 \qquad\qquad \frac{7}{0}-\frac{56}{0}\neq 1$$
$$1\stackrel{\checkmark}{=}1 \qquad\qquad \text{4 is an extraneous solution.}$$

The solution set is $\{3\}$. The graph in Figure 4.17 may be used to reinforce this solution graphically.

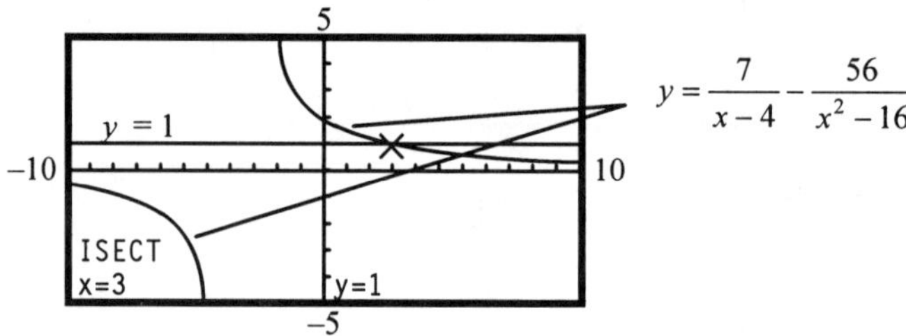

Figure 4.17

PROGRESS CHECK 3

Solve $\frac{5}{x-3}-\frac{30}{x^2-9}=1$. ■

Example 4: Solving a Fractional Equation That Leads to a Quadratic Equation

Solve $\frac{2}{x}+\frac{10}{x^2}=1$.

Solution: Figure 4.18 indicates that the graphs of

$$y=\frac{2}{x}+\frac{10}{x^2} \text{ and } y=1$$

intersect at two points where $x \approx -2.3$ or $x \approx 4.3$. So we can anticipate that these two numbers are approximate solutions for the equation in question. To find the exact values of these roots by algebraic methods, we first remove fractions by multiplying both sides of the equation by the LCD, which is x^2 in this equation. Thus, if $x \neq 0$,

$$x^2\left(\frac{2}{x}+\frac{10}{x^2}\right) = x^2 - 1$$
$$2x + 10 = x^2$$
$$0 = x^2 - 2x - 10.$$

Then, by the quadratic formula

$$x = \frac{-(-2) \pm \sqrt{(-2)^2 - 4(1)(-10)}}{2(1)} = \frac{2 \pm \sqrt{44}}{2} = \frac{2 \pm 2\sqrt{11}}{2} = 1 \pm \sqrt{11}.$$

The restriction $x \neq 0$ does not affect the proposed solution, and both numbers check in the original equation. Thus, the solution set is $\{1 \pm \sqrt{11}\}$. Observe that $1+\sqrt{11} \approx 4.3$ and $1-\sqrt{11} \approx -2.3$, so these roots are in agreement with the approximate solutions that we obtained graphically using Figure 4.18.

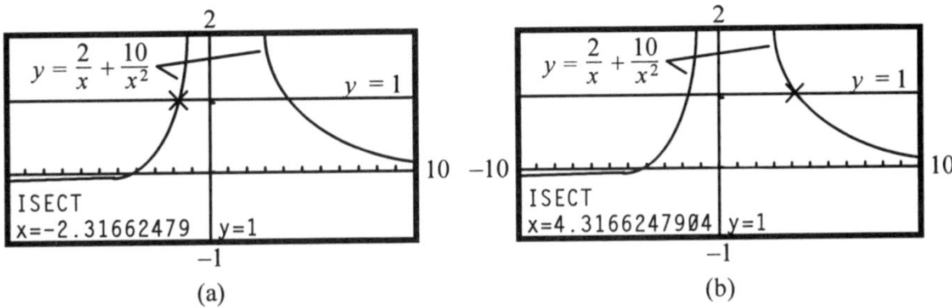

Figure 4.18

PROGRESS CHECK 4

Solve $\frac{3}{x^2} + \frac{4}{x} = 1$. ■

As mentioned in Section R.5, it is sometimes necessary to convert a formula to a form that is more efficient for a particular problem. The methods of this section may be used to rearrange formulas that contain fractions.

Example 5: Solving for a Specified Variable

Solve the formula $T = \frac{32m_1m_2}{m_1 + m_2}$ for m_1.

Solution: The general approach is to first clear fractions and then isolate all terms containing the specified variable on one side of the equation, where it can be factored out.

$$T = \frac{32m_1m_2}{m_1 + m_2}$$

$T(m_1 + m_2) = 32m_1m_2$	Multiply both sides by $m_1 + m_2$.
$Tm_1 + Tm_2 = 32m_1m_2$	Distributive property
$Tm_2 = 32m_1m_2 - Tm_1$	Subtract Tm_1 from both sides.
$Tm_2 = m_1(32m_2 - T)$	Factor out m_1.
$\frac{Tm_2}{32m_2 - T} = m_1$	Divide both sides by $32m_2 - T$.

PROGRESS CHECK 5

Solve $t = \frac{v - v_0}{v}$ for v. ■

One type of application that involves equations with fractions is called a **work problem**. The goal in such problems is to find the time needed to complete a job when two or more people or machines work together. The basis for analyzing work problems is to assume a *constant work rate*. **If a job requires t units of time to complete, then $1/t$ of the job is completed in one unit of time.** For example, if a pump can empty a storage tank in 4 hours, then (assuming a constant pumping rate) it empties 1/4 of the tank for each hour of pumping.

Example 6: Solving a Work Problem

One pipe can fill a swimming pool in 4 hours. A second pipe can fill the pool in 12 hours. How long will it take to fill the pool if both pipes operate together?

Solution: Let x represent the number of hours required to fill the pool if both pipes operate together. Then apply the basic principle outlined above.

First pipe:	This pipe fills the pool in 4 hours, so it fills 1/4 of the pool in 1 hour.
Second pipe:	This pipe fills the pool in 12 hours, so it fills 1/12 of the pool in 1 hour.
Together:	Both pipes operating together fill the pool in x hours, so they fill $1/x$ of the pool in 1 hour.

Now set up an equation as follows.

$$\underbrace{\text{Part done by first pipe in 1 hour}} + \underbrace{\text{Part done by second pipe in 1 hour}} = \underbrace{\text{Part done by both pipes in 1 hour}}$$

$$\frac{1}{4} + \frac{1}{12} = \frac{1}{x}$$

To solve this equation multiply both sides by the LCD, $12x$.

$$12x\left(\frac{1}{4}+\frac{1}{12}\right) = 12x\left(\frac{1}{x}\right)$$
$$3x + x = 12$$
$$4x = 12$$
$$x = 3$$

It takes the two pipes 3 hours to fill the pool when they are in operation simultaneously. To check this answer numerically, note that in 3 hours the first pipe does $3\cdot(1/4)$ of the job, while the second pipe does $3\cdot(1/12)$ of the job. Because $3/4 + 1/4 = 1$, the whole pool is filled, and the solution checks. Figure 4.19 shows a graph that reaffirms that 3 is the solution of the equation that was set up to solve the problem.

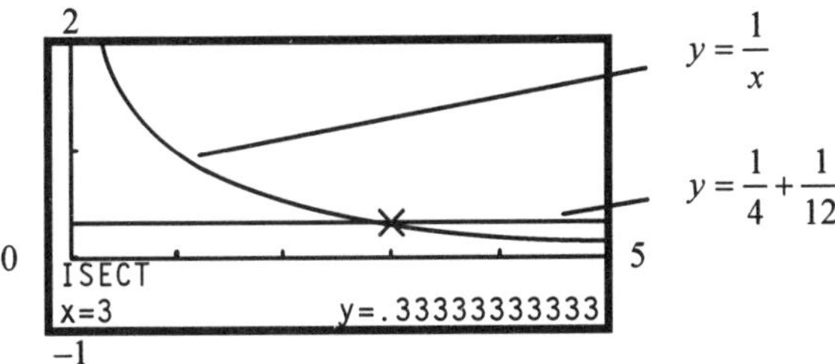

Figure 4.19

PROGRESS CHECK 6

One robot assembly line (A) can fill an order in 20 minutes. Another (B) can fill the order in 30 minutes. How long would it take the two lines working together to fill the order? (Assume that the two machines do not interfere with one another.) ■

Example 7: Solving an Average Speed Problem

Solve the problem in the section introduction on page 333.

Solution: First review Example #10 of Section 3.3. Now if we let x represent the minimum speed that the driver must average for the second lap to qualify, then

$$100 = \frac{2d}{\frac{d}{50} + \frac{d}{x}}$$

$$100\left(\frac{d}{50} + \frac{d}{x}\right) = 2d$$

$$2d + \frac{100d}{x} = 2d$$

$$\frac{100d}{x} = 0.$$

Since $100d$ cannot equal 0, the equation has no solution. This means that it is not possible to qualify, no matter how fast the driver goes on the second lap! To illustrate why this answer makes sense, note that the distance around the track is not important in the above equations, so let us suppose d equals 100 miles. Then the driver is required to travel 200 miles in 2 hours or less to qualify. However, the first lap took 2 hours to complete because the average speed was only 50 mi/hour, so the driver cannot qualify.

To view this solution graphically, let y represent the average speed and observe that

$$y = \frac{2d}{\frac{d}{50} + \frac{d}{x}} = \frac{100xd}{xd + 50d} = \frac{100x}{x + 50}.$$

Then Figure 4.20 illustrates that the graph of $y = 100x/(x + 50)$ approaches, but does not touch, the graph of $y = 100$.

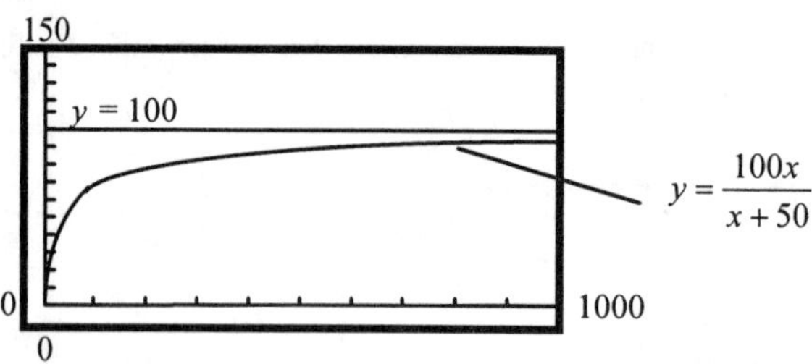

Figure 4.20

PROGRESS CHECK 7

A race car driver must average at least 120 mi/hour for two laps around a track, to qualify for the finals. Because of mechanical trouble the driver is only able to average 90 mi/hour for the first lap. What minimum speed must the driver average for the second lap, to qualify for the finals? ■

EXPLORE 4.2

1. The graph of $y = (ax+b)/(cx+d)$ approaches the horizontal line $y = k$, as $|x|$ increases, where k depends on the values of a and c. Examine the graphs of these functions and determine how you can find k if you know a and c.
   a. $y = \frac{6x+1}{2x-4}$ **b.** $y = \frac{6x+1}{-2x-4}$ **c.** $y = \frac{2x-5}{4x+2}$
2. Equation solvers built into calculators depend on the user supplying a reasonable initial guess at the solution. For instance, a solver will not find the solution to $3/(3-x) = 4/3$ if the initial guess is 5. Look at the graph of $y = 3/(3-x) - 4/3$ and explain why. What will happen if the initial guess is 4? 2? 3.1? 2.9?
3. When the calculator is set to draw in connected mode, it will light a dot at each pixel that is on the graph and connect consecutive points. For instance, if the graph is undefined at $x = 2$, and consecutive points on the screen (as revealed by tracing) are at $x = 1.9048$ and $x = 2.0635$, then the calculator will not "know" there is a break, so it will (incorrectly) connect these points. However, if one of the trace points DOES correspond to $x = 2$, the calculator will not draw any dot there (since y is undefined), and it will NOT connect the points nearest to the break. So if you set the range of x values in such a way that the cursor lands on the break, you will get a correct unconnected picture at that point. The x-values for the pixels are determined by the minimum and maximum settings for x.
   a. Draw the graph of $y = 1/(x-2)$ with several settings of the x range, some of which make a correct graph and some of which do not.
   b. On some calculators making the Trace function hit the break can often be accomplished through a choice such as Zoom Decimal. Explain why this choice works.

EXERCISES 4.2

In Exercises 1-26 solve and check each equation.

1. $\frac{2x-5}{3} - \frac{x}{4} = \frac{1}{2}$ **2.** $\frac{3}{7} - \frac{n-5}{14} = \frac{11n}{14}$

3. $\frac{5}{3a} - \frac{7}{4a} = \frac{5}{6}$ **4.** $\frac{2+x}{6x} - 2 = \frac{3}{5x}$

5. $\frac{9y-5}{7} = \frac{2y-4}{3}$ **6.** $\frac{k}{k+2} = \frac{2}{3}$

7. $\frac{x+4}{x+1} = \frac{x+2}{x+3}$ **8.** $\frac{z-1}{z-2} = \frac{z+1}{z-3}$

9. $\frac{1}{n-3} + \frac{2}{3-n} = \frac{1}{2}$ **10.** $\frac{2}{2-x} + \frac{x}{x-2} = 1$

11. $\frac{4}{x+1} - \frac{3}{x-1} = \frac{-6}{x^2-1}$

12. $\frac{2}{a^2+a}=\frac{1}{a}-\frac{2}{a+1}$ 13. $\frac{3}{s+3}-\frac{s}{3-s}=\frac{s^2+9}{s^2-9}$

14. $\frac{1}{s-2}-\frac{3}{s+2}=\frac{4}{s^2-4}$

15. $\frac{3c}{c+2}-2=\frac{2c-3}{2c-1}$ 16. $\frac{3}{t}+\frac{2}{t-1}=\frac{12}{t^2-t}$

17. $1+\frac{12}{c^2-4}=\frac{3}{c-2}$ 18. $3-\frac{t-4}{t-3}=\frac{4t-1}{2t+3}$

19. $\frac{3}{k+2}-2k=\frac{-5}{2k+4}$ 20. $\frac{5x}{x-2}-\frac{4x}{2x-7}=3$

21. $\frac{3}{x}+\frac{15}{x^2}=1$ 22. $\frac{3}{x^2}-2=\frac{1}{x}$

23. $1-90{,}000x^{-2}=0$ 24. $x-15x^{-1}=14$

25. $2x+9x^{-1}=19$ 26. $x^{-2}+3x^{-1}-10=0$

In Exercises 27–40 solve each formula for the letter indicated.

27. $\frac{W_1}{W_2}=\frac{L_2}{L_1}$ for L_1 28. $\frac{d_1}{d_2}=\frac{F_2}{F_1}$ for d_2

29. Solve $w=\frac{1-p}{p}$ for p.

30. Solve $p=\frac{1}{1+w}$ for w.

31. $t=\frac{v-v_0}{a}$ for v 32. $I=\frac{E}{R+r}$ for r

33. Solve $P=\frac{S}{1+ni}$ for i.

34. Solve $S=\frac{P}{1-nd}$ for d.

35. $S=\frac{a-rL}{1-r}$ for r 36. $I=\frac{nE}{R+nr}$ for n

37. $Z=\frac{Z_1Z_2}{Z_1+Z_2}$ for Z_2 38. $\frac{E}{e}=\frac{R+r}{r}$ for r

39. $\frac{1}{f}=\frac{1}{a}+\frac{1}{b}$ for a 40. $\frac{1}{R}=\frac{1}{R_1}+\frac{1}{R_2}$ for R_2

41. Babylonian mathematics texts from about 3000 B.C. contain problems asking for a number that, when added to its reciprocal, gives a specified sum. They worked out a general procedure for such problems. Try this one. What number when added to its reciprocal equals 5? Give the answers in radical form and in decimal approximation.

42. The sum of a number and its reciprocal is $\frac{29}{10}$. What are the numbers?

43. A man left one-half of his property to his wife, one third to his son, and the remainder, which was $10,000, to his daughter. How much money did the man leave?

44. The harmonic mean H of two numbers a and b is given by $H=\frac{2ab}{a+b}$.
   a. Solve for b.
   b. If $H=8$ and $a=6$, find b.

45. If the total cost of producing n units of a product consists of $1,000 in fixed costs and $10 per unit, then the average cost per unit A is given by $A=\frac{10n+1{,}000}{n}$. How many units should be produced for the average cost to be $20 per unit?

46. In Exercise 45, how many units should be produced for the average cost to be $15 per unit?

47. A telephone company charges 25 cents per minute plus an 80 cent surcharge for a typical call made with a calling card. Therefore, the real cost (y) of a call per minute is given by $y=\frac{80+25x}{x}$, where x is the length of the call in minutes.
   a. What is the total cost for a 2 minute call? What is the real average cost per minute for a 2 minute call?
   b. The longer the call, the less per minute its average cost is. At what length call does the average cost drop to 26 cents per minute?
   c. If the cost per minute is rounded to the nearest cent for billing purposes, at what point does the call really cost 25 cents per minute?

48. A computer printer claims to print 6 pages per minute, but it takes 3 minutes to warm up before the first page will print.
   a. Write a formula for the real speed in pages per minute (y) if you have to turn the machine on before you print x pages.
   b. Graph the function and describe how y changes as x increases for $x>0$.
   c. How many pages must you print before the average speed reaches 5 pages per minute?

49. Coin Sorter A can process a sack of coins in 20 minutes; sorter B can process the sack in 30 minutes. How long would it take the two machines working together to process the sack of coins?

50. One card sorter can process a deck of punched cards in 1 hour; another can sort the deck in 2 hours. How long would it take the two sorters together to process the cards?

51. A pump can empty a tank in 100 minutes. Show algebraically that two such pumps working together will cut the time in half.

52. A swimming pool can be filled through one pipe in 16 hours and through a second pipe in 12 hours. It can be emptied through a drain in 8 hours. If the drain is accidentally left open while both pipes are turned on, how long does it take to fill the pool?

53. A driver drives 20 miles at 20 mi/hour. How fast must the driver go for the next 20 miles, to bring the average speed for the whole trip up to 30 mi/hour?

54. Because of an earlier accident on the autobahn, it takes a sports car driver 1 hour to travel the first 20 miles of a 40-mile trip. If the maximum possible speed of the car is 120 mi/hour, can the driver go fast enough over the next 20 miles to bring the average speed for the whole trip up to 35 mi/hour?

55. A plane flies 360 mi from A to B with a 50 mi/hour tailwind. It flies back against this wind in twice the time.

a. What is the speed of the plane in still air?
b. How long did it take to fly each way?
c. What is the speed for the round trip?

56. Two small children who run at the same speed are playing on a moving sidewalk in an airport. The sidewalk is 1/4 mi long and moves at 1 mi/hour. One child starts at each end, and they run toward each other. They meet in 3 minutes. What is their running speed on regular ground?

THINK ABOUT IT 4.2

1. People who do a lot of probability calculations often use shortcuts for solving some equations. For example, to solve the equation $1/(x-1) = 2/3$ for x, they first notice that the top plus the bottom on the left equals x. To get a new simpler proportion, they say, "Replace each fraction by its (top plus its bottom) over its top." This would yield $x = 5/2$ immediately.

a. Check that 5/2 is the solution to the original equation.
b. Solve $3/(3-x) = 4/3$ by first replacing each fraction by its "top minus bottom over top."
c. Solve $p/(1+p) = 1/4$ using "top over (bottom minus top)."

2. Here are several versions of a classic puzzle.

a. The weight of a brick is half a pound plus half the weight of a brick. How many pounds does the brick weigh?
b. The weight of a brick is 1/4 of a pound plus 1/4 of a brick. How many pounds does the brick weight?
c. The brick weighs 4/5 of a pound plus 4/5 of a brick. How many pounds does the brick weight?
d. This is the problem stated in general. The weight w of a brick is a certain fraction f of a pound plus that fraction of a brick. How many pounds does the brick weight? (Solve for w in terms of f.)
e. For which version of this problem is the answer that the brick weighs 10 lb?

3. Show that in Exercise 51 the value 100 was not necessary to the problem. Show that two similar machines working together will cut the time in half regardless of how long they take working separately.

4. Consider the following statement: "If two fractions are equal and have equal numerators, they also have equal denominators." Now consider the following equation, which we wish to solve for x.

$$\frac{x-1}{x+1}+2=\frac{3x+1}{x+5}$$
$$\frac{x-1}{x+1}+\frac{2(x+1)}{x+1}=\frac{3x+1}{x+5}$$
$$\frac{x-1+2(x+1)}{x+1}=\frac{3x+1}{x+5}$$
$$\frac{3x+1}{x+1}=\frac{3x+1}{x+5}$$

From the above statement we conclude that $x+1=x+5$ or, upon subtracting x from both sides, that $1=5$. What is wrong? What value of x satisfies the equation?

5. Puzzles of all types are popular in magazines and newspapers and have a devout and fanatical following. Mathematical puzzles are no exception. Test your skills on the following problem from the *Mathematical Puzzles of Sam Loyd*, a two-volume series edited by Martin Gardner and published by Dover Publications, © 1960.

How many children are on the carousel?

While enjoying a giddy ride on the carousel, Sammy propounded this problem: "One third of the number of kids riding ahead of me, added to three quarters of those riding behind me gives the correct number of children on this merry-go-round.

How many children were riding the carousel?

● ● ●

4.3 Other Types of Equations

Photo Courtesy of Murphy of Archive Photos, New York

The formula $t = \sqrt{d}/4$ relates the distance d in feet traveled by a free-falling object to the time t of the fall in seconds, disregarding air resistance. If you drop a stone from a bridge that spans a river, and you hear the sound of the splash 3.7 seconds later, then to the nearest foot how far above the water is that particular point on the bridge? The sound created by the splash travels at a speed of about 1100 ft/second.
(See Example 7.)

Objectives

1. Solve radical equations.
2. Solve equations with quadratic form.
3. Solve higher-degree polynomial equations using factoring.

An equation in which the unknown appears under a radical is called a **radical equation**. Example 1 shows how to solve an applied problem that involves a simple radical equation. You will find that this problem is related to (but simpler than) the problem in the section introduction.

Example 1: An Application Involving a Radical Equation

The formula $t = \sqrt{d}/4$ relates the distance d in feet traveled by a free-falling object to the time t of the fall in seconds, disregarding air resistance. If a stone is dropped from a bridge that spans a river and hits the water in 2.3 seconds, then to the nearest foot how far above the water is that particular point on the bridge?

Solution: Replacing t by 2.3 in

$$t = \frac{\sqrt{d}}{4} \text{ gives } 2.3 = \frac{\sqrt{d}}{4}.$$

The resulting equation may be solved by squaring both sides of the equation and solving for d.

$$(2.3)^2 = \left(\frac{\sqrt{d}}{4}\right)^2 \quad \text{Square both sides.}$$

$$(2.3)^2 = \frac{d}{16}$$

$$16(2.3)^2 = d$$

$$84.64 = d$$

To the nearest foot, the stone was dropped from a point on the bridge that is 85 ft above the water.

To check this answer, replace d by 85 in the given formula and determine that

$$t = \frac{\sqrt{85}}{4} \approx 2.3.$$

When $d \approx 85$ ft, $t \approx 2.3$ seconds as specified. Figure 4.21 shows a graphical check that confirms the solution in this example.

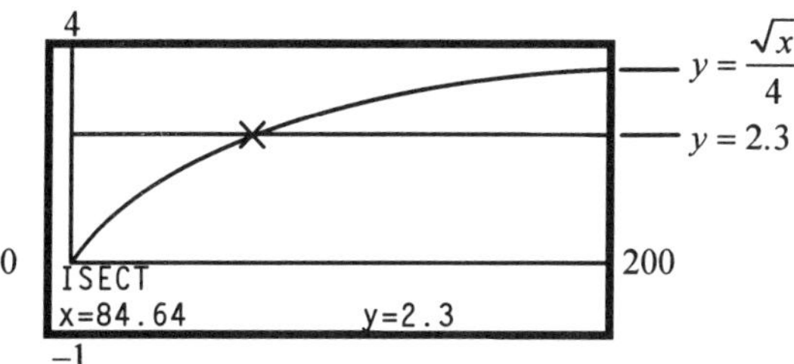

Figure 4.21

PROGRESS CHECK 1
A stone dropped from the recorded height of the Empire State Building in New York City takes about 9.592 seconds to hit the ground. To the nearest foot, what is this recorded height? ■

Example 1 illustrates that the following steps may be used to solve a radical equation.

1. Isolate a radical on one side of the equation.
2. Raise both sides of the equation to the power that matches the index of the radical.
3. Solve the resulting equation and check all solutions in the *original* equation.

The check in step 3 is essential because this procedure is based on the following principle.

Principle of Powers

If P and Q are algebraic expressions, then the solution set of the equation $P = Q$ is a subset of the solution set of $P^n = Q^n$ for any positive integer n.

Thus, every solution of $P = Q$ is a solution of $P^n = Q^n$; however, solutions of $P^n = Q^n$ may or may not be solutions of $P = Q$, so checking *is* necessary. Solutions of $P^n = Q^n$ that do not satisfy the original equation are called **extraneous solutions**, and Example 2 illustrates this possibility.

Example 2: Solving a Radical Equation

Solve $x - \sqrt{x+1} = 1$.

Solution: We can first isolate the radical on one side of the equation.

$$x - \sqrt{x+1} = 1$$
$$x - 1 = \sqrt{x+1}$$

Then the intersection of the graphs of $y = x - 1$ and $y = \sqrt{x+1}$, which is shown in Figure 4.22, indicates that $x - 1$ and $\sqrt{x+1}$ have the same value only when $x = 3$. To confirm this result algebraically we square both sides of the equation $x - 1 = \sqrt{x+1}$ and solve for x.

$$(x-1)^2 = \left(\sqrt{x+1}\right)^2$$
$$x^2 - 2x + 1 = x + 1$$
$$x^2 - 3x = 0$$
$$x(x-3) = 0$$
$$x = 0 \quad \text{or} \quad x = 3$$

Check:

$$0 - \sqrt{0+1} \stackrel{?}{=} 1 \qquad\qquad 3 - \sqrt{3+1} \stackrel{?}{=} 1$$
$$0 - 1 \stackrel{?}{=} 1 \qquad\qquad 3 - 2 \stackrel{?}{=} 1$$
$$-1 \neq 1 \quad \text{extraneous solution} \qquad\qquad 1 \stackrel{\checkmark}{=} 1$$

Only 3 is a solution of the original equation, so the solution set is $\{3\}$.

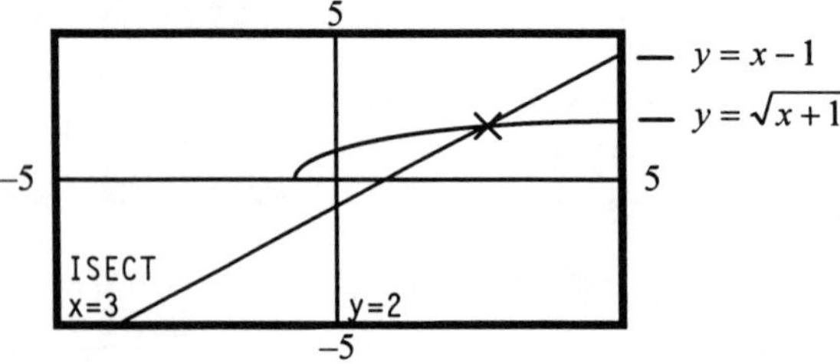

Figure 4.22

PROGRESS CHECK 2
Solve $5+\sqrt{2x-2} = x$. ■

Example 3: Solving a Radical Equation

Solve the equation $\sqrt[3]{x^2+15} = 4$.

Solution: The index of the radical is 3, so we raise both sides of the equation to the third power and then solve for x.

$$\sqrt[3]{x^2+15} = 4$$
$$\left(\sqrt[3]{x^2+15}\right)^3 = 4^3 \quad \text{Cube both sides.}$$
$$x^2 + 15 = 64$$
$$x^2 = 49$$
$$x = \pm 7$$

Check

$$\sqrt[3]{(7)^2+15} \stackrel{?}{=} 4 \qquad\qquad \sqrt[3]{(-7)^2+15} \stackrel{?}{=} 4$$
$$\sqrt[3]{64} \stackrel{?}{=} 4 \qquad\qquad \sqrt[3]{64} \stackrel{?}{=} 4$$
$$4 \stackrel{\checkmark}{=} 4 \qquad\qquad 4 \stackrel{\checkmark}{=} 4$$

Since both solutions check, the solution set is $\{7, -7\}$.

See Figure 4.23 for a graphical solution that supports that 7 and –7 are the two roots of this equation.

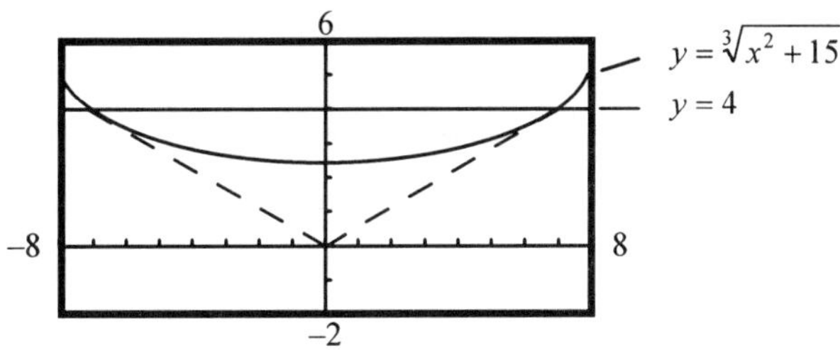

Figure 4.23

PROGRESS CHECK 3

Solve $\sqrt[3]{x^2 - 37} = 3$. ■

Example 4: Solving a Radical Equation with Two Radicals

Solve the equation $\sqrt{x} + \sqrt{x+5} = 5$.

Solution: We first isolate one of the radicals, and then square both sides of the resulting equation.

$$\begin{aligned} \sqrt{x} + \sqrt{x+5} &= 5 \\ \sqrt{x+5} &= 5 - \sqrt{x} \\ \left(\sqrt{x+5}\right)^2 &= \left(5 - \sqrt{x}\right)^2 && \text{Square both sides.} \\ x + 5 &= 25 - 10\sqrt{x} + x \end{aligned}$$

A radical remains, so isolate this radical term and square both sides again.

$$\begin{aligned} -20 &= -10\sqrt{x} \\ 2 &= \sqrt{x} \\ (2)^2 &= \left(\sqrt{x}\right)^2 && \text{Square both sides.} \\ 4 &= x \end{aligned}$$

Check

$$\begin{aligned} \sqrt{4} + \sqrt{4+5} &\stackrel{?}{=} 5 \\ 2 + 3 &\stackrel{?}{=} 5 \\ 5 &\stackrel{\checkmark}{=} 5 \end{aligned}$$

Thus, the solution set is $\{4\}$. The intersection shown in Figure 4.24 may be used to confirm this solution graphically.

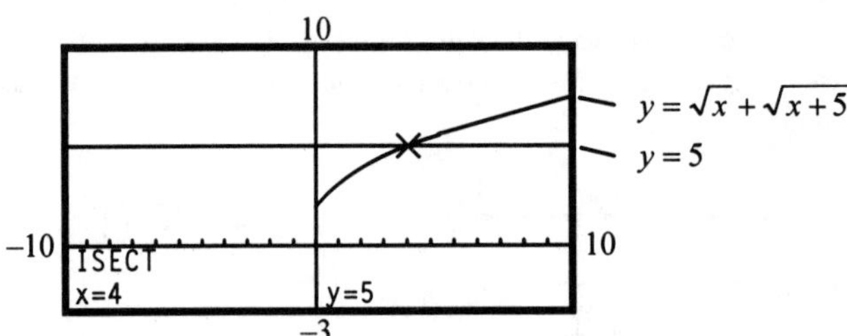

Figure 4.24

PROGRESS CHECK 4
Solve $\sqrt{3x-5}-\sqrt{2x-5}=1$. ∎

Example 5: Solving a Radical Equation

Solve the equation $x+\sqrt{x}-2=0$.

Solution: To solve the equation, we can first isolate the radical.

$$x+\sqrt{x}-2=0$$
$$x-2=-\sqrt{x}$$
$$(x-2)^2=\left(-\sqrt{x}\right)^2 \quad \text{Square both sides.}$$
$$x^2-4x+4=x$$
$$x^2-5x+4=0$$
$$(x-4)(x-1)=0$$
$$x=4 \quad \text{or} \quad x=1$$

Check

$$4+\sqrt{4}-2\stackrel{?}{=}0$$
$$4+2-2\stackrel{?}{=}0$$
$$4\neq 0 \quad \text{extraneous solution}$$

$$1+\sqrt{1}-2\stackrel{?}{=}0$$
$$1+1-2\stackrel{?}{=}0$$
$$0\stackrel{\checkmark}{=}0$$

Therefore, 1 is the only solution of the equation, and the solution set is $\{1\}$. The x-intercept of the graph of $y=x+\sqrt{x}-2$ that is shown in Figure 4.25 is in agreement with this solution.

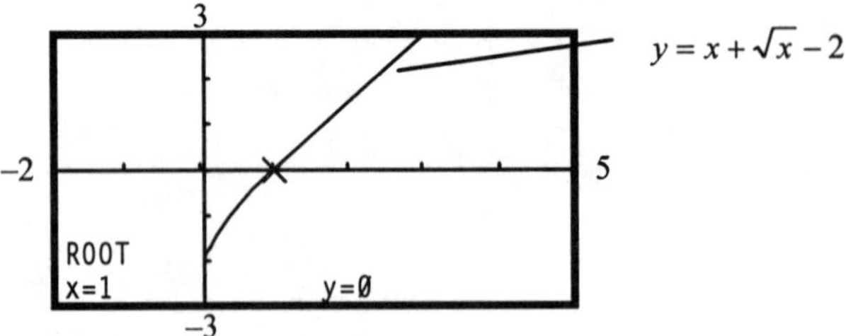

Figure 4.25

PROGRESS CHECK 5
Solve $x-3\sqrt{x}-4=0$. ∎

In Example 5 the solution led to a quadratic equation. Although we often encounter quadratic equations in the process of solving other equations, $x+\sqrt{x}-2=0$ is a special case. Equations that are not themselves quadratic but which are equivalent to equations having the form

$$at^2+bt+c=0 \quad (a\neq 0)$$

are called **equations with quadratic form**. In Example 5, if we let $t=\sqrt{x}$, then $x+\sqrt{x}-2=0$ becomes

$$t^2+t-2=0$$

with solution

$$\begin{aligned}(t+2)(t-1)&=0\\ t=-2 \quad \text{or} \quad t&=1.\end{aligned}$$

Now we can resubstitute $\sqrt{x}$ for t and solve for x.

$$\begin{aligned}\sqrt{x}=-2 \quad &\text{or} \quad \sqrt{x}=1\\ x=4 \quad &\text{or} \quad x=1\end{aligned}$$

As in our solution above, only 1 is a root of the equation. This extraneous root $x=4$ is easy to pick out here because $\sqrt{x}$ cannot be –2. To spot equations with quadratic form, look for the exponent in one term to be double the exponent in another term.

Example 6: Solving an Equation with Quadratic Form

Solve $x^{-2}-4x^{-1}-3=0$.

Solution: First note that

$$x^{-2}-4x^{-1}-3=\left(x^{-1}\right)^2-4\left(x^{-1}\right)-3=0,$$

so if we let $t=x^{-1}$, we have

$$t^2-4t-3=0.$$

By the quadratic formula (see Section 4.1, Example 5).

$$t=2\pm\sqrt{7}.$$

Now we resubstitute x^{-1} for t and solve for x.

$$\begin{aligned}x^{-1}&=2\pm\sqrt{7}\\ x&=\frac{1}{2\pm\sqrt{7}}=\frac{2\pm\sqrt{7}}{-3}=\frac{-2\pm\sqrt{7}}{3}\end{aligned}$$

Check these two solutions in the original equation to confirm that the solution set is $\left\{\left(-2\pm\sqrt{7}\right)/3\right\}$. Since $\left(-2+\sqrt{7}\right)/3\approx 0.22$ and $\left(-2-\sqrt{7}\right)/3\approx -1.55$, these two solutions are in agreement with the x-intercepts in the graph of $y=x^{-2}-4x^{-1}-3$ that is shown in Figure 4.26.

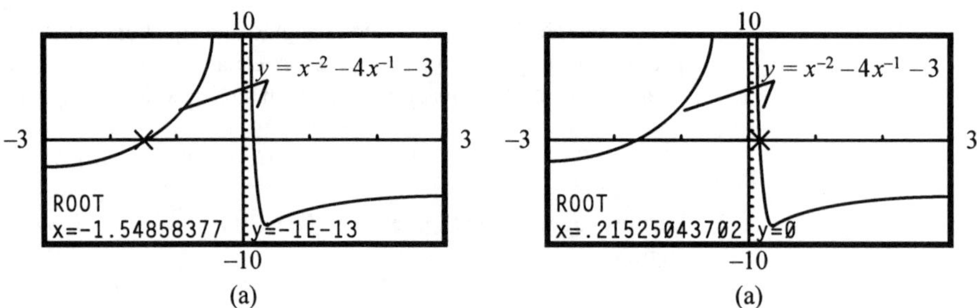

Figure 4.26

PROGRESS CHECK 6
Solve $x^{-2} + 3x^{-1} - 2 = 0$. ■

Example 7: An Application Involving an Equation with Quadratic Form

Solve the problem in the section introduction on page 344.

Solution: The time required for the stone to hit the water is given by $\sqrt{d}/4$. Using $d = rt$, the time required for the sound of the splash to reach you is given by $d/1{,}100$. Since the sound of the splash is heard 3.7 seconds after the stone is released, we have

$$\underbrace{\text{Time for stone to reach water}}_{\dfrac{\sqrt{d}}{4}} \quad \underset{+}{\text{plus}} \quad \underbrace{\text{time for sound of splash to reach you}}_{\dfrac{d}{1{,}100}} \quad \underset{=}{\text{equals}} \quad \underbrace{\text{total elapsed time.}}_{3.7.}$$

A picture of the solution to this equation is shown in Figure 4.27. We can see that if it takes 3.7 seconds for you to hear the sound of the splash, then the stone was dropped from about 198 ft above the water.

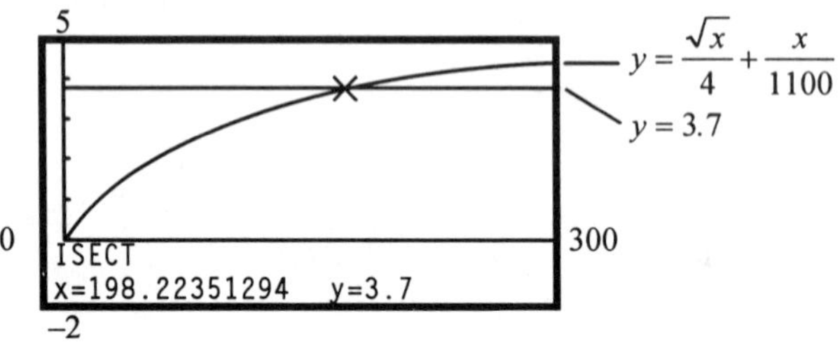

Figure 4.27

To confirm this solution algebraically we can first clear fractions by multiplying both sides of the derived equation by 1,100.

$$275\sqrt{d} + d = 4{,}070$$

Now we choose to solve the resulting equation by treating it as an equation with quadratic form.

$$d + 275\sqrt{d} - 4{,}070 = 0$$
$$z^2 + 275z - 4{,}070 = 0 \quad \text{Let } z = \sqrt{d}\text{; then } z^2 = d.$$

The quadratic formula gives

$$z = \frac{-275 \pm \sqrt{(275)^2 - 4(1)(-4070)}}{2(1)} = \frac{-275 \pm \sqrt{91{,}905}}{2},$$

and using the positive solution for z together with the substitution that $d = z^2$ yields

$$d = z^2 = \left(\frac{-275 + \sqrt{91{,}905}}{2}\right)^2 \approx 198.$$

To the nearest foot, the stone was dropped from a point on the bridge that is 198 ft above the water.

PROGRESS CHECK 7
Redo the problem described in Example 7 assuming that you hear the sound of the splash 2.4 seconds after releasing the stone. ■

Higher-Degree Polynomial Equations

We have considered how to solve linear and quadratic equations that are first- and second-degree polynomial equations. Higher-degree polynomial equations are usually much harder to solve, and we will consider this topic in detail in Chapter 6. For now, however, it is sometimes possible to factor higher-degree polynomials into linear and/or quadratic factors with integer coefficients so that the equations can be solved by using the zero product principle. The idea is the same as with second-degree equations. First, rearrange the equation (if necessary) so that zero is on one side by itself, then factor. Now set each factor equal to zero and obtain solutions by solving the resulting equations.

Example 8: Solving a Polynomial Equation

Solve $x^5 = 10x^3 - 9x$.

Solution: First, rewrite the equation so that one side is 0.

$$\begin{aligned} x^5 &= 10x^3 - 9x \\ x^5 - 10x^3 + 9x &= 0 \end{aligned}$$

Now factor completely.

$$\begin{aligned} x(x^4 - 10x^2 + 9) &= 0 \\ x(x^2 - 9)(x^2 - 1) &= 0 \\ x(x+3)(x-3)(x+1)(x-1) &= 0 \end{aligned}$$

Setting each factor equal to zero gives

$$\begin{array}{lllllllll} x = 0 & \text{or} & x + 3 = 0 & \text{or} & x - 3 = 0 & \text{or} & x + 1 = 0 & \text{or} & x - 1 = 0 \\ & & x = -3 & & x = 3 & & x = -1 & & x = 1. \end{array}$$

Thus, the solution set is $\{0, -3, 3, -1, 1\}$.

Graphical support for this answer may be found by examining the x-intercepts in the graph of $y = x^5 - 10x^3 + 9x$ that is shown in Figure 4.28.

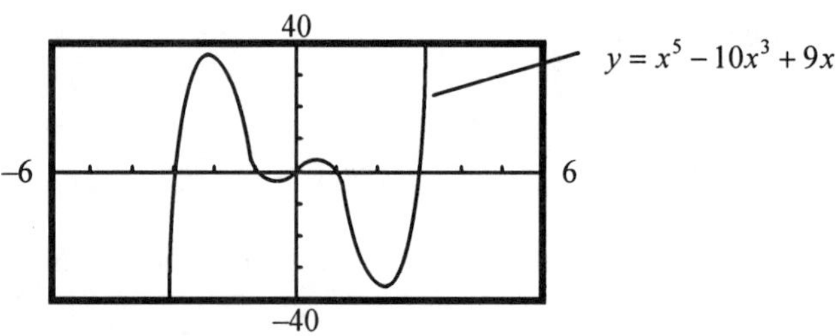

Figure 4.28

PROGRESS CHECK 8

Solve $x^5 + 12x = 7x^3$. ■

EXPLORE 4.3

1. **a.** To solve $x - 1 = \sqrt{x+1}$ graphically, you can look for any intersection points of $y = x - 1$ and $y = \sqrt{x+1}$. Check that there is only one such point. What are the coordinates of that point? What is the solution to the equation?

b. If you square both sides of the equation in part **a** you get $(x-1)^2 = x + 1$. Look at the graphs of $y = (x-1)^2$ and $y = x + 1$. Now, how many points of intersection are there? Are any of these points the same as the point found in part **a**? Compare the solution set of $x - 1 = \sqrt{x+1}$ to the solution set of $(x-1)^2 = x + 1$.

c. Use the results of parts **a** and **b** to answer this question. If you square both sides of an equation to find solutions, why is it necessary to check the resulting solutions in the original equation?

2. This problem examines how a slight change in an equation can change the number of solutions that are real numbers. We use the fact the intersections of graphs may reveal real number solutions.

a. Graphically investigate the solutions to $(x+2)(x-1)^2 = 0$ by looking at the x-intercepts of the graph of $y = (x+2)(x-1)^2$. How many solutions are there? Find the exact values of the solutions.

b. Now consider $(x+2)(x-1)^2 = 1$. How many points of intersection are there between the graphs of $y = (x-2)(x+1)^2$ and $y = 1$? Estimate all the real solutions.

c. Consider $(x+2)(x-1)^2 = -1$. How many real solutions are there?

d. Explain why the equation $(x+2)(x-1)^2 = n$ where n is real, must have at least one real number solution.

3. A common student error is to rewrite $\sqrt{x^2 + 4}$ as $x + 2$..

a. Use a grapher to find all values of x for which

$$\sqrt{x^2 + 4} = x + 2$$

is a true statement.

b. Solve the equation in part **a** to find the answer algebraically.

EXERCISES 4.3

In Exercises 1-36 solve and check each equation.

1. $\sqrt{x+5} = 2$
2. $\sqrt{2x+1} = 7$
3. $\sqrt{x-5} = -3$
4. $\sqrt{x-2} + 3 = 0$
5. $2\sqrt{3x} + 3 = 13$
6. $3\sqrt{2x} - 4 = 5$
7. $\sqrt[4]{2x+1} = \sqrt[4]{3x-5}$

8. $\sqrt[4]{x^2+x+6}=\sqrt[4]{x^2+3x+2}$

9. $\sqrt[3]{x+1}=-2$
10. $\sqrt[3]{x^2-17}=4$

11. $\sqrt{x^2+9}=x+1$
12. $\sqrt{x^2-x+1}=x-1$

13. $\sqrt{8-x^2}+x=0$
14. $\sqrt{x+1}+1=x$

15. $\sqrt{x-2}+x=4$
16. $\sqrt{x+4}-4=x$

17. $2\sqrt{x+1}=2-x$
18. $2\sqrt{x-1}=x-1$

19. $\sqrt{2x^2-2}+x-1=0$

20. $\sqrt{x-4}=5-\sqrt{x+1}$
21. $\sqrt{x-8}+\sqrt{x}=2$

22. $\sqrt{x+5}-9=\sqrt{x-4}$
23. $\sqrt{x}-\sqrt{2x+1}=-1$

24. $\sqrt{4x+2}+\sqrt{2x}-\sqrt{2}=0$

25. $\sqrt{2x}+\sqrt{2x+3}=3$
26. $\dfrac{10}{\sqrt{x-5}}=\sqrt{x-5}-3$

27. $\sqrt{x^2}=x$
28. $\sqrt{(2x-1)^2}=2x-1$

29. $\dfrac{x}{\sqrt{x^2+16}}-1=0$
30. $\dfrac{1}{4}\cdot\dfrac{2x}{2\sqrt{x^2+16}}-\dfrac{1}{5}=0$

31. $\dfrac{1}{\sqrt{2}}=\dfrac{6-a}{\sqrt{5(9+a^2)}}$
32. $\dfrac{2a+3}{\sqrt{5a^2+45}}=1$

33. $x+2\sqrt{x}-3=0$
34. $x-10\sqrt{x}+9=0$

35. $x+23=10\sqrt{x}$
36. $4x+3=8\sqrt{x}$

In Exercises 37–64 solve each equation.

37. $x^4-6x^2+5=0$
38. $x^4+3=4x^2$

39. $4x^4+1=4x^2$
40. $3x^4+14x^2-5=0$

41. $x^4-4x^2+1=0$
42. $x^4-6x^2+4=0$

43. $(x+3)^2-5(x+3)+4=0$

44. $(x-1)^2-(x-1)=0$

45. $4(x-2)^2+9=12(x-2)$

46. $3(x+1)^2+13(x+1)=10$

47. $x^{-2}+x^{-1}=0$
48. $3x^{-2}+1=4x^{-1}$

49. $x^{1/2}+2=3x^{1/4}$
50. $x^{1/3}-1=2x^{-1/3}$

51. $y^3+9y^2+14y=0$
52. $4a^3-13a^2=-3a$

53. $3a^3=3a$
54. $16y^3=9y$

55. $x^2(2x-1)(3x+1)=0$

56. $x^3(1-2x)(4x+5)=0$

57. $x^4+4=5x^2$
58. $x^6=10x^4-9x^2$

59. $3x^3-11x^2=3x-11$

60. $2x^3-x^2-8x+4=0$

61. $x^3-8=0$
62. $x^3+1=0$

63. $x^6-1=0$
64. $x^6=64$

65. Solve for s: $gt=\sqrt{2gs}$.

66. Solve for d: $r=\sqrt[3]{\dfrac{3a}{4\pi d}}$.

67. Solve for y: $x=\sqrt[3]{y+2}$.

68. Solve for g: $t=\pi\sqrt{\dfrac{L}{g}}$.

The formula $t=\sqrt{d}/4$ relates the distance d in feet traveled by a free-falling object to the time t of the fall in seconds disregarding air resistance. Use this relationship to solve Exercises 69 and 70.

69. As you travel along the rim of the Grand Canyon, you stop at a place called Mojave Cliffs. Using a stopwatch, you calculate the length of time it takes for a rock dropped off the edge to hit the bottom of the canyon as 13.7 seconds. Approximate, to the nearest foot, the depth of the Grand Canyon at this point.

70. On a farm there's an old well you'd like to use. You have a bucket, and you're going to buy some rope, but you don't know how much rope to get. You decide to measure the depth of the well by dropping a small stone. If it takes about 1.3 seconds for the stone to hit the water, how deep is the well (to the nearest foot)?

71. The speed in feet per second that a dropped object acquires in falling d feet is given by $v=8\sqrt{d}$. Are there any moments at which the measurement of speed is the same number as the number of feet fallen? If so, find them.

72. It was determined by investigators in New York that an object falling from above hit the sidewalk at 800 ft/sec. Is it reasonable to claim that it fell from a window on the 100th floor of a skyscraper? Assume that each floor occupies about 10 feet vertically. From about what height was the object dropped? Use $v=8\sqrt{d}$.

73. For a certain object thrown upwards, the time in seconds it takes to reach a height of S feet is given by $t = 2 - \dfrac{\sqrt{64 - S}}{4}$. How high is the object when 1 second has elapsed?

74. The surface area of a cone is given by $S = \pi r\sqrt{r^2 + h^2}$, and the volume of the cone is given by $V = \dfrac{1}{3}\pi r^2 h$.

a. What height is needed for a cone with radius 2 cm to have a surface area equal to 25 cm^2?

b. What volume will it have?

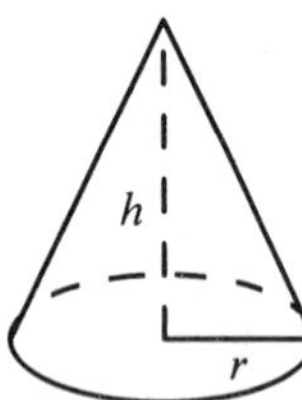

75. The period of a pendulum is the time required for the pendulum to complete one round-trip of motion, that is, one complete cycle. When the period is measured in seconds and the pendulum length l is measured in feet, then a formula for the period is

$$T = 2\pi\sqrt{\frac{l}{32}}.$$

To the nearest hundredth of a foot, what is the length of a pendulum whose period is 1 second?

76. Using the formula from Exercise 75, find the length of a pendulum whose period is 2 seconds.

77. A kind of average called the **root-mean-square** is defined as the square root of the mean of a collection of squares. The formula $A = \sqrt{\dfrac{m^2 + n^2}{2}}$ gives the root-mean-square of two positive numbers, m and n. Find n if $m = 1$ and $A = 5$.

78. Use the formula from Exercise 77 and find n if $m = 2$ and $A = 2\sqrt{5}$.

79. If the perimeter in this right triangle is 14 cm, find x.

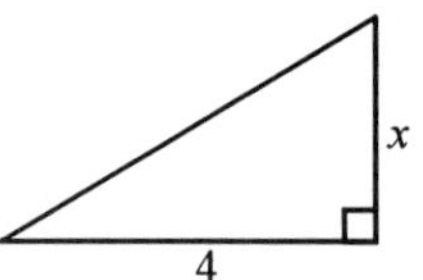

80. The perimeter in a right triangle is 36 m and the height is 3 m longer than the base. Find the base in the triangle.

81. The perimeter of an isosceles right triangle is 12 inches. Find its area.

82. A survey was taken to estimate the percentage of all potential voters who support the president. Because the survey only includes SOME of the potential voters, it has a margin of error, which describes how much it is likely to deviate from the truth. The margin of error in this kind of survey is given by $E = 1.96\sqrt{\dfrac{p - p^2}{n}}$ where n is the number of people interviewed and p is the proportion of those interviewed who said they supported the president. Note that p must be a number from 0 to 1.

a. Use a grapher to illustrate that for any given n, the margin of error is largest when $p = .5$. To do this, draw graphs with several values of n (say, $n = 100$, 500, 1000, 2000 and estimate the coordinates of the maximum point.

b. If you take $p = .5$, about how many people are needed so that the margin of error is .02 or less. This will imply that the survey results probably do not deviate from the truth by more than 2 percent.

THINK ABOUT IT 4.3

1. The velocity in feet/sec of a satellite in orbit at a fixed distance r (in feet) from the Earth is given by $v = \sqrt{\left(14 \times 10^{15}/r\right)}$.

a. The moon travels around the Earth at a speed of about 3300 feet/sec. Use this information and the given formula to estimate (to the nearest 10,000 miles) the distance from the Earth to the moon. There are 5280 feet in a mile.

b. At about what height (to the nearest 1000 miles) should a satellite be placed so that it will orbit the Earth one time in 24 hours?

2. A classic geometry problem asks for the dimensions of a rectangle inscribed in a circle such that the rectangle has maximum possible area. See figure. If the radius is R and x is as shown, the problem leads to the equation $\sqrt{R^2 - x^2} - x^2/\sqrt{R^2 - x^2} = 0$.

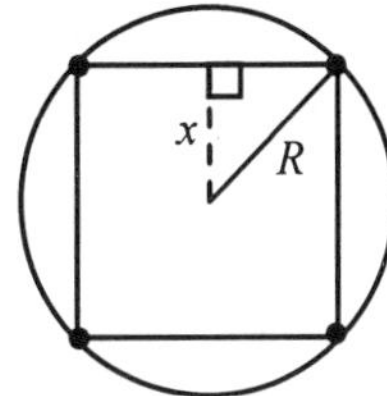

a. Solve this equation for x in terms of R.
b. Then find the area of the rectangle with maximum area.

3. Find all real numbers a for which the distance between $(-1, 1)$ and $(5, a)$ is 10.
4. Solve for x: $4^{1/4} + x^{1/3} = 2/(2-\sqrt{2})$.
5. Explain the apparent paradox.

$a = b$	
$a^2 = b^2$	Square both sides of the equation.
$a^2 - b^2 = 0$	Subtract b^2 from both sides of the equation.
$(a+b)(a-b) = 0$	Factor.
$\dfrac{(a+b)(a-b)}{a-b} = \dfrac{0}{a-b}$	Divide both sides of the equation by the same number.
$a + b = 0$	Subtract b from both sides of the equation.
$a = -b$	

Thus, a equals both b and the negative of b!

● ● ●

4.4 Solving Inequalities Involving Polynomials

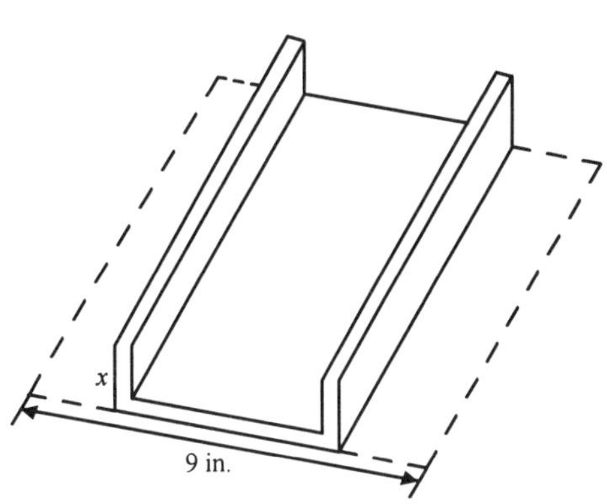

A long strip of galvanized sheet metal 9 in. wide is to be shaped into an open gutter by bending up the edges, as shown in the figure, to form a gutter with a rectangular cross-sectional area. If the area must be at least 9 in.^{2} for the gutter to be useful, then find all possible heights (x) of the gutter. (Ignore the thickness of the metal.)
(See Example 4.)

Objectives

1. Sole quadratic inequalities.
2. Solve polynomial inequalities of degree higher than 2.
3. Solve inequalities involving quotients of polynomials.

Quadratic inequalities are inequalities that may be expressed in the standard forms

$$ax^2 + bx + c > 0, \qquad ax^2 + bx + c \geq 0,$$
$$ax^2 + bx + c < 0, \quad \text{or} \quad ax^2 + bx + c \leq 0,$$

where a, b, and c are real numbers with $a \neq 0$. There are several methods for solving such inequalities, and Example 1 shows a method that is based on the ability to read a graph.

Example 1: Using a Graphical Method

Figure 4.29 shows the graph of $f(x) = x^2 + 2x - 3$. Use this graph to solve each equation or inequality.

a. $x^2 + 2x - 3 = 0$ **b.** $x^2 + 2x - 3 < 0$ **c.** $x^2 + 2x - 3 > 0$ **d.** $x^2 + 2x - 3 \leq 0$

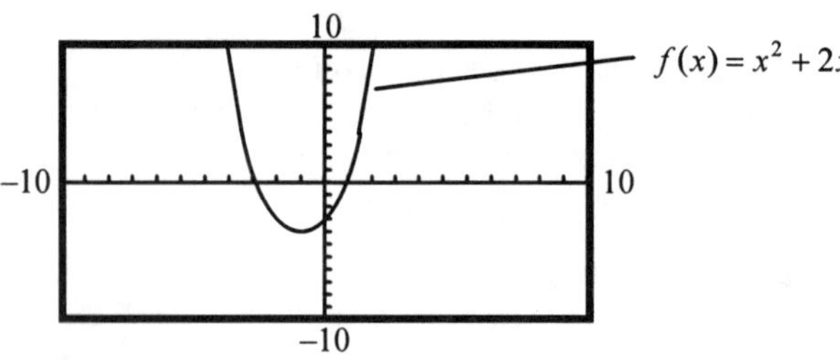

Figure 4.29

Solution:

a. The real number roots of the equation $x^2 + 2x - 3 = 0$ are identical to the x-coordinates of the x-intercepts of the graph of $f(x) = x^2 + 2x - 3$. In Figure 4.29 it appears that the graph is *on the x-axis* when $x = -3$ and $x = 1$, so we check these numbers in the original equation.

$$\text{For } x = -3: \quad (-3)^2 + 2(-3) - 3 = 9 - 6 - 3 = 0$$
$$\text{For } x = 1: \quad (1)^2 + 2(1) - 3 = 1 + 2 - 3 = 0$$

Both apparent solutions check, and the solution set is $\{-3, 1\}$.

b. The y-values are less than zero when the graph is *below the x-axis*. Thus, $x^2 + 2x - 3 < 0$ for $-3 < x < 1$, so the solution set is the interval $(-3, 1)$.

c. The y-values are greater than zero when the graph is *above the x-axis*. Thus, $x^2 + 2x - 3 > 0$ when $x < -3$ or $x > 1$, so the solution set in interval notation is $(-\infty, -3) \cup (1, \infty)$.

d. The y-values are less than or equal to zero when the graph is *on or below the x-axis*. Thus, we need to modify the answer from part **b** to include –1 and 1 in the solution set. So $x^2 + 2x - 3 \leq 0$ for $-3 \leq x \leq 1$, and the solution set is the interval $[-3, 1]$.

PROGRESS CHECK 1

Figure 4.30 shows the graph of $f(x) = x^2 - x - 2$. Use this graph to solve each equation or inequality.

a. $x^2 - x - 2 = 0$ **b.** $x^2 - x - 2 > 0$ **c.** $x^2 - x - 2 < 0$ **d.** $x^2 - x - 2 \geq 0$ ■

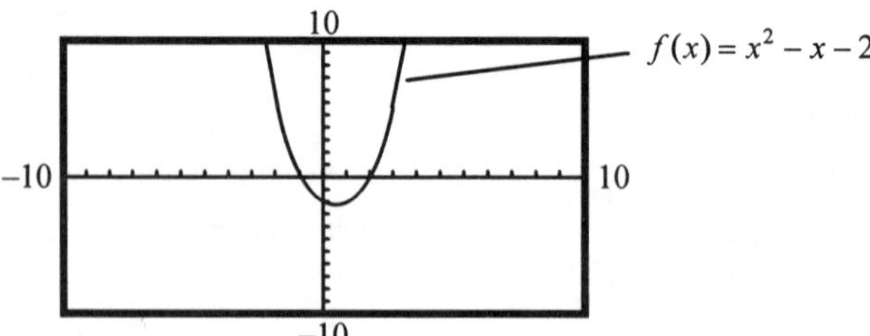

Figure 4.30

When solving quadratic equations or inequalities by the method shown in Example 1, note that the solution set is determined using the criteria described in Figure 4.31.

Quadratic Equation or Inequality	Possible Graph of $y = f(x)$	Interpretation of Solution Set
$ax^2 + bx + c = 0$	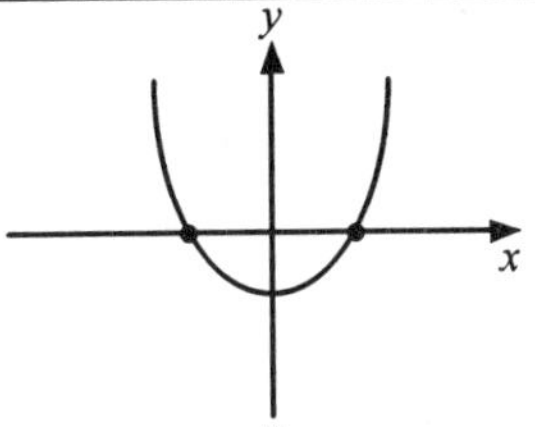	$f(x) = 0$ for all x-values where the graph is *on* the x-axis.
$ax^2 + bx + c < 0$	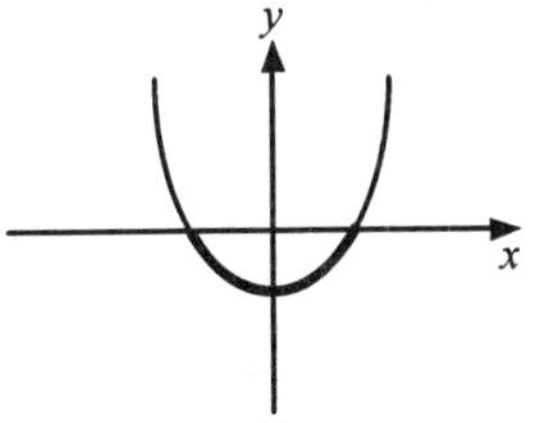	$f(x) < 0$ for all x-values where the graph is *below* the x-axis.
$ax^2 + bx + c > 0$	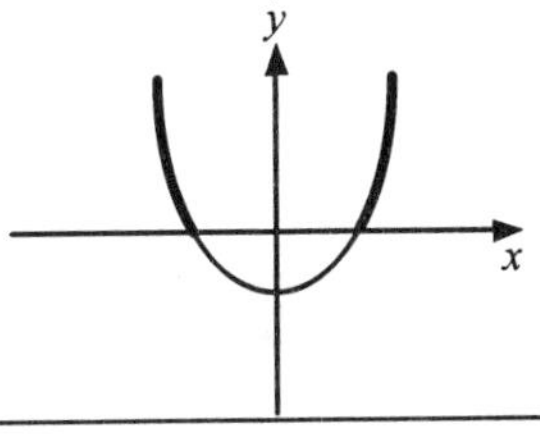	$f(x) > 0$ for all x-values where the graph is *above* the x-axis.

Figure 4.31

The graphing method just described can be used to formulate an algebraic method for solving quadratic inequalities that is also easy to use in practice. First, real numbers that make $ax^2 + bx + c$ equal to zero are called **critical numbers**. Note in Example 1 that the critical numbers are -3 and 1 and that these two critical numbers separate the number line into three intervals, namely, $(-\infty, -3)$, $(-3, 1)$, and $(1, \infty)$. Also, observe that throughout each interval the sign of $x^2 + 2x - 3$ remains the same. Thus an efficient method, called the **test point method**, for solving quadratic inequalities is to find the sign of a convenient number in each of the intervals determined by the critical numbers. By comparing the resulting sign with the inequality in question, we may determine the solution set, as shown in Example 2.

Example 2: Using the Test Point Method

Solve $1 - x^2 \geq 0$ using the test point method.

Solution: First, factor the nonzero side of the inequality.

$$1 - x^2 \geq 0$$
$$(1 + x)(1 - x) \geq 0$$

Because $(1 + x)(1 - x)$ equals 0 when $x = -1$ and when $x = 1$, the critical numbers are -1 and 1. These numbers separate the number line into the intervals $(-\infty, -1)$, $(-1, 1)$, and $(1, \infty)$, and Figure 4.32 shows whether a true statement results when a specific number in each of these intervals is tested.

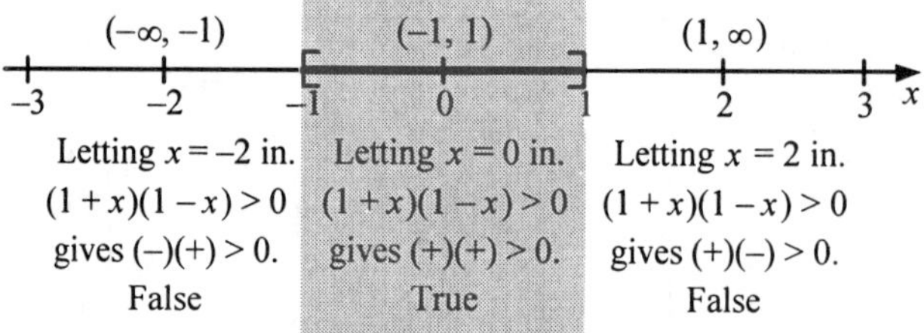

Figure 4.32

All numbers in each interval lead to the same result as the specific number tested, so $(1+x)(1-x)>0$ is a true statement on the interval $(-1, 1)$. Because the inequality in question is greater than or equal to zero, −1 and 1 are also solutions, and the solution set is the interval $[-1, 1]$.

This answer may be checked using the graph in Figure 4.33, which shows that the graph of $f(x)=1-x^2$ is on or above the x axis for $-1\le x\le 1$.

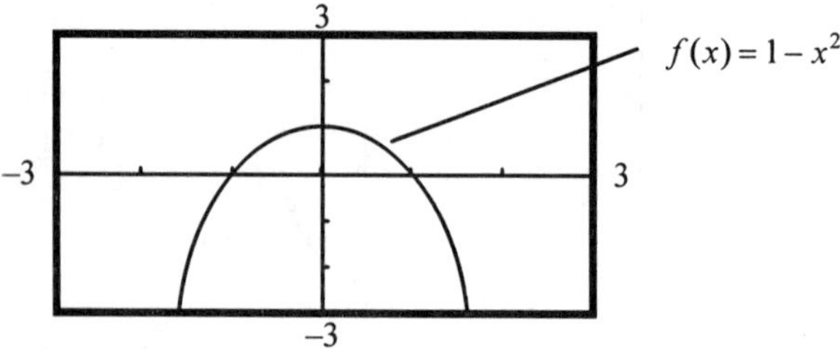

Figure 4.33

Note: In this example specific numbers may be tested in $1-x^2\ge 0$ or in $(1+x)(1-x)\ge 0$. Although it is necessary to determine only the sign that results from the substitution, you may prefer to find the actual number.

PROGRESS CHECK 2

Solve $2x^2-13x-7\ge 0$ using the test point method. ■

In the next example the quadratic formula provides an efficient method for finding the critical numbers that are irrational.

Example 3: Using the Test Point Method and the Quadratic Formula

Solve $x^2-2x-1>0$.

Solution: First, determine the critical numbers by using the quadratic formula to find when x^2-2x-1 equals zero.

$$x=\frac{-(-2)\pm\sqrt{(-2)^2-4(1)(-1)}}{2(1)}=\frac{2\pm\sqrt{8}}{2}=\frac{2\pm 2\sqrt{2}}{2}=1\pm\sqrt{2}$$

The critical numbers are $1+\sqrt{2}$ and $1-\sqrt{2}$, which are approximately equal to 2.4 and −0.4, respectively. Now we may test in the intervals determined by these critical numbers, as shown in Figure 4.34.

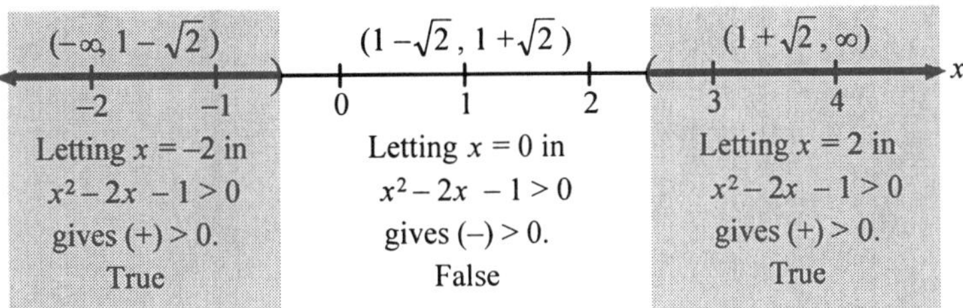

Figure 4.34

The tests show that the inequality $x^2 - 2x - 1 > 0$ is true for all x such that $x < 1 - \sqrt{2}$ or $x > 1 + \sqrt{2}$. Thus the solution set is $(-\infty, 1-\sqrt{2}) \cup (1+\sqrt{2}, \infty)$.

The graph in Figure 4.35 may be used to support this solution since it shows that the graph of $f(x) = x^2 - 2x - 1$ is above the x-axis when x is to the left of $1-\sqrt{2}$ or when x is to the right of $1+\sqrt{2}$.

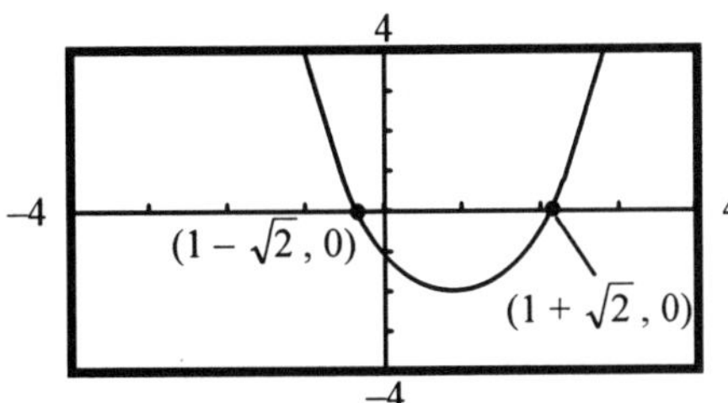

Figure 4.35

PROGRESS CHECK 3
Solve $x^2 - 6x + 4 < 0$. ■

Example 4: Finding a Useful Height for a Gutter

Solve the problem in the section introduction on page 355.

Solution: The sheet metal is 9 in. wide, so bending up edges of length x on each side converts the sheet metal to a gutter with dimensions as shown in Figure 4.36. The rectangular cross-sectional area must be at least 9 in.^{2}, and using $A = lw$ leads to the inequality

$$(9-2x)x \geq 9.$$

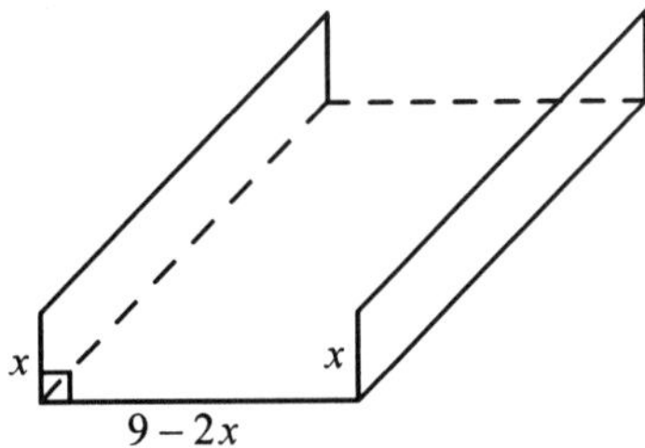

Figure 4.36

Figure 4.37 shows that the graph of $y = (9-2x)x$ intersects or is higher than the graph of $y = 9$ when $1.5 \leq x \leq 3$. So the cross-sectional area for the gutter is at least 9 in.^{2} when the height of the gutter is from 1.5 in. to 3. in.

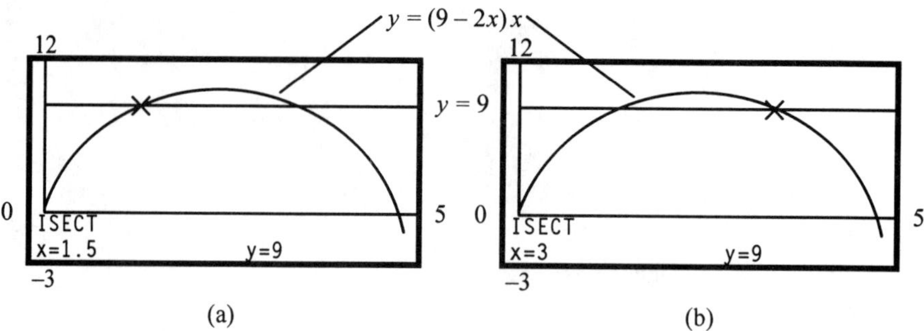

Figure 4.37

An algebraic verification of the proposed solution may be obtained by first converting the derived inequality to the form $ax^2 + bx + c \geq 0$.

$$(9-2x)x \geq 9$$
$$(9-2x)x - 9 \geq 0$$
$$-2x^2 + 9x - 9 \geq 0$$

Now determine the critical numbers by using the quadratic formula (or the factoring method) to determine when $-2x^2 + 9x - 9$ equals 0.

$$x = \frac{-9 \pm \sqrt{9^2 - 4(-2)(-9)}}{2(-2)} = \frac{-9 \pm \sqrt{9}}{-4} = \frac{-9 \pm 3}{-4}$$

Since

$$\frac{-9+3}{-4} = \frac{-6}{-4} = 1.5 \text{ and } \frac{-9-3}{-4} = \frac{-12}{-4} = 3$$

the critical numbers are 1.5 and 3, and we test in the intervals determined by the critical numbers, as shown in Figure 4.38.

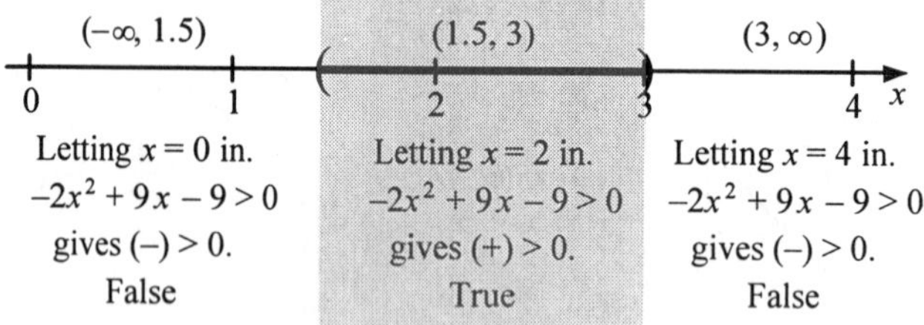

Figure 4.38

The tests show $-2x^2 + 9x - 9 > 0$ if $1.5 < x < 3$. Because the inequality in question is $\geq$, the possible heights for the gutter are from 1.5 in. to 3 in., and the solution set is [1.5 in., 3 in.].

PROGRESS CHECK 4

Redo the problem in Example 4, but assume that the cross-sectional area of the gutter must be at least 7 in.^{2}. ■

The methods for solving quadratic inequalities may be applied to higher-degree polynomial inequalities. This extension is based on the definition that critical numbers are real numbers that make the *polynomial* in question zero, and the principle that throughout each of the intervals determined by the critical numbers, the sign of the polynomial remains the same.

Example 5: Solving a Higher-Degree Polynomial Inequality

Solve $x(2x-3)(x+2)<0$.

Solution: The inequality is already in a workable form; that is, it is factored and one side is zero. Therefore we go straight to the analysis in Figure 4.39. Critical numbers are 0, $3/2$, and -2, since $x(2x-3)(x+2)$ equal zero when $x=0$, $x=3/2$, and $x=-2$.

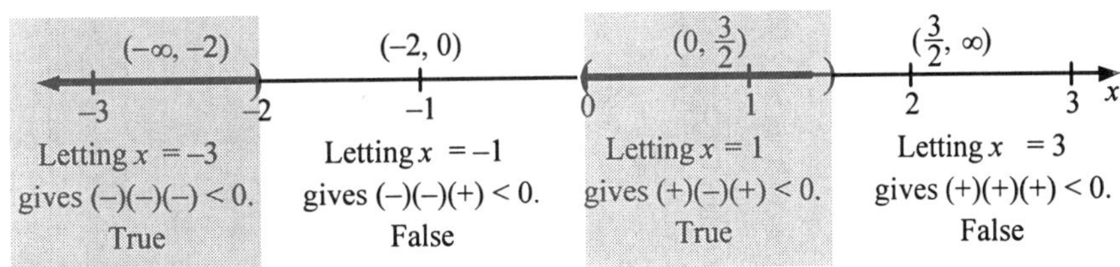

Figure 4.39

Thus, $x(2x-3)(x+2)<0$ is true if $x<-2$ or $0<x<3/2$, and the solution set is $(-\infty,\ -2)\cup(0,\ 3/2)$.

We can see using Figure 4.40 that this solution checks, since the graph of $y=x(2x-3)(x+2)$ appears to be below the x-axis when x is to the left of -2 and when x is between 0 and $3/2$.

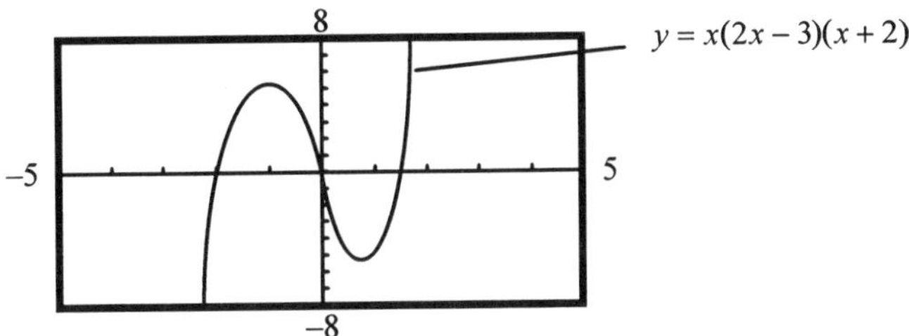

Figure 4.40

PROGRESS CHECK 5

Solve $x(3x+2)(x-2)>0$. ■

Inequalities involving quotients of polynomials may also be solved by our current methods if we define critical numbers in such problems to be real numbers that make either the numerator zero or the denominator zero. When analyzing quotients, it is important to remember that division by zero is undefined, so critical numbers that make the denominator zero are never included in the solution set.

Example 6: Solving a Rational Inequality

For what value of x will $\sqrt{\dfrac{3x-1}{x+1}}$ be a real number?

Solution: For the square root to b a real number, the expression under the radical must be greater than or equal to 0. Thus, we need to solve

$$\frac{3x-1}{x+1} \geq 0.$$

The numerator is zero when $x = 1/3$, and the denominator is zero when $x = -1$. So $1/3$ and -1 are critical numbers, and we may test in the intervals determined by these two numbers, as shown in Figure 4.41.

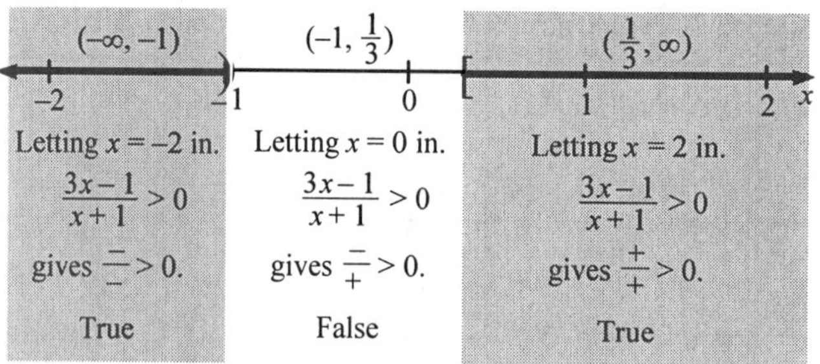

Figure 4.41

See in Figure 4.41 that $(3x-1)/(x+1) > 0$ is true if $x < -1$ or $x > 1/3$. The inequality in question is $\geq$, so we also determine that $1/3$ is included and -1 is excluded from the solution set, because the quotient is 0 when $x = 1/3$ and undefined when $x = -1$. Thus, the solution set is $(-\infty, -1) \cup [1/3, \infty)$.

The graph of $y = \sqrt{(3x-1)/(x+1)}$, which is shown in Figure 4.42, may be used to support this solution because it shows that admissible values for x occur when x is to the left of -1 and when x is to the right of $1/3$. The decision on whether to include the interval endpoints at -1 or $1/3$ in the solution set is made more easily based on the algebraic conditions described above.

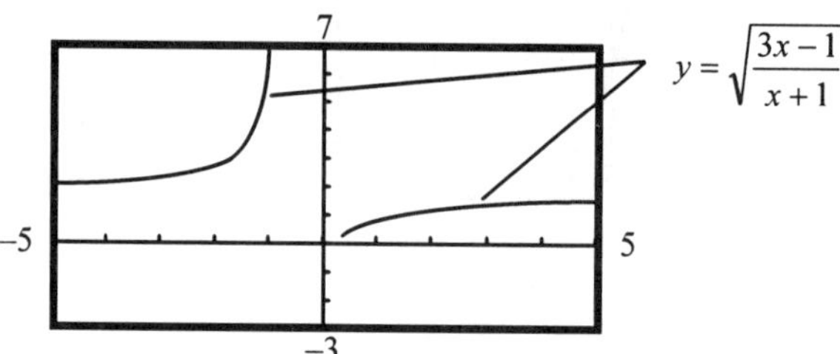

Figure 4.42

PROGRESS CHECK 6

For what values of x will $\sqrt{\dfrac{2x+1}{x-3}}$ be a real number? ■

Example 7: Solving a Rational Inequality

Solve $\dfrac{5x+1}{x+1} \geq 2$.

Solution: First, rewrite the inequality so the right side is zero. Then simplify the resulting expression on the left side into a single fraction.

$$\frac{5x+1}{x+1} \geq 2$$
$$\frac{5x+1}{x+1} - 2 \geq 0$$
$$\frac{5x+1-2(x+1)}{x+1} \geq 0$$
$$\frac{3x-1}{x+1} \geq 0$$

This inequality is solved in Example 6, and in interval notation the solution set is $(-\infty, -1) \cup [1/3, \infty$.

Figure 4.43 shows the graph in dot mode of both $y = (5x+1)/(x+1)$ and $y = 2$. This figure supports the proposed solution because it illustrates that the graph of $y = (5x+1)/(x+1)$ intersects or is higher than the graph of $y = 2$ when $x < -1$ or $x \geq 1/3$.

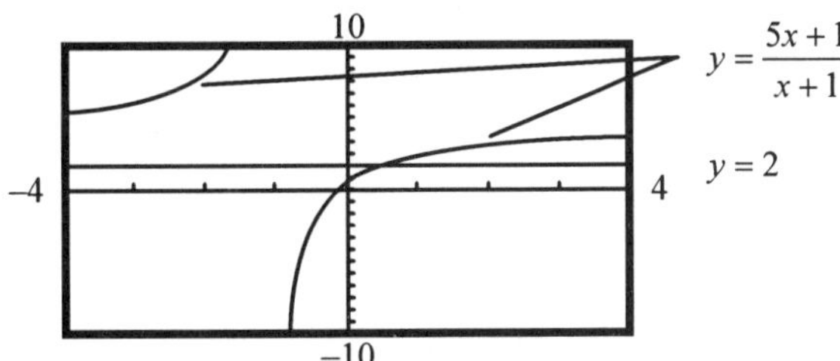

Figure 4.43

PROGRESS CHECK 7

Solve $\frac{2x+3}{3x+2} \leq 1$. ■

EXPLORE 4.4

1. **a.** Describe 3 ways you can use a grapher to solve the inequality $3x^3 < 4x+1$. In method 1, compare the graphs of $y = 3x^3$ and $y = 4x+1$. In method 2, make use of the fact that the given inequality is equivalent to $3x^3 - 4x - 1 < 0$. In method 3, use the graph that results when $y1$ is set equal to $3x^3 < 4x+1$. (See Explore in Section R.7).

b. Solve $3x^3 < 4x+1$ graphically. Round answers to 2 decimal places.

2. Multiplication of both sides by –1 shows that these two inequalities are equivalent.

$$3x^2 - 2x + 4 < 0 \text{ and } -3x^2 + 2x - 4 > 0$$

Confirm that both inequalities have the same solution set by investigating the graphs of the two associated functions.

3. The height of a projectile shot directly up from the ground with an initial velocity of 50 ft/second is given by $y = f(t) = -16t^2 + 50t$, where t is in seconds and y is in feet.

a. Because the graph of f intersects the line $y = 25$ in two points, there are two times when the projectile is 25 feet above the ground. What are these two times? When is the projectile more than 25 feet above the ground?

b. What does it mean that the graph of f intersects the line $y = -25$ in two points?

For the next two explorations, use the fact that graphical methods of solving inequalities also work nicely when the functions involved are not polynomials.

4. The margin of error (m) in a particular survey is given by $m = 0.098\sqrt{p - p^2}$, where p is the proportion of those interviewed who were in favor of a new law. For what values of p will m be between 0 and .04? Round answers to 2 decimal places.

5. You need to borrow \$3,000. You want to pay it back over 4 years, but you can't afford more than \$80 per month as your payment. Use the equal monthly payment formula $E = \Pr/\lfloor 1-(1+r)^{-n} \rfloor$ to find, to the nearest hundredth of a percent, those monthly interest rates (r) that will make the loan possible for you. Recall from Section R.3 that P is the amount borrowed, and that n is the number of months for the loan.

EXERCISES 4.4

In Exercises 1–54, solve each inequality.

1. $(x+1)(x-2) > 0$
2. $(x-3)(x+2) < 0$
3. $y(1-y) \geq 0$
4. $t(1-3t) \leq 0$
5. $x^2 + x - 6 < 0$
6. $x^2 - 4x - 5 > 0$
7. $1 - x^2 \leq 0$
8. $z^2 - 9 \geq 0$
9. $y^2 - 3 > 0$
10. $7 - x^2 < 0$
11. $k^2 > 5k$
12. $2x \leq x^2$
13. $b^2 + b \leq 2$
14. $10 - c^2 > -3c$
15. $2x^2 \geq 7x + 4$
16. $3y^2 + 2 < 5y$
17. $x^2 + 4 \geq 0$
18. $x^2 + 9 \leq 0$
19. $(a-1)^2 \leq 0$
20. $(3x-1)^2 > 0$
21. $x^2 - 4x - 3 > 0$
22. $x^2 - 2x - 5 < 0$
23. $3x^2 \leq 4x + 2$
24. $5x^2 \geq 3x + 4$
25. $2x^2 - x + 3 < 0$
26. $2x^2 - x + 3 > 0$
27. $x(x-1)(x-2) \geq 0$
28. $b(4b+3)(3b-4) > 0$
29. $a(2a-1)^2 < 0$
30. $y^2(1-3y) > 0$
31. $(1-y)(2-y)(3-y)^2 < 0$
32. $k(4k+3)^2(3k-4) \geq 0$
33. $t^3 > 4t$
34. $x^4 \leq 4x^2$
35. $x^4 - 5x^2 \leq -4$
36. $x^3 + 8x^2 + 12x < 0$
37. $c^5 + 9c \geq 10c^3$
38. $a^6 + 9a^2 > 10a^4$
39. $2x^3 - 5x^2 < 2x - 5$
40. $3n^3 - 2n^2 - 6n + 4 > 0$
41. $\dfrac{t-3}{t+2} \leq 0$
42. $\dfrac{x+2}{x-3} > 0$
43. $\dfrac{(x-1)(x+1)}{x} \leq 0$
44. $\dfrac{a(4-a)}{2a+3} \geq 0$
45. $\dfrac{x^2-4}{3-x} \geq 0$
46. $\dfrac{t^2+3t-4}{3t+2} \leq 0$
47. $\dfrac{1}{a} < 2$
48. $\dfrac{2}{x} > 5$
49. $\dfrac{1}{4} \geq \dfrac{7}{7-x}$
50. $\dfrac{1}{y+2} \leq \dfrac{1}{3}$
51. $\dfrac{2m}{m-4} \leq 3$
52. $\dfrac{-x}{2x+1} \geq 1$
53. $\dfrac{3}{x-2} \leq \dfrac{3}{x+3}$
54. $\dfrac{2}{x-1} < \dfrac{1}{x+1}$
55. For what values of x will $\sqrt{4-x^2}$ be a real number?
56. For what values of t will $\sqrt{1.44-t^2}$ be a real number?
57. What is the domain of the function $y = \sqrt{x^2-4}$?
58. What is the domain of the function $y = \sqrt{\dfrac{x+1}{x-2}}$?
59. The height (y) of a projectile that is shot directly up from the ground with an initial velocity 96 ft/second is given by the formula $y = 96t - 16t^2$. During what period is the projectile more than 80 ft off the ground?
60. The height of a projectile that is shot directly up from the ground with an initial velocity of 144 ft/second is given by

$$y = -16t^2 + 144t,$$

where y is measured in feet and t is the elapsed time in seconds. To the nearest hundredth of a second, for what values of t is the projectile more than 250 ft off the ground?

61. In Example 11 of Section 4.1, the equation $R = (100 + x)(99 - 0.5x)$ gave the relation between the revenue for a charter flight and the number of passengers, where x represents the number of passengers above 100. Find all passenger counts that result in a revenue for the flight of (**a**) at least \$10,880 and (**b**) at least \$11,000.

62. In Exercise 87 of Section 4.1, the formula $S = 8x - x^2$ gave the relation between the unit selling price (in dollars) and total sales revenue (S), where $S = 1$ represents sales of \$10,000. What selling prices will yield total sales revenue of (**a**) at least \$150,000 (that is $S \geq 15$) and (**b**) at least \$140,000?

63. What set of real numbers has the following property: the square of the number is smaller than the number itself?

64. What set of real numbers has the following property: the cube of the number is smaller than the number itself?

65. A rectangle has perimeter 16 in. If the length is denoted by x, then the width is denoted by $8 - x$. (See the figure below.) Assume that the length is not shorter than the width.

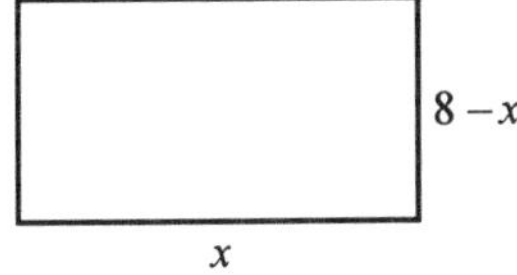

a. For what lengths is the area of the rectangle greater than 10 in.^{2}? (Give the answer exactly using radical notation and also as a decimal to the nearest hundredth.)

b. What length yields the maximum area?

66. The length of a rectangle is 2 ft more than the width, and the area is less than 63 ft^2. What are the possibilities for the length in such a case?

67. For a square metal plate to be useful in the design of a video component, its area must be between 1.08 and 1.12 square inches. If the side length is given by x, this condition gives $1.08 < x^2 < 1.12$. What are the permissible values of x?

68. Under certain conditions the formula $d = 0.045s^2 + 1.1s$ gives the approximate distance in feet that it takes to stop a car that is going s mile per hour. Suppose that because of visibility and road conditions, the safe stopping distance one day is 150 feet. Use a grapher to determine safe speeds under these conditions.

69. This exercise describes a type of laboratory situation that may lead to quadratic models. An animal study investigated the relationship between the dose (x) of a certain drug and weight gain (y) in decagrams. Eight laboratory animals of the same sex, age, and size were selected and randomly assigned to one of eight dosage levels of the given drug. The results are shown in the table. [Data appear in *Applied Regression Analysis* by Kleinbaum and Kupper, Duxbury, 1978.]

Dosage Level (x)	1	2	3	4	5	6	7	8
Weight Gain (y)	1	1.2	1.8	2.5	3.6	4.7	6.6	9.1

Graph these points and note that they appear to satisfy a quadratic model better than a linear one. It can be shown that the parabola defined by $y = 1.13 - 0.41x + 0.17x^2$ does a good job of fitting this data.

a. What values of y does the model predict for $x = 1$, 4, and 8?

b. Find the difference between the predicted and the observed values at each of these three points.

c. From the data in the table we see that for dosage levels 4 to 8 the weight gain is more than 2 decagrams. Now use the quadratic model to determine the dosage levels the model predicts will yield weight gains greater than 2 decagrams. Note that the dosage level must be a whole number.

THINK ABOUT IT 4.4

1. Create a quadratic inequality whose solution set is $[-4, 6]$.
2. Create an inequality involving a quotient that is less than or equal to zero, whose solution set is $(-\infty, -5) \cup [2, \infty)$.
3. If $x^2 + 9 = kx$, what values for k lead to roots that are not real numbers?
4. What set of real numbers satisfies the following condition? When the number and its square are added together, the sum is between 6 and 42.

5. Four square corners are cut from a piece of cardboard 18 in. by 12 in., and the remaining piece is folded to make an open box as shown. What size corners can be cut off so that the volume of the box is at least 160 cubic inches?

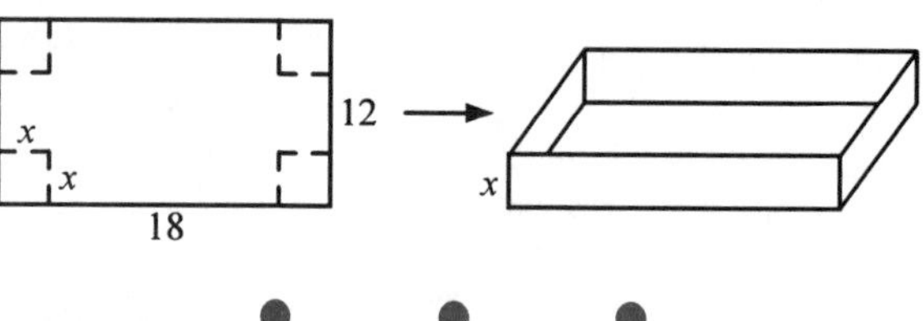

4.5 Absolute Value Equations and Inequalities

Photo Courtesy of Camerique, Inc., The Picture Cube

A projectile fired vertically up from the ground with an initial velocity of 160 ft/second will hit the ground 10 seconds later, and the speed y of the projectile in terms of the elapsed time t is given by

$$y = |160 - 32t|.$$

a. Graph this equation and describe what happens to the speed as t increases.
b. For what values of t is the projectile's speed 48 ft/second?
c. For what values of t is the projectile's speed at least 48 ft/second?

(See Example 8.)

Objectives

1. Solve equations involving absolute value.
2. Solve inequalities of the form $|ax+b| < c$.
3. Solve inequalities of the form $|ax+b| > c$.
4. Write a number interval using inequalities and absolute value.

Both algebraic and geometric perspectives of the absolute value concept are useful for formulating procedures to solve equations or inequalities that involve absolute value. Recall from Section R.1 that the absolute value of a number a is denoted by $|a|$ and is interpreted geometrically as the distance on the number line between a and zero. For instance $|3| = 3$ and $|-3| = 3$, as shown in Figure 4.44.

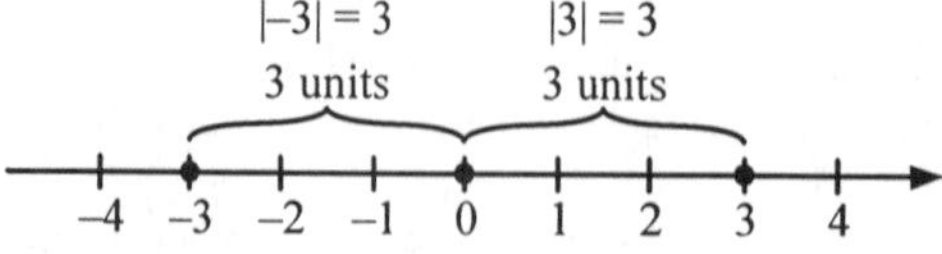

Figure 4.44

We see that if a is a positive number or 0, then $|a| = a$, while if a is a negative number then $|-a| = a$. Thus, the algebraic definition of $|a|$ is

$$|a| = \begin{cases} a, & \text{if } a \geq 0, \\ -a, & \text{if } a < 0. \end{cases}$$

In Example 1 we can now illustrate three main approaches to solving absolute value equations with respect to a simple equation like $|x| = 5$.

Example 1: Solving Equations of the Form $|x| = c$

Solve $|x| = 5$ using **(a)** a number line approach, **(b)** a graphing approach, and **(c)** an algebraic approach.

Solution:

a. To solve $|x| = 5$ using a number line approach, look for two numbers that are 5 units from 0 on the number line. The required numbers are 5 and –5, as shown in Figure 4.45.

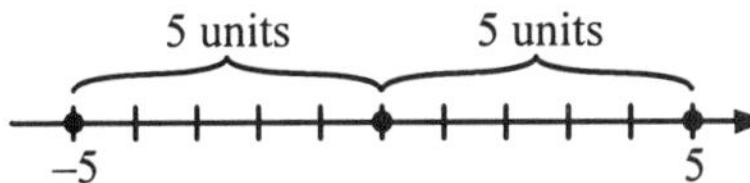

Figure 4.45

b. The x coordinates of the intersection points of the graphs of $y = |x|$ and $y = 5$ give the roots of the equation $|x| = 5$. From Figure 4.46 we can read that these roots are 5 and –5.

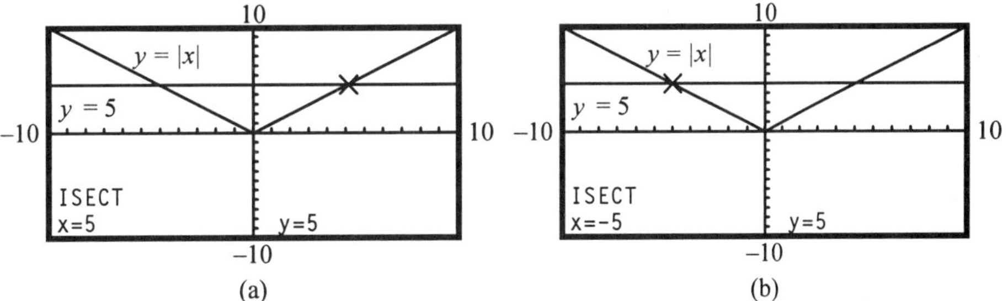

Figure 4.46

c. By the algebraic definition $|x|$ equals either x or $-x$. Setting these two expressions equal to 5 gives

$$x = 5 \text{ or } -x = 5$$

$$\text{so } x = 5 \text{ or } x = -5.$$

Both solutions check and the solution set is $\{5, -5\}$.

PROGRESS CHECK 1

Solve $|x| = 8$ using **(a)** a number line approach, **(b)** a graphing approach, and **(c)** an algebraic approach. ■

Example 1 illustrates that for $c > 0$,

$$|x| = c \text{ implies } x = c \text{ or } x = -c.$$

Other first-degree expressions may replace x in this result to provide a general procedure for solving absolute value equations of the form $|ax+b|=c$.

> **Solution of $|ax+b|=c$**
>
> If $|ax+b|=c$ and $c>0$, then
>
> $$ax+b=c \text{ or } ax+b=-c.$$

To solve $|ax+b|=c$ when $c=0$, note that it is only necessary to solve $ax+b=0$, because 0 is the only number whose absolute value is 0. Also, equations such as $|x|=-3$ have no solution, since the absolute value of a number is never negative; so in general, $|ax+b|=c$ has no solution if $c<0$.

Example 2: Solving Equations of the Form $|ax+b|=c$

Solve $|10x+5|=45$.

Solution: $|10x+5|=45$ is equivalent to

$$\begin{aligned} 10x+5=45 &\quad\text{or}\quad 10x+5=-45 \\ 10x=40 &\quad\text{or}\quad 10x=-50 \\ x=4 &\quad\text{or}\quad x=-5 \end{aligned}$$

Both solutions check (verify this), and the solution set is $\{4,-5\}$. The intersection of the graphs of $y=|10x+5|$ and $y=45$, which is shown in Figure 4.47, is in agreement with this solution.

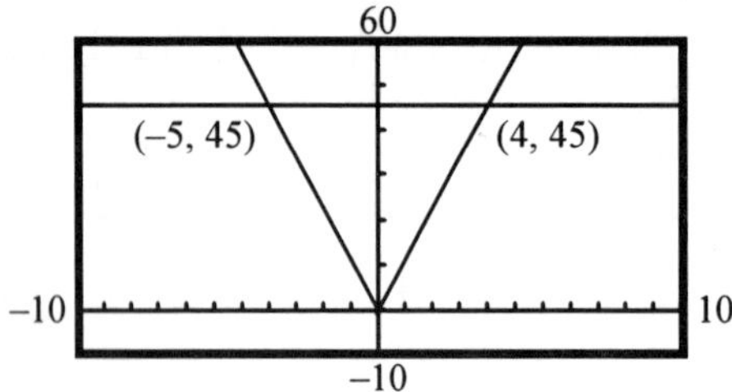

Figure 4.47

PROGRESS CHECK 2

Solve $|14x-21|=63$. ■

The next example shows how to solve absolute value equations of the form $|ax+b|=|cx+d|$.

Example 3: Solving Equations of the Form $|ax+b|=|cx+d|$

Solve $|2x+1|=|3-x|$.

Solution: By definition, $|2x+1|=\pm(2x+1)$ and $|3-x|=\pm(3-x)$. Setting these expressions equal to each other produces two distinct cases:

$$2x+1=3-x \text{ or } 2x+1=-(3-x).$$

Finally, solving each of these equations separately gives

$$\begin{aligned} 2x+1 &= 3-x && \text{or} \quad 2x+1=-(3-x) \\ 3x &= 2 && \text{or} \quad 2x+1=-3+x \\ x &= \frac{2}{3} && \text{or} \quad x=-4. \end{aligned}$$

Thus, the solution set is $\{2/3, -4\}$. A graphical check that supports this solution is shown in Figure 4.48.

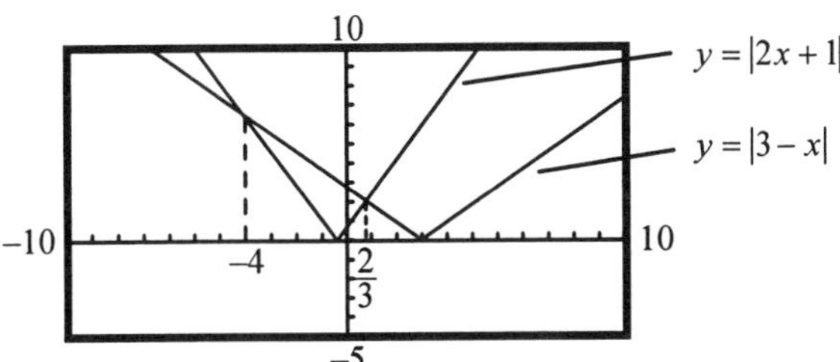

Figure 4.48

[**Note:** The given equation is also satisfied when $-(2x+1)=-(3-x)$ or when $-(2x+1)=3-x$. However, these equations are equivalent to the two we solved and need not be considered.]

PROGRESS CHECK 3

Solve $|7x-8|=|9x+12|$. ■

The idea that absolute value can define a distance on the number line may be extended and used to interpret expressions of the form $|x-a|$, which occur often in higher mathematics. Given any two points a and b, the distance between them on the number line is $|a-b|$, as illustrated in Figure 4.49. Note that $|a-b|=|b-a|$, so the order in the subtraction is not significant. This geometric perspective may be used to analyze equations of the form $[x-\text{constant}]=\text{constant}$, as discussed in Example 4.

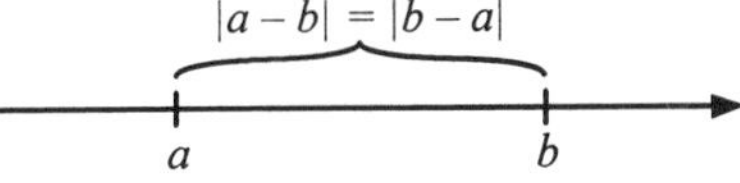

Figure 4.49

Example 4: Solving Equations of the Form $|x-a|=c$.

Solve $|x-2|=5$ using

a. a number line approach **b.** a graphing approach and **c.** an algebraic approach.

Solution:

a. Using a number line approach, $|x-2|$ represents the distance between 2 and some number x on the number line. So when solving $|x-2|=5$, we are looking for two numbers that are 5 units from 2 on the number line. As shown in Figure 4.50, the required numbers are 7 and −3.

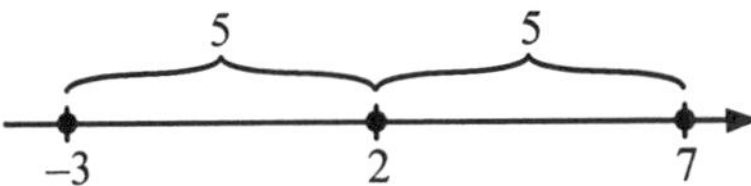

Figure 4.50

b. The graphs of $y = |x-2|$ and $y = 5$ intersect when $x = 7$ and $x = -3$, as illustrated in Figure 4.51. Thus, the solution set is $\{7, -3\}$.

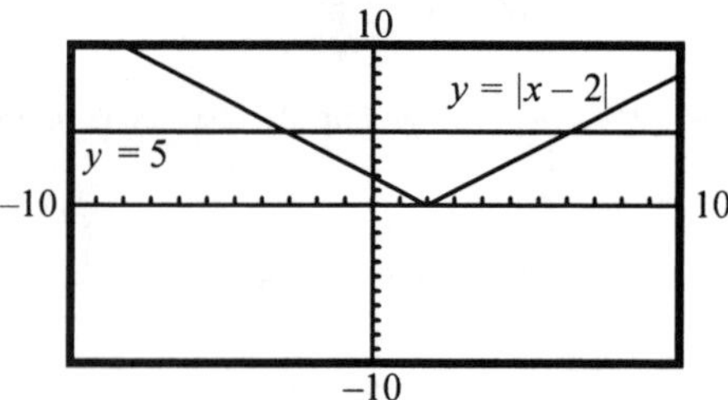

Figure 4.51

c. Algebraically, $|x-2| = 5$ implies

$$\begin{aligned} x-2 &= 5 \quad \text{or} \quad x-2 = -5 \\ x &= 7 \quad \text{or} \quad\quad\ \ x = -3 \end{aligned}$$

Both solutions check and the solution set is $\{7, -3\}$.

PROGRESS CHECK 4
Solve $|x-5| = 7$ using
a. a number line approach, **b.** a graphing approach, and, **c.** an algebraic approach. ■

In many applications involving absolute value, the inequality signs ($<, \le, >, \ge$) express the required relation in a problem. For instance, we can determine when the speed of the projectile described in the section-opening problem is *less than* 48 ft/second by solving.

$$|160 - 32t| < 48$$

To solve such inequalities, first consider that

$$|x| < 3$$

can be solved from a geometric viewpoint by finding all numbers that are less than 3 units from 0 on the number line. This set of numbers is graphed in Figure 4.52 and is expressed in set-builder notation by $\{x\colon\ -3 < x < 3\}$, and in interval notation by $(-3, 3)$. This example illustrates that for $c > 0$

$$|x| < c \text{ implies } -c < x < c,$$

which gives a general procedure for solving inequalities of the form $|ax + b| < c$.

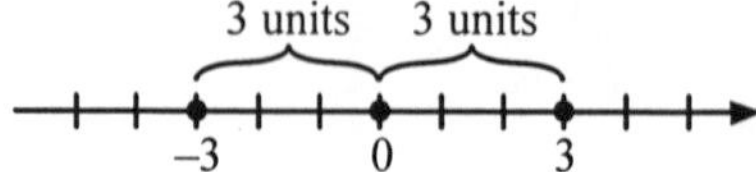

Figure 4.52

Solution of $|ax+b|<c$

> If $|ax+b|<c$ and $c>0$, then
>
> $$-c<ax+b<c.$$

Note that the inequality symbol $\le$ may replace $<$ in our discussion to this point, so by similar reasoning, an inequality such as $|x|\le 3$ implies $-3\le x\le 3$.

Example 5: Solving Inequalities of the Form $|x-a|<c$

Solve $|x-4|<3$ using
a. a number line approach, **b.** a graphing approach, and, **c.** an algebraic approach. ■

Solution:

a. Using a number line approach, $|x-4|<3$ requires that the distance on the number line between 4 and some number x be less than 3 units. Figure 4.53 shows that such an interval starts at 1, ends at 7, and excludes both end points.

$|x-4|<3$ means x is between 1 and 7.

Figure 4.53

b. Figure 4.54 illustrates that the graph of $y=|x-4|$ is *below* the graph of $y=3$ for $1<x<7$, which confirms that the solution set is (1, 7).

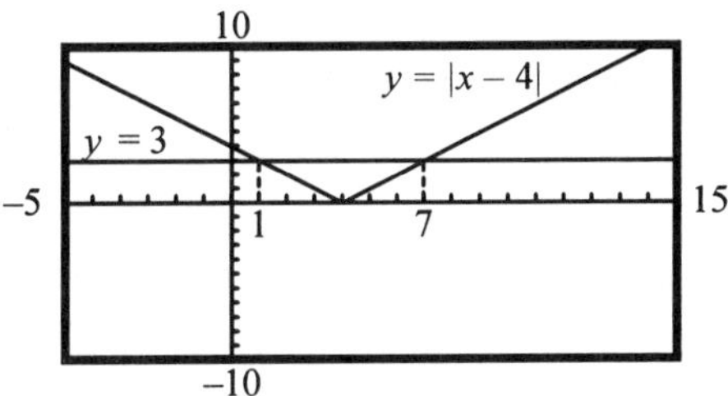

Figure 4.54

c. $|x-4|<3$ implies

$$-3<x-4<3.$$

Now we proceed by obtaining equivalent inequalities using any of the properties given in Section R.7.

$$-3<x-4<3$$
$$1<x<7 \qquad \text{Add 4 to each member.}$$

Thus, any number between 1 and 7 satisfies the given inequality, and the solution set is (1, 7).

PROGRESS CHECK 5

Solve $|x-2|\le 8$ using
a. a number line approach, **b.** a graphing approach, and, **c.** an algebraic approach. ■

Example 6: Solving Inequalities of the Form $|ax+b|<c$

Solve $|3x+5|<11$.

Solution: We rewrite $|3x+5|<11$ as $-11<3x+5<11$ and proceed as follows:

$$-11<3x+5<11$$

$$-16<3x<6 \qquad \text{Subtract 5 from each member.}$$

$$\frac{-16}{3}<x<2. \qquad \text{Divide each member by 3.}$$

Thus, any number between $-16/3$ and 2 satisfies the inequality, and the solution set is $(-16/3,\ 2)$.

To check this solution graphically, observe in Figure 4.55 that the graph of $y=|3x+5|$ is below the graph of $y=11$ when x is between $-16/3$ and 2.

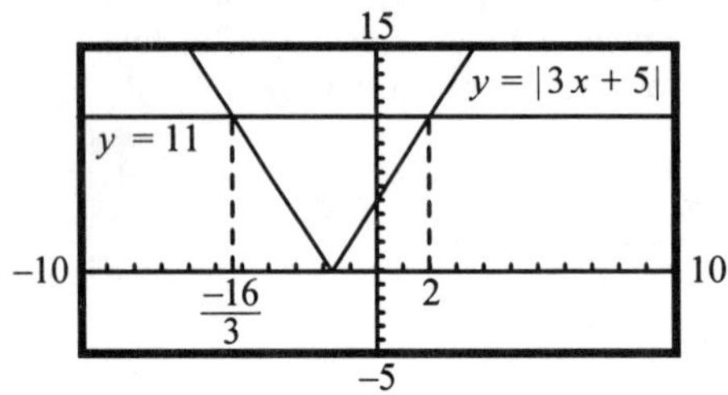

Figure 4.55

PROGRESS CHECK 6

Solve $|5x+7|<12$. ■

To solve absolute value inequalities of the form $|ax+b|>c$, consider that an inequality like

$$|x|>3$$

can be solved by finding all numbers greater than 3 units from 0 on the number line. Figure 4.56 specifies this set of numbers graphically, in set-builder notation, and in interval notation. As suggested by this example, if $c>0$.

$$|x|>c \text{ implies } x<-c \text{ or } x<c.$$

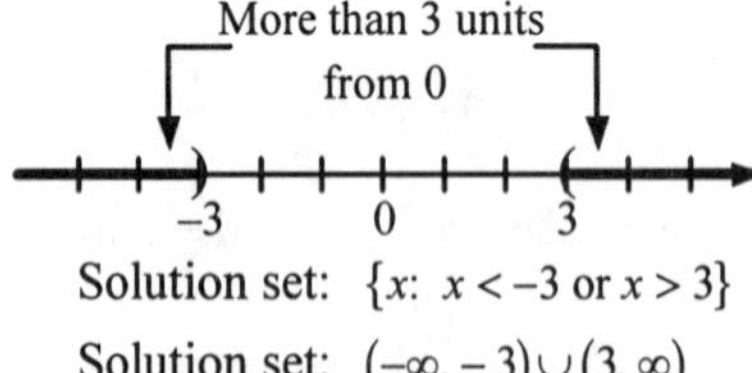

Solution set: $\{x\colon\ x<-3 \text{ or } x>3\}$

Solution set: $(-\infty,-3)\cup(3,\infty)$

Figure 4.56

So inequalities of the form $|ax+b|>c$ may be solved as follows.

Solution of $|ax+b|>c$

> If $|ax+b|>c$ and $c>0$, then
>
> $$ax+b<-c \text{ or } ax+b>c.$$

Example 7: Solving Inequalities of the Form $|x-a|>c$

Solve $|x-4|>3$ using

a. a number line approach, **b.** a graphing approach, and, **c.** an algebraic approach.

Solution:

a. Figure 4.57 illustrates that the numbers located more than 3 units from 4 on the number line are either to the left of 1 or the right of 7, which confirms that the solution set is $(-\infty, 1)\cup(7, \infty)$.

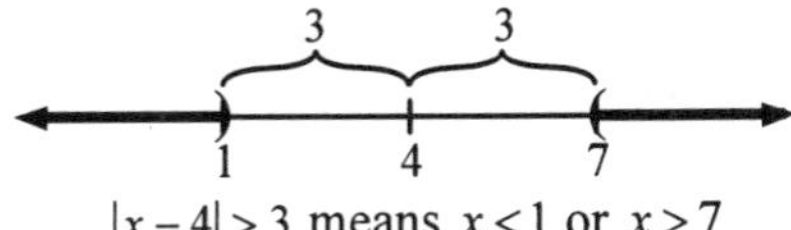

$|x-4|>3$ means $x<1$ or $x>7$

Figure 4.57

b. The graph of $y=|x-4|$ is *above* the graph of $y=3$ when $x<1$ or $x>7$, as can be seen in Figure 4.58. So from this graph we can read that the solution set is $(-\infty, 1)\cup(7, \infty)$.

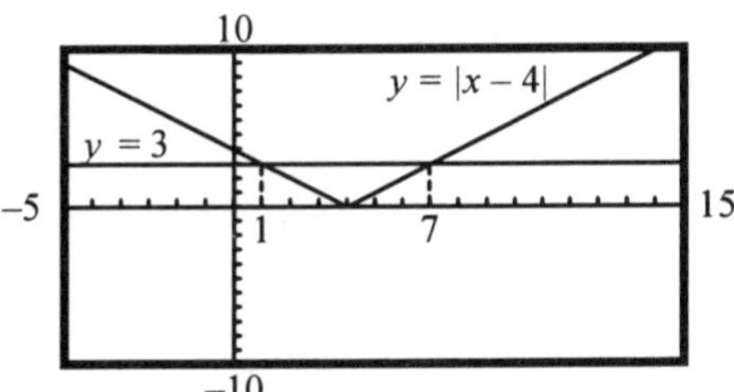

Figure 4.58

c. Algebraically, the given inequality translates to the compound statement

$$x-4<-3 \text{ or } x-4>3.$$

We then add 4 to both sides in both inequalities to get

$$x<1 \text{ or } x>7$$

So the solution set in interval notation is $(-\infty, 1)\cup(7, \infty)$.

PROGRESS CHECK 7

Solve $|x-75|>5$ using

a. a number line approach, **b.** a graphing approach and, **c.** an algebraic approach. ■

Example 8: An Application Involving Absolute Value Equations and Inequalities

Solve the problem in the section introduction on page 366.

Solution:

a. Figure 4.59 shows a graph of $y = |160 - 32t|$ where the variable x is used on a grapher to represent the elapsed time t. The graph shows that the speed of the projectile decreases uniformly from an initial speed of 160 ft/second until it stops momentarily when $t = 5$; then the projectile increases in speed uniformly until it travels at 160 ft/second again, when it hits the ground after 10 seconds have elapsed.

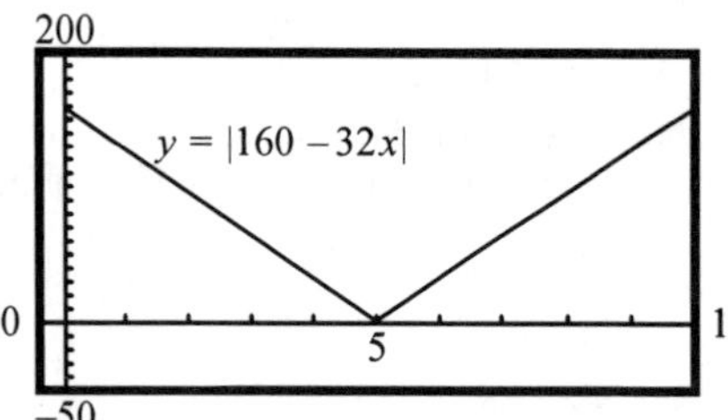

Figure 4.59

b. Replacing y by 48 in the formula $y = |160 - 32t|$ gives $48 = |160 - 32t|$, which implies

$$\begin{aligned} 160 - 32t &= 48 & \text{or} \quad 160 - 32t &= -48 \\ -32t &= -112 & \text{or} \quad -32t &= -208 \\ t &= 3.5 & \text{or} \quad t &= 6.5. \end{aligned}$$

Thus, the projectile attains a speed of 48 ft/second after 3.5 seconds (on its way up) and again (on its way down) when 6.5 second have elapsed. The intersection of the graphs of $y = |160 - 32t|$ and $y = 48$ that is shown in Figure 4.60 reaffirms this solution.

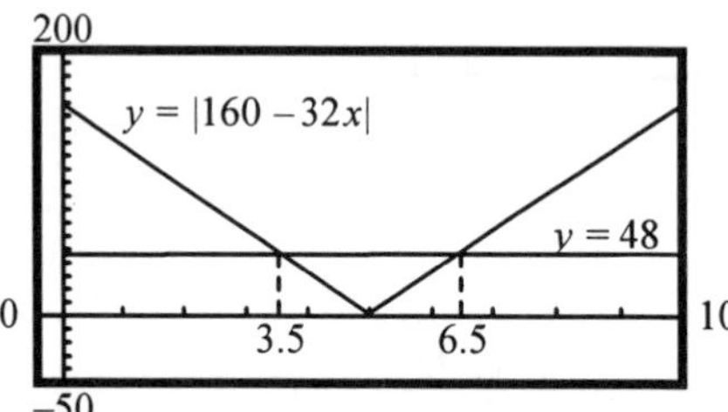

Figure 4.60

c. To find when the speed of the projectile is at least 48 ft/second, we first need to solve $|160 - 32t| \geq 48$, which is equivalent to

$$160 - 32t \leq -48 \text{ or } 160 - 32t \geq 48 .$$

Then solving each of these inequalities gives

$$\begin{aligned} 160 - 32t &\leq -48 & \text{or} \quad 160 - 32t &\geq 48 & & \\ -32t &\leq -208 & -32t &\geq -112 & & \text{Subtract 160 from both sides.} \\ t &\geq 6.5 & t &\leq 3.5 & & \text{Divide by } -32 \text{ and reverse the inequality symbols.} \end{aligned}$$

In the context of the problem, only values for t from 0 to 10 seconds are meaningful so the solution set is

$$[0 \text{ seconds}, 3.5 \text{ seconds}] \cup [6.5 \text{ seconds}, 10 \text{ seconds}].$$

To check this solution graphically, observe in Figure 4.59 that the graph of $y = |160 - 32x|$ intersects or is above the graph of $y = 48$ when $0 \le x \le 3.5$ or when $6.5 \le x \le 10$.

PROGRESS CHECK 8

A projectile fired vertically up from the ground with an initial velocity of 256 ft/second will hit the ground 16 seconds later, and the speed y of the projectile in terms of the elapsed time t is given by $y = |256 - 32t|$. Redo the questions in Example 8 with respect to this projectile. ■

Inequalities involving absolute value are commonly seen in applications in which an interval of numbers is centered around a target number, and all numbers in the interval must be within (or must exceed) a specified distance from this target. The next two examples discuss how to translate such situations to an algebraic model.

Example 9: Translating to an Absolute Value Inequality

The specifications for a component in a solar panel require that the length of this component part be 7.25 cm. When manufacturing this part, a sample is judged acceptable when the length l is not more than 0.04 cm from this specification. Write an inequality involving absolute value that gives all values of l for an acceptable part, and then solve this inequality.

Solution: The expression $|l - 7.25|$ represents the distance between 7.25 and the length of the part. Since this distance must be less than or equal to 0.04, the requested algebraic model is

$$|l - 7.25| \le 0.04.$$

By the methods of this section, this inequality is equivalent to $7.21 \le l \le 7.29$. So the length of an acceptable part measures from 7.21 cm to 7.29 cm.

PROGRESS CHECK 9

An investor buys a certain stock at \$30 per share and wishes to hold onto this stock as long as the stock price p stays within 20 percent of the original purchase price. Write an inequality involving absolute value that gives all values of p for which the investor will continue to own the stock, and then solve this inequality. ■

Example 10: Translating to an Absolute Value Inequality

Write the interval $\{x: \ -3 < x < 7\}$ using inequalities and absolute value.

Solution: The interval between –3 and 7 is 10 units long and centered at 2. As shown in Figure 4.61, we are describing numbers that are within 5 units of 2. This condition means that $-3 < x < 7$ is equivalent to $|x - 2| < 5$.

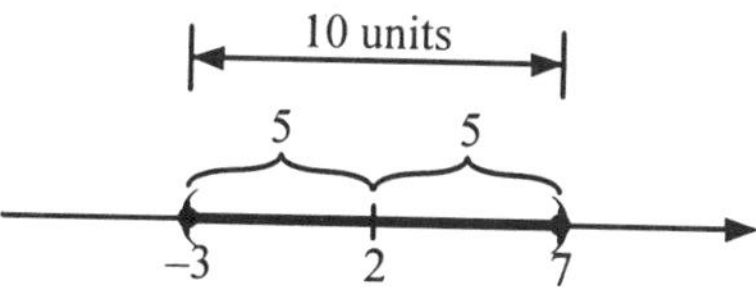

Figure 4.61

PROGRESS CHECK 10

Write the interval $\{x: \ -2 < x < 3\}$ using inequalities and absolute value. ■

EXPLORE 4.5

1. Check that these three approaches all lead to the same graph.

   a. Graph $y = |x|$. **b.** Graph $y = \sqrt{x^2}$. **c.** Graph $y = \begin{cases} -x & \text{if } x < 0 \\ x & \text{if } x \geq 0 \end{cases}$.

   See Section 1.2 for help with graphs of piecewise functions.

2. Use a grapher to solve these inequalities, and round answers to two decimal places.

   a. $|x^2 - 4| < 2$ **b.** $|x^3 - 4| < 2$ **c.** $|x^3 - 4x - 1| < 2$

3. Some graphing calculators have a Shade command. In analyzing the inequality $|x - 4| < 3$, a Shade command can produce a graph like Figure 1, which makes the solution set stand out. Similarly $|x - 4| > 3$ produces a graph like Figure 2. If your grapher has a Shade command, use it to make the solution set $|3x - 4| < 7$ stand out, and express the solution set in interval notation.

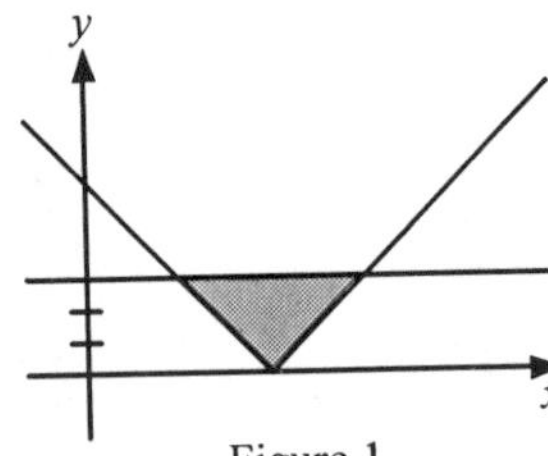

Figure 1

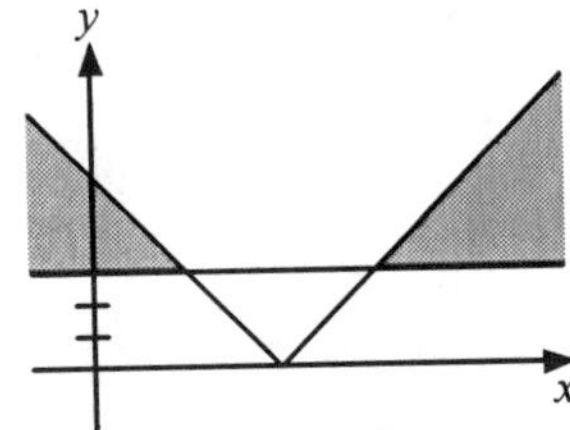

Figure 2

4. A piece of wood in a model is supposed to be 10 inches long. If it is off by E inches, then the relative error is $|E/(10 + E)|$. For what values of E is the relative error less than 1 percent? Round answer to two decimal places.

EXERCISES 4.5

In Exercises 1–8 solve the given equation using **a.** a number line approach, **b.** a graphing approach, and, **c.** an algebraic approach.

1. $|x| = 2$ **2.** $|x| = 1.6$

3. $|x - 1| = 2$ **4.** $|x - 3| = 1.6$

5. $|x - (-2)| = 1$ **6.** $|x - (-4)| = 2$

7. $|x + 8| = 3.1$ **8.** $|x + 1.2| = 8.5$

In Exercises 9–28 solve each equation.

9. $|5x + 10| = 30$ **10.** $|3x - 12| = 24$

11. $\left|\frac{1}{2}x - 9\right| = 8$ **12.** $|3x + 7| = 0$

13. $2|x| + 7 = 10$ **14.** $3|x - 1| + 2 = 11$

15. $|x - 3| = -2$ **16.** $|x + 2| = -5$

17. $|3x - 4| = 1 + x$ **18.** $|-2x + 1| = 7x$

19. $|2x - 1| = x$ **20.** $|-x + 2| = x + 1$

21. $|x - 3| = |x + 1|$ **22.** $|1 - x| = |3x - 1|$

23. $|2x + 1| = |3 + x|$ **24.** $|6x - 4| = |2x - 7|$

25. $|5 - 2x| = |2 - 5x|$ **26.** $|-3x| = |2 - 3x|$

27. $|2 + x| = |2x| + 1$ **28.** $|4 - x| = |x| - 2$

In Exercises 29–36 solve the inequality using **a.** a number line approach, **b.** a graphing approach, and **c.** an algebraic approach.

29. $|x| < 4$ **30.** $|x| \leq \frac{1}{2}$ **31.** $|x - 5| \leq 3$

32. $|x - 3| < 0.01$ **33.** $|x + 1| < 2$

34. $|x + 2| < 1$ **35.** $|x - 5| \geq 3$ **36.** $|x - 0.1| > 0.8$

In Exercises 37–48 solve the given inequality.

37. $|x| > 2$ **38.** $|x| > 0$ **39.** $|3x + 5| < 8$

40. $|6x-7| \le 10$ **41.** $|1-2x| \le 13$

42. $|3-4x| < 5$ **43.** $|2x-1| > 3$

44. $|1-x|+2 \ge 4$ **45.** $|x| < x$

46. $x < |x|$ **47.** $|3x-7| < x$ **48.** $|3x-7| > x$

In Exercises 49–62 write the given interval using inequalities and absolute value.

49. $(-2, 2)$ **50.** $[-7, 7]$

51. $\{x\colon\ 1.1 < x < 1.2\}$ **52.** $\{x\colon\ -0.6 < x < -0.5\}$

53. $(-\infty,\ -2) \cup (2,\ \infty)$ **54.** $(-\infty,\ -0.1) \cup (0.1,\ \infty)$

55. $\{x\colon\ x < 0 \text{ or } x > 6\}$ **56.** $\{x\colon\ x \le -4 \text{ or } x \ge 0\}$

57.

58.

59. x is between 2 and 4.

60. x is between −8 and −1.

61. x is within one-half unit of 3.

62. x is within d units of a.

In Exercises 63 and 64 show a number line solution.

63. $|x-a| < 1$

64. $|y-L| < \in \quad (E > 0)$

65. A projectile fired vertically up from the ground with an initial velocity of 128 ft/second will hit the ground 8 seconds later, and the speed of the projectile in terms of the elapsed time t is given by $y = |128 - 32t|$.

a. Graph this equation and describe what happens to the speed as t increases.

b. For what values of t is the speed 140 ft/sec?

c. For what values of t is the speed less than 80 ft/sec?

d. What is the maximum speed the projectile attains?

66. Redo Exercise 65, but assume that speed is given by $y = |153.6 - 32t|$.

67. A bulb is useful if its brightness b does not differ from 1170 lumens by more than 12.5 lumens. Write an inequality involving absolute value that gives all values of b for an acceptable bulb, and then solve the inequality.

68. In a calculus proof the variable y cannot miss its target L by more than $\in$. Express this condition as an inequality involving absolute value. Show what this condition means on a number line.

69. In statistics the inequality $|p - \hat{p}| \le 0.05$ occurs. Express this inequality in the form $a \le p \le b$.

70. In manufacturing a metal square with side 1 mm, the error in the side must be less than 0.05 mm.

a. Let x be the actual length of the manufactured side and express this requirement using absolute value.

b. Find an inequality that describes the area of the manufactured square.

c. What is the maximum error in the area?

71. From a grapher you see that the root r of an equation is between 1.8940 and 1.9048.

a. Express this interval using the absolute value inequality.

b. What does the answer in part **a** reveal about the maximum error, E, that can be made when the midpoint, M, of the interval is used as an estimate for the root?

c. If E (from part **b**) is less than 0.005, then your estimate is called accurate to 2 decimal places. If E is less than 0.05, then your estimate is accurate to 1 decimal place. To how many decimal places can you say M, as calculated in part **b**, is accurate?

THINK ABOUT IT 4.5

1. Explain how the inequality $|x+3| > 0.1$ may be solved geometrically.
2. Use inequalities and absolute value to describe the set of all real numbers x such that x is within d units of a but not equal to a.
3. Use the property $|a/b| = |a|/|b|$ to solve the inequality

$$\left|\frac{1}{x}\right| > 3.$$

4. a. Compare the graphs of $y = |x+5|$ and $y = |x| + |5|$. Is it true that $|x+5| \le |x| + |5|$ for all x?
   b. Compare the graphs of $y = |x+(-5)|$ and $y = |x| + |-5|$. Is it true that $|x+(-5)| \le |x| + |-5|$ for all x?
   c. Explain by referring to graphs why $|x+a| \le |x| + |a|$ is true for all values of x and a.
5. Sometimes the graphical approach to solving an absolute value equation is much simpler than the algebraic approach. As an illustration solve $|3+x| = 3 - |x|$ graphically and algebraically as described.
   a. Graphical: Find the intersection of $y = |3+x|$ and $y = 3 - |x|$.
   b. Graphical: Rewrite the original equation as $|3+x| + |x| = 3$, then find the intersection of the graphs of the left and right sides.
   c. Algebraic: There are 4 cases depending on whether $|3+x|$ stands for $3+x$ or $-(3+x)$ and whether $|x|$ stand for x or $-x$.

● ● ●

CHAPTER 4 SUMMARY

OBJECTIVES CHECKLIST

Specific chapter objectives are summarized below, along with numbered example problems from the text that should clarify the objectives. If you do not understand any objectives, or do not know how to do the selected problems, then restudy the material.

4.1: Can you:

1. **Solve quadratic equations, using a specified method?** Solve $4x^2 = 12x$ using the factoring method.

   [Example 1]

2. **Find intercepts of graphs of equations of the form $y = ax^2 + bx + c$?** Find the intercepts of the graph of $y = f(x) = x^2 - 2x - 3$.

   [Example 2]

3. **Use the discriminant to determine the number of real roots of a quadratic equation or the number of x-intercepts in the graph of $y = ax^2 + bx + c$?** Use the discriminant to determine the number of real roots of the equation $2x^2 - 5x = 3$.

   [Example 8]

4. **Solve quadratic equations after choosing an efficient method?** Solve the equation $80 = 96t - 16t^2$. Choose an efficient method.

   [Example 9]

5. **Solve applied problems involving quadratic equations?** The height (y) of a projectile that is shot directly up from the ground with an initial velocity of 100 ft/second is given by the formula $y = 100t - 16t^2$, where t is the elapsed time in seconds. To the nearest hundredth of a second, when will the projectile initially attain a height of 115 ft?

   [Example 11]

4.2: Can you:

1. **Solve fractional equations that lead to linear equations?** Solve $\dfrac{2}{3a} = \dfrac{3}{2a-1}$.

   [Example 1]

2. **Solve fractional equations that lead to quadratic equations?** Solve $\dfrac{7}{t-4} - \dfrac{56}{t^2-16} = 1$.

   [Example 3]

3. **Solve formulas that contain fractions for a specified variable?** Solve the formula $T = \dfrac{32m_1m_2}{m_1 + m_2}$ for m_1.

[Example 5]

4. **Solve applied problems involving fractional equations?** One pipe can fill a swimming pool in 4 hours. A second pipe can fill the pool in 12 hours. How long will it take to fill the pool if both pipes operate together?

[Example 6]

4.3: Can you:

1. **Solve radical equations?** Solve $x - \sqrt{x+1} = 1$.

[Example 2]

2. **Solve equations with quadratic form?** Solve $x^{-2} - 4x^{-1} - 3 = 0$.

[Example 6]

3. **Solve higher-degree polynomial equations using factoring?** Solve $x^5 = 10x^3 - 9x$.

[Example 8]

4.4: Can you:

1. **Solve quadratic inequalities?** Solve $x^2 - 2x - 1 > 0$.

[Example 3]

2. **Solve polynomial inequalities of degree higher than 2?** Solve $x(2x-3)(x+2) < 0$.

[Example 5]

3. **Solve inequalities involving quotients of polynomials?** For what values of x will $\sqrt{\dfrac{3x-1}{x+1}}$ be a real number?

[Example 6]

4.5: Can you:

1. **Solve equations involving absolute value?** Solve $|10x + 5| = 45$.

[Example 2]

2. **Solve inequalities of the form $|ax + b| < c$?** Solve $|3x + 5| < 11$.

[Example 6]

3. **Solve inequalities of the form $|ax + b| > c$?** Solve $|x - 4| > 3$ using
   a. an algebraic approach, **b.** a number line approach, and **c.** a graphing approach.

[Example 7]

4. Write a number interval using inequalities and absolute value? Write the interval $\{x: \ -3 < x < 7\}$ using inequalities and absolute value.

KEY CONCEPTS AND PROCEDURES

Section	Key Concepts and Procedures to Review
4.1	• Definitions of a second-degree or quadratic equation • Zero product principle: $a \cdot b = 0$ if and only if $a = 0$ or $b = 0$ • Methods to solve second- and higher-degree equations by factoring • Methods to solve a quadratic equation by using the square root property, by completing the square, and by using the quadratic formula • Quadratic formula: If $ax^2 + bx + c = 0$ and $a \neq 0$, then $x = \dfrac{-b \pm \sqrt{b^2 - 4ac}}{2a}$. • Methods to determine the number of real solutions to a quadratic equation from the discriminant $(b^2 - 4ac)$

- Methods to determine the number of x-intercepts of the graph of a quadratic function from the discriminant $(b^2 - 4ac)$
- Connection between roots and x-intercepts
- Use of grapher to find x-intercepts

4.2
- Methods to solve an equation containing fractions
- With equations that contain fractions, it is important to check solutions in the original equation and reject solutions that result in division by zero.

4.3
- Definitions of radical equations and equations with quadratic form
- Principle of powers
- Methods to solve radical equations and equations with quadratic form
- Methods to solve higher degree polynomial equations by factoring

4.4
- Definition of critical number
- Graphical method to solve inequalities
- Methods to solve inequalities involving polynomials and quotients of polynomials by using critical number analysis

4.5
- Definition of critical number
- Methods to solve absolute value equations and inequalities
- If $|ax+b| = c$ and $c > 0$, then $ax+b = c$ or $ax+b = -c$.
- If $|ax+b| < c$ and $c > 0$, then $-c < ax+b < c$
- If $|ax+b| > c$ and $c > 0$, then $ax+b < -c$ or $ax+b > c$.
- Methods for translating between absolute value inequalities and interval notation.

CHAPTER 4 REVIEW EXERCISES

In Exercises 1–30 solve each equation or inequality.

1. $x^2 = 1 - x$ **2.** $2\sqrt{x+1} = x - 2$

3. $3(x-5)^2 + 12 = 0$ **4.** $x + \sqrt{x} - 12 = 0$

5. $|x| < 5$ **6.** $|x| > 3$

7. $3 - 120{,}000x^{-2} = 0$ **8.** $3x + 3 = 7x^{-1}$

9. $\sqrt{2x+1} - 3 = 0$ **10.** $9x^3 = 25x$

11. $\dfrac{x}{2\sqrt{25+x^2}} - \dfrac{1}{4} = 0$

12. $x(x+3) = x^2 + 3(x+1) - 1$

13. $|5 - x| = 2$ **14.** $|2x - 3| \geq 1$

15. $\dfrac{x+1}{x-1} < 1$ **16.** $\dfrac{x}{2x-5} < 0$

17. $1 - \dfrac{12}{c^2 - 4} = \dfrac{3}{c+2}$ **18.** $|x| - 1 = \dfrac{|x|}{2}$

19. $x(x+2) = 0$ **20.** $x^2 + 3x \geq 4$

21. $b^3 < 9b$ **22.** $3t^3 + 10t^2 - 8t \geq 0$

23. $2x^2 + 5x = 12$ **24.** $(x-2)^2 \geq 0$

25. $a^3 - 5a^2 = a - 5$ **26.** $x^4 + 8 = 6x^2$

27. $\dfrac{x}{3} - \dfrac{5}{3} = 2x^{-1}$ **28.** $\sqrt{x} + \sqrt{x+6} = 6$

29. $\dfrac{2}{x+3} \geq \dfrac{1}{x-1}$ **30.** $\dfrac{x}{x-1} = \dfrac{x-2}{x+3}$

31. Solve for F_1: $\dfrac{d_1}{d_2} = \dfrac{F_2}{F_1}$.

32. Solve for y: $12 - 5y - \dfrac{5y}{x} = \dfrac{y-3}{x}$.

33. Solve for y: $x = \dfrac{2y+1}{1-y}$.

34. Solve for x: $\dfrac{bx+a}{ab} = \dfrac{x^2-1}{ax+b}$.

35. Solve for x: $x^{-1} + a^{-1} = b$.

36. Solve for L: $t = \pi\sqrt{\dfrac{L}{g}}$.

37. What is the domain of the function $y = \sqrt{x^2 + 2x - 3}$?

38. Write the interval $\{x\colon\ 1 < x < 10\}$ using inequalities and absolute value.

39. What number must be added to $x^2 + x$ to make the expression a perfect square?

40. If $2x^2 - 5x = k$, then what value(s) for k lead to each of the following:
a. the roots are equal **b.** one root is 0?

41. A regression model, $y = 20 + 20x - x^2$, relates anxiety level (x) to performance (y), where both variables are measured by special tests invented by the researcher. Use a grapher to graph this equation, and describe what happens to performance as anxiety increases. Estimate to the nearest integer all anxiety levels at which the predicted performance score is more than 100.

42. Find the intercepts of the graph of $y = x^2 - x - 6$.

43. Find a possible equation for this graph.

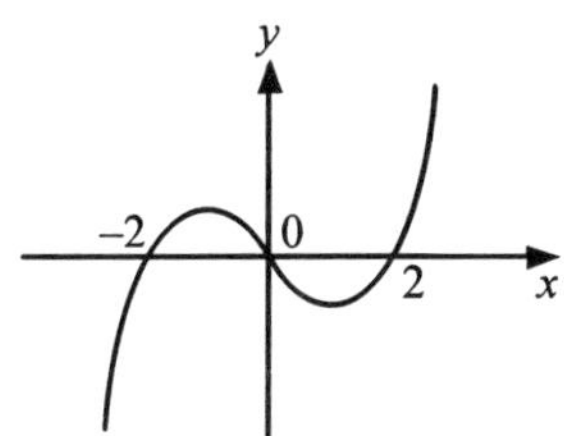

44. The sum of a number and its reciprocal is $25/12$. What are the numbers?

45. One side of a right triangle is 3 ft longer than the other, and the hypotenuse is 15 ft. What is the length of the shortest side in the triangle?

46. A diver jumps from a cliff that is 64 ft above water. Her height above water (y) after t seconds is given by the formula $y = 64 - 16t^2$. When will the diver hit the water? During what period is the diver at least 28 ft above water?

47. An automobile driver travels 30 miles at an average speed of 30 miles per hour. How fast must the driver go for the next 30 miles to bring the average speed for the whole trip up to 40 mi hour?

48. Use the given figure to solve $f(x) \geq 0$.

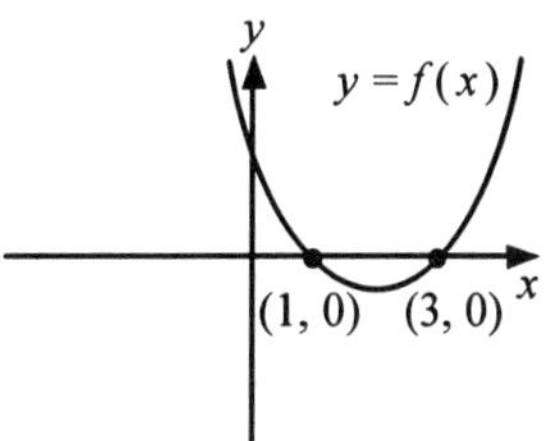

49. If the perimeter in this right triangle is 23 in., find x.

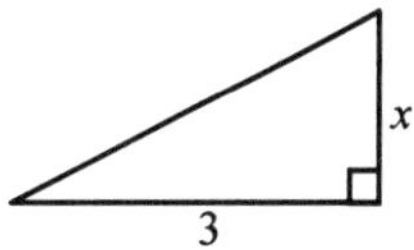

50. A long strip of sheet metal 12 in. wide is to be shaped into an open gutter as shown. If the area of the rectangular cross-section is to be at least 16 in.^{2} for the gutter to be useful, then find all possible heights (x) of the gutter.

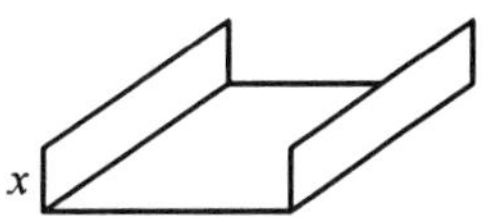

In Exercises 51–60 select the choice that completes the statement or answers the question.

51. If $x^{-2} = -36$, then x equals
a. ± 6 **b.** $\pm 1/6$ **c.** $\pm 6i$ **d.** $\pm i/6$

52. If $\sqrt{x^2 + 3} = 2x$, then x equals
a. both 1 and −1 **b.** only 1
c. only −1 **d.** no solution

53. Pick the statement that describes the set of numbers illustrated

a. $|x - 2| < 1$ **b.** $|x + 2| > 1$
c. $|x - 1| > 2$ **d.** $|x + 1| < 2$

54. If $ax^n - bx = 0$, and $x \neq 0$, then b equals
a. ax^{n-1} **b.** ax^{2n} **c.** a^n **d.** na

55. The value of the discriminant $b^2 - 4ac$ for the equation $2x^2 - x - 5 = 0$ is
a. −39 **b.** 30 **c.** −19 **d.** 41

56. If the discriminant of a quadratic equation equals −4, then the equation has how many real roots?
a. 0 **b.** 1 **c.** 2 **d.** −4

57. The solution set to $x^2 - 6 < x$ is

a. $(-2, 3)$ **b.** $(-\infty, -2) \cup (3, \infty)$
c. $(-\infty, -3) \cup (2, \infty)$
d. $(-3, 2)$

58. If $\frac{1}{a} = \frac{1}{x} - \frac{1}{b}$, then x equals

a. $\frac{ab}{a-b}$ **b.** $\frac{a+b}{ab}$
c. $a+b$ **d.** $\frac{ab}{a+b}$

59. Pick the statement that corresponds to the diagram.

-2 2

a. $|x| < 2$ **b.** $|x| > 2$
c. $|x| \geq 2$ **d.** $|x| \leq 2$

60. If $at = \left(\frac{a}{k} - 1\right)x$, then x equals

a. $\frac{a(kt-1)}{a-1}$ **b.** $\frac{akt}{a-k}$
c. $\frac{akt}{a-1}$ **d.** $\frac{at}{a-k}$

CHAPTER 4 TEST

In Questions 1–12 solve each equation or inequality.

1. $6x^2 + 2x = 3$ **2.** $x - 8\sqrt{x} + 7 = 0$

3. $\frac{3+x}{7x} - 1 = \frac{5}{4x}$ **4.** $\frac{11y}{7} = \frac{y^2+16}{y}$

5. $|4x-7| = |3-x|$ **6.** $3x^3 + 2x = -5x^2$

7. $7(x-5)^2 - 42 = 0$ **8.** $|3x-1| > 5$

9. $t^2 + t < 20$ **10.** $\frac{2}{y} > 3$

11. $\sqrt[3]{x^2+23} = 3$ **12.** $x(x-1)(x+2) \geq 0$

13. Use the given graph to solve $f(x) \leq 0$

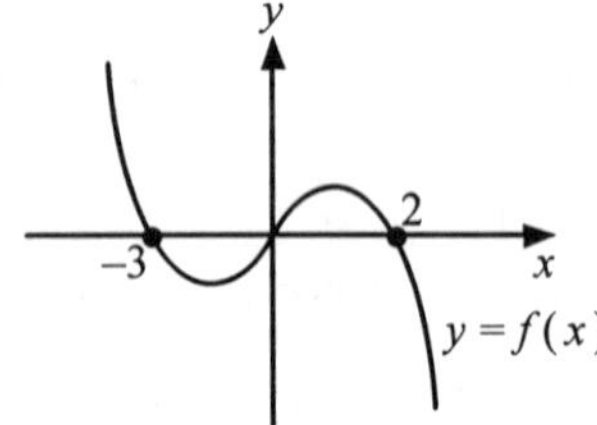

14. Solve for C_1: $\frac{1}{C} = \frac{1}{C_1} + \frac{1}{C_2}$.

15. Write the interval $(1, 9)$ using inequalities and absolute value.

16. The speed (v), in feet per second, that a dropped object acquires in falling d feet is given by $v = 8\sqrt{d}$.

a. If a dropped object falls past a window at a rate of 75 ft/second, from what height above the window was it dropped? Answer to the nearest foot.

b. If the window in part **a** is 30 feet above the ground, how fast will the object be going when it hits the ground? Answer to the nearest whole number.

17. For what values of x will $\sqrt{x^2 + x}$ be a real number?

18. The perimeter of a rectangle is 80 m and the area is 375 m^2. Find the dimensions of the rectangle.

19. If the perimeter in the right triangle below is 12 ft, find x.

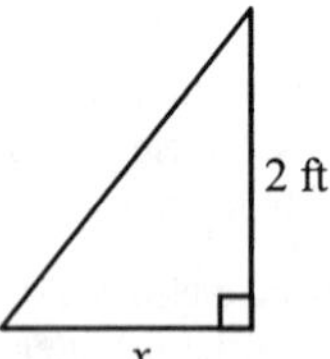

20. The height (y) of a projectile that is shot directly up from the ground with an initial velocity of 80 ft/second is given by the formula $y = 80t - 16t^2$. To the nearest tenth of a second, when will the projectile initially attain a height of 44 ft?

Part 2

Elementary Functions

• • •

Chapter 5

Functions Revisited

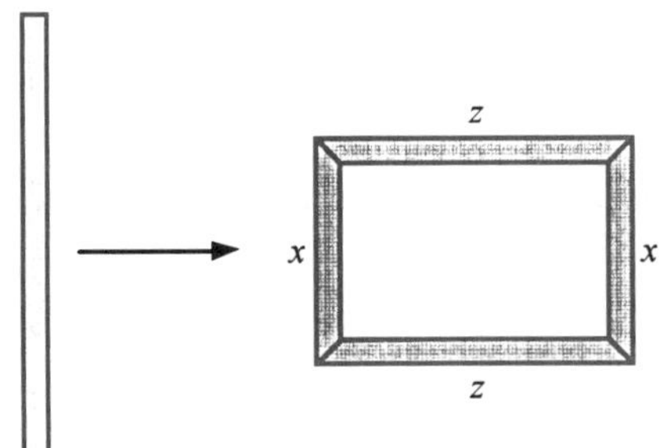

A piece of molding is 46 in. long. You wish to use it to make a rectangular frame of sides x and z units long as shown.

a. Construct a table that shows corresponding values for x, z, and the area of the rectangle. Let x vary from 4 to 24 in. increments of 4.

b. Write a formula that expresses the area of the rectangle as a function of x.

c. Use a grapher to graph the formula from part **b** and describe what happens to the area as x increases. What are the domain and range of this function?

(See Example 1 of Section 5.1.)

In Chapter 1 we introduced the idea of a function. For the next few chapters, this concept will be the central theme as we investigate the behavior of the "elementary functions" studied in calculus. These important functions are called the linear, quadratic, polynomial, rational, exponential, logarithmic, and trigonometric functions. In this chapter we first review some important ideas about functions and graphs, and then extend these concepts to include operations with functions and various methods for measuring the rate of change in a function. One of these methods, the slope of a line, leads to a discussion of linear functions and parallel and perpendicular lines. Finally, we conclude this chapter by considering quadratic functions.

• • •

5.1 Review of Functions and Graphs

Objectives

1. Create a table of values associated with a function.
2. Find a formula that defines a function.
3. Graph a function by the point-plotting method or by using a grapher.
4. Find the domain and range of a function.
5. Determine if a graph or a set of ordered pairs is a function.
6. Evaluate functions using functional notation.

7. Read from a graph of function f the domain, range, function values, values of x for which $f(x) = 0$, $f(x) < 0$, or $f(x) > 0$, and intervals where f is increasing, decreasing, or constant.
8. Prove or classify a function to be even, odd, or neither of these.

We begin by listing, for easy access, the important definitions and rules from Chapter 1. This summary will be followed by some example problems that will further clarify the important ideas. If you find the pace of the review too fast, turn back to Chapter 1, where each topic is discussed in greater detail.

Definition of a Function

A **function** is a rule that assigns to each element x in a set X exactly one element y in a set Y. In this definition, set X is called the **domain** of the function, and the set of all elements of Y that correspond to elements in the domain is called the **range** of the function.

Functional Notation

In this notation we write an equation like $y = x^2 - 5$ as $f(x) = x^2 - 5$. The dependent variable y is replaced by $f(x)$, with the independent variable x appearing in parentheses. The term $f(x)$ is read "f of x" or "f at x" and means the function value (the y value) corresponding to some x value. Similarly, $f(-2)$ is "f of –2" or "f at –2" and means the value of the function when $x = -2$. If $f(x) = x^2 - 5$, we find $f(-2)$ by substituting –2 for x in the equation. Thus, $f(-2) = (-2)^2 - 5 = -1$.

Ordered Pairs

In mathematical notation we represent the correspondence in a function by using ordered pairs. In such pairs the value for the independent variable is listed first, and the value for the dependent variable comes second. For example, in the equation $y = x^2 - 1$, replacing x with 2 gives a y value of 3. Thus (2, 3) is a solution of this equation. Note that (3, 2) is not a solution of the equation.

Functions as Relations

A **relation** is a set of ordered pairs. The set of all first components of the ordered pairs is called the **domain** of the relation. The set of all second components is called the **range** of the relation. A **function** is a relation in which no two different ordered pairs have the same first component.

Graph

The graph of a relation is the set of all points in a coordinate system that correspond to ordered pairs in the **relation**.

Vertical Line Test

Imagine a vertical line sweeping across the graph from left to right. If the vertical line at any position intersects the graph in more than one point, the graph is not a graph of a function.

Even Function

A function f is called an even function when $f(-x) = f(x)$ for all x in the domain of f. The graph of an even function is symmetric about the y-axis.

Odd Function

A function f is called an odd function when $f(-x) = -f(x)$ for all x in the domain of f. The graph of an odd function is symmetric about the origin.

Vertical and Horizontal Shifts

Let c be a positive constant.

1. The graph of $y = f(x) + c$ is the graph of f raised c units.
2. The graph of $y = f(x) - c$ is the graph of f lowered c units.
3. The graph of $y = f(x + c)$ is the graph of f shifted c units to the left.
4. The graph of $y = f(x - c)$ is the graph of f shifted c units to the right.

Reflecting

The graph of $y = -f(x)$ is the graph of f reflected about the x-axis.

Stretching

If $c > 1$, the graph $y = cf(x)$ is the graph of f stretched by a factor of c.

Shrinking

If $0 < c < 1$, the graph of $y = cf(x)$ is the graph of f flattened out by factor of c.

Some Important Functions

In Chapter 1 we encountered a few important types of functions and their graphs. Figure 5.1 summarizes some of the key ideas. Note that the domain and range of each function may be easily read from the graph. The domain is given by the variation of the graph in a horizontal direction, while the range is given by the variation in the vertical direction.

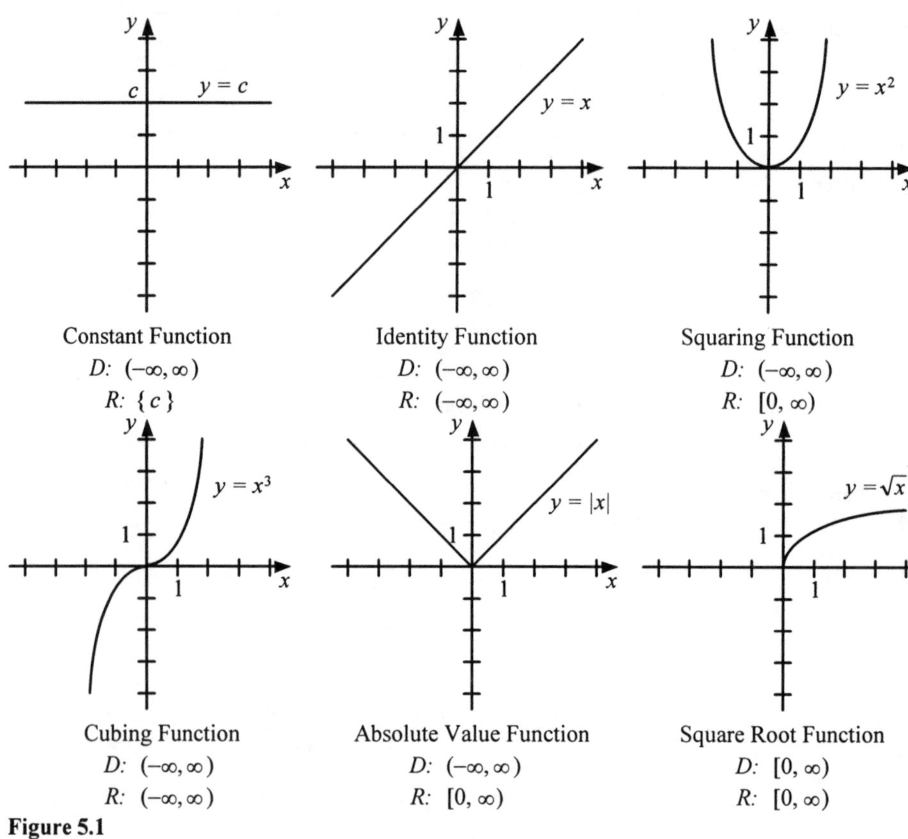

Figure 5.1

Example 1: Analyzing a Relation Numerically, Algebraically, and Graphically

Solve the problem in the chapter introduction on page 385.

Solution:

a. Because the molding is 46 in. long, the perimeter of the rectangle is 46 in., and if $x = 4$ in., as shown in Figure 5.2, then

$$\begin{aligned} P &= 2x + 2z \\ 46 &= 2(4) + 2z \\ 19 &= z. \end{aligned}$$

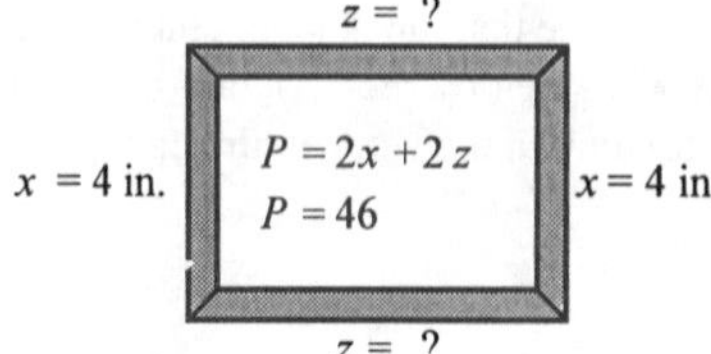

Figure 5.2

Then the area of the rectangle is given by

$$A = xz = 4(19) = 76 \text{ in.}^2.$$

By reasoning in a like way for $x = 8$, 12, 16, 20, and 24, we construct the following table, which shows how x, z, and A are related as x varies from 4 to 24 in increments of 4.

x (inches)	z (inches)	A (square inches)
4	19	76
8	15	120
12	11	132
16	7	112
20	3	60
24	–1 (not meaningful)	–24 (not meaningful)

Note in the table that $x = 24$ does not lead to meaningful values for z and A in the context of the problem.

b. To write a formula that expresses A as a function of x, first express z as a function of x by using the perimeter formulas:

$$P = 2x + 2z$$
$$46 = 2x + 2z$$
$$23 - x = z.$$

Now replacing z by $23 - x$ in the area formula gives

$$A = xz$$
$$= x(23 - x).$$

Thus, $A = x(23 - x)$ expresses A in terms of x.

c. Figure 5.3 shows a graph of $y = x(23 - x)$ where the y axis represents values of the area A. The viewing rectangle $[0, 24]$ by $[0, 150]$ was chosen as reasonable, on the basis of the table of values in part **a**. We see that as x increases, so does the area, for a while (until $x = 11.5$); then as x increases, the area decreases and approaches a value of 0 (when $x = 23$).

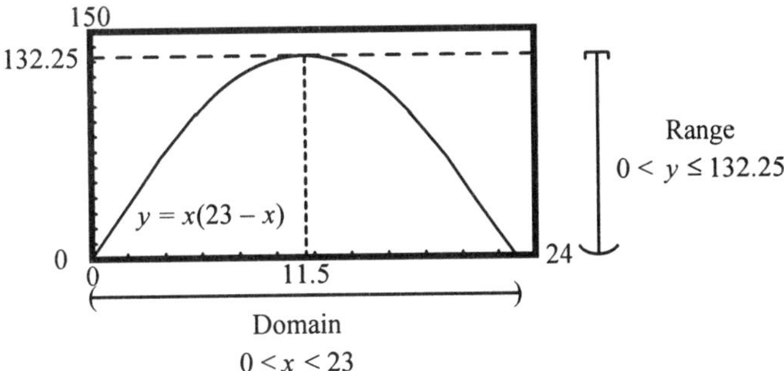

Figure 5.3

Both x and $23 - x$ must be positive, so the domain is the interval $(0, 23)$. This interval is in agreement with the meaningful variation in the horizontal direction of the graph in Figure 5.3. To find the range, you may use the Zoom and Trace features (or a Function Maximum feature, if available) to determine that $x \approx 11.50$ and $y \approx 132.25$ at the highest point on the graph. The lowest y value is 0, but $y > 0$ because of the context. Thus, the range, which is the set of possible values for the area, is the interval $(0, 132.25]$.

PROGRESS CHECK 1

A farmer encloses an area by connecting 500 m of fencing in the shape of a rectangle of sides x and z units.

a. Construct a table that shows corresponding values of x, z and the area of the rectangle. Let x vary from 40 to 280 m in increments of 40.

b. Write a formula that expresses the area of the rectangle as a function of x.

c. Graph the formula from part **b** and describe what happens to the area as x increases. What are the domain and range of this function? ■

Example 2: Finding Domain and Range Algebraically and Graphically

Determine the domain and range of $y = \sqrt{1-x^2}$.

Solution: For the output of $y = \sqrt{1-x^2}$ to be a real number, x must satisfy $1-x^2 \geq 0$. As shown in Example 2 of Section 4.4, $1-x^2$ is nonnegative in the interval $[-1, 1]$, which is the domain. The corresponding y values vary from 0 (when $x = \pm 1$) to 1 (when $x = 0$), so the range is the interval $[0, 1]$. An effective alternative method is to first graph the function as shown in Figure 5.4 and then determine the domain and range by examining the graph. With this graphical approach, analysis of the equation $y = \sqrt{1-x^2}$ is required to determine that $(-1, 0)$ and $(1, 0)$ are endpoints of the graph.

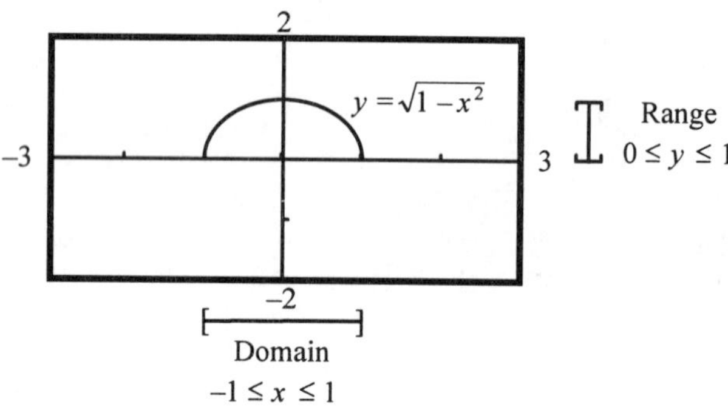

Figure 5.4

Caution: On a calculator, remember to enter the function input (argument) inside of parentheses when the input is more than one term. For instance, the sequence

$$\sqrt{} 1-x^2 \text{ means } \sqrt{1}-x^2,$$

while

$$\sqrt{} (1-x^2) \text{ means } \sqrt{1-x^2}.$$

PROGRESS CHECK 2

Find the domain and range of $y = \sqrt{x-8}$. ■

Example 3: Graphing a Piecewise Function

Graph the function defined as follows:

$$y = \begin{cases} x+1 & \text{if} \quad 0 \le x \le 2 \\ 3 & \text{if} \quad x > 2 \end{cases}$$

Indicate the domain and range on the graph.

Solution: This function assigns to any value of x between 0 and 2 (inclusive) the number that is one greater than the x value. For example, when $x = 1$, $y = 2$, and when $x = 1/2$, $y = 3/2$. This function also assigns the number 3 to each value of x greater than 2. Thus, as x increases, the graph begins as a line segment from (0, 1) to (2, 3), and then the graph proceeds horizontally from (2, 3) to the right, as shown in Figure 5.5. From this graph we can observe that the domain is $[0, \infty)$ and that the range is $[1, 3]$.

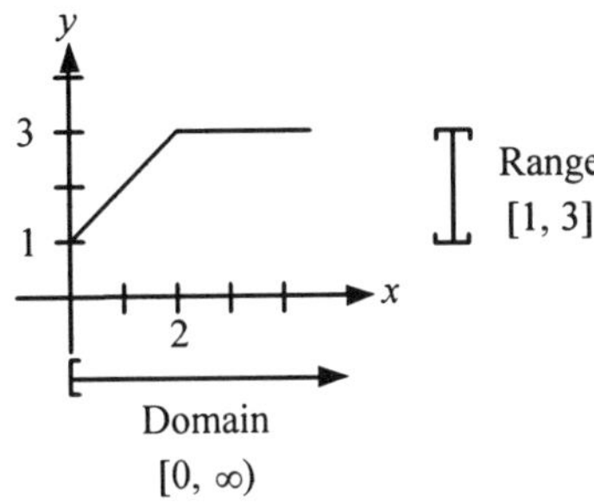

Figure 5.5

Technology Link

It is possible to use some graphing calculators to graph a piecewise function. Many calculators include the ability to enter inequality symbols (often through a key or menu marked Test) so that the following expression may be entered in the equation list to graph the function in this example.

$$y1 = (x+1)(0 \le x)(x \le 2) + 3(x > 2)$$

When graphing piecewise functions it is recommended that the calculator be set in Dot mode, not connected mode, to prevent the calculator from connecting points that are in separate pieces of the graph.

PROGRESS CHECK 3

Graph the function defined as follows:

$$y = \begin{cases} -3 & \text{if} \quad x < 2 \\ x+1 & \text{if} \quad x \ge 2 \end{cases}$$

Indicate the domain and range on the graph. ■

Example 4: Determining Functions and Finding Domain and Range

Determine if the relation is a function. Also specify the domain and range.

a. $\{(2, 1), (3, 1), (4, -1)\}$ **b.** $\{(1, 2), (1, 3), (-1, 4)\}$

Solution:

a. This relation which is diagrammed in Figure 5.6, is a function because the first components in the ordered pairs are always different. The domain is $\{2, 3, 4\}$, and the range is $\{1, -1\}$.

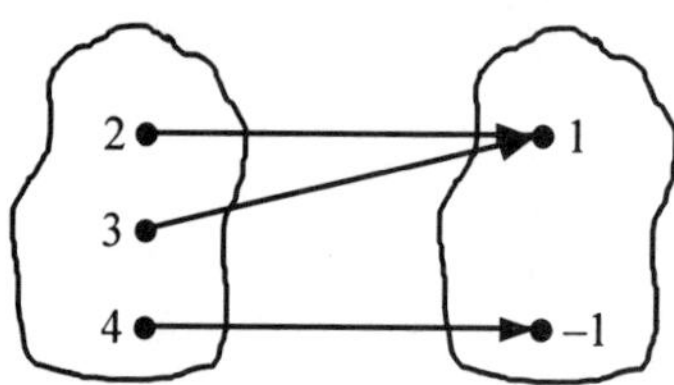

Figure 5.6
Function

b. Figure 5.7 shows a diagram for this relation, which is not a function because the number 1 is the first component in two ordered pairs. The domain is $\{1,\ -1\}$, and the range is $\{2,\ 3,\ 4\}$.

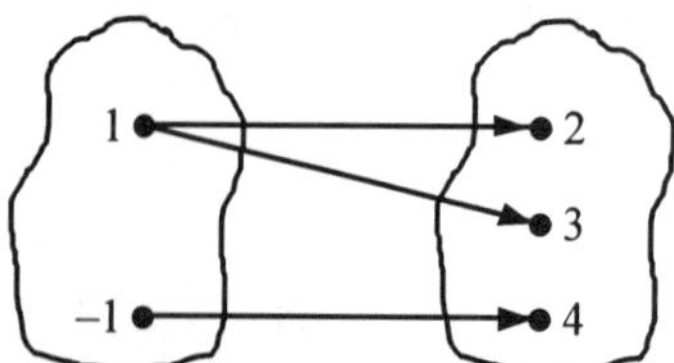

Figure 5.7
Not a Function

PROGRESS CHECK 4

Determine if the relation is a function. Also specify the domain and the range.

a. {(0, 1), (0, 2), (0, 3)} **b.** {(1, 0), (2, 0), (3, 0)} ■

Example 5: Using Functional Notation

If $y = f(x) = 7 + \frac{1}{x}$, find $f(1), f\left(\frac{1}{3}\right), f(0)$, and $f(a)$.

Solution: In each case replace all occurrence of x by the number inside the parentheses, and then simplify.

$$y_{\text{when } x=1} = f(1) = 7 + \frac{1}{1} = 8$$

$$y_{\text{when } x=1/3} = f\left(\frac{1}{3}\right) = 7 + \frac{1}{1/3} = 10$$

$$y_{\text{when } x=0} = f(0) = 7 + \frac{1}{0} = \text{undefined}$$

$$y_{\text{when } x=a} = f(a) = 7 + \frac{1}{a}$$

PROGRESS CHECK 5

If $y = f(x) = \frac{x}{x-5}$ find $f(10), f(5), f\left(\frac{1}{2}\right)$, and $f(b)$. ■

Example 6: Reading a Graph

Consider the graph of $y = f(x)$ in Figure 5.8.

a. What is the domain? **b.** What is the range?

c. Is f even, odd or neither? **d.** Find $f(0)$.

e. Solve $f(x) = 0$.

f. Solve $f(x) > 0$.

g. Find the open intervals on which the function is increasing or decreasing.

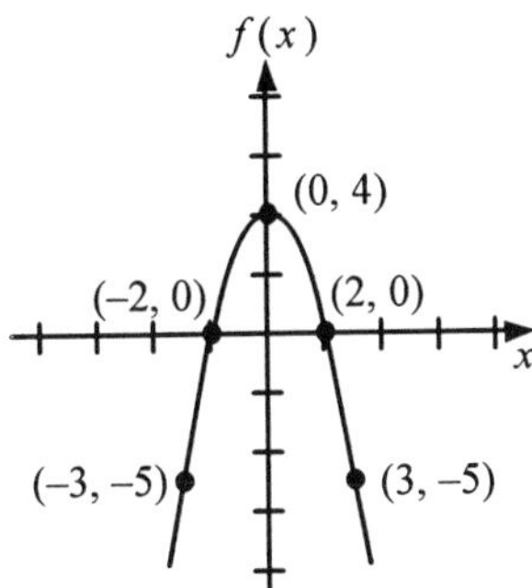

Figure 5.8

Solution:

a. The x values in the figure keep going both to the left and to the right. Thus, the domain is the set of all real numbers.

b. The maximum y value is 4, and the graph extends indefinitely in the negative y direction, so there is no minimum value. Thus, the range is $(-\infty, 4]$.

c. The graph is symmetric with respect to the y-axis, so f is an even function.

d. The notation $f(0)$ means that we want the y value when $x = 0$. The ordered pair (0, 4) tells us that $y = 4$ when $x = 0$. Thus, $f(0) = 4$.

e. Here $f(x)$ or y is 0, and we look at the ordered pairs $(-2, 0)$ and $(2, 0)$. Thus $f(x) = 0$ when $x = -2$ or 2.

f. The y values are greater than zero when the graph is above the x-axis. Thus, $f(x) > 0$ when $-2 < x < 2$, so the solution set is the interval $(-2, 2)$.

g. From left to right, the graph rises to the point (0, 4) and falls thereafter. So the function increases for the interval $(-\infty, 0)$ and decreases for the interval $(0, \infty)$.

PROGRESS CHECK 6

Answer the questions in Example 6 for the graph of $y = f(x)$ in Figure 5.9. ■

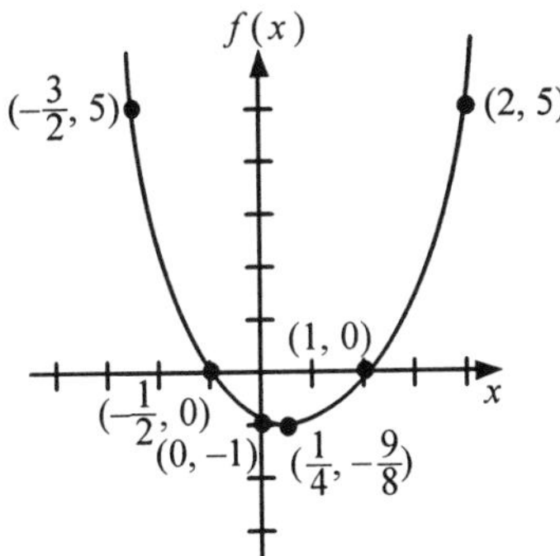

Figure 5.9

Example 7: Graphing Functions Using Translation and Reflection

Use the graph of $y = |x|$ to graph each of the following functions.

a. $f(x) = |x - 2|$ **b.** $f(x) = |x + 2| - 1$ **c.** $f(x) = 2 - |x|$

Solution: The graph of the absolute function $y = |x|$ is ∨ shaped. The rules for shifting and reflecting tells us to move this basic shape as follows, see Figure 5.10.

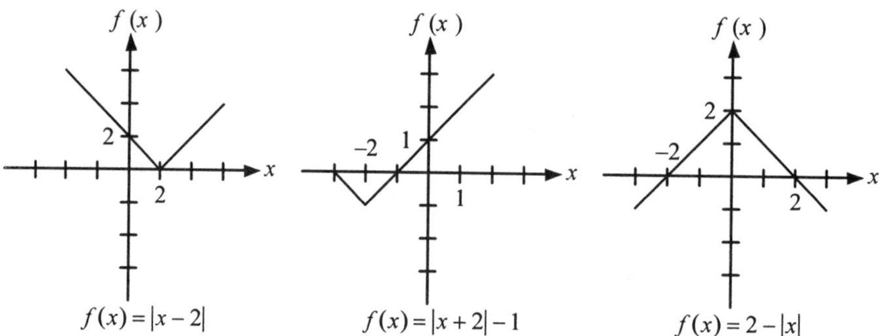

Figure 5.10

a. The constant 2 is subtracted before the absolute value rule is applied, which moves the $\vee$ 2 units to the right.

b. In this case we move the basic shape 2 units to the left and 1 unit down.

c. Because of the negative sign in $-|x|$, we reflect the basic $\vee$ shape about the x-axis. Now we raise our result up 2 units, since adding 2 gives $2-|x|$.

PROGRESS CHECK 7

Use the graph of $y=|x|$ to graph each of the following functions.

a. $f(x)=|x|-2$ **b.** $f(x)=-|x+3|$ **c.** $f(x)=|x-1|-4$ ■

Example 8: Using Symmetry to Graph a Function

Is the function $f(x)=-2/x$ an even function, an odd function, or neither? Figure 5.11 is the graph of this function for $x \geq 0$. Complete the graph and check your result with a grapher.

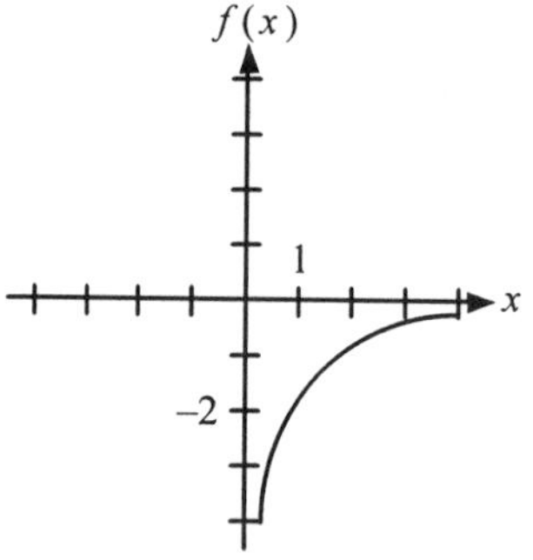

Figure 5.11

Solution: $f(x)=-2/x$ and $f(-x)=-2/-x=2/x=-f(x)$. Since $f(-x)=-f(x)$, f is an odd function, so f is symmetric about the origin. Reflecting the given curve about both the x- and y-axes (in either order) gives the completed graph in Figure 5.12. This result checks with the display in a grapher, as shown in Figure 5.13.

PROGRESS CHECK 8

Is the function $f(x)=-\dfrac{1}{x^2}$ an even function, an odd function, or neither? Figure 5.14 is the graph of this function for $x \geq 0$. Complete the graph and check your result with a grapher. ■

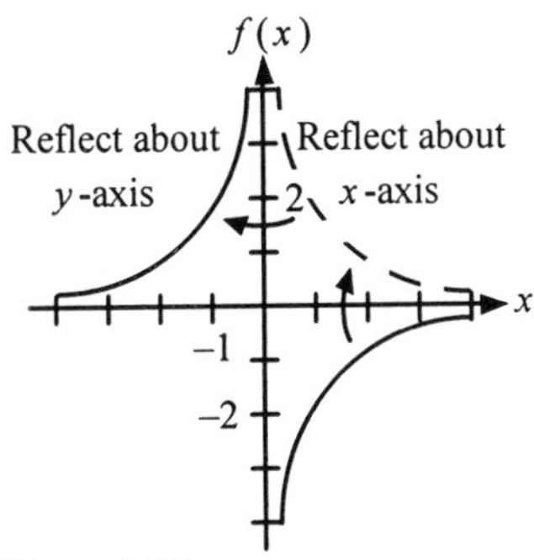

Figure 5.12

Figure 5.13

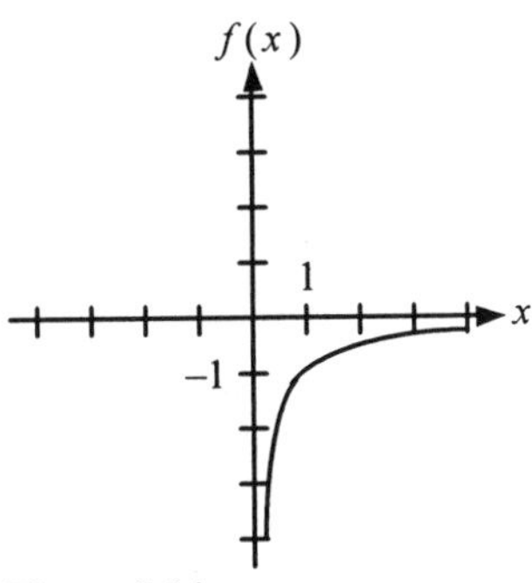

Figure 5.14

EXPLORE 5.1

1. A graphing calculator omits points wherever a function output is not real. To illustrate, graph $y = \sqrt{x}$ and describe what coordinates are shown when you trace where x is negative.
2. Note how important parentheses are in indicating the argument of a square root function. Compare the graphs in each set of functions, and write out without using parentheses the equations that are being graphed.
   a. $y1 = \sqrt{\ } x - 7$ **b.** $y1 = \sqrt{\ } ((x-4)/x)$
   $y2 = \sqrt{\ } (x-7)$ $y2 = \sqrt{\ } (x - 4/x)$
3. If your calculator can use inequality operations, find out what value it assigns to a true statement, such as $8 > 2$, and to a false statement, such as $8 > 20$. Then graph $y = x^2(x > -2)$ both in Connected mode and in Dot mode, and explain any difference between the two graphs.
4. The appearance of a graph may be influenced significantly both by the viewing window and the graphing mode (connected dots or separate dots). First graph $y = x^2$ for [−5, 5] by [−1, 10] using Connected mode. Next look at the graph of $y = x^2$ with each of the following viewing windows and modes, and write an answer to the related question.

Viewing Window	Mode	Question
[−5, 5] by [−1, 10]	Separate	Why do the dots look far apart at the top of the graph and close together near the bottom?
[−1, 1] by [−1, 10]	Connected	Why does the graph look flat instead of curved near the lowest point? Why is the graph made up of horizontal line segments instead of being a smooth curve? Why do these segments get shorter as you move away from the lowest point on the graph?
[−1, 1] by [−1, 10]	Separate	Why does the graph look about the same in both modes?

EXERCISES 5.1

1. The total cost of producing a certain product consists of paying \$600 per month rent plus \$7 per unit for material and labor. Let x equal the number of units produced in a month, and c represent the total cost.
   a. Construct a table that shows corresponding values for x and c. Let x vary from 100 to 700 in increments of 100.
   b. Write a formula that expresses c as a function of x.
   c. Use a grapher to graph the formula from part **b**, and describe what happens to the cost as x increases. What are the domain and range of this function?

2. The federal income tax (t) for a single person depends on the person's taxable income (i). For taxable income between \$10,000 and \$20,000 inclusive, the tax is \$1,504 plus 15 percent of the excess over \$10,000.
   a. Construct a table that shows corresponding values for t and i. Let i vary from 10,000 to 20,000 in increments of 1,000.
   b. Write a formula that expresses t as a function of i.
   c. Use a grapher to graph the formula from part **b** and describe what happens to the tax as i increases. What are the domain and range of this function?

3. The area (A) of a square depends on its perimeter (p).
   a. Construct a table that shows corresponding values for p and A. Let p vary from 12 to 40 in increments of 4.
   b. Write a formula that expresses A as a function of p.
   c. Use a grapher to graph the formula from part **b** and describe what happens to the area as p increases. What are the domain and range of this function?

4. The area (A) of a circle depends on the diameter (d).
   a. Construct a table that shows corresponding values for d and A. Let d vary from 1 to 7 in increments of 1.
   b. Write a formula that expresses A as a function of d.
   c. Use a grapher to graph the formula from part **b** and describe what happens to the area as d increases. What are the domain and range of this function?

5. The sum of two nonnegative numbers (x and z) is 43.8. You wish to study the relation between these numbers and their product, P.
   a. Construct a table that shows corresponding values for x, z, and P. Let x vary from 1 to 7 in increments of 1.
   b. Write a formula that expresses P as a function of x.
   c. Use a grapher to graph the formula from part **b**, and describe what happens to the product as x increases. What are the domain and range of this function?

6. The sum of two nonnegative numbers (x and z) is 43.8. You wish to study the relation between these numbers and their difference, D, where $D = x - z$.
   a. Construct a table that shows corresponding values for x, z, and D. Let x vary from 1 to 7 in increments of 1.
   b. Write a formula that expresses D as a function of x.
   c. Use a grapher to graph the formula from part **b**, and describe what happens to the difference as x increases. What are the domain and range of this function?

7. A wire that is 60 cm long is bent into the shape shown, where one vertical side is twice the length of the other. You wish to study the relation between the shortest side and the area, A.
   a. Construct a table that shows corresponding values for x, $2x$, z and A. Let x vary from 2 to 20 in increments of 2.
   b. Write a formula that expresses A as a function of x.
   c. Use a grapher to graph the formula from part **b**, and describe what happens to the area as x increases. What are the domain and range of this function?

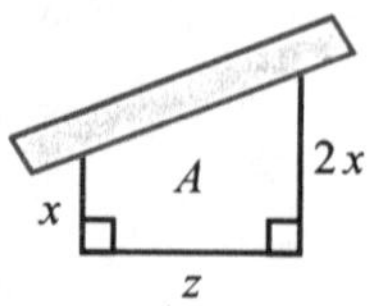

8. Redo Exercise 7, with one vertical side three times the length of the other.

9. The perimeter of the right triangle in the figure is 30 in. We wish to investigate the relation between the lengths of the sides and the area.

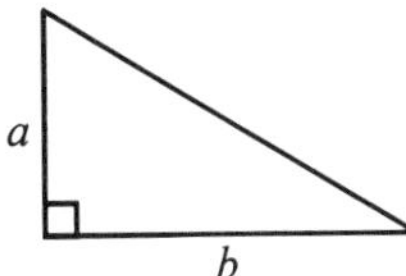

a. Construct a table that shows corresponding values for both sides, the hypotenuse, and the area. Let a vary from 1 to 17 in increments of 4.
b. Write a formula that expresses the area as a function of a.
c. Use a grapher to graph the formula from part **b**, and describe what happens to the area as a increases. What are the domain and range of this function?

10. Redo Exercise 9 if the perimeter is 40 in.

In Exercises 11–22 find the domain and range of the function.

11. $y = \sqrt[3]{x}$ 12. $y = x^3$

13. $y = \dfrac{1}{x-2}$ 14. $y = \dfrac{1}{x+1}$

15. $y = \dfrac{1}{\sqrt{x}}$ 16. $y = \sqrt{x}$

17. $y = \sqrt{9-x^2}$ 18. $y = \sqrt{9-x}$

19. $y = 5$ 20. $y = \sqrt{x^2-9}$

21. $\{(-2,2),(-1,1),(0,0)\}$

22. $\{(3,6),(4,7),(5,8)\}$

In Exercises 23–32 graph the functions. In each case indicate the domain and range on the graph.

23. $y = \begin{cases} 2 & \text{if } x > 1 \\ -2 & \text{if } x \le -1 \end{cases}$ 24. $y = \begin{cases} x & \text{if } x > 0 \\ 1 & \text{if } x = 0 \\ 2 & \text{if } x < 0 \end{cases}$

25. $f(x) = \begin{cases} 1 & \text{if } x > 0 \\ 0 & \text{if } x = 0 \\ -1 & \text{if } x < 0 \end{cases}$

26. $f(x) = \begin{cases} 1-x & \text{if } 0 \le x < 1 \\ x-1 & \text{if } 1 \le x < 2 \\ 1 & \text{if } 2 \le x < 4 \end{cases}$

27. $y = \begin{cases} x & \text{if } x \le 0 \\ -x & \text{if } x > 0 \end{cases}$ 28. $y = \begin{cases} x-1 & \text{if } 0 \le x \le 3 \\ 2 & \text{if } x > 3 \end{cases}$

29. $f(x) = \begin{cases} x^2+2 & \text{if } -4 \le x < 1 \\ 4-x^3 & \text{if } 1 \le x \le 3 \end{cases}$

30. $h(x) = \begin{cases} 2x & \text{if } 0 \le x < 3 \\ x^2 & \text{if } x \ge 3 \end{cases}$

31. $f(x) = \begin{cases} \sqrt{4-x^2} & \text{if } |x| \le 2 \\ \sqrt{x^2-4} & \text{if } |x| > 2 \end{cases}$

32. $g(x) = \begin{cases} \sqrt{9-x^2} & \text{if } 0 \le x < 3 \\ -\sqrt{x-3} & \text{if } x \ge 3 \end{cases}$

In Exercises 33–40 determine if the relation is a function. Also, specify the domain and range.

33. $\{(1,-1),(2,5),(3,1)\}$

34. $\{(-1,1),(5,2),(1,3)\}$

35. $\{(-2,2),(2,-2),(2,0)\}$

36. $\{(0,2),(-2,2),(2,-2)\}$

37. $\{(1,5),(1,6),(1,7)\}$ 38. $\{(5,1),(6,1),(7,1)\}$

39. $\{(-2,4),(2,4)\}$ 40. $\{(4,2),(4,-2)\}$

41. Which of the graphs in the figure represent the graph of a function?

a.

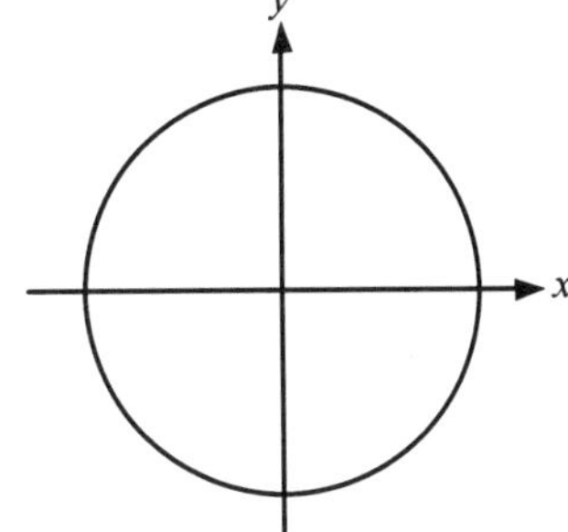

b.

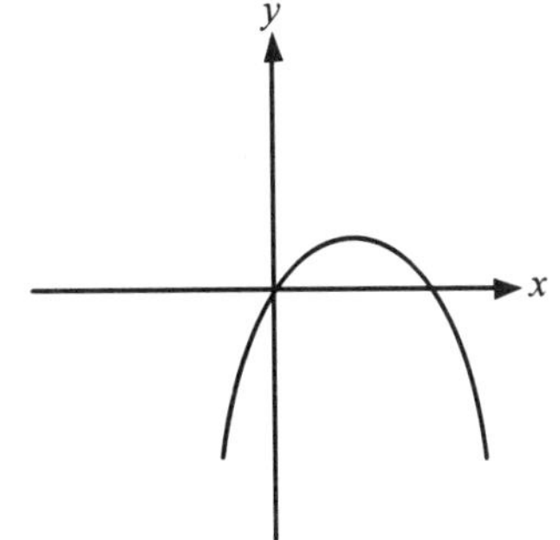

c.

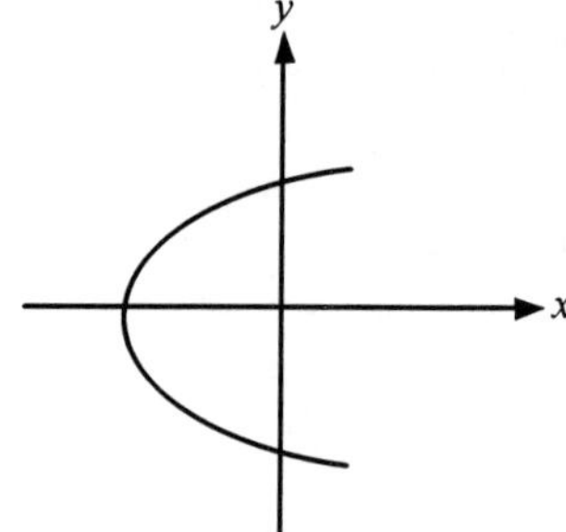

d.

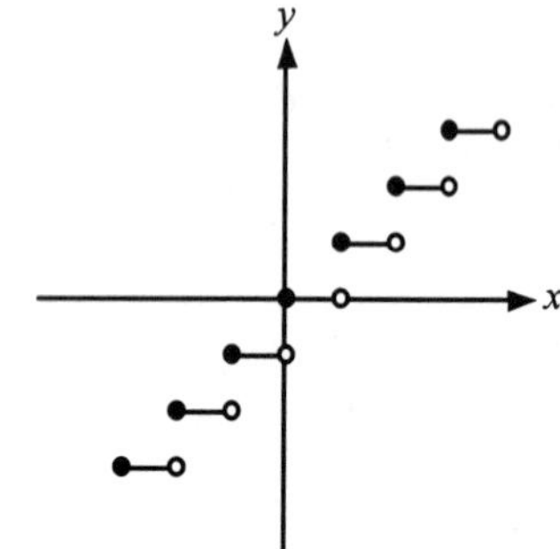

42. Find the domain and range of the function shown.

a.

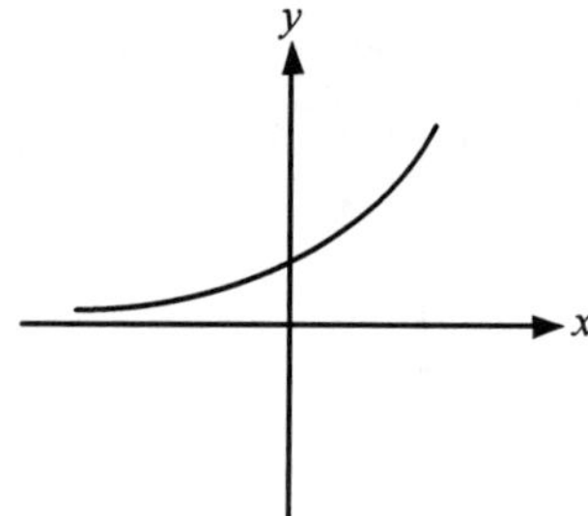

b.

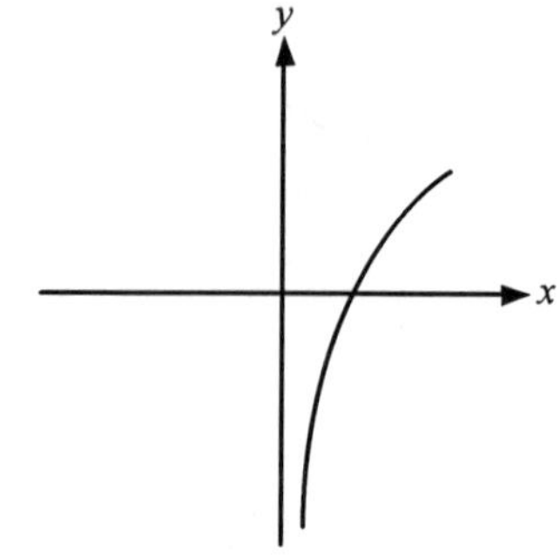

c.

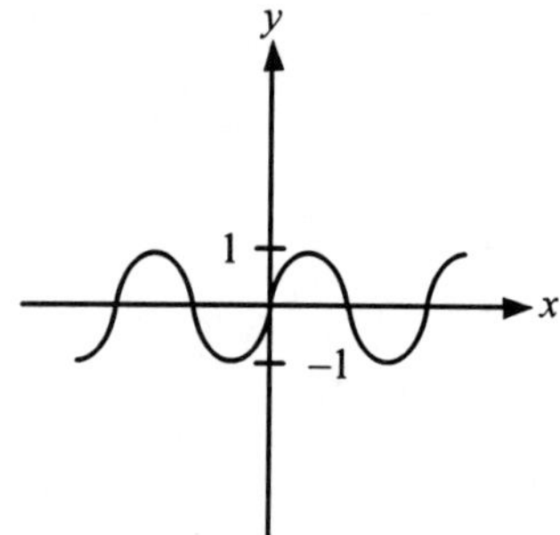

d.

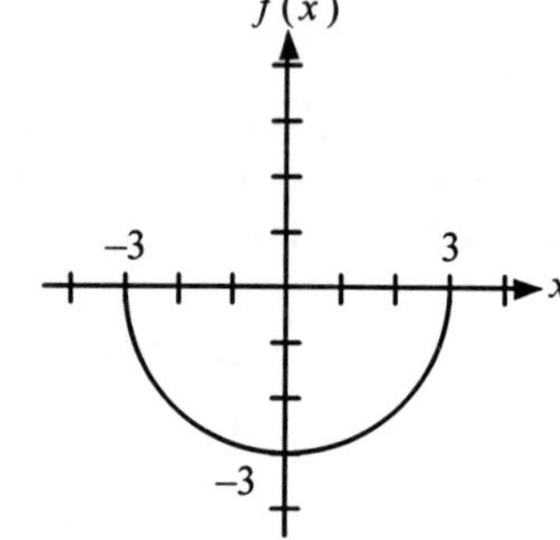

43. If $f(x) = x^3 - x + 1$, find $f(2)$, $f(-1)$, and $f(a)$.

44. If $f(x) = 2x^2 - x + 3$, find $f(3)$, $f(-4)$, and $f\left(\frac{1}{2}\right)$.

45. If $f(x) = -x^2 + x$, find $f(1)$, $f\left(-\frac{1}{2}\right)$, and $f(c)$.

46. If $f(x) = -x^2$, find $f(3)$, $f(-3)$, and $f(h)$.

47. If $g(x) = \frac{1}{x} - 5$, find $g(0)$, $g(-1)$, and $g(x_0 + h)$.

48. If $g(x) = 3 - \frac{2}{x}$, find $g(0)$, $g(-1)$, and $g(a + h)$.

49. If $h(t) = \frac{t+2}{t-1}$, find $h(0)$, $h(1)$, and $h(a)$.

50. If $f(r) = \frac{2r-5}{3r+1}$, find $f(0)$, $f\left(-\frac{1}{3}\right)$, and $f(0.5)$.

51. If $f(x) = 2x - 1$ and $g(x) = x^2$, find

a. $f(0) + g(2)$ **b.** $f(-1) + g(-4)$

c. $4g(-3) + 3f(1)$ **d.** $2f(5) - 5g(-2)$

e. $g(-1) \cdot f(2)$ **f.** $\frac{g(5)}{f(0.5)}$

52. If $f(x) = \sqrt{x}$,

a. Does $f(9+16) = f(9) + f(16)$?

b. Does $f(a+b) = f(a) + f(b)$ for all a and b?

In Exercises 53–56 consider the given graph, then answer these questions.

a. What is the domain? **b.** What is the range?

c. Is f even, odd, or neither?

d. Find $f(0)$. **e.** Solve $f(x) = 0$.

f. Solve $f(x) > 0$.

g. Find the open intervals on which the function is increasing, decreasing, or constant.

53.

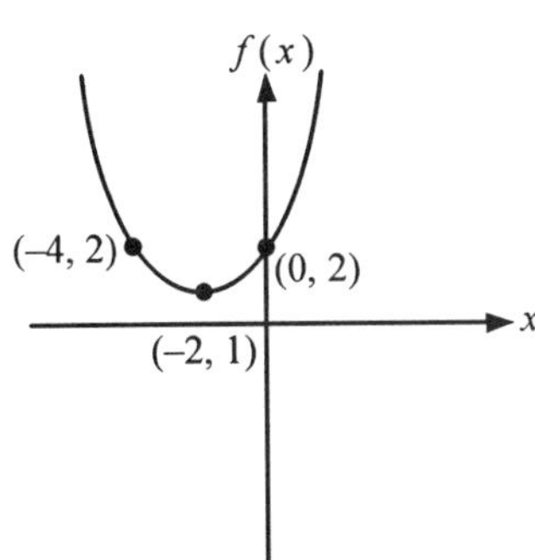

54.

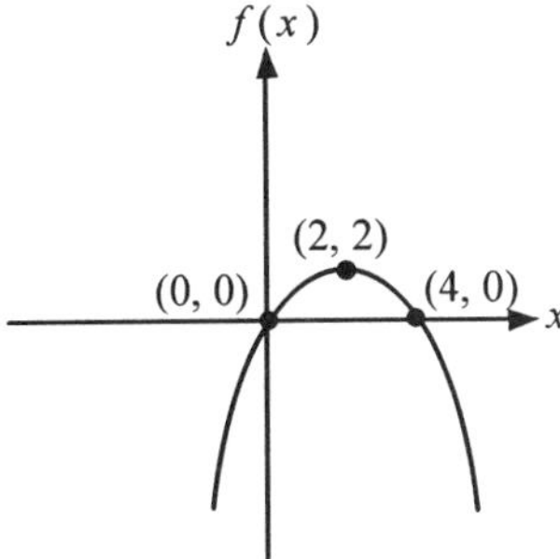

55.

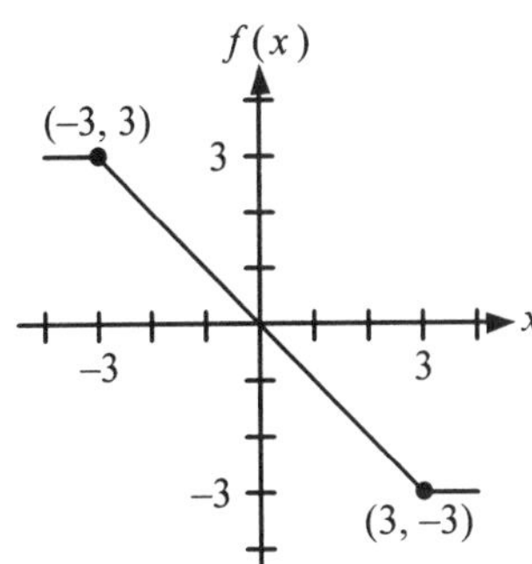

56.

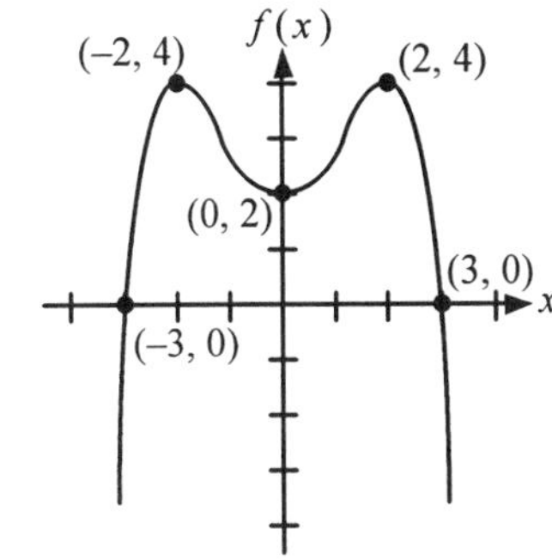

57. Use the graph of $y = x^2$ to graph

a. $f(x) = (x-2)^2$ **b.** $f(x) = x^2 - 2$

c. $f(x) = (x+2)^2 + 2$

58. Use the graph of $y = \sqrt{x}$ to graph

a. $f(x) = \sqrt{x+5}$ **b.** $f(x) = \sqrt{x} + 5$

c. $f(x) = 5 - \sqrt{x}$

In Exercises 59–62 shift the appropriate figure, and graph each function.

59. $y = |x+3|$ **60.** $y = (x+3)^2$

61. $y = \sqrt{x+3}$ **62.** $y = (x+3)^3$

In Exercises 63–70 rearrange the appropriate figure, and graph each function.

63. $y = x^3 + 2$ **64.** $y = 2 - x^2$

65. $y = \sqrt{x - 13}$ **66.** $y = 11 + |x|$

67. $y = -|x+12|$ **68.** $y = \sqrt{x} - 13$

69. $y = (x-2)^2 + 3$ **70.** $y = -x^3$

In Exercises 71–78 classify the function as even, odd, or neither, and then complete the graph of f. The figure shows the graph for $x \geq 0$. Check your result with a grapher.

71. $f(x) = x^2 + 3$

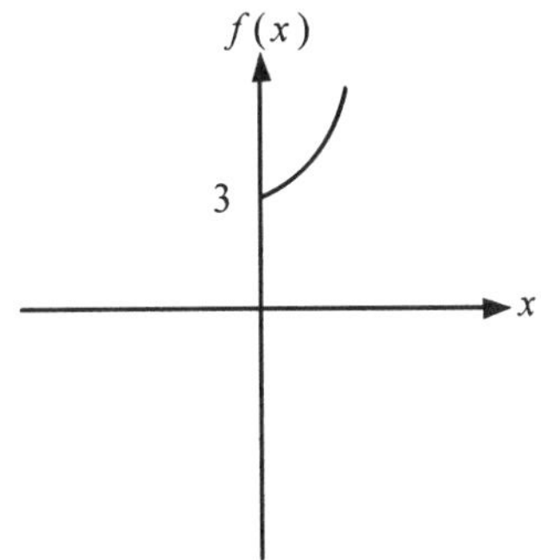

72. $f(x) = x^3 + 3$

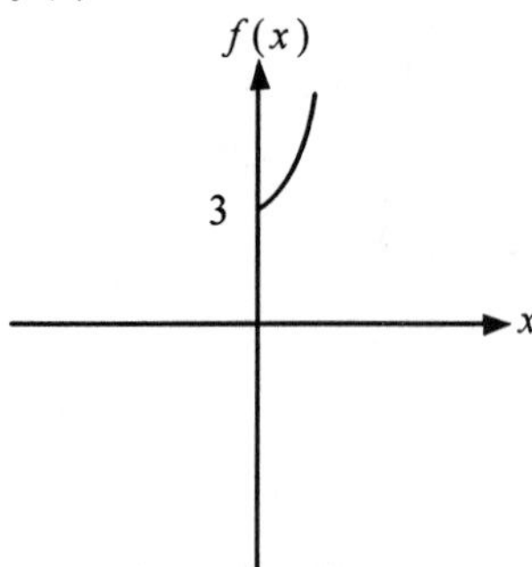

73. $f(x) = -x^2$

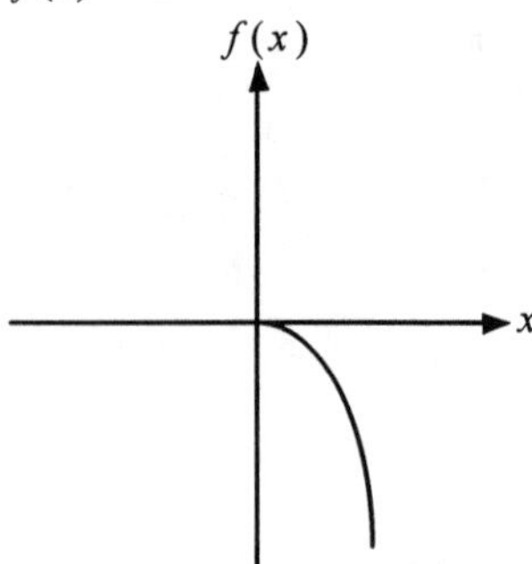

74. $f(x) = -x^3$

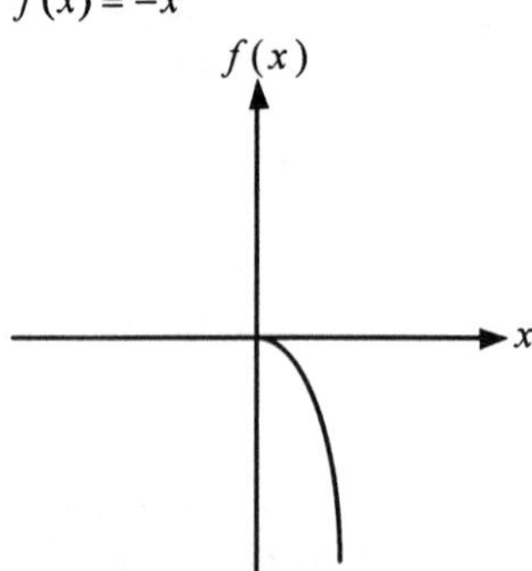

75. $f(x) = \dfrac{5x^2}{1+x^2}$

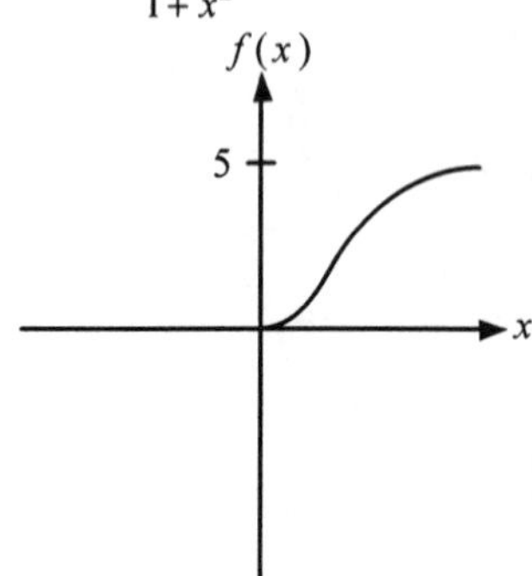

76. $f(x) = \dfrac{10x}{1+x^2}$

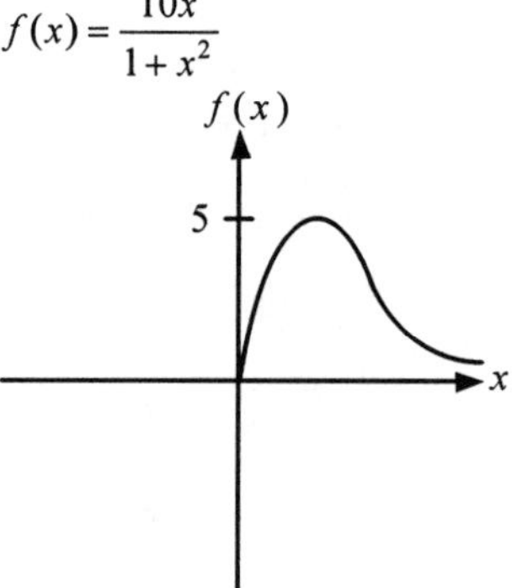

77. $f(x) = -|x|$

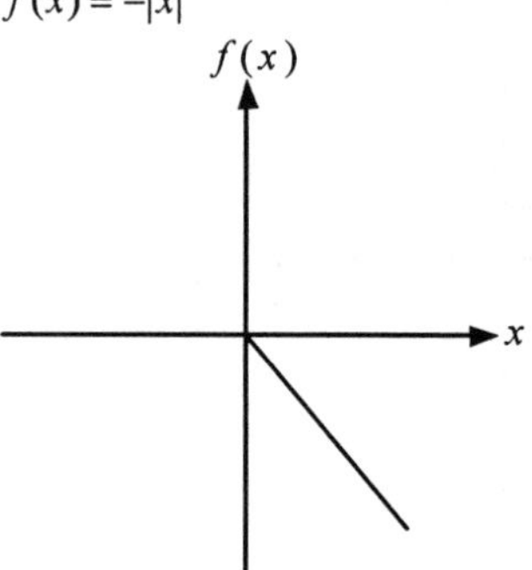

78. $f(x) = \sqrt{4-x^2}$

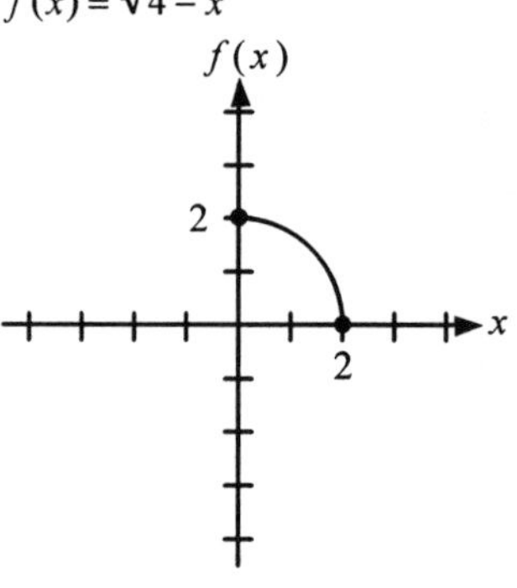

THINK ABOUT IT 5.1

1. Describe a realistic situation that would produce the following graph.

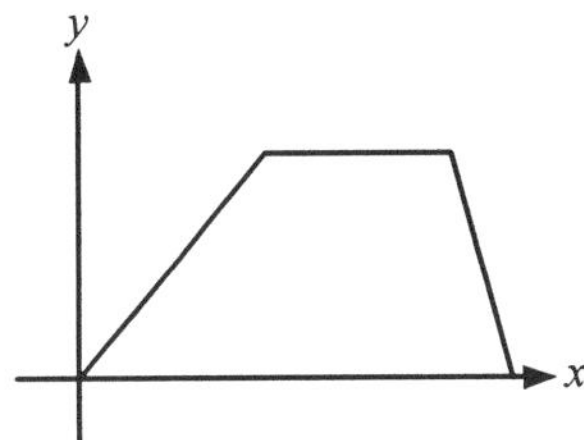

2. The combined federal, state, and local tax rates for a couple place them in the 40 percent tax bracket, so they invest significantly in a tax-free money market fund. Write a formula that expresses the equivalent taxable yield (t) for this couple's investment as a function of the tax-free yield (x). Then use this formula to find the equivalent taxable yield for this investment if the money market fund is currently paying 5.31 percent tax-free.
3. It is often useful to formulate (at least mentally) a verbal description of a function. For example, the function defined by the equation $f(x) = (x-1)^3$ may be expressed in words as follows: For each real number, subtract 1 from it, and then cube the result. Write verbal descriptions for the following functions.

   a. $f(x) = 2x$ **b.** $f(x) = 2$ **c.** $f(x) = x^2 - 1$ **d.** $f(x) = (x-1)^2$

   e. $f(x) = 1/x^2$ **f.** $f(x) = -\sqrt{x}$ **g.** $f(x) = -|x+3|$ **h.** $f(x) = (x+2)^3 - 1$
4. Use the given information about function f to draw a possible graph. Domain: $(-\infty, \infty)$ Range: $[-4, \infty)$

$$f(0) = 5 \quad f(1) = 0 \quad f(5) = 0$$

   f is decreasing over $(-\infty, 3)$, and increasing over $(3, \infty)$.
5. Explain the conditions under which a phone book is a function. Describe the domain and the range.

● ● ●

5.2 Operations with Functions

Photo Courtesy of **Reuters/Jim Bourg** of Archive Photos, New York

An oil leak is spreading over a plane surface in the shape of a circle. The radius of this circle is increasing at a constant rate of 5 cm/second, so the radius of this spill t seconds after the start of the leak may be expressed by $r = g(t) = 5t$. If function f expresses the area of this circular spill as a function of r so that $A = f(r) = \pi r^2$, find and interpret $(f \circ g)(t)$.
(See Example 8.)

Objectives

1. Add, subtract, multiply, and divide two functions.
2. Find the composite function of two functions.
3. Interpret a function h as the composition of simpler functions f and g so that $h(x) = (f \circ g)(x)$.

We may add, subtract, multiply, and divide functions to form new functions. For example, suppose that an item that costs \$5 to manufacture sells for \$7. The total cost of x items is given by the function

$$C(x) = 5x.$$

The total sales revenue from x items is given by

$$S(x) = 7x.$$

Since profit is the difference between sales and cost, we form the profit function as the difference of the two other functions.

$$\begin{aligned} P(x) &= S(x) - C(x) \\ &= 7x - 5x \\ &= 2x \end{aligned}$$

This example illustrates how to find a difference of two functions, and in general, $f+g$, $f-g$, $f \cdot g$ and f/g are defined as follows.

Sum, Difference, Product, and Quotient of Two Functions

If f and g are functions, then:

Sum of f and g: $(f+g)(x) = f(x) + g(x)$

Difference of f and g: $(f-g)(x) = f(x) - g(x)$

Product of f and g: $(f \cdot g)(x) = f(x) \cdot g(x)$

Quotient of f and g: $\left(\dfrac{f}{g}\right)(x) = \dfrac{f(x)}{g(x)},\ g(x) \neq 0$

The domain of the resulting function is the intersection of the domains of f and g. The domain of the quotient function excludes any x for which $g(x) = 0$. In effect, the definitions say that for any x at which both functions are defined, we combine the y values.

Example 1: Combining Functions Algebraically

If $f(x) = x$ and $g(x) = \sqrt{x}$, find $(f+g)(x)$, $(f-g)(x)$, $(f \cdot g)(x)$, and $(f/g)(x)$. In each case give the domain of the resulting function.

Solution: Using the definitions, we have

$$\begin{aligned} (f+g)(x) &= f(x) + g(x) = x + \sqrt{x} \\ (f-g)(x) &= f(x) - g(x) = x - \sqrt{x} \\ (f \cdot g)(x) &= f(x) \cdot g(x) = x\sqrt{x} \\ \left(\frac{f}{g}\right)(x) &= \frac{f(x)}{g(x)} = \frac{x}{\sqrt{x}} \text{ or } \sqrt{x}. \end{aligned}$$

The domain of f is the set of all real numbers, and the domain of g is $[0, \infty)$. The domain of $f+g$, $f-g$, and $f \cdot g$ is the set of numbers common to both domains, which is $[0, \infty)$. In the quotient function $x \neq 0$. Thus, the domain of f/g is the interval $(0, \infty)$.

Technology Link

A graphing calculator can be used to illustrate the concept of a combined function. For instance, to illustrate the sum of f and g in this example, first enter expressions for $y1$, $y2$, and $y3$ as shown in Figure 5.15(a). Then activate the graphing routine using the standard window to see all three graphs, as in Figure 5.15(b).

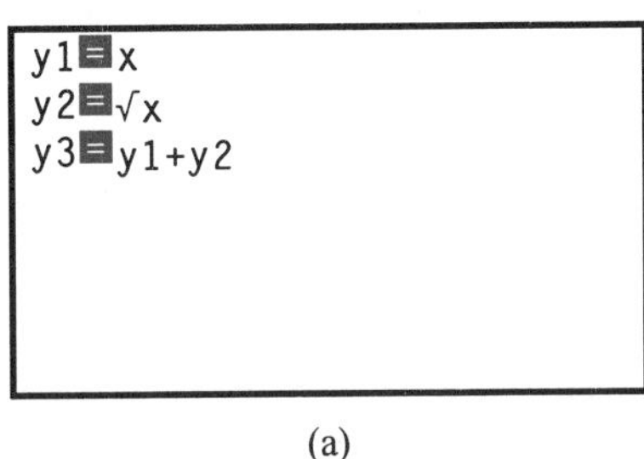

(a)

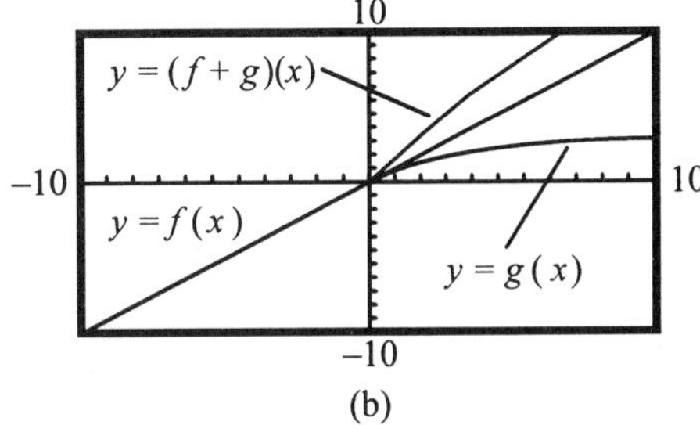

(b)

Figure 5.15

Recall from Section 1.1 that you may turn off and turn on the graph of an equation. You should continue the analysis in this example by making $y3$ the only selected function, and then graphing so that only the graph of $y = (f+g)(x)$ appears in the display.

PROGRESS CHECK 1

If $f(x) = \sqrt{x-1}$ and $g(x) = -x$, find $(f+g)(x)$, $(f-g)(x)$, $(f \cdot g)(x)$, and $(f/g)(x)$. In each case given the domain of the resulting function. ■

Example 2: Evaluating a Combined Function

If $f(x) = 5x$ and $g(x) = \dfrac{9x-4}{x}$, find $(f \cdot g)(2)$.

Solution: $(f \cdot g)(2) = f(2) \cdot g(2)$ so find $f(2)$ and $g(2)$.

$$\begin{aligned} f(x) &= 5x & g(x) &= \frac{9x-4}{x} \\ f(2) &= 5(2) & g(2) &= \frac{9(2)-4}{2} \\ &= 10 & &= 7 \end{aligned}$$

Then,

$$(f \cdot g)(2) = f(2) \cdot g(2) = 10(7) = 70.$$

Note: An alternative solution method is to find $(f \cdot g)(2)$ by first finding $(f \cdot g)(x)$. However, when this method is used, care must be taken to make sure that the function input belongs to the domain of the combined function. For instance, based on the functions defined in this example, $(f \cdot g)(0)$ is undefined; but this result may not be apparent if the simplified form of $f(x) \cdot g(x)$ is used.

PROGRESS CHECK 2

If f and g are defined as in Example 2, then find $(f-g)(10)$. ■

Example 3: Combining Functions Numerically

If $f = \{(0, 1), (1, 2), (2, 3), (3, 4)\}$ and $g = \{(2, 0), (3, 1), (4, -5), (5, -3)\}$, find $f+g$, $f-g$, $f \cdot g$ and f/g.

Solution: Both functions are defined at $x = 2$ and $x = 3$. Thus,

$$f+g = \{(2, 3+0), (3, 4+1)\} = \{(2, 3), (3, 5)\}$$
$$f-g = \{(2, 3-0), (3, 4-1)\} = \{(2, 3), (3, 3)\}$$
$$f \cdot g = \{(2, 3 \cdot 0), (3, 4 \cdot 1)\} = \{(2, 0), (3, 4)\}$$
$$\frac{f}{g} = \left\{\left(3, \frac{4}{1}\right)\right\} = \{(3, 4)\}; \text{ since } \frac{3}{0} \text{ is undefined, we do not list } \left(2, \frac{3}{0}\right).$$

PROGRESS CHECK 3

If $f = \{(0, 1), (1, 2), (2, 3), (3, 4)\}$, and $g = \{(-2, 2), (-1, 0), (1, 0), (2, -2)\}$, find $f+g$, $f-g$, $f \cdot g$ and f/g. ■

There is another important way of combining functions that has no counterpart in arithmetic. The operation is called **composition**. Basically, composition is a substitution that causes a "chain reaction" in which two functions are applied in succession.

Consider the problem of determining the area of a square whose perimeter is 20 in. You would probably reason as follows: If the perimeter is 20 in., the side is 5 in.; the area is then 25 in.^{2}. Given a specific value for the perimeter (P), we first determine the side (s) by using the function

$$s = \frac{P}{4}.$$

The result of this step is then substituted in the function

$$A = s^2$$

to determine the area. Let us put these formulas in functional notation and use the interpretation of a function as a rule. Consider Figure 5.16, in which $s = f(P) = P/4$ and $A = g(s) = s^2$. The output of the f rule feeds the input of the g rule. Thus, we first apply the f function to determine s and then apply the g function to determine A. The area is given by the function

$$A = g[f(P)].$$

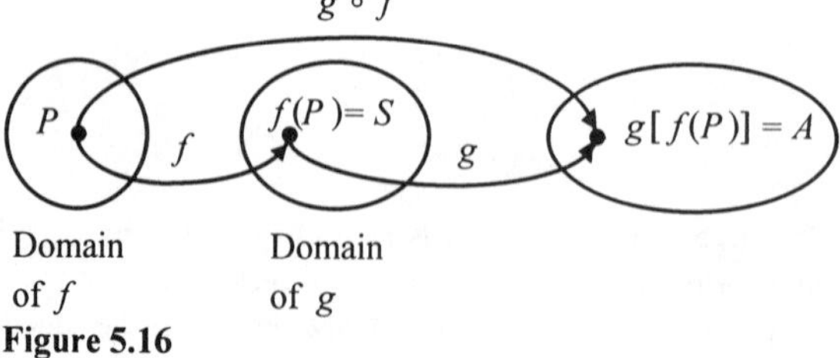

Figure 5.16

This function is said to be a composition of f and g. By substitution, we have

$$A = g[f(P)] = g\left(\frac{P}{4}\right) = \left(\frac{P}{4}\right)^2 = \frac{P^2}{16},$$

and this formula gives area as a function of perimeter. The symbol "$\circ$" is also used for composition. We could write

$$A = (g \circ f)(P).$$

In general, the composite functions of functions f and g are defined as follows:

$$(g \circ f)(x) = g[f(x)]$$
$$(f \circ g)(x) = f[g(x)].$$

Example 4: Finding Composite Functions Algebraically

If $f(x) = 3x - 5$ and $g(x) = x^2$, find the following.

a. $(f \circ g)(x)$ **b.** $(g \circ f)(x)$

Solution:

a. $(f \circ g)(x) = f[g(x)]$ By definition.
$= f(x^2)$ Replace $g(x)$ by x^2.
$= 3x^2 - 5$ Apply the f rule.

b. $(g \circ f)(x) = g[f(x)]$ By definition.
$= g(3x - 5)$ Replace $f(x)$ by $3x - 5$.
$= (3x - 5)^2$ Apply the g rule.
$= 9x^2 - 30x + 25$ Expand.

Note: In this example $(f \circ g)(x)$ does not equal $(g \circ f)(x)$, and this result holds except for special classes of functions. Thus, it is not true that $(f \circ g)(x)$ equals $(g \circ f)(x)$ for all functions f and g, so the order in which the functions are applied is usually important.

Technology Link

A graphing calculator may be used to show that $(f \circ g)(x) \neq (g \circ f)(x)$ in this example. Figure 5.17 shows expressions to enter for $y1$, $y2$, $y3$, and $y4$. Note that $y1 = f(x)$, $y2 = g(x)$, $y3 = (f \circ g)(x)$, and $y4 = (g \circ f)(x)$.

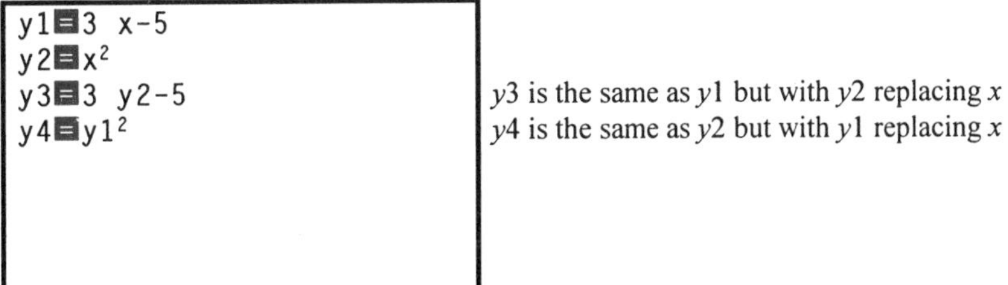

y1=3 x-5
y2=x²
y3=3 y2-5
y4=y1²
y3 is the same as y1 but with y2 replacing x
y4 is the same as y2 but with y1 replacing x

Figure 5.17

By making $y3$ the only selected function, the graph of $y3 = (f \circ g)(x)$ may be obtained, as shown in Figure 5.18(a). Similarly, selecting only $y4$ gives the graph of $y4 = (g \circ f)(x)$ shown in Figure 5.18(b). Because the graphs are different, it is apparent that the order in which the functions are applied is important in this example.

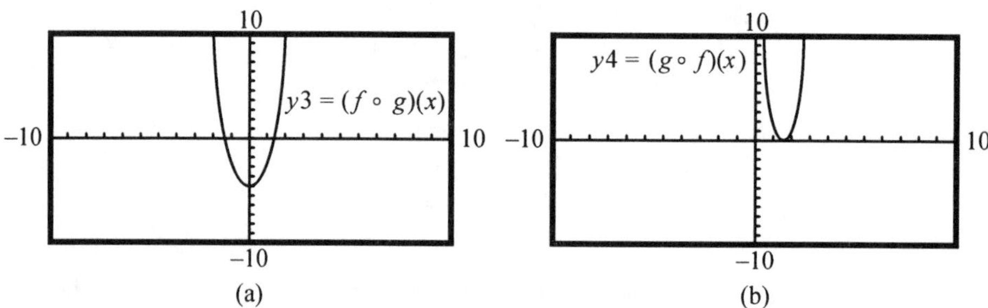

Figure 5.18

PROGRESS CHECK 4
If $f(x) = x^2 - 4$ and $g(x) = 2x + 5$, find the following.
a. $(f \circ g)(x)$ **b.** $(g \circ f)(x)$ ■

A number x must satisfy two requirements to belong to the domain of $g \circ f$.

1. Since we first apply the f function, x must be in the domain of f.
2. The output from step 1, $f(x)$, must be in the domain of g.

Thus, the domain of $g \circ f$ is the set of all real numbers x in the domain of f for which $f(x)$ is in the domain of g. This set consists of all the values in the domain of f for which $g \circ f$ is defined.

Example 5: Finding Domains of Composite Functions

If $f(x) = \sqrt{x}$ and $g(x) = x^2$, find $(g \circ f)(x)$ and $(f \circ g)(x)$. Indicate the domain of each function.

Solution: Find an expression for each composite function and then determine its domain.

$$(g \circ f)(x) = g[f(x)] = g(\sqrt{x}) = (\sqrt{x})^2 = x$$
Domain of f: $[0, \infty)$
$(g \circ f)(x) = x$ is defined for all real numbers.

Thus, the domain of $g \circ f$ is $[0, \infty)$.

$$(f \circ g)(x) = f[g(x)] = f(x^2) = \sqrt{x^2} \text{ or } |x|$$
Domain of g: $(-\infty, \infty)$
$(f \circ g)(x) = |x|$ is defined for all real numbers.

Thus, the domain of $f \circ g$ is the set of all the real numbers.

PROGRESS CHECK 5
If $f(x) = 1 - \sqrt{x}$ and $g(x) = x + 4$, find $(g \circ f)(x)$ and $(f \circ g)(x)$. Indicate the domain of each function. ■

Example 6: Finding Composite Functions Numerically

If $f = \{(0,1), (1,2), (2,3), (3,4)\}$ and $g = \{(2,0), (3,1), (4,-5), (5,-3)\}$, find $f \circ g$ and $g \circ f$. Indicate the domain of each composite function.

Solution: To determine $f \circ g$, we first apply the g function. As illustrated below, the range elements 0 and 1 are in the domain of f. Thus, $f \circ g = \{(2,1), (3,2)\}$ and the domain of $f \circ g$ is $\{2, 3\}$.

$$2 \xrightarrow{g} \quad 0 \xrightarrow{f} 1$$
$$3 \rightarrow \quad 1 \rightarrow 2$$
$$4 \rightarrow -5$$
$$5 \rightarrow -3$$

To find $g \circ f$, we first apply the *f* function. As illustrated below, the range elements 2, 3, and 4 are in the domain of *g*. Thus, $g \circ f = \{(1, 0), (2, 1), (3, -5)\}$ and the domain of $g \circ f = \{1, 2, 3\}$.

$$0 \xrightarrow{f} 1$$
$$1 \rightarrow 2 \xrightarrow{g} \quad 0$$
$$2 \rightarrow 3 \rightarrow \quad 1$$
$$3 \rightarrow 4 \rightarrow -5$$

Caution: To apply the functions in the correct order, note that the function labels are read from right to left. Because we normally read in the reverse direction, it is easy to make an error by overlooking this convention.

PROGRESS CHECK 6

If $f = \{(0, 1), (1, 2), (2, 3), (3, 4)\}$ and $g = \{(-2, 2), (-1, 0), (1, 0), (2, -2)\}$, find $f \circ g$ and $g \circ f$. Indicate the domain of each composite function. ■

Example 7: Graph Reading and Combining Functions

Use the graphs of *f* and *g* in Figure 5.19 to evaluate each expression.

a. $(f + g)(2)$ **b.** $(f \circ g)(3)$

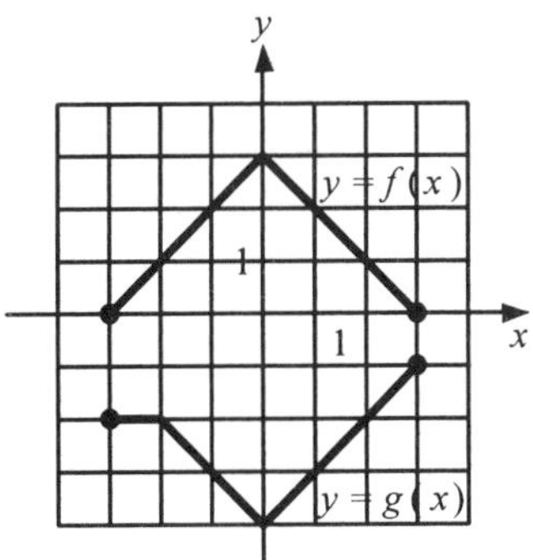

Figure 5.19

Solution:

a. To find $f(2)$ and $g(2)$ requires finding the *y* value in each function when $x = 2$. Since (2, 1) is in *f* and (2, -2) is in *g*, we know $f(2) = 1$ and $g(2) = -2$. Then

$$(f + g)(2) = f(2) + g(2) = 1 + (-2) = -1.$$

b. $(f \circ g)(3) = f[g(3)] = f(-1) = 2$, where $g(3) = -1$ and $f(-1) = 2$ are read from the graphs of *g* and *f*, respectively.

PROGRESS CHECK 7
Evaluate each expression using Figure 5.19.
a. $(f-g)(-2)$ **b.** $(g\circ f)(3)$ ■

Example 8: Spreading of an Oil Leak

Solve the problem in the section introduction on page 401.

Solution: By definition $(f\circ g)(t)=f[g(t)]$, and in this problem $g(t)=5t$. Thus,

$$(f\circ g)(t)=f[g(t)]=f(5t)=\pi(5t)^2=25\pi t^2$$

To interpret this result, if function *f* expresses *A* in terms of *r*, and function *g* expresses *r* in terms of *t*, then the function $f\circ g$ expresses *A* in terms of *t*. Thus, the area *A* of this circular spill *t* seconds after the start of the leak is given by

$$A=(f\circ g)(t)=25\pi t^2.$$

PROGRESS CHECK 8
An oil leak is spreading over a plane surface in the shape of a circle. The radius of this circle is increasing at a rate of 7 cm/second, so that radius of this spill *t* seconds after the start of the leak may be expressed by $r=g(t)=7t$. If function *f* expresses the circumference of this circular spill as a function of *r* so that $C=f(r)=2\pi r$, find and interpret $(f\circ g)(t)$. ■

In calculus it is often important to recognize a function as the composition of simpler functions. Although more than one answer may be possible in such problems, the next example shows a useful way for viewing an expression that involves a power.

Example 9: Interpreting a Function as a Composite

If $h(x)=(2x+3)^3$, find simpler functions *f* and *g* so that $h(x)=(f\circ g)(x)$.

Solution: Because $(f\circ g)(x)=f[g(x)]$, the *g* rule is applied first, and we will refer to *g* as the "inner" function. Try to let $g(x)$ equal an expression that is set off by a grouping symbol, particularly when such expressions may be written as the base in an exponential expression. From this perspective, we note that $h(x)$ equals the third power of $2x+3$; so, if we define the inner function *g* by

$$g(x)=2x+3$$

and let *f* be the cubing function defined by

$$f(x)=x^3,$$

then

$$f[g(x)]=f(2x+3)=(2x+3)^3=h(x).$$

Thus, $(f\circ g)(x)=h(x)$, as desired.
Note: It is also true that $(f\circ g)(x)=h(x)$, if $g(x)=2x$ and $f(x)=(x+3)^3$, and other choices are also possible. However, viewing $(2x+3)^3$ as the third power of $2x+3$ is the type of straightforward and useful analysis that is desired in these problems.

PROGRESS CHECK 9
If $h(x)=\sqrt{10-x}$, find simpler functions *f* and *g* so that $h(x)=(f\circ g)(x)$. ■

EXPLORE 5.2

1. Given $y1 = 2x - 1$ and $y2 = 3 - 2x$, use the calculator to graph each of the following functions. Are there any real numbers for which the functions in parts a–d are not defined? Is this evident from the graphs? For each graph find y when $x = 6$.
   a. $y3 = y1 + y2$ **b.** $y4 = y1 - y2$
   c. $y5 = y1 \cdot y2$ **d.** $y6 = \dfrac{y1}{y2}$
2. Given $y1 = \sqrt{5+x}$ and $y2 = \sqrt{5-x}$, use the calculator to graph each of the following functions. Are there any real numbers for which the functions in parts a–d are not defined? Is this evident from the graphs? For each graph find y when $x = 4$.
   a. $y3 = y1 + y2$ **b.** $y4 = y1 - y2$
   c. $y5 = y1 \cdot y2$ **d.** $y6 = \dfrac{y1}{y2}$
3. Let $y1 = 5x + 1$ and $y2 = x^2$. Graph each of the following and explain why the resulting graphs make sense. For each graph find y when $x = 6$.
   a. $y3 = y1^2$ **b.** $y4 = 5y2 + 1$

EXERCISES 5.2

In Exercises 1–12 find $(f+g)(x)$, $(f-g)(x)$, $(f \cdot g)(x)$, $\left(\dfrac{f}{g}\right)(x)$, $(f \circ g)(x)$ and $(g \circ f)(x)$. Indicate the domain of each function.

1. $f(x) = 2x$; $g(x) = x - 1$
2. $f(x) = -2x$; $g(x) = x + 2$
3. $f(x) = 4x^2 - 5$; $g(x) = -x + 3$
4. $f(x) = 1 + 2x - x^2$; $g(x) = -3x + 5$
5. $f(x) = x^2$; $g(x) = 1$
6. $f(x) = 5$; $g(x) = 10$
7. $f(x) = |x|$; $g(x) = x - 3$
8. $f(x) = x^{-1}$; $g(x) = x$
9. $f(x) = x^2 - 1$; $g(x) = \sqrt{x+1}$
10. $f(x) = \sqrt{x}$; $g(x) = -x^2$
11. $f = \{(0, 2), (1, 3), (2, 4), (3, 5)\}$
    $g = \{(2, -1), (3, 0), (4, 1), (5, 2)\}$
12. $f = \{(-3, 5), (0, 1), (2, 6), (4, 11)\}$
    $g = \{(0, 0), (2, 2), (5, 5), (-3, -3)\}$

Given $f(x) = x^2 - 4x + 5$ and $g(x) = 2x - 3$.

13. Find $(f \cdot g)(3)$. 14. Find $(f - g)(2)$

Given $f(x) = 2x^2 - 3x + 2$ and $g(x) = 5 - 2x$.

15. Find $(f + g)(-1)$. 16. Find $\left(\dfrac{f}{g}\right)(1)$.

Given $f(x) = 3x^2 - x - 1$ and $g(x) = 5x - 14$.

17. Find $\left(\dfrac{f}{g}\right)(2)$. 18. Find $(f + g)(4)$.

Given $f(x) = x^2 + 2$ and $g(x) = x^2 - 4x + 5$.

19. Find $(f + g)(3)$. 20. Find $(f - g)(-3)$.

21. Find $(f \cdot g)(3)$. 22. Find $\left(\dfrac{f}{g}\right)(-3)$.

23. If $f = \{(1, 3), (2, -2), (3, 5)\}$ and $g = \{(2, -1), (3, 4), (4, 0)\}$, find
    a. $(f + g)(2)$ **b.** $(f \cdot g)(3)$
    c. $(g \cdot f)(1)$ **d.** $(g \circ f)(1)$
24. If $f = \{(1, 5), (-2, 4), (-3, 0)\}$ and $g = \{(0, 2), (-3, 1), (-5, -1)\}$, find

a. $\left(\frac{f}{g}\right)(-3)$ b. $\left(\frac{g}{f}\right)(-3)$

c. $(f \circ g)(-3)$ d. $(g \circ f)(-3)$

In Exercises 25–32 use the graphs of *f* and *g* given below to evaluate each expression.

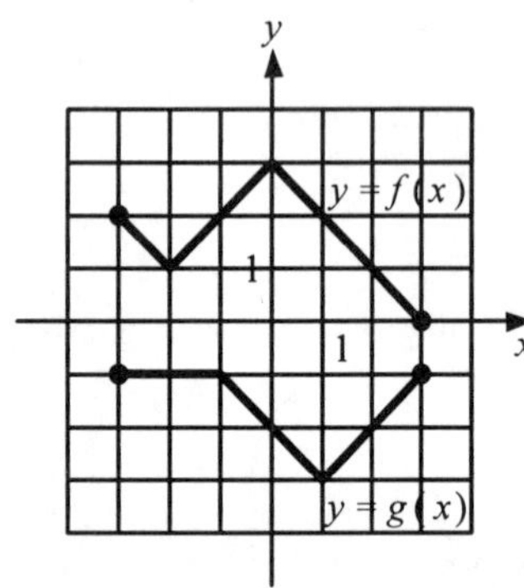

25. $(f+g)(1)$ 26. $(f-g)(2)$

27. $(f \cdot g)(-2)$ 28. $\left(\frac{f}{g}\right)(0)$

29. $(f \circ g)(1)$ 30. $(g \circ f)(1)$

31. $(g \circ f)(0)$ 32. $(f \circ g)(2)$

33. If $f(x)=9x-5$ and $g(x)=\frac{1}{9}x+\frac{5}{9}$, find $(f \circ g)(x)$ and $(g \circ f)(x)$.

34. If $f(x)=x^3-5$ and $g(x)=\sqrt[3]{x+5}$, find $(f \circ g)(x)$ and $(g \circ f)(x)$.

35. If $g(u)=u^3$ and $u=f(x)=5x-4$, find $(g \circ f)(x)$.

36. If $g(u)=\sqrt{u}$ and $u=f(x)=2x^2+9$, find $(g \circ f)(x)$.

37. If $f(r)=\pi r^2$ and $r=g(t)=6t$, find $(f \circ g)(t)$.

38. If $f(r)=\frac{4}{3}\pi r^3$ and $r=g(t)=\frac{5}{2}t$, find $(f \circ g)(t)$.

In Exercises 39–48 find simpler functions *f* and *g* so that $h(x)=(f \circ g)(x)$. (See Example 9.)

39. $h(x)=(4x-1)^3$ 40. $h(x)=\left(x^2+1\right)^2$

41. $h(x)=\sqrt{\frac{x-1}{x+1}}$ 42. $h(x)=(5x-10)^{-4}$

43. $h(x)=\sqrt[3]{2x+1}$ 44. $h(x)=\sqrt{1-x^2}$

45. $h(x)=2(3-x)^4$ 46. $h(x)=\left|x^3-8\right|+5$

47. $h(x)=(1-x)^3+6(1-x)^2$

48. $h(x)=\frac{1}{3x-5}$

49. If $h(x)=\sqrt{2x^2+2}$ and $f(x)=\sqrt{x}$, find $g(x)$ so that $h(x)=(f \circ g)(x)$.

50. If $h(x)=3x^2-1$ and $g(x)=x^2$, find $f(x)$ so that $h(x)=(f \circ g)(x)$.

51. If $f(x)=2x-3$ and $g(x)=5x+b$, find b so that $(f \circ g)(x)=(g \circ f)(x)$.

52. If $f(x)=ax+b$ and $g(x)=5x-2$, find conditions on a and b so that $(f \circ g)(x)=(g \circ f)(x)$.

In Exercises 53–56 find $(f \circ g \circ h)(x)$

53. $f(x)=2x$; $g(x)=x^2$; $h(x)=x-2$

54. $f(x)=x-2$; $g(x)=2x$; $h(x)=x^2$

55. $f(x)=\sqrt{x}$; $g(x)=\frac{1}{x}$; $h(x)=1-x$

56. $f(x)=\frac{1}{x}$; $g(x)=1-x$; $h(x)=\sqrt{x}$

57. Show that the sum of two odd functions is an odd function.

58. Show that the product of two odd functions is an even function.

59. Show that the product of an even function and an odd function is an odd function.

60. If *f* and *g* are odd functions, is $f \circ g$ an odd function or an even function? Verify your answer.

61.
   a. There are 1760 yd. in 1 mi. Write a function that converts miles to yards.
   b. Write a function that converts yards to feet.
   c. Use parts **a** and **b** and composition to construct a function that converts miles to feet.

62.
   a. Express the diagonal of a square as a function of the side.
   b. Express the side of a square as a function of the perimeter.
   c. Use parts **a** and **b** and composition to construct a function that expresses the diagonal of a square as a function of the perimeter.

63. Helium is pumped into a spherical balloon. The radius of this sphere is increasing at a constant speed of 0.5 cm/second so that the radius of this balloon t seconds after the start of inflation may be expressed by $r = g(t) = \frac{1}{2}t$. If function f expresses the volume of this balloon as a function of r so that $V = f(r) = \frac{4}{3}\pi r^3$, find and interpret $(f \circ g)(t)$.

64. An oil leak is spreading over a plane surface in the shape of a circle. If the radius of this circle is increasing at a constant rate of 8 cm/second, express the area A of this circular spill as a function of time t, where t represents the time in seconds from the beginning of the leak.

65. The cost of producing a full shipment of bracelets is given by $C = f(h) = 340h + 200$, where h is the current hourly pay rate at the factory. For the next five years the hourly pay rate will go up 5 percent per year. Thus $g(t) = h(1.05)^t$ gives the hourly pay rate after t years.

a. Find and interpret $(f \circ g)(t)$.

b. Find and interpret $(f \circ g)(4)$.

c. Find and interpret $(f \circ g)(4)$ if $h = 10$.

66. The cost of producing a full shipment of watches is given by $C = f(h) = 500h + 300$, where h is the current hourly pay rate at the factory. For the next five years the hourly pay rate will go up 3 percent per year. Thus $g(t) = h(1.03)^t$ gives the hourly pay rate after t years.

a. Find and interpret $(f \circ g)(t)$.

b. Find and interpret $(f \circ g)(5)$.

c. Find and interpret $(f \circ g)(5)$ if $h = 12$.

THINK ABOUT IT 5.2

1. If $f(x) = \sqrt{x-7}$ and $g(x) = \sqrt{2-x}$, consider the domains of the functions and explain why it does not make sense to form $(f+g)(x)$.

2. If $f(x) = \sqrt{x-7}$ and $g(x) = 2 - x^2$, explain why it does not make sense to form $(f \circ g)(x)$.

3. **a.** Give two examples of functions for which $(f \circ g)(x)$ does *not* equal $(g \circ f)(x)$.

b. Give two examples of functions for which $(f \circ g)(x)$ does equal $(g \circ f)(x)$.

4. If $f(x) = 4x - 9$ and $(f \circ g)(x) = x$ for all values of x, find $g(x)$.

5. Use the graphs below and graph $y = (f+g)(x)$.

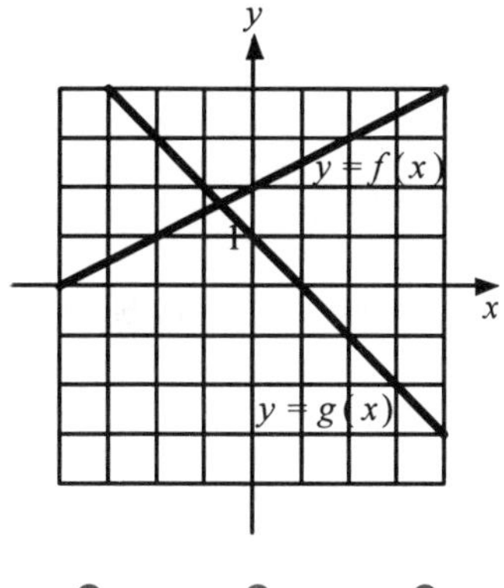

● ● ●

5.3 Slope and Rate of Change

Photo Courtesy of Camerique, Inc./The Picture Cube

A company has determined that the cost in dollars of producing x units per week of a certain product is approximated by

$$C = f(x) = 3{,}000 + 50x - 0.1x^2$$

a. Find in simplest form an expression for the average rate of change of this cost function as the number of units produced varies from x to $x+h$ with $h \neq 0$.
b. Evaluate this expression for $x = 74$, $h = 1$, and interpret the result.

(See Example 7.)

Objectives

1. Find and interpret the slope of a line.
2. Find the average rate of change of a function for a given interval.

An understanding of the process of change is vital to the investigation of the laws of nature, for everything is constantly changing in such significant respects as size, position, and temperature. Consequently, an important topic in mathematics is finding the rate at which one variable changes with respect to another. For example:

A couple is interested in the interest rate on their bank deposit so that their investment will yield a maximum return. An interest rate describes how the value of an investment changes with respect to time.

An economist is interested in knowing the rate at which the demand for a product will change with respect to the price charged for the product.

A motorist is interested in knowing her speed—the rate of change of the distance she is traveling—with respect to time.

A manager needs to know the rate at which the total cost of a production is changing with respect to each additional unit produced.

Social scientists are groping for a way to analyze the accelerating rate of social processes, which crowds an increasing number of events into an arbitrary interval of time. For example, in *Future Shock* (Random House, 1970, p. 2) Alvin Toffler argues that "unless man quickly learns to control the rate of change in his personal affairs as well as in society at large, we are doomed to a massive adaptational breakdown." This disease of change is labeled "future shock."

A health official helps prevent an epidemic by analyzing the rate at which an infectious disease (such as flu) might spread in a certain community.

These examples illustrate the necessity of analyzing change. The most important tool in this analysis is calculus. In fact, it is this mathematical ability of calculus to analyze movement and change that makes it the principal link between practical science and mathematical thought. To begin, we will avoid more complicated functions and study only functions that always change at the same rate, so that their graphs are straight lines. To find the rate of change in such cases, pick *any* two points on the given line, calculate the change in the dependent variable, and divide it by the change in the independent variable. This ratio is called the *slope of the line*.

Definition of Slope

If (x_1, y_1) and (x_2, y_2) are any two distinct points on a nonvertical line, as shown in Figure 5.20, then

$$\text{slope} = m = \frac{\Delta y}{\Delta x} = \frac{y_2 - y_1}{x_2 - x_1}.$$

(**Note:** The symbol Δ is the Greek capital letter delta. The symbol is used to indicate a change.)

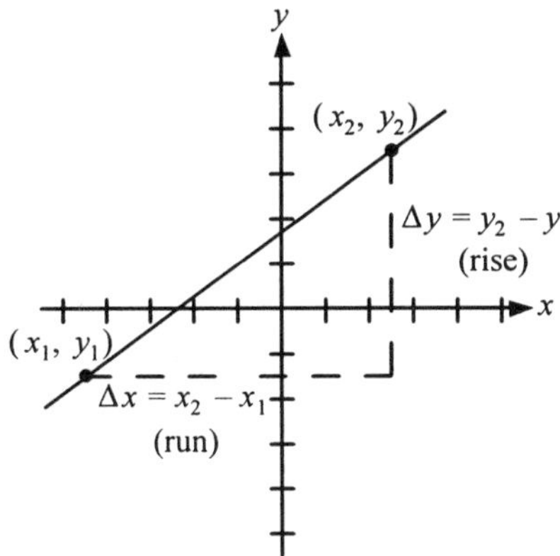

Figure 5.20

The slope of a line is a fundamental notion in analytic geometry as a measure of the inclination or steepness of a line. In this context we often refer to Δy as the "rise" and Δx as the "run," so

$$\text{slope} = \frac{\text{rise}}{\text{run}}.$$

With this interpretation, it is natural to use the idea of slope to analyze such concrete situations as the steepness of a ramp, the pitch of a roof, or the inclination of a ski slope.

Example 1: Using the Definition of Slope

Find the slope of the line through the given points.

a. (2, 1) and (5, 4) **b.** (−1, −1) and (3, −6)

Solution:

a. If we label (2, 1) as point 1, then $x_1 = 2$, $y_1 = 1$, $x_2 = 5$ and $y_2 = 4$. As illustrated in Figure 5.21, the slope formula then gives

$$m = \frac{\Delta y}{\Delta x} = \frac{y_2 - y_1}{x_2 - x_1} = \frac{4-1}{5-2} = \frac{3}{3} = 1.$$

A slope of 1 means that as x increases 1 unit, y increases 1 unit. Notice that the slope is unaffected by the way in which we label the points. If we label (5, 4) as point 1, then $x_1 = 5$, $y_1 = 4$, $x_2 = 2$, and $y_2 = 1$, so

$$m = \frac{\Delta y}{\Delta x} = \frac{y_2 - y_1}{x_2 - x_1} = \frac{1-4}{2-5} = \frac{-3}{-3} = 1.$$

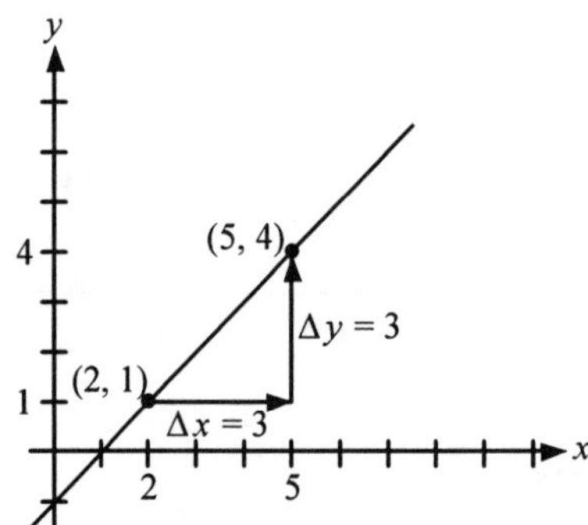

Figure 5.21

b. We let $x_1 = -1$, $y_1 = -1$, $x_2 = 3$, and $y_2 = -6$ and substitute in the slope formula.

$$m = \frac{\Delta y}{\Delta x} = \frac{y_2 - y_1}{x_2 - x_1} = \frac{-6-(-1)}{3-(-1)} = \frac{-5}{4}$$

A slope of $-5/4$ means that as x increases 4 units, y decreases 5 units, as shown in Figure 5.22. Alternately, observe that a slope of $-5/4$ means that as x increases 1 unit, y decreases 1.25 units. In general, note that a slope of m means that as x increases 1 unit, y changes m units.

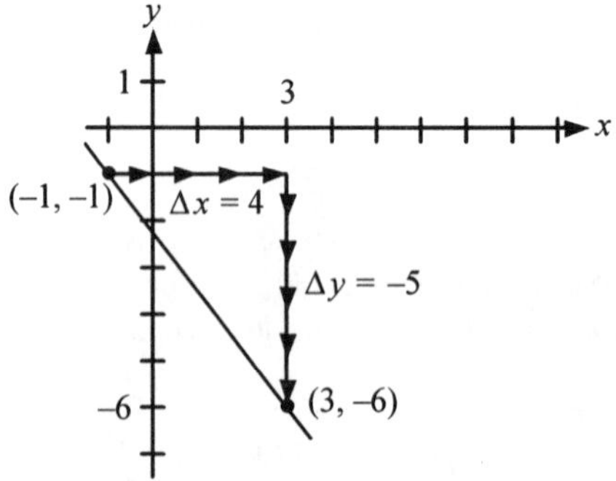

Figure 5.22

PROGRESS CHECK 1

Find the slope of the line through the given points.

a. (2, 1) and (3, 4) **b.** (−2, 1) and (3, −1) ■

The special cases discussed in the next example occur when the slope formula is applied to lines that are horizontal or vertical.

Example 2: Finding the Slope of Horizontal or Vertical Lines

Find the slope of the line containing the following pairs of points.

a. (–1, 4) and (5, 4) **b.** (3, 2) and (3, – 4)

Solution:

a. The line through the given points is horizontal, as shown in Figure 5.23(a). If we let $x_1 = -1$, $y_1 = 4$, $x_2 = 5$, and $y_2 = 4$, then

$$m = \frac{\Delta y}{\Delta x} = \frac{y_2 - y_1}{x_2 - x_1} = \frac{4-4}{5-(-1)} = \frac{0}{6} = 0.$$

The slope of every horizontal line is zero since the numerator of the slope ratio (Δy) is always zero.

b. Figure 5.23(b) shows that the line through the given points is vertical. The slope formula is not meaningful in this case, since letting $x_1 = 3$, $y_1 = 2$, $x_2 = 3$, and $y_2 = -4$ gives

$$m = \frac{\Delta y}{\Delta x} = \frac{y_2 - y_1}{x_2 - x_1} = \frac{-4-2}{3-3} = \frac{-6}{0}$$

The slope of every vertical line is undefined, since the denominator of the slope ratio (Δx) is always zero.

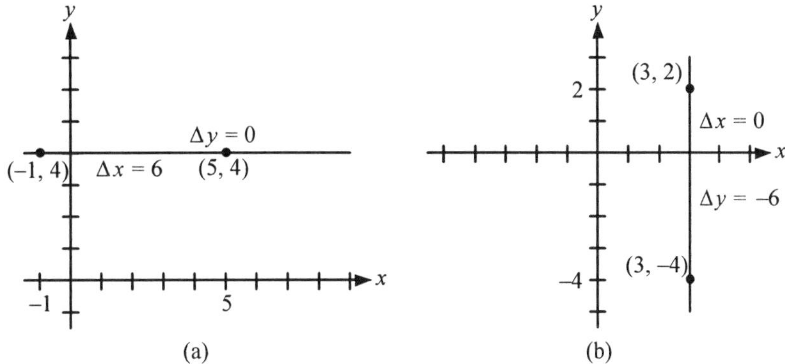

Figure 5.23

PROGRESS CHECK 2

Find the slope of the line through the given points.

a. (–2, – 4) and (–2, – 3) **b.** (0, 1) and (3, 1) ■

Examples 1 and 2 have considered lines with slopes that are positive, negative, zero, or undefined. Important features that are associated with these cases are summarized in Figure 5.24, to help you see some general principles concerning the slope of a line.

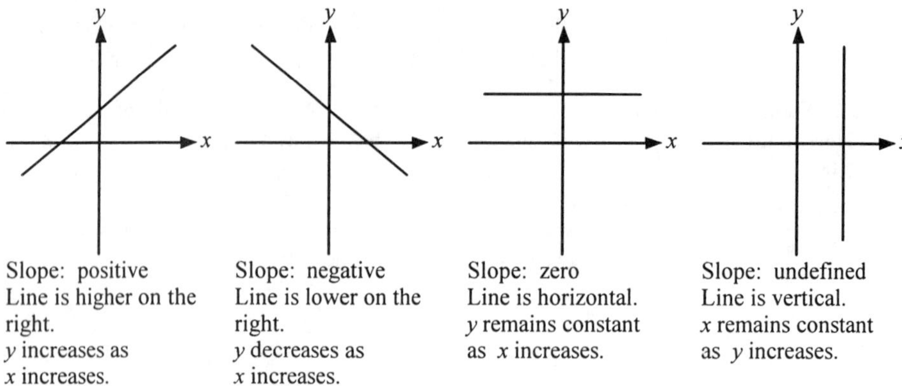

y
x
Slope: positive
Line is higher on the right.
y increases as x increases.
Slope: negative
Line is lower on the right.
y decreases as x increases.
Slope: zero
Line is horizontal.
y remains constant as x increases.
Slope: undefined
Line is vertical.
x remains constant as y increases.

Figure 5.24

The next example considers how to interpret the slope concept in an applied problem.

Example 3: Interpreting Slope in an Applied Problem

A ball thrown vertically up from a roof with an initial velocity of 128 ft/second will continue to climb for 4 seconds, and its velocity is 96 ft/second when $t = 1$ second, and 32 ft/second when $t = 3$ seconds. Find and interpret the slope of the line that describes this linear relationship.

Solution: Figure 5.25 shows the graph of the relation where t replaces x as the independent variable, and v replaces y as the dependent variable. We let $t_1 = 1$, $v_1 = 96$, $t_2 = 3$, and $v_2 = 32$ and substitute in the slope formula.

$$m = \frac{\Delta v}{\Delta t} = \frac{v_2 - v_1}{t_2 - t_1} = \frac{32 - 96}{3 - 1} = \frac{-64}{2} = -32$$

A slope of –32 means that for every additional second (up to 4 seconds) the velocity decreases by 32 ft/second.

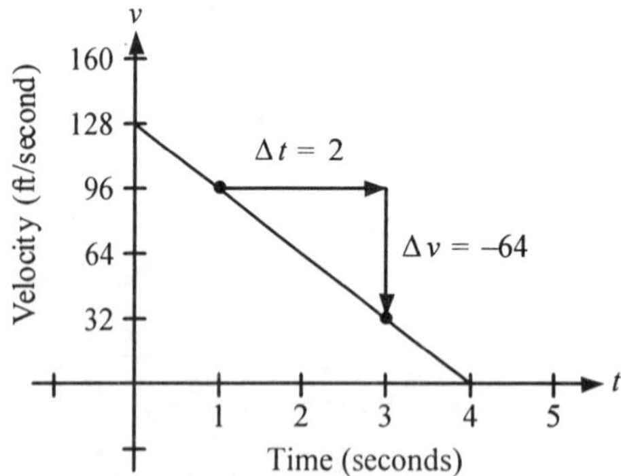

Figure 5.25

In general, the ratio of the vertical unit to the horizontal unit reveals the meaning of slope in any given application. For instance, in this example we have (ft/second)/second so velocity decreases at a rate of 32 ft per second every second.

PROGRESS CHECK 3
A printing shop charges \$20 for 500 copies of a flyer and \$25 for 750 copies. If the relation between the number of copies (x) and the cost (y) graphs as a line, calculate and interpret the slope.
■

In finding the slope of a function, we compute a ratio that can be interpreted as an *average value*. What makes linear relationships relatively easy to analyze is that they always change at the same rate. Thus, we obtain the same average no matter which two points we pick on the line. A function need not always change at the same rate for us to apply the concept of an average rate of change, but we must be careful in our interpretation of this measure. For example, if you drive 165 mi in 3 hours, your average speed is 165/3, or 55 mi *per* hour. This measure does not require or even mean that you go 55 mi *each* hour. It means that over the 3-hour period you average that speed; for another time interval the average speed would probably be different. Since the average varies with the interval chosen, it is important to remember that the average rate of change of a nonlinear relationship has meaning only if we indicate the interval in which we compute the average.

Example 4: Finding an Average Rate of Change

The graph in Figure 5.26 shows the distance of a car from its starting point as a function of elapsed time.

a. The points (1, 30) and (4, 180) belong to the graph. What do these points reveal about the relation?

b. Compute the slope of the line that contains the points (1, 30) and (4, 180) and interpret the result.

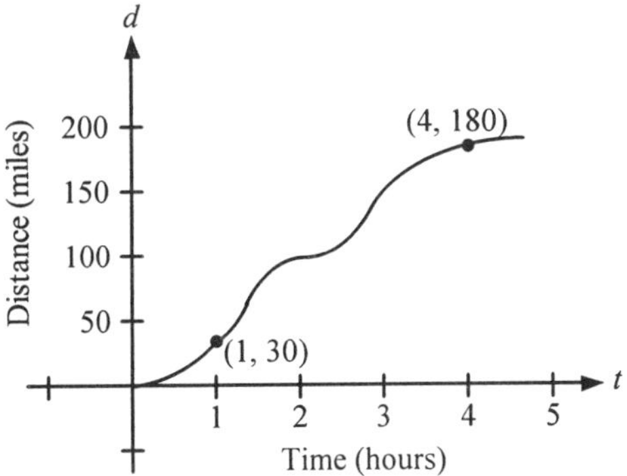

Figure 5.26

Solution:

a. (1, 30) reveals that the car is 30 miles from its starting place after 1 hour, and (4, 180) reveals that the car is 180 miles from its starting place after 4 hours.

b. The slope formula yields

$$m = \frac{\Delta d}{\Delta t} = \frac{180-30}{4-1} = \frac{150}{3} = 50$$

Since the vertical units are *miles* and the horizontal units are *hours*, this slope means that over the interval from 1 to 4 hours, the car's average speed is 50 miles/hour.

PROGRESS CHECK 4

According to United Nations estimates, the world population was 4.5 billion in 1980 and 5.3 billion in 1990. Find and interpret the slope of the line that passes through these two data points. ■

Abstractly, one can describe the average rate of change of a function over the interval $[a, b]$ as the slope of the line passing through the points $(a, f(a))$ and $(b, f(b))$. Much use is made of this concept in calculus.

Example 5: Finding an Average Rate of Change

If $f(x) = 3x^2 + 1$, find the average rate of change of the function as x varies

a. from $a = 0$ to $b = 2$ **b.** from $a = 1$ to $b = 4$

Solution:

a. Since $f(x) = 3x^2 + 1$, we have $f(0) = 3(0)^2 + 1 = 1$ and $f(2) = 3(2)^2 + 1 = 13$. Thus,

$$m = \frac{\Delta y}{\Delta x} = \frac{f(b) - f(a)}{b - a} = \frac{13 - 1}{2 - 0} = 6.$$

The average rate of change of the function $f(x) = 3x^2 + 1$ as x varies from 0 to 2 is 6. The function is graphed in Figure 5.27. Note that from the calculations above, (0, 1) and (2, 13) are points of the graph, and $(\Delta y/\Delta x)$ is the slope of the line joining them.

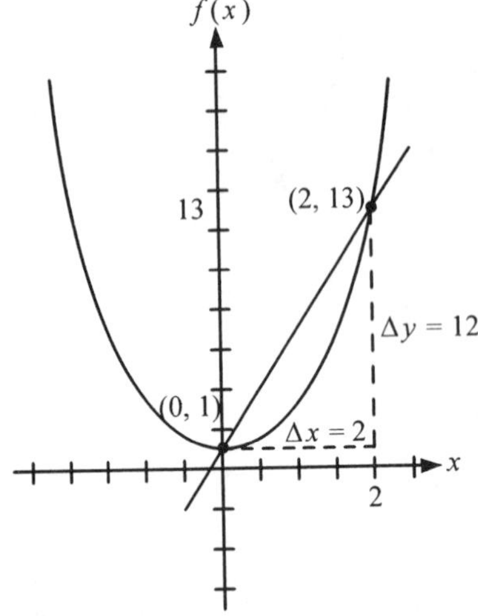

Figure 5.27

b. Since $f(x) = 3x^2 + 1$, we have $f(1) = 3(1)^2 + 1 = 4$ and $f(4) = 3(4)^2 + 1 = 49$. Thus

$$m = \frac{\Delta y}{\Delta x} = \frac{f(b) - f(a)}{b - a} = \frac{49 - 4}{4 - 1} = 15.$$

In the interval from 1 to 4, the average rate of change of the function is 15.

PROGRESS CHECK 5

If $f(x) = 3 + \sqrt{x}$, find the average rate of change of the function as x varies from $a = 0$ to $b = 4$.

■

In many applications it is most useful to determine how a function is changing at a particular point, say x_1. For this purpose, we define the average rate of change in terms of the constant x_1 and the variable Δx. Since $\Delta x = x_2 - x_1$, we have $x_2 = x_1 + \Delta x$, and we may write

$$m = \frac{\Delta y}{\Delta x} = \frac{f(x_2) - f(x_1)}{x_2 - x_1} = \frac{f(x_1 + \Delta x) - f(x_1)}{\Delta x}.$$

This concept is illustrated in Figure 5.28. To simplify algebraic computation with this formula, Δx is usually represented by *h*, and the subscript x_1, is dropped. We therefore have the following important definition.

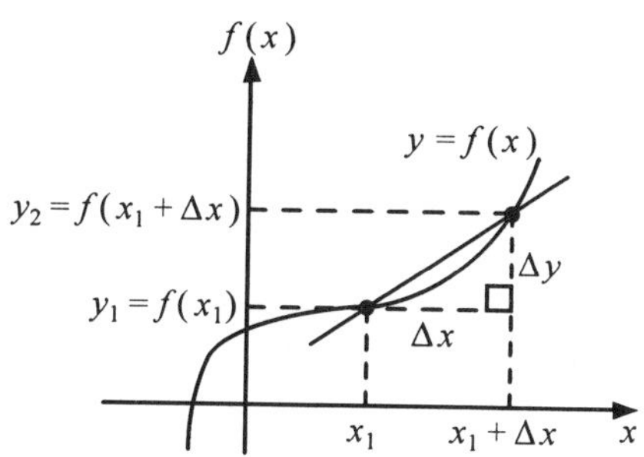

Figure 5.28

Average Rate of Change from x to $x + h$

> The average rate of change of a function *f* as *x* varies from *x* to $x + h$ is
>
> $$\frac{\Delta y}{\Delta x} = \frac{f(x+h) - f(x)}{h}, \text{ with } h \neq 0.$$

Computing this ratio is an essential consideration (and a major stumbling block for students) in calculus. You can probably see on an intuitive level that if we make $|h|$ very small, the above ratio becomes a good approximation for the rate of change of the function at *x*.

Example 6: Using a Formula for Average Rate of Change

If $f(x) = 3x^2 + 1$,

a. Find the formula for the average rate of change of this function from *x* to $x + h$.

b. Evaluate this expression for $x = 1$, $h = 3$, and interpret the result.

Solution:

a. If $f(x) = 3x^2 + 1$, we have

$$f(x+h) = 3(x+h)^2 + 1 = 3x^2 + 6xh + 3h^2 + 1.$$

Then

$$\begin{aligned}\frac{f(x+h)-f(x)}{h} &= \frac{(3x^2+6xh+3h^2+1)-(3x^2+1)}{h}\\ &= \frac{3x^2+6xh+3h^2+1-3x^2-1}{h}\\ &= \frac{6xh+3h^2}{h}\\ &= 6x+3h.\end{aligned}$$

b. Evaluating this expression for $x=1$ and $h=3$, we have

$$\frac{\Delta y}{\Delta x} = 6(1)+3(3) = 15.$$

In the interval from $x=1$ to $x+h=4$, the average rate of change of the function is 15. (See Example 5b.)

PROGRESS CHECK 6

If $f(x)=8-x^2$, then

a. find the formula for the average rate of change of this function from x to $x+h$ and

b. evaluate this expression for $x=4$, $h=1$, and interpret the result. ■

Example 7: Using a Formula for Average Rate of Change

Solve the problem in the section introduction on page 412.

Solution:

a. If $C=f(x)=3{,}000+50x-0.1x^2$,

$$\begin{aligned}f(x+h) &= 3{,}000+50(x+h)-0.1(x+h)^2\\ &= 3{,}000+50x+50h-0.1x^2-0.2xh-0.1h^2.\end{aligned}$$

Then

$$\begin{aligned}f(x+h)-f(x) &= (3{,}000+50x+50h-0.1x^2-0.2xh-0.1h^2)-(3{,}000+50x-0.1x^2)\\ &= 50h-0.2xh-0.1h^2.\end{aligned}$$

Finally,

$$\begin{aligned}\frac{f(x+h)-f(x)}{h} &= \frac{50h-0.2xh-0.1h^2}{h}\\ &= 50-0.2x-0.1h.\end{aligned}$$

b. Evaluating this expression for $x=74$ and $h=1$ gives

$$\frac{\Delta C}{\Delta x} = 50-0.2(74)-0.1(1) = 35.1.$$

In the interval from $x=74$ to $x+h=75$, the average rate of change of the cost function is 35.1. In other words, when 74 units have been produced, then the cost of the next unit is about \$35.10.

PROGRESS CHECK 7

Answer the questions in Example 7, assuming that the cost function is approximated by $C = f(x) = 2{,}500 + 70x - 0.2x^2$. ■

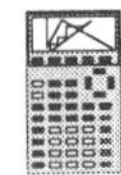

EXPLORE 5.3

1. a. Graph $y = 0.5x + 3$, then find the coordinates of any two points on the line. Calculate the slope of the line using these two points.
   b. Repeat part **a** for the line $y = 0.5x + 7$, and for the line $y = 0.5x - 2$.
   c. Describe the relationship among the three graphs from parts **a** and **b**.
2. Because they don't change direction very fast, many curved graphs look straight when you zoom in enough. That is, each tiny piece of the curve is almost straight. Such a curve is said to exhibit **local linearity**. For any tiny "almost straight" piece, you can approximate a slope by picking two points on it and calculating the change in y divided by the change in x. Of course, both changes will be very small.
   a. Graph $y = x^2 - 3$, zoom in several times near the point (3, 6), use Trace to find the coordinates of two points on this "line," then calculate the slope.
   b. Repeat part **a** for the graph of $y = \sqrt{45 - x^2}$.
   c. The graph of $y = |x - 3| + 6$ has a sharp corner at $(3, 6)$. Repeatedly zooming in will never result in a graph of one straight line through that point. Thus, this curve does not exhibit local linearity at the point $(3, 6)$. Find the slopes of the two different lines that appear when you zoom in on $(3, 6)$.
3. a. Graphing calculators often have a special Draw feature for drawing a vertical line or a horizontal line. If your calculator has such a feature, draw both a vertical line and a horizontal line that pass through the point (3, 6).
   b. If your calculator has a feature that allows you to turn individual points on and off, turn off the point of intersection of the two lines from part **a**.
4. Some calculators have a Draw feature that allows you to draw a line between two points on a graph. If you have such a feature, draw the curve $y = x^2$ then connect the points (−1, 1) and (3, 9) with a straight line. You may need to adjust the viewing window so that the coordinates of the traced points include these points.

EXERCISES 5.3

In Exercises 1–10 find the slope of the line determined by each pair of points.

1. $(1, 2)$ and $(3, 4)$ **2.** $(3, -1)$ and $(5, -7)$

3. $(5, 1)$ and $(-2, -3)$ **4.** $(-1, 2)$ and $(0, 0)$

5. $(4, -4)$ and $(2, 7)$ **6.** $(-5, -1)$ and $(-2, -4)$

7. $(-1, -3)$ and $(2, -3)$ **8.** $(-2, 0)$ and $(5, 0)$

9. $(1, 3)$ and $(1, -1)$ **10.** $(-3, 2)$ and $(-3, -2)$

In Exercises 11–14 find two ordered pairs that satisfy the equation and use them to find the slope of the line.

11. $y = -x + 1$ **12.** $y = 4x - 5$

13. $y = \frac{1}{2}x + 3$ **14.** $y = -\frac{5}{4}x - 2$

15. Approximate the slope of each line in the figure below.

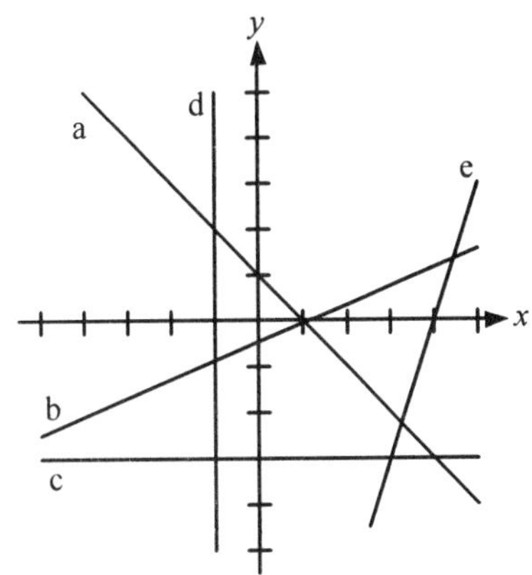

16. Approximate the slope of each line in the figure below.

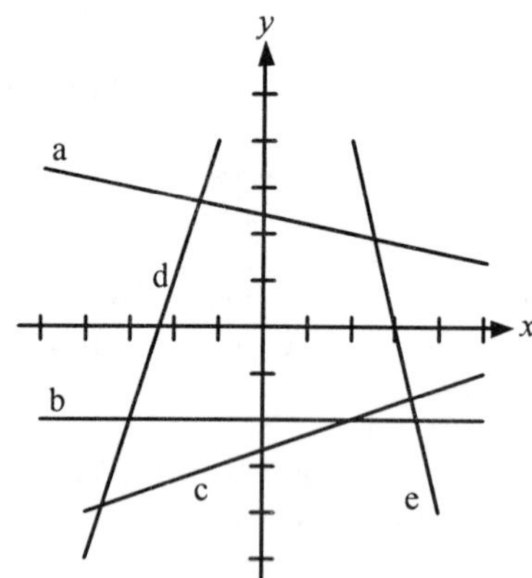

17. On a piece of graph paper, draw lines through the point $(1, 2)$ with the following slopes:

a. 2 **b.** -3 **c.** $-\frac{1}{4}$

d. $\frac{2}{3}$ **e.** 0

18. On a piece of graph paper, draw lines through the point $(-2, -1)$ with the following slopes:

a. 1 **b.** -4 **c.** $-\frac{1}{3}$

d. $\frac{1}{3}$ **e.** $-\frac{3}{2}$

In Exercises 19–26 find the rate of change of each linear relationship by calculating the slope. In each case interpret the meaning of the slope.

19. An object that weighs 24 lb on Earth weighs 4 lb on the moon; a different object that weights 42 lb on Earth weighs 7 lb on the moon.

20. The freezing point is 32° Fahrenheit compared to 0° Celsius; the boiling point is 212° Fahrenheit compared to 100° Celsius.

21. A ball thrown vertically up from the ground with an initial velocity of 320 ft/second will continue to climb for 10 seconds; its velocity is 256 ft/second when $t = 2$ seconds and 64 ft/second when $t = 8$ seconds.

22. A projectile fired vertically up from the ground with an initial velocity of 224 ft/second will continue to climb for 7 seconds; its velocity is 96 ft/second when $t = 4$ seconds.

23. The cost of 50 brochures is \$11; the cost of 100 brochures is \$17.

24. The total cost of manufacturing 100 units of a certain product is \$450; the cost of manufacturing 200 units is \$600.

25. A moving company charges \$50 to move a certain machine 5 mi and \$71 to move the same machine 8 mi.

26. The monthly cost of renting a computer is \$400 if it is used for 5 hours, and \$500 if it is used for 7 hours.

In Exercises 27–30 find the average rate of change of each function in the given interval.

27. $f(x) = x^2$ from $a = 1$ to $b = 3$

28. $y = \sqrt{x}$ from $a = 4$ to $b = 9$

29. $g(x) = 1 - x - x^2$ from $a = -2$ to $b = -1$

30. $f(x) = x^3 + 1$ from $a = 0$ to $b = 2$

In Exercises 31–40 compute $\frac{f(x+h) - f(x)}{h}$ with $h \neq 0$ to determine the average rate of change of the function from x to $x + h$.

31. $f(x) = x^2$ **32.** $f(x) = 2x^2 + 3$

33. $f(x) = 4 - x^2$ **34.** $f(x) = x^2 + 3x + 5$

35. $f(x) = 2x$ **36.** $f(x) = 4x - 1$

37. $f(x) = 1$

38. $f(x) = \sqrt{x}$ (**Note:** Rationalize the numerator.)

39. $f(x) = \frac{1}{x}$ **40.** $f(x) = x^3$

41. Determine the average rate of change of the area of a circle with respect to its radius, as r varies from r to $r + h$.

42. Two cyclists bike for 7 hours to reach a friend's house. The figure shows the distance of the cyclists from their starting point over this time.

a. What distance did they cover during the first 4 hours? Find their average speed over that interval.
b. What happened during the fifth hour?
c. What was their average speed for the whole trip?

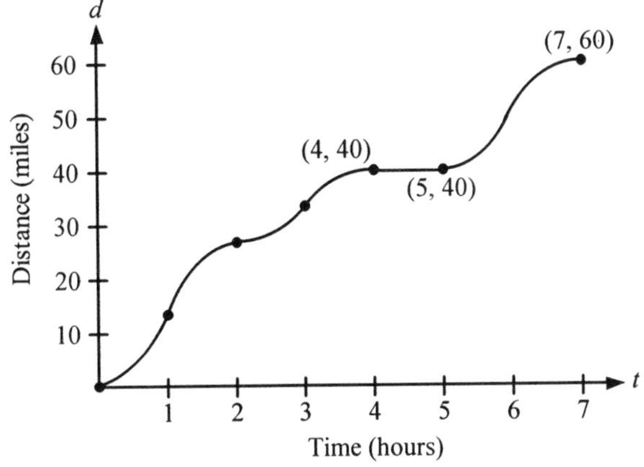

43. According to the Official U.S. Census, the population of California was 23.67 million in 1980, and 29.76 million in 1990. Find and interpret the slope of the line that passes through these two data points.

44. Some liquid is draining into a cylindrical container, but not at a constant rate. The table shows the height of the liquid in the container at various times.

x, elapsed time (minutes)	0	10	20	30	40	50	60
y, height (cm)	0	60	90	105	115	120	124

a. Graph these points.
b. Find and interpret the slope of the line connecting (0, 0) and (10, 60).
c. Find and interpret the slope of the line connecting (50, 120) and (60, 124).
d. Describe the manner in which the liquid is entering the container.

THINK ABOUT IT 5.3

1. Give an example of two points that determine a line with the indicated slope.
   a. $m = 5/2$ **b.** $m = -1/3$ **c.** $m = 0$ **d.** m is undefined
2. Find two ordered pairs that satisfy $y = ax + b$ and use them to find the slope of the line determined by this equation.
3. The distance (d) in feet traveled by a freely falling body in t seconds is given by the formula $d = 16t^2$. What is the average velocity of the body in the interval from $t_1 = 1$ to $t_2 = 1 + h$? What is the average velocity if $h = 0.1$? 0.01? 0.001? Intuitively, what do you think we say the velocity of the body is at $t = 1$?
4. A company has determined that the cost of producing x units per week of a certain product is given by

$$C = f(x) = 1{,}900 + 40x - 0.1x^2.$$

   Find in simplest form an expression for the average rate of change of this cost function, as the number of units produced varies from x to $x + h$ with $h \neq 0$. Evaluate this expression for $x = 89$, $h = 1$, and interpret the result.
5. **a.** For the function $f(x) = x^2 + 5$, find the average rate of change as x varies from 1 to 2, then do the same as x varies from 2 to 3. Compute the average of these two answers.
   b. Find the average rate of change as x changes from 1 to 3. Is this the same answer you got in part **a**?
   c. Find the average rate of change for f, as x varies from 1 to 2, then do the same as x varies from 2 to 5. Compute the average of these two answers, then find the average rate of change as x changes from 1 to 5. Do you get the same answer?
   d. For this function, is it true that the average of the average rates of change as x varies from a to b and from b to c will equal the average rate of change as x varies from a to c? Justify your conclusion, and explain what determines whether the answers are the same or not.

• • •

5.4 Linear Functions

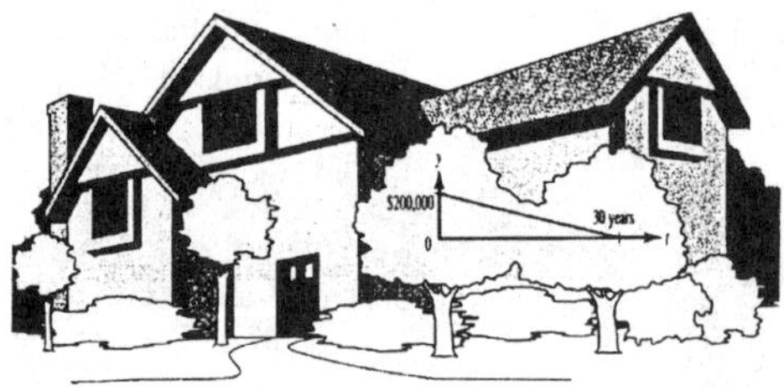

A real-life version of the Monopoly® game for small investors is the purchase of a single family home for use as rental property. An important benefit of this investment is that one can *depreciate* the cost of the building (but not the land) for tax benefits, while the home (hopefully) *appreciates* in value. Usually the investment depreciates linearly by the straight-line method, which means the cost of the building is deducted equally over a specified number of years, as defined in the tax laws. If such an investment depreciates linearly from $200,000 to $0 in 30 years, then how much is the investment worth (for tax purposes) after 7 years?
(See Example 7.)

Objectives

1. Write an equation for a line, given its slope and a point on the line or given two points on the line.
2. Find the slope and y-intercept, given an equation for the line.
3. Graph and write an equation for a line, given the slope and y-intercept.
4. Find the equation of a linear function f, given two ordered pairs in f.
5. Solve applied problems involving linear functions
6. Solve problems involving the equations of parallel and perpendicular lines.

The straight-line method of computing depreciation that is considered in the section-opening problem involves a linear function, which is among the most widely applied functions in mathematics. Such functions are characterized by a rate of change that remains constant (note the regular decrease in the value of the depreciating rental property), so that from a geometric viewpoint their graphs are straight lines. In algebraic terms, linear functions are first-degree polynomial functions, so the following definition applies.

Definition of Linear Function

A function of the form

$$f(x) = ax + b$$

where a and b are real numbers with $a \neq 0$, is called a **linear function**.

When we are working with linear functions, two important formulas for an equation of a line, called the *point-slope equation* and the *slope-intercept equation* are often used. Therefore, we begin by developing these formulas.

Consider any nonvertical line L with slope m that passes through a point (x_1, y_1) as shown in Figure 5.29. (We eliminate vertical lines, since their slope is undefined.) Then, any other point (x, y) with $x \neq x_1$ is on L if and only if the slope of the line segment joining (x, y) and (x_1, y_1) is m, that is, if and only if

$$\frac{y - y_1}{x - x_1} = m.$$

If we write this equation in the form

$$y - y_1 = m(x - x_1),$$

then (x_1, y_1) also satisfies the equation, which means that the points on L match the solutions of $y - y_1 = m(x - x_1)$. We summarize this result as follows.

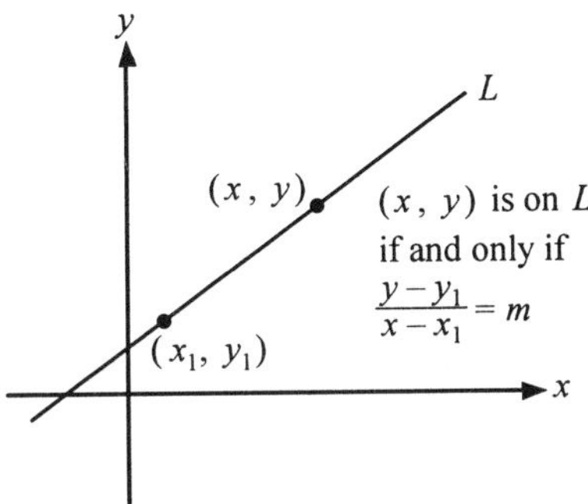

Figure 5.29

Point-Slope Equation

An equation of the line with slope m passing through (x_1, y_1) is

$$y - y_1 = m(x - x_1).$$

This equation is called the **point-slope** form of the equation of a line.

Example 1: Writing an Equation Given the Slope and a Point

Find an equation for the line that contains the point (1, 3) and whose slope is 2. Write the answer in the form $y = ax + b$.

Solution: We are given that $x_1 = 1$, $y_1 = 3$, and $m = 2$. Substituting these numbers in the point-slope equation, we have

$$\begin{aligned} y - 3 &= 2(x - 1) \\ y - 3 &= 2x - 2 \\ y &= 2x + 1. \end{aligned}$$

PROGRESS CHECK 1

Find an equation of the line through (−1, 0) with slope 3. Write the answer in the form $y = ax + b$.

■

Example 2: Writing an Equation Given Two Points

Find an equation for the line that contains the points (1, −2) and (4, 5). Write the answer in the form $y = ax + b$.

Solution: First, we find the slope

$$m = \frac{y_2 - y_1}{x_2 - x_1} = \frac{5-(-2)}{4-1} = \frac{7}{3}$$

Now use the point-slope equation with $m = 7/3$ and either $(1, -2)$ or $(4, 5)$ for (x_1, y_1). Using $x_1 = 4$ and $y_1 = 5$, we have

$$y - 5 = \frac{7}{3}(x-4),$$

which is converted to the form requested as follows:

$$y - 5 = \frac{7}{3}x - \frac{28}{3}$$
$$y = \frac{7}{3}x - \frac{13}{3}.$$

PROGRESS CHECK 2

Find an equation for the line that contains the points (–2, 2) and $(3, -2)$. Write the answer in the form $y = ax + b$.

■

There are many ways of writing the equation of a line. The point-slope equation is used extensively for finding the equation of a line, but it is not very helpful for graphing lines or interpreting linear relationships when the equation is known. To develop a form that is useful in such cases, consider Figure 5.30 and note that any nonvertical line *L* must cross the y-axis at some point. The *x*-coordinate of this point is zero and we will label the *y*-coordinate *b*. This point (0, *b*), where the line crosses the *y*-axis, is called the ***y*-intercept**. If we apply the point-slope equation in the case when (x_1, y_1) is the *y*-intercept, we find that (*x*, *y*) is on *L* if and only if

$$y - b = m(x - 0)$$

or

$$y = mx + b.$$

To emphasize the geometric aspects of our discussion, we state this result as follows.

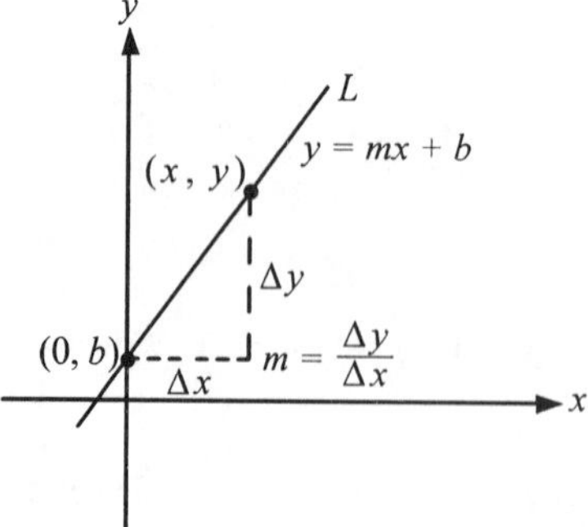

Figure 5.30

Slope-Intercept Equation

> The graph of the equation
>
> $$y = mx + b$$
>
> is a line with slope m and y-intercept $(0, b)$. The equation $y = mx + b$ is called the **slope-intercept** form of the equation of a line.

Note that a linear function f may be defined by an equation of the form $y = f(x) = mx + b$ with $m \neq 0$, so that it follows form the above result that the graph of a linear function is a line.

Example 3: Finding the Slope and the y-Intercept

Find the slope and the y-intercept of the line defined by the equation $2x + 3y = 6$.

Solution: First, transform the equation to the form $y = mx + b$.

$$2x + 3y = 6$$
$$3y = -2x + 6$$
$$y = -\frac{2}{3}x + 2$$

Matching this equation to the form $y = mx + b$, we conclude that

$$m = -\frac{2}{3} \qquad b = 2.$$

Thus, the slope is $-2/3$, and the y-intercept is $(0, 2)$.

Technology Link

To check these results on a grapher, we need to rewrite the equation $2x + 3y = 6$ so that y is expressed as a function of x. One possible equation is $y = (-2/3)x + 2$ (as shown above, while another is $y = (6 - 2x)/3$. Thus, Figure 5.31 shows a graph of either of these equations, and this graph is in agreement with the conclusion that the y-intercept is $(0, 2)$ and the slope is $-2/3$.

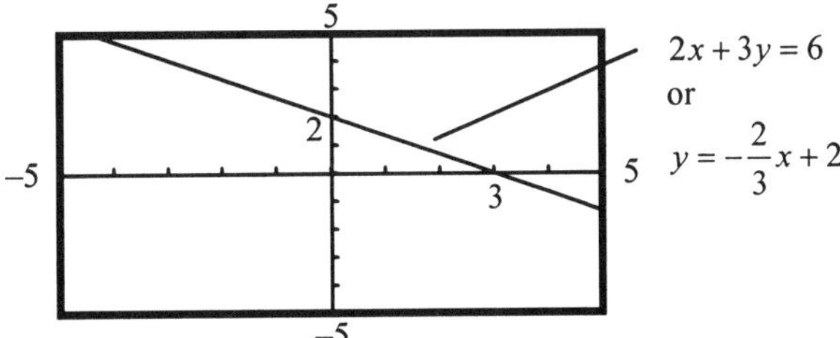

Figure 5.31

PROGRESS CHECK 3
Find the slope and the y-intercept of the line defined by the equation $3x - 4y = 5$. ■

Example 4: Graphing Lines Using Slope-Intercept Form

Graph the line whose slope is $-\frac{2}{3}$ and whose y-intercept is $(0, 4)$. Also, find an equation for the line.

Solution: We are given the y-intercept, so we know that one point on the line is (0, 4). To find another point, we may interpret a slope of $-2/3$ to mean that when x increases 3 units, y decreases 2 units. By starting at (0, 4) and going 3 units to the right and 2 units down, we obtain a second point on the line at (3, 2). Drawing a line through these two points produces the graph in Figure 5.32. By substituting $m = -2/3$ and $b = 4$ in the slope-intercept equation $y = mx + b$, we determine that

$$y = -\frac{2}{3}x + 4$$

is an equation for the line.

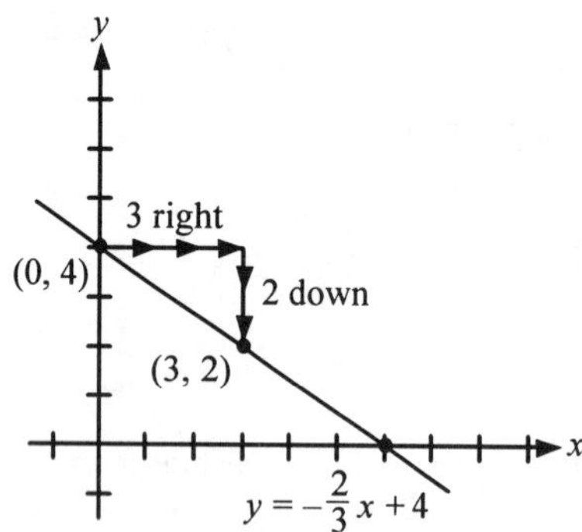

Figure 5.32

PROGRESS CHECK 4

Graph the line whose slope is $\frac{3}{4}$ and whose y-intercept is $(0, -3)$. Also, find an equation for the line. ■

Example 5: Defining a Linear Function *f* Using an Equation

Find the equation that defines the linear function f if $f(1) = 6$ and $f(-1) = 2$.

Solution: If $f(1) = 6$, then when $x = 1$, $y = 6$. Similarly, if $f(-1) = 2$, then when $x = -1$, $y = 2$. Thus, (1, 6) and (–1, 2) are points on the graph. First, calculate the slope.

$$m = \frac{y_2 - y_1}{x_2 - x_1} = \frac{6-2}{1-(-1)} = 2$$

Now use the point-slope form with one of the points, say (1, 6), as follows:

$$\begin{aligned} y - y_1 &= m(x - x_1) \\ y - 6 &= 2(x - 1) \\ y - 6 &= 2x - 2 \\ y &= 2x + 4. \end{aligned}$$

The equation that defines the function f is

$$f(x) = 2x + 4.$$

To check this answer numerically, observe that if $f(x) = 2x + 4$,

$$f(1) = 2(1) + 4 = 6,$$

and

$$f(-1) = 2(-1) + 4 = 2.$$

PROGRESS CHECK 5
Find the equation that defines the linear function *f* if $f(-2) = 3$ and $f(4) = 0$. ■

Example 6: Defining a Linear Relationship Between Temperature Scales

Find an equation that defines the linear relationship between Fahrenheit and Celsius temperatures if the freezing point of water is 32° Fahrenheit and 0° Celsius, and the boiling point is 212° Fahrenheit and 100° Celsius.

Solution: Arbitrarily, let F be the independent variable (like x) and C be the dependent variable (like y). Then, find the slope.

$$m = \frac{\Delta C}{\Delta F} = \frac{100 - 0}{212 - 32} = \frac{100}{180} = \frac{5}{9}$$

Observe that a slope of 5/9 in this relation means that as the Fahrenheit temperature increases 9 degrees, the Celsius temperature increases 5 degrees. Now use the point-slope equation with C replacing y and F replacing x. By selecting $C_1 = 0$ and $F_1 = 32$, we have

$$C - 0 = \frac{5}{9}(F - 32),$$

so an equation that defines the linear relationship is

$$C = \frac{5}{9}(F - 32).$$

(**Note:** If the roles of F and C are reversed in step 1, then $m = \Delta F/\Delta C$, and the resulting equation is $F = (9/5)C + 32$.)

PROGRESS CHECK 6
Find an equation that defines the linear relationship between age (x) and minimum target heart rate (y) when exercising. The recommended targets in beat per minute are 140 at age 20 and 112 at age 60. ■

Example 7: Using the Straight -Line Method of Depreciation

Solve the problem in the section introduction on page 424.

Solution: If we let y represent the depreciated value of the building, then we are given that $y = 200{,}000$, when $t = 0$, and $y = 0$ when $t = 30$. From these corresponding values of y and t, we calculate the slope.

$$m = \frac{\Delta y}{\Delta t} = \frac{200{,}000 - 0}{0 - 30} = -\frac{20{,}000}{3} \qquad (\approx -\$6{,}667)$$

Since $(0,\ 200{,}000)$ is the y-intercept and $-20{,}000/3$ is the slope, the equation of the line is

$$y = -\frac{20,000}{3}t + 200,000.$$

When $t = 7$, we have

$$y = -\frac{20,000}{3}(7) + 200,000 = \$153,333.33.$$

Thus, the value of the building (for tax purposes) after 7 years is \$153,333. Note that the depreciation benefit in each year is about \$6,667.

PROGRESS CHECK 7
If a microcomputer depreciates linearly from \$3,240 to \$0 in 5 years, how much is the microcomputer worth (for tax purposes) after 3 years?

In some applications it is important to be able to recognize a linear relationship from a table of data. Because a linear function is characterized by a rate of change that is constant, we can spot an exact linear relationship as follows: **In any table of a linear function $y = f(x)$, a fixed change in x produced a constant difference between the corresponding y-values.**

Example 8: Recognizing a Linear Relation

The monthly cost y for x hours of usage of an "online" computer service company is shown in the following table.

Time (in hours) x	1	2	3	4	5
Cost (in dollars) y	14.75	19.55	24.35	29.15	33.95

Is y a linear function of x? If yes, find an equation that relates x and y.

Solution: As x increases by 1, observe that each y-value after the first may be found by adding 4.80 to the previous y-value, as shown below.

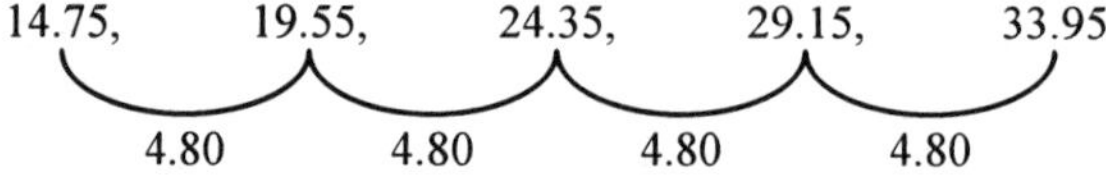

Therefore, a fixed change in x of 1 produces a constant difference of 4.80 between the corresponding y-values, so the relation is linear.

To find an equation for this relation, first observe that

$$\frac{\Delta y}{\Delta x} = \frac{4.80}{1} = 4.80,$$

so the slope is 4.8. Then, using the point-slope equation with one of the points, say (1, 14.75), yields

$$\begin{aligned} y - y_1 &= m(x - x_1) \\ y - 14.75 &= 4.8(x - 1) \\ y - 14.75 &= 4.8x - 4.8 \\ y &= 4.8x + 9.95 \end{aligned}$$

The relation is defined by $y = 4.8x + 9.95$, which means the company charges \$4.80 an hour on top of a \$9.95 monthly membership fee.

PROGRESS CHECK 8

An accountant creates a depreciation schedule for a purchase of office equipment, as shown below. Book values are in dollars and state the value at the beginning of the year given. ■

Year (x)	1	2	3	4	5
Book value (y)	20,000	12,000	7,200	4,320	2,592

Is y a linear function of x? If yes, find an equation that relates x and y.

Parallel and Perpendicular Lines

We often work with parallel and perpendicular lines in analytic geometry, and the slope concept may be used to establish conditions that parallel and perpendicular lines must satisfy.

Conditions for Parallel and Perpendicular Lines

1. Two nonvertical lines are parallel if and only if their slopes are equal.
2. Two nonvertical lines are perpendicular if and only if the product of their slopes is –1.

Intuitively, it is easy to comprehend that parallel lines have the same slope or inclination, and the proof of statement 1 is outlined in Exercise 70. We can establish statement 2 by using the definition of slope and elementary geometry. Consider Figure 5.33, in which L_1 is perpendicular to L_2, and two right triangles have been formed as shown. From the definition of slope, the length of BC (denoted $\overline{BC}$) is m_1, while $\overline{CD} = -m_2$ (since m_2 is negative). Then since L_1 is perpendicular to L_2, both α_1 and α_2 are complementary to θ_1, so

$$\alpha_1 = \alpha_2 \text{ and } \theta_1 = \theta_2$$

and the triangles are similar. In such triangles, corresponding side lengths are proportional, so

$$\frac{\overline{BC}}{\overline{AC}} = \frac{\overline{AC}}{\overline{CD}} \text{ or } \frac{m_1}{1} = \frac{1}{-m_2}.$$

If then follows that $m_1 \cdot m_2 = -1$, which establishes that if L_1 and L_2 are nonvertical perpendicular lines, then the product of their slopes is –1. By reversing each of the steps given, we can also show that if $m_1 \cdot m_2 = -1$, then L_1 and L_2 are perpendicular, thereby establishing statement 2.

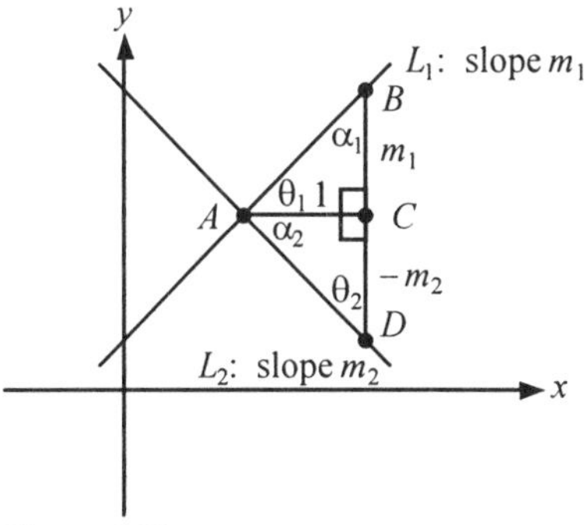

Figure 5.33

Example 9: Finding Equations of Parallel or Perpendicular Lines

Find an equation for the line passing through the point (–4, 1) that is

a. parallel to $5x-2y=3$ **b.** perpendicular to $5x-2y=3$.

Solution: See Figure 5.34

a. First, find the slope of the given line by changing the equation to the form $y=mx+b$.

$$\begin{aligned} 5x-2y&=3\\ -2y&=-5x+3\\ y&=\frac{5}{2}x-\frac{3}{2} \end{aligned}$$

By inspection, the slope of this line is $5/2$. Since the slopes of parallel lines are equal, we want an equation for the line through (–4, 1) with slope $5/2$. Using the point-slope equation, we have

$$\begin{aligned} y-1&=\frac{5}{2}(x-(-4))\\ y&=\frac{5}{2}x+11. \end{aligned}$$

b. As shown in part **a**, the slope of $5x-2y=3$ is $5/2$. It follows from statement 2 that the slope m of every line perpendicular to the given line must satisfy $(5/2)m=-1$, so $m=-2/5$. Then we find an equation for the line through (–4, 1) with slope $-2/5$ as follows:

$$\begin{aligned} y-y_1&=m(x-x_1)\\ y-1&=-\frac{2}{5}(x-(-4))\\ y&=-\frac{2}{5}x-\frac{3}{5}. \end{aligned}$$

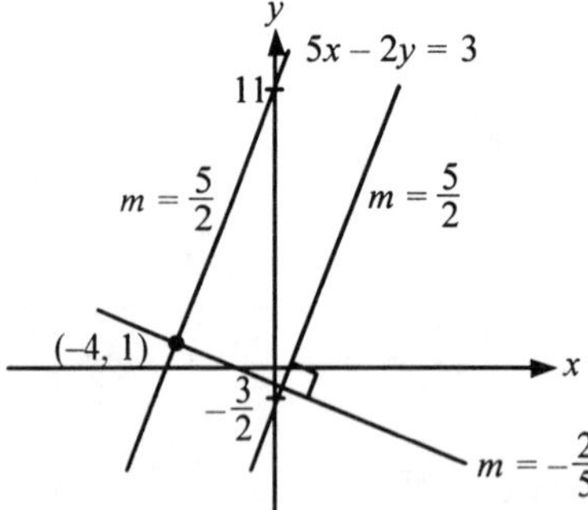

Figure 5.34

Technology Link

To check visually that lines are perpendicular on a grapher, it is necessary to use the Square Setting feature to correct the built-in angular distortion present in the standard display. For instance, Figure 5.35 shows a graphical check of the solution in part **b** using both Zoom Standard and Zoom Square. Observe that the lines appear to be perpendicular only when the Square Setting feature was used.

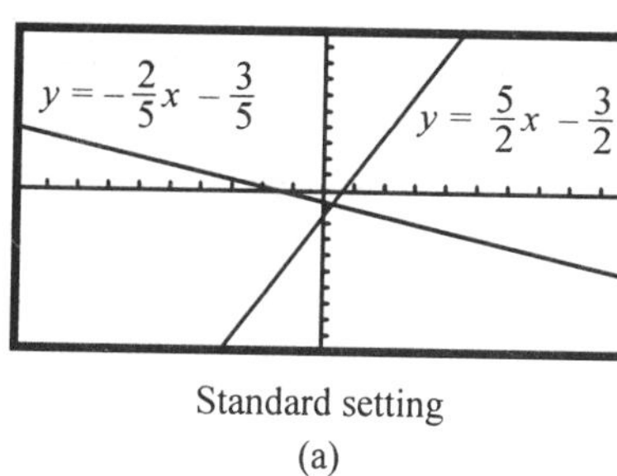

Standard setting
(a)

Square setting
(b)

Figure 5.35

PROGRESS CHECK 9

Find an equation for the line through the point (3, 4) that is

a. parallel to $x - 2y = 1$ and **b.** perpendicular to $x - 2y = 1$. ■

EXPLORE 5.4

1. Graph the equations $y = 3x + b$ for these values of b, and describe the effect of the value of b on the graph.

   a. 4 **b.** 2 **c.** 0 **d.** –2 **e.** $\sqrt{12}$

2. Graph the equations $y = mx + 3$ for these values of m, and describe the effect of the value of m on the graph.

   a. 4 **b.** 2 **c.** 0 **d.** –2 **e.** $\sqrt{12}$

3. Very often in research, related pairs of data are recorded. For example, in medical research we might record the amount of saturated fat a person eats in a week and the person's cholesterol level at the end of the week. In sports, we might record the date and the new world record each time a new record is set. When such ordered pairs are plotted they may seem to almost be in some familiar pattern, such as a line or a parabola. Then we can try to find an equation for this shape, which we can use as a mathematical model of the relationship. Any procedure for finding an equation for a graph that mimics data points is called **Regression**. Many graphing calculators have regression procedures built in (usually as part of a statistics menu). If the data points come close to making a **Straight Line**, and you want an equation of that line, you choose a procedure called **Linear Regression**. There are many ways to do linear regression, each of which may result in slightly different equations. All of them do a good job of mimicking the pattern of points; they just each define "good" in a slightly different way. The most common procedure uses a criterion for "good" that is called **Least Squares**.

   a. If your calculator performs least squares linear regression, use it to find an equation for a line that mimics the points given below. Round off the constants involved to 2 decimal places.

x	1	2	4	7
y	1	3	5	10

   The figure shows the data points with the regression line drawn through them, as produced by a Draw command in a typical statistics menu.

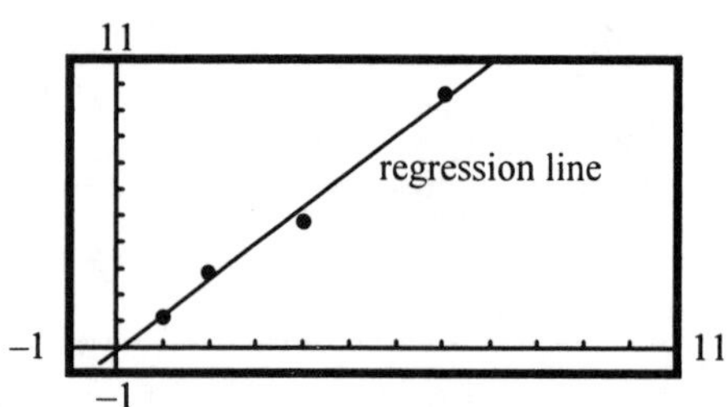

b. In linear regression the calculator also displays a value called r, the **correlation coefficient**. The sign of r matches the sign of the slope of the regression line, and the absolute value of r (which can be from 0 to 1) measures how closely the regression line fits the pattern of points. A perfect fit yields $|r| = 1$; the poorest fit yields 0. Determine r for the data given above.

4. Apply the least squares linear regression procedure to the data given in Exercise 71 on the men's mile record. Round off the slope to 4 decimal places and the intercept to 2 decimal places.

5. The 1994 Statistical Abstract of the U.S. gives the following data for the number of foreigners who visited the U.S. for pleasure (in thousands). Apply regression analysis to this data to find a linear equation that closely models the pattern of the given points.

Year	No. of Visitors
1985	6,609
1986	7,342
1987	8,887
1988	10,821
1989	12,115
1990	13,418
1991	14,734
1992	16,450

EXERCISES 5.4

In Exercises 1–6 find an equation for the line that has the given slope m and passes through the given point.

1. $m=5;\ (4, 3)$ **2.** $m=-4;\ (2, -1)$

3. $m=\frac{1}{2};\ (-2, 0)$ **4.** $m=-\frac{3}{4};\ (-5, 2)$

5. $m=-\frac{1}{2};\ (0, 0)$ **6.** $m=0;\ (1, 2)$

In Exercises 7–10 find an equation for the line that contains the given points.

7. $(3, 2)$ and $(4, 1)$ **8.** $(-4, 3)$ and $(-2, 5)$

9. $(-3, -3)$ and $(-4, -2)$

10. $(2, -3)$ and $(3, -3)$

In Exercises 11–20 find the slope and the y-intercept of the line determined by the following equations.

11. $y=x+7$ **12.** $y=-\frac{2}{3}x-5$

13. $y=5x$ **14.** $y=-x$

15. $y=-2$ **16.** $y=0$

17. $3y+2x=-2$ **18.** $3x+5y=2$

19. $6x-y=-7$ **20.** $2x-7y=-4$

In Exercises 21–26 graph the line whose slope and y-intercept are given, and find an equation for the line.

21. $m=\frac{1}{2};\ (0, 3)$ **22.** $m=5;\ (0, 1)$

23. $m = -3;\ (0, -1)$

24. $m = -\frac{3}{4};\ (0, -2)$

25. $m = 0;\ (0, -5)$

26. $m = 0;\ (0, 0)$

In Exercises 27–32 find the equation that defines the function if *f* is a linear function or a constant function.

27. $f(3) = 0$ and $f(0) = -2$

28. $f(0) = -1$ and $f(2) = 0$

29. $f(-1) = -3$ and $f(2) = -7$

30. $f(-6) = -1$ and $f(-2) = -5$

31. $f(-3) = 1$ and $f(0) = 1$

32. $f(4) = -6$ and $f(-1) = -6$

In Exercises 33–36 determine the given function value; *f* is a linear function or a constant function.

33. $f(3) = 1$ and $f(5) = -3$; find $f(4)$.

34. $f(-2) = 0$ and $f(1) = 6$; find $f(0)$.

35. $f(-4) = -1$ and $f(-1) = -3$; find $f(0)$.

36. $f(-5) = 2$ and $f(1) = 2$; find $f(5)$.

37. A moving company charges $46 to move a certain machine 10 mi, and $58 to move the same machine 30 mi.
    a. Find an equation that defines this relationship if it is linear.
    b. What will it cost to move the machine 25 mi?
    c. What is the minimum charge for moving the machine?
    d. What is the rate for each mile the machine is moved?

38. The total cost of producing a certain item consists of paying rent for the building and paying a fixed amount per unit for material. The total cost is $250 if 10 units are produced and $330 if 30 units are produced.
    a. Find an equation that defines this relationship if it is linear.
    b. What will it cost to produce 100 units?
    c. How much is paid in rent?
    d. What is the cost of the material for each unit?

39. A ball thrown vertically down from a building has a velocity of -252 ft/second when $t = 1$ second, and -316 ft/second when $t = 3$ seconds.
    a. Find an equation that defines this relationship if it linear.
    b. Find the velocity of the ball when $t = 4$ seconds.
    c. What is the initial velocity of the ball?
    d. What is the significance of whether the velocity is positive or negative?

40. A spring that is 24 in. long is compressed to 20 in. by a force of 16 lb and to 15 in. by a force of 36 lb.
    a. Find an equation that defines this relationship if it is linear.
    b. What is the length of the spring if a force of 28 lb is applied?
    c. How much force is needed to compress the spring to 10 in?

41. a. Use the two equalities 0 meters = 0 inches and 1 meter = 39.37 inches to find a linear relationship between inches (*I*) and centimeters (*C*). Remember that 1 meter is 100 centimeters.
    b. Interpret the slope of the relationship you find.
    c. Graph *I* as a function of *C*.
    d. Use the relation to convert 24.8 cm to inches.

42. a. Find a linear relationship between pounds (*P*) and kilograms (*K*) if 1 pound equals 453.59 grams. Remember that 1 kilogram equals 1,000 grams.
    b. Interpret the slope of the relationship you find.
    c. Use the relation to convert 180 pounds to kilograms.

43. a. Two students invent their own temperature scales called the Smith scale and the Jones scale. Find a linear relationship between them if 0° Smith = 100° Jones, and 100° Smith = 50° Jones.
    b. Interpret the slope of the relation.
    c. Convert 32° Smith to Jones degrees.

44. a. Use the fact that 55 mi/hour equals 88.5 km/hour to find a linear relation between the two systems.
    b. Use the relation you found to convert 65 mi/hour to km/hour.
    c. Express 120 km/hour in miles per hour.

45. A St. Bernard dog has been gaining the same amount of weight every week since it was adopted at age 8 weeks, when it weighed 6 pounds. At age 48 weeks it weighs 132 pounds.

a. Find a linear relationship between age in weeks and weight.
b. To the nearest week, when did the puppy first weigh more than 45 pounds?

46. An office copier depreciates linearly in value from \$11,000 to \$1,000 in 8 years. Express this relationship with a function that expresses the value after t years have elapsed. To the nearest year, when does the value first fall below \$5,000?

In Exercises 47–60 determine if the values shown in each table represent a linear relation. If so, express y as a function of x.

47.

x	1	2	3	4	5
y	12.1	12.8	13.5	14.2	14.9

48.

x	1	3	5	7	9
y	23	19	15	11	7

49.

x	0	6	9	14	15
y	6	18	24	34	36

50.

x	0	2	4	6	8
y	1	5	17	37	65

In Exercises 51–62 find an equation for the line passing through the given point that is
a. parallel to the given line
b. perpendicular to the given line

51. $y = x - 2;\ (1, 2)$ **52.** $y = 3x + 1;\ (-2, -1)$

53. $y = -2x + 7;\ (3, 0)$ **54.** $y = -x;\ (-4, 2)$

55. $x + 7y = 2;\ (0, 0)$ **56.** $3x - 2y = 5;\ (0, -5)$

57. $5x - 3y = 2;\ (1, -3)$ **58.** $4x - y = -8;\ (-1, -4)$

59. $3x + 12y = 10;\ (5, 2)$

60. $x + 5y = -3;\ (-2, 3)$

61. $2y - 3x + 1 = 0;\ \left(\frac{1}{2}, -\frac{1}{3}\right)$

62. $4x - y - 1 = 0;\ \left(-\frac{3}{4}, -\frac{1}{2}\right)$

63. Show that the points $A(-4, -2)$, $B(-3, 2)$, and $C(1, 1)$ are the vertices of a right triangle, using the concept of slope.

64. Show that the points $A(-7, -7)$, $B(-1, 1)$, and $C(2, 5)$ lie on a straight line using the concept of slope.

65. If the line through $(-2, -3)$ and $(a, 4)$ is perpendicular to the graph of $y = 4 - x$, find a.

66. Write an equation for the line with y-intercept $(0, 1)$ and parallel to the graph of $y = x$.

67. If two nonvertical lines $A_1x + B_1y + C_1 = 0$ and $A_2x + B_2y + C_2 = 0$ are parallel, then show that $A_1B_2 = A_2B_1$.

68. Solve $y - y_1 = m(x - x_1)$ for y, and determine the y-intercept of the line.

69. For the function $f(x) = mx + b$ compute $\frac{f(x+h) - f(x)}{h}$ with $h \neq 0$, and determine the average rate of change of the function from x to $x + h$. Interpret the meaning of your result.

70. Use the illustration and the fact that L_1 is parallel to L_2 if and only if $\theta_1 = \theta_2$ to prove the following statements.

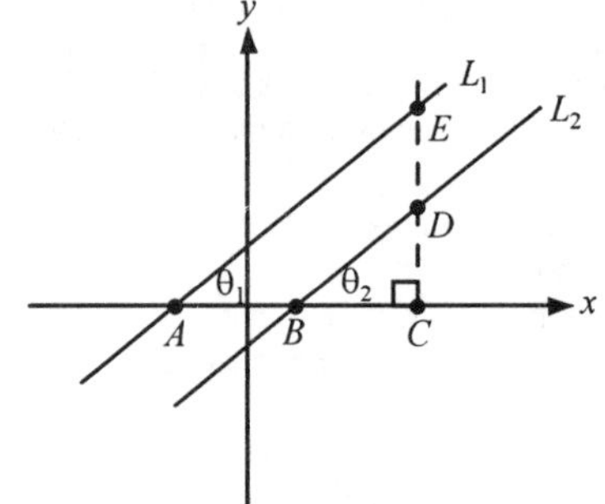

a. If two nonvertical lines are parallel, then their slopes are equal.
b. If two nonvertical lines have equal slopes, then the lines are parallel.

71. Linear functions are often used to approximate the relationship between two variables so that predictions can be made about future events or trends. The history of the record for the 1-mi run is an interesting case in point. Following are a table and graph that show the evolution of the record in this event. Many people are surprised to see that the times have not been leveling off. Instead, they have dropped in a rather predictable linear pattern.

Time	Runner	Year
4:36.5	Richard Webster, Britain	1865
4:29	William Chinnery, Britain	1868
4:28.8	W. C. Gibbs, Britain	1868
4:26	Walter Slade, Britain	1874
4:24.5	Walter Slade, Britain	1875
4:23.2	Water George, Britain	1880
4:21.4	Walter George, Britain	1882
4:19.4	Walter George, Britain	1882
4:18.4	Walter George, Britain	1884
4:18.2	Fred Bacon, Scotland	1894
4:17	Fred Bacon, Scotland	1895
4:15.6	Thomas Conneff, U.S.	1895
4:15.4	John Paul Jones, U.S.	1911
4.14.6	John Paul Jones, U.S.	1913
4:12.6	Norman Taber, U.S.	1915
4:10.4	Paavo Nurmi, Finland	1923
4:09.2	Jules Ladourmegue, France	1931
4:07.6	Jack Lovelock, New Zealand	1933
4:06.8	Glen Cunningham, U.S.	1934
4:06.4	Sydney Wooderson, Britain	1937
4:06.2	Gunder Haegg, Sweden	1942
4:06.2	Arne Andersson, Sweden	1942
4:04.6	Gunder Haegg, Sweden	1942
4:02.6	Arne Andersson, Sweden	1943
4:01.6	Arne Andersson, Sweden	1944
4:01.4	Gunder Haegg, Sweden	1945
3:59.4	Roger Bannister, Britain	1954
3:58	John Landy, Australia	1954
3:57.2	Derek Ibbotson, Britain	1957
3:54.5	Herb Elliott, Australia	1958
3:54.4	Peter Snell, New Zealand	1962
3:54.1	Peter Snell, New Zealand	1964
3:53.6	Michel Jazy, France	1965
3:51.3	Jim Ryun, U.S.	1966
3:51.1	Jim Ryun, U.S.	1967
3:50	Filbert Bayi, Tanzania	1975
3:49.4	John Walker, New Zealand	1975
3:48.9	Sebastian Coe, Britain	1979
3:48.8	Steve Ovett, Britain	1980
3:48.5	Sebastian Coe, Britain	1981
3:48.4	Steve Ovett, Britain	1981
3:48.3	Sebastian Coe, Britain	1981
3:46.3	Steve Cram, Britain	1985
3:44.4	Noureddine Morceli, Algeria	1993

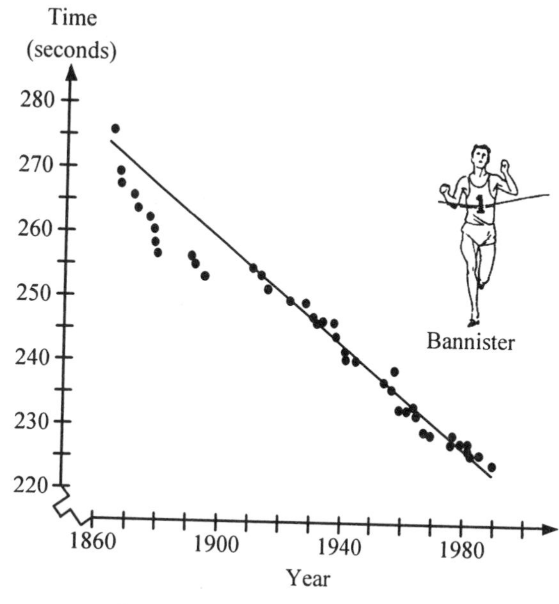

a. The line shown in the figure nicely "fits" the points and can be used to predict with some reliability what will happen in the future in this event. This line passes through the points (1900, 258) and (1950, 239). Write an equation for this line. Interpret the slope.

b. Use the answer to part **a** and predict the time for the mile run in the year 2010.

c. Use the answer to part **a** and predict the year in which the record will be 3:39.

72. The 1994 Statistical Abstract of the U.S. gives the following data for the monthly average number of subscribers for Cable TV (in thousands).

a. Find the equation of a line that fits the data well, by passing through the points (1987, 41,200) and (1992, 54,300).

Year	Number of Subscribers
1985	35,500
1986	38,200
1987	41,200
1988	44,200
1989	47,500
1990	50,520
1991	52,600
1992	54,300
1993	56,300

b. Use the answer from part **a** and predict the number of subscribers in 2000.

c. Use the answer from part **a** and predict the year the number of subscribers will reach 70 million.

THINK ABOUT IT 5.4

1. Create a triangle from the intersection of three lines with positive slope such that the origin is inside the triangle. Write an equation for each of the three lines and display your solution graphically.
2. Line L_1 passes through $(1, -2)$ and $(4, 3)$, and line L_2 passes through $(-5, k)$ and $(-2, 1)$.
   a. Find k if L_1 is parallel to L_2. **b.** Find k if L_1 is perpendicular to L_2.
3. If y is a linear function of x, is x also a linear function of y? Justify your conclusion algebraically.
4. The following figure shows five lines whose equations are as follows:
   a. $y = ax$ **b.** $y = bx + c$ **c.** $y = dx + e$ **d.** $y = mx + k$ **e.** $y = px + n$
   If $d = n = 0$, $b = m$, and $a < k < p$, then match each equation to the corresponding line.

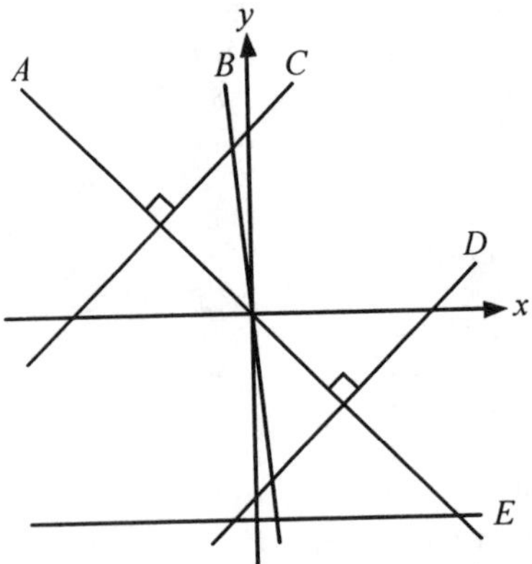

5. Two points A and B are said to be *symmetric about the line* L if line segment AB is perpendicular to L and points A and B are equidistant from L [see illustration (a)]. Use this definition to show in illustration (b) that the points (a, b) and (b, a) are symmetric about the line $y = x$. (**Note:** This exercise demonstrates why the graphs of inverse functions are symmetric about the line $y = x$. This concept will be considered in Section 7.1.)

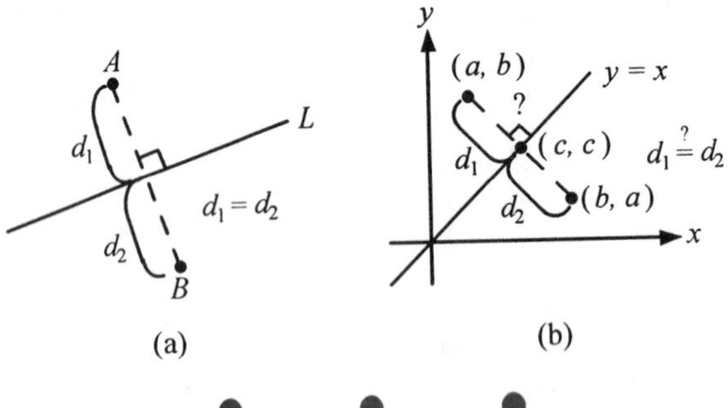

• • •

5.5 Quadratic Functions

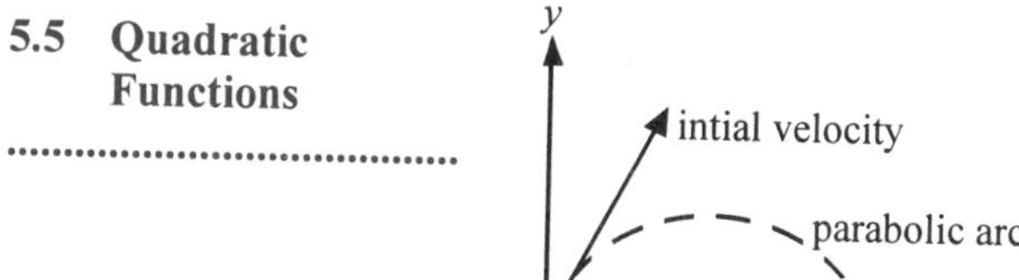

In the seventeenth century Galileo studied the motion of projectiles to help determine the path taken by a shell shot from a cannon. He found that the path formed part of a parabola, as shown in the figure. The actual formula for the trajectory is

$$y = y_0 + k_1 x - k_2 x^2$$

where y is the vertical distance the projectile has traveled, x is the horizontal distance, y_0 is the initial height, and k_1 and k_2 are constants that depend on the initial velocity or on the angle at which the projectile is launched with respect to the horizontal. If a cannon fires a shell from 4 ft above ground with an initial velocity of 160 ft/second, and the cannon is aimed at a 60° angle, then the trajectory of the shell is approximated by $y = 4 + \sqrt{3}x - .0025x^2$.

a. Find the horizontal distance traveled by the shell before it hits the ground. Answer to the nearest tenth of a foot.

b. To the nearest foot, what the maximum height reached by the shell?

(See Example 6.)

Objectives

1. Graph quadratic functions, indicating the vertex and intercepts of the graph, and the range of the function.
2. Solve applied problems involving quadratic functions.
3. Solve quadratic inequalities using a graphing approach.

The trajectory of a projectile is a parabolic path, as discussed in the section-opening problem. Such curves are characteristic of the graphs of second-degree polynomial functions, which are called quadratic functions.

Definition of Quadratic Function

A function of the form

$$f(x) = ax^2 + bx + c,$$

where a, b, and c are real numbers with $a \neq 0$, is called a **quadratic function**.

To see the essential properties of a quadratic function, consider the graphs of $f(x) = x^2 - 4x + 3$ (Figure 5.36) and $f(x) = 4x - 2x^2$ (Figure 5.37), which were obtained on a grapher.

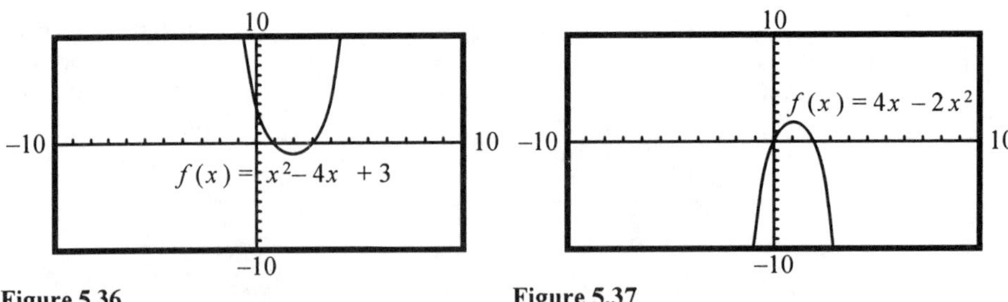

Figure 5.36 **Figure 5.37**

The graphs in both instances are curves that are called **parabolas**. It can be shown that the graph of every quadratic function is a parabola, and we may use some features of parabolas to graph quadratic functions efficiently. First, note in the graph of $f(x) = x^2 - 4x + 3$ in Figure 5.36 that the parabola has a minimum turning point and opens up like a cup. This occurs whenever the coefficient of x^2 is a positive number. If the coefficient of x^2 is a negative number, then the graph turns at the highest point on the graph, as shown in the graph of $f(x) = 4x - 2x^2$ in Figure 5.37. We now know some important facts about the graph of a quadratic form.

Graph of a Quadratic Function

For the function defined by $y = ax^2 + bx + c$ with $a \neq 0$

1. The graph is a parabola.
2. If $a > 0$, the parabola opens upward and turns at the lowest point on the graph.
3. If $a < 0$, the parabola opens downward and turns at the highest point on the graph.

For both graphing and applications purposes, it is important to develop a way to locate the turning point of the parabola, which is called the **vertex**. Consider Figure 5.38, which illustrates that the vertex in the graph of a quadratic function is located on a vertical line called the **axis of symmetry** of the parabola. This line divides the parabola into two segments such that if we make a fold on the line, the two halves will coincide.

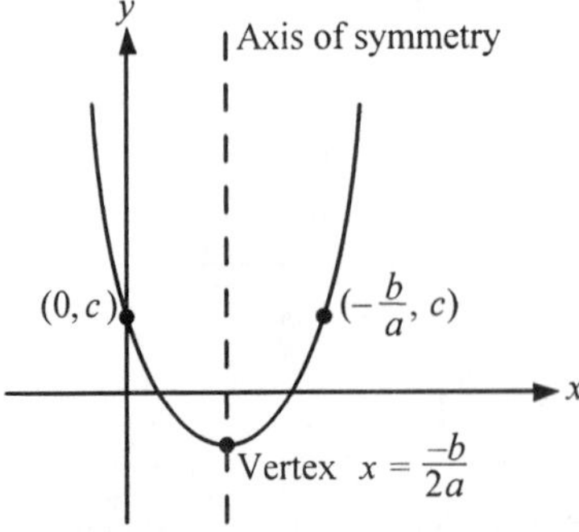

Figure 5.38

To find the equation of the axis of symmetry, note that this line is halfway between any pair of points on the parabola that have the same y-coordinate. The easiest pair of points to analyze is the pair of points on the graph of $y = ax^2 + bx + c$ whose y-coordinate is c. Therefore, we replace y by c in this equation and solve for x.

$$\begin{aligned}
y &= ax^2 + bx + c & &\\
c &= ax^2 + bx + c & &\text{Replace } y \text{ by } c.\\
0 &= ax^2 + bx & &\text{Subtract } c \text{ from both sides.}\\
0 &= x(ax + b) & &\text{Factor the nonzero side.}\\
x = 0 \quad &\text{or} \quad ax + b = 0 & &\text{Set each factor equal to 0.}\\
&\qquad ax = -b & &\\
&\qquad x = \frac{-b}{a} & &
\end{aligned}$$

The x-coordinate halfway between 0 and $-b/a$ is given by

$$x = \frac{0 + (-b/a)}{2},$$

so the equation of the axis of symmetry is

$$x = \frac{-b}{2a}.$$

Because the vertex lies on the axis of symmetry, the x-coordinate of the vertex is $-b/2a$. We may find the y-coordinate of the vertex by finding $f(-b/2a)$, which is the y value when $x = -b/2a$, and we have established the following formula.

Vertex Formula

> The vertex of the graph of $y = f(x) = ax^2 + bx + c$, with $a \neq 0$, is located at
>
> $$\left(\frac{-b}{2a}, f\left(\frac{-b}{2a}\right)\right).$$

The last important feature of the graph of a quadratic function that we consider is the coordinates of the points where the curve crosses the axes. The point where the parabola crosses the y-axis (that is, the y-intercept) can be found by substituting 0 for x in the equation $y = ax^2 + bx + c$.

$$\begin{aligned}
y &= a(0)^2 + b(0) + c\\
&= c
\end{aligned}$$

Thus the y-intercept is always $(0,\ c)$.

To find the points where the parabola crosses the x-axis (that is, the x-intercepts), we substitute 0 for y in the equation $y = ax^2 + bx + c$. Thus, the x-coordinates of the x-intercepts are found by solving the equation $ax^2 + bx + c = 0$.

Example 1: Graphing a Quadratic Function

Graph the function defined by $f(x) = x^2 - 7x + 6$ and indicate

a. the coordinates of the x- and y-intercepts

b. the equation of the axis of symmetry

c. the coordinates of the maximum or minimum point

d. the range of the function

Solution:

a. To find the x-intercepts, set $f(x) = 0$ and solve the resulting equation.

$$x^2 - 7x + 6 = 0$$
$$(x-6)(x-1) = 0$$
$$x - 6 = 0 \quad \text{or} \quad x - 1 = 0$$
$$x = 6 \qquad\qquad x = 1$$

x-intercepts: (1,0) and (6,0); y-intercept: (0,6) since $c = 6$

b. Axis of symmetry:

$$x = \frac{-b}{2a} = \frac{-(-7)}{2(1)} = \frac{7}{2}.$$

c. Since $a > 0$, we have a minimum point. The x-coordinate of the minimum point is $7/2$, since the minimum point lies on the axis of symmetry. We find the y-coordinate of the minimum point by finding the y value when $x = 7/2$.

$$f(x) = x^2 - 7x + 6$$
$$f\left(\frac{7}{2}\right) = \left(\frac{7}{2}\right)^2 - 7\left(\frac{7}{2}\right) + 6$$
$$= \frac{49}{4} - \frac{49}{2} + 6$$
$$= \frac{49}{4} - \frac{98}{4} + \frac{24}{4}$$
$$= -\frac{25}{4}$$

Minimum point: $(7/2, -25/4)$

d. The graph is the parabola shown in Figure 5.39. As illustrated, the range is the interval $[-25/4, \infty)$.

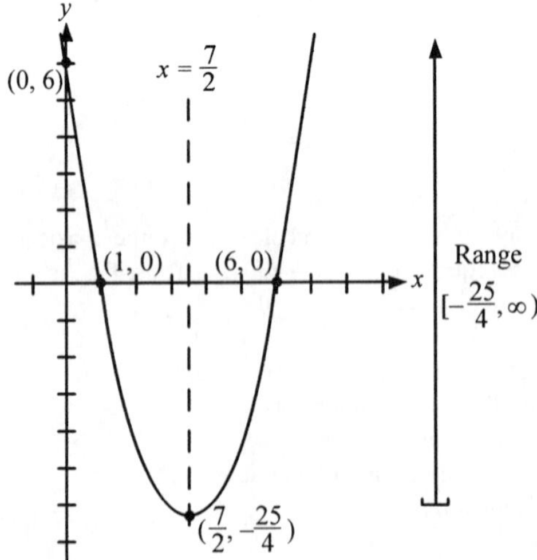

Figure 5.39

Technology Link

The results in this example are not difficult to check with a grapher. For instance, if we graph the given function as shown in Figure 5.40 and utilize the Trace feature, we can obtain approximations for the intercepts and vertex that support the exact answers we determined algebraically. In addition, many calculators have built-in operations for estimating an x-intercept, a maximum point, or a minimum point. You should learn to use these features, if they are available.

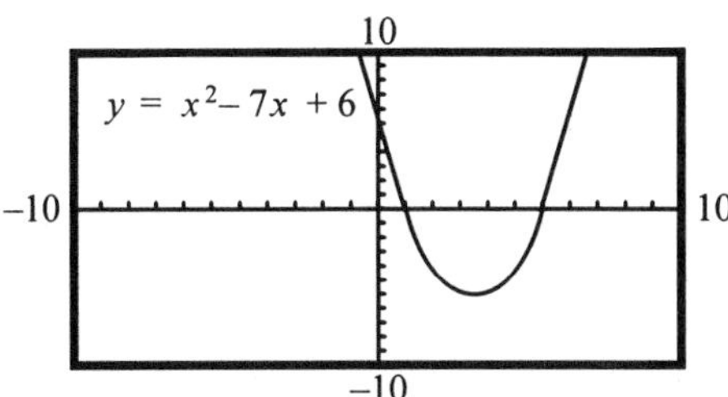

Figure 5.40

PROGRESS CHECK 1

Redo the problems in Example 1 for the function defined by $f(x) = x^2 - 6x + 8$. ■

Example 2: Graphing a Quadratic Function

Graph the function defined by $f(x) = 3 - x - x^2$ and indicate

a. the coordinates of the x- and y-intercepts
b. the equation of the axis of symmetry
c. the coordinates of the maximum or minimum point
d. the range of the function

Solution:

a. To find the x-intercepts, set $y = 0$ and solve the resulting equation:

$$3 - x - x^2 = 0$$

By the quadratic formula (with $a = -1$, $b = -1$, and $c = 3$) we have

$$\begin{aligned} x &= \frac{-(-1) \pm \sqrt{(-1)^2 - 4(-1)(3)}}{2(-1)} \\ &= \frac{1 \pm \sqrt{13}}{-2}. \end{aligned}$$

x-intercepts: $\left((1+\sqrt{13})/-2, 0\right)$ and $\left((1-\sqrt{13})/-2, 0\right)$, which are about $(-2.3, 0)$ and $(1.3, 0)$
y-intercept: (0,3), since $c = 3$.

b. Axis of symmetry:

$$x = \frac{-b}{2a} = \frac{-(-1)}{2(-1)} = -\frac{1}{2}$$

c. Since $a < 0$, we have a maximum point. The x-coordinate of the maximum point is $-1/2$, since the maximum point lies on the axis of symmetry. We find the y-coordinate of the maximum point by finding $f(-1/2)$, the value of the function when $x = -1/2$.

$$f(x) = 3 - x - x^2$$
$$f\left(-\frac{1}{2}\right) = 3 - \left(-\frac{1}{2}\right) - \left(-\frac{1}{2}\right)^2$$
$$= 3 + \frac{1}{2} - \frac{1}{4}$$
$$= \frac{13}{4}$$

Maximum point: $(-1/2, 13/4)$

d. By drawing a parabola through the vertex and the intercepts, we obtain the graph of f in Figure 5.41. From the graph, we read that the range is the interval $(-\infty, 13/4]$. Figure 5.42 shows a graph of $y = 3 - x - x^2$ from a grapher. Check that this graph is in agreement with our results in this example.

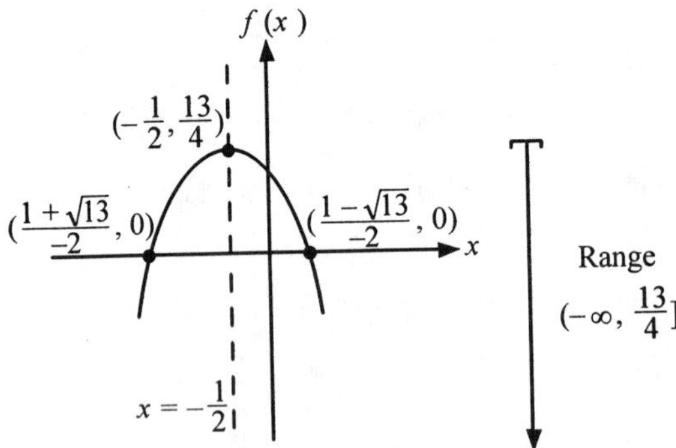

Figure 5.41

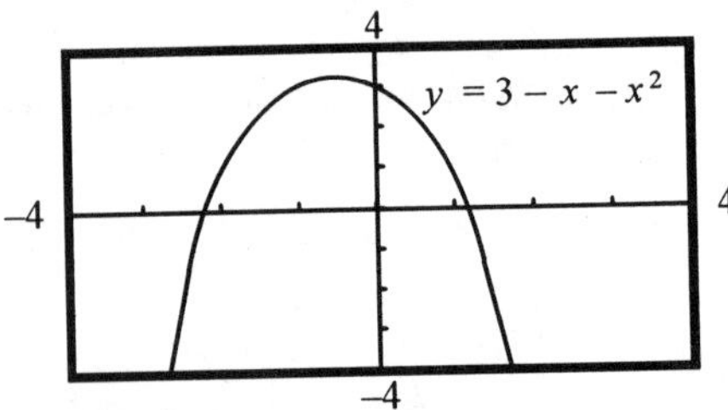

Figure 5.42

PROGRESS CHECK 2

Redo the problems in Example 2 for the function defined by $f(x) = 3 + 4x - x^2$. ■

The vertex of a parabola can be found more efficiently in some cases by employing the graphing techniques discussed in Section 1.3. For instance, in Example 1c, of that section we graphed $y = (x+1)^2 - 2$ by moving the graph of $y = x^2$, 1 unit to the left and 2 units down, which placed the vertex of the parabola at $(-1, -2)$. More generally, any quadratic function may be placed in the standard form.

$$f(x) = a(x - h)^2 + k \quad (\text{with } a \neq 0)$$

by completing the square, and in this form our graphing techniques tell us that the graph of f is the graph of $y = ax^2$ (a parabola) shifted so the vertex is the point (h,k) and the axis of symmetry is the vertical line $x = h$ (see Figure 5.43). Example 3 illustrates this result using the functions analyzed in Examples 1 and 2.

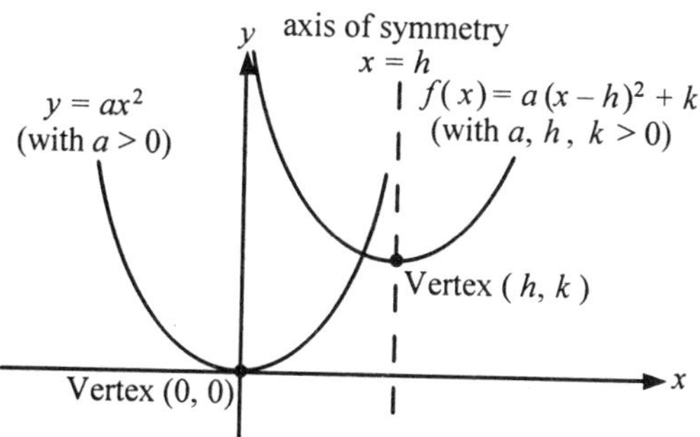

Figure 5.43

Example 3: Writing a Quadratic Function in Standard Form

Determine the vertex and axis of symmetry of the graph of f by matching the function to the form $f(x) = a(x-h)^2 + k$.

a. $f(x) = x^2 - 7x + 6$ **b.** $f(x) = 3 - x - x^2$

Solution:

a. First, write the equation as

$$f(x) = (x^2 - 7x \quad) + 6.$$

As discussed in Section 4.1, we complete the square for $x^2 + bx$ by adding $(b/2)^2$, which is the square of one-half of the coefficient of x. Half of -7 is $-7/2$ and $(-7/2)^2 = 49/4$. Now inside the parentheses both add $49/4$ (so we can complete the square) and subtract $49/4$ [so $f(x)$ does not change], and proceed as follows:

$$\begin{aligned} f(x) &= \left(x^2 - 7x + \frac{49}{4} - \frac{49}{4}\right) + 6 \\ &= \left(x^2 - 7x + \frac{49}{4}\right) - \frac{49}{4} + 6 \\ &= \left(x - \frac{7}{2}\right)^2 - \frac{25}{4}. \end{aligned}$$

Matching this result to the form $f(x) = a(x-h)^2 + k$, we determine $h = 7/2$ and $k = -25/4$, so the vertex of the parabola is $(7/2, -25/4)$, and the axis of symmetry is $x = 7/2$.

b. To complete the square, the coefficient of x^2 must be 1, so first factor out -1 from the x terms and then proceed as in part **a**.

$$\begin{aligned} f(x) &= 3 - x - x^2 \\ &= -1(x^2 + x) + 3 \\ &= -1\left(x^2 + x + \frac{1}{4} - \frac{1}{4}\right) + 3 \\ &= -1\left(x^2 + x + \frac{1}{4}\right) + (-1)\left(-\frac{1}{4}\right) + 3 \\ &= -1\left(x + \frac{1}{2}\right)^2 + \frac{13}{4} \end{aligned}$$

Once again, we match this equation to the form $f(x) = a(x-h)^2 + k$, which gives $h = -1/2$ and $k = 13/4$. Thus, the vertex is $(-1/2, 13/4)$, and the axis of symmetry is $x = -1/2$.

PROGRESS CHECK 3

Determine the vertex and axis of symmetry of the graph of *f* by matching the equation to the form $f(x) = a(x-h)^2 + k$.

a. $f(x) = x^2 + 6x + 8$ **b.** $f(x) = 1 + x - x^2$ ■

The next three examples illustrate how quadratic functions can be applied in practical situations. Notice, in particular, the significance of the maximum or minimum value, which is an important topic in applied mathematics.

Example 4: Finding a Maximum Area

A manager wants to fence in a parking lot of which one side is bounded by a building, but because of budget constraints, only a total of 300 ft of fencing is available. What are the dimensions of the largest rectangular parking lot that can be enclosed with the available fencing?

Solution: First, draw a sketch of the situation (Figure 5.44). If we let *x* represent the length of one side of the lot, the opposite side also has length *x*. The length of the side opposite of the building is then $300-2x$. We wish to maximize the area of this rectangular region, which is given by the following quadratic function.

$$\text{Area} = x(300 - 2x)$$
$$A = 300x - 2x^2$$

Since $a < 0$, we have a maximum point. Axis of symmetry:

$$x = \frac{-b}{2a} = \frac{-300}{2(-2)} = \frac{-300}{-4} = 75.$$

Length of the side opposite the building:

$$300 - 2x = 300 - 2(75) = 300 - 150 = 150.$$

The area is a maximum when the dimensions are 75 ft by 150 ft.

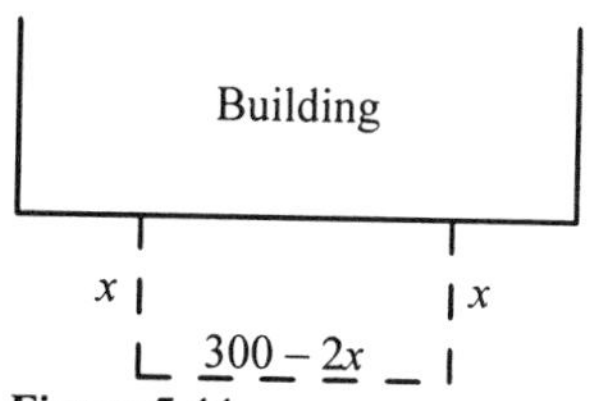

Figure 5.44

As mentioned in Example 1, some calculators have a built-in operation that finds a highest point of a graph in an interval. For example, on one such calculator the Function Maximum operation may be utilized to display the highest point for a graph of the equation in this example, as shown in Figure 5.45. This display is further evidence that the maximum area is attained when $x = 75$.

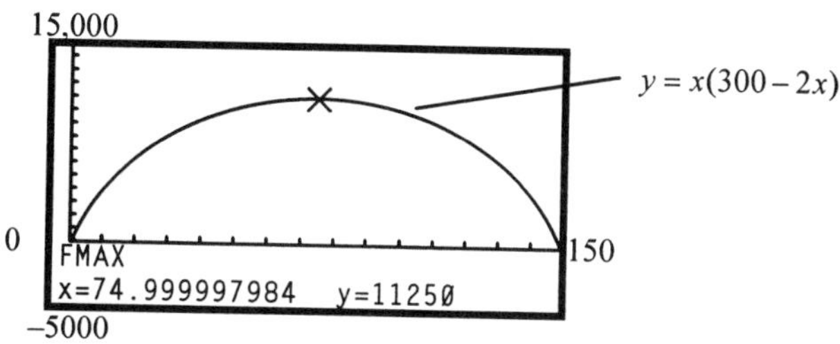

Figure 5.45

We will use the Function Maximum or Minimum feature occasionally in the text because this operation gives very accurate answers. But keep in mind that repeated use of Zoom and Trace is a workable alternative, if this special feature is not available on your calculator.

PROGRESS CHECK 4

A rectangular field is adjacent to a river and is to have fencing on three sides because the side on the river requires no fencing. If 500 yd of fencing are available, what are the dimensions of the rectangular section with largest area that can be enclosed with the available fencing? ■

Example 5: A Projectile with Up-Down Motion

The height (y) of a projectile shot vertically up from the ground with an initial velocity of 144 ft/second is given by the formula $y = 144t - 16t^2$. Graph the function defined by this formula and indicate

a. when the projectile will strike the ground.

b. when the projectile attains its maximum height

c. the maximum height

Solution:

a. When the projectile hits the grounds, the height (y) of the projectile is 0. Thus, we want to find the values of t for which $144t - 16t^2$ equals 0.

$$144t - 16t^2 = 0$$

$$16t(9 - t) = 0$$

$$16t = 0 \quad \text{or} \quad 9 - t = 0$$

$$t = 0 \qquad\qquad t = 9$$

The projectile will hit the ground 9 seconds later.

b. Since $a < 0$, we have a maximum point. Axis of symmetry:

$$t = \frac{-b}{2a} = \frac{-(144)}{2(-16)} = \frac{-144}{-32} = 4.5$$

The projectile reaches its highest point when $t = 4.5$ seconds.

c. We can find the maximum height by finding the value of y when $t = 4.5$ (or $t = 9/2$) seconds.

$$\begin{aligned} y = f(t) &= 144t - 16t^2 \\ f\left(\frac{9}{2}\right) &= 144\left(\frac{9}{2}\right) - 16\left(\frac{9}{2}\right)^2 \\ &= 72 \cdot 9 - 16\left(\frac{81}{4}\right) \\ &= 324 \end{aligned}$$

Thus, the projectile reaches a maximum height of 324 ft, and Figure 5.46 shows the graph of this function.

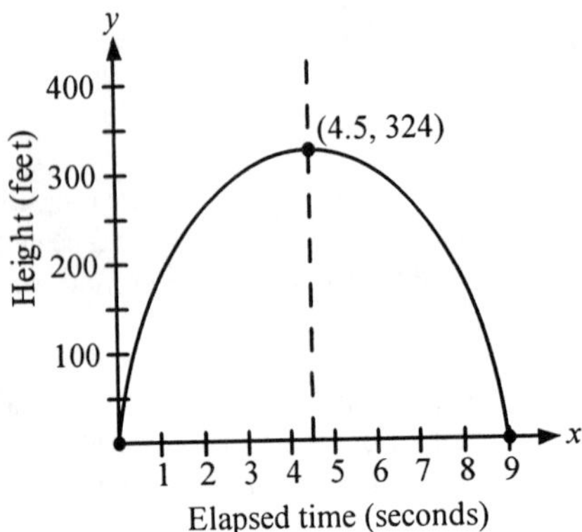

Figure 5.46

PROGRESS CHECK 5

The height (y) in feet of a projectile shot vertically up from the ground with an initial velocity of 128 ft/second is given by $y = 128t - 16t^2$. Answer the questions in Example 5 with respect to this formula. ■

Example 6: Trajectory of a Projectile

Solve the problem in the section introduction on page 439.

Solution:

a. We are given that the path of the shell is part of the parabola defined by $y = 4 + \sqrt{3}x - 0.0025x^2$. When the projectile hits the ground, its vertical position (y) is zero, and replacing y by 0 in the given formula leads to

$$0 = 4 + \sqrt{3}x - 0.0025x^2.$$

Then applying the quadratic formula to the resulting equation in which $a = -0.0025$, $b = \sqrt{3}$ and $c = 4$ gives

$$x = \frac{-\sqrt{3} \pm \sqrt{(\sqrt{3})^2 - 4(-0.0025)(4)}}{2(-0.0025)} = \frac{-\sqrt{3} \pm \sqrt{3.04}}{-0.005}.$$

To the nearest tenth, the two roots are

$$x_1 = \frac{-\sqrt{3} + \sqrt{3.04}}{-0.005} \approx -2.3 \text{ or } x_2 = \frac{-\sqrt{3} - \sqrt{3.04}}{-0.005} \approx 695.1.$$

In the context of the problem, only positive solutions are meaningful. Therefore, the projectile travels a horizontal distance of 695.1 ft before it hits the ground.

b. The shell reaches its maximum height when

$$x = \frac{-b}{2a} = \frac{-\sqrt{3}}{2(-0.0025)} \approx 346.41.$$

To find the maximum height, replace x by 346.41 in the given formula.

$$\begin{aligned} y &= 4 + \sqrt{3}x - 0.0025x^2 \\ y &= 4 + \sqrt{3}(346.41) - 0.0025(346.41)^2 \\ &= 304 \end{aligned}$$

Thus, the shell reaches a maximum height of about 304 ft. Figure 5.47 shows a check of the results of this example on a grapher.

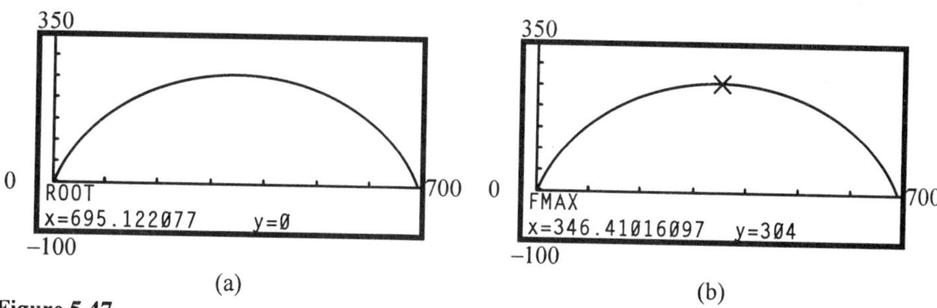

(a) (b)

Figure 5.47

Note: You should distinguish between the two uses of a parabola in Examples 5 and 6. In Example 5 the path of the projectile is vertical (straight up and down), and a parabola results when the height is graphed as a function of time. In Example 6 the parabola actually shows the path of the projectile.

PROGRESS CHECK 6

If a baseball is hit from 3.5 ft above home plate with an initial velocity of 80 ft/second, and the ball is initially propelled at a 60° angle with respect to the horizontal, then the trajectory of the ball is approximated by $y = 3.5 + \sqrt{3}x - 0.01x^2$, where y gives the height (in feet) of the ball when the ball is x feet from home plate.

a. When the ball hits the ground, how far is it from home plate?

b. To the nearest tenth of a foot, what is the maximum height reached by the ball? ■

At this point you should be proficient at graphing quadratic functions. Recall that it is easy to solve quadratic equations or inequalities with the aid of such graphs. Figure 5.48 reviews the graphical method for solving quadratic equations or inequalities that was introduced in Section 4.4.

Quadratic Equation or Inequality	Possible Graph of $y = f(x)$	Interpretation of Solution Set
$ax^2 + bx + c = 0$	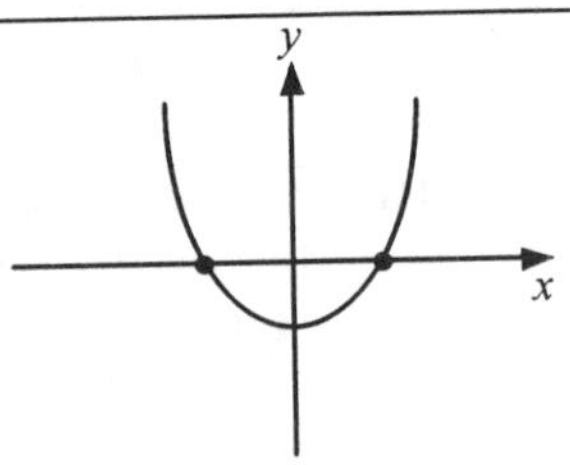	$f(x) = 0$ for all x-values where the graph is *on* the x-axis.
$ax^2 + bx + c < 0$	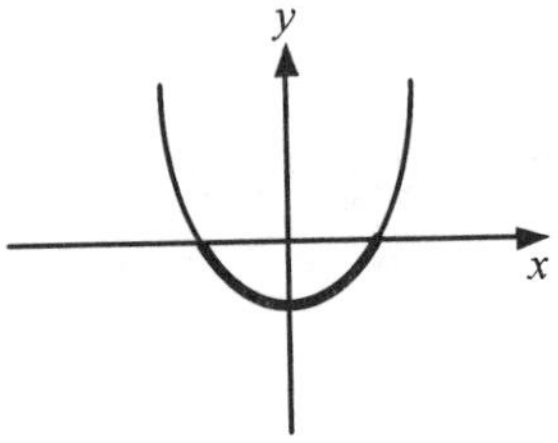	$f(x) < 0$ for all x-values where the graph is *below* the x-axis.
$ax^2 + bx + c > 0$	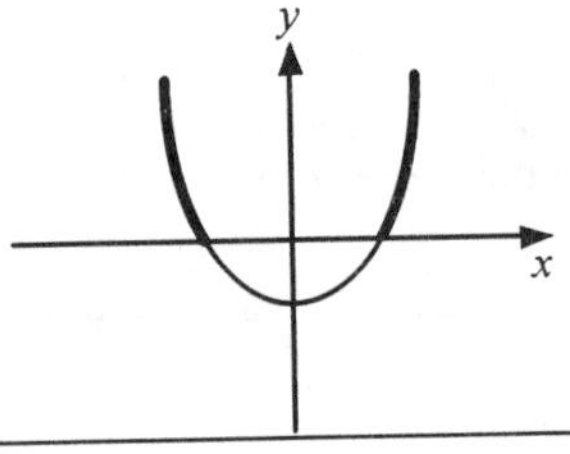	$f(x) > 0$ for all x-values where the graph is *above* the x-axis.

Figure 5.48

Example 7: Solving a Quadratic Inequality Graphically

Solve $x^2 - 6 > x$ using a graphical method.

Solution: $x^2 - 6 > x$ in the form $ax^2 + bx + c > 0$ is $x^2 - x - 6 > 0$. Therefore we want to sketch the graph of $y = x^2 - x - 6$ with particular emphasis on the x-intercepts and on whether the parabola has a maximum or minimum turning point.

Vertex: Since $a > 0$, the graph has a minimum turning point and opens up. The actual coordinates of the vertex need not be found.

***x*-intercepts:** To find the x-intercepts, set $y = 0$ and solve the resulting equation.

$$0 = x^2 - x - 6$$
$$0 = (x-3)(x+2)$$
$$x - 3 = 0 \quad \text{or} \quad x + 2 = 0$$
$$x = 3 \qquad\qquad x = -2$$

x intercepts: (3,0) and (−2, 0)

Now sketch Figure 5.49. To solve $f(x) > 0$, we specify all x values for which the parabola is *above* the x axis. Thus, we read the solution set from the graph to be $(-\infty, -2) \cup (3, \infty)$.

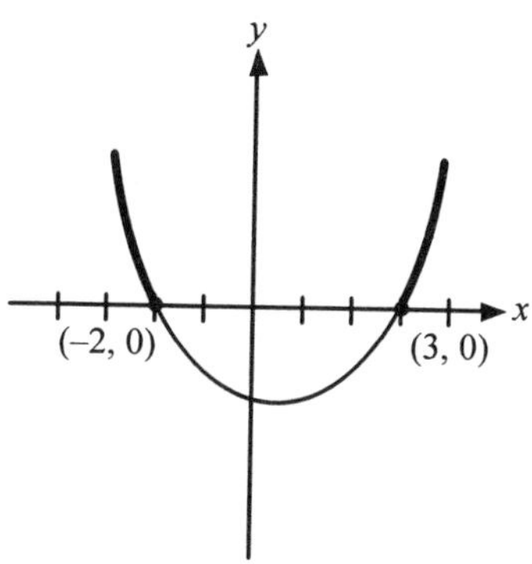

Figure 5.49

Technology Link

A graphical method of solving quadratic inequalities is usually the method of choice because it is easy to obtain a graph of a quadratic function on a grapher. To illustrate, using the inequality in this example, consider Figure 5.50. It is apparent that $x^2 - x - 6$ is positive for x values that are to the right of 3, or for x values that are to the left of –2. We can check that the graph intersects the x-axis when $x = 3$ and $x = -2$ as follows.

For $x = 3$: $y = 3^2 - 3 - 6 = 0$
For $x = -2$: $y = (-2)^2 - (-2) - 6 = 0$

Both apparent x-intercepts check, so $x^2 - x - 6 > 0$ is true when $x < -2$ or $x > 3$.

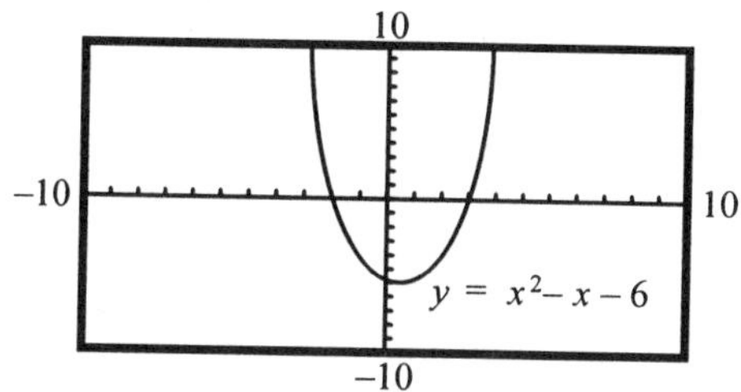

Figure 5.50

PROGRESS CHECK 7
Solve $x^2 + 2x \geq 15$ using a graphical method. ■

Example 8: Solving a Quadratic Inequality Graphically

Solve $x^2 + 2 < 0$ using a graphical method.

Solution: Figure 5.51 shows a graph of $y = x^2 + 2$. We see that the graph is never below the x-axis, so $x^2 + 2$ is never negative. Therefore, there are no values of x for which the inequality is a true statement, and the solution set is $\varnothing$.

PROGRESS CHECK 8
Solve $3x^2 + 2x + 2 > 0$ using a graphical method. ■

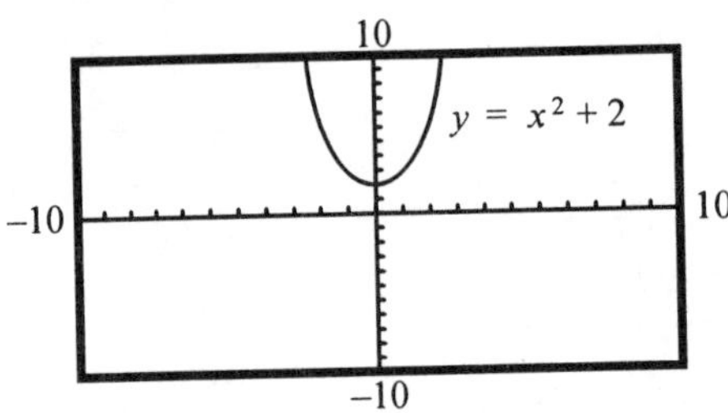

Figure 5.51

EXPLORE 5.5

1. The graph of the quadratic function $f(x) = ax^2 + bx + c$ is a parabola. Investigate the effects of the constants *a*, *b*, and *c* on the graph.
   a. Set *a* and *b* equal to 1, then look at graphs for $c = -5, -1, 1,$ and 5. What seems to be the effect of *c* on the graph?
   b. Set *b* and *c* equal to 1, then look at the graphs for $a = -5, -1, 1,$ and 5. What seems to be the effect of *a* on the graph?
   c. Set *a* and *c* equal to 1, then look at the graphs for $b = -5, -1, 1,$ and 5. What seems to be the effect of *b* on the graph?
2. The 1994 Statistical Abstract of the U.S. gives the following data for the attendance at NCAA Women's college basketball games (in thousands).

Year	Attendance
85	2072
87	2156
88	2325
89	2502
90	2777
91	3013
92	3397

   a. Graph these points and notice that they appear to curve like a parabola.
   b. Assume that the vertex of the parabola is the point (85, 2072), and the point (92, 3397) is on the graph. What is the resulting equation?
   c. If your calculator performs quadratic regression, apply that technique to this set of points and determine the equation of the parabola.
3. The problem that opens this section concerns the parabolic path of a trajectory. A graphing calculator with a Parametric Mode offers another approach to this topic. By the laws of physics the path of a projectile can be analyzed into separate vertical and horizontal components. For instance, when a stone is thrown horizontally from the lip of a cliff at 4 feet per second, it falls in a parabolic arc. Its horizontal position is given by $x = 4t$, where *t* is elapsed time in seconds. That is, each second it is 4 feet further out from the cliff side. Its vertical position is given by $-16t^2$. That is, it falls $16t^2$ feet in *t* seconds. Parametric graphing allows you to enter these 2 equations independently, then the calculator plots the (*x*, *y*) points corresponding to different values of *t*. If your calculator has parametric graphing mode, use the following steps to graph the path of the stone for 5 seconds.
   1. Set calculator to Parametric mode.
   2. Let $x_{1t} = 4t$
      Let $y_{1t} = -16t^2$

3. For the viewing window settings let $t_{min} = 0$ and $t_{max} = 5$.
Let $t_{step} = 0.1$ (to follow the stone every tenth of a second).
Let $x_{min} = 0$ and $x_{max} = 20$.
Let $y_{min} = -500$ and $y_{max} = 0$.

After graphing the path of the stone, use Trace and create a table that shows how it moves every tenth of a second.

EXERCISES 5.5

In Exercises 1–20 graph the quadratic function defined by the equation. On the graph indicate

a. the coordinates of the x- and y-intercepts
b. the equation of the axis of symmetry
c. the coordinates of the maximum or minimum point
d. the range of the function

1. $y = x^2 - 3x + 2$
2. $y = x^2 + 6x + 5$
3. $f(x) = x^2 - 4$
4. $g(x) = 3x^2 - 3$
5. $y = x^2 - 2x$
6. $f(x) = 2x^2 - 3x$
7. $g(x) = 4x - x^2$
8. $y = 1 - x^2$
9. $y = x^2 + 2x + 1$
10. $y = x^2 - 4x + 4$
11. $f(x) = -x^2 + x + 2$
12. $f(x) = -x^2 + 3x - 4$
13. $y = x^2 - 5$
14. $y = 2 - x^2$
15. $y = 5 - 4x^2$
16. $y = 2x^2 - 3$
17. $f(x) = 2x^2 - x - 1$
18. $y = 3x^2 - 5x + 1$
19. $y = 3 + 6x - 2x^2$
20. $g(x) = -3x^2 + 5x + 2$

In Exercises 21–30 determine the vertex and the axis of symmetry of the graph of f by matching the function to the form $f(x) = a(x-h)^2 + k$.

21. $f(x) = (x+2)^2 - 3$
22. $f(x) - 2(x-1)^2 + 5$
23. $f(x) = 3x^2 - 4$
24. $f(x) = 3(x-4)^2$
25. $f(x) = x^2 - 2x + 2$
26. $f(x) = x^2 + 6x + 5$
27. $f(x) = 3x^2 - 9x + 1$
28. $f(x) = 2x^2 + 4x - 5$
29. $f(x) = 2 + x - x^2$
30. $f(x) = 1 - 3x - 4x^2$

In Exercises 31–50 solve the quadratic inequality

31. $x^2 - 4x - 5 > 0$
32. $x^2 + x - 6 < 0$
33. $x^2 < 4x + 5$
34. $x^2 + 5 < 6x$
35. $-x^2 + x + 2 \le 0$
36. $-10 \ge 3x - x^2$
37. $x^2 \ge 4$
38. $x^2 - 5 \le 0$
39. $x^2 < 3x$
40. $0 > 5x - 10x^2$
41. $2x^2 + x \le 6$
42. $4 + x > x^2$
43. $3x^2 - 2x < 5$
44. $5x^2 \ge 4x - 3$
45. $x^2 + 2x > -1$
46. $4x^2 \le 12x - 9$
47. $x^2 - 5x + 7 < 0$
48. $x^2 + 1 \ge 0$
49. $1 - x^2 \ge 0$
50. $-2x^2 + x > 4$

51. If the cannon described in Example 6 fires at a 45° angle with the same initial velocity of 160 ft/second, then the trajectory of the shell is approximated by $y = 4 + x - 0.00125x^2$.
    a. Find the horizontal distance traveled by the shell before it hits the ground. Answer to the nearest tenth of a foot.
    b. To the nearest foot, what is the maximum height reached by the shell?

52. The height of a ball thrown directly up from a roof 144 ft high with an initial velocity of 128 ft/second is given by the formula $y = 144 + 128t - 16t^2$. What is the maximum height attained by the ball? When does the ball hit the ground?

53. The height of a projectile shot vertically upward from the ground with an initial velocity of 96 ft/second is given by the formula $y = 96t - 16t^2$. What is the maximum height attained by the projectile? When does the projectile hit the ground?

54. The total sales revenue for a company is estimated by the formula $S = 8x - x^2$, where x is the unit selling price (in dollars) and S is the total sales revenue (1 unit

equals \$10,000). What should be the unit selling price if the company wishes to maximize the total revenue?

55. In a 120-volt line having a resistance of 10 ohms, the power W in watts when the current I is flowing is given by the formula $W = 120I - 10I^2$. What is the maximum power that can be delivered in this circuit?

56. Find two positive numbers whose sum is 8 and whose product is a maximum.

57. Find two positive numbers whose sum is 20 and whose product is a maximum.

58. What positive number exceeds its square by the largest amount?

59. Find two positive numbers whose sum is 20 such that the sum of their squares is a minimum.

60. The sum of the base and altitude of a triangle is 12 in. Find each dimension if the area is to be a maximum.

61. A rectangular field is adjacent to a river and is to have fencing on three sides because the side of the river requires no fencing. If 400 yd of fencing are available, what are the dimensions of the largest rectangular section that can be enclosed with the available fencing?

62. A farmer wants to make a rectangular enclosure along the side of a barn and then divide the enclosure into two pens with a fence constructed at a right angle to the barn. If 300 ft of fencing are available, what are the dimensions of the largest section that can be enclosed with the available fencing?

63. A sheet of metal 12 in. wide and 20 ft long is to be made into a gutter by turning up the same amount of material on each edge at right angles to the base. Determine the amount of material that should be turned up to maximize the volume that the gutter can carry.

64. Find the maximum possible area of a rectangle with a perimeter of 100 ft.

65. A bus tour charges a fare of \$10 per person and carries 200 people per day. The manager estimates that she will lose 10 passengers for each \$1 increase in fare. Find the most profitable fare for her to charge.

66. An airline offers a charter flight at a fare of \$100 per person if 100 passengers sign up. For *each* passenger above 100, the fare for *each* passenger is reduced by 50 cents. What is the maximum total revenue that the airline can obtain if the plane has 200 seats?

67. The figure shown has a perimeter of 60 inches.
    - **a.** Find a formula for its area.
    - **b.** Find the value of x that maximizes the area.
    - **c.** Find the maximize area to 2 decimal places.

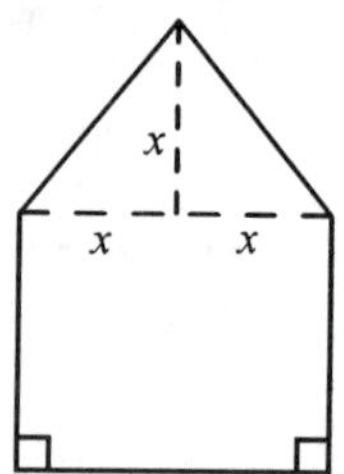

68. The figure shown has a perimeter of 60 inches. The curve is a semicircle.
    - **a.** Find a formula for its area.
    - **b.** Find the value of x that maximizes the area.
    - **c.** Find the maximum area to 2 decimal places.

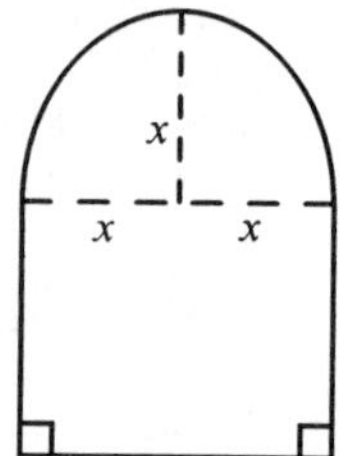

69. The figure below shows that for any value of x the vertical distance between the line and the parabola is given by the distance in their y values.
    - **a.** Write a formula for this vertical distance.
    - **b.** Find any points on the parabola and the line where the vertical distance is 3.
    - **c.** Find the points on the line and parabola that minimize the vertical distance.
    - **d.** Find the minimum distance.

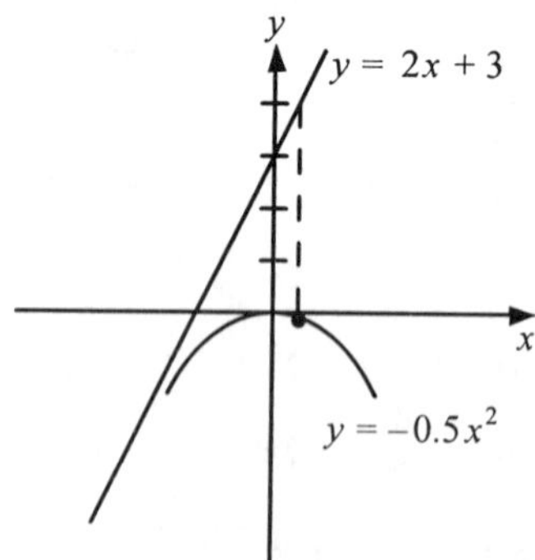

70. Sometimes mathematical models are used in business to ascertain the effect of certain decisions on sales. One example (given in *Applied Linear Statistical*

Models, by Neter and Wasserman, Irwin 1974) describes a study done for a chain of cafeterias, to analyze the relation between the number of self-service coffee dispensers in the cafeteria line and sales of coffee. The example presents the quadratic model $y = 503.35 + 78.94x - 3.97x^2$, where x is defined only from 0 to 7. In this model x represents the number of dispensers, and y is sales in hundreds of gallons. In this study no cafeteria had more than 7 coffee dispensers, but if the model continues to hold for greater numbers, what number of dispensers would yield maximum sales? [**Caution:** even if it were reasonable to extend the model until the maximum sales **volume** is reached, it would not be reasonable to extend the model beyond that. It is always risky to extend a model to data points outside the range of observed data. Extension beyond observations is called **extrapolation**; extrapolation is always chancy.]

THINK ABOUT IT 5.5

1. By trial and error find the value of c that places the vertex of the graph of $y = x^2 + 6x + c$ on the x-axis. Explain algebraically why this value of c produces the desired result.
2. Find an equation that defines the quadratic function f with the given properties.
   a. The graph of f passes through the origin and has its vertex at $(1, -4)$.
   b. The intercepts of the graph of f are (1, 0), (3, 0), and $(0, -2)$.
3. Create a quadratic inequality with the given solution set.
   a. $[-2, 5]$ **b.** $(-\infty, 0) \cup (3, \infty)$ **c.** $(-\infty, \infty)$ **d.** $\varnothing$
4. If f is a linear function and g is a quadratic function, is the function $f \circ g$ linear, quadratic, or neither? Also, answer this question for the function $g \circ f$. In both cases prove that your answer is correct.
5. By completing the square, place the quadratic function defined by $f(x) = ax^2 + bx + c$ with $a \neq 0$ in the form

$$f(x) = a(x-h)^2 + k$$

and show that the vertex is at

$$\left(\frac{-b}{2a}, \frac{4ac - b^2}{4a}\right).$$

● ● ●

CHAPTER 5 SUMMARY

OBJECTIVES CHECKLIST

Specific chapter objectives are summarized below along with numbered example problems from the text that should clarify the objectives. If you do not understand any objectives, or do not know how to do the selected problems, then restudy the material.

5.1: Can you:

1. **Create a table of values associated with a function?** A piece of molding is 46 in. long. You wish to use it to make a rectangular frame of sides x and z units, as shown.

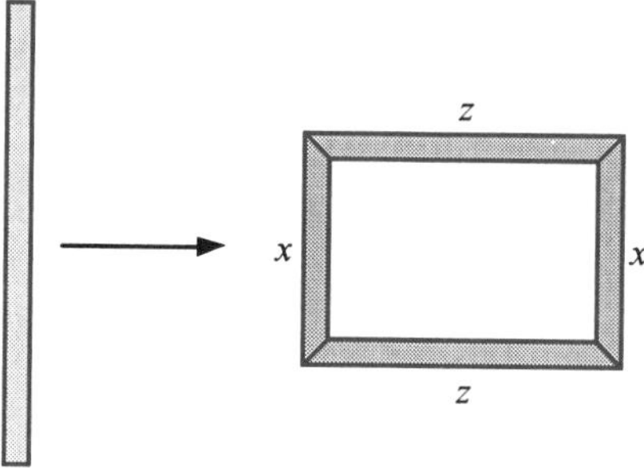

Construct a table that shows corresponding values for x, z, and the area of the rectangle. Let x vary from 4 to 24 in increments of 4.

[Example 1a]

2. **Find a formula that defines a function?** For the molding in the previous problem, write a formula that expresses the area of the rectangle as a function of x.

[Example 1b]

3. **Graph a function by the point-plotting method or by using a grapher?** Use a grapher to graph the formula from the pervious problem, and describe what happens to the area as x increases. What are the domain and range of this function?

[Example 1c]

4. **Find the domain and range of a function?** Determine the domain and range of $y = \sqrt{1-x^2}$.

[Example 2]

5. **Determine if a graph or a set of ordered pairs is a function?** Determine if the relation is a function? Also specify the domain and range.

   a. $\{(2, 1), (3, 1), (4, -1)\}$ **b.** $\{(1, 2), (1, 3), (-1, 4)\}$

[Example 4]

6. **Evaluate functions using functional notation?** If $y = f(x) = 7 + 1/x$, find $f(1)$, $f(1/3)$, $f(0)$, and $f(a)$.

[Example 5]

7. **Read from a graph of function *f* the domain, range, function values, values of *x* for which $f(x) = 0$, $f(x) < 0$, or $f(x) > 0$, and intervals where *f* is increasing, decreasing, or constant?** Consider the graph of $y = f(x)$ in the figure.

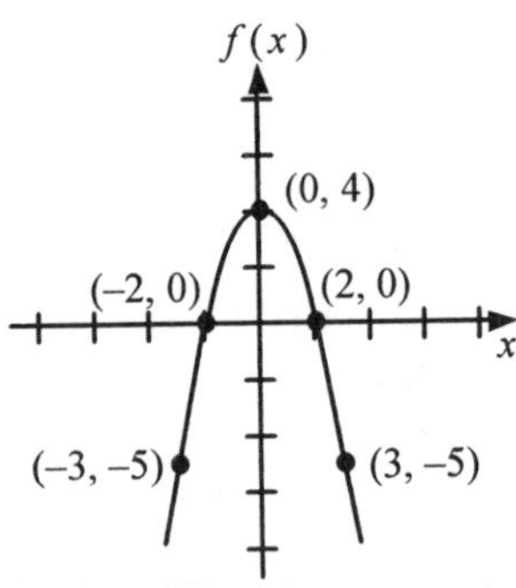

   a. What is the domain of *f*? **b.** What is the range of *f*?
   c. Is *f* even, odd, or neither? **d.** Find $f(0)$.
   e. Solve $f(x) = 0$. **f.** Solve $f(x) > 0$.
   g. Find the open intervals on which the function is increasing and decreasing

[Example 6]

8. **Prove or classify a function to be even, odd, or neither of these?** Is the function $f(x) = -2/x$ an even function, an odd function, or neither? The figure is the graph of this function for $x \geq 0$. Complete the graph and check your result with a grapher.

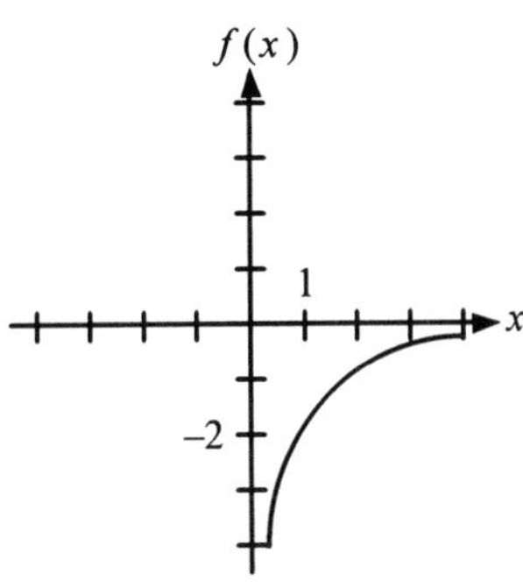

[Example 8]

5.2: Can you:

1. **Add, subtract, multiply, and divide two functions?** If $f(x) = x$ and $g(x) = \sqrt{x}$, find $(f+g)(x)$, $(f-g)(x)$, $(f \cdot g)(x)$, and $(f/g)(x)$. In each case give the domain of the resulting function.

   [Example 1]

2. **Find the composite function of two functions?** If $f(x) = 3x - 5$ and $g(x) = x^2$, find the following.
   a. $(f \circ g)(x)$ **b.** $(g \circ f)(x)$

   [Example 4]

3. **Interpret a function *h* as the composition of simpler functions *f* and *g* so that $h(x) = (f \circ g)(x)$?** If $h(x) = (2x+3)^3$, find simpler functions f and g so that $h(x) = (f \circ g)(x)$.

   [Example 9]

5.3: Can you:

1. **Find and interpret the slope of a line?** A ball thrown vertically up from a roof with an initial velocity of 128 ft/second continues to climb for 4 seconds, and its velocity is 96 ft/second, when $t = 1$ second, and 32 ft/second when $t = 3$. Find and interpret the slope of the line that describes this linear relationship.

   [Example 3]

2. **Find the average rate of change of a function for a given interval?** If $f(x) = 3x^2 + 1$,
   a. Find the formula for the average rate of change of this function from x to $x+h$.
   b. Evaluate this expression for $x = 1$, $h = 3$ and interpret the result.

   [Example 6]

5.4: Can you:

1. **Write an equation for a line, given its slope and a point on the line or given two points on the line?** Find an equation for the line that contains the point (1, 3) and whose slope is 2. Write the answer in the form $y = ax + b$.

   [Example 1]

2. **Find the slope and the y-intercept, given an equation for the line?** Find the slope and the y-intercept of the line defined by the equation $2x + 3y = 6$.

   [Example 3]

3. **Graph and write an equation for a line, given the slope and *y*-intercept?** Graph the line whose slope is $-2/3$ and whose y-intercept is $(0, 4)$. Also, find an equation for the line.

   [Example 4]

4. **Find the equation of a linear function *f*, given two ordered pairs in *f*?** Find the equation that defines the linear function *f* if $f(1) = 6$ and $f(-1) = 2$.

   [Example 5]

5. **Solve applied problems involving linear functions?** Find an equation that defines the linear relationship between Fahrenheit and Celsius temperatures if the freezing point of water is 32° Fahrenheit and 0° Celsius, and the boiling point is 212° Fahrenheit and 100° Celsius.

[Example 6]

6. **Solve problems involving the equations of parallel and perpendicular lines?** Find an equation for the line passing through the point (–4, 1) that is
   a. parallel to $5x - 2y = 3$ **b.** perpendicular to $5x - 2y = 3$

[Example 9]

5.5: Can you:

1. **Graph quadratic functions indicating the vertex and intercepts of the graph, and the range of the function?** Graph the function defined by $f(x) = x^2 - 7x + 6$ and indicate
   a. the coordinates of the x- and y-intercepts
   b. the equation of the axis of symmetry
   c. the coordinates of the maximum or minimum point
   d. the range of the function

[Example 1]

2. **Solve applied problems involving quadratic functions?** A manager wants to fence in parking lot, one side of which is bounded by a building. Because of budget constraints, however, only a total of 300 ft of fencing is available. What are the dimensions of the largest rectangular parking lot that can be enclosed with the available fencing?

[Example 4]

3. **Solve quadratic inequalities using a graphing approach?** Solve $x^2 - 6 > x$ using a graphical method.

[Example 7]

KEY CONCEPTS AND PROCEDURES

Section **Key Concepts and Procedures to Review**

5.1

- Rule definition of a function
- Ordered pair definitions of a relation and a function
- Definitions of domain, range, even function, odd function, and the graph of a function
- The graph of an even function is symmetric about the y-axis
- The graph of an odd function is symmetric about the origin
- Methods to graph variations of a familiar function by using vertical and horizontal shifts, reflecting, stretching, and shrinking
- Vertical line test
- Functional notation
- Domain, range, and graph of a constant function, and the identity, squaring, cubing, absolute value, and square root functions

5.2

- Definitions of $f + g$, $f - g$, $f \cdot g$, and f/g for two functions f and g
- The symbol " $\circ$ " denotes the operation of composition. The composite functions are $(f \circ g)(x) = f[g(x)]$ and $(g \circ f)(x) = g[f(x)]$.
- Methods to determine the domain of $f + g$, $f - g$, $f \cdot g$, f/g, $f \circ g$, $g \circ f$

5.3

- Slope formula: $m = \dfrac{\Delta y}{\Delta x} = \dfrac{y_2 - y_1}{x_2 - x_1}$ $(x_2 \neq x_1)$
- Average rate of change from x to $x + h$: $\dfrac{\Delta y}{\Delta x} = \dfrac{f(x+h) - f(x)}{h}$, $h \neq 0$

5.4

- Definition of linear function
- Point-slope equation: $y - y_1 = m(x - x_1)$
- Slope-intercept equation: $y = mx + b$
- For two lines with slopes m_1 and m_2:
  parallel lines: $m_1 = m_2$
  perpendicular lines: $m_1 \cdot m_2 = -1$
- In a table of a linear function $y = f(x)$, a fixed change in x produces a constant difference between the corresponding y-value.

5.5

- Definitions of quadratic function, quadratic inequality, and axis of symmetry
- Methods to graph a quadratic function
- Axis of symmetry formula: $x = -b/2a$
- Vertex formula: $(-b/2a,\ f(-b/2a))$
- The graph of $f(x) = a(x-h)^2 + k$ (with $a \neq 0$) is the graph of $y = ax^2$ (a parabola) shifted so the vertex is $(h,\ k)$ and the axis of symmetry is $x = h$.
- Methods to solve a quadratic inequality

CHAPTER 5 REVIEW EXERCISES

In Exercises 1 and 2 use $f(x) = 2x + 1$ and $g(x) = \sqrt{x}$.

1. Find $3f(-2) - 4g(9)$.

2. Find $(g - f)(4)$.

3. What is the domain of the function $y = \dfrac{1}{\sqrt{4 - x^2}}$?

4. What is the domain of $y = \dfrac{3x - 1}{4x + 2}$?

5. What is the range of $y = 2x - x^2$?

6. Determine the vertex and axis of symmetry of the graph of $f(x) = 3(x+5)^2 - 1$.

7. Write the equation of the line with an x-intercept of 3 and a y-intercept of –2.

8. If $g(x) = 3x - 4$, for what value of x does $g(x) = 0$?

9. Is the set of ordered pairs $\{(-3,\ 0),\ (0,\ 0),\ (1,\ 0)\}$ a function?

10. Show that $(f \circ f)(x) = x$ if $f(x) = 3 - x$.

In Exercises11–16 graph each function.

11. $y = -\sqrt{x}$ **12.** $y = |x - 4|$

13. $f(x) = 5 - 2x$ **14.** $y = 2$

15. $y = (x-1)^2 + 2$ **16.** $g(x) = x - x^2$

In Exercises 17 and 18 solve each inequality.

17. $x^2 + x \geq 6$ **18.** $x^2 > x - 1$

In Exercises 19–22 consider the function

$$f(x) = \begin{cases} 1 & \text{if } x > 2 \\ -x & \text{if } x \leq -2. \end{cases}$$

19. Find $f(-3)$. **20.** Graph the function.

21. What is the domain?

22. Solve $f(x) = 4$.

23. Line L_1 passes through $(-2,\ -2)$ and $(0,\ 3)$, and line L_2 passes through $(-4,\ k)$ and $(2,\ -1)$.

a. Find k if L_1 is parallel to L_2.

b. Find k if L_1 is perpendicular to L_2.

24. Is the function $f(x) = -2/x^2$ even, odd, or neither? The following figure is the graph of this function for $x \geq 0$. Complete the graph.

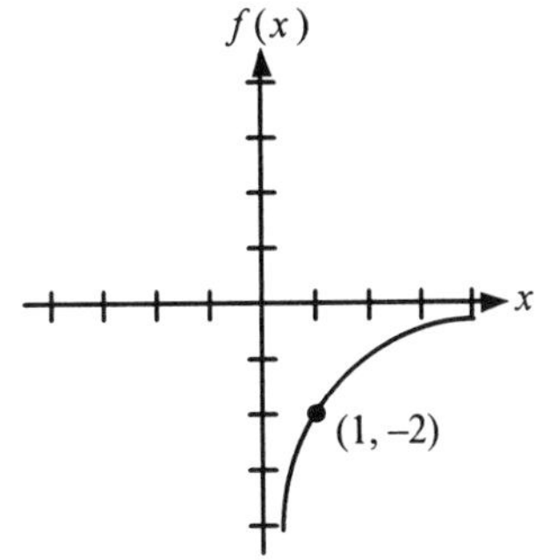

25. If the parabola $y = x^2 - 8x + k$ has its vertex on the x-axis, what is the value of k?

26. Find the x-intercepts of the graph of the function $f(x) = x^2 - x - 5$. Plot the approximate position of these points on the x-axis.

In Exercises 27–30 compute in simplest form $\frac{\Delta y}{\Delta x} = \frac{f(x+h) - f(x)}{h}$, $h \neq 0$ to determine the average rate of change of the function from x to $x + h$.

27. $f(x) = 3x^2 - 1$

28. $f(x) = -3$

29. $f(x) = 3 - x$

30. $f(x) = \frac{1}{x+1}$

31. Find the equation that defines the linear function f if $f(-1) = 6$ and $f(2) = -3$.

32. Give three ordered pairs in f if $f(x) = 7$.

33. Write an equation for the line that passes through the origin and is parallel to the line joining $(5, -1)$ and $(-2, 3)$.

34. If $f(x) = x^2$ and $g(x) = -1/x$, find $(f \circ g)(x)$. What is the domain of $f \circ g$?

35. Write a different function that has the same domain and range as the function $\{(1, 4), (2, 5)\}$.

36. If $h(x) = 3(1-x)^2$, find simpler functions f and g so that $h(x) = (f \circ g)(x)$.

In Exercises 37–44 use the following figure.

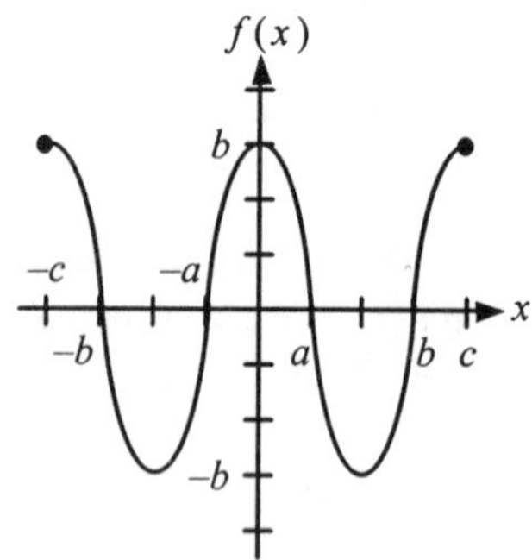

37. What is the domain?

38. What is the range?

39. Is f even, odd, or neither?

40. Find $f(b)$.

41. Solve $f(x) = b$.

42. Solve $f(x) < 0$.

43. Graph $y = -\frac{1}{2} f(x)$.

44. Graph $y = f(x - c)$.

45. Express the area of an equilateral triangle as a function of its side length (x).

46. Which of these tables represent linear functions?

(a)

x	y
0	12
1	10
2	8
3	6
4	4

(b)

x	y
0	3.5
1	3.6
2	3.7
3	3.8
4	3.9

(c)

x	y
0	1
1	2
2	4
3	7
4	11

(d)

x	y
0	8
1	9
2	10
3	9
4	8

47. Express the monthly earnings (e) of a salesperson in terms of the case amount (a) of merchandise sold if the salesperson earns \$500 per month plus 9 percent commission on sales above \$2,000.

48. Show that the largest rectangular field that can be enclosed by a given length of fence is a square.

49. A manager owns a parking lot that she wants to enclose with a fence. One side of the lot is bounded by a building, so no fencing is required there. There is to be a 16-ft opening in the front of the fence, which is opposite the building. If side fencing costs \$1 per foot and front fencing costs \$1.50 per foot, what are the dimensions of the largest rectangular lot that can be fenced for \$300?

50. A salesperson receives a monthly salary plus a commission on sales. She earned \$1,050 during a month in which her sales totaled \$5,000; the next month, she sold \$6,000 worth of merchandise and earned \$1,150.
   a. Find the equation that defines this linear relationship.
   b. How much will she earn if she sells \$9,000 worth of merchandise for the month?
   c. What is her monthly salary?
   d. What is her rate of commission?

51. A projectile shot vertically upward has a velocity of 192 ft/second at $t = 4$ seconds, and 96 ft/second when $t = 7$ seconds.
    a. Find the equation that defines this linear relationship.
    b. Find the velocity of the projectile when $t = 9$ seconds.
    c. What is the initial velocity of the projectile (velocity when $t = 0$)?
    d. When is the velocity 0? What is the physical significance of this?

52. A computer manufacturer sells 6,000 units per week of its \$3,000 model? Because of a backlog of computers, a rebate program is introduced. It is estimated that each \$100 decrease in price will result in the sale of 300 more units. The weekly revenue R is thus given by the formula $R = (6,000 + 300x)(3,000 - 100x)$, where x is the number of \$100 discounts offered. What rebate should be given to produce maximum revenue?

In Exercises 53–62 select the choice that completes the statement.

53. A rectangular field is completely enclosed by x feet of fencing. If the length of the rectangular region is twice the width, then the area of the field as a function of x equals

    a. $\frac{x^2}{18}$ b. $\frac{x^2}{72}$ c. $\frac{x^2}{16}$ d. $\frac{x^2}{36}$

54. The slope of the line $3x - 2y = 6$ is

    a. 3 b. −3 c. $\frac{3}{2}$ d. $-\frac{3}{2}$

55. A quadratic function with no x-intercepts in its graph is

    a. $y = x^2 - 4x + 5$ b. $y = x^2 - 5x + 4$
    c. $y = x^2 + 5x - 4$ d. $y = x^2 + 4x - 5$

56. The equation of the line that passes through the point (−3, 1) and is perpendicular to $y = \frac{1}{2}x + 4$ is

    a. $y = \frac{1}{2}x - \frac{1}{4}$ b. $y = \frac{1}{2}x + \frac{5}{2}$
    c. $y = 2x + 7$ d. $y = -2x - 5$

57. If $f = \{(0, -3)\ (-1, -2), (2, -1)\}$ and $g = \{(-1, 2)\ (-2, 1), (-3, 3)\}$, then $(g \circ f)(-1)$ equals

    a. −1 b. 1 c. 2 d. −2

58. The range of the function $f(x) = 4 - 4x^2$ is

    a. $(-\infty, 0]$ b. $(-\infty, \infty)$
    c. $[-4, \infty)$ d. $(-\infty, 4]$

59. The axis of symmetry for $y = 2x^2 + 4x - 1$ is

    a. $y = 1$ b. $x = 2$
    c. $x = -1$ d. $y = 4$

60. If $f(1) = 6$ and $f(-1) = 2$, and f is a linear function, then $f(7)$ equals

    a. 11 b. 9
    c. 14 d. 18

61. If $f(x) = (x-1)^2$, then $\frac{f(x+h) - f(x)}{h}$ with $h \neq 0$ in simplest form is

    a. $2x$ b. $2h$
    c. $2x + h - 2$ d. $2x + h$

62. The graph of an even function is symmetric with respect to the

    a. x-axis b. y-axis c. origin

CHAPTER 5 TEST

1. If $f(x) = \frac{2}{x-9}$, find the domain of f.

2. What is the range of the function defined by $y = 6 - |x|$?

3. Graph $y = (x+2)^2 - 1$.

4. Graph $y = \begin{cases} x & \text{if } x \geq 0 \\ 2 & \text{if } x < 0. \end{cases}$

5. Is the function defined by $f(x) = 3 - x^2$ even, odd, or neither of these?

6. If $f = \{(0, -3), (1, 2), (-1, 0)\}$ and $g = \{(0, 5), (1, -4), (2, 0)\}$, find $f + g$.

7. Find $(g \circ f)(x)$ if $f(x) = 3x^2 - 1$ and $g(x) = 2x + 5$.

8. If $h(x) = \sqrt{25 - x^2}$, find simpler functions f and g so that $h(x) = (f \circ g)(x)$.

9. Find the domain and range of the function graphed below.

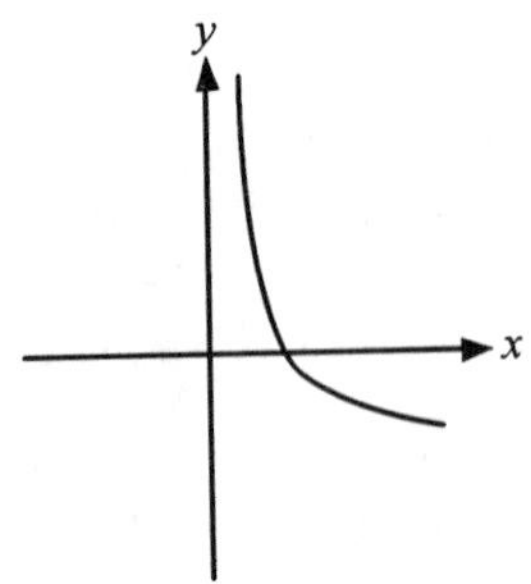

10. A rectangular area of sides *x* and *y* units is enclosed using 800 ft of fencing. Find a formula for the enclosed area as a function of *x*.

11. The total cost of manufacturing 100 units of a certain product is \$560, while the cost of manufacturing 250 units is \$740. If the relation between cost and units produced is linear, calculate the slope to determine the rate of change in this relation. Also, interpret the meaning of the slope.

12. Compute $\frac{f(x+h)-f(x)}{h}$ with $h \neq 0$ in simplest form if $f(x) = x^2 - 5x + 3$.

13. Find the slope and the *y*-intercept of the line given by $6x + 2y = 5$.

14. Graph the line whose slope is $-3/2$ and whose *y*-intercept is $(0, 2)$. Also, find an equation for the line.

15. Write an equation for the line passing through (–4, 1) that is perpendicular to the line given by $y = -\frac{1}{2}x + 3$.

16. Find the equation that defines the linear function *f* if $f(0) = -3$ and $f(5) = 0$.

17. Find the coordinates of the *x*-intercepts of the graph of $g(x) = 2x^2 - x - 3$.

18. Find the range of the function defined by $y = 6x - x^2$.

19. Solve $x^2 - 6 \leq 2x$.

20. The height (*y*) of a projectile shot vertically upward from the ground with an initial velocity of 128 ft/second is given by the formula $y = 128t - 16t^2$. What is the maximum height attained by the projectile?

• • •

Chapter 6

Polynomial and Rational Functions

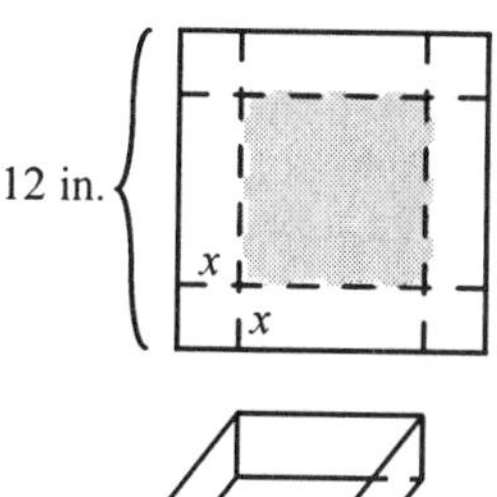

A manufacturer wants to convert a square piece of cardboard with side length 12 in. into an uncovered box, by cutting a small square of cardboard from each corner and bending up the sides as shown.

a. Find a formula for the volume V of the box in terms of the length (x) of the side of the square that is cut from each corner.

b. What is the domain of this volume function?

c. Use a grapher to graph this volume function and to estimate to two decimal places the length of the corner to cut out to obtain the box with maximum volume.

(See Example 9 of Section 6.1.)

In this chapter we continue with the theme of a function and discuss many ideas associated with a large and important class of functions called polynomial functions. Basically, a polynomial function is characterized by terms of the form:

$$(\text{real number})\,x^{\text{nonnegative integer}}$$

so that a polynomial function may contain terms such as 4, $2x$, $-3x^2$, $\sqrt{5}x^3$ and so on. More technically, a **polynomial function of degree *n*** is a function of the form

$$y = P(x) = a_n x^n + a_{n-1}x^{n-1} + \cdots + a_1 x + a_0 \quad (a_n \neq 0),$$

where n is a nonnegative integer and a_n, a_{n-1}, $\ldots$, a_1, and a_0 are real-number constants. For example, the function $P(x) = 5x^3 + 2x^2 - 1$ is a polynomial function of degree 3 in which $a_3 = 5$, $a_2 = 2$, $a_1 = 0$, and $a_0 = -1$. Since n must be a nonnegative integer, note that functions with terms such as $x^{1/2}$ (or $\sqrt{x}$) and x^{-2} (or $1/x^2$) are not polynomial functions. Although $y = 1/x^2$ is not a polynomial function, it can be expressed as a quotient of two polynomials and is called a **rational function**. This chapter concludes by showing how to graph rational functions.

● ● ●

6.1 Zeros and Graphs of Polynomial Functions

Objectives

1. Determine if a function is a polynomial function.
2. Write a polynomial function, given its degree and zeros.
3. Find the zeros of a polynomial function in factored form.
4. Use translation, reflection, stretching, and shrinking to graph certain variations of $y = x^n$.
5. Graph a polynomial function, by using information obtained from the intercepts or by using a grapher.
6. Write an equation for a polynomial function, using information obtained from the intercepts.

The most widely applied family of functions is the polynomial functions. From the definition of a polynomial function given in the chapter introduction, you should recognize that we have already discussed the following polynomial functions.

1. The constant function $y = P(x) = c$ (with $c \neq 0$) is a polynomial function of degree 0.
2. The linear function $y = P(x) = mx + b$ (with $m \neq 0$) is a polynomial function of degree 1.
3. The quadratic function $y = P(x) = ax^2 + bx + c$ (with $a \neq 0$) is a polynomial function of degree 2.

The next example will help you recognize other polynomial functions.

Example 1: Recognizing a Polynomial Function

Determine whether or not the function is a polynomial function. If yes, state the degree.

a. $f(x) = 4x^5 - 2x^{1/2} - 3$ **b.** $f(x) = 4x^5 - \frac{1}{2}x^2 - 3$

Solution:

a. Every term in a polynomial in a single variable, say x, may be written in the form ax^n, where $n = 0, 1, 2, 3, 4, 5, \ldots$, and a is a real number constant. Thus, $f(x) = 4x^5 - 2x^{1/2} - 3$ is not a polynomial function because of the term $-2x^{1/2}$.

b. $f(x) = 4x^5 - (1/2)x^2 - 3$ is a polynomial function. The degree of this polynomial function is 5, because the highest-degree term $4x^5$ has degree 5. Observe that $-3 = -3x^0$, so this constant term has degree 0.

PROGRESS CHECK 1

Determine whether or not the function is a polynomial function. If yes, state the degree.

a. $f(x) = 50 - x^6$ **b.** $f(x) = 50 + x^{2.5}$ ■

In this section we are mainly concerned with graphing polynomial functions and finding their zeros. A **zero** of a function f is a value of x for which $f(x) = 0$. For example, the zeros of the function $f(x) = x^2 - 5x + 4$ are 1 and 4, since

$$f(1) = (1)^2 - 5(1) + 4 = 0$$

and

$$f(4) = (4)^2 - 5(4) + 4 = 0.$$

Zeros is merely a new name applied to a familiar concept. Depending on the frame of reference, consider three closely related names applied to the preceding example.

1. The **solutions** or **roots** of the **equation** $x^2 - 5x + 4 = 0$ are 1 and 4.
2. The **zeros** of the **function** $f(x) = x^2 - 5x + 4$ are 1 and 4.

3. The **x-intercepts** of the **graph** of the function $y = x^2 - 5x + 4$ are (1, 0) and (4, 0) as shown in Figure 6.1.

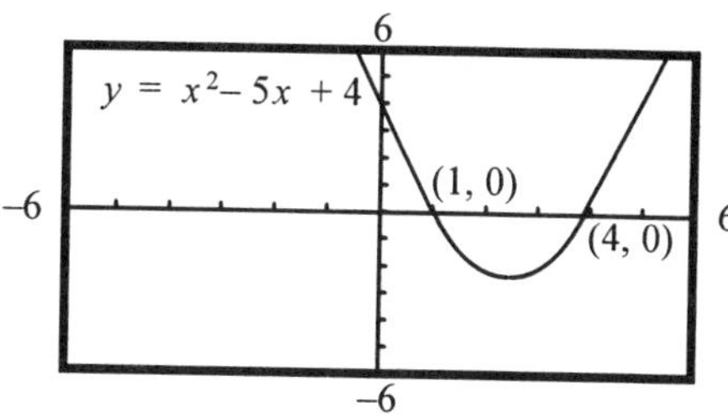

Figure 6.1

Finding zeros for polynomial functions of degrees greater than 2 using algebraic methods is often difficult. There is no easy formula (like the quadratic formula) that can be used to produce them, and numerical methods that require a grapher or a computer are frequently used to approximate such zeros. Nevertheless, there are some theorems you should be aware of concerning zeros of polynomial functions. The first one we consider is called the *factor theorem.*

Factor Theorem

If b is a zero of the polynomial function $y = P(x)$, then $x - b$ is a factor of $P(x)$; conversely, if $x - b$ is a factor of $P(x)$, then b is as zero of $y = P(x)$.

We can illustrate this theorem using $P(x) = x^2 - 5x + 4$. Observe that the zeroes of this function are 1 and 4, as shown above, while $x^2 - 5x + 4$ factors as $(x-1)(x-4)$. The factor theorem indicates that if 1 and 4 are zeroes of $y = P(x)$, then $x - 1$ and $x - 4$ are factors of $P(x)$; conversely, if $x - 1$ and $x - 4$ are factors of $P(x)$, then 1 and 4 are zeros of $y = P(x)$. The following two examples illustrate the usefulness of the factor theorem (which will be proven formally in the next section).

Example 2: Using the Factor Theorem

Write a polynomial function (in factored form) of degree 3 with zeros of 2, –4, and 3.

Solution: If 2, –4, and 3 are zeros of $y = P(x)$, then by the factor theorem, $x - 2$, $x - (-4)$, and $x - 3$ are factors of $P(x)$. Thus, a possible polynomial function is $P(x) = (x-2)(x+4)(x-3)$.

PROGRESS CHECK 2

Write a polynomial function (in factored form) of degree 3 with zeros of –5, 0, and 1. ■

Example 3: Using the Factor Theorem

If $P(x) = x(x+2)(x-5)^2$, what are the zeros of the function?

Solution: Since x, $x + 2$, and $x - 5$ are factors of $P(x)$, we conclude by the factor theorem that 0, –2, and 5 are zeros of the function. Remember that from a slightly different viewpoint we can also conclude that 0, –2, and 5 are roots or solutions of the equation $x(x+2)(x-5)^2 = 0$, and that the graph of $P(x) = x(x+2)(x-5)^2$ intersects the x-axis at 0, –2, and 5, as shown with the aid of a grapher in Figure 6.2.

PROGRESS CHECK 3

If $P(x) = 4x^2(x-1)(x+3)$, what are the zeros of the function? ■

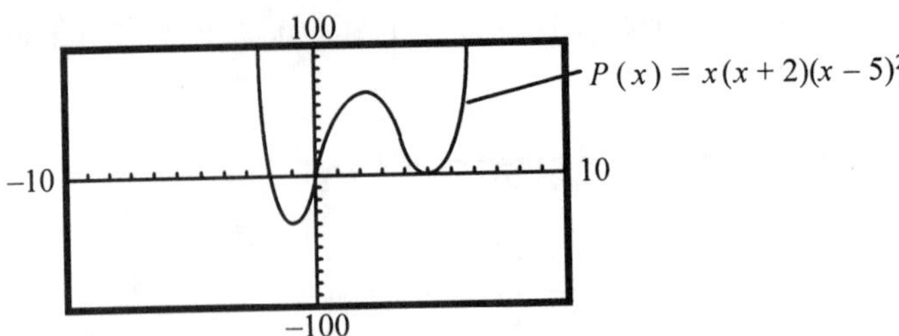

Figure 6.2

Note in Example 3 that the function defined by $P(x) = x(x+2)(x-5)^2$ is shorthand for $P(x) = x(x+2)(x-5)(x-5)$. In this case we have four factors but only three zeros, with the zero of 5 being repeated from two different factors. It is useful to adopt a convention that indicates how many factors of $P(x)$ result in the same zero. Using this convention in this example, we say 5 is a **zero of multiplicity** 2, since there are two factors of $x-5$. Similarly, for $P(x) = (x-3)^4$ we say 3 is a zero of multiplicity 4. In general, the multiplicity of zero b is given by the highest power of $x-b$ that is a factor of $P(x)$.

Example 4: Finding the Multiplicity of a Zero

If $P(x) = x^4(x-2)(x+7)^3$, find the zeros of the function as well as the multiplicity of each zero. What is the degree of the polynomial function?

Solution: $P(x)$ contains only three different factors: x, $x-2$, and $x+7$. Thus, 0, 2, and −7 are the three distinct zeros of the function. However, since x appears as a factor four times, 0 is a zero of multiplicity 4. Similarly, $(x+7)^3$ means −7 is a zero of multiplicity 3, while $(x-2)^1$ means 2 is a zero of multiplicity 1 (called a **simple zero**). The degree of the polynomial function is given by the sum of the multiplicities of the zeros. Thus, the polynomial function is of degree 8.

PROGRESS CHECK 4

If $P(x) = (x-4)^2(x+1)$, find the zeros of the function as well as the multiplicity of each zero. What is the degree of the polynomial function? ■

Knowing the multiplicity of a real-number zero is useful when graphing polynomial functions. Before we see how, let us first consider the graphs of polynomial functions of the form

$$y = x^n$$

in which y equals a power of x. We have already graphed such functions for $n = 1$, 2, and 3, and the graphs of $y = x$, $y = x^2$, and $y = x^3$ are shown in Figure 6.3 for reference purposes.

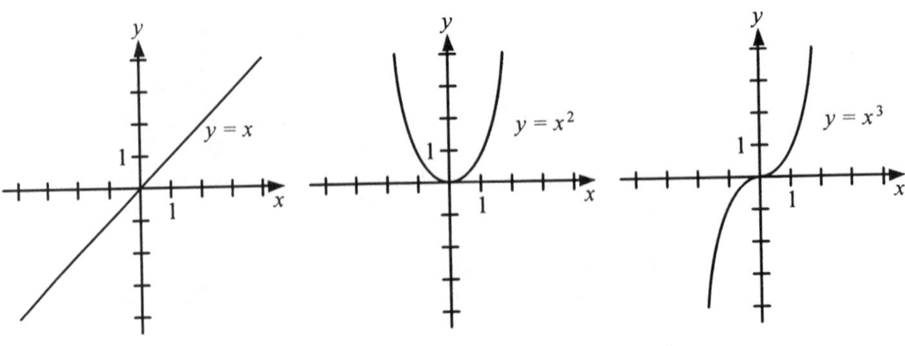

Figure 6.3

For larger values of n, the graph of $y = x^n$ is similar to the graph of $y = x^2$ if n is even, and it is similar to the graph of $y = x^3$ is n is odd. To see this, consider Figure 6.4. Note that when n is even, the graph is symmetric with respect of the y-axis and passes through $(-1, 1)$, $(0, 0)$, and $(1, 1)$. When n is odd, the graph is symmetric with respect to the origin and passes through $(-1, -1)$, $(0, 0)$, and $(1, 1)$. In both bases as the exponent increases, the graph becomes flatter on the interval $[-1, 1]$ and rises or falls more quickly for $|x| > 1$. To graph certain variations of $y = x^n$, we may use the graphing techniques discussed in Section 1.3, as shown in the next example.

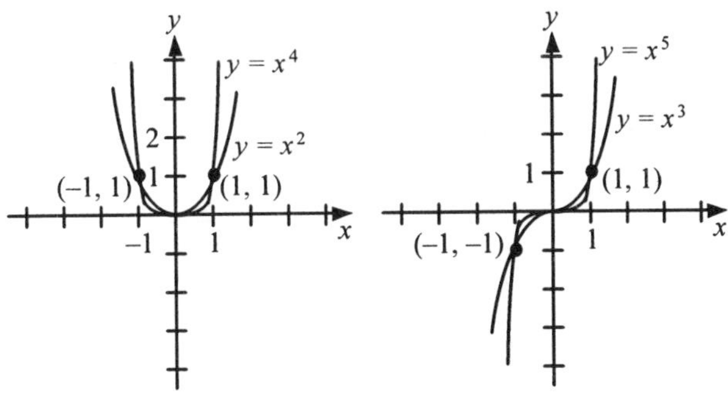

Figure 6.4

Example 5: Graphing Variations of $y = x^4$

Use the graph of $f(x) = x^4$ to graph each function.

a. $y = -x^4$ **b.** $y = 3x^4$ **c.** $y = (x+1)^4 - 2$

Solution:

a. The graph of $y = -f(x)$ is the graph of $y = f(x)$ reflected about the x-axis. In Figure 6.5(a) the graph of $y = x^4$ has been reflected about the x-axis to obtain the graph of $y = -x^4$.

b. If $f(x) = x^4$, then $y = 3x^4 = 3f(x)$. To graph $y = 3f(x)$, we triple each y value in f and stretch the graph of f by a factor of 3. Thus, the graph in Figure 6.5(b) is the graph of $y = 3x^4$.

c. The graphs of $y = f(x+c)$ and $y = f(x) + c$ are horizontal and vertical translations of the graph of f, respectively. To graph $y = (x+1)^4 - 2$, we translate the graph of $y = x^4$, 1 unit to the left and 2 units down, to obtain the graph in Figure 6.5(c).

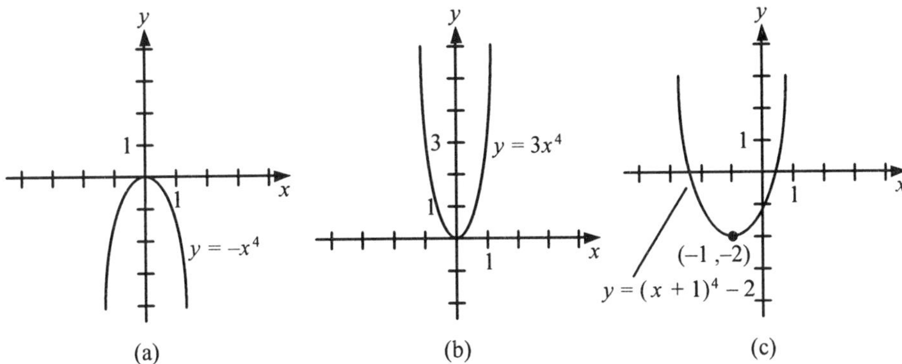

Figure 6.5

PROGRESS CHECK 5
Use the graph of $f(x) = x^4$ to graph each function.

a. $y = -\frac{1}{4}x^4$ **b.** $y = (x-2)^4 - 10$ ■

We have already indicated that the real-number zeros of $y = P(x)$ also give the x-intercepts in the graph of $y = P(x)$. In the case of $y = ax^n$, note that the only real-number zero is 0; and that when n is odd, the graph crosses the x-axis at $x = 0$, while the graph turns around and stays on the same side of the x-axis if n is even. This behavior occurs because x values change sign as x passes through 0, so y changes sign when n is odd, and y keeps the same sign when n is even. This type of analysis is applicable to all x-intercepts and indicates that the multiplicity of each real-number zero reveals whether or not the graph crosses the x-axis at such intercepts, according to the following theorem.

Graph of $y = P(x)$ near x-intercepts

> If b is a real-number zero with multiplicity n of $y = P(x)$, then the graph of $y = P(x)$ crosses the x-axis at $x = b$ if n is odd, while the graph turns around and stays on the same side of the x-axis at $x = b$ if n is even.

To use this theorem, we need to express $P(x)$ in factored form. Then the x-intercepts (or real-number zeros) can be obtained from the factor theorem, while the behavior of the graph at an x-intercept, say $(b, 0)$, can be determined from the multiplicity of zero b [or the highest power of $x = b$ that is a factor of $P(x)$]. We illustrate this theorem in the next three examples. The smooth, unbroken types of curves shown in the examples are characteristic of the graphs of polynomial functions.

Example 6: Using Intercepts to Graph a Polynomial Function

Graph $y = (x+1)(x-2)^2$ on the basis of information obtained from the intercepts.

Solution: By setting $x = 0$, we determine that the y-intercept is $(0, 4)$. Since $x+1$ is a factor with an odd exponent, we conclude that $(-1, 0)$ is an x-intercept at which the graph crosses the x-axis. Since $(x-2)^2$ is a factor with an even exponent, we conclude that $(2, 0)$ is an x-intercept at which the graph touches the x-axis and then turns around. We also note that as x gets very large, so does y; while as x increases in magnitude in the negative direction, y becomes very small. Figure 6.6(a) illustrates our results so far. Finally, since the function is a polynomial function, we draw a smooth, unbroken curve that satisfies the conditions in Figure 6.6(a), and we obtain the graph in Figure 6.6(b).

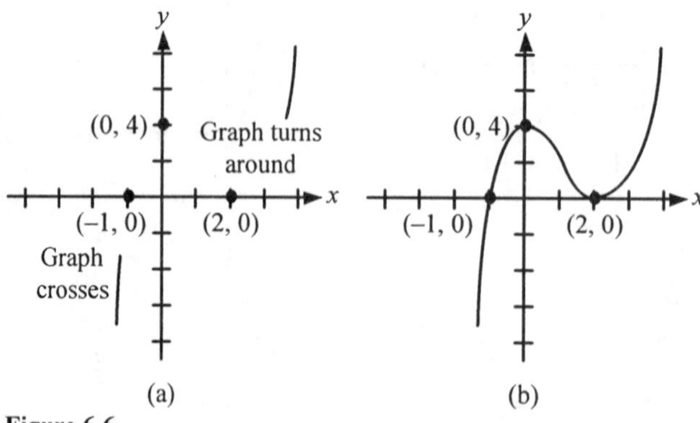

Figure 6.6

PROGRESS CHECK 6

Graph $y = (x-1)^2(x+2)$ on the basis of information obtained from the intercepts. ■

With the aid of the polynomial graphs drawn to this point, we now discuss some important properties of the graphs of polynomial functions.

Continuity

The graph of a polynomial function is a continuous curve. This property guarantees that polynomial graphs cannot have breaks such as those illustrated in Figure 6.7.

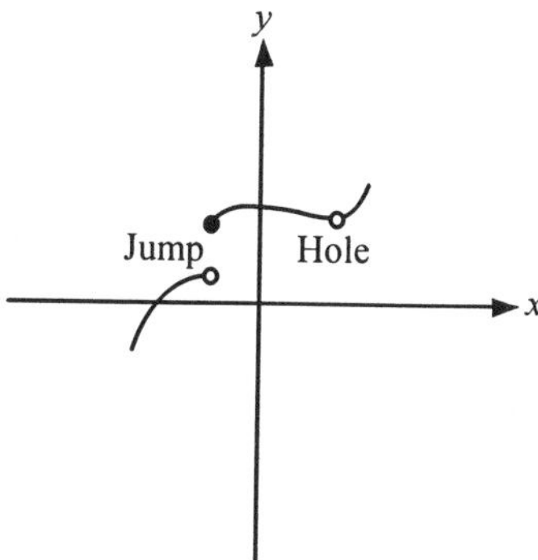

Figure 6.7

Turning Points

Points at which the graph changes from rising to falling, or vice versa, are called **turning points**, and it can be shown that a polynomial function of degree n has at most $n-1$ turning points. For example, Figure 6.8 shows the two turning points for the third-degree polynomial graphed in Example 6. Note that it is possible for a polynomial function of degree n to have fewer than $n-1$ turning points. For instance, $y = x^3$ is a polynomial function of degree 3 with no turning points. All turns in polynomial graphs are rounded turns, so a sharp turn (or corner), as illustrated in Figure 6.9, cannot occur in these graphs.

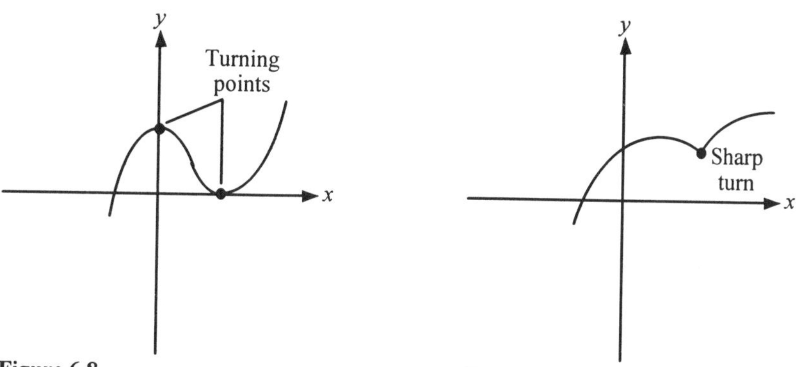

Figure 6.8 **Figure 6.9**

Behavior for Large $|x|$

The graph of the nth-degree polynomial function

$$y = a_n x^n + \cdots + a_1 x + a_0$$

resembles the graph of $y = a_n x^n$ when $|x|$ is large. This behavior occurs because for x values far from the origin, the leading term $a_n x^n$ is much larger than the sum of all other terms in the polynomial. Thus, the behavior of a polynomial graph for large $|x|$ depends on whether n is even or odd, and on the sign of a_n, as specified in Figure 6.10.

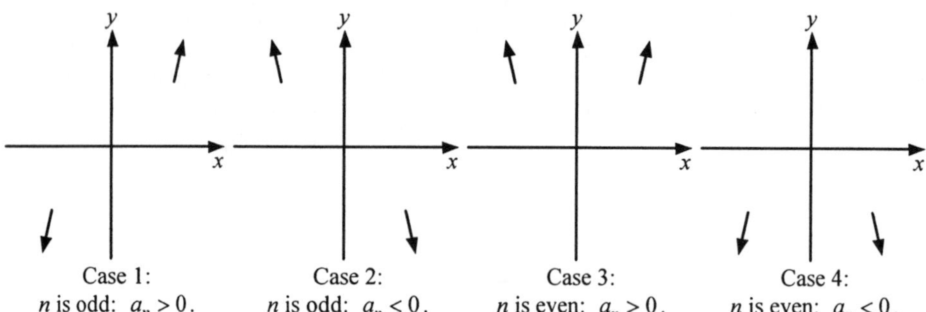

Figure 6.10

Example 7: Using Intercepts to Graph a Polynomial Function

Graph $y = x^5 - 4x^3$ on the basis of information obtained from the intercepts.

Solution: By setting $x = 0$, we determine that (0, 0) is a y- (and x-) intercept. To find x-intercepts, we next factor the polynomial as follows:

$$y = x^5 - 4x^3 = x^3(x^2 - 4) = x^3(x+2)(x-2).$$

Since x^3, $x+2$, and $x-2$ are factors, we conclude that the x-intercepts are (0, 0), (–2, 0), and (2, 0). All of these intercepts are derived from factors with odd exponents, so the graph crosses the x-axis at each of these points. By using this information and observing that the graph behaves like $y = x^5$ (or case 1 in Figure 6.10) for large values of $|x|$, we draw the graph in Figure 6.11. Keep in mind that our current methods produce only a rough sketch of the graph. A more detailed graph of $y = x^5 - 4x^3$ that is obtained on a grapher is given in Figure 6.12. Note that the graph indicates turning points when $x \approx \pm 1.55$ and an unanticipated wavy pattern between the turning points. Determining such subtleties is best left to displaying functions on a grapher or to analysis in a calculus course.

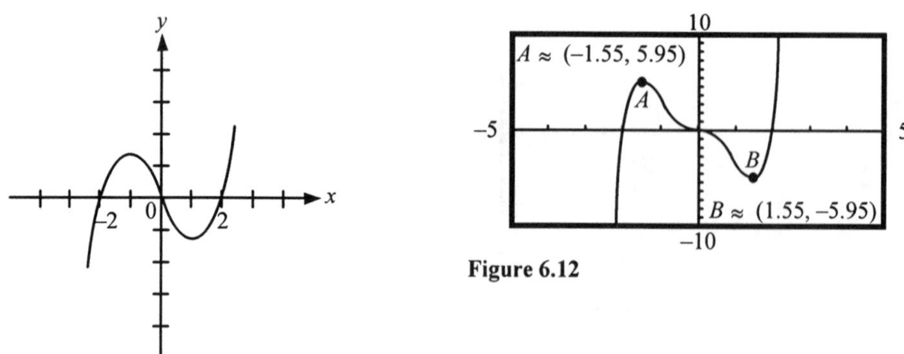

Figure 6.12

Figure 6.11

PROGRESS CHECK 7

Graph $y = x^4 - 9x^2$ on the basis of information obtained from the intercepts. ■

Example 8: Writing an Equation for a Polynomial Graph

Find an equation for the third-degree polynomial function that is graphed in Figure 6.13.

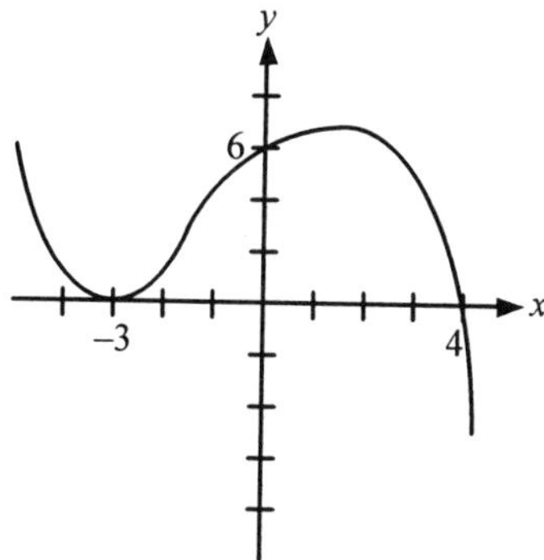

Figure 6.13

Solution: Let $y = P(x)$ define the third-degree polynomial function pictured. Because the graph touches the x-axis at $x = -3$ and then turns around, $(x+3)^2$ is a factor of $P(x)$. Since the graph crosses the x-axis at $x = 4$, another factor of $P(x)$ is $x - 4$. Thus, the equation fits the form

$$y = k(x+3)^2(x-4).$$

To find k, observe that the y-intercept is $(0,\ 6)$ so

$$\begin{aligned} 6 &= k(0+3)^2(0-4) \\ 6 &= -36k \\ -\frac{1}{6} &= k. \end{aligned}$$

So, an equation for the function is

$$y = -\frac{1}{6}(x+3)^2(x-4).$$

Graph this function with the aid of a grapher to check this result.

PROGRESS CHECK 8

Find an equation for the third-degree polynomial function that is graphed in Figure 6.14. ■

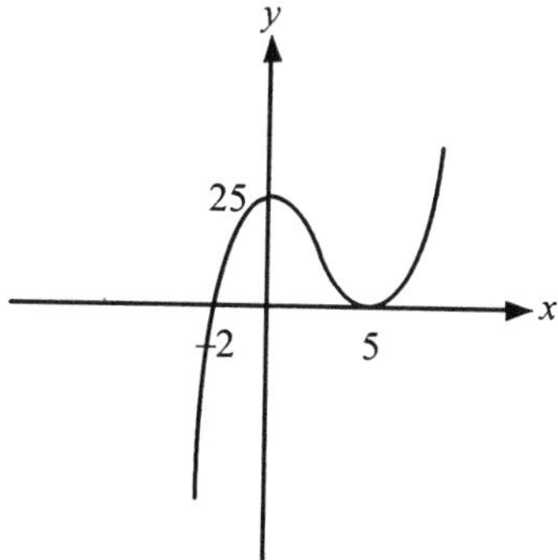

Figure 6.14

Among the simple applications of third-degree polynomial functions are problems involving the volume of a box. The next example illustrates a problem of this type.

Example 9: Volume of a Box

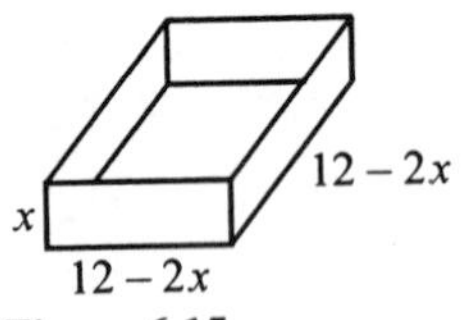

Figure 6.15

Solve the problem in the chapter introduction on page 463.

Solution:

a. When a square of side length x is cut from each corner, then x is the height of the box, and $12-2x$ is both the length and the width of the box, as shown in Figure 6.15. Using $V = lwh$, the volume of the box is given by

$$V = (12-2x)(12-2x)x,$$

or

$$V = x(12-2x)^2.$$

b. In the context of the problem, both x and $12-2x$ must be positive, so the domain is the set of real numbers between 0 and 6, which is (0 in., 6 in.) in interval notation.

c. Figure 6.16 shows a complete graph of $y = x(12-2x)^2$. However, the given formula is meaningful only for $0 < x < 6$, so we regraph the equation using this domain restriction, as shown in Figure 6.17. To find the value of x that yields the maximum volume, you may use the Zoom and Trace features (or a Function Maximum feature, if available) to determine that $x \approx 2.00$ and $y \approx 128$ at the highest point on the graph. So the maximum volume is obtained when a 2.00 in. square is cut from each corner.

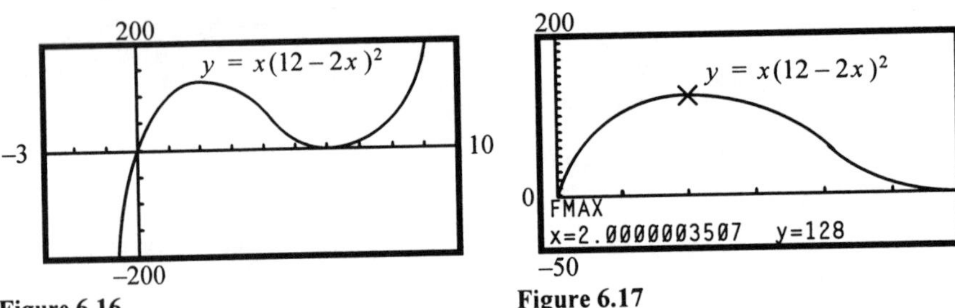

Figure 6.16 **Figure 6.17**

Note: To graph $y = x(12-2x)^2$ using the theorems in this section, it is useful to factor out –2 from each factor of $12-2x$, and then rewrite the equation as $y = 4x(x-6)^2$. In this form it is apparent that the graph crosses the x-axis at $x = 0$ and that the graph turns when it intersects the x-axis at $x = 6$.

PROGRESS CHECK 9

Redo the questions in Example 9 assuming that the manufacturer starts with a rectangular piece of cardboard that measures 18 in. by 12 in. ■

EXPLORE 6.1

1. Graphs of polynomial functions are smooth continuous curves with no breaks, where the number of turning points is limited by the degree of the polynomial. By trying different values for the constants, complete this table. Be sure to include some negative values.

Polynomials and Turning Points

Degree	Polynomial Form ($a \neq 0$)	Maximum Number of Turning Points
1	$ax+b$	0
2	ax^2+bx+c	
3	ax^3+bx^2+cx+d	
4	$ax^4+bx^3+cx^2+dx+e$	

What seems to be the relation between the degree and the maximum number of turning points?

2. The graph of a third-degree polynomial function, $y=ax^3+bx^2+cx+d$, cannot have more than two turning points. Thus, there are a limited number of basic shapes for third-degree polynomial functions. They are shown in the figures below.

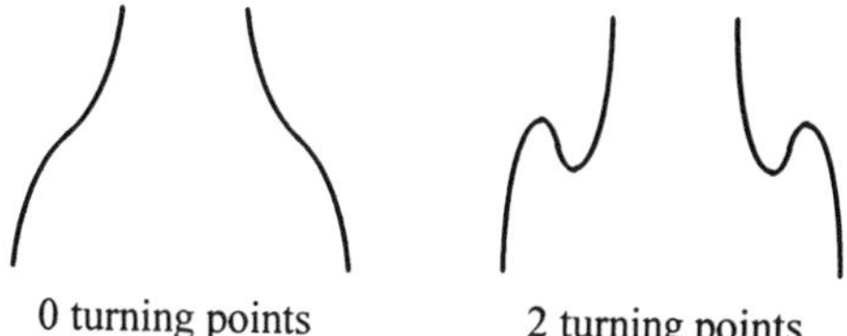

0 turning points 2 turning points

 a. By trying different values for a, b, c, and d, find formulas that yield graphs that have each of the given shapes.
 b. Sketch all the possible shapes of a fourth-degree polynomial, using the fact that a fourth-degree polynomial cannot have more than 3 turning points. Find formulas that yield graphs that have each of the given shapes.

3. By the "global" shape of a graph we mean the overall shape that appears as you repeatedly zoom out. The global shape of a polynomial graph is determined by its degree. For instance, all second-degree polynomials have parabolic graphs, essentially the shape of $y=x^2$ or $y=-x^2$. Similarly, all third-degree polynomials look like one of the shapes in the figure above but if you zoom out, the "middle" behavior fades, and the global shape is simply like $y=x^3$ or $y=-x^3$. Check this behavior by graphing $y=x^3-8x^2-10x+2$ with the following viewing windows.
 a. $[-3, 10]$ by $[-10, 10]$
 b. $[-5, 12]$ by $[-200, 100]$
 c. $[-10, 15]$ by $[-1000, 500]$
 d. $[-100, 100]$ by $[-10{,}000, 10{,}000]$

4. Some graphing calculators have a built-in operation that finds an x-intercept of a graph in a specified interval. This operation works best when the graph actually crosses the x-axis somewhere in the interval. But, curiously, it may fail with some cases that you could do mentally with ease. This is particularly the case for zeros of even multiplicity, where the graph turns on the x-axis. If your calculator has such a feature, illustrate this point by using this feature to find the x-intercepts of the graph of $P(x)=(x-2)^4$. You may get some type of error message.

EXERCISES 6.1

In Exercises 1-10 is the function a polynomial function? If yes, state the degree.

1. $f(x)=x^3+\sqrt{5}x-3$
2. $f(x)=x^3+5\sqrt{x}-3$
3. $f(x)=4x-3$
4. $f(x)=7+4x-x^2$
5. $f(x)=x^2+x^{1/2}-1$
6. $f(x)=x^{-3}+2x^2-x+3$
7. $f(x)=\dfrac{1}{x}$
8. $f(x)=3^{-1}$

9. $f(x) = \pi$ 10. $f(x) = x^{100} - 1$

In Exercises 11–16 write a polynomial function (in factored form) with the given degree and zeros.

11. degree 3; zeros are 1, 2 and 3

12. degree 3, zeros are $\sqrt{2}$, $-\sqrt{2}$, and 0

13. degree 4; zeros are –4, 0, i, and $-i$

14. degree 5; zeros are 5, $2 \pm i$, and $1 \pm \sqrt{3}$

15. degree 4; 1 and 5 are both zeros of multiplicity 2

16. degree 6; –2 is a zero of multiplicity 1; 0 is a zero of multiplicity 2; and 5 is a zero of multiplicity 3

In Exercises 17–24 find the zeros of each polynomial function. In each case state the degree of the polynomial function and the multiplicity of each zero.

17. $P(x) = (x+3)^2$ 18. $P(x) = x(x-2)^3$

19. $P(x) = (x+4)(x+\sqrt{2})(x-\sqrt{2})$

20. $P(x) = 4x^3(x-1)^5(x+6)$

21. $P(x) = 3x^2(2-x)$ 22. $P(x) = (x-4)(2x-1)$

23. $P(x) = \dfrac{1}{2}x^4$

24. $P(x) = -x(x-1)^2\left(x + \dfrac{1}{3}\right)^3$

In Exercises 25–28 use the graph of $y = x^3$ to graph the given functions.

25. $y = x^3 - 1$ 26. $y = 1 - x^3$

27. $y = (x-2)^3$ 28. $y = (x+2)^3 - 5$

In Exercises 29–32 use the graph of $y = x^4$ to graph the given functions.

29. $y = (x+2)^4$ 30. $y = x^4 - 4$

31. $y = 2 - (x-1)^4$ 32. $y = -\dfrac{1}{2}x^4$

In Exercises 33–36 use the graph of $y = x^5$ to graph the given functions.

33. $y = -x^5$ 34. $y = 2x^5$

35. $y = (x+1)^5 + 4$ 36. $y = 1 - (x-1)^5$

In Exercises 37–40 use the graph of $y = x^6$ to graph the given functions.

37. $y = \dfrac{1}{4}x^6$ 38. $y = -x^6$

39. $y = 2 - x^6$ 40. $y = (x+1)^6 + 3$

In Exercises 41–50 graph the functions on the basis of information obtained from the intercepts.

41. $y = (x-2)(x+1)$ 42. $y = (x-2)(x+1)^2$

43. $y = (x-2)^2(x+1)$ 44. $y = (x-2)^2(x+1)^2$

45. $y = (x^2-1)(x^2-4)$ 46. $y = x^4 - 3x^2 + 2$

47. $y = x^3 - 2x^2 - 3x$ 48. $y = x^3 - 4x$

49. $y = x^2 - x^3$ 50. $y = x^3 + x^2$

In Exercises 51–56 find an equation of the lowest possible degree for the polynomial function that is graphed.

51.

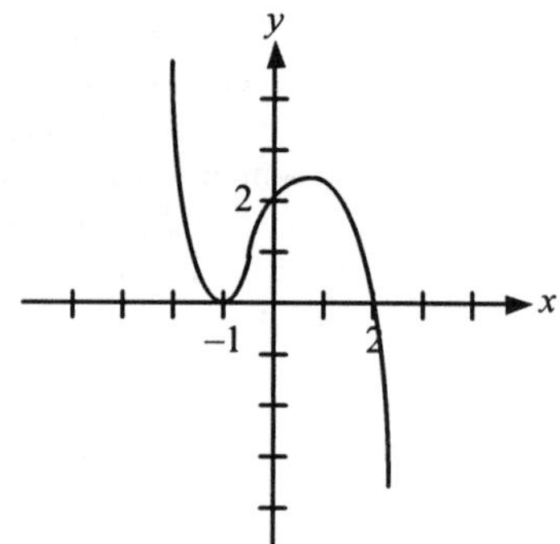

52.

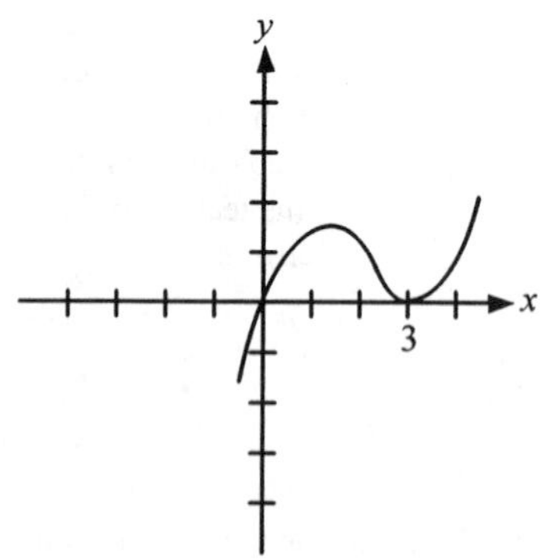

53.

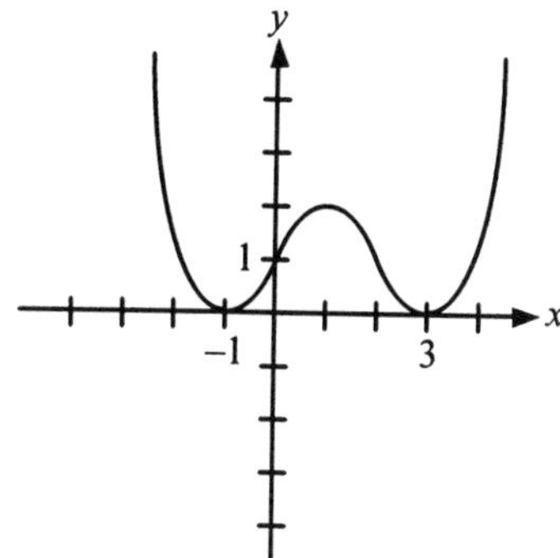

54.

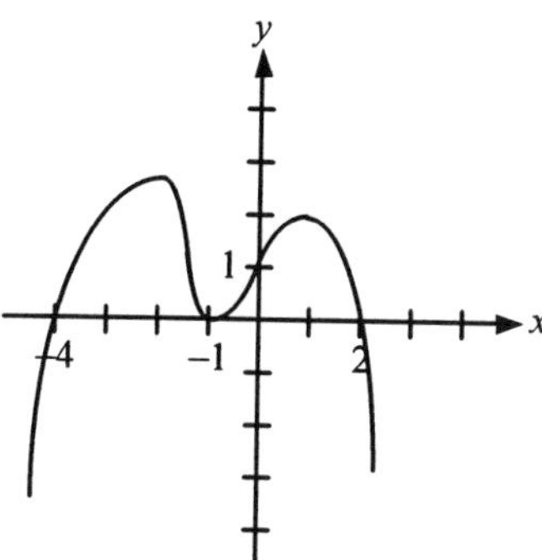

55.

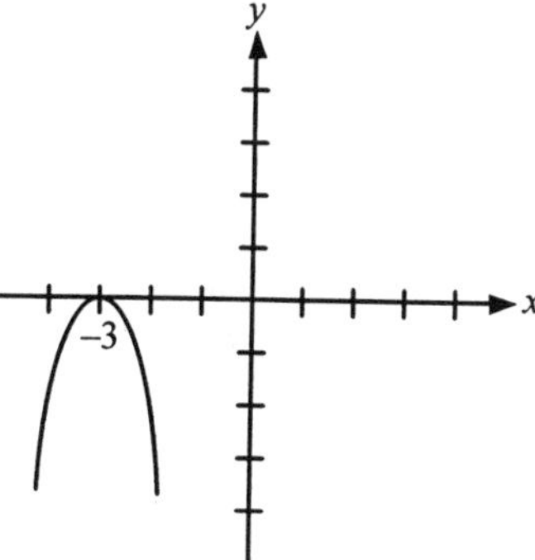

56.

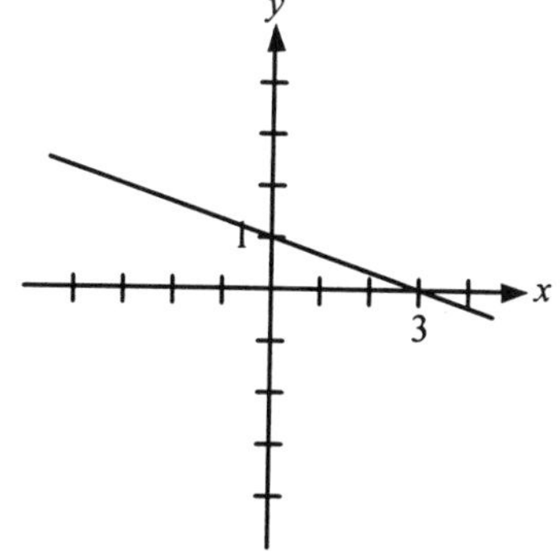

57. A square piece of cardboard with side length 10 in. is to be converted into an uncovered box, as described in Example 9.

a. Find a formula for the volume V of the box in terms of the length (x) of the side of the square that is cut from each corner.

b. What is the domain of this volume function?

c. Use a grapher to graph this volume function and to estimate to two decimal places the length of the corner to cut out to obtain the box with maximum volume. What fraction of the side length is cut out to create the box with maximum volume?

58. Redo Exercise 57, but assume that the original piece of cardboard is square with a side length of 8 in.

59. Here are three different boxes that can be constructed from cardboard.

a. In each case find the value of x that gives the box with maximum volume, and calculate the maximum value. Assume that the dimensions are in inches.

1. Open box: $V(x) = (24 - 2x)(12 - 2x)x$

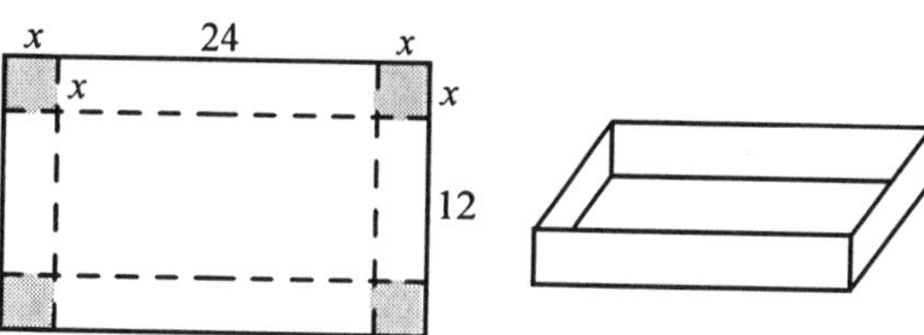

2. Reinforced end: $V(x) = (24 - 4x)(12 - 2x)x$

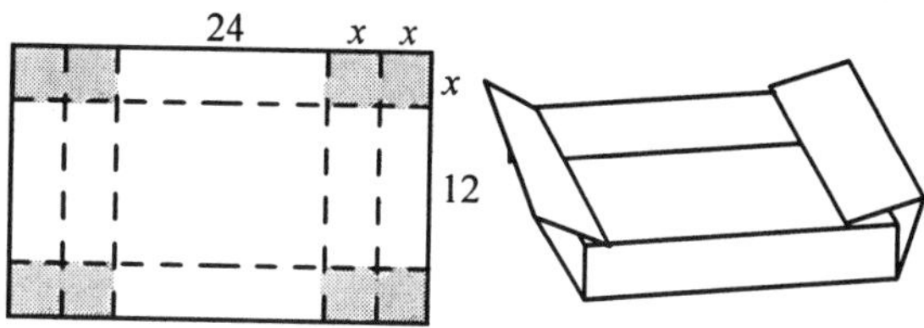

3. With top: $V(x) = (12 - x)(12 - 2x)x$

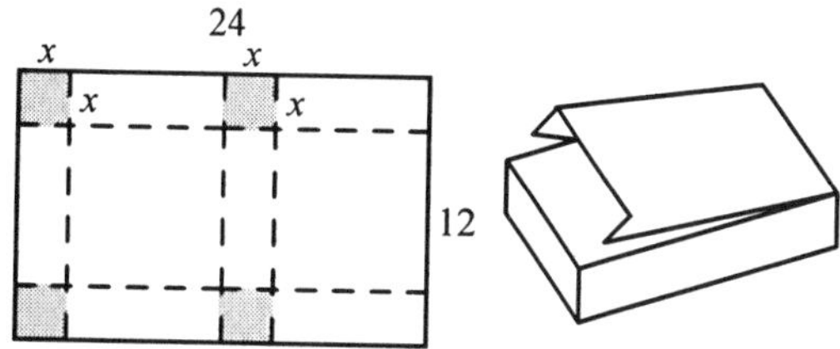

b. Are there any values of x for which two of the boxes have the same volume?

60. A square piece of tin, 12 inches on each side, is to be bent into a 3-edged pan, as shown, after square corners of side x are cut out. What size corner should be cut out so that the pan has maximum volume?

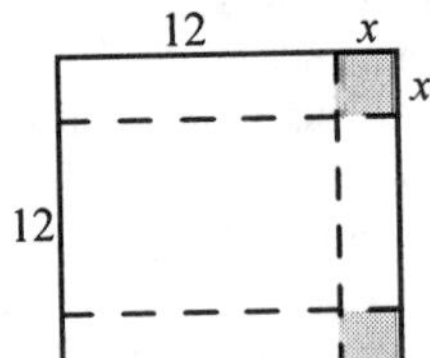

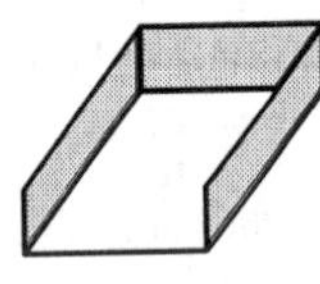

61. The volume of a cube is supposed to be 125 cm^3. If there is an error of x cm in each side, then the volume is $(5+x)^3$. For what values of x will the volume be between 120 and 130 cm^3?

62. Redo Exercise 61, if the volume must be less than 1 cm^3 in error from 125 cm^3.

THINK ABOUT IT 6.1

1. Since polynomial functions have smooth continuous graphs, there are no sharp turns or corners. In contrast, the graph of the absolute value function does change direction abruptly. Some interesting graphs therefore result from applying absolute value to polynomial functions. Graph each of these pairs of functions, and describe how the second is related to the first. Zoom in on any turning points of these graphs and describe the behavior of the graph there.
   - **a.** $y = x^2 - 3$; $y = |x^2 - 3|$
   - **b.** $y = x^3 + 3x^2 - x - 3$; $y = |x^3 + 3x^2 - x - 3|$
   - **c.** $y = x^3 - 3$; $y = |x^3 - 3|$
2. Create a third-degree polynomial inequality with the given solution set.
   - **a.** $(-\infty, -2] \cup [1, 5]$
   - **b.** $(-2, 3) \cup (3, \infty)$
3. What can't a cubic polynomial have exactly 1 turning point? Hint: consider the end behavior. Why can't a fourth-degree polynomial have exactly two turning points? What seems to be the relation between the degree of the polynomial and the possible numbers of turning points?
4. A famous problem in the history of mathematics dates to the early sixteenth century in Italy, when there was much competition to be able to solve cubic equations. Niccolo Tartaglia posed this problem. *To divide the number 8 into two parts such that the result of multiplying the product of those parts by their difference is maximal.*
   - **a.** If you let the two parts be x and $8 - x$ check that the problem leads to the function $f(x) = x(8-x)[x-(8-x)]$. Find the coordinates of the maximum point to three decimal places. [Only solutions between 0 and 8 are sensible.] Give the two parts of 8, each accurate to 3 decimal places.
   - **b.** Tartaglia said that the solution was found as: *Halve the number 9; the square of that half augmented by a third of that square will be equal to the square of the difference of the two parts.* Show that these words lead to the equation $64/3 = (2x-8)^2$. Solve this quadratic equation and confirm that the solutions found in parts **a** and **b** are the same.

   Source: Tikhomirov, *Stories about Maxima and Minima*
5. A square piece of cardboard, s inches per side, has corners of side x cut out, so that it can be folded into an open box.
   - **a.** Express the volume of the resulting open box as a function of s and x.
   - **b.** Graph the volume function for several values of s, including $s = 18$, 12, 6, and 3. For each value of s, find the value of x that maximizes the volume. What seems to be the relation between s and the maximizing value of x?

• • •

6.2 Polynomial Division and Theorems About Zeros

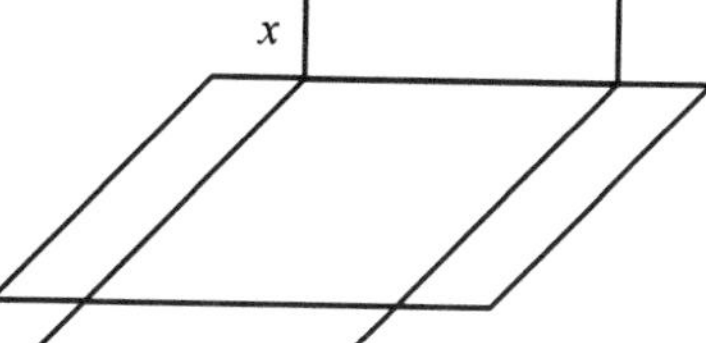

A rectangular piece of cardboard 18 in. by 12 in. is to be used to make an open-top box with a volume of 160 in.^{3}, by cutting a small square from each corner and bending up the sides as shown. As determined in Progress Check 9 of Section 6.1, the volume of the box is given by $V = (18-2x)(12-2x)x$. So the side length x for each square cutout may be found by solving $160 = (18-2x)(12-2x)x$, which is equivalent to

$$x^3 - 15x^2 + 54x - 40 = 0.$$

It is apparent that 1 in. is a solution, since replacing x by 1 yields $1-15+54-40=0$, and $x = 1$ in. is meaningful in the context of the problem. Find all other possible side lengths for the cutout.
(See Example 6.)

Objectives

1. Divide one polynomial by another using long division.
2. Divide a polynomial by a polynomial of the form $x-b$, using synthetic division.
3. Evaluate a polynomial using the remainder theorem.
4. Find all zeros of a third-degree or fourth-degree polynomial function when one or two zeros, respectively, are apparent.
5. Use the conjugate-pair theorems to find zeros or write equations for polynomial functions.

When we attempt to find exact values for the zeros of polynomial functions, we can sometimes spot some simple answers. For instance, in the section-opening problem we need to find the zeros for

$$P(x) = x^3 - 15x^2 + 54x - 40.$$

You may be able to spot a zero in this case because one of the zeros is a simple number, namely 1. Alternately, a graph of $y = P(x)$ on a grapher will reveal that 1 is a zero, because the graph has an x-intercept at $x = 1$ (along with other integer values), as shown in Figure 6.18.

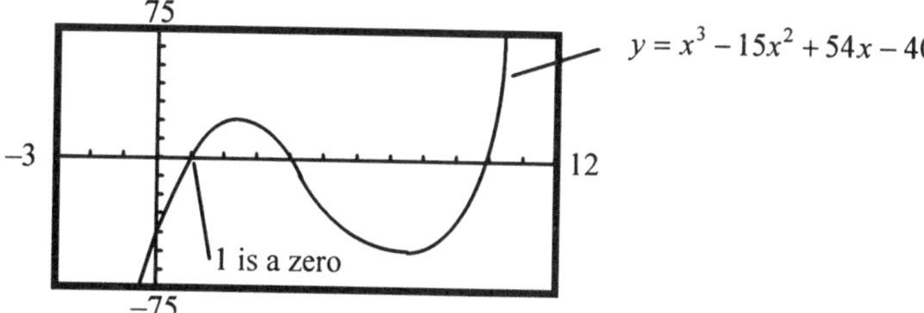

Figure 6.18

Once it is known that 1 is a zero of a polynomial function $y = P(x)$, then the factor theorem indicates that $x-1$ is a factor of $P(x)$. Therefore, $P(x) = x^3 - 15x^2 + 54x - 40$ may be written as

$$P(x) = (x-1)Q(x),$$

where $Q(x)$ is a second-degree polynomial. To find $Q(x)$, which may be used to find the exact values of the remaining zeros, we may use the long-division method for dividing polynomials that is illustrated below. As you study this procedure, note the names of the key components in such a division.

$$
\begin{array}{rrl}
 & x^2 - 14x + 40 & \leftarrow \text{Quotient} \\
\text{Divisor} \rightarrow x - 1 & \overline{)\, x^3 - 15x^2 + 54x - 40} & \leftarrow \text{Dividend} \\
\text{Subtract:} & \underline{x^3 - x^2} & \\
 & -14x^2 + 54x - 40 & \\
\text{Subtract:} & \underline{-14x^2 + 14x } & \\
 & 40x - 40 & \\
\text{Subtract:} & \underline{40x - 40} & \\
 & 0 & \leftarrow \text{Remainder}
\end{array}
$$

Then, dividend = (divisor)(quotient) + remainder, so

$$x^3 - 15x^2 + 54x - 40 = (x-1)\left(x^2 - 14x + 40\right).$$

From this result, it follows that

$$P(x) = (x-1)(x-4)(x-10),$$

which implies that 1, 4, and 10 are the zeros of the function. Observe that the graph in Figure 6.18 supports this conclusion, because the graph appears to have x-intercepts when $x = 1$, $x = 4$, and $x = 10$.

The long-division method for dividing polynomials is useful in situations other than finding zeros of polynomial functions algebraically. Therefore, a detailed description of this procedure is given next.

Long Division of Polynomials

1. Arrange the terms of the dividend and the divisor with descending powers. *If a lower power is absent in the dividend, write 0 as its coefficient.*
2. Divide the first term of the dividend by the first term of the divisor to obtain the first term of the quotient.
3. Multiply the entire divisor by the first term of the quotient, and subtract this result from the dividend.
4. Use the remainder as the new dividend, and repeat the above procedure until the remainder is of lower degree than the divisor.

Example 1: Using Long Division

Divide $2x^4 + 6x^3 - 5x^2 - 1$ by $2x^2 - 3$. Express the answer in the form

$$\frac{\text{dividend}}{\text{divisor}} = \text{quotient} + \frac{\text{remainder}}{\text{divisor}}.$$

Solution: The division is performed below. Note that $0x$ is inserted to help align like terms vertically, and that the division process stopped when the remainder $9x-4$ was of lower degree than the divisor $2x^3-3$.

$$
\begin{array}{lrl}
 & x^2+3x-1 & 2x^4/2x^2 = x^2,\ 6x^3/2x^2 = 3x. \\
 & & \text{and } -2x^2/2x^2 = -1 \\
2x^2-3 \,\big) & 2x^4+6x^3-5x^2+0x-1 & \\
\text{subtract} & \underline{2x^4 \qquad\quad -3x^2} & \text{This line is } x^2(2x^2-3). \\
 & 6x^3-2x^2+0x-1 & \\
\text{subtract} & \underline{6x^3 \qquad\quad -9x} & \text{This line is } 3x(2x^2-3). \\
 & -2x^2+9x-1 & \\
\text{subtract} & \underline{-2x^2 \qquad +3} & \text{This line is } -1(2x^2-3). \\
 & 9x-4 &
\end{array}
$$

The answer in the form requested is

$$\frac{2x^4+6x^3-5x^2-1}{2x^2-3} = x^2+3x-1+\frac{9x-4}{2x^2-3}.$$

Technology Link

One way to check this answer on a grapher is to let

$$y1 = \frac{2x^4+6x^3-5x^2-1}{2x^2-3}$$

$$y2 = x^2+3x-1+\frac{9x-4}{2x^2-3}$$

$$y3 = \frac{y1}{y2}.$$

Then graph only $y3$. The quotient of identical expressions is 1, so check for a graph that is a horizontal line where y is fixed at 1.

PROGRESS CHECK 1

Divide $9x^4+5x^2+x+3$ by $3x-1$. Answer in the form requested in Example 1. ■

The division of any polynomial by a polynomial of the form $x-b$ is of theoretical and practical importance. This division may be performed by a shorthand method called **synthetic division**. Consider the arrangement for dividing $2x^3+5x^2-1$ by $x-2$.

$$
\begin{array}{rl}
 & 2x^2+9x^2+18 \\
x-2 \,\big) & 2x^3+5x^2+\ 0x-\ 1 \\
 & \underline{\textcircled{2}x^3-4x^2} \\
 & \qquad 9x^2+\ 0x-\ 1 \\
 & \qquad \underline{\textcircled{9}x^2-18x} \\
 & \qquad\qquad 18x-\ 1 \\
 & \qquad\qquad \underline{\textcircled{18}x-36} \\
 & \qquad\qquad\qquad 35
\end{array}
$$

When the polynomials are written with terms in descending powers of x, there is no need to write all the x's. Only the coefficients are needed. Also, notice that the encircled coefficients entailed needless writing. Using only the necessary coefficients, we may abbreviate this division as follows:

$$\begin{array}{r|rrrr} & 2 & 9 & 18 & \\ \hline -2 & 2 & 5 & 0 & -1 \\ & & -4 & -18 & -36 \\ \hline & & 9 & 18 & 35. \end{array}$$

If we bring down 2 as the first entry in the bottom row, all the coefficients of the quotient appear. The arrangement may then be shorted to

$$\begin{array}{r|rrrrl} -2 & 2 & 5 & 0 & -1 & \leftarrow \text{coefficients of dividend (row 1)} \\ & \downarrow & -4 & -18 & -36 & \text{(row 2)} \\ \hline \text{coefficients of quotient} \rightarrow & 2 & 9 & 18 & 35 & \leftarrow \text{remainder (row 3)} \end{array}$$

Finally, if we replace –2 by 2, which is the value of b, we may change the sign of each number in row 2 and add at each step instead of subtracting. The final arrangement for synthetic division is then as follows:

$$\begin{array}{r|rrrrl} 2 & 2 & 5 & 0 & -1 & \text{(row 1)} \\ & & 4 & 18 & 36 & \text{(row 2)} \\ \hline & 2 & 9 & 18 & 35 & \rightarrow \text{remainder (row 3)} \\ & \downarrow & \downarrow & \downarrow & & \\ \text{quotient} \rightarrow & 2x^2 + & 9x + & 18 & & \end{array}$$

Use this example as a basis for understanding the synthetic division procedure outlined next.

Synthetic Division

To divide a polynomial $P(x)$ by $x - b$:

1. Form row 1 by writing the coefficients of the terms in the dividend $P(x)$. The dividend must be written in descending powers, and 0 must be entered as the coefficient of any missing term. Write the value of b to the left of these coefficients.
2. Bring down the first dividend entry as the first coefficient in the quotient.
3. Multiply this quotient coefficient by b. Place the result under the next number in row 1, and then add.
4. Repeat the procedure in step 3 until all entries in row 1 have been used.
5. The last number in the bottom row is the remainder. The other numbers in the bottom row are, from left to right, the coefficients of descending powers of the quotient.

It is important to remember that synthetic division applies only to division by a polynomial of the form $x - b$. Because this divisor is a first-degree polynomial, the degree of the polynomial in the quotient is always one less than the degree of the polynomial in the dividend.

Example 2: Using Synthetic Division

Use synthetic division to divide $x^5 - 1$ by $x - 1$. Express the result in the form

dividend = (divisor)(quotient) + remainder.

Solution: We use 0's as the coefficients of the missing x^4, x^3, x^2, and x terms. We divide by $x - 1$ so that $b = 1$ (not -1). The arrangement is as follows:

$$\begin{array}{r|rrrrrr} 1 & 1 & 0 & 0 & 0 & 0 & -1 \\ & & 1 & 1 & 1 & 1 & 1 \\ \hline & 1 & 1 & 1 & 1 & 1 & 0 \leftarrow \text{remainder} \end{array}$$

quotient $\longrightarrow$ $x^4 + x^3 + x^2 + x + 1.$

The answer in the form requested is

$$x^5 - 1 = (x - 1)(x^4 + x^3 + x^2 + x + 1) + 0.$$

PROGRESS CHECK 2

Use synthetic division to divide $4x^3 - 2x^2 + 3x - 1$ by $x - 2$. Express the result in the form requested in Example 2. ■

Example 3: Using Synthetic Division

Use synthetic division to divide $4x^3 - x^2 + 2$ by $x + 3$. Express the result in the form

dividend = (divisor)(quotient) + remainder.

Solution: We use 0 as the coefficient of the missing x term. We divide by $x + 3$ or $x - (-3)$, so that $b = -3$ (not 3). The arrangement is as follows:

$$\begin{array}{r|rrrr} -3 & 4 & -1 & 0 & 2 \\ & & -12 & 39 & -117 \\ \hline & 4 & -13 & 39 & -115 \leftarrow \text{remainder} \end{array}$$

quotient $\longrightarrow$ $4x^2 - 13x + 39.$

The answer in the form requested is

$$4x^3 - x^2 + 2 = (x + 3)(4x^2 - 13x + 39) - 115.$$

PROGRESS CHECK 3

Use synthetic division to divide $3x^3 + 10x^2 - 6x + 8$ by $x + 4$. Express the result in the form requested in Example 3. ■

In Example 3 we found that the function

$$P(x) = 4x^3 - x^2 + 2$$

may be written as

$$P(x) = (x + 3)(4x^2 - 13x + 39) - 115.$$

When $x = -3$, the factor $x + 3$ equals 0; thus,

$$\begin{aligned} P(-3) &= 0 - 115 \\ &= -115. \end{aligned}$$

The value of the function when $x = -3$ is the same as the remainder obtained when $P(x)$ is divided by $x - (-3)$. This discussion suggests the following theorem.

Remainder Theorem

> If a polynomial $P(x)$ is divided by $x - b$, the remainder is $P(b)$.

To prove this theorem, let $Q(x)$ and r represent the quotient and remainder when $P(x)$ is divided by $x - b$. Then

$$\begin{aligned} \text{dividend} &= (\text{divisor})(\text{quotient}) + \text{remainder} \\ P(x) &= (x - b)Q(x) + r. \end{aligned}$$

This statement is true for all values of x. If $x = b$, then

$$\begin{aligned} P(b) &= (b - b)Q(b) + r \\ &= 0 \cdot Q(b) + r. \end{aligned}$$

Thus,

$$P(b) = r.$$

We know that $P(b)$ may be found by substituting b for x in the function. The remainder theorem provides an alternative method. That is, we find $P(b)$ by determining the remainder when $P(x)$ is divided by $x - b$. Since this remainder may be obtained by synthetic division, this approach is often simpler than direct substitution. Later in this section, other advantages of the remainder theorem will be discussed.

Example 4: Using the Remainder Theorem

If $P(x) = 3x^5 + 5x^4 + 7x^3 - 4x^2 + x - 24$, find $P(-2)$ by

a. direct substitution and **b.** the remainder theorem.

Solution:

a. By direct substitution, we have

$$\begin{aligned} P(-2) &= 3(-2)^5 + 5(-2)^4 + 7(-2)^3 - 4(-2)^2 + (-2) - 24 \\ &= -96 + 80 - 56 - 16 - 2 - 24 \\ &= -114. \end{aligned}$$

b. By the remainder theorem, we have

$$\begin{array}{r|rrrrrr} -2 & 3 & 5 & 7 & -4 & 1 & -24 \\ & & -6 & 2 & -18 & 44 & -90 \\ \hline & 3 & -1 & 9 & -22 & 45 & -114. \end{array}$$

Since the remainder is -114, $P(-2) = -114$.

PROGRESS CHECK 4

If $P(x) = 2x^4 - 5x^3 + 11x^2 - 3x - 5$, find $P(-1/2)$ by

a. direct substitution and **b.** the remainder theorem. ■

In the previous section, the factor theorem was introduced. We restate this theorem next for reference purposes, and then prove it with the aid of the remainder theorem. Example 5 will then show an advantage of using the remainder theorem method to find $P(b)$.

Factor Theorem

> If b is a zero of the polynomial function $y = P(x)$, then $x - b$ is a factor of $P(x)$; conversely, if $x - b$ is a factor of $P(x)$, then b is a zero of $y = P(x)$.

We now prove the first part of this theorem. As in the proof of the remainder theorem, let $Q(x)$ and r represent the quotient and remainder when $P(x)$ is divided by $x - b$. Then

$$P(x) = (x - b)Q(x) + r.$$

By the remainder theorem, $r = P(b)$, so we have

$$P(x) = (x - b)Q(x) + P(b).$$

If b is a zero of $y = P(x)$, then $P(b) = 0$. Thus,

$$P(x) = (x - b)Q(x) + 0,$$

and $x - b$ is a factor of $P(x)$. The proof of the second part of this theorem is left as Exercise 69.

Example 5: Using the Factor and Remainder Theorems

If $P(x) = x^3 - 2x^2 - 5x + 6$, then is $x - 1$ a factor of $P(x)$? If yes, then factor $P(x)$ completely.

Solution: By the factor theorem, $x - 1$ is a factor of $P(x)$ if and only if $P(1) = 0$. We may find $P(1)$ by using direct substitution or by using the remainder theorem.

Direct substitution method: $P(1) = 1^3 - 2(1)^2 - 5(1) + 6 = 0$

Remainder theorem method:

$$\begin{array}{r|rrrr} 1 & 1 & -2 & -5 & 6 \\ & & 1 & -1 & -6 \\ \hline & 1 & -1 & -6 & 0 \leftarrow P(1) \end{array}$$

Both methods show $P(1) = 0$, so $x - 1$ is a factor of $P(x)$. However, observe that the remainder theorem method is especially helpful here because it also reveals that the quotient is $x^2 - x - 6$. Thus, $P(x)$ factors as

$$P(x) = (x - 1)(x^2 - x - 6).$$

For this polynomial we can factor further, since $x^2 - x - 6$ factors as $(x - 3)(x + 2)$. So the complete factorization is

$$P(x) = (x - 1)(x - 3)(x + 2).$$

To check this result on a grapher, note that applying the factor theorem to this factorization gives that 1, 3, and –2 are zeros of $y = P(x)$, so the graph of $y = P(x)$ should intersect the x-axis at these numbers. These intersections are apparent in the graph of $y = P(x)$ shown in Figure 6.19.

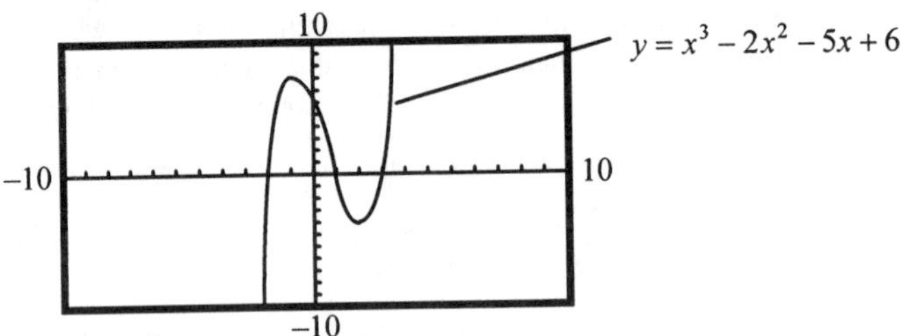

Figure 6.19

PROGRESS CHECK 5
If $P(x) = x^3 + 3x^2 - 10x - 24$, then is $x + 2$ a factor of $P(x)$? If yes, then factor $P(x)$ completely. ■

Now that we have developed methods for dividing polynomials and established the remainder and factor theorems, we can return to the main topic of finding zeros for polynomial functions. In this section we present three theorems that are fundamental to the theory of finding such zeros. You need a good grasp of number systems to understand these statements thoroughly and it may be helpful if you review the relationships among the various sets of numbers, which are given in Sections R.1 and 3.6.

Fundamental Theorem of Algebra

> Every polynomial function of degree $n \geq 1$ with complex number coefficients has at least one complex zero.

Although the proof of this theorem is beyond the scope of this book, we can prove an important corollary of this theorem that answers a basic question: How many zeros are there for a polynomial function of degree n?

Number of Zeros Theorem

> Every polynomial function of degree $n \geq 1$ has exactly n complex zeros, where zeros of multiplicity k are counted k times.

For instance, we have seen that $P(x) = x^4(x-2)(x+7)^3$ is a polynomial function of degree 8 with distinct zeros of 0, 2 and –7 (see Example 4 of Section 6.1). However, the multiplicities of these zeros are 4, 1, and 3, respectively, so that if we take into account the idea of multiplicity, then this 8th-degree polynomial function has exactly eight zeros. More generally, we can prove the above theorem by first noting that if $y = P(x)$ is a polynomial function of degree $n \geq 1$ with leading coefficient a_n, then the fundamental theorem of algebra guarantees that $y = P(x)$ has at least one complex zero, say c_1. By the factor theorem, since c_1 is a zero, $x - c_1$ is a factor of $P(x)$, so

$$P(x) = (x - c_1)Q_1(x),$$

where polynomial $Q_1(x)$ has degree $n-1$ and leading coefficient a_n. If the degree of $Q_1(x)$ is at least 1, then once again the fundamental theorem of algebra guarantees that $y = Q_1(x)$ has at least one zero, say c_2, so as above,

$$Q_1(x) = (x - c_2)Q_2(x),$$

where polynomial $Q_2(x)$ has degree $n-2$ and leading coefficient a_n. Then combining results, we have

$$P(x) = (x - c_1)(x - c_2)Q_2(x).$$

This process is continued for a total of *n* times until $Q_n(x) = a_n$. Thus, $P(x)$ can be factored into *n* linear factors and written as

$$P(x) = a_n(x - c_1)(x - c_2)\cdots(x - c_n)$$

so that, by the factor theorem, $y = P(x)$ has n zeros: $c_1, c_2, \ldots, c_n$. Furthermore, no other number, say *c*, distinct from $c_1, c_2, \ldots, c_n$ can be a zero since

$$P(c) = a_n(c - c_1)(c - c_2)\cdots(c - c_n) \neq 0,$$

because none of the factors is zero. Thus, every polynomial function of degree $n \geq 1$ has exactly *n* (not necessarily distinct) complex zeros.

Example 6: Using a Known Root to Find All Roots

Solve the problem in the section introduction on page 477.

Solution: We need to find all roots of the equation

$$x^3 - 15x^2 + 54x - 40 = 0,$$

given that 1 is a solution. Observe that this equation is a third-degree polynomial equation, so there are three (not necessarily distinct) roots. If we let $P(x)$ equal the polynomial on the left side of the equal sign, the solution at $x = 1$ implies that $x - 1$ is a factor of $P(x)$. We can then use synthetic division to find another factor of $P(x)$ as follows:

$$\begin{array}{r|rrrr} \underline{1} & 1 & -15 & 54 & -40 \\ & & 1 & -14 & 40 \\ \hline & 1 & -14 & 40 & 0. \end{array}$$

As expected, the remainder is 0, so $P(1) = 0$, and 1 is a solution. Because the remainder is 0, the division above indicates that $P(x)$ factors as

$$x^3 - 15x^2 + 54x - 40 = (x - 1)(x^2 - 14x + 40)$$

so we get

$$x^3 - 15x^2 + 54x - 40 = 0$$
$$(x - 1)(x^2 - 14x + 40) = 0.$$

The remaining roots are the solutions of $x^2 - 14x + 40 = 0$, and for this equation the factoring method applies.

$$\begin{aligned} x^2 - 14x + 40 &= 0 \\ (x-4)(x-10) &= 0 \\ x - 4 = 0 \quad &\text{or} \quad x - 10 = 0 \\ x = 4 \quad & \quad\quad x = 10 \end{aligned}$$

Thus, the three roots are 1, 4, and 10 (check Figure 6.18 again). In the context of the problem, 1 and 4 are meaningful answers, but it is not possible to cut out a square with side length 10 in. from each corner. Thus, the only other possible side length for the cutout is 4 in.

PROGRESS CHECK 6

If the box described in Example 6 is to have a volume of 224 in.^{3}, then the side length x for each square cutout may be found by solving $(18-2x)(12-2x)x = 224$, which is equivalent to

$$x^3 - 15x^2 + 54x - 56 = 0.$$

One solution to this problem is $x = 2$ in. Find all other possible side lengths for the cutout. ■

Example 7: Using a Known Zero to Find All Zeros

If 2 is a zero of $P(x) = x^3 - 2x^2 + 3x - 6$, find the other zeros.

Solution: Since $y = P(x)$ is a polynomial function of degree 3, there are exactly three (not necessarily distinct) zeros. We are given that 2 is one of these zeros, and if 2 is a zero of $y = P(x)$, then by the factor theorem, $x - 2$ is a factor of the polynomial. We can then use synthetic division to find another factor of $P(x)$ as follows:

$$\begin{array}{r|rrrr} 2 & 1 & -2 & 3 & -6 \\ & & 2 & 0 & 6 \\ \hline & 1 & 0 & 3 & 0. \end{array}$$

The coefficients of another factor of $P(x)$ are given in the bottom row of this synthetic division. Thus,

$$P(x) = (x-2)(x^2+3).$$

The remaining zeros of $y = P(x)$ are the solutions of $x^2 + 3 = 0$.

$$\begin{aligned} x^2 + 3 &= 0 \\ x^2 &= -3 \\ x &= \pm i\sqrt{3} \end{aligned}$$

Thus, the other two zeros are $i\sqrt{3}$ and $-i\sqrt{3}$. Figure 6.20 helps to confirm that the only real zero of this function is at $x = 2$.

PROGRESS CHECK 7

If 1 is a zero of $P(x) = x^3 - x^2 - 15x + 15$, find the other zeros. ■

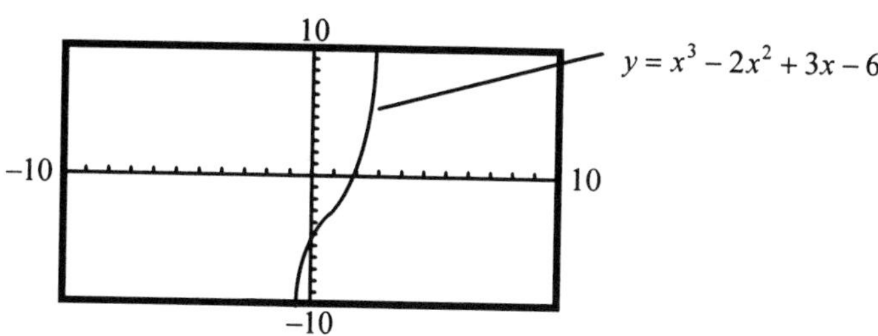

Figure 6.20

Note in Example 7 that the zeros $i\sqrt{3}$ and $-i\sqrt{3}$ are conjugates of each other. In general, the following theorems show that when certain restrictions are placed on the coefficients of a polynomial, then zeros of polynomial functions always occur in conjugate pairs.

Conjugate-Pair Theorems

1. If a complex number $a + bi$ is a zero of a polynomial function of degree $n \geq 1$ with *real*-number coefficients, then it conjugate $a - bi$ is also a zero.
2. Let a, b, and c be rational numbers. If an irrational number of the form $a + b\sqrt{c}$ is a zero of a polynomial function of degree $n \geq 1$ with *rational*-number coefficients, then $a - b\sqrt{c}$ is also a zero.

The restrictions on the coefficients of the polynomial are crucial conditions in these theorems. For example $P(x) = x - \sqrt{2}$ has only one zero, namely $\sqrt{2}$. Note that the conjugate $-\sqrt{2}$ is not also a zero, and that the second conjugate-pair theorem does not apply to $P(x) = x - \sqrt{2}$, because the polynomial does not have rational-number coefficients.

We can prove the first conjugate-pair theorem by using the properties of complex conjugates discussed in Section 3.6 (which you may need to review). We start with a polynomial function of degree $n \geq 1$ with real-number coefficients that we may write as

$$P(x) = a_n x^n + a_{n-1} x^{n-1} + \cdots + a_1 x + a_0.$$

Now if $z = a + bi$ is a zero of $y = P(x)$, then $P(z) = 0$, so

$$0 = a_n z^n + a_{n-1} z^{n-1} + \cdots + a_1 z + a_0.$$

We then take the conjugate of both sides of the equation (noting that the conjugate of 0 is 0) and proceed as follows:

$$\begin{aligned}
0 &= \overline{a_n z^n + a_{n-1} z^{n-1} + \cdots + a_1 z + a_0} \\
&= \overline{a_n z^n} + \overline{a_{n-1} z^{n-1}} + \cdots + \overline{a_1 z} + \overline{a_0} && \text{(conjugate of a sum = sum of conjugates)} \\
&= \overline{a_n}\,\overline{z^n} + \overline{a_{n-1}}\,\overline{z^{n-1}} + \cdots + \overline{a_1}\,\overline{z} + \overline{a_0} && \text{(conjugate of a product = product of conjugates)} \\
&= a_n \overline{z^n} + a_{n-1} \overline{z^{n-1}} + \cdots + a_1 \overline{z} + a_0 && \text{(conjugate of a real number = the real number)} \\
&= a_n \left(\overline{z}\right)^n + a_{n-1} \left(\overline{z}\right)^{n-1} + \cdots + a_1 \overline{z} + a_0. && \text{(conjugate of a power = power of the conjugate)}
\end{aligned}$$

This last equation tells us that $P(\overline{z}) = 0$, so $\overline{z}$ is also a zero of $y = P(x)$, which proves the theorem.

Example 8: Using the Conjugate-Pair Theorems If $-2i$ and $2+\sqrt{7}$ are zeros of the polynomial function $P(x)=x^4-4x^3+x^2-16x-12$, find the other zeros.

Solution: $P(x)$ has a rational-number coefficients. Thus, both conjugate-pair theorems apply, and if $-2i$ and $2+\sqrt{7}$ are zeros, then $2i$ and $2-\sqrt{7}$ are also zeros. There are no other zeros, since the polynomial has degree 4.

PROGRESS CHECK 8
If $4-2i$ and $\sqrt{3}$ are zeros of the polynomial function $P(x)=x^4-8x^3+17x^2+24x-60$ find the other zeros. ■

Example 9: Using the Conjugate-Pair Theorems Write a polynomial function (in factored form) of the lowest possible degree that has rational coefficients and zeros of 0, $1-i$, and $\sqrt{2}$.

Solution: Since the polynomial has rational coefficients, we utilize both conjugate-pair theorems. Thus, in addition to the three zeros given, $1+i$ and $-\sqrt{2}$ are also zeros. The lowest possible degree for the polynomial is then degree 5, and in factored form one possibility is

$$P(x)=x\left(x-\sqrt{2}\right)\left(x+\sqrt{2}\right)(x-(1+i))(x-(1-i)).$$

PROGRESS CHECK 9
Write a polynomial function (in factored form) of the lowest possible degree that has rational coefficients and zeros of 2, $\sqrt{2}$, and $2i$. ■

In conclusion, let us discuss briefly why the fundamental theorem of algebra is so central to the theory of polynomial equations. During the historical growth of equation solving, the solution to certain types of equations continued to force mathematicians to develop new types of numbers. For example, if an equation uses just positive integers (for example, $x+5=1$), it may be necessary to leave the positive integers for a solution. In the case of $x+5=1$, an extension to –4 and the negative integers is needed. Integers were not enough for equations such as $2x+5=0$, so rational numbers were required. Similarly, $x^2-5=0$ brings an extension to irrational (or real) numbers, while the equation $x^2+5=0$ can be solved only by introducing imaginary (or complex) numbers. However, the process stops here. The fundamental theorem of algebra guarantees that every polynomial equation involving complex numbers can be solved using only complex numbers. No extension of the number system is necessary. The mathematician Carl Friedrich Gauss was the first to prove this theorem. He did so in his doctoral thesis in 1799, when he was 20.

EXPLORE 6.2

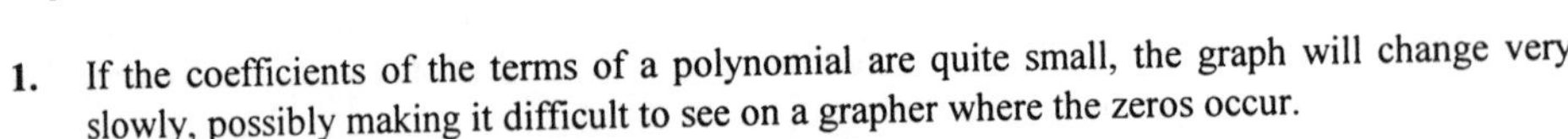

1. If the coefficients of the terms of a polynomial are quite small, the graph will change very slowly, possibly making it difficult to see on a grapher where the zeros occur.
   a. Demonstrate this by examining the graph of

$$y=P(x)=0.01(x-1)^3+0.02x^2.$$

In a standard viewing window, the graph seems to run along the x-axis, but because this polynomial has degree 3, it cannot actually touch the x-axis at more than 3 points. Use a grapher to determine the number of real zeros of $y=P(x)$. Estimate all real zeros to 2 decimal places.

   b. Use a grapher to determine the number of real zeros of $y=P(x)$, and estimate them to 2 decimal places.

$$y=P(x)=0.01(x+1)^3+0.02(x-1)^2-0.05$$

2. Sometimes, for mathematical convenience, polynomials are used to approximate functions that are not polynomials. Such approximations are usually good only for some limited domain. For instance, the function $f(x)=\sqrt{4-x^2}$, whose graph is a semicircle, is not a polynomial.
   a. The polynomial $P(x)=-0.533x^2+2.159$ does a reasonable job of approximating $f(x)$ over the domain $[-2, 2]$. Use a grapher to graph both $y=f(x)$ and $y=P(x)$.
   b. Graph the function $y=f(x)-P(x)$. For how many points on the given domain do the 2 graphs have exactly the same value? To 3 decimal places, what is the maximum error over the interval $[-2, 2]$ in using $P(x)$ to approximate $f(x)$?
   c. Repeat parts **a** and **b** taking $P(x)=-0.077x^4-0.191x+2$.
3. Some calculators have a Polynomial Regression feature. It usually appears on a menu from which you select the degree of the regression model. You may see a choice such as "cubic regression," or "*P*3 regression." For instance, cubic regression finds the equation of a third-degree polynomial that comes close to the set of given data points. In the special case where the number of points is exactly 1 more than the degree of the polynomial, the resulting equation will have no error, and its graph will actually hit each of the given points. If you have this feature, use it to find the equation of
   a. a quadratic function whose graph includes $(3, 0)$, $(5, 0)$, and $(0, 4)$.
   b. a cubic function whose graph includes $(-1, 0)$, $(2, 0)$, $(1, 3)$, and $(0, 2)$.

EXERCISES 6.2

In Exercises 1–6 perform the indicated divisions by using long division. Express the answer in the form

$$\frac{\text{dividend}}{\text{divisor}}=\text{quotient}+\frac{\text{remainder}}{\text{divisor}}.$$

1. $\dfrac{x^2-5}{x+1}$
2. $\dfrac{x^2+3x-4}{x-3}$
3. $\dfrac{3x^4-5x^2+7}{x^2+2x+1}$
4. $\dfrac{2x^4-8x^3-7x^2+1}{2x^2-5}$
5. $\dfrac{x^3+1}{x(x-1)}$
6. $\dfrac{x^3-5}{(x-1)^2}$

In Exercises 7–12 perform the indicated divisions by using long division. Express the result in the form dividend = (divisor)(quotient) + remainder.

7. $(x^2+7x-2)\div(x+5)$
8. $(x^2-4)\div(x-1)$
9. $(6x^3-3x^2+14x-7)\div(2x-1)$
10. $(4x^3+5x^2-10x+4)\div(4x-3)$
11. $(3x^4+x-2)\div(x^2-1)$
12. $(2x^3-3x^2+10x-5)\div(x^2+5)$

In Exercises 13–22 perform the indicated division by using synthetic division. Express the result in the form dividend = (divisor)(quotient) + remainder.

13. $(x^3-5x^2+2x-3)\div(x-1)$
14. $(x^3+x^2-9x-6)\div(x-3)$
15. $(2x^3+9x^2-x+14)\div(x+5)$
16. $(3x^3-5x^2-18x+9)\div(x+2)$
17. $(7+6x-2x^2-x^3)\div(x+3)$
18. $(4-x+3x^2-x^3)\div(x-4)$
19. $(2x^3+x-5)\div(x+1)$
20. $(3x^4-x^2+7)\div(x+3)$
21. $(x^4-16)\div(x-2)$
22. $(x^3+27)\div(x-3)$

In Exercises 23–30 find the given function value by **a.** direct substitution and **b.** the remainder theorem.

23. $P(x)=x^3-4x^2+x+6$; find $P(2)$ and $P(3)$.

24. $P(x) = 2x^3 + 5x^2 - x - 7$; find $P(4)$ and $P(-3)$.

25. $P(x) = 2x^4 - 7x^3 - x^2 + 4x + 11$; find $P(2)$ and $P(-2)$.

26. $P(x) = x^4 - 5x^3 - 4x^2 + 17x + 15$; find $P(-1)$ and $P(5)$.

27. $P(x) = 2x^4 + 5x^3 - 20x - 32$; find $P(-2)$ and $P(2)$.

28. $P(x) = 2x^4 - x^3 + 2x - 1$; find $P\left(\frac{1}{2}\right)$ and $P\left(-\frac{1}{2}\right)$.

29. $P(x) = 6x^4 + 2x^3 - 5x - 5$; find $P\left(\frac{1}{3}\right)$ and $P\left(-\frac{1}{3}\right)$.

30. $P(x) = 3x^4 + x^2 - 7x + 2$; find $P(0.1)$ and $P(-0.1)$.

31. If $P(x) = x^3 - 2x^2 + 2x + 5$, then is $x + 1$ a factor of $P(x)$? If yes, then factor $P(x)$ completely.

32. If $P(x) = x^3 - 3x^2 + 8x - 12$, then is $x - 2$ a factor of $P(x)$? If yes, then factor $P(x)$ completely.

33. If $P(x) = x^3 - 5x^2 + 2x + 8$, then is $x - 4$ a factor of $P(x)$? If yes, then factor $P(x)$ completely.

34. If $P(x) = x^3 + 9x^2 + 24x + 20$, then is $x + 5$ a factor of $P(x)$? If yes, then factor $P(x)$ completely.

35. If $P(x) = x^3 + 3x^2 + x + 6$, then is $x - 2$ a factor of $P(x)$? If yes, then factor $P(x)$ completely.

36. If $P(x) = x^3 + x^2 + 2x + 6$, then is $x + 3$ a factor of $P(x)$? If yes, then factor $P(x)$ completely.

37. A rectangular piece of paper 8 in. by 10 in. is to be used to make an open-top box with volume 48 in.^{3}, as described in Example 6 of this section. If the side for each cutoff corner is given by x, then the volume is given by $V = (8 - 2x)(10 - 2x)x$. Check that $x = 1$ in. is one solution, then find all other possible solutions.

38. Redo Exercise 37 if the volume is 24 in.^{3}, given that $x = 3$ in. is one solution.

39. The rectangular open metal figure shown is made from 12 rods. The total amount of wire used for the edges is 40 inches.
  a. Confirm that $a + b + c = 10$.
  b. Suppose that height (b) is 2 inches more than the length (a). Confirm that the volume is then given by
  $$V = a(8 - 2a)(a + 2).$$
  c. If $a = 1$, then the volume of the figure is 18 in.^{3}. Is there any other choice for a that also yields volume 18 in.^{3}?

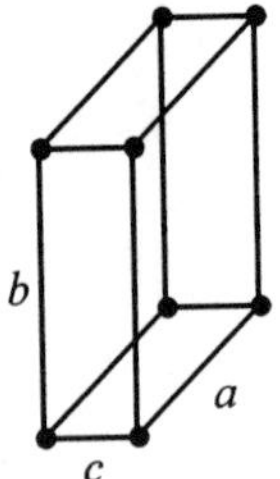

40. Redo Exercise 39 part **c**, assuming that $V = 32$ in.^{3}. In this case $a = 2$ is one solution. Are there any others?

41. Given that -2 is a zero of $P(x) = x^3 + 2x^2 + x + 2$, find all other zeros.

42. Given that -3 is a zero of $P(x) = x^3 + 3x^2 - 5x - 15$, find all other zeros.

In Exercises 43-56 one or more zeros (is/are) given for each of the following polynomial functions. Find the other zeros.

43. $P(x) = x^2 - 4x - 3$; $2 + \sqrt{7}$

44. $P(x) = 3x^2 + 4x - 2$; $\dfrac{-2 - \sqrt{10}}{3}$

45. $P(x) = x^2 + 2$; $i\sqrt{2}$

46. $P(x) = 3x^2 - 2x + 1$; $\dfrac{1}{3} - \dfrac{i\sqrt{2}}{3}$

47. $P(x) = 2x^3 - 11x^2 + 28x - 24$; $\dfrac{3}{2}$ and $2 + 2i$

48. $P(x) = x^4 + 13x^2 - 48$; $\sqrt{3}$ and $-4i$

49. $P(x) = 2x^3 + 5x^2 + 4x + 1$; -1

50. $P(x) = 3x^3 - 2x^2 - 10x + 4$; 2

51. $P(x) = x^4 - 6x^3 + 7x^2 + 12x - 18$; 3 is a zero of multiplicity 2

52. $P(x) = 2x^4 - 5x^3 + 11x^2 - 3x - 5$; 1 and $-\dfrac{1}{2}$

53. $P(x) = x^4 - 4$; $\sqrt{2}$

54. $P(x) = x^4 - 14x^2 + 45$; $-\sqrt{5}$

55. $P(x) = x^4 - 1; i$ **56.** $P(x) = x^4 - 16; 2i$

In Exercises 57–62 write a polynomial function (in factored form) with rational coefficients of the lowest possible degree with the given zeros.

57. 2 and $\sqrt{3}$ **58.** i and 0

59. 0, $2+3i$, and $4-\sqrt{3}$ **60.** -5, $-\sqrt{5}$, and $5i$

61. 3, -3 and $2i$ **62.** $3i$ and $1+\sqrt{2}$

63. If -1 is a zero of $y = x^3 - x^2 - 10x - 8$, find all intercepts and then graph the function.

64. Graph the function in Exercise 51 on the basis of information obtained from the intercepts.

In Exercises 65–68 $y = P(x)$ is a polynomial function with real coefficients. Answer true or false.

65. Every polynomial function of degree 3 has at least one real zero.

66. Every polynomial function of degree 4 has at least one real zero.

67. Every polynomial function of degree $n \geq 1$ has exactly n (not necessarily different) real zeros.

68. Every polynomial function of degree $n \geq 1$ has exactly n (not necessarily different) complex zeros.

69. Show that if $x - b$ is a factor of the polynomial $P(x)$, then b is a zero of $y = P(x)$.

THINK ABOUT IT 6.2

1. The dividend is x^2, the quotient is $x-4$, and the remainder is 16. What is the divisor?
2. For what value of k is $x-2$ a factor of $2x^4 - 5x^3 + kx^2 - 5x + 2$?
3. Write a polynomial function (in factored form) with *real* coefficients of lowest possible degree with zeros of 0, $-\sqrt{3}$, and $3i$.
4. If $f(x) = ax^3 + bx^2 + cx + d$, where a, b, c, and d are real numbers with $a \neq 0$, then explain why the graph of f must have at least one x-intercept.
5. A polynomial function of degree n can't have more than n zeros. Thus its graph cannot hit the x-axis more than n times.
   - **a.** Explain why the graph of a polynomial function of degree n cannot hit some other horizontal line more than n times. Hint: If $P(x)$ has degree n, what degree does $P(x) - c$ have, where c is a real constant?
   - **b.** Explain why no section of a polynomial graph can be horizontal.

● ● ●

6.3 Additional Theorems About Zeros

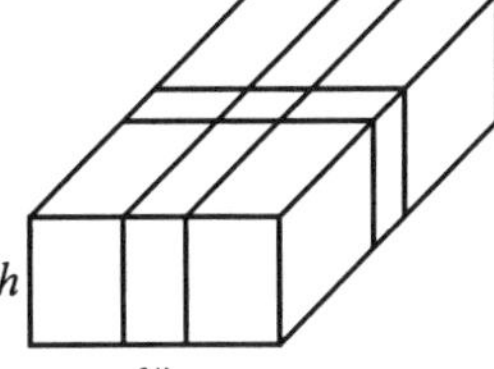

The volume of a rectangular packaging carton is to be 30 ft^3. If the length and width are respectively 3 ft and 1 ft greater than the height, show that only one set of dimensions is possible for the given carton. What are these dimensions?
(See Example 4.)

Objectives

1. List the possible rational zeros of a polynomial function with integer coefficients.
2. Determine the maximum number of positive and negative real zeros of a polynomial function.
3. Find all the zeros of a polynomial function, when given a function with a sufficient number of rational zeros to allow all zeros to be determined.
4. Use the location theorem to verify the existence of a zero of a function in a given interval.
5. Approximate a zero of a function in a given interval to the nearest tenth.

In the pervious section we saw that we can find all the zeros of a third-or fourth-degree polynomial function when one or two zeros, respectively, are apparent. The following theorem can help us locate some apparent answers because it may be used to obtain a list of *possible* rational zeros of a polynomial function with integer coefficients. We emphasize the word *possible*. The zeros may all be either irrational or complex. This theorem states only that if there are rational zeros, they satisfy the requirement indicated.

Rational-Zero Theorem

> If p/q, a rational number in lowest terms, is a zero of the polynomial function with integer coefficients
>
> $$P(x) = a_n x^n + a_{n-1}x^{n-1} + \cdots + a_1 x + a_0 \quad (a_n \neq 0),$$
>
> then p is an integral factor of the constant term a_0, and q is an integral factor of the leading coefficient a_n.

To prove this theorem, first note that if p/q is a zero of $y = P(x)$, then

$$a_n(p/q)^n + a_{n-1}(p/q)^{n-1} + \cdots + a_1(p/q) + a_0 = 0.$$

If we multiply both sides of this equation by q^n, we have

$$a_n p^n + a_{n-1}p^{n-1}q + \cdots + a_1 pq^{n-1} + a_0 q^n = 0.$$

Next, we subtract $a_0 q^n$ from both sides of the equation, yielding

$$a_n p^n + a_{n-1}p^{n-1}q + \cdots + a_1 pq^{n-1} = -a_0 q^n,$$

and finally, we factor the common factor p from each term on the left to obtain

$$p\left(a_n p^{n-1} + a_{n-1}p^{n-2}q + \cdots + a_1 q^{n-1}\right) = -a_0 q^n.$$

Since p is a factor of the left-hand side of the equation, p must also be a factor of $-a_0 q^n$. But by hypothesis, p/q is in lowest terms, so p and q have no common factor, and p is not a factor of q^n. Thus, p is a factor of a_0.

To prove that q is a factor of a_n, we rewrite our second equation as

$$q\left(a_{n-1}p^{n-1} + a_{n-2}p^{n-2}q + \cdots + a_0 q^{n-1}\right) = -a_n q^n$$

and reason as above. In practice, the rational-zero theorem is not difficult to apply.

Example 1: Finding Possible Rational Zeros

List the possible rational zeros of the function $P(x) = 2x^3 - x^2 - 6x - 3$.

Solution: The constant term a_0 is –3. The possibilities for p are the integers that are factors of 3. Thus,

$$p = \pm 3,\ \pm 1.$$

The leading coefficient a_n is 2. The possibilities for q are the integers that are factors of 2. Thus,

$$q = \pm 2,\ \pm 1.$$

The possible rational zeros p/q are then

$$3, \frac{3}{2}, 1, \frac{1}{2}, -\frac{1}{2}, -1, -\frac{3}{2}, -3.$$

PROGRESS CHECK 1
List the possible rational zeros of the function $P(x) = 5x^3 - 7x^2 - 8x + 4$. ■

Since we are often faced with a large list of possible rational zeros, we need some help in narrowing down the possibilities. Consider the theorem that follows, which helps us chip away at the problem. Note that it is a statement about real zeros. Thus, it applies to zeros that are rational or irrational.

Descartes' Rule of Signs

The maximum number of positive real zeros of the polynomial function $y = P(x)$ is the number of changes in sign of the coefficients in $P(x)$. The number of changes in sign of the coefficients in $P(-x)$ is the maximum number of negative real zeros. In both cases, if the number of zeros is not the maximum number, then it is less than this number by a multiple of 2.

We illustrate this theorem in the following example.

Example 2: Finding Zeros of a Third-Degree Polynomial Function

Find the zeros of $P(x) = 2x^3 + 5x^2 + 4x + 1$.

Solution: Steps **a** to **d** outline a systematic approach to finding the zeros algebraically.

a. Use Descartes' rule of signs to determine the maximum number of positive and negative real zeros.

$$P(x) = 2x^3 + 5x^2 + 4x + 1.$$

All the coefficients in $P(x)$ are positive. There are no positive real zeros since there are no changes in sign.

$$\begin{aligned} P(-x) &= 2(-x)^3 + 5(-x)^2 + 4(-x) + 1 \\ &= -2x^3 + 5x^2 - 4x + 1 \end{aligned}$$

There are three sign changes in $P(-x)$. At most, there may be three negative real zeros. By decreasing this number by 2, we determine that one negative real zero is the only other possibility.

b. Determine the possible rational zeros. The constant term a_0 is 1. The possibilities for p are the integers that are factors of 1. Thus,

$$p = \pm 1.$$

The leading coefficient a_n is 2. The integers that are factors of 2 give the possibilities for q. Thus,

$$q = \pm 2, \ \pm 1.$$

The possible rational zeros p/q are then

$$1,\ \frac{1}{2},\ -\frac{1}{2},\ -1.$$

From part **a** we eliminate the positive possibilities, leaving

$$-\frac{1}{2},\ -1.$$

c. Use synthetic division to test whether one of the possibilities, say –1, is a zero.

$$\begin{array}{r|rrrr} -1 & 2 & 5 & 4 & 1 \\ & & -2 & -3 & -1 \\ \hline & 2 & 3 & 1 & 0 \end{array}$$

Since the remainder is zero, $P(-1)=0$. Thus, –1 is a zero of the function.

d. The factor theorem states that if –1 is a zero, $x-(-1)$ or $x+1$ is a factor. The coefficients of the other factor are given in the bottom row of the synthetic division. That is,

$$2x^3+5x^2+4x+1=(x+1)(2x^2+3x+1).$$

We now find other zeros by setting the factor $2x^2+3x+1$ equal to 0. The big advantage is that after finding a zero we may lower by 1 the degree of the equation we are solving. When we reach a second-degree equation, as in this case, we may apply the quadratic formula or (sometimes) the factoring method.

$$\begin{aligned} 2x^2+3x+1&=0 \\ (2x+1)(x+1)&=0 \\ 2x+1=0 \qquad & x+1=0 \\ x=-\frac{1}{2} \qquad & x=-1 \end{aligned}$$

Thus the zeros are –1, –1, and $-1/2$. Remember that it is possible for a number to be counted as a zero more than once, and in this case we say –1 is a zero of multiplicity 2. Based on the *x*- and *y*-intercepts and the multiplicity of each zero, the function is graphed in Figure 6.21.

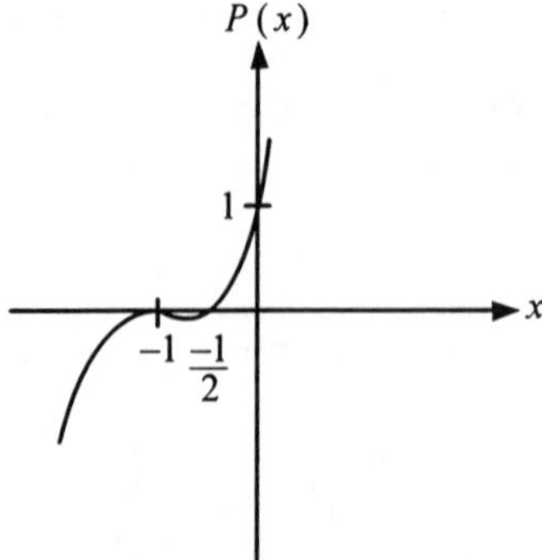

Figure 6.21

Technology Link

With a grapher it is possible to begin the analysis in this problem by first graphing the function in the standard viewing window, as shown in Figure 6.22, and then refining this graph so the x-intercepts are more apparent, as shown in Figure 6.23. This method will help you to locate quickly potential rational-number zeros.

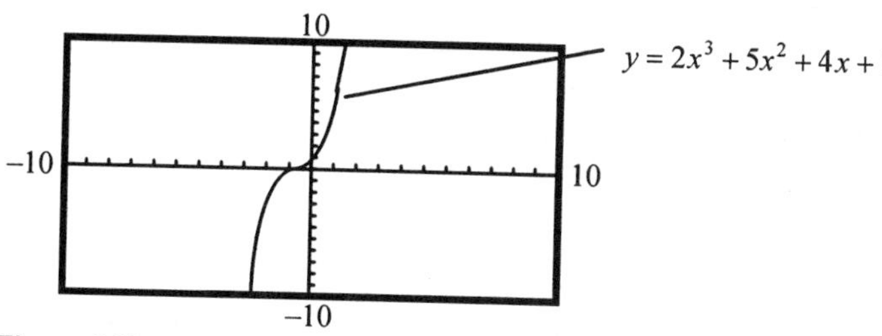

Figure 6.22

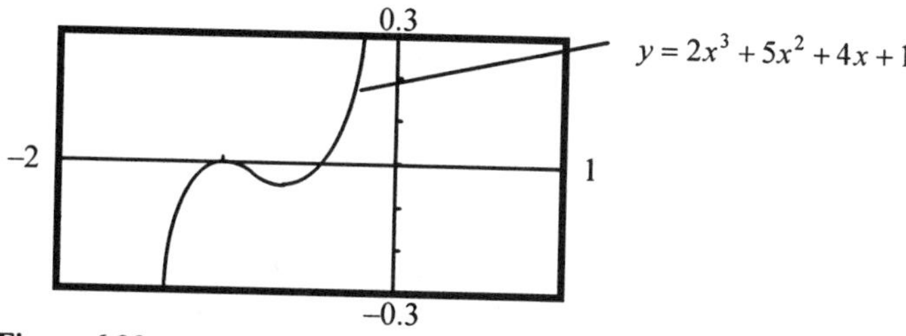

Figure 6.23

PROGRESS CHECK 2
Find the zeros of $P(x) = 2x^3 - 7x^2 + 8x - 3$. ■

For the next example we add another useful rule, which permits us to establish upper and lower bounds on the zeros of a polynomial function.

Upper and Lower Bounds Theorem

1. If we divide a polynomial $P(x)$ synthetically by $x - b$, where $b > 0$ and all the numbers in the bottom row have the same signs, then there is no zero greater than b. We say that b is an **upper bound** for the zeros of $y = P(x)$.
2. If we divide a polynomial $P(x)$ synthetically by $x - c$, where $c < 0$ and the numbers in the bottom row alternate in sign, then there is no zero less than c. We say that c is a **lower bound** for the zeros of $y = P(x)$.

For the purposes of these tests, zero may be denoted as +0 or –0.

A rationale for these tests is contained in the next example.

Example 3: Finding Zeros of a Fourth-Degree Polynomial Function

Find the zeros of $P(x) = 2x^4 - 5x^3 + 11x^2 - 3x - 5$.

Solution: A systematic algebraic approach to obtaining the zeros is given in steps **a** to **e**.

a. Use Descartes' rule of signs to determine the maximum number of positive and negative real zeros.

$$P(x) = 2x^4 - 5x^3 + 11x^2 - 3x - 5$$

There are three sign changes in $P(x)$. Thus, the number of positive real zeros is three or one.

$$\begin{aligned} P(-x) &= 2(-x)^4 - 5(-x)^3 + 11(-x)^2 - 3(-x) - 5 \\ &= 2x^4 + 5x^3 + 11x^2 + 3x - 5 \end{aligned}$$

There is one sign change in $P(-x)$, so Descartes' rule of signs guarantees exactly one negative *real* zero in this case.

b. Determine the possible rational zeros. The constant term a_0 is –5. The possibilities for p are the integers that are factors of –5. Thus,

$$p = \pm 5,\ \pm 1.$$

The leading coefficient a_n is 2. The integral factors of 2 give the possibilities for q. Thus,

$$q = \pm 2,\ \pm 1.$$

The possible rational zeros are then

$$5,\ \frac{5}{2},\ 1,\ \frac{1}{2},\ -\frac{1}{2},\ -1,\ -\frac{5}{2},\ -5.$$

c. We test negative zeros first, because part **a** indicated that there is, at most, one negative zero. If we find a negative zero, we then switch to the positive possibilities. Pick –1, a number in the middle of the negative choices. If it is not a zero, it may be a lower bound that eliminates other possibilities.

–1	2	–5	11	–3	–5
		–2	7	–18	21
	2	–7	18	–21	16

The remainder is 16, so –1 is not a zero. We tested a negative possibility, and the numbers in the bottom row alternate in sign. Thus, there is no zero less than –1, as specified in statement 2 of the above theorem. To see why, note that if we test a negative choice, say c, greater in absolute value than –1, then the numbers in the bottom row of the synthetic division will continue to alternate in sign with respective numbers of greater absolute value (after the first entry) than our previous bottom row $(2,\ -7,\ 18,\ -21,\ 16)$. Thus $P(c) > 16$ when $c < -1$, so –1 is a lower bound for the zeros of $y = P(x)$. This eliminates $-5/2$ and –5, so try $-1/2$.

$-\frac{1}{2}$	2	–5	11	–3	–5
		–1	3	–7	5
	2	–6	14	–10	0

Since the remainder is zero, $P(-1/2) = 0$. Thus, $-1/2$ is a zero of the function.

d. Since $-1/2$ is a zero, $x - (-1/2)$ is a factor, and we write

$$2x^4 - 5x^3 + 11x^2 - 3x - 5 = \left(x + \frac{1}{2}\right)\left(2x^3 - 6x^2 + 14x - 10\right).$$

We now try to find the values of x for which $2x^3 - 6x^2 + 14x - 10$ equals 0. We found the one negative zero, so we switch to the positive possibilities. Pick 1. If it is not a zero, the numbers in the bottom row in the synthetic division may have the same sign. Such a result would establish 1 as an upper bound for the zeros (thereby eliminating 5/2 and 5), since any synthetic division by $x-b$, where $b>1$, would continue to produce bottom-row entries that have the same sign and even larger absolute values (after the first entry) than those that result from the division involving $b=1$.

$$\begin{array}{r|rrrr} \underline{1} & 2 & -6 & 14 & -10 \\ & & 2 & -4 & 10 \\ \hline & 2 & -4 & 10 & 0 \end{array}$$

Therefore, 1 is a zero of the function.

e. We now use the quadratic formula to complete the solution.

$$2x^2-4x+10=0$$

$$x^2-2x+5=0$$

$$x=\frac{-(-2)\pm\sqrt{(-2)^2-4(1)(5)}}{2(1)}=\frac{2\pm\sqrt{-16}}{2}=\frac{2\pm 4i}{2}=1\pm 2i$$

The zeros are $-1/2$, 1, $1+2i$, and $1-2i$.

A graph of this function which illustrates the two real-number zeros, is shown in Figure 6.24.

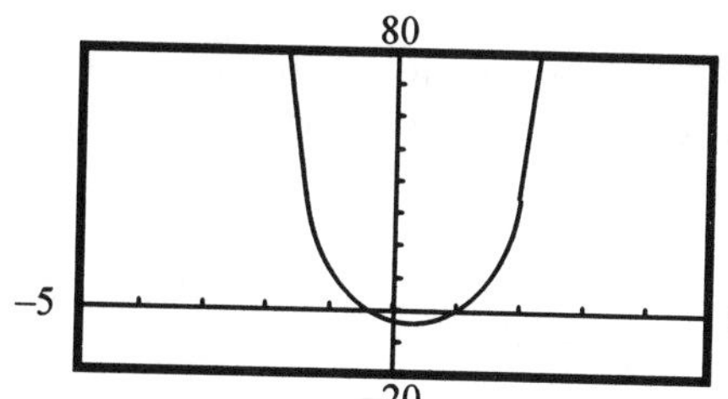

$y=2x^4-5x^3+11x^2-3x-5$

Figure 6.24

Technology Link

Some calculators have a built-in feature that may be used to solve polynomial equations. For instance, Figure 6.25 shows the coefficient entry screen and the solution screen associated with solving the equation in this example, on a particular calculator with a Polynomial Root-Finding feature. Observe that the calculator output displays the complex number $a+bi$ in the form (a, b).

```
a4X^4+...+a1X+a0=0
 a4=2
 a3=-5
 a2=11
 a1=-3
 a0=-5
```

(a)

```
a4X^4+...+a1X+a0=0
 x1=(1,2)
 x2=(1,-2)
 x3=(1,0)
 x4=(-.5,0)
```

(b)

Figure 6.25

PROGRESS CHECK 3

Find the zeros of $P(x)=2x^4-9x^3+42x-20$.

Example 4: Solving a Volume Problem

Solve the problem in the section introduction on page 491.

Solution: If x represents the height of the carton, then $x+3$ and $x+1$ represent the length and width, respectively. Since the formula for the volume is $V = lwh$ and the volume of the carton is 30 ft^3, we have

$$(x+3)(x+1)x = 30$$
$$x^3 + 4x^2 + 3x = 30$$
$$x^3 + 4x^2 + 3x - 30 = 0.$$

The possible positive rational zeros are then

30, 15, 10, 6, 5, 3, 2, 1.

We can determine that 2 is a solution either by graphing the associated function, as shown in Figure 6.26, or by testing possibilities with the aid of the upper bound theorem. The synthetic division that confirms that 2 is a solution is shown next.

$$\begin{array}{r|rrrr} 2 & 1 & 4 & 3 & -30 \\ & & 2 & 12 & 30 \\ \hline & 1 & 6 & 15 & 0 \end{array}$$

The two other roots are then determined by solving $x^2 + 6x + 15 = 0$. However, since the discriminant $(b^2 - 4ac)$ is less than 0, these roots are not real numbers. Thus, the only real-number solution is 2, and the box has unique dimensions of 5 ft by 3 ft by 2 ft.

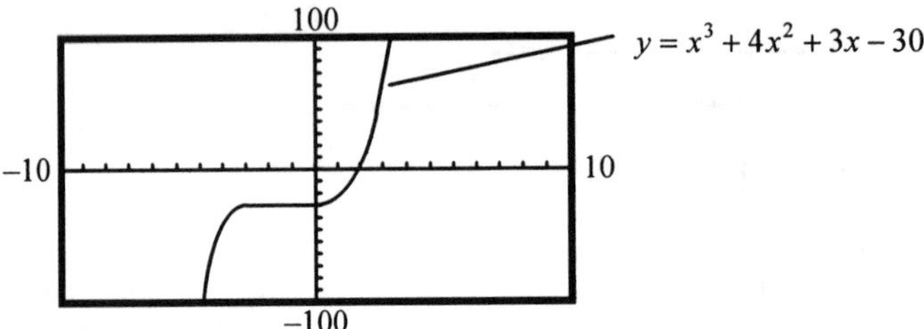

Figure 6.26

PROGRESS CHECK 4

The volume of a rectangular box is to be 350 in.^3. If the width is 2 in. greater than the height, and the length is twice the height, show that only one set of dimensions is possible for the given box. What are these dimensions? ■

Up to this point we have limited our discussion to finding rational zeros of polynomial functions of degree greater than 2. Unfortunately, irrational zeros of higher-degree polynomial functions are usually very difficult to find algebraically. Complicated formulas are available for polynomial functions of degrees 3 and 4, and it can be shown that no formula exists for degree 5 and greater. In such cases, numerical methods that utilize a grapher or a computer are employed to approximate irrational zeros, and this topic is usually considered in detail in a course in numerical analysis. For our purposes, we need only illustrate the following two theorems which enable us to approximate zeros of polynomial functions.

Intermediate Value Theorem for Polynomials

> Let $y = P(x)$ be a polynomial function. If a and b are real numbers with $a < b$, and if d is any number between $P(a)$ and $P(b)$, inclusive, then there is at least one number c in the interval $[a, b]$ such that $P(c) = d$.

This result is easy to see geometrically. Consider Figure 6.27 and keep in mind that the graph of a polynomial function is an unbroken curve. Through any number d between $P(a)$ and $P(b)$, inclusive, a horizontal line may be drawn that must intersect the graph of $y = P(x)$ at least once in the interval $[a, b]$. Then, the x-coordinate of an intersection point is a number c in the interval $[a, b]$ such that $P(c) = d$.

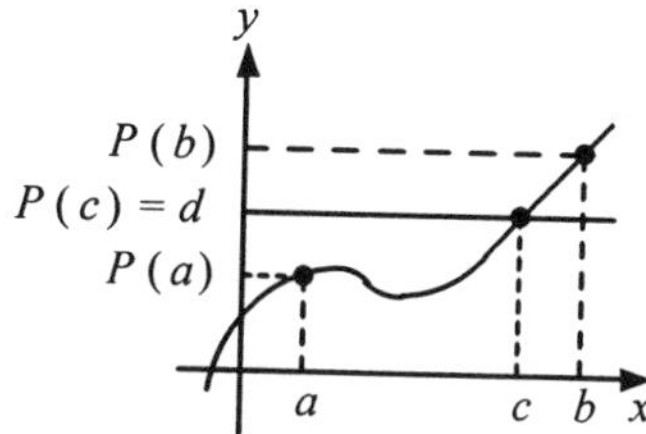

Figure 6.27

The intermediate value theorem is used in many ways when analyzing polynomial functions, and a special case of this theorem, called the *location theorem*, may be used to approximate zeros of such functions. When $P(a)$ and $P(b)$ are opposite in sign, then 0 is between $P(a)$ and $P(b)$ so the intermediate value theorem guarantees that there is at least one number c in the interval (a, b) such that $P(c) = 0$. Thus, $y = P(x)$ has at least one real zero between a and b, which leads to the following result.

Location Theorem

> Let $y = P(x)$ be a polynomial function. If $P(a)$ and $P(b)$ have opposite signs, then $y = P(x)$ has at least one real zero between a and b.

The geometric interpretation of this theorem is shown in Figure 6.28. Since the graph is on different sides of the x-axis at a and b, and no breaks are possible, the graph must cross the x-axis between a and b, guaranteeing at least one real zero in this interval.

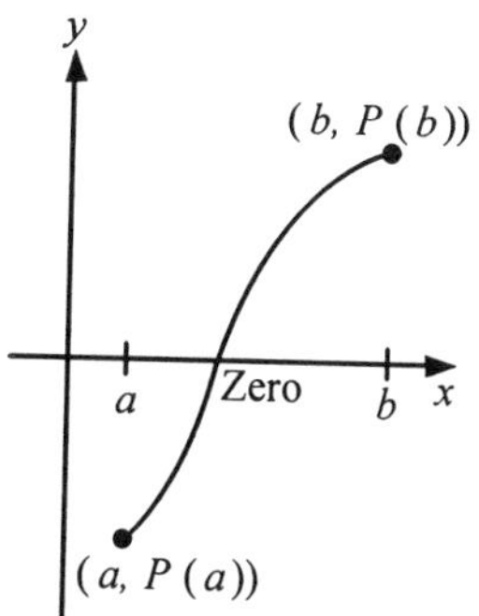

Figure 6.28

Example 5: Using the Location Theorem

Verify that $P(x) = 3x^3 + 16x^2 - 8$ has a zero between –5 and –6.

Solution: By direct substitution or synthetic division, we determine that $P(-5) = 17$ and $P(-6) = -80$. Since $P(-5)$ and $P(-6)$ have opposite signs, the location theorem assures us of at least one real zero between –5 and –6. This zero can be seen in the graph of the function that is shown in Figure 6.29.

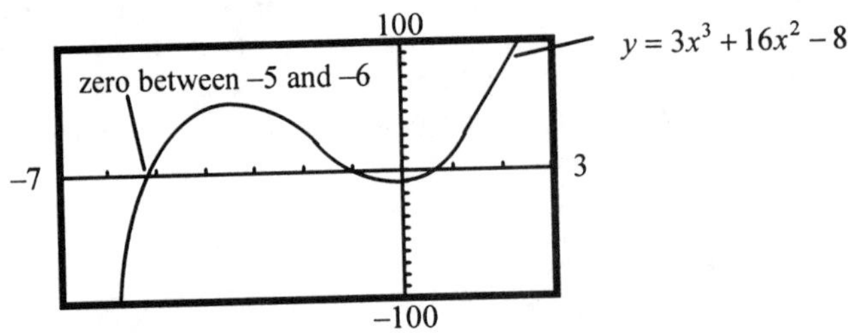

Figure 6.29

PROGRESS CHECK 5

Verify that $P(x) = 3x^3 + 16x^2 - 8$ has zero between 0 and 1. ■

Once we have determined an interval that contains a zero, the job is to keep narrowing the interval until we obtain a sufficient degree of accuracy. The most efficient methods for narrowing the interval are developed in calculus. A simple and relatively effective alternative is basically to halve the interval each time until we attain a specified accuracy.

Example 6: Approximating a Zero of a Polynomial Function

To the nearest tenth, approximate the zero of the function $P(x) = 3x^3 + 16x^2 - 8$ that is between –5 and –6.

Solution: By using the location theorem and a calculator, and by successively halving the intervals, we determine the following.

Calculation	Comment: A Zero is Between
$P(-5) = 17$	
$P(-6) = -80$	–5 and –6
$P(-5.5) \approx -23$	–5 and –5.5
$P(-5.25) \approx -1.1$	–5 and –5.25
$P(-5.13) \approx 8.1$	–5.13 and –5.25
$P(-5.19) \approx 3.6$	–5.19 and –5.25

Thus, to the nearest tenth, a zero is –5.2. Figure 6.30 shows a check of this answer on a grapher.

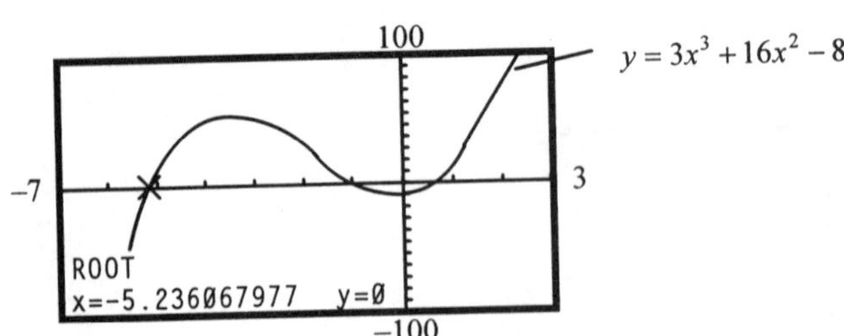

Figure 6.30

Note: Exercise 14 in this section asks you to find the exact value of this zero by using the previous methods of this section.

PROGRESS CHECK 6

To the nearest tenth, approximate the zero of the function $P(x) = 3x^3 + 16x^2 - 8$ that is between 0 and 1.

■

EXPLORE 6.3

1. Some calculators have a special feature for finding all the zeros of polynomial functions (including imaginary ones). The degree (or order) of the polynomial is entered, and then just the coefficients. As with the feature that finds the x-intercepts of any graphs, this polynomial zero-finding algorithm sometimes fails when zeros have even multiplicity. If your calculator has such a feature, illustrate this point by finding the zeros of $P(x) = x^4 - 8x^3 + 24x^2 - 32x + 16$. Although the exact zero is 2 with multiplicity 4, you may get complex answers with the real part close to 2, but with a non-zero (but small) imaginary part. Similarly, although the exact zeros of $P(x) = x^4 + 2x^2 + 1$ are i and $-i$, you may get complex answers with the real part close to zero and the imaginary part close to 1 or −1. If your calculator has this feature, try it on the 2 given polynomials, then make up an example of your own that will cause the machine to fail.
2. The calculator often estimates a zero of a function f by an iterative process that repeats a certain set of steps until the answer is "close enough." (If the answer doesn't become close enough after a certain number of tries, the calculator displays some kind of error message.) There are various criteria for deciding when an answer is close enough. Here are three common stopping criteria:

Method 1. For each new estimate, x_{new}, of the zero evaluate $f(x_{\text{new}})$, and finally stop when the absolute value of $f(x_{\text{new}})$ is smaller than some agreed-upon value, often denoted by epsilon, $(\in)$. For instance, stop when $|f(x_{\text{new}})| < 0.001$.

Method 2. For each new sequential pair of estimates, x_1 and x_2 of the zero, evaluate $|x_1 - x_2|$ and finally stop when this expression is smaller than some agreed-upon value. For instance, stop when $|x_1 - x_2| < 0.001$.

Method 3. For each new sequential pair of estimates, x_1 and x_2 of the zero, evaluate $|x_1 - x_2|/(1 + |x_2|)$ and finally stop when this expression is smaller than some agreed-upon value. For instance, stop when $|x_1 - x_2|/(1 + |x_2|) < 0.001$.

a. Suppose a calculator is estimating a zero of $f(x) = x^3 - 4x^2 + 4$ by an iterative process that gives the following sequence of estimates. Use each of the three methods above to determine the answer given by the calculator if the value of $\in$ is 0.001.

$x_1 = 1.5000$	$x_5 = 1.2188$	$x_9 = 1.1895$	$x_{13} = 1.1942$
$x_2 = 0.7500$	$x_6 = 1.1719$	$x_{10} = 1.1924$	$x_{14} = 1.1940$
$x_3 = 1.1250$	$x_7 = 1.1953$	$x_{11} = 1.1938$	$x_{15} = 1.1939$
$x_4 = 1.3125$	$x_8 = 1.1836$	$x_{12} = 1.1946$	$x_{16} = 1.19394$

b. What would the answers be if $\in$ is 0.0001?

EXERCISES 6.3

In Exercises 1–4 list the possible rational zeros of the function.

1. $P(x) = x^3 + 8x^2 - 10x - 20$

2. $P(x) = x^4 - 3x^3 + x^2 - x - 6$

3. $P(x) = 4x^4 - 3x^3 + x^2 - x - 6$

4. $P(x) = 3x^3 + 7x^2 + 8$

In Exercises 5–8 use Descartes' rule of signs to determine the maximum number of positive and negative real zeros.

5. $P(x) = x^3 - 2x^2 + x - 3$

6. $P(x) = 5x^3 + x^2 + 4x + 1$

7. $P(x) = 2x^4 + 5x^3 - x^2 - 3x - 7$

8. $P(x) = x^8 + 1$

In Exercises 9–18 find the zeros of the function.

9. $P(x) = x^3 - x^2 - 10x - 8$

10. $P(x) = x^3 + 6x^2 + 11x + 6$

11. $P(x) = 3x^3 + x^2 + 15x + 5$

12. $P(x) = 4x^3 - x^2 - 28x + 7$

13. $P(x) = 8x^3 - 12x^2 + 6x - 1$

14. $P(x) = 3x^3 + 16x^2 - 8$

15. $P(x) = 4x^4 - 5x^3 - 2x^2 - 3x - 10$

16. $P(x) = x^4 - 6x^3 + 7x^2 + 12x - 18$

17. $P(x) = 2x^4 + 7x^3 + 25x^2 + 47x + 18$

18. $P(x) = 6x^4 + 11x^3 - 63x^2 - 7x + 5$

In Exercises 19–22 answer the following questions with respect to the given functions

a. What are all the zeros of the function?
b. How many zeros are the rational numbers? Name them.
c. How many zeros are real numbers? Name them.
d. How many zeros are imaginary numbers? Name them.

19. $P(x) = x^3 - x^2 - 2x - 12$

20. $P(x) = x^3 + 7x^2 + 12x + 6$

21. $P(x) = 3x^4 + 5x^3 - 23x^2 - 35x + 14$

22. $P(x) = 10x^4 + 35x^3 - 73x^2 + 7x - 15$

23. The volume of a rectangular box is 105 in.^{3}. If the width is 2 in. greater than the height, and the length is 1 in. greater than twice the height, show that only one set of dimensions is possible for the given box. What are these dimensions?

24. The volume of a rectangular box is 20 ft^3. If the box has a square base and the height is 3 ft greater than the measure of the edges of the base, show that only one set of dimensions is possible for the given box. What are these dimensions?

25. A rectangular piece of cardboard 12 in. by 18 in. is to be used to make an open-top box, by cutting a small square from each corner and bending up the sides. If the volume of the box is to be 216 in.^{3}, then find the length of the side of the square that is to be cut from each corner.

26. A rectangular sheet of tin 20 cm by 24 cm has identical squares cut out from each corner. The sides are then turned up to form an open box. Find the side length for each square cutout, given that the volume is to be 640 cm^3.

In Exercises 27–30 use the location theorem and verify each statement.

27. $P(x) = 3x^3 + 16x^2 - 8$ has a zero between 0 and –1.

28. $P(x) = 4x^3 - x^2 - 28x + 7$ has a zero between 2 and 3.

29. $P(x) = x^4 - 6x^3 + 7x^2 + 12x - 18$ has a zero between 1 and 2 and another zero between –1 and –2.

30. $P(x) = 6x^4 + 11x^3 - 63x^2 - 7x + 5$ has a zero between 0 and 1 and another zero between –4 and –5.

In Exercises 31–36 approximate the zero of the function in the given interval to the nearest tenth.

31. $P(x) = x^3 + x - 1$; a zero is between 0.6 and 0.7.

32. $P(x) = x^3 + x^2 - 10x - 10$; a zero is between –3.1 and –3.2.

33. $P(x) = x^3 - 3x + 1$; a zero is between 0 and 1.

34. $P(x) = x^3 - 3x + 1$; a zero is between 1 and 2.

35. $P(x) = 2x^3 - 5x^2 - 3x + 9$; a zero is between –1 and –2.

36. $P(x) = 2x^3 - 5x^2 - 3x + 9$; a zero is between 2 and 3.

THINK ABOUT IT 6.3

1. Find a polynomial function that has 1 zero between 2 and 3. Find a polynomial function that has 2 zeros between 2 and 3. Find a polynomial function that has 4 zeros between 2 and 3. What is the minimum degree possible for a polynomial function that has 10 zeros between 2 and 3?
2. If $P(x) = x^3 + px + q$, determine the nature of the zeros when
   a. p and q are both positive and
   b. p is positive and q is negative.
3. Consider the function defined by $f(x) = x^n + a$.
   a. If n is a positive even integer and a is a negative real number, find in terms of n the number of zeros of f that are not real numbers.
   b. Find in terms of n the number of zeros of f that are not real numbers, if n is a positive odd integer and a is a positive real number.
4. Use the rational-zero theorem and the function $P(x) = x^2 - 2$ to show that $\sqrt{2}$ is not a rational number.
5. Show that $\sqrt[5]{7}$ is not a rational number, by using the rational-zero theorem.

● ● ●

6.4 Rational Functions

Photo Courtesy of ***Janice Fullman***, *The Picture Cube*

A regional telephone company charges 25 cents per minute plus an 80-cent surcharge for a typical call made with its calling card. Therefore, the average cost (y) of a call per minute is given by

$$y = \frac{25x + 80}{x},$$

where x is the length of the call in minutes, and y is measured in cents.

a. What is the total cost for a 10-minute call? What is the average cost per minute for a 10-minute call?

b. Graph the function and describe how y changes as x increases for $x > 0$. What is the significance of the horizontal asymptote?

(See Example 4.)

Objectives

1. Determine any vertical and horizontal asymptotes for rational function and graph the function.
2. Determine any vertical and slant asymptotes for a rational function and graph the function.
3. Graph rational functions that involve common factors.

In the section-opening problem, the formula for the average cost per minute of a certain type of telephone call is

$$y = \frac{25x + 80}{x}.$$

This equation is an example of a formula that involves a quotient of two polynomials. In general, a function of the form

$$y = \frac{P(x)}{Q(x)},$$

where $P(x)$ and $Q(x)$ are polynomials with $Q(x) \neq 0$, is called a **rational function**. Some additional examples of rational functions are

$$y = \frac{1}{x},\ y = \frac{2x^2 + x - 3}{x^2},\ \text{and } y = \frac{x^2 - 1}{x + 2}.$$

The behavior of a rational function often differs dramatically from that of a polynomial function. This difference may be seen easily by comparing the graph of a rational function to the smooth unbroken curves that characterize polynomial functions.

Consider the rational function $y = 1/x$. Since division by zero is undefined, x cannot equal zero. Thus, the graph of this function does not intersect the line $x = 0$ (the y-axis). We can, however, let x approach zero and consider x values as close to zero as we wish. From the following tables, note that as the x values squeeze in on zero, $|y|$ becomes larger.

x	1	0.5	0.1	0.01	0.001
$y = \frac{1}{x}$	1	2	10	100	1,000

x	−1	−0.5	−0.1	−0.01	−0.001
$y = \frac{1}{x}$	−1	−2	−10	−100	−1,000

Figure 6.31 shows the behavior of $y = 1/x$ in the interval $-1 \leq x \leq 1$. The vertical line $x = 0$ that the curve approaches, but never touches, is called a **vertical asymptote**. We may use the following rule to determine if the graph of a rational function has any vertical asymptotes.

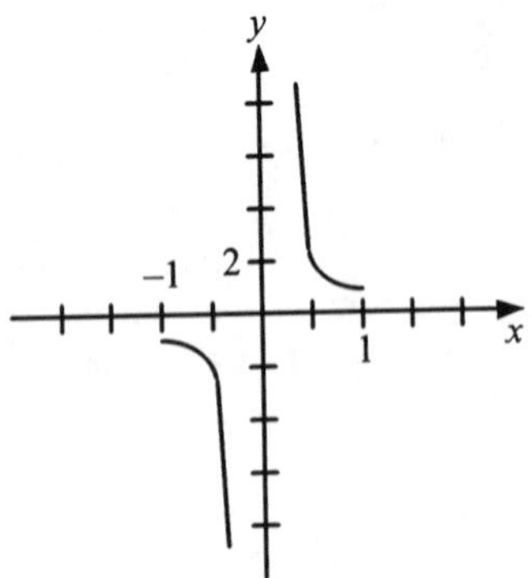

Figure 6.31

Vertical Asymptotes

> If $P(x)$ and $Q(x)$ have no common factors, then the graph of the rational function $y = P(x)/Q(x)$ has as a vertical asymptote the line $x = a$ for each real number a at which $Q(a) = 0$.

Initially we will consider only functions where $P(x)$ and $Q(x)$ have no common factors, so the rational function is in lowest terms. In this case the rule above indicates that a vertical asymptote occurs whenever the denominator is zero.

Example 1: Finding Vertical Asymptotes

Find any vertical asymptotes of the function

$$y = \frac{x+1}{(x-3)(x+2)}.$$

Solution: The polynomial in the denominator

$$Q(x) = (x-3)(x+2)$$

equals 0 when x is 3 or -2. Thus, there are two vertical asymptotes: $x = 3$ and $x = -2$. **Note:** As a visual aid, the graph of this function is shown in Figure 6.32. By the end of this section, we will have developed sufficient techniques for you to be able to draw this graph.)

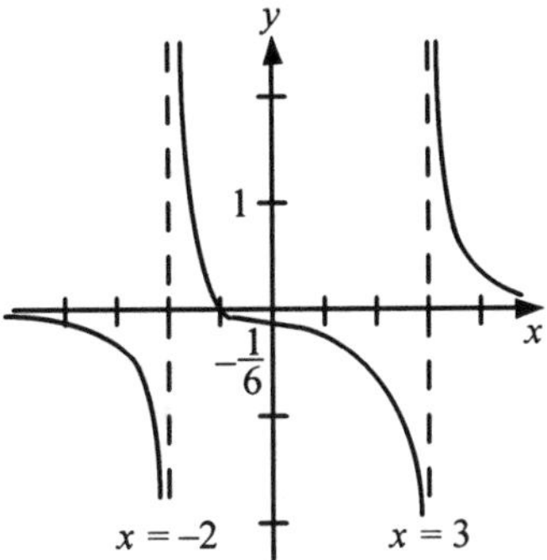

Figure 6.32

PROGRESS CHECK 1

Find any vertical asymptotes of $y = \dfrac{3x}{x^2 - 4}$. ■

To complete the graph of $y = 1/x$, we must consider the behavior of the function as $|x|$ becomes larger (this is called the end behavior of the function). It is not difficult to see that $1/x$ squeezes in on zero from the positive side when x takes on larger positive values. Similarly, $1/x$ squeezes in on zero from the negative side when x increases in magnitude in the negative direction. Thus, the curve gets closer to the line $y = 0$, and the x-axis is a **horizontal asymptotes**. Figure 6.33 shows a graph of $y = 1/x$.

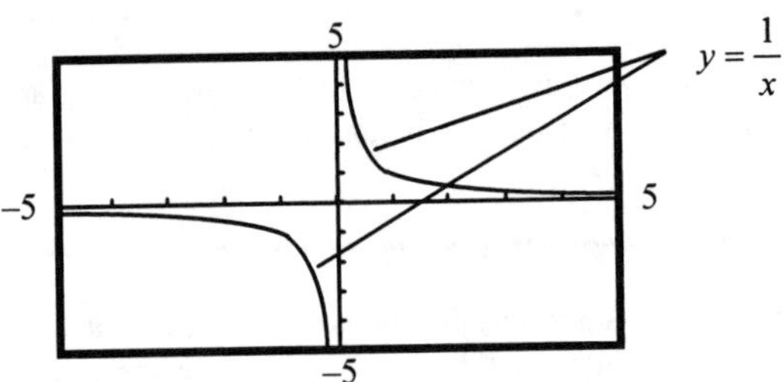

Figure 6.33

Identifying any vertical and horizontal asymptotes is very important. These lines, together with plotting the intercepts and a few points, enable us to graph a rational function. A technique for determining horizontal asymptotes is included in the following examples.

Example 2: Graphing a Rational Function

If $y = \frac{x-1}{x+1}$, find any asymptotes and graph the function.

Solution: Steps **a** to **d** outline a systematic approach to graphing this function.

a. The polynomial in the denominator $x+1$ equals zero when x is -1. Thus, $x=-1$ is a vertical asymptote.

b. To determine any horizontal asymptotes, we change the form of the function by dividing each term in the numerator and the denominator by the highest power of x in the expression.

$$y = \frac{x-1}{x+1} = \frac{\frac{x}{x}-\frac{1}{x}}{\frac{x}{x}+\frac{1}{x}} = \frac{1-\frac{1}{x}}{1+\frac{1}{x}} \quad \text{(assuming that } x \neq 0\text{)}$$

Now as $|x|$ gets larger, $1/x$ approaches zero. Thus, y approaches $(1-0)/(1+0)$, and $y=1$ is a horizontal asymptote.

c. By setting $x=0$, we determine that $(0, -1)$ is the y-intercept. Similarly, by setting $y=0$, we determine that $(1, 0)$ is the x-intercept.

d. The vertical asymptote divides the x-axis into two regions. The intercepts are two points to the right of $x=-1$. If we plot a couple of points to the left of the vertical asymptote, say $(-2, 3)$ and $(-3, 2)$, we may complete the graph (see Figure 6.34).

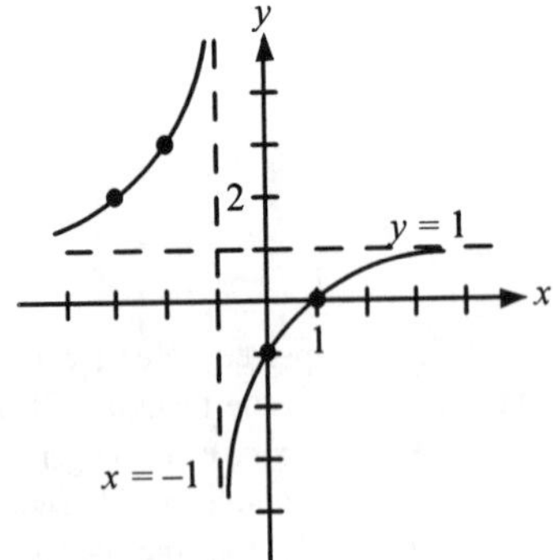

Figure 6.34

When graphing rational functions on a grapher, extra care is often needed to determine an appropriate viewing window and to decide whether Connected Mode or Dot Mode gives a more recognizable graph. For instance, Figure 6.35 shows a graph of the function in this example using the standard viewing window and Connected Mode. Note that the calculator erroneously draws an almost vertical line at about $x = -1$ because it is set to connect points that are in separate pieces of the graph.

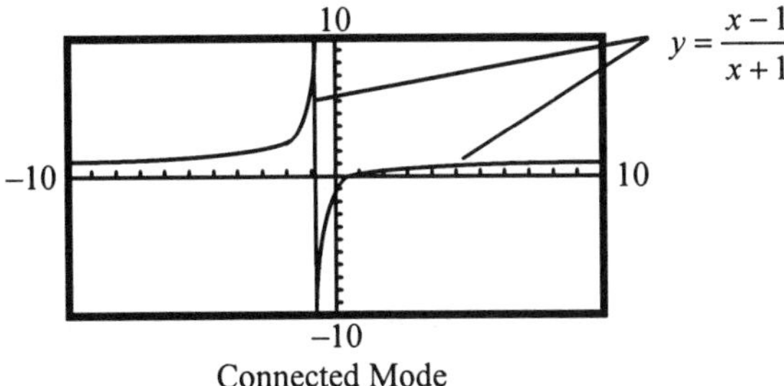

Figure 6.35

One way to make the graph more readable may be to switch the calculator to Dot Mode and redraw the graph in the viewing window as shown in Figure 6.36. Now the graph is centered about the two asymptotes, and only the calculated points for the function are displayed. You should be prepared to experiment with different viewing windows and graphing modes when graphing a rational function, until you obtain a picture you can interpret. However, under no circumstances should your hand-drawn graph show solid lines at vertical asymptotes, like those that may appear on a calculator screen.

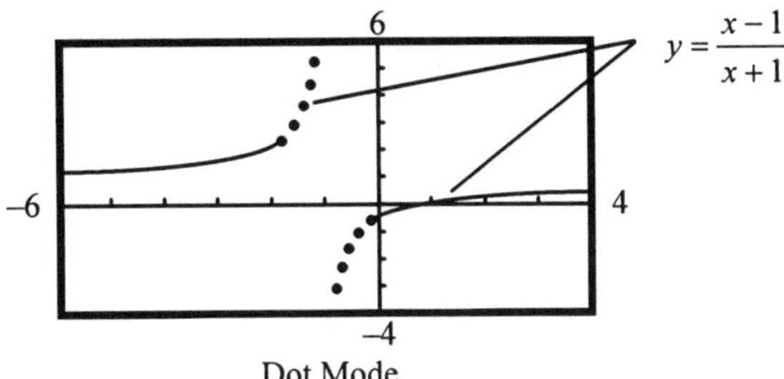

Figure 6.36

PROGRESS CHECK 2

If $y = \dfrac{4x-5}{2x+6}$, find any asymptotes and graph the function. ■

In Example 2 an important question is unanswered. We know that as $|x|$ becomes large, y approaches, but never quite reaches, 1. How do we know that y is not 1 when $|x|$ is small? We answer this question by determining if there is any value of x for which

$$\frac{x-1}{x+1} = 1.$$

Solving this equation, we have

$$\begin{aligned} x - 1 &= x + 1 \\ -1 &= 1 \quad \text{false.} \end{aligned}$$

The equation has no solution, and we may conclude that the curve never crosses the horizontal asymptote. Example 3 illustrates why this possibility must be considered.

Example 3: Graphing a Rational Function

If $y = \dfrac{2x^2 + x - 3}{x^2}$, find any asymptotes and graph the function.

Solution: A systematic algebraic approach to obtaining the graph is given in steps **a** to **e**.

a. The polynomial in the denominator equals zero when x is zero. Thus, $x = 0$ is a vertical asymptote.

b. To determine any horizontal asymptotes, we use the procedure from Example 2. Dividing each term in the numerator and the denominator by x^2 gives

$$y = \frac{\frac{2x^2}{x^2} + \frac{x}{x^2} - \frac{3}{x^2}}{\frac{x^2}{x^2}} = \frac{2 + \frac{1}{x} - \frac{3}{x^2}}{1} \quad (x \neq 0).$$

As $|x|$ gets larger, $1/x$ and $-3/x^2$ approach 0. Thus, y approaches 2, and $y = 2$ is a horizontal asymptote.

c. To determine if the curve ever crosses the horizontal asymptote, find out if there are any values of x for which

$$\frac{2x^2 + x - 3}{x^2} = 2.$$

Solving this equation for x, we have

$$\begin{aligned} 2x^2 + x - 3 &= 2x^2 \\ x - 3 &= 0 \\ x &= 3. \end{aligned}$$

The curve crosses the asymptote at (3, 2).

d. Since x cannot be zero, there is no y-intercept. After setting $y = 0$, we solve the equation $2x^2 + x - 3 = 0$ to find the x-intercepts $(1, 0)$ and $(-3/2,\ 0)$.

e. To determine the behavior of the curve before it drops and starts approaching 2, we plot a couple of points to the right of $x = 3$, say $(4,\ 33/16)$ and $(5,\ 52/25)$. Additional points may always be plotted. Figure 6.37 shows the graph of the function.

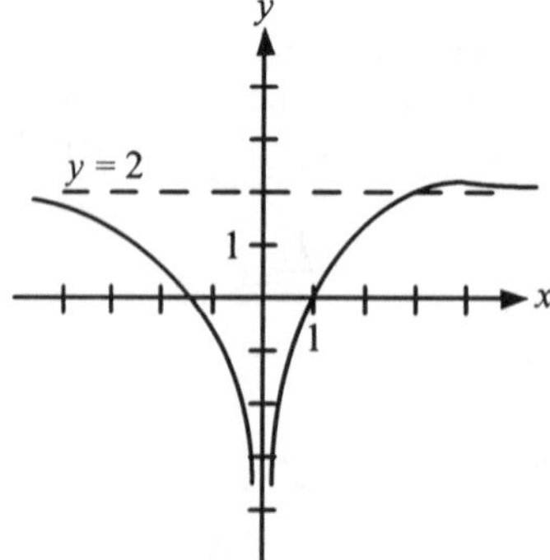

Figure 6.37

Technology Link

Figure 6.38 shows a complete graph of the function in this example on a grapher. Observe that it is not easy to recognize in this graph that the curve crosses the horizontal asymptote at (3, 2) and then approaches $y = 2$ from above. However, this behavior is apparent if we adjust the viewing window and also graph $y = 2$ as shown in Figure 6.39.

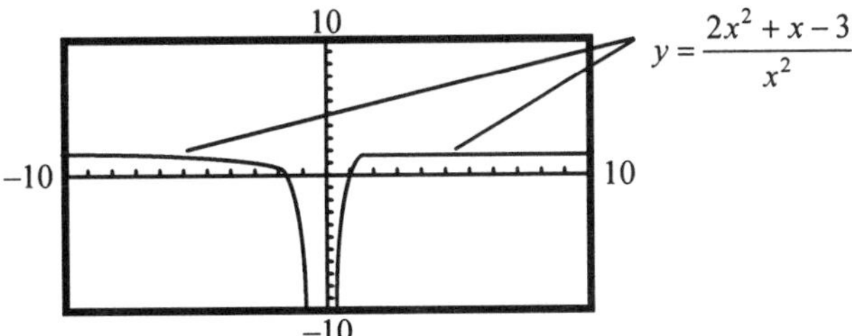

Figure 6.38

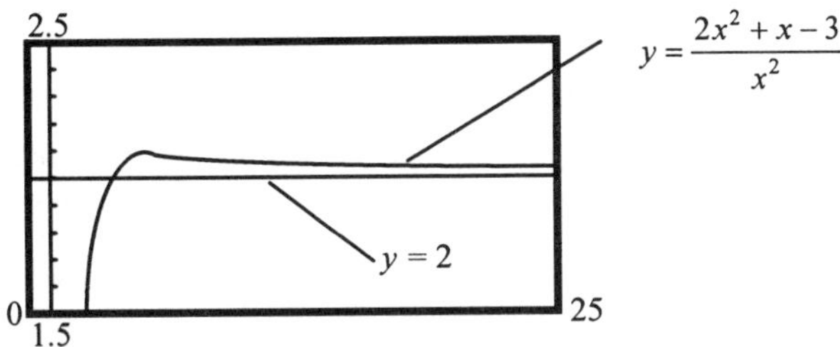

Figure 6.39

PROGRESS CHECK 3

If $y = \dfrac{2x^2 - 2}{x^3}$, find any asymptotes and graph the function. ■

There are two theorems, about asymptotes that can simplify our work. The first theorem indicates the behavior of the graph near any vertical asymptote. It is based on the fact that even powers of positive and negative numbers have the same sign (positive), while odd powers of positive and negative numbers are opposite in sign.

Graph of $y = P(x)/Q(x)$ Near Vertical Asymptotes

If $P(x)$ and $Q(x)$ have no common factors, and if $(x - a)^n$ is a factor of $Q(x)$, where n is the largest positive integer for which this statement is true, then:

1. The graph of $y = P(x)/Q(x)$ goes in opposite directions about the vertical asymptote $x = a$ when n is odd.
2. The graph of $y = P(x)/Q(x)$ goes in the same direction about the vertical asymptote $x = a$ when n is even.

For instance, in Example 2 the graph goes to positive infinity on the left side of $x = -1$ and to negative infinity on the right side (see Figure 6.34). We can predict this type of behavior since $Q(x) = (x + 1)^1$, and the odd exponent indicates that the graph goes in opposite directions about the vertical asymptote. In Example 3, however, $Q(x) = x^2$, and the even exponent indicates that the graph goes in the same direction (to negative infinity) about the vertical asymptote ($x = 0$), as shown in Figure 6.37.

The second theorem about asymptotes enables us to pick out horizontal asymptotes almost by inspection. This theorem may be derived by using the procedure shown in Examples 2 and 3 for determining a horizontal asymptote, and applying it to the general form for a rational function given below.

Horizontal Asymptote Theorem

The graph of the rational function

$$y = \frac{P(x)}{Q(x)} = \frac{a_n x^n + a_{n-1}x^{n-1} + \cdots + a_0}{b_m x^m + b_{m-1}x^{m-1} + \cdots + b_0},$$

where $a_n,\ b_m \neq 0$, has

1. a horizontal asymptote at $y = 0$ (the x-axis) if $n < m$
2. a horizontal asymptote at $y = a_n/b_m$ if $n = m$
3. no horizontal asymptote if $n > m$

For instance, the graph of $y = 1/x$ satisfies the first case, since the numerator is a zero-degree polynomial while the denominator is a first-degree polynomial. Since the higher-degree polynomial is in the denominator, $y = 0$ is a horizontal asymptote. Examples 2 and 3 illustrate the second case, in which both polynomials are of the same degree. Thus, the horizontal asymptotes for

$$y = \frac{1x - 1}{1x + 1} \text{ and } y = \frac{2x^2 + x - 3}{1x^2}$$

are $y = 1/1 = 1$ and $y = 2/1 = 2$, respectively. To illustrate another example of the second case, we solve the section-opening problem next.

Example 4: Average Cost of Using a Calling Card Solve the problem in the section introduction on page 503.

Solution:

a. The telephone company charges 25 cents per minute plus a 80-cent surcharge, so the total cost c of a 10-minute call is

$$C = 25(10) + 80 = 330 \text{ cents or } \$3.30.$$

The formula $y = (25x + 80)/x$ gives the average cost (y) of a call per minute. For a 10-minute call,

$$y = \frac{25(10) + 80}{10} = 33$$

Thus, the average cost per minute for a 10-minute call is 33 cents per minute.

b. To graph $y = (25x + 80)/x$ for $x > 0$, first determine any asymptotes.

Vertical asymptote: The polynomial in the denominator equals zero when x is zero. Thus, $x = 0$ is a vertical asymptote.

Horizontal asymptote: The polynomial in the numerator has the same degree (degree 1) as the polynomial in the denominator. So the horizontal asymptote for

$$y = \frac{25x + 80}{1x}$$

is $y = 25/1 = 25$. To determine if the curve ever crosses the horizontal asymptote, we set $y = 25$ and solve

$$25 = \frac{25x + 80}{x}.$$

This equation is equivalent to $25x = 25x + 80$, which is never true. Therefore the curve does not cross the horizontal asymptote.

Figure 6.40 gives a graph of this function that shows the curve approaching these asymptotes. Observe that as x increases for $x > 0$, the average cost per minute always decreases. However, y decreases rapidly at first, and then y levels off and approaches $y = 25$. Thus, the horizontal asymptote is significant because it reveals that as the length of the call increases, the average cost per minute approaches 25 cents, which is the charge per minute for the call.

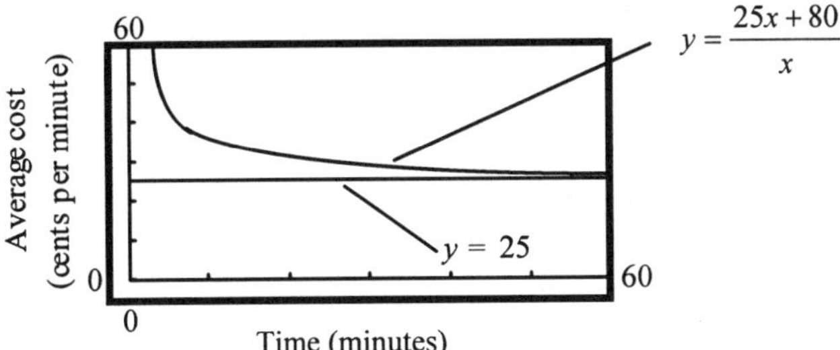

Figure 6.40

PROGRESS CHECK 4

Redo the problem in Example 4 given that the company charges 30 cents per minute plus a 50-cent surcharge. The average cost per minute is then given by $y = \frac{30x + 50}{x}$. ■

When there are no horizontal asymptotes as described in the third case of the horizontal asymptote theorem, other techniques may be employed. In particular, if $n = m + 1$ (that is, if the degree of the numerator is one more than the degree of the denominator), then the graph has a slant or oblique asymptote, as discussed in the next example.

Example 5: Graphing a Rational Function with a Slant Asymptote

If $f(x) = \frac{x^2 - 1}{x + 2}$ find any asymptotes and graph the function.

Solution: Steps **a** to **d** outline a systematic approach to graphing this function.

a. The polynomial in the denominator equals zero when x is -2. Thus, $x = -2$ is a vertical asymptote. Also, since $Q(x) = (x + 2)^1$, $Q(x)$ is an odd power of $x + 2$, so the graph of f goes in opposite directions about $x = -2$.

b. Because the higher-degree polynomial is in the numerator, there is no horizontal asymptote. However, we can change the form of the equation by dividing $x^2 - 1$ by $x + 2$ as follows:

$$\begin{array}{r} x-2 \\ x+2\overline{)x^2 \qquad -1} \\ \underline{x^2+2x} \\ -2x-1 \\ \underline{-2x-4} \\ 3. \end{array}$$

Thus,

$$f(x)=\frac{x^2-1}{x+2}=x-2+\frac{3}{x+2}.$$

Now as $|x|$ gets larger, $3/(x+2)$ approaches 0, and y approaches $x-2$. Thus, the graph of f approaches the oblique (neither horizontal nor vertical) line $y=x-2$, and we call this line a **slant asymptote** of the graph of f.

c. By setting $x=0$, we may determine that $(0, -1/2)$ is the y-intercept. Similarly, by setting $f(x)=0$, we may determine that (1, 0) and (–1, 0) are x-intercepts.

d. Using the above information and plotting one point to the left of the vertical asymptote, say (–3, – 8), we draw the graph in Figure 6.41.

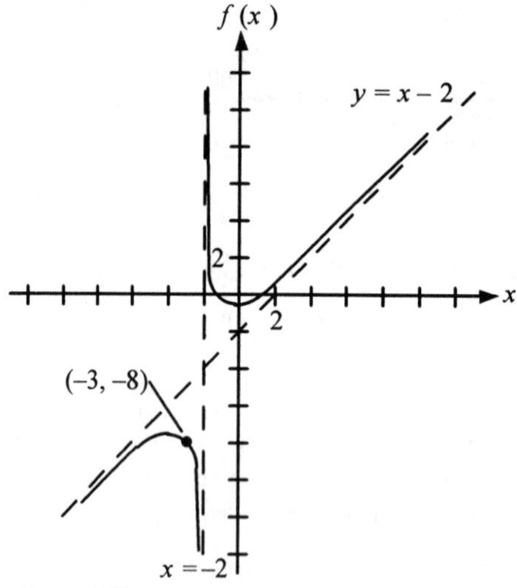

Figure 6.41

Use a grapher to check the results in this example.

PROGRESS CHECK 5

If $f(x)=\dfrac{x^2-4}{x-1}$, find any asymptotes and graph the function. ■

Example 6: Using the Least Amount of Fencing

A classic type of optimization problem asks for the dimensions that would require the least amount of fencing to enclose a rectangular region with fixed area. To illustrate, assume that a farmer plans to make a rectangular enclosure with area 500 ft^2 along the side of a barn, so that no fencing is needed along the barn. Use a grapher to determine to two decimal places the dimensions that would require the least amount of fencing.

Solution: First, draw a sketch of the situation, as in Figure 6.42. Since the area is fixed at 500 ft^2, if x represents the length of the fencing opposite the barn, then $500/x$ represents the length of the fencing perpendicular to the barn. So the length y of the required fencing is given by

$$\begin{aligned} y &= x + 2\left(\frac{500}{x}\right) \\ &= x + \frac{1000}{x}. \end{aligned}$$

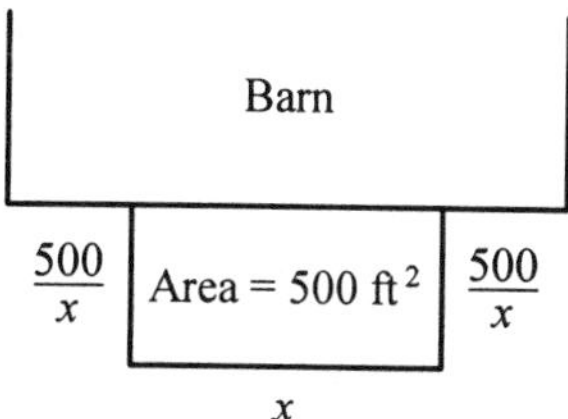

Figure 6.42

From the form of the equation, it is apparent that the graph has a vertical asymptote at $x = 0$ and a slant asymptote at $y = x$. Figure 6.43 gives a graph of this function for $x > 0$ that shows the curve approaching these asymptotes. To find the lowest point in the graph you may use the Zoom and Trace features (or a Function Minimum feature, if available) to determine that $x \approx 31.62$ at the minimum point. The other side dimension is given by $500/x$, so this side length is

$$\frac{500}{31.62} \approx 15.81.$$

Thus, the least amount of fencing is required when the length of the fencing opposite the barn is 31.62 ft, and the length of the fencing perpendicular to the barn is 15.81 ft.

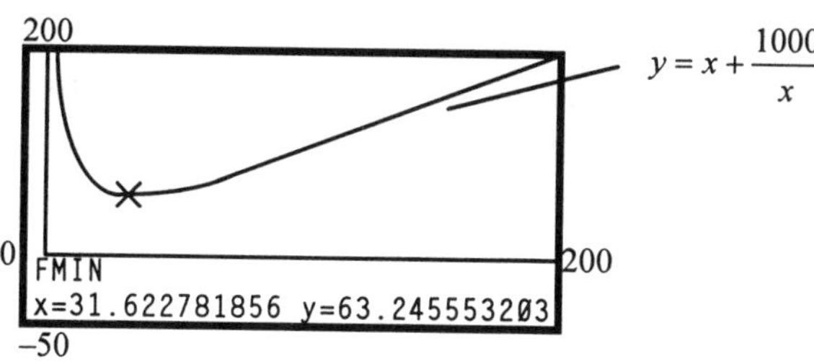

Figure 6.43

Note: In a calculus course, algebraic methods are developed for finding the exact value of x that minimizes the amount of fencing. In this problem, that dimension is $x = \sqrt{1000}$ ft.

PROGRESS CHECK 6

Redo the problem in Example 6, but assume that fencing is needed on all four sides (including the side along the barn). ∎

In our statements about rational functions, we have assumed that $P(x)$ and $Q(x)$ had no common factors. If the same factor does appear in the numerator and the denominator, we can divide it out as long as we keep track of all values that make the factor zero. The next example illustrates such a case.

Example 7: Graphing a Rational Function with Common Factors

Graph the function $f(x) = \dfrac{x^2 - 4}{x - 2}$.

Solution: By factoring the numerator, we rewrite the expression as

$$f(x) = \frac{x^2 - 4}{x - 2} = \frac{(x+2)(x-2)}{x-2}.$$

If $x = 2$, the expression becomes $0/0$, which is undefined. If $x \neq 2$, we can divide out common factors and obtain $x + 2$. Thus, we can rewrite the expression as follows:

$$f(x) = \frac{x^2 - 4}{x - 2} = \frac{(x+2)(x-2)}{x-2} = \begin{cases} x+2 & \text{if } x \neq 2 \\ \text{undefined} & \text{if } x = 2. \end{cases}$$

As shown in Figure 6.44, the graph is the set of all points on the line given by $f(x) = x + 2$ except (2, 4).

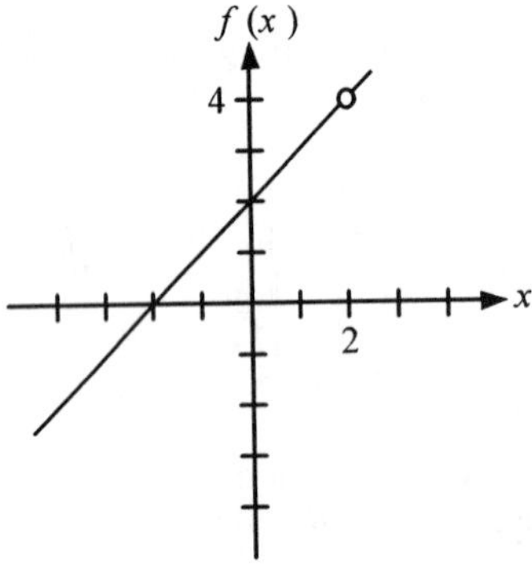

Figure 6.44

Technology Link

On a grapher the graph of $y = (x^2 - 4)/(x - 2)$ is indistinguishable from the graph of $y = x + 2$, unless a trace point coincides with (2, 4). So be careful in your graph to account for all values that make the denominator zero.

PROGRESS CHECK 7

Graph the function $f(x) = \dfrac{x^2 - x}{x - 1}$. ■

Finally, here is an outline of the procedure for graphing rational functions with the form $y = P(x)/Q(x)$.

1. If $P(x)$ and $Q(x)$ have common factors, we can divide them out on the following conditions:
   a. If the degree of the common factor in the numerator is greater than or equal to the degree of the common factor in the denominator, then any value of x that makes the common factor zero produces a hole in the graph, as shown in Example 7.
   b. If the degree of the common factor in the numerator is less than the degree of the common factor in the denominator, then any value of x that makes the common factor zero produces a vertical asymptote.

2. If $P(x)$ and $Q(x)$ have no common factors, find any vertical asymptotes by solving $Q(x) = 0$. Determine the behavior of the graph near any vertical asymptote by using the theorem given in this section.
3. Find any horizontal asymptotes by using the horizontal asymptote theorem or by dividing each term by the highest power of x and determining the value approached by y as $|x|$ gets larger. If the degree of the numerator is one more than the degree of the denominator, then determine a slant asymptote by dividing $P(x)$ and $Q(x)$ and setting y equal to the quotient.
4. Check if the curve crosses the horizontal asymptote for small values of x by finding if there are any values of x for which $P(x)/Q(x)$ equals the value of the horizontal asymptote. There's no need to check for crossings of vertical asymptotes—they can't happen.
5. Find any x-intercepts by setting $y = 0$ and solving $P(x) = 0$. Find any y-intercepts by setting $x = 0$ and evaluating the expression.
6. Plot additional points as needed and draw the graph. Use a grapher to check all of your results.

EXPLORE 6.4

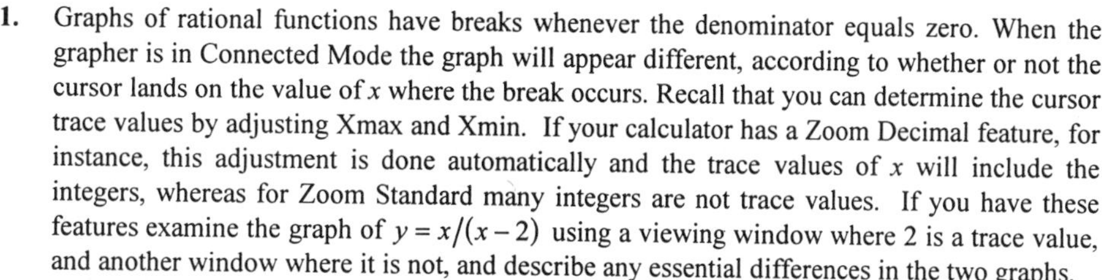

1. Graphs of rational functions have breaks whenever the denominator equals zero. When the grapher is in Connected Mode the graph will appear different, according to whether or not the cursor lands on the value of x where the break occurs. Recall that you can determine the cursor trace values by adjusting Xmax and Xmin. If your calculator has a Zoom Decimal feature, for instance, this adjustment is done automatically and the trace values of x will include the integers, whereas for Zoom Standard many integers are not trace values. If you have these features examine the graph of $y = x/(x-2)$ using a viewing window where 2 is a trace value, and another window where it is not, and describe any essential differences in the two graphs.
2. Because the equation of a rational function involves fractions, special care is needed when you enter the equation. The following sequences of key strokes produce different graphs. In each case write out in ordinary algebraic notation the function that corresponds to these sets of key strokes, and use a grapher to graph each function. In each case give the domain of the function.

$$y = 1/x - 1$$
$$y = 1/(x-1)$$
$$y = x + 2/x - 4$$
$$y = (x+2)/x - 4$$
$$y = (x+2)/(x-4)$$

3. The global view of the graph of a rational function (what you see if you zoom out far enough) is highly influenced by the highest-degree terms in the top and bottom polynomials. For instance, if you zoom out far enough, the graph of $y = \left(x^3 + 2x^2 - 4x + 5\right)/(x+3)$ will look like the graph of $y = x^2$ because $x^3/x = x^2$. The lower-degree terms are important when the absolute value of x is small (the part of the graph nearer the origin), but once the absolute value of x is large enough, the terms with lower degree contribute less to the shape of the graph. Graph the following curves using Zoom Standard, then zoom out and describe the basic shape of the graph.

a. $y = \dfrac{x^3 + 2x^2 - 4x + 5}{x+3}$ **b.** $y = \dfrac{x^3 - x + 2}{x^2 + 2}$

c. $y = \dfrac{9x^4 - x^2 + 5}{x^4 + x^3 + 2}$ **d.** $y = \dfrac{-x^3 - 3}{x+4}$

Note: You can control the amount that a calculator zooms when you press Zoom Out or Zoom In by setting the zoom factors. For instance, if you set the x zoom factors to 10, then zooming out one time will multiply xMin and xMax by 10.

EXERCISES 6.4

In Exercises 1–4 find any vertical asymptotes of the function.

1. $y=\dfrac{x+7}{2x-3}$ **2.** $y=\dfrac{x^2+1}{x(x+4)}$

3. $y=\dfrac{2}{x^2-5x-6}$ **4.** $y=\dfrac{x-3}{x^2+x}$

In Exercises 5–24 determine any asymptotes and graph the function.

5. $y=\dfrac{-1}{x}$ **6.** $y=\dfrac{3}{x^2}$ **7.** $y=\dfrac{2}{x+1}$

8. $y=\dfrac{-5}{3x+2}$ **9.** $y=\dfrac{x-2}{x+2}$

10. $y=\dfrac{x+3}{x-5}$ **11.** $y=\dfrac{2x-5}{3x+5}$ **12.** $y=\dfrac{4x+1}{2x+7}$

13. $y=\dfrac{2}{(x-1)^2}$ **14.** $y=\dfrac{2}{x^2-1}$

15. $y=\dfrac{1}{(x+1)(x-4)}$ **16.** $y=\dfrac{x-1}{x^2-4}$

17. $y=\dfrac{x^2-1}{x^3}$ **18.** $y=\dfrac{x}{x^2+1}$

19. $y=\dfrac{x^2-x-6}{x^2}$ **20.** $y=\dfrac{3x^2+x-2}{2x^2}$

21. $y=\dfrac{x^2+1}{x}$ **22.** $y=\dfrac{(x-1)^2}{x}$

23. $y=\dfrac{x^2-9}{x+2}$ **24.** $y=\dfrac{x^2-2x-3}{x+3}$

In Exercises 25–34 graph the given function, with particular emphasis on any value of x for which $P(x)=Q(x)=0$.

25. $h(x)=\dfrac{x(x+1)}{x}$ **26.** $y=x\dfrac{(x+1)}{x+1}$

27. $y=\dfrac{x^2-4}{x+2}$ **28.** $y=\dfrac{x^2-x}{x}$

29. $g(x)=\dfrac{(2x+1)(x-3)(x+2)}{(x-3)(x+2)}$

30. $f(x)=\dfrac{(x-1)(x-2)(x-3)}{(x-1)(x-2)}$

31. $f(x)=\dfrac{x}{x(x+1)}$ **32.** $y=\dfrac{x}{x^2-2x}$

33. $y=\dfrac{x-1}{(x-1)^2}$ **34.** $y=\dfrac{x+1}{(x+1)^2}$

35. The cost to rent a recording studio is a flat fee of \$100 plus \$1 per minute. Thus the average cost per minute for x minutes use of the studio is given by $y=\dfrac{100+x}{x}$.

a. What is the total cost for a 3-hour recording session? What is the average cost per minute for a 3-hour session?

b. Graph the function and describe how y changes as x increases for $x>0$. What is the significance of the horizontal asymptote.

36. Redo Exercise 35 if the cost is a \$75 flat fee plus \$10 per hour.

37. Assume that a farmer plans to make a rectangular enclosure with area 900 ft^2 along the side of a barn, so that no fencing is needed along the barn. Use a grapher to determine to 2 decimal places the dimensions that would require the least amount of fencing.

38. Redo Exercise 37 but assume that fencing is needed on all four sides (including the side along the barn). Does your solution imply that the yard is square?

Exercises 39 and 40 demonstrate numerically that the behavior of a rational function is determined mainly by the highest-degree terms in the top and bottom as the absolute value of x grows. In each exercise complete these tables and find a simple formula that expresses the relationship between the emerging patterns of x and y values.

39. $y=\dfrac{3x^2-x+1}{x+1}$

x	0	10	100	1,000	10,000	100,000
y						

40. $y = \dfrac{2x^2 - x + 1}{-x + 5}$

x	0	10	100	1,000	10,000	100,000
y						

In some applications the cost increases as the process goes on. For instance, in mining it is less expensive to extract a ton of ore when a mine is new and the ore is easy to reach; it is more expensive when the mine is old and it is difficult to reach the ore. In deep drilling it is less expensive to drill the first foot than the 2000th foot. Situations such as these may be modeled with rational functions, as illustrated by Exercises 41 and 42.

41. Let $y = \dfrac{5000}{500 - x}$, for $0 < x \le 300$, where x is the number of feet drilled, and y is the cost in dollars for drilling the xth foot.
   a. How much does it cost to drill the first foot? the 100th? the 300th?
   b. Make a table for the integer values $x = 1, 2, 3, \ldots, 10$ and calculate the total expense for drilling 10 feet.

42. Let $y = \dfrac{60}{15 - 0.2x}$, for $0 < x \le 7$, where x is the number of tons of ore (in thousands) removed from a mine, and y is the time (in days) it takes to extract the x thousandth ton.
   a. How long does it take to extract the first 1000 tons of ore?
   b. Make a table for the integer values $x = 1, 2, 3, 4, 5$ and calculate the total time to extract 5 thousand tons of ore.
   c. When the time to extract a thousand tons of ore reaches 20 days, the project stops. How many tons can be extracted before this limit is reached?

In many applications it is desirable to have a mathematical model that equals zero when $x = 0$, then increases to some maximum value, and then decreases to zero. Rational functions can be useful in such situations, as illustrated in Exercises 43 and 44.

43. **a.** Explain why $f(x) = \dfrac{P(x)}{Q(x)}$, where $P(x) = x(24 - x)$ and $Q(x) \ne 0$, will equal zero when $x = 0$, and $x = 24$.
   b. Compare the graphs for these 2 functions, and determine the value of x for which they achieve maximum value.

$$y = \frac{x(24 - x)}{x + 1}$$
$$y = \frac{x(24 - x)}{x + 5}$$

   c. Choose a denominator $Q(x)$ for $f(x) = \dfrac{x(24 - x)}{Q(x)}$ so that $f(x)$ will achieve a maximum at about $x = 10$.

44. **a.** Explain why $f(x) = \dfrac{P(x)}{Q(x)}$, where $P(x) = x(60 - x)$ and $Q(x) \ne 0$, will equal zero when $x = 0$, and $x = 60$.
   b. Compare the graphs for these 2 functions, and determine the value of x for which they achieve maximum value.

$$y = \frac{x(60 - x)}{x^2 + 1}$$
$$y = \frac{x(60 - x)}{x^2 + 5}$$

   c. Choose a denominator $Q(x)$ for $f(x) = \dfrac{x(60 - x)}{Q(x)}$ so that $f(x)$ will achieve a maximum at about $x = 5$.

45. Some "pure" orange juice is diluted by adding water. At first, a 64-ounce jug contains just 32 ounces of pure juice.
   a. What percent of the solution is juice after 4 ounces of water have been added? After 8 ounces have been added?
   b. Write a formula that gives y the percent of juice in the solution after x ounces of water have been added. What is the domain of this function?
   c. Graph the formula from part **b** over the domain from part **b**.
   d. At what point is the solution 80% juice? 50%? 40%?
   e. Explain algebraically and realistically why the graph has a horizontal asymptote.
   f. How big a container is needed if the juice is diluted down to a 10% solution?

46. Redo Exercise 45, but assume that at the beginning the jug contains a mixture of 9 ounces of orange juice and 1 ounce of water.

47. In economics there are various "laws of diminishing returns." In most contexts such a law implies that as one variable increases, the rate of growth of a second one increases for a while but then decreases. A graph that behaves this way is shown in this figure:

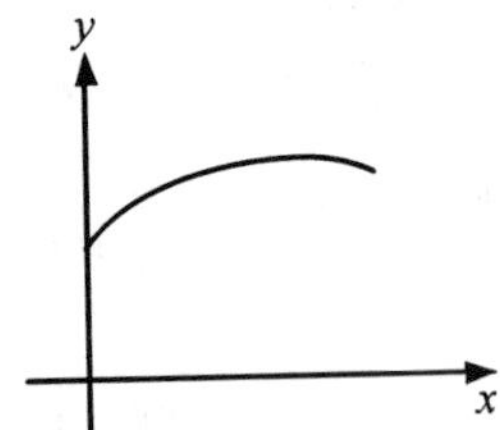

a. Complete the given formula for such a graph by finding the value for k so that the graph includes the points (0, 50) and (100, 75).

$$y = \frac{x(150 - x)}{x + k} + 50.$$

b. Graph the formula from part **a** over the interval [0, 100], and find its maximum value.

48. The formula $f(x) = \frac{1}{\pi(1 + x^2)}$ is called the Cauchy density function. It is used in problems that involve finding the tangents of angles picked at random. Find the maximum value of f, and describe any asymptotes. For what value of x is $f(x) = 0.25$?

49. In the figure shown, the two slanted sides are equal. Find the least perimeter that will enclose an area equal to 100 square feet. Does your solution lead to an equilateral triangle?

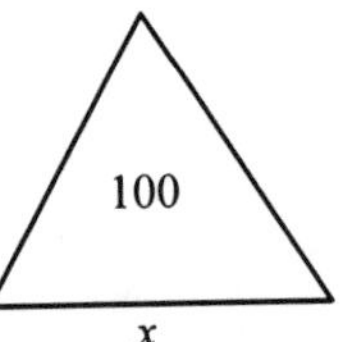

50. In a math journal, Brian Bolt described the effect of some factors on the flight of a golf ball hit from a tee. He showed that the velocity U of the ball is roughly proportional to the velocity V of the clubhead at the moment the clubhead hits the ball. That is, $U = kV$. The constant of proportionality k was shown to be $M(1+e)/(M+m)$, where M is the mass of the clubhead, e is a constant that depends mostly on the elastic quality of the ball, and m is the mass of the ball. The author analyzed the effect of the weight of the clubhead on the velocity of the ball. For this purpose he used 0.7 as a representative value of e, and 46g as a typical mass for a ball. This gives the equation $k = 0.7M/(M+46)$.

a. Graph this function, and describe what happens to k as the mass of the clubhead increases.

b. A typical value for M is 200 g. What is the corresponding value of k?

c. How much would k increase if the mass of the club were increased from 200 g to 1000 g? (Of course, it would be much harder to swing the club.)

[Source: *UMAP Journal*, vol. 4, no. 1., 1983]

THINK ABOUT IT 6.4

1. Graphs of real data often exhibit apparently asymptotic behavior. One may then try to explain why the data have this behavior. The two graphs shown illustrate this phenomenon.

a. Figure 1 appears in the 1993 edition of *Vital Signs*, published by the Worldwatch Institute. These worldwide data were compiled by the Motor Vehicle Manufacturers Association. Describe how the number of people per automobile changed from 1950 onward. Discuss some reasons why this curve may be leveling off.

b. Figure 2 appears in *Environmental Studies* 2nd edition,. by D. Botkin and E. Keller. The figure is based on data compiled by the World Bank. Total fertility rate is the average number of children expected to be born to a woman during her lifetime. Describe how the total fertility rate changes with respect to individual income. Discuss some reasons why this curve may be leveling off.

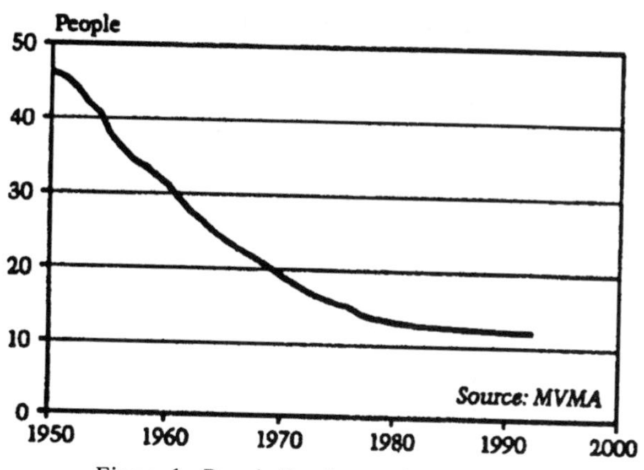

Figure 1: People Per Automobile, 1950–92

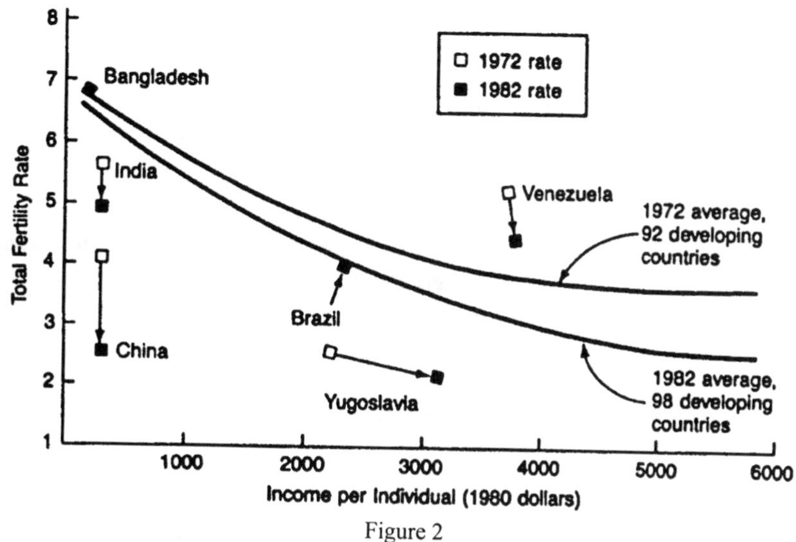

Figure 2

2. What is the difference between the graphs of $y = (x^2 - 1)/(x - 1)$ and $y = x + 1$?
3. If $f(x) = 4/(x^2 - 1)$, what is the range of f?
4. Give two examples of an equation that defines a rational function such that the vertical asymptote is $x = -3$ and the horizontal asymptote is $y = 5/2$.
5. a. Explain why the graph of a rational function cannot cross a vertical asymptote.
   b. We have seen that it is possible for the graph of a rational function to cross its horizontal asymptote. Do you think a graph may cross its oblique asymptote? Explain.
   c. Graph $y = x^3/(x^2 - 1)$ and check that your explanation in part **b** is in agreement with this graph.

CHAPTER 6 SUMMARY

OBJECTIVES CHECKLIST

Specific chapter objectives are summarized below along with numbered example problems from the text that should clarify the objectives. If you do not understand any objectives, or do not know how to do the selected problems, then restudy the material.

6.1: Can you:

1. **Determine if a function is a polynomial function?** Determine whether the function is a polynomial function. If yes, state the degree.
   a. $f(x) = 4x^5 - 2x^{1/2} - 3$ **b.** $f(x) = 4x^5 - \frac{1}{2}x^2 - 3$

   [Example 1]

2. **Write a polynomial function, given its degree and zeros?** Write a polynomial function (in factored form) of the third degree with zeros of 2, –4, and 3.

   [Example 2]

3. **Find the zeros of a polynomial function that is given in factored form?** If $P(x) = x^4(x-2)(x+7)^3$, find the zeros of the function as well as the multiplicity of each zero. What is the degree of the polynomial function?

   [Example 4]

4. **Use translation, reflection, stretching and shrinking to graph certain variations of $y = x^n$?** Use the graph of $y = x^4$ to graph each function.
   a. $y = -x^4$ **b.** $y = 3x^4$ **c.** $y = (x+1)^4 - 2$

   [Example 5]

5. **Graph a polynomial function on the basis of information obtained from the intercepts, or by using a grapher?** Graph $y = (x+1)(x-2)^2$ on the basis of information obtained from the intercepts.

   [Example 6]

6. **Write an equation for a polynomial function on the basis of information obtained from the intercepts?** Find an equation for the third-degree polynomial function that is graphed in the figure.

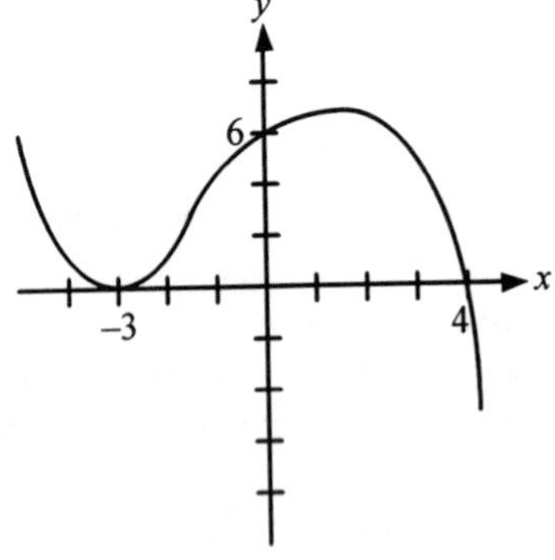

   [Example 8]

6.2: Can you:

1. **Divide one polynomial by another using long division?** Divide $2x^4 + 6x^3 - 5x^2 - 1$ by $2x^2 - 3$. Express the answer in the form

$$\frac{\text{dividend}}{\text{divisor}} = \text{quotient} + \frac{\text{remainder}}{\text{divisor}}.$$

[Example 1]

2. **Divide a polynomial by a polynomial of the form $x - b$ using synthetic division?** Use synthetic division to divide $x^5 - 1$ by $x - 1$. Express the result in the form

$$\text{dividend} = (\text{divisor})(\text{quotient}) + \text{remainder}.$$

[Example 2]

3. **Evaluate a polynomial using the remainder theorem?** If

$$P(x) = 3x^5 + 5x^4 + 7x^3 - 4x^2 + x - 24,$$

find $P(-2)$ by
a. direct substitution and **b.** the remainder theorem.

[Example 4]

4. **Find all zeros of a third- or fourth-degree polynomial function when one or two zeros, respectively, are apparent?** If 2 is a zero of $P(x) = x^3 - 2x^2 + 3x - 6$, find the other zeros.
[Example 7]

5. **Use the conjugate-pair theorems to find zeros or write equations for polynomial functions?** If $-2i$ and $2 + \sqrt{7}$ are zeros of the polynomial function $P(x) = x^4 - 4x^3 + x^2 - 16x - 12$, find the other zeros.

[Example 8]

6.3: Can you:

1. **List the possible rational zeros of a polynomial function with integer coefficients?** List the possible rational zeros of the function $P(x) = 2x^3 - x^2 - 6x - 3$.

[Example 1]

2. **Determine the maximum number of positive and negative real zeros of a polynomial function?** Use Descartes' rule of signs to determine the maximum number of
a. positive and **b.** negative real zeros of $P(x) = 2x^3 + 5x^2 + 4x + 1$.
[Example 2, step a]

3. **Find all the zeros of a polynomial function when given a function with a sufficient number of rational zeros so that all zeros can be determined?** Find the zeros of $P(x) = 2x^4 - 5x^3 + 11x^2 - 3x - 5$.

[Example 3]

4. **Use the location theorem to verify the existence of a zero of a function in a given interval?** Verify that $P(x) = 3x^3 + 16x^2 - 8$ has a zero between –5 and –6.

[Example 5]

5. **Approximate a zero of a function in a given interval to the nearest tenth?** To the nearest tenth, approximate the zero of the function $P(x) = 3x^3 + 16x^2 - 8$ that is between –5 and –6.
[Example 6]

6.4: Can you:

1. **Determine any vertical and horizontal asymptotes for a rational function and graph the function?** If $y = (x - 1)/(x + 1)$, find any asymptotes and graph the function.

[Example 2]

2. **Determine any vertical and slant asymptotes for a rational function and graph the function?** If $f(x) = (x^2 - 1)/(x + 2)$, find any asymptotes and graph the function.

[Example 5]

3. **Graph rational functions that involve common factors?** Graph the function $f(x) = (x^2 - 4)/(x - 2)$.

[Example 7]

KEY CONCEPTS AND PROCEDURES

Section	Key Concepts and Procedures to Review
6.1	• Definition of a polynomial function of degree n • Definitions of a zero of a function and the multiplicity of a zero • Factor theorem • Theorem about graphing polynomial functions by knowing the multiplicity of real-number zeros • Properties of polynomial graphs concerning continuity, turning points, and behavior for large $\lvert x \rvert$
6.2	• Long-division procedure for the division of polynomials • Synthetic division procedure for the division of a polynomial by $x - b$ • Remainder theorem • Fundamental theorem of algebra • Number of zeros theorem • Theorem about when complex zeros come in conjugate pairs • Theorem about when irrational zeros of the form $a \pm b\sqrt{c}$ come in conjugate pairs
6.3	• Rational-zero theorem • Descartes' rule of signs • Upper and lower bounds theorem • Intermediate value theorem • Location theorem
6.4	• Definition of a rational function • Methods to determine vertical, horizontal, or slant asymptotes • Horizontal asymptote theorem • Theorem about the behavior of a graph near any vertical asymptote(s) • Outline of the procedure for graphing rational functions

CHAPTER 6 REVIEW EXERCISES

In Exercises 1–6 indicate if the given function is a polynomial function

1. $f(x) = \dfrac{2x - 1}{3}$
2. $y = \dfrac{2x - 1}{x}$
3. $y = x^2 + x^{-1} + 2$
4. $y = x^2 + x + 2^{-1}$
5. $y = \sqrt{x}$
6. $y = \sqrt{2}$

In Exercises 7–12 find the zeros of the function.

7. $y = 2x^2 - 3x$
8. $f(x) = 3 - 5x$
9. $g(x) = x^3 - 2x^2 + x - 2$
10. $y = x^2(x + 7)(x - 4)$
11. $f(x) = \dfrac{2x^2 - 5x + 1}{x^3}$
12. $y = \dfrac{1}{x^2 - 1}$

In Exercises 13–22 graph each function.

13. $y = \dfrac{-2}{3x^2}$
14. $y = \dfrac{3}{2x - 3}$
15. $y = \dfrac{x + 1}{x}$
16. $y = \dfrac{x + 1}{2}$

17. $y = \dfrac{x^2 + 2x}{x}$ 18. $y = \dfrac{5x^2 + 4}{4x^2 - 1}$

19. $y = (x + 2)(x - 1)^2$ 20. $y = x^2(x + 7)(x - 4)$

21. $y = x^3 - 3x^2 - 4x$ 22. $y = x^4 - x^2$

23. Determine the vertical asymptotes for the function $y = \dfrac{3x + 1}{x^2 + 2x}$.

24. Determine the horizontal asymptote for the function $y = \dfrac{2x^2 + 3}{3x^2 + x - 2}$. Does the graph cross the horizontal asymptote? If yes, find the point at which it crosses.

25. What is the difference between the graphs of $y = \dfrac{x^2 - 4}{x - 2}$ and $y = x + 2$?

26. Write a polynomial function (in factored form) with rational coefficients of the lowest possible degree with zeros of $-i$, $\sqrt{2}$, and 0.

27. Write a polynomial function (in factored form) of degree 4 with zeros of 5, –2, 0, and 3.

28. If $f(x) = 2x^4 - x + 7$, find $f(-2)$ by
    a. direct substitution and
    b. the remainder theorem

29. Divide $2x^4 - x + 7$ by $x^2 + 2$. Express the result in the form dividend = (divisor)(quotient) + remainder.

30. The dividend is x^2, the quotient is $x + 3$, and the remainder is 9. What is the divisor?

31. List the possible rational zeros of the function $P(x) = 3x^3 - 7x^2 + x - 2$.

32. Use Descartes' rule of signs to determine the maximum number of
    a. positive and
    b. negative real zeros of $P(x) = x^4 - 3x^3 + x^2 - x + 1$.

33. If 2 is a zero of $P(x) = 2x^3 - 4x^2 - 6x + 12$, find the other zeros.

34. Describe how to graph $y = (x + 1)^4 - 3$ using the graph of $y = x^4$.

35. If $P(x) = 7x^3(x + 2)^2$, find the zeros of the function and state the multiplicity of each zero.

36. If 1 is a root of $x^3 - 2x^2 - 5x + 6 = 0$, what is the quadratic equation that can be used to find the other two roots?

37. If b is a zero of $P(x) = x^3 + ax^2 + ax + 1$, show that $1/b$ is also a zero.

38. Find an equation for the third degree polynomial shown.

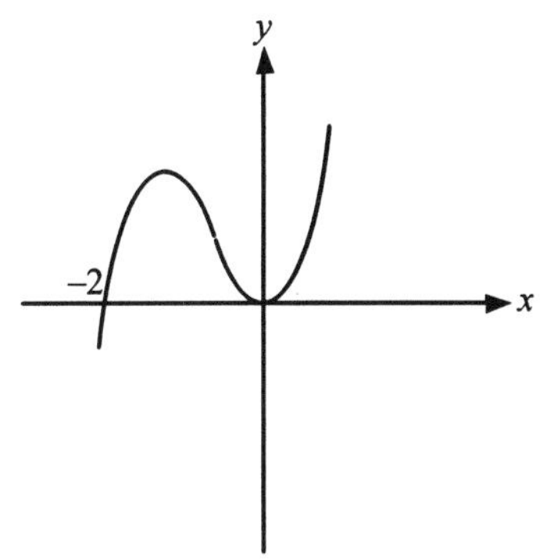

39. Verify that $P(x) = x^3 - 3x^2 - 1$ has a real zero between 3 and 4, by using the location theorem.

40. Approximate the zero of $P(x) = x^3 - 2x - 7$ that is between 2 and 3, to the nearest tenth.

In Exercises 41–50 select the choice that answers the question or completes the statement.

41. The degree of $P(x) = x^2(x + 3)(x - 3)^4$ is
    a. 6 **b.** 7 **c.** 8 **d.** 16

42. If $P(x) = x^2(x + 3)(x - 3)^4$, what is the multiplicity of the zero of –3?
    a. 1 **b.** 2 **c.** 4 **d.** 5

43. The graph of $y = \dfrac{x^2 + 1}{x^3}$ lies in quadrants
    a. 1 and 2 **b.** 1 and 3
    c. 2 and 4 **d.** 3 and 4

44. If $x^3 - 2x^2 + ax + 9$ is exactly divisible by $x + 3$ then a equals
    a. –10 **b.** –12 **c.** 7 **d.** 16

45. The remainder when $2x^9 + 5x^4 + 3$ is divided by $x - 1$ is
    a. 0 **b.** 6 **c.** –4 **d.** 10

46. The vertical asymptote for $y = \dfrac{x - 1}{x - 2}$ is
    a. $x = 1$ **b.** $y = 1$ **c.** $y = 2$ **d.** $x = 2$

47. Which one of the following is a rational zero of $P(x) = 2x^3 + 19x^2 + 37x + 14$?

a. $-\frac{1}{2}$ **b.** $\frac{7}{2}$ **c.** $\frac{5}{2}$ **d.** $-\frac{3}{2}$

48. According to Descartes' rule of signs, the maximum number of negative real zeros for $P(x) = 3x^3 + x^2$ is

a. none **b.** one **c.** two **d.** three

49. If $\frac{2}{5}$ is a zero of $y = ax^3 + bx^2 + cx + d$, in which a, b, c, and d are integers with $a \neq 0$, then 5 must be a factor of

a. a **b.** b **c.** c **d.** d

50. The function $y = x^3 - 3x - 3$ has exactly one positive real zero in which one of the following intervals?

a. [0, 1] **b.** [1, 2] **c.** [2, 3] **d.** [3, 4]

CHAPTER 6 TEST

1. True or false: The function $y = 1 - x^3$ is a polynomial function.

2. What is the degree of $P(x) = 3x^2(x-1)(x+2)^4$?

3. Write a polynomial function of degree 3 with zeros of –2, –1, and 4.

4. Describe how to graph $y = (x-3)^4 + 2$ using the graph of $y = x^4$.

5. Verify that $P(x) = x^3 - x^2 + 5$ has a real zero between –1 and –2 by using the location theorem.

6. True or false: The graph of $y = \dfrac{x^2 - 3x}{x}$ has a vertical asymptote at $x = 0$.

7. Divide $\dfrac{x^3 - 1}{x(x+1)}$. Express the answer in the form

$$\frac{\text{dividend}}{\text{divisor}} = \text{quotient} + \frac{\text{remainder}}{\text{divisor}}.$$

8. What is the remainder when $5x^4 - 3x + 12$ is divided by $x^2 - 2$?

9. What is the quotient when $2x^4 - x^2 + 5$ is divided by $x + 3$?

10. If $P(x) = x^3 - 3x^2 + x + 9$, find $P(-2)$ by

a. direct substitution and
b. the remainder theorem.

11. If $P(x) = 4x^2(x+1)^3$, find the zeros of the function and state the multiplicity of each zero.

12. Graph $y = x^4 - 4x^2$ based on information obtained from the intercepts.

13. If –5 is a zero of $P(x) = x^3 + 5x^2 - 6x - 30$, then find the other zeros for the function.

14. By the rational-zero theorem what are the possible rational zeros for $P(x) = 3x^3 + 2x^2 - 3x - 2$?

15. By Descartes' rule of signs what is the maximum number of negative real zeros of the function $P(x) = x^3 + 4x^2 - x + 5$?

16. Find the zeros of $P(x) = 2x^3 + 3x^2 + 6x + 9$.

17. Find any vertical asymptotes for the graph of $y = \dfrac{2x - 7}{x^2 - 6x}$.

18. Find all points at which the graph of $y = \dfrac{x^2 - 4}{x^3}$ crosses its horizontal asymptote.

19. Find the slant asymptote for the graph of $y = \dfrac{x^2 - 4x - 5}{x - 2}$.

20. Graph $y = \dfrac{2x - 5}{x + 3}$.

● ● ●

Chapter

7

Exponential and Logarithmic Functions

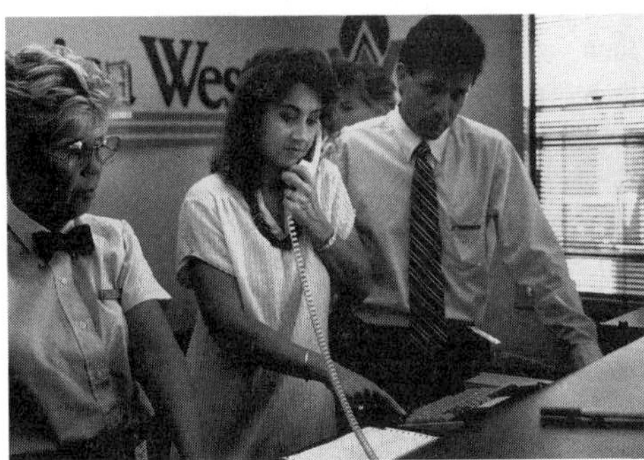

*Photo Courtesy of **Laima Druskis** of Stock Boston, Boston, MA*

Certain airline employees accept a 20 percent cut in hourly wage to avoid layoffs. What percent raise will then be needed to return these employees to their original hourly wage?
(See Example 8 of Section 7.1)

Exponential functions and their inverses, the logarithmic functions, are important rules for analyzing a wide variety of relationships. Compound interest, population growth, radioactive decay, pH, decibels, and heat loss (from a home or murder victim) are but some of the applications we consider in this chapter. Whereas a typical polynomial function looks like $y = x^2$, a typical exponential function looks like $y = 2^x$. The explosive growth of exponential functions such as $y = 2^x$ can be quite amazing, and the introductory problem to Section 7.2 addresses this issue. But first, we begin with a general discussion of inverse functions, because important relationships between exponential and logarithmic functions are easily seen once we establish some ways in which all inverse functions are related.

● ● ●

7.1 Inverse Functions

Objectives

1. Find the inverse of a function and its domain and range.
2. Determine if a function has an inverse function.
3. Graph $y = f^{-1}(x)$ from the graph of $y = f(x)$.
4. Find an equation that defines f^{-1}.
5. Determine whether two functions are inverses of each other.
6. Use inverse function concepts in applications.

Consider the following tables, which illustrate ordered pairs that belong to the function $f(x)=x^3$ and ordered pairs that belong to the function $g(x)=\sqrt[3]{x}$.

x	$f(x)=x^3$	Ordered Pairs in f
1	$1^3=1$	(1, 1)
2	$2^3=8$	(2, 8)
–2	$(-2)^3=-8$	(–2, –8)
3	$3^3=27$	(3, 27)
–3	$(-3)^3=-27$	(–3, –27)

x	$g(x)=\sqrt[3]{x}$	Ordered Pairs in g
1	$\sqrt[3]{1}=1$	(1, 1)
8	$\sqrt[3]{8}=2$	(8, 2)
–8	$\sqrt[3]{-8}=-2$	(–8, –2)
27	$\sqrt[3]{27}=3$	(27, 3)
–27	$\sqrt[3]{-27}=-3$	(–27, –3)

Observe that functions f and g are related and have reverse assignments. For example, $f(3)=27$ and $g(27)=3$. Two functions with exactly reverse assignments are called **inverse functions** of each other, and $f(x)=x^3$ and $g(x)=\sqrt[3]{x}$ are examples of inverse functions. The special symbol f^{-1} is used to denote the inverse of function f, so

$$\text{if } f(x)=x^3 \quad \text{then } f^{-1}(x)=\sqrt[3]{x}$$
$$\text{and if } f(x)=\sqrt[3]{x}, \quad \text{then } f^{-1}(x)=x^3.$$

The reverse assignments of f and f^{-1} mean that when (a, b) belongs to a function, then (b, a) belongs to its inverse function. Because the components in the ordered pairs are reversed, the domain of f equals the range of f^{-1}, and the range of f equals the domain of f^{-1}.

Example 1: Finding an Inverse Function Numerically

If $f=\{(7,\,-2),\,(9,\,1),\,(-3,\,2)\}$, find f^{-1}. Find and compare the domain and range of the two functions.

Solution: Reversing assignment gives

$$f^{-1}=\{(-2,\,7),\,(1,\,9),\,(2,\,-3)\}.$$

The set of all first components give the domain of a function, and the set of all second components gives the range. Therefore,

$$\text{domain of } f = \text{range of } f^{-1} = \{-3,\,7,\,9\}$$
$$\text{range of } f = \text{domain of } f^{-1} = \{-2,\,1,\,2\}$$

As expected, f and f^{-1} interchange their domain and range.

PROGRESS CHECK 1

If $f=\{(-1,\,1),\,(-2,\,2),\,(-3,\,3)\}$, find f^{-1}. Find and compare the domain and the range of the two functions. ■

Not all functions have an inverse function. For instance, if a function f is given by

$$\{(1,\,5),\,(2,\,5),\,(3,\,5)\}$$

then reversing assignments produces

$$\{(5,\ 1),\ (5,\ 2),\ (5,\ 3)\}.$$

The resulting relation is not a function because the number 5 is the first component in more than one ordered pair. Because the second component in f becomes the first component in the inverse relation, the following method may be used in determining whether f has an inverse function.

One-to-One Function

> A function is **one-to-one** when each x value in the domain is assigned a different y value so that no two ordered pairs have the same second component. If f is one-to-one, then f has an inverse function, and if f is not one-to-one, then f does not have an inverse function.

Example 2: Testing Numerically for an Inverse Function

If $f = \{(-6,\ 6),\ (0,\ 0),\ (6,\ 6)\}$, does f have an inverse function?

Solution: Because 6 appears as the second component in two ordered pairs, f is not a one-to-one function, and f does not have an inverse function?

PROGRESS CHECK 2
If $f = \{(-3,\ 3),\ (0,\ 0),\ (3,\ -3)\}$ does f have an inverse function? ■

It is easy to recognize the graph of a one-to-one function because none of its points can have the same y-coordinate. Therefore, the graph of a one-to-one function cannot contain two or more points that lie on the same horizontal line. This feature is often summarized in the horizontal line test.

Horizontal Line Test

> Imagine a horizontal line sweeping down the graph of a function f. If the horizontal line at any position intersects the graph in more than one point, then f is not a one-to-one function, and f does not have an inverse function.

Example 3: Testing Graphically for an Inverse Function

Which functions graphed in Figure 7.1 have an inverse function?

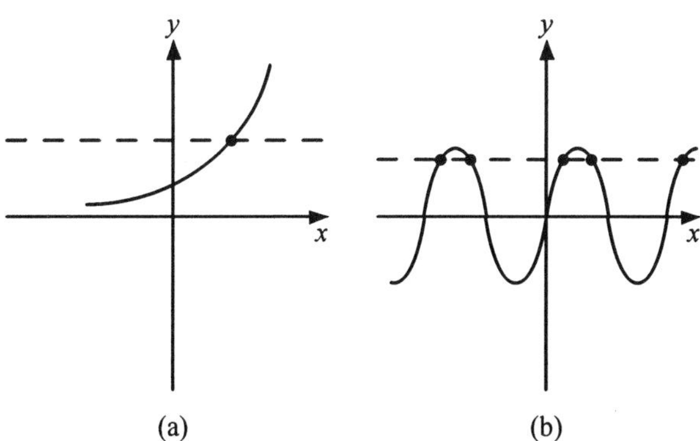

Figure 7.1

Solution:

a. This function has an inverse function because no horizontal line intersects the graph at more than one point.

b. This function does not have an inverse function, since there is a horizontal line that intersects the graph at more than one point.

PROGRESS CHECK 3

Which functions graphed in Figure 7.2 have an inverse function?

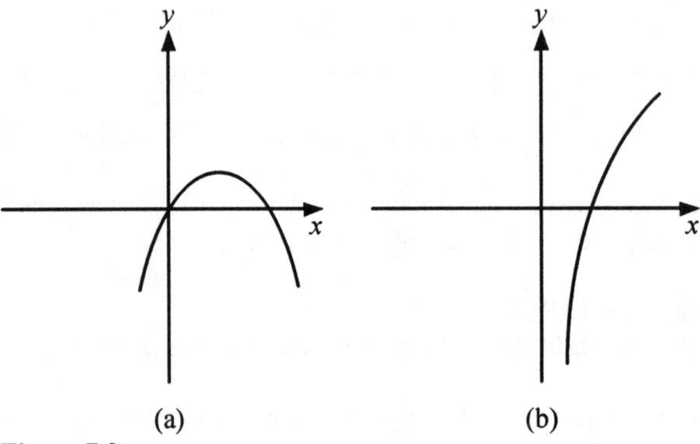

Figure 7.2 ■

When a function f has an inverse function, then there is a simple geometric method for drawing f^{-1}. Because the x-and y-coordinates change places, the graphs of inverse functions are related in that each one is the reflection of the other across the line $y = x$. Figure 7.3 shows this relationship between the square root and squaring functions for $x \geq 0$.

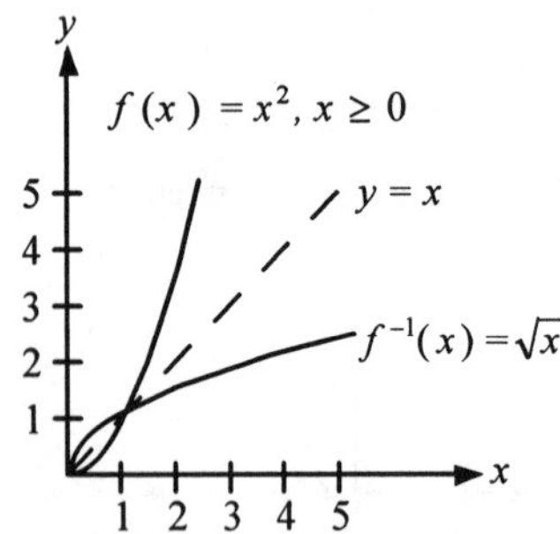

Figure 7.3

Example 4: Finding an Inverse Function Graphically

Consider the graph of $y = f(x)$ in Figure 7.4. Use this graph to sketch $y = f^{-1}(x)$.

Solution: To graph $y = f^{-1}(x)$, we reflect the curve for $y = f(x)$ about the line $y = x$. Both graphs are shown in Figure 7.5. Note that the ordered pairs (–1, 1/3) and (1, 3) from f become (1/3, – 1) and (3, 1) in f^{-1}.

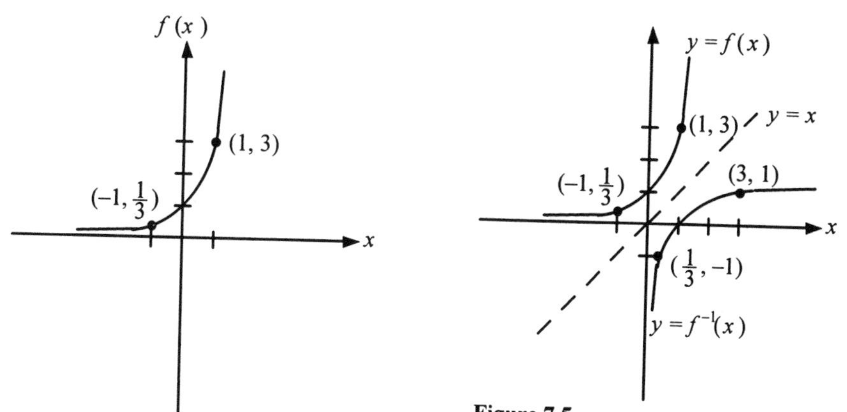

Figure 7.4

Figure 7.5

PROGRESS CHECK 4

Use the graph of $y = f(x)$ in Figure 7.6 to graph $y = f^{-1}(x)$.

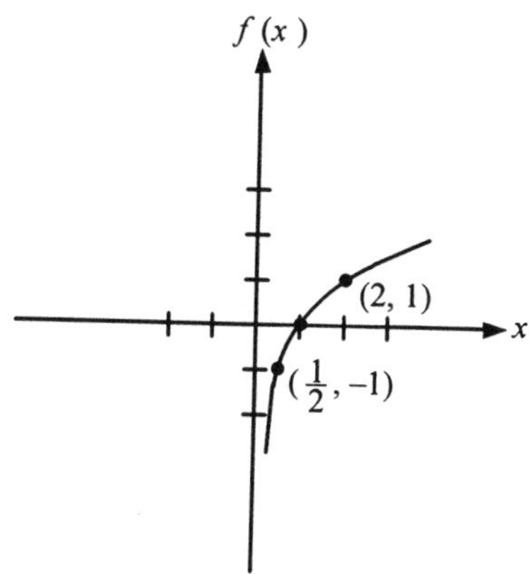

Figure 7.6

■

When *f* is a one-to-one function that is defined by an equation, then it is sometimes possible to find an equation that defines f^{-1}. Because *f* and f^{-1} make reverse assignments, we will reverse the roles of *x* and *y* in the equation that defines *f* and try to solve for *y* to obtain an equation that defines f^{-1}. For example, the inverse of the function defined by

$$y = x^3$$

is defined by

$$x = y^3.$$

Then taking the cube root of both sides of the resulting equation gives

$$y = \sqrt[3]{x}.$$

Thus, if $f(x) = x^3$, then $f^{-1}(x) = \sqrt[3]{x}$. The method just discussed is incorporated in the detailed procedure that follows for finding an equation that defines an inverse function.

To Find an Equation Defining f^{-1}

1. Start with a one-to-one function $y = f(x)$ and interchange x and y in this equation.
2. Solve the resulting equation for y, and then replace y by $f^{-1}(x)$.
3. Define the domain of f^{-1} to be equal to the range of f.

Example 5: Finding an Inverse Function Algebraically

Find $f^{-1}(x)$ for each function. If the given function is not one-to-one, so that no inverse function exists, state this. Also, evaluate $f^{-1}(2)$ for each function if it is defined.

a. $f(x) = \dfrac{7-x}{3}$ **b.** $f(x) = \sqrt{x+1}$ **c.** $f(x) = x^2$

Solution:

a. The function $f(x) = (7-x)/3$, or $f(x) = -(1/3)x + 7/3$, graphs as a straight line (see Figure 7.7), so the function is one-to-one and the inverse is a function. To find an equation for f^{-1}, write the equation for f as

$$y = \frac{7-x}{3}$$

and then proceed as follows:

$$x = \frac{7-y}{3} \qquad \text{Interchange } x \text{ and } y.$$

$$\left.\begin{aligned} 3x &= 7 - y \\ y &= -3x + 7 \end{aligned}\right\} \quad \text{Solve for } y.$$

$$f^{-1}(x) = -3x + 7. \qquad \text{Replace } y \text{ by } f^{-1}(x).$$

The domain of f^{-1} is equal to the range of f, which is the set of all real numbers.

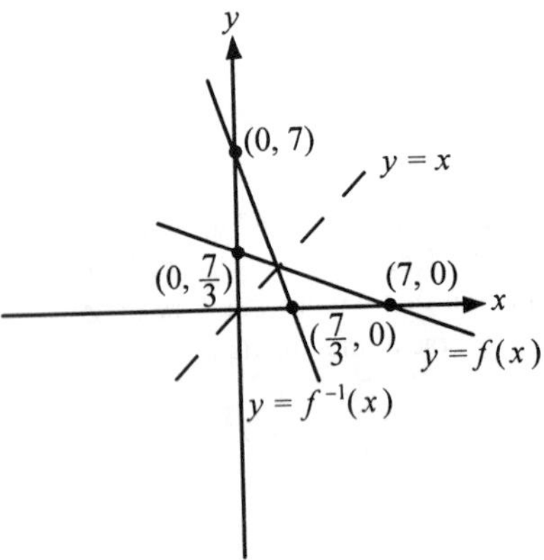

Figure 7.7

To find $f^{-1}(2)$, replace x by 2 in the equation for f^{-1}.

$$f^{-1}(2) = -3(2) + 7 = 1$$

b. The graph of $f(x) = \sqrt{x+1}$ in Figure 7.8 shows the function is one-to-one and the range is the interval $[0, \infty)$. We start with $y = f(x)$, so first write the equation as

$$y = \sqrt{x+1}.$$

Then interchange x and y and solve for y.

$$x = \sqrt{y+1}$$
$$x^2 = y+1$$
$$x^2 - 1 = y$$

Now replace y by $f^{-1}(x)$, so

$$f^{-1}(x) = x^2 - 1.$$

Finally, although $y = x^2 - 1$ is defined for any real number, we need to match the domain of f^{-1} to the range of f so that the functions will be inverses of each other. Thus, we restrict x as follows:

$$f^{-1}(x) = x^2 - 1, \quad x \geq 0.$$

From this formula we evaluate

$$f^{-1}(2) = 2^2 - 1 = 3.$$

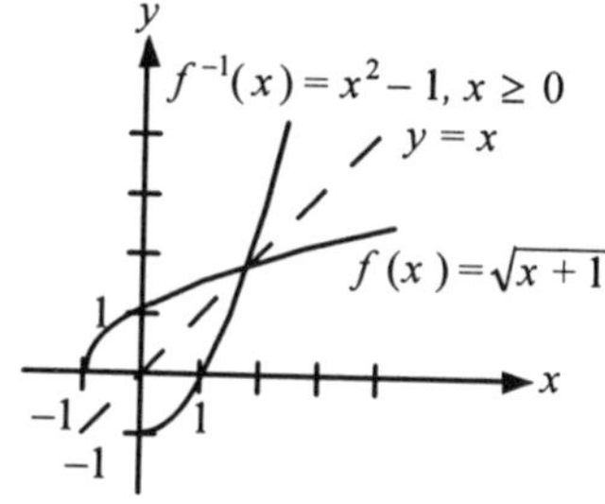

Figure 7.8

c. The squaring function $f(x) = x^2$ is not a one-to-one function (as can be seen by applying the horizontal line test to its graph in Figure 7.9). Thus, f does not have an inverse function.

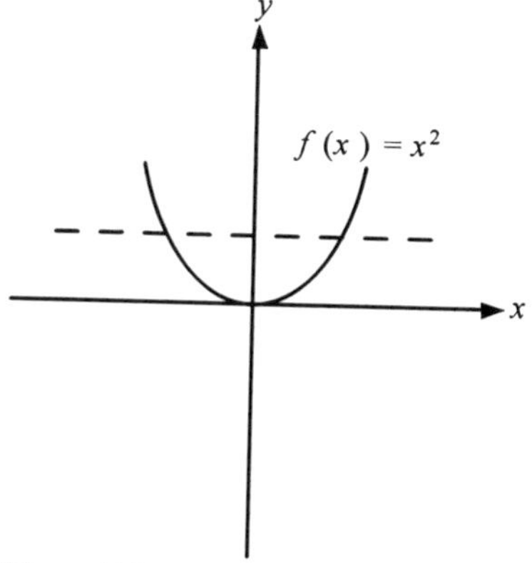

Figure 7.9

Caution: Remember that f^{-1} is the special symbol that denotes the inverse function of function f. Do *not* interpret the -1 in this symbol as an exponent.

Technology Link

If you have found a formula for the inverse of a function f, then it is straightforward to use a grapher to display the graphs of f, f^{-1}, and the line $y = x$ at the same time. This picture can serve as a visual check on the algebra, since the graphs of f and f^{-1} should appear as reflections about the line $y = x$. In Figure 7.10 we show a typical example using the functions from Example 5**b**. You should invoke the Zoom Square feature to correct the built-in angular distortion present in the standard display.

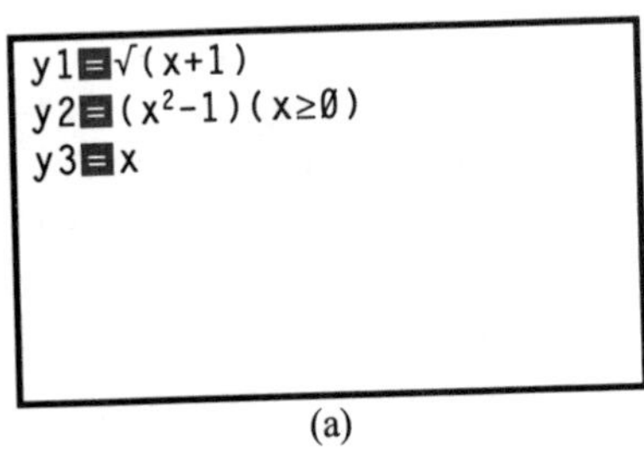

(a)

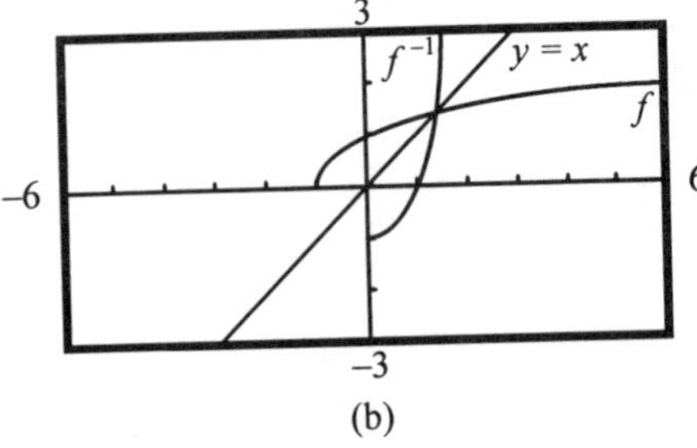

(b)

Figure 7.10

PROGRESS CHECK 5

Find $f^{-1}(x)$ for each function. If no inverse function exists, state this. Also, evaluate $f^{-1}(3)$ for each function if it is defined.

a. $f(x) = 3x + 2$ **b.** $f(x) = |x|$ **c.** $f(x) = \sqrt{x - 4}$ ■

To this point we have always worked with equations that give a direct relationship between x and y. However, when analyzing inverse functions it may be more useful to express both x and y in terms of a third variable, which is called a **parameter**. Common parameters are the variable t (often for time) and the variable θ (for an angle measure). Associated with each value of a parameter t, there is an x value and a y value, as defined by the parametric equations

$$x = f(t) \text{ and } y = g(t).$$

Many graphing calculators have a feature called **Parametric mode**, in which it is convenient to enter expressions for $f(t)$ and $g(t)$ and then draw the graph relating x and y. In this mode it is easy to reverse x and y pairs to graph the inverse of a given function, as demonstrated in the next example.

Example 6: Using Parametric Mode to Graph f^{-1}

Use a grapher set in Parametric mode to graph $f(x) = x^3 + x + 1$, its inverse, and the line $y = x$.

Solution: Set the calculator to Parametric mode. In this mode the domain variable is t instead of x. Then enter the equations as shown in Figure 7.11(a). Observe that the graphs of f and f^{-1} are symmetrical about the line $y = x$, as expected. You should also take the time to check numerically that the two functions are inverses, by using the Trace feature and the Arrow keys.

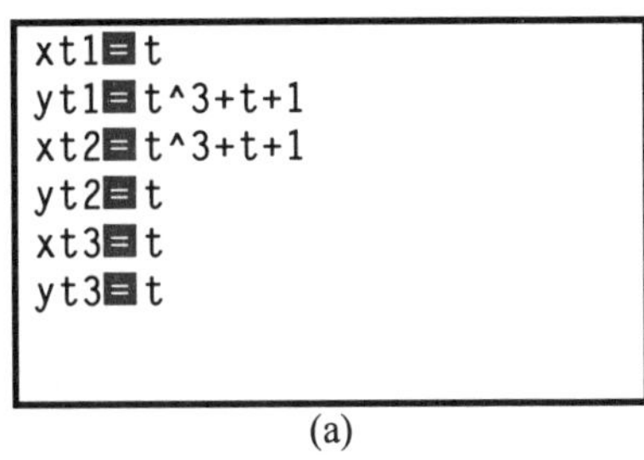

(a)

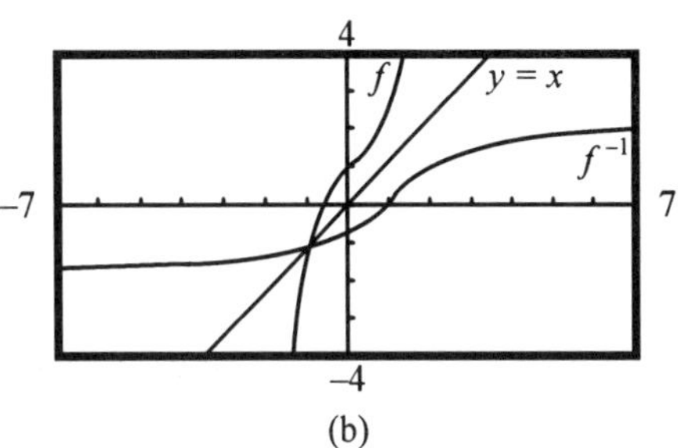

(b)

Figure 7.11

Note: Observe that *xt*1 and *yt*1 define (*x*, *y*) pairs for $y = x^3 + x + 1$, *xt*2 and *yt*2 define (*x*, *y*) pairs for $x = y^3 + y + 1$, and *xt3* and *yt3* define (*x*, *y*) pairs for $y = x$. The calculator draws the graph by giving *t* values as determined in the window settings and computing the corresponding (*x*, *y*) pairs for the graphs. This example illustrates that the concept of an inverse function may be applied even when we cannot find an equation that defines f^{-1} because the algebra involved is difficult or impossible. For instance, try finding an equation that defines f^{-1} if $f(x) = x^3 + x + 1$.

PROGRESS CHECK 6

Use a grapher set in parametric mode to graph $f(x) = x^3 + 3x - 2$, its inverse, and the line $y = x$.

■

An important concept associated with inverse functions is that each function "undoes" the other. Therefore, applying *f* and f^{-1}, one after the other, to a meaningful input *x* produces *x* as the output. For instance, if $f(x) = x^3$, then 2 is a meaningful input and $f^{-1}(x) = \sqrt[3]{x}$. Observe that

$$f(2) = 8 \text{ and } f^{-1}(8) = 2, \text{ so } f^{-1}[f(2)] = 2.$$

Recall from Section 5.2 that applying two functions in succession is called a composition of the functions, and inverse functions may be defined in terms of this operation as follows.

Definition of Inverse Functions

Two functions *f* and *g* are said to be inverses of each other provided that

$$(f \circ g)(x) = f[g(x)] = x \text{ for all } x \text{ in the domain of } g$$

and

$$(g \circ f)(x) = g[f(x)] = x \text{ for all } x \text{ in the domain of } f.$$

This definition may be used to prove that two functions are inverse functions, as shown next.

Example 7: Using Composition to Verify Inverses

Verify that $f(x) = 5x - 4$ and that $g(x) = (1/5)x + (4/5)$ are inverses of each other, using the composition definition of inverse function.

Solution: We first show that $f[g(x)] = x$ for all real numbers (which is the domain of *g*).

$$\begin{aligned} f[g(x)] &= f\left(\frac{1}{5}x + \frac{4}{5}\right) \\ &= 5\left(\frac{1}{5}x + \frac{4}{5}\right) - 4 \\ &= x + 4 - 4 \\ &= x \end{aligned}$$

Next we verify that $g[f(x)] = x$ for all real numbers (which is the domain of f).

$$\begin{aligned} g[f(x)] &= g(5x - 4) \\ &= \frac{1}{5}(5x - 4) + \frac{4}{5} \\ &= x - \frac{4}{5} + \frac{4}{5} \\ &= x \end{aligned}$$

Thus, f and g are inverses of each other.

Note: The fact that f and f^{-1} must satisfy $f\left[f^{-1}(x)\right] = x$ provides an alternative method for finding an equation for f^{-1} that does not involve all the symbol switching of our previous method. For example, to redo Example 5b, we start with $f(x) = \sqrt{x+1}$ and find $f\left[f^{-1}(x)\right]$ by replacing x by $f^{-1}(x)$ to obtain

$$f\left[f^{-1}(x)\right] = \sqrt{f^{-1}(x) + 1}.$$

Then by definition, $f\left[f^{-1}(x)\right] = x$ equals x, so

$$\begin{aligned} x &= \sqrt{f^{-1}(x) + 1} \\ x^2 &= f^{-1}(x) + 1 \\ x^2 - 1 &= f^{-1}(x), \end{aligned}$$

which agrees with our previous result. Of course, the restriction $x \geq 0$ applies, as discussed in Example 5b. If you feel comfortable working with functional notation, then this method may be helpful.

PROGRESS CHECK 7

Verify that $f(x) = 4x - 1$ and $g(x) = (x + 1)/4$ are inverses of each other, using the composition definition of inverse functions. ■

The fact that inverse functions "undo" each other is the basis for one type of application considered in Example 8.

Example 8: Using an Inverse to Undo a Function

Solve the problem in the chapter introduction on page 525.

Solution: A 20 percent cut in hourly wage means that the new wage is 80 percent of the original wage. Therefore, the reduced wage is given by

$$y = f(x) = 0.8x.$$

The rule for f^{-1} gives the formula to "undo *f*," and the inverse of the function defined by $y = 0.8x$ is defined by

$$x = 0.8y.$$

Solving this equation for *y* then yields

$$y = \frac{x}{0.8} \text{ or } y = 1.25x.$$

The offsetting formula $y = f^{-1}(x) = 1.25x$ indicates that a 25 percent increase is needed to return the workers to their original hourly wage.
Note: To check this result, observe that

$$f^{-1}[f(x)] = f^{-1}(0.8x) = x.$$

Thus, *f* and f^{-1} undo each other to produce the original hourly wage.

PROGRESS CHECK 8
An investor's portfolio loses one-third of its value during the first year. For the following year, what percent increase is needed for the portfolio to return to its original value? ■

Another type of application involving inverse functions is based on the fact that inverse functions contain the same information in different forms. Which function is more useful depends on what is given and what is to be found. In this context certain problems are viewed as inverse problems, as illustrated below.

Inverse Problems
→ *Problem:* Find the Celsius temperature for a given Fahrenheit temperature.
→ *Problem:* Find the Fahrenheit temperature for a given Celsius temperature.

The inverse problems of converting between two temperature scales are analyzed further in the next example.

Example 9: Using an Inverse to Express a Relation Conveniently

The function $y = f(x) = (5/9)(x - 32)$ converts degrees Fahrenheit (*x*) to degrees Celsius (*y*). Find an equation for the inverse function. What formula does the inverse function represent?

Solution: The equation $y = f(x) = (5/9)(x - 32)$ defines a linear function, so an inverse function exists and an equation defining f^{-1} may be found as follows.

$$y = \frac{5}{9}(x - 32) \quad \text{Start with } y = f(x).$$

$$x = \frac{5}{9}(y - 32) \quad \text{Interchange } x \text{ and } y.$$

$$\left.\begin{aligned} \frac{9}{5}x &= y - 32 \\ \frac{9}{5}x + 32 &= y \end{aligned}\right\} \quad \text{Solve for } y.$$

$$\frac{9}{5}x + 32 = f^{-1}(x) \quad \text{Replace } y \text{ by } f^{-1}(x).$$

Thus, if $y = f(x) = (5/9)(x - 32)$, then $y = f^{-1}(x) = (9/5)x + 32$. In the context of the problem, it is stated that the formula given in function *f* converts degrees Fahrenheit (x) to degrees Celsius (y). Therefore, the formula in the inverse function converts degrees Celsius (x) to degrees Fahrenheit (y).

PROGRESS CHECK 9

The function $y = f(x) = x^2,\ x > 0$, gives the formula for the area (y) of a square in terms of the side length (x). Find an equation for the inverse function. What formula does the inverse function represent? ■

EXPLORE 7.1

1. In examining a function on a graphing calculator to see if it has an inverse, a part of the graph may appear horizontal when it really is not. Examine the graph of $y = \sqrt[3]{x+1}$ using Zoom Standard, then look more closely at the interval from 1 to 2. Is the graph horizontal? Use Trace to help decide. How can you know algebraically that the graph is not horizontal? Why does the graph seem to be made of horizontal segments?
2. Graph $y = x^3 + x + 1$, then use the graph and Trace to estimate $f^{-1}(-1)$ and $f^{-1}(2)$.
3. Sometimes we restrict the domain of a function *f* to intervals for which *f* is increasing (or decreasing) so that f^{-1} will exist over the restricted domain. Use a grapher to estimate any intervals between –2 and 2 where *f* is increasing.
   a. $f(x) = (x-1)^2$ **b.** $f(x) = 4x^3 - 3x^2 - 6x + 1$
4. If two functions *f* and *g* are inverses, then both compositions $f \circ g$ and $g \circ f$ must be identically equal to *x*. Thus the graph of $f \circ g$ and $g \circ f$ will each be the line $y = x$. Graph both $f \circ g$ and $g \circ f$ for these functions and determine whether they are inverses of one another. On a grapher let $y1 = f(x)$, $y2 = g(x)$, $y3 = f[g(x)]$, and $y4 = g[f(x)]$, then determine if the graphs of *y*3 and *y*4 are the line $y = x$. See Section 5.2 for more on graphing composition.
   a. $f(x) = x^3 + 4;\ g(x) = \sqrt[3]{x-4}$ **b.** $f(x) = x^2 - 3;\ g(x) = \sqrt{x} + 3$
5. Explore the properties of graphing in parametric mode by examining the parametric graph of $y = x^2$. Let $xt1 = t$, and $yt1 = t^2$.
   a. Describe what happens when you use Zoom Standard. Check the viewing window settings after you use Zoom Standard. Why don't you get a complete graph of the parabola? What happens when you set $t\text{Min} = -10$?
   b. Set the graphing format to DrawDot to investigate the result of altering *t*Step. Try different values for *t*Step, such as .1, .5, 1, and 2.
   c. Set *t*Step = .1 then use Trace to see what happens to each of *t*, *x*, and *y* as you press the cursor.

EXERCISES 7.1

In Exercises 1–6, find the inverse of the function. Also, find the domain and range of both functions.

1. {(1, 7), (2, 8), (3, 9)}
2. {(4, – 1), (8, – 2), (12, – 3)}
3. {(–2, 10), (0, 0), (5, – 1)}
4. {(–7, 1), (–4, 11), (1, 13)}
5. {(4, 4), (5, 5), (6, 6)}
6. $\left\{(1, 1), \left(2, \frac{1}{2}\right), \left(\frac{1}{2}, 2\right)\right\}$

In Exercises 7–10 determine if the inverse of the function is a function.

7. {(1, 4), (2, 4), (3, 4)}
8. {(–4, 4), (–5, 5), (–6, 6)}
9. {(0, 0), (5, 0.2), (0.2, 5)}
10. {(–1, 1), (0, 0), (1, 1)}

In Exercises 11–18 determine if the inverse of the function graphed is a function.

11.

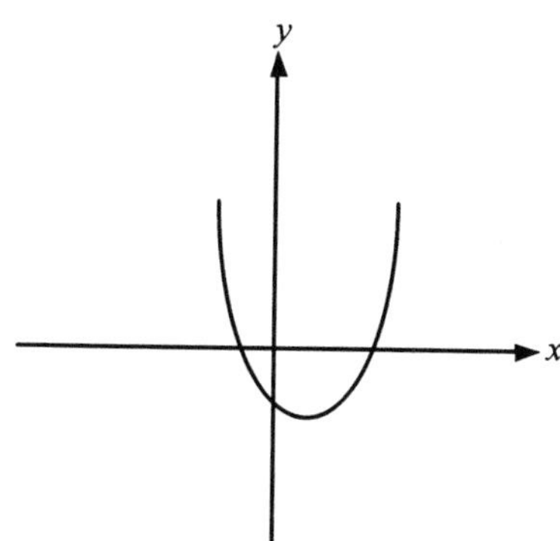

12.

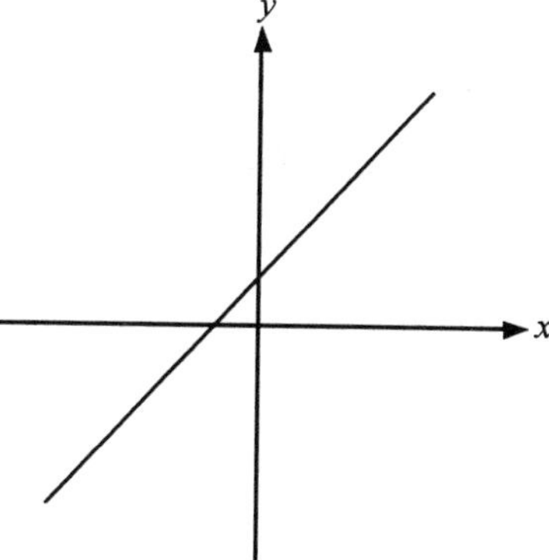

13.

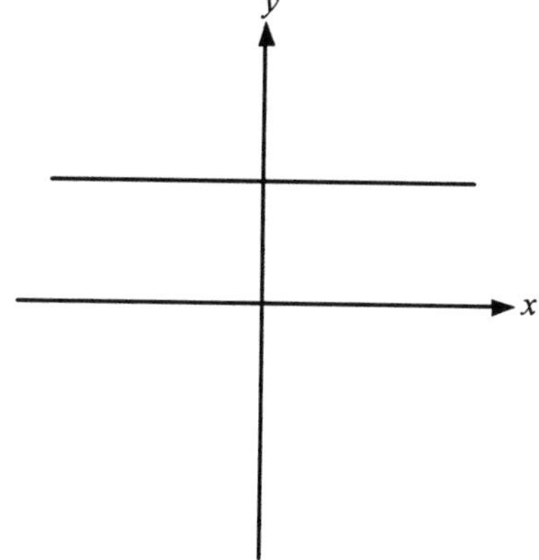

14.

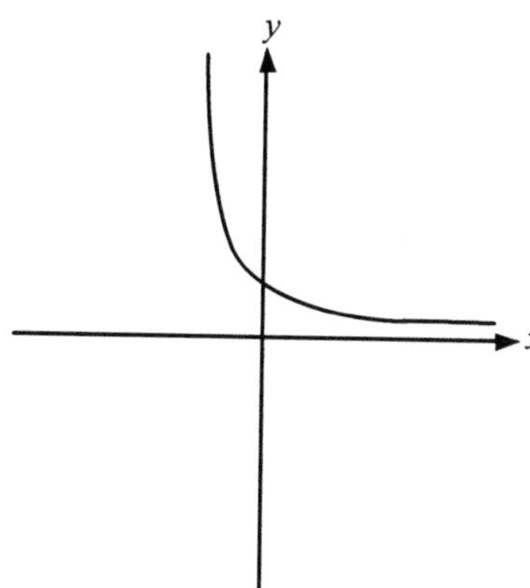

15.

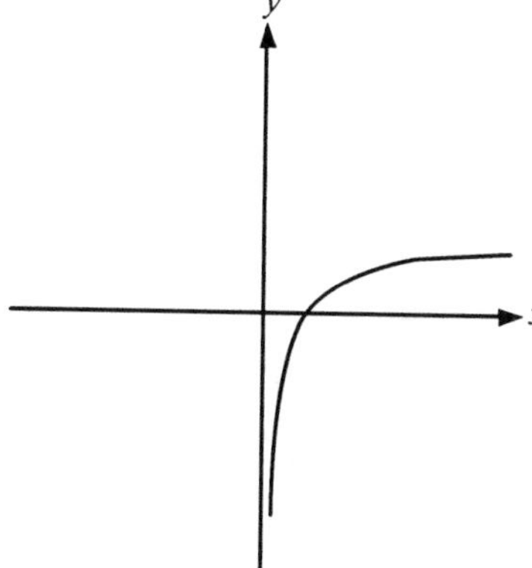

16.

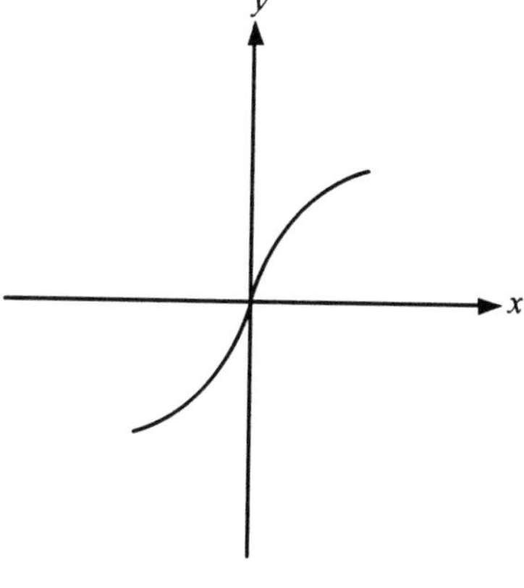

17.

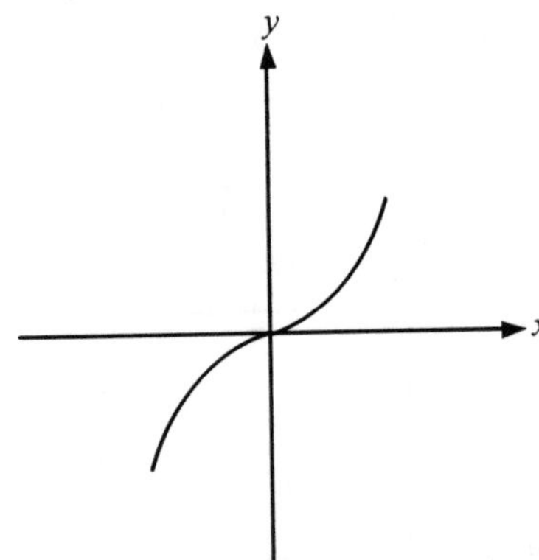

18.

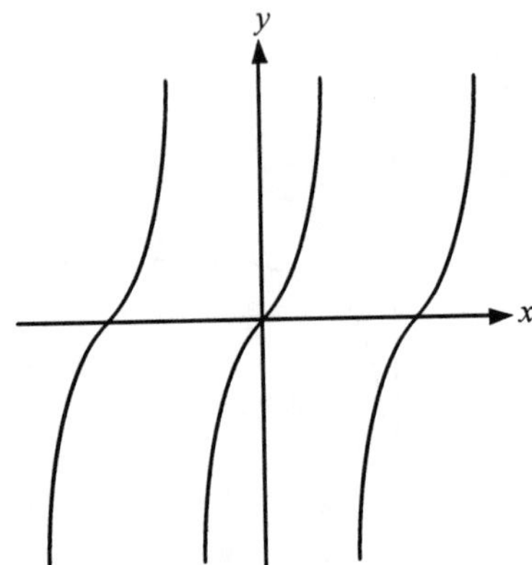

In Exercises 19–24 use the graph of $y = f(x)$ to graph $y = f^{-1}(x)$. If no inverse function exists, state this.

19.

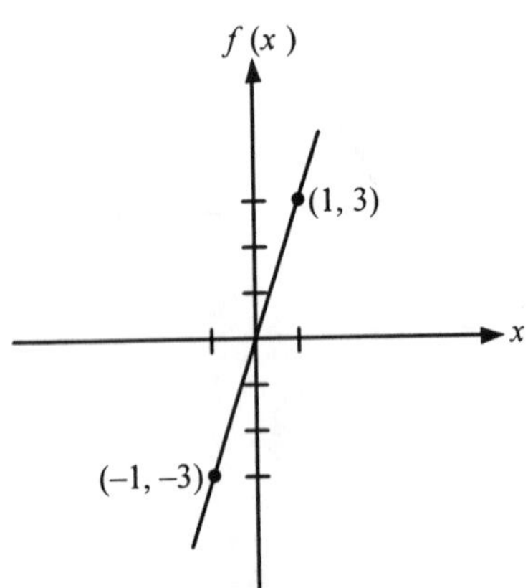

20.

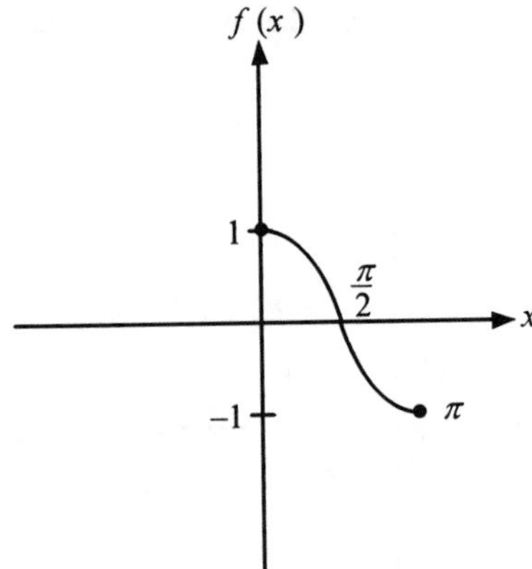

21.

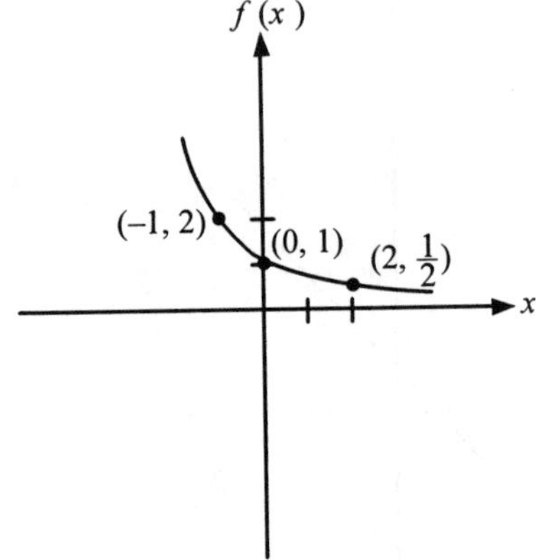

22.

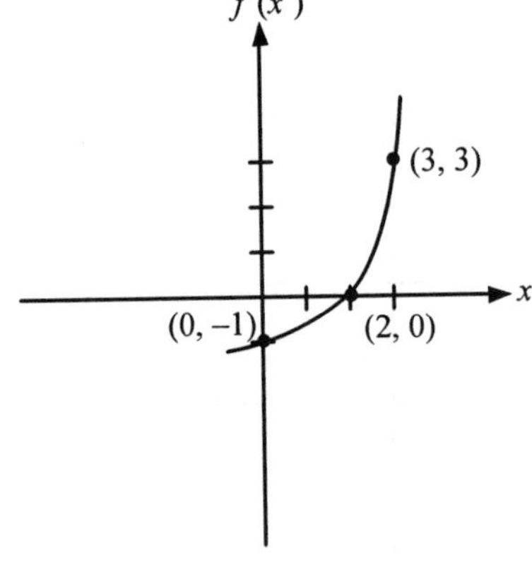

23.

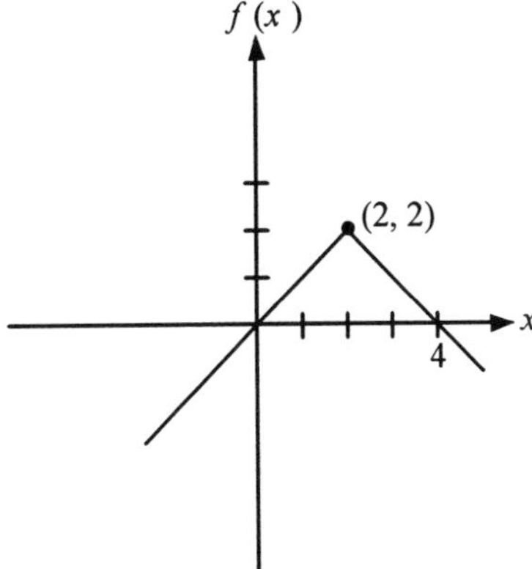

24.

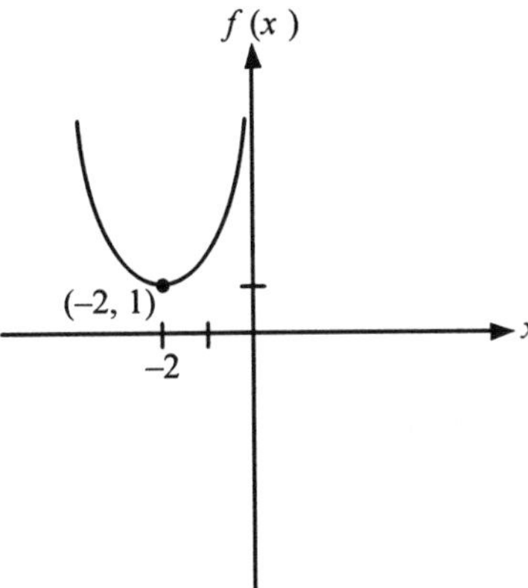

In Exercises 25–42 find the inverse function and graph both the function and its inverse function on the same set of axes. Evaluate the inverse function at $x = 2$ if it is defined. If the given function is not one-to-one, so that no inverse function exists, state this.

25. $y = x - 10$

26. $g(x) = -\frac{1}{2}x + 5$

27. $f(x) = \frac{1}{3}x - \frac{7}{3}$

28. $y = \frac{4-x}{11}$

29. $y = x^2$

30. $y = x^3$

31. $f(x) = 2$

32. $f(x) = |x| - x$

33. $y = \frac{1}{x}$

34. $f(x) = \frac{1}{x+3}$

35. $h(x) = \sqrt[3]{x-4}$

36. $g(x) = \sqrt{x}$

37. $f(x) = x^2 - 4,\ x \ge 0$

38. $g(x) = 1 - x^2,\ x < 0$

39. $y = \sqrt{x+3}$

40. $f(x) = 2 + \sqrt{x}$

41. $f(x) = \sqrt{1-x^2}$

42. $h(x) = \sqrt{1-x^2},\ x \ge 0$

In Exercises 43–48 use a grapher set in parametric mode to graph *f*, its inverse, and the line $y = x$. Indicate whether or not the inverse relation is a function.

43. $f(x) = x^3 + 2x - 1$

44. $f(x) = x^3 - x + 3$

45. $f(x) = \sqrt[3]{x^2 - 9}$

46. $f(x) = \sqrt[3]{x^2 + 9}$

47. $f(x) = \frac{x^2 - 4}{x^2 + 1}$

48. $f(x) = \frac{x^2 + 4}{x^2 + 1}$

In Exercises 49–58 verify that the given functions are inverses of each other by using the composition definition of inverse function.

49. $f(x) = x + 5,\ g(x) = x - 5$

50. $f(x) = 3x - 2,\ g(x) = \frac{1}{3}x + \frac{2}{3}$

51. $f(x) = \sqrt[3]{x},\ g(x) = x^3$

52. $f(x) = \sqrt[3]{x} + 2,\ g(x) = (x-2)^3$

53. $f(x) = \frac{1}{x+4},\ g(x) = \frac{1-4x}{x}$

54. $f(x) = \frac{1}{x+2},\ g(x) = \frac{1}{x} - 2$

55. $f(x) = \sqrt{x+1},\ x \ge -1,\ g(x) = x^2 - 1,\ x \ge 0$

56. $f(x) = \sqrt{2x-4},\ x \ge 2,\ g(x) = \frac{x^2+4}{2},\ x \ge 0$

57. If $f(x) = \frac{1}{x}$, show that *f* is its own inverse. That is, show $f[f(x)] = x$.

58. If $f(x) = \frac{2x+1}{3x-2}$, show that *f* is its own inverse. That is, show $f[f(x)] = x$.

59. What percent raise is required to restore a 10 percent wage cut? Round the answer to the nearest hundredth of a percent.

60. A store owner sets the retail price of a shirt 100 percent higher than the cost to the store. By what percent can this retail price be cut before the store loses money on the sale?

61. A manager's weekly salary is cut by 15 percent during a financial crisis and then raised by 15 percent when the crisis is over. Let c and r be functions that define the cut and the raise in salary, respectively, and show that r is not the inverse of c. That is, show $(r \circ c)(x) \neq x$. What is the percent gain or loss in the manager's weekly salary?

62. Voters cut a school budget by 10% one year, then raise it 10% the next year. Let c and r be functions that define the cut and raise, and show that r is not the inverse of c. What is the percent gain or loss in the budget?

63. The function $y = x^3$, $x > 0$, gives the formula for the volume (y) of a cube in terms of the side length (x). Find the inverse function. What formula does this function represent?

64. The function $y = \frac{5}{9}x + \frac{45967}{180}$ converts degrees Fahrenheit (x) to degrees Kelvin (y). Find the inverse function. What does the inverse function represent?

THINK ABOUT IT 7.1

1. a. Give two graphical examples of an odd function that is not a one-to-one function.
   b. If f is an even function, then does f^{-1} exist? Explain.
2. If function g is the inverse of function f, then the rule in g must "undo" the rule in f so that $(g \circ f)(x) = x$. Find the inverse function for each of the following functions by undoing in words the rule in f. For example, if $f(x) = 5x - 2$, then the rule in f is "for each real number, multiply it by 5, and then subtract 2 from the result." In words, the rule to undo f says "for each real number, add 2 to it, and then divide the result by 5," so f^{-1} is defined by $f^{-1}(x) = (x+2)/5$.

   a. $f(x) = 2x - 7$ b. $f(x) = \frac{x+5}{4}$ c. $f(x) = \frac{x}{4} + 5$

   d. $f(x) = -x + 4$ e. $f(x) = (x+2)^3$ f. $f(x) = \sqrt[3]{x-6}$
3. Explain the different purposes of the horizontal and vertical line tests.
4. Explain why a function that is increasing over the set of real numbers must have an inverse function.
5. Every linear function has an inverse function. What is the inverse of the function $f(x) = mx + b$, $m \neq 0$?

● ● ●

7.2 Exponential Functions

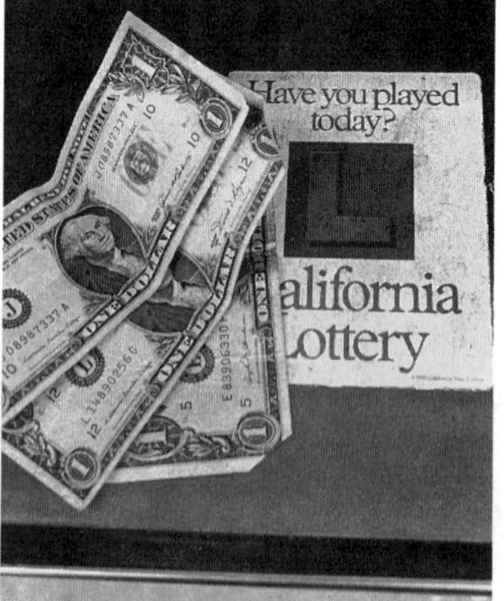

*Photo Courtesy of **Debbie and Steve Takatsuno**, The Picture Cube*

If you win the state lottery and are offered \$1 million or a penny that doubles in value on each day in the month of November, which option should you accept? Answer the following questions to find out.

a. Write a formula that gives the value (in cents) of the penny at the end of x days.

b. What is the value of the penny by the end of the month? Which offer is better?

(See Example 1).

Objectives

1. Determine function values for an exponential function.
2. Graph an exponential function.
3. Solve exponential equations using $b^x = b^y$ implies $x = y$.
4. Find the base in the exponential function $y = b^x$ given an ordered pair in the function.
5. Solve applied problems involving exponential functions.

The section-opening problem, with slight variations in wording, often appears in puzzle sections of newspapers and magazines. We begin by solving this problem because it provides an interesting and revealing introduction to the concept of an exponential function.

Example 1: Finding a Formula for an Exponential Function

Solve the problem in the section introduction on page 540.

Solution:

a. Let y represent the value (in cents) of the penny. Then

$y = 1$ when $x = 0$ days
$y = 1 \cdot 2 = 2$ when $x = 1$ day
$y = 2 \cdot 2 = 2^2 = 4$ when $x = 2$ days
$y = 2^2 \cdot 2 = 2^3 = 8$ when $x = 3$ days
$y = 2^3 \cdot 2 = 2^4 = 16$ when $x = 4$ days
and in general
$y = 2^x$,

where x represents the number of complete days that have elapsed.

b. By the end of November, when $x = 30$ days,

$$y = 2^{30} = 1{,}073{,}741{,}824 \text{ cents}$$
$$= \$10{,}737{,}418.24.$$

Thus the penny is worth more than $10 million by the end of the month, and the penny option is the better offer.

Note: Often the results of calculations involving exponential expressions result in extremely large (or small) numbers that are displayed in scientific notation format. For instance, on some calculators 2^{30} is given as 1.073742×10^9.

PROGRESS CHECK 1

A scientist has 1 g of a radioactive element which is disappearing through radioactive decay. Careful observations reveal that for every hour that passes, the quantity of this element that still remains is one-half the amount that was present at the beginning of that hour. (Scientists say the "half-life" of the element is 1 hour.)

a. Find a formula showing the amount of the element present in x hours.

b. Approximately how much of this element is left at the end of 1 day? ■

Example 1 requires that 2^x have meaning only for nonnegative integer values of x because no change is considered to have occurred until an entire time period has elapsed. However, for the exponential function defined by $y = 2^x$, we wish 2^x to be meaningful for *all* real values of x. Up to this point in the text, there is no difficulty interpreting 2^x where x is a rational number. For instance,

$$2^4 = 16,$$
$$2^0 = 1,$$
$$2^{-3} = \frac{1}{2^3} = \frac{1}{8},$$
$$2^{2/3} = \sqrt[3]{2^2} = \sqrt[3]{4} \approx 1.59.$$

A precise definition for 2^x, where x is irrational, is given in higher mathematics, and the Power key on a calculator may be used to approximate such expressions. For example, the display in Figure 7.12 shows $2^\pi \approx 8.82$ and $2^{\sqrt{2}} \approx 2.67$. Observe that a result like $2^\pi \approx 8.82$ is a sensible value, since $2^\pi \approx 2^{3.14}$, which is a little larger than 2^3, or 8. Thus, an expression like 2^x is meaningful for all real numbers x, and it can be shown that all previous laws of exponents (as first listed in Section R.3) are valid for all real number exponents. An exponential function with a positive base other than 1 may now be defined.

```
2^π
                8.82497782708
2^√2
                2.66514414269
```

Figure 7.12

Exponential Function

The function f defined by

$$f(x) = b^x,$$

with $b > 0$ and $b \neq 1$, is called the **exponential function with base *b*.**

The restriction $b \neq 1$ is made because $1^x = 1$ for all values of x, so f is a constant function in this case. Only positive bases are used in the defining equation because expressions such as $(-4)^{1/2}$ and 0^{-2} are not real numbers.

Example 2: Evaluating an Exponential Function

If $f(x) = 9^x$, find $f(3)$, $f(-2)$, and $f(5/2)$. Also approximate $f(\sqrt{2})$ to the nearest hundredth.

Solution: Using $f(x) = 9^x$ gives

$$f(3) = 9^3 = 729$$
$$f(-2) = 9^{-2} = \frac{1}{9^2} = \frac{1}{81}$$
$$f\left(\frac{5}{2}\right) = 9^{5/2} = \left(\sqrt{9}\right)^5 = 3^5 = 243.$$

By calculator,

$$f\left(\sqrt{2}\right) = 9^{\sqrt{2}} = 22.36 \text{ (to the nearest hundredth).}$$

PROGRESS CHECK 2

If $f(x)=8^x$, find $f(2)$, $f(-1)$, and $f(2/3)$. Also approximate $f(\sqrt{3})$ to the nearest hundredth. ■

In the next two examples, we investigate the effect of the value of b on the behavior of $y=b^x$.

Example 3: Graphing an Exponential Function with $b>1$

Sketch the graphs of $g(x)=2^x$ and $h(x)=4^x$ on the same coordinate system. Indicate domain and range on the graph.

Solution: Generate a table of values by replacing x with integer values from, say, –3 to 3.

x	–3	–2	–1	0	1	2	3
2^x	$\frac{1}{8}$	$\frac{1}{4}$	$\frac{1}{2}$	1	2	4	8
4^x	$\frac{1}{64}$	$\frac{1}{16}$	$\frac{1}{4}$	1	4	16	64

By graphing these solutions and drawing a smooth curve through them, we sketch the graphs of g and h, which are shown in Figure 7.13.

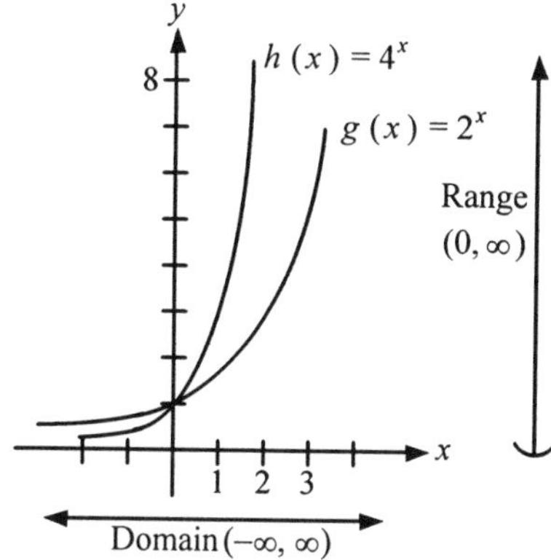

Figure 7.13

For each graph the domain is the set of all real numbers, and the range is the set of positive real numbers. Observe that both graphs are increasing over the whole domain, and that the graph with the larger value of b increases faster. The behavior of these functions is called exponential growth. Notice that in an exponential table, a fixed change in x produces a **constant ratio** between the corresponding y values. When the fixed change in x is 1, then the ratio of the y values is the base of the exponential function.

Technology Link

The appearance of an exponential graph on a grapher display is noticeably influenced by the choice of viewing window. In Figure 7.14 are displayed two different viewing window versions of the graphs in Figure 7.13.

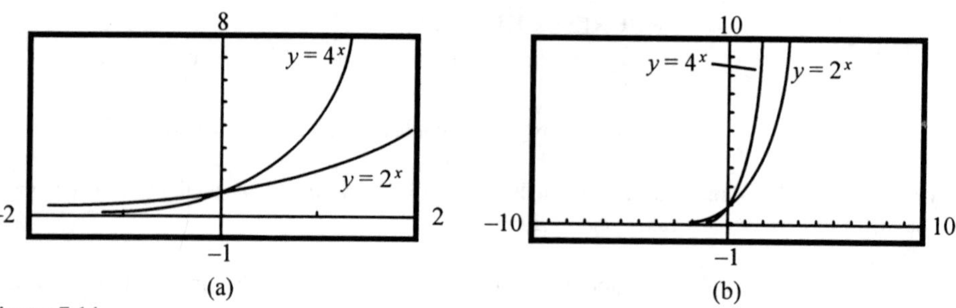

Figure 7.14

PROGRESS CHECK 3

Sketch the graphs of $g(x) = 3^x$ and $h(x) = 8^x$ on the same coordinate system. Indicate domain and range on the graph. ■

Example 4: Graphing Exponential Functions with $0 < b < 1$

Sketch the graphs of $g(x) = (1/2)^x$ and $h(x) = (1/4)^x$ on the same coordinate system. Indicate domain and range on the graph.

Solution: Construct a table of values, as follows, and then graph the equations, as shown in Figure 7.15.

x	−3	−2	−1	0	1	2	3
$\left(\frac{1}{2}\right)^x$	8	4	2	1	$\frac{1}{2}$	$\frac{1}{4}$	$\frac{1}{8}$
$\left(\frac{1}{4}\right)^x$	64	16	4	1	$\frac{1}{4}$	$\frac{1}{16}$	$\frac{1}{64}$

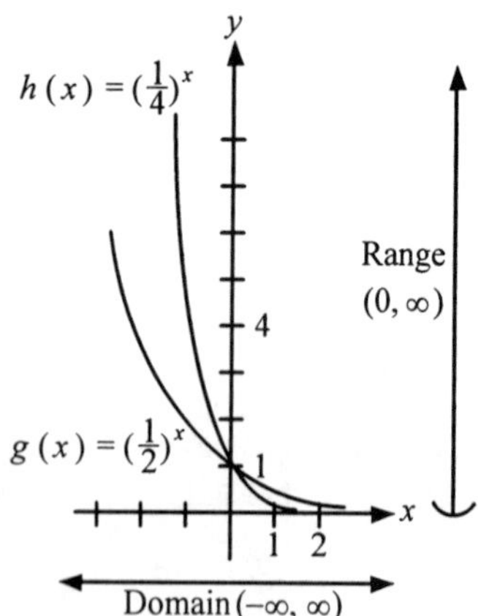

Figure 7.15

Observe that taking b between 0 and 1 yields a decreasing exponential function, and that the graph with the smaller value of b decreases faster. These graphs illustrate **exponential decay**. The domain and range are the same as in the previous example.

Note: The equation $y = (1/2)^x$ is equivalent to $y = 2^{-x}$, since

$$\left(\frac{1}{2}\right)^x = \frac{1}{2^x} = 2^{-x}$$

Therefore, exponential decay may also be described by equations of the form $y = b^{-x}$ with $b > 1$.

PROGRESS CHECK 4

Sketch the graphs of $g(x) = (1/3)^x$ and $h(x) = (0.8)^x$ on the same coordinate system. Indicate domain and range on the graph. ■

Several important features of the exponential function with base b were illustrated in the graphs in Examples 3 and 4. These properties are summarized in the next box.

Properties of $f(x) = b^x$ (with $b > 0$, $b \neq 1$) and its Graph

1. If $b > 1$, then as x increases, y increases; a quantity is growing exponentially.
2. If $0 < b < 1$, then as x increases, y decreases; a quantity is decaying exponentially.
3. The y-intercept is always (0, 1), and there are no x-intercepts.
4. The x-axis is a horizontal asymptote for the graph of f.
5. The domain of f is $(-\infty, \infty)$, and the range of f is $(0, \infty)$.

We can graph certain variations of the function $y = b^x$ by using the graphing techniques given in Section 1.3. Example 5 illustrates such a case.

Example 5: Applying Graphing Techniques to Exponential Functions

Graph $y = 1 - 2^{-x}$. What is the range of this function?

Solution: We begin by noting that 2^{-x} is equivalent to $(1/2)^x$, so we start with the graph of $y = (1/2)^x$ in Figure 7.13. We also note that $1 - 2^{-x} = -2^{-x} + 1$. To graph $y = -2^{-x} + 1$, we first reflect the graph of $y = 2^{-x}$ about the x-axis to obtain the graph of $y = -2^{-x}$ [see Figure 7.16(a)]. Then we raise the graph in Figure 7.16(a) up 1 unit, since the constant 1 is added after the exponential term. The completed graph is shown in Figure 7.16(b). From this graph we read that the range is the interval $(-\infty, 1)$.

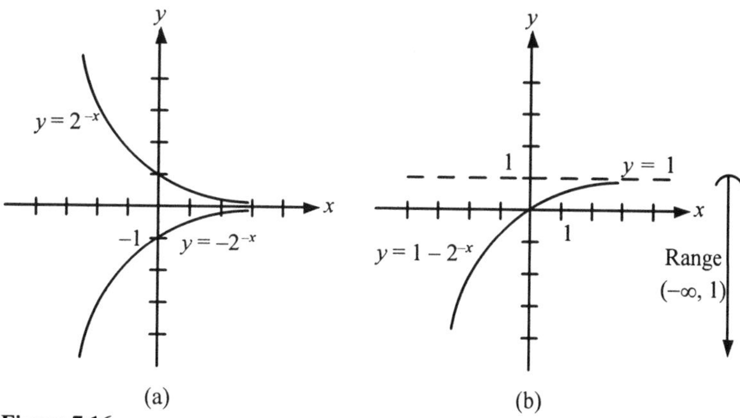

Figure 7.16

PROGRESS CHECK 5

Graph $y = 1 - \left(\frac{1}{2}\right)^x$ and indicate its range on the graph. ■

In many problems involving an exponential function, we need to solve an **exponential equation**, which is an equation that has a variable in an exponent. In certain cases it is not difficult to write both sides of the equation in terms of the same base. Then because an exponential function takes on each value in its range exactly once, the following principles applies.

Equation-Solving Principle

> If b is a positive number other than 1, then
>
> $$b^x = b^y \text{ implies } x = y.$$

This principle indicates that when certain exponential expressions with the same base are equal, then the equation can be solved by equating exponents, as illustrated in Examples 6 and 7.

Example 6: Completing an Ordered Pair Solution

If $(x,\ 4)$ is a solution of the equation $y = 8^x$, find x.

Solution: We need to find the value of x that satisfies the equation $8^x = 4$. In many cases it is hoped that through an understanding of exponents, you can determine the solution by inspection. If not, then try to rewrite the expressions in terms of a common base. In this case since $8 = 2^3$ and $4 = 2^2$ we solve the equation as follows:

$$\begin{aligned} 8^x &= 4 \\ (2^3)^x &= 2^2 \\ 2^{3x} &= 2^2 \\ 3x &= 2 \quad \text{Since } b^x = b^y \text{ implies } x = y. \\ x &= \frac{2}{3} \end{aligned}$$

Thus, the solution set is $\{2/3\}$.

Technology Link

Figure 7.17 shows how a grapher may be used to estimate the solution in this example. To obtain this display the equations $y1 = 8^x$ and $y2 = 4$ were graphed in the viewing window shown and the ISECT (intersection) operation was used to estimate the point of intersection. (If your calculator does not have this feature, then the Zoom and Trace features may be used to estimate this point). Observe that the displayed x-coordinate of the point of intersection is in agreement with the solution $x = 2/3$. On advantage of this method is that it may be applied to solve equations such as $8^x = 5$, in which it is difficult to rewrite expressions in terms of a common base. Algebraic methods for solving such equations involve logarithms, so this topic is not considered until Section 7.5.

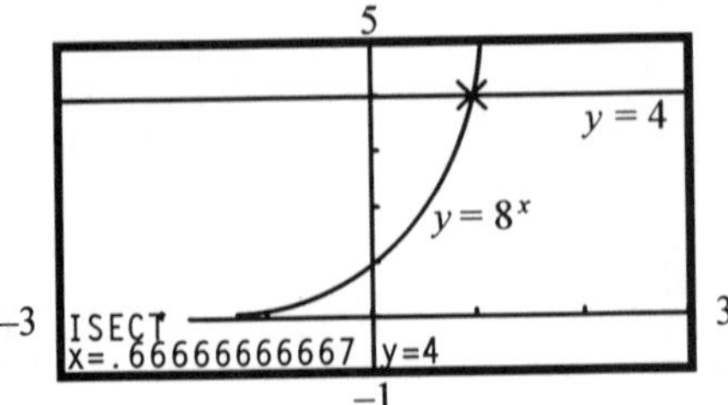

Figure 7.17

PROGRESS CHECK 6

If $\left(x, \frac{1}{6}\right)$ is a solution of the equation $y = 36^x$, find x. ■

Example 7: Solving an Exponential Equation

Solve $32^x = \left(\frac{1}{4}\right)^{x+1}$.

Solution: To rewrite this equation so both sides are expressed with the same base, recognize that $32 = 2^5$ and $\frac{1}{4} = \frac{1}{2^2} = 2^{-2}$. Then

$$\begin{aligned} 32^x &= \left(\frac{1}{4}\right)^{x+1} \\ \left(2^5\right)^x &= \left(2^{-2}\right)^{x+1} \\ 2^{5x} &= 2^{-2x-2} \\ 5x &= -2x - 2 && \text{Since } b^x = b^y \text{ implies } x = y. \\ 7x &= -2 \\ x &= -\frac{2}{7} \end{aligned}$$

Thus, the solution set is $\{-2/7\}$. See Figure 7.18 for a graphical solution that supports this answer.

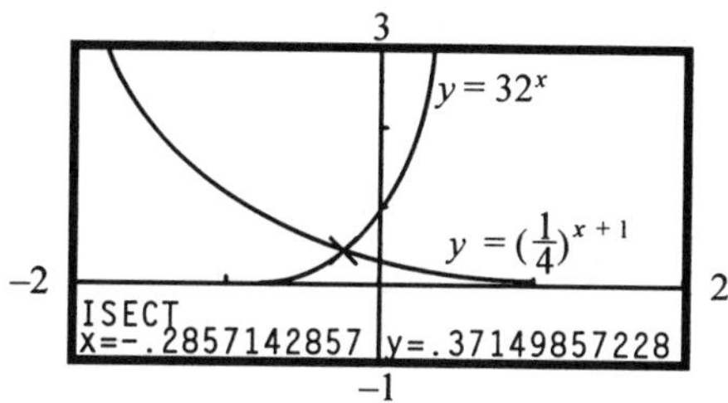

Figure 7.18

PROGRESS CHECK 7

Solve $27^x = \left(\frac{1}{3}\right)^{x-1}$. ■

Another type of problem associated with analyzing exponential functions involves finding the base of an exponential function when given one ordered pair in the function.

Example 8: Finding the Base of an Exponential Function

Find the base of the exponential function $y = b^x$ that contains the point $(-1/3,\ 1/4)$.

Solution: Replacing x by $-1/3$ and y by $1/4$ in the equation $y = b^x$ gives

$$\frac{1}{4} = b^{-1/3}.$$

To find b, use an extension of the principle of powers and raise both sides of this equation to the reciprocal power of $-1/3$, which is -3, and then simplify.

$$\left(\frac{1}{4}\right)^{-3} = \left(b^{-1/3}\right)^{-3} \quad \text{Raise both sides to the reciprocal power.}$$
$$4^3 = b^1$$
$$64 = b$$

Since $64^{-1/3} = (1/64)^{1/3} = \sqrt[3]{1/64} = \frac{1}{4}$, the solution checks; so the base of the exponential function containing the given point is 64.

PROGRESS CHECK 8
Find the base of the exponential function $y = b^x$ that contains the point (4, 2). ■

The next three examples consider applications in which a quantity either grows or decays exponentially.

Example 9: Finding a Formula for Compound Interest

$1,000 is invested at 6 percent interest compounded annually.

a. Describe in functional notation the value of the investment at the end of t years.
b. Construct a table that shows corresponding values for t and $f(t)$. Let t vary from 0 to 3 in increments of 1.
c. How much is the investment worth after 4 years?

Solution:

a. The $1,000 amounts to $\$1,000(1+0.06)$, or $1,060, at the end of 1 year. The principal during the second year is $1,000(1+0.06)$. Thus, the amount at the end of 2 years is $1,000(1+0.06)(1+0.06)$, or $1,000(1+0.06)^2$. In general, the value of the investment after t years is

$$f(t) = 1,000(1.06)^t$$

b. Using the formula from part **a**, we create the following table:

t	0	1	2	3
$f(t)$	1,000	1,060	1,123.60	1,191.02

c. Evaluating the function when $t = 4$, we have

$$f(4) = 1,000(1.06)^4$$
$$= \$1,262.48.$$

PROGRESS CHECK 9
$2,000 is invested at 5 percent interest compounded annually.

a. Describe in functional notation the value of the investment at the end of t years.
b. Construct a table that shows corresponding values for t and $f(t)$. Let t vary from 0 to 4 in increments of 1.
c. How much is the investment worth in 5 years? ■

The procedure from Example 9 can be generalized to obtain the following formula for the compounded amount A when an original principal P is compounded annually for t years at annual interest rate r:

$$A = P(1+r)^t$$

The next example makes use of this formula.

Example 10: Finding an Interest Rate

At what interest rate compounded annually must a sum of money be invested if it is to double in 9 years?

Solution: When an original principal P has doubled, then $A = 2P$. Replace A by $2P$ and t by 9 in the formula for interest compounded annually, and solve for r.

$$2P = P(1+r)^9$$
$$2 = (1+r)^9$$
$$2^{1/9} = 1 + r \qquad \text{Raise both sides to the reciprocal power.}$$
$$2^{1/9} - 1 = r$$

By calculator, $2^{1/9} - 1 \approx 0.0800597$, so the required interest rate is about 8.01 percent.

PROGRESS CHECK 10

At what interest rate compounded annually must a sum of money be invested if it is to triple in 16 years?

■

Example 11: Finding a Formula for Depreciation

A company purchases a new machine for $3,000. The value of the machine depreciates at a rate of 10 percent each year.

a. Find a formula (rule) showing the value (y) of the machine at the end of x years.

b. Construct a table that shows corresponding values for x and y. Let x vary from 0 to 3 in increments of 1.

c. What is the machine worth after 4 years?

Solution:

a. If the machine depreciates 10 percent each year, at the end of a year the machine is worth 90 percent of the value at which it began the year. Thus, the value is $3,000(0.9) after 1 year; 3,000(0.9)(0.9), or $3{,}000(0.9)^2$, after 2 years; and

$$y = 3{,}000(0.9)^x \qquad \text{after } x \text{ years.}$$

b. Using the formula from part a, we get the following table

x	0	1	2	3
y	3,000	2,700	2,430	2,187

c. Evaluating the function when $x = 4$, we have

$$y = 3.000(0.9)^4$$
$$= \$1{,}968.30.$$

Note: In the previous example the formula $A = P(1+r)^t$ was used to analyze exponential growth, where r is the growth rate. To analyze exponential decay, this formula may be modified to

$$A = A_0(1-r)^t$$

where A is the amount at time t, A_0 is the initial amount, and r is the decay rate. From this viewpoint, the depreciation formula in this example is given by

$$y = 3{,}000(1-0.10)^x = 3{,}000(0.9)^x.$$

PROGRESS CHECK 11

A rented property is valued at \$50,000 in January 1996. For tax purposes it is depreciated at a rate of 3.5 percent each year.

a. Find a formula showing the value (y) of the property after x years have elapsed.

b. Construct a table that shows corresponding values for x and y. Let x vary from 0 to 4 in increments of 1.

c. What is the property valued at in January 2003? ■

In this section the applications describe situations in which no change is considered to have occurred until an entire time period has elapsed. Thus, the \$1,000 investment in Example 9 is compounded annually and does not change until the end of the year. For this reason, the domain of these functions is the set of nonnegative integers. In Section 7.6 we discuss phenomena that change continuously. The domain of the function in these applications is usually the set of nonnegative real numbers.

EXPLORE 7.2

1. Investigate the behavior of the family of functions $y = b^x$ as the value of b changes. In each case describe important features that remain unchanged as well as those features that change.
   - **a.** Use various values of $b > 1$.
   - **b.** Use various values of b between 0 and 1.
   - **c.** We have not defined the exponential function for negative values of b. What will your calculator draw for $y = (-2)^x$? Use several different windows in this exploration, including Zoom Decimal if available.
2. Any exponential function $y = b^x$ with $b > 1$ will eventually grow very rapidly and will surpass any polynomial function.
   - **a.** $y = x^2$ and $y = 2^x$ intersect in 3 points. Find them. (**Caution:** A calculator Solver can only find one point of intersection at a time; it will find the one closest to your initial guess.) When does $y = 2^x$ permanently surpass $y = x^2$?
   - **b.** When does $y = 2^x$ permanently surpass $y = x^6$?
   - **c.** Compare the graphs of $y = 2^x + 2^{-x}$ and $y = 2^{|x|}$. Do they intersect? Explain how you can be sure.
3. Exponential growth is characterized by constant doubling time. For each given base b, a particular change in x will cause y to double. Examine the graphs of these exponential functions to estimate their doubling time.
   - **a.** $y = 1.5^x$
   - **b.** $y = 1.0185^x$ (This graph approximates current world population growth, where x is in years).
   - **c.** $y = 5^x$
4. Use a grapher to solve these equations; give solutions to the nearest tenth.
   - **a.** $2^x = 5$
   - **b.** $0.4^x = 3$
   - **c.** $3^{2x} = 2$
   - **d.** $-4^{3x} = -5$
   - **e.** $3^x = x^2$
   - **f.** $4^x = x^4$
5. Your grapher may be able to do **exponential regression**. This feature finds the formula for an exponential function that comes as close as possible to a given set of points. Use exponential regression to find a formula of the form $y = ab^x$ for a curve that approximates the pattern given by these points. (The correct formula is $y = 4447.148(1.018)^x$ with the constants rounded to 3 decimal places.) To see an accurate graph of the regression curve, use a Draw Regression feature that maintains all the decimal places in the coefficients.

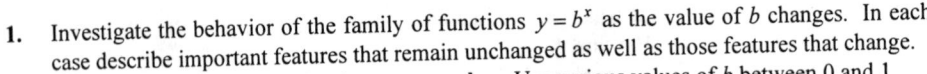

x	3	4	5	6	7	8	9	10
y	4686	4768	4852	4937	5024	5113	5203	5293

For this data, x represents years elapsed since 1980, and y is the world population in millions.
   - **a.** Use this formula to predict the world population in 2000.
   - **b.** According to this formula, when will the population be double its 1990 size?

EXERCISES 7.2

In Exercises 1–4 evaluate each function for the given element of the domain. Approximate irrational number results to the nearest hundredth.

1. $f(x)=4^x$; $f(2)$; $f(-1)$; $f\left(\frac{3}{2}\right)$; $f(\sqrt{2})$

2. $g(x)=9^x$; $g(0)$; $g\left(-\frac{1}{2}\right)$; $g(3)$; $g(\sqrt{2})$

3. $h(x)=\left(\frac{2}{3}\right)^x$; $h(4)$; $h(-3)$; $h(0)$; $h(\pi)$

4. $f(x)=8^{-x}$; $f(1)$; $f(-2)$; $f\left(\frac{2}{3}\right)$; $f(\pi)$

In Exercises 5–20 solve each equation

5. $3^{x-1}=3^5$ 6. $2^{2x}=2^6$ 7. $9^x=3$

8. $8^x=2$ 9. $2^x=\frac{1}{8}$ 10. $6^x=\frac{1}{36}$

11. $\left(\frac{3}{4}\right)^x=\frac{16}{9}$ 12. $\left(\frac{2}{3}\right)^x=\frac{27}{8}$

13. $3^{-x}=\frac{1}{81}$ 14. $2^{3-x}=1$ 15. $4^{x+3}=\sqrt{2}$

16. $9^{2x-1}=\sqrt[3]{3}$ 17. $4^x=8^{1-x}$ 18. $27^x=9^{x+2}$

19. $\left(\frac{1}{16}\right)^{x-1}=32^x$ 20. $\left(\frac{1}{64}\right)^{2-x}=\left(\frac{1}{32}\right)^{2x}$

In Exercises 21–32 fill in the missing component of each of the ordered pairs that makes the pair a solution of the given equation. Also, choose a convenient scale for the vertical axis and graph each function.

21. $y=2^x$; $(-3,\ )$, $(\ ,1)$, $\left(\ ,\frac{1}{2}\right)$, $(2,\ )$

22. $y=\left(\frac{1}{2}\right)^x$; $(0,\ )$, $(2,\ )$, $(\ ,2)$, $\left(\ ,\frac{1}{2}\right)$

23. $y=\left(\frac{1}{10}\right)^x$; $(\ ,0.01)$, $(\ ,1)$, $(-1,\ )$, $(-2,\ )$

24. $g(x)=4^{-x}$; $(2,\ )$, $(\ ,0.5)$, $(-0.5,\ )$, $(\ ,4)$

25. $y=4^x$; $\left(\frac{1}{2},\ \right)$, $(\ ,1)$, $(1,\ )$, $\left(\ ,\frac{1}{2}\right)$

26. $y=-5^x$; $(0,\ )$, $(1,\ )$, $(-1,\ )$, $(\ ,-0.04)$

27. $h(x)=\left(\sqrt{2}\right)^x$; $(0,\ )$, $(-2,\ )$, $(1,\ )$, $(\ ,2)$

28. $g(x)=\left(\frac{1}{\sqrt{3}}\right)^x$; $(0,\ )$, $(-2,\ )$, $\left(\ ,\frac{1}{3}\right)$, $(\ ,9)$

29. $f(x)=3(2)^x$; $(-1,\ )$, $(0,\ )$, $(1,\ )$, $(2,\ )$

30. $f(x)=100(1.06)^x$; $(-1,\ )$, $(0,\ )$, $(1,\ )$, $(2,\ )$

31. $y=3^x+3^{-x}$; $(-1,\ )$, $(0,\ )$, $(1,\ )$, $(2,\ )$

32. $y=2^x-2^{-x}$; $(-1,\ )$, $(0,\ )$, $(1,\ )$, $(\ ,0)$

In Exercises 33–42 graph each function. Also determine the domain and range of the function in each case.

33. $f(x)=3^{-x}$ 34. $y=-3^x$

35. $y=1-3^{-x}$ 36. $y=3^x-1$

37. $y=2^{x-2}$ 38. $y=2^{2-x}$

39. $h(x)=-\left(\frac{1}{2}\right)^{-x}$ 40. $y=\left(\frac{1}{2}\right)^{x+3}$

41. $y=2^x+2^{-x}$ 42. $f(x)=2^{|x|}$

In Exercises 43–52 find the base of the exponential function $y=b^x$ that contains the given point.

43. $(1, 4)$ 44. $(2, 9)$ 45. $(-1, 3)$

46. $\left(-2,\frac{1}{25}\right)$ 47. $\left(\frac{1}{2},4\right)$ 48. $(0.5, 10)$

49. $\left(\frac{2}{3},\frac{1}{4}\right)$ 50. $\left(\frac{3}{2},27\right)$ 51. $\left(-\frac{1}{3},2\right)$

52. $\left(-\frac{2}{3},\frac{1}{9}\right)$

53. Show that if $f(x)=b^x$ is the exponential function with base b, then $f(x_1+x_2)=f(x_1)\cdot f(x_2)$.

54. Show that if $f(x)=b^x$ then $f(x_1-x_2)=\dfrac{f(x_1)}{f(x_2)}$.

55. A biologist has 500 cells in a culture at the start of an experiment. Hourly readings indicate that the number of cells is doubling every hour.

a. Find a formula (rule) showing the number of cells present at the end of t hours.
b. How many cells are present at the end of 4 hours?
c. At the end of how many hours are 32,000 cells present?

56. A sheet of paper is 0.005 in. thick. Every time it is folded in half, the thickness doubles.
a. Find a formula showing the thickness after n folds.
b. How thick is it after 6 folds?
c. How many folds are needed for the thickness to exceed 1 inch?

57. \$100 is invested at 5 percent compounded annually.
a. Find a formula showing the value of the investment at the end of t years.
b. Construct a table that shows corresponding values for t and y. Let t vary from 0 to 4 in increments of 1.
c. How much is the investment worth at the end of 3 years?

58. The market value of a company's investment is \$100,000. This value increases at a rate of 10 percent each year.
a. Describe in functional notation the market value $f(t)$ of the investment t years from now.
b. Construct a table that shows corresponding values for t and $f(t)$. Let t vary from 0 to 4 in increments of 1.
c. What should be the market value of the investment 2 years from now?

59. A company purchases a machine for \$10,000. The value of the machine depreciates at a rate of 20 percent each year.
a. Find a formula to show the value of the machine after t years.
b. What will be the value of the machine 3 years from now?

60. After purchasing it for \$110,000 a landlord depreciates a rental house at a rate of 12 percent each year.
a. Find a formula to show the value of the property after t years.
b. What will be the value of the property after 6 years?

61. A new element has a half-life of 30 minutes; that is, if x oz of the element exist at a given time, $\frac{x}{2}$ oz exist 30 minutes later. The other half disintegrates into another element. Suppose that 1 g of the element is present 2 hours after the start of an experiment.
a. Find a formula showing the amount of the element present t hours after the start of the experiment.
b. How much of the element was present 1 hour after the start of the experiment?

62. A biologist grows a colony of a certain kind of bacteria. It is found experimentally that $N = N_0 3^t$, where N represents the number of bacteria present at the end of t days, N_0 is the number of bacteria present at the start of the experiment. Suppose that there are 153,000 bacteria present at the end of 2 days.
a. How many bacteria were present at the start of the experiment?
b. How many bacteria are present at the end of 4 days?
c. At the end of how many days are 459,000 bacteria present?

63. In 1776 a family lent the American government \$450,000. The loan was not repaid, and in 1990 the descendants sued for repayment.
a. If the loan had been issued at 6 percent simple annual interest, what amount would have been due in 1990? Use $A = P(1+rt)$.
b. What if it had been issued at 6 percent interest compounded annually?

64. You purchase a home for \$150,000 and take out a 30-year mortgage. If the annual inflation rate remains fixed at 5 percent and the value of the home keeps pace with inflation, then what is the value of your home by the time the mortgage is paid off?

65. At what interest rate compounded annually must a sum of money be invested if it is to double in 5 years?

66. At what interest rate compounded annually must a sum of money be invested if it is to triple in 10 years?

67. At what annual rate must a population grow if it is to double in 40 years?

68. At what annual rate must a population grow if it is to double in 100 years?

69. The average cost for photovoltaic modules (cells that convert light to electricity) has been dropping. A graph of the data suggests that a decreasing exponential function would be a reasonable model for the pattern. See the following figure. A formula that matches these data values pretty closely is $y = 53.8(0.84)^x$, where x is the number of years elapsed since 1975, and y is the average cost of a PV cell in dollars per watt. These data are taken from *Vital Signs 1993* (Worldwatch Institute). Use the

formula to estimate the average cost of a PV module in 2005.

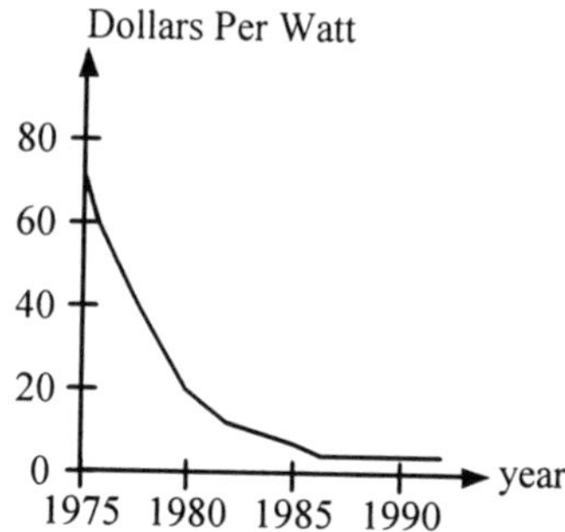

70. The table shows the observed density of a colony of Salmonella bacteria as it grows in a laboratory medium. An exponential function that approximates the growth pattern is $y = 1.06 \times 10^8 (1.86)^x$, where x is the number of hours elapsed since the first observation, and y is the density of the culture in cells per ml. Estimate the density after 10 hours, assuming that this model will still hold.

x	1	2	3	4
y	1.9×10^8	3.6×10^8	6.9×10^8	1.3×10^9

x	5	6	7	8
y	2.5×10^9	4.7×10^9	8.5×10^9	1.4×10^{10}

In Exercises 71–74 the table was generated by using an exponential function. Recall that in any exponential table a fixed change in x produces a constant ratio between the corresponding y values.

a. Find b, the value of the ratio when the change in x is 1.

b. Use the value of b and the value of y when $x = 0$ to find an equation of the form $y = ab^x$ that will generate the given table. Confirm that the formula is correct by regenerating the table.

71.

x	0	1	2	3	4	5
y	1000	2000	4000	8000	16,000	32,000

72.

x	0	1	2	3	4	5
y	1000	500	250	125	62.5	31.25

73.

x	0	2	4	6	8	Hint: Solve $10b^2 = 30$
y	10	30	90	270	810	

74.

x	0	3	6	9	12	Hint: Solve
y	80	20	5	1.25	0.3125	$80b^3 = 20$

In Exercises 75–80 examine the given table and decide if the table represents a linear function, an exponential function, or neither. For the linear and exponential cases, find a formula that will generate the tabled values. Confirm that the formula is correct by regenerating the table.

75.

x	0	1	2	3	4
y	1.2301	1.4602	1.6903	1.9204	2.1505

76.

x	0	1	2	3	4
y	5.2	5.0	4.8	4.6	4.4

77.

x	0	1	2	3	4
y	1.2	1.44	1.728	2.0736	2.48832

78.

x	0	1	2	3	4
y	5.1	2.04	0.816	0.3264	0.13056

79.

x	0	1	2	3	4
y	1.2	1.3	1.6	2.1	2.8

80.

x	0	1	2	3	4
y	5.5	5.4	5.1	4.6	3.9

In Exercises 81 and 82 determine the exponential function whose graph is shown.

81.

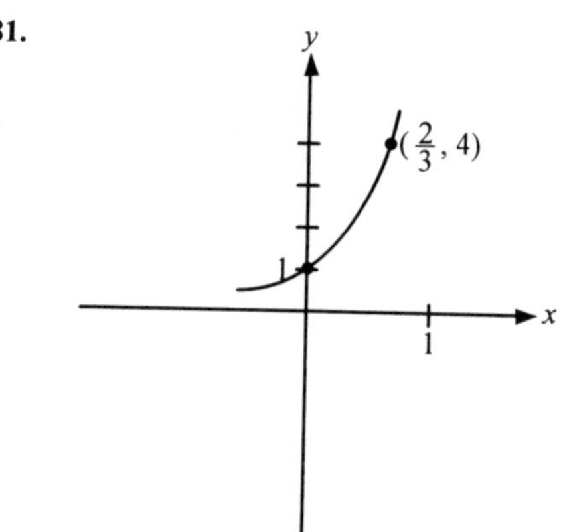

82.

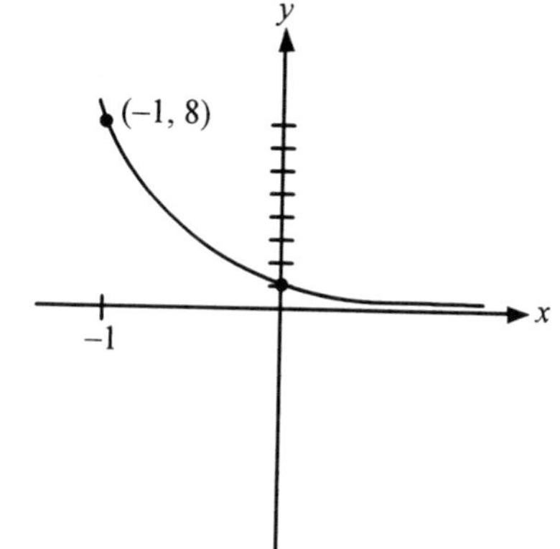

THINK ABOUT IT 7.2

1. Without using a calculator, place the following numbers in their correct numerical order: $5^{\sqrt{2}}$, 5^{π}, 0.04, $\sqrt[3]{5}$, $(1/5)^{1.3}$. Justify your answer in terms of the graph of $y = 5^x$.
2. Graph $y = (1/5)^x$, $y = (1/3)^x$, $y = 3^x$, and $y = 5^x$ on the same coordinate system. Use these graphs to explain why $a^x = b^x$ implies $a = b$, if a and b are positive real numbers with $x \neq 0$.
3. There is often more than one type of function whose graph will contain certain given points.
   - **a.** Given the points (0, 1) and (1, 4), find a linear function, a quadratic function, and an exponential function whose graphs contain both points.
   - **b.** For each function in part **a**, find y when $x = 2$.
4. **a.** If $f(x) = 10^x$, find $f^{-1}(100)$. **b.** If $f(x) = 8^x$, find $f^{-1}(4)$.
   c. If $f(x) = 9^x$, find $f^{-1}(3)$.
5. **a.** Graph $f(x) = 10^x$ and then use this graph to explain why the inverse of *f* is a function.
   b. Use the graph of $y = f(x)$ in part **a** to graph $y = f^{-1}(x)$.
   c. Find the domain and range of *f*. **d.** Find the domain and range of f^{-1}.

● ● ●

7.3 Logarithmic Functions

*Photo Courtesy of **Glenn Kulbako**, The Picture Cube*

The most widely used unit in communications engineering is the decibel (dB), named after Alexander Graham Bell. The decibel gain G for an amplifier is defined by the equation

$$G = 10\log\frac{P_2}{P_1},$$

where P_1 is the input power and P_2 is the output power. When rating an amplifier from a radio or public address system, P_1 assumes the arbitrary reference level value of 6 milliwatts (or 0.006 watt). What is the gain in decibels of a 60-watt amplifier?
(See Example 9.)

Objectives

1. Convert from the exponential form $b^L = N$ to the logarithmic form $\log_b N = L$ and vice versa.
2. Determine the value of the unknown in an expression of the form $\log_b N = L$
3. Determine the common logarithm and antilogarithm of a number by using a calculator.
4. Graph logarithmic functions.
5. Solve applied problems involving logarithmic functions.

From Section 7.2 we know that if a penny doubles in value each day, then the formula $V = 2^t$ gives the value (in cents) of the penny after t days have elapsed. This formula is easy to apply to find V for a given value of t. However, to solve the inverse problem of finding how much time is required for V to grow to a given value, it is more useful to have a formula that gives t as a function of V. To write such a formula we must introduce the concept of a logarithm.

The **logarithm** (abbreviated **log**) of a number is the *exponent* to which a fixed base is raised to obtain the number. The exponential statement $2^3 = 8$ is written in logarithmic form as $\log_2 8 = 3$, and we say that 3 is the logarithm to the base 2 of 8. In general, the key relation between exponential form and logarithmic form is expressed in the following definition.

Definition of Logarithm

> If b and N are positive numbers with $b \neq 1$, then
>
> $$\log_b N = L \text{ is equivalent to } b^L = N.$$

In this definition it is important to observe that a logarithm is an exponent, as the following diagram emphasizes.

logarithm or exponent

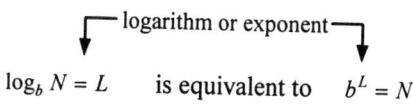

$$\log_b N = L \quad \text{is equivalent to} \quad b^L = N$$

With the aid of this definition we may now solve $V = 2^t$ for t, since

$$2^t = V \text{ is equivalent to } \log_2 V = t.$$

Examples 1 and 2 give further illustrations of converting between exponential form and logarithmic form using this definition.

Example 1: Converting Equations from Exponential to Logarithmic Form

Write $4^2 = 16$ and $b^r = s$ in logarithmic form.

Solution: Since $b^L = N$ implies $\log_b N = L$,

$$4^2 = 16 \text{ may be written as } \log_4 16 = 2$$
$$b^r = s \text{ may be written as } \log_b s = r.$$

PROGRESS CHECK 1
Write $10^2 = 100$ and $a^x = 4$ in logarithmic form. ■

Example 2: Converting Equations from Logarithmic to Exponential Form

Write $\log_2 \frac{1}{8} = -3$ and $\log_a u = v$ in exponential form.

Solution: Since $\log_b N = L$ implies $b^L = N$,

$$\log_2 \frac{1}{8} = -3 \text{ may be written as } 2^{-3} = \frac{1}{8}$$
$$\log_a u = v \text{ may be written as } a^v = u.$$

PROGRESS CHECK 2
Write $\log_9 3 = \frac{1}{2}$ and $\log_b x = y$ in exponential form. ■

In expressions of the form $\log_b N = L$, the value of L, N, and b may be determined provided two of these numbers are known, as shown next.

Example 3: Finding an Unknown in $\log_b N = L$

Determine the value of the unknown in each expression.

a. $\log_3 1 = y$ **b.** $\log_9 3 = n$ **c.** $\log_{1/2} x = -2$ **d.** $\log_b 8 = -3$

Solution: In each case it is helpful to first convert (at least mentally) the expression from logarithmic form to exponential form.

a. $\log_3 1 = y$ is equivalent to $3^y = 1$. Because 3 raised to the zero power is 1,

$$y = \log_3 1 = 0.$$

b. $\log_9 3 = n$ is equivalent to $9^n = 3$. If the solution, $n = \frac{1}{2}$, is not apparent in this form, then rewrite both sides of the equation in terms of base 3 and equate exponents as shown below.

$$\begin{aligned} 9^n &= 3 \\ (3^2)^n &= 3^1 \\ 3^{2n} &= 3^1 \\ 2n &= 1 \qquad \text{Since } b^x = b^y \text{ implies } x = y. \\ n &= \frac{1}{2} \end{aligned}$$

Thus, $n = \log_9 3 = \frac{1}{2}$.

c. $\log_{1/2} x = -2$ is equivalent to $\left(\frac{1}{2}\right)^{-2} = x$. Thus,

$$x = \left(\frac{1}{2}\right)^{-2} = \left(\frac{2}{1}\right)^2 = 4$$

d. $\log_b 8 = -3$ is equivalent to $b^{-3} = 8$. To find b, raise both sides of the equation to the reciprocal power of -3 (which is $-\frac{1}{3}$) to get

$$(b^{-3})^{-1/3} = 8^{-1/3},$$

and then simplify to obtain

$$b = 8^{-1/3} = \frac{1}{8^{1/3}} = \frac{1}{\sqrt[3]{8}} = \frac{1}{2}$$

Thus, $b = \frac{1}{2}$.

PROGRESS CHECK 3

Determine the value of the unknown in each expression.

a. $\log_4 4^3 = n$ **b.** $\log_8 16 = x$ **c.** $\log_{10} x = -3$ **d.** $\log_b 8 = \frac{3}{2}$ ■

In applications of logarithms, two specific bases are most prevalent: base 10 or **common logarithms** and base e or **natural logarithms**. It is standard notation to write a common logarithm as $\log N$ (with base 10 being understood) and a natural logarithm as $\ln N$ (with base e being understood). Thus, a typical graphing calculator has two different logarithm keys that are labeled LOG and LN. The next two examples show how to work with common logarithms on a calculator. Logarithms to the base e are used extensively in calculus and will not be considered until Section 5.6 where the irrational number e is developed.

Example 4: Finding A Common Logarithm

Evaluate log 12.3 to four decimal places.

Solution: To approximate log 12.3 with a graphing calculator simply press

[LOG] 12.3 [ENTER].

The display that results in Figure 7.19 shows $\log 12.3 = 1.0899$ to four decimal places.

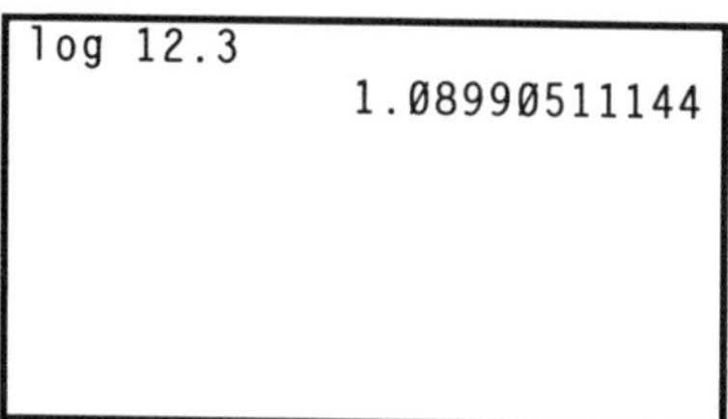

Figure 7.19

PROGRESS CHECK 4
Evaluate log 3.21 to four decimal places.

■

The next example asks that the answer be rounded off to a specified number of significant digits. Note that all digits, except the zeros that are used to indicate the position of the decimal point, are **significant digits**.

Example 5: Finding an Antilogarithm

If $\log N = -1.8$, find N to three significant digits.

Solution: $\log N = -1.8$ is equivalent to $10^{-1.8} = N$. Thus, N may be found as shown in Figure 7.20.

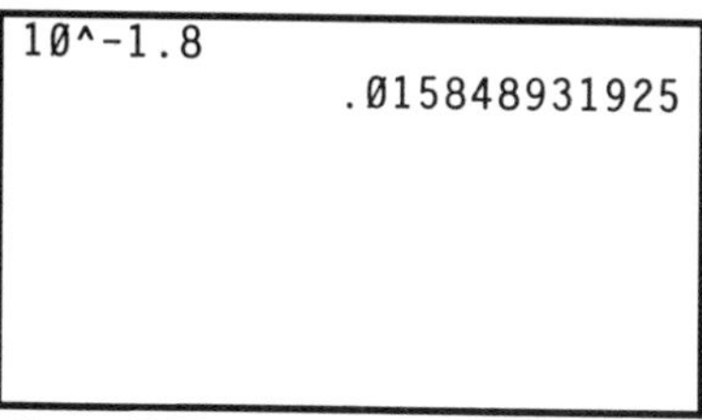

Figure 7.20

The display in Figure 7.20 shows $N = 0.0158$ to three significant digits. Observe that the zeros in this answer are not significant digits, because the number is 158 ten-thousandths, and the zeros are written to indicate the correct position of the decimal point.
Note: In science undoing a common logarithm is usually called taking an antilogarithm. In this convention,

$$\text{antilog}\, x = 10^x$$

PROGRESS CHECK 5
If $\log N = -4.2$, find N to three significant digits.

■

Inverse functions for exponential functions may now be defined using logarithms, and these functions may be evaluated and graphed by calculator.

$y = b^x$	Start with $y = f(x)$.
$x = b^y$	Interchange x and y.
$y = \log_b x$	Solve for y using the definition of logarithm.
$f^{-1}(x) = \log_b x$	Replace y by $f^{-1}(x)$.

Thus, $y = b^x$ and $y = \log_b x$ are inverse functions. Because the base in an exponential function must be a positive number other than 1, this same restriction applies to a logarithmic function.

To see another important restriction, consider that

$$y = \log_b x \text{ is equivalent to } x = b^y$$

Since a positive base raised to any power is positive, it is also necessary to require that x be positive. In other words, we must incorporate in the definition of a logarithmic function that we may only take the logarithms of positive numbers.

Logarithmic Function

If b and x are positive numbers with $b \neq 1$, then the function f defined by

$$f(x) = \log_b x$$

is called the **logarithmic function with base *b*.**

In Example 6 we compare the graphs of a logarithmic function and its inverse.

Example 6: Graphing a Logarithmic Function and Its Inverse

Graph $y = \log_2 x$. Then sketch the graphs of $y = \log_2 x$ and $y = 2^x$ on the same coordinate system, and describe how the graphs are related.

Solution: To graph $y = \log_2 x$, first construct a table of values, as shown below. To generate this table, rewrite

$$y = \log_2 x \text{ as } x = 2^y,$$

and then replace y with integer values from -3 to 3.

x	$\frac{1}{8}$	$\frac{1}{4}$	$\frac{1}{2}$	1	2	4	8
y	-3	-2	-1	0	1	2	3

By graphing these solutions and drawing a smooth curve through them, we obtain the graph of $y = \log_2 x$ shown in Figure 7.21.

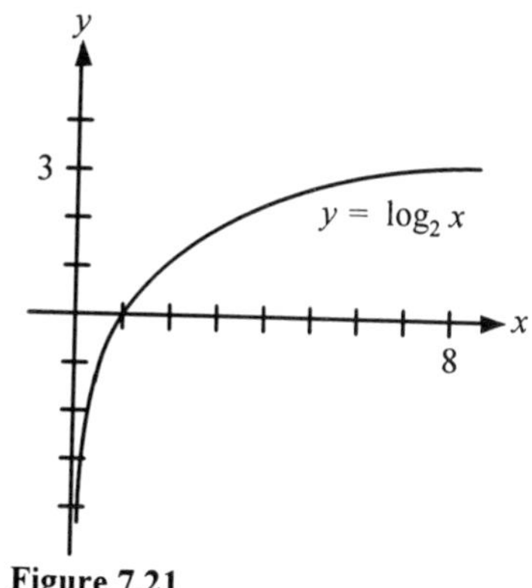

Figure 7.21

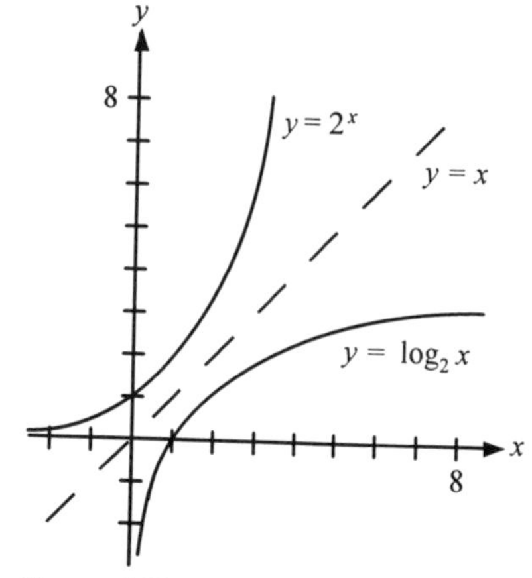

Figure 7.22

To relate this graph to the graph of its inverse $y = 2^x$, sketch both graphs on the same coordinate system, as in Figure 7.22. As expected, the graphs are related in that each is the reflection of the other about the line $y = x$.

Technology Link

One way to obtain the graph of $f(x) = \log_2 x$ on a graphing calculator is to use the **change of base formula**

$$\log_b x = \frac{\log_a x}{\log_a b}$$

which will be proved in Section 5.5. Based on this formula,

$$f(x) = \log_2 x = \frac{\log x}{\log 2} \text{ or } f(x) = \log_2 x = \frac{\ln x}{\ln 2}.$$

So Figure 7.23(a) shows an expression that may be entered to graph *f*. Figure 7.23(b) gives the graph that results in the viewing window $[-2, 8]$ by $[-4, 4]$.

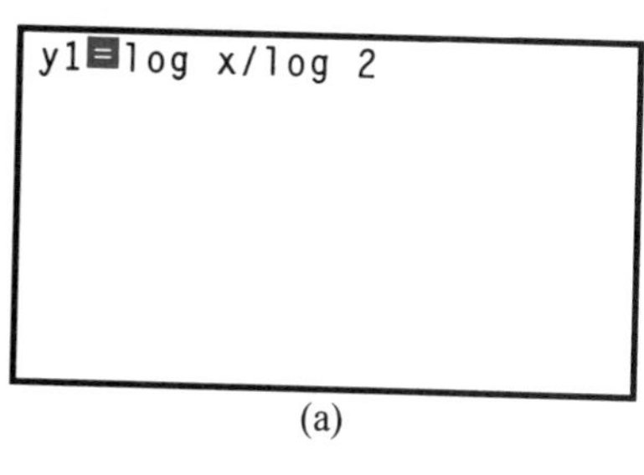

(a)

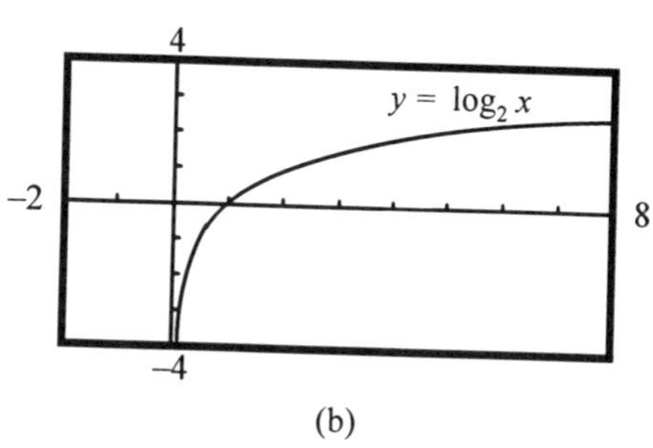

(b)

Figure 7.23

PROGRESS CHECK 6

Graph $y = \log_{1/2} x$. Then sketch the graphs of $y = \log_{1/2} x$ and $y = \left(\frac{1}{2}\right)^x$ on the same coordinate system, and describe how the graphs are related.

■

Two typical logarithm functions were graphed in Example 6 and "Progress Check" Exercise 6:

$$y = \log_2 x \text{ in which } b > 1,$$
$$y = \log_{1/2} x \text{ in which } 0 < b < 1.$$

In general, the graphs of $y = \log_b x$ and $y = b^x$ for these two cases are as shown in Figure 7.24. From these graphs some properties of the logarithmic function with base b are apparent.

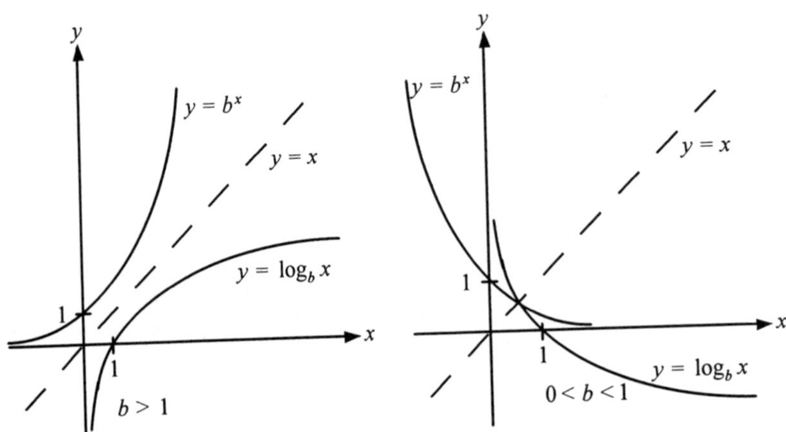

Figure 7.24

Properties of $f(x) = \log_b x$ (with $b > 0$, $b \neq 1$) and Its Graph

1. The domain of $y = \log_b x$ (which is the range of $y = b^x$) is $(0, \infty)$.
2. The range of $y = \log_b x$ (which is the domain of $y = b^x$) is $(-\infty, \infty)$.
3. The graph of $y = \log_b x$ has the x-intercept (1,0), and the y-axis is a vertical asymptote.
4. The graph of $y = \log_b x$ is the reflection of the graph of $y = b^x$ about the line $y = x$.
5. If $b > 1$, then as x increases, y increases. If $0 < b < 1$, then as x increases, y decreases.

Graphs of common logarithmic functions are easily shown by using the LOG key on a graphing calculator. An exploration involving such functions is discussed in Example 7.

Example 7: Using a Grapher to Explore Common Logarithmic Functions

Graph the functions $y1 = f(x) = \log x$, $y2 = g(x) = 3 + \log x$, and $y3 = h(x) = \log(x + 3)$ on $[-3, 5]$ by $[-2, 4]$. Use the Trace feature and explain how the graph of g and the graph of h are related to the graph of f.

Solution: The appropriate calculator displays for defining and picturing the functions are given in Figure 7.25.

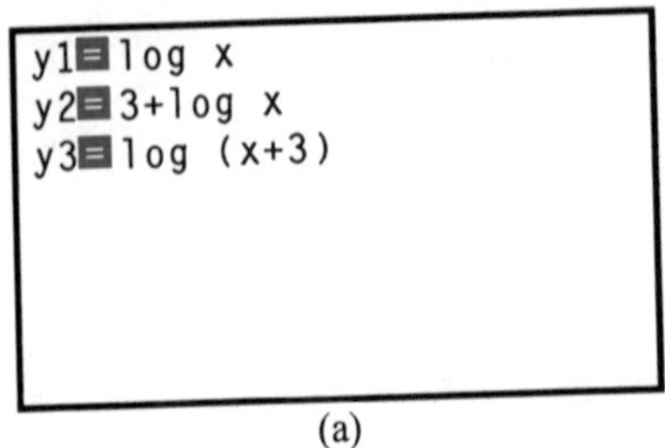

(a)

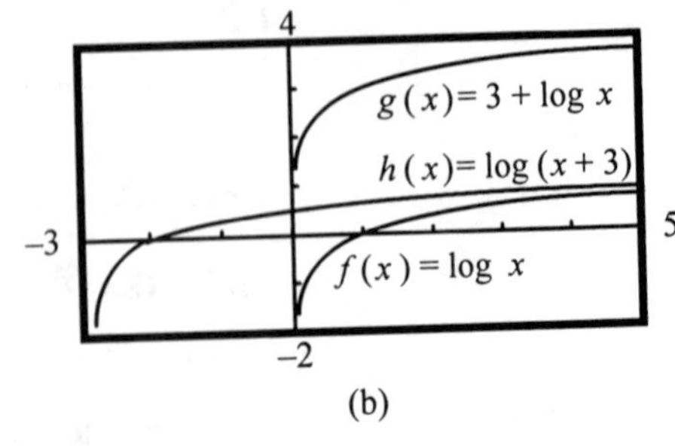

(b)

Figure 7.25

Use of the Trace feature suggests that the graph of *g* is the graph of *f* shifted 3 units up, while the graph of *h* is the graph of *f* shifted 3 units to the left. In fact these observations may be confirmed by the graphing techniques given in Section 1.3.

PROGRESS CHECK 7

Graph $y1 = f(x) = \log x$, $y2 = g(x) = -\log x$, and $y3 = h(x) = \log(x-2)$ on $[-1, 7]$ by $[-3, 3]$. Use the Trace feature and explain how the graph of *g* and the graph of *h* are related to the graph of *f*.

■

Example 7 illustrated that the graphing techniques of Section 1.3 may be used to graph certain variations of $y = \log x$. These techniques apply to any logarithmic function, and the next example illustrates the case of a horizontal shift of $y = \log_2 x$.

Example 8: Applying Graphing Techniques to Logarithmic Functions

Graph $y = \log_2(x+3)$. What is the domain of the function?

Solution: We start with the graph of $y = \log_2 x$ in Figure 7.21. To graph $y = \log_2(x+3)$, we note that x has been replaced by $x+3$, and the logarithmic rule is then applied to $x+3$. Such a replacement means that the graph of $y = \log_2(x+3)$ is the graph of $y = \log_2 x$ shifted 3 units to the left. The completed graph is shown in Figure 7.26, and we read from this graph that the domain of $y = \log_2(x+3)$ is $(-3, \infty)$. Without the graph, we may determine the domain by noting that logarithms are only defined for positive numbers. Thus, $x+3>0$, so $x>-3$.

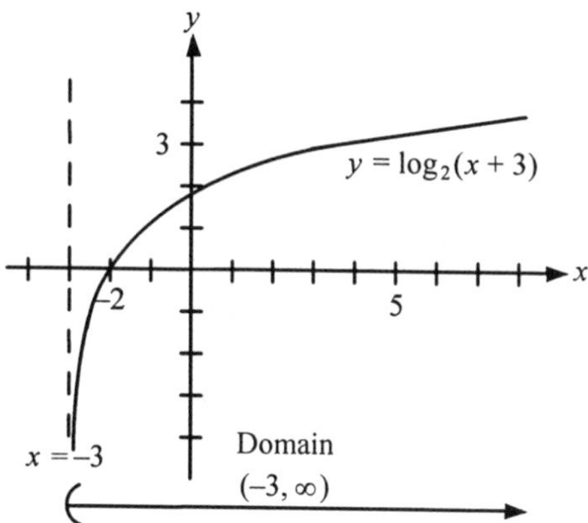

Figure 7.26

Technology Link

Figure 7.27 shows a complete graph of $y = \log_2(x+3)$ on a graphing calculator. To obtain this graph we set $y1 = \log(x+3)/\log 2$.

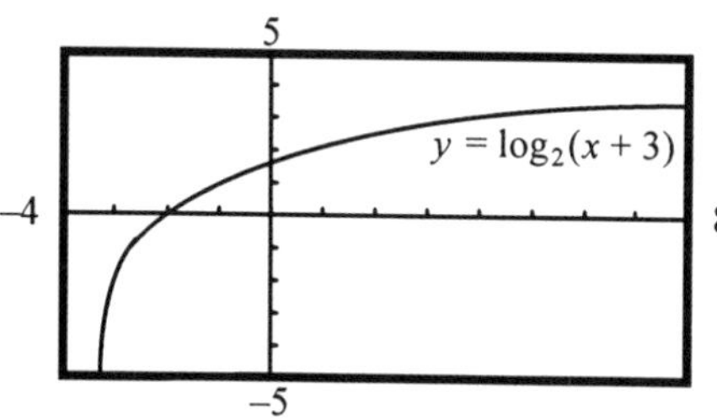

Figure 7.27

PROGRESS CHECK 8

Graph $y = \log_3(x+2)$. What is the domain of the function? ■

In the remaining examples we consider applications in which a relationship is analyzed by using a logarithmic function.

Example 9: Finding the Decibel Gain of an Amplifier

Solve the problem in the section introduction on page 554.

Solution: Replace P_2 by 60 and P_1 by 0.006 in the given formula and simplify by calculator.

$$\begin{aligned} G &= 10\log\frac{P_2}{P_1} \\ &= 10\log\frac{60}{0.006}. \\ &= 40 \end{aligned}$$

Thus, the decibel gain of a 60-watt amplifier is 40 dB.

PROGRESS CHECK 9

The loudness L of a sound in decibels is defined by the equation $L = 10\log(I/I_0)$ where I is the intensity of the sound and I_0 is an arbitrary reference level of 10^{-12} watts per square meter, which is the intensity of the faintest sound that can be heard. In a normal conversation the intensity of the sound is about 10^{-6} watts per square meter. How loud in decibels is this sound? ■

You are probably aware of the concept of an acid, since it is associated with such familiar and diverse topics as indigestion, shampoo, swimming pools, and car batteries. In chemistry a logarithm is used to define pH (hydrogen potential), which is a convenient measure of the acidity of a solution. Briefly, let's see why pH is defined as a logarithm.

When an atom of hydrogen loses an electron, it is called a hydrogen ion. Since this atom is positively charged, the hydrogen ion is symbolized H^+. The electrical imbalance in an ion greatly affects chemical reactions, and the concentration of hydrogen ions (symbolized $[H^+]$) in a solution determines its acidity. However, hydrogen ion concentrations are very small numbers. For example, a weak acid solution might have a concentration (measured in moles/liter) of only 1 part H^+ in 10,000 or $1/10{,}000$, or 10^{-4}. When writing such small numbers, the exponential form is obviously helpful. In 1909 a further simplification, pH notation, was introduced. With this notation only the exponent is considered, as follows:

$$\begin{aligned} \text{If } [H^+] &= 10^{-4} \text{ then pH} = 4 \\ [H^+] &= 10^{-8.3} \text{ then pH} = 8.3 \\ [H^+] &= 10^{-x} \text{ then pH} = x. \end{aligned}$$

Since pH is defined as an exponent, we determine pH by finding a logarithm. That is, from

$$[H^+] = 10^{-\text{pH}}$$

we may write

$$-\text{pH} = \log_{10}\left[\text{H}^+\right]$$
$$\text{pH} = -\log\left[\text{H}^+\right]$$

Thus, pH is defined as the negative of the logarithm of the hydrogen ion concentration. A solution with $\text{pH} = 7$ is called neutral. In acids $\text{pH} < 7$; bases or alkalies have $\text{pH} > 7$.

Example 10: Finding pH given $\left[\text{H}^+\right]$

The hydrogen ion concentration of distilled water is 10^{-7}.

a. What is the pH of distilled water?

b. If a solution has double the hydrogen ion concentration of distilled water, then find its pH to the nearest tenth.

Solution:

a. For distilled water, $\left[\text{H}^+\right] = 10^{-7}$, so

$$\text{pH} = -\log\left[\text{H}^+\right] = -\log 10^{-7} = -(-7) = 7.$$

So the pH of distilled water is 7.

b. Double the hydrogen ion concentration of distilled water may be represented as 2×10^{-7}. Then

$$\text{pH} = -\log\left[\text{H}^+\right] = -\log\left(2 \times 10^{-7}\right) \approx 6.7.$$

To the nearest tenth, the pH of the solution is 6.7. Note that as $\left[\text{H}^+\right]$ increases, the solution becomes more acidic and the pH becomes lower.

Technology Link

Remember to take advantage of the scientific notation capabilities of a calculator when finding pH. For instance in part b the expression $-\log\left(2 \times 10^{-7}\right)$ may be entered on a graphing calculator with the keystroke sequence

[(-)] [LOG] 2 [EE] [(-)] 7 [ENTER].

PROGRESS CHECK 10

The hydrogen ion concentration in a sample of nitric acid is 4.8×10^{-4}.

a. What is the pH of this sample?

b. If the sample is diluted so the concentration of hydrogen ions is halved, then find its pH to the nearest tenth.

■

Example 11: Finding $\left[\text{H}^+\right]$ given pH

The pH of tomato juice is about 4.5. Determine $\left[\text{H}^+\right]$ to two significant digits.

Solution: Replace pH by 4.5 in the formula for pH and solve for $\log\left[\text{H}^+\right]$.

$$\text{pH} = -\log\left[\text{H}^+\right]$$
$$4.5 = -\log\left[\text{H}^+\right]$$
$$-4.5 = \log\left[\text{H}^+\right]$$

Then $\log[H^+] = -4.5$ is equivalent to $10^{-4.5} = [H^+]$, so $[H^+]$ may be found by evaluating $10^{-4.5}$ on a calculator. The result for this calculation shows $[H^+] = 3.2 \times 10^{-5}$ (to two significant digits).

PROGRESS CHECK 11

To two significant digits determine $[H^+]$ for a sample of seawater whose pH is 8.7. ■

EXPLORE 7.3

1. Investigate the behavior of the family of functions $y = \log(cx)$ as the value of c changes. In each case describe important features that remain unchanged as well as those features that change.
   - **a.** Let $c = 1$, 10, 100, and 1000. Use Trace to check the vertical distance between curves.
   - **b.** Let $c = -1, -10, -100$, and -1000. What is the domain in these cases?
   - **c.** Describe the graph of $y = \log|cx|$. Use several values of c.
2. The equation $t = 768.5 + 125.6 \log P$ is based on the current growth rate of the world population, and this equation expresses the year t as a function of the world population P.
   - **a.** Make a table of ordered pairs from $P = 5$ billion to $P = 12$ billion in increments of 1 billion. Round off values for t to one decimal place.
   - **b.** Use scales [5 billion, 12 billion] by [1950, 2100] and draw a graph of this equation.
   - **c.** Analyze the interval between years in which the world population is adding a billion people and discuss what is happening to this interval.
3. Use the calculator to find these (common) logarithms, and describe the relation that is being illustrated.
   - **a.** $\log 2519$ **b.** $\log 251.9$ **c.** $\log 25.19$
   - **d.** $\log 2.519$ **e.** $\log 0.2519$ **f.** $\log 0.02519$
4. Log functions are used in applications where growth occurs slowly. In general the function $y = k \log x$ at some point grows more slowly than any power function $y = x^p$. This means that beyond a certain value of x the log graph remains below the graph of the power function, and falls further and further below it as x increases.
   - **a.** Compare the graphs of $y = \log x$ and $y = \sqrt{x}$. Confirm that the log graph is below the square root graph for all positive values of x.
   - **b.** Compare the graphs of $y = 3\log x$ and $y = \sqrt{x}$. For what values of x does the log graph fall below the square root graph and stay there permanently?
5. Use the grapher in parametric mode to graph $y = \log_2 x$ and its inverse. Check that the graph of $y = 2^x$ is the same as the graph of the inverse of $y = \log_2 x$.

EXERCISES 7.3

In Exercises 1-10 express in logarithm form.

1. $3^2 = 9$ **2.** $2^3 = 8$ **3.** $\left(\frac{1}{2}\right)^2 = \frac{1}{4}$

4. $\left(\frac{1}{3}\right)^3 = \frac{1}{27}$ **5.** $4^{-2} = \frac{1}{16}$

6. $\left(\frac{1}{2}\right)^{-3} = 8$ **7.** $25^{1/2} = 5$ **8.** $8^{-1/3} = \frac{1}{2}$

9. $7^0 = 1$ **10.** $10^{-4} = 0.0001$

In Exercises 11–20 express in exponential form.

11. $\log_5 5 = 1$ **12.** $\log_2 32 = 5$

13. $\log_{1/3} \frac{1}{9} = 2$ **14.** $\log_{1/2} \frac{1}{16} = 4$

15. $\log_2 \frac{1}{4} = -2$ **16.** $\log_{1/4} 4 = -1$

17. $\log_{49} 7 = \frac{1}{2}$ **18.** $\log_{27} \frac{1}{3} = -\frac{1}{3}$

19. $\log_{100} 1 = 0$ 20. $\log_{10} 0.001 = -3$

In Exercises 21–30 find the value of each expression.

21. $\log_3 9$ 22. $\log_2 16$ 23. $\log_4 4$

24. $\log_5 1$ 25. $\log_3 \frac{1}{3}$ 26. $\log_{10} 0.01$

27. $\log_3 \sqrt{3}$ 28. $\log_9 3$ 29. $\log_{10}(\log_5 5)$

30. $\log_2(\log_4 2)$

In Exercises 31–40 determine the value of the unknown by inspection or by writing in exponential form.

31. $\log_2 8 = y$ 32. $\log_4 2 = y$

33. $\log_5 x = -1$ 34. $\log_{10} x = -5$

35. $\log_b 125 = 3$ 36. $\log_b 10 = \frac{1}{2}$

37. $\log_{10} 10^3 = y$ 38. $\log_{10} 10^{2.4} = y$

39. $\log_b b = 1$ 40. $\log_b 1 = 0$

In Exercises 41-44 evaluate the logarithm to four decimal places

41. $\log 4.56$ 42. $\log 65.4$

43. $\log 218.7762$ 44. $\log(2.4 \times 10^{18})$

In Exercises 45–48 find N to three significant digits.

45. $\log N = -2.5$ 46. $\log N = 2.5$

47. $\log N = 3.14$ 48. $\log N = -3.14$

In Exercises 49–60 graph the function.

49. $y = \log_{10} x$ 50. $y = \log_3 x$

51. $y = \log_{1/4} x$ 52. $y = \log_{1/2} x$

53. $y = -\log_4 x$ 54. $y = -\log_{1/4} x$

55. $y = 1 - \log_2 x$ 56. $y = \log_2(-x)$

57. $y = \log_2(1 + x)$ 58. $y = 1 + \log_2 x$

59. $y = |\log_{10} x|$ 60. $y = \log_{10}|x|$

In Exercises 61–64 use a grapher to graph all three functions. Use the Trace feature and explain how the graphs of g and h are each related to the graph of f.

61. $f(x) = \log x$ $g(x) = 2 + \log x$
$h(x) = \log(x - 2)$

62. $f(x) = \log x$ $g(x) = -3 + \log x$
$h(x) = \log(x + 3)$

63. $f(x) = \log x$ $g(x) = 5 - \log x$
$h(x) = \log(5 - x)$

64. $f(x) = \log x$ $g(x) = -3 - \log x$
$h(x) = \log(-3 - x)$

In Exercises 65–72 use a grapher to graph the given function. Give the domain.

65. $y = \log_4(x + 1)$ 66. $y = \log_{40}(x + 1)$

67. $y = \log_{1/2}(x - 5)$ 68. $y = \log_{1/10}(x - 10)$

69. $f(x) = \log_{10}(2x - 1)$ 70. $g(x) = \log_2(x^2 - 9)$

71. $f(x) = \log_{10}|x - 1|$ 72. $f(x) = \log_{10} \sqrt{x + 1}$

73. Use the relationship $G = 10\log\frac{P_2}{P_1}$ to determine the gain in decibels of a 40-watt amplifier. Let P_1 be the standard reference power of 0.006 watts.

74. Use the relationship $L = 10\log\frac{I}{I_0}$ to determine the loudness in decibels of a sound whose intensity is 10 watts per square meter. (This is a commonly accepted value for the threshold of pain.) I_0 is an arbitrary reference value of 10^{-12} watts per square meter.

In Exercises 75 and 76 use the relationship:

$$\text{difference in loudness} = 10\log\frac{P_2}{P_1}.$$

75. Determine the difference in loudness (in decibels) between sounds 1 and 2 if the power ratio $\left(\frac{P_2}{P_1}\right)$ between the sounds is
a. 1 **b.** 10 **c.** 100 **d.** 1,000
As the power ratio is progressively multiplied by 10, how does the loudness in decibels increase?

76. In a normal conversation the power ratio $\frac{P_2}{P_1}$ between the highest and lowest sound volume is about 300 to 1. What is the range of speech in decibels? (**Note:** The decibel scale for loudness ranges from 0 dB for a sound at the threshold of hearing to about 140 dB for an airplane engine. Normal conversation is rated at about 60 dB.)

In Exercises 77–80 determine the pH of the solution with the given hydrogen ion concentration.

77. $\left[H^+\right]=10^{-7}$ (water)

78. $\left[H^+\right]=4.2\times10^{-3}$ (5-percent vinegar)

79. $\left[H^+\right]=2.5\times10^{-4}$ (orange juice)

80. $\left[H^+\right]=3.3\times10^{-8}$ (swimming pool water)

In Exercises 81–84 determine the hydrogen ion concentration of the solution with the given pH.

81. pH = 0 (pure sulfuric acid)

82. pH = 8.9 (seawater)

83. pH = 2.1 (stomach juices)

84. pH = 11.3 (household ammonia)

85. Some growth patterns have a generally logarithmic shape. The following table displays data that follow the growth of one person. A model that fits this pattern reasonably well is $y = 1.07 + 3.75\log x$.

Growth of a person

x	Age in years	3	6	9	12	15
y	Height in ft	3	3.75	4.5	5	5.8

x	Age in years	18	21	24	27
y	Height in ft	6.1	6.15	6.17	6.18

Use this logarithmic model to estimate the person's height at age 10.

86. In applied statistics it is sometimes preferable to work with linear relations. When the data initially exhibit an exponential relation they may be transformed by using logarithms to produce a more linear pattern.

a. The following table shows days of training (X) and Performance Score (Y) for 10 sales trainees. The data appear in the text *Applied Linear Statistical Models* (Neter and Wasserman, Irwin, 1990). Graph the points (X, Y) and note that the relation is not linear, but (somewhat) exponential.

X:	1	1	2	2	3	3
Y:	45	40	60	62	75	81

X:	4	5	5	5
Y:	115	150	145	148

b. Complete a new table replacing Y by $Y' = \log Y$ as shown next:

X:	1	1	2	2	3	3
Y':	1.65321	1.60206				

X:	4	5	5	5
Y':				

c. Graph the new points (X, Y') and confirm that the relation is more linear in appearance.

THINK ABOUT IT 7.3

1. For the logarithmic function defined by $y = \log_b x$, explain why the restriction $b \neq 1$ is necessary for this equation to define a function.
2. By calculator, find $10^{\log_{10} 3}$ and $10^{\log_{10} 5}$. Generalize to $b^{\log_b x}$, and explain why these answers make sense.
3. The Richter scale rating for an earthquake of intensity I equals $\log_{10}(I/I_0)$, where I_0 is a standard reference level number. How many times stronger is a quake rated at 6 than a quake rated at 2:
4. Recall the equation $\left[H^+\right]=10^{-pH}$ defines the pH of a solution.
   a. Show that when the pH is doubled, the hydrogen ion concentration is squared.
   b. In general, what happens to the hydrogen ion concentration when the pH is multiplied by n?
5. Determine the domain for each of these functions. Describe the pattern that evolves.
   a. $\log x$ **b.** $\log\log x$ **c.** $\log\log\log x$ **d.** $\log\log\log\log x$

• • •

7.4 Properties of Logarithms

The loudness L of a sound in decibels is defined by the equation $L = 10\log(I/I_0)$, where I is the intensity of the sound and I_0 is an arbitrary reference level of 10^{-12} watts per square meter, which is the intensity of the faintest sound that can be heard.

a. Show that $L = 10\log(I/10^{-12})$ and $L = 10\log I + 120$ are equivalent formulas for finding the loudness of a sound.

b. The intensity of the sound of riveting is about 10^{-2} watts per square meter. Find the loudness of this sound in decibels using both formulas to illustrate that the formulas are equivalent.

(See Example 6.)

Objectives

1. Use properties of logarithms to express certain log statements in terms of simpler logarithms or expressions.
2. Use properties of logarithms to convert certain statements involving logarithms to a single logarithm with coefficient 1.
3. Use properties of logarithms in applications.

Because a logarithm is an exponent, properties of logarithms follow from exponent properties. Three key exponent properties from Section R.3 that are the basis for the product, quotient, and power rules for logarithms may be stated as follows.

1. $b^m \cdot b^n = b^{m+n}$ Product property
2. $\dfrac{b^m}{b^n} = b^{m-n}$ Quotient property
3. $(b^m)^n = b^{mn}$ Power-to-a-power property

To each of these exponent properties, there corresponds a logarithm property, as stated next.

Product, Quotient, and Power Properties of Logarithms

If b, x, and y are positive numbers with $b \neq 1$, and k is any real number, then

1. $\log_b xy = \log_b x + \log_b y$ Product property
2. $\log_b \dfrac{x}{y} = \log_b x - \log_b y$ Quotient property
3. $\log_b x^k = k\log x$ Power-to-a-power property

To use the product property of exponents to prove the product property of logarithms, let

$$x = b^m \text{ and } y = b^n,$$

and observe that the respective logarithmic forms of these statements are

$$\log_b x = m \text{ and } \log_b y = n.$$

By the product rule of exponents,

$$x \cdot y = b^m \cdot b^n = b^{m+n},$$

and converting $xy = b^{m+n}$ to logarithmic form gives

$$\log_b xy = m + n.$$

Finally, substituting $\log_b x$ for m and $\log_b y$ for n yields the property

$$\log_b xy = \log_b x + \log_b y.$$

The quotient and power properties of logarithms may be established in similar ways (and these proofs are requested in Exercises 53 and 54).

One use of these properties is to convert certain logarithms to a sum, difference, or product involving simpler logarithms, as illustrated in Example 1.

Example 1: Converting to Expressions Involving Simpler Logarithms

Express each logarithm as a sum, difference, or product involving simpler logarithms.

a. $\log_b(4 \cdot 5)$ **b.** $\log_2\left(\frac{3}{7}\right)$ **c.** $\log_4\sqrt{5}$ **d.** $\log_b\left(\frac{xy}{z}\right)$ **e.** $\log_{10} 2\pi\sqrt{\frac{L}{g}}$

Solution:

a. $\log_b(4 \cdot 5) = \log_b 4 + \log_b 5$ Product property

b. $\log_2\left(\frac{3}{7}\right) = \log_2 3 - \log_2 7$ Quotient property

c. $\log_4\sqrt{5} = \log_4(5)^{1/2} = \frac{1}{2}\log_4 5$ $\sqrt{s} = s^{1/2}$; Power property

d. $\log_b\left(\frac{xy}{z}\right) = (\log_b x + \log_b y) - \log_b z$ Product and Quotient properties

e. $\log_{10} 2\pi\sqrt{\frac{L}{g}} = \log_{10} 2 + \log_{10}\pi + \log_{10}\sqrt{\frac{L}{g}}$ Product property

$= \log_{10} 2 + \log_{10}\pi + \frac{1}{2}\log_{10}\left(\frac{L}{g}\right)$ $\sqrt{\frac{L}{g}} = \left(\frac{L}{g}\right)^{1/2}$; Power property

$= \log_{10} 2 + \log_{10}\pi + \frac{1}{2}(\log_{10} L - \log_{10} g)$ Quotient property

Caution: Properties have been stated for the logarithm of a product, a quotient, and a power, but *not* for a *sum* or *difference*. In particular,

$\log_b(x + y)$ may *not* be replaced by $\log_b x + \log_b y$
and $\log_b(x - y)$ may *not* be replaced by $\log_b x - \log_b y$.

PROGRESS CHECK 1

Express each logarithm as a sum, difference, or product involving simpler logarithms.

a. $\log_b(3 \cdot 8)$ **b.** $\log_3\left(\frac{2}{5}\right)$ **c.** $\log_5\sqrt[3]{2}$ **d.** $\log_b\left(\frac{x}{2y}\right)$ **e.** $\log_{10}\frac{2\pi}{\sqrt{mn}}$ ■

Two additional properties that are often used to simplify logarithmic expressions are the direct result of the exponent laws $b^1 = b$ and $b^0 = 1$. Converting these expressions to logarithmic form gives the result that if b is a positive number other than 1, then

$$\log_b b = 1 \text{ and } \log_b 1 = 0.$$

For instance, $\log_{10} 10 = 1$ and $\log_2 1 = 0$. In the remaining examples, these two properties will be used to further simplify logarithmic expressions whenever applicable.

Example 2: Simplifications Involving $\log_b b = 1$ and $\log_b 1 = 0$

Express each logarithm as a sum, difference, or product involving simpler logarithms.

a. $\log_5 5x$ **b.** $\log_{10} 10^3$ **c.** $\log_b \frac{1}{4}$

Solution:

a. $\log_5 5x = \log_5 5 + \log_5 x$ Product property

$= 1 + \log_5 x$ $\log_b b = 1$

b. $\log_{10} 10^3 = 3\log_{10} 10$ Power property

$= 3(1)$ $\log_b b = 1$

$= 3$

c. $\log_b \frac{1}{4} = \log_b 1 - \log_b 4$ Quotient property

$= 0 - \log_b 4$ $\log_b 1 = 0$

$= -\log_b 4$

PROGRESS CHECK 2

Express each logarithm as a sum, difference, or product involving simpler logarithms.

a. $\log_{10} 10x$ **b.** $\log_4 4^2$ **c.** $\log_b \frac{1}{b}$ ■

Example 3: Finding Relationships Among Logarithm Values

If $\log_b 2 = m$ and $\log_b 3 = n$ express each of the following in terms of m and/or n.

a. $\log_b 24$ **b.** $\log_b \frac{1}{9}$

Solution:

a. Using factors of 2 and 3, we write 24 as $2^3 \cdot 3$. Then

$$\log_b 24 = \log_b\left(2^3 \cdot 3\right) = \log_b 2^3 + \log_b 3 = 3\log_b 2 + \log_b 3 = 3m + n.$$

b. Writing 9 as 3^2 and applying logarithm properties gives

$$\log_b\left(\frac{1}{3^2}\right) = \log_b 1 - \log_b 3^2 = 0 - \log_b 3^2 = -2\log_b 3 = -2n.$$

PROGRESS CHECK 3

If $\log_b 2 = m$ and $\log_b 3 = n$, express each of the following in terms of m and/or n.

a. $\log_b \frac{1}{2}$ **b.** $\log_b 72$ ■

There are two more important properties of logarithms that are often useful. We have been saying all along that a logarithm is an exponent. One way to express this idea is to say that

$$b^{\log_b x} = x$$

for all positive values of b and x. This property is merely a convenient way of restating the definition that $\log_b x$ is the exponent to which b is raised to obtain x. Similarly, since x is the exponent to which b is raised to obtain b^x, it follows that

$$\log_b b^x = x.$$

From a different viewpoint, these properties are a direct consequence of the inverse relation between exponential and logarithmic functions. That is, if we let $f(x) = b^x$, then $f^{-1}(x) = \log_b x$. From the definition of inverse functions in Section 7.1 we have

$$f\left[f^{-1}(x)\right] = f^{-1}[f(x)] = x.$$

Thus,

$$f\left[f^{-1}(x)\right] = b^{\log_b x} = x$$

and

$$f^{-1}[f(x)] = \log_b b^x = x.$$

Because of this relation, these two properties are referred to as **inverse properties**.

Inverse Properties

If b is a positive number with $b \neq 1$, then

1. $\log_b b^x = x$,
2. $b^{\log_b x} = x$, for $x > 0$.

Example 4: Using Inverse Properties

Simplify each expression.

a. $2^{\log_2 8}$ b. $\log_{10} 10^p$

Solution:

a. For $x > 0$, $b^{\log_b x} = x$, so $2^{\log_2 8} = 8$. We can verify this answer since $\log_2 8 = 3$ so that $2^{\log_2 8} = 2^3 = 8$.

b. Since $\log_b b^x = x$, we have $\log_{10} 10^p = p$. We also could reason

$$\log_{10} 10^p = p \log_{10} 10 = p(1) = p.$$

PROGRESS CHECK 4

Simplify each expression.

a. $\log\left(m \times 10^k\right)$ b. $10^{\log 1000}$ ■

To this point, logarithm properties have been used mainly to convert a single log statement to a sum, difference, or product that involved simpler logarithms. Depending on the application, it may be more useful to convert sums and differences of logarithms to a single logarithm with coefficient 1. This type of conversion is considered in the next example.

Example 5: Converting to a Single Logarithm

Express as a single logarithm with coefficient 1.

a. $\log_{10} 9 + \log_{10} 2$ b. $\log_5 x - \log_5 7$ c. $3\log_b x + 2\log_b y$

d. $\frac{1}{2}(\log_{10} x - 3\log_{10} y)$

Solution:

a. $\log_{10} 9 + \log_{10} 2 = \log_{10}(9 \cdot 2)$ — Product property

$= \log_{10} 18$

b. $\log_5 x - \log_5 7 = \log_5(x / 7)$ — Quotient property

c. $3\log_b x + 2\log_b y = \log_b x^3 + \log_b y^2$ — Power property

$= \log_b x^3 y^2$ — Product property

d. $\frac{1}{2}(\log_{10} x - 3\log_{10} y) = \frac{1}{2}(\log_{10} x - \log_{10} y^3)$ — Power property

$= \frac{1}{2}(\log_{10}(x/y^3))$ — Quotient property

$= (\log_{10}(x/y^3))^{1/2}$ or $\log_{10}\sqrt{x/y^3}$ — Power property

PROGRESS CHECK 5

Express as a single logarithm with coefficient 1.

a. $\log_b 12 - \log_b 4$ **b.** $\log_2 x + \log_2(x-1)$ **c.** $2\log_{10} x + \frac{1}{2}\log_{10} 9$

d. $\frac{1}{2}[\log_{10} L - \log_{10} g]$ ■

Applications of logarithm properties generally involve converting an expression to a form that is more useful for a particular problem. For example, converting to a single logarithm with coefficient 1 is particularly useful in the next section for solving logarithmic equations. The next two examples illustrate cases in which converting to a sum or difference of simpler logarithms is desirable.

Example 6: Using Logarithm Properties to Establish Alternate Formulas

Solve the problem in the section introduction on page 567.

Solution:

a. The two expressions for L are equivalent since

$$10\log\left(\frac{I}{10^{-12}}\right) = 10(\log I - \log 10^{-12}) \quad \text{Quotient property}$$
$$= 10[\log I - (-12)] \quad \log_b b^x = x$$
$$= 10\log I + 120$$

b. For the sound of riveting, $I = 10^{-2}$. Using $L = 10\log\left(\frac{I}{10^{-12}}\right)$ gives

$$L = 10\log\left(\frac{10^{-2}}{10^{-12}}\right) = 10\log 10^{10} = 10(10) = 100\text{dB}.$$

Using $L = 10\log I + 120$ gives

$$L = 10\log 10^{-2} + 120 = 10(-2) + 120 = 100\text{dB}.$$

Both formulas show that riveting has a loudness of 100 dB.

PROGRESS CHECK 6

In the previous section the formula $\mathrm{pH} = -\log[\mathrm{H}^+]$ was given to define the pH of a solution.

a. Show that $\mathrm{pH} = \log\dfrac{1}{[\mathrm{H}^+]}$ is equivalent to $\mathrm{pH} = -\log[\mathrm{H}^+]$.

b. For a sample of acid rain, $[\mathrm{H}^+] = 3.7 \times 10^{-4}$. Find the pH of the sample using both formulas to illustrate that the formulas are equivalent. ■

Example 7: Converting an Exponential Relation to a Linear Relation

In some applications, for mathematical simplicity, researchers try to reexpress the relation between two variables using a *linear* function. For instance, $y = 10 \cdot 2^x$ is an exponential relation typical of population growth, so y is *not* a linear function of x. However, show that by taking logarithms of both sides of $y = 10 \cdot 2^x$ you can express $\log y$ as a linear function of x. Graph this line, assuming $x \geq 0$.

Solution: We begin by using the fact that if two expressions are equal then their logarithms are also equal.

$$
\begin{aligned}
y &= 10 \cdot 2^x \\
\log y &= \log(10 \cdot 2^x) && \text{Apply common logarithms to each side.} \\
&= \log 10 + \log 2^x && \text{Product property} \\
&= \log 10 + x \log\ 2 && \text{Power property}
\end{aligned}
$$

Then $\log 10 = 1$ and $\log 2 \approx 0.3010$, so the relation is given by

$$\log y = 1 + 0.3010x.$$

This result shows that $\log y$ is a linear function of x, with slope $= 0.3010$ and intercept (0, 1) where the horizontal axis represents x and the vertical axis represents $\log y$. The graph is shown in Figure 7.28.

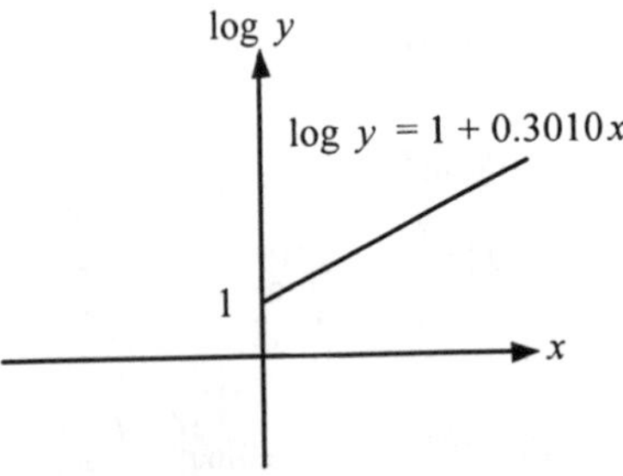

Figure 7.28

Note: Special graph paper, called **semilogarithmic graph paper**, may be used when the horizontal axis represents x and the vertical axis represents $\log y$.

PROGRESS CHECK 7

In a study involving exponential decay, researchers used the model $y = 1000 \cdot 2^{-x}$. Express $\log y$ as a linear function of x and graph the line assuming $x \geq 0$. ■

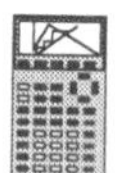

EXPLORE 7.4

1. Compare the graphs of these logarithmic functions to the graph of $y = \log x$ and describe any relations between them. Confirm the relation by referring to the product and quotient properties of logarithms. In general, what is the relation between the graphs of $y = \log x$ and $y = \log cx$, where c is any positive constant?
   a. $y = \log 10x$ b. $y = \log 100x$ c. $y = \log(x/10)$ d. $y = \log(x/100)$
2. Check graphically that the properties of logarithms hold for these functions.
   a. Show that the graph of $y = \log 5x$ appears to be the same as the graph of $y = \log 5 + \log x$.
   b. Show that the graph of $y = \log(x/50)$ appears to be the same as the graph of $y = \log x - \log 50$.
   c. Show that the graph of $y = \log x^3$ appears to be the same as the graph of $y = 3\log x$.
3. a. Verify graphically that $\log(3+x)$ is not equal to $\log 3 + \log x$ by showing that $y = \log(3+x)$ and $y = \log 3 + \log x$ have different graphs.
   b. Verify graphically that $\log(4-x)$ is not equal to $\log 4 - \log x$ by showing that $y = \log(4-x)$ and $y = \log 4 - \log x$ have different graphs.
4. a. Graph $y = \log x^2$. What is the domain?
   b. Graph $y = 2\log x$. What is the domain?
   c. What restrictions are necessary to say that $\log x^2$ is equal to $2\log x$?

EXERCISES 7.4

In Exercises 1–26 express each logarithm as a sum, difference, or product of simpler logarithms.

1. $\log_{10}(7 \cdot 5)$
2. $\log_b xyz$
3. $\log_6\left(\frac{3}{5}\right)$
4. $\log_b\left(\frac{x}{y}\right)$
5. $\log_5(11)^{16}$
6. $\log_b x^{16}$
7. $\log_2 \sqrt{3}$
8. $\log_b \sqrt{x}$
9. $\log_{10} \sqrt[5]{16}$
10. $\log_b \sqrt[5]{x}$
11. $\log_4(4^2 \cdot 3^3)$
12. $\log_b x^2y^3$
13. $\log_b \sqrt{xy}$
14. $\log_b \sqrt{\frac{x}{y}}$
15. $\log_b \sqrt[3]{x^5}$
16. $\log_b(x)^\pi$
17. $\log_b \sqrt[4]{\frac{xy^2}{z}}$
18. $\log_{10} \sqrt{s(s-a)(s-b)(s-c)}$
19. $\log_6 6x$
20. $\log_4 4^5 x$
21. $\log_b \frac{1}{10}$
22. $\log_b b^3$
23. $\log_2 8x^2$
24. $\log_3 27x^6$
25. $\log_2(2^{3/5})$
26. $\log_2(4^{1/3})$

27. If $\log_b 3 = m$ and $\log_b 4 = n$ express each of the following in terms of m and/or n.
   a. $\log_b 12$ b. $\log_b 6$ c. $\log_b \frac{1}{36}$
28. If $\log_a 2 = x$ and $\log_a 3 = y$, express each of the following in terms of x and/or y.
   a. $\log_a 27$ b. $\log_a 72$
   c. $\log_a \sqrt{\frac{1}{3}}$ d. $\log_a \sqrt[5]{\frac{2}{3}}$

In Exercises 29–36 simplify each expression.

29. $10^{\log 100}$
30. $10^{\log(1/10)}$
31. $\log_{10}(m \times 10^k)$
32. $3^{\log_3 2}$
33. $2^{\log_2 15 - \log_2 5}$
34. $b^{r \log_b N}$
35. $\frac{n}{x} b^{n \log_b x}$
36. $b^{\log_b x + \log_b y}$

In Exercises 37–50 express each statement as a single logarithm with coefficient 1.

37. $\log_2 3 + \log_2 4$
38. $\log_b x + \log_b y$

39. $\log_4 20 - \log_4 5$ 40. $\log_b y - \log_b x$

41. $2\log_7 3$ 42. $3\log_b z$

43. $\frac{1}{2}\log_{10} 9 - 3\log_{10} 2$ 44. $3\log_b x - \frac{1}{2}\log_b z$

45. $\frac{1}{3}\log_b x + \frac{2}{3}\log_b y$ 46. $2\log_b x + \log_b(x+y)$

47. $\log_b(x^2-1) - \log_b(x+1)$

48. $\log_b(y-2) + \log_b y - 2\log_b x$

49. $\frac{1}{2}[\log_b x - (5\log_b y - 3\log_b z)]$

50. $\frac{1}{2}[(\log_b x - 5\log_b y) - 3\log_b z]$

51. Show that $\log_b\left(\frac{1}{a}\right) = -\log_b a$.

52. If $f(x) = 10^x$ and $f^{-1}(x) = \log_{10} x$, verify that $f^{-1}[f(x)] = x$ for all x in the domain of f.

Prove the following properties of logarithms; state any restrictions placed on x, y, b, and k.

53. $\log_b \frac{x}{y} = \log_b x - \log_b y$

54. $\log_b (x)^k = k\log_b x$

55. One property of sound measured on the decibel scale is called sound intensity, and a formula for calculating the intensity level (L) in decibels (dB) is

$$L = 10\log(I) + 90,$$

where I is the intensity measured in ergs per square centimeter per second.

a. Show that the given formula is equivalent to

$$L = 10\log\left(\frac{I}{10^{-9}}\right),$$

which is another common version of this formula.

b. Find L when $I = 100$.

c. Find I when $L = 120$ decibels.

56. Another characteristic of sound that is measured in decibels is called the sound pressure. One formula for the pressure level is

$$L = 20\log(p) + 74,$$

where p is measured in dynes per square centimeter.

a. Show that the formula is equivalent to

$$L = 20\log\left(\frac{p}{10^{-3.7}}\right).$$

b. Find L when $p = 2\ \text{dynes/cm}^2$.

c. Find p when $L = 0$ decibels.

57. A useful equation in statistics is called the *log-linear model* because the logarithm of one variable is expressed as a linear function of the other variable. Such models are often used to investigate the effects of various health hazards on the survival time of patients. An example of a log-linear model is $\log y = ax + b$.

a. Solve this equation for x.

b. Find x when $a = 0.4$, $b = 0$, and $y = 16$.

c. Solve the given equation for y.

d. Find y when $a = 0.4$, $b = 0$, and $x = 3$.

58. In the statistical study of life expectancy, a simple but useful mathematical model is $\log p = -mt$, where p represents the fraction of some original population that is still alive after time t. The constant m is called the "hazard rate" or the "force of mortality."

a. Solve this equation for p.

b. Solve this equation for t.

c. Assume that $m = 0.1$ and $t = 4$. Find p to the nearest hundredth.

d. Assume that $m = 0.1$ and $p = 0.5$. Find t to the nearest hundredth.

In Exercises 59–62 use the given equation and express $\log y$ as a linear function of x.

59. $y = 100 \cdot 3^x$ 60. $y = 10 \cdot 3^{-2x}$

61. $y = k \cdot b^x$ 62. $y = c \cdot 10^{ax^2+bx}$

63. Time-study experts make use of a mathematical model called a **learning curve**, which describes the time it takes a production line to make a complex product (such as an airplane). It attempts to describe the fact that the time decreases as the workers get used to the project. This model was based on the observed phenomenon that the labor time to make product number 2^{n+1} is (approximately) 80% of the time needed to make product number 2^n. An example is in the following table.

x Unit Number	y Hours of labor
1	40,000
2	32,000
4	25,600
8	20,480
16	16,384

This pattern is described by $y = 40{,}000(0.80)^{\log_2 x}$. Show that by taking logarithms (base 2) of both sides you can express $\log y$ as a linear function of $\log x$. (Special graph paper, called **log-log paper** may be used when the horizontal axis represents $\log x$ and the vertical axis represents $\log y$.)

64. Keeping the previous exercise in mind, find an equation that describes the relation in this table.

x	1	2	4	8	16
y	64	32	16	8	4

THINK ABOUT IT 7.4

1. In words, the product property of logarithms states that the logarithm of a product is equal to the sum of the logarithms of the factors. Give a verbal description for the quotient and power properties of logarithms.
2. Use the fact that $x = b^{\log_b x}$ and $y = b^{\log_b y}$ to prove the product property of logarithms.
3. Liquid *X* has 100 times the concentration of hydrogen ions as liquid *Y*. What is the relation of the pH of *X* to the pH of *Y*?
4. Give specific counterexamples to *disprove* each of the following statements.
   a. $\log_b xy = (\log_b x)(\log_b y)$ **b.** $(\log_b x)(\log_b y) = \log_b x + \log_b y$
   c. $\log_b \dfrac{x}{y} = \dfrac{\log_b x}{\log_b y}$ **d.** $\dfrac{\log_b x}{\log_b y} = \log_b x - \log_b y$
   e. $\log_b x^k = (\log_b x)^k$ **f.** $(\log_b x)^k = k \log_b x$
5. Examine the following line of reasoning: $3 > 2$. If we multiply both sides of the inequality by $\log_{10}(1/2)$, we have

$$3\log_{10}\left(\frac{1}{2}\right) > 2\log_{10}\left(\frac{1}{2}\right)$$

$$\log_{10}\left(\frac{1}{2}\right)^3 > \log_{10}\left(\frac{1}{2}\right)^2 .$$

$$\log_{10}\left(\frac{1}{8}\right) > \log_{10}\left(\frac{1}{4}\right)$$

Thus,

$$\frac{1}{8} > \frac{1}{4}.$$

Our conclusion is incorrect. What went wrong?

● ● ●

7.5 Exponential and Logarithmic Equations

Photo Courtesy of Archive Photos, New York

The doubling time of an exponentially increasing quantity is the time required for the quantity to double its size or value. Use the fact that the population of the United States is currently growing at a rate of about 0.71 percent per year to determine the current doubling time (to the nearest year) for the U.S. population. The required formula is

$$P = P_0(1+r)^t$$

where P is the population t years from now, P_0 is the current U.S. population, and r is the annual growth rate.
(See Example 3.)

Objectives

1. Solve exponential equations by using logarithms.
2. Apply the change of base formula.
3. Solve logarithmic equations.

In Section 7.2 we solved exponential equations in which it was not too hard to rewrite the expressions in terms of a common base. For instance, $9^x = 1/27$ was written as $3^{2x} = 3^{-3}$, so $x = -3/2$. Since this procedure is limited, we now consider a general approach that uses the following principle, which is based on the fact that a logarithmic correspondence is a one-to-one function.

Equation-Solving Principle

If x, y, and b are positive real numbers with $b \neq 1$, then

1. $x = y$ implies $\log_b x = \log_b y$, and conversely,
2. $\log_b x = \log_b y$ implies $x = y$.

We can therefore solve **exponential equations** (the unknown is in the exponent) by taking the logarithm of both sides of the equation, as shown in the following examples.

Example 1: Solving an Exponential Equation

Solve for x: $3^x = 8$. Give the solution to four significant digits.

Solution: To obtain a picture of the solution to this equation, we can look for the intersection of the graphs of $y = 3^x$ and $y = 8$. As shown in Figure 7.29, $3^x = 8$ when $x \approx 1.893$. This solution can be confirmed algebraically by taking the common logarithm of both sides of $3^x = 8$ and applying the power property of logarithms.

$$3^x = 8$$
$$\log 3^x = \log 8 \quad \text{Apply common logarithms to each side}$$
$$x \log 3 = \log 8 \quad \text{Power property}$$
$$x = \frac{\log 8}{\log 3}$$
$$= 1.89278926\ldots \quad \text{By calculator}$$

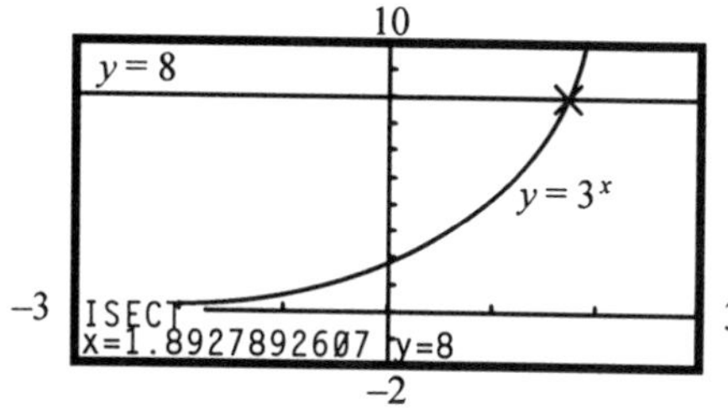

Figure 7.29

Thus, to four significant digits the solution set is $\{1.893\}$.

PROGRESS CHECK 1
Solve for t: $2^t = 9$.. Give the solution to four significant digits. ■

Example 1 illustrates a general approach for solving exponential equations.

To Solve Exponential Equations Using Logarithms

1. Take the logarithm to the same base of both sides of the equation.
2. Simplify by applying the property $\log_b x^k = k \log_b x$.
3. Solve the resulting equation using previous equation-solving methods.

Example 2: Solving an Exponential Equation

Solve for x: $3^{x+2} = 5^{2x-1}$. Give the solution to four significant digits.

Solution: Figure 7.30 shows that 3^{x+2} and 5^{2x-1} have the same value when $x \approx 1.795$, and algebraic verification of this result is shown next.

$$3^{x+2} = 5^{2x-1}$$
$$\log 3^{x+2} = \log 5^{2x-1} \quad \text{Apply common logarithms to each side}$$
$$(x+2)\log 3 = (2x-1)\log 5 \quad \text{Power property}$$
$$x\log 3 + 2\log 3 = 2x\log 5 - \log 5 \quad \text{Distributive property}$$
$$x\log 3 - 2x\log 5 = -2\log 3 - \log 5 \quad \text{Equivalent equation grouping } x$$
$$x(\log 3 - 2\log 5) = -2\log 3 - \log 5 \quad \text{Factoring}$$
$$x = \frac{-2\log 3 - \log 5}{\log 3 - 2\log 5} \approx 1.795$$

To four significant digits, the solution set is $\{1.795\}$.

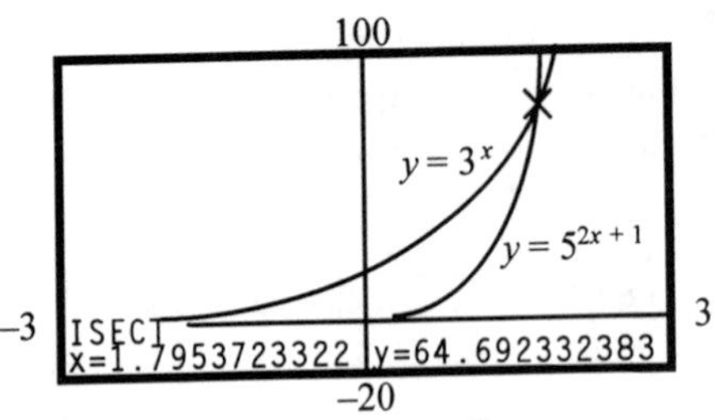

Figure 7.30

PROGRESS CHECK 2
Solve for x: $5^x = 3^{x+1}$. Give the solution to four significant digits. ■

Example 3: Finding Doubling Time in Exponential Growth

Solve the problem in the section introduction on page 576.

Solution: When the current U.S. population has doubled, then $P = 2P_0$. Replace P by $2P_0$ and r by 0.71 percent (or 0.0071) in the given formula and solve for t.

$$P = P_0(1+r)^t$$

$$2P_0 = P_0(1+.0071)^t \quad \text{Replace } P \text{ by } 2P_0 \text{ and } r \text{ by } 0.0071$$

$$2 = (1.0071)^t$$

$$\log 2 = \log(1.0071)^t \quad \text{Apply common logarithms to each side}$$

$$\log 2 = t\log(1.0071) \quad \text{Power property}$$

$$\frac{\log 2}{\log 1.0071} = t$$

By calculator, $t \approx 97.9725$, so the current doubling time for the U.S. population is about 98 years. Observe that the intersection of the graphs of $y = 1.0071^x$ and $y = 2$ that is shown in Figure 7.31 supports this estimate.

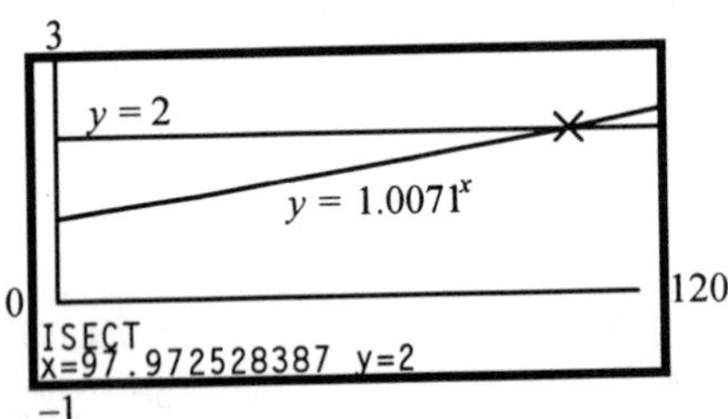

Figure 7.31

PROGRESS CHECK 3
What is the doubling time (to the nearest year) for the population of Greece given that the current annual growth rate is about 0.06 percent? ■

Depending on the application, it may be more convenient to write a logarithmic statement in a certain base, and the change-of-base formula that was given in Section 7.3 may be derived using our current methods for solving exponential equations. To express $\log_b x$ in terms of a different base, say a, first recall that

$$y = \log_b x \quad \text{is equivalent to} \quad b^y = x.$$

By taking the logarithm to the base a of both sides of $b^y = x$, we have

$$\log_a b^y = \log_a x$$
$$y \log_a b = \log_a x$$
$$y = \frac{\log_a x}{\log_a b}.$$

Then replacing y by $\log_b x$ yields the formula

$$\log_b x = \frac{\log_a x}{\log_a b}.$$

Because calculators have a Log key, converting to base 10 logarithms is often useful, as illustrated next.

Example 4: Using the Change-of-Base Formula

Use logarithms to the base 10 to determine each logarithm to four significant digits.

a. $\log_2 7$ **b.** $\log_5 0.043$

Solution: Convert to common logarithms using the change-of-base formula.

a. $\log_2 7 = \frac{\log 7}{\log 2} \approx 2.807$ **b.** $\log_5 0.043 = \frac{\log 0.043}{\log 5} \approx -1.955$

Note: When solving equations such as $3^x = 8$, students sometimes begin by writing

$$x = \log_3 8$$

but then are stumped. The continuation of this line of reasoning uses the change-of-base formula, so

$$x = \log_3 8 = \frac{\log 8}{\log 3} \approx 1.893.$$

Compare this alternative method with the solution shown in Example 1.

PROGRESS CHECK 4

Evaluate each logarithm to four significant digits.

a. $\log_4 70$ **b.** $\log_{1/3} 0.45$ ■

The change-of-base formula was introduced in Section 7.3 so that a grapher could be used to quickly draw graphs for equations of the form $y = \log_b x$. The next example reviews this important use of the change-of-base formula.

Example 5: Graphing $y = \log_b x$ by Calculator

Use a grapher to graph $f(x) = \log_2 x$ and $g(x) = \log_{1/3} x$.

Solution: Using the change-of-base formula to convert to common logarithms gives

$$f(x) = \log_2 x = \frac{\log x}{\log 2} \text{ and } g(x) = \log_{1/3} x = \frac{\log x}{\log(1/3)}$$

Figure 7.32 shows expressions that may be entered to graph f and g along with their graphs in the viewing window $[-1,6]$ and $[-4,4]$.

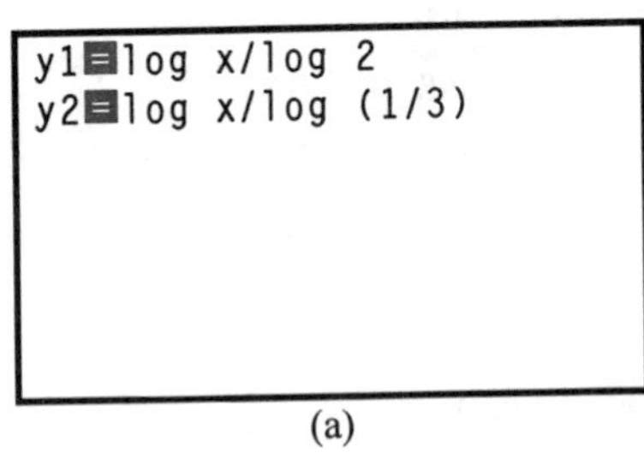

(a)

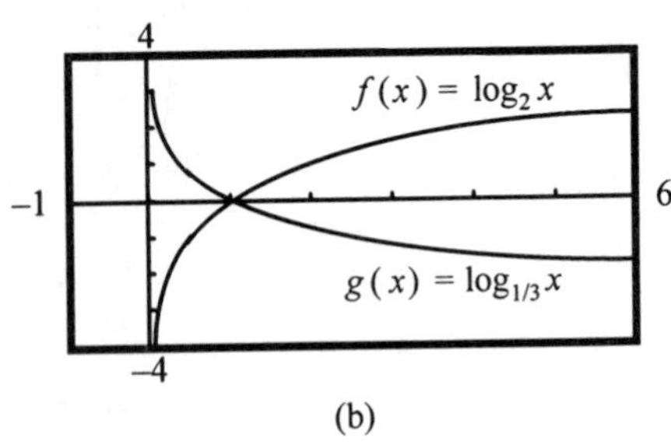

(b)

Figure 7.32

PROGRESS CHECK 5
Use a grapher to graph $f(x) = \log_3 x$ and $g(x) = \log_{1/2} x$. ■

A **logarithmic equation** (the unknown is in the log statement) is sometimes solved by using the principle that $\log_b x = \log_b y$ implies $x = y$. For instance

$$\text{if } \log(3x-8) = \log x\text{, then } 3x-8 = x\text{, and } x = 4.$$

If the equation contains only one log statement, we often solve by changing to exponential form. Thus

$$\text{if } \log(x+7) = 2\text{, then } x+7 = 10^2\text{, and } x = 93.$$

In some cases we obtain the single log statement by applying a property of logarithms, as shown in the next two examples. It is important to remember that logarithms are not defined for negative numbers or zero. Therefore, in the solution of logarithmic equations, it is necessary to check answers in the original equation and accept only solutions that result in the logarithms of positive numbers.

Example 6: Solving a Logarithmic Equation

Solve for x: $\log_3(x^2-4) - \log_3(x+2) = 2$.

Solution: Figure 7.33 shows that $\log_3(x^2-4) - \log_3(x+2) = 2$ when $x = 11$. To obtain this display you can enter

$$y1 = \frac{\log(x^2-4)}{\log 3} - \frac{\log(x+2)}{\log 3} \text{ and } y2 = 2.$$

An algebraic solution to this equation relies on the quotient property of logarithms as shown below.

$$\begin{aligned} \log_3(x^2-4) - \log_3(x+2) &= 2 \\ \log_3\left(\frac{x^2-4}{x+2}\right) &= 2 \qquad \text{Quotient property of logarithms} \\ \log_3(x-2) &= 2 \\ 3^2 &= x-2 \\ 11 &= x \end{aligned}$$

If we substitute 11 in our original equation, both log expressions are defined. Thus, the solution set is $\{11\}$.

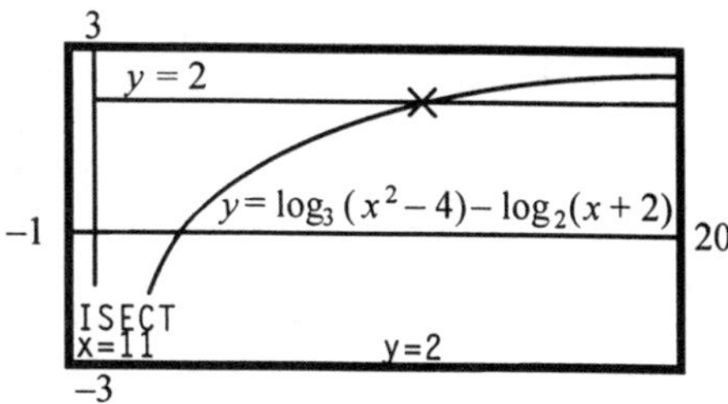

Figure 7.33

PROGRESS CHECK 6

Solve for x: $\log_4(x^2 - 9) - \log_4(x - 3) = 3$. ■

Example 7: Solving a Logarithmic Equation

Solve for x: $\log_2(x - 3) = 2 - \log_2 x$.

Solution: First rewrite the given equation as

$$\log_2 x + \log_2(x - 3) = 2$$

so that we may apply the product property of logarithms to obtain

$$\log_2[x(x - 3)] = 2.$$

Now change from logarithmic form to exponential form and solve

$$\begin{aligned} 2^2 &= x(x - 3) \\ 4 &= x^2 - 3x \\ 0 &= x^2 - 3x - 4 \\ 0 &= (x - 4)(x + 1) \end{aligned}$$

$$x - 4 = 0 \qquad x + 1 = 0$$

$$x = 4 \qquad x = -1$$

A check in the original equation shows that 4 is a solution, while −1 is extraneous, because the domain of $\log_2 x$ is the set of positive real numbers. Thus, the solution set is $\{4\}$. See Figure 7.34 for a graphical solution that supports this answer.

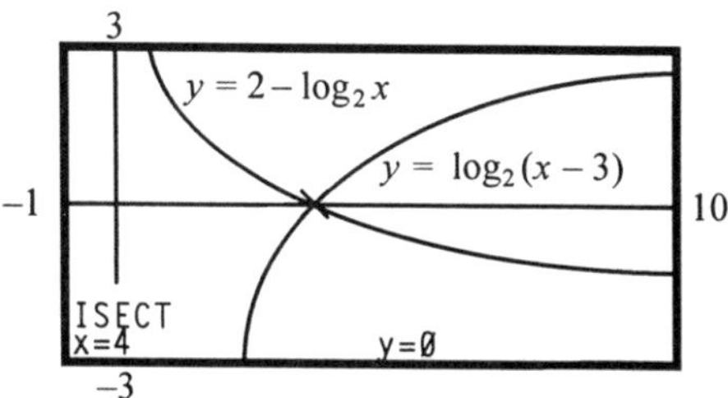

Figure 7.34

PROGRESS CHECK 7

Solve for x: $\log_3(x + 6) = 3 - \log_3 x$. ■

EXPLORE 7.5

1. a. To the nearest tenth, solve $2^x = 3^{x-1}$ graphically.
   b. To the nearest hundredth, solve $2^x = 3^{x-1}$ using the Solver feature.
   c. Find the exact solution to $2^x = 3^{x-1}$ in terms of logarithms.
   d. Solve $b^x = a^{x-1}$ for x in terms of logarithms. Assume a and b are positive numbers other than 1.
2. a. Graph $y = \log_{1/2} x$. b. Graph $y = -\log_2 x$.
   c. How are the graphs in parts **a** and **b** related? Explain this relationship in terms of the change of base formula.
3. Compare the graphs of $y = \log_b x$ for various values of b. Describe those characteristics of the graph that change when b changes, as well as those characteristics that do not change.
4. a. Find any points of intersection of $y = (1/2)^x$ and $y = \log_{1/2} x$.
   b. Check that the graphs of $y = 2^x$ and $y = \log_2 x$ have no points of intersection.
   c. Approximately (to 2 decimal places), what is the largest value of b for which the graphs of $y = b^x$ and $y = \log_b x$ intersect?
5. Some equations that mix exponential, logarithmic, and polynomial expressions are impossible to solve algebraically, and are best solved approximately by graphical or numerical methods. Use a grapher to solve these two examples.
   a. $5^x = x + 1$ b. $\log x = (x-6)^2$

EXERCISES 7.5

In Exercises 1–14 use logarithms to solve the equation.

1. $2^x = 10$ 2. $5^x = 100$ 3. $4^{x+1} = 17$

4. $8^{x-1} = 25$ 5. $2^{-x} = 9$ 6. $3^{2x} = 5$

7. $(0.7)^t = 0.3$ 8. $(1.05)^t = 3$

9. $2^x = 3^{2x+1}$ 10. $3^{2x-1} = 4^{x+2}$

11. $5^{x-2} = 6^{2x}$ 12. $10^{1-x} = 5^{-x}$

13. $6{,}000 = 2{,}000(1.03)^{2t}$

14. $1{,}000 = 100\left(1 + \dfrac{0.05}{4}\right)^{4t}$

In Exercises 15–22 use logarithms to the base 10 to determine each logarithm.

15. $\log_2 9$ 16. $\log_4 31$ 17. $\log_3 5$

18. $\log_5 3$ 19. $\log_{1/2} 19$ 20. $\log_{1/3} \dfrac{2}{3}$

21. $\log_4 0.012$ 22. $\log_6 0.735$

23. If $\log_b a = 3$, find $\log_a b$.

24. Simplify $\log_x 5 \cdot \log_5 x$.

In Exercises 25–34 use a grapher to graph the given functions. Describe the domain of the function.

25. $f(x) = \log_4 x$ 26. $f(x) = \log_5 x$

27. $f(x) = \log_{1/4} x$ 28. $f(x) = \log_{1/5} x$

29. $y = \log_2(x-8)$ 30. $y = \log_3(x+5)$

31. $y = 3\log_2(5x^2)$ 32. $y = -2\log_5(x^2+1)$

33. $y = \log|x|$ 34. $y = |\log x|$

In Exercises 35–64 solve the logarithmic equation.

35. $\log x = 2$ 36. $\log x = 0$

37. $5\log x = 5$ 38. $2\log x = \log 2$

39. $\log(1-x) = -1$ 40. $\log(-x) = \dfrac{1}{2}$

41. $\log(2x-5) = \log x$ 42. $\log x = \log(1-x)$

43. $2\log x = \log 8$ 44. $3\log x = \log(3x)$

45. $1 + \log x = \log 5$ 46. $\log 2 + \log 3 = \log x$

47. $\log_4 x + \log_4 2 = 1$ 48. $\log_3 x - \log_3 4 = 2$

49. $\log_2(x^2-1) = 3$ 50. $\log_5(x^2+9) = 2$

51. $\log_6(x+1)+\log_6 x=1$

52. $\log_2(x+1)+\log_2(x+4)=2$

53. $\log_2(x-2)+\log_2 x=3$

54. $\log x=1-\log(x-3)$

55. $\log_3(x-4)=2-\log_3(x+4)$

56. $\log_2(x+1)=3-\log_2(x-1)$

57. $\log(x+6)-2\log x=0$

58. $2\log x-\log(x+2)=0$

59. $\log(x-4)+\log(3x-4)=\log 11$

60. $\log_5(5x-6)+\log_5(x-1)=\log_5 4$

61. $\log_4(x-1)-\log_4(x+3)=\log_4 x$

62. $\log(3x-2)=1+\log(x+4)$

63. $\log_b 2x=\log_b 4x-\log_b 2$

64. $\log_7(x^2-x)=\log_7 x+\log_7(x-1)$

65. How long will it take for money invested at 4.5 percent compounded annually to double?

66. How long will it take for money invested at 5 percent compounded annually to double?

In Exercises 67–70 use the following information. What is the doubling time (to the nearest year) for the populations given, given their current annual growth rate? (Source: *1995 World Population Sheet*; Population Reference Bureau)

67. World; 1.5 percent
68. Greece; 0.04 percent
69. Syria; 3.5 percent
70. Japan; 0.3 percent

71. The U.S. Population has been growing recently at about 1 percent annually. In 1990 the population was about 248.7 million.
   a. To the nearest year, when will the population reach 300 million?
   b. In about what year will the U.S. population be double the 1990 number?

72. The population of Canada has been growing recently at about 0.8 percent annually. In 1990 the population was about 26.6 million.
   a. To the nearest year, when will the population reach 30 million?
   b. In about what year will the Canadian population be double the 1990 number?

73. What annual growth rate (to the nearest tenth) will cause a population to double in 5 years?

74. What annual growth rate (to the nearest tenth) will cause a population to double in 25 years?

75. Geologists construct mathematical models that relate properties of the soil to the age of the soil. In one such study by Milan Pavich and Nataša Vidic, for a particular region of the earth (the Sava River Valley in Slovenia), the equation $y=-17.0+3.2\log x$ was found to given an approximate relation between the age (x) of the soil in years and the thickness (y) of the soil in meters. Use this equation to estimate the age (to the nearest thousand years) of soil that is 6 meters thick. The model is valid for soils up to about 2 million years old. (Source: *Geophysical Monograph 78*, 1993, American Geophysical Union)

76. A chemical reaction that reaches equilibrium is characterized by a number K called the **equilibrium constant** for that reaction. In one kind of reaction (like that in a battery cell) the equation $E=\frac{0.0592}{2}\log K$ gives the cell potential in volts.
   a. In one reaction between copper and tin $E=0.183$ volts. Find the equilibrium constant.
   b. In a reaction between iodine and tin $E=0.381$ volts. Find K.

THINK ABOUT IT 7.5

1. If $A=P(1+r)^t$, express t in terms of the common logarithms of A, P, and $1+r$.
2. If $\log_b x=a$, find $\log_{1/b} x$ in terms of a.
3. Solve each equation.
   a. $x^2 10^x=10^x$ b. $\log(\log x)=1$ c. $2^{2x}-20=2^x$
4. Just as squaring both sides of an equation can introduce extraneous roots, certain operations with logarithms can alter the solution set of an equation. Solve $\log_3(x-5)^2=2$ by two methods and compare solution sets.
   Method 1: Use the definition of logarithm to get $3^2=(x-5)^2$. Then solve this quadratic equation.

Method 2: Use the power property of logarithms to get $2\log_3(x-5)=2$. Divide both sides by 2, and solve the resulting logarithmic equation.

Which method gives the correct solution set to the original equation?

5. In the equation for population growth, $P = P_0(1+r)^t$ does doubling the annual growth rate (r) cut the doubling time in half?
   a. Compare the doubling time for $r=5$ and $r=10$ percent.
   b. Compare the doubling time for $r=10$ and $r=20$ percent.
   c. What value of r has exactly half the doubling time of $r=5$ percent?

● ● ●

7.6 More Applications and the Number *e*

Photo Courtesy of Archive Photos, New York

Newton's law of cooling states that when a warm body is placed in colder surroundings at temperature t_a, the temperature (T) of the body at time t is given by

$$T - t_a = D_0 e^{kt},$$

where D_0 is the initial difference in temperature and k is a constant. Because of an ice storm, there is a loss of power in a home heated to 68° F. If the outside temperature remains fixed at 28°F and the temperature in the house drops from 68 to 64°F in 1 hour, when will the temperature in the house be down to 50°F? (See Example 11.)

Objectives

1. Use the compound-interest formula for an investment compounded n times per year.
2. Use the compound-interest formula for an investment compounded continuously.
3. Graph exponential and logarithmic functions involving the number e.
4. Solve exponential and logarithmic equations involving the number e.
5. Solve applied problems involving continuous growth or decay.

The formula $A = P(1+r)^t$ was used in Section 7.2 to analyze investments that are compounded once a year. However, if the interest is computed more frequently and added to the principal, then the amount grows at a faster rate as the additional interest earns interest. For instance, if \$6,000 is invested at 8 percent compounded annually, then this investment will be worth \$8,815.97 after 5 years. (Use $A = P(1+r)^t$ and check this result.) The effect of changing this investment to compounding twice per year is to increase the compounded amount by \$65.50, as can be seen from Example 1.

Example 1: Compounding *n* Times per Year

\$6,000 is invested at 8 percent compounded semiannually.

a. Find a formula showing the value of the investment at the end of t years.
b. How much is the investment worth after 5 years?

Solution:

a. Since the investment is compounded semiannually, the interest is determined two times per year, and the interest rate for each of these periods is 8 percent/2, or 0.04. Then,

$$A = 6{,}000(1.04) \quad \text{when} \quad t = \frac{1}{2} \text{ year},$$
$$A = 6{,}000(1.04)^2 \quad \text{when} \quad t = 1 \text{ year},$$
$$A = 6{,}000(1.04)^4 \quad \text{when} \quad t = 2 \text{ years}.$$

and in general, the compounded amount after t years is given by

$$A = 6{,}000(1.04)^{2t}.$$

b. When $t = 5$

$$\begin{aligned} A &= 6{,}000(1.04)^{2(5)} \\ &= 6{,}000(1.04)^{10} \\ &= \$8881.47 \end{aligned}$$

PROGRESS CHECK 1

\$9,000 is invested at 5 percent compounded semiannually.

a. Find a formula showing the value of the investment at the end of *t* years.

b. How much is the investment worth after 12 years? ■

The procedure for Example 1 can be generalized to obtain the following compound-interest formula.

Compound-Interest Formula

> The compounded amount A when an original principal P is compounded n times per year for t years at annual interest rate r is given by
>
> $$A = P\left(1 + \frac{r}{n}\right)^{nt}.$$

Observe that the above formula simplifies to $A = P(1+r)^t$ for interest compounded annually, since $n = 1$ in this case.

Example 2: Compounding *n* Times per Year

\$100 is invested at 5 percent interest compounded quarterly. How much will it amount to in 3 years?

Solution: Substituting $P = 100$, $r = 0.05$, $n = 4$, and $t = 3$ in the formula above, we have

$$\begin{aligned} A &= 100\left(1 + \frac{0.05}{4}\right)^{4(3)} \\ &= 100(1.0125)^{12} \\ &= \$116.08 \end{aligned}$$

PROGRESS CHECK 2

\$2,000 is invested in an IRA account by a college student on her 17th birthday. If the account grows at 8.5 percent compounded daily, what will be the value of this account on her 65th birthday? Use $n = 365$, and round to the nearest dollar. ■

When an investment is compounded n times per year, then we have observed that the compounded amount increases as n increases. However, there is a limit to this growth. To illustrate, fix P, r, and t at 1 in the compound-interest formula to obtain

$$A = 1\left(1+\frac{1}{n}\right)^{n(1)} = \left(1+\frac{1}{n}\right)^{n}.$$

The result gives the compounded amount when \$1 is invested at 100 percent interest for 1 year and is compounded n times. Now consider how A changes as the frequency of compounding is increased, as shown in the following table.

Type of Compounding	Number of Conversions (n)	Compounded Amount
Annually	1	$A = \left(1+\frac{1}{1}\right)^{1} = 2$
Semiannually	2	$A = \left(1+\frac{1}{2}\right)^{2} = 2.25$
Quarterly	4	$A = \left(1+\frac{1}{4}\right)^{4} \approx 2.441\ldots$
Monthly	12	$A = \left(1+\frac{1}{12}\right)^{12} \approx 2.613\ldots$
Daily	365	$A = \left(1+\frac{1}{365}\right)^{365} \approx 2.714\ldots$
Hourly	8,760	$A = \left(1+\frac{1}{8{,}760}\right)^{8{,}760} \approx 2.718\ldots$

Notice that A increases by a small amount as the conversion period changes from daily to hourly, and more frequent conversions lead to even smaller changes in A. In higher mathematics it is shown that as n gets larger, $(1+1/n)^n$ gets closer to an irrational number that is denoted by the letter e. (The choice of e as the symbol for this number was made by the Swiss mathematician Leonhard Euler in about 1728.)

To six significant digits,

$$e \approx 2.71828..$$

When n increases without bound, we say that the investment is compounded continuously. If the \$1 investment analyzed above is compounded in this manner, then it grows to \$$e$ by the end of the year. A base e exponential function therefore describes investments that are compounded continuously, and a general formula that allows for interest rates and principals that are more practical than 100 percent and \$1 is stated next.

Continuous-Compounding Formula

> The compounded amount A when an original principal P is compounded continuously for t years at annual interest rate r is given by
>
> $$A = Pe^{rt}.$$

To derive this formula, we start with the compound-interest formula involving n conversions per year and let $x = n/r$ so that $r/n = 1/x$ and $n = xr$. Then

$$A = P\left(1+\frac{r}{n}\right)^{nt} \text{ may be written as } A = P\left[\left(1+\frac{1}{x}\right)^{x}\right]^{rt}.$$

The number e is the value approached by $(1+1/x)^x$ as x increases without bound. Therefore, when an investment is compounded continuously

$$A = P\left[(1+1/x)^x\right]^{rt} \text{ may be replaced by } A = Pe^{rt},$$

which establishes the continuous-compounding formula.

Example 3: Compounding Continuously

\$2,000 is invested at 6 percent compounded continuously. How much will the investment be worth in 10 years?

Solution: Substituting $P = 2{,}000$, $r = 0.06$, and $t = 10$ in the formula $A = Pe^{rt}$ gives

$$A = 2{,}000e^{(0.06)(10)}$$
$$= 2{,}000e^{0.6}$$

To evaluate powers of e, most calculators make e^x the second function associated with the [LN] key. So a typical keystroke sequence to evaluate $2{,}000e^{0.6}$ is

2,000 [×] [2nd] [e^x] 0.6 [ENTER].

The result of this computation shows that the compounded amount is \$3,644.24.

PROGRESS CHECK 3

\$6,000 is invested at 5 percent compounded continuously. How much will the investment be worth in 7 years?

■

On the graphing calculator the keystroke sequence [2nd] [e^x] is also used to obtain the graph of $y = e^x$ and others related to it. Three graphs of this type are shown in the next example.

Example 4: Graphing an Exponential Function Involving Base *e*

Graph each function using a grapher.

a. $y = e^x$ **b.** $y = e^{-x}$ **c.** $y = 256\left(1 - e^{-x/8}\right);\ 0 \le x \le 50$

Solution: The functions may be defined on a grapher as shown in Figure 7.35.

```
y1=e^x
y2=e^-x
y3=256(1-e^(-x/8))
```

Figure 7.35

a. The graph of $y = e^x$ resembles the graph of $y = 2.7^x$, so we can anticipate that the graph increases rapidly for $x > 0$ and that the x-axis is a horizontal asymptote for the graph. These features are confirmed in the graph of $y = e^x$ shown in Figure 7.36.

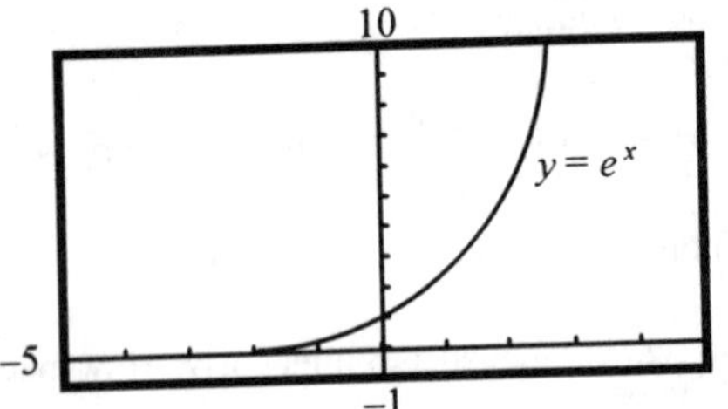

Figure 7.36

b. Recall that exponential decay may be described by equations of the form $y = b^{-x}$ with $b > 1$. Therefore, the graph of $y = e^{-x}$ illustrates exponential decay, as shown in Figure 7.37.

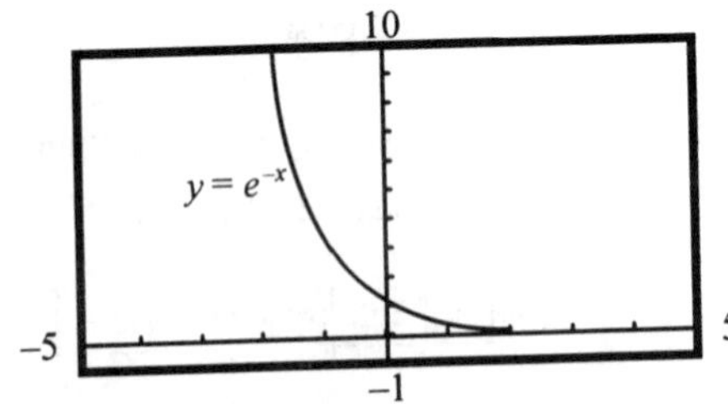

Figure 7.37

c. As directed, first restrict x to the interval [0, 50]. To set an appropriate vertical range for the graph, observe that when $x = 0$, $y = 0$, and that as x gets large, $e^{-x/8}$ approaches 0, so y approaches 256. Consequently we choose [0, 300] to define the vertical range and then obtain the graph of $y = 256\left(1 - e^{-x/8}\right)$ in Figure 7.38.

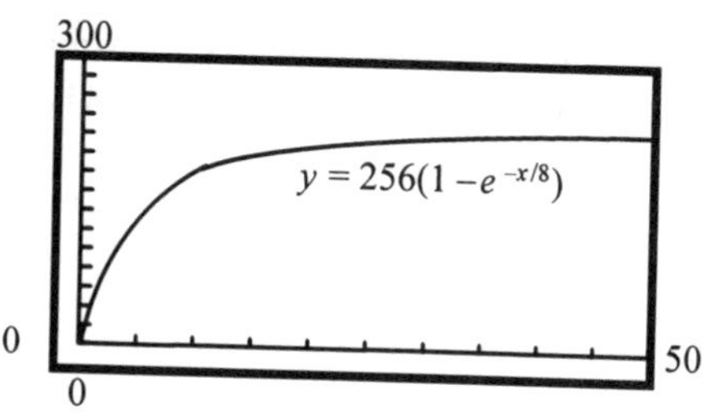

Figure 7.38

Note: The equation in part c is an example of an exponential function that fits the general form

$$y = L(1 - e^{kt}).$$

There are many applications associated with this mathematical model, and Example 12 and Exercises 83–88 consider a few applied problems of this type.

PROGRESS CHECK 4

Graph each function using a grapher.

a. $y = e^{0.5x}$ **b.** $y = 4e^{-x}$ **c.** $y = 175(1 - e^{-0.2x})$; $0 \le x \le 50$ ■

The positioning of the e^x function above the LN key on most calculators once again suggests the inverse relation between exponential and logarithmic functions. The inverse of the function defined by $y = e^x$ is defined by $x = e^y$, which is equivalent to $y = \log_e x$. As mentioned in Section 7.3 logarithms to the base e are called **natural logarithms**, and $\log_e x$ is usually abbreviated as $\ln x$. Thus,

$y = e^x$ and $y = \ln x$ are inverse functions.

The graphs of $y = e^x$ and $y = \ln x$ are shown in Figure 7.39. Observe that the graphs are symmetric about the line $y = x$ (as are all pairs of inverse functions), and they behave as exponential and logarithmic functions with base b that satisfy $b > 1$ (since $e > 1$).

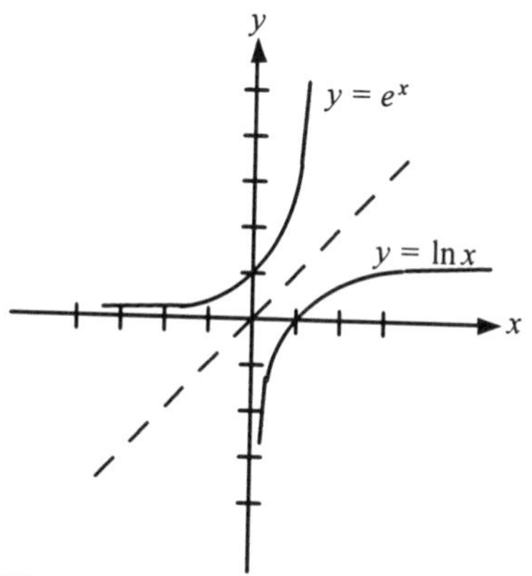

Figure 7.39

Graphs of natural logarithm functions are easily shown by using the LN key on a graphing calculator. An exploration involving such functions is discussed in Example 5.

Example 5: Using a Grapher to Explore Natural Logarithm Functions

Graph $y = \ln x^3$ and $y = 3\ln x$ in the same display using the standard viewing window. How are the graphs related? Use logarithm properties to explain the basis for this relation.

Solution: The appropriate calculator displays for defining and picturing the functions are given in Figure 7.40.

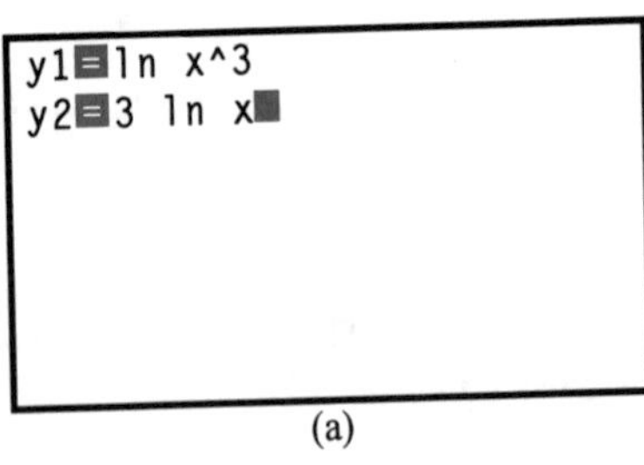

(a)

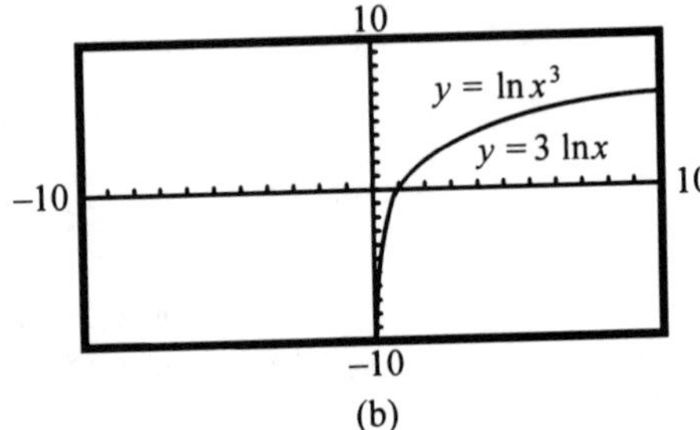

(b)

Figure 7.40

It appears that the graph of $y = \ln x^3$ is the same as the graph of $y = 3\ln x$. In fact this observation may be confirmed on the basis of the power property of logarithms given in Section 7.4. That is, it follows from $\log_b x^k = k \log_b x$, that $\ln x^3 = 3\ln x$.

PROGRESS CHECK 5

Graph $y = \ln 100x$ and $y = \ln 100 + \ln x$ in the same display using the standard viewing window. How are the graphs related? Use logarithm properties to explain the basis for this relation. ■

Example 5 illustrates that the logarithm properties given in Section 7.4 may be used to establish properties of natural logarithms. For instance, stating the product, quotient, and power properties in terms of base *e* logarithms results in the following properties.

Properties of Natural Logarithms

If x and y are positive numbers and k is any real number, then

1. $\ln xy = \ln x + \ln y$ Product property
2. $\ln \frac{x}{y} = \ln x - \ln y$ Quotient property
3. $\ln x^k = k \ln x$. Power property

Furthermore, the inverse properties $\log_b b^x = x$ and $b^{\log_b x} = x$ (if $x > 0$) become

$$\ln e^x = x \quad \text{and} \quad e^{\ln x} = x \quad (\text{if } x > 0)$$

when stated in terms of base *e*, while

$$\ln e = 1 \quad \text{and} \quad \ln 1 = 0.$$

In the remaining examples, properties of natural logarithms are used together with the equation-solving methods from Section 7.5 to solve exponential and logarithmic equations that involve the number e. Among these, the last five examples will demonstrate applications that involve this topic.

Example 6: Solving an Exponential Equation Involving *e*

Solve $e^{-0.32t} = 0.5$. Give the solution to four significant digits.

Solution: Because this equation involves e, we employ natural logarithms in the solution.

$$e^{-0.32t} = 0.5$$

$$\ln e^{-0.32t} = \ln 0.5 \qquad \text{Apply natural logarithms to each side.}$$

$$-0.32t = \ln 0.5 \qquad \text{Inverse property } \ln e^x = x.$$

$$t = \frac{\ln 0.5}{-0.32}$$

$$= 2.1660849\ldots$$

Thus, to four significant digits the solution set is $\{2.166\}$. See Figure 7.41 for a graphical solution that supports this answer.

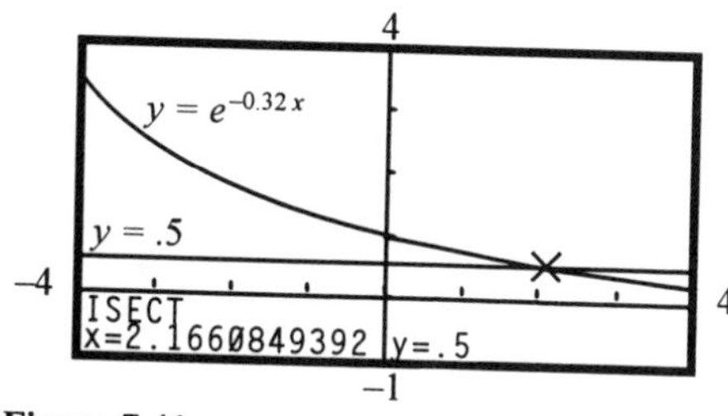

Figure 7.41

PROGRESS CHECK 6
Solve $e^{0.14t} = 2$. Give the solution to four significant digits.

■

Example 7: Solving a Logarithmic Equation Involving *e*

Solve each equation. Give exact answers and also approximate the solution to four significant digits.

a. $\ln x = 1.9$ **b.** $\ln(x+1) = 2 - \ln x$

Solution:

a. If $\ln x = 1.9$, then $x = e^{1.9}$, and the solution set is $\{e^{1.9}\}$. By calculator $e^{1.9} = 6.686$ to four significant digits, and this approximate answer is supported by the graphical solution shown in Figure 7.42(a).

b. Applying logarithmic properties leads to a quadratic equation, as shown below.

$$\ln(x+1) = 2 - \ln x$$

$$\ln x + \ln(x+1) = 2$$

$$\ln[x(x+1)] = 2$$

$$x(x+1) = e^2$$

$$x^2 + x - e^2 = 0$$

Now using the quadratic formula with $a = 1$, $b = 1$, and $c = -e^2$ gives

$$x = \frac{-1 \pm \sqrt{1^2 - 4(1)(-e^2)}}{2(1)} = \frac{-1 \pm \sqrt{4e^2 + 1}}{2}.$$

A check in the original equation shows that $\left(-1+\sqrt{4e^2+1}\right)/2$ is a solution (use your calculator), while $\left(-1-\sqrt{4e^2+1}\right)/2$ is extraneous (why?). Thus, the solution set is $\left\{\left(-1+\sqrt{4e^2+1}\right)/2\right\}$.

To four significant digits $x = 2.264$, and this solution is in agreement with the solution of the system of equations graphed in Figure 7.42(b).

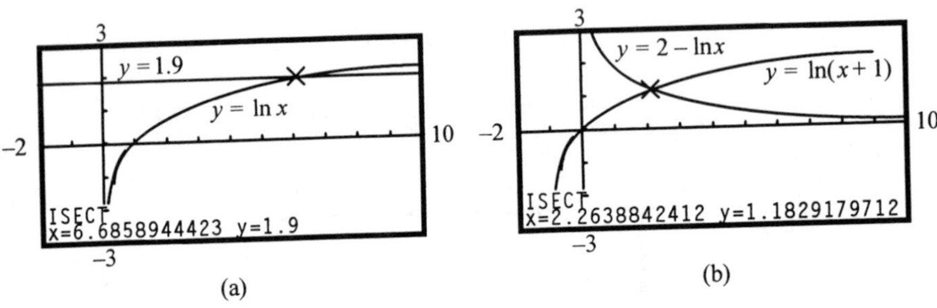

Figure 7.42

PROGRESS CHECK 7
Solve each equation. Give exact answers and also approximate solutions to four significant digits.

a. $\ln x = -2.5$ **b.** $\ln(x+1) = -\ln(x-1)$ ■

Example 8: Finding Doubling Time in Continuous Growth

An amount is invested at 7 percent compounded continuously. In how many years does the amount double?

Solution: When the original principal P has doubled, then $A = 2P$. Substitute $2P$ for A and 0.07 for r in the formula $A = Pe^{rt}$ and solve for t.

$$A = Pe^{rt}$$

$$2P = Pe^{0.07t}$$

$$2 = e^{0.07t}$$

$$\ln 2 = \ln e^{0.07t} \quad \text{Apply natural logarithms to each side.}$$

$$\ln 2 = 0.07t \quad \text{Inverse property } \ln e^x = x$$

$$\frac{\ln 2}{0.07} = t$$

By calculator, $t \approx 9.9021026$ so the required doubling time is about 9.9 years.

PROGRESS CHECK 8
An amount is invested at 3 percent compounded continuously. In how many years does the amount double? ■

For analysis of exponential growth or decay at a continuous rate for physical quantities, the formula $A = Pe^{rt}$ is expressed more generally as

$$A = A_0e^{kt},$$

where A is the amount at time t, A_0 is the initial amount, and k is the growth or decay constant. The constant k is positive when describing growth and negative when describing decay.

The formula above applies when the rate of change of some quantity is proportional to the amount present.

Example 9: Population Growth

If there is no restriction on food and living space, the rate of growth of a population of living organisms is proportional to the size of the population. Thus, the formula $A = A_0e^{kt}$ is appropriate. Assume that in the absence of hunters, a certain animal population in New York State triples every 11 years.

a. Find the size of a population that initially numbers 500 after 5 years.

b. In how many years will the population number 5,000?

Solution:

a. We know that $A_0 = 500$. Thus, our formula

$$A = A_0e^{kt}$$

becomes

$$A = 500e^{kt}.$$

We can find k since we know that $A = 1{,}500$, when $t = 11$.

$$1{,}500 = 500e^{k(11)}$$

$$3 = e^{11k}$$

$$\ln 3 = \ln e^{11k} \qquad \text{Apply natural logarithms to each side.}$$

$$\ln 3 = 11k \qquad \text{Inverse property } \ln e^x = x$$

$$k = \frac{\ln 3}{11} \approx 0.10$$

The formula is then

$$A = 500e^{0.1t}.$$

Substituting $t = 5$, we have

$$A = 500e^{0.1(5)}$$

$$= 500e^{0.5}$$

$$A \approx 824.$$

b. Substituting $A = 5{,}000$ in our formula, we have

$$5{,}000 = 500e^{0.1t}$$
$$10 = e^{0.1t}$$
$$\ln 10 = \ln e^{0.1t} \quad \text{Apply natural logarithms to each side.}$$
$$\ln 10 = -.1t \quad \text{Inverse property } \ln e^{x} = x$$
$$t = \frac{\ln 10}{0.1} \approx 23 \text{ years}$$

PROGRESS CHECK 9

Answer questions **a** and **b** in Example 9 assuming that the population triples every 20 years. ■

Example 10: Half-Life Physicists tell us that the half-life of radium is 1,620 years; that is, a given amount (A) of radium decomposes to the amount $A/2$ in 1,620 years. Radium (like all radioactive substances) decays at a rate proportional to the amount present. Thus, the formula $A = A_0e^{kt}$ is appropriate. If a research center owns 1 g of radium, how much radium will it have in 100 years?

Solution: We know that $A_0 = 1g$; thus, our formula

$$A = A_0e^{kt}$$

becomes

$$A = 1e^{kt}.$$

We can find k since we know that $A = 0.5$ when $t = 1{,}620$.

$$0.5 = e^{k(1{,}620)}$$
$$\ln 0.5 = \ln e^{1{,}620k} \quad \text{Apply natural logarithms to each side.}$$
$$\ln 0.5 = 1{,}620k \quad \text{Inverse property } \ln e^{x} = x$$
$$k = \frac{\ln 0.5}{1{,}620} \approx -0.00043$$

The formula is then

$$A = e^{-0.00043t}.$$

Substituting $t = 100$, we have

$$A = e^{-0.00043(100)}$$
$$= e^{-0.043}$$
$$A \approx 0.96\text{g}.$$

PROGRESS CHECK 10

The half-life of radon 222 is 3.82 days. If a given amount is present on Monday, Jan. 1 at noon, what percent of the original is still present on Friday, Jan. 5 at noon? ■

Example 11: Newton's Law of Cooling

Solve the problem in the section introduction on page 584.

Solution: We know that $t_a = 28$ and $D_0 = 68 - 28 = 40$. Thus, or formula becomes

$$T - 28 = 40e^{kt}.$$

We can find k since we know that $T = 64$ when $t = 1$.

$$64 - 28 = 40e^{k(1)}$$
$$0.9 = e^k$$
$$\ln 0.9 = \ln e^k \quad \text{Apply natural logarithms to both sides.}$$
$$\ln 0.9 = k \quad \text{Inverse property } \ln e^x = x$$
$$k \approx -0.105$$

(*Note*: k depends on the insulation in the house.) The formula is then

$$T - 28 = 40e^{-0.105t}.$$

To find the number of hours needed for the temperature to fall to 50°F, we let $T = 50$ and solve for t.

$$50 - 28 = 40e^{-0.105t}$$
$$0.55 = e^{-0.105t}$$
$$\ln 0.55 = \ln e^{-0.105t} \quad \text{Apply natural logarithms to both sides.}$$
$$\ln 0.55 = -0.105t \quad \text{Inverse property } \ln e^x = x$$
$$t = \frac{\ln 0.55}{-0.105}$$
$$t \approx 5.7 \text{ hours}$$

To check this answer graphically, it is instructive to first solve the formula in this example for T to obtain $T = 40e^{-0.105t} + 28$. The graph of this function then displays exponential decay where T is approaching the outside temperature of 28°F. Figure 7.43 shows that the intersection of this graph with the graph of $T = 50$ reaffirms that it takes about 5.7 hours for the temperature in the house to drop to 50°F.

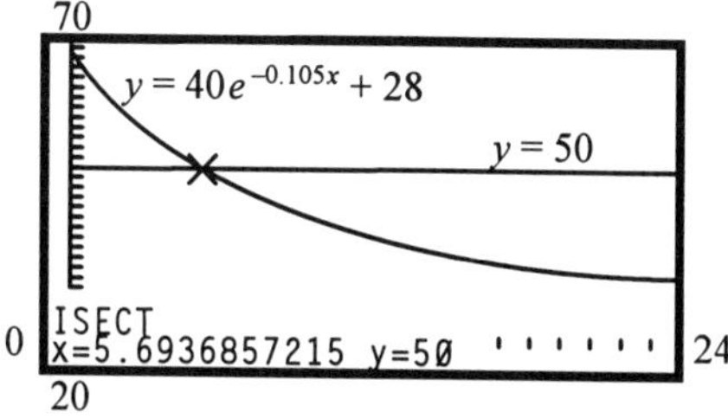

Figure 7.43

PROGRESS CHECK 11

Redo the problem in Example 11 but assume that the temperature in the house drops from 68° to 62°F in 1 hour, and that the outside temperature remains fixed at 18°F. ■

Example 12: Skydiving

When an object falls through a medium that offers resistance (such as air or water) it may approach a limiting velocity. The following pair of equations is typical of those that describe the velocity of a skydiver who jumps from an airplane. For the first 10 seconds the jumper is in free fall with velocity given by

$$v = 256\left(1 - e^{-t/8.06}\right), \quad 0 < t < 10.$$

Then the parachute opens, and from that time until the jumper hits the ground 20 seconds later the velocity is given by

$$v = 16 + 166e^{20-2t}, \quad 10 \le t < 30.$$

a. Show a graph of the velocity for $0 \le t < 30$, and describe its general behavior.
b. Approximately what is the fastest the jumper falls?
c. About how long after the chute opens does the velocity slow to 18 ft/second ?

Solution:

a. The graph of the velocity may be obtained on a graphing calculator in piecewise fashion as

$$y = 256\left(1 - e^{-x/8.06}\right)(x < 10) + \left(16 + 166e^{20-2x}\right)(x \ge 10).$$

The graph is shown in Figure 7.44.

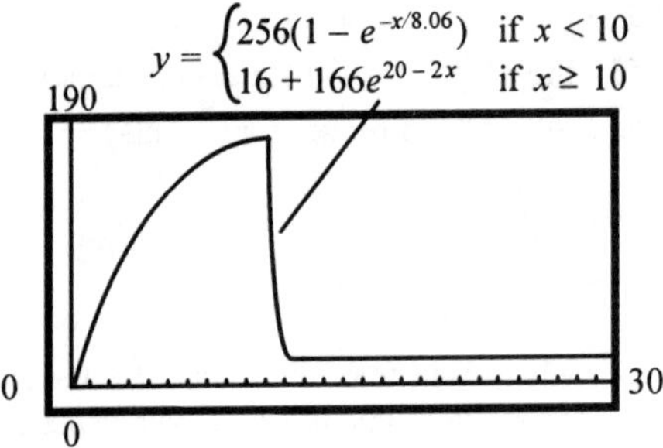

Figure 7.44

We observe that the velocity increases over the first 10 seconds of free fall from 0 to about 182 ft/second, then decreases rapidly at first when the parachute opens, then more slowly, and levels off at about 16 ft/second.

b. The maximum velocity occurs at $t = 10$ seconds (when the parachute opens). Since

$$\begin{aligned} 16 + 166e^{20-2(10)} &= 16 + 166e^{0} \\ &= 182 \end{aligned}$$

the maximum velocity is 182 ft/second. You should check this answer graphically by using the Zoom and Trace features of a calculator to estimate the highest point in the graph in Figure 7.44.

c. From the graph we see that we must set the equation for the second piece equal to 18.

$$\begin{aligned} 18 &= 16 + 166e^{20-2t} \\ 2 &= 166e^{20-2t} \\ 2/166 &= e^{20-2t} \\ \ln(2/166) &= \ln e^{20-2t} \\ \ln(2/166) &= 20 - 2t \\ t &= \frac{20 - \ln(2/166)}{2} \approx 12.2 \end{aligned}$$

The velocity slows to 18 ft/second about 12.2 seconds after the jump, or about 2.2 seconds after the chute opens.

PROGRESS CHECK 12

An object falls from rest towards the earth meeting air resistance that results in the velocity equation $v = 4(1 - e^{-8t})$ where t is in seconds and v is in feet per second.

a. Graph the velocity for $0 \le t < 3$, and describe its behavior.
b. Approximately what is the limiting velocity?
c. About how long does it take to reach a velocity of 3 ft/second? ■

In this section we discussed in detail only a few of the basic applications of the number e. However, keep in mind that the scope and power of the number e is remarkable. For example, consider the following relationships.

1. When the power is turned off, the electric current in a circuit does not vanish instantly but rather decreases exponentially.
2. The intensity of sunlight decreases exponentially as a diver descends further into ocean depths.
3. When a drug is injected into the bloodstream, the drug drains out of the body exponentially. Analyzing this relationship is important in medicine, for it helps the doctor plan a series of injections that will maintain the prescribed level of some drug (such as an antibiotic) in a patient's bloodstream.
4. Atmospheric pressure decreases exponentially as altitude above sea level increases.

The list could easily be continued. In calculus you will probably consider some of these applications.

EXPLORE 7.6

1. Any exponential function can be expressed in terms of one with base e. To illustrate, find by trial and error the constant k (to the nearest tenth) that makes the graph of $y = e^{kt}$ match the graph of $y = 2^t$. Confirm your answer by using natural logarithms to find the exact value of k.
2. This exercise explores the formula given in this section in the discussion of continuous growth. Graph $y = (1 + 1/x)^x$ for $x > 0$ and check that as x increases, the value of y approaches e. What is the smallest integer value of x for which y exceeds 2.71?
3. The number e is used in one of the most famous equations in mathematics, $e^{i\pi} + 1 = 0$, which was first proved by Euler. This is an impressive equation because it involves some fundamental quantities that have no obvious relation to one another. If your calculator can represent complex numbers, evaluate the expression $e^{i\pi}$ to confirm that it equals -1.
4. Many calculators include **logarithmic regression** in the regression menu. This feature finds the coefficients for the "best fitting" function of the form $y = a + b\ln x$ for a given set of points. If your calculator has this capability, apply it to the data of Exercise 85 in Section 7.3, then compare this regression model to the one from that exercise, which gave a model using base 10 logarithms. Estimate height at age 10 by both models. Use appropriate Draw features to show the given points and the regression curve.

EXERCISES 7.6

In Exercises 1-8 first find a formula that shows the value of the investment at the end of t years, then find the value of the investment at the end of the number of years given.

		Amount Invested	Annual Interest Rate (percent)	Compounded	Length of Time
1.	**a.**	\$1,000	6	semiannually	3 years
	b.	\$5,000	12	quarterly	2 years
2.	**a.**	\$3,000	5.4	monthly	1 year
	b.	\$2,000	5	daily	5 years
3.	**a.**	\$1,500	3.5	daily	2 years
	b.	\$2,000	10	daily	10 years
4.	**a.**	\$25,000	4	quarterly	3 years
	b.	\$12,000	4.5	monthly	10 years
5.	**a.**	\$1,000	6	continuously	5 years
	b.	\$2,000	10	continuously	10 years
6.	**a.**	\$10,000	24	continuously	1 month
	b.	\$3,000	6.4	continuously	6 quarters
7.	**a.**	\$5,000	9	monthly	15 months
	b.	\$5,000	9	continuously	15 months
8.	**a.**	\$6,000	12	quarterly	18 months
	b.	\$6,000	12	continuously	18 months

9. Which becomes more valuable after 20 years, \$1 invested continuously at 10 percent or \$2 invested continuously at 5 percent?

10. You are going to invest \$10,000 for 3 years. Which offer is best?
a. 10 percent compounded monthly
b. 10.7 percent compounded annually
c. 9.8 percent compounded continuously

11. Assume that in a specific culture the number of bacteria (A) present after t hours is given by $A = 1000e^{0.15t}$. How many bacteria are present after $1/2$ hour?

12. The amount (A) of a certain radioactive element remaining after t minutes is given by $A = A_0e^{-0.03t}$. How much of the element would remain after 1 hour if 50 g were present initially?

In Exercises 13–20 use a grapher to graph the given function. Indicate the domain and range.

13. $y = e^{2x}$ **14.** $y = e^{3x}$

15. $y = e^{-2x}$ **16.** $y = e^{-3x}$

17. $y = 50\left(1 - e^{-x/2}\right);\ 0 \le x \le 20$

18. $y = 20\left(3 - e^{-x/10}\right);\ 0 \le x \le 30$

19. $y = e^{-x^2}$ **20.** $y = e^{-x^2/2}$

In Exercises 21–28 use a grapher to display both graphs in the same window using the standard viewing window. Describe how the graphs are related and use logarithm properties to explain the basis for the relation.

21. $y = \ln x^5 \quad y = 5\ln x$ **22.** $y = \ln x^{-3} \quad y = -3\ln x$

23. $y = \ln 5x \quad y = \ln 5 + \ln x$

24. $y = \ln\dfrac{3}{x} \quad y = \ln 3 - \ln x$

25. $y = \ln 3x^2 \quad y = \ln 3 + 2\ln x$

26. $y = \ln 2x^3 \quad y = \ln 2 + 3\ln x$

27. $y = \ln\dfrac{x^2}{3} \quad y = 2\ln x - \ln 3$

28. $y = \ln\dfrac{5x}{3} \quad y = \ln 5 + \ln x - \ln 3$

In Exercises 29–38 solve the given equation. Give the solution to four significant digits.

29. $e^{-0.21t} = 0.2231$ **30.** $e^{-0.36t} = 0.4628$

31. $e^{-3.52t} = 0.3753$ **32.** $e^{-4.71t} = 0.1246$

33. $e^{0.13t} = 1.364$ **34.** $e^{0.47t} = 3.413$

35. $e^{0.06t} = 2.175$ **36.** $e^{0.05t} = 1.831$

37. $e^{2k} = 2$ **38.** $e^{-k} = 0.25$

In Exercises 39–50 solve the given equation for x, and

a. give the exact value and

b. give the approximate value to four significant digits.

39. $\ln x = 7.4$ **40.** $\ln x = 0.56$

41. $\ln x = -3.8$ **42.** $\ln x = -0.4$

43. $\ln x + \ln 3 = 6$ **44.** $\ln x = 3 + \ln 2$

45. $2\ln x = 3 + 2\ln 5$ **46.** $3\ln x = 2 + 3\ln 2$

47. $\ln(x+2) = 3 - \ln x$ **48.** $\ln(x+3) = 4 - \ln x$

49. $\ln(x+1) + \ln(x-1) = 0$

50. $\ln(x+1) - \ln(x-1) = 0$

In Exercises 51–56 at which rate of interest compounded continuously must money be invested to grow as indicated? (Round to the nearest hundredth of a percent.)

51. Double in 7 years **52.** Double in 10 years

53. Triple in 10 years **54.** Triple in 7 years

55. Quadruple in 10 years

56. Quadruple in 7 years

57. About how long does it take an investment to double in value if it is invested at 5 percent interest compounded continuously?

58. About how long does it take an investment to double in value if it is invested at 4 percent interest compounded continuously?

59. How much would a person have to invest today at 6 percent compounded continuously in order to have \$1,000 1 year from today? (*Note*: An important consideration of an economist is how much a dollar x years from now is worth today.)

60. How much would a person have to invest today at 6 percent compounded continuously in order to have \$100,000 30 years from today?

61. An amount is invested at 6 percent compounded monthly. In how many years does the amount triple?

62. How long will it take for \$20,000 to grow to \$30,000 at 10 percent compounded quarterly?

63. If left unchecked, a certain animal population triples in size every 15 years.

a. If this population initially includes 200 members, about how many will it include in 10 years?

b. In how many years will the population number 1000?

64. If left unchecked, a certain animal population doubles in size every 10 years.

a. If this population initially includes 50 members, about how many will it include in 12 years?

b. In how many years will the population number 500?

65. The number of bacteria in a culture at the start of an experiment is about 10^5. Four hours later the population has increased to about 10^7.

a. Approximate the size of the population after 6 hours.

b. When was the number of bacteria 250,000?

66. In 1960 the population of a certain town was 1,000, and in 1970 it was 4,000. What population can the city planning commission expect in 1994 if this growth rate continues?

67. Physicists tell us that the half-life of an isotope of strontium is 25 years. If a research center own 10 g of this isotope, how much will it have in 10 years?

68. A radioactive substance decays from 10 g to 6 g in 5 days. Find the half-life of the substance.

69. Specimens in geology and archaeology are dated by considering the exponential decay of a particular radioactive element found in the specimen. This technique is very reliable because radioactive disintegration is not affected by conditions such as pressure and temperature, which change the rate of ordinary chemical reactions. For dating to an age of about 60,000 years, the researcher considers the concentration of carbon 14, an isotope of carbon. In the cells of all living plants and animals, there is a fixed ratio of carbon 14 to ordinary stable carbon. However, when the plant or animal dies, the carbon 14 decreases according to the law of exponential decay.

a. The half-life of carbon 14 is about 5,600 years. Determine k, the decay constant, for this element.

b. What percentage of the carbon 14 is left in a specimen of bone that is 2,000 years old?

(*Note*: The radiocarbon dating method was developed in 1947 by Dr. Williard Libby, who received the Nobel prize in chemistry for his discovery. For specimens older than 60,000 years, too much carbon 14 has disintegrated for an accurate measurement. In such cases the researcher considers the concentration of other radioactive substances. For example, long periods of geologic time are frequently determined by measuring the presence of uranium 238, which has a half-life of about 4.5 billion years.)

70. One method of telling when a buried layer of soil was on the earth's surface is to use radioactive dating methods to measure the presence of Beryllium 10 $\left({}^{10}\text{Be}\right)$ in the soil sample. It is assumed that the Beryllium was deposited over the years at a constant rate through rainfall. In a recent research paper by M. Pavich and N. Vidic, the equation estimating the age of the soil was $t = -(1/\lambda)\ln(1-(\lambda N/q))$ where t gives the age of the soil in years, λ is the decay constant of ${}^{10}\text{Be}$, N is the amount of Beryllium in atoms per cm^2, and q is the rate of deposition of ${}^{10}\text{Be}$ in hundreds of cm of rainfall per year.

a. Solve this equation for N.

b. Find N when $t = 32{,}000$, $\lambda = -4.33\times10^{-7}$, and $q = 1.2\times10^{6}$.

71. The decay constant for carbon 14 is -0.000121. Approximately what percent of the initial carbon 14 remains in a fossil that is 10,000 years old?

72. The decay constant for carbon 14 is -0.000121. Approximately what percent of the initial carbon 14 remains in a fossil that is 100 years old?

73. The great physicist Ernest Rutherford was among the first to use radioactive decay to date the age of the earth. After analyzing the decay of radium and uranium in a pitchblende rock, he dramatically declared (about 100 years ago) to a geology professor at Cambridge University that he was sure the rock in his hand was 700 million years old. The decay constant for uranium 238 is -1.55×10^{-10}. What percentage of an initial amount of uranium 238 remains after 700 million years?

74. The most widely used method for radiometric dating employed by geologists today involves the radioactive decay of potassium 40, which has a decay constant of -5.55×10^{-10}. What percentage of an initial amount of potassium 40 remains after 700 million years?

75. A nuclear accident at Chernobyl in Ukraine in 1986 released dangerous amounts of the radioactive isotope cesium 137, which has a decay constant of -0.023. Approximately what percent of the cesium released in 1986 will remain in the year 2000?

76. The Chernobyl accident in 1986 also released dangerous amounts of strontium 90, which has a decay constant of -0.024. Approximately what percent of the strontium released in 1986 will remain in the year 2000?

77. A coroner examines the body of a murder victim at 9 A.M. and determines its temperature to be 88°F. An hour later the body temperature is down to 84°F. If the temperature of the room in which the body was found is 68°F, and if the victim's body temperature was 98°F at the time of death, approximate the time of the murder.

78. The temperature of a six-pack of beer (bought from a distributor on a hot summer day) is 90°F. The beer is placed in a refrigerator with a constant temperature of 40°F. If the beer cools to 60°F in 1 hour, when will the beer reach the more thirst-quenching temperature of 45°F?

79. A thermometer has been stored at 75°F. After 5 minutes outside it says 65°F. After another 5 minutes it says 60°F. What is the outside temperature?

80. A turkey which is at room temperature (68°F), is placed in a 350°F oven. After 10 minutes the temperature of the turkey is 90°F. About how long will it take the turkey to reach 300°F?

81. The analysis of growth given limited resources often involves equations containing e. A common form is called the **logistic growth function**. One of the characteristic forms of a logistic growth function is given by:

$$y = \frac{L}{1+be^{-kt}}.$$

To see what a typical logistic growth curve looks like, assume that $L = 100$, $b = 99$, and $k = 2$, and plot the graph of y. Construct a table for $t = 0, 1, 2, 3, 4, 5, 6$. In the given equation, y represents the size of the growing entity, and t represents elapsed time.

82. In 1920 two scientists, Pearl and Reed, published a formula that described the growth of the U.S. population over the years 1790 to 1910. Their formula is $N = \dfrac{197{,}273{,}000}{1+e^{-0.0314t}}$, where N is the population and t is the number of years elapsed since 1914 (t is negative for years before 1914). Their formula was also exactly on target for the years 1790, 1850, and 1910. What population does their formula yield for each of these years? Round to the nearest thousand.

Many phenomena exhibit limiting behavior that may be modeled by $y = L(1 - e^{kt})$. Exercises 83–88 illustrate this relationship.

83. An object falls from a great height and meets air resistance in such a way that its velocity in ft/second at time t is given by $v = 4(1 - e^{-8t})$.

a. Graph the velocity over the interval 0 to 3 seconds. Describe the general behavior of this graph.
b. Find the velocity when $t = 0.25$ seconds.
c. Approximately, what is the fastest this object will ever move?

84. A stone falls through a liquid in such a way that its velocity at time t is given by $v = 1.5(1 - e^{-2t})$, where v is in ft/second.

a. Graph the velocity over the interval 0 to 3 seconds. Describe the general behavior of this graph.
b. Find the velocity when $t = 1$ second.
c. Approximately, what is the fastest the stone will ever move?

85. Suppose that as an election campaign progresses, the percentage of voters who say they will vote for some previously unknown candidate increases and then "levels off" to the "final" percentage.

a. Find values for L and k, so that the final percentage of votes is 35 percent, and the vote percentage grows from 0 to 30 percent in 8 weeks.
b. About what percentage of voters are "for" the candidate 2 weeks into the campaign?
c. About how long does it take the candidate to have 15 percent of the vote?

86. Suppose that the continuous administration of an intravenous drug to a patient causes the amount of a chemical in the blood to rise to only a certain level because the body speeds up the rate of elimination of the chemical, in response to the increased level of the drug.

a. Find values for L and k, so that the limiting blood level is 4.5 units, and the level climbs from 0 to 4.4 in 10 hours.
b. About how long will it take the blood level to reach 2.25 units?
c. What is the blood level after 12 hours?

87. Determine the constants in $y = L(1 - e^{kt})$ so that the graph begins at the origin, passes through the point (1, 50) and then rises towards a limiting value of 100. For what value of t will $y = 75$?

88. Determine the constants in $y = L(1 - e^{kt})$ so that the graph begins at the origin, passes through the point (1, 1) and then rises towards a limiting value of 100. For what value of t will $y = 10$?

It is sometimes surprising to look into an old math book and see what applications were of interest in the past. The next two exercises are based on those in a book from 1924. The book is called *Calculus Made Easy: Being a very-simplest introduction to those beautiful methods of reckoning which are generally called by the terrifying names of the DIFFERENTIAL CALCULUS and the INTEGRAL CALCULUS* by S. P. Thompson, (MacMillan).

89. The damping on a telephone line can be ascertained from the relation $i = i_0 e^{-\beta\ell}$ where i is the strength, after t seconds, of a telephonic current of initial strength i_0. ℓ is the length of the line in kilometers. For the Franco-English submarine cable laid in 1910, $\beta = 0.0114$. Find the damping at the other end of the cable (40 kilometers).

90. The intensity I of a beam of light which has passed through a thickness ℓ of some transparent medium is $I = I_0 e^{-K\ell}$, where I_0 is the initial intensity of the beam and K is a "constant of absorption." Determine K if a beam has its intensity diminished by 18% in passing through 10 cm of a certain medium. What thickness will cut the intensity in half?

91. If a population has an annual birth rate of b babies per 1000 population, and an annual death rate of d people per 1000 population, then the size of the population at time t is given by $P = P_0 e^{\frac{b-d}{1000}t}$.

a. Describe what happens to the population if the death rate is equal to the birth rate.
b. What happens if the birth rate is greater than the death rate? How long will it take the population to double in size?
c. What happens if the birth rate is less than the death rate? How long will it take the population to be cut in half?

92. In one model for the spread of an epidemic, an equation gives the number of people in a population who remain susceptible to becoming infected as time goes on. This number will drop towards zero for the kind of disease a person can only get one time. Thus, the equation will have the form $x = x_0 e^{-kt}$. Graph this function for $k = 0.7$. If t is time in weeks, and there are originally 1000 people in the population, how long will it take before there are fewer than 100 susceptible people in the population?

93. In the seventeenth century many mathematicians investigated functions that could be expressed as infinite series. In 1668, Nicolaus Mercator worked out $\ln(1+x)=\frac{x}{1}-\frac{x^2}{2}+\frac{x^3}{3}-\frac{x^4}{4}+\cdots$. Use this formula to estimate ln 1.1 to four decimal places. How many terms are needed to achieve this degree of accuracy?

THINK ABOUT IT 7.6

1. From the following answers, identify *all* choices for k that result in the condition described.
   a. $0<k<1$ **b.** $k>1$ **c.** $k=0$ **d.** $k=1$
   e. $k<0$
   (i) $A=A_0e^{kt}$ describes growth (ii) $A=A_0(k)^t$ describes growth
   (iii) $A=A_0e^{kt}$ describes decay.
2. If an amount is invested at p percent compounded continuously, then a convenient rule of thumb for estimating the number of years required for the amount to double is

$$\text{doubling time} \approx \frac{70}{p}.$$

   For example, an amount will double in about 10 years at 7 percent, and in about 5 years at 14 percent. Explain the mathematical basis for this rule of thumb.
3. Given $A=A_0e^{kt}$, find an equation that expresses half-life as a function of the decay constant, k.
4. Simplify each expression algebraically.
   a. $\left(\frac{e^x+e^{-x}}{2}\right)^2-\left(\frac{e^x+e^{-x}}{2}\right)^2$
   b. $\left(\frac{e^2-1}{4e}+\frac{e^2+1}{2e}-e\right)-\left(\frac{e^2-1}{4e}-\frac{e^2+1}{2e}-\frac{1}{e}\right)$
5. Solve the following equations which are equations with quadratic form.
   a. $e^{2x}+4e^x=32$ **b.** $(\ln x)^2=\ln x^2$ **c.** $(\ln x)^2-3\ln x+1=0$

● ● ●

CHAPTER 7 SUMMARY

OBJECTIVES CHECKLIST

Specific chapter objectives are summarized below along with numbered example problems from the text that should clarify the objectives. If you do not understand any objectives, or do not know how to do the selected problems, then restudy the material.

7.1: Can you:

1. **Find the inverse of a function and its domain and range?** If $f=\{(7,-2),(9,1),(-3,2)\}$, find f^{-1}. Find and compare the domain and range of the two functions.
   [Example 1]
2. **Determine if a function has an inverse function?** If , $f=\{(-6,6),(0,0),(6,6)\}$ does f have an inverse function?
   [Example 2]
3. **Graph $y=f^{-1}(x)$ from the graph of $y=f(x)$?** Consider the graph of $y=f(x)$ in Figure 7.45. Use this graph to sketch $y=f^{-1}(x)$.
   [Example 4]

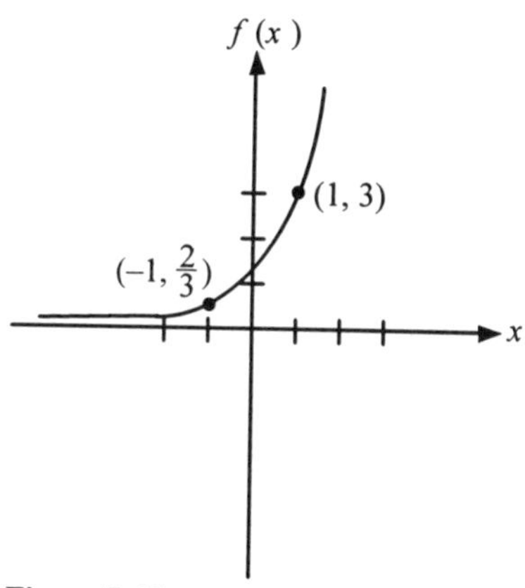

Figure 7.45

4. **Find an equation that defines f^{-1}?** Find $f^{-1}(x)$ for $f(x) = 7 - x/3$.

[Example 5a]

5. **Determine whether two functions are inverses of each other?** Verify that $f(x) = 5x - 4$ and $g(x) = (1/5)x + 4/5$ are inverses.

[Example 7]

6. **Use inverse function concepts in applications?** Certain airline employees accept a 20 percent cut in hourly wage to avoid layoffs. What percent raise will then be needed to return these employees to their original hourly wage?

[Example 8]

7.2: Can you:

1. **Determine function values for an exponential function?** If $f(x) = 9^x$, find $f(3)$, $f(-2)$, and $f(5/2)$. Also approximate $f(\sqrt{2})$ to the nearest hundredth.

[Example 2]

2. **Graph an exponential equation?** Sketch the graphs of $g(x) = 2^x$ and $h(x) = 4^x$ on the same coordinate system. Indicate domain and range on the graph.

[Example 3]

3. **Solve exponential equations using $b^x = b^y$ implies $x = y$?** Solve $32^x = (1/4)^{x+1}$.

[Example 7]

4. **Find the base in the exponential function $y = b^x$ given an ordered pair in the function?** Find the base of the exponential function $y = b^x$ that contains the point $(-1/3,\ 1/4)$.

[Example 8]

5. **Solve applied problems involving exponential functions?** A company purchases a new machine for \$3,000. The value of the machine depreciates at a rate of 10 percent each year. Find a formula (rule) showing the value (y) of the machine at the end of x years.

[Example 11a]

7.3: Can you:

1. **Convert from the exponential form $b^L = N$ to the logarithmic form $\log_b N = L$, and vice versa?** Write $4^2 = 16$ and $b^r = s$ in logarithmic form.

[Example 1]

2. **Determine the value of the unknown in expressions of the form $\log_b N = L$?** Determine the value of the unknown: $\log_9 3 = n$.

[Example 3]

3. **Determine the common logarithm and antilogarithm of a number by using a calculator?** If $\log N = -1.8$, find N to three significant digits.
[Example 5]

4. **Graph logarithmic functions?** Graph $y = \log_2 x$. Then sketch the graphs of $y = \log_2 x$ and $y = 2^x$ on the same coordinate system, and describe how the graphs are related.
[Example 6]

5. **Solve applied problems involving logarithmic functions?** The decibel gain G for an amplifier is defined by the equation $G = 10 \log P_2/P_1$, where P_1 is the input power and P_2 is the output power. Assume P_1 is the arbitrary reference value of 6 milliwatts (or 0.006 watt), and find the gain in decibels of a 60-watt amplifier.
[Example 9]

7.4: Can you:

1. **Use properties of logarithms to express certain log statements in terms of simpler logarithms or expressions?** Express $\log_b(xy/z)$ as a sum, difference, or product involving simpler logarithms.
[Example 1d]

2. **Use properties of logarithms to convert certain statements involving logarithms to a single logarithm with coefficient 1?** Express $3\log_b x + 2\log_b y$ as a single logarithm with coefficient 1.
[Example 5c]

3. **Use properties of logarithms in applications?** The intensity of the sound of riveting is about 10^{-2} watts per square meter. Find the loudness of this sound in decibels. Use the formulas given in the section-opening problem.
[Example 6]

7.5: Can you:

1. **Solve exponential equations by using logarithms?** Solve $3^x = 8$.
[Example 1]

2. **Apply the change of base formula?** Use logarithms to the base 10 to determine $\log_2 7$.
[Example 4a]

3. **Solve logarithmic equations?** Solve $\log_3(x^2 - 4) - \log_3(x + 2) = 2$.
[Example 6]

7.6: Can you:

1. **Use the compound-interest formula for an investment compounded *n* times per year?** \$100 is invested at 5 percent interest compounded quarterly. How much will it amount to in 3 years?
[Example 2]

2. **Use the compound-interest formula for an investment compounded continuously?** \$2,000 is invested at 6 percent compounded continuously. How much will the investment be worth in 10 years?
[Example 3]

3. **Graph exponential and logarithmic functions involving the number *e*?** Use a grapher to graph $y = e^x$.
[Example 4a]

4. **Solve exponential and logarithmic equations involving the number *e*?** Solve $\ln x = 1.9$.
[Example 7a]

5. **Solve applied problems involving continuous growth or decay?** Assume that in the absence of hunters a certain animal population triples every 11 years. Use the formula $A = A_0e^{kt}$, and find the size of a population that initially numbers 500 after 5 years.

[Example 9a]

KEY CONCEPTS AND PROCEDURES

Section	Key Concepts and Procedures to Review
7.1	• Definition of a one-to-one function and inverse functions • The special symbol f^{-1} is used to denote the inverse function of *f* • Methods to determine if the inverse of a function is a function, and how to find f^{-1}, if it exists • Horizontal line test • Procedures to use a grapher in parametric mode • *f* and f^{-1} interchange their domain and range. • The graphs of *f* and f^{-1} are reflections of each other across the line $y = x$.
7.2	• Definition of the exponential function with base *b* • For $b > 0$, b^x is increasing; for $0 < b < 1$, b^x is decreasing. • All previous laws of exponents hold for real number exponents. • For $f(x) = b^x$ with $b > 0$, $b \neq 1$: Domain: $(-\infty, \infty)$ Horizontal asymptote: *x*-axis Range: $(0, \infty)$ *y*-intercept: (0, 1) • If $b > 0$, $b \neq 1$, then $b^x = b^y$ implies $x = y$. • Methods from Section 1.3 to graph variations of $f(x) = b^x$
7.3	• Definitions of a logarithm, a common logarithm, and the logarithmic function with base *b* • $\log_b N = L$ is equivalent to $b^L = N$. • For $b > 0$, $b \neq 1$, and $x > 0$, $y = \log_b x$ if and only if $x = b^y$ and for $f(x) = \log_b x$: Domain: $(0, \infty)$ Vertical asymptote: *y*-axis Range: $(-\infty, \infty)$ *x*-intercept: (1, 0) • For $b > 1$, $y = \log_b x$ is an increasing function; for $0 < b < 1$, $y = \log_b x$ is a decreasing function. • The logarithmic function $y = \log_b x$ and the exponential function $y = b^x$ are inverse functions. Consequently, **a.** The functions interchange their domain and range. **b.** The graphs of the functions are symmetric about the line $y = x$. • Methods to graph $y = \log_b x$ using a grapher • Methods from Section 1.3 to graph variations of $f(x) = \log_b x$
7.4	• Properties of logarithms (for $b, x, y > 0$, $b \neq 1$ and *k* any real number): **1.** $\log_b xy = \log_b x + \log_b y$ **2.** $\log_b(x/y) = \log_b x + \log_b y$ **3.** $\log_b x^k = k \log_b x$ **4.** $\log_b b = 1$ **5.** $\log_b 1 = 0$ **6.** $\log_b b^x = x$ **7.** $b^{\log_b x} = x$

7.5
- If $x, y, b > 0$, with $b \neq 1$, then $x = y$ implies $\log_b x = \log_b y$, and $\log_b x = \log_b y$ implies $x = y$.
- Change of base formula: $\log_b x = \frac{\log_a x}{\log_a b}$
- Methods to solve exponential and logarithmic equations

7.6
- Definition of a natural logarithm
- As n gets larger and larger, $(1+1/n)^n$ gets closer and closer to an irrational number that is denoted by the letter e. To six significant digits, $e \approx 2.71828$
- Compound-interest formula: $A = P(1+\frac{r}{n})^{nt}$
- Continuous growth or decay formula: $A = A_0 e^{kt}$
  For $k > 0$, this formula describes growth; for $k < 0$ it describes decay.
- $y = \ln x$ and $y = e^x$ are inverse functions.
- Graphs of $y = \ln x$ and $y = e^x$
- $\ln e^x = x$ and, if $x > 0$, $e^{\ln x} = x$

CHAPTER 7 REVIEW EXERCISES

In Exercises 1–10 solve each equation for x. Use a calculator or tables where necessary.

1. $\log_5 5^7 = x$
2. $2^{-x} = 3$
3. $\ln 8 = \ln x - \ln 4$
4. $\log 0.123 = x$
5. $x^{0.1} = 5.67$
6. $x = e^{\ln a}$
7. $\log x = -2.4157$
8. $5^{x-3} = 1$
9. $e^{3x} = 4$
10. $\log(3x-2) - \log(x+4) = 1$

In Exercises 11 and 12 express the sum or difference as a single logarithm with coefficient 1.

11. $2\ln x + \frac{1}{3}\ln y$
12. $\log_a(x+h) - \log_a x$

In Exercises 13 and 14 express each as the sum or difference of simpler logarithms.

13. $\log_b \sqrt{\frac{x^3}{y}}$
14. $\ln\sqrt[3]{xy^2}$

15. What is the domain of $f(x) = \ln(3x-7)$?

16. Express in logarithmic form $4^{-1} = \frac{1}{4}$.

17. Express in exponential form: $\log_2 8 = 3$.

18. Simplify $\frac{n}{x}e^{n\ln x}$.

19. Sketch on the same axes the graphs of $f(x) = \left(\frac{1}{2}\right)^x$ and $g(x) = \log_{1/2} x$. What are the domain and the range of each function?

20. True or false: if false, give a specific counterexample.
$$\frac{\log_b x}{\log_b y} = \log_b x - \log_b y$$

21. If $y = \left(\frac{1}{4}\right)^x$, fill in the missing component of each of the following ordered pairs:
$$(3,\ ), (\ , 4), (\ , 1), (\tfrac{3}{2},\ ), (-2,\ ).$$

22. If \$500 is invested at 6 percent compounded annually, find a formula for the value (y) of the investment at the end of t years.

23. \$3,000 is invested at 5 percent compounded continuously. How much will the investment be worth in 4 years?

24. A radioactive substance decays from 10 g to 8 g in 11 days. Find the half-life of the substance.

25. If $f = \{(3,-1), (4,-2)\}$, find f^{-1}. Determine the domain and range of each function.

26. If $g(x) = 5^x$, find $g(\sqrt{2})$ to the nearest hundredth.

27. If $\log_{10} M = K$, find 100^K in terms of M.

28. Evaluate $10(1-e^{-x})$ when $x = 3.5$.

29. Use $\log_b 2 = x$ and $\log_b 3 = y$ to find $\log_b \sqrt[3]{6}$ in terms of x and y.

30. If $f(x) = \log_b x$, find $f\left(\frac{1}{b}\right)$.

31. If $h(x) = \log_4 x$, find $h(1)$.

32. If $g(x) = 9^x$, find $g\left(-\frac{3}{2}\right)$.

33. If $f(x) = 5^x - 2$, find the domain and range of f.

In Exercises 34–42 graph the given function,

34. $y = \ln x$ **35.** $y = 4^x$ **36.** $y = 10 - 10^{-x}$

37. $y = -\log_3(x-2)$ **38.** $y = e^x$

39 $y = \log_5 x$ **40.** $y = e^{-2x}$

41. The inverse of $f(x) = \log(x+3)$

42. The inverse of $f(x) = -x^3 - 2x - 2$

43. Which one of these functions has an inverse function?

a.

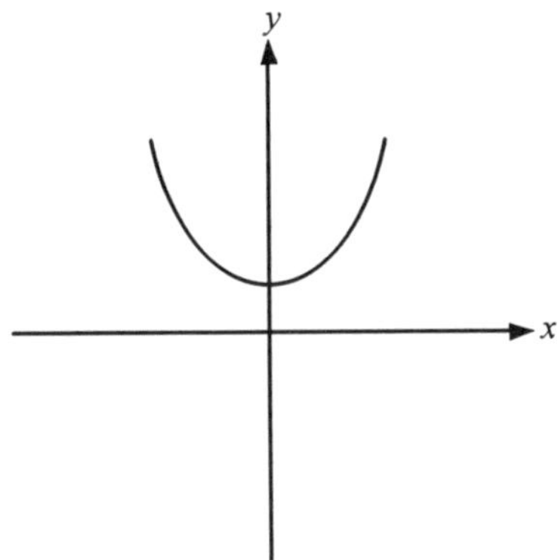

b.

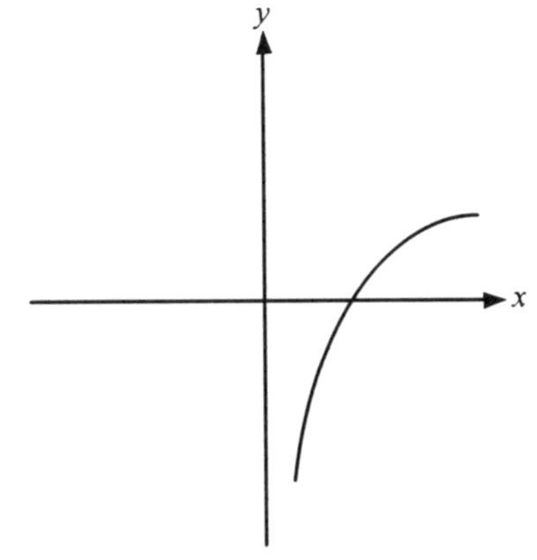

44. Solve: $9^x = \sqrt{3}$. **45.** Solve: $\left(\frac{1}{2}\right)^x = 2^{x+3}$.

46. Solve $a^{x-4} = \left(\frac{1}{a}\right)^x$ for x if a is a positive real number other than 1.

47. If $\log_b\left(\frac{1}{10}\right) = -\frac{1}{2}$, find b.

48. Find the base of the exponential function $y = b^x$ that contains the point $\left(\frac{3}{2}, 8\right)$.

49. Evaluate $\log_9 3 - \frac{1}{2}\log_5 1 + 2\log_3 27$.

50. Evaluate $\log_2 36 + \log_2 \frac{4}{9}$.

51. Solve $50 = 35 + 40e^{10k}$ for k by using natural logarithms.

52. Use logarithms to base 10 to evaluate $\log_4 9$.

53. Show that if $y = r + ce^{-kt}$, then $t = \frac{1}{k}\ln\frac{c}{y-r}$.

54. A common formula for pH is $\text{pH} = \log_{10}\frac{1}{[\text{H}^+]}$. Show that this formula is equivalent to the one given in this chapter. That is, show $\log_{10}\frac{1}{[\text{H}^+]} = -\log_{10}[\text{H}^+]$.

55. Liquid X has 10 times the concentration of hydrogen ions as liquid Y. What is the relation of the pH of X to the pH of Y?

56. If $\log_{10} 9! = x$, find $\log_{10} 10!$

57. If $g(x) = \frac{1}{2}x - \frac{5}{2}$, find

a. $g^{-1}(x)$ and **b.** $g^{-1}(-1)$.

58. Is the inverse of the function $f = \{(3,1), (0,0), (1,3)\}$, also a function?

59. If $f(x) = -\ln x$, find $f^{-1}(x)$. Specify the domain of f^{-1}.

60. If $f(x) = \sqrt{x} - 1$, find $f^{-1}(x)$. Specify the domain of f^{-1}.

61. Verify by the composition definition of inverse functions that $f(x)=\frac{1}{x+1}$ and $g(x)=\frac{1-x}{x}$ are inverses of each other.

62. Is the function $f(x)=x^3-3$ a one-to-one function? Is the inverse of *f* also a function?

In Exercises 63–72 select the choice that answers the question or completes the statement.

63. Which function is an exponential function?

a. $y=x^{-3}$ **b.** $y=x^{1/3}$
c. $y=3^{-1}$ **d.** $y=-3^x$

64. The inverse of the function $y=2x$ is

a. $y=-2x$ **b.** $y=\frac{1}{2}x$
c. $y=-\frac{1}{2}x$ **d.** $y=\frac{1}{2x}$

65. The graph of $f(x)=\log_3 x$ lies in quadrants

a. 1 and 2 **b.** 1 and 3
c. 1 and 4 **d.** 2 and 4

66. If $\log_{10} x^2=a$, then $\log_{10} 10x$ equals

a. $\frac{a+2}{2}$ **b.** $10a$
c. $a+1$ **d.** a^2

67. An expression equivalent to $\frac{\ln 27}{3}$ is

a. $\ln 9$ **b.** $-\ln 27$ **c.** $\ln 3$ **d.** $\ln 24$

68. An expression equivalent to $\frac{\log_{10} x}{\log_5 x}$ is

a. $\log_2 x$ **b.** 2
c. $\log_5 10$ **d.** $\log_{10} 5$

69. If $4^x=\sqrt{8}$, then x equals

a. $\frac{3}{4}$ **b.** $-\frac{2}{3}$ **c.** $\frac{3}{2}$ **d.** $\frac{1}{3}$

70. If $x^a=y^{a+1}$, and *x* and *y* are positive numbers other than 1, then *a* equals

a. $\frac{\log x}{\log x-\log y}$ **b.** $\frac{\log y}{\log x-\log y}$
c. $\frac{\log y-\log x}{\log x}$ **d.** $\frac{\log x+\log y}{\log y}$

71. If $f(x)=\ln x$, then $f(ab)$ equals

a. $f(a)\cdot f(b)$ **b.** $f(a+b)$
c. $f(a)+f(b)$ **d.** $f(a)-f(b)$

72. If $A/2=Ae^{-kt}$, then *k* equals

a. $\frac{\ln\frac{1}{2}}{t}$ **b.** $t\cdot\ln\frac{1}{2}$ **c.** $\ln\frac{2}{t}$ **d.** $\frac{\ln 2}{t}$

CHAPTER 7 TEST

1. If $(-2,a)$ is a solution of the equation $y=\left(\frac{1}{3}\right)^x$, find a.

2. Solve for x: $4^x=\frac{1}{8}$.

3. If $f(x)=10^x-3$, what is the range of *f*?

4. A company purchases a machine for \$8,000. If the value of the machine depreciates at a rate of 30 percent each year, what is the value of the machine 5 years after purchase?

5. **a.** Express in logarithmic form: $8^{2/3}=4$.
b. Express in exponential form: $\log_{10} 0.01=-2$.

6. If $f(x)=\log(x+2)$, what is the domain of *f*?

7. Graph $y=\log_2(x-1)$.

8. The hydrogen ion concentration $\left[H^+\right]$ in a sample of grapefruit juice is 5.7×10^{-4}. Use $pH=-\log_{10}\left[H^+\right]$ to determine the pH of this sample.

9. Express $\log_b\left(5\sqrt{x}\right)$ as the sum or difference of simpler logarithms.

10. If $\log_b 7=x$ and $\log_b 3=y$, express $\log_b\left(\frac{7}{9}\right)$ in terms of *x* and *y*.

11. Express $3\log_b 4-\log_b 32$ as a single logarithm with coefficient 1.

12. Graph $y=5\cdot2^{x-1}$.

13. Solve: $3^{x+2}=72$.

14. Solve: $\log_4 x + \log_4 (x-3) = 1$.

15. Evaluate $\log_4 25$ to four significant digits.

16. If $f(x) = x^3 - 5$, find $f^{-1}(x)$

17. Use the composition definition of inverse functions and verify that $f(x) = 7x - 2$ and $g(x) = \frac{1}{7}x + \frac{2}{7}$ are inverses of each other.

18. Given that the population of China is now growing at an annual rate of about 1.1 percent, find the time (to the nearest year), that it will take the population of China to double.

19. A 2-year certificate of deposit (CD) is purchased for \$4,000. If the CD pays 6 percent compounded monthly, then what is the value of the CD at maturity?

20. How long will it take for money invested at 9 percent compounded continuously to triple?

● ● ●

Chapter

8

Trigonometric Functions of Real Numbers

Photo Courtesy of J. C. LeJeune, Stock Boston

The **latitude** of the location of a point on the Earth's surface specifies the angle north or south of the equator between the location and the plane of the Earth's equator. Find, to the nearest 10 minutes, the latitude of Chicago, Illinois, which is 2,890 mi north of the equator. Assume that the Earth is a sphere of radius 3,960 mil.
(See Example 4 of Section 8.1.)

In Chapter 2 we initially viewed trigonometry in terms of right triangles, and then expanded our coverage to the general angle definitions of the trigonometric functions. We now consider the modern concept of the trigonometric functions of real numbers, which is the approach most useful in calculus. We begin by considering the radian measure of an angle. **Radians** are used extensively in calculus and are the link that makes a cohesive unit of trigonometry.

● ● ●

8.1 Radians

Objectives

1. Use $\theta = s/r$ to find either the central angle, the intercepted arc, or the radius, given measures for two variables in the relationship.
2. Convert from degree measure of an angle to radian measure, and vice versa.
3. Determine the area of a sector of a circle, given the radius and the central angle.
4. Find linear velocity, given an angular velocity and a radius.
5. Find angular velocity, given a linear velocity and a radius.

Up to now we have measured angles in degrees. However, in many applications of these functions, and in calculus, a different angle measure,—called a **radian**,—is more useful. We define the radian measure of an angle by first placing the vertex of the angle at the center of a circle. Let s be the length of the intercepted arc, and let r be the radius. Then

$$\theta = \frac{s}{r}$$

is the radian measure of the angle (see Figure 8.1). Equivalently, an angle measuring 1 radian intercepts an arc equal in length to the radius of the circle (see Figure 8.2). In plane geometry it is shown that *s varies directly as r*, so that we can find the radian measure of θ by using a circle of *any* radius.

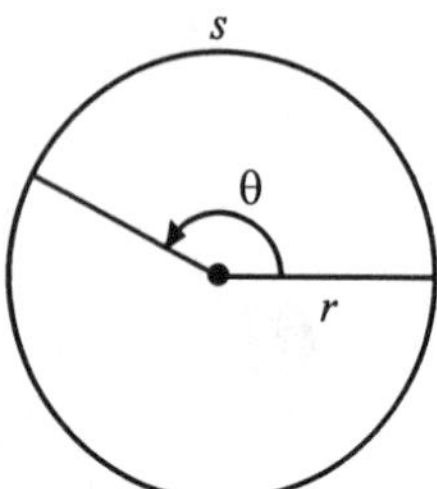

Figure 8.1

s = r
1 radian
r

Figure 8.2

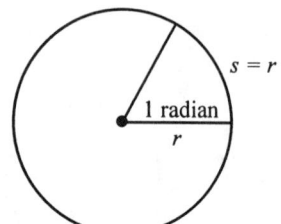

Example 1: Finding the Radian Measure of an Angle

A central angle in a circle of radius 3 in. intercepts an arc of 12 in. Find the radian measure of the angle.

Solution: Substituting $r = 3$ in. and $s = 12$ in. in the formula

$$\theta = \frac{s}{r},$$

we have

$$\theta = \frac{12 \text{ in.}}{3 \text{ in.}} = 4.$$

Thus, the radian measure of the angle is 4. Note that r and s are measured in the same unit, which divides out in the ratio. Thus, the radian measure of an angle is a number without dimension. Although the word "radian" is often added, an angle measure with no units means radian measure.

PROGRESS CHECK 1
In a circle a central angle intercepts an arc equal in length to one-half the circumference of the circle. What is the measure of the central angle in radians?
(**Hint:** Recall that $C = 2\pi r$.) ■

We may find the relation between degrees and radians by considering the measure of an angle that makes one complete rotation. In degrees, the measure of the angle is 360. Since the circumference of a circle is $2\pi r$, the radian measure is

$$\theta = \frac{s}{r} = \frac{2\pi r}{r} = 2\pi.$$

Thus,

$$360° = 2\pi \text{ radians}.$$

From this relation we derive the following conversion rules between degrees and radians.

Degree-Radian Conversion Rules

Degrees to radians formula:

$$1° = \frac{\pi}{180} \text{ radian} \approx 0.0175 \text{ radians}$$

Radians to degree formula:

$$1 \text{ radian} = \frac{180°}{\pi} \approx 57.3°.$$

Example 2: Converting from Degrees to Radians

Express 30°, 45°, and 270°, in terms of radians.

Solution: Using the degrees to radians formulas, we have

$$30° = 30 \cdot 1° = 30 \cdot \frac{\pi}{180} = \frac{\pi}{6}$$
$$45° = 45 \cdot 1° = 45 \cdot \frac{\pi}{180} = \frac{\pi}{4}$$
$$270° = 270 \cdot 1° = 270 \cdot \frac{\pi}{180} = \frac{3\pi}{2}.$$

PROGRESS CHECK 2
Express 90°, 100°, and 315° in terms of radians. ■

Example 3: Converting from Radians to Degrees

Express $\frac{\pi}{3}$, π, and $\frac{7\pi}{5}$ radians in terms of degrees.

Solution: Using the radians to degrees formula, we have

$$\frac{\pi}{3} = \frac{\pi}{3} \cdot 1 = \frac{\pi}{3} \cdot \frac{180°}{\pi} = 60°$$
$$\pi = \pi \cdot 1 = \pi \cdot \frac{180°}{\pi} = 180°$$
$$\frac{7\pi}{5} = \frac{7\pi}{5} \cdot 1 = \frac{7\pi}{5} \cdot \frac{180°}{\pi} = 252°.$$

PROGRESS CHECK 3
Express $\frac{\pi}{9}$, $\frac{2\pi}{5}$, and $\frac{5\pi}{2}$ radians in terms of degrees. ■

Example 4: Finding Latitude

Solve the problem in the chapter introduction on page 611.

Solution: A sketch of the situation in the problem is shown in Figure 8.3. Angle θ specifies the latitude for Chicago, and θ in radians is given by

$$\theta = \frac{s}{r} = \frac{2890 \text{ mi}}{3960 \text{ mi}} = \frac{289}{396}.$$

To convert this result to degrees, we multiply by $180°/\pi$ to obtain

$$\theta = \frac{289}{396} \cdot \frac{180°}{\pi} \approx 41.814°,$$

as shown in Figure 8.4. Finally, to convert to degree-minute format multiply the decimal portion of angle θ by 60 minutes, as shown in the last four lines of Figure 8.4. To the nearest 10 minutes, the decimal portion rounds to 50 minutes, so the latitude of Chicago is 41°50′ N.

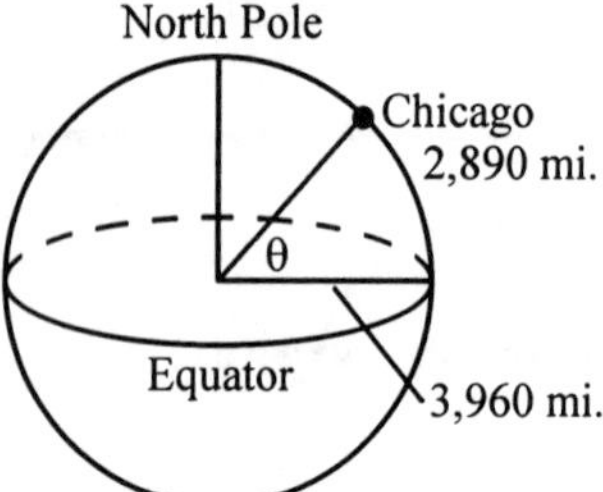

Figure 8.3

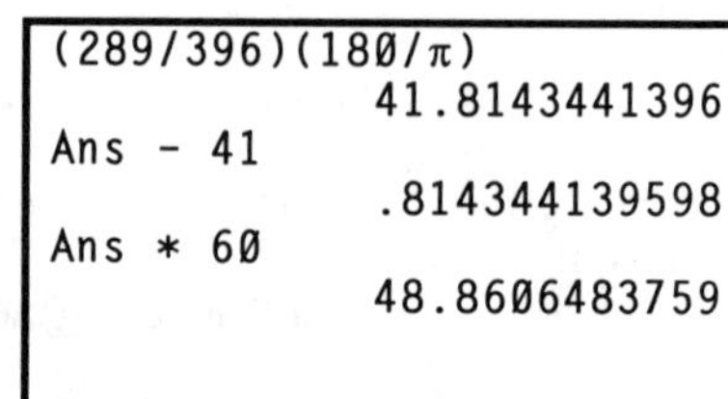

```
(289/396)(180/π)
                41.8143441396
Ans - 41
                .814344139598
Ans * 60
                48.8606483759
```

Figure 8.4

PROGRESS CHECK 4

Find, to the nearest 10 minutes, the latitude of Baltimore, Maryland, which is 2,720 mi north of the equator. Assume that the earth is a sphere of radius 3960 mi. ■

To illustrate another application of radians, consider the problem of determining the area of the shaded section in Figure 8.5. If you do not remember from geometry, notice intuitively that the area (A) varies directly as the central angle (θ). That is,

$$A = k\theta, \text{ where } k \text{ is a constant.}$$

We may find k, since we know that $A = \pi r^2$ when the angle (θ) is a complete rotation of 2π radians. Thus,

$$\pi r^2 = k \cdot 2\pi$$
$$\frac{1}{2}r^2 = k.$$

The formula (when θ is expressed in radians) is then

$$A = \frac{1}{2}r^2\theta.$$

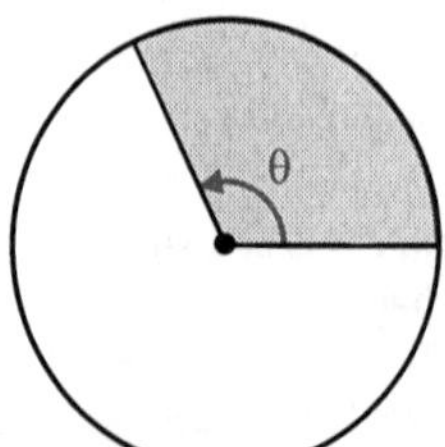

Figure 8.5

Example 5: Finding the Area of a Sector of a Circle

In a circle of radius 12 in., find the area of a sector whose central angle is 120°.

Solution: To use the formula, we must first convert the angle to radians.

$$120° = 120 \cdot \frac{\pi}{180} = \frac{2\pi}{3}$$

Now, substituting $r = 12$ and $\theta = 2\pi/3$ in the formula

$$A = \frac{1}{2} r^2 \theta,$$

we have

$$A = \frac{1}{2}(12)^2 \cdot \frac{2\pi}{3} = 48\pi \text{ in.}^2.$$

PROGRESS CHECK 5

Find the area of a sector of a circle whose central angle is 135° and whose radius is 8 ft. ■

One of the most important applications of radians concerns linear and angular velocity. If a point P moves along the circumference of a circle at a constant speed, then the linear velocity v is given by the formula

$$v = \frac{s}{t},$$

where s is the distance traveled by the point (or the length of the arc traversed), and t is the time required to travel this distance. During the same time interval, the radius of the circle connected to point P swings through θ angular units. Thus, the angular velocity ω (Greek lowercase omega) is given by

$$\omega = \frac{\theta}{t}.$$

We can relate linear and angular velocity through the familiar formula

$$s = \theta r.$$

Dividing both side of this equation by t, we obtain

$$\frac{s}{t} = \frac{\theta}{t} r.$$

Substituting v for s/t and ω for θ/t yields

$$v = \omega r.$$

Thus, the linear velocity is equal tot he product of the angular velocity and the radius.

Since $s = \theta r$ is valid only when θ is measured in radians, the angular velocity ω must be expressed in radians per unit of time when using the formula $v = \omega r$. However, in many common applications, the angular velocity is expressed in revolutions per minute. For these problems remember to convert ω to radians (rad) per unit of time through the relationship $1 \text{ rpm} = 2\pi \text{rad/minute}$, before using the formula.

Example 6: Angular Velocity and Linear Velocity

A circular saw blade 12 in. in diameter rotates at 1,600 rpm. Find

a. the angular velocity in radians per second and

b. the linear velocity in inches per second at which the teeth would strike a piece of wood.

Solution:

a. Since $1 \text{ rpm} = 2\pi \text{rad/minute}$, we have

$$\omega = 1{,}600 \text{ rpm} = 1{,}600\left(2\pi \frac{\text{rad}}{\text{minute}}\right) = 3{,}200\pi \frac{\text{rad}}{\text{minute}}.$$

To convert to rad/second, we use 1 minute = 60 seconds, as follows:

$$\omega = 3{,}200\pi \frac{\text{rad}}{\text{minute}} \frac{1 \text{ minute}}{60 \text{ seconds}} = \frac{160\pi}{3} \frac{\text{rad}}{\text{second}}.$$

b. Since ω is expressed in radians per unit of time, we find v by the formula $v = \omega r$, with $r = 6$ in.

$$\begin{aligned} v &= \omega r \\ &= \frac{160\pi}{3} \frac{\text{rad}}{\text{second}} \cdot 6 \text{ in.} \\ &= 320\pi \frac{\text{in.}}{\text{second}} \end{aligned}$$

Remember that the radian measure of an angle is a number without dimension, so the word *radian* does not appear in the units for linear velocity.

PROGRESS CHECK 6

A phonograph record 7 in. in diameter is being played at 45 rpm.

a. Find the angular velocity in radians per minute.

b. Find the linear velocity in inches per minute of a point on the circumference of the record. ■

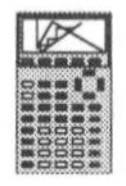

EXPLORE 8.1

1. To convert angles from degree measure to radian measure, you multiply by the conversion factor $\pi/180$ because an angle of 180 degrees is equal to an angle of π radians. Therefore, when you use a calculator to convert degrees to radians, the result is often a decimal that is an approximation to the correct value. It is helpful to recognize the decimal approximations for certain common fractions and multiples of π, so that you can also write the exact value. For instance, if you use a calculator to multiply 90° by $\pi/180$, the calculator shows 1.570796... It is convenient to recognize this value as an approximation of $\pi/2$.
By experimenting with a calculator, complete the following table showing the fraction of π approximated by the given decimal.

Decimal	Fraction or multiple of π
3.141592..	π
1.570796..	$\pi/2$
6.283185..	____
1.047197..	____
0.523598..	____
0.785398..	____
0.261799..	____
2.094395..	____
2.356194..	____

2. The values for the trigonometric functions of angles depend on the system used to measure the angles. For instance, the sine of 30 degrees is not the same as the sine of 30 radians. So you must know how to set your calculator to the correct mode of angle measurement. This is usually done through a Mode key. On some calculators, you can also use a degree symbol or a radian symbol to indicate the units.
   a. Use your calculator to find the sine of 30 degrees and the sine of 30 radians, each to four decimal places.
   b. As part **a** shows, in general it is not true that the sine of D degrees equals the sine of D radians. But, it is possible to find certain angles A, such that the sine of A degrees does equal the sine of A radians. Find one such nonzero angle, to the nearest hundredth.
3. In the study of chaos theory you explore the behavior of functions in which you use the current "output" as the next "input." Some functions tend toward a single value, some fluctuate periodically among a limited set of values, and some wildly fluctuate with no apparent pattern. [Source; R. Devaney, *Chaos, Fractals, and Dynamics*, Addison-Wesley]
   a. To examine the behavior of the function $y = \cos x$, pick any starting value for x, and get its cosine, then get the cosine of that result, and then the cosine of that result, etc. Keep doing this until you can describe what is happening. Does it matter if the calculator is in degree or radian mode?
   b. Examine the behavior of the sine function and the tangent function.

EXERCISES 8.1

In Exercises 1–8 complete the table by replacing each question mark with the appropriate number.

	The Radius Is	The Intercepted Arc Is	The Central Angle Is
1.	20 ft	100 ft	?
2.	8 in.	?	3
3.	?	56 yd	7
4.	5.2 m	5.2 m	?
5.	?	4.8 m	$\frac{1}{2}$
6.	11 ft	?	3.2
7.	1 unit	5 units	?
8.	1 unit	π units	?

In Exercises 9–28 express each angle in radian measure.

9.	30°	**10.**	45°	**11.**	60°	**12.**	90°
13.	120°	**14.**	135°	**15.**	150°	**16.**	180°
17.	210°	**18.**	225°	**19.**	240°	**20.**	270°
21.	300°	**22.**	315°	**23.**	330°	**24.**	360°
25.	200°	**26.**	75°	**27.**	20°	**28.**	162°

In Exercises 29–40 express each angle in degree measure.

29.	$\frac{\pi}{3}$	**30.**	$\frac{\pi}{4}$	**31.**	$\frac{2\pi}{9}$	**32.**	$\frac{\pi}{6}$
33.	$\frac{2\pi}{3}$	**34.**	$\frac{11\pi}{6}$	**35.**	$\frac{7\pi}{9}$	**36.**	$\frac{13\pi}{10}$

37. $\frac{12\pi}{5}$ **38.** $\frac{2\pi}{15}$ **39.** $\frac{7\pi}{18}$ **40.** $\frac{\pi}{12}$

In Exercises 41–46 find, to the nearest degree, the number of degrees in each angle. (**Note:** Use $\pi \approx 3.14$.)

41. 2 **42.** 1 **43.** 4

44. 6 **45.** 5.8 **46.** 3.5

47. Refer to Example 4 of Section 8.1. Find, to the nearest 10 minutes, the latitude of Miami, Florida, which is about 1,762 miles north of the equator.

48. Refer to Example 4 of Section 8.1. Find, to the nearest 10 minutes, the latitude of Bellingham, Washington, which is about 3,366 miles north of the equator.

49. **a.** Starting at the equator, if a ship sails north at 20 miles per hour, then to the nearest degree, what will its latitude be in 24 hours?
b. Find the ship's average speed in degrees per hour.

50. **a.** Starting at the equator, if a ship sails south at 25 miles per hour, then to the nearest degree, what will its latitude be in 24 hours?
b. Find the ship's average speed in degrees per hour.

51. If you consider Earth to be a sphere of radius 3,960 miles, then, to the nearest mile, how many miles north of the equator is latitude 1 degree?

52. If you consider Earth to be a sphere of radius 6,373 kilometers, then, to the nearest kilometer, how many kilometers north of the equator is latitude 1 degree?

For Exercises 53 and 54: When a ship travels at 4 **knots**, that means it travels 4 **nautical miles per hour**. The original idea of a nautical mile was 1 minute of arc along the equator. For ordinary calculations, the nautical mile can be considered equal to 1.15 mi.

53. If a ship travels due north from latitude 0° to latitude 2° in 12 hours, find its average speed in knots, to the nearest whole number.

54. If a ship travels from latitude 10° to latitude 15° in 24 hours, find its average speed in knots, to the nearest whole number.

55. Find the area of a sector of a circle whose central angle is $\frac{\pi}{4}$ and whose radius is 10 in.

56. Find the area of a sector of a circle whose central angle is 120° and whose radius is 5 in.

57. In a circle of radius 2 ft, the arc length of a sector is 8 ft. Find the area of the sector.

58. In a circle of radius 5 ft, the arc length of a sector is 3 ft. Find the area of the sector.

59. The radius of a wheel is 20 in. When the wheel moves 110 in., through how many radians does a point on the wheel turn? How many revolutions are made by the wheel?

60. The radius of a wheel is 16 in. Find the number of radians through which a point on the circumference turns when the wheel moves a distance of 2 ft. How many revolutions are made by the wheel?

61. What is the angular velocity in radians/second of an object that moves along the circumference of a circle of radius 9 in. with a linear velocity of 54 in./second?

62. What is the angular velocity in radians/second of the minute hand of a clock?

63. A phonograph record 12 in. in diameter is being played at $33\frac{1}{3}$ rpm.

a. Find the angular velocity in radians/minute.
b. Find the linear velocity in inches/minute of a point on the circumference of the record.

64. A pulley 20 in. in diameter makes 400 rpm.
a. Find the angular velocity in radians/second.
b. Find the speed in inches/minute of the belt that drives the pulley.
(**Note:** The speed of a point on the circumference of the pulley is the same as the speed of the belt.)

65. In a dynamo, an armature 12 in. in diameter makes 1,500 rpm. Find the linear velocity in inches/second of the tip of the armature.

66. What is the linear velocity in inches/second of an object that moves along the circumference of a circle of radius 12 in. with an angular velocity of 7 rad/second?

67. A train is traveling at 120 mi/hr (176 ft/second) on wheels 40 in. in diameter. Find the angular velocity of the wheels in radians per minute.

68. A car is traveling at 60 mi/hr (88 ft/second) on tires 30 in. in diameter. Find the angular velocity of the tires in radians/second.

69. Arcs of circles are used in many styles of molding that can be found in homes. Consider the figure shown here. If $\overline{AB} = \overline{DE} = \frac{1}{2}\overline{BC} = \frac{1}{2}\overline{CD}$ and $\overline{AB} = \frac{1}{2}$ in., what is the length of the curved portion of the molding?

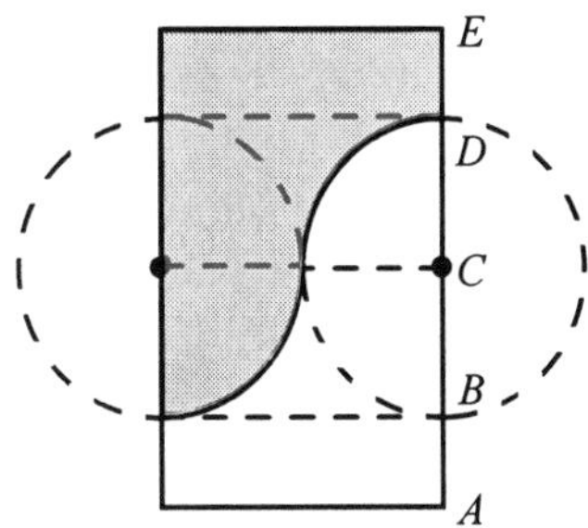

70. Refer to Exercise 69. If the length of $\overline{AB}$ is $\frac{3}{4}$ in., what is the length of the curved portion of the molding?

THINK ABOUT IT 8.1

1. **a.** If a central angle of α radians intercepts an arc of length a in a circle of radius b, express b in terms of α and a.
   b. In a circle a central angle intercepts an arc equal in length to 5/8 of the diameter of the circle. What is the measure of the central angle in radians?
2. Express $74°25'12''$ in radian measure to four significant digits.
3. In the history of sailing it was a major challenge to determine longitude at sea accurately, until good clocks were invented that would work on a ship. Such clocks were not readily available until about the time of the American Revolution. In determining longitude the idea was to compare noon on the ship at sea (when the sun was at its highest point) to the time on a clock that indicated noon at the port from which the ship had sailed. Thus, while at sea, the clock had to be able to tell what the time was in the home port. This ability was of great importance because military and commercial success depended on being able to navigate the open sea. [For a fascinating history of the competition to achieve this capability, read *Longitude* by Dava Sobel, Walker and Co., 1995.]
   a. Since Earth takes 24 hours to go from one noon to the next at a given place, Earth rotates through 360° in 24 hours. Through how many degrees does Earth rotate in 1 hour?
   b. At the equator (where the radius of Earth is 3,960 miles), to the nearest mile, how many miles are there in 15° of longitude?
   c. At 60° N latitude how many miles are there in 15° of longitude? [Hint: Find the radius of the latitude circle at 60° N.]
4. The formula $s = \theta r$ may be used to estimate certain distances when θ is a small angle (to about 10°). For example, consider the given illustration. An observer measures the angle from Earth to opposite ends of the sun's diameter to be 0.53°. Since θ is a small angle, the arc length s is a good approximation for the diameter d. If the distance from Earth to the sun is 93,000,000 mi, determine the diameter of the sun to two significant digits.

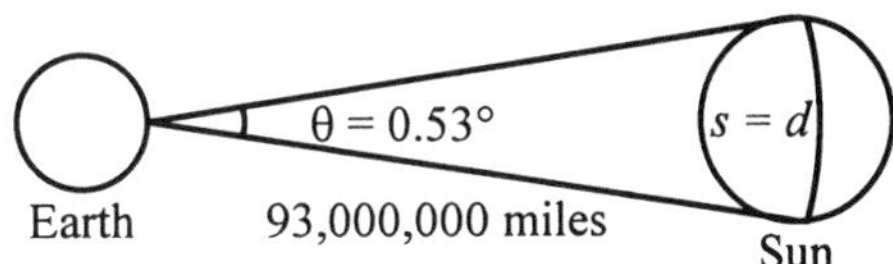

5. In architecture, arcs of circles are basic to the construction of many beautiful designs. For example, consider the quatrefoil in this illustration. If the side of the square measures 4 in., determine

a. the perimeter of the quatrefoil and
b. the area of the quatrefoil (the colored portion of the figure.)

• • •

8.2 Trigonometric Functions of Real Numbers

*Photo Courtesy of **Glen Kulbako**, The Picture Cube*

The monthly revenue R in dollars from sales of a sun screen product in the New York region during a certain year is approximated by

$$R = 6{,}500 - 5{,}000\cos(0.5236t),$$

where t is the number of months that have elapsed from February 1. Following this model, estimate the monthly revenue for this product for March and for August of the year in question. (See Example 8.)

Objectives

1. Determine the exact value of any trigonometric function of an arc length that terminates at one of the axes.
2. Determine the exact value of any trigonometric function of an arc length with a reference arc of $\pi/6$, $\pi/4$, or $\pi/3$.
3. Use a calculator to approximate a trigonometric value of a real number.
4. Solve applied problems involving the evaluating of a trigonometric function of a real number.

The revenue for a company from sales of a seasonal product, which is considered in the section-opening problem, provides one example of the many common events that repeat over definite periods of time. The waves broadcast by a radio station; the rhythmic motion of the heart; alternating electric current; weather-related issues such as air pollution; and the economic pattern of expansion, retrenchment, recession, and recovery—all these occur in cycles. The trigonometric functions, also repetitive, are very useful in analyzing such periodic phenomena. However, the independent variable in these applications is the time, not the angle. Thus, we need to define the trigonometric functions in terms of real numbers, not degrees. Since radians measure angles in terms of real numbers, they are the starting point for our discussion of the trigonometry of real numbers.

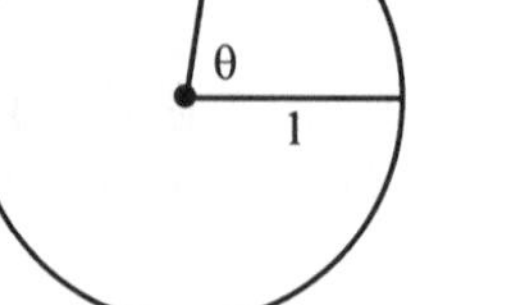

Figure 8.6

Consider Figure 8.6, in which the central angle θ is measured in radians. It is convenient to label the radius of the circle as 1 unit so that $r = 1$. Then

$$\theta = \frac{s}{r} = \frac{s}{1}$$
$$\theta = s.$$

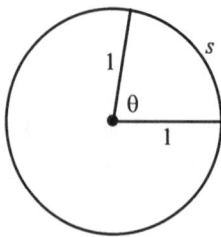

In a unit circle the same real number measures both the central angle θ and the intercepted arc s. Thus, we may base the definitions of the trigonometric functions on either an angle or an arc length. The results are valid for both interpretations. Since angles are meaningless in periodic phenomena, it is useful to emphasize the interpretation of a real number as the measure of an arc length s. We liberally interpret arc length as the distance traveled by a point as it moves around the unit circle, repeating its behavior every 2π units.

To illustrate more forcefully the correspondence between real numbers and arc lengths, consider a unit circle with its center at the origin of the Cartesian coordinate system. The equation of this circle is $x^2 + y^2 = 1$. Through the point (1, 0) and parallel to the y-axis we draw a real number line, labeled s. The zero point of s coincides with the point (1, 0) on the circle. Units are marked off in the same scale as the y-axis (see Figure 8.7).

If the positive half of s is wrapped around the circle counterclockwise, and the negative half of s is wrapped around the circle clockwise, we establish a one-to-one correspondence between real numbers and arc lengths of the circle (see Figure 8.8). Thus, each numbered point on s coincides with exactly one point on the circle, and the length of an arc may be read from this curved s-axis. We now relate this discussion to trigonometry.

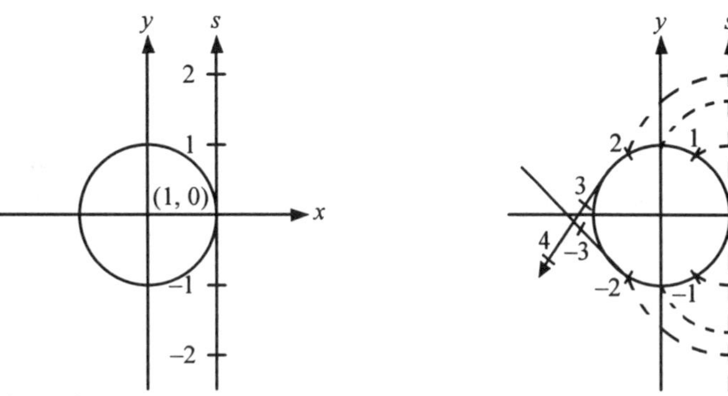

Figure 8.7 **Figure 8.8**

Definition of the Sine and Cosine Functions

> Consider a point (x, y) on the unit circle $x^2 + y^2 = 1$ at arc length s from (1, 0). We define the cosine of s to be the x-coordinate of the point, and the sine of s to be the y-coordinate.
>
> $$\cos s = x$$
> $$\sin s = y$$

Note in Figure 8.9 that our definitions are consistent with the definitions of sine and cosine in terms of the sides and hypotenuse of a right triangle. The domain of both the sine and cosine functions is the set of all real numbers, since the arc length s is determined by wrapping a real number line around the unit circle. The x- and y-coordinates in a unit circle vary between −1 and 1 (inclusive). Thus, for the range we have

$$-1 \le \cos s \le 1$$
$$-1 \le \sin s \le 1.$$

The remaining trigonometric functions are defined as follows:

Name of Function	Abbreviation	Ratio
Tangent of s	$\tan s$	$y/x = \sin s/\cos s \quad (x \neq 0)$
cotangent of s	$\cot s$	$x/y = 1/\tan s \quad (y \neq 0)$
secant of s	$\sec s$	$1/x = 1/\cos s \quad (x \neq 0)$
cosecant of s	$\csc s$	$1/y = 1/\sin s \quad (y \neq 0)$

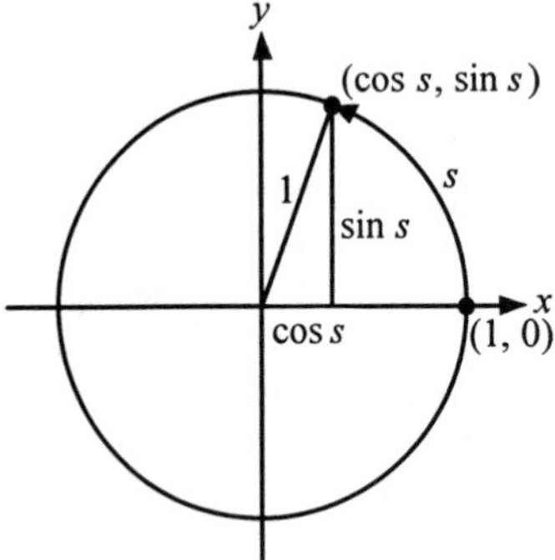

Figure 8.9

We evaluate trigonometric functions of real numbers by using these definitions or by using a calculator. Once again, the calculator method is most efficient for evaluating a trigonometric expression. Just set the calculator for Radian mode and use the appropriate function keys, as discussed in Section 2.1. However, this shortcut bypasses many important ideas in trigonometry. So let us consider how to evaluate the trigonometric functions from their definitions, as we did in Section 2.3.

Using the above definitions, we find the values of the trigonometric functions by determining the rectangular coordinates (x, y) of points on the unit circle. For certain real numbers these coordinates are easy to find. For example, let us determine the values of the trigonometric functions of zero. The coordinates for an arc length of zero are (1, 0) (see Figure 8.10). Thus,

$$\sin 0 = y = 0 \quad \leftarrow \text{reciprocals} \rightarrow \quad \csc 0 = \frac{1}{y} = \frac{1}{0} \text{ undefined}$$

$$\cos 0 = x = 1 \quad \leftarrow \text{reciprocals} \rightarrow \quad \sec 0 = \frac{1}{x} = \frac{1}{1} = 1$$

$$\tan 0 = \frac{x}{y} = \frac{0}{1} = 0 \quad \leftarrow \text{reciprocals} \rightarrow \quad \cot 0 = \frac{x}{y} = \frac{1}{0} \text{ undefined.}$$

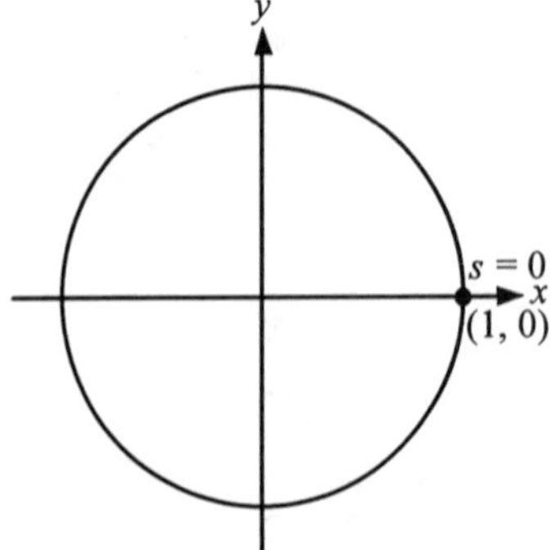

Figure 8.10

We can repeat this procedure for other arc lengths that terminate at one of the axes. Since the circumference of a circle or radius r is $2\pi r$, the circumference of a unit circle is 2π. The x- or y-axes may then intersect the unit circle at arc lengths of 0, $\pi/2$, π, $3\pi/2$, and so on. In the next example we consider how to find an exact trigonometric value when the arc length is $\pi/2$, π, or $3\pi/2$.

Example 1: Evaluating Trigonometric Functions by Definition

Use the definition of the trigonometric functions to find each function value.

a. $\sin\frac{\pi}{2}$ **b.** $\cos\pi$ **c.** $\sec\frac{3\pi}{2}$

Solution: Figure 8.11 shows the points on the unit circle that are assigned to $\pi/2$, π, and $3\pi/2$.

a. The point (0, 1) is assigned to $\pi/2$, and $\sin s = y$, so $\sin \pi/2 = 1$.

b. The point (–1, 0) is assigned to π, and $\cos s = x$, so $\cos \pi = -1$.

c. The point $(0, -1)$ is assigned to $3\pi/2$, and $\sec s = 1/x$, so $\sec(3\pi)/2 = 1/0$, which is undefined.

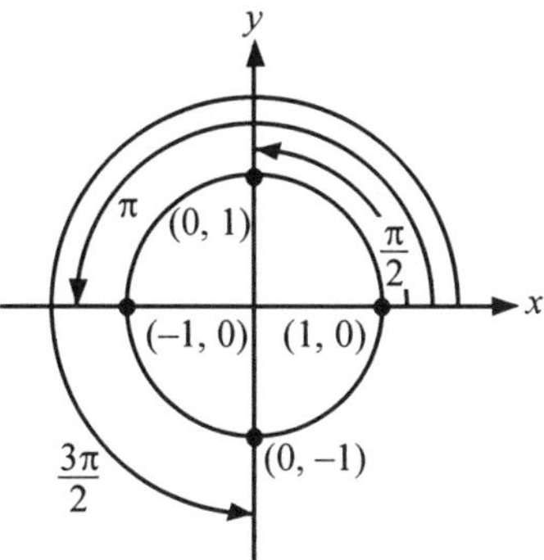

Figure 8.11

Technology Link

You should check these trigonometric evaluations by calculator, using the following guidelines on calculator usage.

1. Be sure to set the calculator for Radian mode.
2. Most calculators have a key labeled π that should be used to evaluate expressions involving π, as in this example.
3. You will often need to use parentheses around the argument of the trigonometric function. For instance,

$$\sin \pi/2 \text{ means } \frac{\sin \pi}{2},$$

$$\text{while } \sin(\pi/2) \text{ means } \sin\frac{\pi}{2}.$$

4. You may use

$$\csc s = \frac{1}{\sin s}, \ \sec s = \frac{1}{\cos s}, \text{ or } \cot s = \frac{1}{\tan s}$$

to evaluate cosecant, secant, or cotangent expressions, respectively.

PROGRESS CHECK 1

Use the definition of the trigonometric functions to find each function value.

a. $\sin\frac{3\pi}{2}$ **b.** $\tan\pi$ **c.** $\csc\frac{\pi}{2}$ **d.** $\cot\frac{\pi}{2}$ ■

Trigonometric evaluations involving 0, $\pi/2$, π, and $3\pi/2$ appear often in trigonometry. It is, therefore, useful to use the methods shown in Example 1 to create the following table, which lists the exact values of all the trigonometric functions of these special numbers. Note that tabular values match our results for 0°, 90°, 180° and 270° from Section 2.3.

s	**sin s**	**csc s**	**cos s**	**sec s**	**tan s**	**cot s**
0	0	undefined	1	1	0	undefined
$\pi/2$	1	1	0	undefined	undefined	0
π	0	undefined	−1	−1	0	undefined
$3\pi/2$	−1	−1	0	undefined	undefined	0

Other numbers terminate at one of the axes, but their trigonometric values are the same as one of the four listed. For example, the trigonometric values of 2π are the same as the trigonometric values of 0, since both numbers are assigned the point (1, 0) on the unit circle. The numbers 4π, 6π, -2π, and -2π are also assigned this point. A basic fact in our development is that in laying off a length 2π, we pass around the circle and return to our original point. Thus, the x- and y-coordinates repeat themselves at intervals of length 2π, and for any trigonometric function f we have

$$f(s+2\pi k)=f(s), \quad \text{where } k \text{ is an integer.}$$

For example,

$$f(4\pi)=f(0+2\pi(2))=f(0)$$
$$f(-2\pi)=f(0+2\pi(-1))=f(0).$$

This observation is very important because it means that if we determine the values of the trigonometric functions in the interval $[0,\ 2\pi)$, we know their values for all real s.

Example 2: Using $f(s+2\pi k)=f(s)$

Find $\sin 7\pi$, $\cos 12\pi$, $\tan\frac{11\pi}{2}$, $\cot\frac{17\pi}{2}$, $\sec(-5\pi)$, and $\csc\frac{-5\pi}{2}$.

Solution: Since the values of the trigonometric functions repeat themselves at multiples of 2π, we have

$$\sin 7\pi=\sin(\pi+6\pi)=\sin[\pi+2\pi(3)]=\sin\pi=0$$
$$\cos 12\pi=\cos(0+12\pi)=\cos[0+2\pi(6)]=\cos 0=1$$
$$\tan\frac{11\pi}{2}=\tan 5\tfrac{1}{2}\pi=\tan\left(\frac{3\pi}{2}+4\pi\right)=\tan\left[\frac{3\pi}{2}+2\pi(2)\right]=\tan\frac{3\pi}{2} \quad \text{undefined}$$
$$\cot\frac{17\pi}{2}=\cot 8\tfrac{1}{2}\pi=\cot\left(\frac{\pi}{2}+8\pi\right)=\cot\left[\frac{\pi}{2}+2\pi(4)\right]=\cot\frac{\pi}{2}=0$$
$$\sec(-5\pi)=\sec[\pi+(-6\pi)]=\sec[\pi+2\pi(-3)]=\sec\pi=-1$$
$$\csc\left(\frac{-5\pi}{2}\right)=\csc(-2\tfrac{1}{2}\pi)=\csc\left[\frac{3\pi}{2}+(-4\pi)\right]=\csc\left[\frac{3\pi}{2}+2\pi(-2)\right]=\csc\frac{3\pi}{2}=-1.$$

Check these results by calculator.

PROGRESS CHECK 2

Find $\sin\frac{7\pi}{2}$, $\cos 13\pi$, $\tan\frac{-5\pi}{2}$, $\cot\frac{-5\pi}{2}$, $\sec 8\pi$, and $\csc\frac{15\pi}{2}$. ■

A **trigonometric identity** is a statement that is true for all real numbers for which the expressions are defined. We simplify our work by developing identities that relate a trigonometric function of a negative number to the same function of a positive number. The symmetry of the unit circle makes these identities easy to derive. Consider Figures 8.12 and 8.13, which illustrate the symmetry for two possible values of *s*. Note that the numbers *s* and –*s* are assigned the same *x*-coordinate, so

$$\cos(-s) = \cos s.$$

The *y*-coordinates differ only in their sign. Thus,

$$\sin(-s) = -\sin s.$$

The remaining functions are ratios of the sine and cosine, so it follows that

$$\begin{aligned}\tan(-s) &= -\tan s\\ \cot(-s) &= -\cot s\\ \sec(-s) &= \sec s\\ \csc(-s) &= -\csc s.\end{aligned}$$

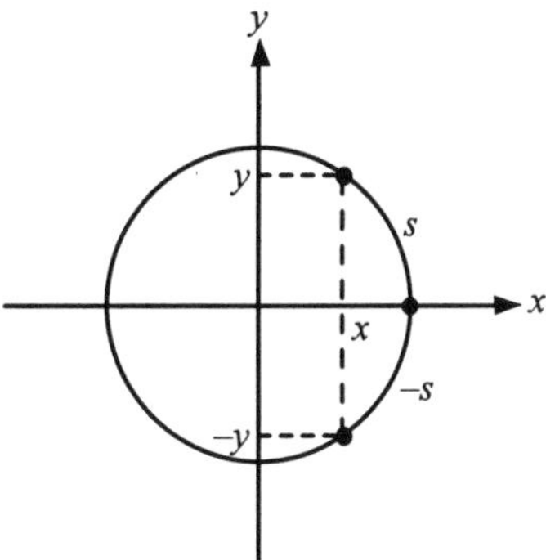

Figure 8.12

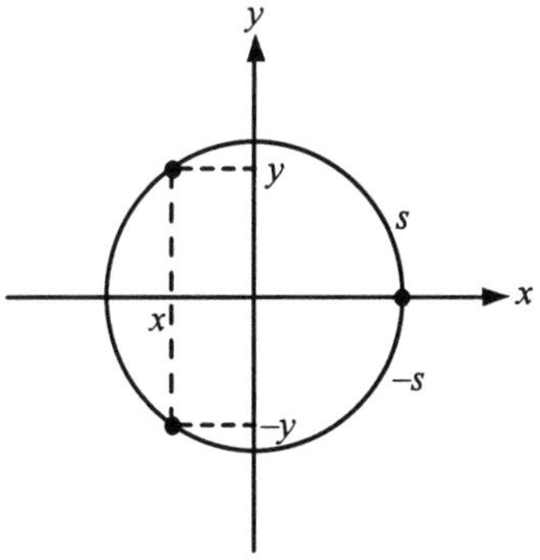

Figure 8.13

Example 3: Using Negative Angle Identities

Find $\cos\frac{-3\pi}{2}$ and $\csc\frac{-5\pi}{2}$, using negative angle identities..

Solution: $\cos(-s) = \cos s$ and $\csc(-s) = -\csc s$, so

$$\begin{aligned}\cos\left(\frac{-3\pi}{2}\right) &= \cos\frac{3\pi}{2} = 0\\ \csc\left(\frac{-5\pi}{2}\right) &= -\csc\frac{5\pi}{2} = -\csc 2\frac{1}{2}\pi = -\csc\left(\frac{\pi}{2}+2\pi\right)\\ &= -\csc\frac{\pi}{2} = -1.\end{aligned}$$

You should confirm these results with a calculator.

PROGRESS CHECK 3

Find $\sin\frac{-\pi}{2}$ and $\sec -3\pi$, using negative angle identities. ■

Besides arc lengths that terminate at one of the axes, the numbers $\pi/4$, $\pi/3$, and $\pi/6$ are also considered special numbers. These numbers correspond to angles of 45°, 60°, and 30°, respectively, and in Section 2.1 we derived their exact trigonometric values. These results are given in the following table and repeated on the endpaper at the back of the book.

s	$\sin s$	$\csc s$	$\cos s$	$\sec s$	$\tan s$	$\cot s$
$\frac{\pi}{3}$	$\frac{\sqrt{3}}{2}$	$\frac{2}{\sqrt{3}}$	$\frac{1}{2}$	2	$\sqrt{3}$	$\frac{1}{\sqrt{3}}$
$\frac{\pi}{4}$	$\frac{1}{\sqrt{2}}$	$\sqrt{2}$	$\frac{1}{\sqrt{2}}$	$\sqrt{2}$	1	1
$\frac{\pi}{6}$	$\frac{1}{2}$	2	$\frac{\sqrt{3}}{2}$	$\frac{2}{\sqrt{3}}$	$\frac{1}{\sqrt{3}}$	$\sqrt{3}$

The above results can also be obtained by unit circle derivations. Let us illustrate in the case of $s = \pi/4$. Since $\pi/4$ is halfway between 0 and $\pi/2$, $\pi/4$ is assigned the midpoint of the arc joining the points (1, 0) and (0, 1) on the unit circle (see Figure 8.14. Thus, the x- and y-coordinates are equal (that is, $x = y$). Since the coordinates. of any point on the circle satisfy the equation $x^2 + y^2 = 1$, we have

$$\begin{aligned} x^2 + x^2 &= 1 \\ x^2 &= \frac{1}{2} \\ x &= \frac{1}{\sqrt{2}} \text{ or } -\frac{1}{\sqrt{2}} \end{aligned}$$

Since x and y are positive in the first quadrant, we have

$$x = y = \frac{1}{\sqrt{2}}.$$

The trigonometric values of $\pi/4$ are then

$$\begin{aligned} &\sin\frac{\pi}{4} = y = \frac{1}{\sqrt{2}} && \leftarrow \text{reciprocals} \rightarrow && \csc\frac{\pi}{4} = \frac{1}{\sin(\pi/4)} = \sqrt{2} \\ &\cos\frac{\pi}{4} = x = \frac{1}{\sqrt{2}} && \leftarrow \text{reciprocals} \rightarrow && \sec\frac{\pi}{4} = \frac{1}{\cos(\pi/4)} = \sqrt{2} \\ &\tan\frac{\pi}{4} = \frac{\sin(\pi/4)}{\cos(\pi/4)} = 1 && \leftarrow \text{reciprocals} \rightarrow && \cot\frac{\pi}{4} = \frac{\cos(\pi/4)}{\sin(\pi/4)} = 1. \end{aligned}$$

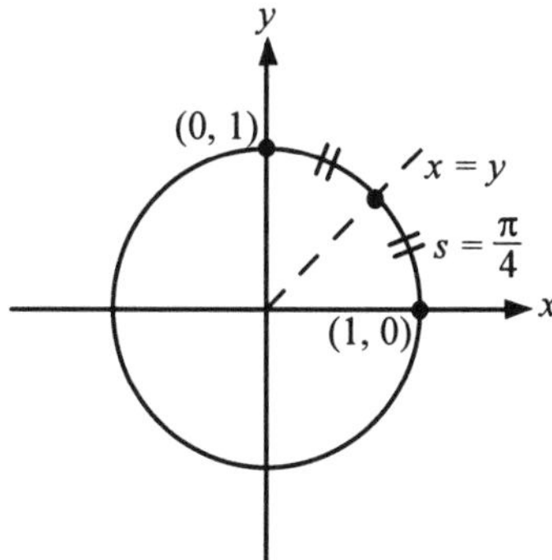

Figure 8.14

Example 4: Evaluations Involving $\pi/3$, $\pi/4$ or $\pi/6$

Find the exact values of $\sin\frac{-\pi}{3}$, $\cos\frac{-\pi}{4}$, $\sin\frac{9\pi}{4}$, and $\tan\frac{13\pi}{3}$.

Solution: Using the table just developed and our previous procedure, we have

$$\sin\left(-\frac{\pi}{3}\right) = -\sin\frac{\pi}{3} = -\frac{\sqrt{3}}{2}$$

$$\cos\left(-\frac{\pi}{4}\right) = \cos\frac{\pi}{4} = \frac{1}{\sqrt{2}}$$

$$\sin\frac{9\pi}{4} = \sin 2\frac{1}{4}\pi = \sin\left(\frac{\pi}{4} + 2\pi\right) = \sin\frac{\pi}{4} = \frac{1}{\sqrt{2}}$$

$$\tan\frac{13\pi}{3} = \tan 4\frac{1}{3}\pi = \tan\left(\frac{\pi}{3} + 4\pi\right) = \tan\frac{\pi}{3} = \sqrt{3}.$$

To check these results by calculator, compare the decimal approximation from the calculator for a particular trigonometric evaluation to the approximation for the corresponding radical value. Such a check is shown in Figure 8.14 for the evaluation of $\sin(9\pi/4)$.

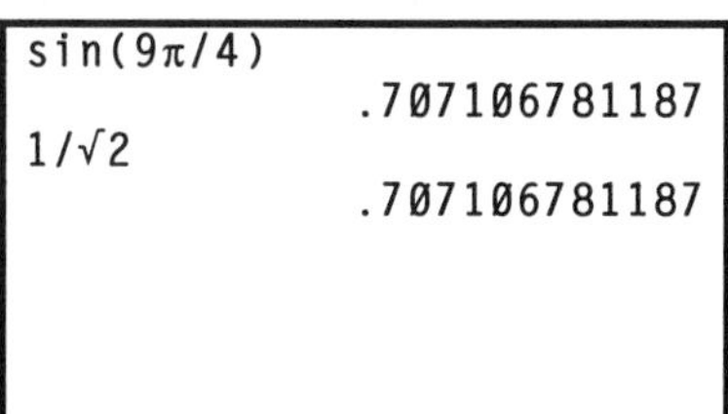

Figure 8.14

PROGRESS CHECK 4

Find the exact values of $\tan\frac{-\pi}{6}$ and $\sin\frac{7\pi}{3}$. ■

A basic fact in our development has been that in laying off a length 2π, we pass around the unit circle and return to our original point. Thus, if we determine the values of the trigonometric functions in the interval $[0, 2\pi)$, we know their values for all real s. Consider Figure 8.15, which illustrates how the symmetry of the circle may be used to further simplify the evaluation.

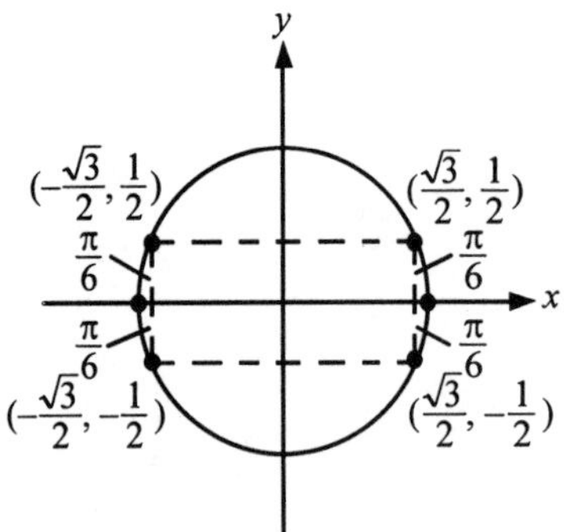

Figure 8.15

Note that the coordinates of the point assigned to $\pi/6$ differ only in sign from the coordinates assigned to $\pi - \pi/6 = 5\pi/6$, $\pi + \pi/6 = 7\pi/6$, and $2\pi - \pi/6 = 11\pi/6$. Thus, for a specific trigonometric function, say sine, we have

$$\sin\frac{\pi}{6} = \left|\sin\frac{5\pi}{6}\right| = \left|\sin\frac{7\pi}{6}\right| = \left|\sin\frac{11\pi}{6}\right| = \frac{1}{2}.$$

The number $\pi/6$ is called the **reference number** for $5\pi/6$, $7\pi/6$, and $11\pi/6$. In general, we determine the reference number for s (denoted by s_R) by finding the shortest positive arc length between the point on the circle assigned to s and the x-axis. This discussion indicates that the trigonometric values of s and s_R are related as follows.

Trigonometric Values of s and s_R

> Any trigonometric function of s is equal in absolute value to the same numbed function of its reference number s_R.

We determine the correct sign by considering the function definitions together with the sign of x and y in the four quadrants. The chart in Figure 8.16 indicates the signs of the functions in the various quadrants (and this diagram corresponds to our previous chart in Figure 2.25).

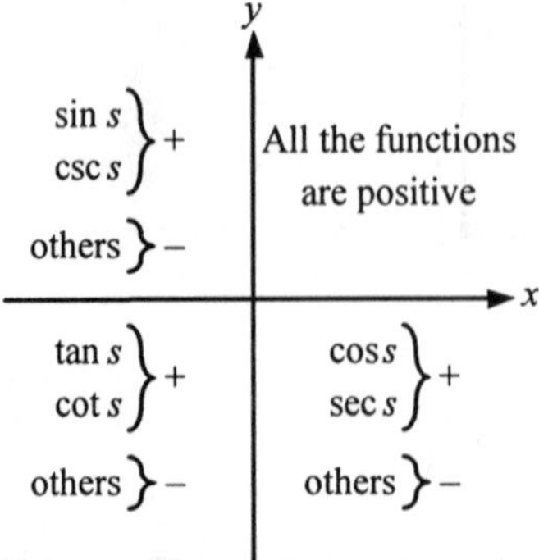

Figure 8.16

Example 5: Evaluations Involving $\pi/3$, $\pi/4$ or $\pi/6$ Reference Numbers

Find the exact value of $\sin\frac{7\pi}{4}$.

Solution: First, determine the reference number.

$$s_R = 2\pi - \frac{7\pi}{4} = \frac{\pi}{4} \quad \text{(See Figure 8.17.)}$$

Second, determine $\sin(\pi/4)$.

$$\sin\frac{\pi}{4} = \frac{1}{\sqrt{2}}$$

Third, determine the correct sign. The point assigned to $7\pi/4$ is in Q_4, where the value of the sine function is negative. Therefore,

$$\sin\frac{7\pi}{4} = -\frac{1}{\sqrt{2}}.$$

You should check this result by calculator.

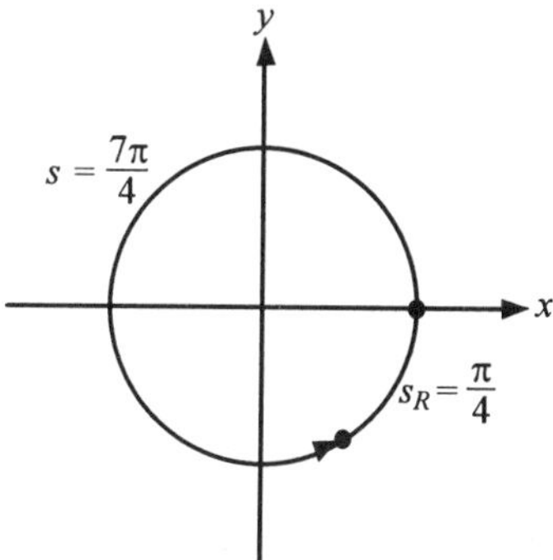

Figure 8.17

PROGRESS CHECK 5

Find the exact value of $\cos\frac{11\pi}{6}$. ∎

Example 6: Evaluations Involving $\pi/3$, $\pi/4$ or $\pi/6$ Reference Numbers

Find the exact value of $\cos\frac{-10\pi}{3}$.

Solution: First, simplify the expression to the form cos s, where $0 \le s < 2\pi$.

$$\cos\left(\frac{-10\pi}{3}\right) = \cos\frac{10\pi}{3} = \cos 3\frac{1}{3}\pi = \cos\left(\frac{4\pi}{3} + 2\pi\right) = \cos\frac{4\pi}{3}$$

Second, determine s_R.

$$s_R = \frac{4\pi}{3} - \pi = \frac{\pi}{3} \quad \text{(See Figure 8.18.)}$$

Third, determine $\cos(\pi/3)$.

$$\cos\frac{\pi}{3} = \frac{1}{2}$$

Fourth, determine the correct sign. The point assigned to $4\pi/3$ is in Q_3, where the value of the cosine function is negative. Therefore,

$$\cos\left(\frac{-10\pi}{3}\right) = -\frac{1}{2}.$$

Check this result by calculator.

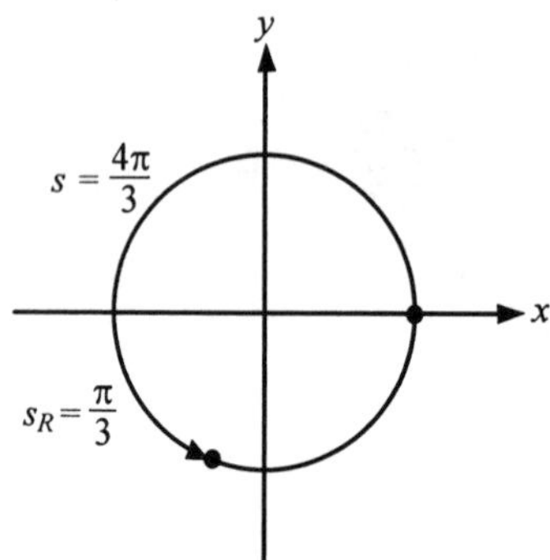

Figure 8.18

PROGRESS CHECK 6

Find the exact value of $\sin\frac{-17\pi}{3}$. ■

Arc lengths that terminate on an axis or that have reference numbers of $\pi/3$, $\pi/4$, or $\pi/6$ may be considered special cases because exact values may be found for trigonometric functions of these numbers. For other numbers, we can easily approximate their function values by using a calculator, as shown in the next two examples.

Example 7: Approximating Trigonometric Values by Calculator

Find the approximate value of each expression. Round off answers to four decimal places.

a. $\sin 2$ **b.** $\sec\frac{8\pi}{5}$

Solution: See Figure 8.19.

a. $\sin 2 \approx 0.9093$. Note that a calculator interprets sin 2 to mean the sine of 2 radians or the sine of 2 degrees, depending on the mode setting for the unit of angle measure.

b. Using $\sec(8\pi/5) = 1/\cos(8\pi/5)$ yields $\sec(8\pi/5) \approx 3.2361$,

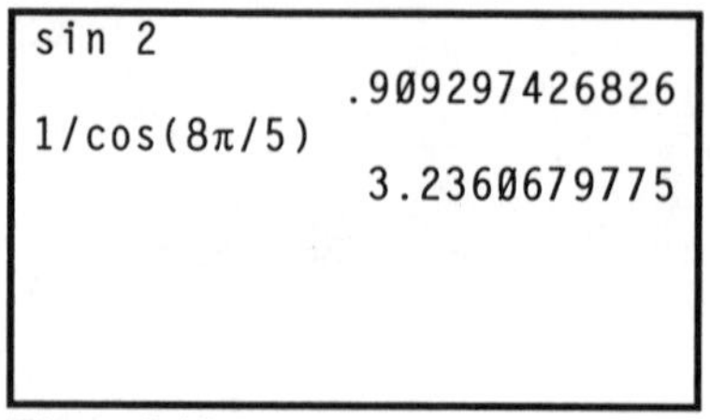

```
sin 2
                .909297426826
1/cos(8π/5)
                3.2360679775
```

Figure 8.19

PROGRESS CHECK 7

Find the approximate value of each expression. Round off answers to four decimal places.

a. $\cos 5$ **b.** $\cot\dfrac{9\pi}{7}$ ■

Example 8: Revenue from Sales of a Seasonal Product

Solve the problem in the section introduction on page 620.

Solution: By the end of March, 2 months have elapsed from February 1, and replacing t by 2 in the given formula yields

$$R = 6{,}500 - 5{,}000\cos(0.5236 \cdot 2) \approx \$4{,}000.$$

In a similar way, 7 months elapse from February 1 to the end of August, and replacing t by 7 gives

$$R = 6{,}500 - 5{,}000\cos(0.5236 \cdot 7) \approx \$10{,}830.$$

Thus, the estimates from the formula are that monthly revenue for March was \$4,000, and the monthly revenue for August was \$10,830.

PROGRESS CHECK 8

On May 2, the tide in a certain harbor is described by $y = 12 + 5.3\cos(0.5067t)$, where y is the height of the water in feet and t is the time that has elapsed from 12 midnight on May 1. To one decimal place, estimate the water level in this harbor on May 2 at 7 a.m. and at 11 p.m. ■

EXPLORE 8.2

1. To familiarize yourself with the trigonometric capabilities of your calculator, evaluate each of these expressions, and determine when you need to use parentheses.
   a. $\sin 3\pi$ **b.** $(\sin 3)\pi$ **c.** $\sin(5+\pi)$ **d.** $(\sin 5)+\pi$
   e. $\sin(\pi/2)$ **f.** $(\sin \pi)/2$ **g.** $\sin \pi^2$ **h.** $(\sin \pi)^2$
2. In this chapter we take the domain of the trigonometric functions to be the set of real numbers (not necessarily interpreted as angles). Therefore, it is sensible to form the composition of trigonometric functions with other functions. Use a calculator to evaluate each of these expressions to 3 decimal places.
   a. $\sin(\log 51)$ **b.** $\log(\sin 51)$ **c.** $\sin(\cos 100)$ **d.** $\cos(\sin 100)$
   e. $\tan[\cos(\sin 100)]$
3. Recall that the fraction $0/0$ is undefined. But, in calculus, we must often determine the behavior of a fraction when both the numerator and denominator get very close to 0.
   a. Complete the following table and describe what happens to the value of the given fractions as x approaches (but does not reach) zero.

x	$\sin x$	$\dfrac{\sin x}{x}$	$\dfrac{\sin^2 x}{x^2}$	$\dfrac{\sin^2 x}{x}$	$\dfrac{\sin x}{x^2}$
.5					
.2					
.1					
.01					
.001					

   b. Graph each of the four fractional functions from part **a**, and describe the behavior of the graph as x approaches zero. What does the calculator give as the value of each function when x is exactly equal to zero?

EXERCISES 8.2

In Exercises 1–30, find the *exact* function value.

1. $\cos 4\pi$ **2.** $\sin 100\pi$ **3.** $\sin 5\pi$

4. $\cos 11\pi$ **5.** $\tan(-2\pi)$ **6.** $\cot(-36\pi)$

7. $\cos\left(-\frac{\pi}{2}\right)$ **8.** $\sin\left(-\frac{3\pi}{2}\right)$ **9.** $\sec\frac{5\pi}{2}$

10. $\tan\frac{19\pi}{2}$ **11.** $\csc\left(\frac{-7\pi}{2}\right)$ **12.** $\cos\left(\frac{-9\pi}{2}\right)$

13. $\sin\left(\frac{11\pi}{2}\right)$ **14.** $\sin\left(\frac{-97\pi}{2}\right)$

15. $\cos\left(-\frac{\pi}{3}\right)$ **16.** $\sin\left(-\frac{\pi}{4}\right)$ **17.** $\tan\frac{9\pi}{4}$

18. $\cos\frac{9\pi}{4}$ **19.** $\sin\frac{13\pi}{6}$ **20.** $\sec\frac{17\pi}{4}$

21. $\cot\frac{7\pi}{3}$ **22.** $\sin\frac{19\pi}{3}$ **23.** $\cos\left(\frac{-13\pi}{6}\right)$

24. $\sin\left(\frac{-13\pi}{3}\right)$ **25.** $\csc\left(\frac{-31\pi}{3}\right)$

26. $\cos\left(\frac{-25\pi}{4}\right)$ **27.** $\cot\frac{55\pi}{3}$

28. $\tan\left(\frac{-73\pi}{6}\right)$ **29.** $\sin\left(\pi-\frac{4\pi}{3}\right)$

30. $\cos\left(2\pi-\frac{13\pi}{6}\right)$

In Exercises 31–50, find the *exact* function value. Use reference numbers.

31. $\cos\frac{7\pi}{6}$ **32.** $\sin\frac{5\pi}{3}$ **33.** $\tan\frac{4\pi}{3}$

34. $\sec\frac{11\pi}{6}$ **35.** $\sin\frac{7\pi}{4}$ **36.** $\cos\frac{2\pi}{3}$

37. $\csc\frac{5\pi}{6}$ **38.** $\sin\frac{3\pi}{4}$ **39.** $\cos\frac{5\pi}{3}$

40. $\cot\frac{5\pi}{4}$ **41.** $\sin\left(\frac{-7\pi}{6}\right)$ **42.** $\cos\left(\frac{-4\pi}{3}\right)$

43. $\sec\left(\frac{-2\pi}{3}\right)$ **44.** $\tan\left(\frac{-7\pi}{4}\right)$ **45.** $\cos\frac{11\pi}{4}$

46. $\sin\frac{8\pi}{3}$ **47.** $\cot\frac{23\pi}{6}$ **48.** $\csc\frac{15\pi}{4}$

49. $\sin\left(\frac{-20\pi}{3}\right)$ **50.** $\cos\left(\frac{-23\pi}{6}\right)$

In Exercises 51–70 find the approximate function value. Round off answers to four decimal places.

51. cos 2 **52.** sin 3 **53.** tan 4

54. sec 5 **55.** sin 5.41 **56.** csc 2.23

57. cot 3.71 **58.** cos 1.84 **59.** sin(−6.07)

60. tan(−1.69) **61.** cos 11.73

62. cot 9.61 **63.** csc(−7.57)

64. sin(−10.25) **65.** $\sin\frac{2\pi}{5}$

66. $\tan\frac{9\pi}{5}$ **67.** $\cos\frac{3\pi}{7}$ **68.** $\sin\frac{4\pi}{9}$

69. $\sec\left(\frac{-7\pi}{8}\right)$ **70.** $\cos\left(\frac{-7\pi}{10}\right)$

71. The monthly revenue R in dollars from sales of ice cream for a sweet shop in Montpelier, Vermont is approximated by

$$R = 4200 + 2100\cos(0.5236t),$$

where t is the number of months that have elapsed from August 1. Use this model to estimate to the nearest hundred dollars, the total revenue for the two month period August and September, and the total revenue for the two-month period January and February.

72. The monthly number of kilowatt hours (E) of electricity sold by a particular utility company is approximated by

$$E = 384.5 + 89.5\cos(.5236t),$$

where t is the number of months that have elapsed from January 1. Use this model to approximate the total number of kilowatt hours sold in the 2-month period March and April, and the total number of kilowatt hours sold in the 2-month period September and October.

73. The analysis of oscillations in physics may involve trigonometric equations. For instance, the motion of a swinging pendulum can be described by such equations. Suppose that a body oscillates according to the equation

$$x = 6.0\cos\left(3\pi t + \frac{\pi}{3}\right),$$

where x represents the horizontal displacement from a reference point in meters, and t is the elapsed time in seconds. Because the cosine function periodically decreases and increases, the moving body periodically gets closer to and farther away from the reference point.

a. Find the horizontal displacement of the body when $t = 0$ and when $t = 2$.

b. What is the maximum possible value of $\cos\left(3\pi t + \frac{\pi}{3}\right)$? What is the maximum possible horizontal displacement of this body?

74. When you wiggle a rope to make a wave travel along it, you create what physicists call a **transverse wave**. If you look at a particular point along the rope, it will have different displacements at different times. For example, the equation

$$y = 10\sin[\pi(0.01x - 2.00t)]$$

gives the displacement y (in centimeters) of a rope at a point x centimeters from the end at time t seconds.

a. Find the displacements of the rope 10 cm from the end when $t = 2, 2.25, 2.5,$ and 3.

b. What is the maximum possible value of $\sin[\pi(0.01x - 2.00t)]$? What is the maximum possible displacement of this rope?

75. The average daily temperature T in degrees Fahrenheit at the surface of the ground in Athens, Georgia, over the course of a year is approximated fairly well by the equation

$$T = 61.3 + 17.9\cos\left(\frac{2\pi}{365}t\right),$$

where t is in days, and $t = 0$ represents July 1. [Source for Exercises 75 and 76: *Mathematical Modeling and Cool Buttermilk in the Summer*, Corbitt and Edwards, 1979 Yearbook, NCTM]

a. Estimate the average temperature (to the nearest degree) on July 1.

b. Estimate the average temperature (to the nearest degree) on January 1.

76. An equation similar to the one in Exercise 75 gives the temperature T and time t in a cellar that is x cm below the surface of the ground. For a location where the average temperature is about 16°C and the extremes are 5° and 27°, the equation is

$$T = 16 + 11e^{-.00706x}\cos\left(\frac{2\pi}{k}t - .00706x\right).$$

In this equation, t is in seconds ($t = 0$ is 12 a.m. July 1), and $k = 365 \times 24 \times 3600$, the number of seconds in a year.

a. Note that $\frac{\pi}{.00706} = 445$. Use this relationship to simplify the equation for a cellar of depth $x = 445$ cm.

b. Find the temperature of the cellar in part **a** on July 1 (when $t = 0$).

c. Find the temperature of the cellar in part **a** on January 1 (when $t = \frac{k}{2}$).

THINK ABOUT IT 8.2

1. **a.** Explain why there is no real number s such that $\sin s > 1$.

b. Classify each of the six trigonometric functions as either an even function or an odd function. Give reasons for your answers.

2. The following formulas (derived from calculus) may be used to compute the values of $\cos x$ and $\sin x$.

$$\cos x = 1 - \frac{x^2}{2!} + \frac{x^4}{4!} - \frac{x^6}{6!} + \cdots$$

$$\sin x = x - \frac{x^3}{3!} + \frac{x^5}{5!} - \frac{x^7}{7!} + \cdots,$$

where $n! = 1 \cdot 2 \cdots n$ (for example, $3! = 1 \cdot 2 \cdot 3$). Use the first three terms of these formulas to approximate the following expressions. Compare your results with the values given by calculator.

a. sin 1 **b.** cos 1 **c.** cos 0 **d.** sin 0.5

3. Determine the coordinates of the point on the unit circle assigned to $\pi/6$. (**Hint:** The method is similar to the one given for $\pi/4$. The arc in the circle from $-\pi/6$ to $\pi/6$ is equal to the arc from $\pi/6$ to $\pi/2$.)
4. Consider the unit circle in the accompanying figure, in which each of the six trigonometric functions can be represented as a line segment. For example, since $\overline{OC} = 1$ in right triangle *OAC*, we have

$$\sin\theta = \frac{\overline{AC}}{\overline{OC}} = \frac{\overline{AC}}{1} = \overline{AC}.$$

Notice that we obtained the desired line segment by selecting a right triangle where the denominator in the defining ratio is 1. Determine the line segments representing the five remaining trigonometric functions in the figure.

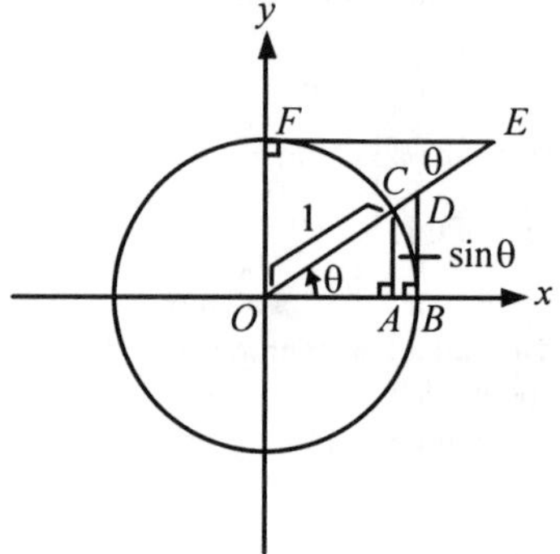

5. The trigonometric functions are sometimes called "circular" functions because their values are determined by points on a circle. We can invent other periodic functions based on different geometric shapes. In this exercise we invent "square" functions as follows.
   a. Construct a square with center at the origin and vertices as shown in the figure.

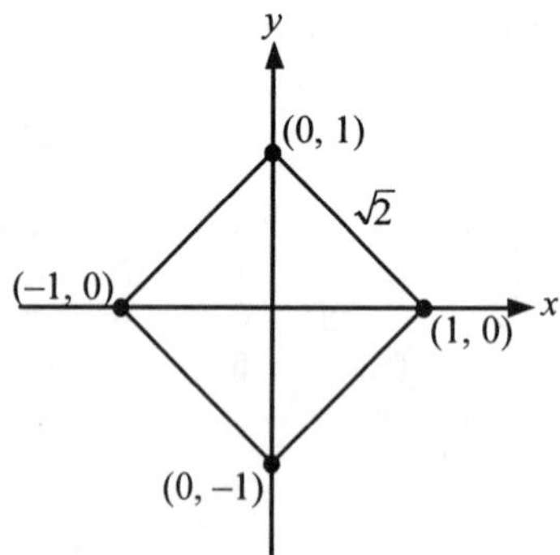

   b. Define $\sin \text{sq}(t)$ and $\cos \text{sq}(t)$ to be the y- and x-coordinates, respectively, of the point reached by traveling along the square for a distance t in a counterclockwise direction, starting from the point $(1, 0)$. For example, when $t = \sqrt{2}$ the point is $(0, 1)$.
   c. Use the definitions of $\sin \text{sq}(t)$ and $\cos \text{sq}(t)$ to complete the following table.

t	sin sq(t)	cos sq(t)
0		
$\sqrt{2}$		
$2\sqrt{2}$		
$3\sqrt{2}$		
$\sqrt{2}/2$		
$\sqrt{2}/3$		
$\sqrt{2}/4$		
1		

d. Explain why $|\sin \text{sq}(t)| + |\cos \text{sq}(t)| = 1$ for all t.

● ● ●

8.3 Graphs of Sine and Cosine Functions

Photo Courtesy of ***Crady von Pawlak*** *of Archive Photos*

Based on the orbit of Earth around the sun, the number of hours of daylight during each day in a specific location demonstrates cyclic behavior. This behavior may be modeled by a sine function whose specific equation depends on the latitude of the location. For the city of Houston, Texas, an equation that may be used to approximate the number of hours of daylight (y) during the xth day of the year is

$$y = 12 + 2.3\sin\left[\frac{2\pi}{365}(x - 80)\right].$$

a. Graph this function over the interval from $x = 1$ (January 1) to $x = 365$ (December 31) with the aid of a grapher.

b. What are the amplitude and the period of the function?

c. According to your graph, which day of the year has the minimum number of hours of daylight in Houston?

(See Example 10)

Objectives

1. Find the amplitude and period for functions of the form $y = a\sin bx$ and $y = a\cos bx$, and graph the function for a given interval.
2. Write the equation of the form $y = a\sin bx$ or $y = a\cos bx$ that corresponds to a given graph.
3. Graph sine and cosine functions that involve horizontal translations or vertical translations.
4. Solve applied problems involving the evaluating of a trigonometric function of a real number.

A picture or graph of the sine and cosine functions helps us understand their cyclic behavior. This insight is crucial because this behavior is the basis for the use of these functions to model periodic events (such as the number of hours of daylight in Houston, which is considered in the section-opening problem). In terms of notation we use the xy-plane and associate the arc length values with points on the horizontal or x-axis. This means that we are now using x, instead of s, to represent the independent variable, which is an arc length. The vertical axis, labeled the y-axis, is used to represent the function values.

We begin with a numeric approach, by considering the values of the sine function as we lay off on the x-axis a length 2π and pass around the unit circle. The following table indicates some values of the sine function on this typical interval.

$y = \sin x$ (for $0 \le x \le 2\pi$)													
x	0	$\frac{\pi}{6}$	$\frac{\pi}{3}$	$\frac{\pi}{2}$	$\frac{2\pi}{3}$	$\frac{5\pi}{6}$	π	$\frac{7\pi}{6}$	$\frac{4\pi}{3}$	$\frac{3\pi}{2}$	$\frac{5\pi}{3}$	$\frac{11\pi}{6}$	2π
y	0	0.5	0.87	1	0.87	0.5	0	–0.5	–0.87	–1	–0.87	–0.5	0

If we plot these points and join them with a smooth curve, we obtain the graph in Figure 8.20, which describes the essential characteristics of the sine function during one cycle. The plot of $y = \sin x$ starts at the origin, attains a maximum at one-fourth of the cycle length, returns to zero halfway through the cycle, attains a minimum at the three-quarter point, and returns to zero at the end of the cycle. Each time we lay off a length 2π, we pass around the circle and repeat this behavior. Thus, the graph of $y = \sin x$ weaves continuously through cycles in both directions, as illustrated in Figure 8.21.

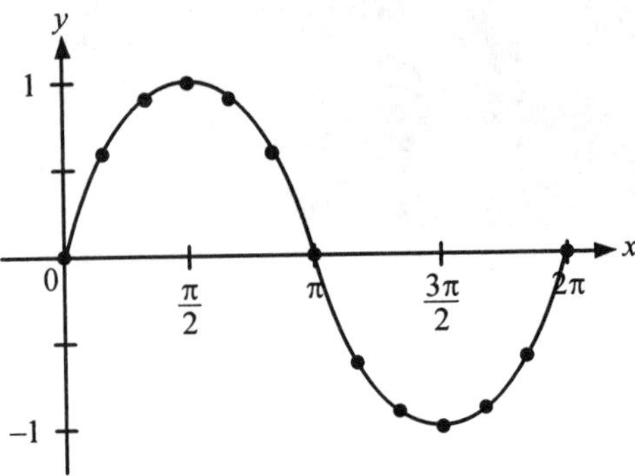

Figure 8.20

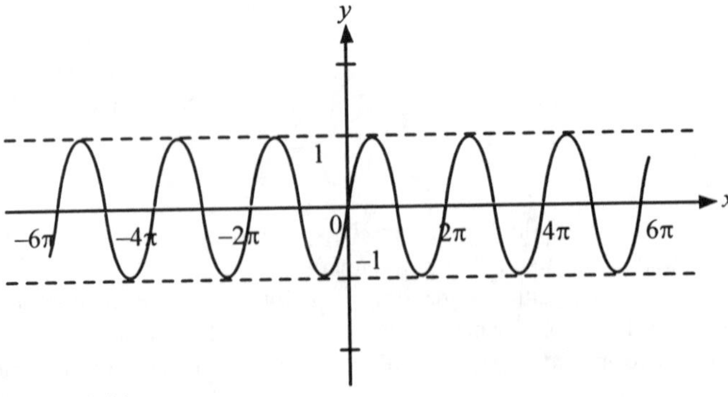

Figure 8.21

The graph in Figure 8.21 is easy to generate on a graphing utility. The main steps are to set the angle mode to *radian*, enter $y1 = \sin x$ in the equation editor, and use $[-6\pi, 6\pi]$ by $[-2, 2]$ to define a viewing window. To mimic Figure 8.21, set the x scale to π and the y scale to 1, and then graph the function as shown in Figure 8.22. Note that the standard viewing window $[-10, 10]$ by $[-10, 10]$ is not well suited for graphing $y = \sin x$. Therefore, many graphing calculators have a built-in feature that automatically sets a window that is suitable for many basic trigonometric graphs. For instance, Figure 8.23 shows the result of graphing $y = \sin x$ using the Zoom Trig feature on a particular Texas Instruments calculator. You should learn how to use this type of feature if it is available on your calculator.

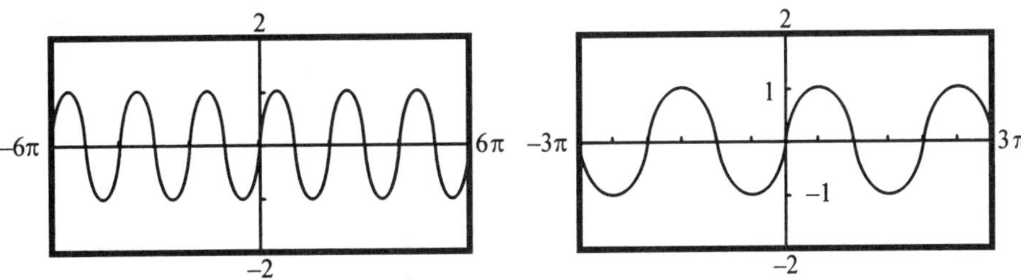

Figure 8.22 **Figure 8.23**

When we examine our graphs of the sine function we see that the function repeats its values on intervals of length 2π. For this reason, the sine function is said to be *periodic*. We define a periodic function as follows.

Periodic Function

A function f is periodic if

$$f(x) = f(x + p)$$

for all x in the domain of f. The smallest positive number p for which this is true is called the **period** of the function.

This definition applies to the sine function since

$$\sin x = \sin(x + 2\pi),$$

where 2π is the smallest positive constant for which this type of statement is true. Thus the sine function is periodic, with period 2π.

This information is critical for investigating the graph of the family $y = \sin bx$ as b changes for $b > 0$. Such an exploration is conducted with the aid of a grapher in Example 1.

Example 1: Exploring the Family $y = \sin bx$

Let $f(x) = \sin x$, $g(x) = \sin 2x$, and $h(x) = \sin \frac{1}{2}x$.

a. Graph f, g, and h on a grapher using $[-2\pi, 2\pi]$ by $[-4, 4]$ to determine the viewing window.
b. Use the graphs in part **a** to determine the period of each function.
c. Describe how to obtain the period of g and the period of h by using the period of $f(x) = \sin x$.

Solution:

a. The requested graphs are shown in Figure 8.24.

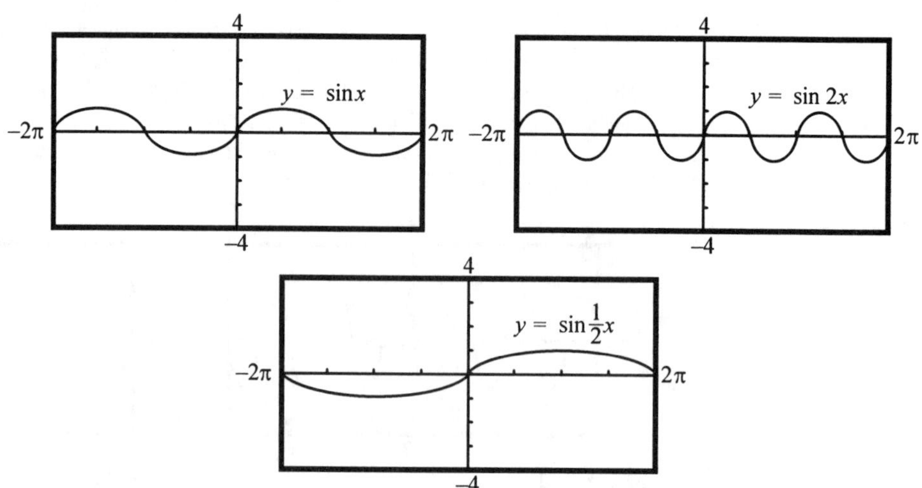

Figure 8.24

b. We read from the graphs that the smallest intervals on which the graphs complete one full cycle are 2π for $f(x) = \sin x$, π for $g(x) = \sin 2x$, and 4π for $h(x) = \sin(1/2)x$. Thus, the periods for f, g, and h are the 2π, π, and 4π, respectively.

c. The period of $g(x) = \sin 2x$ may be found by taking the period of $f(x) = \sin x$ and dividing it by 2. That is,

$$\text{period of } g = \frac{2\pi}{2} = \pi.$$

Similarly, taking the period of $f(x) = \sin x$ and dividing it by 1/2 gives the period of $h(x) = \sin(1/2)x$. Thus,

$$\text{period of } h = \frac{2\pi}{1/2} = 4\pi.$$

PROGRESS CHECK 1

Let $g(x) = \sin 4x$ and $h(x) = \sin \pi x$.

a. Graph g and h using $[-2\pi, 2\pi]$ by $[-4, 4]$ to determine the viewing window.

b. Use the graphs in part **a** to determine the period of each function.

c. Describe how to obtain the period of g and the period of h by using the period of $f(x) = \sin x$.

The result in Example 1 suggests the period of $y = \sin bx$ for $b > 0$ is $2\pi/b$. This result makes sense analytically because $y = \sin bx$ completes one full cycle as bx ranges from 0 to 2π. Since

$$bx = 0 \text{ when } x = 0$$

and

$$bx = 2\pi \text{ when } x = \frac{2\pi}{b},$$

it follows that the period of $y = \sin bx$ for $b > 0$ is $2\pi/b$.

Example 2: Graphing a Sine Function Using Its Period

State the period and graph $y = \sin\frac{x}{3}$ for $-p \le x \le p$, where p is the period of the function.

Solution: Since $\sin(x/3)$ is equivalent to $\sin(\frac{1}{3}x)$, we determine the period by substituting 1/3 for b in the formula for the period.

$$\text{Period} = \frac{2\pi}{b} = \frac{2\pi}{1/3} = 6\pi$$

The length of one cycle is 6π, so we draw two cycles to graph the function on the interval $[-6\pi, 6\pi]$, as shown in Figure 8.25. You should confirm this result with a grapher.

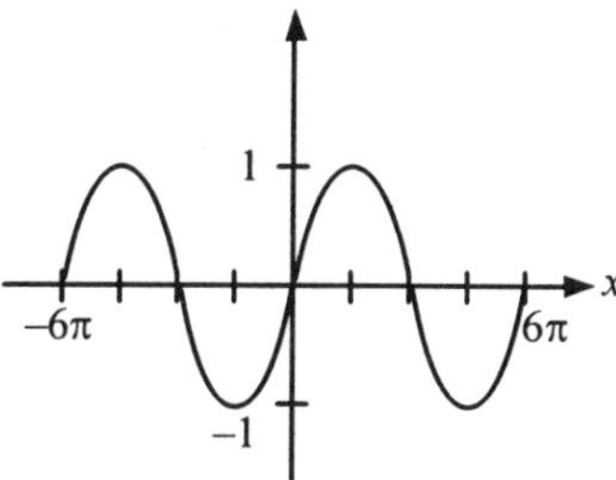

Figure 8.25

PROGRESS CHECK 2

State the period and graph $y = \sin\frac{x}{5}$ for $-p \le x \le p$, where p is the period of the function. ■

Two other key characteristics of the graph of $y = \sin x$ are that the y values oscillate from -1 to 1, and that the x-axis serves as a *midline* that runs halfway between these minimum and maximum values. When the midline of the graph of a periodic function is the x-axis, then the maximum y value is called the *amplitude* of the function. Thus, the amplitude of $y = \sin x$ is 1.

We can use this information to sketch functions of the form $y = a\sin bx$, where a represents a real number. These functions are similar to $y = \sin bx$ in that they have the same basic shape and period, but they may differ by having different amplitudes. For example, to sketch the graph of $y = 3\sin x$, we obtain values for sin x and multiply these values by 3. Since the greatest y value that sin x attains is 1, the greatest y value that $3\sin x$ attains is 3. Thus, the amplitude of $y = 3\sin x$ is 3.

In general, since the greatest value that sin x attains is 1, the greatest value that a sin x attains is $|a|$. Thus, the amplitude of $y = a\sin bx$ is $|a|$. Figure 8.26 compares the graphs of $y = \sin x$ and $y = 3\sin x$ on $[0, 2\pi]$.

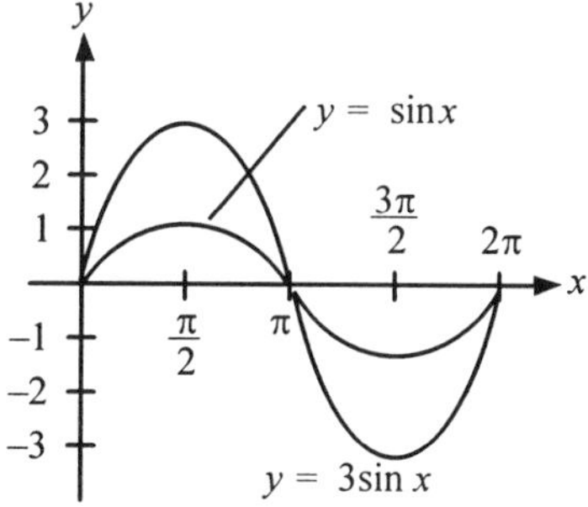

Figure 8.26

Note that we can anticipate the relation between these two graphs from our work with graphing techniques in Section 1.3, because the graph of $y = c \cdot f(x)$ with $c > 0$ is the graph of $y = f(x)$ stretched or flattened out by a factor of c.

Example 3: Graphing a Sine Function Using Its Amplitude and Period

State the amplitude and the period and sketch the graph of $y = 2\sin 3x$ for $0 \le x \le 2\pi$.

Solution: First, determine the amplitude and period.

$$\text{Amplitude} = |a| = |2| = 2$$
$$\text{Period} = \frac{2\pi}{b} = \frac{2\pi}{3}$$

If the curve completes one cycle on the interval $[0,\ 2\pi/3]$, the curve completes three cycles on the given interval. Note that b gives the number of cycles on the interval $[0,\ 2\pi]$. Since the amplitude is 2, the curve oscillates between a maximum value of 2 and a minimum value of –2. The graph is sketched in Figure 8.27, and you should check this result with a grapher.

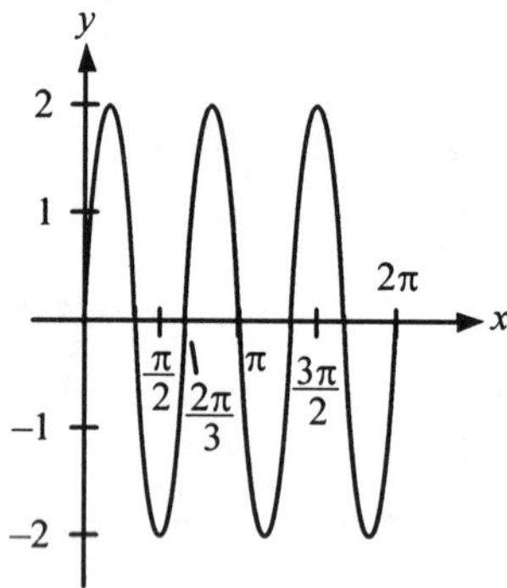

Figure 8.27

PROGRESS CHECK 3

State the amplitude and the period and sketch the graph of $y = 5\sin 2x$ for $0 \le x \le 2\pi$. ∎

Example 4: Graphing $y = a \sin bx$ Where $a < 0$

State the amplitude and the period and sketch the graph of $y = -\sin \pi x$ for $0 \le x \le 2\pi$.

Solution: First, determine the amplitude and period.

$$\text{Amplitude} = |a| = |-1| = 1$$
$$\text{Period} = \frac{2\pi}{b} = \frac{2\pi}{\pi} = 2$$

If the curve completes one cycle every 2 units, the curve completes slightly more than three cycles on the interval $[0,\ 2\pi]$. Since the amplitude is 1, the curve oscillates between a maximum value of 1 and a minimum value of -1. Because a is negative, we obtain the graph by reflecting the graph of $y = \sin \pi x$ about the x-axis, as shown in Figure 8.28.

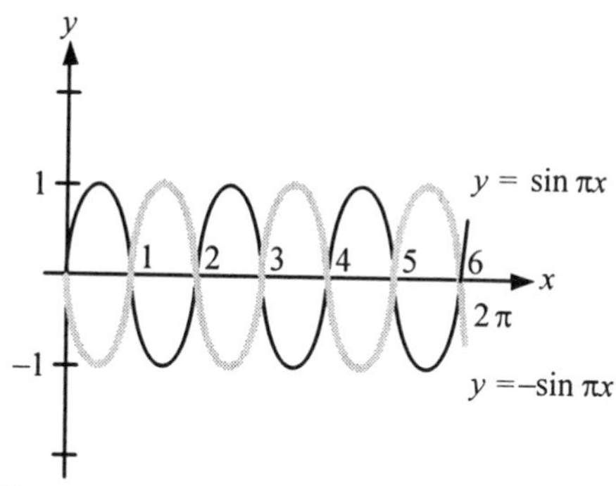

Figure 8.28

Observe that this result is in agreement with the graph in Figure 8.29, which was obtained by graphing $y = -\sin \pi x$ on a grapher.

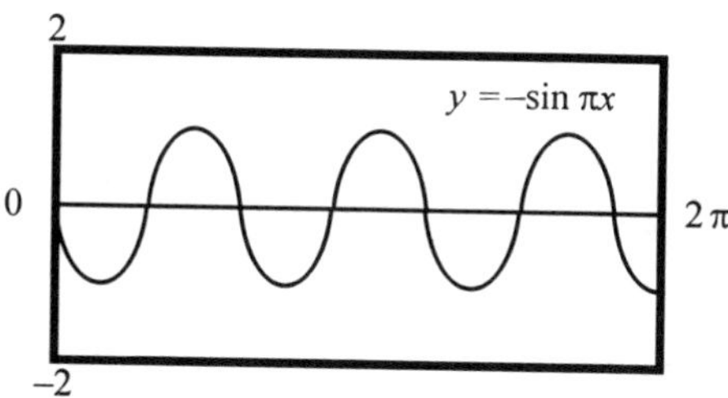

Figure 8.29

PROGRESS CHECK 4

State the amplitude and the period and sketch the graph of $y = -3\sin 2\pi x$ for $0 \le x \le 2\pi$. ■

The graph of the cosine function has the same essential characteristics as the graph of the sine function. That is, the amplitude of the cosine function is 1 and the period is 2π. This is evidenced by the following table, which indicates some values of the cosine function on the interval $[0, 2\pi]$.

$y = \cos x$ (for $0 \le x \le 2\pi$)													
x	0	$\frac{\pi}{6}$	$\frac{\pi}{3}$	$\frac{\pi}{2}$	$\frac{2\pi}{3}$	$\frac{5\pi}{6}$	π	$\frac{7\pi}{6}$	$\frac{4\pi}{3}$	$\frac{3\pi}{2}$	$\frac{5\pi}{3}$	$\frac{11\pi}{6}$	2π
y	1	0.87	0.5	0	–0.5	–0.87	–1	–0.87	–0.5	0	0.5	0.87	1

If we plot these points and join them with a smooth curve, we obtain the graph shown in Figure 8.30. This graph demonstrates that the cosine function completes one cycle on the interval $[0, 2\pi]$ and attains a maximum value of 1. Like that of the sine function, this graph can be reproduced indefinitely in both directions to obtain as much of the graph of the cosine function as desired (see Figure 8.31).

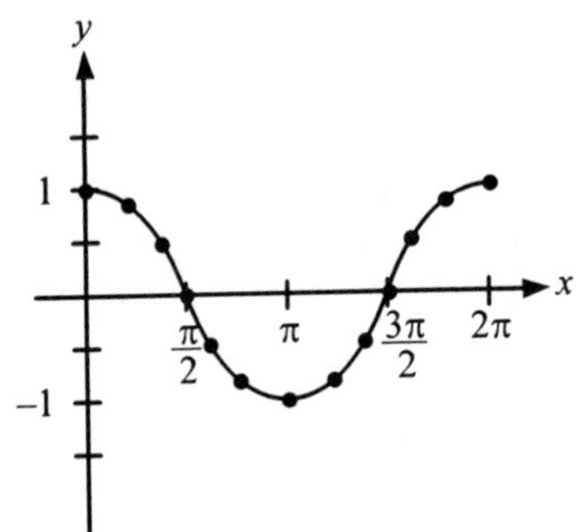

Figure 8.30

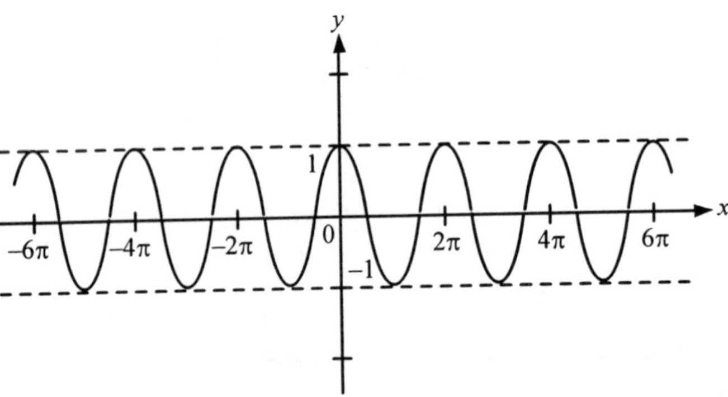

Figure 8.31

The great similarity between the graphs of the sine and cosine functions should be apparent. In fact, if we shift the graph of the cosine function $\pi/2$ units to the right, the resulting graph is the sine function. Thus, the only difference between the two graphs is that one curve leads the other by $\pi/2$. That is, $\cos(x - \pi/2) = \sin x$.

We graph functions of the form $y = a\cos bx$ in a manner similar to graphing $y = a\sin bx$; that is, we find the amplitude by computing $|a|$ and the period by using $2\pi/b$. The difference is that the graph of $y = a\cos bx$ attains a maximum or minimum height at $x = 0$.

Example 5: Graphing $y = a \cos bx$ Where $a > 0$

State the amplitude and the period and sketch the graph of $y = \frac{1}{2}\cos 2x$ for $0 \le x \le 2\pi$.

Solution: First, determine the amplitude and period.

$$\text{Amplitude} = |a| = \left|\frac{1}{2}\right| = \frac{1}{2}$$

$$\text{Period} = \frac{2\pi}{b} = \frac{2\pi}{2} = \pi$$

If the curve completes one cycle on $[0, \pi]$, the curve will complete two cycles on $[0, 2\pi]$. Since the amplitude is $1/2$, the curve oscillates between a maximum value of $1/2$ and a minimum value of $-1/2$, as shown in Figure 8.32. This result should be checked using a grapher.

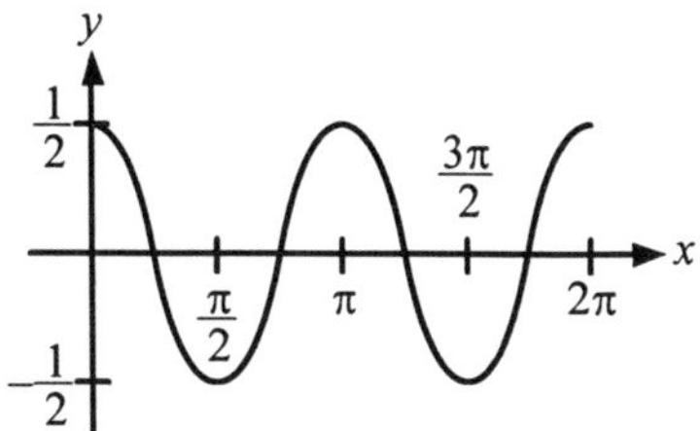

Figure 8.32

PROGRESS CHECK 5

State the amplitude and the period and sketch the graph of $y = 10\cos 4x$ for $0 \le x \le 2\pi$. ■

Example 6: Graphing $y = a \cos bx$ Where $a < 0$

State the amplitude and the period and sketch one cycle of the graph of $y = -4\cos 10x$.

Solution: First, determine the amplitude and period.

$$\text{Amplitude} = |a| = |-4| = 4$$

$$\text{Period} = \frac{2\pi}{b} = \frac{2\pi}{10} = \frac{\pi}{5}$$

The curve completes one cycle on the interval $[0, \pi/5]$ and attains a maximum value of 4 and a minimum value of −4. Since a is a negative number, we obtain the graph shown in Figure 8.33 by starting and ending the graph at a minimum point. You should use a grapher to check this result.

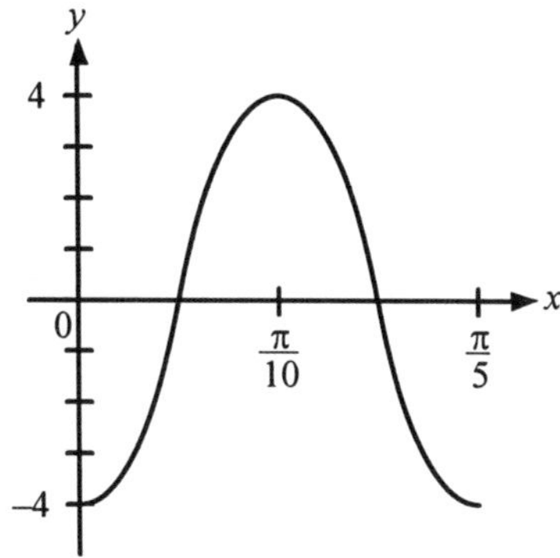

Figure 8.33

PROGRESS CHECK 6

State the amplitude and the period and sketch one cycle of the graph of $y = -\frac{3}{2}\cos\frac{x}{4}$. ■

Example 7: Determining an Equation That Fits a Graph

Find an equation for the curve with the single cycle shown in Figure 8.34. The equation should be written in the form $y = a\sin bx$ or $y = a\cos bx$.

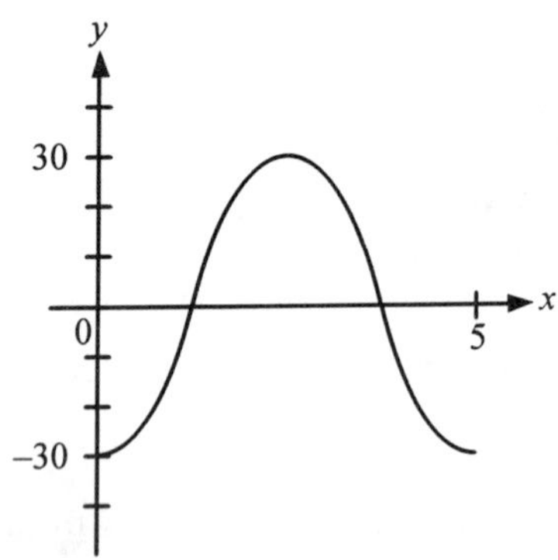

Figure 8.34

Solution: Since the cycle shown starts at a minimum y value, the form of the equation is $y = a\cos bx$ with $a < 0$. The amplitude is 30 and $a < 0$, so $a = -30$. Since the period is 5, we have

$$\frac{2\pi}{b} = 5 \text{ so } b = \frac{2\pi}{5}.$$

Thus, an equation for the graph is $y = -30\cos\left(\frac{2\pi}{5}x\right)$. You should graph the equation on a grapher to check this result.

PROGRESS CHECK 7
Redo the question in Example 7 for the curve with the single cycle shown in Figure 8.35. ■

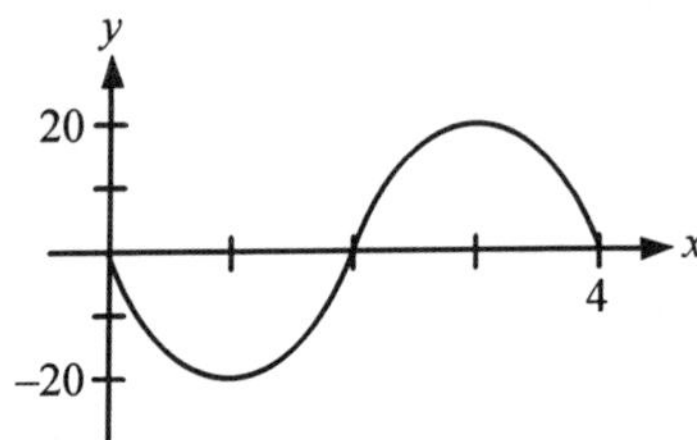

Figure 8.35

Vertical and Horizontal Shifts Involving Sine and Cosine

To model cyclic events in many applied problems, it is necessary to use vertical or horizontal translations of sine or cosine curves. To shift the midline of such curves off the x-axis, we use a vertical translation. Recall from Section 1.3 that if $d > 0$, then the graph of $y = f(x) + d$ is the graph of f raised d units, while the graph of $y = f(x) - d$ is the graph of f lowered d units. Thus, the midline of the functions $y = d + a\sin bx$ and $y = d + a\cos bx$ is the horizontal line $y = d$.

Example 8: Using a Vertical Translation

Graph one cycle of the function $y = 3 + 2\sin 4x$. Indicate the amplitude, period, and midline.

Solution: First, observe that the amplitude and period of $y = 2\sin 4x$ are 2 and $\pi/2$, respectively. So we graph one cycle of this function as shown in black in Figure 8.36. Then we graph in color $y = 3 + 2\sin 4x$ by raising the graph of $y = 2\sin 4x$ up 3 units where the graph oscillates about a midline of $y = 3$. Knowledge of the midline, the amplitude, and the period is crucial for setting an appropriate viewing window for such graphs, and you should try to use a grapher to duplicate the results in this example.

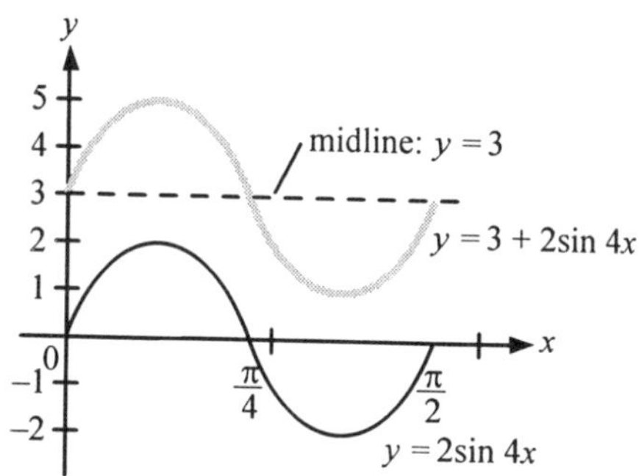

Figure 8.36

PROGRESS CHECK 8

Graph one cycle of the function $y = 800 - 400\cos\frac{\pi}{6}x$. Indicate the amplitude, period, and midline.

■

A horizontal shift is created when the constant c is not zero in a function of the form $y = d + a\sin(bx + c)$. To illustrate, consider the equation $y = \sin(x - \pi/2)$. As $x - \pi/2$ ranges from 0 to 2π, the curve completes one sine wave.

$$x - \frac{\pi}{2} = 0 \text{ when } x = \frac{\pi}{2}$$
$$x - \frac{\pi}{2} = 2\pi \text{ when } x = \frac{5\pi}{2}$$

Thus, the function completes one cycle in the interval from $\pi/2$ to $5\pi/2$. The period of the function is 2π and the amplitude is 1. For comparison, one cycle of the graphs of $y = \sin x$ and $y = \sin(x - \pi/2)$ is given in Figure 8.37.

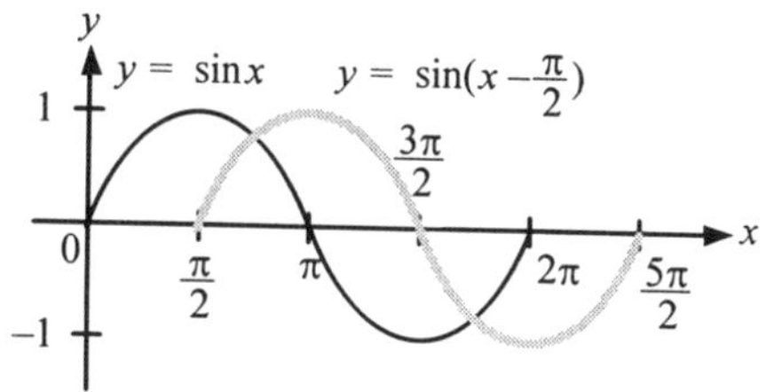

Figure 8.37

The complete graph of $y = \sin(x - \pi/2)$ is sketched by repeating the cycle shown in Figure 8.37 to the left and to the right. Notice that the graph of $y = \sin(x - \pi/2)$ may be obtained by shifting the graph of $y = \sin x$ to the right $\pi/2$ units. The sine wave then starts a cycle at $\pi/2$ instead of 0, and we call $\pi/2$ the **phase shift**. This horizontal shift can be anticipated from our work in Section 1.3, where we found that the graph of $y = f(x - c)$ with $c > 0$ is the graph of f shifted c units to the right.

In general, the constant c in the function $y = d + a\sin(bx + c)$ causes a shift of the graph of $y = d + a\sin bx$. The shift is of distance $|c/b|$ and is to the left if $c > 0$ and to the right if $c < 0$. The phase shift is given by $-c/b$. Similar remarks holds for functions of the form $y = d + a\cos(bx + c)$.

Example 9: Using a Horizontal Translation

Graph one cycle of the function $y = 3\cos\left(2x + \frac{\pi}{2}\right)$. Indicate the amplitude, period, and phase shift.

Solution: Determine the amplitude and period.

$$\text{Amplitude} = |a| = |3| = 3$$
$$\text{Period} = \frac{2\pi}{b} = \frac{2\pi}{2} = \pi$$

The function completes one cosine cycle as $2x + \pi/2$ varies from 0 to 2π.

$$2x + \frac{\pi}{2} = 0 \text{ when } x = -\frac{\pi}{4}$$
$$2x + \frac{\pi}{2} = 2\pi \text{ when } x = \frac{3\pi}{4}$$

Thus, the function completes one cycle in the interval from $-\pi/4$ to $3\pi/4$ (see Figure 8.38). This interval checks with the computed period, since $3\pi/4 - (-\pi/4) = \pi$. A cycle starts at $-\pi/4$, so $-\pi/4$ is the phase shift. We may verify the phase shift, since

$$\frac{-c}{b} = \frac{-\pi/2}{2} = -\frac{\pi}{4}.$$

All of these results should be checked by graphing $y = 3\cos(2x + \pi/2)$ on a grapher.

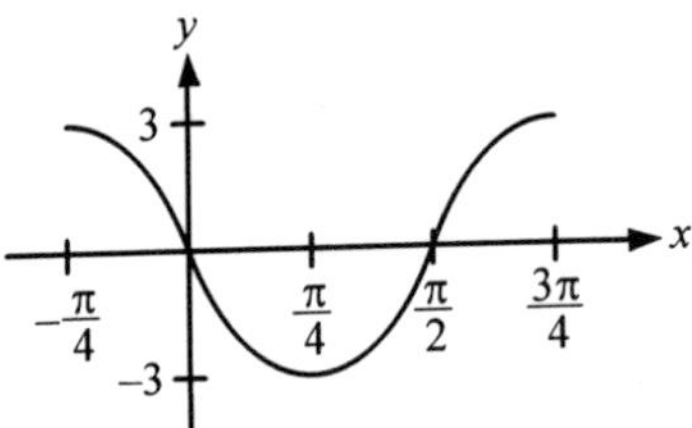

Figure 8.38

PROGRESS CHECK 9

Graph one cycle of the function $y = 4\sin\left(\frac{x}{4} - \frac{\pi}{2}\right)$. Indicate the amplitude, period, and phase shift. ∎

Before solving the section-opening problem, it is useful to consolidate our results about graphing sine and cosine curves.

Graphs of Sine and Cosine Functions

For the graphs of $y = d + a\sin(bx + c)$ and $y = d + a\cos(bx + c)$ where $a \neq 0$ and $b > 0$:

1. The **amplitude** is $|a|$.
2. The **period** is $2\pi/b$.
3. The **phase shift** is $-c/b$.
4. The **midline** is $y = d$.

Example 10: Number of Hours of Daylight in a Day

Solve the problem in the section introduction on page 635.

Solution:

a. For the given equation

$$y = 12 + 2.3\sin\left[\frac{2\pi}{365}(x - 80)\right],$$

$a = 2.3$, $b = 2\pi/365$, $c = -80$, and $d = 12$. Because the curve oscillates 2.3 units about the midline $y = 12$, it is sensible to let $9 \le y \le 15$ when creating the viewing window for the graph. Given the context of the problem, it is also sensible to let $1 \le x \le 365$. Then, Figure 8.39 shows a graph of the given equation in the viewing window just described.

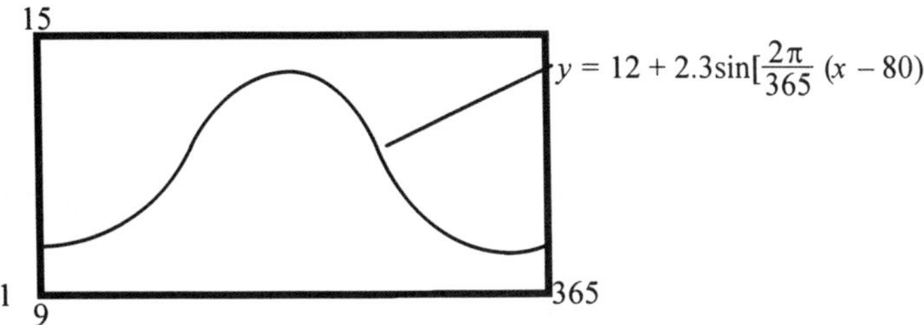

Figure 8.39

b. Since $a = 2.3$, and $b = 2\pi/365$,

$$\text{Amplitude} = |a| = |2.3| = 2.3$$

$$\text{Period} = \frac{2\pi}{b} = \frac{2\pi}{2\pi/365} = 365$$

Observe that the graph in Figure 8.39 supports these results.

c. Figure 8.40 shows that the given function has a minimum point at about $x = 354$, $y = 9.7$. (If your calculator does not have a Function Minimum feature, then you can use Zoom and Trace to determine this point.) Thus, the day of the year in Houston with the minimum number of hours of daylight is the 354th day, which is December 20.

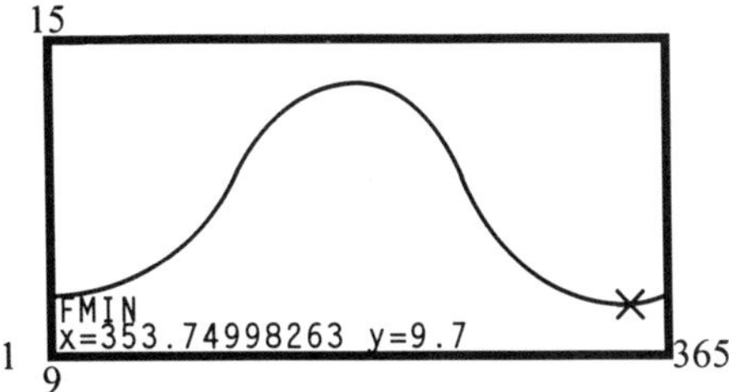

Figure 8.40

PROGRESS CHECK 10

Based on the rhythmic motion of her heart, a certain woman's blood pressure y is given by $y = 105 + 20\sin(140\pi t)$, where t is the elapsed time in minutes.

a. Graph this function on the interval [0 minutes, 0.1 minutes].

b. What are the maximum (systolic) and minimum (diastolic) readings for her blood pressure?

c. Find the women's heart rate in beats per minute. ■

Amplitude, period, midline, and phase shift are important considerations when we are analyzing any periodic phenomena. For example, let us briefly discuss two familiar concepts: musical sounds and radio waves. Musical sounds are caused by regular vibrations that have a definite period. On an electronic instrument called an **oscilloscope**, which changes sounds to electrical impulses and then to light waves, the sound from a tuning fork has the shape illustrated in Figure 8.41(a). The period of the wave depends on the pitch of the sound. With higher notes the pitch or frequency of the sound increases, producing a wave that has a smaller period [Figure 8.41(b)]. Human beings can detect frequencies between about 50 and 15,000 vibrations (cycles) per second. However, some animals, such as the bat, can hear frequencies as high as 120,000 hertz (cycles per second). The amplitude of the wave depends on the intensity of the sound. Since human beings hear best at a frequency of about 3,500 hertz, the loudness of a sound depends on both intensity and frequency. Although most musical sounds are very complex [such as the sound produced by the piano in Figure 8.41(c), the French mathematician Joseph Fourier, in about 1800, showed that any periodic function is the sum of simple sine functions. Thus, all these sounds an be graphed and analyzed by some combination of sine waves. This analysis is indispensable in the design of sound recording and reproducing equipment.

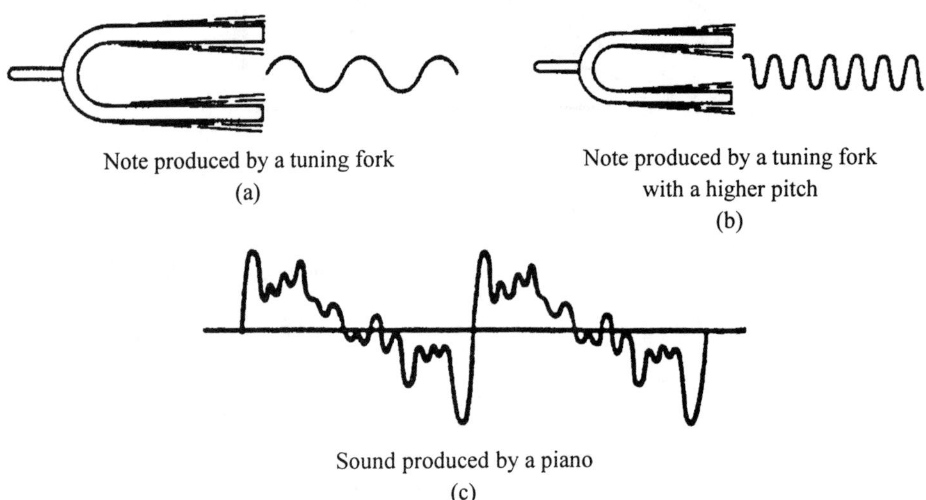

Note produced by a tuning fork
(a)

Note produced by a tuning fork
with a higher pitch
(b)

Sound produced by a piano
(c)

Figure 8.41

Radio waves, used to transmit information at the speed of light (186,000 mi/second), are produced by oscillations in an electric current. Each current cycle produces a single radio wave. Special electronic equipment is needed to produce the alternating current, since ordinary stations broadcast between 500,000 and 1,500,000 radio waves per second. Each station is licensed to broadcast a fixed number of radio waves per second, called its **frequency**. You receive the station's program by adjusting or tuning your set to accept this frequency.

Information is imposed on a radio wave as follows: A carrier wave [Figure 8.42(a)] is produced at a transmitting station, which makes the licensed number of cycles per second. This carrier wave is then modulated by the program current from the broadcasting site. In amplitude-modulated (AM) broadcasting, the amplitude of the carrier is made to vary according to the message; the wavelength remains constant [see Figure 8.42(b)]. With frequency modulation (FM), the amplitude of the wave remains constant and the wavelength varies [see Figure 8.42(c)].

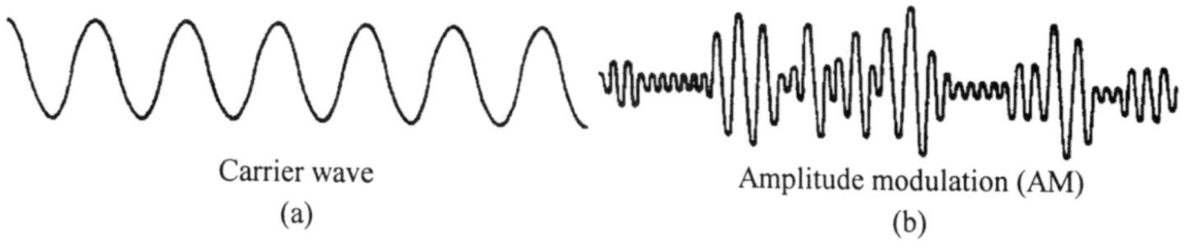

Carrier wave
(a)

Amplitude modulation (AM)
(b)

Frequency modulation (FM)
(c)

Figure 8.42

FM broadcasting is superior to AM in that it produces better fidelity of sound and is relatively free from static and interference. However, AM stations are more numerous since they are less expensive and have a greater broadcasting range. In television an AM signal is used to transmit the picture, but an FM signal carries the sound.

EXPLORE 8.3

1. Investigate how the graph for each family of functions changes as the specified parameter changes. In each case describe important features that are shared and important features that change. Use a Zoom Trigonometric setting if one is available on your calculator.
  a. The family $y = a\sin x$ as a changes. **b.** The family $y = \sin bx$ as b changes.
  c. The family $y = \sin(x + c)$ as c changes. **d.** The family $y = d + \sin x$ as d changes.

2. **a.** Graph $y = |\sin x|$, and describe the range. What is the relationship between the graph of $y = \sin x$ and $y = |\sin x|$?
  b. Without a calculator, sketch the graphs of $y = \cos x$ and $y = |\cos x|$. Confirm your sketch with a calculator.
  c. Without a calculator, sketch the graph of $y = \sin \pi$. Confirm your sketch with a calculator.

3. If your calculator supports parametric graphing, set $x_1 = \cos t$, and $y_1 = \sin t$. Then set the viewing window so that t ranges from -2π to 2π, while x and y range from -2 to 2. Describe the resulting graph, and explain why it makes sense. Check the appearance of the graph using a Zoom Square Setting.

4. **a.** It is possible to define trigonometric functions of imaginary numbers. If you have a calculator that handles imaginary numbers, confirm that $\sin(i) \approx 1.1752i$, and $\cos(i) \approx 1.5431$. Does it appear that $\sin^2 i + \cos^2 i = 1$?
  b. The sine of a real number cannot be greater than 1, but the sine of an imaginary number can be greater than 1. If your calculator does calculations with imaginary numbers, find $\sin^{-1} 2$, and determine approximately what imaginary number z has $\sin z = 2$.

EXERCISES 8.3

In Exercises 1–12 indicate the amplitude, period, and midline, and sketch the curve for $-p \le x \le p$, where p is the period of the function.

1. $y = 2\sin x$

2. $y = 3\cos 2x$

3. $y = -3\cos x$

4. $y = -4\sin\frac{1}{3}x$

5. $y = 2\sin 3x$

6. $y = \frac{1}{2}\cos 4x$

7. $y = -\cos 18x$ **8.** $y = -6\sin\frac{x}{4}$

9. $y = 10\cos(\pi x) + 4$

10. $y = 110\sin(120\pi x) - 110$

11. $y = 5 + \sin 2x$ **12.** $y = 4 - 2\sin x$

In Exercises 13–24 indicate the amplitude, period, and midline, and sketch the curve for $0 \le x \le 2\pi$.

13. $y = 3\cos 4x$ **14.** $y = 2\cos\frac{x}{4}$

15. $y = -\sin\frac{x}{2}$ **16.** $y = -3\sin 2x$

17. $y = \frac{1}{2}\sin 3x$ **18.** $y = 1.5\cos\frac{1}{3}x$

19. $y = \sin\frac{\pi}{2}x$ **20.** $y = 2\cos \pi x$

21. $y = 100 + 2\sin\frac{\pi}{4}x$ **22.** $y = -100 - 2\sin\frac{\pi}{3}x$

In Exercises 25–36 indicate the amplitude, period, midline, and phase shift, and sketch one cycle of the function.

25. $y = \sin\left(x + \frac{\pi}{2}\right)$ **26.** $y = 2\sin(x - \pi)$

27. $y = \cos\left(x - \frac{\pi}{4}\right)$ **28.** $y = 3\cos\left(x + \frac{\pi}{3}\right)$

29. $y = \frac{1}{2}\cos\left(2x + \frac{\pi}{4}\right)$ **30.** $y = -\sin\left(\frac{x}{2} - \pi\right)$

31. $y = -\cos\left(\frac{x}{4} + \frac{\pi}{2}\right)$ **32.** $y = \sin(x - 1)$

33. $y = \sin(\pi x - \pi)$

34. $y = 1.2\cos\left(2\pi x - \frac{\pi}{2}\right) - 3$

35. $y = 30 + 30\sin\left(\frac{\pi}{2}x - \frac{\pi}{2}\right)$

36. $y = -40 - 40\cos\left(\frac{\pi}{3}x + \frac{\pi}{3}\right)$

In Exercises 37–44 find an equation for the curves with the given single cycle. The equations should be written in the form $y = a\sin bx$ or $y = a\cos bx$.

37.

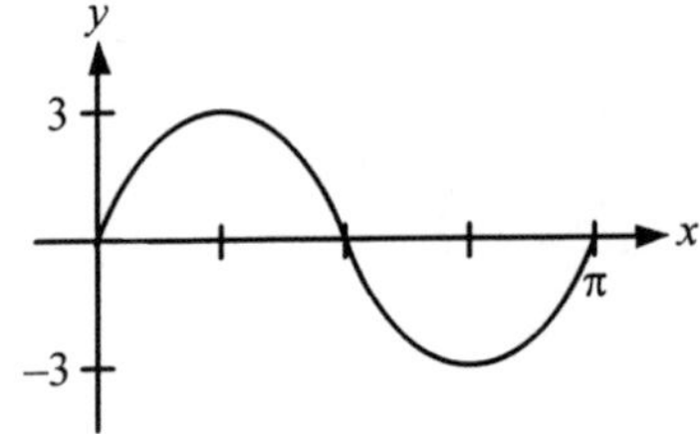

38.

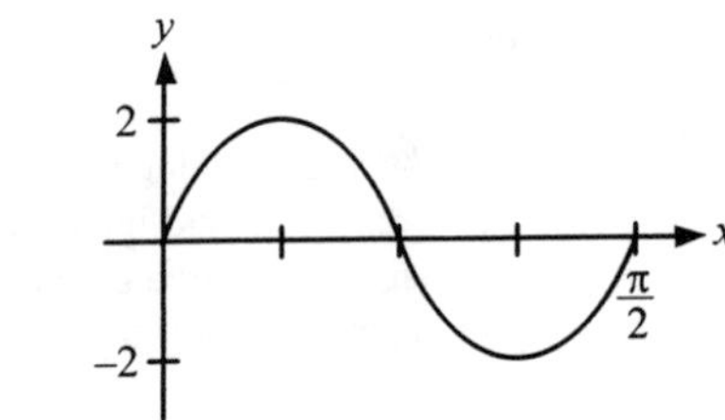

39.

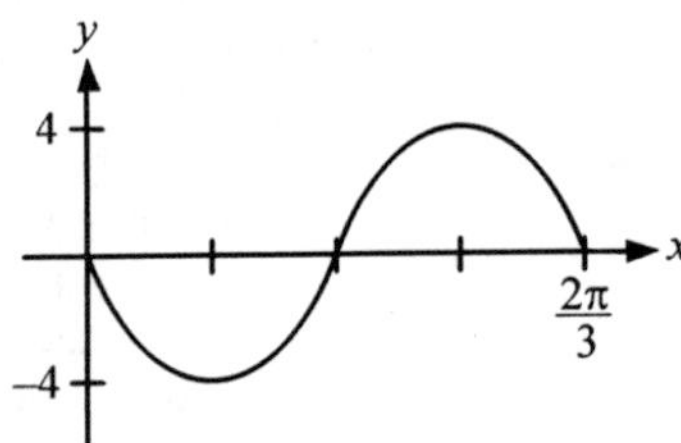

40.

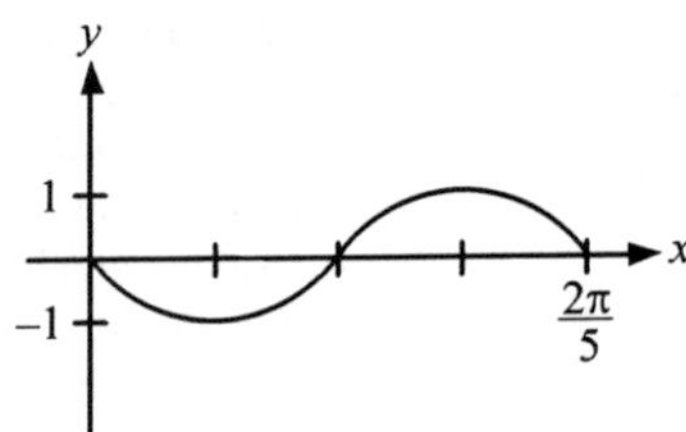

41.

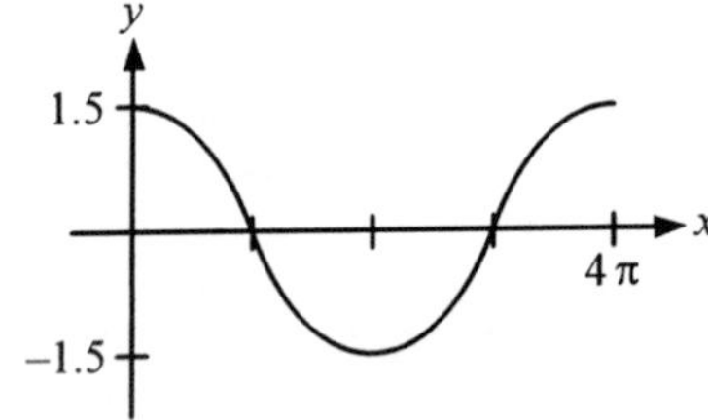

42.

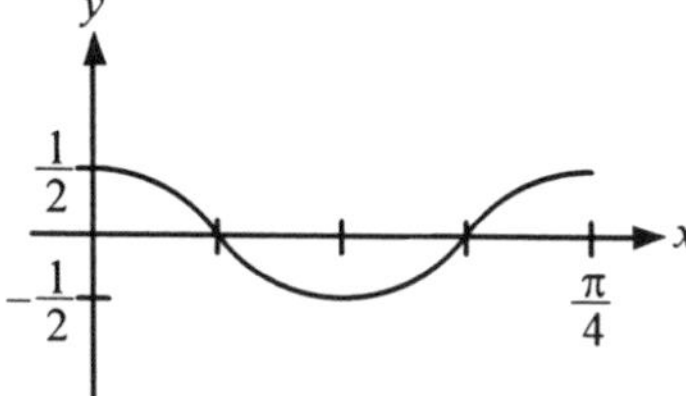

43.

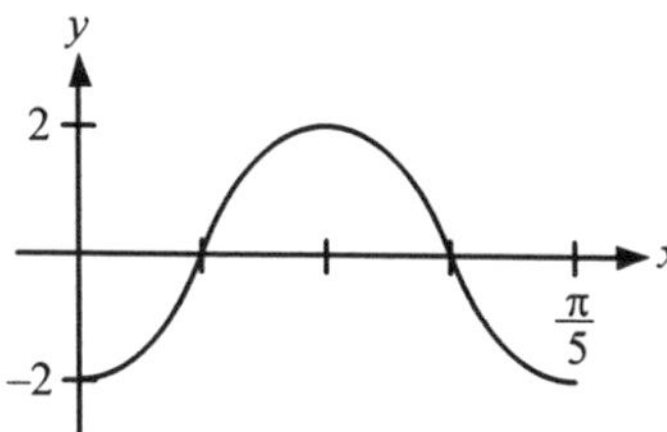

44.

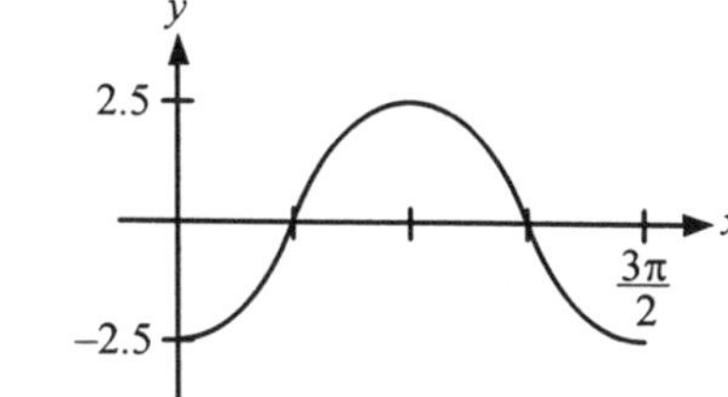

In Exercises 45–50 let $M =$ the maximum value of y and let $m =$ the minimum value of y, then use the facts that $d = \frac{M+m}{2}$, and $a = \frac{M-m}{2}$ to find an equation of the form $y = d + a\sin(x)$, for the given conditions. Graph your equation to confirm that the graph meets the given conditions.

	Maximum	Minimum
45.	20	14
46.	5	0
47.	0	−10
48.	−10	−20
49.	50	−10
50.	3	−9

In Exercises 51–56 find an equation for the curve with the given single cycle.

51.

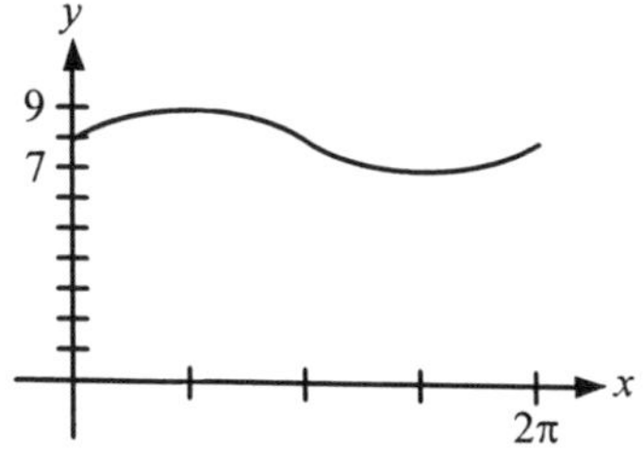

52.

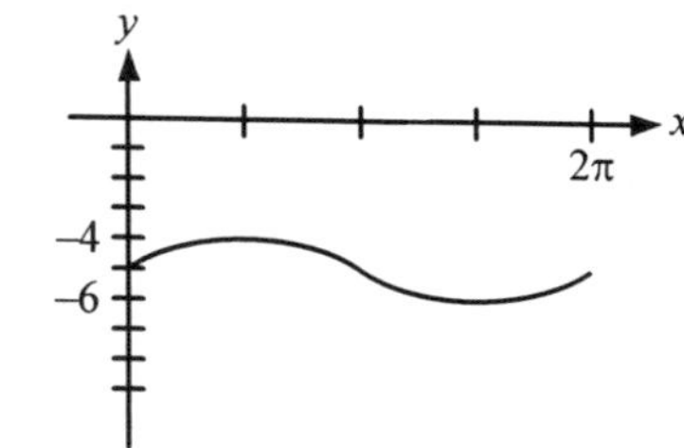

53.

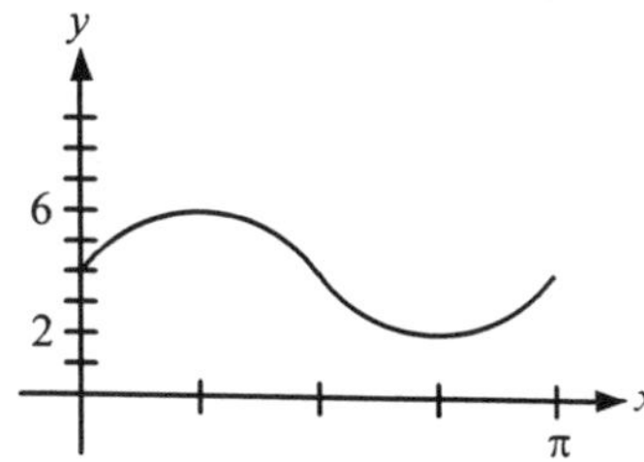

54.

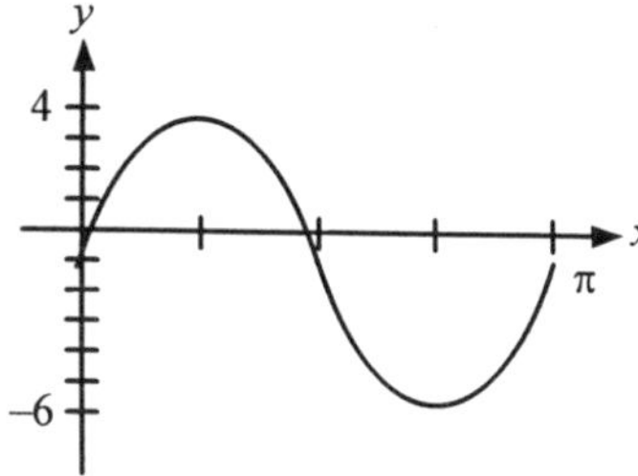

55.

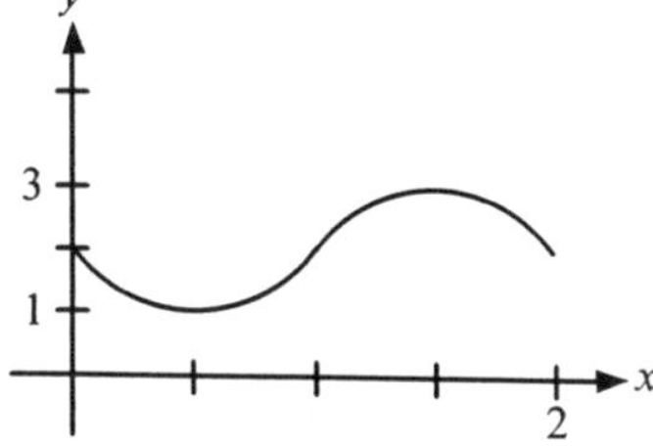

56.

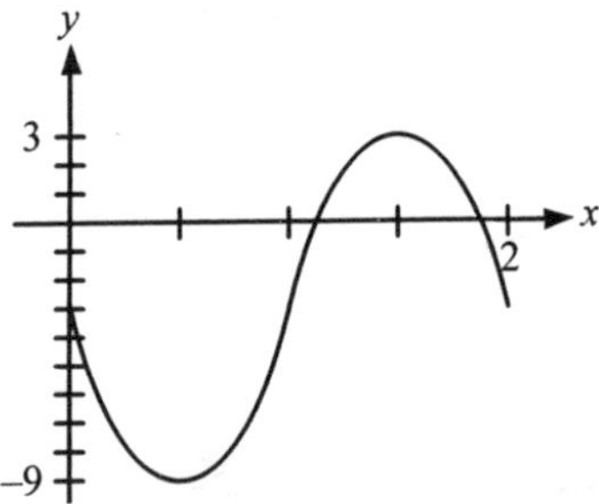

57. Refer to Example 10 of Section 8.3. For the city of Key West, Florida, an equation that may be used to approximate the length in hours (y) of the xth day is:

$$y = 12 + 1.65\sin\left[\frac{2\pi}{365}(x - 80)\right].$$

a. Graph this function over the interval from $x = 1$ (January 1) to $x = 365$ (December 31), with the aid of a grapher.
b. What are the amplitude and period of the function?
c. Find the day that has the maximum number of hours of daylight, and estimate the number of hours of daylight on that day.

58. Refer to Example 10 of Section 8.3. For the city of Nome, Alaska, an equation that may be used to approximate the length in hours (y) of the xth day is

$$y = 12.5 + 8.5\sin\left[\frac{2x}{365}(x - 80)\right].$$

a. Graph this function over the interval from $x = 1$ (January 1) to $x = 365$ (December 31) with the aid of a grapher.
b. What are the amplitude and period of the function?
c. Find the day that has the maximum number of hours of daylight, and estimate the number of hours of daylight on that day.

59. The monthly revenue R in dollars from sales of ice cream for a sweet shop in Montpelier, Vermont is approximated by

$$R = 4{,}200 + 2{,}100\cos\left(\frac{2\pi}{12}x\right),$$

where $x = 0$ represents July.

a. Graph this function over the interval from $x = 0$ to $x = 12$.
b. What are the amplitude and the period of the function?
c. According to this graph, which month has the minimum revenue?

60. The monthly number of kilowatt hours (E) of electricity sold by a particular utility company is approximated by

$$E = 384.5 + 89.5\cos\left(\frac{2\pi}{12}x\right),$$

where $x = 0$ represents January.

a. Graph this function over the interval from $x = 0$ to $x = 12$.
b. What are the amplitude and the period of the function?
c. According to this graph, which month has the minimum usage?

61. Suppose a swinging object oscillates according to the equation

$$y = 6.0\cos\left(3\pi x + \frac{\pi}{3}\right),$$

where y represents the horizontal displacement from a reference point in meters, and x is the elapsed time in seconds.

a. Graph this function over the interval from $x = 0$ to $x = \frac{2}{3}$.
b. What are the amplitude and the period of the function?
c. What is the maximum possible horizontal displacement of this body? At what time in the cycle does it occur?

62. The equation

$$y = 10\sin[\pi(0.01x - 2.00t)]$$

gives the displacement y (in centimeters) of a wiggling rope at a point x centimeters from the end at time t seconds.

a. Let $x = 100$ cm, and graph this function over the interval from $t = 0$ to $t = 2$ seconds, in order to visualize the behavior of the rope 100 cm from the end.
b. What are the amplitude and the period of the function?
c. At what point over the interval from $t = 0$ to $t = 1$ does the rope achieve maximum displacement?

63. The average daily temperature y in degrees Fahrenheit at the surface of the ground in Athens, Georgia, over the course of a year is approximated fairly well by the equation

$$y = 61.3 + 17.9\cos\left(\frac{2\pi}{365}x\right),$$

where x is in days and $t = 0$ represents July 1.

a. Graph this function over the interval from $x = 0$ to $x = 365$.

b. What are the amplitude and the period of the function?

c. In what month is the average daily temperature at a minimum?

64. In a certain location, the equation

$$y = 16 + 11e^{-.00706d}\cos\left(\frac{2\pi}{k}x - .00706d\right).$$

approximates the temperature in a storage cellar d cm below ground level at time x. The variable x is given in seconds, where $x = 0$ represents 12 a.m. July 1. The constant $k = 365 \times 24 \times 3600$, the number of seconds in a year

a. Let $d = 445$ cm and graph this function over the interval from 0 to k, in order to visualize the pattern of temperature change over 1 year in a cellar of depth 445 cm.

b. What are the amplitude and the period of the function?

c. Approximate, to the nearest degree, the warmest and coldest temperatures reached in this cellar.

65. A clock is mounted in a square frame as shown. The second hand is 5 in. long. The tip of the second hand traces the circumference of a circle.

a. Beginning when the second hand is on the 12, and every 10 seconds after that, for one entire revolution, calculate the distance from the tip of the second hand to the left edge of the frame, and complete the table shown.

time (t):	0	10	20	30	40	50	60
distance (d):	5						

b. Use the table to graph d as a function of t.

c. Find a formula that expresses d as a function of t.

66. Answer the questions posed in Exercise 65, this time measuring the distance from the tip of the second hand to the top edge of the frame.

THINK ABOUT IT 8.3

1. In each case describe how the graph of g may be obtained from the graph of f.
   a. $f(x) = \sin x$, $g(x) = -\sin x$
   b. $f(x) = \cos x$, $g(x) = 4\cos x$
   c. $f(x) = 4\cos x$, $g(x) = 4\cos(x + \pi)$
   d. $f(x) = 3\sin \pi x$, $g(x) = 3\sin(\pi x - \pi)$
2. a. Give two examples of an equation of the form $y = \sin(bx + c)$ whose graph passes through $(\pi/4, 0)$.
   b. Give two examples of an equation of the form $y = \cos(bx + c)$ whose graph passes through $(1, 0)$.
3. Find the smallest nonnegative value of a for which the graph of $y = \sin x$ is symmetric about the line $x = a$. Confirm your answer graphically.
4. Use this sunrise and sunset data for the city of Honolulu, Hawaii, to find a sine function that approximates the number of hours of daylight on the xth day of the year.

Day	**Sunrise (a.m.)**	**Sunset (p.m.)**
Mar. 21	6:35	6:43
June 21	5:50	7:16
Sept. 21	6:20	6:29
Dec. 21	7:05	5:55

5. Sine waves may be used to represent sound waves. Since the loudness of sound can change with time, a graph is needed with changing amplitude.
   a. Investigate the graphs for the given functions, and describe what is happening to the loudness.

      (1) $y = 0.9^x \sin x$ (2) $y = 1.1^x \sin x$ (3) $y = 5e^{-x^2/100}\sin 2x$
   b. Find an equation that will produce a sine wave similar to the one shown here.

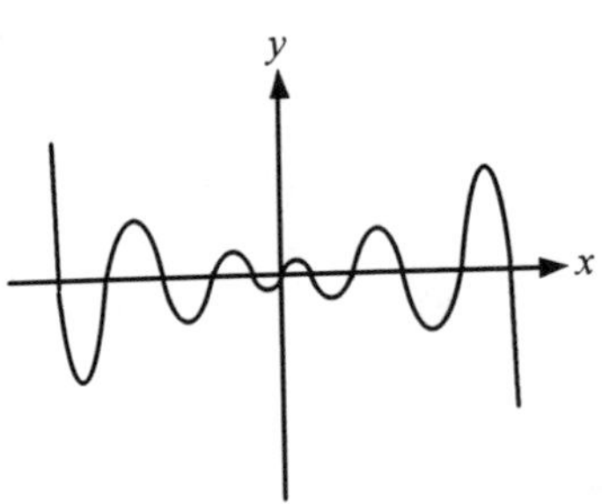

c. When statisticians analyze time series graphs, they may use a model that superimposes periodic behavior on a linear trend. For example, ice cream sales fluctuate periodically with the seasons, but over the last several years sales have generally been rising. Graph $y = .2x + \sin x$, using several viewing windows, and confirm that it shows the kind of behavior described above. Invent an equation that will superimpose periodic fluctuations on a decreasing linear trend.

● ● ●

8.4 Graphs of Other Trigonometric Functions

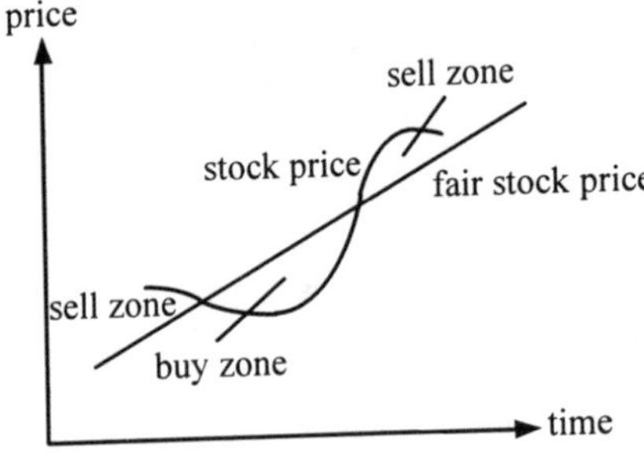

In a value-oriented mutual fund, portfolio managers try to invest in well-established companies whose stocks are currently undervalued as their share price oscillates in value. This investment approach can be illustrated graphically as shown in the figure

If a manager projects that the price y in dollars for a share of stock in a particular company may be modeled by

$$y = 62 + 0.3x - 5\sin\frac{\pi}{7}x,$$

where x is the elapsed time in months from January 1, then graph this equation simultaneously with the equation for the fair stock price, using a grapher. Let x range over the 2-year period that starts at January 1.
(See Example 5. Source: *Twentieth Century Mutual Funds–Conservative Equity Fund Investing*)

Objectives

1. Graph tangent, cotangent, secant, and cosecant functions.
2. Graph $f(x) = g(x) + h(x)$, where g or h is trigonometric, using addition of ordinates or using a grapher.

The objectives listed above indicate that in this section we will learn to sketch graphs of the trigonometric functions other than sine and cosine, and that we will discus a graphing technique known as *addition of ordinates*. Problems that involve this latter topic are handled most efficiently by using a grapher, and we will rely on such an aid to solve the problem on value-oriented investing given in the section opener. We begin by discussing how to graph $y = \tan x$.

Although the tangent function is periodic, its behavior differs dramatically from that of the sine and cosine. To see this difference, compare the graph of $y = \tan x$ to the smooth, weaving curves of these functions. First, we use a numeric approach and construct the following table, which lists some values of the tangent function on the interval $[0,\ 2\pi]$.

$y = \tan x = \sin x/\cos x$ (for $0 \le x \le 2\pi$)														
x	0	$\frac{\pi}{6}$	$\frac{\pi}{3}$	$\frac{\pi}{2}$	$\frac{2\pi}{3}$	$\frac{5\pi}{6}$	π	$\frac{7\pi}{6}$	$\frac{4\pi}{3}$	$\frac{3\pi}{2}$	$\frac{5\pi}{3}$	$\frac{11\pi}{6}$	2π
y	0	0.6	1.7	und.	–1.7	–0.6	0	0.6	1.7	und.	–1.7	–0.6	0

Unlike the sine and cosine, the tangent function is not defined for all real numbers. That is, $y = \tan x = \sin x/\cos x$ is undefined when $\cos x = 0$. Thus, we must exclude from the domain of this function $\pi/2$, $3\pi/2$, and any x for which $x = (\pi/2) + k\pi$ (k any integer). As x approaches $\pi/2$, $\sin x$ approaches 1, and $\cos x$ approaches 0. This means that $|\tan x|$ becomes very large as x gets close to $\pi/2$. On the basis of the preceding discussion and the fact that $\tan(-x) = -\tan x$, a portion of the graph $y = \tan x$ is presented in Figure 8.43.

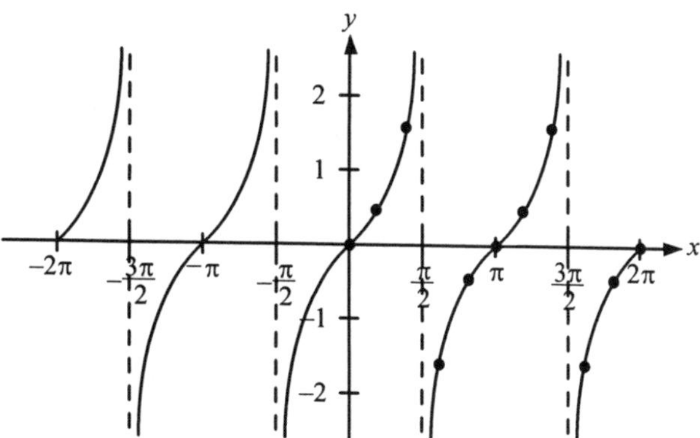

Figure 8.43

It is easy to use a grapher to check the result in Figure 8.43, and graphs for $y = \tan x$ are displayed in Dot mode and in Connected mode in Figures 8.4 and 8.5, respectively. Observe in Figure 8.5 that the calculator erroneously draws almost vertical lines near the vertical asymptotes because it is set to connect points that are in separate pieces of the graph. You should be prepared to experiment with different viewing windows and graphing modes, when graphing trigonometric functions with vertical asymptotes, until you obtain a picture you can interpret. However, under no circumstances should your hand-drawn graph show solid lines at vertical asymptotes like those that may appear on a calculator screen.

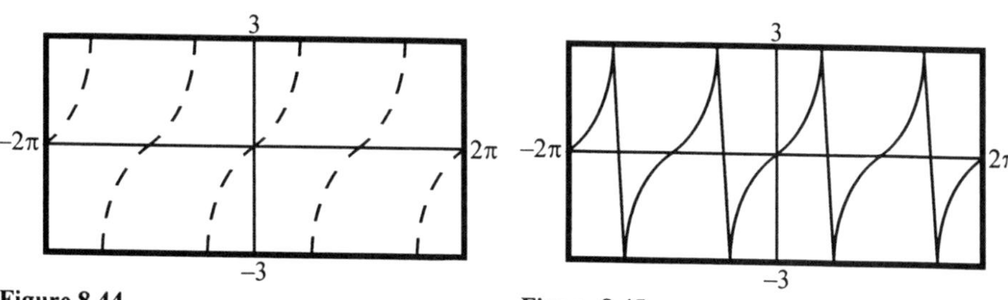

Figure 8.44

Figure 8.45

The graphs of $y = \tan x$ may be used to contrast the tangent function to the sine and cosine functions with regard to several basic features.

Comparison of the Sine, Cosine, and Tangent Functions

1. The domain of the sine and cosine functions is the set of all real numbers. The tangent function excludes $x = (\pi/2) + k\pi$ (k any integer).
2. The range of the sine and cosine functions is $[-1, 1]$. The range of the tangent function is the set of all real numbers.
3. The sine and cosine are periodic with period 2π. Careful consideration of Figure 8.43 shows that the tangent function repeats its value every π units. Thus, the period of the tangent functions is π, and the period of $y = \tan bx$ $(b > 0)$ is π/b.

The other three trigonometric functions may be graphed as the reciprocals of the sine, cosine, or tangent. That is,

$$\csc x = \frac{1}{\sin x}, \quad \sec x = \frac{1}{\cos x}, \quad \cot x = \frac{1}{\tan x}.$$

In Figures 8.46–8.48 we first graph the sine, cosine, and tangent as black curves. After obtaining the reciprocals of various y values, we then graph in color the cosecant, secant, and cotangent. You should use a grapher to check each of the graphs by using the reciprocal identities and entering $1/\sin x$ for $\csc x$, and so on. Observe from these graphs the following relations between a function and its reciprocal function.

1. As one function increase, the other decreases, and vice versa.
2. The two functions always have the same sign.
3. When one function is zero, the other is undefined.
4. When the value of the function is 1 or –1, the reciprocal function has the same value.

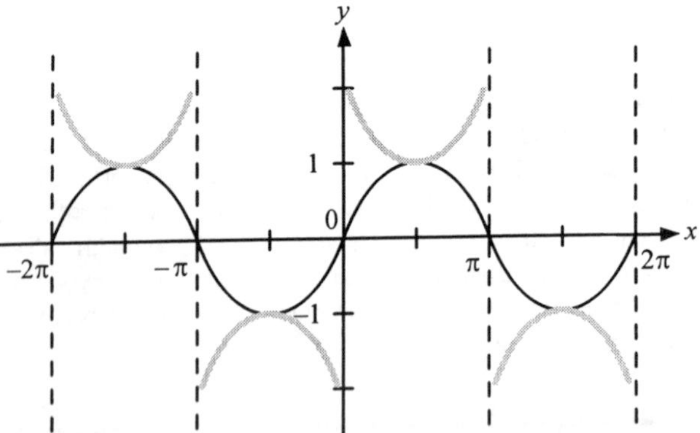

Figure 8.46
$y = \csc x$

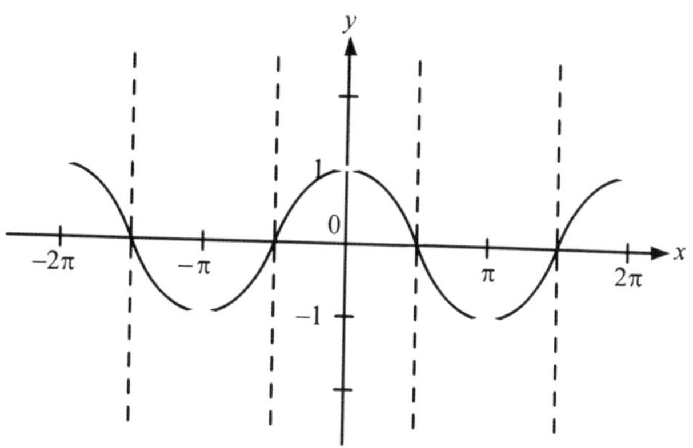

Figure 8.47
$y = \sec x$

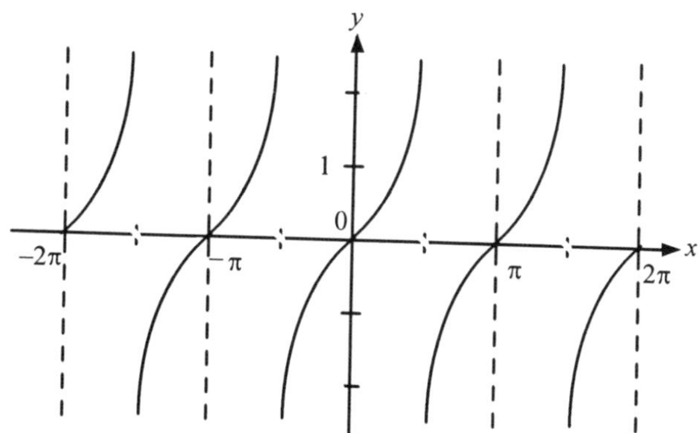

Figure 8.48
$y = \cot x$

Example 1: Graphing a Tangent Function Using a Translation

Graph one cycle of the function $y = \tan\left(x + \frac{\pi}{4}\right)$.

Solution: We start with the graph of $y = \tan x$ in Figure 8.43. To graph $y = \tan(x + \pi/4)$, we note that x has been replaced by $x + \pi/4$, so by our graphing techniques from Section 1.3, the graph of $y = \tan(x + \pi/4)$ is the graph of $y = \tan x$ shifted $\pi/4$ units to the left. Figure 8.49 shows the graph of one cycle on the interval that contains $x = 0$. Check this result on a grapher.

PROGRESS CHECK 1
Graph one cycle of the function $y = \cot(x + \pi/2)$.

■

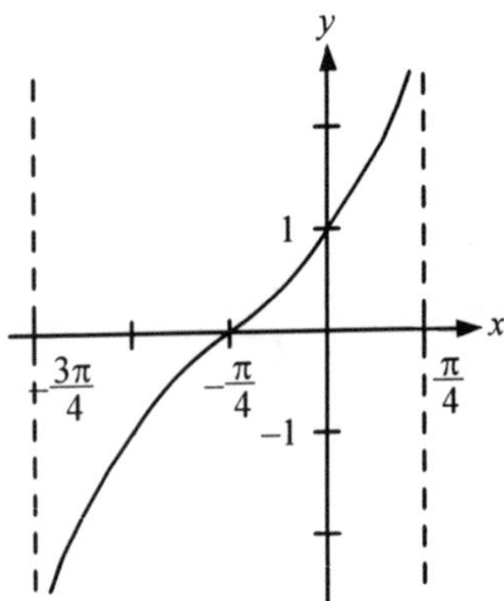

Figure 8.49

Example 2: Using a Reflection and a Translation

Graph one cycle of the function $y = -\tan x - 1$.

Solution: To graph $y = -\tan x - 1$, first reflect the graph of $y = \tan x$ about the x-axis to obtain the graph of $y = -\tan x$ [see Figure 8.50(a)]. Then we lower the graph in Figure 8.50(a) down 1 unit, because the constant 1 is subtracted from $-\tan x$. The completed graph is shown in Figure 8.50(b). You should use a grapher to check this result.

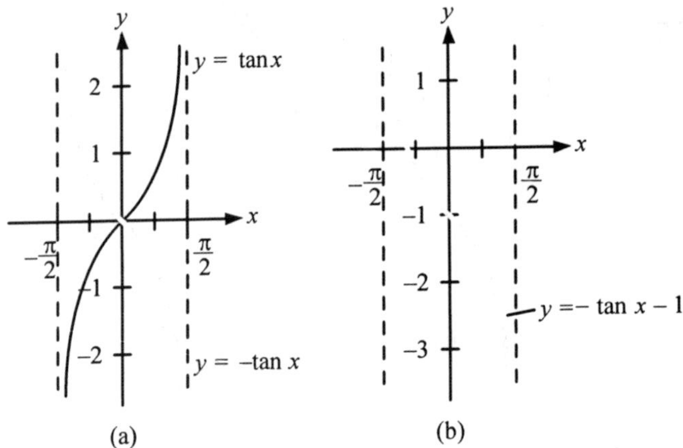

Figure 8.50

PROGRESS CHECK 2

Graph one cycle of the function $y = 2 - \tan x$. ■

Example 3: Graphing a Secant Function

Graph one cycle of the function $y = \sec 2\pi x$.

Solution: Since $\sec 2\pi x = 1/\cos 2\pi x$, we first sketch one cycle of the graph of $y = \cos 2\pi x$ to use as a reference. The amplitude and period for $y = \cos 2\pi x$ are both 1, and the black curve in Figure 8.51 shows the graph of this function. Then by obtaining the reciprocals of the various y values, we graph one cycle of $y = \sec 2\pi x$, as shown in the figure. This result checks with the display in a grapher that is shown in Figure 8.52.

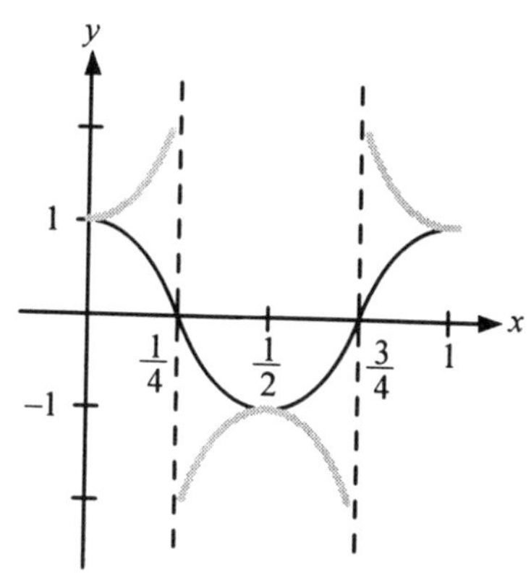

Figure 8.51

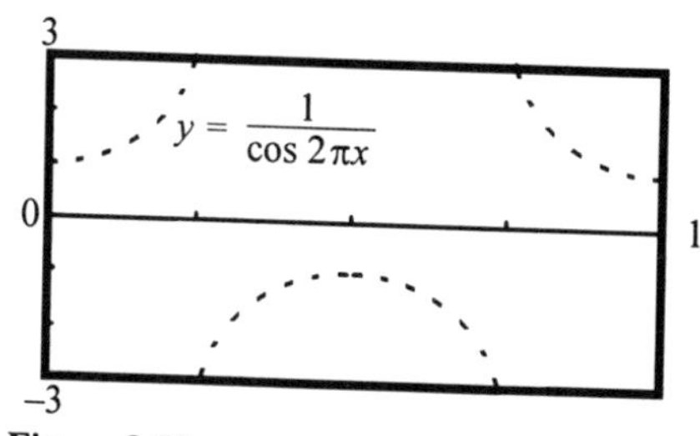

Figure 8.52

PROGRESS CHECK 3
Graph one cycle of the function $y = 2\sec \pi x$.

■

Addition of Ordinates

Graphing by vertical shifts as in Example 2 is a special case of graphing equations of the form $f(x) = g(x) + h(x)$ by addition of ordinates (or y values) at each x value in the domain of f. To illustrate the idea involved in this method, we will graph such an equation in Example 4 without the aid of a grapher. Note that if either g or h is a trigonometric function, then the graph is usually drawn by determining at least all points in the graph of f associated with intercepts, maximum points, or minimum points in the graphs of either g or h.

Example 4: Graphing by Addition of Ordinates

Graph $y = x + \sin x$, by using addition of ordinates.

Solution: Figure 8.53 shows the graphs of $g(x) = x$ and $h(x) = \sin x$ drawn on the same coordinate system. On $[0, 2\pi]$ the points in the graph of $y = x + \sin x$ associated with intercepts, maximum points, or minimum points in the graphs of g and h are determined below.

x	$g(x) = x$	$h(x) = \sin x$	$y = x + \sin x$
0	0	0	$0 + 0 = 0$
$\frac{\pi}{2}$	$\frac{\pi}{2}$	1	$\frac{\pi}{2} + 1 \approx 2.6$
π	π	0	$\pi + 0 = \pi$
$\frac{3\pi}{2}$	$\frac{3\pi}{2}$	-1	$\frac{3\pi}{2} - 1 \approx 3.6$
2π	2π	0	$2\pi + 0 = 2\pi$

By plotting $(0, 0)$, $(\pi/2, \pi/2 + 1)$, (π, π), $(3\pi/2, 3\pi/2 - 1)$, and $(2\pi, 2\pi)$ and drawing a smooth curve through these points, we obtain the graph of $y = x + \sin x$ in Figure 8.53. Note that it is useful to view this graph as the graph of $y = \sin x$ oscillating about a rising midline of $y = x$. You should try to use a grapher to duplicate the results shown in Figure 8.53.

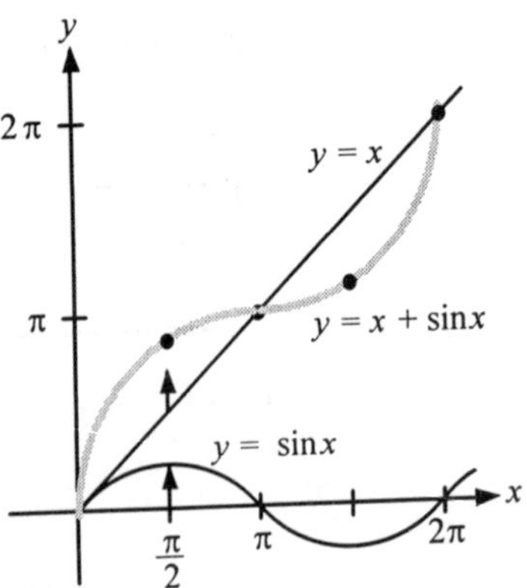

Figure 8.53

PROGRESS CHECK 4

Graph $y = 2 + 0.5x - \sin x$ by using addition of ordinates. ■

Example 4 illustrates that graphing by addition or ordinates may involve considerable work with computing function values and then plotting points. Fortunately, in practice, a grapher can accurately plot many such points in a few seconds, and we will use this method to draw the graphs in the next two examples.

Example 5: Value-Oriented Investing

Solve the problem in the section introduction on page 654.

Solution: The equation that is given for the anticipated stock price,

$$y = 62 + 0.3x - 5\sin\frac{\pi}{7}x,$$

has a graph that oscillates about a rising midline

$$y = 62 + 0.3x,$$

which shows the manager's estimate for the fair price of the stock. To obtain a meaningful graph for these models over the 24-month period that starts at January 1, we use [0, 24] by [55, 80] to define a viewing window, and then graph these equations as shown in Figure 8.54.

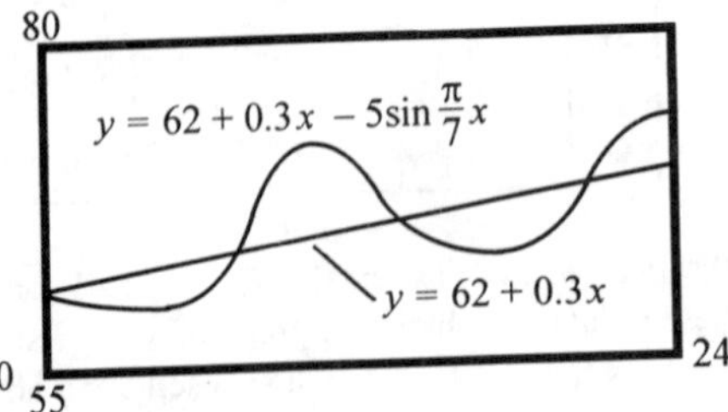

Figure 8.54

PROGRESS CHECK 5

In a certain wildlife area, the population of rabbits is given by $y = 90 + 3x + 25\sin\frac{\pi x}{6}$, where y is the number of rabbits per square mile, and x is the elapsed time in months for January 1. Graph this equation simultaneously with the equation for the midline, using a grapher. Let x range over the 3-year period that starts at January 1.

■

Example 6: Finding the Period and the Amplitude

Use a grapher to graph $y = \sin x - \cos x$ and to find the period and the amplitude of this function.

Solution: A graph of $y = \sin x - \cos x$ is shown in Figure 8.55 and it is apparent that the graph is a sine wave. To determine the period and the amplitude, we can use the Function Maximum feature (or Zoom and Trace) to find that the coordinates of the two relative maximum points in the displayed graph are located at about (2.356, 1.414) and (−3.927, 1.414). The midline in the graph is the x-axis, so the amplitude is given by the maximum y value. Thus

$$\text{amplitude} \approx 1.414.$$

The period, which is the length of one cycle, is the difference between the x-coordinates at these two maximum points, so

$$\text{period} \approx 2.356 - (-3.927) \approx 6.283.$$

You should note that $6.283 \approx 2\pi$, so the period in terms of π is given by 2π. This result makes sense for the period of $y = \sin x - \cos x$ because the period of both $y = \sin x$ and $y = \cos x$ is 2π. In general, if $f(x) = g(x) + h(x)$ where g and h are periodic functions, then the period of f is the least common multiple of the periods of g and h.

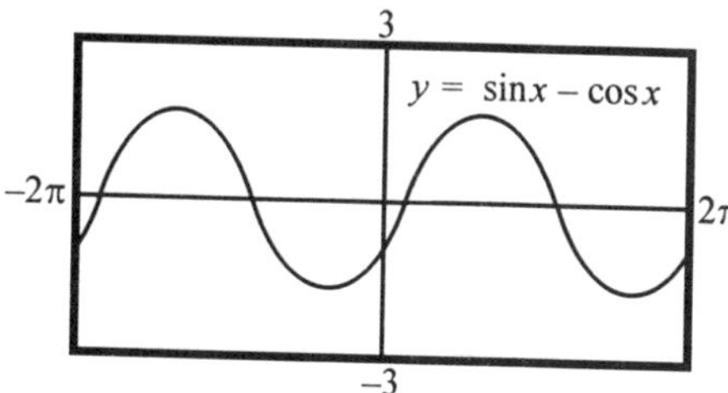

Figure 8.55

Note: In general, the equation $y = a\sin x + b\cos x$ may be written in the form $y = A\sin(x + c)$. Thus, you can anticipate that the graph of $y = a\sin x + b\cos x$ is a sine graph. For this example, an identity derived in Section 8.8 can be used to confirm that $\sin x - \cos x = \sqrt{2}\sin(x - \pi/4)$.

PROGRESS CHECK 6

Use a grapher to graph $y = 2\sin 2x + \cos 2x$ and to find the amplitude and the period of this function.

■

EXPLORE 8.4

1. Explore the family of graphs for $y = \tan(ax)$ where a is a constant, and determine the period as a function of a. For example, try $a = 1$, 2, and 4.
2. **a.** Use a grapher to examine the graph of $y = |\tan x|$ over the interval $x = -\pi$ to $x = \pi$.
   b. Does this graph have a vertical asymptote at $x = \pi/2$?
   c. Evaluate y for $x = \pi/2 \pm 0.0000001$, and explain why you get these answers.
3. If a sine function and a cosine function have the same period, then their sum may be expressed as some sine function.

a. Check graphically that $y = 3\sin 2x + \cos(2x + \pi)$ appears to be a sine function.

b. Estimate the period, amplitude, and phase shift of the graph in part **a**, then determine a, b, and c so that the graph of $y = a\sin(bx + c)$ is the same as the graph in part **a**.

EXERCISES 8.4

In Exercises 1–4 complete the table for the function and then sketch the curve from these points on the interval $[0, 2\pi]$. Confirm your result with a grapher.

x	0	$\frac{\pi}{6}$	$\frac{\pi}{3}$	$\frac{\pi}{2}$	$\frac{2\pi}{3}$	$\frac{5\pi}{6}$	π	$\frac{7\pi}{6}$	$\frac{4\pi}{3}$	$\frac{3\pi}{2}$	$\frac{5\pi}{3}$	$\frac{11\pi}{6}$	2π
y													

1. $y = \cot x$
2. $y = \sec x$
3. $y = \csc x$
4. $y = \cot 2x$

In Exercises 5–8 use Figures 8.47–8.49 or a grapher to determine the domain, the range, and the period of the function.

5. $y = \cot x$
6. $y = \sec x$
7. $y = \csc x$
8. $y = \cot 2x$

In Exercises 9–14 complete the table by determining if the function is increasing or decreasing in the interval. Use graphs to determine your answers.

		$\left(0, \frac{\pi}{2}\right)$	$\left(\frac{\pi}{2}, \pi\right)$	$\left(\pi, \frac{3\pi}{2}\right)$	$\left(\frac{3\pi}{2}, 2\pi\right)$
9.	$\sin x$				
10.	$\cos x$				
11.	$\tan x$				
12.	$\cot x$				
13.	$\sec x$				
14.	$\csc x$				

In Exercises 15–24 graph one cycle of the given functions. State the period.

15. $y = -\cot\left(x + \frac{\pi}{2}\right)$
16. $y = \tan\left(x - \frac{\pi}{4}\right)$
17. $y = \tan 2x$
18. $y = \cot\frac{1}{2}x$
19. $y = 3\csc\frac{x}{2}$
20. $y = \sec 2x$
21. $y = -\sec\left(x + \frac{\pi}{4}\right)$
22. $y = 2\csc \pi x$
23. $y = \csc(\pi x - \pi)$
24. $y = \sec\left(2\pi x - \frac{\pi}{2}\right)$

In Exercises 25–34 graph one cycle of the given functions. Use vertical shifts. State the period.

25. $y = -3 + \cos x$
26. $y = 2 + \sin x$
27. $y = -\sin x - 1$
28. $y = -\cos x - 2$
29. $y = 1 - 2\cos 2x$
30. $y = 2 + 3\sin 4x$
31. $y = -1 + \csc \pi x$
32. $y = 1 + \sec x$
33. $y = 2 - \tan x$
34. $y = -\cot x - 1$

In Exercises 35–40 use a grapher or graph by the method of addition of ordinates.

35. $y = x - \sin x$
36. $y = x + \cos x$
37. $y = \cos x - x$
38. $y = -x + \sin x$
39. $y = .02x^2 + \sin x$ for $-15 \le x \le 15$
40. $y = .02x^2 - \cos x$ for $-15 \le x \le 15$

In Exercises 41–46 use a grapher to graph the function and to find the period and amplitude.

41. $y = \sin x + \cos x$
42. $y = \cos x - \sin x$
43. $y = 2\cos x + \sin x$
44. $y = 2\sin x - \cos x$
45. $y = \sin x - \cos 2x$
46. $y = \cos x + \sin 2x$
47. The number of reported cases of a disease in a large city generally decreased over a period of 5 years, but it fluctuated seasonally, as described approximately by the equation

$$y = 20{,}000 - 90x + 400\cos\left[\frac{2\pi}{12}(x - 3)\right]$$

where $x = 0$ represents January 1 of the first year.

a. Use a grapher to graph this function for the 5-year interval from $x = 0$ to $x = 60$ months.

b. During the second year, which month had the highest number of cases reported? To the nearest 100, how many cases were reported that month?

48. For a five-year interval, the quarterly sales revenue (y) for a seasonal product is approximated by the equation

$$y = 100 + 25x + \sin\left[\frac{2\pi}{12}(x-5)\right],$$

where $x = 0$ represents the first quarter of the first year. The revenue y is given in thousands of dollars.

a. Use a grapher to graph this function for the 5-year interval from $x = 0$ to $x = 20$.

b. During the third year, which quarter had the highest sales revenue? To the nearest $1,000, what was the revenue that quarter?

49. The following equation appeared in the analysis of an electric circuit. The variables x and y represent time in seconds and current in amperes, respectively.

$$y = 4.47\sin(40x - 1.11) + 4e^{-10x}$$

This expresses the current as the sum of a sinusoidal term and an exponential term. The exponential term becomes very small very quickly, and becomes negligible for practical purposes; it is called the **transient** term. The remaining sinusoidal term is called the **steady state** term.

a. Use a grapher to graph this function over the interval from 0 to 1. Use a viewing window [0, 1] by [−5, 7] for a reasonable view.

b. Estimate the maximum value of the function during the first cycle, and during the second cycle.

c. At what value of x does the difference between the given function and the steady state part become less than .1? Less than .01?

50. In the study of electric circuits, it is sometimes useful to describe the way that an oscillation dies out. The following type of equation describes such a "dying" oscillation.

$$y = e^{-ax}[k_1\cos(bx) + k_2\sin(bx)]$$

a. Use a grapher to graph the specific case where $a = .1$, $b = 1$, $k_1 = 1$ by $k_2 = 2$ for a reasonable view of the oscillation.

b. At what value of x does the amplitude of the dying wave become smaller than .01?

THINK ABOUT IT 8.4

1. **a.** Describe how we may translate the graph of $y = \csc x$ to obtain the graph of $y = \sec x$.

b. What is the value of c closest to zero such that

$$\csc(x + c) = \sec x$$

for all values of x for which the expressions are defined?

2. In each case find how many times the graphs of f and g intersect when drawn on the same coordinate system. Confirm your answer graphically.

a. $f(x) = \sec$, $g(x) = (1/2)x$ **b.** $f(x) = \csc x$, $g(x) = (1/2)x$

3. Find the first 5 positive values of x for which the graphs of $y = x$ and $y = \tan x$ intersect. Estimate the x values to three decimal places. How many solutions does the equation $\tan x = x$ have?

4. Solve each inequality.

a. $\tan x > 0$ **b.** $\cot x < 0$

5. The function $y = [\sin(x^2 + x^3) + \tan(x^4 + x^2)]/[1 - \cos(x + 2x^5)]$ was mentioned on an electronic bulletin board for math teachers in a discussion about the appropriate use of graphers.

a. Graph $y = [\sin(x^2 + x^3) + \tan(x^4 + x^2)]/[1 - \cos(x + 2x^5)]$. What appears to be the y-intercept of this graph?

b. Evaluate y when $x = 0$. Is this the same answer you got in part **a**? Does this graph have a y-intercept?

● ● ●

8.5 Inverse Trigonometric Functions

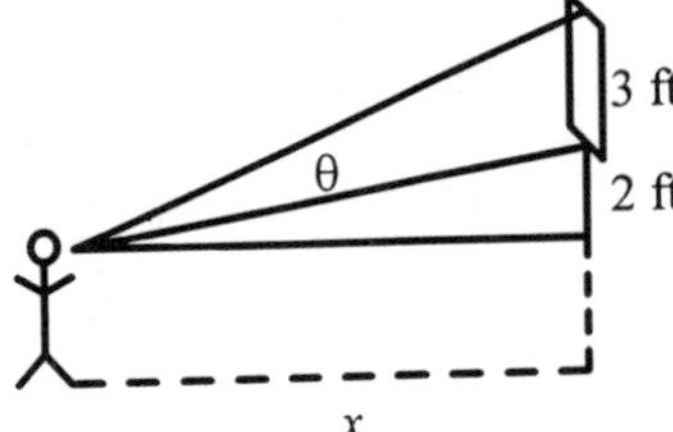

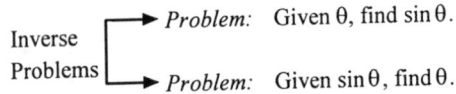

At an art gallery, a painting that is 3 ft high is mounted so that its base is 2 ft above the anticipated eye level of a viewer.

a. If a viewer is standing x ft from the wall, as shown in the diagram, then express the viewing angle θ in terms of x (**Hint:** θ may be expressed as the difference between two inverse trigonometric functions.)

b. Use a grapher to graph the relation in part **a**.

c. How far back should a viewer stand so that the pictures appears to be the largest? In other words, what value for x produces the maximum value for θ?

(See Example 3.)

Objectives

1. Evaluate expressions involving the inverse trigonometric functions.
2. Graph an expression that involves an inverse trigonometric function.
3. Evaluate a composition of functions that involves an inverse trigonometric function.
4. Simplify a trigonometric function of an inverse trigonometric function.
5. Rewrite a trigonometric expression using a given substitution.
6. Use the inverse trigonometric function properties when they apply.

The problem in the section opener seeks to find an expression for the viewing angle θ. Recall from Section 2.1 that when the problem is to find θ, it is useful to view the following types of problems as inverse problems.

Inverse Problems
- *Problem:* Given θ, find $\sin\theta$.
- *Problem:* Given $\sin\theta$, find θ.

Based on this perspective, the function that reverses the assignments of the sine function is called the **inverse sine function**, and, in general, the functions that reverse the assignments of the trigonometric functions are called the **inverse trigonometric functions**. Before we investigate these functions by introducing the inverse sine function, you should consider carefully four facts about any inverse functions that we learned in Section 7.1.

1. Two functions with exactly reverse assignments are inverse functions.
2. The domain of a function f is the range of its inverse function, and the range of f is the domain of its inverse.
3. A function is one-to-one when each x value in the domain is assigned a different y value so that no two ordered pairs have the same second component. We define the inverse of f, denoted f^{-1}, only when f is a one-to-one function.
4. The graphs of inverse functions are symmetric about the line $y = x$.

Now consider the graph of $y = \sin x$ in Figure 8.56. Because the sine function is periodic, many x values are assigned the same y value. For example,

$$\sin 0 = \sin\pi = \sin 2\pi = \sin(-\pi) = 0.$$

Thus, the inverse of the sine function is not a function. The so-called **inverse sine function** is defined by restricting the domain of $y = \sin x$ to the interval $[-\pi/2,\ \pi/2]$. Note in Figure 8.56 that each x value is assigned a different y value in this limited version of the sine function.

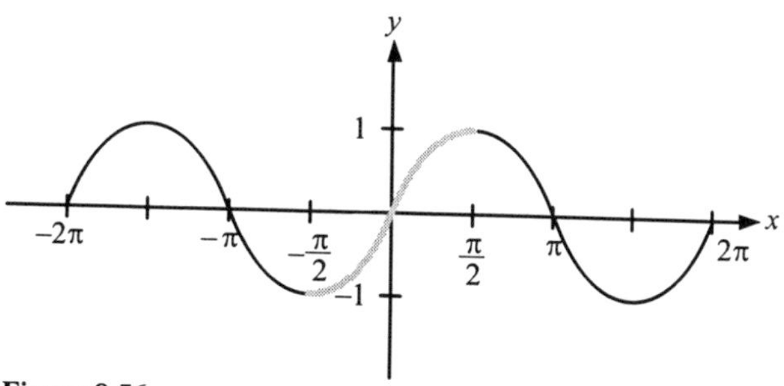

Figure 8.56

We find the rule for the inverse sine function by interchanging the x and y variables. Thus, the inverse of the function defined by

$$y = \sin x \quad -\frac{\pi}{2} \le x \le \frac{\pi}{2}$$

is defined by

$$x = \sin y \quad -\frac{\pi}{2} \le y \le \frac{\pi}{2},$$

which is written in inverse notation as

$$y = \arcsin x \text{ or } y = \sin^{-1} x.$$

For example, since $\sin(\pi/2) = 1$, we write

$$\frac{\pi}{2} = \arcsin 1 \text{ or } \frac{\pi}{2} = \sin^{-1} 1.$$

Both expressions are read "$\pi/2$ is the arc (or number) whose sine is 1," and note that the -1 in the function name $\sin^{-1}$ is not an exponent. Because a function and its inverse interchange their domain and range, for the inverse sine function the domain is the interval $[-1,\ 1]$ and the range is $[-\pi/2,\ \pi/2]$. The following definition sums up the key aspects of our discussion.

Inverse Sine Function

> The inverse sine function, denoted by **arcsin** or $\mathbf{sin^{-1}}$, is defined by
>
> $$y = \arcsin x \ \textbf{ if and only if } \ x = \sin y,$$
>
> where $-1 \le x \le 1$ and $-\pi/2 \le y \le \pi/2$.

Most graphing calculators have a key labeled $\sin^{-1}$ for the inverse sine function so it is easy to obtain a picture for $y = \sin^{-1} x$, as shown in Figure 8.57. To interpret this graph it is helpful to understand that the graph of $y = \sin^{-1} x$ may be obtained by reflecting the graph of $y = \sin x$ that is limited to $-\pi/2 \le x \le \pi/2$ about the line $y = x$, as shown in Figure 8.58. From this graph it is then apparent that the domain of $y = \sin^{-1} x$ is $[-1, 1]$, and the range is $[-\pi/2, \pi/2]$.

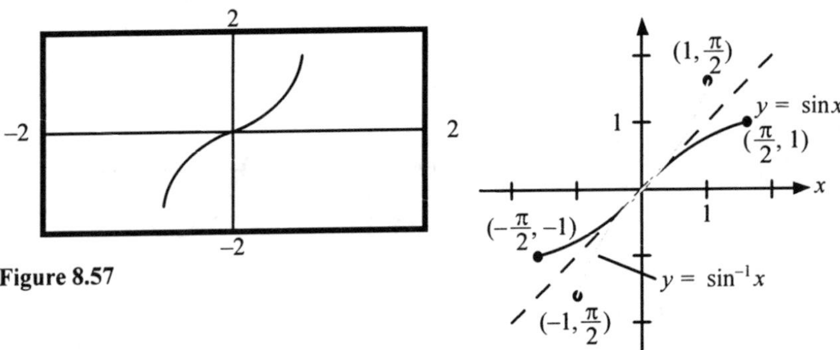

Figure 8.57

Figure 8.58

Example 1: Evaluating the Inverse Sine Function

Evaluate each expression. Specify exact values, where possible.

a. $\arcsin\frac{1}{2}$ **b.** $\sin^{-1}(0.5509)$ **c.** $\arcsin 2$

Solution: On the basis of our work in Section 8.2, it will be possible to specify exact values when answers involve 0, $\pi/6$, $\pi/4$, $\pi/3$, or $\pi/2$.

a. Let $\arcsin(1/2) = y$; then $1/2 = \sin y$. We solve this equation and find the number y in the interval $[-\pi/2, \pi/2]$ whose sine is $1/2$. Since $\sin(\pi/6) = 1/2$,

$$\arcsin\frac{1}{2} = \frac{\pi}{6}.$$

The first four lines in Figure 8.59 show a calculator check of this result.

```
sin⁻¹(1/2)
              .523598775598
π/6
              .523598775598
sin⁻¹(-Ø.55Ø9)
             -.583442252547
```

Figure 8.59

b. We need to use a calculator to determine that

$$\sin^{-1}(-0.5509) \approx -0.58,$$

as shown in the last two lines in Figure 8.59.

c. The expression arcsin 2 has no value because $y = \arcsin 2$ implies $2 = \sin y$ and this equation has no solution. Check this result by calculator. You should obtain either an error message or a complex number result that has meaning only in advanced mathematics.

PROGRESS CHECK 1

Evaluate each expression. Specify exact values, where possible.

a. $\sin^{-1}\left(-\dfrac{\sqrt{3}}{2}\right)$ **b.** $\arcsin \pi$ **c.** $\sin^{-1}(0.8109)$ ■

Through similar considerations we may define inverses for the other five trigonometric functions. For example, Figure 8.60 shows that $y = \cos x$ is one-to-one and assumes all its range values for $0 \le x \le \pi$. Thus, we may define the inverse function, $y = \arccos x$, as graphed, with domain $[-1, 1]$ and range $[0, \pi]$ from this limited version of the cosine function. Similarly, Figure 8.61 shows that we use $y = \tan x$ for $-\pi/2 < x < \pi/2$ to define $y = \arctan x$ with the set of all real numbers for its domain and the interval $(-\pi/2, \pi/2)$ for its range. The remaining inverse functions are used less frequently, so we just list their respective domains and ranges in the following summary. A consideration of the graphs of $y = \cot x$, $y = \sec x$, and $y = \csc x$ would show the respective intervals to be suitable (though arbitrary) choices.

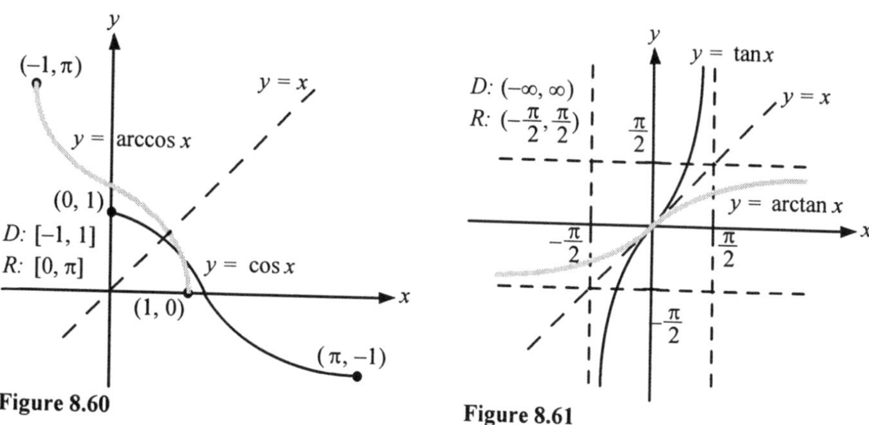

Figure 8.60 **Figure 8.61**

Inverse Trigonometric Functions

Function	Domain	Range
$y = \arcsin x$	$[-1, 1]$	$[-\pi/2, \pi/2]$
$y = \arccos x$	$[-1, 1]$	$[0, \pi]$
$y = \arctan x$	$(-\infty, \infty)$	$(-\pi/2, \pi/2)$
$y = \operatorname{arccot} x$	$(-\infty, \infty)$	$(0, \pi)$
$y = \operatorname{arcsec} x$	$(-\infty, -1] \cup [1, \infty)$	$[0, \pi/2) \cup [\pi, 3\pi/2)$
$y = \operatorname{arccsc} x$	$(-\infty, -1] \cup [1, \infty)$	$[0, \pi/2) \cup (-\pi, -\pi/2]$

There are two areas concerning inverse trigonometric functions in which there is no general agreement, so take note.

1. In terms of notation, an inverse trig function is sometimes capitalized and written as Arcsin x or $\text{Sin}^{-1}x$. With this convention, $\arcsin x$ and $\sin^{-1} x$ do not represent functions but represent the set of all numbers whose sine is x with no restriction placed on the range values.

2. The inverse secant and cosecant functions are sometimes assigned different range intervals depending on the application of the functions. The intervals chosen above are usually selected to obtain simpler derivative and integration formulas in calculus.

Example 2: Evaluating the Inverse Cosine or Tangent Function

Evaluate each expression. Specify exact answers, where possible.

a. arccos 0 **b.** $\cos^{-1}\left(-\frac{1}{2}\right)$ **c.** arctan(−1) **d.** $\tan^{-1} 2$

Solution: On the basis of our work in Section 8.2, it is possible to specify exact values when answers involve 0, $\pi/6$, $\pi/4$, $\pi/3$, $\pi/2$, or π. You should check each of the results that follow by using your calculator.

a. Let $y = \arccos 0$; then $0 = \cos y$. We seek the number y in the interval $[0, \pi]$ whose cosine is 0. Since $\cos\pi = 0$, $\arccos 0 = \pi$.

b. Let $y = \cos^{-1}(-1/2)$; then $-1/2 = \cos y$. We solve this equation and find the number y in the interval $[0, \pi]$ whose cosine is $-1/2$. Since $\cos(\pi/3) = 1/2$, we have

$$\cos\left(\pi - \frac{\pi}{3}\right) = \cos\frac{2\pi}{3} = -\frac{1}{2}.$$

Thus, $\cos^{-1}(-1/2) = 2\pi/3$.

c. Let $y = \arctan(-1)$; then $-1 = \tan y$. We solve this equation and find the number y in the interval $(-\pi/2, \pi/2)$ whose tangent is -1. Since $\tan(-\pi/4) = -1$,

$$\arctan(-1) = \frac{\pi}{4}.$$

d. By calculator, $\tan^{-1} 2 \approx 1.11$.

PROGRESS CHECK 2

Evaluate each expression. Specify exact answers, where possible.

a. arctan 0 **b.** $\cos^{-1}\left(\frac{1}{\sqrt{2}}\right)$ **c.** $\cos^{-1}\left(-\frac{1}{\sqrt{2}}\right)$ **d.** $\tan^{-1}(-5)$ ■

Example 3: Maximizing a Viewing Angle

Solve the problem in the section introduction on page 664.

Solution:

a. Consider the simplified sketch of the problem in Figure 8.62. Observe that $\theta = \alpha - \beta$ where $\tan\alpha = 5/x$ and $\tan\beta = 2/x$. So

$$\alpha = \tan^{-1}\frac{5}{x},\ \beta = \tan^{-1}\frac{2}{x},\ \text{and}\ \theta = \tan^{-1}\frac{5}{x} - \tan^{-1}\frac{2}{x}.$$

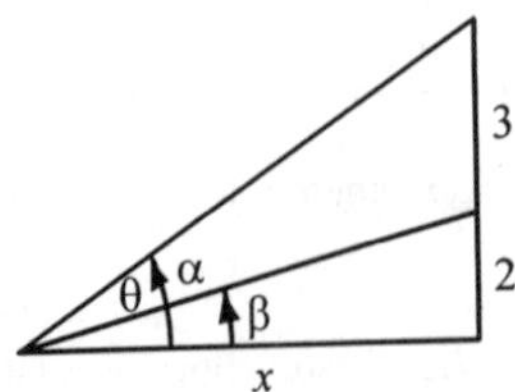

Figure 8.62

b. The equation that defines θ in terms of x is graphed in Figure 8.63.

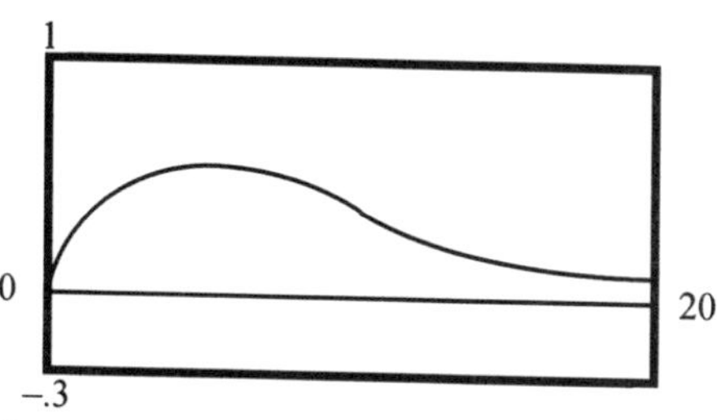

Figure 8.63

c. By using the Function Maximum feature (or Zoom and Trace), we can determine that the maximum point in the graph of this relation is reached at about (3.16, 0.44), as shown in Figure 8.64. Thus, when x is about 3.2 ft, the viewing angle reaches a maximum value of about 0.44 radians (or 25°).

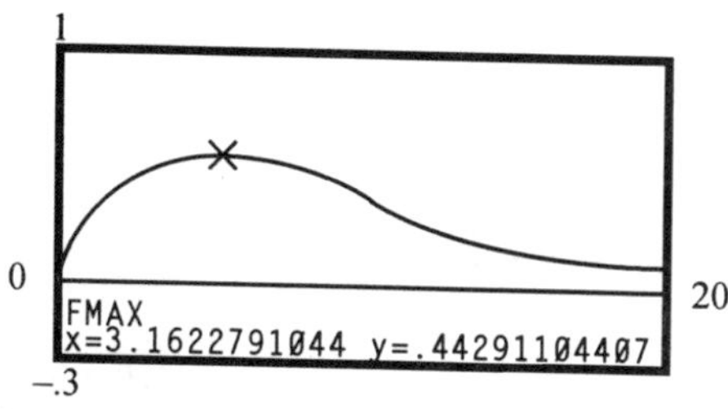

Figure 8.64

PROGRESS CHECK 3

Redo the problem in Example 3, assuming that the painting is 5 ft high and its base is hung 3 ft above the anticipated eye level of a viewer. ■

Example 4: Evaluating a Composition of Functions

Evaluate each expression. Find exact answers where possible.

a. $\sin(\arccos 0)$ **b.** $\tan\left(\sin^{-1}\frac{4}{5}\right)$

Solution:

a. First, determine arcos 0. Since $\cos(\pi/2) = 0$, we have

$$\frac{\pi}{2} = \arccos 0.$$

Now replace arcos 0 by $\pi/2$ in the original expressions.

$$\sin(\arccos 0) = \sin\frac{\pi}{2} = 1$$

b. It is useful to interpret $\sin^{-1}(4/5)$ as the measure of an angle in a right triangle. Let $\theta = \sin^{-1}(4/5)$ and sketch the triangle in Figure 8.65. The length of the side opposite θ is 4 and the hypotenuse has length 5. The remaining side length is found by the Pythagorean relationship.

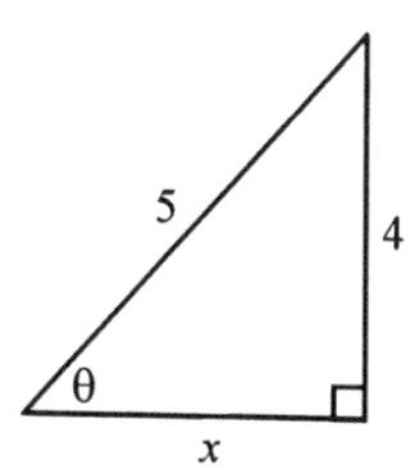

Figure 8.65

$\theta = \sin^{-1}\frac{4}{5}$

$$x = \sqrt{5^2 - 4^2} = 3$$

Thus,

$$\tan\theta = \tan\left(\sin^{-1}\frac{4}{5}\right) = \frac{4}{3}.$$

Note: It is not difficult to evaluate a composition of functions with a calculator, and Figure 8.66 shows a calculator display that confirms the results in this example.

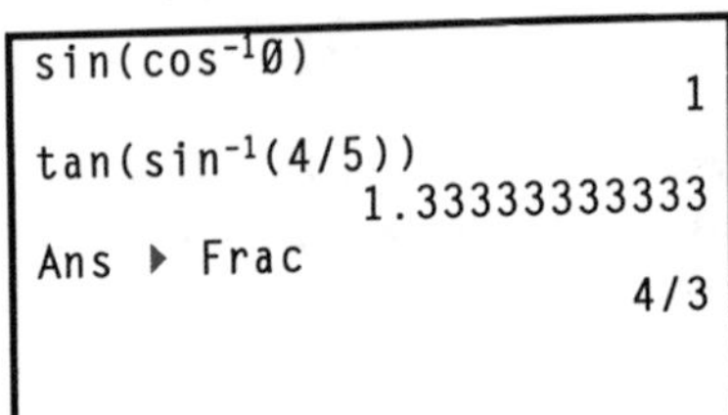

Figure 8.66

PROGRESS CHECK 4

Evaluate each expression. Find exact answers where possible.

a. $\cos(\arcsin(-1))$ **b.** $\sin\left(\tan^{-1}\frac{15}{8}\right)$ ■

The right-triangle method used to evaluate $\tan(\sin^{-1}(4/5))$ in Example 4b can be extended to solve a type of problem that occurs in calculus. The next two examples illustrate this important technique.

Example 5: Simplifying a Composition of Functions

Simplify $\sin(\cos^{-1} x)$.

Solution: Let $\theta = \cos^{-1} x$. Then $\cos\theta = x$, where $0 \le \theta \le \pi$. First, assume that x is positive and sketch the triangle in Figure 8.67. The length of the side adjacent to θ is x, and the hypotenuse has length 1. We find the opposite side length a to be $\sqrt{1-x^2}$ using the Pythagorean relation.

$$x^2 + a^2 = 1^2$$
$$a^2 = 1$$
$$a = \sqrt{1-x^2}$$

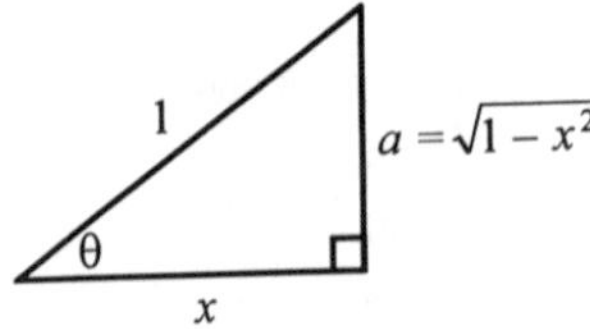

Figure 8.67

Now sketch Figure 8.68 with the aid of this triangle, where θ may be in either Q_1 or Q_2 because $0 \le \theta \le \pi$.

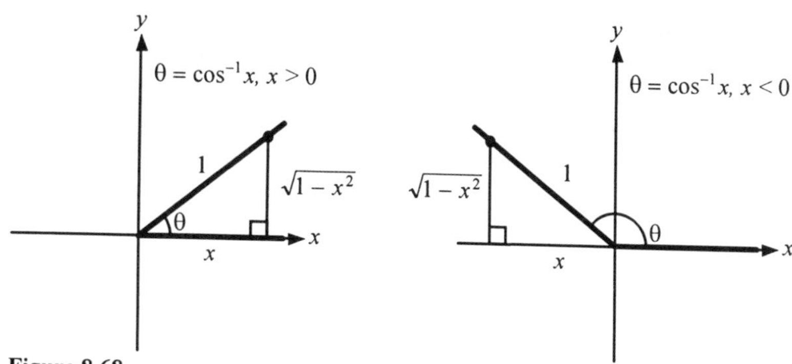

Figure 8.68

Then for all θ in $[0, \pi]$,

$$\sin\theta = \frac{\sqrt{1-x^2}}{1},$$

so

$$\sin(\cos^{1} x) = \sin\theta = \sqrt{1-x^2}.$$

PROGRESS CHECK 5

Simplify $\cos\left(\tan^{-1}\frac{x}{2}\right)$.

■

Example 6: Using a Substitution to Rewrite an Expression

If $x = \sin\theta$, where $-\pi/2 \le \theta \le \pi/2$, write $-\cot\theta - \theta$ as a function of x.

Solution: If $x = \sin\theta$, where $-\pi/2 \le \theta \le \pi/2$, then $\theta = \arcsin x$. We can represent $\cot\theta$ by first assuming that $x > 0$ and sketching the triangle in Figure 8.69. The length of the side opposite θ is x, the hypotenuse has length 1, and the remaining side length is found by the Pythagorean relation (as in Example 5) to be $\sqrt{1-x^2}$. Now sketch Figure 8.70 with the aid of this triangle, where θ may be in either Q_1 or Q_4, because $-\pi/2 \le \theta \le \pi/2$. Then $\cot\theta = \sqrt{1-x^2}/x$, so

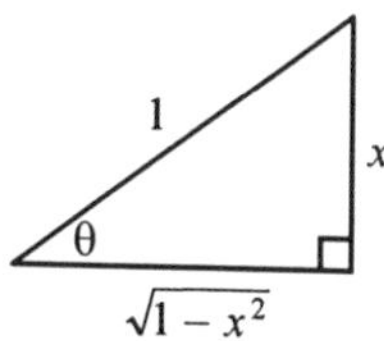

Figure 8.69

$$-\cot\theta - \theta = \frac{-\sqrt{1-x^2}}{x} - \arcsin x.$$

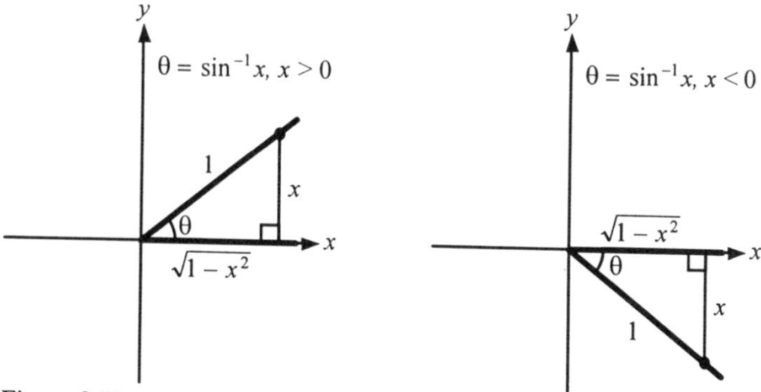

Figure 8.70

PROGRESS CHECK 6

If $x = 3\sec\theta$, where $0 \le \theta < \pi/2$ or $\pi \le \theta < 3\pi/2$, write $3\tan\theta - 3\theta$ as a function of x. ■

Inverse Properties

As discussed in Section 7.1, inverse functions offset each other; so that if f and g are inverses, then

$$f[g(x)] = x \text{ for all } x \text{ in the domain of } g$$

and

$$g[f(x)] = x \text{ for all } x \text{ in the domain of } f.$$

Applying this definition to the restricted sine, cosine, and tangent functions and their inverses yields the following properties.

$$\sin(\sin^{-1} x) = x \quad \text{for} \quad -1 \le x \le 1$$
$$\sin^{-1}(\sin x) = x \quad \text{for} \quad -\pi/2 \le x \le \pi/2$$
$$\cos(\cos^{-1} x) = x \quad \text{for} \quad -1 \le x \le 1$$
$$\cos^{-1}(\cos x) = x \quad \text{for} \quad 0 \le x \le \pi$$
$$\tan(\tan^{-1} x) = x \quad \text{for} \quad \text{all } x$$
$$\tan^{-1}(\tan x) = x \quad \text{for} \quad -\pi/2 < x < \pi/2$$

The next example shows that knowing the intervals for which these properties hold is essential to applying the properties.

Example 7: Monitoring the Domain of the Inverse Properties

Evaluate each expression, if possible.

a. $\sin^{-1}\left[\sin\left(\frac{\pi}{4}\right)\right]$ **b.** $\cos^{-1}(\cos 2\pi)$ **c.** $\sin(\sin^{-1} 2)$

Solution:

a. Because $\pi/4$ is in the interval $[-\pi/2, \pi/2]$, we use $\sin^{-1}(\sin x) = x$ to obtain $\sin^{-1}[\sin(\pi/4)] = \pi/4$. To check, note that $\sin(\pi/4) = 1/\sqrt{2}$ and $\sin^{-1}(1/\sqrt{2}) = \pi/4$.

b. Because 2π is not in the interval $[0, \pi]$, we cannot use $\cos^{-1}(\cos x) = x$. Instead, since $\cos 2\pi = 1$, we have

$$\cos^{-1}(\cos 2\pi) = \cos^{-1} 1 = 0.$$

c. Because 2 is not in the interval $[-1, 1]$, we cannot use $\sin(\sin^{-1} x) = x$. There is no value for $\sin(\sin^{-1} 2)$, since 2 is not in the domain of $y = \sin^{-1} x$.

PROGRESS CHECK 7

Evaluate each expression, if possible.

a. $\tan^{-1}\left[\tan\frac{\pi}{6}\right]$ **b.** $\cos(\cos^{-1} 2\pi)$ **c.** $\tan(\tan^{-1} 2)$ ■

EXPLORE 8.5

1. a. Use the viewing window [–4, 4] by [–4, 4] to graph $y = \sin^{-1} x$, and determine the domain and range.
   b. Use the viewing window [–4, 4] and [–4, 4] to graph $y = \cos^{-1} x$, and determine the domain and range.
   c. Use graphs to estimate the solution to the equation $\sin^{-1} x = \cos^{-1} x$.
2. a. Estimate the solution to $\cos^{-1} x = \cos x$ to 3 decimal places.
   b. Estimate the solution to $\cos x = x$ to 3 decimal places.
   c. Why do the equations in parts **a** and **b** have the same solution?
3. a. By definition, $\sin^{-1}(\sin x) = x$. This implies that the graphs of $y = \sin^{-1}(\sin x)$ and $y = x$ are the same. Are they? Explain what happened; what are the domain and range of the inverse sine function?
   b. By definition, $\sin(\sin^{-1} x) = x$. This implies that the graphs of $y = \sin(\sin^{-1} x)$ and $y = x$ are the same. Are they? Explain what happened.
   c. Graph $y = x^2 - \sin(\sin^{-1}(x^2 - 1))$, and explain the result.
   d. Graph $y = x^2 - \sin^{-1}(\sin(x^2 - 1))$, and explain the result.

EXERCISES 8.5

In Exercises 1–30 evaluate the expression. Specify exact values where possible.

1. $\arccos\frac{1}{2}$ **2.** $\arccos\left(-\frac{1}{2}\right)$

3. $\sin^{-1}\left(-\frac{\sqrt{3}}{2}\right)$ **4.** $\arcsin\frac{\sqrt{3}}{2}$

5. arctan 1 **6.** $\arctan \pi$

7. $\arcsin(-1)$ **8.** $\arcsin\left(\frac{1}{\pi}\right)$

9. $\arcsin\frac{1}{\sqrt{2}}$ **10.** $\tan^{-1}\sqrt{3}$

11. $\tan^{-1}(-\sqrt{3})$ **12.** $\arccos\left(-\frac{1}{\sqrt{2}}\right)$

13. $\arcsin 0.3124$ **14.** $\sin^{-1}(-0.3124)$

15. $\cos^{-1}(-0.5509)$ **16.** arccos 0.5509

17. arctan 1.758 **18.** $\tan^{-1}(-1.758)$

19. $\cos^{-1}\left(\frac{4}{5}\right)$ **20.** $\arctan\frac{3}{2}$

21. $\cos(\arcsin 0)$ **22.** $\sin(\arccos 1)$

23. $\sin\left(\cos^{-1}\frac{\sqrt{3}}{2}\right)$ **24.** $\cos\left[\sin^{-1}\left(-\frac{1}{2}\right)\right]$

25. $\cot(\arccos 0)$ **26.** $\tan^{-1}(\cos 0)$

27. $\arcsin(\tan 0)$ **28.** $\csc(\arcsin 1)$

29. $\tan[\arcsin(-0.5518)]$

30. $\csc(\arccos 0.0129)$

In Exercises 31–40 simplify the expression.

31. $\sin\left(\tan^{-1}\frac{3}{4}\right)$ **32.** $\cos\left[\arcsin\left(-\frac{1}{3}\right)\right]$

33. $\tan\left(\arcsin\frac{12}{13}\right)$ **34.** $\cot\left(\cos^{-1}\frac{8}{17}\right)$

35. $\cos(\tan^{-1} x)$ **36.** $\cot(\sin^{-1} x)$

37. $\cot(\arcsin x)$ **38.** $\sin(\arctan x)$

39. $\tan\left(\arcsin\frac{x-2}{3}\right)$ **40.** $\sec\left(\tan^{-1}\frac{x}{2}\right)$

In Exercises 41–48 write the expression as a functions of x. Use inverse functions and/or right triangles.

41. θ if $x = \sin\theta$, where $-\frac{\pi}{2} \le \theta \le \frac{\pi}{2}$

42. θ if $x = \tan\theta$, where $-\frac{\pi}{2} < \theta < \frac{\pi}{2}$

43. θ if $x - 2 = 3\sec\theta$, where $0 \le \theta \le \frac{\pi}{2}$ or $\pi \le \theta < \frac{3\pi}{2}$

44. θ if $x = a\sin\theta$, where $-\frac{\pi}{2} \le \theta \le \frac{\pi}{2}$

45. $\frac{-1}{4\sin\theta}$ if $x = 2\tan\theta$, where $-\frac{\pi}{2} < \theta < \frac{\pi}{2}$

46. $2\sec\theta\tan\theta$ if $x = 2\sec\theta$, where $0 \le \theta < \frac{\pi}{2}$ or $\pi \le \theta < \frac{3\pi}{2}$

47. $-\frac{1}{2}\csc\theta$ if $2(x+2) = \sec\theta$, where $0 \le \theta < \frac{\pi}{2}$ or $\pi \le \theta < \frac{3\pi}{2}$

48. $\theta + \sin\theta\cos\theta$ if $x - 1 = \sin\theta$, where $-\frac{\pi}{2} \le \theta \le \frac{\pi}{2}$

In Exercises 49–52 solve the formula for θ.

49. $m \cdot \sin\theta = 1$, where $-\frac{\pi}{2} \le \theta \le \frac{\pi}{2}$

50. $\tan\theta = \frac{b}{a}$, where $-\frac{\pi}{2} < \theta < \frac{\pi}{2}$

51. $T = \frac{2V_0\sin\theta}{g}$, where $-\frac{\pi}{2} \le \theta \le \frac{\pi}{2}$

52. $V_x = V\cos\theta$, where $0 \le \theta \le \pi$

In Exercises 53–58 use the inverse properties on page 672 to determine what values of x make each statement true.

53. $\sin^{-1}(\sin x) = x$

54. $\sin(\sin^{-1} x) = x$

55. $\cos^{-1}(\cos x) = x$

56. $\cos(\cos^{-1} x) = x$

57. $\tan(\arctan x) = x$

58. $\arctan(\tan x) = x$

In Exercises 59–68 evaluate each expression.

59. $\cos(\arccos 0)$

60. $\cos^{-1}(\cos 0)$

61. $\sin^{-1}[\sin(-1)]$

62. $\sin[\arcsin(-1)]$

63. $\cos^{-1}\left[\cos\left(\frac{3\pi}{2}\right)\right]$

64. $\sin^{-1}(\sin 2\pi)$

65. $\tan\left[\tan^{-1}(-1)\right]$

66. $\tan^{-1}[\tan(-\pi)]$

67. $\cos\left(\cos^{-1} 2\right)$

68. $\sin\left(\sin^{-1} 2\pi\right)$

69. Sketch the graph of $y = \operatorname{arcsec} x$.

70. Sketch the graph of $y = \operatorname{arccot} x$.

71. Sketch the graph of $y = \arcsin(x-1)$.

72. Sketch the graph of $y = \arccos(x+2)$.

73. Consider the illustration below.

a. Write θ as a function of a.

b. Write θ as a function of b.

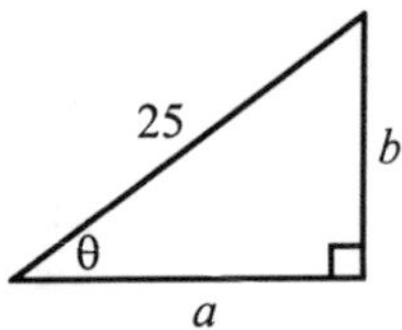

74. Consider the illustration below. Write θ as the difference between two inverse tangent expressions.

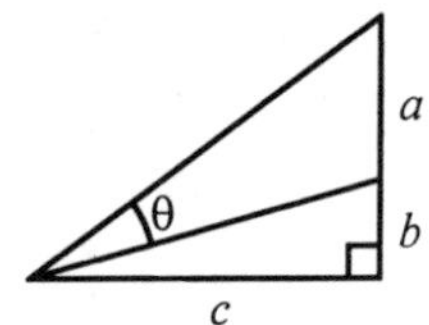

75. Consider the illustration below. The area of the region enclosed by the graphs of $y = \frac{1}{\sqrt{1-x^2}}$, $y = 0$, $x = 0$ and $x = \frac{\sqrt{2}}{2}$ is shown in calculus to be given by $\arcsin\left(\frac{\sqrt{2}}{2}\right) - \arcsin 0$. Find this area.

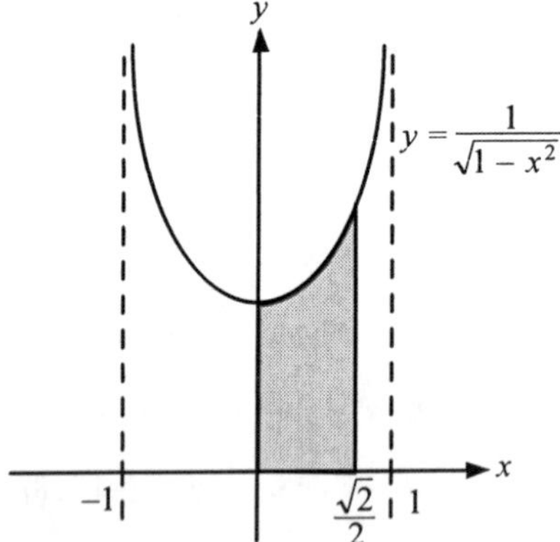

76. Consider the illustration below. The area of the region enclosed by the graphs of $y = \frac{1}{x^2+1}$, $y = 0$, $x = -1$,

and $x = 1$ is shown in calculus to be given by $\arctan 1 - \arctan(-1)$. Find this area.

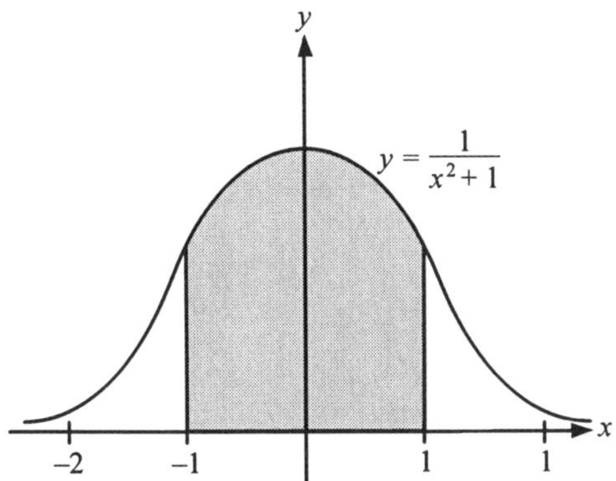

77. As shown in the figure, the length s of the shadow cast by a pole that is t feet tall depends on θ, the angle of inclination of the sun.
    a. Express s as a function of θ. State the domain and range of this function.
    b. Express θ as a function of s. State the domain and range of this function.

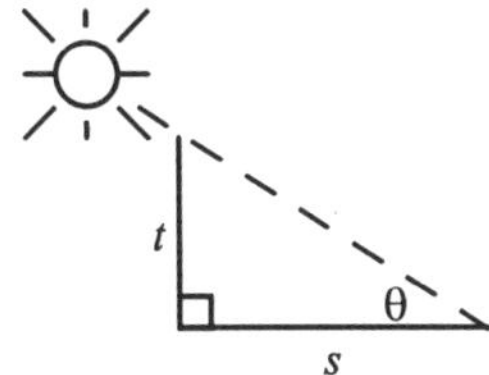

78. The height s of the right triangle shown in the figure depends on the acute angle θ.
    a. Express the height as a function of θ. Give the domain and range of this function.
    b. Express the area of the triangle as a function of θ. Give the domain and range of this function.
    c. Express θ as a function of the area. Give the domain and range of this function.

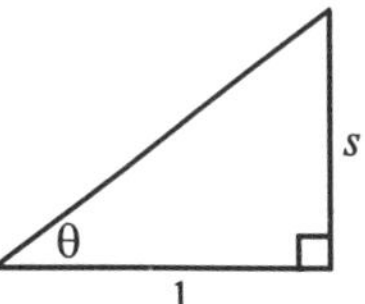

79. Solve $\arcsin x = \arccos\frac{2}{3}$ without the aid of a calculator.

80. Solve $\arcsin x = \arctan\frac{5}{4}$ without the aid of a calculator.

81. The equation $y = .001\sin(880\pi t)$ may be used to describe a sound wave in air created by a vibrating tuning fork used to tune a musical instrument. The vibration of the tuning fork causes the displacement y in inches of individual air molecules at time t seconds.
    a. The given equation expresses y as a function of t. Find the period and the amplitude of this function.
    b. Graph one-quarter of one cycle of this function.
    c. For the interval covered in part **b**, express t as a function of y.

82. When a weight on a spring bobs up and down, the displacement of the weight from its resting position for a short time t can be approximated by $y = D\cos\sqrt{\frac{k}{m}}t$, where k and m are constants that depend on the material and construction of the particular spring, and D is the original displacement. Over one-half of a cycle, no displacements are repeated, so this function has an inverse. Use this fact to express t as a function of y over this interval.

THINK ABOUT IT 8.5

1. Give an example of a value for x that shows that the given equation is *not true* for all values of x for which the expressions are defined.
    a. $\tan^{-1} x = (\sin^{-1} x)/(\cos^{-1} x)$
    b. $\sin^{-1} x = 1/(\sin x)$
    c. $\text{arccot}\, x = 1/(\arctan x)$
    d. $\text{arccot}\, x = \arctan(1/x)$
    e. $\arccos(\cos x) = x$
    f. $\cos(\cos^{-1} x) = x$
2. Explain why the calculator value for $\sin^{-1}(\sin 2) = 1.14159$, and not 2. What is the relationship between π and 1.14159?
3. a. Show that $\arcsin(-x) = -\arcsin x$.
    b. If $f(x) = \arcsin x$, is f an even function, an odd function, or neither?
    c. Show that $\arccos(-x) = \pi - \arccos x$.
4. a. Show that $\arctan(1/2) + \arctan(1/3) = \pi/4$.
    b. In 1706, John Machin determined 100 correct decimal places for π with the aid of the relation

$$\frac{\pi}{4} = 4\arctan\left(\frac{1}{5}\right) - \arctan\left(\frac{1}{239}\right).$$

Use identities to show that this relation is true.

5. Use the cofunction identity $\cos[\pi/2 - \theta] = \sin\theta$, which is true for all values of θ, and show that

$$\arcsin x + \arccos x = \frac{\pi}{2}$$

for all x in $[-1, 1]$. In terms of right triangles, explain the interpretation of this identity if $0 < x < 1$.

● ● ●

8.6 Trigonometric Equations

Photo Courtesy of **Steven Baratz**, The Picture Cube

On May 2 the tide in a certain harbor is described by

$$y = 12 + 5.3\cos(0.5067t),$$

where y is the height of the water in feet and t is the time that has elapsed from 12 midnight on May 1. To the nearest minute, find all times on May 2 when the water level is at a height of 14 ft.
(See Examples 1 and 6.)

Objectives

1. Find the approximate solution to trigonometric equations by using graphical methods.
2. Use reference numbers to solve trigonometric equations, specifying exact solutions where possible.
3. Use factoring or the square root property to solve certain trigonometric equations.
4. Solve trigonometric equations involving functions of multiple angles.

Recall from Section 2.4 that equations involving trigonometric expressions commonly appear in two types: identities and conditional equations. An identity is an equation such as

$$\tan\theta = \frac{\sin\theta}{\cos\theta}$$

that is true for all values of θ for which expressions are defined. This topic is discussed in detail in the next three sections. Conditional trigonometric equations differ from identities in that they are true only for certain values of the unknown. For instance, in the section-opening problem we need to find all values of t in the interval $0 \le t \le 24$ that satisfy

$$14 = 12 + 5.3\cos(0.5067t),$$

and we will find that there are just four such solutions. Now that we have discussed in detail how to graph trigonometric functions, it is instructive to first show how we can approximate these four solutions by using graphical methods.

Example 1: Tidal Water Height—A Graphing Approach

Solve the problem in the section introduction on page 676, using graphical methods.

Solution: To obtain a picture of the solution to the given equation, we first graph $y1 = 14$ and $y2 = 12 + 5.3\cos(0.5067x)$ on a grapher that is set to Radian Mode, as shown in Figure 8.71. Observe that a viewing window of [0, 24], by [6, 18] was chosen because the problem requests all solutions for the 24-hour interval associated with May 2, and because this cosine graph oscillates about a midline of $y = 12$ with an amplitude of 5.3.

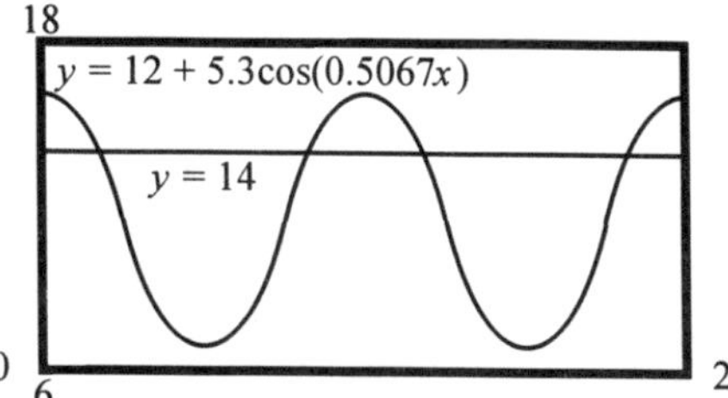

Figure 8.71

We see the two graphs intersect at four points, so there are four solutions in the interval shown. To estimate the earliest solution, we can use the Intersection operation (or Zoom and Trace) to find that $x \approx 2.3364$ for the intersection point shown in Figure 8.72. By the same method, you can determine that the x coordinates of the other three intersection points are 10.0638, 14.7366, and 22.4640. Thus, to the nearest minute, the water level in this harbor is at a height of 14 ft at 2:20 a.m., 10:04 a.m., 2:44 p.m. and 10:28 p.m.

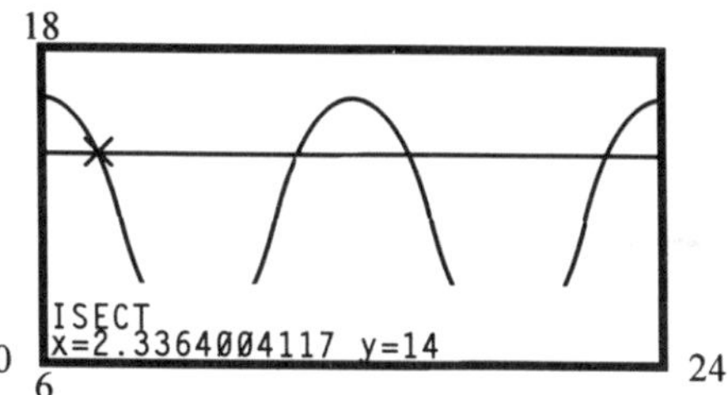

Figure 8.72

PROGRESS CHECK 1

Use the formula from Example 1 and find all times on May 2 in this harbor when the water level is at a height of 11 ft. Use a graphical method. ■

The solutions in Example 1 may be confirmed algebraically by extending the methods that were introduced in Section 2.4. Before considering the problem, it is helpful to first develop the necessary procedures using simpler equations. The next two examples illustrate how to find solutions for $0 \le x < 2\pi$ when exact solutions may be specified.

Example 2: Finding Exact Solutions Where $0 \le x < 2\pi$

Solve $\sin x = -1/2$ for $0 \le x < 2\pi$. Specify exact solutions.

Solution: First, determine the quadrant that contains the point assigned to x.

$$\sin x = -\frac{1}{2}, \text{ which is a negative number.}$$

The point assigned to x could be in either Q_3 or Q_4, since the sine function is negative in both quadrants.

Second, determine the reference number. Discard the sign on the function value and use the inverse sine function to determine that

$$\sin^{-1}\frac{1}{2} = \frac{\pi}{6}.$$

Therefore, the reference number is $\pi/6$.

Third, determine the appropriate values of x (see Figure 8.73).

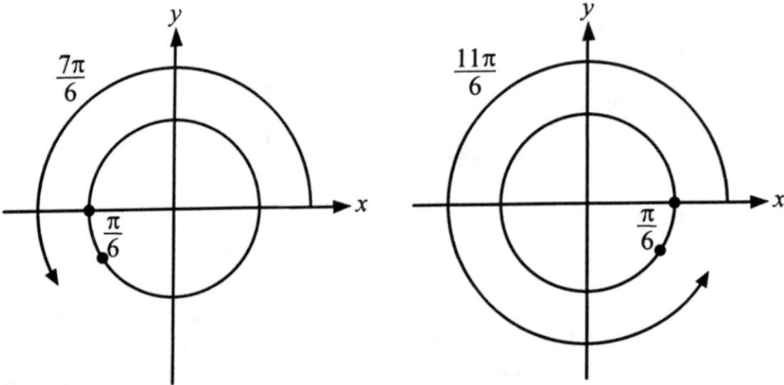

Figure 8.73
$\pi + \pi/6 = 7\pi/6$ is the number in Q_3 with a reference number of $\pi/6$.
$2\pi - \pi/6 = 11\pi/6$ is the number in Q_4 with a reference number of $\pi/6$.

Therefore, $7\pi/6$ and $11\pi/6$ make the equation a true statement, and the solution set in the interval $[0, 2\pi)$ is $\{7\pi/6, 11\pi/6\}$. Figure 8.74 provides graphical evidence that supports this result.

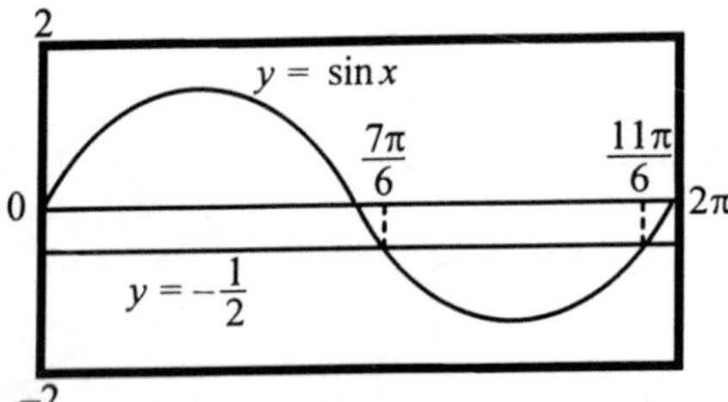

Figure 8.74

PROGRESS CHECK 2

Solve $\cos x = -1/\sqrt{2}$ for $0 \le x < 2\pi$. Specify exact solutions. ■

When solving certain trigonometric equations, it may be necessary to first solve the equations for the functions of x in the problem. Factoring, the square root property, or identities may be needed in such cases. To illustrate, the next example shows a case in which factoring is useful.

Example 3: Using Factoring

Solve the equation $2\sin x\cos x - \sin x = 0$ in the interval $[0,\ 2\pi)$.

Solution: First, factor out the common factor sin x.

$$2\sin x\cos x - \sin x = 0$$
$$\sin x(2\cos x - 1) = 0$$

Note that we have found two factors whose product is zero. Hence, the original equation will be satisfied whenever either factor is zero, and we treat each factor separately from this point on.

First factor $\sin x = 0$ The sine function is 0 when $x = 0$ and $x = \pi$.
Second factor $2\cos x - 1 = 0$
$\cos x = 1/2$

1. Since cos x is positive, the point assigned to x could be in Q_1 or Q_4.
2. Since $\cos^{-1}(1/2) = \pi/3$, the reference is $\pi/3$.
3. $\pi/3$ is the number in Q_1 with a reference number of $\pi/3$ (Figure 8.75.)

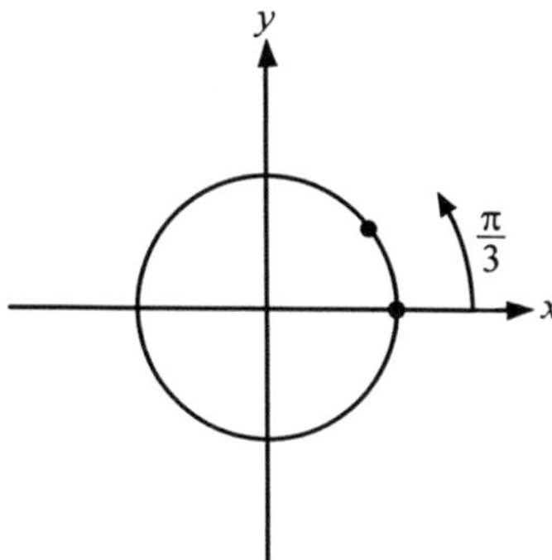

Figure 8.75

$2\pi - \pi/3 = 5\pi/3$ is the number in Q_4 with a reference number of $\pi/3$ (Figure 8.76.)

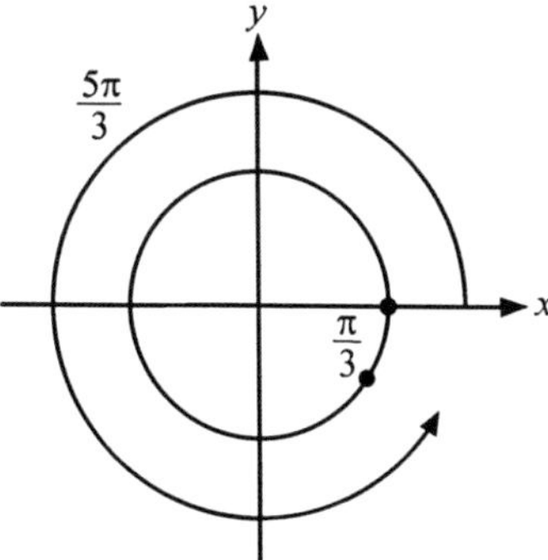

Figure 8.76

Thus, in the interval $[0,\ 2\pi)$ the solution set is $\{0,\ \pi/3,\ \pi,\ 5\pi/3\}$. Observe that the four x-intercepts in the graph in Figure 8.77 on $[0,\ 2\pi)$ are in agreement with these four solutions.

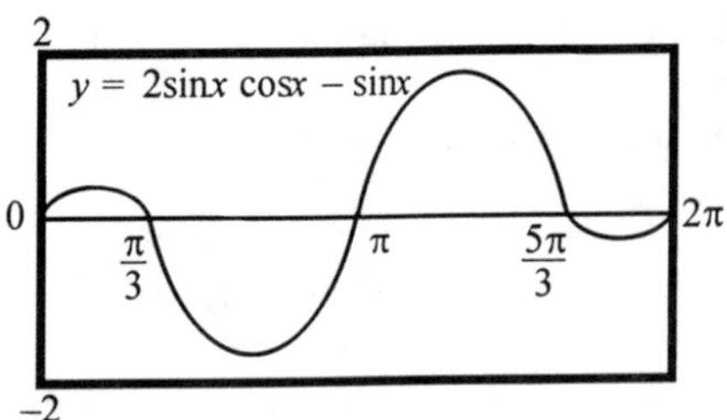

Figure 8.77

PROGRESS CHECK 3
Solve $2\sin^2 x = \sin x$ for $0 \le x < 2\pi$. ■

A solution in Q_1, Q_3 and Q_4 was illustrated in Examples 2 and 3. Observe that solutions in each of the quadrants are determined as follows:

Quadrant	Solution
1	reference number
2	$\pi -$ reference number
3	$\pi +$ reference number
4	$2\pi -$ reference number

Once we have the solutions of a trigonometric equation that are between 0 and 2π, we can determine all the solutions, since the trigonometric functions are periodic. In laying off a length 2π, we pass around the unit circle and return to our original point. Thus, we generate all the solutions to an equation by adding multiples of 2π to the solutions that are in the interval $[0,\ 2\pi)$.

Example 4: Approximating All Solutions to a Trigonometric Equation

Approximate all the solutions to $4\cos x + 1 = 0$.

Solution: First, solve the equation for cos x.

$$4\cos x + 1 = 0$$
$$\cos x = -\frac{1}{4} = -0.2500$$

Second, determine the quadrant that contains the point assigned to x.

$\cos x = -0.2500$, which is a negative number.

The point assigned to x could be in either Q_2 or Q_3, since the cosine function is negative in both quadrants.
Third, determine the reference number. Discard the sign on the function value and use the inverse cosine function to determine that

$$\cos^{-1} 0.25 \approx 1.32.$$

Therefore, the reference number is 1.32.
Fourth, determine the appropriate values of x (Figure 8.78).

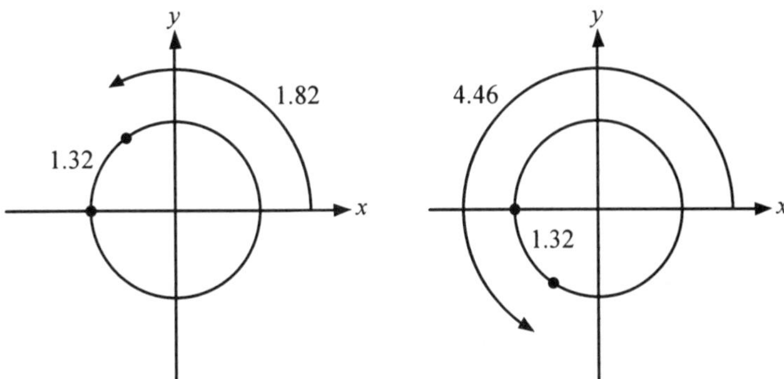

Figure 8.78
$3.14 - 1.32 = 1.82$ is the number in Q_2 with a reference number of 1.32.
$3.14 + 1.32 = 4.46$ is the number in Q_3 with a reference number of 1.32.

Thus, the formulas,

$$1.82 + k2\pi \text{ and } 4.46 + k2\pi,$$

where k is an integer, generate all the solutions to the equation, and the solution set is $\{x\colon\ x = 1.82 + k2\pi \text{ or } x = 4.46 + k2\pi,\ k \text{ any integer}\}$. A graphical check that supports this result for $-4\pi \leq x \leq 4\pi$ is shown in Figure 8.79.

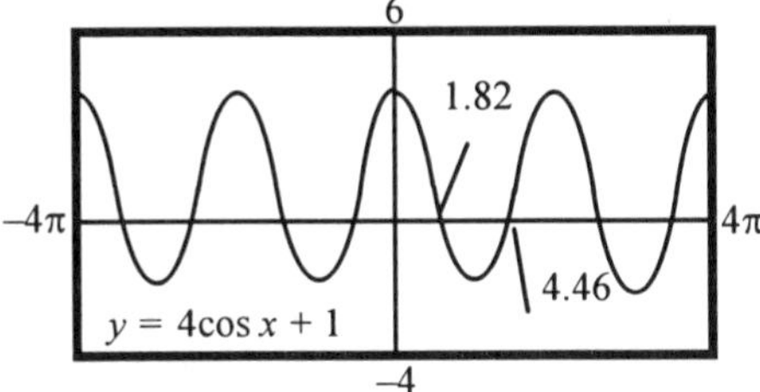

Figure 8.79

PROGRESS CHECK 4
Approximate all the solutions to $5\sin x + 2 = 0$. ■

The last main idea we need to develop, in order to show an algebraic solution to the section-opening problem, is a method for solving trigonometric equations that involve multiple angles. Example 5 illustrates a basic procedure for solving an equation of this type.

Example 5: Solving Equations Involving Functions of Multiple Angles

Solve $\sin 3x = \dfrac{1}{\sqrt{2}}$ for $0 \leq x < 2\pi$.

Solution: Since the function $y = \sin 3x$ completes three cycles for $0 \leq x < 2\pi$, we can anticipate that $\sin 3x = 1/\sqrt{2}$ has three times as many solutions as $\sin x = 1/\sqrt{2}$ for this interval. Figure 8.80 confirms this prediction since it shows that $\sin 3x = 1/\sqrt{2}$ has six solutions in the interval $[0,\ 2\pi)$.

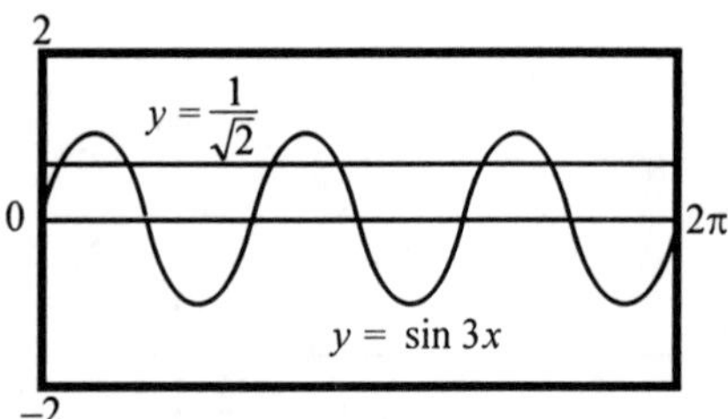

Figure 8.80

To solve this equation algebraically we initially solve the equation for $3x$. Then the formulas for $3x$ will imply formulas for x, as shown in Step 3 below.

1. Since $\sin 3x$ is positive, the point assigned to $3x$ could be in Q_1 or Q_2.
2. Since $\sin^{-1}(1/\sqrt{2}) = \pi/4$, the reference number is $\pi/4$.
3. Determine the appropriate values of $3x$ and then of x (Figure 8.81).

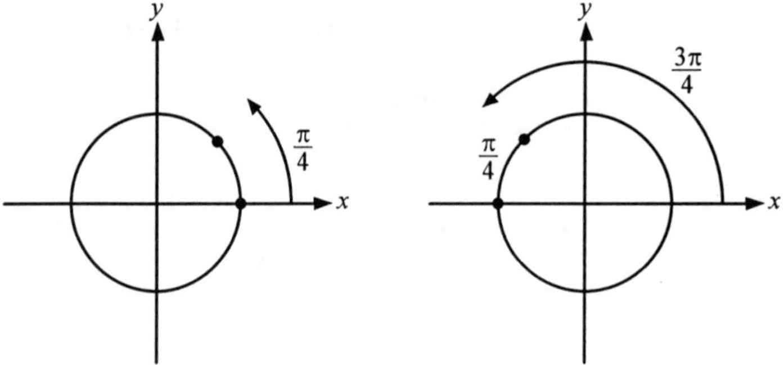

Figure 8.81

$\pi/4$ is the number in Q_1 with a reference number of $\pi/4$
$\pi - \pi/4 = 3\pi/4$ is the number in Q_2 with a reference number of $\pi/4$.

Thus,

$$3x = \frac{\pi}{4} + k2\pi \text{ or } 3x = \frac{3\pi}{4} + k2\pi,$$

from which we have

$$x = \frac{\pi}{12} + k\frac{2\pi}{3} \text{ or } x = \frac{\pi}{4} + k\frac{2\pi}{3}.$$

4. To obtain solutions in the interval $[0, 2\pi)$, use 0, 1, and 2 as replacements for k.

$$\frac{\pi}{12} + (0)\frac{2\pi}{3} = \frac{\pi}{12} \qquad \frac{\pi}{4} + (0)\frac{2\pi}{3} = \frac{\pi}{4}$$
$$\frac{\pi}{12} + (1)\frac{2\pi}{3} = \frac{3\pi}{4} \qquad \frac{\pi}{4} + (1)\frac{2\pi}{3} = \frac{11\pi}{12}$$
$$\frac{\pi}{12} + (2)\frac{2\pi}{3} = \frac{17\pi}{12} \qquad \frac{\pi}{4} + (2)\frac{2\pi}{3} = \frac{19\pi}{12}$$

Thus $\{\pi/12,\ \pi/4,\ 3\pi/4,\ 11\pi/12,\ 17\pi/12,\ 19\pi/12\}$ is the solution set for $0 \le x < 2\pi$.

PROGRESS CHECK 5

Solve $\cos 2x = -\dfrac{1}{2}$ for $0 \le x < 2\pi$. ■

Example 6: Tidal Water Height—An Algebraic Approach

Solve the problem in the section introduction on page 676 by using algebraic methods.

Solution: Replacing y by 14 in the given equation yields

$$14 = 12 + 5.3\cos(0.5067t)$$

which is equivalent to

$$\cos(0.5067t) = \frac{2}{5.3}.$$

The cosine function is positive in Q_1 and Q_4, and the reference number is $\cos^{-1}(2/5.3)$. The numbers in Q_1 and Q_4 with this reference number are $\cos^{-1}(2/5.3)$ and $2\pi - \cos^{-1}(2/5.3)$, respectively. Thus,

$$0.5067t = \cos^{-1}(2/5.3) + k2\pi \quad \text{or} \quad 0.5067t = 2\pi - \cos^{-1}(2/5.3) + k2\pi$$

from which we have

$$t = \frac{\cos^{-1}(2/5.3) + k2\pi}{0.5067} \quad \text{or} \quad t = \frac{2\pi - \cos^{-1}(2/5.3) + k2\pi}{0.5067}.$$

Finally to obtain solutions in the interval [0, 24), use 0 and 1 as replacements for k to obtain 2.3364, 10.0638, 14.7366, and 22.4640. By converting these answers to standard times that are rounded to the nearest minute, we find that the water level in this harbor is at a height of 14 ft at 2:20 a.m., 10:04 a.m., 2:44 p.m., and 10:28 p.m.

PROGRESS CHECK 6

Use the formula from Example 6 and find all times on May 2 in this harbor when the water level is at a height of 11 ft. Use an algebraic method. ■

EXPLORE 8.6

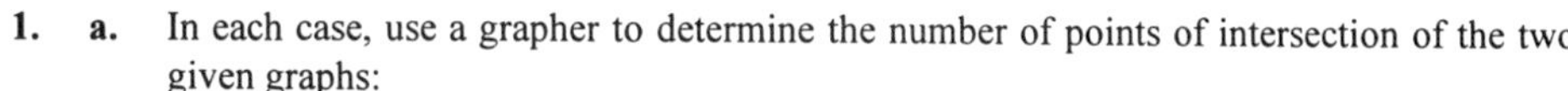

1. **a.** In each case, use a grapher to determine the number of points of intersection of the two given graphs:

i. $y = \sin x$, $y = 2x$ **ii.** $y = \sin x$, $y = 0.5x$ **iii.** $y = \sin x$, $y = 0.1x$

b. Determine a value of $c > 0$ such that $y = \sin x$ and $y = cx$ intersect in eleven points.

c. Is it possible to determine a value to $c > 0$ such that $y = \sin x$ and $y = cx$ intersect in five points?

2. **a.** Use a grapher to estimate the solutions to this nonlinear system. Find solutions to three decimal place accuracy.

$$y = 4\cos x$$
$$2x^2 + y^2 = 16$$

b. Use a grapher to estimate, to the nearest tenth, the solutions of this equation in the interval $[-2\pi, 2\pi]$.

$$\csc x = \frac{1}{2}x$$

Note: your eye can fool you; check that your answer makes algebraic sense.

c. Solve $2\cos x = x^3 - 2x$. Find solutions to the nearest tenth.

3. A problem that was discussed on an electronic bulletin board for math teachers said that the function $y = \sin(\ln x)$ has more than four zeros in the interval (0, 1].

a. Graph this function, and decide if that statement is correct.

b. Confirm any zeros you find algebraically. **Hint:** What is $\ln e^{-\pi}$?

EXERCISES 8.6

In Exercises 1–20 solve for x in the interval $[0, 2\pi)$. Specify exact solutions where possible.

1. $\sin x = \frac{\sqrt{3}}{2}$ **2.** $\sin x = -\frac{\sqrt{3}}{2}$

3. $\cos x = -\frac{1}{2}$ **4.** $\tan x = 1$

5. $\tan x = -1$ **6.** $\sec x = 2$

7. $\sin x = 0.1219$ **8.** $\sin x = -0.1219$

9. $\tan x = -3.145$ **10.** $\tan x = 3.145$

11. $\sqrt{2}\cos x = 1$ **12.** $\csc x - \sqrt{2} = 0$

13. $\sin x + 2 = 0$ **14.** $\cos x - 3 = 0$

15. $3\tan x - 1 = 0$ **16.** $2\tan x + 7 = 0$

17. $2\cot x + 3 = 0$ **18.** $\frac{\cos x}{4} = \frac{1}{100}$

19. $\sec x = -5$ **20.** $\frac{3\sin x}{5} = \frac{3}{8}$

In Exercises 21–36 solve for x in the interval $[0, 2\pi)$. Specify exact solutions, where possible.

21. $2\sin^2 x - 1 = 0$ **22.** $3\tan^2 x - 1 = 0$

23. $\cos^2 x = \cos x$ **24.** $\sin^3 x = \sin x$

25. $2\cos x \sin x - \cos x = 0$

26. $\tan x \cos x = \tan x$ **27.** $\tan^2 x + 4\tan x - 21 = 0$

28. $2\sin^2 x - 5\sin x = 3$ **29.** $\cos 3x = 0$

30. $\sin 4x = 1$ **31.** $\sin 2x = \frac{1}{2}$

32. $\sin\frac{1}{2}x = \frac{1}{2}$ **33.** $\tan\frac{1}{3}x = 1$

34. $\csc\frac{1}{4}x = 2$ **35.** $\sin^2 2x = \sin 2x$

36. $\cot^2 2x + 2\cot 2x + 1 = 0$

In Exercises 37–48 find all solutions to the equation. Specify exact solutions, where possible.

37. $4\sin x - 1 = 1$ **38.** $10\cos x - 2 = 0$

39. $4\csc x + 9 = 0$ **40.** $5\cot x + 3 = 0$

41. $3\tan x - 5 = 4$ **42.** $5\tan x - 3 = 4$

43. $\frac{\cos(2x) - .2}{.5} = 1$ **44.** $\frac{5 - \sin(2x)}{3} = 1.5$

45. $4\cos(3x) + 1 = 6$ **46.** $5 - \sin(3x) = 3.2$

47. $\sin 2x(2\sin x - 1) = 0$

48. $\cos 2x(2\cos x + 1) = 0$

49. The average daily temperature T in degrees Fahrenheit at a certain location is approximated fairly well by the equation

$$T = 61.3 + 17.9\cos\left(\frac{2\pi}{365}t\right),$$

where t is in days and $t = 0$ represents July 1. Find all days in the year (assume 365 days) for which the average daily temperature is between 70.5 and 71.5 degrees.

50. Suppose a pendulum oscillates according to the equation

$$x = 6.0\cos\left(3\pi t + \frac{\pi}{3}\right),$$

where x represents the horizontal displacement from a reference point in meters, and t is the elapsed time in seconds. Use a grapher to find, to the nearest tenth of a second, the first time the displacement is 5 meters.

51. The monthly revenue R in dollars from sales of ice-cream for a sweet shop in Montpelier, Vermont is approximated by

$$R = 4{,}200 + 2{,}100\cos(0.5236t)$$

where $t = 0$ represents August. Use a grapher to determine in which months the revenue is more than \$5,000.

52. The monthly number of kilowatt hours (E) of electricity sold by a particular utility company is approximated by

$$E = 384.5 + 89.5\cos(.5236t),$$

where $t = 0$ represents January. Use a grapher to determine in which months the utility sold at least 425 kilowatt hours.

53. A formula for the horizontal distance traveled by a projectile, neglecting air resistance, is

$$d = \frac{1}{32}V^2 \sin 2\theta,$$

where θ and V measure the angle of elevation and the initial velocity in feet/second of the projectile, respectively. If a professional field goal kicker boots a football with an initial velocity of 76 feet/second and the ball travels 180 ft, find θ to the nearest degree. (**Note:** There are two possible solutions.)

54. As the bottom of a ladder is pulled away from a wall, the top slides toward the floor with increasing speed. This means that the angle between the ladder and the floor shrinks with increasing speed. See the figure. The rate of speed at which the angle shrinks before the top loses contact with the wall is given in radians per second by

$$v = -\frac{k}{L\sin\theta},$$

where k is the speed in ft per second at which the bottom moves away from the wall, and L is the length of the ladder in feet.
Suppose for a particular 41-ft ladder, the bottom is being pulled away from the wall at 10 ft/second. At the moment the ladder loses contact with the wall, $v = -.6585$ radians per second. Find to the nearest hundredth, the angle θ at which the ladder loses contact with the wall. [Source: P. Scholten and A. Simpson, "The Falling Ladder Paradox," *The College Mathematics Journal*, Jan. 1996].

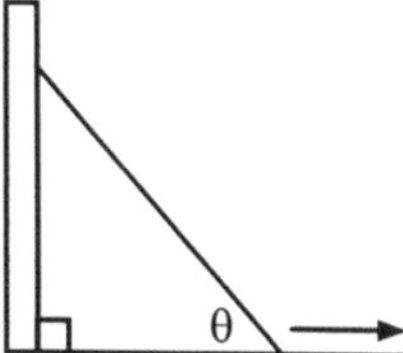

THINK ABOUT IT 8.6

1. Explain in terms of the unit circle definitions of the trigonometric functions why the solution set of the equation $\sec t = -1/2$ is $\varnothing$.
2. Is $\sin(x - (\pi/2)) = \cos x$ a conditional equation or an identity? Explain why.
3. Find all solutions to $\sin(\ln x) = 0$. **Hint:** See Explore Exercise 3 on page 684.
4. Solve for x in the interval $[0,\ 2\pi)$. Use the quadratic formula, $\cos^2 x + \cos x - 1 = 0$.
5. Solve $\sin 3x = \sin x$ in the interval $[0,\ 2\pi)$ with the aid of the sum-to-product formula

$$\sin A - \sin B = 2\cos\frac{A+B}{2}\sin\frac{A-B}{2}.$$

● ● ●

8.7 Fundamental Identities

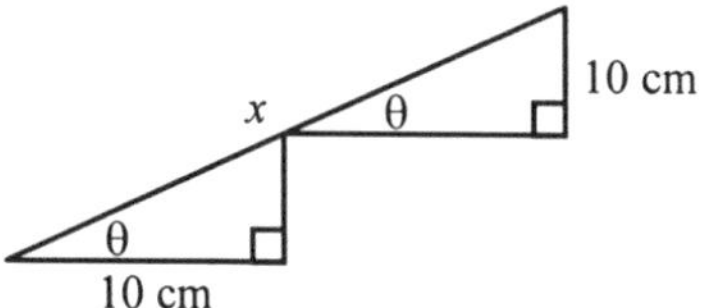

For the type of brace that is shown in the diagram, the length of the brace x is a function of θ. Show that

$$x = \frac{10(\sin\theta + \cos\theta)}{\sin\theta + \cos\theta}.$$

(See Example 8.)

Objectives

1. Transform a given trigonometric expression and show that it is identical to another given expression.
2. Verify that certain trigonometric equations are identities.

3. Use the graph of a trigonometric expression to make a conjecture about how the expression simplifies, and then confirm this conjecture algebraically.
4. Use a trigonometric substitution to simplify the square root of a quadratic expression.
5. Use identities to evaluate various trigonometric expressions, given the value of one trigonometric function of x and the interval containing x.
6. Use the fundamental identities to help solve certain trigonometric equations.

Trigonometry is characterized by many formulas that may be used to simplify trigonometric expressions. For example, the formula

$$\cos(-x) = \cos x$$

made it easier to evaluate the cosine of a negative number. In Section 8.4 we used the formula

$$\csc x = \frac{1}{\sin x}$$

to study the behavior of the cosecant function, and Example 8 shows that this formula may be used in conjunction with

$$\sec x = \frac{1}{\cos x}$$

to solve the section-opening problem. These formulas are examples of **trigonometric identities**. That is, they are true for all values of x for which the expressions are defined.

Some basic trigonometric identities are easy to develop. Since all the trigonometric functions may be defined in terms of the sine and/or cosine, these functions are interrelated. This enables us to change the form of many trigonometric expressions to an expression that is either simpler or more useful for a particular problem. For convenience, we list below six fundamental identities that are used often.

Six Fundamental Identities

Identity 1 $\csc x = \dfrac{1}{\sin x}$ or $\sin x = \dfrac{1}{\csc x}$ or $\sin x \csc x = 1$

Identity 2 $\sec x = \dfrac{1}{\cos x}$ or $\cos x = \dfrac{1}{\sec x}$ or $\cos x \sec x = 1$

Identity 3 $\cot x = \dfrac{1}{\tan x}$ or $\tan x = \dfrac{1}{\cot x}$ or $\tan x \cot x = 1$

Identity 4 $\tan x = \dfrac{\sin x}{\cos x}$

Identity 5 $\cot x = \dfrac{\cos x}{\sin x}$

Identity 6 $\sin^2 x + \cos^2 x = 1$ or $\sin^2 x = 1 - \cos^2 x$ or $\cos^2 x = 1 - \sin^2 x$

Except for identity 6, these statements follow directly from the definition of the trigonometric functions. Graphical support for identity 6 can be seen in Figure 8.82, which shows a graph of $y = (\sin x)^2 + (\cos x)^2$ on a grapher. It appears that the graph is the same as the graph of $y = 1$, and further evidence of this result may be obtained by using the Trace feature to display points in the graph.

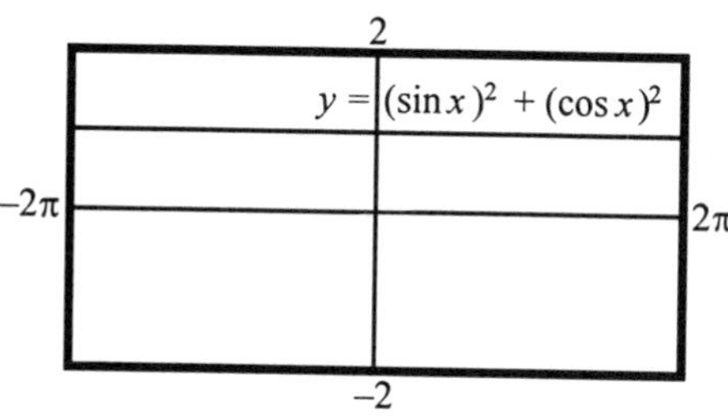

Figure 8.82

To confirm this identity algebraically, recall that by definition

$$x = \cos s \text{ and } y = \sin s.$$

Any point (x, y) on the unit circle must satisfy the equation $x^2 + y^2 = 1$. Thus, by substitution we have

$$(\cos s)^2 + (\sin s)^2 = 1.$$

For convenience, this statement is usually written with x as the independent variable. Thus,

$$\sin^2 x + \cos^2 x = 1.$$

Example 1: Verifying that Expressions are Identical

Show that the expression $\sin x \cot x \sec x$ is identical to 1.

Solution:

$$\begin{aligned} \sin x \cot x \sec x &= \sin x\left(\frac{\cos x}{\sin x}\right)\sec x && \text{Identity 5} \\ &= \sin x \frac{\cos x}{\sin x}\frac{1}{\cos x} && \text{Identity 2} \\ &= 1 && \text{Simplify} \end{aligned}$$

Thus, $\sin x \cot x \sec x = 1$ for all values of x at which the expressions are defined. You can graph

$$y1 = (\sin x)\left(\frac{\cos x}{\sin x}\right)\left(\frac{1}{\cos x}\right)$$

and

$$y2 = 1$$

in the same display on a grapher to check this result. You should see that both equations graph as a horizontal line that is fixed at 1.

PROGRESS CHECK 1

Show that $\dfrac{\tan x}{\sin x}$ is identical to $\sec x$. ∎

Example 2: Verifying an Identity

Prove the identity $\tan^2 x + 1 = \sec^2 x$.

Solution:

$$\begin{aligned} \tan^2 x + 1 &= \frac{\sin^2 x}{\cos^2 x} + 1 && \text{Identity 4} \\ &= \frac{\sin^2 x + \cos^2 x}{\cos^2 x} \\ &= \frac{1}{\cos^2 x} && \text{Identity 6} \\ &= \sec^2 x && \text{Identity 2} \end{aligned}$$

Thus, $\tan^2 x + 1 = \sec^2 x$ is an identity. Graphical support for this identity may be seen in Figure 8.83, which shows that the graphs of $y1 = (\tan x)^2 + 1$ and $y2 = 1/(\cos x)^2$ are the same graph.

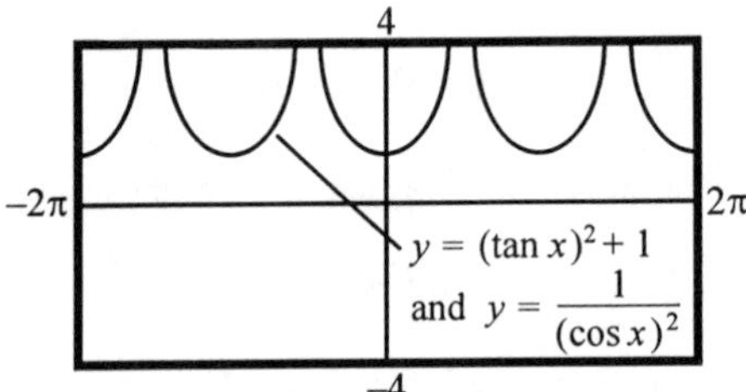

Figure 8.83

Note: Recall that you may also test for equivalent expressions on a grapher by graphing the quotient of the expressions in question. This quotient will be 1 for equivalent expressions throughout the domain of this quotient function. To illustrate, the identity in this example may be checked by defining $y1$ and $y2$ as stated above and then graphing only $y3 = y1/y2$, as shown in Figure 8.84.

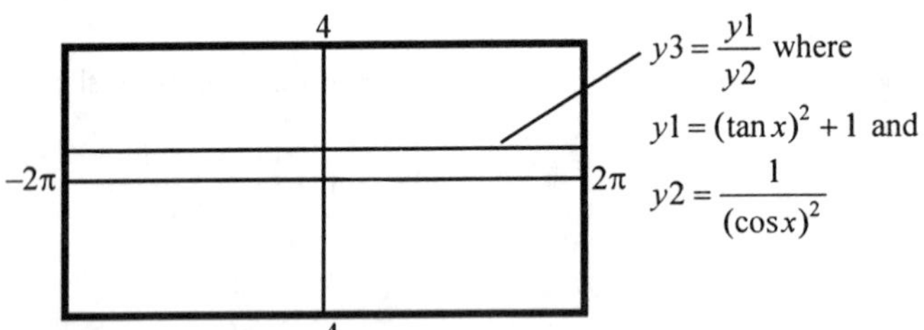

Figure 8.84

This graph, which is a horizontal line where y is fixed at 1, supports our results. This method is recommended when you find it hard to interpret the display that results from graphing the expressions on both sides of the proposed identity.

PROGRESS CHECK 2

Prove the identity $\cot^2 x + 1 = \csc^2 x$. ■

The first two examples discussed how a grapher may be used to check that two expressions are identical. A grapher may also be used to suggest a simpler form for a trigonometric expression, as shown in the next example.

Example 3: Discovering and Confirming an Identity

Graph $f(x)=\dfrac{\sec x-\cos x}{\tan x}$ and make a conjecture about an identity involving $f(x)$. Then, confirm this conjecture algebraically.

Solution: Figure 8.85 shows a graph of the given function. This graph appears to be the same as the graph of $y=\sin x$, so we suspect that

$$\frac{\sec x-\cos x}{\tan x}=\sin x$$

is an identity. To check algebraically, observe that

$$\frac{\sec x-\cos x}{\tan x}=\frac{(1/\cos x)-\cos x}{\sin x/\cos x}=\frac{1-\cos^2 x}{\sin x}=\frac{\sin^2 x}{\sin x}=\sin x\,.$$

This result confirms the relation we saw from the graph.

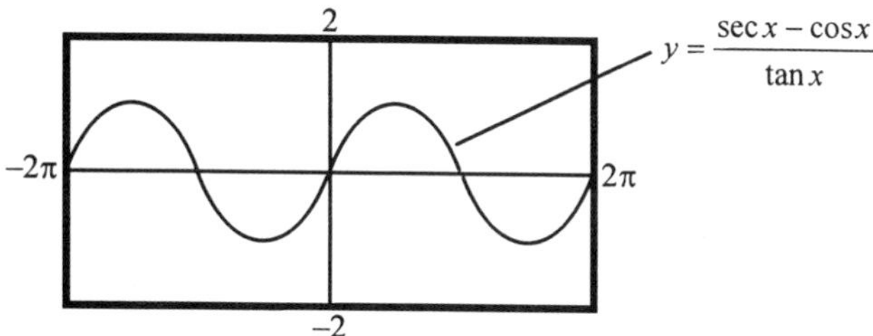

Figure 8.85

PROGRESS CHECK 3

Graph $f(x)=\dfrac{\csc x-\sin x}{\cot x}$ and make a conjecture about an identity involving $f(x)$. Then, confirm this conjecture algebraically. ■

In Example 2 we established the identity $\tan^2 x+1=\sec^2 x$, and in the associated Progress Check exercise you were asked to confirm that $\cot^2 x+1$ is identical to $\csc^2 x$. These two identities are used often, and we need to add them to our list of basic identities.

Identity 7 $\tan^2 x+1=\sec^2 x$

Identity 8 $\cot^2 x+1=\csc^2 x$

The eight identities listed are commonly referred to as the fundamental identities, and among these, identities 6, 7, and 8 are called the Pythagorean identities. In the next example we include an alternative solution that effectively employs identity 7.

Example 4: Using Pythagorean Identities

Show that the equation $\cos^2 x(1+\tan^2 x)=1$ is an identity.

Solution:

$$\begin{aligned}
\cos^2 x(1+\tan^2 x) &= \cos^2 x\left(1+\frac{\sin^2 x}{\cos^2 x}\right) && \text{Identity 4}\\
&= \cos^2 x\cdot 1+\cos^2 x\cdot\frac{\sin^2 x}{\cos^2 x} && \\
&= \cos^2 x+\sin^2 x && \text{Simplify}\\
&= 1 && \text{Identity 6}
\end{aligned}$$

Alternative Solution

$$\begin{aligned}
\cos^2 x(1+\tan^2 x) &= \cos^2 x(\sec^2 x) && \text{Identity 7}\\
&= \cos^2 x\left(\frac{1}{\cos^2 x}\right) && \text{Identity 2}\\
&= 1
\end{aligned}$$

Thus, $\cos^2 x(1+\tan^2 x)=1$ is an identity.

PROGRESS CHECK 4

Verify the identity $\dfrac{\cot^2 x+1}{\cot^2 x}=\sec^2 x$. ■

In general, there is no standard procedure for working with identities. In fact, a given identity can usually be proved in several ways. However, these suggestions should be helpful.

Guidelines for Proving Identities

1. Change the more complicated expression in the identity to the same form as the less complicated expression. If both expressions are complicated, you might try to change them both to the same expression.
2. If you are having difficulty, change all functions to sines and cosines. This procedure might necessitate more algebra in some instances, but it will provide a direct approach to the problem. Gradually try to make use of the other trigonometric functions.
3. It is often helpful to rewrite expressions by adding fractions, simplifying complex fractions, or factoring expressions.
4. Do *not* attempt to prove an identity by treating it as an equation and using the associated techniques, for this involves assuming what you want to prove.

Example 5: Converting to Sines and Cosines

Show that the expression $\dfrac{\tan^2 x-1}{\tan^2 x+1}$ is identical to the expression $\sin^2 x-\cos^2 x$.

Solution:

$$\begin{aligned}
\frac{\tan^2 x - 1}{\tan^2 x + 1} &= \frac{(\sin^2 x/\cos^2 x) - 1}{(\sin^2 x/\cos^2 x) + 1} && \text{Identity 4} \\
&= \frac{\cos^2 x[(\sin^2 x/\cos^2 x) - 1]}{\cos^2 x[(\sin^2 x/\cos^2 x) + 1]} && \\
&= \frac{\sin^2 x - \cos^2 x}{\sin^2 x + \cos^2 x} && \left.\right\} \text{Simplify the complex fraction.} \\
&= \frac{\sin^2 x - \cos^2 x}{1} && \text{Identity 6} \\
&= \sin^2 x - \cos^2 x
\end{aligned}$$

Thus,

$$\frac{\tan^2 x - 1}{\tan^2 x + 1} = \sin^2 x - \cos^2 x$$

is an identity.

PROGRESS CHECK 5

Verify the identity $\dfrac{\cot^2 x - 1}{\cot^2 x + 1} = \cos^2 x - \sin^2 x$. ■

Example 6: Converting to Sines and Cosines

Prove the identity $\dfrac{\cos x \csc x}{\cot^2 x} = \tan x$.

Solution:

$$\begin{aligned}
\frac{\cos x \csc x}{\cot^2 x} &= \frac{\cos x(1/\sin x)}{\cos^2 x/\sin^2 x} && \text{Identities 1 and 5} \\
&= \frac{\sin^2 x(\cos x/\sin x)}{\sin^2 x(\cos^2 x/\sin^2 x)} && \\
&= \frac{\sin x \cos x}{\cos^2 x} && \left.\right\} \text{Simplify the complex fraction.} \\
&= \frac{\sin x}{\cos x} && \\
&= \tan x && \text{Identity 4}
\end{aligned}$$

Thus, the given equation is an identity.

PROGRESS CHECK 6

Show that the expression $\dfrac{\sin x \sec x}{\tan^2 x}$ is identical to the expression $\cot x$. ■

Example 7: Using Factoring

Verify the identity $\sin x - \cos^2 x \sin x = \sin^3 x$.

Solution:

$$\begin{aligned}\sin x - \cos^2 x \sin x &= \sin x(1-\cos^2 x) && \text{Factor}\\ &= \sin x(\sin^2 x) && \text{Identity 6}\\ &= \sin^3 x\end{aligned}$$

PROGRESS CHECK 7

Verify the identity $\dfrac{\sin^2 x}{1+\cos x} = 1-\cos x$. ■

Because calculators do not have keys for the cotangent, secant, and cosecant functions, it may be convenient to write certain formulas that arise in applied problems by using only the sine, cosine, or tangent functions. We will make such a conversion as we derive the formula that is required in the section-opening problem.

Example 8: Using Identities to Derive a Formula

Solve the problem in the section introduction on page 685.

Solution: From the sketch of the brace that is shown in Figure 8.86, we see that

$$\sec\theta = \frac{a}{10}, \text{ so } a = 10\sec\theta,$$

and

$$\csc\theta = \frac{b}{10}, \text{ so } b = 10\csc\theta.$$

The length of the brace x is the sum of a and b. Thus,

$$x = 10\sec\theta + 10\csc\theta.$$

To convert to the requested formula, which involves only the sine and cosine functions, replace $\sec\theta$ with $1/\cos\theta$ and replace $\csc\theta$ with $1/\sin\theta$, to obtain

$$x = 10\left(\frac{1}{\cos\theta}\right) + 10\left(\frac{1}{\sin\theta}\right) = \frac{10\sin\theta + 10\cos\theta}{\sin\theta\cos\theta} = \frac{10(\sin\theta + \cos\theta)}{\sin\theta\cos\theta}.$$

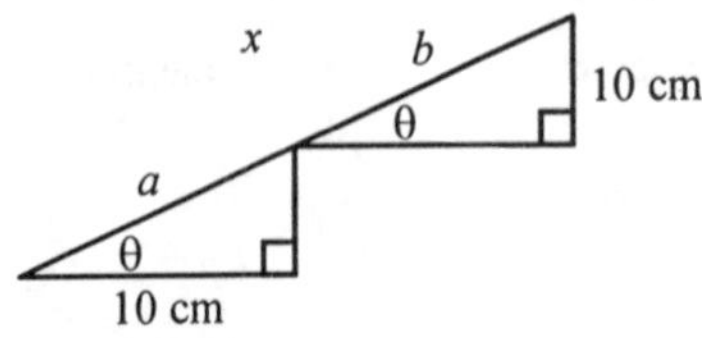

Figure 8.86

PROGRESS CHECK 8

Show that the formula derived in Example 8 may also be written as $x = \dfrac{10(1+\tan\theta)}{\sin\theta}$. ■

In calculus a trigonometric substitution may be used to simplify the square root of a quadratic expression. In such problems the Pythagorean identities are used in the forms $\cos^2\theta = 1-\sin^2\theta$, $\sec^2\theta = \tan^2\theta + 1$ and $\tan^2\theta = \sec^2\theta - 1$.

Example 9: Using Identities and Trigonometric Substitution

Eliminate the radical in the expression $\sqrt{9-x^2}$ by substituting $3\sin\theta$ for x, where $-\frac{\pi}{2} \le \theta \le \frac{\pi}{2}$.

Solution: Replace x by $3\sin\theta$ as directed and then simplify, as follows.

$$\begin{aligned}\sqrt{9-x^2} &= \sqrt{9-(3\sin\theta)^2}\\ &= \sqrt{9-9\sin^2\theta}\\ &= \sqrt{9(1-\sin^2\theta)}\\ &= \sqrt{9\cos^2\theta} && \text{Identity 6}\\ &= 3\cos\theta\end{aligned}$$

Note that the restriction $-\pi/2 \le \theta \le \pi/2$ allows us to replace $\sqrt{\cos^2\theta}$ with $\cos\theta$ in the solution.

PROGRESS CHECK 9

Eliminate the radical in the expression $\sqrt{x^2+16}$ by substituting $4\tan\theta$ for x where $-\frac{\pi}{2} \le \theta \le \frac{\pi}{2}$.

In Section 2.3 we considered a procedure for using a given trigonometric value to find other trigonometric values of that angle. We now show an alternative method for solving such problems that uses identities.

Example 10: Evaluating Functions Using Identities

If $\sin x = \frac{3}{5}$ and $\frac{\pi}{2} < x < \pi$, find

a. $\cos x$ **b.** $\tan x$ **c.** $\sec x$

Solution:

a. Since $\pi/2 < x < \pi$, where $\cos x < 0$, the identity $\sin^2 x + \cos^2 x = 1$ when solved for $\cos x$ becomes

$$\cos x = -\sqrt{1-\sin^2 x} = -\sqrt{1-\left(\frac{3}{5}\right)^2} = -\sqrt{\frac{16}{25}} = -\frac{4}{5}.$$

b. $\tan x = \dfrac{\sin x}{\cos x} = \dfrac{3/5}{-4/5} = \dfrac{3}{-4}$ **c.** $\sec x = \dfrac{1}{\cos x} = \dfrac{1}{-4/5} = \dfrac{5}{-4}$

PROGRESS CHECK 10

If $\cos x = \frac{5}{13}$ and $\frac{3\pi}{2} < x < 2\pi$, find

a. $\sin x$ **b.** $\tan x$ **c.** $\csc x$ ■

Our final example illustrates how an analytic solution to a trigonometric equation can depend on a knowledge of identities.

Example 11: Solving Trigonometric Equations Using Fundamental Identities

Solve $2\cos^2 x = 1 - \sin x$ in the interval $[0,\ 2\pi)$.

Solution: Figure 8.87 indicates that the graphs of $y = 2\cos^2 x$ and $y = 1 - \sin x$ intersect at three points in the interval $[0,\ 2\pi)$, so we can anticipate three solutions to the equation in question. To find these solutions algebraically, use the identity $\cos^2 x = 1 - \sin^2 x$ to rewrite the equation in terms of only $\sin x$, and then solve the resulting equation by the methods shown in Section 8.6.

$$2\cos^2 x = 1 - \sin x$$
$$2(1 - \sin^2 x) = 1 - \sin x$$
$$2 - 2\sin^2 x = 1 - \sin x$$
$$0 = 2\sin^2 x - \sin x - 1$$
$$0 = (2\sin x + 1)(\sin x - 1)$$
$$2\sin x + 1 = 0 \quad \text{or} \quad \sin x - 1 = 0$$
$$\sin x = -\frac{1}{2} \quad \text{or} \quad \sin x = 1$$

In the interval $[0,\ 2\pi)$, $\sin x = -1/2$ when $x = 7\pi/6$ and $x = 11\pi/6$ (see Example 2 of Section 8.6), while $\sin x = 1$ when $x = \pi/2$. Thus, the solution set in $[0,\ 2\pi)$ is $\{\pi/2,\ 7\pi/6,\ 11\pi/6\}$.

You should look at Figure 8.87 once again to confirm that this solution set is reasonable.

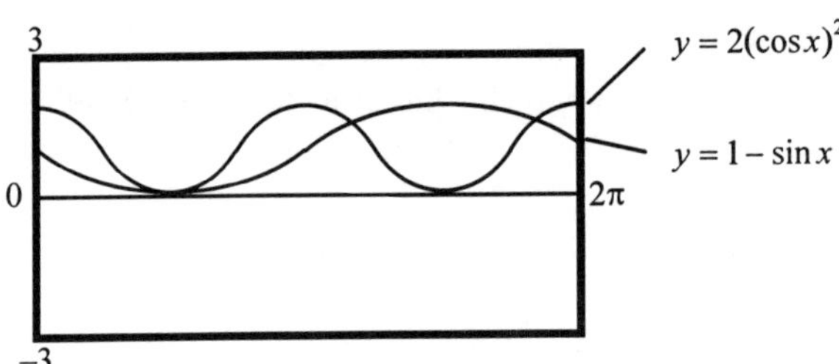

Figure 8.87

PROGRESS CHECK 11

Solve $\tan x + \cot x = 2\csc x$ in the interval $[0,\ 2\pi)$. ■

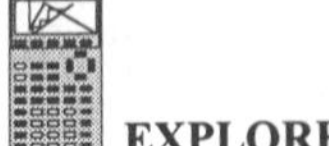

EXPLORE 8.7

1. Use a calculator to gather numerical evidence for or against the claim that these equations are identities. By trying several values of *x* in each case, determine which of these equations may be true for all values of *x*.
   a. $\sin 2x = 2\sin x$ **b.** $\cos 2x = 2\cos x$
   c. $\sin 2x = 2\sin x\cos x$ **d.** $\cos 2x = 2\sin x\cos x$
   e. $\sin 2x = \cos^2 x - \sin^2 x$ **f.** $\cos 2x = \cos^2 x - \sin^2 x$
2. If an equation $f(x) = g(x)$ is an identity, then the graphs of $y_1 = f(x)$ and $y_2 = g(x)$ will be the same. Furthermore, the graph of $y_3 = y_1/y_2 = f(x)/g(x)$ will be a horizontal line at $y = 1$. Use these graphical techniques to determine which of the equations in the previous problem are identities.
3. Use a grapher to support the claim that these equations are identities,
   a. $\sin^2 x = \dfrac{1 - \cos 2x}{2}$ **b.** $\cos^2 x = \dfrac{1 + \cos 2x}{2}$

EXERCISES 8.7

In Exercises 1–20 transform each first expression and show that it is identical to the second expression.

1. $\sin x \sec x$; $\tan x$
2. $\tan x \csc x$; $\sec x$
3. $\sin^2 x \csc x$; $\sin x$
4. $\tan x \cot^2 x$; $\cot x$
5. $\cos x \tan x$; $\dfrac{1}{\csc x}$
6. $\sec x \cot x$; $\dfrac{1}{\sin x}$
7. $\cos x \tan x \csc x$; 1
8. $\cot x \sec x \sin x$; 1
9. $\sin^2 x \cot^2 x$; $\cos^2 x$
10. $\tan^2 x \csc^2 x$; $\sec^2 x$
11. $\sin^2 x \sec^2 x$; $\tan^2 x$
12. $\dfrac{\cot x}{\csc x}$; $\cos x$
13. $\dfrac{\cos^2 x}{\cot^2 x}$; $\sin^2 x$
14. $\dfrac{\sin x \sec x}{\tan x}$; 1
15. $\dfrac{\cot x \tan x}{\sec x}$; $\cos x$
16. $1+\cot^2 x$; $\csc^2 x$
17. $(1-\cos x)(1+\cos x)$; $\dfrac{1}{\csc^2 x}$
18. $\sin x(\csc x - \sin x)$; $\cos^2 x$
19. $\dfrac{\cos^2 x}{1+\sin x}$; $1-\sin x$
20. $\sin^4 x - \cos^4 x$; $\sin^2 x - \cos^2 x$

In Exercises 21–40 prove that the equation is an identity.

21. $\dfrac{1}{\sin^2 x} - 1 = \cot^2 x$
22. $\csc x \sin x - \dfrac{1}{\sec^2 x} = \sin^2 x$
23. $\dfrac{1+\tan^2 x}{\csc^2 x} = \tan^2 x$
24. $\dfrac{1+\tan^2 x}{\tan^2 x} = \csc^2 x$
25. $\sin x \tan x + \cos x = \sec x$
26. $\tan x \csc^2 x - \tan x = \cot x$
27. $\dfrac{\tan^2 x - \sin^2 x}{\tan^2 x} = \sin^2 x$
28. $\dfrac{\sec^2 x + \csc^2 x}{\sec^2 x} = \csc^2 x$
29. $\dfrac{\csc x}{\tan x + \cot x} = \cos x$
30. $\dfrac{\sec x + \csc x}{1+\tan x} = \csc x$
31. $(1-\sin x)(\sec x + \tan x) = \cos x$
32. $\sec x \csc x - 2\cos x \csc x + \cot x = \tan x$
33. $\sin x \cos^3 x + \cos x \sin^3 x = \sin x \cos x$
34. $\cos^4 x + 2\cos^2 x \sin^2 x + \sin^4 x = 1$
35. $(\sin x + \cos x)^2 + (\sin x - \cos x)^2 = 2$
36. $\dfrac{1}{\sec^3 x \cos^4 x} = \sec x$
37. $\sec x - \cos x = \sin x \tan x$
38. $\sec^4 x - \tan^4 x = 2\sec^2 x - 1$
39. $\dfrac{\sin x}{\csc x} + \dfrac{\cos x}{\sec x} = 1$
40. $\dfrac{\csc x}{\sin x} - \dfrac{\cot x}{\sin x} = 1$

In Exercises 41–46 use a grapher to graph the given expression, and make a conjecture about an identity involving $f(x)$. Then confirm the identity algebraically, if possible.

41. $f(x) = \dfrac{1-\cos^2(x)}{\cos x \tan x}$
42. $f(x) = \dfrac{1-\sin^2 x}{\sin x \cos x}$
43. $f(x) = (\sin x - \tan x)(\sin x + \tan x) + \cos^2 x + \dfrac{1}{\cos^2 x}$
44. $f(x) = \cos^2 x\left(1 - \dfrac{1}{\cos^4 x}\right) + \sin^2 x\left(1 + \dfrac{1}{\cos^2 x}\right)$
45. $f(x) = \sin x \cos x$
46. $f(x) = \cos^2 x - \sin^2 x$

In Exercises 47–56 eliminate the radical in the expression by using the given substitution, where θ is restricted to the interval specified.

47. $\sqrt{1-x^2}$; $x = \sin\theta$, $\left[-\dfrac{\pi}{2}, \dfrac{\pi}{2}\right]$

48. $\sqrt{1+x^2}$; $x=\tan\theta$, $\left(-\frac{\pi}{2},\frac{\pi}{2}\right)$

49. $\sqrt{x^2+9}$; $x=3\tan\theta$, $\left(-\frac{\pi}{2},\frac{\pi}{2}\right)$

50. $\sqrt{25-x^2}$; $x=5\sin\theta$, $\left[-\frac{\pi}{2},\frac{\pi}{2}\right]$

51. $\sqrt{x^2-1}$; $x=\sec\theta$, $\left[0,\frac{\pi}{2}\right)\cup\left[\pi,\frac{3\pi}{2}\right)$

52. $\sqrt{(x-3)^2-1}$; $x-3=\sec\theta$, $\left[0,\frac{\pi}{2}\right)\cup\left[\pi,\frac{3\pi}{2}\right)$

53. $\sqrt{1-(x-2)^2}$; $x-2=\sin\theta$, $\left[-\frac{\pi}{2},\frac{\pi}{2}\right]$

54. $\sqrt{\left(16+x^2\right)^3}$; $x=4\tan\theta$, $\left(-\frac{\pi}{2},\frac{\pi}{2}\right)$

55. $\sqrt{\left(4x^2-9\right)^3}$; $2x=3\sec\theta$, $\left[0,\frac{\pi}{2}\right)\cup\left[\pi,\frac{3\pi}{2}\right)$

56. $\sqrt{4-9(x-2)^2}$; $3(x-2)=2\sin\theta$, $\left[-\frac{\pi}{2},\frac{\pi}{2}\right]$

In Exercises 57–60 use identities to find the given function value if $\sin x=\frac{12}{13}$ and $\frac{\pi}{2}<x<\pi$.

57. $\csc x$ 58. $\cos x$ 59. $\tan x$ 60. $\cot x$

In Exercises 61–64 use identities to find the given function value if $\cos x=a$ and $\pi<x<\frac{3\pi}{2}$.

61. $\sin x$ 62. $\csc x$ 63. $\sec x$ 64. $\tan x$

In Exercises 65–68 use identities to find the values of the remaining trigonometric functions of x.

65. $\sin x=\frac{8}{17}$, $\cos x>0$

66. $\cos x=-\frac{3}{5}$, $\sin x<0$

67. $\sec x=-\frac{4}{3}$, $\sin x>0$

68. $\tan x=2$, $\cos x<0$

In Exercises 69–72 express the given function of x in terms of $\sin x$.

69. $\cos x$ 70. $\tan x$ 71. $\sec x$ 72. $\csc x$

In Exercises 73–76 express the given function of x in terms of $\cos x$.

73. $\sin x$ 74. $\cot x$ 75. $\sec x$ 76. $\csc x$

In Exercises 77–90 solve each equation for $0\le x<2\pi$ through the aid of the *fundamental identities.*

77. $\sin x=\cos x$ 78. $\sqrt{3}\sin x-\cos x=0$

79. $\sin x=\tan x$ 80. $2\sin x-\tan x=0$

81. $\sin^2 x+\cos^2 x=\tan x$

82. $\tan^2 x-\sec^2 x=\cos x$

83. $\sin x+\cos x\tan x=1$

84. $\tan x+4\cot x=5$ 85. $2\sin x-\csc x=1$

86. $1-\cos^2 x=\sin x$ 87. $2\cos^2 x=3\sin x+3$

88. $2\sin^2 x=\cos x+1$ 89. $\sec^2 x+5\tan x+4=0$

90. $\tan^2 x-3\sec x=9$

91. One approach to analyzing the surface area of a cylinder involves inscribing regular polygonal solids inside the cylinder. The diagram shown is useful in this analysis. Note that the length x is a function of θ, where $0°<\theta<90°$.

a. Show that $x=\sqrt{\sin^2\theta+\left(1-\cos^2\theta\right)}$.

b. Show that $x=\sqrt{2-2\cos\theta}$.

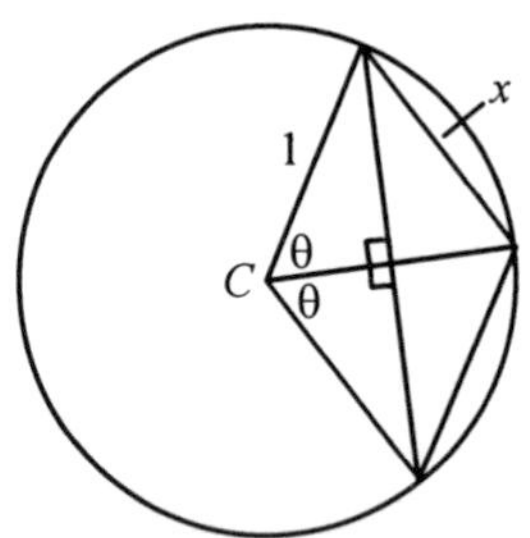

92. In the study of polarized light, light is passed through two sheets of polarized material. As the top sheet is rotated through an angle θ, the intensity I of the light that is transmitted through both sheets changes. The relationship that describes this process was discovered in 1809 by Etienne Louis Malus, and is now called the Law of Malus. The Law is given by the equation $I=M-M\tan^2\theta\cos^2\theta$, where M is the maximum intensity transmitted. Show that this law may also be expressed as $I=M\cos^2\theta$.

For Exercises 93 and 94 refer to the diagram shown. This diagram is used in the analysis of satellite data between a satellite S at height h above the Earth and a point R on the Earth's surface. The angle at the horizon H is a right angle, and r represents the radius of the Earth. [Source: *Space Mathematics*, NASA, 1985].

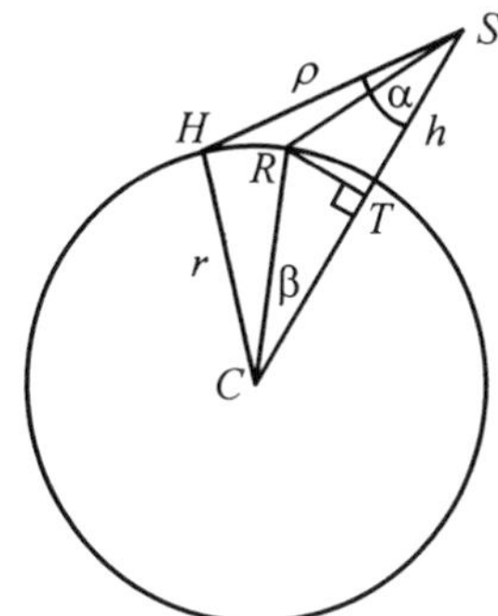

93. Use fundamental identities and the fact that $\tan\alpha = \dfrac{RT}{CS - CT}$ to do parts **a** and **b**.

a. Show that $\tan\alpha = \dfrac{r\sin\beta}{(r+h) - r\cos\beta}$.

b. Divide the numerator and denominator from part **a** by $r+h$ to show that $\tan\alpha = \dfrac{\sin\rho\sin\beta}{1-\sin\rho\sin\beta}$.

94. Show that the relationship given in part **b** of Exercise 93 can also be written as

$$\tan\alpha = \frac{1}{\csc\rho\csc\beta - 1}.$$

95. A major application of calculus is to describe the rate of change of functions. The following expression gives the rate of change of the function $y = \tan x$.

$$\text{rate of change of } \tan x = \frac{\cos x(\cos x) - \sin x(\sin x)}{(\cos x)^2}$$

Show that this expression simplifies to $\sec^2 x$

96. The following expression gives the rate of change of the function $y = \cot x$.

$$\text{rate of change of } \cot x = \frac{\sin x(\sin x) - \cos x(\cos x)}{(\sin x)^2}$$

Show that this expression simplifies to $-\csc^2 x$.

THINK ABOUT IT 8.7

1. Find all values of x in $[0, 2\pi)$ that make the given equation a true statement.
   - **a.** $\sin x = \sqrt{1-\cos^2 x}$
   - **b.** $\sin x = -\sqrt{1-\cos^2 x}$
   - **c.** $\sec x = -\sqrt{1+\tan^2 x}$
   - **d.** $\cot x = \sqrt{\csc^2 x - 1}$
2. Use the graph of $y = \tan^2 x$ that follows to graph $y = \sec^2 x$.

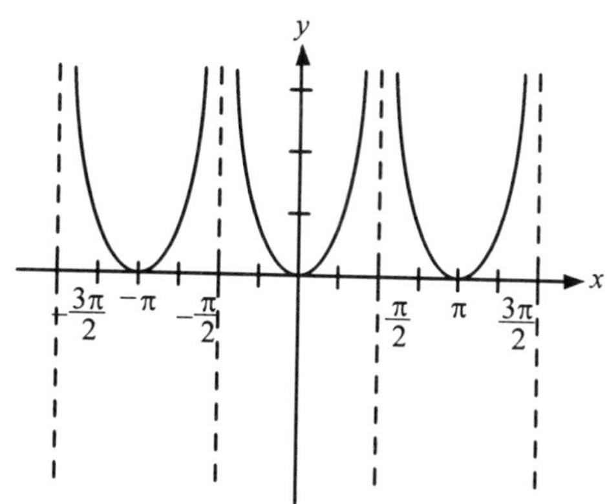

3. **a.** If $\log\sin a = b$, express $\log\csc a$ in terms of b.
   b. Verify the identity:

$$-\ln(\sec x - \tan x) = \ln(\sec x + \tan x).$$

4. Verify the identity: $\dfrac{\sec x + 1}{\tan x} = \dfrac{1}{\csc x - \cot x}$.

5. The figure below, called the **function hexagon**, arranges the six trigonometric functions of x so that many basic identities are easily generated, as shown in the following exercises.

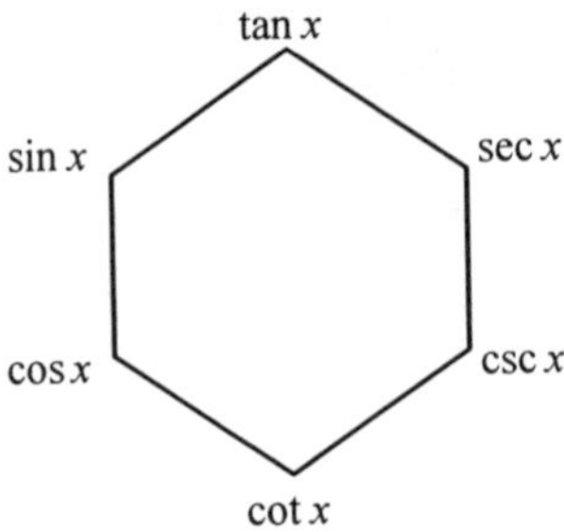

a. Describe in geometric terms how reciprocal functions are positioned in the hexagon.

b. Any function of x is identical to the product of the two functions of x on either side of it. For example $\sin x = \cos x \tan x$ is an identity. Create another example of this type and verify that the equation is an identity.

c. The identities $\tan x = \sin x / \cos x$ and $\tan x = \sec x / \csc x$ illustrate quotient identities that may be read from the arrangement. How is this done? State and prove two identities of this type for $\sin x$.

d. Another type of identity involves the product of three functions of x in alternate positions around the hexagon. Give two specific examples and state the general rule for this type of identity.

● ● ●

8.8 Sum and Difference Formulas

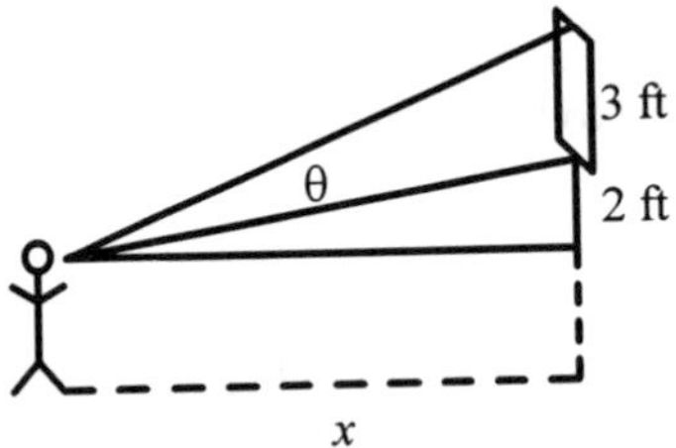

At an art gallery, a painting that is 3 ft high is mounted so that its base is 2 ft above the anticipated eye level of a viewer.

a. If a viewer is standing x ft from the wall, as shown in the diagram, then use the difference formula for the tangent to show that

$$\theta = \tan^{-1}\left(\frac{3x}{x^2+10}\right).$$

b. Use a grapher to graph the relation in part **a**.

c. How far back should a viewer stand so that the picture appears to be the largest? In other words, what value for x produces the maximum value for θ?

(See Example 7. Note that the problem of maximizing the viewing angle θ in this situation was also considered in Example 3 of Section 8.5. However, for that problem we chose to express θ as the difference between two inverse tangent functions in part **a**.)

Objectives

1. Use the sum or difference formulas to verify certain other identities.
2. Use the sum or difference formulas to find exact function values of certain angles or numbers.
3. Use the sum or difference formulas to evaluate certain trigonometric expressions involving two angles, given function values of these angles.
4. Investigate a relation with the aid of a sum or difference formula.

In the section-opening problem we are asked to use the identity for the tangent of the difference of two angles to determine a formula that expresses the viewing angle θ in terms of the viewer's distance x from the wall. Trigonometric expressions such as $\tan(\alpha-\beta)$, $\sin(x+h)$, and $\cos(\pi-x)$ that involve the sum or difference of two angles or numbers occur often, so we will now develop identities for analyzing such expressions.

We begin by considering the formula for the cosine of the sum of two numbers. First, observe that

$$\cos\left(\pi+\frac{\pi}{2}\right)=\cos\frac{3\pi}{2}=0$$

and

$$\cos\pi+\cos\frac{\pi}{2}=-1+0=-1$$

so

$$\cos\left(\pi+\frac{\pi}{2}\right)\neq\cos\pi+\cos\frac{\pi}{2}.$$

We see that $\cos(x_1+x_2)$ is not identical to $\cos x_1+\cos x_2$. Instead, the identity for the cosine of the sum of two numbers is more involved, and is given by

$$\cos(x_1+x_2)=\cos x_1\cos x_2-\sin x_1\sin x_2.$$

This formula is usually memorized as a verbal rule.

The cosine of a sum is the cosine of the first times the cosine of the second minus the sine of the first times the sine of the second.

The derivation of this identity is quite long, but once established, this formula yields many important results. Consider the coordinates of the points on the unit circle in Figure 8.88. The arc lengths from $(1, 0)$ to $[\cos(s_1+s_2),\ \sin(s_1+s_2)]$ and from $[\cos(-s_1),\ \sin(-s_1)]$ to $(\cos s_2,\ \sin s_2)$ are both of length s_1+s_2. Equal arcs in a circle subtend equal chords. Thus,, by the distance formula we have

$$\sqrt{[\cos(s_1+s_2)-1]^2+[\sin(s_1+s_2)-0]^2}=\sqrt{(\cos s_2-\cos s_1)^2+[\sin s_2-(-\sin s_1)]^2}\,.$$

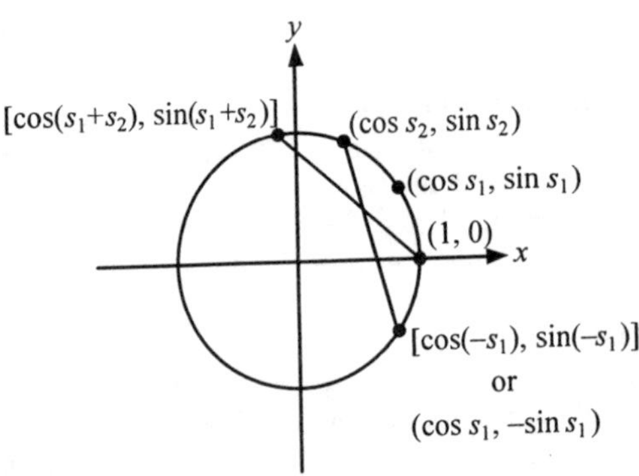

Figure 8.88

After squaring both sides to remove radicals and performing the indicated operations, we obtain

$$\cos^2(s_1+s_2)-2\cos(s_1+s_2)+1+\sin^2(s_1+s_2)=\cos^2 s_2-2\cos s_1\cos s_2+\cos^2 s_1$$
$$+\sin^2 s_2+2\sin s_1\sin s_2+\sin^2 s_1.$$

Because $\sin^2 s+\cos^2 s=1$, we regroup the terms as follows:

$$\left[\sin^2(s_1+s_2)+\cos^2(s_1+s_2)\right]-2\cos(s_1+s_2)+1=\left(\sin^2 s_2+\cos^2 s_2\right)+\left(\sin^2 s_1+\cos^2 s_1\right)$$
$$-2\cos s_1\cos s_2+2\sin s_1\sin s_2.$$

Substituting 1 for the appropriate expressions, we have

$$1-2\cos(s_1+s_2)+1=1+1-2\cos s_1\cos s_2+2\sin s_1\sin s_2-2\cos(s_1+s_2)$$
$$-2\cos(s_1+s_2)=-2\cos s_1\cos s_2+2\sin s_1\sin s_2\cos(s_1+s_2)$$
$$\cos(s_1+s_2)=\cos s_1\cos s_2-\sin s_1\sin s_2.$$

For convenience, we let x represent the independent variable. Thus,

$$\cos(x_1+x_2)=\cos x_1\cos x_2-\sin x_1\sin x_2.$$

Example 1: Using the $\cos(x_1+x_2)$ Formula

Verify the identity $\cos(2\pi+x)=\cos x$.

Solution: Using the formula for the cosine of the sum of two numbers gives

$$\cos(2\pi+x)=\cos 2\pi\cos x-\sin 2\pi\sin x$$
$$=1\cdot\cos x-0\cdot\sin x$$
$$=\cos x.$$

■

PROGRESS CHECK 1

Verify the identity $\cos\left(\dfrac{3\pi}{2}+x\right)=\sin x$.

The identity for the cosine of a sum of two numbers leads directly to an identity for the cosine of a difference. We merely rewrite $\cos(x_1-x_2)$ as $\cos[x_1+(-x_2)]$ and use the formula for the cosine of a sum.

$$\cos[x_1 + (-x_2)] = \cos x_1 \cos(-x_2) - \sin x_1 \sin(-x_2)$$

Since $\cos(-x_2) = \cos x_2$ and $\sin(-x_2) = -\sin x_2$, we have

$$\cos[x_1 + (-x_2)] = \cos x_1 \cos x_2 - \sin x_1(-\sin x_2)$$

or

$$\cos(x_1 - x_2) = \cos x_1 \cos x_2 + \sin x_1 \sin x_2.$$

Example 2: Using the $\cos(x_1 - x_2)$ Formula

Verify the identity $\cos\left(\frac{\pi}{2} - x\right) = \sin x$.

Solution: We use the formula for $\cos(x_1 - x_2)$ and replace x_1 with $\pi/2$ and x_2 with x.

$$\begin{aligned}\cos\left(\frac{\pi}{2} - x\right) &= \cos\frac{\pi}{2}\cos x + \sin\frac{\pi}{2}\sin x \\ &= 0 \cdot \cos x + 1 \cdot \sin x \\ &= \sin x\end{aligned}$$

As in the previous section, you can obtain visual evidence for an identity by overlaying the graphs that are associated with the expressions on each side of the equal sign and checking for identical graphs. Figure 8.89 shows such a check for the identity in this example.

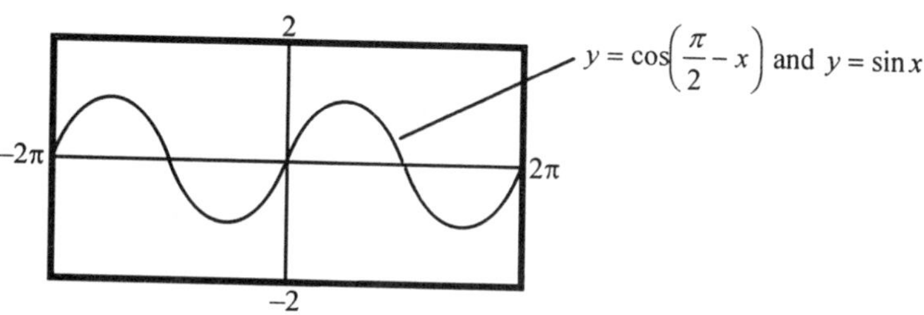

Figure 8.89

PROGRESS CHECK 2

Verify the identity $\cos(2\pi - x) = \cos x$.

■

To this point, only identities involving the cosine function have been developed. We will now show how the formula for the sine of the sum of two numbers is usually derived. This formula can then be used to derive a list of identities concerning the sine function, and the exercises ask you to verify many of these formulas.

In Example 2, we proved

$$\cos\left(\frac{\pi}{2} - x\right) = \sin x.$$

Replacing x by $x_1 + x_2$ gives

$$\sin(x_1 + x_2) = \cos\left[\frac{\pi}{2} - (x_1 + x_2)\right] = \cos\left[\left(\frac{\pi}{2} - x_1\right) - x_2\right].$$

If we now use the formula for the cosine of the difference of two numbers, we obtain

$$\sin(x_1 + x_2) = \cos\left[\left(\frac{\pi}{2} - x_1\right) - x_2\right] = \cos\left(\frac{\pi}{2} - x_1\right)\cos x_2 + \sin\left(\frac{\pi}{2} - x_1\right)\sin x_2.$$

By the identity in Example 2 we know that $\cos(\pi/2 - x_1) = \sin x_1$. In Exercise 64 of this section you are asked to verify that $\sin(\pi/2 - x_1) = \cos x_1$. The formula for the sine of the sum of two numbers is then

$$\sin(x_1 + x_2) = \sin x_1 \cos x_2 + \cos x_1 \sin x_2.$$

Once again, such a formula is usually best memorized in terms of words.

The sine of a sum is the sine of the first times the cosine of the second plus the cosine of the first times the sine of the second.

This formula leads directly to the following identity, whose verification is left to the exercises.

$$\sin(x_1 - x_2) = \sin x_1 \cos x_2 - \cos x_1 \sin x_2$$

Applications of these two formulas are shown in the next two examples, and Example 4 is an application from calculus.

Example 3: Using the $\sin(x_1 - x_2)$ Formula

Verify the identity $\sin(2\pi - x) = -\sin x$.

Solution: Use the formula for the sine of the difference of two numbers.

$$\begin{aligned}\sin(2\pi - x) &= \sin 2\pi \cos x - \cos 2\pi \sin x \\ &= 0 \cdot \cos x - 1 \cdot \sin x \\ &= -\sin x\end{aligned}$$

PROGRESS CHECK 3

Verify the identity $2\sin\left(x - \frac{\pi}{6}\right) = \sqrt{3}\sin x - \cos x$. ■

Example 4: Simplifying a Difference Quotient

If $f(x) = \sin x$, show that

$$\frac{f(x+h) - f(x)}{h} = \sin x\left(\frac{\cos h - 1}{h}\right) + \cos x\left(\frac{\sin h}{h}\right).$$

Solution: If $f(x) = \sin x$, we have

$$f(x+h) = \sin(x+h) = \sin x \cos h + \cos x \sin h.$$

Then

$$\frac{f(x+h)-f(x)}{h}=\frac{\sin x\cos h+\cos x\sin h-\sin x}{h}$$
$$=\frac{\sin x(\cos h-1)+\cos x\sin h}{h}$$
$$=\sin x\left(\frac{\cos h-1}{h}\right)+\cos x\left(\frac{\sin h}{h}\right).$$

PROGRESS CHECK 4

If $f(x)=\cos x$, show that $\dfrac{f(x+h)-f(x)}{h}=\cos x\left(\dfrac{\cos h-1}{h}\right)-\sin x\left(\dfrac{\sin h}{h}\right)$. ■

Because $\tan x=\sin x/\cos x$, the identities for $\sin(x_1+x_2)$ and $\cos(x_1+x_2)$ may be used to derive an identity for $\tan(x_1+x_2)$. The result and the formula for $\tan(x_1-x_2)$ follow. The verification of these identities is requested in Exercises 53 and 56.

$$\tan(x_1+x_2)=\frac{\tan x_1+\tan x_2}{1-\tan x_1\tan x_2}$$
$$\tan(x_1-x_2)=\frac{\tan x_1-\tan x_2}{1+\tan x_1\tan x_2}$$

In the next two examples we consider some numerical applications of the sum and difference identities.

Example 5: Finding the Exact Value of Certain Trigonometric Expressions

Find the exact value of $\tan 15°$.

Solution: The initial objective is to write 15° as a sum or a difference of two special angles whose exact trigonometric values are known. By observing that $15°=60°-45°$ and using the formula for $\tan(x_1-x_2)$, we have

$$\tan 15°=\tan(60°-45°)$$
$$=\frac{\tan 60°-\tan 45°}{1+\tan 60°\tan 45°}$$
$$=\frac{\sqrt{3}-1}{1+\sqrt{3}(1)}.$$

To simplify this result, rationalize the denominator.

$$\tan 15°=\frac{\sqrt{3}-1}{1+\sqrt{3}}\cdot\frac{1-\sqrt{3}}{1-\sqrt{3}}$$
$$=\frac{\sqrt{3}-3-1+\sqrt{3}}{1-3}$$
$$=2-\sqrt{3}$$

PROGRESS CHECK 5

Find the exact value of $\tan 75°$. ■

Example 6: Evaluating Certain Trigonometric Expressions Involving Two Angles

If $\sin x_1 = -\frac{1}{2}$, where $\pi < x_1 < \frac{3\pi}{2}$, and $\cos x_2 = \frac{\sqrt{3}}{2}$, where $0 < x_2 < \frac{\pi}{2}$, find.

a. $\cos x_1$ **b.** $\sin x_2$ **c.** $\cos(x_1 + x_2)$ **d.** $\sin(x_1 - x_2)$

Solution:

a. Since $\pi < x_1 < 3\pi/2$, where $\cos x_1$ is negative, we have

$$\cos x_1 = -\sqrt{1-\sin^2 x_1} = -\sqrt{1-\left(-\frac{1}{2}\right)^2}$$
$$= -\sqrt{\frac{3}{4}} = -\frac{\sqrt{3}}{2}.$$

b. Since $0 < x_2 < \pi/2$, where $\sin x_2$ is positive, we have

$$\sin x_2 = \sqrt{1-\cos^2 x_2} = \sqrt{1-\left(\frac{\sqrt{3}}{2}\right)^2} = \sqrt{\frac{1}{4}} = \frac{1}{2}.$$

c. $\cos(x_1 + x_2) = \cos x_1 \cos x_2 - \sin x_1 \sin x_2$

$$= \left(-\frac{\sqrt{3}}{2}\right)\left(\frac{\sqrt{3}}{2}\right) - \left(-\frac{1}{2}\right)\left(\frac{1}{2}\right)$$
$$= -\frac{3}{4} - \left(-\frac{1}{4}\right)$$
$$= -\frac{1}{2}$$

d. $\sin(x_1 - x_2) = \sin x_1 \cos x_2 - \cos x_1 \sin x_2$

$$= \left(-\frac{1}{2}\right)\left(\frac{\sqrt{3}}{2}\right) - \left(-\frac{\sqrt{3}}{2}\right)\left(\frac{1}{2}\right)$$
$$= -\frac{\sqrt{3}}{4} - \left(-\frac{\sqrt{3}}{4}\right)$$
$$= 0$$

PROGRESS CHECK 6

If $\sin\alpha = \frac{3}{5}$ and $\cos\beta = -\frac{12}{13}$, where α is a first-quadrant angle and β is a second-quadrant angle, then find

a. $\sin(\alpha+\beta)$ **b.** $\cos(\alpha+\beta)$ ■

In our final example we will investigate the relation that is described in the section-opening problem with the aid of a sum or difference formula.

Example 7: Maximizing a Viewing Angle

Solve the problem in the section introduction on page 698.

Solution:

a. Consider the simplified sketch of the problem in Figure 8.90. Observe that $\theta = \alpha - \beta$, $\tan\alpha = 5/x$, and $\tan\beta = 2/x$. So

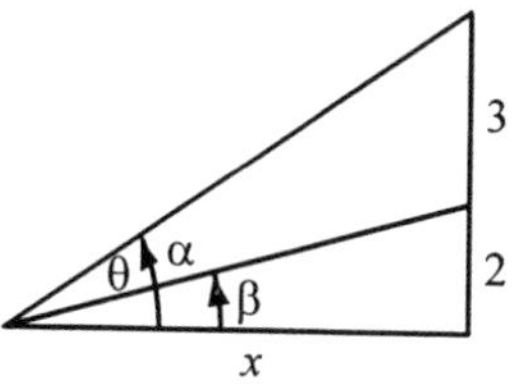

Figure 8.90

$$
\begin{aligned}
\tan\theta &= \tan(\alpha - \beta) \\
&= \frac{\frac{5}{x} - \frac{2}{x}}{1 + \frac{5}{x}\cdot\frac{2}{x}} \\
&= \frac{3x}{x^2 + 10}
\end{aligned}
$$

and

$$\theta = \tan^{-1}\left(\frac{3x}{x^2 + 10}\right).$$

b. The equation $\theta = \tan^{-1}(3x/(x^2 + 10))$ is graphed in Figure 8.91.

c. By using the Function Maximum feature (or Zoom and Trace), we can determine that the maximum point in the graph of this relation is reached at about (3.16, 0.44), as shown in Figure 8.92. Thus, when x is about 3.2 ft, the viewing angle reaches a maximum value of about 0.44 radians (or 25°).

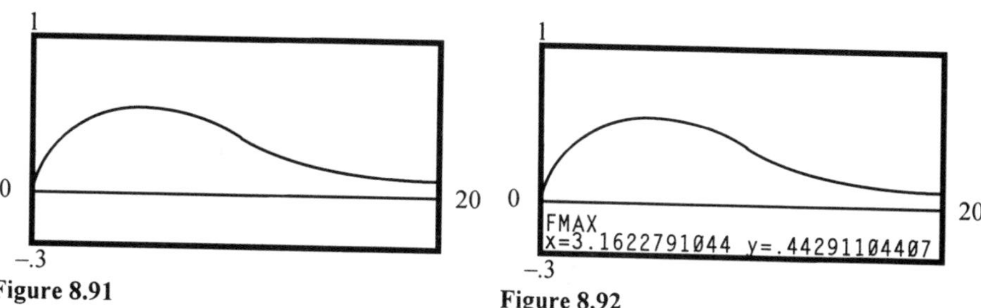

Figure 8.91 **Figure 8.92**

PROGRESS CHECK 7

Redo the problems in Example 7, assuming that the painting is 5 ft high and that its base is hung 3 ft above the anticipated eye level of a viewer. In part a use the difference formula for the tangent to express θ in terms of x.

In summary, the six sum or difference identities developed in this section are listed below in consolidated form. Read the upper signs for the sum formula and the lower signs for the difference formula.

Sum or Difference Formulas

$$
\begin{aligned}
\sin(x_1 \pm x_2) &= \sin x_1 \cos x_2 \pm \cos x_1 \sin x_2 \\
\cos(x_1 \pm x_2) &= \cos x_1 \cos x_2 \mp \sin x_1 \sin x_2] \\
\tan(x_1 \pm x_2) &= \frac{\tan x_1 \pm \tan x_2}{1 \mp \tan x_1 \tan x_2}
\end{aligned}
$$

EXPLORE 8.8

1. a. Many students assume that $\sin(a + b)$ is identically equal to $\sin a + \sin b$. Use a calculator to complete this table, and describe what the table reveals about this assumption.

a	b	$\sin(a+b)$	$\sin a + \sin b$
0	0		
0	−6		
10	0		
1	2		
1	−1		

b. How many counterexamples are necessary to prove that an equation is not an identity?

2. **a.** Check by calculator to see which of these statements are true:

i. $\sin(1+2) = \sin 1 \cos 2 - \cos 1 \sin 2$ ii. $\sin(1+2) = \sin 1 \cos 2 + \cos 1 \sin 2$

iii. $\cos(1+2) = \cos 1 \cos 2 + \sin 1 \sin 2$ iv. $\sin(1+2) = \cos 1 \cos 2 - \sin 1 \sin 2$

b. On the basis of your investigation in part **a**, guess what might be the right-hand side of these two identities:

i. $\sin(a+b) =$ ii $\cos(a+b) =$

EXERCISES 8.8

In Exercises 1–6 verify the identity. Use the formulas for the cosine of either the sum or difference of two numbers

1. $\cos\left(\frac{\pi}{2} + x\right) = -\sin x$ **2.** $\cos(\pi + x) = -\cos x$

3. $\cos(\pi - x) = -\cos x$ **4.** $\cos\left(\frac{3\pi}{2} - x\right) = -\sin x$

5. $\cos\left(x - \frac{\pi}{2}\right) = \sin x$ **6.** $\cos(x - \pi) = -\cos x$

In Exercises 7–12 verify the identity. Use the formulas for the sine of either the sum or difference of two numbers.

7. $\sin\left(\frac{\pi}{2} - x\right) = \cos x$ **8.** $\sin\left(\frac{\pi}{2} + x\right) = \cos x$

9. $\sin(\pi + x) = -\sin x$ **10.** $\sin(\pi - x) = \sin x$

11. $\sin\left(\frac{3\pi}{2} - x\right) = -\cos x$

12. $\sin\left(\frac{3\pi}{2} + x\right) = -\cos x$

In Exercises 13–16 verify the identity. Use the formulas for the tangent of either the sum or difference of two numbers.

13. $\tan(\pi + x) = \tan x$ **14.** $\tan(2\pi + x) = \tan x$

15. $\tan(2\pi - x) = -\tan x$ **16.** $\tan(\pi - x) = -\tan x$

In Exercises 17–20 verify the cofunction identity.

17. $\sin(90° - \pi) = \cos\theta$ **18.** $\cos(90° - \pi) = \sin\theta$

19. $\sec(90° - \pi) = \csc\theta$ **20.** $\csc(90° - \pi) = \sec\theta$

In Exercises 21–26 verify the identity.

21. $\sqrt{2}\sin\left(x - \frac{\pi}{4}\right) = \sin x - \cos x$

22. $\sqrt{2}\sin\left(x + \frac{\pi}{4}\right) = \sin x + \cos x$

23. $\cos\left(x + \frac{\pi}{3}\right) = \frac{1}{2}\left(\cos x - \sqrt{3}\sin x\right)$

24. $\cos\left(x - \frac{\pi}{6}\right) = \frac{1}{2}\left(\sqrt{3}\cos x + \sin x\right)$

25. $\tan\left(x + \frac{\pi}{4}\right) = \frac{1 + \tan x}{1 - \tan x}$

26. $\tan\left(x - \frac{\pi}{4}\right) = \frac{\tan x - 1}{\tan x + 1}$

27. If $f(x) = 4\sin x$, show that
$$\frac{f(x+h) - f(x)}{h} = 4\sin x\left(\frac{\cos h - 1}{h}\right) + 4\cos x\left(\frac{\sin h}{h}\right).$$

28. If $f(x) = 8\cos x$, show that
$$\frac{f(x+h) - f(x)}{h} = 8\cos x\left(\frac{\cos h - 1}{h}\right) - 8\sin x\left(\frac{\sin h}{h}\right).$$

29. If $f(x) = \tan x$, show that
$$\frac{f(x+h) - f(x)}{h} = \frac{\tan h}{h} \cdot \frac{\sec^2 x}{1 - \tan x \tan h}.$$

30. If $f(x)=3\tan x$, show that

$$\frac{f(x+h)-f(x)}{h}=\frac{\tan h}{h}\cdot\frac{3\sec^2 x}{1-\tan x\tan h}.$$

In Exercises 31–38 find exact function values.

31. $\sin 75°$ **32.** $\cos 15°$ **33.** $\cos 255°$

34. $\tan 105°$ **35.** $\sin\left(\frac{\pi}{12}\right)$ **36.** $\cos\left(\frac{5\pi}{12}\right)$

37. $\tan\left(\frac{13\pi}{12}\right)$ **38.** $\sin\left(\frac{11\pi}{12}\right)$

39. If $\sin x_1=\frac{1}{2}$, where $0<x_1<\frac{\pi}{2}$, and $\cos x_2=-\frac{\sqrt{3}}{2}$, where $\frac{\pi}{2}<x<\pi$, find

a. $\cos x_1$ **b.** $\sin x_2$
c. $\cos(x_1+x_2)$ **d.** $\sin(x_1-x_2)$

40. If $\cos x_1=\frac{3}{5}$, where $\frac{3\pi}{2}<x_1<2\pi$, and $\sin x_2=-\frac{5}{13}$, where $\frac{3\pi}{2}<x_2<2\pi$, find

a. $\sin x_1$ **b.** $\cos x_2$
c. $\cos(x_1-x_2)$ **d.** $\sin(x_1+x_2)$

41. If $\sin x_1=\frac{1}{2}$, where $0<x_1<\frac{\pi}{2}$, and $\cos x_2=-\frac{1}{2}$, where $\pi<x_2<\frac{3\pi}{2}$, find

a. $\cos x_1$ **b.** $\sin x_2$
c. $\cos(x_1+x_2)$ **d.** $\sin(x_1-x_2)$

42. If $\cos x_1=\frac{\sqrt{5}}{5}$, where $0<x<\frac{\pi}{2}$, and $\sin x_2=\frac{\sqrt{5}}{5}$, where $\frac{\pi}{2}<x_2<\pi$, find

a. $\sin x_1$ **b.** $\cos x_2$
c. $\cos(x_1-x_2)$ **d.** $\sin(x_1+x_2)$

In Exercises 43–48 find exact function values if $\sin\alpha=\frac{12}{13}$ and $\cos\beta=-\frac{7}{25}$, where α is a first-quadrant angle and β is a second-quadrant angle.

43. $\sin(\alpha+\beta)$ **44.** $\sin(\alpha-\beta)$

45. $\cos(\alpha-\beta)$ **46.** $\cos(\alpha+\beta)$

47. $\tan(\alpha+\beta)$ **48.** $\tan(\alpha-\beta)$

In Exercises 49–52 find exact function values. Use a sum or difference formula and right triangles.

49. $\cos\left(\sin^{-1}\frac{3}{5}-\cos^{-1}\frac{24}{25}\right)$

50. $\cos\left(\arctan\frac{4}{3}+\arcsin\frac{3}{5}\right)$

51. $\sin\left[\arctan\left(-\frac{4}{3}\right)-\arccos\frac{12}{13}\right]$

52. $\sin\left[\sin^{-1}\frac{8}{17}+\cos^{-1}\left(-\frac{12}{13}\right)\right]$

In Exercises 53 and 54 use $\tan x=\frac{\sin x}{\cos x}$ to verify the identity.

53. $\tan(x_1+x_2)=\frac{\tan x_1+\tan x_2}{1-\tan x_1\tan x_2}$

54. $\tan\left(\frac{\pi}{2}+x\right)=-\cot x$

In Exercises 55 and 56 verify the identity. Use addition formulas, writing x_1-x_2 as $x_1+(-x_2)$.

55. $\sin(x_1-x_2)=\sin x_1\cos x_2-\cos x_1\sin x_2$

56. $\tan(x_1-x_2)=\frac{\tan x_1-\tan x_2}{1+\tan x_1\tan x_2}$

In Exercises 57 and 58 verify the identity. Use addition formulas, writing $2x$ as $x+x$.

57. $\cos 2x=\cos^2 x-\sin^2 x$

58. $\sin 2x=2\sin x\cos x$

59. Show that $\sin(x+y)+\sin(x-y)=2\sin x\cos y$ is an identity.

60. Show that

$$\sin a+\sin b=2\sin\frac{a+b}{2}\cos\frac{a-b}{2}$$

is an identity. **[Hint:** Start with the identity in Exercise 59, and let $x=\frac{a+b}{2}$ and $y=\frac{a-b}{2}$.]

61. If $\tan(x+y)=2$ and $\tan x=1$, find $\tan y$.

62. Show that angle C equals the sum of angles A and B in this diagram by showing that $\tan C=\tan(A+B)$.

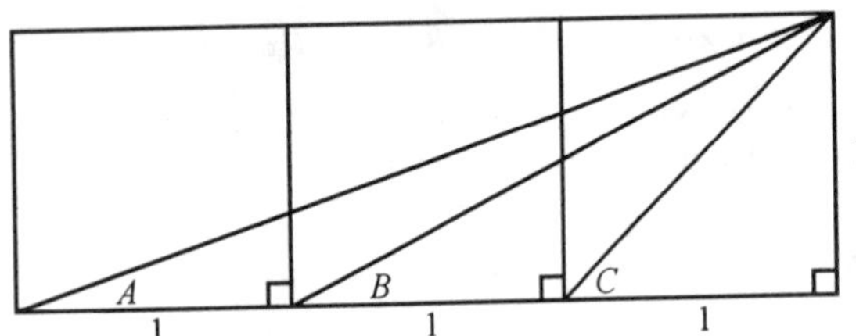

63. Show that $\cos(-x) = \cos x$ by expanding $\cos(0 - x)$ using the formula for the cosine of a difference. This result shows that the cosine function is an even function.

64. Show that $\sin\left(\frac{\pi}{2} - x\right) = \cos x$. (**Hint:** Write $\cos x$ as $\cos\left[\frac{\pi}{2} - \left(\frac{\pi}{2} - x\right)\right]$ and use the formula for the cosine of the difference of two numbers.)

65. A 10 ft high billboard is mounted on the roof of a 50 ft building, as shown in the sketch. The viewing angle θ is a function of the distance x from the building.

   a. Show that $\theta = \tan^{-1}\left(\frac{10x}{x^2 + 3{,}000}\right)$.

   b. Use a grapher in Degree mode to graph the relation in part **a**.

   c. What value of x produces the maximum value for θ?

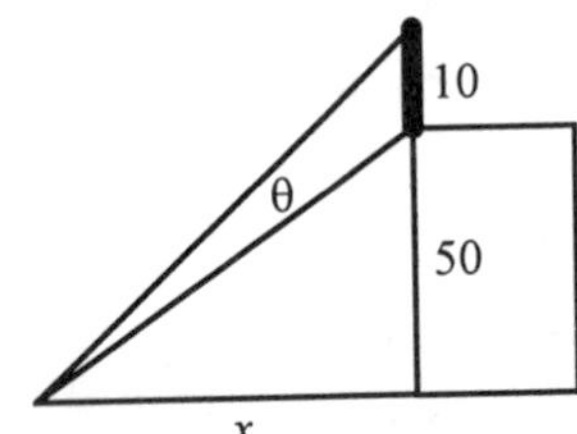

66. Suppose the billboard in Exercise 65 is rebuilt so that it is 15 ft high.

   a. Show that $\theta = \tan^{-1}\left(\frac{15x}{x^2 + 3{,}250}\right)$.

   b. Use a grapher in Degree mode to graph the relation in part **a**.

   c. What value of x produces the maximum value for θ?

67. In this sketch of a mounted painting in an art gallery, the eye level of the viewer is between the top and bottom of the painting. This implies that the viewing angle increases as the viewer approaches the wall.

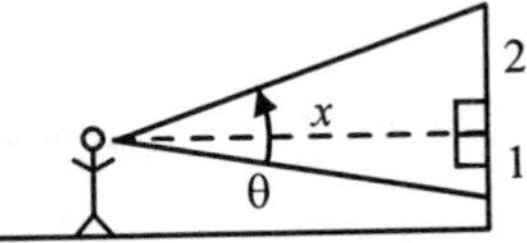

   a. Show that $\theta = \tan^{-1}\left(\frac{3x}{x^2 - 2}\right)$.

   b. Use a grapher in Degree mode to graph the relation in part **a**.

   c. Confirm from the diagram that when $x = \sqrt{2}$ ft, then $\theta = 90°$, and note that when x is less than $\sqrt{2}$, then θ is between 90° and 180°. Why will the graph in part **b** not show this result properly? [**Hint:** What is the range of the inverse tangent function?] For instance, when $x = 1$ ft, the equation in part **a** yields $\theta = -71.6°$, and not some angle between 90° and 180°. What is the correct value of θ when $x = 1$?

68. Suppose that the painting in Exercise 67 is lowered $\frac{1}{2}$ ft, and that the same viewer observes it.

   a. Show that $\theta = \tan^{-1}\left(\frac{3x}{x^2 - 1.5^2}\right)$.

   b. Use a grapher in Degree mode to graph the relation in part **a**.

   c. Confirm that when $x = 1.5$ ft, then $\theta = 90°$. Find θ when $x = 1$ ft. [Read the discussion in Exercise 67, if necessary.]

THINK ABOUT IT 8.8

1. **a.** Give an example of specific values for x_1 and x_2 such that $\cos(x_1 + x_2)$ does not equal $\cos x_1 + \cos x_2$. Verify your answer without using a calculator.

   b. Give an example of specific values for x_1 and x_2 such that $\cos(x_1 + x_2)$ does equal $\cos x_1 + \cos x_2$. Verify your answer without using a calculator.

2. Determine the exact value of the expression $\sin(3/10)\pi\cos(1/20)\pi - \cos(3/10)\pi\sin(1/20)\pi$.

3. Show that $\sin(A + B) = \sin C$ if A, B, and C are the angle measures in a triangle.

4. There are many ways to establish the sum and difference formulas for sine and cosine. Here is one unusual way to derive the formula for $\cos(A - B)$. It involves drawing the two angles, A and B, in two different arrangements, finding a certain distance both times, and setting the two results equal to one another. [Source: E. J. McShane, "The Addition Formulas for the Sine and Cosine", in *Selected Papers on Precalculus*, MA]

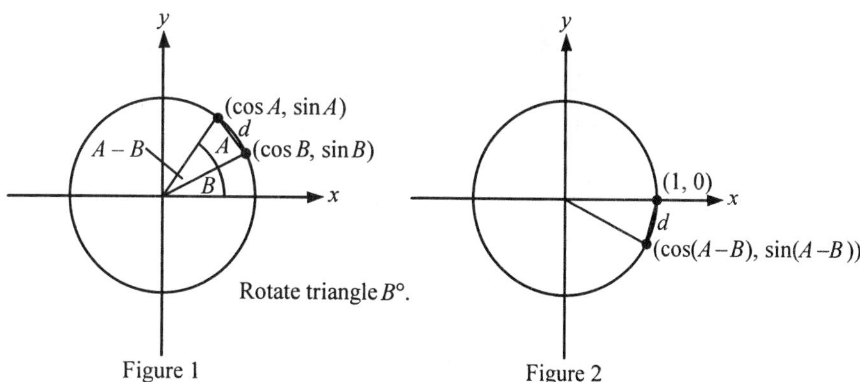

Figure 1 Figure 2

a. From Figure 1 show that $d^2 = (\cos A - \cos B)^2 + (\sin A - \sin B)^2$. Square both terms and simplify the result.

b. From Figure 2 show that $d^2 = (\cos(A - B) - 1)^2 + \sin^2(A - B)$. Square the first term and simplify the result.

c. Equate the results of parts **a** and **b**, and solve for $\cos(A - B)$.

5. Use the diagram below and show that if α, β, and $\alpha + \beta$ are acute angles, then

$$\sin(\alpha + \beta) = \sin\alpha\cos\beta + \cos\alpha\sin\beta.$$

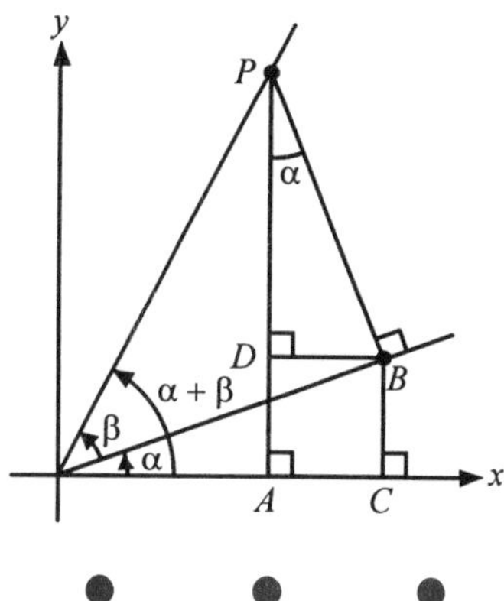

• • •

8.9 Multiple-Angle Formulas

Photo Courtesy of **Lora E. Askinazi**, The Picture Cube

A formula for the horizontal distance traveled by a projectile, neglecting air resistance, is

$$d = V^2 \frac{\sin\theta\cos\theta}{16},$$

where θ and V measure the angle of elevation and the initial velocity in ft/second of the projectile, respectively. Use a multiple-angle formula to show that an alternate form of this formula is

$$d = \frac{1}{32}V^2 \sin 2\theta.$$

(See Example 2.)

Objectives

1. Evaluate functions involving double-angles or half-angles, given a trigonometric function of x and the interval containing x.
2. Verify certain identities by using the double-angle, half-angle, or power reduction formulas.
3. Use double-angle formulas to solve certain trigonometric equations.
4. Use the half-angle formulas to find exact function values of certain angles or numbers.
5. Use the double-angle or half-angle identities to simplify certain trigonometric expressions.
6. Derive multiple-angle formulas for $\sin 3x$, $\cos 3x$, $\sin 4x$, or $\cos 4x$.

The question posed in the section-opening problem requires that a multiple-angle identity be used to show that two formulas concerning projectile motion are equivalent.

Multiple-angle formulas refer to identities for $\sin kx$, $\cos kx$, and $\tan kx$, where k is a positive rational number. The most useful formulas of this type result for $k = 2$ and are called the double-angle formulas.

Double-Angle Formulas

$$\sin 2x = 2\sin x\cos x$$

$$\cos 2x = \cos^2 x - \sin^2 x = 2\cos^2 x - 1 = 1 - 2\sin^2 x$$

$$\tan 2x = \frac{2\tan x}{1 - \tan^2 x}$$

To prove these formulas, we use the addition identities from the preceding section, rewriting $2x$ as $x + x$. For example, a formula for $\cos 2x$ may be derived as follows.

$$\begin{aligned}\cos 2x &= \cos(x + x) = \cos x\cos x - \sin x\sin x \\ &= \cos^2 x - \sin^2 x\end{aligned}$$

There are two alternative forms that express $\cos 2x$ in terms of $\cos x$ or in terms of $\sin x$. By solving the identity $\sin^2 x + \cos^2 x = 1$ for $\sin^2 x$ it follows that $\sin^2 x = 1 - \cos^2 x$. Thus,

$$\begin{aligned}\cos 2x &= \cos^2 x - \sin^2 x\\ &= \cos^2 x - \left(1-\cos^2 x\right)\\ &= 2\cos^2 x - 1.\end{aligned}$$

Similarly, because $\cos^2 x = 1-\sin^2 x$, we have

$$\begin{aligned}\cos 2x &= \cos^2 x - \sin^2 x\\ &= \left(1-\sin^2 x\right) - \sin^2 x\\ &= 1-2\sin^2 x.\end{aligned}$$

Formulas for sin $2x$ and tan $2x$ may be derived by similar methods, and these derivations are requested in Exercises 83 and 84.

Example 1: Using Double-Angle Formulas to Evaluate Functions

If $\sin x = \frac{3}{5}$ and $\frac{\pi}{2} < x < \pi$, find

a. cos $2x$ **b.** sin $2x$ **c.** tan $2x$

Solution:

a. Because sin x is known, use the formula that expresses cos $2x$ in terms of sin x.

$$\cos 2x = 1-2\sin^2 x = 1-2\left(\frac{3}{5}\right)^2 = \frac{7}{25}$$

b. To find sin $2x$, both sin x and cos x must be known. First, determine $\cos x = -4/5$ as shown in Example 10 of Section 8.7 (or use the methods of Section 2.3). Then,

$$\sin 2x = 2\sin x\cos x = 2\left(\frac{3}{5}\right)\left(-\frac{4}{5}\right) = -\frac{24}{25}.$$

c. sin $2x$ and cos $2x$ are now known, so the easiest way to find tan $2x$ is to use a fundamental identity.

$$\tan 2x = \frac{\sin 2x}{\cos 2x} = \frac{-24/25}{7/25} = -\frac{24}{7}$$

If sin $2x$ or cos $2x$ were unknown, then we could use the double-angle formula for tan $2x$. From $\sin x = 3/5$ and $\cos x = -4/5$, it follows that $\tan x = -3/4$. Then,

$$\tan 2x = \frac{2\tan x}{1-\tan^2 x} = \frac{2(-3/4)}{1-(-3/4)^2} = \frac{-24}{16-9} = -\frac{24}{7}.$$

PROGRESS CHECK 1

If $\cos x = \frac{5}{13}$ and $\frac{3\pi}{2} < x < 2\pi$, find

a. cos $2x$ **b.** sin $2x$ **c.** tan $2x$ ■

Example 2: Projectile Motion

Solve the problem in the section introduction on page 710.

Solution: The formula that is given is

$$d = V^2 \frac{\sin\theta\cos\theta}{16}$$

In this formula the factor $\sin\theta\cos\theta$ suggests the double-angle identity for $\sin 2\theta$, which is $\sin 2\theta = 2\sin\theta\cos\theta$. So

$$d = V^2 \frac{\sin\theta\cos\theta}{16} = V^2\left(\frac{2\sin\theta\cos\theta}{2\cdot 16}\right) = \frac{1}{32}V^2\sin 2\theta,$$

which establishes the alternative formula stated in the question.

PROGRESS CHECK 2

The relation in Example 2 is sometimes described by

$$d = \frac{V^2}{16}\sin A\sin B,$$

where A is the angle of elevation of the projectile and $B = 90° - A$. Show that an alternate form of this formula is

$$d = \frac{1}{32}V^2\sin 2A.$$

■

Example 3: Using Double-Angle Formulas to Verify an Identity

Verify the identity $\cos^2 x = \dfrac{1+\cos 2x}{2}$.

Solution: Because the left-hand member of the identity is $\cos^2 x$, use the formula that expresses cos 2*x* in terms of cos *x*.

$$\begin{aligned}\frac{1+\cos 2x}{2} &= \frac{1+\left(2\cos^2 x - 1\right)}{2}\\ &= \frac{2\cos^2 x}{2}\\ &= \cos^2 x\end{aligned}$$

PROGRESS CHECK 3

Verify the identity $\dfrac{2\tan x}{1+\tan^2 x} = \sin 2x$ ■

Example 4: Using Double-Angle Identities to Solve an Equation

Solve $\cos 2x + \cos x = 0$ for $0 \le x < 2\pi$.

Solution: The solutions of the equation $\cos 2x + \cos x = 0$ are the same as the *x*-coordinates of the *x*-intercepts of the graph of $y = \cos 2x + \cos x$. So Figure 8.93 shows that the equation has three solutions in the interval $[0, 2\pi)$. To find these solutions algebraically, use an identity for cos 2*x* to rewrite the equation in terms of only cos *x*, and then solve the resulting equation by the methods shown in Section 8.6.

$$\cos 2x + \cos x = 0$$
$$\left(2\cos^2 x - 1\right) + \cos x = 0$$
$$2\cos^2 x + \cos x - 1 = 0$$
$$(2\cos x - 1)(\cos x + 1) = 0$$
$$2\cos x - 1 = 0 \quad \text{or} \quad \cos x + 1 = 0$$
$$\cos x = \frac{1}{2} \quad \text{or} \quad \cos x = -1$$

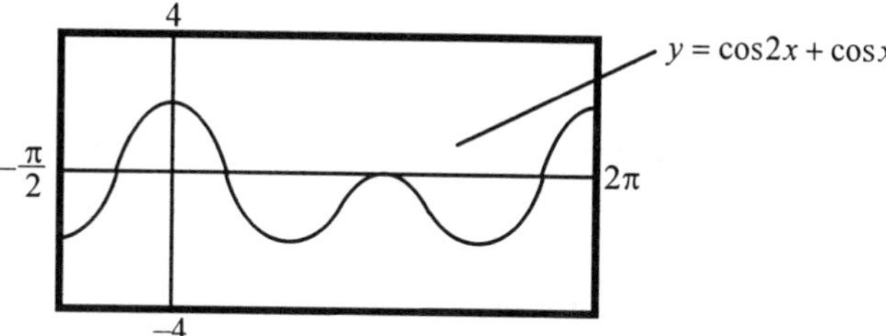

Figure 8.93

In the interval $[0, 2\pi)$, $\cos x = 1/2$ when $x = \pi/3$ and $x = 5\pi/3$ (see Example 3 of Section 8.6), while $\cos x = -1$ when $x = \pi$. Thus, the solution set in $[0, 2\pi)$ is $\{\pi/3, 5\pi/3, \pi\}$. You should look at Figure 8.93 once again to confirm that this solution set is reasonable.

PROGRESS CHECK 4

The x-coordinates of the relative maximum and minimum points in the graph in Figure 8.93 are the solutions of the equation $-2\sin 2x - \sin x = 0$. Find these x-coordinates in the interval $[0, 2\pi)$. ■

The identity in Example 3 may also be derived by solving for $\cos^2 x$ in the identity

$$\cos 2x = 2\cos^2 x - 1.$$

Similarly, solving for $\sin^2 x$ in the identity $\cos 2x = 1 - 2\sin^2 x$ yields

$$\sin^2 x = \frac{1 - \cos 2x}{2}.$$

Thus, the following identities, called power reduction formulas, are alternative forms of $\cos 2x$ formulas.

Power Reduction Formulas

$$\sin^2 x = \frac{1 - \cos 2x}{2}$$
$$\cos^2 x = \frac{1 + \cos 2x}{2}$$

An important application of these formulas in calculus is the simplification of powers of sine and cosine, as shown in the next example.

Example 5: Using Power Reduction Formulas

Verify the identity $\cos^4 x = \frac{3}{8} + \frac{1}{2}\cos 2x + \frac{1}{8}\cos 4x$.

Solution: Rewrite $\cos^4 x$ in terms of $\cos^2 x$, use the power reduction formula for $\cos^2 x$, and then perform the indicated operations.

$$\begin{aligned}\cos^4 x &= \left(\cos^2 x\right)^2 \\ &= \left(\frac{1+\cos 2x}{2}\right)^2 \\ &= \frac{1}{4}\left(1+2\cos 2x+\cos^2 2x\right)\end{aligned}$$

Now $\cos^2 2x = (1+\cos 4x)/2$ by the power reduction formula for $\cos^2 x$, with x replaced by $2x$. Thus,

$$\begin{aligned}\frac{1}{4}\left(1+2\cos 2x+\cos^2 2x\right) &= \frac{1}{4}\left(1+2\cos 2x+\frac{1+\cos 4x}{2}\right) \\ &= \frac{1}{4}+\frac{1}{2}\cos 2x+\frac{1}{8}+\frac{1}{8}\cos 4x \\ &= \frac{3}{8}+\frac{1}{2}\cos 2x+\frac{1}{8}\cos 4x.\end{aligned}$$

Note that the right-hand member of the identity is simpler, in the sense that the *fourth power* of $\cos x$ is expressed in terms of the *first powers* of the cosine of multiples of x.

PROGRESS CHECK 5

Verify the identity $\sin^2 4x \cos^2 4x = \frac{1}{8}(1-\cos 16x)$. ■

A second group of multiple-angle identities, are the half-angle formulas. In these formulas, which follow, the correct sign is determined by the quadrant in which $x/2$ lies for identities involving the plus or minus sign.

Half-Angle Formulas

$$\sin\frac{x}{2} = \pm\sqrt{\frac{1-\cos x}{2}} \qquad \cos\frac{x}{2} = \pm\sqrt{\frac{1+\cos x}{2}}$$

$$\tan\frac{x}{2} = \pm\sqrt{\frac{1-\cos x}{1+\cos x}} = \frac{1-\cos x}{\sin x} = \frac{\sin x}{1+\cos x}$$

The derivations of the identities for $\sin(x/2)$ and $\cos(x/2)$ are applications of the power reduction formulas, with x replaced by $x/2$. Thus,

$$\sin^2\frac{x}{2} = \frac{1-\cos 2(x/2)}{2} \quad \text{implies} \quad \sin\frac{x}{2} = \pm\sqrt{\frac{1-\cos x}{2}}$$

$$\cos^2\frac{x}{2} = \frac{1+\cos 2(x/2)}{2} \quad \text{implies} \quad \cos\frac{x}{2} = \pm\sqrt{\frac{1+\cos x}{2}}.$$

Question 5 in the "Think About It" exercises will lead you through derivations of the formulas for $\tan(x/2)$.

Example 6: Using Half-Angle Formulas

Find the exact values of $\sin 165°$ and $\cos 165°$.

Solution: $165° = 1/2(330°)$, and we know that $\cos 330° = \sqrt{3}/2$ by the methods of Section 2.3 or 8.2. Because $165°$ lies in Q_2, where $\sin\theta > 0$ and $\cos\theta < 0$, choose the positive square root to find $\sin 165°$ and the negative square root to find $\cos 165°$ in the half-angle formulas.

$$\sin 165° = \sin\frac{330°}{2} = \sqrt{\frac{1-\cos 330°}{2}}$$
$$= \sqrt{\frac{1-\sqrt{3}/2}{2}} = \frac{\sqrt{2-\sqrt{3}}}{2}$$
$$\cos 165° = \cos\frac{330°}{2} = -\sqrt{\frac{1+\cos 330°}{2}}$$
$$= -\sqrt{\frac{1+\sqrt{3}/2}{2}} = \frac{-\sqrt{2+\sqrt{3}}}{2}$$

PROGRESS CHECK 6

Find the exact values of $\sin\frac{\pi}{12}$ and $\cos\frac{\pi}{12}$. ■

Example 7: Simplifying Expressions Using Multiple-Angle Formulas

Use the double-angle or half-angle identities to write each expression as a single function of a multiple angle.

a. $\dfrac{\sin 6t}{1+\cos 6t}$ **b.** $\sin 5\theta\cos 5\theta$ **c.** $5\cos^2\frac{x}{8} - 5\sin^2\frac{x}{8}$

Solution:

a. The given expression suggests the half-angle identity

$$\tan\frac{x}{2} = \frac{\sin x}{1+\cos x}.$$

Replacing x with $6t$, we have

$$\frac{\sin 6t}{1+\cos 6t} = \tan\frac{6t}{2} = \tan 3t.$$

b. Use the double-angle identity $\sin 2x = 2\sin x\cos x$, with $x = 5\theta$, as follows.

$$\sin 5\theta\cos 5\theta = \left(\frac{1}{2}\cdot 2\right)\sin 5\theta\cos\theta$$
$$= \frac{1}{2}(2\sin 5\theta\cos 5\theta)$$
$$= \frac{1}{2}\sin(2\cdot 5\theta)$$
$$= \frac{1}{2}\sin 10\theta$$

c. Factor out the common factor 5, and then apply the double-angle identity $\cos 2x = \cos^2 x - \sin^2 x$, with $x/8$ in place of x.

$$\begin{aligned} 5\cos^2 \frac{x}{8} - 5\sin^2 \frac{x}{8} &= 5\left(\cos^2 \frac{x}{8} - \sin^2 \frac{x}{8}\right) \\ &= 5\cos\left(2 \cdot \frac{x}{8}\right) \\ &= 5\cos\frac{x}{4} \end{aligned}$$

PROGRESS CHECK 7

Use the double-angle or half-angle identities to write each expression as a single function of a multiple angle.

a. $\cos^2 3x - \sin^2 3x$ **b.** $10\sin 3\theta \cos 3\theta$ **c.** $\dfrac{1-\cos 4t}{\sin 4t}$ ■

Besides the double-angle and half-angle formulas, there are other multiple-angle identities that are sometimes useful. The next example shows how we may derive a triple-angle formula.

Example 8: Deriving Other Multiple-Angle Formulas

Derive a formula for $\cos 3x$ in terms of $\cos x$.

Solution: Use the formula for the cosine of the sum of two angles, rewriting $3x$ as $2x + x$. Then express the result in terms of $\cos x$ by using the double-angle formulas for $\cos 2x$ and $\sin 2x$ and replacing $\sin^2 x$ with $1 - \cos^2 x$.

$$\begin{aligned} \cos 3x &= \cos(2x + x) \\ &= \cos 2x \cos x - \sin 2x \sin x \\ &= \left(2\cos^2 x - 1\right)\cos x - (2\sin x \cos x)\sin x \\ &= 2\cos^3 x - \cos x - 2\cos x \sin^2 x \\ &= 2\cos^3 x - \cos x - 2\cos x\left(1 - \cos^2 x\right) \\ &= 2\cos^3 x - \cos x - 2\cos x + 2\cos^3 x \\ &= 4\cos^3 x - 3\cos x \end{aligned}$$

Thus, $\cos 3x = 4\cos^3 x - 3\cos x$ is a triple-angle formula for $\cos 3x$. You should test this result on a grapher by checking that $y = \cos 3x$ and $y = 4\cos^3 x - 3\cos x$ have identical graphs.

PROGRESS CHECK 8

Show that $\sin 4x = 8\cos^3 x \sin x - 4\sin x \cos x$ is an identity. ■

Finally, the major identities we have considered are summarized below. We have derived many of these formulas, while the verifications of others have been left to the exercises. For reference purposes, we include on this list the *product-to-sum formulas* and *sum-to-product formulas* that have occasional uses in higher mathematics. These formulas are derived from the sum or difference formulas, and you should be able to specify these formulas by consulting a reference source if the need arises.

Summary of Trigonometric Identities

Fundamental Identities

$\csc x = \dfrac{1}{\sin x}$ $\quad \sec x = \dfrac{1}{\cos x}$ $\quad \cot x = \dfrac{1}{\tan x}$

$\tan x = \dfrac{\sin x}{\cos x}$ $\quad \cot x = \dfrac{\cos x}{\sin x}$

$$\left.\begin{aligned} &\sin^2 x + \cos^2 x = 1 \\ &\tan^2 x + 1 = \sec^2 x \\ &\cot^2 x + 1 = \csc^2 x \end{aligned}\right\} \textbf{Pythagorean Identities}$$

Negative Angle Formulas

$\sin(-x) = -\sin x \quad \cos(-x) = \cos x$

$\tan(-x) = -\tan x$

Double-Angle Formulas

$\sin 2x = 2\sin x \cos x$

$\tan 2x = \dfrac{2\tan x}{1 - \tan^2 x}$

$$\begin{aligned} \cos 2x &= \cos^2 x - \sin^2 x \\ &= 2\cos^2 x - 1 \\ &= 1 - 2\sin^2 x \end{aligned}$$

Half-Angle Formulas

$\sin\dfrac{x}{2} = \pm\sqrt{\dfrac{1-\cos x}{2}} \qquad \cos\dfrac{x}{2} = \pm\sqrt{\dfrac{1+\cos x}{2}}$

$\tan\dfrac{x}{2} = \pm\sqrt{\dfrac{1-\cos x}{1+\cos x}} = \dfrac{1-\cos x}{\sin x} = \dfrac{\sin x}{1+\cos x}$

Power Reduction Formulas

$\sin x = \dfrac{1-\cos 2x}{2}, \ \cos^2 x = \dfrac{1+\cos 2x}{2}$

Sum or Difference Formulas

$\sin(x_1 \pm x_2) = \sin x_1 \cos x_2 \pm \cos x_1 \sin x_2$

$\cos(x_1 \pm x_2) = \cos x_1 \cos x_2 \mp \sin x_1 \sin x_2$

$\tan(x_1 \pm x_2) = \dfrac{\tan x_1 \pm \tan x_2}{1 \mp \tan x_1 \tan x_2}$

(**Note:** Read the upper signs for the sum formula and the lower signs for the difference formula.)

Product-to-Sum Formulas

$\sin x \cos y = \dfrac{1}{2}[\sin(x-y) + \sin(x+y)]$

$\sin x \sin y = \dfrac{1}{2}[\cos(x-y) - \cos(x+y)]$

$\cos x \cos y = \dfrac{1}{2}[\cos(x-y) + \cos(x+y)]$

Sum-to-Product Formulas

$\sin A + \sin B = 2\sin\dfrac{A+B}{2}\cos\dfrac{A-B}{2}$

$\sin A - \sin B = 2\cos\dfrac{A+B}{2}\sin\dfrac{A-B}{2}$

$\cos A + \cos B = 2\cos\dfrac{A+B}{2}\cos\dfrac{A-B}{2}$

$\cos A - \cos B = -2\sin\dfrac{A+B}{2}\sin\dfrac{A-B}{2}$

EXPLORE 8.9

1. Some of the equations below are identities and some are not. By trying several values of x in each case, determine which of the equations are definitely not identities, and which may possibly be identities.
   a. $\sin 2x = 2\sin x$ **b.** $\cos 2x = 2\cos x$
   c. $\sin 2x = 2\sin x \cos x$ **d.** $\cos 2x = 2\sin x \cos x$
   e. $\sin 2x = \cos^2 x - \sin^2 x$ **f.** $\cos 2x = \cos^2 x - \sin^2 x$
2. If an equation $f(x) = g(x)$ is an identity, then the graphs of $y_1 = f(x)$ and $y_2 = g(x)$ will be the same. Furthermore, the graph of $y_3 = y_1/y_2 = f(x)/g(x)$ will be a horizontal line at $y = 1$, throughout the domain of the quotient function. Use these graphical techniques to determine which of the equations in the previous problem are identities.
3. Use a grapher to support the claim that these equations are identities.
   a. $\sin^2 x = \dfrac{1-\cos 2x}{2}$ **b.** $\cos^2 x = \dfrac{1+\cos 2x}{2}$

EXERCISES 8.9

In Exercises 1–6 find $\sin 2x$, $\cos 2x$, and $\tan 2x$ by using the double-angle formulas.

1. $\sin x = \frac{4}{5},\ 0 < x < \frac{\pi}{2}$

2. $\cos x = \frac{12}{13},\ \frac{3\pi}{2} < x < 2\pi$

3. $\csc x = -\frac{4}{3},\ \frac{3\pi}{2} < x < 2\pi$

4. $\sec x = -3,\ \pi < x < \frac{3\pi}{2}$

5. $\tan x = \frac{1}{2},\ \pi < x < \frac{3\pi}{2}$

6. $\cot x = -\sqrt{2},\ \frac{\pi}{2} < x < \pi$

In Exercises 7–16 show that the equation is an identity.

7. $\sin^2 x = \frac{1-\cos 2x}{2}$

8. $\frac{\sin 2x}{2\sin x} = \cos x$

9. $\csc x \sin 2x = 2\cos x$

10. $\cos^4 x - \sin^4 x = \cos 2x$

11. $\cos 2x + 2\sin^2 x = 1$

12. $\frac{2\cot x}{\csc^2 x} = \sin 2x$

13. $\frac{1+\cos 2x}{\sin 2x} = \cot x$

14. $\frac{\cos 2x}{\sin x} + \frac{\sin 2x}{\cos x} = \csc x$

15. $\frac{2\tan x}{2-\sec^2 x} = \tan 2x$

16. $\frac{2\cot x}{\cot^2 x - 1} = \tan 2x$

17. **a.** Solve $\cos 2x - \cos x = 0$ for $0 \le x < 2\pi$.
    b. The x-coordinates of the relative maximum and minimum points in the graph for part **a** are the solutions of the equation $\sin x - 2\sin 2x = 0$. Find those x-coordinates in the interval $[0, 2\pi)$.

18. **a.** Solve $\sin 2x - \cos x = 0$ for $0 \le x < 2\pi$.
    b. The x-coordinates of the relative maximum and minimum points in the graph for part **a** are the solutions of the equation $2\cos 2x + \sin x = 0$. Find these x-coordinates in the interval $[0, 2\pi)$.

In Exercises 19–26 use double-angle identities to solve for x in the interval $[0, 2\pi)$.

19. $\sin 2x + \sin x = 0$

20. $\sin 2x = \sin x$

21. $4\sin x \cos x = -1$

22. $4\sin x \cos x = \sqrt{3}$

23. $\cos 2x = \cos x$

24. $\sin x - \cos 2x = 0$

25. $\cos 2x = 2\sin x \cos x$

26. $\sin 2x - \cos 2x = 1$

In Exercises 27–36 use the power reduction formulas to verify each identity.

27. $2\cos^2 4\theta = 1 + \cos 8\theta$

28. $\frac{1}{2}(1-\sin\theta)^2 = \frac{3}{4} - \sin\theta - \frac{1}{4}\cos 2\theta$

29. $\sin^2 x \cos^2 x = \frac{1}{8}(1-\cos 4x)$

30. $\sin^2 2x \cos^2 2x = \frac{1}{8}(1-\cos 8x)$

31. $\sin^4 x = \frac{3}{8} - \frac{1}{2}\cos 2x + \frac{1}{8}\cos 4x$

32. $\cos^4 \frac{x}{2} = \frac{3}{8} + \frac{1}{2}\cos x + \frac{1}{8}\cos 2x$

33. $\sec^2 \frac{x}{2} = \frac{2}{1+\cos x}$

34. $\csc^2 \frac{x}{2} = \frac{2}{1-\cos x}$

35. $\cos^2 \frac{x}{2} = \frac{\sec x + 1}{2\sec x}$

36. $\sin^2 \frac{x}{2} = \frac{\tan x - \sin x}{2\tan x}$

In Exercises 37–40 find $\sin\frac{x}{2}$, $\cos\frac{x}{2}$, and $\tan\frac{x}{2}$ by using the half-angle formulas.

37. $\cos x = \frac{15}{17},\ \frac{3\pi}{2} < x < 2\pi$

38. $\sin x = \frac{3}{5},\ 0 < x < \frac{\pi}{2}$

39. $\csc x = -4,\ \pi < x < \frac{3\pi}{2}$

40. $\tan x = -1,\ \frac{\pi}{2} < x < \pi$

In Exercises 41–46 find the exact value of each expression by using the half-angle formulas.

41. $\sin 105°$ **42.** $\cos 105°$ **43.** $\cos\left(\frac{\pi}{8}\right)$

44. $\sin\left(\frac{\pi}{8}\right)$ **45.** $\tan\left(\frac{\pi}{12}\right)$ **46.** $\tan 165°$

47. Show that $\sin\frac{\pi}{12} = \frac{\sqrt{2-\sqrt{3}}}{2}$.

48. Show that $\cos\frac{\pi}{12} = \frac{\sqrt{2+\sqrt{3}}}{2}$.

49. Show that $\tan\left(\frac{7\pi}{12}\right) = -2-\sqrt{3}$.

50. Show that $\tan 67.5° = \sqrt{2}+1$

In Exercises 51–54 use the figure below and write each expression in terms of a, b, and c.

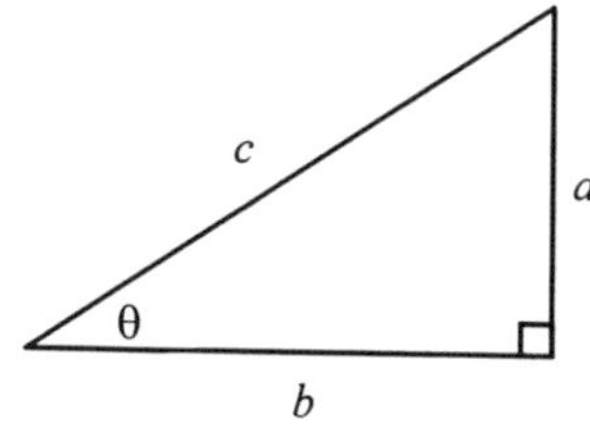

51. $\cos 2\theta$ **52.** $\sin 2\theta$

53. $\tan 2\theta$ **54.** $\tan\left(\frac{\theta}{2}\right)$

In Exercises 55–60 write each expression in terms of a if $\cos x = a$ and $\pi < x < \frac{3\pi}{2}$. Do not simplify radicals.

55. $\sin 2x$ **56.** $\cos 2x$ **57.** $\tan 2x$

58. $\tan\left(\frac{x}{2}\right)$ **59.** $\cos\left(\frac{x}{2}\right)$ **60.** $\sin\left(\frac{x}{2}\right)$

In Exercises 61–70 use the double-angle or half-angle identities to write each expression as a single function of a multiple angle.

61. $2\cos^2 6t - 1$ **62.** $2\sin 8\theta \cos\theta$

63. $\frac{\sin 4x}{1+\cos 4x}$ **64.** $\frac{1-\cos 6x}{\sin 6x}$

65. $\pm\sqrt{\frac{1+\cos 10\theta}{2}}$ **66.** $\frac{2\tan 5t}{1-\tan^2 5t}$

67. $\sin 2x \cos 2x$ **68.** $3\cos^2\frac{x}{4} - 3\sin^2\frac{x}{4}$

69. $4 - 8\sin^2 9t$ **70.** $\frac{6\sin 7\theta}{5+5\cos 7\theta}$

In Exercises 71–82 show that the equation is an identity.

71. $\sin 6x = 2\sin 3x \cos 3x$

72. $\cos 4x = \cos^2 2x - \sin^2 2x$

73. $10\cos^2\frac{x}{2} - 5 = 5\cos x$

74. $6\sin\frac{x}{2}\cos\frac{x}{2} = 3\sin x$

75. $\csc 2x = \frac{1}{2}(\cot x + \tan x)$

76. $\csc 2x = \frac{1}{2}\sec x \csc x$

77. $\cot 2x = \frac{1}{2}(\cot x - \tan x)$

78. $\cot 2x = \frac{\cot^2 x - 1}{2\cot x}$ **79.** $\sec 2x = \frac{1}{2\cos^2 x - 1}$

80. $\sec 2x = \frac{\sec^2 x}{2-\sec^2 x}$ **81.** $\tan\frac{x}{2} = \csc x - \cot x$

82. $\cot\frac{x}{2} = \csc x + \cot x$

83. Use the formula for $\sin(x_1 + x_2)$ to derive the double-angle formula for sin 2x.

84. Use the formula for $\tan(x_1 + x_2)$ to derive the double-angle formula for $\tan 2x$.

85. Derive a formula for $\sin 3x$ in terms of x.

86. Derive a formula for cos 4x in terms of cos x.

87. **a.** Since $\sin 2x = 2\sin x\cos x$, it follows that $\sin t = 2\sin\frac{t}{2}\cos\frac{t}{2}$. Use this relation and the given figure to show that the area of an isosceles triangle is given by $A = \frac{s^2}{2}\sin\theta$. Use the fact that the bisector of the vertex angle is perpendicular to the base, then express the base and height of the triangle in terms of $\sin\frac{\theta}{2}$ and $\cos\frac{\theta}{2}$.

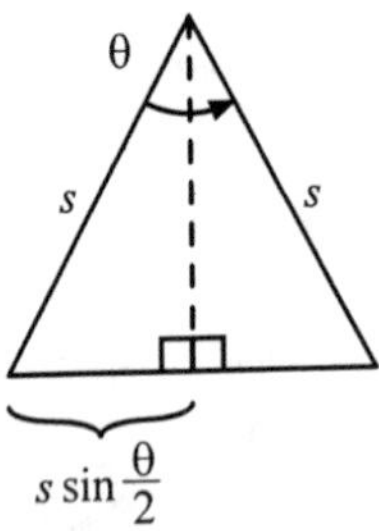

b. For a given value of s, what value of θ yields maximum area?

c. The triangle in part **a** is the cross-section of the prism shown in the figure. If the length of the prism is L, then use the results of part **a** to find a formula for the volume of the prism.

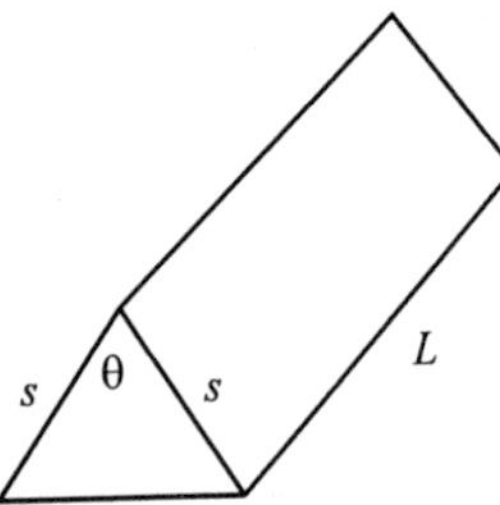

88. Demonstrate that the area of the right triangle shown is given both by $A = \frac{1}{2}\sin 2x \cos 2x$ and by $A = \sin x \cos x\left(1 - 2\sin^2 x\right)$.

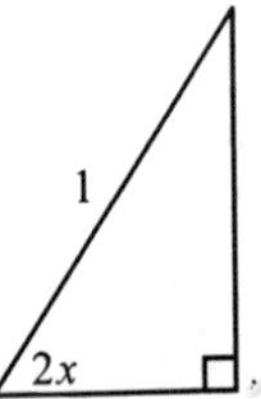

89. One approach to analyzing the surface area of a cylinder involves inscribing regular polygonal solids inside the cylinder. The diagram shown is useful in this analysis. Note that the length x is a function of θ, where $0° < \theta < 90°$.

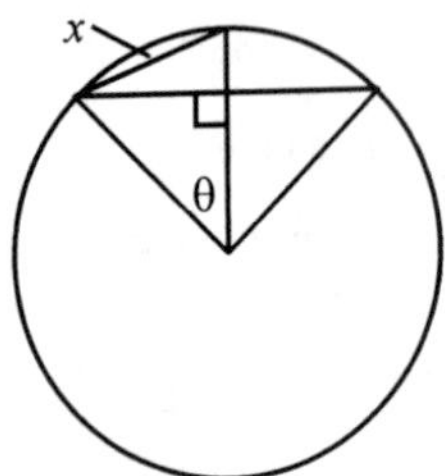

a. First show that $x = \sqrt{2 - 2\cos\theta}$, and then show that $x = \sin\left(\frac{\theta}{2}\right)$.

b. The line x in part a forms the base of a rectangle whose height is h. Use the results of part **a** to find a formula for the area of this rectangle.

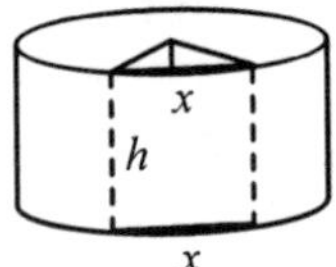

90. In the given sketch of $y = \cos 2x$, as the vertical line moves to the right from $x = 0$ towards $x = \frac{\pi}{4}$, the shaded area increases. When the line is at $x = t$, where $0 \le t \le \frac{\pi}{4}$, the shaded area is given by $A = \sin t \cos t$.

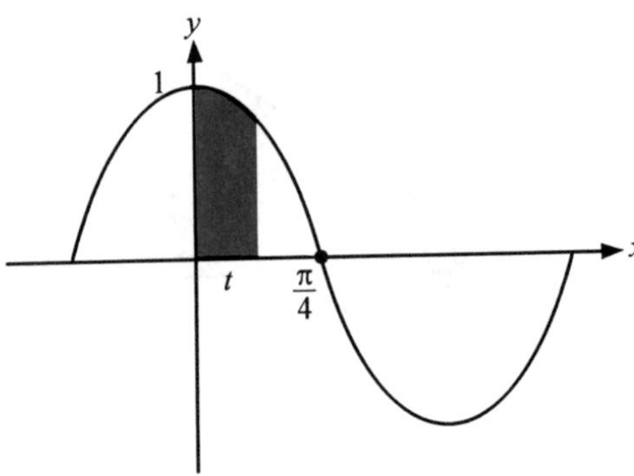

a. Find the shaded area when $t = \frac{\pi}{8}$, and when $t = \frac{\pi}{4}$.

b. Use a double-angle identity to express A as a function of the sine of $2t$. Then use this new expression to confirm the two answers in part *a*.

For Exercises 91–94: One approach to solving equations involves temporary substitution by variables that make the solution easier to find. The substitution $t = \tan\left(\frac{\theta}{2}\right)$ provides an example of this.

91. Suppose we let $t = \tan\left(\frac{\theta}{2}\right)$. Show that $\tan\theta = \frac{2t}{1-t^2}$.

Hint: Use the identity $\tan 2x = \frac{2\tan x}{1 - \tan^2 x}$.

92. Use the results of Exercise 91 to show

a. $\sin\theta = \dfrac{2t}{1+t^2}$ **b.** $\cos\theta = \dfrac{1-t^2}{1+t^2}$

93. Use the results of Exercise 92 to find all solutions to $5\sin\theta - 21\cos\theta = 10$ in the interval $[-\pi, \pi]$. **Hint:** Replace $\sin\theta$ and $\cos\theta$ by the appropriate expressions from Exercise 92 and solve for t. Then use the inverse tangent function to find $\dfrac{\theta}{2}$, and then find θ.

94. Use the results of Exercise 92 to find all solutions to $25\sin\theta - 10\cos\theta = 23$ in the interval $[-\pi, \pi]$.

For Exercises 95 to 98: If two strings on a musical instrument are slightly "out of tune" they vibrate with slightly different periods. Thus, one could be represented by $y = \cos Bt$ and the other by $y = \cos(B+\delta)t$, where δ is a small number. The vibration caused by sounding the two strings at the same time is given by the sum $y = \cos BT + \cos(B+\delta)t$. This vibration displays a *beating phenomenon*, which musicians can hear and use to improve the tuning. Whenever this wave reaches a maximum, there is said to be a *beat*.

95. Use the sum-to-product formula for the sum of two cosines to show that

$$y = \cos Bt + \cos(B+\delta)t$$

can be rewritten as

$$y = \left\{2\cos\left(\frac{\delta}{2}\right)t\right\}\cos\left(B + \frac{\delta}{2}\right)t\,.$$

96. The frequency of the musical note A is 440 cycles per second. Therefore, its period is $\dfrac{1}{440}$ second, and its sound wave could be represented as

$$y = \cos 880\pi t \quad \left(\text{since } \frac{2\pi}{1/440} = 880\pi\right).$$

Suppose another note is vibrating slightly faster with wave given by

$$y = \cos 900\pi t\,.$$

Use the sum-to-product formula for the sum of two cosines to show that

$$y = \cos 880\pi t + \cos(880\pi + 20\pi)t$$

can be rewritten as

$$y = \{2\cos 10\pi t\}\cos 890\pi t\,.$$

97. **a.** Graph $y = \cos 880\pi t$ and $y = \cos 900\pi t$ with a viewing window $\left[0, \dfrac{6}{440}\right]$ by $[-2, 2]$ to see how they gradually get more and more out of phase.

b. Graph the sum $y = \cos(880\pi t) + \cos(900\pi t)$ with the viewing window $[0, .2]$ by $[-2, 2]$, and confirm that the maximum points occur at 0, .1, and .2. The graph on a TI-85 should look like the one shown here. This is the pattern of the beating phenomenon. The beat occurs every $\frac{1}{10}$, at the maximums. The structure of this beating can be seen if you look at the other form of the equation [See Exercise 96]:

$$y = \{2\cos(10\pi)t\}\cos(890\pi t)$$

and interpret the first factor as a varying amplitude, which hits a maximum of 2 with period .1 .

98. **a.** Graph $y = \cos 880\pi t$ and $y = \cos 920\pi t$ with a viewing window $\left[0, \dfrac{6}{440}\right]$ by $[-2, 2]$ to see how they gradually get more and more out of phase.

b. Use the sum-to-product formula for the sum of two cosines to show that $y = \cos 880\pi t + \cos 920\pi t$ can be rewritten as $y = \{2\cos 20\pi t\}\cos 900\pi t$.

c. Graph the sum $y = \cos(880\pi t) + \cos(920\pi t)$ with the viewing window $[0, .2]$ by $[-2, 2]$, and locate the points in this window where the y-value is a maximum. What is the period of this beating function?

THINK ABOUT IT 8.9

1. Use a power reduction formula to graph the given equation.
   a. $y = \sin^2 x$ **b.** $y = \cos^2 x$
2. Explain how the figure below can be used to show geometrically that

$$\tan\frac{\theta}{2} = \frac{\sin\theta}{1+\cos\theta}$$

if θ is an acute angle.

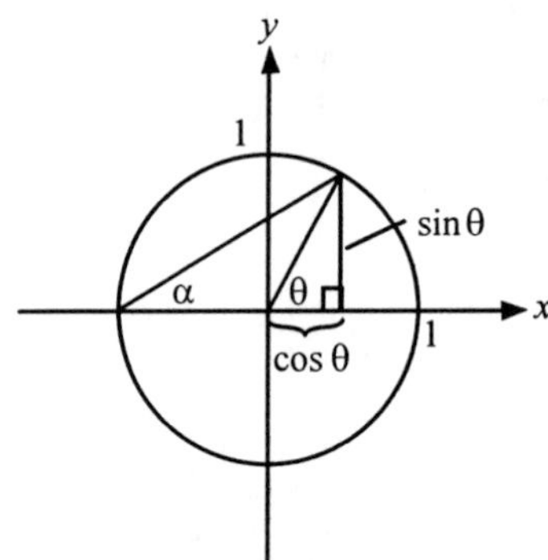

3. Is $\tan 2x$ less than or greater than $2\tan x$, if $0 < x < \pi/4$? Explain.

4. In 1706, John Machin determined 100 correct decimal places for π with the aid of the relation

$$\frac{\pi}{4} = 4\arctan\left(\frac{1}{5}\right) - \arctan\left(\frac{1}{239}\right).$$

Use identities o show that this relation is true.

5. **a.** Use $\tan x = (\sin x)/(\cos x)$ to verify the identity

$$\tan\frac{x}{2} = \pm\sqrt{\frac{1-\cos x}{1+\cos x}}.$$

b. Verify the identity

$$\tan\theta = \frac{\sin 2\theta}{1+\cos 2\theta}.$$

c. Replace θ by $x/2$ in the identity in part b to verify the identity

$$\tan\frac{x}{2} = \frac{\sin x}{1+\cos x}.$$

d. Multiply by $1-\cos x$ in the numerator and the denominator of the right-hand member of the identity in part c to verify the identity

$$\tan\frac{x}{2} = \frac{1-\cos x}{\sin x}.$$

● ● ●

CHAPTER 8 SUMMARY

OBJECTIVES CHECKLIST

Specific chapter objectives are summarized below, along with numbered example problems from the text that should clarify the objectives. If you do not understand any objectives, or do not know how to do the selected problems, then restudy the material.

8.1: Can you:

1. **Use $\theta = s/r$ to find either the central angle, the intercepted arc, or the radius, given measures for two variables in the relationship?** A central angle in a circle of radius 3 in. intercepts an arc of 12 in. Find the radian measure of the angle.

   [Example 1]

2. **Convert from degree measure of an angle to radian measure, and vice versa?** Express $\pi/3$, π, and $7\pi/5$ radians in terms of degrees.

   [Example 2]

3. **Determine the area of a sector of a circle, given the radius and the central angle?** In a circle of radius 12 in. find the area of a sector whose central angle is $120°$.

   [Example 5]

4. **Find linear velocity, given an angular velocity and a radius?** A circular saw blade 12 in. in diameter rotates at 1,600 rpm. Find the linear velocity in inches per second at which the teeth would strike a piece of wood.

   [Example 6b]

5. **Find angular velocity, given a linear velocity and a radius?** A circular saw blade 12 in. in diameter rotates at 1,600 rpm. Find the angular velocity in radians per second.

   [Example 6a]

8.2: Can you:

1. **Determine the exact value of any trigonometric function of an arc length that terminates at one of the axes?** Use the definition of the trigonometric functions to find $\sin(\pi/2)$.

   [Example 1]

2. **Determine the exact value of any trigonometric function of an arc length with a reference arc of $\pi/6$, $\pi/4$, or $\pi/3$?** Find the exact value of $\sin(-\pi/3)$.

   [Example 4]

3. **Use a calculator to approximate a trigonometric value of a real number?** Find the approximate value of sin 2. Round off to four decimal places.

   [Example 7a]

4. **Solve applied problems involving the evaluation of at trigonometric function of a real number?** The monthly revenue R in dollars from sales of a sun screen product in the New York region during a certain year is approximated by

   $$R = 6{,}500 - 5{,}000\cos(0.5236t)$$

   where t is the number of months that have elapsed from February 1. Based on this model, estimate the monthly revenue for this product for March and for August of the year in question.

   [Example 8]

8.3: Can you:

1. **Find the amplitude and period for functions of the form $y = a\sin bx$ and $y = a\cos bx$, and graph the function for a given interval?** State the amplitude and period, and sketch the graph, of $y = 2\sin 3x$ for $0 \le x \le 2\pi$.

   [Example 3]

2. **Write the equation of the form $y = a\sin bx$ or $y = a\cos bx$ that corresponds to a given graph?** Find an equation for the curve with the single cycle shown in the figure. The equation should be written in the form $y = a\sin bx$ or $y = a\cos bx$.

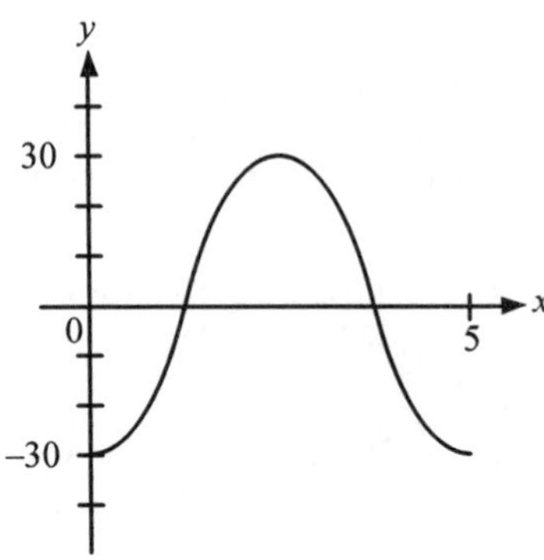

[Example 7]

3. **Graph sine and cosine functions that involve horizontal translations or vertical translations?** Graph one cycle of the function $y = 3 + 2\sin 4x$. Indicate the amplitude, period, and midline.

[Example 8]

8.4: Can you:

1. **Graph tangent, cotangent, secant, and cosecant functions?** Graph one cycle of the function $y = \tan(x + \pi/4)$.

[Example 1]

2. **Graph $f(x) = g(x) + h(x)$, where g or h is trigonometric, by using addition of ordinates or by using a grapher?** Graph $y = x + \sin x$ by using addition of ordinates.

[Example 4]

8.5: Can you:

1. **Evaluate expressions involving the inverse trigonometric functions?** Evaluate $\arcsin 1/2$. Specify the exact value, if possible.

[Example 1a]

2. **Graph an expression that involves an inverse trigonometric function?** At an art gallery, a painting that is 3 ft high is mounted so that its base is 2 ft above the anticipated eye level of a viewer.

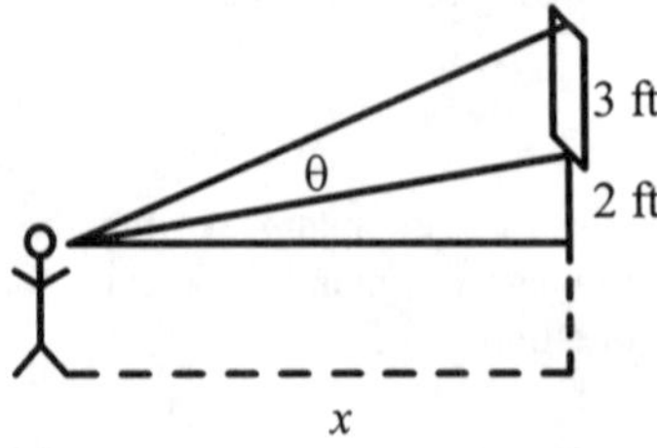

If a viewer is standing x ft from the wall, as shown in the diagram, then express the viewing angle in terms of x, and graph this relation.

[Example 3 a,b]

3. **Evaluate a composition of functions that involves an inverse trigonometric function?** Evaluate $\sin(\arccos 0)$. Find an exact answer, if possible.

[Example 4]

4. **Simplify a trigonometric function of an inverse trigonometric function?** Simplify $\sin(\cos^{-1} x)$.

[Example 5]

5. **Rewrite a trigonometric expression, using a given substitution?** If $x = \sin\theta$, where $-\pi/2 \le \theta \le \pi/2$, write $-\cot\theta - \theta$ as a function of *x*.

[Example 6]

6. **Use the inverse trigonometric function properties, when they apply?** Evaluate $\cos^{-1}(\cos 2\pi)$, if possible.

[Example 7b]

8.6: Can you:

1. **Find the approximate solution to trigonometric equations by using graphical methods?** On May 2 the tide in a certain harbor is described by

$$y = 12 + 5.3\cos(0.5067t),$$

where *y* is the height of the water in feet, and *t* is the time that has elapsed from 12 midnight on May 1. To the nearest minute, find all times on May 2 when the water level is at a height of 14 ft.

[Example 1]

2. **Use reference numbers to solve trigonometric equations, specifying exact solutions where possible?** Solve $\sin x = -1/2$ for $0 \le x < 2\pi$. Specify exact solutions.

[Example 2]

3. **Use factoring or the square-root property to solve certain trigonometric equations?** Solve the equation $2\sin x\cos x = 0$ in the interval $[0, 2\pi)$.

[Example 3]

4. **Solve trigonometric equations involving functions of multiple angles?** Solve $\sin 3x = 1/\sqrt{2}$ for $0 \le x < 2\pi$.

[Example 5]

8.7: Can you:

1. **Transform a given trigonometric expression and show that it is identical to another given expression?** Show that the expression $\sin x \cot x \sec x$ is identical to 1.

[Example 1]

2. **Verify that certain trigonometric equations are identities?** Prove the identity $\tan^2 x + 1 = \sec^2 x$.

[Example 2]

3. **Use the graph of a trigonometric expression to make a conjecture about how the expression simplifies, and then confirm this conjecture algebraically?** Graph $f(x) = (\sec x - \cos x)/(\tan x)$ and make a conjecture about an identity involving $f(x)$. Then, confirm the conjecture algebraically.

[Example 3]

4. **Use a trigonometric substitution to simplify the square root of a quadratic expression?** Eliminate the radical in the expression $\sqrt{9-x^2}$ by substituting $3\sin\theta$ for x, where $-\pi/2 \le \theta \le \pi/2$.

[Example 9]

5. **Use identities to evaluate various trigonometric expressions, given the value of one trigonometric function of x and the interval containing x?** If $\sin x = 3/5$ and $-\pi/2 < x < \pi$, find cos x.

[Example 10]

6. **Use the fundamental identities to help solve certain trigonometric equations?** Solve $2\cos^2 x = 1 - \sin x$ in the interval $[0, 2\pi)$.

[Example 11]

8.8: Can you:

1. **Use the sum or difference formulas to verify certain other identities?** Verify the identity $\cos(2\pi + x) = \cos x$.

[Example 1]

2. **Use the sum or difference formulas to find exact function values of certain angles or numbers?** Find the exact value of $\tan 15°$.

[Example 5]

3. **Use the sum or difference formulas to evaluate certain trigonometric expressions involving two angles, given function values of these angles?** If $\sin x_1 = -1/2$, where $\pi < x_1 < 3\pi/2$, and $\cos x_2 = \sqrt{3}/2$, where $0 < x_2 < \pi/2$, find $\cos(x_1 + x_2)$.

[Example 6c]

4. **Investigate a relation with the aid of a sum or difference formula?** At an art gallery, a painting that is 3 ft high is mounted so that its base is 2 ft above the anticipated eye level of a viewer.

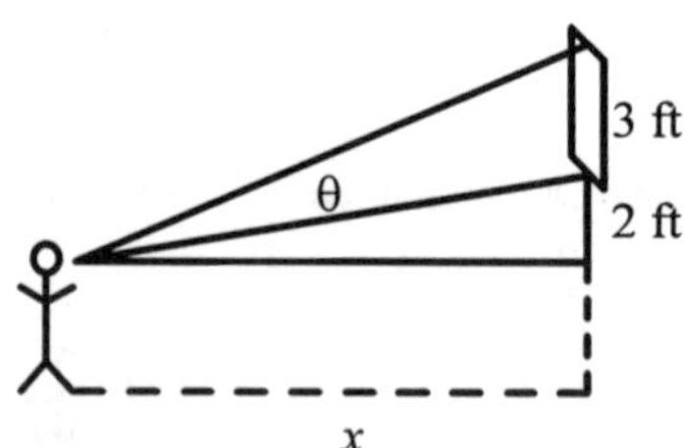

If a viewer is standing x ft from the wall, as shown in the diagram, then use the difference formula for the tangent to show that

$$\theta = \tan^{-1}\left(\frac{3x}{x^2 + 10}\right).$$

[Example 7a]

8.9: Can you:

1. **Evaluate functions involving double-angles or half-angles, given a trigonometric function of x and the interval containing x?** If $\sin x = 3/5$ and $\pi/2 < x < \pi$, find cos $2x$.

[Example 1a]

2. **Verify certain identities by using the double-angle, half-angle, or power reduction formulas?** Verify the identity $\cos^2 x = (1+\cos 2x)/2$.

   [Example 3]

3. **Use double-angle formulas to solve certain trigonometric equations?** Solve $\cos 2x + \cos x = 0$ for $0 \le x < 2\pi$.

   [Example 4]

4. **Use the half-angle formulas to find exact function values of certain angles or numbers?** Find the exact value of $\sin 165°$.

   [Example 6]

5. **Use the double-angle or half-angle identities to simplify certain trigonometric expressions?** Use the double-angle or half-angle identities to write this expression as a single function of a multiple angle.

$$\frac{\sin 6t}{1+\cos 6t}$$

   [Example 7a]

6. **Derive multiple-angle formulas for sin 3*x*, cos 3*x*, sin 4*x* or cos 4*x*?** Derive a formula for cos 3*x* in terms of cos *x*.

   [Example 8]

KEY CONCEPTS AND PROCEDURES

Section	Key Concepts and Procedures to Review
8.1	• Definition of 1 radian • Radian measure formula: $\theta = s/r$ • Degrees to radians formula: $1° = \pi/180$ radians • Radians to degree formula: 1 radian $= 180°/\pi$ • Area formula: $A = \frac{1}{2}r^2\theta$ (θ in radians) • Formula relating linear velocity (v) and angular velocity (ω): $v = \omega r$
8.2	• Definition of the trigonometric functions for a unit circle • Definitions of trigonometric identity and reference number • In a unit circle the same real number measures both the central angle θ and the intercepted arc s. (That is, $\theta = s$.) • For the sine and cosine functions, the domain is the set of all real numbers, and the range is the set of real numbers between –1 and 1, inclusive. • Methods to determine (if defined) approximate trigonometric values for any number and exact values in special cases • For any trigonometric function *f*, we have $f(s+2\pi k) = f(s)$, where *k* is an integer. • Negative angle identities • When using a calculator, be sure to set the calculator for radian mode. • Any trigonometric function of s is equal in absolute value to the same named function of its reference number s_R.
8.3	• Definitions of periodic function, period, amplitude, phase shift, and midline • For $y = d + a\sin(bx+c)$ and $y = d + a\cos(bx+c)$, with $b > 0$, amplitude $= \lvert a \rvert$, period $= 2\pi/b$, phase shift $= -c/b$, midline: $y = d$.

8.4

- Graphs of $y = \tan x$, $y = \cot x$, $y = \sec x$, and $y = \csc x$
- Domain, range, and period for $y = \tan x$, $y = \cot x$, $y = \sec x$, and $y = \csc x$
- Relations between a function and its reciprocal function
- Methods to graph trigonometric functions using horizontal shifting, vertical shifting, reflection, and addition of ordinates

8.5

- The inverse sine function is denoted by arcsin or $\sin^{-1}$. By definition, $y = \arcsin x$ if and only if $x = \sin y$, where $-1 \le x \le 1$ and $-\pi/2 \le y \le \pi/2$. Similar remarks hold for the other inverse trigonometric functions.
- Domain and range of the six inverse trigonometric functions
- Graphs of $y = \arcsin x$, $y = \arccos x$, and $y = \arctan x$
- Right triangle method to simplify a trigonometric function of an inverse trigonometric expression
- Inverse properties involving inverse trigonometric functions

8.6

- Methods to solve certain trigonometric equations (Reference numbers, identities, and a graphing calculator may be involved.)
- Determine solutions between 0 and 2π as follows:

Quadrant	Solution
1	Reference number
2	$\pi-$ reference number
3	$\pi+$ reference number
4	$2\pi-$ reference number

- We generate all the solutions to a trigonometric equation by adding multiples of 2π to the solutions that are in the interval $[0, 2\pi)$.

8.7

- Statements and applications of the **fundamental identities:**

$$\csc x = \frac{1}{\sin x} \qquad \sec x = \frac{1}{\cos x} \qquad \cot x = \frac{1}{\tan x}$$

$$\tan x = \frac{\sin x}{\cos x} \qquad \cot x = \frac{\cos x}{\sin x}$$

$$\left.\begin{aligned} &\sin^2 x + \cos^2 x = 1 \\ &\tan^2 x + 1 = \sec^2 x \\ &\cot^2 x + 1 = \csc^2 x \end{aligned}\right\} \textbf{Pythagorean Identities}$$

- Guidelines for proving identities

8.8

- Statements and applications of the **sum and difference formulas:**

$$\sin(x_1 \pm x_2) = \sin x_1 \cos x_2 \pm \cos x_1 \sin x_2$$

$$\cos(x_1 \pm x_2) = \cos x_1 \cos x_2 \mp \sin x_1 \sin x_2$$

$$\tan(x_1 \pm x_2) = \frac{\tan x_1 \pm \tan x_2}{1 \mp \tan x_1 \tan x_2}$$

(**Note:** Read the upper signs for the sum formula and the lower signs for the difference formula.)

8.9 • Statements and applications of the **double angle, half angle, and power reduction formulas:**

$\sin 2x = 2\sin x\cos x \qquad \cos 2x = \cos^2 x - \sin^2 x$

$\tan 2x = \dfrac{2\tan x}{1-\tan^2 x} \qquad = 2\cos^2 x - 1$

$\qquad = 1 - 2\sin^2 x$

$\sin\dfrac{x}{2} = \pm\sqrt{\dfrac{1-\cos x}{2}} \qquad \cos\dfrac{x}{2} = \pm\sqrt{\dfrac{1+\cos x}{2}}$

$\tan\dfrac{x}{2} = \pm\sqrt{\dfrac{1-\cos x}{1+\cos x}} = \dfrac{1-\cos x}{\sin x} = \dfrac{\sin x}{1+\cos x}$

$\sin^2 x = \dfrac{1-\cos 2x}{2} \qquad \cos^2 x = \dfrac{1+\cos 2x}{2}$

CHAPTER 8 REVIEW EXERCISES

In Exercises 1–10 find the *exact* value of the given expression.

1. $\sin\left(\dfrac{\pi}{3}\right)$ **2.** $\cos\left(-\dfrac{\pi}{4}\right)$ **3.** $\sin 99\pi$

4. $\tan\left(\dfrac{56\pi}{3}\right)$ **5.** $\cot\left(\dfrac{5\pi}{3}\right)$ **6.** $\cos\left(\dfrac{3\pi}{4}\right)$

7. $\arctan(-1)$ **8.** $\arccos\left(-\dfrac{\sqrt{3}}{2}\right)$

9. $\sin\left[\cos^{-1}(-1)\right]$ **10.** $\tan\left(\sin^{-1}\dfrac{2}{3}\right)$

In Exercises 11–20 use a calculator to find the value of the given expression to three significant digits.

11. $\sin 1$ **12.** $\sin 1°$ **13.** $\tan\left(\dfrac{8\pi}{5}\right)$

14. $\cot\left(\dfrac{\pi}{9}\right)$ **15.** $\sec 6$ **16.** $\csc(-4)$

17. $\arccos(-1.11)$ **18.** $\arcsin(-0.4439)$

19. $\cot(\operatorname{arcsec} 1.238)$ **20.** $\tan\left[\sin^{-1}(-0.9563)\right]$

In Exercises 21–24 sketch the graph for $0 \le x \le 2\pi$.

21. $y = -\cos x$ **22.** $y = \sin 2x$

23. $y = \cot x$ **24.** $y = \sec x$

In Exercises 25–30 state the amplitude, the period, the midline, and the phase shift, and sketch one cycle of the graph.

25. $y = \dfrac{1}{2}\cos 3x$ **26.** $y = 2\sin \pi x$

27. $y = -\cos\left(2x + \dfrac{\pi}{2}\right)$ **28.** $y = \sin\left(\pi x - \dfrac{\pi}{4}\right)$

29. $y = 4 - 2\sin\left(\dfrac{1}{2}x\right)$ **30.** $y = -10 + 3\cos(2x)$

In Exercises 31–34 sketch one cycle of the function.

31. $y = 3\sec\left(\dfrac{x}{2}\right)$ **32.** $y = \csc 2x$

33. $y = 1 - \tan x$ **34.** $y = \cot\left(x + \dfrac{\pi}{4}\right)$

35. Graph $y = \sin^{-1} x$ **36.** Graph $y = \arccos(x-1)$

In Exercises 37–42 state the domain and the range of the function.

37. $y = \tan x$ **38.** $y = \cos x$

39 $y = \csc x$ **40.** $y = \sec x$

41. $y = \arctan x$ **42.** $y = \arcsin x$

43. In which quadrant do $y = \sin x$ and $y = \cos x$ both decrease?

In Exercises 44 and 45 find an equation for the curve in the following illustration. The equation should be written in the form $y = d + a\sin bx$ or $y = d + a\cos bx$.

44.

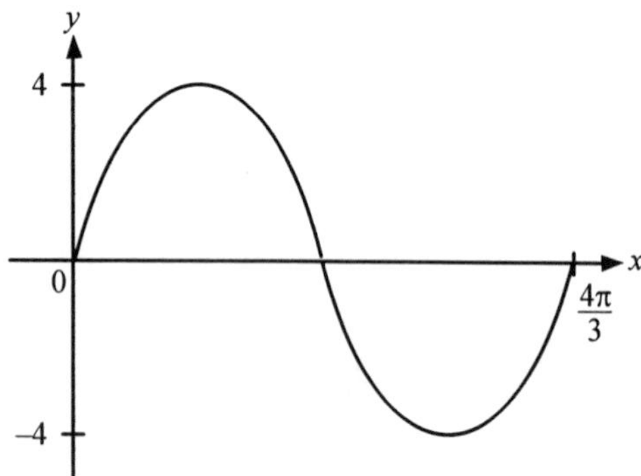

45.

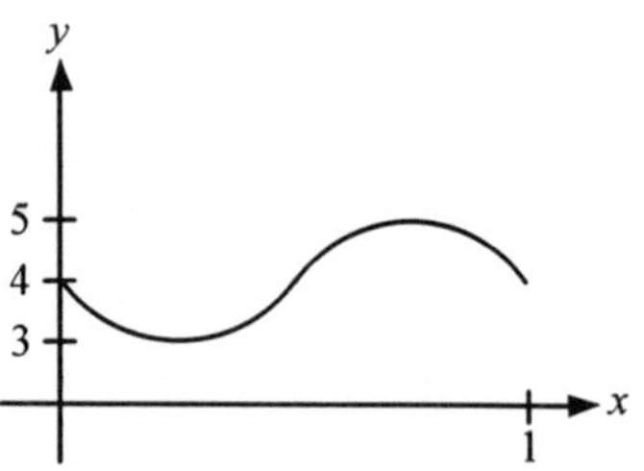

46. Graph $y = 2x + 2\sin 2x$ over the interval $[0,\ 2\pi]$.

47. For the given diagram express θ as a function of x

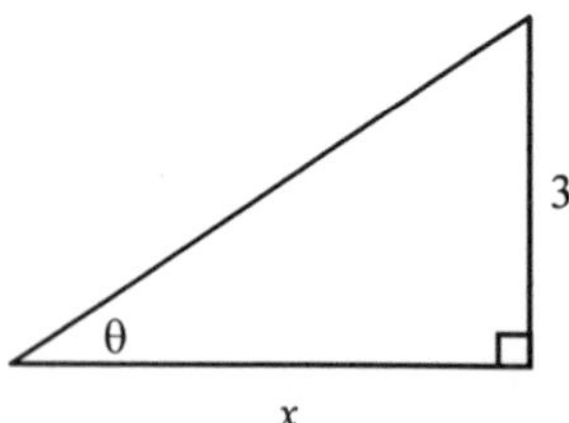

48. The diameter of a globe of the Earth is 12 in. A point on this globe is 4 in north of the equator of the globe. To the nearest 10 minutes, find its latitude.

49. On June 15 the tide in a certain harbor is described by $y = 20 + 8.2\cos(0.5067t)$, where y is the height of the water in feet and t is the time that has elapsed from 12 midnight on June 14. To one decimal place, estimate the water level in this harbor on June 15 at 2 p.m.

50. For the tide described in Exercise 49 find to the nearest minute all times on June 15 when the water level is 22 ft.

In Exercise 51–56 perform the indicated operations and simplify.

51. **a.** Change $210°$ to radians.
b. Change $2\pi/5$ radians to degrees.

52. If $\tan\theta = \frac{1}{2}$ and $\cos\theta < 0$, find $\sin\theta$.

53. For what values of x is $\sin(\sin^{-1}x) = x$ a true statement?

54. Find the area of a sector of a circle of radius 10 in. that is subtended by a central angle of $60°$.

55. In a circle a central angle intercepts an arc equal in length to the diameter to the circle. What is the measure of the central angle in radians?

56. If $y = \arccos\left(\frac{x}{2}\right)$, find x in terms of y.

57. If $x = 3\sin\theta$, where $-\frac{\pi}{2} \le \theta \le \frac{\pi}{2}$, write $\frac{9}{2}\theta - \frac{9}{4}\sin\theta\cos\theta$ as a function of x.

58. A car is traveling at 60 mi/hour (88 ft/second) on tires 32 in. in diameter. Find the angular velocity of the tires in radians/second.

In Exercises 59–76 verify the given identity.

59. $\cot^2 x\sec^2 x = \csc^2 x$

60. $\dfrac{\sin^2 x}{1+\cos x} = 1 - \cos x$

61. $\sin^2 x\tan^2 x = \tan^2 x - \sin^2 x$

62. $\sec\theta\csc\theta = \tan\theta + \cot\theta$

63. $\dfrac{\sin\theta + \tan\theta}{1+\sec\theta} = \sin\theta$ **64.** $\dfrac{\sec x}{\cos x} - \dfrac{\tan x}{\cot x} = 1$

65. $(\sin x - \cos x)^2 = 1 - \sin 2x$

66. $\dfrac{1-\cos 2\theta}{1+\cos 2\theta} = \tan^2\theta$ **67.** $\dfrac{1-\cos 2x}{\sin 2x} = \tan x$

68. $\dfrac{2\cot x}{\csc^2 x - 2} = \tan 2x$ **69.** $2\sin^2 3x = 1 - \cos 6x$

70. $\tan\left(\frac{\theta}{2}\right) = \csc\theta - \cot\theta$

71. $\cot 2x = \cot x - \csc 2x$

72. $\sin(x - \pi) = -\sin x$

73. $\sin 3x = 3\sin x\cos^2 x - \sin^3 x$

74. $\sqrt{3}\sin x + \cos x = 2\sin\left(x + \frac{\pi}{6}\right)$

75. $\sin^4 2x = \frac{3}{8} - \frac{1}{2}\cos 4x + \frac{1}{8}\cos 8x$

76. $\cos(u + v) + \cos(u - v) = 2\cos u\cos v$

In Exercises 77–86 find exact function values if $\cos x = \frac{4}{5}$ and $\frac{3\pi}{2} < x < 2\pi$.

77. $\sin x$ **78.** $\csc x$ **79.** $\sec x$

80. $\tan x$ **81.** $\cos 2x$ **82.** $\sin 2x$

83. $\tan 2x$ **84.** $\tan\left(\frac{x}{2}\right)$ **85.** $\sin\left(\frac{x}{2}\right)$

86. $\cos\left(\frac{x}{2}\right)$

87. Is $\sin\left(\frac{\pi}{2} + x\right) = -\cos x$ a conditional equation or an identity?

88. Find $\sin\left(\frac{\pi}{4} - x\right)$ if $x = \frac{4}{5}$ and $0 < x < \frac{\pi}{2}$.

89. Find $\cos 2x$ if $\sin x = \frac{1}{4}$ and $\frac{\pi}{2} < x < \pi$.

90. If $\cos x = a$, express $\cos 2x$ in terms of a.

91. If $\tan\alpha = a$ and $\tan\beta = 2a$, express $\tan(\beta - \alpha)$ in terms of a.

92. Find the exact value of cos 75° by using a sum formula.

93. Eliminate the radical in the expression $\sqrt{x^2 + 25}$ by substituting $5\tan\theta$ for x, where $-\frac{\pi}{2} < \theta < \frac{\pi}{2}$.

94. Express $8\cos^2\left(\frac{x}{4}\right) - 8\sin^2\left(\frac{x}{4}\right)$ as a single function of a multiple angle.

95. Simplify $\sin(3\pi - x)$.

96. Simplify $\tan 2x\cos 2x\csc x$.

97. Find exactly: $\cos\left(\sin^{-1}\frac{24}{25} - \cos^{-1}\frac{3}{5}\right)$.

98. If $\sin x_1 = -\frac{1}{3}$, where $\pi < x_1 < \frac{3\pi}{2}$, and $\cos x_2 = -\frac{1}{2}$, where $\frac{\pi}{2} < x_2 < \pi$, find $\sin(x_1 - x_2)$.

99. Use the identity for $\sin(x_1 + x_2)$ to derive the identity for $\sin(x_1 - x_2)$.

100. Use the formula for the sine of the sum of two numbers and derive a formula for sin 4x in terms of sin x and cos x.

In Exercises 101–106 solve for x in the interval $[0, 2\pi)$.

101. $\csc x = -2$ **102.** $4\sin^2 x - \sin x - 2 = 0$

103. $2\cos^2 x = 3 - 3\sin x$ **104.** $\tan 3x = 1$

105. $\sin 2x = \sin x$ **106.** $\tan x = \cot x$

In Exercises 107 and 108 find all real number solutions.

107. $3\tan^2\theta - 1 = 0$ **108.** $2\sin\theta\cos\theta - \cos\theta = 0$

In Exercises 109–120 select the choice that completes the statement or answers the question.

109. The expression $\sin x + \frac{\cos^2 x}{\sin x}$ is identical to

a. $\sec x$ **b.** $\csc x$
c. $\cos x$ **d.** 1

110. If $x = \arccos\left(-\frac{1}{2}\right)$, then x equals

a. $\frac{\pi}{3}$ **b.** $\frac{\pi}{6}$ **c.** $\frac{5\pi}{6}$ **d.** $\frac{2\pi}{3}$

111. Which number is not in the range of $y = \sin x$?

a. 1 **b.** $-\frac{1}{2}$ **c.** 2 **d.** 0

112. If $\sin x < 0$ and $\tan x > 0$, then the point assigned to x lies in quadrant

a. 1 **b.** 2 **c.** 3 **d.** 4

113. The equation $\cos(-x) = -\cos x$ is true for

a. all values of x
b. only certain values of x
c. no values of x

114. A function having the period π is

a. $y = 2\sin x$ **b.** $y = \frac{1}{2}\sin x$

c. $y = \sin\frac{1}{2}x$ **d.** $y = \sin 2x$

115. To convert from radians to degrees, we multiply the number of radians by

a. $\frac{180°}{\pi}$ **b.** $\frac{\pi}{90°}$ **c.** $\frac{90°}{\pi}$ **d.** $\frac{\pi}{180°}$

116. The expression $\cos(x-\pi)$ simples to

a. $\cos x$ **b.** $-\cos x$ **c.** $\sin x$ **d.** $-\sin x$

117. If a central angle of a radians intercepts an arc of length b in a circle of radius c, then

a. $a=\frac{b}{c}$ **b.** $a=\frac{\pi}{b}$ **c.** $a=\frac{c}{b}$ **d.** $a=bc$

118. If $f(x)=\cos 3x+\tan 2x$, then $f\left(\frac{\pi}{6}\right)$ equals

a. $\sqrt{3}$ **b.** $1+\sqrt{3}$

c. $\frac{\sqrt{3}}{3}$ **d.** $\frac{3+\sqrt{3}}{3}$

119. If $\log\tan x=a$, then $\log\cot x$ equals

a. $\frac{1}{a}$ **b.** $-a$ **c.** $1-a$ **d.** a

120. Which one of the following is an identity?

a. $\sin x+\cos x=1$ **b.** $\sec x\cdot\csc x=1$

c. $\sin\frac{1}{2}x=\frac{1}{2}\sin x$ **d.** $\cos^2 x=1-\sin^2 x$

CHAPTER 8 TEST

1. What is the range of the function $y=\arcsin x$?

2. What is the domain of the function $y=\tan x$?

3. Find the exact value of $\tan\left(\frac{2\pi}{3}\right)$.

4. Sketch one cycle of the graph of $y=-3\cos 4\pi x$.

5. Sketch one cycle of the graph of $y=\tan\left(x-\frac{\pi}{4}\right)$.

6. Graph $y=\csc x$ for $0\le x\le 2\pi$.

7. Find the amplitude, the period, and the phase shift for the graph of $y=-\sin\left(3x-\frac{\pi}{2}\right)$.

8. **a.** Change 270° to radians.

b. Change $\frac{7\pi}{4}$ radians to degrees.

9. Find the arc length of a sector of a circle of radius 10 cm that is subtended by a central angle of 150°.

10. Find the exact value of $\arcsin\left(-\frac{\sqrt{3}}{2}\right)$.

11. Simplify $\tan\left(\cos^{-1}x\right)$.

12. For what value of x is $\sin^{-1}(\sin x)=x$ a true statement?

13. Show that $\frac{1+\cot^2 x}{\cot^2 x}=\sec^2 x$ is an identity.

14. Simplify $\sec x\cos x-\frac{1}{\csc^2 x}$.

15. Show that $\cos(x-\pi)=-\cos x$ is an identity.

16. Find all real number solutions: $4\sin x+3=1$.

17. Solve for x in the interval $[0,\ 2\pi)$: $2\sin^2 x=1-\cos x$.

18. Use identities to find $\tan x$ if $\cos x=\frac{5}{13}$ and $\frac{3\pi}{2}<x<2\pi$.

19. Show that $\frac{1-\tan^2 x}{1+\tan^2 x}=\cos 2x$ is an identity.

20. Solve for x in the interval $[0,\ 2\pi)$: $\cos 3x=\frac{\sqrt{2}}{2}$.

• • •

Chapter

9

Further Applications of Trigonometry

*Photo Courtesy of **Joe Peaco**/National Park Service/ Corporate Digital Archive*

Two observation towers, *A* and *B*, are located $1\overline{0}$ mi apart. A fire is sighted at point *C*, and the observer in tower *A* measures angle *CAB* to be 80°. At the same time, the observer in tower *B* measures angle *CBA* to be 40°. How far is the fire from tower *A*?
(See Example 1 of Section 9.1).

In this chapter we continue our discussion of trigonometry by considering additional applications of the trigonometric functions with angle measures as domain elements. These applications involve the law of sines, the law of cosines, the area of a triangle, vectors, the trigonometric form of complex numbers, and polar coordinates.

● ● ●

9.1 Law of Sines

Objectives

1. Use the law of sines to solve a triangle, given two angles and one side of the triangle.
2. Determine exact solutions for unknowns associated with certain triangles, assuming exact numbers.
3. Use the law of sines to solve a triangle, given two sides of the triangle and the angle opposite one of them.
4. Determine how many triangles are described by certain conditions.
5. Solve applied problems using the law of sines.

In Section 2.2 we learned to solve right triangles. We now wish to extend our ability to solve triangles, by considering the solution of general triangles, which may or may not be right triangles. This generalization is useful because the analysis of a problem often leads to a triangle that does not contain a 90° angle. For instance, the angles in the triangle that is used to determine the distance to the fire in the chapter-opening problem measure 80°, 40°, and 60°, so right-triangle trigonometry is not useful in this case. Remember, we "solve" a triangle by finding the measures of its three angles and three sides. To accomplish this, at least three of these six values must be known, and one or more must be a side length.

The first technique that we use to solve general triangles is called the **law of sines**. We can derive this law by placing triangle ABC on a rectangular coordinate system so that angle A is in standard position. Figure 9.1(a) and (b) show the result when A is an acute angle and an obtuse angle, respectively. In both cases we draw the altitude of the triangle, CD, and note that its length is y.

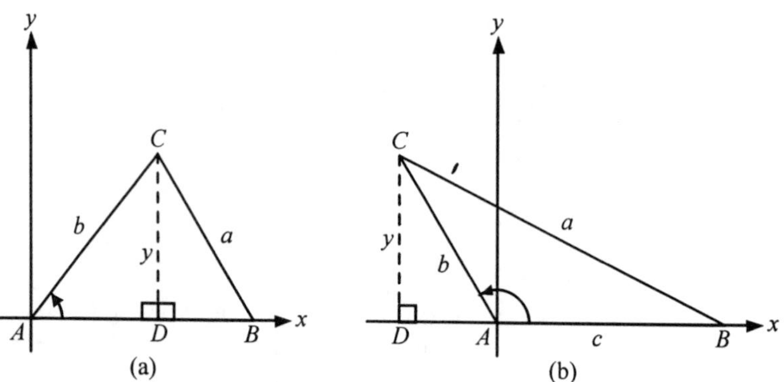

Figure 9.1
(a) A is an acute angle. (b) A is an obtuse angle.

Then by the general definition of the sine function,

$$\sin A = \frac{y}{b}, \text{ so } y = b\sin A.$$

Also, in the right triangle, BDC,

$$\sin B = \frac{y}{a}, \text{ so } y = a\sin B.$$

Setting these two expressions for y equal to each other, we have

$$b\sin A = a\sin B, \text{ so } \frac{\sin A}{a} = \frac{\sin B}{b}.$$

Similarly, by placing angle C in standard position, we can show that

$$\frac{\sin A}{a} = \frac{\sin C}{c}.$$

Combining these results, we have

$$\frac{\sin A}{a} = \frac{\sin B}{b} = \frac{\sin C}{c}.$$

This relationship, **the law of sines**, states the following.

Law of Sines

> The sines of the angles in a triangle are proportional to the lengths of the opposite sides.

Note that if C is a right angle, $\sin C = \sin 90° = 1$, and the law of sines yields the right-triangle relationships

$$\sin A = \frac{a}{c} \text{ and } \sin B = \frac{b}{c}.$$

With the assistance of the law of sines we can now solve the problem in the chapter introduction.

Example 1: Distance to a Fire

Solve the problem in the chapter introduction on page 733.

Solution: First, draw a diagram picturing the data (Figure 9.2). Second, find angle C.

$$\begin{aligned} A + B + C &= 180° \\ 80° + 40° + C &= 180° \\ C &= 60° \end{aligned}$$

Third, we find b by applying the law of sines.

$$\begin{aligned} \frac{\sin B}{b} &= \frac{\sin C}{c} \\ \frac{\sin 40°}{b} &= \frac{\sin 60°}{10} \\ b &= \frac{10 \sin 40°}{\sin 60°} \\ &= 7.4 \end{aligned}$$

Thus, the fire is about 7.4 mi from station A.

Note: Remember that our computed results cannot be more accurate than the data that are given. Guidelines for the desired accuracy in a solution can be found in Section 2.2

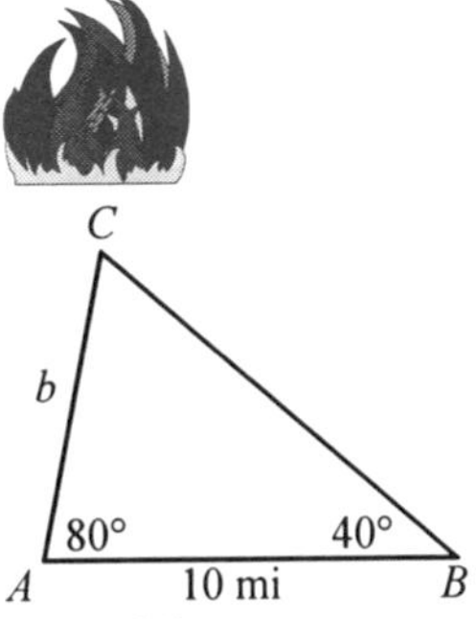

Figure 9.2

PROGRESS CHECK 1

Two surveyors establish a baseline AB on a level field. The surveyor at point A is 375 ft from the surveyor at point B. Each one sights a stake at point C. The surveyor at A measures angle CAB to be 82.3°, while the surveyor at B measures angle CBA to be 65.4°. Find the distance from A to C.

■

The law of sines can be used to solve any triangle in the following two cases:

1. If we know the measures for two angles and one side of the triangle.
2. If we know the measures for two sides of the triangle and the angle opposite one of them.

The following example illustrates how the law of sines can be used to solve a triangle in the first case, in which the measures for two angles and one side of the triangle are known. Note that in computing results, the symbol for equality (=) is generally used, even though the symbol for approximation (≈) may be more appropriate.

Example 2: Solving a Triangle: Angle-Angle-Side Case (AAS)

Approximate the missing parts of triangle ABC in which $A = 35°$, $B = 50°$ and $a = 12$ ft.

Solution: First, sketch Figure 9.3. We can find angle C, since the sum of the angles in a triangle is 180°.

$$\begin{aligned} A + B + C &= 180° \\ 35° + 50° + C &= 180° \\ C &= 95° \end{aligned}$$

Second, we find side length b by applying the law of sines.

$$\begin{aligned} \frac{\sin A}{a} &= \frac{\sin B}{b} \\ \frac{\sin 35°}{12} &= \frac{\sin 50°}{b} \\ b &= \frac{12 \sin 50°}{\sin 35°} \\ &= 16 \text{ ft} \end{aligned}$$

Third, we find c by applying the law of sines.

$$\begin{aligned} \frac{\sin A}{a} &= \frac{\sin C}{c} \\ \frac{\sin 35°}{12} &= \frac{\sin 95°}{c} \\ c &= \frac{12 \sin 95°}{\sin 35°} \\ &= 21 \text{ ft} \end{aligned}$$

Thus, the solution to the triangle is as shown in Figure 9.4.

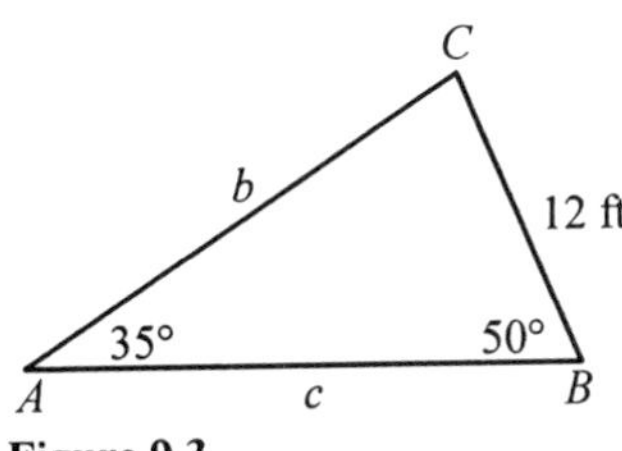

Figure 9.3

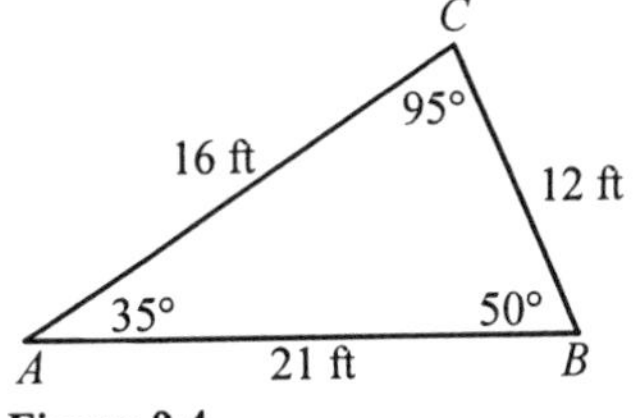

Figure 9.4

PROGRESS CHECK 2
Solve the triangle *ABC* in which $A = 30°$, $B = 40°$ and $a = 2\overline{0}$ ft. ■

For the next example, we will assume that all numbers are exact numbers, and because the angles given measures 45° and 30°, we may determine an exact solution.

Example 3: Using the Law of Sines, Assuming Exact Numbers

In triangle *RST*, $r = 8$, $R = 45°$ and $S = 30°$. Assume exact numbers and find *s*.

Solution: First, sketch Figure 9.5. Then relate *r*, *s*, sin *R*, and sin *S* by the law of sines and solve for *s*.

$$\frac{\sin R}{r} = \frac{\sin S}{s}$$

$$\frac{\sin 45°}{8} = \frac{\sin 30°}{s}$$

$$s = \frac{8\sin 30°}{\sin 45°}$$

$$s = \frac{8 \cdot 1/2}{\sqrt{2}/2}$$

$$s = 4\sqrt{2}$$

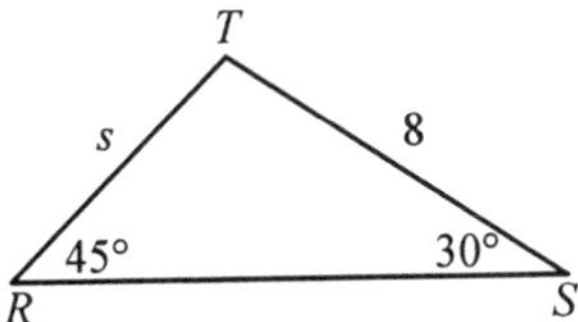

Figure 9.5

PROGRESS CHECK 3
In triangle *PQR*, $R = 120°$, $P = 30°$, and $p = 12$. Assume exact numbers and find *r*. ■

The following examples illustrate how the law of sines can be used to solve a triangle in the second case, in which the measures for two sides of the triangle and the angle opposite one of them are known.

Example 4: Solving a Triangle: Side-Side-Angle Case (SSA)

Solve the triangle *ABC* in which $B = 60°$, $b = 5\overline{0}$ ft, and $c = 3\overline{0}$ ft.

Solution: First, sketch Figure 9.6. We find angle C by applying the law of sines.

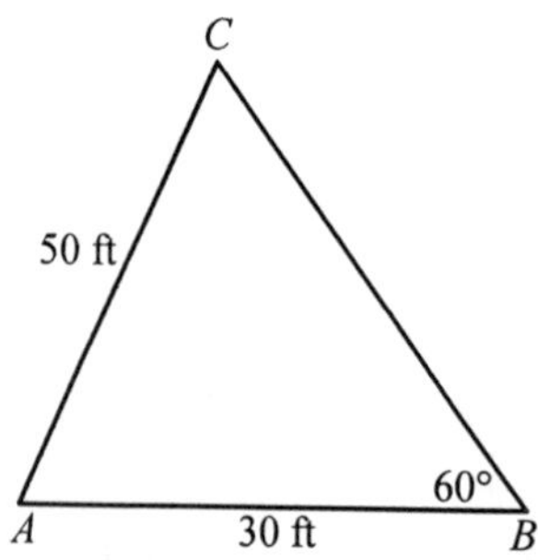

Figure 9.6

$$\frac{\sin B}{b} = \frac{\sin C}{c}$$
$$\frac{\sin 60^\circ}{50} = \frac{\sin C}{30}$$
$$\frac{30(\sin 60^\circ)}{50} = \sin C$$
$$0.5196 = \sin C$$

We now have two possibilities for angle C, since the sine of both first and second quadrant angles is positive.

Case 1 (acute angle in Q_1): $\sin C = 0.5196$

Case 2 (obtuse angle in Q_2): $\sin C = 0.5196$

$$\text{reference angle} = \sin^{-1} 0.5196 = 31.3^\circ \quad \text{(by calculator)}$$

31.3° is the angle in Q_1 with a reference angle of 31.3° [Figure 9.7(a)].
148.7° is the angle in Q_2 with a reference angle of 31.3° [Figure 9.7(b)]

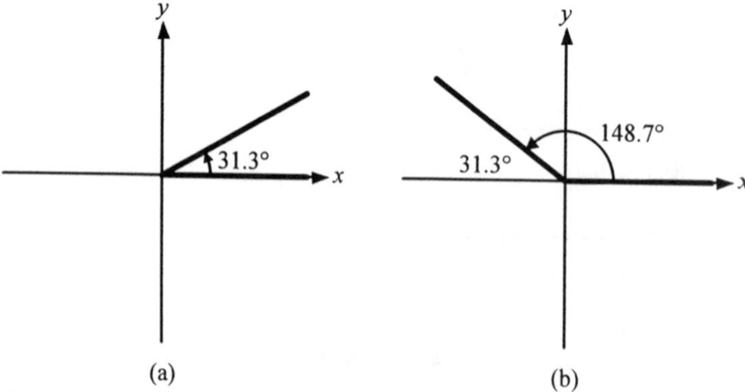

Figure 9.7

$$\text{Therefore, } C = 31.3^\circ \text{ or } C = 148.7^\circ.$$

Second, we find angle A in both of the above cases.

Case 1 $A + B + C = 180°$ Case 2 $A + B + C = 180°$

$A + 60° + 31.3° = 180°$ $B = 60°$ $C = 148.7°$

$A = 88.7°$

In Case 2 we find $B + C = 208.7°$, so regardless of the value of A, the sum of the angles of the triangle exceeds 180°. Therefore, we reject $C = 148.7°$ as a solution.
Third, we find side length a by applying the law of sines.

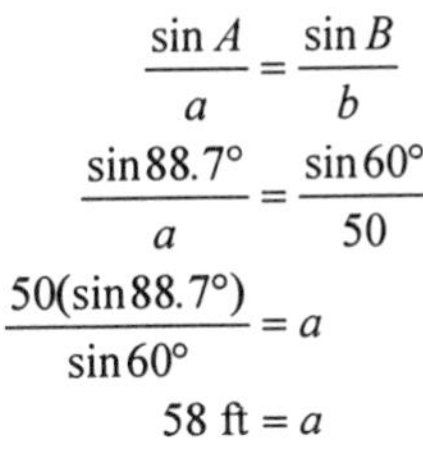

$$\frac{\sin A}{a} = \frac{\sin B}{b}$$

$$\frac{\sin 88.7°}{a} = \frac{\sin 60°}{50}$$

$$\frac{50(\sin 88.7°)}{\sin 60°} = a$$

$$58 \text{ ft} = a$$

Figure 9.8

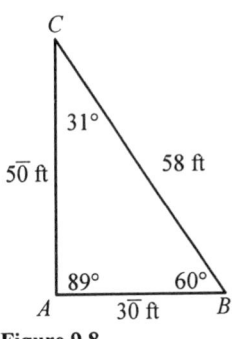

When we round off the angle measures to the nearest degree, the solution to the triangle is as shown in Figure 9.8.

PROGRESS CHECK 4

Approximate the missing parts of the triangle ABC in which $A = 55°$, $a = 75$ ft, and $b = 42$ ft. ■

Example 5: Solving a Triangle: Side-Side-Angle Case (SSA)

Approximate the missing parts of triangle ABC in which $A = 37°20'$, $a = 12.5$ ft and $c = 20.1$ ft.

Solution: First, sketch Figure 9.9. We find angle C by applying the law of sines.

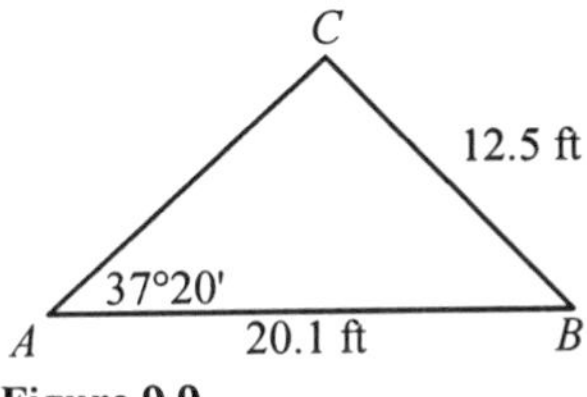

Figure 9.9

$$\frac{\sin A}{a} = \frac{\sin C}{c}$$

$$\frac{\sin 37°20'}{12.5} = \frac{\sin C}{20.1}$$

$$\frac{20.1(\sin 37°20')}{12.5} = \sin C$$

$$0.9753 = \sin C$$

We now have two possibilities.

Case 1 (acute angle in Q_1): $\sin C = 0.9753$

Case 2 (obtuse angle in Q_2): $\sin C = 0.9753$

reference angle $= \sin^{-1} 0.9753 = 77°10'$

$77°10'$ is the angle in Q_1 with a reference angle of $77°10'$ [Figure 9.10(a)].
$102°50'$ is the angle in Q_2 with a reference angle of $77°10'$ [Figure 9.10(b)].

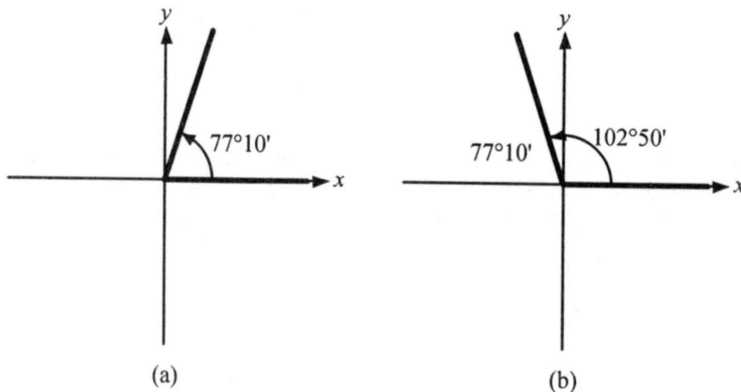

Figure 9.10

$$\text{Therefore, } C = 77°10' \text{ or } C = 102°50'.$$

Second, we find angle B in both of the above cases.

$$\begin{aligned} A + B + C &= 180° \\ 37°20' + B + 77°10' &= 180° \\ B + 114°30' &= 180° \\ B &= 65°30' \end{aligned} \qquad \begin{aligned} A + B + C &= 180° \\ 37°20' + B + 102°50' &= 180° \\ B + 140°10' &= 180° \\ B &= 39°50' \end{aligned}$$

Third, we find side length b by applying the law of sines.

$$\begin{aligned} \frac{\sin A}{a} &= \frac{\sin B}{b} \\ \frac{\sin 37°20'}{12.5} &= \frac{\sin 65°30'}{b} \\ b &= \frac{12.5(\sin 65°30')}{\sin 37°20'} \\ &= 18.8 \text{ ft} \end{aligned} \qquad \begin{aligned} \frac{\sin A}{a} &= \frac{\sin B}{b} \\ \frac{\sin 37°20'}{12.5} &= \frac{\sin 39°50'}{b} \\ b &= \frac{12.5(\sin 39°50')}{\sin 37°20'} \\ &= 13.2 \text{ ft} \end{aligned}$$

The two possible solutions from the given data are shown in Figure 9.11.

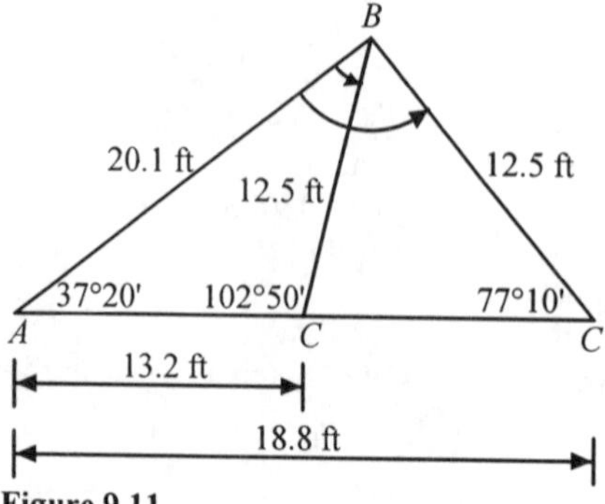

Figure 9.11

PROGRESS CHECK 5

Solve the triangle ABC in which $A = 28°40'$, $a = 162$ ft, and $b = 225$ ft. ■

Note that when we attempt to solve a triangle in which the measures for two sides of the triangle and the angle opposite one of them are given, there may be one triangle that fits the data (as in Example 4), or there may be two triangles that fit the data (as in Example 5). Consequently, this case is called the **ambiguous case** of the law of sines. It is also possible in the ambiguous case that no triangle can be constructed from the data; then we say the data are inconsistent. Figure 9.12 shows conditions that determine the various cases when *a*, *b*, and acute angle *A* are given. In this diagram it is helpful to think that the side of length *a* can swing like a pendulum.

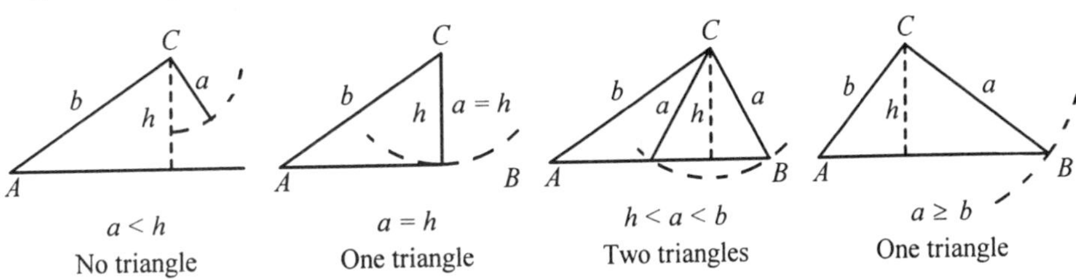

Figure 9.12

$\sin A = \frac{h}{b}$, so $h = b \sin A$.

If A is an obtuse angle, then the possibilities are more obvious, as shown in Figure 9.13.

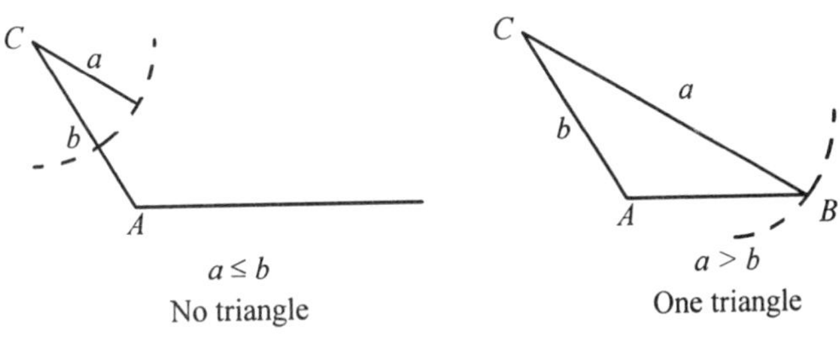

Figure 9.13

It is not recommended that the information in Figures 9.12 and 9.13 be memorized. Instead, in a given problem you should be aware of the various possibilities, make a careful sketch of the given information, and let an analysis based on the law of sines determine the case when the case is not obvious.

Example 6: Finding the Number of Triangles in the SSA Case

Determine if no triangle, one triangle, or two triangles are determined by the conditions that $A = 52°$, $a = 55$, and $b = 75$.

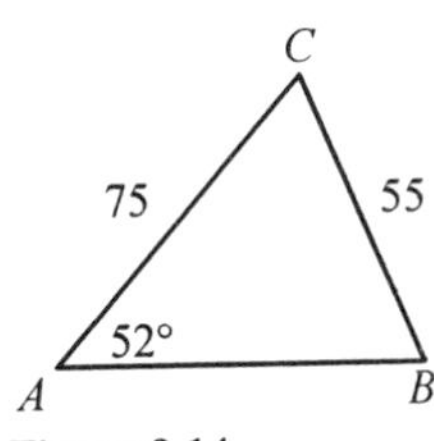

Figure 9.14

Solution: First, sketch Figure 9.14. Then relate *a*, *b*, sin *A* and sin *B* by the law of sines and solve for *B* using the given data.

$$\frac{\sin B}{b} = \frac{\sin A}{a}$$

$$\frac{\sin B}{55} = \frac{\sin 52°}{75}$$

$$\sin B = \frac{55 \sin 52°}{75} \approx 1.0746$$

Because the range of the sine function is $[-1, 1]$, $\sin B$ cannot equal 1.0746 so no triangle exists.

PROGRESS CHECK 6

Determine if no triangle, one triangle, or two triangles are determined by the conditions that $A = 52°$, $a = 45$ and $b = 55$.

■

EXPLORE 9.1

1. In solving general triangles some angles may be acute (between 0° and 90°), and some may be obtuse (between 90° and 180°). Confirm by calculator exploration that the cosine of an acute angle is positive, but the cosine of an obtuse angle is negative. Compare $\cos 89.99°$, $\cos 90.01°$ and $\cos 90°$. Compare the graphs of $y = \cos(90° + x)$ and $y = \cos(90° - x)$ for $0° < x < 90°$.
2. The law of sines states that $\sin B = b(\sin A)/a$. In triangle ABC suppose $A = 30°$, and $a = 10$.
   a. Calculate $\sin A/a$.
   b. Show that 20 is the largest possible value for b so that sin B is not greater than 1. Recall that the sine of an angle in a triangle cannot be greater than 1.
   c. Suppose you take $b = 25$. Then you get $\sin B = 1.25$. Use a calculator to solve $\sin B = 1.25$. Some calculators will yield an imaginary value, and some will yield an error message. Check the response of your calculator. In either case, this is a warning that something is peculiar about the triangle. Try to draw a triangle with $A = 30°$, $a = 10$, and $b = 25$. What is the problem?

EXERCISES 9.1

(**Note:** There are more problems on the law of sines in Exercises 9.2.) In Exercises 1–6 approximate the remaining parts of the triangle for the data given.

1. $A = 30°$, $a = 25$ ft, and $B = 45°$
2. $C = 60°$, $c = 4\overline{0}$ ft, and $A = 80°$
3. $B = 120°$, $C = 40°$, and $a = 55$ ft
4. $C = 135°$, $c = 98$ ft, and $B = 15°$
5. $A = 62°10'$, $a = 31.5$ ft, and $B = 76°30'$
6. $A = 98°30'$, $B = 6°10'$, and $a = 415$ ft

In Exercises 7–16 assume exact numbers and find exact answers.

7. In triangle ABC, $a = 8$, $b = 12$, and $A = 30°$. Find $\sin B$.
8. In triangle ABC, $b = 8$, $c = 10$, and $C = 150°$. Find $\sin B$.
9. In triangle PQR, $p = 9$, $\sin P = \frac{3}{4}$, and $\sin Q = \frac{1}{2}$. Find q.
10. In triangle PQR, $\sin R = 0.6$, $\sin Q = 0.4$, and $q = 14$. Find r.
11. In triangle RST, $\sin R = \frac{1}{4}$ and $\sin S = \frac{7}{8}$. Find $\frac{s}{r}$.
12. In triangle RST, $S = 30°$ and $T = 45°$. Find $\frac{s}{t}$.
13. In triangle ABC, $b = 20$, $B = 45°$, and $C = 30°$. Find c.
14. In triangle ABC, $a = 10$, $A = 30°$, and $B = 60°$. Find b.
15. Find b and c in the figure below.

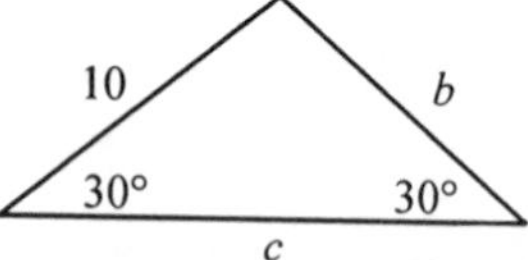

16. Find p and q in the figure below.

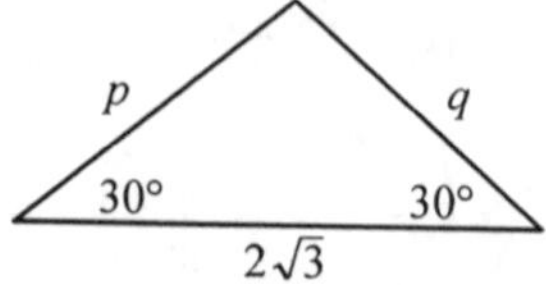

In Exercises 17–22 approximate the remaining parts of the triangle for the data given.

17. $A = 45°$, $a = 8\overline{0}$ ft, and $b = 5\overline{0}$ ft
18. $C = 60°$, $c = 75$ ft, and $a = 45$ ft

19. $B = 30°$, $b = 3\bar{0}$ ft, and $a = 4\bar{0}$ ft

20. $B = 22°$, $b = 78$ ft, and $a = 86$ ft

21. $C = 150°$, $c = 92$ ft, and $b = 69$ ft

22. $C = 105°30'$, $c = 46.1$ ft, and $b = 75.2$ ft

In Exercises 23–28 determine if no triangle, one triangle, or two triangles are determined by the given conditions.

23. $A = 18°$, $a = 15$, $b = 28$

24. $A = 65°$ $a = 18$, $b = 24$

25. $C = 30°$, $b = 16$, $c = 32$

26. $B = 45°$, $a = 26$, $b = 21$

27. $A = 130°$, $a = 14$, $b = 18$

28. $A = 96°$, $a = 15$, $b = 11$

29. Two surveyors establish a baseline AB on a level field. The surveyor at point A is $20\bar{0}$ ft from the surveyor at point B. Each one sights a stake at point C. The surveyor at A measures angle CAB to be $72°30'$, while the surveyor at B measures angle CBA to be $81°20'$. Find the distance from B to C.

30. Engineers wish to build a bridge across a river to joint point A on one side to either point B or point C on the other side. The distance from B to C is $40\bar{0}$ ft, angle ABC is $67°20'$ and angle ACB is $84°30'$. By how many feet does the distance from A to B exceed the distance from A to C?

31. Airport A is $3\bar{0}0$ mi mi due north to airport B. Their radio stations receive a distress signal from a ship located at point C. It is determined that point C is located 54° south of east with respect to airport A, and 76° north of east from airport B. How far is the ship from airport A?

32. Two engineers are located at points A and B on the opposite sides of a hill. They are both able to see a stake at point C, which is at a distance of $8\bar{0}0$ ft from A and $7\bar{0}0$ ft from B. If angle ABC is 25°, find the distance $\overline{AB}$ through the hill.

33. If ABC is a right triangle ($C = 90°$), show that the law of sines simplifies to the right-triangle relationships $\sin A = a/c$ and $\sin B = b/c$.

34. In triangle RST express sin T in terms of r, t, and sin R.

THINK ABOUT IT 9.1

1. In the given illustration find acute angles θ, α, and β to the nearest degree and x and y to two significant digits.

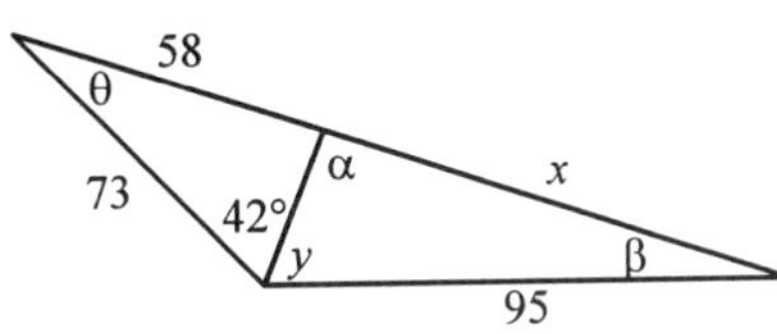

2. The following equations are equivalent forms of an equation that is one of the **Mollweide's check formulas**.

$$\frac{a-b}{c} = \frac{\sin\frac{1}{2}(A-B)}{\cos\frac{1}{2}C}; \quad \frac{b-a}{c} = \frac{\sin\frac{1}{2}(B-A)}{\cos\frac{1}{2}C}$$

These equations relate all the parts in a triangle and are therefore useful for checking solutions when solving triangles. Check the solution given in Example 2 of this section by using the form from above that produces positive results on both sides of the equation. If the same result is not obtained on both sides (with minor allowances for round-off error), then the solution is incorrect.

3. In a triangle if A, a, and b are given and $a = b\sin A$, then use the law of sines to show that the given conditions determine a right triangle.

4. Prove, using the law of sines, that if the measures of two angles of a triangle are equal, then the lengths of the sides opposite these angles are equal.

5. In the figure below use the law of sines to show that if line segments AB and DE are parallel, then $e/b = d/a$. (This problem shows that if a line is drawn parallel to a side of a triangle, then the other two sides are divided proportionately.)

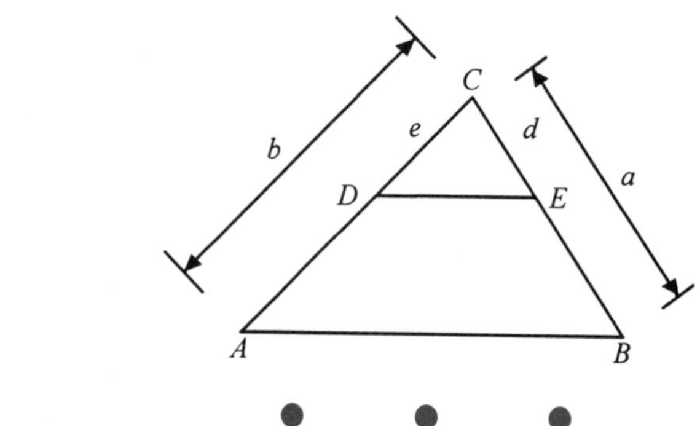

9.2 Law of Cosines and Area of Triangles

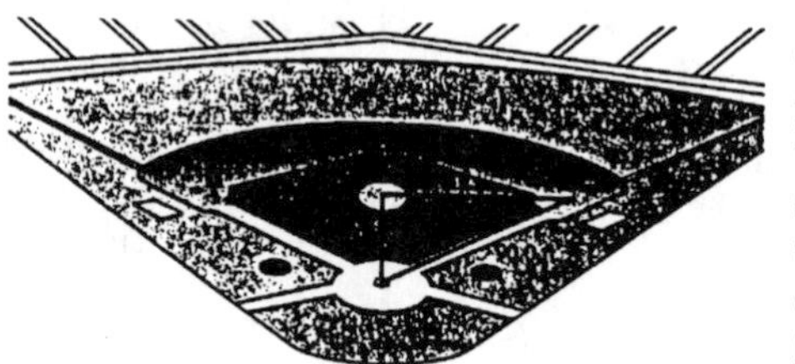

On a ground ball to the first base side, the pitcher is expected to cover first base and beat the batter to this bag. If a baseball diamond is a square that is 90.0 ft long on each side, and if the pitcher's mound is 60.5 ft from home plate on the diagonal from home to second base, then how far is the pitcher's mound from first base? (See Example 1.)

Objectives

1. Use the law of cosines to solve a triangle, given two sides of the triangle and the angle between these two sides.
2. Use the law of cosines to solve a triangle, given the lengths of three sides of the triangle.
3. Solve applied problems using the law of cosines.
4. Solve a problem involving a triangle by determining whether the law of sines or the law of cosines is appropriate for the problem.
5. Find the area of a triangle, given two sides of the triangle and the angle between these two sides.
6. Find the area of a triangle, given the lengths of three sides.

In Section 9.1 we found that the law of sines can be used to solve any triangle in the following two cases:

1. If we know the measures for **two angles** and **one side** of the triangle
2. If we know the measures for **two sides** of the triangle and the **angle opposite** one of them

However, there exist two other cases for which the law of sines cannot be applied. They are:

1. If we know the measures for **two sides** of the triangle and the **angle between** these two sides
2. If we know the measures for the **three sides** of the triangle

An illustration of case 3 is readily available when we try to find the distance from the pitcher's mound to first base in the section-opening problem. Because the pitcher's mound is on the diagonal from home to second base, the angle from the pitcher's mound to home to first base measures one-half of 90°, which is 45°, and this angle is *between* the two known sides. To solve this type of problem and problems involving case 4, we use the law of cosines, which states the following.

Law of Cosines

> In any triangle, the square of any side length equals the sum of the squares of the other two side lengths, minus twice the product of these other two side lengths and the cosine of their included angle.

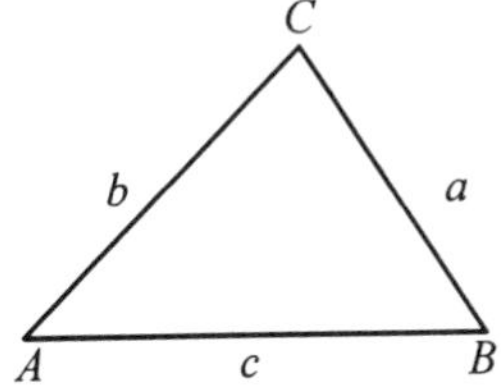

Figure 9.15

Thus, for triangle ABC in Figure 9.15 we have

$$a^2 = b^2 + c^2 - 2bc\cos A$$
$$b^2 = a^2 + c^2 - 2ac\cos B$$
$$c^2 = a^2 + b^2 - 2ab\cos C.$$

If we know the measures for two sides of the triangle and the included angle, we find the third side length by substituting in one of these formulas. After finding this part, we complete the solution by means of the law of sines. To obtain accuracy in the angle measures, the computed side length should be carried in the calculations to at least one more significant digit than stated in the solution.

If the three forms of the law of cosines are solved for the cosine of the angle, we have

$$\cos A = \frac{b^2 + c^2 - a^2}{2bc}$$
$$\cos B = \frac{a^2 + c^2 - b^2}{2ac}$$
$$\cos C = \frac{a^2 + b^2 - c^2}{2ab}.$$

These formulas are used to find the angle measures in a triangle when we know the three side lengths. In this case we do not use the law of sines to complete the solution because results are more accurate when they are computed from the data given.

Before starting the sample problems, let us first derive the law of cosines. Once again, we place triangle ABC on a rectangular coordinate system with angle A in standard position, and consider both an acute and obtuse possibility for angle A, as shown in Figure 9.16(a) and (b). In both cases vertex B obviously has coordinates $(c, 0)$. Also, in both cases the x-coordinate of vertex C is $b \cos A$, and the y-coordinate is $b \sin A$, because $\cos A = x/r$ and $\sin A = y/r$, where r is given by side length b in the triangles. If we now apply the distance formula to find the square of side length a, we have

$$\begin{aligned} a^2 &= (c - b\cos A)^2 + (0 - b\sin A)^2 \\ &= c^2 - 2bc\cos A + b^2\cos^2 A + b^2\sin^2 A \\ &= b^2\left(\sin^2 A + \cos^2 A\right) + c^2 - 2bc\cos A. \end{aligned}$$

Since $\sin^2 A + \cos^2 A = 1$ is an identity, the equation becomes

$$a^2 = b^2 + c^2 - 2bc\cos A,$$

which is one form of the law of cosines. A similar procedure with angles B and C, respectively, placed in standard position yields the other two forms.

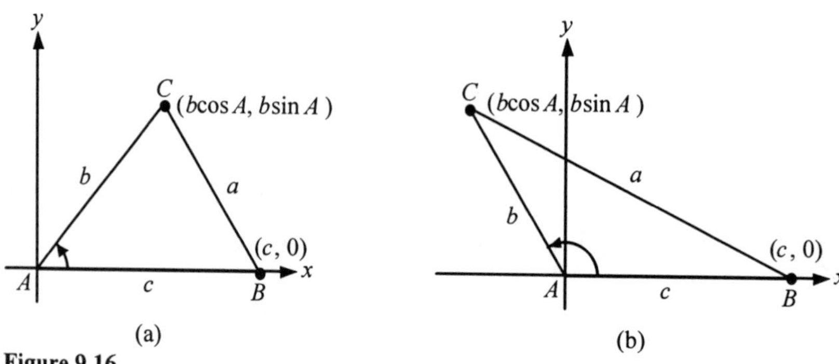

Figure 9.16

Example 1: Distance Between the Pitcher's Mound and First Base

Solve the problem in the section introduction on page 744.

Solution: Consider the simplified sketch of the problem in Figure 9.17. Observe that the angle between the two known sides measures 45° because the pitcher's mound is on the diagonal from home to second base, and the diagonals in a square bisect the right angles in the square. We can now find x by applying the law of cosines, which says that the square of any side length in a triangle equals the sum of the squares of the other two side lengths, minus twice the product of these other two side lengths and the cosine of their included angle.

$$\begin{aligned} x^2 &= (60.5)^2 + (90)^2 - 2(60.5)(90)\cos 45^\circ \\ &= 4{,}059.93 \\ x &= \sqrt{4{,}059.93} \approx 63.7 \end{aligned}$$

Thus, the pitcher's mound is about 63.7 ft from first base, which should give the pitcher an adequate head start in a race to the bag.

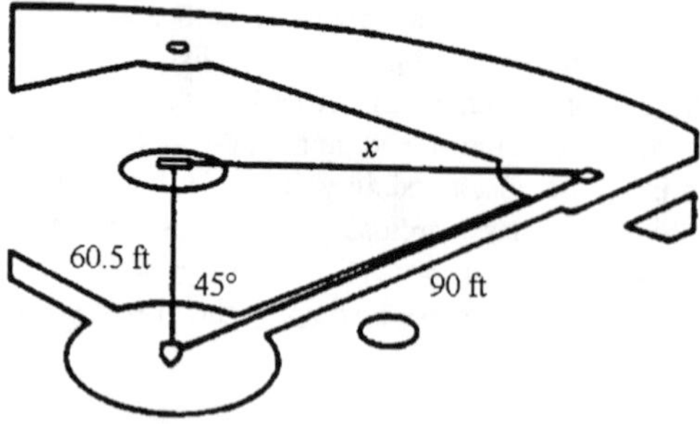

Figure 9.17

PROGRESS CHECK 1

What is the distance between the two islands shown in Figure 9.18? ■

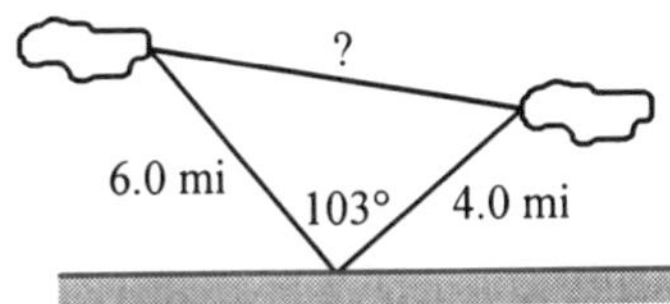

Figure 9.18

Example 2: Solving a Triangle: Side-Angle-Side Case (SAS)

Approximate the missing parts of triangle ABC in which $A = 60°$, $b = 25$ ft, and $c = 42$ ft.

Solution: First, sketch Figure 9.19. We find side length a by applying the law of cosines.

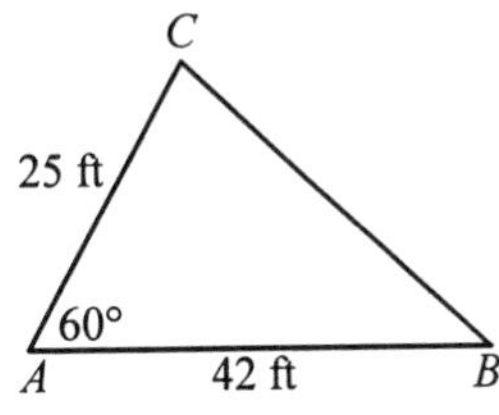

Figure 9.19

$$\begin{aligned} a^2 &= b^2 + c^2 - 2bc\cos A \\ &= (25)^2 + (42)^2 - 2(25)(42)\cos 60° \\ &= 1{,}339 \\ a &= \sqrt{1{,}339} \approx 36.6 \\ a &= 37 \text{ ft} \end{aligned}$$

Second, we find the *smaller* of the remaining angles, angle B, by applying the law of sines. This angle must be acute. (Why?) We use 36.6 for a, for better accuracy.

$$\begin{aligned} \frac{\sin A}{a} &= \frac{\sin B}{b} \\ \frac{\sin 60°}{36.6} &= \frac{\sin B}{25} \\ \frac{25\sin 60°}{36.6} &= \sin B \\ 0.5915 &= \sin B \\ 36° &= B \end{aligned}$$

(**Note:** $\sin B = 0.5915$ is true if $B = 36°$ or if $B = 144°$. We eliminate 144° as a possible solution, since we know that angle B must be acute.)

Third, we find angle C.

$$\begin{aligned} A + B + C &= 180° \\ 60° + 36° + C &= 180° \\ C &= 84° \end{aligned}$$

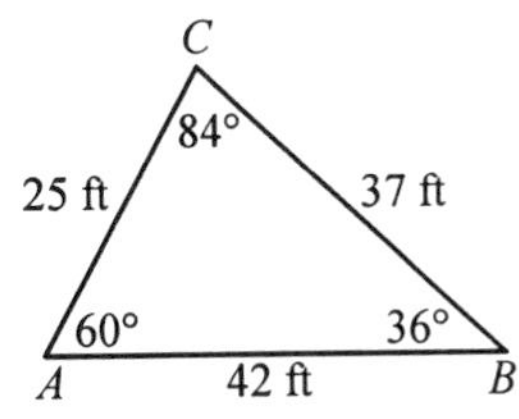

Figure 9.20

Thus, the solution to the triangle is as shown in Figure 9.20. ■

PROGRESS CHECK 2

Solve the triangle ABC in which $B = 48°$, $a = 31$ ft, and $c = 55$ ft.

Example 3: Solving a Triangle: Side-Side-Side Case (SSS)

Approximate the missing parts of triangle ABC in which $a = 23.5$ ft, $b = 44.2$ ft, and $c = 30.1$ ft.

Solution: First, sketch Figure 9.21. We find angle A by applying the law of cosines.

$$\cos A = \frac{b^2 + c^2 - a^2}{2bc}$$
$$= \frac{(44.2)^2 + (30.1)^2 - (23.5)^2}{2(44.2)(30.1)}$$
$$= 0.8672$$
$$A = 29.9°$$

(**Note:** Remember that if the cosine of the angle is positive, the angle is acute; if the cosine of the angle is negative, the angle is in Q_2 and is obtuse.)

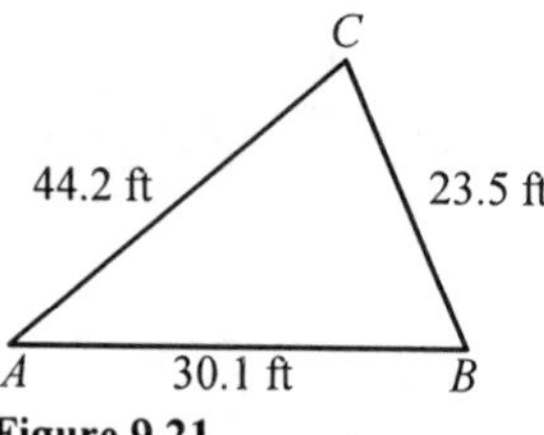

Figure 9.21

Second, we find the smaller of the remaining angles, acute angle *C*, by applying the law of cosines.

$$\cos C = \frac{a^2 + b^2 - c^2}{2ab}$$
$$= \frac{(44.2)^2 + (23.5)^2 - (30.1)^2}{2(44.2)(23.5)}$$
$$= 0.7701$$
$$C = 39.6°$$

Third, we find angle *B*.

$$A + B + C = 180°$$
$$29.9° + B + 39.6° = 180°$$
$$B = 110.5°$$

Thus, the solution to the triangle is as shown in Figure 9.22

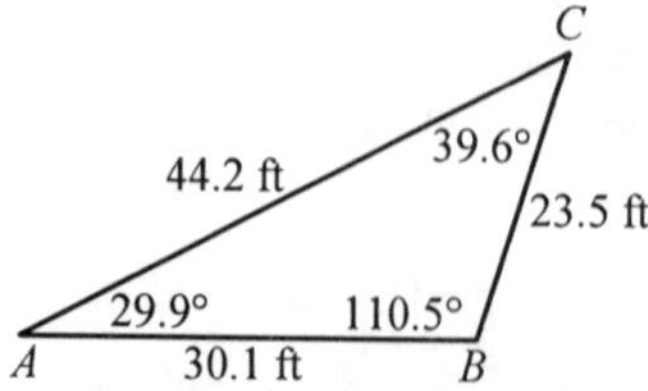

Figure 9.22

PROGRESS CHECK 3

Find the measures of the three angles in the triangle in which $a = 13$ ft, $b = 25$ ft, and $c = 22$ ft. ■

We have now illustrated all four cases for solving a triangle by applying the law of sines or the law of cosines. At this point it is useful to consider problems in which it is necessary to determine which of these two laws is the appropriate law for the problem in question.

Example 4: Choosing the Appropriate Law If parallelogram *ABCD* the lengths of sides *AB* and *AD* are 12 m and 19 m, respectively. If $A = 38°$, find the length of the longer diagonal of the parallelogram. Choose between the law of sines and the law of cosines to solve the problem.

<u>**Solution:**</u> First, sketch Figure 9.23 and note that the longer diagonal is *AC*. In the parallelogram $\overline{AB} = \overline{DC} = 12$ and the sum of angles *A* and *D* is 180°. Thus,

$$A + D = 180°$$
$$38° + D = 180°$$
$$D = 142°.$$

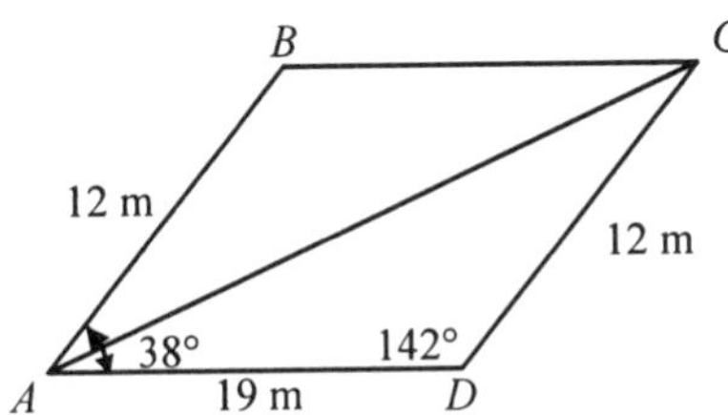

Figure 9.23

We now know two side lengths of a triangle involving diagonal *AC* and the measure of the angle between these two sides. Therefore, we can find the length of the longer diagonal by applying the law of cosines.

$$\left(\overline{AC}\right)^2 = \left(\overline{AD}\right)^2 + \left(\overline{DC}\right)^2 - 2\left(\overline{AD}\right)\left(\overline{DC}\right)\cos D$$
$$\left(\overline{AC}\right)^2 = (19)^2 + (12)^2 - 2(19)(12)\cos 142°$$
$$\left(\overline{AC}\right)^2 = 864.3$$
$$\overline{AC} = 29$$

Thus, the longer diagonal is about 29 m.

PROGRESS CHECK 4
In a parallelogram the shorter diagonal makes angles of 36° and 68° with the sides. If the length of the longer side is 8.0 yd, what is the length of the shorter side? Choose between the law of sines and the law of cosines to solve the problem. ■

Area of a Triangle

The area of a triangle is equal to one-half the product of its base and its altitude, which translates to the formula $A = (1/2)bh$. Because *h* is often unknown, it is useful to develop area formulas that do not require *h*. First, consider Figure 9.24 and note that in both cases by the general definition of the sine function

$$\sin A = \frac{h}{b}, \text{ so } h = b\sin A.$$

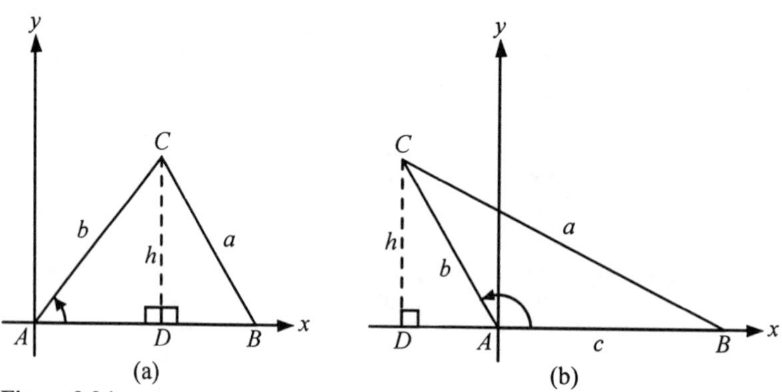

Figure 9.24
(a) A is an acute angle. (b) A is an obtuse angle.

Thus, if we represent the area by K (to avoid ambiguity with angle A), then

$$K = \frac{1}{2}(base)(height)$$
$$K = \frac{1}{2}c(b\sin A)$$
$$K = \frac{1}{2}bc\sin A.$$

Using similar reasoning with angle B and then angle C in standard position produces the formulas

$$K = \frac{1}{2}ac\sin B$$
$$K = \frac{1}{2}ab\sin C.$$

These formulas establish the following theorem.

Area of a Triangle

> The area of a triangle is given by one-half the product of the lengths of two sides and the sine of the angle between these two sides.

Example 5: Finding the Area of a Triangle

Find the area of triangle ABC if $a = 32.4$ cm, $b = 49.2$ cm, and $C = 18.5°$.

Solution: Using the formula containing a, b, and C, we have

$$\begin{aligned} K &= \frac{1}{2}ab\sin C \\ &= \frac{1}{2}(32.4)(49.2)\sin 18.5° \\ &= 253 \text{ cm}^2 \qquad \text{(three significant digits).} \end{aligned}$$

PROGRESS CHECK 5
Find the area of triangle ABC if $a = 14$ ft, $c = 11$ ft and $B = 37°$. ■

Example 6: Proving an Area Formula for a Triangle

Show by using the law of sines that the area K of triangle ABC may be given by

$$K = \frac{c^2 \sin A \sin B}{2\sin C}.$$

Solution: Consideration of the above equation suggests that we start with an area formula that contains the factor $c \sin A$ or the factor $c \sin B$. We select $K = (1/2)bc\sin A$ and note that by the law of sines

$$\frac{\sin B}{b} = \frac{\sin C}{c}, \text{ so } b = \frac{c\sin B}{\sin C}.$$

Then,

$$K = \frac{1}{2}bc\sin A = \frac{1}{2}\left(\frac{c\sin B}{\sin C}\right)c\sin A = \frac{c^2 \sin A \sin B}{2\sin C}.$$

PROGRESS CHECK 6

Show by using the law of sines that the area K of triangle ABC may be given by

$$K = \frac{b^2 \sin A \sin C}{2\sin B}$$

■

When the three side measures of a triangle are known, then the area may be determined by using the following formula, which is named after the mathematician Heron of Alexandria.

Heron's Area Formula

The area K of the triangle with side lengths a, b, and c is

$$K = \sqrt{s(s-a)(s-b)(s-c)}$$

where s is the semiperimeter, which is given by $(1/2)(a+b+c)$.

To derive this formula, start with one of the formulas for the area of a triangle in terms of two side lengths and the included angle, say $K = (1/2)bc\sin A$, and square both sides of the equation to obtain

$$\begin{aligned} K^2 &= \frac{1}{4}b^2c^2\sin^2 A = \frac{1}{4}b^2c^2\left(1-\cos^2 A\right) \\ &= \frac{bc}{2}(1+\cos A)\frac{bc}{2}(1-\cos A). \end{aligned}$$

Then by the law of cosines, $\cos A = \left(b^2+c^2-a^2\right)/(2bc)$, so

$$K^2 = \frac{bc}{2}\left(1+\frac{b^2+c^2-a^2}{2bc}\right)\frac{bc}{2}\left(1-\frac{b^2+c^2-a^2}{2bc}\right).$$

Through algebraic methods (see "Think About It" Question 4), this equation leads to

$$K^2 = \left(\frac{b+c+a}{2}\right)\left(\frac{b+c-a}{2}\right)\left(\frac{a-b+c}{2}\right)\left(\frac{a+b-c}{2}\right).$$

Finally, rewrite each factor so that it contains the expression $(a+b+c)/2$, which represents the semiperimeter s of the triangle. Then, K^2 equals

$$\left(\frac{a+b+c}{2}\right)\left(\frac{a+b+c}{2}-a\right)\left(\frac{a+b+c}{2}-b\right)\left(\frac{a+b+c}{2}-c\right).$$

Thus, $K^2 = s(s-a)(s-b)(s-c)$, so $K = \sqrt{s(s-a)(s-b)(s-c)}$.

Example 7: Using Heron's Area Formula

Find the area of triangle ABC if $a = 3.0$ ft, $b = 4.0$ ft, and $c = 6.0$ ft.

Solution: First, find the semiperimeter s.

$$s = \frac{1}{2}(a+b+c) = \frac{1}{2}(3+4+6) = 6.5$$

Then, Heron's formula gives

$$\begin{aligned} K &= \sqrt{s(s-a)(s-b)(s-c)} \\ &= \sqrt{6.5(6.5-3)(6.5-4)(6.5-6)} \\ &= 5.3 \text{ ft}^2 \qquad \text{(two significant digits).} \end{aligned}$$

PROGRESS CHECK 7

Find the area of triangle ABC is $a = 2.25$ km, $b = 3.07$ km, and $c = 2.08$ km. ■

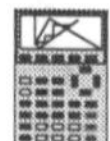

EXPLORE 9.2

1. The sides of a triangle must satisfy the requirement that the sum of the lengths of any two sides must be greater than the length of the third side.
   a. What would go wrong if the sum of the lengths of two of the sides was less than the length of the third side? **Hint:** Try to draw a triangle where one side is 10, and the others are 2.
   b. The law of cosines includes the lengths of all three sides of a triangle. Therefore, something peculiar should happen if the sides do not satisfy the requirement given above. Suppose $a = 10$, $b = 2$, and $c = 2$. Use a calculator to show that $\cos A = -11.5$, then find $A = \cos^{-1}(-11.5)$. Similarly, find B, and C. Some calculators will yield imaginary values for the angles, and some will yield an error message. Check the response of your calculator.
2. The mathematician Ptolemy, who lived in Alexandria, Egypt, about 150 A.D., did much work in developing trigonometry in his book called the *Almagest*. The book includes tables that were like tables of trigonometric functions, but they were neither sine nor cosine tables. Instead, they gave the lengths of chords of a circle. He used a circle of radius 60, as shown in the figure. [Source: A. Aaboe, Episodes from the Early History of Mathematics, Random House, 1964.]

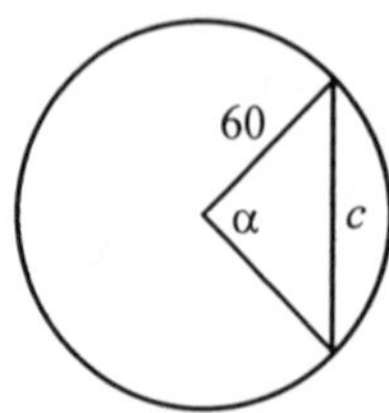

a. Using the definition of sine, show that $(c/2)/60 = \sin(\alpha/2)$. Then show that $c = 120\sin(\alpha/2)$. We will call this value crd α.

b. Use a calculator and the formula from part **a** to complete this table for crd α. Verify that the given values are correct. (These value are used in "Think About It" exercises.)

α	crd α
10°	
20°	
40°	41.0424
60°	
140°	112.7631

3. The 3–4–5 right triangle is very well known. This exercise uses the law of cosines to explore other triangles *ABC* whose sides are consecutive integers. From the law of cosines

$$A = \cos^{-1}\left(\frac{b^2 + c^2 - a^2}{2bc}\right).$$

a. Use this formula to determine angle A for each of these triangles. Note that A is the smallest angle in the triangle. Describe any pattern that you see. What do you think would happen if you continue the table?

a	b	c	A
2	3	4	
3	4	5	
4	5	6	
5	6	7	
6	7	8	
7	8	9	
8	9	10	

b. If you were to continue the table, what would be the smallest value of a for which $A \geq 59°$? Can A ever reach 60°?

c. Why was 1–2–3 omitted from the table?

EXERCISES 9.2

In Exercises 1–6 use the law of cosines.

1. A surveyor at point C sights two points A and B on opposite sides of a lake. If C is 760 ft from A and 920 ft from B, and angle ACB is 96°, how wide is the lake?

2. What is the distance between the two islands in the illustration shown?

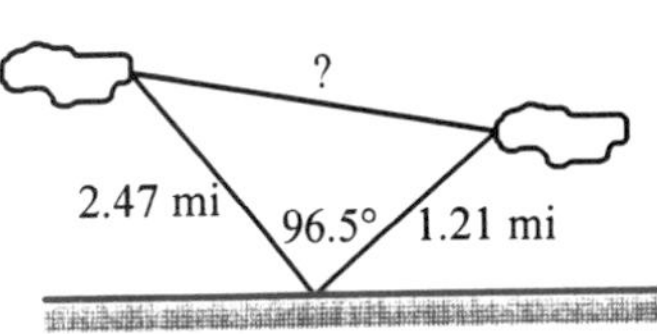

3. In parallelogram *ABCD* the lengths of sides *AB* and *AD* are 21 m and 13 m, respectively. If $A = 52°$, find the length of the longer diagonal of the parallelogram.

4. In parallelogram *ABCD* the length of sides *AB* and *AD* are 6.0 and 8.0 m, respectively. If the length of the shorter diagonal is 5.0 m, find angle *A*.

5. On a football field a kicker is positioned at point *B*, as shown in the figure. The goal post is 24 ft long, and the kicker is 35 yds from the right end of the goal post. Angle *A* is 50°. At what angle *B* must the ball be kicked so that it will cross the goal post in the center (at point *C*)?

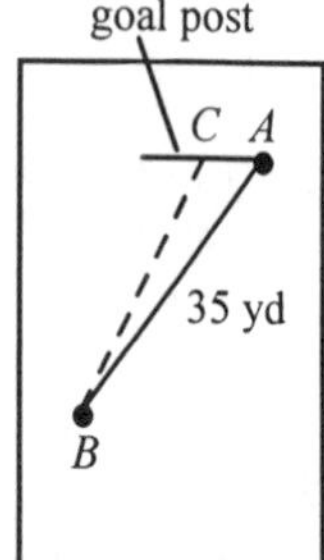

6. On a particular softball field the distance between the bases is $6\overline{0}$ ft, and the pitcher's mound is 43 ft from home plate. Find the distance from the pitcher's mound to first base.

In Exercises 7–16 approximate the remaining parts of triangle *ABC*. The law of cosines will be needed in at least one of the steps.

7. $a = 12$ ft, $b = 15$ ft, and $C = 60°$

8. $a = 2\overline{0}$ ft, $c = 3\overline{0}$ ft, and $B = 30°$

9. $c = 19.2$ ft, $a = 46.1$ ft, and $B = 10°20'$

10. $b = 36$ ft, $c = 75$ ft, and $A = 98°$

11. $b = 11.1$ ft, $a = 19.2$ ft, and $C = 95°40'$

12. $a = 11$ ft, $b = 15$ ft, and $c = 19$ ft

13. $a = 12$ ft, $b = 5.2$ ft, $c = 8.1$ ft

14. $a = 4.9$ ft, $b = 5.3$ ft, and $c = 2.6$ ft

15. $a = 34.4$ ft, $b = 56.1$ ft, and $c = 42.3$ ft

16. $a = 45.0$ ft, $b = 108$ ft, and $c = 117$ ft

In Exercises 17–20 refer to the following triangle and complete each statement by using the law of cosines.

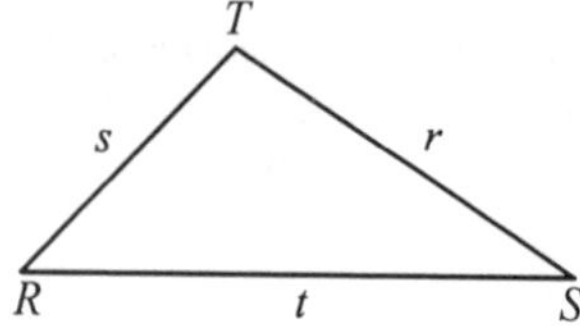

17. $r^2 =$ ______ **18.** $t =$ ______

19. $\cos T =$ ______ **20.** $\cos S =$ ______

In Exercises 21–24 assume exact numbers and find exact answers.

21. In triangle *ABC*, $a = 7$, $b = 5$, and $c = 6$. Find $\cos A$.

22. Find the cosine of the largest angle of the triangle whose sides measure 3, 5, and 6 units.

23. In triangle *RST*, $R = 60°$, $s = 7$ and $t = 4$. Find *r*.

24. Find the perimeter in triangle *ABC* if $A = 120°$, $b = 16$, and $c = 11$.

In Exercises 25–36 the problems are mixed; that is, some use the law of sines, some the law of cosines, and some use both. Solve each triangle.

25. $A = 15°$, $C = 87°$, and $b = 42$ ft

26. $B = 68°$ $C = 72°$, and $a = 18$ ft

27. $a = 4\overline{0}$ ft, $b = 5\overline{0}$ ft, and $C = 120°$

28. $b = 126$ ft, $c = 92.1$ ft, and $A = 72°50'$

29. $B = 111°20'$, $C = 35°40'$, and $a = 142$ ft

30. $c = 127$ ft, $b = 315$ ft, and $A = 162°30'$

31. $a = 4.0$ ft, $b = 2.0$ ft, and $c = 3.0$ ft

32. $A = 95°$, $a = 54$ ft, and $c = 38$ ft

33. $B = 7°10'$, $b = 74.8$ ft, and $c = 92.4$ ft

34. $a = 84.8$ ft, $b = 36.8$ ft, and $c = 76.5$ ft

35. $a = 15\overline{0}$ ft, $b = 175$ ft, and $c = 20\overline{0}$ ft

36. $B = 152°50'$, $b = 13\overline{0}$ ft, and $c = 45.0$ ft

In Exercises 37–40 the solution will require a decision as to whether the law of sines or the law of cosines is appropriate.

37. A ship sails due east for $4\overline{0}$ mi and then changes direction and sails 20° north of east for $6\overline{0}$ mi. How far is the ship from its starting point?

38. One gun is located at point *A*, while a second gun at point *B* is located 5.0 mi directly east of *A*. Form point *A* the direction to the target is 27° north of east. From point *B* the direction to the target is 72° north of east. For what firing range should the guns be set?

39. In a parallelogram the shorter diagonal makes angles of 25° and 72° with the sides. If the length of the shorter side is 15 m, what is the length of the longer side?

40. A and B are two points located on opposite edges of a lake. A third point C is located so that $\overline{AC}$ is 421 ft and $\overline{BC}$ is 376 ft. Angle ABC is measured to be 65.5°. Compute the distance $\overline{AB}$ across the lake.

In Exercises 41–52 find the area of the triangle satisfied by the given conditions.

41. $a = 7.0$ ft, $b = 4.0$ ft, and $C = 30°$

42. $a = 6.0$ ft, $b = 8.0$ ft, and $C = 150°$

43. $a = 12$ m, $c = 14$ m, and $B = 110°$

44. $a = 25$ m, $c = 19$ m, and $B = 70°$

45. $b = 4.74$ km, $c = 3.42$ km, and $A = 21.5°$

46. $b = 51.7$ cm, $c = 55.9$ cm, and $A = 16°50'$

47. $a = 5.0$ ft, $b = 4.0$ ft, and $c = 7.0$ ft

48. $a = 3.0$ m, $b = 4.0$ m, and $c = 3.0$ m

49. $a = 23.0$ m, $b = 14.0$ m, and $c = 18.0$ m

50. $a = 538$ ft, $b = 726$ ft, and $c = 981$ ft

51. $a = 2.51$ km, $b = 1.95$ km, and $c = 2.14$ km

52. $a = 42.56$ cm, $b = 37.83$ cm, and $c = 53.17$ cm

In Exercises 53–56 assume exact number and find exact answers.

53. Find the area of triangle ABC if $A = 120°$, $b = 16$, and $c = 11$.

54. Find the area of an isosceles triangle in which the vertex angle measures 30° and each leg measures 8 units.

55. Find the area of an equilateral triangle in which each side measures 6 units by using $K = \frac{1}{2}ab\sin C$.

56. Find the area of an equilateral triangle in which each side measures 6 units by using Heron's formula.

57. If ABC is a right triangle $(C = 90°)$, show that the law of cosines simplifies to the relationship $c^2 = a^2 + b^2$.

58. Write a formula for the area K of an isosceles triangle in which s and θ measure the legs and vertex angle, respectively.

59. Three circles are tangent externally, as shown in the diagram below. If the diameters of the circles are given by $d_1 = 32.0$ mm, $d_2 = 16.0$ mm, and $d_3 = 18.0$ mm, then find the area of the triangle joining their centers.

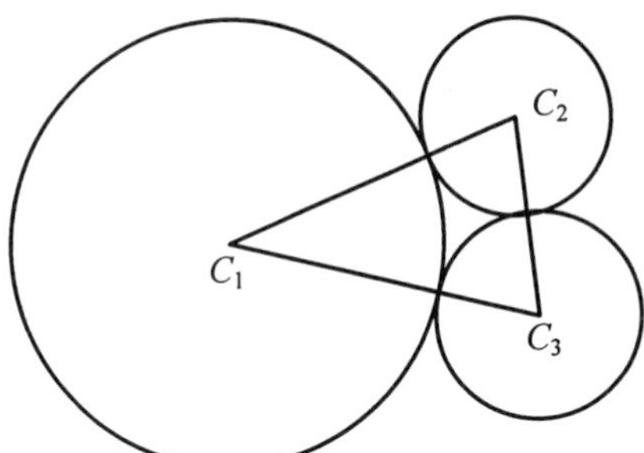

60. Show by using the law of sines that the area K of triangle ABC may be given by

$$K = \frac{a^2 \sin B \sin C}{2 \sin A}.$$

THINK ABOUT IT 9.2

1. **a.** Find the perimeter of the regular pentagon in the given illustration.
b. Find the area of the regular pentagon shown in the figure.

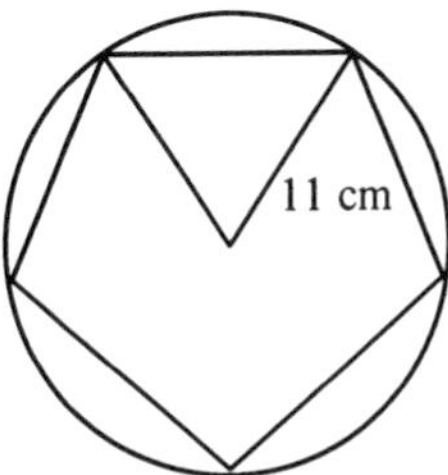

2. Show that the area K of an isosceles triangle in which s and β measure the legs and base angles, respectively, is given by

$$K = \frac{1}{2}s^2 \sin 2\beta.$$

3. Use Heron's formula to show that the area K of an isosceles triangle in which s and b measure the legs and base, respectively, is given by

$$K = \frac{1}{4}b\sqrt{4s^2 - b^2}.$$

4. In the derivation of Heron's formula show that

$$\frac{bc}{2}\left(1+\frac{b^2+c^2-a^2}{2bc}\right)\frac{bc}{2}\left(1-\frac{b^2+c^2-a^2}{2bc}\right)=\left(\frac{b+c+a}{2}\right)\left(\frac{b+c-a}{2}\right)\left(\frac{a-b+c}{2}\right)\left(\frac{a+b-c}{2}\right).$$

5. Using Ptolemy's table of chord lengths (described in Explore 2), it is possible to solve right triangles. First, you must use the fact that if you draw a circle through the three vertices of a right triangle, then the hypotenuse of the triangle is a diameter of that circle. See the figure.

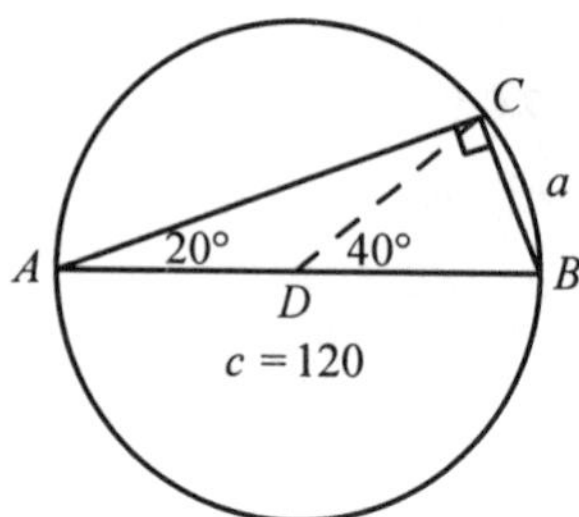

Suppose right triangle ABC with $C = 90°$ has $A = 20°$, and $c = 120$. Then side a is just the chord from Ptolemy's table for the angle $2A = 40°$. (Check that the central angle $\angle CDB$ is $2A$. Note that the two smaller triangles are isosceles.) Thus, $a = \text{crd}\,40°$ and $b = \text{crd}\,140°$. These values were given in Explore 2.

a. Use the tabled values from Explore 2 to solve right triangle ABC.

b. How would the answer to part **a** change if the length of side c were 12 instead of 120?

c. Use Ptolemy's table to solve the right triangle ABC where $A = 10°$ and $a = 30$.

● ● ●

9.3 Analytic Approach to Vectors

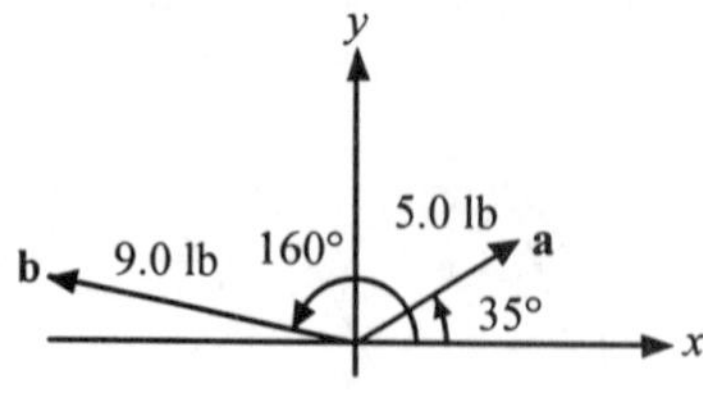

Two forces **a** and **b** act on an eyebolt connector as shown in the diagram. Find the vector sum or resultant **v** of forces **a** and **b** and give the magnitude and direction angle of **v**. (See Example 7.)

Objectives

1. Find the magnitude and direction angle of a vector.
2. Find sums, differences, and scalar multiples of vectors written in component form.
3. Prove vector properties involving vector addition and scalar multiplication.
4. Find sums, differences, and scalar multiples of vectors written as linear combinations of **i** and **j**.
5. Express a vector as a linear combination of **i** and **j** given the components or the magnitude and direction angle of the vector.

The basic question in the section-opening problem was first considered in Example 6 of Section 2.5, where vectors were approached from a geometric viewpoint. In this section we take an analytic approach to vectors and consider definitions and properties of vectors that enable us to analyze vectors using algebraic methods. We have seen that in a geometric approach, a vector is a directed line segment. Two vectors **AB** and **CD** are said to be **equal**, and we write **AB** = **CD**, if and only if they have the same length and direction (see Figure 9.25). This definition means that a vector may be shifted horizontally and vertically to different positions as long as its magnitude and direction are not changed. Thus, if we introduce a coordinate plane and place a vector **v** in this plane, then there are many representatives for **v**, as illustrated in Figure 9.26.

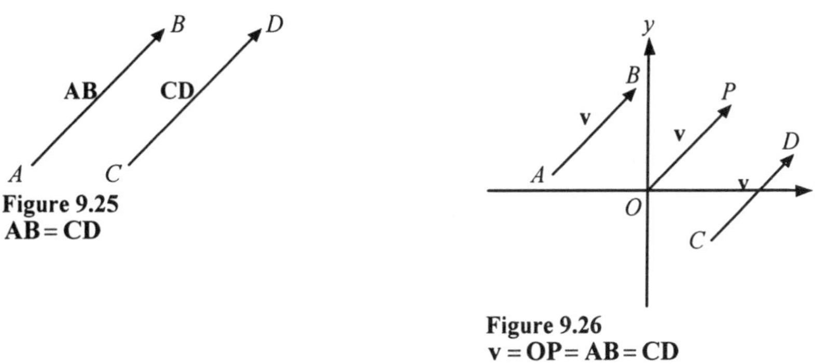

Figure 9.25
AB = CD

Figure 9.26
v = OP = AB = CD

It is usually most convenient to work with the representation of **v** that has its initial point at the origin. Such a vector is called a **position vector**, and vector **OP** in Figure 9.26 is an example of a position vector. A position vector **v** that terminates at the point (v_1, v_2), as shown in Figure 9.27, is called the **vector** $(\mathbf{v}_1, \mathbf{v}_2)$. This vector is written $\langle v_1, v_2 \rangle$ so that vectors and points are not confused, and the numbers v_1 and v_2 are called the **components** of $\langle v_1, v_2 \rangle$. A one-to-one correspondence has now been established between vectors in a plane and ordered pairs of real numbers, and to analyze vectors analytically means to work from the ordered-pair interpretation of a vector. Using the analytic approach, two vectors $\mathbf{v} = \langle v_1, v_2 \rangle$ and $w = \langle w_1, w_2 \rangle$ are **equal** if and only if

$$v_1 = w_1 \text{ and } v_2 = w_2$$

The **magnitude**, **length**, or **norm**, of **v**, symbolized $\|\mathbf{v}\|$, is given by the distance from the origin to the point (v_1, v_2), so

$$\|\mathbf{v}\| = \sqrt{v_1^2 + v_2^2}.$$

The **direction angle** θ of **v** is the smallest positive angle from the positive x-axis to the position vector, so if $v_1 \neq 0$, then

$$\tan\theta = \frac{v_2}{v_1}.$$

When $v_1 = 0$, then $\theta = 90°$ if $v_2 > 0$, and $\theta = 270°$ if $v_2 < 0$. As a special case, the vector $\langle 0, 0 \rangle$ is called the **zero vector** and denoted by a boldface **0**. The magnitude of **0** is 0, and the zero vector has no direction.

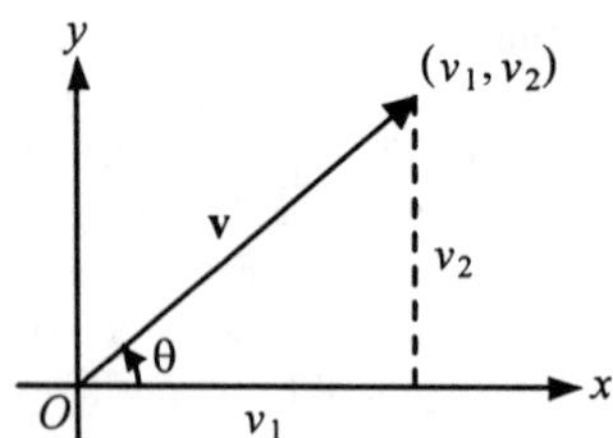

Figure 9.27

Example 1: Finding the Magnitude and Direction Angle

Find the magnitude and the direction angle of each vector.

a. $\mathbf{v} = \langle -1, \sqrt{3} \rangle$ **b.** $w = \langle 0, -3 \rangle$

Solution: (See Figure 9.28).

a. Replacing v_1 by -1 and v_2 by $\sqrt{3}$ in the magnitude formula gives

$$\|\mathbf{v}\| = \sqrt{(-1)^2 + (\sqrt{3})^2} = 2.$$

Now determine the direction angle θ.

$$\tan\theta = \frac{v_2}{v_1} = \frac{\sqrt{3}}{-1} = -\sqrt{3}$$

Since $(-1, \sqrt{3})$ is in Q_2 and the reference angle is 60°,

$$\theta = 180° - 60° = 120°.$$

b. The magnitude is

$$\|\mathbf{w}\| = \sqrt{0^2 + (-3)^2} = 3.$$

By considering the sketch of **w** in Figure 9.28, we conclude that the direction angle is 270°. We may also reasons that $\theta = 270°$ because $w_1 = 0$ and $w_2 < 0$.

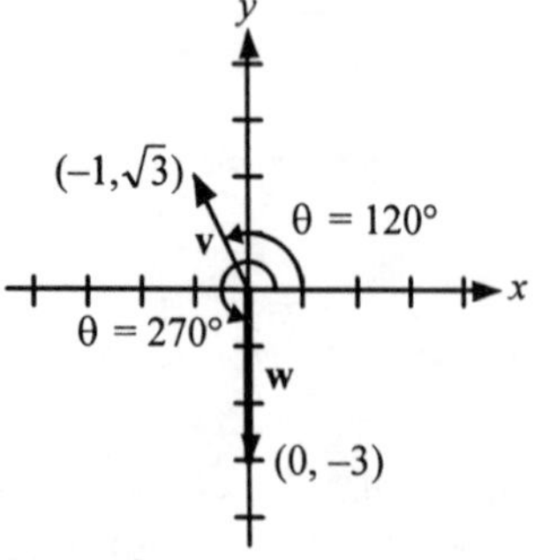

Figure 9.28

PROGRESS CHECK 1
Find the magnitude and direction angle of each vector.
a. $\mathbf{v} = \langle 0,\ 7\rangle$ **b.** $w = \langle 5,\ -5\rangle$ ■

The basic arithmetic operations on vectors, called vector addition and scalar multiplication, are defined below in component form. Note that when discussing vectors, the term **scalar** means a real number. In the sciences scalars are quantities, such as length or temperature, that are completely described by the magnitudes of the quantities.

Vector Addition and Scalar Multiplication

For vectors $\mathbf{v} = \langle v_1,\ v_2\rangle$ and $\mathbf{w} = \langle w_1,\ w_2\rangle$ and scalar k:

$$\mathbf{v} + \mathbf{w} = \langle v_1 + w_1,\ v_2 + w_2\rangle$$
$$k\mathbf{v} = \langle kv_1,\ \mathrm{k}v_2\rangle$$

To reinforce these definitions geometrically, consider Figure 9.29. Note Figure 9.29(a) shows that the component definition of vector addition is in agreement with the parallelogram method of the previous section for finding a vector sum. This figure also shows that the vector sum $\mathbf{v}+\mathbf{w}$ may be found geometrically by placing the initial point of **w** at the endpoint of **v**. Figure 9.29(b) shows the geometric interpretation of the vector $k\mathbf{v}$. The magnitude of $k\mathbf{v}$ is $|k|$ times the magnitude of **v**, and $k\mathbf{v}$ has the same direction as **v** if $k > 0$, and the opposite direction of **v** if $k < 0$. Finally, Figure 9.29(c) shows the geometric interpretation of $k\mathbf{v}$ in component form if $k > 1$.

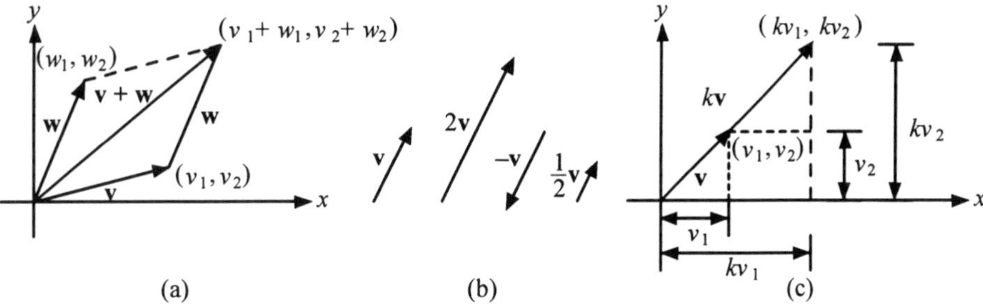

Figure 9.29

Example 2: Vector Addition and Scalar Multiplication

If $\mathbf{v} = \langle -3,\ 4\rangle$ and $\mathbf{w} = \langle 1,\ 5\rangle$, find the following vectors.
a. $\mathbf{v} + \mathbf{w}$ **b.** $5\mathbf{v}$ **c.** $-4\mathbf{w} + 3\mathbf{v}$

Solution: Apply the definitions of vector addition and scalar multiplication.
a. $\mathbf{v} + \mathbf{w} = \langle -3,\ 4\rangle + \langle 1,\ 5\rangle = \langle -3 + 1,\ 4 + 5\rangle = \langle -2,\ 9\rangle$
b. $5\mathbf{v} = 5\langle -3,\ 4\rangle + \langle 5(-3),\ 5(4)\rangle = \langle -15,\ 20\rangle$
c. $-4\mathbf{w} + 3\mathbf{v} = -4\langle 1,\ 5\rangle + 3\langle -3,\ 4\rangle = \langle -4,\ -20\rangle + \langle -9,\ 12\rangle = \langle -13,\ -8\rangle$

PROGRESS CHECK 2
If $\mathbf{v} = \langle 7,\ -2\rangle$ and $\mathbf{w} = \langle -5,\ 8\rangle$, find the following vectors.
a. $\mathbf{v} + \mathbf{w}$ **b.** $3\mathbf{w}$ **c.** $-2\mathbf{v} + 5\mathbf{w}$ ■

Analogous to arithmetic with real numbers, the **negative** or **additive inverse** of $\mathbf{v} = \langle v_1,\ v_2\rangle$ is defined by

$$-\mathbf{v} = \langle -v_1,\ -v_2\rangle \text{ so } -\mathbf{v} = (-1)\mathbf{v}\,.$$

and the **difference** of **w** and **v**, denoted by $\mathbf{w}-\mathbf{v}$ is defined by

$$\mathbf{w}-\mathbf{v}=\mathbf{w}+(-\mathbf{v}).$$

Thus, if $\mathbf{w}=\langle w_1, w_2\rangle$ and $\mathbf{v}=\langle v_1, v_2\rangle$, then

$$\mathbf{w}-\mathbf{v}=\langle w_1-v_1, w_2-v_2\rangle.$$

Figure 9.30 shows how to interpret a difference of two vectors geometrically. The difference $\mathbf{w}-\mathbf{v}$ is the vector from the endpoint of **v** to the endpoint of **w**. We can check this interpretation by the methods for vector addition, since $\mathbf{v}+(\mathbf{w}-\mathbf{v})=\mathbf{w}$.

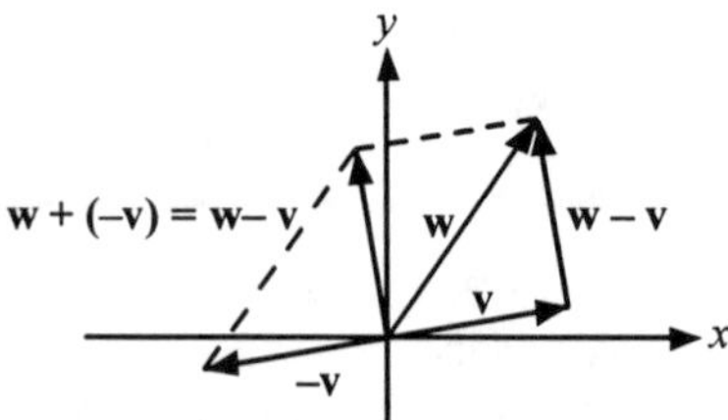

Figure 9.30

Example 3: Negative of a Vector and Vector Subtraction

If $\mathbf{v}=\langle 2, -6\rangle$ an $\mathbf{w}=\langle 1, -4\rangle$, find the following vectors.

a. $-\mathbf{w}$ **b.** $\mathbf{w}-\mathbf{v}$ **c.** $3\mathbf{v}-2\mathbf{w}$

Solution:

a. $-\mathbf{w}=\langle -(1), -(-4)\rangle=\langle -1, 4\rangle$

b. $\mathbf{w}-\mathbf{v}=\langle 1-2, -4-(-6)\rangle=\langle -1, 2\rangle$

c. Because $3\mathbf{v}=\langle 6, -18\rangle$ and $2\mathbf{w}=\langle 2, -8\rangle$, $3\mathbf{v}-2\mathbf{w}=\langle 6, -18\rangle-\langle 2, -8\rangle=\langle 4, -10\rangle$

PROGRESS CHECK 3

If $\mathbf{v}=\langle -1, 3\rangle$ and $\mathbf{w}=\langle 6, 5\rangle$, find the following vectors.

a. $-\mathbf{v}$ **b.** $\mathbf{v}-\mathbf{w}$ **c.** $5\mathbf{w}-4\mathbf{v}$ ■

Many of the familiar laws associated with real numbers carry over to working with vectors, and we list below the fundamental properties of vector addition and scalar multiplications.

Properties of Vector Addition and Scalar Multiplication

If **u**, **v**, and **w** are vectors, and *c* and *d* are scalars (real numbers), then

1. $\mathbf{u}+\mathbf{v}=\mathbf{v}+\mathbf{u}$
2. $(\mathbf{u}+\mathbf{v})+\mathbf{w}=\mathbf{u}+(\mathbf{v}+\mathbf{w})$
3. $\mathbf{u}+\mathbf{0}=\mathbf{u}$
4. $\mathbf{u}+(-\mathbf{u})=\mathbf{0}$
5. $c(\mathbf{u}+\mathbf{v})=c\mathbf{u}+c\mathbf{v}$
6. $c(d\mathbf{u})=(cd)\mathbf{u}$
7. $(c+d)\mathbf{u}=c\mathbf{u}+d\mathbf{u}$
8. $1\mathbf{u}=\mathbf{u}$

In Example 4 we prove analytically the fifth property listed above, and the proofs of the remaining properties are requested in the exercises and in Progress Check 4.

Example 4: Proving a Property of Vectors

Show that $c(\mathbf{u}+\mathbf{v}) = c\mathbf{u} + c\mathbf{v}$, where $\mathbf{u} = \langle u_1,\ u_2\rangle$, $\mathbf{v} = \langle v_1,\ v_2\rangle$ and c is a scalar.

Solution: The verification depends on the distributive property of real numbers that is used in the fourth step of the proof below.

$$\begin{aligned} c(\mathbf{u}+\mathbf{v}) &= c(\langle u_1,\ u_2\rangle + \langle v_1,\ v_2\rangle) \\ &= c\langle u_1+v_1,\ u_2+v_2\rangle \\ &= \langle c(u_1+v_1),\ c(u_2+v_2)\rangle \\ &= \langle cu_1+cv_1,\ cu_2+cv_2\rangle \\ &= \langle cu_1,\ cu_2\rangle + \langle cv_1,\ cv_2\rangle \\ &= c\langle u_1,\ u_2\rangle + c\langle v_1,\ v_2\rangle \\ &= c\mathbf{u} + c\mathbf{v} \end{aligned}$$

PROGRESS CHECK 4

Show that $c(d\mathbf{u}) = (cd)\mathbf{u}$, where $\mathbf{u} = \langle u_1,\ u_2\rangle$, and c and d are scalars. ■

Unit Basis Vectors

A **unit vector** is a vector of length 1. Two special unit vectors, denoted by **i** and **j**, are shown in Figure 9.31 and are defined as follows.

$$\mathbf{i} = \langle 1,\ 0\rangle, \qquad \mathbf{j} = \langle 0,\ 1\rangle$$

These vectors are especially useful because for any vector $\mathbf{v} = \langle v_1,\ v_2\rangle$, we have

$$\begin{aligned} \mathbf{v} &= \langle v_1,\ v_2\rangle = \langle v_1,\ 0\rangle + \langle 0,\ v_2\rangle \\ &= v_1\langle 1,\ 0\rangle + v_2\langle 0,\ 1\rangle = v_1\mathbf{i} + v_2\mathbf{j}. \end{aligned}$$

The vector sum $v_1\mathbf{i} + v_2\mathbf{j}$ is called a linear combination of **i** and **j**; and since every vector in a plane is expressible in this form, **i** and **j** are called **unit basis vectors**. To see a significant benefit of this notation, we restate below the rules for vector addition and scalar multiplication when vectors **v** and **w** are written as linear combinations of **i** and **j**.

$$\begin{aligned} (v_1\mathbf{i}+v_2\mathbf{j}) + (w_1\mathbf{i}+w_2\mathbf{j}) &= (v_1+w_1)\mathbf{i} + (v_2+w_2)\mathbf{j} \\ (v_1\mathbf{i}+v_2\mathbf{j}) - (w_1\mathbf{i}+w_2\mathbf{j}) &= (v_1-w_1)\mathbf{i} + (v_2-w_2)\mathbf{j} \\ c(v_1\mathbf{i}+v_2\mathbf{j}) &= cv_1\mathbf{i} + cv_2\mathbf{j} \end{aligned}$$

Note that operating on vectors in this form is like operating on ordinary algebraic expressions.

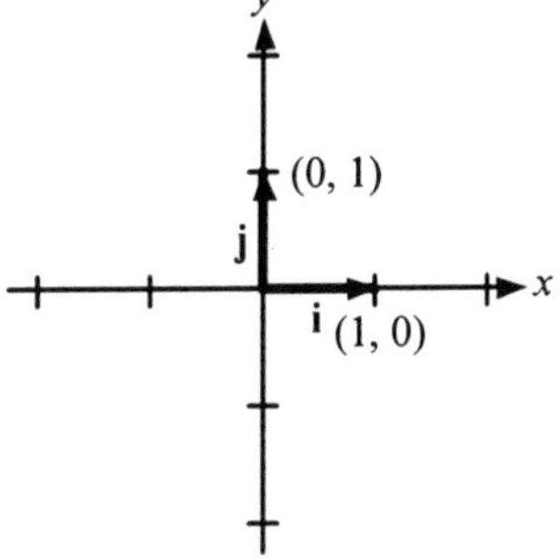

Figure 9.31

Example 5: Vector Operations Using Unit Basis Vectors

If $\mathbf{v} = 3\mathbf{i} - \mathbf{j}$ and $w = -2\mathbf{i} + 5\mathbf{j}$, express the following vectors as linear combinations of **i** and **j**.

a. $\mathbf{w} + \mathbf{v}$ **b.** $\mathbf{v} - \mathbf{w}$ **c.** $2\mathbf{w} - 3\mathbf{v}$

Solution:

a. $\mathbf{w} + \mathbf{v} = (-2\mathbf{i} + 5\mathbf{j}) + (3\mathbf{i} - \mathbf{j}) = \mathbf{i} + 4\mathbf{j}$

b. $\mathbf{v} - \mathbf{w} = (3\mathbf{i} - \mathbf{j}) - (-2\mathbf{i} + 5\mathbf{j}) = 5\mathbf{i} - 6\mathbf{j}$

c. $2\mathbf{w} - 3\mathbf{v} = 2(-2\mathbf{i} + 5\mathbf{j}) - 3(3\mathbf{i} - \mathbf{j}) = (-4\mathbf{i} + 10\mathbf{j}) - (9\mathbf{i} - 3\mathbf{j}) = -13\mathbf{i} + 13\mathbf{j}$

PROGRESS CHECK 5

If $\mathbf{v} = -7\mathbf{i} + \mathbf{j}$ and $\mathbf{w} = 4\mathbf{i} - 5\mathbf{j}$, express the following vectors as linear combinations of **i** and **j**.

a. $2\mathbf{v} + \mathbf{w}$ **b.** $\mathbf{w} - \mathbf{v}$ **c.** $5\mathbf{v} - 8\mathbf{w}$ ■

To express a vector **v** as a linear combination of **i** and **j** when given the magnitude and direction angle of **v**, note in Figure 9.32 that

$$\cos\theta = \frac{v_1}{\|\mathbf{v}\|} \quad \text{implies} \quad v_1 = \|\mathbf{v}\|\cos\theta$$

and

$$\sin\theta = \frac{v_2}{\|\mathbf{v}\|} \quad \text{implies} \quad v_2 = \|\mathbf{v}\|\sin\theta.$$

Thus, $\mathbf{v} = v_1\mathbf{i} + v_2\mathbf{j}$ may be written as

$$\mathbf{v} = \|\mathbf{v}\|(\cos\theta)\mathbf{i} + \|\mathbf{v}\|(\sin\theta)\mathbf{j}.$$

Consideration of Figure 9.32 also indicates why v_1 is called the **horizontal** or x component of $v_1\mathbf{i} + v_2\mathbf{j}$, while v_2 is called the **vertical** or y component of this vector.

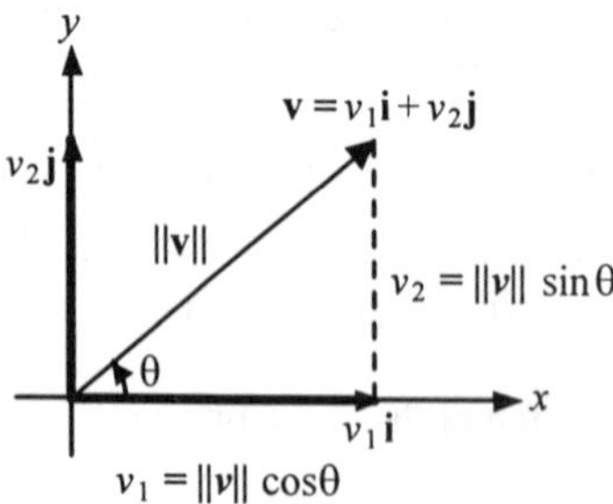

Figure 9.32

Example 6: Expressing a Vector in Terms of i and j

Express **v** as a linear combination of **i** and **j** if the magnitude of **v** is 8 and the direction angle of **v** is 120°.

Solution: Using $\|\mathbf{v}\| = 8$ and $\theta = 120°$ in the above form of **v** gives

$$\begin{aligned}\mathbf{v} &= \|\mathbf{v}\|(\cos\theta)\mathbf{i} + \|\mathbf{v}\|(\sin\theta)\mathbf{j}\\ &= 8(\cos 120^\circ)\mathbf{i} + 8(\sin 120^\circ)\mathbf{j}\\ &= 8\left(-\frac{1}{2}\right)\mathbf{i} + 8\left(\frac{\sqrt{3}}{2}\right)\mathbf{j}\\ &= -4\mathbf{i} + 4\sqrt{3}\mathbf{j}.\end{aligned}$$

PROGRESS CHECK 6

Express **v** as a linear combination of **i** and **j** if the magnitude of **v** is 14 and the direction angle of **v** is 210°. ■

In the next example we redo the basic question in Example 6 of Section 2.5, using our current notation and methods. Many applied problems may be solved in this way.

Example 7: Finding a Resultant Force

Solve the problem in the section introduction on page 756.

Solution: First, sketch Figure 9.33. Then use the methods of Example 6 to write **a** and **b** as linear combinations of **i** and **j**.

$$\begin{aligned}\mathbf{a} &= 5(\cos 35^\circ)\mathbf{i} + 5(\sin 35^\circ)\mathbf{j} = 4.096\mathbf{i} + 2.868\mathbf{j}\\ \mathbf{b} &= 9(\cos 160^\circ)\mathbf{i} + 9(\sin 160^\circ)\mathbf{j} = -8.457\mathbf{i} + 3.078\mathbf{j}\end{aligned}$$

The vector sum **v** of **a** and **b** is given by

$$\begin{aligned}\mathbf{v} &= \mathbf{a} + \mathbf{b}\\ &= (4.096\mathbf{i} + 2.868\mathbf{j}) + (-8.457\mathbf{i} + 3.078\mathbf{j})\\ &= -4.361\mathbf{i} + 5.946\mathbf{j},\end{aligned}$$

and the magnitude of the resultant is

$$\|\mathbf{v}\| = \sqrt{v_1^2 + v_2^2} = \sqrt{(-4.361)^2 + (5.946)^2} = 7.4$$

Now determine the direction angle.

$$\tan\theta = \frac{v_2}{v_1} = \frac{5.946}{-4.361} = -1.363$$

Since (−4.361, 5.946) is in Q_2 and the reference angle (by calculator) is 54°, we conclude that

$$\theta = 180^\circ - 54^\circ = 126^\circ.$$

In summary, the vector sum or resultant is $-4.361\mathbf{i} + 5.946\mathbf{j}$, and the magnitude and direction angle of the resultant are 7.4 lb and 126°, respectively.

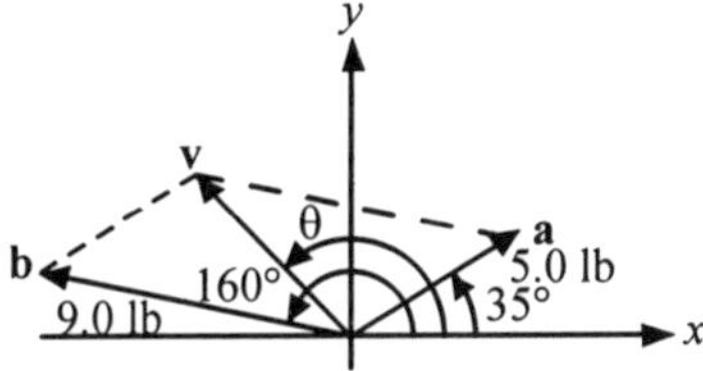

Figure 9.33

Technology Link

Some calculators have a vector capability that includes features on vector operations and features to convert between component form and magnitude-direction angle form. You should learn to use such features if they are available on your calculator.

PROGRESS CHECK 7
Three forces **a**, **b**, and **c** act in the same plane on a tree stump. Force **a** has a magnitude of 25 lb and a direction angle of 82°. Force **b** has a magnitude of 47 lb and a direction angle of 146°. Force **c** has a magnitude of 38 lb and a direction angle of 61°. Find the vector sum or resultant **v** of forces **a**, **b**, and **c** and give the magnitude and direction angle of **v**. ■

EXPLORE 9.3

Some calculators have special vector capabilities, including special vector operations. Others have matrix capability, which can be used for vector manipulation. For the latter type, you may represent a vector as a 1 by 2 matrix. These explorations encourage you to experiment with these features. The behavior of the calculator may depend on the mode settings. In order to enter vectors in component notation, you may need to set the mode to "rectangular."

1. Use the calculator vector command for "norm" or "length" to show that the magnitude of $\langle -1, \sqrt{3} \rangle$ is 2.
2. Use the calculator vector command for conversion to polar form to show that $\langle -1, \sqrt{3} \rangle$ has direction angle 120°.
3. a. Use calculator vector commands to name and store vectors **v** and **w** if $\mathbf{v} = \langle -3, 4 \rangle$ and $\mathbf{w} = \langle 1, 5 \rangle$.
   b. Use calculator vector commands to show that $-4\mathbf{w} + 3\mathbf{v} = \langle -13, -8 \rangle$, given the two vectors in part **a**.
   c. The dot product of two vectors is defined before Exercises 63–66 of this section. Use calculator vector commands to compute $\mathbf{u} \cdot \mathbf{v}$ for the vectors given in part **a**.

EXERCISES 9.3

In Exercises 1–10 find the magnitude and the direction angle of each vector.

1. $\langle -1, 1 \rangle$ 2. $\langle 2, 2 \rangle$ 3. $\langle -\sqrt{3}, -1 \rangle$

4. $\langle 3, -3\sqrt{3} \rangle$ 5. $\langle 0, 1 \rangle$ 6. $\langle -5, 0 \rangle$

7. $\langle 5, -12 \rangle$ 8. $\langle -4, -3 \rangle$ 9. $\left\langle \frac{3}{5}, \frac{4}{5} \right\rangle$

10. $\left\langle -\frac{1}{2}\sqrt{3}, \frac{1}{2} \right\rangle$

In Exercises 11–22 let $\mathbf{v} = \langle 4, -3 \rangle$ and $\mathbf{w} = \langle -5, 1 \rangle$ and find the vectors.

11. $\mathbf{v} + \mathbf{w}$ 12. $\mathbf{w} + \mathbf{v}$ 13. $3\mathbf{w}$

14. $7\mathbf{v}$ 15. $-5\mathbf{v} + 2\mathbf{w}$ 16. $2\mathbf{v} + 3\mathbf{w}$

17. $-\mathbf{w}$ 18. $-\mathbf{v}$ 19. $\mathbf{v} - \mathbf{w}$

20. $\mathbf{w} - \mathbf{v}$ 21. $5\mathbf{w} - 2\mathbf{v}$ 22. $3\mathbf{v} - 4\mathbf{w}$

In Exercises 23–26 sketch vectors corresponding to **v**, **w**, $\mathbf{v} + \mathbf{w}$, $2\mathbf{v}$, $-\mathbf{v}$, and $\mathbf{w} - \mathbf{v}$.

23. $\mathbf{v} = \langle 1, 0 \rangle$, $\mathbf{w} = \langle 0, 1 \rangle$

24. $\mathbf{v} = \langle 0, -2 \rangle$, $\mathbf{w} = \langle -2, 0 \rangle$

25. $\mathbf{v} = \langle 3, 2 \rangle$, $\mathbf{w} = \langle 1, 4 \rangle$

26. $\mathbf{v} = \langle -2, 4 \rangle$, $\mathbf{w} = \langle 5, -1 \rangle$

In Exercises 27–32 express
a. $\mathbf{v} + \mathbf{w}$, b. $\mathbf{v} - \mathbf{w}$, and
c. $5\mathbf{w} - 2\mathbf{v}$ as linear combinations of **i** and **j**.

27. $\mathbf{v} = \mathbf{i} + \mathbf{j}$, $\mathbf{w} = 3\mathbf{i} + 2\mathbf{j}$

28. $\mathbf{v} = \mathbf{i} - \mathbf{j}$, $\mathbf{w} = -\mathbf{i} - \mathbf{j}$

29. $\mathbf{v} = \mathbf{i} - 3\mathbf{j}$, $\mathbf{w} = -5\mathbf{i} + 2\mathbf{j}$

30. $\mathbf{v} = 4\mathbf{i} - \mathbf{j}$, $\mathbf{w} = -\mathbf{i} + 7\mathbf{j}$

31. $\mathbf{v} = \mathbf{i}$, $\mathbf{w} = 5\mathbf{j}$ 32. $\mathbf{v} = -2\mathbf{i}$, $\mathbf{w} = \mathbf{j}$

In Exercises 33–40 express the vectors as linear combinations of **i** and **j**.

33. $\langle 4, 7\rangle$ 34. $\langle -9, -2\rangle$ 35. $\langle 0, -5\rangle$

36. $\langle 3, 0\rangle$

37. magnitude 6, direction angle 180°

38. magnitude 3, direction angle 300°

39. magnitude 25, direction angle 120°

40. magnitude 11, direction angle 201°

In Exercises 41–46 find the horizontal and vertical component of each vector.

41. $2\mathbf{i} - 8\mathbf{j}$ 42. $-\mathbf{i} + 7\mathbf{j}$

43. magnitude 5, direction angle 135°

44. magnitude 17, direction angle 270°

45. magnitude 8, direction angle 330°

46. magnitude 16, direction angle 303°

In Exercises 47–50 find the magnitude and direction angle of **v**.

47. $\mathbf{v} = 3\mathbf{i} - 3\mathbf{j}$ 48. $\mathbf{v} = -\mathbf{i} - \mathbf{j}$

49. $\mathbf{v} = 7\mathbf{j}$ 50. $\mathbf{v} = -10\mathbf{i}$

In Exercises 51–56 prove the given properties (which were started in this section), where $\mathbf{u} = \langle u_1, u_2\rangle$, $\mathbf{v} = \langle v_1, v_2\rangle$, $\mathbf{w} = \langle w_1, w_2\rangle$, and c and d are scalars.

51. $\mathbf{u} + \mathbf{0} = \mathbf{u}$ 52. $\mathbf{u} + (-\mathbf{u}) = \mathbf{0}$

53. $\mathbf{u} + \mathbf{v} = \mathbf{v} + \mathbf{u}$ 54. $1\mathbf{u} = \mathbf{u}$

55. $(\mathbf{u} + \mathbf{v}) + \mathbf{w} = \mathbf{u} + (\mathbf{v} + \mathbf{w})$

56. $(c + d)\mathbf{u} = c\mathbf{u} + d\mathbf{u}$

In Exercises 57–62 prove the following additional properties of vector addition and scalar multiplication.

57. $-\mathbf{u} = (-1)\mathbf{u}$ 58. $0\mathbf{u} = 0$

59. $\|-2\mathbf{v}\| = 2\|\mathbf{v}\|$ 60. $c(\mathbf{u} - \mathbf{v}) = c\mathbf{u} - c\mathbf{v}$

61. Prove the vector **i** is a unit vector.

62. The magnitude of $c\mathbf{v}$ is $|c|$ times the magnitude of **v**. That is, $\|c\mathbf{v}\| = |c|\|\mathbf{v}\|$.

The **dot product** $\mathbf{u} \cdot \mathbf{v}$ of two vectors $\mathbf{u} = \langle u_1, u_2\rangle$ and $\mathbf{v} = \langle v_1, v_2\rangle$ is defined by

$$\mathbf{u} \cdot \mathbf{v} = u_1 v_1 + u_2 v_2$$

Note the dot product of two vectors is a real number (not a vector). For this reason the dot product is sometimes called the **scalar product**. In Exercises 63–66 find $\mathbf{u} \cdot \mathbf{v}$ for the given vectors.

63. $\mathbf{u} = \langle 4, 5\rangle$, $\mathbf{v} = \langle 3, -2\rangle$

64. $\mathbf{u} = \langle 3, -4\rangle$, $\mathbf{v} = \langle -4, -3\rangle$

65. $\mathbf{u} = 3\mathbf{i}$, $\mathbf{v} = 4\mathbf{j}$

66. $\mathbf{u} = \mathbf{i} - \mathbf{j}$, $\mathbf{v} = 2\mathbf{i} + 7\mathbf{j}$

In Exercises 67–72 find the vector sum or resultant **v** of the given vectors and give the magnitude and direction of **v**.

	Vector	*Magnitude*	*Direction*
67.	**a**	5.0 lb	26°
	b	3.0 lb	84°
68.	**c**	6.0 lb	95°
	d	9.0 lb	15°
69.	**c**	12 lb	110°
	d	15 lb	180°
70.	**a**	4.0 lb	90°
	b	3.0 lb	190°
71.	**a**	$1\bar{0}$ lb	42°
	b	$2\bar{0}$ lb	140°
	c	$3\bar{0}$ lb	240°
72.	**a**	5.0 lb	60°
	b	2.0 lb	210°
	c	1.0 lb	270°

In Exercises 73–76 use the methods of this section to solve the problem. (**Note** Alternative methods for solving these problems were shown in section 2.5.)

73. A force of 3.0 lb and a force of 4.0 lb are acting on a body with an angle of 90° between the two forces. Find
    a. the magnitude of the resultant and
    b. the angle between the resultant and the larger force.

74. A ship is headed due east at $2\bar{0}$ knots (nautical miles per hour) while the current carries the ship due south at 5.0 knots. Find
    a. the speed of the ship and
    b. the direction (course) of the ship.

75. An airplane, pointed due west, is traveling 10° north of west at a rate of $\bar{400}$ mi/hr. This resultant course is due to a wind blowing north. Find

a. the velocity of the plane if there were no wind (that is, the vector pointing due west) and

b. the velocity of the wind (that is, the vector pointing due north).

76. A body is acted on by two forces with magnitudes of 78 lb and 42 lb that act at an angle of 25° with each other. Find

a. the magnitude of the resultant and

b. the angle between the resultant and the larger force.

THINK ABOUT IT 9.3

1. Show that if $\mathbf{v}$ is a nonzero vector and $\mathbf{v} = v_1\mathbf{i} + v_2\mathbf{j}$, then $\mathbf{u} = (v_1/\|\mathbf{v}\|)\mathbf{i} + (v_2/\|\mathbf{v}\|)\mathbf{j}$ is a unit vector having the same direction as $\mathbf{v}$.

2. Use the result in Question 1 and find a unit vector having the same direction as the given vector.

a. $\mathbf{v} = 3\mathbf{i} + 4\mathbf{j}$ **b.** $\mathbf{v} = \mathbf{i} - \mathbf{j}$

3. Use the law of cosines and the definition of dot product in Exercises 63–66 to show that if θ is the angle between two nonzero vectors $\mathbf{u} = \langle u_1, u_2 \rangle$ and $\mathbf{v} = \langle v_1, v_2 \rangle$, as shown below, then

$$\mathbf{u} \cdot \mathbf{v} = \|\mathbf{u}\| \, \|\mathbf{v}\| \cos\theta.$$

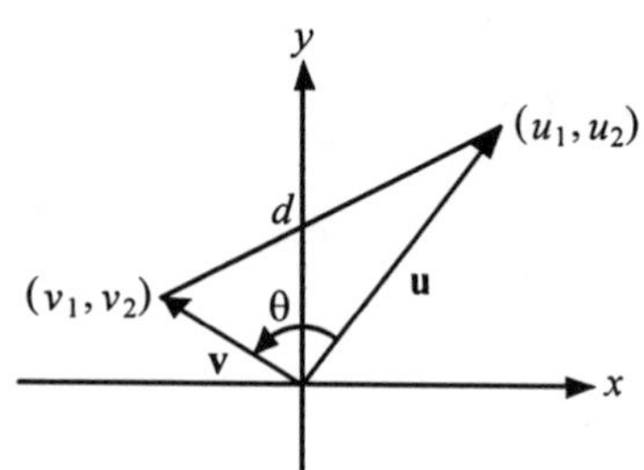

(**Note: The angle between two nonzero vectors u** and **v** is the angle θ, satisfying $0° \le \theta \le 180°$, that is determined by the position vectors of **u** and **v**.)

4. Two nonzero vectors are called **orthogonal** if and only if the angle between them measures 90°.

a. Show that two nonzero vectors **u** and **v** are orthogonal if and only if $\mathbf{u} \cdot \mathbf{v} = 0$.

b. Show that $\mathbf{u} = 2\mathbf{i} + 7\mathbf{j}$ and $\mathbf{v} = -7\mathbf{i} + 2\mathbf{j}$ are orthogonal.

5. Use the definition of dot product in Exercises 63–66 to prove the following properties of the dot product, where $\mathbf{u} = \langle u_1, u_2 \rangle$, $\mathbf{v} = \langle v_1, v_2 \rangle$, $\mathbf{w} = \langle w_1, w_2 \rangle$, and c is a scalar.

a. $\mathbf{u} \cdot \mathbf{v} = \mathbf{v} \cdot \mathbf{u}$ **b.** $\mathbf{v} \cdot \mathbf{v} = \|\mathbf{v}\|^2$

c. $c(\mathbf{u} \cdot \mathbf{v}) = (c\mathbf{u}) \cdot \mathbf{v}$ **d.** $\mathbf{u} \cdot (\mathbf{v} + \mathbf{w}) = \mathbf{u} \cdot \mathbf{v} + \mathbf{u} \cdot \mathbf{w}$

● ● ●

9.4 Trigonometric Form of Complex Numbers

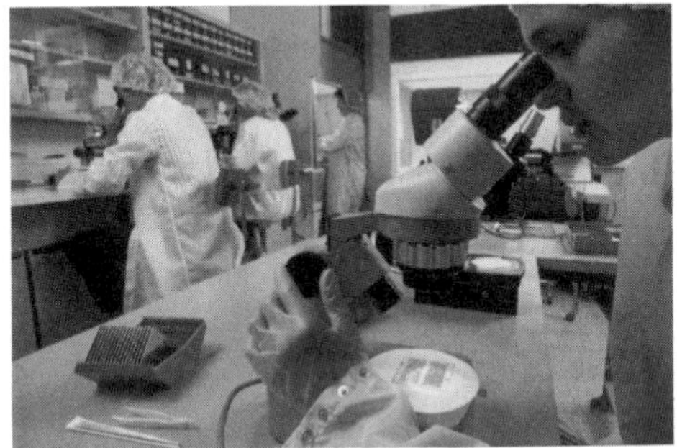

Photo Courtesy of **Eric Roth, Picture Cube**

In an alternating current circuit, the total impedance Z for two impedances Z_1 and Z_2 that are connected in series and in parallel is given by

Series		Parallel
$Z = Z_1 + Z_2$	and	$Z = \dfrac{Z_1 Z_2}{Z_1 + Z_2}.$

Find the complex number in trigonometric form that measures the total impedance if

$$Z_1 = 2.81(\cos 11.7° + i \sin 11.7°) \text{ ohms},$$
$$Z_2 = 3.54(\cos 57.2° + i \sin 57.2°) \text{ ohms},$$

and the given impedances are connected

a. in series and **b.** in parallel.

(See Example 5.)

Objectives

1. Graph a complex number $a+bi$ and find its absolute value.
2. Write a complex number $a+bi$ in trigonometric form.
3. Convert a complex number from trigonometric form to $a+bi$ form.
4. Find the product and quotient of complex numbers in trigonometric form.
5. Use De Moivre's theorem to find powers of complex numbers.
6. Use De Moivre's theorem to find roots of complex numbers and graph these roots.
7. Solve equations of the form $x^n = z$, where n is a positive integer.

Complex numbers were introduced in Section 3.6 from an algebraic perspective. We can now expand this coverage to include both geometric and trigonometric representations of such numbers. These representations will make more apparent why complex numbers can convey more information about a quantity. For example, a complex number is often used in physics to represent a vector, because this representation indicates both the magnitude and the direction of the vector. In electronics complex numbers are used extensively to represent voltage, current, and other electrical quantities, as illustrated in the section-opening problem. This representation is useful since it indicates both the strength and time (or phase) relationships of the quantities.

Each complex number $a+bi$ involves a pair of real numbers, a and b. Graphically, this means that we may represent a complex number as a point in a rectangular coordinate system. The values for a are plotted on the horizontal axis (x-axis) and the values for b are plotted on the vertical axis (y-axis). Thus, the complex number $a+bi$ is represented by the point (a, b) with x-value a and y value b. The plane on which complex numbers are graphed is called the **complex plane**, and in this context the horizontal axis is called the **real axis** and the vertical axis is called the **imaginary axis**. In Figure 9.34, the geometric representations of several complex numbers are shown.

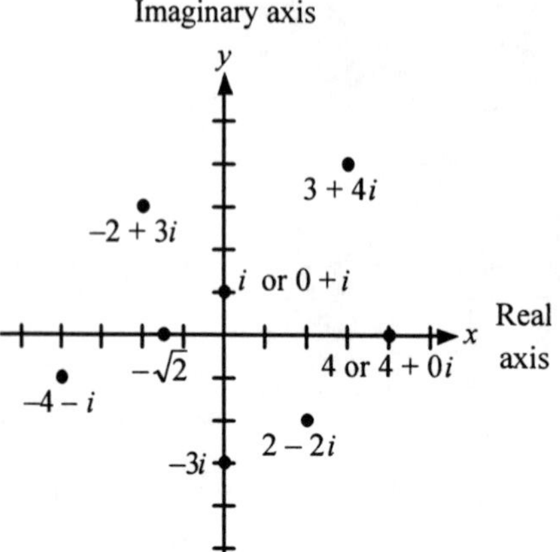

Figure 9.34

The **absolute value** or **modulus** of a complex number $a+bi$, denoted $|a+bi|$, is interpreted geometrically as the distance from the origin to the point (a, b), as shown in Figure 9.35. Thus, algebraically $|a+bi|$ is defined by

$$|a+bi|=\sqrt{a^2+b^2}.$$

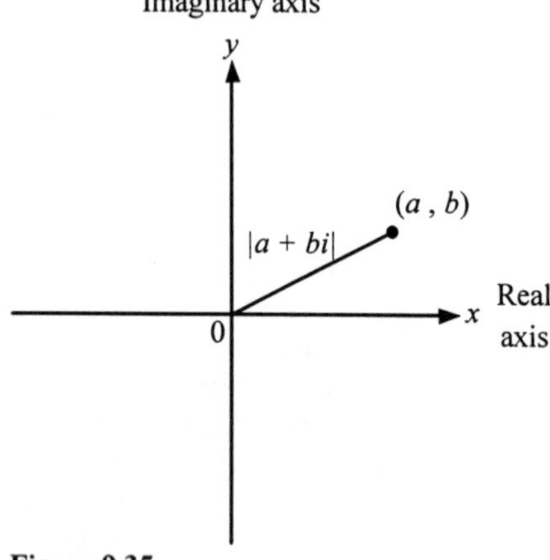

Figure 9.35

Example 1: Graph and Absolute Value of a Complex Number

Graph $-3+2i$ and find $|-3+2i|$.

Solution: The complex number $-3+2i$, which is represented by the point $(-3, 2)$, is graphed in Figure 9.36. Since $a=-3$ and $b=2$,

$$|-3+2i|=\sqrt{(-3)^2+2^2}=\sqrt{13}.$$

Thus, the absolute value of $-3+2i$ is $\sqrt{13}$.

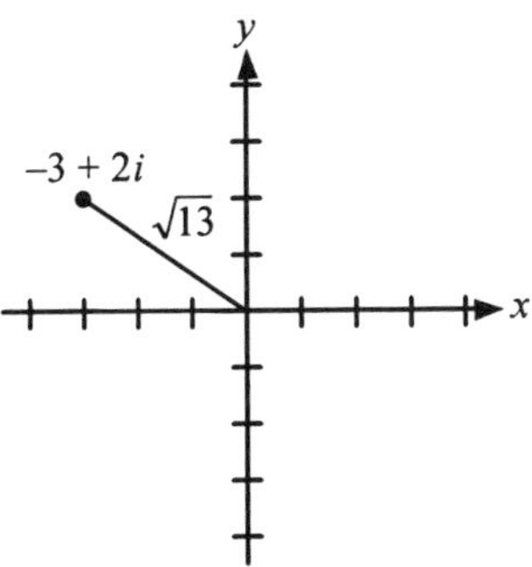

Figure 9.36

PROGRESS CHECK 1
Graph $8-6i$ and find the absolute value of this number. ■

A useful new way of writing a complex number is to interpret the position of its geometric representation in terms of the trigonometric functions. Consider the point corresponding to the nonzero complex number $a+bi$ in Figure 9.37. When we use trigonometry, we obtain the following relationships.

$$\sin\theta = \frac{b}{r} \text{ therefore } \boldsymbol{b = r\sin\theta}$$

$$\cos\theta = \frac{a}{r} \text{ therefore } \boldsymbol{a = r\cos\theta}$$

$$\boldsymbol{r = \sqrt{a^2+b^2}}$$

$$\boldsymbol{\tan\theta = \frac{b}{a}}$$

Thus, we may change the from of the complex number as follows:

$$a+bi = r\cos\theta + (r\sin\theta)i = r(\cos\theta + i\sin\theta).$$

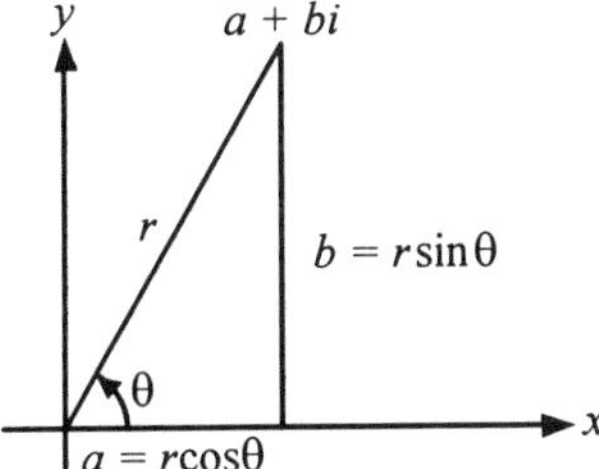

Figure 9.37

In this trigonometric form, r is the absolute value of $a+bi$, and θ is called the **argument** of the complex number. Note that the argument is not unique, since all angles of the form $\theta + k\cdot 360°$, where k is an integer, are coterminal to θ and serve as suitable choices for the argument. In most cases we use the **principal argument** of the complex number, which is the unique choice for θ that satisfies $0° \le \theta < 360°$.

Example 2: Expressing a Complex Number in Trigonometric Form

Write the number $3-3i$ in trigonometric form.

Solution: First, determine the absolute value, r. Since $a=3$ and $b=-3$, we have

$$r=\sqrt{a^2+b^2}=\sqrt{(3)^2+(-3)^2}=\sqrt{18} \text{ or } 3\sqrt{2}.$$

Now determine the argument,

$$\tan\theta=\frac{b}{a}=\frac{-3}{3}=-1.$$

Since $(3,\ -3)$ is in Q_4 and the reference angle is 45°, we conclude that

$$\theta=315^\circ.$$

The number is written in trigonometric form as

$$3\sqrt{2}(\cos 315^\circ+i\sin 315^\circ),$$

and this number is graphed in Figure 9.38. Observe that any angle coterminal with 315° may also be used. For instance, we may write $3\sqrt{2}[\cos(-45^\circ)+i\sin(-45^\circ)]$.

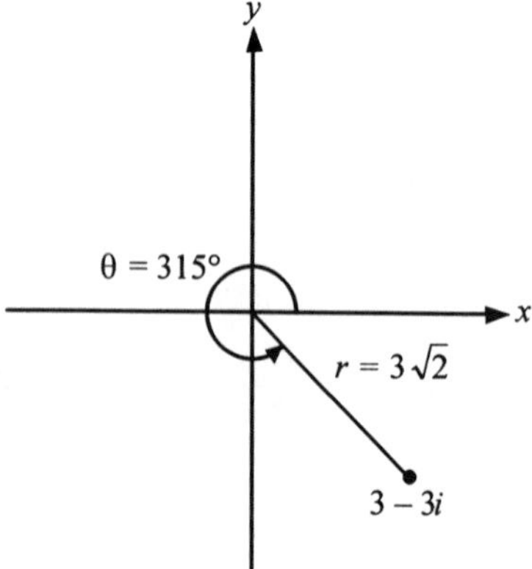

Figure 9.38

Note: In much of the literature dealing with complex numbers and their applications, the expression

$$r(\cos\theta+i\sin\theta)$$

is written as

$$r\operatorname{cis}\theta \text{ or } r\angle\theta.$$

These forms are merely convenient abbreviations, and the expressions are interchangeable.

PROGRESS CHECK 2

Write the number $-\sqrt{3}-i$ in trigonometric form. ■

Example 3: Converting a Complex Number to $a+bi$ Form

Write the number $2(\cos 150° + i\sin 150°)$ in the form $a+bi$.

Solution: Since $\cos 150° = -\sqrt{3}/2$ and $\sin 150° = 1/2$, we have

$$\begin{aligned} 2(\cos 150° + i\sin 150°) &= 2\left[\frac{-\sqrt{3}}{2} + i\left(\frac{1}{2}\right)\right] \\ &= -\sqrt{3} + i. \end{aligned}$$

PROGRESS CHECK 3

Write the number $6(\cos 225° + i\sin 225°)$ in the form $a+bi$. ■

Multiplication and division of complex numbers are easy when the numbers are in trigonometric form. We derive the product rule as follows:

$$\begin{aligned} &r_1(\cos\theta_1 + i\sin\theta_1)\cdot r_2(\cos\theta_2 + i\sin\theta_2) \\ &= r_1 r_2\left[\cos\theta_1\cos\theta_2 + i\cos\theta_1\sin\theta_2 + i\sin\theta_1\cos\theta_2 + i^2\sin\theta_1\sin\theta_2\right]. \end{aligned}$$

Replacing i^2 by -1 and grouping the terms yields

$$r_1 r_2[(\cos\theta_1\cos\theta_2 - \sin\theta_1\sin\theta_2) + i(\sin\theta_1\cos\theta_2 + \cos\theta_1\sin\theta_2)].$$

We now simplify the expressions in the parentheses by applying the formulas for the sine and cosine of the sum of two angles (see Section 8.8) and obtain our final result.

$$r_1(\cos\theta_1 + i\sin\theta_1)\cdot r_2(\cos\theta_2 + i\sin\theta_2) = r_1 r_2[\cos(\theta_1+\theta_2) + i\sin(\theta_1+\theta_2)]$$

In words, the product of two complex numbers is a third complex number. The absolute value is the product of the absolute values of the given numbers, and the argument is the sum of the arguments of the given numbers.

By a similar procedure we can obtain the following quotient rule.

$$\frac{r_1(\cos\theta_1 + i\sin\theta_1)}{r_2(\cos\theta_2 + i\sin\theta_2)} = \frac{r_1}{r_2}[\cos(\theta_1-\theta_2) + i\sin(\theta_1-\theta_2)], \text{ if } r_2 \neq 0$$

In words, the quotient of two complex numbers is a third complex number. The absolute value is the quotient of the absolute values of the given numbers, and the argument is the difference of the arguments of the given numbers.

Example 4: Multiplying and Dividing Complex Numbers

Determine

a. the product $z_1 \cdot z_2$ **b.** the quotient $\dfrac{z_1}{z_2}$ **c.** the quotient $\dfrac{z_2}{z_1}$

of the following complex numbers.

$$z_1 = 3(\cos 72° + i\sin 72°),\ z_2 = 5(\cos 43° + i\sin 43°)$$

Solution:

a. $3(\cos 72° + i\sin 72°)\cdot 5(\cos 43° + i\sin 43°) = 3\cdot 5[\cos(72°+43°) + i\sin(72°+43°)]$
$= 15(\cos 115° + i\sin 115°)$

b. $\dfrac{3(\cos 72° + i\sin 72°)}{5(\cos 43° + i\sin 43°)} = \dfrac{3}{5}[\cos(72°-43°) + i\sin(72°-43°)] = \dfrac{3}{5}(\cos 29° + i\sin 29°)$

c. $\dfrac{5(\cos 43° + i\sin 43°)}{3(\cos 72° + i\sin 72°)} = \dfrac{5}{3}[\cos(43°-72°) + i\sin(43°-72°)] = \dfrac{5}{3}[\cos(-29°) + i\sin(-29°)]$

or

$= \dfrac{5}{3}(\cos 331° + i\sin 331°)$

PROGRESS CHECK 4

Find

a. the product $z_1 \cdot z_2$, **b.** the quotient $\dfrac{z_1}{z_2}$, and

c. the quotient $\dfrac{z_2}{z_1}$ of the complex numbers given by

$$z_1 = 8(\cos 54° + i\sin 54°) \text{ and } z_2 = 4(\cos 90° + i\sin 90°).$$

■

Example 5: Total Impedance in Series and in Parallel

Solve the problem in the section introduction on page 767.

Solution:

a. The total impedance in a series connection is given by

$$Z = Z_1 + Z_2.$$

The values for Z_1 and Z_2 are given in trigonometric form, and we cannot compute a sum in this form. So we first convert each number to $a + bi$ form.

$$Z_1 = 2.81(\cos 11.7° + i\sin 11.7°) \approx 2.7516 + 0.5698i$$
$$Z_2 = 3.54(\cos 57.2° + i\sin 57.2°) \approx 1.9176 + 2.9756i$$

Now we add.

$$\begin{aligned} Z &= Z_1 + Z_2 \\ &= (2.7516 + 0.5698i) + (1.9176 + 2.9756i) \\ &= 4.6692 + 3.5454i. \end{aligned}$$

Finally, converting back to trigonometric form yields

$$Z = 5.86(\cos 37.2° + i\sin 37.2°) \text{ ohms.}$$

b. For a parallel connection, the total impedance is given by

$$Z = \frac{Z_1 Z_2}{Z_1 + Z_2}$$

The trigonometric form of complex numbers is ideal for finding products and quotients, and using the value for $Z_1 + Z_2$ that was determined in part **a** gives

$$\begin{aligned} Z &= \frac{2.81(\cos 11.7° + i\sin 11.7°)\cdot 3.54(\cos 57.2° + i\sin 57.2°)}{5.86(\cos 37.2° + i\sin 37.2°)} \\ &= \frac{2.81(3.54)}{5.86}[\cos(11.7° + 57.2° - 37.2°) + i\sin(11.7° + 57.2° - 37.2°)] \\ &= 1.70(\cos 31.7° + i\sin 31.7°) \text{ ohms.} \end{aligned}$$

Technology Link

Some graphing calculators have a complex number capability. If your calculator can perform operations with complex numbers, then you should redo the problems in Examples 1–5 and compare your results to the text's answers. Be advised that for some calculator models, the complex number

$$r(\cos\theta + i\sin\theta) \text{ is denoted by } (r\angle\theta).$$

To illustrate, Figure 9.39 shows the computations involved in Example 5, using a calculator that works with this notation and is set to a polar complex number display format.

```
(2.81 ∠ 11.7) + (3.54 ∠
57.2)
(5.86277661449 ∠ 37.20...
(2.81 ∠ 11.7) * (3.54 ∠
57.2)/(5.86 ∠ 37.2)
 (1.69750853242 ∠ 31.7)
```

Figure 9.39

PROGRESS CHECK 5

Find the total impedance if impedances Z_1 and Z_2 are connected

a. in series and **b.** in parallel, given that

$Z_1 = 4.09(\cos 24.5° + i\sin 24.5°)$ ohms and $Z_2 = 1.93(\cos 72.8° + i\sin 72.8°)$ ohms ■

When a complex number z is expressed in trigonometric form, it is easy to find powers of z. To establish a pattern that suggests a formula for z^n, we begin by establishing a formula for z^2 when $z = \cos\theta + i\sin\theta$.

$$\begin{aligned} z^2 &= [r(\cos\theta + i\sin\theta)][r(\cos\theta + i\sin\theta)] \\ &= r\cdot r[\cos(\theta + \theta) + i\sin(\theta + \theta)] \\ &= r^2(\cos 2\theta + i\sin 2\theta) \end{aligned}$$

Similarly, by writing z^3 as $z^2 \cdot z$ and applying the product rule, we obtain

$$z^3 = r^3(\cos 3\theta + i\sin 3\theta).$$

Continuing in this way suggests that

$$z^n = r^n(\cos n\theta + i\sin n\theta)$$

is true for every positive integer n, and in fact, this pattern generalizes to the following statement, which is known as De Moivre's theorem.

De Moivre's Theorem

If $r(\cos\theta + i\sin\theta)$ is any complex number and n is any real number, then

$$[r(\cos\theta + i\sin\theta)]^n = r^n(\cos n\theta + i\sin n\theta).$$

Although advanced mathematics is required to prove the above result, De Moivre's formula may be proved for every positive integer n by using mathematical induction (See Section 12.4), and the result may be extended to all integral exponents, as discussed in "Think About It" Question 5.

Example 6: Finding a Power of a Complex Number

Find $(1+i)^{12}$.

Solution: First, write the number in trigonometric form.

$$r = \sqrt{a^2+b^2} = \sqrt{(1)^2+(1)^2} = \sqrt{2}$$

$$\tan\theta = \frac{b}{a} = \frac{1}{1} = 1 \qquad \theta = 45°$$

Thus, $1+i = \sqrt{2}(\cos 45° + i\sin 45°)$. Now De Moivre's theorem tells us that

$$\begin{aligned}\left[\sqrt{2}(\cos 45° + i\sin 45°)\right]^{12} &= \left(\sqrt{2}\right)^{12}[\cos(12\cdot 45°) + i\sin(12\cdot 45°)]\\ &= \left(2^{1/2}\right)^{12}[\cos 540° + i\sin 540°]\\ &= 2^6[(-1)+i(0)]\\ &= -64.\end{aligned}$$

PROGRESS CHECK 6
Find $(1-i)^8$. ■

Analogous to roots of real numbers, an ***n*th root of a complex number** z is a complex number w such that

$$w^n = z,$$

where n is a positive integer. Any complex number has two square roots, three cube roots, four fourth roots, and so on. To find these roots we use De Moivre's theorem. However, to find all of the roots, you must remember that there are many trigonometric representations for the same complex number. For example, since the angle $\theta + k\cdot 360°$ (k any integer) is coterminal to θ, we have

$$\begin{aligned}1+i &= \sqrt{2}(\cos 45° + i\sin 45°) = \sqrt{2}(\cos 405° + i\sin 405°)\\ &= \sqrt{2}(\cos 765° + i\sin 765°), \text{ and so on.}\end{aligned}$$

Thus, the method is to find one root, add 360° to θ, use the new trigonometric representation to find another root, and repeat this procedure until all n roots are found.

Example 7: Find the *n*th Roots of a Number

Find and graph the five fifth roots of 32.

Solution: First, we write 32 in trigonometric form as

$$32(\cos 0° + i\sin 0°).$$

Now applying De Moivre's theorem, we have

$$[32(\cos 0° + i\sin 0°)]^{1/5} = 32^{1/5}\left(\cos\frac{0°}{5} + i\sin\frac{0°}{5}\right)$$
$$= 2(\cos 0° + i\sin 0°) = 2.$$

The first root is 2. To find another root, first add 360° to θ to obtain a different trigonometric representation for 32.

$$32 = 32[\cos(0° + 360°) + i\sin(0° + 360°)]$$

Then by De Moivre's theorem

$$\text{2nd root is } 32^{1/5}\left(\cos\frac{0° + 360°}{5} + i\sin\frac{0° + 360°}{5}\right) = 2(\cos 72° + i\sin 72°).$$

Repeating this procedure, we have

$$\text{3rd root is } 32^{1/5}\left(\cos\frac{0° + 2\cdot 360°}{5} + i\sin\frac{0° + 2\cdot 360°}{5}\right) = 2(\cos 144° + i\sin 144°)$$
$$\text{4th root is } 32^{1/5}\left(\cos\frac{0° + 3\cdot 360°}{5} + i\sin\frac{0° + 3\cdot 360°}{5}\right) = 2(\cos 216° + i\sin 216°)$$
$$\text{5th root is } 32^{1/5}\left(\cos\frac{0° + 4\cdot 360°}{5} + i\sin\frac{0° + 4\cdot 360°}{5}\right) = 2(\cos 288° + i\sin 288°).$$

The five roots are graphed in Figure 9.40, which illustrates that the roots all lie on a circle of radius 2 centered at O and that the arguments of consecutive roots differ by 72°. We do not have to consider arguments outside of the interval [0°, 360°) because the associated graphs will repeat points already obtained.

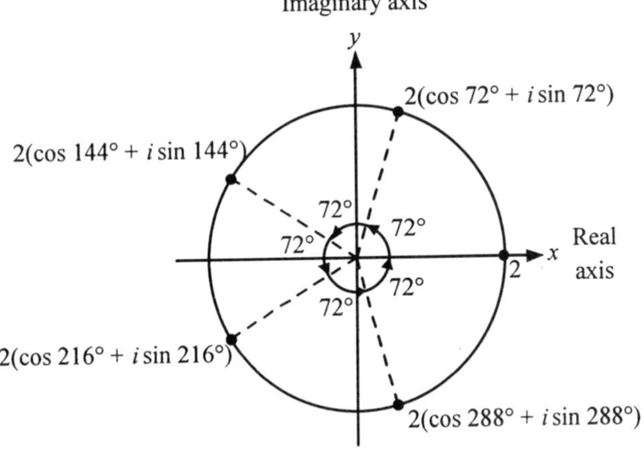

Figure 9.40

PROGRESS CHECK 7

Find and graph all the cube roots of –27. ■

The type of problem considered in Example 7 may be solved more efficiently by generalizing our current methods. Because $\theta + k \cdot 360°$, where k is an integer, generates all angles coterminal to θ, the complex number $z = r(\cos\theta + i\sin\theta)$ may also be represented by

$$z = r[\cos(\theta + k \cdot 360°) + i\sin(\theta + k \cdot 360°)].$$

Then, by De Moivre's theorem, this number may be written as

$$z = \left[r^{1/n}\left(\cos\frac{\theta + k \cdot 360°}{n} + i\sin\frac{\theta + k \cdot 360°}{n}\right)\right]^n.$$

Since $w^n = z$, where n is a positive integer, implies that w is an nth root of z, every complex number of the form

$$r^{1/n}\left(\cos\frac{\theta + k \cdot 360°}{n} + i\sin\frac{\theta + k \cdot 360°}{n}\right)$$

is an nth root of z. Furthermore, we obtain n distinct nth roots of z for $k = 0, 1, 2, \ldots, n-1$. Other integral values for k will lead to arguments that are coterminal to those previously obtained, and no different nth roots will be derived. Thus, we may find all of the nth roots of a complex number by applying the following formula.

***n*th Root Formula**

The n distinct nth roots of the complex number $r(\cos\theta + i\sin\theta)$ are given by

$$\sqrt[n]{r}\left(\cos\frac{\theta + k \cdot 360°}{n} + i\sin\frac{\theta + k \cdot 360°}{n}\right),$$

where $k = 0, 1, 2, \ldots, n-1$.

The geometric interpretation of the above result is that the graphs of the nth roots all lie on a circle of radius $\sqrt[n]{r}$ that is centered at the origin. Also, because the arguments of consecutive roots differ by $360°/n$, the nth roots are equally spaced on this circle.

Example 8: Using the *n*th Root Formula

Find and graph the three cube roots of $-8i$.

Solution: First, we write $-8i$ in trigonometric form as

$$8(\cos 270° + i\sin 270°).$$

By the nth root formula, the three distinct cube roots are given by

$$\sqrt[3]{8}\left(\cos\frac{270° + k \cdot 360°}{3} + i\sin\frac{270° + k \cdot 360°}{3}\right), \text{ where } k = 0, 1, 2,$$

which simplifies to

$$2[\cos(90°+k\cdot 120°)+i\sin(90°+k\cdot 120°)],\text{ where } k=0,\ 1,\ 2.$$

Finally, replacing k by 0, 1, and 2 yields the roots

$$2(\cos 90°+i\sin 90°)=2i$$
$$2(\cos 210°+i\sin 210°)=-\sqrt{3}-i$$
$$2(\cos 330°+i\sin 330°)=\sqrt{3}-i.$$

The points determined by these roots are on a circle of radius 2, as shown in Figure 9.41.

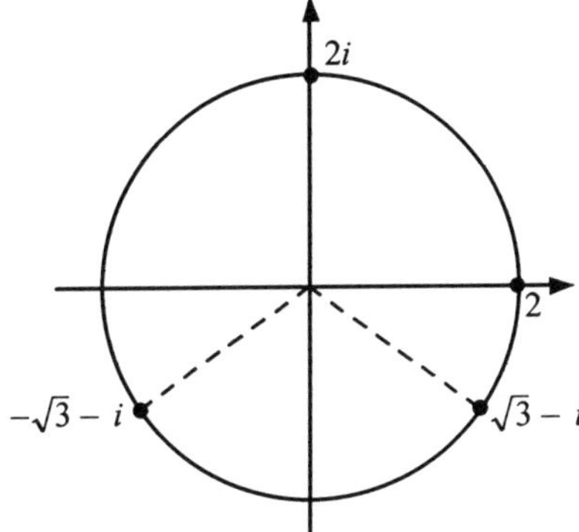

Figure 9.41

PROGRESS CHECK 8
Find and graph the four fourth roots of i. ■

The nth root formula may be used to solve equations that may be written in the form

$$x^n = z,\text{ where } n \text{ is a positive integer,}$$

since the roots of this equation are the n distinct nth roots of z. Note in the next example that the solutions are the sixth roots of 1, which is a special case. The nth roots of 1 are called the ***n*th roots of unity**, and such roots are especially useful in higher mathematics.

Example 9: Solving Equations of the Form $x^n = z$

Find all solutions for $x^6 = 1$.

Solution: To solve $x^6=1$, we find the six 6th roots of 1. In trigonometric form 1 is written as $1(\cos 0°+i\sin 0°)$, so the nth root formula with $n=6$ yields

$$\sqrt[6]{1}\left(\cos\frac{0°+k\cdot 360°}{6}+i\sin\frac{0°+k\cdot 360°}{6}\right),\text{where } k=0,\ 1,\ 2,\ 3,\ 4,\ 5,$$

which simplifies to

$$\cos(0°+k\cdot 60°)+i\sin(0°+k\cdot 60°),\text{ where } k=0,\ 1,\ 2,\ 3,\ 4,\ 5.$$

Then the six roots are:

$$\cos 0° + i\sin 0° = 1 \qquad \text{(for } k = 0\text{)}$$
$$\cos 60° + i\sin 60° = \frac{1}{2} + \frac{\sqrt{3}}{2}i \qquad \text{(for } k = 1\text{)}$$
$$\cos 120° + i\sin 120° = -\frac{1}{2} + \frac{\sqrt{3}}{2}i \qquad \text{(for } k = 2\text{)}$$
$$\cos 180° + i\sin 180° = -1 \qquad \text{(for } k = 3\text{)}$$
$$\cos 240° + i\sin 240° = -\frac{1}{2} - \frac{\sqrt{3}}{2}i \qquad \text{(for } k = 4\text{)}$$
$$\cos 300° + i\sin 300° = \frac{1}{2} - \frac{\sqrt{3}}{2}i \qquad \text{(for } k = 5\text{).}$$

The six roots, which are the six 6th roots of unity, are graphed in Figure 9.42.

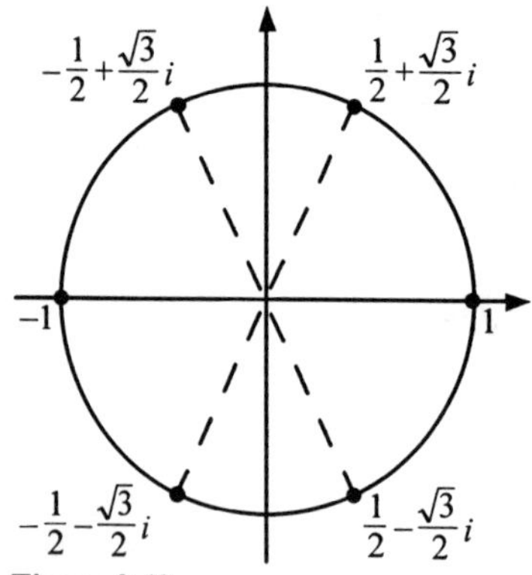

Figure 9.42

PROGRESS CHECK 9

Solve $x^4 + 16 = 0$.

■

EXPLORE 9.4

1. If your calculator has complex number capability, then it can accept complex numbers in $a + bi$ format (called **rectangular form**) or in $r(\cos\theta + i\sin\theta)$ format (called **trigonometric or polar form**). You will need to check the calculator format settings accordingly. Additionally, there is probably a command for converting from one form to the other. This exercise lets you explore these calculator capabilities. See the **Technology Link** in this section of the text for an illustration.
   a. Use Rectangular format to enter and find the sum of $z_1 = 3 + 4i$ and $z_2 = 5 + 3i$, then convert the answer to trigonometric form.
   b. Use Polar format to enter and find the sum of $z_1 = 2(\cos 30° + i\sin 30°)$ and $z_2 = 3(\cos 30° + i\sin 30°)$, then convert the answer to rectangular form.
2. It is possible to define an exponential function with a complex coefficient. Euler's formula states that $e^{i\theta} = \cos\theta + i\sin\theta$. In this formula θ is given in radians. Simplify this equation for the case where $\theta = \pi$ to derive one of the most famous identities in mathematics. If your calculator can operate with complex quantities, compute each of these values two ways, by using the Exponential key and by using the Sine and Cosine keys.
   a. $e^{-\pi i}$ **b.** $e^{(\pi/2)i}$ **c.** $e^{(\pi/5)i}$

EXERCISES 9.4

In Exercises 1–10 graph each complex number and find its absolute value.

1. $4+3i$ **2.** $5-12i$ **3.** $-1+2i$

4. $-2-i$ **5.** $\sqrt{3}-i$ **6.** $\sqrt{2}+\sqrt{2}i$

7. -3 **8.** 1 **9.** $4i$ **10.** $-2i$

In Exercises 11–24 write the number in trigonometric form.

11. 3 **12.** i **13.** $-2i$ **14.** -4

15. $1-i$ **16.** $-1+i$ **17.** $4+3i$

18. $-5-12i$ **19.** $-3-2i$ **20.** $7-4i$

21. $-1+\sqrt{3}i$ **22.** $\sqrt{3}-i$ **23.** $\sqrt{2}-\sqrt{2}i$

24. $-2-2\sqrt{3}i$

In Exercises 25–34 write the number in the form $a+bi$.

25. $3(\cos 90°+i\sin 90°)$ **26.** $5(\cos 0°+i\sin 0°)$

27. $4(\cos 180°+i\sin 180°)$

28. $\sqrt{2}(\cos 270°+i\sin 270°)$

29. $\sqrt{3}(\cos 120°+i\sin 120°)$

30. $2(\cos 210°+i\sin 210°)$

31. $2(\cos 225°+i\sin 225°)$

32. $\sqrt{6}(\cos 315°+i\sin 315°)$

33. $\cos 52°+i\sin 52°$ **34.** $10(\cos 115°+i\sin 115°)$

In Exercises 35–44 find

a. the product $z_1 \cdot z_2$ **b.** the quotient $\frac{z_1}{z_2}$ and

c. the quotient $\frac{z_2}{z_1}$ of the given complex numbers.

35. $z_1 = 2(\cos 52°+i\sin 52°)$, $z_2 = 4(\cos 11°+i\sin 11°)$

36. $z_1 = 6(\cos 7°+i\sin 7°)$, $z_2 = 9(\cos 90°+i\sin 90°)$

37. $z_1 = \cos 90°+i\sin 90°$, $z_2 = \cos 180°+i\sin 180°$

38. $z_1 = \cos 33°+i\sin 33°$, $z_2 = 4(\cos 63°+i\sin 63°)$

39. $z_1 = 3(\cos 131°+i\sin 131°)$, $z_2 = 12(\cos 205°+i\sin 205°)$

40. $z_1 = \cos 270°+i\sin 270°$, $z_2 = 5(\cos 3°+i\sin 3°)$

41. $z_1 = 2(\cos 300°+i\sin 300°)$, $z_2 = \sqrt{2}(\cos 45°+i\sin 45°)$

42. $z_1 = 3\sqrt{2}(\cos 315°+i\sin 315°)$, $z_1 = \cos 0°+i\sin 0°$

43. $z_1 = \cos(-20°)+i\sin(-20°)$,
$z_2 = \cos(-45°)+i\sin(-45°)$

44. $z_1 = 3(\cos 0°+i\sin 0°)$, $z_2 = 4[\cos(-90°)+i\sin(-90°)]$

In Exercises 45–58 use De Moivre's theorem and express the result in the form $a+bi$.

45. $[2(\cos 10°+i\sin 10°)]^3$

46. $[3(\cos 15°+i\sin 15°)]^4$

47. $[4(\cos 135°+i\sin 135°)]^5$

48. $[2(\cos 225°+i\sin 225°)]^6$

49. $(1+i)^8$ **50.** $(-1+i)^6$ **51.** $(1-\sqrt{3}i)^5$

52. $(-\sqrt{3}-i)^7$ **53.** $\left(\frac{\sqrt{2}}{2}+\frac{\sqrt{2}}{2}i\right)^{16}$

54. $\left(\frac{\sqrt{2}}{2}-\frac{\sqrt{2}}{2}i\right)^{14}$ **55.** $(1+\sqrt{3}i)^{-1}$

56. $(1-i)^{-1}$ **57.** $(-\sqrt{2}+\sqrt{2}i)^{-3}$

58. $(-\sqrt{3}+i)^{-4}$

In Exercises 59–70 find and graph all roots. Express the answers in both trigonometric form and in the form $a+bi$.

59. the square roots of $25(\cos 60°+i\sin 60°)$

60. the cube roots of $8(\cos 135°+i\sin 135°)$

61. the fourth roots of $16(\cos 80°+i\sin 80°)$

62. the square roots of $9(\cos 70°+i\sin 70°)$

63. the cube roots of 1 **64.** the fourth roots of -1

65. the square roots of $-i$

66. the square roots of i

67. the fourth roots of $-16i$

68. the cube roots of $8i$ **69.** the fifth roots of $1+\sqrt{3}i$

70. the fifth roots of $1+i$

In Exercises 71–76 find all solutions of the equation. Leave the answers in trigonometric form except for answers that may be expressed exactly in $a+bi$ form.

71. $x^3+8=0$

72. $x^3-27i=0$

73. $x^4=1$

74. $x^5=1$

75. $x^5+1=0$

76. $x^5+32=0$

77. Find and graph the fifth roots of unity.

78. Find and graph the fourth roots of unity.

For Exercises 79 and 80 refer to the formulas given in Example 5. Find the total impedance if impedances Z_1 and Z_2 are connected

a. in series and

b. in parallel, using the given values of Z_1 and Z_2.

79. $Z_1 = 3.14(\cos 12.5° + i\sin 12.5°)$ ohms
$Z_2 = 3.14(\cos 77.5° + i\sin 77.5°)$ ohms

80. $Z_1 = 1.11(\cos 0.5° + i\sin 0.5°)$ ohms
$Z_2 = 9.99(\cos 10.1° + i\sin 10.1°)$ ohms

81. In an electrical circuit, the admittance is the reciprocal of the impedance. If the impedance is $Z = 2.81(\cos 11.7° + i\sin 11.7°)$, what is the admittance?

82. Refer to Exercise 81. If the impedance is $Z = 3.54(\cos 57.2° + i\sin 57.2°)$, what is the admittance?

83. In the study of chaos theory, mathematicians employ functions called "Möbius transformations," which have the form

$$T(z) = \frac{\alpha z + \beta}{\gamma z + \delta},$$

where all the variables are complex. [Source: Devaney, *An Introduction to Chaotic Dynamical Systems*, Addison-Wesley]. Given $\alpha = \cos 45° + i\sin 45°$, $\beta = \cos 90° + i\sin 90°$, $\gamma = \cos 0° + i\sin 0°$, and $\delta = \cos 45° - i\sin 45°$, find $T(\cos 30° + i\sin 30°)$.

84. Refer to Exercise 83. Given $\alpha = \cos 45° + i\sin 45°$, $\beta = \cos 90° + i\sin 90°$, $\gamma = \cos 0° + i\sin 0°$, and $\delta = \cos 45° - i\sin 45°$, find $T(\cos 60° + i\sin 60°)$.

In Exercises 85 and 86 let $z = a + bi$ and establish each result.

85. $|\bar{z}| = |z|$

86. $|z| = \sqrt{z\bar{z}}$

In Exercises 87–90 let $z = r(\cos\theta + i\sin\theta)$ and establish each result.

87. $-z = r[\cos(\theta+\pi) + i\sin(\theta+\pi)]$, where $-z = -1\cdot z$

88. $z^{-1} = r^{-1}[\cos(-\theta) + i\sin(-\theta)]$, where $z^{-1} = \dfrac{1}{z}$ and $z \neq 0$

89. $z^2 = r^2(\cos 2\theta + i\sin 2\theta)$

90. $z^3 = r^3(\cos 3\theta + i\sin 3\theta)$

THINK ABOUT IT 9.4

1. Graph all complex numbers z that satisfy the given condition.
   a. $|z| = 1$ **b.** $z = \bar{z}$
   c. The real part and the imaginary part of z are equal.

2. **a.** Multiply $z = r(\cos\theta + i\sin\theta)$ by i in trigonometric form. Use the result and the diagram below to describe what happens geometrically when a complex number is multiplied by i.

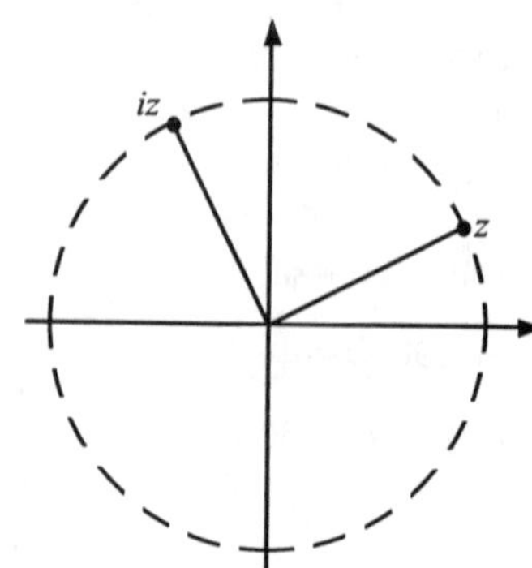

   b. From part **a** it follows that multiplication in the complex plane can describe the rotation of geometric shapes. For instance, a triangle can be rotated 90° around one of the vertices. Under such a transformation the coordinates of the vertices change in a regular manner. This pattern can be shown clearly if we think of the vertices as points in the complex plane. In the given figure we can represent point A as $0+3i$, B as $4+0i$, and C as $0+0i$. After rotation through $90°$, A becomes $A' = -3+0i$, B becomes $B' = 0+4i$ and C becomes $C' = 0+0i$. Multiply each of the original vertices by i and notice that in each case the result gives the location of the vertex after rotation by 90°. Thus, we say that multiplication by i describes rotation through $90°$. This rotation occurs because the magnitude of i is 1, and its principal argument is $90°$.

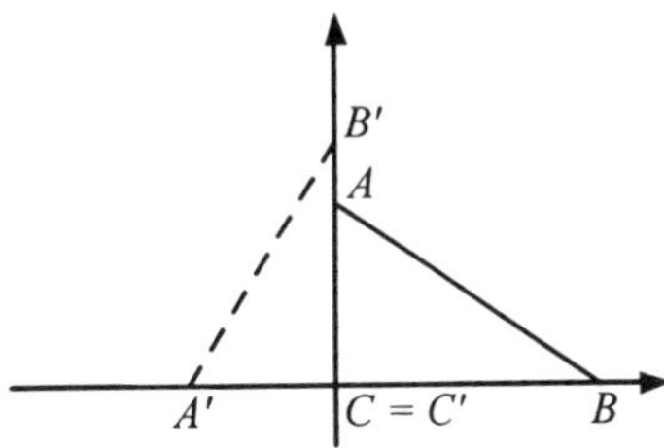

c. What happens to the triangle of part **b**, if the multiplier is a complex number different from i? First, guess what will happen by analyzing the magnitude and direction of the multiplier, then test your guess by doing the appropriate calculation. What happens to the triangle ABC if each vertex is multiplied by $-i$? by $2i$? by $1+i$?

3. Apply the product rule to the expression $[\cos\theta + i\sin\theta][\cos(-\theta) + i\sin(-\theta)]$ and show that $(\cos\theta + i\sin\theta)^{-1} = \cos(-\theta) + i\sin(-\theta)$.

4. **a.** Solve $x^3 - 1 = 0$ using factoring and the quadratic formula. Verify that the methods of this section produce the same roots.

b. Solve $x^3 + 1 = 0$ using factoring and the quadratic formula. Verify that the methods of this section produce the same roots.

5. **a.** Show that $\sin 2\theta = 2\sin\theta\cos\theta$ by applying De Moivre's theorem to $(\cos\theta + i\sin\theta)^2$.

b. Show that $\cos 2\theta = \cos^2\theta - \sin^2\theta$ by applying De Moivre's theorem to $(\cos\theta + i\sin\theta)^2$.

● ● ●

9.5 Polar Coordinates

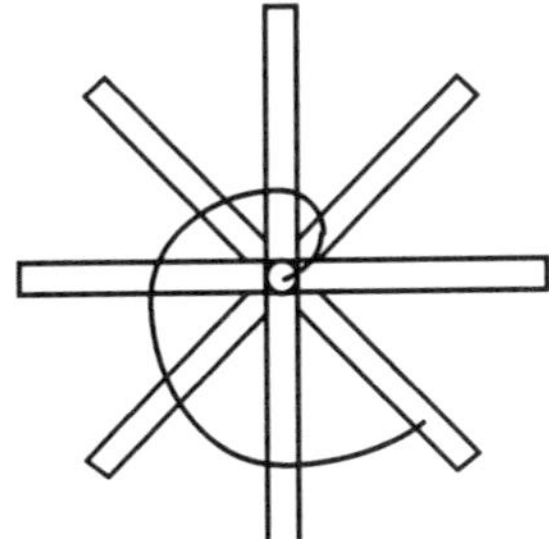

Many curves in mathematics describe paths taken by moving objects. One of the most famous of these is the curve named after the Greek mathematician Archimedes (287–212 BC). He investigated the path followed by an object moving steadily along a stick which itself is rotating at a uniform rate. (See the figure.) This motion results in a spiral-shaped path. In polar coordinate notation, the equation of this path is

$$r = c\theta, \text{ where } c > 0 \text{ is some constant.}$$

Determine c if the stick is rotating once per minute and the object is moving at 5 ft per minute, then graph the spiral for $0 \le \theta \le 3\pi$. (See Example 7.)

Objectives

1. Represent points in the polar coordinate system.
2. Convert between polar and rectangular coordinates.

3. Graph polar equations.
4. Convert equations between polar and rectangular forms.

A coordinate system in a plane is a system that allows you to locate specific points in the plane. To accomplish this location, it is not necessary to have two axes that intersect at a right angle. In the **polar coordinate system**, we begin by fixing one point O to be the origin, or **pole**, for the system. From O we construct an initial ray called the **polar axis**, as shown in Figure 9.43. Then any point P in the plane can be located by assigning it a distance r from the pole and an angle θ measured from the polar axis. If P is located in this way, we say that r and θ are the **polar coordinates** of the point P, and we write $P(r, \theta)$. The angle θ can be given either in degrees or radians. Jacob Bernoulli is credited with the first publication on the use of polar coordinates in 1691. Newton wrote about it earlier, but his work wasn't published until after Bernoulli's.

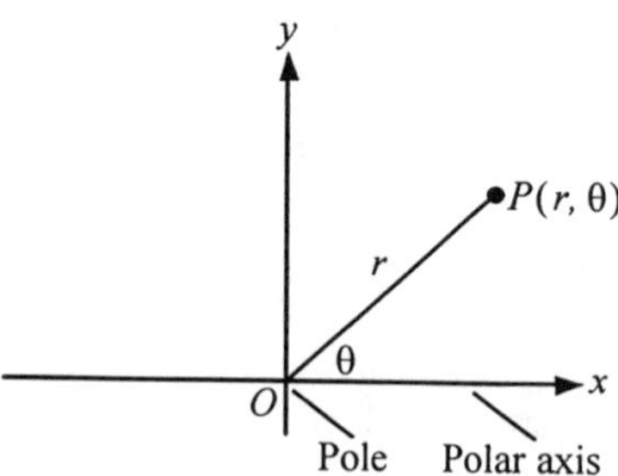

Figure 9.43

In polar coordinates, both r and θ are signed values. A positive value of θ represents an angle measured counterclockwise from the polar axis, and a negative value represents an angle measured clockwise. The point $(-r, \theta)$ is located in the opposite direction from the point (r, θ). These ideas are illustrated in Example 1.

Example 1: Graphing Points Using Polar Coordinates

Graph the points whose polar coordinates are given as follows.

a. $(2, 60°)$ **b.** $(1, -45°)$ **c.** $\left(-3, \dfrac{5\pi}{6}\right)$ **d.** $\left(-\dfrac{1}{2}, -\dfrac{\pi}{4}\right)$ **e.** $(0, \pi)$

Solution:

a. The point $(2, 60°)$ lies at the intersection of a circle with radius 2 and the terminal side of a $60°$ angle. See Figure 9.44(a).
b. The point $(1, -45°)$ lies at the intersection of a circle with radius 1 and the terminal side of a $-45°$ angle, which places it in the 4th quadrant. See Figure 9.44 (b).
c. The angle $5\pi/6$ is in the 2nd quadrant, but since r is negative, the point $(-3, 5\pi/6)$ is in the 4th quadrant. See Figure 9.44 (c).
d. Since $-\pi/4$ is in the 4th quadrant, this point is in the 2nd quadrant. See Figure 9.44 (d).
e. Since $r = 0$, this point is the pole. In general, $(0, \theta)$ is the pole for all values of θ. See Figure 9.44(e).

PROGRESS CHECK 1

Graph the points whose polar coordinates are given as follows.

a. $(3, 45°)$ **b.** $(2, -135°)$ **c.** $\left(-1, \dfrac{7\pi}{6}\right)$ **d.** $\left(-\dfrac{3}{4}, -\dfrac{\pi}{3}\right)$ **e.** $\left(0, \dfrac{\pi}{2}\right)$ ■

Any point in the plane can be located by giving its rectangular or its polar coordinates. From Figure 9.45 it can be seen for the point (x, y) that $\cos\theta = x/r$ and $\sin\theta = y/r$. From these equations we get the following relations between the polar and rectangular coordinates of a point.

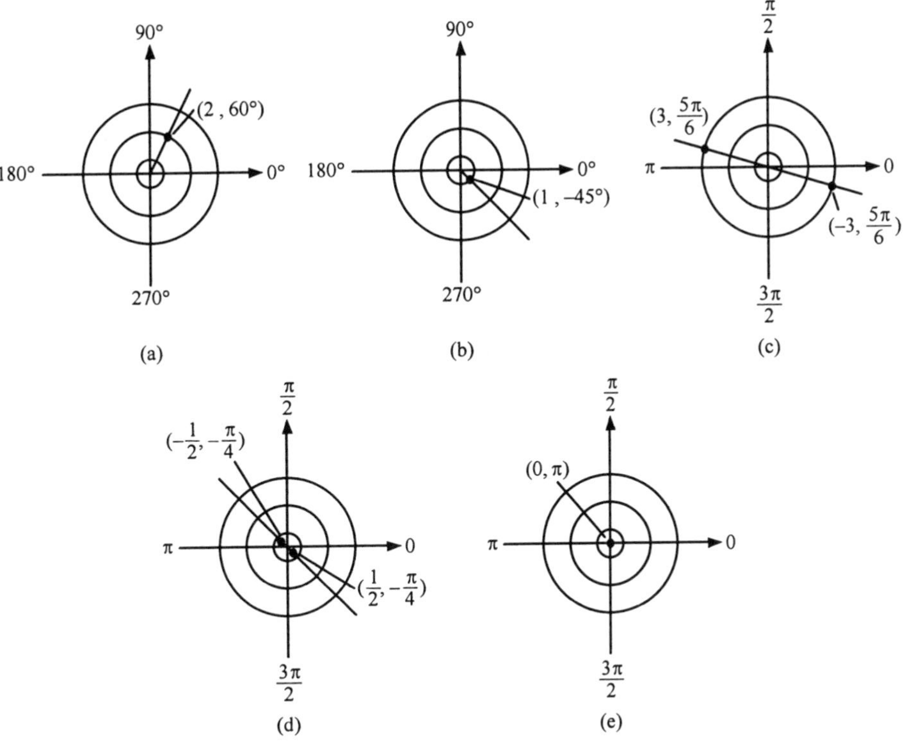

Figure 9.44

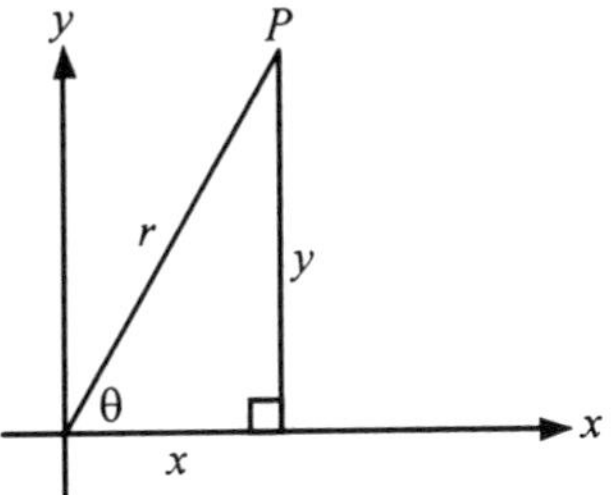

Figure 9.45

Converting Between Polar and Rectangular Coordinates

If the point P has polar coordinates (r, θ) and rectangular coordinates (x, y), then the following relations are true for these coordinates.

$$x = r\cos\theta$$
$$y = r\sin\theta$$
$$x^2 + y^2 = r^2$$
$$\tan\theta = y/x \quad (x \neq 0)$$

Example 2: Converting Between Polar and Rectangular Coordinates

a. Represent the point $(r, \theta) = (2, \pi/4)$ in rectangular coordinates.

b. Represent the point $(x, y) = (-1/2, 1/2)$ in polar coordinates. Use radian measure for the angle.

Solution:

a. Using the conversion relations above, we get

$$x = r\cos\theta \qquad\qquad y = r\sin\theta$$
$$= 2\cos\left(\frac{\pi}{4}\right) \qquad\qquad = 2\sin\left(\frac{\pi}{4}\right)$$
$$= 2\left(\frac{\sqrt{2}}{2}\right) \qquad\qquad = 2\left(\frac{\sqrt{2}}{2}\right)$$
$$= \sqrt{2} \qquad\qquad = \sqrt{2}$$

The rectangular coordinates are $\left(\sqrt{2},\ \sqrt{2}\right)$.

b. We observe that the point is in the 2nd quadrant, and we find an appropriate value of θ. From the conversion relation $\tan\theta = y/x$, we get $\tan\theta = (1/2)/(-1/2) = -1$. Since the point is in the 2nd quadrant, we can take $\theta = 3\pi/4$. Next we use the conversion relation $r^2 = x^2 + y^2$ to calculate r.

$$r^2 = x^2 + y^2 = \left(-\frac{1}{2}\right)^2 + \left(\frac{1}{2}\right)^2 = \frac{1}{2}$$
$$r = \sqrt{\frac{1}{2}} = \frac{\sqrt{2}}{2}$$

Thus, the polar coordinates of the given point are $\left(\sqrt{2}/2,\ 3\pi/4\right)$.

Note: The polar coordinates for a given point are not unique; a point has more than one pair of coordinates in polar notation. For instance, you should confirm that $\left(\sqrt{2}/2,\ 3\pi/4\right)$, $\left(\sqrt{2}/2,\ 11\pi/4\right)$, and $\left(-\sqrt{2}/2,\ 7\pi/4\right)$ all refer to the same point.

PROGRESS CHECK 2

a. Represent the point $(r,\ \theta) = (1,\ \pi/2)$ in rectangular coordinates.

b. Represent the point $(x,\ y) = \left(-1,\ \sqrt{3}\right)$ in polar coordinates. ■

Just as the equation $y = f(x)$ describes the relation between y and x, so the equation $r = f(\theta)$ describes the relation between r and θ. Similarly, we can visualize polar relations by drawing graphs in the polar coordinate system. Three such graphs are illustrated in Examples 3 and 4.

Example 3: Graphing Polar Equations

Sketch the graphs of these equations

a. $r = 2$ **b.** $\theta = 60°$

Solution:

a. This graph consists of all points in the polar plane for which $r = 2$. Since these points are all 2 units from the pole, the graph is a circle of radius 2 with center at the pole. In general, the graph of $r = \text{constant}$ is a circle of radius c with center at the pole. The graph is shown in Figure 9.46(a).

b. This graph consists of all points in the polar plane for which $\theta = 60°$. These points lie on the ray that makes and angle of 60° with the polar axis, and so the graph is a straight line through the pole. In general, the graph of $\theta = \text{constant}$ is the line associated with the terminal side of angle θ. The graph of $\theta = 60°$ is shown in Figure 9.46 (b).

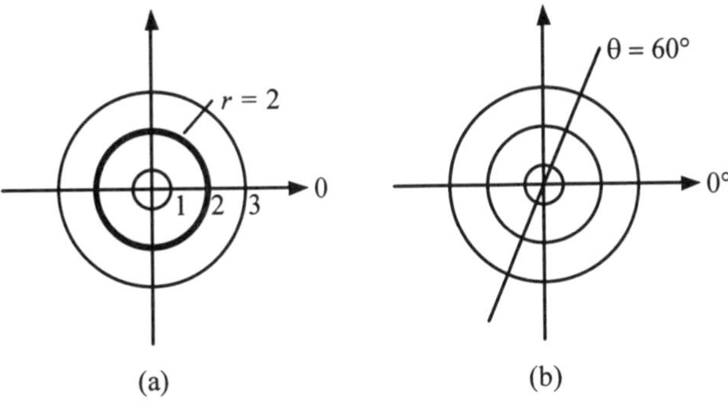

Figure 9.46

PROGRESS CHECK 3

Sketch the graphs of these equations.

a. $r = 1$ **b.** $\theta = 210°$ ■

Example 4: Graphing Polar Equations

Graph the equation $r = 2\cos\theta$.

Solution: The cosine function has period 2π, so that we need only plot points for the interval $0 \le \theta < 2\pi$. Table 9.1 shows some points on the graph.

θ	0	$\frac{\pi}{6}$	$\frac{\pi}{4}$	$\frac{\pi}{3}$	$\frac{\pi}{2}$	$\frac{2\pi}{3}$	$\frac{3\pi}{4}$	$\frac{5\pi}{6}$	π
$2\cos\theta$	2	$\sqrt{3}$	$\sqrt{2}$	1	0	-1	$-\sqrt{2}$	$-\sqrt{3}$	-2

θ	$\frac{7\pi}{6}$	$\frac{5\pi}{4}$	$\frac{4\pi}{3}$	$\frac{3\pi}{2}$	$\frac{5\pi}{3}$	$\frac{7\pi}{4}$	$\frac{11\pi}{6}$	2π
$2\cos\theta$	$-\sqrt{3}$	$-\sqrt{2}$	-1	0	1	$\sqrt{2}$	$\sqrt{3}$	2

Table 9.1

The graph of these points is shown in Figure 9.47. The graph appears to be a circle of radius 1 with center at 1 on the polar axis. We will confirm in Example 5 that this description is correct.

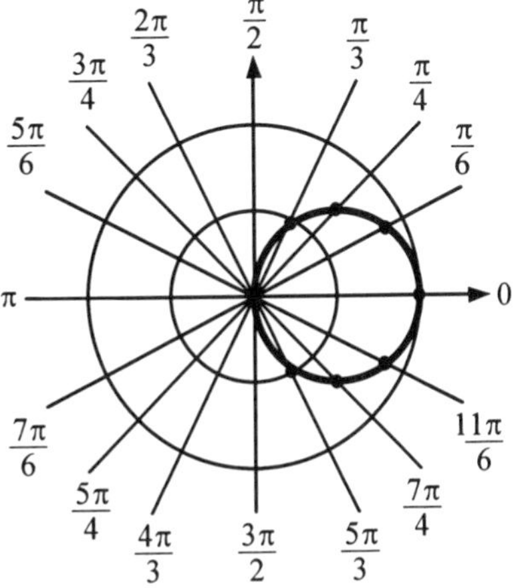

Figure 9.47

Technology Link

1. Many graphing calculators can be set in **polar mode**. In this mode the equation menu allows you to enter r as a function of θ directly. The viewing window settings include minimum and maximum values for θ, as well as an increment for θ. See Figure 9.48 for an illustration of screens used to graph $r = 2\cos\theta$. A Square setting was used to preserve the shape of the circle.

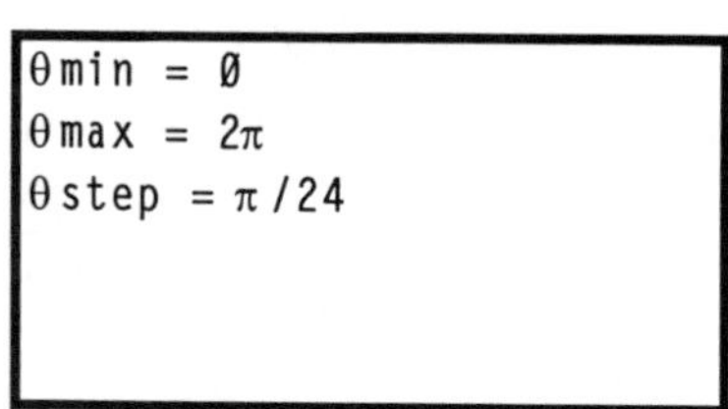

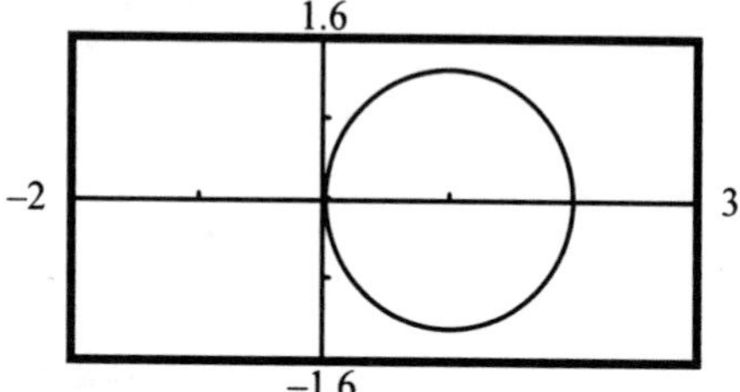

Figure 9.48

2. Parametric mode on a calculator may also be used to graph equations of the form $r = f(\theta)$. Using the conversion equivalence, $x = r\cos\theta$, and the given equation $r = f(\theta)$, we can write $x = f(\theta)\cos\theta$. Similarly, we write $y = f(\theta)\sin\theta$. Using t as the parameter when the calculator is set in Parametric mode, enter the graph equation as follows:

$$xt = f(t)\cos t$$
$$yt = f(t)\sin t$$

For example, to graph $r = 2\cos\theta$, enter

$$xt = 2\cos t\cos t$$
$$yt = 2\cos t\sin t$$

The resulting graph is the same as that in Figure 9.48.

PROGRESS CHECK 4

Graph the equation $y = 2\sin\theta$. ■

In Example 2 we demonstrated that a point in the plane can be represented both with rectangular and polar coordinates. This dual representation extends naturally to the equations that define graphs. A graph may be described by an equation both in terms of rectangular coordinates and in terms of polar coordinates.

Example 5: Converting Equations from Polar to Rectangular Form

Find an equation using rectangular coordinates for each of these polar equations.

a. $r = 2$ **b.** $\theta = 60°$ **c.** $r = 2\cos\theta$

Solution:

a. We use a conversion relation to replace r by an expression involving x and y.

$$r = 2$$
$$\pm\sqrt{x^2 + y^2} = 2 \quad \text{conversion relation, } x^2 + y^2 = r^2$$
$$x^2 + y^2 = 4 \quad \text{squaring both sides}$$

This last equation may be recognized as a circle with radius 2 centered at the origin. Recall that in Example 3 we showed that the graph of $r = 2$ is a circle of radius 2 with center at the pole.

b. Since we need to replace θ by an expression involving x and y, we use the conversion relation $\tan\theta = y/x$.

$$\theta = 60°$$
$$\tan\theta = \tan 60°$$
$$\tan\theta = \sqrt{3}$$
$$\frac{y}{x} = \sqrt{3} \qquad \tan\theta = \frac{y}{x}$$
$$y = \sqrt{3}x$$

This last equation graphs as a line through the origin with slope $\sqrt{3}$. Thus it makes a 60° angle with the x-axis. Recall that in Example 3 we showed the graph of $\theta = 60°$ is this same line.

c. Since the given polar equation involves both r and $\cos\theta$, we use the conversion relation $x = r\cos\theta$.

$$r = 2\cos\theta$$
$$r = 2\frac{x}{r} \quad \text{conversion relation, } \cos\theta = \frac{x}{r}$$
$$r^2 = 2x$$
$$x^2 + y^2 = 2x \quad \text{conversion relation, } r^2 = x^2 + y^2$$

To see that this equation represents a circle, we complete the square

$$x^2 - 2x + y^2 = 0$$
$$x^2 - 2x + 1 + y^2 = 1$$
$$(x-1)^2 + y^2 = 1$$

This relation may be recognized as a circle with center (1, 0) and radius 1.

PROGRESS CHECK 5

Find an equation using rectangular coordinates for each of these polar equations.

a. $r = 3$ **b.** $\theta = \frac{\pi}{4}$ **c.** $r = 2\sin\theta$ ■

Example 6: Converting Equations from Rectangular to Polar Form

Transform $x^2 + y^2 + 6x = 0$ to an equation in polar form.

Solution: Since $x^2 + y^2 = r^2$ and $x = r\cos\theta$, we proceed as follows:

$$x^2 + y^2 + 6x = 0$$
$$r^2 + 6r\cos\theta = 0$$
$$r(r + 6\cos\theta) = 0$$
$$r = 0 \quad r + 6\cos\theta = 0$$
$$r = -6\cos\theta$$

Since $r = -6\cos\theta$ includes the pole, or $r = 0$, the complete relation is expressed by $r = -6\cos\theta$. Note that both equations fit the form for circles in their respective systems.

PROGRESS CHECK 6
Transform $x^2 + y^2 - 6x = 0$ to an equation in polar form. ■

In Example 5 we showed that the equations $r = 2\cos\theta$ and $(x-1)^2 + y^2 = 1$ represent the same circle. For certain types of graphs the polar equation is simpler than the rectangular one and may be preferred. This is typically the case for graphs that spiral, loop, or wind around. Example 7 involves such a graph.

Example 7: The Spiral of Archimedes

Solve the problem in the section introduction on page 781.

Solution: We solve the problem by locating the object when a certain amount of time has elapsed. For convenience, we choose to locate it when 1 minute has elapsed. If the stick is rotating once per minute, then the angle θ must equal 2π when 1 minute has elapsed. Also, when 1 minute has elapsed, the object must be 5 ft from the pole (since it moves at the constant rate of 5 ft per minute). Thus the point $(5, 2\pi)$ must satisfy the equation. Replacing r by 5, and θ by 2π will then determine c.

$$r = c\theta$$
$$5 = c(2\pi)$$
$$c = \frac{5}{2\pi}$$

Therefore, the value of c is $5/2\pi$.

The graph of $r = (5/2\pi)\theta$ is shown in Figure 9.49. Note that the equation $r = c\theta$ states that r, the distance of the point from the pole, is directly proportional to the size of the angle. Thus, as the angle increases, the points are farther from the origin.

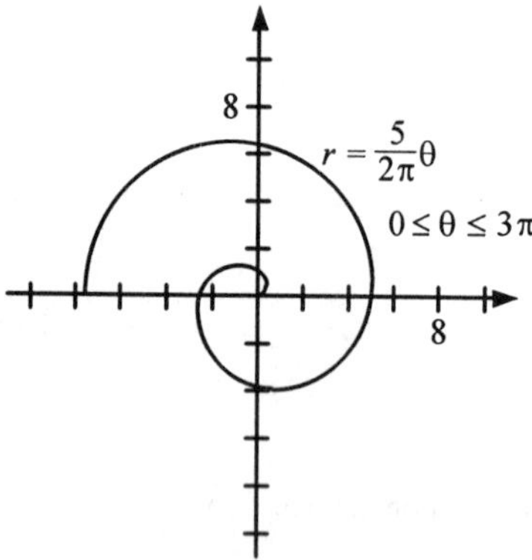

Figure 9.49

PROGRESS CHECK 7
Repeat Example 7 but suppose that the object is now moving along the stick at 3 ft per second.

In Table 9.2 several well-known polar graphs are shown. Typically, they have much simpler equations in polar notation than in rectangular notation. Mathematicians in the seventeenth and eighteenth centuries developed many of these equations, giving them Latin or Greek names suggested by their shapes. The graphs may be drawn conveniently with the aid of a grapher set in Polar or Parametric mode.

Table 9.2 Some well-known polar graphs

Circles (from the Latin word meaning "small ring")

All the following circles have radius $a > 0$.

$r = a$	Center at the origin
$r = \pm 2a\sin\theta$	Center on y-axis
$r = \pm 2a\cos\theta$	Center on x-axis

Three specific cases are illustrated using $a = 1$.

a. $r = 1$

1

b. 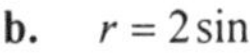 $r = 2\sin\theta$

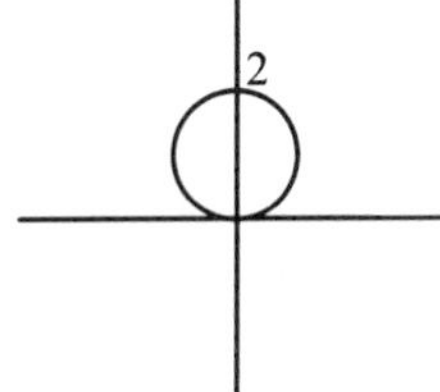

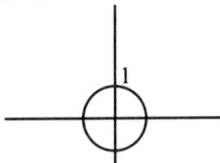

c. $r = 2\cos\theta$

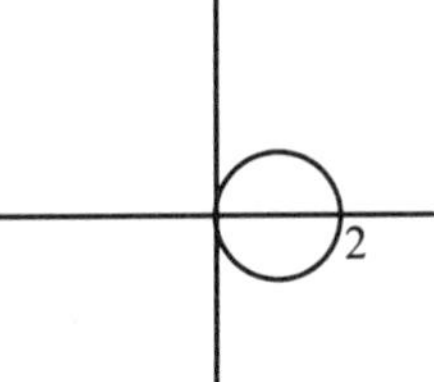

Spirals (from Latin and Greek words meaning "twisted" or "coiled")

$r = a\theta$	Archimedes Spiral
$r^2 = a^2\theta$	Parabolic Spiral

Two specific cases are illustrated using $a = 1$.

a. $r = \theta$

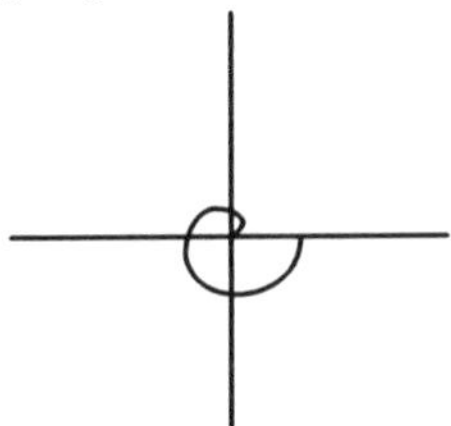

b. $r^2 = \theta$

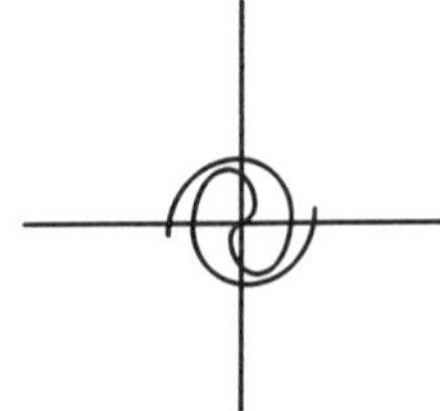

Limaçons (from the Latin word for "snail")

$r = a \pm b\cos\theta$; $r = a \pm b\sin\theta$

The shape depends on the relative size of a and b. When $a < b$ the curve has an inner loop. When $a = b$ the curve is heart-shaped.

Three specific cases are illustrated.

a. $r = 2 + 2\cos\theta$
Cardioid (pointed indentation)

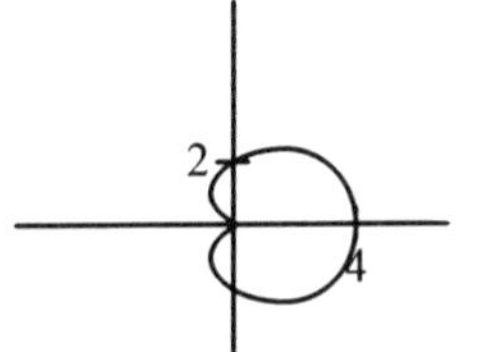

b. $r = 3 + 2\cos\theta$
(Smooth indentation)

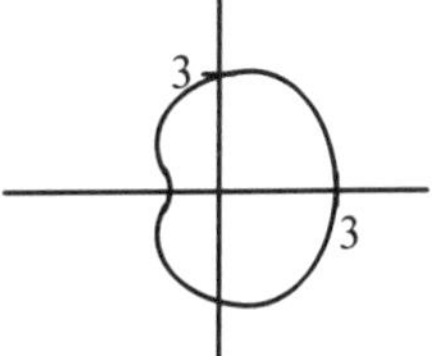

Table 9.2 Some well-known polar graphs (continued)

c. $r = 1 + 2\cos\theta$
(Inner loop)

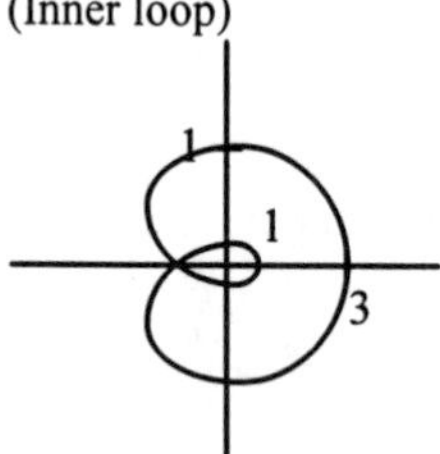

Lemniscates (from a Greek word describing a hair ribbon made of wool from the island of Lemnos)

$r^2 = a^2 \cos 2\theta$; $r^2 = a^2 \sin 2\theta$

Two specific cases are illustrated

a. $r^2 = \cos 2\theta$

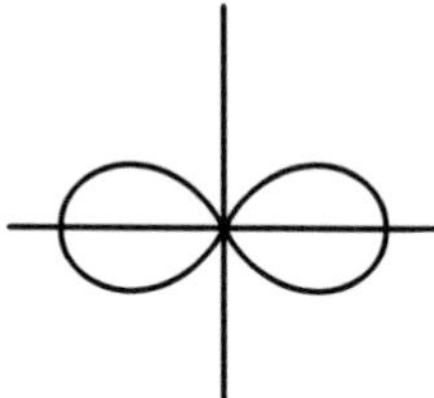

b. $r^2 = \sin 2\theta$

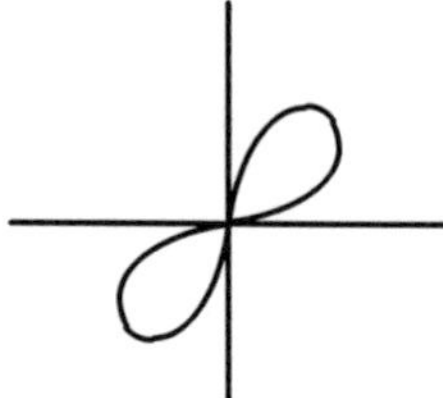

Roses (because of the flower shape)

The constant n determines the number of petals.

$r = a\cos n\theta \quad (n \geq 2)$; $r = a\sin n\theta \quad (n \geq 2)$

Two specific cases are illustrated.

a. $r = \cos 4\theta$
When n is even, there are $2n$ petals.

b. $r = \cos 5\theta$

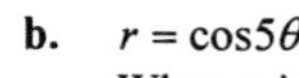

When n is odd, there are n petals.

EXPLORE 9.5

1. By setting a calculator in Polar mode, you can enter equations directly in the form $r = f(\theta)$. The graphs are very sensitive to the window settings. The viewing window settings will include settings for θ as well as for x and y. To explore how changing the window settings affects the graph, look at the graph of $y = \cos(5\theta)$ under these windows.

a.

	minimum	maximum	step
θ	0	2π	$\pi/36$
x	−1	1	.1
y	−1	1	.1

b. Change the θ step to $\pi/12$ and explain how the appearance of the graph changes.

c. Reset θ step to $\pi/36$ and change the θ maximum to π. Explain the resulting graph. What is the result if θ maximum is $\pi/2$?

d. Experiment with various Zoom settings to become familiar with their effects.

2. Explore the Parametric mode feature for graphing in polar coordinates. This approach depends on the basic relations $x = r\cos\theta$ and $y = r\sin\theta$, with r replaced by the appropriate equation. In many calculators the default variable in Parametric mode is t, so you will use t instead of θ.
   a. Graph $r = \cos(5\theta)$ using the following settings.
      Equations: $xt = \cos(5t)\cos(t)$
      $yt = \cos(5t)\sin(t)$
      Window settings: $t\min = 0$; $t\max = 2\pi$
      $x\min = y\min = -1$
      $x\max = y\max = 1$
   b. Graph $r = 3\sin(4\theta)$ using settings of your own choice.
   c. Use a grapher in Parametric mode to graph: $xt = 2\cos t$
      $yt = 2\sin t$
      Using the basic relations given in this exercise, explain why the graph must be a circle with radius 2.
3. In Table 9.2 in this section of the text, several well-known polar graphs are shown. Use a grapher to confirm that the graphs correspond to the given equations.

EXERCISES 9.5

1. Plot the following points in polar coordinates.
   a. $(2, 60°)$ b. $(3, \pi)$
   c. $(1, -30°)$ d. $(0, 150°)$
   e. $\left(0.5, -\frac{2\pi}{3}\right)$ f. $(2, 500°)$

2. Plot the following points in polar coordinates.
   a. $(-1, 90°)$ b. $\left(-2, -\frac{\pi}{6}\right)$
   c. $\left(-\frac{3}{2}, 135°\right)$ d. $(-2, 210°)$
   e. $(-3, -\pi)$ f. $\left(-1, \frac{2\pi}{3}\right)$

In Exercises 3–8 express each point in rectangular coordinates.

3. $(1, \pi)$ 4. $(2, 60°)$ 5. $(-3, 3\pi)$

6. $(-6, 0°)$ 7. $\left(\sqrt{2}, \frac{5\pi}{4}\right)$ 8. $(-\sqrt{3}, -120°)$

In Exercises 9–14 express each point in polar coordinates with $r \geq 0$ and $0° \leq \theta < 360°$.

9. $(-1, 0)$ 10. $(0, -2)$ 11. $(1, -1)$

12. $(-3, -3)$ 13. $(-\sqrt{3}, 1)$ 14. $(-2, 2\sqrt{3})$

15. Express the following points in polar coordinates with $r \geq 0$ and $0° \leq \theta < 360°$ (or $0 \leq \theta < 2\pi$).
   a. $(1, -30°)$ b. $\left(-2, \frac{5\pi}{6}\right)$
   c. $\left(-1, \frac{4\pi}{3}\right)$ d. $(3, 585°)$
   e. $\left(-1, -\frac{7\pi}{4}\right)$ f. $(2, -180°)$

16. Express the following points in polar coordinates with $r \geq 0$ and $0° \leq \theta < 360°$ (or $0 \leq \theta < 2\pi$).
   a. $(1, -60°)$ b. $\left(-1, \frac{\pi}{6}\right)$
   c. $\left(-2, \frac{5\pi}{3}\right)$ d. $(5, 600°)$
   e. $\left(-3, -\frac{\pi}{4}\right)$ f. $(1, -270°)$

17. List three other polar representations for $\left(1, \frac{5\pi}{6}\right)$ with one having $r < 0$.

18. List three other polar representations for $(3, 210°)$ with one having $r < 0$.

In Exercises 19–34 graph each equation in polar coordinates and give its name.

19. $r = 3$ 20. $\theta = \frac{\pi}{4}$ 21. $r = \cos\theta$

22. $r = 3\sin\theta$ 23. $r = -4\sin\theta$

24. $r = -2\cos\theta$ 25. $r = 1 + \sin\theta$

26. $r = 1 - \sin\theta$ **27.** $r = \sin\theta - 1$

28. $r = 1 + \cos\theta$ **29.** $r = 2(1 - \cos\theta)$

30. $r = \cos\theta - 1$ **31.** $r = \cos 2\theta$

32. $r = 3\sin 2\theta$ **33.** $r = 2\sin 3\theta$

34. $r = \cos 4\theta$

In Exercises 35–42 graph each equation in polar coordinates.

35. $r = 2 + \cos\theta$ (limaçon)

36. $r = 3 - \sin\theta$ (limaçon)

37. $r = 1 - 2\sin\theta$ (limaçon, inner loop)

38. $r = 1 + 2\cos\theta$ (limaçon, inner loop)

39. $r^2 = 4\cos 2\theta$ (lemniscate)

40. $r^2 = \sin 2\theta$ (lemniscate)

41. $r = \theta,\ \theta \geq 0$ (spiral) **42.** $r = \theta,\ \theta \leq 0$ (spiral)

In Exercises 43–52 transform each equation to rectangular form.

43. $r = 3$ **44.** $\theta = -\dfrac{\pi}{6}$ **45.** $r\cos\theta = 2$

46. $r\sin\theta = -3$ **47.** $r = -3\sin\theta$

48. $r = 3\cos\theta$ **49.** $r = 2\csc\theta$

50. $r = \sec\theta$ **51.** $r(3\sin\theta + 2\cos\theta) = 1$

52. $r = \sin 2\theta$

In Exercises 53–62 transform each equation to polar form.

53. $x = 2$ **54.** $y = -2$

55. $x^2 + y^2 = 9$ **56.** $y = x$

57. $x^2 + y^2 = x$ **58.** $x^2 + y^2 - 4y = 0$

59. $y = x^2$ **60.** $y = x^3$

61. $x^2 - y^2 = 4$ **62.** $x^2 + y^2 + x = \sqrt{x^2 + y^2}$

63. Starting at the center, a bug is crawling outward along a radius of a revolving phonograph record. The path of the bug is described by $r = c\theta$, where $c > 0$ is some constant. Determine c if the record is spinning at 45 revolutions per minute, and the bug is crawling at 2 in. per minute.

64. If the record in Exercise 63 is spinning at 78 revolutions per minute, determine c.

THINK ABOUT IT 9.5

1. Explain how the figure below can be used to interpret geometrically the graph of the equation $r = 2a\cos\theta$.

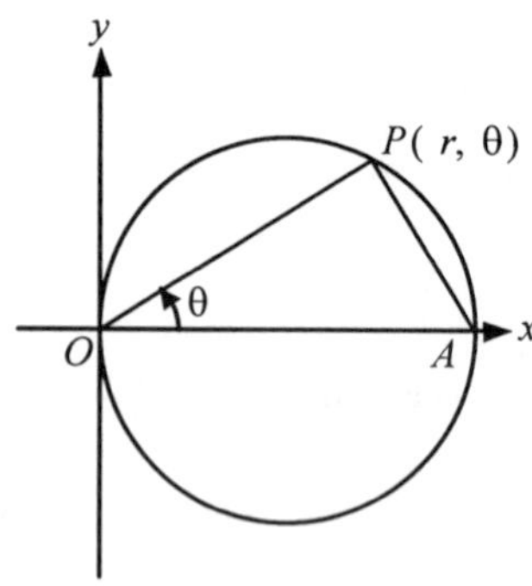

2. **a.** Convert $r = a\sin\theta + b\cos\theta$ to rectangular form. Name the graph of the resulting equation.

b. Graph $r = 2\sin\theta + 2\cos\theta$.

3. Graph $r = \sin\theta$ and $r = \cos\theta$ on the same coordinate system and give a representation for all points of intersection of the two curves. Explain why geometric methods must be included when finding intersection points of two curves that are given in polar form.

4. **a.** Consider the graph of $r = \sin 2\theta$. Is the graph symmetric about the x-axis?

b. If replacing θ by $-\theta$ produces an equivalent polar equation, then the graph of the equation is symmetric about the x-axis. Apply this test to the graph of $r = \sin 2\theta$. Is the test result in agreement with your answer tp part **a**? Explain.

5. Use different polar representations for $(r, -\theta)$ and state two tests for determining symmetry about the x-axis that are different from the test in Question **4b**. Can you conclude from either of these alternative tests that the graph of $r = \sin 2\theta$ is symmetric about the x-axis?

● ● ●

CHAPTER 9 SUMMARY

OBJECTIVES CHECKLIST

Specific chapter objectives are summarized below along with numbered example problems from the text that should clarify the objectives. If you do not understand any objectives, or do not know how to do the selected problems, then restudy the material.

9.1: Can you:

1. **Use the law of sines to solve a triangle, given two angles and one side of the triangle?** Approximate the missing parts of triangle ABC in which $A = 35°$, $B = 50°$, and $a = 12$ ft.
[Example 2]

2. **Determine exact solutions for unknowns associated with certain triangles, assuming exact numbers?** In triangle RST, $r = 8$, $R = 45°$, and $S = 30°$. Assume exact numbers and find s.
[Example 3]

3. **Use the law of sines to solve a triangle, given two sides of the triangle and the angle opposite one of them?** Solve the triangle ABC in which $B = 60°$, $b = 5\overline{0}$ ft, and $c = 3\overline{0}$ ft.
[Example 4]

4. **Determine how many triangles are described by certain conditions?** Determine if no triangle, one triangle, or two triangles are determined by the condition that $A = 52°$, $a = 55$, and $b = 75$.
[Example 6]

5. **Solve applied problems using the law of sines?** Two observation towers, A and B, are located $1\overline{0}$ mi apart. A fire is sighted at point C, and the observer in tower A measures angle CAB to be 80°. At the same time, the observer in tower B measures angle CBA to be 40°. How far is the fire from tower A.
[Example 1]

9.2: Can you:

1. **Use the law of cosines to solve a triangle, given two sides of the triangle and the angle between these two sides?** Approximate the missing parts of triangle ABC in which $A = 60°$, $b = 25$ ft, and $c = 42$ ft.
[Example 2]

2. **Use the law of cosines to solve a triangle, given the lengths of the three sides of the triangle?** Approximate the missing parts of triangle ABC in which $a = 23.5$ ft, $b = 44.2$ ft, and $c = 30.1$ ft.
[Example 3]

3. **Solve applied problems using the law of cosines?** On a ground ball to the first base side, the pitcher is expected to cover first base and beat the batter to this bag. If a baseball diamond is a square that 90.0 ft long on each side, and if the pitcher's mound is 60.5 ft from home plate on the diagonal from home to second base, then how far is the pitcher's mound from first base?
[Example 1]

4. **Solve a problem involving a triangle by determining whether the law of sines or the law of cosines is appropriate for the problem?** In parallelogram $ABCD$ the lengths of the sides AB and AD are 12 m and 19 m, respectively. If $A = 38°$, find the length of the longer diagonal of the parallelogram. Choose between the law of sines and the law of cosines to solve the problem.
[Example 4]

5. **Find the area of a triangle, given two sides of the triangle and the angle between these two sides?** Find the area of triangle ABC if $a = 32.4$ cm, $b = 49.2$ cm, and $C = 18.5°$.

[Example 5]

6. **Find the area of a triangle, given the lengths of three sides?** Find the area of triangle ABC if $a = 3.0$ ft, $b = 4.0$ ft, and $c = 6.0$ ft.

[Example 7]

9.3: Can you:

1. **Find the magnitude and direction angle of a vector?** Find the magnitude and direction angle for $\mathbf{v} = \langle -1, \sqrt{3} \rangle$.

[Example 1a]

2. **Find sums, differences, and scalar multiples of vectors written in component form?** If $\mathbf{v} = \langle -3, 4 \rangle$, and $\mathbf{w} = \langle 1, 5 \rangle$, find $-4\mathbf{w} + 3\mathbf{v}$.

[Example 2c]

3. **Prove vector properties involving vector addition and scalar multiplication?** Show that $c(\mathbf{u} + \mathbf{v}) = c\mathbf{u} + c\mathbf{v}$, where $\mathbf{u} = \langle u_1, u_2 \rangle$, $\mathbf{v} = \langle v_1, v_2 \rangle$, and c is a scalar.

[Example 4]

4. **Find sums, differences, and scalar multiples of vectors written as linear combinations of i and j?** If $\mathbf{v} = 3\mathbf{i} - \mathbf{j}$ and $\mathbf{w} = -2\mathbf{i} + 5\mathbf{j}$, express $2\mathbf{w} - 3\mathbf{v}$ as a linear combination of **i** and **j**.

[Example 5c]

5. **Express a vector as a linear combination of i and j given the components or the magnitude and direction angle of the vector?** Express **v** as a linear combination of **i** and **j** if the magnitude of **v** is 8 and the direction angle of **v** is 120°

[Example 6]

6. **Solve applied problems involving vectors?** Two forces **a** and **b** act on an eyebolt connector as shown in the diagram. Find the vector sum or resultant **v** of forces **a** and **b**, and give the magnitude and direction angle of **v**.

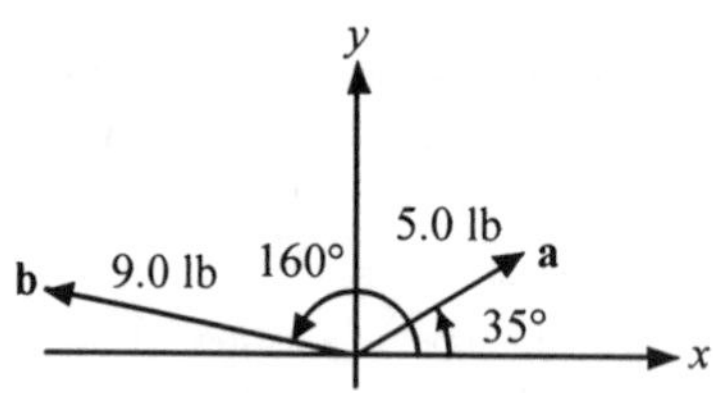

[Example 7]

9.4: Can you:

1. **Graph a complex number $a + bi$ and find its absolute value?** Graph $8 - 6i$ and find the absolute value of this number.

[Example 1]

2. **Write a complex number $a + bi$ in trigonometric form?** Write the complex number $3 - 3i$ in trigonometric form.

[Example 2]

3. **Convert a complex number from trigonometric form to $a + bi$ form?** Write the number $2(\cos 150° + i \sin 150°)$ in the form $a + bi$.

[Example 3]

4. **Find the product and quotient of complex numbers in trigonometric form?** Determine the quotient z_1/z_2 if $z_1 = 3(\cos 72° + i\sin 72°)$ and $z_2 = 5(\cos 43° + i\sin 43°)$.

[Example 4b]

5. **Use DeMoivre's theorem to find powers of complex numbers?** Find $(1+i)^{12}$.

[Example 6]

6. **Use DeMoivre's theorem to find roots of complex numbers and graph these roots?** Find and graph the five fifth roots of 32.

[Example 7]

7. **Solve equations of the form $x^n = z$, where n is a positive integer?** Find all solutions for $x^6 = 1$.

[Example 9]

9.5: Can you:

1. **Represent points in the polar coordinate system?** Graph the point $(2,\ 60°)$.

[Example 1a]

2. **Convert between polar and rectangular coordinates?** Represent the point $(r,\ \theta) = (2,\ \pi/4)$ in rectangular coordinates.

[Example 2a]

3. **Graph polar equations?** Graph the equation $r = 2\cos\theta$.

[Example 4]

4. **Convert equations between polar and rectangular forms?** Find an equation using rectangular coordinates for $r = 2\cos\theta$.

[Example 5]

KEY CONCEPTS AND PROCEDURES

Section | **Key Concepts and Procedures to Review**

9.1

- Law of sines: $\dfrac{\sin A}{a} = \dfrac{\sin B}{b} = \dfrac{\sin C}{c}$
- Guidelines on when to use the law of sines
- When given the measures for two sides of a triangle and the angle opposite one of them, there may be one triangle that fits the data (see Example 4) or there may be two triangles that fit the data (see Example 5). Sometimes no triangle can be constructed from the data; then we say the data are inconsistent.

9.2

- Law of cosines: $a^2 = b^2 + c^2 - 2bc\cos A$
  $b^2 = a^2 + c^2 - 2ac\cos B$
  $c^2 = a^2 + b^2 - 2ab\cos C$
- Guidelines on when to use the law of cosines
- The area of a triangle is given by one-half the product of the lengths of two sides and the sine of the angle between these two sides
- Heron's area formula: $K = \sqrt{s(s-a)(s-b)(s-c)}$ where $s = \dfrac{1}{2}(a+b+c)$

9.3

- Definitions of vector equality, position vector, components of **v**, magnitude and direction angle of **v**, vector addition, scalar multiplication, vector subtraction, and unit vector
- Magnitude (or length) of **v**: $\|\mathbf{v}\| = \sqrt{v_1^2 + v_2^2}$
- The direction angle of **v** may be determined from $\tan\theta = v_2/v_1$ if $v_1 \neq 0$

- Every vector **v** in a plane is expressible in the form $\mathbf{v} = v_1\mathbf{i} + v_2\mathbf{j}$, where $\mathbf{i} = \langle 1, 0\rangle$ and $\mathbf{j} = \langle 0, 1\rangle$. The vector sum $v_1\mathbf{i} + v_2\mathbf{j}$ is called a linear combination of **i** and **j**
- Use $\mathbf{v} = \|\mathbf{v}\|(\cos\theta)\mathbf{i} + \|\mathbf{v}\|(\sin\theta)\mathbf{j}$ to express **v** as a linear combination of **i** and **j** when given the magnitude and direction angle of **v**

9.4

- Definitions of complex plane, real axis, imaginary axis, and the argument and absolute value of a complex number
- Definitions of an *n*th root of a complex number and the *n*th roots of unity
- Graph of a complex number
- In trigonometric form $a + bi$ is written as $r(\cos\theta + i\sin\theta)$
- The absolute value, *r*, is given by $r = |a + bi| = \sqrt{a^2 + b^2}$
- The argument, θ, is given by $\tan\theta = b/a$
- Methods to multiply and divide two complex numbers in trigonometric form
- De Moivre's theorem
- The *n* distinct *n*th roots of the complex number $r(\cos\theta + i\sin\theta)$ are given by

$$\sqrt[n]{r}\left(\cos\frac{\theta + k\cdot 360^\circ}{n} + i\sin\frac{\theta + k\cdot 360^\circ}{n}\right),$$

where $k = 0, 1, 2, \ldots, n-1$

9.5

- Definitions of pole, polar axis, and polar coordinates
- If $r < 0$, we plot (r, θ) by measuring $|r|$ units in a direction opposite to the terminal ray of θ
- Formulas for different representations of a given point:
  $(r, \theta) = (r, \theta + n\cdot 360^\circ)$, where *n* is an even integer
  $(-r, \theta) = (r, \theta + n\cdot 180^\circ)$, where *n* is an odd integer
- Basic graphs:

$$r = a,\ r = \pm 2a\sin\theta,\ r = \pm 2a\cos\theta\text{: circle}$$

$$r = a \pm a\sin\theta,\ r = a \pm a\cos\theta\text{: cardioid}$$

$$r = a\sin n\theta,\ r = a\cos n\theta\text{: rose}\begin{cases} n \text{ loops, if } n \text{ is odd} \\ 2n \text{ loops, if } n \text{ is even}\end{cases}$$

- Formulas relating rectangular coordinates and polar coordinates:
  $x = r\cos\theta,\ y = r\sin\theta,\ x^2 + y^2 = r^2,\ \tan\theta = y/x$

CHAPTER 9 REVIEW EXERCISES

In Exercises 1–6 find the indicated part in triangle *ABC*.

1. Determine *B* if $a = 5.0$, $b = 9.0$, and $c = 6.0$.

2. Determine *C* if $A = 81^\circ$, $a = 11$, and $c = 35$.

3. Determine *A* if $C = 44^\circ 50'$, $B = 86^\circ 20'$, and $a = 62.7$.

4. Determine *B* if $C = 90^\circ$, $b = 5.00$, and $c = 8.00$.

5. Determine c if $C = 120^\circ$, $a = 3.0$, and $b = 4.0$.

6. Determine *b* if $A = 40^\circ$, $B = 60^\circ$, and $c = 6.0$.

In Exercises 7–10 perform the indicated operations. Write the answers in the form $a + bi$ and in trigonometric form.

7. $8(\cos 90^\circ + i\sin 90^\circ) \div (\cos 0^\circ + i\sin 0^\circ)$

8. $2(\cos 73^\circ + i\sin 73^\circ)\cdot 3(\cos 107^\circ + i\sin 107^\circ)$

9. $\left(\frac{\sqrt{2}}{2} + \frac{\sqrt{2}}{2}i\right)^{10}$ **10.** $\sqrt[3]{-1}$ (all roots)

In Exercises 11–14 graph each equation in polar coordinates.

11. $r = -2\sin\theta$ **12.** $r = 2$

13. $r = 2 + 2\cos\theta$ **14.** $r = 2\cos 3\theta$

15. Express the point $(-1, -1)$ in polar coordinates with $r \geq 0$ and $0° \leq \theta < 360°$.

16. Express the point $\left(-2, \frac{7\pi}{4}\right)$ in rectangular coordinates.

17. If $\mathbf{a} = 5(\cos 30°)\mathbf{i} + 5(\sin 30°)\mathbf{j}$, where **i** and **j** are unit basis vectors, then what is the magnitude and direction angle for vector $10\mathbf{a}$?

18. Find all solutions for the equation $x^4 + 4 = 0$

19. Find the area of triangle *ABC* if $a = 16$ m, $c = 11$ m, and $B = 116°$.

20. In triangle *ABC*, $b = 5$, $c = 6$, and $\cos A = -\frac{1}{3}$. Find side length *a*.

21. Transform the equation $r(2\cos\theta - \sin\theta) = 3$ to rectangular form.

22. Transform the equation $x^2 + y^2 = y$ to polar form.

23. Express the roots of $x^3 + 27 = 0$ in the form $r(\cos\theta + i\sin\theta)$.

24. Express the number $-1 - i$ in trigonometric form.

In Exercises 25–28 find the magnitude and direction angle of **v**.

25. $\mathbf{v} = \langle 1, -\sqrt{3} \rangle$ **26.** $\mathbf{v} = \langle 3, 0 \rangle$

27. $\mathbf{v} = -5\mathbf{j}$ **28.** $\mathbf{v} = -2\mathbf{i} + 2\mathbf{j}$

29. If $\mathbf{v} = \langle 8, -1 \rangle$ and $\mathbf{w} = \langle -5, -2 \rangle$, find $\mathbf{v} - 4\mathbf{w}$.

30. Express v as a linear combination of **i** and **j** if the magnitude of **v** is 6 and the direction angle of **v** is 150°.

31. If $\mathbf{v} = 2\mathbf{i} - 7\mathbf{j}$ and $\mathbf{w} = 5\mathbf{i} + 3\mathbf{j}$, express $2\mathbf{v} - \mathbf{w}$ as a linear combination of **i** and **j**.

32. Vector **a** has a magnitude of 7.0 lb and a direction angle of 74°. Vector **b** has a magnitude of 4.0 lb and a direction angle of 195°. Find the vector sum or resultant **v** of vectors **a** and **b**, and give the magnitude and direction angle of **v**.

33. If the outfielders are positioned as shown below, then how far is each of them from third base in case they need to throw out a runner at that base?

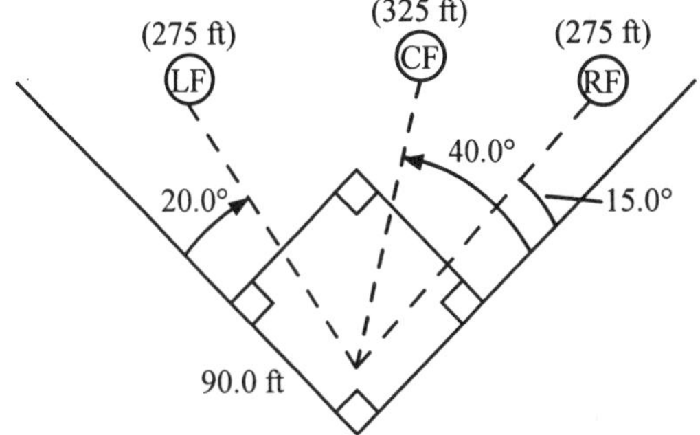

34. Two points, *A* and *B*, are $1\bar{0}0$ yd apart. Point *C* across a canyon is located so that angle *CAB* is 70° and angle *CBA* is 80°. Compute the distance $\overline{BC}$ across the canyon.

35. A draftsman drew to scale (1 in.= $5\bar{0}$ yd) a map of a development that includes a triangular recreation area with sides of lengths 75 yd, 125 yd, and 150 yd. What are the angles of the triangle representing the recreation area on the map?

36. A force of 12 lb and a force of 15 lb are acting on a body with an angle of 90° between the two forces. Find the magnitude of the resultant and the angle between the resultant and the larger force.

37. A force of $4\bar{0}$ lb and a force of $3\bar{0}$ lb act on a body so that their resultant is a force of 38 lb. Find the angle between the two original forces.

38. Vector **A** has a magnitude of 6.0 lb and makes an angle of 55° with the positive *x*-axis. Vector **B** has a magnitude of 2.0 lb and makes an angle of 110° with the positive *x*-axis. Find the resultant (or vector sum) of **A** and **B**.

39. Use the law of cosines to show that if *p*, *q*, and *r* are the side lengths of a triangle and $r^2 = p^2 + q^2$, then triangle *PQR* is a right triangle.

40. Use the law of sines to show that the area *K* of triangle *ABC* may be given by

$$K = \frac{b^2 \sin A \sin C}{2\sin B}.$$

In Exercises 41–50 select the choice that completes the statement or answers the question.

41. The number of triangles satisfying the conditions that $B = 30°$, $a = 57$ ft, and $b = 39$ ft is

a. two **b.** one **c.** none

42. We write $2(\cos 120° + i\sin 120°)$ in the form $a + bi$ as

a. $1 - \sqrt{3}i$ **b.** $-\sqrt{3} + i$
c. $\sqrt{3} - i$ **d.** $-1 + \sqrt{3}i$

43. The graph of $r = 2 + 2\cos\theta$ is a

a. circle **b.** cardioid
c. rose **d.** line

44. The x component of a vector with a magnitude of 25 lb that makes an angle of 67° with the x-axis is

a. 9.8 lb **b.** 12 lb
c. 17 lb **d.** 23 lb

45. An alternate polar representation for $(-1, -45°)$ is

a. (1, 45°) **b.** (1, 135°)
c. (−1, 135°) **d.** (1, 315°)

46. To solve a triangle when we know the measures for two sides of the triangle and the angle between these two sides, we first apply the

a. law of sines **b.** law of cosines

47. $(1 + i)^6$ simplifies to

a. 8 **b.** −8 **c.** $-8i$ **d.** $8i$

48. In triangle ABC, if $\sin A = \frac{3}{4}$ and $\sin B = \frac{1}{2}$, then the ratio of side length a to side length b is

a. 3:2 **b.** 8:3 **c.** 3:1 **d.** 4:3

49. Which illustration shows the graph of $r = \sin 5\theta$?

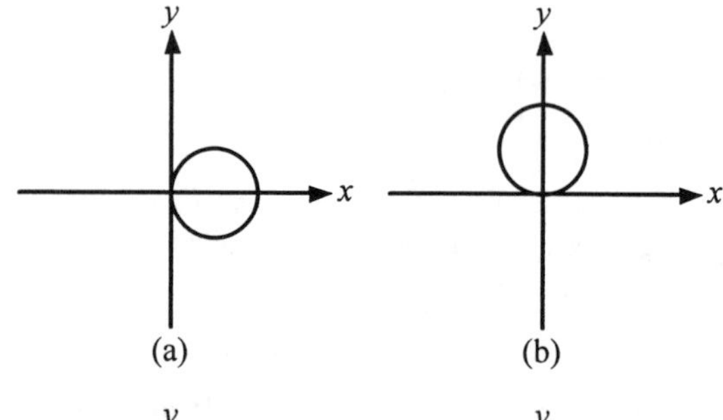

(a) (b)

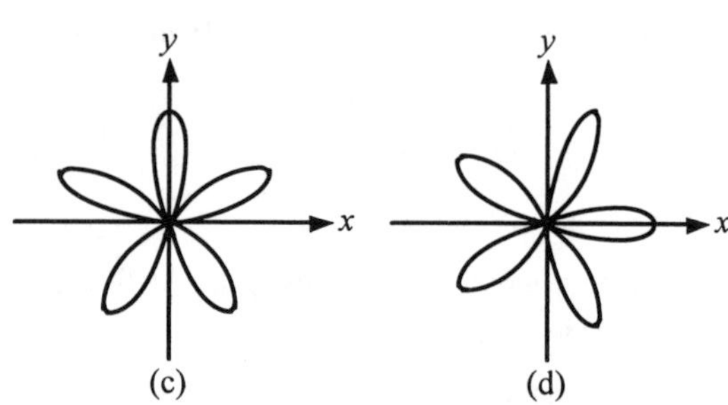

(c) (d)

50. The absolute value of the complex number $-2 + 2i$ is

a. $2\sqrt{2}$ **b.** 2 **c.** 4 **d.** $\sqrt{2}$

CHAPTER 9 TEST

1. True or false: To solve a triangle when given the measures for two sides of the triangle and the angle between these two sides, we first apply the law of cosines.

2. Assume exact numbers and find r in triangle RST if $R = 45°$, $T = 30°$, and $t = 24$.

3. In triangle ABC, $A = 34.5°$, $a = 11.6$ ft and $c = 19.1$ ft. Find the two possible solutions for C.

4. Find the measure of the largest angle in a triangle with side measures of 18.5 m, 15.0 m, and 26.0 m.

5. Solve the triangle ABC in which $B = 76°$, $b = 45$ ft, and $c = 35$ ft.

6. An equilateral triangle is inscribed in a circle with radius 12.0 cm. What is the perimeter of the triangle?

7. Find the area of triangle PQR if $p = 5.6$ cm, $r = 4.1$ cm, and $Q = 48.0°$.

8. If $\mathbf{v} = \langle 3\sqrt{3}, -3\rangle$, find the magnitude and direction angle of $\mathbf{v}$.

9. If $\mathbf{v} = \langle 7, -1\rangle$ and $\mathbf{w} = \langle -3, 5\rangle$, find $2\mathbf{v} - 3\mathbf{w}$.

10. Express $\mathbf{v}$ as a linear combination of $\mathbf{i}$ and $\mathbf{j}$ if the magnitude of $\mathbf{v}$ is 4 and the direction angle of $\mathbf{v}$ is 120°.

11. What is the vertical component of a vector with a magnitude of 56 lb that makes an angle of 37° with the positive x-axis?

12. A force of 28 lb and a force of 12 lb are acting on a body with an angle of 28° between the two forces. Find the magnitude of the resultant force.

13. Write the number $-15+8i$ in trigonometric form.

14. If complex numbers $z_1 = 6(\cos 44° + i\sin 44°)$ and $z_2 = 3(\cos 136° + i\sin 136°)$, write the product $z_1 z_2$ in the form $a+bi$.

15. Find $(1+i)^{10}$.

16. Find the three cube roots of -125.

17. List two other polar representations for $(-3, 225°)$ with one having $r > 0$.

18. Graph $r = 5 + 5\sin\theta$.

19. Express the point $\left(6, \dfrac{2\pi}{3}\right)$ in rectangular coordinates.

20. Transform the equation $x^2 + y^2 - 9x = 0$ to polar form.

● ● ●

Chapter

10

Analytic Geometry: Conic Sections

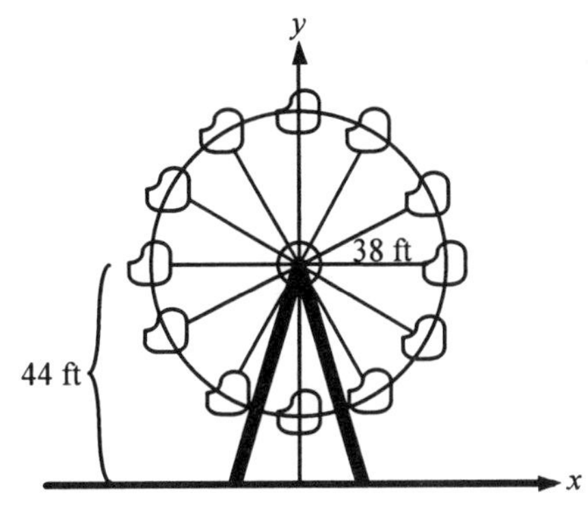

A ferris wheel has a radius of 38 ft. To analyze the trajectory of a cab on this wheel, a representation of the wheel is positioned on a coordinate system as diagrammed.

a. Find an equation of the graph of a point on a cab of this ferris wheel that moves in such a way that it is always 38 ft from the center.

b. To the nearest foot, how high above the ground is this point on the cab when the cab is 15 ft to the right and higher than the center?

(See Example 3 of Section 10.1.)

Analytic geometry bridges the gap between algebra and geometry. By representing the ordered pairs that satisfy some algebraic equation as points in the Cartesian coordinate system, we generate a geometric picture or graph. The fundamental relationship between an equation and its graph was first stated in Chapter 1 and merits repeating here:

> Every ordered pair that satisfies an equation corresponds to a point in its graph, and every point in the graph corresponds to an ordered pair that satisfies the equation.

We have considered some techniques for obtaining the graphs of a wide variety of functions. The next three sections will extend our coverage of graphing to obtain pictures of the solutions of second-degree equations in two variables with general form

$$Ax^2 + Bxy + Cy^2 + Dx + Ey + F = 0,$$

where A, B, and C are not all zero. In addition, this chapter also considers the reverse question to obtaining the graph of an equation:

> If we are given some object or motion from the physical world in geometric terms, can we describe it with an equation?

If so, the powerful methods of algebra can be used to analyze such objects or motions. Thus, the two primary problems of this chapter are finding the graphs of second-degree equations and determining second-degree equations that correspond to certain geometric conditions.

10.1 Introduction to Analytic Geometry and the Circle

Objectives

1. Find the equation in standard form of a circle, given its center and either the radius of the circle or a point on the circle.
2. Find the center and radius of a circle, given its equation, and then graph it.
3. Use the methods of analytic geometry to prove geometric theorems.
4. Find the midpoint of a line segment joining a given pair of points.

Circles, ellipses, hyperbolas, and parabolas are referred to as **conic sections**. This name was first used by Greek mathematicians who discovered that these curves result from the intersections of a cone with an appropriate plane, as shown in Figure 10.1. If the cutting plane does not contain the vertex, then the important conic sections are obtained as follows:

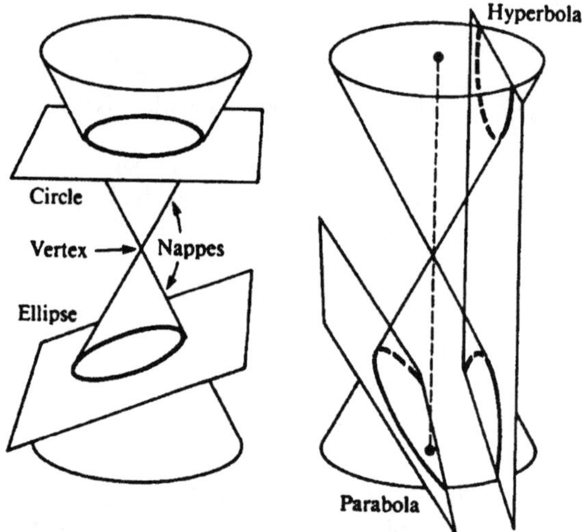

Figure 10.1

Circle: The intersecting plane is parallel to the base of the cone.
Ellipse: The intersecting plane is parallel to neither the base nor the side and intersects only one nappe of the cone.
Hyperbola: The intersecting plane cuts both nappes of the cone.
Parabola: The intersecting plane is parallel to the side of the cone.

Three degenerate cases occur when the cutting plane passes through the vertex. These degenerate conic sections are a point, a line, and a pair of intersection lines.

The conic sections may be analyzed from different viewpoints, and in this chapter we take the analytic geometry approach, in which we derive equations for the conic sections by starting with their geometric definitions. To illustrate, we begin with the definition of a circle that is given in plane geometry.

Definition of a Circle

A **circle** is the set of all points in a plane at a given distance from a fixed point.

From this definition we derive an equation of a circle as follows (see also Figure 10.2):

Let (h,k) be the fixed point, the center of the circle.
Let r be the given distance, the radius of the circle.
Let (x,y) be any point on the circle.

Then by the distance formula, we have

$$\sqrt{(x-h)^2+(y-k)^2}=r$$

or, after squaring both sides of the equation,

$$(x-h)^2+(y-k)^2=r^2.$$

This equation is the standard form of the equation of a circle of radius r with center (h,k).

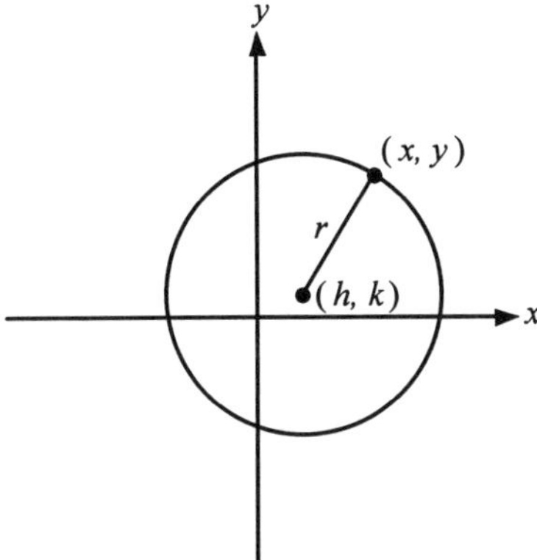

Figure 10.2

Standard Form of the Equation of a Circle

> The standard form of the equation of a circle of radius r with center (h,k) is
>
> $$(x-h)^2+(y-k)^2=r^2$$

Example 1: Finding the Standard Equation of a Circle

Find the equation in standard form of the circle with center at $(2,-1)$ and radius 4.

Solution: Substituting $h=2$, $k=-1$, and $r=4$ in the equation

$$(x-h)^2+(y-k)^2=r^2$$

we have

$$(x-2)^2+[y-(-1)]^2=(4)^2$$
$$(x-2)^2+(y+1)^2=16 \quad .$$

PROGRESS CHECK 1
Find the equation in standard form of the circle with center at (4,−7) and radius 9. ■

Example 2: Finding the Standard Equation of a Circle

Find the equation in standard form of the circle with center at (4,1) that passes through the origin.

Solution: Substituting $h = 4$ and $k = 1$ into the standard equation for a circle, we have

$$(x-4)^2 + (y-1)^2 = r^2.$$

Since the circle passes through the origin, we find r^2 by substituting the coordinates of the point (0,0) in the equation.

$$(0-4)^2 + (0-1)^2 = r^2$$
$$17 = r^2$$

The equation of the circle in standard form is then

$$(x-4)^2 + (y-1)^2 = 17.$$

PROGRESS CHECK 2
Find the equation in standard form of the circle with center at (6,3) that passes through (10,6). ■

Example 3: Trajectory of a Cab on a Ferris Wheel

Solve the problem in the section introduction on page 801.

Solution:

a. The point on the cab of the ferris wheel moves so that the trajectory is the set of all points in a plane that are 38 ft from a fixed point. Thus, the trajectory is a circle with radius 38 ft. Based on the positioning of the coordinate system, as diagrammed in Figure 10.3, the center is at (0,44). So the standard equation for the circle that gives the trajectory is

$$(x-0)^2 + (y-44)^2 = 38^2$$
$$x^2 + (y-44)^2 = 1{,}444.$$

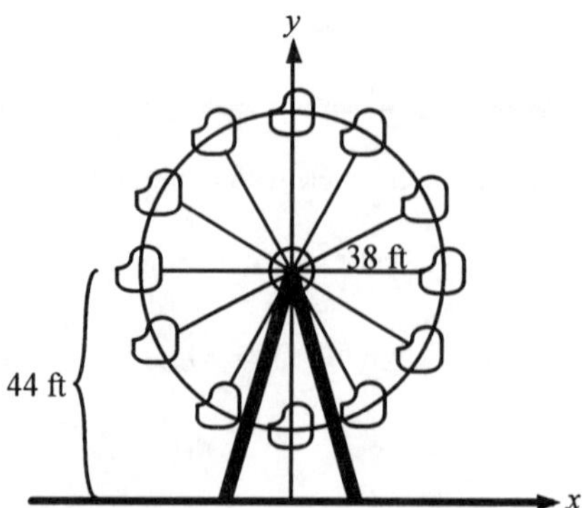

Figure 10.3

b. To find y for a given value of x, we first solve the equation that gives the trajectory for y.

$$\begin{aligned} x^2 + (y-44)^2 &= 1{,}444 \\ (y-44)^2 &= 1{,}444 - x^2 \\ y - 44 &= \pm\sqrt{1{,}444 - x^2} \\ y &= 44 \pm \sqrt{1{,}444 - x^2} \end{aligned}$$

When the cab is 15 ft to the right and higher than the center, then x is replaced by 15, and the radical term is added to give

$$\begin{aligned} y &= 44 + \sqrt{1{,}444 - 15^2} \\ &= 79\text{ft} \qquad \text{(to the nearest foot)}. \end{aligned}$$

The point on the cab is about 79 ft above the ground at the specified point.

Technology Link

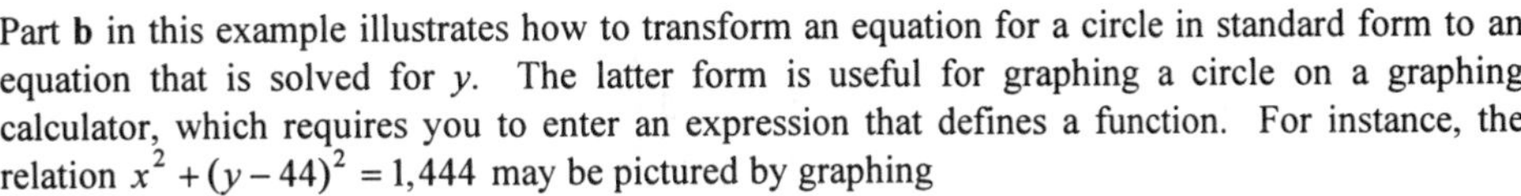

Part **b** in this example illustrates how to transform an equation for a circle in standard form to an equation that is solved for y. The latter form is useful for graphing a circle on a graphing calculator, which requires you to enter an expression that defines a function. For instance, the relation $x^2 + (y-44)^2 = 1{,}444$ may be pictured by graphing

$$y1 = 44 + \sqrt{1{,}444 - x^2} \quad \text{(top semicircle)}$$

and

$$y2 = 44 - \sqrt{1{,}444 - x^2} \quad \text{(bottom semicircle)}$$

in the same display, as shown in Figure 10.4. For this figure, a Square Setting feature was used to ensure that the graph would be round instead of oval.

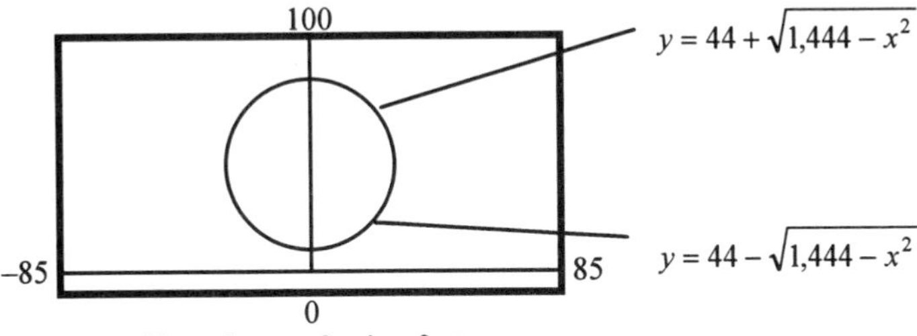

Figure 10.4

PROGRESS CHECK 3

Use the ferris wheel as described in Example 3 to answer the following questions.

a. Find an equation of the graph of a point on a cab of this ferris wheel that moves in such a way that it is always 37 ft from the center.

b. To the nearest foot, how high above the ground is this point on the cab when the cab is 18 ft to the right and lower than the center? ■

Expressing an equation for a circle in standard form is beneficial because the center and the radius of the circle can be identified by inspection of the equation. Once these features are known, it is easy to graph the circle.

Example 4: Finding the Center, Radius, and Graph of a Circle

Find the center and the radius of the circle given by

$$(x+3)^2 + y^2 = 25.$$

Then, sketch the circle.

Solution: To match standard form, view the given equation as

$$[x-(-3)]^2 + (y-0)^2 = (5)^2$$

Then $h=-3$, $k=0$, and $r=5$, so the center is (–3,0) and the radius is 5. Figure 10.5 shows the graph of this circle.

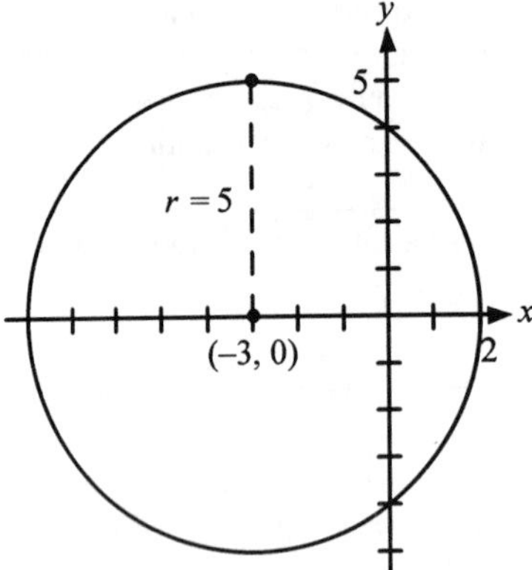

Figure 10.5

Technology Link

Some graphing calculators have a Circle operation that lets you graph a circle by specifying the coordinates of its center and the radius of the circle. If your calculator has this capability, you should use it to graph the circle in the example with center at (–3,0) and radius 5. Note that, depending on the window setting, it may be necessary to invoke a Square Setting feature so that the graph appears round.

PROGRESS CHECK 4
Find the center and radius of the circle given by $x^2 + (y+2)^2 = 9$. Then, sketch the circle. ■

Example 5: Converting to Standard Form

Find the center and radius of the circle given by

$$x^2 + y^2 + 4x - 6y = 12.$$

Then, sketch the circle.

Solution: We must transform the given equation to the standard form.

$$(x-h)^2 + (y-k)^2 = r^2$$

We start by completing the square in the x terms and the y terms. To do this, first group the equation as

$$\left(x^2 + 4x \quad\right) + \left(y^2 - 6y \quad\right) = 12.$$

Now find one-half of the coefficient of $x(2)$. Square it (4) and add the result to both sides of the equation. Similarly, find one-half of the coefficient of $y(-3)$. Square it (9) and add the result to both sides of the equation. Thus, we have

$$\left(x^2 + 4x + 4\right) + \left(y^2 - 6y + 9\right) = 12 + 4 + 9$$
$$(x + 2)^2 + (y - 3)^2 = 25$$

By comparing this equation to the standard form, we determine that the center of the circle is at $(-2,3)$ and that the radius is $\sqrt{25}$, or 5.

Figure 10.6 shows the graph of this circle.

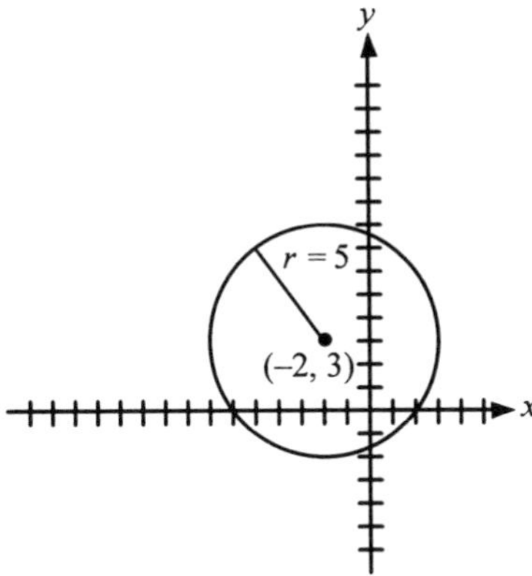

Figure 10.6

PROGRESS CHECK 5

Find the center and the radius of the circle given by $x^2 + y^2 - 8x - 4y - 5 = 0$. Then, sketch the circle. ■

Another aspect of analytic geometry is the proof of geometric theorems by using a coordinate system and algebraic methods, as illustrated by the next example.

Example 6: Proving a Geometry Theorem Using Analytic Geometry

Analytic geometry allows us to use the powerful methods of algebra to analyze geometry problems. To illustrate this, use Figure 10.7 and the concept of slope to show that if a triangle is inscribed in a circle with the diameter as one of its sides, then the triangle is a right triangle.

Solution: If the product of the slopes of line segments PP_1 and PP_2 is -1, then these sides are perpendicular, and the triangle is a right triangle. The product of the slopes is

$$m_1 m_2 = \frac{y-0}{x-(-r)} \cdot \frac{y-0}{x-r} = \frac{y^2}{x^2 - r^2}.$$

Since $x^2 + y^2 = r^2$ is an equation of a circle of radius r with center at the origin, we know that $y^2 = r^2 - x^2$, so

$$m_1 m_2 = \frac{r^2 - x^2}{x^2 - r^2} = -1.$$

Thus, the product of the slopes is -1, so the triangle is a right triangle.

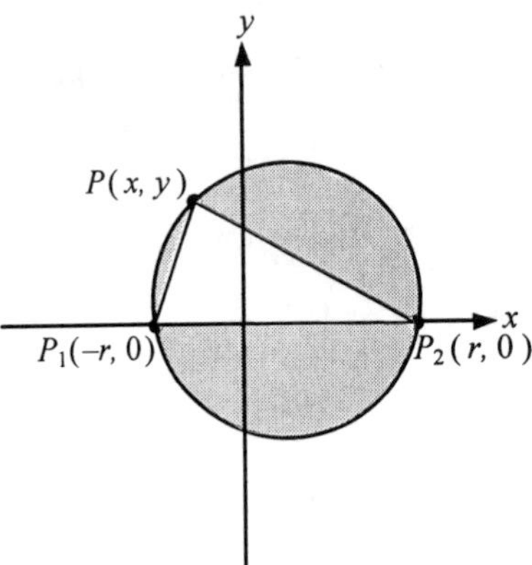

Figure 10.7

PROGRESS CHECK 6
Show by the methods of analytic geometry that the diagonals of a square are perpendicular to each other. ■

Up to now we have relied on distance and slope formulas in our analysis. Another useful formula that is simple to apply is the formula for finding the coordinates of the midpoint of a line segment. To derive this formula, let $P_1(x_1, y_1)$ and $P_2(x_2, y_2)$ be two points in a plane with $x_1 < x_2$ and $y_1 < y_2$, and let $P(x, y)$ be any other point on line segment P_1P_2. We then construct similar triangles P_1RP and PSP_2, as shown in Figure 10.8, and since corresponding side lengths are proportional in such triangles, we know that

$$\frac{x - x_1}{x_2 - x} = \frac{y - y_1}{y_2 - y} = \frac{\overline{P_1P}}{\overline{PP_2}}.$$

If $P(x, y)$ is the midpoint of the segment, then $\overline{P_1P}/\overline{PP_2} = 1$, and we may solve for x and y as follows:

$$\begin{aligned} \frac{x - x_1}{x_2 - x} &= 1 & \frac{y - y_1}{y_2 - y} &= 1 \\ x - x_1 &= x_2 - x & y - y_1 &= y_2 - y \\ 2x &= x_1 + x_2 & 2y &= y_1 + y_2 \\ x &= \frac{x_1 + x_2}{2} & y &= \frac{y_1 + y_2}{2} \end{aligned}$$

These results are valid regardless of the location of P_1 and P_2, and we have the following result.

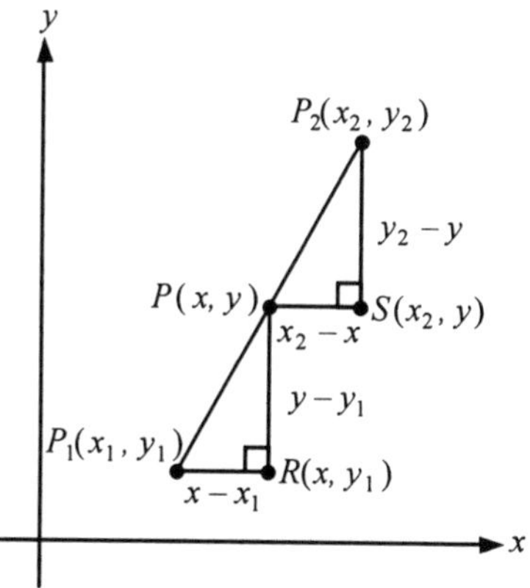

Figure 10.8

Midpoint Formula

> The midpoint of the line segment joining $(x_1 y_1)$ and $(x_2 y_2)$ is
>
> $$\left(\frac{x_1+x_2}{2}, \frac{y_1+y_2}{2}\right)$$

Note that the x-coordinate of the midpoint is simply the average of the x-coordinates of the two endpoints, and a similar relation applies for the y-coordinate of the midpoint. Also, in the above derivation of the midpoint formula, it is easy to extend our result by letting $\overline{P_1P}/\overline{PP_2}$ equal any positive number r and thereby develop a formula for the coordinates of the point that divides a line segment in any given ratio. We will pursue this idea in the **THINK ABOUT IT** exercises.

Example 7: Finding the Midpoint of a Line Segment

Find the midpoint of the line segment joining $(-3,4)$ and $(7,-1)$.

Solution: Using the midpoint formula, the midpoint is

$$\left(\frac{-3+7}{2}, \frac{4+(-1)}{2}\right)=\left(2,\frac{3}{2}\right).$$

PROGRESS CHECK 7

Find the midpoint of the line segment joining $(3,5)$ and $(11,-7)$. ■

Example 8: Proving a Geometry Theorem Using Analytic Geometry

Use analytic geometry to prove that the midpoint of the hypotenuse of any right triangle is equidistant from each of the three vertices.

Solution: We are free to place the specified figure in any convenient position on a coordinate system, so we start by placing the legs of a right triangle on the x- and y-axes, as shown in Figure 10.9. The vertex of the right angle is at the origin, while the other vertices are at the points $A(a,0)$ and $B(0,b)$. By the midpoint formula, the midpoint M of the hypotenuse is

$$\left(\frac{a+0}{2}, \frac{0+b}{2}\right)=\left(\frac{a}{2},\frac{b}{2}\right).$$

Then by the distance formula, the distance between the midpoint and each vertex is as follows:

$$\overline{AM} = \sqrt{\left(a - \frac{a}{2}\right)^2 + \left(0 - \frac{b}{2}\right)^3} = \sqrt{\frac{a^2}{4} + \frac{b^2}{4}}$$

$$\overline{BM} = \sqrt{\left(0 - \frac{a}{2}\right)^2 + \left(b - \frac{b}{2}\right)^2} = \sqrt{\frac{a^2}{4} + \frac{b^2}{4}}$$

$$\overline{CM} = \sqrt{\left(0 - \frac{a}{2}\right)^2 + \left(0 - \frac{b}{2}\right)^2} = \sqrt{\frac{a^2}{4} + \frac{b^2}{4}}$$

Therefore, $\overline{AM} = \overline{BM} = \overline{CM}$, which establishes the given theorem.

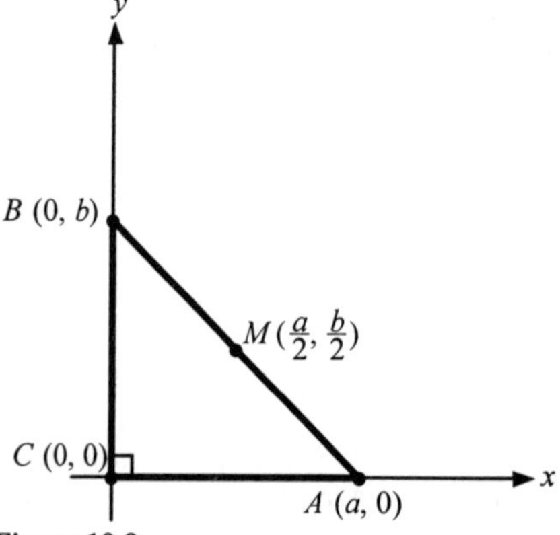

Figure 10.9

PROGRESS CHECK 8

Use analytic geometry to prove that the line segment joining the midpoints of two sides of a triangle is parallel to the third side and is one-half its length. ■

EXPLORE 10.1

1. Graph of circles will not appear round in the standard viewing window. You must either use a Square Setting feature or adjust the x and y limits yourself to achieve the desired shape. Compare the graphs of $y = \pm\sqrt{9 - x^2}$ in the standard viewing window and in another window that you choose to make the graph appear round.
   Note: For circles, it is easier to enter $y2 = -y1$ rather than enter the defining equation for $y2$.
2. Because the equation for a semicircle involves the square root of a difference, you must be careful about the use of parentheses. Describe and explain any differences between the graphs of $y = \sqrt{\ }(9 - x^2)$ and $y = \sqrt{\ }9 - x^2$.
3. Some calculators have special commands for drawing points, lines, and circles. These commands are often found on the Draw menu. Note however, that the Circle Draw command is intended to produce a round figure and may ignore the window settings. To explore this feature, choose the standard viewing window, then draw a circle with radius 6 by entering $y1 = \sqrt{\ }(36 - x^2)$ and $y2 = -y1$. Next, use the Circle Draw feature to draw a circle with center (0,0) and radius 6. Describe the results of this exploration. What happens if you use Square Setting features instead of the standard window?
4. Two distinct circles may or may not intersect. Use the calculator to estimate to two decimal places the coordinates of all points of intersection of these pairs of circles.
   a. $x^2 + y^2 = 49$; $(x - 5)^2 + y^2 = 4$ **b.** $x^2 + y^2 = 49$; $(x - 5)^2 + y^2 = 5$
   c. $x^2 + y^2 = 49$; $(x - 5)^2 + y^2 = 3$

EXERCISES 10.1

In Exercises 1–10, find the equation in standard form for each of the circles from the given information.

1. Center at $(-3,4)$, radius 2
2. Center at $(-2,1)$, radius 3
3. Center at (0,0), radius 1
4. Center at (0,2), radius 5
5. Center at $(-4,-1)$, tangent to the line $x=-1$ (**Note:** A line is tangent to a circle if the two intersect at exactly one point. The tangent is perpendicular to the radius at the point of intersection.)
6. Center at (1,4), tangent to the line $y=-3$
7. Center at $(3,-3)$, passes through the origin
8. Center at $(2,-5)$, passes through $(1,0)$
9. Center at $(-1,-1)$, passes through $(6,2)$
10. Center at $(-4,-3)$, passes through $(-1,-5)$

In Exercises 11–24 find the center and radius of each circle.

11. $(x-2)^2+(y-5)^2=16$
12. $(x+1)^2+(y+4)^2=49$
13. $(x-3)^2+y^2=20$
14. $x^2+(y+2)^2=5$
15. $x^2+y^2=9$
16. $x^2+y^2-4=0$
17. $x^2+y^2-10x-6y=15$
18. $x^2+y^2-4x+6y=23$
19. $x^2+y^2+8x-2y-1=0$
20. $x^2+y^2+4x-9=0$
21. $x^2+y^2+7x+3y+4=0$
22. $x^2+y^2-5x-y-3=0$
23. $4x^2+4y^2+8x-16y-29=0$
24. $2x^2+2y^2-10x-2y-31=0$

In Exercises 25–30 find the midpoint of the line segment joining the given pair of points.

25. $(-2,3)$, $(6,-1)$
26. $(5,-4)$, $(-9,0)$
27. $(7,2)$, $(-7,-9)$
28. $(-1,1)$, $(1,-1)$
29. $(0,0)$, $(0,3)$
30. $(8,-2)$, $(5,-2)$

In Exercises 31–34 use the illustration below to prove the given theorem.

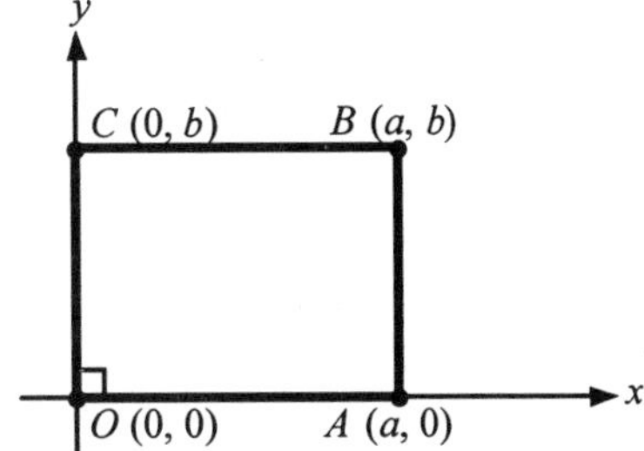

31. The diagonals of a rectangle are equal in length.
32. The diagonals of a rectangle bisect each other.
33. If the diagonals of a rectangle are perpendicular, then the rectangle is a square.
34. The midpoints of the sides of rectangle are the vertices of a rhombus (that is, a parallelogram with all sides equal in length.)

In Exercises 35–36 prove each theorem by the methods of analytic geometry.

35. The diagonals of a parallelogram bisect each other.
36. The line segments joining the midpoints of the opposite sides of any quadrilateral bisect each other.
37. **a.** For the semicircle shown, prove that the area of the inscribed triangle is given by $A=r\sqrt{r^2-x^2}$.
    b. What is the maximum possible area for the inscribed triangle?

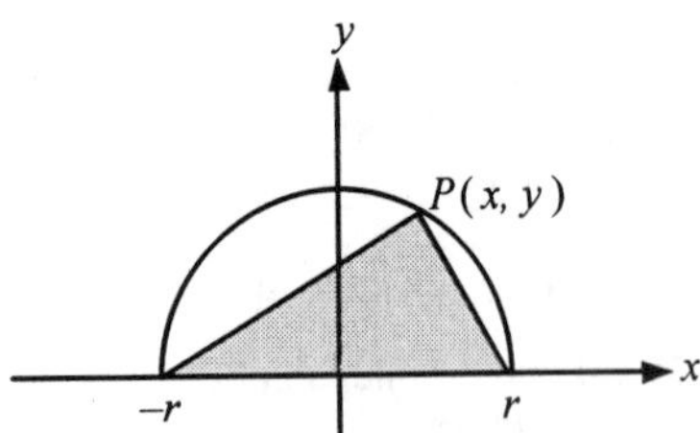

38. a. For the semicircle shown, find a formula for the area of the inscribed rectangle.
b. If $r = 1$, use a calculator to estimate to 3 decimal places the maximum possible area for the inscribed rectangle.

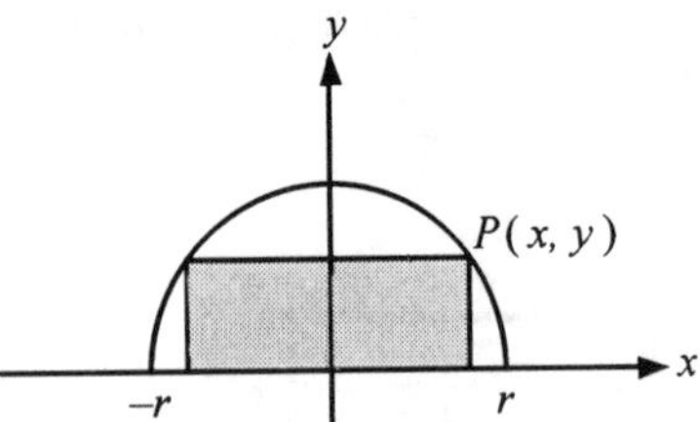

39. The tip of the minute hand of a clock traces a circle with diameter 20 inches. The clock is mounted on a wall with its center 6 feet above the ground.
a. Using the representation of the coordinate axes as shown, find an equation for the circle traced by the tip of the minute hand.
b. How high above the floor is the tip of the minute hand at 3:05 p.m.? **Note:** When the tip of the minute hand is on the 1, it is 5 inches to the right of center.

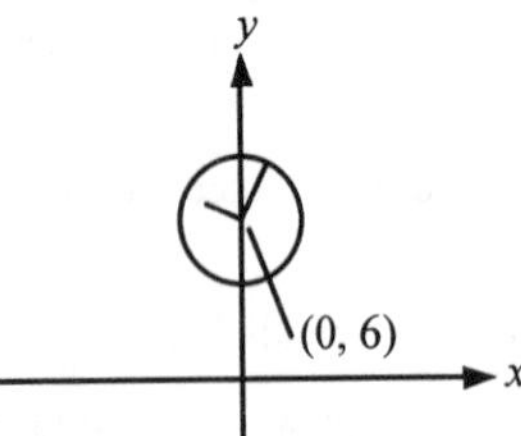

40. For the clock in the previous exercise, the length of the hour hand is 6 inches.
a. Using the representation of the coordinate axes as shown in the previous exercise, find an equation for the circle traced by the tip of the hour hand.
b. How high above the floor is the tip of the hour hand at 1 p.m.? **Note:** At 1 o'clock the tip of the hour hand is 3 inches to the right of center.

41. The cross-section of a tunnel is the semicircle shown in the figure. How high above the road is the tunnel ceiling 30 ft on each side of the center?

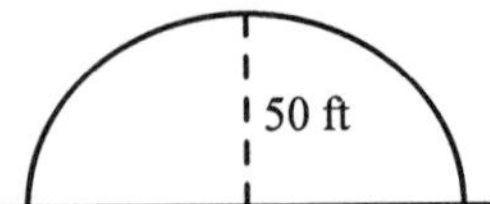

42. Refer to the figure in Exercise 41. How high above the road is the tunnel 25 ft on each side of the center? Given the answer in exact radical form and also approximated to the nearest tenth of a foot.

43. The figure for this exercise shows two circles. Point A is on the larger circle, and it is directly above C, the center of the smaller circle.
a. Find the coordinates of point C.
b. Find the equation of the larger circle.
c. Find the coordinates of point A.
d. Find the distance between A and B.
e. Find the area of triangle ABO.

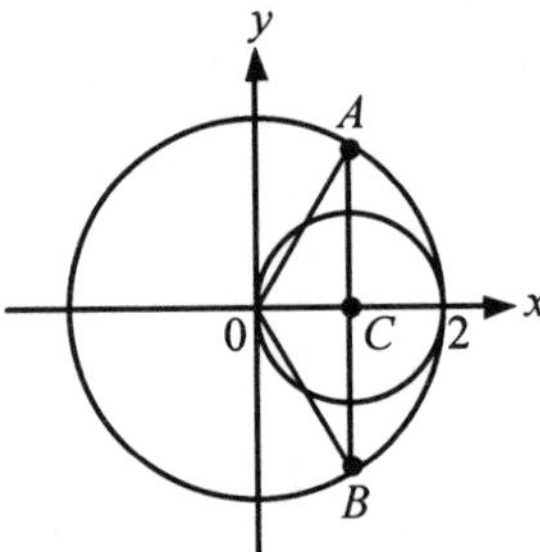

44. For the figure given below, answer the questions in Exercise 43.

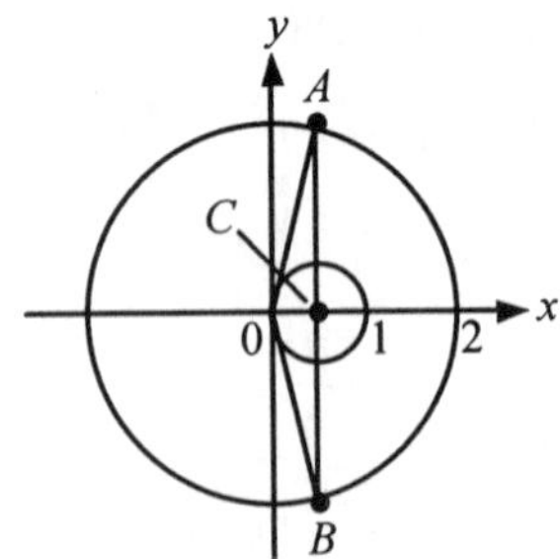

45. The circle shown below has equation $x^2 + y^2 = 2$. Find the area of the shaded square.

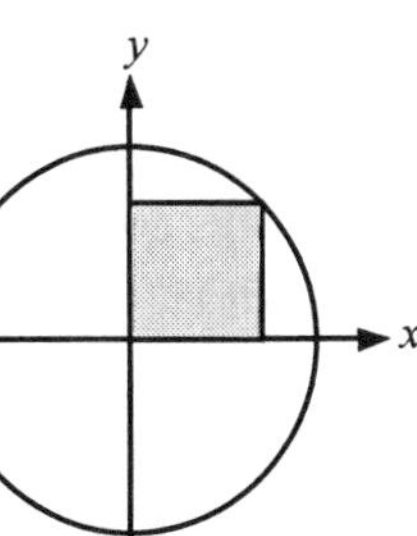

46. The circle shown below has equation $x^2 + y^2 = 1$. Find the are of the shaded square.

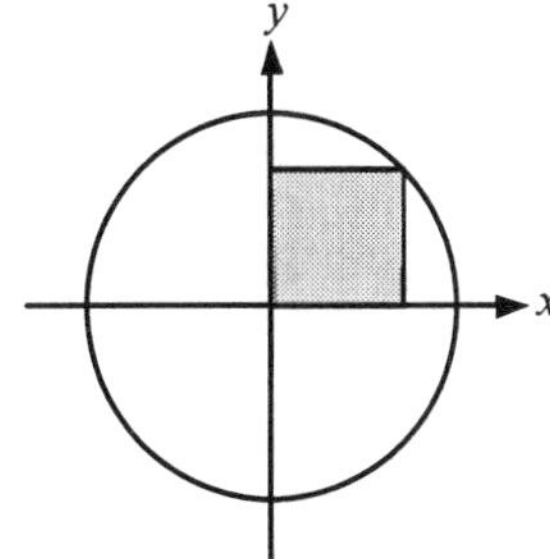

THINK ABOUT IT 10.1

1. Find the equation in standard form of the circle with $P(-4,-3)$ and $Q(4,5)$ at endpoints of a diameter.
2. If the line $y = mx$ passes through the center of the circle $4x^2 + 4y^2 - 4x + 12y - 1 = 0$, find the value of m.
3. Show by the methods of analytic geometry that the sum of the squares of the side lengths of a parallelogram is equal to the sum of the squares of the lengths of the diagonals.
4. **a.** Consider the derivation of the midpoint formula. If $P(x, y)$ is a point that divides the line segment from P_1 to P_2 so that $\overline{P_1P}/\overline{PP_2}$ equals a given positive ratio r, then show that the coordinates of $P(x, y)$ are

$$\left(\frac{x_1 + rx_2}{1+r}, \frac{y_1 + ry_2}{1+r}\right).$$

This formula is called the *point-of-division formula.*

b. Derive the midpoint formula from the point-of-division formula.

5. Solve each problem by using the point-of-division formula.

a. Find the coordinates of the point that divides the line segment from $P_1(1,3)$ to $P_2(8,-4)$ in the ratio $\frac{3}{4}$.

b. Determine the two points of trisection of the line segment joining $(0,-4)$ and $(6,4)$.

● ● ●

10.2 The Ellipse

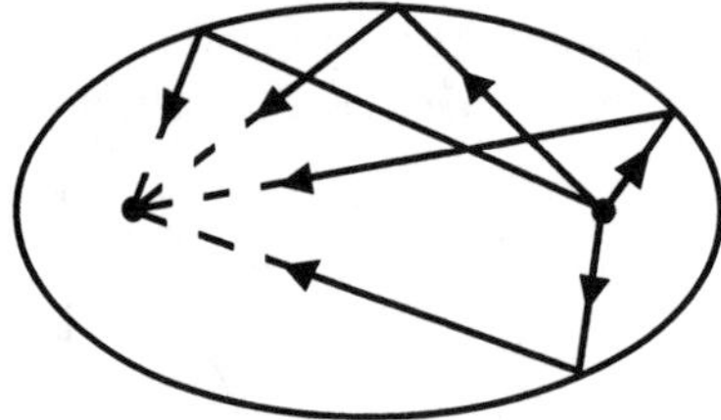

When kidney stones become trapped in the kidneys, they may be dispersed using a lithotripter, which is a machine that uses ultrasound to pulverize the stones into a powder that can easily be excreted from the body. A lithotripter has the shape of an ellipse that is rotated about its major axis. This shape is used because the ellipse has a reflection property that causes any ray or wave that originates at one focus to strike the ellipse and pass through the other focus, as shown in the figure.

When a lithotripter is used, the kidney stones are positioned at one focus, and the device that produces the ultrasound is positioned at the other focus. Because of the reflection property of the ellipse, the waves from the ultrasound are then concentrated on the kidney stones and shatter them.

a. The design for a certain lithotripter is based on rotating the ellipse given by $\left(x^2/4\right)+\left(y^2/9\right)=1$ about its major axis. Graph this ellipse.

b. When a doctor uses the lithotripter described in part **a** to break up kidney stones, what are the coordinate positions that show where to position the patient's kidney stones and where to position the source of the ultrasound?

(See Example 2.)

Objectives

1. Use an equation of an ellipse to graph the relation and to find the coordinates of the center, the foci, and the endpoints of the major and minor axes.
2. Find an equation for an ellipse satisfying various conditions concerning its center, foci, vertices, and major and minor axes.

The ellipse, which is an oval-shaped curve, is defined geometrically as follows.

Definition of an Ellipse

An **ellipse** is the set of all points in a plane the sum of whose distances from two fixed points is a constant.

Each fixed point is called a **focus** of the ellipse, and the two fixed points are called the **foci**. The locations of the foci are important in many applications of the ellipse. For instance, one of the well-known applications of the ellipse is that the Earth moves in an elliptical orbit with the sun at one focus, as shown in Figure 10.10.

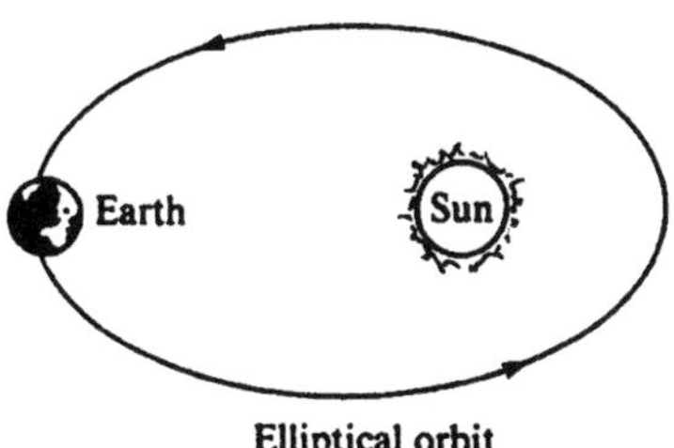

Figure 10.10

To understand the geometric definition, consider Figure 10.11, which shows an ellipse centered at the origin with foci on the x-axis. If we let $P(x,y)$ be any point on the ellipse as shown, then by definition,

$$d_1 + d_2 = \text{positive constant}.$$

In more concrete terms this definition means that an ellipse may be drawn using pencil, string, and tacks, as shown in Figure 10.12. Can you explain the mathematical basis for this method, referring to the definition of the ellipse?

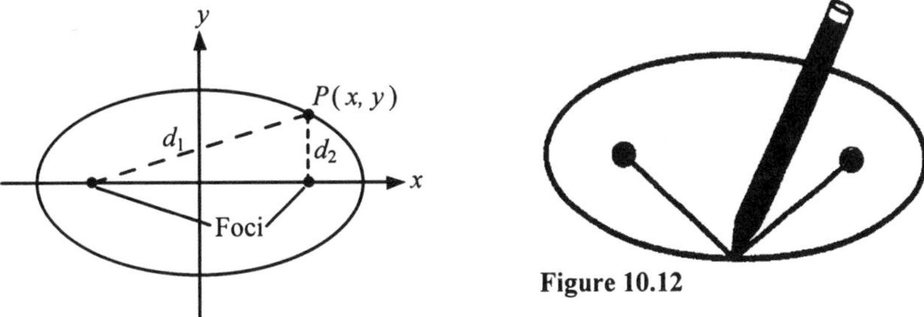

Figure 10.12

Figure 10.11

The answer is that the length of the string is a constant, and tacking down the ends determines two fixed points, which are the foci. The curve traced in Figure 10.12 is thus an ellipse because it is the set of all points in a plane, the sum of whose distances from two fixed points (located at the tacks) is a constant (the length of the string).

Equations for ellipses may be derived from this definition in the following way. For simplicity, let the foci be positioned at $(-c,0)$ and $(c,0)$, and let $2a$ represent the constant sum. If we let $P(x,y)$ be any point on the ellipse, as shown in Figure 10.13, then by definition $d_1 + d_2 = 2a$, so it follows from the distance formula that

$$\sqrt{(x+c)^2 + (y-0)^2} + \sqrt{(x-c)^2 + (y-0)^2} = 2a$$

or

$$\sqrt{(x+c)^2 + y^2} = 2a - \sqrt{(x-c)^2 + y^2}.$$

Now square both sides of the equation and simplify.

$$(x+c)^2 + y^2 = 4a^2 - 4a\sqrt{(x-c)^2 + y^2} + (x-c)^2 + y^2$$

$$xc = a^2 - a\sqrt{(x-c)^2 + y^2}$$

$$a\sqrt{(x-c)^2 + y^2} = a^2 - cx$$

Again square both sides of the equation and simplify.

$$a^2\left[(x-c)^2 + y^2\right] = a^4 - 2a^2cx + c^2x^2$$

$$a^2x^2 - 2a^2cx + a^2c^2 + a^2y^2 = a^4 - 2a^2cx + c^2x^2$$

$$a^2x^2 - c^2x^2 + a^2y^2 = a^4 - a^2c^2$$

$$\left(a^2 - c^2\right)x^2 + a^2y^2 = a^2\left(a^2 - c^2\right)$$

$$\frac{x^2}{a^2} + \frac{y^2}{a^2 - c^2} = 1$$

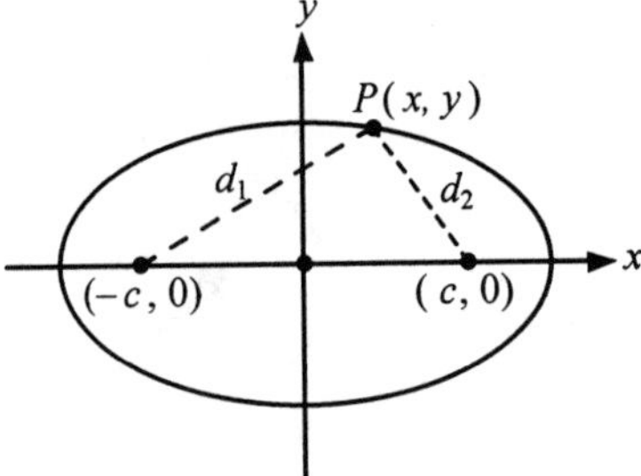

Figure 10.13

Since $d_1 + d_2$ is greater than the distance between the foci, it follows that $a > c$ and $a^2 - c^2 > 0$. If we now define

$$b^2 = a^2 - c^2 \text{ or } a^2 = b^2 + c^2,$$

we obtain the **standard form** of the equation of an ellipse with center at the origin and foci on the x-axis.

$$\frac{x^2}{a^2} + \frac{y^2}{b^2} = 1$$

Consider Figure 10.14. By setting $y = 0$, we find that the x-intercepts are $(-a,0)$ and $(a,0)$. By setting $x = 0$, we find that the y-intercepts are $(0,-b)$ and $(0,b)$. The larger segment from $(-a,0)$ to $(a,0)$ is called the **major axis**; the **minor axis** is the segment from $(0,-b)$ to $(0,b)$. The endpoints of the major axis are called the **vertices of the ellipse**.

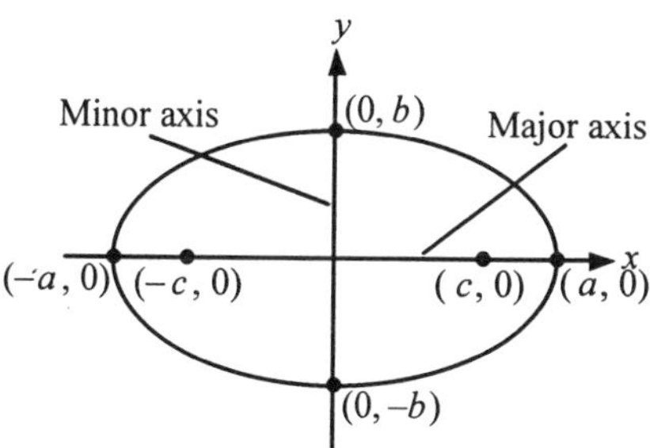

Foci: $(-c,0)$, $(0,-c)$
Vertices: $(0,a)$, $(0,-a)$
Figure 10.14

If the foci are placed on the y-axis at $(0,-c)$ and $(0,c)$, then the **standard form** of the equation of an ellipse is

$$\frac{x^2}{b^2}+\frac{y^2}{a^2}=1$$

where the larger denominator is denoted by a^2. The major axis is then along the y-axis, as shown in Figure 10.15. Note that the foci always lie on the major axis.

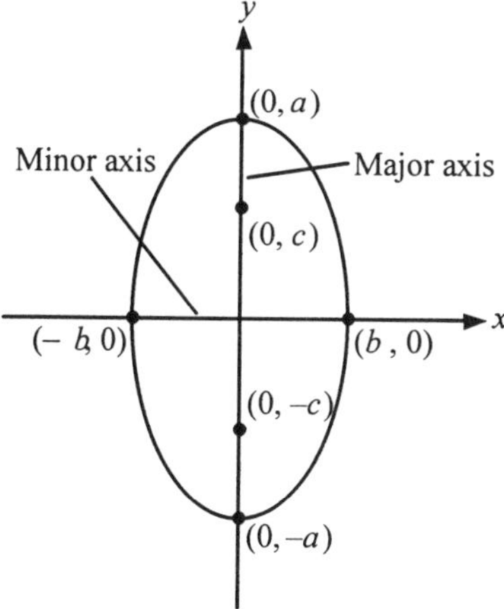

Foci: $(0,c)$, $(0,-c)$
Vertices: $(0,a)$, $(0,-a)$
Figure 10.15

Example 1: Using a Standard Form of an Ellipse

Find the coordinates of the foci and the endpoints of the major and minor axes of the ellipse given by $4x^2+y^2=36$. Also, sketch the ellipse.

Solution: First, divide both sides of the equation by 36 to obtain the standard form.

$$\frac{x^2}{9}+\frac{y^2}{36}=1$$

Then

$$a^2 = 36 \quad a = 6$$
$$b^2 = 9 \quad b = 3$$

To find c, we replace a^2 by 36 and b^2 by 9 in the formula

$$a^2 = b^2 + c^2$$
$$36 = 9 + c^2$$
$$27 = c^2 \quad \text{or } c = \sqrt{27} = 3\sqrt{3}$$

The major axis is along the y-axis since the larger denominator appears in the y term. Thus,

endpoints of major axis: $(0,-6)$, $(0,6)$
endpoints of minor axis: $(-3,0)$, $(3,0)$
coordinates of foci: $(0,-3\sqrt{3})$, $(0,3\sqrt{3})$.

These points are shown in the graph of the ellipse in Figure 10.16.

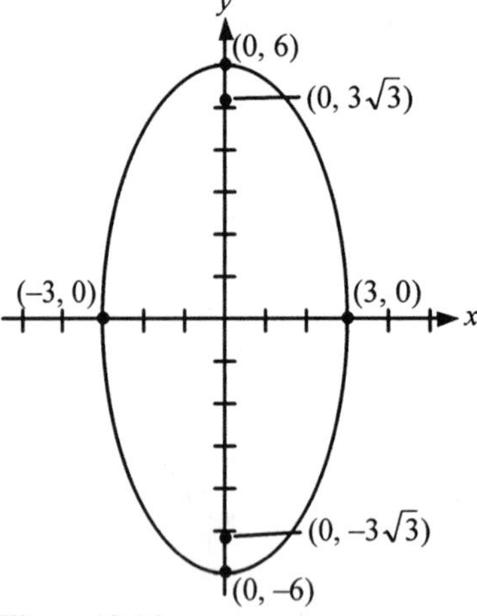

Figure 10.16

Technology Link

Visual support for the results in this example may be obtained with a graphing calculator. The ellipse $4x^2 + y^2 = 36$ may be pictured by graphing

$$y1 = \sqrt{36 - 4x^2} \quad \text{(top semiellipse)}$$

and

$$y2 = -\sqrt{36 - 4x^2} \quad \text{(bottom semiellipse)}$$

in the same display, as shown in Figure 10.17. Once again a viewing window that uses a square setting is recommended, to avoid distortion in the shape of the graph.

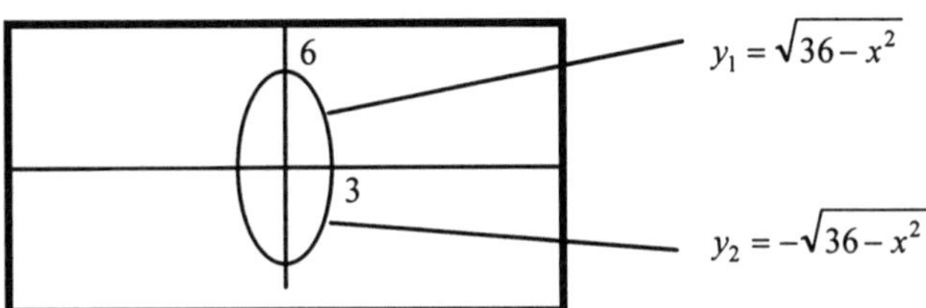

Figure 10.17

PROGRESS CHECK 1

Find the coordinates of the foci and the endpoints of the major and minor axes of the ellipse given by $x^2 + 4y^2 = 16$. Also, sketch the ellipse. ■

Example 2: Dispersing Kidney Stones in a Lithotripter

Solve the problem in the section introduction on page 814.

Solution:

a. The ellipse $(x^2/4) + (y^2/9) = 1$ is in standard form in the case where the larger denominator, a^2, is in the y term, so the major axis is vertical. Then

$$a^2 = 9, \text{ so } a = 3 \text{ and } b^2 = 4, \text{ so } b = 2.$$

Therefore, the endpoints of the major axis are (0, 3) and (0, –3), and the endpoints of the minor axis are (2, 0) and (–2, 0). Drawing an ellipse using these endpoints gives the graph of the equation in Figure 10.18. Use a square viewing window with $y1 = \sqrt{(36 - 9x^2)/4}$ and $y2 = -\sqrt{(36 - 9x^2)/4}$ to check this result on a grapher.

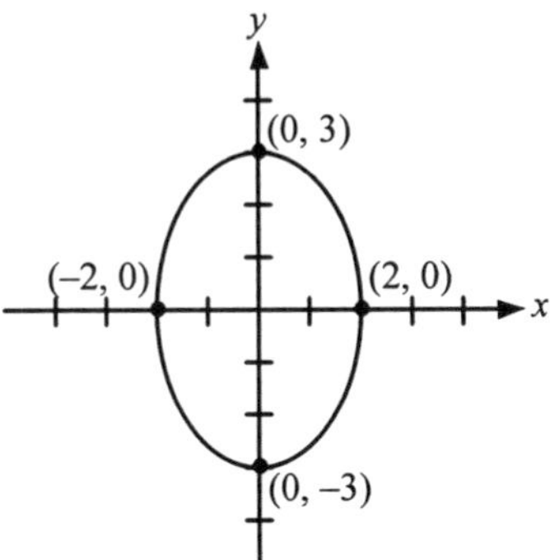

Figure 10.18

b. To disperse kidney stones in a lithotripter, the patient's kidney stones are positioned at the upper focus, and the source of the ultrasound is positioned at the lower focus. To locate these positions, we find c by replacing a^2 by 9 and b^2 by 4 in the equation

$$a^2 = b^2 + c^2$$

$$9 = 4 + c^2$$

$$5 = c^2 \text{ or } c = \sqrt{5}$$

The foci are located on the major axis, which is vertical so the requested coordinate positions are $(0,\sqrt{5})$ and $(0,-\sqrt{5})$.

Note: The reflection property of an ellipse has an interesting application in acoustics. In a room with an elliptical ceiling, a slight noise made at one focus can be heard at the other focus because of this reflection property. However, if you are standing between the foci, you hear nothing. Such rooms are called "whispering galleries," and a famous one is located in the Capitol in Washington, D. C.

PROGRESS CHECK 2
Redo the problems in Example 2, assuming that the design for a certain lithotripter is based on rotating the ellipse given by $4x^2 + y^2 = 4$ about its major axis. ■

Example 3: Finding the Standard Equation of an Ellipse

Find the equation in standard form of the ellipse with center at the origin, one focus at (3, 0), and one vertex at (4, 0).

Solution: Since c is the distance from the center to a focus, c is 3. Since a is the distance from the center to a vertex, a is 4. To find b^2, replace c by 3 and a by 4 in the formula

$$\begin{aligned} a^2 &= b^2 + c^2 \\ (4)^2 &= b^2 + (3)^2 \\ 7 &= b^2 . \end{aligned}$$

Since the major axis is along the x-axis, the standard form is

$$\frac{x^2}{a^2} + \frac{y^2}{b^2} = 1 .$$

Thus,

$$\frac{x^2}{16} + \frac{y^2}{7} = 1$$

is the standard equation of the ellipse.

PROGRESS CHECK 3
Find the equation in standard form of an ellipse with center at the origin, one focus at (2, 0), and one vertex at (5, 0). ■

If the ellipse is centered at the point (h, k), the **standard form** is

$$\frac{(x-h)^2}{a^2} + \frac{(y-h)^2}{b^2} = 1$$

when the major axis is parallel to the x-axis and

$$\frac{(x-h)^2}{b^2}+\frac{(y-k)^2}{a^2}=1$$

when the major axis is parallel to the y-axis. As in the case of the circle, the standard form is obtained by completing the square.

Example 4: Converting to Standard Form

Find the coordinates of the foci and the endpoints of the major and minor axes of the ellipse given by

$$9x^2+4y^2+18x-8y-23=0.$$

Also, sketch the curve.

Solution: First, put the equation in standard form by completing the square.

$$\begin{aligned}(9x^2+18x\quad)+(4y^2-8y\quad)&=23\\ 9(x^2+2x\quad)+4(y^2-2y\quad)&=23\\ 9(x^2+2x+1)+4(y^2-2y+1)&=23+9(1)+4(1)=36\end{aligned}$$

Be careful to add 9(1) and 4(1) to the right side of the equation. Now divide both sides of the equation by 36.

$$\frac{(x+1)^2}{4}+\frac{(y-1)^2}{9}=1$$

From this equation in standard form we conclude that the center is at $(-1,1)$.

$$\begin{aligned}a^2&=9\quad a=3\\ b^2&=4\quad b=2\end{aligned}$$

To find c, we replace a^2 by 9 and b^2 by 4 in the formula

$$\begin{aligned}a^2&=b^2+c^2\\ 9&=4+c^2\\ 5&=c^2 \text{ or } c=\sqrt{5}\cdot\end{aligned}$$

The major axis is parallel to the y-axis, since the larger denominator appears in the y term; a is the distance from the center to the vertices. Thus,

endpoints of major axis: $(-1,-2)$, $(-1,4)$.

The distance from the center to the endpoints of the minor axis is b. Thus,

endpoints of minor axis: $(-3,1)$, $(1,1)$.

The distance from the center to the foci is c. Thus,

$$\text{coordinates of the foci: } \left(-1, 1-\sqrt{5}\right), \left(-1, 1+\sqrt{5}\right).$$

The ellipse is graphed in Figure 10.19.

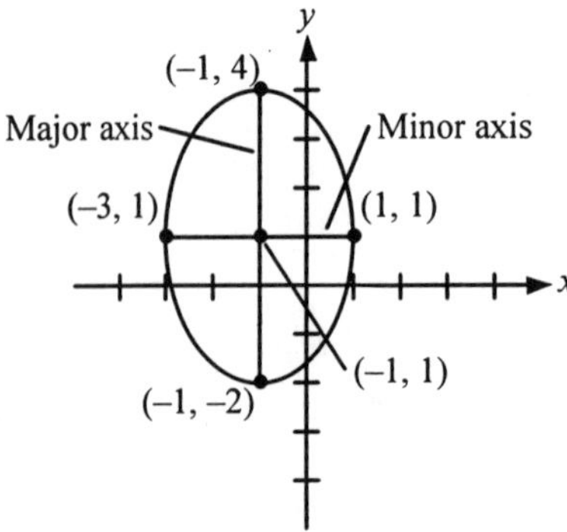

Figure 10.19

Use a square viewing window with $y1 = 1 + \sqrt{\left(36 - 9(x+1)^2\right)/4}$ and $y2 = 1 - \sqrt{\left(36 - 9(x+1)^2\right)/4}$ to check this result on a grapher.

PROGRESS CHECK 4
Find the coordinates of the foci and the endpoints of the major and minor axes of the ellipse given by $16x^2 + 25y^2 - 64x + 50y - 311 = 0$. Also, sketch the curve. ■

EXPLORE 10.2

To graph an ellipse with a grapher, you can enter separate equations for the top and bottom halves, as illustrated in the Technology Link in this section. Note that if the cursor does not land on the x-intercepts, the top and bottom halves may appear disconnected. This disconnection can be avoided by setting the viewing window appropriately—for instance, by using Zoom Decimal if the ellipse has integer intercepts.

1. **a.** Use a grapher to draw several ellipses defined by

$$\frac{x^2}{a^2} + \frac{y^2}{25} = 1$$

for several values of a, such as 1, 2, 5, and 8. Describe the effect of changing the value of a. To avoid distortion of the elliptical shape use a Square Setting feature.

   b. Do a similar exploration for various values of b in the equation: $x^2/25 + y^2/b^2 = 1$.

2. Use a grapher to estimate to two decimal places the points of intersection of these two ellipses.

$$x^2 + 2y^2 = 64$$

$$2x^2 + y^2 = 64$$

3. An elliptical arch is to be built as a roof, as shown. It must meet supports that are 3 feet from the center and 2 meters high.

   a. Show that there is more than one equation for an ellipse that meets these requirements. Hint: In the equation $x^2/a^2 + y^2/b^2 = 1$, let $x = 3$, $y = 2$, and solve for b in terms of a.

b. Using five different values for a, show simultaneously on a grapher five different semiellipses that include the points $(-3,2)$ and $(3,2)$. Why must a be greater than 3?

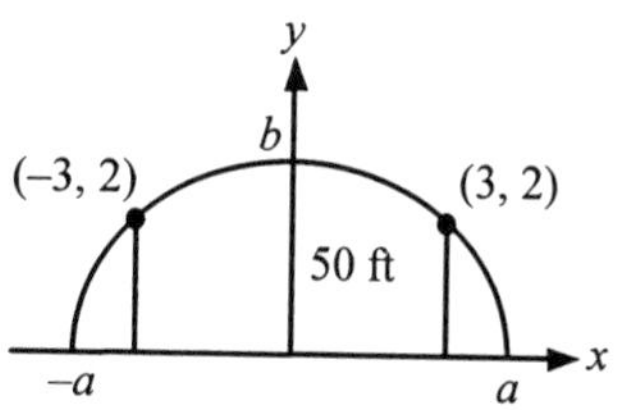

4. Most graphing calculators have a Shade command, which can be used to enhance graphs. Use such a grapher to graph a circle of radius 4 inside the ellipse with semiaxes of lengths 7 and 5. Then use the Shade command to shade the area between the circle and the ellipse. **Hint:** You may have to use some techniques for piecewise graphing.

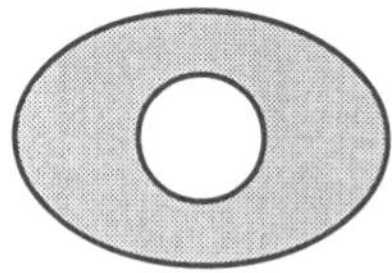

EXERCISES 10.2

In Exercises 1–14 find the coordinates of the foci and the endpoints of the major and minor axes. Sketch each ellipse.

1. $\dfrac{x^2}{25}+\dfrac{y^2}{16}=1$

2. $\dfrac{x^2}{36}+\dfrac{y^2}{100}=1$

3. $\dfrac{x^2}{1}+\dfrac{y^2}{4}=1$

4. $\dfrac{x^2}{49}+\dfrac{y^2}{9}=1$

5. $x^2+9y^2=36$

6. $9x^2+4y^2=144$

7. $4x^2+y^2=1$

8. $x^2+9y^2=25$

9. $\dfrac{(x-1)^2}{9}+\dfrac{(y+3)^2}{25}=1$

10. $\dfrac{(x+2)^2}{100}+\dfrac{(y-2)^2}{64}=1$

11. $4(x-3)^2+25(y+1)^2=100$

12. $25(x+4)^2+16y^2=400$

13. $4x^2+9y^2+8x-54y+49=0$

14. $9x^2+y^2-36x+2y+1=0$

In Exercises 15–24 find the equation in standard form of the ellipse satisfying the conditions.

15. Center at origin, vertex (0, 5), focus (0, 3)

16. Center at origin, vertex $(-10,0)$, focus $(-8,0)$

17. Center at origin, focus (4, 0), length of major axis 12

18. Center at origin, focus $(0,-2)$, length of minor axis 2

19. Center at origin, vertex (3, 0), length of minor axis 4

20. Center at origin, vertex (0, 3), passes through (2, 1)

21. Center at (2, 2), focus $(2,-1)$, vertex $(2,-3)$

22. Center at $(-1,3)$, vertex (3, 3), length of minor axis 6

23. Foci at (5, 3) and $(-1,3)$, length of major axis 10

24. Vertices at (4, 1) and $(-8,1)$, focus $(-4,1)$

In Exercises 25–28 find an equation for the ellipse shown.

25.

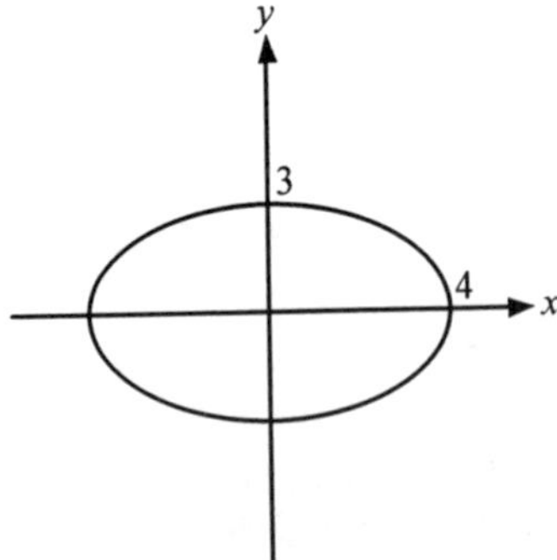

26.

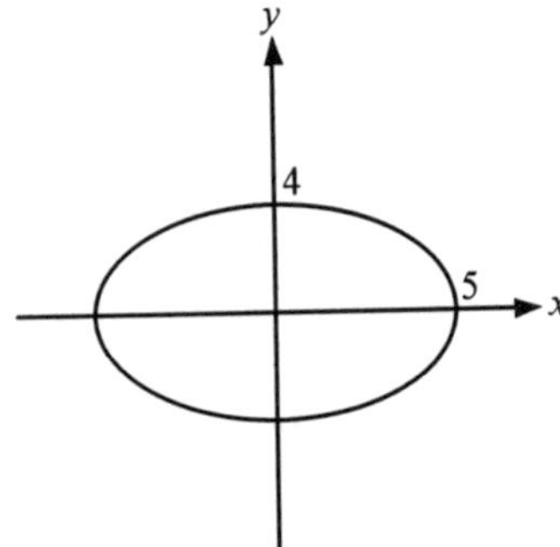

27.

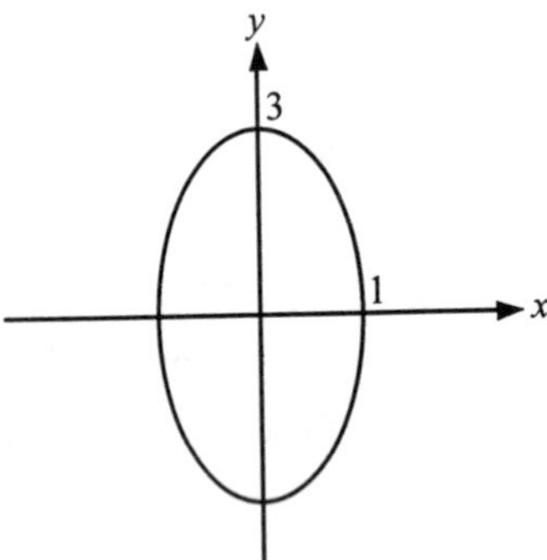

28.

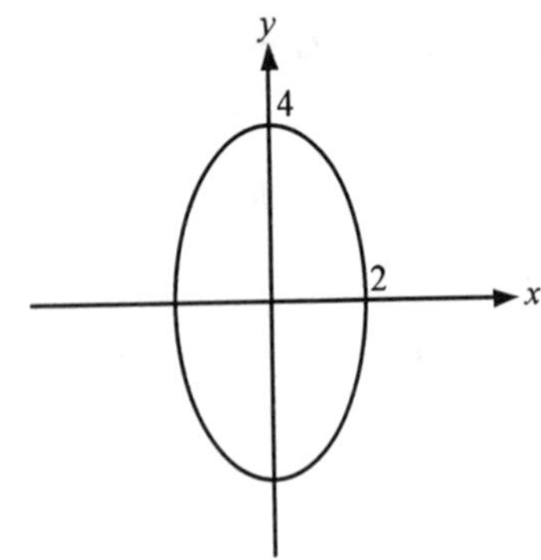

29. Earth travels in an elliptical orbit around the sun, which is located at one of the foci. The longest distance between Earth and the sun is 94,500,000 mi, and the shortest distance is 91,500,000 mi.

a. Find the length of the major axis of the ellipse.

b. Find the distance between the foci of the orbit.

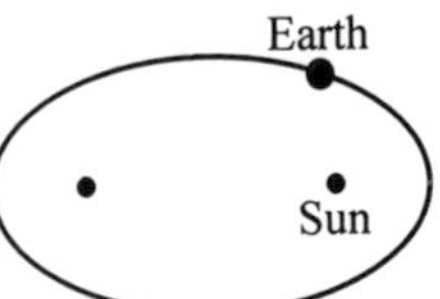

30. Refer to the previous exercise. When Earth is directly "above" the sun, how far is it from the sun?

31. The cross-section at the center of a greenhouse is a semiellipse, as indicated in the sketch. The horizontal span is 12 ft, and the height at the center is 8 ft. To the nearest inch, how high is the ceiling 3 ft to the left of center?

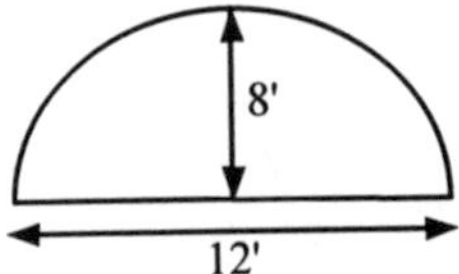

32. For the greenhouse in the previous exercise, how far from the center is the ceiling 5 ft high?

33. An elliptical arch has a height of 10 ft and a span of 10 ft. How far from the center is the height of the arch 5 ft?

34. An elliptical arch has a height of 20 ft and a span of 50 ft. How high is the arch 15 ft each side of the center?

35. A certain whispering gallery is 60 ft wide and 25 ft high at the center. How far from the walls are the foci? See the **NOTE** in Example 2 for a description of a whispering gallery.

36. You want to build a whispering gallery that is 50 ft wide with the foci 4 ft from the walls. What must the height be at the center?

THINK ABOUT IT 10.2

1. Describe the graph of the ellipse $(x^2/a^2)+(y^2/b^2)=1$, if $a=b$. How are the foci related in this case?
2. Refer to the figure below of an ellipse. The ratio c/a is called the **eccentricity** of the ellipse. Eccentricity means "out of roundness." If $c=0$, then the focus is at the center, a equals b, the eccentricity is 0, and the figure is a circle. Recall that the orbits of the planets around the sun are ellipses with the sun at one focus. For instance, the orbit of Venus has eccentricity 0.0068, while Earth's eccentricity is 0.0168, and Mercury's is 0.206.

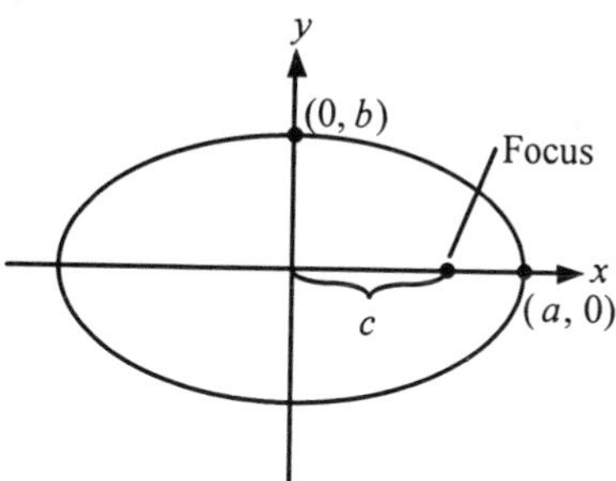

   a. Find the eccentricity of the ellipse whose x-intercepts are (2,0) and $(-2,0)$ and whose y-intercepts are $(0,1)$ and $(0,-1)$.
   b. If the x-intercepts are twice as far from the origin as the y-intercepts, will the eccentricity be the same as that in part **a**?
3. If the area A of the ellipse enclosed by

$$\frac{(x-h)^2}{a^2}+\frac{(y-k)^2}{b^2}=1$$

is given by $A=\pi ab$, find the area enclosed by $5x^2+6y^2+10x-12y-169=0$.
4. Use properties of the ellipse to find an equation for the graph that is the set of all points in a plane the sum of whose distances from (2, 0) and $(-2,0)$ is 10.
5. A line segment joining two points on an ellipse that contains a focus and that is perpendicular to the major axis is called a **focal chord** (or a **latus rectum**) of the ellipse. Find the endpoints for the two focal chords of an ellipse whose equation is $(x^2/a^2)+(y^2/b^2)=1$. What is the length of these chords?

● ● ●

10.3 The Hyperbola

Photo Courtesy of Archive Photos, New York

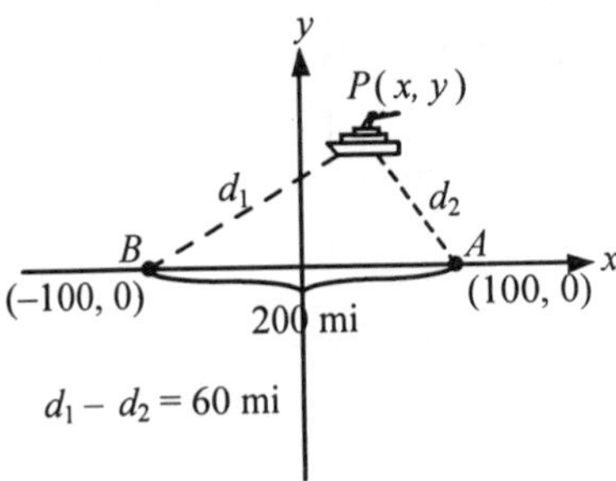

Ships and aircraft commonly use radio navigation systems, which have the advantage of being independent of weather conditions. One such system is called LORAN, an abbreviation of the phrase LOng RAnge Navigation. A Loran transmitter system consists of a pair of synchronized transmitting stations spaced a fixed distance apart. To estimate the position of, say, a ship, a receiver on the ship finds the time difference between the signals received from these transmitters. Because radio waves travel at 186,000 miles per second, this time difference reveals the difference in the distances from the ship to each station. The possible positions that satisfy this condition are then given by a curve known as a **hyperbola**. By switching to another pair of Loran transmitters and repeating the procedure, a second hyperbola is determined, and the intersection of these two curves then gives the position of the ship.

a. Using the Loran navigation system, a navigator on a ship determines that the ship is positioned 60 miles closer to transmitter A than to transmitter B. If the transmitting stations are 200 miles apart, as diagrammed, then find an equation of the branch of the hyperbola that gives all possible positions of the ship.

b. Graph the branch of the hyperbola determined in part **a** on a grapher.

(See Example 4.)

Objectives

1. Use an equation of a hyperbola to graph the relation, determine its asymptotes, and find the coordinates of the center, the foci, and the vertices.
2. Find an equation for a hyperbola satisfying various conditions concerning its center, foci, vertices, and transverse and conjugate axes.

The next conic section we consider is the hyperbola, and its geometric definition resembles the definition of the ellipse.

Definition of a Hyperbola

> A **hyperbola** is the set of all points in a plane the difference of whose distances from two fixed points (**foci**) is a positive constant.

Thus, the distances between the foci and a point on the figure maintain a *constant difference* for a hyperbola and a *constant sum* for an ellipse. To derive the standard form of the equation of a hyperbola with foci on the x-axis and center at the origin, we position the foci at $(-c,0)$ and $(c,0)$.

Let $2a$ represent the positive constant of the definition, and let $P(x,y)$ be any point on the hyperbola, as shown in Figure 10.20.

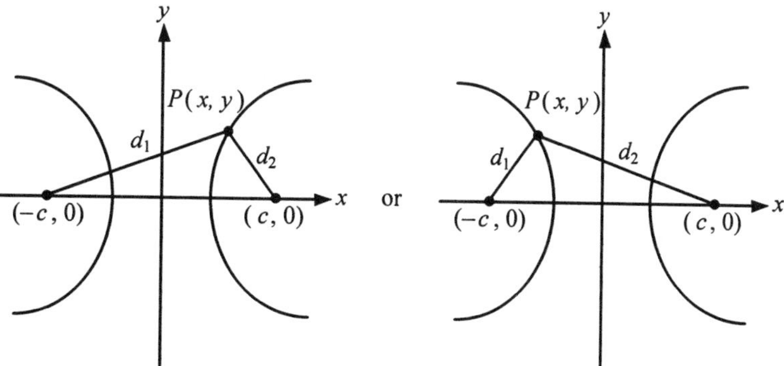

Figure 10.20

Since $d_1 < d_2$ or $d_1 > d_2$, by definition,

$$|d_1 - d_2| = 2a \text{ or } d_1 - d_2 = \pm 2a.$$

Then by the distance formula, we have

$$\sqrt{(x+c)^2 + (y-0)^2} - \sqrt{(x-c)^2 + (y-0)^2} = \pm 2a$$

or

$$\sqrt{(x+c)^2 + y^2} = \pm 2a + \sqrt{(x-c)^2 + y^2}.$$

Now square both sides of the equation and simplify.

$$\begin{aligned}
(x+c)^2 + y^2 &= 4a^2 \pm 4a\sqrt{(x-c)^2 + y^2} + (x-c)^2 + y^2 \\
4cx &= 4a^2 \pm 4a\sqrt{(x-c)^2 + y^2} \\
cx - a^2 &= \pm a\sqrt{(x-c)^2 + y^2}
\end{aligned}$$

Again square both sides and simplify.

$$\begin{aligned}
c^2x^2 - 2cxa^2 + a^4 &= a^2\left[(x-c)^2 + y^2\right] \\
c^2x^2 - 2cxa^2 + a^4 &= a^2x^2 - 2cxa^2 + a^2c^2 + a^2y^2 \\
c^2x^2 - a^2x^2 - a^2y^2 &= a^2c^2 - a^4 \\
(c^2 - a^2)x^2 - a^2y^2 &= a^2(c^2 - a^2) \\
\frac{x^2}{a^2} - \frac{y^2}{c^2 - a^2} &= 1
\end{aligned}$$

If we now define

$$b^2 = c^2 - a^2 \quad \text{or} \quad c^2 = a^2 + b^2,$$

we obtain the **standard form** of a hyperbola with center at the origin and foci on the x-axis.

$$\frac{x^2}{a^2} - \frac{y^2}{b^2} = 1$$

Consider Figure 10.21. By setting $y = 0$, we find that the x-intercepts are $(-a,0)$ and $(a,0)$. The line segment joining these two points is called the **transverse axis**. The endpoints of the transverse axis are called the **vertices of the hyperbola**. By setting $x = 0$, we find that there are no y-intercepts. The line segment from $(0,b)$ to $(0,-b)$ is called the **conjugate axis**.

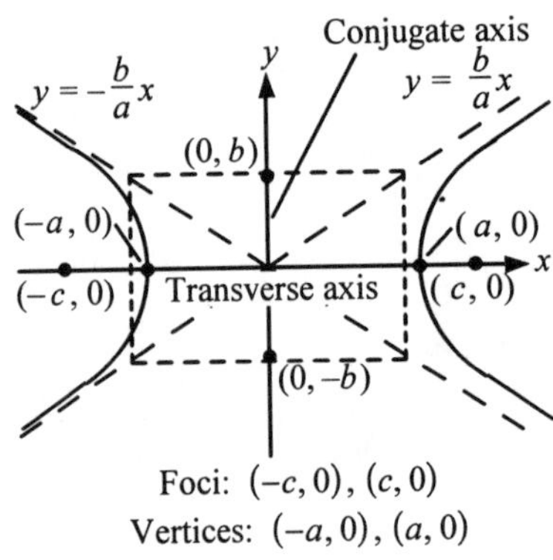

Figure 10.21

To determine the significance of b, we rewrite $(x^2/a^2) - (y^2/b^2) = 1$ as

$$y = \frac{\pm bx}{a}\sqrt{1 - \frac{a^2}{x^2}}.$$

As $|x|$ gets very large, $1 - a^2/x^2$ approaches 1. Thus, the graph of the hyperbola approaches the lines

$$y = \pm\frac{b}{a}x$$

These lines are called the **asymptotes of the hyperbola**. They are a great aid in sketching the curve. As shown in Figure 10.21, the asymptotes are the diagonals of a rectangle of dimensions $2a$ by $2b$.

Example 1: Using a Standard Form of a Hyperbola

For the hyperbola given by $4x^2 - y^2 = 36$, find the coordinates of the vertices and the foci. Also, determine the asymptotes and sketch the curve.

Solution: First, divide both sides of the equation by 36 to obtain the standard form.

$$\frac{x^2}{9} - \frac{y^2}{36} = 1$$

Then

$$a^2 = 9 \quad a = 3$$
$$b^2 = 36 \quad b = 6$$

Note that a^2 is not necessarily the larger denominator. Since a is the distance from the center to the vertices, we have

$$\text{coordinates of the vertices: } (-3,0),(3,0)$$

To find c, we replace a^2 by 9 and b^2 by 36 in the formula

$$c^2 = a^2 + b^2$$
$$= 9 + 36$$
$$c^2 = 45 \quad \text{or} \quad c = \sqrt{45} = 3\sqrt{5}$$

The distance from the center to the foci is c. Thus,

$$\text{coordinates of the foci: } \left(-3\sqrt{5},0\right),\left(3\sqrt{5},0\right)$$

For the asymptotes we have

$$y = \pm\frac{b}{a}x = \pm\frac{6}{3}x = \pm 2x$$

The hyperbola is sketched in Figure 10.22.

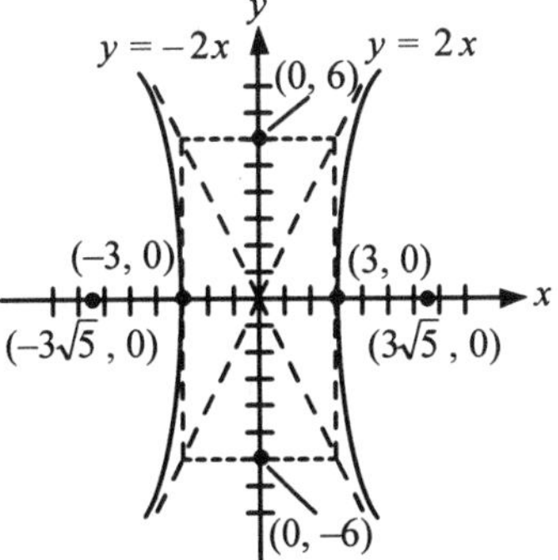

Figure 10.22

PROGRESS CHECK 1

For the hyperbola given by $16x^2 - 9y^2 = 144$, find the coordinates of the vertices and the foci. Also determine the asymptotes and sketch the curve. ■

If the foci are positioned on the *y*-axis at $(0,-c)$ and $(0,c)$, the *standard form* of the equation of a hyperbola is

$$\frac{y^2}{a^2}-\frac{x^2}{b^2}=1$$

In this case the asymptotes are given by $y=\pm ax/b$, as shown in Figure 10.23.

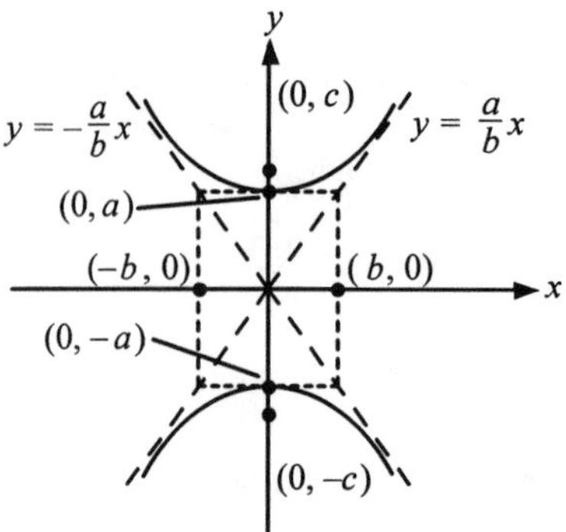

Figure 10.23

Example 2: Using a Standard Form of a Hyperbola

For the hyperbola given by $25y^2-4x^2=100$, find the coordinates of the vertices and the foci. Also, determine the asymptotes and sketch the curve.

Solution: First, divide both sides of the equation by 100 to obtain the standard form.

$$\frac{y^2}{4}-\frac{x^2}{25}=1$$

Then

$$a^2=4 \quad a=2$$
$$b^2=25 \quad b=5$$

Since a is the distance from the center to the vertices, we have

coordinates of the vertices: $(0,2)$, $(0,-2)$.

To find c, we replace a^2 by 4 and b^2 by 25 in the formula

$$\begin{aligned} c^2 &= a^2+b^2 \\ &= 4+25 \\ c^2 &= 29 \quad c=\sqrt{29} \end{aligned}$$

The distance from the center to the foci is c. Thus,

coordinates of foci: $(0,\sqrt{29})$, $(0,-\sqrt{29})$.

For the asymptotes we have

$$y = \pm\frac{a}{b}x = \pm\frac{2}{5}x.$$

The hyperbola is sketched in Figure 10.24.

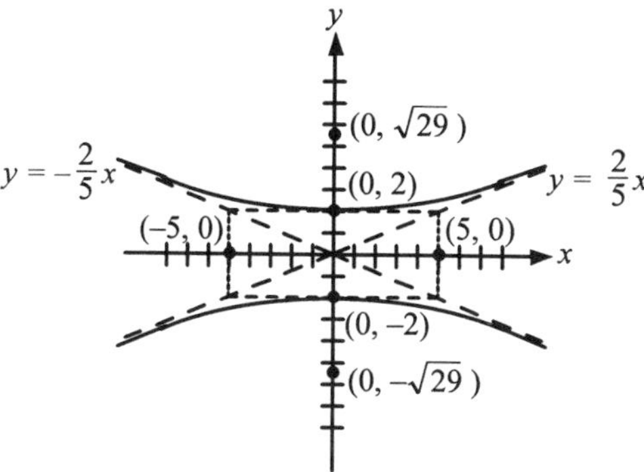

Figure 10.24

A graphing calculator may be used to obtain visual evidence that the graph of a hyperbola approaches it asymptotes. For instance, in this example the hyperbola $25y^2 - 4x^2 = 100$ and its asymptotes may be pictured by graphing

$y1 = \sqrt{(4x^2 + 100)/25}$ (top branch)
$y2 = -\sqrt{(4x^2 + 100)/25}$ (bottom branch)
$y3 = (2/5)x$
and $y4 = (-2/5)x$

in the same display as shown in Figure 10.25. When you have duplicated this graph on your calculator, try using the Trace feature to detect numerically that the difference between the curve and its nearby asymptote approaches 0 as $|x|$ increases.

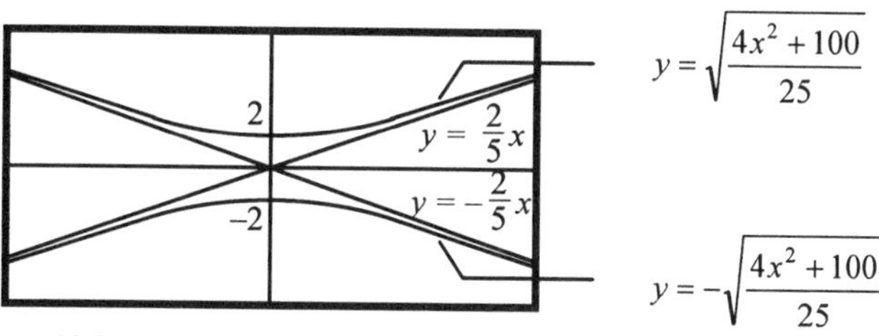

Figure 10.25

PROGRESS CHECK 2
For the hyperbola given by $16y^2 - 25x^2 = 400$, find the coordinates of the vertices and the foci. Also, determine the asymptotes and sketch the curve. ∎

Example 3: Finding the Standard Equation of a Hyperbola

Find the equation in standard form of the hyperbola with center at the origin, one focus at (0,5), and one vertex at (0,3).

Solution: Since c is the distance from the center to a focus, c is 5. Since a is the distance form the center to a vertex, a is 3. To find b^2, we replace c by 5 and a by 3 in the formula

$$c^2 = a^2 + b^2$$
$$(5)^2 = (3)^2 + b^2.$$
$$16 = b^2$$

Since the foci are along the y-axis, the standard form is

$$\frac{y^2}{a^2} - \frac{x^2}{b^2} = 1.$$

Thus, the standard equation of the hyperbola is

$$\frac{y^2}{9} - \frac{x^2}{16} = 1.$$

PROGRESS CHECK 3

Find the equation in standard form of the hyperbola with center at the origin, one focus at (10,0), and one vertex at (6,0). ■

To solve the section-opening problem, it is necessary to recall that a hyperbola is the set of all points in a plane the difference of whose distances from two fixed points (foci) is a positive constant, and that this constant was represented by $2a$ when the equations for the standard forms of a hyperbola were developed.

Example 4: Using the Loran Navigation System

Solve the problem in the section introduction on page 826.

Solution:

a. The possible positions of the ship are given by the set of all points $P(x,y)$ that satisfy $d_1 - d_2 = 60$, as diagrammed in Figure 10.26. Such points $P(x,y)$ lie on one branch of a hyperbola whose foci are located at the two transmitting stations. Since these stations are located at $(100,0)$ and $(-100,0)$, we know that $c = 100$. We also know that the constant difference, which is given by $2a$, equals 60, so $a = 30$. To find b^2, we replace c by 100 and a by 30 in the formula $c^2 = a^2 + b^2$ to obtain

$$100^2 = 30^2 + b^2$$
$$9{,}100 = b^2$$

Since the foci are along the x-axis, the standard form is $\left(x^2/a^2\right) - \left(y^2/b^2\right) = 1$, so the equation for the hyperbola is

$$\frac{x^2}{900} - \frac{y^2}{9100} = 1.$$

Finally, the ship is closer to transmitter A, so its position lies on the right branch of the hyperbola. The vertex associated with this branch is located at (30,0), so this branch may be defined by

$$\frac{x^2}{900}-\frac{y^2}{9100}=1 \quad \text{with} \quad x \geq 30.$$

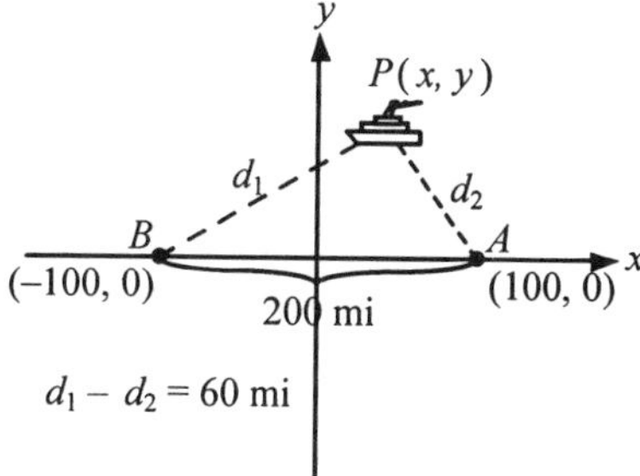

Figure 10.26

b. The branch of the hyperbola determined in part **a** may be pictured on a grapher by entering

$$y1=\sqrt{(91x^2/9-9100)}(x \geq 30)$$

$$\text{and} \quad y1=-\sqrt{(91x^2/9-9100)}(x \geq 30)$$

in the equation list and using [–200,200] by [–200,200] for the viewing window. The graph is shown in Figure 10.27.

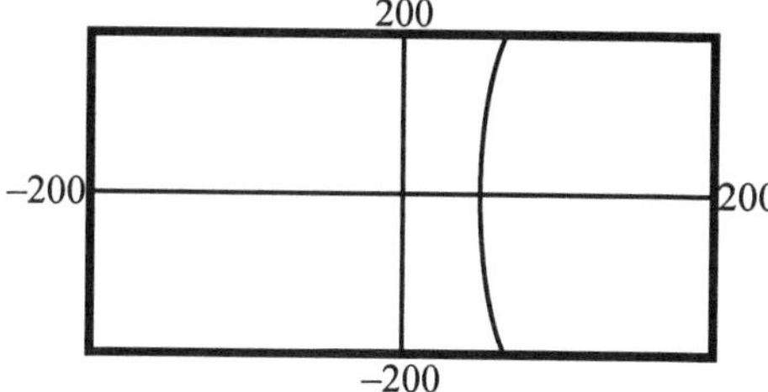

Figure 10.27

PROGRESS CHECK 4

Redo the problems in Example 4, assuming that this ship is positioned 80 miles closer to transmitter B than to transmitter A. ■

If the hyperbola is centered at the point (h,k), the **standard form** is

$$\frac{(x-h)^2}{a^2}-\frac{(y-k)^2}{b^2}=1$$

when the transverse axis is parallel to the x-axis. In this case the asymptotes are given by $y-k=(\pm b/a)(x-h)$. If the transverse axis is parallel to the y-axis, the **standard form** is

$$\frac{(x-k)^2}{a^2}-\frac{(x-h)^2}{b^2}=1$$

and the asymptotes are given by $y-k=(\pm a/b)(x-h)$.

Example 5: Converting to Standard Form

Find the coordinates of the vertices and the foci of the hyperbola given by $16y^2-x^2-32y+4x-4=0$. Also, determine the asymptotes and sketch the curve.

Solution: First, put the equation in standard form by completing the square.

$$\begin{aligned}(16y^2-32y\quad)-(x^2-4x\quad)&=4\\16(y^2-2y\quad)-(x^2-4x\quad)&=4\\16(y^2-2y\quad+1)-(x^2-4x+4\quad)&=4+16(1)+(-1)4\\16(y-1)^2-(x-2)^2&=16\end{aligned}$$

Dividing both sides of the equation by 16, we have

$$\frac{(y-1)^2}{1}-\frac{x-2^2}{16}=1.$$

From this equation in standard form we conclude that the center is at (2,1).

$$\begin{aligned}a^2&=1 \qquad a=1\\b^2&=16 \qquad b=4\end{aligned}$$

The transverse axis is parallel to the y-axis, since the y term is positive; a is the distance form the center to the vertices. Thus,

coordinates of vertices: (2,0) (2,2).

To find c, we replace a^2 by 1 and b^2 by 16 in the formula

$$\begin{aligned}c^2&=a^2+b^2\\&=1+16\\c^2&=17 \quad\text{or}\quad c=\sqrt{17}.\end{aligned}$$

The distance from the center to the foci is c. Thus,

coordinates of foci: $(2,1-\sqrt{17}),\ (2,1+\sqrt{17})$.

The asymptotes are the diagonals of the rectangle of dimensions $2a$ by $2b$ that is centered at (2,1). The equations of these asymptotes are

$$y - k = \pm\frac{a}{b}(x - h)$$

$$y - 1 = \pm\frac{1}{4}(x - 2).$$

The hyperbola is graphed in Figure 10.28.

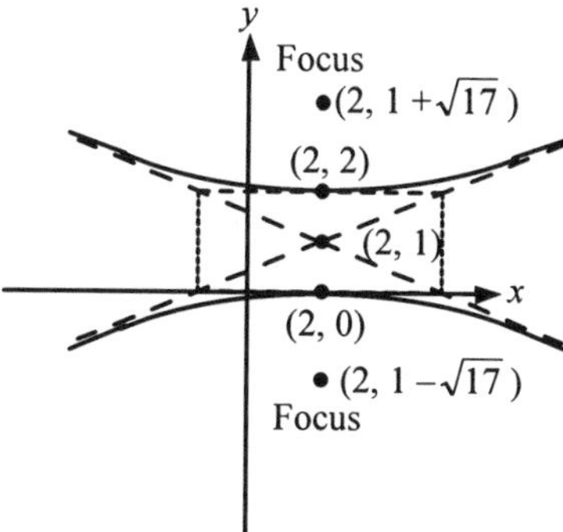

Figure 10.28

You should check these results with a grapher.

PROGRESS CHECK 5

Find the coordinates of the vertices and the foci of the hyperbola given by

$$4y^2 - 9x^2 - 36x - 8y - 68 = 0.$$

Also, determine the asymptotes and sketch the curve. ■

EXPLORE 10.3

1. Use a grapher to graph the hyperbolas given by the following equations. In each case solve for y and graph the top and bottom curves separately. Also note that Zoom Decimal will help get a connected graph in cases where the x component of the vertex is an integer.
   a. $x^2 - y^2 = 6$ **b.** $y^2 - x^2 = 6$
2. The graph of a hyperbola has asymptotes. However, it may appear in some viewing windows of a grapher that the curve intersects an asymptote when in fact this is not the case. You should be able to determine the behavior by zooming in sufficiently. Use a grapher to graph $x^2/4 - y^2/9 = 1$ and its asymptotes $y = \pm 3x/2$, using several different viewing windows, and find a viewing window that gives a good global picture that shows the asymptotic behavior of the graph.
3. In this section it is explained that various location systems like the Loran navigation system depend on finding the intersection of two hyperbolas. The algebra for finding the intersection exactly is covered in Section 10.5, but you can always approximate points of intersection graphically. Use a grapher, and if the hyperbolas intersect, estimate to two decimal places the coordinates of all points of intersection.
   a. $\dfrac{x^2}{4} - \dfrac{y^2}{9} = 1$ and $\dfrac{y^2}{4} - \dfrac{x^2}{9} = 1$ **b.** $\dfrac{x^2}{4} - \dfrac{y^2}{9} = 1$ and $\dfrac{y^2}{9} - \dfrac{x^2}{4} = 1$
4. Two hyperbolas both with center at (0,0), have the major axis on the x-axis, but they have different focus points. Is it possible for them to intersect? If so, find equations for two such hyperbolas. If not, explain why it cannot be done.

EXERCISES 10.3

In Exercises 1-14 find the coordinates of the vertices and the foci, determine the asymptotes, and sketch each curve.

1. $\frac{x^2}{16} - \frac{y^2}{9} = 1$ **2.** $\frac{x^2}{36} - \frac{y^2}{100} = 1$

3. $\frac{y^2}{25} - \frac{x^2}{16} = 1$ **4.** $\frac{y^2}{1} - \frac{x^2}{4} = 1$

5. $x^2 - 9y^2 = 36$ **6.** $9x^2 - 4y^2 = 144$

7. $4y^2 - x^2 = 1$ **8.** $y^2 - 9x^2 = 25$

9. $\frac{(x+2)^2}{9} - \frac{(y-3)^2}{25} = 1$

10. $\frac{(y-1)^2}{64} - \frac{(x-2)^2}{36} = 1$

11. $4(y+1)^2 - 25(x+3)^2 = 100$

12. $25(x+4)^2 - 16y^2 = 400$

13. $4x^2 - 9y^2 + 8x - 54y - 113 = 0$

14. $9y^2 - x^2 - 36y + 2x - 1 = 0$

In Exercises 15–24 find the equation in standard form of the hyperbola satisfying the conditions.

15. Center at origin, focus (0,5), vertex (0,4)

16. Center at origin, focus $(-10,0)$, vertex $(-8,0)$

17. Center at origin, focus (4,0), length of transverse axis 6

18. Center at origin, focus $(0,-2)$, length of conjugate axis 2

19. Center at origin, vertex (3,0), length of conjugate axis 10

20. Center at origin, vertex (0,2), passes through (3,4)

21. Center at $(-3,4)$, focus $(2,-4)$, vertex $(0,-4)$

22. Center at (5,0), vertex (5,6), length of conjugate axis 8

23. Foci at (3,4) and $(3,-2)$, length of transverse axis 4

24. Vertices at (4,1) and $(-10,1)$, focus (7,1)

25. Refer to Example 4 for a description of a Loran navigation system. A navigator on a ship determines that the ship is positioned 50 miles closer to transmitter *A* than to transmitter *B*.

- **a.** If the transmitting stations are 240 miles apart, then find an equation of the branch of the hyperbola that gives all possible positions of the ship.
- **b.** Use a grapher to graph the branch of the hyperbola determined in part **a.**

26. Redo Exercise 25 if the ship is positioned 75 miles closer to transmitter *A* than to transmitter *B*.

27. Points *A* and *B* are 5000 feet apart. The sound of an explosion at *x* reaches *A* 0.5 seconds before it reaches *B*. Thus *A* is closer to the explosion than is *B*. See the diagram.

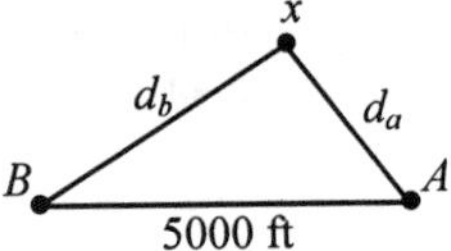

- **a.** Assuming that sound travels at 1100 feet per second, find the difference between the distance from *A* to *X* and the distance from *B* to *X*.
- **b.** Use the results of part **a** to find an equation of the branch of the hyperbola that gives all possible positions of the explosion, and use a grapher to draw the graph of that equation.

28. Points *A* and *B* are underwater and 8000 feet apart. The sound of a whale reaches *A* 0.6 seconds before it reaches *B*.

- **a.** Assuming that sound travels through water at 5000 feet per second, find the difference between the distance from *A* to the whale and the distance from *B* to the whale.
- **b.** Use the results of part **a** to find an equation of the branch of the hyperbola that gives all possible positions of the whale, and use a grapher to draw the graph of that equation.

29. Like the ellipse, the hyperbola has certain reflective properties. Light originating at one focus reflects off the closest branch in the direction of the straight line connecting the other focus to the point of reflection. See the diagram.

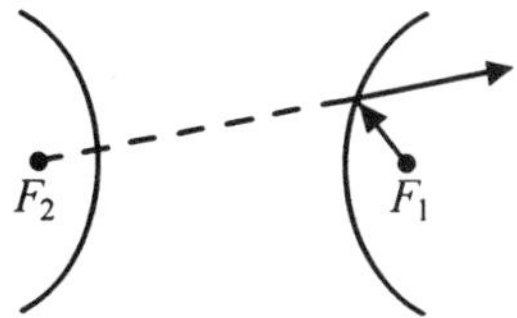

Take the hyperbola defined by $\frac{x^2}{9} - \frac{y^2}{16} = 1$. A beam of light from the focus at (5,0) hits the closest branch at the point in the first quadrant where $x = 4$. Find the slope of the line along which the light is reflected.

30. Repeat the previous problem if the light hits the branch where $x = 5$.

THINK ABOUT IT 10.3

1. Find equations in standard form for the two hyperbolas that satisfy the following conditions: the asymptotes are $y = \pm(2/3)x$, and the length of the transverse axis is 12.

2. If the asymptotes of the hyperbola $(x^2/a^2) - (y^2/b^2) = 1$ are perpendicular, then what is the relation between a and b? Establish the relation with a proof.

3. The names *ellipse*, *parabola*, and *hyperbola* were given to the conic sections by the Greek mathematician Apollonius (c. 262–190 B.C.). His work was entirely geometric but is equivalent to algebraically noting three types of equations for conics.

$$y^2 = px - \frac{px^2}{d} \quad \text{The case of "ellipsis," which means to fall short}$$
$$y^2 = px \quad \text{The case of "parabole," which means to coincide}$$
$$y^2 = px + \frac{px^2}{d} \quad \text{The case of "hyperbole," which means to exceed}$$

The letters p and d refer to lengths of line segments used in the geometric construction of these shapes.

a. Let $p = 4$ and $d = 1$; then by plotting points, draw each of the resulting graphs and confirm that they do give the "correct" shape.

b. Let $p = 1$ and $d = 4$, and repeat part *a*.

c. What value for p and d will make the "ellipse" into a circle?

4. Use properties of the hyperbola to find an equation for the graph that is the set of all points in a plane the difference of whose distances from (4,0) and (−4,0) is 6.

5. A line segment joining two points on a hyperbola that contains a focus and that is perpendicular to the transverse axis is called a **focal chord** (or **latus rectum**) of the hyperbola. Show that $2b^2/a$ is the length of both focal chords of a hyperbola whose equation is $(x^2/a^2) - (y^2/b^2) = 1$.

● ● ●

10.4 The Parabola and Classifying Conic Sections

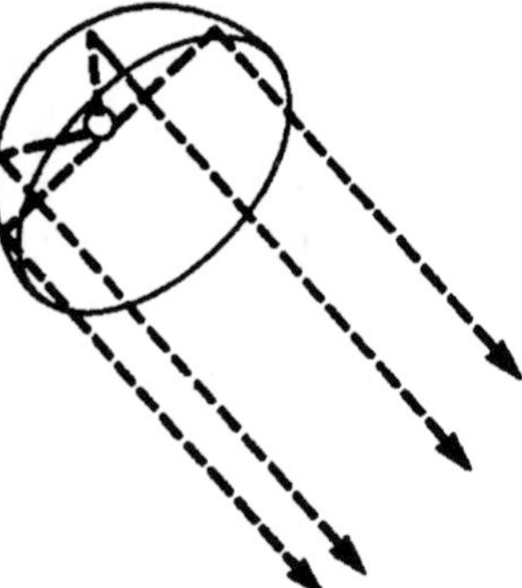

Figure 1

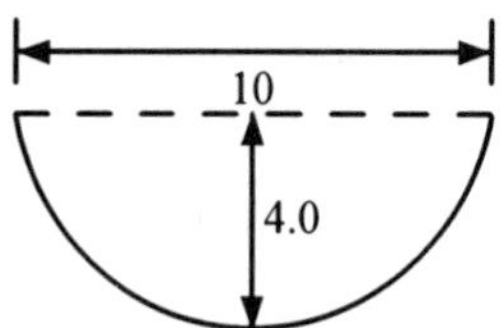

Figure 2

The parabola has an important reflection property. Any ray or wave that originates at the focus and strikes the parabola is reflected parallel to the axis of symmetry (see Figure 1). For this reason, an instrument such as a flashlight or a searchlight uses a parabolic reflection, with the bulb located at the focus. The reflector redirects light that would otherwise be wasted parallel to the axis, so that a straight beam of light is formed.

a. Consider the parabolic reflector in the searchlight shown in Figure 2. Choose axes in a convenient position and determine an equation of the parabola.

b. For the reflector described in part **a**, what is the coordinate position that shows where to place the bulb in the searchlight?

(See Example 4.)

Objectives

1. Use an equation of a parabola to graph the relation, determine the equation of the directrix, and find the coordinates of the vertex and the focus.
2. Find an equation in standard form for a parabola satisfying various conditions concerning its vertex, focus, and directrix.
3. Classify certain equations as defining either a circle, an ellipse, a hyperbola, or a parabola, or as defining a degenerate case of a conic section.

We have already done some work with parabolas in connection with graphic quadratic functions in Section 5.5. We now expand our coverage by considering some geometric properties of this curve and developing the standard forms of equations of parabolas with a horizontal or vertical axis of symmetry. As in the preceding sections, a geometric definition is our starting point.

Definition of a Parabola

> A **parabola** is the set of all points in a plane equidistant from a fixed line (**directrix**) and a fixed point (**focus**) not on the line.

From this definition we drive a general equation for a parabola as follows. For simplicity, let the directrix be the line $x = -p$, and position the focus at $(p,0)$. $P(x,y)$ represents any point on the parabola (see Figure 10.29). Since P_1 and P have the same y-coordinate, the distance d_1 is given by

$$d_1 = |x - (-p)| = |x + p|$$

The distance from P to $(p,0)$ is given by

$$d_2 = \sqrt{(x-p)^2 + (y-0)^2}$$

The geometric condition for a parabola states that

$$d_1 = d_2.$$

Thus,

$$|x+p| = \sqrt{(x-p)^2 + y^2}$$

Squaring both sides of the equation yields

$$(x+p)^2 = (x-p)^2 + y^2.$$

Simplifying this equation, we have

$$x^2 + 2px + p^2 = x^2 - 2px + p^2 + y^2$$

$$\boxed{4px = y^2}$$

This equation is the **standard form** of a parabola with directrix $x = -p$ and focus at $(p, 0)$. The line through the focus that is perpendicular to the directrix is called the **axis of symmetry**. In this case the axis of symmetry is the x-axis. The point on the axis of symmetry that is midway between the focus and the directrix is called the **vertex**. The vertex is the turning point of the parabola. In this case the vertex is the origin. Note that p is the distance from the vertex to the focus and from the vertex to the directrix.

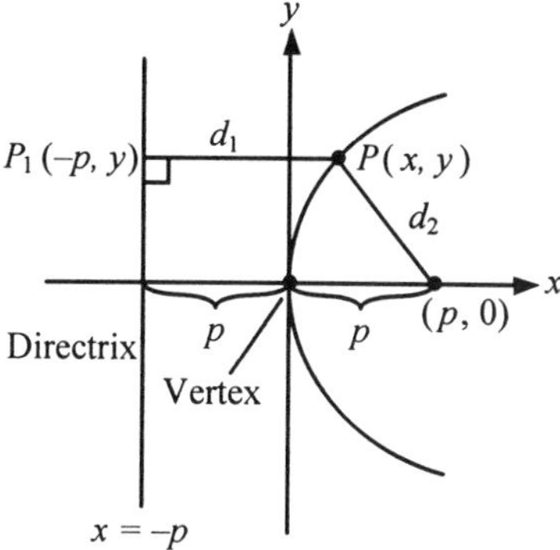

Figure 10.29

Example 1: Using a Standard Form of a Parabola

Find the focus and the directrix of the parabola given by $y^2 = 12x$. Also, sketch the parabola.

Solution: Matching the equation $y^2 = 12x$ to the standard form

$$y^2 = 4px,$$

we write

$$y^2 = 12x = 4(3)x.$$

Thus,

$$p = 3.$$

The focus is on the axis of symmetry (the x-axis) p units to the right of the vertex (the origin). Thus

focus: (3, 0)

The directrix is the line p units to the left of the vertex. Thus,

directrix: $x = -3$.

The parabola is sketched in Figure 10.30. Use a square viewing window with $y1 = \sqrt{12x}$ and $y2 = -\sqrt{12x}$ to check this result with a grapher.

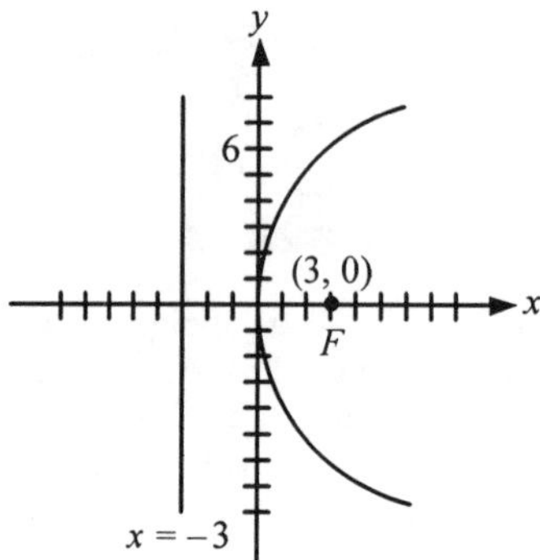

Figure 10.30

PROGRESS CHECK 1
Find the focus and directrix of the parabola given by $y^2 = 16x$. Also sketch the parabola. ■

There are three other possibilities for a parabola with a directrix that is parallel to the x- or y-axis and whose vertex is the origin. These cases are illustrated in Figures 10.31–10.33. In all cases $p > 0$.

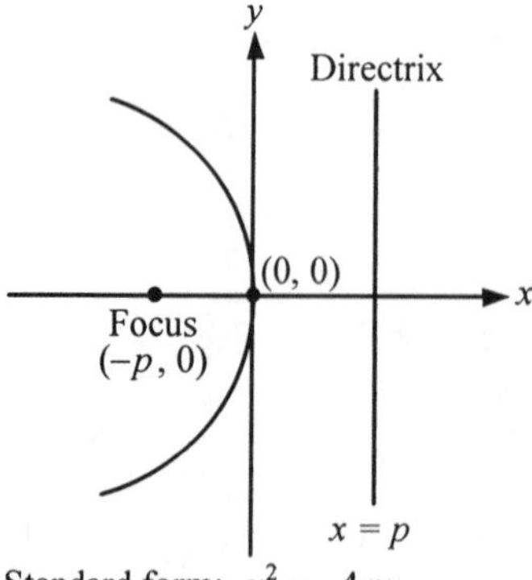

Standard form: $y^2 = -4px$
Axis of symmetry: x-axis
Opens to the left
Figure 10.31 Case 2

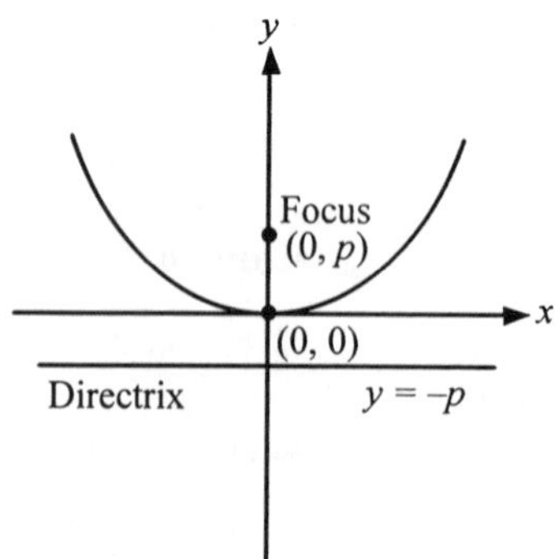

Standard form: $x^2 = 4py$
Axis of symmetry: y-axis
Opens upward
Figure 10.32 Case 3

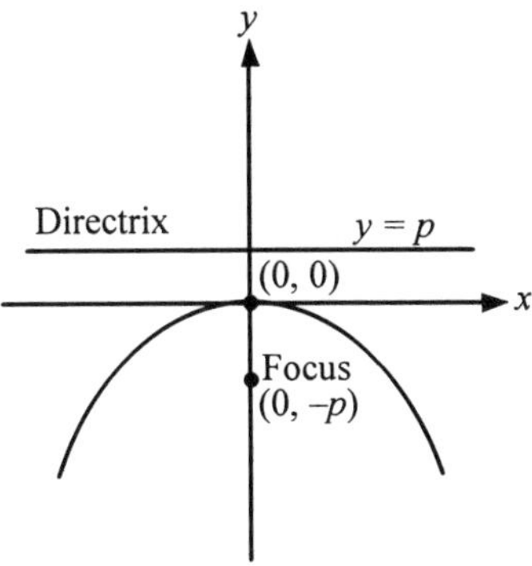

Standard form: $x^2 = -4py$
Axis of symmetry: y-axis
Opens downward
Figure 10.33 Case 4

To summarize, the graphs of the equations $y^2 = \pm 4px$ and $x^2 = \pm 4py$ are parabolas. If the y term is squared, the axis of symmetry is the x-axis. The parabola opens to the right when the coefficient of x is positive and opens to the left when this coefficient is negative. When the x term is squared, the axis of symmetry is the y-axis. The parabola opens upward when the coefficient of y is positive, and opens downward when this coefficient is negative. In all cases, p gives the distance from the vertex to the focus and from the vertex to the directrix.

Example 2: Using a Standard Form of a Parabola

Find the focus and directrix of the parabola given by $x^2 = -6y$. Also, sketch the parabola.

Solution: Matching the equation $x^2 = -6y$ to the standard form in case 4,

$$x^2 = -4py,$$

we write

$$x^2 = -6y = -4\left(\frac{3}{2}\right)y.$$

Thus,

$$p = \frac{3}{2}.$$

The focus is on the axis of symmetry (the y-axis) p units down from the vertex (the origin). Thus,

$$\text{focus: } \left(0, -\frac{3}{2}\right).$$

The directrix is the line p units up from the vertex. Thus,

$$\text{directrix: } y = \frac{3}{2}.$$

Figure 10.34 shows a sketch of the parabola. Check this result with a grapher.

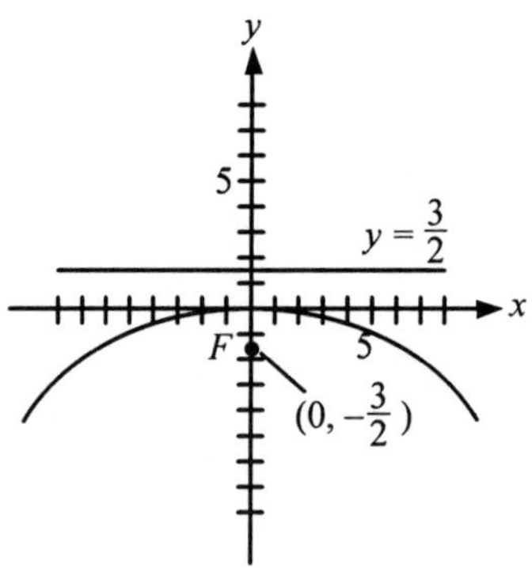

Figure 10.34

PROGRESS CHECK 2

Find the focus and directrix of the parabola given by $y^2 = -10x$. Also sketch the parabola. ■

Example 3: Finding the Standard Equation of a Parabola

A parabola with its vertex at the origin has its focus at $(-5,0)$. Find the equation in standard form of the parabola.

Solution: This is an example of case 2. The term *p*, which represents the distance from the vertex (0, 0) to the focus $(-5,0)$, is 5. To find the standard equation of the parabola, replace *p* by 5 in the form

$$y^2 = -4px$$

to obtain

$$y^2 = -20x.$$

The parabola opens to the left and is graphed in Figure 10.35.

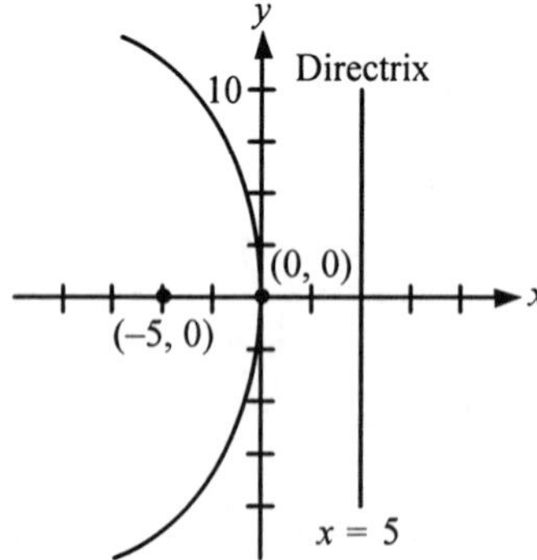

Figure 10.35

PROGRESS CHECK 3

A parabola with its vertex at the origin has its focus at $(0,-2)$. Find the equation in standard form of the parabola. ■

Example 4: Parabolic Reflector in a Searchlight

Solve the problem in the section introduction on page 838.

Solution:

a. First, place the parabola with its vertex at the origin, as shown in Figure 10.36. The standard form for a parabola in this case is $y^2 = -4py$. Since (5, 4) is a point on the parabola, we have

$$5^2 = 4p(4), \text{ so } p = 25/16.$$

Thus, an equation for the parabola is

$$x^2 = 4\left(\frac{25}{16}\right)y$$

$$x^2 = \frac{25y}{4}$$

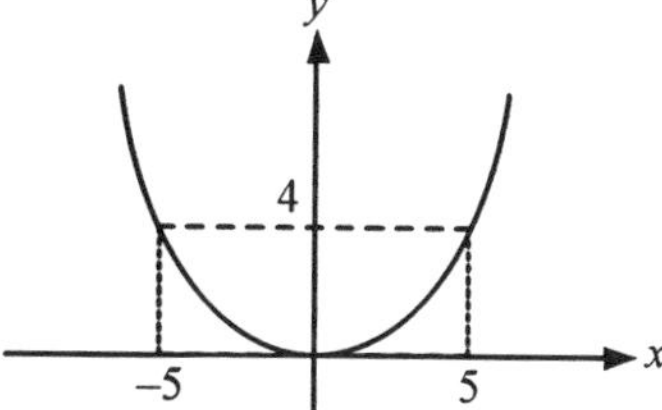

Figure 10.36

b. The bulb in the searchlight should be positioned at the focus, which is on the axis of symmetry p units up from the vertex. The coordinate position of this point is $(0, 25/16)$.

Note: The reflection property of a parabola also causes any ray or wave that comes into a parabolic reflector parallel to the axis of symmetry to be directed to the focus point, as diagrammed in Figure 10.37. Radar, radio antennas, and reflecting telescopes often make use of this principle.

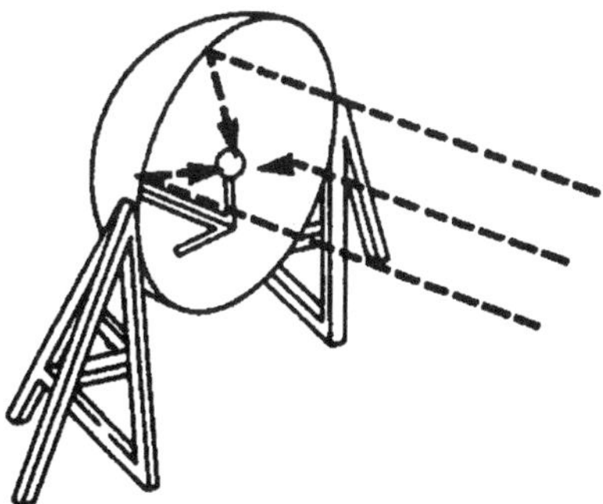

Figure 10.37

PROGRESS CHECK 4

Answer the questions in Example 4 for the parabolic reflector shown in Figure 10.38. ■

If the vertex of the parabola is at the point (h, k) and the directrix is parallel to the x- or y-axis, then there are four possible standard forms for the parabola. These cases are illustrated in Figure 10.39. In all cases, $p > 0$.

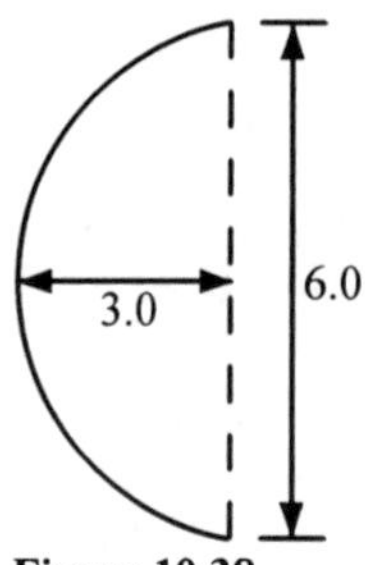

Figure 10.38

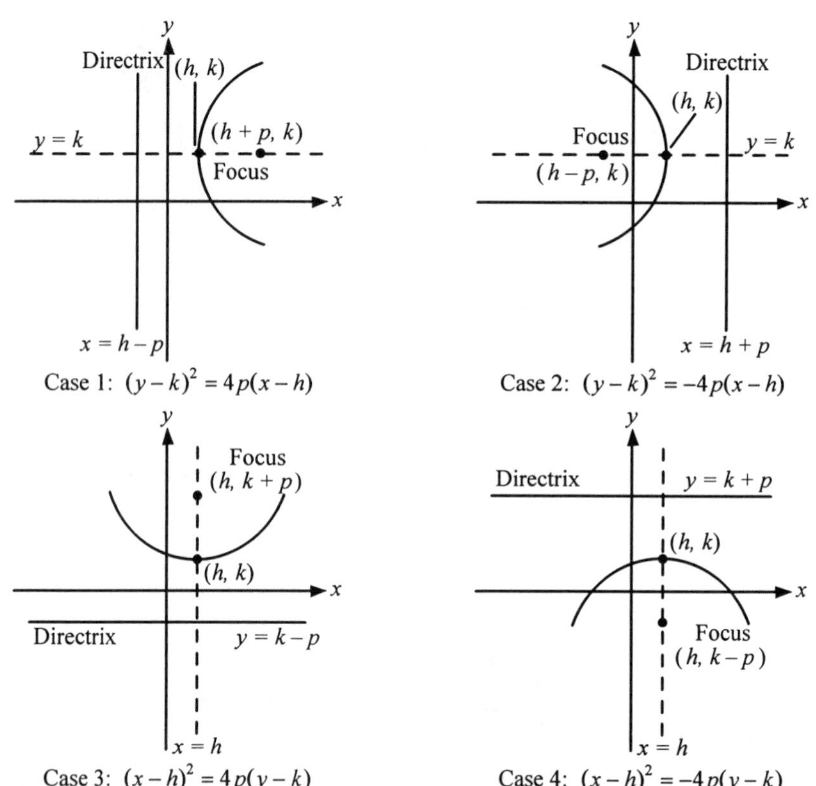

Figure 10.39

Example 5: Converting to Standard Form

For the parabola given by $y = x^2 - 2x$, find the vertex, the focus, and the directrix. Also, sketch the curve.

Solution: First, put the equation in standard form by completing the square,

$$y = \left(x^2 - 2x \quad\right)$$
$$y + 1 = \left(x^2 - 2x + 1\right)$$
$$y + 1 = (x - 1)^2$$

Matching this equation to the form in case 3,

$$(x-h)^2 = 4p(y-k),$$

we conclude that

$$h = 1, \ k = -1, \ 4p = 1, \text{ or } p = \frac{1}{4}.$$

Thus,

$$\text{vertex: } (1,-1).$$

The focus is on the axis of symmetry $(x = 1)$ p unit up from the vertex $(1,-1)$. Thus,

$$\text{focus: } \left(1,-\frac{3}{4}\right).$$

The directrix is the line p units down from the vertex. Thus,

$$\text{directrix: } y = -\frac{5}{4}.$$

The parabola opens upward and is graphed in Figure 10.40.

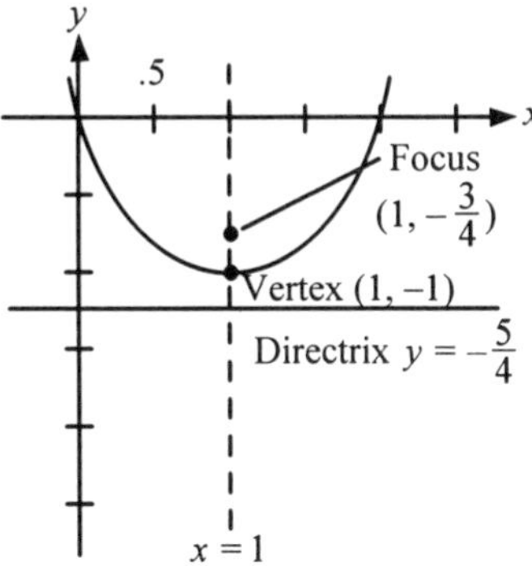

Figure 10.40

Use a viewing window like $[-1,3]$ by $[-2,1]$ to check this result on a grapher.

PROGRESS CHECK 5
For the parabola given by $y = x^2 + 2x - 2$ find the vertex, the focus, and the directrix. Also, sketch the curve. ■

The equations considered in Sections 10.1–10.4 have all been of the general form

$$Ax^2 + Cy^2 + Dx + Ey + F = 0$$

where A and C are not both zero. This equation is called the general form of an equation of a conic section with axis or axes parallel to the coordinate axes. This form omits the case of a second-degree equation with a Bxy term. When Bxy is present, the graph (for the nondegenerate case) represents a conic section that has been rotated.

If a graph is a circle, an ellipse, a hyperbola, or a parabola, then in general form, A and C must satisfy certain conditions that are apparent from the standard forms of these conics. We summarize these conditions below, along with the degenerate possibilities for each case.

Conditions on A and C	**Conic Section**	**Degenerate Possibilities**
$A = C \neq 0$	Circle	A point or no graph at all
$A \neq C$, A and C have the same sign.	Ellipse	A point or no graph at all
A and C have opposite signs.	Hyperbola	Two intersecting straight lines
$A = 0$ or $C = 0$ (but not both)	Parabola	A line, a pair of parallel lines, or no graph at all

Example 6 illustrates the main application of this chart.

Example 6: Identifying a Conic Section

Identify the graph of the following equations if the graph is a circle, an ellipse, a hyperbola, or a parabola.

a. $4x^2 - 9y^2 + 8x - 54y - 113 = 0$ **b.** $2x^2 - x = -3y^2$

Solution: Since the degenerate possibilities have been eliminated, we proceed as follows:

a. The equation $4x^2 - 9y^2 + 8x - 54y - 113 = 0$ is in general form with $A = 4$, and $C = -9$. Since A and C have different signs, the graph of the equation is a hyperbola.

b. To determine the conic given by $2x^2 - x = -3y^2$, we first write the equation in general form as

$$2x^2 + 3y^2 - x = 0.$$

In this case A and C have the same sign, with $A \neq C$. Thus, the graph of the equation is an ellipse.

PROGRESS CHECK 6

Identify the graph of each of the following equations if the graph is a circle, an ellipse, a hyperbola, or a parabola.

a. $2x^2 + 2y^2 - 10x - 2y - 31 = 0$ **b.** $y^2 - 2y - 8x + 50 = 0$ ■

The next example shows the analysis of two equations that lead to degenerate cases. It is interesting to note that many of these cases also result from the intersection of a cone with an appropriate plane, and the degenerate possibilities are called **degenerate conic sections**. It can be shown that the graph of every second-degree equation is a conic section or a degenerate conic section.

Example 7: Identifying a Degenerate Conic Section

Identify the graph of each of the following equations.

a. $2x^2 + y^2 + 1 = 0$ **b.** $x^2 + y^2 + 4x - 6y + 13 = 0$

Solution:

a. The equation $2x^2 + y^2 + 1 = 0$ does not define an ellipse, even though A and C have the same sign, with $A \neq C$. There is no ordered pair of real numbers that satisfies $2x^2 + y^2 = -1$, since $2x^2$ and y^2 are never negative, so their sum can never be -1. Thus, the equation has no graph.

b. On first inspection, this equation seems to define a circle. To make sure, however, we need to complete the square and convert the equation to standard form. By doing so, in this case we obtain

$$(x+2)^2+(y-3)^2=0.$$

Only $x=-2$, $y=3$ satisfies the equation, so the graph is the point $(-2,3)$.

PROGRESS CHECK 7

Identify the graph of each of the following equations.

a. $x^2+y^2+6y+10=0$ **b.** $4y^2-9x^2-36x-8y-32=0$ ■

EXPLORE 10.4

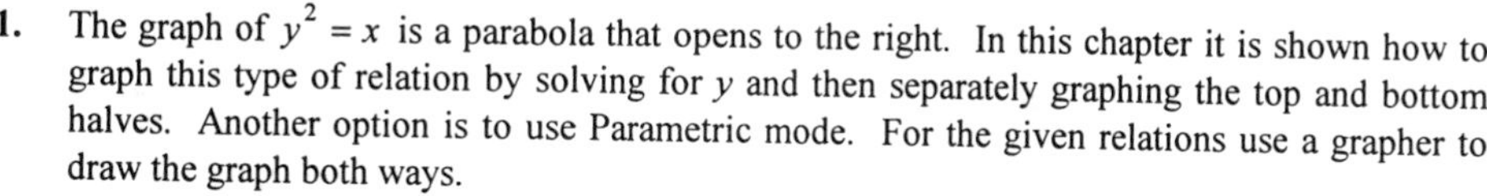

1. The graph of $y^2=x$ is a parabola that opens to the right. In this chapter it is shown how to graph this type of relation by solving for y and then separately graphing the top and bottom halves. Another option is to use Parametric mode. For the given relations use a grapher to draw the graph both ways.

a. Graph $y^2=x$ by graphing separately $y1=\sqrt{x}$ and $y2=-\sqrt{x}$.
Graph $y^2=x$ in Parametric mode by letting $xt1=t^2$ and $yt1=t$.

b. Graph $(y-2)^2=x+9$ by solving for y.
Graph $(y-2)^2=x+9$ in parametric mode by letting $xt1=(t-2)^2-9$ and $yt1=t$.

c. Graph $(y+2)^2=x-4$ by both methods.

2. **a.** As discussed in this section, the value of p in the equation $x^2=4py$ determines the position of a horizontal line called the directrix of the parabola. Examine what happens to the graph of $x^2=4py$ as p increases. Let $p=1$, 2, and 3, and use a grapher to graph both the parabola and its directrix $y=-p$. If desired, use List capabilities to enter the equations. For instance, let $y1=x^2/4\{1,2,3\}$.

b. Similarly, for the graph of $y^2=4px$, the vertical line $x=-p$ is the directrix. Use a grapher to graph the parabola and the directrix when $p=5$. Hint: Try Parametric mode for this exploration.

EXERCISES 10.4

In Exercises 1–14 find the vertex, the focus, and the equation of the directrix, and sketch each curve.

1. $y^2=4x$ **2.** $y^2=-8x$

3. $x^2=-2y$ **4.** $x^2=5y$

5. $y^2+12x=0$ **6.** $y^2-10x=0$

7. $4x^2-3y=0$ **8.** $2x^2+3y=0$

9. $(y+2)^2=4(x+1)$ **10.** $(x-1)^2=-6y$

11. $y=x^2-4x$ **12.** $x=y^2+2y$

13. $x^2-2x-4y-7=0$ **14.** $y^2+4y+3x-8=0$

In Exercises 15–24 find the equation in standard from of the parabola satisfying the conditions.

15. Vertex at origin, focus (0, 3)

16. Vertex at origin, focus (–2,0)

17. Vertex at origin, directrix $x=-\frac{1}{2}$

18. Vertex at origin, directrix $y=\frac{3}{2}$

19. Focus (0, 4), directrix $y=-4$

20. Focus $\left(-\frac{5}{2},0\right)$, directrix $x=\frac{5}{2}$

21. Focus (2, 1), directrix $x = -4$

22. Focus $(-1, -3)$, directrix $y = 5$

23. Vertex (3, 0), focus (3, 3)

24. Vertex (1, 4), directrix $x = \frac{7}{2}$

25. Consider the parabolic reflector shown. Choose axes in a convenient position and determine
- **a.** an equation of the parabola and
- **b.** the location of the focus point.

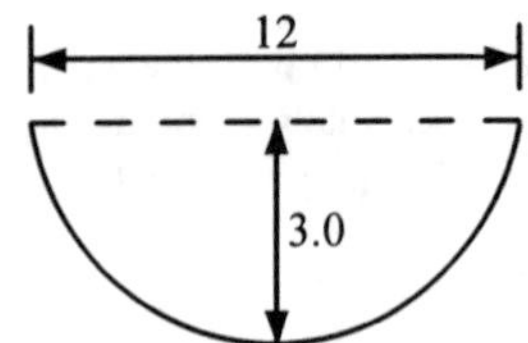

26. Answer the questions in Exercise 25 for the parabolic reflector shown below.

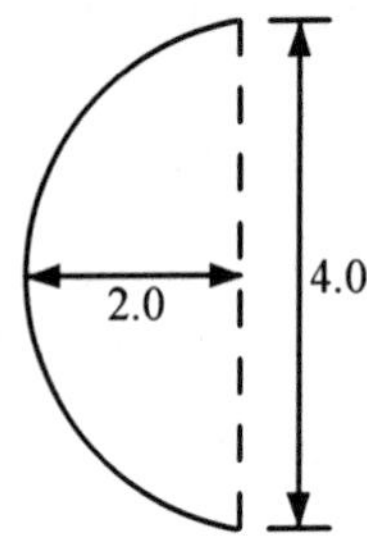

27. In equatorial countries efficient solar cookers are made by placing a cooking surface at the focus of a parabolic reflector, usually coated with aluminum foil.
- **a.** Choose axes in a convenient position and determine an equation for the parabolic sunlight reflector shown in the figure.
- **b.** What is the coordinate position that shows where to place the item to be cooked?

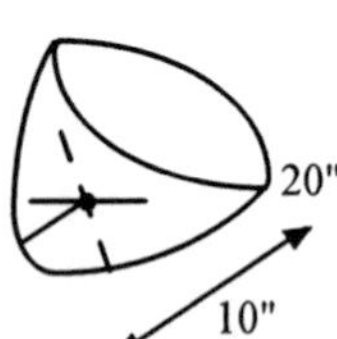

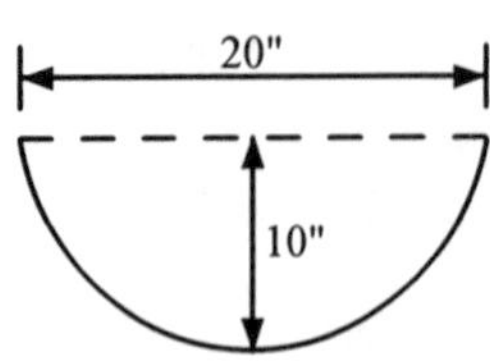

28. A parabolic listening device has a microphone placed at the focus.
- **a.** Choose axes in a convenient position and determine an equation for the parabolic device shown in the figure.
- **b.** What is the coordinate position that shows where to place the microphone?

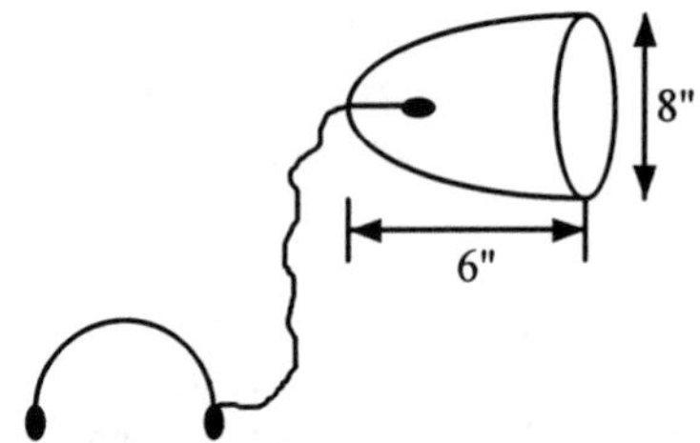

29. A parabolic arch has a height of 25 ft and a span of 40 ft. How high is the arch 8 ft each side of the center?

30. For the arch in the previous exercise, how far from the center is the arch 16 ft high?

31. The great Greek mathematician Archimedes (c225 B.C.) showed that the area inside a parabola (see the figure) equals $\frac{4}{3}$ the area of the triangle. Use his discovery to find the area enclosed by the parabola $y = x^2$ and the line $y = 4$.

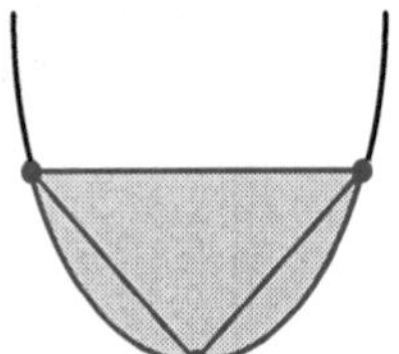

32. If you place the line in the previous exercise at $y = k\,(k > 0)$, what is the resulting area inside the parabola?

In Exercises 33–42 identify the graph of each equation if the graph is a circle, an ellipse, a hyperbola, or a parabola.

33. $y^2 - 3x + 2y + 1 = 0$

34. $9y^2 - x^2 - 36x + 2y - 1 = 0$

35. $4x^2 + 9y^2 + 8x - 54y - 49 = 0$

36. $x^2 + y^2 + 7x + 3y - 4 = 0$

37. $x^2 = y^2 - 4$

38. $3y^2 = 4x - x^2$

39. $3y = 4x - x^2$

40. $5x^2 = 1 - 5y^2$

41. $2y^2 = 9 - 2x^2$

42. $5x - 2y = y^2$

In Exercises 43–52 identify the graph of each equation. Degenerate cases are possible here.

43. $x^2 - y^2 - 4y - 4 = 0$

44. $x^2 + y^2 - 4y + 4 = 0$

45. $x^2 + 2x = 1 - y^2$

46. $x^2 + 2x = y^2 - 1$

47. $x^2 + 2x = 1 - y$

48. $x^2 + 2x = y^2 - 4$

49. $9x^2 + 4y^2 + 18x - 8y + 13 = 0$

50. $4x^2 + 9y^2 + 2x - 4 = 0$

51. $4x^2 + 9y^2 - 2x + 4 = 0$

52. $4y^2 + 3 = 8y$

THINK ABOUT IT 10.4

1. **a.** Find the vertex of the parabola whose equation is $y = ax^2 + bx + c$ with $a \neq 0$.

b. Find the focus and the directrix of the parabola whose equation is $y = ax^2 + bx + c$ with $a \neq 0$.

2. **a.** A line segment joining two points on a parabola that contains the focus and that is perpendicular to the axis of symmetry is called the **focal chord** (or **latus rectum**) of the parabola. Find the length of the focal chord of a parabola whose equation is $y^2 = 4px$.

b. Use the result in part **a** to find equations in standard form for the two parabolas that satisfy the following conditions: the vertex is at the origin, the axis of symmetry is the x-axis, and the length of the focal chord is 8.

3. A graph is the set of all points in a plane whose distance from (2, 0) is always one-half its distance from $x = 8$. Find an equation for this graph, and classify the graph.

4. **a.** If $x^2 + y^2 + Dx + Ey + F = 0$ is the graph of a circle, then find the center and the radius in terms of D, E, and F.

b. State conditions involving D, E, and F that determine whether the equation in part **a** represents a circle or a point, or has no graph.

5. **a.** Match each graph with the equation that illustrates that case.

Graph	Equation
1. Two distinct lines through the origin	a. $x^2 + y^2 = 0$
2. Two distinct parallel lines	b. $x^2 + y^2 = -1$
3. One line through the origin	c. $x^2 - y^2 = 0$
4. A point (the origin)	d. $x^2 = 1$
5. No graph	e. $x^2 = 0$

b. Generalize from the result in part **a** and specify conditions for A, C, and k in the equation

$$Ax^2 + Cy^2 = k, \text{ with } A^2 + C^2 \neq 0,$$

that result in a graph that is a circle, an ellipse, a hyperbola, or each of the special cases listed above.

• • •

10.5 Nonlinear Systems of Equations

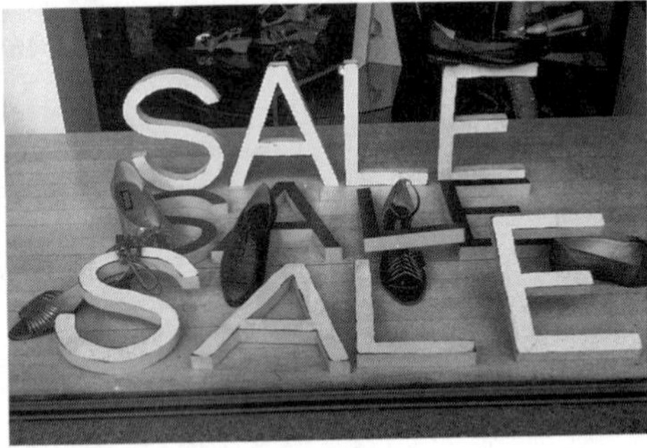

Photo Courtesy of ***R. P. Kingston****, The Picture Cube*

In economics an analysis of the law of supply and demand involves the intersection of two graphs. Basically, as the price for an item increases, the quantity of the product that is supplied increases while the quantity that is demanded decreases. The point at which the supply and demand graphs intersect is called the **point of market equilibrium**.

a. The supply and demand equations in a certain location for a model of sandal priced at p dollars are given by

$$\text{Supply: } q = p^2 + 4p - 60$$
$$\text{Demand: } q = 240 - p$$

Estimate the equilibrium price and the corresponding number of units supplied and demanded by finding the intersection of the graphs of these equations for $p > 0$ on a grapher.

b. Confirm your solution from part **a** by algebraic methods.

(See Example 2.)

Objectives

1. Solve a nonlinear system of equations by the graphing method, the substitution method, or the addition-elimination method.
2. Solve applied problems involving a nonlinear system of equations.

Systems of linear equations in two variables were solved algebraically and graphically in Section 1.5. On the basis of the work with second-degree equations in this chapter, we can now solve certain **nonlinear systems of equations**, which are systems with at least one equation that is not linear. Graphically, solving most of the systems considered in this section will require finding all intersection points of one of the conic sections (parabola, circle, ellipse, hyperbola) with a line or another conic section. The algebraic approach for finding such solutions will rely once again on either the substitution method or the addition-elimination method. As with linear systems, when at least one equation in the system is solved for one of the variables, the substitution method is easy to apply, as illustrated in Example 1.

Example 1: Solving a Nonlinear System Graphically and Algebraically

Solve the system:

$$y = x^2 - 6x + 8 \quad (1)$$
$$y = x + 2 \quad (2)$$

Use graphical and algebraic methods.

Solution: The solutions to this system are easy to picture on a grapher. Equation (1) graphs as a parabola, equation (2) graphs as a line, and these two graphs intersect at (1,3) and (6,8), as shown in Figure 10.41.

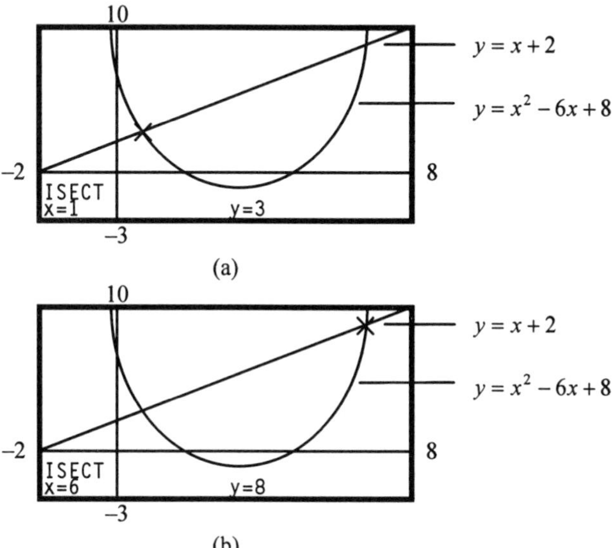

Figure 10.41

To confirm these solutions algebraically, we choose the substitution method and replace y by $x^2 - 6x + 8$ in equation (2) and solve for x.

$$x^2 - 6x + 8 = x + 2$$
$$x^2 - 7x + 6 = 0$$
$$(x - 6)(x - 1) = 0$$
$$x - 6 = 0 \quad \text{or} \quad x - 1 = 0$$
$$x = 6 \qquad\qquad x = 1$$

Thus, the parabola and the line intersect at $x = 1$ and $x = 6$. To find the y coordinates, substitute these numbers in the simpler equation $y = x + 2$.

$$\begin{aligned} y &= x + 2 \\ &= (1) + 2 \\ &= 3 \end{aligned} \quad \text{or} \quad \begin{aligned} y &= x + 2 \\ &= (6) + 2 \\ &= 8 \end{aligned}$$

The solutions, (1,3) and (6,8), should be checked in both original equations (1) and (2).

Note: A line and a conic section may intersect in two, one, or no points, as illustrated in Figure 10.42. Therefore, we can anticipate that a system with one first-degree equation and one second-degree equation will have two, one, or no real solutions.

PROGRESS CHECK 1

Solve the system: $\begin{aligned} x^2 + y^2 &= 25 \\ y &= x - 1 \end{aligned}$. Use graphical and algebraic methods. ■

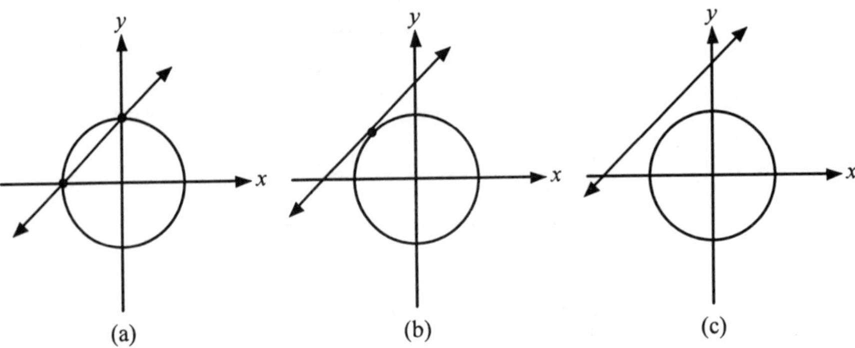

Figure 10.42

Example 2: Point of Market Equilibrium

Solve the problem in the section introduction on page 850.

Solution:

a. The intersection of the graphs of the equations given for supply and demand for $p > 0$ may be displayed on a grapher as shown in Figure 10.43. It appears that the equilibrium price is \$15, and that 225 pairs of sandals are supplied and demanded when the market for this commodity is in equilibrium.

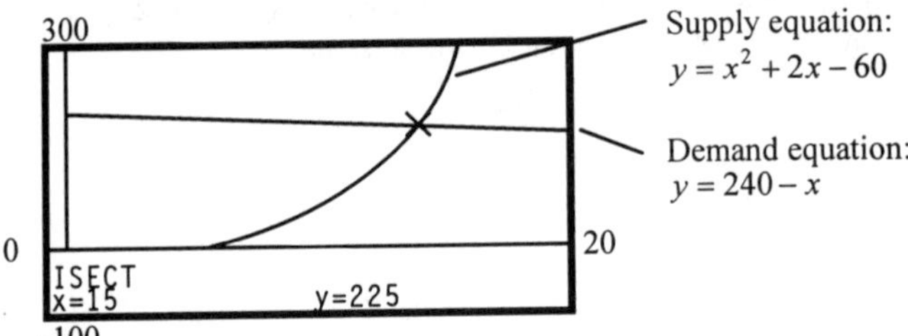

Figure 10.43

b. To confirm the proposed solution algebraically, we can use the substitution method and replace q by $p^2 + 4p - 60$ in the demand equation.

$$p^2 + 4p - 60 = 240 - p$$
$$p^2 + 5p - 300 = 0$$
$$(p + 20)(p - 15) = 0$$
$$p + 20 = 0 \quad \text{or } p - 15 = 0$$
$$p = -20 \quad\quad p = 15$$

In the context of the problem only $p = 15$ is meaningful, so the equilibrium price is \$15. To find q, replace p by 15 in the demand equation, which is the simpler equation.

$$\begin{aligned} q &= 240 - p \\ &= 240 - 15 \\ &= 225 \end{aligned}$$

When this market is in equilibrium, manufacturers will supply 225 pairs of sandals, and consumers will demand 225 pairs of sandals. Thus, we have algebraic confirmation of the results from part **a**.

PROGRESS CHECK 2
Redo the problems in Example 2, assuming that the supply equation is $q = p^2 + 3p - 70$, and the demand equation is $q = 410 - p$. ■

When the two equations in a system both contain an x^2 term and a y^2 term, then the addition-elimination method is often useful, as shown next.

Example 3: Using the Addition-Elimination Method

Find all intersection points of the ellipse $2x^2 + 3y^2 = 5$ and the hyperbola $4y^2 - 3x^2 = 1$.

Solution: The system we need to solve is

$$2x^2 + 3y^2 = 5$$
$$-3x^2 + 4y^2 = 1$$

If we form equivalent equations by multiplying the top equation by 3 and the bottom equation by 2, we can eliminate the x variable.

$$\begin{aligned} 6x^2 + 9y^2 &= 15 \\ \underline{-6x^2 + 8y^2} &\underline{= 2} \\ 17y^2 &= 17 \quad \text{Add the equations} \\ y^2 &= 1 \\ y &= \pm 1 \end{aligned}$$

Now substituting $y = \pm 1$ into $2x^2 + 3y^2 = 5$ gives

$$\begin{aligned} 2x^2 + 3(\pm 1)^2 &= 5 \\ x^2 &= 1 \\ x &= \pm 1 \cdot \end{aligned}$$

Thus, the solutions are $(1,1)$, $(1,-1)$, $(-1,1)$, and $(-1,-1)$. Figure 10.44 shows the ellipse and the hyperbola intersecting at these points.

PROGRESS CHECK 3
Find all intersection points of the ellipse $6x^2 + 7y^2 = 159$ and the hyperbola $3x^2 - y^2 = 39$.

The next two examples provide interesting contrasts between graphical and algebraic methods for solving nonlinear systems. In Example 4, a graphical approach quickly leads to meaningful estimates of the solutions, while the exact solutions are revealed only after some detailed algebraic analysis. However, a clever substitution leads to a simple algebraic solution for the system in Example 5, which is more difficult to interpret graphically.

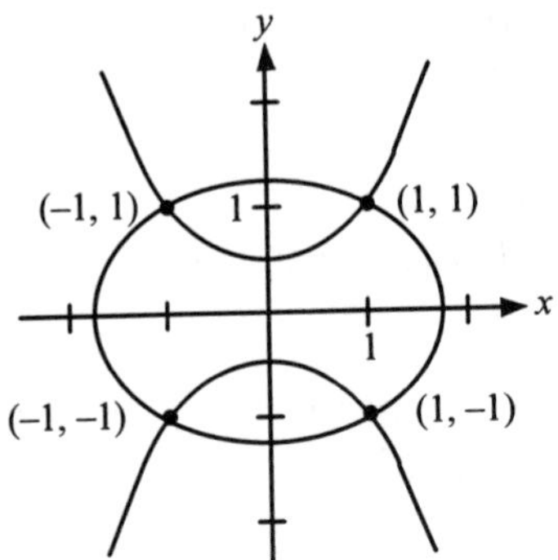

Figure 10.44

Example 4: Contrasting Graphical and Algebraic Methods

Solve the system: $\begin{aligned} x^2 + y^2 &= 1 \\ y &= x^2 \end{aligned}$.

Solution: Both equations in this system have basic graphs, and it is easy to envision the two solutions to the system as shown in Figure 10.45. In addition, estimates of these solutions may be obtained quickly on a graphing calculator. Figure 10.46 shows that the quadrant one solution is about (0.79, 0.62), and the quadrant two solution is then about $(-0.79, 0.62)$, because of the y-axis symmetry of the graphs.

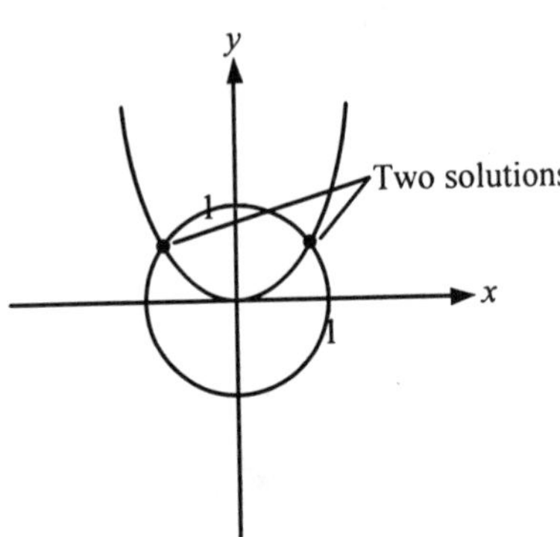

Figure 10.45

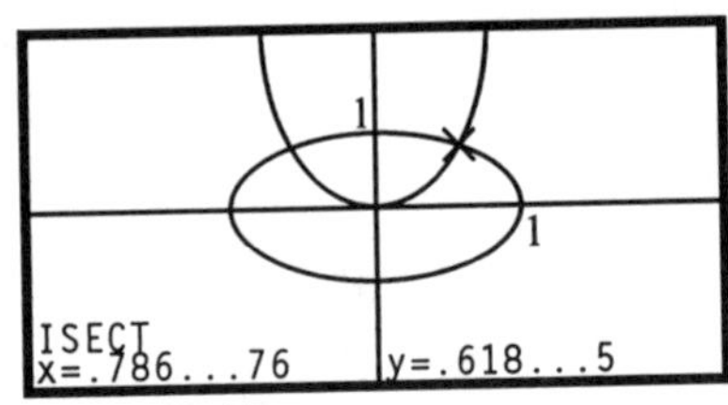

Figure 10.46

To confirm these results algebraically and to determine the exact solutions, we can use the substitution method. Replacing y by x^2 in the equation for the circle gives

$$x^2 + \left(x^2\right)^2 = 1 \text{ or } \left(x^2\right)^2 + x^2 - 1 = 0,$$

which is an equation with quadratic form. As discussed in Section 3.3, if we let $t = x^2$, the equation becomes

$$t^2 + t - 1 = 0.$$

By the quadratic formula,

$$t=\frac{-1\pm\sqrt{(1)^2-4(1)(-1)}}{2(1)}=\frac{-1\pm\sqrt{5}}{2},$$

so

$$x^2=\frac{-1\pm\sqrt{5}}{2} \text{ and } x=\pm\sqrt{\frac{-1\pm\sqrt{5}}{2}}.$$

Since x must be a real number, we eliminate $\pm\sqrt{\frac{-1-\sqrt{5}}{2}}$ and conclude that

$$x=\pm\sqrt{\frac{-1+\sqrt{5}}{2}}.$$

Finally, substituting these two values of x into $y=x^2$ gives us solutions of

$$\left(\sqrt{\frac{-1+\sqrt{5}}{2}},\frac{-1+\sqrt{5}}{2}\right) \text{ and } \left(-\sqrt{\frac{-1+\sqrt{5}}{2}},\frac{-1+\sqrt{5}}{2}\right).$$

You should evaluate the radical expressions involved in these solutions, to confirm that they are approximately $(0.79, 0.62)$ and $(-0.79, 0.62)$.

PROGRESS CHECK 4

Solve the system: $\begin{array}{r} y^2-x^2=4 \\ y=x^2 \end{array}$. ■

Example 5: Contrasting Graphical and Algebraic Methods

Solve the system

$$\begin{aligned} \frac{5}{x}+\frac{2}{y}-1=0 \quad (1) \\ \frac{1}{x}-\frac{3}{y}-7=0 \quad (2) \end{aligned}$$

Solution: If we rewrite the above system as

$$\begin{aligned} 5\left(\frac{1}{x}\right)+2\left(\frac{1}{y}\right)=1 \\ \left(\frac{1}{x}\right)-3\left(\frac{1}{y}\right)=7 \end{aligned}$$

and let $a=1/x$ and $b=1/y$, we obtain

$$\begin{aligned} 5a+2b=1 \\ a-3b=7 \end{aligned},$$

which is a relatively simple system of linear equations. If we now add –5 times the second equation to the first equation, we find

$$17b = -34, \text{ so } b = -2.$$

Then substituting $b = -2$ into equation 2 gives

$$\begin{aligned} a - 3(-2) &= 7 \\ a &= 1 \end{aligned}.$$

Finally, from the definition of a and b we have

$$\frac{1}{x} = 1 \text{ and } \frac{1}{y} = -2$$
$$x = 1 \text{ and } y = -\frac{1}{2}$$

Thus, the solution is $\left(1, -\frac{1}{2}\right)$.

To check this result graphically, solve equation (1) and equation (2) for y to obtain $y = 2x/(x-5)$ and $y = 3x/(1-7x)$, respectively. Then the intersection of the graphs of these rational functions that is shown in Figure 10.47 is in agreement with the proposed solution. Note that the location of the asymptotes for these rational functions was influential in determining the viewing window used in displaying this solution.

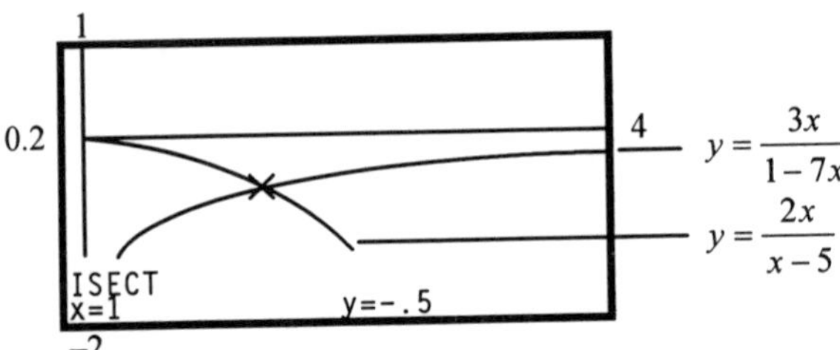

Figure 10.47

PROGRESS CHECK 5

Solve the system $\begin{aligned} \frac{7}{x} - \frac{5}{y} &= 11 \\ \frac{12}{x} + \frac{2}{y} &= 3 \end{aligned}$ ■

EXPLORE 10.5

1. For the circle $x^2 + y^2 = 25$ and the family of parabolas $y = x^2 + c$ use a grapher to determine the values of c for which the two curves have $n = 4$, 3, 2, 1, or 0 points of intersection. Confirm your answers algebraically. Be especially careful to check your results algebraically for the case of 2 points of intersection. Use Zoom features to help see clearly what happens.
2. In about the year 200 B.C., the Greek mathematician Apollonius wrote a major book about conic sections. In fact, it is from this book that we get the names "ellipse," "parabola," and "hyperbola." One of his problems involves constructing the tangent to an ellipse. In this exploration we look at his result and at a related question you can solve by analytic geometry.

a. If you pick a point on the ellipse and draw a tangent line as shown, Apollonius determined that the line hits the x-axis so that $\overline{AH}/\overline{AG} = \overline{BH}/\overline{BG}$. Use his theorem to find where the tangent to $x^2/25 + y^2/9 = 1$ at the point $(4, 9/5)$ hits the x-axis.

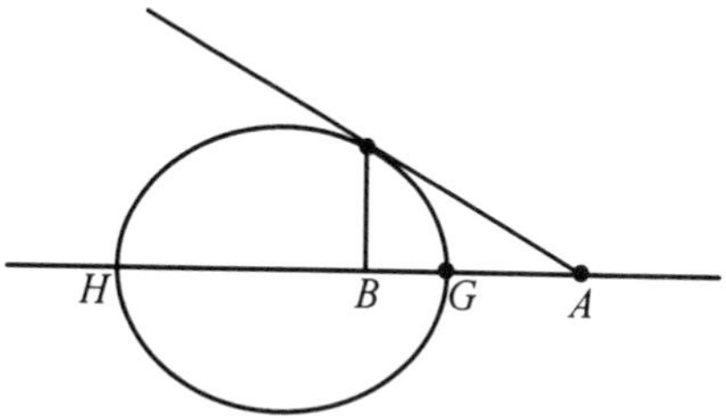

b. Show that the line $4x + 5y = 25$ is the tangent line to the ellipse $x^2/25 + y^2/9 = 1$, by graphing both figures and showing that they have only one point in common. Confirm the point of intersection algebraically.

3. In Section 10.3, the opening example describes the Loran navigation system for determining the position of a ship. This exploration continues that discussion. Suppose there are 3 transmitters, A, B, and C, as shown in the figure.

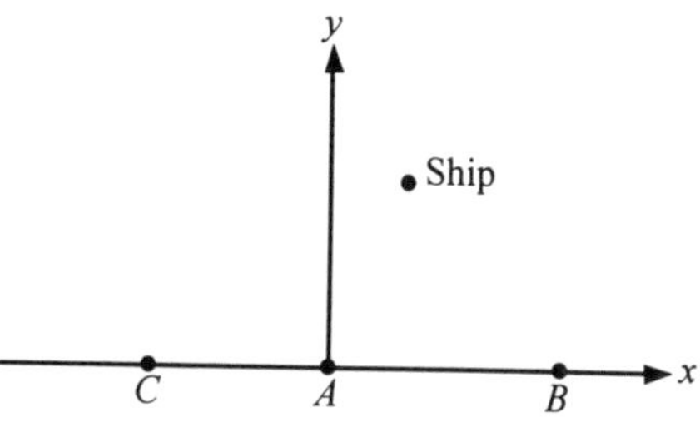

a. The distance between A and B is 200 miles, and the navigator on the ship determines that the ship is 60 miles closer to A than it is to B. Show that the equation for the hyperbola that gives all possible positions of the ship is $(x-100)^2/900 - y^2/9100 = 1$. Use a grapher to draw the branch of this hyperbola that shows the ship is closer to A.

b. The distance from A to C is 120 miles, and the ship is 40 miles closer to C than it is to A. Show that the equation for the hyperbola that gives all possible positions of the ship is $(x+60)^2/400 - y^2/3200 = 1$, and graph the branch of this curve that shows the ship is closer to A.

c. To the nearest mile, find the position of the ship by determining the appropriate intersection of the two hyperbolas in parts **a** and **b**.

EXERCISES 10.5

In Exercises 1-6 use both graphical and algebraic methods to solve the given system of equations.

1. $y = x^2 - 5x + 6$
$y = x - 1$

2. $y = 4 - x^2$
$y = 3 - x$

3. $(x-1)^2 + (y-2)^2 = 9$
$y = (x-1)^2 - 1$

4. $x^2 + y^2 = 9$
$y = x^2 + 3$

5. $y = x^2 + 3$
$y = 2(x-1)^2 + 2$

6. $3x^2 + 4y^2 = 12$
$6x^2 - 4y^2 = 15$

In Exercises 7–34 solve the given system.

7. $y = x^2$
$y = x$

8. $f(x) = 2 - x^2$
$g(x) = -x$

9. $g(x) = 10 - x^2$
$f(x) = 1$

10. $y = x^2 - 2x - 3$
$y = 2x + 2$

11. $y = x^2 - 4$
$y = 4 - x^2$

12. $f(x) = x^2$
$g(x) = 2 - x^2$

13. $x^2 + y^2 = 100$
$3x + y = 10$

14. $x^2 + y^2 = 1$
$y = x + 1$

15. $x^2 + y^2 = 8$
$2x^2 - y^2 = 4$

16. $x^2 - y^2 = 8$
$2x^2 + y^2 = 19$

17. $x^2 + 2y^2 = 40$
$2x^2 + y^2 = 32$

18. $2x^2 + 3y^2 = 5$
$3x^2 = 4y^2 - 1$

19. $xy = 15$
$x + y = 8$

20. $xy = 1$
$x - 2y = 1$

21. $xy = 2$
$x^2 + 2y^2 = 9$

22. $xy = 4$
$x^2 + y^2 = 8$

23. $x^2 + y^2 = 2$
$x = y^2$

24. $x^2 + y^2 = 3$
$x = y^2$

25. $x - y = 2$
$x = y^2 - 4$

26. $y^2 = x + 9$
$x = 3 - 2y - y^2$

27. $3x + y = 7$
$x^2 y = 4$

28. $x = y^3 - 3y^2$
$x - y = -3$

29. $x^2 + y^2 = 5$
$y = x^2$

30. $(x-1)^2 + (y+2)^2 = 5$
$y + 2x = 0$

31. $\dfrac{x^2}{4} + \dfrac{y^2}{9} = 1$
$\dfrac{x^2}{4} - \dfrac{y^2}{9} = 1$

32. $\dfrac{x^2}{4} + y^2 = 1$
$\dfrac{x^2}{4} - \dfrac{y^2}{2} = 1$

33. $\dfrac{3}{x} + \dfrac{2}{y} = 27$
$\dfrac{2}{x} + \dfrac{5}{y} = -1$

34. $\dfrac{5}{x} + \dfrac{3}{y} - 5 = 0$
$\dfrac{4}{x} + \dfrac{1}{y} - 11 = 0$

35. Find all intersection points of the ellipse $x^2 + 2y^2 = 3$ and the hyperbola $3x^2 - y^2 = 2$.

36. Find all intersection points of the semicircle $y = -\sqrt{36 - x^2}$ and the line $y = -x$.

37. Find to the nearest tenth all points where the circle $x^2 + y^2 = 9$ and the ellipse $\dfrac{x^2}{25} + \dfrac{y^2}{4} = 1$ intercept.

38. Find to the nearest hundredth all intersection points of the circle $x^2 + y^2 = 1$ and the parabola $x = y^2$.

39. Find the length of the legs of a right triangle if the area is 36 ft^2 and the hypotenuse is $5\sqrt{6}$ ft.

40. Find the radius of each circle in the illustration below if the combined area of the two circles is 48.5π square inches.

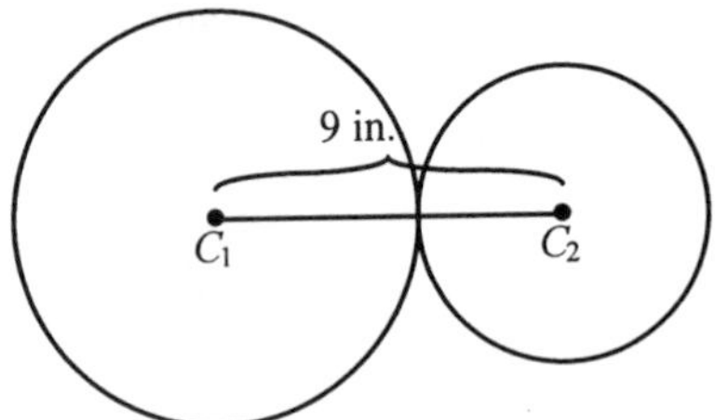

41. Both the parallelogram and the triangle in the illustration below have the areas of 36m^2. Find b and h.

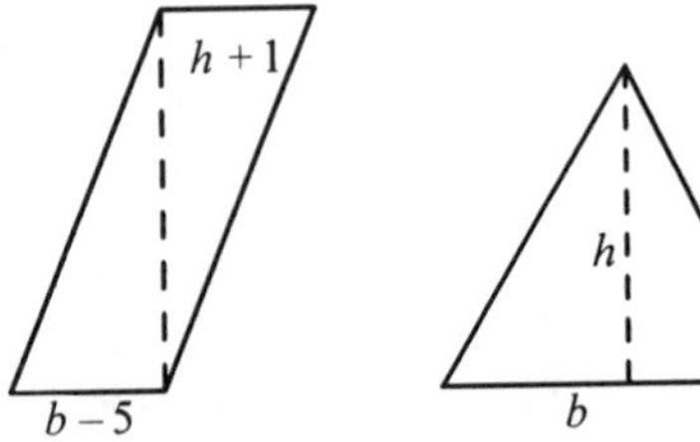

42. If the area and perimeter of a rectangle are denoted by a and p, respectively, then show that the length is given by $\dfrac{p + \sqrt{p^2 - 16a}}{4}$.

43. Greek legend has it that Apollo was responsible for a plague in Delos. The oracle there said that Apollo would be appeased if the people would double the size of the altar, keeping its shape the same. The shape was cubical. The mathematical question then is, "How much longer should each side of the new cube be so that its volume is double the original volume?" This was a difficult problem for Greek geometry. One solution, often attributed to Menaechmus (c. 350 B.C.), depending on finding the intersection of a parabola and a hyperbola. The side of the new alter is given by the x-coordinate of the intersection of the graphs of $y = x^2$ and $xy = 2$. Solve this system and show that multiplying the side (s) of a cube by this solution gives a new cube with double the volume of the old one.

44. Refer to the previous exercise. Suppose the god Apollo got greedy and demanded that the new alter must be 10 times the size of the original. Now the side of the new altar is found by multiplying the original side s by the x-coordinate of the intersection of $y = x^2$ and $xy = 10$. Solve this system and confirm that this solution is correct.

THINK ABOUT IT 10.5

1. Show that a hyperbola defined by the equation $\left(x^2/a^2\right)-\left(y^2/b^2\right)=1$ does not intersect its asymptotes.
2. Find all intersection points of the graphs of $y = \log(x-4)$ and $y = -1 + \log(6x+16)$.
3. A rectangle is inscribed in a circle. If the areas of the rectangle and the circle are 4 m^2 and $(5/2)\pi\text{m}^2$, respectively, then find the dimensions of the rectangle.
4. Find in terms of r all intersection points of the circle $x^2 + y^2 = r^2$ and the hyperbola $y^2 - x^2 = 2$. For what values of r do the curves intersect?
5. Find the slope-intercept equation of the line through (1,1) that intersects the parabola $y = x^2$ in exactly one point.

CHAPTER 10 SUMMARY

OBJECTIVES CHECKLIST

Specific chapter objectives are summarized below along with numbered example problems from the text that should clarify the objectives. If you do not understand any objectives, or do not know how to do the selected problems, then restudy the material.

10.1: Can you:

1. **Find the equation in standard form of a circle, given its center and either the radius of the circle or a point on the circle?** Find the equation in standard form of the circle with center at $(2,-1)$ and radius 4.

   [Example 1]

2. **Find the center and radius of a circle, give its equation and then graph it?** Find the center and radius of the circle given by

$$(x+3)^2 + y^2 = 25.$$

   Then, sketch the circle.

   [Example 4]

3. **Use the methods of analytic geometry to prove geometric theorems?** Use the figure below and the concept of slope to show that if a triangle is inscribed in a circle with the diameter as one of its sides, then the triangle is a right triangle.

   [Example 6]

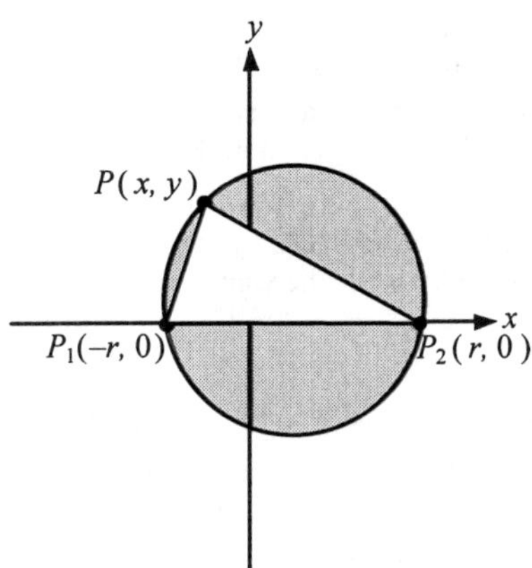

4. **Find the midpoint of a line segment joining a given pair of points?** Find the midpoint of the line segment joining $(-3,\ 4)$ and $(7,\ -1)$.

[Example 7]

10.2: Can you:

1. **Use an equation of an ellipse to graph the relation and to find the coordinates of the center, the foci, and the endpoints of the major and minor axes?** Find the coordinates of the foci and the endpoints of the major and minor axes of the ellipse given by $4x^2 + y^2 = 36$. Also, sketch the ellipse.

[Example 1]

2. **Find an equation for an ellipse satisfying various conditions concerning its center, foci, vertices, and major and minor axes?** Find the equation in standard form of the ellipse with center at the origin, one focus at (3, 0), and one vertex at (4, 0).

[Example 3]

10.3: Can you:

1. **Use an equation of a hyperbola to graph the relation, to determine its asymptotes, and to find the coordinates of the center, the foci, and the vertices?** For the hyperbola given by $4x^2 - y^2 = 36$, find the coordinates of the vertices and the foci. Also, determine the asymptotes and sketch the curve.

[Example 1]

2. **Find an equation for a hyperbola satisfying various conditions concerning its center, foci, vertices, and transverse and conjugate axes?**
Find the equation in standard form of the ellipse with center at the origin, one focus at (0, 5), and one vertex at (0, 3).

[Example 3]

10.4: Can you:

1. **Use an equation of a parabola to graph the relation, to determine the equation of the directrix, and to find the coordinates of the vertex and focus?** Find the focus and the directrix of the parabola given by $y^2 = 12x$. Also, sketch the parabola.

[Example 1]

2. **Find an equation in standard form for a parabola satisfying various conditions concerning its vertex, focus, and directrix?** A parabola with its vertex at the origin has its focus at $(-5,\ 0)$. Find the equation in standard form of the parabola.

[Example 3]

3. **Classify certain equations as defining a circle, an ellipse, a hyperbola, or a parabola, or as defining a degenerate case of a conic section?** Identify the graph of $4x^2 - 9y^2 + 8 - 54y - 113 = 0$ if the graph is a circle, an ellipse, a hyperbola, or a parabola. [Example 6a]

10.5: Can you:

1. **Solve a nonlinear system of equations by the graphing method, the substitution method, or the addition-elimination method?** Solve the system: $\begin{array}{l} y = x^2 - 6x + 8 \\ y = x + 2 \end{array}$ [Example 1]

2. **Solve applied problems involving a nonlinear system of equations?** The supply and demand equations in a certain location for a model of sandal priced at p dollars are given by

$$\text{Supply: } q = p^2 + 4p - 60$$
$$\text{Demand: } q = 240 - p.$$

Estimate the equilibrium price and the corresponding number of units supplied and demanded by finding the intersection of the graphs of these equations for $p > 0$ on a grapher. Confirm your solution by algebraic methods. [Example 2]

KEY CONCEPTS AND PROCEDURES

Section **Key Concepts and Procedures to Review**

10.1

- Fundamental relationship between an equation and its graph: Every ordered pair that satisfies the equation corresponds to a point in its graph, and every point in the graph corresponds to an ordered pair that satisfies the equation
- Definition of a circle
- Standard form of a circle of radius r with center (h, k): $(x-h)^2 + (y-k)^2 = r^2$
- Procedures for proving certain theorems in plane geometry by using a coordinate system and algebraic methods
- Midpoint formula: $\left(\dfrac{x_1 + x_2}{2}, \dfrac{y_1 + y_2}{2}\right)$

10.2

- Definitions of ellipse, foci, major axis, minor axis, and vertices
- Summary for an ellipse:

Foci	**Center**	**Standard Form** $(a > b)$
On x-axis at $(\pm c, 0)$	Origin	$\dfrac{x^2}{a^2} + \dfrac{y^2}{b^2} = 1$
On y-axis at $(0, \pm c)$	Origin	$\dfrac{x^2}{b^2} + \dfrac{y^2}{a^2} = 1$
On major axis parallel to x-axis	(h,k)	$\dfrac{(x-h)^2}{a^2} + \dfrac{(y-k)^2}{b^2} = 1$
On major axis parallel to y-axis	(h,k)	$\dfrac{(x-h)^2}{b^2} + \dfrac{(y-k)^2}{a^2} = 1$

- In general, $a^2 = b^2 + c^2$, where a, b, and c represent the following distances:

a is the distance from the center to the endpoints on the major axis.
b is the distance from the center to the endpoints on the minor axis.
c is the distance from the center to the foci.

10.3

- Definitions of hyperbola, foci, transverse axis, vertices, conjugate axis, and asymptotes of the hyperbola
- Summary for a hyperbola:

Foci	Center	Standard Form	Asymptotes
On x-axis at $(\pm c,0)$	Origin	$\frac{x^2}{a^2}-\frac{y^2}{b^2}=1$	$y=\pm\frac{b}{a}x$
On y-axis at $(0,\pm c)$	Origin	$\frac{y^2}{a^2}-\frac{x^2}{b^2}=1$	$y=\pm\frac{a}{b}x$
On transverse axis parallel to x-axis	(h,k)	$\frac{(x-h)^2}{a^2}-\frac{(y-k)^2}{b^2}=1$	$y-k=\pm\frac{b}{a}(x-h)$
On transverse axis parallel to y-axis	(h,k)	$\frac{(x-k)^2}{a^2}-\frac{(y-h)^2}{b^2}=1$	$y-k=\pm\frac{a}{b}(x-h)$

- In general, $c^2=a^2+b^2$, where a, b, and c represent the following distances:

  a is the distance from the center to the endpoints on the transverse axis.
  b is the distance from the center to the endpoints on the conjugate axis.
  c is the distance from the center to the foci.

10.4

- Definitions of parabola, directrix, focus, axis of symmetry, and vertex
- Summary for a parabola with vertex at the origin: (In all cases p gives the distance form the vertex to the focus and from the vertex to the directrix. See Figures 10.24–10.27.)

Standard Form	Opens	Axis of Symmetry	Focus	Directrix
$y^2=4px$	Right	x-axis	$(p,0)$	$x=-p$
$y^2=-4px$	Left	x-axis	$(-p,0)$	$x=p$
$x^2=4py$	Upward	y-axis	$(0,p)$	$y=-p$
$x^2=-4py$	Downward	y-axis	$(0,-p)$	$y=p$

- Figure 10.39 summarizes the cases when the vertex of the parabola is at the point (h,k) and the directrix is parallel to the x-or y-axis.
- The general form of an equation of a conic section with axis or axes parallel to the coordinate axes is

$$Ax^2+Cy^2+Dx+Ey+F=0,$$

  where A and C are not both zero.
- Chart summarizing the graphing possibilities for the above equation

10.5

- Methods to solve nonlinear systems of equations by the graphing method, by the substitution method, and by the addition-elimination method.

CHAPTER 10 REVIEW EXERCISES

1. Find an equation for a circle with center at the origin and radius 1.

2. Classify the graph defined by each of the following equations.
   a. $9x^2 = 4y^2 - 54x - 16y + 61$
   b. $x^2 + 4x + 8y = 4$
   c. $7x - 2y = 3$
   d. $16x^2 + 25y^2 - 32x - 284 = 0$
   e. $x^2 + 6x = 3 - y^2$

3. Determine the center and radius of the circle given by $x^2 + y^2 + 4x - 6y = 12$.

4. Determine the standard equation of the circle with center at (4, −3) that passes through (0, −1).

5. Find the coordinates of the foci of the ellipse given by $25x^2 + 16y^2 = 400$.

6. What is the equation in standard form of the ellipse with vertices at (3, 2) and (−7, 2) and a minor axis of length 6?

7. Find the equations of the asymptotes of the hyperbola given by $\frac{y^2}{64} - \frac{x^2}{100} = 1$.

8. Determine the standard equation of the hyperbola whose center is at the origin with one focus at (−8, 0) and a transverse axis of length 14.

9. What are the coordinates of the focus point of the parabola given by $x^2 = -7y$?

10. Find the standard equation of the parabola with focus at (5, 1) and directix $x = -1$.

11. Find the standard equation of the circle whose diameter extends from (−2, 2) to (4, 2).

12. Show by the methods of analytic geometry that the midpoints of the sides of a rectangle are the vertices of a quadrilateral whose perimeter is equal to the sum of the lengths of the diagonals of the rectangle.

13. Graph $16(y - 2)^2 - 25(x + 3)^2 = 400$.

14. If the line $y = mx - 1$ passes through the center of the circle $(x - 2)^2 + (y + 3)^2 = 25$, find the value of m.

15. What are the coordinates of the vertices of the hyperbola given by $9(x - 3)^2 - 4(y + 1)^2 = 144$?

16. Find in terms of r all intersection points of the circle $x^2 + y^2 = r^2$ and the hyperbola $x^2 - y^2 = 5$. For what values of r do the curves intersect?.

17. The area and perimeter of a rectangle are 341 m^2 and 84 m, respectively. Find the length and width of the rectangle.

18. Find an equation for this elliptical arch.

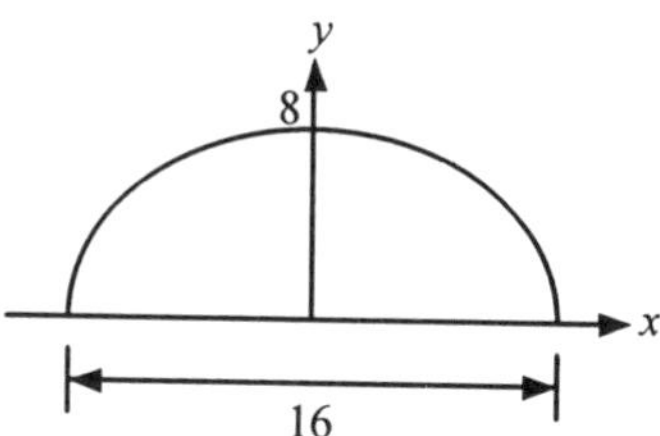

19. What is an equation for this semicircle?

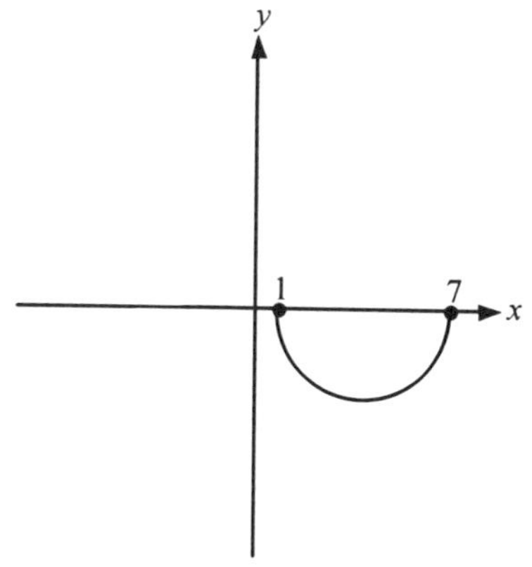

In Exercises 20–25 select the choice that completes the statement or answers the question.

20. The graph of $x^2 = 1 - 2y^2$ is
   a. a circle b. an ellipse
   c. a parabola d. a hyperbola

21. Which conic section is defined as the set of all points in a plane equidistant from a fixed line and a fixed point not on the line?
    a. circle b. ellipse
    c. parabola d. hyperbola

22. If s varies directly as the square of t, then the graph of this relation is a
    a. line b. circle
    c. parabola d. hyperbola

23. Which of the following is the equation of a hyperbola?
    a. $2x^2 + y^2 = 5$ b. $x^2 = 1 - y^2$
    c. $x = y^2 - 4$ d. $x^2 = y^2 - 4$

24. The standard equation of the circle with center at $(1, -3)$ and radius 5 is
    a. $(x+1)^2 + (y-3)^2 = 5$
    b. $(x-1)^2 + (y+3)^2 = 25$
    c. $(x-1)^2 + (y+3)^2 = 5$
    d. $(x+1)^2 + (y-3)^2 = 25$

25. Which illustration shows the graph of $(x-1)^2 + y^2 = 1$?

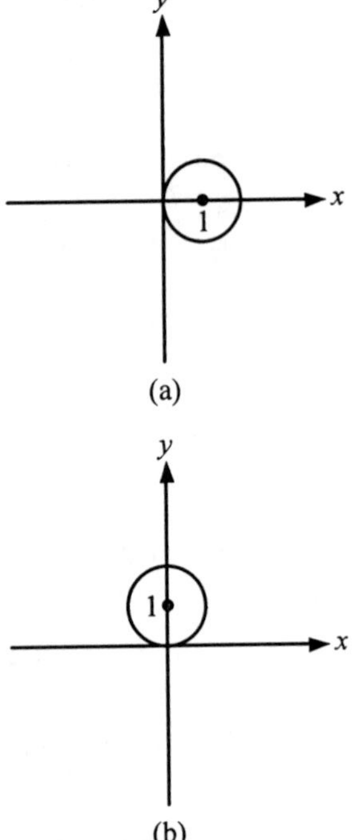

CHAPTER 10 TEST

1. Find an equation for the circle that is the set of all points in a plane that are 5 units from the point $(-4,3)$.

2. Find all intersection points of the parabola $y = x^2 + 5$ and the line $x + y = 7$.

3. Show by the methods of analytic geometry that the diagonals of a square are perpendicular to each other.

4. Find the standard equation of the circle with center $(1,-2)$ that passes through the origin.

5. Find the center and radius of the circle given by $(x+5)^2 + (y-1)^2 = 36$.

6. Find the radius of the circle given by $x^2 + y^2 + 2x - 4y - 27 = 0$.

7. Write the standard equation of the ellipse with center $(-1,1)$, one vertex at (5, 1), and minor axis of length 4.

8. An elliptical arch has a height of 8 ft and a span of 20 ft. How high is the arch 5 ft each side of the center?

9. Graph $\dfrac{(x+3)^2}{25} + \dfrac{y^2}{9} = 1$.

10. Graph $9y^2 - x^2 = 36$.

11. Find the coordinates of the foci of the hyperbola given by $x^2 - 4y^2 = 1$.

12. Find the equation of the asymptotes of the graph of $x^2 - 4y^2 = 1$.

13. Find the vertex of the parabola given by $y^2 + 4y + 5x - 11 = 0$.

14. Write the standard equation for the parabola with vertex at (2, 1) and focus at (2, 4).

15. Graph $(y+1)^2 = 2(x-2)$.

16. Identify the graph of $y^2 = x^2 - 4x + 5$ if the graph is a circle, an ellipse, a hyperbola, or a parabola.

17. Identify the graph of $9x^2 + y^2 - 36x + 2y + 37 = 0$. Degenerate cases are possible here.

18. Find an equation for this hyperbola if the foci are at $(-4,0)$ and $(4, 0)$.

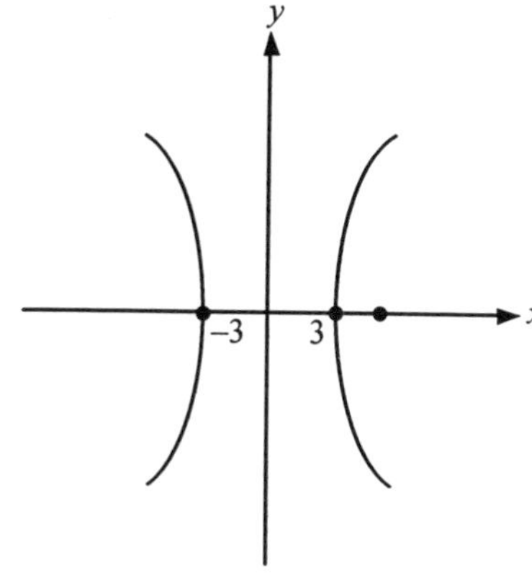

19. Solve the system: $\begin{aligned} x^2 + y^2 &= 16 \\ y^2 - x^2 &= 4 \end{aligned}$

20. For the ellipse shown, find the length of line segment *AB*.

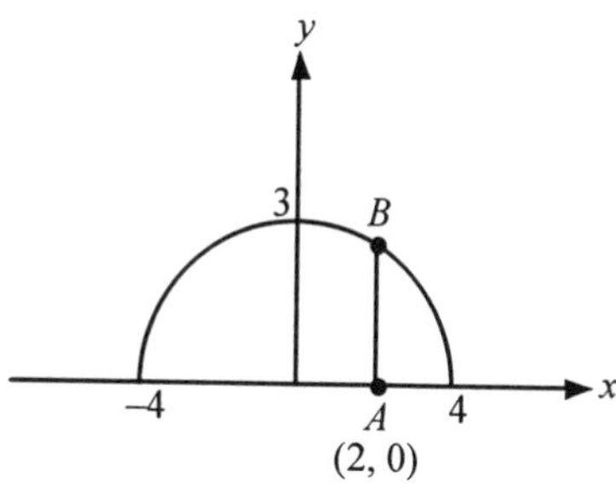

• • •

Chapter 11

Systems of Equations and Inequalities

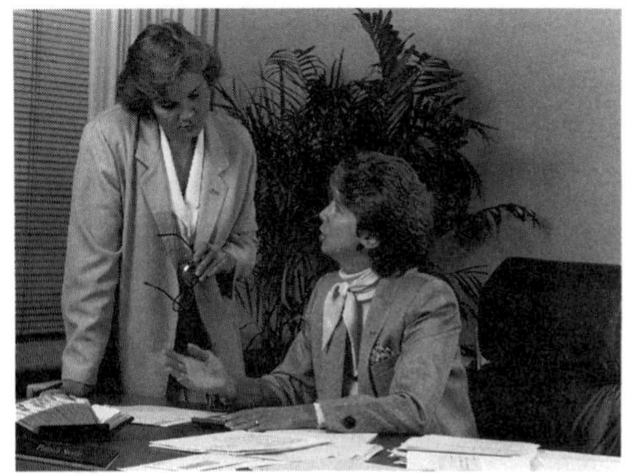

A company is to give a performance bonus to one of its top managers. The income in the manager's division is \$200,000 and the bonus rate is 10 percent. The bonus is based on profit, so it is based on income after both taxes and the bonus have been subtracted from the \$200,000. The taxes amount to 30 percent of the taxable income, with the bonus counting as a tax deduction. What is the manager's bonus and how much is paid in taxes? (See Example 2 of Section 11.1).

Photo Courtesy of Stock Boston, Boston, MA

In the analysis of a problem, we often must take into account many variables and many relationships among the variables. In most cases we deal with **linear equations in *n* variables**, which are equations of the form

$$a_1x_1 + a_2x_2 + \cdots + a_nx_n = c$$

where $a_1, a_2, \ldots, a_n$ and c are real numbers, and $x_1, x_2, \ldots, x_n$ are variables. A set of linear equations is called a **linear system**. In this chapter we first review the addition-elimination method and the substitution method for solving linear systems in two variables, and then we extend these methods to solve linear systems in three variables. Next, we show how to solve linear systems by using Gaussian elimination, Gauss-Jordon elimination, determinants, and matrix algebra. Finally, we discuss systems of linear inequalities. Partial fractions and linear programming are considered as applications of systems of equations and inequalities, respectively.

11.1 Linear Systems in Two Variables and Three Variables

Objectives

1. Solve linear systems in two variables (by the methods of Section 1.5).
2. Solve linear systems in three variables using the addition-elimination method.
3. Some applied problems by setting up and solving a system of linear equations.

Recall from Section 1.5 that the graph of a linear system in two variables consists of two straight lines, and that the solution set for such a system is given geometrically by all the points where the lines intersect. There are three possible cases, and they are illustrated in Figure 11.1.

Case 1 The equations represent two lines that intersect at one point and so have 1 point in common. This system is called **consistent.**
Case 2 The equations represent two distinct lines that are parallel and do not intersect at all, and so have no points in common. This system is called **inconsistent**.
Case 3 Both equations represent the same line and so have all the points of that line in common. This system is called **dependent** (and consistent).

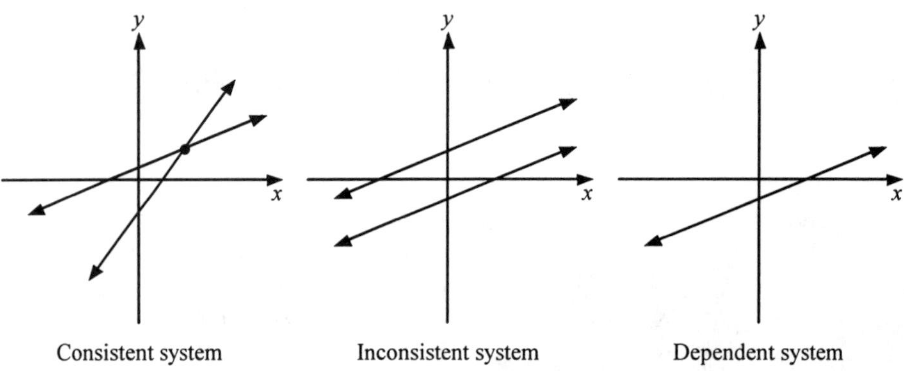

Consistent system Inconsistent system Dependent system

Figure 11.1

The most useful case is usually the one in which the graphs intersect at exactly one point. Example 1 reviews how to find the exact coordinates of such an intersection point by the addition-elimination method.

Example 1: Solving a Linear System in Two Variables by Addition-Elimination

Solve by addition-elimination:

$$2x - 5y = 5 \quad (1)$$
$$4x + 3y = 23. \quad (2)$$

Solution: For this method to result in the elimination of a variable, the coefficients of either x or y must be opposites. For this system, the x variable can be eliminated by multiplying both sides of equation (1) by -2 and then adding the resulting equation to equation (2).

$$\begin{array}{rrl} -2(2x-5y) = -2(5) \rightarrow & -4x + 10y = -10 & (3) \\ & 4x + \;\;3y = \;\;23 & (2) \\ \hline & 13y = \;\;13 & \text{Add the equations} \\ & y = \;\;\;1 & \end{array}$$

To find x, replace y by 1 in equation (1) or equation (2).

$$\begin{array}{rcr} 2x - 5y = 5 & \text{or} & 4x + 3y = 23 \\ 2x - 5(1) = 5 & & 4x + 3(1) = 23 \\ 2x = 10 & & 4x = 20 \\ x = 5 & & x = 5 \end{array}$$

The solution is (5,1). Check this result numerically by replacing x by 5 and y by 1 in both original equations (1) and (2). Figure 11.2 shows a graphical check that confirms this solution.

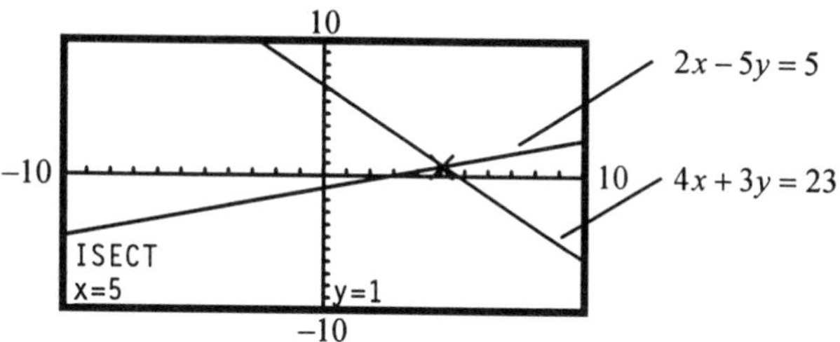

Figure 11.2

PROGRESS CHECK 1

Solve by addition-elimination: $\begin{matrix} 2x+3y=5 \\ 3x+5y=7 \end{matrix}$. ■

A second algebraic method for solving linear systems of equations is called the **substitution method**. This method is most efficient when at least one equation in a linear system is solved for one of the variables. This situation occurs in the chapter-opening problem, where two variables are defined in terms of each other. Thus, they have interlocking solutions that lead naturally to the topic of simultaneous equations.

Example 2: Solving a Linear System in Two variables by Substitution

Solve the problem in the chapter introduction on page 867.

Solution: Let b represent the bonus and t represent the tax. Then write an equation for each in terms of the other.

Tax is 30 percent of income after bonus deduction.

Tax: $\quad t = 0.3 \cdot (200{,}000 - b) \qquad (1)$

Bonus is 10 percent of income after bonus and tax deductions

Bonus: $\quad b = 0.1 \cdot (200{,}000 - b - t) \qquad (2)$

The resulting linear system with parentheses cleared is

$$t = 60{,}000 - 0.3b \qquad (3)$$
$$b = 20{,}000 - 0.1b - 0.1t \qquad (4)$$

Since the equations are solved for a variable, we choose to use the substitution method and replace t by $60{,}000 - 0.3b$ in equation (4).

$$b = 20,000 - 0.1b - 0.1(60,000 - 0.3b)$$
$$b = 20,000 - 0.1b - 6,000 + 0.03b$$
$$1.07b = 14,000$$
$$b = \frac{14,000}{1.07} \approx 13,084$$

Then $t = 0.3(200,000 - b)$, so

$$t = 0.3(200,000 - 13,084) \approx 56,075$$

Thus, to the nearest dollar, the bonus is \$13,084, and the tax is \$56,075. To check this result observe that

$$\text{tax} = 0.3(200,000 - 13,084) \approx 56,075, \text{ and}$$
$$\text{bonus} = 0.1(200,000 - 13,084 - 56,075) \approx 13,084,$$

so the solution checks.

PROGRESS CHECK 2

Redo the problem in Example 2, but assume that the bonus rate is 5 percent, the tax rate is 40 percent, and the income in the manager's division is \$500,000. ■

It is useful to extend our current methods beyond linear systems in two variables, because problems often involve many variables and many relationships among the variables. When linear (first-degree) equations in three variables are required, then we need to deal with equations of the form

$$ax + by + cz = d,$$

where a, b, c, and d are real numbers, and x, y, and z are variables. A solution of such an equation is an **ordered triple** of numbers of the form (x,y,z) that satisfies the equation. Furthermore, the solution set of a system of three linear equations in three variables consists of all ordered triples that satisfy all the equations at the same time. The next example discusses how the addition-elimination method may be extended to solve a linear system in three variables.

Example 3: Solving a Linear System in Three Variables by Addition-Elimination

Solve the system.

$$\begin{aligned} 3x + 2y + z &= -3 \quad (1) \\ 2x + 3y + 2z &= 5 \quad (2) \\ -2x + y - z &= 3 \quad (3) \end{aligned}$$

Solution: The initial goal is to obtain two equations in two variables that may be solved by methods already established. To obtain a first equation, select any pair of equations in the system and use the addition method to eliminate one of the variables. For the given system, z is eliminated simply by adding equations (1) and (3).

$$\begin{array}{rcrcrcr} 3x & + & 2y & + & z & = & -3 \\ -2x & + & y & - & z & = & 3 \\ \hline x & + & 3y & & & = & 0 \quad (4) \end{array}$$

To obtain a second equation, select a different pair of equations in this system and eliminate the *same* variable, z. We choose equations (2) and (3) and eliminate z by multiplying both sides of equation (3) by 2 and adding the result to equation (2)

$$\begin{array}{rrcrcrcrl} & 2x & + & 3y & + & 2z & = & 5 & \\ 2(-2x+y-z)=2(3)\rightarrow & -4x & + & 2y & - & 2z & = & 6 & \\ \hline & -2x & + & 5y & & & = & 11 & (5) \end{array}$$

Equations (4) and (5) can now be used to obtain a linear system in two variables (our initial goal).

$$\begin{array}{rcrcr} x & + & 3y & = & 0 \\ -2x & + & 5y & = & 11 \end{array}$$

To solve this system, we choose to eliminate x and solve for y, as shown next.

$$\begin{array}{rcrrcrcr} 2(x+3y) & = & 2(0)\rightarrow & 2x & + & 6y & = & 0 \\ & & - & 2x & + & 5y & = & 11 \\ \hline & & & & & 11y & = & 11 \\ & & & & & y & = & 1 \end{array}$$

Then substituting 1 for y in equation (4) gives

$$\begin{aligned} x+3(1) &= 0 \\ x &= -3 \end{aligned}$$

Finally, by replacing x by -3 and y by 1 in equation (1), we have

$$\begin{array}{rcrcrcr} 3(-3) & + & 2(1) & + & z & = & -3 \\ & & & & z & = & 4. \end{array}$$

Thus, the solution is $(-3,1,4)$. Check it in the original equations (1), (2), and (3).

Technology Link

Some graphing calculators have a built-in feature for solving linear systems (or simultaneous equations), and you should learn to use this feature if it is available. When this special feature is not available, then most graphing calculators have the ability to solve linear systems by using matrices, and these approaches are discussed in Sections 11.2 and 11.4

PROGRESS CHECK 3

Solve the system

$$\begin{array}{rcrcrcr} 2x & - & y & + & z & = & 7 \\ -x & + & 2y & - & z & = & 6 \\ 2x & - & 3y & - & 2z & = & 9 \end{array}$$

■

The methods of Example 3 illustrate a general procedure for solving a linear system in three variables, which is summarized next.

To Solve a Linear System in Three Variables

1. Select any pair of equations in the system and use the addition method to eliminate one of the variables.
2. Choose a different pair of equations in the system and eliminate the *same* variable by using the addition method again.
3. Use the results of steps **1** and **2** to obtain a linear system in two variables, and solve this system.
4. Substitute the values of the two variables obtained in step **3** into one of the original equations to find the value of the third variable.
5. Check the solution in all three of the original equations.

The next example shows how a linear system in three variables may be used to determine the equation of a parabola that passes through three points (not all in a straight line).

Example 4: Fitting an Equation to a Parabola

The points (1,2), (2,7), and (−1,4) lie on the parabola given by $y = ax^2 + bx + c$. Find *a*, *b*, and *c*, and write an equation for the parabola. Check your result with a grapher.

Solution: Substituting the values $x = 1$ and $y = 2$, $x = 2$, and $y = 7$, and $x = -1$ and $y = 4$, respectively, into the equation $y = ax^2 + bx + c$ gives the following system of equations.

$$\begin{aligned} 2 &= a(1)^2 + b(1) + c & \text{or} \quad a + b + c &= 2 \quad (1) \\ 7 &= a(2)^2 + b(2) + c & 4a + 2b + c &= 7 \quad (2) \\ 4 &= a(-1)^2 + b(-1) + c & a - b + c &= 4 \quad (3) \end{aligned}$$

To solve this system we follow the steps of the general procedure and choose to eliminate *b*.

Step 1 Eliminate *b* using equations (1) and (3)

$$\begin{array}{rcl} a + b + c &=& 2 \quad (1) \\ a - b + c &=& 4 \quad (3) \\ \hline 2a \qquad + 2c &=& 6 \\ a + c &=& 3 \quad (4) \end{array}$$

Step 2 Eliminate *b* again using a different pair of equations, say (2) and (3).

$$\begin{array}{rrcl} & 4a + 2b + c &=& 7 \\ 2(a - b + c) = 2(4) \rightarrow & 2a - 2b + 2c &=& 8 \\ \hline & 6a \qquad + 3c &=& 15 \\ & 2a + c &=& 5 \end{array}$$

Step 3 Solve the linear system in two variables resulting from steps 1 and 2.

$$\begin{array}{rrcl} -1(a + c) = -1(3) \rightarrow & -a - c &=& -3 \\ & 2a + c &=& 5 \\ \hline & a &=& 2 \end{array}$$

Then substituting 2 for *a* in equation (4) yields $c = 1$.

Step 4 Replacing a by 2 and c by 1 in equation (1) gives

$$\begin{aligned} 2 + b + 1 &= 2 \\ b &= -1. \end{aligned}$$

Step 5 The solution to the linear system is $a = 2$, $b = -1$, and $c = 1$. Check it in the original equations (1), (2), and (3).

The solution to the linear system reveals that the equation of the parabola passing through the given points is $y = 2x^2 - x + 1$. Check this result numerically by substituting the coordinates of the three given points in this equation. Figure 11.3 shows a graphical check that confirms this solution.

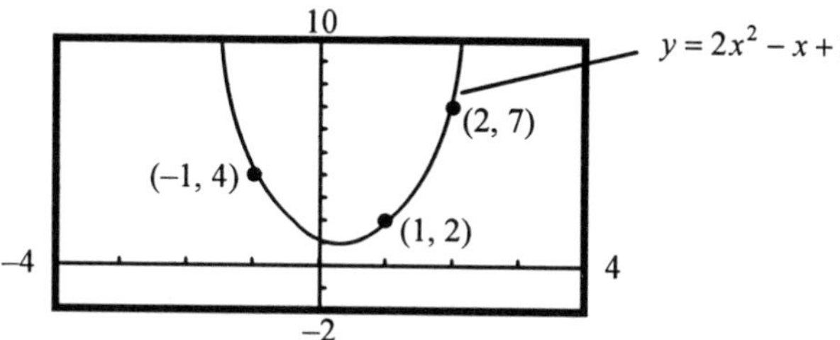

Figure 11.3

PROGRESS CHECK 4
The points $(1,-2)$, $(2,-3)$, and $(3,-6)$ lie on the parabola given by $y = ax^2 + bx + c$. Find a, b, and c and write an equation for the parabola. Check your result with a grapher. ■

When the application of our current methods for solving a linear system in three variables results in a false equation *at any step*, then the system is inconsistent and has no solution. Example 5 illustrates this case.

Example 5: An Inconsistent System

Solve the system.

$$\begin{aligned} 2x + y - z &= 2 \quad (1) \\ x + 2y + z &= 5 \quad (2) \\ x - y - 2z &= -2 \quad (3) \end{aligned}$$

Solution: We follow the steps of the general procedure and choose to eliminate z.

Step 1 Eliminate z using equations (1) and (2).

$$\begin{array}{rcrcrcl} 2x & + & y & - & z & = & 2 \quad (1) \\ x & + & 2y & + & z & = & 5 \quad (2) \\ \hline 3x & + & 3y & & & = & 7 \quad (4) \end{array}$$

Step 2 Eliminate z again using a different pair of equations, say (2) and (3).

$$\begin{array}{rcl} 2(x + 2y + z) = 2(5) \rightarrow & 2x + 4y + 2z = & 10 \\ & x - y - 2z = & -2 \\ \hline & 3x + 3y = & 8 \quad (5) \end{array}$$

Step 3 Solve the linear system in two variables resulting from steps 1 and 2.

$$\begin{array}{rrcrcrcr} & 3x & + & 3y & = & & 7 \\ -1(3x+3y)=-1(8)\rightarrow & -\ 3x & - & 3y & = & - & 8 \\ \hline & & & 0 & = & - & 1 \end{array}$$

The false equation $0=-1$ indicates that the system is inconsistent and has no solution. The solution set for every inconsistent system is Ø.

Note: At any step a false equation implies an inconsistent system. Therefore, if step 1 results in a false equation, then conclude without further work that the system is inconsistent.

PROGRESS CHECK 5

Solve the system.

$$\begin{aligned} x-y-z &= 5 \\ 4x+y+3z &= 8 \\ -2x+2y+2z &= 7 \end{aligned}$$

■

A linear system in three variables may also be a dependent system, and the next example illustrates this case.

Example 6: A Dependent System Solve the system.

$$\begin{aligned} x-y+2z &= 0 \quad (1) \\ -x+4y+z &= 0 \quad (2) \\ -x+2y-z &= 0 \quad (3) \end{aligned}$$

Solution: We follow steps of the general procedure and choose to eliminate x.

Step 1 Eliminate x using equations (1) and (2).

$$\begin{array}{rcrcrcrl} x & - & y & + & 2z & = & 0 & (1) \\ -x & + & 4y & + & z & = & 0 & (2) \\ \hline & & 3y & + & 3z & = & 0 & (4) \end{array}$$

Step 2 Eliminate x again using a different pair of equations.

$$\begin{array}{rcrcrcrl} x & - & y & + & 2z & = & 0 & (1) \\ -x & + & 2y & - & z & = & 0 & (3) \\ \hline & & y & + & z & = & 0 & (5) \end{array}$$

Step 3 Solve the system consisting of equations (4) and (5).

$$\begin{array}{rrcrcr} & 3y & + & 3z & = & 0 \\ -3(y+z)=-3(6)\rightarrow & -3y & - & 3z & = & 0 \\ \hline & & & 0 & = & 0 \end{array}$$

The identity $0=0$ indicates that the system is dependent and that the number of solutions is infinite. In higher mathematics, specifying the solution set for the system in this example is considered.

Note: It is also common to conclude that a system is dependent if *both* step 1 and step 2 in our current methods produce identities. In this case all the equations are equivalent, and the solution set is the set of all ordered triples satisfying any equation in the system. However, if one step produces an identity and the other step produces a false equation, then the system is inconsistent. Remember that a false equation *at any step* implies an inconsistent system.

PROGRESS CHECK 6
Solve the system.

$$\begin{aligned} 3x + 2y - z &= 3 \\ x - y + z &= 1 \\ 5x + z &= 5 \end{aligned}$$

■

As with linear equations in two variables, it is possible to interpret geometrically a linear system in three variables. However, an equation that may be written in the form $ax + by + cz = d$ graphs as a plane in a three-dimensional space. Consider carefully Figure 11.4, which summarizes the geometric solution: either exactly one point, no point, infinitely many points on a line, or infinitely many points in a plane.

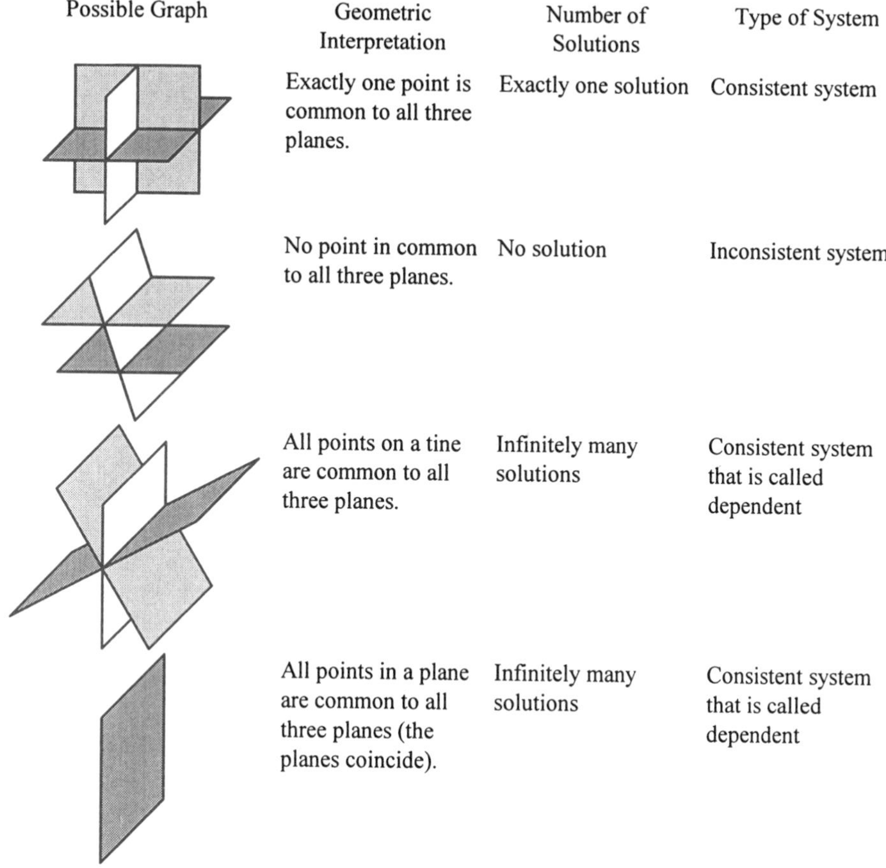

Possible Graph	Geometric Interpretation	Number of Solutions	Type of System
	Exactly one point is common to all three planes.	Exactly one solution	Consistent system
	No point in common to all three planes.	No solution	Inconsistent system
	All points on a tine are common to all three planes.	Infinitely many solutions	Consistent system that is called dependent
	All points in a plane are common to all three planes (the planes coincide).	Infinitely many solutions	Consistent system that is called dependent

Figure 11.4

EXPLORE 11.1

1. Very slight changes in the coefficients in a system of equations can make a huge difference in the solution. Solve these 2 systems and compare the solutions.

System 1

$$\begin{aligned} 1.01x + 1.02y + 1.03z &= 1 \\ 1.02x + 1.03y + 1.04z &= 1 \\ 1.01x + y + z &= 1 \end{aligned}$$

System 2

$$\begin{aligned} 1.01x + 1.01y + 1.03z &= 1 \\ 1.02x + 1.03y + 1.04z &= 1 \\ 1.01x + y + z &= 1 \end{aligned}$$

2. Some calculators have a feature for solving a linear system of equations automatically. The algorithm used by the calculator involves much arithmetic, which makes the calculations subject to possible rounding error.
   - **a.** Check by substitution that the exact solution of the first system in the previous Explore exercise is $x = 100$, $y = -300$, $z = 200$. Then use the system-solving feature of a calculator to see if you get this exact solution.
   - **b.** If you alter the original coefficients by inserting 4 zeros after each decimal point (for example, change 1.01 to 1.00001), you will make the round-off problem even worse. The correct solution is now (1 million, −3 million, 2 million). Compare that to the solution your calculator produces.

EXERCISES 11.1

In Exercises 1–22 solve the given system.

1. $\begin{aligned} 4x + 3y &= 7 \\ x - 3y &= 3 \end{aligned}$

2. $\begin{aligned} x - 2y &= 1 \\ x + 2y &= 3 \end{aligned}$

3. $\begin{aligned} 2x + 3y &= 5 \\ 3x + 2y &= 5 \end{aligned}$

4. $\begin{aligned} 3x - 4y &= 1 \\ 4x - 3y &= 1 \end{aligned}$

5. $\begin{aligned} 2x + 5y &= 7 \\ 4x + 10y &= 6 \end{aligned}$

6. $\begin{aligned} -6x + 3y &= 2 \\ 2x - y &= 1 \end{aligned}$

7. $\begin{aligned} 4x - 6y &= 10 \\ 2x - 3y &= 5 \end{aligned}$

8. $\begin{aligned} 3x - 4y &= 1 \\ 9x - 12y &= 3 \end{aligned}$

9. $\begin{aligned} x + y + z &= 10 \\ x + 2y - z &= 6 \\ -x - 3y + 4z &= 0 \end{aligned}$

10. $\begin{aligned} x + y + z &= 4 \\ 2x - y + 2z &= -1 \\ -3x - y - z &= 0 \end{aligned}$

11. $\begin{aligned} 3x + 2y + z &= 10 \\ 2x - y - 2z &= -6 \\ 5x - 3y + 3z &= 8 \end{aligned}$

12. $\begin{aligned} -x + y + z &= 0 \\ 2x + 3y + 4z &= -8 \\ -4x - 2y + 3z &= -9 \end{aligned}$

13. $\begin{aligned} x + y + z &= 3 \\ 2x + 2y + z &= 4 \\ -x + 3y - z &= 5 \end{aligned}$

14. $\begin{aligned} 2x - y + 4z &= 1 \\ 4x + 2y - 5z &= 2 \\ -2x - 4y + z &= -1 \end{aligned}$

15. $\begin{aligned} 3x + 2y + z &= 10 \\ 2x - y - 2z &= -6 \\ x + 3y + 3z &= 16 \end{aligned}$

16. $$\begin{aligned} x - y + 2z &= 2 \\ 3x - y + 4z &= 0 \\ 2x - y + 3z &= 1 \end{aligned}$$

17. $$\begin{aligned} 3x + 2y + z &= 6 \\ 3x + y - z &= 6 \\ 2x - y - 4z &= 4 \end{aligned}$$

18. $$\begin{aligned} 2x - 2y + 4z &= 0 \\ 3x - 3y + 6z &= 0 \\ -x + y + 3z &= 1 \end{aligned}$$

19. $$\begin{aligned} 2x + y - z &= 3 \\ -x + y + z &= 1 \\ -3x + 2z &= -1 \end{aligned}$$

20. $$\begin{aligned} -2x + 4y - z &= 1 \\ x - 2y + z &= 3 \\ x - 2y + 2z &= 2 \end{aligned}$$

21. $$\begin{aligned} x + 2y - 3z &= 5 \\ -x + y - z &= 2 \\ 3x + 9y - 13z &= 8 \end{aligned}$$

22. $$\begin{aligned} 2x + y - z &= 0 \\ 3x - y + 6z &= 5 \\ -x - 3y + 8z &= 5 \end{aligned}$$

23. A shopper was offered a 20 percent discount on hardback books and a 10 percent discount on paperbacks. The regular price for the purchase was \$254.70, on which the total discount was \$40.44. How much did the shopper spend on each kind of book?

24. A ski shop discounts red-tag items by 30 percent and yellow-tag items by 20 percent. A shopper who bought only red- and yellow-tagged items got a total discount of \$120 on a purchase, bringing the bill down to \$305. After the discount, how much did the shopper spend in each category?

25. One year a couple had total income equal to \$90,000, and their total tax bill came to \$28,000. If the income from their salaries was taxed at 32 percent, and the income from their capital gains was taxed at 28 percent, how much income did they have in each category?

26. A student borrowed a total of \$6,200 one year to pay for tuition and other expenses. The tuition loan had an annual interest rate of 8 percent, and the other loan had an annual interest rate of 10 percent. The combined interest payments for the year came to \$520. Find the amount of each loan.

27. A company gives a bonus to a division manager according to this principle: The bonus rate is 15 percent of the division's after-tax and after-bonus income. The tax, meanwhile, is 28 percent of the division's income after the bonus is paid. If the division's income is \$400,000, to the nearest dollar what is the manager's bonus?

28. A corporation operates in a state that levies a 4 percent tax on the income that remains after paying the federal tax. Meanwhile, the federal tax is 30 percent of the income that remains after paying the state tax. If, during the current year, the corporation has \$1 million in taxable income, determine (to the nearest dollar) the state and federal income tax.

29. A child runs at her top speed for 0.1 mi the "wrong" way along a moving sidewalk in an airport. This takes 3 minutes. When she runs (also at top speed) the "right" way, it takes 1 minute. How fast is the sidewalk moving?

30. Redo Exercise 29 but with the speed of the sidewalk changed. It now takes the child 1.2 minutes running the right way and 2 minutes the wrong way. What is the speed of the moving sidewalk?

31. The points (0,8), (1,5), and (2,4) lie on the parabola given by $y = ax^2 + bx + c$. Find *a*, *b*, and *c*, and write an equation for the parabola. Check your result with a grapher.

32. The points $(-5,-2)$, $(-2,1)$, and $(-1,-2)$ lie on the parabola given by $y = ax^2 + bx + c$. Find *a*, *b*, and *c*, and write an equation for the parabola. Check your result with a grapher.

33. Applying Kirchoff's law to the electric circuit shown leads to the following system

$$\begin{aligned} I_1 + I_2 - I_3 &= 0 \\ 30I_2 + 10I_3 &= 4 \\ 25I_1 + 10I_3 &= 6 \end{aligned}$$

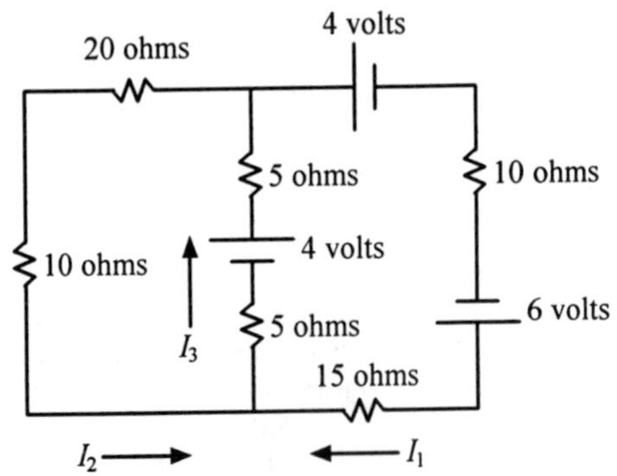

Find the values of currents I_1, I_2, and I_3, in amperes.

34. In the circuit of Exercise 33, if all the resistances are doubled, the resulting system is

$$\begin{aligned} I_1 + I_2 - I_3 &= 0 \\ 60I_2 + 20I_3 &= 4 \\ 50I_1 + 20I_3 &= 6 \end{aligned}$$

Solve the system for the three new currents. How do they compare with the answers in Exercise 33?

35. The following problem is typical of those from algebra textbooks of 100 years ago. (It is from William J. Milne's *High School Algebra*, published by the American Book Company in 1892.) Divide 125 into four parts such that, if the first be increased by 4, the second diminished by 4, the third multiplied by 4, and the fourth divided by 4, all these results will be equal. (**Hint:** Call the 4 parts a, b, c, and $125-a-b-c$. The first equation is then $a+4=b-4$.)

36. Divide 180 into four parts such that if the first be increased by 5, the second diminished by 5, the third multiplied by 5, and the fourth divided by 5, all these results will be equal.

37. The graph of $y=a2^x+b2^{-x}+c$ is shaped something like the graph of $y=x^2$ if a and b have the same sign. What values of a, b, and c will cause the graph to go through the points $(0,-7)$, $\left(4,\frac{1}{32}\right)$, and $\left(-4,\frac{1}{32}\right)$? Check your result with a grapher.

38. The graph of $y=a2^x+b2^{-x}+c$ is shaped something like the graph of $y=x^3$ if a and b have opposite signs. What values of a, b, and c will cause the graph to go through the points $(0,0)$, $(1,3)$, and $(-1,-3)$? Check your result with a grapher.

THINK ABOUT IT 11.1

1. Extend the methods of this section to solve this system of four linear equations in four variables. (**Hint:** First eliminate one variable to get a new system of three linear equations in three variables.)

$$\begin{aligned} w - x + y + z &= -4 \\ 2w - 3x + y - z &= -3 \\ w + 2x - 2y + 3z &= 1 \\ 3w + x + 2y - 3z &= 9 \end{aligned}$$

2. Clay tablets from more than 3,000 years ago show that the mathematicians of ancient Babylonia also solved systems of linear equations, but their approach was different from ours. One of their problems leads to this system.

$$\begin{aligned} \frac{2}{3}x - \frac{1}{2}y &= 500 \quad (1) \\ x + y &= 1{,}800 \quad (2) \end{aligned}$$

The basic idea of their solution is to start with a wrong answer and fix it, so they start with the guess that x and y are equal. From equation 2, this gives $x=y=900$ as a starting point.

a. Letting x and y each equal 900 in the left-hand side of equation (1) gives 150, which is 350 short of the desired 500. Check this.

b. What change in x and y will make up this shortage? Since (by equation 2) every increase of 1 in x must be accompanied by a decrease of 1 in y, then find that when x increases by 1, the left-hand side of equation (1) increases by $7/6$. Check that $(2/3)(1)-(1/2)(-1)=7/6$.

c. To find out how many unit increases (u) will make up the shortage of 350, they solve $(7/6)u = 350$, which gives $u = 300$. Therefore, $x = 900 + 300 = 1{,}200$, and $y = 900 - 300 = 600$. Use the Babylonian method to solve

$$\begin{aligned} \frac{1}{2}x - \frac{1}{4}y &= 350 \\ x + y &= 1{,}000. \end{aligned}$$

(This problem and others like it are described in *A History of Mathematics* by Victor J. Katz (HarperCollins, 1993).]

3. Recall that a system of three linear equations in three variables is inconsistent if there is no point that is simultaneously in all three planes (as shown in Figure 11.4). Draw and describe two other possible graphs of inconsistent systems.
4. Solve for x, y, and z in terms of a, b, and c.

$$\begin{aligned} x + z &= a \\ x + y &= b \\ y + z &= c \end{aligned}$$

5. In Exercises 21 and 22 the systems are dependent and therefore have infinitely many solutions. For each Exercise find three of the solutions.

• • •

11.2 Triangular Form and Matrices

Photo Courtesy of Len Rubenstein, The Picture Cube

A financial planner wants to split a \$400,000 investment into three accounts with annual yields of 6 percent, 9 percent, and 11 percent, respectively, in order to obtain an annual income of \$31,600. Because the risk of losing investment capital increases as the yield increases, the planner wants to invest twice as much money at 6 percent as at 11 percent. How much should be invested in each account? (See Example 2.)

Objectives

1. Solve a linear system by transforming the system to triangular form.
2. Solve a linear system using row operations on matrices.

Systems of three linear equations with three variables, and more complicated linear systems, can always be solved by using a technique known as **Gaussian elimination**. This method is basically the addition-elimination method from Section 11.1, and its systematic nature can be programmed to allow effective computer solutions. To understand this method, first consider the following system of equations that is said to be in **triangular form**.

$$\begin{aligned} 4x + 2y - z &= 2 \\ 3y - 3z &= 6 \\ 5z &= 10 \end{aligned}$$

This system is easy to solve. The third equation, $5z = 10$, tells us that $z = 2$, and then back substitution yields

$$\begin{aligned} 3y - 3(2) &= 6 \\ 3y &= 12 \\ y &= 4 \end{aligned} \qquad \text{and} \qquad \begin{aligned} 4x + 2(4) - (2) &= 2 \\ 4x &= -4 \\ x &= -1. \end{aligned}$$

Thus, the solution is $x = -1$, $y = 4$, and $z = 2$. However, systems of equations rarely start out in triangular form, so we need some procedure for obtaining this form for any given system. The three operations that follow are called the **elementary operations**, and they are used in Gaussian elimination to produce **equivalent systems** (ones with the same solution) until we reach triangular form.

Operations That Produce Equivalent Systems

1. Multiply both sides of an equation by a nonzero number.
2. Add a multiple of one equation to another equation.
3. Interchange the order in which two equations of a system are listed.

Note that we have already used the first two operations in Section 11.1, while the third operation clearly affects only the form of the system, and not the solution. Now consider carefully Example 1, which shows how we can use these operations to change a system into triangular form.

Example 1: Using Elementary Operations on Equations

Solve the system of equations

$$\begin{aligned} 3x - y + 6z &= 1 \\ x + 2y - 3z &= 0 \\ 2x - 3y - z &= -9. \end{aligned}$$

Solution: We want x to appear only in the first equation. It is easier to eliminate x from the other equations when the coefficient of x in the first equation is 1. So, first, we change the order of the equations to

$$\begin{aligned} x + 2y - 3z &= 0 \\ 3x - y + 6z &= 1 \\ 2x - 3y - z &= -9. \end{aligned}$$

We now add -3 times the first equation to the second equation, and we also add -2 times the first equation to the third equation. The result is

$$\begin{aligned} x + 2y - 3z &= 0 \\ -7y + 15z &= 1 \\ -7y + 5z &= -9. \end{aligned}$$

Finally, to eliminate y in the third equation, we add -1 times the second equation to the third equation to obtain

$$\begin{aligned} x + 2y - 3z &= 0 \\ -7y + 15z &= 1 \\ -10z &= -10 \end{aligned}$$

The system is now in triangular form. From the third equation we know that $z=1$, while back substitution gives

$$\begin{aligned} -7y+15(1)&=1 \\ -7y&=-14 \\ y&=2 \end{aligned} \quad \text{and} \quad \begin{aligned} x+2(2)-3(1)&=0 \\ x&=-1. \end{aligned}$$

Thus, the solution is $x=-1$, $y=2$, and $z=1$.

PROGRESS CHECK 1
Solve the system of equations.

$$\begin{aligned} 2x - 3y - 2z &= 9 \\ -x + 2y + z &= -2 \\ x - y + z &= -1. \end{aligned}$$

■

Example 2: Allocating Funds in an Investment

Solve the problem in the section introduction on page 879.

Solution: To find the amount invested at each rate, let

x = amount invested at 6 percent
y = amount invested at 9 percent
and z = amount invested at 11 percent.

Then the three invested amounts sum to \$400,000 so

$$x+y+z=400{,}000. \qquad (1)$$

Since twice as much money is invested at 6 percent as at 11 percent, a second equation is $x=2z$, or

$$x-2z=0. \qquad (2)$$

The sum of the returns from the three investments needs to produce an annual income of \$31,600, so a third equation is

$$0.06x+0.09y+0.11z=31{,}600 \quad (3)$$

To solve the system formed from these three equations, we now add -1 times equation (1) to equation (2), and we also add -0.06 times equation (1) to equation (3). The result is

$$\begin{aligned} x+y+z &= 400{,}000 \\ -y-3z &= -400{,}000 \quad (4) \\ 0.03y+0.05z &= 7{,}600 \quad (5) \end{aligned}$$

Finally, to eliminate y in equation (5), we add 0.03 times equation (4) to equation (5), to obtain

$$\begin{aligned} x+y+z &= 400{,}000 \\ -y-3z &= -400{,}000 \\ -0.04z &= -4{,}400. \end{aligned}$$

The system is now in triangular form. From the third equation we know that $z = \$110,000$, while back substitution yields $y = \$70,000$ and $x = \$220,000$. You should check this solution in the context of the original problem.

PROGRESS CHECK 2

Redo the problem in Example 2, assuming that the annual yields of the three accounts are revised to 5 percent, 8 percent, and 12 percent, respectively. ■

We can improve the procedure in our current method by keeping track of only the constants in the equations and not writing down the variables. The standard notation for such an abbreviation utilizes matrices. A **matrix** is a rectangular array of numbers that is enclosed in brackets (or parentheses) and commonly denoted by a capital letter such as A or B. Each number in the matrix is called an **entry** or **element** of the matrix. There are two matrices associated with the system

$$\begin{aligned} a_1x &+ b_1y + c_1z = d_1 \\ a_2x &+ b_2y + c_2z = d_2 \\ a_3x &+ b_3y + c_3z = d_3. \end{aligned}$$

The **coefficient matrix** consists of the coefficients of x, y, and z and is written as

$$\begin{bmatrix} a_1 & b_1 & c_1 \\ a_2 & b_2 & c_2 \\ a_3 & b_3 & c_3 \end{bmatrix},$$

while the **augmented matrix** that follows includes these coefficients and an additional column (usually separated by a dashed line) that contains the constants on the right side of the equals sign.

$$\left[\begin{array}{ccc:c} a_1 & b_1 & c_1 & d_1 \\ a_2 & b_2 & c_2 & d_2 \\ a_3 & b_3 & c_3 & d_3 \end{array}\right]$$

We now restate in the language of matrices the operations that produce equivalent systems. These operations are called **elementary row operations**, and they can be used to obtain a matrix solution to a system of equations. By analogy to *equivalent* systems of equations, two matrices that can be derived from each other by using one or more of these elementary row operations are called **equivalent matrices**.

Corresponding Operations for Solving a Linear System

Elementary Operations on Equations	Elementary Row Operations on Matrices
1. Multiply both sides of an equation by a nonzero number.	1. Multiply each entry in a row by a nonzero number.
2. Add a multiple of one equation to another.	2. Add a multiple of the entries in one row to another row.
3. Interchange two equations.	3. Interchange two rows.

We display both of these methods in the next example to reinforce the similarities in the methods.

Example 3: Contrasting Equation Form with Matrix Form

Solve the system

$$\begin{aligned} 4x - 2y + z &= 11 \\ x - y + 3z &= 6 \\ x + y + z &= 2 \end{aligned}$$

and show both the matrix form of the system and the corresponding equations.

Solution: We use the elementary operations above and proceed as follows:

Equation	Matrix Form
$\begin{aligned} 4x - 2y + z &= 11 \\ x - y + 3z &= 6 \\ x + y + z &= 2 \end{aligned}$	$\left[\begin{array}{ccc:c} 4 & -2 & 1 & 11 \\ 1 & -1 & 3 & 6 \\ 1 & 1 & 1 & 2 \end{array}\right]$
↓ Interchange equations 1 and 3.	↓ Interchange rows 1 and 3.
$\begin{aligned} x + y + z &= 2 \\ x - y + 3z &= 6 \\ 4x - 2y + z &= 11 \end{aligned}$	$\left[\begin{array}{ccc:c} 1 & 1 & 1 & 2 \\ 1 & -1 & 3 & 6 \\ 4 & -2 & 1 & 11 \end{array}\right]$
↓ Add −1 times the first equation to the second equation	↓ Add −1 times each entry in row 1 to the corresponding entry in row 2.
$\begin{aligned} x + y + z &= 2 \\ -2y + 2z &= 4 \\ 4x - 2y + z &= 11 \end{aligned}$	$\left[\begin{array}{ccc:c} 1 & 1 & 1 & 2 \\ 0 & -2 & 2 & 4 \\ 4 & -2 & 1 & 11 \end{array}\right]$
↓ Add −4 times the first equation to the third equation	↓ Add −4 times each entry in row 1 to the corresponding entry in row 3.
$\begin{aligned} x + y + z &= 2 \\ -2y + 2z &= 4 \\ -6y + 3z &= 3 \end{aligned}$	$\left[\begin{array}{ccc:c} 1 & 1 & 1 & 2 \\ 0 & -2 & 2 & 4 \\ 0 & -6 & -3 & 3 \end{array}\right]$
↓ Add −3 times the second equation to the third equation.	↓ Add −3 each entry in row 2 to the corresponding entry in row 3.
$\begin{aligned} x + y + z &= 2 \\ -2y + 2z &= 4 \\ -9z &= -9 \end{aligned}$	$\left[\begin{array}{ccc:c} 1 & 1 & 1 & 2 \\ 0 & -2 & 2 & 4 \\ 0 & 0 & -9 & -9 \end{array}\right]$

The last row or last equation tells us that $-9z = -9$, so $z = 1$. Then

$$\begin{aligned} -2y + 2(1) &= 4 \\ y &= -1 \end{aligned} \quad \text{and} \quad \begin{aligned} x + (-1) + 1 &= 2 \\ x &= 2. \end{aligned}$$

Thus, the solution is $x = 2$, $y = -1$, and $z = 1$.

PROGRESS CHECK 3
Solve the system

$$\begin{aligned} 2x - 3y + z &= 11 \\ x + 3y + 2z &= 4 \\ 3x - y - 3z &= 4 \end{aligned}$$

and show both the matrix form of the system and the corresponding equations. ■

Example 4: Using Elementary Row Operations on Matrices

Use matrix form to solve the system

$$\begin{aligned} 6x + 10y &= 7 \\ 15x - 4y &= 3. \end{aligned}$$

Solution: The augmented matrix for the system is

$$\left[\begin{array}{cc|c} 6 & 10 & 7 \\ 15 & -4 & 3 \end{array}\right].$$

It is easier to obtain 0 for the second entry in column 1 when the first entry in column 1 is 1, so first multiply each entry in row 1 by $1/6$.

$$\left[\begin{array}{cc|c} 1 & \frac{5}{3} & \frac{7}{6} \\ 15 & -4 & 3 \end{array}\right]$$

Now add -15 times row 1 to row 2.

$$\left[\begin{array}{cc|c} 1 & \frac{5}{3} & \frac{7}{6} \\ 0 & -29 & -\frac{29}{2} \end{array}\right]$$

The last row tells us that $-29y = -(29/2)$, so $y = 1/2$. Then

$$\begin{aligned} x + \frac{5}{3}\left(\frac{1}{2}\right) &= \frac{7}{6} \\ x &= \frac{7}{6} - \frac{5}{6} \\ &= \frac{1}{3} \end{aligned}.$$

Thus, the solution is $\left(\frac{1}{3}, \frac{1}{2}\right)$.

PROGRESS CHECK 4
Use matrix form to solve the system

$$\begin{aligned} 2x - 3y &= 4 \\ 5x + 2y &= 1. \end{aligned}$$

■

For the remaining examples we will use the following convenient abbreviations for the elementary row operations.

Elementary row operation	**Abbreviation**
1. Replace row i by multiplying each entry in row i by k.	1. $R_i \to kR_i$
2. Replace row i by adding k times each entry in row j to the corresponding entry in row i	2. $R_i \to kR_j + R_i$
3. Interchange row i and row j.	3. $R_i \leftrightarrow R_j$

Example 5: Using Elementary Row Operations on Matrices

Use matrix form to solve the system

$$\begin{array}{ccccccccc} a & + & b & + & c & + & d & = & 1 \\ a & - & b & + & c & + & d & = & 1 \\ & & b & + & c & - & d & = & 1 \\ & & b & & & + & d & = & 1. \end{array}$$

Solution: The augmented matrix for the system is

$$\left[\begin{array}{cccc|c} 1 & 1 & 1 & 1 & 1 \\ 1 & -1 & 1 & 1 & 1 \\ 0 & 1 & 1 & -1 & 1 \\ 0 & 1 & 0 & 1 & 1 \end{array}\right].$$

To obtain 0's in the first column after row 1, we need only add −1 times the first row to the second row.

$$\left[\begin{array}{cccc|c} 1 & 1 & 1 & 1 & 1 \\ 0 & -2 & 0 & 0 & 0 \\ 0 & 1 & 1 & -1 & 1 \\ 0 & 1 & 0 & 1 & 1 \end{array}\right] \quad R_2 \to 1R_1 + R_2$$

The second row represents the equation $-2b = 0$, so we already know that $b = 0$. To obtain 0's in the second column after row 2, we first multiply each entry in row 2 by $-1/2$ to make the coefficient of b in the second equation a 1.

$$\left[\begin{array}{cccc|c} 1 & 1 & 1 & 1 & 1 \\ 0 & 1 & 0 & 0 & 0 \\ 0 & 1 & 1 & -1 & 1 \\ 0 & 1 & 0 & 1 & 1 \end{array}\right] \quad R_2 \to -\frac{1}{2}R_2$$

Now add −1 times the second row to the third row as well as to the fourth row.

$$\left[\begin{array}{cccc|c} 1 & 1 & 1 & 1 & 1 \\ 0 & 1 & 0 & 0 & 0 \\ 0 & 0 & 1 & -1 & 1 \\ 0 & 0 & 0 & 1 & 1 \end{array}\right] \quad \begin{array}{l} R_3 \to -1R_2 + R_3 \\ R_4 \to 1R_2 + R_4 \end{array}$$

Row 2 and row 4 tell us that $b = 0$ and $d = 1$, while substitution of these values into the equations corresponding to row 3 and row 1 gives

$$\begin{aligned} c - d &= 1 \\ c - (1) &= 1 \\ c &= 2 \end{aligned} \quad \text{and} \quad \begin{aligned} a + b + c + d &= 1 \\ a + 0 + 2 + 1 &= 1 \\ a &= -2 \end{aligned}$$

Thus, the solution is $a = -2$, $b = 0$, $c = 2$, and $d = 1$.

Technology Link

Matrices may be entered and utilized on most graphing calculators. Among the operations that can usually be performed on matrices with such calculators are the elementary row operations discussed in this section. If your calculator has this capability, then you should redo the row operations described in Examples 3–5 and compare your results to the matrices shown in the text.

PROGRESS CHECK 5
Use matrix form to solve the system

$$\begin{aligned} 4a - 3b + 2d &= 9 \\ a + 3c &= 14 \\ -b + 2d &= 7 \\ 3a + 4d &= 26. \end{aligned}$$

■

There are two additional ideas to consider about solving linear systems using row operations. First, if a row of zeros results in the coefficient portion in any matrix, answer that there is no unique solution to the problem. Such systems are either dependent (if the last column entry is also zero) or inconsistent (if the last column entry is not zero). Second, in our example problems to this point, we stopped when we reached triangular form and completed the solution by back substitution. An alternative method, called **Gauss-Jordan elimination**, is to continue to produce equivalent matrices until we reach a form like

$$\left[\begin{array}{ccc|c} 1 & 0 & 0 & a \\ 0 & 1 & 0 & b \\ 0 & 0 & 1 & c \end{array}\right].$$

From this final form of the matrix, we directly read that the solution is $x = a$, $y = b$, and $z = c$. The matrix above is an example of a **reduced row echelon matrix**, which is defined as follows.

Reduced Row-Echelon Matrix

Matrix A is a reduced row-echelon matrix if and only if

1. Rows containing all 0's (if any exist) occur at the bottom of A.
2. The first nonzero entry in each nonzero row is 1, called a **leading 1**.
3. These leading 1's are positioned further to the right in succeeding lower rows of A.
4. Every column that contains a leading 1 has all 0's above and below that leading 1.

Example 6 shows a systematic procedure that may be used to obtain a reduced row-echelon matrix.

Example 6: Using Gauss-Jordan Elimination Use Gauss-Jordan elimination to solve the system

$$\begin{aligned} 3x + y + z &= 1 \\ x - y + z &= 3 \\ 2x + 2y - z &= -3. \end{aligned}$$

Solution: The augmented matrix for the system is

$$\left[\begin{array}{ccc|c} 3 & 1 & 1 & 1 \\ 1 & -1 & 1 & 3 \\ 2 & 2 & -1 & -3 \end{array}\right].$$

In the Gauss-Jordan elimination method, we transform this matrix to a reduced row-echelon matrix, as shown next.

Row operation	Equivalent matrix	Objective
$R_1 \leftrightarrow R_2$	$\left[\begin{array}{ccc\|c} 1 & -1 & 1 & 3 \\ 3 & 1 & 1 & 1 \\ 2 & 2 & -1 & -3 \end{array}\right]$	1. Obtain a leading 1 in row 1, column 1.
$R_2 \to -3R_1 + R_2$ $R_3 \to -2R_1 + R_3$	$\left[\begin{array}{ccc\|c} 1 & -1 & 1 & 3 \\ 0 & 4 & -2 & -8 \\ 0 & 4 & -3 & -9 \end{array}\right]$	2. Use the leading 1 in row 1, column 1, to get other column 1 entries to be 0.
$R_2 \to \frac{1}{4}R_2$	$\left[\begin{array}{ccc\|c} 1 & -1 & 1 & 3 \\ 0 & 1 & -\frac{1}{2} & -2 \\ 0 & 4 & -3 & -9 \end{array}\right]$	3. Obtain a leading 1 in row 2, column 2.
$R_1 \to R_2 + R_1$ $R_3 \to -4R_2 + R_3$	$\left[\begin{array}{ccc\|c} 1 & 0 & \frac{1}{2} & 1 \\ 0 & 1 & -\frac{1}{2} & -2 \\ 0 & 0 & -1 & -1 \end{array}\right]$	4. Use the leading 1 in row 2, column 2, to get other column 2 entries to be 0.
$R_3 \to -1R_3$	$\left[\begin{array}{ccc\|c} 1 & 0 & \frac{1}{2} & 1 \\ 0 & 1 & -\frac{1}{2} & -2 \\ 0 & 0 & 1 & 1 \end{array}\right]$	5. Obtain a leading 1 in row 3, column 3.
$R_1 \to -\frac{1}{2}R_3 + R_1$ $R_2 \to \frac{1}{2}R_3 + R_2$	$\left[\begin{array}{ccc\|c} 1 & 0 & 0 & \frac{1}{2} \\ 0 & 1 & 0 & -\frac{3}{2} \\ 0 & 0 & 1 & 1 \end{array}\right]$	6. Use the leading 1 in row 3, column 3, to get other column 3 entries to be 0.

From the system of equations corresponding to the reduced row-echelon matrix at the bottom, we determine that the solution is $x = 1/2$, $y = -3/2$, and $z = 1$.

PROGRESS CHECK 6

Use Gauss-Jordan elimination to solve the system

$$\begin{aligned} 3x + y + 2 &= -1 \\ x + 2y + 4z &= 3 \\ -x + 2y - 4z &= -1. \end{aligned}$$

■

EXPLORE 11.2

1. Graphing calculators with matrix capability can perform elementary row operations. If you have such a calculator, perform the following sequence of operations, which puts a matrix into reduced row-echelon form. (This is the same sequence illustrated in Example 6 of this section of the text. We use the abbreviation system defined after Example 4.) After each step, check that matrix A looks like the corresponding one shown in Example 6.
Note: It may be necessary to use a Store command after each operation so that the next operation can be referred to a named matrix. For instance, an instruction similar to

rowSwap(A,1,2) STO B

swaps rows 1 and 2 of matrix A, and saves the resulting matrix as matrix B.

a. Create a 3-by-4 matrix named A, where $A = \begin{bmatrix} 3 & 1 & 1 & 1 \\ 1 & -1 & 1 & 3 \\ 2 & 2 & -1 & -3 \end{bmatrix}$

b. $R_1 \leftrightarrow R_2$ Swap rows 1 and 2.

c. $R_2 \to -3R_1 + R_2$ Replace row 2 by $-3 \times$ row 1 plus row 2.

d. $R_3 \to -2R_1 + R_3$ Replace row 3 by $-2 \times$ row 1 plus row 3.

e. $R_2 \to \frac{1}{4}R_2$ Multiply row 2 by $1/4$.

f. $R_1 \to R_2 + R_1$ Replace row 1 by row 2 plus row 1.

g. $R_3 \to -4R_2 + R_3$ Replace row 3 by $-4 \times$ row 2 plus row 3.

h. $R_3 \to -R_3$ Multiply row 3 by -1.

i. $R_1 \to -\frac{1}{2}R_3 + R_1$ Replace row 1 by $-\frac{1}{2} \times$ row 3 plus row 1.

j. $R_2 \to \frac{1}{2}R_3 + R_2$ Replace row 2 by $\frac{1}{2} \times$ row 3 plus row 2:

2. **a.** Some calculators have a matrix operation that automatically puts a matrix into a special triangular form called row-echelon form, which has all ls on the main diagonal. If your calculator has this operation, apply it to matrix A of the previous problem and confirm that the result is correct.

   b. If your calculator has a matrix operation for reduced row-echelon form, apply it to matrix A of the previous problem and confirm that the result is correct.

EXERCISES 11.2

In Exercises 1–4 solve the given systems of equations and show both the matrix form of the systems and the corresponding equations.

1.
$$\begin{aligned} x + y &= 7 \\ x - y &= -2 \end{aligned}$$

2.
$$\begin{aligned} 2a + 5b &= 1 \\ 3a - 4b &= 13 \end{aligned}$$

3.
$$\begin{aligned} a + b + c &= 2 \\ 3a - b - c &= -1 \\ 2a + 2b - c &= 1 \end{aligned}$$

4.
$$\begin{aligned} 4x - 2y + 2z &= 0 \\ 3x + 2z &= 0 \\ x - 2y &= 0 \end{aligned}$$

In Exercises 5–20 use matrix form to solve the given systems of equations. Use Gaussian elimination with back substitution or Gauss-Jordan elimination.

5.
$$\begin{aligned} 5x + 2y + 17 &= 0 \\ -2z - y + 5 &= 0 \end{aligned}$$

6.
$$\begin{aligned} x - y &= 3 \\ -x - y &= -3 \end{aligned}$$

7.
$$\begin{aligned} 2x - 3y &= 1 \\ -6x + 9y &= -3 \end{aligned}$$

8.
$$\begin{aligned} x - y &= 3 \\ -x + y &= -5 \end{aligned}$$

9.
$$\begin{aligned} x + y &= 4 \\ y + z &= -8 \\ x + z &= 2 \end{aligned}$$

10.
$$\begin{aligned} -a + b &= -1 \\ b - c &= 3 \\ a + c &= -12 \end{aligned}$$

11.
$$\begin{aligned} x + y + z &= 1 \\ x + y - z &= 3 \\ x - y - z &= 5 \end{aligned}$$

12.
$$\begin{aligned} x + y + 3z &= 1 \\ 2x + 5y + 2z &= 0 \\ 3x - 2y - z &= 3 \end{aligned}$$

13.
$$\begin{aligned} x_1 - x_2 + x_3 &= 2 \\ 2x_1 - 3x_2 + 2x_3 &= 6 \\ 3x_1 + x_2 + x_3 &= 2 \end{aligned}$$

14.
$$\begin{aligned} x_1 - x_2 + x_3 &= 3 \\ 3x_1 - x_2 + x_3 &= 1 \\ 2x_1 + 2x_2 - x_3 &= -3 \end{aligned}$$

15.
$$\begin{aligned} 6A + 3B + 2C &= 1 \\ 5A + 4B + 3C &= 0 \\ A + B + C &= 0 \end{aligned}$$

16.
$$\begin{aligned} 2a + 3b - c &= 3 \\ a + b - 3c &= -4 \\ -a - b + 5c &= 8 \end{aligned}$$

17.
$$\begin{aligned} 3x - 2y - z &= -2 \\ 2x + 2y + 2z &= 2 \\ 2x - 3y - 2z &= -4 \end{aligned}$$

18.
$$\begin{aligned} x + y &= 2 \\ y + z &= 1 \\ z + w &= -1 \\ x + w &= 0 \end{aligned}$$

19.
$$\begin{aligned} a + b + c + d &= 0 \\ a - b + 2c + d &= 1 \\ 4a + b + 2c &= 5 \\ 5a + 4c + 2d &= 6 \end{aligned}$$

20.
$$\begin{aligned} x_1 + x_2 + x_3 + x_4 &= 1 \\ x_1 - x_2 + x_3 - x_4 &= -1 \\ -2x_1 + x_2 - x_3 - 2x_4 &= 1 \\ 2x_1 - 2x_2 + 2x_3 + x_4 &= 1 \end{aligned}$$

21. A college is given \$1 million by a rich graduate. The gift is invested and the interest is used for scholarships. The college invests the money in three mutual funds for one year. The Alpha fund earns 22 percent interest for the year, the Beta Fund earns 8 percent that year, and the Gamma fund loses 4 percent. Equal amounts were put into Beta and Gamma, with the rest in Alpha. At the end of the year, the investments are worth a total of \$1,140,200. How much was invested in each fund?

22. A \$2 million investment is distributed among three mutual funds, Delta Fund, Epsilon Fund, and Omicron Fund. Equal amounts are put in Delta and Epsilon, with the remainder in Omicron. At the end of one year, Delta earns 15 percent, Epsilon earns 3 percent, and Omicron loses 18 percent. As a result, the total value of the investment is \$1,954,000. How much was invested in each fund?

23. The points $(-1,-8)$, $(1, 2)$, $(2, 4)$ lie on the parabola $y = ax^2 + bx + c$. Find *a*, *b*, and *c*.

24. Let $f(x) = ax^2 + bx + c$. Find values of *a*, *b*, and *c* such that $f(1) = 0$, $f(2) = 8$ and $f(3) = 22$.

25. If the graph of $y = ax^3 + bx^2 + cx + d$ passes through the points $(1, 1)$, $(2, 1)$, $(3,-11)$, and $(-1,-11)$ find *a*, *b*, *c*, and *d*.

26. The points $(3, 3)$, $(-2,-2)$, and $(1,-1)$ lie on the circle $x^2 + y^2 + Dx + Ey + F = 0$. Find *D*, *E*, and *F*. What are the center and radius of this circle?

27. Find the radius of each circle in the diagram blow.

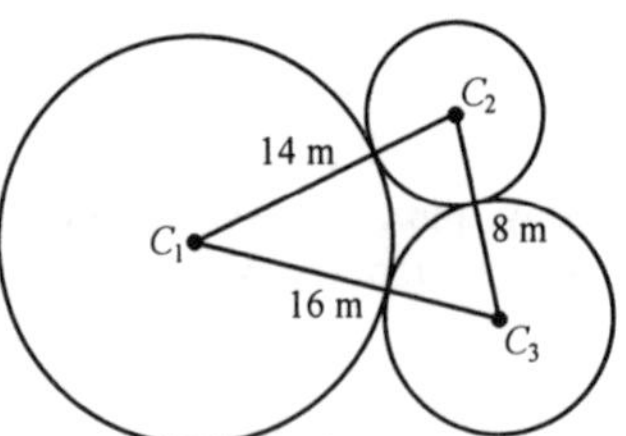

28. A 2,000-seat theater was sold out for a weekend concert. Total receipts on Friday night were \$31,500, when seats in the orchestra, mezzanine, and balcony sold for \$20, \$15, and \$10, respectively. On Saturday, prices were raised to \$25, \$20, and \$10, and total receipts increased to \$39,000. How many seats are in the mezzanine?

29. A used-records shop sells records, CDs and cassettes. The CDs sell for \$8, while the records and cassettes sell for \$4. One week the sales totaled \$1,000. During the week they sold 10 more cassettes than records, and they sold twice as many CDs as records and cassettes combined. How many of each were sold?

30. A gas station sells three grades of gasoline—regular, plus, and super—for \$1.10, \$1.15, and \$1.25 per gallon, respectively. One day sales were \$2,760. On this day as much regular was sold as the other two types of gas combined, and the amount of plus sold was the same as the amount of super. How many gallons of each were sold? If the station's profit is 5¢ per gallon, how much profit did it make for the day?

THINK ABOUT IT 11.2

1. Solve for *x*, *y*, and *z* in terms of the given constants.

$$\begin{aligned} Ax + By + Cz &= D \\ Ex + Fy + Gz &= H \\ Jx + Ky + Lz &= M \end{aligned}$$

2. A linear system of 2 equations in 3 variables is underdetermined. Mathematically, there may be an infinite number of solutions. In a particular application, however, only a limited number of these solutions may be sensible. As illustrated by the following problem, sometimes it is necessary to use a supplementary trial-and-error approach to determine the sensible solutions. At a charity performance a cashier notices that 100 tickets were sold for a total of \$100. The tickets were priced at \$5 for adults, 10¢ for children, and \$1 for senior citizens. Find the number of each type of ticket sold. There may be more than one sensible answer.
[Source: Vermont high school mathematics portfolio assessment seminar, sponsored by the Vermont Institute for Science and Math Teaching (VISMT)]
3. Find in terms of *D* an equation for all of the circles given by $x^2 + y^2 + Dx + Ey + F = 0$ that pass through the points $(1, 3)$ and $(-7,-1)$. Choose arbitrarily two distinct values for *D* and write equations of two different circles passing through these two points.

4. One of the classic stories from the history of mathematics concerns Archimedes (c. 250 B.C.), a great mathematician and scientist of ancient times. Archimedes lived in the Greek city-state of Syracuse, where the king, Hieron, suspected a goldsmith of giving him a gold crown that contained hidden silver. The king referred the problem to Archimedes and asked him to determine, without destroying the crown, the percentage of pure gold in the crown. While taking a bath, Archimedes found the principle needed to solve this problem. What Archimedes noticed in his bath was the fact that when a body is immersed in water, it displaces a volume of water that is equal to the volume of the body. He also knew that bodies of the same weight do not necessarily have the same volume. Using these principles, Archimedes then filled a bucket of water to the brim. Suppose the crown weighed 10 lb and displaced 18 cubic in. of water, and that 10 lb of pure gold and 10 lb of pure silver displaced 15 and 30 in.^{3} of water, respectively. What percent of the crown would you tell King Hieron is made of each metal? **Note** Solve the linear system with two equations in two variables by a method of this section.)
5. In the "Think About it" exercises of Section 4.2, we considered a problem from the *Mathematical Puzzles of Sam Loyd*, a two-volume series edited by Martin Gardner and published by Dover Publications, © 1960. Here's another one for you to try.

How much does each jar hold?

MRS. HUBBARD has invented a clever system for keeping tabs on her jars of blackberry jam. She has arranged the jars in her cupboard so that she has twenty quarts of jam on each shelf. The jars are in three sizes. Can you tell how much each size contains?

• • •

11.3 Determinants and Cramer's Rule

One perfume contains 1 percent essence of rose, while a second contains 1.8 percent essence of rose. How much of each should be combined to make 24 oz of perfume that contains 1.5 percent essence of rose?
(See Example 4.)

Objectives

1. Evaluate the determinant of a 2-by-2 matrix.
2. Solve a linear system in two variables using Cramer's rule.
3. Evaluate the determinant of a 3-by-3 matrix.
4. Solve a linear system in three variables using Cramer's rule.

A linear system with n equations in n variables that is neither inconsistent nor dependent may be solved by using formulas known as Cramer's rule. To understand the derivation and application of these formulas, we start by finding the solution of the general system with two equations in two variables

$$\begin{aligned} a_1x + b_1y &= c_1 \\ a_2x + b_2y &= c_2 . \end{aligned}$$

To solve for x, we multiply each member of the first equation by b_2 and each member of the second equation by $-b_1$, to obtain

$$\begin{aligned} a_1b_2x + b_1b_2y &= c_1b_2 \\ -b_1a_2x - b_1b_2y &= -b_1c_2 . \end{aligned}$$

Now add the two equations to get

$$a_1b_2x - b_1a_2x = c_1b_2 - b_1c_2 .$$

Factoring the left side of the equation, we have

$$(a_1b_2 - b_1a_2)x = c_1b_2 - b_1c_2 .$$

Thus, if $a_1b_2 - b_1a_2 \neq 0$,

$$x = \frac{c_1b_2 - b_1c_2}{a_1b_2 - b_1a_2} .$$

In a similar manner, multiplying each member of the first equation by $-a_2$ and each member of the second equation by a_1, and then adding, leads to

$$y = \frac{a_1c_2 - c_1a_2}{a_1b_2 - b_1a_2} .$$

These formulas may now be used to find x and y whenever $a_1b_2 - b_1a_2 \neq 0$. If $a_1b_2 - b_1a_2 = 0$ there is no unique solution for x and y, and the system is either dependent or inconsistent.

We do not memorize the formulas in this form, since they may be obtained by defining what is called a determinant. Consider the expression $a_1b_2 - b_1a_2$, which is the denominator in the formulas for both x and y, and note that the coefficient matrix of the linear system is

$$A = \begin{bmatrix} a_1 & b_1 \\ a_2 & b_2 \end{bmatrix}.$$

This matrix is an example of a square matrix, which is a matrix having the same number of rows as columns. To each square matrix A there is assigned a unique real number called its determinant and denoted by $|A|$. As suggested above, when A has two rows and two columns, called a 2-by-2 matrix, then we define the value of $|A|$ to be $a_1b_2 - b_1a_2$.

Determinant of a 2-by-2 Matrix

If $A = \begin{bmatrix} a_1 & b_1 \\ a_2 & b_2 \end{bmatrix}$, then the determinant of A is given by

$$|A| = \begin{vmatrix} a_1 & b_1 \\ a_2 & b_2 \end{vmatrix} = a_1b_2 - b_1a_2.$$

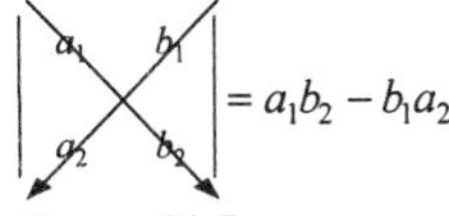

Figure 11.5

The numbers a_1 and b_2 are elements of the principal diagonal, and the numbers b_1 and a_2 are elements of the secondary diagonal. Note in Figure 11.5 that the value of the determinant is the product of the elements of the principal diagonal minus the product of the elements of the secondary diagonal.

Example 1: Determinant of a 2-by-2 Matrix

Evaluate $\begin{vmatrix} 1 & 2 \\ 3 & 4 \end{vmatrix}$.

Solution: Using the definition above gives

$$\begin{vmatrix} 1 & 2 \\ 3 & 4 \end{vmatrix} = 1(4) - 2(3) = 4 - 6 = -2.$$

PROGRESS CHECK 1

Evaluate $\begin{vmatrix} 5 & -2 \\ 1 & 3 \end{vmatrix}$. ■

Example 2: Determinant of a 2-by-2 Matrix

If $A = \begin{bmatrix} 3 & -5 \\ 4 & -7 \end{bmatrix}$, find $|A|$.

Solution: The determinant of matrix A is given by

$$|A| = \begin{vmatrix} 3 & -5 \\ 4 & -7 \end{vmatrix} = 3(-7) - (-5)(4) = -21 + 20 = -1.$$

Technology Link

The determinant of a matrix A may be denoted by $|A|$ or det A. The latter notation is commonly used on calculators with a matrix capability, to offer a function DET that will evaluate the determinant of a matrix. If your calculator has a Determinant function, then you should evaluate the determinants shown in Examples 1 and 2 using this function and compare your results to the text's answers.

PROGRESS CHECK 2

If $A = \begin{bmatrix} 3 & -2 \\ 4 & -3 \end{bmatrix}$, find $|A|$. ■

We now state Cramer's rule for two equations with two variables, which shows how determinants can be used to solve such a system of equations.

Cramer's Rule for 2-by-2 Systems

The solution to the system

$$\begin{aligned} a_1x + b_1y &= c_1 \\ a_2x + b_2y &= c_2 \end{aligned}$$

with $a_1b_2 - b_1a_2 \neq 0$ is

$$x = \frac{D_x}{D} = \frac{\begin{vmatrix} c_1 & b_1 \\ c_2 & b_2 \end{vmatrix}}{\begin{vmatrix} a_1 & b_1 \\ a_2 & b_2 \end{vmatrix}} = \frac{c_1b_2 - b_1c_2}{a_1b_2 - b_1a_2}$$

$$y = \frac{D_y}{D} = \frac{\begin{vmatrix} a_1 & c_1 \\ a_2 & c_2 \end{vmatrix}}{\begin{vmatrix} a_1 & b_1 \\ a_2 & b_2 \end{vmatrix}} = \frac{a_1c_2 - c_1a_2}{a_1b_2 - b_1a_2}$$

In Cramer's rule the formulas for x and y are not difficult to remember. In both cases the determinant in the denominator is formed from the coefficients of x and y. Different determinants are used in the numerators. When solving for x, the coefficients of x, which are a_1 and a_2, are replaced by the constants c_1 and c_2. Similarly, when solving for y, the coefficients of y, which are b_1 and b_2, are replaced by the constants c_1 and c_2.

Example 3: Using Cramer's Rule to Solve a 2-by-2 System

Use Cramer's rule to solve the system of equations

$$\begin{aligned} 3x - 2y &= 27 \\ 2x + 5y &= -1 \end{aligned}$$

Solution: First, evaluate the determinant in the denominator, which consists of the coefficients of x and y:

$$D = \begin{vmatrix} 3 & -2 \\ 2 & 5 \end{vmatrix} = 3(5) - (-2)(2) = 15 + 4 = 19$$

Second, evaluate the determinant in the numerator of the formula for x. Replace the column containing the coefficients of x by the column with the constants on the right side of the equations.

$$D_x = \begin{vmatrix} 27 & -2 \\ -1 & 5 \end{vmatrix} = 27(5) - (-2)(-1) = 135 - 2 = 133$$

Third, evaluate the determinant in the numerator of the formula for y. We replace the column containing the coefficients of y by the column with the constants on the right side of the equations.

$$D_y = \begin{vmatrix} 3 & 27 \\ 2 & -1 \end{vmatrix} = 3(-1) - 27(2) = -3 - 54 = -57$$

Fourth, use Cramer's rule.

$$x = \frac{D_x}{D} = \frac{\begin{vmatrix} 27 & -2 \\ -1 & 5 \end{vmatrix}}{\begin{vmatrix} 3 & -2 \\ 2 & 5 \end{vmatrix}} = \frac{133}{19} = 7$$

$$y = \frac{D_y}{D} = \frac{\begin{vmatrix} 3 & 27 \\ 2 & -1 \end{vmatrix}}{\begin{vmatrix} 3 & -2 \\ 2 & 5 \end{vmatrix}} = \frac{-57}{19} = -3$$

Thus, the solution is $(7,-3)$. Check this solution in the given system.

PROGRESS CHECK 3

Solve by Cramer's rule:
$$\begin{aligned} 5x + 2y &= 1 \\ x - 3y &= 7 \end{aligned}$$
■

To solve the section-opening problem, we recall from Section R.6 that liquid mixture problems are analyzed using

$$\begin{pmatrix}\text{percent of} \\ \text{an ingredient}\end{pmatrix} \cdot \begin{pmatrix}\text{amount of} \\ \text{solution}\end{pmatrix} = \begin{pmatrix}\text{amount of} \\ \text{ingredient}\end{pmatrix}$$

Example 4: Determining a Perfume Mixture

Solve the problem in the section introduction on page 892.

Solution: Let $x =$ amount used in ounces of 1 percent essence of rose
$y =$ amount used in ounces of 1.8 percent essence of rose
and organize the key components in the problem in a chart format.

Solution	Percent of Essence of Rose	·	Amount of Solution (Ounces)	=	Amount of Essence of Rose (Ounces)
First perfume	1		x		$0.01x$
Second perfume	1.8		y		$0.018y$
New perfume	1.5		24		0.015(24), or 0.36

Because x ounces of the first perfume are combined with y ounces of the second perfume to form 24 ounces of the new perfume,

$$x + y = 24. \quad (1)$$

Also, the amount of essence of rose in the new perfume is the sum of the amounts contributed by the two concentrations of perfume, so

$$0.01x + 0.018y = 0.36 \quad (2)$$

To solve the resulting system, we choose to first multiply both sides of equation (2) by 1000 to clear decimals, and then we apply Cramer's rule to the system.

$$\begin{aligned} x + y &= 24 \\ 10x + 18y &= 360 \end{aligned}$$

The three determinants defined in Cramer's rule are

$$D = \begin{vmatrix} 1 & 1 \\ 10 & 18 \end{vmatrix} = 1(18) - 1(10) = 8,$$

$$D_x = \begin{vmatrix} 24 & 1 \\ 360 & 18 \end{vmatrix} = 24(18) - 1(360) = 72,$$

$$D_y = \begin{vmatrix} 1 & 24 \\ 10 & 360 \end{vmatrix} = 1(360) - 24(10) = 120,$$

so

$$x = \frac{D_x}{D} = \frac{72}{8} = 9 \quad \text{and} \quad y = \frac{D_y}{D} = \frac{120}{8} = 15.$$

Thus, the desired perfume is obtained by mixing 9 oz of 1 percent essence of rose with 15 oz of 1.8 percent essence of rose.

PROGRESS CHECK 4

Jewelers often use alloys to add strength, color, or other properties to the metals in jewelry. If a jeweler has two alloys, one 35 percent gold and one 25 percent gold, to be combined by melting them down, how much of each should be used to produce 0.45 oz of 31 percent gold alloy? ■

Cramer's rule may be extended to solve linear systems in three variables; to apply it, however, we must learn to evaluate determinants with three rows and three columns.

We first define the **minor** of an element to be the determinant formed by deleting the row and column containing the given element. For example,

$$\text{minor for } a_1 = \begin{vmatrix} a_1 & b_1 & c_1 \\ a_2 & b_2 & c_2 \\ a_3 & b_3 & c_3 \end{vmatrix} = \begin{vmatrix} b_2 & c_2 \\ b_3 & c_3 \end{vmatrix}$$

$$\text{minor for } b_2 = \begin{vmatrix} a_1 & b_1 & c_1 \\ a_2 & b_2 & c_2 \\ a_3 & b_3 & c_3 \end{vmatrix} = \begin{vmatrix} a_1 & c_1 \\ a_3 & c_3 \end{vmatrix}.$$

Now we define the **cofactor** of an element by considering the row and column position of that element. If an element is in the ith row and the jth column, then the cofactor of the element is given by the product

$$(-1)^{i+j} \cdot \text{(minor of the element)}$$

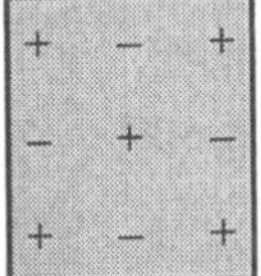

Figure 11.6

In other words, if $i + j$ is even, the cofactor of the element is the same as the minor of the element; if $i + j$ is odd, then the cofactor equals the negative of the minor of the element. In the case of a 3-by-3 determinant, this definition means that the cofactor of an element is found by attaching the sign from the pattern in Figure 11.6 to the minor of that element. Now we can evaluate 3-by-3 (and more complicated) determinants by following these steps.

To Evaluate a Determinant

1. Pick any row or any column of the determinant.
2. Multiply each entry in that row or column by its cofactor.
3. Add the results. This sum is defined to be the value of the determinant.

Remember that to each determinant there corresponds exactly one number as the value of that determinant. This determinant value is independent of the row or column that is chosen in step 1.

Example 5: Determinant of a 3-by-3 Matrix

Evaluate the following determinant by

a. expansion along the first row of the determinant and
b. expansion along the second column of the determinant.

$$\begin{vmatrix} 1 & 1 & 3 \\ 5 & 0 & 2 \\ -2 & 3 & -1 \end{vmatrix}.$$

Solution:

a. The sign pattern in Figure 11.6 for row 1 is +, –, +. Using row 1 and following the steps given on page 896, we have

$$\begin{aligned}
&\begin{vmatrix} 1 & 1 & 3 \\ 5 & 0 & 2 \\ -2 & 3 & -1 \end{vmatrix} \\
&= 1\begin{vmatrix} \not{1} & \not{1} & \not{3} \\ \not{5} & 0 & 2 \\ -\not{2} & 3 & -1 \end{vmatrix} - 1\begin{vmatrix} \not{1} & \not{1} & \not{3} \\ 5 & \not{0} & 2 \\ -2 & \not{3} & -1 \end{vmatrix} + 3\begin{vmatrix} \not{1} & \not{1} & \not{3} \\ 5 & 0 & \not{2} \\ -2 & 3 & -\not{1} \end{vmatrix} \\
&= 1\begin{vmatrix} 0 & 2 \\ 3 & -1 \end{vmatrix} - \begin{vmatrix} 5 & 2 \\ -2 & -1 \end{vmatrix} + 3\begin{vmatrix} 5 & 0 \\ -2 & 3 \end{vmatrix} \\
&= 1[0(-1) - 2(3)] - 1[5(-1) - 2(-2)] + 3[5(3) - 0(-2)] \\
&= 1(-6) - 1(-1) + 3(15) \\
&= -6 + 1 + 45 \\
&= 40
\end{aligned}$$

b. The sign pattern in Figure 11.6 for column 2 is –,+, –. Expanding the determinant by this column gives

$$\begin{aligned}
&\begin{vmatrix} 1 & 1 & 3 \\ 5 & 0 & 2 \\ -2 & 3 & -1 \end{vmatrix} \\
&= -1\begin{vmatrix} \not{1} & \not{1} & \not{3} \\ 5 & \not{0} & 2 \\ -2 & \not{3} & -1 \end{vmatrix} + 0\begin{vmatrix} 1 & \not{1} & 3 \\ \not{5} & \not{0} & \not{2} \\ -2 & \not{3} & -1 \end{vmatrix} - 3\begin{vmatrix} 1 & \not{1} & 3 \\ 5 & \not{0} & 2 \\ -\not{2} & \not{3} & -\not{1} \end{vmatrix} \\
&= -1\begin{vmatrix} 5 & 2 \\ -2 & -1 \end{vmatrix} + 0\begin{vmatrix} 1 & 3 \\ -2 & -1 \end{vmatrix} - 3\begin{vmatrix} 1 & 3 \\ 5 & 2 \end{vmatrix} \\
&= -1[5(-1) - 2(-2)] + 0[\text{not needed}] - 3[1(2) - 3(5)] \\
&= -1(-1) + 0 - 3(-13) \\
&= 1 + 39 \\
&= 40
\end{aligned}$$

Thus, the value of the determinant is 40. As expected, the answer in parts **a** and **b** is the same. Since you have a choice, you may want to evaluate a determinant by picking the row or column with the most 0's. As shown in part **b**, it is unnecessary to evaluate the minor of 0 elements.

PROGRESS CHECK 5

Evaluate the following determinant by

a. expansion along the first column of the determinant and

b. expansion along the second row of the determinant.

$$\begin{vmatrix} 3 & -2 & 1 \\ -2 & 0 & -2 \\ 1 & -2 & 1 \end{vmatrix}$$

■

Cramer's rule extends to systems with three equations and three variables, and the formulas for x, y, and z follow an arrangement similar to that of the system with two equations and two variables. In each case the determinant in the denominator is formed from the coefficients of x, y, and z. Different determinants are used in the numerators. When solving for x, the coefficients of x are replaced by the constants d_1, d_2, and d_3. Similarly, when solving for y or z, the constants d_1, d_2 and d_3 replace the coefficients of the desired variable.

Cramer's Rule for 3-by-3 Systems

The solution to the system

$$\begin{aligned} a_1x + b_1y + c_1z &= d_1 \\ a_2x + b_2y + c_2z &= d_2 \\ a_2x + b_3y + c_2z &= d_3 \end{aligned} \quad \text{with } D = \begin{vmatrix} a_1 & b_1 & c_1 \\ a_2 & b_2 & c_2 \\ a_3 & b_3 & c_3 \end{vmatrix} \neq 0$$

is $x = D_x/D$, $y = D_y/D$, and $z = D_z/D$ where

$$D_x = \begin{vmatrix} d_1 & b_1 & c_1 \\ d_2 & b_2 & c_2 \\ d_3 & b_3 & c_3 \end{vmatrix}, \ D_y = \begin{vmatrix} a_1 & d_1 & c_1 \\ a_2 & d_2 & c_2 \\ a_3 & d_3 & c_3 \end{vmatrix}, \text{ and } D_z = \begin{vmatrix} a_1 & b_1 & d_1 \\ a_2 & b_2 & d_2 \\ a_3 & b_3 & d_3 \end{vmatrix}.$$

Example 6: Using Cramer's Rule to Solve a 3-by-3 System

Use Cramer's rule to solve the system of equations

$$\begin{aligned} 3x - y + 6z &= 1 \\ x + 2y - 3z &= 0. \\ 2x - 3y - z &= -9 \end{aligned}$$

Solution: The determinant in the denominator is formed from the coefficients of x, y, and z. To find the determinant in the numerator for x, we replace the column containing the coefficients of x by the column containing the constants on the right side of the equations.

$$x = \frac{D_x}{D} = \frac{\begin{vmatrix} 1 & -1 & 6 \\ 0 & 2 & -3 \\ -9 & -3 & -1 \end{vmatrix}}{\begin{vmatrix} 3 & -1 & 6 \\ 1 & 2 & -3 \\ 2 & -3 & -1 \end{vmatrix}}$$

$$= \frac{\overbrace{1\begin{vmatrix} 2 & -3 \\ -3 & -1 \end{vmatrix} - 0\begin{vmatrix} -1 & 6 \\ -3 & -1 \end{vmatrix} + (-9)\begin{vmatrix} -1 & 6 \\ 2 & -3 \end{vmatrix}}^{\text{expansion by column 1 because of 0 element}}}{3\begin{vmatrix} 2 & -3 \\ -3 & -1 \end{vmatrix} - \begin{vmatrix} -1 & 6 \\ -3 & -1 \end{vmatrix} + 2\begin{vmatrix} -1 & 6 \\ 2 & -3 \end{vmatrix}} = \frac{70}{-70} = -1$$

The value of the determinant in the denominator for both y and z is -70.
The column containing the constants on the right side of the equations first replaces the coefficients of y (to find y) and then the coefficients of z (to find z).

$$y = \frac{D_y}{D} = \frac{\begin{vmatrix} 3 & 1 & 6 \\ 1 & 0 & -3 \\ 2 & -9 & -1 \end{vmatrix}}{-70}$$

$$= \frac{\overbrace{-1\begin{vmatrix} 1 & -3 \\ 2 & -1 \end{vmatrix} + 0\begin{vmatrix} 3 & 6 \\ 2 & -1 \end{vmatrix} - (-9)\begin{vmatrix} 3 & 6 \\ 1 & -3 \end{vmatrix}}^{\text{expansion by column 2 because of 0 element}}}{-70} = \frac{-140}{-70} = 2$$

$$z = \frac{D_z}{D} = \frac{\begin{vmatrix} 3 & -1 & 1 \\ 1 & 2 & 0 \\ 2 & -3 & -9 \end{vmatrix}}{-70}$$

$$= \frac{\overbrace{1\begin{vmatrix} 1 & 2 \\ 2 & -3 \end{vmatrix} - 0\begin{vmatrix} 3 & -1 \\ 2 & -3 \end{vmatrix} + (-9)\begin{vmatrix} 3 & -1 \\ 1 & 2 \end{vmatrix}}^{\text{expansion by column 3 because of 0 element}}}{-70} = \frac{-70}{-70} = 1$$

The solution is $x = -1,\ y = 2, z = 1$. Check this solution in all three equations in the given system.

PROGRESS CHECK 6
Solve by Cramer's rule.

$$\begin{aligned} x + 2y + 4z &= 3 \\ 3x + y + 2z &= -1 \\ -x + 2y - 4z &= -1 \end{aligned}$$

■

There are three additional points to mention about Cramer's rule. First, if the determinant in the denominator is zero, merely write that there is no unique solution to the problem. Such systems are either inconsistent or dependent. Second, although Cramer's rule extends to systems beyond three equations with three variables, the determinants associated with such systems are usually hard to evaluate without a machine. An easier approach in such cases is to use Gaussian elimination, as discussed in the previous section. Finally, for the purposes of Cramer's rule, zero elements are positioned in determinants to correspond to any missing terms, since a system like

$$\begin{array}{rcrcl} 2x & + & y & = & 1 \\ & & 5y + 4z & = & 1 \\ -3x & + & 2z & = & 1 \end{array} \quad \text{is equivalent to} \quad \begin{array}{rcrcrcl} 2x & + & y & + & 0z & = & 1 \\ 0x & + & 5y & + & 4z & = & 1 \\ -3x & + & 0y & + & 2z & = & 1. \end{array}$$

EXPLORE 11.3

1. Calculators with matrix capability include an operation (often symbolized "det") that computes determinants. If your calculator has this capability, use it to calculate these determinants.

a. $\begin{vmatrix} 5 & -4 \\ -1 & 12 \end{vmatrix}$ **b.** $\begin{vmatrix} 3 & 5 & 7 \\ -0.2 & 3.4 & 6 \\ 2 & 1 & -8 \end{vmatrix}$ **c.** $\begin{vmatrix} 3 & 6 & 9 & -2 \\ 1 & 1 & 1 & 1 \\ 2 & -5 & 3 & 0 \\ -7 & 2 & 2 & -7 \end{vmatrix}$

2. The elements in certain matrices have clear patterns. Patterns that are widely used are given special names. The following two patterned square matrices are useful in statistics as well as other fields of application.

   a. *Vandermonde matrix* In a Vandermonde matrix, the first row is all ls, the second row can be anything, and the rest of the rows are increasing powers of the second row. Confirm that matrix A is a Vandermonde matrix, and compute $|A|$.

$$A = \begin{bmatrix} 1 & 1 & 1 & 1 \\ -1 & 3 & 2 & -2 \\ 1 & 9 & 4 & 4 \\ -1 & 27 & 8 & -8 \end{bmatrix}$$

   b. *Toeplitz Matrix*: A Toeplitz matrix is a square matrix with equal elements on each diagonal, as illustrated. Confirm that matrix B is a Toeplitz matrix, and compute $|B|$.

$$B = \begin{bmatrix} 6 & 2 & 3 & 0 & 0 \\ 4 & 6 & 2 & 3 & 0 \\ 5 & 4 & 6 & 2 & 3 \\ 0 & 5 & 4 & 6 & 2 \\ 0 & 0 & 5 & 4 & 6 \end{bmatrix}$$

EXERCISES 11.3

In Exercises 1–16 evaluate the determinant.

1. $\begin{vmatrix} 1 & 3 \\ 4 & 2 \end{vmatrix}$ **2.** $\begin{vmatrix} -2 & 5 \\ 1 & 3 \end{vmatrix}$

3. $\begin{vmatrix} 1 & -5 \\ 2 & -8 \end{vmatrix}$ **4.** $\begin{vmatrix} 2 & -3 \\ 2 & -5 \end{vmatrix}$

5. $\begin{vmatrix} 5 & -3 \\ 2 & 4 \end{vmatrix}$ **6.** $\begin{vmatrix} -2 & 14 \\ 3 & -3 \end{vmatrix}$

7. $\begin{vmatrix} -3 & 0 \\ 4 & 1 \end{vmatrix}$ **8.** $\begin{vmatrix} 5 & -28 \\ 4 & -10 \end{vmatrix}$

9. $\begin{vmatrix} 2 & 5 \\ -4 & -10 \end{vmatrix}$ **10.** $\begin{vmatrix} 4 & -8 \\ 6 & -12 \end{vmatrix}$

11. $\begin{vmatrix} -2 & 3 & 2 \\ 0 & 7 & 4 \\ 1 & -1 & 3 \end{vmatrix}$ **12.** $\begin{vmatrix} 0 & 2 & -1 \\ 5 & 1 & 0 \\ -3 & 0 & -4 \end{vmatrix}$

13. $\begin{vmatrix} 2 & 1 & 1 \\ 1 & -2 & -1 \\ 3 & 3 & -1 \end{vmatrix}$ **14.** $\begin{vmatrix} 2 & 3 & 2 \\ 1 & -3 & 3 \\ 5 & 1 & -4 \end{vmatrix}$

15. $\begin{vmatrix} 6 & 3 & 11 \\ -3 & 4 & 5 \\ -1 & -2 & 3 \end{vmatrix}$ **16.** $\begin{vmatrix} -1 & 2 & -1 \\ 5 & -2 & -3 \\ 2 & -4 & 2 \end{vmatrix}$

In Exercises 17–20 find $|A|$.

17. $A = \begin{bmatrix} 3 & 8 \\ -8 & 11 \end{bmatrix}$ **18.** $A = \begin{bmatrix} -7 & -9 \\ -10 & -6 \end{bmatrix}$

19. $A = \begin{bmatrix} 2 & -7 & 1 \\ 4 & 1 & 2 \\ 6 & -3 & 3 \end{bmatrix}$ **20.** $A = \begin{bmatrix} 9 & 10 & -2 \\ 4 & 1 & 8 \\ -1 & -7 & 3 \end{bmatrix}$

In Exercises 21–40 solve the given system using Cramer's rule.

21. $\begin{aligned} x + y &= 25 \\ 6x - y &= 3 \end{aligned}$ **22.** $\begin{aligned} 2x + 3y &= 8 \\ 2x - 7y &= -32 \end{aligned}$

23. $\begin{aligned} 5x + 3y &= -2 \\ x - 2y &= -3 \end{aligned}$ **24.** $\begin{aligned} -2x - y &= -5 \\ 5x + 2y &= -17 \end{aligned}$

25. $\begin{aligned} 3x - 2y &= 1 \\ 6x - 4y &= 5 \end{aligned}$ **26.** $\begin{aligned} 2x - 3y &= 1 \\ 6x - 9y &= 3 \end{aligned}$

27. $\begin{aligned} 2x - 5y &= 5 \\ 4x + 3y &= 23 \end{aligned}$ **28.** $\begin{aligned} 4x + 2y &= 2 \\ 6x - 5y &= 27 \end{aligned}$

29. $\begin{aligned} 7x - 2y - 19 &= 0 \\ 3x + 5y + 14 &= 0 \end{aligned}$

30. $\begin{aligned} 6x + 10y - 7 &= 0 \\ 15x - 4y - 3 &= 0 \end{aligned}$

31. $\begin{aligned} x - y + z &= -1 \\ x + y - 3z &= 3 \\ 2x + y - 2z &= 4 \end{aligned}$

32. $\begin{aligned} x + y + z &= 1 \\ x + y - z &= 3 \\ x - y - z &= 5 \end{aligned}$

33. $\begin{aligned} -2x + y - z &= 3 \\ 2x + 3y + 2z &= 5 \\ 3x + 2y + z &= -3 \end{aligned}$

34. $\begin{aligned} x + y - 3z &= -4 \\ -x - y + 5z &= 8 \\ 2x + 3y - z &= 3 \end{aligned}$

35. $\begin{aligned} x - y + z &= 3 \\ 3x + y + z &= 1 \\ 2x + 2y - z &= -3 \end{aligned}$

36. $\begin{aligned} x + y + 3z &= 1 \\ 2x + 5y + 2z &= 0 \\ 3x - 2y - z &= 3 \end{aligned}$

37. $\begin{aligned} 2x + 3y &= 10 \\ 3x + 2z &= 2 \\ 4x + z &= 6 \end{aligned}$

38. $\begin{aligned} x + y + z &= 0 \\ 2x - 5y &= 3 \\ -5y + 3z &= 7 \end{aligned}$

39. $\begin{aligned} 2x - y + 3z &= 1 \\ -x + y - z &= -1 \\ -4x + 2y - 6z &= -2 \end{aligned}$

40. $\begin{aligned} x + y + z &= 1 \\ 2x - 3y - 2z &= -4 \\ 3x - 2y - z &= -2 \end{aligned}$

41. The table below lists the percentage of silver in each of two alloys. How much of each should be used to make 0.80 oz of 75 percent silver?

	percent silver
Alloy 1	68
Alloy 2	78

42. The table below lists the percentage of oil in each of two gasoline-oil mixtures. How much of each should be used to make 2.5 gal of 3 percent oil mixture?

	percent oil
Mix 1	0.5
Mix 2	8

43. A company sells two products, the Terminator and the Avenger. The weight and cost for one of each product is shown in the table.

	Product	
	Terminator	Avenger
cost	\$10	\$20
weight	5 lb	8 lb

On one order the total weight was 6,480 pounds, and the total cost was \$13,900. Determine how many of each item were in the order.

44. A family buys stock in two companies, B&J and Teebear. The cost per share and the gain per share for 1 year are listed in the table below.

	Stock	
	B&J	Teebear
cost per share	$8	$12
gain per share	$0.50	$0.85

At the end of the year, a total original investment of $1000 was worth $1066.50. Determine how many shares of each stock the family bought.

In Exercises 45–48 solve the equations simultaneously to determine the value of the currents I_1, I_2, and I_3 shown in the diagram.

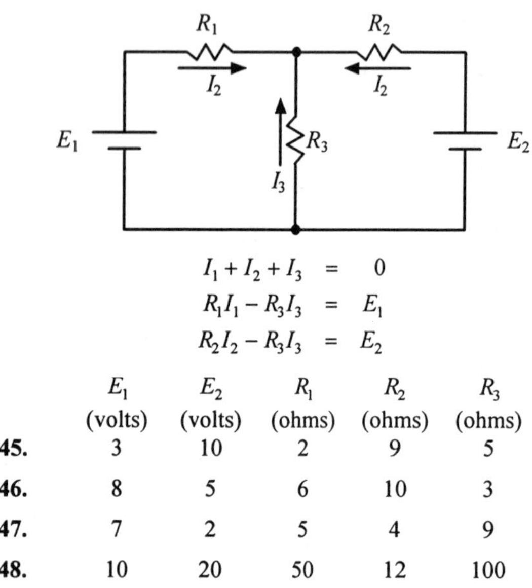

$$
\begin{aligned}
I_1 + I_2 + I_3 &= 0 \\
R_1 I_1 - R_3 I_3 &= E_1 \\
R_2 I_2 - R_3 I_3 &= E_2
\end{aligned}
$$

	E_1 (volts)	E_2 (volts)	R_1 (ohms)	R_2 (ohms)	R_3 (ohms)
45.	3	10	2	9	5
46.	8	5	6	10	3
47.	7	2	5	4	9
48.	10	20	50	12	100

49. As part of a project on nutrition, a student wrote the following table. [Source for values: Sandoz Pharmaceutical Co., *Countdown on Cholesterol*]

Food	Cholesterol (mg)	Total fat (g)	Calories
Wendy's hamburger plain	70	15.0	240
Wendy's french fries (small)	0	12.0	240
Ice cream (vanilla, soft serve)	153	22.5	377

The student imagined a person eating only these foods one day, and calculated that the person's total intake was 656 mg of cholesterol, 168 grams of total fat, and 2914 calories. Determine how many servings of each food the person consumed.

50. The student in the previous exercise designs an alternative diet that is supposed to be healthier. [Source for values: Sandoz Pharmaceutical Co., *Countdown on Cholesterol*]

Food	Cholesterol (mg)	Total fat (g)	Calories
Tuna on rye bread w/mayo	38	14.5	224
Baked potato	0	0.2	220
Yogurt (plain)	14	3.7	70

Assume that a person eats only these foods one day, with a total intake of 218 mg of cholesterol, 80.7 grams of total fat, and 2140 calories. Determine how many servings of each food the person consumes.

THINK ABOUT IT 11.3

1. Consider the system of equations given by

$$
\begin{aligned}
x + y &= b - 1 \\
x + by &= 5
\end{aligned}
$$

a. Use Cramer's rule to solve the system in terms of b.

b. How many solutions are there if $b = 1$?

c. For what value(s) of b does the system have a unique solution?

2. Consider the system of equations given by

$$
\begin{aligned}
ax + 2y + z &= 0 \\
2x - y - 2z &= 0 \\
x + y + 3z &= 0.
\end{aligned}
$$

a. For what value of a will the determinant D of the coefficients of x, y, and z equal 0?
b. When a has the value found in part a, how many solutions are there for the system?
c. For what value(s) of a does the system have a unique solution?
d. What is the unique solution?

3. a. Evaluate $\begin{vmatrix} a_1 & b_1 \\ 0 & b_2 \end{vmatrix}$.

b. Evaluate $\begin{vmatrix} a_1 & b_1 & c_1 \\ 0 & b_2 & c_2 \\ 0 & 0 & c_3 \end{vmatrix}$.

c. Evaluate $\begin{vmatrix} a_1 & b_1 & c_1 & d_1 \\ 0 & b_2 & c_2 & d_2 \\ 0 & 0 & c_3 & d_3 \\ 0 & 0 & 0 & d_4 \end{vmatrix}$.

d. Parts **a–c** show determinants of matrices that are called **upper triangular matrices**. How may the value of this type of determinant be found?

4. The following problems consider three useful properties of determinants that correspond to elementary row operations on matrices.

a. Verify the identity

$$\begin{vmatrix} ka_1 & kb_1 \\ a_2 & b_2 \end{vmatrix} = k\begin{vmatrix} a_1 & b_1 \\ a_2 & b_2 \end{vmatrix}.$$

If each element of a row of matrix A is multiplied by a nonzero real number to form matrix B, then how is $|B|$ related to $|A|$?

b. Verify the identity

$$\begin{vmatrix} a_1 & b_1 & c_1 \\ a_2 & b_2 & c_2 \\ a_3 & b_3 & c_3 \end{vmatrix} = -\begin{vmatrix} a_2 & b_2 & c_2 \\ a_1 & b_1 & c_1 \\ a_3 & b_3 & c_3 \end{vmatrix}.$$

If two rows of matrix A are interchanged to form matrix B, then how is $|B|$ related to $|A|$?

c. Create an example in which you start with a matrix, say A, and add a multiple of the entries in one row to another row to form matrix B. Then, find $|A|$ and $|B|$. What property of determinants is suggested by this example?

5. This section concentrated on the use of determinants to solve systems of linear equations. Determinants have many other applications, as illustrated by the next two problems.

a. If the vertices of a triangle are a points (x_1, y_1), (x_2, y_2), and (x_3, y_3), then the area of the triangle is given by the absolute value of the following determinant.

$$\frac{1}{2}\begin{vmatrix} x_1 & y_1 & 1 \\ x_2 & y_2 & 1 \\ x_3 & y_3 & 1 \end{vmatrix}$$

Use this rule to find the area of the triangle whose vertices are at $(-4,-2)$, $(-3,2)$, and $(1, 1)$.

b. An equation for the line that passes through distinct points (x_1, y_1) and (x_2, y_2) may be written in determinant form as

$$\begin{vmatrix} x & y & 1 \\ x_1 & y_1 & 1 \\ x_2 & y_2 & 1 \end{vmatrix} = 0.$$

Use this rule to find an equation for the line through $(-1,6)$ and $(2,-3)$.

● ● ●

11.4 Solving Systems by Matrix Algebra

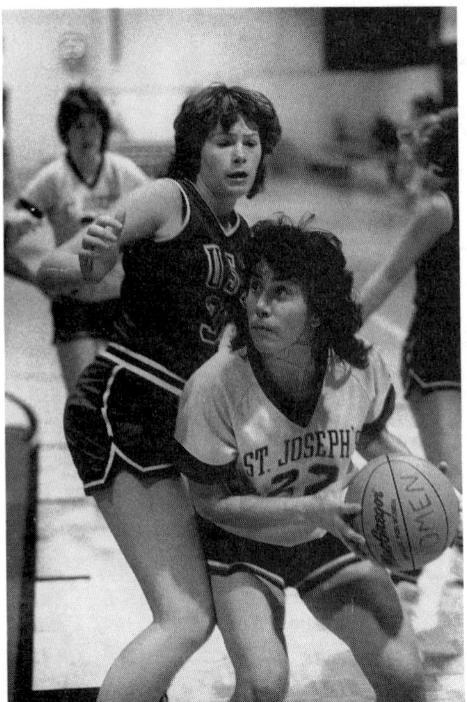

Photo Courtesy of ***Dean Abramson*** *of Stock Boston, Boston, MA*

The 1994 Statistical Abstract of the U.S. gives the following data for the attendance at NCAA women's college basketball games.

Year	Attendance (in thousands)
85	2072
87	2156
88	2325
89	2502
90	2777
91	3013
92	3397

a. A graph of this data reveals a trend that may be approximated by a segment of a parabola, so this relation may be modeled by an equation of the form $y = ax^2 + bx + c$. Find such a model for this data by selecting the data points associated with three years, say 1985, 1988, and 1990, and determining the values of a, b, and c. In your analysis let $x = 0$ correspond to 1980, so that the three data points simplify to (5, 2072), (8, 2325), and (10, 2777), respectively.

b. If the model obtained in part **a** had been used in 1990 to predict attendance for 1992, what would the predicted attendance have been? Compare this prediction to what actually happened.

(See Example 8.)

Objectives

1. Add, subtract, and multiply matrices, and multiply a matrix by a scalar.
2. Find the inverse of a matrix, if it exists.
3. Write a system of linear equations in matrix form $AX = B$ and use A^{-1} to solve the system.

There is one other method for solving a system of linear equations that merits our attention. In Section 11.2 we used elementary row operations on matrices to solve the system. A different matrix method is to develop a matrix algebra that define how to perform various operations with matrices and includes some properties that hold in this system. This approach is beneficial in three ways.

1. Matrices have many applications independent of systems of equations, and the ideas discussed here can serve as an introduction to this important topic. Because of this we develop a little more matrix algebra than is needed to solve a system of equations. More detailed coverage of this topic is usually given in a course in linear algebra.
2. Although the matrix algebra approach usually requires more work than our other methods to solve a given system manually, it is an efficient technique for solving systems such as

$$2x + 3y = b_1$$
$$3x + 5y = b_2$$

where the coefficients of x and y remain fixed but different pairs of constants replace the b's. We consider this problem in Example 7.
3. The matrix capability that is available on most graphing calculators may be used with the matrix algebra approach to give a quick and straightforward method for solving a linear system with n equations and n unknowns.

To begin, we first recall that a **matrix** is a rectangular array of numbers enclosed in brackets (or parentheses) and that each number in the matrix is called an **entry** or **element** of the matrix. We now consider some other basic definitions.

Matrix Definitions

1. A matrix with m rows and n columns is of **dimension** $m \times n$.
2. A **square matrix** is a matrix having the same number of rows as columns.
3. A **zero matrix** is a matrix containing only zero elements. We often denote a zero matrix by a boldface zero: **0**.
4. Two matrices are **equal** if and only if the elements in corresponding positions are equal.
5. To **add** two matrices of the same dimension, add elements in corresponding positions. Only matrices of the same dimension can be added.
6. To multiply a matrix by a real number, multiply each element in the matrix by that number. In this context we call the real number a **scalar**, and we call this operation **scalar multiplication**.
7. $A - B = A + (-B)$ where $-B = (-1)B$.

Now consider Example 1, which illustrate these definitions.

Example 1: Using Matrix Definitions

Answer the following questions given

$$A = \begin{bmatrix} 2 & -1 & 3 \\ 0 & 1 & -2 \end{bmatrix}, \; B = \begin{bmatrix} 0 & 0 \\ 0 & 0 \end{bmatrix}$$

$$C = \begin{bmatrix} 1 & 0 \\ 0 & 1 \end{bmatrix}, \text{ and } D = \begin{bmatrix} 1 & 3 & 6 \\ 2 & 5 & 7 \end{bmatrix}$$

a. What is the dimension of each matrix?

b. Find (if possible) $A + D$.

c. Find (if possible) $A + C$.

d. Find $-A$.

e. If $A + 2X = D$, find X.

Solution:

a. A and D are matrices of dimension 2×3, since both have 2 rows and 3 columns. B and C are square matrices with 2 rows and 2 columns, and are of dimension 2×2. Matrix B is an example of a zero matrix.

b. A and D are both of dimension 2×3, so $A+D$ is defined and is found as follows:

$$\begin{bmatrix} 2 & -1 & 3 \\ 0 & 1 & -2 \end{bmatrix} + \begin{bmatrix} 1 & 3 & 6 \\ 2 & 5 & 7 \end{bmatrix} = \begin{bmatrix} 2+1 & -1+3 & 3+6 \\ 0+2 & 1+5 & -2+7 \end{bmatrix} = \begin{bmatrix} 3 & 2 & 9 \\ 2 & 6 & 5 \end{bmatrix}.$$

c. A and C are not of the same dimension, so $A+C$ is undefined.

d. $-A=(-1)A$, so to find $-A$ we multiply each element in A by -1. Then

$$-A = -1\begin{bmatrix} 2 & -1 & 3 \\ 0 & 1 & -2 \end{bmatrix} = \begin{bmatrix} -2 & 1 & -3 \\ 0 & -1 & 2 \end{bmatrix}.$$

e. Since A and D are of dimension 2×3, matrix X must also have 2 rows and 3 columns. If we let

$$X = \begin{bmatrix} a & b & c \\ d & e & f \end{bmatrix}, \text{ then } 2X = \begin{bmatrix} 2a & 2b & 2c \\ 2d & 2e & 2f \end{bmatrix}, \text{ so}$$

$$\begin{bmatrix} 2 & -1 & 3 \\ 0 & 1 & -2 \end{bmatrix} + \begin{bmatrix} 2a & 2b & 2c \\ 2d & 2e & 2f \end{bmatrix} = \begin{bmatrix} 1 & 3 & 6 \\ 2 & 5 & 7 \end{bmatrix}$$

$$\begin{bmatrix} 2+2a & -1+2b & 3+2c \\ 0+2d & 1+2e & -2+2f \end{bmatrix} = \begin{bmatrix} 1 & 3 & 6 \\ 2 & 5 & 7 \end{bmatrix}.$$

For two matrices to be equal, elements in corresponding positions must be the same. Thus,

$$2+2a=1, \text{ so } a=-\frac{1}{2}; \qquad -1+2b=3, \text{ so } b=2;$$

$$3+2c=6, \text{ so } c=\frac{3}{2}; \qquad 0+2d=2, \text{ so } d=1;$$

$$1+2e=5, \text{ so } e=2; \qquad -2+2f=7, \text{ so } f=\frac{9}{2}.$$

So

$$X = \begin{bmatrix} -\frac{1}{2} & 2 & \frac{3}{2} \\ 1 & 2 & \frac{9}{2} \end{bmatrix}.$$

PROGRESS CHECK 1

Answer the following questions given

$$A = \begin{bmatrix} 2 \\ -5 \end{bmatrix}, \quad B = \begin{bmatrix} -3 \\ 9 \end{bmatrix}, \text{ and } C = \begin{bmatrix} 4 & 0 \end{bmatrix}.$$

a. What are the dimensions of each matrix?

b. Find (if possible) $A+B$.

c. Find (if possible) $A+C$.

d. Find (if possible) $2A-3B$.

e. If $A+2X=B$, find X. ■

Many of the properties from the algebra of real numbers carry over to the algebra of matrices. That is because we have defined our operations to this point in terms of operating on elements that are real numbers and that are in corresponding positions. Some of the more basic properties are as follows.

Properties of Matrices

If A, B, and C are matrices and c and k are scalars (real numbers), then

1. $(A+B)+C = A+(B+C)$
2. $A+B = B+A$
3. $A+0 = 0+A = A$
4. $A+(-A) = (-A)+A = 0$
5. $c(kA) = (ck)A$
6. $c(A+B) = cA+cB$.

These properties provide us with an alternative method for solving the problem in Example 1e that is similar to our previous work with equations. That is, we start with $A+2X = D$ and solve for X as follows:

$$\begin{array}{lclcll} A & + & 2X & = & D & \\ A+(-A) & + & 2X & = & D+(-A) & \text{Add } -A \text{ to both sides.} \\ 0 & + & 2X & = & D-A & \text{Property 4 and subtraction definition} \\ & & 2X & = & D-A & \text{Property 3} \\ & & X & = & \frac{1}{2}(D-A) & \text{Multiply both sides by the scalar } \frac{1}{2}. \end{array}$$

Then

$$X = \frac{1}{2}\left(\begin{bmatrix} 1 & 3 & 6 \\ 2 & 5 & 7 \end{bmatrix} - \begin{bmatrix} 2 & -1 & 3 \\ 0 & 1 & -2 \end{bmatrix}\right)$$

$$= \frac{1}{2}\begin{bmatrix} -1 & 4 & 3 \\ 2 & 4 & 9 \end{bmatrix} = \begin{bmatrix} -\frac{1}{2} & 2 & \frac{3}{2} \\ 1 & 2 & \frac{9}{2} \end{bmatrix}$$

Note that this answer agrees with our previous result.

To this point, our definitions and properties have been straightforward and predictable. Matrix multiplication, however, is unusual and we do *not* multiply matrices by simply multiplying elements in corresponding positions. To understand the usefulness of the different type of product that is defined in matrix multiplication, consider the following matrices.

$$\begin{array}{r} \\ \text{Plant A} \\ \text{Plant B} \\ \text{Plant C} \end{array}\begin{array}{c} \begin{array}{cc} \text{Printers} & \text{Monitors} \end{array} \\ \begin{bmatrix} 70 & 30 \\ 50 & 90 \\ 80 & 10 \end{bmatrix} \end{array} = A \qquad \begin{array}{r} \text{Printers} \\ \text{Monitors} \end{array}\overset{\text{Value}}{\begin{bmatrix} 300 \\ 200 \end{bmatrix}} = B$$

Matrix A summarizes information about the location of computer equipment, while matrix B indicates the value of that equipment. We can find the total value of the equipment at each plant as follows:

	(number of printers)	·	(value per printer)	+	(number of monitors)	·	(value per monitor)	=	total value
Part A	70	·	300	+	30	·	200	=	27,000
Part B	50	·	300	+	90	·	200	=	33,000
Part C	80	·	300	+	10	·	200	=	26,000

and we can represent this result in the following matrix.

$$\begin{matrix} & \text{Value} \\ \text{Plant A} \\ \text{Plant B} \\ \text{Plant C} \end{matrix} \begin{bmatrix} 27{,}000 \\ 33{,}000 \\ 26{,}000 \end{bmatrix} = C.$$

We wish to define matrix multiplication so that $AB = C$. First, note that for this product to make sense, the columns in matrix A (type of computer equipment) must match the rows of matrix B. Thus the product AB of two matrices is defined only when the number of columns in A is the same as the number of rows in B. Also note that the product AB (or matrix C) gets its row description from the rows of A (plant location), and its column description from the column(s) of matrix B(value). Thus, the product AB of two matrices has as many rows as A and as many columns as B. We summarize these results in this diagram.

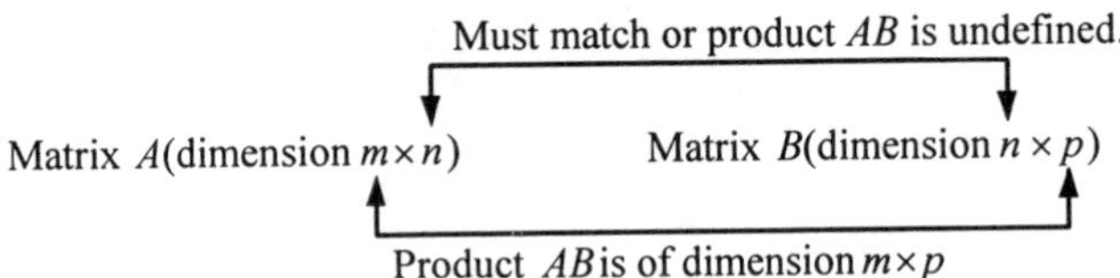

Finally, the element in row 1, column 1, in AB is found by multiplying the entries in the first row of A by the corresponding entries in the first column of B and adding the results. Similarly, the element in row 2, column 1, is found by multiplying the entries in the second row of A by the corresponding entries in the first column of B. This pattern continues, and our observations suggest the following definition for matrix multiplication.

Matrix Multiplication

> If A is an $m \times n$ matrix and B is an $n \times p$ matrix, then the product AB is an $m \times p$ matrix in which the ith row, jth column element of AB is found by multiplying each element in the ith row of A by the corresponding element in the jth column of B and adding the results.

Example 2: Multiplying Two Matrices

Let

$$A = \begin{bmatrix} 3 & 0 & 1 \\ 4 & 4 & 2 \end{bmatrix} \quad B = \begin{bmatrix} 1 & 2 \\ 4 & 0 \\ 7 & 3 \end{bmatrix}$$

Find (if possible) the product AB.

Solution: A is of dimension 2×3, and B is of dimension 3×2. Thus, the product is defined and AB is a 2×2 matrix. Such a matrix has 4 elements, which are determined as follows.

Step 1 (row 1, column 1)

$$\begin{bmatrix} 3 & 0 & 1 \\ 4 & 4 & 2 \end{bmatrix} \begin{bmatrix} 1 & 2 \\ 4 & 0 \\ 7 & 3 \end{bmatrix} \qquad \begin{matrix} \textbf{Computation} \\ 3 \cdot 1 + 0 \cdot 4 + 1 \cdot 7 = 10 \end{matrix}$$

Step 2 (row 1, column 2)

$$\begin{bmatrix} 3 & 0 & 1 \\ 4 & 4 & 2 \end{bmatrix} \begin{bmatrix} 1 & 2 \\ 4 & 0 \\ 7 & 3 \end{bmatrix} \quad 3 \cdot 2 + 0 \cdot 0 + 1 \cdot 3 = 9$$

Step 3 (row 2 column 1)

$$\begin{bmatrix} 3 & 0 & 1 \\ 4 & 4 & 2 \end{bmatrix} \begin{bmatrix} 1 & 2 \\ 4 & 0 \\ 7 & 3 \end{bmatrix} \quad 4 \cdot 1 + 4 \cdot 4 + 2 \cdot 7 = 34$$

Step 4 (row 2, column 2)

$$\begin{bmatrix} 3 & 0 & 1 \\ 4 & 4 & 2 \end{bmatrix} \begin{bmatrix} 1 & 2 \\ 4 & 0 \\ 7 & 3 \end{bmatrix} \quad 4 \cdot 2 + 4 \cdot 0 + 2 \cdot 3 = 14$$

Thus,

$$AB = \begin{bmatrix} 10 & 9 \\ 34 & 14 \end{bmatrix}.$$

PROGRESS CHECK 2

Find (if possible) the product AB, given that

$$A = \begin{bmatrix} 1 & 2 & -1 \\ -3 & 0 & 4 \end{bmatrix} \text{ and } B = \begin{bmatrix} 1 & 2 \\ 3 & 1 \\ 0 & 4 \end{bmatrix}$$

■

Example 3: Multiplying Two Matrices

Let

$$A = \begin{bmatrix} 1 & 2 & 3 & 4 \end{bmatrix} \text{ and } B = \begin{bmatrix} -1 & 4 \\ 0 & 1 \\ 5 & -3 \\ -2 & 0 \end{bmatrix}.$$

Find (if possible)

a. the product AB and **b.** the product BA.

Solution:

a. A is a 1×4 matrix and B is a 4×2 matrix. Thus, the product is defined and is a 1×2 matrix. We determine these two elements as follows:

$$\text{entry in row 1, column 1: } 1(-1)+2(0)+3(5)+4(-2)=6$$
$$\text{entry in row 1, column 2: } 1(4)+2(1)+3(-3)+4(0)=-3$$

Thus,

$$AB=[6 \quad -3].$$

b. Since B is a 4×2 and A is a 1×4 matrix, the number of columns in B does not equal the number of rows in A, so the product BA is undefined.

Note: This example shows that matrix multiplication is not commutative (that is, in general $AB\neq BA$), so we must be careful about the order in which we write matrices when expressing a product. For square matrices of the same dimension we do, however, have an associative property $(AB)C=A(BC)$ and a distributive property $A(B+C)=AB+AC$.

PROGRESS CHECK 3

Find (if possible)

a. the product AB and **b.** the product BA, given that

$$A=\begin{bmatrix}2\\-5\end{bmatrix} \text{ and } B=\begin{bmatrix}3 & -1\\9 & 6\end{bmatrix}.$$

■

Example 4: Expressing a Linear System as a Matrix Equation

Let

$$A=\begin{bmatrix}2 & 3\\3 & 5\end{bmatrix}, \quad X=\begin{bmatrix}x\\y\end{bmatrix}, \quad \text{and} \quad B=\begin{bmatrix}5\\9\end{bmatrix}.$$

Write the system of linear equations represented by $AX=B$.

Solution: A is a 2×2 matrix and X is a 2×1 matrix, so AX is defined and is a 2×1 matrix. The two elements in the product matrix are

$$\begin{matrix}\text{entry in row 1, column 1: } 2x+3y\\ \text{entry in row 2, column 1: } 3x+5y\end{matrix} \quad \text{so } AX=\begin{bmatrix}2x & + & 3y\\3x & + & 5y\end{bmatrix}.$$

Since $AX=B$, we know

$$\begin{bmatrix}2x+3y\\3x+5y\end{bmatrix}=\begin{bmatrix}5\\9\end{bmatrix},$$

so from the definition of equal matrices we obtain the system

$$\begin{aligned}3x+3y &= 5\\3x+5y &= 9.\end{aligned}$$

PROGRESS CHECK 4
Let

$$A=\begin{bmatrix}3 & -1 & 6\\ 1 & 2 & -3\\ 2 & -3 & -1\end{bmatrix},\ X=\begin{bmatrix}x\\ y\\ z\end{bmatrix},\ \text{and } B=\begin{bmatrix}1\\ 0\\ -9\end{bmatrix}.$$

Write the system of linear equations represented by $AX = B$.

Example 4 illustrates that a system of linear equations can be represented as a matrix equation $AX = B$, so if we can solve this matrix equation for X, the result will give us the solution to the system of equations. To solve $AX = B$ for X, you might suggest that we divide both side by A, since that is one approach to solving equations. However, division of matrices is not defined, so this approach doesn't work here. A second approach is to multiply both sides of the equation by $1/A$ (or the multiplicative inverse of A), since in the algebra of real numbers

$$\begin{aligned}\left(\frac{1}{A}\right)Ax &= \left(\frac{1}{A}\right)B\\ 1\cdot x &= \left(\frac{1}{A}\right)B.\\ x &= \left(\frac{1}{A}\right)B\end{aligned}$$

This approach can be extended to matrix algebra if we define matrices that can play the parts of $1/A$ and 1 from the equations above. The necessary definitions for the part of 1 are as follows:

1. The **principal diagonal** of a square matrix consists of the elements in the diagonal extending from the upper-left corner to the lower-right corner.
2. The $n \times n$ **identity matrix** is the square matrix with n rows and n columns with 1's on the principal diagonal and 0's elsewhere. An identity matrix is symbolized by I.

An identity matrix is important because

$$AI = IA = A$$

for any square matrix A. Example 5 illustrates this property.

Example 5: Products Involving an Identity Matrix

If $A=\begin{bmatrix}3 & 1\\ 2 & 4\end{bmatrix}$, verify that $AI = IA = A$.

Solution: Since A is a 2×2 matrix, we use $I=\begin{bmatrix}1 & 0\\ 0 & 1\end{bmatrix}$, which is the 2×2 identity matrix. Then

$$AI=\begin{bmatrix}3 & 1\\ 2 & 4\end{bmatrix}\begin{bmatrix}1 & 0\\ 0 & 1\end{bmatrix}=\begin{bmatrix}3\cdot 1+1\cdot 0 & 3\cdot 0+1\cdot 1\\ 2\cdot 1+4\cdot 0 & 2\cdot 0+4\cdot 1\end{bmatrix}=\begin{bmatrix}3 & 1\\ 2 & 4\end{bmatrix}=A$$

Also,

$$IA=\begin{bmatrix}1 & 0\\ 0 & 1\end{bmatrix}\begin{bmatrix}3 & 1\\ 2 & 4\end{bmatrix}=\begin{bmatrix}1\cdot 3+0\cdot 2 & 1\cdot 1+0\cdot 4\\ 0\cdot 3+1\cdot 2 & 0\cdot 1+1\cdot 4\end{bmatrix}=\begin{bmatrix}3 & 1\\ 2 & 4\end{bmatrix}=A$$

PROGRESS CHECK 5
Verify that $AI = IA = A$ if

$$A = \begin{bmatrix} 5 & -2 & 3 \\ 2 & 4 & -1 \\ 1 & 8 & 6 \end{bmatrix}.$$

We now need a matrix to play the part of $1/A$, so we wish to find a square matrix A^{-1}, called the **multiplicative inverse of *A***, such that

$$A^{-1}A = AA^{-1} = I.$$

It can be shown that A has an inverse if and only if the determinant of A is not zero and that when A^{-1} does exist, it is unique and can be obtained in the following way.

To Find A^{-1}

1. Write the augmented matrix $[A|I]$, where I is the identity matrix with the same dimension as A.
2. Use the elementary row operations shown in Section 11.2 to replace matrix $[A|I]$ with a matrix of the form $[I|B]$.
3. Then A^{-1} is matrix B.

When it is not possible to obtain $[I|B]$ (for instance, you may obtain all 0's in a row in the left portion of the matrix), then A has no inverse, and A is said to be singular. Example 6 illustrates the procedure.

Example 6: Finding the Inverse of a Matrix

If $A = \begin{bmatrix} 2 & 3 \\ 3 & 5 \end{bmatrix}$, find A^{-1} (if it exists).

Solution: First, we form the augmented matrix $[A|I]$.

$$\left[\begin{array}{cc|cc} 2 & 3 & 1 & 0 \\ 3 & 5 & 0 & 1 \end{array}\right]$$

We then convert $[A|I]$ to the form $[I|B]$, using elementary row operations. To obtain 1 for the row 1, column 1, entry, we multiply each element in row 1 by $1/2$.

$$\left[\begin{array}{cc|cc} 1 & \frac{3}{2} & \frac{1}{2} & 0 \\ 3 & 5 & 0 & 1 \end{array}\right]$$

To obtain 0 for the row 2, column 1 entry, we add -3 times each entry in row 1 to the corresponding entry in row 2.

$$\left[\begin{array}{cc|cc} 1 & \frac{3}{2} & \frac{1}{2} & 0 \\ 0 & \frac{1}{2} & -\frac{3}{2} & 1 \end{array}\right]$$

Next, multiply each entry in row 2 by 2, so the row 2, column 2 entry is 1.

$$\left[\begin{array}{cc|cc} 1 & \frac{3}{2} & \frac{1}{2} & 0 \\ 0 & 1 & -3 & 2 \end{array}\right]$$

Finally, add $-(3/2)$ times each entry in row 2 to the corresponding entry in row 1 to obtain 0 for the row 1, column 2 entry.

$$\left[\begin{array}{cc|cc} 1 & 0 & 5 & -3 \\ 0 & 1 & -3 & 2 \end{array}\right]$$

The matrix is now in the form $[I|B]$, so by the given rule, B is the inverse of A. Thus,

$$A^{-1} = \begin{bmatrix} 5 & -3 \\ -3 & 2 \end{bmatrix}.$$

Technology Link

Most graphing calculators have the ability to add, subtract, and multiply matrices, perform scalar multiplication, and find the inverse of a matrix. If your calculator has this capability then you should redo the problems in Examples 1–6 and compare your results to the text's answers. When redoing Examples 1c and 3b, note the type of error message your calculator displays when doing a computation that is undefined because of dimension restrictions. You should also create an example of a matrix whose determinant is zero, and then discover the error message associated with trying to find the inverse of a matrix that is singular.

PROGRESS CHECK 6

If $A = \begin{bmatrix} 7 & 5 \\ 4 & 3 \end{bmatrix}$, find A^{-1} (if it exists). ■

If A^{-1} exists, the matrix equation $AX = B$ can now be solved by multiplying both sides of the equation by A^{-1} and simplifying as follows:

$$\begin{aligned} A^{-1}AX &= A^{-1}B \\ IX &= A^{-1}B \quad (\text{since } A^{-1}A = 1) \\ X &= A^{-1}B \quad (\text{since } IX = X) \end{aligned}$$

Thus, the product $A^{-1}B$ gives the solution to a system of linear equations. In practice, it is usually easier to use row operation to solve a particular linear system than to determine A^{-1} (assuming a calculator with matrix capability is not being used). However, Example 7 considers the problem mentioned in the introduction to this section, in which the matrix algebra method is an efficient approach.

Example 7: Solving Linear Systems Using Matrix Algebra

Solve the system

$$\begin{aligned} 2x + 3y &= b_1 \\ 3x + 5y &= b_2 \end{aligned}$$

where

a. $b_1 = 5$, $b_2 = 9$ **b.** $b_1 = -3$, $b_2 = 0$.

Solution:

a. The given system can be represented by $AX = B$, where

$$A = \begin{bmatrix} 2 & 3 \\ 3 & 5 \end{bmatrix}, \; X = \begin{bmatrix} x \\ y \end{bmatrix}, \text{ and } B = \begin{bmatrix} 5 \\ 9 \end{bmatrix}.$$

If A^{-1} exists, the solution to $AX = B$ is $X = A^{-1}B$. We know from Example 6 that

$$A^{-1} = \begin{bmatrix} 5 & -3 \\ -3 & 2 \end{bmatrix},$$

so

$$\begin{bmatrix} x \\ y \end{bmatrix} = \begin{bmatrix} 5 & -3 \\ -3 & 2 \end{bmatrix}\begin{bmatrix} 5 \\ 9 \end{bmatrix} = \begin{bmatrix} 5(5)+(-3)(9) \\ -3(5)+2(9) \end{bmatrix} = \begin{bmatrix} -2 \\ 3 \end{bmatrix}.$$

Thus, $x = -2$ and $y = 3$.

b. We simply proceed as before but with different elements in matrix *B*.

$$\begin{bmatrix} x \\ y \end{bmatrix} = \begin{bmatrix} 5 & -3 \\ -3 & 2 \end{bmatrix}\begin{bmatrix} -3 \\ 0 \end{bmatrix} = \begin{bmatrix} 5(-3)+(-3)(0) \\ -3(-3)+2(0) \end{bmatrix} = \begin{bmatrix} -15 \\ 9 \end{bmatrix}.$$

Thus, $x = -15$ and $y = 9$. Note that once we know A^{-1}, it is easy to solve the system for any pair of values for b_1 and b_2.

PROGRESS CHECK 7

Solve the system given in Example 7 when $b_1 = 5$ and $b_2 = 6$. ■

A linear system with *n* equations and *n* unknowns may be solved in a fast and straightforward way by using the matrix algebra method in conjunction with a calculator with matrix capability. In the next example we use this approach to solve the application problem that opens this section.

Example 8: Using the Matrix Algebra Method with a Calculator

Solve the problem in the section introduction on page 904.

Solution:

a. Substituting the values $x = 5$ and $y = 2{,}072$, $x = 8$ and $y = 2{,}325$, and $x = 10$ and $y = 2{,}777$ into the equation $y = ax^2 + bx + c$ gives the following system of equations

$$\begin{array}{lcl} 2{,}072 = a(5)^2 + b(5) + c & \text{or} & 25a + 5b + c = 2{,}072 \\ 2{,}325 = a(8)^2 + b(8) + c & & 64a + 8b + c = 2{,}325 \\ 2{,}777 - a(10)^2 + b(10) + c & & 100a + 10b + c = 2{,}777 \end{array}$$

This system can be represented by $AX = B$, where

$$A = \begin{bmatrix} 25 & 5 & 1 \\ 64 & 8 & 1 \\ 100 & 10 & 1 \end{bmatrix}, \; X = \begin{bmatrix} a \\ b \\ c \end{bmatrix}, \text{ and } B = \begin{bmatrix} 2{,}072 \\ 2{,}325 \\ 2{,}777 \end{bmatrix}.$$

The solution to $AX = B$ is $X = A^{-1}B$, so you can solve this system using a calculator with matrix capability by entering matrices A and B and then computing $A^{-1}B$. The result of such a computation reveals that

$$X = A^{-1}B = \begin{bmatrix} 28.333333333 \\ -284 \\ 2783.66666667 \end{bmatrix}.$$

To two decimal places, the solution is $a = 28.33$, $b = -284$, and $c = 2,783.67$, so the equation of the parabola passing through the given points is $y = 28.33x^2 - 284x + 2,783.67$.

b. To find the predicted attendance if the model from part **a** had been used in 1990 to predict the attendance (in thousands) for 1992, it is necessary to replace x by 12 in this formula, since 1980 corresponds to $x = 0$.

$$y = 28.33(12)^2 - 284(12) + 2,783.67 \approx 3,455$$

The model predicts an attendance of about 3,445,000 people. In fact the attendance in 1992 was about 3,397,000 people, so this prediction would have overestimated the actual result by about 58,000 people. Although this may seem like a lot of people, an error of 58,000 with respect to an actual result of 3,397,000 translates to a relative error of less than 2 percent.

PROGRESS CHECK 8

Redo the questions in Example 8 but use the data points associated with the years 1985, 1988, and 1991. ■

EXPLORE 11.4

1. Calculators that have matrix capability can usually do matrix addition and multiplication. If you have such a calculator, enter matrices $A = \begin{bmatrix} 1 & 3 \\ 2 & 5 \end{bmatrix}$, $B = \begin{bmatrix} 3 & -1 \\ -2 & 1 \end{bmatrix}$ and $C = \begin{bmatrix} 2 & 4 \\ 6 & -8 \\ 5 & 1 \end{bmatrix}$.
   - **a.** Calculate $A + B$ and AB. **Note:** If you want to save the answers, you must store them with a new name.
   - **b.** Learn what error message you get if you try to calculate $A + C$ or AC.
2. In general, the *transpose* of an m by n matrix A is the n by m matrix A^T, whose columns are the rows of A, and whose rows are the columns of A.
   - **a.** If your calculator has a menu choice for Transpose (perhaps symbolized as ^{T} or "trns"), use it to produce C^T, the transpose of $C = \begin{bmatrix} 1 & 3 & 5 \\ 2 & 5 & 8 \end{bmatrix}$.
   - **b.** Compute the product $P = C^TC$. What are the dimensions of P?
   - **c.** Explain why for any matrix A, the product A^TA is always defined, but AA^T is not always defined.
3. The *identity matrix* I is a square matrix with ones on the main diagonal and zeros everywhere else. If A is a square matrix, then $AI = A$. The inverse, A^{-1}, of a square matrix A is the matrix for which $AA^{-1} = I$, and also $A^{-1}A = I$. It often takes a lot of calculations to find the inverse of a matrix, so calculators that find the inverse automatically are a great help.
   - **a.** If your calculator computes matrix inverses (often just by pressing the key marked x^{-1}), use it to produce the inverse of $A = \begin{bmatrix} 4 & 1 \\ 3 & 1 \end{bmatrix}$ and save this inverse as matrix B.
   - **b.** Find the products AB and BA, using the matrices from part **a**, and confirm that both products give the identity matrix, $I = \begin{bmatrix} 1 & 0 \\ 0 & 1 \end{bmatrix}$.

4. In some application the entries in a square matrix represent probabilities. For instance, matrix P (called a *transition matrix*) shown below describes a 2-state system when an experiment begins. It says that over the first time period, if the system is in state 1, there is a 20% chance that it will remain there, and an 80% chance that it will switch to state 2. If it is in state 2, there is a 40% chance that it will switch to state 1, and a 60% chance that it will stay in state 2.

$$P = \begin{matrix} & s1 & s2 \\ s1 & .20 & .80 \\ s2 & .40 & .60 \end{matrix}$$

In these applications the *powers* of the transition matrix describe what is likely to happen to the system as time passes.

a. $P^2 = PP$ describes what is likely to happen over 2 periods. Calculate P^2 and interpret the entries.

b. In a similar fashion, $P^3 = PPP$ describes what is likely to happen over 3 periods. Calculate P^3 and interpret the entries.

c. Many such systems eventually reach a "steady state" where the entries of the matrix do not change. Look at higher and higher powers of P, and determine the elements of the steady state matrix. The whole sequence of transition matrices is called a *Markov Chain.*

EXERCISES 11.4

In Exercises 1–4 specify the dimension of each matrix.

1. $\begin{bmatrix} 3 & 0 & 1 & -5 \\ -2 & 8 & 4 & 7 \end{bmatrix}$ **2.** $\begin{bmatrix} 0 & 1 \\ -1 & 0 \end{bmatrix}$

3. $\begin{bmatrix} 6 \\ 4 \end{bmatrix}$ **4.** $\begin{bmatrix} 1 & -3 & 5 \end{bmatrix}$

In Exercises 5–8 find the values of the variables.

5. $\begin{bmatrix} a & b \\ 5 & 4 \end{bmatrix} = \begin{bmatrix} 10 & -3 \\ c & d \end{bmatrix}$ **6.** $\begin{bmatrix} x \\ y \end{bmatrix} = \begin{bmatrix} -1 \\ 4 \end{bmatrix}$

7. $\begin{bmatrix} 1+a & -1-b & 3+c \end{bmatrix} = \begin{bmatrix} 3 & -1 & 0 \end{bmatrix}$

8. $\begin{bmatrix} 2+2a & -1+2b & 3+2c \\ 2d & 1+2e & -2+2f \end{bmatrix} = \begin{bmatrix} 4 & 0 & 1 \\ -2 & 7 & 2 \end{bmatrix}$

In Exercises 9–14 perform the indicated operations.

9. $\begin{bmatrix} 1 & 2 \\ 3 & 1 \end{bmatrix} + \begin{bmatrix} -2 & -5 \\ 1 & 9 \end{bmatrix}$

10. $\begin{bmatrix} \frac{1}{8} \\ \frac{5}{8} \\ \frac{7}{8} \end{bmatrix} - \begin{bmatrix} \frac{3}{8} \\ -\frac{1}{8} \\ \frac{5}{8} \end{bmatrix}$

11. $\frac{1}{2}\begin{bmatrix} 0 & 4 & -2 \\ 8 & -6 & 2 \end{bmatrix}$ **12.** $-3\begin{bmatrix} 1 & 0 \\ 0 & 1 \end{bmatrix}$

13. $2\begin{bmatrix} 3 & 2 \\ 1 & 0 \end{bmatrix} - \begin{bmatrix} -4 & 2 \\ 3 & 5 \end{bmatrix}$

14. $-\frac{1}{2}\begin{bmatrix} 2 & -4 & 6 \end{bmatrix} + \frac{1}{3}\begin{bmatrix} -9 & 3 & 6 \end{bmatrix}$

In Exercises 15–20 find the dimensions of the product AB and the product BA, if they are defined.

15. A is 2×2, B is 2×2 **16.** A is 2×3, B is 2×3

17. A is 2×3, B is 3×2 **18.** A is 4×1, B is 1×4

19. A is 4×1, B is 3×2 **20.** A is 3×3, B is 1×3

In Exercises 21–26 determine the given product.

21. $\begin{bmatrix} 1 & 3 \\ -1 & 0 \end{bmatrix}\begin{bmatrix} 2 & -3 \\ 3 & 1 \end{bmatrix}$ **22.** $\begin{bmatrix} 2 & -3 \\ 3 & 1 \end{bmatrix}\begin{bmatrix} 1 & 3 \\ -1 & 0 \end{bmatrix}$

23. $\begin{bmatrix} 1 & 3 \\ -1 & 0 \end{bmatrix}\begin{bmatrix} 1 & 0 \\ 0 & 1 \end{bmatrix}$ **24.** $\begin{bmatrix} 1 & 3 \\ -1 & 0 \end{bmatrix}\begin{bmatrix} 2 \\ 3 \end{bmatrix}$

25. $\begin{bmatrix} 3 & -4 & 1 \\ 0 & 5 & -3 \\ 2 & -1 & 4 \end{bmatrix}\begin{bmatrix} -1 \\ 2 \\ 1 \end{bmatrix}$

26. $\begin{bmatrix} 1 & 3 & -2 \\ -1 & 0 & 2 \end{bmatrix}\begin{bmatrix} -1 & 3 \\ 2 & 0 \\ 1 & 4 \end{bmatrix}$

In Exercises 27–38 perform the indicated operations (if they are defined) for matrices *A*, *B*, *C*, *D*, and *I*, as defined below.

$A=\begin{bmatrix} 2 & -1 & 3 \\ 0 & 1 & -2 \end{bmatrix}$, $B=\begin{bmatrix} 1 & 2 \\ -3 & -4 \end{bmatrix}$, $C=\begin{bmatrix} -1 & 0 \\ 0 & -1 \end{bmatrix}$,
$D=\begin{bmatrix} 1 & 3 & 6 \\ 2 & 5 & 7 \end{bmatrix}$, $I=\begin{bmatrix} 1 & 0 \\ 0 & 1 \end{bmatrix}$,

27. $A-D$
28. $A-B$
29. AB
30. BA
31. BI
32. IB
33. $2(B+C)$
34. $3D-2A$
35. II
36. $(BC)A$
37. $C(A+D)$
38. $(A+D)C$

In Exercises 39 and 40 write the system of linear equations represented by $AX=B$.

39. $A=\begin{bmatrix} 1 & 1 \\ 5 & -2 \end{bmatrix}$, $X=\begin{bmatrix} x \\ y \end{bmatrix}$, and $B=\begin{bmatrix} 0 \\ 3 \end{bmatrix}$.

40. $A=\begin{bmatrix} 3 & -1 & 5 \\ 1 & 0 & -2 \\ -1 & 1 & -1 \end{bmatrix}$, $X=\begin{bmatrix} x \\ y \\ z \end{bmatrix}$, and $B=\begin{bmatrix} 1 \\ -1 \\ 1 \end{bmatrix}$.

In Exercises 41–50 find the inverse, if it exists, for each matrix.

41. $\begin{bmatrix} -2 & -1 \\ 7 & 3 \end{bmatrix}$
42. $\begin{bmatrix} 1 & 1 \\ 6 & -1 \end{bmatrix}$
43. $\begin{bmatrix} 5 & 3 \\ 1 & -2 \end{bmatrix}$
44. $\begin{bmatrix} -2 & -1 \\ 5 & 2 \end{bmatrix}$
45. $\begin{bmatrix} 3 & -2 \\ 6 & -4 \end{bmatrix}$
46. $\begin{bmatrix} 7 & -2 \\ 3 & 5 \end{bmatrix}$
47. $\begin{bmatrix} 3 & 2 & 1 \\ -8 & 2 & 0 \\ 4 & 1 & 5 \end{bmatrix}$
48. $\begin{bmatrix} 2 & 3 & 0 \\ 3 & 0 & 2 \\ 0 & 4 & 1 \end{bmatrix}$
49. $\begin{bmatrix} 2 & -1 & 3 \\ -1 & 1 & -1 \\ -4 & 2 & -6 \end{bmatrix}$
50. $\begin{bmatrix} 1 & -1 & 1 \\ -1 & 1 & 0 \\ 0 & -1 & 1 \end{bmatrix}$

In Exercises 51–60 write the given system in the form $AX=B$ and use A^{-1} (if it exists) to solve the system. (**Note:** The inverses for these problems are found in Exercises 41–50.)

51. $\begin{aligned} -2x - y &= 2 \\ 7x + 3y &= -1 \end{aligned}$

52. $\begin{aligned} x + y &= 25 \\ 6x - y &= 3 \end{aligned}$

53. $\begin{aligned} 5x + 3y &= -2 \\ x - 2y &= -3 \end{aligned}$

54. $\begin{aligned} -2x - y &= -5 \\ 5x + 2y &= -17 \end{aligned}$

55. $\begin{aligned} 3x - 2y &= 1 \\ 6x - 4y &= 5 \end{aligned}$

56. $\begin{aligned} 7x - 2y &= 19 \\ 3x + 5y &= -14 \end{aligned}$

57. $\begin{aligned} 3x + 2y + z &= 5 \\ -8x + 2y &= -8 \\ 4x + y + 5z &= 14 \end{aligned}$

58. $\begin{aligned} 2x + 3y &= 10 \\ 3x + 2z &= 2 \\ 4y + z &= 6 \end{aligned}$

59. $\begin{aligned} 2x - y + 3z &= 1 \\ -x + y - z &= -1 \\ -4x + 2y - 6z &= -2 \end{aligned}$

60. $\begin{aligned} x - y + z &= 1 \\ -x + y &= -1 \\ y + z &= 2 \end{aligned}$

61. If the inverse of $\begin{bmatrix} 3 & -5 \\ -4 & 7 \end{bmatrix}$ is $\begin{bmatrix} 7 & 5 \\ 4 & 3 \end{bmatrix}$, then solve
$$\begin{aligned} 3x-5y &= b_1 \\ -4x+7y &= b_2 \end{aligned}$$
when
a. $b_1=2$, $b_2=3$, and
b. $b_1=0$, $b_2=-5$.

62. If the inverse of $\begin{bmatrix} 6 & -1 & -5 \\ -7 & 1 & 5 \\ -10 & 2 & 11 \end{bmatrix}$ is $\begin{bmatrix} -1 & -1 & 0 \\ -27 & -16 & -5 \\ 4 & 2 & 1 \end{bmatrix}$,
then solve
$$\begin{aligned} 6x-y-5z &= b_1 \\ -7x+y+5z &= b_2 \\ -10x+2y+11z &= b_3 \end{aligned}$$
when
a. $b_1=1$, $b_2=0$, $b_3=0$ and
b. $b_1=2$, $b_2=-1$, $b_3=-2$.

63. Consider the following matrices.

$$\begin{array}{r} \\ \text{Jennifer} \\ \text{David} \\ \text{Tom} \\ \text{Kelly} \end{array} \begin{array}{c} \begin{array}{cc} \text{Midterm} & \text{Final} \end{array} \\ \begin{bmatrix} 80 & 96 \\ 75 & 83 \\ 94 & 86 \\ 85 & 67 \end{bmatrix} \end{array} = A$$

$$\begin{array}{r} \\ \text{Midterm} \\ \text{Final} \end{array} \begin{array}{c} \begin{array}{cc} \text{System 1} & \text{System 2} \end{array} \\ \begin{bmatrix} 0.5 & 0.4 \\ 0.5 & 0.6 \end{bmatrix} \end{array} = B$$

Find AB and discuss what it means.

64. The following matrices arise in a problem in which 4 people are making drill team flags. There are 3 sizes of flags. Each size has a different number of blue and white stripes. [Source: VISMT]

Number of flags of each size

$$\text{people} \begin{array}{r} \\ \text{Kim} \\ \text{Lee} \\ \text{Pat} \\ \text{Chris} \end{array} \begin{array}{c} \begin{array}{ccc} \text{small} & \text{medium} & \text{large} \end{array} \\ \begin{bmatrix} 0 & 6 & 2 \\ 0 & 3 & 3 \\ 2 & 6 & 2 \\ 8 & 0 & 6 \end{bmatrix} \end{array} = A$$

Number of stripes needed

$$\text{sizes} \begin{array}{r} \\ \text{small} \\ \text{medium} \\ \text{large} \end{array} \begin{array}{c} \begin{array}{cc} \text{blue} & \text{white} \end{array} \\ \begin{bmatrix} 2 & 1 \\ 4 & 3 \\ 10 & 5 \end{bmatrix} \end{array} = B$$

cost per stripe ($)

$$\text{color} \begin{array}{r} \text{blue} \\ \text{white} \end{array} \begin{bmatrix} 1 \\ 2 \end{bmatrix} = C$$

Find each of these products and discuss what they mean.

a. AB **b.** BC **c.** ABC

d. Let $U = \begin{bmatrix} 1 \\ 1 \\ 1 \end{bmatrix}$ and find AU.

e. Let $V = [1 \;\; 1 \;\; 1 \;\; 1]$, and find VA.

Exercises 65 to 68 both lead to a system of linear equations. Solve the system by using the matrix methods of this section.

65. The points in the table suggest a graph like a parabola.

Point	A	B	C	D	E	F
X	0	2	4	6	7	10
Y	1	7	15	40	52	100

a. Plot the points and confirm the parabolic shape.

b. Find the equation of the parabola $y = ax^2 + bx + c$ that includes points A, C, and E.

66. Refer to the table in the previous exercise, and find the equation of the parabola $y = ax^2 + bx + c$ that includes points B, D, and F.

67. A large corporation takes inventory of some computer equipment and reports the number of printers and monitors by location as follows.

	Printers	Monitors
Plant A	70	30
Plant B	50	90
Plant C	80	10

For tax purposes they assign an average dollar value to each printer and to each monitor. As a result they find that the value of the equipment is $28,750, $46,250, and $23,750 at plants A, B, and C, respectively. What dollar value was assigned to printers? To monitors?

68. The following fiscal year, the corporation in Exercise 67 decides to include notebook computers in its inventory, which yields the table shown.

	Printers	Monitors	Notebooks
Plant A	60	40	20
Plant B	55	90	15
Plant C	70	15	30

They find the value of the equipment now to be $75,000, $82,750, and $89,500 at plants A, B, and C, respectively. What dollar value was assigned to the notebooks?

In a very important sense, matrices are generalizations of numbers. Just as numbers with special properties are important, so are matrices with special properties. The next two exercises illustrate some special matrix forms that have wide applicability.

69. A matrix is called *idempotent* if and only if it is equal to its own square; that is, B is idempotent if and only if $BB = B$. Check that each given matrix is idempotent, and calculate its determinant.

$$B = \begin{bmatrix} \frac{1}{2} & \frac{1}{2} \\ \frac{1}{2} & \frac{1}{2} \end{bmatrix}, \quad A = \begin{bmatrix} 1 & 0 \\ 1 & 0 \end{bmatrix}, \quad C = \begin{bmatrix} 1 & 0 & 0 \\ 0 & \frac{2}{3} & \frac{\sqrt{2}}{3} \\ 0 & \frac{\sqrt{2}}{3} & \frac{1}{3} \end{bmatrix}$$

70. The transpose ^{T} of a matrix A is the matrix that results when rows and columns are interchanged. For instance if $A = \begin{bmatrix} 1 & 2 \\ 3 & 4 \end{bmatrix}$, then $A^T = \begin{bmatrix} 1 & 3 \\ 2 & 4 \end{bmatrix}$.

A matrix is called *orthogonal* if and only if its inverse equals its transpose; that is P is orthogonal if and only if $P^{-1} = P^T$. Check that each given matrix is orthogonal, and calculate its determinant.

$$P = \begin{bmatrix} 3/5 & -4/5 \\ 4/5 & 3/5 \end{bmatrix} \qquad A = \begin{bmatrix} -1/\sqrt{2} & 1/\sqrt{2} \\ -1/\sqrt{2} & -1/\sqrt{2} \end{bmatrix}$$

THINK ABOUT IT 11.4

1. Give an example of two matrices A and B of dimension 2×2 for which $AB = 0$ but neither A nor B is the zero matrix.
2. The equation $(A+B)^2 = A^2 + 2AB + B^2$ is an identity if A and B are real numbers. However, in matrix algebra if the square of a matrix, say A, is defined by $A^2 = AA$, then this equation is not necessarily true for square matrices A and B, even though the distributive properties $A(B+C) = AB + AC$ and $(B+C)A = BA + CA$ are true. Explain why.
3. The Trace of a square matrix is the sum of elements on the main diagonal (top left to bottom right). If $A = \begin{bmatrix} a & b \\ c & d \end{bmatrix}$, and $B = \begin{bmatrix} m & n \\ p & q \end{bmatrix}$, show algebraically that the trace of AB equals to the trace of BA.
4. If $A = \begin{bmatrix} a & 0 & 0 \\ 0 & b & 0 \\ 0 & 0 & c \end{bmatrix}$, state conditions on a, b, and c so that A^{-1} exists. Then, find A^{-1}.
5. Consider the following theorem:

   If $A = \begin{bmatrix} a & b \\ c & d \end{bmatrix}$ with $ad - bc \neq 0$, then $A^{-1} = \frac{1}{|A|}\begin{bmatrix} d & -b \\ -c & a \end{bmatrix}$.

   a. Use this theorem to find A^{-1} for $A = \begin{bmatrix} 5 & 3 \\ 1 & -2 \end{bmatrix}$.

   Check that the answer agrees with the result for Exercise 43 of this section.

   b. For A and A^{-1} as specified in the theorem, show that $A^{-1}A = 1$.

● ● ●

11.5 Partial Fractions

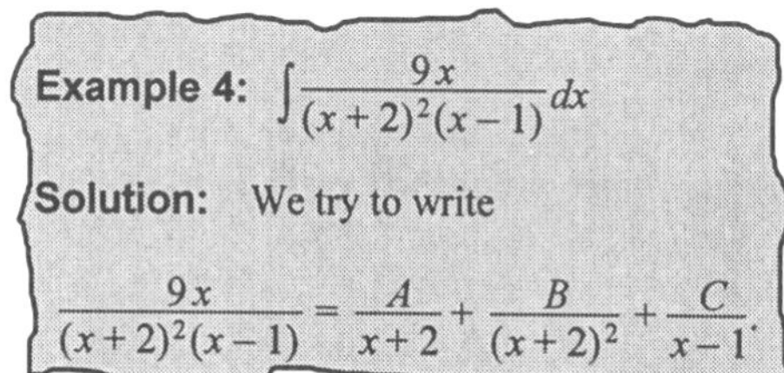

Example 4: $\int \frac{9x}{(x+2)^2(x-1)}dx$
Solution: We try to write
$\frac{9x}{(x+2)^2(x-1)} = \frac{A}{x+2} + \frac{B}{(x+2)^2} + \frac{C}{x-1}$.

The excerpt from a calculus textbook is taken from a section that discusses techniques for integrating rational functions using partial fraction decomposition.

Although the solution to the calculus problem continues, this excerpt is sufficient to illustrate the type of problem we consider in this section. That is, determine the constants A, B, and C so that the equation given in the excerpt is an identity.

(See Example 4.)

Objectives

1. Complete a partial fraction decomposition of $P(x)/Q(x)$ by determining the constants in a given equation that make the equation an identity.
2. Determine a partial fraction decomposition of $P(x)/Q(x)$ that involves distinct linear or quadratic factors of $Q(x)$.
3. Determine a partial fraction decomposition of $P(x)/Q(x)$ that involves repeated linear or quadratic factors of $Q(x)$.

Solving a system of linear equations is a useful component for determining algebraic identities that enable us to analyze rational functions. Recall from Section 6.4 that rational functions fit the form

$$y = \frac{P(x)}{Q(x)} \qquad Q(x) \neq 0,$$

where $P(x)$ and $Q(x)$ are polynomials. In certain situations (especially in calculus), if $P(x)/Q(x)$ is complicated, we try to write this expression as the sum of simpler fractions that are easier to analyze. For example, from the usual addition of fractions we know that

$$\frac{2}{x+2} + \frac{3}{x-1}$$

is identical to

$$\frac{5x+4}{(x+2)(x-1)}.$$

The question is how to work this problem backward and split the more complicated fraction into the sum of the simpler fractions. This sum is called the **partial fraction decomposition** of the expression, and each term in the sum is called a **partial fraction**.

$$\frac{5x+4}{(x+2)(x-1)} \underset{\text{adding}}{\overset{\text{decomposing}}{\rightleftarrows}} \overbrace{\frac{2}{x+2} + \frac{3}{x-1}}^{\text{partial fraction decomposition}}$$

(partial fractions: $\frac{2}{x+2}$ and $\frac{3}{x-1}$)

In the above problem, determining the partial fraction decomposition (assume that it is unknown) amounts to finding constants A and B such that

$$\frac{5x+4}{(x+2)(x-1)} = \frac{A}{x+2} + \frac{B}{x-1}$$

is an identity. To do this, we first clear the expression of fractions by multiplying both sides of the equation by $(x+2)(x-1)$, to obtain

$$5x + 4 = A(x-1) + B(x+2).$$

Now since this equation is an identity, the coefficients of like powers of x must be equal on both sides of the equation. By rewriting the above equation as

$$5x+4=(A+B)x-A+2B$$

and equating coefficients, we obtain the linear system

$$\begin{aligned} A+B &= 5 \\ -A+2B &= 4 \end{aligned}.$$

Adding these two equations yields $3B=9$, so $B=3$. Then it is easy to determine that $A=2$ and that our results check with the original problem.

In this example we have illustrated the method of equating coefficients to determine the constants A and B because it is a powerful method that extends to more complicated cases, and it also gives us additional exposure to systems of linear equations. A shortcut to finding A and B in this particular example is to realize that

$$5x+4=A(x-1)+B(x+2).$$

is an identity and therefore true when $x=1$ and $x=-2$. Substitution of $x=1$ readily yields $9=3B$, so $B=3$; replacing x by -2 gives $-6=-3A$, so $A=2$. However, in the long run this method is limited, and it is most useful only when A and B are coupled with factors that are linear expressions.

Example 1: Completing a Partial Fraction Decomposition

Determine the constants A, B, and C so that

$$\frac{16}{(x+2)(x^2+4)}=\frac{A}{x+2}+\frac{Bx+C}{x^2+4}$$

is an identity.

Solution: First, multiply both sides of the given equation by $(x+2)(x^2+4)$, to obtain

$$16=A(x^2+4)+(Bx+C)(x+2).$$

Then expand the right side of this equation.

$$16=Ax^2+4A+Bx^2+2Bx+Cx+2C$$

Next, group together like powers of x, and then factor.

$$16=(A+B)x^2+(2B+C)x+4A+2c$$

Now equate coefficients of like powers of x and set up a system of linear equations. (Note that $16=0x^2+0c+16$.)

Equating coefficients of x^2:	$A+B=0$
Equating coefficients of x:	$2B+C=0$
Equating constant terms	$4A+2C=16$

Finally, we solve this system by the methods in this chapter to determine that $A=2$, $B=-2$, $C=4$.

Technology Link

In this example, to check algebraically that

$$\frac{16}{(x+2)(x^2+4)}=\frac{2}{x+2}+\frac{-2x+4}{x^2+4}$$

is an identity, you can add the two partial fractions on the right side of this equation to obtain the fraction on the left side. One way to check this result on a grapher is to let

$$\frac{y1+16}{(x+2)(x^2+4)}$$

$$y2=\frac{2}{(x+2)}+\frac{-2x+4}{x^2+4}$$

$$y3=\frac{y1}{y2}.$$

Then graph only $y3$, as shown in Figure 11.7. The quotient of identical expressions is 1, so this graph (which is a horizontal line where y is fixed at 1) supports our results.

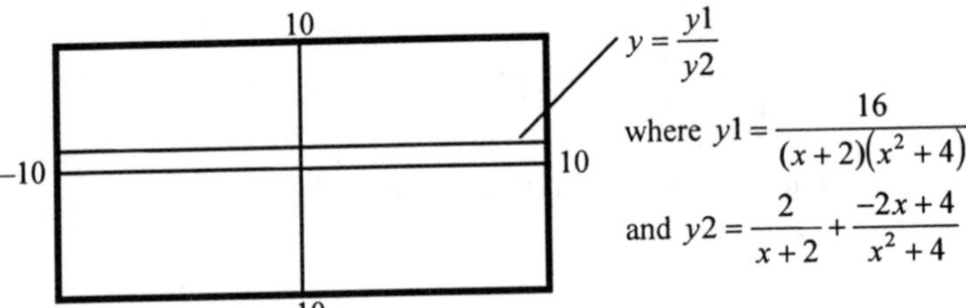

10
−10
10
−10
$y=\frac{y1}{y2}$
where $y1=\frac{16}{(x+2)(x^2+4)}$
and $y2=\frac{2}{x+2}+\frac{-2x+4}{x^2+4}$

Figure 11.7

PROGRESS CHECK 1
Determine constants A and B so that

$$\frac{5x+7}{(x-1)(x+3)}=\frac{A}{x-1}+\frac{B}{x+3}$$

is an identity. ■

To this point, the form of the partial fraction decomposition has been given, and our job has been to determine constants so that the given equation is an identity. In practice, however, you will start only with a rational function

$$y=\frac{P(x)}{Q(x)}\quad Q(x)\neq 0,$$

and you will need to supply the form of the decomposition. To accomplish this, first divide out any common factors of $P(x)$ and $Q(x)$, and then (if necessary) do the following.

1. The degree of $P(x)$ must be less than the degree of $Q(x)$. If it is not, then through long division rewrite the fraction as

$$\frac{P(x)}{Q(x)} = \text{quotient} + \frac{\text{remainder}}{Q(x)},$$

where the remainder term is in proper form.

2. The denominator, $Q(x)$, must be factored into linear factors and/or quadratic factors. The quadratic factors should be **irreducible**, which means that they cannot be factored into linear factors (with real coefficients). Factoring $Q(x)$ in this manner is always possible (theoretically), but it may be very difficult.

Now consider Example 1, which illustrates the form used for distinct linear and quadratic factors. For a linear factor such as $x+2$, we merely put a constant, say A, in the numerator of the partial fraction. For an irreducible quadratic factor such as x^2+4, we just put a linear expression, say $Bx+C$, in the numerator of the partial fraction. In general, we state these rules as follows.

Nonrepeated Linear or Quadratic Factors

Distinct Linear Factors Each linear factor $ax+b$ of $Q(x)$ that is not repeated produces in the decomposition a term of the form

$$\frac{A}{ax+b}.$$

Distinct Quadratic Factors Each irreducible quadratic factor ax^2+bx+c of $Q(x)$ that is not repeated produces in the decomposition a term of the form

$$\frac{Ax+B}{a^2+bx+c}.$$

We illustrate our ideas to this point in the following example.

Example 2: Distinct Linear or Quadratic Factors

Express $\dfrac{x^3+8}{x^3+4x}$ as a sum of partial fractions.

Solution: Since the degree of the numerator is not less than the degree of the denominator, we need long division to determine

$$\frac{x^3+8}{x^3+4x} = 1 + \frac{-4x+8}{x^3+4x}.$$

Now we work with the remainder term. First, factor the denominator and then write the partial fraction decomposition in the form determined by the above rules

$$\frac{-4x+8}{x^3+4x} = \frac{-4x+8}{x(x^2+4)} = \frac{A}{x} + \frac{Bx+C}{x^2+4}$$

Then, multiplying both sides of this equation by $x(x^2+4)$ gives

$$\begin{aligned}-4x+8 &= A(x^2+4)+(Bx+C)x \\ &= Ax^2+4A+Bx^2+Cx \\ &= (A+B)x^2+Cx+4A\ .\end{aligned}$$

Equating the coefficients, we have

$$\begin{aligned}A+B&=0\\ C&=-4\\ 4A&=8\ ,\end{aligned}$$

so

$$A=2\ ,\quad B=-2\ ,\text{ and } C=-4\ .$$

Thus

$$\frac{x^3+8}{x^3+4x}=1+\frac{2}{x}-\frac{-2x-4}{x^2+4}.$$

PROGRESS CHECK 2

Express $\dfrac{x^4+1}{x^3-x}$ as a sum of partial fractions. ■

To handle the cases in which linear or quadratic factors may be repeated in the denominator, we need a more general version of our methods for partial fraction decomposition. These more powerful procedures are as follows.

Partial Fraction Decomposition of $P(x)/Q(x)$

Linear Factors: Each linear factor of $Q(x)$ of the form $(ax+b)^n$ produces in the decomposition a sum of n terms of the from

$$\frac{A_1}{ax+b}+\frac{A_2}{(ax+b)^2}+\cdots+\frac{A_n}{(ax+b)^n}.$$

Quadratic Factors: Each irreducible quadratic factor of $Q(x)$ of the form (ax^2+bx+c) produces in the decomposition a sum of n terms of the form

$$\frac{A_1x+B_1}{ax^2+bx+c}+\frac{A_2x+B_2}{(ax^2+bx+c)^2}+\cdots+\frac{A_nx+B_n}{(ax^2+bx+c)^n}.$$

Note that when a linear or quadratic factor appears just once, then $n=1$ and the above procedures simplify to our previous methods. Also, in practice, we usually denote the constants in the numerators as *A*, *B*, *C*, and so on, instead of using subscript notation.

Example 3: Repeated Linear or Quadratic Factors

Express $\dfrac{1}{x^4-2x^3+x^2}$ as a sum of partial fractions.

Solution: No division is necessary here, so first factor the denominator.

$$x^4 - 2x^3 + x^2 = x^2(x^2 - 2x + 1) = x^2(x-1)^2$$

Now $(x-1)^2$ is a repeated linear factor, and x^2 is also analyzed with the repeated linear factor procedure. Each of these factors produces two terms in the decomposition, as follows:

$$\frac{1}{x^2(x-1)^2} = \frac{A}{x} + \frac{B}{x^2} + \frac{C}{x-1} + \frac{D}{(x-1)^2}.$$

Multiplying both sides of this equation by $x^2(x-1)^2$ gives

$$\begin{aligned} 1 &= Ax(x-1)^2 + B(x-1)^2 + Cx^2(x-1) + Dx^2 \\ &= Ax^3 - 2Ax^2 + Ax + Bx^2 - 2Bx + B + Cx^3 - Cx^2 + Dx^2 \\ &= (A+C)x^3 + (-2A + B - C + D)x^2 + (A - 2B)x + B \quad . \end{aligned}$$

Equating the coefficients, we have

$$\begin{aligned} A + C &= 0 \\ -2A + B - C + D &= 0 \\ A - 2B &= 0 \\ B &= 1, \end{aligned}$$

so

$$A = 2, \quad B = 1, \quad C = -2 \quad \text{and} \quad D = 1.$$

Thus

$$\frac{1}{x^4 - 2x^3 + x^2} = \frac{2}{x} + \frac{1}{x^2} + \frac{-2}{x-1} + \frac{1}{(x=1)^2}.$$

PROGRESS CHECK 3

Express $\dfrac{6x-10}{(x-3)^2}$ as a sum of partial fractions. ■

As mentioned in the section-opening problem, it is sometimes useful in calculus to express a complicated rational function as a sum of simpler rational functions. We have now developed enough procedures to solve the precalculus aspects of this opening problem.

Example 4: A Partial Fraction Decomposition in Calculus

Solve the problem in the section introduction on page 919.

Solution: In the excerpt from the calculus text, observe that the rational function is to be rewritten as

$$\frac{9x}{(x+2)^2(x-1)} = \frac{A}{x+2} + \frac{B}{(x+2)^2} + \frac{C}{x-1}.$$

This form is used because $(x+2)^2$ is a repeated linear factor, and $x-1$ is a distinct linear factor. By multiplying both sides of this equation by $(x+2)^2(x-1)$, we get

$$\begin{aligned} 9x &= A(x+2)(x-1) + B(x-1) + C(x+2)^2 \\ &= Ax^2 + Ax - 2A + Bx - B + Cx^2 + 4Cx + 4C \\ &= (A+C)x^2 + (A+B+4C)x - 2A - B + 4C \quad . \end{aligned}$$

Equating the coefficients gives the system

$$\begin{array}{rcrcrcl} A & + & & + & C & = & 0 \\ A & + & B & + & 4C & = & 9 \\ -2A & - & B & + & 4C & = & 0. \end{array}$$

Finally, we solve this system by the methods of this chapter to determine that the stated equation is an identity when $A=-1$, $B=6$, and $C=1$. Thus, the partial fraction decomposition is

$$\frac{9x}{(x+2)^2(x-1)} = \frac{-1}{x+2} + \frac{6}{(x+2)^2} + \frac{1}{x-1}.$$

PROGRESS CHECK 4

In a calculus text the analysis of a problem involving "up and down motion in resisting media" leads to a partial fraction decomposition of the form

$$\frac{1}{a^2-x^2} = \frac{A}{a-x} + \frac{B}{a+x}.$$

Determine A and B in terms of a so that this equation is an identity. ■

EXPLORE 11.5

In this section of the text, a major goal is to find one algebraic expression that is equivalent to another. This set of exercises explores some ways a graphing calculator can help with this quest.

1. *Graphical Approach* If two expressions $f(x)$ and $g(x)$ are equivalent, then $f(x)/g(x)$ equals 1 (for $g(x) \neq 0$), and the graph of $y = f(x)/g(x)$ is a horizontal line at $y=1$. On the calculator let $y1 = f(x)$ and $y2 = g(x)$, then look at the graph of $y3 = y1/y2$. Use this approach to check the following identities, which appear in Section 11.5.

   a. $\dfrac{16}{(x+2)(x^2+4)} = \dfrac{2}{x+2} + \dfrac{-2x+4}{x^2+4}$

   b. $\dfrac{x^3+8}{x^3+4x} = 1 + \dfrac{2}{x} + \dfrac{-2x-4}{x^2+4}$

   c. $\dfrac{1}{x^4-2x^3+x^2} = \dfrac{2}{x} + \dfrac{1}{x^2} + \dfrac{-2}{x-1} + \dfrac{1}{(x-1)^2}$

2. *Numerical Approach* If two expressions $f(x)$ and $g(x)$ are equivalent, then they will have the same value for any given value of x. Using a calculator, you can easily substitute values of x into both functions to see if you get the same result. This approach is made easier if your calculator has a Table or List capability, which allows you to enter several values of x at one time. Use one of these features to check the identities given in the previous exercise. For instance, evaluate each function for $x=-5$, -3, 0, 2, and 7. Error messages may appear if one element of your list causes a function to be undefined.

EXERCISES 11.5

In Exercises 1–10 determines the constants A, B, C and/or D so that the equation is an identity.

1. $\dfrac{1}{(x+2)(x-2)}=\dfrac{A}{x+2}+\dfrac{B}{x-2}$

2. $\dfrac{1}{(x-1)(x+2)}=\dfrac{A}{x-1}+\dfrac{B}{x+2}$

3. $\dfrac{6}{(x+1)(x^2+1)}=\dfrac{A}{x+1}+\dfrac{Bx+C}{x^2+1}$

4. $\dfrac{2x^3+5x+2}{(x^2+1)(x^2+2)}=\dfrac{Ax+B}{x^2+1}+\dfrac{Cx+D}{x^2+2}$

5. $\dfrac{2x}{(x+2)^2}=\dfrac{A}{x+2}+\dfrac{B}{(x+2)^2}$

6. $\dfrac{3x-10}{(x-1)^2}=\dfrac{A}{x-1}+\dfrac{B}{(x-1)^2}$

7. $\dfrac{x^3-2x^3-4x+3}{(x^2+1)^2}=\dfrac{Ax+B}{x^2+1}+\dfrac{Cx+D}{(x^2+1)^2}$

8. $\dfrac{4x^2+3x-1}{(x^2+x+1)^2}=\dfrac{Ax+B}{x^2+x+1}+\dfrac{Cx+D}{(x^2+x+1)^2}$

9. $\dfrac{-2x+4}{(x-1)^2(x^2+1)}=\dfrac{A}{x-1}+\dfrac{B}{(x-1)^2}+\dfrac{Cx+D}{x^2+1}$

10. $\dfrac{2x^3-5x^2+4x-3}{x^2(x^2+1)}=\dfrac{A}{x}+\dfrac{B}{x^2}+\dfrac{Cx+D}{x^2+1}$

In Exercises 11–24 find the partial fraction decomposition of the given expression.

11. $\dfrac{1}{1-x^2}$

12. $\dfrac{x+2}{x^2-5x+4}$

13. $\dfrac{x}{x^2+2x+1}$

14. $\dfrac{5x+6}{x^3-2x^2}$

15. $\dfrac{x^3}{x^2-4}$

16. $\dfrac{x^2}{x^2-2x+1}$

17. $\dfrac{-12x-6}{x^3+x^2-6x}$

18. $\dfrac{1}{(x+1)(x+2)(x+3)}$

19. $\dfrac{1}{x^3+x}$

20. $\dfrac{x^5-x^3+x-1}{x^4+x^2}$

21. $\dfrac{4x}{x^4-1}$

22. $\dfrac{1}{x^3-1}$

23. $\dfrac{4x^2+x+8}{x^4+3x^2+2}$

24. $\dfrac{4x^3-7x^2+5x-2}{x^4+2x^2+1}$

25. One type of mathematical model for growth when resources are limited is called **logistic growth**. A typical graph of logistic growth is shown in the figure. Sometimes the analysis of this kind of model involves partial fractions. For instance, one such problem leads to the partial fraction decomposition $\dfrac{10}{x(400-x)}=\dfrac{A}{x}+\dfrac{B}{400-x}$. Determine A and B so that the equation is an identity.

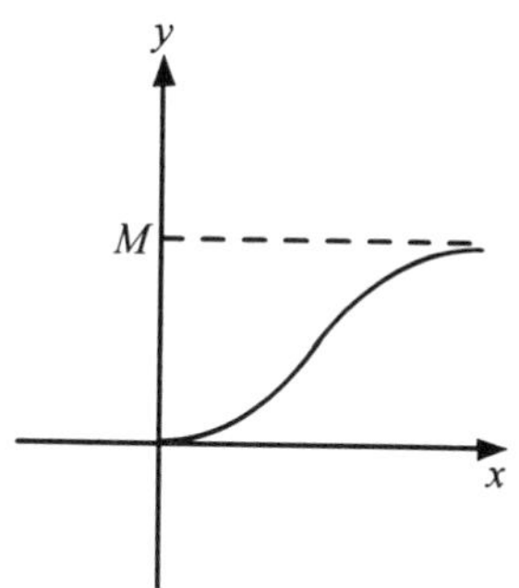

26. In a calculus text the analysis of a problem about a forest regenerating after a fire leads to a partial fraction decomposition of the form $\dfrac{1}{y(M-y)}=\dfrac{A}{y}+\dfrac{B}{M-y}$. Determine A and B in terms of M so that the equation is an identity. [Source: W. C. Ramaly, *Applied Calculus*, Wm. C. Brown, 1995]

Laplace transform Solving a differential equation is a calculus problem with wide application. One popular technique for solving differential equations uses the *inverse Laplace transform* of a function, which involves referring to a published table of functions. To match a rational expression to those in the table, it is sometimes necessary to find the expression's partial fraction decomposition. The next two exercises illustrate this technique.

27. To solve a certain differential equation, the inverse Laplace transform of $f(s)=\dfrac{3s-14}{(s-2)(s-5)}$ is needed.

a. Find the partial fraction decomposition of $f(s)$.

b. The inverse Laplace transform of $\dfrac{1}{s-a}$ is e^{at}.

We can write $L^{-1}\left(\dfrac{1}{s-a}\right)=e^{at}$. Use this information and the partial fraction decomposition of $f(s)$ from part **a** to show that the inverse Laplace transform of $f(s)$ is $\dfrac{8}{3}e^{2t}+\dfrac{1}{3}e^{5t}$.

Note: The transform of a sum is the sum of the transforms. The transform of a constant times a function equals the constant times the transform of the function.

28. To solve a certain differential equation, the inverse Laplace transform of $f(s)=\dfrac{3s}{(s-4)(s+2)}$ is needed.

a. Find the partial fraction decomposition of $f(s)$.

b. Use the partial fraction decomposition of $f(s)$ from part **a** to show that the inverse Laplace transform of $f(s)$ is $2e^{4t}+e^{-2t}$.

THINK ABOUT IT 11.5

1. a. Choose arbitrary values for A, B, and a in the expression

$$\frac{A}{x+a}+\frac{B}{x-a},$$

and then add the fractions.

b. Find the partial fraction decomposition of the answer in part **a** by the methods of this section.

2. Use the shortcut method discussed in the paragraph preceding Example 1 to express

$$\frac{x+2}{x^2-5x+4}$$

as a sum of partial fractions.

3. If x^2 is a factor of the denominator, then decomposition using the repeated linear factor form

$$\frac{A}{x}+\frac{B}{x^2}$$

is preferred to using the distinct quadratic factor form

$$\frac{Ax+B}{x^2}.$$

Start from the bottom form and show that these two forms are equivalent. Discuss why the top form is preferred in terms of the basic goal of partial fraction decomposition.

4. Find the partial fraction decomposition of

$$\frac{bx+c}{(x-a)^2}$$

where a, b, and c are constant.

5. *Generating functions* The coefficients of a polynomial may be used to generate a sequence of numbers. For instance, the unending polynomial $P(s)=1+s+s^2+s^3+\ldots$ gives the sequence $\{1,1,1,\ldots\}$, since all the coefficients are ones. Similarly, the polynomial $Q(s)=1+2s+3s^2+4s^3+\cdots$ gives the sequence $\{1,2,3,4,\ldots\}$. Furthermore, you can check, by performing long division on $1/(1-s)$, that $1/(1-s)=1+s+s^2+s^3+\ldots$, and that $1/(1-s)^2=1+2s+3s^2+4s^3+\ldots$. Thus, we say that $1/(1-s)$ is the *generating function* for

$\{1,1,1,\ldots\}$, and that $1/(1-s)^2$ is the generating function for $\{1,2,3,4,\ldots\}$. Generating functions can be used to solve a type of equation called a *difference equation*, because the solution of a difference equation is a sequence of numbers. Difference equations have great use in probability theory. [Source: S. Goldberg, *Introduction to Difference Equations*, Dover, 1986]. The generating function for the solution to the difference equation $y_{k+1} = 2y_k + 3$ is $Y(s) = \frac{1}{1-2s} + \frac{3s}{(1-s)(1-2s)}$.

a. Find the partial fraction decomposition of the second term of $Y(s)$.

b. Use the result of part a to simplify the expression for $Y(s)$.

c. Use long division on the answer from part b to express $Y(s)$ as an unending polynomial. Confirm that you get $Y(s) = 1 + 5s + 13s^2 + 29s^3 + \ldots$, and compare that to the values of y_k you get from the original generating function if you start with $y_1 = 1$.

● ● ●

11.6 Systems of Linear Inequalities and Linear Programming

Photo Courtesy of Westerman/International Stock

A manufacturer of personal computers makes two types of printers, A and B. To comply with contracts, the company must produce at least 100 type A printers and 200 type B printers each week. A total of 4 labor hours is required to assemble printer A, and a total of 2 labor hours is required to assemble printer B, with a maximum of 1,000 labor hours available each week. If the company can sell all its printers for a profit of \$70 on printer A and \$40 on printer B, how many of each type should be produced weekly for the maximum profit?
(See Example 7.)

Objectives

1. Write a linear inequality that expresses a given relation.
2. Graph the solution set of a linear inequality.
3. Graph the solution set of a system of linear inequalities.
4. Find the maximum and/or minimum values of an objective function subject to given constraints.
5. Solve linear programming problems.

The analysis of certain problems often requires a system of inequalities instead of a system of equations. Example 1 illustrates this point.

Example 1: Translating to a Linear Inequality

A company manufactures two products, A and B. product A costs \$20 per unit to produce, and product B costs \$60 pr unit. The company has \$90,000 to spend on the production. If x represents the number of units produced of product A, and y represents the number of units produced of product B, write an inequality that expresses the restriction placed on the company because of available funds.

Solution: x represents the number of units produced of product A, and each unit costs \$20 to produce. Thus, the company spends $20x$ dollars to produce x units of product A. Similarly, the company spends $60y$ dollars to produce y units of product B. Since the total cost cannot exceed \$90,000 we have

$$20x + 60y \le 90,000.$$

Since the number of units produced cannot be negative, we also have

$$x \ge 0 \text{ and } y \ge 0.$$

Thus, the restriction placed on the company because of available funds is given by

$$20x + 60y \le 90,000 \text{ with } x \ge 0 \text{ and } y \ge 0.$$

PROGRESS CHECK 1

A manufacturer of personal computers makes two types of printers, A and B. A total of 4 labor hours is required to assemble printer A, and a total of 2 labor hours is required to assemble printer B, with a maximum of 1,000 labor hours available each week. If x represents the number of units produced of printer A, and y represents the number of units produced of printer B, write an inequality that expresses the restriction placed on the company because of available labor hours per week. ■

The question in Example 1 is designed to show the usefulness of inequalities in describing practical situations. Capital is only one of many restrictions that a company must consider because of limited resources. For example, machine time, labor hours, and storage space are other considerations. Subject to these restrictions, the company must determine the number of units of each product that should be manufactured for maximum profit. The solution to such problems is found by techniques from a branch of applied mathematics called **linear programming**. To explore this topic further, we first need to develop methods for graphing the solution set to a system of inequalities.

An inequality in two variables such as $y < x$ has infinitely many solutions. To illustrate these solutions graphically, we note that the line $y = x$ separates the plane into two regions. Each region consists of the set of points on one side of the line and is called a **half-plane**. The solution set to $y < x$ is given by the half-plane *below* the line $y = x$, while the solution set to $y > x$ is given by the half-plane *above* $y = x$. We usually indicate the half-place in the solution set through shading, as shown in Figure 11.8, and the sketch of the solution set gives the graph of the inequality. Note that we use either a solid line or a dashed line in the graph, depending on whether the line is included in the solution set.

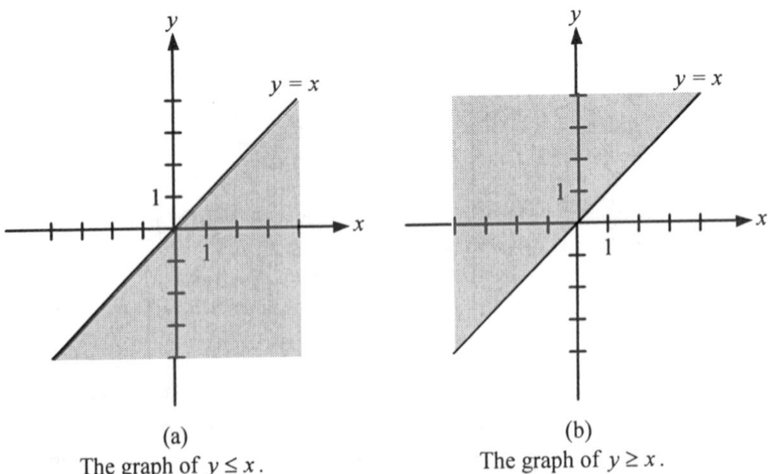

(a) The graph of $y \le x$.

(b) The graph of $y \ge x$.

Figure 11.8

Example 2: Graphing Linear Inequalities

Graph each inequality.

a. $y \le 2x - 1$ **b.** $y > -3x$

Solution:

a. See Figure 11.9. **b.** See Figure 11.10.

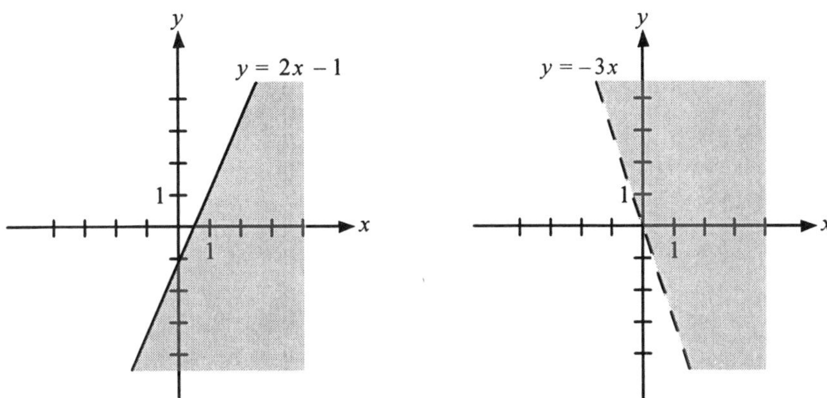

Figure 11.9 **Figure 11.10**

Technology Link

Some graphing calculators have a Shade command that can be used to display the solution set to an inequality in two variables. If your calculator has a Shade command, use it to graph the linear inequalities in this example.

PROGRESS CHECK 2

Graph each inequality.

a. $y \ge -x + 2$ **b.** $y < 2x$ ■

When the inequality is not solved for *y*, you may have difficulty choosing which half-plane to shade in your answer. The easiest way to decide in such cases is usually to pick some **test point** that is not on the line and substitute the coordinates of this point into the inequality. If the resulting statement is true, then shade the half-plane containing the test point. Otherwise, shade the half-plane on the other side of the line from the test point. The origin (0,0) is often a convenient point to use in this test.

Example 3: Graphing a Linear Inequality Using a Test Point

Graph $2x - 3y \ge 6$.

Solution: The origin is not on the line $2x - 3y = 6$, so substitute 0 for *x* and *y* in the above inequality.

$$2(0) - 3(0) \ge 6$$
$$0 \ge 6$$

Since this statement is false, we shade the half plane not containing the origin, as shown in Figure 11.11. If you feel uncomfortable with the test point method, an alternative is to solve the given inequality for *y*. The result $y \le (2/3)x - 2$, tells us to shade the half-plane below the line.

PROGRESS CHECK 3

Graph $3x - y \le 3$. ■

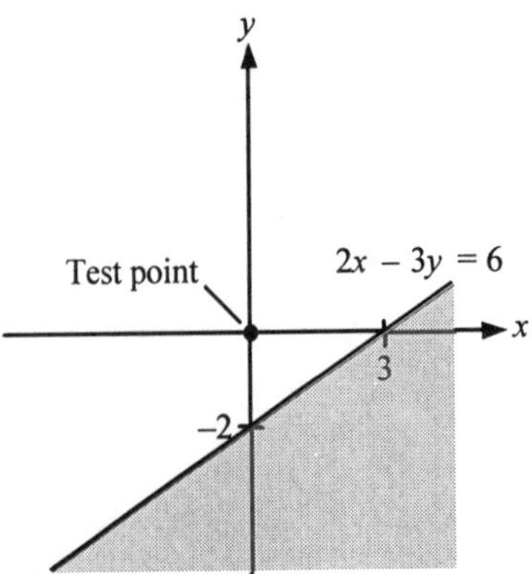

Figure 11.11

The solution set of a system of inequalities is the intersection of the solution sets of all the individual inequalities in the system. The best way to give the solution set is in a graph. To do this, we graph the solutions to each inequality in the system and then shade in the overlap (intersection) of these half-planes.

Example 4: Solving a System of Linear Inequalities

Graph the solution set of the system

$$\begin{aligned} x+y &\le 5 \\ x-2y &\le -4 \end{aligned}$$

Solution: By using the origin in both cases as a test point, we determine that $x+y\le 5$ is satisfied by the points on or below the line $x+y=5$, while $x-2y\le -4$ is satisfied by the points on or above the line given by $x-2y=-4$. The intersection of these two half-planes is shown in Figure 11.12. AS check of this solution that utilizes the shade command on a graphing calculator is displayed in Figure 11.13.

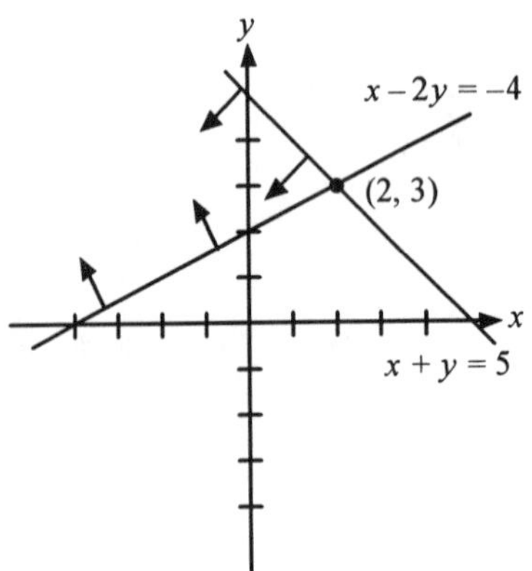

Figure 11.12

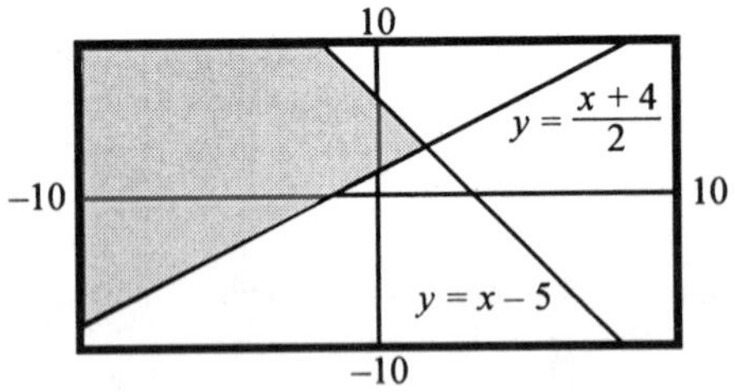

Figure 11.13

PROGRESS CHECK 4
Graph the solution set of the system

$$\begin{aligned} x+y &\ge 3 \\ 2x-y &\ge -4 \end{aligned}$$

■

The **vertices** or **corners** of a region are the intersection points of the bounding sides in the region. For example, the corner in the region defined by the system of inequalities in Example 4 is (2,3), as shown in Figure 11.12. Such corners are important in the solution to linear programming problems, and they can be found by any of the methods we have considered for finding the intersection point for two lines. The next example includes a discussion of this concept.

Example 5: Finding Corners in the Graph of a Solution Set

Graph the solution set of the system

$$\begin{aligned} x + y &\le 6 \\ x - 2y &\le 3 \\ y - x &\le 2 \\ x &\ge 0 \\ y &\ge 0 \end{aligned}$$

Specify the coordinates of any corner in the graph.

Solution: The inequalities $x \ge 0$ and $y \ge 0$ restrict the graph of the solution set to the first quadrant. By using the origin as a test point, we determine that the inequalities specify the region on or below the lines $x + y = 6$ and $y - x = 2$ as well as the region on or above the line $x - 2y = 3$. Figure 11.14 shows the graph of the solution set. To determine the coordinates of the corners, *O*, *A*, *B*, *C*, and *D*, we note that *O* is the origin (0,0); *A* is the *x*-intercept of $x - 2y = 3$, which is (3,0); and *D* is the *y*-intercept of $y - x = 2$, which is (0,2). To find point *B*, we solve the system

$$\begin{aligned} x + y &= 6 \\ x - 2y &= 3 \end{aligned}$$

by adding -1 times the bottom equation to the top equation to get $3y = 3$. Then $y = 1$, so $x = 5$, and point B is (5,1). Finally, the system

$$\begin{aligned} x + y &= 6 \\ y - x &= 2 \end{aligned}$$

solves by adding the two equations together to get $2y = 8$. Then $y = 4$, and $x = 2$, and point *C* is (2,4).

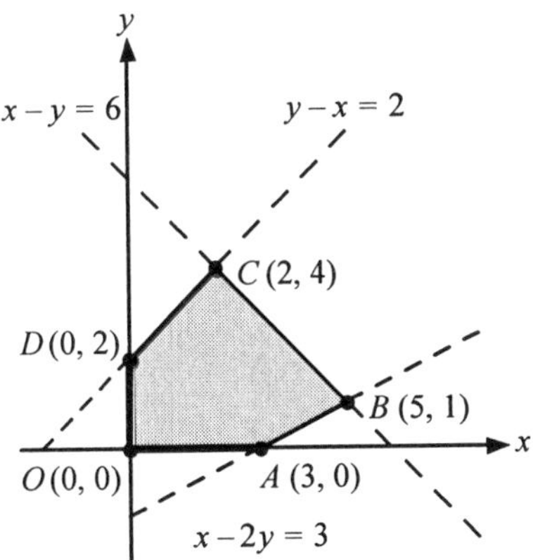

Figure 11.14

PROGRESS CHECK 5
Graph the solution set of the system

$$y \le x + 1$$
$$y \ge -\frac{1}{2}x + 1$$
$$y \ge 4x - 8$$

■

Specify the coordinates of any corner in the graph.

We now return to the topic of linear programming. Basically, linear programming involves finding the maximum or minimum value of a linear function $F = ax + by$, which is called the **objective function**. The variables x and y must satisfy certain inequalities, called **constraints**, that impose restrictions on the solution. Only an ordered pair (x, y) that satisfies all the constraints is a possible or feasible solution, and the set of all feasible solutions is given by the graph of the given system of inequalities (or constraints). Although the set of feasible solutions may seem overwhelming, the corners are the crucial points, as shown in the following theorem.

Corner Point Theorem

> If a linear function $F = ax + by$ assumes a maximum or minimum value subject to a system of linear inequalities, then it does so at a corner or vertex of that system of constraints.

Furthermore, we are guaranteed that a maximum or minimum exists as long as the set of feasible solutions is a **closed** convex polygonal region. The condition that a region be closed merely means that the region must include its boundary lines, while a region is said to be convex provided that the region contains the line segment joining *any* two points in the region. Figure 11.15(a) shows a set of points that is convex, while Figure 11.15(b) shows a nonconvex region.

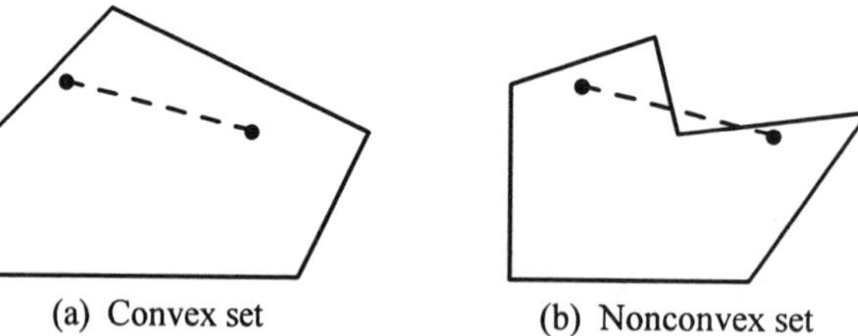

(a) Convex set (b) Nonconvex set
Figure 11.15

Since the solution set of a system of linear inequalities is always a convex region, this condition poses no problem. Thus, the only time we may not be able to find a maximum or minimum value occurs when the set of feasible solutions is unbounded and does not form a closed polygonal region.

We can now consider a typical problem in linear programming.

Example 6: Finding the Maximum Value of an Objective Function

Find the maximum value of $F = 3x - 2y$ subject to the constraints

$$x + y \le 6$$
$$y - x \le 2$$
$$x - 2y \le 3$$
$$x \ge 0$$
$$y \ge 0$$

Solution: The constraints are the system of inequalities in Example 5, so the set of all feasible solutions including all vertices is shown in Figure 11.14. Since the set of feasible solutions is a closed convex polygonal region, we know that a maximum value exists at one of the corners. We need only substitute the coordinates of these vertices into the objective function to determine the correct corner.

Vertices	$F = 3x - 2y$
$O(0,0)$	$3(0) - 2(0) = 0$
$A(3,0)$	$3(3) - 2(0) = 9$
$B(5,1)$	$3(5) - 2(1) = 13$
$C(2,4)$	$3(2) - 2(4) = -2$
$D(0,2)$	$3(0) - 2(2) = -4$

From the table, we see that when $x = 5$ and $y = 1$, the function attains a maximum value of 13. Note that we can also determine from the table that the minimum function value of –4 occurs at (0.2).

PROGRESS CHECK 6
Find the maximum value of $F = 2x + 3y$ subject to the constraints

$$y \le x + 1$$
$$y \ge -\frac{1}{2}x + 1.$$
$$y \ge 4x - 8$$

■

If we look a little closer at Example 6, we can see why any maximum or minimum occurs at a corner. Consider Figure 11.16 which plots the line $F = 3x - 2y$ for various values of F. Note that as F increases, the y-intercepts of the line change, but the slopes remain fixed at $3/2$. The result is a family of parallel lines. The line with the maximum or minimum value of F will then intersect the set of feasible solutions at a corner, as shown in Figure 11.16.

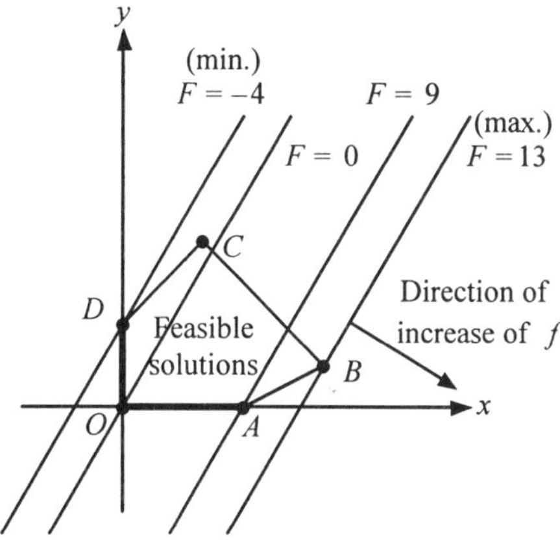

Figure 11.16

The next example considers a linear programming problem in a more practical setting.

Example 7: Maximizing Profit

Solve the problem in the section introduction on page 929.

Solution: If we let x and y represent the number of assembled type A and type B printers, respectively, then the profit function is

$$P = 70x + 40y$$

and the constraints are

$$\begin{aligned} x &\geq 100 \\ y &\geq 200 \\ 4x + 2y &\leq 1{,}000 \end{aligned} .$$

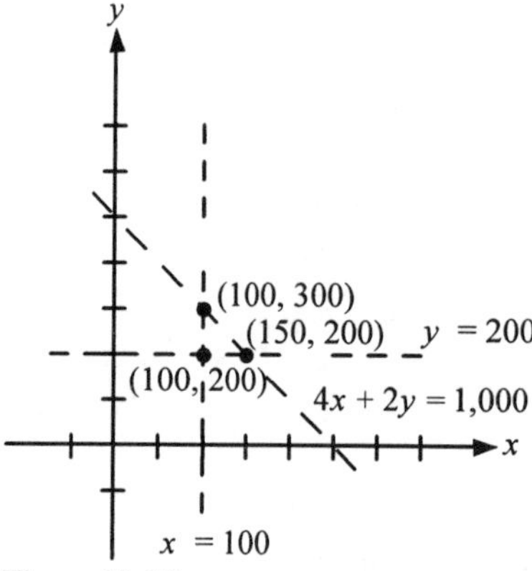

Figure 11.17

The constraints determine the set of feasible solutions shown in Figure 11.17 with vertices (100,200), (150,200), and (100,300). Since the set of feasible solutions is a closed convex polygonal region, we know that a maximum value exists at one of these corners. The value of the profit function at each of these vertices is as follows:

Vertices	$P = 70x + 40y$
(100,200)	$70(100) + 40(200) = 15{,}000$
(150,200)	$70(150) + 40(200) = 18{,}500$
(100,300)	$70(100) + 40(300) = 19{,}000$

Thus, the profit function reaches a maximum value of \$19,000 when 100 type A printers and 300 type B printers are produced each week.

PROGRESS CHECK 7

A manufacturer of personal computers makes two types of disc drives, A and B. A total of 4 labor hours is required to assemble drive A, and a total of 2 labor hours is required to assemble drive B, with a maximum of 160 labor hours available each week. If the manufacturer can produce at most 30 type A drives at a profit of \$40 per drive, and at most 40 type B drives at a profit of \$30 per drive, then find how many of each drives should be produced weekly for the maximum profit. ■

The geometrical method we have discussed is intended as an introduction to linear programming. In most cases the situation being analyzed is quite complicated, and an algebraic method, called the **Simplex method**, is used to solve the problem using computers. Although we have only considered this topic in the context of maximizing profit, other applications are common. Some of these include:

1. Determining the most economical mixture of ingredients that will result in a product with certain minimum requirements.
2. Determining the quickest and most economical route in distributing a product.
3. Determining the most efficient use of industrial machinery.
4. Determining the most effective production schedule.

EXPLORE 11.6

1. Most graphing calculators have a feature that shades specific regions of the screen. Typically they are accessed by a menu choice called Shade. If your calculator has such a feature, use it to create these images.

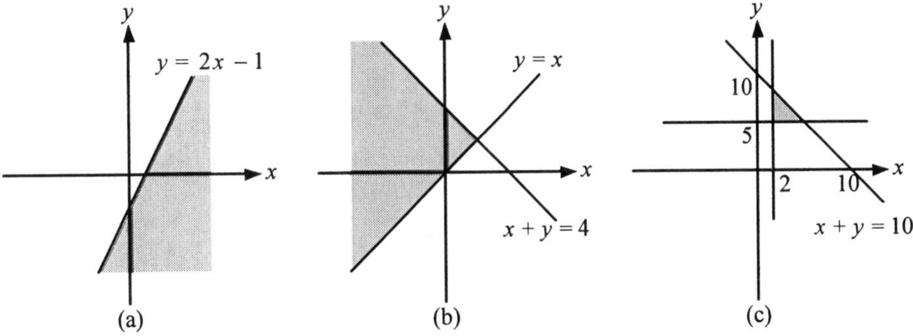

2. In this section we explore the behavior of functions at the corners of polygonal regions. Consider the triangular region bounded by $x = 80$, $y = 150$, and $5x + 4y = 2000$.
   a. Graph the triangular region formed by these three lines, and find the coordinates of the three vertices.
   b. Evaluate the expression $90x + 75y$ at each vertex, and determine which vertex yields the highest value.
   c. Evaluate the expression $100x + 75y$ at each vertex, and determine which vertex yields the highest value.
   d. Determine A and B so that $Ax + By$ has the same value at the two acute vertices of the region.
3. A theorem in this section states that a function that achieves a maximum or minimum over a polygonal region will do so at a corner of the region. For the triangular region given in the previous problem.
   a. Determine the coordinates of the three corners of the region.
   b. Determine the coordinates of any 4 other points inside the triangular region.
   c. Evaluate the expression $90x + 75y$ at each of the points from parts **a** and **b**, and check that the maximum and minimum values occur at the corners.
4. For a region with a curved boundary, there are no vertices. So the point on the boundary that maximizes an expression in x and y may occur anywhere, and different points will maximize different expressions.
   a. Graph the region bounded by the parabola $y = 4 - (x - 2)^2$ and the x-axis.
   b. Find the point on the parabola for which the expression $x + y$ is a maximum.
   c. Find the point on the parabola for which $2x + y$ is a maximum.

EXERCISES 11.6

In Exercises 1–8 graph the given inequalities. Use shading to indicate the graph.

1. $y \le 2x$ **2.** $y > x - 2$ **3.** $y > 1 - 2x$

4. $y \ge 2$ **5.** $x \le 4$ **6.** $x - y > 0$

7. $2x - 4y \ge 8$ **8.** $x + 2y \le 5$

In Exercises 9–16 graph the solution set of the given system of inequalities. Use shading to indicate the graph, and specify the coordinates of any corner in the graph.

9. $x + y \le 7$, $x - 2y \le -8$

10. $x + 2y \ge -12$, $2x - y \ge 1$

11. $y \le 3 - 4x$, $x - y \ge 0$

12. $y \ge 3x + 4$, $x + y \le -2$

13. $x + y \le 4$, $x \ge 0$, $y \ge 0$

14. $x + y \ge 1$, $x - y \le 1$, $x + 2y \le 2$

15. $y \le x + 1$, $y \ge -x + 1$, $y \ge x - 1$, $y \le -x + 3$

16. $x + y \le 8$, $2x + y \le 11$, $-2x + y \le 5$, $x \ge 0$, $y \ge 0$

17. Find the maximum and minimum values of $F = 2x - 3y$ subject to the constraints given in Exercise 13.

18. Find the maximum and minimum values of $F = 6x - 12y$ subject to the constraints given in Exercise 14.

19. Find the values of x and y that yield the maximum value of $F = \frac{1}{2}x + \frac{1}{4}y$ subject to the constraints given in Exercise 15.

20. Find the values of x and y that yield the minimum value of $F = 5x - y$ subject to the constraints given in Exercise 16.

In Exercises 21–22 find the maximum value of P subject to the given constraints.

21. $P = 60x + 40y$
$4x + 2y \le 2{,}000$
$x \ge 100$
$y \ge 200$

22. $P = 120x + 20y$
$15x + 3y \le 12{,}000$
$x \ge 300$
$y \ge 600$

23. Minimize $C = -2x - 3y + 200$ subject to
$3x + 4y \le 180$
$15 \le x \le 40$
$10 \le y \le 30$

24. Maximize $P = 15x + 8y + 9$ subject to
$x - y \le 7$
$x - y \ge 1$
$8 \le x \le 13$

25. A woman has \$20,000 to invest in the stock market. Stock A sells for \$90 per share, stock B costs \$50 per share, and stock C costs \$30 per share. If x, y, and z represent the number of shares that she buys of each stock, respectively, find an inequality that expresses this relationship.

26. A TV manufacturer has available 10,000 labor hours per month. A console requires 7 labor hours, and a portable requires 5 labor hours. If x represents the number of consoles produced, and y represents the number of portables produced, find an inequality that expresses this relationship.

27. Government regulations limit the amount of a pollutant that a company can discharge into the water to A gal/day. If product 1 produces B gal of pollutant per unit, and product 2 produces C gal of pollutant per unit, find an inequality that expresses the restriction placed on this company because of the government regulation.

28. Sad Sam has been informed by his doctor that he would be less sad if he obtained at least the minimum adult requirements of thiamine (2 mg) and of niacin (25 mg) per day. A trip to the supermarket reveals the following facts about his favorite cereals.

Cereal	Thiamine per Ounce (mg)	Niacin per Ounce (mg)
A	1.0	10.0
B	0.7	12.0

If x represents the number of ounces he eats of cereal A, and y represents the number of ounces he eats of cereal B, find two inequalities that express what must be done if he is to satisfy his minimum daily requirement.

29. A manufacturer of personal computers makes two types of disc drives, A and B. To comply with contracts, the company must produce at least 80 type A and 150 type B disc drives each week. A total of 5 labor hours is required to assemble drive A, and a total of 4 labor hours is required to assemble drive B, with a maximum of 2,000 labor hours available each week. If the company can sell all its disc drives for a profit of \$90 on type A and \$75 on type B, how many of each should be produced weekly for the maximum profit?

30. In Exercise 29 suppose the profit is \$100 on type A drives and \$75 on type B. How many of each should be produced weekly for maximum profit?

31. Chris needs at least 9 units of vitamin A and 12 units of vitamin C per day. The table below summarizes information about two of her favorite foods.

Food	Vitamin A per Ounce	Vitamin C per Ounce	Cost per Ounce
A	3 units	3 units	25 cents
B	2 units	4 units	20 cents

How many ounces of each food should Chris eat to satisfy at lowest cost, her minimum requirements for vitamins A and C?

32. Repeat Exercise 31 with food A costing 28¢ per ounce and food B costing 18¢ per ounce.

33. You need to buy some large and some small screws. The small ones each cost 3¢, and the large ones each cost 4¢. You have \$5 to spend, which must cover the screws plus a 5 percent sales tax. You need at least 75 small and 50 large screws. How many of each should you buy to get the maximum number of screws?

34. Use the information given in Exercise 28 and determine the combination of cereals Sam should eat to satisfy his doctor's recommendation at lowest cost, if cereal A costs 15¢ per ounce and cereal B costs 13¢ per ounce.

THINK ABOUT IT 11.6

1. Graph $x^2 + y^2 < 9$.

2. Graph the solution set of the system.

$$y \le 4 - x^2$$
$$y \ge 3x.$$

In Questions 3 and 4 find a system of linear inequalities with the given solution set.

3.

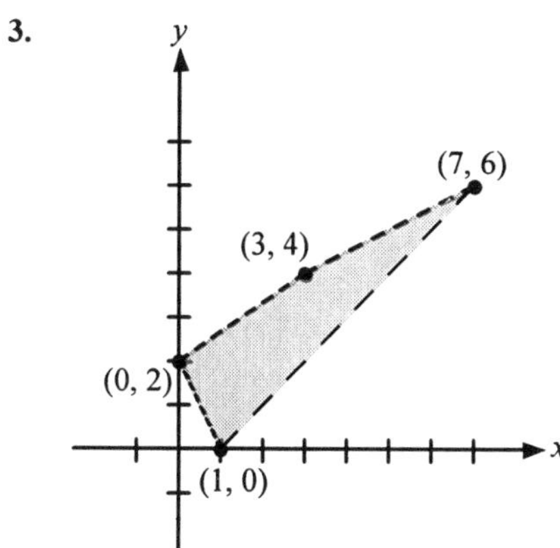

4.

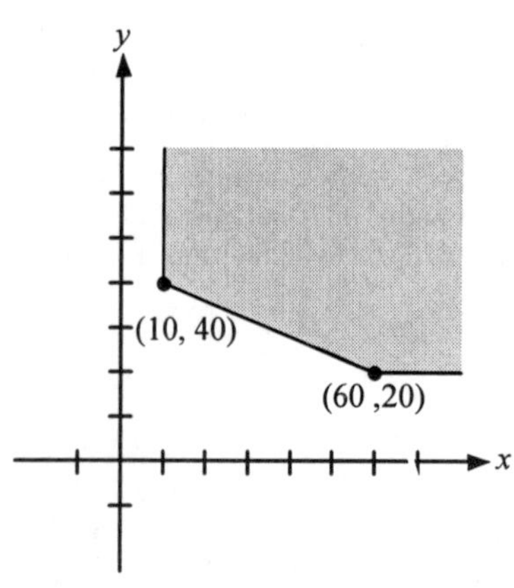

5. Redo the problem in Example 7 of this section, assuming that the company can sell all its printers for a profit of \$80 on printer A and \$40 on printer B. Explain geometrically why the solution is not unique.

● ● ●

CHAPTER 11 SUMMARY

OBJECTIVES CHECKLIST

Specific chapter objectives are summarized below along with numbered example problems from the text that should clarify the objectives. If you do not understand any objectives, or do not know how to do the selected problems, then restudy the material.

11.1: Can you:

1. **Solve linear systems in two variables (by the methods of Section 1.5)?** Solve by addition-elimination: $\begin{aligned} 2x - 5y &= 5 \\ 4x + 3y &= 23 \end{aligned}$

[Example 1]

2. **Solve linear systems in three variables using the addition-elimination methods?** Solve the system $\begin{aligned} 3x + 2y + z &= -3 \\ 2x + 3y + 2z &= 5 \\ -2x + y - z &= 3 \end{aligned}$

[Example 3]

3. **Solve applied problems by setting up and solving a system of linear equations?** The points (1, 2), (2, 7), and (−1, 4) lie on the parabola given by $y = ax^2 + bx + c$. Find a, b, and c, and write an equation for the parabola. Check your result with a grapher.

[Example 4]

11.2: Can you:

1. **Solve a linear system by transforming the system to triangular form?** Solve the system of equations $\begin{aligned} 3x - y + 6z &= 1 \\ x + 2y - 3z &= 0 \\ 2x - 3y - z &= -9 \end{aligned}$

[Example 1]

2. **Solve a linear system using row operations on matrices?** Use matrix form to solve the system $\begin{aligned} 6x + 10y &= 7 \\ 15x - 4y &= 3 \end{aligned}$

[Example 4]

11.3: Can you:

1. **Evaluate the determinant of a 2-by-2 matrix?** Evaluate $\begin{vmatrix} 1 & 2 \\ 3 & 4 \end{vmatrix}$

[Example 1]

2. **Solve a linear system in two variables using Cramer's rule?** Use Cramer's rule to solve the system of equations $\begin{aligned} 3x - 2y &= 27 \\ 2x + 5y &= -1 \end{aligned}$

[Example 3]

3. **Evaluate the determinant of a 3-by-3 matrix?** Evaluate the following determinant by
   a. expansion along the first row of the determinant and
   b. expansion along the second column of the determinant. $\begin{vmatrix} 1 & 1 & 3 \\ 5 & 0 & 2 \\ -2 & 3 & -1 \end{vmatrix}$

[Example 5]

4. **Solve a linear system in three variables using Cramer's rule?** Use Cramer's rule to solve the system of equations $\begin{aligned} 3x - y + 6z &= 1 \\ x + 2y - 3z &= 0 \\ 2x - 3y - z &= -9. \end{aligned}$

[Example 6]

11.4: Can you:

1. **Add, subtract, and multiply matrices, and multiply a matrix by a scalar?** Answer the following questions given

$$A = \begin{bmatrix} 2 & -1 & 3 \\ 0 & 1 & -2 \end{bmatrix},\ B = \begin{bmatrix} 0 & 0 \\ 0 & 0 \end{bmatrix},\ C = \begin{bmatrix} 1 & 0 \\ 0 & 1 \end{bmatrix},\ \text{and}\ D = \begin{bmatrix} 1 & 3 & 6 \\ 2 & 5 & 7 \end{bmatrix}$$

   a. What is the dimension of each matrix?
   b. Find (if possible) $A+D$. **c.** Find (if possible) $A+C$.
   d. Find $-A$. **e.** If $A+2x=D$, find X.

[Example 1]

2. **Find the inverse of a matrix, if it exists?** If $A = \begin{bmatrix} 2 & 3 \\ 3 & 5 \end{bmatrix}$, find A^{-1} (if it exists).

[Example 6]

3. **Write a system of linear equations in matrix form $AX = B$ and use A^{-1} to solve the system?** Solve the system $\begin{aligned} 2x+3y &= b_1 \\ 3x+5y &= b_2 \end{aligned}$ when
   a. $b_1 = 5, b_2 = 9$ and **b.** $b_1 = -3$, $b_2 = 0$.

[Example 7]

11.5: Can you:

1. **Complete a partial fraction decomposition of $\dfrac{P(x)}{Q(x)}$ by determining the constants in a given equation that make the equation an identity?** Determine the constants A, B, and C so that $\dfrac{16}{(x+2)(x^2+4)} = \dfrac{A}{x+2} + \dfrac{Bx+C}{x^2+4}$ is an identity.

[Example 1]

2. **Complete a partial fraction decomposition of $\frac{P(x)}{Q(x)}$ that involves distinct linear or quadratic factors of $Q(x)$?** Express $\frac{x^3+8}{x^3+4x}$ as a sum of partial fractions.

[Example 2]

3. **Complete a partial fraction decomposition of $\frac{P(x)}{Q(x)}$ that involves repeated linear or quadratic factors of $Q(x)$?** Express $\frac{1}{x^4-2x^3+x^2}$ as a sum of partial fractions.

[Example 3]

11.6: Can you:

1. **Write a linear inequality that expresses a given relation?** A company manufactures two products, A and B. Product A costs \$20 per unit to produce, and product B costs \$60 per unit. The company has \$90,000 to spend on the production. If x represents the number of units produced of product A, and y represents the number of units produced of product B, write an inequality that expresses the restriction placed on the company because of available funds.

[Example 1]

2. **Graph the solution set of a linear inequality?** Graph each inequality.
   a. $y \le 2x-1$ **b.** $y > -3x$

[Example 2]

3. **Graph the solution set of a system of linear inequalities?** Graph the solution set of the system $\begin{aligned} x + y &\le 5 \\ x - 2y &\le -4. \end{aligned}$

[Example 4]

4. **Find the maximum and/or minimum values of an objective function subject to given constraints?** Find the maximum value of $F = 3x - 2y$ subject to the constraints

$$\begin{aligned} x + y &\le 6 \\ y - x &\le 2 \\ x - 2y &\le 3 \\ x &\ge 0 \\ y &\ge 0. \end{aligned}$$

[Example 6]

5. **Solve linear programming problems?** A manufacturer of personal computers makes two types of printers, A and B. To comply with contracts, the company must produce at least 100 type A printers and 200 type B printers each week. A total of 4 labor hours is required to assemble printer A, and a total of 2 labor hours is required to assemble printer B, with a maximum of 1,000 labor hours available each week. If the company can sell all its printers for a profit of \$70 on printer A and \$40 on printer B, how many of each type should be produced weekly for the maximum profit?

[Example 7]

KEY CONCEPTS AND PROCEDURES

Section	Key Concepts and Procedures to Review
11.1	• Definitions of linear equations with n variables, linear system, inconsistent system, and dependent system • The solution set of a system of linear equations in two variables is the set of all the ordered pairs that satisfy both equations. Graphically this corresponds to the collection of points where the lines intersect.

- Methods to solve a system of equations by the substitution method and by the addition-elimination method

11.2
- Definitions of triangular form, equivalent systems, matrix, entry or element of a matrix, coefficient matrix, augmented matrix, equivalent matrices, and reduced row-echelon matrix
- The following operations are used in Gaussian elimination and Gauss-Jordan elimination to solve a linear system

Elementary Operations on Equations	**Elementary Row Operations on Matrices**
1. Multiply both sides of an equation by a nonzero number	**1.** Multiply each entry in a row by a nonzero number
2. Add a multiple of one equation to another	**2.** Add a multiple of the entries in one row to another row.
3. Interchange two equations.	**3.** Interchange two rows

11.3
- Definitions of square matrix, determinant, minor of an element, and cofactor of an element
- To evaluate a 2-by-2 determinant, use $\begin{vmatrix} a_1 & b_1 \\ a_2 & b_2 \end{vmatrix} = a_1b_2 - b_1a_2$.
- Cramer's rule
- Method to evaluate 3-by-3 (and more complicated) determinants

11.4
- Definitions of dimension, square matrix, zero matrix, equal matrices, scalar, $n \times n$ identity matrix, and the multiplicative inverse of A (symbolized A^{-1})
- Definitions for scalar multiplication and addition, subtraction, and multiplication of matrices
- Six basic properties of matrices
- Only matrices of the same dimension can be added or subtracted
- The product AB of two matrices is defined only when the number of columns in A matches the number of rows in B.
- Methods to determine if A^{-1} exists, and if it does, how to find it
- If A^{-1} exists, the solution to the linear system $AX = B$ is given by $X = A^{-1}B$.

11.5
- Definitions of partial fraction decomposition, partial fraction, and irreducible quadratic factor
- Methods to determine the partial fraction decomposition of $P(x)/Q(x)$, where $P(x)$ and $Q(x)$ are (nonzero) polynomials (Note that the denominator in each partial fraction in the decomposition is either a linear factor, a repeated linear factor, an irreducible quadratic factor, or a repeated irreducible quadratic factor.)

11.6
- Definitions of half-plane, vertices or corners of a region, objective function, constraints, feasible solution, and convex region
- Methods to graph inequalities in two variables
- Methods to solve a system of inequalities
- The solution set of a system of inequalities is the intersection of the solution sets of all the individual inequalities in the system
- Methods to find the maximum and/or minimum value of an objective function subject to given constraints
- Corner point theorem

CHAPTER 11 REVIEW EXERCISES

In Exercises 1–10 solve the given system of linear equations, using the method indicated.

1. $\begin{aligned} 2x+3y&=-17 \\ 3x-y&=2 \end{aligned}$ (addition-elimination)

2. $\begin{aligned} y&=-3x+2 \\ y&=2x-3 \end{aligned}$ (substitution)

3. $\begin{aligned} y&=\frac{1}{2}x-3 \\ 3x-7y-18&=0 \end{aligned}$ (graphing)

4. $\begin{aligned} 3x+y&=2 \\ -3x+y&=-2 \end{aligned}$ (Cramers rule)

5. $\begin{aligned} x+y&=10 \\ -2x-5y&=-11 \end{aligned}$ (row operations on matrices)

6. $\begin{aligned} 5x+3y&=1 \\ 3x+2y&=0 \end{aligned}$ (matrix algebra)

7. $\begin{aligned} x-y+z&=5 \\ y-z&=0 \\ z&=-1 \end{aligned}$ (substitution)

8. $\begin{aligned} x+y+z&=2 \\ x+y-z&=4 \\ 2x-y-4z&=-5 \end{aligned}$ (row operations on matrices)

9. $\begin{aligned} x+z&=10 \\ y+2z&=3 \\ -x+y&=-2 \end{aligned}$ (Cramers rule)

10. $\begin{aligned} x+3y+3z&=5 \\ x+4y+3z&=2 \\ x+3y+4z&=8 \end{aligned}$ (matrix algebra)

In Exercises 11–16 solve each system of equations.

11. $\begin{aligned} x&=3y-5 \\ y&=4x-2 \end{aligned}$

12. $\begin{aligned} 5x-4y-11&=0 \\ 2x+7y+13&=0 \end{aligned}$

13. $\begin{aligned} 2x-y+z&=3 \\ x-3y+5z&=4 \\ x+y-2z&=0 \end{aligned}$

14. $\begin{aligned} x+y-z&=2 \\ 2x-y+5z&=3 \\ -x-y+z&=-1 \end{aligned}$

15. $\begin{aligned} -x+y+z&=0 \\ x-y+z&=0 \\ x+y-z&=0 \end{aligned}$

16. $\begin{aligned} x+y+z&=1 \\ 2x+2y+2z&=2 \\ x-y-z&=0 \end{aligned}$

In Exercises 17–20 evaluate each determinant.

17. $\begin{vmatrix} -2 & 3 \\ 1 & 4 \end{vmatrix}$

18. $\begin{vmatrix} 1 & -6 & 2 \\ 2 & 2 & -3 \\ 3 & -4 & 1 \end{vmatrix}$

19. $\begin{vmatrix} 3 & -3 & 1 \\ -1 & 9 & -3 \\ 5 & -6 & 2 \end{vmatrix}$

20. $\begin{vmatrix} 1 & 0 & 1 & 2 \\ 0 & 0 & -1 & 0 \\ -2 & 3 & 0 & -1 \\ 1 & 2 & 3 & 0 \end{vmatrix}$

In Exercises 21 and 22 verify the given identity.

21. $\begin{vmatrix} a & b \\ c & d \end{vmatrix} = -\begin{vmatrix} c & d \\ a & b \end{vmatrix}$

22. $\begin{vmatrix} ka & b \\ kc & d \end{vmatrix} = k\begin{vmatrix} a & b \\ c & d \end{vmatrix}$

In Exercises 23–30 find (if defined) the specified matrix given

$$A=\begin{bmatrix} 1 & -2 & 3 \\ -1 & 0 & 4 \end{bmatrix},\quad B=\begin{bmatrix} -3 & 0 \\ 0 & -3 \end{bmatrix},\quad C=\begin{bmatrix} 4 & 5 \\ 3 & 4 \end{bmatrix},\quad D=\begin{bmatrix} 3 & -1 \\ 0 & 1 \\ 2 & 5 \end{bmatrix}$$

23. $B-C$ **24.** BC **25.** CD

26. DC **27.** $3C$ **28.** C^{-1}

29. $AD+C$ **30.** $2A-3D$

31. Graph $2x-y\le 7$. Use shading to indicate the graph.

32. Graph the solution set of the system
$$\begin{aligned} x+4y&\le 17 \\ 4x+3y&\ge 16 \end{aligned}$$

33. Maximize $P=50x+30y$ subject to
$$\begin{aligned} 5x+2y&\le 1{,}000 \\ x&\ge 50 \\ y&\ge 100 \end{aligned}$$

34. Minimize $C=3x-2y+20$ subject to
$$\begin{aligned} x+y&\le 8 \\ y-x&\le 6 \\ x-2y&\le 2 \\ x&\ge 0 \\ y&\ge 0 \end{aligned}$$

35. Find a, b, and c so that the parabola $y=ax^2+bx+c$ passes through the points (1, 3) (−1, 9) and (2, 6).

36. Find the intersection point of the line $y=\frac{3}{4}x-\frac{1}{4}$ and the line whose x- and y-intercepts are $(6, 0)$ and $(0, 4)$.

37. A student took out three loans at different rates: a car loan at 8 percent annual interest, a tuition loan at 6 percent, and a loan for travel expenses at 10 percent. The total interest due in one year was $960. The car loan was $1000 less than the tuition loan. The car and travel loans combined were $1000 more than the tuition loan. Find the size of each loan.

38. A company sells 2 perfumes in 2-ounce bottles, Xcape and Ynot. The price and the profit for one of each bottle are shown in the table.

	Perfume	
	Xcape	Ynot
Price	$30	$75
Profit	$20	$60

For one sales period, total sales were $2,803,500, with total profit of $2,121,000. How many bottles of each item were sold?

39. If the vertices of a triangle are at points (x_1y_1), (x_2y_2), and (x_3y_3), then the area of the triangle is given by the absolute value of the following determinant.

$$\frac{1}{2}\begin{vmatrix} x_1 & y_1 & 1 \\ x_2 & y_2 & 1 \\ x_3 & y_3 & 1 \end{vmatrix}$$

Use this rule to find the area of the triangle whose vertices are at $(0,-3)$, $(-2,-1)$ and $(1,4)$.

40. A company manufactures two types of fertilizer, A and B. A bag of fertilizer A is made from 4 units of nitrogen and 1 unit of potassium, while a bag of fertilizer B is made from 2 units of nitrogen and 3 units of potassium. Raw materials in stock are limited to 600 units of nitrogen and 260 units of potassium. If the company can sell all its fertilizer for profits of $5 per bag of fertilizer A and $4 per bag of fertilizer B, how many of each type should be produced for the maximum profit?

41. True or false: $\begin{bmatrix} 1 & 2 \\ 3 & 4 \end{bmatrix}=\begin{bmatrix} 3 & 4 \\ 1 & 2 \end{bmatrix}$

42. True or false: $\begin{vmatrix} 1 & 2 \\ 3 & 4 \end{vmatrix}=-\begin{vmatrix} 3 & 4 \\ 1 & 2 \end{vmatrix}$

In Exercises 43 and 44 determine the constants A, B, C, and/or D so that the equation is an identity.

43. $\dfrac{-4x-5}{(x+3)^2}=\dfrac{A}{x+3}+\dfrac{B}{(x+3)^2}$

44. $\dfrac{-x^3-x^2-8x+2}{(x^2+3)(x^2-2)}=\dfrac{Ax+B}{x^2+3}+\dfrac{Cx+D}{x^2-2}$

In Exercises 45 and 46 find the partial fraction decomposition of the given expression.

45. $\dfrac{x^2}{x^2-9}$

46. $\dfrac{2x^3+8x}{x^4+4x^2+4}$

CHAPTER 11 TEST

1. Evaluate $\begin{vmatrix} 2 & -3 & 1 \\ -1 & 4 & -7 \\ 3 & 0 & 2 \end{vmatrix}$.

2. If $A=\begin{bmatrix} -3 & 4 \\ -5 & 7 \end{bmatrix}$, find A^{-1} if it exists.

3. Find BA if $B=\begin{bmatrix} 3 & 0 & -5 \\ -4 & 2 & 6 \end{bmatrix}$ and $A=\begin{bmatrix} 8 & 1 \\ 10 & -2 \\ 7 & 0 \end{bmatrix}$.

4. If $A=\begin{bmatrix} 2 & -1 \\ 3 & 2 \end{bmatrix}$ and $B=\begin{bmatrix} 0 & 5 \\ 7 & -3 \end{bmatrix}$, find $2B-A$.

5. Express $\dfrac{4x+6}{x^2-4}$ as a sum of partial fractions.

6. Use Cramer's rule to find the solution for x of the system

$$\begin{aligned} x - 2y - z &= 1 \\ 2x + y - z &= 2. \\ -x + 3y + z &= 1 \end{aligned}$$

7. Solve the system in Exercise 6 by using row operations on matrices.

8. A manager invests a total of $150,000. The investment is split between a bank CD yielding

6 percent interest and a stock that pays a 9 percent dividend. If the total annual income from the investment is \$12,300, then how much is invested in the bank CD?

9. The points $(1,-1)$, $(2, 3)$ and $(-1,9)$ lie on the parabola given by $y = ax^2 + bx + c$. Find *a*, *b*, and *c*.

10. Find the maximum value of $P = 7x + 4y$ subject to the constraints

$$\begin{aligned} x + y &\le 18 \\ 2x + y &\le 24 \\ x &\ge 0, \quad y \ge 0 \end{aligned}$$

● ● ●

Chapter

12

Discrete Algebra and Probability

Photo Courtesy of *The Picture Cube*

An amount of \$2,000 is invested at 6 percent annual interest, payable on the anniversary of the deposit. The formula

$$a_n = 2{,}000(1.06)^n$$

gives the value of the deposit after n complete years. Find the value of a_n for each of the years, 1, 2, 3, 4, 5. What kind of sequence is this?
(See Example 6 of Section 12.1.)

Up to now we have dealt primarily with situations that are continuous in nature and that are analyzed by using the set of real numbers. However, in this chapter we discuss topics in which the positive integers play an important defining role because of the step-by-step or discrete nature of the processes in these problems. Throughout this chapter, as we discuss sequences, series, mathematical induction, the binomial theorem, and counting techniques, it is important to note that in our formulas, n is restricted to the positive integers. Finally, we conclude this chapter with an introduction to probability, since it is a natural outgrowth of our discussion of counting techniques.

12.1 Sequences

Objectives

1. Find any term in a sequence when given a formula for the nth term of the sequence.
2. Graph a sequence with the aid of a grapher.
3. Find a formula for the general term a_n in a given arithmetic sequence.
4. Find a formula for the general term a_n in a given geometric sequence.
5. Determine if a sequence is an arithmetic sequence, a geometric sequence, or neither.
6. Solve applied problems involving arithmetic sequences, geometric sequences, or the Fibonacci sequence.

In everyday speech the word *sequence* means the coming of one thing after another in a fixed order. For instance, if you buy an unassembled piece of furniture, you may find in the accompanying directions that a sequence of illustrations shows the correct method for assembly. In mathematics the concept of a sequence is usually first studied in connection with number sequences, which are sets of numbers arranged in a definite order. For example, the collection of even positive integers

$$2,4,6,8\ldots$$

is a sequence. Each number is the sequence is called a **term** of the sequence. We usually write a sequence as

$$a_1,a_2,a_3\ldots,a_n,\ldots,$$

where the subscript gives the term number, and a_n represents the general or *n*th term. We also denote a sequence by $\{a_n\}$. In the above sequence, 2 is the first term, a_1; 4 is the second term, a_2, and so on. A sequence with a first and last term is called a **finite sequence**; a sequence with an infinite number of terms is called an **infinite sequence**.

Any term in a sequence is an assignment of a_n to *n*. Thus, we can think of a sequence as a function. For example, the sequence of even positive integers is given by

$$a(n)=2n,\ n=1,2,3,\ldots.$$

Note that *a* is a function name (just like *f*), and the domain of a is the set of positive integers. We can convert from this form and list the terms of the sequence by substituting the positive integers for *n* in the given rule.

The graph of this function is given in Figure 12.1 and consists of the points

$$(1,2),\ (2,4),\ (3,6),\ldots$$

Note that we do not connect the points and that we are limited to plotting a finite number of points that suggest the behavior of the function.

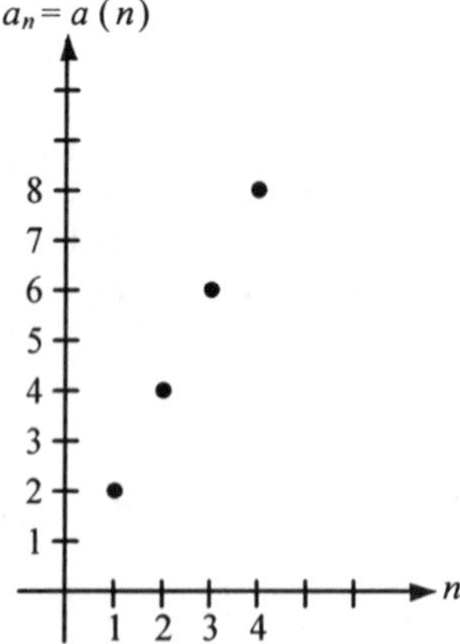

Figure 12.1

		Subscript Notation		Functional Notation		Terms of Sequence or Range Elements
First Term	=	a_1	=	$a(1)$	$=2(1)=$	2
Second Term	=	a_2	=	$a(2)$	$=2(2)=$	4
Third Term	=	a_3	=	$a(3)$	$=2(3)=$	6
$\vdots$		$\vdots$		$\vdots$	$\vdots$	$\vdots$
General or nth Term	=	a_n	=	$a(n)$	$=2(n)=$	$2n$
$\vdots$		$\vdots$		$\vdots$	$\vdots$	$\vdots$

The functional interpretation of sequence leads to the following definitions.

Definition of a Sequence

A **sequence** is a function whose domain is the set of positive integers $1,2,3\ldots$. The functional values or range elements are called the **terms** of the sequence.

The word *sequence* normally means infinite sequence, and this definition is consistent with that understanding. In the examples and exercise that follow, we stick with the custom of using subscript notation, instead of functional notation, when giving a sequence.

Example 1: Finding Terms in a Sequence

Write the first four terms of the sequence given by $a_n = 3n-1$; also find a_{25}.

Solution: Substituting $n = 1,2,3,4$ in the rule for a_n gives

$$a_1 = 3(1)-1 = 2, \qquad a_2 = 3(2)-1 = 5,$$
$$a_3 = 3(3)-1 = 8, \qquad a_4 = 3(4)-1 = 11.$$

For the 25th term, a_{25}, we have

$$a_{25} = 3(25)-1 = 74.$$

PROGRESS CHECK 1
Write the first four terms in the sequence given by $a_n = 5n-2$; also find a_{50}. ■

Example 2: Graphing a Sequence Using a Grapher

Graph the sequence $a_n = 5n-2$ with the aid of a grapher.

Solution: Since the domain in a sequence is the set of positive integers, the graph of a sequence is limited to points in either quadrant 1 or quadrant 4, or points on the positive x-axis. So we begin by using the viewing window $[0,10]$ by $[-50,50]$, to graph $y = 5x-2$, where x can be any real number, as shown in Figure 12.2. Then, the graph of $a_n = 5n-2$ is just the points on the graph of $y = 5x-2$ that correspond to positive integer replacements for x.

In Figure 12.3, colored dots have been placed on the points

$$(1,3),\ (2,8),\ (3,13),\ldots,(10,48)$$

and these points show the graph of $a_n = 5n-2$. For sequence graphs remember that points are not connected and that we are limited to showing a finite number of points that suggest the behavior of the function.

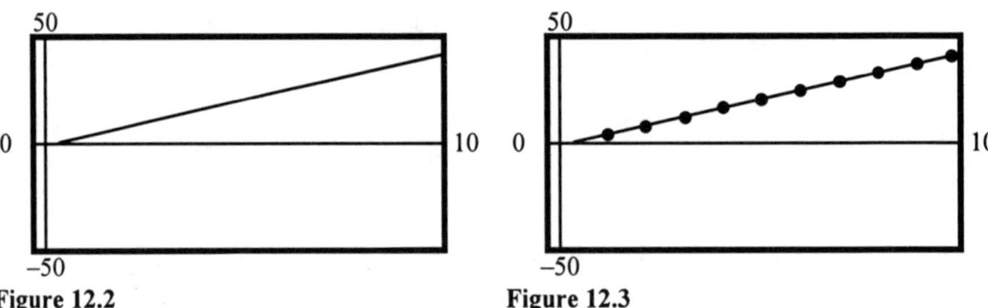

Figure 12.2 **Figure 12.3**

Note: Some graphing calculators have special features, such as Sequence mode, Dot mode, or Zoom Integer, that are helpful for displaying a sequence graph. For instance, Figure 12.4 shows a graph of the sequence in this example, using a graphing calculator that is set in both Sequence mode and Dot mode. You should learn to use such features, if they are available on your calculator.

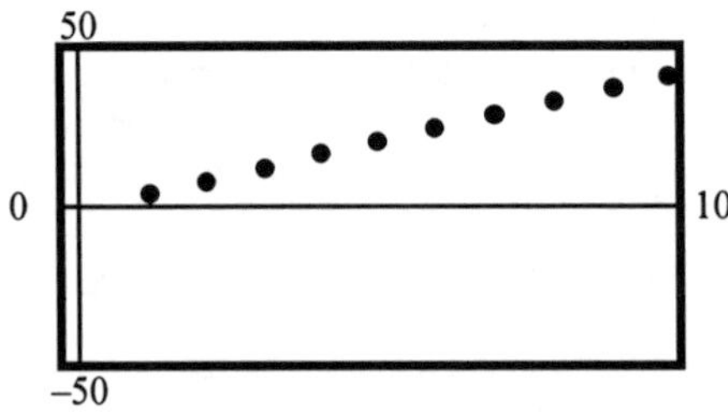

Figure 12.4

PROGRESS CHECK 2

Graph the sequence $a_n = \frac{1}{n}$ with the aid of a calculator. ■

Two special types of sequences that have many applications are called arithmetic sequences and geometric sequences. To illustrate an arithmetic sequence, consider the sequence

$$4, 7, 10, 13, \ldots.$$

Observe that each term after the first may be found by adding 3 to the previous term, as shown below.

$$\underbrace{4, \; 7}_{3}, \; \underbrace{\; 10}_{3}, \; \underbrace{\; 13}_{3}, \; \ldots$$

Therefore, the sequence is called an arithmetic sequence, and the number 3 is called the common difference in this sequence, as specified in the following definitions.

Arithmetic Sequence

An **arithmetic sequence** is a sequence of numbers in which each number after the first is found by adding a constant to the preceding term. This constant is called the **common difference** and is symbolized by d.

A formula for the general term in an arithmetic sequence with first term a_1 and common difference d may be found by observing the following pattern in such a sequence.

$$\begin{array}{lllll} \text{1st term,} & \text{2nd term,} & \text{3rd term,} & \ldots, \text{nth term,} & \ldots \\ a_1, & a_1 + 1d, & a_1 + 2d, & \ldots, a_1 + (n-1)d, & \ldots \end{array}$$

Thus, the nth term is given by $a_1 + (n-1)d$, and we have the following formula for a_n.

Formula for *n*th Term of an Arithmetic Sequence

The nth term of an arithmetic sequence is given by

$$a_n = a_1 + (n-1)d$$

where a_1 is the first term and d is the common difference.

Example 3: Finding the *n*th Term of an Arithmetic Sequence

Find a formula for the general term a_n in the arithmetic sequence

$$3, 7, 11, \ldots$$

What is the 17th term in the sequence?

Solution: The first term is $a_1 = 3$ and the common difference is $d = 7 - 3 = 4$. Substituting these numbers in the above formula gives

$$\begin{aligned} a_n &= 3 + (n-1)4 \\ &= 4n - 1. \end{aligned}$$

To find the 17th term, a_{17}, replace n by 17. Thus,

$$a_{17} = 4(17) - 1 = 67.$$

Note: Figure 12.5 illustrates that the graph of the arithmetic sequence $a_n = 4n - 1$ is just the points on the graph of the linear function $f(x) = 4x - 1$ that correspond to positive integer replacements for x. This graph is visual evidence that an arithmetic sequence is the same as a linear function with slope d in which the independent variable is limited to the positive integers.

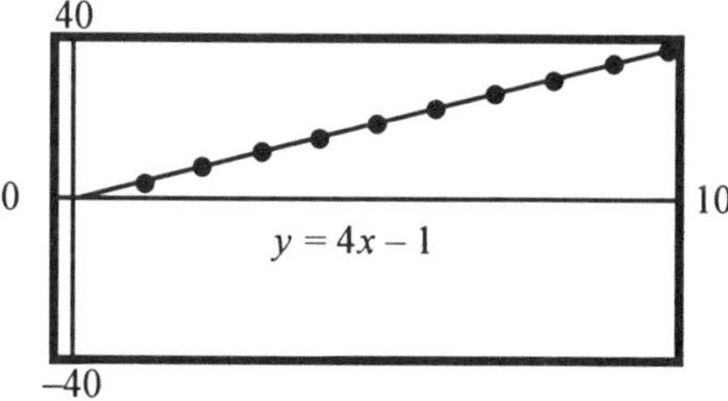

Figure 12.5

PROGRESS CHECK 3

Find a formula for the general term a_n in the arithmetic sequence 5,12,19,26,.... What is the 27th term in this sequence? ■

In an arithmetic sequence, the terms in the sequence maintain a *common difference*. When the terms in a sequence maintain a *common ratio*, then the sequence is called a geometric sequence, as given in the following definitions.

Geometric Sequence

> A **geometric sequence** is a sequence of numbers in which each number after the first number is found by multiplying the preceding term by a constant. This constant is called the **common ratio** and is symbolized by r.

To illustrate, the following sequences are geometric sequences with common ratio r as shown.

$$3,9,27,81\ldots \qquad r=\frac{9}{3}=\frac{27}{9}=\frac{81}{27}=3$$

$$1,1.05,(1.05)^2,(1.05)^3,\ldots \qquad r=\frac{1.05}{1}=\frac{(1.05)^2}{1.05}=\frac{(1.05)^3}{(1.05)^2}=1.05.$$

A formula for the general term in a geometric sequence with first term a_1 and common ratio r may be found by observing the following pattern in such a sequence.

1st term, ②nd term, ③rd term, . . . , ⓝth term, . . .

a_1, $\quad a_1 r^{1}$, $\quad a_1 r^{2}$, $\quad \ldots, a_1 r^{n-1}$, $\quad \ldots$

Thus, a_1r^{n-1} specifies the general term, and we have the following formula for a_n in a geometric sequence.

Formula for *n*th Term in a Geometric Sequence

> The nth term in a geometric sequence is given by
>
> $$a_n = a_1r^{n-1},$$
>
> where a_1 is the first term and r is the common ratio.

Example 4: Finding the *n*th Term of a Geometric Sequence

Find the formula for the general term a_n in the geometric sequence

$$24,12,6\ldots$$

What is the 7th term in the sequence?

Solution: The first term is $a_1 = 24$ and the common ratio is $r=\frac{12}{24}=\frac{1}{2}$. Substituting these values in the above formula gives

$$a_n = 24\left(\frac{1}{2}\right)^{n-1}.$$

To find the 7th term a_7, replace n by 7. Thus,

$$\begin{aligned} a_7 &= 24\left(\frac{1}{2}\right)^{7-1} \\ &= 24\left(\frac{1}{2}\right)^{6} \\ &= 24\left(\frac{1}{64}\right) = \frac{3}{8} \end{aligned}$$

Note: Figure 12.6 illustrates that the graph of the geometric sequence $a_n = 24(1/2)^{n-1}$ is just the points on the exponential function $f(x) = 24(1/2)^{x-1}$ that correspond to positive integer replacements for x. This graph is visual evidence that a geometric sequence with $r > 0$ is the same as an exponential function with base r in which the independent variable is limited to the positive integers.

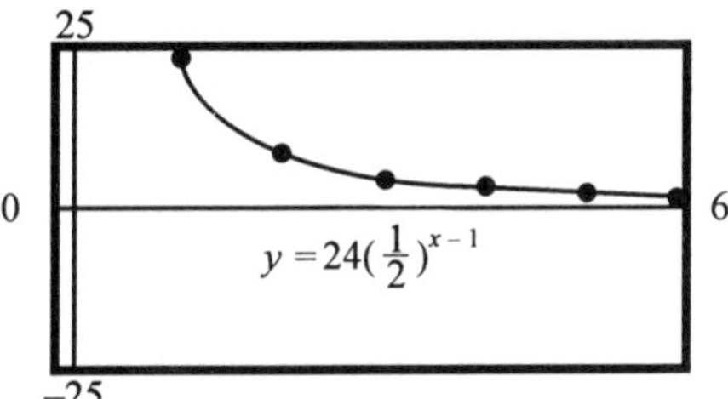

Figure 12.6

PROGRESS CHECK 4

Find a formula for the general term a_n in the geometric sequence $3, 15, 75, 375, \ldots$. What is the 7th term in this sequence? ■

When given a sequence of numbers, we must be able to classify the sequence as arithmetic, geometric, or neither, as considered in the next example.

Example 5: Classifying Sequences

State whether the sequence is an arithmetic sequence, a geometric sequence, or neither. For any arithmetic sequence, state the common difference and write the next two terms. For any geometric sequence, state the common ratio and write the next two terms.

a. $2, 6, 18, 54 \ldots$ **b.** $1, 8, 15, 22 \ldots$ **c.** $1, 4, 9, 16, \ldots$

Solution:

a. Each term after the first is found by multiplying the preceding term by 3, as illustrated below.

$$2, \quad 6, \quad 18, \quad 54, \quad \ldots.$$
$$2 \cdot 3 = 6 \qquad 6 \cdot 3 = 18 \qquad 18 \cdot 3 = 54$$

Therefore, the sequence is geometric, with common ratio 3. The next two terms in the sequence are 162 and 486.

b. Each term after the first is found by adding 7 to the preceding term, as shown next.

$$1, \quad 8, \quad 15, \quad 22, \quad \ldots.$$
$$1 + 7 = 8 \qquad 8 + 7 = 15 \qquad 15 + 7 = 22$$

Therefore, the sequence is arithmetic, with common difference 7. The next two terms in the sequence are 29 an 36.

c. The terms in this sequence do not have a common difference, and they do not have a common ratio. Therefore, the sequence is neither arithmetic nor geometric.

PROGRESS CHECK 5

Answer the questions in Example 5 for each sequence.

a. $1.125, 1.25, 1.375, 1.5, \ldots$ **b.** $1, \frac{1}{2}, \frac{1}{3}, \frac{1}{4}, \ldots$ **c.** $27, 9, 3, 1\ldots$ ■

The mathematics of finance contains many applications of sequences. One such application is considered in the chapter-opening problem.

Example 6: Compound Interest

Solve the problem in the chapter introduction on page 947.

Solution: Substituting $n = 1,2,3,4,5$ in the rule for a_n gives

$$a_1 = 2{,}000(1.06)^1 = \$2{,}120$$
$$a_2 = 2{,}000(1.06)^2 = \$2{,}247.20$$
$$a_3 = 2{,}000(1.06)^3 \approx \$2{,}382.03$$
$$a_4 = 2{,}000(1.06)^4 \approx \$2{,}524.95$$
$$a_5 = 2{,}000(1.06)^5 = \$2{,}676.45$$

In this sequence each term after the first is found by multiplying the preceding term by 1.06, so the sequence is geometric with common ratio 1.06.

Note: You should recognize that the formula for the above sequence follows from our work with compound interest in the chapter on exponential functions. This insight once again demonstrates the relation between geometric sequences and exponential functions.

PROGRESS CHECK 6

If office equipment is purchased for \$20,000 and depreciated at a rate of 40 percent each year, then the formula

$$a_n = 12{,}000(0.6)^{n-1}$$

gives the depreciated value of this equipment after n complete years. Find the value of a_n for each of the years $1,2,3,4,5$. What kind of sequence is this? ■

Arithmetic and geometric sequences by no means exhaust the list of important sequences. As an example, consider the set of numbers

$$1, 1, 2, 3, 5, 8, 13, 21, 34, 55, 89, \ldots,$$

which is called the **Fibonacci sequence**. The rule that generates these numbers is as follows: Start with 1 and 1, and thereafter each term is the sum of the two previous terms. In symbols we write

$$a_1 = 1, \quad a_2 = 1, \quad a_n = a_{n-1} + a_{n-2} \quad \text{for } n \geq 3.$$

Leonardo Fibonacci, who was a mathematician in the Middle Ages, introduced this sequence innocently enough in his book on arithmetic and algebra, as the answer to the following problem (paraphrased): A pair of rabbits produces a pair of baby rabbits by the end of their second month and every month thereafter. Each new pair of rabbits does the same. If no rabbits die, how many pairs of rabbits are there at the beginning of each month? The answer for the first seven months (and take the time to analyze why) is

Beginning of month number	1	2	3	4	5	6	7
Numbers of pairs	1	1	2	3	5	8	13

What is special about this sequence? Why is a mathematics journal, *The Fibonacci Quarterly*, devoted exclusively to issues connected with these numbers? The answer is that the numbers in the Fibonacci sequence keep appearing in the analysis of topics in such diverse areas as mathematics (naturally), physics, chemistry, art, and music, as well as appearing with surprising regularity in nature. Most of these applications are a bit involved, but it is not difficult to consider some of the fascinating ways in which Fibonacci numbers occur in nature.

On pine cones, pineapples, sunflowers, and other growths that appear on the tip of a stem, there are two distinct sets of spirals. As shown in the diagram of a small sunflower in Figure 12.7, one set radiates clockwise, and the other counterclockwise. The number of spirals in each set is a Fibonacci number. In a small sunflower (as in this figure) there are 21 clockwise and 34 counterclockwise spirals. Some common counts for such growths are:

pine cone	5, 8	small sunflower	21, 34
pineapple	8, 13	large sunflower	34, 55
large pineapple	13, 21	giant sunflower	55, 89.

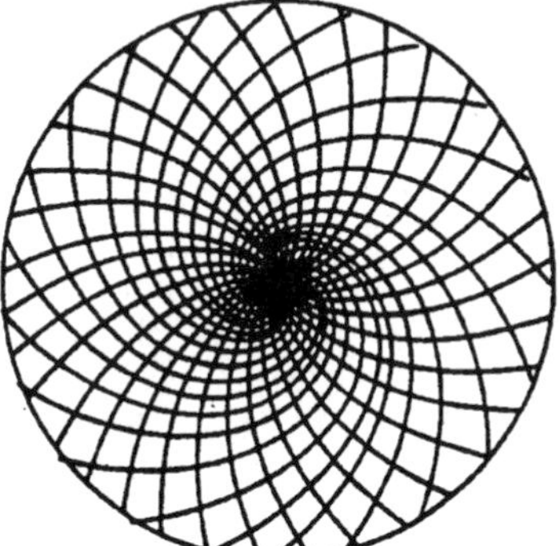

Figure 12.7

Note that each pair of numbers gives consecutive terms in the Fibonacci sequence.

These numbers also occur when considering the spiral leaf growth pattern of many trees. For example, consider the cyclic arrangements in the position of the leaf buds on the sample from an oak tree in Figure 12.8. Note that there are 5 buds to a complete cycle. In addition, each cycle contains 2 revolutions of the spiral. Both numbers are terms of the Fibonacci sequence. Similar remarks are true of other trees (such as elm and birch trees) and certain types of bushes. These examples just begin to scratch the surface of the immense volume of material dealing with the Fibonacci sequence.

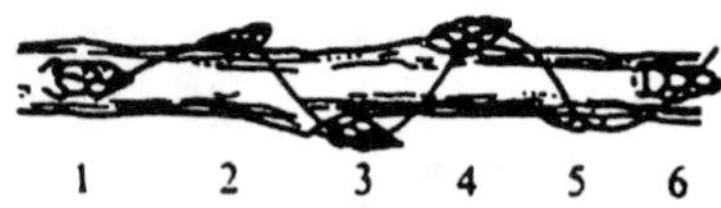

Figure 12.8

EXPLORE 12.1

Some graphing calculators have a Sequence mode, which makes the analysis of sequences easier. In Sequence mode, you can define a sequence and then generate its terms automatically, and graphs of sequences automatically come out as a pattern of unconnected dots. The first four explorations do not assume that your machine has a sequence mode, but if it does, learn how to take advantage of its features.

1. The List feature of a calculator is helpful in calculating the terms of a sequence. Recall that on many calculators the List capability is activated by the use of braces. For instance, this sequence of steps will evaluate $a_n = 3n - 2$ for $n = 1,2,3$ and 4.

$\{1,\ 2,\ 3,\ 4\} \rightarrow N$
$\{1\ 2\ 3\ 4\}$
$3N - 2$
$\{1,\ 4,\ 7,\ 10\}$

On some calculators, only certain variable names, such as L_1, may be used to store lists. On such a machine the sequence of steps would look like this.

$\{1,\ 2,\ 3,\ 4\} \rightarrow L_1$
$\{1\ 2\ 3\ 4\}$
$3L_1 - 2$
$\{1,\ 4,\ 7,\ 10\}$

Use your calculator to calculate and display the first 5 terms of these sequences.

a. $a_n = 2n + 1$ **b.** $a_n = n^2 - 7$

2. **a.** The sequence $a_n = 2n + 1$ is an arithmetic sequence. The function $f(x) = 2x + 1$ is a linear function. With the aid of a grapher, draw the graphs of these two functions, then describe any important ways in which the graphs are the same, and any ways they differ. Explain the relationship between *d*, the common difference for the sequence, and *m*, the slope of the line.

   b. The sequence $a_n = 2^n$ is a geometric sequence. The function $f(x) = 2^x$ is an exponential function. With the aid of a grapher, draw the graphs of these two functions, then describe any important ways in which the graphs are the same, and any ways in which they differ. Explain the relationship between *r*, the common ratio for the sequence, and *b*, the base of the exponential function.

3. The Technology Link in Section 12.1 explains how to interpret continuous graphs for applications involving sequences. For a calculator that does not have a special Sequence mode, this is probably the simplest approach. However, if desired, you can produce dot graphs of sequences by adjusting the viewing window property. Here is one approach based on a special function, "fractional part," which is built into most calculators. This menu choice, often abbreviated "fpart," produces the fractional part of a number. For instance, $\text{fpart}(-1.385) = -.385$.

   a. Set the grapher to Dot mode, then graph $y = \text{fpart}(x)$ in the standard viewing window. Explain the resulting graph. What is the value of *y* when *x* is an integer? When *x* is not an integer?

b. Set the grapher to Dot mode, then graph $y = \text{fpart}(x)$ in a Zoom Decimal viewing window. Explain the resulting graph.

c. Set the grapher to Dot mode. Using the approach explained in Section 1.2 for graphing piecewise functions, enter:

$$y = (2x-3)(\text{fpart}(x) == 0)(x > 0),$$

and graph using the Zoom Decimal setting. Explain why this results in the graph of the sequence $a_n = 2n-3$.

Note: The relations == and > are found on the "test" menu. The relation == means "equals."

d. Use the approach in part **c** to graph these sequences:

$$a_n = n^2 - 1$$
$$a_n = 2n + 1$$

4. You can use the programming capability of a graphing calculator to help with calculation of the *n*th term of a sequence. The following steps describe a simple program for calculating the *n*th term of $a_n = 2n+1$. You can adapt it for your own calculator.

Outline	Specific Example (TI)
1. Ask user to input *n*.	**1.** Prompt N
2. Calculate a_n	**2.** $2N+1 \to A$
3. Display the result.	**3.** Disp A

5. Calculators that have special sequence operations can define and generate a sequence automatically. For instance, to generate the first 5 terms of the sequence $a_n = 2^n$, you use an instruction like $\text{seq}(2^n, n, 1, 5, 1)$. The arguments in the parentheses indicate the formula, the variable name, the starting and stopping index values, and the increment between steps. If your calculator has this capability, use it to generate the first 5 terms of $a_n = 2^n$. Explore what happens if you change the variable from *n* to *x*, and enter the sequence instruction in the graph editor. You would enter something like $y = \text{seq}(2^x, x, 1, 5, 1)$. Explain the graph that appears.

EXERCISES 12.1

In Exercises 1–10 write the first four terms in the sequence with general terms as given; also find the indicated term.

1. $a_n = 2n-1$; a_{50}

2. $a_n = 7n+2$; a_{25}

3. $a_n = n^2$; a_7

4. $a_n = \dfrac{1}{n}$; a_{33}

5. $a_n = \dfrac{(-1)^{n+1}}{n}$; a_{10}

6. $a_n = \dfrac{(-1)^n}{n}$; a_{10}

7. $a_n = \dfrac{1}{2^n}$; a_5

8. $a_n = \dfrac{3^{n-1}}{4^{n-1}}$; a_5

9. $a_n = \dfrac{n(n+1)}{2}$; a_8

10. $a_n = \dfrac{1}{6}n(n+1)(2n+1)$; a_6

In Exercises 11–20 graph the given sequence with the aid of a grapher.

11. $a_n = 4n-3$

12. $a_n = 2-n$

13. $a_n = n^2+2$

14. $a_n = \dfrac{n^2+2}{n+1}$

15. $a_n = \frac{4}{n}$
16. $a_n = -\frac{3}{n}$
17. $a_n = n^3 - n^2$
18. $a_n = \sqrt[3]{n^2}$
19. $a_n = n(n+2)$
20. $a_n = \left(1+\frac{1}{n}\right)^n$

In Exercises 21–30 the sequences are arithmetic sequences. Find the formula for the general term a_n, and the value of the indicated term in the sequence.

21. $3, 7, 11, \ldots$; 40th term
22. $2, 9, 16, \ldots$; 50th term
23. $9, 4, -1, \ldots$; 5th term
24. $2, -2, -6, \ldots$; 10th term
25. $1.1, 1.4, 1.7, \ldots$; a_{25}
26. $3.15, 3.10, 3.05, \ldots$; a_{18}
27. $\frac{1}{3}, \frac{5}{3}, 3, \ldots$; a_{30}
28. $\frac{10}{7}, \frac{1}{7}, -\frac{8}{7}, \ldots$; a_9
29. $4, \frac{35}{8}, \frac{19}{4}, \ldots$; a_8
30. $\frac{13}{4}, \frac{9}{2}, \frac{23}{4}, \ldots$; a_{20}

In Exercises 31–40 the sequences are geometric sequences. Find the formula for the general term a_n and the value of the indicated term in the sequence.

31. $2, 6, 18, \ldots$; 7th term
32. $2, -6, 18, \ldots$; 7th term
33. $1, -\frac{1}{2}, \frac{1}{4}, \ldots$; 6th term
34. $1, \frac{1}{2}, \frac{1}{4}, \ldots$; 6th term
35. $\sqrt{2}, 2, \sqrt{2}, \ldots$; a_8
36. $1, 1.04, (1.04)^2, \ldots$; a_{10}
37. $6, 4, \frac{8}{3}, \ldots$; a_7
38. $1, 0.1, 0.01, \ldots$; a_{10}
39. $1, -\frac{1}{3}, \frac{1}{9}, \ldots$; a_6
40. $1, \frac{3}{4}, \frac{9}{16}, \ldots$; a_5

In Exercises 41–50 state whether the sequence is an arithmetic sequence, geometric sequence or neither. For any arithmetic sequence state the common difference and write the next two terms. For any geometric sequence state the common ratio and write the next two terms

41. $2, 7, 12, 17, 22, \ldots$
42. $-2, 4, -8, 16, -32, \ldots$
43. $1, \frac{1}{2}, \frac{1}{3}, \frac{1}{4}, \frac{1}{5}, \ldots$
44. $1, -4, -9, -14, -19, \ldots$
45. $1, 1.02, (1.02)^2, (1.02)^3, (1.02)^4, \ldots$
46. $\frac{1}{2}, \frac{2}{3}, \frac{3}{4}, \frac{4}{5}, \frac{5}{6}, \ldots$
47. $3, 0.3, 0.03, 0.003, 0.0003, \ldots$
48. $1.01, 1.02, 1.03, 1.04, 1.05, \ldots$
49. $1, \frac{1}{2^2}, \frac{1}{3^2}, \frac{1}{4^2}, \frac{1}{5^2}, \ldots$
50. $1, \frac{1}{2}, \frac{1}{2^2}, \frac{1}{2^3}, \frac{1}{2^4}, \ldots$
51. Write the first four terms in the arithmetic sequence in which $a_1 = 4$ and $d = \frac{1}{2}$.
52. Write the first four terms in the geometric sequence in which $a_1 = 4$ and $r = \frac{1}{2}$.
53. Which term of the geometric sequence $18, 9 \ldots$ is $\frac{9}{16}$?
54. Which terms of the arithmetic sequence $7, 4 \ldots$ is -68?
55. Find the first term in an arithmetic sequence in which $a_{17} = 39$ and $d = 2$.
56. The first term of a geometric sequence is 8, and the fourth term is 27. Find the common ratio.
57. Find two different values of x so that $-\frac{2}{3}, x, -\frac{25}{54}$ are in geometric sequence.
58. If, in an arithmetic sequence $a_1 = 80$, $d = -3$, and $a_n = 26$, find n.
59. If, in an arithmetic sequence, $a_6 = 15$ and $a_{12} = 24$, find a_1 and d.
60. If, in an arithmetic sequence $a_{15} = \frac{22}{3}$ and $a_{30} = \frac{91}{6}$, find a_{40}.
61. An amount of \$1,000 is invested at 8 percent annual interest, payable on the anniversary of the deposit. The formula $a_n = 1000(1.08)^n$ gives the value of the deposit after n complete years. Find the value of a_n for each of the years 1, 2, 3, 4, 5. What kind of sequence is this?
62. According to the "double-declining balance" method of depreciation, the book value of office equipment with a useful life of 10 years declines at 20 percent per year. Find the value of equipment purchased for

\$60,000 at the end of 2, 4, 6, 8, and 10 years. For this equipment, $a_n = 60{,}000(0.8)^n$ gives the value after n years. What kind of sequence is this?

63. Sequences are important in the theory of limits in calculus, where the object is to see if the terms of a sequence are approaching some fixed value. When this happens, that value is called the *limit* of the sequence. In these examples, write the first five terms of the given sequence, and guess the limit.

a. $a_n = \frac{n}{n+1}$ **b.** $a_n = \frac{1}{n}$

c. $a_n = \frac{1}{n^3 - 5n^2 + 11n - 6}$

d. $a_n = \frac{1}{n^2}$ **e.** $a_n = 2 - \frac{1}{3^n}$

f. $a_n = 1 - 0.6^n$

64. Here is a sequence, called the Galileo sequence, with a clear pattern.

$$\frac{1}{3}, \frac{1+3}{5+7}, \frac{1+3+5}{7+9+11}, \ldots$$

a. Follow the given pattern and write the next term.

b. Simplify each term as much as possible, to reveal a remarkable result.

65. Some sequences are clearly defined even though there is no obvious formula. For instance, what are the first five terms of the sequence of prime numbers? That is, $a_n = n$th prime number. Is this sequence arithmetic, geometric, or neither?

66. Prime numbers are positive integers that have exactly two factors. Thus, 3 is prime because it has exactly two factors, namely, 3 and 1. In a similar way, we can construct a sequence of positive integers that have exactly three factors.

	Factors
$a_1 = 4$	1, 2, 4
$a_2 = 9$	1, 3, 9
$a_3 = 25$	1, 5, 25

What is the next number in this sequence? Is this sequence arithmetic, geometric, or neither?

67. State whether the sequence is an arithmetic sequence, geometric sequence, or neither. Find the common difference for any arithmetic sequence and the common ratio for any geometric sequence.

a. $\log 2, \log 4, \log 8, \log 16$

b. $\log 2, \log 4, \log 6, \log 8, \ldots$

c. $\log 2, \log 4, \log 16, \log 256, \ldots$

68. Same as 67, but for these sequences.

a. $\log 3, \log 9, \log 27, \log 81, \ldots$

b. $\log 3, \log 9, \log 81, \log 6561, \ldots$

c. $\log 3, \log 6, \log 9, \log 27, \ldots$

69. **a.** Write the first fifteen terms of the Fibonacci sequence.

b. The Fibonacci sequence appears in the family tree of a male bee. A male bee develops from an unfertilized egg and has only one parent—his mother. A female bee has both a mother and a father. The number of ancestors of a male bee in each successive generation is therefore the numbers of the Fibonacci sequence, as shown below.

Number of Ancestors
1
2
3

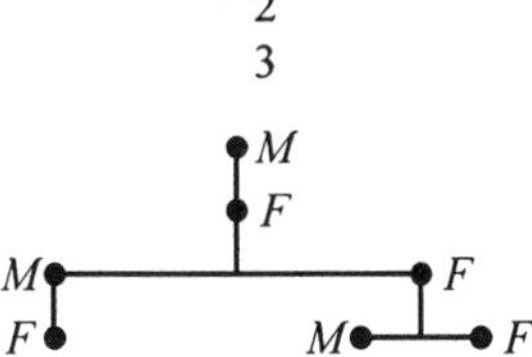

Continue the diagram for the next three generations and verify that the number of ancestors in these generations is given by the next three terms in the Fibonacci sequence.

In Exercises 70–72, let F_n represent the nth term of the Fibonacci sequence.

70. Each positive integer can be written as the sum of Fibonacci numbers, using none of them more than one time. For instance $5 = F_3 + F_4 = 2 + 3$. Write the integers 10, 20, 30 and 40 this way.

71. The ratio of 2 consecutive Fibonacci numbers approaches a limiting value as n increases. For example $\frac{F_4}{F_3} = 1.5$, while $\frac{F_{10}}{F_9} = \frac{55}{34} \approx 1.6176$. Let us define a new sequence where $a_n = \frac{F_{n+1}}{F_n}$. Evaluate a_n for $n = 9$, 10, 11, and 12. Give each answer accurate to 4 decimal places. The terms of this sequence should appear to be getting closer and closer to some limiting value. It can be shown that this limiting value is the so-called golden ratio, which is $\frac{1}{2}(1+\sqrt{5})$.

72. The sum of the squares of the first n Fibonacci numbers equals the product of the nth and $n+1st$ Fibonacci numbers. For instance, the sum of the squares of the first 5 Fibonacci numbers is given by $1^2+1^2+2^2+3^2+5^2=40$, which is the same as 5×8, the product of the 5th and 6th terms. Use this relationship to find the sum of the squares of the first 10 Fibonacci numbers two different ways.

For more information about Fibonacci the man and this sequence of integers see *The History of Mathematics*, by David Burton, Wm. C. Brown, 1991.

73. A generalized Fibonacci sequence is one that starts with two numbers other than $a_1=1$ and $a_2=1$, but then proceeds in the usual way. For instance 8, 15, 23, 38, 61,... is a generalized Fibonacci sequence. Note, in this particular case, that 8 and 15 have no common factors, and that the sequence includes both prime and composite terms. Mathematicians have tried to find generalized Fibonacci sequences with special properties. For instance, can you start with two numbers that have no common factor but then produce a sequence that contains no prime terms? If so what are the two smallest starting values that will work? In this case, the answer is yes, and the two smallest starting numbers are

$a_1 = 1{,}059{,}683{,}225{,}053{,}915{,}111{,}058{,}165{,}$
$141{,}686{,}995$ and

$a_2 = 1{,}786{,}772{,}701{,}928{,}802{,}632{,}268{,}715{,}$
$130{,}455{,}793$

[Source for this sequence: *The Penguin Dictionary of Curious and Interesting Numbers* by David Wells, Penguin, 1986]

a. Prove that a_3 is not prime.

b. Suppose, in a generalized Fibonacci sequence, that a_1 and a_2 are both even integers. Will a_n for $n>2$ ever be a prime number? Justify your answer.

74. A sequence may be given by a rule rather than a formula. Consider the Farey sequence of order n, named for John Farey (1766–1826), an English mathematician. This sequence consists of all positive fractions between $\frac{0}{n}$ to $\frac{n}{n}=\frac{1}{1}$ arranged in increasing numerical order. The top integer cannot be larger than the bottom integer, and the fraction must be in lowest terms. Let the terms of a new sequence a_n be the number of fractions in the Farey sequence of order n. For the first three terms we get

$a_1=2$, since the Farey sequence of order 1 is $\frac{0}{1}, \frac{1}{1}$

$a_2=3$, since the Farey sequence of order 2 is $\frac{0}{2}, \frac{1}{2}, \frac{1}{1}$

$a_3=5$, since the Farey sequence of order 3 is $\frac{0}{3}, \frac{1}{3}, \frac{1}{2}, \frac{2}{3}, \frac{1}{1}$.

Find the Farey sequences of order 4 and 5, then determine a_4 and a_5.

THINK ABOUT IT 12.1

Guessing the formula or pattern for a given sequence is a common puzzle. It is never possible *just by looking at the first several numbers* to be certain that you have the correct formula, because for any given finite set of terms, there are an unlimited number of different formulas that will work.

a. Check, for instance, that the sequence beginning 1, 2, 4,... can be generated both by $a_n=2^{n-1}$ and by $a_n=0.5n^2-0.5n+1$. In each case find the next term in the sequence.

b. Find two different formulas that will generate the sequence beginning 1, 3, 9,.... Find the next term in each sequence.

c. This old puzzle is solved most quickly by young children. What is the next term in this sequence of letters? How many more terms can you determine? O, T, T, F, F,...

d. Guess the next term in this sequence. 0,0,0,0,0,... It was generated by $a_n=n(n-1)(n-2)(n-3)(n-4)(n-5)$. What is the next term according to this formula?

2. The English economist Thomas Malthus, in 1798, wrote a famous essay called "An Essay on the Principle of Population," in which he claimed that the population increases geometrically, while food supply increases arithmetically. Discuss what this means. Why did readers find this proposal so alarming?

3. Is 5,000 a term in the arithmetic sequence 1, 4, 7,...? Explain your answer.
4. a. If a, b, and c are the first three terms in an arithmetic sequence, express b in terms of a and c.
   b. If a, b, and c are the first three terms in a geometric sequence, express b in terms of a and c.
   c. If a, b, and c are three consecutive terms in an arithmetic sequence, b is called the **arithmetic mean** of a and c. If a, b, and c are three consecutive terms in a geometric sequence, b is called the **geometric mean** of a and c. Find the arithmetic mean and the geometric mean of 2 and 8.
5. a. Find three numbers in arithmetic sequence whose sum is 60 and whose product is 5,120.
   b. Find three numbers in geometric sequence whose sum is 52 and whose product is 1,728.

● ● ●

12.2 Series

*Photo Courtesy of **Bob Thomas**/RTS/Popperfoto, of Archive Photos, New York*

If 512 players enter a single elimination tournament (one loss and you're out), determine the number of matches needed to declare a winner by considering the following two approaches:

a. The number of matches from each round form a geometric sequence. Write down this sequence and determine the sum.
b. Instead of reasoning from the first round to the end of the tournament, try a favorite technique in problem solving—reason from the end result. Write an equivalent question to "How many matches were played?" that transforms this problem into one that is really simple. Now, how many matches are needed in this tournament?

(See Example 3.)

Objectives

1. Find the sum of an indicated number of terms in an arithmetic sequence.
2. Find the sum of an indicated number of terms in a geometric sequence.
3. Write a series given in sigma notation in the expanded form, and determine the sum.
4. Write a series given in expanded form, using sigma notation.
5. Solve applied problems involving series.

Associated with any sequence

$$a_1, a_2, \ldots, a_n, \ldots$$

is a **series**

$$a_1 + a_2 + \ldots + a_n + \ldots$$

which is the sum of all the terms in the sequence. The series associated with arithmetic and geometric sequences are called, respectively, an **arithmetic series** and a **geometric series**. In this section we consider only series with a *finite* number of terms. The meaning of a series with an infinite number of terms is more complex. Infinite series are discussed briefly in the next section and in detail in calculus.

We begin with a classic story from the history of mathematics that suggests the procedure for finding the sum of an arithmetic series. Carl Friedrich Gauss dominated mathematics in the nineteenth century and is considered, along with Archimedes and Newton, to be in a special class among the great mathematicians. He contributed profoundly to astronomy, physics (especially electromagnetic theory), non-Euclidean geometry, number theory, probability and statistics, and function theory. Reportedly, he demonstrated his imaginative insight to problem solving at the early age of ten when his class at school was assigned some "busy work." Their task—add all the numbers from 1 to 100. To the teacher's amazement, Gauss quickly scribbled the correct result: 5,050. How did he do it? Well, as is often the case when an ingenious method is used, the answer is easily obtained. Gauss noticed that the 100 numbers could be arranged in 50 pairs that all add up to 101, as follows:

$$\begin{array}{cccccccc} 1 \;+ & 2 \;+ & 3 \;+ & 4 \;+ & \cdots \;+ & 48 \;+ & 49 \;+ & 50 \;+ \\ 100 \;+ & 99 \;+ & 98 \;+ & 97 \;+ & \cdots \;+ & 53 \;+ & 52 \;+ & 51 \\ \hline 101 \;+ & 101 \;+ & 101 \;+ & 101 \;+ & \cdots \;+ & 101 \;+ & 101 \;+ & 101 \end{array}$$

Thus, the result is simply 50(101) or 5,050.

By a similar method of adding pairs of terms with the same sum, we can develop the formula for the sum, denoted S_n, of the first n terms of an arithmetic progression. In general form an arithmetic series can be written as

$$S_n = a_1 + (a_1 + d) + (a_1 + 2d) + \cdots + (a_n - 2d) + (a_n - d) + a_n .$$

By reversing the order of the terms, we can also write this series as

$$S_n = a_n + (a_n - d) + (a_n - 2d) + \cdots + (a_1 + 2d) + (a_1 + d) + a_1$$

If we now add term by term the equivalent expressions for S_n, we have n pairs, which all add up to $a_1 + a_n$. Thus,

$$2S_n = n(a_1 + a_n),$$

so that the general formula is

$$S_n = \frac{n}{2}(a_1 + a_n).$$

This formula is used when we know a_1 and a_n. However, a_n is often not given. If we replace a_n by $a_1 + (n-1)d$, the above formula becomes

$$\begin{aligned} S_n &= \frac{n}{2}\{a_1 + [a_1 + (n-1)d]\} \\ &= \frac{n}{2}[2a_1 + (n-1)d] \end{aligned}$$

We summarize our results as follows.

Arithmetic Series Formulas

The sum of the first n terms of an arithmetic series is given by

$$S_n = \frac{n}{2}(a_1 + a_n)$$

or

$$S_n = \frac{n}{2}[2a_1 + (n-1)d].$$

Example 1: Finding the Sum of an Arithmetic Sequence

Find the sum of the first 21 terms of the arithmetic sequence $4, 7, 10, \ldots$.

Solution: Here $a_1 = 4$, $d = 3$, and $n = 21$. Thus,

$$\begin{aligned} S_n &= \frac{n}{2}[2a_1 + (n-1)d] \\ S_{21} &= \frac{21}{2}[2(4) + (21-1)3] \\ &= \frac{21}{2}(68) = 714 \end{aligned}$$

PROGRESS CHECK 1

Find the sum of the first 250 positive integers. ■

We now develop a formula for the sum of a given number of terms in a geometric sequence. In general form a geometric series with n terms can be written as

$$S_n = a_1 + a_1 r + a_1 r^2 + \cdots + a_1 r^{n-1}.$$

Multiplying both sides of this equation by r gives

$$rS_n = a_1 r + a_1 r^2 + \cdots + a_1 r^{n-1} + a_1 r^n$$

Subtracting the second equation from the first, we obtain

$$\begin{aligned} S_n - rS_n &= a_1 - a_1 r^n \\ S_n(1-r) &= a_1 - a_1 r^n \\ S_n &= \frac{a_1 - a_1 r^n}{1-r}, \text{ for } r \neq 1. \end{aligned}$$

This formula is used when we know a_1, r, and n. If we know the nth or last term in the series, a different formula is used. Since $a_n = a_1 r^{n-1}$, we have $a_1 = a_n / r^{n-1}$. This replacement in the above formula gives

$$S_n = \frac{a_1 - \left(a_n / r^{n-1}\right) r^n}{1-r} = \frac{a_1 - a_n r}{1-r}.$$

To summarize, we have derived the following formulas.

Geometric Series Formulas

The sum of the first n terms of a geometric series with $r \neq 1$ is given by

$$S_n = \frac{a_1 - a_1 r^n}{1-r}$$

or

$$S_n = \frac{a_1 - a_n r}{1-r}.$$

Example 2: Finding the Sum of a Geometric Sequence

Find the sum of the first seven terms of the geometric sequence $1, \frac{1}{2}, \frac{1}{4}, \ldots$.

Solution: Here $a_1 = 1$, $r = \frac{1}{2}$, and $n = 7$. Thus

$$S_n = \frac{a_1 - a_1 r^n}{1-r}$$

$$S_7 = \frac{1 - 1\left(\frac{1}{2}\right)^7}{1 - \frac{1}{2}}$$

$$= \frac{1 - \frac{1}{128}}{\frac{1}{2}} = \frac{\frac{127}{128}}{\frac{1}{2}} = \frac{127}{64}$$

PROGRESS CHECK 2

Find the sum of the first eight terms of the geometric sequence 3,12,48,... ■

Example 3: Number of Matches in a Tennis Tournament

Solve the problem in the section introduction on page 961.

Solution:

a. In the first round, the 512 players are paired up two at a time, so 256 matches are needed. The 256 winners then advance to the second round, which will require 128 matches. Each subsequent round continues to require half the number of matches as the preceding round, so the number of matches from each round forms the following geometric sequence.

256, 128, 64, 32, 16, 8, 4, 2, 1

It is easy enough to just add up these numbers, but we can also find the sum by using the second of our formulas for the sum of a geometric series.

$$S_n = \frac{a_1 - a_n r}{1-r} = \frac{256 - 1\left(\frac{1}{2}\right)}{1 - \frac{1}{2}} = \frac{512 - 1}{2 - 1} = 511$$

Thus, 511 matches are needed in the tournament.

b. When the tournament is *over*, there are 511 losers and 1 winner. Each loser lost exactly one match so the question "How many matches are played?" can be found by answering "How many losers are there?" With this approach it is easy to see that 511 matches are needed.

PROGRESS CHECK 3

Redo the problem in Example 3, assuming that 128 players enter the tournament. ■

A series is a sum, and it is convenient to use the Greek letter Σ, read "sigma," to mean *add*. By this convention, called **sigma notation**, we write

$$S_n = a_1 + a_2 + \cdots + a_n$$

so that $S_n = \sum_{i=1}^{n} a_i$ means to add the terms that result from replacing i by 1, then 2,..., then n. The use of the letter i in this notation is arbitrary, so that

$$\sum_{i=1}^{n} a_i, \quad \sum_{j=1}^{n} a_j, \quad \sum_{k=1}^{n} a_k$$

all represent the same series.

Example 4: Simplifying a Series in Sigma Notation

Write the series $\sum_{i=1}^{4}(2i+3)$ in expanded form and determine the sum.

Solution:

$$\begin{aligned}\sum_{i=1}^{4}(2i+3) &= (2\cdot 1+3)+(2\cdot 2+3)+(2\cdot 3+3)+(2\cdot 4+3)\\ &= 5 + 7 + 9 + 11\\ &= 32\end{aligned}$$

PROGRESS CHECK 4

Write the series $\sum_{i=1}^{5}(i^2-1)$ in expanded form and determine the sum. ■

Example 5: Simplifying a Series in Sigma Notation

Write the series $\sum_{j=2}^{7} 3^j$ in expanded form and determine the sum.

Solution:

$$\sum_{j=2}^{7} 3^j = 3^2 + 3^3 + 3^4 + 3^5 + 3^6 + 3^7.$$

The series is a geometric series with $a_1 = 3^2$, $a_n = 3^7$, and $r = 3$. Thus

$$\begin{aligned}S_n &= \frac{a_1 - a_n r}{1-r} = \frac{3^2 - (3^7)3}{1-3}\\ &= \frac{9 - 6{,}561}{-2} = 3{,}276\end{aligned}$$

Therefore $\sum_{j=2}^{7} 3^j = 3{,}276$.

PROGRESS CHECK 5

Write the series $\sum_{i=3}^{8} 2^i$ in expanded form and determine the sum. ■

Example 6: Expressing a Series in Sigma Notation

Find an expression for the general term and write the series $2+5+8+11+14+17$ in sigma notation.

Solution: In most cases it is hoped that you can predict the general term by inspection. For this series it is

$$a_i = 3i - 1$$

and since there are six terms, we have

$$\sum_{i=1}^{6}(3i-1) = 2+5+8+11+14+17.$$

If you could not predict the general term in this case, you could note the series is an arithmetic series with $a_1 = 2$ and $d = 3$, so that

$$\begin{aligned} a_n &= a_1 + (n-1)d \\ &= 2 + (n-1)3 \\ &= 3n - 1. \end{aligned}$$

PROGRESS CHECK 6

Find an expression for the general term and write the series $7+12+17+22+27+32+37$ in sigma notation. ■

Arithmetic series and geometric series are useful in many applied problems, and the next example illustrates an application of an arithmetic series.

Example 7: Distance Traveled by a Free-Falling Body

A free-falling body that starts from rest drops about 16 ft the first second, 48 ft the second second, 80 ft the third second, and so on. About how many feet does a parachutist drop during the first 10 seconds of free fall?

Solution: The drop in feet for the parachutist during the first 10 seconds of the free fall is given by the sum of the first 10 terms of the sequence

$$16, 48, 80, \ldots \;.$$

This sequence is arithmetic, with $a_1 = 16$ and $d = 32$. Let $n = 10$ and use the formula that gives S_n in terms of a_1, d, and n.

$$\begin{aligned} S_n &= \frac{n}{2}[2a_1 + (n-1)d] \\ S_{10} &= \frac{10}{2}[2(16) + (10-1)32] \\ &= \frac{10}{2}(320) = 1{,}600 \end{aligned}$$

Thus, a parachutist drops about 1,600 ft during the first 10 seconds of free fall.

PROGRESS CHECK 7

From the sequence given in Example 7, about how many feet does a parachutist drop during the first 15 seconds of free fall? ■

We now consider one of the important applications of geometric series. Sooner that you might imagine, one (or more) of your former classmates will probably call you up and ask you to invest in your future. The plan—purchase an annuity. An **annuity** is any series of equal payments made at equal time intervals. There are different types of annuities. Here we discuss an **ordinary annuity**, in which payments are made at the end of each time period that coincides with a point at which interest is converted. As an example, suppose that at the end of each year for 5 years you invest \$100, as shown below, in an account that pays 6 percent interest compounded annually.

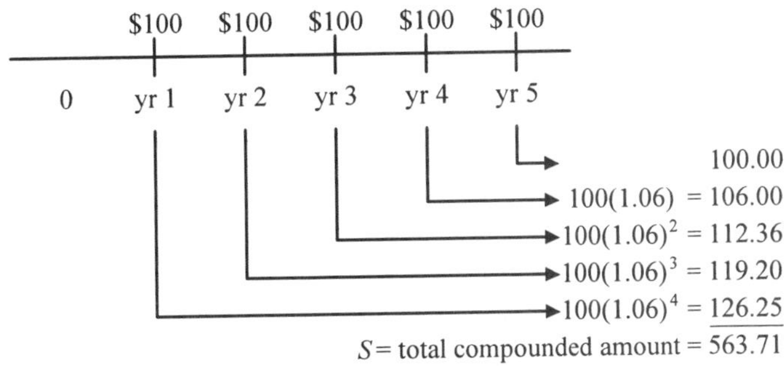

Each deposit earns interest for a different period of time. Thus, we compute separately the compounded amount of each \$100 investment by the methods from Section 7.2, and add the results to obtain the total compounded amount. This sum is called the amount of the annuity and denoted S. The amount of this annuity (or the value of this series of equal periodic payments) is \$563.71.

Treating each \$100 investment separately is cumbersome. It is more efficient to note that

$$S = 100 + 100(1.06) + 100(1.06)^2 + 100(1.06)^3 + 100(1.06)^4$$

is a geometric series with $a_1 = 100$, $r = 1.06$, and $n = 5$. The formula for the sum of a geometric series

$$S = \frac{a_1 - a_1 r^n}{1 - r}$$

is usually written for these problems as

$$\boxed{S = \frac{a_1(r^n - 1)}{r - 1}}$$

so that for this series we have

$$\begin{aligned} S &= \frac{100[(1.06)^5 - 1]}{1.06 - 1} \\ &= \frac{100[1.338226 - 1]}{0.06} \\ &= \frac{33.8226}{0.06} = \$563.71. \end{aligned}$$

This answer coincides with our previous result and is easier to obtain.

Example 8: Amount of an Ordinary Annuity

At the end of each year, you deposit \$1,800 in an account that pays 7 percent interest compounded annually. How much is in the account after 20 deposits have been made?

Solution: The amount S in the account is the sum of 20 terms as expressed by

$$S = 1{,}800 + 1{,}800(1.07) + 1{,}800(1.07)^2 + \ldots + 1{,}800(1.07)^{19}.$$

This series is a geometric series with $a_1 = 1{,}800$, $r = 1.07$, and $n = 20$. Thus

$$\begin{aligned} S_n &= \frac{a_1(r^n - 1)}{r - 1} \\ S_{20} &= \frac{1{,}800(1.07^{20} - 1)}{1.07 - 1} \approx \$73{,}791.89 \text{ (by calculator)}. \end{aligned}$$

PROGRESS CHECK 8

At the end of each month, you deposit \$200 in an account that is compounded at a rate of 1 percent each month. How much is in the account after 20 years (which means 240 deposits have been made)?

■

Finally, in the mathematics of finance it is useful to generalize our discussion and derive the formula for the amount of an annuity—the most efficient way to determine S. Refer to the specific examples, just given to clarify any steps you do not understand. If we let

p = periodic payment of the annuity
i = interest rate per period
n = number of payment periods
S = amount of the annuity

then

$$S = p + p(1+i) + p(1+i)^2 + \cdots + p(1+i)^{n-1}.$$

This series is a geometric series with n terms, in which $r = 1 + i$ and $a_1 = p$. Substituting these values in the formula

$$S = \frac{a_1(r^n - 1)}{r - 1}$$

gives

$$S = \frac{p\left[(1+i)^n - 1\right]}{(1+i)-1} = \frac{p\left[(1+i)^n - 1\right]}{i}.$$

This formula is usually written as

$$S = p\left[\frac{(1+i)^n - 1}{i}\right].$$

To simplify the evaluation of this expression, tables (which may be found in most business math books) gives the values of

$$\frac{(1+i)^n - 1}{i}$$

for common values of i and n. Then we need only multiply the appropriate table value by p (the periodic payment) to determine S (the amount of the annuity).

Since a series is a sum of the terms of a sequence, calculators with sequence capability can be used to analyze series as well.

EXPLORE 12.2

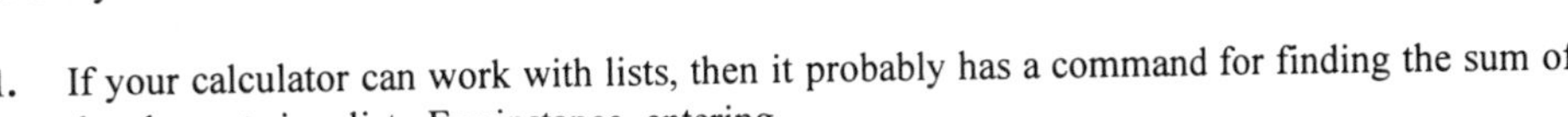

1. If your calculator can work with lists, then it probably has a command for finding the sum of the elements in a list. For instance, entering

   sum{1, 5, 8}

   results in 14. Use the sum feature to compute the sum of the first 5 terms of the sequence $a_n = n^3$.
2. If your calculator has special sequence commands they may be combined in expressions such as Sum Seq $(2^n, n, 1, 5, 1)$, which finds the sum $\sum_{i=1}^{5} 2^i$.
   - **a.** If your calculator has this capability, use it to find the sum of the first 100 terms of $\sum_{i=1}^{n} 1/i$. How many terms are necessary for the sum to exceed 8?
   - **b.** Repeat the exploration of part **a** for $\sum_{i=1}^{n} 1/i^2$.
3. In Section 12.2 appear formulas for the sum of the first n terms of an arithmetic series. Use these formulas and the programming capability of your calculator to write a program that will compute the desired sum after you input a_1, d, and n. Then use the program to find the sum of the first 21 terms of the arithmetic sequence 4, 7, 10, ...
4. In Section 12.2 appear formulas for the sum of the first n terms of a geometric series. Use these formulas and the programming capability of your calculator to write a program that will compute the desired sum after you input a_1, r, and n. Then use the program to find the sum of the first 7 terms of the geometric sequence 1, 1/2, 1/4....

EXERCISES 12.2

In Exercises 1–10 find the sum of the indicated number of terms of the following arithmetic and geometric sequences.

1. 3, 7, 11,...; 10 terms

2. 9, 4, -1,...; 12 terms

3. 2, 6, 18,...; 7 terms

4. 2, -6, 18,...; 7 terms

5. 1, 1.04, $(1.04)^2$, ...; 5 terms

6. $1, 0.1, 0.01, \ldots$; 9 terms

7. $1, \frac{3}{4}, \frac{9}{16}, \ldots$; 6 terms

8. $1.1, 1.4, 1.7, \ldots$; 8 terms

9. $3.15, 3.10, 3.05, \ldots$; 11 terms

10. $\frac{13}{4}, \frac{9}{2}, \frac{23}{4}, \ldots$; 10 terms

In Exercises 11–20 write the series in expanded form and determine the sum.

11. $\sum_{i=1}^{10} i$ 12. $\sum_{i=1}^{4} i^2$ 13. $\sum_{j=1}^{6} (-1)^j$

14. $\sum_{j=1}^{3} \frac{(-1)^{j+1}}{j}$ 15. $\sum_{k=1}^{12} (2k-1)$

16. $\sum_{k=1}^{20} (6k+1)$ 17. $\sum_{i=2}^{5} \left(\frac{1}{2}i+3\right)$

18. $\sum_{j=2}^{8} 2^j$ 19. $\sum_{k=1}^{5} \left(\frac{1}{3}\right)^{k-1}$ 20. $\sum_{i=1}^{4} \left(2+3^i\right)$

In Exercises 21–30 find an expression for the general term and write the series in sigma notation. (**Note:** A given series can be expressed in sigma notation in more than one way.)

21. $3+4+5+\cdots+10$ 22. $1+4+9+16+25$

23. $\frac{1}{2}+\frac{1}{4}+\frac{1}{8}+\frac{1}{16}$ 24. $\frac{1}{2}-\frac{1}{4}+\frac{1}{8}-\frac{1}{16}$

25. $\frac{1}{2}+\frac{2}{3}+\frac{3}{4}+\frac{4}{5}+\frac{5}{6}$ 26. $\frac{2}{3}+\frac{3}{5}+\frac{4}{7}+\frac{5}{9}+\frac{6}{11}$

27. $x+x^2+x^3+x^4+x^5+x^6$

28. $x+x^3+x^5+x^7$

29. $1+2^3+3^3+\cdots+n^3$

30. $1+2^3+3^3+\cdots+(n-1)^3$

31. If 16 basketball teams enter a single elimination tournament (one loss means the team is eliminated), determine the number of games needed to declare a winner.

32. If 64 racquet ball players enter a single elimination tournament (losing one match means the person is eliminated), determine the number of matches needed to declare a winner.

33. What is the sum of the first n even positive integers?

34. What is the sum of the first n odd positive integers?

35. A free-falling body that starts from rest drops about 16 ft the first second, 48 ft the second second, 80 ft the third second, and so on. How many feet does the object fall in 20 seconds?

36. On the moon, a free-falling body starting from rest drops about 2.6 ft the first second, 7.8 ft the second second, 13.0 ft the third second, and so on. How many feet does the object fall in 20 seconds?

37. In many winter celebrations such as Kwanza or Hanukkah, an increasing number of candles are burned each day for a period of time. For instance, over the 8 days of Hanukkah 2, 3, 4, 5, 6, 7, 8, and then 9 candles are used. Express the total number of candles as a series and find its sum. Is the series arithmetic, geometric, or neither?

38. A grandfather clock chimes as many times as the hour. How many times does it chime in striking the hours 1 through 12?

39. The principle of an annuity might help one to stop smoking. Suppose one saves the price of a package of cigarettes (let's say \$1.50) each day of the year (use 365 days). At the end of the year, the money is deposited in an account paying 9 percent interest compounded annually. How much is in the account after 10 deposits have been made?

40. At the end of each year, a woman deposits \$1,000 in an account that pays 6 percent interest compounded annually. How much is in the account after 8 deposits have been made?

41. A person makes a purchase for \$6,000, paying \$1,500 down and agreeing to pay at the end of each year 8 percent interest on the unpaid balance, plus an additional \$900 to reduce the principal.
   - **a.** Show that the loan will be paid off after 5 years.
   - **b.** Write out the five-term series that gives the total of the interest payments. What kind of series is this? What is the total amount of interest paid?

42. A person makes a purchase of \$6,000, paying \$1,200 down and agreeing to pay at the end of each year 8 percent interest on the unpaid balance, plus an additional \$800 to reduce the principal.
   - **a.** Show that the loan will be paid off after 6 years.
   - **b.** Write out the six-term series that gives the total of the interest payments. What kind of series is this? What is the total amount of interest paid?

43. A certain ball always rebounds $\frac{2}{3}$ as far as it falls. If the ball is dropped from a height of 9 ft, how far up and down has it traveled when it hits the ground for the seventh time?

44. A certain ball always rebounds 1/2 as far as it falls. If the ball is thrown 10 ft into the air, how far up and down has it traveled when it hits the ground for the sixth time?

45. To help finance their son's college education, a couple invests $1,000 on his birthday each year, starting with his fifth birthday. If the money is placed in an account that pays 6 percent interest compounded annually, how much is in the account on his 18th birthday? Assume that the last payment is made on his 17th birthday. How much of the total is interest?

46. Redo Exercise 45, but assume that they start investing on his first birthday.

47. What is the total number of your ancestors in the seven generations that immediately precede you? (**Note:** One generation back is your parents, two generations back is your grandparents, and so on.)

48. In this famous riddle the correct answer is 1 (why?), but how many were coming *from* St. Ives? Express the answer as a series. Is it arithmetic, geometric, or neither?

As I was going to St. Ives,
I met a man with seven wives.
Each wife had seven sacks,
Each sack had seven cats,
Each cat had seven kits:
Kits, cats, sacks and wives,
How many were going to St. Ives?

49. You accept a position at a salary of $20,000 for the first year, with an increase of $1,000 per year after that. After how many years will your total earnings equal $513,000?

50. You accept a position at a salary of $10,000 for the first year, with an increase of $1,000 per year each year thereafter. How many years will you have to work for your total earnings to equal $231,000.

51. Refer to the figure and express the total area as the sum of individual rectangular areas. Is this series arithmetic, geometric, or neither?

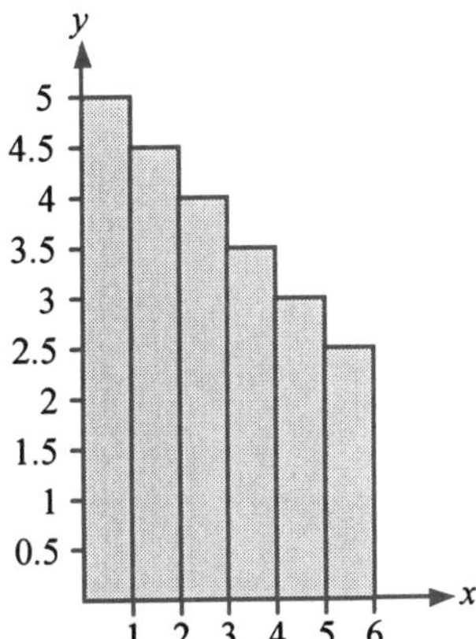

52. The area under the parabola $y = x^2$ may be estimated by the sum of the areas of the 10 rectangles shown in the figure. Write out the 10-term series that represents the sum of the areas of the rectangles and find the sum. (**Hint:** The base of each rectangle is 0.1; the height is given by x^2.) Is this series arithmetic, geometric, or neither?

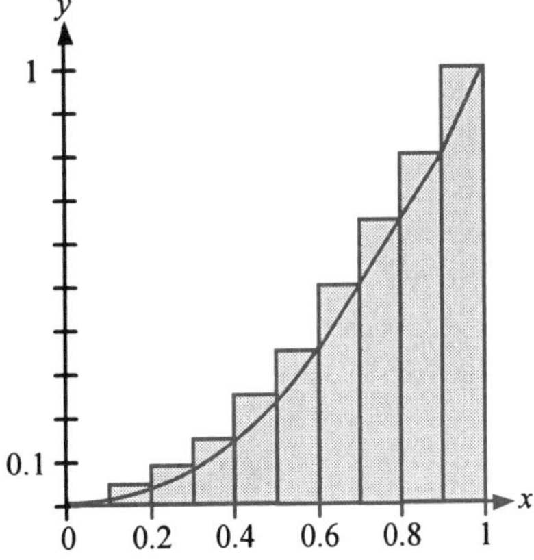

THINK ABOUT IT 12.2

1. Find the sum of the first 250 positive integers in each of the following ways.
   a. Use the method attributed to Gauss in the beginning of this section.
   b. Use the method that was applied when deriving the formula for S_n.
   c. Use the formula $S_n = (n/2)(a_1 + a_n)$

2. Prove each property of series.
   a. $\sum_{i=1}^{n} ca_i = c\sum_{i=1}^{n} a_i$ b. $\sum_{i=1}^{n}(a_i - b_i) = \sum_{i=1}^{n} a_i - \sum_{i=1}^{n} b_i$

3. Find the arithmetic sequence in which the sum of the first n terms is given by $S_n = n^2 + 3n$. What is the 50th term in this arithmetic sequence?
4. At the beginning of each month, you deposit \$100 in an account that pays 6 percent interest compounded monthly. Explain why the amount in this account at the end of 1 year is given by:

$$100\left(1+\frac{0.06}{12}\right)^1 + 100\left(1+\frac{0.06}{12}\right)^2 + \cdots + 100\left(1+\frac{0.06}{12}\right)^{12}.$$

   Use the methods of this section to find this sum.
5. Let $a_1, a_2, \ldots$ be the Fibonacci sequence. For successive values of n, find $a_1 + a_2 + \cdots + a_n$. Compare the sums with the terms of the sequence and determine the formula for the sum of the first n terms in the Fibonacci sequence.

● ● ●

12.3 Infinite Geometric Series

Photo Courtesy of K. H. Photo, International Stock

A system for removing pollutants from kerosene involves passing the kerosene repeatedly through a filtering system. Suppose that each time the kerosene passes through the filters some pollutant is removed and stored in a special tank. In a certain application each pass through the filters removes 10 percent as much pollutant as the previous pass. Suppose the first pass produces 18 gal of pollutant. What size holding tank is large enough to hold all the pollutant this application will produce? (See Example 4.)

Objectives

1. Find the sum of certain infinite geometric series.
2. Use an infinite geometric series to express a repeating decimal as the ratio of two integers.
3. Solve applied problems involving infinite geometric series.
4. Use graphical methods to find the sum of an infinite geometric series.

A question that has intrigued mathematicians for some time is whether it is possible to assign a number as the sum of certain infinite series in some meaningful way. To explore this problem, consider the series

$$\frac{1}{2}+\frac{1}{4}+\frac{1}{8}+\frac{1}{16}+\cdots+\left(\frac{1}{2}\right)^n+\cdots$$

and examine the behavior of S_n (the sum of the first n terms in the series) as n increases from 1 to 6.

$$S_1=\frac{1}{2}$$
$$S_2=\frac{1}{2}+\frac{1}{4}=\frac{3}{4}$$
$$S_3=\frac{1}{2}+\frac{1}{4}+\frac{1}{8}=\frac{7}{8}$$
$$S_4=\frac{1}{2}+\frac{1}{4}+\frac{1}{8}+\frac{1}{16}=\frac{15}{16}$$
$$S_5=\frac{1}{2}+\frac{1}{4}+\frac{1}{8}+\frac{1}{16}+\frac{1}{32}=\frac{31}{32}$$
$$S_6=\frac{1}{2}+\frac{1}{4}+\frac{1}{8}+\frac{1}{16}+\frac{1}{32}+\frac{1}{64}=\frac{63}{64}$$

It appears that as n gets large, S_n approaches, but never equals, 1. In fact, we can get S_n as close to 1 as we wish merely by taking a sufficiently large value for n. It is in this sense that we are going to assign the number 1 as the "sum" of the series. That is, to say 1 is the sum of the above infinite series is to say that as n gets larger, S_n converges to or closes in on 1.

On an intuitive level, observe that if a series is to converge to a sum, then the terms in the series must approach 0 as n gets larger. This condition occurs in infinite geometric series with $|r|<1$ (or equivalently, with $-1<r<1$), so we will analyze this type of series. A formula for the sum of the first n terms of a geometric series is

$$S_n=\frac{a_1-a_1r^n}{1-r}.$$

Now if $|r|<1$, as n gets larger a_1r^n approaches 0, so that S_n, closes in on $a_1/(1-r)$. Thus, we have the following important result.

Sum of an Infinite Geometric Series

> An infinite geometric series with $|r|<1$ converges to the value or sum
>
> $$S=\frac{a_1}{1-r}.$$

Example 1: Finding the Sum of an Infinite Geometric Series

Find the sum of the infinite geometric series

$$1+\frac{1}{3}+\frac{1}{9}+\frac{1}{27}+\cdots.$$

Solution: Here $a_1=1$ and $r=1/3$. Since the common ratio is between -1 and 1, we can assign a sum to the series. Substituting in the formula

$$S=\frac{a_1}{1-r},$$

we have

$$S = \frac{1}{1-\frac{1}{3}} = \frac{1}{\frac{2}{3}} = \frac{3}{2}.$$

PROGRESS CHECK 1

Find the sum of the infinite geometric series $3+1.5+0.75+\ldots$. ■

Example 2: Finding the Sum of an Infinite Geometric Series

Evaluate $\sum_{i=1}^{\infty}(0.2)^{i+1}$.

Solution: In expanded form the series is

$$\sum_{i=1}^{\infty}(0.2)^{i+1} = (0.2)^2 + (0.2)^3 + (0.2)^4 + \ldots.$$

This series is an infinite geometric series with $a_1 = (0.2)^2 = 0.04$ and $r = 0.2$. A sum S can be assigned to this series, since $|r| < 1$ and

$$S = \frac{a_1}{1-r} = \frac{0.04}{1-0.2} = 0.05.$$

PROGRESS CHECK 2

Evaluate $\sum_{i=1}^{\infty}(-3/4)^{i-1}$. ■

In Section R.1 we saw that every repeating decimal is a rational number and can be expressed as the quotient of two integers. One method for finding the appropriate fraction is to use an infinite geometric series. For example, the repeating decimal $0.\overline{3}$ (or equivalently $0.333\ldots$) can be written as

$$0.3 + 0.03 + 0.003 + \cdots.$$

This series is an infinite geometric series with $r = (0.03/0.3) = 0.1$ and $a_1 = 0.3$, so

$$S = \frac{a_1}{1-r} = \frac{0.3}{1-(0.1)} = \frac{0.3}{0.9} = \frac{1}{3}.$$

Thus, $0.\overline{3}$ is equivalent to $1/3$.

Example 3: Converting a Repeating Decimal to a Ratio of Integers

Express the repeating decimal $7.\overline{54}$ as the ratio of two integers.

Solution: The repeating decimal $7.\overline{54}$ can be written as

$$7 + 0.54 + 0.0054 + 0.000054 + \cdots.$$

The series

$$0.54 + 0.0054 + 0.000054 + \cdots$$

is an infinite geometric series with $a_1 = 0.54$ and $r = 0.01$, so

$$S = \frac{a_1}{1-r} = \frac{0.54}{1-(0.01)} = \frac{0.54}{0.99} = \frac{54}{99} = \frac{6}{11}.$$

Adding the first term in the original series, 7, to $6/11$ gives

$$7 + \frac{6}{11} = \frac{77}{11} + \frac{6}{11} = \frac{83}{11}.$$

Thus, $7.\overline{54}$ is equivalent to $83/11$.

PROGRESS CHECK 3

Express the repeating decimal $3.\overline{7}$ as the ratio of two integers. ■

By using the concepts developed in this section, we can now solve the applied problem that opens this section.

Example 4: Removing Pollutants from Kerosene

Solve the problem in the section introduction on page 972.

Solution: The amount S of pollutant being removed and stored is given by

$$S = 18 + 18(0.1) + 18(0.1)^2 + \ldots$$

This series is an infinite geometric series with $r = 0.1$, so S can be determined, since $|r| < 1$. Replacing a_1 by 18 and r by 0.1 in the formula for S gives

$$S = \frac{a_1}{1-r} = \frac{18}{0.9} = 20.$$

Thus, the holding tank must be able to accommodate 20 gal of pollutant.

PROGRESS CHECK 4

In a certain filtration system, wastewater is collected in a holding tank so that each time a liquid is passed through this system, only 75 percent as much wastewater is collected as in the previous pass. If 12 gal of wastewater are collected the first time a liquid is passed through this system, how many gallons of wastewater must the holding tank be able to accommodate? ■

A graphing calculator can provide visual evidence in a graph that certain infinite geometric series converge to a sum, which can be read from the graph. The next example illustrates a method to obtain such a graph in connection with the section-opening problem.

Example 5: Visualizing a Converging Infinite Geometric Series

Use graphical methods to solve the problem in the section introduction on page 972.

Solution: As in Example 4, the amount S of pollutant is given by the infinite geometric series

$$S = 18 + 18(.01) + 18(0.1)^2 + \ldots$$

A formula for the sum of the first n terms of this series is

$$S_n = \frac{a_1 - a_1 r^n}{1-r},$$

and letting $a_1 = 18$ and $r = 0.1$ in this formula yields

$$S_n = \frac{18 - 18(0.1)^n}{1 - 0.1} = 20 - 20(0.1)^n.$$

To see what happens to S_n as n increases without bound, graph $y = 20 - 20(0.1)^x$ and $y = 20$, as shown in Figure 12.9. Observe that the curve approaches the line $y = 20$, so it appears that the line $y = 20$ is a horizontal asymptote. Since the graph of $S_n = 20 - 20(0.1)^n$ is just the points on the line $y = 20 - 20(0.1)^x$ that correspond to positive integer replacements for x, we conclude that the sum of the infinite series above is 20. Note that algebraic analysis, as in Example 4, is required to confirm that the holding tank must be able to accommodate 20 gal of pollutant.

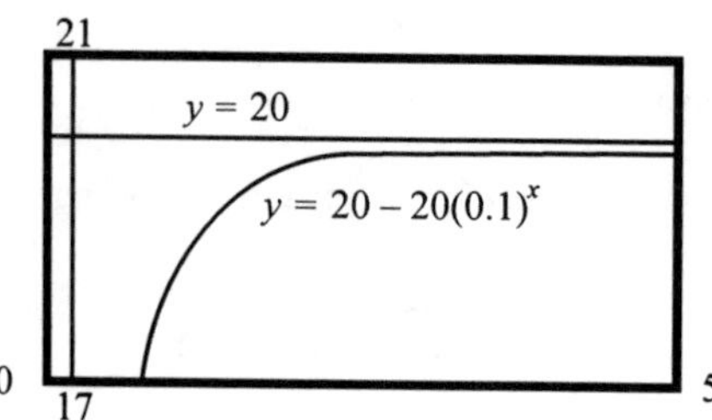

Figure 12.9

PROGRESS CHECK 5

Use graphical methods to find the sum of the infinite geometric series $1 + \frac{1}{3} + \frac{1}{9} + \frac{1}{27} + \ldots$. Confirm your result algebraically. ■

EXPLORE 12.3

Some infinite series converge to a finite sum, and some do not. In a particular case, a graphing calculator may help you decide whether or not an infinite series converges, but the calculator cannot prove the case one way or the other. Algebraic techniques are required for proof.

1. Geometric series are not the only infinite series that converge. With calculus, for instance, you can prove that the series whose terms are the reciprocals of the squares of the positive integers converges.
   a. With the aid of a calculator, estimate to the nearest hundredth the sum of this series.
   $1/1^2 + 1/2^2 + 1/3^2 + 1/4^2 + \ldots$
   Often it is easier to show that a series converges than it is to find exactly the sum it converges to. It can be proved that the series above converges to $\pi^2/6$.
   b. Estimate to the nearest hundredth the sum of this related series, which uses just the odd integers.

$$\frac{1}{1^2} + \frac{1}{3^2} + \frac{1}{5^2} + \frac{1}{7^2} + \ldots$$

   It can be proved that this series converges to $\pi^2/8$.
   c. Why is it reasonable to guess from the results of parts **a** and **b** that $1/2^2 + 1/4^2 + 1/6^2 + \ldots$ converges to $\pi^2/24$?
2. A famous infinite series that does not converge is the harmonic series: $1 + 1/2 + 1/3 + 1/4 + \ldots$. If enough terms are added, the sum will surpass any positive value you choose. Determine, with the aid of a calculator, how many terms are needed for the sum to exceed each of these values:
   a. 2 **b.** 3 **c.** 4 **d.** 10

3. Isaac Newton (1642–1727) made major mathematical breakthroughs in dealing with a particular kind of infinite series called **power series** which are like infinite degree polynomials. For instance, he worked with the equation

$$\frac{1}{1+x}=1-x+x^2-x^3+x^4-x^5\ldots$$

which follows from long division of 1 by $1+x$.

   a. Compare the values of the left-and right-hand expressions when $x=1/4$. What happens as you include more and more terms on the right side?

   b. The power series above converges only for certain values of x. For other values, nonsense results. Compare the values of the left- and right-hand expressions when $x=1$, and when $x=-1$.

4. Write a program for your calculator to find the sum of an infinite geometric series, given a_1 and r.

EXERCISES 12.3

In Exercises 1–10 find the sum of the infinite geometric series.

1. $1+\frac{1}{2}+\frac{1}{4}+\frac{1}{8}+\ldots$ 2. $1+\frac{1}{5}+\frac{1}{25}+\frac{1}{125}+\ldots$

3. $\frac{1}{3}-\frac{1}{9}+\frac{1}{27}-\frac{1}{81}+\ldots$

4. $2-1+\frac{1}{2}-\frac{1}{4}+\ldots$

5. $3+\frac{3}{10}+\frac{3}{100}+\frac{3}{1000}+\ldots$

6. $5+\frac{5}{3}+\frac{5}{9}+\frac{5}{27}+\ldots$ 7. $\frac{3}{2}-1+\frac{2}{3}-\frac{4}{9}+\ldots$

8. $\frac{1}{2}-\frac{1}{3}+\frac{2}{9}-\frac{4}{27}+\ldots$

9. $5+0.5+0.05+0.005+\ldots$

10. $11+1.1+0.11+\ldots$

In Exercises 11–16 evaluate the given sum.

11. $\sum_{i=1}^{\infty}(0.3)^i$ 12. $\sum_{i=1}^{\infty}\left(\frac{1}{3}\right)^i$ 13. $\sum_{i=1}^{\infty}\left(-\frac{1}{3}\right)^{i-1}$

14. $\sum_{k=1}^{\infty}\left(-\frac{2}{3}\right)^{i-1}$ 15. $\sum_{j=1}^{\infty}2\left(\frac{1}{3}\right)^{j+1}$

16. $\sum_{i=1}^{\infty}3\left(\frac{1}{4}\right)^{i+1}$

In Exercises 17–26 express each repeating decimal as the ratio of two integers.

17.	$0.\overline{2}$	**18.**	$0.\overline{9}$	**19.**	$0.\overline{07}$
20.	$0.0\overline{7}$	**21.**	$0.\overline{321}$	**22.**	$0.\overline{6332}$
23.	$5.\overline{9}$	**24.**	$4.\overline{81}$	**25.**	$2.14\overline{3}$
26.	$2.\overline{143}$				

In Exercises 27–32 use graphical methods to find the sum of the infinite geometric series. Confirm your result algebraically.

27. $27+2.7+0.27+0.027+\ldots$

28. $12-2.4+0.48-0.096+\ldots$

29. $4+4(0.8)+4(0.8)^2+\ldots$

30. $5+5(0.9)+5(0.9)^2+\ldots$

31. $(-1)-(-1)(0.5)+(-1)(0.5)^2-(-1)(0.5)^3+\ldots$

32. $100-100(0.9)+100(0.9)^2-100(0.9)^3+\ldots$

33. A system for purifying DNA involves repeatedly rinsing a batch of plant cells in a cleansing solution. Each rinse removes 60 percent as much impurity as the previous rinse did. Suppose the first rinse removes 10 milligrams of impurities. What is the maximum amount of impurities that can be removed by repeated rinsing?

34. Redo Exercise 33, if the system is improved so that each rinse removes 80 percent of the impurities.

35. A pendulum is swinging back and forth, but each diminishing swing takes 0.999 times as long as the previous one. If the first swing takes 2 seconds, what is the total time elapsed before the pendulum stops? Approximate this time as the sum of an infinite geometric series.

36. A certain ball always rebounds $\frac{2}{3}$ as far as it falls. If the ball is dropped from a height of 9 ft, how far up and down has it traveled before it comes to rest? Approximate this distance as the sum of an infinite geometric series.

37. For what values of x does the series

$$(x+1)+2(x+1)^2+4(x+1)^3+\ldots$$

converge to a sum? What is this sum?

38. For what values of x does the series

$$1+x+x^2+x^3+\ldots$$

converge to a sum? What is this sum?

39. A game consists of tossing a coin until you get heads. The appearance of a head (H) is called a success(S). The appearance of a tail (T) is called a failure (F). Here are all the possible outcomes and their probabilities.

Outcome	*Probability*
Heads first appears on 1st toss: H	$\frac{1}{2}$
Heads first appears on 2nd toss: TH	$\frac{1}{2}\cdot\frac{1}{2}=\frac{1}{2^2}$
Heads first appears on 3rd toss: TTH	$\frac{1}{2}\cdot\frac{1}{2}\cdot\frac{1}{2}=\frac{1}{2^3}$
⋮	⋮
Heads first appears on nth toss: $\underbrace{\text{TTT...TH}}_{n-1\text{ T's}}$	$\frac{1}{2^n}$
⋮	⋮

Show that the sum of the probabilities of all the possible outcomes is a geometric series whose sum is 1. (**Note:** In probability theory the sum of the probabilities of all possible outcomes of an experiment is always 1.)

40. A game consists of rolling a die until you get a 6. The appearance of a 6 is called a success (S). The appearance of any other number is called a failure (F). Here are all the possible outcomes for such a game.

Outcome	*Probability*
6 first appears on 1st roll: S	$\frac{1}{6}$
6 first appears on 2nd rolls: FS	$\frac{5}{6}\cdot\frac{1}{6}=\frac{5}{6^2}$
6 first appears on 3rd roll: FFS	$\frac{5}{6}\cdot\frac{5}{6}\cdot\frac{1}{6}=\frac{5^2}{6^3}$
⋮	⋮
6 first appears on nth roll $\underbrace{\text{FFF...FS}}_{n-1\text{ F's}}$	$\frac{5^{n-1}}{6^n}$
⋮	⋮

Show that the sum of the probabilities of all the possible outcomes is a geometric series whose sum is 1.

41. The curve drawn in the figure below is a segment of a parabola. The largest triangle has the same base and vertex as the curve. Each smaller triangle is constructed in the same way, resulting in a sequence of polygons with more and more sides, whose perimeter approaches the shape of the parabola. Archimedes showed that the areas of the successive polygons are given by this sequence, where A represents the area of the largest triangle.

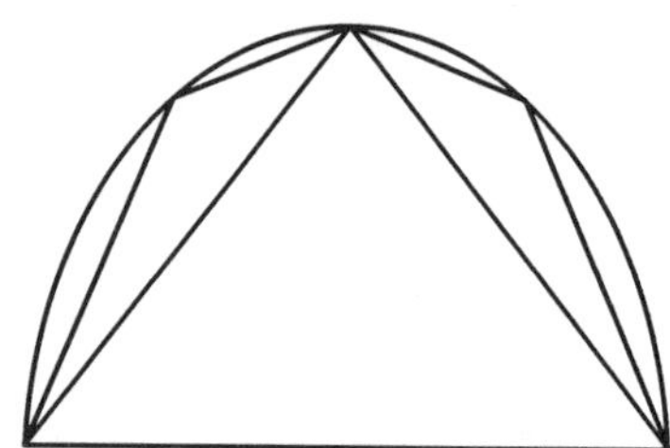

Polygon	**Area**
1st polygon	A
2nd polygon	$A+\frac{1}{4}A$
3rd polygon	$A+\frac{1}{4}A+\frac{1}{4^2}A$
4th polygon	$A+\frac{1}{4}A+\frac{1}{4^2}A+\frac{1}{4^3}A$
⋮	⋮
nth polygon	$A+\frac{1}{4}A+\frac{1}{4^2}A+\ldots+\frac{1}{4^{n-1}}A$

Archimedes argued that the sum of the infinite series gives the area of the parabola in terms of the area (A) of the largest triangle. Discover Archimedes formula by finding the sum of this geometric series. This is the earliest recorded example of the summation of an infinite series. (**Hint:** Factor out A from the series before finding the sum.)

42. Here is a classic problem as discussed in W.W. Sawyer's Mathematician's Delight (Penguin Books, 1943): "If a ton of seed potatoes will produce a crop of 3 tons, which can either be consumed or used again as seed, how much must a gardener buy, if his family wants to consume a ton of potatoes every year forever?" [**Hint:** For the first year's harvest, $\frac{1}{3}$ ton of seed planted now will be enough. For the second year's harvest, $\frac{1}{9}$ to be planted now will be enough. (Why?)]

a. Write a geometric series for the total amount of seed potatoes needed.

b. Find the sum of the series to answer the question.

THINK ABOUT IT 12.3

1. Explain in terms of infinite geometric series why $\sum_{n=1}^{\infty} 1/x^n$ converges to a sum if $|x| > 1$. What is this sum?

2. The given figure shows the start of an infinite sequence of squares that is generated by using the midpoints of the sides of the first square as the vertices of the second square, and so on. If the side length of the initial square is 4 cm, then find the sum of the areas of all the squares. What is the sum of all the perimeters?

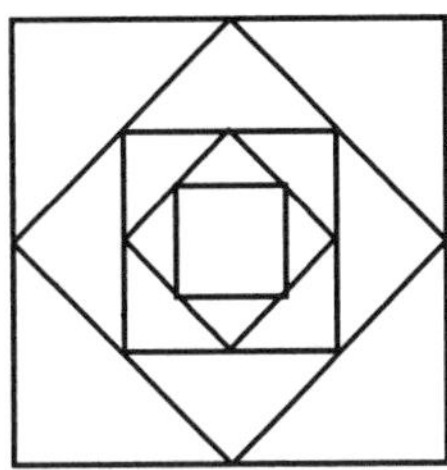

3. If $a_1 + a_2 + \ldots + a_n + \ldots$ is an infinite geometric series with $|r| < 1$, what is the common difference in the sequence $\log|a_1|, \log|a_2|, \ldots, \log|a_n|, \ldots$?

4. **a.** If S denotes the sum of an infinite geometric series with $|r| < 1$, and S_n denotes the sum of the first n terms in this series, write a formula for $S - S_n$.

b. Use the result in part **a** and find the least number of terms in the series $1 + \frac{1}{3} + \frac{1}{9} + \ldots$ that should be summed so that the difference between S and S_n is less than $\frac{1}{1{,}000}$.

5. Consider the series $1 - 1 + 1 - 1 + 1 - 1 + \ldots$. What is the sum of this series? The problem can be approached from three viewpoints.

Method 1 The sum is 0 since $(1-1) + (1-1) + (1-1) + \ldots = 0 + 0 + 0 + \ldots = 0$

Method 2 The sum is 1 since $1 + (-1+1) + (-1+1) + \ldots = 1 + 0 + 0 + \ldots = 1$

Method 3 The sum is $\frac{1}{2}$ since the series is a geometric series with $r = -1$, so $S = \frac{a_1}{1-r} = \frac{1}{1-(-1)} = \frac{1}{2}$. Therefore $0 = 1 = \frac{1}{2}$. What is wrong?

● ● ●

12.4 Mathematical Induction

Photo Courtesy of **John Colletti**, The Picture Cube

An important formula that has been derived in this chapter is

$$a_1 + a_1r + a_1r^2 + \ldots + a_1r^{n-1} = \frac{a_1 - a_1r^n}{1-r}.$$

a. What is this equation the formula for?
b. Prove this formula using mathematical induction.
c. You accept a management position at a salary of \$30,000 for the first year, with an increase of 5 percent per year each year thereafter. Use the formula above to find your total earnings at the end of 10 years.

(See Example 3.)

Objectives

1. Use mathematical induction to prove that a statement is true for all positive integers.
2. Use mathematical induction to prove that a statement is true for all integers greater than or equal to a particular integer.

Certain statements are true for every positive integer, and these results are quite useful in mathematics. We usually prove these statements about positive integers by a method called *mathematical induction.* The word *induction* is used because in many cases we can guess the general formula by looking at some specific cases. For example, consider the following sums of odd integers.

$$\begin{aligned} 1 &\quad \text{or } 1^2 \\ 1 + 3 &= 4 \text{ or } 2^2 \\ 1 + 3 + 5 &= 9 \text{ or } 3^2 \\ 1 + 3 + 5 + 7 &= 16 \text{ or } 4^2 \\ 1 + 3 + 5 + 7 + 9 &= 25 \text{ or } 5^2 \end{aligned}$$

From these additions we strongly suspect (or inductively conclude) that the sum of the first n odd integers always equals n^2 or, equivalently, that

$$1 + 3 + 5 + \cdots + (2n-1) = n^2$$

for every positive integer value of n. However, the validity of this formula in some specific cases does not guarantee that the rule will not break down for some higher number of odd integers. We can never prove the principle for *all* positive integers by checking a finite number of cases. Thus, mathematicians developed a technique called "proof by mathematical induction," which is based on the following principle.

Principle of Mathematical Induction

If a given statement S_n concerning positive integer n is true for $n = 1$, and if its truth for $n = k$ implies its truth for $n = k + 1$, then S_n is true for every positive integer n.

According to this principle, we can prove that a statement (or formula) is true for every positive integer value of n by showing two things.

1. Prove the statement is true for $n = 1$.
2. Prove that if the statement is true for any positive integer k, then it is also true for the next highest integer $k + 1$.

Proof by mathematical induction is similar to lining up a row of dominoes so that when the first is knocked over it causes the second domino to fall, the second knocks down the third, and so on, as illustrated in Figure 12.10. In Step 1 we knock down the first domino, and by proving Step 2 we show that the desired chain reaction then takes place.

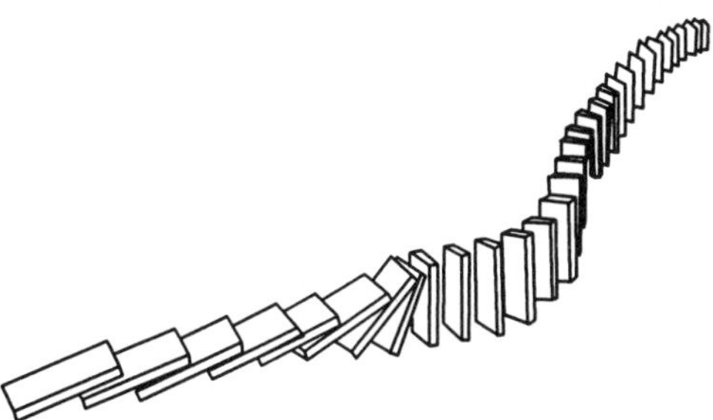

Figure 12.10

By using mathematical induction, we can now continue our discussion concerning the sum of the first n odd integers, and we can prove that this sum always equals n^2. We show how in Example 1.

Example 1: Sum of the First *n* Odd Integers

Use mathematical induction to prove that

$$1 + 3 + 5 + \cdots + (2n - 1) = n^2$$

is true for every positive integer n.

Solution: We prove the statement by mathematical induction with the following two steps.

Step 1 Replacing n by 1 in the formula gives

$$2(1) - 1 = 1^2,$$

which is true. Step 1 is the easy part of the proof.

Step 2 Replace n by k, and write what it means for the formula to work for any positive integer k. Call this equation (1). It is the induction assumption.

$$1 + 3 + 5 + \cdots + (2k - 1) = k^2 \qquad (1)$$

Now replace n by $k + 1$, and write what it means for the formula to work for the next highest integer $k + 1$. Call this equation (2).

$$1 + 3 + 5 + \cdots + (2k - 1) + [2(k + 1) - 1] = (k + 1)^2 \qquad (2)$$

We now show that equation (1) implies equation (2). That is, we assume that equation (1) is true and use this information to prove that equation (2) is true. We accomplish this in either of the following ways.

Method 1 We start with equation (1) and add the $k+1$ term to both sides of this equation. Then

$$\begin{aligned} 1+3+\cdots+(2k-1)+[2(k+1)-1] &= k^2+[2(k+1)-1] \\ &= k^2+2k+1 \\ &= (k+1)^2 \end{aligned}$$

which is equation (2). Thus, equation (1) implies equation (2).

Method 2 We start with the left side of equation (2) and derive the other side with the aid of equation (1). Then

$$\begin{aligned} &1+3+\cdots+(2k-1) \quad +[2(k+1)-1] \\ &= [1+3+\cdots+(2k-1)] \quad +[2(k+1)-1] \\ &= \underbrace{\qquad k^2 \qquad}_{\text{induction assumption}} \quad +[2(k+1)-1] \\ &= k^2+2k+1 \\ &= (k+1)^{22} \end{aligned}$$

which shows that equation (1) implies equation (2).

The proof is complete. Step 1 shows that the statement is true for $n=1$. Step 2 guarantees that the statement is true for the next highest integer 2, then 3, and so on, with no positive integer escaping the chain reaction.

PROGRESS CHECK 1

Use mathematical induction to prove that $2+4+6+\ldots+2n=n(n+1)$ is true for every positive integer n. (This result shows that the sum of the first n even integers always equals $n(n+1)$.) ■

Example 2: Using Mathematical Induction to Prove an Exponent Property

Prove that if a and b are real numbers, then $(ab)^n = a^n b^n$ for every positive integer n.

Solution: We prove the statement by mathematical induction with the following two steps.

Step 1 Replacing n by 1 in the statement gives

$$(ab)^1 = a^1 b^1,$$

which is true.

Step 2 When $n=k$, the statement becomes

$$(ab)^k = a^k b^k. \qquad (1)$$

When $n=k+1$, the statement becomes

$$(ab)^{k+1} = a^{k+1} b^{k+1}. \quad (2)$$

We now use equation (1) to establish equation (2) as follows:

$$\begin{aligned}(ab)^{k+1} &= (ab)^k \cdot (ab) \\ &= \left(a^k b^k\right)\cdot(ab) \qquad \text{By equation (1)} \\ &= \left(a^k \cdot a\right)\cdot\left(b^k \cdot b\right) \\ &= a^{k+1}\cdot b^{k+1}\end{aligned}$$

The proof is complete. Note that it is important to state clearly what you need to prove (the "$k+1$" statement), and what you assume (the "k" statement) so you can prove it.

PROGRESS CHECK 2

Prove that if a and b are real numbers with $b \neq 0$, then $(a/b)^n = a^n/b^n$ for every positive integer n.

■

Example 3: Total Earnings for a Job

Solve the problem in the section introduction on page 980.

Solution:

a. This equation is the formula for the sum of the first n terms in a geometric series.

b. The following two steps are needed to prove this statement by mathematical induction.

Step 1 Replacing n by 1 in the formula gives

$$a_1 = \frac{a_1 - a_1 r^1}{1-r},$$

which is true since

$$\frac{a_1 - a_1 r^1}{1-r} = \frac{a_1(1-r)}{1-r} = a_1 .$$

Step 2 When $n = k$, the statement becomes

$$a_1 + a_1 r + a_1 r^2 + \ldots + a_1 r^{k-1} = \frac{a_1 - a_1 r^k}{1-r} \qquad (1)$$

When $n = k+1$, the statement becomes

$$a_1 + a_1 r + a_1 r^2 + \ldots + a_1 r^{k-1} + a_1 r^k = \frac{a_1 - a_1 r^{k+1}}{1-r} \qquad (2)$$

We now choose to start with the left side of equation (2) and derive the other side with the aid of equation (1). So

$$\begin{aligned}&a_1 + a_1 r + a_1 r^2 + \ldots + a_1 r^{k-1} + a_1 r^k \\ &= \left[a_1 + a_1 r + a_1 r^2 + \ldots + a_1 r^{k-1}\right] + a_1 r^k \\ &= \underbrace{\frac{a_1 - a_1 r^k}{1-r}}_{\text{induction assumption}} + a_1 r^k \\ &= \frac{a_1 - a_1 r^k + a_1 r^k - a_1 r^{k+1}}{1-r} = \frac{a_1 - a_1 r^{k+1}}{1-r}\end{aligned}$$

which shows that equation (1) implies equation (2). Thus, the proof is complete.

c. The total earnings E at the end of 10 years is given by $E = 30{,}000 + 30{,}000(1.05) + 30{,}000(1.05)^2 + \ldots + 30{,}000(1.05)^9$. Then using the formula that was just established with $a_1 = 30{,}000$, $r = 1.05$, and $n = 10$ yields

$$E = \frac{30,000 - 30,000(1.05)^{10}}{1 - 1.05} \approx 377,336.78.$$

At the end of 10 years, the total earnings for this job are \$377,336.78.

PROGRESS CHECK 3

An important formula that has been derived in this chapter is

$$a_1 + (a_1 + d) + \ldots + [a_1 + (n-1)d] = \frac{n}{2}[2a + (n-1)d].$$

a. What is this equation the formula for?

b. Prove this formula using mathematical induction.

c. You accept a management position at a salary of \$30,000 for the first year, with an increase of \$1,500 per year each year thereafter. Use the formula above to find your total earnings at the end of 10 years. ■

Some statements are not true for *all* positive integers, but instead are valid only from some particular integer on up. We prove a statement for all integers greater than or equal to some particular integer (say q) as follows:

1. By direct substitution show the statement is true for $n = q$.

2. Show that if the statement is true for any positive integer $k \geq q$, then it is also true for the next highest integer $k+1$.

Example 4 illustrates this case.

Example 4: Using Mathematical Induction to Prove an Inequality

Prove that $2^n > n + 1$ for every positive integer $n \geq 2$.

Solution: To prove the statement by mathematical induction, we need two steps. In Step 1, however, we cannot start at $n = 1$ because 2^1 is not greater than $1+1$. Thus, as indicated in the problem, we start at $n = 2$.

Step 1 Replacing n by 2 in the inequality gives

$$2^2 > 2 + 1$$

which is true.

Step 2 We need to show that

$$2^k > k + 1 \text{ implies } 2^{k+1} > (k+1) + 1$$

or, equivalently, that

$$2^k > k + 1 \text{ implies } 2^{k+1} > k + 2.$$

Since $2^k \cdot 2 = 2^{k+1}$, we start with $2^k > k + 1$ and multiply both sides by 2 as follows:

$$\begin{aligned} 2^k &> k + 1 \\ 2 \cdot 2^k &> 2(k+1) \\ 2^{k+1} &> 2k + 2 \end{aligned}$$

Since $k \geq 2$, it follows that $2k+2$ is greater than $k+2$. Thus,

$$2^{k+1} > 2k+2 > k+2$$

and the proof is complete.

PROGRESS CHECK 4

Prove that $3n < 3^n$ for every positive integer $n \geq 2$. ■

The principle of mathematical induction is a very powerful method of proof that can be applied to a wide variety of mathematical statements. The main difficulty is that some ingenuity is often needed in deciding how to establish the "$k+1$" statement from the *assumption* of the "k" statement. Note that we accomplished this with different approaches in the example problems after illustrating these approaches in Example 1. To be specific, in Examples 2 and 3 we started with one side of the "$k+1$" statement and derived the other side with the aid of the "k" statement. However, in Example 4 we started with the "k" statement and did something to it (we multiplied both sides by 2) to establish the "$k+1$" statement. It will be easier for you to decide how to proceed in Step 2 if you are aware of both methods. Eventually, Step 2 becomes manageable through the experience gained in considering many problems.

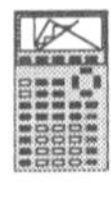

EXPLORE 12.4

1. Greek mathematicians of the fifth Century B.C. classified numbers by geometric shape. For instance, the numbers below are called the *triangular numbers*, because they are related to equilateral triangles, as shown.

   $t_1 = 1$ $t_2 = 3$

   $t_3 = 6$ $t_4 = 10$

   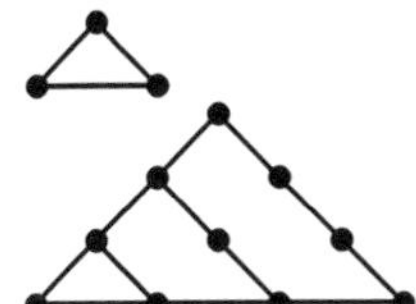

   a. What is the fifth triangular number, t_5?
   b. What number must be added to t_k to get t_{k+1}?
   c. Guess a formula for the *n*th triangular number, t_n.
   d. Use mathematical induction to prove that your formula is correct.

2. The square numbers are the ones whose dots can be formed into squares, as shown.

   $s_1 = 1$ $s_2 = 4$

   $s_3 = 9$ $s_4 = 16$

   a. What is the fifth square number?
   b. Referring to the triangular numbers from the previous exercise calculate each of these sums:

   $$t_1 + t_2,\ t_2 + t_3,\ t_3 + t_4.$$

   c. Prove by mathematical induction that the *n*th square number is the sum of the *n*th triangular number and the previous triangular number. That is, show that $s_n = t_n + t_{n-1}$.

3. Prove by mathematical induction that $1^3 + 2^3 + 3^3 + \ldots + n^3 = t_n^2$, where t_n is the *n*th triangular number as defined in Exploration 1.

EXERCISES 12.4

In Exercises 1–10 prove by mathematical induction that the following formulas are true for every positive integer value of n.

1. $1+2+3+\cdots+n=\dfrac{n(n+1)}{2}$

2. $\dfrac{1}{2}+\dfrac{2}{2}+\dfrac{3}{2}+\cdots+\dfrac{n}{2}=\dfrac{n(n+1)}{4}$

3. $\dfrac{2}{3}+\dfrac{4}{3}+\dfrac{6}{3}+\cdots+\dfrac{2n}{3}=\dfrac{n(n+1)}{3}$

4. $3+6+9+\cdots+3n=\dfrac{3n(n+1)}{2}$

5. $4+8+12+\cdots+4n=2n(n+1)$

6. $2+6+10+\cdots+(4n-2)=2n^2$

7. $2+2^2+2^3+\cdots+2^n=2(2^n-1)$

8. $3+3^2+3^3+\cdots+3^n=\dfrac{3(3^n-1)}{2}$

9. $1^3+2^3+3^3+\cdots+n^3=\dfrac{n^2(n+1)^2}{4}$

10. $1^2+2^2+3^2+\cdots+n^2=\dfrac{n(n+1)(2n+1)}{6}$

In Exercises 11–20 prove the statement by mathematical induction.

11. $2^n > n$ for all positive integers n.

12. $3^n < 3^{n+1}$ for all positive integers n.

13. $n^2 < 2^n$ for all positive integers $n \geq 5$.

14. If a is a real number, then $(a^2)^n = a^{2n}$ for all positive integers n.

15. If the x's are positive real numbers, then $\log(x_1 x_2 \cdots x_n)=\log x_1+\log x_2+\cdots+\log x_n$ for all positive integers $n \geq 2$.

16. If P dollars are invested at r percent compounded annually, then at the end of n years the compounded amount is given by $A_n = P(1+r)^n$.

17. If a is a real number such that $0<a<1$, then $0<a^n<1$ for all positive integers n.

18. n^2+n is divisible by 2 for all positive integers n.

19. If $x \neq a$ then $x^n - a^n$ is divisible by $x-a$ for positive integers n. (**Hint:** Add and subtract xa^k to $x^{k+1}-a^{k+1}$, then group terms and factor.)

20. If $x \neq y$, then $x^{2n}-y^{2n}$ is divisible by $x-y$ for all positive integers n. (**Note:** See Exercise 19.)

21. The formula $(n-1)+(n-2)+\ldots+1=n\dfrac{n-1}{2}$ gives the total number of "clinks" needed for each person in a group of n people to clink glasses with each other person.
    a. Prove this formula using mathematical induction.
    b. Use the formula from part **a** to find the total number of clinks for a group of 20 people.

22. This exercise describes a series of geometric figures. To begin, take an equilateral triangle with sides of length s. Thus, its perimeter is $3s$. At Stage 2, a new triangle is inscribed inside the first triangle by connecting the midpoints, as shown. Since the new sides are half the previous ones, after two stages the sum of both perimeters is $3s+\dfrac{3s}{2}=\dfrac{9s}{2}$. At the next stage a new triangle is inscribed the same way inside the previous smallest one. Let P_n represent the total of all the perimeters after n stages. Thus, $P_1 = 3s$, and $P_2=\dfrac{9s}{2}$.
    a. Find P_3
    b. The formula
    $$3s+\frac{3s}{2}+\frac{3s}{4}+\frac{3s}{8}+\ldots+\frac{3s}{2^{n-1}}=3s\frac{2^n-1}{2^{n-1}}$$ gives the total perimeter after n stages. Prove this formula by mathematical induction.
    c. Use the formula from part **b** to find the sum of the perimeters after 10 stages.

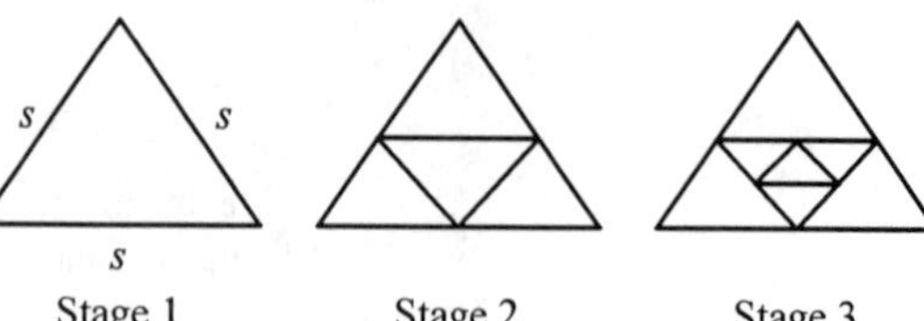

THINK ABOUT IT 12.4

1. Show by mathematical induction that the sum of n even integers is an even integer for $n \geq 2$.
2. Show by mathematical induction that

$$\sum_{i=1}^{n} ca_i = c\sum_{i=1}^{n} a_i.$$

3. The following sequence is an example of a sequence that is defined *recursively*: the first term is 1; thereafter, each term is one more than twice the preceding term. In symbols this translates to $a_1 = 1$ and $a_n = 2a_{n-1} + 1$ for $n \geq 2$. Find a_2, a_3, a_4 and a_5 and then make a conjecture about a formula for a_n in terms of n. Verify that your formula is correct by using mathematical induction.
4. **a.** Simplify each expression.

$$1-\frac{1}{2}; \left(1-\frac{1}{2}\right)\left(1-\frac{1}{3}\right);$$
$$\left(1-\frac{1}{2}\right)\left(1-\frac{1}{3}\right)\left(1-\frac{1}{4}\right)$$

   b. Make a conjecture about the simplification of the following product, and then prove it by using mathematical induction.

$$\left(1-\frac{1}{2}\right)\left(1-\frac{1}{3}\right)\cdots\left(1-\frac{1}{n}\right).$$

5. The following problems analyze false statements to reinforce that proof by mathematical induction is a two-step process, and that neither step by itself is sufficient in such proofs.
   a. Let S_n be the statement that for any positive integer n, $n^2 + n + 11$ is a prime number. Verify that S_1 is true. Show that S_n is true for $n = 1,2,\ldots,9$ but false for $n = 10$.
   b. Let S_n be the statement that for any positive integer n, $n = n + 5$. Show that if S_n is true for any positive integer k, then it is also true for the next highest integer $k+1$. Then, show that S_n is a false statement by showing that S_1 is false.

● ● ●

12.5 Binomial Theorem

*Photo Courtesy of **Reuters/Jeff Mitchell** of Stock Boston, Boston, MA*

A survey about political beliefs found that 64 percent of the national news media describe themselves as "moderates." Given this result, if six members of the media are polled at random, then the probability that exactly three will describe themselves as "moderates" is given by the fourth term in the expansion of $(0.64 + 0.36)^6$. Find this probability to the nearest hundredth of a percent.
(See Example 9. *Source*: Times Mirror Center for People and the Press Survey, 1995.)

Objectives

1. Expand $(a+b)^n$ using Pascal's triangle.
2. Expand $(a+b)^n$ using the binomial theorem.
3. Evaluate binomial coefficients in the form $\binom{n}{r}$ for given values of *n* and *r*.
4. Expand $(a+b)^n$ using the binomial theorem and binomial coefficients.
5. Find the *r*th term in the expansion of $(a+b)^n$.

A leading candidate for the most common mistake in algebra appears when students are asked to find the product or expansion of $(x+y)^2$. Too frequently they write the expansion as x^2+y^2, which leaves out the middle term $2xy$. The problem recurs when expanding other expressions of the form $(x+y)^n$, where *n* is a positive integer, since there are usually many more terms than just x^n and y^n. We now discuss a method for finding these middle terms that is much easier than repeated multiplication.

We start by trying to find some patterns in the following expansions of the powers of $a+b$. You can verify each expansion by direct multiplication.

$$\begin{aligned}
(a+b)^1 &= a+b && \text{(2 terms)}\\
(a+b)^2 &= a^2+2ab+b^2 && \text{(3 terms)}\\
(a+b)^3 &= a^3+3a^2b+3ab^2+b^3 && \text{(4 terms)}\\
(a+b)^4 &= a^4+4a^3b+6a^2b^2+4ab^3+b^4 && \text{(5 terms)}\\
(a+b)^5 &= a^5+5a^4b+10a^3b^2+10a^2b^3+5ab^4+b^5 && \text{(6 terms)}
\end{aligned}$$

Observe that in all the above cases the expansion of $(a+b)^n$ behaved as follows:

1. The number of terms in the expansion is $n+1$. For example in the expression $(a+b)^5$, we have $n=5$, and there are six terms in the expansion.
2. The first term is a^n, and the last term is b^n.
3. The second term is $na^{n-1}b$, and the *n*th term is nab^{n-1}.
4. The exponent of *a* decreases by 1 in each successive term, while the exponent of *b* increases by 1.
5. The sum of the exponents of *a* and *b* in any term is *n*.

Thus, if we assume that this pattern continues when *n* is positive integer greater than 5 (and it can be proved that it does), we only need to find a method for determining the constant coefficients, if we wish to expand such expressions. Our observations to this point may be summarized in the following expansion formula for positive integral powers of $a+b$.

$$\begin{aligned}
(a+b)^n &= a^n + na^{n-1}b + (\text{constant})a^{n-2}b^2\\
&\quad + (\text{constant})a^{n-3}b^3 + \cdots + nab^{n-1} + b^n
\end{aligned}$$

You may think of this expansion formula as an informal version of the binomial theorem. A complete statement of the binomial theorem includes a method for obtaining the constant coefficients, and we will do this later. However, for our present purpose, we have arranged the constant coefficients in the above expansions of $(a+b)^n$ for $n \le 5$ in the triangular array shown in the following chart.

Powers of $a+b$		Constant Coefficients (Pascal's Triangle)
$(a+b)^0$	Row 0	1
$(a+b)^1$	Row 1	1 1
$(a+b)^2$	Row 2	1 2 1
$(a+b)^3$	Row 3	1 3 3 1
$(a+b)^4$	Row 4	1 4 6 4 1
$(a+b)^5$	Row 5	1 5 10 10 5 1
•	•	.
•	•	.
•	•	.

The triangular array of numbers that specifies the constant coefficients in the expansions of $(a+b)^n$ for $n=1,2,\ldots$ is called **Pascal's triangle**. Note that except for the 1's, each entry in Pascal's triangle is the sum of the two numbers on either side of it in the preceding row as diagrammed in the chart. We are now ready to try some problems.

Example 1: Expanding a Binomial Using Pascal's Triangle

Expand $(x+y)^8$ using Pascal's triangle.

Solution: In our expansion formula we substitute *x* for *a*, *y* for *b*, and 8 for *n*. To determine the constant coefficients, we must continue the pattern in the above chart until we have specified the entries in row 8 of Pascal's triangle. To do this, we start at row 5 and proceed as follows:

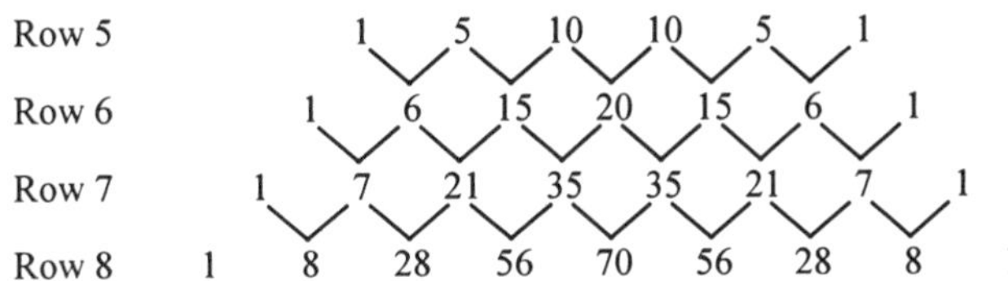

Thus, the expansion of $(x+y)^8$ is

$$(x+y)^8 = x^8 + 8x^7y + 28x^6y^2 + 56^5y^3 + 70x^4y^4 + 56x^3y^5 + 28x^2y^6 + 8xy^7 + y^8.$$

PROGRESS CHECK 1
Expand $(x+h)^9$ using Pascal's triangle. ■

Example 2: Expanding a Binomial Using Pascal's Triangle

Expand $(x+2y)^4$ using Pascal's triangle.

Solution: We substitute *x* for *a*, 2*y* for *b*, and 4 for *n* in our expansion formula. From Pascal's triangle we determine the coefficients of the five terms in the expansion as 1, 4, 6, 4, and 1, respectively. Thus, we have

$$\begin{aligned}(x+2y)^4 &= x^4 + 4x^3(2y) + 6x^2(2y)^2 + 4x(2y)^3 + (2y)^4 \\ &= x^4 + 8x^3y + 24x^2y^2 + 32xy^3 + 16y^4.\end{aligned}$$

PROGRESS CHECK 2

Expand $(x+3y)^5$ using Pascal's triangle. ■

Example 3: Expanding a Binomial Using Pascal's Triangle

Expand $(2x-y)^3$ using Pascal's triangle.

Solution: First, rewrite $(2x-y)^3$ as $\left[2x+(-y)\right]^3$. In our expansion formula we now substitute $2x$ for a, $-y$ for b, and 3 for n. From Pascal's triangle we determine the coefficients of the four terms in the expansion as 1, 3, 3, and 1, respectively. Thus, we have

$$\begin{aligned}(2x-y)^3 &= (2x)^3+3(2x)^2(-y)+3(2x)(-y)^2+(-y)^3\\ &= 8x^3y-12x^2y+6xy^2-y^3.\end{aligned}$$

Note that when the binomial is the difference of two terms, the terms in the expansion alternate in sign.

PROGRESS CHECK 3

Expand $(3x-2y)^4$ using Pascal's triangle. ■

By now you should have the feel of the basic patterns in the binomial expansion, so let's add the final step—a formula for the constant coefficients. We have been using Pascal's triangle to generate these constants. However, this approach is often impractical, since constructing the triangle can be time-consuming, particularly for large values of n.

The formula for the coefficients can be quite efficient and follows a nice pattern. First, though, it is helpful to introduce **factorial** notation. For any positive integer n, the symbol $n!$ (read "n factorial") means the product $n\cdot(n-1)\cdot(n-2)\cdot\dots\cdot 3\cdot 2\cdot 1$. For example,

$$6! = 6\cdot 5\cdot 4\cdot 3\cdot 2\cdot 1 = 720.$$

We also define $0! = 1$. Using factorials, we now state the binomial theorem.

Binomial Theorem

For any positive integer n,

$$\begin{aligned}(a+b)^n &= a^n+\frac{n}{1!}a^{n-1}b+\frac{n(n-1)}{2!}a^{n-2}b^2+\cdots\\ &+\frac{n(n-1)(n-2)\cdots(n-r+1)a^{n-r}b^r}{r!}+\cdots+b^n.\end{aligned}$$

The binomial theorem may be proved by mathematical induction, and we consider such a proof in the exercises. You will find the application of this theorem to be very systematic if you consider the following example carefully.

Example 4: Expanding a Binomial Using the Binomial Theorem

Expand $(x+y)^5$ by the binomial theorem.

Solution: In our statement of the binomial theorem we substitute x for a and y for b. We also note that in this case $n=5$. The expansion goes as follows:

$$(x+y)^5 = x^5 + \frac{5}{1!}x^4y^1 + \frac{5\cdot 4}{2!}x^3y^2 + \frac{5\cdot 4\cdot 3}{3!}x^2y^3 + \frac{5\cdot 4\cdot 3\cdot 2}{4!}xy^4 + y^5$$
$$= x^5 + 5x^4y + 10x^3y^2 + 10x^2y^3 + 5xy^4 + y^5.$$

Technology Link

The ability to compute $n!$ is a standard feature on graphing calculators, and you should learn how to access this feature on your model. Because $n!$ increases very rapidly, many calculators cannot compute beyond 69!. Determine 69! on your calculator and then explain why computing 70! results in an error message on many calculators.

PROGRESS CHECK 4

Expand $(x+2)^6$ by the binomial theorem. ■

In Example 4 there are a few patterns in determining the coefficients that you should note.

1. The coefficients 1, 5, 10, 10, 5, 1 are symmetrical; that is, they increase and then decrease in the same manner. This symmetry cuts our job in half, since we need determine the coefficients only as far as the middle term to obtain the entire list.
2. The coefficients go

$$1,\ \frac{5}{1!},\ \frac{5\cdot 4}{2!},\ \frac{5\cdot 4\cdot 3}{3!},\ \text{etc.}$$

so, in general, as

$$1,\ \frac{n}{1!},\ \frac{n(n-1)}{2!},\ \frac{n(n-1)(n-2)}{3!},\ \text{etc.}$$

Each new coefficient (after n) can be generated from the previous coefficient. Just insert the next lowest integer as a factor in the numerator, and use the next highest factorial in the denominator. It might also help if you note that in each coefficient the denominator is the factorial of the exponent for y in that term.

3. When computing each coefficient it is possible to divide out common factors so that the expression is easy to evaluate. For example,

$$\frac{5\cdot 4\cdot 3\cdot 2}{4!} = \frac{5\cdot 4\cdot 3\cdot 2}{4\cdot 3\cdot 2\cdot 1} = \frac{5}{1} = 5.$$

With a little practice you will find this method to be quite straightforward. The idea is to balance the different benefits that Pascal's triangle and the formula method have to offer. Here is a problem in which using the formula method is a lot easier than constructing Pascal's triangle.

Example 5: Finding the First *n* Terms in a Binomial Expansion

Write the first four terms in the expansion of $(x-2)^{14}$.

Solution: First, rewrite $(x-2)^{14}$ as $[x+(-2)]^{14}$. Here $n=14$, $a=x$, and $b=-2$. Then by the binomial theorem the first four terms are

$$x^{14} + \frac{14}{1!}x^{13}(-2)^1 + \frac{14\cdot 13}{2!}x^{12}(-2)^2 + \frac{14\cdot 13\cdot 12}{3!}x^{11}(-2)^3,$$

which simplifies to

$$x^{14} - 28x^{13} + 364x^{12} - 2{,}912x^{11}.$$

PROGRESS CHECK 5

Write the first four terms in the expansion of $(x-y)^{10}$. ■

To state the binomial theorem in its most efficient form, we need one more refinement in our methods. Note in the statement of the binomial theorem that the constant coefficient of the term containing b^r for $0 < r \leq n$ is

$$\frac{n(n-1)\cdots(n-r+1)}{r!}.$$

For instance, in the expansion of $(x+y)^5$ in Example 4, the constant coefficient of the term containing y^4 is

$$\frac{5 \cdot 4 \cdot 3 \cdot 2}{4!}$$

or

$$\frac{5(5-1)(5-2)(5-4+1)}{4!}.$$

In its present form this formula is awkward to use, and we may restate the formula more efficiently as follows:

$$\begin{aligned}\frac{n(n-1)\cdots(n-r+1)}{r!} &= \frac{n(n-1)\cdots(n-r+1)}{r!} \cdot \frac{(n-r)!}{(n-r)!} \\ &= \frac{n!}{r!(n-r)!}\end{aligned}$$

If we now follow customary notation and denote the constant coefficient of the term containing b^r by $\binom{n}{r}$, then we have developed that a formula for $\binom{n}{r}$ is

$$\binom{n}{r} = \frac{n!}{r!(n-r)!}.$$

In addition, $\binom{n}{0} = \binom{n}{n} = 1$, and we may obtain these results with our formula since by definition $0! = 1$. Our discussion has led us to the definition of a binomial coefficient.

Binomial Coefficient

Let r and n be nonnegative integers with $r \leq n$. Then the symbol $\binom{n}{r}$ is defined by

$$\binom{n}{r} = \frac{n!}{r!(n-r)!}.$$

Each of the numbers $\binom{n}{r}$ is called a **binomial coefficient**.

As you might expect, binomial coefficients are entries in Pascal's triangle, and the next example illustrates this point.

Example 6: Connecting Binomial Coefficients to Pascal's Triangle

Evaluate the following binomial coefficients and compare the results to the entries in Pascal's triangle:

$$\binom{4}{0}, \binom{4}{1}, \binom{4}{2}, \binom{4}{3}, \binom{4}{4}.$$

Solution: By applying the formula in the definition, we have

$$\binom{4}{0} = \frac{4!}{0!(4-0)!} = \frac{4!}{0!4!} = 1$$
$$\binom{4}{1} = \frac{4!}{1!(4-1)!} = \frac{4!}{1!3!} = \frac{4 \cdot 3 \cdot 2 \cdot 1}{1 \cdot (3 \cdot 2 \cdot 1)} = 4$$
$$\binom{4}{2} = \frac{4!}{2!(4-2)!} = \frac{4!}{2!2!} = \frac{4 \cdot 3 \cdot 2 \cdot 1}{(2 \cdot 1)(2 \cdot 1)} = 6$$
$$\binom{4}{3} = \frac{4!}{3!(4-3)!} = \frac{4!}{3!1!} = \frac{4 \cdot 3 \cdot 2 \cdot 1}{(3 \cdot 2 \cdot 1) \cdot 1} = 4$$
$$\binom{4}{4} = \frac{4!}{4!(4-4)!} = \frac{4!}{4!0!} = 1$$

By direct comparison, we note that the binomial coefficients 1, 4, 6, 4, and 1 match the entries in row 4 of Pascal's triangle. This example illustrates the general result that the binomial coefficients

$$\binom{n}{0}, \binom{n}{1}, \binom{n}{2}, \ldots, \binom{n}{n}$$

are precisely the numbers in the nth row of Pascal's triangle.

Technology Link

It is easy to evaluate binomial coefficients using a calculator with a factorial capability. For instance, since $\binom{18}{7} = \frac{18!}{7!11!}$, we can determine that $\binom{18}{7} = 31{,}824$, as shown in the first two lines of the display in Figure 12.11

```
18!/(7!*11!)
                31824
18 nCr 7
                31824
```

Figure 12.11

In addition, most graphing calculators have a menu choice that evaluates binomial coefficients directly. In probability applications, the binomial coefficient $\binom{n}{r}$ gives the number of combinations of n objects taken r at a time (as discussed in the next section), so the menu choice for binomial coefficients is often labeled ${}_nC_r$ instead of $\binom{n}{r}$. The last two lines of the display in Figure 12.11 show the computation of $\binom{18}{7}$ on a calculator that uses ${}_nC_r$ notation.

PROGRESS CHECK 6

Evaluate the following binomial coefficients and compare the results to the entries in Pascal's triangle.

$$\binom{6}{0}, \binom{6}{1}, \binom{6}{2}, \binom{6}{3}, \binom{6}{4}, \binom{6}{5}, \binom{6}{6}$$

■

By using binomial coefficients and sigma notation, the binomial theorem may be stated in a very efficient way.

Binomial Theorem

For any positive integer n,

$$(a+b)^n = \sum_{r=0}^{n} \binom{n}{r} a^{n-r} b^r = \binom{n}{0} a^n + \binom{n}{1} a^{n-1} b + \binom{n}{2} a^{n-2} b^2 + \cdots + \binom{n}{n} b^n.$$

We now redo the expansion from Example 1, using this version of the theorem.

Example 7: Using the Binomial Theorem and Binomial Coefficients

Expand $(x+y)^8$ using the binomial theorem and binomial coefficients.

Solution: Applying the binomial theorem with $x = a$, $y = b$, and $n = 8$ yields

$$(x+y)^8 = \binom{8}{0} x^8 + \binom{8}{1} x^7 y + \binom{8}{2} x^6 y^2 + \binom{8}{3} x^5 y^3 + \binom{8}{4} x^4 y^4$$
$$+ \binom{8}{5} x^3 y^5 + \binom{8}{6} x^2 y^6 + \binom{8}{7} x y^7 + \binom{8}{8} y^8.$$

To evaluate the binomial coefficients, we first note that $\binom{8}{0}=\binom{8}{8}=1$. Then we use the symmetry in the coefficients and the given formula to obtain

$$\binom{8}{1}=\binom{8}{7}=\frac{8!}{1!7!}=8, \qquad \binom{8}{2}=\binom{8}{6}=\frac{8!}{2!6!}=28,$$
$$\binom{8}{3}=\binom{8}{5}=\frac{8!}{3!5!}=56, \qquad \binom{8}{4}=\frac{8!}{4!4!}=70$$

Thus, the desired expansion is

$$(x+y)^8 = x^8+8x^7y+28x^6y^2+56x^5y^3+70x^4y^4+56x^3y^5+28x^2y^6+8xy^7+y^8.$$

Note: The symmetry in the binomial coefficients follows from the identity

$$\binom{n}{r}=\binom{n}{n-r}.$$

You are asked to verify this property in Exercise 48.

PROGRESS CHECK 7

Expand $(x+h)^9$ using the binomial theorem and binomial coefficients. ■

With our current methods it is now easy to write any single term of a binomial expansion without producing the rest of the terms. Note in the binomial theorem that the exponent of b is 1 in the second term, 2 in the third term, and, in general, $r-1$ in the rth term. Since the sum of the exponents of a and b is always n, the exponent above a in the rth term must be $n-(r-1)$. Finally, the binomial coefficient of the term containing b^{r-1} is $\binom{n}{r-1}$. These observations lead to the following formula.

rth Term of the Binomial Expansion

The rth term of the binomial expansion of $(a+b)^n$ is $\binom{n}{r-1}a^{n-(r-1)}b^{r-1}$.

Example 8: Finding a Specified Term in a Binomial Expansion

Find the 14th term in the expansion of $(3x-y)^{15}$.

Solution: In this case $n=15$, $r=14$, $a=3x$, and $b=-y$. By the above formula, the 14th term is

$$\begin{aligned}\binom{15}{13}(3x)^2(-y)^{13} &= 105(9x^2)(-y)^{13}\\ &= -945x^2y^{13}\end{aligned}.$$

PROGRESS CHECK 8

Find the seventh term in the expansion of $(x-4)^{10}$. ■

Example 9: Political Beliefs

Solve the problem in the section introduction on page 987.

Solution: The requested probability is the fourth term in the expansion of $(0.64+0.36)^6$. In this case, $a=0.64$, $b=0.36$, $n=6$, and $r=4$ (so $r-1=3$). By the above formula, the fourth term is

$$\binom{6}{3}(0.64)^3(0.36)^3 = 20(0.64)^3(0.36)^3$$
$$\approx 0.2446118093 \quad .$$

To the nearest hundredth of a percent, the probability that six members of the media are polled and exactly three describe themselves as "moderates" is 24.46 percent.

PROGRESS CHECK 9

The survey described in Example 9 also found that 40 percent of the public describe themselves as "moderates." Given this result, if five members of the public are polled at random, then the probability that exactly two will describe themselves as "moderates" is given by the third term in the expansion of $(0.6+0.4)^5$. Find this probability to the nearest hundredth of a percent. ■

EXPLORE 12.5

1. As mentioned in the Technology Links in Section 12.5, most calculators have special keys or menu choices for calculating factorials and binomial coefficients. Learn how to use these features, and use them to confirm that $\binom{52}{13} = \dfrac{52!}{13!39!}$.
2. The mathematician Euler made many contributions to the field of number theory. A result he proved (often called Fermat's theorem) is that for any integer a and any prime p, $a^p - a$ is divisible by p. For instance, when $a=3$ and $p=5$, $a^p - a$ equals $243-3=240$, which is divisible by 5. Euler's proof depended on noticing that if p is prime, then each binomial coefficient $\binom{p}{k}$ is divisible by p for $k=1,2,\ldots,p-1$. Confirm that this statement is correct for several cases including $p=3$, 5, and 7.
3. A great early contribution by Isaac Newton was the extension of the binomial theorem to fractional and negative exponents. In the statement of the binomial theorem in this section, we wrote that the general expression for the binomial coefficient is

$$\frac{n(n-1)(n-2)\cdots(n-r+1)}{r!} .$$

Newton decided that he could follow the pattern in part **a** even with fractional and negative values in the binomial coefficients. For example,

$$\binom{\frac{1}{2}}{3} = \frac{\frac{1}{2}(\frac{1}{2}-1)(\frac{1}{2}-2)}{1\cdot 2\cdot 3} = \frac{(\frac{1}{2})(-\frac{1}{2})(-\frac{3}{2})}{6} = \frac{3}{48} = \frac{1}{16},$$
$$\binom{-1}{3} = \frac{-1(-1-1)(-1-2)}{1\cdot 2\cdot 3} = \frac{-1(-2)(-3)}{6} = \frac{-6}{6} = -1.$$

Use this technique to evaluate $\binom{\frac{1}{2}}{2}$ and $\binom{-2}{3}$.

4. By analogy with Pascal's triangle, Leibniz formed the harmonic triangle, shown below.
   a. By exploring the triangle, determine how each entry is computed, then write out the next two rows.
   b. What is the relationship between the entries in the harmonic triangle and the corresponding entries in Pascal's triangle?

$$\begin{array}{ccccccccc} & & & & \frac{1}{1} & & & & \\ & & & \frac{1}{2} & & \frac{1}{2} & & & \\ & & \frac{1}{3} & & \frac{1}{6} & & \frac{1}{3} & & \\ & \frac{1}{4} & & \frac{1}{12} & & \frac{1}{12} & & \frac{1}{4} & \\ \frac{1}{5} & & \frac{1}{20} & & \frac{1}{30} & & \frac{1}{20} & & \frac{1}{5} \end{array}$$

EXERCISES 12.5

In Exercises 1–12 expand each expression by the binomial theorem. Use both Pascal's triangle and the formula methods in determining the constant coefficients.

1. $(x+y)^6$ 2. $(x+y)^7$ 3. $(x-y)^5$

4. $(x-y)^4$ 5. $(x+h)^4$ 6. $(x+h)^3$

7. $(x-1)^7$ 8. $(y+1)^5$ 9. $(2x+y)^3$

10. $(x-2y)^6$ 11. $(3c-4d)^4$ 12. $(4c+3d)^3$

In Exercises 13–16 write the first four terms in the expansion of the given expression.

13. $(x+y)^{15}$ 14. $(x-y)^{10}$

15. $(x-3y)^{12}$ 16. $(x+3)^{17}$

In Exercises 17 and 18 evaluate each of the binomial coefficients and compare the results with the entries in Pascal's triangle.

17. $\binom{3}{0}, \binom{3}{1}, \binom{3}{2}, \binom{3}{3}$

18. $\binom{5}{0}, \binom{5}{1}, \binom{5}{2}, \binom{5}{3}, \binom{5}{4}, \binom{5}{5}$

In Exercises 19 and 20 express the entries in the given row of Pascal's triangle using the notation $\binom{n}{r}$.

19. Row 6 20. Row 2

In Exercises 21–30 evaluate each of the binomial coefficients

21. $\binom{5}{3}$ 22. $\binom{6}{4}$ 23. $\binom{9}{4}$

24. $\binom{10}{3}$ 25. $\binom{10}{0}$ 26. $\binom{9}{9}$

27. $\binom{20}{2}$ 28. $\binom{25}{24}$ 29. $\binom{18}{8}$

30. $\binom{15}{5}$

In Exercises 31–40 write the indicated term of the expansion.

31. Second term of $(x+y)^{18}$

32. Fifth term of $(x-y)^7$

33. 12th term of $(3x-y)^{13}$

34. Sixth term of $(x+3y)^{10}$

35. Sixth term of $\left(x^2+1\right)^9$

36. 11th term of $\left(2x^3-1\right)^{14}$

37. Third term of $\left(2-\sqrt{x}\right)^5$

38. Fourth term of $\left(\sqrt{x}+3\right)^6$

39. Middle term of $\left(3x^2+2y^3\right)^6$

40. Middle term of $\left(x^{-1}+y^{-1}\right)^{10}$

41. Find the coefficient of x^7y^4 in the expansion of $(x+y)^{11}$.

42. Find the coefficient of $x^{15}y^5$ in the expansion of $(x-y)^{20}$.

43. Find the coefficient of the term containing x^7 in the expansion of $(x-2y)^{10}$.

44. Find the term with the variable factor x^5 in the expansion of $(2x-y)^{15}$.

45. Find the value of $(1.2)^4$ by writing it in the form $(1+0.2)^4$ and using the binomial theorem.

46. Find the value of $(0.97)^3$ by writing it in the form $(1-0.03)^3$ and using the binomial theorem.

47. Show that $\binom{n}{n-1}=n$.

48. Show that $\binom{n}{r}=\binom{n}{n-1}$.

49. If 80 percent of the marbles in a huge container are red, and 50 marbles are drawn out at random, then the probability that exactly 40 of those drawn will be red is given by the 41st term in the expansion of $(0.20+0.80)^{50}$. Find this probability to the nearest hundredth of a percent.

50. In an algebra class 20 of the 30 students are female. If a committee of five of these students is picked at random, then the probability that the committee contains exactly three males is given by the fourth term of the expansion of $\left(\frac{2}{3}+\frac{1}{3}\right)^5$. Find this probability to the nearest hundredth of a percent.

51. When a fair coin is tossed 100 times, the probability of getting from 49 to 51 heads is given by the sum of the 50th, 51st, and 52nd terms of $\left(\frac{1}{2}+\frac{1}{2}\right)^{100}$. Calculate this probability to the nearest hundredth of a percent.

52. In an algebra class, 20 of the 30 students are female. If a committee of five of these students is picked at random, then the probability that the committee contains more males than females is given by the sum of the last three terms of the expansion of $\left(\frac{2}{3}+\frac{1}{3}\right)^5$. Calculate this probability to the nearest hundredth of a percent.

THINK ABOUT IT 12.5

1. Expand $\left(x^2+x-1\right)^4$ by using the binomial theorem.

2. **a.** Use the binomial expansion of $(1+1)^n$ to show that

$$\binom{n}{0}+\binom{n}{1}+\binom{n}{2}+\cdots+\binom{n}{n}=2^n.$$

   b. What is the sum of the entries in row 12 of Pascal's triangle?

3. If $f(x)=x^n$, where n is a positive integer, show that

$$\frac{f(x+h)-f(x)}{h}=nx^{n-1}+\frac{n(n-1)}{2!}x^{n-2}h+\cdots+h^{n-1}.$$

4. **a.** Show that $\binom{4}{3}+\binom{4}{2}=\binom{5}{3}$.

b. Consider the positions in Pascal's triangle of the binomial coefficients in part **a**. What feature of Pascal's triangle does the statement illustrate?

c. The property we have considering states that for all positive integers k and r with $r \le k$,

$$\binom{k}{r}+\binom{k}{r-1}=\binom{k+1}{r}.$$

Prove this property by using the binomial coefficient formula.

5. Below we use mathematical induction and the property shown in Question 4 to prove the binomial theorem $\left[\text{that is, } (a+b)^n=\sum_{r=0}^{n}\binom{n}{r}a^{n-r}b^r\right]$. Fill in the missing steps (denoted by the letters $\boldsymbol{a}-\boldsymbol{i}$ along with appropriate justifications.

Step 1 Replacing n by 1 gives **a**, which is true.

Step 2 When $n=k$ the statement becomes

$$(a+b)^k=\binom{k}{0}a^k+\binom{k}{1}a^{k-1}b+\cdots+\binom{k}{k-1}ab^{k-1}+\binom{k}{k}b^k. \quad (1)$$

When $n=k+1$ the statement becomes

b (2)

We now use equation (1) to establish equation (2). To begin, multiply both sides of equation (1) by $(a+b)$ and then continue as follows:

$$(a+b)^{k+1}=a\left[\binom{k}{0}a^k+\binom{k}{1}a^{k-1}b+\cdots+\binom{k}{k-1}ab^{k-1}+\binom{k}{k}b^k\right]+b[\mathbf{c}]$$

$$=\binom{k}{0}a^{k+1}+\binom{k}{1}a^kb+\cdots+\binom{k}{k-1}a^2b^{k-1}+\binom{k}{k}ab^k+\mathbf{d}$$

$$=\binom{k}{0}a^{k+1}+\left[\binom{k}{1}+\binom{k}{0}\right]a^kb+\cdots+[\mathbf{e}]ab^k+\binom{k}{k}b^{k+1}$$

$$=(\mathbf{f})\,a^{k+1}+(\mathbf{g})a^kb+\cdots+(\mathbf{h})\,ab^k+(\mathbf{i})\,b^{k+1}$$

which is equation (2). Thus, equation (1) implies equation (2), and the proof by induction is complete.

• • •

12.6 Counting Techniques

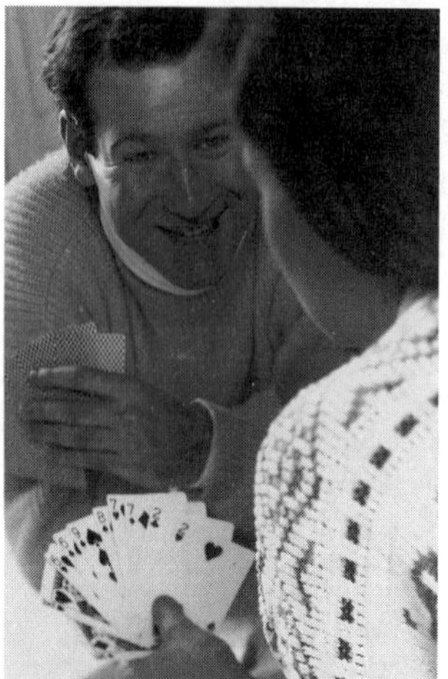

Photo Courtesy of **Mark Bolster/International Stock**

How many different five-card poker hands can be dealt from a deck of 52 cards? (See Example 7).

Objectives

1. Determine the number of ways two (or more) events taken together can occur.
2. Determine the number of distinct permutations of n objects in the case when all the objects are different.
3. Determine the number of permutations of n objects taken r at a time.
4. Determine the number of distinct permutations of n objects when certain members in the set of objects are indistinguishable.
5. Determine the number of combinations of n objects taken r at a time.
6. Solve applied problems by choosing the appropriate counting techniques and applying the associated formulas.

When making certain decisions, it is often important to analyze the possibilities associated with the situation. For example, if your college offers two sections of a required math course and three sections of a required English course, then how many schedules are possible in enrolling for these courses? We can answer this question by constructing a **tree diagram**, as shown in Figure 12.12.

Math Course	English Course	Possible Schedule
	E_a	M_aE_a
M_a	E_b	M_aE_b
	E_c	M_aE_c
	E_a	M_bE_a
M_b	E_b	M_bE_b
	E_c	M_bE_c

Figure 12.12

We start by listing the two math sections, and then we branch out from these points to the three English sections. Reading each branch from left to right, we determine that there are six schedules, as shown in Figure 12.12. Note that the number of possible schedules is the product of the number of math choices and the number of English choices. This example illustrates the following key principle in counting.

Fundamental Counting Principle

If event A can occur in a ways, and following this, event B can occur in b ways, then the two events taken together can occur in $a \cdot b$ ways.

Furthermore, this principle extends to three or more events, as considered in Example 1.

Example 1: Using the Fundamental Counting Principle

In how many ways can a three-item true-false test be answered? Include a tree diagram listing the possibilities.

Solution: There are three questions, so we start by making three boxes.

$$\square \quad \square \quad \square$$

Since there are two choices (true or false) for each question, we now enter a 2 in each box and then multiply as indicated in the fundamental counting principle.

$$\boxed{2} \cdot \boxed{2} \cdot \boxed{2} = 8$$

Thus, there are eight ways to answer the test. As shown in Figure 12.13, we can systematically list these choices by writing T and F under the first question and then branching to T and F for each successive question. Following each branch from left to right then gives the possibilities shown.

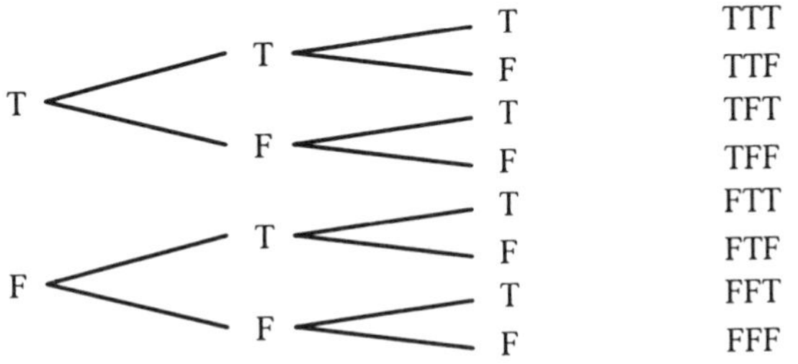

Figure 12.13

PROGRESS CHECK 1

At a community college, students are classified according to their degree program (A.A, A.S, A.A.S) and gender (male, female). How many classifications are possible? Include a tree diagram listing the possibilities. ■

Example 2: Using the Fundamental Counting Principle

Find the number of different batting orders that are possible for a baseball team that starts nine players. (Assume the team has only nine players.)

Solution: The lead-off batter may be any of the nine players. However, there are only eight choices for the second position (since the lead-off batter cannot be used again,) seven choices for the third position, and so on. By the fundamental counting principle, this gives

$$9 \cdot 8 \cdot 7 \cdot 6 \cdot 5 \cdot 4 \cdot 3 \cdot 2 \cdot 1 = 362{,}880$$

different batting orders.

PROGRESS CHECK 2

How many orders of finish (excluding ties) are possible in a race among eight people? ■

The question in Example 2 illustrates an important type of counting problem. A **permutation** of a collection of objects or symbols is an arrangement without repetition of these objects, in which order is important. Thus, in Example 2 we say that there are 362,880 permutations or orderings of the 9 players on a baseball team. In general, when ordering n objects we have n choices for the first position, $n-1$ choices for the second position, $n-2$ choices for the third position, and so on. By using the fundamental counting principle and recalling factorial notation (see Section 12.5), we then conclude the following:

Permutations of a Set of n Objects

> The number of permutations of n different objects using all of them is
>
> $$n(n-1)(n-2)\cdots 3\cdot 2\cdot 1 = n!$$

Example 3: Finding the Number of Permutations

In how many ways can four candidates be arranged on a ballot?

Solution: Since each candidate is listed only once and the order or arrangement is significant, we need to determine the number of permutations of the four candidates. From the rule above, this number is

$$4! = 4\cdot 3\cdot 2\cdot 1 = 24.$$

PROGRESS CHECK 3

In how many ways can the five nominees for "Best Actress" be introduced at the Oscar award ceremony? ■

In many cases it is useful to compute the number of permutations of n objects using only some of the objects in each permutation. For instance, by the fundamental counting principle, the number of four-letter permutations of the 26 letters in the alphabet is $26\cdot 25\cdot 24\cdot 23 = 358{,}800$. We refer to this problem as the **number of permutations of 26 objects taken 4 at a time**, and symbolize this as ${}_{26}P_4$. Note that

$${}_{26}P_4 = \underbrace{26\cdot 25\cdot 24\cdot 23}_{\text{4 factors}}$$

so that to evaluate ${}_{26}P_4$ we multiply four consecutive integers that start at 26 and then decrease. This example suggests the following rule.

Permutation Formula

> If ${}_nP_r$ represents the number of permutations of n objects taken r at a time, then
>
> $${}_nP_r = \underbrace{n(n-1)(n-2)\cdots(n-r+1)}_{r\text{ factors}} \text{ where } r \le n$$

Note in this formula that n is the starting factor, while r indicates the number of factors. When the number of factors is large, it is convenient to evaluate ${}_nP_r$ by applying the formula derived below and using a calculator with a Factorial key.

$$_nP_r = n(n-1)(n-2)\cdots(n-r+1)$$
$$= \frac{n(n-1)(n-2)\cdots(n-r+1)[(n-r)\cdots 2\cdot 1]}{[(n-r)\cdots 2\cdot 1]}$$

so

$$_nP_r = \frac{n!}{(n-r)!}$$

Example 4: Permutations of n Objects Taken r at a Time

In how many ways can nine members from a hockey team be assigned to six different starting positions?

Solution: Since position is important and repetition cannot occur, the number is

$$_9P_6 = \underbrace{9\cdot 8\cdot 7\cdot 6\cdot 5\cdot 4}_{6\text{ factors}}.$$

Using our alternative formula, the evaluation is

$$_9P_6 = \frac{9!}{(9-6)!} = \frac{9!}{3!} = \frac{9\cdot 8\cdot 7\cdot 6\cdot 5\cdot 4\cdot 3\cdot 2\cdot 1}{3\cdot 2\cdot 1} = 60{,}480.$$

Technology Link

Most graphing calculators have a menu item labeled $_nP_r$ for finding the number of permutations of n objects taken r at a time. If your calculator has the capability, then you should redo the problem in this example and compare your result to the text's answer.

PROGRESS CHECK 4

In how many ways can a chairperson and a secretary be selected for a scholarship committee if there are 10 committee members? ■

When we cannot distinguish between certain members in our collection of objects, we must revise our permutation formula. For instance, how many distinct permutations can be made from the letters of the word ERROR? If we could distinguish among the R's by calling them, say, R_1, R_2 and R_3, then there would be **5!**, or 120, permutations of the five letters. Since there are **3!**, or 6, orderings of the three R's, we could group the 120 possibilities into 20 groups in which the six arrangements in each group are not distinguishable when we drop the subscripts. Thus, the number of distinct arrangements of the letters in ERROR is 20, which is 5!÷ 3!. This example suggests the following rule for analyzing this type of problem.

Distinguishable Permutations Formula

The number of distinguishable permutations of n objects taken all at a time, when n_1 are of one kind, n_2 are of a second kind, . . ., and n_k are of a kth kind, where $n_1 + n_2 + \cdots + n_k = n$ is,

$$\frac{n!}{n_1!n_2!\cdots n_k!}.$$

Example 5: Finding the Number of Distinguishable Permutations

How many distinct permutations can be made from the letters of the word Beginning?

Solution: The word BEGINNING has 9 letters with duplications of 3 N's, 2 I's, and 2 G's. By the given theorem the number of distinct permutations is

$$\frac{9!}{3!2!2!} = 15,120.$$

PROGRESS CHECK 5

How many distinct permutations can be made from the letters of the word STATISTICS? ■

We now consider one last, but very important, type of counting problem. A distinguishing characteristic of a permutation is that the order of the objects is significant. However, in many situations order does not matter. For example, if we want to know the number of ways in which two people can be hired from five job applicants, then selecting individuals A and B is the same as selecting B and A. What is important here is only who gets the jobs. A **combination** of a collection of objects or symbols is a selection without repetition in which the order of selection does not matter. In our example problem we therefore need to determine the number of combinations of five applicants when taken two at a time. To answer this, we first note that if order does matter, there are

$${}_5P_2 = 5 \cdot 4 = 20$$

possibilities. However, since the two positions can be ordered in **2!**, or 2, ways, only half of these possibilities are different when we disregard order. Thus, there are $20 \div 2$, or 10, combinations of applicants who can be hired. This example illustrates that we analyze combinations as follows:

Combination Formula

If ${}_nC_r$ represents the number of combinations of n objects taken r at a time, then

$${}_nC_r = \frac{{}_nP_r}{r!} = \frac{n!}{(n-r)!r!}.$$

Note that the formula for ${}_nC_r$ matches the formula for the binomial coefficient $\binom{n}{r}$, which we considered in Section 12.5. For this reason, $\binom{n}{r}$ is often used in place of ${}_nC_r$ to denote the number of combinations of n objects taken r at a time.

Example 6: Combinations of *n* Objects Taken *r* at a Time

From a class of 16 students an instructor asks 3 students to write their homework on the board. How many student selections are possible?

Solution: Order does not matter and repetition is not possible, so we need to determine the number of combinations of 16 students taken 3 at a time.

$${}_{16}C_3 = \frac{{}_{16}P_3}{3!} = \frac{\overbrace{16 \cdot 15 \cdot 14}^{\text{3 factors}}}{3!} = 560$$

An alternative evaluation is

$$_{16}C_3 = \frac{16!}{(16-3)!3!} = \frac{16!}{13!3!} = \frac{16 \cdot 15 \cdot 14}{3!} = 560 .$$

Technology Link

Most graphing calculators have a menu item labeled ${}_nC_r$ for finding the number of combinations of n objects taken r at a time. If your calculator has this capability, then you should redo the problem in this example and compare your result to the text's answer.

PROGRESS CHECK 6
Two cocaptains are selected from a soccer team with 15 players. How many cocaptain selections are possible? ■

Several methods have been shown in this section for counting the number of ways an event can occur. These methods include the multiplication rule contained in the fundamental counting principle, the permutation rules, and the combination rule. To decide if a counting problem involves permutations, combinations, or neither, you need to consider carefully whether or not order is significant and repetition is possible. Then you can solve the problem using the appropriate theorems.

Example 7: Choosing a Counting Method—Poker Hands

Solve the problem in the section introduction on page 1000.

Solution: When a hand of cards is dealt in poker, repetition in the cards is not possible, and the order of the cards in the hand does not matter. Therefore, a poker hand is a combination of 52 cards taken 5 at a time. Using

$${}_nC_r = \frac{n!}{(n-r)!r!} \text{ yields } {}_{52}C_5 = \frac{52!}{47!5!} = 2{,}598{,}960$$

So the number of possible poker hands is 2,598,960.

PROGRESS CHECK 7
In how many ways can six taxi drivers be assigned to seven taxi cabs? ■

EXPLORE 12.6

In Section 12.6 various tools for counting are developed. Most calculators have built-in algorithms for computing permutations and combinations. Common designations for these procedures are ${}_nP_r$ and ${}_nC_r$, respectively. Note that ${}_nC_r$ is another way to designate the binomial coefficient, $\binom{n}{r}$. It is convenient to learn how to use these features of your calculator.

1. The value of ${}_nP_r$ can be unexpectedly large. Use a calculator to determine the values of the following permutations. If your calculator gives the result in scientific notation, rewrite it in standard notation.
   a. ${}_{15}P_r$ for $r = 1,2,3,\ldots,15$ b. ${}_nP_{12}$ for $n = 12,13,14,\ldots,17$
2. a. For a given choice of n and r, the value of ${}_nP_r$ cannot be smaller than the value of ${}_nC_r$. Why not? Can they be equal?
   b. Find particular values for n and r so that ${}_nP_r$ is 6 times larger than ${}_nC_r$. Is it possible for ${}_nP_r$ to be 10 times larger than ${}_nC_r$? Justify your answer.
3. a. Let $f(x) = x!$, and explain why f is a function. What is its domain? It is very tricky to get your calculator to display a graph of $y = f(x) = x!$. Can you do it?
   b. A key feature of the factorial function is that $f(n) = nf(n-1)$. Confirm that this relationship holds when $n = 3$, 4, and 5. Does it hold when $n = 1$? For what values of n does the relationship hold?

4. The gamma function $\Gamma(x)$ is related to the factorial function. It has similar properties, but is defined for all positive real numbers, not just integers. Three properties of the gamma function, are that $\Gamma(x)=(x-1)\Gamma(x-1)$, that $\Gamma(n)=(n-1)!$, and that $\Gamma\left(\frac{1}{2}\right)=\pi^{1/2}$. Using these properties, find $\Gamma(4)$, $\Gamma(5)$, $\Gamma(3/2)$, $\Gamma(5/2)$. CAS systems often have the Gamma function built it. If you have such a system, use it to confirm these results.

EXERCISES 12.6

In Exercises 1–10 evaluate the expression

1. $6!$ 2. $(6-3)!$ 3. $\dfrac{9!}{5!4!}$

4. $\dfrac{100!}{98!}$ 5. ${}_5P_4$ 6. ${}_5C_4$

7. ${}_8P_5$ 8. ${}_4P_2$ 9. ${}_{10}C_1$

10. ${}_{25}C_{22}$

11. One reason 0! is defined as 1 is that this definition helps validate our permutation and combination formulas when r equals n or 0. Use $0!=1$ when evaluating each of the following.
    a. ${}_5P_5$ **b.** ${}_5C_5$ **c.** ${}_5P_0$

In Exercises 12–15 answer the question using the fundamental counting principle.

12. In how many ways can a two-item true-false test be answered? Include a tree diagram listing the possibilities.

13. How many head-tail orderings are possible when a coin is flipped three times? Include a tree diagram listing the possibilities.

14. In how many ways can a student choose 1 of 4 math courses and 1 of 3 business courses?

15. In how many ways can a student choose 2 or 3 computer courses and 3 of 4 elective course?

In Exercises 16–19 answer the question by using the permutation theorems.

16. How many distinct permutations can be made from the letters of the world NUMBER?

17. How many distinct permutations can be made from the letters of the world MISSISSIPPI?

18. In how many ways can eight members from a basketball team be assigned to five different starting positions?

19. How many orders of finish (excluding ties) are possible in a race among six people?

In Exercises 20–23 answer the question by using the combination theorems.

20. In how many ways can we obtain four volunteers from ten people?

21. In how many ways can three elective courses be selected from eight choices?

22. How many different 13-card bridge hands can be dealt from a deck of 52 cards?

23. How many subcommittees of four can be made from a 12-person committee?

In Exercises 24–40 consider carefully whether or not order is significant and repetition is possible. Then answer the question using an appropriate theorem.

24. In how many ways can the 12 members of a jury be seated in the jury box?

25. How many 12-member juries can be selected from 18 potential jurors?

26. In how many ways can 2 cards be selected from a deck of 52 cards if the first card is replaced before the second card is chosen? How many ways are possible if the first card is not replaced?

27. In how many ways can four red flags and two white flags be arranged one above another?

28. Find the number of different batting orders that are possible for a baseball team that starts nine players if the pitcher must bat last. (Assume the team has only nine players.)

29. In a group of ten people, how many handshakes are possible if each person shakes hands once with the other members of the group?

30. In how many ways can the four quadrants in the Cartesian coordinate system be colored if four different colors are used?

31. In how many ways can three of the same math book and two of the same English book be arranged on a shelf?

32. In how many ways can a five-item multiple-choice test with choices a, b, c, and d be answered?

33. How many subcommittees of 3 Republicans and 3 Democrats can be selected from a 10-member committee that includes 6 Republicans and 4 Democrats?

34. In how many ways can we divide 10 people into 2 groups of 5?

35. In how many ways can we divide 10 people into 5 groups of 2?

36. On a test you are asked to answer 4 problems out of 7. How many choices are possible?

37. How many 3-digit odd numbers can be written from the digits 1, 2, 3, 4, and 5 without repetition in the digits.

38. A soccer league consists of six teams. If each team plays two games with each of the other teams, find the total number of games played.

39. In how many ways can we introduce, one by one, the five nominees for an award?

40. How many line segments can be drawn joining seven points, if no three of the points lie on the same line?

41. Solve for n: ${}_nP_2 = 20$.

42. Solve for n: ${}_nC_2 = 66$.

43. Use ${}_nC_r = \dfrac{n!}{(n-r)!r!}$ and prove that ${}_nC_r = {}_nC_{n-r}$.

44. Use ${}_nC_r = {}_nC_{n-r}$, and solve ${}_nC_{16} = {}_nC_4$ for n.

45. If ${}_nC_a = {}_nC_b$ where $a \neq b$ and a, b, and n are positive integers, then

a. $n = a + b$ **b.** $n = a - b$

c. $n = ab$ **d.** $n = \dfrac{a}{b}$

46. If n is a positive integer greater than 1, then

a. ${}_nC_n > {}_nC_{n-1}$ **b.** ${}_nC_n = {}_nC_{n-1}$

c. ${}_nC_n < {}_nC_{n-1}$

THINK ABOUT IT 12.6

1. Create a word problem not discussed in this section that may be solved by computing the given expression.

a. $6!$ **b.** ${}_6P_3$ **c.** ${}_6C_3$

2. In how many different orders may four people stand in line? How about a circle? In general, what is the number of distinguishable circular arrangements of n people?

3. By using sigma notation and combinations, the binomial theorem may be efficiently stated as follows:

For any positive integer n,

$$(a+b)^n = \sum_{r=0}^{n} {}_nC_r a^{n-r} b^r$$

Expand $(a+b)^4$ using this formula.

4. Expand $(x-2y)^5$ using the formula in Question 3.

5. **a.** The rth term in the expansion of $(a+b)^n$ is ${}_nC_{r-1}a^{n-(r-1)}b^{r-1}$. Use this formula and find the third term in the expansion of $(x+y)^6$.

b. Find in simplest form the fifth term in the expansion of $\left(x^2+x^{-1}\right)^8$ by using the formula in part **a**.

• • •

12.7 Probability

From a class of 24 students, an instructor randomly selects 3 students to write their homework on the board. If Mary, Jamila, and Allan are students in this class, then what is the probability they are the three people selected?
(See Example 3.)

Objectives

1. Find the probability of a single event.
2. Find the probability that an event will not occur, using the formula for P (not E).
3. Find the probability of event A and event B both occurring if A and B are independent.
4. Find the probability of event A or event B occurring if A and B are mutually exclusive.
5. Find the probability of event A or event B occurring if A and B are not mutually exclusive.

Once we can analyze the number of ways in which certain events can occur, a natural and important question is to ask what the chances are of these events taking place. Probability theory is used to assign a number between 0 and 1, inclusive, that indicates how certain we are that an event will occur. On this scale, impossible events are assigned a probability of 0, and events that must take place are assigned a probability of 1. Before we begin to develop some of the laws for determining probabilities, it is important to understand that there are two varieties of probability statements. To illustrate this, consider the following predictions.

> There is an 80 percent chance that interest rates will go down.
> There is a 60 percent chance that the Bears will win.
> There is a 50 percent chance that a flipped coin will land heads.
> There is a 25 percent chance of correctly guessing both answers on a two-item true-or false test.

The first two statements are examples of subjective statements, in that the stated number is a reflection of (hopefully) an expert's opinion and cannot really be verified. In such cases a different expert will often predict a different number. However, the last two statements are examples of **theoretical probability**, in that the numbers are determined by the laws of probability, and everyone should obtain the same theoretical result. When we say there is a 50-percent chance that a flipped coin will land heads, we note that in practice we rarely obtain 50 heads in 100 flips of a coin, but theoretically

$$\frac{\text{number of heads}}{n \text{ tosses}} \xrightarrow{\text{approaches}} 50 \text{ percent}$$

as n gets larger. This idea that probability is a ratio suggests the following definition for the probability of an event.

Definition of the Probability of Event E

> If an event E can occur in s ways out of a total of T equally likely outcomes, than the probability of E, denoted $P(E)$, is
>
> $$P(E) = \frac{s}{T} = \frac{\text{number of ways } E \text{ can occur}}{\text{total number of equally likely outcomes}}.$$

Note that this definition assures us that if E cannot occur, $P(E)=0$, while if E is certain, then $P(E)=1$. Now consider Examples 1–4, which illustrate this definition.

Example 1: Finding the Probability of an Event

Find the probability of selecting a spade from a regular deck of 52 cards.

Solution: There are 52 cards with the same chance of being picked. Among these are 13 spades. Thus,

$$P(\text{spade}) = \frac{13}{52}, \text{ or } \frac{1}{4}, \text{ or } 0.25, \text{ or } 25 \text{ percent.}$$

Note that fractional answers are simplified (if possible) and that probabilities often take the form of decimals and percents.

PROGRESS CHECK 1

Find the probability of selecting a red queen from a regular deck of 52 cards. ■

For a given experiment, the set of all possible outcomes is called the **sample space** of the experiment. For some probability problems it is useful to specify the sample space, as shown in Example 2. In other cases it is not practical to list all the possibilities, and Examples 3 and 4 consider such problems.

Example 2: Using a Sample Space

Two dice are rolled. What is the probability that the sum of the outcomes on the dice is 7?

Solution: There are 6 equally likely outcomes on each die, so by the fundamental counting principle, there are 6×6 or 36 outcomes to consider. By using a tree diagram and associating each outcome on one die with each possible outcome on the other, we obtain the sample space in Figure 12.14. As encircled in Figure 12.14 there are 6 outcomes that produce a sum of 7, so

$$P(\text{sum } 7) = \frac{6}{36}, \text{ or } \frac{1}{6}.$$

1,1	1,2	1,3	1,4	1,5	(1,6)
2,1	2,2	2,3	2,4	(2,5)	2,6
3,1	3,2	3,3	(3,4)	3,5	3,6
4,1	4,2	(4,3)	4,4	4,5	4,6
5,1	(5,2)	5,3	5,4	5,5	5,6
(6,1)	6,2	6,3	6,4	6,5	6,6

Figure 12.14

PROGRESS CHECK 2

Two children are playing a game in which each player displays one symbol for either rock, paper, or scissors. What is the probability that at least one player will display the symbol for scissors? Assume that rock, paper, and scissors have the same chance of being displayed. ■

Example 3: Using Combinations

Solve the problem in the section introduction on page 1008.

Solution: The instructor may select ${}_{24}C_3$, or 2,024, equally likely groups. Only one of these combinations is Mary, Jamila, and Allan. Thus, the chances are $1/2,024$.

PROGRESS CHECK 3

From a group of 8 contestants, 4 will be selected randomly for prizes. If Vickie, Ellen, Justin, and Brian are among the contestants, then what the is the probability that they will be the winners? ■

Example 4: Using Combinations and the Fundamental Counting Principle

From a committee of 6 men and 6 women, a 3-person subcommittee is randomly chosen. What is the probability that the subcommittee will contain 2 men and 1 woman?

Solution: There are ${}_{12}C_3$, or 220, equally likely subcommittees. Among these are ${}_6C_2 \cdot {}_6C_1$, or 90, subcommittees with 2 of 6 men and 1 of 6 women. Thus, the chances are $90/220$, or $9/22$.

PROGRESS CHECK 4

From a committee of 8 men and 5 women, a 4-person subcommittee is chosen. What is the probability that the subcommittee will contain 2 men and 2 women? ■

To determine probabilities involving more than one event, it is useful to add to our current methods probability formulas that enable us to analyze "and," "or," and "not" statements. We first illustrate the "not" formula. On one roll of a die it is easy to determine that

$$P(4) = \frac{1}{6} \text{ while } P \text{ (not 4)} = \frac{5}{6}.$$

Note that $P(4) + P(\text{not } 4) = 1$ or, equivalently, $P(\text{not } 4) = 1 - P(4)$. We generalize this observation as follows:

Formula for $P(\text{not } E)$

The probability that event E will not occur is given by

$$P(\text{not } E) = 1 - P(E).$$

Example 5: Using the Relation Between $P(E)$ and $P(\text{not } E)$

If you guess randomly at the answers on a two-item true-false test, what is the probability that you will not get both answers right?

Solution: From a tree diagram (see Figure 12.15), the possibilities in the sample space are

$$R,R \quad R,W \quad W,R \quad W,W,$$

where the question 1 outcome is listed as the first component in each pair. Because you guess randomly, $P(R) = P(W)$. This implies that the four possible outcomes in the tree diagram are equally likely.

Then

$$P(\text{both right}) = \frac{1}{4} \text{ while } P(\text{not both right}) = 1 - \frac{1}{4} = \frac{3}{4}.$$

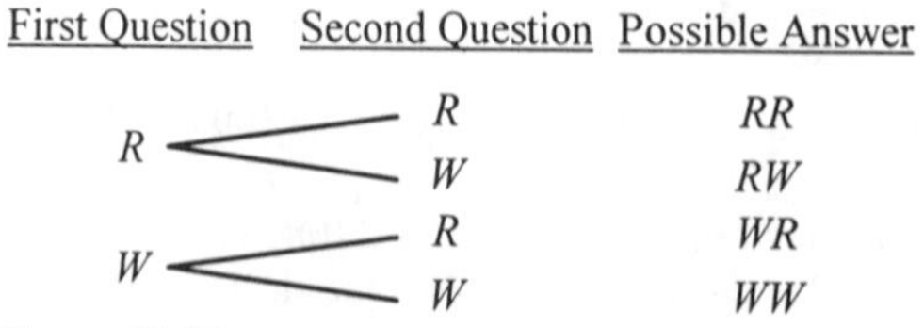

First Question	Second Question	Possible Answer
R	R	RR
	W	RW
W	R	WR
	W	WW

Figure 12.15

PROGRESS CHECK 5
Two dice are rolled. What is the probability that the sum of the outcomes on the dice will not be 11? ■

In Example 5 we can determine the chances of being right on both questions without listing all the outcomes. Note that

$$P(\text{question 1 right}) = \frac{1}{2} \text{ and } P(\text{question 2 right}) = \frac{1}{2},$$

$$\text{while } P(\text{both right}) = \frac{1}{4}.$$

Thus,

$$P(\text{both right}) = P(\text{question 1 right}) \cdot P(\text{question 2 right}).$$

In general, we say two events are **independent** if the outcome of one does not affect the outcome of the other, and we have the following theorem for independent events.

Probability of Independent Events

> If A and B are independent events, than
>
> $$P(A \text{ and } B) = P(A) \cdot P(B).$$
>
> Furthermore, this rule extends to three or more independent events.

Example 6: Finding the Probability of a Sequence of Two Independent Events

Three pills and five capsules are in a box. Two of the pills cause drowsiness, while four of the casuals remedy this side effect. What is the probability that a person who swallows a pill and a capsule will be drowsy? What are the chances of not being drowsy?

Solution: For a person to be drowsy, two independent events must occur as follows:

$$\begin{aligned} P(\text{drowsy}) &= P(\text{pill causes drowsiness and capsule is not a remedy}) \\ &= \frac{2}{3} \times \frac{1}{5} \\ &= \frac{2}{15} \end{aligned}$$

Also,

$$P(\text{not drowsy}) = 1 - P(\text{drowsy}) = 1 - \frac{2}{15} = \frac{13}{15}.$$

PROGRESS CHECK 6
In a certain population 10 percent of the people have type B blood. If two members of this population are chosen at random, what is the probability that both will be type B? What is the probability that neither will be type B? ■

Finally, we need a probability formula for "or" statements. Two events that cannot occur at the same time are called **mutually exclusive**, and for such events the "or" formula is simply

$$P(A \text{ or } B) = P(A) + P(B).$$

For example, on one selection from a regular deck of cards

$$\begin{aligned} P(\text{jack or queen}) &= P(\text{jack}) + P(\text{queen}) \\ &= \frac{4}{52} \quad + \frac{4}{52} \\ &= \frac{8}{52}, \text{or } \frac{2}{13}. \end{aligned}$$

When A and B have outcomes in common, we must subtract this overlapping from the sum, so that we won't count these outcomes twice. Thus, the general "or" formula is

$$P(A \text{ or } B) = P(A) + P(B) - P(A \text{ and } B).$$

For example,

$$\begin{aligned} P(\text{jack or spade}) &= P(\text{jack}) + P(\text{spade}) - \overbrace{P(\text{jack or spade})}^{\text{jack of spades}} \\ &= \frac{4}{52} \quad + \frac{13}{52} \quad - \frac{1}{52} \\ &= \frac{16}{52}, \text{or } \frac{4}{13} \end{aligned}$$

To summarize, the "or" formulas are as follows:

Formulas for *P*(*A* or *B*)

> For any events A and B,
>
> $$P(A \text{ or } B) = P(A) + P(B) - P(A \text{ and } B).$$
>
> When A and B are mutually exclusive, this formula simplifies to
>
> $$P(A \text{ or } B) = P(A) + P(B).$$

Example 7: Using the Formula for *P*(*A* or *B*)

A computer is programmed to generate randomly a number from 1 to 100.

a. What is the probability that the computer will generate a number that is less than 20 or greater than 50?

b. What is the probability that the computer will generate a number that is greater than 50 or a multiple of 10?

Solution:

a. No number is both less than 20 and simultaneously greater than 50, so these two events are mutually exclusive. Therefore, the probability that the computer will generate a number that is less than 20 (event A) or greater than 50 (event B) is

$$P(A \text{ or } B) = P(A) + P(B)$$
$$= \frac{19}{100} + \frac{50}{100} = \frac{69}{100}.$$

b. Five numbers are both greater than 50 and simultaneously multiples of 10, namely 60, 70, 80, 90, and 100. So the probability that the computer will generate a number that is greater than 50 (event A) or a multiple of 10 (event B) is

$$P(A \text{ or } B) = P(A) + P(B) - P(A \text{ and } B)$$
$$= \frac{50}{100} + \frac{10}{100} - \frac{5}{100}$$
$$= \frac{55}{100} = \frac{11}{21}.$$

PROGRESS CHECK 7

A card is drawn from a regular deck of 52 cards. Find the probability that the card drawn is

a. a picture card or a 10 **b.** a picture card or a heart ■

In this section we have calculated probabilities by constructing a list of all possible outcomes, by using the combination rule, and by using probability formulas. For a given problem, you will need to decide whether using more than one of these methods is helpful. For instance, to solve the problem in the next example, students often find it easier to answer the "and" question by formula, and easier to answer the "or" question by considering the list of equally likely outcomes.

Example 8: Contrasting Methods to Find a Probability

We flip a coin and roll a die. Find the following probabilities by listing the equally likely outcomes, and also by using the "and" and "or" formulas.

a. tails on the coin *and* 5 on the die **b.** tails on the coin *or* 5 on the die

Solution:

a. There are 2 equally likely outcomes for the coin and six for the die. So by the fundamental counting principle, we must list 2×6, or 12, outcomes. By using a tree diagram, we obtain the outcomes

$$\begin{matrix} H1 & H2 & H3 & H4 & H5 & H6 \\ T1 & T2 & T3 & T4 & T5 & T6 \end{matrix}.$$

The outcome $T5$ represents tails on the coin and 5 on the die, so the chances for this event are $1/12$. We obtain this answer by formula as follows:

$$P(T \text{ and } 5) = P(T) \cdot P(5) = \frac{1}{2} \cdot \frac{1}{6} = \frac{1}{12}.$$

b. The outcomes $T1$, $T2$, $T3$, $T4$,$T5$, $T6$, and $H5$ all contain either a T or a 5, so the probability of tails on the coin or 5 on the die is $7/12$. This answer is obtained by formula as follows:

$$P(T \text{ or } 5) = P(T) + P(5) - P(T \text{ and } 5)$$
$$= \frac{1}{2} + \frac{1}{6} - \frac{1}{12}$$
$$= \frac{7}{12}.$$

PROGRESS CHECK 8

A boy and a girl simultaneously call out a number from 1 to 5 at random. Find the probability for each of the following by listing the equally likely outcomes, and also by using the "and," and "or," and "not" formulas.

a. Both call an even number. **b.** Either one of them calls 3, while the other does not. ■

EXPLORE 12.7

One of the most famous problems in elementary probability is the so-called "birthday" problem. It asks "In a gathering of n people, what are the chances that at least two people have the same birthday (month and day)?" Many students find the answer surprising. Exploration 1 leads to a solution of the birthday problem. For the simplest case, assume that a year has 365 days, and that no one has a birthday on Feb. 29.

1. **a.** If two people are picked at random, the probability that they have different birthdays is $365/365 \times 364/365$. Explain why this computation is correct. What is the probability that the two people have the same birthday?
   b. If three people are picked at random, the probability that they have three different birthdays is $365/365 \times 364/365 \times 363/365$. Explain why this computation is correct. What is the probability that the three people do not have three different birthdays? What is the probability that at least two of the three people have the same birthday?
   c. If four people are picked at random, what is the probability that at least two of them have the same birthday?
   d. Create a table for $n = 1$ to 40 which shows the probabilities if n people are picked at random that at least two of them have the same birthday. What is the smallest value of n for which the probability of a shared birthday is more than 50 percent? Repeat the question for more than 90 percent.

One way to solve probability questions, when you don't have the mathematical tools to compute the answer exactly, is to act out the situation and see what happens. This procedure is called simulation, and it is an important method in the analysis of complicated probability questions. Often the simulation depends on using random numbers. Most calculators have a feature (often called "rand") that generates random numbers. These numbers are decimal values, r, where $0 \le r < 1$. For instance, by entering rand, you may get 0.943597. The built-in algorithm generates the numbers so that all possible decimals values are equally likely. Exploration 2–5 investigate the behavior and application of the Rand function.

2. Use your calculator to generate 10 random values between 0 and 1, and store them in a list. If you wish, use a sort command to put them in increasing order.
3. The Integer function (Int on most calculators) truncates a number and produces its integer part. Apply the Integer function to the 10 values from part **a**. Why are all the answers the same?
4. If you start with r between 0 and 1, multiply it by 2, add 1, and then apply the Integer function, what are the only possible values you can get for an answer? (**Hint:** Try $r = 0.0123$, and $r = 0.9521$.) Explain how you can use these integer values to simulate the tossing of a fair coin. Use the calculator to simulate the tossing of a fair coin 60 times, record the outcomes, and compute the percentage of times the outcome was heads. You may prefer to write a program to do these 60 tosses automatically.
5. If you start with r between 0 and 1, multiply it by 6, add 1, and then apply the Integer function, what are the only possible values you can get for an answer? Explain how you can use these integer values to stimulate the rolling of a fair die. Use the calculator to simulate the rolling of a fair die 60 times, and record the outcomes. What percentage of the time should a fair die come up 5? What percentage of the time did your simulated die come up 5?

EXERCISES 12.7

1. A committee consists of seven men and five women. If the chairperson is selected at random, find the probability that the chairperson will be a woman.

2. If a fair die is rolled once, find the probability of obtaining each of the following.
   a. a 3 **b.** an outcome less than 3
   c. a 7 **d.** an outcome less than 7

3. A card is drawn from a regular deck of 52 cards. Find the probability that the card drawn will be
   a. the queen of spades
   b. a heart **c.** a picture card
   d. not a red queen **e.** an ace or a king
   f. neither a jack nor a queen
   g. a spade or a 7 **h.** neither a spade nor a 7

4. A student is selected at random from a group in which 30 percent are freshmen, 25 percent are sophomores, 35 percent are juniors, and 10 percent are seniors. Find the probability of selecting each of the following.
   a. a senior **b.** not a senior
   c. a freshman or a sophomore
   d. neither a freshman nor a junior

5. If a woman's chance of winning a price is $\frac{2}{19}$, then what is her chance of not winning?

6. If the probability of playing a game three times and winning all three times is $\frac{27}{64}$, what is the probability of playing the game three times and not winning at least once?

7. Two dice are rolled. What is the probability that the sum of the outcomes on the dice will be 5?

8. On two rolls of a pair of dice, what is the probability that at least once the sum of the outcomes on the dice will be 5?

9. A boy and a girl are playing a game in which they simultaneously call out a number from 1 through 3. Find the probability for each of the following.
   a. Both call an odd number.
   b. Both call the same number.
   c. One of them calls 2, while the other does not.

10. What is the probability that on four flips of a coin the result will be four tails?

11. Four pills and three capsules are in a box. One of the pills causes drowsiness, while two of the capsules remedy this side effect. What is the probability that a person who swallows a pill and a capsule will be drowsy?

12. From the letters *a*, *b*, *c*, and *d*, three letters are selected at random without replacement. What is the probability that a person will draw the word *bad* in the correct order of spelling?

13. From 10 tickets, 3 different winners are to be chosen. If Chris, Amy, and Pat each holds a ticket, what is the probability they are the 3 winners?

14. If five nominees with different last names are introduced randomly, what is the probability that they will be introduced in alphabetical order?

15. What is the probability that a family with three children contains two boys and one girl? (Assume boy and girl are equally likely.)

16. From a committee of six men and four women, a three-person subcommittee is randomly chosen. What is the probability that the committee will contain two women and one man?

17. If your pocket contains seven nickels and five quarters, and you randomly select three coins, what is the probability of obtaining exactly 35¢ to pay for a newsletter?

18. On a multiple-choice test with four choices, what is the probability that you will guess randomly at the answers on three questions and get two out of three right?

19. Of two independent events, A and B, the probability that A will occur is $\frac{1}{4}$, while the probability that B will occur is $\frac{1}{3}$. What is the probability that neither event will occur?

20. If 40 percent of the population has type A blood, find the following probabilities for two blood donors chosen at random.
    a. Both are type A.
    b. Neither is type A.
    c. 1 out of 2 is type A.

THINK ABOUT IT 12.7

1. Which of the following numbers *cannot* be a probability? Explain why.

$$0,\ 1,\ -1,\ \frac{4}{3},\ 30\%,\ 300\%,\ 1.01,\ \frac{31}{365}$$

2. True or false: If events A and B are mutually exclusive, then A and B are independent events. Explain your answer.
3. Create a word problem that may be solved by computing $3(1/6)(5/6)^2$.
4. The *odds against an event* E are given by the ratio $P(\text{not } E){:}P(E)$, or the fraction $P(\text{not } E)/P(E)$. Answer each question using this definition.
   a. When two dice are rolled, what are the odds against the sum of the outcomes on the dice being 7?
   b. If the odds against winning in a certain game are $b{:}a$, what is the probability of winning?
5. What is the probability of being dealt a poker hand that is any type of flush (that is, five cards all of the same suit)?

● ● ●

CHAPTER 12 SUMMARY

OBJECTIVES CHECKLIST

Specific chapter objectives are summarized below along with numbered example problems from the text that should clarify the objectives. If you do not understand any objectives, or do not know how to do the selected problems, then restudy the material.

12.1: Can you:

1. **Find any term in a sequence, when given a formula for the *n*th term of the sequence?** Write the first four terms of the sequence given by $a_n = 3n - 1$; also find a_{25}.

   [Example 1]

2. **Graph a sequence with the aid of a grapher?** Graph the sequence $a_n = 5n - 2$ with the aid of a grapher.

   [Example 2]

3. **Find a formula for the general term a_n in a given arithmetic sequence?** Find a formula for the general term a_n in the arithmetic sequence 3, 7, 11,...

   [Example 3]

4. **Find a formula for the general term a_n in a given geometric sequence?** Find a formula for the general term a_n in the geometric sequence 24, 12, 6,....

   [Example 4]

5. **Determine if a sequence is an arithmetic sequence, a geometric sequence, or neither?** State whether the sequence is an arithmetic sequence, a geometric sequence, or neither. For any arithmetic sequence, state the common difference and write the next two terms. For any geometric sequence, state the common ratio and write the next two terms.

   2, 6, 18, 54,...

   [Example 5a]

6. **Solve applied problems involving arithmetic sequences, geometric sequences, or the Fibonacci sequence?** An amount of \$2,000 is invested at 6 percent annual interest, payable on the anniversary of the deposit. The formula

$$a_n = 2{,}000(1.06)^n$$

gives the value of the deposit after n complete years. Find a_n for each of the years 1, 2, 3, 4, 5. What kind of sequence is this?

[Example 6]

12.2: Can you:

1. **Find the sum of an indicated number of terms in an arithmetic sequence?** Find the sum of the first 21 terms of the arithmetic sequence 4, 7, 10,….

   [Example 1]

2. **Find the sum of an indicated number of terms in a geometric sequence?** Find the sum of the first seven terms of the geometric sequence 1, $1/2$, $1/4$,….

   [Example 2]

3. **Write a series given in sigma notation in its expanded form, and determine the sum?** Write the series $\sum_{j=2}^{7} 3^j$ in expanded form and determine the sum.

   [Example 5]

4. **Write a series given in expanded form using sigma notation?** Find an expression for the general term and write the series $2+5+8+11+14+17$ in sigma notation.

   [Example 6]

5. **Solve applied problems involving series?** A free-falling body that starts from rest drops about 16 ft the first second, 48 ft the second second, 80 feet the third second, and so on. About how many feet does a parachutist drop during the first 10 seconds of free fall?

   [Example 7]

12.3: Can you:

1. **Find the sum of certain infinite geometric series?** Find the sum of the infinite geometric series $1+1/3+1/9+1/27+\cdots$.

   [Example 1]

2. **Use an infinite geometric series to express a repeating decimal as the ratio of two integers?** Express the repeating decimal $7.\overline{54}$ as the ratio of two integers.

   [Example 3]

3. **Solve applied problems involving infinite geometric series?** A system for removing pollutants from kerosene involves passing the kerosene repeatedly through a filtering system. Suppose that each time the kerosene passes through the filters some pollutant is removed and stored in a special tank. In a certain application each pass through the filters removes 10 percent as much pollutant as the previous pass. Suppose the first pass produces 18 gal of pollutant. What size holding tank is large enough to hold all the pollutant this application will produce?

   [Example 4]

4. **Use graphical methods to find the sum of an infinite geometric series?** Use graphical methods to solve the previous problem.

   [Example 5]

12.4: Can you:

1. **Use mathematical induction to prove that a statement is true for all positive integers?** Use mathematical induction to prove that

$$1+3+5+\ldots+(2n-1)=n^2$$

   is true for all positive integers.

   [Example 1]

2. **Use mathematical induction to prove that a statement is true for all integers greater than or equal to a particular integer?** Prove that $2^n > n+1$ for every positive integer $n \geq 2$.
[Example 4]

12.5: Can you:

1. **Expand $(a+b)^n$ using Pascal's triangle?** Expand $(x+y)^8$ using Pascal's triangle.
[Example 1]

2. **Expand $(a+b)^n$ using the binomial theorem?** Expand $(x+y)^5$ using the binomial theorem.
[Example 4]

3. **Evaluate binomial coefficients in the form $\binom{n}{r}$ for given values of n and r?** Evaluate the following binomial coefficients and compare the results to the entries in Pascal's triangle.

$$\binom{4}{0}, \binom{4}{1}, \binom{4}{2}, \binom{4}{3}, \binom{4}{4}.$$

[Example 6]

4. **Expand $(a+b)^n$ using the binomial theorem and binomial coefficients?** Expand $(x+y)^8$ using the binomial theorem and binomial coefficients.
[Example 7]

5. **Find the rth term in the expansion of $(a+b)^n$?** Find the 14th term in the expansion of $(3x-y)^{15}$.
[Example 8]

12.6: Can you:

1. **Determine the number of ways two (or more) events taken together can occur?** In how many ways can a three-item true-false test be answered? Include a tree diagram listing the possibilities.
[Example 1]

2. **Determine the number of distinct permutations of n objects in the case when all the objects are different?** Find the number of different batting orders that are possible for a baseball team that starts nine players. (Assume the team has only nine players.)
[Example 2]

3. **Determine the number of distinct permutations of n objects taken r at a time?** In how many ways can nine members from a hockey team be assigned to six different starting positions?
[Example 4]

4. **Determine the number of distinct permutations of n objects when certain members in the set of objects are indistinguishable?** How many distinct permutations can be made from the letters of the word BEGINNING?
[Example 5]

5. **Determine the number of combinations of n objects taken r at a time?** From a class of 16 students an instructor asks 3 students to write their homework on the board. How many student selections are possible?
[Example 6]

6. **Solve applied problems by choosing the appropriate counting techniques and applying the associated formulas?** How many different 5-card poker hands can be dealt from a deck of 52 cards?

[Example 7]

12.7: Can you:

1. **Find the probability of a single event?** Find the probability of selecting a spade from a regular deck of 52 cards.

[Example 1]

2. **Find the probability that an event will not occur, using the formula for *P* (not *E*)?** If you guess randomly at the answers on a two-item true-false test, what is the probability that you will not get both answers right?

[Example 5]

3. **Find the probability of event *A* and *B* occurring if *A* and *B* are independent?** Three pills and five capsules are in a box. Two of the pills cause drowsiness, while four of the capsules remedy this side effect. What is the probability that a person who swallows a pill and a capsule will be drowsy? What are the chances that the person won't be drowsy?

[Example 6]

4. **Find the probability of event *A* or event *B* occurring if *A* and *B* are mutually exclusive?** A computer is programmed to generate randomly a number from 1 to 100. What is the probability that the computer will generate a number that is less than 20 or greater than 50?

[Example 7a]

5. **Find the probability of event *A* or event *B* occurring if *A* and *B* are not mutually exclusive?** A computer is programmed to generate randomly a number from 1 to 100. What is the probability that the computer will generate a number that is greater than 50 or a multiple of 10?

[Example 7b]

KEY CONCEPTS AND PROCEDURES

Section **Key Concepts and Procedures to Review**

12.1

- Definitions of sequence, terms (of the sequence), arithmetic sequence, common difference, geometric sequence, common ratio, and Fibonacci sequence
- Formulas for the *n*th term:
  arithmetic sequence: $a_n = a_1 + (n-1)d$
  geometric sequence: $a_n = a_1 r^{n-1}$
  Fibonacci sequence: $a_1 = 1$, $a_2 = 1$, $a_n = a_{n-1} + a_{n-2}$ for $n \geq 3$

12.2

- Definitions of series, arithmetic series, and geometric series
- Formulas for the sum of the first *n* terms:
  arithmetic series: $S_n = \frac{n}{2}(a_1 + a_n)$ or $S_n = \frac{n}{2}[2a_1 + (n-1)d]$
  geometric series: $S_n = \frac{a_1 - a_1 r^n}{1-r}$ or $S_n = \frac{a_1 - a_n r}{1-r}$
- The Greek letter Σ (read "sigma") is used to simplify the notation involved with series. We define sigma notation as follows:

$$\sum_{i=1}^{n} a_i = a_1 + a_2 + \ldots a_n$$

12.3
- An infinite geometric series with $|r| < 1$ converges to the value or sum S given by $S = \dfrac{a_1}{1-r}$

12.4
- Principle of mathematical induction
- We prove by mathematical induction that a statement is true for all positive integers by doing the following:
  1. Show by direct substitution that the statement (or formula) is true for $n = 1$
  2. Show that if the statement is true for any positive integer k, then it is also true for the next highest integer $k+1$
- Two methods for establishing the "$k+1$" statement from the assumption of the "k" statement
- Prove that a statement is true for all integers greater than or equal to integer q

12.5
- Binomial theorem: For any positive integer n,

$$(a+b)^n = a^n + \frac{n}{1!}a^{n-1}b + \frac{n(n-1)}{2!}a^{n-2}b^2 + \cdots + \frac{n(n-1)(n-2)\ldots(n-r+1)}{r!}a^{n-r}b^r + \cdots + b^n,$$

or an alternate form is

$$(a+b)^n = \sum_{r=0}^{n}\binom{n}{r}a^{n-r}b^r$$

- The binomial coefficient $\binom{n}{r}$ is defined by

$$\binom{n}{r} = \frac{n!}{r!(n-r)!}$$

- The rth term of the binomial expansion of $(a+b)^n$ is

$$\binom{n}{r-1}a^{n-(r-1)}b^{r-1}$$

- The symbol $n!$ (read "n factorial") means the product $n \cdot (n-1) \cdots 3 \cdot 2 \cdot 1$
- Pascal's triangle

12.6
- Definitions of permutation and combination
- Method to construct a tree diagram
- Fundamental counting principle
- Permutation formulas:
  n objects, n at a time: ${}_nP_n = n!$
  n objects, r at a time: ${}_nP_r = \underbrace{n(n-1)\cdots(n-r+1)}_{r \text{ factors}} = \dfrac{n!}{(n-r)!}$
- Formula for distinguishable permutations (if certain objects are all alike)

- Combination formulas:
  n objects, r at a time: ${}_nC_r = \frac{{}_nP_r}{r!} = \frac{n!}{(n-r)!r!}$
- The order of the objects is important in a permutation but does not matter in a combination

12.7
- Definitions of probability of an event, sample space, independent events, and mutually exclusive events
- The probability of an event is a number between 0 and 1, inclusive, with impossible events assigned a probability of 0 and events that must take place assigned a probability of 1
- Probability formulas
  1. $P(\text{not } E) = 1 - P(E)$
  2. $P(A \text{ and } B) = P(A) \cdot P(B)$ (if A and B are independent events)
  3. a. $P(A \text{ or } B) = P(A) + P(B) - P(A \text{ and } B)$ (for any events A and B)
     b. $P(A \text{ or } B) = P(A) + P(B)$ (if A and B are mutually exclusive)

CHAPTER 12 REVIEW EXERCISES

1. Write the first four terms of the sequence given by $a_n = (-1)^n 2^{n-1}$; also find a_6.

2. Determine a_2, a_4, and a_n so that the following sequence is
   a. an arithmetic sequence and
   b. a geometric sequence.

   $$5,\ a_2,\ 45,\ a_4, \ldots, a_n, \ldots$$

3. Find the sum of the first 70 positive integers.

4. Write the series $\sum_{i=2}^{7} 2^{i-1}$ in expanded from and determine the sum.

5. Write the series $1 + \left(\frac{1}{2}\right)^2 + \left(\frac{1}{3}\right)^2 + \left(\frac{1}{4}\right)^2 + \left(\frac{1}{5}\right)^2$ in sigma notation.

6. Find the sum of the infinite geometric series $1 - \frac{1}{4} + \frac{1}{16} - \frac{1}{64} + \cdots$.

7. By mathematical induction, prove that the sum of the first n terms of an arithmetic series is given by

   $$a + (a+d) + (a+2d) + \cdots + [a + (n-1)d] = \frac{n}{2}[2a + (n-1)d]$$

8. Express the repeating decimal $3.\overline{4}$ as the ratio of two integers. (Use infinite geometric series.)

9. a. If a, b, and c are the first three terms in an arithmetic sequence, express c in terms of a and b.
   b. If a, b, and c are the first three terms in a geometric sequence, express c in terms of a and b.

10. \$2,000 is split up into eight prizes by a lottery system in which each award is \$10 less than the preceding award. How much money is awarded as first prize?

11. Evaluate $\sum_{k=1}^{\infty} \frac{1}{3^k}$.

12. Find the 20th term in the arithmetic sequence 3, 7, 11,....

13. Evaluate $\sum_{k=1}^{5} k^2$.

14. What is the sum of the first n odd positive integers?

15. In how many different orders may seven people stand in a line? How about a circle? In how many different orders can four men and three women stand in a line so that each woman is between two men?

16. How many line segments are needed to connect in all possible ways the nine points on the circle shown?

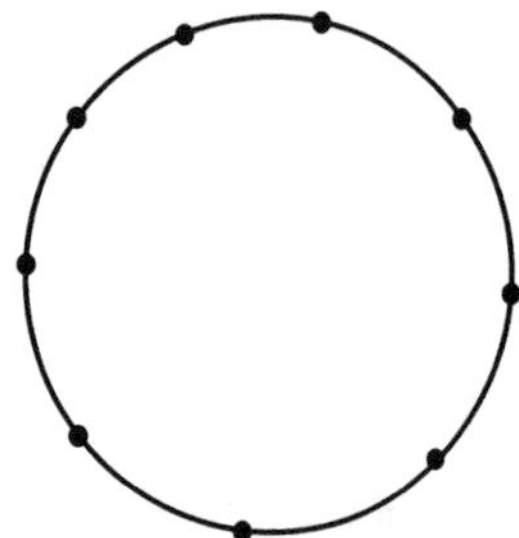

17. The first term in an arithmetic sequence is a_1, the nth term is a_n, the common difference is d, and the number of terms is n. Express d in terms of a_1, a_n, and n.

18. Find the sum of the infinite geometric series $3^{-1}+3^{-2}+3^{-3}+\cdots$.

19. Is $\sum_{i=0}^{n-1}(2i+1)$ equal to $\sum_{k=1}^{n}(2k-1)$?

20. Simplify ${}_nC_{n-1}$.

21. If ${}_nC_{12}={}_nC_3$, find n.

22. Expand $(x+h)^6$ by the binomial theorem.

23. A jack, a queen, and a king are face down on a table. If two of the cards are turned over at random, what is the probability that one of them will be a king?

24. If the probability of winning a game is a/b, what is the probability of not winning the game?

25. How many permutations are there of 7 things taken 3 at a time?

26. How many combinations are there of 7 things taken 3 at a time?

27. If ${}_nP_r=720$ and ${}_nC_r=120$, find n and r.

28. Write the 10th term in the expansion of $(2x-y)^{11}$.

29. Write the 4th term in the expansion of $\left(\sqrt{x}+1\right)^7$.

30. Prove by mathematical induction that the following formula is true for all positive integer values of n.

$$1+5+9+\cdots+(4n-3)=n(2n-1).$$

31. If 10 percent of the population has type B blood, then what is the probability that two blood donors chosen at random will be type B?

32. If one card is selected from a regular deck of 52 cards, then what is the probability this card will be a club or a picture card?

33. Write the first four terms in the expansion of $(1-x)^{15}$.

34. A slot machine contains three independent wheels. On each wheel a star, a lemon, a grape, or a cherry may appear. How many outcomes are possible on the slot machine?

In Exercises 35–44 select the choice that complete the statement or answers the question.

35. The sequence $\sqrt{2}$, $\sqrt{3}$, $\sqrt{4}$, $\sqrt{5}$ is
   a. an arithmetic sequence
   b. a geometric sequence
   c. neither an arithmetic nor a geometric sequence

36. If the first three terms in a geometric sequence are b^2, b^x and b^8, then x equals
   a. 4 **b.** $4\sqrt{2}$ **c.** $2\sqrt{2}$ **d.** 5

37. The next term in the geometric sequence $\sqrt[4]{2}$, $\sqrt{2}$, $\sqrt[4]{8}$,... is
   a. 2 **b.** $2\sqrt{2}$ **c.** 4 **d.** 8

38. Which statement is true?
   a. ${}_{50}C_{10}<{}_{50}C_{40}$ **b.** ${}_{50}C_{10}={}_{50}C_{40}$
   c. ${}_{50}C_{10}>{}_{50}C_{40}$

39. Which number *cannot* be a probability?
   a. 1 **b.** 0 **c.** 2 **d.** $\frac{1}{2}$

40. How many telegraphic characters can be made using three dots and two dashes in each character?
   a. 10 **b.** 8 **c.** 32 **d.** 25

41. The coefficient of the x^9y^2 term in the binomial expansion of $(x+y)^{11}$ is
   a. 55 **b.** 9 **c.** 11 **d.** 165

42. The expression $\sum_{n=1}^{5}c$ (where c is a constant) simplifies to
   a. c **b.** $5c$ **c.** cn **d.** $c+5$

43. The sum of the infinite geometric series

$$5+\frac{5}{3}+\frac{5}{3^2}+\cdots \text{ is}$$

   a. $\frac{25}{3}$ **b.** $\frac{45}{8}$ **c.** $\frac{15}{2}$ **d.** $\frac{125}{9}$

44. The fifth term in the binomial expansion of $(a-b)^7$ is
   a. $35a^3b^4$ **b.** $-35a^3b^4$
   c. $-21a^2b^5$ **d.** $21a^2b^5$

CHAPTER 12 TEST

1. State whether the sequence is an arithmetic sequence, a geometric sequence, or neither: 1, 0.1, 0.01, 0.001,....

2. What is the 70th term of the arithmetic sequence 5, 9, 13,...?

3. Find the formula for the general term a_n of the geometric sequence 27, −18, 12,....

4. Find the sum of the first 30 terms of the arithmetic sequence 2, 7, 12,.....

5. Evaluate $\sum_{n=1}^{6} 100(1.08)^{n-1}$.

6. A certain ball always rebounds $\frac{1}{2}$ as far as it falls. If the ball is dropped from a height of 8 ft, how far up and down has it traveled when it hits the ground for the eighth time?

7. Find the sum of the series $7-\frac{7}{4}+\frac{7}{16}-\frac{7}{64}+\cdots$.

8. Express the repeating decimal $2.\overline{16}$ as the ratio of two integers. (Use infinite geometric series.)

9. Expand $(x-3y)^4$ by the binomial theorem.

10. Write the first four terms in the expansion of $(y+2)^{11}$.

11. Find the sixth term in the expansion of $(x^2-1)^7$.

12. How many seven-digit telephone numbers are possible if the first digit cannot be 0?

13. In how many ways can four taxi drivers be assigned to six taxicabs?

14. On a test, you are asked to answer five problems out of eight. How many choices are possible?

15. Two dice are rolled. What is the probability that the sum of the outcomes on the dice will be 11?

16. What is the probability that a family with three children contains at least one girl?

17. A card is drawn at random from a regular deck of 52 cards. Find the probability that the card drawn will either be a black card or a face card.

18. Prove by mathematical induction: For every positive integer n,

$$10+20+30+\cdots+10n=5n(n+1).$$

Appendix

A.1 Approximate Numbers

Measurements of one kind or another are essential to both scientific and nontechnical work. Weight, distance, time, volume, and temperature are only a few of the quantities for which measurements are required. We measure through the medium of numbers and arbitrary units, such as the foot (British system) or the meter (metric system). These numbers are approximations and can only be as accurate as the measuring instruments allow. For example, if we measure the height of a man to be 68 in., we are saying that his height to the nearest inch is 68 in. This means that his exact height (h) is somewhere in the interval $67.5 \text{ in.} \leq h < 68.5 \text{ in.}$ Using a difference measuring device, we might record the height to be 68.2 in., so that $68.15 \text{ in.} \leq h < 68.5 \text{ in.}$.

The exact height is contained in these intervals because the measurements are estimated by rounding off to some decimal place. We round off by considering the digit in the next place to the right of the desired decimal place. If this digit is less than 5, the digit in the desired decimal place remains the same; if the digit is 5 or greater, the digit in the desired decimal place is increased by one. The digits to the right of the desired decimal place are then dropped and replaced by zeros, if the zeros are needed to maintain the position of the decimal point. Rounding off is best understood by considering the following table, which indicates how we round off a measurement of 265.307 ft to various places.

Precision Desired	Measurement of 265.307 Ft Is Given as
Nearest 100 ft	300 ft
Nearest 10 ft	270 ft
Nearest foot	265 ft
Nearest tenth of a foot	265.3 ft
Nearest hundredth of a foot	265.31 ft

The **precision** of a number refers to the place at which we are rounding off. A number rounded off to the nearest hundredth is more precise than a number rounded off to the nearest tenth, and so on.

Another important consideration is the number of **significant digits** in our approximation. All digits, except the zeros that are required to indicate the position of the decimal point, are significant. The following table illustrates this concept.

Approximate Number	Number of Significant Digits
15.13	Four
0.44	Two
0.003	One; the number is *three* thousandths, and only the 3 is significant. The zeros are needed to indicate the position of the decimal point.
0.0196	Three
307	Three; the zero is *not* used to indicate the position of the decimal point. Zeroes between nonzero digits are always significant.
50.007	Five
8.0	Two; the zero is significant because it shows that the number is rounded off to the nearest tenth. The zero is not required to write the number 8 and should not be written unless it is significant.
0.2600	Four
230	Two; we do not have sufficient information to determine whether the number is rounded off in the tens place or to the nearest unit. Unless it is stated otherwise, we assume that the number to be rounded off in the tens place, so the zero is needed to indicate the position of the decimal point.
2,000	One
$2{,}00\bar{0}$	Four; the bar is used to avoid ambiguity by indicating that the number is rounded off at the digit below the bar.
$2{,}\bar{0}00$	Two

The **accuracy** of a number is determined by the number of significant digits in the number. Thus, 21 (two significant digits) is a more accurate number than 0.07 (one significant digit). It is important to distinguish between the accuracy and the precision of a number. For example, consider the following pairs of numbers.

$$21 \text{ and } 0.07 \begin{cases} 21 \text{ is more accurate, since it contains two significant digits} \\ 0.07 \text{ is more precise, since it is written to the nearest hundredth} \end{cases}$$

$$123 \text{ and } 12.3 \begin{cases} \text{Same accuracy, since both contain three significant digits} \\ 12.3 \text{ is more precise, since it is written to the nearest tenth} \end{cases}$$

$$171 \text{ and } 59 \begin{cases} 171 \text{ is more accurate, since it contains three significant digits} \\ \text{Same precision, since both are written to the nearest integer.} \end{cases}$$

$$517 \text{ and } 1.703 \begin{cases} 1.703 \text{ is more accurate, since it contains four significant digits} \\ 1.703 \text{ is more precise, since it is written to the nearest thousandth} \end{cases}$$

Operations with Approximate Numbers

When computing with approximate numbers, we should express our results to a precision or accuracy that is appropriate to the data. We establish guidelines for the result by considering the error involved in computing with such numbers. For example, suppose that $x \approx 123$ and $y \approx 4.27$. Adding these numbers, we have $123 + 4.27 = 127.27$. However, because of rounding off, we know that the exact values of x and y are in the intervals

$$122.5 \le x < 123.5$$
$$4.265 \le y < 4.275.$$

Adding the minimum values to find the minimum possible sum, and adding the maximum values to find the maximum possible sum, we have

$$122.5 + 4.265 \le x + y < 123.5 + 4.275$$
$$126.765 \le x + y < 127.775$$

To the nearest integer, the minimum sum is 127, an the maximum sum is 128. Thus, the precision in the result may not be given to more than the nearest integer, and there is even a chance of error if we are that precise with our answer. This example motivates us to the following guideline.

Rounding Off in Addition or Subtraction

When **adding** or **subtracting** approximate numbers, perform the operation and then round off the result to the **precision** of the least precise number.

Thus, if $x \approx 123$ and $y \approx 4.27$, then $x + y \approx 123 + 4.27 = 127.27 \approx 127$. We round off to the nearest integer because 123, the less precise number, is rounded off to the nearest integer.

Before adding or subtracting, it is permissible to round off the more precise numbers, since the extra digits are not considered when we round off our result. If this is done, the numbers should be rounded off to one place beyond that of the least precise number.

Example 1

Add the approximate numbers: 12.79, 4.3131, 46.2, 9.618.

Solution: 46.2 is the least precise number, so before adding we may round off the other numbers to the nearest hundredth.

$$\begin{array}{r} 12.79 \\ 4.31 \\ 46.20 \\ \underline{9.62} \\ 72.92 \end{array}$$

Rounding off to the nearest tenth, the result is 72.9.

Example 2

Perform the indicated operation for the approximate numbers: $0.396 - 17.6 + 150$.

Solution: 150 is the least precise number. We assume that 150 is rounded off in the tens place, so before performing the operations, we round off the other numbers to the nearest integer. Thus, we have $0 - 18 + 150 = 132$. Rounding off in the tens place, the result is 130.

To determine the possible error when multiplying or dividing approximate numbers, consider the product of $x \approx 12$ and $y \approx 4.27$.

$$x \cdot y \approx 12(4.27) = 51.24$$

Because of rounding off, we know that the exact values of x and y are in the intervals

$$11.5 \le x < 12.5$$
$$4.265 \le y < 4.275$$

Multiplying the minimum values to find the minimum possible product, and the maximum values to find the maximum possible product, we have

$$(11.5)(4.265) \le xy < (12.5)(4.275)$$
$$49.0475 \le xy < 53.4375$$

To two significant digits, the minimum product is 49 and the maximum product is 53. Thus, the accuracy of the result may not be given to more than two significant digits, and even that accuracy may be wrong. This example motivates us to the following guideline.

Rounding Off in Multiplication or Division

> When **multiplying** or **dividing** approximate numbers, perform the operation and then round off the result to the **accuracy** of the least accurate number.

Thus, if $x \approx 12$ and $y \approx 4.27$, then $x \cdot y \approx 12(4.27) = 51.24 \approx 51$. We round off to two significant digits because 12, the less accurate number, has two significant digits.

Before multiplying or dividing, it is permissible to round off the more accurate numbers to one or more significant digit than the least accurate number. If you are using a calculator, you may find it easier to key in the number without rounding off. Either way, you will usually obtain the same final result.

Since finding a power or root of a number involves multiplication, we are concerned with accuracy in such problems. The result should contain the same number of significant digits as the given approximate number.

Example 3

Multiply the approximate numbers: $(11.5)(0.042649)$.

Solution: 11.5 is the less accurate number, so before multiplying we may round off the other number to four significant digits:

$$(11.5)(0.04265) = 0.490475.$$

Rounding off to three significant digits, the result is 0.490.

Example 4

Perform the indicated operations for the approximate numbers.

$$\frac{(215.1)(7.63)}{0.0002}.$$

Solution: 0.0002 is the least accurate number. We round off the other numbers to two significant digits and then perform the operation.

$$\frac{220(7.6)}{0.0002} = 8,360,000$$

Rounding off to one significant digit the result is 8,000,000.

Example 5 Find the approximate value of $\sqrt{79.1}$.

Solution: Using a calculator, the initial result is $\sqrt{79.1} \approx 8.893818$. Since 79.1 contains three significant digits, the final result is 8.89.

Exact Numbers

Although most of the numbers in scientific work are approximate, there are some numbers that are exact. Exact numbers are often the result of a counting process or a definition, and are commonly seen in formulas. The following are examples of situations in which the numbers are exact.

1. There are 25 students in the class (counting process).
2. Of 100 light bulbs tested, 4 were defective (counting process).
3. There are 60 minutes in 1 degree (definition).
4. A bank pays 6 percent interest (definition).
5. $P = 4s$ (formula for the perimeter of a square).
6. $A = \frac{1}{2}ab$ (formula for the area of a triangle).

Since these number are exact, no error is involved when using them in computations. Only approximate numbers must be considered in determining the accuracy or precision of a result.

EXERCISES A.1

In Exercises 1–20 determine the number of significant digits in the approximate number.

1. 123.14 **2.** 0.59 **3.** 0.02

4. 0.000491 **5.** 904 **6.** 30.03

7. 22.0 **8.** 0.78000

9. 670 **10.** 80,000 **11.** 5.40

12. 540 **13.** 0.054 **14.** 54.0

15. 504 **16.** 540.0 **17.** $5{,}40\overline{0}$

18. $5{,}4\overline{0}0$ **19.** 0.010 **20.** 0.100

In Exercises 21–32 round off the numbers
a. to four significant digits and
b. to two significant digits.

21. 12.3456 **22.** 46.054 **23.** 59,372

24. 72,488 **25.** 29,697 **26.** 59,999

27. 84.096 **28.** 10.998 **29.** 0.068547

30. 0.0034581 **31.** 0.0020006

32. 0.39999

In Exercises 33–44 determine the approximate number in each pair that is
a. more accurate and **b.** more precise.

33. 423, 0.004 **34.** 18, 0.700

35. 402, 40.2 **36.** 312, 98

37. 6.430, 0.304 **38.** 0.9, 12.0

39. 2,000 0.0002 **40.** 1,230, 3,200

41. 9,080, 48.2 **42.** $4{,}\overline{0}00$, $4\overline{0}$

43. $3\overline{0}0$, $30\overline{0}$ **44.** $8{,}88\overline{0}$, 8.008

In Exercises 45–64 perform the indicated operations for the approximate numbers.

45. $16.27 + 2.1515 - 4.3$ **46.** $15 - 3.043 + 1031$

47. $0.002 - 11.0 + 3.59$ **48.** $0.200 + 4.6 + 120$

49. $0.3047 + 0.8 + 0.092 + 0.69$

50. $214.32 + 1{,}000 + 0.75 + 3.2$

51. $(2.1)(0.57892)$ **52.** $(0.08)(489.0)$

53. $(16.4)(2{,}001)$

54. $(7.00)(0.094265)$

55. $408 \div 0.002$

56. $602 \div 0.200$

57. $\dfrac{(799.4)(8.6)}{0.04}$

58. $\dfrac{(0.460)(126.1)}{2{,}200}$

59. $9.714 + (2.3)(0.812)$

60. $(4.50)(0.1234) - 5.8$

61. $\dfrac{4.07}{0.06} + (0.8715)(20)$

62. $5(0.0861) - \dfrac{0.90}{123}$

63. $\sqrt{95.0}$

64. $\sqrt{0.040}$

● ● ●

A.2 Graphs on Logarithmic Paper

Sometimes we need to analyze data that range over a large collection of values. In addition, the analysis often requires us to find a formula that fits a set of data. Here we study a scheme that is useful in such situations.

On logarithmic graph paper the axes are laid off at distances proportional to the logarithms of numbers. Consider Figure A.1, in which we construct an axis by marking on the paper the following numbers:

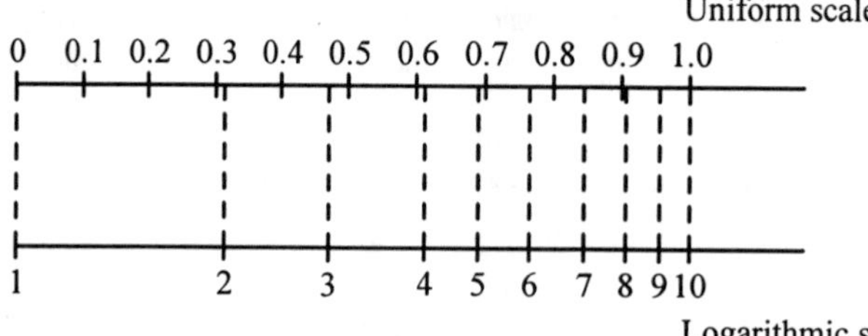

Figure A.1

$\log_{10} 1 = 0$	$\log_{10} 6 = 0.778$
$\log_{10} 2 = 0.301$	$\log_{10} 7 = 0.845$
$\log_{10} 3 = 0.477$	$\log_{10} 8 = 0.903$
$\log_{10} 4 = 0.602$	$\log_{10} 9 = 0.954$
$\log_{10} 5 = 0.699$	$\log_{10} 10 = 1$ etc.

On the logarithmic scale we use 1 as shorthand for $\log_{10} 1$. Since $\log_{10} x$ is undefined for $x \le 0$, only positive numbers may be marked off on logarithmic paper. Using this paper, we may obtain accuracy in a graph in which the variables range over a large collection of values.

Logarithmic paper is especially useful when graphing functions of the form $y = ax^m$, which are called **power functions**. By taking the logarithm to the base 10 of a power function, we have

$$\begin{aligned} \log y &= \log ax^m \\ &= \log a + m \log x \end{aligned}$$

If we let $\log y = Y$, $\log x = X$, and $\log a = b$, the equation becomes

$$Y = b + mX.$$

Thus, on logarithmic paper the graph of a power function is a straight line with slope m and Y-intercept, b. It also follows that if the graph of experimental data is a straight line on logarithmic paper, the variables are related by an equation of the form $y = ax^m$. The slope m is measured with a ruler, and the coefficient a is the intercept on the vertical axis at $x = 1$.

Example 1 Use logarithmic paper to graph $y = 3x^2$.

Solution: First, construct a table of corresponding values of x and y.

x	1	2	3	4	5
y	3	12	27	48	75

As shown in Figure A.2, the graph is a straight line. Using a ruler you may verify that $m = \Delta Y/\Delta X = 2$. The intercept at $x = 1$ is 3. These readings agree with the constants in the equation $y = 3x^2$. In this example the y-axis is scaled in two cycles from $\log 1 = 0$ to $\log 10 = 1$ to $\log 100 = 2$. Logarithmic paper may contain any number of cycles. A cycle ranges between any two numbers whose logarithms are consecutive integers.

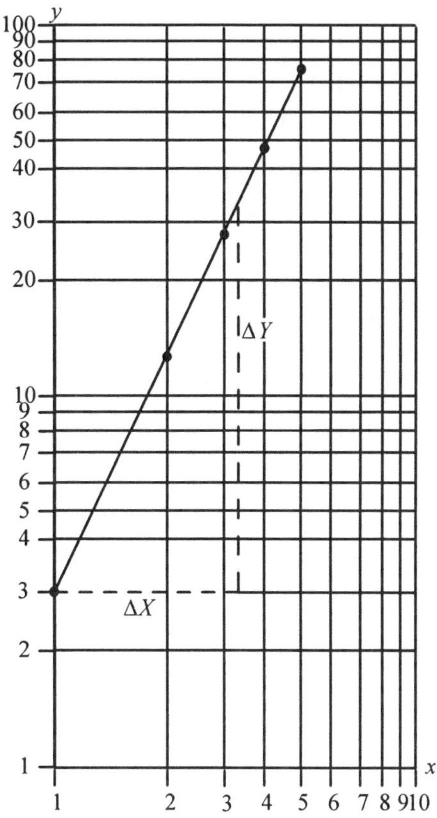

Figure A.2

Example 2 Determine the equation relating the variables in the following table

x	1.0	1.5	2.0	2.5	3.0	4.0	5.0	6.0	7.0	9.0
y	16	21	24	27	29	33	38	41	44	50

Solution: Consider Figure A.3. The vertical axis requires one cycle that ranges from $\log 10 = 1$ to $\log 100 = 2$. Since the graph is approximately a straight line, the equation relating the variables is a power function. Using a ruler, $m = \Delta Y / \Delta X \approx 0.5$. The intercept at $x = 1$ is about 17. Thus, the power function is $y \approx 17x^{0.5}$ or $y \approx 17\sqrt{x}$.

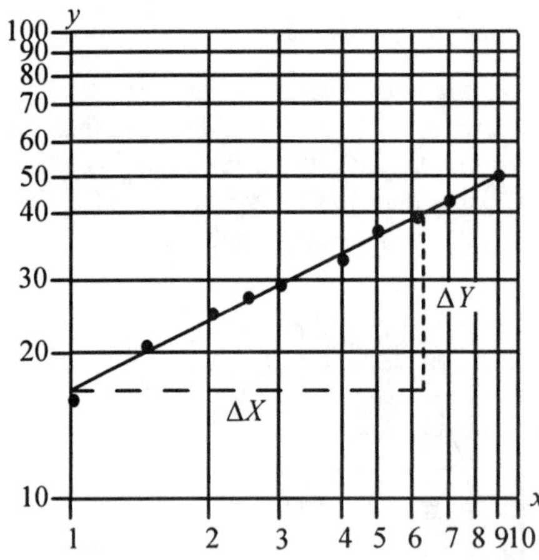

Figure A.3

On semilogarithmic graph paper one axis is scaled as on logarithmic paper, and the other axis is scaled as on Cartesian coordinate paper. This paper is used when only one variable ranges over a large collection of values. Semilogarithmic paper is especially useful when graphing exponential functions, since the graph is a straight line.

Example 3 Use semilogarithmic paper to graph $y = 2^x$.

Solution: First, construct a table of corresponding values of *x* and *y*.

x	-2	-1	0	1	2	3
y	0.25	0.5	1	2	4	8

The vertical axis requires two logarithmic cycles. The horizontal axis is equally spaced and is scaled with positive and negative numbers. As shown in Figure A.4, the graph is a straight line.

Although the graphs in the examples are all straight lines, other curves may result. We obtain a straight line only when graphing a power function on logarithmic paper, and an exponential function on semilogarithmic paper.

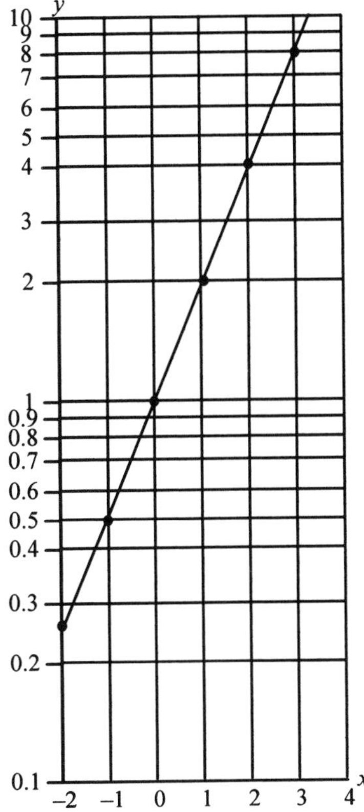

Figure A.4

EXERCISES A.2

In Exercises 1–6 graph the functions on logarithmic paper. Also, find the slope and the intercept at $x = 1$.

1. $y = x^2$ **2.** $y = 2x^3$ **3.** $y = 4\sqrt{x}$

4. $y = 5x^{2/3}$ **5.** $xy = 1$ **6.** $x^2 y = 10$

In Exercises 7–10 use logarithmic paper to determine the equation relating the variables in the given tables.

7.

x	1	2	3	4	5	6	7
y	7.0	5.5	4.8	4.3	4.0	3.7	3.5

8.

x	1.0	2.0	3.0	4.0	5.0	6.0	7.0
y	5.0	1.4	.66	.39	.26	.19	.14

9.

x	1	3	5	7	9	25	30	70
y	3.0	6.0	8.6	11	13	25	28	49

10.

x	1.5	4.5	7.5	11	16	20	30
y	2.1	8.4	16	25	40	53	88

In Exercises 11–14 graph the functions on semilogarithmic paper.

11. $y = 2^{-x}$ **12.** $y = 5^x$

13. $y = 0.5(3)^x$ **14.** $y = 2(3^{-x})$

Answers to Progress Check Exercises

Progress Check R.1

1. $\frac{2}{3}$ **2. a.** None of these **b.** Irrational, real **c.** Rational, real

3. a. Multiplication identity property **b.** Commutative property of multiplication
c. Associative property of multiplication **d.** Distributive property

4. a. $<$ **b.** $=$ **c.** $>$

5. a. -44 **b.** $\frac{5}{9}$ **c.** -8

6. a. -4 **b.** $-\frac{76}{9}$ **c.** 3.6

7. a. -56 **b.** 14.1 **c.** 81

8. a. -125 **b.** -64 **c.** 64

9. a. 6 **b.** $-\frac{21}{11}$ **10.** -72

11. a. $-\frac{4}{7}$ **b.** 5 **12.** \$235,488.15, \$557,222.03, \$1,287,203.63, \$2,906,651.72

Progress Check R.2

1. a. -17 **b.** 247 **2.** $-\frac{3}{2}$

3. a. $-4x^3, x^2, -3x, 7$ **b.** $-\frac{7x}{2}$

4. a. $-7x$ **b.** Cannot be simplified **c.** $3p$ **d.** $5x^2 + 5x$

5. a. $2x - 4y$ **b.** $5x + 4$ **6.** 6π ft

7. a. $1{,}351 \text{ ft}^3$ **b.** 10,133 gal **8. a.** $p + 5$ **b.** $0.08(ab)$

9. a. The sum of *a* times *b* and *c* **b.** The quotient of *a* and 5 more than *a*

Progress Check R.3

1. **a.** $3^5 = 243$ **b.** $3^8 = 6{,}561$ **c.** $4^3 \cdot 2^3 = 512$ **d.** $3^2 = 9$

e. $\frac{1}{3^2} = \frac{1}{9}$ **2.** **a.** $144x^7$ **b.** $\frac{2y}{3x^3}$

3. **a.** -5 **b.** -1 **c.** 4.72

4. **a.** $\frac{1}{7}$ **b.** $\frac{1}{81}$ **c.** $\frac{125}{8}$ **5.** \$365.36

6. **a.** $\frac{1}{256}$ **b.** $\frac{1}{16x^6}$ **c.** $\frac{125}{8x^3}$ **d.** $\frac{y^4}{6}$

e. $\frac{27}{x^8y^2}$ **7.** **a.** 16 **b.** $\frac{27}{x^8y^2}$

8. **a.** a^{2x} **b.** 3^{2n} **c.** x^n

9. **a.** 6.15×10^8 **b.** -0.9×10^{-1} **10.** **a.** 9,200,000,000 **b.** -0.0000027

11. 26.1%

Progress Check R.4

1. **a.** $-6x^4 + 2x^3 - 18x^2$ **b.** $21y^2 - 3xy$ **2.** $-3nx + \frac{1}{2}x$

3. **a.** $21y^2 - 34y + 8$ **b.** $a^{2x} - 2a^x - 15$ **4.** $4x^3 + 15x^2 + 7x - 6$ **5.** $20y^2 - 31y + 22$

6. Revenue = (number of units sold) $\cdot$ (price per unit), where number of units sold $= 20{,}000 + 700x$, and price per unit $= 3{,}000 - 100x$; $60{,}000{,}000 + 100{,}000x - 70{,}000x^2$

7. **a.** $k^2 - 16k + 64$ **b.** $x^2 + 2xh + h^2$ **c.** $9y^2 - 25$

8. **a.** $4x^2 + 28x + 49$ **b.** $4x^2 - 49$ **c.** $9m^2 - 30mn + 25n^2$ **9.** $6x + 3h$

Progress Check R.5

1. $\{-11\}$ **2.** $\left\{-\frac{9}{2}\right\}$ **3.** $\left\{\frac{3}{4}\right\}$ **4.** $\varnothing$

5. $r = 0.08$ **6.** 2.7 kg ≈ 5.9 lb **7.** $y = mx + b$ **8.** $g = \frac{2t - 2S}{t^2}$

9. **a.** $x = \frac{2y + 30}{5}$ **b.** $y = \frac{5x - 30}{2}$

Progress Check R.6

1. \$1,056 **2.** 73 and 74 **3.** 63°

4. \$150,000 at 10%; \$50,000 at 6% **5.** $\frac{3}{20}$ hr or 9 minutes

6. 3.75 lb of 30%; 1.25 lb of 50% **7.** 183 minutes

Progress Check R.7

1. **a.** $(-\infty,4]$

b. $(-3/2,\infty)$

2. **a.** $(-\infty,2]$

3. $(-\infty,-65^\circ \text{ Celsius})$

4. $(0,\infty)$

5. $\varnothing$

6. $(2, 14]$

7. $[61, 101)$

8. $(-\infty,-2]\cup[20/3,\infty)$

Progress Check 1.1

1. $C=9n+2.95$; $D{:}\{1,2,3,\ldots\}$; $R{:}\{\$11.95, \$20.95, \$29.95,\cdots\}$

2. $(-2,-7)$

3. **a.** No; $D{:}\{-2,2\}$; $R{:}\{1,0\}$ **b.** Yes; $D{:}\{-2,2,1\}$; $R{:}\{1,-2\}$

4.

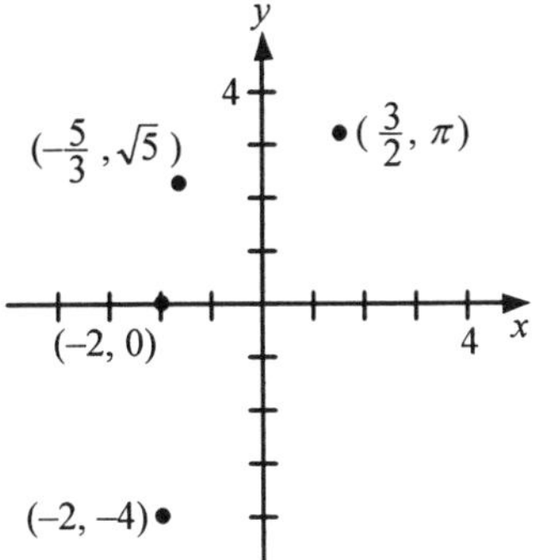

5.

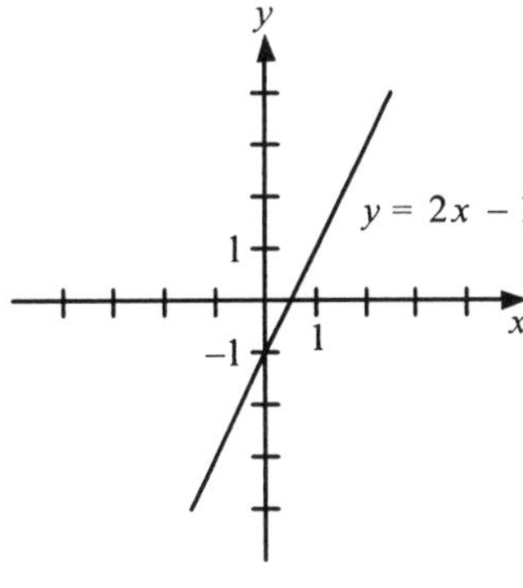

6.

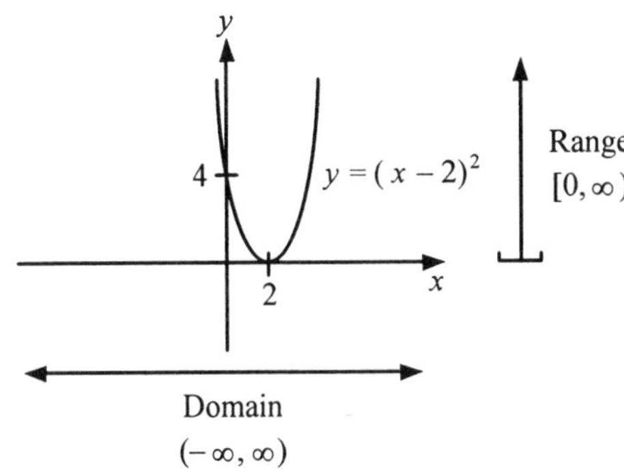

7.

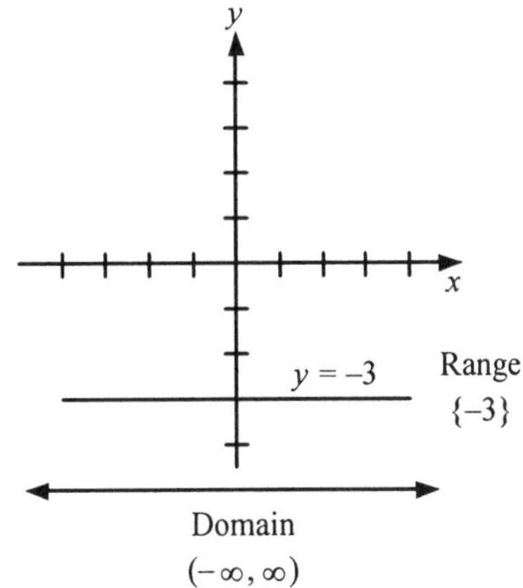

8.

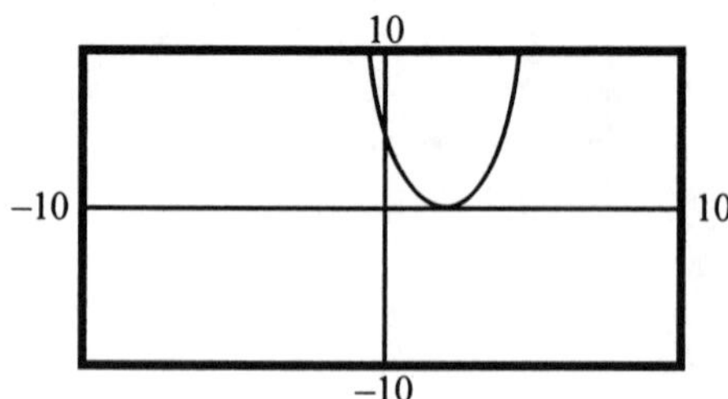

9. Figure a shows a complete graph

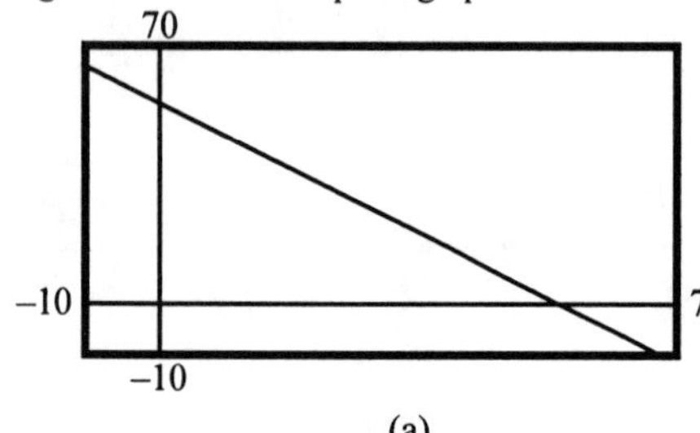

(a)

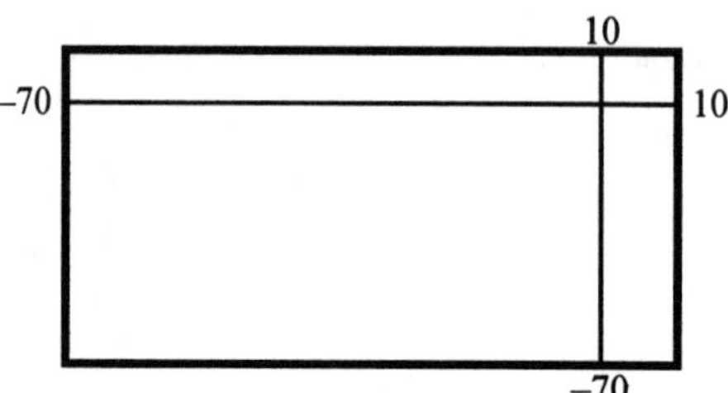

(b)

10.

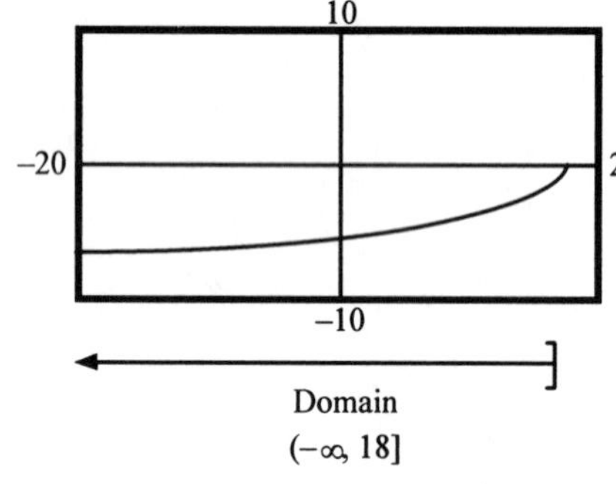

Range
$(-\infty, 0]$

Domain
$(-\infty, 18]$

11.

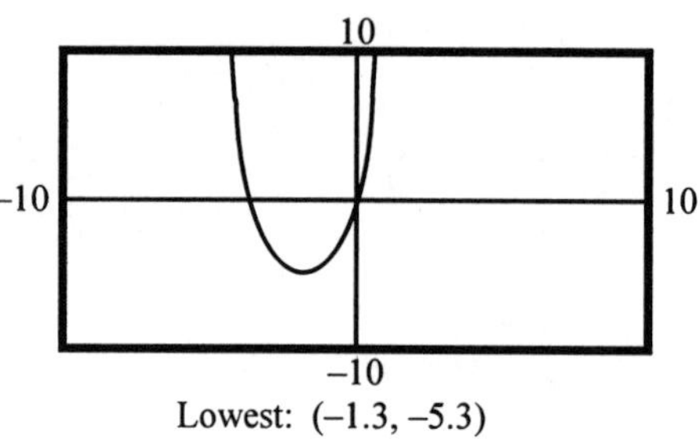

Lowest: (−1.3, −5.3)

12. $A = 2x\sqrt{4-x^2}$; $D{:}(0,2)$; $R{:}(0,4]$

Progress Check 1.2

1. $f(4) = 4$; $f(20) = -300$; $f(-5) = -50$

2. $f(6) = 75{,}036.52$; in six years the value of this art work will be about \$75,000.

3. -14

4. $f(a+b) = 3(a+b) = 3a + 3b$
   $f(a) = 3a$
   $f(b) = 3b$
   Since $3a + 3b = 3a + 3b$, $f(a+b) = f(a) + f(b)$.

5. $2x + h + 3$

6. $e = \begin{cases} \$600 & \text{if } \$0 \le a \le \$10{,}000 \\ \$600 + 0.08(a - \$10{,}000) & \text{if } a > \$10{,}000 \end{cases}$; D: $\{a: a \ge \$0\}$ R: $\{e: e \ge \$600\}$

7. **a.** 2 **b.** Undefined **c.** 0

8.

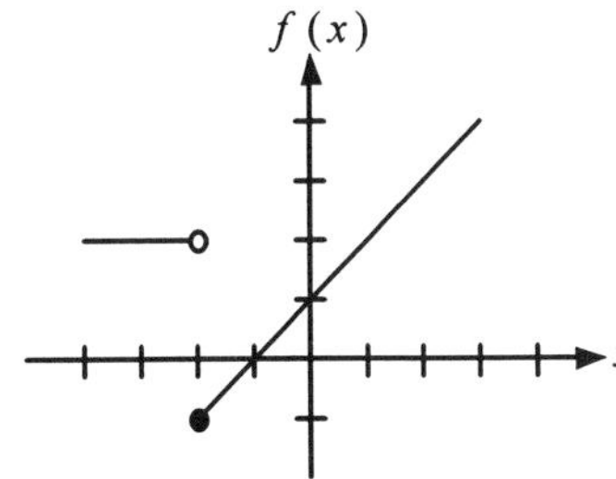

9. Domain: Set of all real numbers except -2

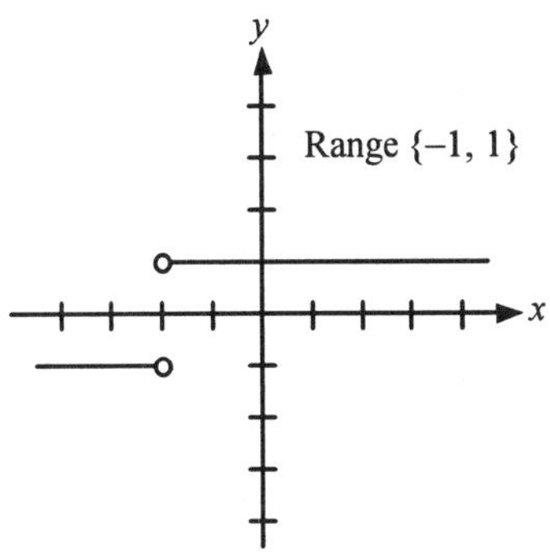

10. **a.** $(-\infty,\infty)$ **b.** $(-\infty,4]$ **c.** 3 **d.** $\{-1,3\}$

e. $(-\infty,-1)\cup(3,\infty)$ **f.** $(-1,3)$ **g.** $(-1,0)$, $(3,0)$, $(0,3)$

h. Increasing for $(-\infty,1)$; decreasing for $(1,\infty)$

Progress Check 1.3

1.

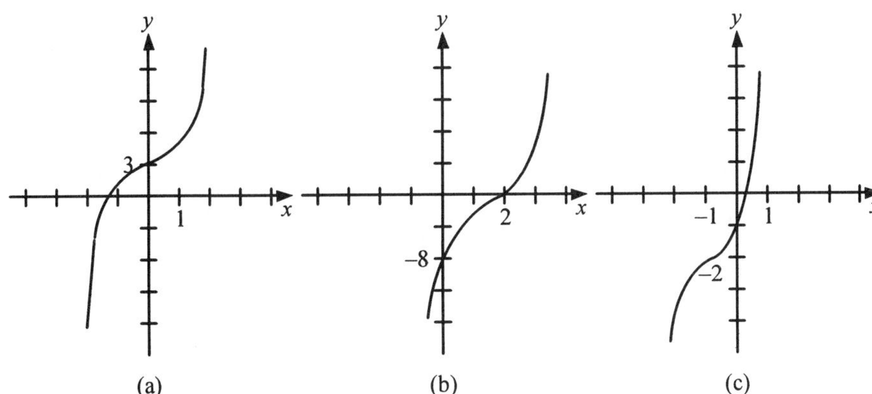

2. **a.**

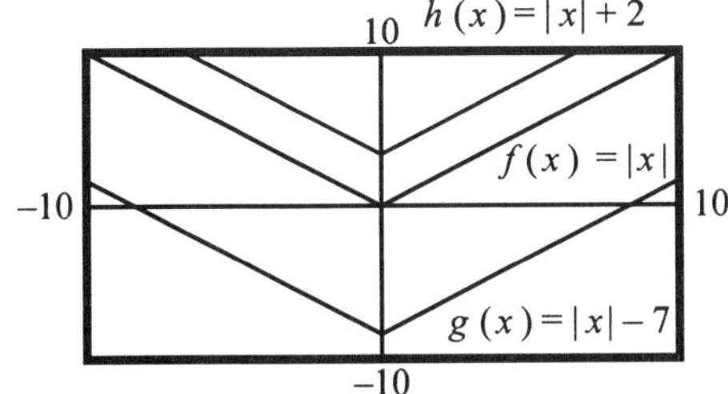

b. The graph of g is the graph of f shifted 7 units down; the graph of h is the graph of f shifted 2 units up.

c. Possible trace points are (0,0) in f, $(0,-7)$ in g, and (0,2) in h. The descriptions appear to be accurate.

3.

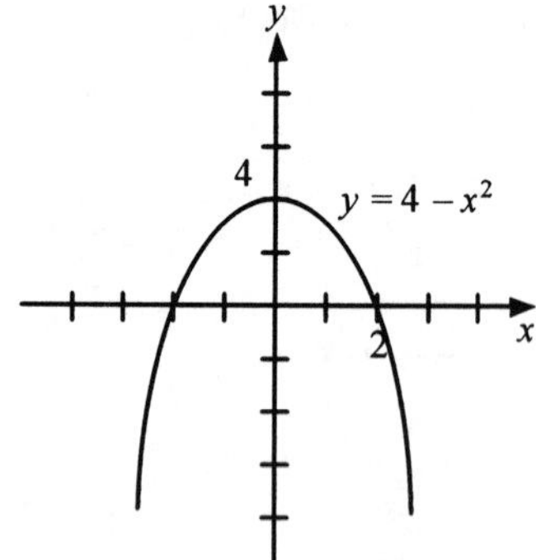

4.

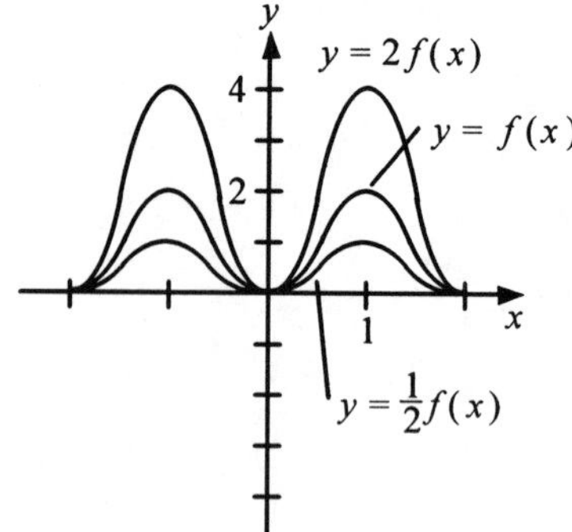

5.

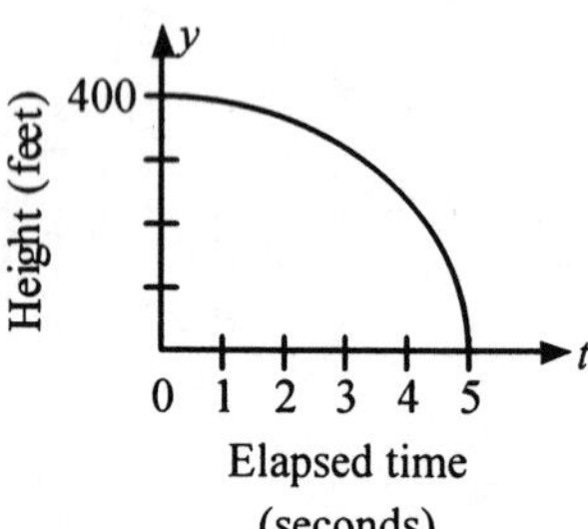

6.

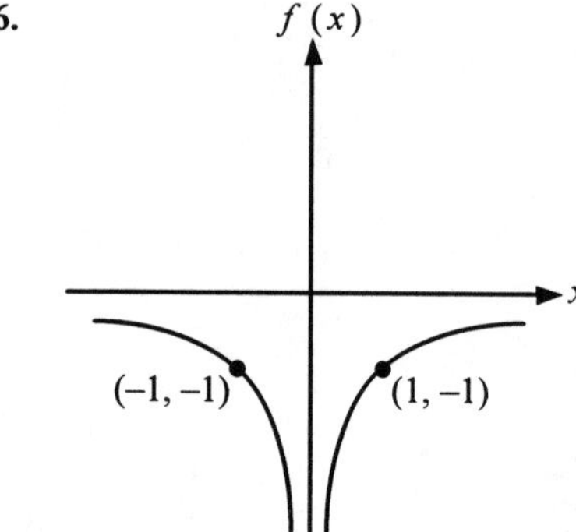

7. $f(x) = x^2$ and $f(-x) = (-x)^2 = x^2 = f(x)$

8.

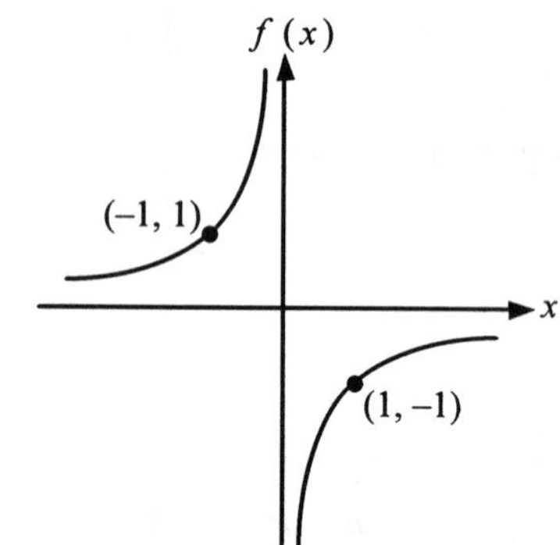

9. $f(x) = x^3$ and $f(-x) = (-x)^3 = -x^3 = -f(x)$

10. f is odd

a.

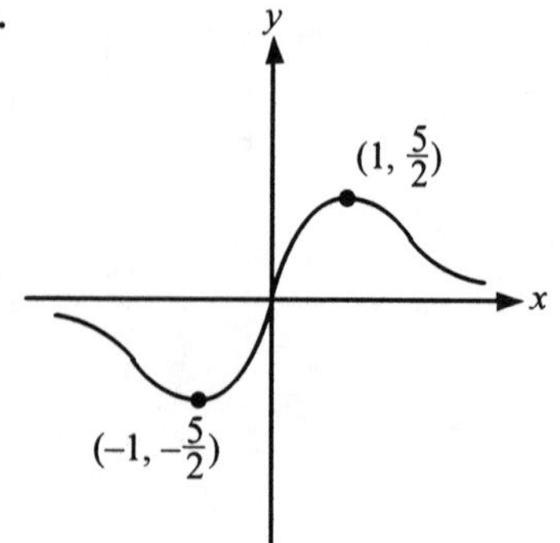

b.

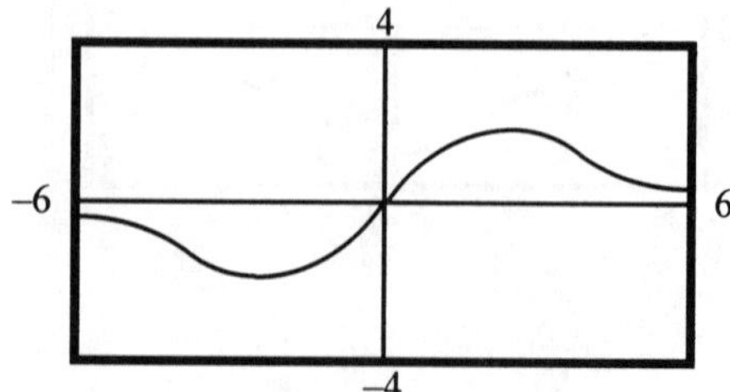

Progress Check 1.4

1. a. 4 b. 3

2. $\sqrt{45}$

3. The length of the two equal sides is $\sqrt{32}$. The length of the other side is 8. The Pythagorean theorem holds, so it is a right triangle.

4.

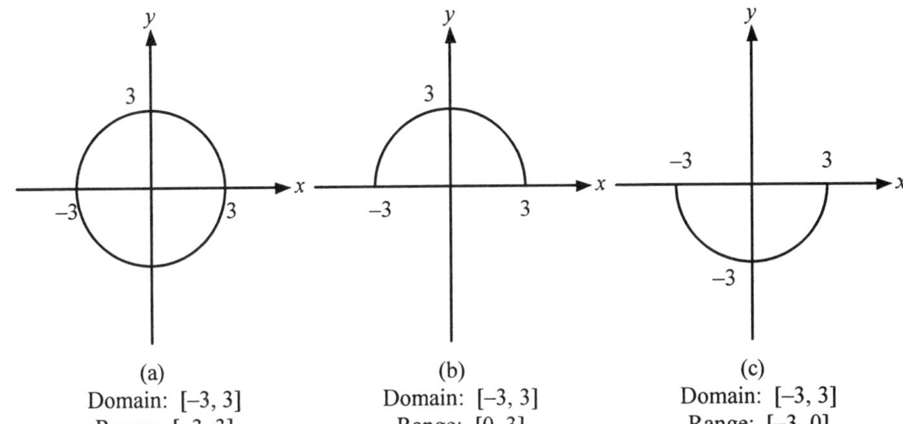

5. 14.3 ft

6. **b** and **c** are functions; **a** and **d** are not functions.

7. a. No b. Yes

Progress Check 1.5

1. a. $y = 0.2x$; $y = 150 + 0.05x$

b.

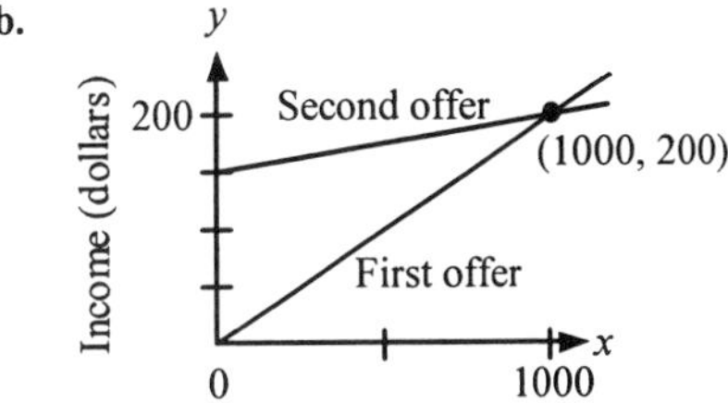

c. For anticipated sales above \$1,000, choose 1st offer; for anticipated sales below \$1,000, choose 2nd offer; for anticipated sales of \$1,000, choose either offer.

2.

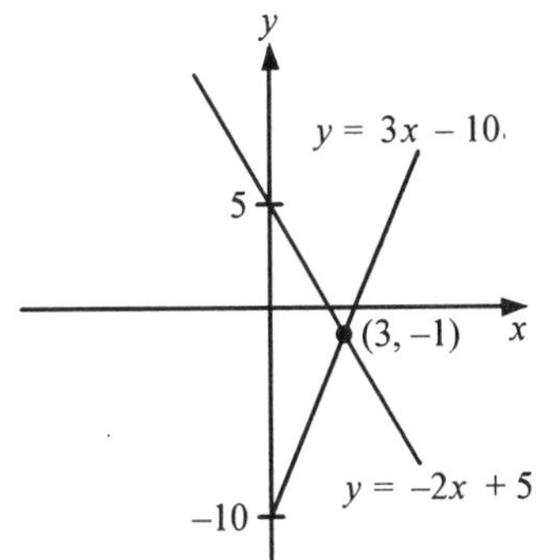

3. (10, 11)

4. 30

5. (9, 5)

6. (7,–10)

7. a. Dependent system: $\left\{(x, y): y = -\frac{1}{2}x + 5\right\}$

b. Inconsistent system; Ø

8. \$12,500 at 8.5%; \$37,500 at 10.5% **9.** **a.** $\{-0.4\}$ **b.** $(-\infty,-0.4)$

Progress Check 1.6

1. **a.** $C = kd$; $k = \pi$ **b.** $T = kv$; k depends on the tax rate of the particular tax district.

2. $y = \left(\frac{3}{4}\right)x$; 9 **3.** **a.** $W_m = \left(\frac{1}{6}\right)W_e$ **b.**

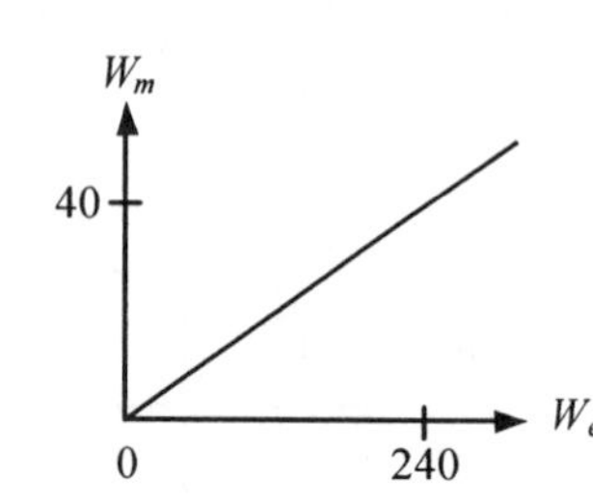

c. 23 lb

4. $y = \frac{65}{x}$; 65 **5.** 180 rpm **6.** $\frac{15,625}{16}$ **7.** 0.24 years (about 88 days)

8. The gravitational attraction drops to $\frac{1}{16}$ of its original value.

Progress Check 2.1

1. $\sin\beta = \frac{4}{5}$, $\cos\beta = \frac{3}{5}$, $\tan\beta = \frac{4}{3}$, $\cot\beta = \frac{3}{4}$, $\sec\beta = \frac{5}{3}$, $\csc\beta = \frac{5}{4}$

2. $\sin\theta = \frac{\sqrt{21}}{5}$, $\tan\theta = \frac{\sqrt{21}}{2}$, $\cot\theta = \frac{2}{\sqrt{21}}$, $\sec\theta = \frac{5}{2}$, $\csc\theta = \frac{5}{\sqrt{21}}$ **3.** 65

4. $\sin 60° = \frac{\sqrt{3}}{2}$, $\cot 30° = \sqrt{3}$, $\csc 30° = 2$, $\tan 45° = 1$ **5.** $\sin 56°$

6. **a.** 1.8807 **b.** 1.0440 **c.** 0.9899 **7.** 344

8. **a.** 74.2°, 74°10′ **b.** 43.7°, 43°40′

Progress Check 2.2

1. $A = 32°$, $b = 56$ ft, $c = 66$ ft **2.** $B = 80°20'$, $a = 1.26$ ft, $b = 7.41$ ft

3. $a = 14.7$ ft, $A = 50°50'$, $B = 39°10'$ **4.** 23.8 ft **5.** 1.190 ft

6. 78.5°

Progress Check 2.3

1. $\sin\theta = -\frac{1}{\sqrt{10}}$, $\csc\theta = -\sqrt{10}$, $\cos\theta = -\frac{3}{\sqrt{10}}$, $\sec\theta = -\frac{\sqrt{10}}{3}$, $\tan\theta = \frac{1}{3}$, $\cot\theta = 3$

2. $\csc\theta = -\frac{4}{3}$, $\cos\theta = -\frac{\sqrt{7}}{4}$, $\sec\theta = -\frac{4}{\sqrt{7}}$, $\tan\theta = \frac{3}{\sqrt{7}}$, $\cot\theta = \frac{\sqrt{7}}{3}$

3. $\sin 180° = 0$, $\csc 180°$ is undefined, $\cos 180° = -1$, $\sec 180° = -1$, $\tan 180° = 0$, $\cot 180°$ is undefined

4. **a.** 1 **b.** 0 **5.** $-\frac{\sqrt{3}}{2}$ **6.** $\frac{1}{\sqrt{2}}$

7. $-\frac{1}{\sqrt{3}}$ **8** **a.** 0.9932 **b.** −1.0785 **9.** 15 mi

Progress Check 2.4

1. $\{30°, 330°\}$ 2. $\{120°, 240°\}$ 3. $\{209°30', 330°30'\}$ 4. $\{121.0°, 301.0°\}$

5. $\{\theta : \theta = 109°30' + k360° \text{ or } \theta = 250°30' + k360°,\ k \text{ any integer}\}$

6. 35.7° or 54.3°

Progress Check 2.5

1. a. 51 lb b. 62° 2. a. 565 mi b. 82.4° south of east

3. 15 lb 4. a. 457 mi/hour due east b. 48 mi/hour due south

5. v, $2\overline{0}$ lb; h, 37 lb 6. $R = 6.6$ lb, $\theta = 96°$

Progress Check 3.1

1. a. $4y^2(2y+1)$ b. $(x+3)(2x+5)$ 2. $A = \pi r(r+s)$

3. a. $3s(2s-3)$ b. $3\sin\theta(2\sin\theta - 3)$ 4. a. $(x+2)(2x+3)$ b. $(x-6)(x^2-6)$

5. $(x-5)(x+4)$ 6. a. $(2t+3)(2t-5)$ b. $(2\tan\theta + 3)(2\tan\theta - 5)$

7. a. $(2\cos\theta - 1)^2$ b. $(2c-1)^2$ 8. $(4x+9)(3x-2)$ 9. $3y^2(2y+1)(y-3)$

10. 225; factorable

Progress Check 3.2

1. a. $(n+4)(n-4)$ b. $(10xy+7z)(10xy-7z)$

2. a. $(1+2\sin\theta)(1-2\sin\theta)$ b. $(1+2s)(1-2s)$

3. a. $y = 625 - 4t^2$ b. $y = (25+2t)(25-2t)$ c. When $t = 12.5$, $y = [25+2(12.5)][25-2(12.5)] = 50(0) = 0$

4. a. $(2y-1)(4y^2+2y+1)$ b. $(x^3+3y^2)(x^6-3x^3y^2+9y^4)$

5. a. $5(x-2)(x^2+2x+4)$ b. $(4m^2+9n^4)(2m+3n^2)(2m-3n^2)$ 6. $16(2+t)(2-t)$

7. $6(\cos\theta - 2)(\cos\theta - 6)$ 8. $(y+1)(y^2-y+1)(y-1)(y^2+y+1)$ 9. $(c+2+d)(c+2-d)$

Progress Check 3.3

1. a. $\frac{1}{2x-1}, x \neq \frac{1}{2}$ b. $x+c, x \neq c$ 2. $-\frac{1}{3x+2}$ 3. $\frac{x-1}{x(x+3)}$

4. $\frac{3(n+5)}{n-5}$ 5. 1 6. $3x(x+5)$ 7. $\frac{b^2+4b+2}{(b+1)(b+2)(b+3)}$

8. $\frac{2y^2+3y-15}{y(y-5)}$ 9. a. $\frac{x}{x-1}$ b. $\frac{-3x^2-3xh-h^2}{x^3(x+h)^3}$ 10. 37.5 mi/hour

11. a. $\frac{2\tan\theta+6}{(\tan\theta+2)(\tan\theta+4)}$ b. $\frac{\sin\theta}{\sin\theta+1}$

Progress Check 3.4

1. a. 10 b. Not real c. –5 d. 5 e. 3
2. a. 4 b. Not real c. –4 d. $\frac{1}{4}$ e. 2
3. a. 27 b. 9 c. $-\frac{1}{8}$ d. 32
4. 16.2 percent
5. a. $7^{1/2}$; $\sqrt{7}$ b. $\frac{1}{2x^2}$ c. $5^{1/4}$; $\sqrt[4]{5}$ d. $\frac{y^{5/2}}{x^{9/5}}$; $\frac{\sqrt{y^5}}{\sqrt[5]{x^9}}$ e. $\frac{y^{1/3}}{x^{7/6}}$; $\frac{\sqrt[3]{y}}{\sqrt[6]{x^7}}$
6. a. $\frac{3x-1}{2x^{3/2}}$; $\frac{3x-1}{2\sqrt{x^3}}$ b. $\frac{3x+1}{2x^{1/2}}$; $\frac{3x+1}{2\sqrt{x}}$
7. a. $\frac{6(3x^2+4)}{x^3}$ b. $\frac{3x+1}{2x^{1/2}}$; $\frac{3x+1}{2\sqrt{x}}$

Progress Check 3.5

1. a. $4\sqrt{2}$ b. $2\sqrt[3]{4}$
2. a. 11 b. y^3 c. $\sqrt[6]{5}$ d. $\sqrt[3]{x^2}$
3. a. $7xy^2\sqrt{2xy}$ b. $xy^2\sqrt[5]{x^2y^2}$
4. a. $\frac{\sqrt{10}}{5}$ b. $\frac{\sqrt[3]{10}}{2}$ c. $\frac{\sqrt{3xy}}{3x}$
5. $\frac{0.3\sqrt{n}}{n}$
6. a. $\frac{74\sqrt{3}}{5}$ b. $-7\sqrt[3]{2}$ c. $16x\sqrt{3y}$
7.

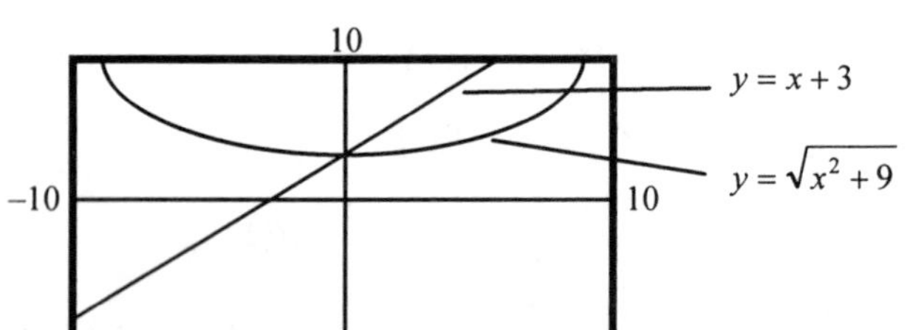

The graphs are different, so $\sqrt{x^2+9}$ is not identical to $x+3$. These two expressions are equal only when $x = 0$.

8. a. $7\sqrt{2}$ b. $54\sqrt[5]{35y^2}$ c. $25-x$
9. a. $4-24\sqrt{x}+36x$ b. -11
10. a. $\frac{\sqrt{30}}{10}$ b. 10 c. $\frac{3\sqrt{2}}{2}$
11. a. $\frac{\sqrt{2}}{2}$ b. $\frac{3\sqrt{x}}{x}$ c. $\frac{3\sqrt{5}}{5}$ d. $\frac{\sqrt{35x}}{7x}$
12. $6+2\sqrt{5}$
13. $\frac{1}{\sqrt{x}+1}$
14. $\frac{4R^2-2h^2}{\sqrt{4R^2-h^2}}$ or $\frac{(4R^2-2h^2)\sqrt{4R^2-h^2}}{4R^2-h^2}$

Progress Check 3.6

1. **a.** $3i$ **b.** $i\sqrt{7}$ **c.** $2i\sqrt{2}$ **d.** $2i\sqrt{3}$

2. **a.** $-2i$ **b.** $7+3i$ **c.** $-7+4i$

3. **a.** -12 **b.** $23+28i$ **c.** $-33+56i$

4. $\left[(-5-3i)+5\right]^2+9=(-3i)^2+9=9i^2+9=-9+9=0$

5. Not a solution

6. $\frac{19}{25}+\frac{17}{25}i$

7. $5+2i$ ohms

8. $\frac{1}{2}+\frac{1}{2}i$

9. **a.** -1 **b.** i

Progress Check 4.1

1. $\{0,-7\}$

2. $(-4,0)$, $(2,0)$, $(0,-8)$

3. **a.** $\{\pm 3\sqrt{2}\}$ **b.** $\{-4\pm i\sqrt{7}\}$

4. $\{6\pm\sqrt{30}\}$

5. $\{1\pm\sqrt{2}\}$

6. $\left\{\frac{4}{5}\right\}$

7. $\left\{\frac{3\pm i\sqrt{7}}{2}\right\}$

8. One

9. $\{3,2\}$

10. 1.17 seconds

11. **a.**

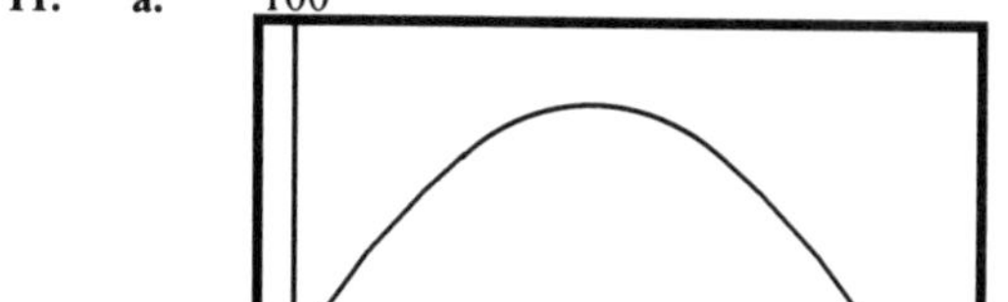
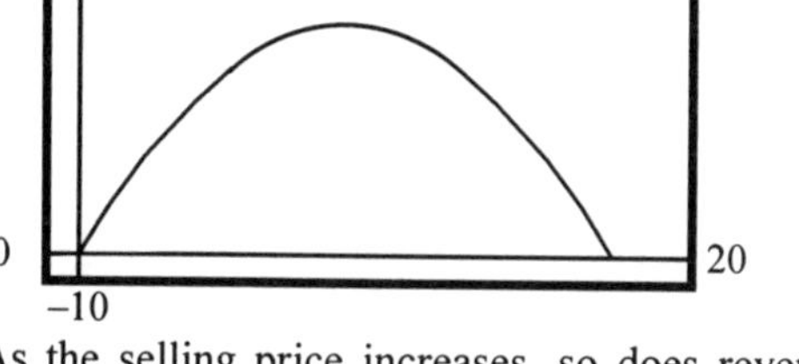

As the selling price increases, so does revenue, for a while; then as price increases, revenue falls off because of lower sales volume.

b. $x\approx \$8$, $x\approx \$9$ **c.** $x=\$8$, $x=\$9$

Progress Check 4.2

1. $\left\{\frac{1}{2}\right\}$

2. $\varnothing$

3. $\{2\}$

4. $\{2\pm\sqrt{7}\}$

5. $v=\frac{-v_0}{t-1}$

6. 12 minutes

7. 180 mi/hour

Progress Check 4.3

1. 1472 ft

2. $\{9\}$

3. $\{8,-8\}$

4. $\{3,7\}$

5. $\{16\}$

6. $\left\{\frac{3\pm\sqrt{17}}{4}\right\}$

7. 86 ft

8. $\{0,2,-2,\sqrt{3},-\sqrt{3}\}$

Progress Check 4.4

1. **a.** $\{-1,2\}$ **b.** $(-\infty,-1)\cup(2,\infty)$ **c.** $(-1,2)$ **d.** $(-\infty,-1]\cup[2,\infty)$

2. $\left(-\infty,-\frac{1}{2}\right]\cup[7,\infty)$ **3.** $\left(3-\sqrt{5},3+\sqrt{5}\right)$ **4.** $\left[1\text{ in.},\frac{7}{2}\text{in.}\right]$ **5.** $\left(-\frac{2}{3},0\right)\cup(2,\infty)$

6. $\left(-\infty,-\frac{1}{2}\right]\cup(3,\infty)$ **7.** $\left(-\infty,-\frac{2}{3}\right)\cup[1,\infty]$

Progress Check 4.5

1. **a.**

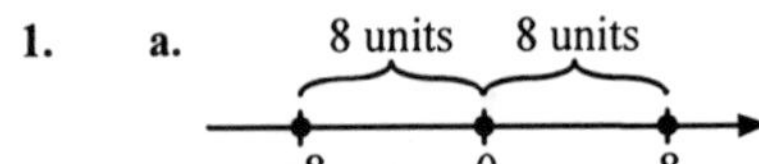

b.

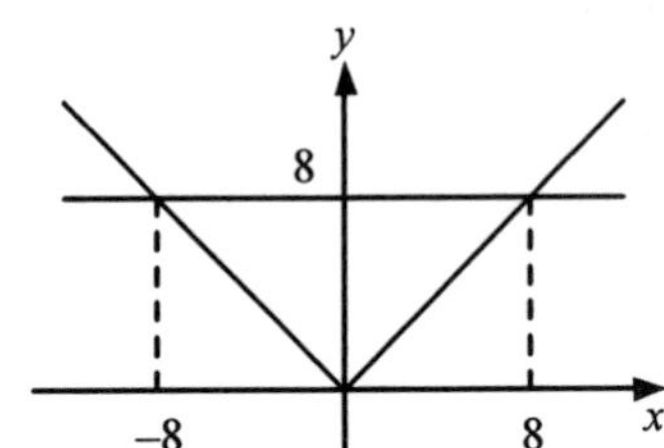

c. $\{8,-8\}$

2. $\{6,-3\}$ **3.** $\left\{-10,-\frac{1}{4}\right\}$

4. **a.**

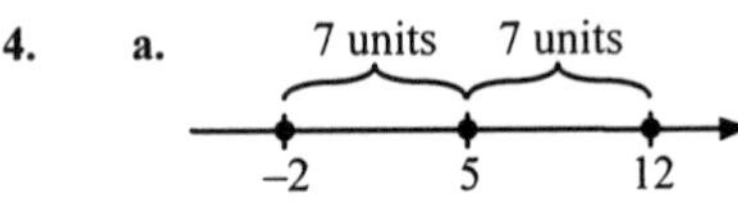

b.

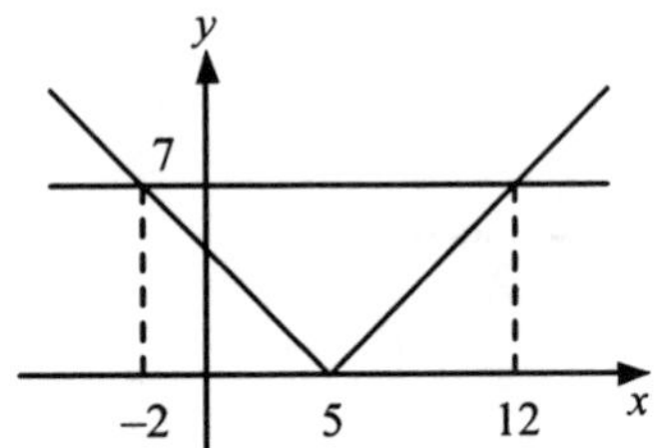

c. $\{-2,12\}$

5. **a.**

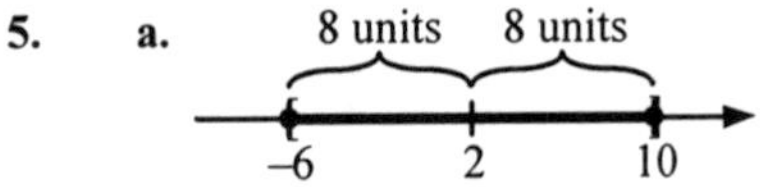

b.

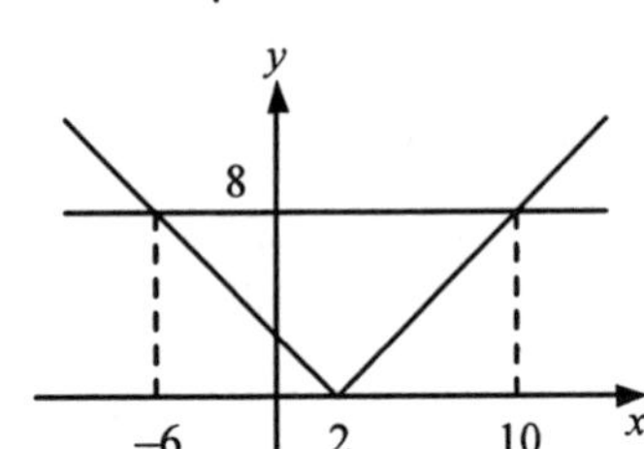

c. $[-6,10]$

6. $\left(-\frac{19}{5},1\right)$

7. **a.**

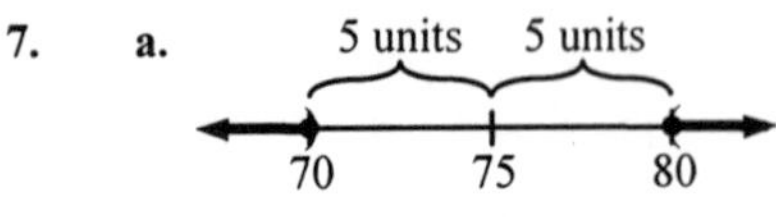

b.

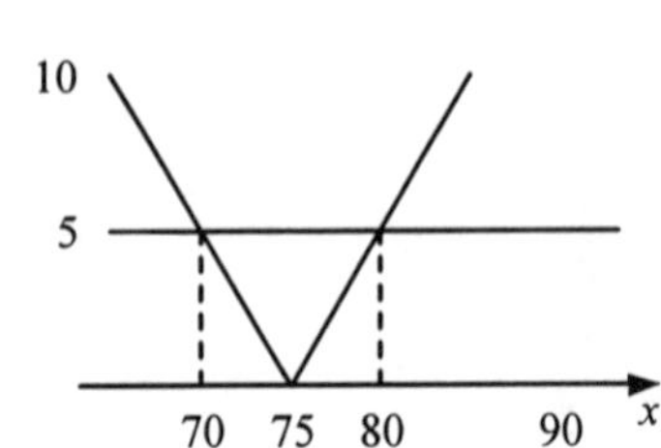

c. $(-\infty,70)\cup(80,\infty)$

8. **a.**

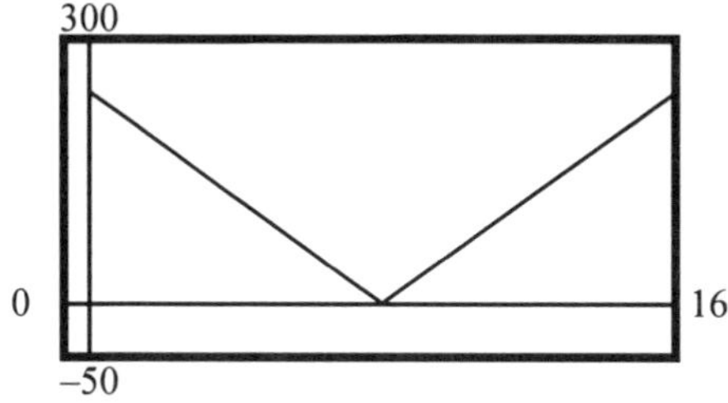

The speed decreases uniformly from 256 ft/second until it stops momentarily when $t = 8$ seconds; then the speed increases uniformly until it hits the ground at 216 ft/second when $t = 16$ seconds.

b. {6.5 seconds; 9.5 seconds}

c. [0 seconds, 6.5 seconds] ∪ [9.5 seconds, 16 seconds]

9. $\{x\colon |x - 0.5| < 2.5\}$

Progress Check 5.1

1. **a.**

x	40	80	120	160	200	240	280	
z	210	170	130	90	50	10	−30	(not meaningful)
A	8,400	13,600	15,600	14,400	10,000	2,400	−8,400	(not meaningful)

b. $A = x(250 - x)$

c.

17,000
15,625
0
0 125 250 300

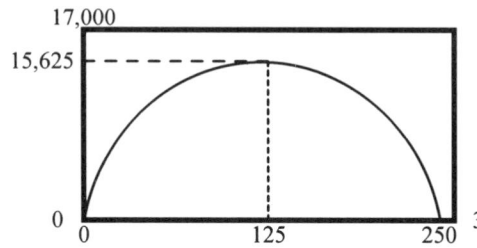

An x increases to 125, A increases to 15,625; then as x increases from 125 to approach 250, A decreases to approach 0. Domain: $(0, 250)$; Range $(0, 15{,}625]$

2. D: $[8, \infty)$; R: $[0, \infty)$

3.

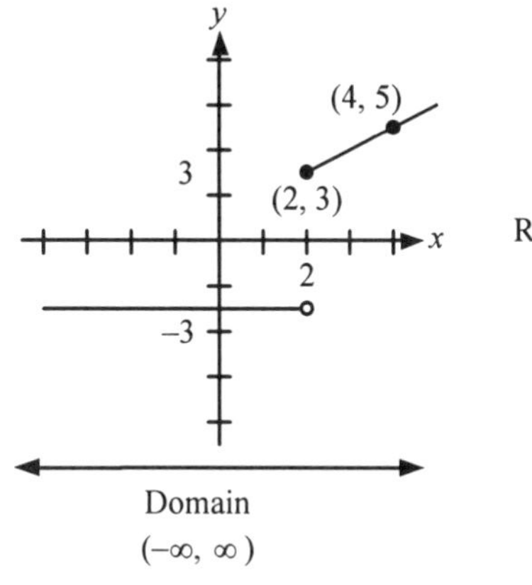

Range = $\{-3\} \cup [3, \infty)$

4. **a.** Not a function: $D = \{0\}$; $R = \{1, 2, 3\}$

b. Function; $D = \{1, 2, 3\}$; $R = \{0\}$

5. 2, undefined, $-\frac{1}{9}$, $\frac{b}{b-5}$

6. **a.** $(-\infty, \infty)$ **b.** $\left[-\frac{9}{8}, \infty\right)$ **c.** Neither **d.** -1

e. $\left\{-\frac{1}{2}, 1\right\}$ **f.** $\left(-\infty, -\frac{1}{2}\right)\cup(1, \infty)$ **g.** Decreasing: $\left(-\infty, \frac{1}{4}\right)$; Increasing: $\left(\frac{1}{4}, \infty\right)$

7. **a.**

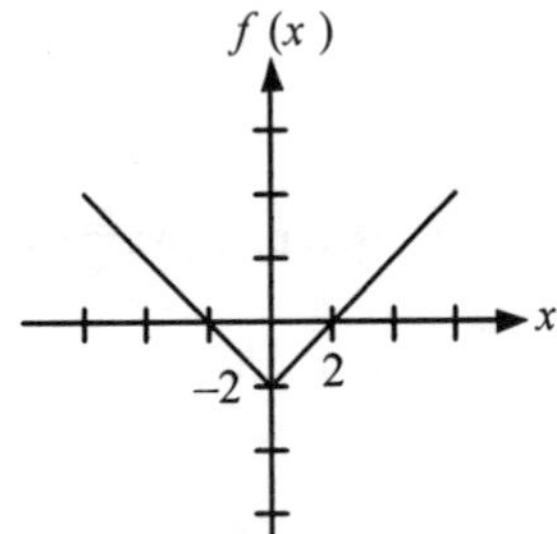

b.

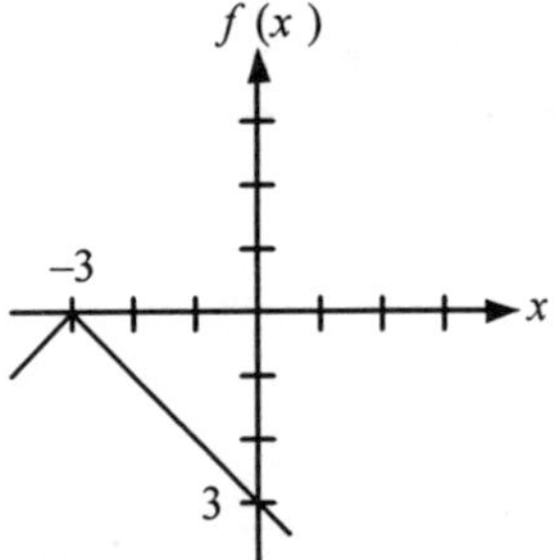

c.

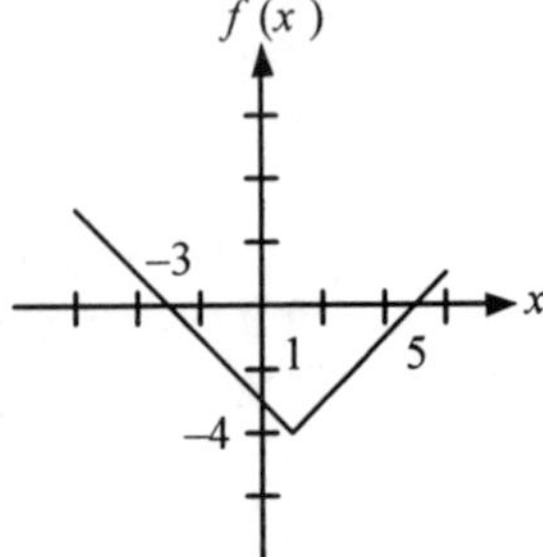

8.

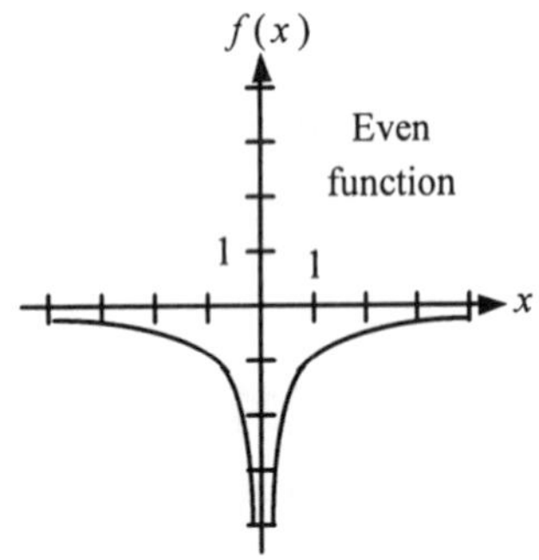

Progress Check 5.2

1. $(f+g)(x)=\sqrt{x-1}-x$, $(f-g)(x)=\sqrt{x-1}+x$, $(f\cdot g)(x)=-x\sqrt{x-1}$; $\left(\frac{f}{g}\right)(x)=-\frac{\sqrt{x-1}}{x}$; domain in all cases is $[1,\infty)$.

2. 41.4

3. $f+g=\{(1,2),(2,1)\}$, $f-g=\{(1,2),(2,5)\}$, $f\cdot g=\{(1,0),(2,-6)\}$, $\frac{f}{g}=\left\{\left(2, -\frac{3}{2}\right)\right\}$

4. **a.** $4x^2+20x+21$ **b.** $2x^2-3$

5. $(g\circ f)(x)=5-\sqrt{x}$, D: $[0,\infty)$; $(f\circ g)(x)=1-\sqrt{x+4}$, D: $[-4,\infty)$

6. $f\circ g=\{(-2,3),(-1,1),(1,1)\}$, D: $\{-2,-1,1\}$; $g\circ f=\{(0,0),(1,-2)\}$; D: $\{0,1\}$

7. **a.** 3 **b.** -4

8. $C = (f \circ g)(t) = 14\pi t$ is the formula for the circumference of the spill t seconds after the start of the leak.

9. $g(x) = 10 - x$, $f(x) = \sqrt{x}$

Progress Check 5.3

1. **a.** 3 **b.** $-\dfrac{2}{5}$ **2.** **a.** undefined **b.** 0

3. Slope $= 0.02$; cost is increasing at 2¢ per copy

4. Slope $= 80{,}000{,}000$; world population increased by an average rate of about 80 million people per year from 1980 to 1990.

5. $\dfrac{1}{2}$

6. **a.** $-2x - h$ **b.** –9; In the interval from 4 to 5, the average rate of change of the function is –9.

7. **a.** $70 - 0.4x - 0.2h$ **b.** 40.2; when 74 units have been produced, then the cost of the next unit is about $40.20.

Progress Check 5.4

1. $y = 3x + 3$ **2.** $y = -\dfrac{4}{5}x + \dfrac{2}{5}$ **3.** Slope $= \dfrac{3}{4}$; y-intercept is $\left(0, -\dfrac{5}{4}\right)$

4.

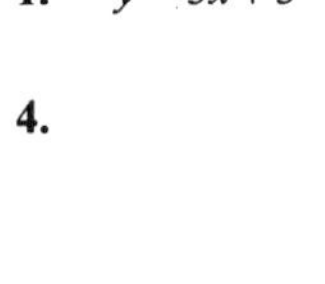

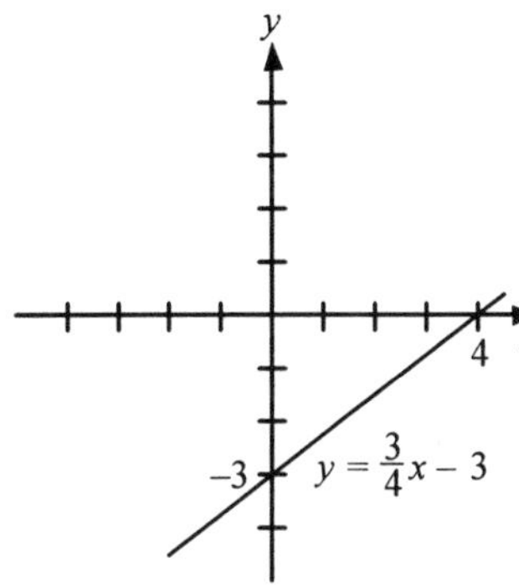

5. $f(x) = -\dfrac{1}{2}x + 2$

6. $y = -0.7x + 154$

7. $1,296

8. No

9. **a.** $y = \dfrac{1}{2}x + \dfrac{5}{2}$ **b.** $y = -2x + 10$

Progress Check 5.5

1.

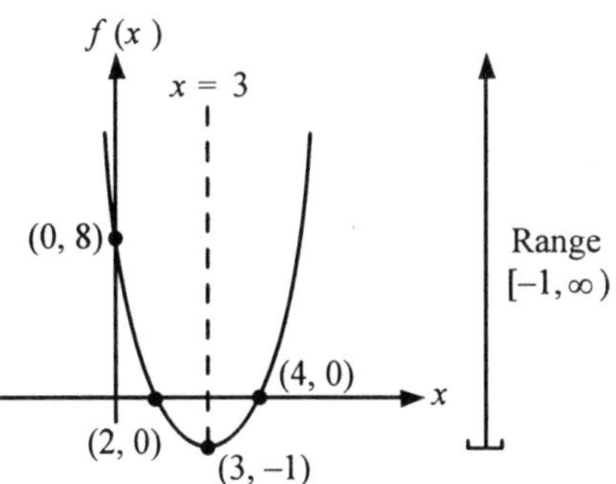

2.

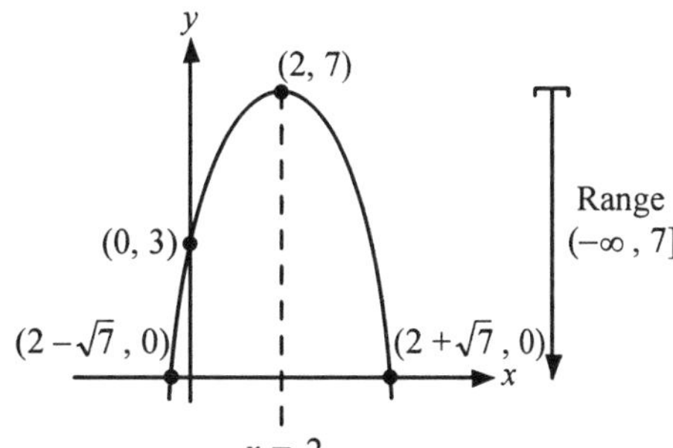

3. **a.** $(-3, -1)$; $x = -3$ **b.** $\left(\dfrac{1}{2}, \dfrac{5}{4}\right)$; $x = \dfrac{1}{2}$ **4.** 125 yd by 250 yd

5. **a.** 8 seconds **b.** 4 seconds **c.** 256 ft

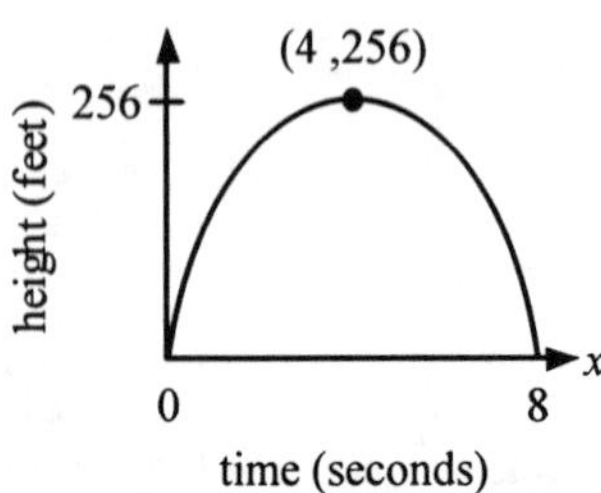

6. **a.** 175.2 ft **b.** 78.5 ft **7.** $(-\infty,-5]\cup[3,\infty)$ **8.** $(-\infty,\infty)$

Progress Check 6.1

1. **a.** Yes; degree 6 **b.** No **2.** $P(x)=x(x+5)(x-1)$ **3.** 0, 1, –3

4. 4 is a zero of multiplicity 2, and −1 is a zero of multiplicity 1; degree 3

5. **a.**

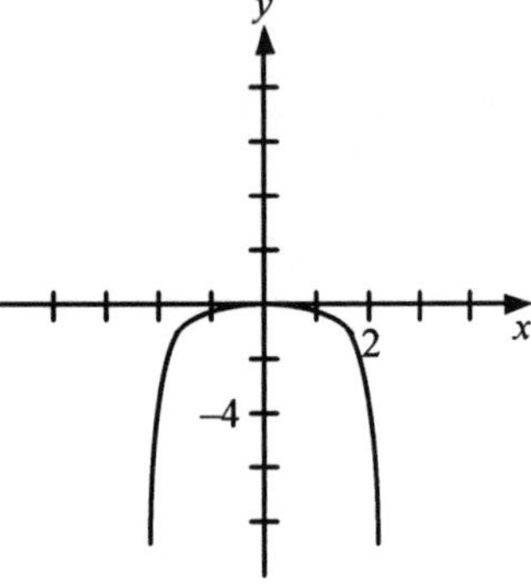

b.

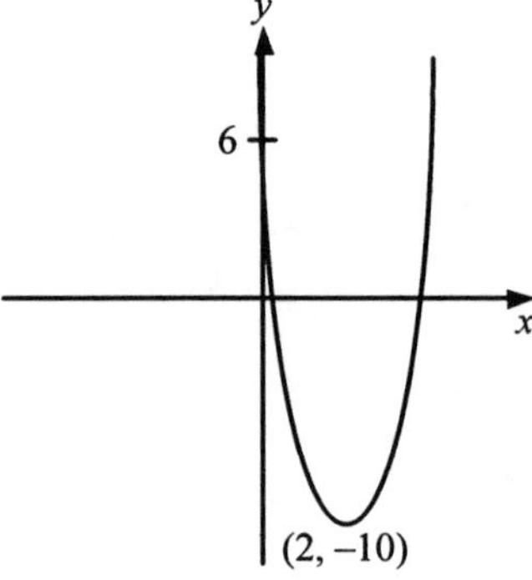

6.

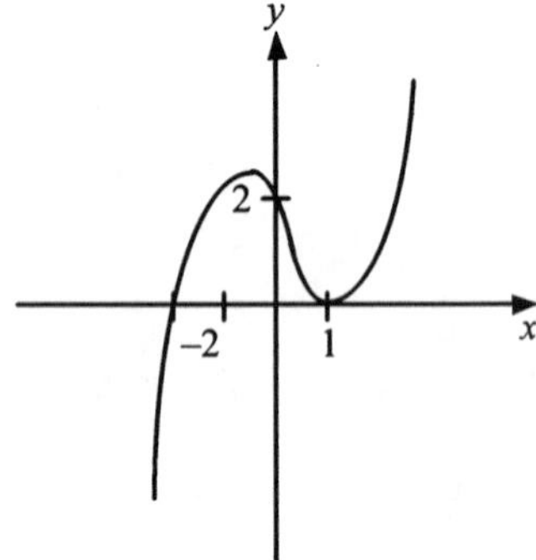

7.

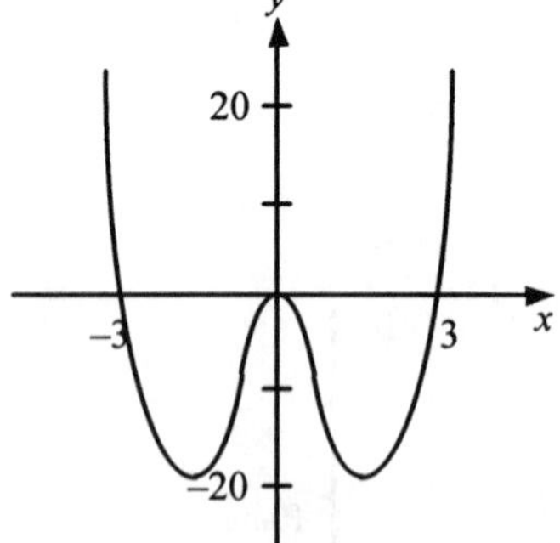

8. $y=\frac{1}{2}(x+2)(x-5)^2$

9. **a.** $V = (18 - 2x)(12 - 2x)x$ **b.** (0 in., 6 in.) **c.** 2.35 in.

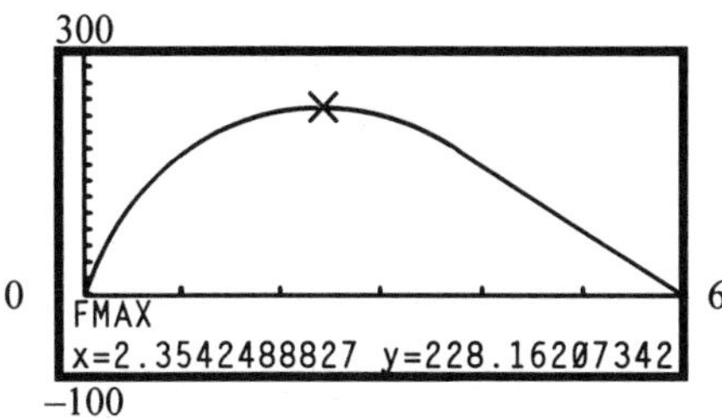

Progress Check 6.2

1. $\dfrac{9x^4 + 5x^2 + x + 3}{3x - 1} = 3x^3 + x^2 + 2x + 1 + \dfrac{4}{3x - 1}$

2. $4x^3 - 2x^2 + 3x - 1 = (x - 2)(4x^2 + 6x + 15) + 29$

3. $3x^3 + 10x^2 - 6x + 8 = (x + 4)(3x^2 - 2x + 2) + 0$

4. $P\left(-\dfrac{1}{2}\right) = 0$

5. Yes; $P(x) = (x + 2)(x - 3)(x + 4)$

6. $x \approx 2.73$ in.

7. $\pm\sqrt{15}$

8. $4 + 2i$; $-\sqrt{3}$

9. $P(x) = (x - 2)(x - \sqrt{2})(x + \sqrt{2})(x - 2i)(x + 2i)$

Progress Check 6.3

1. $4, 2, 1, \dfrac{4}{5}, \dfrac{2}{5}, \dfrac{1}{5}, \dfrac{-1}{5}, \dfrac{-2}{5}, \dfrac{-4}{5}, -1, -2, -4$

2. $\dfrac{3}{2}$, 1 (multiplicity 2)

3. $-2, \dfrac{1}{2}, 3+i, 3-i$

4. $2x(x + 2)x = 350$, so $2x^3 + 4x^2 - 350 = 0$. Then 5 is a root, while the reduced equation $2x^2 + 14x + 70 = 0$ has no real number solution. Unique dimensions: 10 in. by 7 in. by 5 in.

5. $P(0) = -8$, $P(1) = 11$; since $P(0)$ and $P(1)$ have opposite signs, the location theorem guarantees at least one zero between 0 and 1.

6. 0.7

Progress Check 6.4

1. $x = 2$, $x = -2$

2.

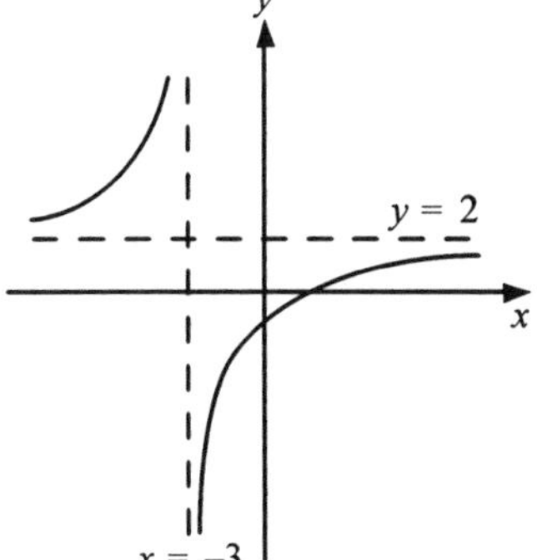

Vertical asymptote: $x = -3$
Horizontal asymptote: $y = 2$

3.

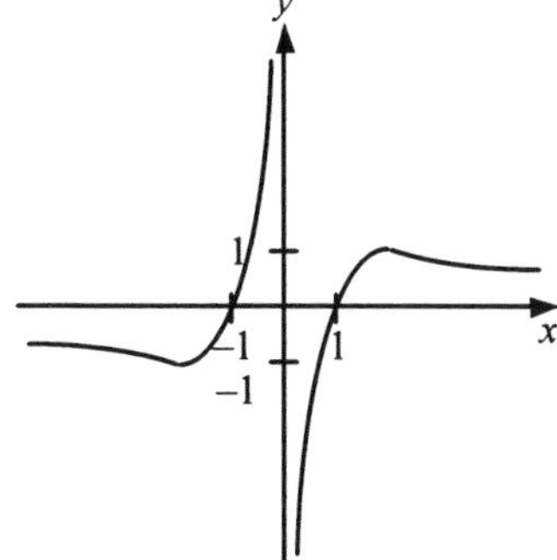

Vertical asymptote: $x = 0$
Horizontal asymptote: $y = 0$

4. **a.** \$3.50; 35¢ per minute

b.

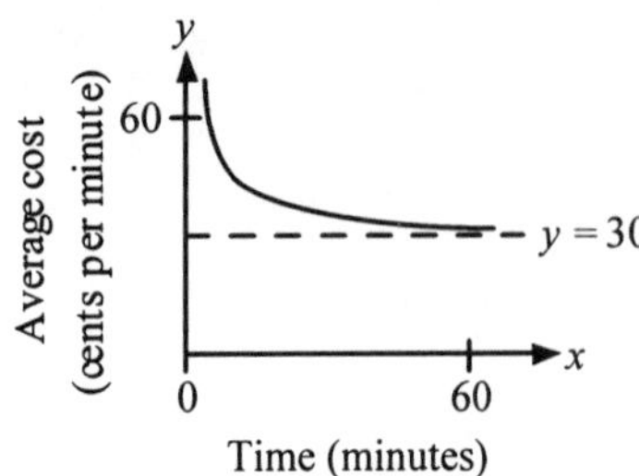

As x increases, y decreases rapidly at first. Then y levels off and approaches $y = 30$. The horizontal asymptote reveals that as the length of the call increases, the average cost per minute approaches the charge per minute for the call.

5.

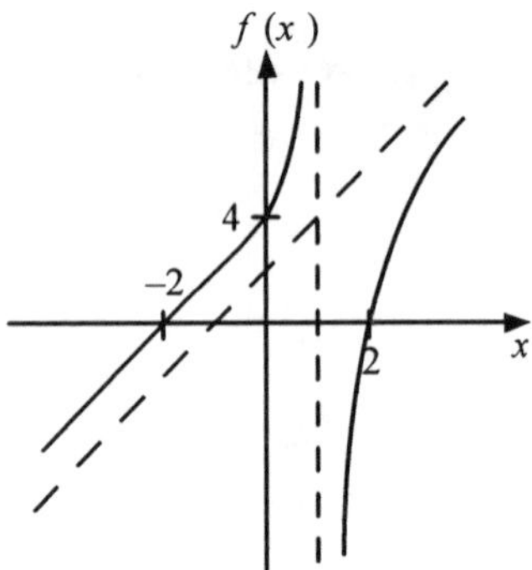

Vertical asymptote: $x = 1$
Slant asymptote: $y = x + 1$

6. 22.36 ft by 22.36 ft

7.

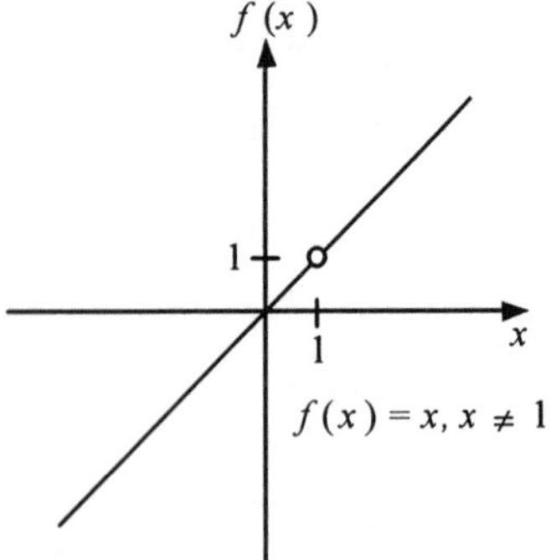

Progress Check 7.1

1. $f^{-1} = \{(1,-1), (2,-2), (3,-3)\}$; domain of f = range of $f^{-1} = \{-1,-2,-3\}$; range of f = domain of $f^{-1} = \{1,2,3\}$

2. Yes

3. **a.** No **b** Yes

4.

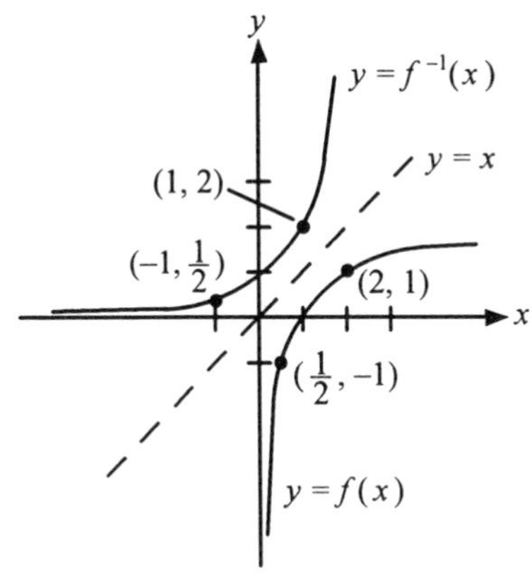

5. **a.** $f^{-1}(x) = \frac{x-2}{3}$; $f^{-1}(3) = \frac{1}{3}$

   b. f does not have an inverse function

   c. $f^{-1}(x) = x^2 + 4$, $x \geq 0$; $f^{-1}(3) = 13$

6.

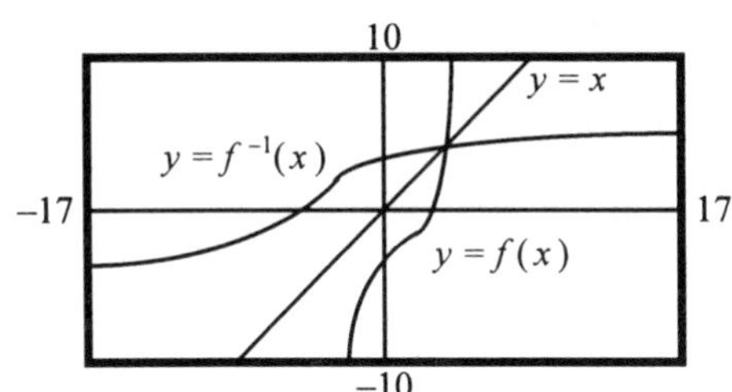

7. $f[g(x)] = 4\left[\frac{x+1}{4}\right] - 1 = x$

   $g[f(x)] = \frac{[4x-1]+1}{4} = x$

8. 50%

9. $y = \sqrt{x}$; $x > 0$. This formula gives the side length of square (y) in terms of the area (x).

Progress Check 7.2

1. **a.** $y = \left(\frac{1}{2}\right)^x$ **b.** $5.96 \times 10^{-8} g$

2. $f(2) = 64$; $f(-1) = \frac{1}{8}$; $f\left(\frac{2}{3}\right) = 4$; $f\left(\sqrt{3}\right) = 36.66$

3.

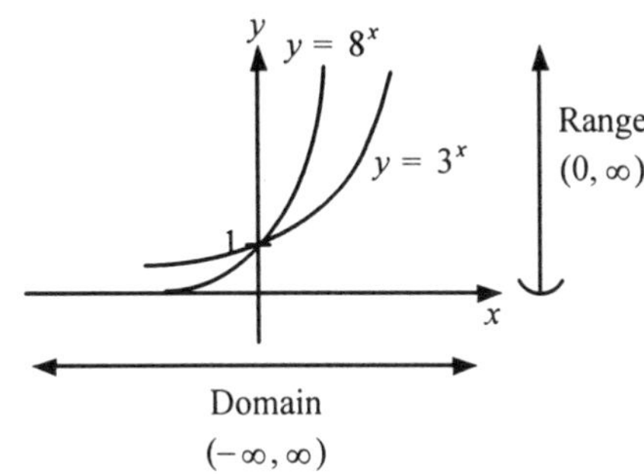

4.

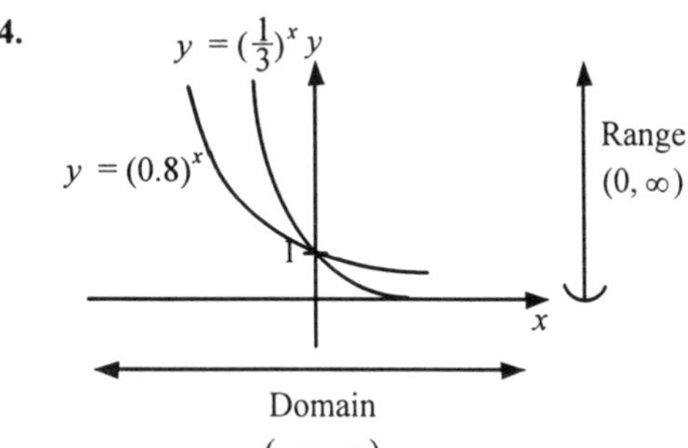

5.

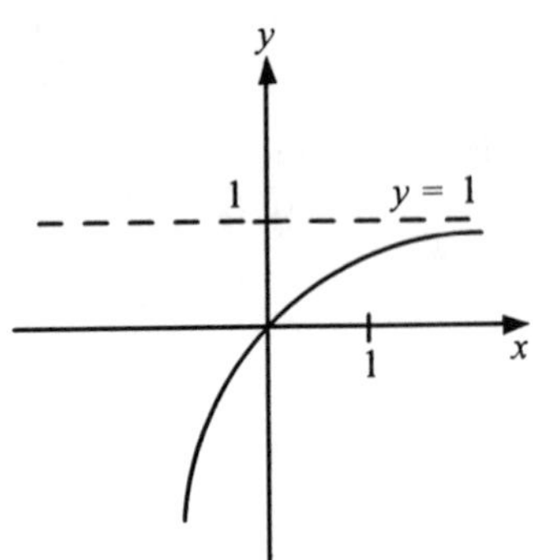

6. $\left\{-\frac{1}{2}\right\}$

7. $\left\{\frac{1}{4}\right\}$

8. $\sqrt[4]{2}$

9. **a.** $f(t) = 2{,}000(1.05)^t$

b.

t	$f(t)$
0	2000
1	2100
2	2205
3	2315.25
4	2431.01

c. \$2,552.56

10. 7.11%

11. **a.** $y = 50{,}000(0.965)^x$

b.

x	y
0	50,000
1	48,250
2	46,561.25
3	44,931.61
4	43,359.00

c. \$38,963.79

Progress Check 7.3

1. $\log_{10} 100 = 2$; $\log_a 4 = x$

2. $9^{1/2} = 3$; $b^y = x$

3. **a.** 3 **b.** $\frac{4}{3}$ **c.** $\frac{1}{1000}$ **d.** 4

4. 0.5065

5. 6.31×10^{-5} or 0.0000631

6.

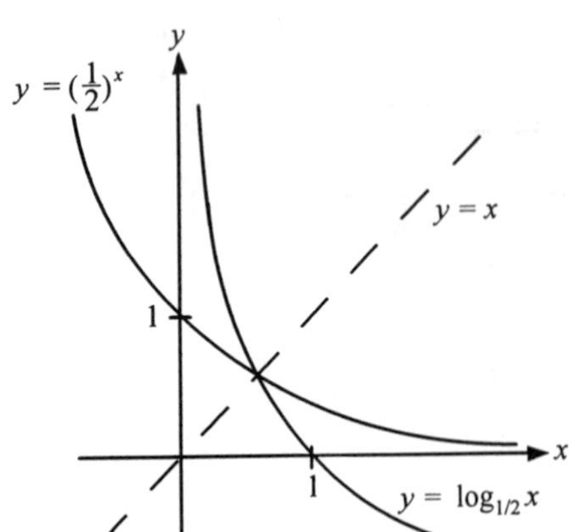

7.

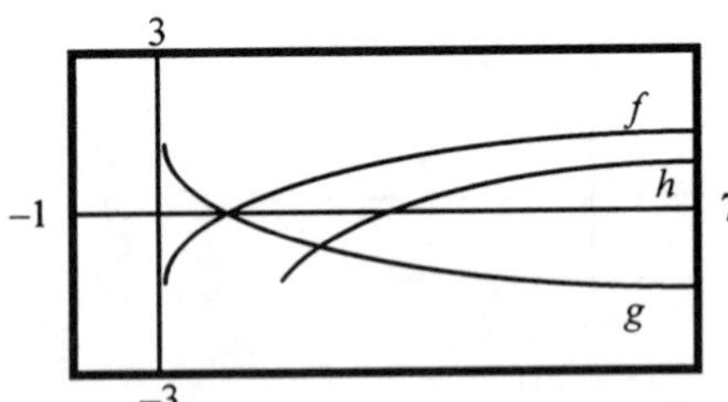

The graph of g is the graph of f reflected about the x = axis; the graph of h is the graph of f shifted 2 units to the right.

8.

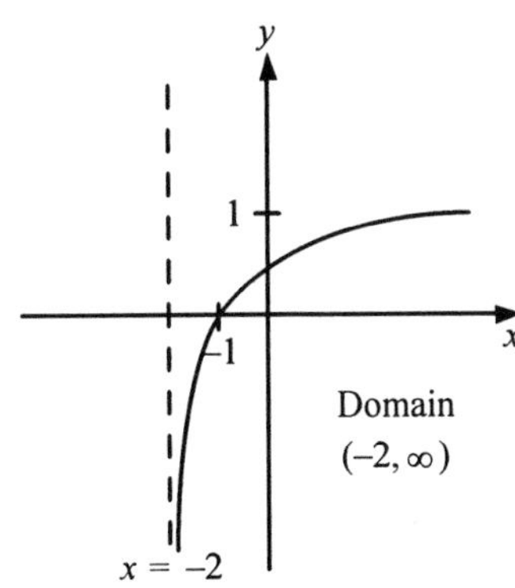

9. 60 dB

10. **a.** 3.3 **b.** 3.6

11. 2.0×10^{-9}

Progress Check 7.4

1. **a.** $\log_b 3 + \log_b 8$ **b.** $\log_3 2 - \log_3 5$ **c.** $\frac{1}{3}\log_5 2$ **d.** $\log_b x - \log_b 2 - \log_b y$

e. $\log_{10} 2 + \log_{10} \pi - \frac{1}{2}(\log_{10} m + \log_{10} n)$

2. **a.** $1 + \log_{10} x$ **b.** 2 **c.** -1

3. **a.** $-m$ **b.** $3m + 2n$

4. **a.** $\log m + k$ **b.** 1000

5. **a.** $\log_b 3$ **b.** $\log_2(x^2 - x)$ **c.** $\log_{10} 3x^2$ **d.** $\log_{10}\sqrt{\frac{L}{g}}$

6. **a.** $\log\frac{1}{[\mathrm{H}^+]} = \log 1 - \log[\mathrm{H}^+] = 0 - \log[\mathrm{H}^+] = -\log[\mathrm{H}^+]$ **b.** $\mathrm{pH} = -\log(3.7 \times 10^{-4}) = 3.4$; $\mathrm{pH} = \log\left(\frac{1}{3.7 \times 10^{-4}}\right) = 3.4$

7.

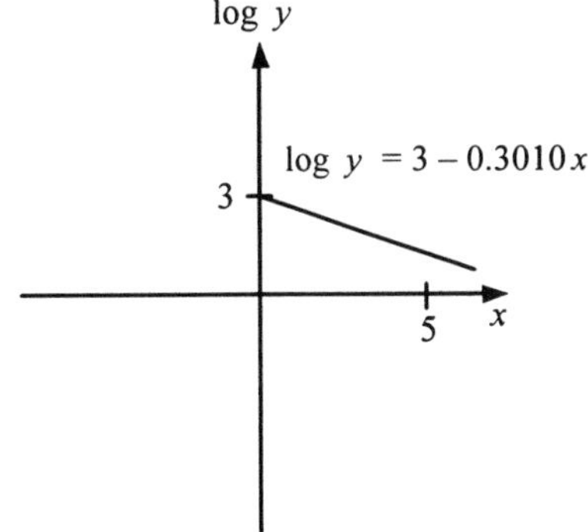

Progress Check 7.5

1. {3.170} **2.** {2.151} **3.** 1,156 years

4. **a.** 3.065 **b.** 0.7268

5.

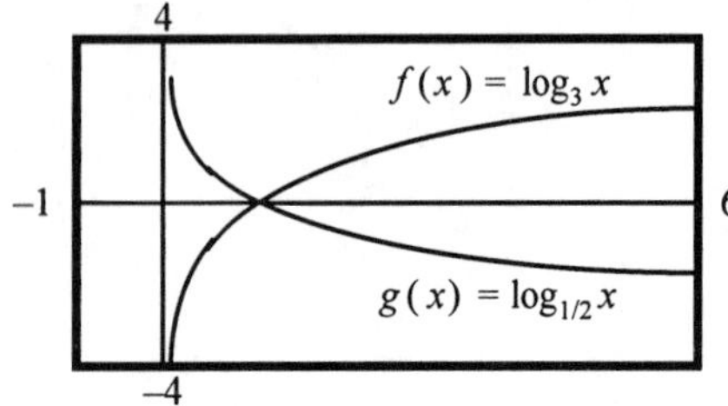

6. {61}

7. {3}

Progress Check 7.6

1. **a.** $A = 9{,}000(1.025)^{2t}$ **b.** \$16,278.53

2. \$118,235

3. \$8,514.41

4.

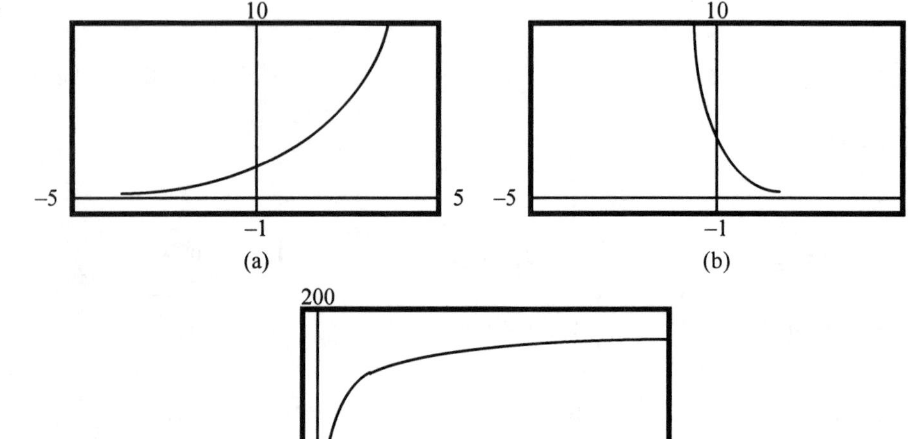

(a) (b)

(c)

5.

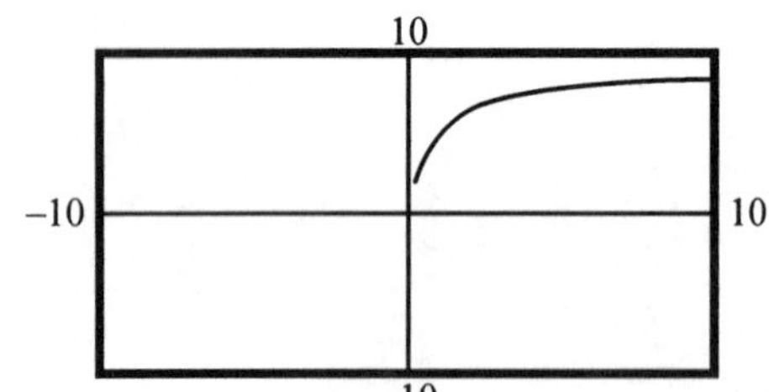

The graphs appear to be the same; the product property $\log_b xy = \log_b x + \log_b y$ implies $\ln 100x = \ln 100 + \ln x$.

6. {4.951}

7. **a.** $\{e^{-2.5}\}$; {0.0821} **b.** $\{\sqrt{2}\}$; {1.414}

8. 23.1

9. **a.** 658 **b.** 42 years

10. 40.4%

11. 3.5 hours

12.

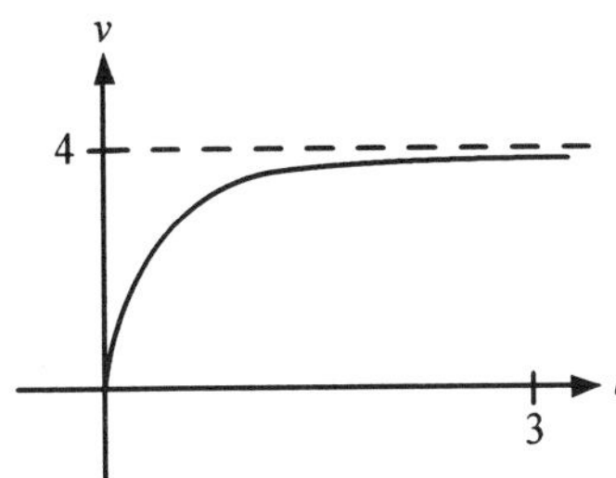

b. 4 ft/sec

c. 0.173 sec
The velocity increases rapidly over the first half-second of free fall to about 3.9 ft/sec. Then the velocity increases very slowly to approach a limiting velocity of 4 ft/sec.

Progress Check 8.1

1. π **2.** $\frac{\pi}{2}, \frac{5\pi}{9}, \frac{7\pi}{4}$ **3.** 20°, 72°, 450° **4.** 39°20′ N

5. $24\pi\,\text{ft}^2$ **6.** **a.** 90π rad/minute **b.** 315π in./minute

Progress Check 8.2

1. **a.** −1 **b.** 0 **c.** 1 **d.** 0

2. −1, −1, undefined, 0, 1, −1 **3.** −1, −1 **4.** $-\frac{1}{\sqrt{3}}, \frac{\sqrt{3}}{2}$

5. $\frac{\sqrt{3}}{2}$ **6.** $\frac{\sqrt{3}}{2}$ **7.** **a.** 0.2837 **b.** 0.7975

8. 7.1 ft at 7 AM; 15.2 ft at 11 PM

Progress Check 8.3

1. **a.**

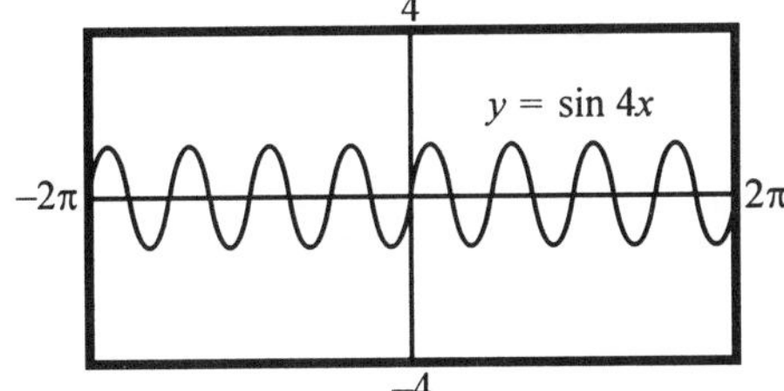

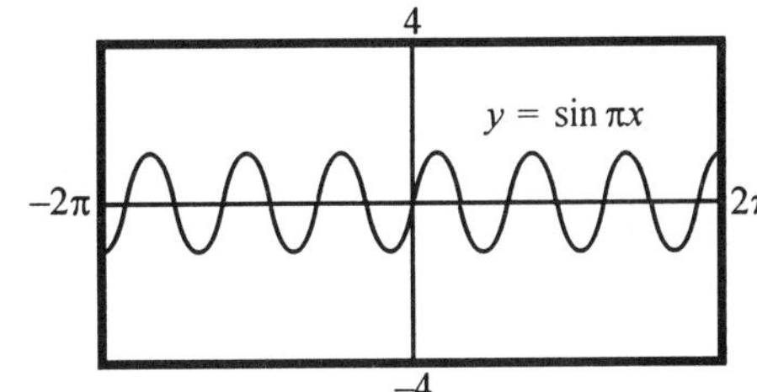

b. Period of $g = \frac{\pi}{2}$; period of $h = 2$

c. The period of $g(x) = \sin 4x$ is the period of $f(x) = \sin x$, which is 2π, divided by 4. The period of $h(x) = \sin \pi x$ is the period of $f(x) = \sin x$ divided by π.

2.

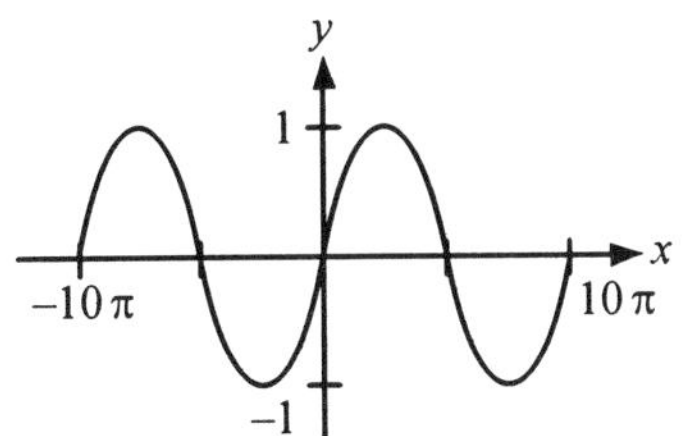

3.

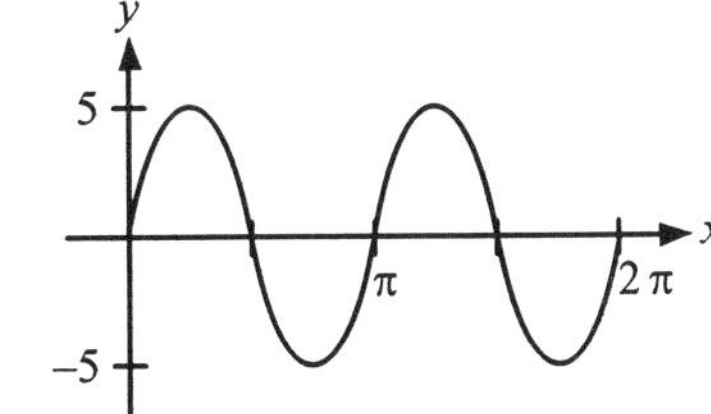

Amplitude: 5
Period: π

4.

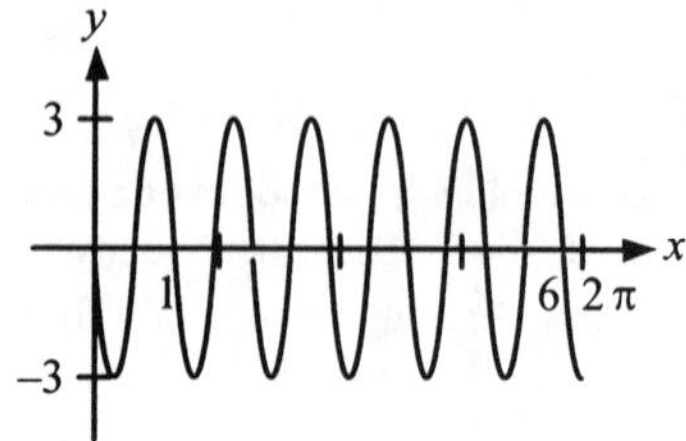

Amplitude: 3
Period: 1

5.

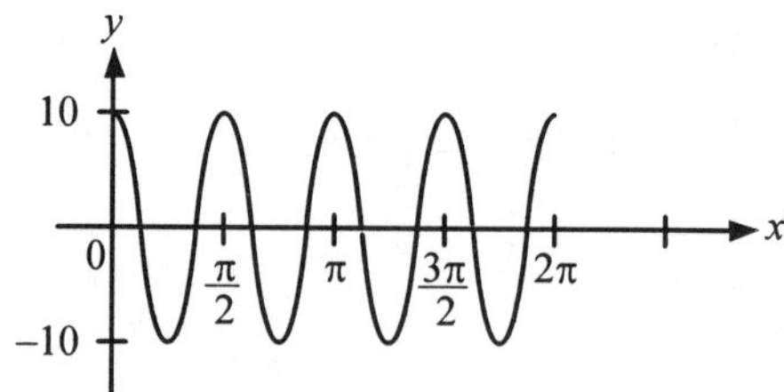

Amplitude: 10
Period: $\frac{\pi}{2}$

6.

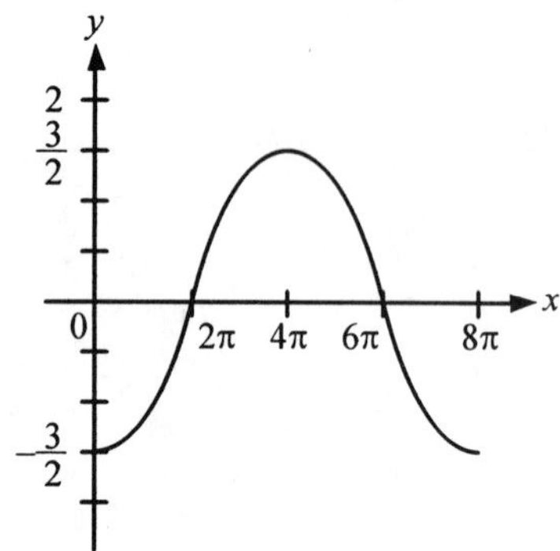

Amplitude: $\frac{3}{2}$

Period: 8π

7. $y = -20\sin\frac{\pi}{2}x$

8.

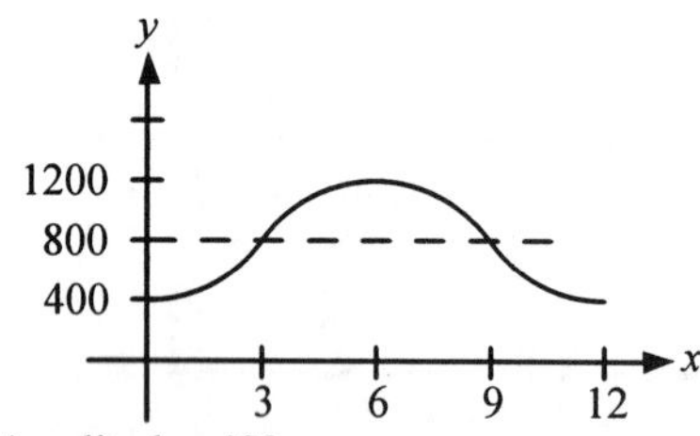

Amplitude: 400
Period: 12
Midline: $y = 800$

9.

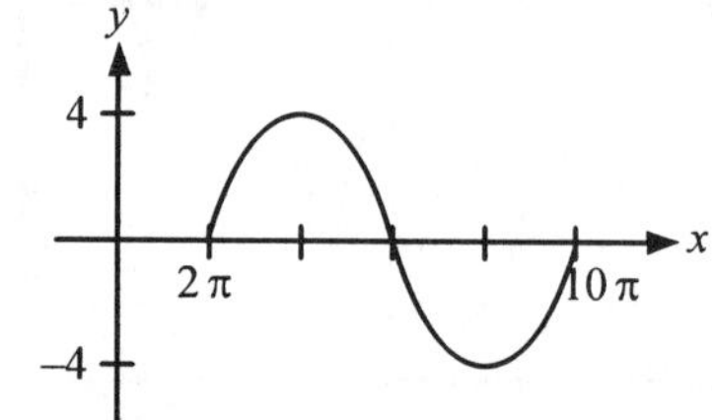

Amplitude: 4
Period: 8π
Phase shift: 2π

10. a.

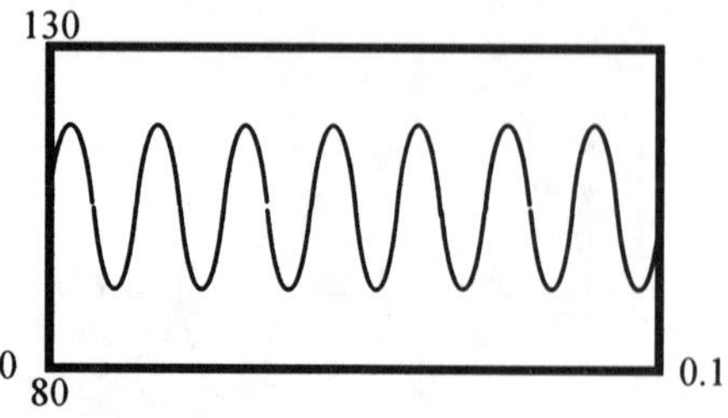

b. Maximum: 125; minimum 85
c. 70 beats/minute

Progress Check 8.4

1.

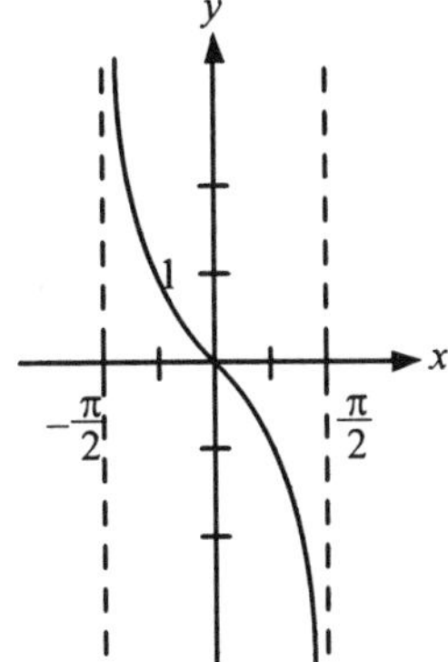

2.

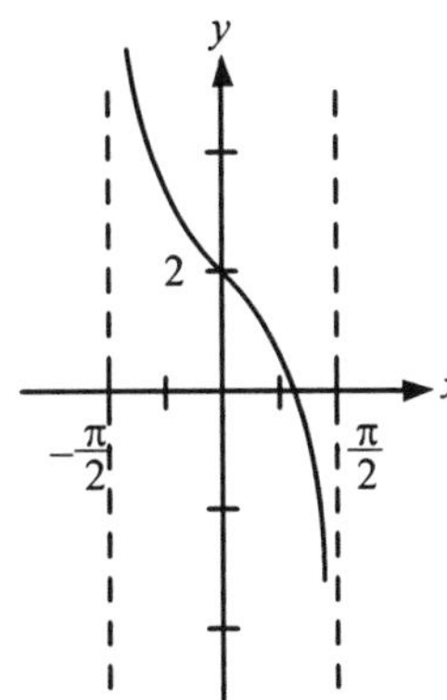

3.

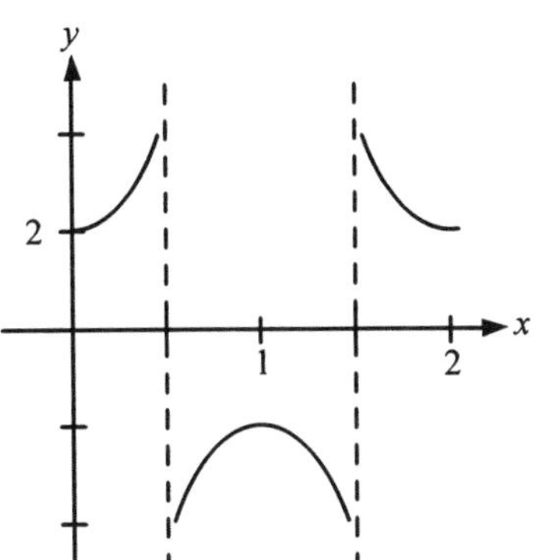

4.

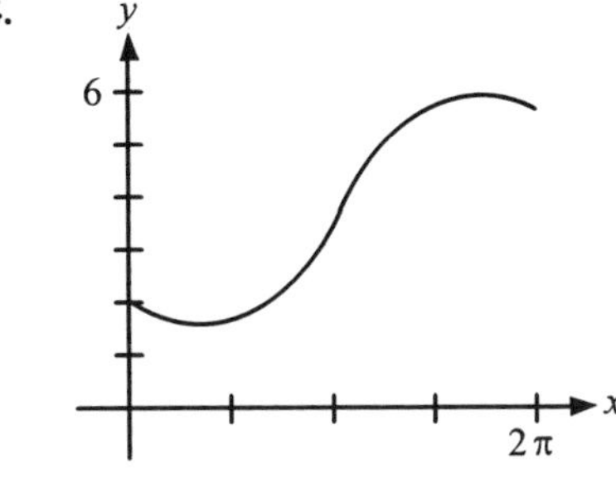

5.

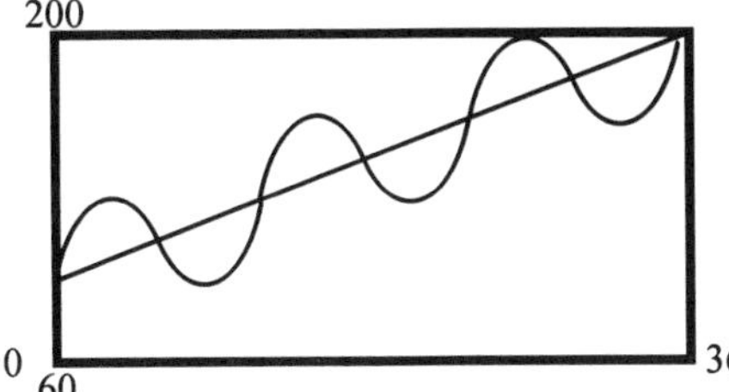

6.

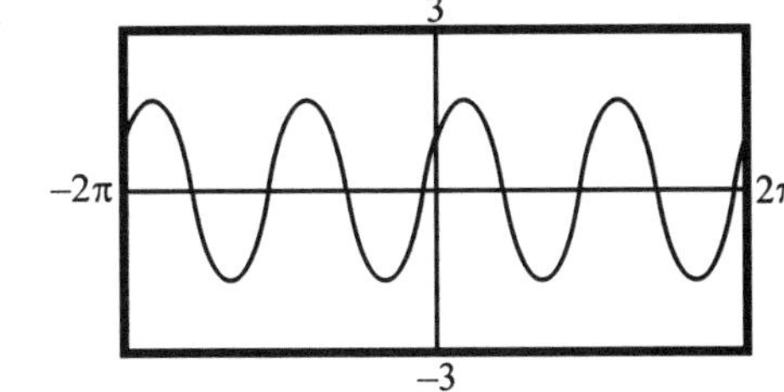

Amplitude: 3.695
Period: $\pi = 3.142$

Progress Check 8.5

1. **a.** $-\frac{\pi}{3}$ **b.** No value **c.** 0.95

2. **a.** 0 **b.** $\frac{\pi}{4}$ **c.** $\frac{3\pi}{4}$ **d.** −1.37

3. **a.** $\theta = \tan^{-1}\left(\frac{8}{x}\right) - \tan^{-1}\left(\frac{3}{x}\right)$

b.

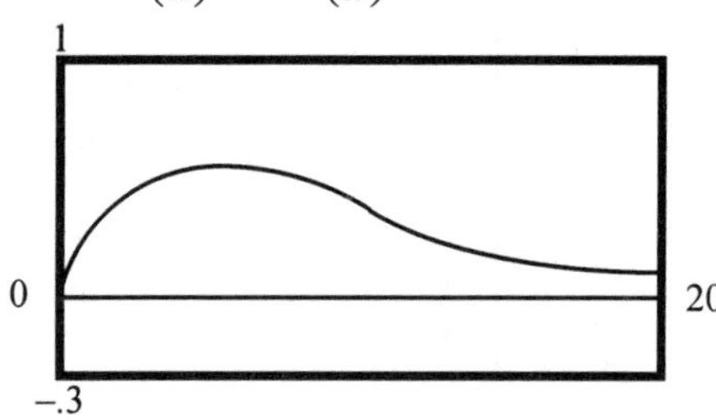

c. 4.9 ft

4. **a.** 0 **b.** $\frac{15}{17}$

5. $\frac{2}{\sqrt{4+x^2}}$

6. $\sqrt{x^2-9} - 3\text{arcsec}\left(\frac{x}{3}\right)$

7. **a.** $\frac{\pi}{6}$ **b.** No value **c.** 2

Progress Check 8.6

1. 3:28 AM, 8:56 AM, 3:52 PM, 9:20 PM

2. $\left\{\frac{3\pi}{4}, \frac{5\pi}{4}\right\}$

3. $\left\{0, \frac{\pi}{6}, \frac{5\pi}{6}, \pi\right\}$

4. $\{x\colon\ x = 3.55 + k2\pi \text{ or } x = 5.87 + k2\pi,\ k \text{ any integer}\}$

5. $\left\{\frac{\pi}{3}, \frac{2\pi}{3}, \frac{4\pi}{3}, \frac{5\pi}{3}\right\}$

6. 3:28 AM, 8:56 AM, 3:52 PM, 9:20 PM

Progress Check 8.7

1. $\frac{\tan x}{\sin x} = \frac{\sin x/\cos x}{\sin x} = \frac{1}{\cos x} = \sec x$

2. $\cot^2 x + 1 = \frac{\cos^2 x}{\sin^2 x} + 1 = \frac{\cos^2 x + \sin^2 x}{\sin^2 x} = \frac{1}{\sin^2 x} = \csc^2 x$

3.

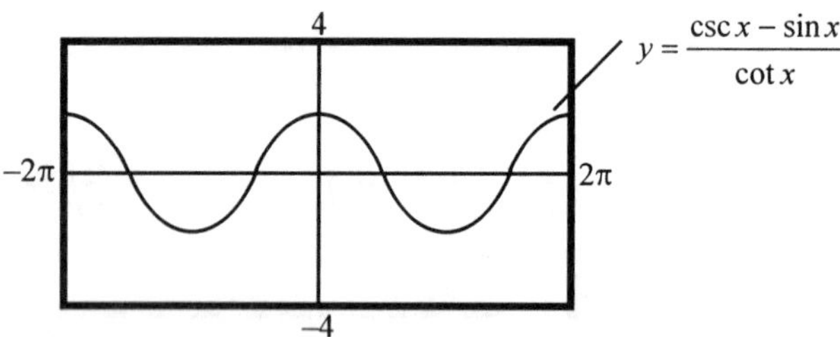

Conjecture: $\frac{\csc x - \sin x}{\cot x} = \cos x$

$$\frac{\csc x - \sin x}{\cot x} = \frac{(1/\sin x) - \sin x}{\cos x/\sin x} = \frac{1-\sin^2 x}{\cos x} = \frac{\cos^2 x}{\cos x} = \cos x$$

4. $\frac{\cot^2 x + 1}{\cot^2 x} = \frac{\csc^2 x}{\cot^2 x} = \frac{1/\sin^2 x}{\cos^2 x/\sin^2 x} = \frac{1}{\cos^2 x} = \sec^2 x$

5. $\frac{\cot^2 x - 1}{\cot^2 x + 1} = \frac{\left(\cos^2 x/\sin^2 x\right) - 1}{\left(\cos^2 x/\sin^2 x\right) + 1} = \frac{\cos^2 x - \sin^2 x}{\cos^2 x + \sin^2 x} = \frac{\cos^2 x - \sin^2 x}{1} = \cos^2 x - \sin^2 x$

6. $\frac{\sin x \sec x}{\tan^2 x} = \frac{\sin x(1/\cos x)}{\sin^2 x/\cos^2 x} = \frac{\sin x \cos x}{\sin^2 x} = \frac{\cos x}{\sin x} = \cot x$

7. $\frac{\sin^2 x}{1+\cos x} = \frac{1-\cos^2 x}{1+\cos x} = \frac{(1+\cos x)(1-\cos x)}{1+\cos x} = 1 - \cos x$

8. $x = \dfrac{10(\sin\theta + \cos\theta)}{\sin\theta\cos\theta} = \dfrac{10(\sin\theta/\cos\theta + \cos\theta/\cos\theta)}{\sin\theta\cos\theta/\cos\theta} = \dfrac{10(\tan\theta + 1)}{\sin\theta}$

9. $4\sec\theta$

10. a. $-\dfrac{12}{13}$ b. $-\dfrac{12}{5}$ c. $-\dfrac{13}{12}$

11. $\left\{\dfrac{\pi}{3}, \dfrac{5\pi}{3}\right\}$

Progress Check 8.8

1. $\cos\left(\dfrac{3\pi}{2} + x\right) = \cos\dfrac{3\pi}{2}\cos x - \sin\dfrac{3\pi}{2}\sin x = 0\cdot\cos x - (-1)\sin x = \sin x$

2. $\cos(2\pi - x) = \cos 2\pi\cos x + \sin 2\pi\sin x = 1\cdot\cos x + 0\cdot\sin x = \cos x$

3. $2\sin\left(x - \dfrac{\pi}{6}\right) = 2\left(\sin x\cos\dfrac{\pi}{6} - \cos x\sin\dfrac{\pi}{6}\right) = 2\left(\sin x\cdot\dfrac{\sqrt{3}}{2} - \cos x\cdot\dfrac{1}{2}\right) = \sqrt{3}\sin x - \cos x$

4. $\dfrac{f(x+h) - f(x)}{h} = \dfrac{\cos x\cos h - \sin x\sin h - \cos x}{h} = \dfrac{\cos x\cos h - \cos x}{h} - \dfrac{\sin x\sin h}{h} = \cos x\left(\dfrac{\cos h - 1}{h}\right) - \sin x\left(\dfrac{\sin h}{h}\right)$

5. $\sqrt{3} + 2$

6. a. $-\dfrac{16}{65}$ b. $-\dfrac{63}{65}$

7. a. $\theta = \tan^{-1}\left(\dfrac{5x}{x^2 + 24}\right)$ b.

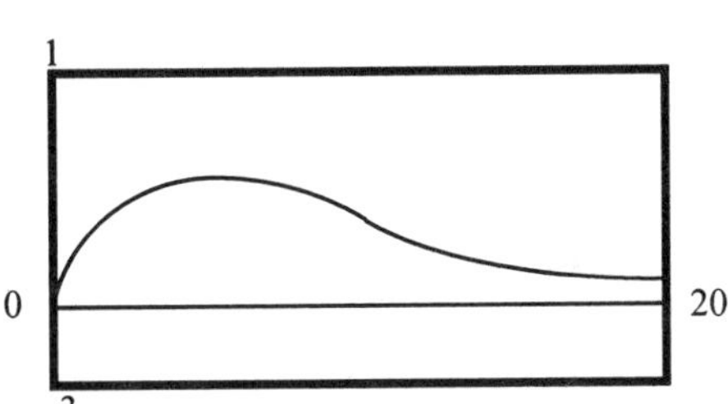

c. 4.9 ft

Progress Check 8.9

1. a. $-\dfrac{119}{169}$ b. $-\dfrac{120}{169}$ c. $\dfrac{120}{119}$

2. $d = \dfrac{V^2}{16}\sin A\sin B = \dfrac{V^2}{16}\sin A\sin(90° - A) = \dfrac{V^2}{16}\sin A\cos A = \dfrac{V^2}{16}\cdot\dfrac{2\sin A\cos A}{2} = \dfrac{1}{32}V^2\sin 2A$

3. $\dfrac{2\tan x}{1 + \tan^2 x} = \dfrac{2\tan x}{\sec^2 x} = \dfrac{2\sin x/\cos x}{1/\cos^2 x} = 2\sin x\cos x = \sin 2x$

4. 0, 1.82, π, 4.46

5. $\sin^2 4x\cos^2 4x = \dfrac{1 - \cos(2\cdot 4x)}{2}\cdot\dfrac{1 + \cos(2\cdot 4x)}{2} = \dfrac{1 - \cos^2 8x}{4} = \dfrac{1}{4}\left[1 - \left(\dfrac{1 + \cos(2\cdot 8x)}{2}\right)\right] = \dfrac{1}{4}\left[\dfrac{1}{2} - \dfrac{\cos 16x}{2}\right]$

$= \dfrac{1}{8}(1 - \cos 16x)$

6. $\sin\dfrac{\pi}{12} = \dfrac{\sqrt{2 - \sqrt{3}}}{2}$; $\cos\dfrac{\pi}{12} = \dfrac{\sqrt{2 + \sqrt{3}}}{2}$

7. a. $\cos 6x$ b. $5\sin 6\theta$ c. $\tan 2t$

8. $\sin 4x = \sin(2 \cdot 2x) = 2\sin 2x \cos 2x = 2(2\sin x \cos x)(2\cos^2 x - 1) = 8\cos^3 x \sin x - 4\sin x \cos x$

Progress Check 9.1

1. 638 ft **2.** $C = 110°$, $b = 26$ ft, $c = 38$ ft **3.** $12\sqrt{3}$

4. $B = 27°$, $C = 98°$, $c = 91$ ft

5. $B = 41°50'$, $C = 109°30'$, $c = 318$ ft or $B = 138°10'$, $C = 13°10'$, $c = 76.6$ ft **6.** Two triangles

Progress Check 9.2

1. 7.9 mi **2.** $b = 41$ ft, $A = 34°$, $C = 98°$ **3.** $A = 31°$, $B = 87°$, $C = 62°$

4. 5.4 yd **5.** 46 ft^2 **6.** $K = \frac{1}{2}ab\sin C = \frac{1}{2}\left(\frac{b\sin A}{\sin B}\right)b\sin C = \frac{b^2 \sin A \sin C}{2\sin B}$

7. 2.34 km^2

Progress Check 9.3

1. a. 7; 90° **b.** $5\sqrt{2}$; 315°

2. a. $\langle 2, 6\rangle$ **b.** $\langle -15, 24\rangle$ **c.** $\langle -39, 44\rangle$

3. a. $\langle 1, -3\rangle$ **b.** $\langle -7, -2\rangle$ **c.** $\langle 34, 13\rangle$

4. $c(d\mathbf{u}) = c(d\langle u_1, u_2\rangle) = c\langle du_1, du_2\rangle = \langle cdu_1, cdu_2\rangle = cd\langle u_1, u_2\rangle = (cd)\mathbf{u}$

5. a. $-10\mathbf{i} - 3\mathbf{j}$ **b.** $11\mathbf{i} - 6\mathbf{j}$ **c.** $-67\mathbf{i} + 45\mathbf{j}$

6. $-7\sqrt{3}\mathbf{i} - 7\mathbf{j}$ **7.** $-17.06\mathbf{i} + 84.27\mathbf{j}$; 86 lb; 101°

Progress Check 9.4

1.

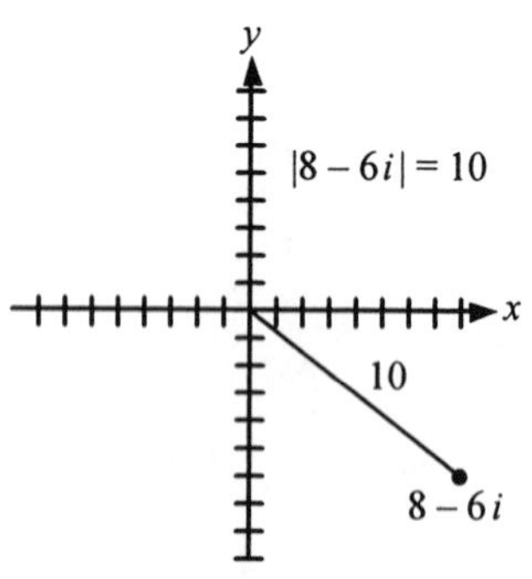

2. $2(\cos 210° + i\sin 210°)$

3. $-3\sqrt{2} - 3\sqrt{2}i$

4. a. $32(\cos 144° + i\sin 144°)$ **b.** $2[\cos(-36°) + i\sin(-36°)]$

c. $\frac{1}{2}(\cos 36° + i\sin 36°)$

5. a. $5.56(\cos 39.5° + i\sin 39.5°)$ ohms **b.** $1.42(\cos 57.8° + i\sin 57.8°)$ ohms

6. 16

7. $3(\cos 60° + i\sin 60°) = \frac{3}{2} + \frac{3\sqrt{3}}{2}i$;
$3(\cos 180° + i\sin 180°) = -3$;
$3(\cos 300° + i\sin 300°) = \frac{3}{2} - \frac{3\sqrt{3}}{2}i$

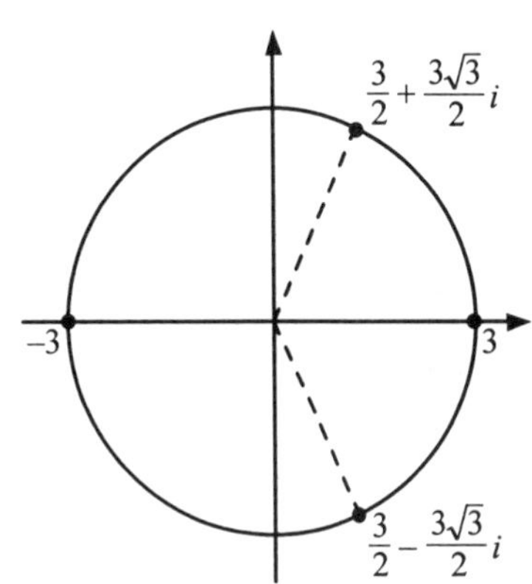

8. $\cos 22.5° + i\sin 22.5° \approx 0.9239 + 0.3827i$
$\cos 112.5° + i\sin 112.5° \approx -0.3827 + 0.9239i$
$\cos 202.5° + i\sin 202.5° \approx -0.9239 - 0.3827i$
$\cos 292.5° + i\sin 292.5° \approx 0.3827 - 0.9239i$

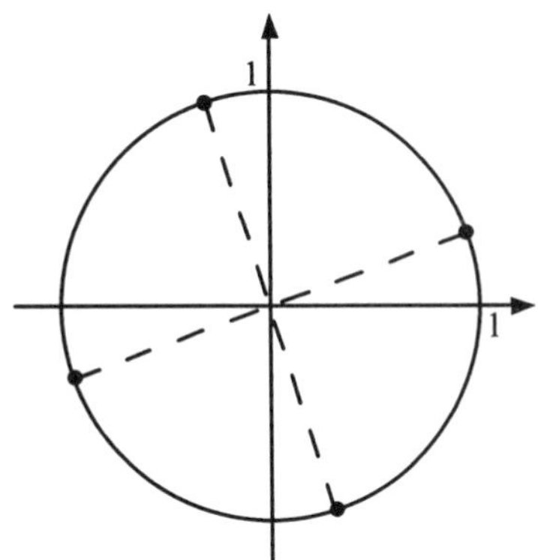

9. $\sqrt{2} + \sqrt{2}i,\ -\sqrt{2} + \sqrt{2}i,\ -\sqrt{2} - \sqrt{2}i,\ \sqrt{2} - \sqrt{2}i$

Progress Check 9.5

1. a.

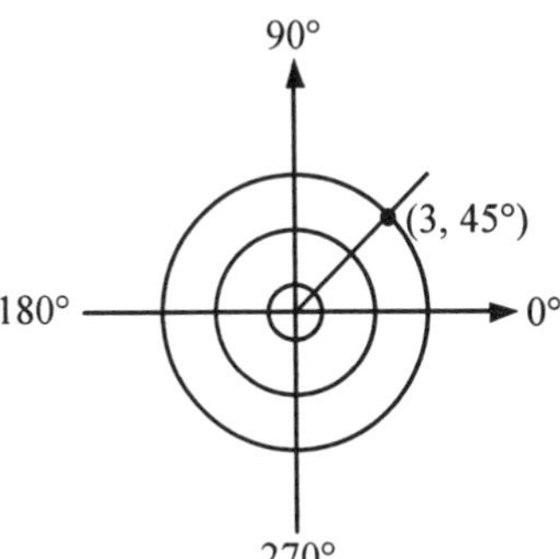

b.

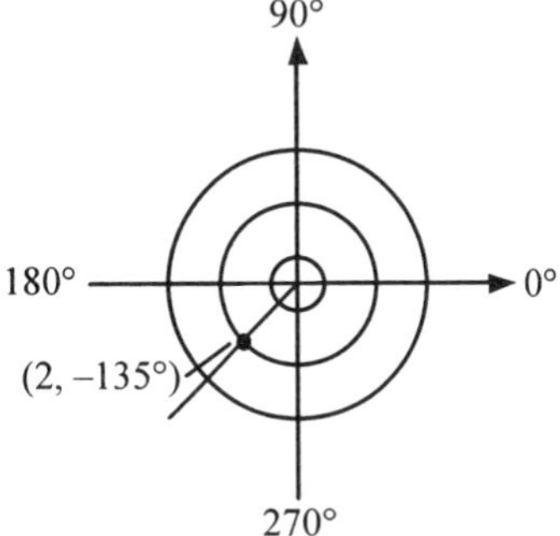

c.

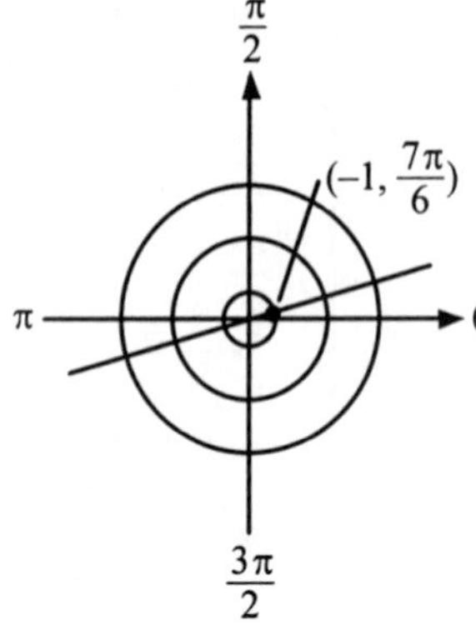

d.

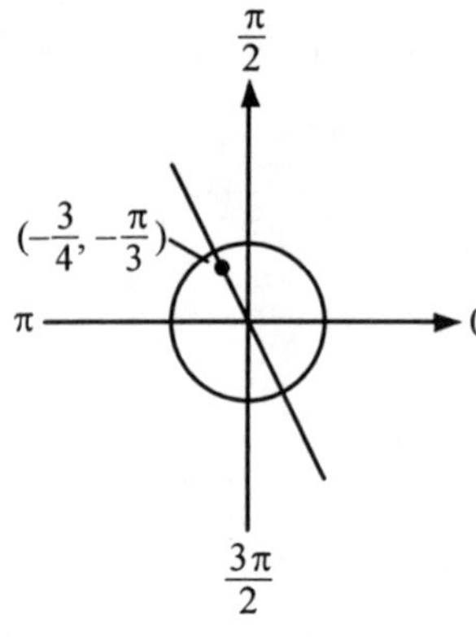

e.

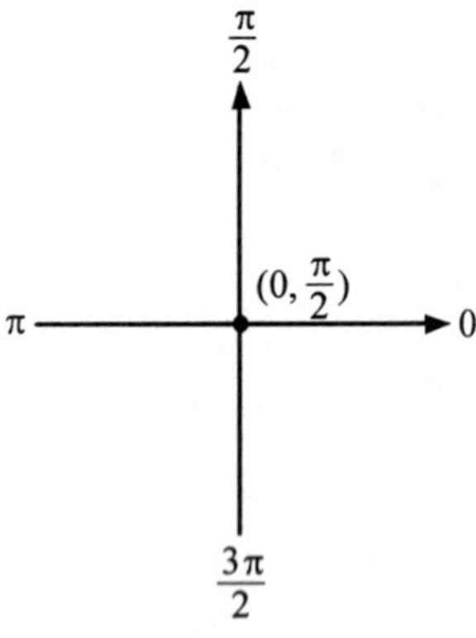

2. **a.** $(0, 1)$ **b.** $\left(2, \frac{2\pi}{3}\right)$

3. **a.**

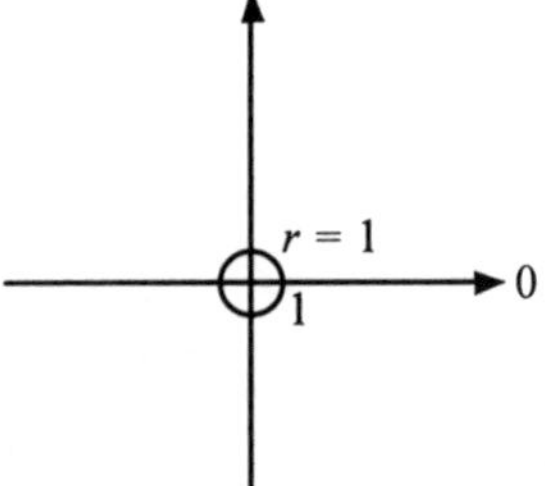

b.

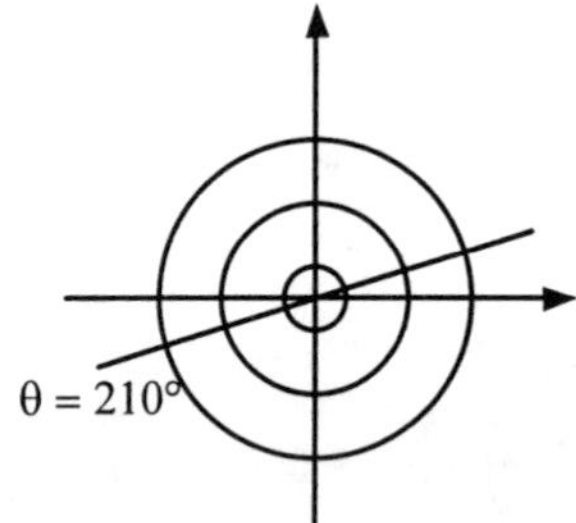

4.

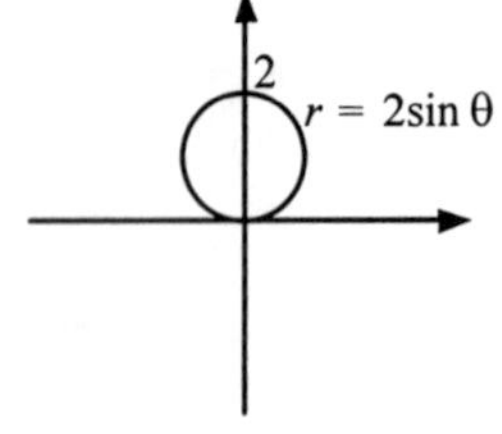

5. **a.** $x^2 + y^2 = 9$

b. $y = x$

c. $x^2 + y^2 = 2y$ or $x^2 + (y-1)^2 = 1$

6. $r = 6\cos\theta$ **7.** $r = \dfrac{3}{2\pi}\theta$

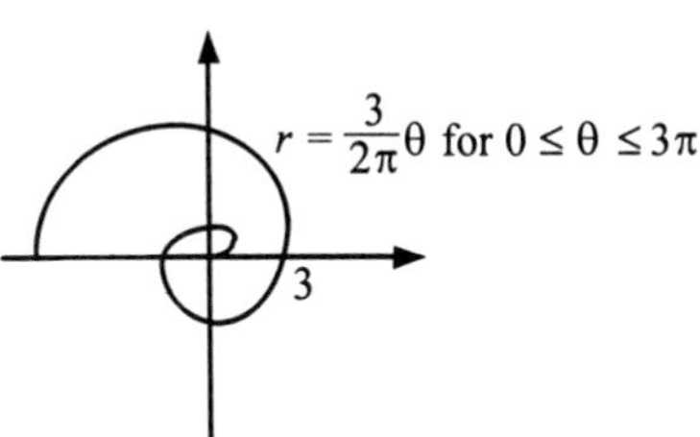

Progress Check 10.1

1. $(x-4)^2+(y+7)^2=81$ **2.** $(x-6)^2+(y-3)^2=25$ **3.** **a.** $x^2+(y-44)^2=1{,}369$ **b.** 12 ft

4.

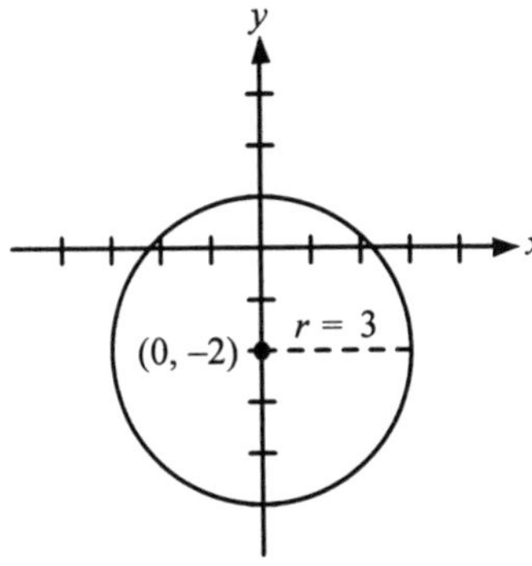

Center (0, –2); radius 3

5.

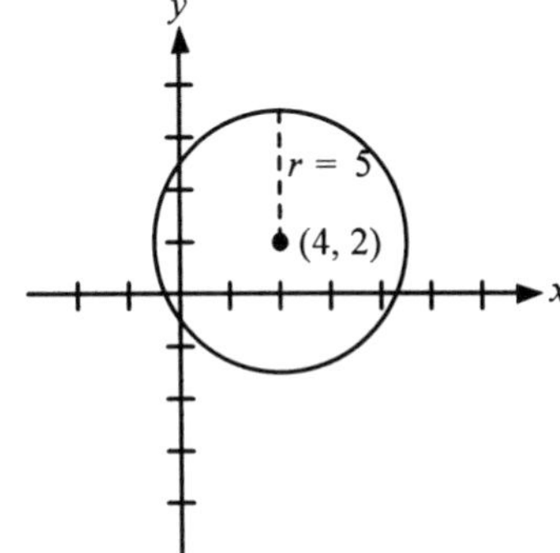

Center (4, 2); radius 5

6. Position square $OABC$ with vertices $O(0,0)$, $A(a,0)$, $B(a,a)$ and $C(0,a)$. Then $m_{OB}=1$ and $m_{AC}=-1$. Because $m_{OB}\cdot m_{AC}=-1$, the diagonals are perpendicular to each other.

7. $(7,-1)$

8. If C and D are the midpoints of OB and AB, then $C=\left(\frac{b}{2},\frac{c}{2}\right)$ and $D=\left(\frac{a+b}{2},\frac{c}{2}\right)$. Since OA and CD both have slope 0, CD is parallel to OA. Also, $\overline{OA}=a$ and $\overline{CD}=\left|\frac{a+b}{2}-\frac{b}{2}\right|=\frac{a}{2}$, so $\overline{CD}$ is one-half of $\overline{OA}$.

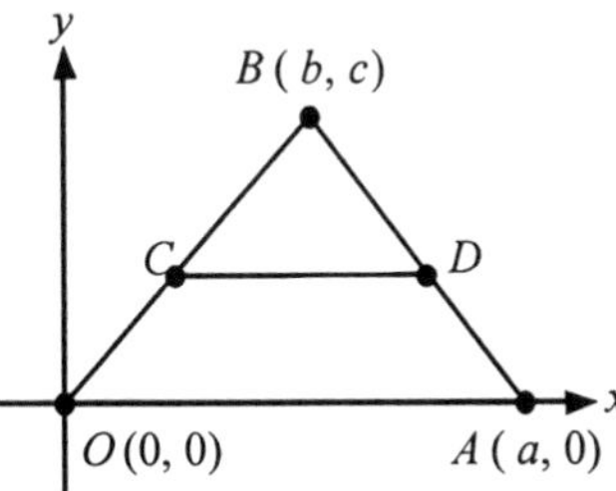

Progress Check 10.2

1.

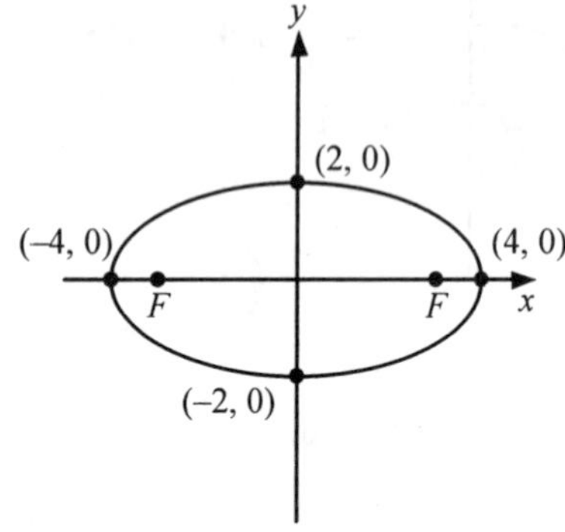

Foci $(2\sqrt{3},0)$, $(-2\sqrt{3},0)$

2. a.

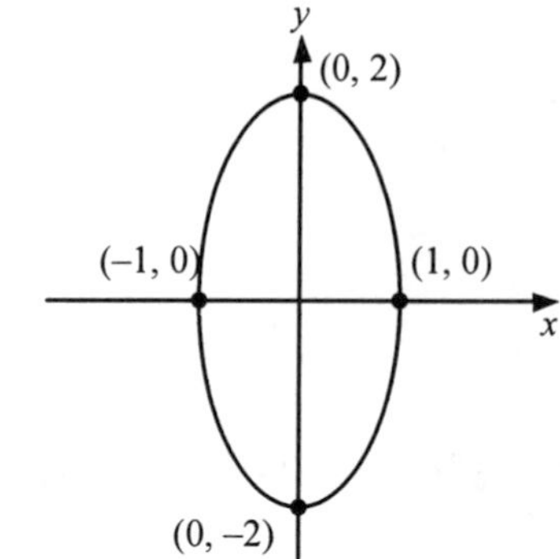

b. $(0,\sqrt{3})$, $(0,-\sqrt{3})$

3. $\dfrac{x^2}{25}+\dfrac{y^2}{21}=1$

4.

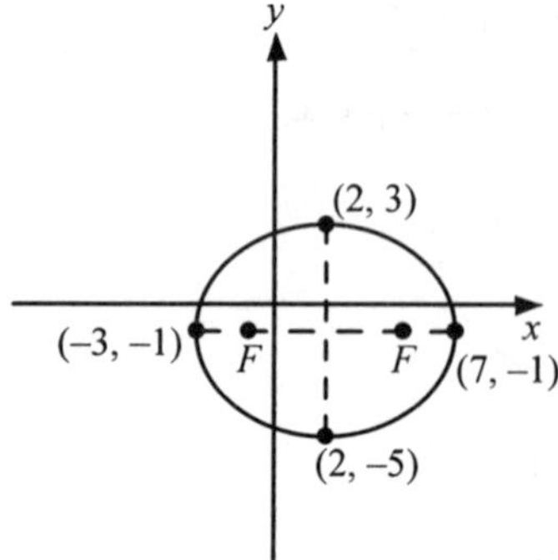

Center: $(2,-1)$

Foci: $(-1,-1)$, $(5,-1)$

Progress Check 10.3

1.

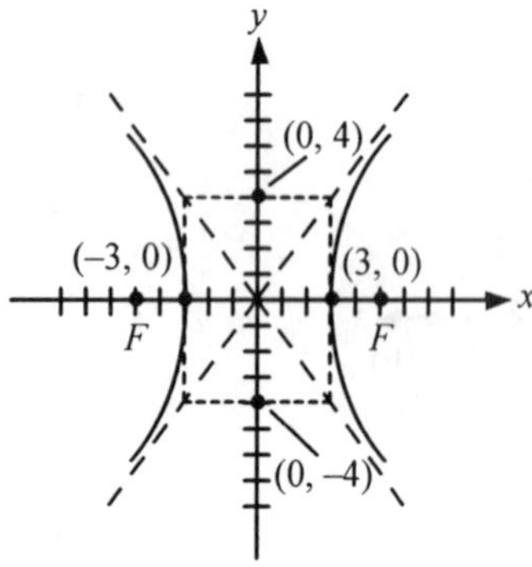

Foci: $(5,0)$, $(-5,0)$

Asymptotes: $y=\pm\dfrac{4}{3}x$

2.

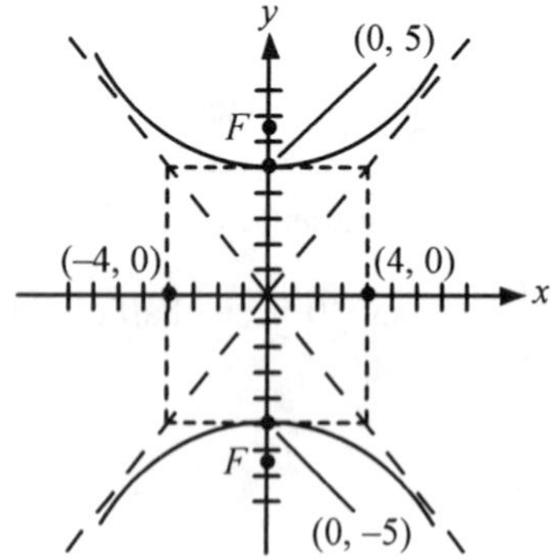

Foci: $(0,\sqrt{41})$, $(0,-\sqrt{41})$

Asymptotes: $y=\pm\dfrac{5}{4}x$

3. $\dfrac{x^2}{36}-\dfrac{y^2}{64}=1$

4. **a.** $\dfrac{x^2}{1600}-\dfrac{y^2}{8400}=1$; $x \le -40$

b.

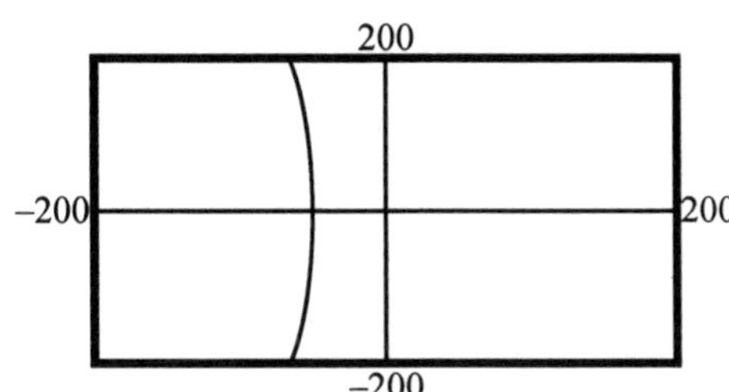

5.

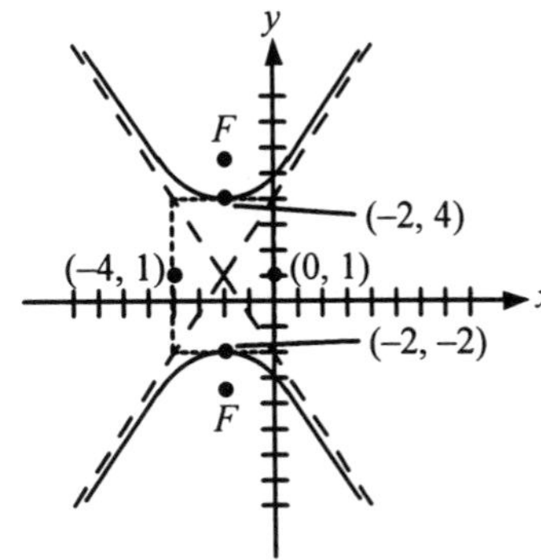

Foci: $\left(-2,1+\sqrt{13}\right)$, $\left(-2,1-\sqrt{13}\right)$

Center: $(-2,1)$

Asymptotes: $y-1=\pm\dfrac{3}{2}(x+2)$

Progress Check 10.4

1.

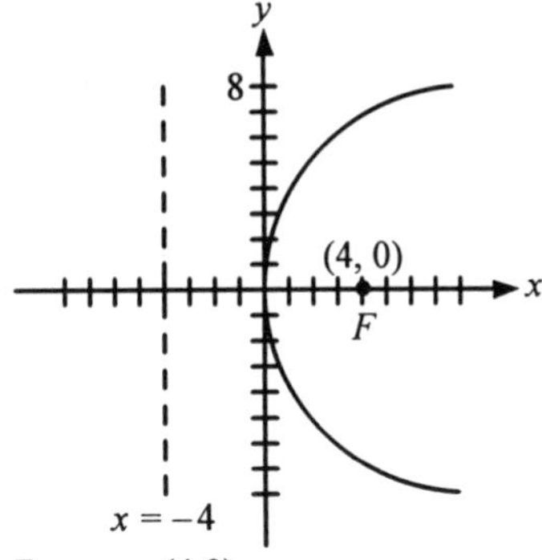

Focus: $(4,0)$
Directrix: $x=-4$

2.

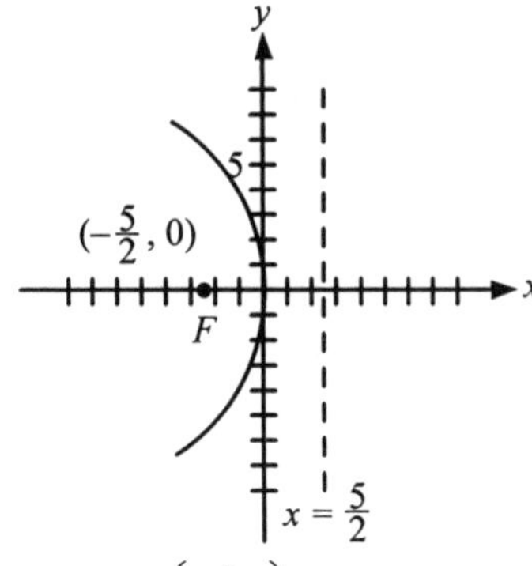

Focus: $\left(-\dfrac{5}{2},0\right)$

Directrix: $x=\dfrac{5}{2}$

3. $x^2=8y$

4. **a.** $y^2=3x$ if the vertex is at $(0,0)$ **b.** $\left(\dfrac{3}{4},0\right)$

5. 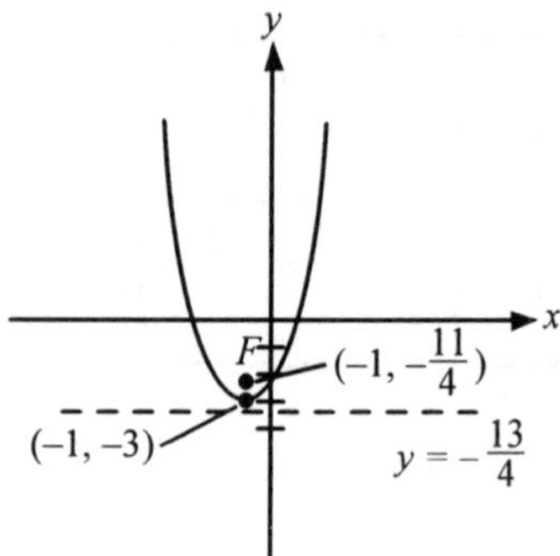

Vertex: $(-1,-3)$

Focus: $\left(-1,-\dfrac{11}{4}\right)$

Directrix: $y=-\dfrac{13}{4}$

6. **a.** Circle
   b. Parabola

7. **a.** No graph
   b. Two intersecting straight lines: $y-1=+\frac{3}{2}(x+2)$

Progress Check 10.5

1. 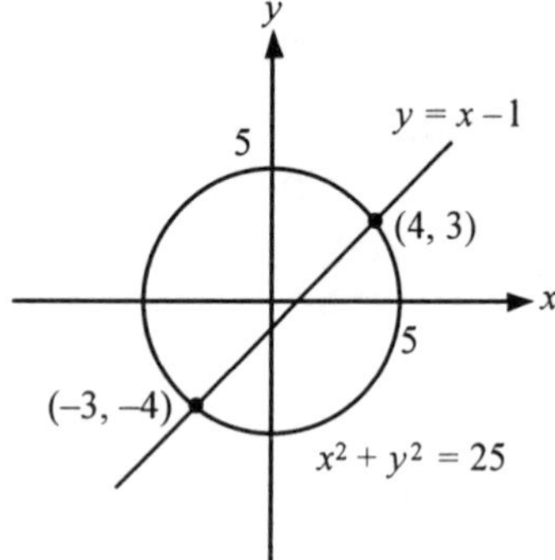

2. **a.** 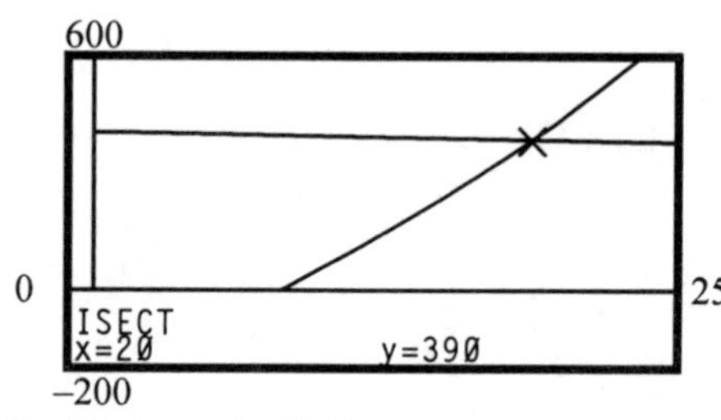

Equilibrium price: $20

Quantity supplied and demanded: 390 pairs of sandals

b. $p^2+3p-70=410-p$ implies $p=20$ if $p>0$. Then, $q=410-20=390$.

3. $(4,3)$, $(4,-3)$, $(-4,3)$, $(-4,-3)$

4. $\left(\sqrt{\dfrac{1+\sqrt{17}}{2}},\dfrac{1+\sqrt{17}}{2}\right)\approx(1.60,\ 2.56)$

$\left(-\sqrt{\dfrac{1+\sqrt{17}}{2}},\dfrac{1+\sqrt{17}}{2}\right)\approx(-1.60,\ 2.56)$

5. $\left(2,-\dfrac{2}{3}\right)$

Progress Check 11.1

1. $(4,-1)$
2. Bonus $=\$14{,}563$; tax $=\$194{,}175$
3. $(8,5,-4)$
4. $a=-1,\ b=2,\ c=-3;\ y=-x^2+2x-3$
5. Ø, inconsistent
6. Dependent

Progress Check 11.2

1. $x=8,\ y=5,\ z=-4$
2. $40,000 at 5%, $340,000 at 8%, and $20,000 at 12%
3. $x=3,\ y=-1,\ z=2$
4. $\left(\dfrac{11}{19},-\dfrac{18}{19}\right)$
5. $a=2,\ b=3,\ c=4,\ d=5$

6. $x = -1, y = \frac{1}{2}, z = \frac{3}{4}$

Progress Check 11.3

1. 17 **2.** -1 **3.** $(1,-2)$

4. 0.27 oz of 35% gold, 0.18 oz of 25% gold **5.** -8 **6.** $x = -1, y = \frac{1}{2}, z = \frac{3}{4}$

Progress Check 11.4

1. **a.** A is 2×1, B is 2×1, C is 1×2. **b.** $\begin{bmatrix} -1 \\ 4 \end{bmatrix}$ **c.** Undefined

d. $\begin{bmatrix} 13 \\ -37 \end{bmatrix}$ **e.** $\begin{bmatrix} -5/2 \\ 7 \end{bmatrix}$ **2.** $\begin{bmatrix} 7 & 0 \\ -3 & 10 \end{bmatrix}$

3. **a.** Undefined **b.** $\begin{bmatrix} 11 \\ -12 \end{bmatrix}$ **4.** $\begin{aligned} 3x - y \quad + 6z &= 1 \\ x \quad + 2y - 3z &= 0 \\ 2x - 3y - z &= -9 \end{aligned}$

5. $\begin{bmatrix} 5 & -2 & 3 \\ 2 & 4 & -1 \\ 1 & 8 & 6 \end{bmatrix} \begin{bmatrix} 1 & 0 & 0 \\ 0 & 1 & 0 \\ 0 & 0 & 1 \end{bmatrix} = \begin{bmatrix} 5 & -2 & 3 \\ 2 & 4 & -1 \\ 1 & 8 & 6 \end{bmatrix}$ and $\begin{bmatrix} 1 & 0 & 0 \\ 0 & 1 & 0 \\ 0 & 0 & 1 \end{bmatrix} \begin{bmatrix} 5 & -2 & 3 \\ 2 & 4 & -1 \\ 1 & 8 & 6 \end{bmatrix} = \begin{bmatrix} 5 & -2 & 3 \\ 2 & 4 & -1 \\ 1 & 8 & 6 \end{bmatrix}$, so $AI = IA = A$.

6. $\begin{bmatrix} 3 & -5 \\ -4 & 7 \end{bmatrix}$ **7.** $(7,-3)$

8. **a.** $y = 24.17x^2 - 229.83x + 2617$

b. About 3,340,000 people, which underestimates the actual result by about 57,000 people.

Progress Check 11.5

1. $A = 3, B = 2$ **2.** $\frac{x^4 + 1}{x^3 - x} = x - \frac{1}{x} + \frac{1}{x+1} + \frac{1}{x-1}$

3. $\frac{6x - 10}{(x-3)^2} = \frac{6}{x-3} + \frac{8}{(x-3)^2}$

4. $A = \frac{1}{2a}, B = \frac{1}{2a}$

Progress Check 11.6

1. $4x + 2y \le 1{,}000$ with $x \ge 0$ and $y \ge 0$

2. a.

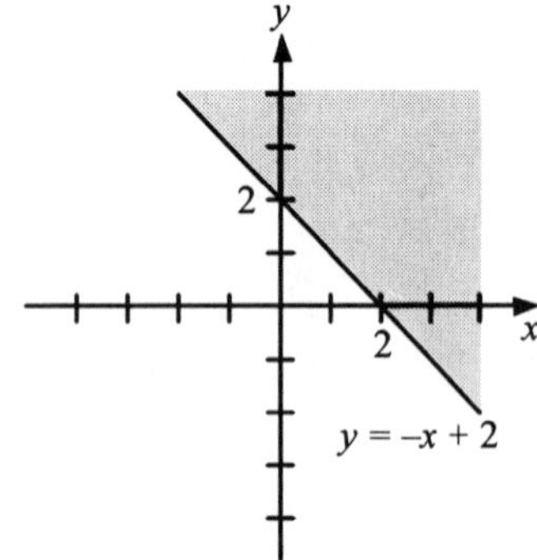

b.

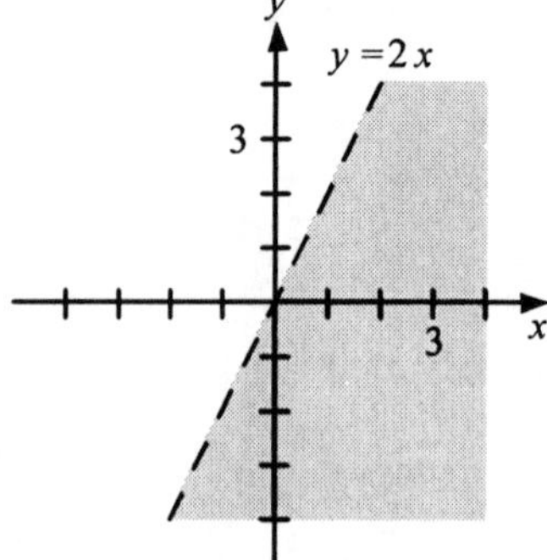

3.

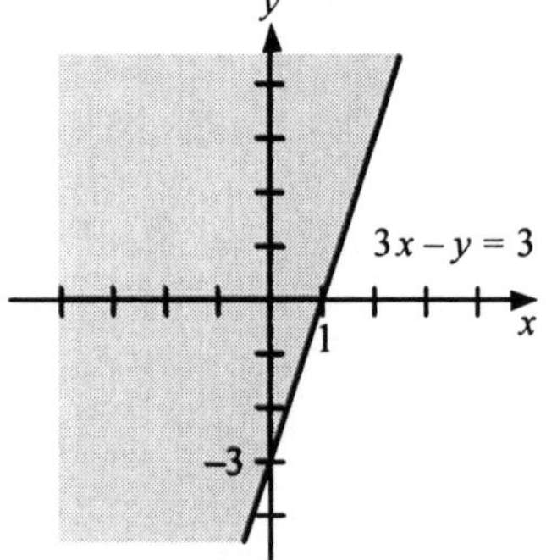

4.

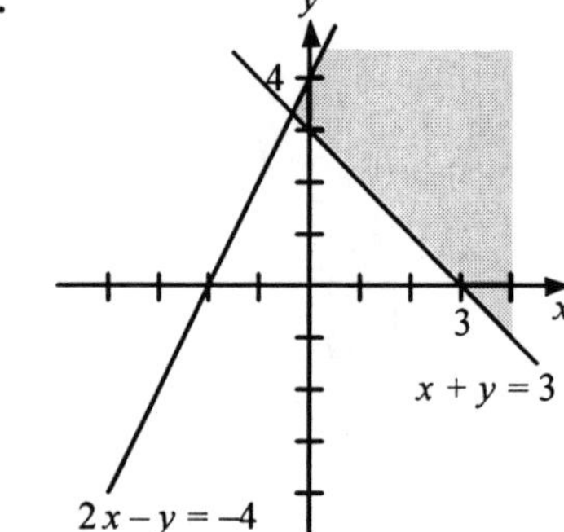

5.

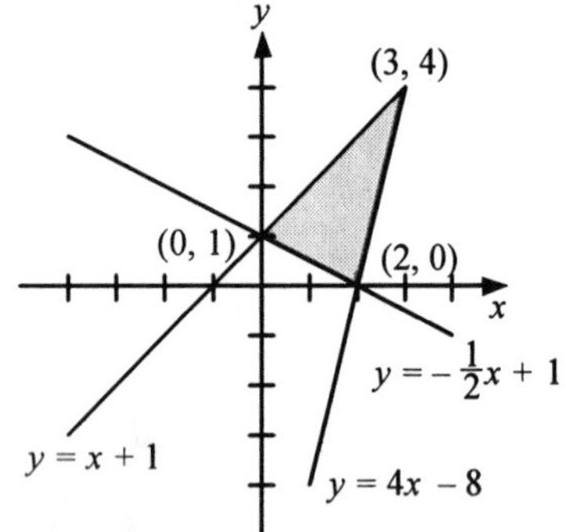

6. 18

7. 20 type A, 40 type B, max profit: $2,000

Progress Check 12.1

1. 3, 8, 13, 18; 248

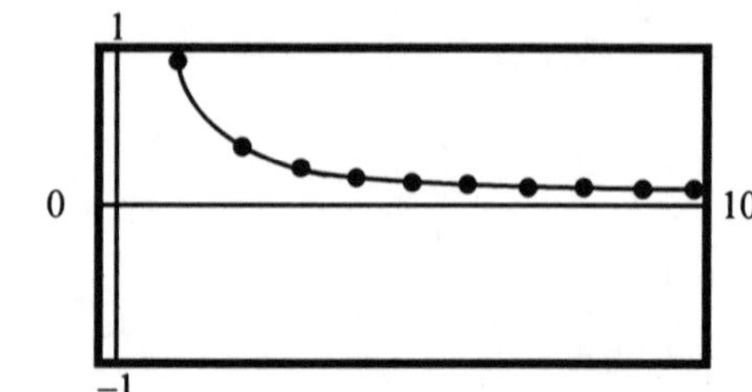

3. $a_n = 7n - 2$; 187

4. $a_n = 3(5)^{n-1}$; 46,875

5. **a.** Arithmetic; 0.125; 1.625, 1.75 **b.** Neither **c.** Geometric; $\frac{1}{3}$; $\frac{1}{3}$, $\frac{1}{9}$

6. \$12,000, \$7,200, \$4,320, \$2,592, \$1,555.20; geometric

Progress Check 12.2

1. 31,375 **2.** 65,535 **3.** **a.** 64, 32, 16, 8, 4, 2, 1; $S_n = \dfrac{64-1(1/2)}{1-(1/2)} = 127$

b. 127 matches are needed because there must be 1 winner and 127 losers, where each loser lost 1 match.

4. $0+3+8+15+24=50$ **5.** $2^3+2^4+2^5+2^6+2^7+2^8=504$ **6.** $a_i = 5i+2$; $\sum_{i=1}^{7}(5i+2)$

7. 3,600 ft **8.** \$197, 851.07

Progress Check 12.3

1. 6 **2.** $\frac{4}{7}$ **3.** $\frac{34}{9}$ **4.** 48 gal

5.

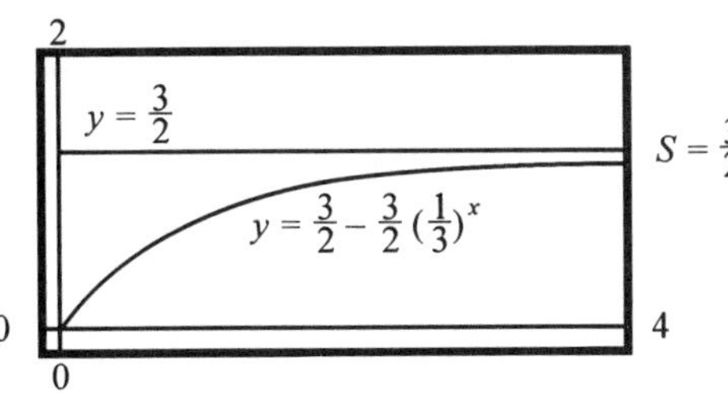

$$S = \frac{1}{1-(1/3)} = \frac{3}{2}$$

Progress Check 12.4

1. Step 1: $2(1) \stackrel{?}{=} 1(1+1)$; $2=2$

Step 2: $\underbrace{2+4+\cdots+2k}_{\text{induction assumption}} + 2(k+1)$

$$= k(k+1) + 2(k+1)$$
$$= (k+1)(k+2)$$
$$= (k+1)[(k+1)+1]$$

2. Step 1: $\left(\frac{a}{b}\right)^1 \stackrel{?}{=} \frac{a^1}{b^1}$; $\frac{a}{b} = \frac{a}{b}$

Step 2:

$$\left(\frac{a}{b}\right)^{k+1} = \left(\frac{a}{b}\right)^k\left(\frac{a}{b}\right)$$
$$= \frac{a^k}{b^k}\left(\frac{a}{b}\right) \quad \text{(induction assumption)}$$
$$= \frac{a^{k+1}}{b^{k+1}}$$

3. **a.** This equation is the formula for the sum of the first n terms in an arithmetic series.

b. Step 1: If $n=1$, $a_1 = \frac{1}{2}\left[2a_1 + (1-1)d\right]$, which is true.

Step 2:

$$\underbrace{a_1 + (a_1+d) + \cdots + \left[a_1 + (k-1)d\right]}_{\text{induction assumption}} + (a_1 + kd)$$

$$= \frac{k}{2}\left[2a_1 + (k-1)d\right] + (a_1 + kd)$$

$$= ka_1 + \frac{k^2 d}{2} - \frac{kd}{2} + a_1 + kd$$

$$= \frac{k^2 d + kd + 2ka_1 + 2a_1}{2}$$

$$= \frac{kd(k+1) + 2a_1(k+1)}{2}$$

$$= \frac{k+1}{2}(2a_1 + kd)$$

c. \$367,500

4. Step 1: If $n=2$, $3(2) < 3^2$, which is true.

$3k < 3^k$ (induction assumption)

Step 2:

$3 \cdot 3k < 3 \cdot 3^k$

$9k < 3^{k+1}$

$3(k+1) < 3^{k+1}$ (since $3(k+1) < 9k$ for $k \ge 2$)

Progress Check 12.5

1. $x^9 + 9x^8h + 36x^7h^2 + 84x^6h^3 + 126x^5h^4 + 126x^4h^5 + 84x^3h^6 + 36x^2h^7 + 9xh^8 + h^9$

2. $x^5 + 15x^4y + 90x^3y^2 + 270x^2y^3 + 405xy^4 + 243y^5$

3. $81x^4 - 216x^3y + 216x^2y^2 - 96xy^3 + 16y^4$

4. $x^6 + 12x^5 + 60x^4 + 160x^3 + 240x^2 + 192x + 64$

5. $x^{10} + 10x^9y + 45x^8y^2 - 120x^7y^3$

6. 1, 6, 15, 20, 15, 6, 1; the binomial coefficients 1, 6, 15, 20, 15, 6, 1 match the entries in row 6 of Pascal's triangle.

7. $x^9 + 9x^8h + 36x^7h^2 + 84x^6h^3 + 126x^5h^4 + 126x^4h^5 + 84x^3h^6 + 36x^2h^7 + 9xh^8 + h^9$

8. $860.160x^4$

9. 34.56%

Progress Check 12.6

1. 6;

A.A — M — A.A, Male
A.A — F — A.A, Female
A.S — M — A.S, Male
A.S — F — A.S, Female
A.A.S — M — A.A.S, Male
A.A.S — F — A.A.S, Female

2. 40,320

3. 120

4. 90

5. 50,400

6. 105

7. 5,040

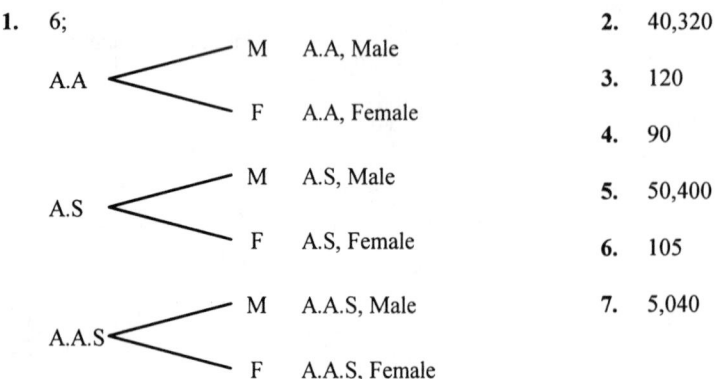

Progress Check 12.7

1. $\frac{1}{26}$

2. $\frac{5}{9}$

3. $\frac{1}{70}$

4. $\frac{56}{143}$

5. $\frac{17}{18}$

6. 0.01; 0.81

7. **a.** $\frac{4}{13}$ **b.** $\frac{11}{26}$

8.

1,1	1,2	1,3	1,4	1,5
2,1	2,2	2,3	2,4	2,5
3,1	3,2	3,3	3,4	3,5
4,1	4,2	4,3	4,4	4,5
5,1	5,2	5,3	5,4	5,5

a. $\frac{4}{25}$; P (boy calls even and girl calls even) $= \frac{2}{5} \cdot \frac{2}{5} = \frac{4}{25}$

b. $\frac{8}{25}$; P (boy calls 3, and girl does not call 3) $+P$ (girl calls 3, and boy does not call 3) $= \left(\frac{1}{5} \cdot \frac{4}{5}\right) + \left(\frac{1}{5} \cdot \frac{4}{5}\right) = \frac{8}{25}$.

● ● ●

Answers to Odd Numbered Problems

Exercises R.1

1. $0.8\overline{0}$ **3.** $0.\overline{45}$ **5.** $6.1\overline{6}$

7. $1.\overline{428571}$ **9.** $\frac{2}{9}$ **11.** $\frac{321}{999}=\frac{107}{333}$

13. $\frac{271}{99}$ **15.** $6\left(\text{or } \frac{6}{1}\right)$ **17.** $\frac{1{,}929}{900}=\frac{643}{300}$

19. Real number, rational number, integer

21. None of these

23. Real number, rational number, integer

25. Real number, rational number, integer

27. Real number, irrational number

29. Real number, rational number

31. Commutative property of addition

33. Associative property of multiplication

35. Multiplication identity property

37. Commutative property of multiplication

39. Distributive property

41. Associative property of multiplication

43. Commutative property of addition

45. Commutative property of multiplication

47. Multiplication inverse property

49. Commutative property of addition

51. $<$ **53.** $<$ **55.** $<$ **57.** $<$

59. $<$ **61.** $>$ **63.** -7

65. 5 **67.** -2 **69.** $\frac{14}{5}$

71. -0.54 **73.** -4.1 **75.** -66

77. 2 **79.** $-\frac{1}{32}$ **81.** $\frac{20}{9}$

83. 0 **85.** -0.04 **87.** 10.8

89. -2 **91.** -14 **93.** 6

95. -27 **97.** -16 **99.** $-\frac{1}{16}$

101. -126 **103.** -40 **105.** -70

107. **a.** $\frac{5}{4}$ **b.** 3.74 **c.** $\frac{3}{5}$ **d.** 11

109. 2
2.594
2.705
2.717
2.718

111. \$3,720.59
\$6,799.13
\$12,217.62

113. 896.2
413.1
135.5
89.2

115. $a(b+c)$ [total area] $= ab$ [area 1] $+ ac$ [area 2]

117. $y=1$, x any real number; $x=0$, y any real number

119. False, $\frac{1}{2}$ **121.** False, 1 **123.** False, $\frac{1}{2}$

125. True **127.** False, 0

Exercises R.2

1. -3 **3.** -32 **5.** -2

7. 10 **9.** 64 **11.** -52

13. -30 **15.** -12 **17.** -15

19. 14 **21.** $\frac{1}{2}$ **23.** 1

25. $-\frac{1}{2}$ **27.** 2 terms: $2n$, $4w$

29. 3 terms: $x^2, -3x, -2$ **31.** 1 term: $\frac{1}{2}bh$

33. $19a$ **35.** $14a-8b$ **37.** $3xy+5cd$

39. $-3x^3+8y^2$ **41.** $2x-5y$

43. $-16a+11b$ **45.** $10x-7y$

47. $2x^3+13x^2y^2+9xy^3-3x^3y$

49. $7y^3-3y^2+4y+6$ **51.** b^3-3b^2+b+5

53. $12a+14$ **55.** $2a+5b$ **57.** 8

59. $12.71x$ **61.** $3.14x^2+12x+7.065$

63. **a.** 302 cubic feet **b.** 2,270 gals

65. $0.045t$ **67.** $c+1$

69. $p+t=p+0.04p=1.04p$

71. $a+mn$

73. The sum of *x* and the product of four and *y*

75. The quotient of *a* and the product of *b* and *c*

Exercises R.3

1. $3^5=243$ **3.** 1 **5.** $\frac{1}{9}$

7. .00204 **9.** 576 **11.** $\frac{3}{4}$

13. $4^3=64$ **15.** $(-3)^5=-243$

17. $\frac{1}{2^3}=\frac{1}{8}$ **19.** $2^3=8$ **21.** $\frac{1}{3^2}=\frac{1}{9}$

23. $\frac{3}{4}$ **25.** x^7 **27.** x^{12}

29. $4p^2$ **31.** $-125y^{15}z^3$ **33.** c^6

35. $\frac{9}{x}$ **37.** $\frac{1}{(x+y)^3}$ **39.** $-27x^4$

41. x^2 **43.** $\frac{1}{t^2}$ **45.** $\frac{25}{y^2}$

47. $\frac{1}{x}$ **49.** $\frac{-8}{x^3}$ **51.** $\frac{9}{(x+h)^2}$

53. $\frac{a^2x^2}{4}$ **55.** $\frac{2}{a^2}$ **57.** $\frac{-x^4}{4y^5z^8}$

59. $\frac{x^4}{27}$ **61.** $\frac{xz^2}{y^2}$ **63.** $\frac{a^4}{8x^{10}}$

65. 2^{x+y} **67.** 2^{1-n} or $\frac{1}{2^{n-1}}$

69. 5^{2x^2} **71.** y^{7a} **73.** $(a-b)^{x+y}$

75. 1 **77.** x^{1+x} **79.** y^{6x^2}

81. $\frac{1}{y}$ **83.** x^{ap-1} **85.** $\frac{x^a}{y}$

87. $(1-x)^{a+2}$ **89.** \$257.49 **91.** 4.2×10^1

93. 3.4251×10^4 **95.** 5.9×10^{12}

97. 1.26×10^6 **99.** 5.3×10^{-23} **101.** 92,000

103. 42.1 **105.** 580,000,000

107. 6,280,000,000,000,000,000

109. .000000000000000000000003

111. 5.58×10^8 miles **113.** 3.6×10^{-25} grams

115. 1 trillion = 1×10^{12}
1 million = 1×10^6
10^6 times

117. **a.** 3.35×10^{25}
b. .0000002 %

Exercises R.4

1. $2x-2y$ **3.** $-5x^4+5x^3+5x^2$

5. $-8x^2yz+2xy^2z-14xyz^2$

7. $p^4q^2+p^2q^4$ **9.** $4n+3x$

11. $2x+h$ 13. $6x^3+8x^2-10x$

15. $a^2+7a+12$ 17. $x^2+2x-35$

19. x^2-16 21. $6x^2-11x+4$

23. $12t^2+28ct-5c^2$ 25. k^2-4k+4

27. $4x^2+12xy+9y^2$ 29. x^3-3x^2+3x-1

31. $y^3+y^2-21y+4$ 33. x^3-y^3

35. $-6y^3+5y^2+3y-2$

37. $x^2-2xy+y^2-2x+2y+1$

39. $x^4-2x^2y^2+y^4$ 41. y^2-9

43. $25n^2-49$ 45. $36x^2-y^2$

47. a^2+2a+1 49. $x^2-14x+49$

51. $9c^2+30c+25$ 53. $4-28x+49x^2$

55. $25x^2+40xy+16y^2$ 57. $100a^2-100ab+25b^2$

59. $x^2-2xy+y^2-2x+2y+1$

61. $x^{2n}+7x^n+10$ 63. $z^{2a}+6z^a+9$

65. $x^{2a}-y^{2b}$ 67. $a^{2bx}+2+a^{-2bx}$

69. $x^{3n}+y^{3n}$

71. $\dfrac{x^2+2xh+h^2+1-x^2-1}{h}=\dfrac{2xh+h^2}{h}=2x+h$

73. $3x^2+3xh+h^2$ 75. $2x+h+2$

77. $810{,}000-2{,}500x^2$
The gross monthly income would constantly decrease, as seen by the second term. On the 18th cut, there would be no revenue.

79. **a.** $225-x^2$ **b.** $x=0$

81. $20x+25$

Exercises R.5

1. $\{3\}$ 3. $\{6\}$ 5. $\{-1\}$

7. $\{3\}$ 9. $\{2\}$ 11. $\{-6\}$

13. Set of all real numbers 15. $\emptyset$

17. $\{0\}$ 19. $\{8\}$ 21. $\{0\}$

23. $\{5\}$ 25. $\{-5\}$ 27. $\{3\}$

29. $\left\{\dfrac{36}{7}\right\}$ 31. 32 33. 40

35. 8 37. 11 39. 6

41. $m=\dfrac{F}{a}$ 43. $a=\dfrac{2d}{t^2}$ 45. $m=\dfrac{P}{gh}$

47. $b=3A-a-c$ 49. $L=\dfrac{2s}{n}-a$

51. $t=\dfrac{a-p}{pr}$ 53. $R^2=\dfrac{A}{\pi}+r^2$

55. $r=\dfrac{a-S}{-S}$ 57. $D=An-A$

59. $r=\dfrac{2E-IR}{2I}$ 61. $x=\dfrac{a}{2}$

63. $x=3az-1-t$ 65. $x=\dfrac{c-b}{4}$

67. $x=\dfrac{a+5b}{2}$

69. **a.** $x=4-2y$ **b.** $y=\dfrac{4-x}{2}$

71. **a.** $x=-y-2$ **b.** $y=-x-2$

73. **a.** $x=\dfrac{6-4y}{3}$ **b.** $y=\dfrac{6-3x}{4}$

75. **a.** $x=\dfrac{2y-50}{3}$ **b.** $y=\dfrac{3x+50}{2}$

77. **a.** $w=5h-200$ **b.** $w=130$ pounds

79. 1.3 seconds 81. 61.75 in.^{2}

Exercises R.6

1. 1,100 cans, 3,300 cans, 1,110 cans

3. \$8,500 5. \$5,450 7. \$1,200

9. 2 11. −5 13. 45

15. 40 degrees, 60 degrees, 80 degrees

17. 67.5 degrees 19. 35 degrees, 145 degrees

21. \$18,000 @12%; \$36,000 @9%

23. **a.** \$271,000 @5%; \$104,000 @12%
b. \$129,000 @5%; \$246,000 @12%

25. 4 hours 27. 21 miles 29. 6.5 hours

31. A: 4.5 quarts; B: 13.5 quarts

33. 38 lb @35%; 57 lb @10%

35. 20 quarts **37.** 12.5 lb **39.** 9 hours

41. 44 ft **43.** 95 **45.** 324 gal

47. 92 mi **49.** 25 qt

51. **a.** 15 ft **b.** $\frac{12}{5}$ ft **c.** $\frac{25}{4}$ mi

Exercises R.7

1. $(8,\infty)$

0 2 8

3. $(2,\infty)$

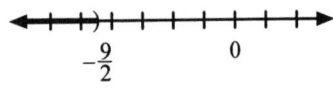

0 2

5. $(-\infty,1]$

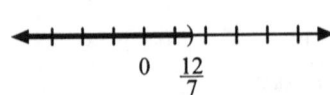

0 1

7. $\left(-\infty,-\frac{9}{2}\right)$

$-\frac{9}{2}$ 0

9. $(-\infty,-1]$

−1 0

11. $(-\infty,-6)$

−6 0 2

13. $\left(-\infty,\frac{12}{7}\right)$

0 $\frac{12}{7}$

15. $(-2,\infty)$

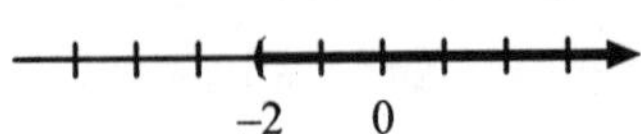

17. $(-\infty,-3]$

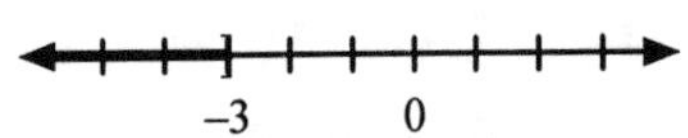

19. $(-8,\infty)$

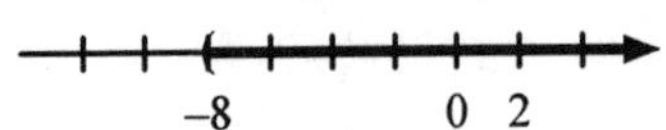

21. Set of all real numbers

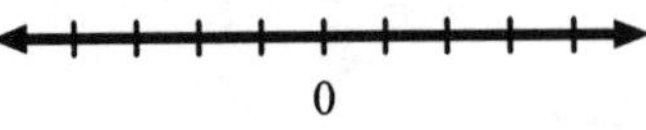

23. $(0,\infty)$

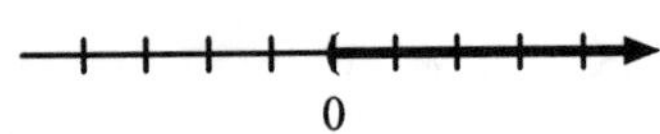

25. $\varnothing$

27. $(-\infty,0]$

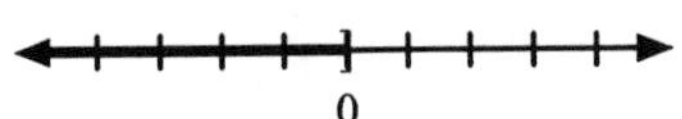

29. $\varnothing$

31. $\{x{:}1 \le x < 4\}$

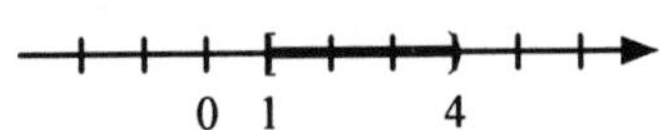

33. $(-\infty,3]$

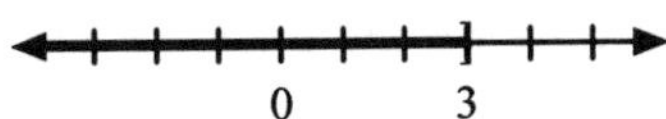

35. $(-3,0)$; $\{x{:} -3 < x < 0\}$

37. $\{x{:}\ x \ge 0\}$

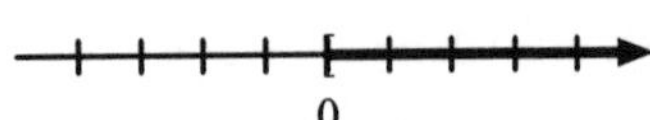

39. $(-\infty,-2)\cup(2,\infty)$; $\{x{:}\ x < -2 \text{ or } x > 2\}$

41. $(2,6)$

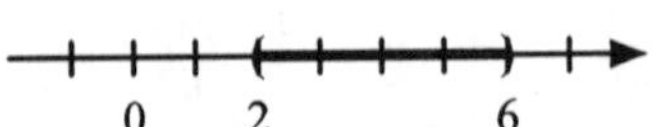

43. $(1,2)$

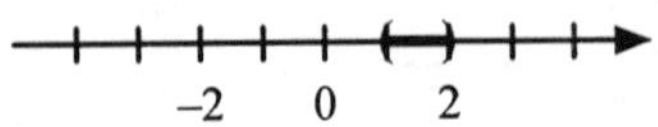

45.  $[-5, 7]$

47. $\left[\frac{3}{2}, \frac{5}{2}\right]$

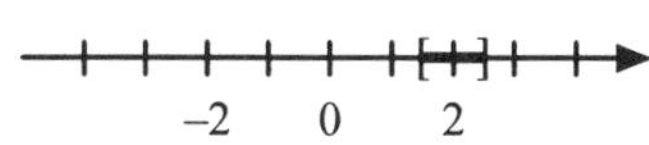

49. $\left(\frac{7}{4}, 4\right]$

0 $\frac{7}{4}$ 4

51. $(-\infty, 1) \cup (6, \infty)$

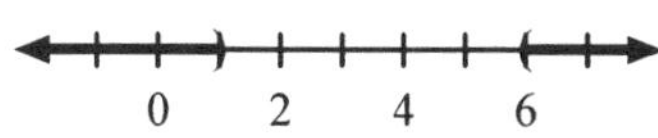

53. $(-\infty, -3) \cup (-1, \infty)$

–3 –1 1 3

55. $\left(\frac{4}{3}, \infty\right)$

–2 0 2

57. $\left(-\infty, \frac{1}{4}\right) \cup \left(\frac{3}{4}, \infty\right)$

–1 0 1

59. $[24{,}500 \text{ mi}, 25{,}500 \text{ mi})$; $[24{,}855 \text{ mi}, 24{,}865 \text{ mi})\infty$

61. $(0^\circ, 40^\circ)$

63. $\{x: x > 11{,}500, \text{ where } x \text{ is an integer}\}$

65. $-37.97^\circ < F < 673.84^\circ$

67. $[71, 91)$

69. Purchases over \$300 would be needed to pay for the card and save over \$25.

Chapter R Review Exercises

1. **a.** 7 **b.** $\frac{1}{2}$

3. $-\frac{5}{7}$ **5.** 9 **7.** $3x-3$

9. $1.\overline{18}$ **11.** $3x-1$

13. **a.** Commutative property of addition
b. Distributive property

15. Real, rational **17.** –2

19. 0.000040 **21.** $81x^6$ **23.** 2^{x+1}

25. 3^{2x^2} **27.** $\frac{1}{81}$ **29.** 3^{1-n}

31. $\frac{27}{x^8 y}$ **33.** $2x^3y - 3x^2y^2 + 4x^2y$

35. $s - 2c^2$ **37.** $1 - 2k + k^2$

39. $x^{2a} - 2x^a y^b + y^{2b}$ **41.** $a^3 + b^3$

43. $\{2\}$ **45.** $[0, \infty)$ **47.** $\{100\}$

49. $(1, \infty)$ **51.** $\varnothing$ **53.** $\frac{7}{2} > x > \frac{1}{2}$

55. $r = \frac{C}{2\pi}$ **57.** $x = \frac{n-b}{a}$

59. $y = x - z - 10P$ **61.** False, $\sqrt{2}$

63. False, 0 **65.** True **67.** False, –1

69. True **71.** 678.6 cubic inches

73. 82 **75.** \$522 **77.** $R = \frac{E}{I}$

79. 51.3 mi **81.** \$85,106 **83.** a

85. d **87.** a **89.** c **91.** a

Chapter R Test

1. Real number, rational number, integer

2. Distributive property

3. –220 **4.** $\frac{1}{81}$ **5.** $\frac{a^3x^6}{1{,}000}$

6. $\frac{1}{x^2}$ **7.** $\{-6\}$ **8.** $\left(-\infty, \frac{2}{3}\right]$

9. $(-\infty, \infty)$ **10.** $\left\{\frac{10}{3}\right\}$ **11.** $15x^2 + 16x - 7$

12. $-11x + 2$ **13.** $-\frac{1}{2}$ **14.** 5

15. $[3,8)$

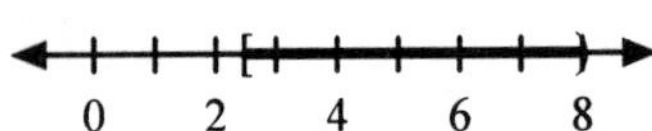

16. $x = \dfrac{5-2y}{3}$ **17.** $c = \dfrac{2A - hb}{h}$

18. $2x + h$ **19.** 1.72×10^8 **20.** $882

Exercises 1.1

1. $A = s^2$: $(0,\infty)$, $(0,\infty)$

3. $s = P/4$; $(0,\infty)$, $(0,\infty)$

5. $e = 28n$; $[0,\infty)$, $[0,\infty)$

7. $g = 294t$, $[0,\infty)$, $[0,\infty)$

9. $c = 5x + 400$; $\{x: x \ge 0$ where x is an integer$\}$
$\{400, 405, 410, \ldots\}$

11. $(-2,-5)$, $(1,4)$

13. $\{(-2,6), (-1,5), (0,4), (1,3), (2,2)\}$ answers will vary

15. $\{(-2,-1), (-1,-2), (0,-1), (1,2), (2,7)\}$ answer will vary

17. $(0, 7), \left(\frac{7}{3}, 0\right), (-5, 22), \left(\frac{2}{3}, 5\right)$

19. 5

21. Function, $\{1,2,3\}$, $\{-3,0,1\}$

23. Function, $\{2,3,4\}$, $\{-1,0\}$

25. Not a function, $\{2\}$, $\{1,2,3\}$

27. Not a function **29.** Function

31.

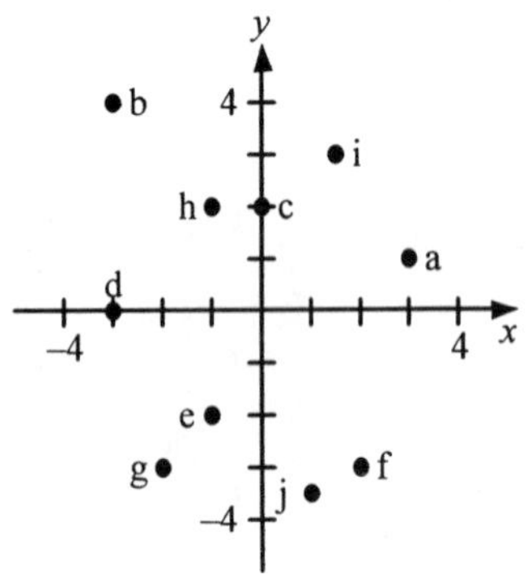

33.

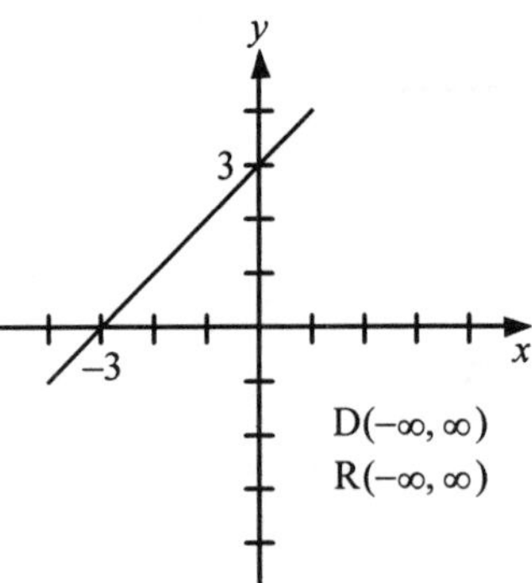

35.

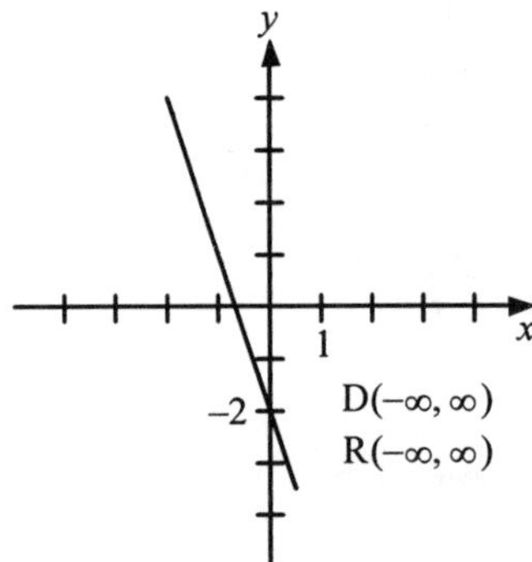

37.

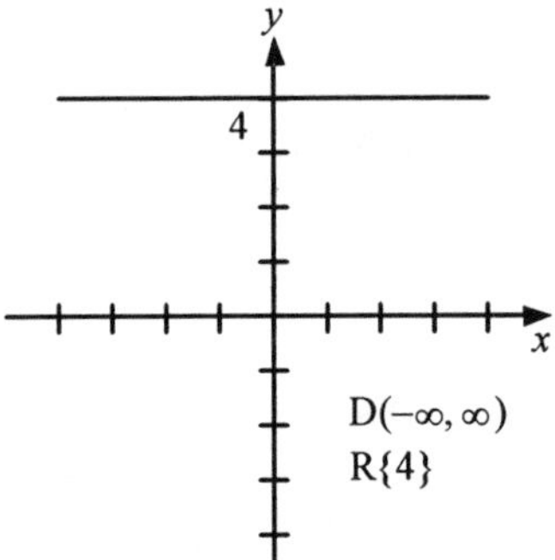

39.

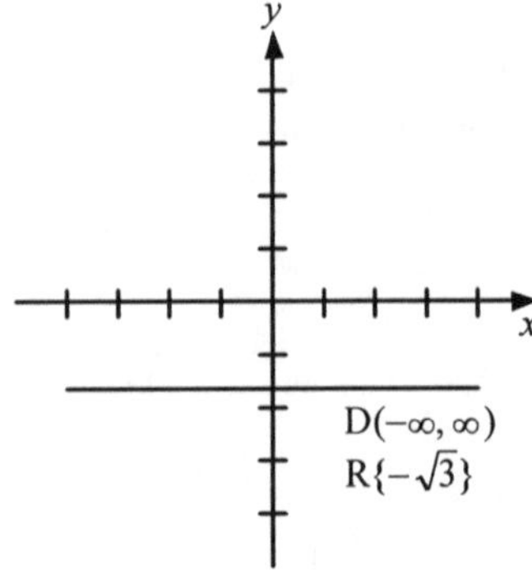

41.

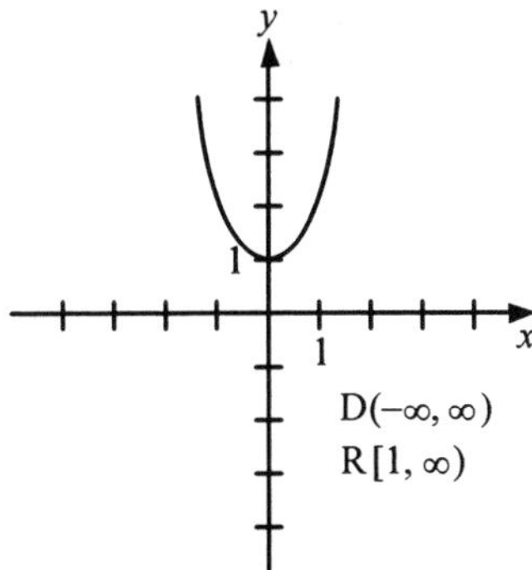

43.

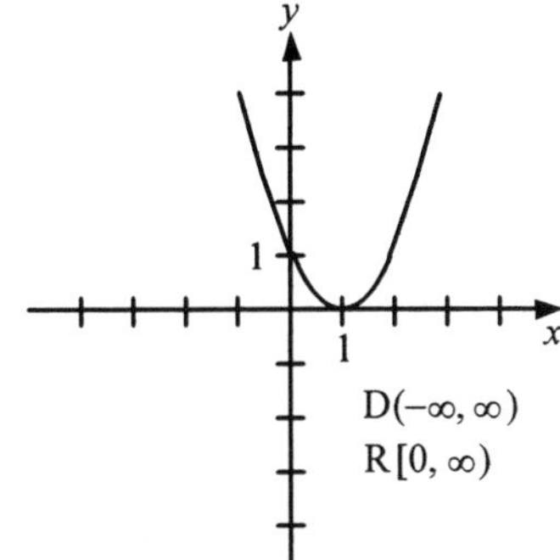

45.

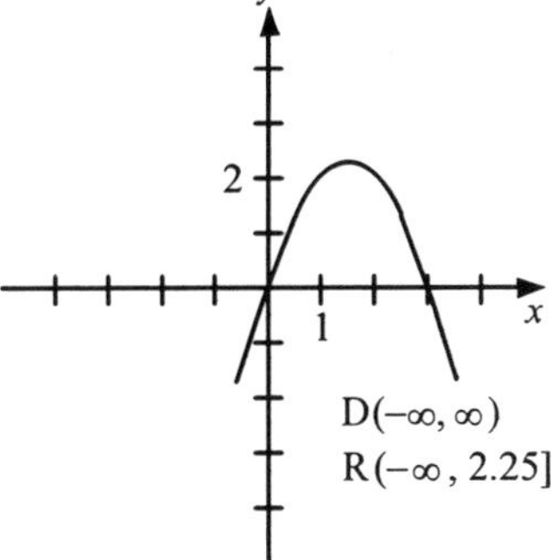

47.

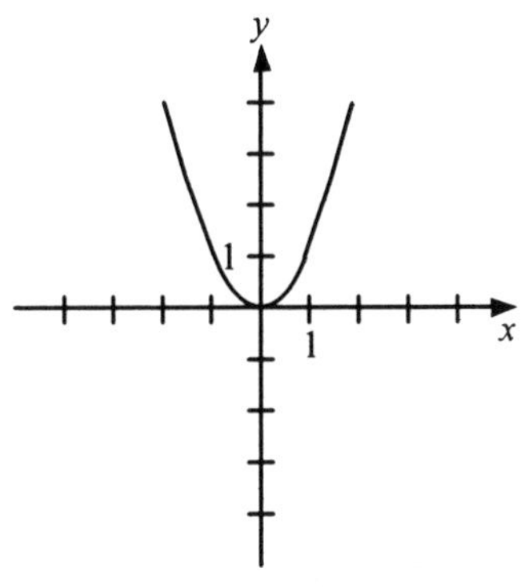

49.

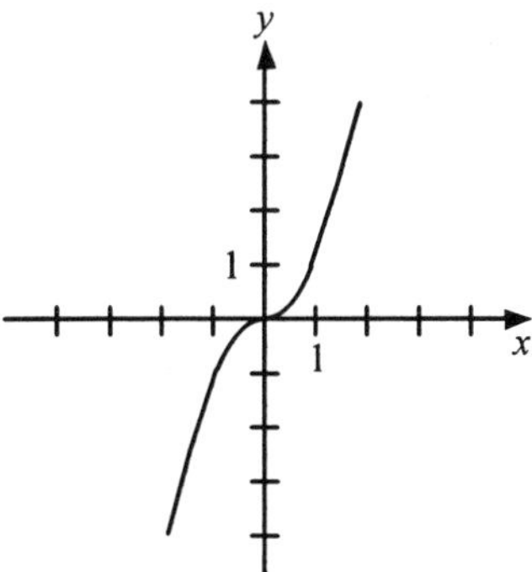

51.

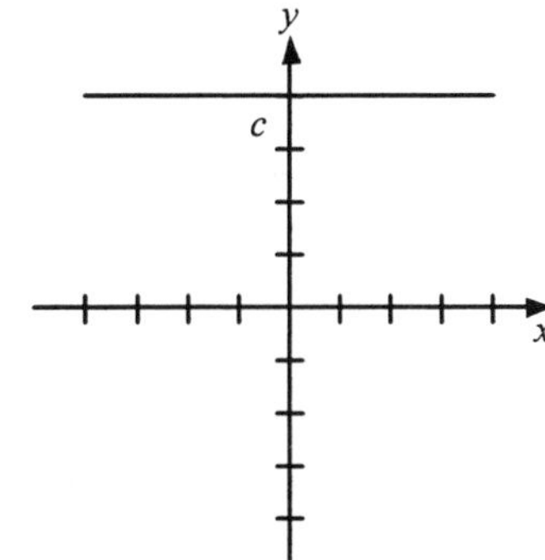

53.

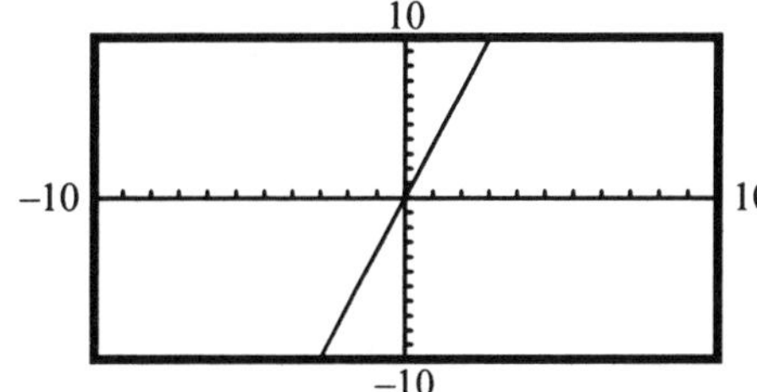

55.

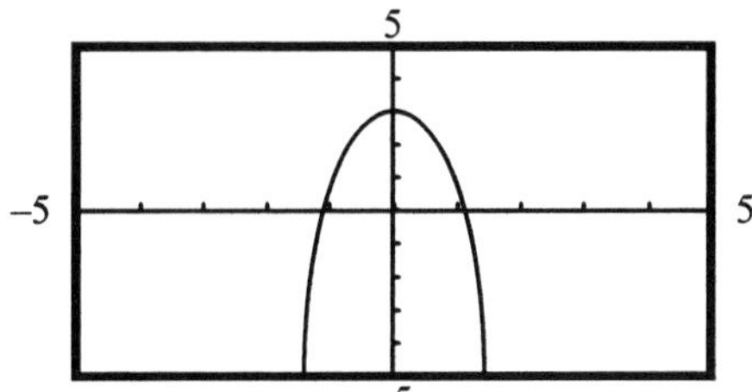

57.

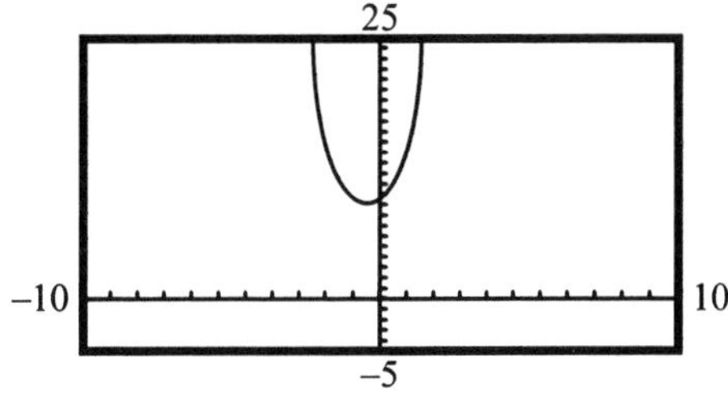

59.

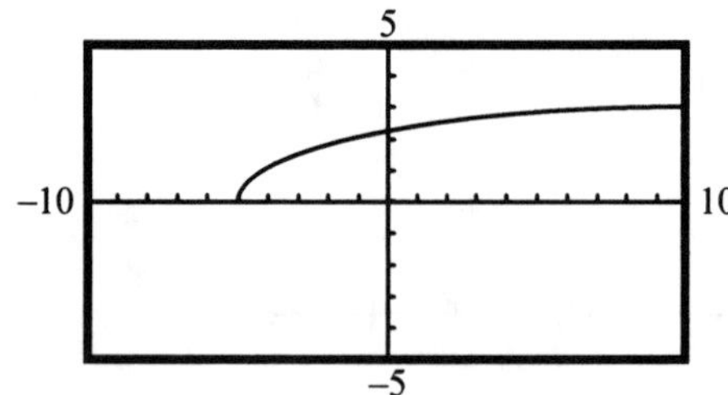

61.

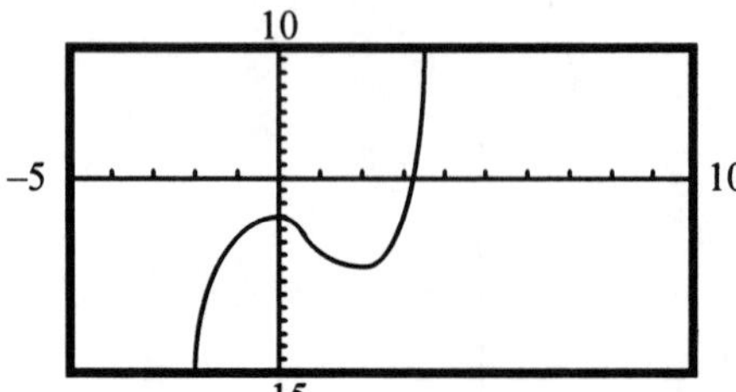

63.

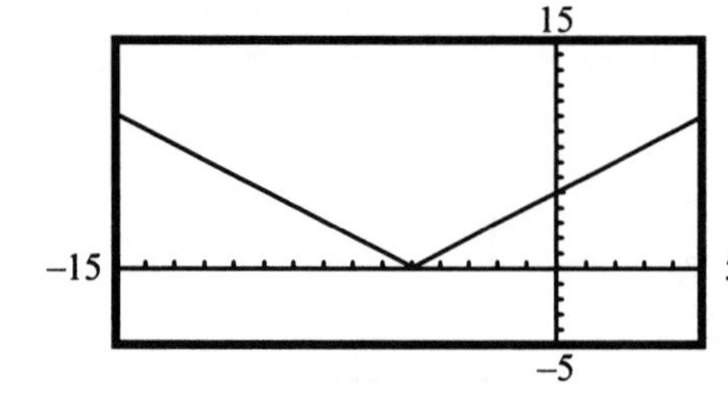

65. b **67.** a

69. $[-1, 15]$ by $[-1, 10]$ **71.** $[-5, 20]$ by $[-5, 20]$

73. $[-6, \infty)$; $[0, \infty)$ **75.** $[3, \infty)$; $[5, \infty)$

77. $(-\infty, \infty)$; $[2, \infty)$ **79.** $(-\infty, \infty)$; $\{7\}$

81. $x \neq -1,\ y \neq 0$ **83.** $(2.5, 3.3)$

85. $(-0.2, -2.2)$ **87.** $(3.3, 20.0)$

89.

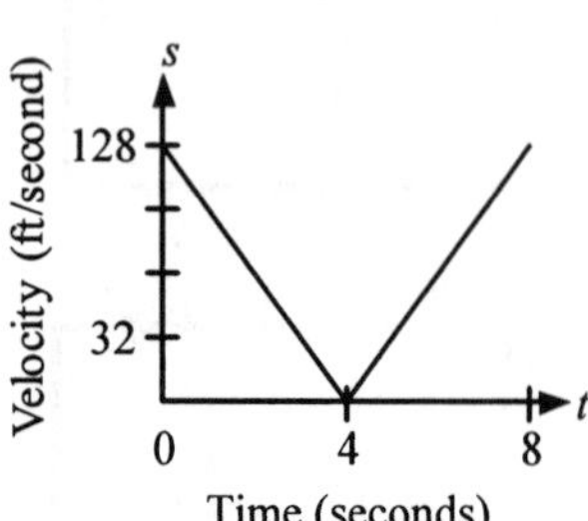

D: $[0, 8]$
R: $[0, 128]$

91. $a = 10{,}000 - 50n$; $[0, 200]$ $[0, 10{,}000]$

93. $t = 31,\ 172 + .36(i - 115{,}000)$; $[115{,}000, 250{,}000]$, $[31{,}172, 79{,}772]$

95. $A = 3(8 - x)$; $(0, 8)$, $(0, 24)$

97. $A = 2x\sqrt{9 - x^2}$; $(0, 3)$, $(0, 9]$

Exercises 1.2

1. $-2, -3, 2$ **3.** $11, 5, -3$

5. $19, 4, 7$ **7.** $\frac{2}{5} - 2$, undefined

9. $5, 5, 5$ **11.** $3a + 3$

13. **a.** 3 **b.** −1 **c.** No

15. $f(50) = \$48{,}764.58$, which is the value of the banjo today

17. $f(1) = 5.1$ m above water; $f(1.42) = .12$ m above water

19. $M(2) = 3$ prime; $M(3) = 7$ prime; $M(5) = 31$ prime

21. **a.** 2 **b.** −10 **c.** 10
d. −4 **e.** 3 **f.** $\frac{1}{2}$
g. 4

23. **a.** −7 **b.** 3 **c.** −4
d. $\frac{20}{3}$

25. **a.** 3 **b.** 4 **c.** 6
d. no, $7 \neq 6$ **e.** $a + 1$ **f.** $b + 1$
g. $a + b + 1$. **h.** no, $a + b + 2 \neq a + b + 1$

27. $a + b + 4 \neq a + 4 + b + 4$

29. $a^3 + 3a^2b + 3ab^2 + b^3 \neq a^3 + b^3$

31. **a.** $2(x + h)^2 - 1$ **b.** $4xh + 2h^2$
c. $4x + 2h$

33. $1 + 2x + h$ **35.** 7

37. 0 **39.** $-2x - h$

41. $t = f(p) = 400 + .28(p - 1000)$

43. $c = f(n) = \begin{cases} 5.25 & \text{if } 0 \le n \le 12 \\ 5.25 + .28(p - 1000) & \text{if } 12 < n \le 48 \end{cases}$

45. $s = f(x) = \begin{cases} 0.28x & \text{if} \quad 0 \le x \le 3400 \\ 952 + 0.03x & \text{if} \quad 3{,}400 < x \le 13{,}100 \\ 1345 + 0.06x & \text{if} \quad x > 13{,}100 \end{cases}$

47. **a.** −2 **b.** −1 **c.** undefined

49. **a.** −1 **b.** 0 **c.** 1
d. −1

51. **a.** 25 **b.** 25 **c.** undefined
d. 30 **e.** 25

53. **a.** 0 **b.** 0 **c.** 1
d. 1 **e.** undefined
f. assigns an integer value to each number based on the integer less than or equal to the number

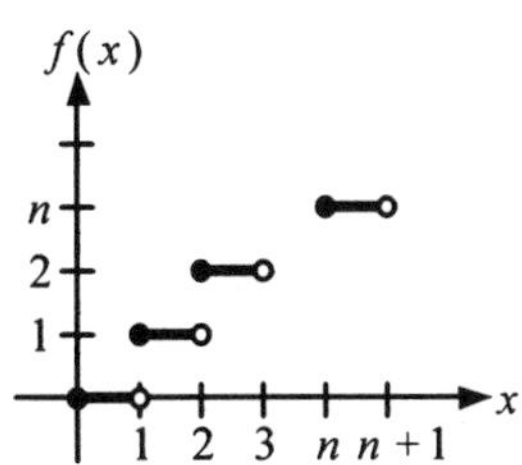

55.

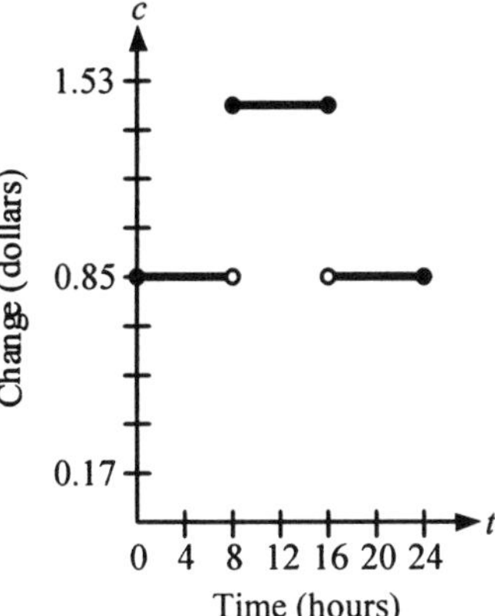

57.

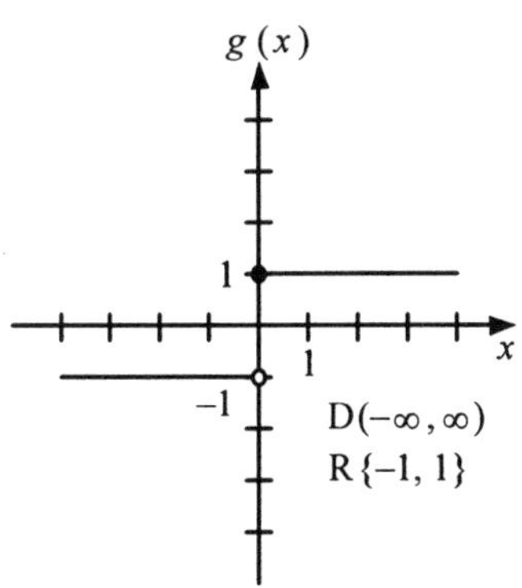

59.

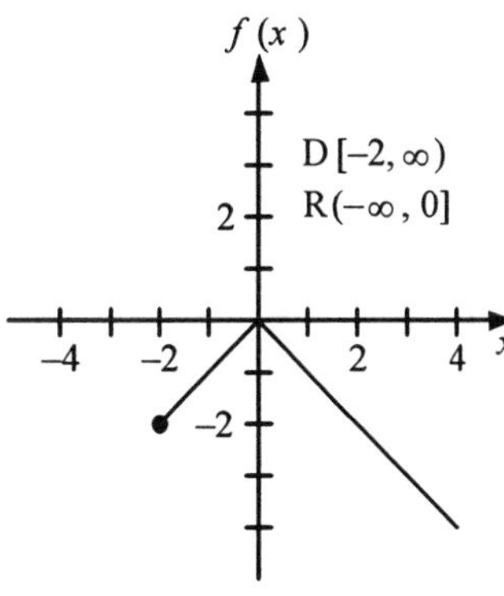

61.

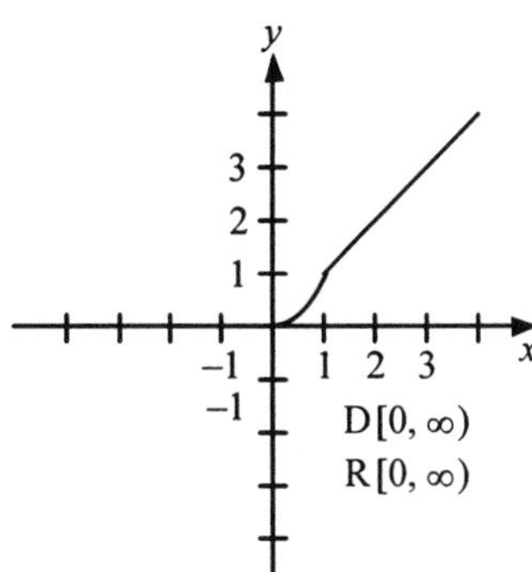

63.

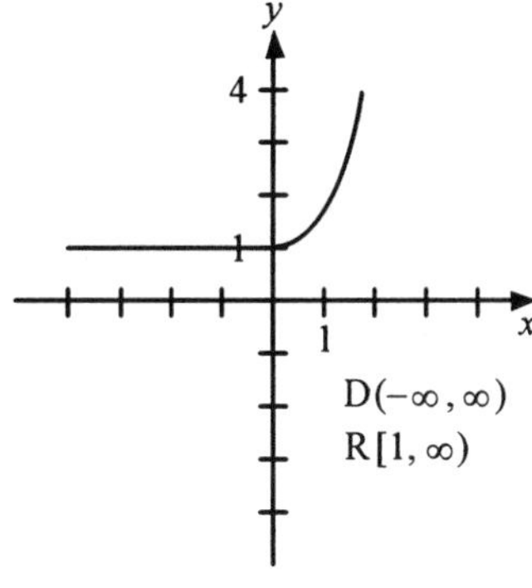

65.

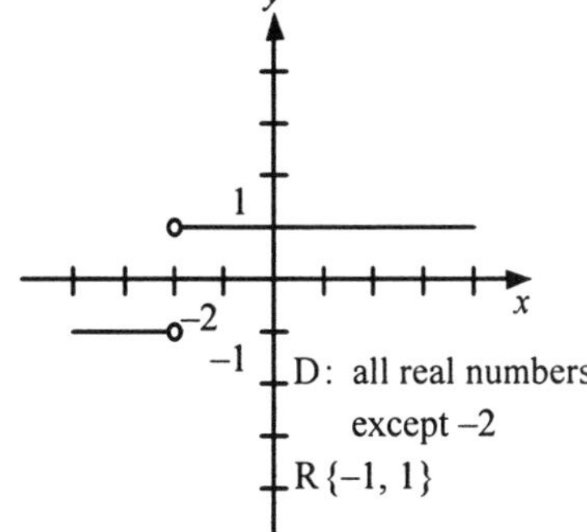

67.

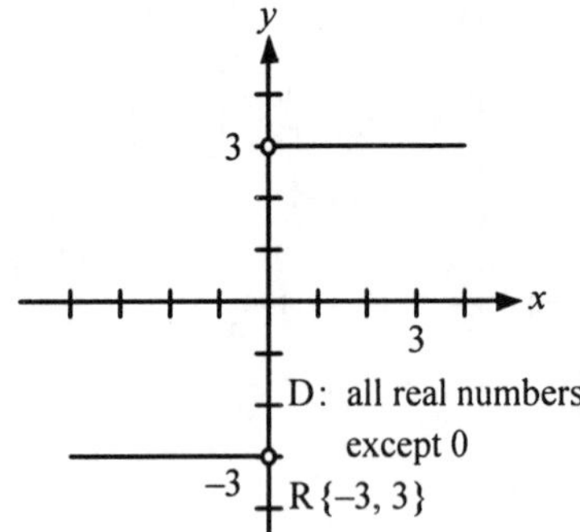

69. **a.** $(-\infty, \infty)$ **b.** $(0, \infty)$ **c.** true
d. none **e.** true

71. **a.** $(-\infty, \infty)$ **b.** $(-\infty, 4]$ **c.** $-5; 0$
d. $\{3\}$ **e.** $\{1, 5\}$ **f.** $(1, 5)$
g. $(-\infty, 1) \cup (5, \infty)$
h. $(0, 6)$ **i.** $(0, -5)$, $(1, 0)$, $(5, 0)$
j. inc. $(-\infty, 3)$; dec. $(3, \infty)$

73. **a.** 0 **b.** 1 **c.** −2 **d.** −4
e.

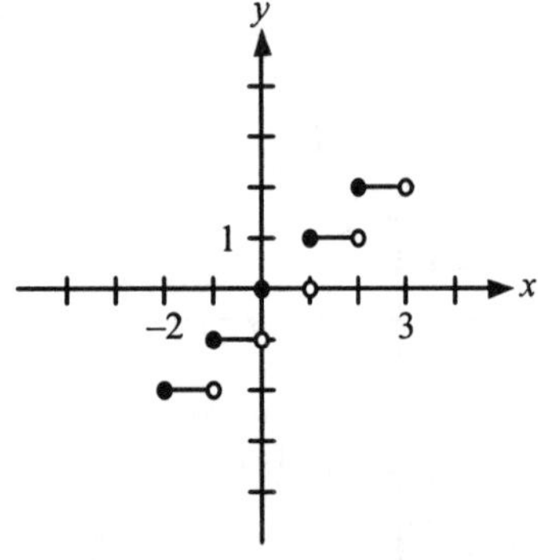

75. $p = 0.32 + 0.23[x]$ if $x > 0$ and x is not an integer;
$p = 0.32 + 0.23(x - 1)$ if $x \geq 1$ and x is an integer;

Exercises 1.3

1.

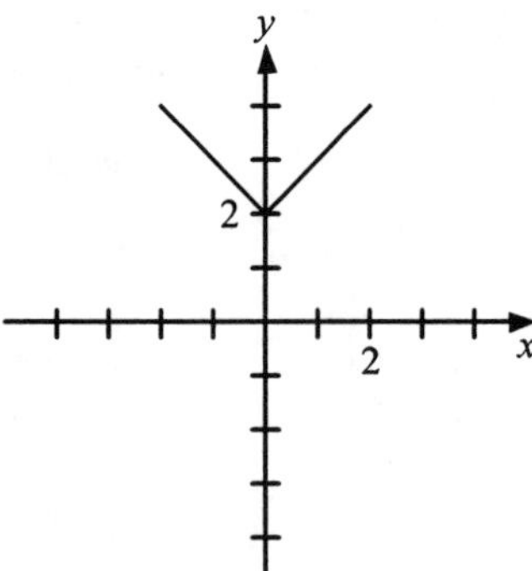

3.

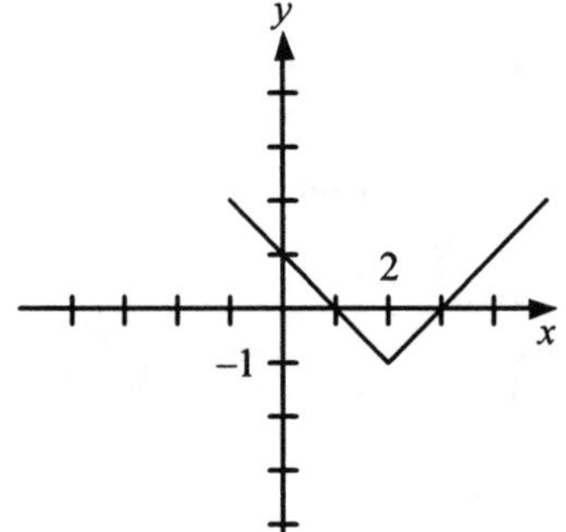

5.

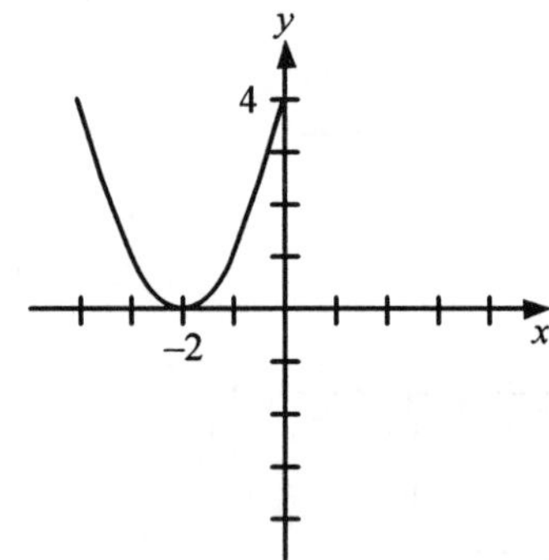

7.

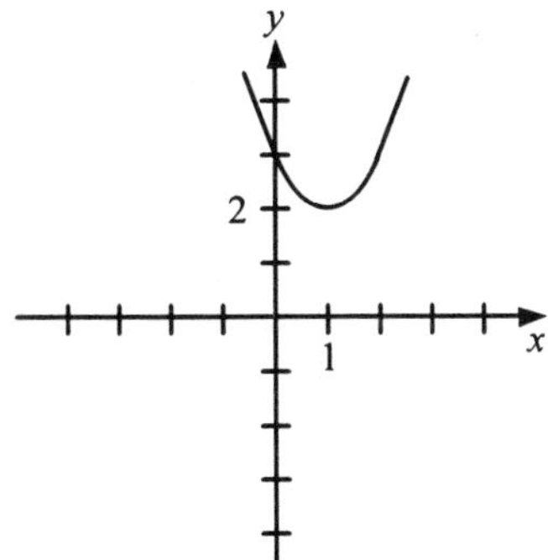

9.

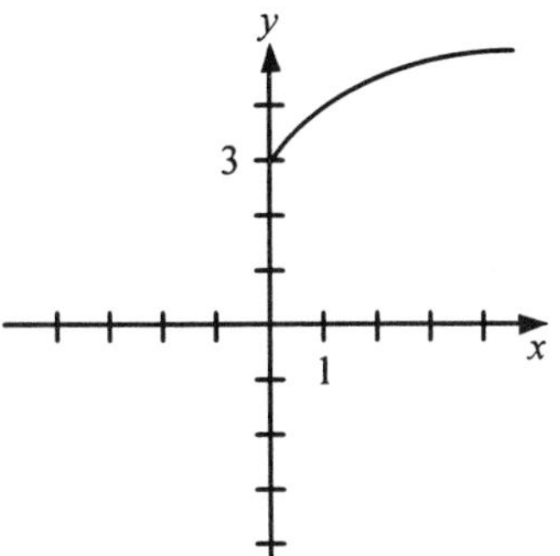

11.

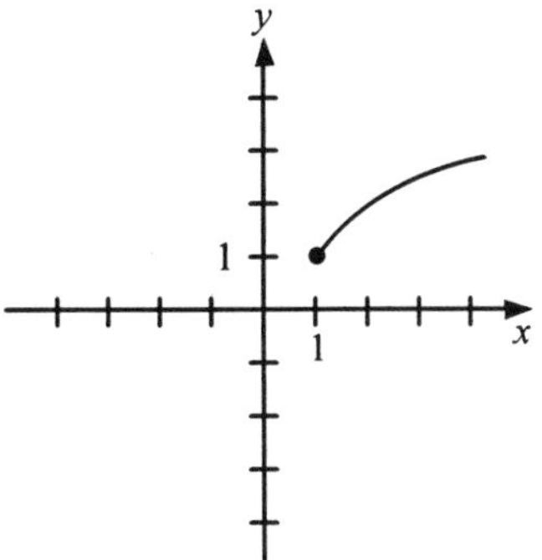

13.

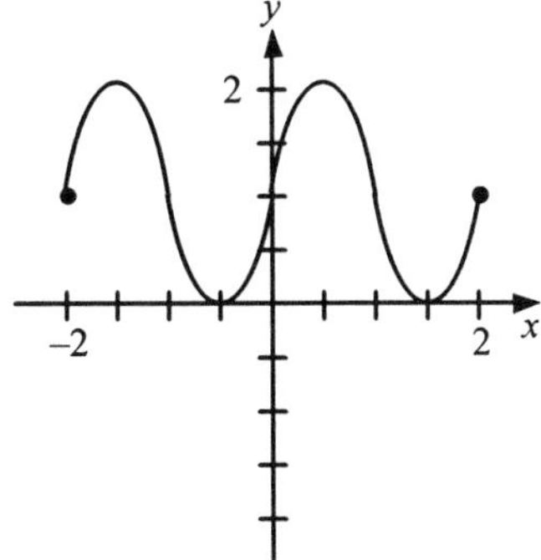

15. The graph in **b** is shifted 20 units to the left of the graph in **a**. The graph in **c** is raised 15 units above the graph values in **a**.

17. $y = (x-3)^2$

19. $y = (x+3)^2 - 1$

21. $y = f(x) + 1$

23. $y = f(x+2) - 1$

25.

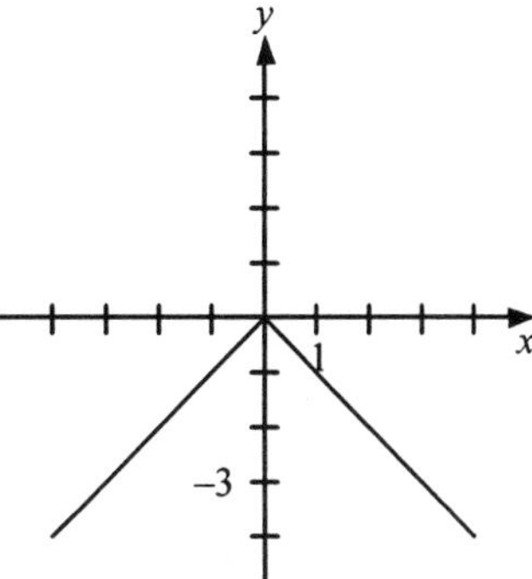

27.

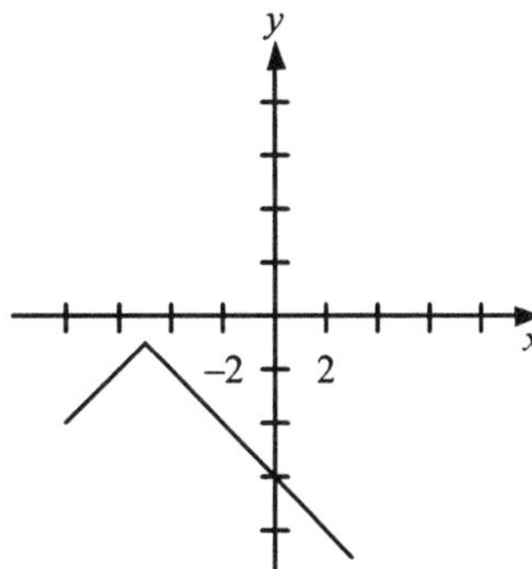

29.

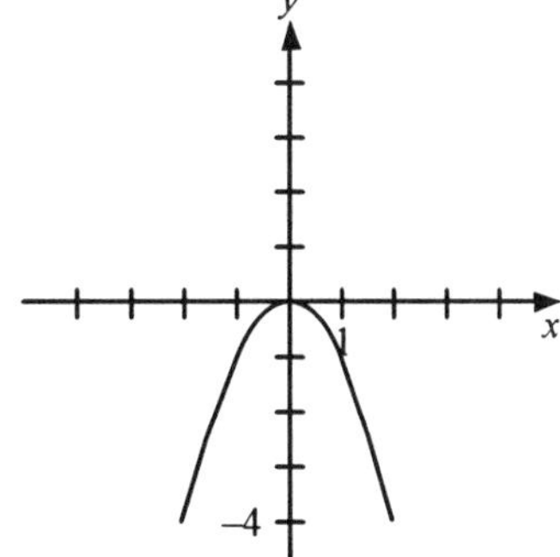

31.

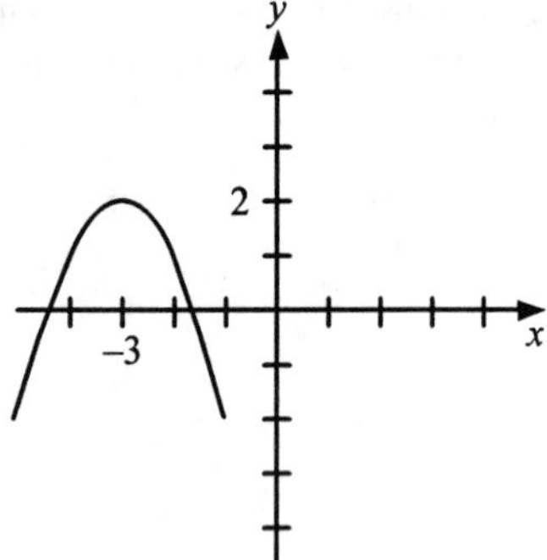

33.

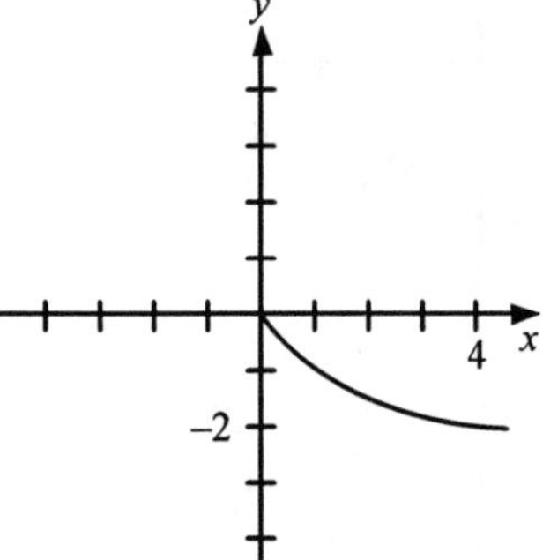

35.

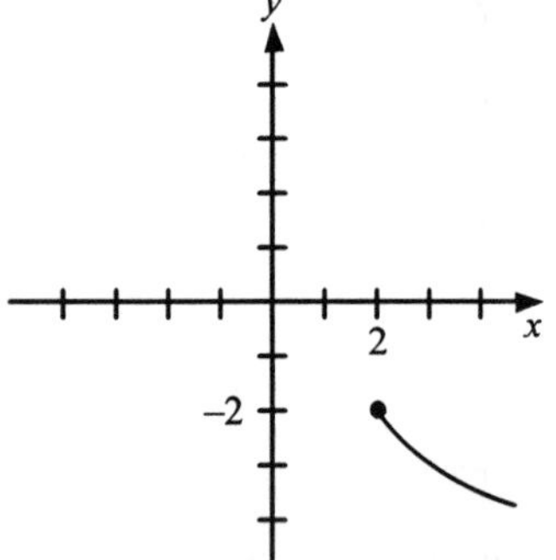

37. $y = -(x-2)^3 + 2$ **39.** $y = -(x-4)^3$

41.

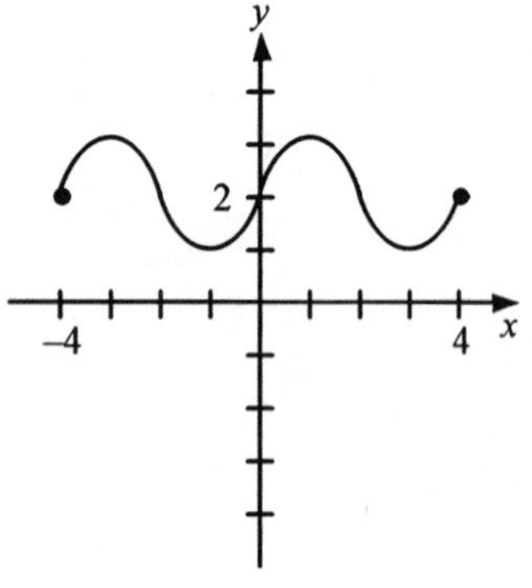

43.

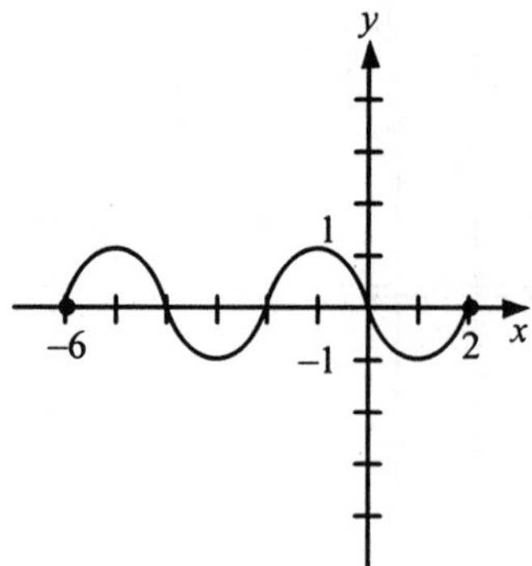

45. $y = g(x-2)$ **47.** $y = -g(x)$

49.

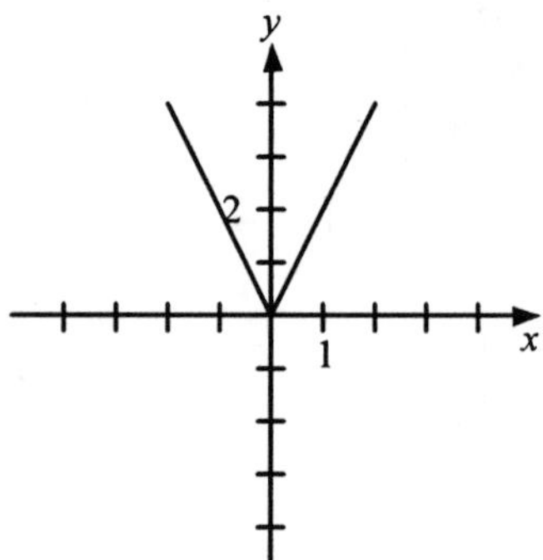

51.

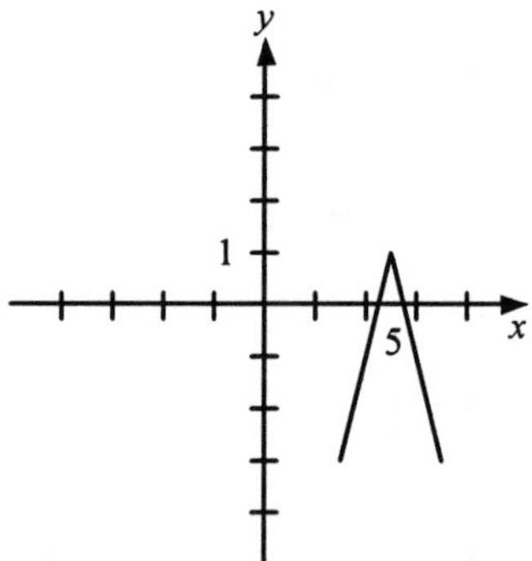

53. $[0, 40]$ by $[0, 50]$ is one choice

55.

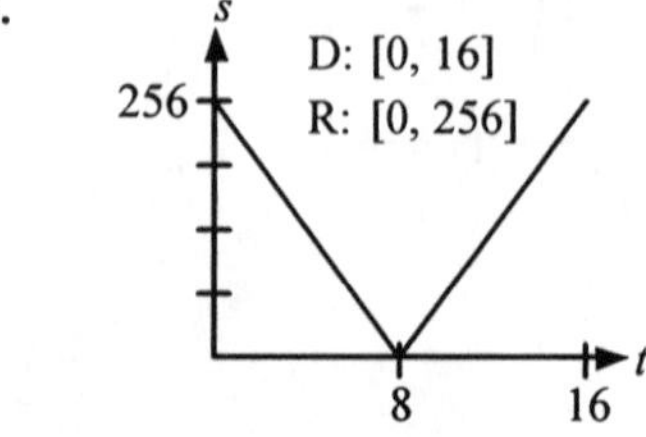

57. **a.**

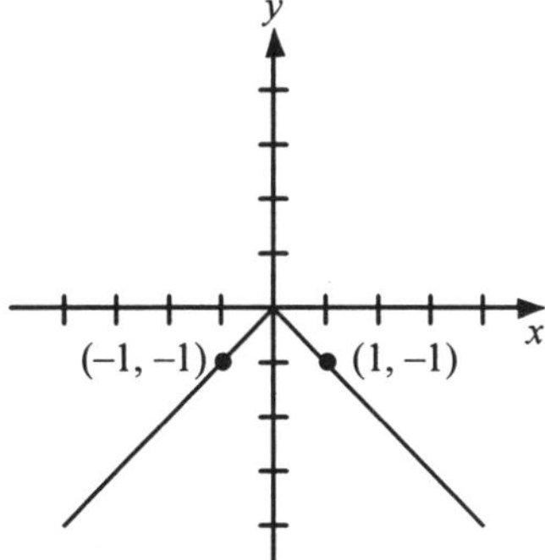

b.

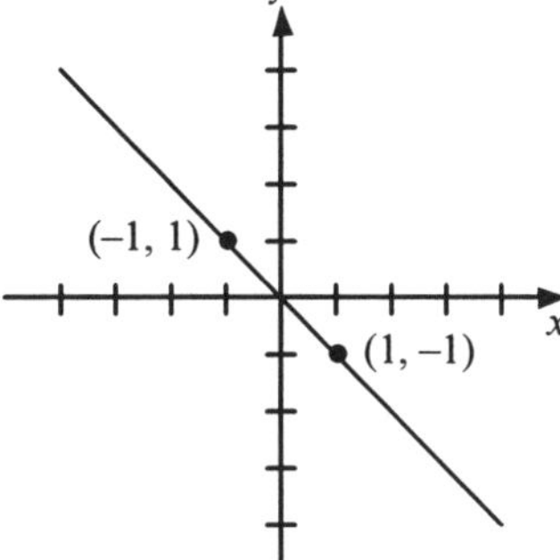

59. **a.**

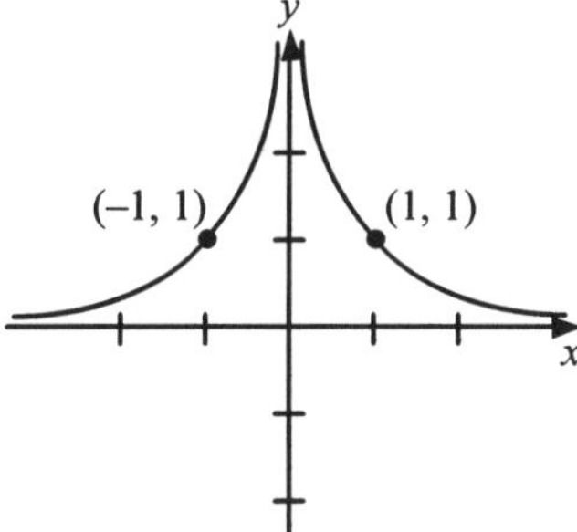

b.

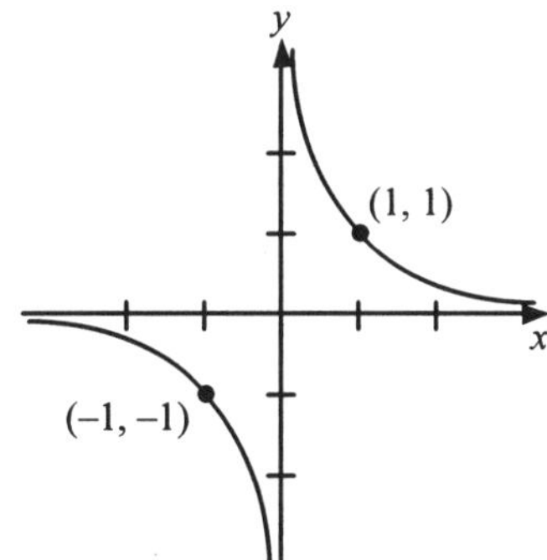

61. Even **63.** Odd **65.** Even

67.

$$\begin{aligned} f(x) &= |3x| \\ f(-x) &= |3(-x)| = |3x| \\ f(x) &= f(-x) \end{aligned}$$

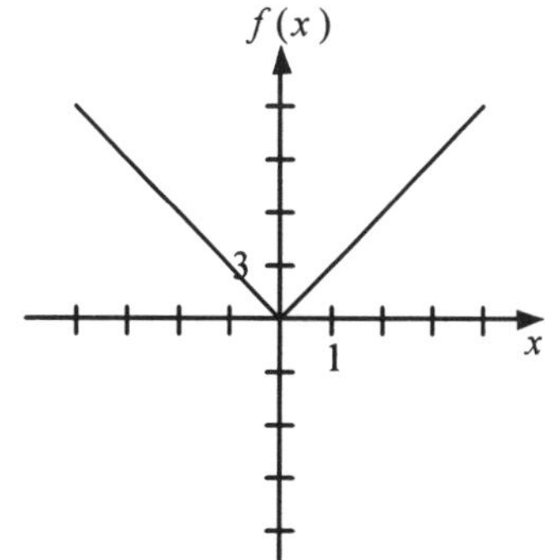

69.

$$\begin{aligned} f(x) &= 7 \\ f(-x) &= 7 \\ f(x) &= f(-x) \end{aligned}$$

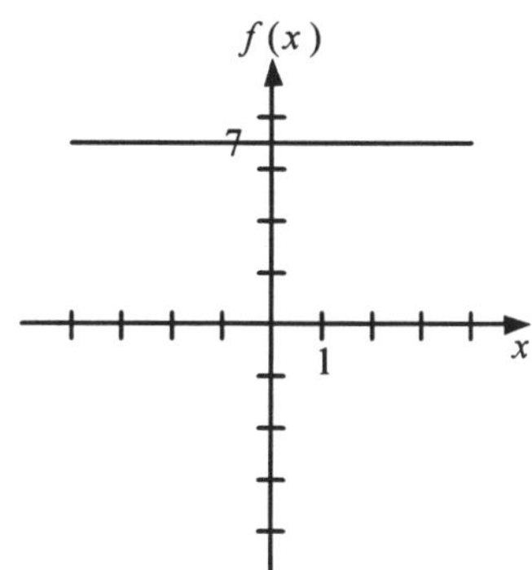

71.

$$\begin{aligned} f(-x) &= 3(-x) = -3x \\ -f(x) &= -(3x) = -3x \\ f(-x) &= -f(x) \end{aligned}$$

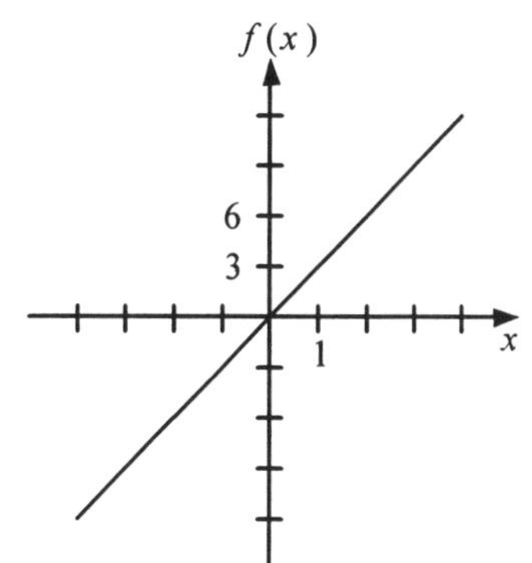

73. $f(-x)=\dfrac{-2(-x)^3-(-x)}{(-x)^2}=\dfrac{2x^3+x}{x^2}$

$-f(x)=-\dfrac{-2x^3-x}{x^2}=\dfrac{2x^3+x}{x^2}$

$f(-x)=-f(x)$

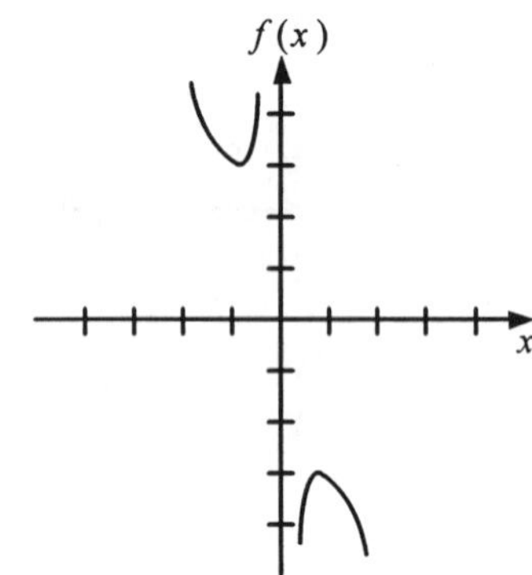

75. Even | 77. Neither

79. **a.** graphs $y=1$ **b.** graphs $y=-1$
c. odd

81. **a.** neither **b.** odd

Exercises 1.4

1. 2 | 3. 4 | 5. 13

7. $\sqrt{40}$ or $2\sqrt{10}$ | 9. $\sqrt{2}$

11. 10 | 13. $(-1,-1)$; 16

15. $d_1=\sqrt{40}$, $d_2=\sqrt{90}$, $d_3=\sqrt{130}$; since $(d_3)^2=(d_1)^2+(d_2)^2$, the Pythagorean theorem ensures a right triangle.

17. $\dfrac{\sqrt{34}}{2}$ | 19. 9π

21. $\left(1, 2\sqrt{2}\right)$; infinite

23. **a.**

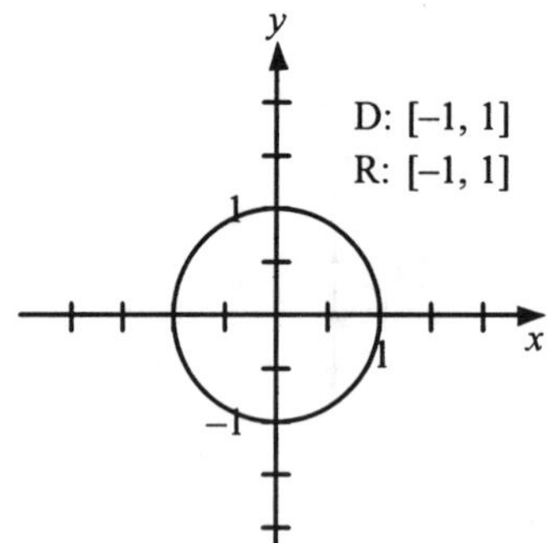

b.

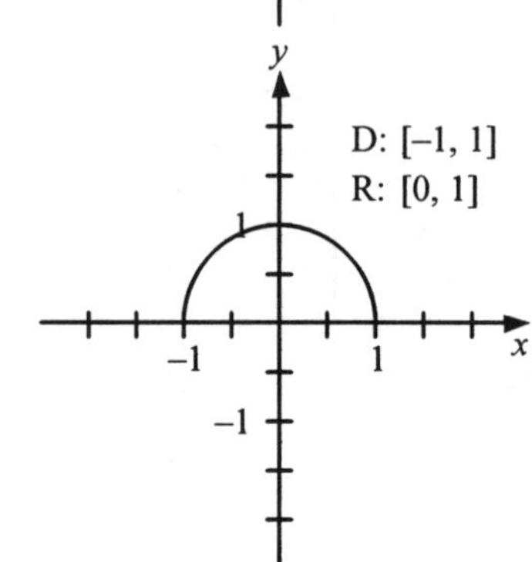

c.

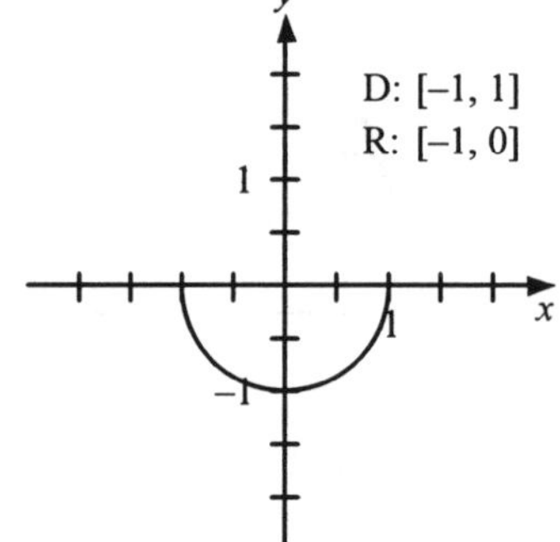

25. **a.**

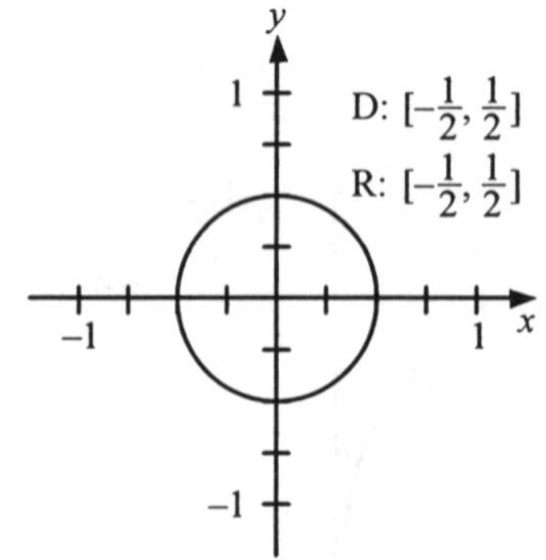

b.

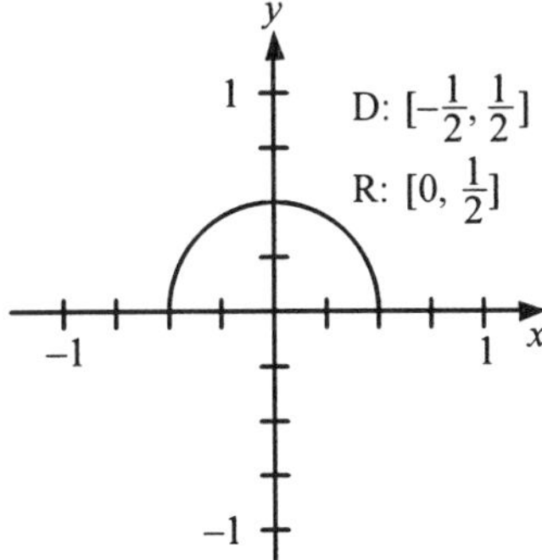

c.

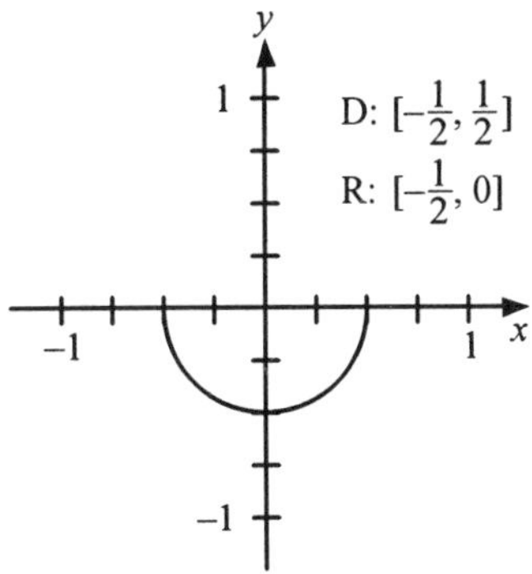

27. **a.**

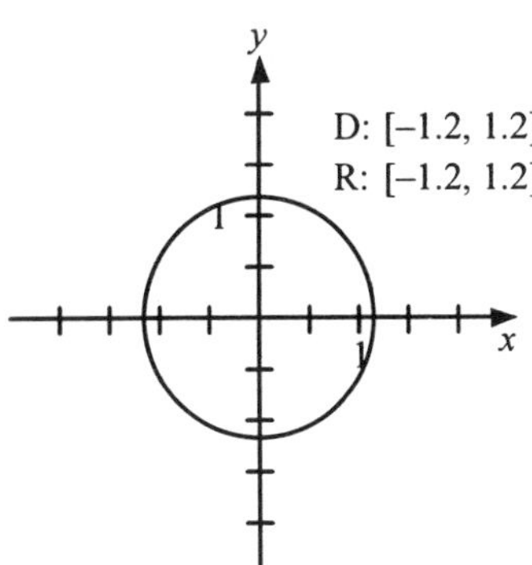

b.

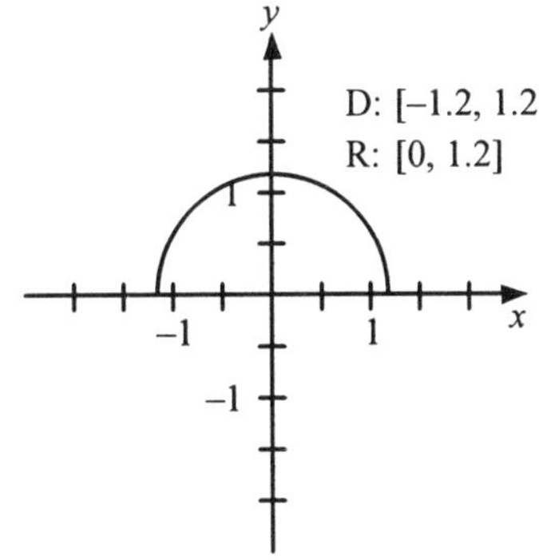

c.

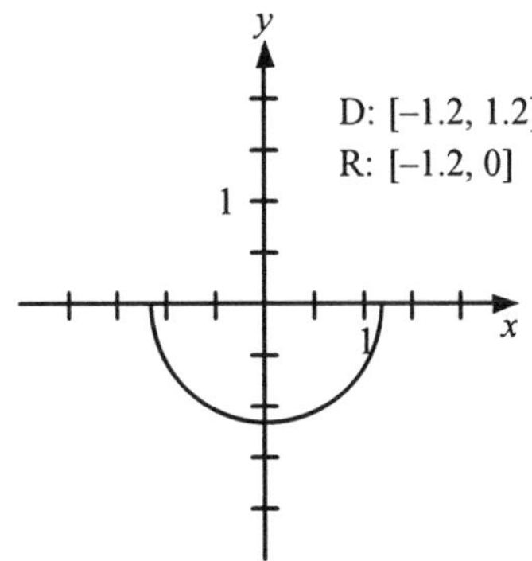

29. **a.** $x^2 + y^2 = 9$ **b.** $\overline{AB} = 4\sqrt{2}$; $\overline{AC} = 2\sqrt{3}$

c. $4\sqrt{2}$

31. **a.** Suspension bar provides the radius as the bob swings back and forth. $y = -\sqrt{324 - x^2}$

b. 12.7 in.

33. $y = \pm\sqrt{16 - 2x^2}$, D: $\left[-\sqrt{8}, \sqrt{8}\right]$, R: $\left[-4, 4\right]$

Intercepts: $(\pm 2.8, 0)$, $(0, \pm 4.0)$

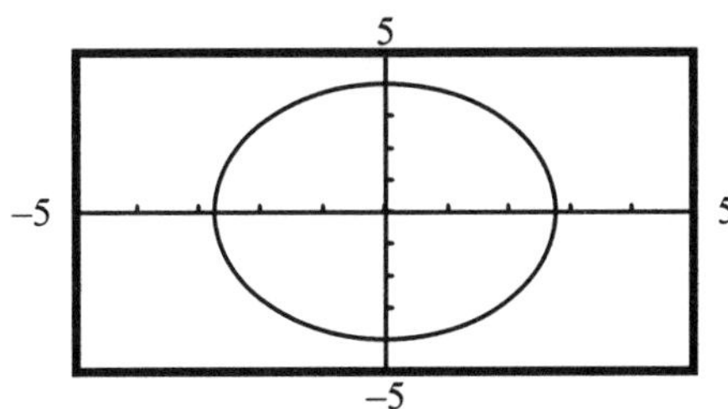

35. Function **37.** Not a function

39. Not a function **41.** Function

43. Function

45. Function, $y = 9$

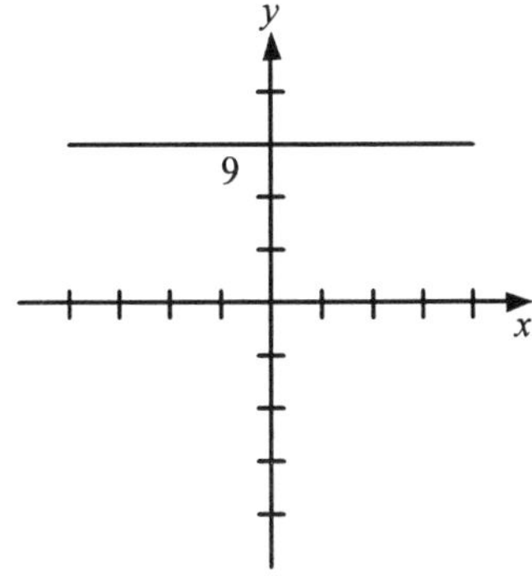

47. Function, $y = 9 - x$

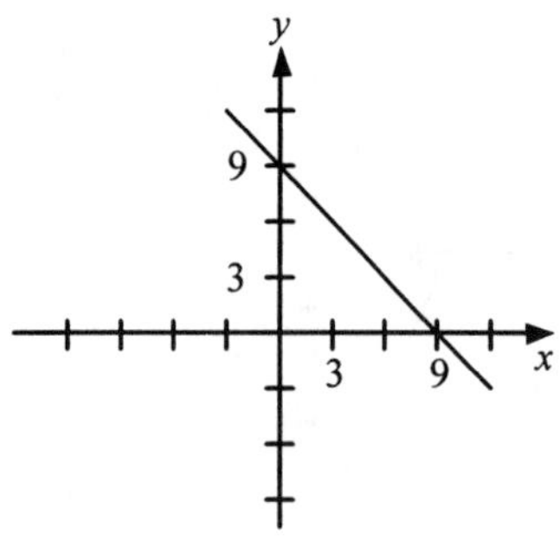

49. Not a function, $y = \pm\sqrt{9 - x^2}$

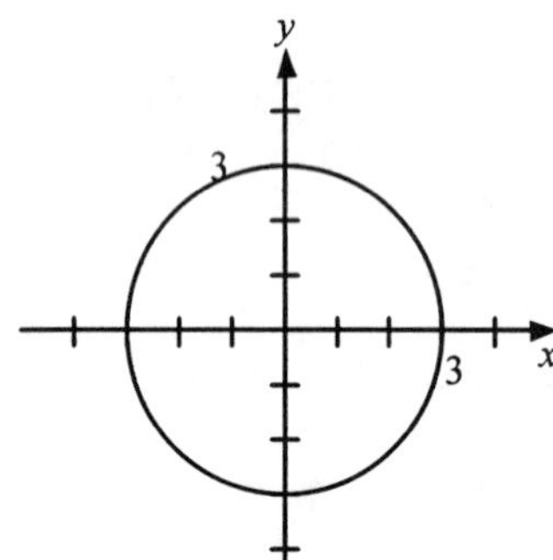

51. Function, $y = 9 - x^3$

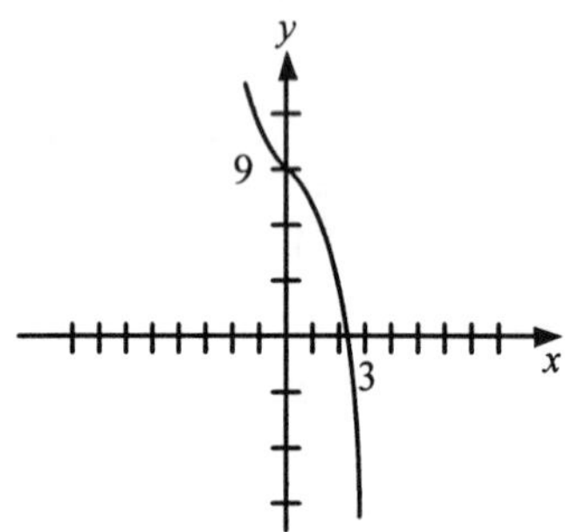

53. Function, $y = \sqrt[3]{9 - x^2}$

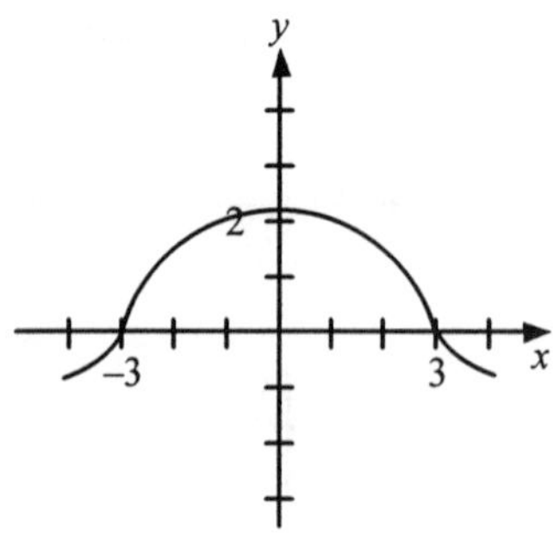

Exercises 1.5

1. $(-2, -4)$ **3.** $\left(-\frac{3}{2}, -\frac{1}{2}\right)$

5. $(0, 4)$ **7.** $(-6, -6)$ **9.** Dependent

11. $(4, 21)$ **13.** $(-1, 1)$

15. No solution, inconsistent

17. $(5, 1)$ **19.** $\left(\frac{67}{41}, -\frac{155}{41}\right)$

21. 46, 24 **23.** 140°, 40° **25.** 200 g

27. $v_0 = 100$ ft/second; $a = -32$ ft/second^2

29. $F_1 = 12$ lb; $F_2 = 6$ lb

31. 2 gals of 8 percent; 6 gals of 6 percent

33. \$8,000 at 6 percent, \$2,000 at 11 percent

35. **a.** $y = 0.15x$; $y = 0.10x + 60$

b.

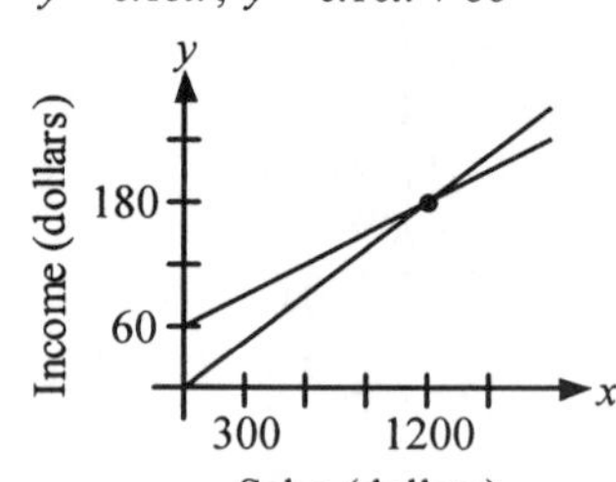

c. If you can see more than \$1,200 each week, the second offer would be better

37. $p = \dfrac{d - b}{a - c}$

39. **a.** $\{-1.7\}$ **b.** $(-1.7, \infty)$

41. **a.** $\{0\}$ **b.** $[0, \infty)$

43. **a.** Ø **b.** Ø

Exercises 1.6

1. $P = ks$; $k = 3$ **3.** $T = kp$

5. $V = \dfrac{k}{P}$ **7.** $y = \dfrac{7}{3}x$; $\dfrac{70}{3}$

9. $y = \dfrac{3}{4}x^2$; 12 **11.** $y = \dfrac{72}{x}$; 3

13. $y = \dfrac{24}{x^3}$; $\dfrac{8}{9}$ **15.** 60

17. 49 **19.** 4 in. **21.** 2,000,000 tons

23. 3 in. **25.** 600 rpm

27. $\frac{500\pi}{3}$ cubic units **29.** 181 lb.

31. Exposure time in multiplied by 4.

33. $\frac{5}{3}$ atm

35. Force is multiplied by 24.

37. **a.**

	A	**B**	**C**
1.	385	692	923
2.	692	1246	1662

b. $A=\left(\frac{5}{26}\right)P$ $B=\left(\frac{9}{26}\right)P$; $C=\left(\frac{6}{13}\right)P$

39. **a.** $P=\frac{0.25w}{s^2}$

b. $P=\frac{30}{s^2}$

Domain $=(0,1]$

Range $=[30,\infty)$

The pressure on the floor varies directly as a fourth of the weight of the woman and inversely as the area of the square where the heel touches the floor.

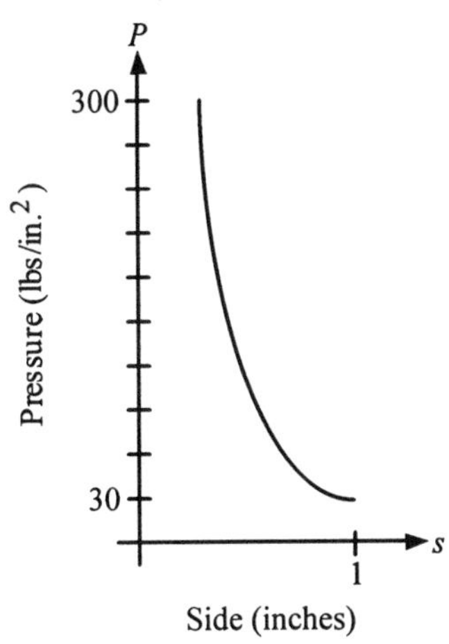

41. $y=\frac{8}{x}$; $(1,8)$; $(4,2)$

Chapter 1 Review Exercises

1. $(0,4)$, $\left(\frac{5}{2},-1\right)$, $(-1,6)$, $(2,0)$, $(3,-2)$

3. Yes **5.** Yes **7.** $\left(\frac{5}{3},\frac{2}{3}\right)$

9. $y=\frac{5}{12}x$; $\frac{25}{3}$

11. D: Set of all real numbers except -4; R: set of all real numbers except 0

13. $A=\frac{d^2}{2}$ **15.** Yes **17.** No

19. Yes **21.** Odd function

23.

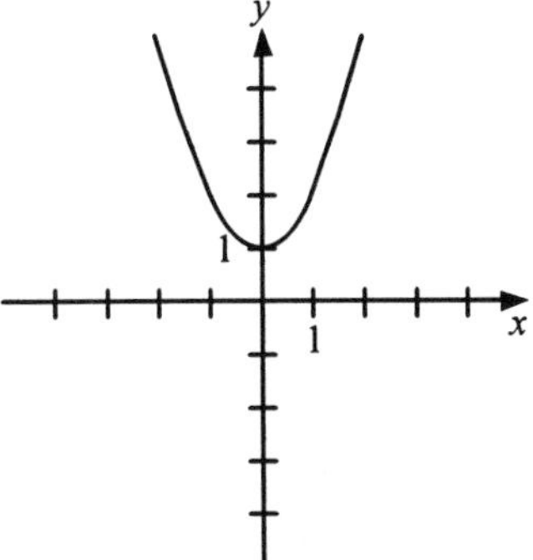

25.

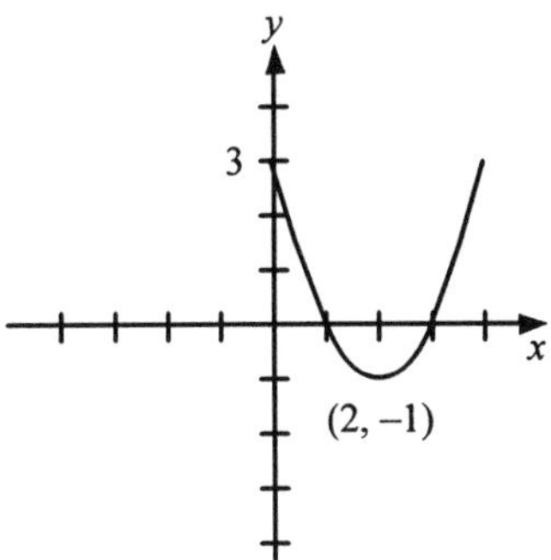

27. $\frac{9}{4}$ **29.** $A=\frac{\pi d^2}{4}$

31.

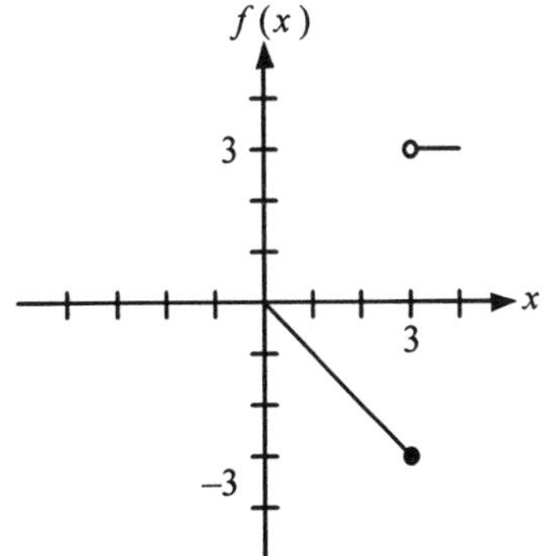

33. $\{y: y = 3 \text{ or } -3 \le y \le 0\}$

35.

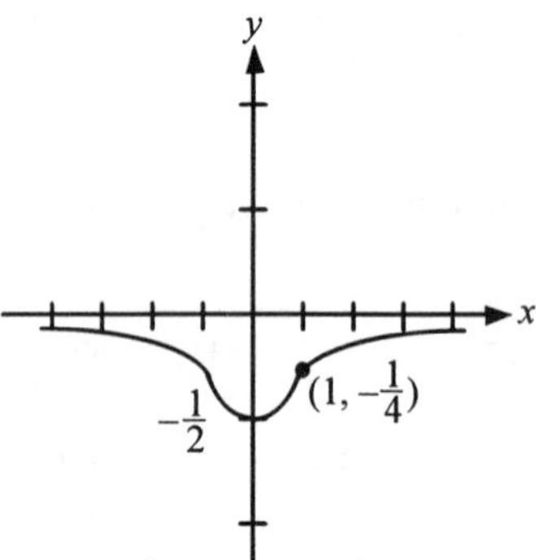

37.

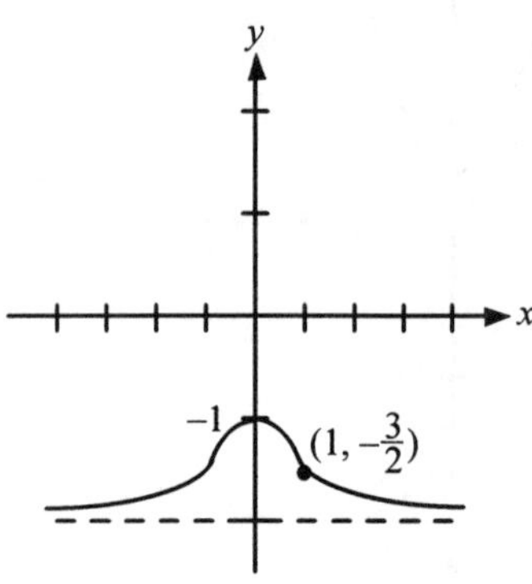

39.

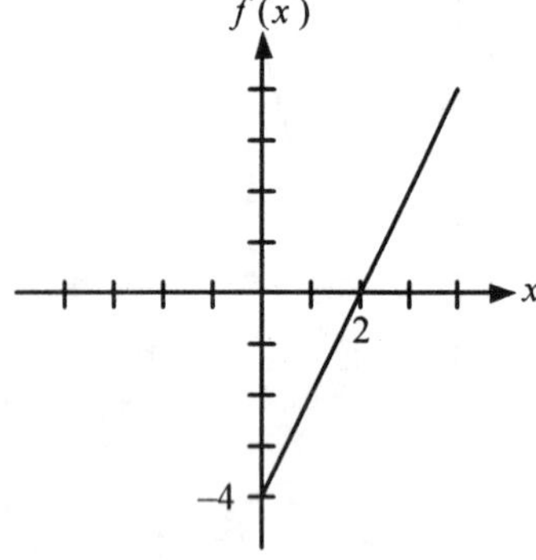

41.

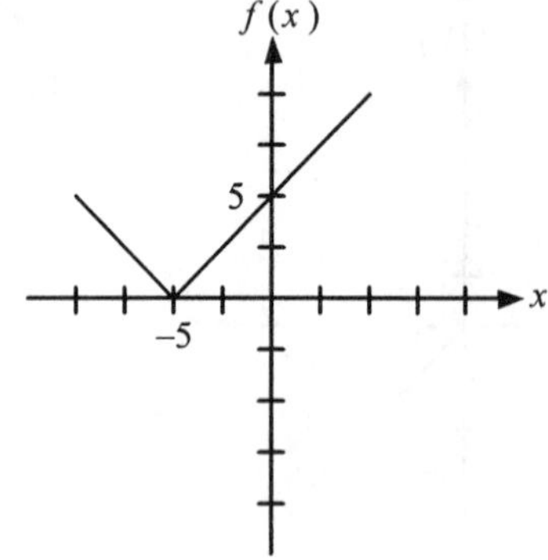

43. Even function

45. **a.** (3) **b.** (1) **c.** (2)

47. $\left[-\frac{\pi}{2}, \frac{\pi}{2}\right]$ **49.** 0 **51.** $(0, 1]$

53. $(0, 0)$ **55.** $6x + 3h$ **57.** $(-\infty, 1]$

59. **a.** $y = 5{,}000 + 2x$; $y = 8{,}000 + x$
b. 3,000 pkgs; machine B
c. 1,667 pkgs

61. Yes

63. $e = \begin{cases} 500 & \text{if } 0 \le a \le 2{,}000 \\ 500 + .09(a - 2{,}000) & \text{if } a > 2{,}000 \end{cases}$

65. $[-c, c]$ **67.** Even function

69. $\{-c, 0, c\}$

71.

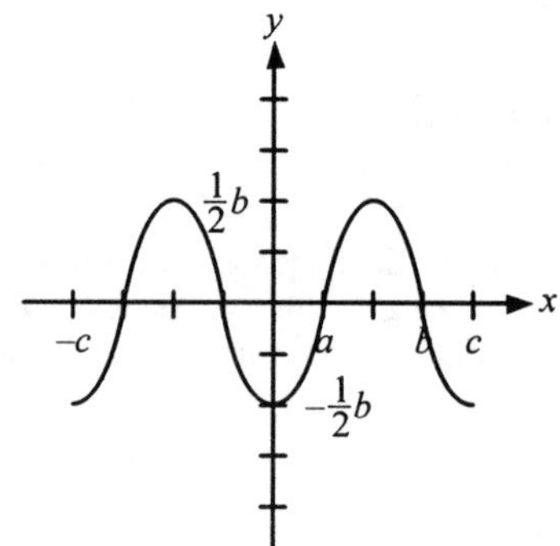

73. $(-b, 0)$, $(-a, 0)$, $(a, 0)$, $(b, 0)$, $\left(0, -\frac{1}{2}b\right)$

75. **a.** D: $(-\infty, \infty)$; R: $[-1, 1]$
b. D: $[-3, 3]$; R: $[-3, 0]$

77. b **79.** d **81.** b

83. a **85.** a

Chapter 1 Test

1. Set of all real numbers except 9

2. $(-\infty, 6]$ **3.** 38 **4.** 19

5. $A = x(400 - x)$ **6.** 10π

7. Yes **8.** $\left(\frac{17}{9}, \frac{7}{9}\right)$

9. Domain: $[-5, 5]$; range: $[0, 5]$; $y = \sqrt{25 - x^2}$

10.

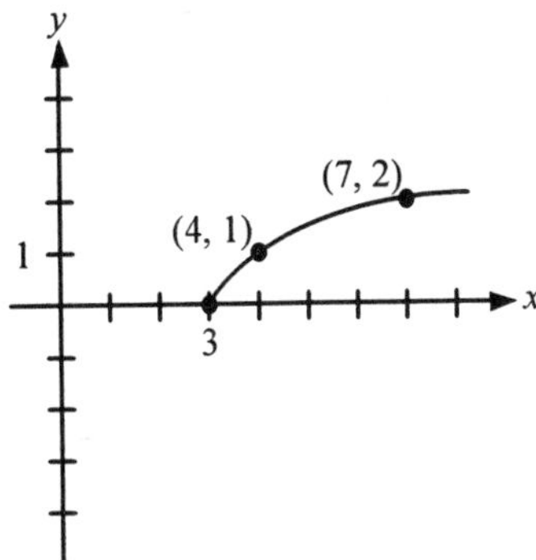

11.

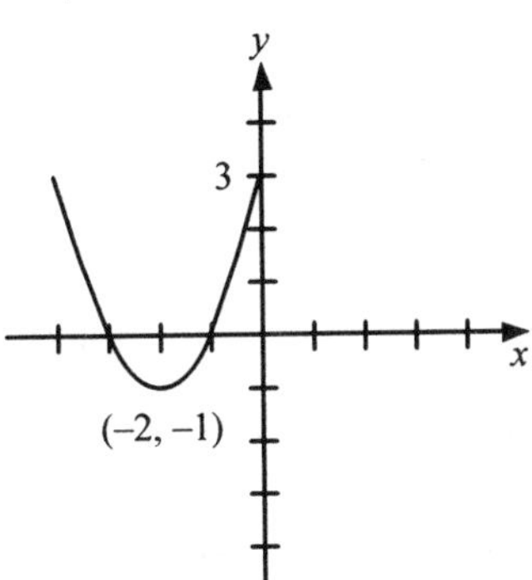

12.

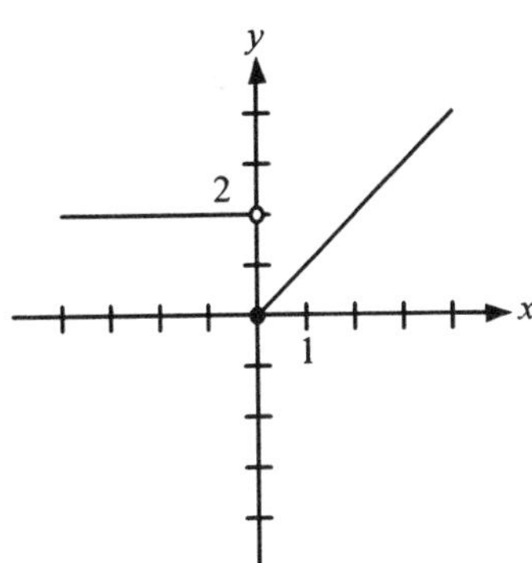

13. Even **14.** $\frac{3}{2}$ **15.** No

16. \$15,000 at 8.5%; \$45,000 at 10.5%

17. $\{1\}$

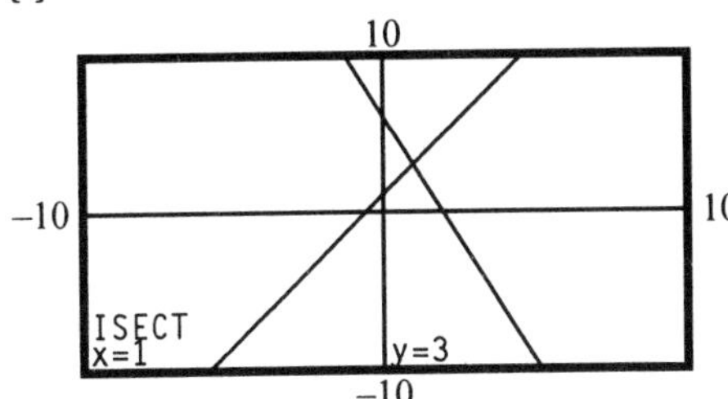

18. $y = 24$ **19.** 6.4 in.

20. Exposure time is divided by 9.

Exercises 2.1

1. $\sin A = \cos B = \frac{5}{13}$, $\cos A = \sin B = \frac{12}{13}$,
$\tan A = \cot B = \frac{5}{12}$, $\cot A = \tan B = \frac{12}{5}$,
$\sec A = \csc B = \frac{13}{12}$, $\csc A = \sec B = \frac{13}{5}$

3. $\sin\theta = \cos\beta = \frac{4}{5}$, $\cos\theta = \sin\beta = \frac{3}{5}$, $\tan\theta = \cot\beta = \frac{4}{3}$,
$\cot\theta = \tan\beta = \frac{3}{4}$, $\sec\theta = \csc\beta = \frac{5}{3}$, $\csc\theta = \sec\beta = \frac{5}{4}$

5. $\frac{x}{y}$ **7.** $\frac{40}{9}$ **9.** $\frac{y}{x}$

11. $\sec X$ or $\csc Z$ **13.** 3

15. $\frac{5}{7}$ **17.** 1

19. $\csc\theta = 2$, $\cos\theta = \frac{\sqrt{3}}{2}$, $\sec\theta = \frac{2}{\sqrt{3}}$, $\tan\theta = \frac{1}{\sqrt{3}}$,
$\cot\theta = \sqrt{3}$

21. $\tan\theta = \frac{5}{3}$, $\sin\theta = \frac{5}{\sqrt{34}}$, $\csc\theta = \frac{\sqrt{34}}{5}$, $\cos\theta = \frac{3}{\sqrt{34}}$,
$\sec\theta = \frac{\sqrt{34}}{3}$

23. $\cot\theta = \frac{1}{2}$, $\sin\theta = \frac{2}{\sqrt{5}}$, $\csc\theta = \frac{\sqrt{5}}{2}$, $\cos\theta = \frac{1}{\sqrt{5}}$,
$\sec\theta = \sqrt{5}$

25. 20 **27.** 36 **29.** $\cos A = \frac{12}{13}$

31. $\cos 73^\circ$ **33.** $\cot 8^\circ 30'$ **35.** $\sec 21.9^\circ$

37. No **39.** No **41.** Yes

43. $2\sqrt{3}$, 4 (hypotenuse)

45. $\frac{\sqrt{3}}{2}$ **47.** $\frac{2}{\sqrt{3}} = \frac{2\sqrt{3}}{3}$

49. $\frac{1}{\sqrt{3}} = \frac{\sqrt{3}}{3}$ **51.** $\frac{1}{\sqrt{2}} = \frac{\sqrt{2}}{2}$

53. 0.9925 **55.** 0.1228 **57.** 5.396

59. 1.287 **61.** 0.2130 **63.** 0.0017

65. 27.06 **67.** 60° **69.** 60°

71. 45° **73.** 45°00′ or 45.0°

75. 38°20′ or 38.3° **77.** 55°50′ or 55.8°

79. 52°30′ or 52.6° **81.** 3°40′ or 3.6°

83. 85°50′ or 85.8° **85.** 69 ft

87. **a.** 44.25° **b.** 48.6°

Exercises 2.2

1. $b = 87$ ft, $c = 1\bar{0}0$ ft, $B = 60°$

3. $a = 13$ ft, $b = 7.5$ ft, $B = 30°$

5. $a = 8.1$ ft, $b = 24$ ft, $A = 19°$

7. $a = 2\bar{0}$ ft, $b = 14$ ft, $B = 35°$

9. $b = 85.1$ ft, $c = 121$ ft, $B = 44.5°$

11. $a = 379$ ft, $b = 497$ ft, $A = 37°20'$

13. $b = 975$ ft, $c = 975$ ft, $A = 1°50'$

15. $b = 14$ ft, $A = 24°$, $B = 66°$

17. $c = 1.4$ ft, $A = 45°$, $B = 45°$

19. $a = 23.1$ ft, $A = 62°30'$, $B = 27°30'$

21. 25 ft **23.** 52 yd **25.** 89 ft

27. $4\bar{0}0$ ft **29.** 280 ft **31.** 10°

33. 22 ft **35.** 120 in. **37.** 16 ohms, 15°

39. 55°, 55°, 70° **41.** 88 ft

43. 480 m **45.** 1.4 in.

Exercises 2.3

	sin θ	csc θ	cos θ	sec θ	tan θ	cot θ
1.	$-\frac{4}{5}$	$-\frac{5}{4}$	$\frac{3}{5}$	$\frac{5}{3}$	$-\frac{4}{3}$	$-\frac{3}{4}$
3.	$\frac{5}{13}$	$\frac{13}{5}$	$-\frac{12}{13}$	$-\frac{13}{12}$	$-\frac{5}{12}$	$-\frac{12}{5}$
5.	$-\frac{4}{5}$	$-\frac{5}{4}$	$\frac{3}{5}$	$\frac{5}{3}$	$-\frac{4}{3}$	$-\frac{3}{4}$
7.	$\frac{1}{\sqrt{2}}$	$\sqrt{2}$	$\frac{1}{\sqrt{2}}$	$\sqrt{2}$	1	1
9.	$\frac{2}{\sqrt{5}}$	$\frac{\sqrt{5}}{2}$	$-\frac{1}{\sqrt{5}}$	$-\sqrt{5}$	-2	$-\frac{1}{2}$

11. Q_4 **13.** Q_3 **15.** Q_2

	sin θ	csc θ	cos θ	sec θ	tan θ	cot θ
17.	$-\frac{3}{5}$	$-\frac{5}{3}$	$-\frac{4}{5}$	$-\frac{5}{4}$	$\frac{3}{4}$	$\frac{4}{3}$
19.	$-\frac{\sqrt{3}}{2}$	$-\frac{2}{\sqrt{3}}$	$\frac{1}{2}$	2	$-\sqrt{3}$	$-\frac{1}{\sqrt{3}}$
21.	$-\frac{3}{4}$	$-\frac{4}{3}$	$-\frac{\sqrt{7}}{4}$	$-\frac{4}{\sqrt{7}}$	$\frac{3}{\sqrt{7}}$	$\frac{\sqrt{7}}{3}$
23.	$\frac{1}{3}$	3	$\frac{\sqrt{8}}{3}$	$\frac{3}{\sqrt{8}}$	$\frac{1}{\sqrt{8}}$	$\sqrt{8}$
25.	-1	-1	0	undef	undef	0
27.	0	undef	1	1	0	undef
29.	0	undef	-1	-1	0	undef

31. 1 **33.** 1 **35.** undefined

37. $-\frac{1}{2}$ **39.** $\frac{2}{\sqrt{3}}$ **41.** -1

43. $\frac{1}{2}$ **45.** $-\frac{1}{\sqrt{2}}$ **47.** $\frac{2}{\sqrt{3}}$

49. $-\sqrt{3}$ **51.** $-\frac{1}{\sqrt{2}}$ **53.** $-\frac{1}{\sqrt{2}}$

55. $-\frac{\sqrt{3}}{2}$ **57.** $\sqrt{3}$ **59.** 1

61. $\frac{2}{\sqrt{3}}$ **63.** 1 **65.** -0.5299

67. 3.487 **69.** 1.923 **71.** -0.8557

73. -0.5354 **75.** 0.0494 **77.** -1.360

79. -0.9983 **81.** -0.6111 **83.** 0.3346

85. 0.1564 **87.** 1.150 **89.** 0.4848

91. -0.6787 **93.** 1.881

95. **a.** $\frac{16\sin 70°}{\sin 40°} = 24.39$ ft

b. $\frac{8}{\cos 70°} + 1 = 24.39$ ft

97. 8.9 mi **99.** 165 ft

Exercises 2.4

1. $\{60°, 300°\}$ **3.** $\{135°, 315°\}$

5. $\{45°, 135°\}$ **7.** $\{150°, 210°\}$

9. $\{30°, 210°\}$ **11.** $\{225°, 315°\}$

13. $\{60°, 240°\}$ **15.** $\{240°, 300°\}$

17. $\{210°, 330°\}$ **19.** $\{45°, 225°\}$

21. $\{7°00', 173°00'\}$ **23.** $\{120°50', 239°10'\}$

25. $\{22°10', 202°10'\}$ **27.** $\{161°00', 199°00'\}$

29. $\{72°20', 252°20'\}$ **31.** $\{101°20', 281°20'\}$

33. $\{131°50', 228°10'\}$ **35.** $\{74°00', 254°00'\}$

37. $\{206°20', 333°40'\}$ **39.** $\varnothing$

41. 405°, 765°, 1,125°, –315°, –675°

43. 382°10′, 742°10′, 1,102°10′, –337°50′, –697°50′

45. 480°, 840°, 1,200°, –240°, –600°

47. –390°, –750°, –1,120°, 330°, 690°

49. –460°50′, –820°50′, –1,180°50′, 259°10′, 619°10′

51. $\{\theta\colon \theta = 45°+k360° \text{ or } \theta = 135°+k360°\}$

53. $\{\theta\colon \theta = 199°30' + k360° \text{ or } \theta = 340°30' + k360°\}$

55. $\{\theta\colon \theta = 113°30' + k360° \text{ or } \theta = 246°30' + k360°\}$

57. $\varnothing$ **59.** $\{\theta\colon \theta = 180°+k360°\}$

Note: In Exercises 41–50 other solutions are possible; in Exercises 51–60 k may be any integer.

61. 20°, 70° **63.** 33°

Exercises 2.5

1. **a.** 13 lb **b.** 23°

3. **a.** 18 lb **b.** 34°

5. **a.** 16 mi/hour **b.** 18°

7. **a.** 510 mi **b.** 81°

9. **a.** 13° **b.** 310 mi/hour

11. **a.** 340 mi/hour **b.** 91 mi/hour

13. 34 lb **15.** 608 lb **17.** 26 lb

19. **a.** $4\overline{0}$ lb **b.** 12°

21. **a.** 321 lb **b.** 5°50′

23. 23 lb **25.** 30°, 56°, 86°

27. v, $5\overline{0}$ lb; h, 87 lb **29.** v, 8.2 lb; h, 16 lb

31. $R = 7.1$ lb, $\theta = 47°$ **33.** $R = 22$ lb, $\theta = 149°$

35. $R = 24$ lb, $\theta = 196°$

Chapter 2 Review Exercises

1. 460°, 820°, 1,180°, –260°, –620°

3. $\frac{1}{a}$ **5.** cot 68° **7.** $\frac{\sqrt{21}}{5}$

9. $\frac{1}{\sqrt{2}}$ **11.** False **13.** True

15. False **17.** $\frac{1}{2}$ **19.** 3

21. $\{\theta\colon \theta = 71°30' + k360° \text{ or } \theta = 251°30' + k360°\}$, k any integer

23. $\varnothing$ **25.** 2 and $\sqrt{3}$ cm.

27. $A = 43°$, $b = 48$ ft, $c = 66$ ft

29. Q_4 **31.** 3 **33.** 3.8 ft^2

35. 35°20′ **37.** 190 yd **39.** 4°

41. 77 ft **43.** 116°

45. $R = 7.3$ lb, $\theta = 68°$ **47.** b

49. a **51.** c **53.** b **55.** d

Chapter 2 Test

1. $\frac{\sqrt{3}}{2}$ **2.** $\frac{8}{15}$ **3.** $\frac{4}{\sqrt{41}}$

4. $-\frac{1}{\sqrt{2}}$ **5.** –1.471

6. 510°, 870°, –210° **7.** 15°

8. $\{139°, 221°\}$

9. $\{\theta\colon \theta = 135°+k360° \text{ or } \theta = 315°+k360°\}$, k any integer

10. $-\frac{2}{\sqrt{13}}$ **11.** True **12.** 1

13. 78 ft **14.** 19.5° **15.** Q_3

16. 34 lb **17.** $\{120°, 240°\}$

18. $A = 39°$, $B = 51°$, $a = 28$ ft

19. 6.6 m **20.** 39 lb

Exercises 3.1

1. $y(x+z)$ **3.** $4(2x-3)$ **5.** $b(1-b)$

7. $5x^2y(x-2y)$ **9.** $3ax(3ax+1)$

11. $x(2xz^2+4y-5xy^2)$ **13.** $(x+1)(x+2)$

15. $(t-4)(7t-1)$ **17.** $(a-c)(b+d)$

19. $(x-5)(x+2)$ **21.** $(x+y)(a+5)$

23. $(y+2)(3x+1)$ **25.** $(x-2)(4x+3)$

27. $(a-3)^2$ **29.** $(x^2+1)(x^3+5)$

31. $(x+7)(x^2-7)$ **33.** $(a+4)(a+1)$

35. $(x-7)(x-2)$ **37.** $(y+15)(y-2)$

39. $(c-5)(c+2)$ **41.** $(x-3)^2$

43. $(t+5)^2$ **45.** $(x+5y)(x+3y)$

47. $(a+4b)(a-b)$ **49.** $(7y-6)(y-1)$

51. $(3x+1)(x-2)$ **53.** $(3k-2)(2k-1)$

55. $(3a-2)^2$ **57.** $(9b+2)(b-3)$

59. $(12x-5y)(x-2y)$ **61.** $3(k-4)(k+2)$

63. $6(t+4)(t-2)$ **65.** $3x(x-3)(x-1)$

67. $y(3y-5)(3y-2)$ **69.** $8xy(3x+2y)(x+y)$

71. $7(\sin\theta+\cos\theta)$ **73.** $7\sin\theta(3\sin\theta-2)$

75. $3\cos\theta\tan\theta(5\tan\theta-7\cos\theta)$

77. $(\sin\theta-5)(\cos\theta+\tan\theta)$

79. $(\tan\theta-2)(\tan\theta-1)$

81. $(\tan\theta-5)(\tan\theta-4)$

83. $(3\sin\theta+1)(\sin\theta-3)$

85. $(3\sin\theta+2)(2\sin\theta+1)$

87. $(5\sin\theta-12)(4\sin\theta+1)$

89. $(4\cos\theta-\tan\theta)(\cos\theta-2\tan\theta)$

91. $(2\sin\theta-1)^2$ **93.** 1, factorable

95. 13, not factorable **97.** 81, factorable

99. 985, not factorable **101.** $S=2b(b+2a)$

103. $A=\dfrac{3\pi R^2}{4}$ **105.** $0.91x$; less

107. $1.32x$; a 32 percent increase of the original price

109. $(2x+11)(x-5)=0$

Exercises 3.2

1. $(n+3)(n-3)$ **3.** $(6x+1)(6x-1)$

5. $(5p+7q)(5p-7q)$ **7.** $(x+y+2)(x+y-2)$

9. $(6r^3+k^2)(6r^3-k^2)$ **11.** $(y+3)(y^2-3y+9)$

13. $(x-1)(x^2+x+1)$ **15.** $(ab+c)(a^2b^2-abc+c^2)$

17. $(x^2+4)(x^4-4x^2+16)$

19. $(x+1)(x^2+8x+19)$ **21.** $x(1+y)(1-y)$

23. $16(1+t)(1-t)$ **25.** $(1+t^2)(1+t)(1-t)$

27. $3(7x-3)(-x-3)$ **29.** $t(t+1)(t^2-t+1)$

31. $2y^2(y-2)(y^2+2y+4)$

33. $\pi(R+r)(R-r)$ **35.** $3(3n^2+n+1)$

37. $6x(x-3)(x-1)$

39. $(x^2+1)(x+1)(x^2-x+1)$

41. $4x^2(x+6)(x-6)$

43. $(x+1)(x^2-x+1)(x-1)(x^2+x+1)$

45. $(x^4+y^4)(x^2+y^2)(x+y)(x-y)$

47. $(x+2+y)(x+2-y)$

49. $(x+y+z)(x-y-z)$

51. $(\sin\theta+1)(\sin\theta-1)$

53. $(2\tan\theta+5\sin\theta)(2\tan\theta-5\sin\theta)$

55. $7(\sin\theta+3)(\sin\theta-3)$

57. $\cos\theta(\cos\theta+1)(\cos\theta-1)$

59. $(\tan^2\theta+1)(\tan\theta+1)(\tan\theta-1)$

61. $(\sin\theta-1)(\sin^2\theta+\sin\theta+1)$

63. $4a(a+b)(a-b)$

65. **a.** $(c+d)^2$

b. c d / c / d

67. $\pi(R+r)(R-r)$

69. **a.** $y = 5329 - 441t^2$

b. $y = (73+21t)(73-21t)$

c. $\left[73+21\left(\frac{73}{21}\right)\right]\left[73-21\left(\frac{73}{21}\right)\right] = 146(0) = 0$

d. 1533 ft/second $\approx$ 1045 mi / hour

Exercises 3.3

1. $\frac{4}{11}$ **3.** $\frac{1}{3b+1}$ **5.** -1

7. $\frac{-1}{ax}$ **9.** $\frac{a+1}{a-4}$ **11.** $\frac{3x(2x-5)}{2(x-4)}$

13. x^2+ax+a^2 **15.** $\frac{y}{4x^2}$

17. $\frac{2}{a(a+3)}$ **19.** $\frac{(x-1)^2}{(x+1)^2}$ **21.** $\frac{1}{x}$

23. $\frac{n-1}{n-3}$ **25.** $\frac{3b(b+4)}{3b+4}$

27. $\frac{(x+1)(x-1)^2}{7x}$ **29.** 1

31. $(x+3)^2(x-3)$ **33.** $x(x+2)(x-2)$

35. $(x-2)(x+1)(x+3)$ **37.** $6x^2(x-1)^2$

39. $\frac{3a}{xy}$ **41.** $\frac{x-1}{x+2}$ **43.** $\frac{3n-2}{n-6}$

45. $\frac{9a}{5}$ **47.** $\frac{5k^2-3k+2}{k^3}$

49. $\frac{2x+47}{10x}$ **51.** $\frac{6x+11}{(2x-3)(2x+3)}$

53. $\frac{1}{x-y}$ **55.** $\frac{65}{72}$ **57.** $\frac{x+1}{x-1}$

59. $\frac{n^2+n+2}{n^2-n-2}$ **61.** $\frac{y+x}{x}$ **63.** $\frac{x^2+3x-4}{4}$

65. $\frac{1}{y-x}$ **67.** $\frac{-1}{2h}$ **69.** $x+a-1$

71. $\frac{7y-4}{y^2-1}$ **73.** 0 **75.** $\frac{-1}{x(x+h)}$

77. $\frac{1}{(x+1)(x+h+1)}$ **79.** $\frac{x^2+xh-1}{x(x+h)}$

81. $y+x$ **83.** $3+\cos\theta$

85. $\sin\theta+\cos\theta$ **87.** $\frac{9}{25}$

89. $\frac{-\sin\theta}{1+\sin\theta}$ **91.** $\frac{1}{\tan\theta}$ **93.** -3

95. 1 **97.** $\frac{\sin\theta+7}{(2\sin\theta-1)(\sin\theta+2)}$

99. $\frac{1+2\sin\theta}{3\sin\theta-1}$ **101.** $\frac{\tan\theta}{1+2\tan\theta}$ **103.** $f=\frac{ad}{a+d}$

105. $\frac{m-y}{y}$ **107.** 24 mi/hour

109. $\frac{2s_g s_h}{s_h+s_g}$ **111.** $V=\frac{h}{3}\left(a^2+ab+b^2\right)$

Exercises 3.4

1. 7 **3.** 2

5. Not a real number **7.** -2

9. 5 **11.** -3 **13.** 2

15. $\frac{5}{3}$ **17.** 5 **19.** -2

21. 27 **23.** 32 **25.** 9

27. $\frac{27}{8}$ **29.** 4 **31.** -27

33. $\frac{1}{5}$ **35.** $x^{3/5}$; $\sqrt[5]{x^3}$ **37.** $b^{1/3}$; $\sqrt[3]{b}$

39. $\frac{1}{y^{3/4}}$; $\frac{1}{\sqrt[4]{y^3}}$ **41.** $x^{1/3}$; $\sqrt[3]{x}$

43. $\frac{1}{x^{7/6}}$; $\frac{1}{\sqrt[6]{x^7}}$ **45.** $2xy^2$

47. $\frac{a}{2b^3}$ **49.** $\frac{x^{1/2}y^2}{z^{2/3}}$; $\frac{y^2\sqrt{x}}{\sqrt[3]{z^2}}$

51. $x+1$

53. $(x-2)^{1/2}$; $\sqrt{x-2}$

55. $\dfrac{1}{(a-b)^{1/4}}$; $\dfrac{1}{\sqrt[4]{a-b}}$

57. $6u+3u^{9/2}$; $6u+3\sqrt{u^9}$

59. $\dfrac{2x^{1/2}-1}{4}$; $\dfrac{2\sqrt{x}-1}{4}$

61. $\dfrac{y^{5/2}}{x}$; $\dfrac{\sqrt{y^5}}{x}$

63. $\dfrac{125x^3}{343y^{24}}$

65. $\dfrac{4}{9}x^{5/2}y^{7/12}$; $\dfrac{4}{9}\sqrt{x^5}\sqrt[12]{y^7}$

67. $\dfrac{2x-1}{2x^{3/2}}$; $\dfrac{2x-1}{2\sqrt{x^3}}$

69. $\dfrac{x+1}{x-1}$

71. $\dfrac{x-2}{2x^{3/2}}$; $\dfrac{x-2}{2\sqrt{x^3}}$

73. $\dfrac{x-2}{2x^{3/2}}$; $\dfrac{x-2}{2\sqrt{x^3}}$

75. $\dfrac{1+x^3}{x^5}$

77. $\dfrac{4(3x^4+2)}{x^5}$

79. $\dfrac{3(x-5)}{x^{2/3}}$

81. $\dfrac{4x^{3/2}+1}{2x^{1/2}}$

83. $\dfrac{x(x^2-5x-1)}{(x-5)^{1/2}}$

85. $\sin\theta$

87. $\cos^4\theta$

89. $\sin\theta+\cos\theta$

91. 25.2 percent

93. 2.0

95. Earth: 1 year; Mercury: 0.24 years; Jupiter: 11.86 years

97. $A=6s^{2/3}$ or $A=6\sqrt[3]{s^2}$; 952 in.^{2}

99.
   a. 9
   b. In this case, geometric mean < arithmetic mean, since $9<15$.

Exercises 3.5

1. $6\sqrt{3}$

3. $10\sqrt{2}$

5. $-2\sqrt[3]{4}$

7. $2x^3y^2\sqrt{3x}$

9. $xy\sqrt[4]{9x^2y^3}$

11. $\sqrt[3]{7}$

13. $\sqrt{5}$

15. $\sqrt[4]{y^2}$

17. $\sqrt[4]{8(x-1)^3}$

19. $y\sqrt[3]{9y}$

21. $6\sqrt{3}$

23. $\dfrac{\sqrt[4]{10}}{2}$

25. $\dfrac{3\sqrt{x}}{x}$

27. $\dfrac{\sqrt[3]{xy^2}}{y}$

29. $\dfrac{x\sqrt{2}}{2}$

31. $11\sqrt{2}$

33. $5\sqrt{3}$

35. $-7\sqrt{2}$

37. $-14\sqrt{11}$

39. $13\sqrt{3}$

41. $7\sqrt{2}$

43. $\dfrac{2\sqrt{6}}{3}$

45. $\dfrac{37\sqrt{6}}{6}$

47. $9\sqrt[3]{3}$

49. $\dfrac{-29\sqrt[3]{3}}{3}$

51. $8\sqrt{2x}$

53. $(2y-6x)\sqrt{5x}$

55. $\dfrac{5x\sqrt{2xy}}{2}$

57. $\left(\dfrac{2}{y}-\dfrac{2}{x^2}+\dfrac{5x}{4}\right)\sqrt{2xy}$

59. $\left(\dfrac{1}{y}-1\right)\sqrt[4]{xy^3}$

61. $2\sqrt{3}$

63. $2\sqrt{2}$

65. -4

67. $3\sqrt{5}$

69. -1

71. $29-4\sqrt{30}$

73. $3x\sqrt{10}$

75. $2xy\sqrt[3]{3xy}$

77. $\dfrac{\sqrt{6x}}{x}$

79. $\dfrac{\sqrt[4]{56x}}{2}$

81. $\dfrac{2\sqrt{3}}{3}$

83. $\dfrac{\sqrt{3}-3}{3}$

85. $3-2\sqrt{2}$

87. $\dfrac{x-\sqrt{xy}}{x-y}$

89. $\dfrac{\sqrt{x^2-1}}{x+1}$

91. $\dfrac{1}{\sqrt{x}+2}$

93. $\dfrac{1}{\sqrt{x+h}+\sqrt{x}}$

95. $\dfrac{2}{\sqrt{2x+2h+1}+\sqrt{2x+1}}$

97. $1{,}200\sqrt{5}$

99. $2x+6$

101. $\dfrac{2\sqrt{x+1}-1}{2\sqrt{x+1}}=\dfrac{2x+2-\sqrt{x+1}}{2x+2}$

103. $\sqrt{a^2+b^2}$

105. $\dfrac{\pi(4R^2-2h^2)}{\sqrt{4R^2-h^2}}=\dfrac{\pi(4R^2-2h^2)\sqrt{4R^2-h^2}}{4R^2-h^2}$

107. $\dfrac{-1}{x^2\sqrt{x^2+1}}=\dfrac{-\sqrt{x^2+1}}{x^2(x^2+1)}$

109. Equal when $x = 0$

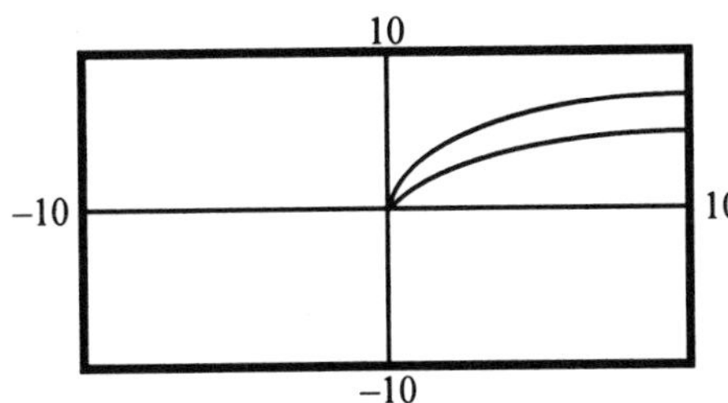

111. Equal when $x = 0$ or $x = -10$

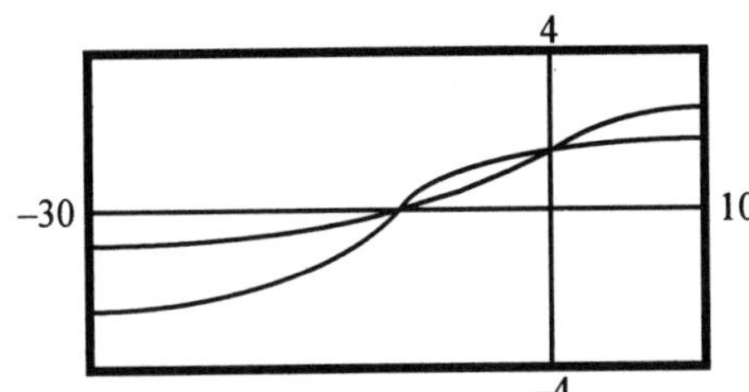

113. Equal when $x = 0$

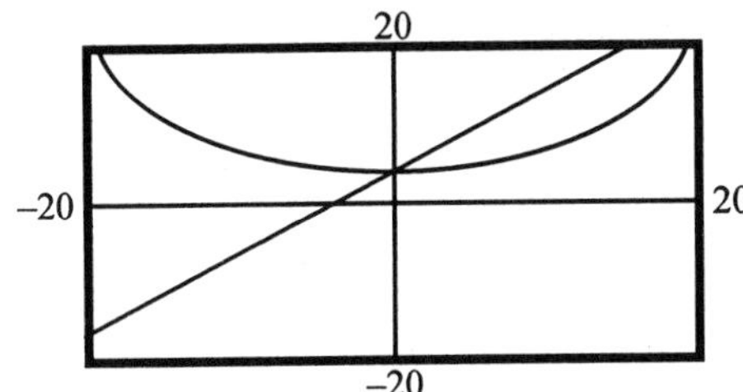

115. $10\sqrt{\sin\theta}$ **117.** $\dfrac{\sqrt{\tan\theta}}{\tan\theta}$ **119.** $\tan\theta$

121. $1 - \tan\theta$ **123.** $3\sqrt{17}$

125. $\pi\sqrt{2}$ seconds; 4.44 seconds

127. **a.** $e = \dfrac{k\sqrt{n}}{n}$

b. $k = 0.09$; $e = 1.8$ percent

129. $L = \dfrac{\sqrt{2A\sqrt{3}}}{3}$; 2

Exercises 3.6

1. $5i$ **3.** $8i$ **5.** $i\sqrt{22}$

7. $i\sqrt{3}$ **9.** $2 - 8i\sqrt{2}$ **11.** $2i$

13. $2i\sqrt{2}$ **15.** $\dfrac{1}{4} - \dfrac{\sqrt{23}}{4}i$

17. $-2 + i$ **19.** $-\dfrac{1}{5} - \dfrac{\sqrt{14}}{5}i$

21. $0 - 3i$ **23.** $0 + 0i$ **25.** $-3 - i$

27. $5 - 6i$ **29.** $-6 + 0i$ **31.** $-4 + 0i$

33. $10 - 10i$ **35.** $13 + 0i$ **37.** $-3 - 4i$

39. $\dfrac{80}{9} - 2i$ **41.** $(5i)^2 + 25 = -25 + 25 = 0$

43. $\left(-2i\sqrt{2}\right)^2 + 8 = -8 + 8 = 0$

45. $(1 - i)^2 - 2(1 - i) + 2 = 1 - 2i + i^2 - 2 + 2i + 2 = 0$

47. $2\left(\dfrac{1}{4} + \dfrac{i\sqrt{23}}{4}\right)^2 - \left(\dfrac{1}{4} + \dfrac{i\sqrt{23}}{4}\right) + 3$

$= \dfrac{1}{8} + \dfrac{i\sqrt{23}}{4} - \dfrac{23}{8} - \dfrac{1}{4} - \dfrac{i\sqrt{23}}{4} + 3 = 0$

49. $(2i)^4 - 16 = 16i^4 - 16 = 16(1) - 16 = 0$

51. Solution **53.** Solution

55. Not a solution **57.** Solution

59. Not a solution **61.** $3 - 4i$

63. $-i$ **65.** -7 **67.** $\dfrac{1}{5} - \dfrac{2}{5}i$

69. $0 + \dfrac{2}{5}i$ **71.** $-\dfrac{4}{17} - \dfrac{1}{17}i$

73. $\dfrac{7}{10} + \dfrac{1}{10}i$ **75.** $\dfrac{1}{3} - \dfrac{2\sqrt{2}}{3}i$

77. $0 + i$ **79.** $\dfrac{1}{2} - \dfrac{1}{2}i$ **81.** $-i$

83. $-i$ **85.** 1 **87.** i

89. **a.** $\overline{(1+i) + (2-3i)} = \overline{3 - 2i} = 3 + 2i$;

$\overline{1+i} + \overline{2-3i} = (1 - i) + (2 + 3i) = 3 + 2i$

b. $\overline{(1+i)(2-3i)} = \overline{5 - i} = 5 + i$;

$\overline{1+i} \cdot \overline{2-3i} = (1 - i)(2 + 3i) = 5 + i$

c. $\overline{(1+i)^2} = \overline{2i} = -2i$; $\left(\overline{1+i}\right)^2 = (1 - i)^2 = -2i$

91. $\overline{z + w} = \overline{(a+c) + (b+d)i} = (a + c) - (b + d)i$

$= (a - bi) + (c - di) = \overline{z} + \overline{w}$

93. if z is a real number, $z = a + 0i$; so $\overline{z} = a - 0i$ and $\overline{z} = z$.

95. $6-4i$ ohms 97. $111+52i$ volts

99. $6+6i$ amperes

Chapter 3 Review Exercises

1. $2^{3/2}$ 3. $4\sqrt{3}$ 5. $4-2\sqrt{3}$

7. $2\pi\sqrt{3}$ 9. $5\sqrt{3}$ 11. $2R\sqrt{3}$

13. $\dfrac{3\sqrt{10}}{10}$ 15. $x^{1/6}$ 17. $\dfrac{3+\sqrt{6}}{3}$

19. $\dfrac{2}{h}$ 21. $\cos^2\theta-4\sin\theta$

23. $\dfrac{2-x}{2}$ 25. $a^{2x}-4a^x+3$

27. $4x^3+6x^2h+4xh^2+h^3$

29. $\dfrac{x+5}{x-1}$ 31. $\dfrac{\cos\theta-1}{\cos\theta+1}$ 33. $\dfrac{2x}{3y^3}$

35. $\dfrac{5-k}{k-1}$ 37. $\dfrac{6\sin^2\theta-5}{15\sin\theta}$ 39. $\dfrac{s-1}{s+1}$

41. $\dfrac{-2}{(x+1)(x+h+1)}$ 43. $\dfrac{\sin\theta-3}{\sin\theta+6}$

45. $\dfrac{x+1}{x^{n+1}}$ 47. $\dfrac{x-y}{x+y}$

49. $\dfrac{1}{\tan\theta(\tan\theta+1)}$ 51. $x+a+1$

53. $\dfrac{-1}{ax}$ 55. $\dfrac{3x-3}{2x^{1/2}}$ 57. $\dfrac{2x^2}{\sqrt{1-x^2}}$

59. $\dfrac{-4}{u^2\sqrt{u^2+4}}=\dfrac{-4\sqrt{u^2+4}}{u^2(u^2+4)}$

61. $-i$ 63. $2i$

65. $\dfrac{9}{13}+\dfrac{6}{13}i$ 67. $(\cos\theta+4)^2$

69. $t^3(1+x)(1-x)$ 71. $(x+1)(x^2+1)$

73. $2\sin\theta\cos\theta(\sin\theta-5)(\sin\theta-4)$

75. $(a+b)(a^2-ab+b^2)$
$=a^3-a^2b+ab^2+a^2b-ab^2+b^3=a^3+b^3$

77. $72x^2y^2z^3$

79. $\dfrac{-1}{\sqrt{x}\sqrt{x+h}\left(\sqrt{x}+\sqrt{x+h}\right)}$

81. $\dfrac{1}{(x+2)\sqrt{x-2}}$

83. $(1+i)^2-2(1+i)+2=1+2i+i^2-2-2i+2=0$

85. $\dfrac{2x}{x-1}$

87. $\dfrac{(x-1)^3-(x+1)\cdot 3(x-1)^2}{(x-1)^6}$
$=\dfrac{(x-1)-3(x+1)}{(x-1)^4}=\dfrac{-2x-4}{(x-1)^4}=\dfrac{-2(x+2)}{(x-1)^4}$

89. $A=\dfrac{\sqrt{3}s^2(k^2-1)}{4k^2}$ 91. 76.82 percent

93. d 95. c 97. b 99. d

Chapter 3 Test

1. $\dfrac{y-4}{y}$ 2. $\dfrac{x}{x+a}$ 3. 22

4. $x-2\sqrt{ax}+a$ 5. $\dfrac{13\sqrt{14}}{7}$

6. $\dfrac{\sin^2\theta-\cos^2\theta}{\cos\theta\sin\theta}$ 7. $\dfrac{x+7}{1-x}$

8. $\dfrac{x-4}{6(3x+2)}$ 9. $\dfrac{10}{3}$

10. $\dfrac{(x-1)(x-2)}{(x+2)(x+1)}$ 11. $x(x+1)(x-1)$

12. $(3x-2y)(9x^2+6xy+4y^2)$

13. $2(5\sin\theta+3)(2\sin\theta-1)$

14. $t(2t+3)^2$ 15. 13 16. $a^{9/10}$, $\sqrt[10]{a^9}$

17. $6+2\sqrt{3}$ 18. $18x^2y^3$ 19. $\dfrac{y}{x^{-1/2}}$

20. 120 mi/hour

Exercises 4.1

1. $\{0,-5\}$ **3.** $\{0,5\}$ **5.** $\{2,-2\}$

7. $\{4,-2\}$ **9.** $\{4,-1\}$ **11.** $\left\{5,\frac{1}{3}\right\}$

13. $\left\{1,\frac{2}{3}\right\}$ **15.** $\left\{2,-\frac{1}{2}\right\}$ **17.** $\{-5,1.5\}$

19. x – int.: $(5,0)$, $(-1,0)$; y – int.: $(0,-5)$

21. x – int.: $(2,0)$; y – int.: $(0,4)$

23. x – int.: $\left(\frac{2}{3},0\right)$, $\left(-\frac{3}{2},0\right)$; y – int.: $(0,-6)$

25. $\{\pm 3\}$ **27.** $\{\pm 5i\}$ **29.** $\{\pm 2i\sqrt{2}\}$

31. $\{3,7\}$ **33.** $\{-1\pm i\}$ **35.** $\{7\pm i\sqrt{7}\}$

37. $\{1\pm i\}$ **39.** $\left\{\frac{-1\pm\sqrt{13}}{2}\right\}$

41. $\left\{\frac{2\pm\sqrt{10}}{3}\right\}$ **43.** $\left\{\frac{-1\pm i\sqrt{14}}{5}\right\}$

45. $\{-4,-1\}$ **47.** $\{3\}$ **49.** $\left\{\frac{1\pm\sqrt{17}}{2}\right\}$

51. $\left\{\frac{1\pm\sqrt{19}}{3}\right\}$ **53.** $\left\{\frac{1\pm i\sqrt{2}}{3}\right\}$ **55.** $\left\{-1,\frac{3}{4}\right\}$

57. none **59.** one double root

61. two **63.** two **65.** none

67. $\{3,-2\}$ **69.** $\left\{\frac{\pm 2i\sqrt{5}}{5}\right\}$ **71.** $\left\{0,\frac{2}{3}\right\}$

73. $\{1\pm\sqrt{5}\}$ **75.** $\{60,9\}$ **77.** $\{0,10\}$

79. 37.4 mph **81.** 7 **83.** 3.8 in., 7.8 in.

85. 10 units

87. \$5, \$3; The low unit price produces high sales volume, and the higher unit price produces lower sales volume.

89. $\frac{\sqrt{10}}{4}$ seconds

91. $x=\frac{7\pm\sqrt{29}}{2}$ inches and both solutions are meaningful in the context of the problem.

93. 8 ohms, 24 ohms

95.

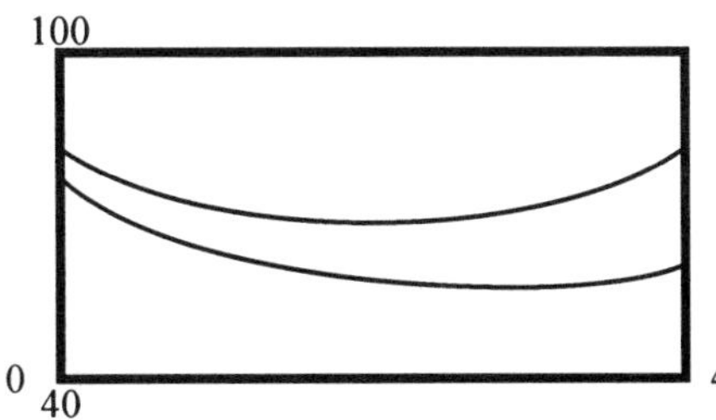

a. In each case the approval rating drops as soon as the term begins. During peacetime the approval rating levels off and begins to increase sooner than during wartime. In each case the approval rating is rising near the end of the term.

b. Graphically: 1.0 years, 3.4 years, 0.3 years

Algebraically: $2.18-\sqrt{\frac{4.52}{3.22}}$ years ,

$2.18+\sqrt{\frac{4.52}{3.22}}$ years

$3.11-\sqrt{\frac{26.92}{3.45}}$ years

c. Peacetime: at $x=0$

d. Wartime: at $x=3.1$ years

Exercises 4.2

1. $\left\{\frac{26}{5}\right\}$ **3.** $\left\{-\frac{1}{10}\right\}$ **5.** $\{-1\}$

7. $\left\{-\frac{5}{2}\right\}$ **9.** $\{1\}$ **11.** $\emptyset$

13. $\emptyset$ **15.** $\{1\}$ **17.** $\{1\}$

19. $\frac{-2\pm\sqrt{15}}{2}$ **21.** $\frac{3\pm\sqrt{69}}{2}$ **23.** $\{\pm 300\}$

25. $\left\{9,\frac{1}{2}\right\}$ **27.** $L_1=\frac{L_2W_2}{W_1}$ **29.** $p=\frac{1}{w+1}$

31. $v=v_0+at$ **33.** $i=\frac{S-P}{Pn}$ **35.** $r=\frac{S-a}{S-L}$

37. $Z_2=\frac{ZZ_1}{Z_1-Z}$ **39.** $a=\frac{fb}{b-f}$

41. $\frac{5 \pm \sqrt{21}}{2}$; 4.8, .21 **43.** $60,000

45. 100 units

47. **a.** \$1.30; \$0.65 **b.** 80 minutes
c. The cost per minute is 25.5 cents when x is 160 minutes. After 160 minutes the cost per minute is 25 cents when rounded to the nearest cent.

49. 12 minutes

51. $\frac{1}{100} + \frac{1}{100} = \frac{1}{x}$
$x + x = 100$
$x = 50$ minutes, which is half the time

53. 60 mph

55. **a.** 150 mph **b.** 1.8 hr; 3.6 hr
c. $133\frac{1}{3}$ mph

Exercises 4.3

1. $\{-1\}$ **3.** $\emptyset$ **5.** $\left\{\frac{25}{3}\right\}$

7. $\{6\}$ **9.** $\{-9\}$ **11.** $\{4\}$

13. $\{-2\}$ **15.** $\{3\}$ **17.** $\{0\}$

19. $\{-3,1\}$ **21.** $\emptyset$ **23.** $\{0,4\}$

25. $\left\{\frac{1}{2}\right\}$ **27.** $[0,\infty)$ **29.** $\emptyset$

31. $\{-9,1\}$ **33.** $\{1\}$ **35.** $\{27 \pm 10\sqrt{2}\}$

37. $\{\pm 1, \pm\sqrt{5}\}$ **39.** $\left\{\frac{\pm\sqrt{2}}{2}\right\}$ **41.** $\left\{\pm\sqrt{2 \pm \sqrt{3}}\right\}$

43. $\{1,-2\}$ **45.** $\left\{\frac{7}{2}\right\}$ **47.** $\{-1\}$

49. $\{16,1\}$ **51.** $\{0,-2,-7\}$ **53.** $\{0,1,-1\}$

55. $\left\{0, \frac{1}{2}, -\frac{1}{3}\right\}$ **57.** $\{2,-2,1,-1\}$

59. $\left\{1,-1,\frac{11}{3}\right\}$ **61.** $\{2,-1 \pm i\sqrt{3}\}$

63. $\left\{1,-1,\frac{1 \pm i\sqrt{3}}{2},\frac{-1 \pm i\sqrt{3}}{2}\right\}$

65. $s = \frac{gt^2}{2}$ **67.** $y = x^3 - 2$ **69.** 3,003 ft

71. Yes, at $d = 0$ and 64.

73. 48 ft **75.** 0.81 ft **77.** 7

79. $\frac{21}{5}$ cm **81.** 6.18 in^2

Exercises 4.4

1. $(-\infty,-1)\cup(2,\infty)$ **3.** $[0,1]$

5. $(-3,2)$ **7.** $(-\infty,-1]\cup[1,\infty)$

9. $(-\infty,-\sqrt{3})\cup(\sqrt{3},\infty)$ **11.** $(-\infty,0)\cup(5,\infty)$

13. $[-2,1]$ **15.** $\left(-\infty,-\frac{1}{2}\right]\cup[4,\infty)$

17. $(-\infty,\infty)$ **19.** $\{1\}$

21. $(-\infty,2-\sqrt{7})\cup(2+\sqrt{7},\infty)$

23. $\left[\frac{2-\sqrt{10}}{3},\frac{2+\sqrt{10}}{3}\right]$

25. $\emptyset$ **27.** $[0,1]\cup[2,\infty)$

29. $(-\infty,0)$ **31.** $(1,2)$

33. $(-2,0)\cup(2,\infty)$ **35.** $[-2,-1]\cup[1,2]$

37. $[-3,-1]\cup[0,1]\cup[3,\infty)$

39. $(-\infty,-1)\cup\left(1,\frac{5}{2}\right)$ **41.** $(-2,3]$

43. $(-\infty,-1]\cup(0,1]$ **45.** $(-\infty,-2]\cup[2,3)$

47. $(-\infty,0)\cup\left(\frac{1}{2},\infty\right)$ **49.** $(-\infty,-21]\cup(7,\infty)$

51. $(-\infty,4)\cup[12,\infty)$ **53.** $(-3,2)$

55. $[-2, 2]$ **57.** $(-\infty,-2]\cup[2,\infty)$

59. (1 second, 5 seconds)

61. **a.** [128, 170] **b.** [135, 163]

63. $(0,1)$

65. **a.** $(4-\sqrt{6}$ in, $4+\sqrt{6}$ in$)$, (1.55 in., 6.45 in.)
b. 4 in.

67. $\left(\dfrac{3\sqrt{3}}{5}\text{ in.}, \dfrac{2\sqrt{7}}{5}\text{ in.}\right)$

69. **a.**

X	1	4	8
Y	.89	2.21	8.73

b. Differences are −0.11, −0.29, and −0.37, respectively.

c. $[4,\infty)$

Exercises 4.5

1. **a.** 2 units 2 units

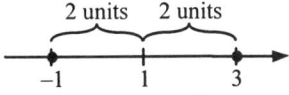

−2 0 2

b.

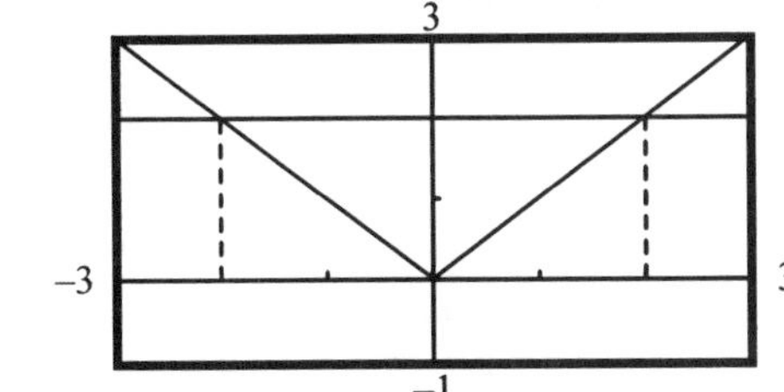

c. $\{-2,2\}$

3. **a.** 2 units 2 units

−1 1 3

b.

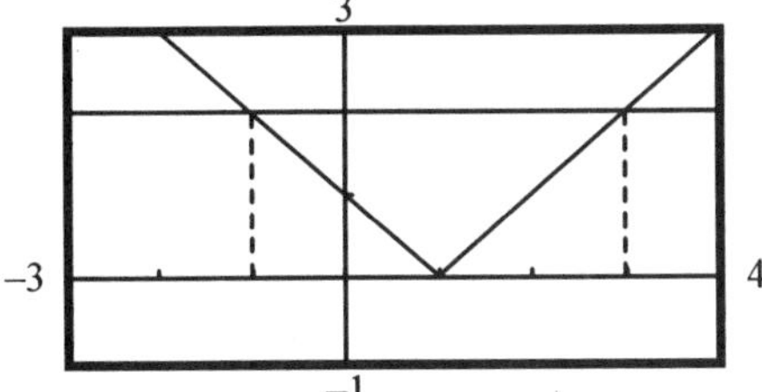

c. $\{-1,3\}$

5. **a.** 1 unit 1 unit

−3 −2 −1

b. 4

−4 3

−2

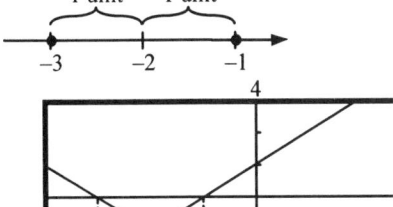

c. $\{-3,-1\}$

7. **a.**

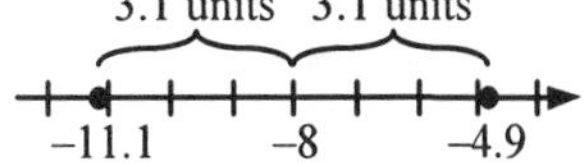

b.

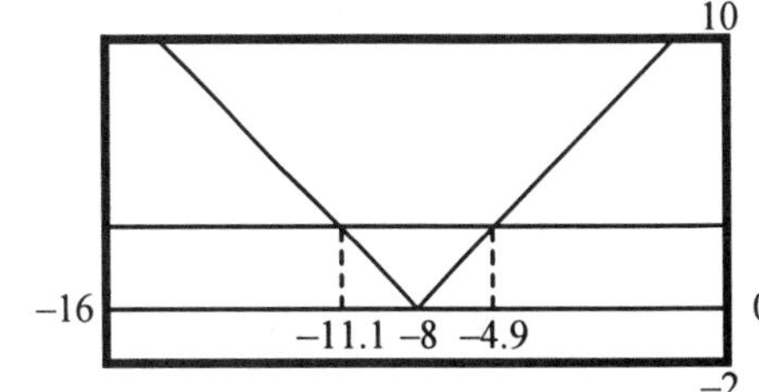

c. $\{-11.1,-4.9\}$

9. $\{-8,4\}$ **11.** $\{34,2\}$ **13.** $\left\{\dfrac{3}{2},-\dfrac{3}{2}\right\}$

15. $\emptyset$ **17.** $\left\{\dfrac{3}{4},\dfrac{5}{2}\right\}$ **19.** $\left\{1,\dfrac{1}{3}\right\}$

21. $\{1\}$ **23.** $\left\{2,-\dfrac{4}{3}\right\}$ **25.** $\{-1,1\}$

27. $\left\{-\dfrac{1}{3},1\right\}$

29. **a.**

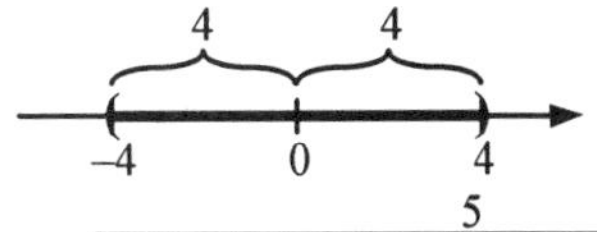

b.

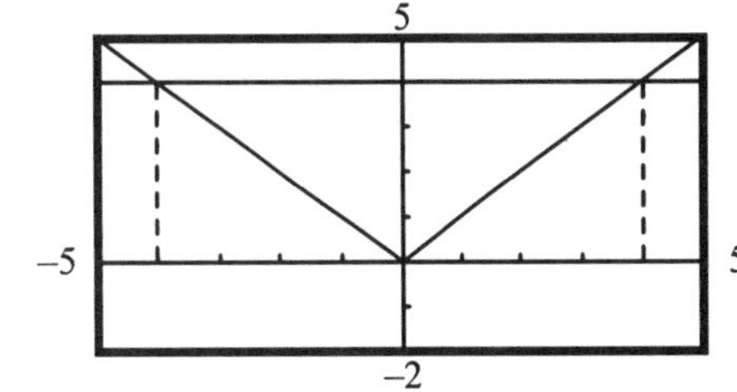

c. $(-4,4)$

31. **a.**

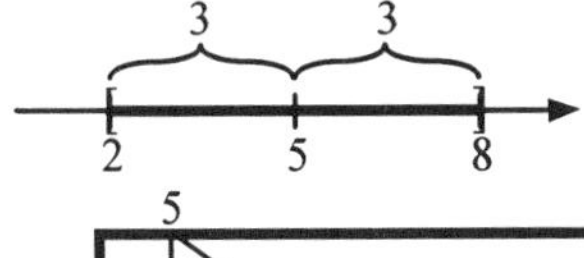

b.

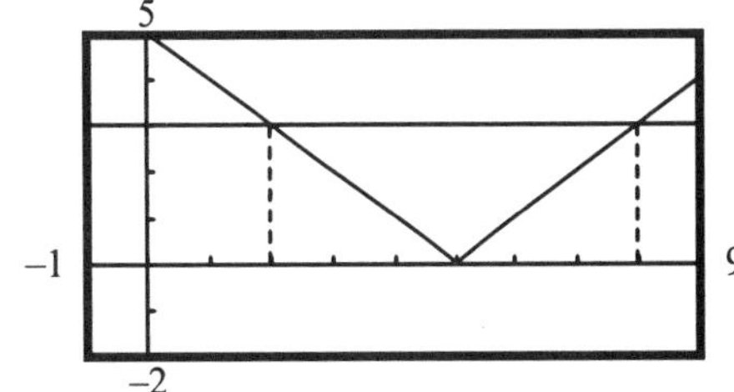

c. $[2,8]$

33. **a.**

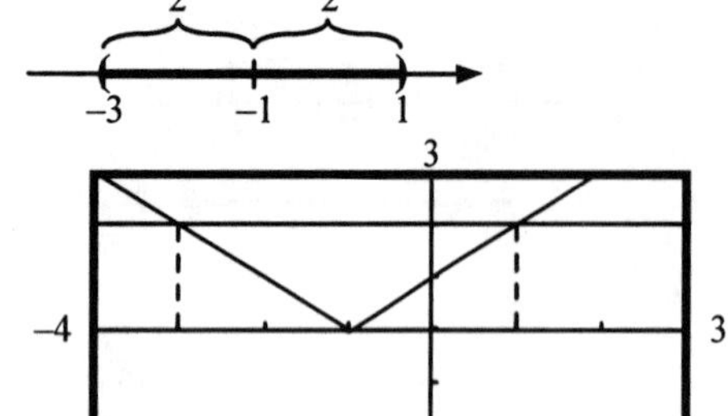

b.

c. $(-3,1)$

35. **a.**

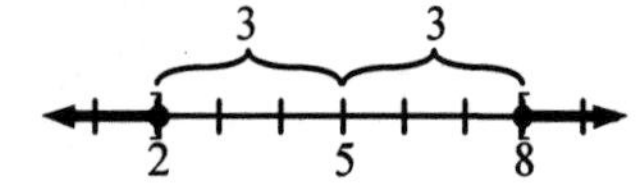

b.

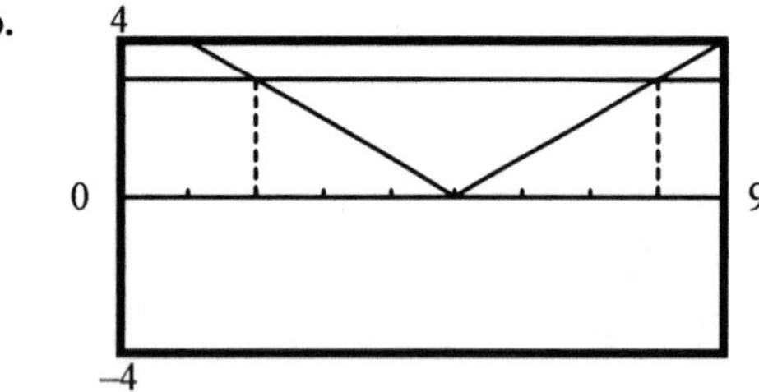

c. $(-\infty,2]\cup[8,\infty)$

37. $(-\infty,-2)\cup(2,\infty)$ **39.** $\left(-\frac{13}{3},1\right)$

41. $[-6,7]$ **43.** $(-\infty,-1)\cup(2,\infty)$

45. $\emptyset$ **47.** $\left(\frac{7}{4},\frac{7}{2}\right)$ **49.** $\{x:|x|<2\}$

51. $\{x:|x-1.15|<0.05\}$ **53.** $\{x:|x|>2\}$

55. $\{x:|x-3|>3\}$ **57.** $\{x:|x|<5\}$

59. $\{x:|x-3|<1\}$ **61.** $\left\{x:|x-3|<\frac{1}{2}\right\}$

63.

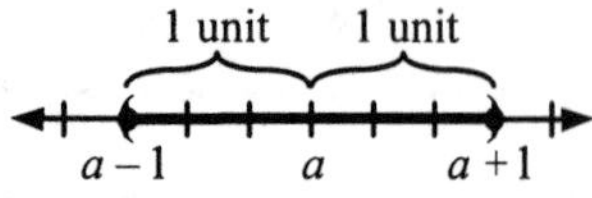

65. **a.** At t increases during the first 4 seconds, the speed decreases and momentarily equals zero. Then the speed increases as t increases during the next four seconds.

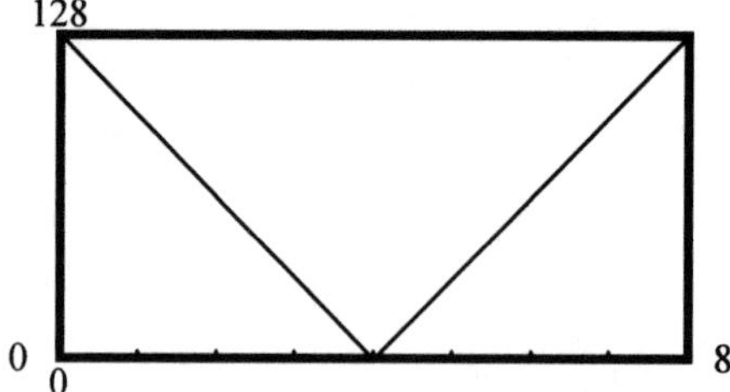

b. $\emptyset$ **c.** (1.5 seconds, 6.5 seconds)
d. 128 ft/second

67. $\{b:|b-1170|\le 12.5\}$; $[1157.5,\ 1182.5]$

69. $\hat{p}-0.05\le p\le \hat{p}+0.05$

71. **a.** $\{r:|r-1.8994|<0.0054\}$
b. If $M=1.8994$ is used as an estimate of r, then the maximum error is 0.0054.
c. 1 decimal place

Chapter 4 Review Exercises

1. $\left\{\frac{-1\pm\sqrt{5}}{2}\right\}$ **3.** $\{5-2i,5+2i\}$

5. $(-5,5)$ **7.** $\{\pm 200\}$ **9.** $\{4\}$

11. $\left\{\frac{5\sqrt{3}}{3}\right\}$ **13.** $\{3,7\}$ **15.** $(-\infty,1)$

17. $\{5\}$ **19.** $\{0,-2\}$

21. $(-\infty,-3)\cup(0,3)$ **23.** $\left\{\frac{3}{2},-4\right\}$

25. $\{1,-1,5\}$ **27.** $\{6,-1\}$

29. $(-3,1)\cup[5,\infty)$ **31.** $F_1=\frac{F_2 d_2}{d_1}$

33. $y=\frac{x-1}{x+2}$ **35.** $x=\frac{a}{ab-1}$

37. $(-\infty,-3]\cup[1,\infty)$ **39.** $\frac{1}{4}$

41. As anxiety starts to increase, performance increases until a point is reached where it momentarily levels off. As anxiety continues to increase after this point, performance starts to decrease $(6,14)$

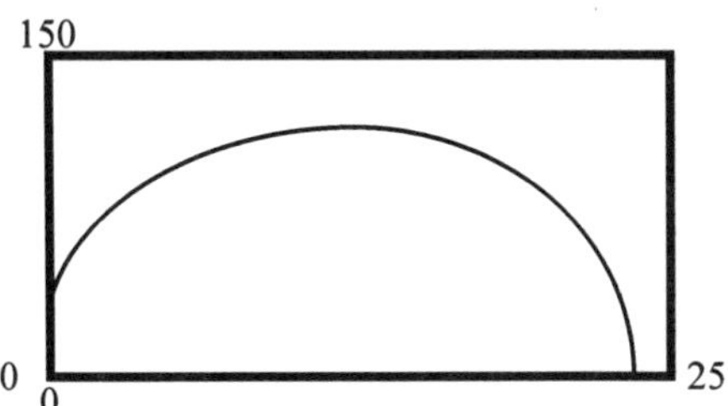

43. $y=x^3-4x$ **45.** 9 ft **47.** 60 mph

49. $\frac{391}{40}$ **51.** d. **53.** c.

55. d. **57.** a. **59.** a.

Chapter 4 Test

1. $\left\{\frac{-1\pm\sqrt{19}}{6}\right\}$ **2.** $\{1,49\}$

3. $\left\{-\frac{23}{24}\right\}$ **4.** $\{\pm 2\sqrt{7}\}$ **5.** $\left\{2,\frac{4}{3}\right\}$

6. $\left\{0,-1,-\frac{2}{3}\right\}$ **7.** $\{5\pm\sqrt{6}\}$

8. $\left(-\infty,-\frac{4}{3}\right)\cup(2,\infty)$ **9.** $(-5,4)$

10. $\left(0,\frac{2}{3}\right)$ **11.** $\{-2,2\}$

12. $[-2,0]\cup[1,\infty)$ **13.** $[-3,0]\cup[2,\infty)$

14. $C_1=\frac{CC_2}{C_2-C}$ **15.** $\{x: |x-5|<4\}$

16. **a.** 88 feet **b.** 44 ft/second

17. $(-\infty,-1]\cup[0,\infty)$ **18.** 15 m, 25 m

19. 4.8 ft **20.** 0.6 second

Exercises 5.1

1. **a.**

x	100	200	300	400	500	600	700
c	1300	2000	2700	3400	4100	4800	5500

b. $c=7x+600$

c. As x increases, the cost increases uniformly and the relation is described by a line. Domain: $[0,\infty)$; Range: $[600,\infty)$. (Technically $D=\{0, 1, 2, 3, \ldots\}$ and $R=\{600, 607, 614, 621, \ldots\}$)

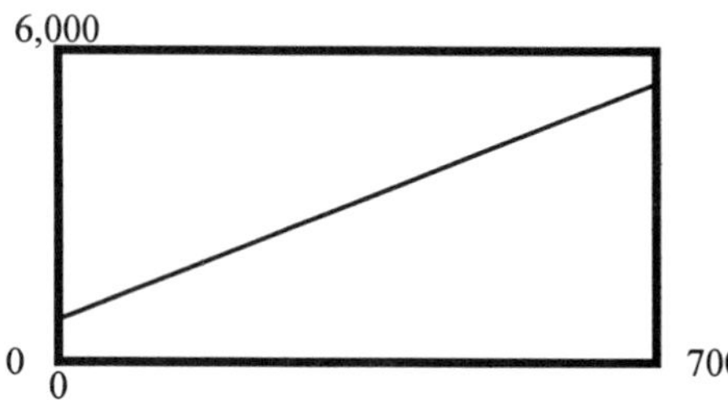

3. **a.**

p	12	16	20	24	28	32	36	40
A	9	16	25	36	49	64	81	100

b. $A=\left(\frac{p}{4}\right)^2=\frac{p^2}{16}$

c. As p increases, the area increases and the relation is described by a segment of a parabola. Domain: $(0,\infty)$; Range: $(0,\infty)$.

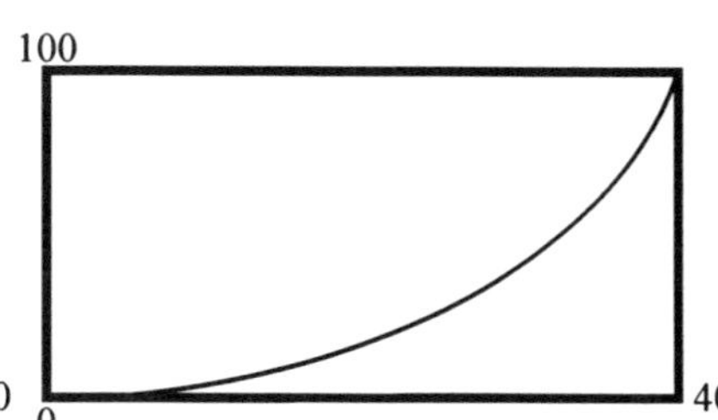

5. **a.**

x	1	2	3	4	5	6	7
z	42.8	41.8	40.8	39.8	38.8	37.8	36.8
p	42.8	83.6	122.4	159.2	194	226.8	257.6

b. $p=x(43.8-x)$

c. As x increases to 21.9, p increases to 479.61; then as x increases from 21.9 to 43.8, p decreases to 0. Domain: $[0, 43.8]$; Range: $[0, 479.61]$.

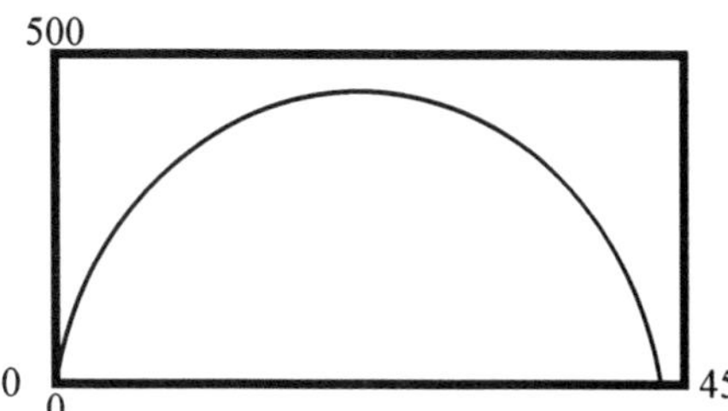

7. **a.**

x	2	4	6	8	10	12	14	16	18	20
$2x$	4	8	12	16	20	24	28	32	36	40
z	54	48	42	36	30	24	18	12	6	0
A	162	288	378	432	450	432	378	288	162	0

b. $A = 1.5x(60 - 3x)$

c. As x increases to 10, A increases to 450; then as x increases from 10 to 20, A decreases to approach 0. Domain: $(0, 20)$; Range: $(0, 450]$.

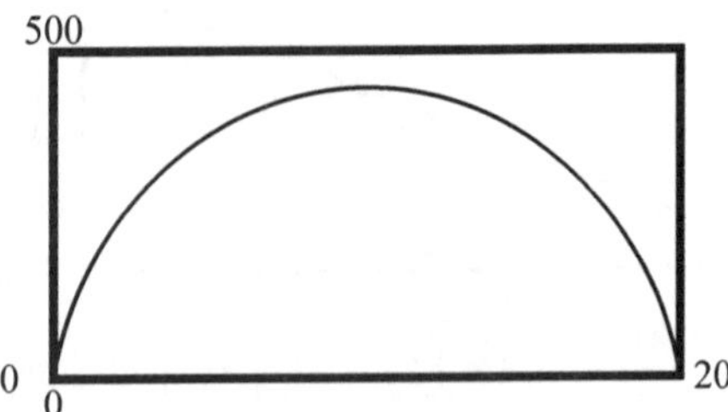

9. **a.**

a	1	5	9	13	17
b	$\frac{420}{29}$	12	$\frac{60}{7}$	$\frac{60}{17}$	$-\frac{60}{13}$ (not meaningful)
A	$\frac{210}{29}$	30	$\frac{270}{7}$	$\frac{390}{17}$	$-\frac{510}{13}$ (not meaningful)

b. $A = \dfrac{15a(15 - a)}{30 - a}$

c. As a increases to about 8.79, A increases to about 38..60; then as x increases from 8.79 to 15, A decreases to approach 0. Domain: $(0, 15)$; Range: $(0, 38.60]$.

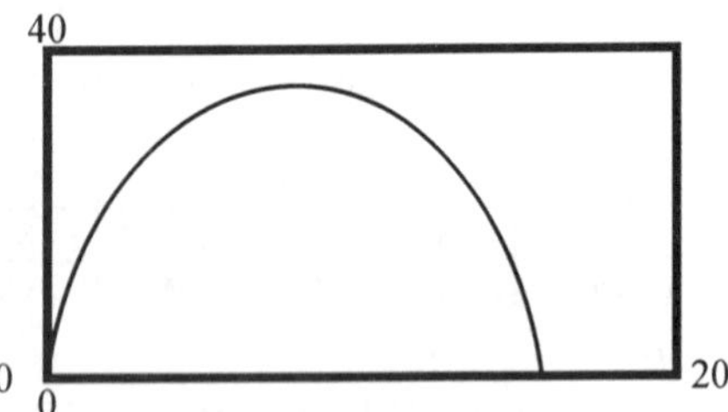

11. D: $(-\infty, \infty)$; R: $(-\infty, \infty)$

13. D: set of all real numbers except 2; R: set of all real numbers except 0

15. D: $(0, \infty)$; R: $(0, \infty)$

17. D: $[-3, 3]$; R: $[0, 3]$

19. D: $(-\infty, \infty)$; R: $\{5\}$

21. D: $\{-2, -1, 0\}$; R: $\{2, 1, 0\}$

23.

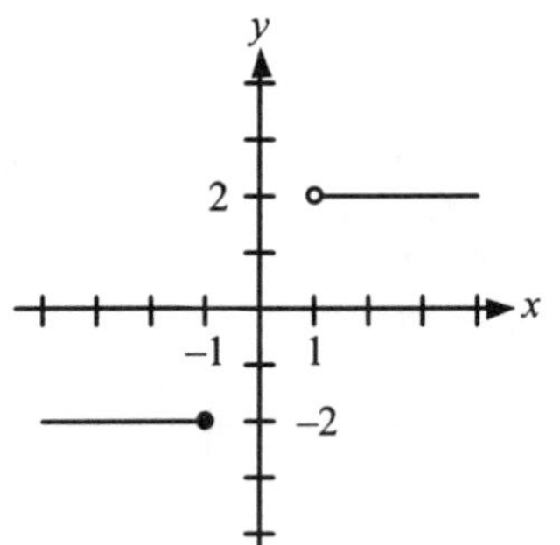

D: $(-\infty, -1] \cup (1, \infty)$; R: $\{2, -2\}$

25.

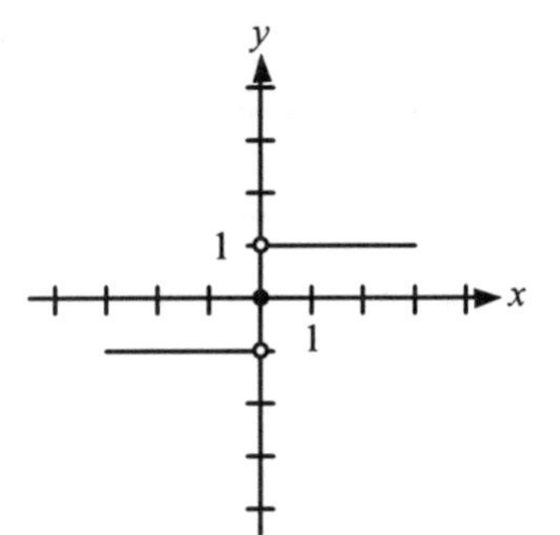

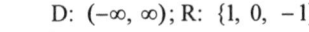

D: $(-\infty, \infty)$; R: $\{1, 0, -1\}$

27.

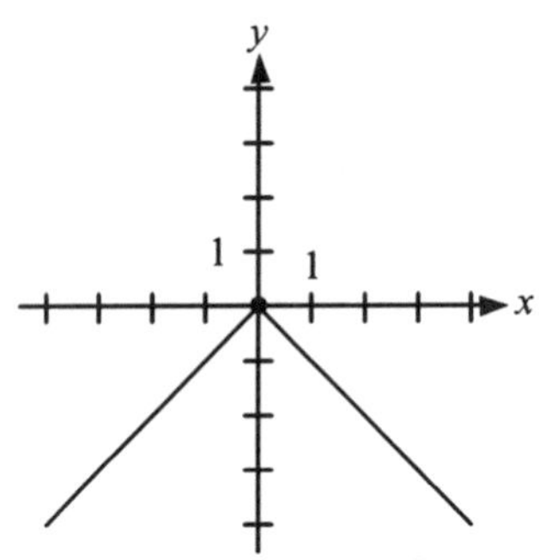

D: $(-\infty, \infty)$; R: $(-\infty, 0]$

29.

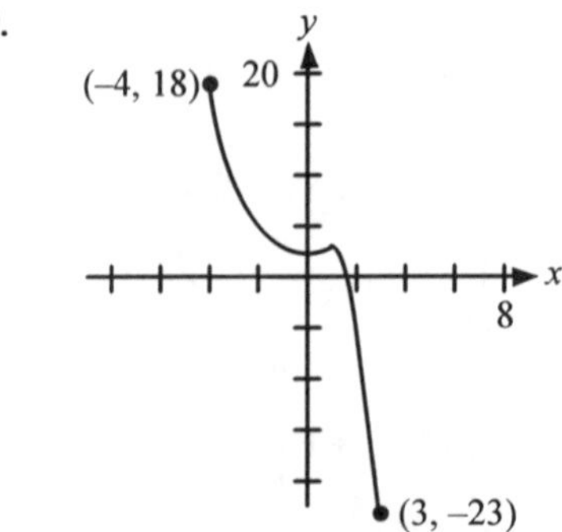

D: $[-4, 3]$; R: $[-23, 18]$

31.

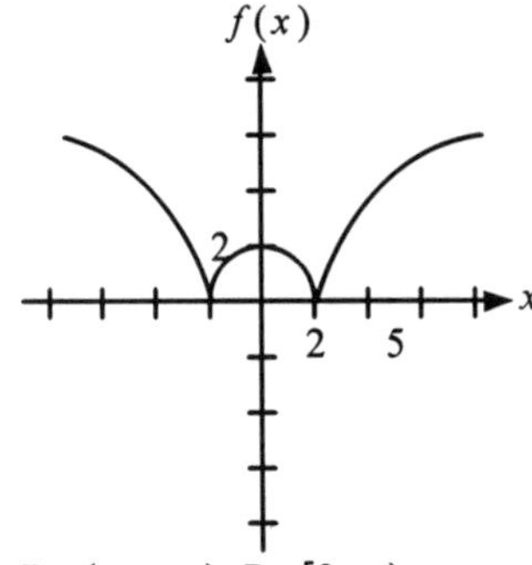

D: $(-\infty, \infty)$; R: $[0, \infty)$

33. Function; D: $\{1, 2, 3\}$; R: $\{-1, 5, 1\}$

35. Not a function; D: $\{-2, 2\}$; R: $\{2, -2, 0\}$

37. Not a function; D: $\{1\}$; R: $\{5, 6, 7\}$

39. Function; D: $\{-2, 2\}$; R: $\{4\}$

41. b, d

43. 7, 1, $a^3 - a + 1$

45. $0, -\frac{3}{4}, -c^2 + c$

47. Undefined, –6, $\frac{1}{x_0 + h} - 5$

49. –2, Undefined, $\frac{a+2}{a-1}$

51. **a.** 3 **b.** 13 **c.** 39
d. –2 **e.** 3 **f.** Undefined

53. **a.** $(-\infty, \infty)$ **b.** $[1, \infty)$ **c.** Neither
d. 2 **e.** $\varnothing$ **f.** $(-\infty, \infty)$
g. Increasing: $(-2, \infty)$; decreasing: $(-\infty, -2)$

55. **a.** $(-\infty, \infty)$ **b.** $[-3, 3]$ **c.** Odd
d. 0 **e.** $\{0\}$ **f.** $(-\infty, 0)$
g. Decreasing: $(-3, 3)$

57. **a.**

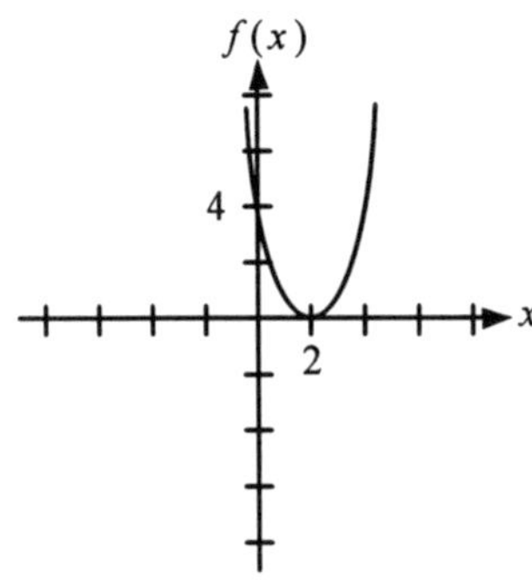

b.

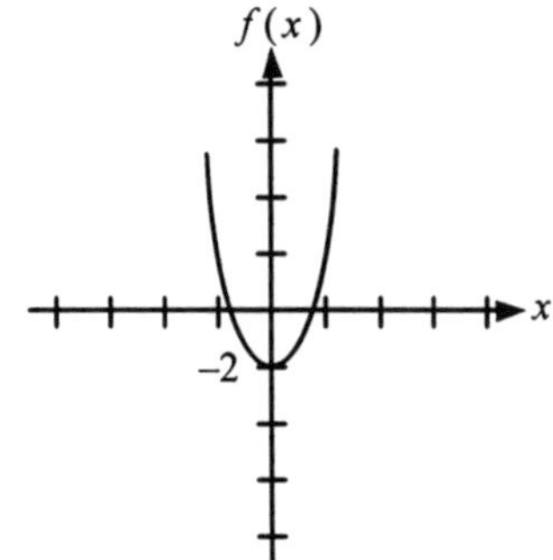

c.

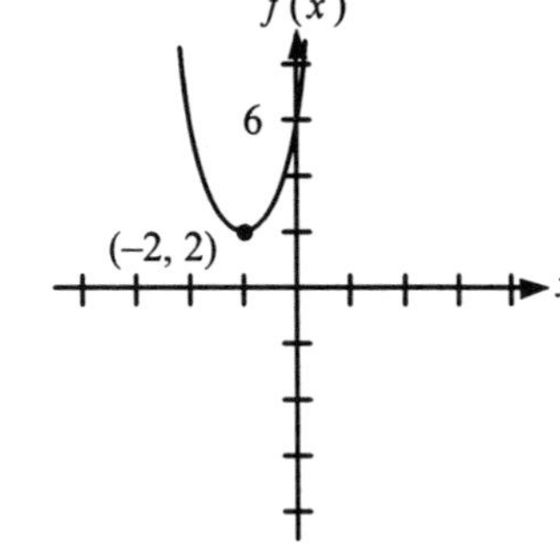

59.

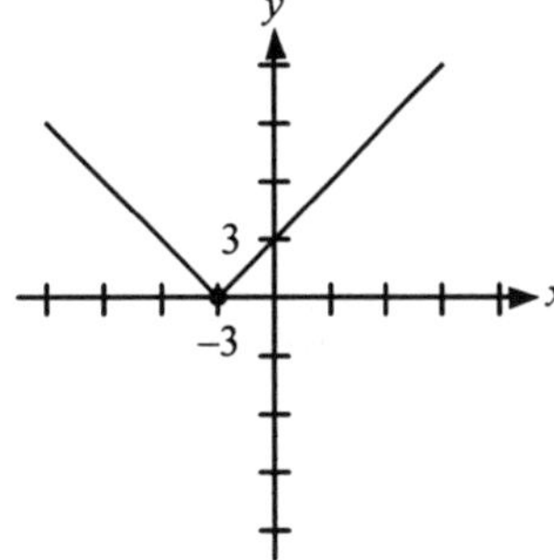

61.

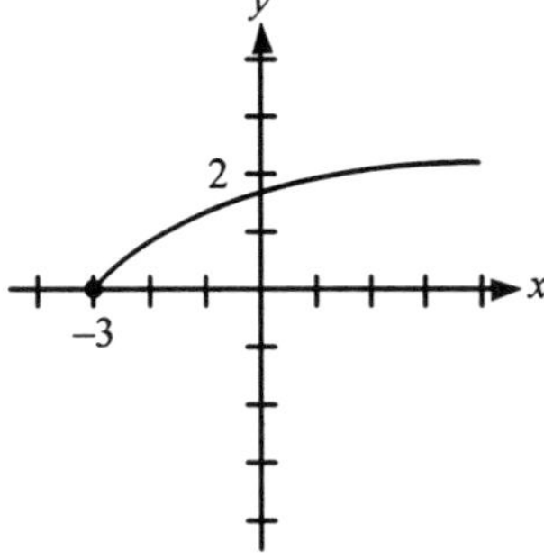

63.

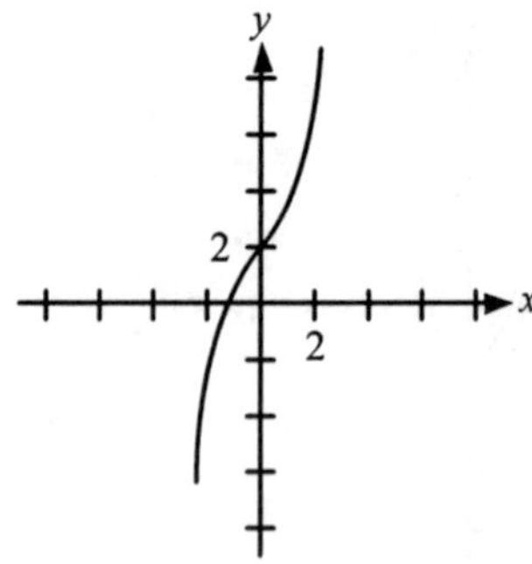

65.

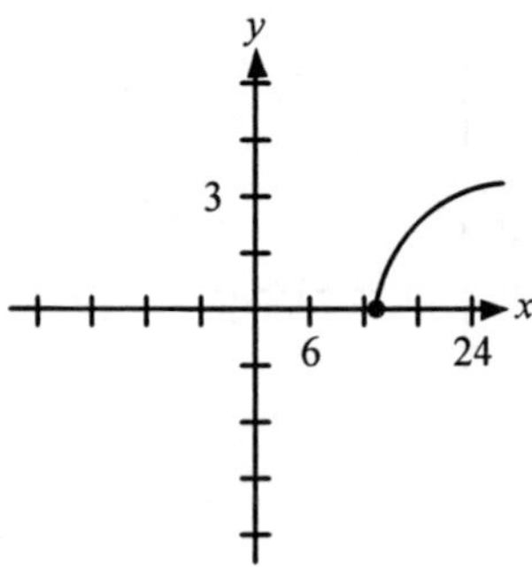

67.

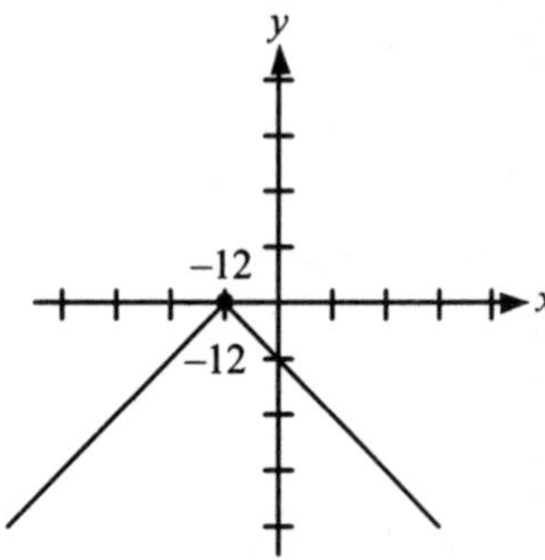

69.

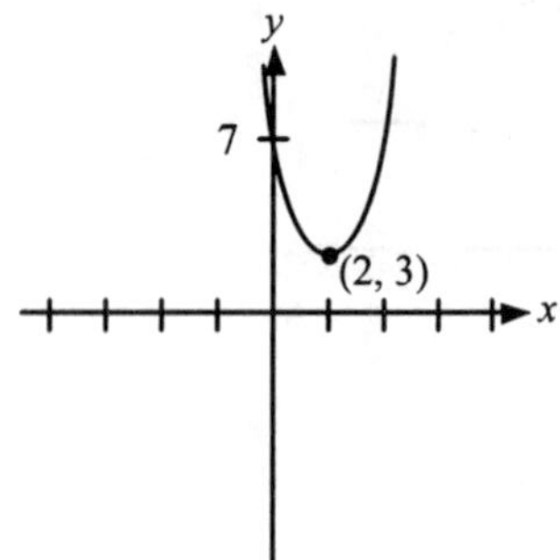

71.

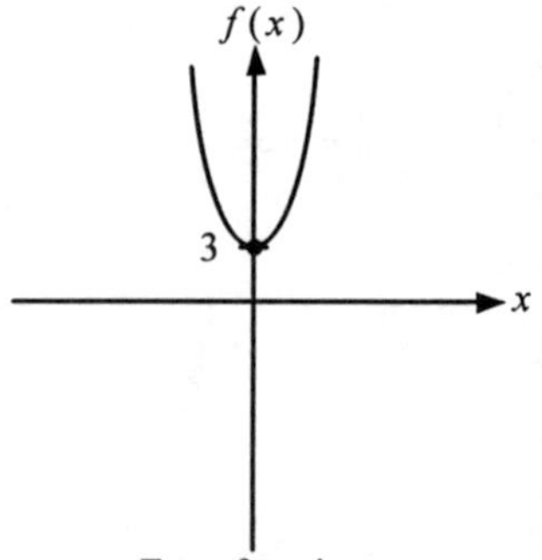

Even function

73.

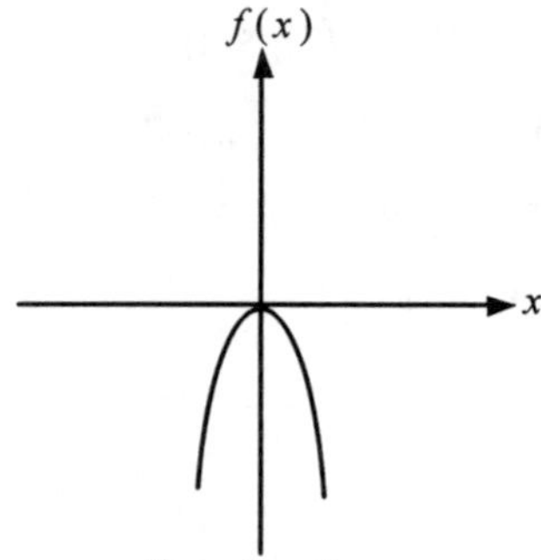

Even function

75.

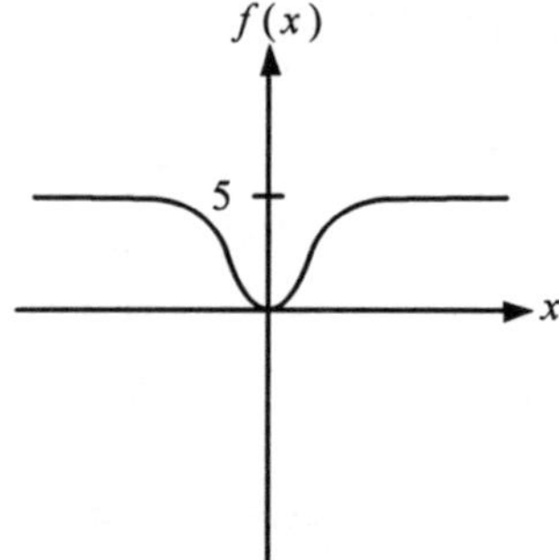

Even function

77.

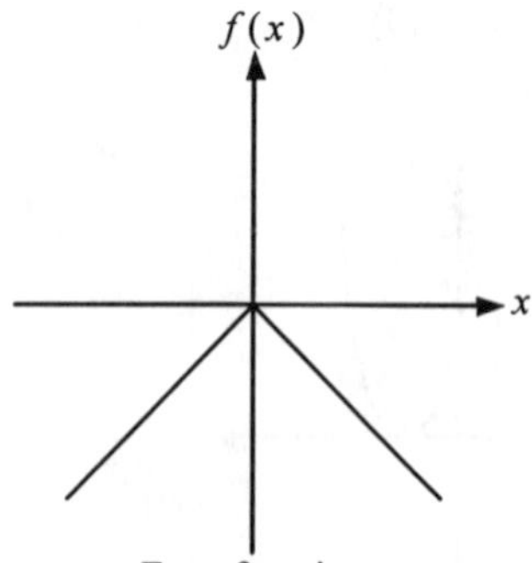

Even function

Exercises 5.2

1. $(f+g)(x)=3x-1$, $(f-g)(x)=x+1$, $(f\cdot g)(x)=2x^2-2x$, $\left(\dfrac{f}{g}\right)(x)=\dfrac{2x}{x-1}$, $(f\circ g)(x)=2x-2$, $(g\circ f)(x)=2x-1$; $D_{f/g}$ is the set of all real numbers except 1, otherwise the domain is $(-\infty, \infty)$.

3. $(f+g)(x)=4x^2-x-2$, $(f-g)(x)=4x^2+x-8$, $(f\cdot g)(x)=-4x^3+12x^2+5x-15$, $\left(\dfrac{f}{g}\right)(x)=\dfrac{4x^2-5}{-x+3}$, $(f\circ g)(x)=4x^2-24x+31$, $(g\circ f)(x)=-4x^2+8$; $D_{f/g}$ is the set of all real numbers except 3, otherwise the domain is $(-\infty, \infty)$.

5. $(f+g)(x)=x^2+1$, $(f-g)(x)=x^2-1$, $(f\cdot g)(x)=x^2$, $\left(\dfrac{f}{g}\right)(x)=x^2$, $(f\circ g)(x)=1$, $(g\circ f)(x)=1$; domain in all cases is $(-\infty, \infty)$.

7. $(f+g)(x)=|x|+x-3$, $(f-g)(x)=|x|-x+3$, $(f\cdot g)(x)=|x|(x-3)$, $\left(\dfrac{f}{g}\right)(x)=\dfrac{|x|}{x-3}$, $(f\circ g)(x)=|x-3|$, $(g\circ f)(x)=|x|-3$; The domain of $\left(\dfrac{f}{g}\right)(x)$ is all real numbers except 3. The domain in all other cases is all real numbers.

9. $(f+g)(x)=x^2-1+\sqrt{x+1}$, $(f-g)(x)=x^2-1-\sqrt{x+1}$, $(f\cdot g)(x)=(x^2-1)\sqrt{x+1}$, $\left(\dfrac{f}{g}\right)(x)=(x-1)\sqrt{x+1}$, $(f\circ g)(x)=x$, $(g\circ f)(x)=|x|$; $D_{f+g}=D_{f-g}=D_{f\cdot g}=D_{f\circ g}=[-1, \infty)$, $D_{f/g}=(-1, \infty)$, $D_{g\circ f}=(-\infty, \infty)$

11. $f+g=\{(2, 3), (3, 5)\}$, $f-g=\{(2, 5), (3, 5)\}$, $f\cdot g=\{(2, -4), (3, 0)\}$, $\dfrac{f}{g}=\{(2, -4)\}$, $f\circ g=\{(3, 2), (4, 3), (5, 4)\}$, $g\circ f=\{(0, -1), (1, 0), (2, 1), (3, 2)\}$; $D_{f+g}=D_{f-g}=D_{f\cdot g}=\{2, 3\}$, $D_{f/g}=\{2\}$, $D_{f\circ g}=\{3, 4, 5\}$, $D_{g\circ f}=\{0, 1, 2, 3\}$

13. 6 **15.** 14 **17.** $-\dfrac{9}{4}$

19. 13 **21.** 22

23. **a.** –3 **b.** 20 **c.** Undefined **d.** 4

25. –1 **27.** –1 **29.** 2 **31.** –1

33. $(f\circ g)(x)=x$, $(g\circ f)(x)=x$

35. $(g\circ f)(x)=(5x-4)^3$

37. $(f\circ g)(t)=36\pi t^2$

39. $g(x)=4x-1$; $f(x)=x^3$

41. $g(x)=\dfrac{x-1}{x+1}$; $f(x)=x^{1/2}$

43. $g(x)=2x+1$; $f(x)=\sqrt[3]{x}$

45. $g(x)=3-x$; $f(x)=2x^4$

47. $g(x)=1-x$; $f(x)=x^3+6x^2$

49. $g(x)=2x^2+2$ **51.** –12

53. $(f\circ g\circ h)(x)=2(x-2)^2$

55. $(f\circ g\circ h)(x)=\sqrt{\dfrac{1}{1-x}}$

57.
$$(f+g)(-x)=f(-x)+g(-x)=-f(x)-g(x)$$
$$=-[f(x)+g(x)]=-(f+g)(x)$$

59. Let f be an even function and g be an odd function. Then

$$(f\cdot g)(-x)=f(-x)\cdot g(-x)=f(x)\cdot[-g(x)]$$
$$=-[f(x)\cdot g(x)]=-(f\cdot g)(x)$$

so the product is an odd function.

61. **a.** $y=1{,}760m$ **b.** $f=3y$
c. $f=3(1{,}760m)=5{,}280m$

63. $(f\circ g)(t)=\dfrac{1}{6}\pi t^3$; $V=(f\circ g)(t)=\dfrac{1}{6}\pi t^3$ gives the volume of the balloon t seconds after the start of inflation.

65. **a.** $(f\circ g)(t)=340h(1.05)^t+200$; this gives the cost of a full shipment of bracelets t years from now.
b. $(f\circ g)(4)=413.27h+200$; this gives the cost of a full shipment of bracelets 4 years from now.
c. $(f\circ g)(4)=4{,}332.7$; 4 years from now it will cost $4,332.70 to produce a full shipment of bracelets.

Exercises 5.3

1. 1 **3.** $\frac{4}{7}$ **5.** $-\frac{11}{2}$ **7.** 0

9. Undefined **11.** -1 **13.** $\frac{1}{2}$

15. **a.** -1 **b.** $\frac{1}{2}$ **c.** 0
d. Undefined **e.** 3

17.

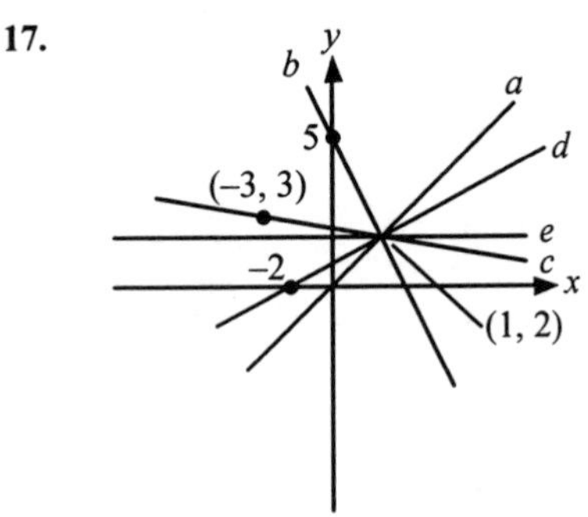

19. $\frac{1}{6}$; for each 6-lb increase of the earth weight of an object, the moon weight increases 1 lb.

21. -32; for each second (up to 10 seconds) the velocity decreases by 32 ft/second.

23. 0.12; for each brochure the cost increases \$0.12.

25. 7; for each mile the cost increases \$7.

27. 4 **29.** 2 **31.** $2x+h$

33. $-2x-h$ **35.** 2 **37.** 0

39. $-\frac{1}{x(x+h)}$ **41.** $2\pi r+\pi h$

43. Slope = 609,000; population of California increased by an average rate of about 609,000 people per year from 1980 to 1990.

Exercises 5.4

1. $y=5x-17$ **3.** $y=\frac{1}{2}x+1$

5. $y=-\frac{1}{2}x$ **7.** $y=-x+5$

9. $y=-x-6$ **11.** 1; $(0,7)$

13. 5; $(0,0)$ **15.** 0; $(0,-2)$

17. $-\frac{2}{3}$; $\left(0,-\frac{2}{3}\right)$ **19.** 6; $(0,7)$

21.

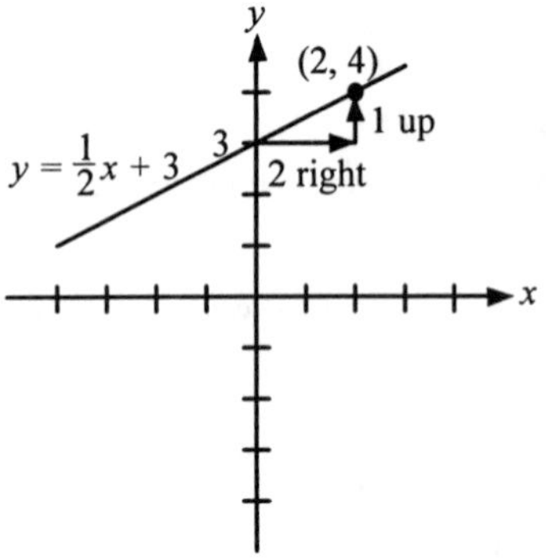

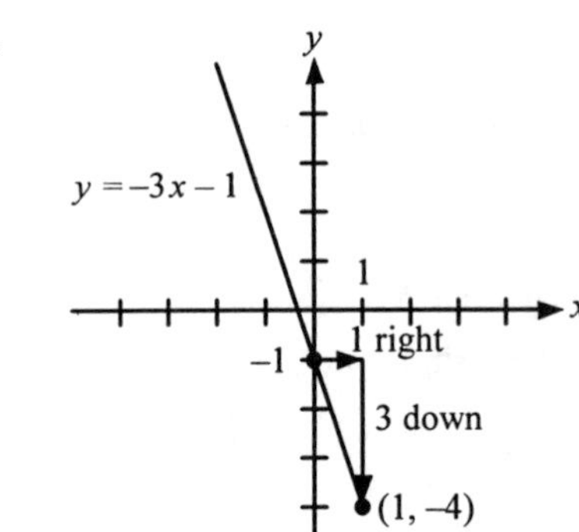

23.

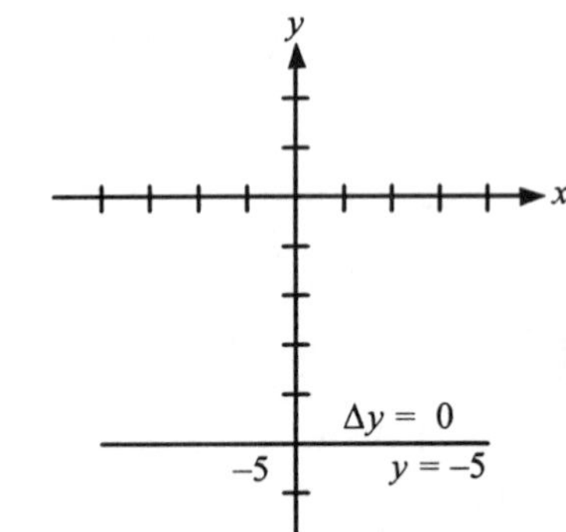

25.

27. $f(x)=\frac{2}{3}x-2$ **29.** $f(x)=-\frac{4}{3}x-\frac{13}{3}$

31. $f(x)=1$ **33.** -1 **35.** $-\frac{11}{3}$

37. **a.** $y=0.6x+40$ **b.** \$55
c. \$40 **d.** \$0.60

39. **a.** $v=-32t-220$
b. -348 ft/second

c. -220 ft/second

d. + ball rising, – ball falling

41. a. $m = .3937$; there are .3937 cm per inch

b.

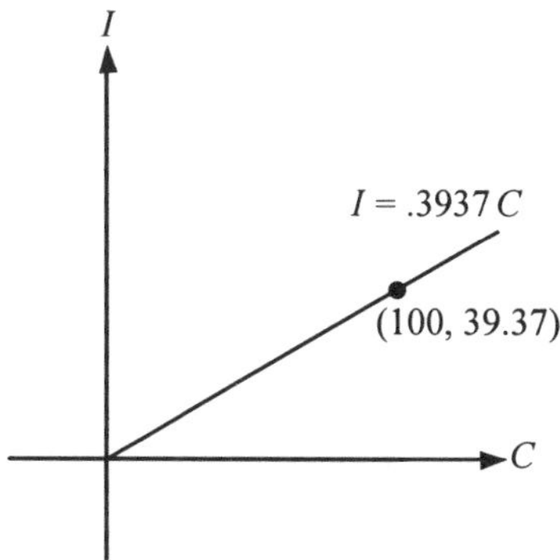

c. 9.76 in.

43. a. $J = -\frac{1}{2}S + 100$

b. The Jones temperature drops $\frac{1}{2}$ degree when the Smith temperature increases by 1 degree

c. 84° Jones

45. a. $W = 3.15A - 19.2\ (A \geq 8)$

b. 21 weeks

47. linear; $y = 11.4 + .7x$

49. linear; $y = 6 + 2x$

51. a. $y = x + 1$ b. $y = -x + 3$

53. a. $y = -2x + 6$ b. $y = \frac{1}{2}x - \frac{3}{2}$

55. a. $y = -\frac{1}{7}x$ b. $y = 7x$

57. a. $y = \frac{5}{3}x - \frac{14}{3}$ b. $y = -\frac{3}{5}x - \frac{12}{5}$

59. a. $y = -\frac{1}{4}x + \frac{13}{4}$ b. $y = 4x - 18$

61. a. $y = \frac{3}{2}x - \frac{13}{12}$ b. $y = -\frac{2}{3}x$

63. $m_{AB} = 4$, $m_{BC} = -\frac{1}{4}$; since $m_{AB} \cdot m_{BC} = -1$, the segments are perpendicular, ensuring a right triangle.

65. 5

67. $m_1 = -\frac{A_1}{B_1}$, $m_2 = -\frac{A_2}{B_2}$; since the lines are parallel, $-\frac{A_1}{B_1} = -\frac{A_2}{B_2}$, so $A_1B_2 = A_2B_1$.

69. m; the rate of change is a constant.

71. a. $y = -0.38x + 980$ Each year the record drops about $\frac{4}{10}$ sec.

b. 3:36 c. 2003

Exercises 5.5

1.

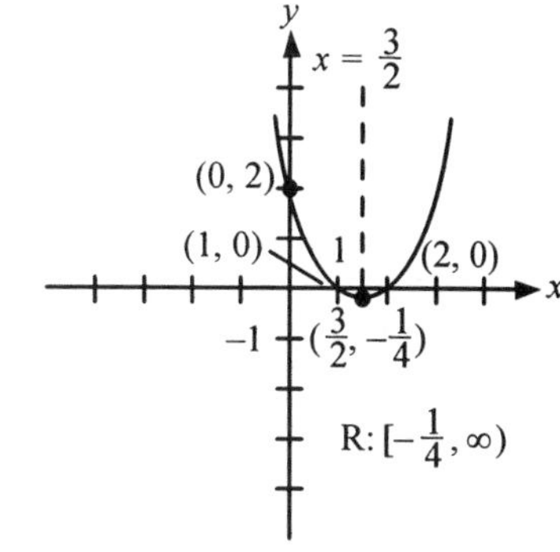

3.

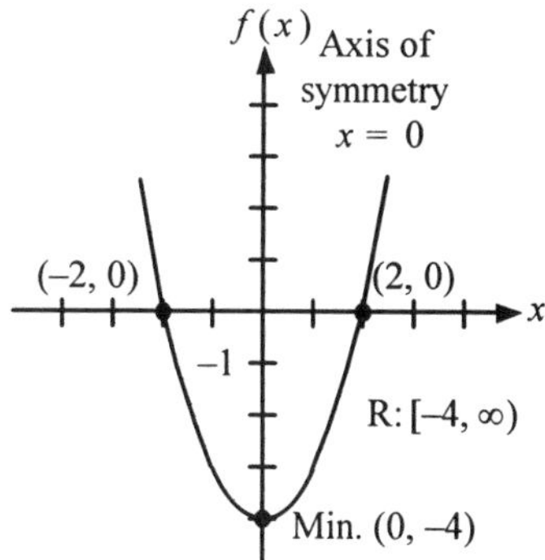

5.

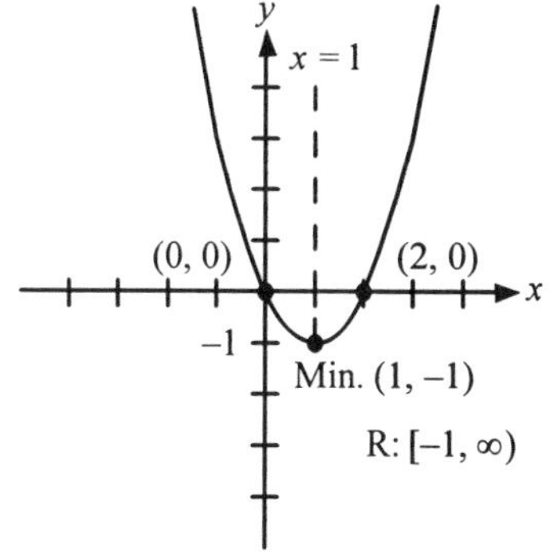

7.

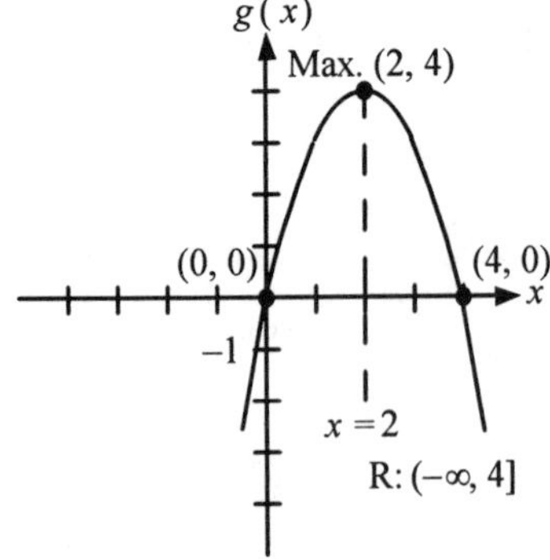

9.

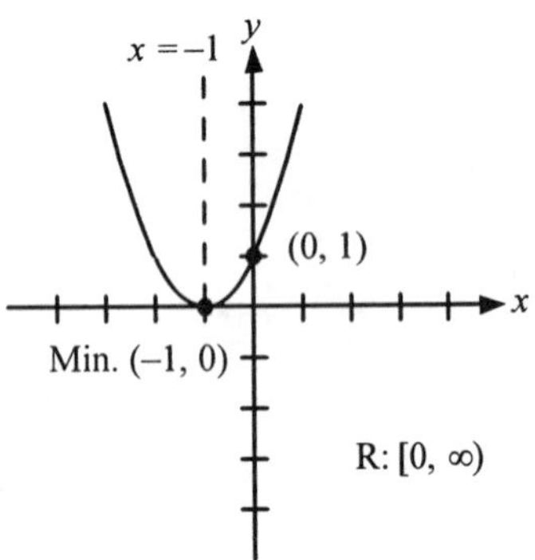

11.

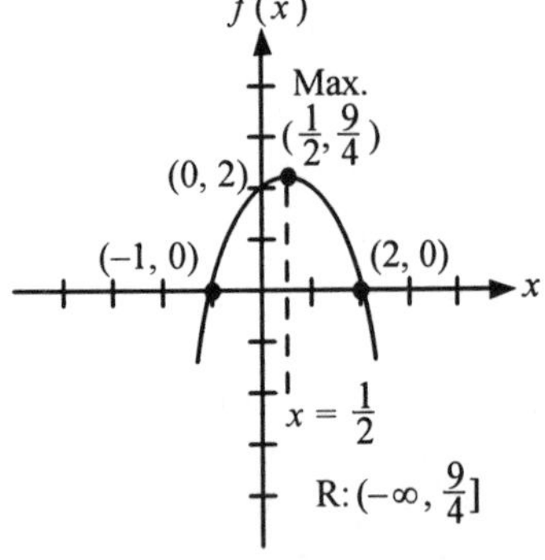

13.

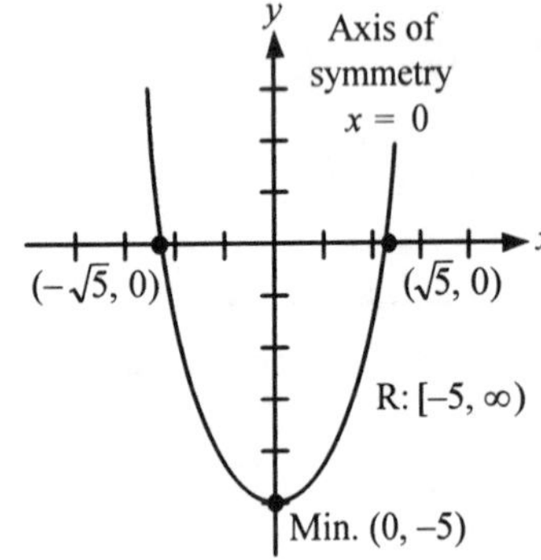

15.

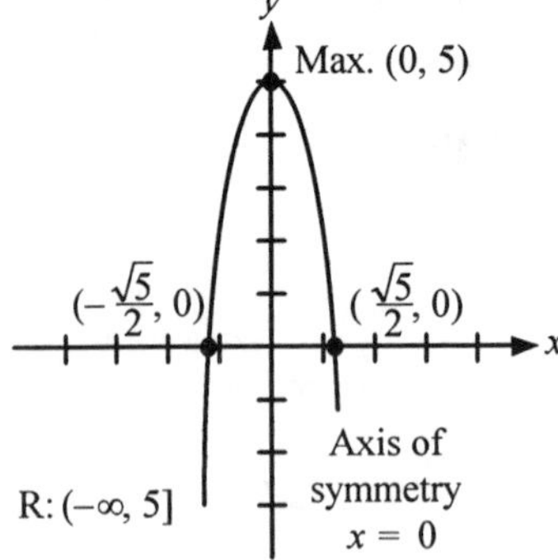

17.

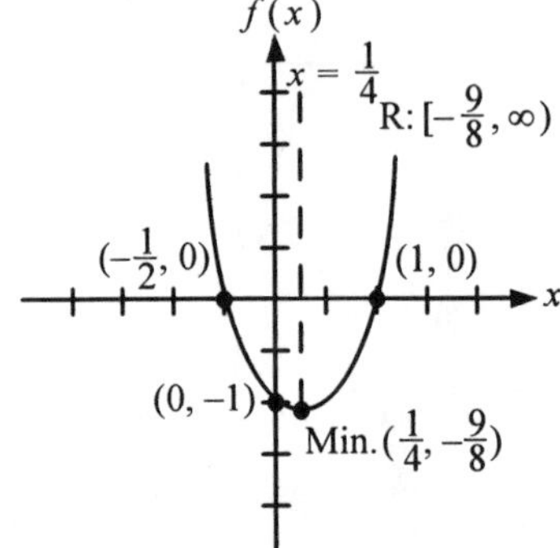

19.

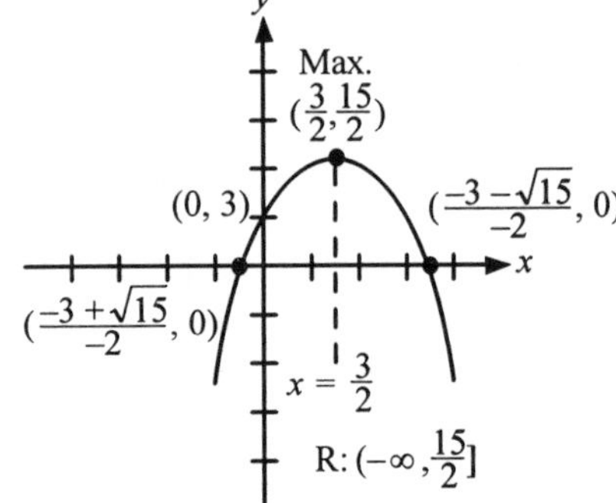

21. $(-2, -3)$; $x = -2$ **23.** $(0, -4)$; $x = 0$

25. $(1, 1)$; $x = 1$ **27.** $\left(\frac{3}{2}, -\frac{23}{4}\right)$; $x = \frac{3}{2}$

29. $\left(\frac{1}{2}, \frac{9}{4}\right)$; $x = \frac{1}{2}$ **31.** $(-\infty, -1) \cup (5, \infty)$

33. $(-1, 5)$ **35.** $(-\infty, -1] \cup [2, \infty)$

37. $(-\infty, -2] \cup [2, \infty)$ **39.** $(0, 3)$

41. $\left[-2, \frac{3}{2}\right]$ **43.** $\left(-1, \frac{5}{3}\right)$

45. Set of all real numbers except −1

47. ∅ **49.** $[-1, 1]$

51. **a.** 804.0 ft **b.** 204 ft

53. 144 ft; 6 seconds **55.** 360 watts

57. 10, 10 **59.** 10, 10

61. 200 yd by 100 yd **63.** 3 in.

65. $15

67. **a.** $A = x^2 + 2x\left(30 - x - \sqrt{2}x\right)$

b. $\dfrac{60}{2 + 4\sqrt{2}} \approx 7.84$

c. 235.08 in^2

69. **a.** $d = 2x + 3 + 0.5x^2$

b. $x = -4$ and $x = 0$; Pts: $(-4, 5)$, $(-4, 8)$ and $(0, 3)$, $(0, 0)$

c. $(-2, -1)$ and $(-2, -2)$

d. 1

Chapter 5 Review Exercises

1. −21 **3.** (−2, 2) **5.** $(-\infty, 1]$

7. $y = \dfrac{2}{3}x - 2$ **9.** Yes

11.

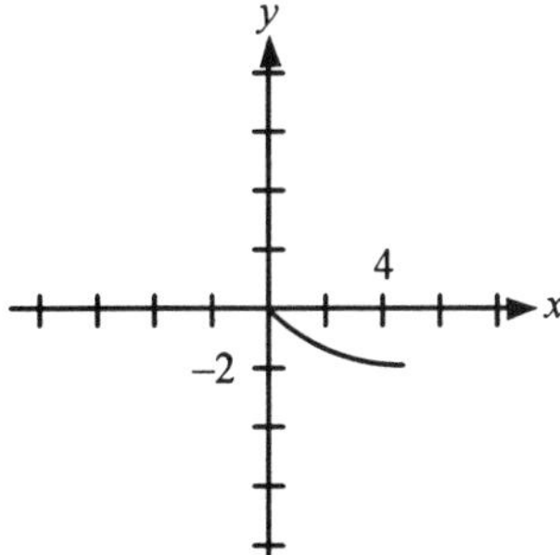

13.

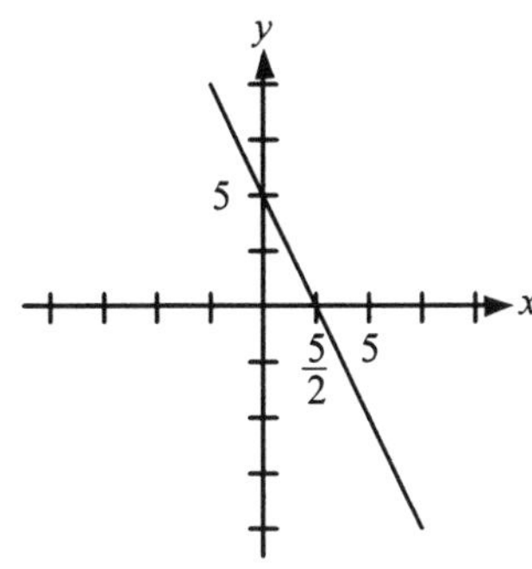

15.

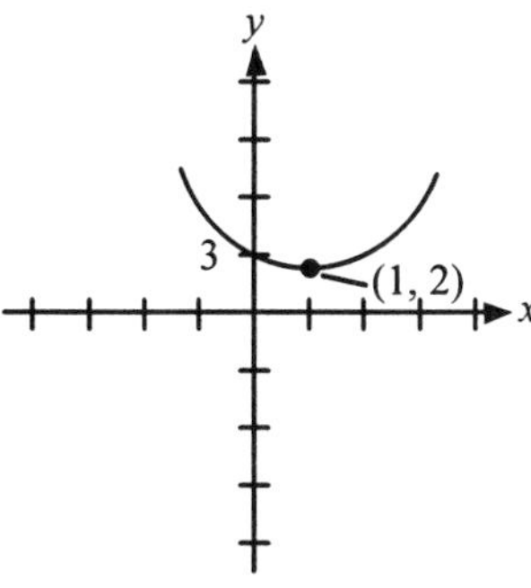

17. $(-\infty, -3] \cup [2, \infty)$ **19.** 3

21. $(-\infty, -2] \cup (2, \infty)$

23. **a.** −16 **b.** $\dfrac{7}{5}$

25. 16 **27.** $6x + 3h$ **29.** −1

31. $f(x) = -3x + 3$ **33.** $y = -\dfrac{4}{7}x$

35. {(1, 5), (2, 4)} **37.** $[-c, c]$

39. Even function **41.** $\{-c, 0, c\}$

43.

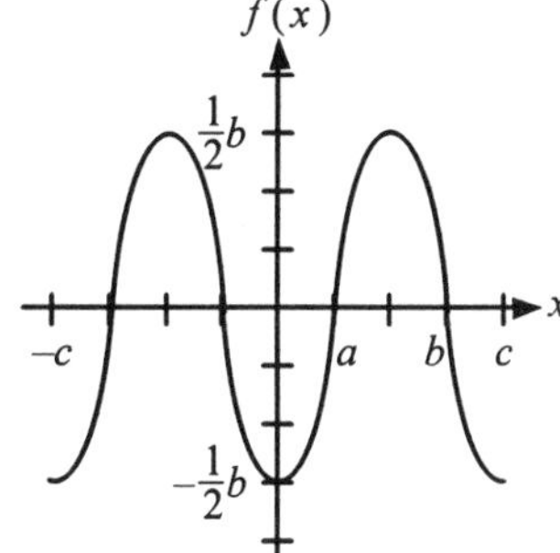

45. $A = \dfrac{\sqrt{3}x^2}{4}$

47. $e = \begin{cases} 500 & \text{if } 0 \le a \le 2{,}000 \\ 500 + 0.09(a - 2{,}000) & \text{if } a > 2{,}000 \end{cases}$

49. 108 ft (front) by 81 ft

51. **a.** $v = -32t + 320$

b. 32 ft/second **c.** 320 ft/second

d. 10 seconds; Projectile reaches its highest point when $t = 10$ seconds.

53. a **55.** a **57.** b

59. c **61.** c

Chapter 5 Test

1. All real numbers except 9

2. $(-\infty, 6]$

3.

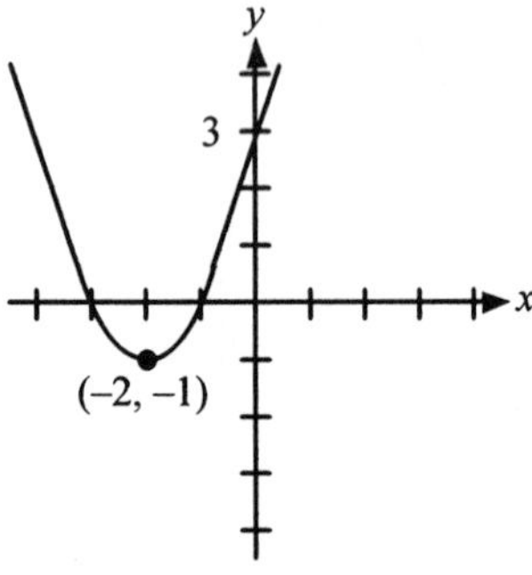

4.

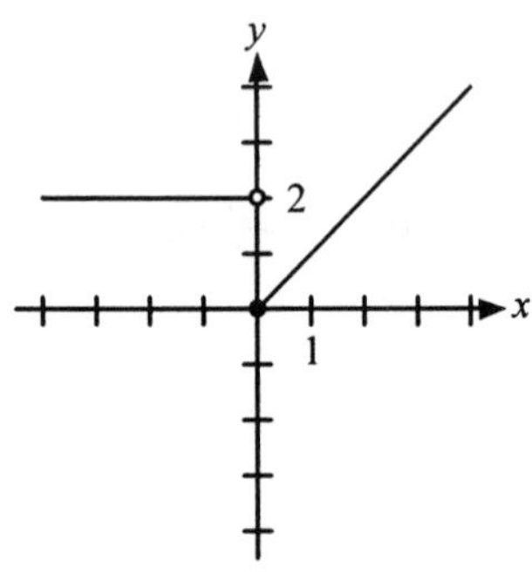

5. Even **6.** $\{(0, 2), (1, -2)\}$

7. $(g \circ f)(x) = 6x^2 + 3$

8. $g(x) = 25 - x^2$; $f(x) = \sqrt{x}$

9. D: $(0, \infty)$; R: $(-\infty, \infty)$

10. $A = x(400 - x)$

11. 1.2; for each unit cost increases \$1.20.

12. $2x + h - 5$ **13.** -3; $\left(0, \dfrac{5}{2}\right)$

14.

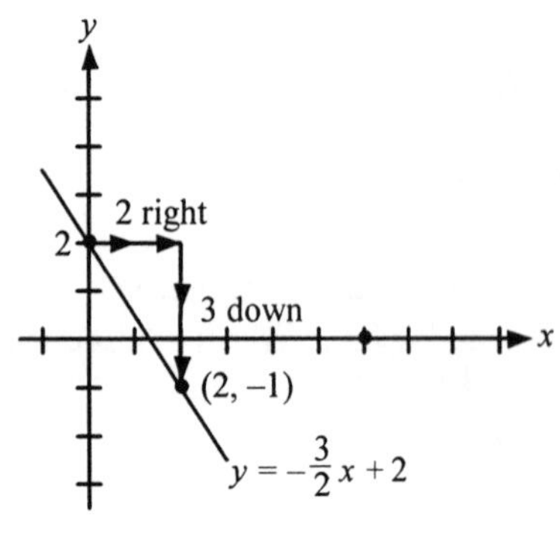

15. $y = 2x + 9$ **16.** $f(x) = \dfrac{3}{5}x - 3$

17. $(-1, 0), \left(\dfrac{3}{2}, 0\right)$ **18.** $(-\infty, 9]$

19. $\left[1-\sqrt{7}, 1+\sqrt{7}\right]$ **20.** 256 ft

Exercises 6.1

1. Yes; 3 **3.** Yes; 1 **5.** No

7. No **9.** Yes; 0

11. $P(x) = (x-1)(x-2)(x-3)$

13. $P(x) = x(x+4)(x-i)(x+i)$

15. $P(x) = (x-1)^2(x-5)^2$

17. Degree 2; -3 is a zero of multiplicity 2

19. Degree 3; -4, $\sqrt{2}$, and $-\sqrt{2}$ are zeros of multiplicity 1.

21. Degree 3; 0 is a zero of multiplicity 2; 2 is a zero of multiplicity 1.

23. Degree 4; 0 is a zero of multiplicity 4

25.

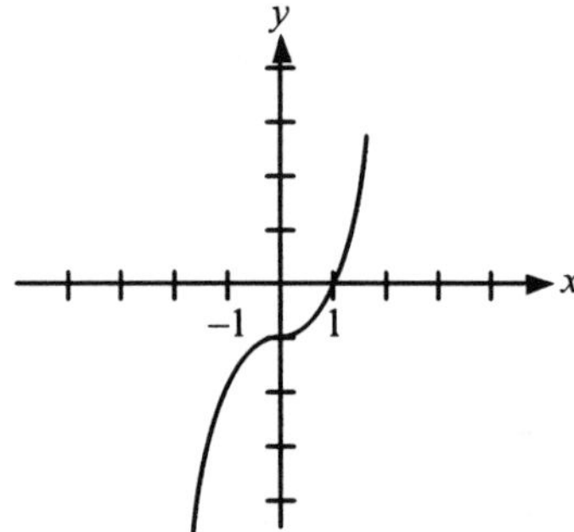

27.

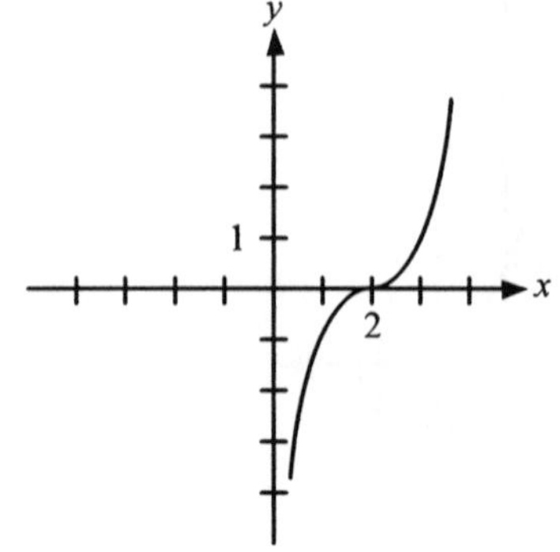

29.

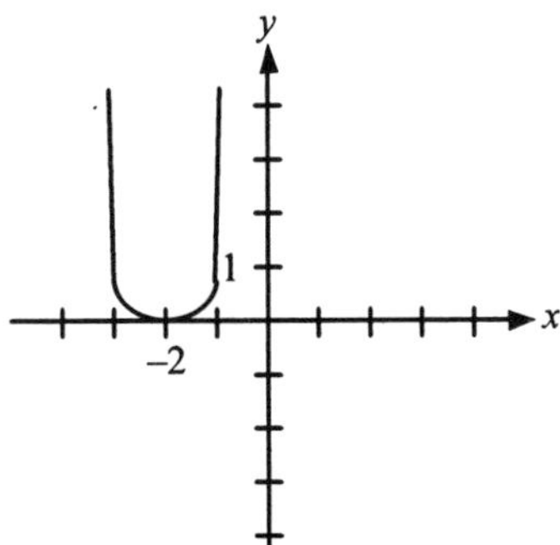

31.

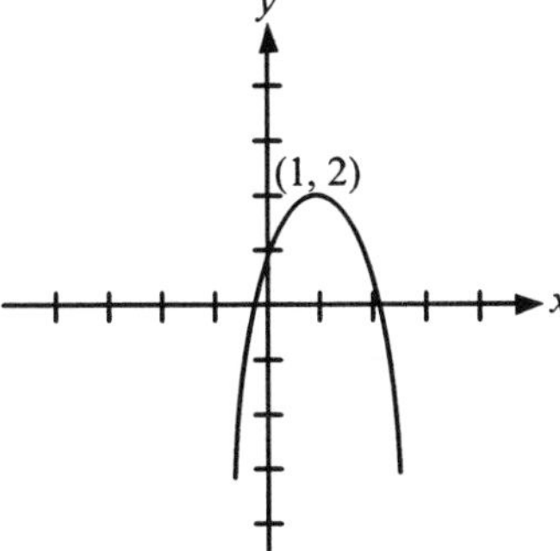

33.

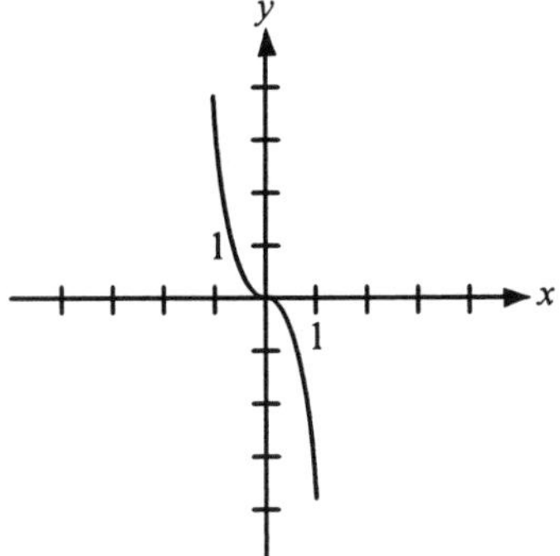

35.

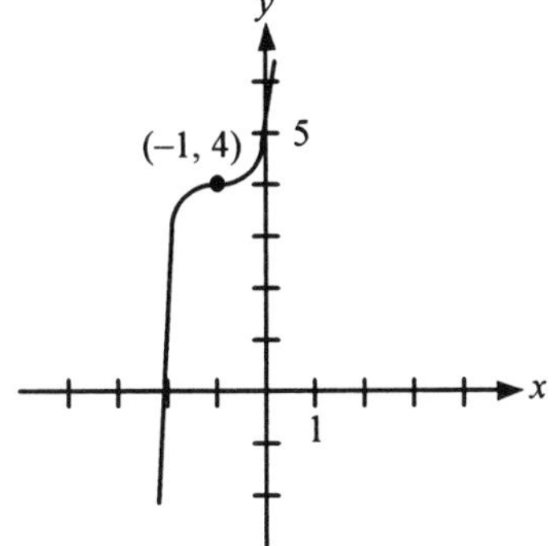

37.

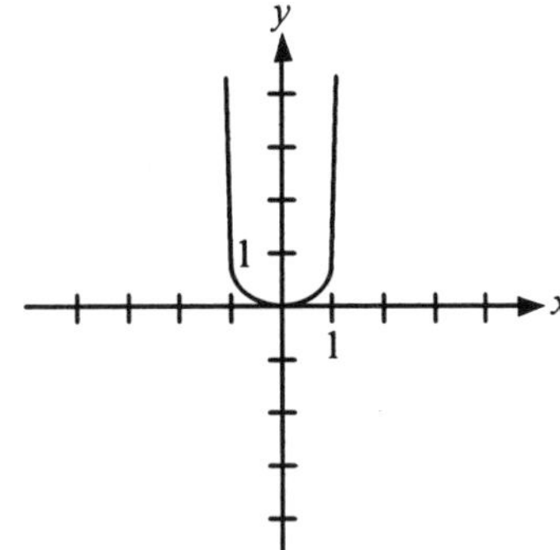

39.

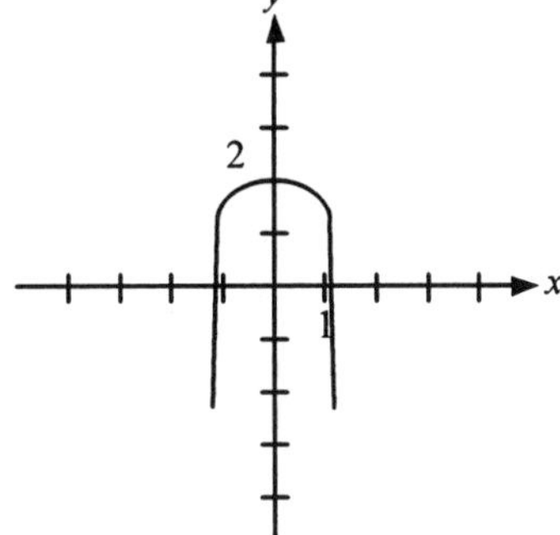

41.

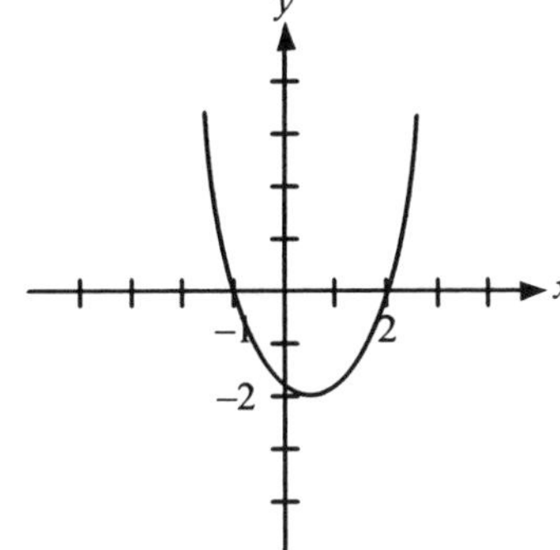

43.

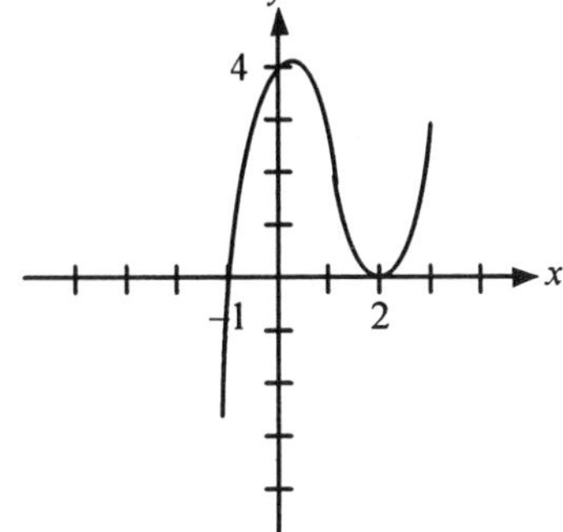

45.

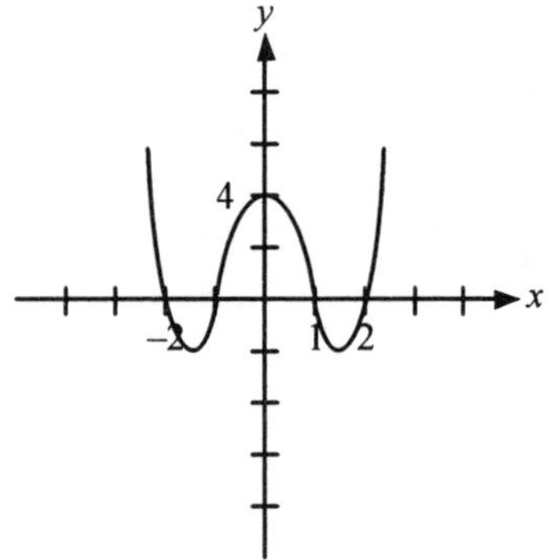

47.

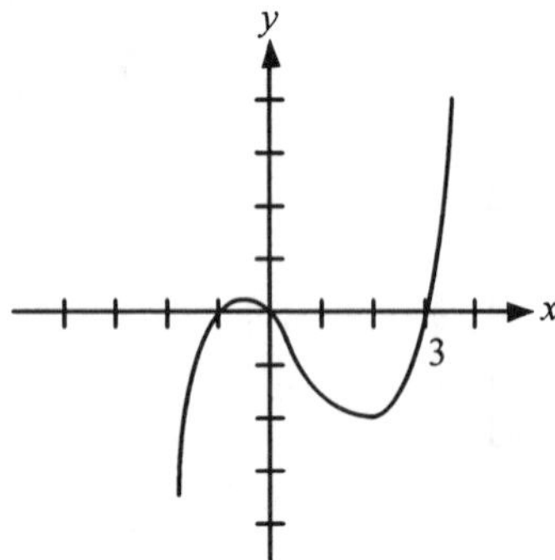

49.

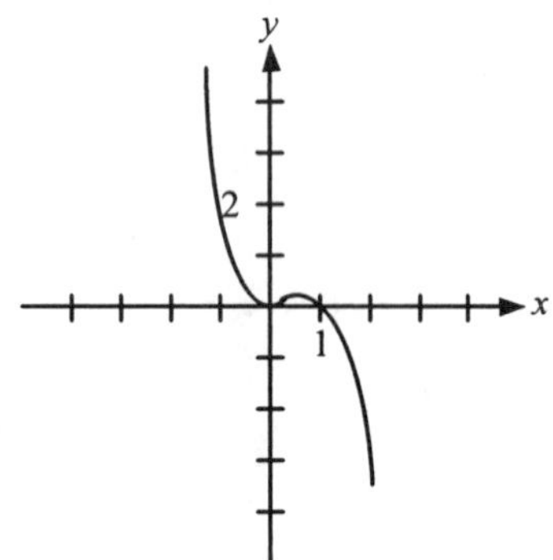

51. $y = -(x+1)^2(x-2)$ 53. $y = \frac{1}{9}(x+1)^2(x-3)^2$

55. $y = -k(x+3)^2$ where $k > 0$

57. a. $V = x(10-2x)^2$

b. Domain: $(0, 5)$ c. $\frac{5}{3}; \frac{1}{6}$

59. a. 1. 2.54; 332.55 in.^{3}

2. 2; 256 in.^{3}

3. 2.54; 166.28 in.^{3}

b. For $x = 4$, boxes 2 and 3 each have volume 128 in.^{3}

61. $-.068 < x < .066$; $\sqrt[3]{120} - 5 < x < \sqrt[3]{130} - 5$

Exercises 6.2

1. $\frac{x^2-5}{x+1} = x - 1 + \frac{-4}{x+1}$

3. $\frac{3x^4 - 5x^2 + 7}{x^2 + 2x + 1} = 3x^2 - 6x + 4 + \frac{-2x+3}{x^2+2x+1}$

5. $\frac{x^3+1}{x(x-1)} = x + 1 + \frac{x+1}{x(x-1)}$

7. $x^2 + 7x - 2 = (x+5)(x+2) - 12$

9. $6x^3 - 3x^2 + 14x - 7 = (2x-1)(3x^2+7) + 0$

11. $3x^4 + x - 2 = (x^2-1)(3x^2+3) + x + 1$

13. $x^3 - 5x^2 + 2x - 3 = (x-1)(x^2 - 4x - 2) - 5$

15. $2x^3 + 9x^2 - x + 14 = (x+5)(2x^2 - x + 4) - 6$

17. $7 + 6x - 2x^2 - x^3 = (x+3)(-x^2 + x + 3) - 2$

19. $2x^3 + x - 5 = (x+1)(2x^2 - 2x + 3) - 8$

21. $x^4 - 16 = (x-2)(x^3 + 2x^2 + 4x + 8) + 0$

23. 0, 0 25. −9, 87 27. 0, 0

29. $-\frac{176}{27}, -\frac{10}{3}$

31. Yes; $P(x) = (x+1)(x^2 - 3x + 5)$

33. Yes; $P(x) = (x-4)(x+1)(x-2)$

35. Not factorable 37. $x = 2$ in.

39. a. perimeter $= 4a + 4b + 4c = 40$, so $a + b + c = 10$

b. $b = a + 2$; $c = 8 - 2a$; $V = a(a+2)(8-2a)$

c. $a = \frac{1+\sqrt{37}}{2} \approx 3.54$ in.

41. $i, -i$ 43. $2 - \sqrt{7}$

45. $-i\sqrt{2}$ 47. $2 - 2i$

49. $-\frac{1}{2}$ and -1 (multiplicity 2)

51. $\pm\sqrt{2}$

53. $\pm i\sqrt{2},\ -\sqrt{2}$

55. $1,\ -1,\ -i$

57. $P(x)=(x-2)(x-\sqrt{3})(x+\sqrt{3})$

59. $P(x)=x(x-(4-\sqrt{3}))(x-(4+\sqrt{3}))\cdot(x-(2+3i))(x-(2-3i))$

61. $P(x)=(x-3)(x+3)(x-2i)(x+2i)$

63.

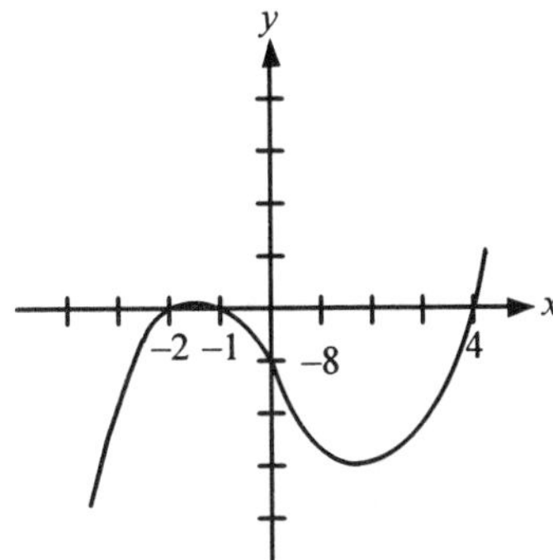

65. True

67. False

69. Since $x-b$ is a factor of $P(x)$, we know $P(x)=(x-b)Q(x)$. Then $P(b)=(b-b)Q(b)=0$, so b is a zero of $y=P(x)$.

Exercises 6.3

1. $\pm 20,\ \pm 10,\ \pm 5,\ \pm 4,\ \pm 2,\ \pm 1$

3. $\pm 6,\ \pm 3,\ \pm 2,\ \pm\frac{3}{2},\ \pm 1,\ \pm\frac{3}{4},\ \pm\frac{1}{2},\ \pm\frac{1}{4}$

5. Positive: 3; negative: 0

7. Positive: 1; negative: 3

9. 4, −1, −2

11. $-\frac{1}{3},\ \pm i\sqrt{5}$

13. $\frac{1}{2}$ (multiplicity 3)

15. $-1,\ 2,\ \frac{1\pm i\sqrt{79}}{8}$

17. $-2,\ -\frac{1}{2},\ \frac{-1\pm i\sqrt{35}}{2}$

19. **a.** $3,\ -1\pm i\sqrt{3}$ **b.** One; 3
c. One; 3
d. Three; $3+0i$, $-1+\sqrt{3}i$, $-1-\sqrt{3}i$

21. **a.** $-2,\ \frac{1}{3},\ \pm\sqrt{7}$ **b.** Two; $-2,\ \frac{1}{3}$
c. Four; $-2,\ \frac{1}{3},\ \sqrt{7},\ -\sqrt{7}$
d. Four; $-2+0i,\ \frac{1}{3}+0i,\ \sqrt{7}+0i,\ -\sqrt{7}+0i$

23. $105=x(x+2)(2x+1)$, so $2x^3+5x^2+2x-105=0$. Then 3 is a root, while the reduced equation $2x^2+11x+35=0$ has no real number solutions. Unique dimensions: 7 in. by 5 in. by 3 in.

25. 3 in. or $6-3\sqrt{2}$ in.

27. $P(0)=-8$, $P(-1)=5$; since $P(0)$ and $P(-1)$ have opposite signs, the location theorem guarantees at least one real zero between 0 and −1.

29. $P(1)=-4$; $P(2)=2$; since $P(1)$ and $P(2)$ have opposite signs, the location theorem guarantees at least one real zero between 1 and 2. Similarly, since $P(-1)=-16$ and $P(-2)=50$, there is at least one real zero between −1 and −2.

31. 0.7

33. 0.3

35. −1.3

Exercises 6.4

1. $x=\frac{3}{2}$

3. $x=6,\ x=-1$

5.

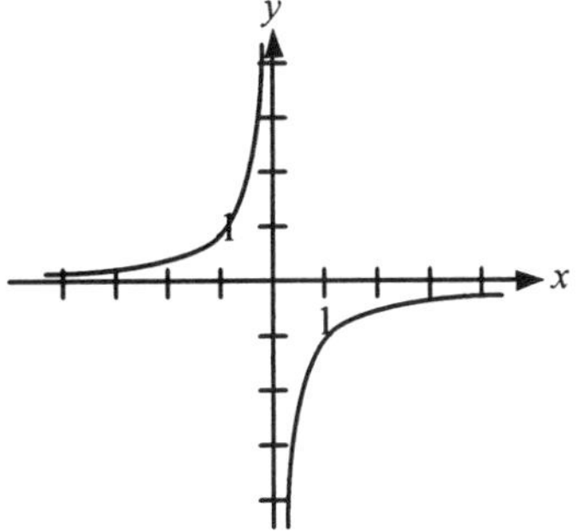

Vertical asymptote: $x=0$
Horizontal asymptote: $y=0$

7.

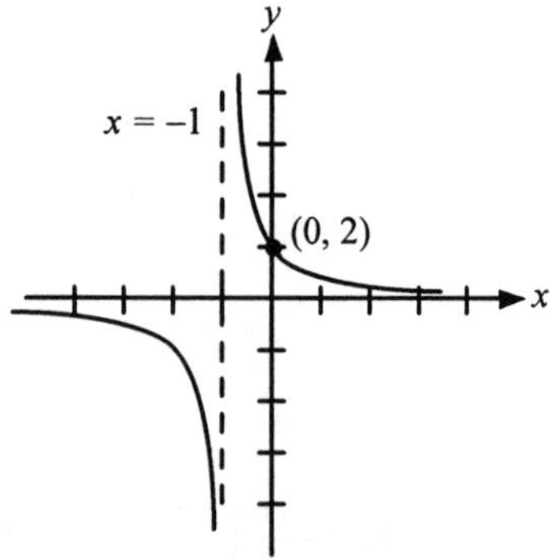

Vertical asymptote: $x = -1$
Horizontal asymptote: $y = 0$

9.

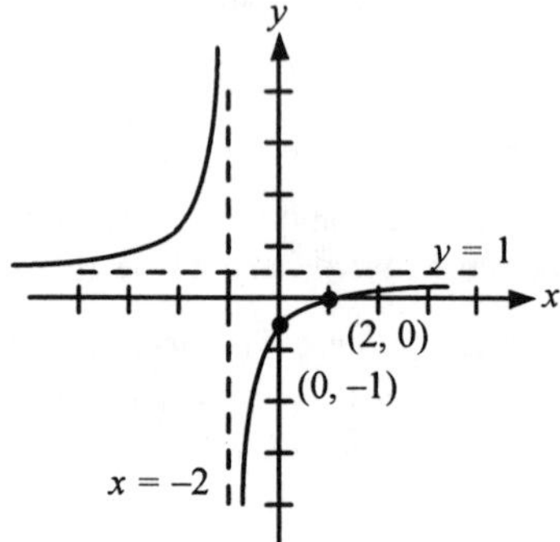

Vertical asymptote: $x = -2$
Horizontal asymptote: $y = 1$

11.

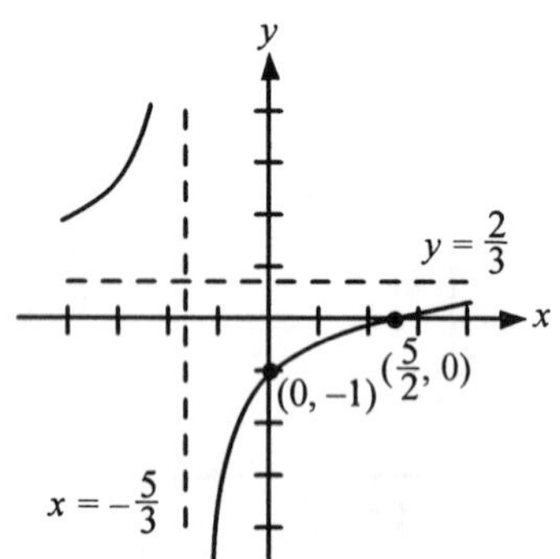

Vertical asymptote: $x = -\frac{5}{3}$

Horizontal asymptote: $y = \frac{2}{3}$

13.

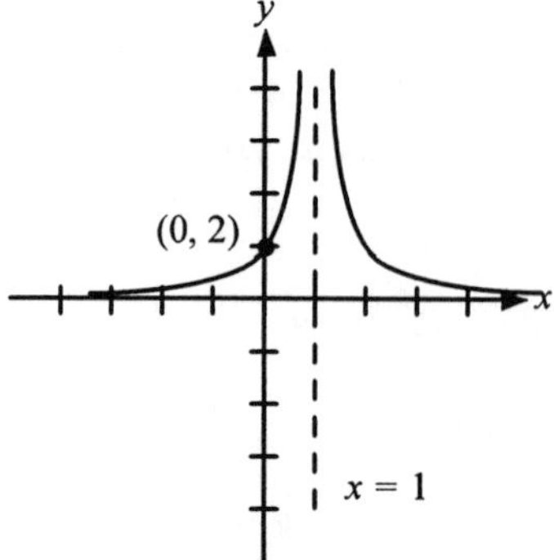

Vertical asymptote: $x = 1$
Horizontal asymptote: $y = 0$

15.

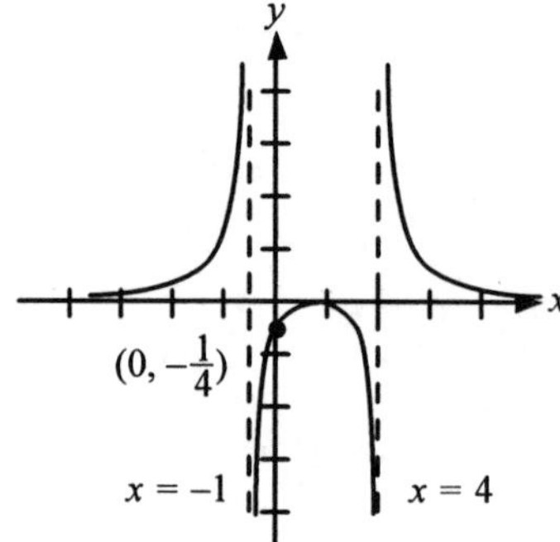

Vertical asymptotes: $x = -1$
$x = 4$
Horizontal asymptote: $y = 0$

17.

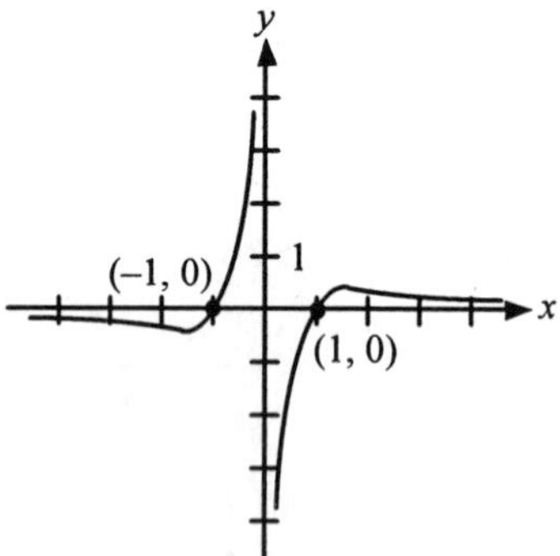

Vertical asymptote: $x = 0$
Horizontal asymptote: $y = 0$

19.

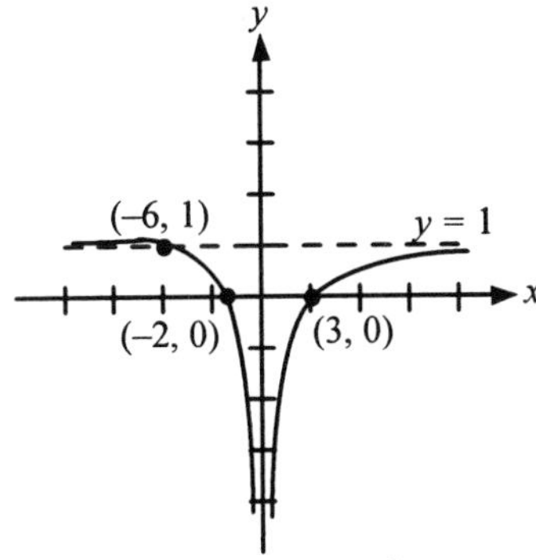

Vertical asymptote: $x = 0$
Horizontal asymptote: $y = 1$

21.

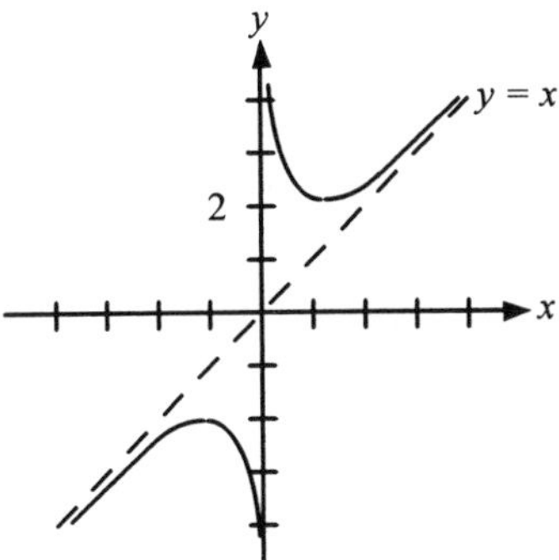

Vertical asymptote: $x = 0$
Horizontal asymptote: none

23.

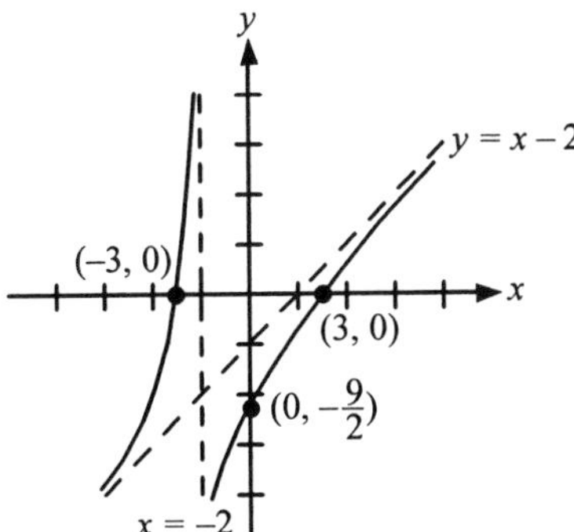

Vertical asymptote: $x = -2$
Horizontal asymptote: none

25.

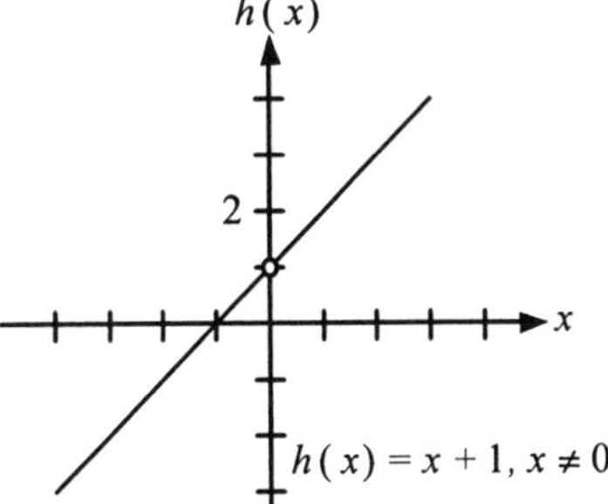

27.

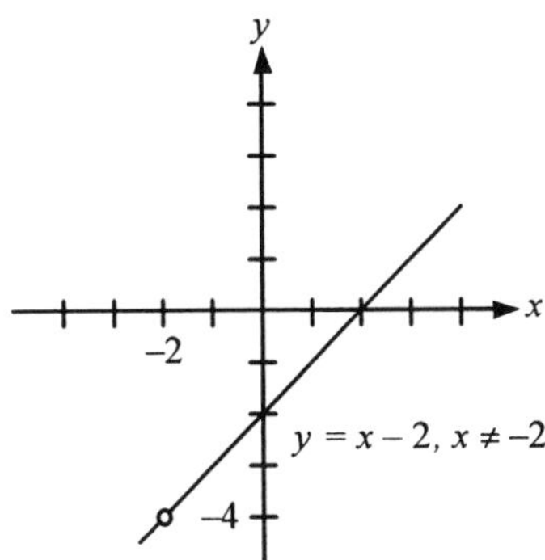

29.

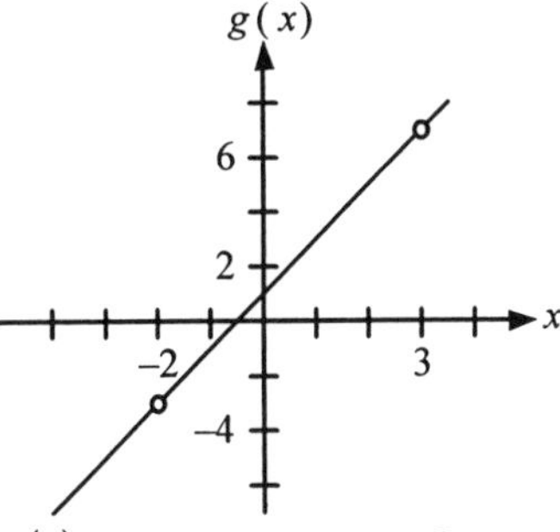

$g(x) = 2x + 1,\ x \neq 3,\ x \neq -2$

31.

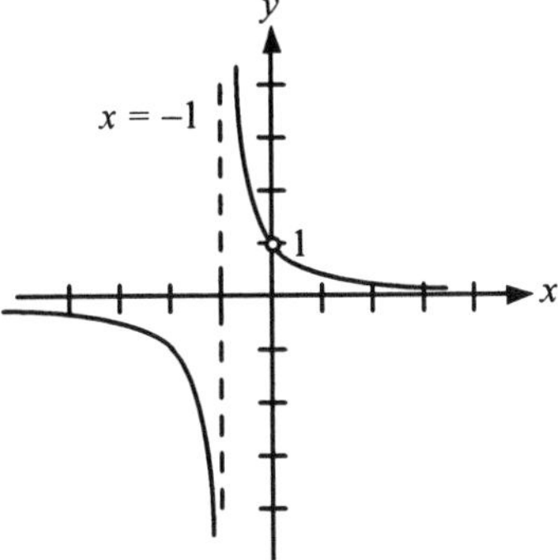

$f(x) = \dfrac{1}{x+1},\ x \neq 0$
Vertical asymptote: $x = -1$
Horizontal asymptote: $y = 0$

33.

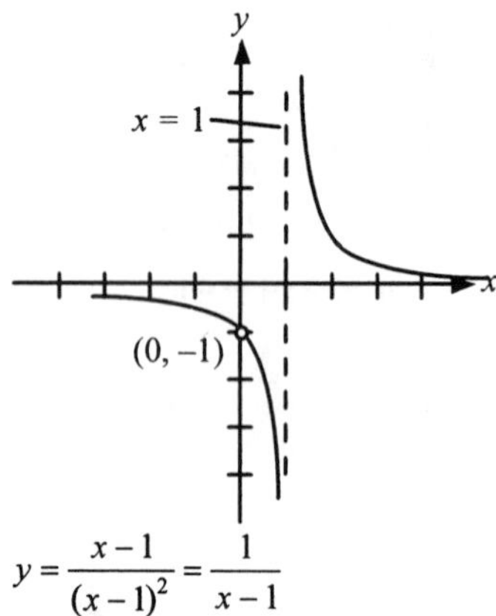

$y = \dfrac{x-1}{(x-1)^2} = \dfrac{1}{x-1}$

Vertical asymptote: $x = 1$

Horizontal asymptote: $y = 0$

35. **a.** \$280; \$1.56

b.

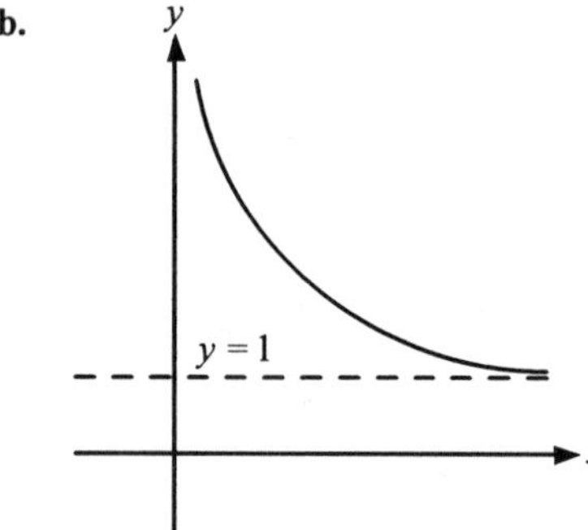

y decreases as x increases. The average cost per minute approaches \$1.

37. 21.2 by 42.4

39.

x	0	10	100	1,000	10,000	100,000
y	1.00	26.45	296.05	2,996.00	29,996.00	299,996.00

$y = 3x$

41. **a.** \$10.02, \$12.50, \$25

b.

x	1	2	3	4	5
cost	10.02	10.04	10.06	10.08	10.10

x	6	7	8	9	10
cost	10.12	10.14	10.16	10.18	10.20;

total = \$101.12

43. **a.** The numerator is zero, and the denominator is not zero, so $f(x) = 0$.

b. max for $\dfrac{x(24-x)}{x+1}$ occurs at $x = 4$;

max for $\dfrac{x(24-x)}{x+5}$ occurs at $x = 7.04$.

c. One possibility is $Q(x) = x + 25$.

45. **a.** 88.89%; 80%

b. $y = \dfrac{32}{32+x} \cdot 100 \quad D\text{: } [0, 32]$

c.

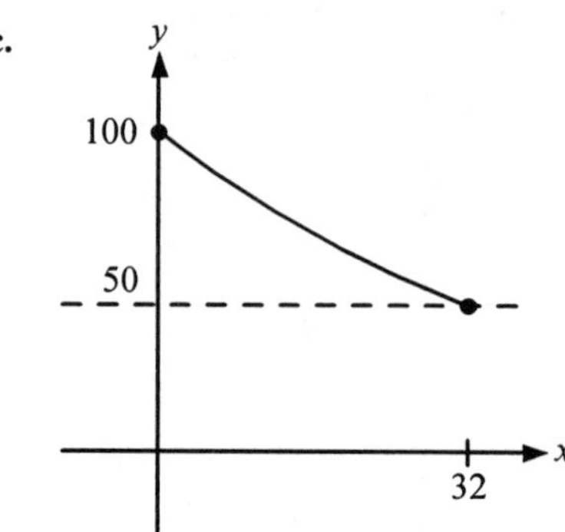

d. 8 oz. of water; 32 oz. of water; not possible in a 64 oz jug

e. The percentage of juice can only approach zero because there is always some juice.

f. 320 ounce

47. **a.** $k = 100$

b.

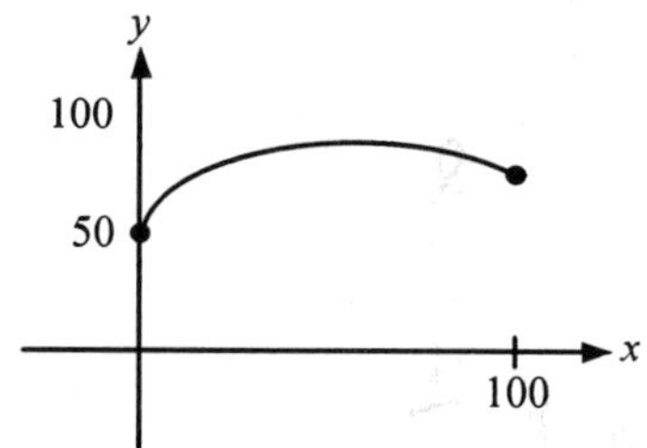

maximum value $= 83.77$ when $x = 58.11$

$y = \dfrac{x(150-x)}{x+100} + 50$

49. Least perimeter found by minimizing

$P = 2\sqrt{\left(\dfrac{x}{2}\right)^2 + \left(\dfrac{200}{x}\right)^2}$; least perimeter = 45.59 ft.

This is an equilateral triangle. The exact value of the side is $\dfrac{20}{\sqrt[4]{3}}$ ft.

Chapter 6 Review Exercises

1. Yes **3.** No **5.** No

7. $0, \dfrac{3}{2}$ **9.** $2, i, -i$ **11.** $\dfrac{5 \pm \sqrt{17}}{4}$

13.

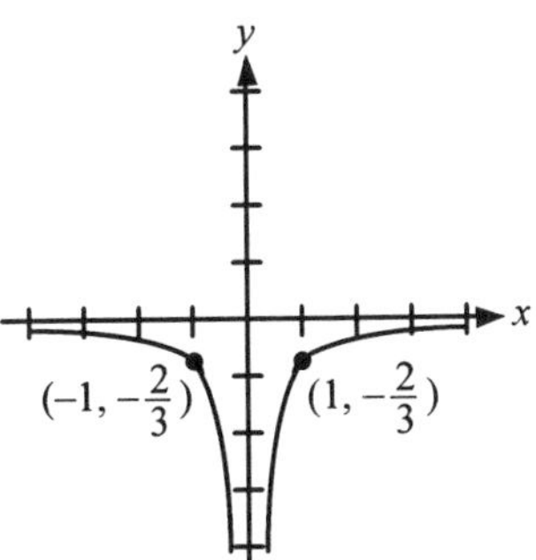

15.

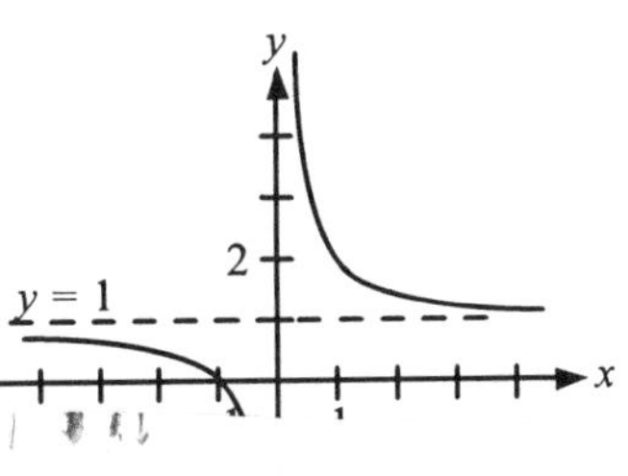

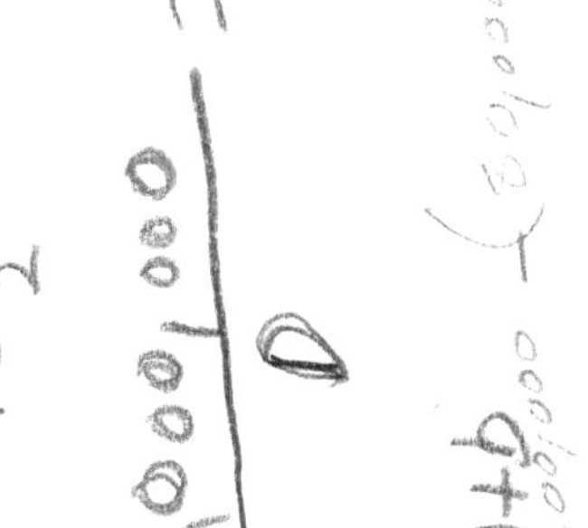

21.

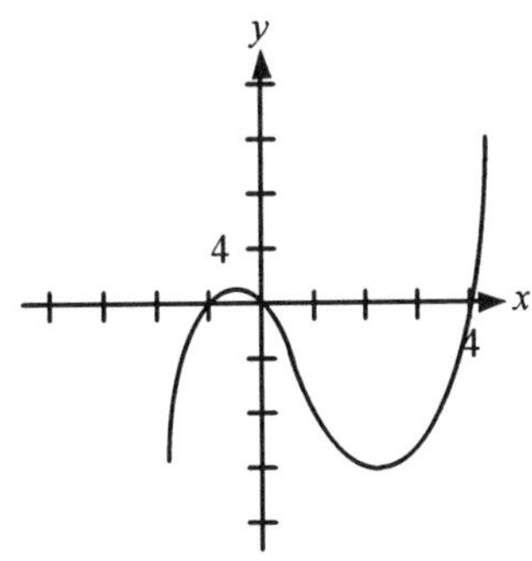

23. $x = 0,\ x = -2$

25. For $x \neq 2$, the same straight line is the graph for both functions. At $x = 2$ the graph of $y = x + 2$ contains the point (2, 4). Since $\dfrac{x^2 - 4}{x - 2}$ is undefined at $x = 2$, the graph of this function shows an open circle at (2, 4).

27. $P(x) = x(x-5)(x+2)(x-3)$

29. $2x^4 - x + 7 = \left(x^2 + 2\right)\left(2x^2 - 4\right) + (-x + 15)$

31. $\pm 2,\ \pm 1,\ \pm\dfrac{2}{3},\ \pm\dfrac{1}{3}$

33. $\pm\sqrt{3}$

35. 0 is a zero of multiplicity 3; –2 is a zero of multiplicity 2.

37. Since b is a zero, $P(b) = b^3 + ab^2 + ab + 1 = 0$. Then,

$$P\left(\frac{1}{b}\right) = \frac{1}{b^3} + \frac{a}{b^2} + \frac{a}{b} + 1 = \frac{1 + ab + ab^2 + b^3}{b^3}$$
$$= \frac{0}{b^3} = 0$$

39. $P(3) = -1$ and $P(4) = 15$. Since $P(3)$ and $P(4)$ have opposite signs, the location theorem guarantees at least one real zero between 3 and 4.

41. b **43.** b **45.** d

47. a **49.** a

Chapter 6 Test

1. True **2.** Degree 7

3. $P(x) = (x+2)(x+1)(x-4)$

4. Translate the graph of $y = x^4$ right 3 units and up 2 units.

5. $P(-1) = 3$, $P(-2) = -7$; since $P(-1)$ and $P(-2)$ have opposite signs, the location guarantees at least one real zero between −1 and −2.

6. False

7. $\frac{x^3-1}{x^2+x} = x - 1 + \frac{x-1}{x^2+x}$

8. $-3x + 32$

9. $2x^3 - 6x^2 + 17x - 51$

10. **a.** −13 **b.** −13

11. 0 is a zero of multiplicity 2; −1 is a zero of multiplicity 3.

12.

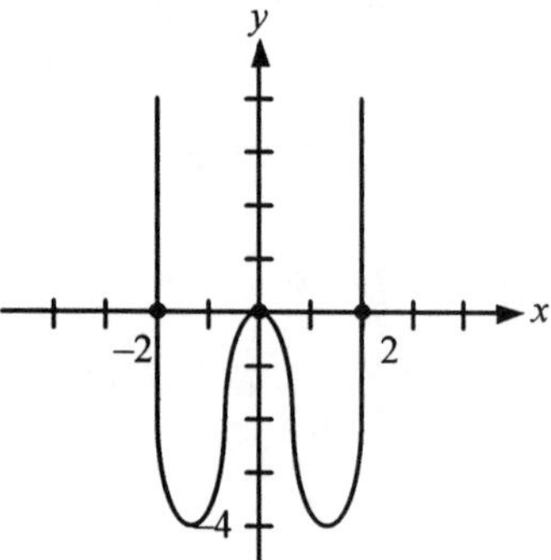

13. $\pm\sqrt{6}$

14. $\pm 2, \pm 1, \pm\frac{2}{3} \pm\frac{1}{3}$

15. 1

16. $-\frac{3}{2}$, $\pm i\sqrt{3}$

17. $x = 0$, $x = 6$

18. $(2,0)$, $(-2,0)$

19. $y = x - 2$

20.

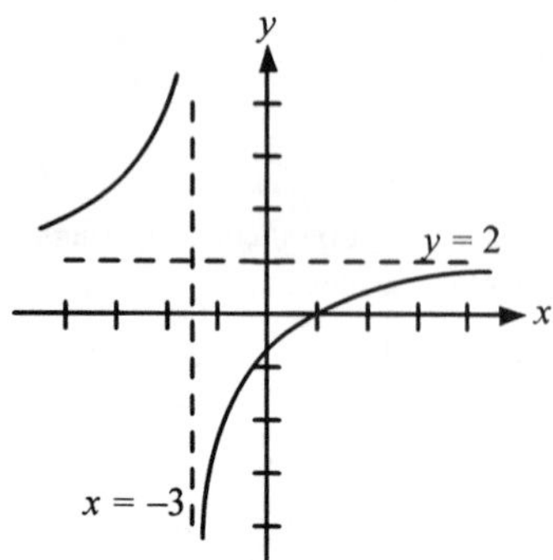

Exercises 7.1

1. $f^{-1} = \{(7,1),(8,2)(9,3)\}$; $D_{f^{-1}} = \{7,8,9\}$;
$R_{f^{-1}} = \{1,2,3\}$; $D_f = \{1,2,3\}$; $R_f = \{7,8,9\}$

3. $f^{-1} = \{(10,-2),(0,0)(-1,5)\}$; $D_{f^{-1}} = \{10,0,-1\}$;
$R_{f^{-1}} = \{-2,0,5\}$; $D_f = \{-2,0,5\}$; $R_f = \{10,0,-1\}$

5. $f^{-1} = \{(4,4),(5,5)(6,6)\}$;
$D_{f^{-1}} = R_{f^{-1}} = D_f = R_f = \{4,5,6\}$

7. No **9.** Yes **11.** No

13. No **15.** Yes **17.** Yes

19.

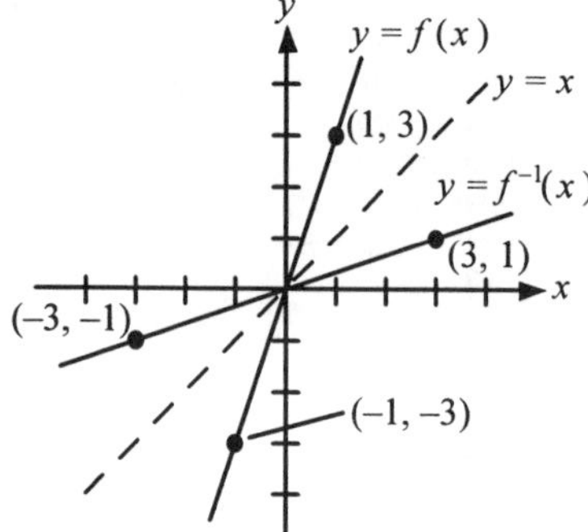

21.

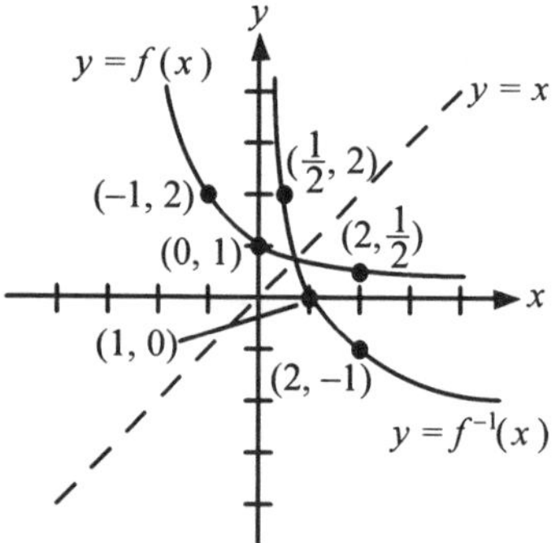

23. No inverse function exist.

25. $f^{-1}(2) = 12$

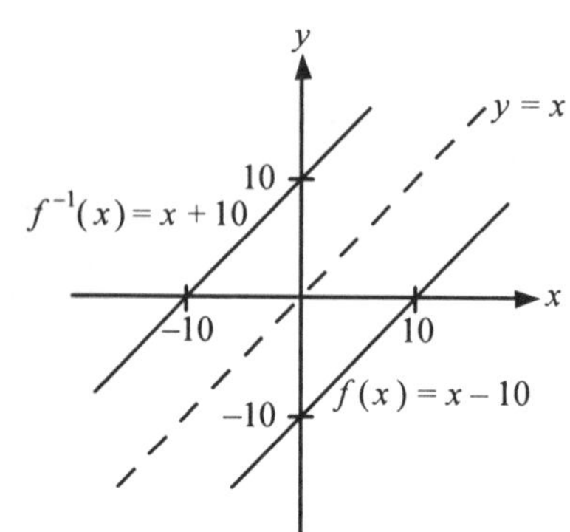

27. $f^{-1}(2) = 13$

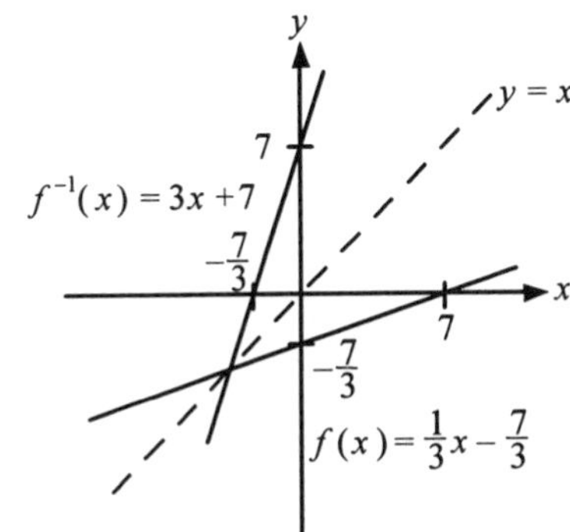

29. $y = x^2$ is not a one-to-one function, so no inverse function exists.

31. $f(x) = 2$ is not a one-to-one function, so no inverse function exists.

33.

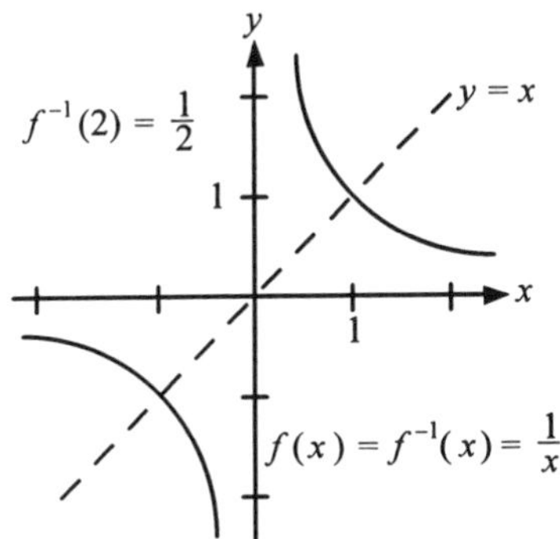

35. $h^{-1}(2) = 12$

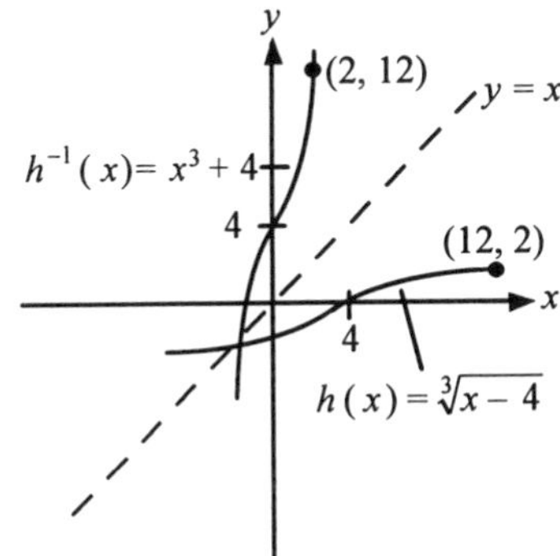

37. $f^{-1}(2) = \sqrt{6}$

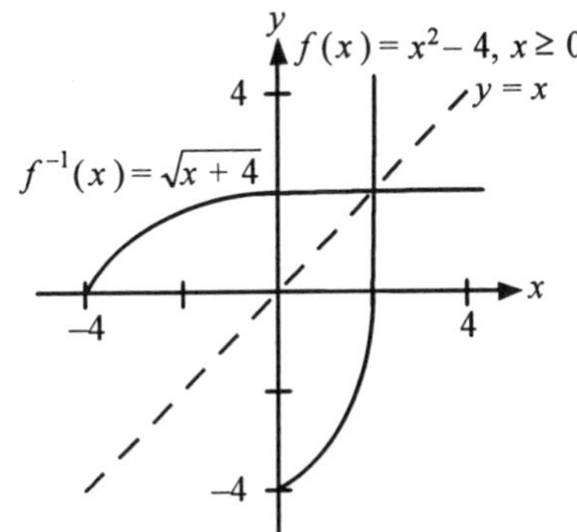

39. $f^{-1}(2) = 1$

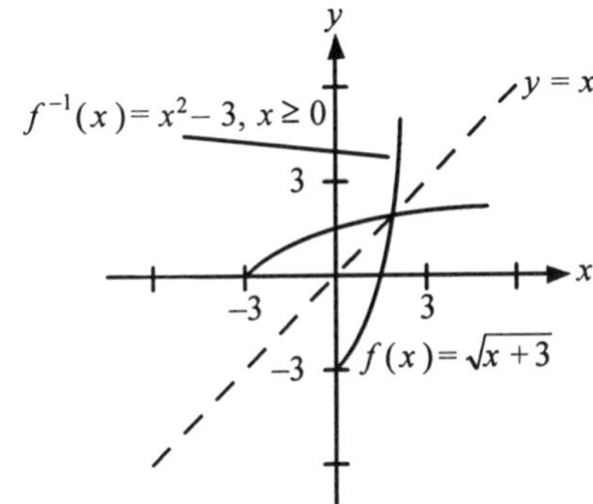

41. $f(x) = \sqrt{1 - x^2}$ is not a one-to-one function, so no inverse function exists

43. Function

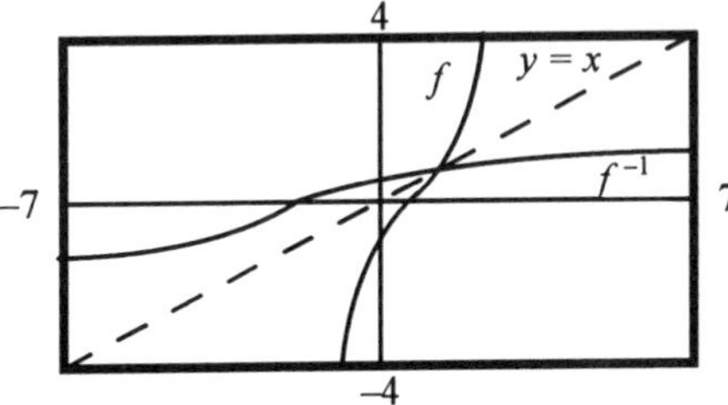

45. Inverse is not a function.

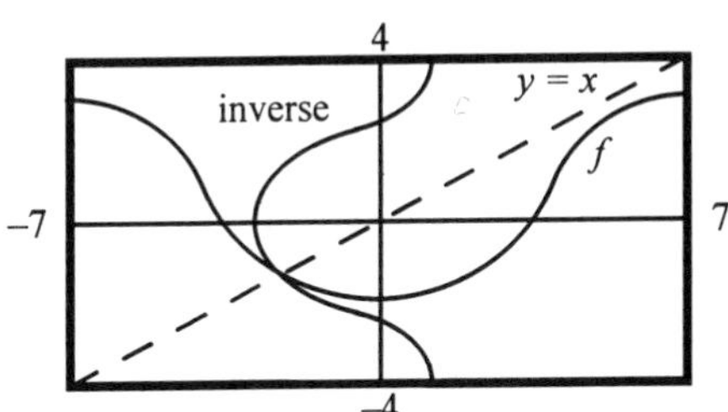

47. Inverse is not a function

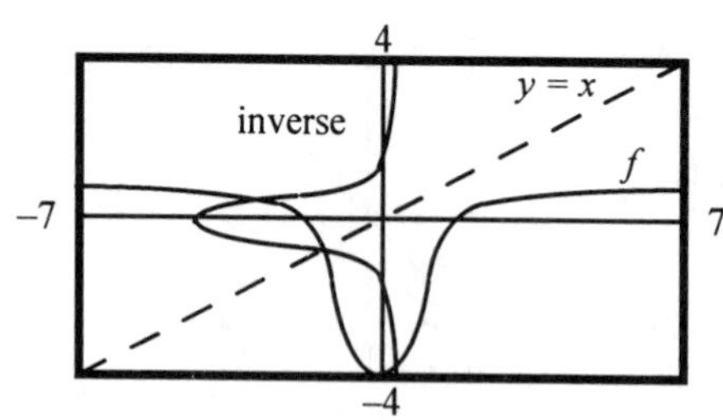

49. $f[g(x)] = (x-5)+5 = x$
$g[f(x)] = (x+5)-5 = x$

51. $f[g(x)] = \sqrt[3]{x^3} = x$
$g[f(x)] = \left(\sqrt[3]{x}\right)^3 = x$

53. $f[g(x)] = \dfrac{1}{[(1-4x)/x]+4} = \dfrac{x}{1-4x+4x} = x$;
$g[f(x)] = \dfrac{1-4[1/(x+4)]}{1/(x+4)} = \dfrac{(x+4)-4}{1} = x$

55. $f[g(x)] = \sqrt{(x^2-1)+1} = \sqrt{x^2} = x$ (since $x \geq 0$);
$g[f(x)] = \left(\sqrt{x+1}\right)^2 - 1 = (x+1)-1 = x$

57. $f[f(x)] = \dfrac{1}{(1/x)} = x$

59. 11.11%

61. $(r \circ c)(x) = 1.15(0.85x) = .9775x \neq x$; 2.25% loss

63. $y = \sqrt[3]{x}$, $x > 0$. This is the formula for the side length (y) of a cube in terms of the area (x).

Exercises 7.2

1. 16; $\frac{1}{4}$; 8; 7.10 **3.** $\frac{16}{81}$; $\frac{27}{8}$; 1; 0.28

5. $\{6\}$ **7.** $\left\{\frac{1}{2}\right\}$ **9.** $\{-3\}$

11. $\{-2\}$ **13.** $\{4\}$ **15.** $\left\{-\frac{11}{4}\right\}$

17. $\left\{\frac{3}{5}\right\}$ **19.** $\left\{\frac{4}{9}\right\}$

21.

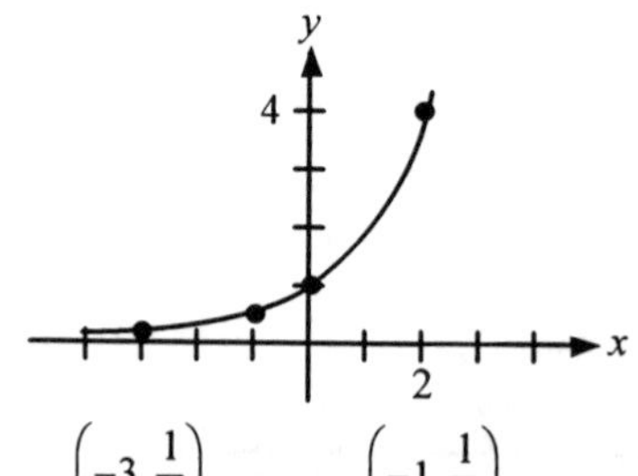

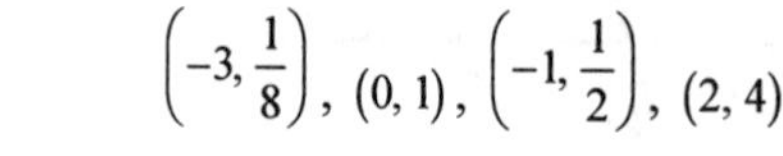

$\left(-3, \frac{1}{8}\right)$, $(0, 1)$, $\left(-1, \frac{1}{2}\right)$, $(2, 4)$

23.

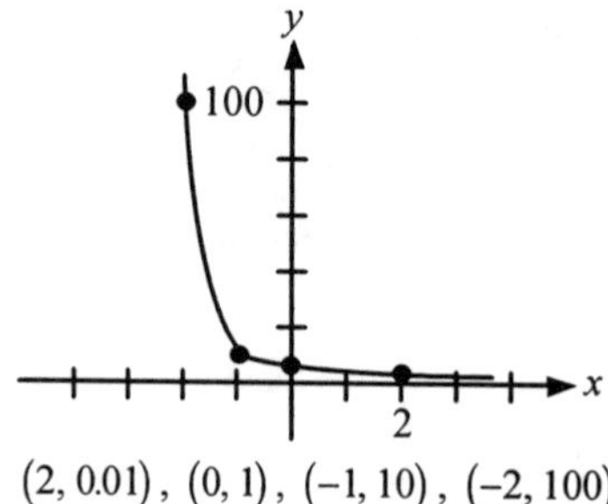

$(2, 0.01)$, $(0, 1)$, $(-1, 10)$, $(-2, 100)$

25.

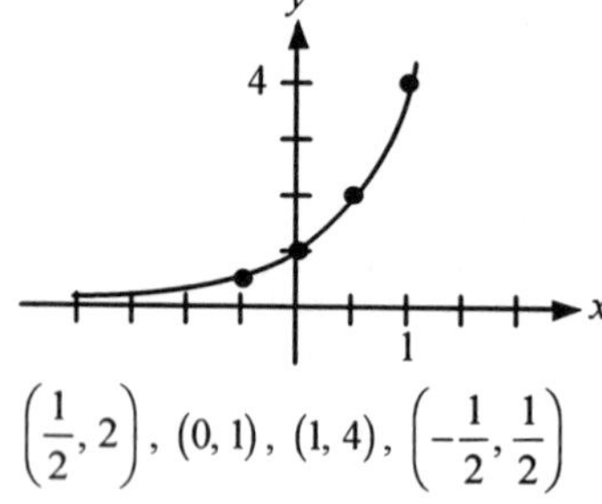

$\left(\frac{1}{2}, 2\right)$, $(0, 1)$, $(1, 4)$, $\left(-\frac{1}{2}, \frac{1}{2}\right)$

27.

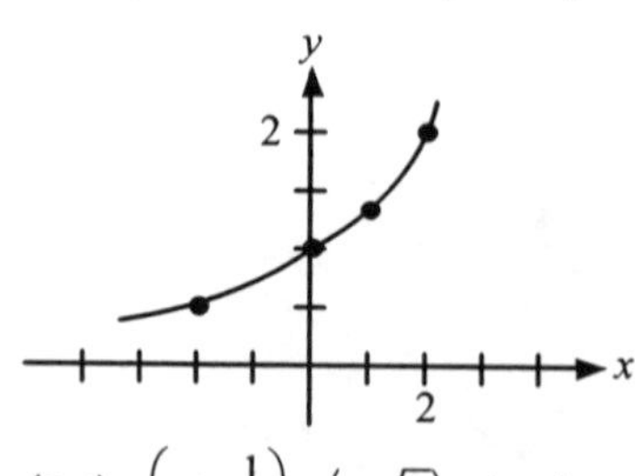

$(0, 1)$, $\left(-2, \frac{1}{2}\right)$, $\left(1, \sqrt{2}\right)$, $(2, 2)$

29.

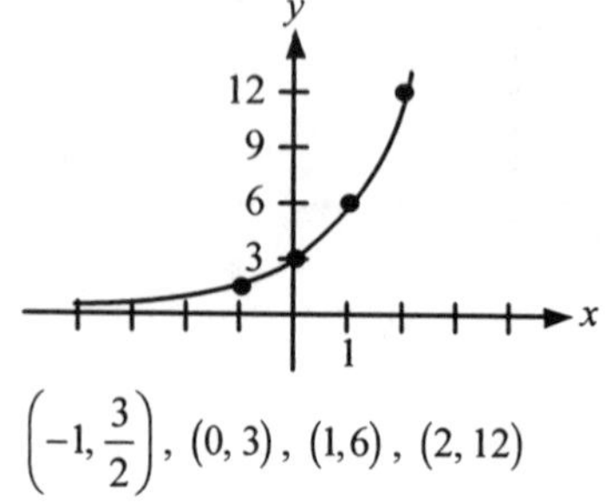

$\left(-1, \frac{3}{2}\right)$, $(0, 3)$, $(1, 6)$, $(2, 12)$

31. 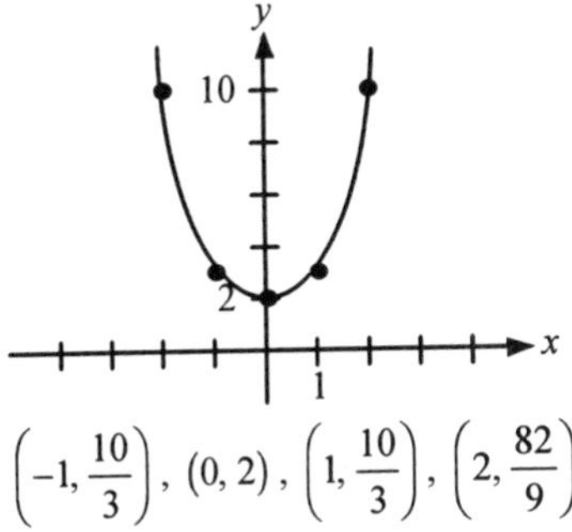

$\left(-1, \frac{10}{3}\right)$, $(0, 2)$, $\left(1, \frac{10}{3}\right)$, $\left(2, \frac{82}{9}\right)$

33. 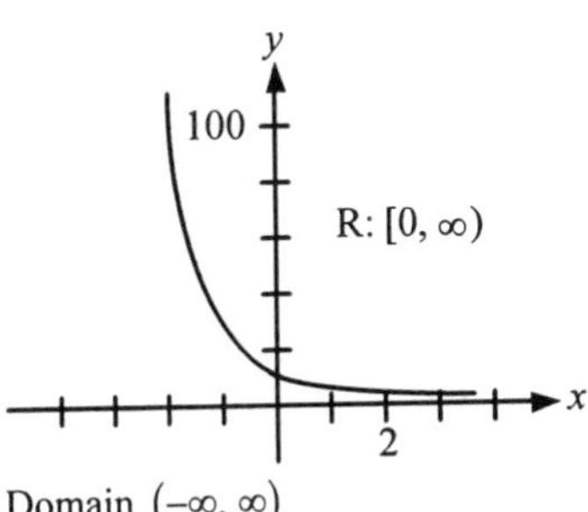

35. 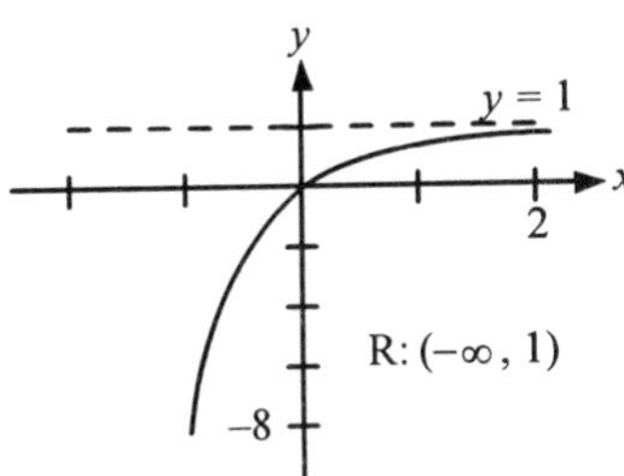

37. 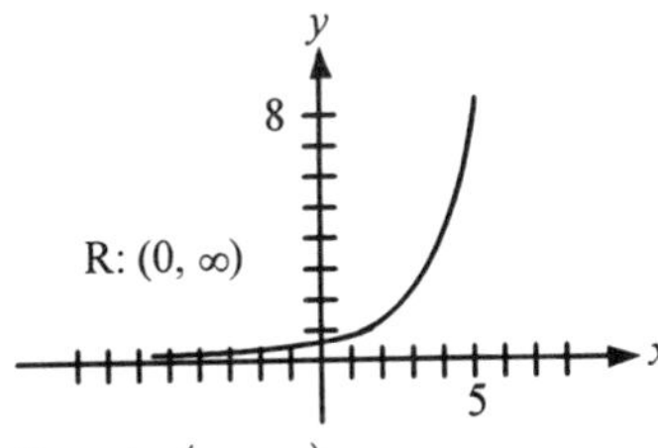

39. 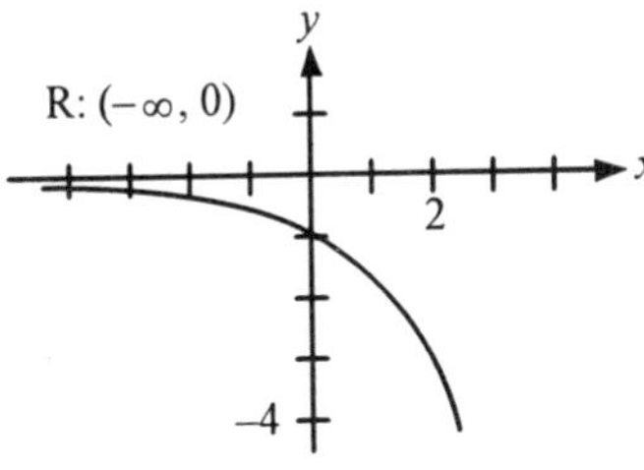

41. 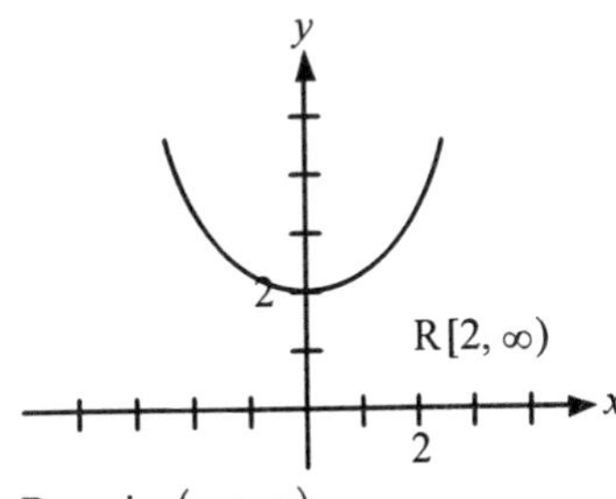

43. 4 45. $\frac{1}{3}$ 47. 16

49. $\frac{1}{8}$ 51. $\frac{1}{8}$

53. If $f(x) = b^x$, then

$f(x_1 + x_2) = b^{x_1 + x_2} = b^{x_1} \cdot b^{x_2} = f(x_1)f(x_2)$

55. a. $y = 500(2)^t$ b. 8000
c. 6 hours

57. a. $y = 100(1.05)^t$
b.

t	0	1	2	3	4
y	100	105	110.25	115.76	121.55

59. a. $f(t) = 10{,}000(0.8)^t$
b. \$5,120

61. a. $y = 16\left(\frac{1}{2}\right)^{2t}$ b. 4g

63. a. \$6,228,000 b. $\$1.17 \times 10^{11}$

65. 14.87% 67. 1.75% 69. 29¢ per watt

71. a. $b = 2$ b. $y = 1000(2)^x$

73. a. $b = \sqrt{3}$ b. $y = 10(\sqrt{3})^x$

75. linear $y = 1.2301 + 0.2301x$

77. exponential $y = (1.2)^{x+1}$

79. neither 81. $y = 8^x$

Exercises 7.3

1. $\log_3 9 = 2$ 3. $\log_{1/2}\left(\frac{1}{4}\right) = 2$

5. $\log_4\left(\frac{1}{16}\right) = -2$ 7. $\log_{25} 5 = \frac{1}{2}$

9. $\log_7 1 = 0$ **11.** $5^1 = 5$

13. $\left(\frac{1}{3}\right)^2 = \frac{1}{9}$ **15.** $2^{-2} = \frac{1}{4}$

17. $49^{1/2} = 7$ **19.** $100^0 = 1$

21. 2 **23.** 1 **25.** −1

27. $\frac{1}{2}$ **29.** 0 **31.** 3

33. $\frac{1}{5}$ **35.** 5 **37.** 3

39. $b > 0,\ b \neq 1$ **41.** .6590

43. 2.3400 **45.** .00316 **47.** 1380

49.

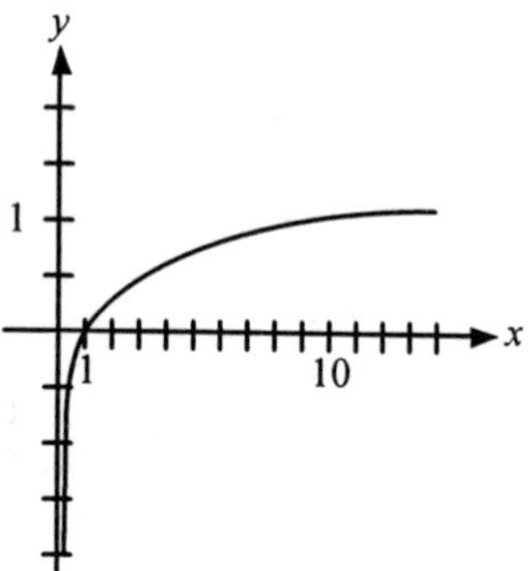

51.

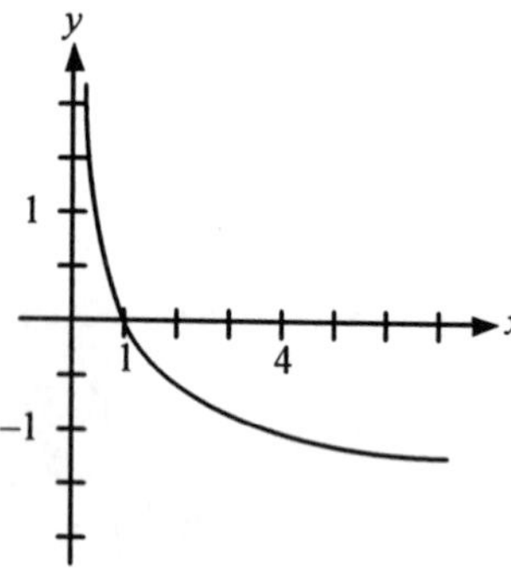

53.

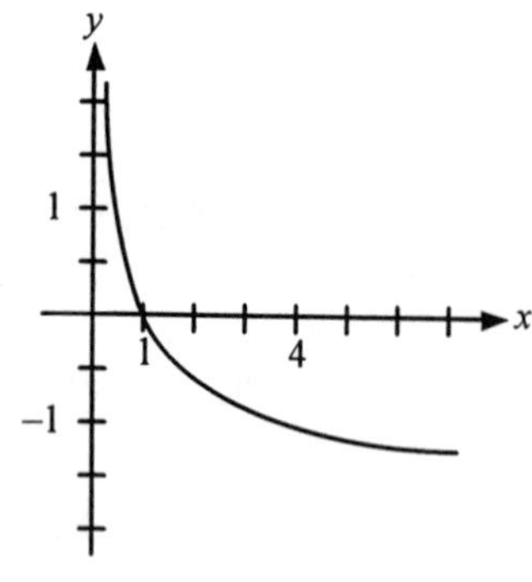

55.

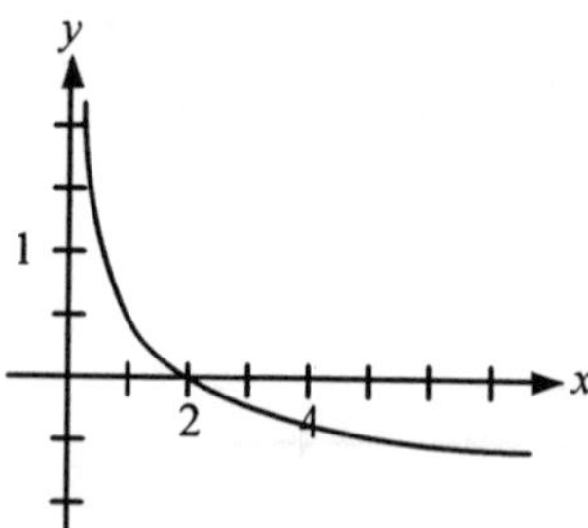

57.

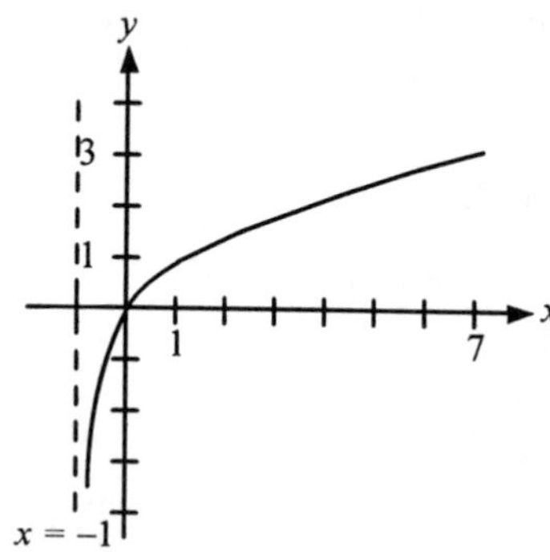

59.

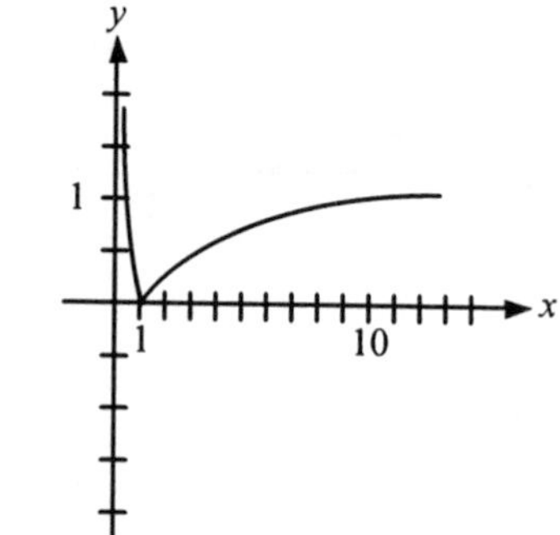

61. $g(x)$ is 2 units higher; $h(x)$ is shifted 2 units to the right.

63. $g(x)$ is reflected about the x-axis and raised 5 units. $h(x)$ is shifted to the right 5 units and reflected about the y-axis.

65.

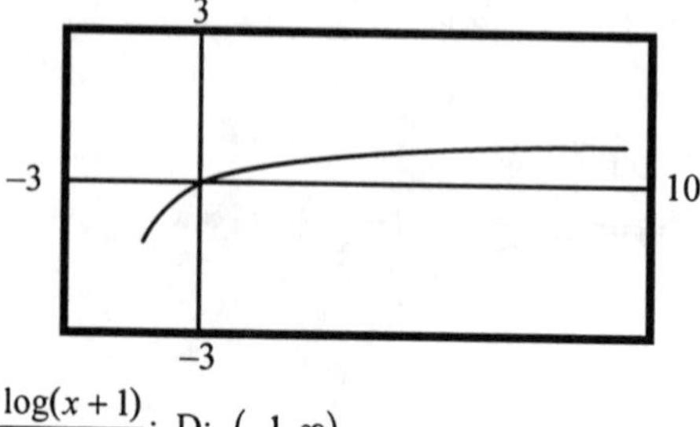

$\frac{\log(x+1)}{\log 4}$; D: $(-1, \infty)$

67.

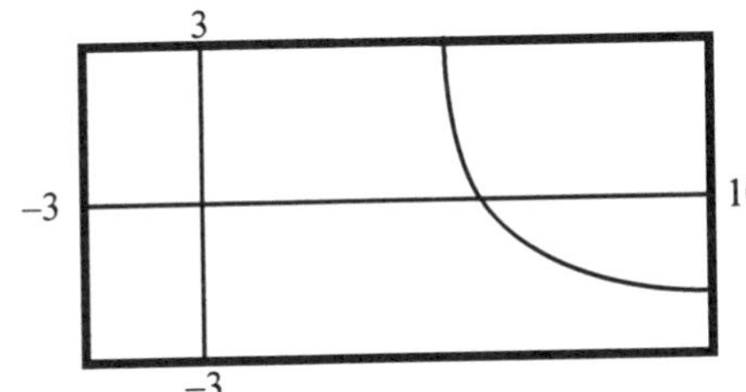

$\dfrac{\log(x-5)}{\log(1/2)}$; D: $(5, \infty)$

69.

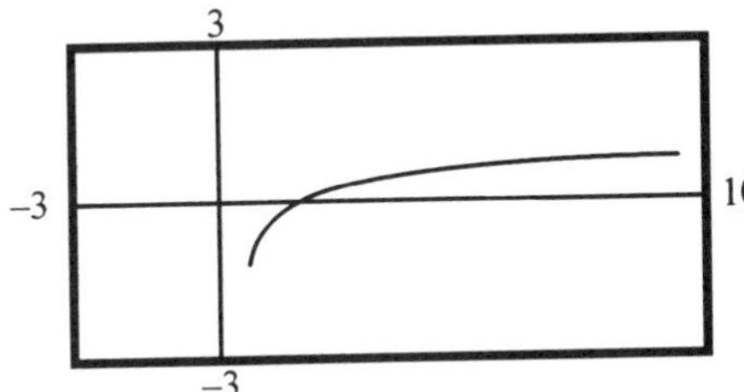

$\log(2x-1)$; D: $\left(\dfrac{1}{2}, \infty\right)$

71.

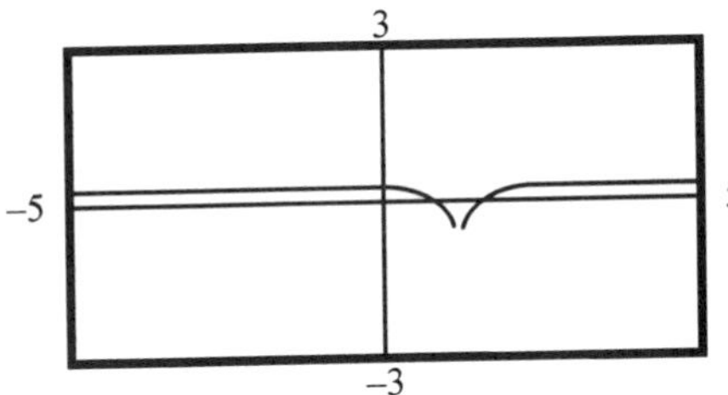

$\log|x-1|$

Domain: set of all real numbers except 1

73. 38.2

75. **a.** 0 **b.** 10 **c.** 20
d. 30; if the power ratio is multiplied by 10, add 10 dB to loudness

77. 7 **79.** 3.6 **81.** 1

83. 7.9×10^{-3} **85.** 4.82 ft

Exercises 7.4

1. $\log_{10} 7 + \log_{10} 5$ **3.** $\log_6 3 - \log_6 5$

5. $16 \log_5 11$ **7.** $\dfrac{1}{2}\log_2 3$

9. $\dfrac{1}{5}\log_{10} 16$ **11.** $2 + 3\log_4 3$

13. $\dfrac{1}{2}(\log_b x + \log_b y)$ **15.** $\dfrac{5}{3}\log_b x$

17. $\dfrac{1}{4}(\log_b x + 2\log_b y - 3\log_b z)$

19. $1 + \log_6 x$ **21.** $-\log_b 10$

23. $3 + 2\log_2 x$ **25.** $\dfrac{3}{5}$

27. **a.** $m+n$ **b.** $m+.5n$ **c.** $-n-2m$

29. 100 **31.** $k + \log_{10} m$ **33.** 3

35. $\dfrac{n}{x}x^n = nx^{n-1}$ **37.** $\log_2 12$

39. $\log_4 4$ or 1 **41.** $\log_7 9$

43. $\log_{10}\dfrac{3}{8}$ **45.** $\log_b \sqrt[3]{xy^2}$

47. $\log_b(x-1)$ **49.** $\log_b \sqrt{\dfrac{xz^3}{y^5}}$

51. $\log_b\left(\dfrac{1}{a}\right) = \log_b a^{-1} = -\log_b a$

53. Let $x = b^m$ and $y = b^n$, then $m = \log_b x$ and $n = \log_b y$. $\dfrac{x}{y} = \dfrac{b^m}{b^n} = b^{m-n}$; thus

$\log_b \dfrac{x}{y} = m - n = \log_b x - \log_b y$.

55. **a.**
$$\begin{aligned} L &= 10\log(I) + 90 \\ &= 10\log(I) + 90\log(10) \\ &= 10[\log(I) + 9\log(10)] \\ &= 10[\log(I) - \log 10^{-9}] \\ &= 10\log\left(\frac{I}{10^{-9}}\right) \end{aligned}$$

b. 110 dB

c. $1000\dfrac{\text{ergs}}{\text{cm}^2\ \text{sec}}$ or $1000\dfrac{\text{dynes}}{\text{cm}^2}$

57. **a.** $x = \dfrac{\log y - b}{a}$

b. 3.01 **c.** $y = 10^{ax+b}$ **d.** 15.85

59. $\log y = 2 + .4771x$

61. $\log y = \log k + x\log b = c + wx$, where $\log k$ and $\log b$ are constants

63. $\log_2 y = 15.2877 - .3219\log_2 x$

Exercises 7.5

1. $\{3.322\}$ **3.** $\{1.044\}$ **5.** $\{-3.170\}$

7. $\{3.376\}$ **9.** $\{-0.7304\}$ **11.** $\{-1.631\}$

13. $\{18.58\}$ **15.** 3.170 **17.** 1.465

19. −4.248 **21.** −3.190 **23.** $\frac{1}{3}$

25.

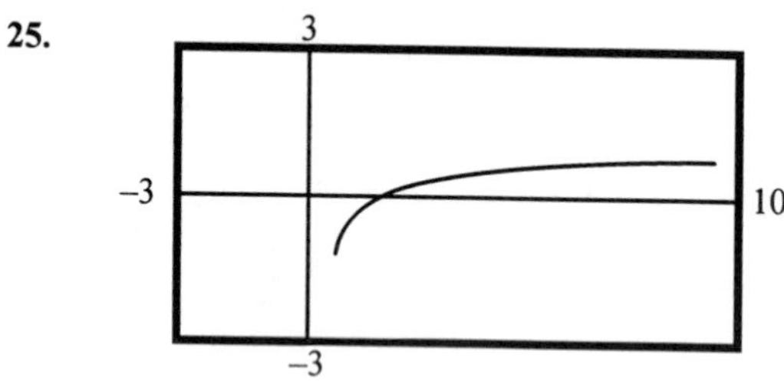

Domain: $(0, \infty)$

27.

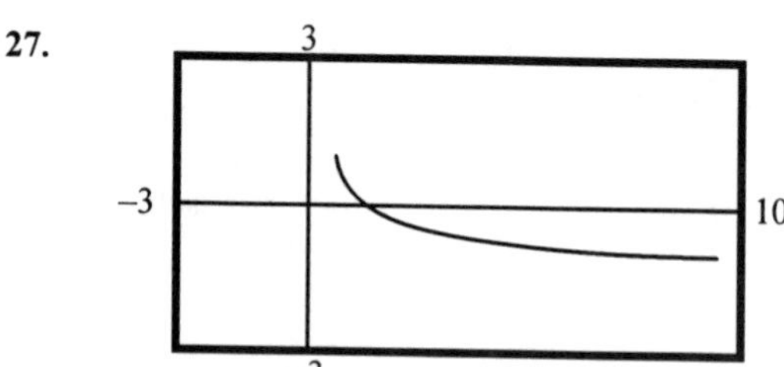

Domain: $(0, \infty)$

29.

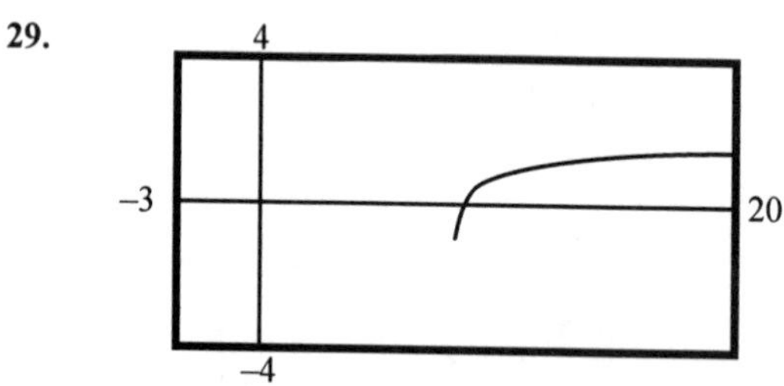

Domain: $(8, \infty)$

31.

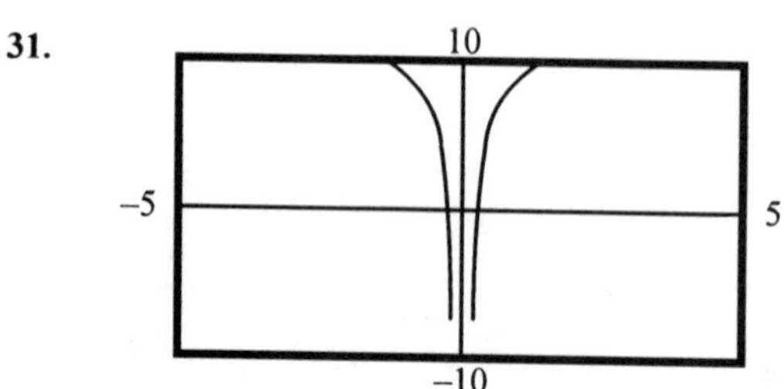

Domain: set of all real numbers except 0

33.

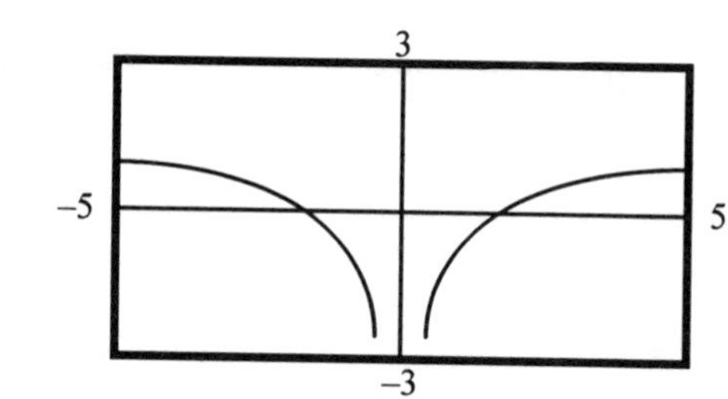

Domain: set of all real numbers except 0

35. $\{100\}$ **37.** $\{10\}$ **39.** $\left\{\frac{9}{10}\right\}$

41. $\{5\}$ **43.** $\{2\sqrt{2}\}$ **45.** $\left\{\frac{1}{2}\right\}$

47. $\{2\}$ **49.** $\{\pm 3\}$ **51.** $\{2\}$

53. $\{4\}$ **55.** $\{5\}$ **57.** $\{3\}$

59. $\{5\}$ **61.** $\varnothing$ **63.** $(0, \infty)$

65. 15.7 years **67.** 47 years **69.** 20 years

71. **a.** 2009 **b.** 2,060

73. 14.9% **75.** 15,399,000

Exercises 7.6

1. **a.** $A = 1000(1.03)^{2t}$; \$1194.05

b. $A = 5000(1.03)^{4t}$; \$6333.85

3. **a.** $A = 1500\left(1 + \frac{0.35}{365}\right)^{365t}$; \$1608.76

b. $A = 2000\left(1 + \frac{.1}{365}\right)^{365t}$; \$5435.82

5. **a.** $A = 1000e^{.06t}$; \$1349.86

b. $A = 2000e^{.1t}$; \$5436.56

7. **a.** $A = 5000(1.0075)^{12t}$; \$5593.01

b. $A = 5000e^{.09t}$; \$5595.01

9. \$1 invested continuously at 10 percent

11. 1078

13.

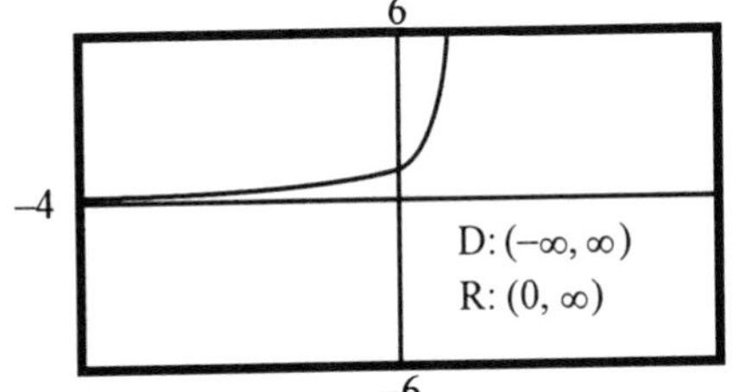

15.

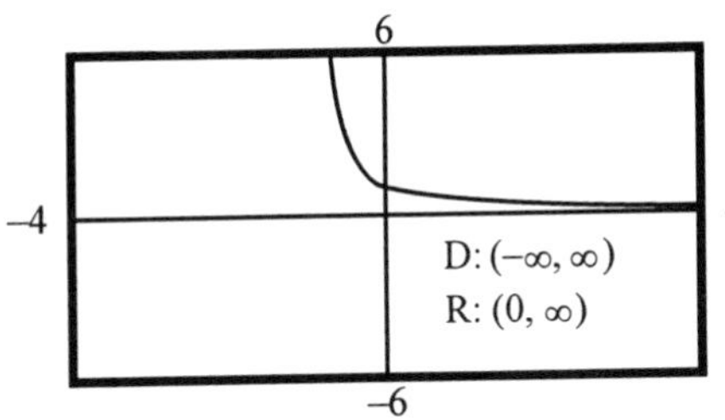

17.

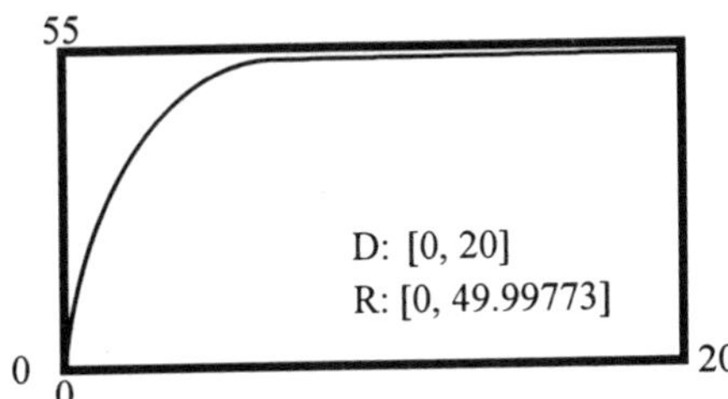

19.

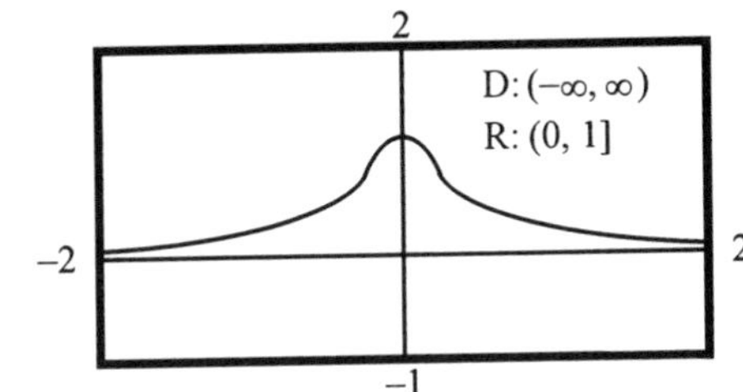

21.

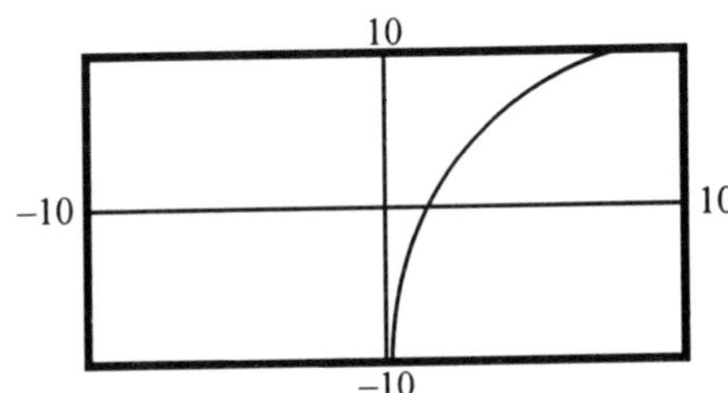

Same graphs since $\log_b x^k = k \log_b x$

23.

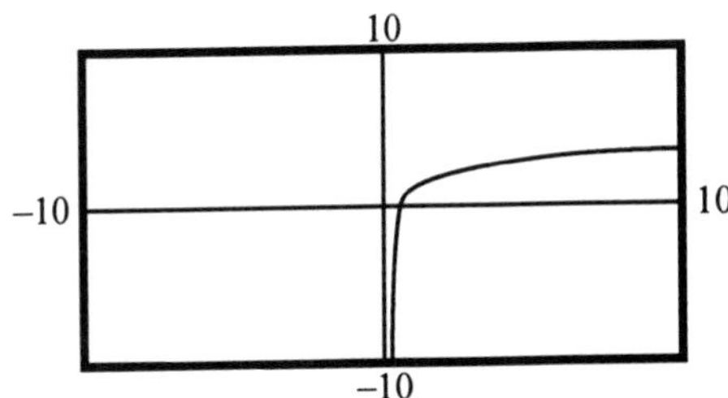

Same graphs since $\log_b kx = \log_b k + \log_b x$

25.

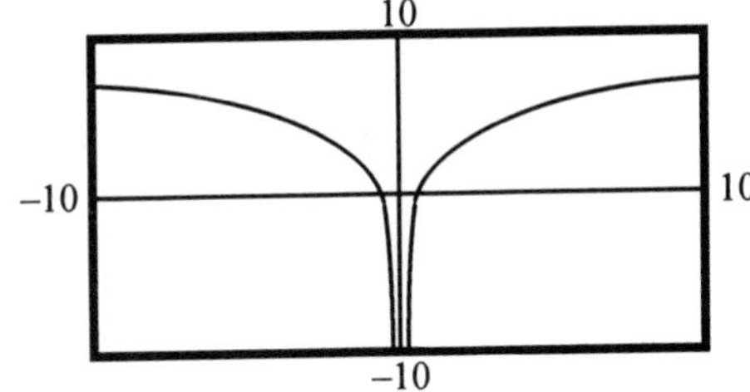

Same graphs since $\log_b kx^c = \log_b k + c \log_b x$

27.

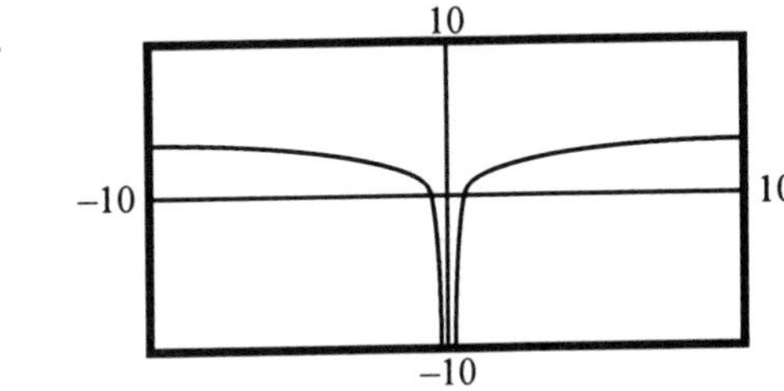

Same graphs since $\ln \frac{x^k}{c} = k \ln x - \ln c$

29. 7.144 **31.** .2784 **33.** 2.388

35. 12.95 **37.** .3466

39. **a.** $e^{7.4}$ **b.** 1636

41. **a.** $e^{-3.8}$ **b.** .0224

43. **a.** $\frac{e^6}{3}$ **b.** 134.5

45. **a.** $5e^{3/2}$ **b.** 22.41

47. **a.** $\frac{-2 + \left(4 + 4e^3\right)^{1/2}}{2}$

b. 3.592

49. **a.** $2^{1/2}$ **b.** 1.414

51. 9.90% **53.** 10.99% **55.** 13.86%

57. 13.9 years **59.** \$941.76 **61.** 18.36 years

63. **a.** 416 **b.** 21.97 years

65. **a.** 100,000,000 **b.** .80 hrs.

67. 7.6*g*

69. **a.** −0.000124 **b.** 78%

71. 29.8% **73.** 89.7% **75.** 72.5%

77. 7:10 a.m. **79.** 55 degrees

81. $y = \dfrac{100}{1 + 99e^{-2t}}$

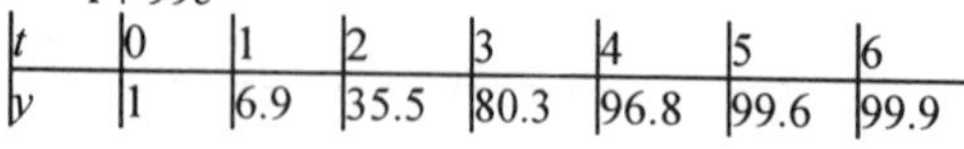

t	0	1	2	3	4	5	6
y	1	6.9	35.5	80.3	96.8	99.6	99.9

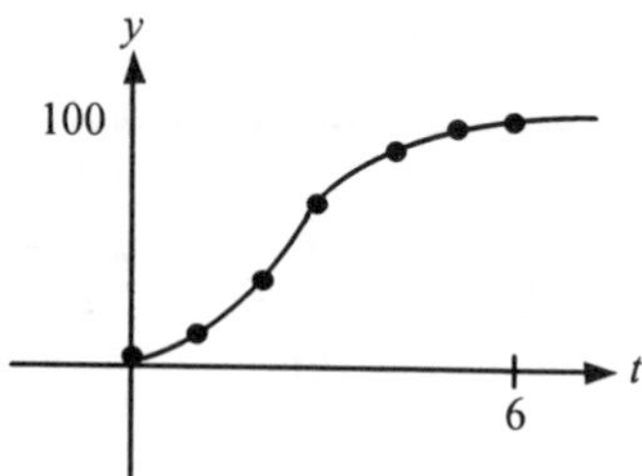

83. **a.** Velocity increases quickly then levels out.

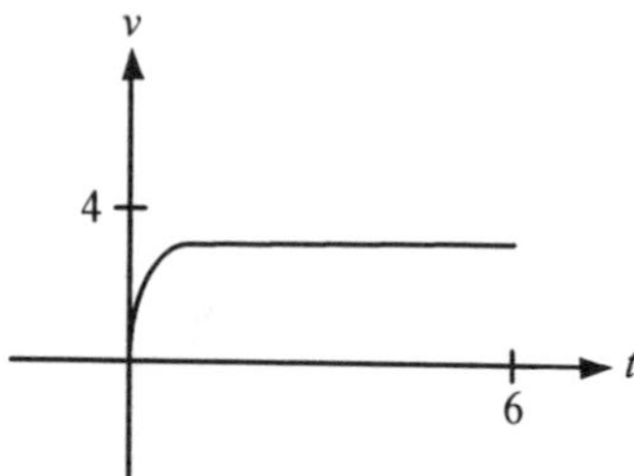

b. $v = 3.46$ ft/second **c.** 4 ft/second

85. **a.** $L = 35,\ k = -.243$
b. 13.5% **c.** 2.3 weeks

87. **a.** $L = 100,\ k = -.693$
b. $t = 2$

89. $i = .63i_0$

91. **a.** no growth

b. population increases; $t = 1{,}000\dfrac{\ln 2}{b-d}$

c. population decreases; $t = 1{,}000\dfrac{\ln 5}{b-d}$

93. .0953; 3 terms

Chapter 7 Review Exercises

1. $\{7\}$ **3.** $\{32\}$ **5.** $\{3.43 \times 10^7\}$

7. $\{0.00384\}$ **9.** $\{.462\}$ **11.** $\ln\left(x^2 \cdot \sqrt[3]{y}\right)$

13. $\frac{1}{2}[3\log_b x - \log_b y]$

15. $\left(\frac{7}{3}, \infty\right)$ **17.** $2^3 = 8$

19.

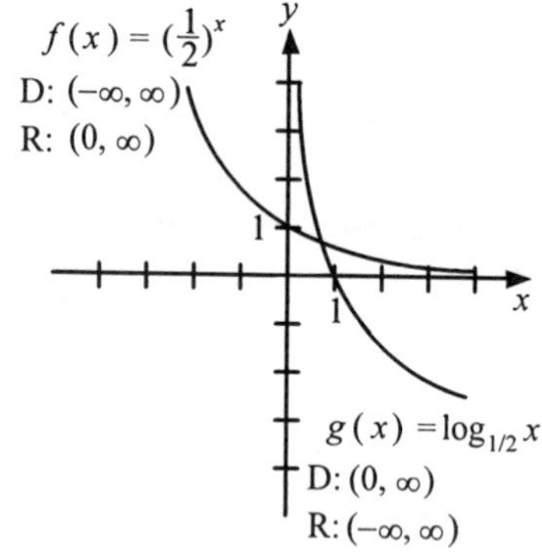

21. $\left(3, \frac{1}{64}\right)$, $(-1, 4)$, $(0, 1)$, $\left(\frac{3}{2}, \frac{1}{8}\right)$, $(-2, 16)$

23. \$3,664.21

25. $f^{-1} = \{(-1, 3), (-2, 4)\}$; $D_f = R_{f^{-1}} = \{3, 4\}$;
$D_{f^{-1}} = R_f = \{-1, -2\}$

27. M^2 **29.** $\frac{1}{3}(x + y)$ **31.** 0

33. Domain: $(-\infty, \infty)$; range: $(-2, \infty)$

35.

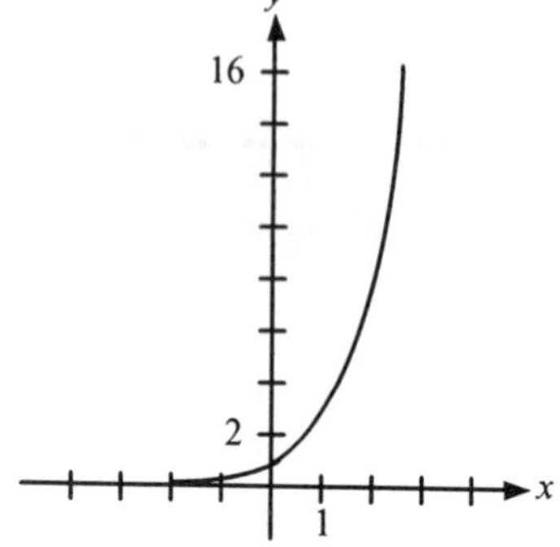

37. 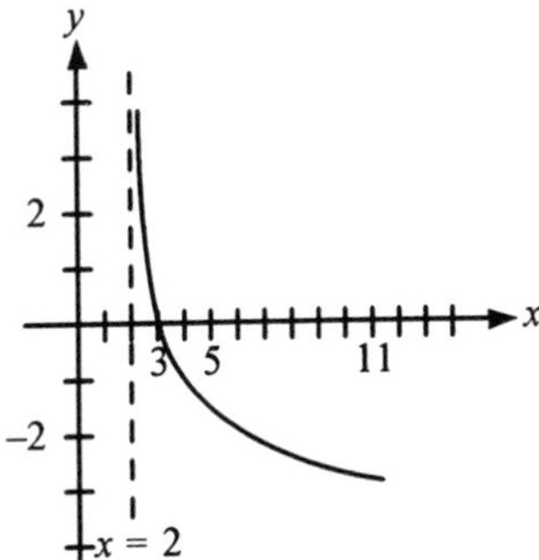

39. 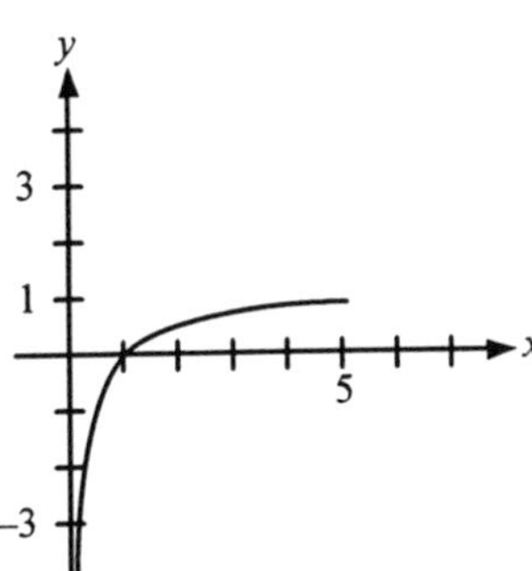

41. 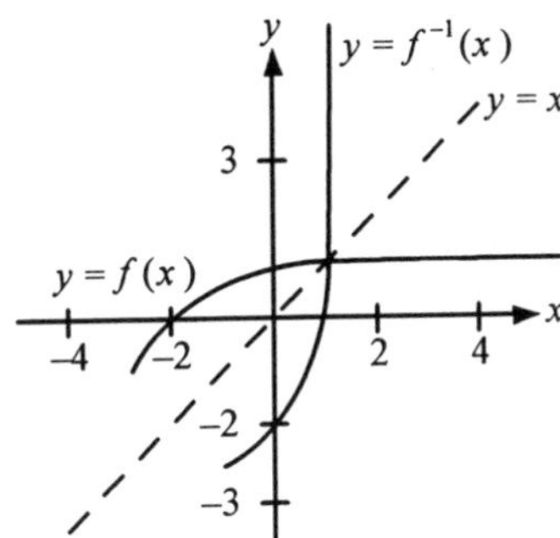

43. b

45. $\left\{-\frac{3}{2}\right\}$

47. 100

49. $\frac{13}{2}$

51. −0.098

53. $y = r + ce^{-kt}$; $\frac{y-r}{c} = e^{-kt}$; $-kt = \ln\frac{y-r}{c}$;

$$t = \frac{-1}{k}\ln\frac{y-r}{c} = \frac{1}{k}\ln\left(\frac{y-r}{c}\right)^{-1} = \frac{1}{k}\ln\frac{c}{y-r}$$

55. (pH of X) = (pH of Y) − 1

57. **a.** $g^{-1}(x) = 2x + 5$
b. 3

59. $f^{-1}(x) = e^{-x}$; $(-\infty, \infty)$

61. $f[g(x)] = \frac{1}{[(1-x)/x]+1} = \frac{x}{1-x+x} = x$;
$g[f(x)] = \frac{1-[1/(x+1)]}{1/(x+1)} = \frac{(x+1)-1}{1} = x$

63. d

65. c

67. c

69. a

71. c

Chapter 7 Test

1. 9

2. $\left\{-\frac{3}{2}\right\}$

3. $(-3, \infty)$

4. $1,344.56

5. **a.** $\log_8 4 = \frac{2}{3}$ **b.** $10^{-2} = 0.01$

6. $(-2, \infty)$

7. 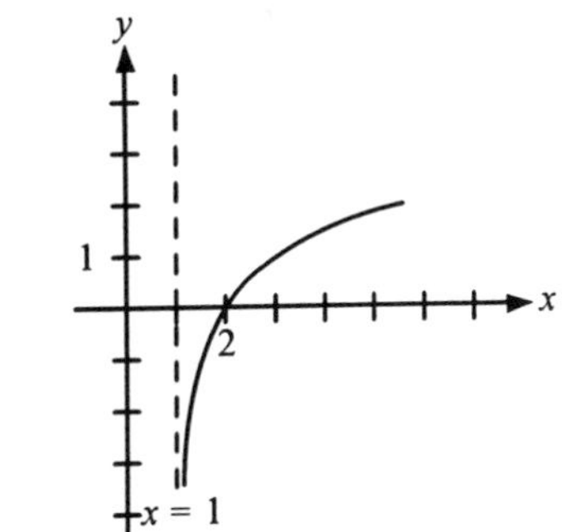

8. 3.2

9. $\log_b 5 + \frac{1}{2}\log_b x$

10. $x - 2y$

11. $\log_b 2$

12. 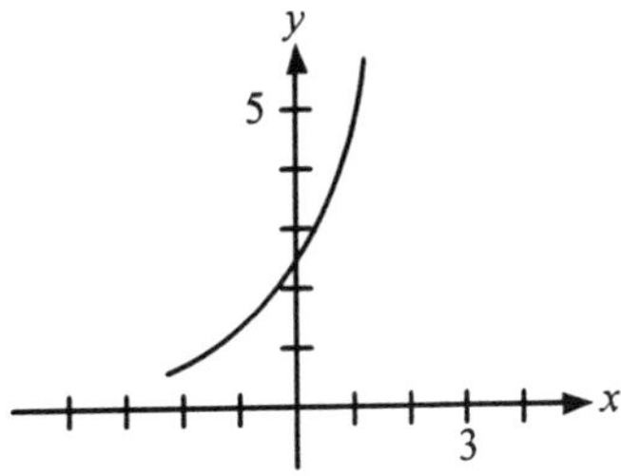

13. {1.893}

14. {4}

15. 2.322

16. $f^{-1}(x) = \sqrt[3]{x+5}$

17. $f[g(x)] = 7\left(\frac{1}{7}x + \frac{2}{7}\right) - 2 = x$;
$g[f(x)] = \frac{1}{7}(7x-2) + \frac{2}{7} = x$

18. about 63 years **19.** \$4,508.64

20. 12.2 years

Exercises 8.1

1. 5 **3.** 8 yd **5.** 9.6 m

7. 5 **9.** $\frac{\pi}{6}$ **11.** $\frac{\pi}{3}$

13. $\frac{2\pi}{3}$ **15.** $\frac{5\pi}{6}$ **17.** $\frac{7\pi}{6}$

19. $\frac{4\pi}{3}$ **21.** $\frac{5\pi}{3}$ **23.** $\frac{11\pi}{6}$

25. $\frac{10\pi}{9}$ **27.** $\frac{\pi}{9}$ **29.** 60°

31. 40° **33.** 120° **35.** 140°

37. 432° **39.** 70° **41.** 115°

43. 229° **45.** 332° **47.** 25°30′N

49. **a.** 7°N **b.** 0.3 deg/hr

51. 69 mi **53.** 10 knots **55.** $\frac{25\pi}{2}$ in.^{2}

57. 8 ft^2 **59.** $\frac{11}{2}, \frac{11}{4\pi}$ **61.** 6 rad/second

63. **a.** $\frac{200\pi}{3}$ rad/minute
b. 400π in./minute

65. 300π in./second **67.** 6,336 rad/minute

69. π in.

Exercises 8.2

1. 1 **3.** 0 **5.** 0

7. 0 **9.** Undefined **11.** 1

13. –1 **15.** $\frac{1}{2}$ **17.** 1

19. $\frac{1}{2}$ **21.** $\frac{1}{\sqrt{3}}$ **23.** $\frac{\sqrt{3}}{2}$

25. $-\frac{2}{\sqrt{3}}$ **27.** $\frac{1}{\sqrt{3}}$ **29.** $-\frac{\sqrt{3}}{2}$

31. $-\frac{\sqrt{3}}{2}$ **33.** $\sqrt{3}$ **35.** $-\frac{1}{\sqrt{2}}$

37. 2 **39.** $\frac{1}{2}$ **41.** $\frac{1}{2}$

43. –2 **45.** $-\frac{1}{\sqrt{2}}$ **47.** $-\sqrt{3}$

49. $-\frac{\sqrt{3}}{2}$ **51.** –0.4161 **53.** 1.158

55. –0.7664 **57.** 1.566 **59.** 0.2116

61. 0.6702 **63.** –1.042 **65.** 0.9511

67. 0.2225 **69.** –1.082

71. \$11,300; \$4,500

73. **a.** 3 meters; 3 meters
b. 6; 6 meters

75. **a.** 79° **b.** 43°

Exercises 8.3

1.

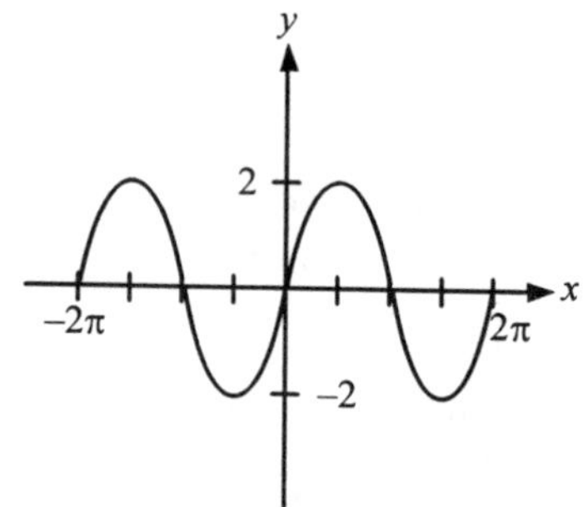

Amplitude: 2
Period: 2π
Midline: $y = 0$

3.

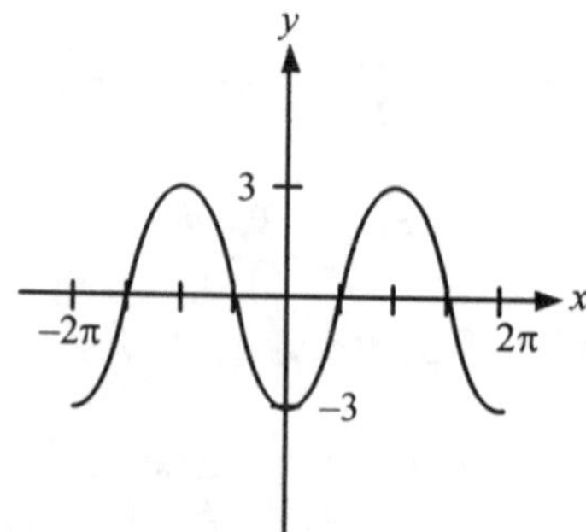

Amplitude: 3
Period: 2π
Midline: $y = 0$

5.

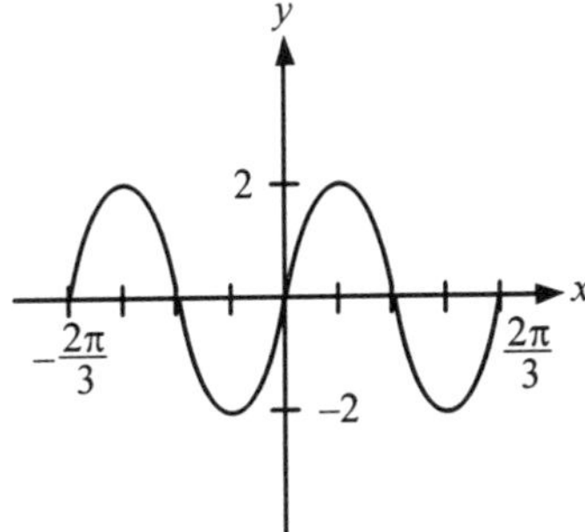

Amplitude: 2
Period: $\frac{2\pi}{3}$
Midline: $y = 0$

7.

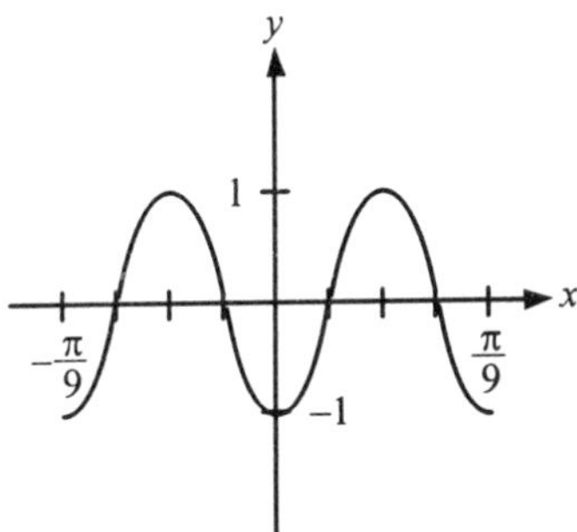

Amplitude: 1
Period: $\frac{\pi}{9}$
Midline: $y = 0$

9.

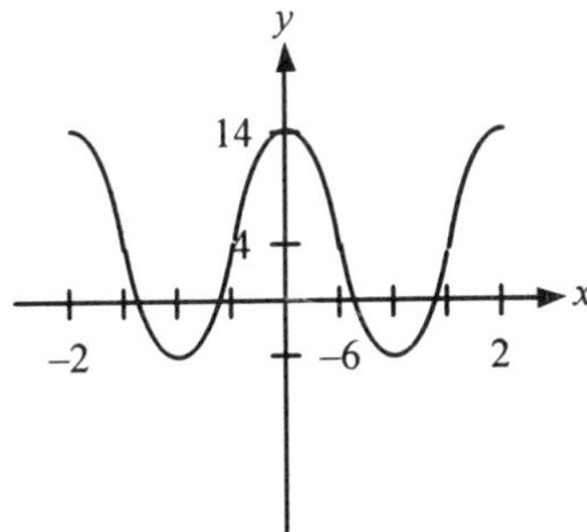

Amplitude: 10
Period: 2
Midline: $y = 4$

11.

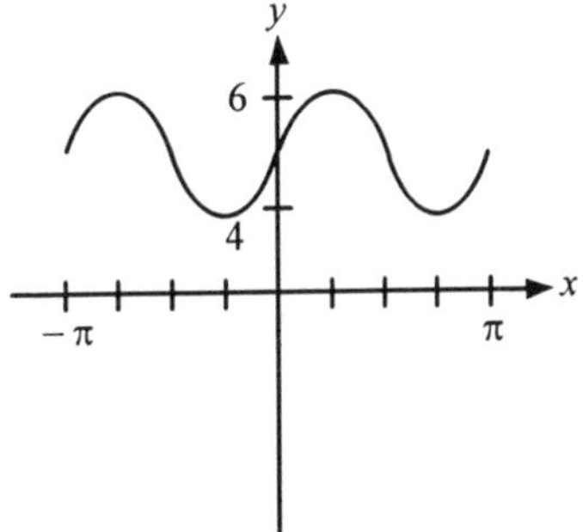

Amplitude: 1
Period: π
Midline: $y = 5$

13.

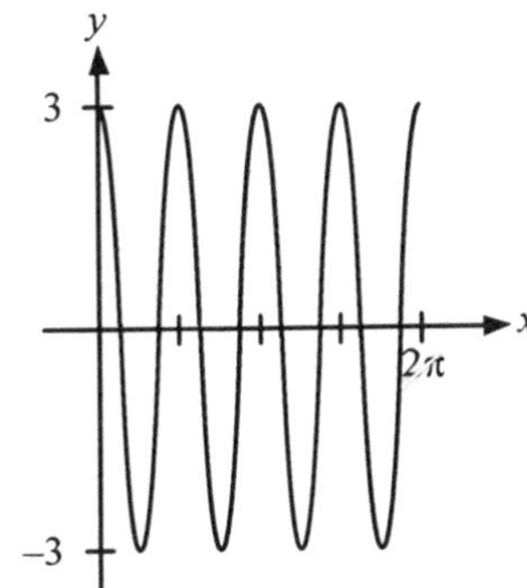

Amplitude: 3
Period: $\frac{\pi}{2}$
Midline: $y = 0$

15.

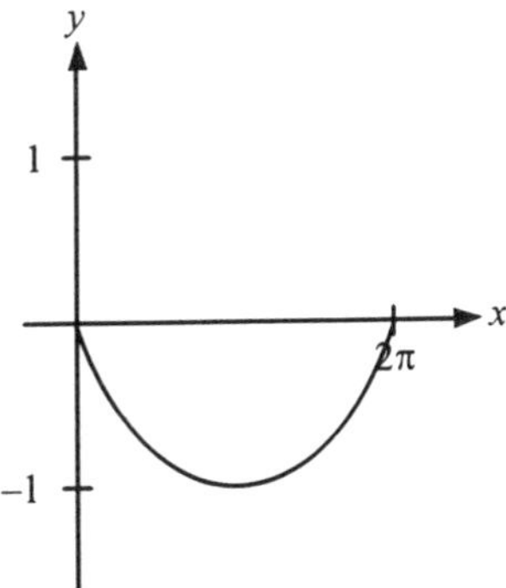

Amplitude: 1
Period: 4π
Midline: $y = 0$

17.

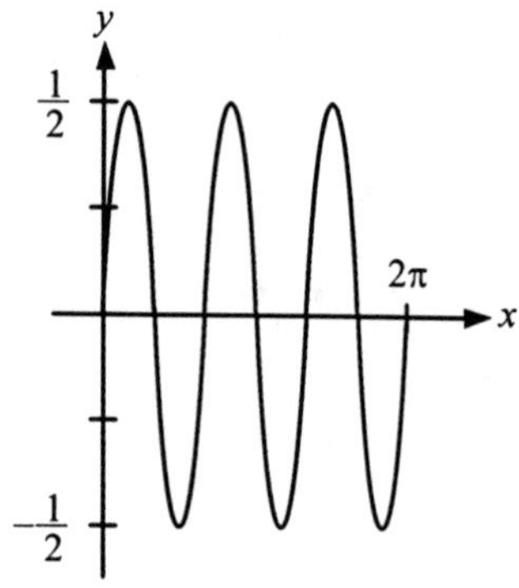

Amplitude: $\frac{1}{2}$

Period: $\frac{2\pi}{3}$

Midline: $y = 0$

19.

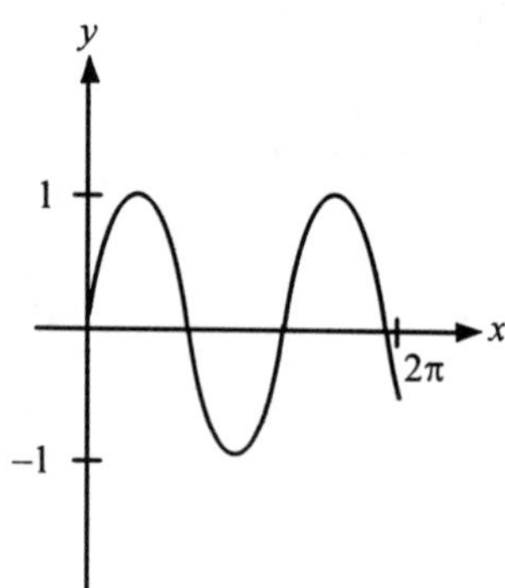

Amplitude: 1
Period: 4
Midline: $y = 0$

21.

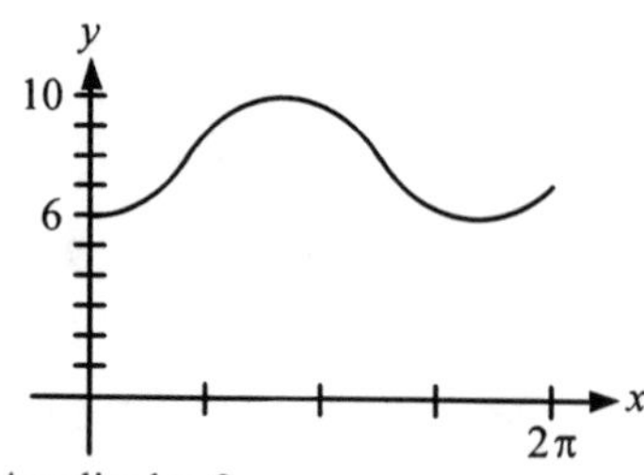

Amplitude: 2
Period: 6
Midline: $y = 8$

23.

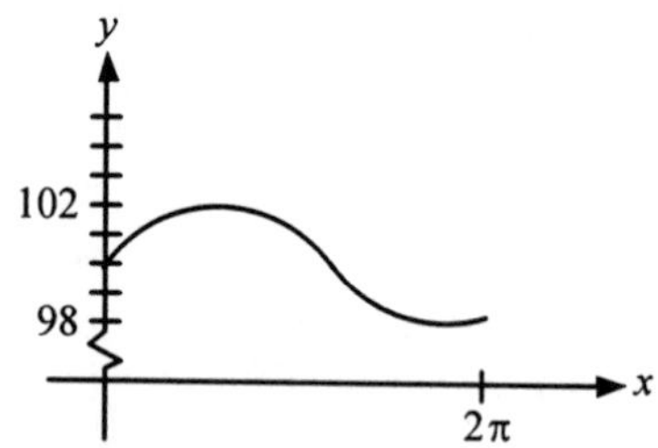

Amplitude: 2
Period: 8
Midline: $y = 100$

25.

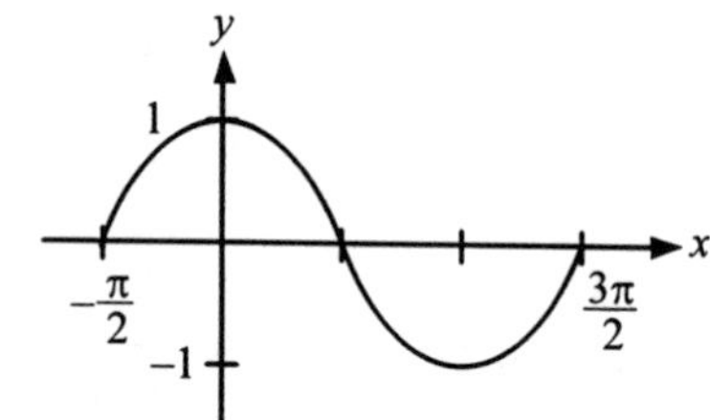

Amplitude: 1
Period: 2π

Phase shift: $-\frac{\pi}{2}$

Midline: $y = 0$

27.

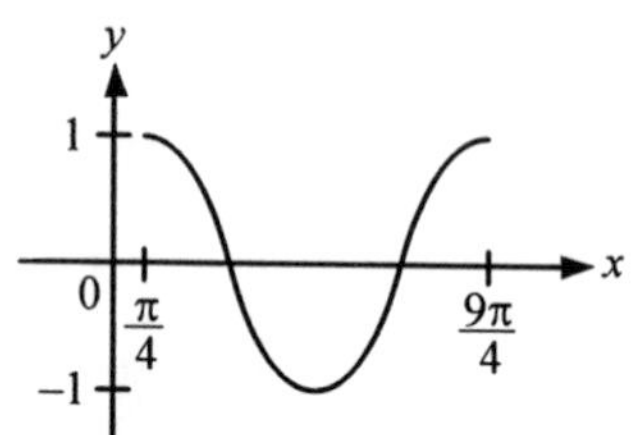

Amplitude: 1
Period: 2π

Phase shift: $\frac{\pi}{4}$

Midline: $y = 0$

29.

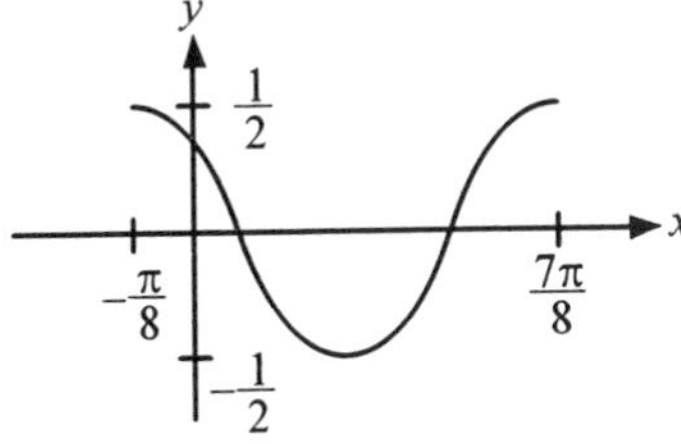

Amplitude: $\frac{1}{2}$

Period: π

Phase shift: $-\frac{\pi}{8}$

Midline: $y = 0$

31.

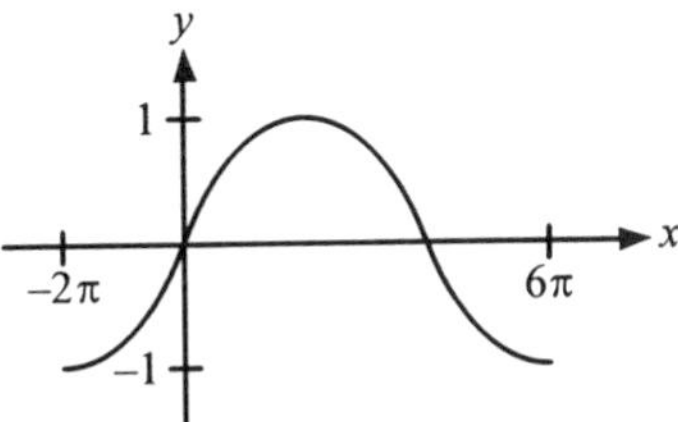

Amplitude: 1
Period: 8π
Phase shift: -2π
Midline: $y = 0$

33.

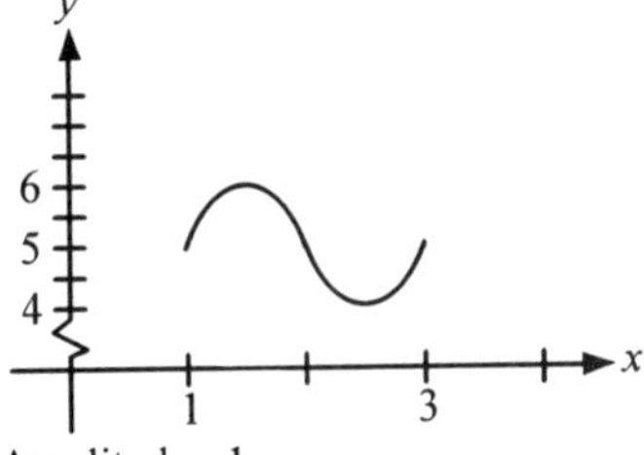

Amplitude: 1
Period: 2
Phase shift: 1
Midline: $y = 5$

35.

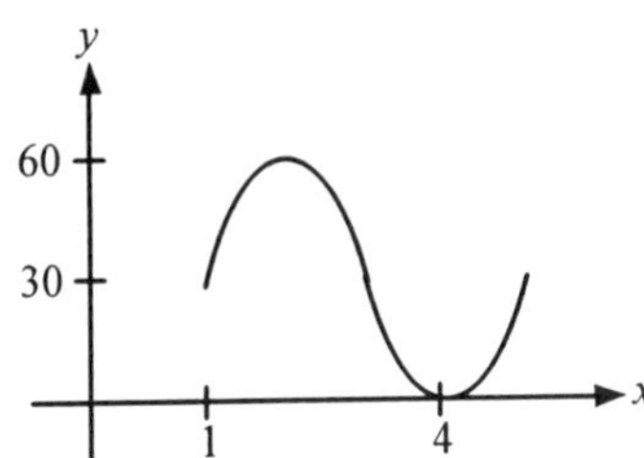

Amplitude: 30
Period: 4
Phase shift: 1
Midline: $y = 30$

37. $y = 3\sin 2x$ **39.** $y = -4\sin 3x$

41. $y = 1.5\cos\frac{1}{2}x$ **43.** $y = -2\cos 10x$

45.

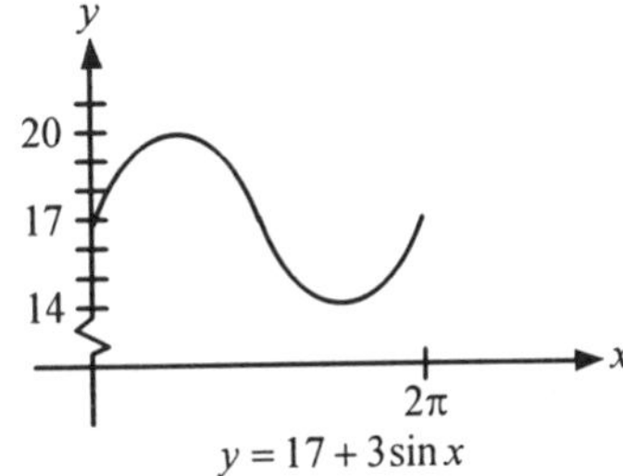

$y = 17 + 3\sin x$

47.

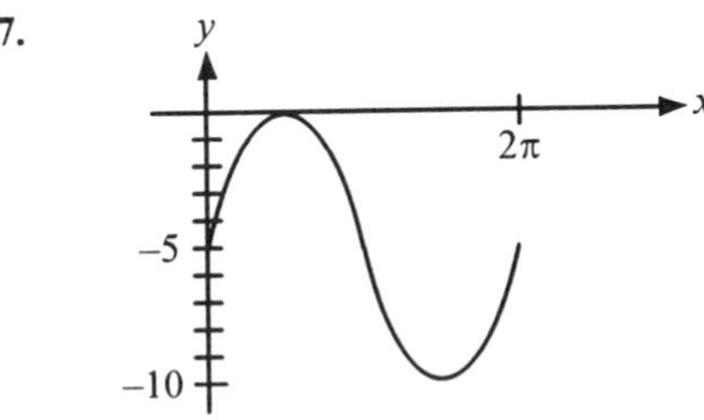

$y = -5 + 5\sin x$

49.

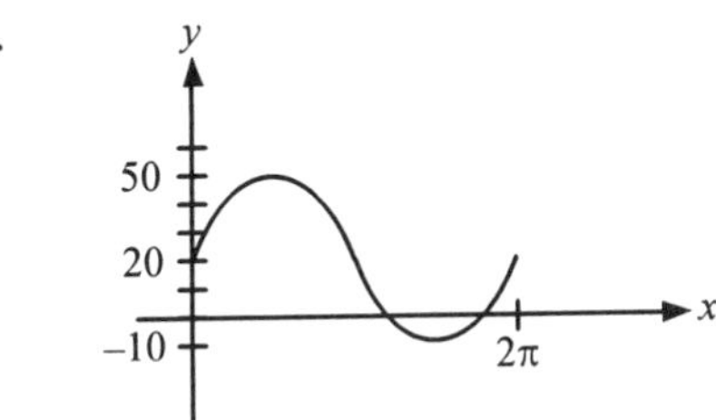

$y = 20 + 30\sin x$

51. $y = 8 + \sin x$ **53.** $y = 4 + 2\sin 2x$

55. $y = 2 - \sin \pi x$

57. **a.**

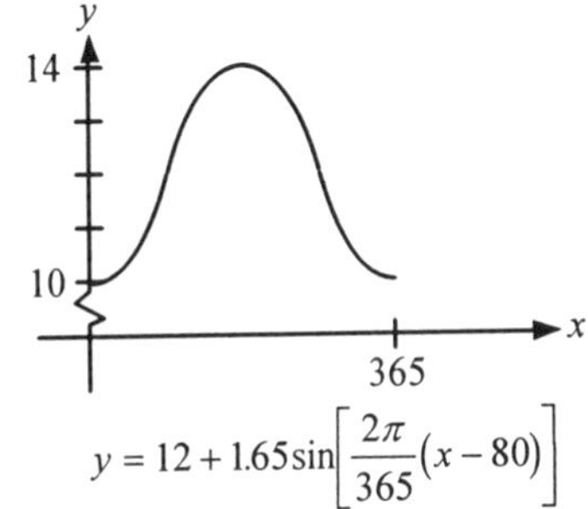

$y = 12 + 1.65\sin\left[\frac{2\pi}{365}(x - 80)\right]$

b. Amplitude $= 1.65$; Period $= 365$ days

c. Days number 171 and 172 have about 13.65 hours of daylight. (About June 20 or June 21)

59. **a.**

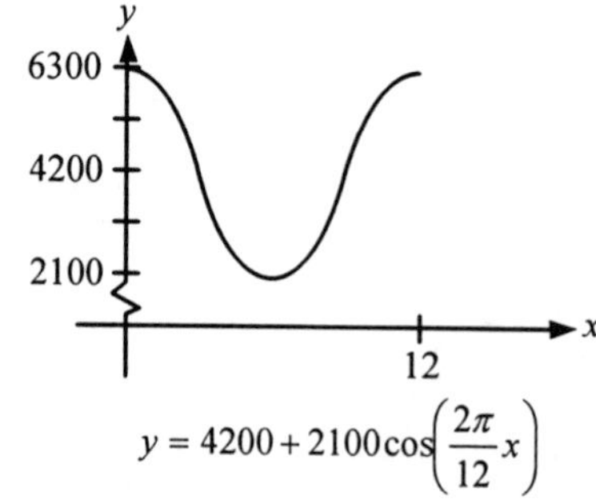

$$y = 4200 + 2100\cos\left(\frac{2\pi}{12}x\right)$$

b. Amplitude $= 2{,}100$; Period $= 12$ months

c. Month 6, (January)

61. **a.**

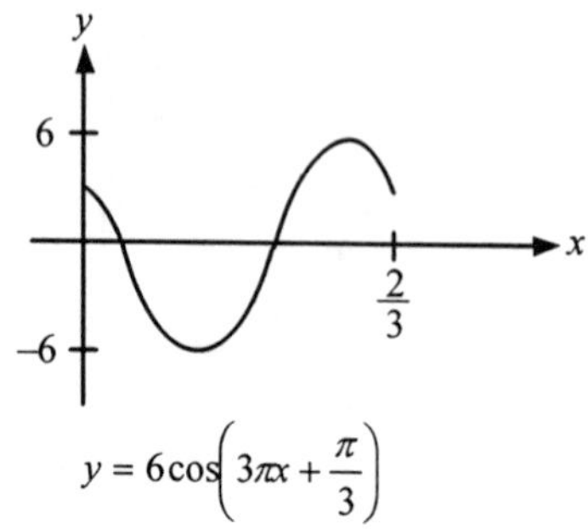

$$y = 6\cos\left(3\pi x + \frac{\pi}{3}\right)$$

b. Amplitude $= 6$; Period $= \frac{2}{3}$

c. Displacement is 6 meters twice each cycle; at $x = \frac{2}{9}$ and $x = \frac{5}{9}$ seconds.

63. **a.**

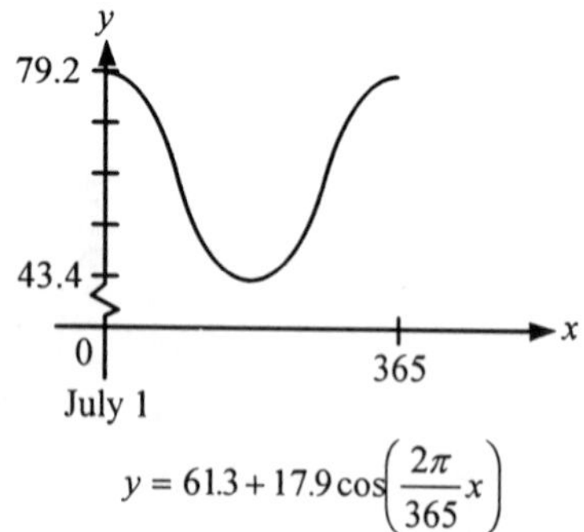

$$y = 61.3 + 17.9\cos\left(\frac{2\pi}{365}x\right)$$

b. Amplitude $= 17.9$; Period $= 365$ days

c. December

65. **a.**

t	0	10	20	30
d	5	$5+\frac{5}{2}\sqrt{3}$	$5+\frac{5}{2}\sqrt{3}$	5
t	40	50	60	
d	$5-\frac{5}{2}\sqrt{3}$	$5-\frac{5}{2}\sqrt{3}$	5	

b.

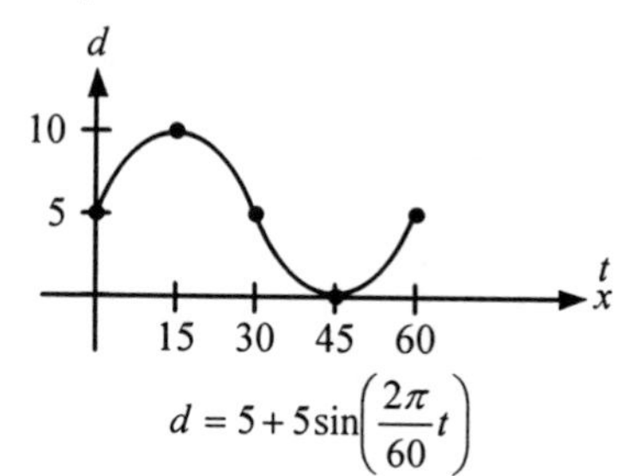

$$d = 5 + 5\sin\left(\frac{2\pi}{60}t\right)$$

c. $d = 5 + 5\sin\left(\frac{2\pi}{60}t\right)$

Exercises 8.4

1.

x	0	$\frac{\pi}{6}$	$\frac{\pi}{3}$	$\frac{\pi}{2}$	$\frac{2\pi}{3}$	$\frac{5\pi}{6}$	π
$\cot x$	und.	1.7	0.6	0	−0.6	−1.7	und.
x	$\frac{7\pi}{6}$	$\frac{4\pi}{3}$	$\frac{3\pi}{2}$	$\frac{5\pi}{3}$	$\frac{11\pi}{6}$	2π	
$\cot x$	1.7	0.6	0	−0.6	−1.7	und.	

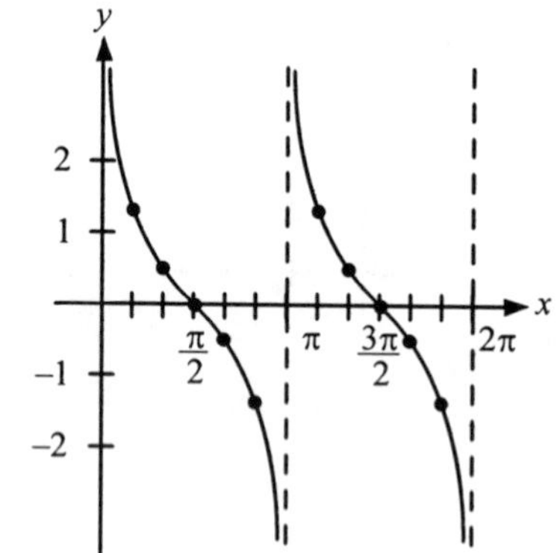

3.

x	0	$\frac{\pi}{6}$	$\frac{\pi}{3}$	$\frac{\pi}{2}$	$\frac{2\pi}{3}$	$\frac{5\pi}{6}$	π
$\csc x$	und.	2	1.2	1	1.2	2	und.
x	$\frac{7\pi}{6}$	$\frac{4\pi}{3}$	$\frac{3\pi}{2}$	$\frac{5\pi}{3}$	$\frac{11\pi}{6}$	2π	
$\csc x$	−2	−1.2	−1	−1.2	−2	und.	

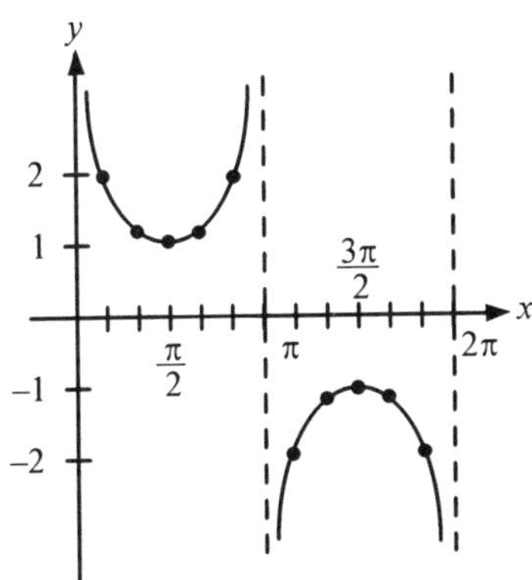

5. D: set of all real numbers except $k = k\pi$ (k any integer); R: $(-\infty, \infty)$; period: π

7. D: set of all real numbers except $x = k\pi$ (k any integer); R: $(-\infty, -1] \cup [1, \infty)$; period: 2π

9. Inc., dec., dec., inc.

11. Always increasing

13. Inc., inc., dec., dec.

15.

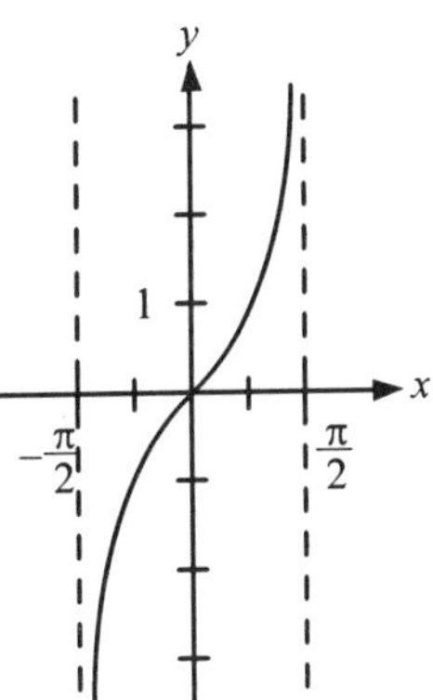

17.

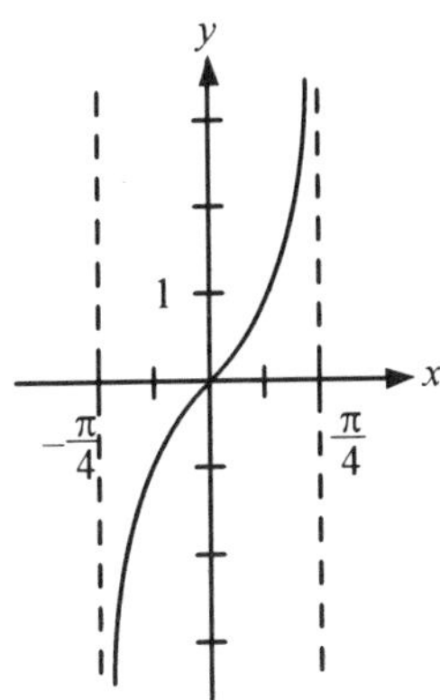

19.

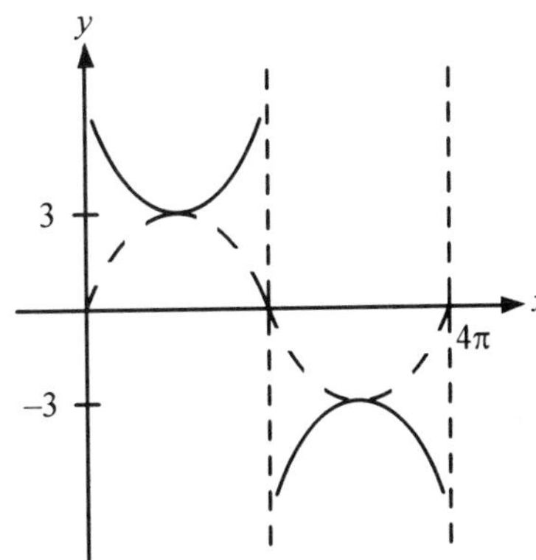

21.

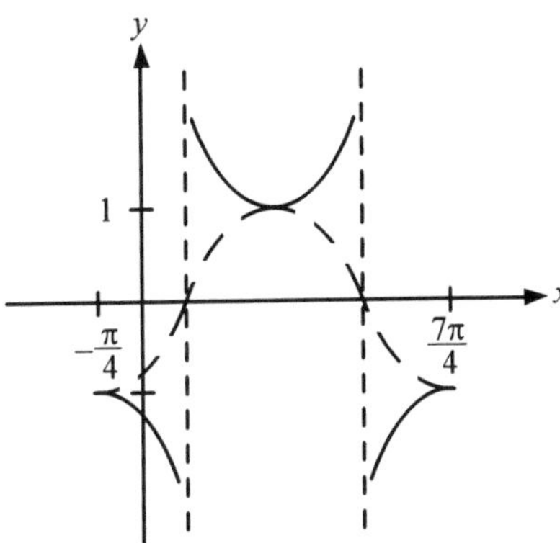

23.

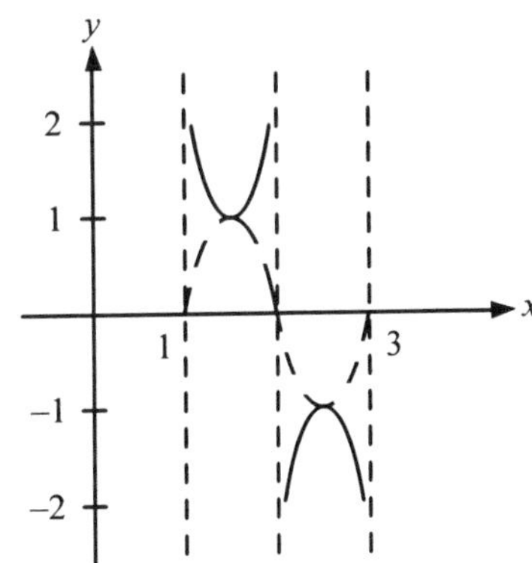

25.

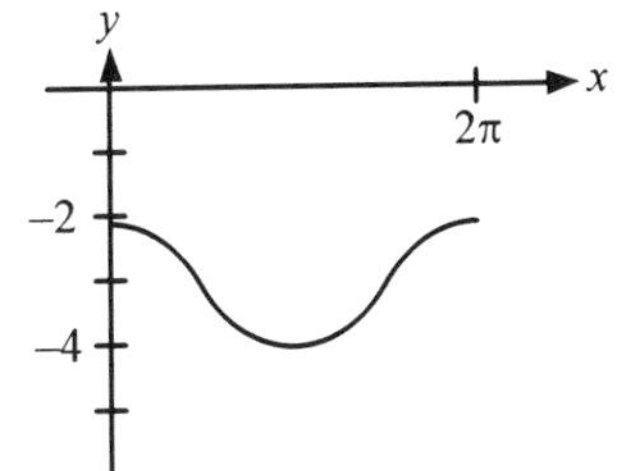

27.

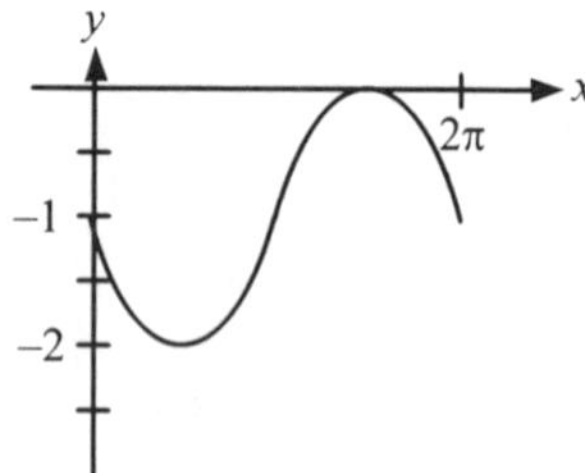

29.

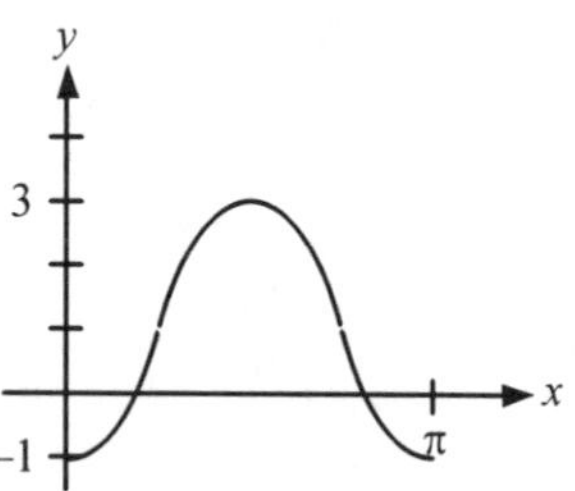

31.

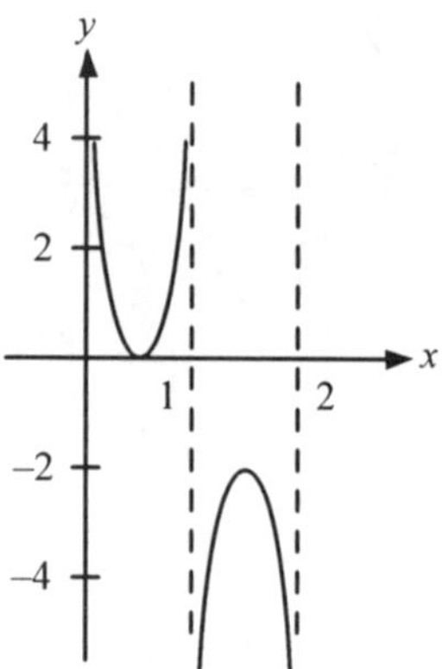

33.

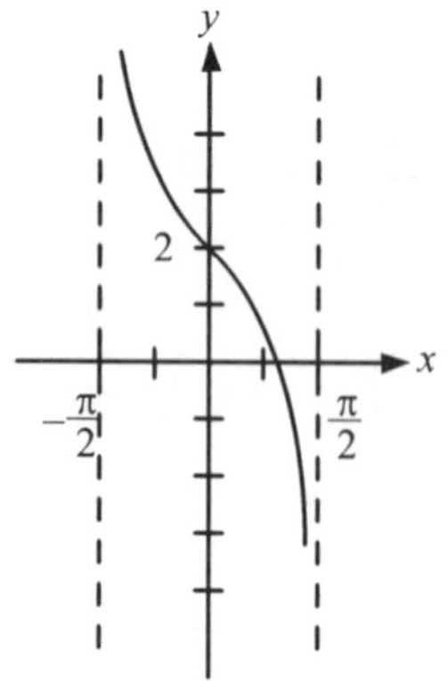

35.

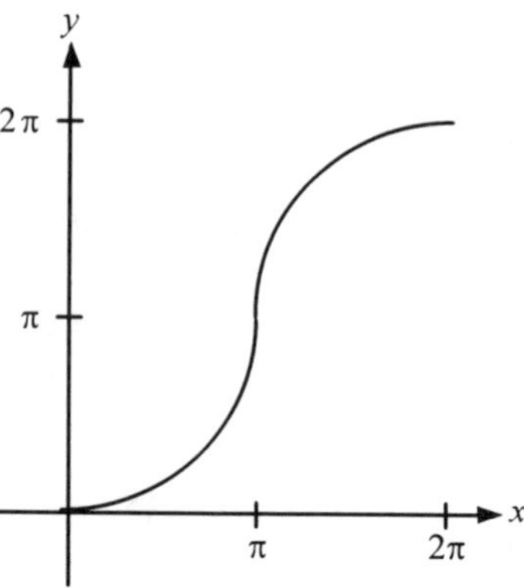

37.

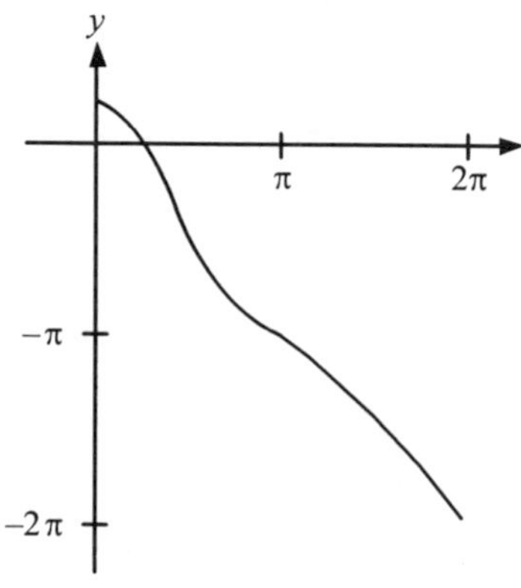

39.

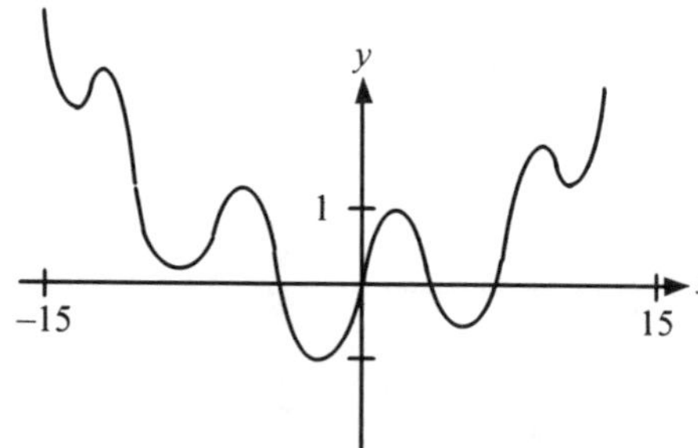

41.

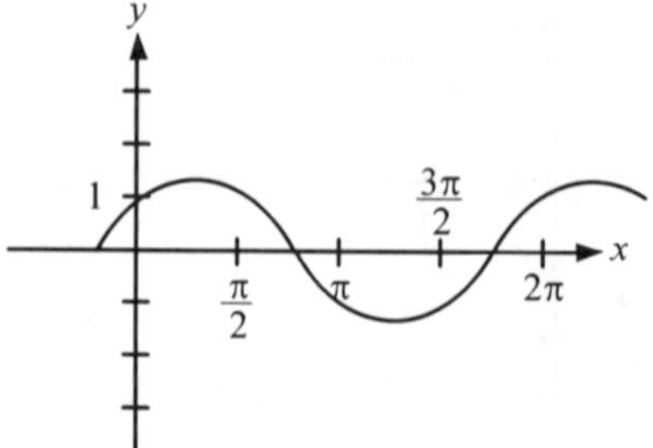

Amplitude: 1.4
Period: 2π

43.

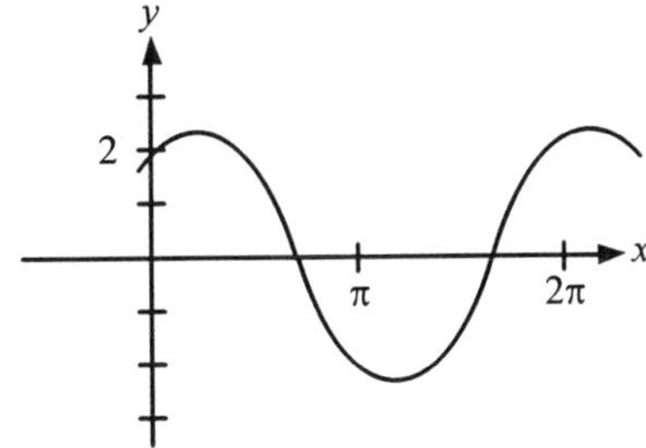

Amplitude: 2.24
Period: 2π

45.

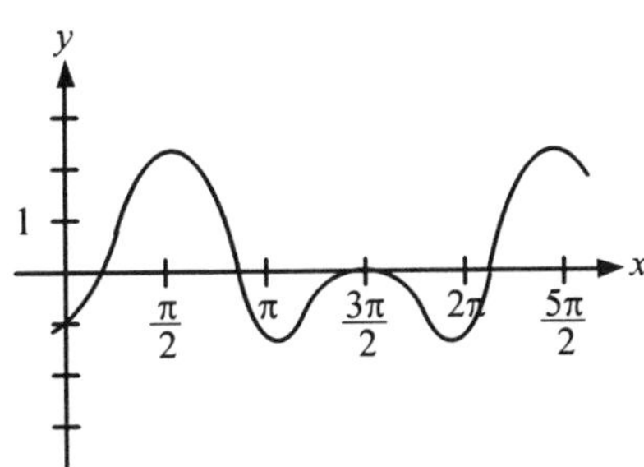

Amplitude: 1.56
Period: 2π

47. **a.**

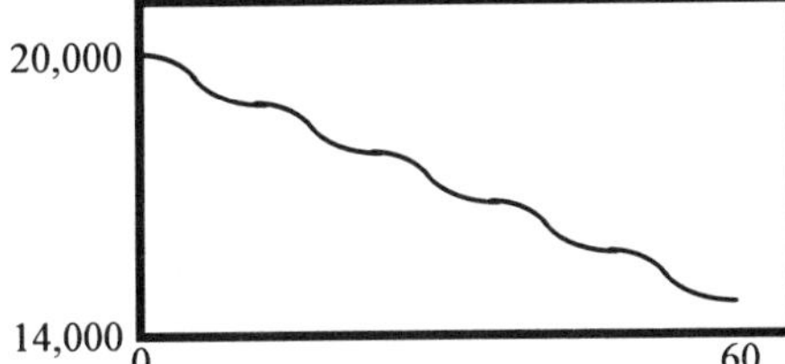

b. February; 19,100

49. **a.**

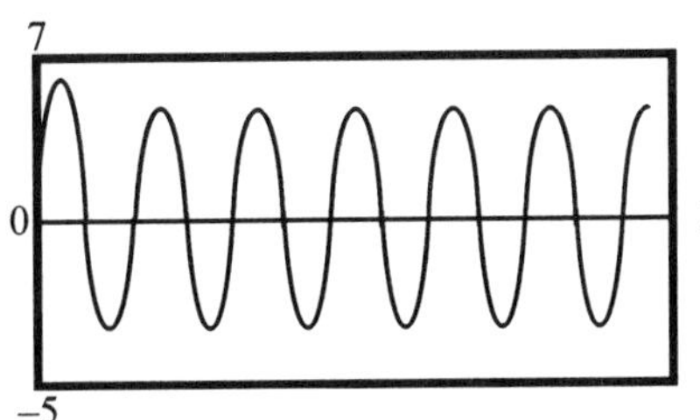

b. 6.55; 4.90

c. $4e^{-10x} < .1$; $x > .37$; $4e^{-10x} < .01$; $x > .60$

Exercises 8.5

1. $\frac{\pi}{3}$ **3.** $-\frac{\pi}{3}$ **5.** $\frac{\pi}{4}$

7. $-\frac{\pi}{2}$ **9.** $\frac{\pi}{4}$ **11.** $-\frac{\pi}{3}$

13. 0.32 **15.** 2.15 **17.** 1.05

19. 0.64 **21.** 1 **23.** $\frac{1}{2}$

25. 0 **27.** 0 **29.** –0.6616

31. $\frac{3}{5}$ **33.** $\frac{12}{5}$ **35.** $\frac{1}{\sqrt{x^2+1}}$

37. $\frac{\sqrt{1-x^2}}{x}$ **39.** $\frac{x-2}{\sqrt{9-(x-2)^2}}$

41. $\arcsin x$ **43.** $\operatorname{arcsec}\left(\frac{x-2}{3}\right)$

45. $-\frac{\sqrt{x^2+4}}{4x}$ **47.** $-\frac{x+2}{\sqrt{4(x+2)^2-1}}$

49. $\theta = \arcsin\left(\frac{1}{m}\right)$ **51.** $\theta = \arcsin\left(\frac{Tg}{2v_0}\right)$

53. $\left[-\frac{\pi}{2}, \frac{\pi}{2}\right]$ **55.** $[0, \pi]$ **57.** $(-\infty, \infty)$

59. 0 **61.** –1 **63.** $\frac{\pi}{2}$

65. –1 **67.** No value

69.

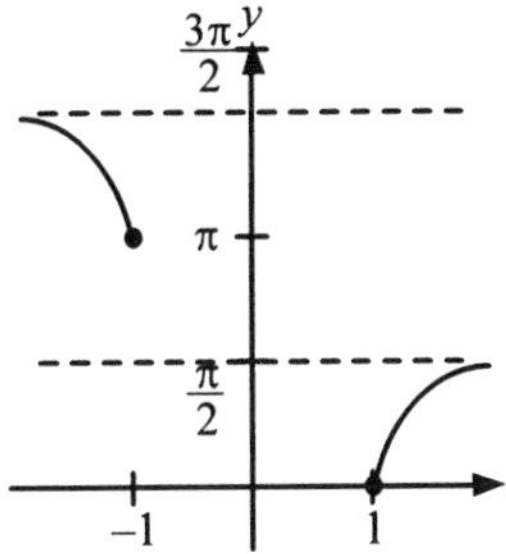

71.

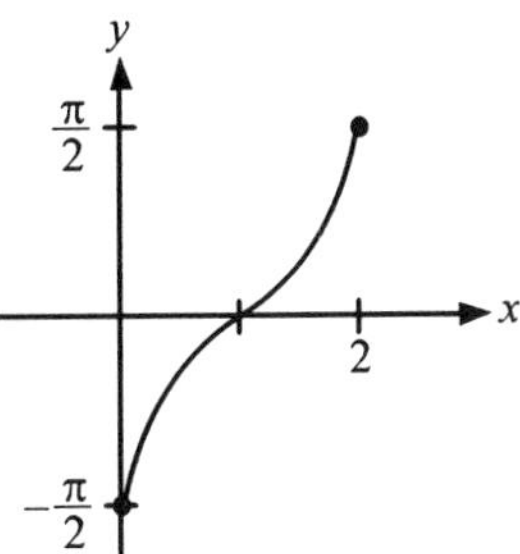

73. **a.** $\theta = \arccos\left(\frac{a}{25}\right)$

b. $\theta = \arcsin\left(\frac{b}{25}\right)$

75. $\dfrac{\pi}{4}$

77. $s = t\cot\theta$, D: $(0°, 90°)$, R: $(0, \infty)$

79. $\dfrac{\sqrt{5}}{3}$

81. **a.** $\dfrac{1}{440}$ second; .001 in.

b.

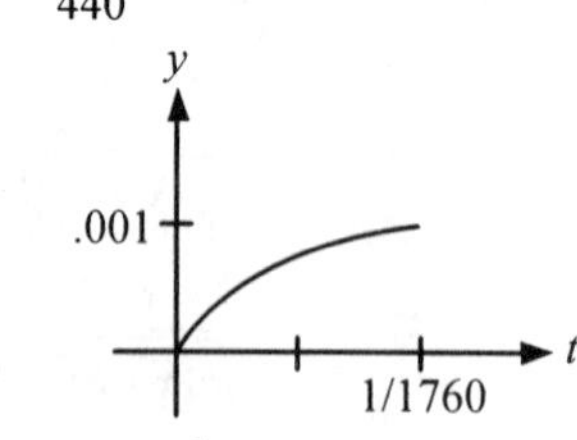

c. $t = \dfrac{\sin^{-1}(1000y)}{880\pi}$

Exercises 8.6

1. $\left\{\dfrac{\pi}{3}, \dfrac{2\pi}{3}\right\}$ **3.** $\left\{\dfrac{2\pi}{3}, \dfrac{4\pi}{3}\right\}$

5. $\left\{\dfrac{3\pi}{4}, \dfrac{7\pi}{4}\right\}$ **7.** $\{0.12, 3.02\}$

9. $\{1.88, 5.02\}$ **11.** $\left\{\dfrac{\pi}{4}, \dfrac{7\pi}{4}\right\}$

13. $\emptyset$ **15.** $\{0.32, 3.46\}$

17. $\{2.55, 5.70\}$ **19.** $\{1.77, 4.51\}$

21. $\left\{\dfrac{\pi}{4}, \dfrac{3\pi}{4}, \dfrac{5\pi}{4}, \dfrac{7\pi}{4}\right\}$ **23.** $\left\{0, \dfrac{\pi}{2}, \dfrac{3\pi}{2}\right\}$

25. $\left\{\dfrac{\pi}{6}, \dfrac{\pi}{2}, \dfrac{5\pi}{6}, \dfrac{3\pi}{2}\right\}$ **27.** $\{1.25, 1.71, 4.39, 4.85\}$

29. $\left\{\dfrac{\pi}{6}, \dfrac{\pi}{2}, \dfrac{5\pi}{6}, \dfrac{7\pi}{6}, \dfrac{3\pi}{2}, \dfrac{11\pi}{6}\right\}$

31. $\left\{\dfrac{\pi}{12}, \dfrac{5\pi}{12}, \dfrac{13\pi}{12}, \dfrac{17\pi}{12}\right\}$

33. $\left\{\dfrac{3\pi}{4}\right\}$ **35.** $\left\{0, \dfrac{\pi}{4}, \dfrac{\pi}{2}, \pi, \dfrac{5\pi}{4}, \dfrac{3\pi}{2}\right\}$

37. $\left\{x\!: x = \dfrac{\pi}{6} + k2\pi \text{ or } x = \dfrac{5\pi}{6} + k2\pi, k \text{ any integer}\right\}$

39. $\{x\!: x = 3.60 + k2\pi \text{ or } x = 5.82 + k2\pi, k \text{ any integer}\}$

41. $\{x\!: x = 1.249 + k\pi, k \text{ any integer}\}$

43. $\{x\!: x = 0.398 + k\pi \text{ or } x = 2.744 + k\pi, k \text{ any integer}\}$

45. $\emptyset$

47. $\left\{x\!: x = k\dfrac{\pi}{2} \text{ or } x = \dfrac{\pi}{6} + k2\pi \text{ or } x = \dfrac{5\pi}{6} + k2\pi,\right.$ $\left. k \text{ any integer}\right\}$

49. Days 57, 58, 59 and 306, 307, 308; August 27, 28, 29 and May 2, 3, 4

51. 0, 1, 2, 10, 11, 12; May through October

53. 43° and 47°

Exercises 8.7

41. $f(x) = \sin x$

43. $f(x) = 2$, when $\cos x \neq 0$

45. $f(x) = \dfrac{\sin 2x}{2}$

47. $\cos\theta$ **49.** $3\sec\theta$ **51.** $\tan\theta$

53. $\cos\theta$ **55.** $27\tan^3\theta$ **57.** $\dfrac{13}{12}$

59. $-\dfrac{12}{5}$ **61.** $-\sqrt{1-a^2}$ **63.** $\dfrac{1}{a}$

65. $\csc x = \dfrac{17}{8}$, $\cos x = \dfrac{15}{17}$, $\sec x = \dfrac{17}{15}$, $\tan x = \dfrac{8}{15}$, $\cot x = \dfrac{15}{8}$

67. $\cos x = -\dfrac{3}{4}$, $\sin x = \dfrac{\sqrt{7}}{4}$, $\csc x = \dfrac{4\sqrt{7}}{7}$, $\tan x = -\dfrac{\sqrt{7}}{3}$, $\cot x = -\dfrac{3\sqrt{7}}{7}$

69. $\pm\sqrt{1-\sin^2 x}$ **71.** $\pm\dfrac{1}{\sqrt{1-\sin^2 x}}$

73. $\pm\sqrt{1-\cos^2 x}$ **75.** $\dfrac{1}{\cos x}$

77. $\left\{\dfrac{\pi}{4}, \dfrac{5\pi}{4}\right\}$ **79.** $[0, \pi]$ **81.** $\left\{\dfrac{\pi}{4}, \dfrac{5\pi}{4}\right\}$

83. $\left\{\dfrac{\pi}{6}, \dfrac{5\pi}{6}\right\}$ **85.** $\left\{\dfrac{\pi}{2}, \dfrac{7\pi}{6}, \dfrac{11\pi}{6}\right\}$

87. $\left\{\dfrac{7\pi}{6}, \dfrac{3\pi}{2}, \dfrac{11\pi}{6}\right\}$ **89.** $\{1.84, 2.20, 4.98, 5.34\}$

91. **a.**

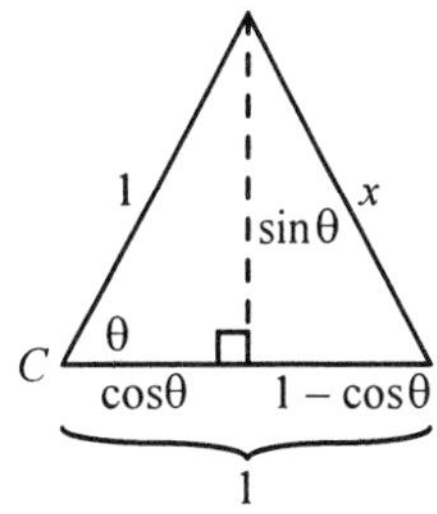

$\sin^2\theta + (1-\cos\theta)^2 = x^2$ by Pythagorean Theorem

Exercises 8.8

31. $\dfrac{\sqrt{2}+\sqrt{6}}{4}$ **33.** $\dfrac{\sqrt{2}-\sqrt{6}}{4}$

35. $\dfrac{\sqrt{6}-\sqrt{2}}{4}$ **37.** $2-\sqrt{3}$

39. **a.** $\dfrac{\sqrt{3}}{2}$ **b.** $\dfrac{1}{2}$

c. -1 **d.** $-\dfrac{\sqrt{3}}{2}$

41. **a.** $\dfrac{\sqrt{3}}{2}$ **b.** $-\dfrac{\sqrt{3}}{2}$

c. 0 **d.** $\dfrac{1}{2}$

43. $\dfrac{36}{325}$ **45.** $\dfrac{253}{325}$ **47.** $-\dfrac{36}{323}$

49. $\dfrac{117}{125}$ **51.** $-\dfrac{63}{65}$

65. **b.**

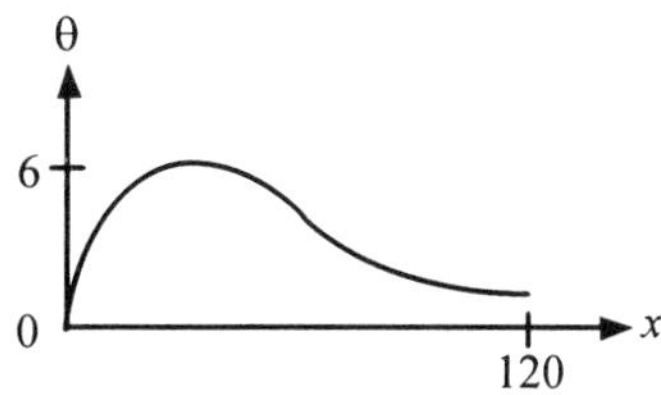

c. 54.77 ft

67. **b.**

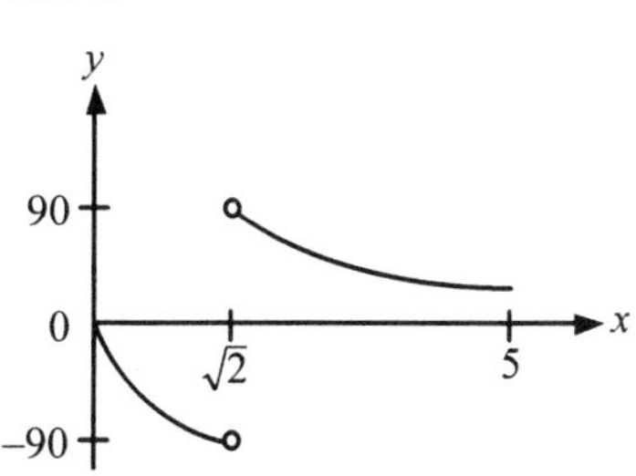

c. $-71.6° + 180° = 108.4°$

Exercises 8.9

1. $\sin 2x = \dfrac{24}{25},\ \cos 2x = -\dfrac{7}{25},\ \tan 2x = -\dfrac{24}{7}$

3. $\sin 2x = -\dfrac{3\sqrt{7}}{8},\ \cos 2x = -\dfrac{1}{8},\ \tan 2x = 3\sqrt{7}$

5. $\sin 2x = \dfrac{4}{5},\ \cos 2x = \dfrac{3}{5},\ \tan 2x = \dfrac{4}{3}$

17. **a.** $\left\{0, \dfrac{2\pi}{3}, \dfrac{4\pi}{3}\right\}$ **b.** $\left\{0, \dfrac{\pi}{3}, \pi, \dfrac{5\pi}{3}\right\}$

19. $\left\{0, \dfrac{2\pi}{3}, \pi, \dfrac{4\pi}{3}\right\}$ **21.** $\left\{\dfrac{7\pi}{12}, \dfrac{11\pi}{12}, \dfrac{19\pi}{12}, \dfrac{23\pi}{12}\right\}$

23. $\left\{0, \dfrac{2\pi}{3}, \dfrac{4\pi}{3}\right\}$ **25.** $\left\{\dfrac{\pi}{8}, \dfrac{5\pi}{8}, \dfrac{9\pi}{8}, \dfrac{13\pi}{8}\right\}$

37. $\sin\left(\dfrac{x}{2}\right) = \dfrac{\sqrt{17}}{17},\ \cos\left(\dfrac{x}{2}\right) = -\dfrac{4\sqrt{17}}{17},\ \tan\left(\dfrac{x}{2}\right) = -\dfrac{1}{4}$

39. $\sin\left(\dfrac{x}{2}\right) = \dfrac{\sqrt{8+2\sqrt{15}}}{4},\ \cos\left(\dfrac{x}{2}\right) = -\dfrac{\sqrt{8-2\sqrt{15}}}{4},$

$\tan\left(\dfrac{x}{2}\right) = -4-\sqrt{15}$

41. $\dfrac{\sqrt{2+\sqrt{3}}}{2}$ **43.** $\dfrac{\sqrt{2+\sqrt{2}}}{2}$

45. $2-\sqrt{3}$

47. $\sin\dfrac{\pi}{12} = \sin\dfrac{\pi/6}{2} = \sqrt{\dfrac{1-\cos(\pi/6)}{2}} = \sqrt{\dfrac{1-(\sqrt{3}/2)}{2}}$

$= \sqrt{\dfrac{2-\sqrt{3}}{4}} = \dfrac{\sqrt{2-\sqrt{3}}}{2}$

49. $\tan\dfrac{7\pi}{12} = \tan\dfrac{7\pi/6}{2} = \dfrac{1-\cos(7\pi/6)}{\sin(7\pi/6)}$

$= \dfrac{1-(-\sqrt{3}/2)}{-1/2} = -2-\sqrt{3}$

51. $\dfrac{b^2-a^2}{c^2}$ **53.** $\dfrac{2ab}{b^2-a^2}$

55. $-2a\sqrt{1-a^2}$ **57.** $\dfrac{-2a\sqrt{1-a^2}}{2a^2-1}$

59. $-\sqrt{\dfrac{1+a}{2}}$ **61.** $\cos 12t$

63. $\tan 2x$ **65.** $\cos 5\theta$

67. $\frac{1}{2}\sin 4x$ **69.** $4\cos 18t$

85. $\sin 3x = 3\sin x - 4\sin^3 x$

87. **b.** 90° **c.** $V = \frac{Ls^2}{2}\sin\theta$

89. **b.** $A = h\sin\frac{\theta}{2}$ **91.** Replace x by $\frac{\theta}{2}$.

93. $\frac{10t - 21 + 21t^2}{1+t^2} = 10;\ t = \{-2.1937, 1.2846\};$

$\frac{\theta}{2} = \{.9093, -1.1431\};\ \theta = \{1.82, -2.29\}$

Chapter 8 Review Exercises

1. $\frac{\sqrt{3}}{2}$ **3.** 0 **5.** $-\frac{\sqrt{3}}{3}$

7. $-\frac{\pi}{4}$ **9.** 0 **11.** 0.841

13. –3.08 **15.** 1.04 **17.** No value

19. 1.37

21.

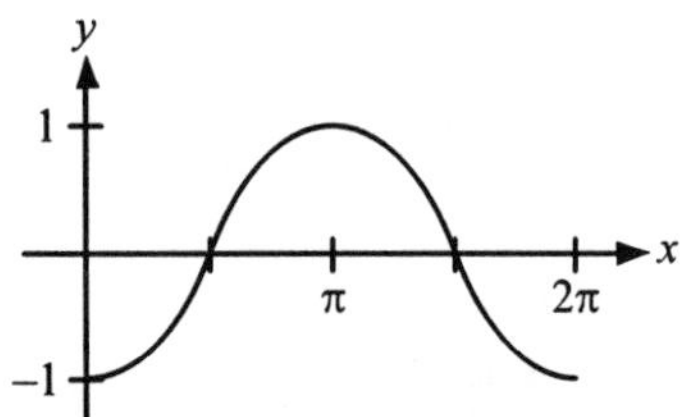

23.

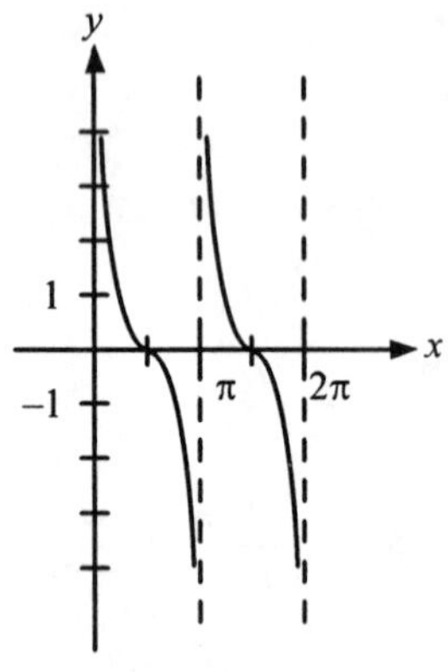

25.

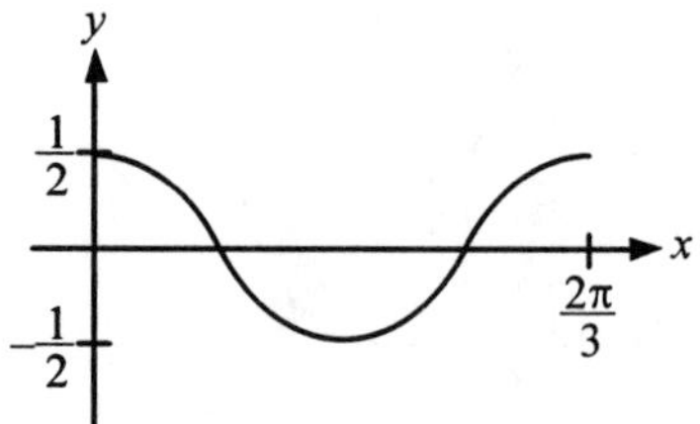

Amplitude: $\frac{1}{2}$

Period: $\frac{2\pi}{3}$

Phase shift: 0

Midline: $y = 0$

27.

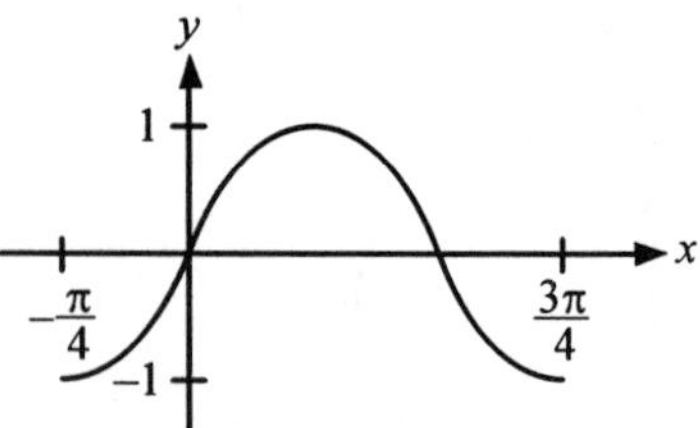

Amplitude: 1

Period: π

Phase shift: $-\frac{\pi}{4}$

Midline: $y = 0$

29.

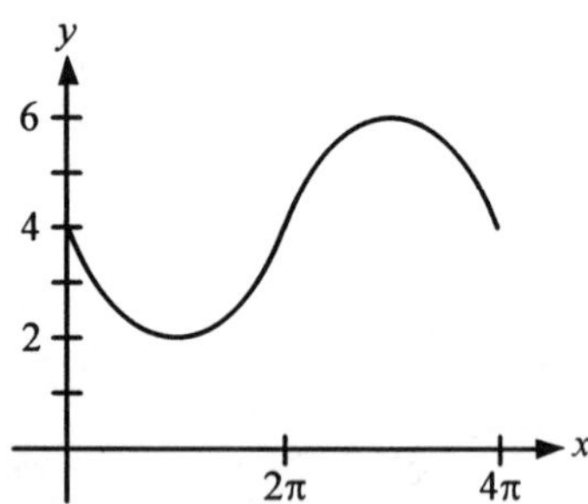

Amplitude: 2

Period: 4π

Phase shift: 0

Midline: $y = 4$

31.

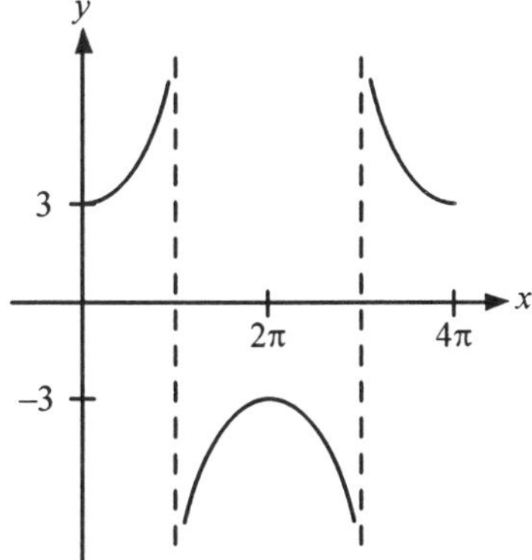

33.

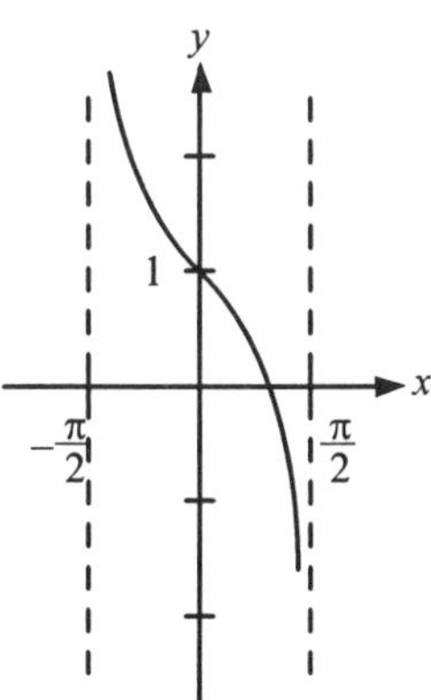

35.

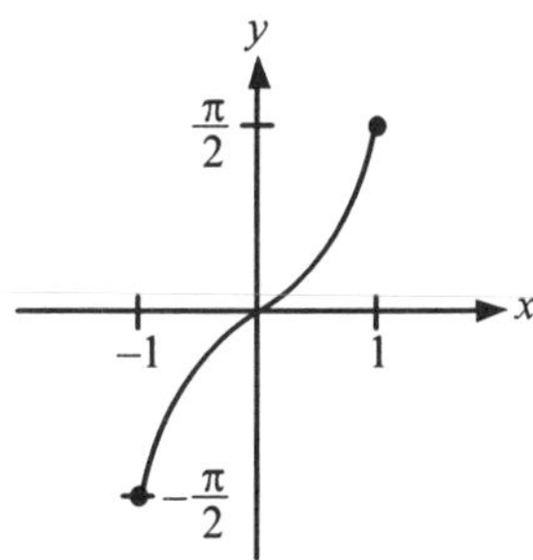

37. Domain: set of all real numbers except $x = \frac{\pi}{2} + k\pi$ (k any integer); range: $(-\infty, \infty)$

39. Domain: set of all real numbers except $x = k\pi$ (k any integer); range: $(-\infty, -1] \cup [1, \infty)$

41. Domain: $(-\infty, \infty)$; range: $\left(-\frac{\pi}{2}, \frac{\pi}{2}\right)$

43. Q_2 **45.** $y = 4 - \sin(2\pi x)$

47. $\theta = \tan^{-1}\frac{3}{x}$ **49.** 25.7 ft

51. **a.** $\frac{7\pi}{6}$ **b.** 72°

53. $[-1, 1]$ **55.** 2

57. $\frac{9}{2}\arcsin\left(\frac{x}{3}\right) - \frac{x\sqrt{9-x^2}}{4}$

77. $-\frac{3}{5}$ **79.** $\frac{5}{4}$ **81.** $\frac{7}{25}$

83. $-\frac{24}{7}$ **85.** $\frac{\sqrt{10}}{10}$

87. Conditional equation

89. $\frac{7}{8}$ **91.** $\frac{a}{1+2a^2}$ **93.** $5\sec\theta$

95. $\sin x$ **97.** $\frac{117}{125}$

99.
$$\begin{aligned}\sin(x_1 - x_2) &= \sin[x_1 + (-x_2)]\\ &= \sin x_1 \cos(-x_2) + \cos x_1 \sin(-x_2)\\ &= \sin x_1 \cos x_2 + \cos x_1(-\sin x_2)\\ &= \sin x_1 \cos x_2 - \cos x_1 \sin x_2\end{aligned}$$

101. $\left\{\frac{7\pi}{6}, \frac{11\pi}{6}\right\}$ **103.** $\left\{\frac{\pi}{6}, \frac{5\pi}{6}, \frac{\pi}{2}\right\}$

105. $\left\{0, \frac{\pi}{3}, \pi, \frac{5\pi}{3}\right\}$

107. $\left\{\theta\colon\ \theta = \frac{\pi}{6} + k\pi \text{ or } \theta = \frac{5\pi}{6} + k\pi,\ k \text{ any integer}\right\}$

109. b **111.** c **113.** b

115. a **117.** a **119.** b

Chapter 8 Test

1. $\left[-\frac{\pi}{2}, \frac{\pi}{2}\right]$

2. Set of all real numbers except $x = \frac{\pi}{2} + k\pi$ (k any integer)

3. $-\sqrt{3}$

4.

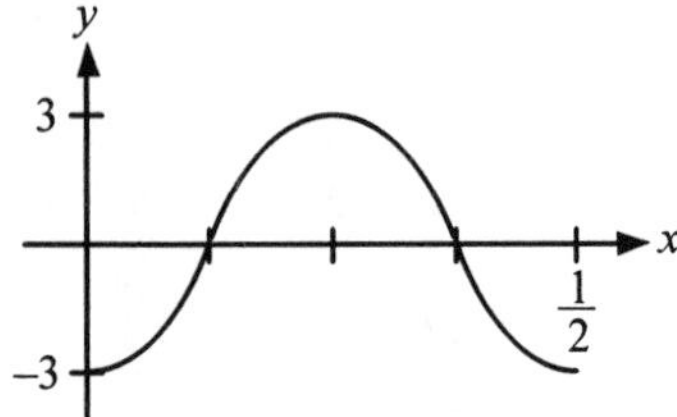

5.

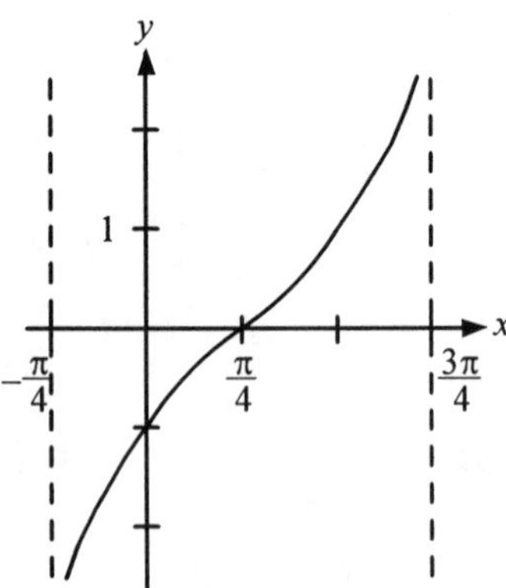

6.

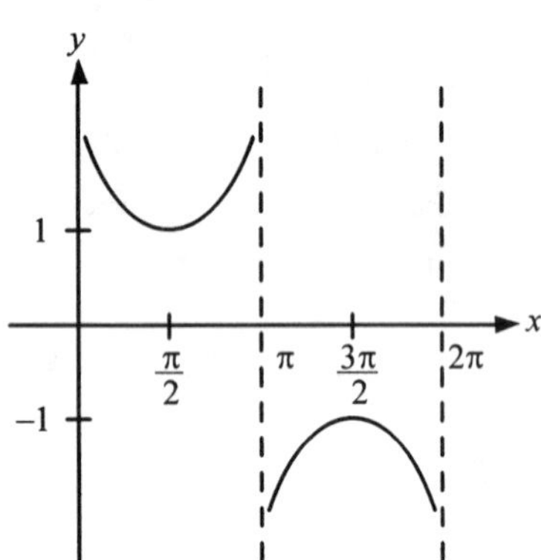

7. Amplitude: 1; period: $\dfrac{2\pi}{3}$; phase shift: $\dfrac{\pi}{6}$

8. **a.** $\dfrac{3\pi}{2}$ **b.** 315°

9. $\dfrac{25\pi}{3}$cm

10. $-\dfrac{\pi}{3}$

11. $\dfrac{\sqrt{1-x^2}}{x}$

12. $\left[-\dfrac{\pi}{2}, \dfrac{\pi}{2}\right]$

13. $\dfrac{1+\cot^2 x}{\cot^2 x} = \dfrac{\csc^2 x}{\cot^2 x} = \dfrac{1/\sin^2 x}{\cos^2 x/\sin^2 x} = \dfrac{1}{\cos^2 x} = \sec^2 x$

14. $\cos^2 x$

15. $\cos(x-\pi) = \cos x \cos \pi + \sin x \sin \pi$
$= (-1)\cos x + (0)\sin x = -\cos x$

16. $\left\{x\colon\ x = \dfrac{7\pi}{6} + k2\pi \text{ or } x = \dfrac{11\pi}{6} + k2\pi,\ k \text{ any integer}\right\}$

17. $\left\{0, \dfrac{2\pi}{3}, \dfrac{4\pi}{3}\right\}$

18. $-\dfrac{12}{5}$

19. $\dfrac{1-\tan^2 x}{1+\tan^2 x} = \dfrac{1-\left(\sin^2 x/\cos^2 x\right)}{1+\left(\sin^2 x/\cos^2 x\right)} = \dfrac{\cos^2 x - \sin^2 x}{\cos^2 x + \sin^2 x}$
$= \dfrac{\cos^2 x - \sin^2 x}{1} = \cos 2x$

20. $\left\{\dfrac{\pi}{12}, \dfrac{7\pi}{12}, \dfrac{3\pi}{4}, \dfrac{5\pi}{4}, \dfrac{17\pi}{12}, \dfrac{23\pi}{12}\right\}$

Exercises 9.1

1. $b = 35$ ft, $c = 48$ ft, $C = 105°$

3. $b = 140$ ft, $c = 1\overline{0}0$ ft, $A = 20°$

5. $b = 34.6$ ft, $c = 23.5$ ft, $C = 41°20'$

7. $\dfrac{3}{4}$

9. 6

11. $\dfrac{7}{2}$

13. $10\sqrt{2}$

15. $b = 10$, $c = 10\sqrt{3}$

17. $c = 110$ ft, $B = 26°$, $C = 109°$

19. $c = 57$ ft, $A = 42°$, $C = 108°$ or $c = 12$ ft, $A = 138°$, $C = 12°$

21. $a = 26$ ft, $A = 8°$, $B = 22°$

23. Two triangles

25. One triangle

27. No triangle

29. 433 ft

31. 95 mi

33. Since $\sin C = \sin 90° = 1$, $\dfrac{\sin A}{\sin 90°} = \dfrac{a}{c}$ becomes $\sin A = \dfrac{a}{c}$ and $\dfrac{\sin B}{\sin 90°} = \dfrac{b}{c}$ becomes $\sin B = \dfrac{b}{c}$.

Exercises 9.2

1. 1,300 ft

3. 31 m

5. 5.4°

7. $c = 14$ ft, $A = 49°$, $B = 71°$

9. $b = 27.4$ ft, $A = 162°30'$, $C = 7°10'$

11. $c = 23.1$ ft, $A = 55°50'$, $B = 28°30'$

13. $A = 128°$, $B = 20°$, $C = 32°$

15. $A = 37°40'$, $B = 93°30'$, $C = 48°50'$

17. $r^2 = s^2 + t^2 - 2st\cos R$

19. $\cos T = \dfrac{r^2 + s^2 - t^2}{2rs}$

21. $\dfrac{1}{5}$ **23.** $\sqrt{37}$

25. $a = 11$ ft, $c = 43$ ft, $B = 78°$

27. $c = 78$ ft, $A = 26°$, $B = 34°$

29. $b = 243$ ft, $c = 152$ ft, $A = 33°00'$

31. $A = 104°$, $B = 29°$, $C = 47°$

33. $a = 165$ ft, $A = 164°00'$, $C = 8°50'$ or $a = 17.4$ ft, $A = 1°40'$, $C = 171°10'$

35. $A = 46°30'$, $B = 58°00'$, $C = 75°30'$

37. 99 mi **39.** 34 m **41.** 7.0 ft^2

43. 79 m^2 **45.** 2.97 km^2 **47.** 9.8 ft^2

49. 126 m^2 **51.** 2.02 km^2

53. $44\sqrt{3}$ square units **55.** $9\sqrt{3}$ square units

57. Since $\cos C = \cos 90° = 1$, $c^2 = a^2 + b^2 - 2ab\cos 90°$ becomes $c^2 = a^2 + b^2$.

59. 195 mm^2

Exercises 9.3

1. $\sqrt{2}$; 135° **3.** 2; 210° **5.** 1; 90°

7. 13; 293° **9.** 1; 53° **11.** $\langle -1, -2\rangle$

13. $\langle -15, 3\rangle$ **15.** $\langle -30, 17\rangle$ **17.** $\langle 5, -1\rangle$

19. $\langle 9, -4\rangle$ **21.** $\langle -33, 11\rangle$

23.

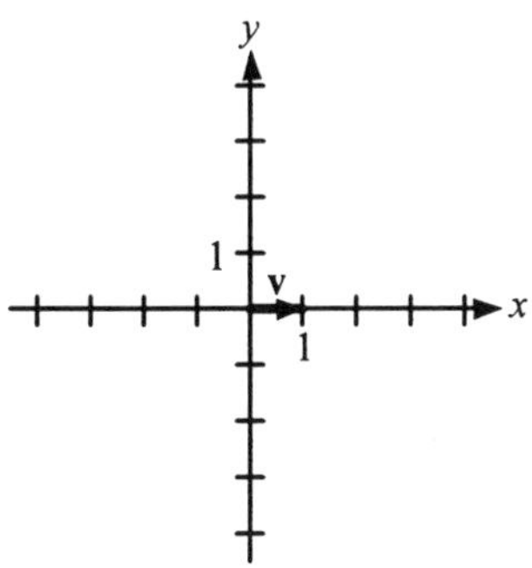

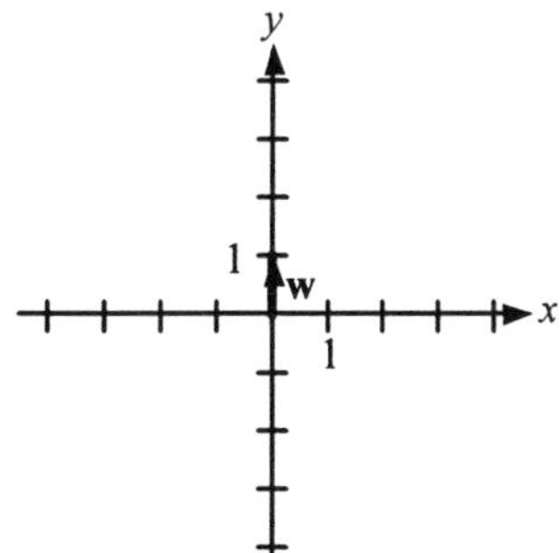

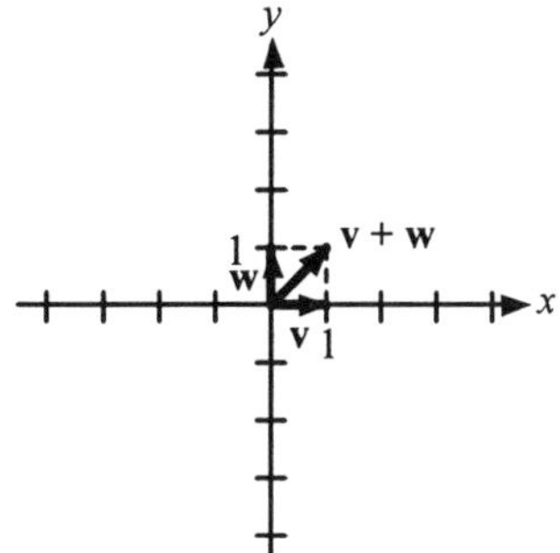

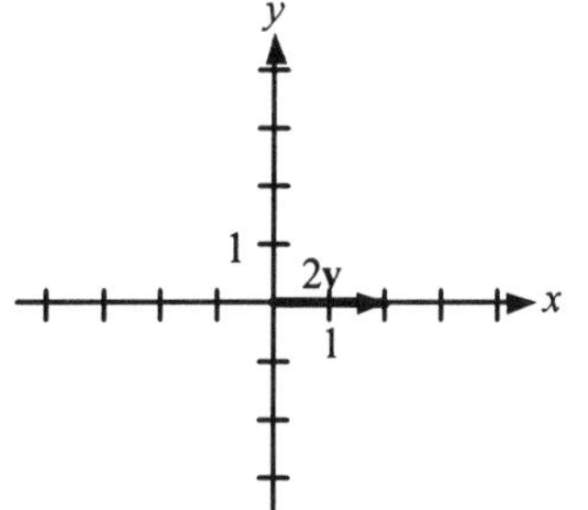

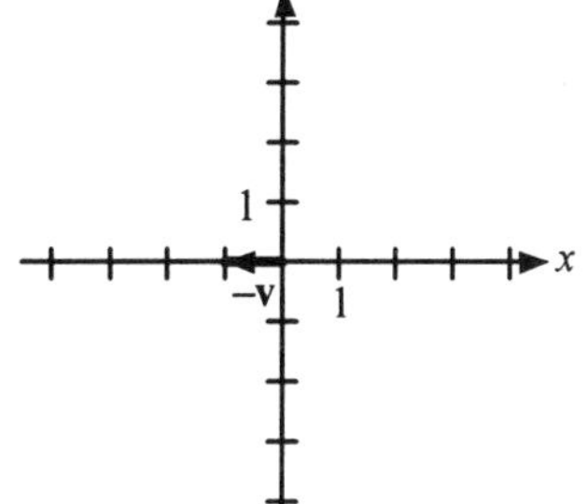

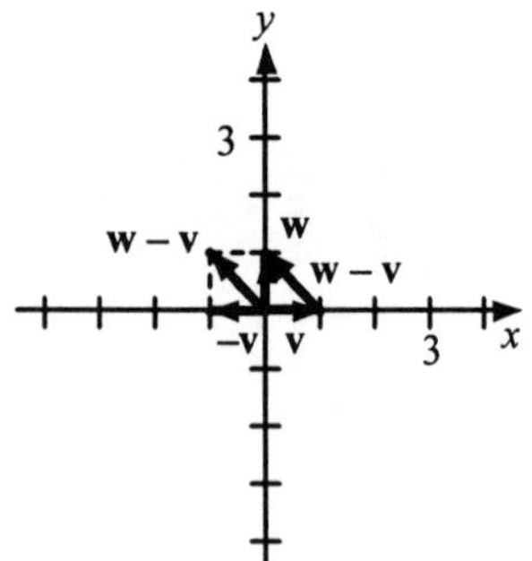

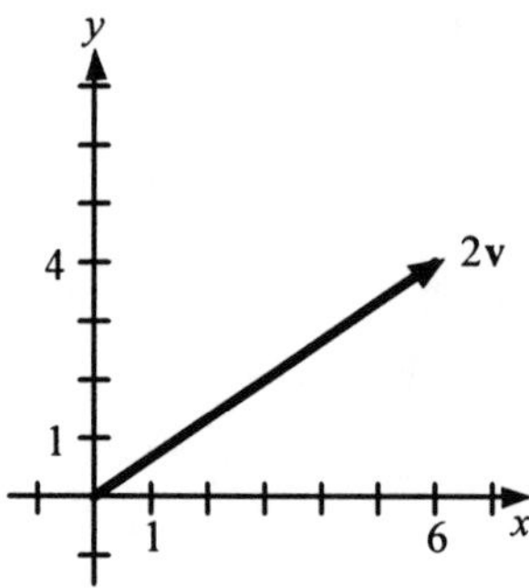

25.

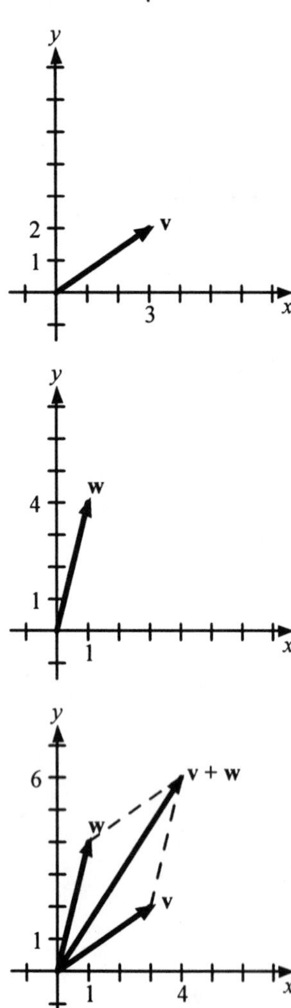

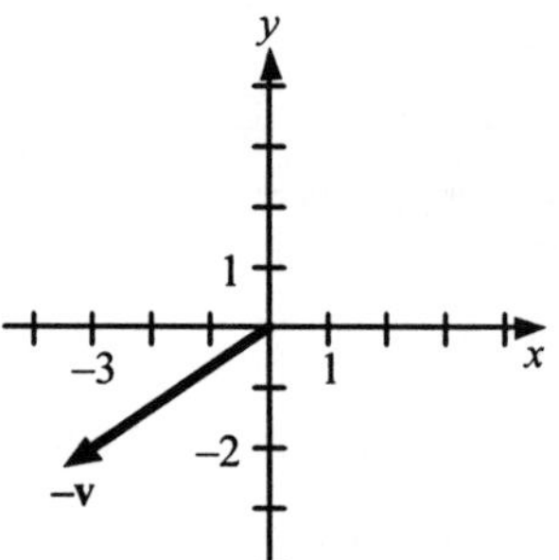

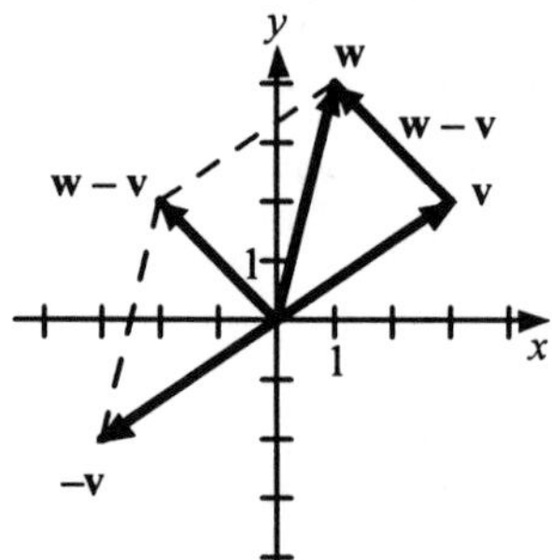

27. **a.** $4\mathbf{i}+3\mathbf{j}$ **b.** $-2\mathbf{i}-\mathbf{j}$ **c.** $13\mathbf{i}+8\mathbf{j}$

29. **a.** $-4\mathbf{i}-\mathbf{j}$ **b.** $6\mathbf{i}-5\mathbf{j}$ **c.** $-27\mathbf{i}+10\mathbf{j}$

31. **a.** $\mathbf{i}+5\mathbf{j}$ **b.** $\mathbf{i}-5\mathbf{j}$ **c.** $-2\mathbf{i}+25\mathbf{j}$

33. $4\mathbf{i}+7\mathbf{j}$ **35.** $-5\mathbf{j}$ **37.** $-6\mathbf{i}$

39. $-\frac{25}{2}\mathbf{i}+\frac{25}{2}\sqrt{3}\mathbf{j}$ **41.** h, 2; v, −8

43. h, $\frac{-5\sqrt{2}}{2}$; v, $\frac{5\sqrt{2}}{2}$ **45.** h, $4\sqrt{3}$; v, −4

47. $3\sqrt{2}$; 315° **49.** 7; 90°

51. $\mathbf{u}+\mathbf{0}=\langle u_1, u_2\rangle+\langle 0, 0\rangle=\langle u_1+0, u_2+0\rangle=\langle u_1, u_2\rangle=\mathbf{u}$

53. $\mathbf{u}+\mathbf{v}=\langle u_1, u_2\rangle+\langle v_1, v_2\rangle=\langle u_1+v_1, u_2+v_2\rangle$
$=\langle v_1+u_1, v_2+u_2\rangle=\mathbf{v}+\mathbf{u}$

55.
$$\begin{aligned}(\mathbf{u}+\mathbf{v})+\mathbf{w}&=(\langle u_1, u_2\rangle+\langle v_1, v_2\rangle)+\langle w_1, w_2\rangle\\&=\langle u_1+v_1, u_2+v_2\rangle+\langle w_1, w_2\rangle\\&=\langle (u_1+v_1)+w_1, (u_2+v_2)+w_2\rangle\\&=\langle u_1+(v_1+w_1), u_2+(v_2+w_2)\rangle\\&=\langle u_1, u_2\rangle+\langle v_1+w_1, v_2+w_2\rangle\\&=\mathbf{u}+(\mathbf{v}+\mathbf{w})\end{aligned}$$

57. $-\mathbf{u}=\langle -u_1, -u_2\rangle=\langle(-1)u_1, (-1)u_2\rangle=(-1)\langle u_1, u_2\rangle$
$=(-1)\mathbf{u}$

59. $\|-2\mathbf{v}\|=\|\langle -2v_1, -2v_2\rangle\|=\sqrt{(-2v_1)^2+(-2v_2)^2}$
$=\sqrt{4v_1^2+4v_2^2}=\sqrt{4}\sqrt{v_1^2+v_2^2}=2\|\mathbf{v}\|$

61. $\|\mathbf{i}\|=\|\langle 1, 0\rangle\|=\sqrt{1^2+0^2}=1$, so $\mathbf{i}$ is a unit vector.

63. 2 **65.** 0

67. $4.808\mathbf{i}+5.175\mathbf{j}$; 7.1 lb; 47°

69. $-19.10\mathbf{i}+11.28\mathbf{j}$; 22 lb; 149°

71. $-22.89\mathbf{i}-6.434\mathbf{j}$; 24 lb; 196°

73. **a.** 5.0 lb **b.** 37°

75. **a.** 390 mi/hour due west
b. 69 mi/hour due north

Exercises 9.4

1.

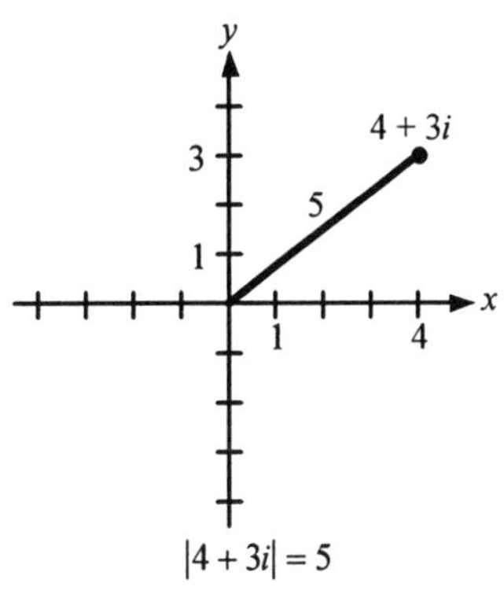

$|4+3i|=5$

3.

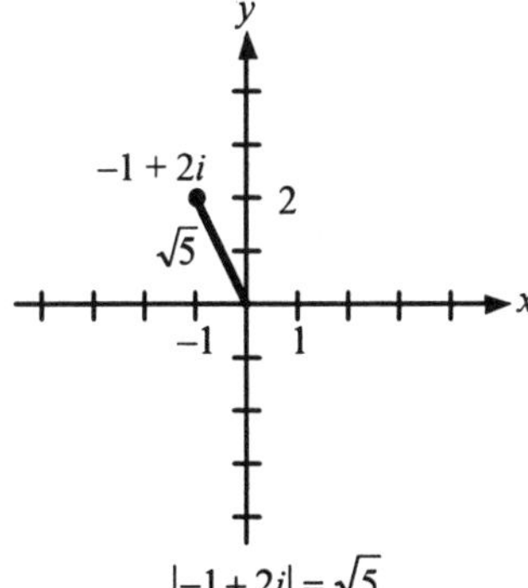

$|-1+2i|=\sqrt{5}$

5.

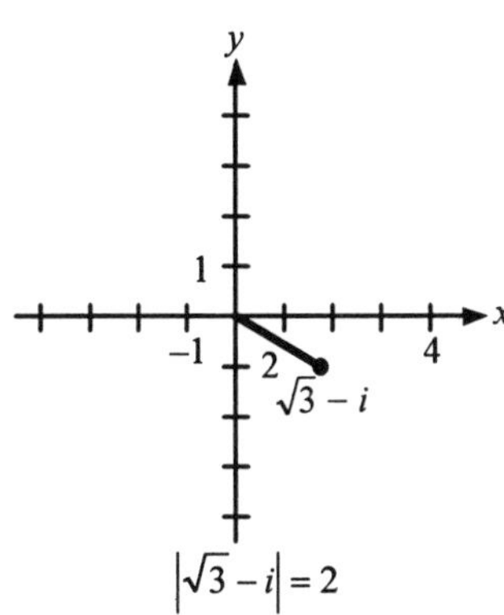

$|\sqrt{3}-i|=2$

7.

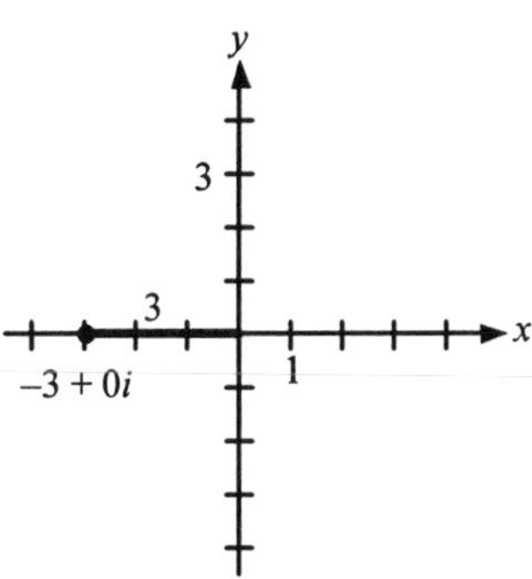

$|-3+0i|=3$

9.

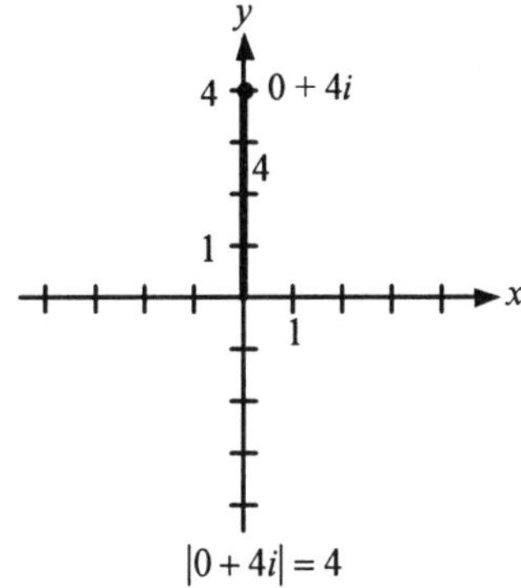

$|0+4i|=4$

11. $3(\cos 0° + i\sin 0°)$ **13.** $2(\cos 270° + i\sin 270°)$

15. $\sqrt{5}(\cos 315° + i\sin 315°)$

17. $5(\cos 36°50' + i\sin 36°50')$

19. $\sqrt{13}(\cos 213°40' + i\sin 213°40')$

21. $2(\cos 120° + i\sin 120°)$

23. $2(\cos 315° + i\sin 315°)$

25. $0+3i$ **27.** $-4+0i$

29. $-\frac{\sqrt{3}}{2}+\frac{3}{2}i$ **31.** $-\sqrt{2}-\sqrt{2}i$

33. $0.6157+0.7880i$

35. **a.** $8(\cos 63° + i\sin 63°)$
b. $\frac{1}{2}(\cos 41° + i\sin 41°)$
c. $2[\cos(-41°) + i\sin(-41°)]$

37. **a.** $\cos 270° + i\sin 270°$
b. $\cos(-90°) + i\sin(-90°)$
c. $\cos 90° + i\sin 90°$

39. **a.** $36(\cos 336° + i\sin 336°)$
b. $\frac{1}{4}[\cos(-74°) + i\sin(-74°)]$
c. $4(\cos 74° + i\sin 74°)$

41. **a.** $2\sqrt{2}(\cos 345° + i\sin 345°)$
b. $\sqrt{2}(\cos 255° + i\sin 255°)$
c. $\frac{\sqrt{2}}{2}[\cos(-255°) + i\sin(-255°)]$

43. **a.** $\cos(-65°) + i\sin(-65°)$
b. $\cos 25° + i\sin 25°$
c. $\cos(-25°) + i\sin(-25°)$

45. $4\sqrt{3}+4i$ **47.** $512\sqrt{2}-512\sqrt{2}i$

49. $16+0i$ **51.** $16+16\sqrt{3}i$

53. $1+0i$ **55.** $\frac{1}{4}-\frac{\sqrt{3}}{4}i$ **57.** $\frac{\sqrt{2}}{16}-\frac{\sqrt{2}}{16}i$

59. $5(\cos 30° + i\sin 30°) = \frac{5\sqrt{3}}{2}+\frac{5}{2}i$,
$5(\cos 210° + i\sin 210°) = -\frac{5\sqrt{3}}{2}-\frac{5}{2}i$

61. $2(\cos 20° + i\sin 20°) \approx 1.8794 + 0.6840i$,
$2(\cos 110° + i\sin 110°) \approx -0.6840 + 1.8794i$,
$2(\cos 200° + i\sin 200°) \approx -1.8794 - 0.6840i$,
$2(\cos 290° + i\sin 290°) \approx 0.6840 - 1.8794i$

63. $\cos 0° + i\sin 0° = 1+0i$,
$\cos 120° + i\sin 120° = -\frac{1}{2}+\frac{\sqrt{3}}{2}i$,
$\cos 240° + i\sin 240° = -\frac{1}{2}-\frac{\sqrt{3}}{2}i$

65. $\cos 135° + i\sin 135° = -\frac{\sqrt{2}}{2}+\frac{\sqrt{2}}{2}i$,
$\cos 315° + i\sin 315° = \frac{\sqrt{2}}{2}-\frac{\sqrt{2}}{2}i$

67. $2(\cos 67.5° + i\sin 67.5°) \approx 0.7654 + 1.8478i$,
$2(\cos 157.5° + i\sin 157.5°) \approx -1.8478 + 0.7654i$,
$2(\cos 247.5° + i\sin 247.5°) \approx -0.7654 - 1.8478i$,
$2(\cos 337.5° + i\sin 337.5°) \approx 1.8478 - 0.7654i$

69. $\sqrt[5]{2}(\cos 12° + i\sin 12°) \approx 1.1236 + 0.2388i$,
$\sqrt[5]{2}(\cos 84° + i\sin 84°) \approx 0.1201 + 1.1424i$,
$\sqrt[5]{2}(\cos 156° + i\sin 156°) \approx -1.0494 + 0.4672i$,
$\sqrt[5]{2}(\cos 228° + i\sin 228°) \approx -0.7686 - 0.8536i$,
$\sqrt[5]{2}(\cos 300° + i\sin 300°) \approx 0.5743 - 0.9948i$

71. $1+\sqrt{3}i$, $-2+0i$, $1-\sqrt{3}i$

73. $1+0i$, $0+i$, $-1+0i$, $0-i$

75. $\cos 36° + i\sin 36°$, $\cos 108° + i\sin 108°$, $-1+0i$,
$\cos 252° + i\sin 252°$, $\cos 324° + i\sin 324°$

77.

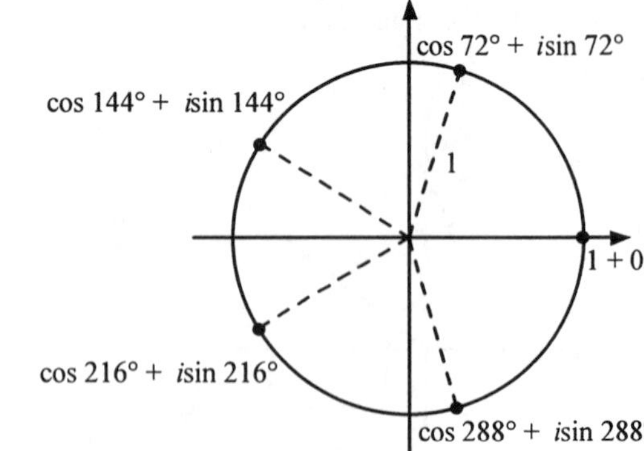

79. **a.** $5.30(\cos 45° + i\sin 45°)$ ohms
b. $1.86(\cos 45° + i\sin 45°)$ ohms

81. $0.36[\cos(-11.7°) + i\sin(-11.7°)]$ or
$0.36(\cos 11.7° + i\sin 11.7°)$

83. $1.25(\cos 90° + i\sin 90°)$ or $1.25i$

85. $|z| = |a+bi| = \sqrt{a^2+b^2}$ and
$|\bar{z}| = |a-bi| = \sqrt{a^2+(-b)^2} = \sqrt{a^2+b^2}$, so $|\bar{z}| = |z|$.

87. $-z = -1 \cdot z = (\cos\pi + i\sin\pi)\cdot r(\cos\theta + i\sin\theta)$
$= r[\cos(\theta+\pi) + i\sin(\theta+\pi)]$

89. $z^2 = (r\cos\theta + i\sin\theta)\cdot(r\cos\theta + i\sin\theta)$
$= r\cdot r[\cos(\theta+\theta) + i\sin(\theta+\theta)]$
$= r^2(\cos 2\theta + i\sin 2\theta)$

Exercises 9.5

1.

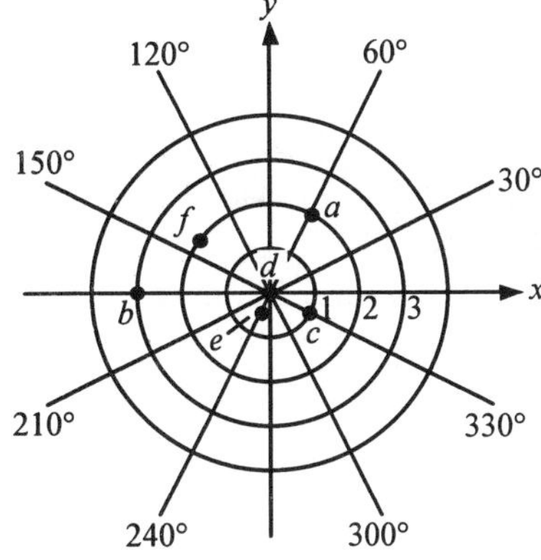

3. $(-1, 0)$ **5.** $(3, 0)$ **7.** $(-1, -1)$

9. $(1, 180°)$ **11.** $(\sqrt{2}, 315°)$ **13.** $(2, 150°)$

15. **a.** $(1, 330°)$ **b.** $\left(2, \frac{11\pi}{6}\right)$ **c.** $\left(1, \frac{\pi}{3}\right)$

d. $(3, 225°)$ **e.** $\left(1, \frac{5\pi}{4}\right)$ **f.** $(2, 180°)$

17. Some possibilities are: $\left(-1, -\frac{\pi}{6}\right)$, $\left(1, \frac{17\pi}{6}\right)$, $\left(1, \frac{29\pi}{6}\right)$

19.

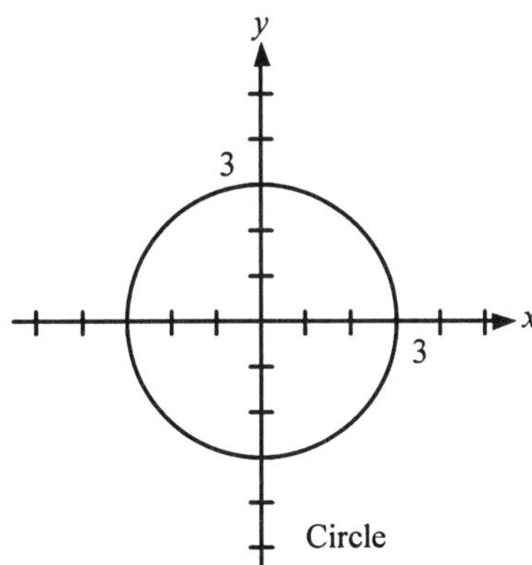

21.

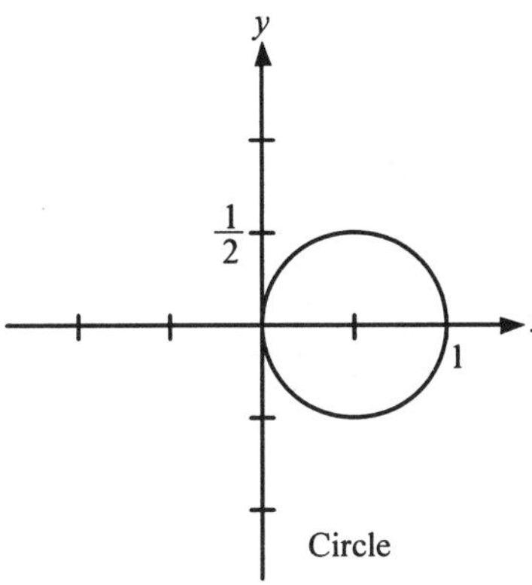

23.

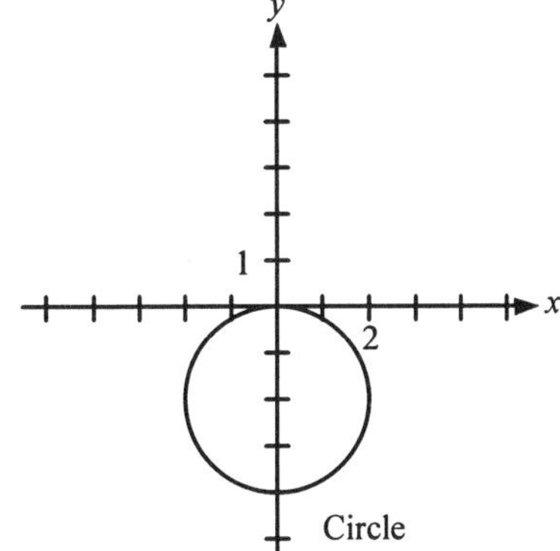

25.

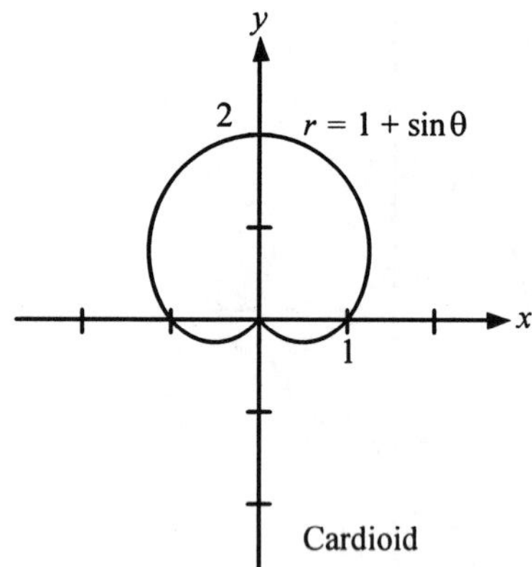

27.

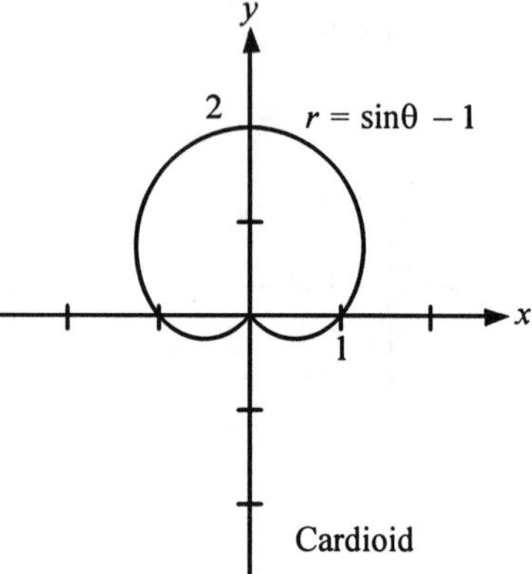

29.

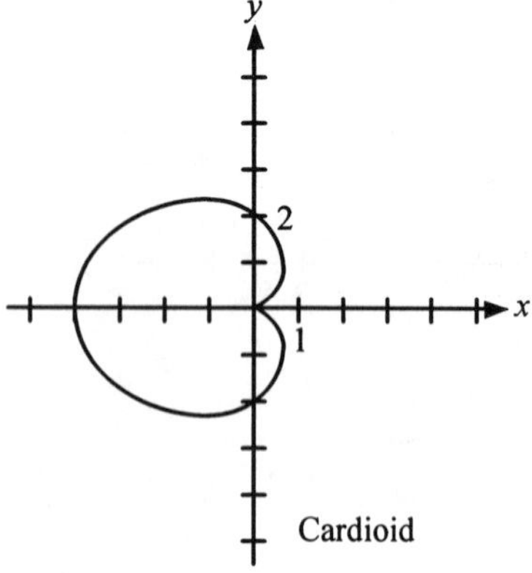

31.

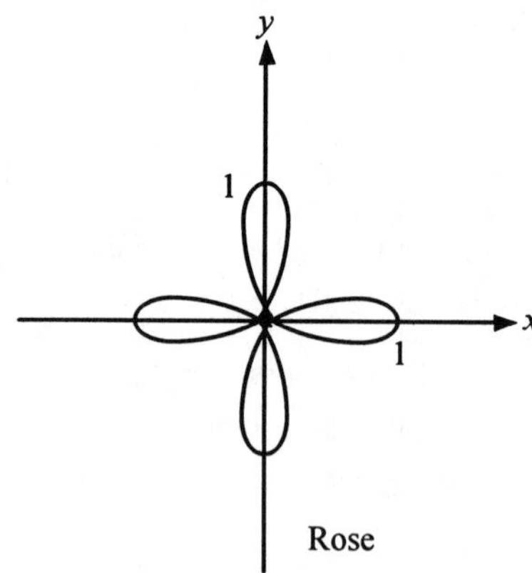

33.

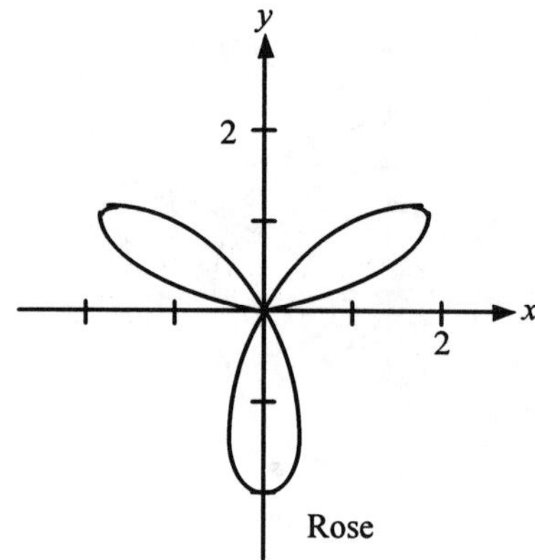

35.

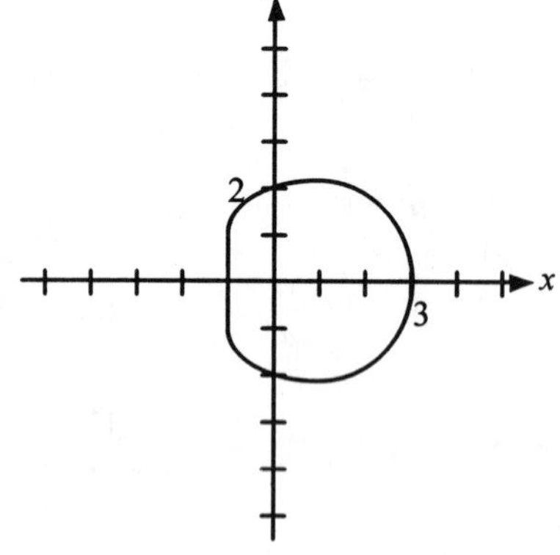

37.

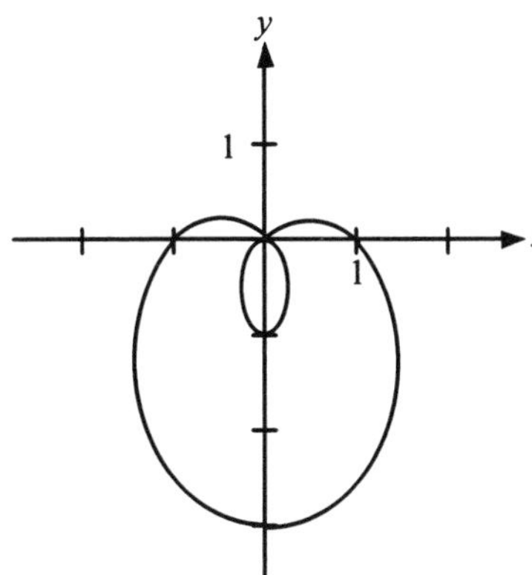

39.

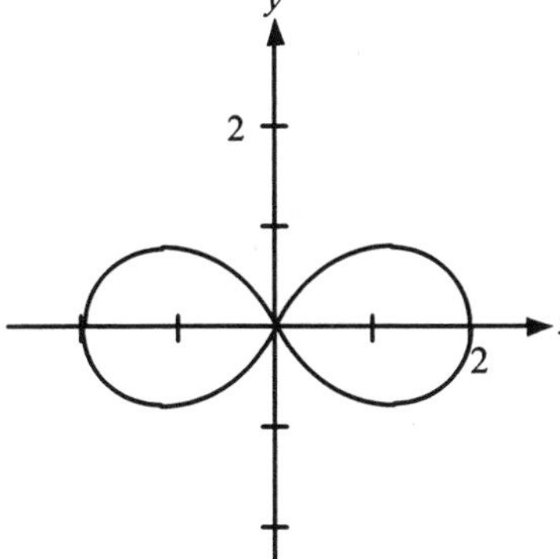

41.

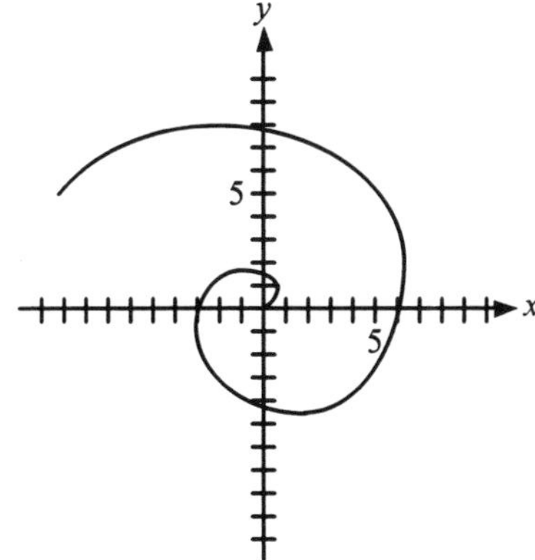

43.	$x^2 + y^2 = 9$	**45.**	$x = 2$
47.	$x^2 + y^2 = -3y$	**49.**	$y = 2$
51.	$3y + 2x = 1$	**53.**	$r = 2\sec\theta$
55.	$r = 3$	**57.**	$r = \cos\theta$
59.	$r = \tan\theta\sec\theta$	**61.**	$r^2 = 4\sec 2\theta$

63. $c = \dfrac{1}{45\pi}$

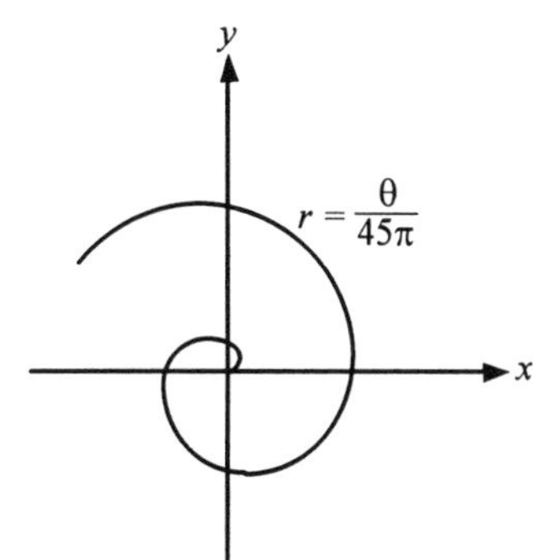

Chapter 9 Review Exercises

1. 109° **3.** 48°50′ **5.** $\sqrt{37} \approx 6.1$

7. $8(\cos 90° + i\sin 90°)$; $0 + 8i$

9. $\cos 450° + i\sin 450°$; $0 + i$

11.

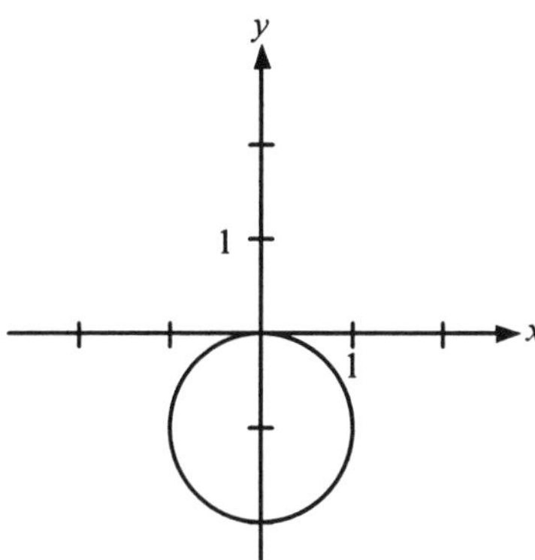

13.

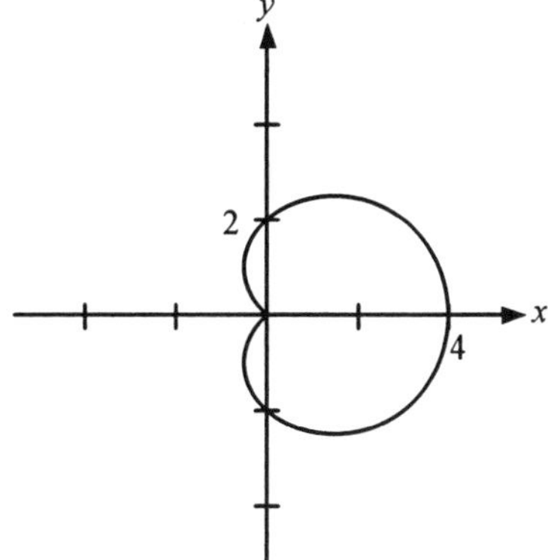

15. $(\sqrt{2}, 225°)$ **17.** 50; 30°

19. 79 m^2 **21.** $2x - y = 3$

23. $3(\cos 60° + i\sin 60°)$, $3(\cos 180° + i\sin 180°)$, $3(\cos 300° + i\sin 300°)$

25. 2; 300° **27.** 5; 270°

29. $\langle 28, 7\rangle$ **31.** $-\mathbf{i} - 17\mathbf{j}$

33. LF: 193 ft; CF: 276 ft; RF: 266 ft

35. 30°, 56°, 94° **37.** 116°

39. By the law of cosines $r^2 = p^2 + q^2 - 2pq\cos R$. If $r^2 = p^2 + q^2$, then $2pq\cos R = 0$, so $\cos R = 0$ and $R = 90°$. Thus, PQR is a right triangle.

41. a **43.** b **45.** b

47. c **49.** c

Chapter 9 Test

1. True **2.** $24\sqrt{2}$

3. 68.8° or 111.2° **4.** 101.3°

5. $C = 49°$, $A = 55°$, $a = 38$ ft

6. 62.4 cm **7.** 8.5 cm^2 **8.** 6; 330°

9. $\langle 23, -17\rangle$ **10.** $-2\mathbf{i} + 2\sqrt{3}\mathbf{j}$

11. 34 lb **12.** 39 lb

13. $17(\cos 152° + i\sin 152°)$

14. $-18 + 0i$ **15.** $32i$

16. $\frac{5}{2} + \frac{5\sqrt{3}}{2}i$, $-5 + 0i$, $\frac{5}{2} - \frac{5\sqrt{3}}{2}i$

17. $(3, 45°)$, $(-3, -135°)$

18.

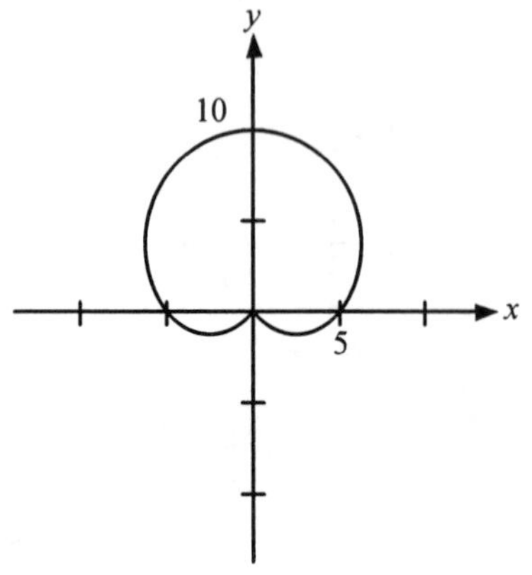

19. $\left(-3, 3\sqrt{3}\right)$ **20.** $r = 9\cos\theta$

Exercises 10.1

1. $(x+3)^2 + (y-4)^2 = 4$

3. $x^2 + y^2 = 1$

5. $(x+4)^2 + (y+1)^2 = 9$

7. $(x-3)^2 + (y+3)^2 = 18$

9. $(x+1)^2 + (y+1)^2 = 58$

11. $(2,5)$; 4 **13.** $(3,0)$; $2\sqrt{5}$

15. $(0,0)$; 3 **17.** $(5,3)$; 7

19. $(-4,1)$; $3\sqrt{2}$ **21.** $\left(-\frac{7}{2}, -\frac{3}{2}\right)$; $\frac{\sqrt{42}}{2}$

23. $(-1,2)$; $\frac{7}{2}$ **25.** $(2,1)$

27. $\left(0, -\frac{7}{2}\right)$ **29.** $\left(0, \frac{3}{2}\right)$

31. $\overline{OB} = \sqrt{a^2 + b^2}$; $\overline{AC} = \sqrt{a^2 + b^2}$; thus $\overline{OB} = \overline{AC}$

33. $m_{OB} = \frac{b}{a}$ and $m_{AC} = -\frac{b}{a}$. If the diagonals are perpendicular, $\left(\frac{b}{a}\right)\left(-\frac{b}{a}\right) = -1$, so $a^2 = b^2$, which implies $a = b$. Thus, the rectangle is a square.

35. Midpoint of OB is $\left(\frac{a+b}{2}, \frac{c}{2}\right)$; midpoint of AC is $\left(\frac{a+b}{2}, \frac{c}{2}\right)$; thus, OB and AC bisect each other.

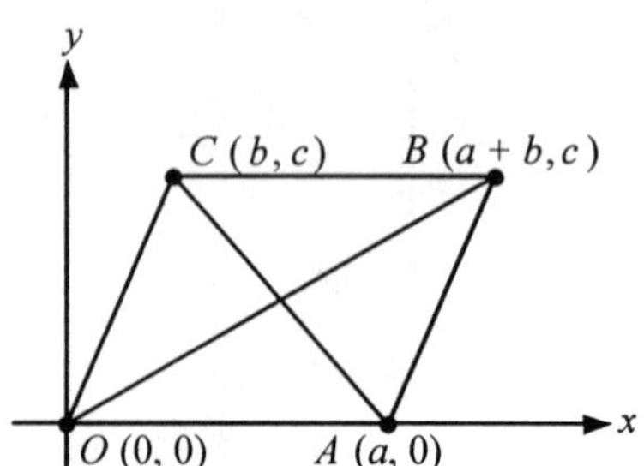

37. **a.** $A = \frac{1}{2}(2r)\sqrt{r^2 - x^2} = r\sqrt{r^2 - x^2}$

b. r^2

39. **a.** $x^2 + (y - 72)^2 = 100$

b. $72 + 5\sqrt{3} = 80.66$ in.= 6 ft 8.66 in.

41. 40 ft

43. **a.** $(1,0)$ **b.** $x^2 + y^2 = 4$ **c.** $\left(1, \sqrt{3}\right)$

d. $2\sqrt{3}$ **e.** $\sqrt{3}$

45. 1

Exercises 10.2

1.

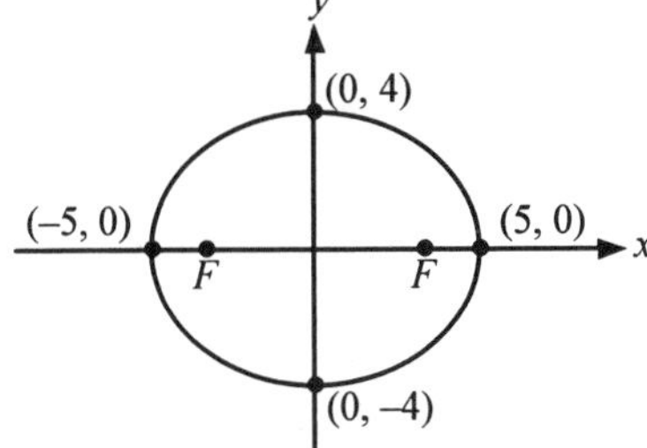

Foci: $(3, 0)$, $(-3, 0)$

3.

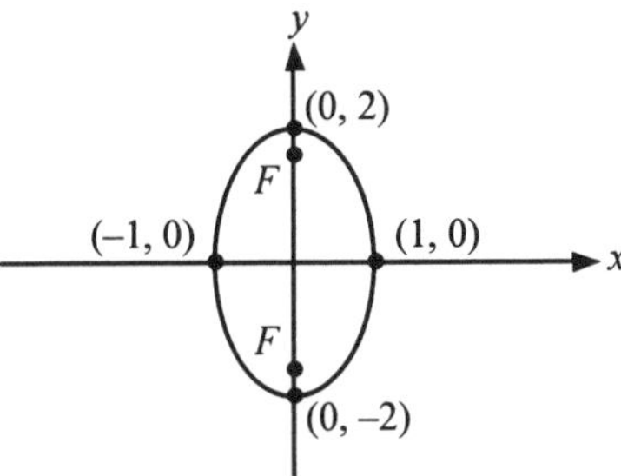

Foci: $\left(0, \sqrt{3}\right)$, $\left(0, -\sqrt{3}\right)$

5.

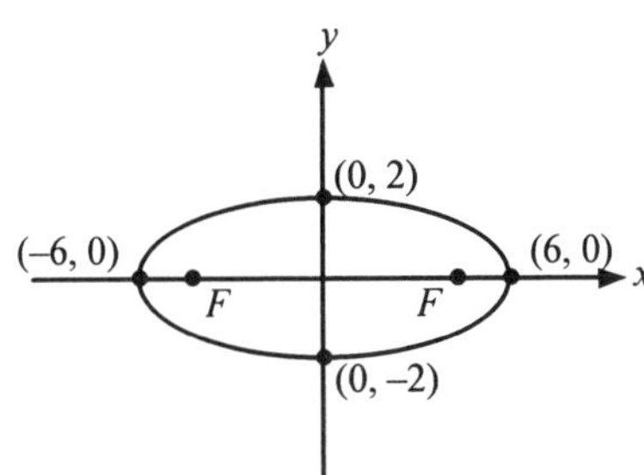

Foci: $\left(4\sqrt{2}, 0\right)$, $\left(-4\sqrt{2}, 0\right)$

7.

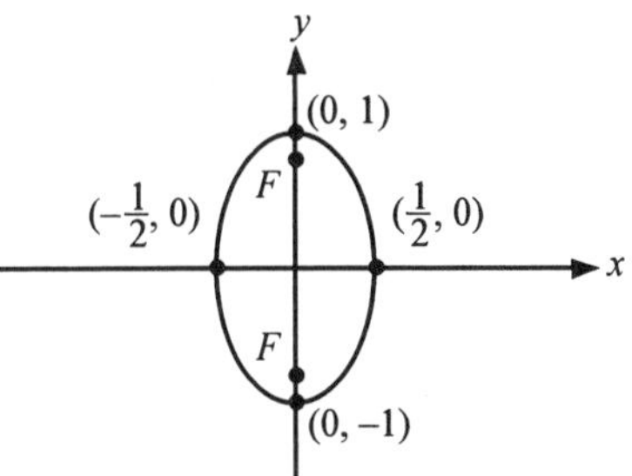

Foci: $\left(0, \dfrac{\sqrt{3}}{2}\right)$, $\left(0, -\dfrac{\sqrt{3}}{2}\right)$

9.

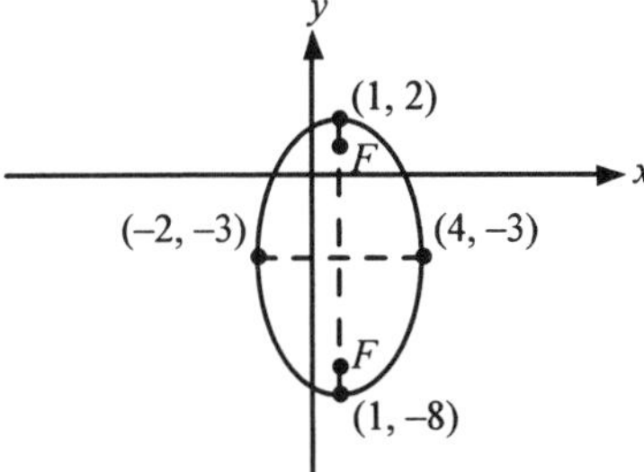

Foci: $(1, 1)$, $(1, -7)$

Center: $(1, -3)$

11.

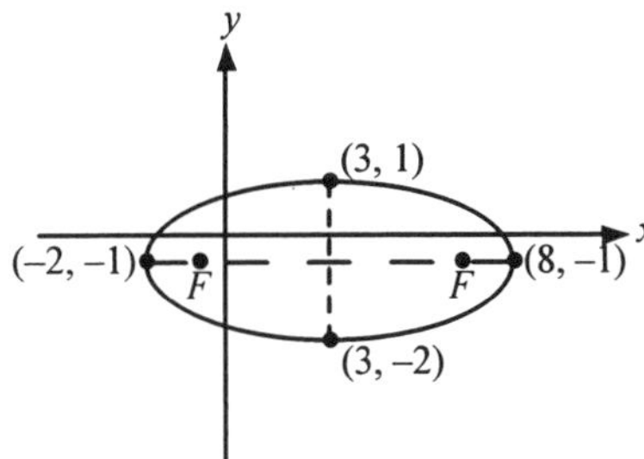

Foci: $\left(3 - \sqrt{21}, -1\right)$, $\left(3 + \sqrt{21}, -1\right)$

Center: $(3, -1)$

13.

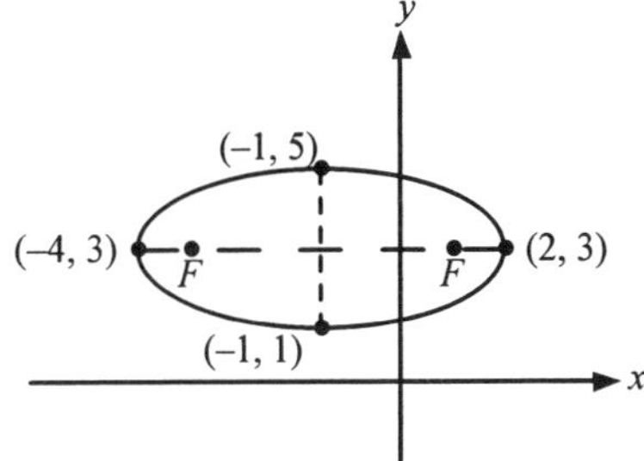

Foci: $\left(-1 - \sqrt{5}, 3\right)$, $\left(-1 + \sqrt{5}, 3\right)$

Center: $(-1, 3)$

15. $\dfrac{x^2}{16}+\dfrac{y^2}{25}=1$ **17.** $\dfrac{x^2}{36}+\dfrac{y^2}{20}=1$

19. $\dfrac{x^2}{9}+\dfrac{y^2}{4}=1$ **21.** $\dfrac{(x-2)^2}{16}+\dfrac{(y-2)^2}{25}=1$

23. $\dfrac{(x-2)^2}{25}+\dfrac{(y-3)^2}{16}=1$

25. $\dfrac{x^2}{16}+\dfrac{y^2}{9}=1$

27. $\dfrac{x^2}{1}+\dfrac{y^2}{9}=1$; $9x^2+y^2=9$

29. **a.** 186,000,000 mi **b.** 3,000,000 mi

31. $\sqrt{48}$ ft ≈ 6 ft 11 in. **33.** $\sqrt{18.75}$ ft ≈ 4 ft 4 in.

35. 13.42 ft

Exercises 10.3

1.

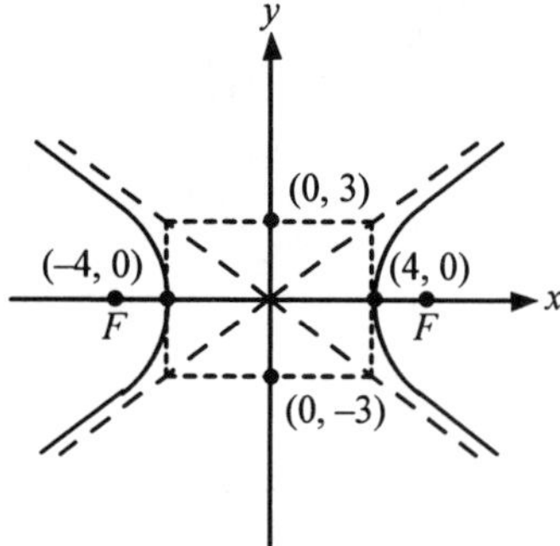

Foci: (5, 0), (–5,0)

Asymptotes: $y=\pm\dfrac{3}{4}x$

3.

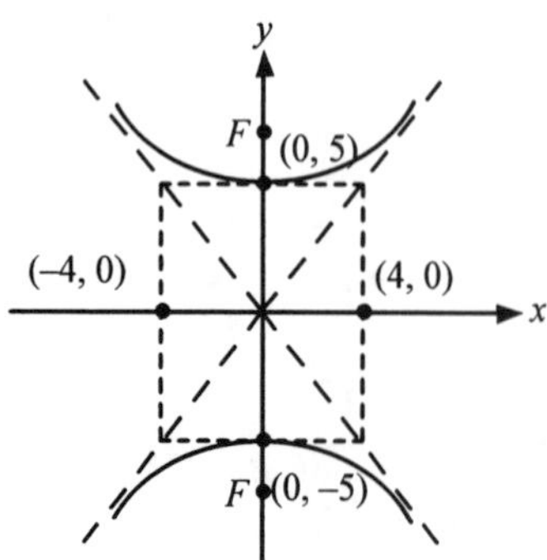

Foci: $(0,\sqrt{41})$, $(0,-\sqrt{41})$

Asymptotes: $y=\pm\dfrac{5}{4}x$

5.

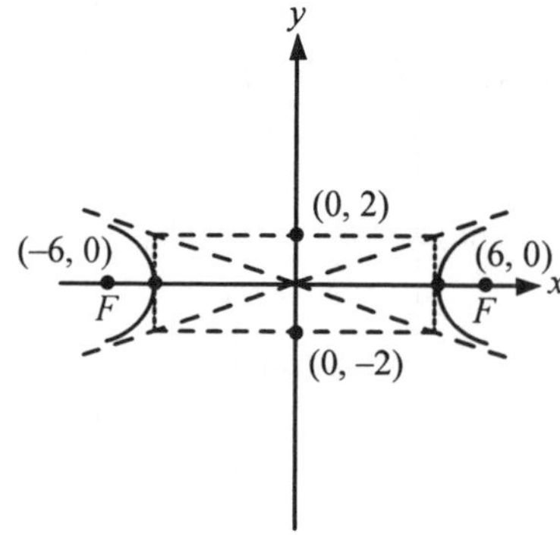

Foci: $(\sqrt{40},0)$, $(-\sqrt{40},0)$

Asymptotes: $y=\pm\dfrac{1}{3}x$

7.

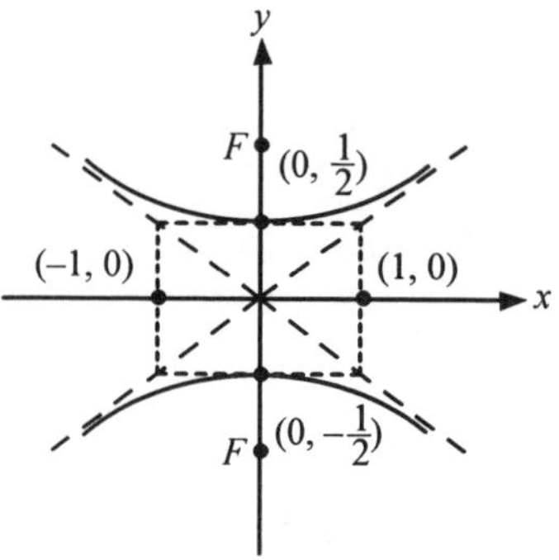

Foci: $\left(0,\dfrac{\sqrt{5}}{2}\right)\left(0,-\dfrac{\sqrt{5}}{2}\right)$

Asymptotes: $y=\pm\dfrac{1}{2}x$

9.

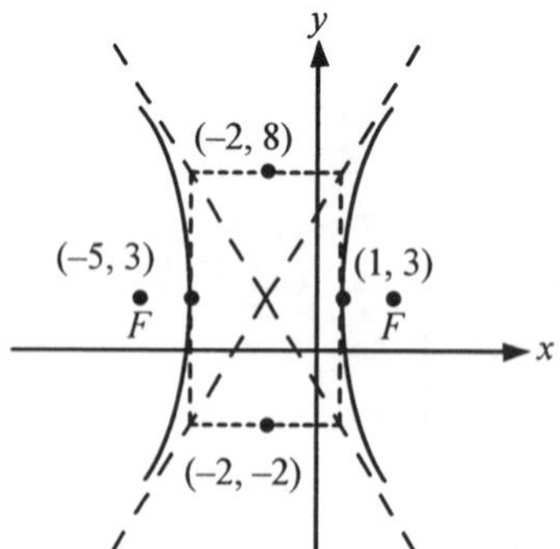

Foci: $(-2+\sqrt{34},3),(-2-\sqrt{34},3)$

Center: $(-2,3)$

Asymptotes: $y-3=\pm\dfrac{5}{3}(x+2)$

11.

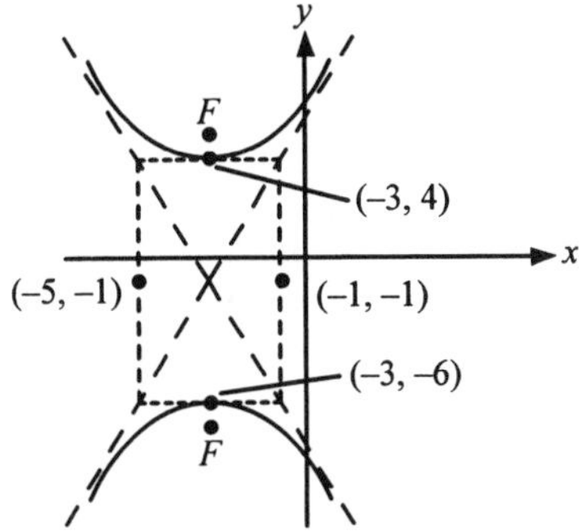

Foci: $\left(-3,\ -1+\sqrt{29}\right)$, $\left(-3,\ -1-\sqrt{29}\right)$

Center: $(-3,\ -1)$

Asymptotes: $y+1=\pm\frac{5}{2}(x+3)$

13.

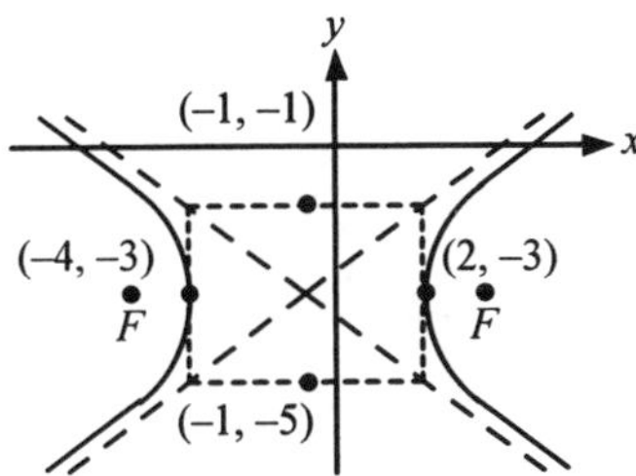

Foci: $\left(-1+\sqrt{13},\ -3\right)$, $\left(-1-\sqrt{13},\ -3\right)$

Center: $(-1,\ -3)$

Asymptotes: $y+3=\pm\frac{2}{3}(x+1)$

15. $\frac{y^2}{16}-\frac{x^2}{9}=1$ **17.** $\frac{x^2}{9}-\frac{y^2}{7}=1$

19. $\frac{x^2}{9}-\frac{y^2}{25}=1$ **21.** $\frac{(x+3)^2}{9}-\frac{(y+4)^2}{16}=1$

23. $\frac{(y-1)^2}{4}-\frac{(x-3)^2}{5}=1$

25. **a.** $\frac{x^2}{625}-\frac{y^2}{13775}=1$; $y=\pm\sqrt{13775\left(\frac{x^2}{625}-1\right)}$

$(x \ge 25)$; Focus at $(120,\ 0)$

b.

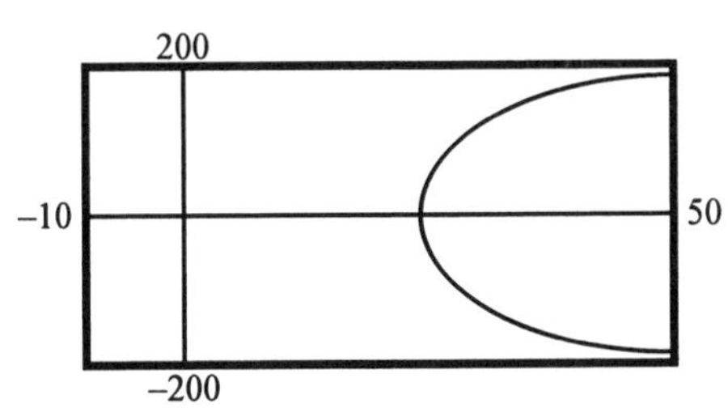

27. **a.** 550 ft

b. $y=\sqrt{6{,}174{,}375\left(\frac{x^2}{75{,}625}-1\right)}$

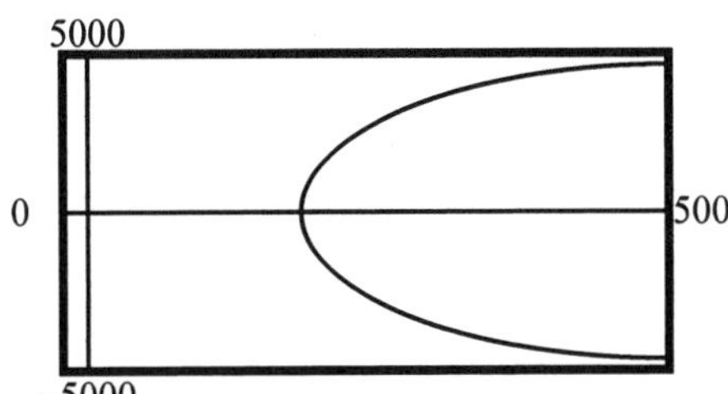

29. $\frac{4\sqrt{7}}{27}$

Exercises 10.4

1.

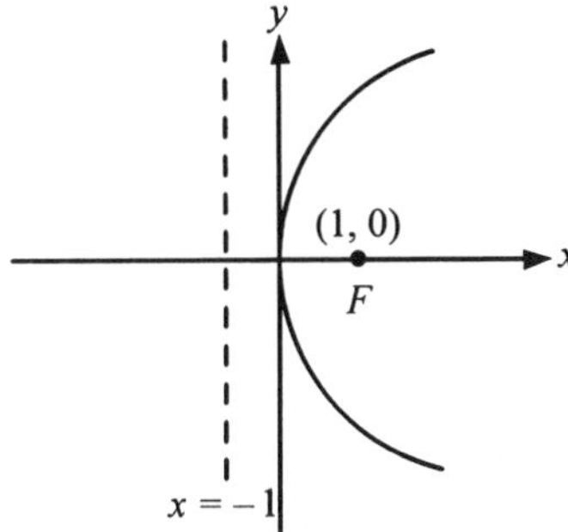

Vertex: $(0,\ 0)$

Focus: $(1,\ 0)$

Directrix: $x=-1$

3.

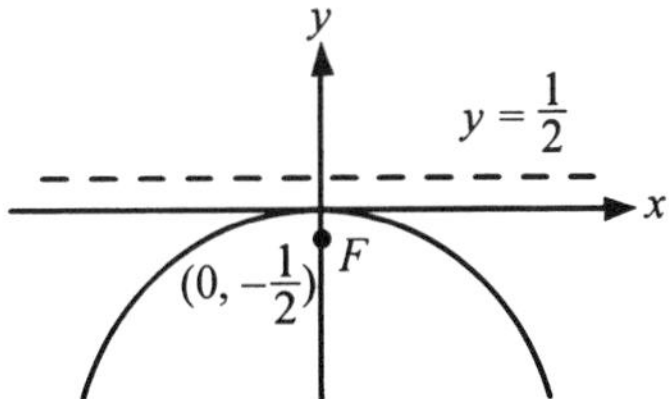

Vertex: $(0,\ 0)$

Focus: $\left(0,\ -\frac{1}{2}\right)$

Directrix: $y=\frac{1}{2}$

5.

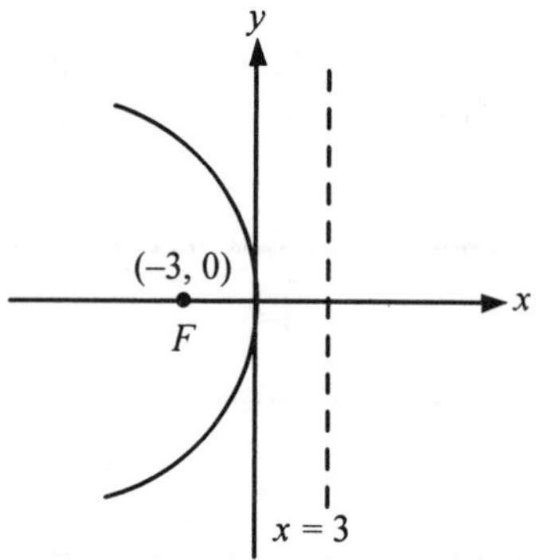

Vertex: (0, 0)
Focus: (–3, 0)
Directrix: $x = 3$

7.

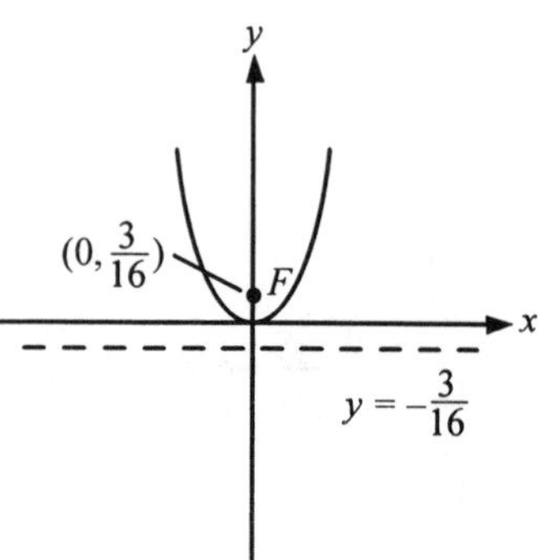

Vertex: (0, 0)
Focus: $\left(0, \frac{3}{16}\right)$
Directrix: $y = -\frac{3}{16}$

9.

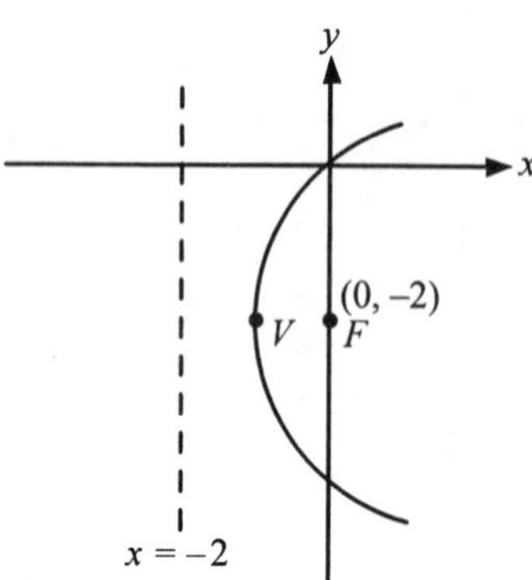

Vertex: (–1, – 2)
Focus: (0, – 2)
Directrix: $x = -2$

11.

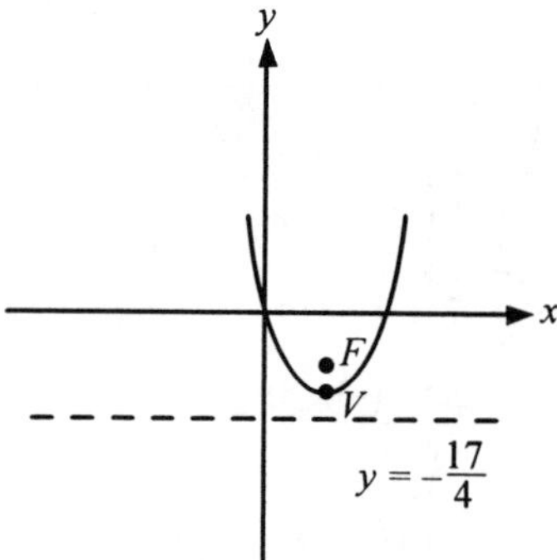

Vertex: (2, – 4)
Focus: $\left(2, -\frac{15}{4}\right)$
Directrix: $y = -\frac{17}{4}$

13.

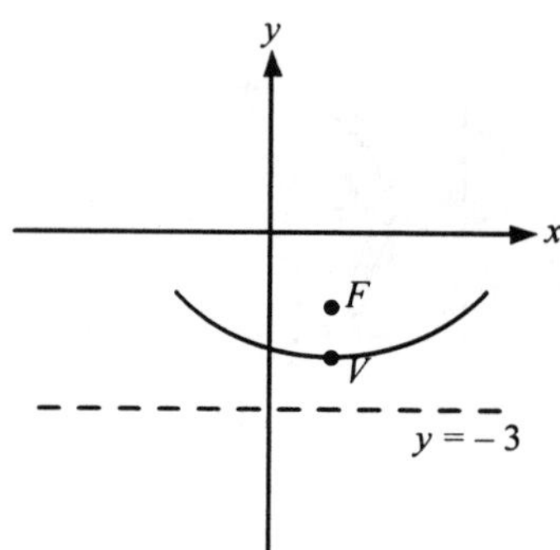

Vertex: (1, – 2)
Focus: (1, – 1)
Directrix: $y = -3$

15. $x^2 = 12y$ **17.** $y^2 = 2x$ **19.** $x^2 = 16y$

21. $(y-1)^2 = 12(x+1)$ **23.** $(x-3)^2 = 12y$

25. **a.** $x^2 = 12y$ if the vertex is at (0, 0)
b. (0, 3)

27. **a.** $x^2 = 10y$ if the vertex is at (0, 0)
b. $\left(0, \frac{5}{2}\right)$

29. 21 ft

31. Triangle: Area = 8; Parabola: Area $= \left(\frac{4}{3}\right)(8) = \frac{32}{3}$

33. Parabola **35.** Ellipse **37.** Hyperbola

39. Parabola **41.** Circle

43. Two intersecting straight lines: $y+2=\pm x$

45. Circle **47.** Parabola

49. Point: $(-1, 1)$ **51.** No graph

Exercises 10.5

1.

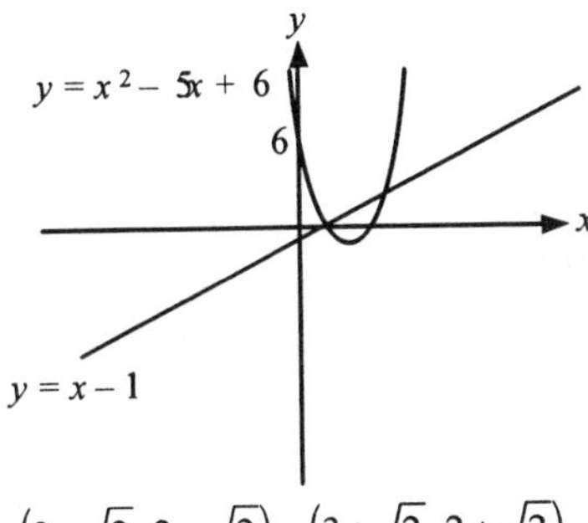

$\left(3-\sqrt{2}, 2-\sqrt{2}\right)$, $\left(3+\sqrt{2}, 2+\sqrt{2}\right)$

3.

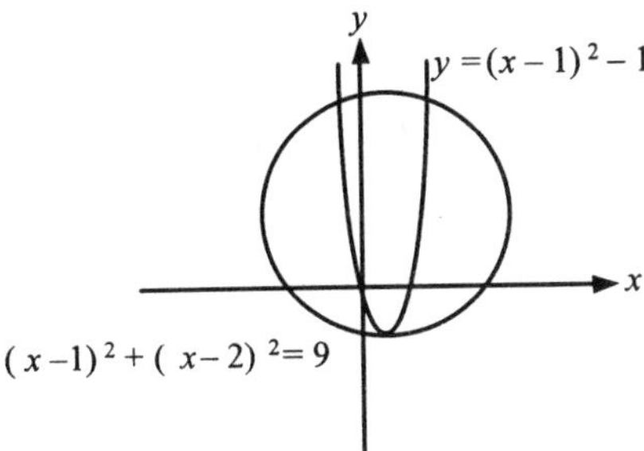

$\left(1+\sqrt{5}, 4\right)$, $\left(1-\sqrt{5}, 4\right)$, $(1, -1)$

5.

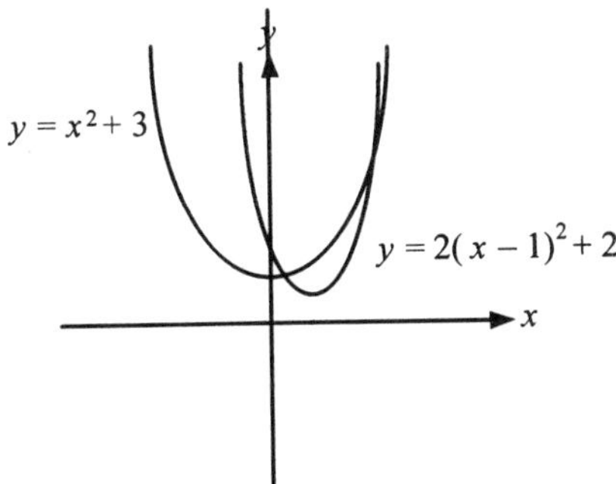

$\left(2-\sqrt{3}, 10-4\sqrt{3}\right)$, $\left(2+\sqrt{3}, 10+4\sqrt{3}\right)$

7. $(0, 0), (1, 1)$ **9.** $(3, 1), (-3, 1)$

11. $(2, 0), (-2, 0)$ **13.** $(0, 10), (6, -8)$

15. $(2, 2), (2, -2), (-2, 2), (-2, -2)$

17. $\left(2\sqrt{2}, 4\right)$, $\left(2\sqrt{2}, -4\right)$, $\left(-2\sqrt{2}, 4\right)$, $\left(-2\sqrt{2}, -4\right)$

19. $(5, 3), (3, 5)$

21. $(1, 2), (-1, -2), \left(2\sqrt{2}, \frac{\sqrt{2}}{2}\right), \left(-2\sqrt{2}, -\frac{\sqrt{2}}{2}\right)$

23. $(1, 1), (1, -1)$ **25.** $(0, -2), (5, 3)$

27. $(1, 4), (2, 1), \left(-\frac{2}{3}, 9\right)$

29. $\left(\sqrt{\frac{-1+\sqrt{21}}{2}}, \frac{-1+\sqrt{21}}{2}\right)$, $\left(-\sqrt{\frac{-1+\sqrt{21}}{2}}, \frac{-1+\sqrt{21}}{2}\right)$

31. $(2, 0), (-2, 0)$ **33.** $\left(\frac{1}{7}, -\frac{1}{3}\right)$

35. $(1, 1), (1, -1), (-1, 1), (-1, -1)$

37. $(2.4, 1.7), (2.4, -1.7), (-2.4, 1.7), (-2.4, -1.7)$

39. $4\sqrt{6}$ ft by $3\sqrt{6}$ ft

41. $b=9$ m, $h=8$ m

43. $y=x^2$; $xy=2$
By substitution we get $x\left(x^2\right)=2$, or $x^3=2$. Thus the solution is $\sqrt[3]{2}$. New volume is $\left(\sqrt[3]{2}s\right)^3=2s^3$ or $2\cdot$ (old volume).

Chapter 10 Review Exercises

1. $x^2+y^2=1$ **3.** $(-2, 3); 5$

5. $(0, 3), (0, -3)$ **7.** $y=\pm\frac{4}{5}x$

9. $\left(0, -\frac{7}{4}\right)$ **11.** $(x-1)^2+(y-2)^2=9$

13.

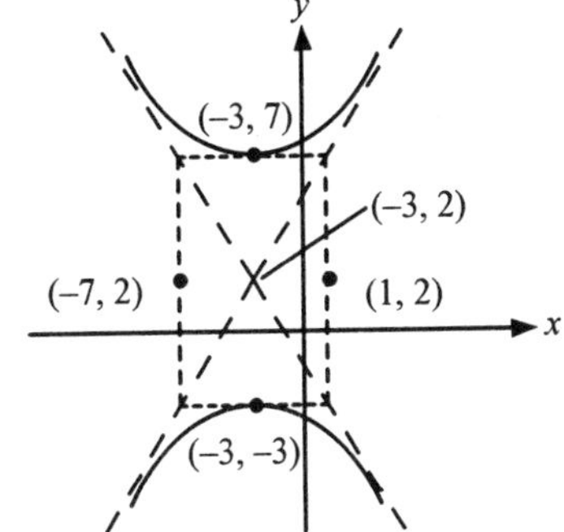

15. $(-1,\ -1),\ (7,\ -1)$

17. Length = 31, Width = 11

19. $y=-\sqrt{9-(x-4)^2}$ **21.** c

23. d **25.** a

Chapter 10 Test

1. $(x+4)^2+(y-3)^2=25$

2. $(1,\ 6),\ (-2,\ 9)$

3. Position square $OABC$ with vertices $O(0,\ 0)$, $A(a,\ 0)$, $B(a,\ a)$, and $C(0,\ a)$. Then $m_{OB}=1$ and $m_{AC}=-1$. Because $m_{OB}\cdot m_{AC}=-1$, the diagonals are perpendicular to each other.

4. $(x-1)^2+(y+2)^2=5$

5. $(-5,\ 1);\ 6$ **6.** $4\sqrt{2}$

7. $\dfrac{(x+1)^2}{36}+\dfrac{(y-1)^2}{4}=1$

8. $4\sqrt{3}$ ft

9.

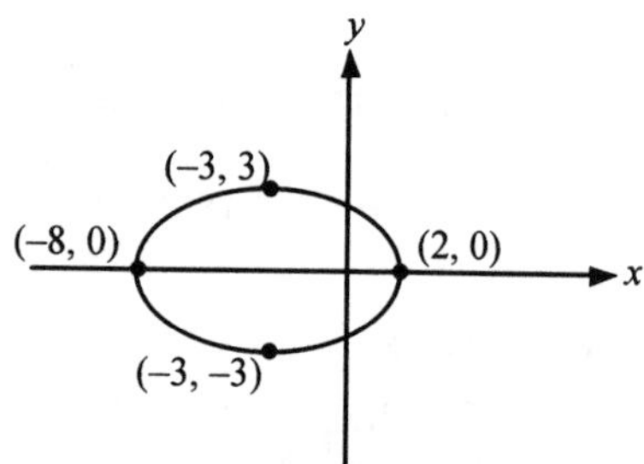

10.

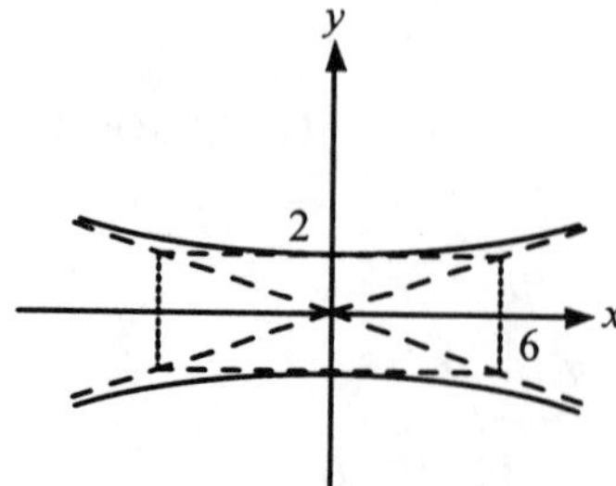

11. $\left(\dfrac{\sqrt{5}}{2},\ 0\right),\ \left(-\dfrac{\sqrt{5}}{2},\ 0\right)$

12. $y=\pm\dfrac{1}{2}x$ **13.** $(3,\ -2)$

14. $(x-2)^2=12(y-1)$

15.

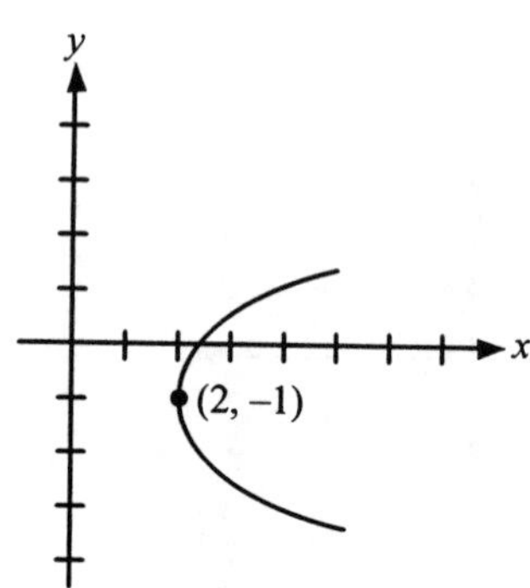

16. Hyperbola **17.** Point: $(2,\ -1)$

18. $\dfrac{x^2}{9}-\dfrac{y^2}{7}=1$

19. $\left(-\sqrt{6},-\sqrt{10}\right),\ \left(-\sqrt{6},\sqrt{10}\right),\ \left(\sqrt{6},-\sqrt{10}\right),\ \left(\sqrt{6},\sqrt{10}\right)$

20. $\dfrac{3\sqrt{3}}{2}$

Exercises 11.1

1. $\left(2,\ -\dfrac{1}{3}\right)$ **3.** $(1,\ 1)$ **5.** Inconsistent $\varnothing$

7. Dependent $\left(x,\ \dfrac{2x-5}{3}\right)$

9. $(8,\ 0,\ 2)$ **11.** $(1,\ 2,\ 3)$ **13.** $(-1,\ 2,\ 2)$

15. Dependent system; Infinite Number of solutions

17. Dependent system; Infinite Number of solutions

19. Inconsistent $\varnothing$ **21.** Inconsistent $\varnothing$

23. \$105 for paperbacks; \$149.70 for hardback

25. \$70,000 salary; \$20,000 capital gains

27. \$38,989 Bonus; \$101,083 Tax

29. $\dfrac{1}{30}$ mi/min = 2 mi/hr

31. $a=1,\ b=-4,\ c=8,\ y=x^2-4x+8$

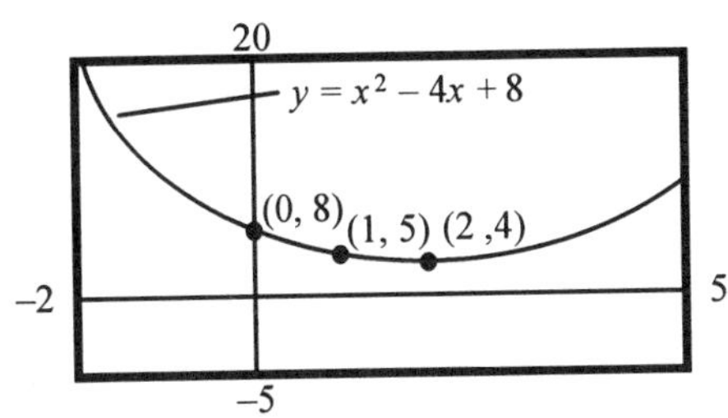

33. $\frac{2}{13}$ amp; $\frac{4}{65}$ amp; $\frac{14}{65}$ amp

35. 16, 24, 5, 80 37. $a = .5,\ b = .5,\ c = -8$

Exercises 11.2

1. $\left(\frac{5}{2}, \frac{9}{2}\right)$ 3. $a = \frac{1}{4},\ b = \frac{3}{4},\ c = 1$

5. (−27, 59)

7. No unique solution (dependent)

9. $x = 7,\ y = -3,\ z = -5$

11. $x = 3,\ y = -1,\ z = -1$

13. $x_1 = 2,\ x_2 = -2,\ x_3 = -2$

15. $A = \frac{1}{2},\ B = -1,\ C = \frac{1}{2}$

17. No unique solution (inconsistent)

19. $a = 2,\ b = -1,\ c = -1,\ d = 0$

21. \$601,000 Alpha; \$199,500 Beta; \$199,500 Gamma

23. $a = -1,\ b = 5,\ c = -2$

25. $a = -1,\ b = 0,\ c = 7,\ d = -5$

27. $r_1 = 11$ m, $r_2 = 3$ m, $r_3 = 5$ m

29. 20 records, 30 cassettes, 100 CD's

Exercises 11.3

1. −10 3. 2 5. 26

7. −3 9. 0 11. −52

13. 17 15. 254 17. 97

19. 0 21. (4, 21) 23. (−1, 1)

25. No solution, inconsistent

27. (5, 1) 29. $\left(\frac{67}{41}, -\frac{155}{41}\right)$

31. $x = 1,\ y = 2,\ z = 0$ 33. $x = -3,\ y = 1,\ z = 4$

35. $x = \frac{1}{2},\ y = -\frac{3}{2},\ z = 1$

37. $x = 2,\ y = 2,\ z = -2$ 39. No unique solution

41. .24 oz. of 68% silver; .56 oz. of 78% silver

43. 920 Terminators; 235 Avengers

45. $I_1 = -\frac{8}{73}$ ampere; $I_2 = \frac{55}{73}$ ampere; $I_3 = -\frac{47}{73}$ ampere

47. $I_1 = \frac{73}{101}$ ampere; $I_2 = -\frac{35}{101}$ ampere;
$I_3 = -\frac{38}{101}$ ampere

49. 5 burgers, 4 french fries, 2 ice cream

Exercises 11.4

1. 2×4 (read 2 by 4)

3. 2×1 (read 2 by 1)

5. $a = 10,\ b = -3,\ c = 5,\ d = 4$

7. $a = 2,\ b = 0,\ c = -3$

9. $\begin{bmatrix} -1 & -3 \\ 4 & 10 \end{bmatrix}$ 11. $\begin{bmatrix} 0 & 2 & -1 \\ 4 & -3 & 1 \end{bmatrix}$

13. $\begin{bmatrix} 10 & 2 \\ -1 & -5 \end{bmatrix}$

15. Both *AB* and *BA* are 2×2

17. *AB* is 2×2, *BA* is 3×3

19. Both *AB* and *BA* are undefined.

21. $\begin{bmatrix} 11 & 0 \\ -2 & 3 \end{bmatrix}$ 23. $\begin{bmatrix} 1 & 3 \\ -1 & 0 \end{bmatrix}$ 25. $\begin{bmatrix} -10 \\ 7 \\ 0 \end{bmatrix}$

27. $\begin{bmatrix} 1 & -4 & -3 \\ -2 & -4 & -9 \end{bmatrix}$ 29. Undefined

31. B 33. $\begin{bmatrix} 0 & 4 \\ -6 & -10 \end{bmatrix}$ 35. I

37. $\begin{bmatrix} -3 & -2 & -9 \\ -2 & -6 & -5 \end{bmatrix}$ 39. $\begin{aligned} x + y &= 0 \\ 5x - 2y &= 3 \end{aligned}$

41. $\begin{bmatrix} 3 & 1 \\ -7 & -2 \end{bmatrix}$ 43. $\frac{1}{13}\begin{bmatrix} 2 & 3 \\ 1 & -5 \end{bmatrix}$

45. A^{-1} does not exist. **47.** $\begin{bmatrix} \frac{5}{47} & \frac{-9}{94} & \frac{-1}{47} \\ \frac{20}{47} & \frac{11}{94} & \frac{-4}{47} \\ \frac{-8}{47} & \frac{5}{94} & \frac{11}{47} \end{bmatrix}$

49. A^{-1} does not exist. **51.** $(5, -12)$

53. $(-1, 1)$

55. The inverse of A does not exist; no unique solution; inconsistent

57. $(1, 0, 2)$

59. The inverse of A does not exist; no unique solution; dependent

61. **a.** $(29, 17)$ **b.** $(-25, -15)$

63. $\begin{bmatrix} 88 & 89.6 \\ 79 & 79.8 \\ 90 & 89.2 \\ 76 & 74.2 \end{bmatrix}$ This represents the grade where the midterm and the final were given equal weight followed by the grade where the midterm was weighted .4 and the final was weighted .6 of the grade.

65. **a.**

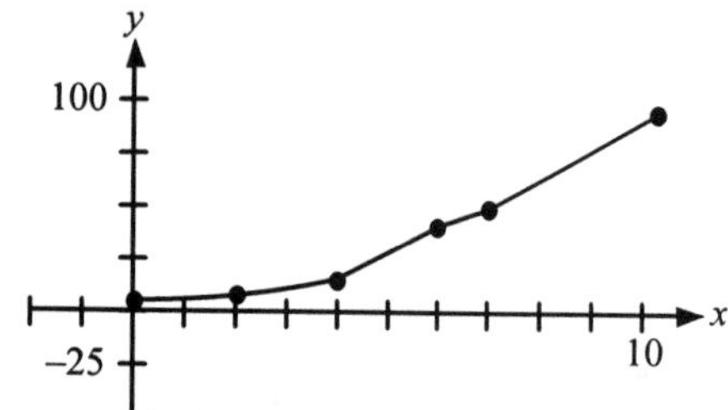

b. $y = \frac{53}{42}x^2 - \frac{65}{42}x + 1$

67. printer = \$250; monitor = \$375

69. $\det[B] = 0$; $\det[A] = 0$; $\det[C] = 0$

Exercises 11.5

1. $A = -\frac{1}{4}, B = \frac{1}{4}$ **3.** $A = 3, B = -3, C = 3$

5. $A = 2, B = -4$

7. $A = 1, B = -2, C = -5, D = 5$

9. $A = -2, B = 1, C = 2, D = 1$

11. $\frac{1/2}{1+x} + \frac{1/2}{1-x}$ **13.** $\frac{1}{x+1} + \frac{-1}{(x+1)^2}$

15. $x + \frac{2}{x+2} + \frac{2}{x-2}$ **17.** $\frac{1}{x} + \frac{-3}{x-2} + \frac{2}{x+3}$

19. $\frac{1}{x} + \frac{-x}{x^2+1}$ **21.** $\frac{1}{x+1} + \frac{1}{x-1} + \frac{-2x}{x^2+1}$

23. $\frac{x+4}{x^2+1} + \frac{-x}{x^2+2}$ **25.** $A = \frac{1}{40}$; $B = \frac{1}{40}$

27. **a.** $f(s) = \frac{8/3}{s-2} + \frac{1/3}{s-5}$

b. $L^{-1}\left(\frac{8}{3} \cdot \frac{1}{s-2}\right) = \frac{8}{3}e^{2t}$; $L^{-1}\left(\frac{1}{3} \cdot \frac{1}{s-5}\right) = \frac{1}{3}e^{5t}$

Exercises 11.6

1.

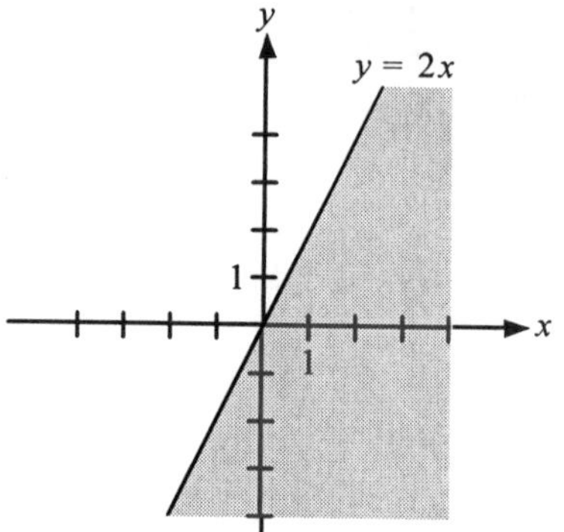

3.

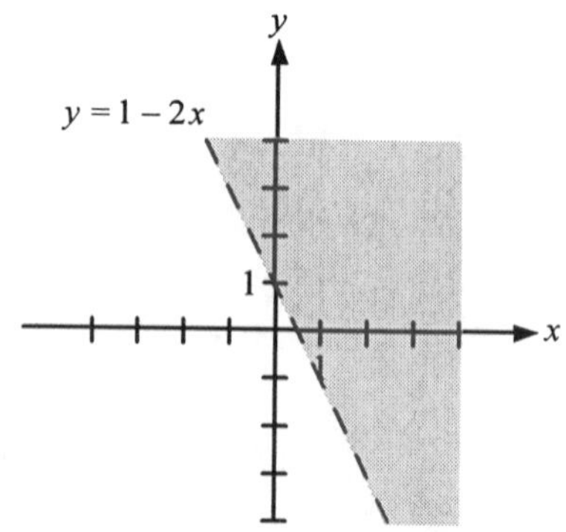

5.

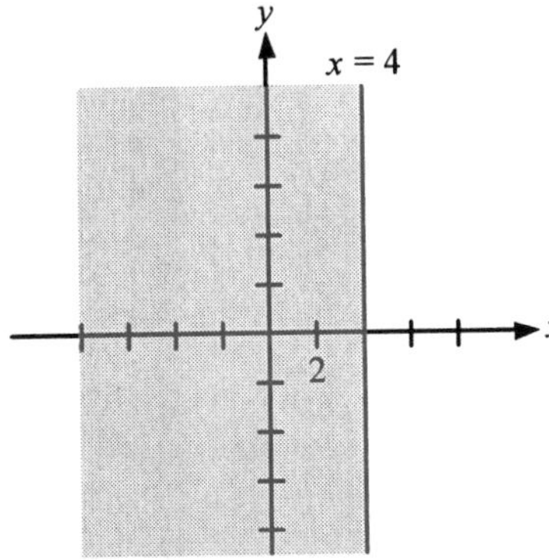

7.

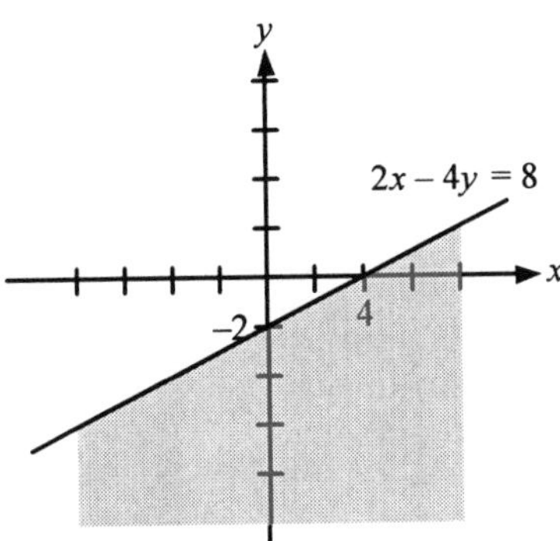

9.

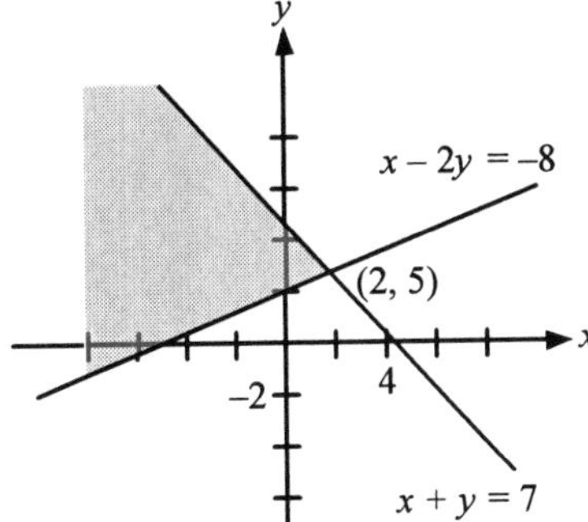

11.

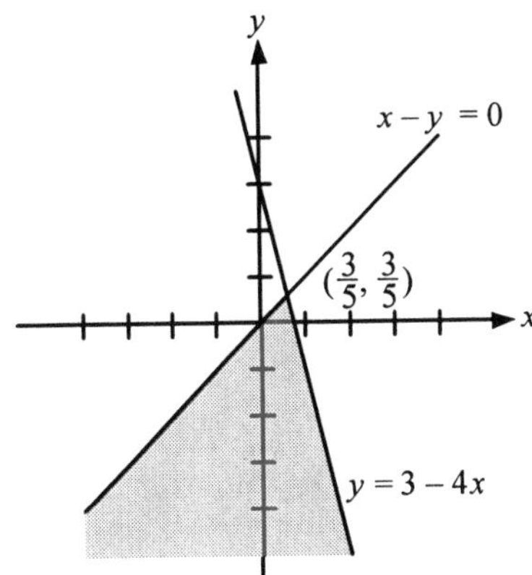

13.

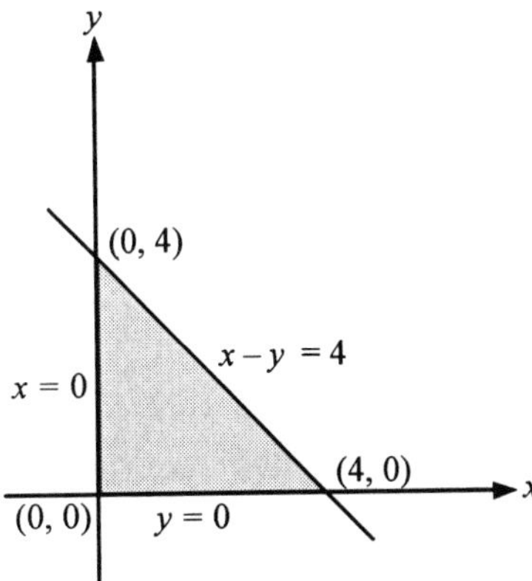

15.

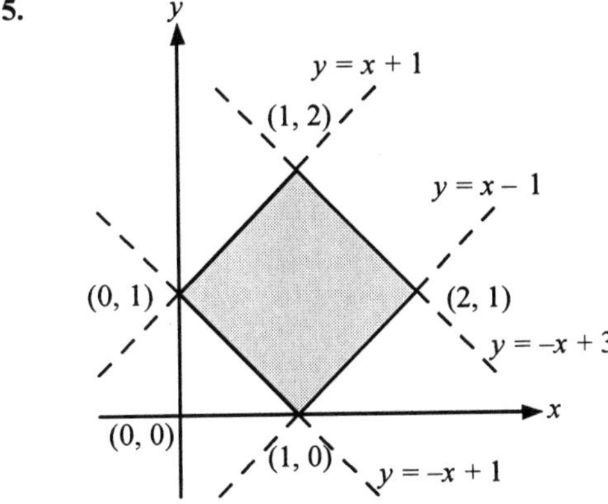

17. Max: 8; min: −12

19. (2, 1)

21. 38,000

23. $C = 70$ at $x = 20,\ y = 30$

25. $0 \le 90x + 50y + 30z \le 20{,}000;\ x,\ y,\ z \ge 0$

27. $0 \le Bx + Cy \le A;\ x,\ y \ge 0$

29. 80 type A, 400 type B, max profit: $37,200

31. A: 2 oz; B: $\frac{3}{2}$ oz; $C = .25x + .20y$, $3x + 2y \ge 9$; $3x + 4y \ge 12$; $x,\ y \ge 0$

33. 92 small; 50 large

Chapter 11 Review Exercises

1. $(-1,\ -5)$

3. $(6, 0)$

5. $(13,\ -3)$

7. $x = 5,\ y = -1,\ z = -1$

9. $x = 15,\ y = 13,\ z = -5$

11. $(1, 2)$

13. $x = 1,\ y = -1,\ z = 0$

15. $x = 0,\ y = 0,\ z = 0$

17. -11

19. 0

21. $-\begin{vmatrix} c & d \\ a & b \end{vmatrix} = -(cb - da) = ad - bc = \begin{vmatrix} a & b \\ c & d \end{vmatrix}$

23. $\begin{bmatrix} -7 & -5 \\ -3 & -7 \end{bmatrix}$ **25.** Undefined

27. $\begin{bmatrix} 12 & 15 \\ 9 & 12 \end{bmatrix}$ **29.** $\begin{bmatrix} 13 & 17 \\ 8 & 25 \end{bmatrix}$

31.

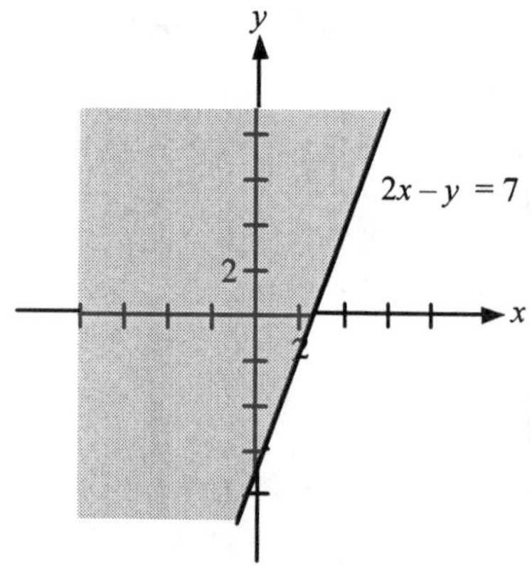

33. $P = 13{,}750$ at $x = 50,\ y = 375$

35. $a = 2,\ b = -3,\ c = 4$

37. Car: \$5000; Tuition: \$6,000; Travel: \$2,000

39. 8 **41.** False **43.** $A = -4,\ B = 7$

45. $1 + \dfrac{-3/2}{x+3} + \dfrac{3/2}{x-3}$

Chapter 11 Test

1. 61 **2.** $\begin{bmatrix} -7 & 4 \\ -5 & 3 \end{bmatrix}$ **3.** $\begin{bmatrix} -11 & 3 \\ 30 & -8 \end{bmatrix}$

4. $\begin{bmatrix} -2 & 11 \\ 11 & -8 \end{bmatrix}$ **5.** $\dfrac{1/2}{x+2} + \dfrac{7/2}{x-2}$

6. $x = -5$ **7.** $x = -5,\ y = 2,\ z = -10$

8. \$40,000 **9.** $a = 3,\ b = -5,\ c = 1$

10. 90

Exercises 12.1

1. 1, 3, 5, 7; 99 **3.** 1, 4, 9, 16; 49

5. $1,\ -\frac{1}{2},\ \frac{1}{3},\ -\frac{1}{4};\ -\frac{1}{10}$

7. $\frac{1}{2},\ \frac{1}{4},\ \frac{1}{8},\ \frac{1}{16};\ \frac{1}{32}$ **9.** 1, 3, 6, 10; 36

11.

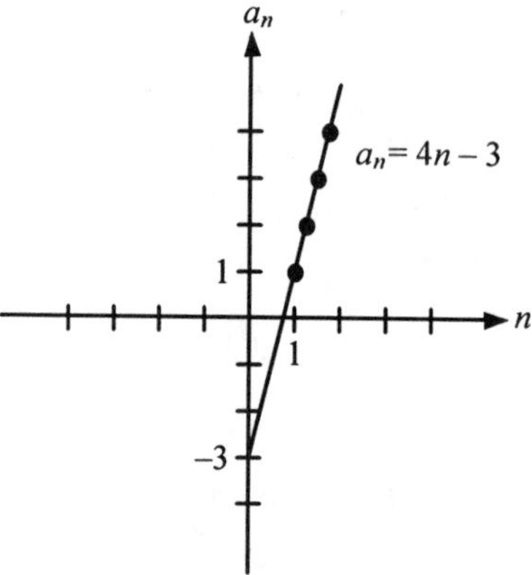

13.

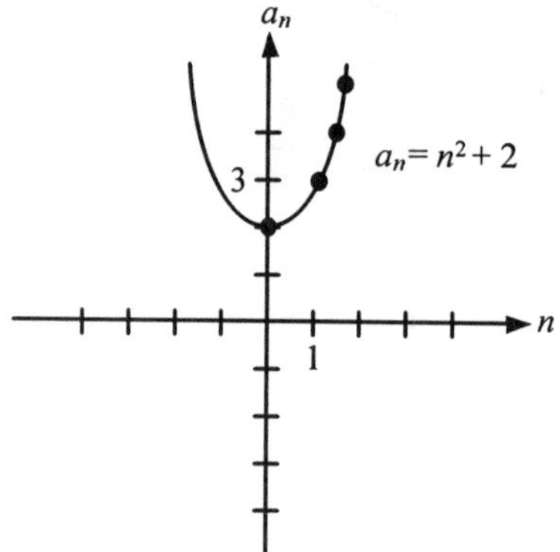

15.

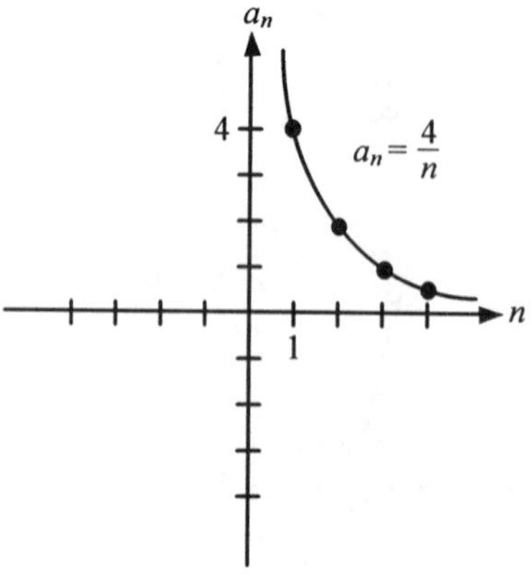

17.

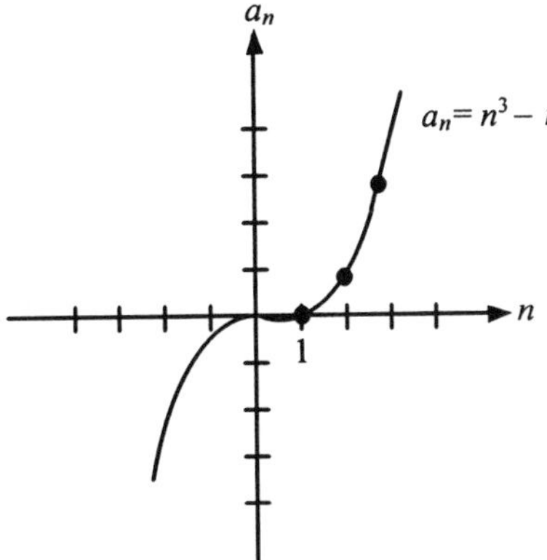

19.

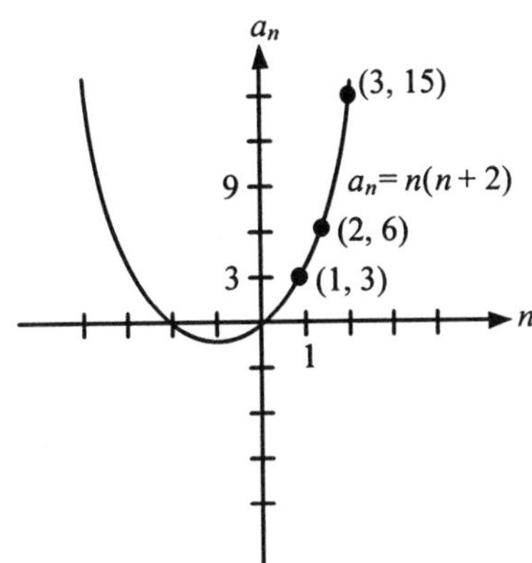

21. $a_n = 4n - 1$; 159 **23.** $a_n = -5n + 14$; -11

25. $a_n = 0.3n + 0.8$; 8.3 **27.** $a_n = \frac{4}{3}n - 1$; 39

29. $a_n = \frac{3}{8}n + \frac{29}{8}$; $\frac{53}{8}$ **31.** $a_n = 2(3)^{n-1}$; 1,458

33. $a_n = \left(-\frac{1}{2}\right)^{n-1}$; $-\frac{1}{32}$ **35.** $a_n = \sqrt{2}\left(\sqrt{2}\right)^{n-1}$; 16

37. $a_n = 6\left(\frac{2}{3}\right)^{n-1}$; $\frac{128}{243}$ **39.** $a_n = \left(-\frac{1}{3}\right)^{n-1}$; $-\frac{1}{243}$

41. AP; $d = 5$; 27, 32 **43.** Neither

45. GP; $r = 1.02$; $(1.02)^5$, $(1.02)^6$

47. GP; $r = 0.1$; 0.00003, 0.000003

49. Neither **51.** 4, $\frac{9}{2}$, 5, $\frac{11}{2}$

53. 6 **55.** 7 **57.** $\frac{5}{9}$ or $-\frac{5}{9}$

59. $a_1 = \frac{15}{2}$, $d = \frac{3}{2}$

61. \$1080; \$1166.40; \$1259.71; \$1360.49; \$1469.33; geometric

63. **a.** $\frac{1}{2}, \frac{2}{3}, \frac{3}{4}, \frac{4}{5}, \frac{5}{6}$
The terms appear to be approaching 1.

b. $1, \frac{1}{2}, \frac{1}{3}, \frac{1}{4}, \frac{1}{5}$
The terms appear to be approaching 0.

c. $1, \frac{1}{4}, \frac{1}{9}, \frac{1}{22}, \frac{1}{49}$
The terms appear to be approaching 0.

d. $1, \frac{1}{4}, \frac{1}{9}, \frac{1}{16}, \frac{1}{25}$
The terms appear to be approaching 0.

e. $\frac{5}{3}, \frac{17}{9}, \frac{53}{27}, \frac{161}{81}, \frac{485}{243}$
The terms appear to be approaching 2.

f. 0.4, 0.64, 0.784, 0.8704, 0.92224
The terms appear to be approaching 1.

65. The first five terms of the sequence of prime numbers are 2, 3, 5, 7, 11. This sequence is neither arithmetic nor geometric, since there is no common difference or ratio between terms.

67. **a.** Arithmetic sequence; $d = \log 2$
b. Neither **c.** Geometric sequence; $r = 2$

69. **a.** 1, 1, 2, 3, 5, 8, 13, 21, 34, 55, 89, 144, 233, 377, 610

b.

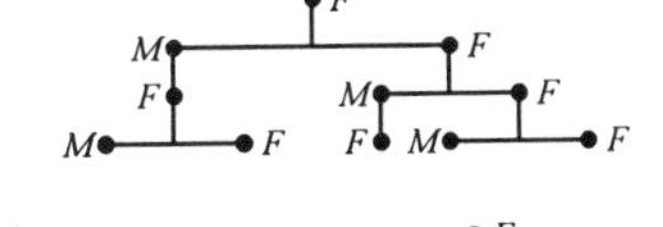

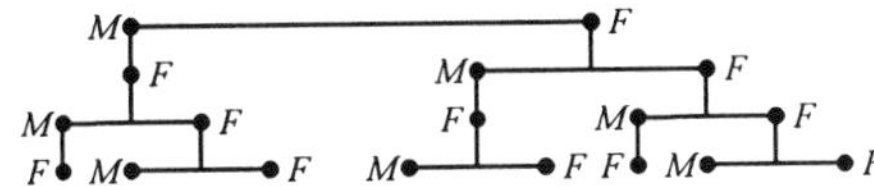

71. $a_9 = \frac{55}{34} = 1.6176$; $a_{10} = \frac{89}{55} = 1.6182$;
$a_{11} = \frac{144}{89} = 1.6180$; $a_{12} = \frac{233}{144} = 1.6181$

73. **a.** Since the last digit of a_3 is 8, z is an even number and so is not prime.
b. No, the sum of two even integers is always even.

Exercises 12.2

1. 210 **3.** 2,186 **5.** About 5.42

7. $\dfrac{3,367}{1,024}$ **9.** 31.9

11. $1+2+3+4+\cdots+9+10=55$

13. $-1+1-1+1-1+1=0$

15. $1+3+5+7+\cdots+21+23=144$

17. $4+\dfrac{9}{2}+5+\dfrac{11}{2}=19$

19. $1+\dfrac{1}{3}+\dfrac{1}{9}+\dfrac{1}{27}+\dfrac{1}{81}=\dfrac{121}{81}$

21. $\sum_{i=3}^{10} i$ **23.** $\sum_{i=1}^{4}\dfrac{1}{2^i}$ **25.** $\sum_{i=1}^{5}\dfrac{i}{i+1}$

27. $\sum_{i=1}^{6} x^i$ **29.** $\sum_{i=1}^{n} i^3$ **31.** 15

33. $n(n+1)$ **35.** 6,400 ft

37. $2+3+4+\cdots+9=44$; arithmetic $d=1$

39. \$8,318.13

41. **a.** $\dfrac{6,000-1,500}{900}=5$

b. $360+288+216+144+72=\$1,080$; arithmetic, $d=-72$

43. $\dfrac{3389}{81}$ ft

45. $\sum_{i=1}^{13} 1,000(1+0.06)^i$. There is \$20,015.07 in the account. \$7,015.07 is interest.

47. 254 **49.** 18 years

51. $5+4.5+4+3.5+3+2.5=22.5$; arithmetic, $d=-0.5$

Exercises 12.3

1. 2 **3.** $\dfrac{1}{4}$ **5.** $\dfrac{10}{3}$

7. $\dfrac{9}{10}$ **9.** $\dfrac{50}{9}$ **11.** $\dfrac{3}{7}$

13. $\dfrac{3}{4}$ **15.** $\dfrac{1}{28}$ **17.** $\dfrac{2}{9}$

19. $\dfrac{7}{99}$ **21.** $\dfrac{107}{333}$ **23.** 6

25. $\dfrac{643}{300}$

27.

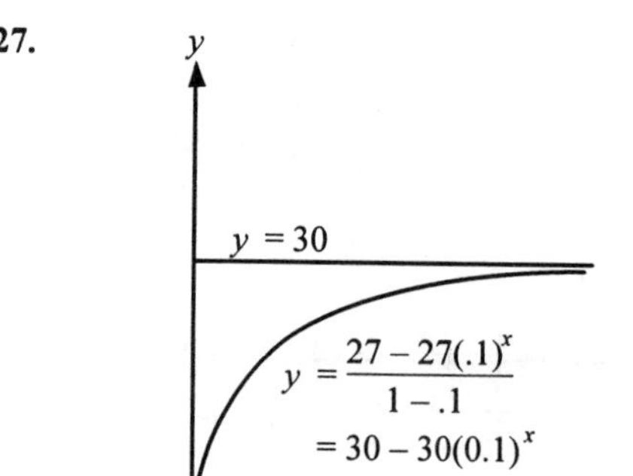

29.

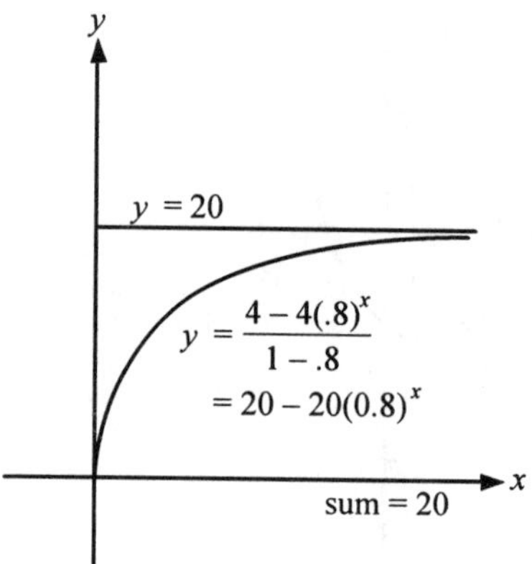

31.

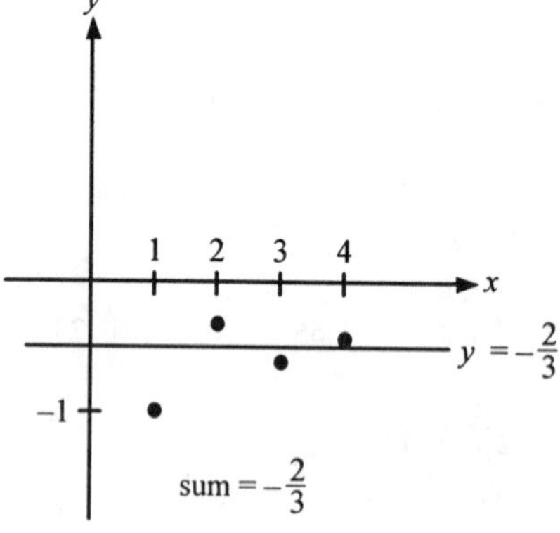

33. 25 mg

35. 2,000 seconds = $33\dfrac{1}{3}$ minutes

37. $-\dfrac{3}{2}<x<-\dfrac{1}{2}$; $\dfrac{x+1}{-2x-1}$

39. $S=\dfrac{1/2}{1-1/2}=1$ **41.** $S=\dfrac{4}{3}A$

Exercises 12.4

1. *Step 1*: $1 \stackrel{?}{=} \frac{1(1+1)}{2}$; $1=1$

Step 2: $\underbrace{1+2+3+\cdots+k}_{\text{induction assumption}}+(k+1)$

$= \frac{k(k+1)}{2} + (k+1)$

$= (k+1)\left(\frac{k}{2}+1\right)$

$= \frac{(k+1)(k+2)}{2}$

$= \frac{(k+1)[(k+1)+1]}{2}$

3. *Step 1*: $\frac{2(1)}{3} \stackrel{?}{=} \frac{1(1+1)}{3}$; $\frac{2}{3}=\frac{2}{3}$

Step 2: $\underbrace{\frac{2}{3}+\frac{4}{3}+\cdots+\frac{2k}{3}}_{\text{induction assumption}}+\frac{2(k+1)}{3}$

$= \frac{k(k+1)}{3} + \frac{2(k+1)}{3}$

$= \frac{(k+1)(k+2)}{3}$

$= \frac{(k+1)[(k+1)+1]}{3}$

5. *Step 1*: $4(1) \stackrel{?}{=} 2(1)(1+1)$; $4=4$

Step 2: $\underbrace{4+8+\cdots+4k}_{\text{induction assumption}}+4(k+1)$

$= 2k(k+1) + 4(k+1)$

$= 2(k+1)(k+2)$

$= 2(k+1)[(k+1)+1]$

7. *Step 1*: $2^1 \stackrel{?}{=} 2(2^1-1)$; $2=2$

Step 2: $\underbrace{2+2^2+\cdots+2^k}_{\text{induction assumption}}+2^{k+1}$

$= 2(2^k-1) + 2^{k+1}$

$= 2^{k+1}-2+2^{k+1}$

$= 2(2^{k+1})-2$

$= 2(2^{k+1}-1)$

9. *Step 1*: $1^3 \stackrel{?}{=} \frac{1^2(1+1)^2}{4}$; $1=1$

Step 2: $\underbrace{1^3+2^3+\cdots+k^3}_{\text{induction assumption}}+(k+1)^3$

$= \frac{k^2(k+1)}{4} \quad (k+1)^3$

$= (k+1)^2\left[\frac{k^2}{4}+(k+1)\right]$

$= (k+1)^2\left[\frac{k^2+4k+4}{4}\right]$

$= \frac{(k+1)^2(k+2)^2}{4}$

$= \frac{(k+1)^2[(k+1)+1]^2}{4}$

11. *Step 1*: If $n=1$, $2^1>1$, which is true.

Step 2: $2^k>k$ (induction assumption)

$2\cdot 2^k > 2\cdot k$

$2^{k+1} > 2k$

$2^{k+1} > k+1$ (since $2k \ge k+1$, for $k \ge 1$)

13. *Step 1*: If $n=5$, $5^2<2^5$, which is true.

Step 2: $k^2<2^k$ (induction assumption)

$k^2+(2k+1) < 2^k+(2k+1)$

$(k+1)^2 < 2^k+2k+k$ (since $k>1$)

$= 2^k+3k$

$< 2^k+k^2$ (since $k>3$)

$< 2^k+2^k$ (induction assumption)

$= 2(2^k) = 2^{k+1}$

15. *Step 1*: If $n=2$, $\log(x_1\cdot x_2)=\log x_1+\log x_2$, which is true.

Step 2: $\log(x_1x_2\cdots x_nx_{n+1})$

$= \log[(x_1x_2\cdots x_n)x_{n+1}]$

$= \underbrace{\log(x_1x_2\cdots x_n)}_{\text{induction assumption}}+\log x_{n+1}$

$= (\log x_1+\log x_2+\cdots+\log x_n)+\log x_{n+1}$

$= \log x_1+\log x_2+\cdots+\log x_n+\log x_{n+1}$

17. *Step 1*: If $n = 1$, $0 < a < 1$, which is given.
Step 2: $0 < a_k < 1$ (induction assumption)

$$a \cdot 0 < a \cdot a^k < a \cdot 1$$
$$0 < a^{k+1} < a$$
$$0 < a^{k+1} < 1 \text{ (since } 0 < a < 1)$$

19. *Step 1*: True for $n = 1$, since $x - a$ is divisible by $x - a$.
Step 2: $x^{k+1} - a^{k+1}$

$$= x^{k+1} - xa^k + xa^k - a^{k+1}$$
$$= x(x^k - a^k) + a^k(x - a)$$

Then $a^k(x-a)$ is obviously divisible by $x - a$, while $x(x^k - a^k)$ is divisible by $x - a$ since $x - a$ divides $x^k - a^k$ by the induction assumption. Thus, $x^{k+1} - a^{k+1}$ is divisible by $x - a$.

21. **a.** *Step 1*: If $n = 2$, $(2-1) = \frac{2(2-1)}{2}$; $1 = 1$

Step 2:
$$[(k+1)-1] + [(k+1)-2] + [(k+1)-3] + \cdots + 1$$
$$= k + \underbrace{(k-1) + (k-2) + \cdots + 1}_{\text{induction assumption}}$$
$$= k + \frac{k(k-1)}{2}$$
$$= \frac{2k + k(k-1)}{2}$$
$$= \frac{k(2 + k - 1)}{2}$$
$$= \frac{k(k+1)}{2}$$

b. $\frac{20(19)}{2} = 190$

Exercises 12.5

1. $(x+y)^6 = x^6 + 6x^5y + 15x^4y^2 + 20x^3y^3 + 15x^2y^4 + 6xy^5 + y^6$

3. $(x-y)^5 = x^5 - 5x^4y + 10x^3y^2 - 10x^2y^3 + 5xy^4 - y^5$

5. $(x+h)^4 = x^4 + 4x^3h + 6x^2h^2 + 4xh^3 + h^4$

7. $(x-1)^7 = x^7 - 7x^6 + 21x^5 - 35x^4 + 35x^3 - 21x^2 + 7x - 1$

9. $(2x+y)^3 = 8x^3 + 12x^2y + 6xy^2 + y^3$

11. $(3c-4d)^4 = (3c)^4 + 4(3c)^3(-4d) + 6(3c)^2(-4d)^2 + 4(3c)(-4d)^3 + (-4d)^4 = 81c^4 - 432c^3d + 864c^2d^2 - 768cd^3 + 256d^4$

13. $(x+y)^{15} = x^{15} + 15x^{14}y + 105x^{13}y^2 + 455x^{12}y^3 + \cdots$

15. $(x-3y)^{12} = x^{12} + 12x^{11}(-3y) + 66x^{10}(-3y)^2 + 220x^9(-3y)^3 + \cdots$
$= x^{12} - 36x^{11}y + 594x^{10}y^2 - 5{,}940x^9y^3 + \cdots$

17. 1, 3, 3, 1; the binomial coefficients 1, 3, 3, 1 match the entries in row 3 of Pascal's triangle.

19. $\binom{6}{0}, \binom{6}{1}, \binom{6}{2}, \binom{6}{3}, \binom{6}{4}, \binom{6}{5}, \binom{6}{6}$

21. 10 **23.** 126 **25.** 1

27. 190 **29.** 43,758 **31.** $18x^{17}y$

33. $-702x^2y^{11}$ **35.** $126x^8$ **37.** $80x$

39. $4{,}320x^6y^9$ **41.** 330 **43.** -960

45. $(1.2)^4 = 1 + 4(0.2) + 6(0.2)^2 + 4(0.2)^3 + (0.2)^4 = 2.0736$

47. $\binom{n}{n-1} = \frac{n!}{(n-1)![n-(n-1)]!} = \frac{n!}{(n-1)!} = \frac{n(n-1)!}{(n-1)!} = n$

49. 13.98 percent **51.** 23.56 percent

Exercises 12.6

1. 720 **3.** 126 **5.** 120

7. 6,720 **9.** 10

11. **a.** $\frac{5!}{0!} = 120$ **b.** $\frac{5!}{(0!5!)} = 1$

c. $\frac{5!}{(5!0!)} = 1$

13. 8

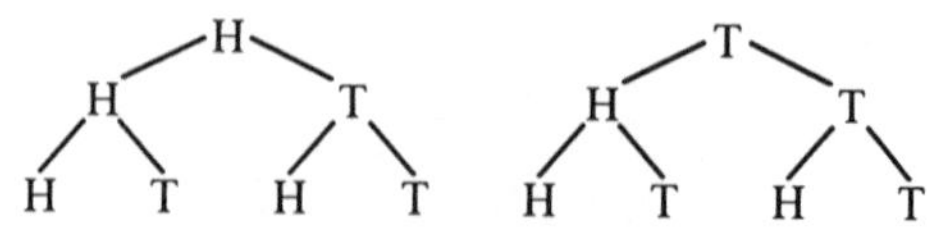

HHH HHT HTH HTT THH THT TTH TTT

15. 12 **17.** 34,650 **19.** 720

21. 56 **23.** 495 **25.** 18,564

27. 15 **29.** 45 **31.** 10

33. 80 **35.** 113,400 **37.** 36

39. 120 **41.** 5

43. ${}_nC_{n-r} = \dfrac{n!}{[n-(n-r)]!(n-r)!} = \dfrac{n!}{r!(n-r)!} = {}_nC_r$

45. a

Exercises 12.7

1. $\frac{5}{12}$

3. **a.** $\frac{1}{52}$ **b.** $\frac{1}{4}$ **c.** $\frac{3}{13}$
d. $\frac{25}{26}$ **e.** $\frac{2}{13}$ **f.** $\frac{11}{13}$
g. $\frac{4}{13}$ **h.** $\frac{9}{13}$

5. $\frac{17}{19}$ **7.** $\frac{1}{9}$

9. **a.** $\frac{4}{9}$ **b.** $\frac{1}{3}$ **c.** $\frac{4}{9}$

11. $\frac{1}{12}$ **13.** $\frac{1}{120}$ **15.** $\frac{3}{8}$

17. $\frac{21}{44}$ **19.** $\frac{1}{2}$

Chapter 12 Review Exercises

1. −1, 2, −4, 8; 32 **3.** 2, 485

5. $\sum_{i=1}^{5} \frac{1}{i^2}$

7. *Step 1*: If $n=1$, $a = \frac{1}{2}[2a+(1-1)d]$, which is true.

Step 2: $\underbrace{a+(a+d)+\cdots+[a+(k-1)d]}_{\text{induction assumption}}+(a+kd)$

$$= \frac{k}{2}[2a+(k-1)d] \quad +(a+kd)$$
$$= ka + \frac{k^2d}{2} - \frac{kd}{2} + a + kd$$
$$= \frac{k^2d + kd + 2ka + 2a}{2}$$
$$= \frac{kd(k+1)+2a(k+1)}{2}$$
$$= \frac{k+1}{2}(2a+kd)$$

9. **a.** $c = 2b - a$ **b.** $c = \frac{b^2}{a}$

11. $\frac{1}{2}$ **13.** 55

15. 5,040; 720; 144 **17.** $d = \frac{a_n - a_1}{n-1}$

19. Yes **21.** 15 **23.** $\frac{2}{3}$

25. 210 **27.** $n=10$, $r=3$

29. $35x^2$ **31.** 0.01

33. $1-15x+105x^2-455x^3$

35. c **37.** a **39.** c

41. a **43.** c

Chapter 12 Test

1. Geometric sequence **2.** 281

3. $a_n = 27\left(-\frac{2}{3}\right)^{n-1}$ **4.** 2,235

5. 733.5929 **6.** $\frac{191}{8}$ ft

7. $\frac{28}{5}$ **8.** $\frac{214}{99}$

9. $x^4 - 12x^3y + 54x^2y^2 - 108xy^3 + 81y^4$

10. $y^{11} + 22y^{10} + 220y^9 + 1320y^8$

11. $-21x^4$ **12.** 9,000,000

13. 360 **14.** 56 **15.** $\frac{1}{18}$

16. $\frac{7}{8}$ **17.** $\frac{8}{13}$

18. *Step 1*: If $n=1$, $10(1) = 5(1)(1+1)$, which is true.

Step 2: $\underbrace{10+20+\cdots+10k}_{\text{induction assumption}}+10(k+1)$

$$= \quad 5k(k+1) \quad +10(k+1)$$
$$= 5(k+1)(k+2)$$
$$= 5(k+1)[(k+1)+1]$$

Exercises A.1

1.	Five	**3.**	One	**5.**	Three
7.	Three	**9.**	Two	**11.**	Three
13.	Two	**15.**	Three	**17.**	Four
19.	Two				

21. **a.** 12.35 **b.** 12

23. **a.** 59,370 **b.** 59,000

25. **a.** $29{,}7\bar{0}0$ **b.** $3\bar{0}{,}000$

27. **a.** 84.10 **b.** 84

29. **a.** 0.06855 **b.** 0.069

31. **a.** 0.002001 **b.** 0.0020

33. **a.** 423 **b.** 0.004

35. **a.** Same accuracy **b.** 40.2

37. **a.** 6.430 **b.** Same precision

39. **a.** Same accuracy **b.** 0.0002

41. **a.** Same accuracy **b.** 48.2

43. **a.** $30\bar{0}$ **b.** $30\bar{0}$

45.	14.1	**47.**	−7.4	**49.**	1.9
51.	1.2	**53.**	32,800	**55.**	200,000
57.	200,000	**59.**	11.6	**61.**	90
63.	9.75				

Exercises A.2

1. $y = x^2$ is a power function with $m = 2$ and $a = 1$. The graph is a straight line with slope 2. The intercept at $x = 1$ is 1.

3. $y = 4\sqrt{x}$ is a power function with $m = \frac{1}{2}$ and $a = 4$. The graph is a straight line with slope $\frac{1}{2}$. The intercept at $x = 1$ is 4.

5. $xy = 1$, so $y = \frac{1}{x} = x^{-1}$. The function is a power function with $m = -1$ and $a = 1$. The graph is a straight line with slope −1. The intercept at $x = 1$ is 1.

7. $y = 7.0x^{-0.35}$ **9.** $y = 3.0x^{0.65}$

11. The function $y = 2^{-x}$ is an exponential function and its graph on semilogarithmic paper is a straight line with negative slope.

13. The function $y = 0.5(3^x)$ is an exponential function and its graph on semilogarithmic paper is a straight line with positive slope.

● ● ●

INDEX

INDEX

Trigonometry

General Angle Definitions If θ is an angle in standard position, and if (x, y) is any point on the terminal ray of θ [except (0, 0)], then

$\sin\theta = \frac{y}{r} \leftarrow \text{reciprocals} \rightarrow \csc\theta = \frac{r}{y}$

$\cos\theta = \frac{x}{r} \leftarrow \text{reciprocals} \rightarrow \sec\theta = \frac{r}{x}$

$\tan\theta = \frac{y}{x} \leftarrow \text{reciprocals} \rightarrow \cot\theta = \frac{x}{y}.$

Note: $r = \sqrt{x^2 + y^2}$

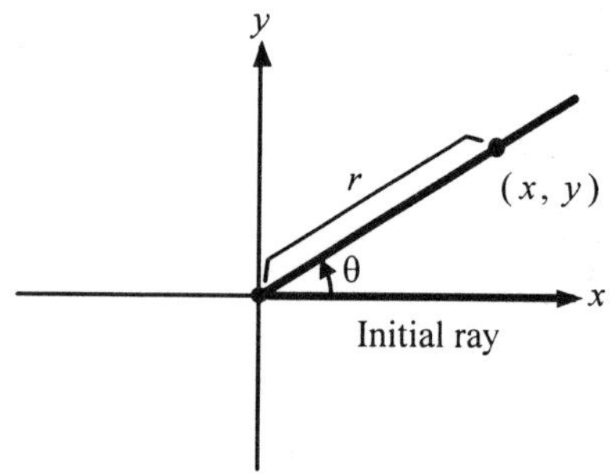

Unit Circle Definitions Consider a point (x, y) on the unit circle $x^2 + y^2 = 1$ at arc length s from (1, 0). Then

$\sin s = y \leftarrow \text{reciprocals} \rightarrow \csc s = \frac{1}{y}$

$\cos s = x \leftarrow \text{reciprocals} \rightarrow \sec s = \frac{1}{x}$

$\tan s = \frac{y}{x} \leftarrow \text{reciprocals} \rightarrow \cot s = \frac{x}{y}.$

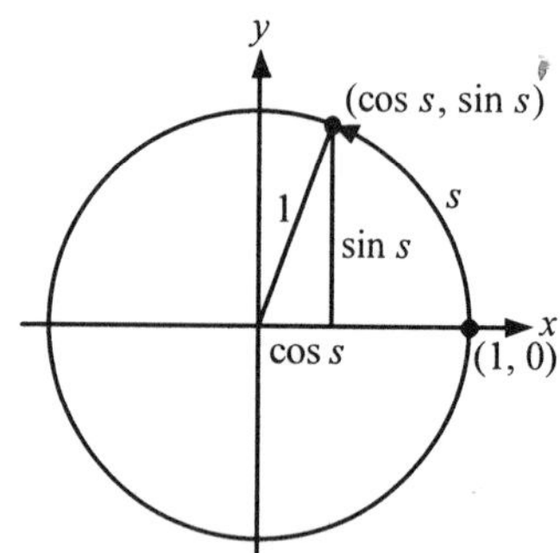

Right Triangle Definitions If θ is an acute angle in a right triangle, as shown below, then

$\sin\theta = \frac{\text{opposite}}{\text{hypotenuse}} \leftarrow \text{reciprocals} \rightarrow \csc\theta = \frac{\text{hypotenuse}}{\text{opposite}}$

$\cos\theta = \frac{\text{adjacent}}{\text{hypotenuse}} \leftarrow \text{reciprocals} \rightarrow \sec\theta = \frac{\text{hypotenuse}}{\text{adjacent}}$

$\tan\theta = \frac{\text{opposite}}{\text{adjacent}} \leftarrow \text{reciprocals} \rightarrow \cot\theta = \frac{\text{adjacent}}{\text{opposite}}.$

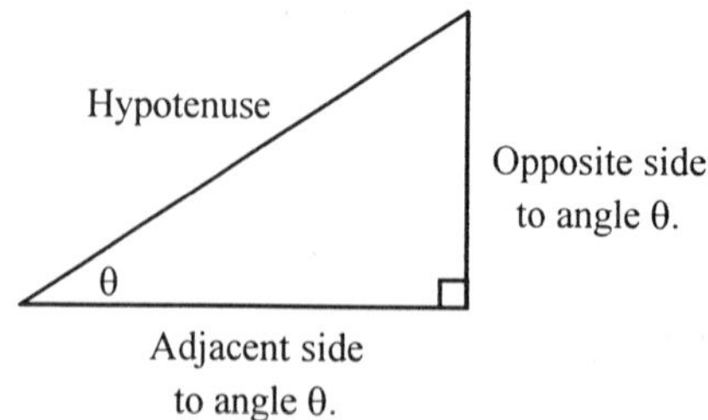

Signs of the Trigonometric Ratios

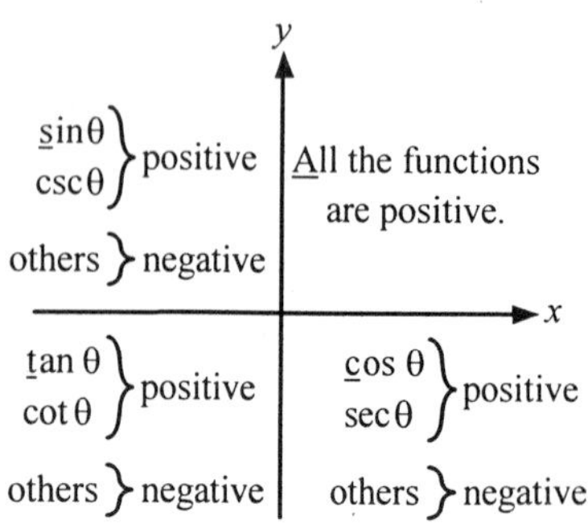